K. Kajimura H. Hayakawa (Eds.)

Advances in Superconductivity III

Proceedings of the 3rd International Symposium on Superconductivity (ISS '90), November 6-9, 1990, Sendai

ISTEC

With 1546 Figures

Springer-Verlag Tokyo Berlin Heidelberg New York
London Paris Hong Kong Barcelona

Dr. Koji Kajimura
Electrotechnical Laboratory
Umezono, Tsukuba
305 Japan

Prof. Dr. Hisao Hayakawa
School of Engineering
Nagoya University
Chikusa-ku, Nagoya
464 Japan

ISBN 4-431-70071-4 Springer-Verlag Tokyo Berlin Heidelberg New York
ISBN 3-540-70071-4 Springer-Verlag Berlin Heidelberg New York Tokyo
ISBN 0-387-70071-4 Springer-Verlag New York Berlin Heidelberg Tokyo

Printed in Hong Kong

Printing/Binding: Excel Graphic Arts, Hong Kong

Foreword

Since the discovery of high temperature superconductors, many new materials have been invented. In the last year, several new materials were also discovered, but their critical temperatures are still below 100K. Precise physical and chemical work has made tremendous progress in the theoretical and experimental study of physical properties and carrier state characterizations. The de Haas van Alphen effect measurement showed the existence of a Fermi surface in YBCO. Flux dynamics is a well-known new problem in which flux creep and irreversibility line features are especially important for a fundamental understanding of the critical current and flux pinning. Flux pinning centers which are intentionally added using non-superconducting precipitates, neutrons, and protons, etc. increase critical currents to practical levels. The analysis of electric and magnetic properties are expected to reveal the pinning mechanism and also to further application development.

As for wires and bulks, many melt-like sintering techniques have improved the material performance of critical current densities. A new seeding Quench-Melt-Growth technique enlarged crystal size and increased the repulsion force. These melting processes, in conjunction with a mechanical strength improvement have been effectively introduced into wire fabrication in order to realize kilometer range wires and will put the oxide wires to practical use. Where thin film is concerned, when many fabrication methods had been developed using the assistance effect of activated oxygen such as ozone and oxygen radicals, a high current density of $10^6 A/cm^2$ at 77K was reported. For further improvement, substrate materials should be considered as the most important problem regarding lattice constant matching. The progress of thin film technology is accelerating the efforts to develop electronics devices. High frequency passive elements are very promising applications because of their surface loss advantage. Superconducting antenna, resonators, etc. have shown higher performances than those obtained by non-superconducting devices. Great efforts have been devoted to developing a tunnel junction using hetero-epitaxial techniques, which have resulted in the hystererisis of current voltage characteristics at 77K. The confirmation of intensity oscillation in RHEED signal indicates a new technology of layer-by-layer epitaxy for creating new atomic layer materials. A new device, a superconducting flux flow transistor operating at 40GHz, has been developed. As for systems applications, magnetic levitation, superconducting propulsion, and energy systems applications have been intensively studied for future use.

As mentioned above, superconductivity research has entered the second phase of study on application feasibility, although the fundamental research in new oxide materials and physics is still important. The world's first human levitation using about 250 superconductor pieces, fabricated by Melt Powder Melt Growth, was the most exciting exhibition. It was a fundamental experiment in magnetic levitation, but it indicates that a new application field will be opened with aggressive, but inspired research.

SHOJI TANAKA
Vice President, ISTEC

Preface

Advances in Superconductivity III is the proceedings of the Third International Symposium on Superconductivity (ISS '90) held in Sendai, Japan, November 6-9, 1990, under the cosponsorship of the International Superconductivity Technology Center (ISTEC), Miyagi Prefecture, Sendai City, Tohoku Economic Federation, and the Industrial Vitalization Center for Tohoku. Approximately 710 participants, including 55 from 12 overseas countries, attended the Symposium. One of the characteristic features of ISS '88-'90 was to offer a forum where the world's scientists, engineers, business administrators, and entrepreneurs — in the spirit of international cooperation and friendly rivalry — present, exchange, and discuss their views and findings on superconductivity and its future applications. In fact, in ISS '90, as many as 319 papers, including 137 from industry, 85 from universities, 53 from the Superconductivity Research Laboratory (SRL), and 44 from public institutes, were presented.

The volume begins with the Organization of ISS '90 and a list of its Committees. The papers from two "Special Plenary Lectures" made at the opening of the Symposium by J. Clarke (University of California, Berkeley) and by M. Murakami (SRL) and four "Plenary Lectures" are arranged at the beginning of the volume in the order in which they were presented in the Symposium program. The contents of this volume are arranged as (1) physics, (2) chemistry, (3) flux pinning, (4) wires, tapes, and bulks, (5) preparation and properties of thin films, and (6) application to devices and systems. Invited papers, which are distinguished by their length (six-eight pages) from the contributed papers (four pages), are arranged at the beginning of sections or subsections.

Remarkable progress was reported at the Symposium in the improvement of bulk materials, wires and tapes, especially in strong flux pinning in MPMG-processed bulk YBaCuO, which, in the near future, will open doors to applications in high-performance electrical machines. Considerable efforts have been made to develop high-quality thin films, including hetero-epitaxial films, in order to realize electronic devices.

The editors are indebted to many people for invaluable assistance at various stages of the preparation of the Symposium. The International Advisory Committee helped us in selecting the invited speakers. The Program and Editorial Committee selected invited papers, arranged all the invited and contributed papers, and refereed the contributed papers with the assistance of many scientists who participated in the Symposium. We would like to acknowledge all the people who helped to make the Symposium a success.

Finally, we are very grateful to the authors who cooperated by carefully preparing their manuscripts for *Advances in Superconductivity III*.

KOJI KAJIMURA
HISAO HAYAKAWA

Organization of ISS '90

Sponsored by:
International Superconductivity Technology Center

Co-sponsored by:
Miyagi Prefecture
City of Sendai
Tohoku Economic Federation
Industrial Vitalization Center for Tohoku

Supported by:
Ministry of International Trade and Industry Agency of Industrial Science and Technology
Tohoku Bureau of International Trade and Industry
Science and Technology Agency
New Energy and Industrial Technology Development Organization
Ceramic Society of Japan
The Chemical Society of Japan
The Cryogenic Society of Japan
The Federation of Electric Power Companies
The Institute of Electrical Engineers of Japan
The Institute of Electronics Information and Communication Engineers, Japan
The Iron and Steel Institute of Japan
Japan Chemical Industry Association
The Japan Electrical Manufactures' Association
Japan Electronic Industry Development Association
Japan Fine Ceramics Association
The Japan Institute of Metals
The Japan Iron and Steel Federation
The Japan Key Technology Center
Japan Mining Industry Association
The Japan Society of Applied Physics
The Japanese Electric Wire & Cable Maker's Association
The Magnetic Society of Japan
Miyagi Industry Association
Physical Society of Japan
Research and Development Association for Future Electron Devices
Sendai Convention Bureau
Tohoku Intelligent Cosmos Promotion Council

Organizing Committee

Chairman:	
G. Hiraiwa	President, ISTEC
Members:	
J. Aoi	Vice President, ISTEC
K. Fueki	Professor, Faculty of Science and Technology Science University of Tokyo
S. Homma	Governor, Miyagi Prefectural Government
T. Ishii	Mayor, City of Sendai
H. Kashiwagi	Director-General, Electrotechnical Laboratory
T. Masumoto	Director-Professor, IMR, Tohoku University
Y. Muto	Emeritus Professor, Tohoku University
S. Nakajima	Chairman of Department of Physics, Tokai University, School of Science
S. Saito	Chairman, The Society of Non-Traditional Technology
M. Sugiura	Director-General, AIST, MITI
T. Tamakawa	Chiriman, The Industrial Vitalization Center for Tohoku
S. Tanaka	Vice President, ISTEC
D. Teranishi	Director-General, Tohoku Bureau of International Trade and Industry
T. Yamamoto	Vice President, ISTEC
S. Yamamura	Member, The Japan Academy

International Advisory Committee

B. Batlogg	AT & T Bell Laboratories
M.R. Beasley	Stanford University
P. Chaudhari	IBM T.J. Watson Research Center
C.W. Chu	University of Houston
J. Clarke	University of California-Berkeley
R.C. Dynes	AT & T Bell Laboratories
J.E. Evetts	University of Cambridge
D.K. Finnemore	AMES Laboratory
K. Fossheim	Norwegian Institute of Technology
K. Fueki	Science University of Tokyo
W. Klose	Karlsruhe Nuclear Research Center
P. Komarek	Karlsruhe Nuclear Research Center
D.C. Larbalestier	University of Wisconsin
W.Y. Liang	University of Cambridge
M.B. Maple	University of California-San Diego
K.A. Müller	IBM Zürich
Y. Muto	Tohoku University
S. Nakajima	Tokai University
B. Raveau	University of Caen
J.M. Rowell	Conductus Inc.
A.W. Sleight	Oregon State University
M. Suenaga	Brookhaven National Laboratory

Steering Committee

Chairman:	
S. TANAKA	Vice President, ISTEC
Members:	
K. ARATAME	Tohoku Electric Power Co., Inc.
Y. FURUTO	The Furukawa Electric Co., Ltd.
E. MASADA	The University of Tokyo
T. MORIYA	City of Sendai
I. NAKAMURA	Miyagi Prefectural Government
O. OHTSUKI	Fujitsu Laboratories Ltd.
M. SHIMAKITA	Mitsubishi Heavy Industry, Ltd.
Y. SYONO	Tohoku University
K. UCHINOKURA	The University of Tokyo
K. WASA	Matsushita Electric Industrial Co., Ltd.
A. YAMAZAKI	Tohoku Bureau of International Trade and Industry
S. YOSHII	Ube Industries, Ltd.

Program and Editorial Committee

Chairman:	
K. KAJIMURA	Director, Electrotechnical Laboratory
Members:	
H. ABE	Oki Electric Industry Co., Ltd.
K. ANDO	Toshiba Corporation
Y. BANDO	Kyoto University
J. FUJIE	Railway Technical Research Institute
H. FUKUYAMA	The University of Tokyo
H. HAYAKAWA	Nagoya University
H. IGARASHI	NEC Corporation
I. IGUCHI	University of Tsukuba
N. KOBAYASHI	Tohoku University
T. KOBAYASHI	Osaka University
N. KOSHIZUKA	SRL-ISTEC
M. KOYANAGI	Electrotechnical Laboratory
S. MATSUDA	Hitachi Ltd.
T. MATSUSHITA	Kyushu Institute of Technology
M. NAKAO	SANYO Electric Co., Ltd.
K. SATO	Mitsubishi Electric Corp.
K. SAWANO	Nippon Steel Corporation
Y. SYONO	Tohoku University
K. TADA	Sumitomo Electric Industries, Ltd.
T. TANAKA	Central Research Institute of Electric Power Industry
Y. TANAKA	The Furukawa Electric Co., Ltd.
T. TOMINAGA	Chubu Electric Power Co., Inc.
S. UCHIDA	The University of Tokyo
H. YAMAUCHI	SRL-ISTEC

General Affairs Committee

Chairman:	
Y. Syono	Professor, Tohoku University
Members:	
A. Gonoi	Miyagi Prefectural Government
H. Hirama	The Industrial Vitalization Center for Tohoku
S. Komagata	Tohoku Bureau of International Trade and Industry
Y. Matsuo	Matsushita Electric Ind., Ltd.
K. Morooka	Miyagi Industrial Association
K. Satoh	Tohoku Electric Power Co., Inc.
K. Sato	City of Sendai
T. Shiono	Showa Electric Wire & Cable Co., Ltd.
Y. Sugino	Sendai Convention Bureau
Y. Takeoka	Nippon Steel Corporation
N. Tsubouchi	NEC Corporation
K. Watanabe	Tohoku University

Table of Contents

2.4 Magnetic Resonance and Mössbauer Studies

2.5 Optical Properties

2.6 Magnetic Properties

3 Chemistry

3.1 Material Synthesis

3.2 Structural Analysis

3.3 Substitution and Doping Effects

3.4 Phase Diagrams

4.2 Flux Line Lattice and Flux Distribution

4.3 Electromagnetic Properties

5 Tape and Bulk

5.1 Wires and Tapes

5.2 Bulk Materials

5.3 Others

6 Preparation and Properties of Thin Films

6.1 YBCO and Related Films

6.2 BSCCO and Related Films

7.2 Systems

1 Plenary Lectures

High-T_c Thin-Film Microelectronics

JOHN CLARKE, DU QUAN, M.J. FERRARI, J.J. KINGSTON, A.H. MIKLICH, and F.C. WELLSTOOD

Department of Physics, University of California, Berkeley, CA 94720, USA
Materials Sciences Division, Lawrence Berkeley Laboratory, University of California, Berkeley, CA 94720, USA
M.S. Colclough, Conductus, Inc., 969 West Maude Avenue, Sunnyvale, CA 94086, USA

ABSTRACT

A complete microelectronics technology based on high-transition temperature superconducting thin films requires two key components: a reproducible Josephson weak link and a multilayer process for interconnects. Although many kinds of weak link have been studied, none is yet satisfactory for circuits involving more than a few elements. On the other hand, a viable interconnect technology has been developed in which two $YBa_2Cu_3O_{7-x}$ films may be either electrically isolated or connected via a superconducting contact. The junctions and interconnect technology have been used to produce magnetometers involving a flux transformer with a multiturn input coil coupled to a dc Superconducting QUantum Interference Device.

KEY WORDS: Junction, interconnect, SQUID, flux transformer

INTRODUCTION

In the four years since the discovery of the high transition temperature (T_C) oxide superconductors [1], there has been remarkable progress in the technology for fabricating thin films of these materials, particularly in the case of $YBa_2Cu_3O_{7-x}$ (YBCO). However, the ability to produce good single layer films is far from being sufficient to enable one to construct electronic circuits. A complete superconducting microelectronic technology also requires Josephson junctions [2] and interconnects; the latter involve superconducting contacts and electrically insulated crossovers between two superconducting layers. In this paper we briefly review the status of junctions and interconnects, and discuss their application to magnetometers based on dc Superconducting QUantum Interference Devices (SQUIDs) [3]. Needless to say, in such a brief review, we have been able to touch on only a fraction of all the work that has been published.

JOSEPHSON JUNCTIONS

The development of a controllable junction process remains the single greatest challenge for the creation of a successful high-T_C microelectronics technology. At this point, no well-defined Josephson tunnel junction - that is, with a current-voltage (I-V) characteristic exhibiting both a discontinuous switching to a well-defined energy gap and a high subgap resistance - has been fabricated from the liquid nitrogen temperature superconductors. On the other hand, a variety of structures with nonhysteretic I-V characteristics have been investigated, and we briefly describe the main types.

Grain boundary junctions. The first thin film junctions were formed from naturally occurring grain boundaries [4]. The junctions can be obtained by patterning a microbridge, typically a few micrometers in width and length, in a film of YBCO, $Bi_2Sr_2CaCu_2O_{8+y}$ (BSCCO) or $Tl_2CaBa_2Cu_2O_{8+z}$ (TCBCO) with grain sizes of the order of 1μm. Suitable c-axis oriented YBCO films may be grown, for example, on (100) MgO substrates where the lattice mismatch is relatively large. Although the a- and b-axes have a preferential alignment along the cubic axis of the substrate, some grains are rotated by 45°; the fraction of such grains can be controlled by the growth conditions [5, 6]. The 45° grain boundaries so formed provide a region of lowered critical current density, and thus form one or more junctions in the microbridge. Junctions with relatively low critical current density ($\lesssim 10^5$ A cm^{-2}) can exhibit [5] almost ideal resistively shunted junction (RSJ) behavior [7], while those with higher critical current densities exhibit behavior consistent with flux flow and flux creep [5]. The nature of the junction remains elusive; the boundary layer between the weak links may be normal, semiconducting, insulating (albeit with a linear quasiparticle resistance) or superconducting with a reduced energy gap.

Grain boundary junctions on bicrystals. A more controlled grain boundary junction has been devised by Dimos *et al.* [8]. In this technique, a $SrTiO_3$ substrate is cut and fused together to form a bicrystal with a misorientation that can be chosen at will, and the microbridge is deposited across the interface. Thus, the grains on the two sides of the boundary have a predictable relative orientation θ, and the critical current has a relatively reproducible dependence on θ. As with naturally occurring grain boundary junctions, the I-V characteristics for lower critical currents are close to the predictions of the RSJ model.

Step-edge microbridges. Simon *et al.* [9] have fabricated microbridges by etching a sharp step in the substrate and depositing a film across the step with a thickness less than the step height. The important feature of this process is the formation of a sharp step edge so that the length of the junction is made as short as possible. Once again, the details of the junction structure are not entirely clear, but transmission electron microscopy suggests that a grain boundary is formed at the step edge. These junctions exhibit a supercurrent at and above 77K, and have I_oR products of several millivolts at 4.2K (I_o is the critical current and R the resistance).

Other "weakened" structures. Various other techniques have been devised to weaken the superconductivity over a restricted length. In one process [10], a YBCO film is deposited over a thin Al stripe which "poisons" the superconductor and yields a weak link structure. In another method [11], one focuses high energy ions on to a microbridge [12] or uses an electron-beam lithographically patterned mask to define the area exposed to an unfocussed ion beam [13]. It is likely that the irradiation destroys the superconductivity in most of the film, leaving narrow superconducting filaments connecting the banks of unirradiated film. In yet another technique [14], controlled electrical pulses were applied to a microbridge at 77K, producing a weakened region that, again, possibly consisted of narrow superconducting threads.

Normal metal bridges. Schwarz and coworkers [15] fabricated superconductor-normal metal-superconductor (SNS) junctions by depositing a strip of Au across a narrow slit (~100nm) in a YBCO film. Provided there is a good proximity effect between the YBCO and the gold, one would expect a substantial supercurrent for bridges as long as $10\xi_N$ or more, where $\xi_N = \hbar v_F/2\pi k_B T \sim 15$ nm at 77K is the decay length [16] (v_F is the Fermi velocity). In fact, a supercurrent was observed, but only up to about 16K, implying that the Josephson coupling energy was extremely small. Subsequently, other measurements [17] have shown that the proximity effect is extremely weak along the direction of the c-axis, so that the weak coupling is not surprising.

Planar junctions. Rogers *et al.* [18] fabricated planar YBCO-PrBCO-YBCO junctions in which the PrBCO layer, typically 50nm thick, is nonsuperconducting. The YBCO films were c-axis oriented, and some junctions supported a supercurrent at temperatures as high as 65K. This type of junction has much to commend it, since the normal layer can have a much higher resistance than a noble metal, and the structure takes advantage of the natural epitaxy of the materials. However, it is now clear that c-axis oriented films yield a very weak proximity effect [17], and it is quite likely that the observed supercurrent arose from YBCO filaments extending through the normal layer.

Edge junctions. The first high-T_c edge junctions were made by Laibowitz *et al.* [19]. A film of c-axis oriented YBCO is deposited, followed by an insulating layer which is not necessarily epitaxial. The films are patterned into a narrow line, part of which is removed with an ion mill to leave an edge. After an "oxyfluoridation" process, a second film of YBCO is deposited and makes contact to the edge of the first film. Some of these edge junctions support supercurrents, although the yield is low. However, as with the PrBCO barriers, it is likely that the supercurrents were carried by YBCO filaments rather than by tunneling through a barrier. A variant on this technique has been made by Gao *et al.* [20], who deposited a PrBCO film followed by a YBCO counterelectrode over the tapered edge of the first YBCO film. All three films were c-axis oriented, so that supercurrent flowed along the ab-axes of the YBCO films. The yield was high, and for a PrBCO thickness of 6nm, the I_0R product was as high as 8mV at 4.2K.

$BaKBiO_3$ junctions. Before leaving the subject of junctions, we mention briefly the results obtained by Dynes and coworkers [21] on the isotropic material $BaKBiO_3$, which has T_c ~30K. By pressing a single crystal against a thin film, they achieved a classic quasiparticle tunneling curve as well as a Josephson current. Thus, one may be able to make thin film tunnel junctions with hysteretic behavior using this material.

In concluding this section, we note that although much useful information has been gained from the study of grain boundary junctions, the lack of control of their properties and location makes them unsuitable for use in circuits. Grain boundaries made deliberately on bicrystals or step edges offer a reasonable degree of reproducibility, and are of interest for circuits requiring a few junctions, but are not extendible to more complicated circuits. The structures that are weakened by poisoning, irradiation or electrical pulsing tend to be rather extended in the direction of current flow, possibly leading to a nonsinusoidal current-phase relation. Again, they may be useful for circuits with a few junctions, but are unlikely to find their way into complex circuits. The SNS junctions deserve further study now that a-axis films are available and one could hope for a stronger proximity effect. These junctions could, in principle, become reasonably reproducible, but the low I_0R product arising from the low resistivity of the metal is likely to be a drawback in most applications. However, one obtains a higher resistance using PrBCO barriers, and, in configurations such as edge junctions in which the current flows along the a-axis, there is a good chance that a successful technology will emerge. Finally, if one can fabricate $BaKBiO_3$ thin film junctions, these may be attractive for computer elements operated at (say) 15K in a closed-cycle refrigerator.

INTERCONNECTS

We turn now to a discussion of interconnects, which involve superconducting connections and electrical isolation between superconducting layers. To minimize interdiffusion between layers, it is probably essential that such films be deposited using *in situ* techniques, and YBCO is at present the superconductor of choice. The constraints on the insulating layer are demanding, and include (i) epitaxial or highly oriented growth on both the substrate and on YBCO; (ii) low interdiffusion and chemical reactivity with the substrate and YBCO at temperatures as high as 750°C; (iii) good coverage of and adhesion to YBCO and the substrate, particularly at edges and irregularities; (iv) the ability to support the growth of high quality YBCO films; and (v) high resistivity ρ at temperatures below T_c. The obvious choices for the insulator are much the same as for substrates: $SrTiO_3$, MgO, YSZ or $LaAlO_3$.

Kingston *et al.* [22] demonstrated a successful crossover technology using laser ablated films on MgO substrates, with $SrTiO_3$ as the insulator. The nominal substrate temperatures during deposition were 730°C and 690°C for YBCO and $SrTiO_3$, respectively, with an O_2 pressure of 190 mTorr in both cases. The films were patterned with stainless steel or etched Si shadow masks pressed against the substrate. The vacuum chamber was opened between depositions to allow the targets and masks to be changed. The first layer was an H-shaped YBCO film, and the second was a $SrTiO_3$ film, typically 0.4nm thick, that covered the crossbar of the H. The upper YBCO film consisted of two strips perpendicular to the crossbar. Figures 1(b) and (c) show the resistance R vs. temperature T for the lower and upper YBCO films; for comparison, Fig. 1(a) shows R vs. T for a single layer YBCO film. We see that the transition temperature and width are not significantly degraded in the trilayer. Figure 1(d) shows R vs. T measured between the two YBCO films, through the $SrTiO_3$

layer. The resistance at 77K, ~$10^8\Omega$, corresponds to a resistivity of 4 x $10^9\Omega$cm. This value is much higher than that needed for most applications. The excellent electrical characteristics of the crossover are due to the fact that each film grows epitaxially, as demonstrated by high resolution transmission electron microscopy [23]. To investigate the properties of superconducting contacts, the same workers also fabricated cross-strips of YBCO with no intervening insulating layer. Despite the need to expose the first film to the atmosphere to change masks, there was no difficulty in obtaining contacts with high critical currents.

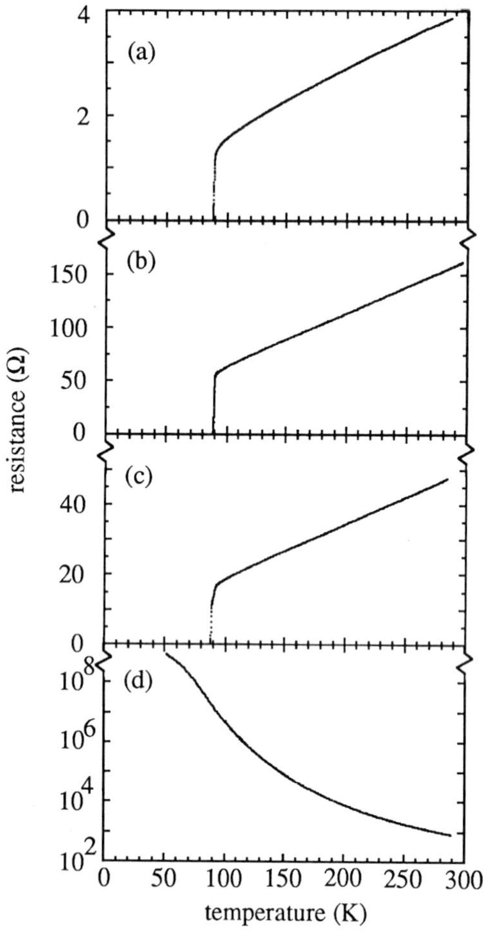

Fig. 1. Resistance (R) vs. temperature for (a) single-layer YBCO film deposited on MgO; (b) lower YBCO film of trilayer; (c) upper YBCO film of trilayer; (d) $SrTiO_3$ insulating layer measured between the upper and lower YBCO films [22].

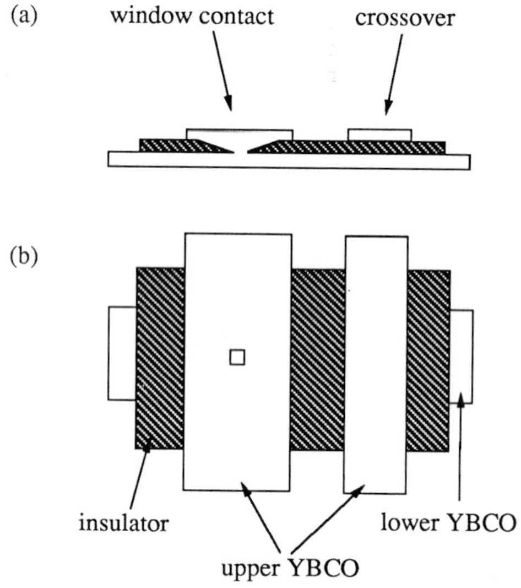

Fig. 2. (a) Cross-sectional view and (b) top view of window contact (left) and crossover (right) [24].

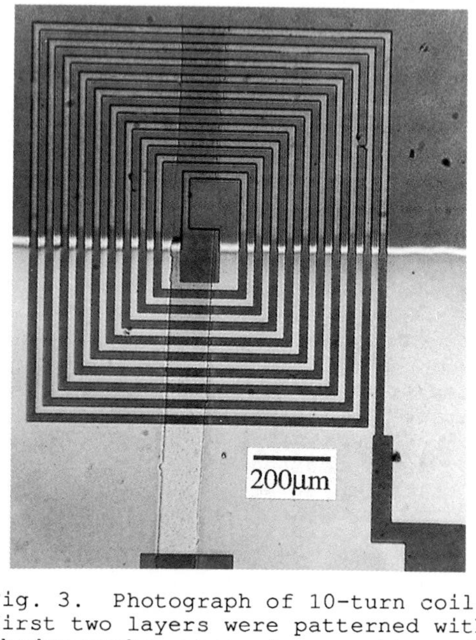

Fig. 3. Photograph of 10-turn coil; first two layers were patterned with shadow masks, upper layer patterned photolithographically [34].

Although shadow masks may be satisfactory for small numbers of devices with not-too-narrow linewidths, it is clear that a process in which each layer is patterned photolithographically is preferable for most applications. A suitable process has been developed by Kingston *et al.* [24]. There are two particular problems to be overcome when one replaces shadow masks with photolithographic patterning. First, while a shadow mask produces a line with rounded edges, photolithography followed by an etch tends to yield sharp edges that are more difficult to insulate. Second, exposure of YBCO films to photoresist leaves a layer of contamination that often prevents the epitaxial growth of subsequent layers. Both problems can be ameliorated by etching the surface in a 2% solution of Br in methanol. Figure 2 shows a window contact and an insulating crossover. After the first YBCO film is patterned, it is etched in the Br solution prior to the deposition of the $SrTiO_3$ layer. The window in the $SrTiO_3$ is patterned by means of an ion mill and a resist mask. An important consideration is the need to bevel the edges of the insulator to support the epitaxial growth of the upper YBCO film. The bevel was obtained by exposing the resist somewhat out-of-focus and subsequently ion-milling the sample. No Br etch is used prior to the deposition of the upper YBCO film; the ion mill apparently leaves a sufficiently clean, albeit somewhat damaged, YBCO surface. The crossover is made by a similar process to that used for the contact, with the omission of the patterning of the window.

It is evident that the basic technology for interconnects now exists. However, for some applications, the dielectric constant of $SrTiO_3$ may be far too high and other insulators may be required; to this end, Kingston *et al.* [25] have produced satisfactory crossovers using YSZ.

DC SQUIDS

Using the junction and interconnect technology described above, a number of groups have begun to produce more complicated structures. The first of these was a dc SQUID, made by Koch *et al.* [4] from YBCO, which relied on grain boundary junctions. The device was in the form of a square washer with an inner hole of side $d \approx 50\mu m$ with an estimated inductance $L = 1.25\mu_o d \approx 60pH$. Subsequently, numerous other workers [26] have described grain boundary SQUIDs made from YBCO, and similar devices made from TCBCO [27, 28] and BSCCO [29] have also been reported. Recently, SQUIDs with more controlled junctions have been fabricated, notably grain boundary junctions on bicrystals [30] and edge junctions [19, 20]. Although it is not possible to review the properties of all these devices here, they do share one common feature, namely a high level of 1/f noise at low frequencies. The conventional figure of merit is the intrinsic noise energy per unit bandwidth $\varepsilon(f) = S_\Phi(f)/2L$, where $S_\Phi(f)$ is the spectral density of the flux noise. The quietest YBCO SQUID reported appears to be that of Gross *et al.*, [30] which had an estimated inductance of 60pH. At 77K, the noise energy was about 1.2×10^{-28} JHz^{-1} at 10Hz, improving approximately as 1/f to about 1.2×10^{-30} JHz^{-1} at 10kHz. The latter noise energy is the best yet reported for a SQUID at 77K and is close to the value predicted for the white noise $\varepsilon \approx 9k_BTL/R$ [31], where R is the resistance of each junction. However, at 77K, a TCBCO SQUID [27] with an estimated inductance of 80 pH was quieter at low frequencies, with a noise energy of about 5×10^{-29} JHz^{-1} at 10Hz; the noise energy fell off more slowly than 1/f, reaching about 10^{-29} JHz^{-1} at 1kHz.

At frequencies above 100 Hz, the noise energy obtained in the best devices is better than that in some commercially available Nb rf SQUIDs operated at 4.2K. However, the high level of 1/f noise remains a serious limitation at lower frequencies. Koch and coworkers have reported that this noise is not reduced by standard double modulation techniques [32], yet it appears not to arise from the motion of flux lines in the body of the SQUID. It is to be hoped that greater insight into this problem will be gained in the near future.

FLUX TRANSFORMERS AND MAGNETOMETERS

Although, as we have seen, progress has been made with thin film dc SQUIDs, it is important to realize that these devices are not particularly sensitive as magnetometers. The reason is simple: the need for low inductance reduces the area of the hole in the SQUID to typically $2500\mu m^2$. Even with a flux focussing factor of (say) 5, a flux noise $S_\Phi^{1/2}(1Hz)$ of 10^{-4} Φ_o $Hz^{-1/2}$ corresponds to a magnetic field sensitivity of 16 pT $Hz^{-1/2}$, a value that is not competitive with flux gate magnetometers. Thus, to achieve a high magnetic field sensitivity, it is essential to couple the SQUID to a superconducting flux transformer. These transformers consist of a multiturn input coil with inductance L_i, tightly coupled to the SQUID, connected to a pickup loop with a much larger area and inductance $L_p \approx L_i$. A single-turn input coil, for example that described by Oh *et al.*, [33] yields only a modest improvement in magnetic field sensitivity because L_p and L_i are not matched.

Multiple-turn coils and flux transformers have been made and tested by Wellstood and coworkers [34-36, 24],using their interconnect technology described above. In the early coils [34], the first two layers were patterned with shadow masks and the third layer with photolithography and an Ar ion mill etch. A photograph of a completed coil is shown in Fig. 3. The first layer was a YBCO strip, about 100μm wide, stretching from one side of the chip to the other. The second layer, $SrTiO_3$, covered approximately 2mm of the length of the YBCO strip. The entire chip was then covered with YBCO, which was subsequently patterned to produce either a 10-turn or a 19-turn coil. In the patterning process, parts of the YBCO strip between the turns of the coil were removed so that the turns were not shorted; however, a portion of this strip provided a superconducting connection between the pickup coil and the innermost turn of the input coil. The coil was approximately 1mm across, and the best sample had a transition temperature of 82K and a critical current at 77K of 1.4mA, corresponding to a critical current density of 2×10^4 Acm^{-2} in the turns of the coil. Subsequently, Kingston *et al.* [24] fabricated smaller coils 0.25mm across, in which all three layers were patterned photolithographically in the configuration of Fig. 4(a). A window contact was made between the first YBCO layer and the innermost turn of the coil, and between the outer end of the first YBCO film and one end of the YBCO pickup loop. A photograph of a completed coil appears in Fig. 5. The best of these microfabricated coils had a transition temperature of 87K and a critical current of 135μA at 82K (5μV measurement criterion).

Both large and small coils have been integrated into flux transformers by patterning the last layer to contain a 10 x $10mm^2$ pickup loop as well as the input coil [see Fig. 4(b)]. The performance of the flux transformers was tested by coupling the input coil to a SQUID. In the first measurements, a

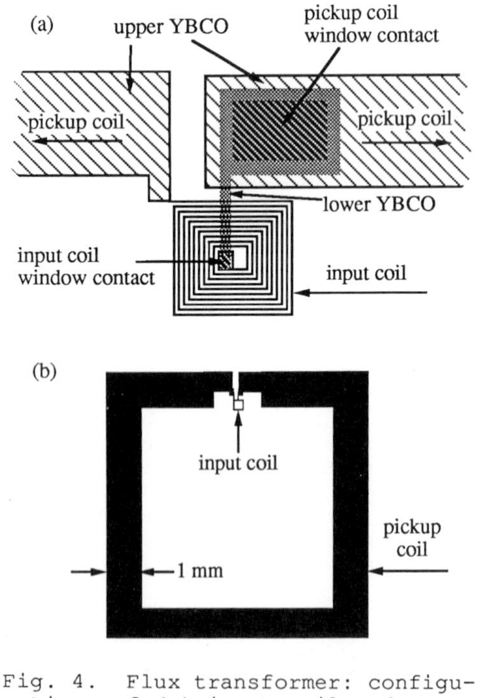

Fig. 4. Flux transformer: configurations of (a) input coil and connections to pickup loop, and (b) pickup loop (not to scale) [24].

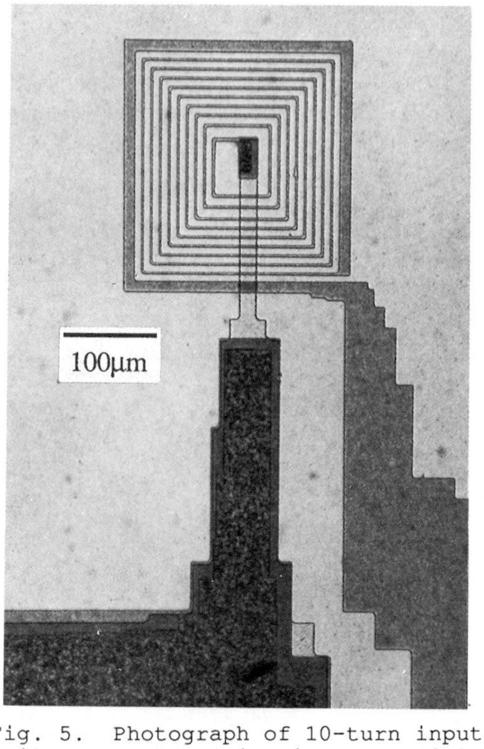

Fig. 5. Photograph of 10-turn input coil patterned as in Fig. 4(a) [24].

large transformer was coupled to a Nb-based SQUID [35, 36]. The magnetic field response was enhanced over that of the bare SQUID by a factor of 9.5. The noise was limited by 1/f noise in the transformer; the measured noise referred to the pickup loop was 0.9pT $Hz^{-1/2}$ at 1Hz and 60K.

Subsequently, both the large and small transformers were coupled to a TCBCO SQUID [28]. The large transformer had a gain of 7.5 and operated at temperatures up to 79K. The small transformer had a gain of 8.7 and ceased to operate at 25K. However, above 25K, it was found that the transformer contained a small resistance (~0.4 mΩ) which probably arose from a localized defect. The measured field sensitivities are shown in Fig. 6. In each case, external noise was observed, indicating that shielding of external magnetic field fluctuations was inadequate. The magnetic field sensitivities with the large transformer at 38K and the small transformer at 4.2K were very similar, about 3pT $Hz^{-1/2}$ at 10Hz and 0.35pT $Hz^{-1/2}$ at 1kHz. In both cases, the sensitivity was limited by noise in the SQUID; a quieter high-T_c SQUID should produce an improvement in the magnetic field sensitivity. If such an improvement can indeed be achieved, high-T_c magnetometers may well find their way into applications within one or two years.

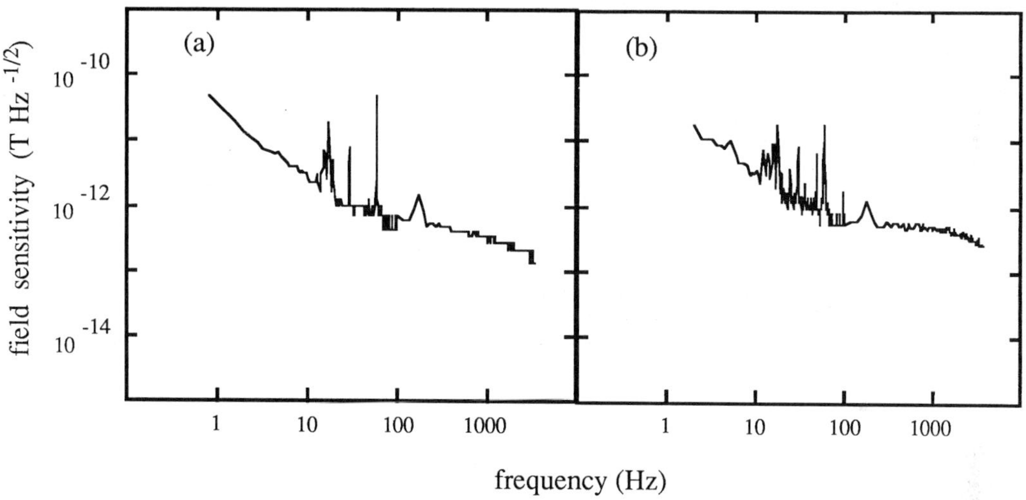

Fig.6 Field sensitivities B_N (f) of the magnetometer with (a) large flux transformer at 38 K and (b) small transformer at 4.2 K. The measured flux-locked bandwidths were ~ 3.5 kHz and 3.8 kHz respectively [28].

ACKNOWLEDGMENTS

We thank A.H. Cardona, L.C. Bourne, W.L. Olson and M.M. Eddy of Superconductor Technologies, Inc. for providing the TCBCO SQUID used in our own work. A.H.M. thanks the Hertz Foundation for a fellowship. The work was supported by Conductus, Inc., the California Competitive Technology Program and by the Director, Office of Energy Research, Office of Basic Energy Sciences, Materials Sciences Division of the U.S. Department of Energy under contract number DE-AC03-76SF00098.

REFERENCES

1. Bednorz JG, Muller KA (1986) Z Phys.B 64: 189
2. Josephson BD (1962) Phys.Lett.1: 251
3. For a review, see, for example, Clarke J (1989) In: Weinstock H, Nisenoff M (eds) Superconducting Electronics. Springer-Verlag, Berlin, p 87
4. Koch RH, Umbach CP, Clark RJ, Chaudhari P, Laibowitz RB (1987) Appl.Phys. Lett.51: 200
5. Russek SE, Lathrop DK, Moeckly BH, Buhrman RA, Shin DH, Silcox J (1990) Appl.Phys.Lett.57: 1155
6. Newman N, private communication

7. McCumber DE (1968) J.Appl.Phys.39: 3113; Stewart WC (1968) Appl.Phys.Lett.12: 277
8. Dimos D, Chaudhari P, Mannhart J, LeGoues FK (1988) Phys.Rev.Lett.61: 219
9. Daly KP, Dozier WD, Burch JF, Coons SB, Hu R, Platt CE, Simon RW (unpublished)
10. Simon RW, Burch JF, Daly KP, Dozier WD, Hu R, Lee AE, Luine JA, Manasevit HM, Platt CE, Schwarzbek SM, St.John D, Wire MS, Zani MJ (1990) Proceedings of the Conference on the Science and Technology of Thin Film Superconductors, Denver, April 1990
11. White AE, Short KT, Dynes RC, Leir AFJ, Anzlowar M, Baldwin KW, Polakos PA, Fulton TA, Dunkleberger LN (1988) Appl.Phys.Lett.53: 1010
12. Zani MJ, Luine JA, Simon RW, Davidheiser RA (unpublished)
13. Cui G, Zhang Y, Hermann K, Buchal Ch, Schubert J, Zander W. Braginski AE, Heiden C (1990) LT-19 Satellite Conference on High Temperature Superconductors, Cambridge, England, August 1990
14. Robbes D. Miklich AH, Kingston JJ, Lerch Ph, Wellstood FC, Clarke J (1990) Appl.Phys.Lett.56: 2240
15. Schwarz DB, Mankiewich PM, Howard RE, Jackel LD, Straughn BL, Burhart EG, Dayem AH (1989) IEEE Trans.Magn.MAG-25: 1298
16. Clarke J (1969) Proc.Roy.Soc.A 208: 447
17. See, for example, Akoh H, Camerlingo C, Takata S (1990) Appl.Phys.Lett.56: 1487; Lee M, Lew D, Eom CB, Geballe TH, Beasley MR (1990) Appl.Phys.Lett.57: 1152
18. Rogers CT, Inam A, Hegde MS, Dutta B, Wu XD, Venkatesan T (1989) Appl.Phys.Lett.55: 2032
19. Laibowitz RB, Koch RH, Gupta A, Koren G, Gallagher WJ, Foglietti V, Oh B, Viggiano JM (1990) Appl.Phys.Lett.56: 686
20. Gao J, Aarnink WAM, Gerritsma GJ, Veldhuis D, Rogalla H (1990) Proceedings of the Applied Superconductivity Conference, Snowmass, Colorado, Sept. 24-28, 1990 (to be published)
21. Dynes RC (1990) Proceedings of LT-19, Brighton, England, August 1990
22. Kingston JJ, Wellstood FC, Lerch Ph, Miklich AH, Clarke J (1990) Appl.Phys.Lett.56: 189
23. Tidjani ME, Gronsky R, Kingston JJ, Wellstood FC, Clarke J (1990) (unpublished)
24. Kingston JJ, Wellstood FC, Du Q, Clarke J (1990) Proceedings of the Applied Superconductivity Conference, Snowmass, Colorado, Sept. 24-28, 1990 (to be published)
25. Kingston JJ, Wellstood FC, Clarke J (unpublished)
26. See, for example, Nakane H, Tarutani Y, Nishino T, Yamada H, Kawabe U (1987) Jap.J.Appl.Phys.26: L1925; Häuser B, Diegel M, Rogalla H (1988) Appl.Phys.Lett.52: 844; Katoh Y, Asahi M, Asano H, Michikami O (1988) Jap.J.Appl.Phys.27: L1110; Takeuchi I, Tsai JS, Tsuge H, Matsukara N (1988) Jap.J. Appl.Phys.27: 2265; Golovashkin AI, Gudkov AL, Krasnosvobadtsev SI, Kuzmin LS, Likharev KK, Maslennikov YV, Pashkin YA, Pechen EV, Snigirev OV (1989) IEEE Trans.Magn.MAG-25: 943; Irie A, Sashara H, Yamashita T, Kurosawa H, Yamane H, Hirai T (1990) Proceedings of the Applied Superconductivity Conference, Snowmass, Colorado, Sept. 24-28, 1990 (to be published)
27. Koch RH, Gallagher WJ, Bumble B, Lee WY (1989) Appl.Phys.Lett.54: 951; Foglietti V, Koch RH, Gallagher WJ, Oh B, Bumble B, Lee WY (1989) Appl.Phys.Lett.54: 2259
28. Miklich AH, Wellstood FC, Kingston JJ, Clarke J, Colclough MS, Cardona AH, Bourne LC, Olson WL, Eddy MM (1990) Proceedings of the Applied Superconductivity Conference, Snowmass, Colorado, September 24-28, 1990 (to be published)
29. Face DW, Graybeal JM, Orlando TP, Rudman DA (1990) Appl.Phys.Lett.56: 1493
30. Gross R. Chaudhari P, Kawasaki M, Ketchen MB, Gupta A (1990) Appl.Phys.Lett.57: 727
31. Tesche CD, Clarke J (1977) J.Low Temp.Phys.29: 301
32. Koch RH, private communication
33. Oh B, Koch RH, Gallagher WJ, Foglietti V, Koren G, Gupta A, Lee WY (1990) Appl.Phys.Lett.56: 2575
34. Wellstood FC, Kingston JJ, Clarke J (1990) Appl.Phys.Lett.56: 2336
35. Wellstood FC, Kingston JJ, Ferrari MJ, Clarke J (in press) Appl.Phys.Lett.
36. Ferrari MJ, Kingston JJ, Wellstood FC, Clarke J (1990) (unpublished)

Flux Pinning in Y-Ba-Cu-O Superconductor Prepared by Melt-Powder-Melt-Growth Process

M. MURAKAMI, T. OYAMA, H. FUJIMOTO, S. GOTOH, K. YAMAGUCHI, T. TAKATA, Y. SHIOHARA, N. KOSHIZUKA, and S. TANAKA

Superconductivity Research Laboratory, International Superconductivity Technology Center, 10-13, Shinonome 1-chome, Koto-ku, Tokyo, 135 Japan

ABSTRACT

The YBaCuO prepared by the Melt-Powder-Melt-Growth (MPMG) process contains fine Y_2BaCuO_5 (211) inclusions embedded in the $YBa_2Cu_3O_x$ matrix. Flux pinning force can be increased with increasing the volume fraction of the 211 inclusions. While MPMG processed $Y_{1.5}Ba_{2.25}Cu_{3.25}O_x$ which contains 25 volume % 211 inclusions exhibits a Jc value exceeding $3x10^4 A/cm^2$ at 77K and 1T, $YBa_2Cu_3O_x$ which contains no 211 exhibit a Jc value of $3x10^3 A/cm^2$ at the same condition. Flux creep can also be reduced significantly by the 211 doping. These results support that the 211 inclusions can work as pinning centers. Theoretical estimation of flux pinning force due to the 211 inclusions was in good agreement with the experimental results. We also present the results of microstructural observation for MPMG processed YBaCuO with an emphasis on structural defects which can work as pinning centers.

KEYWORDS: flux pinning, critical current density, $YBa_2Cu_3O_x$, Y_2BaCuO_5, MPMG

INTRODUCTION

For most applications of superconductors, critical current densities of 10^5 - 10^6 A/cm^2 are required in the presence of significant magnetic fields. Although, Jc values of oxide superconductors had been much lower than the required level, recent developments in processing made it possible to achieve Jc values exceeding 10^4 A/cm^2 at 77K and 1T in bulk YBaCuO and BiSrCaCuO tape[1-4]. The Jc values should be raised by an order of magnitude to find practical application.

It is known that the introduction of structural inhomogeneity which can work as pinning centers is needed to improve Jc values in type II superconductors. However, the identification of pinning centers is still under investigation. Larbalestier et al.[5] have proposed that oxygen defects can act as pinning centers. H. Kupfer et al.[6] also suggested that the pinning center in neutron-irradiated YBaCuO was oxygen deficient region. However, the optimization of oxygen defects is highly difficult and as pointed out by Larbalestier, intragranular weak-links are occasionally introduced in oxygen deficient sample, which is undesirable.

Matsushita[7] and Murakami[8] have proposed that non-superconducting precipitates can work as pinning centers in oxide superconductors, even when their size is much larger than the coherence length. Although, it is apparent that Jc values can be increased by doping Y_2BaCuO_5 (211) particles into the $YBa_2Cu_3O_7$ (123) matrix[9], the question whether the 211 itself can work as pinning center or not is still open since the structural defects produced by the 211 doping may be responsible for the improvement in the Jc values[10].

In this paper, we present how the critical current characteristics are affected by the 211 doping. We then propose that the 211 can help increase flux pinning force.

EXPERIMENTAL

The YBaCuO samples were prepared by the Melt-Powder-Melt-Growth (MPMG) process whose details are reported elsewhere[11-13]. Here, the special care was paid to the oxygenation so that the oxygen content of the 123 matrix is close to 7.0, since oxygen defects affect flux pinning significantly as pointed out by Kupfer and Larbalestier. Using the MPMG process, it is possible to change the volume fraction of the 211 particles.

Physical properties related to flux pinning such as J_c, flux creep, irreversibility line are compared among the MPMG grown YBaCuO crystals with different 211 contents.

Microstructural observation was performed using an optical microscope and a transmission electron microscope.

RESULTS

Comparison of physical properties

Figure 1 shows magnetic field dependence of J_c for three MPMG grown YBaCuO crystals with different 211 contents at 77K. The J_c values were estimated from magnetization hysteresis based on the critical state model. It is apparent that the J_c values can be improved by increasing the volume fraction of the 211.

Figure 2 shows time decay of magnetization hysteresis for two different MPMG grown YBaCuO crystals with and without the 211. Flux creep rate can be reduced by the 211 doping.

Figure 3 shows the irreversibility line for YBaCuO single crystal and MPMG grown YBaCuO containing the 211 inclusions when the magnetic field was applied parallel to the c axis[14]. The line was determined from the peak of imaginary component of AC susceptibility in the presence of DC magnetic field. The irreversibility line can be shifted toward higher H-T region by the 211 doping.

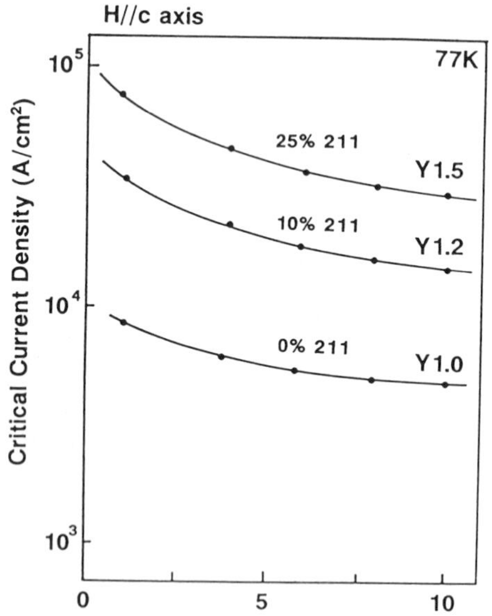

Fig. 1. Magnetic field dependence of J_c for three different MPMG grown YBaCuO crystals with different 211 contents.

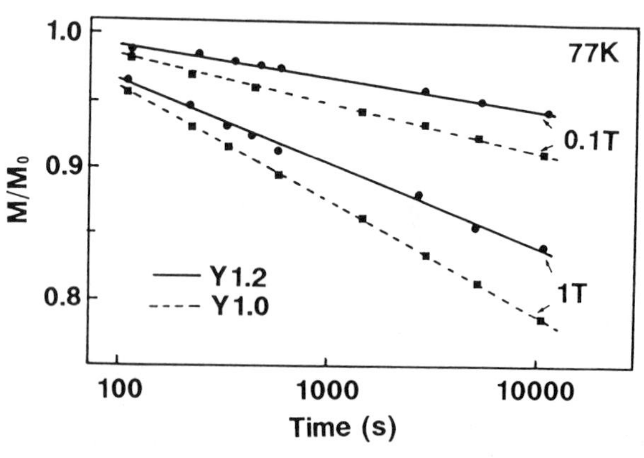

Fig. 2. Time decay of magnetization hysteresis for two different MPMG grown YBaCuO crystals with and without 211.

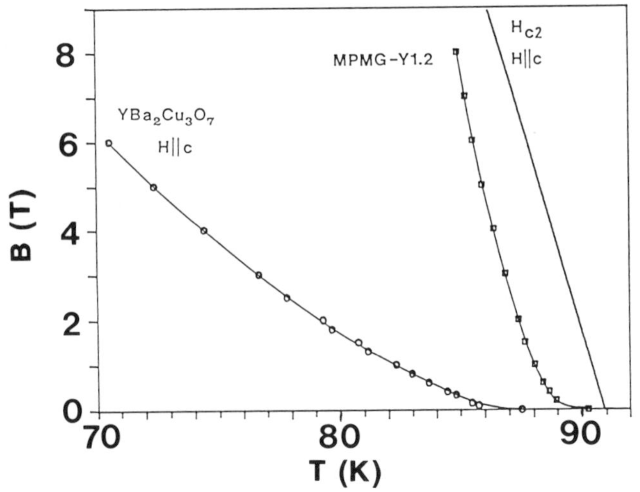

Fig. 3. The irreversibility lines for YBaCuO single crystal and MPMG grown YBaCuO with 211.

Microstructure of MPMG grown YBaCuO crystals[15)]

Figure 4 shows transmission electron micrographs for MPMG grown YBaCuO crystals with and without the 211 particles. The structure was observed from (001) direction. Dense twin structure is observed in both samples. It is also notable that the matrix in the vicinity of the 211 particle is not heavily distorted. Figure 5 is a higher resolution electron micrograph of the interface between the 211 and the 123 matrix. The defect-free 123 lattice can be clearly observed, indicating that the density of defects is not markedly enhanced by the 211 doping.

However, the above observation is not suitable for observing the stacking defects along the c axis. Therefore, the observation was also performed from the direction perpendicular to the c axis. As shown in Fig. 6, there exists a number of stacking faults in the both samples with and without the 211 inclusions. The density of the stacking faults did not vary appreciably between these two samples. We believe that such stacking faults are created during solidification process since they can be observed regardless of the presence of the 211.

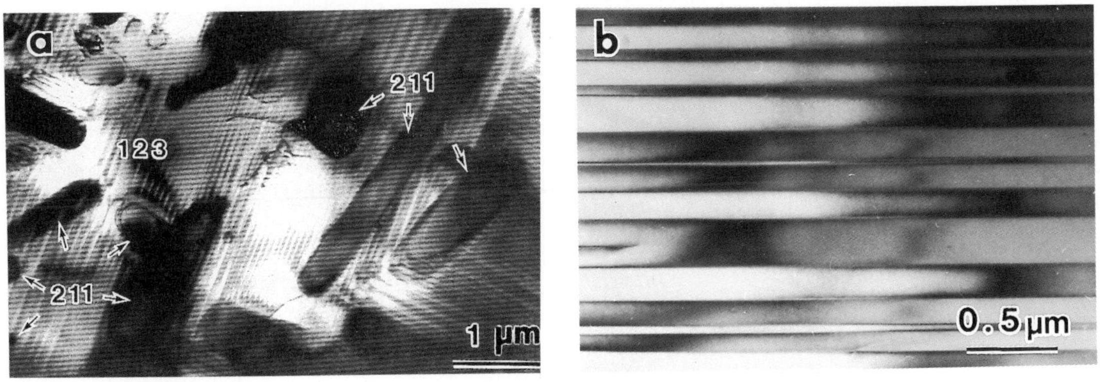

Fig. 4. Transmission electron micrographs for MPMG grown YBaCuO crystals (a) with and (b) without 211.

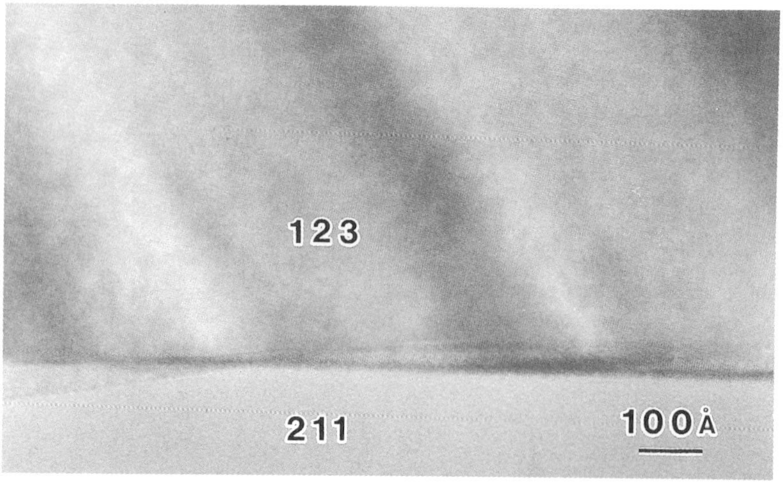

Fig. 5. High resolution transmission electron micrograph of the interface between the 211 and the 123 matrix.

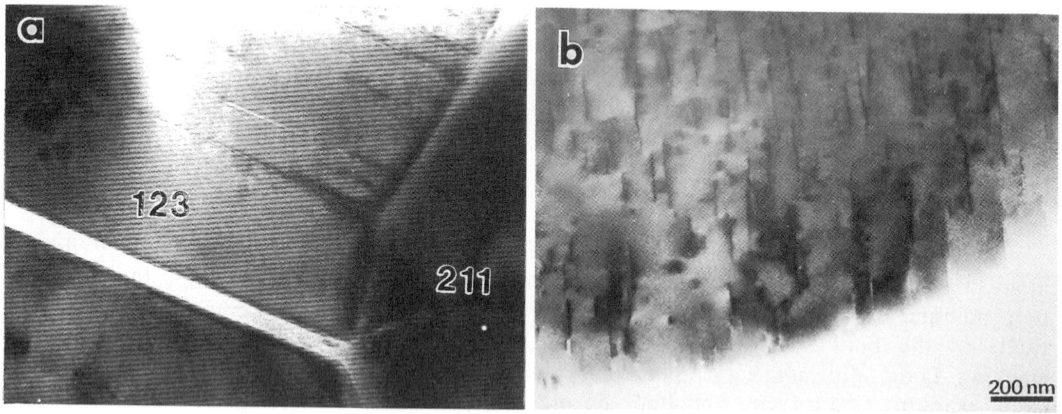

Fig. 6. Transmission electron micrographs for MPMG grown YBaCuO crystals (a) with and (b) without 211 observed from the direction perpendicular to the c axis.

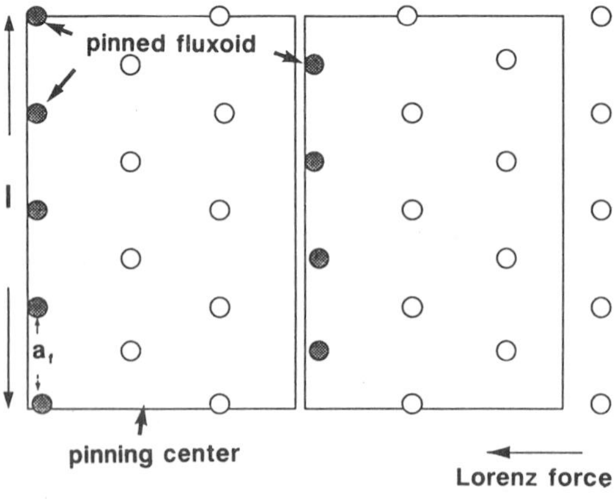

Fig. 7. Schematic illustration of the interaction between flux line lattice and a large non-superconducting particle.

DISCUSSION

All the above observations support that the dispersion of 211 inclusions into the 123 matrix can contribute to the increase in Jc. However, there remains the question whether the 211 itself can contribute to flux pinning or the structural defects produced by the 211 doping are responsible for the improvement in flux pinning force.

According to the microscopic observation, no appreciable change was observed in the density of defects in the matrix by the 211 doping. The YBaCuO crystals grown by the MPMG process contains stacking faults along the (001) plane whether they contain the 211 or not. The relatively large J_c values in the sample without the 211 will be explained by the presence of these stacking faults. It may also be concluded that the improvement in J_c values after the 211 doping is mainly attributable to the flux pinning due to the 211 inclusions themselves. However, as pointed out by many researchers[10,16], the size of the 211 is much larger than that of the coherence length, which will reduce the effectiveness of the 211 inclusions as the pinning centers. On the contrary, it has been reported that the large precipitates can work as pinning centers in conventional superconductors[17].

Generally speaking, any inhomogeneity can work as a pinning center. The problem is how effective it is. In the following discussion we estimate how large J_c values can be expected by the 211 inclusions observed in the MPMG grown crystals.

Estimation of the pinning force

The elementary pinning energy U_p, which is the energy gained per unit interaction between the fluxoid and the non-superconducting particle of the 211 is given by

$$U_p = (H_c^2/8\pi)\,\pi\xi^2 l$$

where H_c is the thermodynamical critical field, ξ is the Ginzburg-Landau coherence length, and l is the size of the 211. The elementary pinning force f_p is then obtained by

$$f_p = U_p/2\xi = (H_c^2/16)\,\xi l$$

In order to obtain the bulk pinning force per unit volume F_p, we need to take account of the density and size of the 211 and the elastic constant of the flux line lattice (FLL). It is known that in oxide superconductors, the penetration depth is relatively large and the coherence length is fairly small. This indicates that the FLL of oxide superconductors is soft. In such case, each fluxoid can be placed at the energy minimum or the pinning site, and therefore the direct summation becomes a good approximation for obtaining F_p and F_p can be obtained as

$$F_p = N(l/a_f)\, f_p$$

where N is the density of the 211 inclusions and a_f is the flux lattice spacing. l/a_f is the number of fluxoids pinned by the 211 in size l (see Fig. 7). J_c is obtained by the relation $F_p = J_c H/10$ (in practical units, e. g. J_c in A/cm^2, H in Oe and F_p in dyn), therefore

$$J_c = 10F_p/H$$

Now, if we know the average size and the density of the 211, the increase in J_c due to the 211 can be estimated and tabulated in Table 1. According to the optical microscopic observation, l is in the range 1 to 2.5 microns and the volume fraction (Nl^3) varies from 0 to 0.25. The calculated values are in good agreement with the experimental values.

Table 1. Calculated Jc values based on the flux pinning due to the 211 inclusions.

volume fraction of 211	diameter (μm)	critical current density (10^4A/cm^2, 77K) B (T)	0.2	0.4	0.6	0.8	1.0
0.10	2.5		2.2	1.6	1.3	1.1	1.0
0.25	2.5		5.6	4.0	3.2	2.8	2.5
0.4	2.0		8.9	6.3	5.2	4.5	4.0
0.1	1.0		5.6	4.0	3.2	2.8	2.5

The Jc value in the sample without the 211 is attributable to the presence of stacking faults. In the samples containing the 211, the increase in Jc values due to the 211 is added.

In conclusion, we can attribute the observed Jc values to the flux pinning due to stacking faults and the 211.

CONCLUSIONS

MPMG grown YBaCuO crystals contain stacking faults regardless of the presence of the 211. Therefore Jc values in the sample without the 211 is attributable to the stacking faults. The increase in Jc values due to the 211 doping is then attributed to the flux pinning by the 211 inclusions themselves.

REFERENCES

1) S. Jin, T. H. Tiefel, R. C. Sherwood, M. E. Davis, R. B. van Dover, G. W. Kammlott, R. A. Fastnacht and H. D. Keith: *Appl. Phys. Lett.* **52** (1988) 2074.
2) K. Salama, V. Selvamanickam, L. Gao and K. Sun: *Appl. Phys. Lett.* **54** (1989) 2352.
3) M. Murakami, M. Morita, K. Doi and K. Miyamoto: *Jpn. J. Appl. Phys.* **28** (1989) 1189.
4) K. Sato, T. Hikita, H. Mukai, T. Masuda, M. Ueyama, H. Hitotsuyanagi, T. Mitusui and M. Kawashima: *Proc. ISS 89* (Springer-Verlag, Tokyo 1990) p. 335.
5) M. Daeumling, J. M. Seuntjens and D. C. Larabalestier: *Nature* **346** (1990) 332.
6) H. Kupfer et al.: Paper presented at ASC 90, to appear in *IEEE Trans. Mag-27* (1991).
7) T. Matsushita, B. Ni, M. Murakami, M. Morita, K. Miyamoto, K. Sawano and S. Matsuda: *Proc. MT-11* (Elsevier, London, 1990) p. 1460.
8) M. Murakami, S. Gotoh, H. Fujimoto, N. Koshizuka and S. Tanaka: Proc. LT 19 Satellite Conf. to appear in *Superconductor Science and Technology*(1991).
9) M. Murakami, S. Gotoh, N. Koshizuka, S. Tanaka, T. Matsushita, S. Kambe and K. Kitazawa: *Cryogenics* **30** (1990) 390.
10) S. H. Wang, Z. X. Li, D. X. Pang, M. Suenaga, D. O. Welch, S. Jin and T. H. Tiefel: *Physica C* **168** (1990) 185.
11) H. Fujimoto, M. Murakami, S. Gotoh, N. Koshizuka, T. Oyama, Y. Shiohara and S. Tanaka: *Proc. ISS 89* (Springer-Verlag, Tokyo, 1990) p. 285.
12) M. Murakami: *Mod. Phys. Lett. B* **4** (1990) 163.
13) M. Murakami, T. Oyama, H. Fujimoto, S. Gotoh, Y. Shiohara, N. Koshizuka and S. Tanaka: to appear in *Proc. ICMC 90.*
14) L T. Sagdahl, T. Lagreid, K. Fossheim, M. Murakami, H. Fujimoto, S. Gotoh, K. Yamaguchi, H. Yamauchi, N. Koshizuka and S. Tanaka: submitted to *Physica C.*
15) K. Yamaguchi, M. Murakami, H. Fujimoto, S. Gotoh, N. Koshizuka and S. Tanaka: *Jpn. J. Appl. Phys.* **29** (1990) L1428.
16) D. C. Larbalestier et al.: Paper presented at ASC 90, to appear in IEEE Trans. Mag-27 (1991).
17) R. I. Coote, J. E. Evetts and A. E. Campbell: *Canad. J. Phys.* **50** (1972) 421.

Electronic Raman Scattering in High Temperature Superconductors

M.V. Klein, F. Slakey, D. Reznik, J.P. Rice, and D.M. Ginsberg

Department of Physics, Materials Research Laboratory, and Science and Technology Center for Superconductivity, University of Illinois at Urbana-Champaign, 104 S. Goodwin Ave., Urbana IL 61801, USA

ABSTRACT

Electronic Raman scattering (ERS) in the high temperature superconductors is dominated by a broad continuum that does not fit any of the conventional effects in metals. We review the normal state properties of this continuum in Yttrium-Barium-Copper-Oxide (YBCO) and Bismuth-Strontium-Calcium-Copper-Oxide (BSCCO) and present recent results on YBCO from 0-1000 cm^{-1}. The results are conveniently parametrized using a modified relaxational form suggested by marginal Fermi liquid theory. At low temperatures the continuum in T_c = 90K YBCO redistributes into a broad maximum suggestive of a 2Δ gap feature. In T_c = 60K YBCO, this feature exists above T_c, perhaps indicating precursor fluctuations of a pairing nature.

KEY WORDS: Gap, light scattering, continuum

INTRODUCTION

In traditional metals and in semiconductors doped to be metallic electronic Raman scattering (ERS) reveals excitations due to: (1) Clean-limit kinematics [For temperature $T \ll T_{Fermi}$ $R''(\omega)$, the imaginary part of the response function rises linearly from $\omega = 0$ to a cutoff near $\omega = qv_{Fermi}$, where the wave-vector q is approximately given by the reciprocal of the optical penetration depth.]. (2) Impurity scattering [$R''(\omega)$ takes the relaxational form proportional to $\omega\Gamma/(\omega^2 + \Gamma^2)$.]. (3) Bulk plasmons [$R''(\omega)$ has a peak at the plasma frequency.]. (4) Crystal-field excitations due to localized f-electrons. (5) Magnetic excitations such as magnons. ERS in the high temperature superconductors is dominated by a broad continuum that does not fit any of the above cases. Here we will cite evidence for this continuum over a broad energy range and will present recent results from our group on its behavior for values of Raman frequency shift ω of almost 0 to 1000 cm^{-1} in both normal and superconducting states of $YBa_2Cu_3O_{7-\delta}$ (YBCO).

OVERVIEW OF CONTINUUM

To our knowledge the first reference to the continuum using Raman scattering was by Bozovic et al.[1], who found that it extended from $\omega = 0$ to beyond 4000 cm^{-1} (1/2 eV) in a YBCO film. That the continuum might be universal in the cuprate hole superconductors (and even extend to the non-cuprate $Ba_xK_{1-x}BiO_3$) can be seen dramatically in the work of Sugai, Enomoto, and Murakami.[2] These authors contrast the nearly-constant ω-dependence of the continuum in YBCO, $Bi_2Sr_2CaCu_2O_{8+y}$ (BSCCO) and $La_{2-x}Sr_xCuO_4$ with the strong, broad peak due to two-magnon Raman scattering seen in the anti-ferromagnetic insulator "parent" phase. Data showing that the continuum extends beyond 8000 cm^{-1} have been recently published by Staufer, Hackl, and Müller [3]. They point out that the measured photon-scattering cross-section, corrected for instrumental response, should be multiplied by a factor $[\omega_L/(\omega_L - \omega)]$, where ω_L is the frequency of the laser photon, in order to give a quantity proportional to $S(\omega)$, the Raman spectral density. The factor in square brackets corrects the fall-off seen in the Sugai et al. results at larger ω; the result is that $S(\omega)$ is essentially constant from 0 to 8000 cm^{-1}.

Recognizing that the constancy of the continuum did not fit any of the ordinary-metal behaviors sketched in the Introduction, Varma et al. [4], conjectured that the carriers in the cuprate superconductors formed a "marginal Fermi liquid" (MFL). In a MFL, the quasi-particle self-energy $\Sigma(q,\omega)$ is proportional to ω for small ω, instead of to ω^2, as in an ordinary Fermi liquid, and two-particle polarizations or response functions $R(\omega,T)$ can be expressed in relaxational form with a damping factor Γ approximately proportional to the greater of ω and T. Such a response was suggested earlier by Collins et al., [5] in their analysis of infrared results, and, at their suggestion, by our group [6] as a way of describing the Raman results.

NORMAL-STATE RESPONSE OF 90K YBCO

Since the twinned nature of ordinary single crystals can introduce strains and complicate analysis of spectra, we have chosen untwinned single crystals of YBCO[7] for the Raman work to be described here[8]. Appropriate annealing treatments produced T_c's either close to 90K or close to 60K. Raman spectra were taken in the pseudo-backscattering geometry with incident photon polarization essentially along [110] and scattered photon polarization either along [110] for A_{1g} symmetry or along [110] for B_{1g} symmetry. Here, we assume that tetragonal Raman selection rules apply. This has been shown to be approximately the case, especially for the continuum. Laser heating was minimized by using low power densities (10W/cm^2); temperatures were verified by using Stokes to anti-Stokes ratios of the spectral densities. We present our results for $R''(\omega)$, the imaginary part of the response function. According to the fluctuation-dissipation theorem, $R'' = S(\omega)\,[1 - \exp(-\omega/T)]$, when ω and T are expressed in the same units. Thus $R'' \approx \omega S/T$ for small ω/T, and $R'' \approx S$ for large ω/T. $R''(\omega)$ must be an odd function of ω; ordinarily it is proportional to ω for small ω.

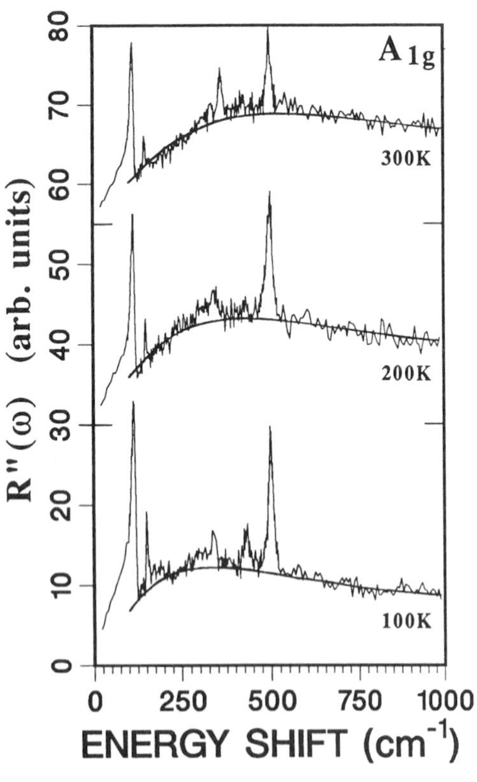

Fig. 1. Imaginary part of response function in A_{1g} symmetry at 100, 200, and 300 K. Zeros of upper curves are at tic marks.

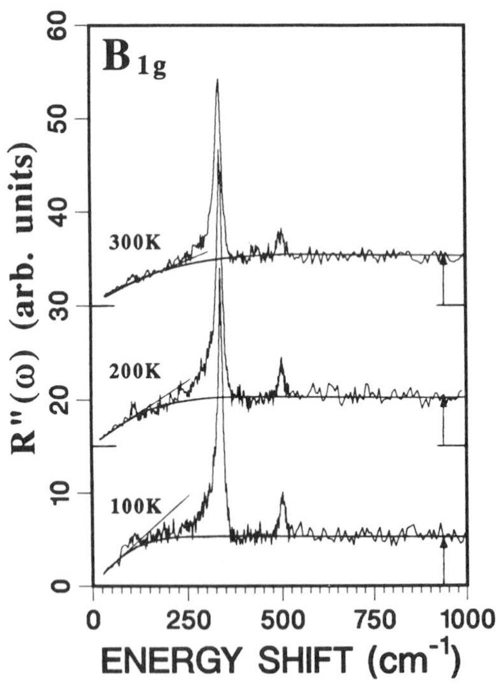

Fig. 2. Imaginary part of response function in B_{1g} symmetry at 100, 200, and 300 K. Zeros of upper curves are at tic marks.

Results for the normal state of the 90K sample are shown in Figs. 1 and 2. One notes that there are several sharp phonon features, on top of which is the electronic continuum. Note that the B_{1g} spectra rise linearly to essentially a constant value. The initial slopes are almost exactly proportional to $1/T$. This follows from the independence of S on ω and T and is consistent with MFL theory. The A_{1g} spectra have broad maxima in the 300-500 cm^{-1} region and hence appear to be qualitatively different from the B_{1g} spectra. These differences may be quantitative rather than qualitative, as all spectra may be fit to the same general expression (smooth solid lines) of modified relaxational form:

$$R'' \propto \omega\Gamma/(\omega^2 + \Gamma^2), \text{ with } \Gamma = \Gamma(\omega,T) = [(\alpha\omega)^2 + (\beta T)^2]^{1/2} \qquad (1)$$

The parameters α and β are of order unity when Γ, ω, and T are expressed in the same units. They are still weakly dependent on temperature.[8] Equation (1) is qualitatively consistent with MFL theory.

The success of the fits is evidence that Eq. (1) provides a convenient parametrization, and is not necessarily a "proof" of MFL theory. As evidence that the values for the scattering or damping rate Γ may have additional physical significance, we have used them to calculate the electrical resistivity according to

$$\rho = 4\pi\Gamma(0,T)/\omega_p^2, \qquad (2)$$

where ω_p is the screened plasma frequency, assumed equal to 1.4 eV.[9] The results are shown in Fig. 3. Different weighting of anisotropic scattering rates might lead to different numerical values for the Γ's appearing in different Raman symmetries or in transport. In particular, the Raman spectra might sample some excitations associated with localized states that would not contribute to transport.[10]

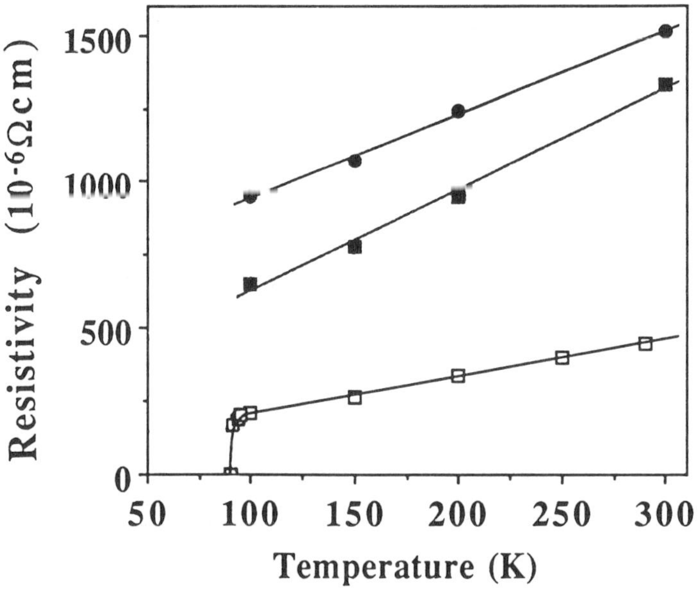

Fig. 3. Open squares: measured resistivity for single crystal of T_c = 90K YBCO. Solid circles and squares represent calculated resistivities using $\Gamma(0,T)$ from fits to A_{1g} and B_{1g} spectra, respectively, and using a value of 1.4 eV for the screened plasma frequency.

RESPONSE OF 90K YBCO IN SUPERCONDUCTING STATE

As the temperature T is lowered below T_c , samples of YBCO with $T_c \approx 90K$ show the flat continuum [in S(ω)] redistributing into gap-like peaks that might be interpreted as giving a measure of 2Δ. A linear-in-ω region persists below these peaks. The spectra differ in peak position when viewed in tetragonal A_{1g} or B_{1g} scattering geometry.[11,12] Recent results on untwinned single crystals are shown in Figs. 4 and 5. The maximum at low temperature in the A_{1g} continuum is at about 350 cm^{-1}; that in the B_{1g} continuum is at about 530 cm^{-1}. If these peaks represent values of 2Δ, then the corresponding values of $2\Delta/k_BT_c$ are about 5.6 and 8.5, respectively. Values for this ratio of 5.2 and 4.95 have been obtained by the Stuttgart group by analysis of renormalization of the position[13] and width[14] of the 330 cm^{-1} Raman-active phonon using strong-coupling BCS theory and electron-phonon coupling constants from electronic structure calculations. It should be pointed out that sharp renormalization is shown by other phonons as the temperature is lowered below T_c; a Landau theory can explain this more general phenomenon as well as the renormalization of the frequency of the 330 cm^{-1} mode.[15]

The linear dependence on ω is apparent in Figs. 4 and 5 for small ω, even at the lowest temperatures. This persistence of excitations of energy well below k_BT_c remains puzzling. It is not an artifact of a poor surface layer; it remains after the surface is chemically etched. Such a low energy linear term is also seen at low temperatures in samples of $Bi_2Sr_2CaCu_2O_8$,[16,17] which readily cleave on Bi-O planes, revealing a fresh surface. If this linear dependence indeed were to persist to ω = 0, one would expect a linear term in the low temperature specific heat as a function of temperature. One group of experts believes that experiments show that there is no such term.[18] Other experts believe that there may be a linear term.[19,20]

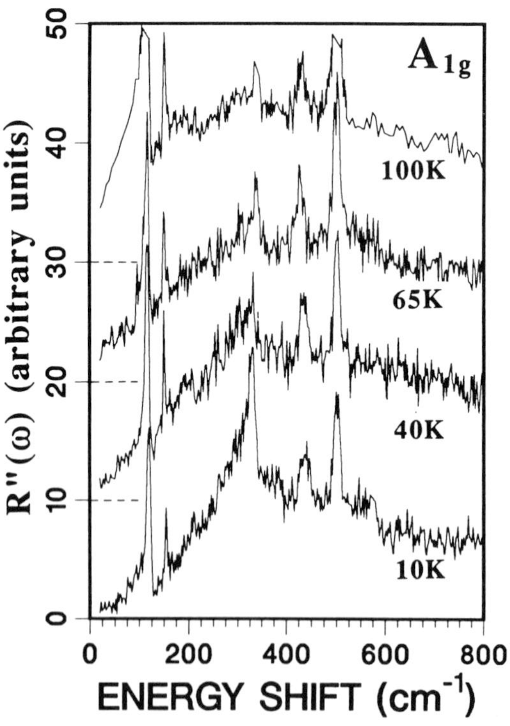

Fig. 4. Imaginary part of response function in A_{1g} symmetry at 100, 65, 40 and 10 K. Zeros of upper curves are at tic marks.

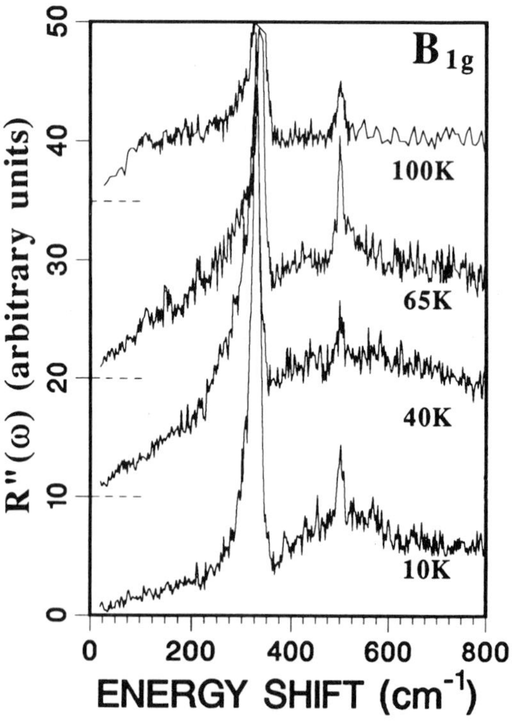

Fig. 5. Imaginary part of response function in B_{1g} symmetry at 100, 65, 40, and 100 K. Zeros of upper curves are at tic marks.

BEHAVIOR OF THE CONTINUUM IN 60K YBCO

We have studied the continuum in an untwinned single crystal of YBCO with $T_c \approx 60K$.[21] At room temperature the spectral function $S(\omega)$ is essentially independent of ω below 800 cm^{-1}. At low temperatures broad gap-like peaks are found, but they do not scale with T_c as one would expect for a BCS superconductor. This can be seen in the 10K data of Fig.6. [At such a low temperature there is essentially no difference between $S(\omega)$ and the response function $R''(\omega)$.] The A_{1g} spectrum has a broad maximum at about 530 cm^{-1}; this should be contrasted with the sharper maximum at 350 cm^{-1}

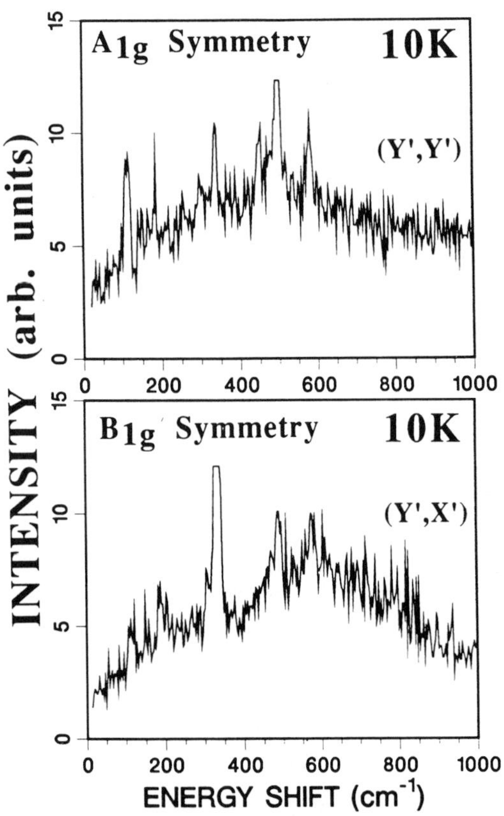

Fig. 6. Spectra in A_{1g} and B_{1g} symmetries of T_c = 60K YBCO. Phonons have been truncated to clarify the background.

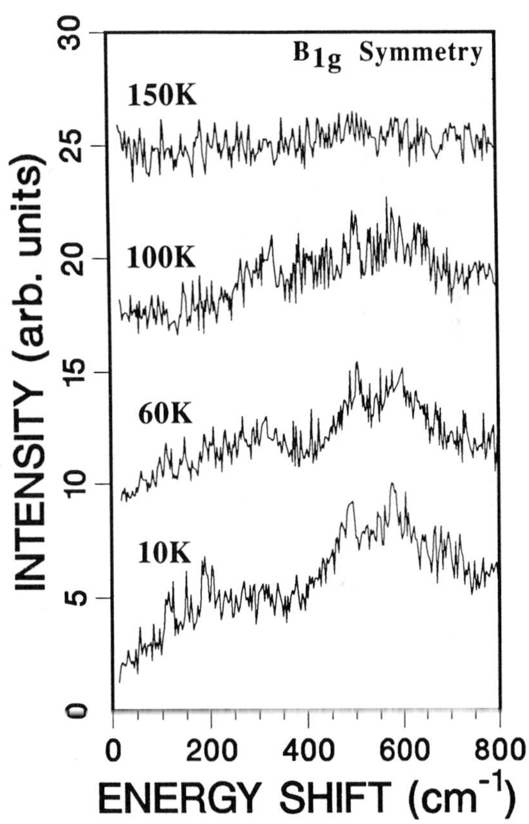

Fig. 7. Temperature-dependent B_{1g} spectra with phonon features subtracted.

of the T_c = 90K sample in Fig. 4. The B_{1g} spectrum has a maximum at the same position and with the same width as that of the T_c = 90K sample in Fig. 5, but there is now a shoulder at 200 cm^{-1}.

Evolution of $S(\omega)$ is shown for the B_{1g} spectra of the 60K sample in Fig. 7. One sees that the broad peak at 550 cm^{-1} starts to form already at 100K and is fully present at T_c. This energy cannot be assigned to the gap in the superconducting state; it must relate to another excitation. A feature near this energy was also found in the normal state of reduced-T_c YBCO using infrared reflectivity measurements.[22] One possibility is that Cooper pairs form already in the normal state and condense into a superfluid at T_c. Such a picture is highly controversial. A great deal of additional data will be needed in order to fully characterize the behavior of gap-like features in reduced-T_c high temperature superconductors.

ACKNOWLEDGMENTS

This work was supported by the National Science Foundation under grants DMR 8715103, 8920538, 9714555 and received general support through the Science and Technology Center for Superconductivity under NSF DMR 8809854.

REFERENCES

1. Bozovic I, Kirillov D, Kapitulnik A, Char K, Hahn MR, Beasley MR, Geballe TH, Kim YH, Heeger AJ (1987) Phys. Rev. Letters 59: 2219
2. Sugai S, Enomoto Y, Murakami T (1989) Solid State Commun. 72: 1193
3. Staufer T, Hackl R, Müller P (1990) Solid State Commun. 75: 975
4. Varma CM, Littlewood PB, Schmitt-Rink S, Abrahams E, Ruckenstein AE (1989) Phys. Rev. Letters 63: 1996
5. Collins RT, Schlesinger Z, Holtzberg F, Chaudari P, Feild C (1989) Phys. Rev. B 39: 6571
6. Klein MV, Cooper SL, Slakey F, Rice JP, Bukowski ED, Ginsberg DM (1989) In: Fukuyama H, Maekawa S, Malozemoff AP (eds) Strong correlations and superconductivity. Springer, Berlin, Heidelberg, p 226
7. Rice JP, Ginsberg DM, Rabin MW, Vandervoort KG, Crabtree GW, Claus H (1990) Phys. Rev. B 41: 6532
8. Slakey F, Klein MV, Rice JP, Ginsberg DM (1990) preprint submitted to Phys. Rev. B
9. Bozovic I (1990) Phys. Rev. B 42: 1969
10. Phillips JC, private communication
11. Hackl R, Gläser W, Müller P, Einzel D, Andres K (1988) Phys. Rev. B 38: 7133
12. Cooper SL, Slakey F, Klein MV, Rice JP, Bukowski ED, Ginsberg DM (1988) Phys. Rev. B 38: 11934
13. Thomsen C, Cardona M, Friedl B, Rodriguiz CO, Mazin EE, Anderson OK (1990) Solid State Commun. 75: 219
14. Friedl B, Thomsen C, Cardona M (1990) Phys. Rev. Letters 65: 915
15. Salje E (1990) Phil. Mag. Letters 62: 277
16. Yamanaka A, Kimura T, Minami F, Inoue K, Takekawa S (1988) Jpn. J. Appl. Phys. 27: L1902
17. Slakey F, Klein MV, Bukowski ED, Ginsberg DM (1990) Phys. Rev. B 41: 2109
18. Phillips NE, Fisher RA, Gordon JE, Kim S, Stacy AM, Crawford MK, McCarron EM (1990) Phys. Rev. Letters 65: 357
19. Ginsberg DM, private communication
20. Loram J, private communication
21. Slakey F, Klein MV, Rice JP, and Ginsberg DM (1990) Phys. Rev. B 42: 2643
22. Orenstein J, Thomas GA, Millis AJ, Cooper SL, Rapkine DH, Timusk T, Schneemeyer LF, Waszczak JV (1990) Phys. Rev. B 42: 6342

Fabrication and Superconducting Properties of $YBa_2Cu_3O_7/PrBa_2Cu_3O_7$ Artificial Superlattices

Jean-Marc Triscone[2], Louis Antognazza[1], Olivier Brunner[1], Laurent Miéville[1], Michael Karkut[3], Thierry Boichat[1], Oliver Eibl[4], and Øystein Fischer[1]

[1] Département de la Matière Condensée, University of Geneva, 24 Quai E.-Ansermet, 1211 Geneva 4, Switzerland
[2] Department of Applied Physics, Stanford University, Stanford CA 94305, USA
[3] Laboratoire de Chimie Minérale B, University of Rennes I, av. du Général Leclerc, Cedex 35042, France
[4] Siemens Research Laboratories, Otto Hahn Ring 6, D-8000 Munich, Germany

ABSTRACT

$YBa_2Cu_3O_7/PrBa_2Cu_3O_7$ (YBCO/PrBCO) superlattices grown by single target dc planar magnetron sputtering are used as a model system to investigate flux motion in high T_c superconductors. In parallel field (parallel to the a-b plane) we find for multilayers containing thin individual YBCO layers (<50Å) separated by PrBCO layers thicker than 24Å that the typical broadening of the resistive transitions is strongly reduced. In perpendicular field we show that the activation energy for flux motion is proportional to the YBCO thickness (for PrBCO thicknesses>24Å). These results allow the estimation of the correlation length for YBCO along the vortices to be about 450 Å .

KEY WORDS: HTS superlattices, epitaxial growth, flux motion.

INTRODUCTION.

The growth of superlattices of High T_c Superconductors (HTS), materials already naturally layered, is motivated by the many possibilities offered by artificial structures. Among these possibilities is the ability to make model systems allowing the physical properties to be investigated in a systematic way.

In the so called "123" material YBCO and PrBCO are particularly interesting superlattice constituents. The reason is that PrBCO has the same structure as YBCO, about the same lattice parameters [1], but unlike YBCO it is neither superconducting nor even conducting[2].

We have shown that sandwiching ultrathin YBCO layers in between insulating PrBCO layers gives access to the properties of ultrathin YBCO layers [3], properties which cannot be easily obtained in single layer studies because of the substrate-film mismatch. In a more recent work[4] we have shown that these superconducting-insulating superlattice systems allow the investigation of the remarkable properties of the mixed state in HTS. More precisely we have shown that the activation energy for flux motion with the magnetic field parallel to the c-axis can be systematically changed by modifying the individual thickness of YBCO and PrBCO in the multilayers.

In this paper we summarize our results on the flux motion in parallel and perpendicular fields for our YBCO/PrBCO multilayer and discuss the implications of these results.

PREPARATION DETAILS AND CHARACTERIZATION.

The preparation details for single films and multilayers can be found respectively in ref. 5, and 6. We use for the deposition of the superlattices a UHV system whose typical base pressure is $<10^{-7}$ Torr. The multilayers are magnetron sputtered onto (100) MgO or $SrTiO_3$ at 650-700°C in a mixture of Ar and O_2 ($Ar+O_2$=300mTorr; O_2=20mTorr). The two magnetron sputtering guns having stoichiometric targets of either PrBCO or YBCO are placed 180° apart in the system and a computer controlled stepping motor positions the samples in front of the desired gun for a certain time. When a total thickness of about 1200Å is reached, the deposition stops and the samples are slowly cooled in O_2[5].

X-ray diffraction studies[6,7] show that this process results in highly oriented c-axis films both on (100) MgO and $SrTiO_3$. These films are oriented both in the c-direction and in the a,b plane. X-ray satellite peaks arising from the artificial modulation are clearly visible and their intensities are consistent with very little mixing between the YBCO and PrBCO layers[7]. We recently carried out transmission electron microscopy (TEM) investigations of the multilayers[8] in order to further characterize the microstructure of the layers. Fig. 1 shows a TEM cross section of a 9-unit cells/ 9 unit cells YBCO/PrBCO film deposited on (100) $SrTiO_3$. Analysis of the TEM pictures show that the interfaces between the YBCO and PrBCO layers are very sharp and no evidence of interdiffusion between the Y and Pr planes could be seen on the TEM images. Additionally the TEM studies show that the (001) planes pass continuously from regions with perfect substrate surfaces to regions containing a larger density of defects. A detailed study can be found in ref.8 .

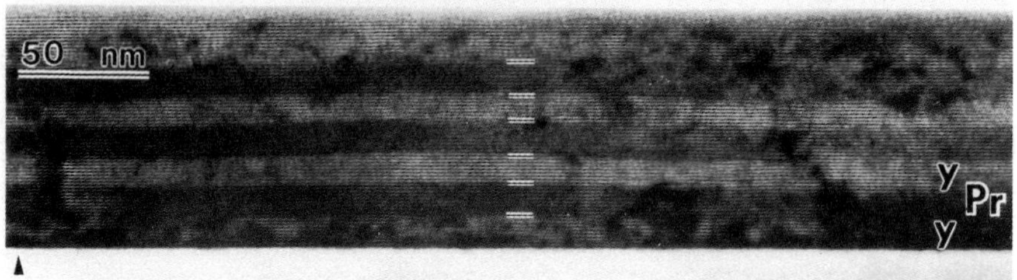

Fig. 1. TEM cross section of a 9-unit cells/ 9 unit cells YBCO/PrBCO film deposited on (100) $SrTiO_3$. The sharp contrast between the YBCO and PrBCO layers suggest very little mixing or interdiffusion at the interfaces.

FLUX MOTION IN YBCO/PrBCO SUPERLATTICES

The T_c behavior of YBCO/PrBCO multilayers as a function of both the YBCO and the PrBCO thickness has been the primary subject of other publications (see for example ref.3 and 9). Here we shall focused our attention on the flux motion in these artificial structures.

In High T_c Superconductors thermal fluctuations play an important role. In the mixed state unusual properties are observed, essentially because of the possibility for single flux line or flux bundle to be thermally activated. Perhaps the most direct manifestation of these unusual properties is the observed broadening of the resistive transition when a magnetic field is applied[10] instead of the shift observed in classical superconductor. In the "visible" part of the transition this broadening corresponds to a flux flow, i.e. a viscous motion of the vortices as in the tail of the transition, say for $\rho < 0.01\rho_n$, the observed behavior is generally understood as a thermally activated flux flow[11] (TAFF). In the latter case each individual flux line is pinned in a potential well of depth U and the measured dissipation is related to the activation of individual flux line.

Fig.2 shows the resistive transitions in a magnetic field parallel and perpendicular to the layers (parallel and perpendicular to the a,b plane) for: a YBCO thin film, a 24Å/12Å , a 24Å/24Å, and a 24Å/144Å YBCO/PrBCO multilayer. The different curves are for fields of 0, 1, 3, 6, and 9T. In parallel field we see that as the insulating PrBCO layer thickness is increased the multilayers become progressively less and less sensitive to the application of a magnetic field. The remaining very small broadening observed for e) and g) could be due to the fact that the layers are slightly wavy[8] and thus a perpendicular component of the magnetic field is always present even when the field is aligned at best. We think that the insensitivity to the magnetic field, which also extends to the tail of the transition[12], reflects the progressive suppression of flux flow in these artificial structures. The

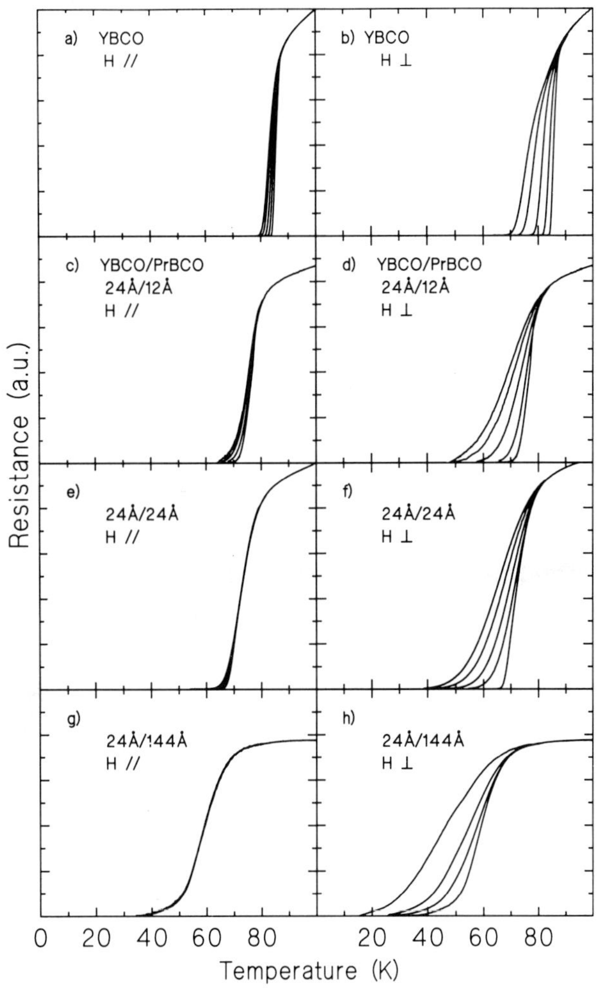

Fig. 2. Resistive transitions in a magnetic field parallel and perpendicular to the layers (parallel and perpendicular to the a,b plane) for a YBCO thin film, a 24Å/12Å , a 24Å/24Å, and a 24Å/144Å YBCO/PrBCO multilayer. The different curves are for fields of 0, 1, 3, 6, and 9T (the 6T curve is missing in h).

reason is that it is energetically more favorable for the magnetic flux to sit in an insulating material. As long as the flux is in the PrBCO layers there will be no normal vortex core in the YBCO layers and thus no conventional flux flow. Because of the large penetration depth there will be probably some shielding currents in the planes but the experiments show that this does not lead to significant dissipation. As the separation between vortex rows is about 60Å at 9T the vortex should start to penetrate the YBCO layers when d-YBCO is >60Å and this is exactly what is observed[13]. Note that to restore a broadening comparable to YBCO a layer thickness of about 200Å is necessary[13].

When the field is applied perpendicular to the a, b plane, the resistive transitions in field display grossly the typical behavior of YBCO, i.e. a broadening of the transition when the field is applied. However, in the tail of the transitions where TAFF is expected to occur important differences are observed[4](these differences are not visible on Fig.2 because of the scale used). Typically for $\rho < 0.01\rho_n$,the resistivity ρ can be written as:

$$\rho(T,H)=\rho_o \exp(-U(T,H)/k_BT)$$

where U is the activation energy for flux motion which is normally both field and temperature dependent.

The standard way to measure the activation energy is to plot the log of the resistivity as a function of 1/T. For $\rho < 0.01\rho_n$ a straight line behavior is expected (if the temperature dependence of U is not too important) whose slope U' is directly related to the activation energy. In principle the relation between U' and U depends on the temperature dependence of U. However, if $U(t)=U(0)(1-t)^q$ ($t=T/T_c$) and q=1, U' is strictly equal to U(0) and in this case the slope of the Arrhenius plots is directly the zero temperature activation energy and no correction is necessary[4]. We have demonstrated in ref. 4 that the temperature dependence of U for these multilayers is $U(t)=U(0)(1-t)^q$ with q very close to 1. This means that the slope of the Arrhenius plots is not very different from U(0) and we did not make any correction for the temperature dependence. Fig. 3 shows Arrhenius plots of the resistivity (logρ as a function of 1/T) for a YBCO thin film (top), a 96Å/96Å multilayer, and (bottom) a 24Å/96Å multilayer. The curves are for different fields 0.5, 1, 3, 6, and 9T. As can be seen immediately the activation energy decreases as the individual YBCO thickness in the multilayer decreases. Additionally, we find that the ratio between the activation energy of the 96Å/96Å and the 24Å/96Å sample is about 4 for each value of the magnetic field.

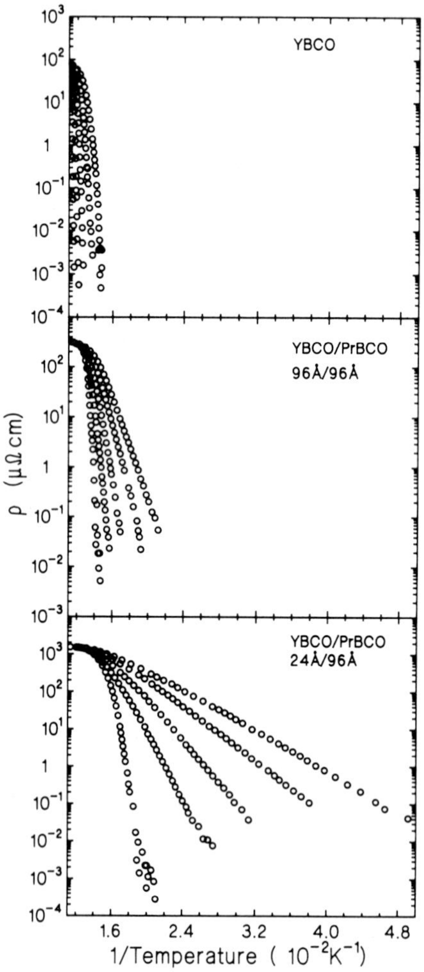

Fig. 3. Logρ as a function of 1/T for a YBCO thin film (top), a 96Å/96Å multilayer (middle), and (bottom) a 24Å/96Å YBCO/PrBCO multilayer. The curves are for different fields 0.5, 1, 3, 6, and 9T.

This result can be understood if we write U as :

$$U=p(\mu oH_c/2)V_c$$

where p is the fraction of the condensation energy density and V_C, the volume of flux involved in the activation process. Now V_C can be written[11] as $V_C=R_C^2L_C$ where R_C^2 is the correlation area in the plane perpendicular to the flux line and L_C the correlation length along the vortex. What is happening is that if the PrBCO layer thickness is large enough to magnetically decouple the YBCO layers (this is so if d-PrBCO>24Å [4]) then we have "pancake" vortices in each YBCO layer and L_C must be limited to d-YBCO. That means that $L_C(96Å/96Å)=4L_C(24Å/96Å)$ and therefore the activation energy must scale with d-YBCO.

Fig. 4 demonstrates this essential result for different multilayers with thick PrBCO layers (either 96Å or 144Å) and various YBCO thicknesses ranging from 24Å to 264Å and for different values of the magnetic field. As can be seen U' scales with d-YBCO for each magnetic field. This result proves that each YBCO layer behaves independently and since the activation energy for pure YBCO is known we can extract L_C for pure YBCO. We find that L_C should be between 400 and 500Å for YBCO.

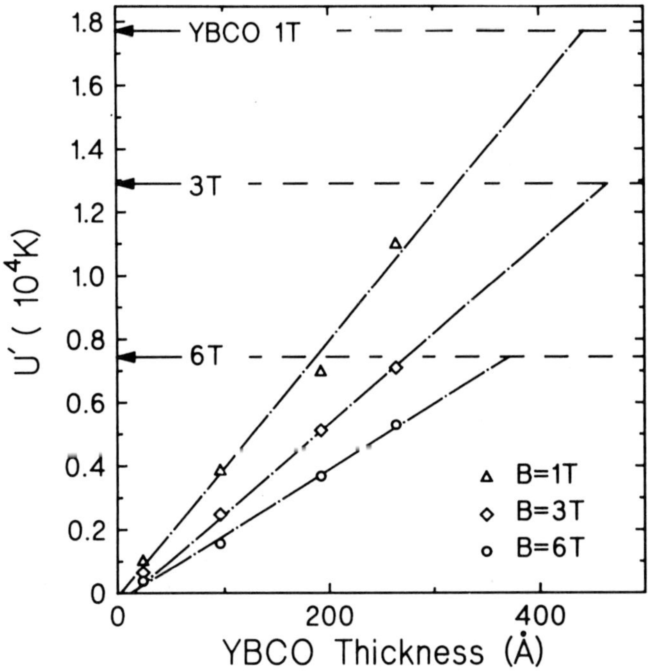

Fig. 4. Activation energy derived from the Arrhenius plots for YBCO/PrBCO multilayers having thick (96 or 144Å) insulating PrBCO layers as a function of the YBCO thickness in the multilayer.

Additionally these results have some implication on the nature of the activation process. If one believes that the activation energy is determined by the pinning of each flux line, these results imply that the pinning centers are very densely and homogenously distributed in the material. In fact since U is proportional to d-YBCO, even for d-YBCO=24Å the implication seems to be that there must be one pinning site per unit cell. If the pinning is due to defects, these results imply a very large number of defects. Another possibility could be that the pinning is due to the twin boundaries which are parallel to the c-axis and would give an activation energy proportional to d-YBCO. A third possibility is that the activation process is related to the formation of dislocation pairs in the flux line lattice. This explanation is supported by the fact that we found that the magnetic field dependence of U' is not a power law but that U'/d-YBCO=-αlogH+β [4] where α and β are constants. Both the log dependence and the value of α and β were predicted by Feigel'man et al. in a theory of collective flux pinning[14] for an activation process related to the formation of dislocation pairs in the flux line lattice. This last idea is discussed in more detail in ref. 4 and these different issues are currently under investigation.

In summary we have shown that high quality YBCO/PrBCO superlattices can be grown by magnetron sputtering and used as a model system to study flux motion in HTS. When the magnetic field is parallel to the a-b plane we find for multilayers containing thin individual YBCO layers (<50Å) separated by thick enough (>24Å) PrBCO layers that the typical broadening of the resistive transitions is strongly reduced showing that the flux flow induced dissipation can be suppressed or considerably decreased in these artificial structures. When the field is applied perpendicular to the a, b plane we show that the activation energy for flux motion is proportional to the YBCO thickness (for YBCO thicknesses<300Å, and PrBCO thicknesses>24Å). These results allow to estimate the correlation length for YBCO along the vortices to be about 450 Å and suggest that there should be at least one pinning center per YBCO unit cell .

ACKNOWLEDGEMENTS

One of us (J.-M.T.) would like to thank Dr. Glen Carey for a careful reading of the manuscript. We also want to thank E. Perreard and A. Stettler for their expert assistance.

REFERENCES

1. See, for instance, Hulliger F., Ott H.R. (1987) Z. Phys. B 67: 291.
2. See, for instance, Dalichaouch Y., Torikachvili M.S., Early E.A, Lee B.W., Seaman C.L., Yang K.N., Zhou H., Maple M.B. (1988) Sol. Statt. Com. 65: 1001.
3. Triscone J.-M., Fischer Ø., Brunner O., Antognazza L., Kent A.D., and Karkut,M.G. (1990) Phys. Rev. Lett. 64: 804 .
4. Brunner O., Antognazza L., Triscone J.-M., Miéville L., Fischer Ø. (submitted to) Phys. Rev. Lett.
5. Triscone J.-M., Karkut M.G., Brunner O., Antognazza L., Decroux, Fischer Ø. (1989) Physica C 158: 293.
6. Triscone J.-M., Karkut M.G., Antognazza L., Brunner O., Fischer Ø. (1989) Phys. Rev. Lett. 63: 1016.
7. Triscone J.-M., Fischer Ø., Antognazza L., Brunner O., Kent A.D., Miéville L., Karkut M.G. (to appear in) "Science and Technology of Thin Film Superconductors 2", edited by R.D. McDonnel, and S.A. Wolf.
8. Eibl O., Hoenig H.E., Triscone J.-M., Fischer Ø., Antognazza L., Brunner O. (submitted to) Physica C.
9. Antognazza L., Triscone J.-M., Brunner O., Karkut M.G., Fischer Ø. (1990), Physica B 165-166: 471.
10. See, for instance, Iye Y., Tamegei T., Takeya H., Takei H. (1987) Jpn. J. Appl. Phys. 27 : L1850..
11. Palstra T.T.M., Batlogg B., van Dover R.B., Schneemeyer L.F., Waszczak J.V., (1990) Phys. Rev. B 41: 6621.
12. Fischer Ø., Triscone J.-M., Brunner O., Antognazza L., Affronte M., Miéville L. (to appear in the) "Proceedings of NATO Science Forum 90", Biarritz, France, September 1990.
13. Brunner O., Triscone J.-M., Antognazza L., Karkut M.G., Fischer Ø. (to appear in the) "Proceedings of the E-MRS Spring Meeting", Strasbourg 1990.
14. Feigel'man M.V., Geshkenbein V.B., Larkin A.I. (1990) Physica C 167: 177.

A 4.2K Josephson Digital Computer

SUSUMU TAKADA

Electrotechnical Laboratory, 1-4, Umezono 1-chome, Tsukuba, Ibaraki, 305 Japan

ABSTRACT

A 4-bit Josephson digital computer, ETL-JC1, consisting of Nb superconducting tunnel junctions has been designed and fabricated to investigated the feasibility of a Josephson computer system. The ETL-JC1 consists of four Josephson LSI chips including 1-kbit ROM and 1-kbit RAM. The computer system was designed based on a RISC (reduced instruction set computer) architecture. These four Josephson LSI chips were made by integrating Nb/AlOx/Nb tunnel junctions. The ETL-JC1 was embodied by mounting these four Josephson LSI chips on the nonmagnetic printed circuit board. The 128-step computer program composed with 10-bit instruction codes sufficiently enough to construct general computer program was installed in the 1-kbit ROM chip. In the experiment, the computer functions have been verified by executing the computer program in the ETL-JC1 at a low repetition frequency. The ETL-JC1 consisting of 22,000 Josephson tunnel junctions consumed a total power of 6.2 mW including regulator junctions on the power lines.

KEY WORDS: Josephson tunnel junction, Nb superconductor, logic circuit, memory, computer

INTRODUCTION

Up to date there are many efforts to develop Josephson computer technology using Nb and NbN tunnel junctions. Josephson processors[1,2,3] and memories[4,5] with a LSI complexity have been reported by several groups. However, a Josephson computer composed of the processor and the memory chips, in which the computer program can be executed, has not yet been reported after the work of Josephson cross-sectional model by the IBM group[6].

In the Electrotechnical Laboratory the key technology for the Josephson computer has been developed under the national project from fiscal 1981 to 1989 sponsored by Agency of Industry and Science Technology, MITI, Japan. The key technology covers Nb tunnel junction integration process, LSI logic circuit, memory chips of 1-kbit ROM(read-only memory) and 1-kbit RAM(random access memory), CAD(computer aided design) system for logic circuits, and multi-phase power supply system. A logic gate family based on a current injection logic gate, 4JL(four Josephson junction logic), has been developed to compose LSI logic circuits including combination logic and sequential logic[7]. The 1-kbit RAM chip has been developed using a new type of Josephson RAM cell, named the variable threshold memory cell. This 1-kbit RAM has been confirmed fully to access all bits for writing and reading without any bit failures[8]. The 1-kbit ROM (read only memory) chip also has been developed by employing a plane-like two-junction dc-SQUID, in which the 128-step program consisting of 10-bit instruction codes was installed[9].

In order to investigate the feasibility of the Josephson computer system, a 4.2K multichip Josephson computer named ETL-JC1 (Electrotechnical Laboratory Josephson computer no. 1) has been designed and fabricated based on the key technology developed[10]. The computer functions in the ETL-JC1 has been verified successfully by executing the 128-step computer program. This paper describes the ETL-JC1.

DESIGN OF THE ETL-JC1

A block diagram of the ETL-JC1 is shown in Fig.1. The computer circuit is divided into four Josephson LSI chips named RALU(register & arithmetic logic unit), SQCU(sequence control unit), IROU(instruction read-only memory unit) and DRAU(data random access memory unit). The RALU chip decodes the 10-bit instruction code and executes arithmetic logic operation following the instruction code. The RALU chip generates the skip control signal for the sequence control following the state of accumulator(ACC) and carry flag register(C). The SQCU chip controls the sequence of the computer program execution by 7-bit program address data. The 128-step computer program was installed in the 1,280-bit ROM plane on the IROU chip to operate the ETL-JC1. Every 10-bit instruction code is read from the IROU chip through the 10-bit instruction bus line and is decoded in both RALU and SQCU chips. The DRAU chip consists of 1-kbit RAM and address decoder. The 4-bit data are accessed in the 1-kbit RAM plane by the RALU chip. All circuits are driven by the two-phase power supply. Shaded blocks in Fig.1 are driven by phase 1 and non-shaded blocks are phase 2.

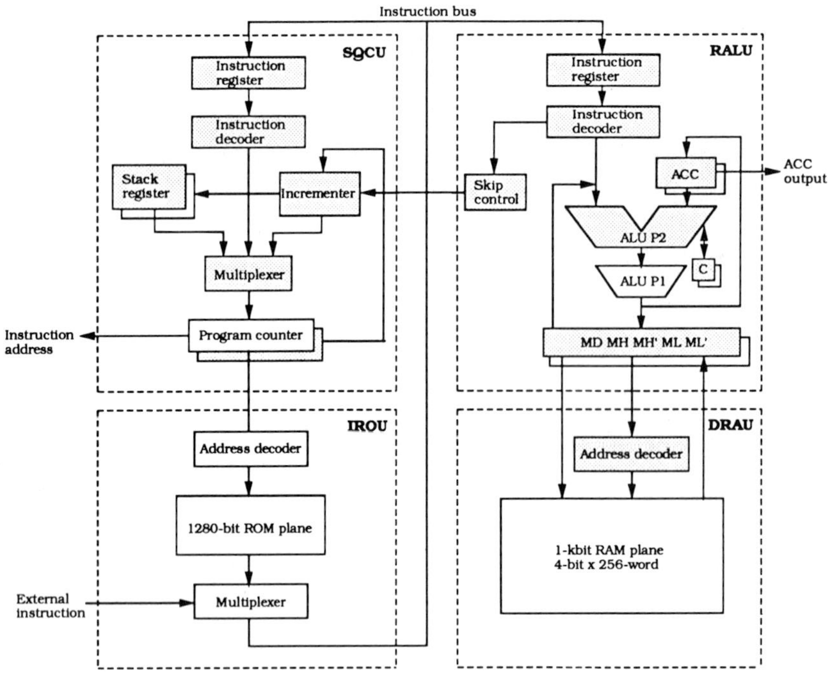

Fig. 1. A block diagram of the ETL-JC1.

Main feature of the ETL-JC1 is listed in Table I. A RISC (reduced instruction set computer) architecture is employed to design the computer system, which can execute all instructions within one clock cycle. A Harvard architecture was used to separate the 4-bit data bus line from the 10-bit instruction bus line. The random logic circuits were designed using dual-rail logic which enables to construct combination logic circuits employing both OR-gates and AND-gates. In order to store the data temporarily in the registers, a data latch circuit was composed

Table I. Main feature of the ETL-JC1

Architecture	RISC (reduced instruction set computer) architecture, Harvard architecture
Arithmetic logic unit	4-bit data(ripple carry)
Register	4-bit data x 6 (ACC, MD, ML, ML', MH and MH'), 7-bit address data x 2 (two-level stack register), 10-bit code x 2 (instruction)
Instruction ROM	10-bit x 128-step(program address), 2-junction SQUID cell
Data RAM	4-bit x 256-word(data address), Variable threshold memory cell
Functions	16 arithmetic logic(add, subtract, OR, XOR), 9 sequence control(skip, subroutine call/return), 2 memory control(write/read)
Logic gate	Four-junction logic (4JL) gate family
Data latch	Direct-coupled data-latch
External input/output	10-bit instruction code, 2-bit external control/4-bit ACC, 7-bit program address
Power	Tow-phase power supply, split power supply
Number of logic gates	3,658
Number of junctions	22,000 (Nb/AlOx/Nb tunnel junction)

of the 4JL gates without conventional superconducting inductance loop. The data latch circuit holds and transfer the data from one phase power to other phase power. This circuit is used at the interface between non-shaded circuit blocks and shaded ones in Fig.1. The logic function NOT is performed by the data latch circuit, which was placed between the combination logic circuits. An OR-type address decoder is implemented in both memory chips of ROM and RAM instead of the conventional AND-decoder in order to reduce the number of logic gates and to make a delay time short. On the basis of RISC architecture a set of instruction codes was provided, in which 27 kinds of basic instruction codes including arithmetic logic functions, conditional skips, write/read access for RAM, jump, shifts, moves, and subroutine call/return were included. These instruction codes are sufficiently enough to make any computer program. In order to start the program execution it is necessary to initialize the registers of ACC, carry flag register and general purpose registers of MD, ML, ML', MH, and MH'. This is done by feeding the external instruction codes into the IROU chip. In order to verify the program execution, the 4-bit data in the ACC and the 7-bit program address are monitored.

FABRICATION

Figure 2 shows a schematic cross section of the Josephson LSI chips. Josephson LSI circuits were fabricated on a 2-inch silicon wafer. The silicon wafer is used only as a substrate and whose electrical properties does not used in the Josephson LSI circuits. All LSI chips have been fabricated using Nb/AlOx/Nb tunnel junctions with 3-μm design rules. A tri-layer sandwich of Nb/AlOx/Nb was fabricated by a sequential dc-magnetron sputtering method combining thermal oxidation process for the tunnel barrier. Small Josephson junctions in the circuit were provided by this large sandwich using a reactive ion etching(RIE) process. A thin MgO film layer was inserted to protect SiO and Nb layer in the RIE process, since the etching rate of the MgO is very small. In order to improve the characteristics of the tunnel junctions, Nb underlayer method was used in the process. The Josephson critical current density was chosen to be 560 A/cm^2 in all LSI chips, which makes a maximum gate current of the 4JL gate 200 μA. The current density was controlled by the oxidation time for the tunnel barrier formation. The resistors were integrated using Pd thin film, which has a sheet resistance of 2 Ω. The two-level wiring layers were made by using Nb film and Pb-In alloy film. 8-level photomasks were used to fabricate RALU, SQCU and IROU chips, and 9 levels for DRAU chip.

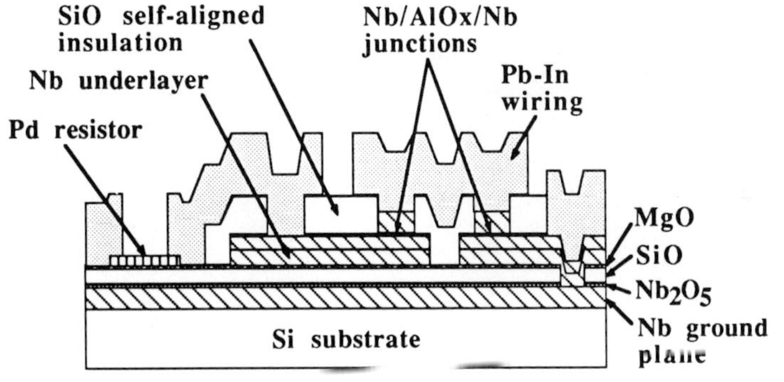

Fig. 2. Schematic cross section of the LSI chips.

LSI CHIPS OF THE ETL-JC1

Figure 3(a) shows the RALU chip fabricated using Nb/AlOx/Nb tunnel junctions. 1273 4JL gates are integrated on the 4.3mm x 5mm chip. A standard cell method was applied to make the logic circuit of this chip. The layout of the standard logic cells was done by using the CAD system. The standard logic cells were placed and connected each other between the two-phase power lines as shown in the figure. This RALU chip was fabricated in 1988 and tested in liquid helium. The exactly same RALU chip was used in the present ETL-JC1 experiment. It should be noted that the RALU chip was completely operated without any failure during the ETL-JC1 experiment. During this period the chip has been stored in atmosphere at room temperature and thermally cycled between room temperature and 4.2K more than ten times. Therefore the Josephson LSI chips fabricated with the present process reveals high stability and reliability.

The SQCU chip shown in Fig.3(b) was also designed by the CAD system and fabricated using Nb tunnel junctions. 593 4JL-gates were integrated on the 3mm x 3.1mm chip.

Figure 3(c) shows the IROU chip, in which the 128-step program was installed. The chip size is 3.5mm x 5mm. Plane-like dc-SQUIDs were employed to make ONE- and ZERO-ROM cells. In the ONE ROM cell the damping

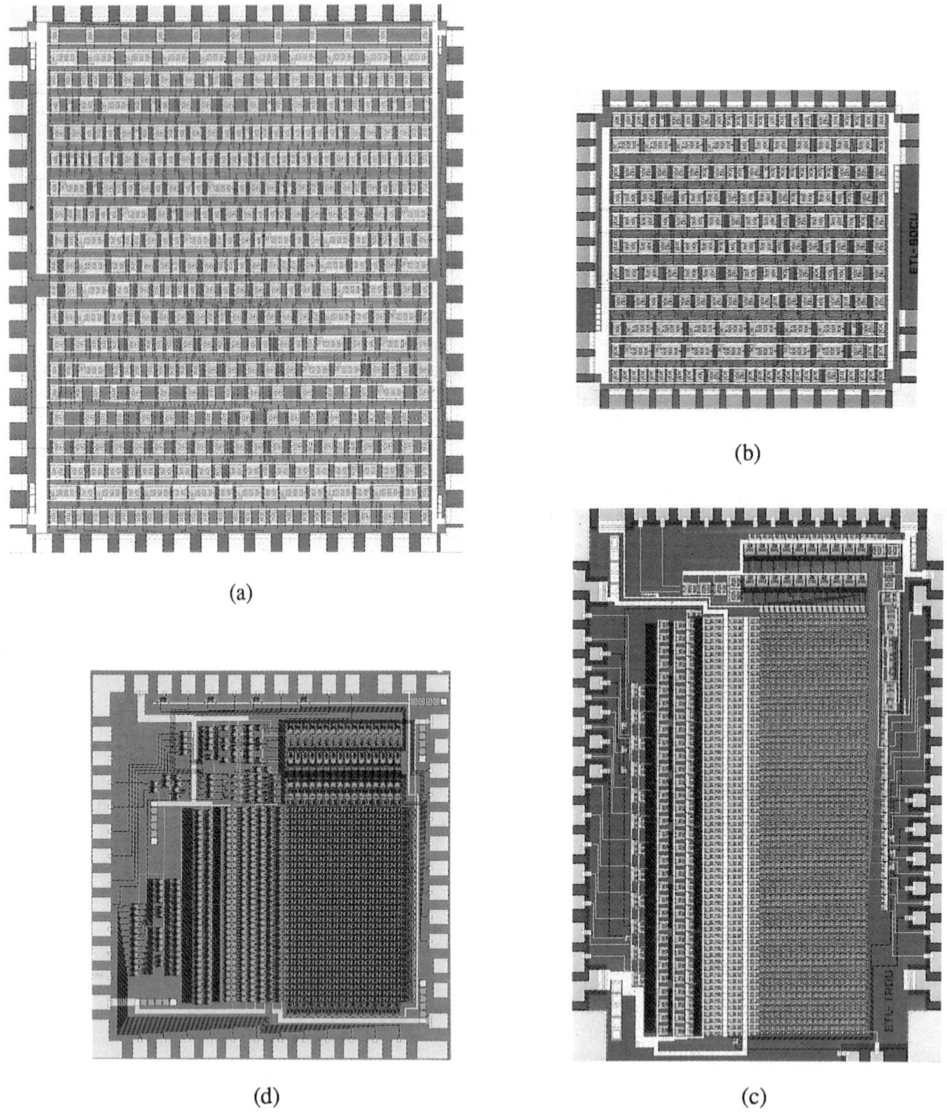

Fig. 3. Photographs of the LSI chips: (a) RALU, (b) .SQCU, (c) IROU, and (d) DRAU.

resistor was used to suppress the resonance in the dc-SQUID. In the ZERO ROM cell the superconducting shunt was carried out not to generate the output voltage using the top wiring layer. Then, the operating margin was improved in the ROM cell plane. The 128-step 10-bit instruction codes were installed in the 1,280-bit ROM plane with two 10-bit columns x 64 lines. The source program was first written by mnemonic expression in the computer (VAX11/780), and was translated to the 10-bit instruction codes. These instruction codes were transferred to the logic CAD system to make automatic layout of ROM cells in the 1,280-bit ROM plane.

Figure 3(d) shows DRAU chip of the 1-kbit Josephson RAM (random access memory). The chip size is 3.7mm x 3.8mm. A set of 256 words with 4-bit data can be written and read out in the RAM chip. 100% memory cell access experiment was confirmed in the chip. The RAM plane consists of the variable threshold memory cells. This cell was designed based on the asymmetric dc-SQUID. The cell has the great advantages of a simple structure and a stable operation by adopting a unique sense circuit instead of the conventional one. The cell size was 30μm x 72μm. The cell itself has no sense gate. The data are stored with a single quantum magnetic flux. In the DRAU chip the set and reset operation is performed by logic gates without any large superconducting closed loop. By this method, the drive current on the bit line can be kept constant without fluctuations due to the quantum magnetic flux, resulting in a wide operating margin in the RAM chip. By introducing the re-writing circuit of the feedback loop consisting of the logic gates, non destructive operation was achieved.

EXPERIMENTAL RESULTS AND DISCUSSION

To construct the ETL-JC1, four LSI chips were connected each other on a nonmagnetic printed circuit board as shown in Fig.4. Each chip was wire-bonded on a 24mm x 24mm chip carrier. Each chip carrier was mounted on the printed circuit board with a 68-pin socket. The printed circuit board was fixed on the end of a plastic pipe as shown at the bottom of the figure. The board size is 80mm x 100mm. Chip-to-chip connection among four chips was carried out by using copper-metal strip lines on both front and back sides of the printed circuit board. The output signals of the 4-bit data in the ACC register and the 7-bit program address in the program counter were monitored to check the program execution.

Fig. 4. The ETL-JC1.

The liquid helium dewar was shielded by four-fold 1 mm thick permalloy metals to reduce an environmental static magnetic field to the chips. Moreover, a double permalloy metal cylinder was placed to cover the printed circuit board in liquid helium. All components of the chip carriers, sockets, screws, and printed circuit board were made of nonmagnetic materials.

The experimental operation of the ETL-JC1 was carried out in liquid helium. The two-phase power was supplied to drive all chips with a repetition frequency of a few kilohertz. The 128-step program execution was verified by monitoring the 4-bit data in the ACC of the RALU chip and 7-bit program address in the SQCU chip. Figure 5 shows an example of the program execution in the ETL-JC1. These data coincide completely with the simulated

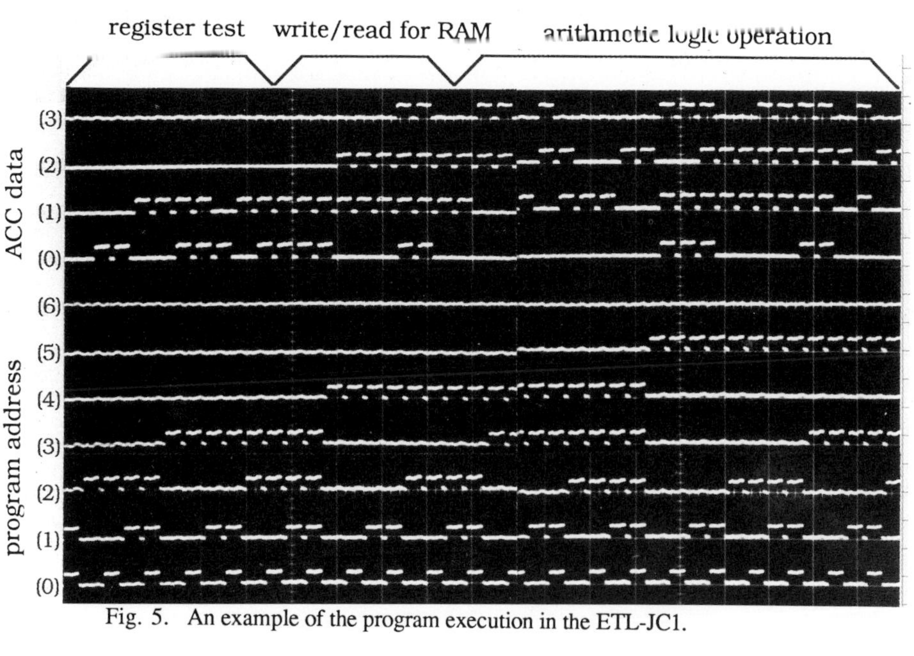

Fig. 5. An example of the program execution in the ETL-JC1.

one by the logic CAD. For all 128-step program operation no failure appeared. Incorrect program operations due to the magnetic flux trap caused by the power current was not observed in the ETL-JC1. The operating margin was obtained to be more than ±10 % for the stable computer operation. A total power dissipation in the computer was 6.2 mW, which includes more than 22,000 single Josephson junctions.

The performance of the ETL-JC1 was carefully investigated by the CAD system. The critical delay time was found in the circuits which include the instruction register, the instruction decoder, and the skip control in the RALU chip, the incrementer, the multiplexer and the program counter in the SQCU chip, and the address decoder and the ROM plane in the IROU chip. This critical path restricts a machine cycle of this computer system. The critical path was evaluated to have a delay time of 1092psec due to the logic gate delay by taking account of the signal delays in the CPU, the ROM, and the RAM at 75% bias current. 1GIPS(giga-instruction per second) operation can be expected with a single CPU of this system. This evaluation was done by the logic gate delay and does not include propagation delay and timing margin of the two-phase power. The first Josephson multichip computer ETL-JC1, consisting of superconducting tunnel junctions, has been demonstrated successfully. In the next step in order to realize high frequency operation of the Josephson computer system high speed packaging technology and cryogenic circumstance should be developed.

REFERENCES

1. Nakagawa H, Kosaka S, Kurosawa I, Aoyagi M, Hamazaki Y, Okada Y, Takada S(1989) Extended Abstracts of ISEC'89: 387
2. Hatano Y, Yano S, Mori H, Yamada H, Hirano M, and Kawabe U (1989) Extended Abstracts of ISEC'89: 375
3. Kotani S, Inoue A, Imamura T, and Hasuo S (1990) Digest of Technical Papers of 1990 IEEE International Solid-State Circuits Conference: 148
4. Kurosawa I, Nakagawa H, Kosaka S, Aoyagi M, Takada S (1989) IEEE J. Solid-State Circuits 24: 1034
5. Nagasawa S, Wada Y, Hidaka M, Tsuge H, Ishida I, and Tahara S (1989) IEEE J. Solid-State Circuits 24: 1363
6. Ketchen MB,.Herrell DJ, and.Anderson CJ (1985) J. Appl. Phys.57: 2550
7. Nakagawa H, Kosaka S, Kawamura H, Kurosawa I, Aoyagi M, Hamazaki Y, Okada Y, and Takada S (1989) IEEE J. Solid-State Circuits 24: 1076
8. Kurosawa I, Nakagawa H, Aoyagi M, Kosaka S, and Takada S (1990) Digest of Technical Papers of VLSI Circuits Symposium: 67
9. Aoyagi M, Nakagawa H, Kurosawa I, Kosaka S, Okada Y, Hamazaki Y, and Takada S (1990) IEEE J. Solid-State Circuits 25: 971
10. Nakagawa H, Kurosawa I, Aoyagi M, Kosaka S, Hamazaki Y, Okada Y, and Takada S (1991) IEEE Trans. on applied superconductivity (submitted)

United States Progress in Superconducting Magnetic Energy Storage (SMES)

ROBERT B. SCHAINKER

Energy Storage Program, Electric Power Research Institute, Palo Alto, California, USA

ABSTRACT

In the United States research and development has been sponsored since 1970 on applying the physical state of superconductivity to the storage of energy; namely, superconducting magnetic energy storage (SMES). This paper summarizes the progress made in this endeavor, culminating in the present day with the completion of component tests and two competing designs for a SMES pilot plant. The contract to construct and test one of these designs is to initiate in early 1991. The two competing designs were developed by teams of engineers and scientists at Bechtel National Inc. and Ebasco Services Inc. The pilot plant is currently rated for two discharge levels: 10 MW for 2 hours and 400 MW for 100 seconds. The paper also presents a brief description of the pilot plant, its construction schedule and test plan. The sponsors for the pilot plant efforts are the Electric Power Research Institute and the US Department of Defense.

KEY WORDS: SMES, Superconducting Magnetic Energy Storage

INTRODUCTION

Variations in electric power demand are experienced by all electric utilities. Most variations are cyclic, with periods ranging from a few seconds to diurnal cycles, which are the most troublesome because of the magnitude and rate of power variation that may occur in a 24 hour period. Power level changes depend on the specific types of load associated with a given utility's service territory; however, these power level changes are often greater than 50% to 200% of the night-time minimum load conditions existing at any given utility. Since night-time base-loaded power plants are highly efficient and utilize solid fuels (e.g., coal or nuclear fuel) that are relatively inexpensive compared to liquid and/or gas fuels used by relatively inefficient day-time peaking and intermediate power plants, the use of storage power plants can save large amounts of money, if the storage plant has a relatively high round-trip efficiency and is charged with the less expensive night-time energy.

In the United States at present, the only storage power plant widely used is the pumped hydroelectric power plant, which relies on the use of night-time electricity to pump water to a high elevation and later allow this water to drive hydro turbines to produce electricity. The round-trip efficiency of this type of storage plant is about 75%. Although about 10 to 15% of the US capacity could cost-effectively be in the form of storage power plants, at present only about 3% (20 GW's) of the US capacity is in the form of pumped hydroelectric plants. These units have been of great value to those utilities fortunate enough to have the necessary topographic, geologic, and other required conditions; namely, an available location with adequate water and sites for upper and lower reservoirs with a differential height of several hundred feet or more. The US is now running out of such favorable sites, and when they do exist, the environmental permit process is often fraught with insurmountable hurdles.

Several other storage power plant options do exist, and the US Electric Power Research Institute (EPRI) is currently performing R&D on a number of them; namely, compressed air energy storage, battery energy storage, and superconducting magnetic energy storage (SMES).

SMES is a relatively recent energy storage option with many possible benefits for electric utilities. SMES plants would enable utilities to store electric energy directly (i.e., without converting it into another energy form). Thus, round-trip energy storage efficiency would be greater than about 95%. Basically, off-peak electricity would be converted to dc electricity and fed into an electromagnetic coil wound with superconducting wire (see Exhibits 1 and 2). It is via dc electricity flowing through an inductor that produces a dc magnetic field that the energy is stored. The amount of stored energy is equal to one-half the product of the inductance and the square of the dc current. Since the coil is superconducting, it has essentially zero resistance; thus, any current induced into the coil by a charging dc voltage will continue to flow indefinitely, until a discharging resistance (i.e., a utility load) is connected to the coil. The method to apply charging voltage or discharging resistance to the circuit is accomplished by solid-state ac-dc-ac converter switches. SMES plants will switch easily between different power levels and between charging and discharging, all with minimal losses and within about 20 milliseconds. The magnetic field created by the SMES coil is dc as opposed to the ac magnetic fields from ac transmission lines. The dc magnetic field follows the charge and discharge energy levels. At present, there is no known deleterious consequence of a dc magnetic field; however, extensive basic research is taking place on this topic.

An alternative way to understand the storage of energy in a SMES plant is by analogy with a more familiar and older storage technology, the flywheel. A flywheel at rest is like a SMES plant without current. To energize a flywheel, a torque (i.e., a "voltage" pressure) is applied to rotate it. As the flywheel rotates faster and faster, the energy stored is equal to the product of one-half the moment of inertia (i.e., "inductance") and to the square of its angular velocity (i.e., "current"). To the extent that there is no resistance in the flywheel bearings, and windage losses are zero, the flywheel will continue to rotate indefinitely, until some "resistance" is applied to the flywheel shaft, which produces output energy in the form of work or heat.

SMES COMPARED TO OTHER STORAGE ALTERNATIVES

Insights gained from recent generation planning studies have shown that SMES and Battery Energy Storage (BES) plants compete with each other for the short time duration storage plant market (i.e., they are "peaking" storage technologies as compared to compressed air energy storage (CAES) and pumped hydro (PH) plants, which are "intermediate" and "base" storage technologies respectively). This insight is mainly due to the cost for each hour of storage, which is about \$200/kW for SMES and BES plants, as compared to less than \$20/kW for CAES and PH plants. The economic niche for SMES and BES plants is likely in the range of less than 3 hours per discharge. Although SMES and BES plants have virtually the same very high ramp rate (i.e., 20 msec.), SMES plants do have important unique advantages over BES plants: 20% higher efficiency; better performance for very short, high power discharge duty; and higher expected life due to the absence of corrosion at cryogenic temperatures. However, BES plants do not have any magnetic fringe field and they are generally easier to site due to their modularity.

UNITED STATES ACTIVITIES ON SMES

The SMES commercialization process in the US has been following the classic approach of four major types of work efforts:

- Basic research and development;
- Pilot plant design, construction and testing;
- Demonstration plant construction and operation;
- Commercial-scale plant offerings.

The basic R&D occurred in the 1970's, principally by the University of Wisconsin and Los Alamos National Laboratory, and to a smaller extent by EPRI. The University of Wisconsin work was funded by the Wisconsin Electric Utility Research Program and the US Department of Energy. The Los Alamos National Laboratory work was sponsored by the US Department of Energy. It culminated in the early 1980's with an important SMES verification and proof-of-principle project with Bonneville Power Administration (BPA), which resulted in a working plant rated at 10 MW for 3 seconds. This plant was constructed and satisfactorily tested for a network stability application at the BPA Tacoma, Washington, substation.

A new activity planned to be sponsored by EPRI is the comprehensive quantification of the benefits associated with SMES applied to electric utility systems. The method of approach is to perform utility specific investigations from both the generation system and the transmission system viewpoint. This effort is starting in the latter part of 1990 and is expected to continue until mid 1992. It will capitalize on past results from battery economic studies and will quantify the benefits for the very short discharge market (i.e., less than the 60 second discharge level) and the longer discharge market (i.e., the 30 minute to 3 hour discharge level).

The pilot plant effort started in 1988 with funding from the US Department of Defence (DOD) and EPRI. It is expected to continue until 1996 when the testing by utility representatives is complete. The demonstration plant effort will likely occur in the late 1990's, depending upon the outcome and cost projections for the pilot plant effort. The commercial-scale plant offerings will likely occur in the early 2000's, depending of course on the outcome from the demonstration plant effort.

What makes the SMES commercialization path in the US rather unique is that the pilot plant effort will concurrently verify the design assumptions, performance, and cost for the very short and longer discharge market niches cited above. The very short (less than 60 seconds) discharge level addresses the frequency regulation, area regulation, VAR control and/or transmission stability market. The longer (30 minute to 3 hours) discharge level addresses the spinning reserve, load-following, and/or load-leveling market. These two market niches can be explored by one SMES pilot plant because the only component that changes is the relatively inexpensive ac-dc-ac power conditioner, which can "rather easily" be designed to reconfigure itself in the field to satisfy two power levels (kW), while keeping constant the maximum energy stored (kWh) in the more expensive coil construction effort.

Also of significant pragmatic and financial importance in the US is the market niche envisioned by the US military for the same very short discharge level as that for the utility market niche. SMES has a unique capability to satisfy very short bursts of power requirements at very high power levels.

This US military market niche has accelerated the SMES commercialization process in the US by enabling the near-term funding for most of the pilot plant work efforts. Identifying and implementing this synergy to cosponsor the pilot plant effort to meet U.S. electricity utility and US military technology needs, has been a very effective component in both the EPRI and US DOD strategy to accelerate the development of the SMES technology.

UNITED STATES SMES PILOT PLANT

The largest on-going SMES program in the US is the EPRI-DOD funded program to build a pilot plant. For the present design of the pilot plant, two discharge levels can be tested: 10 MW for 2 hours and 400 MW for 100 seconds. The overall purpose for this program is to design and build a pilot plant facility that will verify the design and cost assumptions prior to building a demonstration and/or commercial scale plant.

Individual goals of this program are as follows:

- Demonstrate technical performance
- Demonstrate power/energy management capability
- Evaluate integrated system design trade-offs
- Assess plant reliability, maintenance, and safety
- Evaluate potential environmental impacts

The EPRI-DOD pilot plant program is divided into three phases:

- Phase 1: Pilot plant preliminary design and cost; and testing of critical components (e.g., the superconductor cable).

- Phase 2: Pilot plant final design and cost; final site selection; construction; plant start-up; and preliminary testing for a six month EPRI-DOD performance evaluation period.

- Phase 3: Pilot plant testing, over about a three year period, headed up by EPRI for a wide variety of electric utility load and transmission needs; such as VAR control, frequency regulation, area control, transmission stability, spinning reserve, load-following, and load-leveling.

The Phase 1 effort cost about $ 32 million and ended in mid-1990. Its work efforts carried the design of many components from a conceptual stage to the detailed design stage, based on full-scale tests of major plant components such as the conductor (see Exhibit 3) and the support struts. Two different viable design approaches have been developed: one led by a Bechtel team, the other lead by an

Ebasco team. Each uses the solenoid design approach with the radially outward magnetic forces retained by a subsurface trench wall. Fortunately, there do not appear to be any show stoppers associated with either design. In early 1991, one team will be selected to build and test their plant design -- an effort costing about $150M spread over 5 years.

The Phase 2 effort is planned to commence at the beginning of 1991 and end in 1996. It is expected to cost about $150 million. The Phase 3 effort is expected to be a multi-million dollar effort, based on contributions from EPRI and utility members of the SMES User's Group. This last phase of the program is expected to end in 1999. Non-US electric utilities may wish to be a part of the test program. Based on an agreement between the US DOD and EPRI, such international participation is to be arranged with EPRI. The initialization fee from non-US utilities is $5 million and the annual fee is $0.1 million.

As a result of the Phase 1 effort, the major components are the superconducting cable, structure/dewar, refrigerator and ac-dc-ac power conditioning system. Each of these plant components is only briefly described below, since contractor details are proprietary and the proposals for the Phase 2 effort are in the process of being evaluated.

In each competing design the superconducting cable is made from a niobium-titanium alloy in the form of wire filaments, a few microns in diameter, extruded within a copper sheath. Such wire is manufactured from a number of domestic and foreign vendors, principally for use in medical diagnostic imaging machines and atomic particle accelerators. At present, the amperage carrying capacity of commercially available filaments exceeds 400,000 amps/sq.cm at 1.8K temperature. In mid-1990, General Dynamics (under contract from the US DOD, via Bechtel) satisfactorily tested the world's highest rated superconducting cable at 200,000 amps, dc (see Exhibit 3). This capability exceeded previous commercial grade superconductors by a factor of four. The outside diameter of this cable is only 2-1/4 inches.

The structure/dewar supports the superconducting cable and transmits its magnetic forces (both radially outward and axially compressive) from the cable to the coil trench walls via struts, while maintaining the cryogenic temperature environment required by the superconducting filaments. The weight of the magnetic coil is also supported by struts, all of which are fabricated from fiberglass reinforced epoxy plastic, which is a non-electric, low heat-conducting material. During cool-down, each component of the SMES coil and structure will contract at its characteristic rate. To accommodate the resulting stresses, a number of acceptable design approaches have been developed.

The refrigerator is a set of compressors, heat exchangers and expanders designed to provide the superconducting coil with 1.8K helium at atmospheric pressure. This temperature is chosen because it is the optimum value for the highest current in the superconductor while satisfying heat transfer requirements and liquid helium characteristics. The SMES refrigeration system requirements are well within the state-of-art used in atomic particle accelerator coils and utilizes components from highly reliable liquid-air plant refrigerators.

The power conditioning system is an ac-dc-ac converter, similar to those used for high voltage dc transmission substations. To charge or discharge the coil, the voltage across the coil is made positive or negative, respectively. When the unit is in standby, independent of storage level, the voltage across the coil is held to zero causing the current to remain constant. The state-of-art for the solid-state electronic components in the converter, for both silicon controlled rectifier (SCR) and gate turn-off (GTO) devices, is so rapid that full power can be achieved, for either charge or discharge, within one cycle of 60 Hertz; i.e., within 20 milliseconds. This extremely fast performance can be achieved while host utility requirements are maintained for both power factor and frequency harmonics. The above characteristics allow a SMES plant to function in a wide variety of modes: load-leveling, spinning reserve, VAR control, voltage support, frequency regulation, area regulation, network stability, and black start.

Though most experts believe the many components of a SMES plant can be made to operate in an integrated, successful manner, there are still many engineering design details to be worked out and verified. Recent experience with very large superconducting facilities does give confidence that SMES plants can work safely and successfully in a utility environment.

CONCLUSION

In the US SMES is increasingly being recognized as a potentially important option for electricity supply and transmission management. This results from (1) increased appreciation of the importance and value of energy storage technologies for power system management; (2) the unique characteristics of SMES: fast response (within 20 msec), high dispatch efficiency (> 95%), and long life (> 30 years); (3) successful progress on component tests and system design resulting from the 10 MW-2 hr/400 MW-100 sec EPRI-DOD pilot plant program; (4) improved superconducting materials for higher temperature applications, and (5) technical readiness of all plant components and subsystems.

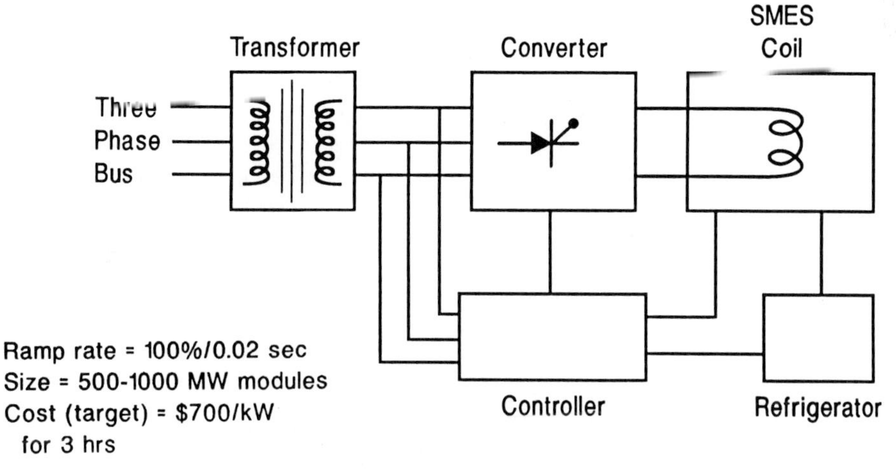

Exhibit 1. SMES Schematic

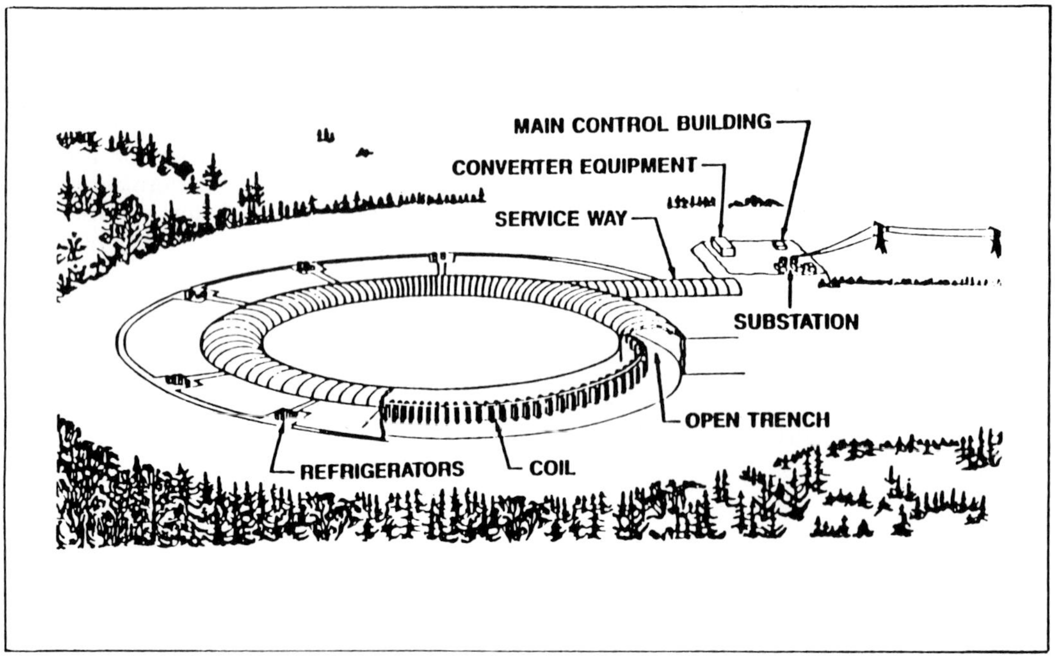

Exhibit 2. Superconducting Magnetic Energy Storage Pilot Plant (10 MW-2 Hr), Aerial View

Exhibit 3. Bechtel 200,000 Amp Superconductor (Outside Diameter = 2.25 inch)

2 Physics

2.1 Theoretical Studies

Superconducting Properties in Layered Cuprate Oxides

M. Tachiki, T. Koyama, and S. Takahashi

Institute for Materials Research, Tohoku University, Sendai, 980 Japan

ABSTRACT

It is probable in layered cuprate oxides such as $YBa_2Cu_3O_7$ and $Bi_2Sr_2CaCu_2O_8$ that superconductivity is generated in the CuO_2 layers and superconductivity in the other layers is induced by the proximity effect. According to the present theoretical study, it was found that unusual properties observed in the experiments of tunneling conductance, optical conductivity, and nuclear magnetic resonance in the oxides are explained on a basis of this model. The single flux line structure and thus the flux line lattice structure in the cuprate oxides are strongly influenced from their layer structure. The anomalous structure of the flux line is important for the flux pinning in the oxides.

KEY WORDS : tunneling conductance, optical conductivity, Knight shift, Lawrence-Doniach model, flux line state, flux pinning

INTRODUCTION

All the high-T_c superconducting oxides which have been found up to now includes the CuO_2 layers. Many experimental results imply that CuO_2 layers are responsible for the high-T_c superconductivity. The CuO_2 layers and the other layers are alternately stacked in the oxides. The coherence length in the direction perpendicular to the layers is comparable or shorter than the lattice period. Therefore the spatial variation of the superconducting order parameter is expected to play an important role for their physical properties. Unusual properties observed in the tunneling conductance [1-4], the optical conductivity [5-8], and the nuclear magnetic resonance [9-14] are shown to be explained using this layered model. The nature of the flux lines and the pinning of the flux lines in the layer structure are also discussed.

We consider $YBa_2Cu_3O_7$ as an example of the layered cuprate oxides [15]. The crystal of $YBa_2Cu_3O_7$ is constructed by a stacking of the two CuO_2 layers and the one CuO_3 chain layer in the unit cell as shown in Fig.1. In Fig.1, the Y and Ba ions are neglected, since the electronic levels of both the ions do not appear near the Fermi level. The resonant photoemission experiments [16] show that the electronic state near the Fermi level is mainly of the p electron character of oxygen.
The p electron state mixes with the d electron state of the copper ions. In this paper, we describe the electronic state near the Fermi level by the freedom of the p orbits of oxygens. The transfer integrals between the neighboring oxygens in the CuO_2 layers are large. The transfer integrals in the CuO_3 chain layer are also large and comparable to those in the CuO_2 layer. The transfer integrals between the layers are much smaller than those of the interlayers. It is shown that the Fermi surfaces calculated using this model well reproduce those obtained by the angle resolved photoemission experiments [17], if suitable values for the transfer integrals are used. We assume the superconducting pairing interaction acting between the oxygen p-holes in the CuO_2 layers. The CuO_3 chain layers are originally in a normal state and the superconductivity in these layers is induced by the proximity effect. Using the above model, we calculate the physical quantities of $YBa_2Cu_3O_7$ in the following.

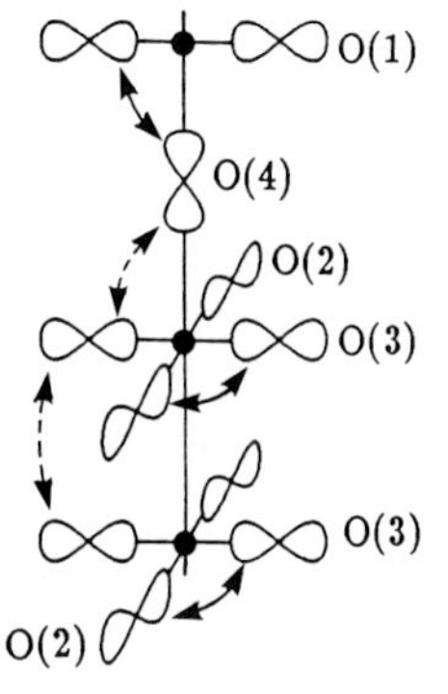

Fig.1 Definition of the oxygen sites in the unite cell of $YBa_2Cu_3O_7$.

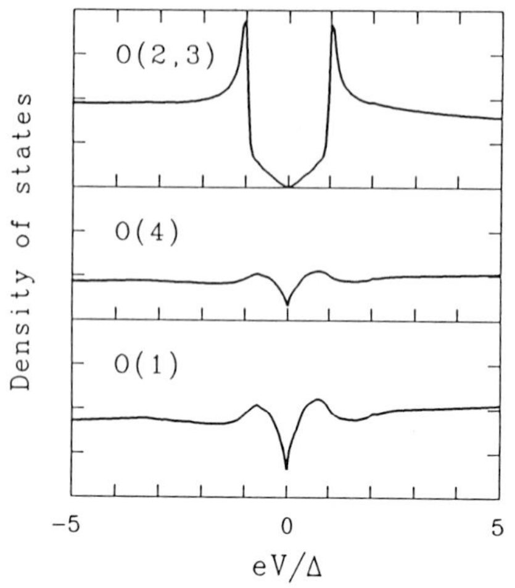

Fig.2 Densities of states. Figures (a), (b), and (c) show the densities of states at the oxygen sites O(2,3), O(4) and O(1), respectively.

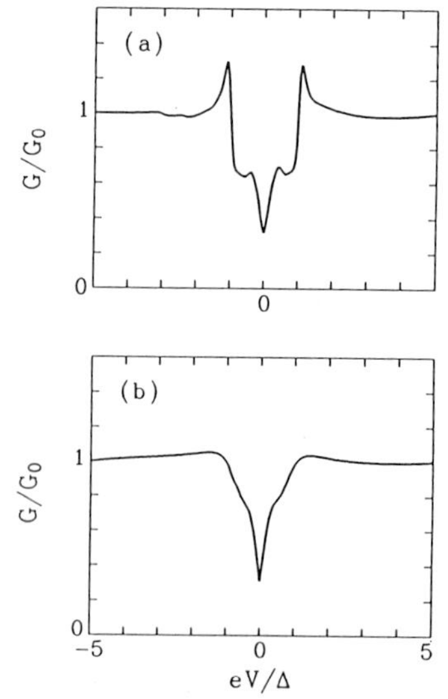

Fig.3 Normalized tunneling conductance as a function of the normalized bias voltage. G_0 is the tunneling conductance in the normal state. (b) shows the tunneling conductance when a pair breaking effect is included in the tunneling conductance in (a).

LOCAL DENSITY OF STATES AND TUNNELING CONDUCTANCE

Using the above model, we first calculate Green's function for the m-th layer $G^{ij}_m(\mathbf{k},\omega)$ which is the i,j component of Green's function with a two by two matrix in the Nambu space. If we define the spectral function of Green's function by $\rho^{ij}_m(\mathbf{k},\omega) = -(1/\pi)\mathrm{Im}G^{ij}_m(\mathbf{k},\omega)$, the local density of sates at the m-th layer is expressed by [18]

$$\rho_m(\omega) = \sum_{\mathbf{k}} \rho^{11}_m(\mathbf{k},\omega). \tag{1}$$

In Green's function the self-energy of holes which originates from the scattering of holes due to the spin and charge fluctuations is included. Figure 2 shows the calculated result of the density of states at each layer. As seen in Fig.2, the large energy gap opens at the site O(2,3) in the CuO_2 layer and the small gap does at the sites O(1) and O(4) in the CuO_3 chain layer. This small gap arises from the superconductivity induced in the CuO_3 chain layers by the proximity effect.

When the electron tunneling process is diffusive in a junction with a superconductor and a normal metal, the tunneling conductance is proportional to the density of states at the energy shifted from the Fermi level by the applied bias voltage. Usually, we measure some average of the local densities of states using the tunnel junction, since the layers are distributed at the junction interface. Figures 3 (a) and (b) show the tunneling conductance as a function of the bias voltage in the case that the CuO_2 layer is less distributed. The large gap mainly originates from the CuO_2 layer and the small gap inside the large gap mainly comes from the CuO_3 chain layers. In Fig.3(b) a pair breaking effect which may be present near the junction interface is included. The peak structure inside the gap in Fig.3(a) changes to the shoulder in Fig.3(b) by a smearing effect. These structures seen in Figs. 3(a) and 3(b) have been observed in the tunnel junctions of $YBa_2Cu_3O_7$ and normal metals [1-4].

OPTICAL CONDUCTIVITY

The optical conductivity gives us information on the particle-hole excitations and the superconducting energy gap at the CuO_2 and CuO_3 layers. Let us take the b axis along the CuO_3 chain in $YBa_2Cu_3O_7$, and the a axis perpendicular to the chain in the layer. For the light polarized along the a axis, one probes only the optical conductivity of the CuO_2 layers, since the light cannot excite electrons in the chains. On the other hand, for the light polarized the b axis, one probes the optical conductivities from both the CuO_2 layers and

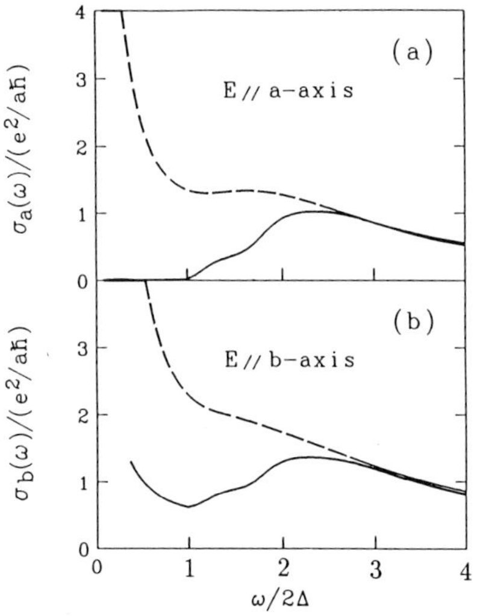

Fig.4 Optical conductivities as a function of frequency. The solid and dashed curves indicate the optical conductivities in the superconducting and normal states, respectively.

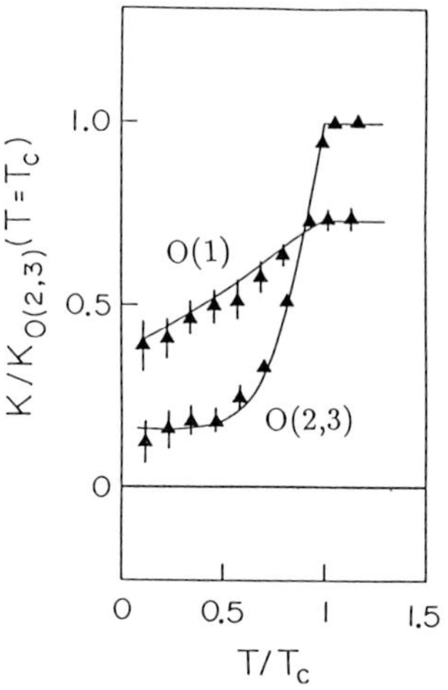

Fig.5 Knight shift as function of temperature. Solid curves indicate the theoretical values and the triangles indicate the experimental values obtained by Takigawa et al.[13]

the CuO_3 chain layers. The optical conductivity of the m-th layer is expressed by [15]

$$\sigma_m(\omega) = \frac{\pi}{2}e^2\sum_{\mathbf{k}} v_m^2(\mathbf{k})\frac{1}{\omega}\int_{-\infty}^{\infty} d\nu \mathrm{Tr}\Big[\varrho_m(\mathbf{k},\nu)\varrho_m(\mathbf{k},\nu-\omega)\Big]\Big(\tanh\frac{\nu}{2T} - \tanh\frac{\nu-\omega}{2T}\Big), \tag{2}$$

$v_m(\mathbf{k})$ being the velocity of hole in the m-th layer. The optical conductivities along the a and b axes are, respectively, given by $\sigma_a(\omega)$ and $\sigma_b(\omega) = \sigma_a(\omega)+\sigma_c(\omega)$, where $\sigma_a(\omega)$ and $\sigma_c(\omega)$ are the optical conductivities of the CuO_2 and CuO_3 layers. Figures 4(a) and 4(b) show the optical conductivities $\sigma_a(\omega)$ and $\sigma_b(\omega)$ respectively. The dashed curves indicate the optical conductivity in the normal state. The peaks at $\omega/2\Delta \sim 0$ correspond to the Drude conductivity. The superconducting energy gap with 2Δ is seen in the spectrum of $\sigma_a(\omega)$. Although the optical conductivity $\sigma_b(\omega)$ has a drastic decrease around 2Δ, it has a finite value even below 2Δ. The finite value of $\sigma_b(\omega)$ below 2Δ is a contribution from $\sigma_c(\omega)$, which has no gap structure even in the superconducting state and behaves like a normal conductivity. The frequency dependence of $\sigma_a(\omega)$ and $\sigma_b(\omega)$ in Fig.4 is qualitatively in agreement with that of the optical conductivity measured by Schlesinger et al. using untwinned single crystals of $YBa_2Cu_3O_7$ [5].

KNIGHT SHIFT

To add another information about the intensity of superconductivity at each layer, we calculate the Knight shift of oxygen ^{17}O. The Knight shift of ^{17}O at the site i is given by

$$K_{O(i)} = \alpha\chi_S^{(i)} + K_0, \tag{3}$$

where $\chi_S^{(i)}$ is the local spin susceptibility of the p holes on the oxygen at the site i, and α and K_0 are constants. The first term at r.h.s. of Eq.(3) denotes the contribution from the spin susceptibility and the second term denotes the remaining part which includes the contribution from the orbital susceptibility. The values of α and K_0 are assumed to be common to the oxygens $^{17}O(2,3)$ and $^{17}O(1)$. In the calculation of the temperature dependence of $\chi_S^{(i)}$, we use the superconducting order parameter whose temperature dependence is expressed by an interpolation formula $\Delta(T) = \Delta(0)\tanh\big[\beta\sqrt{(T_c-T)/T}\big]$.

The normalized experimental values of the Knight shift of $YBa_2Cu_3O_7$ are shown by the solid triangles in Fig.5. If we take $2\Delta(0)/T_c$=6.5, β=1.74 which is the BCS value, and $K_0/\{\alpha\chi_S^{(2,3)}(T{=}T_c)\}$=−0.2, the calculated values which are shown by the solid curves in Fig.5 are in good agreement with the experimental ones [13]. The above large value of $2\Delta(0)/T_c$ is consistent with the values obtained by other experiments. The large value of $2\Delta(0)/T_c$ leads to a rapid decrease of $K_{O(2,3)}$ below T_c compared with that in the weak coupling BCS theory. On the other hand, $K_{O(1)}$ mildly decreases, since the superconductivity induced by the proximity effect in the CuO_3 chain layer is weak.

A SINGLE FLUX LINE IN LAYERED SUPERCONDUCTORS

When the coupling between layers is sufficiently weak in a layered superconductor, the superconductor may be well described phenomenologically by the Lawrence-Doniach model[19,20]. In this model the superconductor is composed of two dimensional sheets stacking with a distance D and the current flowing across the layers is induced by the Josephson effect. In this section we study the structure of a single flux line, based on this model, in a field parallel to the layers. We consider here a superconductor with inequivalent layers, keeping the high-T_c oxides in mind, that is, the superconductor is composed of a periodic array of superconducting and normal (or weakly superconducting) layers. We take the z-axis in the direction perpendicular to the layers. An external field is applied along the x-axis which is parallel to the layers. Then assuming $B_y=B_z=0$ for the internal field $\mathbf{B}$ and choosing the gauge $A_x=0$ for the vector potential $\mathbf{A}$, we approximate the free energy functional in the following form which is valid for the high κ case, neglecting the spatial dependence of the absolute value of the superconducting order parameter in the generalized Lawrence-Doniach model[21],

$$F = \int dyD\sum_{\ell}[Q_\ell(y)^2/(8\pi\Lambda_\ell)^2 + (\frac{c}{2e})(1-\cos P_{\ell,\ell+1}(y))/(4\pi\Gamma^2_{\ell,\ell+1}D^2)]$$

$$+\int dydz\frac{1}{8\pi}B_x(y,z)^2, \tag{4}$$

where

$$Q_l(y) = \frac{c}{2e}\partial_y\varphi_\ell(y) - A_y(y,z), \tag{5}$$

$$P_{\ell,\ell+1}(y) = \varphi_{\ell+1}(y) - \varphi_\ell(y) - \frac{2e}{c}\int_\ell^{\ell+1} dzA_z(y,z), \tag{6}$$

with $\varphi_\ell(y)$ being the phase of the superconducting order parameter at a point y on the ℓ-th layer. In Eq.(4) Λ_ℓ is the fictitious penetration depth defined on the ℓ -th layer, $m_\ell c^2/(8\pi e^2\Delta_\ell^2)$, and $\Gamma_{\ell,\ell+1}$ is a constant with the dimension of length, $M_{\ell,\ell+1}c^2/(8\pi e^2\Delta_\ell\Delta_{\ell+1})$, where m_ℓ is the effective mass on the ℓ -th layer, Δ_ℓ is the absolute value of the order parameter and $M_{\ell,\ell+1}$ is the inverse coupling constant between the ℓ -th and (ℓ +1)-th layers with the dimension of mass. The variation of the free energy functional (4) with respect to $\mathbf{A}$ yields the equations

$$B_\ell(y) - B_{\ell-1}(y) = DQ_\ell(y)^2/\Lambda_\ell^2, \tag{7}$$

$$\partial_y B_\ell(y) = -(\frac{c}{2e})\sin P_{\ell,\ell+1}(y)/(\Gamma^2_{\ell,\ell+1}D) \tag{8}$$

Here $B_\ell(y)$ is the internal magnetic field in between the ℓ-th and $(\ell+1)$-th layers,

$$B_x(y,z) = \sum_\ell[\theta(z-z_\ell) - \theta(z-z_{\ell+1})]B_\ell(y), \tag{9}$$

where $\theta(z)$ is the step function, that is, $B_x(y,z)$ is constant in between adjacent layers for fixed y in this model. Eliminating the vector potential in $Q_\ell(y)$ and $P_{\ell,\ell+1}(y)$ in Eqs. (7) and (8), we obtain the equation for $B_\ell(y)$ from Eq.(7) and (8) for the single flux line state,

$$\partial_y\sin^{-1}[\frac{2e}{c}D\Gamma^2_{\ell,\ell+1}\partial_y B_\ell(y)] + \frac{2e}{c}[\Lambda^2_{\ell+1}B_{\ell+1}(y) - (\Lambda^2_{\ell+1}+\Lambda^2_\ell+D^2)B_\ell(y)$$

$$+\Lambda^2_{\ell-1}B_{\ell-1}(y)]/D = -2\pi\delta_{\ell,0}\delta(y). \tag{10}$$

Here we assumed that the single flux line is located at $y = 0$ between the zeroth layer ($\ell = 0$) and its neighboring first layer ($\ell = 1$). Let us now show the numerical results. We numerically solved Eq.(10) for the superconductors with a periodic array of two successive superconducting and two successive normal layers as shown in Fig.6. In this case the parameters $\Gamma_{\ell,\ell+1}$ and Λ_ℓ can be sorted, respectively, into three independent parameters, Γ_{SS}, Γ_{SN} and Γ_{NN}, and into two independent parameters, Λ_S and Λ_N. For example, $\Gamma_{\ell,\ell+1}$ is equal to Γ_{SN} when the ℓ-th layer is superconducting and the $(\ell+1)$-th layer is normal, etc.. By the definitions we have the inequalities, $\Gamma_{NN} > \Gamma_{SN} > \Gamma_{SS}$, and $\Lambda_N > \Lambda_S$. As shown in Fig.6 the flux line can take three inequivalent positions (S-S, S-N and N-N) in this superconductor, if we assume that the single flux line is located in between adjacent layers. Figures 7-9 show typical results for the field distribution, $B_\ell(y)$, for three different values of the parameter, $\gamma^2_{NN} \equiv \Gamma^2_{NN}/\Lambda^2_S$ in the case of $\Gamma^2_{NN} \gg \Gamma^2_{SN} > \Gamma^2_{SS}$. The values of the other parameters were chosen as $\gamma^2_{SS} \equiv \Gamma^2_{SS}/\Lambda^2_S = 30$, $\gamma^2_{SN} \equiv \Gamma^2_{SN}/\Lambda^2_{SN} = 40$, $\lambda^2_N \equiv \Lambda^2_N/\Lambda^2_S = 2$ and $D/\Lambda_S = 0.1$ in these three cases. As seen in Figs.7-9, the structure of the single flux line shows clear difference, depending on where the flux line is located in the layer structure. The intensity of the magnetic field near the center of the flux line in the S-S case is larger than that in the N-N case. Consequently the extension of the magnetic field along the layers is larger in the N-N case than in the S-S case, since the total flux should be quantized to the value, $hc/2e$, in both cases. As increasing the value of γ_{NN}^2, which corresponds to that the coupling between normal layers is weakened, the flux line in the N-N case spreads out along the layers over an extremely wide range. In contrast the flux line sitting in between the superconducting layers is not so changed for the variation of the value of γ_{NN}^2.

Table I The self-energy of the single flux line in the unit of $(c/2e)^2/(4\Lambda_S^2)$. f_{SS} and f_{NN} indicate the self-energy when the flux lines are situated between the S-layers and N-layers, respectively.

γ_{NN}^2	f_{SS}	f_{NN}	$f_{SS} - f_{NN}$
200	0.561	0.300	0.261
600	0.508	0.188	0.320
3000	0.455	0.094	0.361

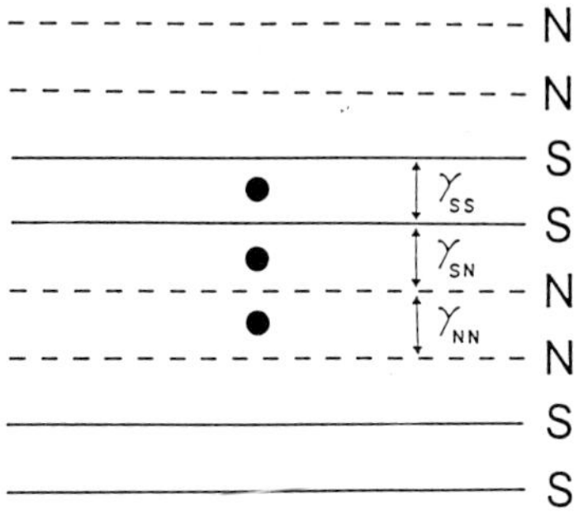

Fig.6 The structure of the layered superconductor. The black circles indicate the three inequivalent positions of the flux line parallel to the layers.

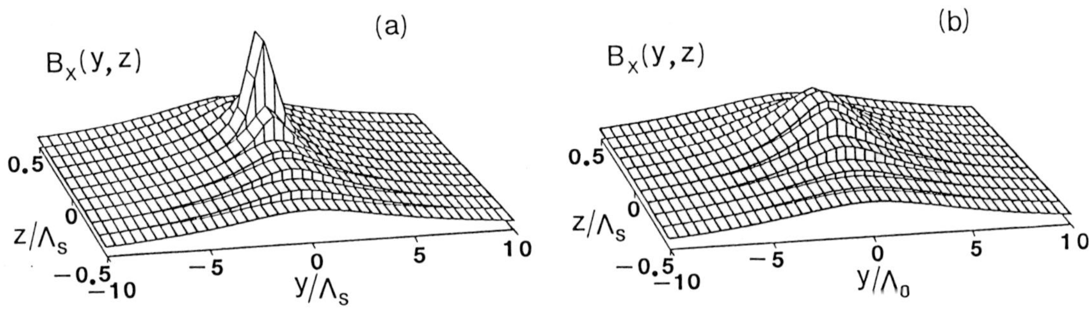

Fig.7 Field distribution of the single flux line for $\gamma_{NN}^2 = 200$ in the (a) S-S and (b) N-N cases.

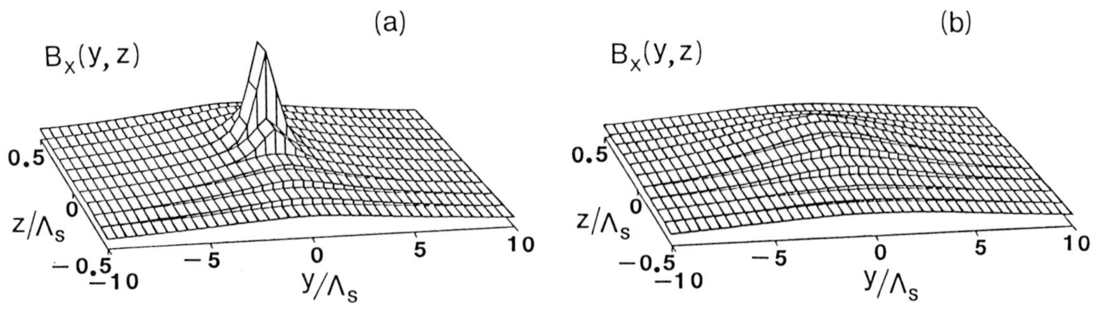

Fig.8 Field distribution of the single flux line for $\gamma_{NN}^2 = 600$ in the (a) S-S and (b) N-N cases.

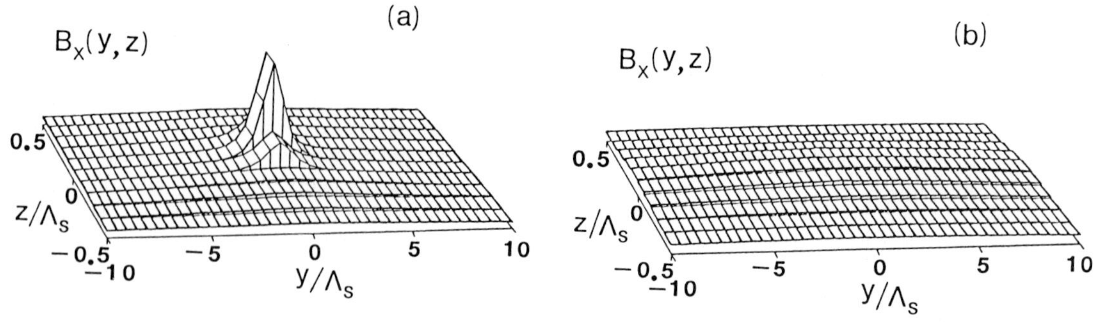

Fig.9 Field distribution of the single flux line for $\gamma_{NN}^2 = 3000$ in the (a) S-S and (b) N-N cases.

FLUX PINNING

As shown in ref.[21], the self-energy of the single flux line (the free energy of the single flux line state) can be expressed as

$$F_{\text{self}} = \frac{\phi_0}{8\pi} B_0(0) + O(d). \tag{12}$$

Here the second term represents the correction of $O(d)$ originating from the layer structure. The self-energy is, thus, approximately proportional to the intensity of the internal magnetic field at the center of flux line. Therefore, the results for $B_t(y)$ given in Figs.7-9 indicate that the single flux line situated in between normal layers is much stabler than that in between superconducting layers. We listed the calculated values of the self-energy including the second term in Eq.(12) in Table I in the cases given in Figs.7-9. This self-energy difference causes the intrinsic pinning force for the flux line moving in the direction perpendicular to the layers. As seen in Table I the difference between the self-energies in the S-S and N-N cases increases as the value of γ_{NN}^2 increases. Therefore, in superconductors containing extremely weak superconducting layers the strong pinning force for the single flux line appears along the direction perpendicular to the layers. In the mixed state with more than one flux line the flux lines are expected to sit on the weakly superconducting layers when the number of the flux lines is not so large. In this region we expect the strong pinning effect arising from the layer structure itself for the flux line lattice, since the pinning force for each flux line additively works [21-25].

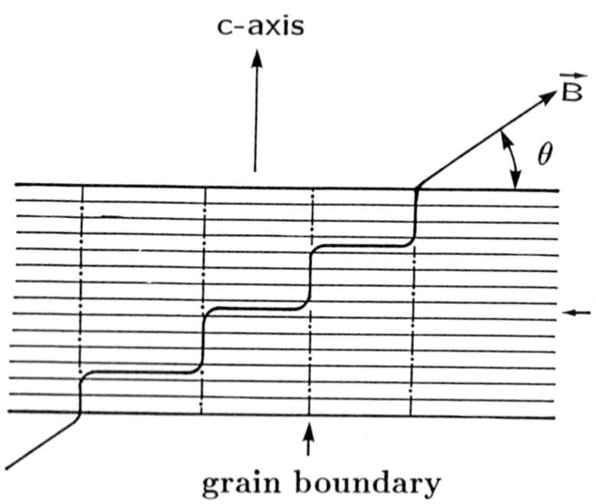

Fig.10 Penetration of a flux line in a sample. The bold solid line indicates a pass of the flux-line core. The horizontal solid lines indicate the CuO_2 layers and the vertical dashed lines indicate imperfections such as grain boundaries.

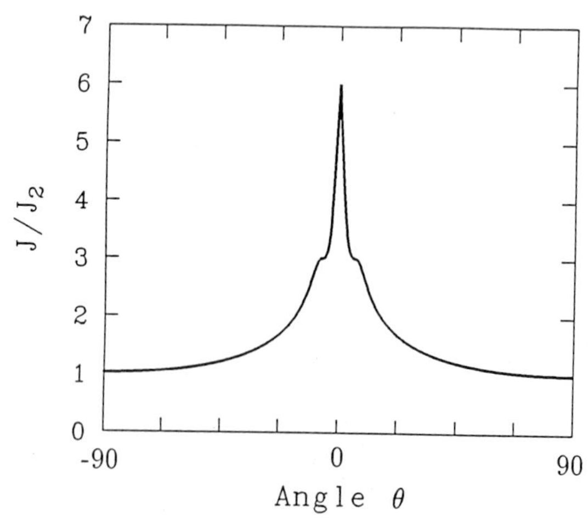

Fig.11 Angular dependence of the critical current density as a function of tilt angle.

Figure 10 shows the core of the flux line penetrating a sample. When an external magnetic field is tilted from the layer above a critical angle, jogs of the flux line shown in Fig.10 may appear[22-25]. We assume that the motion of the flux line along the c axis is pinned by the intrinsic pinning which was discussed above and the motion of the jogs in the plane of the layers is pinned by some imperfections such as grain boundaries. Then, if we consider a configuration of measurement in which the current flows perpendicular to the magnetic field in the plane, the critical current is expected to be strongly dependent on the tilt angle of the magnetic field from the plane [22,26,27]. When we select a set of values for the creation energy of the jogs and the pinning force density, the angular dependence of the critical current density J_c is calculated as shown in Fig.11. The angle θ denotes the tilt angle of the applied magnetic field from the plane of the layers. The sharp peak of J_c around θ=0 is related to the intrinsic pinning and the shoulder of J_c is related to the pinning of the motion of the jogs. The behavior of the angular dependence of J_c is similar to that obtained by Adrian using $YBa_2Cu_2O_{7-\delta}$ films of high quality [28].

REFERENCES

1. Geerk J, Xi X X, Linker G (1988) Z. Phys. **B73**: 329.
2. Gurvitch M, Valles Jr. JM, Cucolo AM, Dynes RC, Garno JP, Schneemeyer LF, Waszczak JV (1989) Phys.Rev.Lett. **63**: 1008.
3. Fournel A, Oujia I, Sorbier I, Noel H, Levet JC, Potel M, Gougeon P (1988) Europhys. Lett. **6**: 653.
4. Takeuchi I, Tsai JS, Shimakawa Y, Manako T, Kubo Y (1989) Physica **C158**: 83.
5. Schlesinger Z, Collins RT, Holtzberg F, Feild C, Blanton SH, Welp U, Crabtree GW, Fang Y, Liu JZ (1990) Phys. Rev. Lett. **64**: 801.
6. Thomas GA, Orenstein J, Rapkine DH, Capizzi M, Millis AJ, Bhatt RN, Schneemeyer LF, Waszczak JV (1988) Phys. Rev. Lett. **61**: 1313.
7. Kamarás K, Herr SL, Porter CD, Tache N, Tanner DB, Etemad S, Venkatesan T, Chase E, Inam A, Wu XD, Hegde MS, Dutta B (1990) Phys. Rev. Lett. **64**: 84.
8. Timusk T, Reedyk M, Hughes R, Bonn DA, Garrett JD, Greedan JE, Stager CV, Tanner DB, Gao F, Herr SL, Kamaras K, Thomas GA, Cooper SL, Orenstein J, Schneemeyer LF, Millis AJ (1989) Physica **C162-164**: 841.
9. Warren Jr. WW, Walstedt RE, Brennert GF, Espinosa GP, Remeika JP (1987) Phys.Rev.Lett. **59**: 1860.
10. Kitaoka Y, Hiramatsu S, Kondo T, Asayama K (1988) J. Phys. Soc. Jpn. **57**: 30.
11. Imai T, Shimizu T, Yasuoka H, Ueda Y, Kosuge K (1988) J. Phys. Soc. Jpn. **57**: 2280.
12. P.C. Hammel, M. Takigawa, R.H. Heffner, Z. Fisk, K.C. Ott (1989) Phys.Rev.Lett. **63**: 1992.
13. Takigawa M, Hammel PC, Heffner RH, Fisk Z, Ott KC, Thompson JD (1989) Physica **C162-164**: 853.
14. Barrett SE, Durand DJ, Pennington CH, Slichter CP, Friedmann TA, Rice JP, Ginsberg DM (1990) Phys. Rev. **B41**: 6283.
15. Takahashi S, Tachiki M (1990) Physica **C170**: 505.
16. Takahashi T, Matsuyama M, Katayama-Yoshida H, Okabe Y, Hosaya S, Seki K, Fujimoto H, Sato M , Inokuchi H, Nature **334**: (1988)691.
17. Campuzano JC, Jennings G, Faiz M, Beaulaigue L, Veal BW, Liu JZ, Paulikas AP, Vandervoort K, Claus H, List RS, Arko AJ, . Bartlett RJ (1990) Phys. Rev. Lett. **64**: 2308.
18. Tachiki M, Takahashi S, F. Steglich, H. Adrian (1990) Z.Phys. **B80**: 161.
19. Lawrence W, Doniach S (1971) In: Kanda E (eds) in *Proc. 12th Int. Cof. on Low Temperature Physics.* Academic Press of Japan, Kyoto, p.361
20. Klemm RA, Luther A, Beasley MR (1975) Phys. Rev. **B12**: 877.
21. Koyama T, Takezawa N, Tachiki M, to be published in Physica C
22. Tachiki M, Takahashi S (1989) Solid State Commun. **70**: 291, *ibid.* **72**: 1083.
23. B.I. Ivlev, N.B. Kopnin (1990) Phys. Rev. Lett. **64**: 1828.
24. Chakravatry S, Ivlev BI, Ovchinnikov YN (1990) Phys. Rev. Lett. **64**: 3187.
25. Feinberg D, Villard C (1990) Phys. Rev. Lett. **65**: 919.
26. Roas B, Schultz L, Saemann-Ischenko G (1990) Phys. Rev. Lett. **64**: 479.
27. Raffy H, Labdi S, Laborde O, Monceau P (1990) Physica **B165-166**: 1423.
28. Adrian H, private communications.

Gauge Field Theory of the High-T_c Superconductors

N. NAGAOSA

Department of Applied Physics, University of Tokyo, Tokyo, 113 Japan

ABSTRACT

The uniform resonating valence bond state is investigated in terms of the gauge field, which represents the local constraint and the fluctuation of the spin chirality. The two contradictory pictures of the normal state of the high-T_c superconductors, i.e., the doped Mott insulator and the metal with a large fermi surface, are unified. The transport properties are governed by the bosons and we obtain the T-linear resistivity and temperature dependent Hall constant which are in agreement with the experiments.

KEY WORDS: RVB, gauge field, spin-charge separation, transport

INTRODUCTION

There are two contradictory pictures of the normal state of the high-T_c superconductors. One is the doped Mott insulator and the other is the metal with a large fermi surface which satisfies Luttinger's theorem. In the former picture the charge carriers are the doped holes (whose concentration is denoted by x below), while the electrons carry current in the latter. Some experiments, e.g., positive Hall constant roughly proportional to $1/x$, support the former and other experiments, e.g., a large fermi surface detected in the angle-resolved photoemission spectra, support the latter.

In this paper we propose the uniform resonating valence bond state [1-3] as a promising candidate for a model of the normal state of high-T_c superconductors, investigating its properties in terms of the gauge field which restores the local constraint neglected in the mean field theory. The physical electron operator $C_{i,\sigma}$ is decomposed into $f_{i,\sigma}b_i^\dagger$. The mean field order parameters are $\chi_{ij} =< f_{i,\sigma}^\dagger f_{j,\sigma} >$ and $B_{ij} =< b_i^\dagger b_j >$ which are spatially uniform in the uniform RVB state. The effective Lagrangian which describes the low energy excitation of this state is given by [2,3]:

$$L = \sum_{i,\sigma} f_{i,\sigma}^*(\partial_\tau - a_0 - A_0 - \mu_F) f_{i,\sigma} + \sum_{<ij>} J e^{ia_{ij}} f_{i,\sigma}^* f_{j,\sigma} + c.c.$$

$$+ \sum_i b_i^*(\partial_\tau - a_0 - \mu_B) b_i + \sum_{<ij>} \tilde{t} e^{ia_{ij}} b_i^* b_j + c.c., \tag{1}$$

where b, b^* and f_σ, f_σ^* represent the bosons and fermions, respectively. The chemical potentials μ_F, μ_B enforce $< b_i^\dagger b_i >= x$ and $\sum_\sigma < f_{i,\sigma}^\dagger f_{i,\sigma} >= 1 - x$. The time component of the gauge field a_0 is nothing but the Lagrange multiplier. There are symmetric and antisymmetric combinations of the two phases of the RVB order parameters χ_{ij} and B_{ij}. The symmetric one is lower in energy which is nothing but the space component a_{ij} of the gauge field. In the large N mean field theory $\tilde{t}$ is the original hopping matrix element t of the physical electron, and the Bose-Einstein(BE) condensation temperature $T_{BE}^{(0)} \approx 4\pi\tilde{t}x$ is very high. Below this $T_{BE}^{(0)}$, the system behaves as a fermi liquid. In reality, however, N=2 and the fluctuations are important. It has been discussed that $\tilde{t}$ is subject to the downward renormalization from t to J (exchange interaction) because the hole distorts the spin configuration around itself considerably [4]. Therefore T_{BE} becomes of order $4\pi Jx \sim 1000K$ which is still higher than the temperature of interest. In this paper, we discuss that the inelastic scattering of the bosons by the gauge field suppresses T_{BE} down to $\sim 100K$ which at the same time explains the T-linear resistivity and the temperature dependent Hall constant. As will be discussed later, we identify T_{BE} with the superconducting T_c. In the continuum limit, the Lagrangian given in eqs.(1) and (2) is written as

$$L=\sum_\sigma \int \mathrm{d}^2\vec{r}\Big[f_\sigma^*(\vec{r})(\partial_\tau - a_0 - A_0 - \mu_F) f_\sigma(\vec{r}) + f_\sigma^*(\vec{r})\frac{1}{2m_F}(i\nabla - \vec{a} - \vec{A})^2 f_\sigma(\vec{r})\Big]$$

$$+\int \mathrm{d}^2\vec{r}\Big[b^*(\vec{r})(\partial_\tau - a_0 - \mu_B) b(\vec{r}) + b^*(\vec{r})\frac{1}{2m_B}(i\nabla - \vec{a})^2 b(\vec{r})\Big], \tag{2}$$

where the lattice constant is put to be unity. The energy is measured in units of J because $1/m_B$ and $1/m_F$ are considered to be of order J from the above discussion. In terms of the gauge field $\vec{a}$, the spin chirality density is expressed as $\nabla\times\vec{a}$ whose integral over an area enters as the phase factor of the quantum mechanical amplitude for the particle (fermion or boson) going around the loop enclosing that area. The external electromagnetic field $\vec{A}$ transforms as $\vec{A}\to\vec{A}+\nabla\theta$ with the gauge transformation $C_\sigma \to C_\sigma e^{i\theta}$. This gauge transformation is realized by $f_\sigma \to f_\sigma e^{i\alpha\theta}$, $b\to b e^{i(\alpha-1)\theta}$ for arbitrary α, and the coupling of the system with $\vec{A}$ is specified for each α but the gauge invariant quantities are not affected by the choice of α. In eq.(2) we choose $\alpha=1$.

If we integrate $\vec{a}$ and a_0 first in eq.(2), we obtain the constraint $j_{F\mu}+j_{B\mu}=0$ with $\mu=0,x,y$ because the coupling is written as $a_\mu(j_{F\mu}+j_{B\mu})$. Instead we integrate the fermions and bosons first to obtain the effective action S_g for the gauge field as

$$S_g=\sum_{\mu\nu}\sum_{q\omega_n}\Big[\Pi_{F\mu\nu}(q,\omega_n)(a_\mu(q,\omega_n)+A_\mu(q,\omega_n))(a_\nu(-q,-\omega_n)+A_\nu(-q,-\omega_n))$$

$$+\Pi_{B\mu\nu}(q,\omega_n)a_\mu(q,\omega_n)a_\nu(-q,-\omega_n)\Big], \tag{3}$$

where $\Pi_{F(B)}$ is the fermion (boson) current-current correlation function. In the transverse gauge, the time component (0) and the transverse spatial component ($\perp$) are decoupled, and the square is completed for each component in eq.(3).

$$S_g=\sum_{\beta=0,\perp}\sum_{q\omega_n}(\Pi_{F\beta}+\Pi_{B\beta})\left(a_\beta+\frac{\Pi_{F\beta}}{\Pi_{F\beta}+\Pi_{B\beta}}A_\beta\right)^2+\frac{\Pi_{F\beta}\Pi_{B\beta}}{\Pi_{F\beta}+\Pi_{B\beta}}A_\beta^2. \tag{4}$$

$-\Pi_{F\beta}A_\beta/(\Pi_{F\beta}+\Pi_{B\beta})$ is the shift of the average of a_β , and the gauge field which the fermion (boson) system feels on the average is $\Pi_{B\beta}A_\beta/(\Pi_{F\beta}+\Pi_{B\beta})$ $(-\Pi_{F\beta}A_\beta/(\Pi_{F\beta}+\Pi_{B\beta}))$. As will be discussed below, this screening effect restores the local constraint and leads to the composition laws for various physical quantities. The coefficient of the square is nothing but the inverse of the propagator of the gauge field. The important low lying excitation comes from the transverse component whose propagator D^T is given by

$$D^T=(\Pi_{F\perp}+\Pi_{B\perp})^{-1}=\left(\frac{|\omega_n|}{q}+\chi_d q^2\right)^{-1}, \tag{5}$$

where $\chi_d=\chi_F+\chi_B$ is the sum of the Landau diamagnetic susceptibility of the fermions and bosons. It should be noticed that only the gauge invariant quantities have physical meaning, and these quantities can be represented by a closed loop in the space-time Feynman path formalism [2]. Of particular importance are the current-current correlation functions Π_F,Π_B , which should be determined self-consistently taking into account the interaction with the gauge field, and the Green function of the physical electron $G_\sigma(\vec{r},\tau)=-<T_\tau f_\sigma(\vec{r},\tau)b^\dagger(\vec{r},\tau)f_\sigma^\dagger(0)b(0)>$. The boson polarization $\Pi_B(\vec{r},\tau)$ is expressed by the integral over the closed path passing the two space-time points $(\vec{0},0)$ and $(\vec{r},\tau)$ as follows:

$$\Pi_B(\vec{r},\tau)=\int D\vec{r}_1(t)\left\langle \exp\left[-\frac{m_B}{2}\int_0^{2\tau}dt\dot{\vec{r}}_1(t)^2+i\Phi[\vec{r}_1]\right]\right\rangle_{\vec{a}}. \tag{6}$$

Here $\vec{r}_1(0)=\vec{r}_1(2\tau)=\vec{0}$, $\vec{r}_1(\tau)=\vec{r}$, $\Phi[\vec{r}_1]=\oint\vec{a}(\vec{r}_1(t))\cdot\dot{\vec{r}}_1(t)dt$ and $<\ \ >_{\vec{a}}$ means the average over the gauge field fluctuations. Since the important contributions come from $\omega<T^{3/2}<<T$ because $T<<J$, we approximate the transverse gauge field by a static but spatially varying "magnetic" field $h=\nabla\times\vec{a}$ and $\Phi[\vec{r}_1]$ becomes $\Phi[\vec{r}_1]=\int\int h dA$ which is the flux penetrating the area A enclosed by the path $\vec{r}_1(t)$ projected onto the real space. The fluctuation of this "magnetic" field h is given by

$$<h_q h_{-q}>=\int d\omega\frac{1}{e^{\omega/T}-1}q^2\mathrm{Im}D^T(q,\omega)\cong\frac{T}{\chi_d}, \tag{7}$$

for $q<q_0$ with $q_0\approx(T/\chi_d)^{1/3}$. Thus we envision a spatially random h which varies on a scale q_0^{-1}. When the linear dimension of the area A for the integral eq.(6) is larger than q_0^{-1}, the integral for $\Phi[\vec{r}_1]$ is the sum

of many independent flux fluctuations , i.e., $\Phi[\vec{r}_1] = \sum_{i=1}^{N} \Phi_i$. In this case, $< \Phi[\vec{r}_1]^2 >= N < \Phi_i^2 >$ where the number of the independent flux fluctuations N is roughly Aq_0^2 and $< \Phi_i^2 >$ is estimated as

$$< \Phi_i^2 > \cong q_0^{-4} \int_0^{q_0} q dq \frac{T}{\chi_d} \cong q_0^2 \frac{T}{\chi_d}. \tag{8}$$

Averaging $e^{i\Phi[\vec{r}_1]}$ over $\vec{a}$ gives the Debye-Waller factor $e^{-<\Phi[\vec{r}_1]^2>/2}$, and we obtain

$$\Pi_B(\vec{r},\tau) = \int D\vec{r}_1(t) \exp\left[-\frac{m_B}{2}\int_0^{2\tau} dt \dot{\vec{r}}_1(t)^2 - \frac{gA}{\lambda^2}\right], \tag{9}$$

where $g = 1/m_B\chi_d$ is the dimensionless coupling constant and $\lambda = (2\pi/m_BT)^{1/2}$ is the thermal length of the boson. For the unperturbed trajectories, the first term of the exponent in eq.(9) is of order of 1, and $A \sim A_0 = r(\tau/m_B)^{1/2} < \lambda^2$. Therefore, for $g << 1$ the first term dominates the second and we obtain

$$\Pi_B(\vec{r},\tau) = \Pi_B^{(0)}(\vec{r},\tau) \exp[-gTm_Br(\tau/m_B)^{1/2}], \tag{10}$$

where $\Pi_B^{(0)}(\vec{r},\tau) = \tau^{-2}\exp[-2m_Br^2/\tau]$. Thus putting the typical $r \sim (\tau/m_B)^{1/2}$ into eq.(10), the mean free time τ_B and the mean free path l_B of the boson are read as $\tau_B^{-1} \cong gT$ and $l_B \cong \lambda/g$. In the opposite limit $g >> 1$, the loops with areas of order $A \sim \lambda^2/g$ contribute. The reduction in the area A means that the almost retracing paths give dominant contributions. Even if we restrict to the exactly retraceable paths so that $\vec{r}_1(\tau+t) = \vec{r}_1(\tau - t)$ for $0 < t < \tau$, we still have $\Pi_B(\vec{r},\tau)$ which is the Green function G_B of the boson with the mass $2m_B$ and its thermal length $\lambda' = \lambda/\sqrt{2}$. In general, we can choose an arbitrary function $f(t)$ and construct a path by $\vec{r}_1(\tau+t) = \vec{r}_1(\tau - f(t))$ which is also retraceable when projected onto the real space. Therefore Π_B is not exactly G_B, but the characteristic length, i.e., the mean free path l_B, still remains of order $\lambda' \sim \lambda$. Interpolating these results for $g << 1$ and $g >> 1$, we conclude $l_B \sim \lambda/\min(1,g)$, which is the characteristic magnitude of $|\vec{r}|$, the width of the Feynman tube $\sim \lambda/(1+g)$ and the area $A \sim \lambda^2/(1+g)$. Therefore the boson resistivity $\rho_B \sim m_BT\min(1,g)/x$ which remains $T-$linear irrespective of the $T-$dependence of $g = 1/m_B\chi_d$ as long as $g > 1$. The response of the boson system to the magnetic field H, however, is reduced because H always appears in the form of the flux $\phi = HA$ and A is reduced. For example, since σ_{xy} is proportional to H, the Hall constant R_H^B is reduced from the weak coupling value x^{-1} to $x^{-1}(1+g)^{-1}$. The Landau diamagnetic susceptibility χ_B is related to H^2 and hence reduced by the factor $(1+g)^{-2}$. If we write $\chi_B = T_{BE}/m_BT$, T_{BE} is also reduced by $(1+g)^{-2}$ from $4\pi Jx \sim 1000K$.

The similar discussions can be done for Π_F and G_σ. For Π_F the width of the Feynman tube is of order k_F^{-1} which is much smaller than q_0^{-1}, and the calculation of the Debye-Waller factor is modified from the boson case. We obtain the mean free time $\tau_F \cong (T/\chi_d)^{4/3}E_F$ and the mean free path $l_F \cong v_F\tau_F$. It should be noted that $l_Fk_F >> 1$ and the interaction between the fermion and the gauge field is always in the weak coupling regime in contrast to the boson case. For the physical electron Green function G_σ we obtain its lifetime τ_{in} near the quasi-particle peak as $\tau_{in}^{-1} \cong (T/\chi_d)^{2/3}(m_Bm_F^2)^{-1/3}$ in the weak coupling limit $g << 1$.

We now discuss the physical quantities of the total system [2]. The dynamics of the fermions and bosons are correlated by the constraint or equivalently through the coupling with the gauge field. The observable current J of the electron is $J_F = -J_B$. The fermion (boson) current $J_F(J_B)$ is $\Pi_F(\Pi_B)$ times the average gauge field $\Pi_{B\beta}A_\beta/(\Pi_{F\beta}+\Pi_{B\beta})$ $(-\Pi_{F\beta}A_\beta/(\Pi_{F\beta}+\Pi_{B\beta}))$ which the fermions (bosons) feel. Then we obtain $J = J_F = -J_B = \Pi A$ with $\Pi = \Pi_F\Pi_B/(\Pi_F+\Pi_B)$ which can be derived also from the last term on the righthand side of eq.(4). From this, the conductivity σ of the total system is $\sigma_F\sigma_B/(\sigma_F+\sigma_B)$ which means that the resistivity ρ is the sum $\rho_B+\rho_F$. From the above discussion, $\rho_B >> \rho_F$ and $\rho \cong \rho_B \cong m_BT\min(1,g)/x$. The dimensionless coupling constant g decreases when approaching the superconducting T_c because $\chi_d(T)$ diverges toward T_c. Well above T_c in the normal phase, g is estimated to be of order unity from the theoretical point of view and $g > 1$ is suggested from the analysis of the Hall constant below. Therefore $\rho \cong Tm_B/x$ which explains the experiment. When the magnetic field is applied, the fermions (bosons) feel the magnetic field $\chi_BH/(\chi_F+\chi_B)$ $(-\chi_FH/(\chi_F+\chi_B))$ and the Hall constant R_H is given by

$$R_H = \frac{\chi_FR_H^B + \chi_BR_H^F}{\chi_F+\chi_B}, \tag{11}$$

where R_H^B is given above. R_H^B is the Hall constant for the fermion system which depends on the details of the band structure but its magnitude is expected to be smaller than x^{-1} and we aproximate R_H by $R_H \cong x^{-1}\chi_F/(m_B^{-1}+\chi_F+\chi_B)$. Well above T_c, $\chi_F >> \chi_B$ and the T-dependence of χ_F, which is expected from the pairing fluctuation of the fermions, can explain the decreasing R_H as a function of T. Experimentally, the high temperature limit of R_H is reduced from x^{-1} by a factor of 2 to 4 depending on the material which can be regarded as $1+g$. The Landau diamagnetic susceptibility χ is $\chi_B\chi_F/(\chi_F+\chi_B)$ which is approximately χ_B when $\chi_F >> \chi_B$. Because the reduction factor for χ_B is the square of that for R_H, we expect an order of magnitude reduction of χ_B and hence T_{BE}, which results in $T_{BE} \sim 4\pi Jx/10 \sim 100K$.

The physical electron Green function $G_\sigma(\vec{k}, \Omega)$ is given by the convolution of the Green functions of the fermion and boson in the mean field approximation. Taking into account the interaction with the gauge field we obtain the following results. The spectral function $\mathrm{Im}G_\sigma(\vec{k}, \Omega)$ consists of a continuum for $\Omega < -|\mu_B| - (|\vec{k}| - k_F)^2/2m_F$ and a peak at $\Omega = \vec{k}^2/2m_F - \mu_F$. The width of the peak is equal to $\max(\Omega^{2/3}J^{1/3}, (Tm_B)^{1/2}m_F^{-1})$. $(Tm_B)^{1/2}m_F^{-1}$ comes from the momentum distribution of the bosons while $\Omega^{2/3}J^{1/3}$ from the coupling with the gauge field. This width violates the Landau criterion for the quasi-particle. If the bosons condense, however, this width becomes $\max(\Omega^2, T^2)$ and the system is a fermi liquid. On the other hand, the location of the peak in $\vec{k}$ space for small $|\Omega|$ is determined by the large "spinon" fermi surface which satisfies Luttinger's theorem.

We conclude by outlining a scenario whereby T_{BE} can be identified with the superconducting T_c. The inelastic scattering by the gauge field gives rise to the pair-breaking effect and the transition temperature of the fermion pairing is reduced from its mean field value $T_D^{(0)}$ to a lower value T_D as in the case of the Bose-Einstein temperature T_{BE}. When either Π_F or Π_B shows the Meissner effect, $\chi_d q^2$ in eq.(5) is replaced by the superfluidity density ρ_S of the fermions or the bosons. This stiffens up the gauge field and the reduction of the transition temperature disappears. Therefore Bose-Einstein condensation and the pairing of the fermions occur simultaneously at $T_c = \max(T_D, T_{BE})$ as long as T_c is below $\min(T_D^{(0)}, T_{BE}^{(0)})$. This explains why the non-fermi liquid behaviour is seen in the normal phase of the concentration region with high-T_c. In the over-doped case, however, $T_{BE} > T_D^{(0)} = T_c$ and we have the fermi liquid behavior, i.e., $\rho \propto T^2$ and negative R_H, between these two temperatures. When the fluctuations are taken into account, the transition at T_{BE} becomes only a cross-over with the exponential growth of the correlation length ξ of the boson as $e^{T_{BE}/T}$ below T_{BE}.

The results reported here are obtained in collaboration with Patrick Lee, and I acknowledge him for discussion and the hospitality at MIT.

REFERENCES

1. Anderson P.W. (1987) Science 235: 1196
2. Nagaosa N, Lee P.A. (1990) Phys. Rev. Lett. 64: 2450 and unpublished
3. Ioffe L.B., Larkin A.I. (1989) Phys. Rev. B39: 8988
4. Kane C.N., Lee P.A., Read.N (1989) Phys. Rev. B39: 6880

Superconductivity due to Interband Pair-Transfer Interaction Divergently Enhanced by a Filled Band

K. YAMAJI and T. KINOSHITA

Electrotechnical Laboratory, 1-4, Umezono 1-chome, Tsukuba, 305 Japan

ABSTRACT

An expression of T_c is derived for a two-band model which has an interband pair-transfer interaction between the conductive and a completely filled second bands. The interband ladder diagram process enhances the effect of the pair-transfer interaction, remarkably reducing the effective on-site Coulomb energy and leading even to high-T_c superconductivity with a negative U. The isotope effect is shown to weaken with increasing T_c.

KEY WORDS: Two-band model, pair-transfer interaction, ladder diagram, expression of T_c, negative U

INTRODUCTION

We have been looking for the mechanism of high-T_c superconductivity in the multiple-band nature of cuprates. Between two tight-binding bands consisting of different-type atomic orbitals, $\phi_1(\mathbf{r})$ and $\phi_2(\mathbf{r})$, on identical atomic sites, there arises an off-diagonal Coulomb integral on one atomic site, called exchange-like inegral, which is defined by $K \equiv \int d\mathbf{r}\int d\mathbf{r}'\phi_1{}^*(\mathbf{r})\phi_1{}^*(\mathbf{r}')(e^2/|\mathbf{r}-\mathbf{r}'|)\phi_2(\mathbf{r}')\phi_2(\mathbf{r})$. Due to a finite value of K ~ 0.5-1 eV, there exists an interband pair-transfer interaction through which BCS-like pairs consisting of electrons with spin up and down in one band are scattered between the two bands. This process promotes a BCS-type superconductivity [1,2]. The effect of this interaction was found to be remarkably enhanced by a process expressed by ladder diagrams [3]. This enhancement is considered to be particularly strong when the two bands cross the Fermi level and their Fermi surfaces have a nesting property with each other [3,4]. In this Communication we show that the enhancement is also important even in Kondo's original case where the second band does not cross the Fermi energy. For this purpose, a simple solvable model is treated. The obtained expression of T_c reveals that the pair-transfer interaction can strongly reduce the effective on-site Coulomb energy by the help of the ladder-diagram process and promote superconductivity, even leading to high T_c in the extreme case. This effect may be relevant in oxide, organic and transition-metal superconductors. With increase of T_c the isotope effect is clearly seen to weaken.

MODEL AND LADDER DIAGRAM

The Hamiltonian of our two-band model is defined by

$$H = \sum_{j=1}^{2}\sum_{\mathbf{k}\sigma} \varepsilon_{j\mathbf{k}} c^{\dagger}_{j\mathbf{k}\sigma} c_{j\mathbf{k}\sigma} + \sum_{j=1}^{2} \frac{U_j}{N} \sum_{\mathbf{k}_1\mathbf{k}_2\mathbf{k}_3\mathbf{k}_4} \sum_{\sigma\sigma'} \delta_{\mathbf{k}_1+\mathbf{k}_2,\mathbf{k}_3+\mathbf{k}_4} c^{\dagger}_{j\mathbf{k}_1\sigma} c^{\dagger}_{j\mathbf{k}_2\sigma'} c_{j\mathbf{k}_3\sigma'} c_{j\mathbf{k}_4\sigma}$$

$$+ \sum_{\mathbf{k}_1\mathbf{k}_2\mathbf{k}_3\mathbf{k}_4} \delta_{\mathbf{k}_1+\mathbf{k}_2,\mathbf{k}_3+\mathbf{k}_4} \left[\frac{K}{N} (c^{\dagger}_{1\mathbf{k}_1\uparrow} c^{\dagger}_{1\mathbf{k}_2\downarrow} c_{2\mathbf{k}_3\downarrow} c_{2\mathbf{k}_4\uparrow} + \mathrm{H.c.}) + \frac{U'}{N} \sum_{\sigma\sigma'} c^{\dagger}_{1\mathbf{k}_1\sigma} c^{\dagger}_{2\mathbf{k}_2\sigma'} c_{2\mathbf{k}_3\sigma'} c_{1\mathbf{k}_4\sigma} \right], \quad (1)$$

where $c_{jk\sigma}$ ($c_{jk\sigma}^\dagger$) is the annihilation (creation) operator of the electron in the j-th band with wave vector $\mathbf{k}$ and spin σ; N is the number of the electronic sites; U_j is the on-site Coulomb energy for the j-th band; U' is the interband Coulomb energy between the two bands; we call the interaction with the exchange-like integral K as the coupling constant by the name of exchange-like interaction or interband pair-transfer interaction. We use the convention $k_B = \hbar = 1$. As the irreducible vertex transferring a BCS pair from the second band to the first, there appear interband ladder diagrams with the advent of the pair-transfer interaction in the diagrammatic perturbation expansion in powers of the interaction terms in eq. (1) [3]. The ladder diagrams are summed up to

$$\Gamma_{12}(-k, k+q) = (K/N)/\mathrm{Re}\{[1 - (U'+K)\Pi_{12}(\mathbf{q}, \omega_\nu)][1 - (U'-K)\Pi^*_{12}(\mathbf{q}, \omega_\nu)]\}, \quad (2)$$

where Re{ } denotes the real part; k and q stand for $(\mathbf{k},\omega_n)$ and $(\mathbf{q}, \omega_\nu)$, respectively; $\omega_n = \pi T(2n+1)$ and $\omega_\nu = 2\pi T\nu$ where T is the temperature and n and ν are integers; this irreducible part links a BCS state consisting of $(2,k+q,\uparrow)$ and $(2,-k-q,\downarrow)$ to another pair state of $(1,-k,\uparrow)$ and $(1,k,\downarrow)$. $\Pi_{12}(\mathbf{q}, \omega_\nu)$ is the interband polarization function given by $\Pi_{12}(\mathbf{q}, \omega_\nu) = \Sigma_{\mathbf{k}}[f(\varepsilon_{1\mathbf{k}}) - f(\varepsilon_{2\mathbf{k+q}})]/(i\omega_\nu + \varepsilon_{2\mathbf{k+q}} - \varepsilon_{1\mathbf{k}})$, where $f(\varepsilon)$ is the electron distribution function. This is positive at least for $\omega_\nu = 0$ where the absolute value of this function is maximized. Therefore eq. (2) means that the effective matrix element Γ_{12} for the pair transfer between the two bands is strongly enhanced from the bare value K/N by the denominator, especially when the factors in the square bracket approach zero. This remarkable denominator is confirmed by Entin-Wohlman and Imry by an ordinary RPA treatment, in which it is obtained as due to screening (actually anti-screening) of the external spin density [5]. This effect has the potential to enhance the effect of the pair transfer process in all the cases where $(U'+K)\Gamma_{12}$ becomes close to unity.

SIMPLE MODEL AND EXPRESSION OF T_c

In this Communication we would like to stress that the remarkable enhancement of Γ_{12} in eq. (2) is also possible, even in Kondo's original two-band situation [1] where the second band does not cross the Fermi level, only if $(U'+K)\Gamma_{12}$ closes to unity. To demonstrate this, we treat a very simple model with two bands illustrated in Fig. 1. Band 1 defined by $\varepsilon_{1\mathbf{k}} \equiv \varepsilon_{\mathbf{k}}$ is assumed to be rather narrow and half-filled. We choose zero level of energy at the Fermi energy ε_F, as shown in Fig. 1. Band 2 is assumed completely flat lying at $-E_2 < 0$. Then we get as a first approximation

$$\Pi_{12}(\mathbf{q}, i\omega_\nu) \cong N/2E_2, \quad (3) \qquad \Gamma \equiv \Gamma_{12} \cong K/\{[1 - (U'+K)/2E_2][1 - (U'-K)/2E_2]\}. \quad (4)$$

When $(U'+K)/2E_2$ approaches unity, Γ is so enhanced that superconductivity should occur. As in the previous paper [4], superconducting transition temperature T_c is determined as the temperature where unity is reached by the largest eigenvalue of the matrix **A** defined by the following matrix elements:

$$A(1k, 1k') = - (T/N)[U - V_{BCS}\theta(\omega_D-|\varepsilon_{\mathbf{k}}|)\theta(\omega_D - |\varepsilon_{\mathbf{k}'}|)]G(1k')G(1,-k'), \quad (5a)$$

$$A(1k, 2k') = - (T/N)\Gamma G(2k')G(2,-k'), \quad (5b) \qquad A(2k, 1k') = - (T/N)\Gamma G(1k')G(1,-k'), \quad (5c)$$

$$A(2k, 2k') = - (T/N)U_2G(2k')G(2,-k'), \quad (5d)$$

where 1k, e.g., denotes $(\mathbf{k}, \omega_n)$ of the band 1 electron; $U = U_1$; we have added in eq. (5a) the BCS attractive interaction with its coupling constant V_{BCS} due to electron-phonon

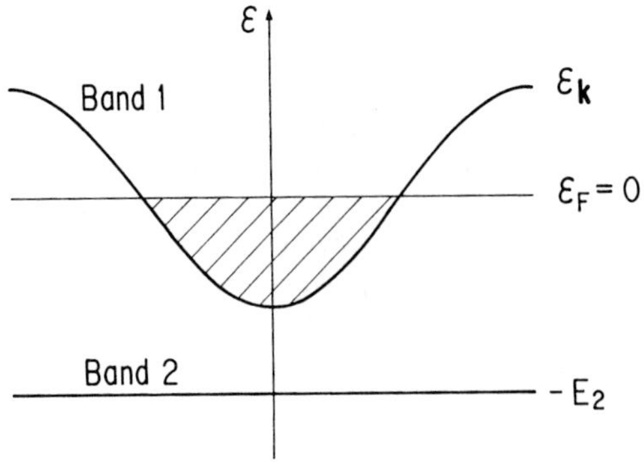

Fig. 1 Simple two-band model. The hatched part denotes occupied part. Zero level of energy is chosen at the Fermi level ε_F.

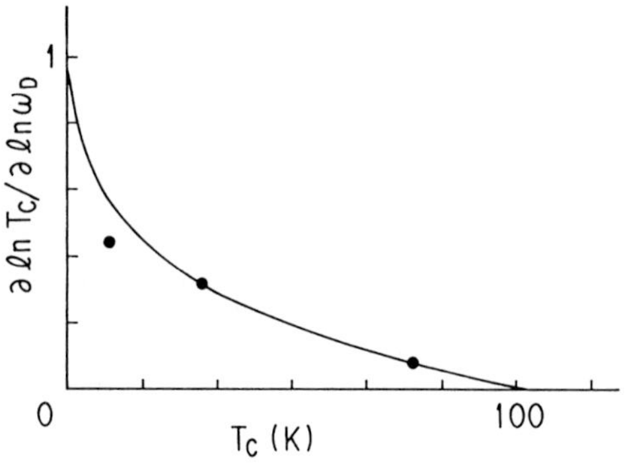

Fig. 2 Curve for $\partial\ln T_c/\partial\ln\omega_D$ vs T_c defined by eqs. (6) and (9). Closed circles denotes observed values for BPBO, $La_{1.85}Sr_{0.15}CuO_4$ and YBCO.

interaction working in the energy range $|\varepsilon_k| < \omega_D$ = Debye energy; $\theta(\ )$ is the step function; $G(1k) = 1/(i\omega_n-\varepsilon_k)$; $G(2k) = 1/(i\omega_n+E_2)$. This problem is solved in a good approximation. Assuming a constant state density N(0) per site for band 1, T_c is obtained as

$$T_c = 1.13\ \omega_D \exp[\ -1/(\lambda - \mu^*)\ ], \qquad (6)$$

$$\mu^* = \mu/[1 + \mu\ \ln(D/\omega_D)], \quad (7) \qquad \mu = N(0)[\ U - \Gamma^2/(2E_2 + U_2)\ ], \quad (8)$$

where $\lambda = N(0)V_{BCS}$ and D is the average half width of band 1.

This result for T_c had been obtained in [2], although the enhancement of K to Γ was not taken into account and Γ was set to be K. It is slightly different from the result in [5] in that the term $\Gamma^2/(2E_2+U_2)$ in eq. (8) is subtracted from $N(0)\mu^*$ in [5] but not from U as in our result.

When U'+K is close to E_2, Γ is divergently magnified in eq. (6) so that μ and μ^* decrease, becoming even negative and possibly leading to high T_c. Negative μ means negative U. This result can be interpreted as due to the reduction of U caused by the virtual transfer of the pair between the two bands. Even if $\lambda = 0$, superconductivity can occur, when μ is negative, with $T_c = 1.13D\exp(1/\mu)$. With increase of the role of negative μ in driving superconductivity, the isotope effect weakens, as is seen from

$$2\alpha \equiv \partial\ln T_c/\partial\ln\omega_D = 1 - [\mu^*/(\lambda - \mu^*)]^2. \qquad (9)$$

In oxide superconductors the value for this derivative is reported to be 0.44, 0.32 and 0.08 with T_c values 11, 33 and 92 K for BPBO, $La_{1.85}Sr_{0.15}CuO_4$ [6] and YBCO [7], respectively. In Fig. 3 these values are plotted in the 2α vs T_c plane and fitted by the curve defined by eqs. (6) and (9). The parameter values $\omega_D = 110\ k_B$ and $\lambda = 0.14$ are chosen as common parameters. Although these values are not so realistic, good fitting result is encouraging.

DISCUSSION

The obtained expression of T_c is derived for the case of weak coupling. For the cases of intermediate and strong coupling where Γ in eq. (4) approaches divergence, validity of eq. (6) is not assured. We expect, however, it remains qualitatively correct even in those cases, since such is often the case in other problems. In eqs. (5) we have neglected for simplicity the intraband ladder diagrams which enhances the effect of U. Therefore, U should be taken as a quantity enhanced by this process. Then, whether μ in eq. (8) is actually decreased or not depends on the features of the two-band system. Since the enhancement of Γ can be greater than that of U, there is an ample parameter region where μ is decreased from the bare value of N(0)U.

It has been established that the stably observed high-T_c conductivity in cuprates occurs in the CuO_2 plane. The Fermi-liquid-like characteristics of the electronic system in this plane are becoming more and more clear. There appeared also observations of the multiple bands crossing the Fermi level, although it is not yet known if these lie on one CuO_2 plane or not. We assume that the Cu $d_{x^2-y^2}$ - O p_σ antibonding band provides the Fermi-level-crossing band after the SDW ordering melts. Although the possibility that the oxygen $2p_{\pi*}$ band crosses the Fermi level was pointed out by quantum chemists [8], experimental evidence is not yet obtained. Another possibility of the second band is the Cu d_{z^2} band appearing above the Fermi level; such a possibility is suggested by a recent XAS experiment [9]. Thus the possibility of two ε_F-crossing bands in the CuO_2 plane is still controversial. However, even if the second band is submerged below ε_F, the present result demonstrates that the two-band feature can play an essential role in driving high T_c. The two-band model of the present type was originally proposed by Kondo for La and other transition-metal superconductors. We have to examine the newly found enhancement of the effect of the exchange-like interaction due to the ladder diagrams. There is another family of systems in which the present effect may work, i.e., organic superconductors. These systems have another band within 1 eV from ε_F in the conduction band.

REFERENCES

1. Kondo J (1963) Prog.Theor.Phys. 29: 1-9
2. Yamaji K, Abe S (1987) J.Phys.Soc.Jpn. 56: 4237-4240
3. Yamaji K (1988) Solid State Commun. 64: 1157-1160
4. Yamaji K (1990) J.Phys.Soc.Jpn. 59: 677-691
5. Entin-Wohlman O, Imry Y (1989) Phys.Rev.B 40: 6731-6738
6. Batlogg B, Cava RJ, Stavola M (1988) Phys.Rev.Lett. 60: 754-754
7. Leary KJ, zur Loye HC, Keller SW, Faltens TA, Ham WK (1987) Phys.Rev.Lett. 59: 1236-1239
8. Guo Y, Langlois JM, Goddard WA III (1988) Science 239: 896-899; Asai Y (1989) J.Phys.Soc.Jpn. 58: 3264-3269
9. Bianconi A, Castrucci P, Fabrizi A, Pompa M, Flank AM, Lagarde P, Katayama-Yoshida H, Calestani G (1990) In: Bednorz JG, Müller KA (eds) Earlier and Recent Aspects of Superconductivity. Springer-Verlag Berlin, Heidelbeg, pp 407-421

Mott-Hubbard Band and Charge Transfer Gap in La_2CuO_4

Mitsuo Kataoka

Institute for Materials Research, Tohoku University, Aoba-ku, Sendai, 980 Japan

ABSTRACT

The Mott-Hubbard (MH) band and the charge transfer gap in La_2CuO_4 is investigated theoretically. The calculation of the band structure is made for the 5-orbital system of the CuO_4 plane which is composed by the usual CuO_2 plane plus O_2 at the two apical O sites. The orbitals are assumed to be all subjected to the MH splitting. We apply Hubbard's approximation to the Coulomb interactions between electrons in same orbital states and the Hartree-Fock approximation to the Coulomb interactions between electrons in different orbital states. The obtained result shows that about one hole enters into $d(3z^2-r^2)$ (with the z axis perpendicular to the CuO_2 plane) rather than $d(x^2-y^2)$, and that the calculated value of the charge transfer gap is close to the observed value 1.5 eV.

KEY WORDS: Mott-Hubbard band, charge transfer gap, band structure, La_2CuO_4

INTRODUCTION

One of the major characteristics of the electronic state in the perovskite-family, high-T_c cuprates is the occurrence of the Mott-Hubbard (MH) splitting in electronic bands due to strong Coulomb correlations. To understand such the characteristics, many authors have concentrated their attention on the CuO_2 plane parallel to the x-y plane, and studied the orbital system which consists of $3d^0(x^2-y^2)$ on Cu at the origin (numbered 0), $2p^1(x)$ on O in the x direction (numbered 1) and $2p^2(y)$ on O in the y direction (numbered 2) [1]. Despite many theoretical investigations, however, the true electronic state has not been fully clarified. For example, this 3-orbital system can not describe observed effects of $2p^3(z)$ and $2p^4(z)$ on two oxygens at the apical sites in the z-direction (numbered 3 and 4) [2], and of $3d^0(3z^2-r^2)$ [3]. The purpose of this paper is to study theoretically the MH bands in La_2CuO_4, which becomes a superconductor under hole doping, by extending the 3-orbital model.

The ingredient which we concern ourselves with is the CuO_4 plane. This plane is composed of the CuO_2 plane given above and O^3 and O^4 at the two apical sites. The orbitals relevant to our purpose are $d^0(x^2-y^2)$, $d^0(3z^2-r^2)$, $p^1(x)$, $p^2(y)$, and $\{ p^3(z)+p^4(z) \}/\sqrt{2}$. This 5-orbital system is assumed to accomodate 1 hole (in the undoped system). In the previous paper [4], we calculated the electronic structure for the case where only either $d^0(x^2-y^2)$ or $d^0(3z^2-r^2)$ is subjected to the MH splitting. The result showed that about one hole enters into $d^0(3z^2-r^2)$ rather than $d^0(x^2-y^2)$. To confirm this result, we calculate in this paper the electronic structure of the 5-orbital system by allowing the MH splitings of all orbitals. This situation of the MH splitting is possible for large on-site Coulomb interactions of both Cu and O much larger than the transfer integral as in La_2CuO_4.

MODEL HAMILTONIAN AND FORMATION OF THE MOTT-HUBBARD BANDS

For the present 5-orbital system, the Hamiltonian is written as

$$H = H_1 + H_2 , \tag{1}$$

$$H_1 = \sum_i \sum_d \sum_\sigma \varepsilon_d a^{d+}_{i\sigma} a^d_{i\sigma} + \sum_i \sum_p \sum_\sigma \varepsilon_p a^{p+}_{i\sigma} a^p_{i\sigma} + \frac{1}{2} U_{Cu} \sum_i \sum_{dd'} \sum_{\sigma\sigma'} n^d_{i\sigma} n^{d'}_{i\sigma'} (1 - \delta_{dd'}\delta_{\sigma\sigma'}) + \frac{1}{2} U_O \sum_i \sum_p \sum_\sigma n^p_{i\sigma} n^p_{i-\sigma} + V \sum_{\{id,jp\}} \sum_{\sigma\sigma'} n^d_{i\sigma} n^p_{j\sigma'} , \quad (2)$$

$$H_2 = \sum_{\{id,jp\}} \sum_\sigma (t^{dp}_{ij} a^{d+}_{i\sigma} a^p_{j\sigma} + h.c.) + \sum_{\{ip,jp'\}} \sum_\sigma t^{pp'}_{ij} a^{p+}_{i\sigma} a^{p'}_{j\sigma} , \quad (3)$$

where d =(x^2-y^2), $(3z^2-r^2)$, and p, p'= x, y, and z denote $d^0(x^2-y^2)$, $d^0(3z^2-r^2)$, $p^1(x)$, $p^2(y)$, and $\{p^3(z)+p^4(z)\}/\sqrt{2}$, respectively, $a_{i\sigma}{}^{d(p)}$ and $n_{i\sigma}{}^{d(p)}$ are, respectively, the annihilation and number operators of an electron with spin σ in the orbital state d(p) on an ion in the i-th unit cell, $\varepsilon_{d(p)}$ is the energy level of the d(p) orbital, t's are transfer integrals, U's and V are respectively the on-site and intersite Coulomb interactions, δ is Kronecker's delta, and $\{\cdots\}$ means pairs of orbitals in neighboring sites.

The values of the parameters in eqs. (2) and (3) have been estimated by fitting the tight binding bands to the calculated augmented-plane-wave bands [5] and by some ab-initio calculations [6]. In the previous [4] and present papers, the following values (in units of eV) are used:

$$\varepsilon_{(x^2-y^2)} = -52.33, \quad \varepsilon_{(3z^2-r^2)} = -49.12, \quad \varepsilon_x = \varepsilon_y = -19.94, \quad \varepsilon_z = -13.19,$$

$$t^{(x^2-y^2),x} = -1.38, \quad t^{(3z^2-r^2),x} = -0.80, \quad t^{(3z^2-r^2),z} = -0.76, \quad t^{x,y} = -0.33,$$

$$U_{Cu} = 10, \quad U_O = 4, \text{ and } V = 2.$$

When the Hartree-Fock (HF) approximation is made to Coulomb interactions between electrons in different orbital states, H_1 is rewritten as

$$H_1 = \sum_i \sum_d \sum_\sigma E_d a^{d+}_{i\sigma} a^d_{i\sigma} + \sum_i \sum_d \sum_\sigma E_p a^{p+}_{i\sigma} a^p_{i\sigma} + \frac{1}{2} U_{Cu} \sum_i \sum_d \sum_\sigma n^d_{i\sigma} n^d_{i-\sigma} + \frac{1}{2} U_O \sum_i \sum_p \sum_\sigma n^p_{i\sigma} n^p_{i-\sigma} , \quad (4)$$

$$E_{(x^2-y^2)} = \varepsilon_{(x^2-y^2)} + U_{Cu} n_{(3z^2-r^2)} + V(2n_x + 2n_y + n_z),$$
$$E_{(3z^2-r^2)} = \varepsilon_{(3z^2-r^2)} + U_{Cu} n_{(x^2-y^2)} + V(2n_x + 2n_y + n_z),$$
$$E_x = E_y = \varepsilon_x + 2V\{ n_{(x^2-y^2)} + n_{(3z^2-r^2)}\}, \quad (5)$$
$$E_z = \varepsilon_z + V\{ n_{(x^2-y^2)} + n_{(3z^2-r^2)}\} ,$$

where $n_{d(p)}$ represents the expectation value of $\Sigma_\sigma\, n_{i\sigma}{}^{d(p)}$ in the ground state. We apply Hubbard's approximation [7] to the Hamiltonian given by eqs. (1), (3), and (4).

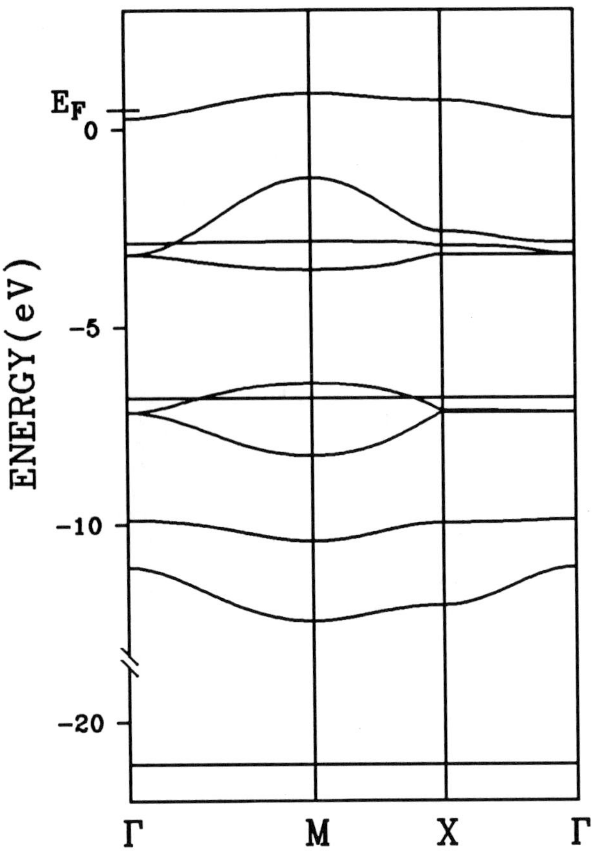

Fig. 1. Electronic energy band structure of the CuO_4 plane in La_2CuO_4. The calculation has been made for the orbital system of $d^0(x^2-y^2)$, $d^0(3z^2-r^2)$, $p^1(x)$, $p^2(y)$, and $\{ p^3(z)+p^4(z) \}/\sqrt{2}$ which are all subjected to the MH splitting. The unit cell and the notation of the high symmetry points Γ, M, and X are the same as in [4].

The Green's functions which we are interested in are

$$G^{\nu\mu}_{ij,\sigma}(E) = \langle\langle a^{\nu}_{i\sigma};\ a^{\mu+}_{j\sigma} \rangle\rangle_E ,$$

$$\Gamma^{\nu\mu}_{ij,\sigma}(E) = \langle\langle n^{\nu}_{i-\sigma}a^{\nu}_{i\sigma};\ a^{\mu+}_{j\sigma}\rangle\rangle_E ,$$

with ν, μ = d, p, and energy E. (See [7] for their definitions.) Their equations of motion involve a different type of Green's function such as $\langle\langle n^{\nu}_{i-\sigma}a^{\nu}_{\ell\sigma};a^{\mu+}_{j\sigma}\rangle\rangle$. This type of Green's function is replaced by $(n_{\nu}/2)G^{\nu\mu}_{\ell j,\sigma}(E)$ in order to break off the sequence of equations of motion. By use of G's and Γ's thus obtained, we calculate electronic bands together with n_{ν}'s. The obtained n_{ν}'s should be equal to those in eq. (5) by requirement of self-consistency. The self-consistent solutions are found by iterating the calculation with the use of the calculated (or initially chosen) values of n_{ν}'s. It is noted that the present calculation gives rise to the MH splittings in all 5 orbitals.

RESULTS AND CONCLUSION

The iteration of the calculation leads us to two self-consistent solutions. The band structure corresponding to the solution with the lower total energy is shown in Fig. 1. In this figure, the two bands lying at about -21 eV and -12 eV are the lower and upper MH bands originating mainly from $d^0(x^2-y^2)$. The two bands lying at about -10 eV and 0.5 eV are the MH bands originating mainly from $d^0(3z^2-r^2)$. On the other hand, the two groups of 3 bands just below and above -5 eV are the MH bands originating mainly from the three p orbitals. The fermi level ε_F cuts the upper MH band of $d^0(3z^2-r^2)$. However, ε_F may be moved into the gap below this band by an improved calculation to explain the insulating behavior of La_2CuO_4.

Also in the present calculation, the upper MH band of $d^0(3z^2-r^2)$, rather than $d^0(x^2-y^2)$, can accommodate a large part of the one hole. The reason for this is explained as follows: The transfer integral between $d^0(x^2-y^2)$ and $p^1(x)$ (or $p^2(y)$) is significantly large, being able to smear a large charge polarization between these orbitals by the intersite Coulomb interaction V. In contrast to the case of $d^0(x^2-y^2)$, the mixing between $d^0(3z^2-r^2)$ and the top valence band is very small, since the latter is largely of the $\{ p^1(x)-p^2(y) \}$ character. This fact allows an electronic transfer from $d^0(3z^2-r^2)$ to $p^1(x)$ and $p^2(y)$. Once the O^3 and O^4 sites are charged up, the energy level of $d^0(3z^2-r^2)$ is pushed above ε_F by an increase of the HF potential at the Cu site.

The resultant indirect gap between the bottom of the upper MH band of $d^0(3z^2-r^2)$ and the top of the valence band is calculated to be 1.49 eV for our choice of the parameter values. This gap value is larger by about 0.5 eV than that calculated in the previous paper. These values are, however, still comparable to the charge transfer gap 1.5 eV observed by optical measurements [8].

Extra doped holes will enter into the $p^1(x)$ and $p^2(y)$ orbitals, since the top valence band consists mainly of these orbitals. This result is consistent with the fact that the density of states at ε_F has predominantly oxygen 2p character in the metallic states realized by hole doping [9]. It is noted, however, that those hole states are the upper MH band states of the p orbitals and the oxygens can be weaky magnetic in the present calculation.

In conclusion, our calculation shows that one hole in the 5-orbital system of La_2CuO_4 enters into the upper MH band of $d^0(3z^2-r^2)$. This holds also for the MH splitting of all 5 orbitals. Comparisons of this calculated result with experiments are left for the future.

REFERENCES

1. Emery VJ (1987) Phys. Rev. Lett. 58: 2794.
2. Maruyama C, Mōri N, Yomo S, Takagi H, Uchida S, and Tokura Y (1989) Nature 339: 293.
3. Bianconi A, De Santis M, Flank AM, Fontaine A, Lagarde P, Marcelli A, Katayama-Yoshida H, and Kotani A (1988) Physica C153-155: 1760.
4. Kataoka M and Ikebe M Prog. Theor. Phys. Suppl. to be published.
5. Mila F (1988) Phys. Rev. B38: 11358.
6. Schluter M, Hybertsen MS, Christensen NE (1988) Physica C153-155: 1217; Stechel EB and Jennion DR (1988) Phys. Rev. B38: 8873.
7. Hubbard J (1963) Proc. Roy. Soc. (london) A276: 238.
8. Tajima S, Tanaka S, Ido T, and Uchida S Proc. 2'nd Int. Symp. on Superconductivity, Tsukuba, Japan 1989.
9. Nücker N, Fink J, Fuggle JC, Durham PJ, and Temmerman WM (1988) Phys. Rev. B37: 5158.

Electronic States of Metal (Ti,V,Cr) Oxides

KIMICHIKA FUKUSHIMA[1] and HIROHIKO ADACHI[2]

[1] Toshiba Research and Development Center, 4-1, Ukishima-cho, Kawasaki-ku, Kawasaki, 210 Japan
[2] Hyogo University of Teacher Education, Yashiro-cho, Kato-gun, Hyogo, 673-14 Japan

ABSTRACT

Electronic states of metal (Ti, V, Cr) oxides were calculated by using the spin-polarized DV-Xα molecular orbital method. We adopted artificial model clusters, TiO_6^{9-}, VO_6^{8-} and CrO_6^{7-}. For the metal ion $3d^1$ configuration, we first calculated the usual case, where the lattice distortion is small. The metal 3d orbitals split into e_g and t_{2g} levels, and d_{xz}, d_{yz} and d_{xy} orbitals are partially occupied. We second calculated the case with some possibility, where the metal-oxygen distance in the ab plane is elongated. The energy levels formed by metal 3d shift lower with increasing atomic number of metal atom. The main component of the highest occupied molecular orbital is d_{xy}, and states formed by d_{xy} are half-filled. The situation is rather similar to Cu oxides, though O 2p hole is more important in the latter.

KEY WORDS: transition metal oxides, Ti, V, Cr, electronic structure, molecular orbital calculation, spin-polarized DV-Xα method

INTRODUCTION

Since the discovery of high-T_c copper oxide superconductors [1], intensive theoretical [2-5] and experimental efforts have been made to clarify the physical and chemical properties of these materials. Copper oxides have octahedrons or modified structures with a central copper atom and oxygen ligands. The octahedron is deformed by the Jahn-Teller effect, and the Cu-O distance in the ab plane is shorter than that in the c axis. In the insulating phase, the configuration of Cu ion is $3d^9$, and $d_{x^2-y^2}$ orbitals are half-filled. The energy levels of Cu 3d and O 2p orbitals are almost same, and these orbitals are strongly mixed. In the metallic phase, charge carriers flow in the two-dimensional CuO_2 plane.

With regard to other transition metal oxides, the electronic states have not yet been fully understood. If it is possible to find the similar electronic structure as mentioned above in the other materials, we may expect that these materials would become high-T_c superconductors. The molecular orbital method is a useful tool to search such new materials, and we have calculated the electronic states of transition metal (Ti, V, Cr) oxides. We adopted MO_6(M=Ti, V, Cr) clusters as a model. For the $3d^1$ configuration of transition metals, the distortion is usually small. We first consider the case where the M-O distances are same both in the ab plane and in the c axis. There is some possibility that the M-O distance in the ab plane is elongated in the crystal field produced by the other ions. We next discuss the case where the M-O distance in the ab plane is longer than the one in the c axis. Computations were carried out by the spin-polarized discrete variational (DV) Xα molecular orbital method [6].

This paper is organized as follows: We first present the computational method and model clusters. We next report the energy level diagrams and discuss the similarity and dissimilarity of electronic structures.

COMPUTATIONAL PROCEDURE

We calculated electronic structures by the use of the spin-polarized DV-Xα method [6], which solves the following one-electron equation

$$H(\mathbf{r})\phi_i(\mathbf{r}) = \varepsilon_i\phi_i(\mathbf{r}), \quad (1)$$

where $\phi_i(\mathbf{r})$ is the electron wave function and ε_i is the orbital energy. Hamiltonian is given by

$$H(\mathbf{r}) = -(1/2)\nabla^2 + V_N(\mathbf{r}) + V_c(\mathbf{r}) + V_{xc}(\mathbf{r}). \quad (2)$$

Here, $V_N(\mathbf{r})$ is the nuclear attractive potential and $V_c(\mathbf{r})$ is the electronic Coulomb potential. The Slater's exchange-correlation potential has the form

$$V_{xc}(\mathbf{r}) = -3\alpha\,[(3/8\pi)\rho(\mathbf{r})]^{1/3}. \quad (3)$$

In the self-consistent iterations, the charges are shared with each atom by using the Mulliken population analysis [7]. The LCAO basis functions used are 1s-4p for 3d metals and 1s-2p for O. These atomic orbitals are obtained by numerically solving the atomic Hartree-Fock-Slater equation.

The metal oxides considered here have a perovskite lattice structure. We use model clusters MO_6 (M=Ti, V, Cr) with the form of octahedron, and the nominal charges of clusters were taken to be -9, -8 and -7, respectively. The model cluster is illustrated in Fig.1, where a metal atom is denoted by the solid circle with symbol M, and six ligand O atoms are represented by the open circles. O atoms in the ab plane are indicated by symbol O_I, and apex O atoms in the c axis by symbol O_{II}. For Cu oxides, M-O distance R_I in the ab plane is shorter than the distance R_{II} in the c axis. On the other hand, the lattice distortion is usually small for transition metal oxides with the ion d^1 configuration. We first took R_I=R_{II}=2.0Å and calculated TiO_6^{9-} with O_h symmetry. There is some possibility that the distance R_I is elongated in the crystal field produced by the other ions. For this case, calculations were carried out for the artificial deformed octahedron with distances of R_I= 2.1Å and R_{II}= 1.9Å . We calculated these clusters in the symmetry of C_{2v}.

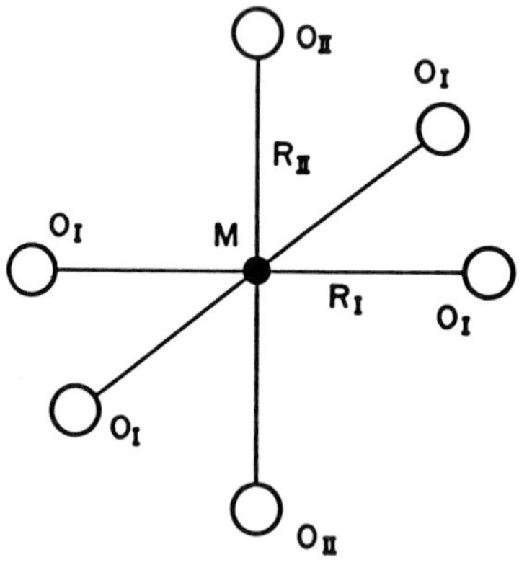

Fig. 1 Schematic diagram of MO_6 (M=Ti, V, Cr) model cluster.

RESULTS AND DISCUSSION

First of all, calculations were performed for the octahedron of TiO_6^{9-} which has the same Ti-O distances in the ab plane and in the c axis. Ti has the highest 3d orbital energy among free atoms of Ti, V and Cr. Figure 2 shows the energy level diagram for TiO_6, and it is compared with Ti 3d and O 2p levels. For the TiO_6 cluster, the levels from $6a_{1g\alpha}$ to $1t_{1g\alpha}$ are bonding and nonbonding orbitals composed mainly of O 2p component and are completely occupied. Here, α and β refer to up and down spins, respectively. The levels from $2t_{2g\alpha}$ to $4e_{g\beta}$ are antibonding orbitals, whose main component is Ti 3d. The atomic Ti 3d orbitals split into $2t_{2g}$ and $4e_g$ levels with both up and down spins in TiO_6 cluster. The highest occupied molecular orbital (HOMO) is $2t_{2g\alpha}$. The components of $2t_{2g\alpha}$ level are d_{xz}, d_{yz} and d_{xy} of Ti (96 %), and O 2p (4%). Although each d_{xz}, d_{yz} and d_{xy} orbital has a local two-dimensional-like character, they do not form a distinct two-dimensional system as a whole unlike the Cu3d-O2p network in Cu oxides.

Next, we calculated TiO_6^{9-} cluster elongated in the ab plane, whose level diagram is depicted on the left hand of Fig. 3. The bonding and nonbonding levels from $14a_{1\alpha}$ to $9b_{1\beta}$ for the distorted octahedron correspond to $6a_{1g\alpha}$ to $1t_{1g\alpha}$ for the undistorted one, and the main component of these orbitals is O 2p. The antibonding levels from $2t_{2g\alpha}$ to $4e_{g\beta}$ consisting mainly of Ti 3d for the undistorted octahedron split into $3a_{2\alpha}$ to $23a_{1\beta}$ in the distorted case. The energy difference between the top of the bonding and nonbonding states and the bottom of the antibonding states is 7.2 eV. HOMO is $3a_{2\alpha}$, whose components are Ti $3d_{xy}$ (96 %) and O 2p (4%) with up spin. The $3a_{2\beta}$ level consisting mainly of Ti $3d_{xy}$ with down spin is above $10b_{1\alpha}$ and $10b_{2\alpha}$ levels. The states whose main component is Ti $3d_{xy}$ are half-filled and have a gap with 1.1 eV. This suggests that Ti $3d_{xy}$ forms a two-dimensional electron system and the metal-insulator transition occurs. This electronic structure is rather similar to the Cu oxide case, where HOMO consists mainly of O 2p and Cu $3d_{x^2-y^2}$, though the role of O 2p is more important in the latter.

The charge distribution in TiO_6^{9-} analyzed by Mulliken populations is shown in Table 1. The orbital populations on the left hand are for up spin, and those in the parenthesis on the right hand are for down spin. Ti 3d orbital is occupied by about one electron with up spin. Since each atomic orbital takes the non-

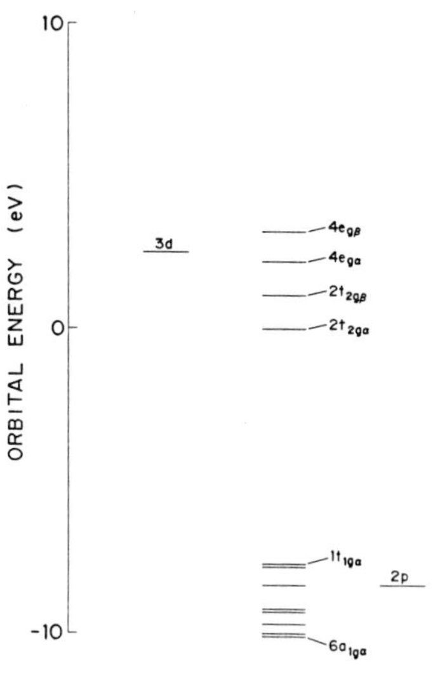

Fig.2 Level diagram for the undistorted TiO_6^{9-} octahedron.

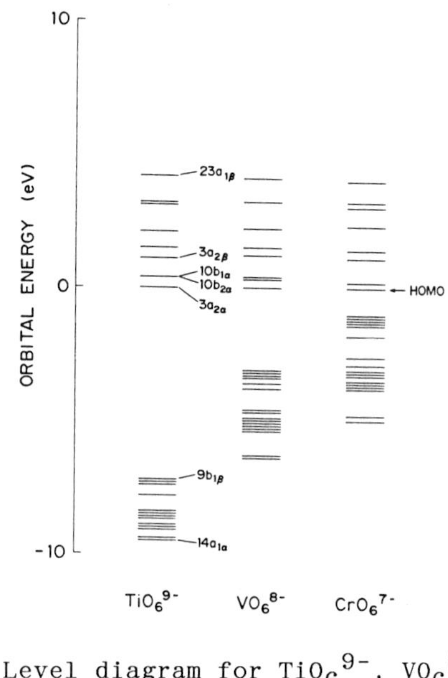

Fig.3 Level diagram for TiO_6^{9-}, VO_6^{8-} and CrO_6^{7-} elongated in the ab plane.

Table 1 Charge distribution in TiO_6^{9-} elongated in the ab plane.

Orbital and ion	Mulliken population
Ti -3d	1.384 (0.351)
O_I -2p	2.933 (2.935)
O_{II}-2p	2.900 (2.917)
Ti ion	+ 1.954
O_I ion	- 1.845
O_{II} ion	- 1.787

integer population, gross atomic charges obtained by summing the orbital populations deviate somewhat from formal ion charges. The charge of Ti is +1.95, and those of O_I and O_{II} are -1.85 and -1.79, respectively.

We calculated the other oxide clusters including V and Cr atoms. The level diagram for VO_6^{8-} and CrO_6^{7-} are shown in the middle and on the right of Fig. 3, respectively with that for TiO_6^{9-} on the left. These three clusters have the similar level structures and the main component of HOMO is commonly metal $3d_{xy}$. The main difference in the level structure is that the energy separation between the top of bonding and nonbonding states and the bottom of antibonding states are 7.2, 3.2 and 1.0 eV for TiO_6, VO_6 and CrO_6, respectively. As atomic number increases, the 3d levels are lowered and covalent interaction between metal 3d and oxygen 2p becomes stronger. As a result, the charge transfer increases as is seen from the fact that charges of Ti, V and Cr ions are +2.0, +2.1 and +2.0, (the starting charges in the self-consistent iteration are +3, +4 and +5), respectively.

CONCLUSION

We have calculated electronic states of transition metal (Ti, V, Cr) oxides by means of the spin-polarized DV-Xα molecular orbital method. For the metal ion $3d^1$ configuration, the lattice distortion is usually small. In this case, 3d orbitals split into e_g and t_{2g} levels, and d_{xz}, d_{yz} and d_{xy} orbitals are partially occupied. There is some possibility that the metal-oxygen distance in the ab plane is elongated in the crystal field produced by the other ions. For this case, we investigated by using artificial model clusters, TiO_6^{9-}, VO_6^{8-} and CrO_6^{7-} with the form of octahedron elongated in the ab plane. The metal 3d levels lower with increasing atomic number of the metal atoms. The highest occupied molecular orbital (HOMO) is d_{xy} with up spin. The states formed by d_{xy} are half-filled and have a gap. The result suggests that d_{xy} orbitals form a two-dimensional electron system, and the metal-insulator transition occurs. The situation is rather similar to the Cu oxide case, though O 2p hole is more important in the latter.

REFERENCES

1. Bednorz JG and Müller KA (1986) Z Phys B64 189.
2. Varma CM, Schmitt-Rink S and Abrahams E (1987) Solid State Commun 62 681.
3. Anderson PW (1987) Science 235 1196.
4. Fukushima K and Sato H (1989) phys status solidi (b) 153 K141.
5. Fukushima K and Sato H (1990) in: Ishiguro T and Kajimura K (eds) Advances in Superconductivity II. (Proceedings of the 2nd International Symposium on Superconductivity.) Springer. Tokyo p.455.
6. Adachi H, Tsukada M and Satoko C (1976) J Phys Soc Japan 45 875.
7. Mulliken RS (1955) J Chem Phys 23 1833.

Superconductivity in a System with Intermediate Statistics

VLADIMIR L. SAFONOV

Department of Molecular Physics, Kurchatov Institute of Atomic Energy, 123182 Moscow, USSR

ABSTRACT

Assuming that the system of quasiparticles obeys parastatistics, the superconducting state of pairs of quasiparticles is considered. The description is based on the Anderson's spin analogy. It is shown that in the weak coupling limit the ratio of energy gap to the critical temperature is less than that in the Fermi system. Some physical consequences of the model are discussed.

KEY WORDS: quasiparticle, parastatistics, spin analogy

INTRODUCTION

A concept of quasiparticles is a very useful tool for description of excitations in solids. Quasiparticles characterize a collective motion of a set of interacting particles (electrons, ions) in crystal and alongside with the energy can carry the electric charge and the magnetic moment. In many respects quasiparticles behave as usual particles but they live in the initially anisotropic space and this fact of their biography imposes some specificity on their characters. So far as the symmetry of the crystalline structure is much more lower than the spherical one the law of conservation of angular momentum is not valid in these conditions. Strictly speaking one has no foundation to introduce such a quantum variable as spin for quasiparticle (as it is classically introduced for particle in standard course of quantum mechanics [1]). In this connection a question about a kind of quantum statistics for given quasiparticles is of interest.

It should be noted that in addition to well known Fermi-Dirac and Bose-Einstein statistics one can consider intermediate statistics known as parastatistics [1] in which the number of particles in a definite energy state can reach some whole number p (the case p=1 corresponds to Fermi statistics and p=∞ to Bose one). There are no reasons in principle to forbid the possibility of parastatistics for quasiparticles in solids. Moreover parastatistics have no restriction in contrast to so called "fractional" statistics [3] existing in two dimensions.

The possibility of the parastatistics for quasiparticles in solids can be explained on the base of the following analogy between objects in real and momentum spaces. The population of the elementary cell for Fermi particles in the **k**-space is analogous to the population for levels of two-level system in the real space. The same analogy exists for Bose particles and levels of harmonic oscillator. Indeed, the equidistant multilevel system in the real space is the analogue to an ideal gas of quasiparticles with parastatistics having the same number p as the total number of levels of the system. An example of such system is a spin S in a magnetic field (p=2S). So the spin operators can be considered for description of quasiparticles with parastatistics.

In the present paper the conditions of condensation in the **k**-space are analyzed for weakly interacting quasiparticles with parastatistics. The part of the obtained results can be considered as the generalization of results of BCS theory to the case of quasiparticles with intermediate statistics.

CONDENSATE IN AN IDEAL GAS

The equilibrium distribution function n_p for an ideal gas of quasiparticles with parastatistics can be represented in the form

$$n_p(x) = S - S\cdot B_S(S\cdot x) \, , \tag{1}$$

where $S \equiv p/2$,

$$B_S(S\cdot x) \equiv 1 - \frac{1}{S}\left(\frac{1}{\exp(x) - 1} - \frac{2S+1}{\exp[(2S+1)x] - 1} \right) \tag{2}$$

is the Brillouin function and $x \equiv (E_{\mathbf{k}} - \mu)/T$, $E_{\mathbf{k}}$ is the energy of quasiparticle with the wave vector $\mathbf{k}$, T is the temperature, μ is the chemical potential (the following system of units is used: $\hbar=k_B=1$). In the low temperature limit

$$\lim_{T\to 0} n_p(x) = \begin{cases} p & E_{\mathbf{k}} < \mu_F \, , \\ 0 & E_{\mathbf{k}} > \mu_F \, , \end{cases} \tag{3}$$

where μ_F is the analog of the Fermi energy.

If the total number of quasiparticles $N \le p$, then all quasiparticles will occupy the lowest energy level at $T = 0$. The process of condensation begins at $T = T_p$, when the chemical potential comes to the first excited energy level of the system. In the simplest model with $E_{\mathbf{k}} = k^2/2m$, where m is the mass of quasiparticle, one can obtain

$$T_p \simeq 2\pi\cdot\left\{ N \Big/ \zeta(\tfrac{3}{2})[1 - (p+1)^{-1/2}] \right\}^{-2/3} \cdot L^{-2} m^{1} \, , \tag{4}$$

where L is the characteristic linear size of the sample and $\zeta(x)$ is the zeta function.

Note that at $p \to \infty$ the formula (4) is reduced to the well known expression for the temperature of Bose condensation [2]. An estimation of (4) at $N = p$, $m \sim m_e$ and $L \sim 100$ Å gives $T_p \sim 0.3\, p^{2/3}$ [K] .

A PAIR CONDENSATE

To describe the superconducting properties of pairs of quasiparticles we shall use the famous Anderson's spin analogy [4] with the Hamiltonian

$$H = -\sum_{\mathbf{k}} 2\varepsilon_{\mathbf{k}} S^z_{\mathbf{k}} - V \sum_{\mathbf{k},\mathbf{q}} (S^x_{\mathbf{k}} S^x_{\mathbf{q}} + S^y_{\mathbf{k}} S^y_{\mathbf{q}}) \, , \tag{5}$$

where $\varepsilon_{\mathbf{k}} \equiv E_{\mathbf{k}} - \mu_F$, $V > 0$ is the amplitude of pair interaction within the energy interval $|\varepsilon_{\mathbf{k}}| \le \omega_*$, ω_* is the characteristic energy of virtual excitations (for example, ω_* is the Debye energy for phonons). The fictious spin operators $\mathbf{S}_{\mathbf{k}}$ describe the population of pairs of quasiparticles with wave vectors $\mathbf{k}$ and $-\mathbf{k}$. Note that for $S = 1/2$ the Hamiltonian (5) is entirely equivalent to the pair Hamiltonian of BCS

theory. Because $\langle S_{\mathbf{k}}\rangle = S\cdot B_S(S\cdot h_{\mathbf{k}} / \tilde{T})$, where $h_{\mathbf{k}}$ is the effective "magnetic" field and $\tilde{T}$ is the temperature corresponding to the pair subspace of the system, the pair of quasiparticles obeys the same statistics as their components (see eq.(1)). So the Hamiltonian (5) for $S > 1/2$ can be considered as the generalization of BSC Hamiltonian to the case of intermediate statistics.
Using mean field approximation, one can write $\langle S^x_{\mathbf{k}}\rangle = \langle S_{\mathbf{k}}\rangle \sin\theta_{\mathbf{k}}$ and $\langle S^z_{\mathbf{k}}\rangle = \langle S_{\mathbf{k}}\rangle \cos\theta_{\mathbf{k}}$, where $\theta_{\mathbf{k}}$ is the fictious angle in the **k**-space. Because the ratio of the mean transverse and mean longitudinal spin components is equal to the ratio of correspondent effective fields, one can obtain the following equations:

$$\mathrm{tg}\theta_{\mathbf{k}} = V \sum_{\mathbf{q}} \langle S_{\mathbf{q}}\rangle \sin\theta_{\mathbf{q}} / \varepsilon_{\mathbf{k}} , \qquad (6a)$$

$$\langle S_{\mathbf{q}}\rangle = S\cdot B_S\{ 2S\cdot[\varepsilon^2_{\mathbf{q}} + (V \sum_{\mathbf{q}_1} \langle S_{\mathbf{q}_1}\rangle \sin\theta_{\mathbf{q}_1})^2]^{1/2} / \tilde{T} \} . \qquad (6b)$$

The solution $\sin\theta_{\mathbf{k}} = 0$ for all **k** describes the normal state.

Denoting $\mathrm{tg}\theta_{\mathbf{k}} \equiv \Delta / \varepsilon_{\mathbf{k}}$, one has

$$\sin\theta_{\mathbf{k}} = \Delta / \tilde{\varepsilon}_{\mathbf{k}} , \quad \cos\theta_{\mathbf{k}} = \varepsilon_{\mathbf{k}} / \tilde{\varepsilon}_{\mathbf{k}} \quad \text{and} \quad \tilde{\varepsilon}_{\mathbf{k}} \equiv (\Delta^2 + \varepsilon^2_{\mathbf{k}})^{1/2} .$$

In the limit of low temperatures ($\tilde{T} = T = 0$) we have $\langle S_{\mathbf{k}}\rangle = S$ and the eq.(6a) after transition from summing to integration acquires the form

$$1 = \frac{1}{2} p\lambda \cdot \int_{-\omega_*}^{\omega_*} \frac{d\varepsilon}{(\Delta^2+\varepsilon^2)^{1/2}} , \qquad (7)$$

where $\lambda \equiv V\cdot\rho_F$ is the parameter of coupling, ρ_F is the density of states on the Fermi level μ_F ($\omega_* \ll \mu_F$). The solution of (7) is

$$\Delta = \omega_* / \mathrm{sh}(1/p\lambda) \simeq 2\omega_*\cdot\exp(-1/p\lambda) . \qquad (8)$$

The latter approximate equation is valid at $p\lambda \ll 1$.

The difference between the energy of the system in the normal and in the superconducting states can be represented by the energy of critical magnetic field. In the weak coupling limit one obtains

$$H_c = 2(\pi p\rho_F)^{1/2}\cdot\Delta . \qquad (9)$$

The eqs.(6a,b) are simplified in the vicinity of the phase transition temperature and reduced to

$$1 = SV \sum_{\mathbf{k}} B_S(2S\, \varepsilon_{\mathbf{k}} / \tilde{T}_c) / \varepsilon_{\mathbf{k}} . \qquad (10)$$

An analysis of statistical sums shows that $\tilde{T}_c = 2\cdot T_c$, where T_c is the real crytical temperature of the system. So the eq.(10) can be rewritten in the form

$$1 = 2S\rho_F V \cdot F(\omega_* / T_c) \quad , \qquad (11)$$

where

$$F(\xi) \equiv \int_0^{\xi} d\eta \; B_S(S\eta)/\eta \quad .$$

The solution of eq.(11) at $p\lambda \ll 1$ is:

$$T_c \simeq \phi(p) \cdot \omega_* \exp(-1/p\lambda) \quad , \qquad (12)$$

where $\phi(p) \sim 1$. Dividing the eq.(8) on the eq.(12) we obtain the famous ratio $2\Delta/T_c$ having no characteristic parameters of the model. The final results are represented in the following table

p	1	2	3	4	5	6	7	8	9
$2\Delta/T_c$	3,53	2,72	2,22	1,89	1,64	1,46	1,31	1,19	1.09

It is interesting to mention that the ratio $2\Delta/T_c$ for several "classical" superconductors is less than 3.5 , namely [5]: Ru (2.9±0.3), Zr (2.7±0.3), Tl (2.8±0.2), La (1.5-3,2). However there is an explanation to the fact of diminishing of $2\Delta/T_c$ due to magnetic impurities. Recently [6] $2\Delta/T_c \simeq 1.7$ has been obtained for YBCO . But even in this case the small value of $2\Delta/T_c$ can be interpreted as an influence of magnetic fluctuations in YBCO [7].

According to the analogy with the Fermi system, the ratio $2\Delta/T_c$ will increase with the increasing of $p\lambda$ in the range of $p\lambda \geq 1$ (the strong coupling limit).

Thus we have obtained that the energy gap, critical magnetic field and critical temperature Δ , H_c , $T_c \propto \exp(-1/p\lambda)$ are greater in the system with parastatistics than the same values in the Fermi system, but the ratio $2\Delta/T_c$ and the coherent length $\xi \propto 1/\Delta$ are smaller.

The detailed calculations will be published elsewhere [8].

REFERENCES

1. Landau LD, Lifshits EM (1974) Quantum mechanics. Nauka. Moscow
2. Isihara A (1971) Statistical physics. Academic Press. New York - London
3. Canright GS, Girvin SM (1990) Science 247: 1197-1205
4. Anderson PW (1959) Phys Rev 112: 1900-1916
5. Lynton E A (1969) Superconductivity. Methuen and Co.Ltd. London
6. Kuzmany H, Sumetsberger B, Matus M, Zachmann G (1990) Europhys Lett **12**: 703-708
7. Antsygina TN, Slyusarev VA (1990) Fiz Nizk Temp **16**: 942-945
8. Safonov VL (1990) Superconductivity: Physics, Chemistry, Technics **3**: No.3 (in press)

2.2 Electron Spectroscopy

Oxygen K-Absorption Spectroscopy of $La_{2-x}Sr_xCuO_4$ (x=0.0-0.5), $YBa_2Cu_3O_{7-\delta}$ (δ=0.0-0.7), and $Nd_{2-y}Ce_yCuO_4$ (y=0.0-0.2)

TAKASHI TAKAHASHI[1], HIROYOSHI MATSUYAMA[1], TAKANORI WATANABE[1], HIROSHI KATAYAMA-YOSHIDA[1], SHIGERU SATO[1], NOBUHIRO KOSUGI[2], AKIRA YAGISHITA[3], SHIN-ICHI SHAMOTO[4], and MASATOSHI SATO[4]

[1] Department of Physics, Tohoku University, Sendai, 980 Japan
[2] Division of Molecular Engineering, Kyoto University, Kyoto, 606 Japan
[3] Photon Factory, KEK, Tsukuba, 305 Japan
[4] Department of Physics, Nagoya University, Nagoya, 464 Japan

ABSTRACT

Oxygen K-absorption spectra were measured for La-, Y-, and Nd-system high-Tc superconductors. The spectrum of insulator for La- and Y-systems has a small structure about 1 eV above the absorption edge, being assigned to the $Cu3d^{10}$ state. Hole-doping produces a new structure just at the absorption edge, whose intensity increases with hole concentration. On the other hand, the intensity of the $Cu3d^{10}$ state gradually decreases. This indicates that the hole-doping brings about a transfer of weight of electronic states from the $Cu3d^{10}$ states to the mid-gap states at the Fermi level. No discontinuous change was observed in the spectrum between the superconductor and the normal metal for La-system, suggesting that the electronic structure of the high-Tc superconductor is essentially the same as that of normal metal. Oxygen K-absorption spectrum of Nd-system exhibits almost no change against Ce doping, suggesting that the O2p orbital is not directly involved in the high-Tc mechanism for the electron-doped superconductor.

Key words: electronic structure, x-ray absorption, doping effect, mid-gap state

INTRODUCTION

The existence of electronic band(s) which cross(es) the Fermi level has been experimentally established by angle-resolved photoemission spectroscopy [1-3], which suggests a close similarity in the electronic structure between the superconductor and the normal metal. The orbital-symmetry of the electronic states at the Fermi level has been studied by polarized x-ray absorption and electron-energy-loss (EELS) spectroscopies [4-6], which showed that the electronic states at the Fermi level has a dominant $O2p_{x,y}$ nature. The next step to approach the high-Tc mechanism is to elucidate the origin of the electronic states in the vicinity of the Fermi level produced by hole-doping. In this paper, we report an x-ray absorption spectroscopy on the La-, Y-, and Nd-system high-Tc superconductors to study the hole-concentration dependence of the electronic structure near the Fermi level. We found that the hole-doping does not necessarily cause a rigid shift of the electronic states relative to the Fermi level, but brings about a transfer of weight of the electronic states from the unoccupied Cu3d ($3d^{10}$) states to mid-gap states in the vicinity of the Fermi level produced by hole-doping.

EXPERIMENTAL

Ceramic samples for the x-ray absorption measurements were prepared by sintering mixture of high-purity oxide or carbonate powders with the respective molar ratio of the composition. $La_{2-x}Sr_xCuO_4$ samples with high Sr-doping ($x>0.3$) were prepared by annealing under high-pressure oxygen atmosphere. The temperature-dependence of resistivity for the La-systems is shown in Fig. 1. As for $YBa_2Cu_3O_{7-\delta}$, the hole concentration (δ) was changed by quenching the samples in air or nitrogen atmosphere from various temperatures from 300 to 900°C. Figure 2 shows the result of the Meissner-effect measurement of the $YBa_2Cu_3O_{7-\delta}$ samples, from which we suppose that the sample denoted by (a), which has the highest Tc and the largest Meissner volume, has the oxygen content of $\delta = 0.0$, the sample (g) has that of $\delta = 0.5$, and the samples (b)-(f) have the intermediate contents between the two compounds while the sample (h), which was quenched in nitrogen atmosphere, has the composition of $\delta = 0.7$. As for $Nd_{2-x}Ce_xCuO_4$ (x=0.04-0.25), all the samples were prepared in a reduced atmosphere and only the sample with x=0.15 shows the superconductivity with the Tc of 20K.

X-ray absorption measurements were performed at the Photon Factory, National Laboratory for High Energy Physics (KEK), Japan using a total-photoemission-yield method. The total energy resolution was about 1 eV. In order to obtain a clean surface for photoemission measurement, we scraped the surface by a diamond file in the spectrometer just before the measurement and kept it at room temperature. No surface degradation was observed throughout the measurements.

RESULTS AND DISCUSSION

X-ray absorption measures an optical transition from a core level to unoccupied states above the Fermi level. Different from inverse photoemission, an x-ray absorption spectrum represents a partial density of states of a certain atomic orbital because of the selection rule ($\Delta \ell = \pm 1$). In the case of oxygen K absorption, the spectrum represents a partial density of O2p states.

Figure 3 shows oxygen K absorption spectra of $YBa_2Cu_3O_{7-\delta}$ with various hole (oxygen) concentration. The sample denoted by (a) is a superconductor with the Tc = 90K while the sample (h) does not show superconductivity. The oxygen content (δ) and therefore the hole concentration decrease from (a) of δ =0.0 to (h) of δ =0.7. As shown in Fig. 3, the superconductor (a) has a prominent peak at the absorption edge of 528 eV while the insulator (h) has a shoulder at 529 eV. We should remark that the spectral change from (h) to (a) is not a shift of a peak but is a growth of a new peak at 528 eV together with gradual diminishment of a peak at 529 eV. This behavior is most clearly seen in the spectra of (g) and (f). We should also note that the spectrum is not shifted as a whole by the change of hole concentration, as seen from a stationary feature of a main band at 536 eV. The shoulder at 529 eV in the spectrum of (h) is attributed to $Cu3d^{10}$ state (upper Hubbard band) since the insulator has a charge-transfer gap of about 2eV between a lower-lying O2p state and an upper-lying Cu3d state. The experimental result in Fig. 3 indicates that the hole doping introduces new electronic states in the charge-transfer gap and at the same time it diminishes the population of $Cu3d^{10}$ states, in other words, the hole doping causes a transfer of weight of electronic states from the upper Hubbard band to mid-gap states introduced by hole doping. It is noted that the behavior of the mid-gap states is very similar to that of deep impurity states created in the energy gap of semiconductor by doping. Kuiper et al. have already reported a similar experimental result on

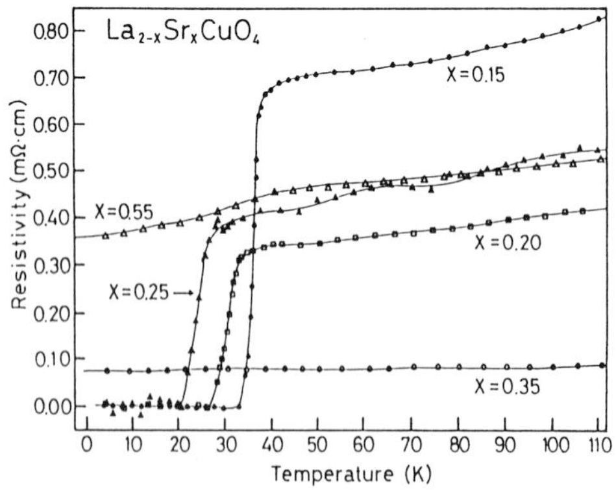

Fig.1 Temperature dependence of resistivity for $La_{2-x}Sr_xCuO_4$

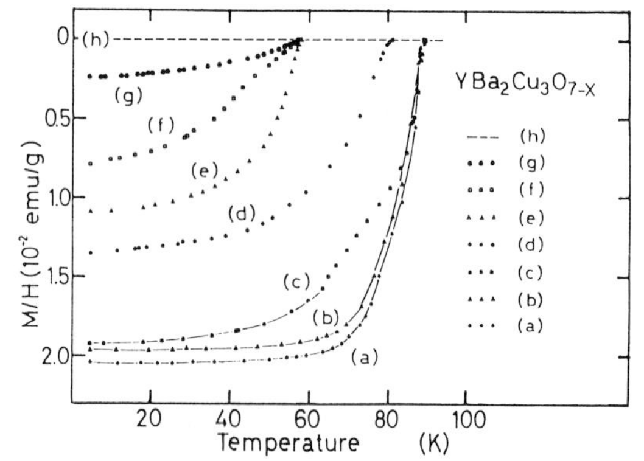

Fig.2 Temperature dependence of magnetic susceptibility for $YBa_2Cu_3O_{7-\delta}$

$YBa_2Cu_3O_{7-\delta}$, but did not explicitly mention this point [7].

Figure 4 shows the result for $La_{2-x}Sr_xCuO_4$ (x=0.0-0.55). Transfer of the electronic states from the $Cu3d^{10}$ states to mid-gap states by hole doping is also seen for $La_{2-x}Sr_xCuO_4$ although it is not so clear as for $YBa_2Cu_3O_{7-\delta}$ (Fig. 4). A recent high-energy-resolution EELS measurement shows more clearly the transfer of the electronic states [8]. Another interesting point in Fig. 4 is the spectral change in the heavy-doping region since $La_{2-x}Sr_xCuO_4$ changes from superconductor to "normal metal" around x=0.3. This problem is directly related to a long-standing problem in the high-Tc superconductor whether the electronic states near the Fermi level is regarded as the Fermi-liquid states or not. The series of oxygen K absorption spectra in Fig. 4 shows no discontinuous change even when crossing the border between the superconductor and the normal metal (x=0.25-0.3). Transfer of the electronic states from the upper Hubbard band to the mid gap states is still seen in the heavy doped region. This indicates that the electronic states near the Fermi level has a close similarity between the superconducting phase and the normal metal phase, suggesting that the electronic states near the Fermi level in the high-Tc superconductor is described in terms of the Fermi-liquid states.

The most essential problem in the electron-doped high-Tc superconductor is what atomic orbital accommodates the doped electrons. Figure 5 shows a series of oxygen K absorption spectra of $Nd_{2-x}Ce_xCuO_4$ with various Ce (electron) concentration (x=0.04-0.25). Among these samples, only that of x=0.15 shows superconductivity with Tc=20K. All the spectra have a small peak at about 529 eV, whose intensity shows almost no change against the Ce (electron) concentration. This peak is attributed to the $Cu3d^{10}$ states, because photon energy of 528 eV corresponds to an optical transition from the O1s core level to the electronic states just at the Fermi level since the binding energy of the O1s core level is the same between the hole doped superconductor and the electron doped one [9]. Observed stationary feature of the

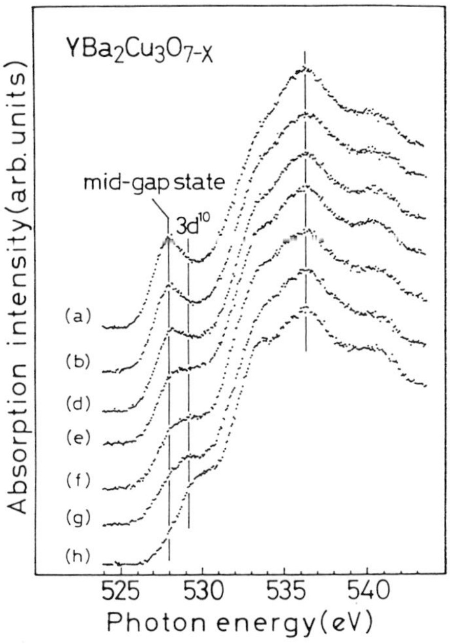

Fig.3 Oxygen K absorption spectra for $YBa_2Cu_3O_{7-\delta}$

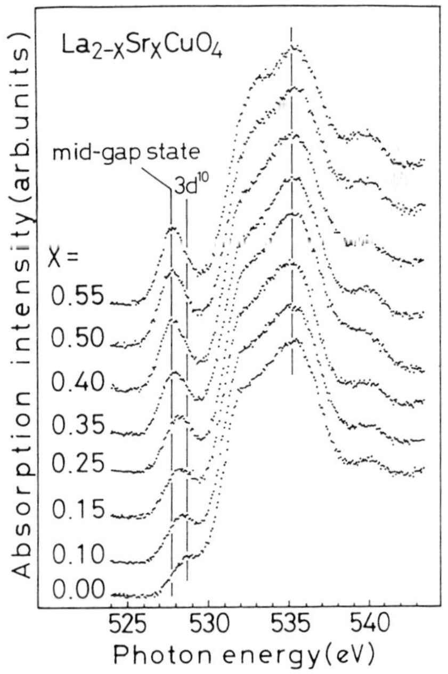

Fig.4 Oxygen K absorption spectra for $La_{2-x}Sr_xCuO_4$

oxygen K absorption spectrum against the Ce (electron) doping as shown in Fig. 5 makes a sharp contrast with the case of the hole doped high-Tc superconductors (Figs. 3 and 4). This means that the O2p state near the Fermi level is not affected by the Ce doping, suggesting that the O2p state is not essential to the occurrence of superconductivity in the electron doped system.

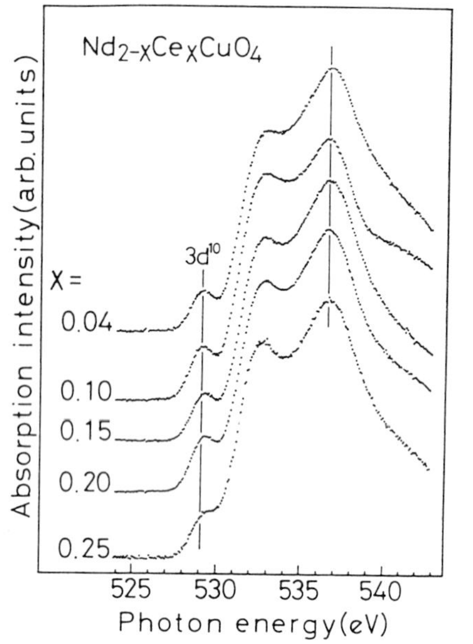

Fig.5 Oxygen K absorption spectra for Nd2-xCexCuO4

CONCLUSION

We have measured oxygen K absorption spectra for La2-xSrxCuO4 , YBa2Cu3O7-δ, and Nd2-y-CeyCuO4 with various hole or electron concentrations. The result shows that hole doping brings about a transfer of weight of the electronic states from the Cu3d^{10} states (upper Hubbard band) to the mid-gap states produced by hole doping. The oxygen K absorption measurement for a wide range of hole concentration from insulator to normal metal showed that no discontinuous change occurs across the border between the superconductor and the normal metal, suggesting that the electronic state near the Fermi level in the high-Tc superconductor is described essentially in terms of the Fermi-liquid states. The experimental result for the electron doped system suggests that the O2p atomic orbital is not directly involved in the high-Tc mechanism.

ACKNOWLEDGEMENT

We thank S. Hosoya and M. Tanaka for their collaboration in characterization of the samples. This work was supported by a grant from the Ministry of Education, Culture and Science of Japan.

REFERENCES

1. T. Takahashi, H. Matsuyama, H. Katayama-Yoshida, Y. Okabe, S. Hosoya, K. Seki, H. Fujimoto, and H. Inokuchi; Nature 334 (1988)691: Phys. Rev. B 39 (1989) 6636.
2. R. Manzke, T. Buslaps, R. Claessen, M. Skibowski, and J. Fink; Physica C 162-164 (1989) 1381.
3. C.G. Olson, R. Liu, A.B. Yang, D.W. Lynch, A.J. Arko, R.S. List, B.W. Viel, Y.C. Chang, P.X. Jiang,and A.P. Paulikas; Science 245 (1989) 731.
4. F.J. Himpsel, G.V. Chandrashekhar, A.B. McLean, and M.W. Shafer; Phys. Rev. B 38 (1988) 11946.
5. N. Nucker, H. Romberg, X.X. Xi, J. Fink, B. Gegenheimer, and Z.X. Zhao; Phys. Rev. B 39 (1989) 6619.
6. H. Matsuyama, T. Takahashi, H. Katayama-Yoshida, T. Kashiwakura, Y. Okabe, S. Sato, N. Kosugi, A. Yagishita, K. Tanaka, H. Fujimoto, and H. Inokuchi; Physica C 160 (1989) 567.
7. P. Kuiper, G. Kruizinga, J. Ghijsen, M. Grioni, P.J.W. Weijs, F.M.F. de Groot, G.A. Sawatzky, H. Verweij, L.F. Feiner, and H. Petersen; Phys. Rev. B 38 (1988) 6483.
8. H. Romberg, M. Alexander, N. Nucker, P. Adelmann, and J. Fink; Phys. Rev. B 42 (1990) in press.
9. T. Takahashi, H. Katayama-Yoshida, and H. Matsuyama; Z. Phys. B 78 (1990) 343.

Electron Energy Loss Spectroscopy in $Ba_2YCu_3O_7$ Superconductor

KENSUKE SHIRAISHI and HIROSHI ITOH

Japan Atomic Energy Research Institute, Takasaki Radiation Chemistry Research Establishment, Watanuki-machi, Takasaki, Gunma, 370-12 Japan

ABSTRACT

A single phased Ba2YCu307 ceramic was milled with 3 keV Ar-ion and electron energy loss spectra in transmission were measured for about 500 to 1000 eV range with incident electron energy of 120 keV. In the oxygen K-edge, pre-edge peak was generally observed at 528.5 eV. The pre-edge peak shifted to 529.5 eV when incident electrons were illuminated from the direction almost exactly parallel to the a-axis and the spectrum was acquired with collection angle of 1.9 mrad.

KEY WORDS: Ba2YCu307 ceramic, ion-milling, EELS, oxygen K-edge, microstructure

INTRODUCTION

Electron energy loss spectroscopy (EELS) of the Ba2YCu307 has been studied for understanding the mechanism of superconductivity of the material[1-7], and the fine structure in the oxygen K-edge is analyzed in terms of the electronic structure of the superconductor, especially the unoccupied density of states near Fermi level. The pre-edge in the oxygen K-edge is characterized to the superconductivity; profile of the pre-edge region is changed by reducing the oxygen content in the superconducting material[3,4,6,7]. Furthermore, the pre-edge structure is reported to be different in the direction of the momentum transfer of incident electrons[1,2,4,5]. Nevertheless, the mechanism of the superconductivity in the oxide system including the Ba2YCu307 has not yet been well understood, and more reliable experimental data are required for the crystallographic orientation dependence of the EELS in the oxide system.

The crystallographic orientation dependent EELS is expected to be readily obtained with a ion-milled ceramic, providing that radiation damage induced by the milling is within a tolerable limit. It is obvious that defect clusters are produced by irradiation of heavy-ions during the milling. On the other hand, high-energy electrons also displace, at least, oxygen atoms from their original lattice site. Threshold energies for the production of visible defects were estimated to be 152 and 131 keV in directions near the a- and b-axes, respectively[8]. This implies that the defects are expected not to be produced during illumination of electrons having energies lower than 131 keV unless the threshold energy for a direction near to c-axis is lower than 131 keV. Thus, the electron irradiation damage can be essentially eliminated when EELS is carried out with incident electron of 120 keV. Furthermore, the characteristic scattering angle for the oxygen K-edge at around 530 keV is about 2.2 mrad for the 120 keV incident electrons[1]. Then, the orientation dependence of EELS is examined by changing the collection angle.

This paper describes crystallographic orientation dependent EELS for an ion-milled superconducting Ba2YCu307-x ceramic.

EXPERIMENTAL

The material used in this experiment was a single phased Ba2YCu307-x ceramic having the superconducting transition temperature of 92.5 K[9]. The thin disk of 3 mm in diameter was dimpled to a thickness of about 0.02 mm, and then milled at room temperature to perforation with 3 keV Ar-ions incident on both the surface under an angle of 10 degree.

EELS was carried out in imaging mode of 104 in magnification with Gatan model 666 attached to a JEM 2000FX II operating at 120 kV. Crystallographic orientation dependence was examined with tilting the specimen and/or changing objective aperture of 1.9, 3.7 and 7.1 mrad in corresponding collection angle.

RESULTS AND DISCUSSION

In the ion-milled specimen, single crystalline area suitable for EELS is readily obtained as shown in Fig. 1; defect clusters of a few nm in diameter produced by ion-milling is observed as black dots in the photograph. Whereas defect clusters may appear during fairly intensive electron illumination for EELS, the electron-irradiation damage is small compared to the damage induced by ion-milling. Furthermore, the spectrum is not appreciably changed after several series of EELS measurements under similar conditions.

Fig. 1 Microstructure for the ion-milled specimen after a series of EELS measurement.

Figure 2 shows an electron energy loss spectrum of 470 - 970 eV region; the spectrum was measured in the area between twin boundaries shown in Fig. 1 and incident electrons were illuminated from a direction nearly parallel to c-axis. In addition to the oxygen K-edge at about 530 eV, the Ba-M4,5 edges and Cu-L2,3 edges are seen at around 800 and 950 eV, respectively. The detailed structure of the oxygen K-edge in the spectra measured with collection angles of 1.9 and 3.7 mrad is presented in Fig. 3. Appreciable difference is observed between the spectra given in the figure, and the width in the loss energy between the oxygen main- and pre-peak was generally 8 eV. Then, the pre-peak is considered to corre-

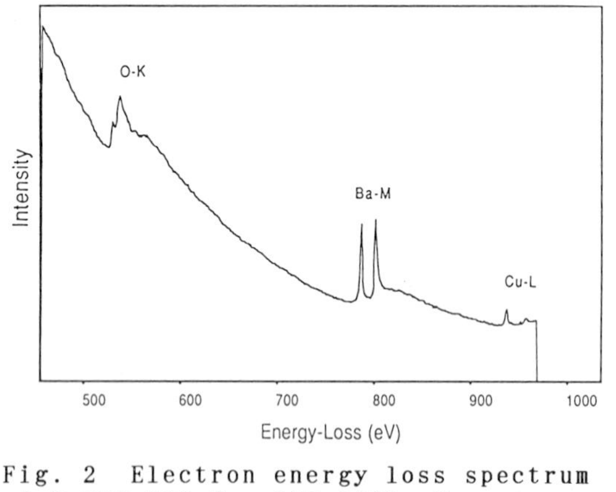

Fig. 2 Electron energy loss spectrum of Ba2YCu3O7 for 500-1000 eV region.

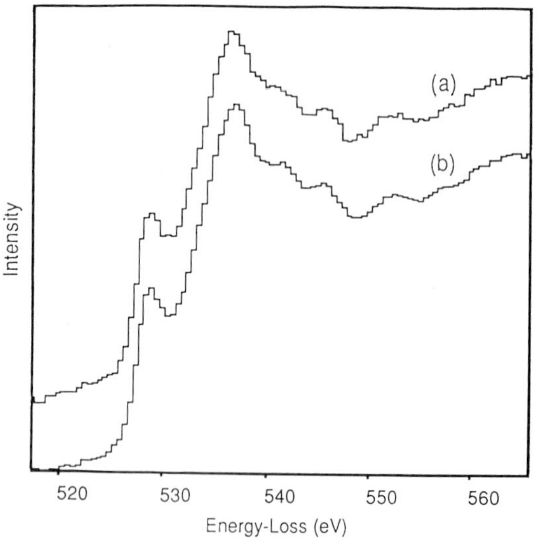

Fig. 3 Oxygen K-edge fine structure measured with collection angles of 1.9 (a) and 3.7 mrad(b).

spond to the mean value of the binding energies of the O 1s levels of the four different O sites of 528.5 eV[5]. When the pre-peak is assigned to 528.5 eV and then the main oxygen peak to 536.5 eV, the Ba-M4,5 peaks are observed at 786 and 801 eV and the Cu-L2,3 peaks at 936 and 956 eV. In addition, the change in shape of the Cu-L2,3 edges with changing the collection angle from 3.7 to 7.1 mrad is also not apparent in this experiment.

The fine structure in the oxygen K-edge is presented in Fig. 4 as a function of the direction of the incident electrons; the spectra were measured with a collection angle of 1.9 mrad and normalized at the main peak of 536.5 eV. When the specimen was illuminated from the direction almost exactly parallel to the a-axis, the oxygen pre-edge peak was observed at 529.5 eV (a), whereas the peak appeared at 528.5 eV when the EELS was performed with incident electrons from the direction almost parallel to the a-axis with g=(003) (b) and far away from the a-axis with (001) system reflections (c).

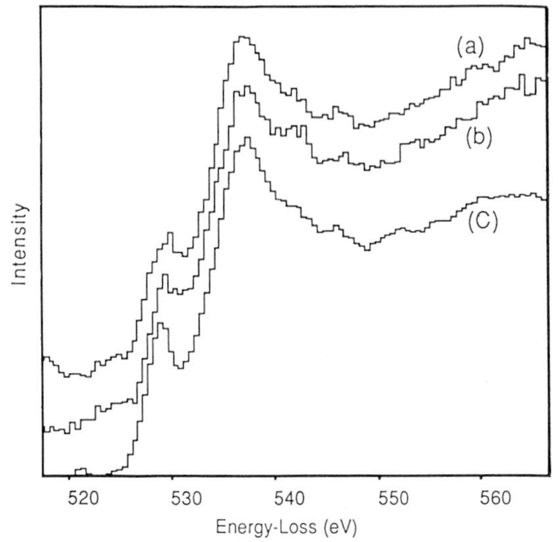

Fig. 4 Orientation-dependent oxygen K-edge fine structure with 1.9 mrad in collection angle; electrons were illuminated from the direction
(a) almost exactly parallel to,
(b) almost parallel to and
(c) away from the a-axis.

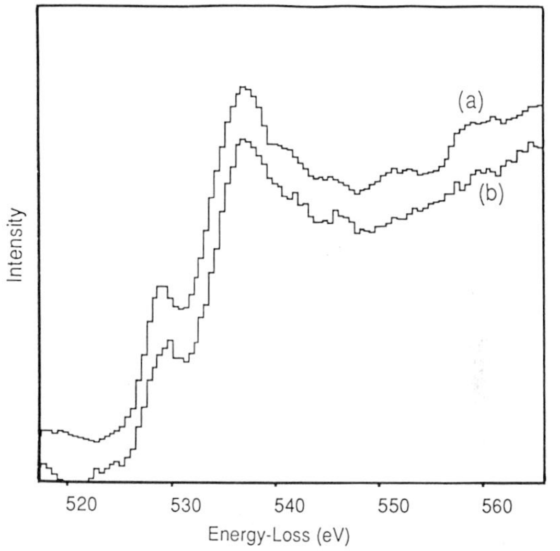

Fig. 5 Change with collection angle in the oxygen K-edge fine structure observed from the direction almost exactly parallel to the a-axis.

The shift of the pre-edge is also observed by changing the collection angle from 1.9 to 3.7 mrad as shown in Fig. 5. The shift of the pre-peak between 529.5 and 528.5 ev is considered to indicate the anisotropy of momentum transfer of the incident electrons along the a- and c-axes.

Figure 6 shows the microstructure for the area where the EELS was carried out to obtain the spectra presented in Figs. 4 and 5; the photograph was took with g=(003) and objective aperture corresponding to 3.7 mrad in collection angle. The lattice image of 1.1 nm spacing is observed in the photograph, although defect clusters were induced by ion-milling. The results indicate that crystallographic orientation dependence of EELS can be examined with ion-milled Ba2YCu3O7 superconductive ceramic.

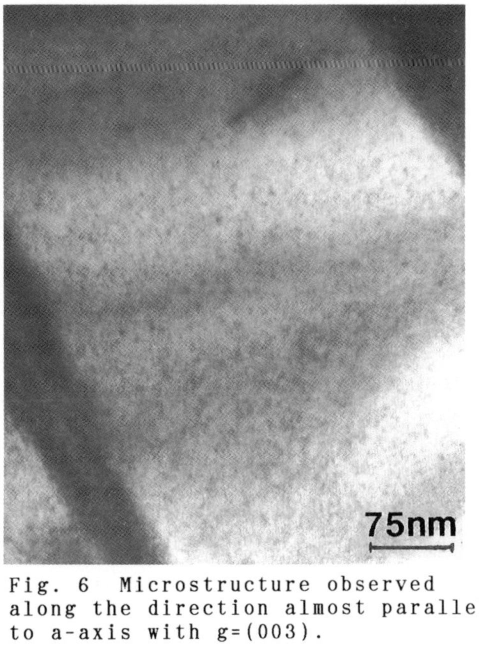

Fig. 6 Microstructure observed along the direction almost parallel to a-axis with g=(003).

SUMMARY

The fine structure in the oxygen K-edge of the superconducting $Ba_2YCu_3O_7$ ceramic was observed by EELS with incident energy of 120 keV. In addition to the main peak at 536.5 eV, pre-edge peak was clearly seen at 528.5 eV in the ion-milled specimen, although small defect clusters were observed in the specimen. The pre-peak shifted to 529.5 eV when incident electrons were illuminated from the direction almost exactly parallel to the a-axis and measured with a collection angle of 1.9 mrad. The pre-peak shifted back to 528.5 eV when the specimen was tilted away from the a-axis and/or measured with larger collection angle of 3.7 mrad. The results indicate that orientation dependent oxygen K-edge fine structure can be investigated with ion-milled superconductive $Ba_2YCu_3O_7$ ceramic.

REFERENCES

1. Batson PE, Chisholm MF (1988) J. Electron Microscopy Technique 8: 311-315
2. Batson PE, Chisholm MF (1988) Phys. Rev. B37: 635-637
3. Nucker N, Fink J, Fuggle JC, Durham PJ, Temmerman WM (1988) Phys. Rev. B37: 5158-5163
4. Nucker N, Fink J, Fuggle JC, Durham PJ, Temmerman WM (1988) Physica C153-155: 119-120
5. Nucker N, Romberg H, Xi XX, Fink J, Gegenheimer B, Zhao ZX (1989) Phys. Rev. B39: 6619-6629
6. Shindo D, Hiraga K, Hirabayashi M, Kikuchi M, Shono Y, Furuno S, Houjou K, Soga T, Otsu H (1989) J. Electron Microsc. 38: 155-157
7. Shindo D, Hiraga K, Tokiwa A, Soga T, Houjou K, Furuno S, Otsu H, Nakajima S, Hirabayashi M, Kikuchi M, Shono Y (1989) Jpn. J. Appl. Phys. 28: L1197-L1200
8. Kirk MA, Baker, MC, Liu JZ, Lam DJ, Weber HW (1988) In: Brodsky MB, Dynes RC, Kitazawa K, Tuller HL (eds) High Temperature Superconductor, Mater. Res. Soc. Pittsburgh pp209-214
9. Shiraishi K, Itoh H, Yoda O (1988) Jpn. J. Appl. Phys. 27: L2339-L2341

EELS and XPS Study of $Y_{1-x}Ca_xBa_2Cu_4O_8$ (x=0~0.1)

KAZUKI YOSHIMURA, YASUHARU KODAMA, and SAKAE TANEMURA

Government Industrial Research Institute, Nagoya, 1-1, Hirate-cho, Kita-ku, Nagoya, 462 Japan

ABSTRACT

The electronic structures of polycrystalline $Y_{1-x}Ca_xBa_2Cu_4O_8$ (x=0~0.1), which were synthesized under high oxygen pressure, have been investigated using EELS (Electron Energy Loss Spectroscopy) including HREELS (High Resolution Electron Energy Loss Spectroscopy) and XPS (X-ray Photoelectron Spectroscopy). EELS and XPS spectra of $Y_{1-x}Ca_xBa_2Cu_4O_8$ (x=0.0, 0.05 and 0.1) show no significant difference among three samples, and these spectra are quite similar to the spectra of $YBa_2Cu_3O_{7-y}$. In EELS spectra structures due to interband transition, plasmon excitation and core excitation are observed. The combination of EELS and XPS measurements enable us to obtain the information about the basic electronic structures of $Y_{1-x}Ca_xBa_2Cu_4O_8$ (x=0.0, 0.05 and 0.1).

KEY WORDS: $YBa_2Cu_4O_8$, $Y_{1-x}Ca_xBa_2Cu_4O_8$, EELS, HREELS, XPS

INTRODUCTION

$YBa_2Cu_4O_8$ (124 phase) was firstly observed as a type of planar defect in $YBa_2Cu_3O_{7-y}$ (123 phase) by high-resolution transmission electron microscopy [1]. Then syntheses of bulk 124 phase at high oxygen partial pressure [2-4] and at ambient pressure [5-8] were reported. The study on the crystal structure of this compound by x-ray [9, 10] diffraction revealed that 124 phase had double Cu-O chains along the b axis, while 123 phase had single Cu-O chains. The superconducting critical temperature of 124 phase is about 80 K, but recently it has been reported that Ca partial substitution for Y results in increasing of the critical temperature of the 124 phase up to 90K [11].

The 124 phase shows an excellent thermal stability in oxygen content up to high temperature in air [2], which is important advantage in a practical use. So far the wide range electronic structure of 124 phase has not been investigated, while the Fermi edge structure was intensively studied by UPS [12]. In this paper the results of EELS and XPS measurements for the sample of polycrystalline $Y_{1-x}Ca_xBa_2Cu_4O_8$ (x=0.0, 0.05 and 0.1) which we successfully synthesized with good crystallinity [13] are reported.

EXPERIMENTAL

Samples of polycrystalline $Y_{1-x}Ca_xBa_2Cu_4O_8$ (x=0.0, 0.05 and 0.1) were synthesized under high oxygen pressure. Stoichiometric amounts of Y_2O_3, $CaCO_3$, $BaCO_3$, and CuO powders were mixed and calcined at 880°C for 24 hrs and at 900°C for 48 hrs with intermediate grinding. The calcined powders were confirmed to be mixture of the 123 phase and CuO by powder X-ray diffractometry. They were pressed into pellets and reacted at 1010°C for 6 hrs under 40 MPa in oxygen partial pressure by using an O_2-HIP apparatus made by Kobe Steel Co. Ltd. Reacted powders were identified to be the 124 phase by X-ray diffractometry. All peaks could be indexed by the 124 structure for the sample of x=0.0. From sharp separation between (017) and (111) peaks, it is confirmed that the sample had good crystallinity. Although x=0.05 and 0.1 had broader peaks than x=0.0 and these samples had very weak CuO peaks in powder neutron diffraction patterns, it showed good crystal quality concerned with 124 phase by x-ray diffraction pattern. No peaks related to the 123 phase and $Y_2Ba_4Cu_7O_{15-y}$ (247 phase) were observed in these samples.

XPS and EELS measurements were made by using VG Scientific ADES400 system. Twin anode X-ray source was used in XPS measurements. Two kinds of electron source were used in EELS measurements. One is conventional non-monochromatized electron gun (PHI 14-50) and monochromatized electron gun (VSW EM50). The latter enable us to obtain HREELS spectra.
The pressure during the measurement in the vacuum chamber was maintained at $5x10^{-8}$ Pa. The samples were scratched by a diamond file before each measurements for surface cleaning. All measurements were performed at room temperature.

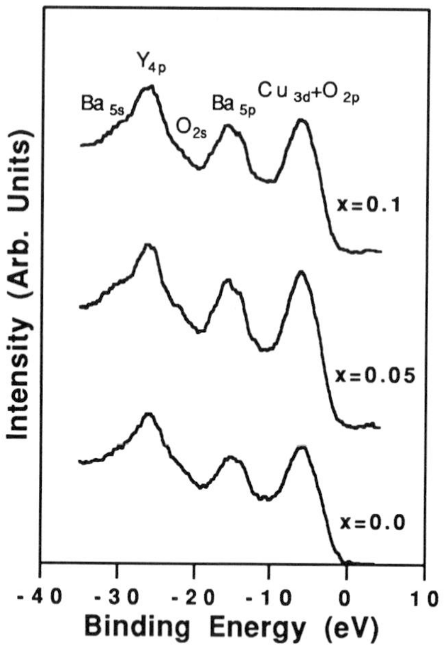

Fig. 1 XPS spectra of $Y_{1-x}Ca_xBa_2Cu_4O_8$ (x=0.0, 0.05 and 0.1) in the valence band and shallow core region.

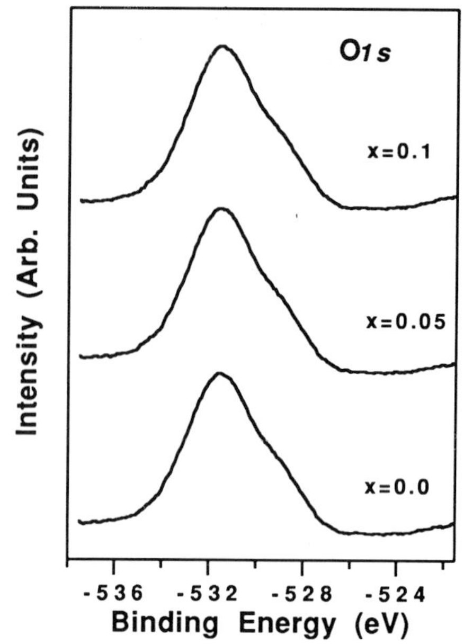

Fig. 2 O*1s* XPS spectra of $Y_{1-x}Ca_xBa_2Cu_4O_8$ (x=0.0, 0.05 and 0.1)

RESULTS AND DISCUSSION

Figure 1 shows the XPS spectra of $Y_{1-x}Ca_xBa_2Cu_4O_8$ (x=0.0, 0.05 and 0.1) in valence band and shallow core region. Al Kα line was used as an exciting source. The structure between -10 eV and 0 (=EF) is the valence band, which mainly consists of Cu(*3d*)and O(*2p*)state. The structure between -20 eV and -10 eV arises from Ba(*5p*)state. The structure below -20 eV mainly consists of Y(*4p*) core with slight structure of O(*2s*) state (the upper side) and Ba(*5s*) states (the lower side). The peak position of Y(*4p*) state is 26.2 eV. These spectra have no significant difference among each samples with different x values and all spectra are quite similar to those of $YBa_2Cu_3O_{7-y}$ [14]. It means that the valence band structure and shallow core states of 124 phase is almost the same as that of 123 phase.

O(*1s*) spectra of $Y_{1-x}Ca_xBa_2Cu_4O_8$ (x=0.0, 0.05 and 0.1) are shown in Fig. 2, respectively. These spectra also quite resemble to each other and that of $YBa_2Cu_3O_{7-y}$ [14, 15]. The energy position of main peak is 521.6 eV, which corresponds to O^{1-} states. But Gourieux *et al.* reported that some contamination peak, especially CO_3 peak appears at -532.3 eV [16]. Hence there is a possibility that this O(*1s*) spectrum is affected by contamination although the sample was scratched in vacuum.

Figure 3 is the EELS spectra of $YBa_2Cu_4O_8$. Incident angle and scattered angle were 20° respectively with respect to the sample normal. This spectra was measured by non-monochromatized electron gun and incident electron energy (Ep) was 1600 eV. The lower part is measured data and the upper part is its second derivative which makes clear the fine structures, respectively. 8 peaks labeled as peak A through H are observed. Their energy position is given in Table I. These spectra shapes with respect to both non-derivative form and second derivative form are quite similar to that of polycrystalline $YBa_2Cu_3O_{7-y}$ which was reported by Ando *et al.* [17]. Identified energy position in Fig. 3 are given in Table I with that of the EELS spectrum of $YBa_2Cu_3O_{7-y}$ [17]. Most of peak positions show good similarity except peak A. There is also a possibility that the energy position of peak A is affected by contamination.

Peak A through F arise from interband transitions. On the other hand the peak G at 24.9 eV is due to plasmon excitation, because this structure is the most dominant structure in the non-derivative spectrum. Using free electron approximation, volume plasmon energy was calculated to be 25.2 eV for 123 phase [17], while it was identified at 24.9 eV. The theoretical volume plasmon energy of 124 phase is 25.9 eV using the lattice constant determined by Marsh *et al.* [9]. In our spectrum it is observed in 24.9 eV, which is a little lower than expected energy. It should be noted that in each case (Ref. 17 and our experiments) the measurements were performed for polycrystalline samples. It means these EELS spectra were integrated over the momentum. So the dispersion of plasmon may make the plasmon peak broad and disturb the precise determination of plasmon energy. Peak F at 29.9 eV is due to the transition between Y(*4s*) state and conduction band.

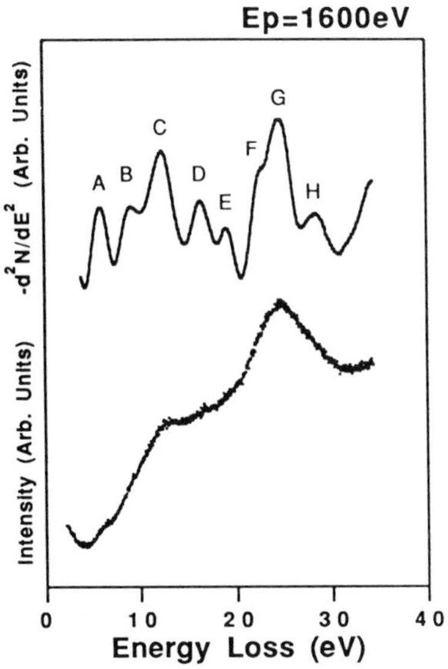

Fig. 3 EELS spectrum of $YBa_2Cu_4O_8$ (the lower part) and its second derivative (the upper part)

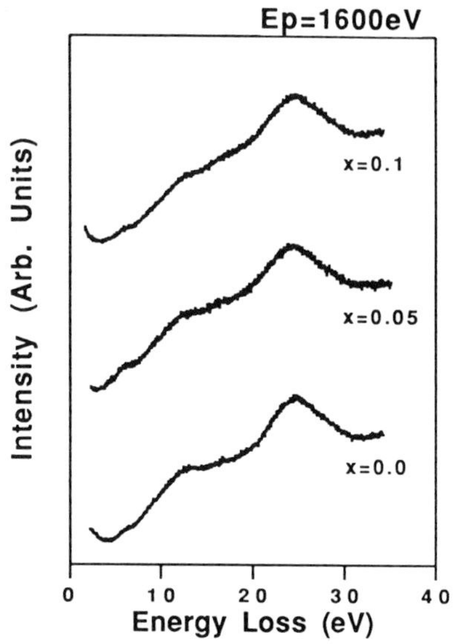

Fig. 4 EELS spectra of $Y_{1-x}Ca_xBa_2Cu_4O_8$ (x=0.0, 0.05 and 0.1).

Figure 4 shows the comparison of the EELS spectra of $Y_{1-x}Ca_xBa_2Cu_4O_8$ (x=0.0, 0.05 and 0.1) . There were no significant difference among them. It indicates that the basic electronic structures of these samples have almost no dependence on the Ca content.

Figure 5 is the EELS spectrum of $YBa_2Cu_4O_8$ in the energy region of 90 - 120 eV. In this energy region the structure due to core excitation are observed. Because core states are structureless, these core transitions well reflect the final empty states (conduction band). This spectrum is quite similar to that of 123 phase [12] and this indicates that the structure of the conduction band of 123 phase and 124 phase are also quite similar. Peak positions of the spectra and their assignment are summarized in Table I.

To observe the more detail of electronic structures, HREELS measurements were carried out. Figure 6 shows the HREELS spectra of $Y_{1-x}Ca_xBa_2Cu_4O_8$ (x=0.0, 0.05 and 0.1) . Incident angle and scattered angle were 30° respectively with respect to the sample normal and incident energy were 5 eV. These spectra exhibits a strong inelastic scattering tail. Theoretically these tail structures are determined by surface conductivity [18]. Intensity of elastic peak and loss structure are inversely proportional to the surface conductivity. To perform the quantitative data analysis of these spectra, the measurement of temperature dependence of the intensity of these tail structures is necessary, which we are now measuring. Weak additional structure is observed in Fig. 6 around 400 meV. These structure are also observed in the HREELS spectrum of 123 phase [19], but the origin of this structure have not been clear. The measurement of temperature dependence of the loss intensity will also be helpful to make clear the origin of this structure.

Table I Peak positions and assignments for the spectrum of 123 phase [17] and that of 124 phase (Fig. 3)

Peak	123 phase* Energy (eV)	124 phase Energy (eV)	Assignment
A	4.2	5.8	
B	8.8	9.2	
C	12.5	12.5	interband transition
D	16.8	16.5	(valence band --> conduction band)
E	19.2	19.1	
F	22.9	22.7	
G	24.9	24.9	volume plasmon
H	29.1	28.5	Y(*4p*) --> conduction band
I	93.2	93.8	Ba($4d_{5/2}$) --> conduction band
J	97.2	97.1	Ba($4d_{3/2}$) --> conduction band
K	104.9	106.8	Ba($4d_{5/2}$) and Ba($4d_{3/2}$) --> conduction band
L	120.1	118.2	Cu(*3s*) --> conduction band

* Ref. 17

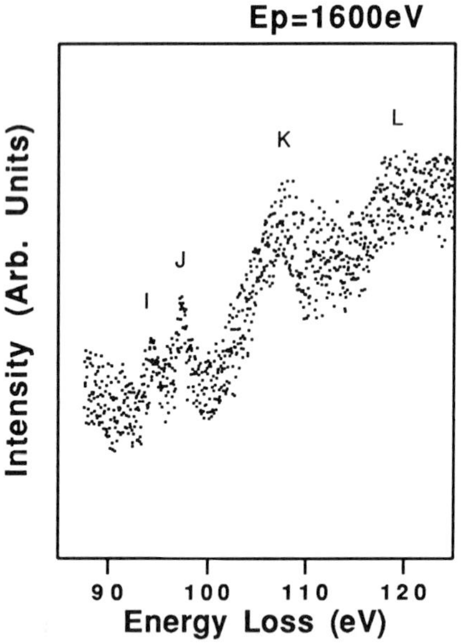

Fig. 5 EELS spectrum of $YBa_2Cu_4O_8$ in the energy region of 90 - 120 eV.

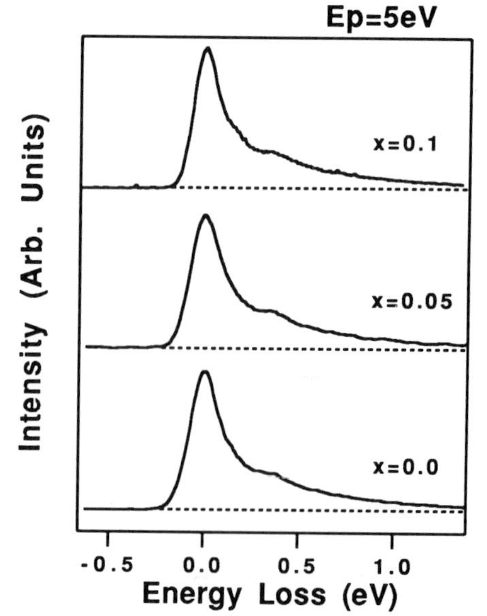

Fig. 6 HREELS spectra of $Y_{1-x}Ca_xBa_2Cu_4O_8$ (x=0.0, 0.05 and 0.1).

CONCLUSIONS

By using EELS and XPS the structure of valence band and conduction band of $Y_{1-x}Ca_xBa_2Cu_4O_8$ (x=0.0, 0.05 and 0.1) are investigated. These results indicate that the electronic structure of $Y_{1-x}Ca_xBa_2Cu_4O_8$ (x=0.0, 0.05 and 0.1) are quite similar to that of $YBa_2Cu_3O_{7-y}$.

ACKNOWLEDGEMENT

We are grateful to Drs. Yamada and Matsumoto (National Research Institute for Metals) for preparing sample powders.

REFERENCES

1. Zaudbergen HW, Gronsky R, Wang K, Thomas G (1988) NATURE **331** : 596
2. Karpinski J, Kaldis E, Jilek E, Rusiecki S, Bucher B (1988) NATURE **336** : 660
3. Morris DE, Nickel JH, Wei JYT, Asmar NG, Scott JS, Scheven UM, Hultgren CT, Markelz AG, Post JE, Heaney PJ, Veblen DR, Hazen RM (1989) Phys. Rev. B**39** : 7347
4. Y.Yamada, M.Yata, Y.Kaieda, H.Irie and T.Matsumoto: Jpn. J. Appl. Phys. **28**(1989) L797
5. CavaRJ, Krajewski JJ, Peck Jr.WF, Batlogg B, Rupp Jr.LW, Fleming RM, James ACWP, Marsh P (1989) NATURE **338 :** 328
6. Kourtakis K, Robbin M, Gallagher PK, Tiefel T (1989) J. Mater. Res. **4** : 1289
7. Balachandran U, Biznek ME, Tomlins GW, Veal BW, Poeppel RB (1990) Physica C **165 :** 335
8. Murakami H, Yaegashi S, Nishino J, Shiohara Y, Tanaka S (1990) Jpn. J. Appl. Phys. **29 :** L445
9. Marsh P, Fleming RM, Mandich ML, DeSantolo AM, Kwo J, Hong M, Martinez-Miranda LJ (1988) NATURE **334** : 141
10. Hazen RM, Finger LW, Morris E (1989) Appl. Phys. Lett. **54** : 1057
11. Miyatake T, Gotoh S, Koshizuka N, Tanaka S (1989) NATURE **341 :** 41
12. Itti R, Miyatake T, Ikeda K, Yamaguchi K, Koshizuka N, Tanaka S (1990) Phys. Rev. **B41** : 9559
13. Kodama Y, Tanemura S, Ikeda T : *submitting to Jpn. J. Appl. Phys.*
14. Yoshimura K, Nogawa S, Tanemura S (1989) Proceeding of SPIE **948** *"High-Tc Superconductivity:Thin Films and Devices"* : 99
15. Horn S, Cai J, Shaheen S A, Jeon Y, Croft M, Chang CL, den Boer ML (1987) Phys. Rev. **B36** : 3895
16. Gourieux T, Krill G, Maurer M, Ravet F, Menny A, Tolentino H, Fontaine A (1988) Phys. Rev. **B37** : 7516
17. Ando A, Saiki K, Ueno K, Koma A (1988) Jpn. J. Appl. Phys. **27** : L304
18. Jacobi K, Sarma DD, Geng P, Simmons C T, Kaidl G (1988) Phys. Rev. **B38** : 863
19. Sarma DD, Prabhakaran K, Rao CNR (1988) Physica **C153-155** : 151

Superconductivity and Electronic State of Nd-Ce-Cu-O System

YUKO YOKOYAMA[1], TOSHIKAZU KATAYAMA[1], HIROYUKI OYANAGI[1],
YUICHI HASUMI[2], and YOSHIKAZU NISHIHARA[1]

[1] Electrotechnical Laboratory, Tsukuba, Ibaraki, 305 Japan
[2] Faculty of Science & Technology, Meiji University, Kawasaki, 214 Japan

Effects of Ce-doping of reduction on the superconductivity and electronic state of $Nd_{2-x}Ce_xCuO_{4-\delta}$ have been investigated by measurements of magnetization and X-ray absorption near-edge structure of the Cu K-edge for the specimens annealed at various oxygen pressures and temperatures. The superconducting onset temperature and dia-magnetization have a peak at Ce content x=0.15. We find that the electron density estimated from the edge shift of the Cu K edge absorption also shows a peak around 0.15. It is concluded that effect of Ce doping changes below and above x=0.15. Below x~0.15 Ce-doping increases the electron density at Cu site and above that vice versa. The result suggests that the electronic state changes above and below x~0.15 in this system.

KEY WORDS: Nd-Ce-Cu-O, magnetization, X-ray absorption, reduction

Introduction

Most of the high Tc superconducting oxides such as $La_{2-x}Sr_xCuO_{4-\delta}$ and $YB_2C_3O_{7-\delta}$ as far found are p-type in which electron holes are responsible for electric conduction. On the contrary, $Nd_{2-x}Ce_xCuO_{4-\delta}$ is n-type superconductor [1] where the major carriers are electron. In this system, superconductivity appears after the annealing under reducing atmosphere in a narrow Ce concentration range [1]. Many experiments such as measurements of magnetization, electric conductivity, photoemittion, X-ray absorption have been studied [2-5]. However, systematic experiments by changing reducing condition and Ce concentration are very few. We have studied effects of the Ce doping and the annealing under the reducing atmosphere.

In this paper, the results on the sample preparation, magnetization measurement and X-ray absorption experiment of $Nd_{2-x}Ce_xCuO_{4-\delta}$ ($0.10 \leq x \leq 0.20$) annealed with various oxygen partial pressures and temperatures and the discussion about the effect of Ce doping on the electronic state of Cu are given.

Experimental

Samples of $Nd_{2-x}Ce_xCuO_{4-\delta}$ were prepared by solid state reaction from Nd_2O_3, CeO_2 and CuO with 99.99% purity. Stoichiometric mixture in air at 950°C for 20 hours and pressed into pellets. The pellets were sintered in air at 1100°C for 10 hours and then cooled to room temperature. After sintering, the samples were then reduced at 900°C in a flowing Ar (4N pure) or at 1050°C in a flowing mixture gas of Ar/O_2 with oxygen partial pressures ranging from 10^{-4} to 1 atm. After reduction, the samples were rapidly cooled by the removal of them from the furnace under the flowing Ar. Figure 1 shows the X-ray diffraction patterns for samples of $Nd_{1.85}Ce_{0.15}CuO_{4-\delta}$ annealed in various oxygen partial pressure P_{O_2} at 1050°C for 20 hours. In the as-grown surface of the

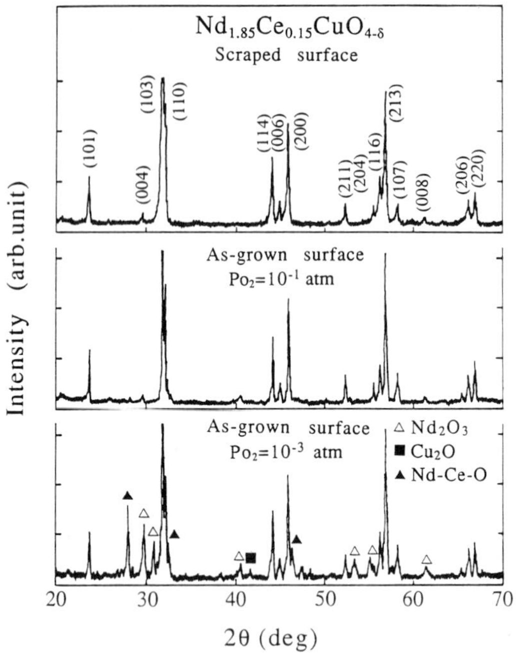

Fig. 1. X-ray diffraction patterns for as-grown and scraped surfaces of the $Nd_{1.85}Ce_{0.15}CuO_{4-\delta}$ pellets annealed at 1050°C in $Po_2=10^{-1}$ and 10^{-3} atm.

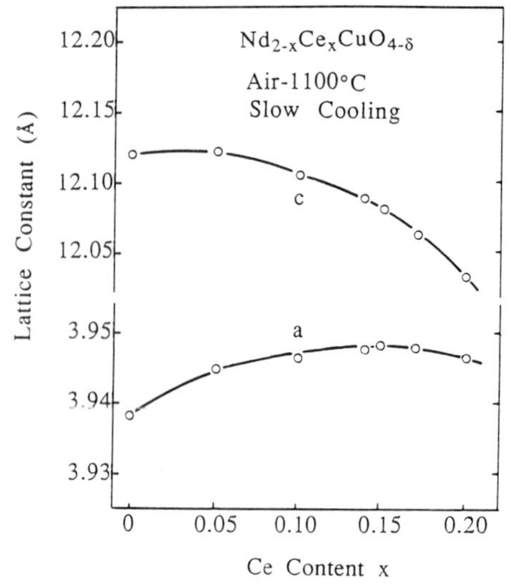

Fig. 2. Lattice parameters a and c of T' phase $Nd_{2-x}Ce_xCuO_{4-\delta}$ as a function of Ce content x. These values were corrected by using an internal silicon standard (a=5.43088Å).

pellet annealed at Po_2 below 10^{-2} atm impurities such as $Nd_2Ce_2O_7$ [6], Nd_2O_3 and Cu_2O were observed together with the superconducting T′ phase as shown in Fig.1. However, in scraped surface where the as-grown surface layer was taken away, there was no phase due to impurities. The measurements of magnetization and X-ray absorption arc done after taking away this surface layer. Figure 2 shows the Ce content dependence of lattice parameters for sintered pellets. These values are nearly equal to the results of Peng et al.[7]. We measured the Po_2 dependence of the lattice parameters for $Nd_{1.85}Ce_{0.15}CuO_{4-\delta}$ specimen, however we could not obtain the systematic change due to Po_2.

The magnetization was measured by a SQUID magnetometer as a function of temperatures in the range of 5~40 K on two different procedures (a) and (b) as follows, (a) the sample was cooled to 5 K at zero-field, after that, a magnetic field was applied, and then the magnetization M(ZFC) due to the shielding current was measured, (b) the sample was cooled from the temperature higher than the critical temperature in a magnetic field and the magnetization due to the Meissner effect M(FC) was measured. In our measurements demagnetization correction is not done. X-ray absorption experiments were performed using synchrotron radiation from the 2.5 GeV storage ring at the Photon Factory.

Results and Discussion

We measured the M(FC) and M(ZFC) in a magnetic field H of 10 Oe for $Nd_{2-x}Ce_xCuO_{4-\delta}$ reduced in flowing Ar (4N pure) for 24 hours at 900°C. Figures 3 (a) and (b) show the temperature dependence of M(FC) and the Ce content dependence of the M(FC) and M(ZFC) at 5 K. M(FC) shows the maximum value at x=0.15, but the volume fraction is smaller than 20% of perfect diamagnetization. In order to increase the volume fraction,

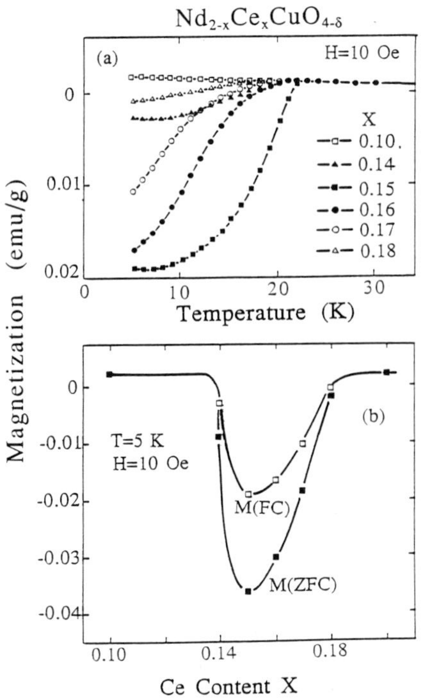

Fig. 3. (a) Temperature dependence of M(FC) in H=10 Oe for $Nd_{2-x}Ce_xCuO_{4-\delta}$ annealed in Ar at 900°C for 24 hours, (b) the Ce content dependence of M(FC) and M(ZFC) at 5 K.

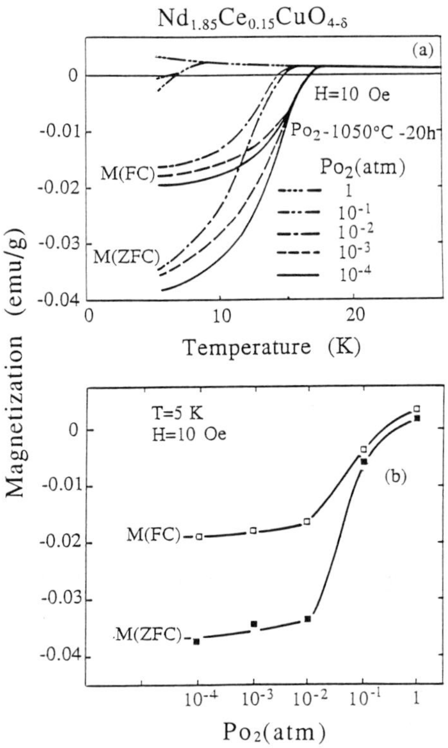

Fig. 4. (a) Temperature dependence of M(FC) and M(ZFC) in H=10 Oe for $Nd_{1.85}Ce_{0.15}CuO_{4-\delta}$ annealed in various Po_2 at 1050°C for 20 hours, (b) the Po_2 dependence of M(FC) and M(ZFC) at 5 K.

we tried the annealing for various conditions by changing annealing temperatures, annealing times and oxygen partial pressures, however we could not obtain a clear increase of Meissner volume. Superconducting transition is broad especially for the specimens with $x \geq 0.16$ where the Meissner magnetizations do not saturate even at 5 K. The results may be ascribed to the inhomogeneous oxygen distribution, as previously discussed [7,8]. The peak position of M(FC) shifts to the lower Ce content region by 0.01 from that of Liang et al. [3].

In order to investigate the reducing effect of superconductivity, we measured the M(FC) and M(ZFC) in H=10 Oe for the specimens of x=0.15 reduced at 1050°C in the Ar/O_2 gas with the oxygen partial pressure Po_2 from 10^{-4} atm to 1 atm. Figures 4 (a) and (b) show the temperature dependence of the M(FC) and M(ZFC) and the Po_2 dependence of those at T=5 K. Superconducting transition temperature and superconducting magnetization increase with decrease of Po_2 and saturate at $Po_2 \geq 10^{-3}$ atm.

Cu K near-edge absorption spectra were measured for these specimens. Figure 5 shows the difference spectra for $Nd_{2-x}Ce_xCuO_{4-\delta}$ ($0.10 \leq x \leq 0.15$) annealed at 900°C in flowing Ar. Each spectrum is obtained by subtraction of the data for $Nd_2CuO_{4-\delta}$ as a standard, which can thus single out the effect of Ce doping on the Cu valence. Details of the X-ray

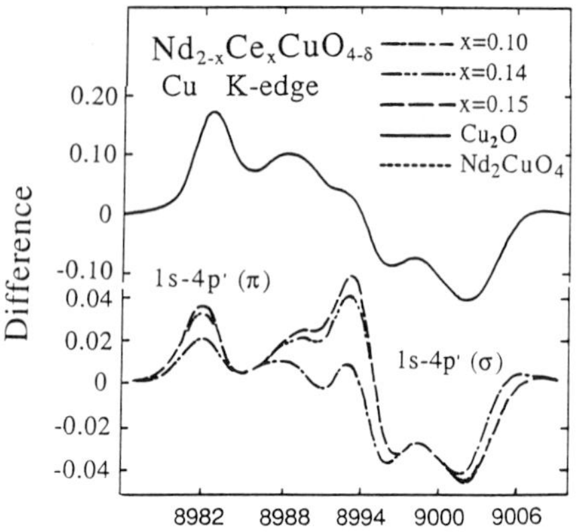

Fig. 5. Difference spectra of Cu K near-edge absorption for $Nd_{2-x}Ce_xCuO_{4-\delta}$ ($0.10 \leq x \leq 0.15$) annealed in flowing Ar and Cu_2O subtracted by absorption spectrum for $Nd_2CuO_{4-\delta}$.

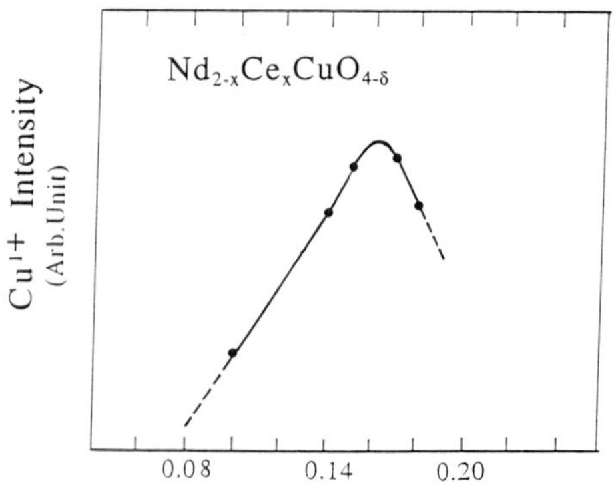

Fig. 6. Ce content dependence of the Cu^{1+} intensity estimated from the Cu K near-edge absorption spectra for $Nd_{2-x}Ce_xCuO_{4-\delta}$ annealed in flowing Ar.

absorption experiments will be published elsewhere [9]. Ce-doping induces new peaks at 8982 eV and 8994 eV which grow with the decrease of P_{O_2}. These new peaks are ascribed to the 1s-4p* (π) transition, which indicate the increase of Cu^{1+}.

Figure 6 shows the Ce content dependence of the intensity of Cu^{1+} component estimated from the peak at 8982 eV. In this figure the results for $0.17 \leq x \leq 0.18$ annealed at 900°C in flowing Ar are displayed together. This figure points out that Cu^{1+} intensity increases as increasing Ce-doping, but at high Ce-doping, the intensity decreases inversely as increasing Ce-doping. Fron the charge nutrality, when Ce remains in Ce^{4+}, we expect that the carrier density simply increases with increasing Ce content. However, above x~0.15 the d density at Cu site decreases. Though the effect of reduction also increases the electron density, the Ce doping has another effect different from the reduction. The electronic state in this system may be different above and below x~0.15.

References

1. Tokura Y, Takagi H and Uchida S (1989) Nature **337:** 345.
2. Takagi H, Uchida S and Tokura Y (1989) Phys. Rev. Lett. **62**: 1197.
3. Liang G, Chen J and Croft M (1989) Phys. Rev. B **40**: 2646.
4. Uji S, Simbo M and Aoki H (1989) Jpn. J. Appl. Phys. **28:** L804.
5. Tranquada J, Heald A, Moodenbaugu A, Liang G and Croft M (1989) Nature **337**: 720.
6. Lopez-Morales ME, Savoy J and Grant PM (1989) Solid State Comm. **71**: 1079.
7. Peng JL, Shelton RN and Radousky HB (1990) Phys. Rev. B **41**: 187.
8. Tarascon J-M, Wang E, L. Green LH, Bagley BG, Hull, D'Egidio SM, Miceli PF, Wang ZZ, Jing TW, Clayhold J, Brawner D and Ong NP (1989) Phys. Rev. B **40**: 4494.
9. Oyanagi H, Yokoyama Y, Yamaguchi H, Kuwahara Y, Katayama T and Nishihara Y (1990) to be published in Phys. Rev. B **40**.

Unoccupied Electronic Structure of $Bi_2Sr_2Ca_{1-x}Y_xCu_2O_8$ Studied by Inverse Photoemission

T. WATANABE, S. SUZUKI, T. TAKAHASHI, H. KATAYAMA-YOSHIDA, and S. SATO

Department of Physics, Tohoku University, Sendai, 980 Japan

ABSTRACT

Inverse photoemission spectroscopy has been performed for $Bi_2Sr_2Ca_{1-x}Y_xCu_2O_8$ single crystal (x=0.0) and ceramics samples (x=0.0, 0.3 and 0.5) to investigate the unoccupied electronic structure. Angle-resolved measurement on a single crystal shows that there are two bands crossing the Fermi level along the Γ-X direction in the Brillouin zone. Angle-integrated measurements were done for ceramics samples where the hole concentration are varied by substituting Y^{3+} for Ca^{2+}. It was found that the intensity at the Fermi level gradually decreases as reducing the hole concentration while no detectable shift of the spectrum relative to the Fermi level was observed. These experimental results indicate that there are at least two Fermi surfaces in 2212-Bi-system-high -Tc superconductor and the electronic states near the Fermi level are a kind of imputiry states produced by hole doping.

KEY WORDS: electronic structure, band structure, inverse photoemission, doping effect, imputiry state

INTRODUCTION

The electronic structure of high-Tc superconductor has been gradually revealed by high-energy spectroscopies such as photoemission and x-ray absorption. Angle-resolved photoemission [1-3] has succeeded in directly observing dispersive energy band(s) crossing the Fermi level, which has been regarded as a direct evidence for existence of the Fermi-liquid states in high-Tc superconductors. However, it is true that there is a serious discrepancy among the three angle-resolved photoemission experiments on $Bi_2Sr_2CaCu_2O_8$ [1-3] in the number of bands crossing the Fermi level. Takahashi et al.[1] observed two dispersive bands intersecting the Fermi level while Manzke et al.[2] and Olson et al.[3] reported only one although the energy position of the band observed by the latter two research groups is apparently different from each other. This problem is directly related to that of the shape of the Fermi surface(s) in the high-Tc superconductor. In order to get an insight to this problem, we have performed an angle-resolved inverse photoemission spectroscopy on Bi2212 system. Inverse photoemission spectroscopy is a complemental experimental technique to photoemission; photoemission spectroscopy gives an information on occupied electronic states while inverse photoemission probes unoccupied electronic states.

The origin of the Fermi-liquid states is an essential problem not only for understanding the electronic structure but also for approaching the high-Tc mechanism. We have studied the hole-concentration dependence of the electronic structure near the Fermi level by inverse photoemission. In Bi2212 system, the hole concentration and as a result the superconducting transition temperature (Tc) are controlled by substituting Y^{3+} atoms for Ca^{2+}[4].

EXPERIMENTAL

A single crystal used in this study was grown by a floating zone method. It has a mirror-like smooth plane of $3x9mm^2$with about 0.5 mm thickness. Its single-crystalinity was confirmed by x-ray diffraction and resistivity measurement showed that it becomes superconductive at 85K. Ceramics samples $Bi_2Sr_2Ca_{1-x}Y_xCu_2O_8$ were prepared by sintering mixtures of high-purity Bi_2O_3, $SrCO_3$, $CaCO_3$, Y_2O_3, and CuO powders with respective molar ratios. The Tc's of ceramic samples were 76, 88, and 47 K for x=0.0, 0.3, and 0.5, respectively.

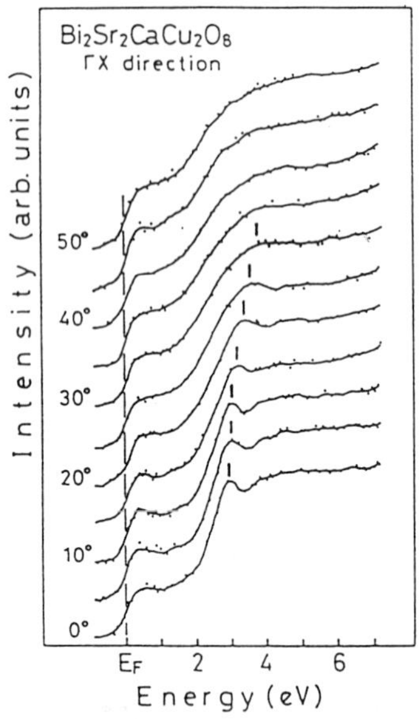

Fig. 1 (a)

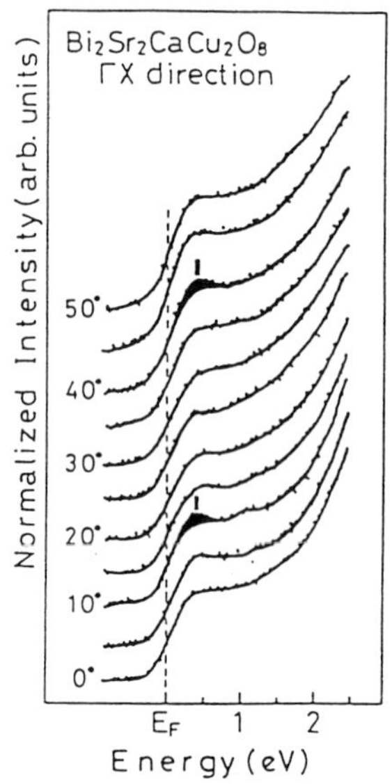

Fig. 1 (b)

Fig. 1 Angle-resolved inverse photoemission spectra of $Bi_2Sr_2CaCuO_8$ measured for Γ-X direction in the Brillouin zone. (a) and (b) show a wide-energy range and the vicinity of the Fermi level, respectively. Shaded areas in (b) show two maxima in the inverse photo-emission intensity at the Fermi level.

Inverse photoemission measurements were done with a home-made angle-resolved inverse photoemission spectrometer[5] which has a Pierce-type electron gun with a BaO dispenser cathode and a Geiger-Muller type counter. The typical electron current impinging on the sample surface was 6μA. Outcoming photons were focused by a concave mirror into the GM counter which detects photons with energy of 9.5 ± 0.3 eV by a bandpass filter of a SrF_2 window and iodine gas in it. Angle-resolved measurement was made by rotating the single crystal with respect to the direction of electron beams. The angular resolution was about 2°. Clean surfaces for measurement were obtained by peeling off surface layers with Scotch tape for a single crystal and by scraping the surface with a diamond file for ceramics. The measurement was done at room temperature under ultrahigh vacuum better than 5×10^{-10} Torr. Degradation of the sample surface, probably due to damage by electron beams, was detected after two or three days. All the spectra for the present interpretation were recorded before the degradation of the sample surface was detected. The Fermi level of the sample was referred to that of a silver film deposited on each sample.

RESULTS AND DISCUSSION

Angle-Resolved Inverse Photoemission Study of $Bi_2Sr_2CaCu_2O_8$

Figure 1 shows angle-resolved inverse photoemission spectra of $Bi_2Sr_2CaCu_2O_8$ measured for a high-symmetry direction in the Brillouin zone Γ X; Figure 1(a) shows the wide-energy-range spectra and Fig. 1(b) enlarges the vicinity of the Fermi-level. In Fig. 1(a), a dispersive feature from 3 eV at $\theta = 0°$ to 4 eV at $\theta = 30°$ is clearly seen. In a previous angle-resolved inverse photoemission study on $Bi_2Sr_2CaCu_2O_8$ [6], such a dispersive feature was not observed. Although the reason for this discrepancy is not clear at present, we would like to stress the higher energy and angular resolutions in the present study. Although this dispersive feature can be correlated to one of calculated bands in the band structure calculation, we tentatively assign it as a surface-image-potential state. A surface-image-potential state has been often observed in inverse photoemission measurements of metals and semiconductors [7] and is an extrinsic electronic state produced by a mirror image of an incoming electron. The observed dispersion in Fig. 1 (a) represents the energy dispersion of the parallel component of the surface-image-potential state which can be described as an almost free electron.

In Fig. 1, we find a substantial inverse photoemission intensity at the Fermi level for all the polar angles from 0° to 50°. This also shows a sharp contrast with the previous angle-resolved inverse photoemission study [6] where the inverse photoemission intensity at the Fermi level is

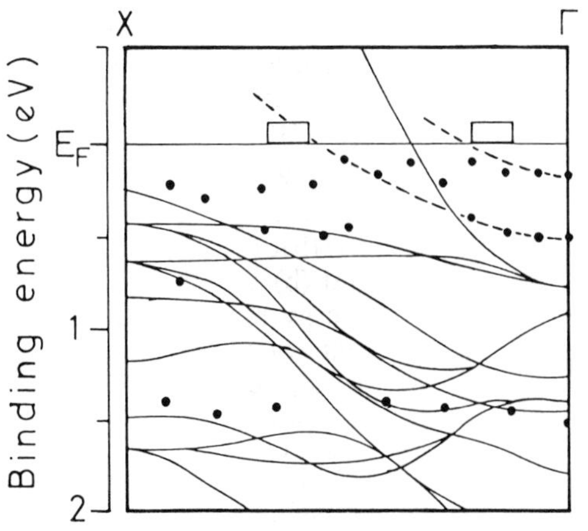

Fig. 2 Experimental band structure of $Bi_2Sr_2CaCu_2O_8$. Rectangles indicate crossing points of bands observed in the present angle-resolved inverse photoemission. Filled circles are the result from an angle-resolved photoemission [1] and broken lines show a possible connection of the two experimental results. A band structure calculation [8] is also shown by thin solid lines for comparison.

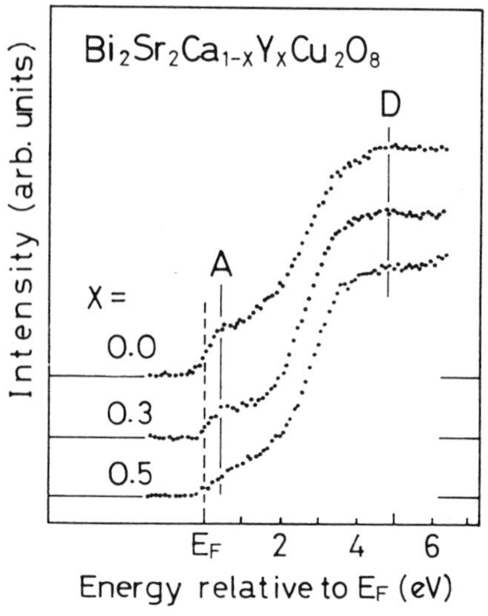

Fig. 3 Inverse photoemission spectra of $Bi_2Sr_2Ca_{1-x}Y_xCu_2O_8$ ceramics (x=0.0, 0.3, and 0.5). Note that the inverse photoemission intensity at the Fermi level (A) gradually decreases as decreasing the hole concentration (in other words, increasing the Y content) without a detectable rigid shift of the spectrum as a whole.

almost zero at θ =0°and increases monotonically up to 50°. The reason for this discrepancy is not clear at present, but it is noted that the behavior shown in Fig. 1 was observed for all three single crystals which we measured. And also the observed clear dispersion around 3-4 eV shown in Fig. 1(a) warrants that we certainly saw a single crystal surface. There are some possible causes which give such a background in an inverse photoemission spectrum; surface imperfections which produce surface states near the Fermi level, incommensurate superstructure of BiO plane which folds band dispersions into a small Brillioun zone, and a strong contribution from phonon upon an optical transition which weakens the selection rule of direct transition. When considering that the crystal surface was relatively insensitive to residual gases and the band folding effect is in general not so strong, we tentatively attribute the observed background to a phonon-assisted indirect transition. This may suggest a strong electron-phonon coupling in the high-Tc superconductor.

In contrast with the previous angle-resolved inverse photoemission measurement [6], we found a characteristic variation of inverse photoemission intensity near the Fermi level. Figure 1(b) shows that the intensity has two maxima at 10° and 40° polar angles. We think that at these two points energy bands cross the Fermi level. The points are shown by rectangles in a band structure scheme in Fig. 2, together with the result from the angle-resolved photoemission (filled circles)[1] and a band structure calculation (thin solid lines)[8]. The present angle-resolved inverse photoemission shows a qualitative agreement with the angle-resolved photoemission result by Takahashi et al.[1] in that there are two bands crossing the Fermi level, though the crossing points seem to be a little different between the two experiment. It may be useful here to compare the difference among the three angle-resolved photoemission experiments in order to check the validity of the present angle-resolved inverse photoemission result. Takahashi et al.[1] reported two bands (named A and B by them) crossing the Fermi level while Manzke et al.[2] and Olson at al.[3] observed only one. When comparing the observed binding energy of the bands at the Γ point, we find that the band observed by Manzke et al. corresponds to one of the two bands (band A) reported by Takahashi et al. and the band reported by Olson et al. corresponds to the other band (band B). These experimental results including the present angle-resolved inverse photoemission indicate that there are two bands which cross the Fermi level, in other words, there are two types of Fermi surfaces in the Bi2212 high-Tc superconductor. The discrepancy between the experiment and the band structure calculation requires further experimental and theoretical studies.

Inverse Photoemission Study of $Bi_2Sr_2Ca_{1-x}Y_xCu_2O_8$ Ceramics

Figure 3 shows (angle-integrated) inverse photoemission spectra from $Bi_2Sr_2Ca_{1-x}Y_xCu_2O_8$ ceramics (x=0.0, 0.3, and 0.5). In Bi2212 system, we can control the hole concentration and as a result the Tc by substituting trivalent Y atoms for divalent Ca [4]. We find in Fig. 3 that as the hole concentration decreases (in other words, the Y content increase) the inverse photoemission intensity near the Fermi level gradually decreases with no detectable shift of the spectrum as a whole. A similar experimental result has been obtained from angle-resolved [9] and angle-integrated [10] photoemission and x-ray absorption experiments [11] of $Bi_2Sr_2Ca_{1-x}Y_xCu_2O_8$. All these experimental results indicate that the hole-doping does not necessarily cause a rigid shift of the density of electronic states relative to the Fermi level but creates new electronic states near the Fermi level.

CONCLUSION

We performed an angle-resolved inverse photoemission spectroscopy to study the electronic states near the Fermi level. The inverse photoemission intensity at the Fermi level shows two maxima between the center (Γ point) and the boundary (X point) of the Brillouin zone, suggesting that there are two bands crossing the Fermi level in support with a previous angle-resolved photoemission [1]. Inverse photoemission spectroscopy on $Bi_2Sr_2Ca_{1-x}Y_xCu_2O_8$ ceramics indicates that the hole-doping produces new electronic states near the Fermi level.

ACKNOWLEDGEMENT

We thank A. Yamanaka and F. Minami at Hokkaido University and S. Takekawa at National Institute for Inorganic Materials for supplying Bi2212 single crystals. This work was supported by a grant from the Ministry of Education, Culture and Science of Japan

REFERENCES

1. T.Takahashi, H.Matsuyama, H.Katayama-Yoshida, Y. Okabe, S.Hosoya, K.Seki, H.Fujimoto, M.Sato, and H.Inokuchi; Nature 334(1988)691; Phys. Rev.B39 (1989) 6636.
2. R.Manzke, T.Buslaps, R.Claessen, M.Skibowski, and J.Fink ; Physica C 162-164 (1989) 1381.
3. C.G.Olson, R.Liu, A.B.Yang, D.W.Lynch, A.J.Arko, R.S.List, B.W.Veal,Y.C.Chang, P.X.Jiang, and A.P.Paulikas ; Science 245 (1989) 731.
4. T.Tamegai, K.Koga, and K.Suzuki ; Jpn.J.Appl.Phys.28 (1989) 112.
5. H.Ohsawa, T.Takahashi, T.Kinoshita, Y.Enta, H.Ishii, and T.sagawa ; Solid State Commun. 61 (1987) 347.
6. R.Claessen, R.Manzke, H.Carstensen, B.Burandt, T.Buslaps, and M.Skibowski ; Phys. Rev. B39 (1989) 73.
7. N.V.Smith ; Rep.Prog.Phys. 51 (1988) 1227-1294.
8. S.Massidda, Jaejun Yu, and A.J.Freeman;Physica C152 (1988) 251.
9. T.Takahashi, H.Matsuyama, H.Katayama-Yoshida, K.Seki, K.Kamiya, and H.Inokuchi ; Physica C170 (1990) 416.
10. R.Itti, F.Munakata, K.Ikeda, N.Koshizuka, and S.Tanaka ; Physica B165 -166 (1990) 1223.
11. H.Matsuyama, T.Takahashi, H.Katayama-Yoshida, T.Kashiwakura, Y.Okabe, S.Sato, N.Kosugi, A.Yagishita, K.Tanaka, H.Fujimoto, and H.Inokuchi ; Physica C160 (1989) 567.

Photoemission Study of $Bi_2Nd_{0.5}Ca_{2.5}Cu_2O_y$ and $Bi_2(La,Pr)_{0.5}Ca_{2.5}Cu_2O_y$ Superconductors

KAZUTO IKEDA, FUMIO MUNAKATA, RITTAPORN ITTI, TAKASHI KAWANO, H. YAMAUCHI, and NAOKI KOSHIZUKA

Superconductivity Research Laboratory, International Superconductivity Technology Center, 10-13, Shinonome 1-chome, Koto-ku, Tokyo, 135 Japan

ABSTRACT

X-ray photoelectron spectra (XPS) of Cu $2p$ have been studied for oxygen- and air-sintered samples of $Bi_2(La,Pr)_{0.5}Ca_{2.5}Cu_2O_y$ and $Bi_2Nd_{0.5}Ca_{2.5}Cu_2O_y$. We estimated the number of Cu $3d$ electrons, n_d, and charge transfer energy, Δ, from these results. We point out the possibility that the difference of T_c between these compounds with the same doped hole concentrations is associated with the difference of n_d. The role of the Cu $3d$ - O$2p$ charge transfer in superconductivity is discussed.

KEY WORDS: $Bi_2Nd_{0.5}Ca_{2.5}Cu_2O_y$, $Bi_2(La,Pr)_{0.5}Ca_{2.5}Cu_2O_y$, photoemission

INTRODUCTION

Since the discovery of high T_c superconductors of $Bi_2Sr_2Ca_{n-1}Cu_nO_{2n+4}$ (n = 1, 2, 3) [1,2], a lot of studies have been made on these Bi cuprates. The superconducting transition temperatures, T_c, of 5 - 20 K [1,3], 84 K [2], and 105 K [2] were reported for n = 1 (2201 phase), 2 (2212 phase) and 3 (2223 phase) materials, respectively. Inoue *et al.* found a new family of Sr - free Bi copper - based superconductors in Bi-La - Ca - Cu - O system. [4]

The crystallographic structure of $Bi_2Nd_{0.5}Ca_{2.5}Cu_2O_y$ and $Bi_2(La,Pr)_{0.5}Ca_{2.5}Cu_2O_y$ is the same as that of $Bi_2Sr_2CaCu_2O_y$; there are two CuO_5 pyramids in the unit cells. They have following properties: (a) T_c of the oxygen - sintered sample is higher than that of air - sintered sample and (b) the oxidation state of Cu, p, of the oxygen - sintered sample is higher than that of air - sintered sample for both compounds. (c) These compounds sintered in the same atmosphere have different T_c although the p value are same. [5]

In this paper, we report on a study of the Cu $3d$ - O$2p$ charge - transfer [6-8] carried out by the measurement of the Cu $2p$ core - level X - ray photoelectron spectra in Bi cuprates. Electronic parameters characterizing the charge - transfer are calculated from the spectra. The relationship between T_c and the degree of charge - transfer is discussed.

EXPERIMENTAL

Samples were prepared by solid - state reactions of Bi_2O_3, $CaCO_3$, CuO, and Nd_2O_3 (or La_2O_3 and Pr_6O_{11}) powders, whose purities were higher than 99.9%. [9,10] In this report, "sintered in air" means "sintered in 21% oxygen/nitrogen mixed gas - flow".

Powder X-ray diffraction indicated that samples were of single phase. We determined T_C by means of AC-magnetization measurement using LAKESHORE model 7000 AC susceptmeter. The oxidation states of Cu, p, were determined by iodometric titration.[7]

X-ray photoelectron spectroscopy (XPS) was performed at room temperature using VG ESCALAB Mk II spectrometer. Al K_α radiation was used as an X-ray source. In order to expose clean surfaces, samples were scraped *in-situ* with a diamond file[11] equipped in a preparation chamber. The cleanness of surface was judged by O1s spectra. Oxygen 1s spectra of the samples had a single peak, which indicates superconducting phase surface.

RESULTS AND DISCUSSION

Figure 1 shows Cu 2p core-level XPS spectra of $Bi_2Nd_{0.5}Ca_{2.5}Cu_2O_y$ and $Bi_2(La,Pr)_{0.5}Ca_{2.5}Cu_2O_y$ sintered in air and oxygen. The intensity ratios, I_s/I_m, of the satellite peak ($2p^5 3d^9$) to main peak ($2p^5 3d^{10}\underline{L}$) and splittings, ΔE_{ms}, of these peaks were obtained from the spectra after subtraction of the back ground to determine the difference of Cu valency between samples. Table I shows I_s/I_m and ΔE_{ms} with experimental errors. It is thought that the Cu valency changes from divalent to more monovalent state with decrease of the I_s/I_m value from the fact there is no satellite structure on Cu 2p spectrum of Cu_2O. Considering the I_s/I_m values, Cu valency of $Bi_2(La,Pr)_{0.5}Ca_{2.5}Cu_2O_y$ seems to be more monovalent like than that of $Bi_2Nd_{0.5}Ca_{2.5}Cu_2O_y$. This indicates that the charge transfer from O 2p to Cu 3d state is likely to take place in $Bi_2(La,Pr)_{0.5}Ca_{2.5}Cu_2O_y$ than in $Bi_2Nd_{0.5}Ca_{2.5}Cu_2O_y$.

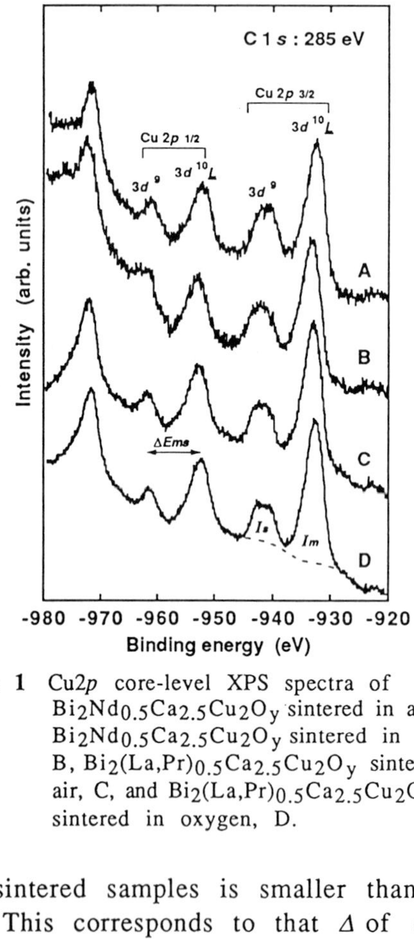

Figure 1 Cu2p core-level XPS spectra of $Bi_2Nd_{0.5}Ca_{2.5}Cu_2O_y$ sintered in air, A, $Bi_2Nd_{0.5}Ca_{2.5}Cu_2O_y$ sintered in oxygen, B, $Bi_2(La,Pr)_{0.5}Ca_{2.5}Cu_2O_y$ sintered in air, C, and $Bi_2(La,Pr)_{0.5}Ca_{2.5}Cu_2O_y$ sintered in oxygen, D.

As shown in Table I, I_s/I_m of the oxygen-sintered samples is smaller than that of air-sintered samples for both compounds. This corresponds to that Δ of oxygen-sintered sample is smaller than that of air-sintered sample. For each compound, T_C of oxygen-sintered sample is higher than that of air-sintered sample (see Table II). This increase of T_C is interpreted by the increase in the copper oxidation state, p, which is measured by means of iodometric titration.[12] The concentration of holes doped to a CuO_5 pyramid from excess oxygens is proportional to p. Therefore, I_s/I_m difference between the oxygen- and air-sintered samples indicates that Δ decreases with increasing doped-holes for each compound. This is in agreement with that of Rao *et al.* [6] This means that the hole doping from excess oxygens has an influence on the Cu $3d$ - O $2p$ charge transfer.

Table I Relative intensities, I_s/I_m, and splittings, ΔE_{ms}, of the main and satellite peaks in the Cu $2p$ XPS core-level spectra. Charge - transfer energies, Δ, on - site Coulomb energies between Cu $2p$ core and $3d$ electrons, U_{cd}, transfer integrals, T, and numbers of Cu $3d$ electrons, n_d, are also given. Δ, U_{cd}, T, and n_d are calculated from I_s/I_m and ΔE_{ms} on the basis of a model of van der Laan *et al.* [13]

samples	I_s/I_m	ΔE_{ms} (eV)	Δ (eV)	U_{cd} (eV)	T (eV)	n_d
$Bi_2Nd_{0.5}Ca_{2.5}Cu_2O_y$ (air)	0.39±0.02	8.9±0.1	0.56±0.15	8.1±0.2	2.4	9.45±0.01
$Bi_2Nd_{0.5}Ca_{2.5}Cu_2O_y$ (O_2)	0.37±0.02	8.7±0.1	0.51±0.15	7.8±0.2	2.4	9.45±0.01
$Bi_2(La,Pr)_{0.5}Ca_{2.5}Cu_2O_y$ (air)	0.31±0.02	9.0±0.1	0.04±0.15	7.7±0.2	2.4	9.49±0.01
$Bi_2(La,Pr)_{0.5}Ca_{2.5}Cu_2O_y$ (O_2)	0.30±0.02	9.2±0.1	-0.10±0.15	7.8±0.2	2.4	9.49±0.01

It has been reported that the strong Coulomb correlation nearly localizes Cu $3d$ electrons, and that holes doped from the excess oxygens are dominantly at O $2p$ state. [11] The copper - based superconductors without conducting curriers are the charge - transfer type insulators [11] as shown in Figure 2.

We estimated the electronic structure parameters and the number of Cu $3d$ electrons of these superconductors based on the model of van der Laan *et al.* [13] According to their analysis, the charge transfer energy, Δ , and on-site Coulomb energy, U_{cd}, between Cu $2p$ core and $3d$ electrons are obtained as functions of the Cu $3d$ to O $2p$ transfer integral, T, by using the experimental values of I_s/I_m and ΔE_{ms}. This result is shown in Table I. Since the value of T can not be uniquely determined from the experimental results, we used a T value for $Bi_2(SrCa)_3Cu_2O_y$ in Ref. 8 in this estimation. The fact that n_d of $Bi_2(La,Pr)_{0.5}Ca_{2.5}Cu_2O_y$ is larger than that of $Bi_2Nd_{0.5}Ca_{2.5}Cu_2O_y$ is thought to be due to the smaller Δ value of $Bi_2(La,Pr)_{0.5}Ca_{2.5}Cu_2O_y$ than that of $Bi_2Nd_{0.5}Ca_{2.5}Cu_2O_y$.

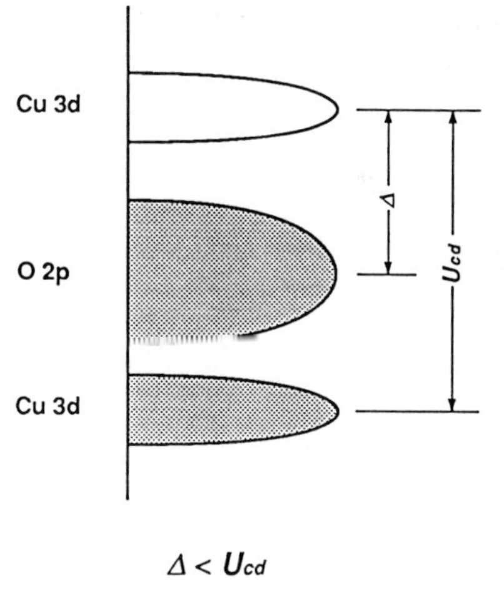

Figure 2 Schematic band structure of the charge - transfer type insulators.

The relationship between T_C and the doped - hole concentration per a CuO_5 pyramid was so far discussed for several superconductors. [5,12] However, the difference in T_C between $Bi_2Nd_{0.5}Ca_{2.5}Cu_2O_y$ and $Bi_2(La,Pr)_{0.5}Ca_{2.5}Cu_2O_y$ can not be explained by the doped hole concentration only because they are in the same oxidation state of Cu (see Table II). This XPS analysis suggests that T_C depends on the degree of the charge transfer from O $2p$ to Cu $3d$ state as well as the doped hole concentration.

The number of Cu $3d$ electrons, n_d, which reflect the degree of Cu $3d$ - O$2p$ charge transfer might be one of the key parameters determining T_C together with the hole numbers per a CuO_5 pyramid in the copper - based superconductors.

Table II Superconducting transition temperatures, T_C, from AC magnetization. Copper oxidation states, p, determined from iodometric titration.

samples	T_C (K)	p	n_d
$Bi_2Nd_{0.5}Ca_{2.5}Cu_2O_y$ (air)	47	2.05±0.01	9.45±0.01
$Bi_2Nd_{0.5}Ca_{2.5}Cu_2O_y$ (O_2)	54	2.08±0.01	9.45±0.01
$Bi_2(La,Pr)_{0.5}Ca_{2.5}Cu_2O_y$ (air)	52	2.05±0.01	9.49±0.01
$Bi_2(La,Pr)_{0.5}Ca_{2.5}Cu_2O_y$ (O_2)	63	2.08±0.01	9.49±0.01

REFERENCES

[1] J. Akimitsu, A. Yamazaki, H. Sawa and H. Fujiki: Jpn. J. Appl. Phys. 26, L2080 (1987).
[2] H. Maeda, Y. Tanaka, M. Fukutomi and T. Asano: Jpn. J. Appl. Phys. 27, L209 (1988).
[3] H. Sawa, H. Fujiki, K. Tomimoto, and J. Akimitsu: Jpn. J. Appl. Phys. 27, L830 (1988).
[4] O. Inoue, S. Adachi, Y. Takahashi, H. Hirono, and S. Kawashima: Jpn. J. Appl. Phys. 28, L778 (1989).
[5] T. Den, and J. Akimitsu: Jpn. J. Appl. Phys. 28, L193 (1989) .
[6] C. N. R. Rao, G. R. Rao, M. K. Rajuumon, and D. D. Sarma: Phys. Rev. B42, 1026 (1990).
[7] F. Munakata, T. Kawano, H. Yamauchi, and Y. Inoue: (to be published in Phys. Rev. B).
[8] A. Fujimori, E. Takayama-Muromachi, Y. Uchida, and B. Okai: Phys. Rev. B35, (1987) 8814; Solid State Commun. 63, 857 (1987).
[9] T. Kawano, F. Munakata, A. Nozaki, T. Ikemachi, H. Yamauchi, and S. Tanaka: Physica C 165, 67 (1990).
[10] T. Kawano, F. Munakata, A. Nozaki, T. Ikemachi, H. Yamauchi, and S. Tanaka: Physica C 165, 143 (1990) .
[11] A. Fujimori, S. Takekawa, E. Takayama-Muromachi, Y. Uchida, A. Ono, T. Takahashi, Y. Okabe, and H. Katayama-Yoshida: Phys. Rev. B39, 2255 (1989).
[12] T. Tamegai, K. Koga, K. Suzuki, M. Ichihara, F. Sakai, and Y. Iye: Jpn. J. Appl. Phys. 28, L112 (1989).
[13] G. van der Laan, C. Westra, C. Haas and G. A. Sawatzky: Phys. Rev. B23, 4369 (1981).

Photoelectron Spectroscopic Study of Zn/Cu Substituted $La_{1.85}Sr_{0.15}(Cu_{1-x}Zn_x)O_y$

R. ITTI, S. IKEGAWA, K. IKEDA, H. YAMAUCHI, N. KOSHIZUKA, and S. TANAKA

Superconductivity Research Laboratory, International Superconductivity Technology Center,
10-13 Shinonome 1-chome, Koto-ku, Tokyo, 135 Japan

ABSTRACT

A photoelectron spectroscopic study of Zn/Cu substituted polycrystalline $La_{1.85}Sr_{0.15}(Cu_{1-x}Zn_x)O_y$ (x=0 - 0.04) is reported. As x increases, Tc of this system decreases and superconductivity disappears for x>0.3. The reason for these occurrences is not yet clear. We have performed XPS and UPS measurements to investigate the respective core levels and the valence band, especially the density of states at the Fermi level for samples with different x's at room temperature. The results showed that i) Zn substituted for Cu was divalent, ii) the core level binding energies of La, Sr, Cu, and O were not affected by substitution, iii) a finite DOS at the Fermi level was observed, not only for x=0 sample, but also for samples with x up to 0.04, which is not superconducting down to 4.2K. This is interesting in the sense that for a cuprate oxide, one may observe a Fermi edge with finite magnitude on samples with nearly the same carrier concentration independent of whether the sample is superconducting.

KEYWORDS: Photoelectron Spectroscopy, $La_{1.85}Sr_{0.15}CuO_4$, Substitution

INTRODUCTION

Chemical substitution is a method widely adopted to elucidate the mechanism of high-Tc superconductivity. For cuprate oxides, Zn is one of the elements mostly used because it is normally divalent (Zn^{2+}) with a $3d^{10}$ electronic configuration, i.e. non-magnetic, while a divalent Cu^{2+} is magnetic. Zn substitution is not expected to alter the carrier density, but to affect the spin correlation through the spin vacancies on the Zn sites.

From the transport and structural studies of $La_{1.85}Sr_{0.15}(Cu_{1-x}Zn_x)O_y$ [1-3], it is shown that Tc decreases rapidly with a small amount of Zn substitution and superconductivity disappears for x ~ 0.03. The lattice parameters also change upon substitution in this range. As the Zn content increases, the magnitude of the Hall coefficient decreases. This can not be understood by a simple picture because the carrier density estimated from a chemical analysis remains constant.[3] At present, there is no clear explanation for the mechanism that results in Tc suppression and the disappearance of superconductivity upon this kind of substitution in this system.

The purpose of this study is to investigate the electronic structure of Zn substituted $La_{1.85}Sr_{0.15}CuO_y$ system by XPS and UPS measurements. The effects of substitution on the core level binding energies, i.e. the change in chemical environments, and on the valence band structure, especially the density of states at the Fermi level, were studied.

EXPERIMENTAL

The samples used in this study were polycrystalline prepared by a standard solid state reaction method.[3] High quality polycrystalline $La_{1.85}Sr_{0.15}(Cu_{1-x}Zn_x)O_y$ (x=0-0.04) samples with no secondary phases were used in this study. Measurements were performed in an ultra high vacuum system equipped with an Al x-ray source (hv=1486.6eV) for XPS and a noble gas discharge lamp for UPS. The base pressure of the system during measurements was typically $2x10^{-10}$ Torr. The sample surface was prepared by abrading with a diamond file *in-situ* before each measurement which was typically less than 30 minutes. The surface quality was verified by the absence of extrinsic oxygen peaks in the XPS spectra.

RESULTS AND DISCUSSION

A. XPS and Inner Core Levels

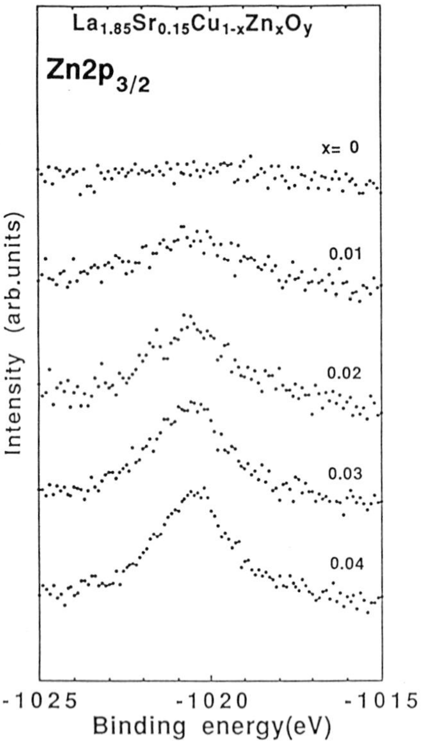

Fig.1 $Zn2p_{3/2}$ core level XPS spectra from $La_{1.85}Sr_{0.15}(Cu_{1-x}Zn_x)O_y$

As shown in Fig.1, the Zn $2p_{3/2}$ XPS peak appeared at about 1020.5eV with intensity increasing in proportion to the substituted Zn composition, x. The result indicates that the substituted Zn occupies the sites in a metallic (or covalent) environment rather than an ionic one, because the observed binding energy is lower than the ~1021.7eV value for $Zn2p_{3/2}$ in ZnO. Actually, this value is even lower than that of the ~1021.5eV value of Zn metal. The result indicates that Zn is divalent but with strong metallic character which may reflect the relatively good mobility of the carriers in the (Cu/Zn)-O planes even with Zn substitution. This agrees well with the fact that the transport property of $La_{1.85}Sr_{0.15}(Cu_{1-x}Zn_x)O_y$ was metallic at high temperature.[3] From the XPS results, and considering that the ionic radii of La^{3+} (1.03Å) and Sr^{2+}(1.18Å) are large while the values are nearly equal for Zn^{2+}(0.73Å) and Cu^{2+}(0.71Å), it is concluded that substituted Zn entered the Cu sites, however, further evidence from neutron-diffraction studies will have to be performed.

Core level XPS spectra for other elements were also investigated as a function of Zn content, x. Though the results are not shown here, it is concluded that the core level binding energies were not affected by Zn substitution in this range. We will show here that this can be understood by a method called "bond valence sum".[4] It is pointed out that the bond valence sum is a method useful for the understanding of the core level shifting in any system with varying compositions.[5] In this method, the bond valence sum of an i ion V_i is determined by the expression.

$$V_i = \Sigma_j S_{ij} = \Sigma_j [\exp((r_0 - r_{ij})/0.37)]$$

where r_0 is empirically determined for each i-j ion pair[4] and r_{ij} is the bond length between i and surrounding j ion. It is obvious that V_i is affected directly by the change of r_{ij}, i.e. the bond length. In general, for any system, a change in composition (e.g. by substitution, oxygen reduction) usually results in the change of r_{ij}'s. Thus one would expect the change in the V_i's of the composite elements. Physically, the change in V_i should reflect the change of the chemical environment which in turn should be observable as a shift of the core level binding energy in XPS. However, if the change in composition only results in small changes of the bond lengths, or the changes of the bond lengths are such that the corresponding calculated bond valence, S_{ij} compensates each other, there will be a negligible change in V_i. In this case, one will observe essentially no shift of core level binding energies.

For $La_{1.85}Sr_{0.15}(Cu_{1-x}Zn_x)O_y$, the fact that the core levels do not shift as Zn content x is varied can be understood by the fact that although the bond lengths change, the bond valence sums are not effectively altered by the substitution in this range. In contrast to this system, it is noted that for the $Bi_2Sr_2Ca_{1-x}Y_xCu_2O_y$ system, as x is varied from 0 to 1, the core levels of all the composite elements shift significantly. The results were successfully explained by the change in the bond valence sums as the result of the change of the bond lengths upon varying x in the system.[5] It is also interesting to note that Cava et al.[6] have calculated the bond valence sums of the $YBa_2Cu_3O_y$ system as a function of oxygen content y and pointed out a close relation between the change of the critical temperature and the bond valence sum of the in-plane Cu. Although there would be some uncertainty of the physical meaning on the absolute value of the bond valence sum, it is plausible that its relative change as the system varied is physically indicative.

B. UPS and Valence Band Electronic Structure

Figure 2 shows the UPS spectra in the valence band region of $La_{1.85}Sr_{0.15}(Cu_{1-x}Zn_x)O_y$ with different x's. The spectrum for x=0, i.e. the superconducting composition, is essentially the same as previous reported results. Upon substitution, no distinct change could be observed, possibly indicating that only a small but critical change of the electronic structure occurs. Figure 3 shows the UPS spectra in the vicinity of the Fermi level. It is noted that not only for the superconducting, x=0 composition, but also for the Zn substituted samples, a weak edge indicating a finite density of states(DOS) at the Fermi level could be observed. Considering that all of the samples are metallic at high temperature, the observed DOS should reflect the metallic transport property of the samples. However, it is noted that since the samples with x>0.2 became semiconducting at low temperature and exhibited no superconductivity, the nature of the observed DOS at the Fermi level in the UPS spectra should be different between super and non-super compositions. However, to elucidate the origins of these states appears to be difficult. Low temperature UPS measurements and structural studies would be considerably suggestive.

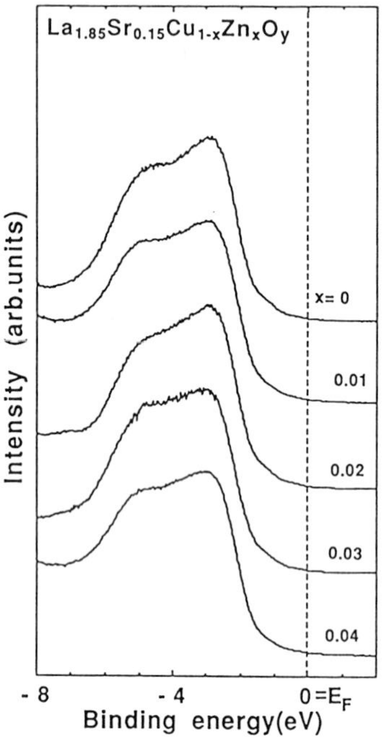

Fig.2 Valence band UPS spectra from $La_{1.85}Sr_{0.15}(Cu_{1-x}Zn_x)O_y$

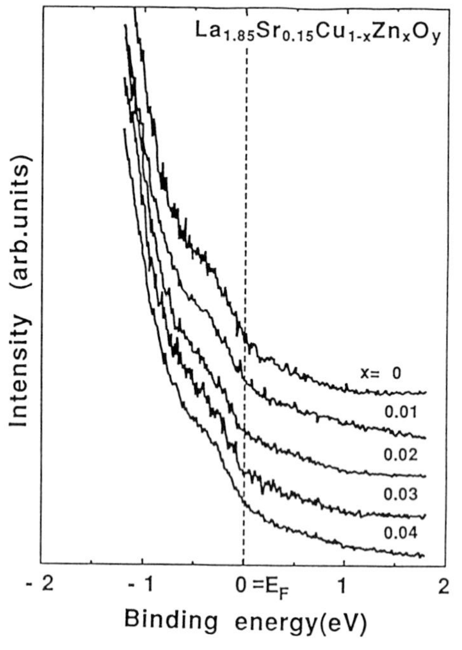

Fig.3 Enlargement of the UPS spectra in the Fermi level (E_F) region. Peak at about -0.3eV and finite DOS at E_F could be observed for all samples.

CONCLUSION

XPS and UPS measurements to investigate the core levels and the valence band of $La_{1.85}Sr_{0.15}(Cu_{1-x}Zn_x)O_y$ with different x's were performed. The results indicate that Zn substituted for Cu is divalent with a metallic environment. The core level binding energies of the constituent elements are not affected by substitution. The bond valence sum seems to be useful for the discussion of core level shifting. In the UPS spectra, a finite DOS at the Fermi level was observed not only for superconducting samples, but also for non-superconducting samples. Certainly, the origin of the DOS of the super and non-superconductors observed should be different. The results indicate that for a substituted cuprate oxide, a finite DOS at the Fermi level can be observed even though the sample is not a superconductor.

REFERENCES

[1] Tarascon JM, Greene LH, Barboux P, McKinnon WR, Hull GW, Orland TP, Dellin KA, Foner S, McNiff EJ Jr (1987) Phys.Rev.B 36: 8393

[2] Xiao G, Bakhshai A, Cieplak MZ, Tesanovic Z, Chien CL (1989) Phys.Rev.B 39: 315

[3] Ikegawa S, Yamashita T, Sakurai T, Itti R, Yamauchi H, Tanaka S :to be published in Phys.Rev.B.

[4] Brown ID, Altermatt D (1985) Acta Crystallogr.B 41: 244

[5] Itti R, Munakata F, Ikeda K, Koshizuka N, Tanaka S (1990) Physica B 165&166: 1223 and unpublished results

[6] Cava RJ, Hewat AW, Hewat EA, Batlogg B, Marezio M, Rabe KM, Krajewski JJ, Peck WF Jr, Rupp LW Jr (1990) Physica C 165: 419

2.3 Transport and Thermal Properties

The Physical Effect of Reduction on The Hall Effect and Superconducting Properties of $Nd_{2-x}Ce_xCuO_{4-\delta}$

N.A. FORTUNE, K. MURATA, M. ISHIBASHI, Y. YOKOYAMA, and Y. NISHIHARA

Electrotechnical Laboratory, 1-4, Umezono 1-chome, Tsukuba, 305 Japan

ABSTRACT

To better understand the need for heat treatment in a reducing atmosphere to induce superconductivity in $Nd_{2-x}Ce_xCuO_{4-\delta}$, we have systematically varied the degree of reduction in a series of ceramic samples. Surprisingly large, systematic changes occur in thermodynamic and transport properties with small changes in degree of reduction and oxygen stoichiometry. Both the Meissner effect and the apparent electron carrier density strongly decrease with decreasing reduction for x=0.15. For x=0.17 and 0.18, we find that the Hall coefficient changes sign below 75 K for superconducting samples.

KEY WORDS: Nd-Ce-Cu-O, Hall effect, Meissner effect, oxygen stoichiometry, reduction

INTRODUCTION

As grown samples of $Nd_{2-x}Ce_xCuO_{4-\delta}$ are not superconducting; superconductivity appears only after heat treatment in a reducing atmosphere[1-4]. Why reduction is required, however, has remained unexplained. To better understand the physical role of reduction, we have varied the degree of reduction in ceramic samples of $Nd_{2-x}Ce_xCuO_{4-\delta}$ by varying the partial pressures of oxygen $P(O_2)$ and argon P(Ar) in a reducing atmosphere while keeping the total pressure constant. Ceramic samples have been studied instead of single crystals because of their relative ease in synthesis and reduction and the tendency of single crystals to exhibit Ce concentration gradients [5]. For x=0.15, we show that increasing the degree of reduction strongly increases the apparent electron carrier density and the superconducting Meissner fraction. We also find that the ability to produce a single phase superconducting sample with no residual resistivity strongly depends on the reducing conditions. For x=0.17 and 0.18, the sign of Hall effect depends on the temperature and degree of reduction.

EXPERIMENTAL

Unreduced samples of $Nd_{2-x}Ce_xCuO_{4-\delta}$ were prepared by solid state reaction from 99.99% purity Nd_2O_3, CeO_2 and CuO. A stoichiometric mixture of starting materials was ground thoroughly and heated in an Al_2O_3 crucible at 950 ^{o}C for 20 hours in air. The resulting powders were reground and pressed into pellets. The pellets were sintered in air at 1100 ^{o}C for 10 hours and then cooled to room temperature over a period of 10 hours. Trace amounts of $Nd_2Ce_2O_7$ and unreacted CuO present in the pre-sintered calcined powder are absent in the slowly cooled scraped sintered pellet, which has the expected tetragonal T' structure with lattice parameters (a = 3.947 Å and c = 12.073 Å for x=0.15).

After sintering, the pellets were reduced by one of two different methods. In method A, pellets were reduced for 24 hours at 930 ^{o}C for 24 hours in a pure argon (99.99%-99.9999%) atmosphere. While pellets reduced for 20-24 hours exhibited significantly larger superconducting Meissner fractions than those annealed for 5-10 hours, no further increase in the superconducting Meissner fraction was observed by longer annealing times up to 200 hours. X-ray diffraction on sintered pellets with x=0.15 and x=0.18 reduced in pure argon at $T \geq$ 1000 ^{o}C show evidence of surface impurities and partial sample decomposition, while pellets reduced at T $\leq$ 950 ^{o}C show no such evidence. In method B, pellets were reduced for 20 hours at 1050 ^{o}C in an 1 atm Ar-O_2 gas mixture with $P(O_2)$ varying from 5×10^{-4} to 10^{-1} atmospheres, followed by quenching to room temperature in the same reducing atmosphere. X-ray diffraction on annealed scraped pellets again show no evidence for impurities, second phases or sample decomposition. By successive thermogravimetric (TGA) measurements (5 ^{o}C/min up to 1000 ^{o}C) on a 167.5 mg sintered unreduced sample in pure argon(99.9995%), pure oxygen(99.99%) and pure argon, we find a reversible change in mass of 0.15±0.05 mg at 950 ^{o}C due to reduction in argon, equivalent to a change in oxygen off-stoichiometry δ = 0.023±0.08. The oxygen loss does not begin until 875 ^{o}C in argon, but oxygen reabsorption begins at temperatures as low as 600 ^{o}C, as previously observed [6].

After thinning (≅100 microns), six Au contacts in a rectangular Hall bar pattern were evaporated onto the samples, then annealed for 1 hour at 300 °C under flowing argon, reducing the contact resistances to less than 1 ohm[7]. Hall effect and resistivity measurements were performed continuously as the temperature of the samples was slowly varied (≅5 K/hour). The Hall voltage was measured using a 5 tesla magnet, with field reversal to eliminate the longitudinal resistance component. Spurious voltages arising from thermoelectric effects were eliminated using the ohmicity of the Hall voltage and current levels were checked for the absence of self heating. Resistance measurements and Hall voltage measurements were made with two different pairs of contacts for each sample. After elimination of spurious voltages, the measured Hall coefficients (R_H) between different pairs agree within the scatter of the data, while the resistivities agree within 1-10%. No corrections were made for the packing fraction (81±1%), which is independent of the degree of reduction.

RESULTS AND DISCUSSION

For x=0.15, we reduced sintered pellets in pure argon at 930 °C(method A) and in oxygen partial pressures $P(O_2)$ = 5x10^{-4}, 10^{-2} and 10^{-1} atm at 1050 °C(method B). The onset T_c's for each sample (determined from the resistivity and the Meissner effect) are shown in Fig. 1a. In Fig. 1b we show the field cooled (Meissner) and zero field cooled (shielding) magnetization in a field of 10 Oe for the same samples. The corresponding Meissner and shielding volume fractions at 5K in an applied magnetic field of 10 Oe for the sample reduced in pure argon are 12% and 24%, respectively.

Although only very small changes in δ occur with reduction, we find surprisingly large systematic changes occur in the apparent carrier density N (α $1/R_H$). As shown in Fig. 2, the measured value of N at 100 K increases from 0.6 ±0.05 carriers/unit cell for $P(O_2)$= 1×10^{-1} atm to 0.9 ± 0.15 for$P(O_2)$ =5x10^{-4} atm to 1.45 ± 0.10 carriers/unit cell for the sample reduced in pure argon. R_H is negative over the entire temperature range for all samples, in agreement with previous measurements on ceramic samples [1], c-axis oriented thin films [8], and measurements at 77 and 300 K on a single crystal [9]. A previous measurement on a ceramic $Nd_{1.85}Ce_{0.15}CuO_{4-\delta}$ sample annealed for 10 hours under Ar-O_2 gas mixture flow with $P(O_2)$= 1×10^{-4} atm found 1.2 n-type carriers/unit cell at 100 K[1], while measurement on an c-axis oriented epitaxial thin film found 2.0 n-type carriers/unit cell at 100 K [8]. Comparison between our results and those on oriented thin films [8] suggests that averaging over any crystalline anisotropy may affect the degree of temperature dependence but does not change the sign of R_H from that measured for magnetic fields normal to the Cu-O plane. The positive Hall coefficient and large increase in magnitude with decreasing temperature observed in a recent measurement of single crystals [10] of $Nd_{1.85}Ce_{0.15}CuO_{4-\delta}$ is not observed here.

The temperature variation of the resistivity for the same samples is shown in Fig. 3. Only the sample reduced with $P(O_2)$= 5×10^{-4} atm exhibits a zero resistance state above 4.2 K. For the most highly reduced sample (pure argon atmosphere), the superconducting onset remains at 20 K, but has a large residual resistivity. A similar dependence of the resistivity below T_c has been reported [11] for a sample reduced in CO_2 for 1 hour at 900 °C. Interestingly, the residual resistivity disappeared for samples annealed at 1000 or 1100 °C, where partial sample decomposition occurs. The absence of such a zero resistance state above 4.2 K in the most highly reduced sample indicates that the achievement of zero resistance depends on the reducing conditions. While the sensitivity of the resistivity to degree of reduction is consistent with the existence of large local variations in local oxygen concentration for non optimally reduced samples, our TGA results suggest that annealing after reduction at temperatures just below 600 K might reduce any inhomogeneities in the oxygen distribution and sharpen the resistive transitions.

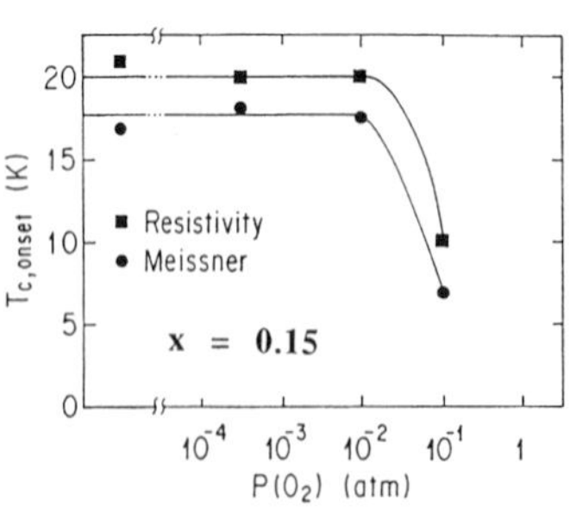

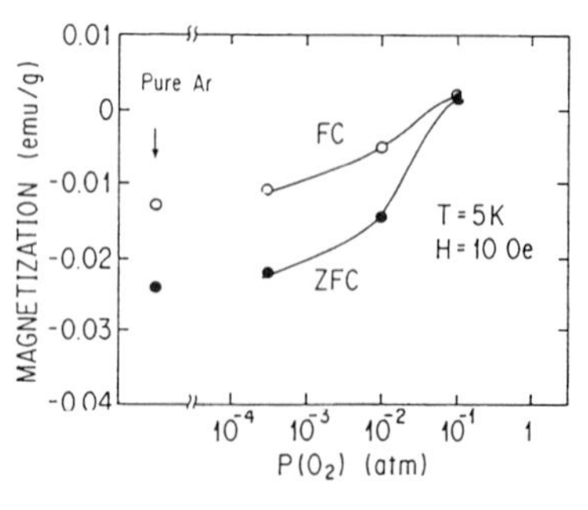

Figure 1: $P(O_2)$ dependence of (a) T_c from magnetization and resistivity measurements and (b) the field cooled and zero field cooled (shielding effect) magnetization in a field of 10 Oe at 5 K for x=0.15.

How can our results for x=0.15 be explained?If the physical effect of reduction is only to increase the number of carriers by $2\Delta\delta$, then much smaller changes in the Hall coefficient and superconducting Meissner fraction due to reduction would be expected. The anomalously large changes in the apparent carrier density could be due to the removal of oxygen atoms from "key sites" which when occupied, strongly suppress the mobile carrier density in the Cu-O planes. In fact, recent experimental work has shown that the oxygen atoms that are removed during heat treatment come from oxygen sites within the O_2 plane that separates CuO_2 planes or from interstitial apical oxygen sites not occupied in the perfect stoichiometric crystal [6]. The former possibility has been independently suggested from "quantum percolation" theory [12], which predicts that the percentage of the sample that is locally superconducting will scale with the number of O vacancies bridging Ce atoms. X-ray absorption measurements [13] by some of us also imply a large increase in the electron density at the copper sites with increasing degree of reduction. The Hall effect variations could also be due to the presence of a very small number of highly mobile holes in addition to the majority electron carriers. Choosing between these alternatives will require (1) accurate knowledge not only of δ before reduction but also the change in δ and the distribution of oxygen vacancies/excess oxygen atoms for each degree of reduction or (2) independent evidence for a minority band of much higher mobility holes. Nevertheless, it is difficult to understand how an electron-hole two band model could explain the observed variations in Meissner effect and resistivity with degree of reduction, while the "key site" model is consistent with all three experiments.

Finally, we present a few preliminary results for $x \neq 0.15$. For x=0.17, the resistivity (not shown) and temperature dependence of the Hall coefficient for $P(O_2)$=1 atm indicates semiconducting behavior, as shown in Fig. 4. Heat treatment in a partially reducing atmosphere ($P(O_2)$=10-3 atm) leads to a change in the sign of R_H below 75 K. The resistivity for this partially reduced sample, shown in Fig. 5, indicates two T_c onsets, perhaps due to large variations in local oxygen concentration. The quantitative dependence of R_H for an x=0.18 sample with $P(O_2)$=10-3 atm is nearly identical to that for the x=0.17 sample. From previous measurements [1] at 300 and 80 K, it has been thought that R_H becomes positive (p-type) for $x \geq 0.17$. We find here, however, that the sign (and magnitude) of R_H is strongly temperature dependent in superconducting samples.

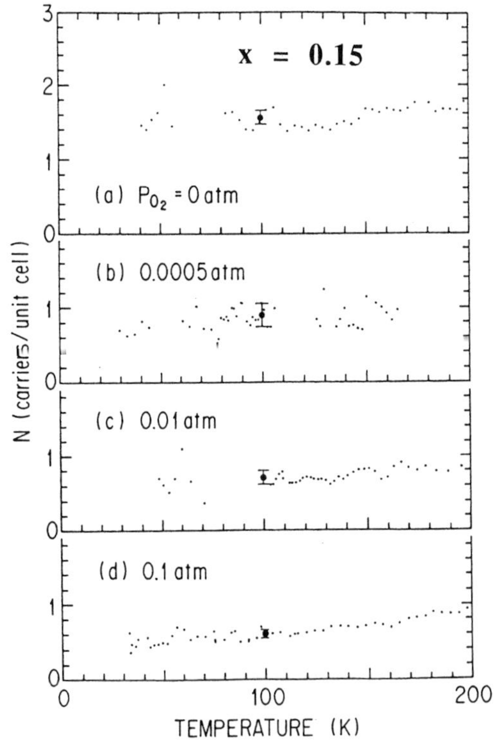

Figure 2: Apparent electron carrier density N as a function of O_2 partial pressure $P(O_2)$ during reduction for x=0.15. (a) $P(O_2)$ = 0 atm, (b) $P(O_2) = 5\times10^{-4}$ atm, (c) $P(O_2) = 1\times10^{-2}$ atm, (d) $P(O_2) = 1\times10^{-1}$ atm. $P(O_2) + P(Ar)$ = 1 atm.

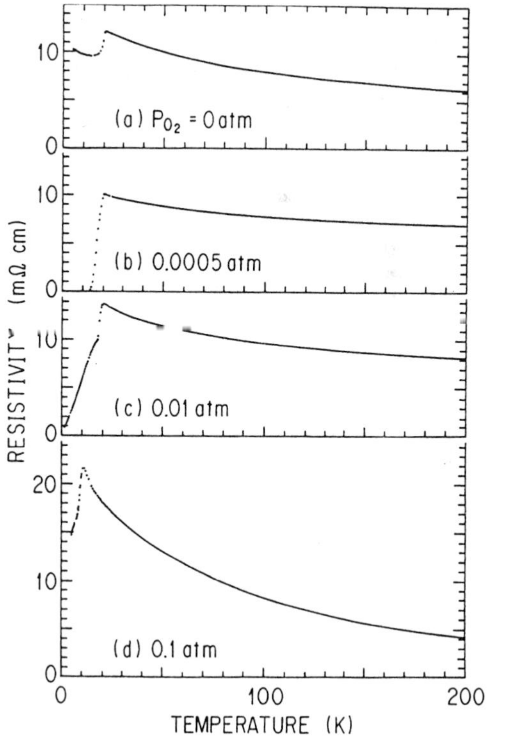

Figure 3: Resistance versus temperature as a function of degree of reduction for the same samples as in Figs. 1 and 2.

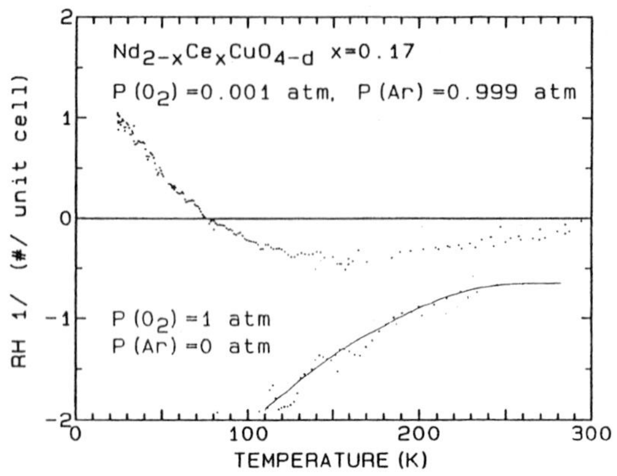

Figure 4: Temperature dependent Hall coefficient as a function of degree of reduction for x=0.17.

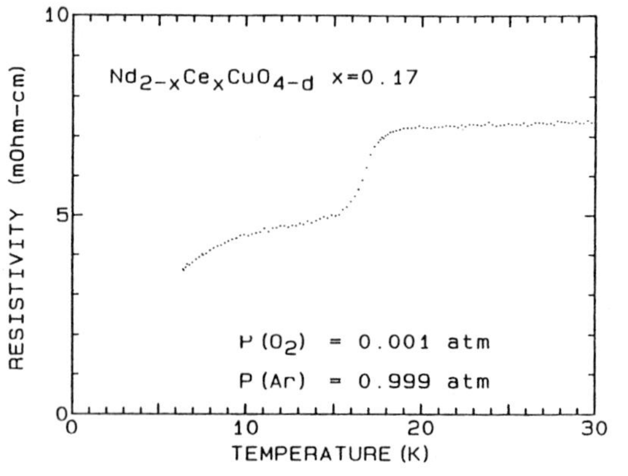

Figure 5:Resistivity for the partially superconducting $P(O_2) = 1\times10^{-3}$ atm x=0.17 sample shown in Fig. 4.

In conclusion, we have observed large, systematic changes in Meissner fraction, Hall coefficient and resistivity with increasing degree of reduction in $Nd_{2-x}Ce_xCuO_{4-\delta}$. Further understanding of the different physical roles of cerium doping and oxygen reduction in this material will require more precise determinations of δ before and after reduction and in the distribution of oxygen atoms/vacancies due to reduction.

REFERENCES

1. Takagi H, Uchida S and Tokura Y (1989) Phys. Rev. Lett. 62: 1197
2. Tokura Y, Takagi H and Uchida S (1989) Nature 337: 345
3. Hidaka Y and Suzuki M (1989) Nature 338: 635
4. Tarascon J-M, Wang E, Greene LH, Bagley BG, Hull GW, D'Egidio SM, Miceli PF, Wang ZZ, Jing TW, Clayhold J, Brawner D, and Ong NP (1989) Phys. Rev. B 40: 4494
5. Hidaka Y (1990) In: Ishiguro T and Kajimura K (eds) Advances in Superconductivity II. Springer-Verlag, Tokyo, p. 229-233
6. Wang E, Tarascon J-M, Greene LH, Hull GW and McKinnon WR (1990) Physical Review B 41: 658
7. Nichols TR, Murata K, Yokoyama Y and Nishihara Y (1990) Bulletin of the Electrotechnical Laboratory 54: 609
8. Hayashi S, Hirochi K, Adachi H, Kohiki S, Hatta S, Setsune K, Kirao T and Wasa K (1990) In: Ishiguro T and Kajimura K (eds) Advances in Superconductivity II. Springer-Verlag, Tokyo, pp. 949-952
9. Y. Tokura, private communication
10. Jing TW, Wang ZZ, Chien TR, Ong NP, Tarascon JM, Wang E (1990) In: Ishiguro T and Kajimura K (eds) Advances in Superconductivity II. Springer-Verlag, Tokyo,pp. 499-504
11. Takayama-Muromachi E, Izumi F, Uchida Y, Kato K and Asano H (1990) Physica C 159: 634
12. Phillips JC (1990) Phys. Rev. B (RC) 41: 850
13. Oyanagi H, Yokoyama Y, Yamaguchi H, Kuwahara Y, Katayama T and Nishihara Y (1990) Phys.Rev. B, to be published

Properties of the Cubic Superconductor $Ba_{0.6}K_{0.4}BiO_3$

N. SAVVIDES[1], R. WINN[1], C. ANDRIKIDIS[1], S.J. COLLOCOTT[1], K.-H. MÜLLER[1], and E.R. VANCE[2]

[1] CSIRO Division of Applied Physics, Sydney, 2070 Australia
[2] ANSTO, Sydney, 2234 Australia

ABSTRACT

AC susceptibility, magnetotransport properties and specific heat of single phase $Ba_{0.6}K_{0.4}BiO_3$ ceramic, with $T_c \simeq 31$ K, are reported. The complex AC susceptibility, $\chi = \chi' + i\chi''$, shows both intragrain and intergrain components characteristic of a granular superconductor, where the positions of the χ'' intra- and intergranular peaks are strongly field dependent. From the onset of the resistive superconducting transition in DC magnetic fields we determine values for the slope of the upper critical field, $dH_{c2}/dT = -0.87$ T K^{-1}, the zero-temperature critical field, $H_{c2}(0) = 18$ T, and the coherence length, $\xi(0) = 43$ Å. Specific heat, C_p, measurements from 0.5 to 20 K show no evidence of a term that has a linear temperature dependence or an upturn in C_p/T.

KEY WORDS: AC susceptibility, specific heat, upper critical field, coherence length, $Ba_{0.6}K_{0.4}BiO_3$.

INTRODUCTION

The occurrence of superconductivity in the $Ba_{1-x}K_xBiO_3$ system, $x \simeq 0.35 - 0.50$, provides the opportunity to investigate the nature of superconductivity in a perovskite oxide which does not contain copper. Unlike the layered Cu-O based superconductors which are highly anisotropic in their properties, this compound is isotropic in the superconducting state as it crystallizes in a simple cubic perovskite structure [1]. Optimum normal-state and superconducting properties occur for potassium dopant levels of $x = 0.4$ giving $T_c \simeq 30$ K. Hinks [2] recently reviewed the current state of knowledge of the nature of superconductivity in cubic $Ba_{1-x}K_xBiO_3$.

In this paper measurements are reported of the dependence of AC susceptibility on the AC field amplitude (0.001–50 G rms), the resistive transition in DC fields up to 7 T, and the specific heat, C_p, from 0.5 to 20 K of single phase $Ba_{0.6}K_{0.4}BiO_3$ ceramic. We analyse the susceptibility data using a critical state model for granular superconductors developed by Müller [3]. The magnetotransport measurements are used to determine the upper critical field, $H_{c2}(0)$, and the coherence length, $\xi(0)$.

EXPERIMENTAL TECHNIQUES

The $Ba_{0.6}K_{0.4}BiO_3$ ceramic material was prepared by the melt-processing technique of Hinks *et al.* [4]. X-ray diffraction confirmed the material to be single phase, with a lattice constant of $a = 4.285$ Å.

The real and imaginary parts of the complex AC susceptibility, $\chi = \chi' + i\chi''$, were measured at 144 Hz in a conventional mutual inductance susceptometer using AC magnetic field amplitudes in the range 0.001–50 G rms, with the field applied along the long axis of the specimen [5]. The cylindrical specimen (diameter 5mm, length 13mm) was held fixed in the centre of one of a pair of secondary coils wound oppositely and connected in series to achieve null balance. High permeability shields, placed around the glass dewar, were used to reduce the Earth's field to about 0.003 G in the measurement axis. The sample was zero-field cooled, then measurements were taken with the temperature increasing from 4.2 K to above T_c. A lock-in amplifier was used to measure χ' and χ'' from the out-of-balance voltage signal. The correct phase was set to null χ'' at 4.2 K for low fields ($\mu_0 H \simeq 0.01$ G).

The DC magnetic field dependence of the resistive superconducting transition was determined using fields up to 7 T. The specific heat, C_p was measured in a ^{3}He-^{4}He semi-adiabatic heat-pulse calorimeter from 0.5 to 20 K [6].

RESULTS AND DISCUSSION

AC Susceptibility

The AC susceptibility data (not corrected for demagnetization effects) are shown in Fig. 1. The diamagnetic onset, determined at the lowest field, occurs at $T_c = 31.0 \pm 0.2$ K, while the midpoint $T_c(10-90\%) = 28.9$ K and $\Delta T_c = 2.8$ K.

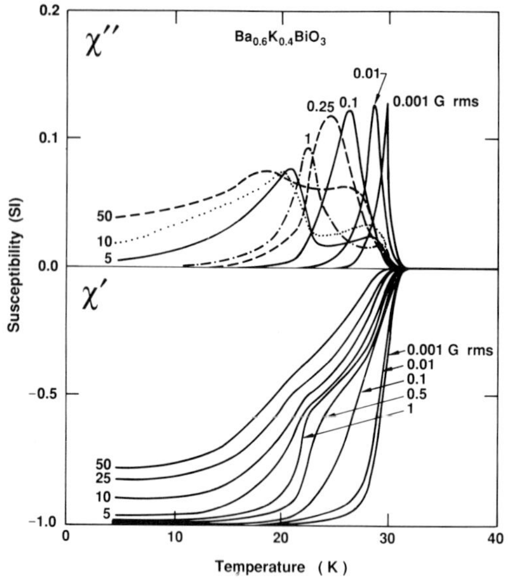

Figure 1. AC susceptibilities χ' and χ'' as a function of temperature for different values of the rms AC magnetic field, $\mu_0 H$. The χ'' curves show an intragrain peak just below T_c and an intergrain peak at a lower temperature.

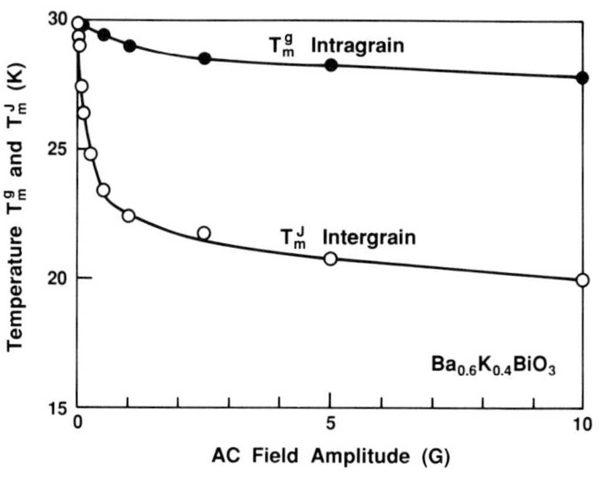

Figure 2. The χ''-peak temperatures T_m^J (intergrain) and T_m^g (intragrain) as function of the AC field amplitude.

The overall behavior of the susceptibility data is typical of weakly-coupled superconductors [3]. The real part, χ', of the susceptibility has a sharp drop just below T_c which is assigned to intragranular shielding followed by a more gradual drop due to intergranular shielding. For the lowest field used, 0.001 G, χ' reaches a value of -1 (i.e. 100% shielding) at about 20 K . Near T_c at strong AC fields ($\sim$ 50 G) flux freely penetrates the grains and grain boundaries causing a region of $\chi' \simeq 0$.

The imaginary part, χ'', shows two distinct peaks that reflect bulk hysteresis losses incurred due to flux motion in the grains and in the grain boundary regions of the granular microstructure. As the temperature increases towards T_c flux lines first penetrate the intergrain regions, where flux pinning is weak, then the intragrain regions (the grains) where flux pinning is stronger. χ'' reaches a peak as the flux fully penetrates each region in turn. The intergrain peak is located at the lower temperature, T_m^J, while the intragrain peak is located at a higher temperature, T_m^g. As the amplitude of the AC field increases the peak position moves to lower temperatures where correspondingly stronger pinning forces must be overcome in order to cause full flux penetration.

Müller [3] recently developed a critical state model of flux dynamics, which adequately explains the observed behavior, by considering flux movement in (Josephson) weak links formed between grains, and movement of Abrikosov vortices in the superconducting grains. A striking feature of the susceptibility data of Fig. 1 is that the height of the χ'' intragrain peak is greater than is usually seen in the Cu-O based superconductors. This is because the intragrain pinning in the $Ba_{1.6}K_{0.4}BiO_3$ specimen is weaker than in the Cu-O based superconductors, causing the intragrain peak to appear even in weak fields. Also, the large average grain size ($\sim$ 30μm) of the specimen favours the appearance of a strong χ'' peak [3].

Figure 2 shows χ''-peak temperatures T_m^J and T_m^g as a function of AC field amplitude. In contrast to the $YBa_2Cu_3O_7$ superconductor, we do not observe a linear dependence of T_m on applied AC field [3]. A non-linear dependence is also seen in the Bi-based high-T_c superconductors, and a detailed analysis of these data, which includes a flux creep correction to the critical state model, is presented elsewhere [7].

Resistivity and Upper Critical Field

In contrast to previous reports of an increasing normal state resistance, $R(T)$, with decreasing temperature [4,8], our specimen has a weak metallic behavior with $R(T)$ decreasing with temperature to a residual value below 50 K [5]. The onset of the resistive superconducting transition and the shape of the $R(T)$ curve depend strongly on the measuring current [5,8]. Figure 3 shows the behavior of the resistive superconducting transition in DC magnetic fields; the measuring current density was $\simeq$ 2 mA cm^{-2}. The onset of superconductivity is systematically suppressed to lower temperatures by the increasing field. The small amount of broadening, which is typical of traditional type II superconductors but in sharp contrast to that of Cu-O based high-T_c superconductors, is consistent with a nearly isotropic critical field expected for the cubic perovskite crystal structure.

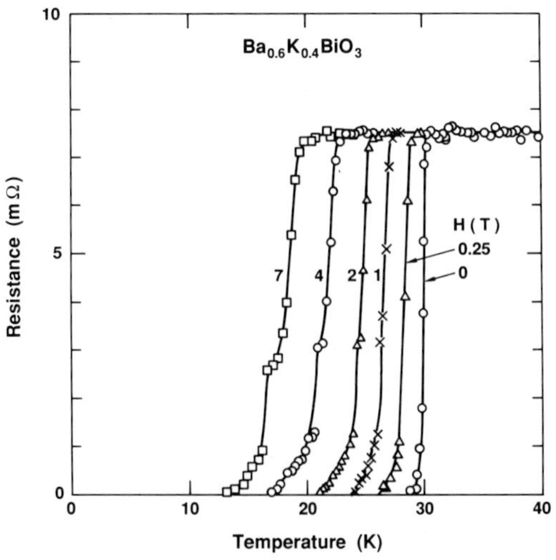

Figure 3. Temperature dependence of the resistive superconducting transition in DC magnetic fields up to 7 T.

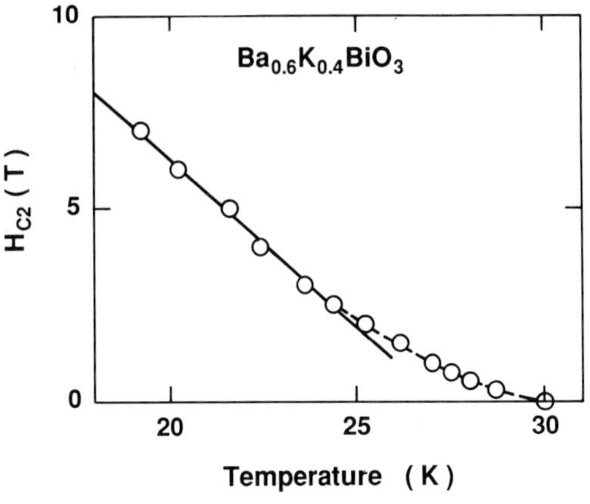

Figure 4. Upper critical field, H_{c2}, versus temperature. The slope given by the straight line, $dH_{c2}/dT = -0.87$ T K^{-1}.

Figure 4 shows the temperature dependence of the upper critical field, $H_{c2}(T)$, determined from the 90% onset of the transition. Close to T_c the data show an upward curvature of H_{c2} with decreasing temperature similar to the behavior seen in high-T_c superconductors. For fields greater than about 2 T there is an approximately linear relation with slope $dH_{c2}/dT = -0.87$ T K^{-1}. Using the Werthamer-Helfand-Hohenberg expression [9]

$$H_{c2}(0) = -0.69\, T_c\, [dH_{c2}/dT], \tag{1}$$

we determine the upper critical field, $H_{c2}(0) = 18$ T. The superconducting coherence length at zero temperature, $\xi(0)$, can be evaluated from $H_{c2}(0)$ by use of the standard expression from Ginsburg-Landau theory, in the dirty limit,

$$H_{c2}(0) = \Phi_0/2\pi\xi(0)^2 \tag{2}$$

where Φ_0 is the flux quantum, giving $\xi(0) = 43$ Å. This value agrees well with a value of 46 Å determined from the data of Welp *et al.* [8]. The data are summarized in Table 1.

Specific Heat

Recent interest in the low temperature specific heat, C_p, of the ceramic oxide superconductors with the perovskite structure, has focused on the coefficient, γ_0, of the linear term in C_p, and the upturn in C_p/T [10]. In particular, there has been much discussion as to the origin of the linear term: whether it is intrinsic or extrinsic to the superconducting state, whether the presence of Cu-O planes is significant, and whether it is a universal feature of oxide superconductors. Low-temperature specific heat studies of the cubic perovskite $Ba_{0.6}K_{0.4}BiO_3$ [6] should assist in resolving the uncertainties that surround γ_0 in these materials.

Specific heat results are plotted in Figure 5 as C_p/T versus T^2 and no upturn in C_p/T is evident at the lowest temperatures. From 2 to 12 K the data are well represented by

$$C_p = 0.060T + 0.278T^3 + 0.001257T^5 \quad \text{mJ/(molK)} \tag{3}$$

with a percent rms deviation for the fit of 2.3%. A lower limit of 2 K was chosen for the fit, as below this temperature the increased scatter in the experimental C_p data affects adversely the fitting procedure. Re-fitting the C_p data from 2 to 12 K with γ_0 set equal to zero, i.e. a two parameter fit, gives no improvement in the %rms deviation. On this basis we conclude that γ_0 is in all probability zero, with an upper limit of 0.060 mJ/(mol K^2), in agreement with previous C_p studies [11]. A linear term in C_p may thus not be a universal feature of the ceramic oxide superconductors. From the coefficient of the T^3 term a limiting value for the Debye temperature of $Ba_{0.6}K_{0.4}BiO_3$ of 324 ± 4 K is obtained.

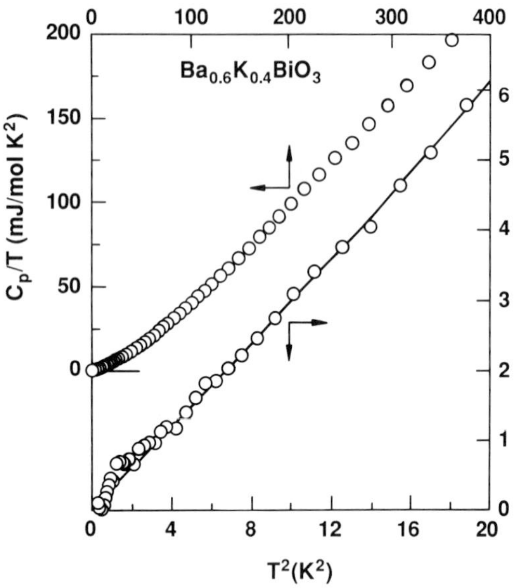

Figure 5. C_p/T versus T^2. The solid curve is given by equation (3) in the text.

Table 1

Physical properties of $Ba_{0.6}K_{0.4}BiO_3$

Transition temperature	$T_c = 31$ K
Normal state resistivity	$\rho_{300} \simeq 12$ mohm cm
Lattice constant	$a = 4.285$ Å
Upper critical field	$dH_{c2}/dT = -0.87$ T K^{-1}
	$H_{c2}(0) = 18$ T
Coherence length	$\xi(0) = 43$ Å
Debye temperature	$\theta_D = 324 \pm 4$ K

CONCLUSION

Measurements are reported on a bulk polycrystalline specimen of the cubic superconductor $Ba_{0.6}K_{0.4}BiO_3$. The complex AC susceptibility shows a behavior typical of a granular superconductor, and both the intragrain and the intergrain components are sensitive to the AC magnetic field amplitude. The resistive superconducting transition is systematically suppressed to lower temperatures with increasing DC magnetic field like in conventional superconductors. The slope of the upper critical field, dH_{c2}/dT, has a value of -0.87 T K^{-1}, from which an upper critical field $H_{c2}(0) = 18$ T and a coherence length $\xi(0) = 43$Å are determined. The specific heat shows no evidence of an upturn in C_p/T and γ_0, the coefficient of the linear term in C_p, is probably zero.

Acknowledgement: This work was partially supported by Metal Manufactures Limited and Department of Industry Technology and Commerce GIRD grant No.15017.

REFERENCES

1. L.F. Mattheis, E.M. Gyorgy and D.W. Johnson Jr., *Phys. Rev. B* **37** (1988) 3745; R.J. Cava, B. Batlogg, J.J. Krajewski, R.C. Farrow, L.W. Rupp Jr., A.E. White, K.T. Short, W.F. Peck Jr. and T.V. Kometani, *Nature* **332** (1988) 814; S. Pei, J.D. Jorgensen, B. Dabrowski, D.G. Hinks, D.R. Richards, A.W. Mitchell, J.M. Newsam, S.K. Sinha, D. Vaknin and A.J. Jacobson, *Phys. Rev. B* **41** (1990) 4126.
2. D.G. Hinks, *MRS Bulletin* **15** (1990) 55.
3. K.-H. Müller, *Physica C* **159** (1989) 717.
4. D.G. Hinks, A.W. Mitchell, Y. Zheng, D.R. Richards and B. Dabrowski, *Appl. Phy. Lett.* **54** (1989) 1585.
5. N. Savvides, S.J. Collocott, C. Andrikidis, K.-H. Müller and E.R. Vance, *Physica C* **171** (1990).
6. S.J. Collocott, N. Savvides and E.R. Vance, *Phys. Rev. B* **42** (1990) 4794.
7. K.-H. Müller and N. Savvides, to be published.
8. U. Welp, W.K. Kwok. G.W. Crabtree, H. Claus, K.G. Vandervoort, B. Dabrowski, A.W. Mitchell, D.R. Richards, D.T. Marx and D.G. Hinks, *Physica C* **156** (1988) 27.
9. N.R. Werthamer, E. Helfand and P.C. Hohenberg, *Phys. Rev.* **147** (1966) 295.
10. R.A. Fisher, J.E. Gordon and N.E. Phillips, *J. Superconductivity* **1** (1988) 231; A. Junod, *Physical Properties of High Temperature Superconductors*, edited by D.M. Ginsberg (World Scientific, Singapore, 1990), Vol.2, p.1.
11. S.E. Stupp, M.E. Reeves, D.M. Ginsberg, D.G. Hinks, B. Dabrowski and K.G. Vandervoort, *Phys. Rev. B* **40** (1989) 10878; M.F. Hundley, J.D. Thompson and G.H. Kwei, *Solid State Commun.* **70** (1989) 1155.

Preparation, Pressure Effect, and Structural Phase Transition in $Ba_{1-x}K_xBiO_3$ and $Ba_{1-x}Rb_xBiO_3$

AKIRA IYO[1], ATSUO YAMADA[1], YUKIO OIJI[1], AKIHIKO TAKAHASHI[1], AKIRA INOUE[1], HIROSHI WATANABE[1], HIROMOTO UWE[1], TUNETARO SAKUDO[1], and KEIZO MURATA[2]

[1] Institute of Applied Physics, University of Tsukuba, Tsukuba, Ibaraki, 305 Japan
[2] Electrotechnical Laboratory, Tsukuba, Ibaraki, 305 Japan

ABSTRACT

Characteristic temperatures for the preparation of $Ba_{1-x}K_xBiO_3$ and $Ba_{1-x}Rb_xBiO_3$ in the normal-pressure atmosphere have been determined. In atmosphere of N_2 gas, the compound is found to be stable only above a certain temperature, below which the atmosphere should be replaced by O_2 gas. However, above another temperature in the O_2 atmosphere, another phase separation is found. The structural-phase-transition temperature has been determined by measuring the temperature dependence of the elastic compliance. The temperature of the flopping of the rotation axes of the oxygen octahedra from <110> to <111> is found for $Ba_{1-x}K_xBiO_3$. In superconducting $Ba_{1-x}K_xBiO_3$ and $Ba_{1-x}Rb_xBiO_3$, both of the Tc and the normal-state conductivity increase with pressure. In $BaPb_{1-x}Bi_xO_3$, both of them decrease with pressure to the contrary. Therefore, it is suggested that Tc is connected to the density of the states at the Fermi energy which manifests itself in the conductivity.

KEY WORDS: $Ba_{1-x}K_xBiO_3$, $Ba_{1-x}Rb_xBiO_3$, phase separation, pressure effect, structural phase transition

INTRODUCTION

$Ba_{1-x}K_xBiO_3$ has the highest superconducting transition temperature(Tc=30K) reported for any oxide superconductor not containing copper. The distinctive features of this system are the absence of magnetic order and two dimensional conduction planes which are believed to cause a high Tc in the copper-oxide superconductors. Although, since the discovery of this system, many experiments have been carried out, the experimental results do not always agree between them, one of the reasons for which might be due to the difficult sample preparation. There are two preparation methods for this system; one is the method of the synthesis in a sealed quartz tube introduced by Cava et al.[1] and the other is the two-step synthesis method brought up by Hinks et al.[2] The difference of the oxygen pressure for the sample preparation seems to affect the physical properties of the sample.

In our previous study of superconductivity and the pressure effect for $Ba_{1-x}K_xBiO_3$,[3] the preparation method of Cava et al. was employed. In this paper, results of elastic measurements for those samples are presented for the structural phase transition temperature. Since, in the method of Hinks et al., thermal measurements including the thermogravimetric analysis (TGA) and the differential thermal analysis(DTA) are easy to be achieved by utilizing the material reaction in normal pressure of gas, the phase separation temperature in oxygen[4] and the solidus and liquidus temperatures for $Ba_{1-x}Rb_xBiO_3$ have been determined. Furthermore, we have found a phase separation temperature in nitrogen atmosphere, below which the material makes a phase separation into the $BaBiO_3$ and the more concentrated $Ba_{1-x}Rb_xBiO_3$ phases. After determining these characteristic temperatures, preparation of the sample has successfully done and the pressure effect of the resistivity has been measured.

SAMPLE PREPARATION AND EXPERIMENTAL

$Ba_{1-x}K_xBiO_3$ samples were prepared by the method of the synthesis in a sealed quartz tube. $Ba_{1-x}Rb_xBiO_3$ samples were prepared on the basis of a phase diagram which will be mentioned after in detail. Stoichiometric mixture of BaO_2, Rb_2CO_3, and Bi_2O_3, which was ground in dry helium, was placed in a Pt crucible and sintered at ca 890 C for 3 hours in flowing nitrogen. Several samples were melted above 930-950C. On the cooling process after the firing, the atmosphere should be replaced by O_2 from N_2 at a certain temperature T_N and cooled in the furnace below a certain temperature T_O at an appropriate cooling rate. Then the sample was annealed at 400 C for several hours. In this preparing process, the temperature T_N at which atmosphere should be replaced by O_2, and the cooling rate after replacing atmosphere were very important to get good quality samples. In this experiment, the T_N was 545 C for x=0.4 and cooling rate was about 5-10 C/min.

The resistivity was measured by the four probe method and the pressure dependence of resistivity was measured up to 12 kbar in a clamp cell. The elastic property was measured by a composite-bar method using a 0 X-cut quartz resonator.[5]

RESULTS AND DISCUSSION

In Fig.1, the (220) X-ray diffraction peaks are shown for various samples as a function of the temperature at which the atmosphere in the sample preparation was replaced from N_2 to O_2 gas for $Ba_{0.6}Rb_{0.4}BiO_3$. As the temperature is lowered from 565 C to 545 C, the (220) peak moves slightly upwards, and returns again downwards at 525 C and 505 C, where the phases of shorter lattice parameters and of $BaBiO_3$ appear as shown by shaded lines. It is noted here that the lattice parameter is shorter in more concentrated samples. Therefore, this result shows that the phase separation temperature is 545K, above which the material is separated into a less-concentrated phase and another unknown Rb-concentrated one. Below the temperature, the material is separated into $BaBiO_3$ and the more concentrated one. The samples should not be cooled below this phase separation temperature in N_2 gas.

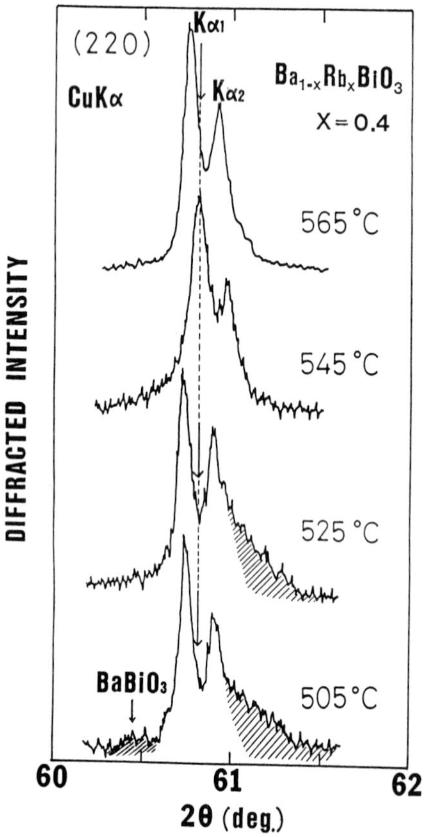

Fig. 1. The (220) X-ray diffraction peaks for various samples as a function of the temperature at which the atomosphere in the sample preparation wsa replaced from N_2 to O_2.

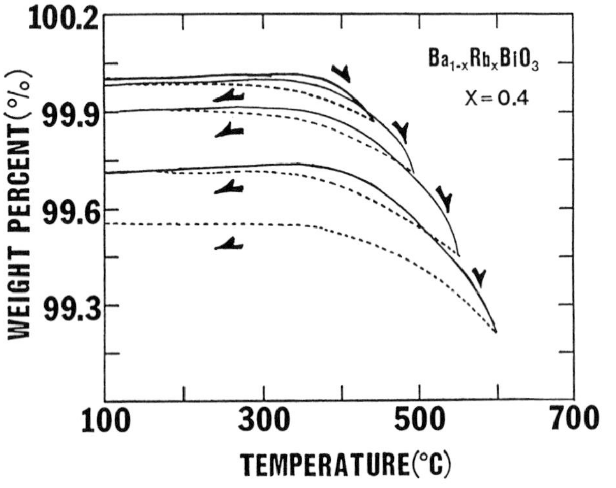

Fig. 2. The results of TGA. The sample(x=0.4) was heated to successively higher temperature in O_2. Heating and cooling curves are solid and dashed lines, respectively.

The results of the TGA in Fig.2 shows that, for the sample which experienced a.. ve the temperature of 500 C in O_2 gas, the weight does not recover again. The sample separates $RbBiO_2$ and Ba-Rb-Bi-O compound with the Rb concentration less than the starting composition.[4] The solidus and liquidus temperatures in Ar were determined by DTA for increasing and decreasing temperatures, respectively. These characteristic temperatures necessary for the sample preparation are summarized as a function of Rb concentration in Fig.3. The temperature T_N of the phase separation in N_2 decreases with the Rb concentration, while the T_O of the phase separation in O_2 is almost constant. There is an unstable temperature region both in atmospheres of N_2 and O_2. Therefore, the sample should be passed through the region as fast as possible in O_2 atmosphere.

Figure 4 shows the temperature dependence of the resistivity for $Ba_{.6}Rb_{.4}BiO_3$ as a function of the pressure. With increasing the pressure, the resistivity decreases and the superconducting temperature Tc increases. For $Ba_{1-x}K_xBiO_3$, Tc and the normal state conductivity increases with pressure: For $BaPb_{1-x}Bi_xO_3$, both of them decrease with pressure to the contrary.[3] These result suggests that Tc corresponds to the density of states at the Fermi energy which is sensitive to the pressure.

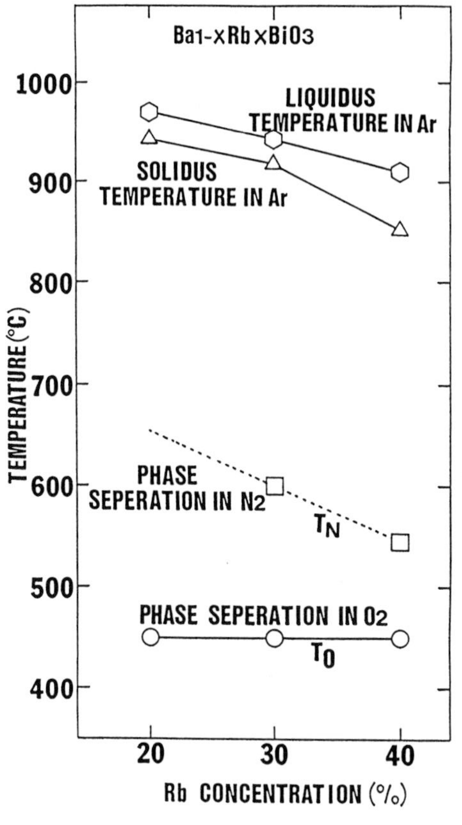

Fig. 3. The temperature T_N and T_O of the phase separations in N_2 and in O_2, respectively for $Ba_{1-x}Rb_xBiO_3$. The solidus and liquidus temperatures in Ar were determined by DTA.

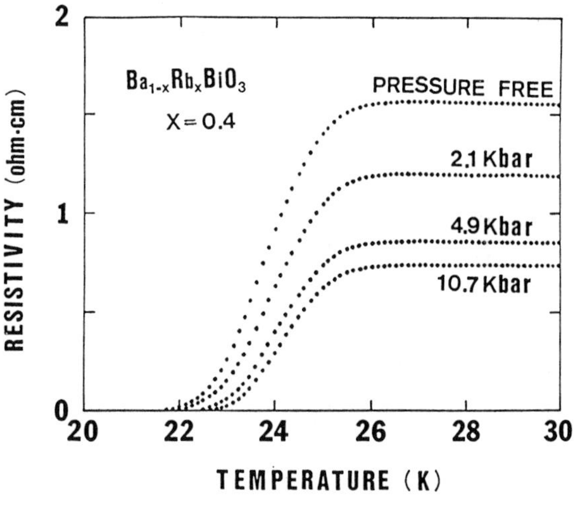

Fig. 4. Temperature dependence of the resistivity for $Ba_{0.6}Rb_{0.4}BiO_3$ at various pressures.

Figure 5 shows the temperature dependence of the resonance frequency corresponding to the elastic constant from 5K to 673K. For the sample of x=0, a clear elastic anomaly is found at 437K, which is the first order transition temperature due to the flopping of the rotation axes of the oxygen octahedra from <110> to <111>.[6] This temperature slightly decreases with increasing the K substitution on the contrary to the result by a neutron scattering experiment[6] which shows a rapid decrease of the temperature for the samples by the two step synthesis method. The difference between these experiments may come from the different sample preparation method. The cubic to rhombohedral transition is found for the sample of x=0.15 and partly for x=0.10 in Fig.5. These temperatures for the phase transition are plotted in Fig.6, where the phase boundary between the superconductive and the insulator phases was determined previously.[3]

In summary, we have determined the characteristic temperatures necessary for the sample preparation of $Ba_{1-x}Rb_xBiO_3$ in the atmosphere of normal pressure. The pressure application for the $Ba_{1-x}Rb_xBiO_3$ prepared by this method increases Tc and the conductivity as in the case for $Ba_{1-x}K_xBiO_3$ prepared by the synthesis in a fused quartz. We also found the structural phase transition temperature for $Ba_{1-x}K_xBiO_3$ prepared by the synthesis in a fused quartz which exhibits a different K-concentration dependence from that in the sample of the two step synthesis.

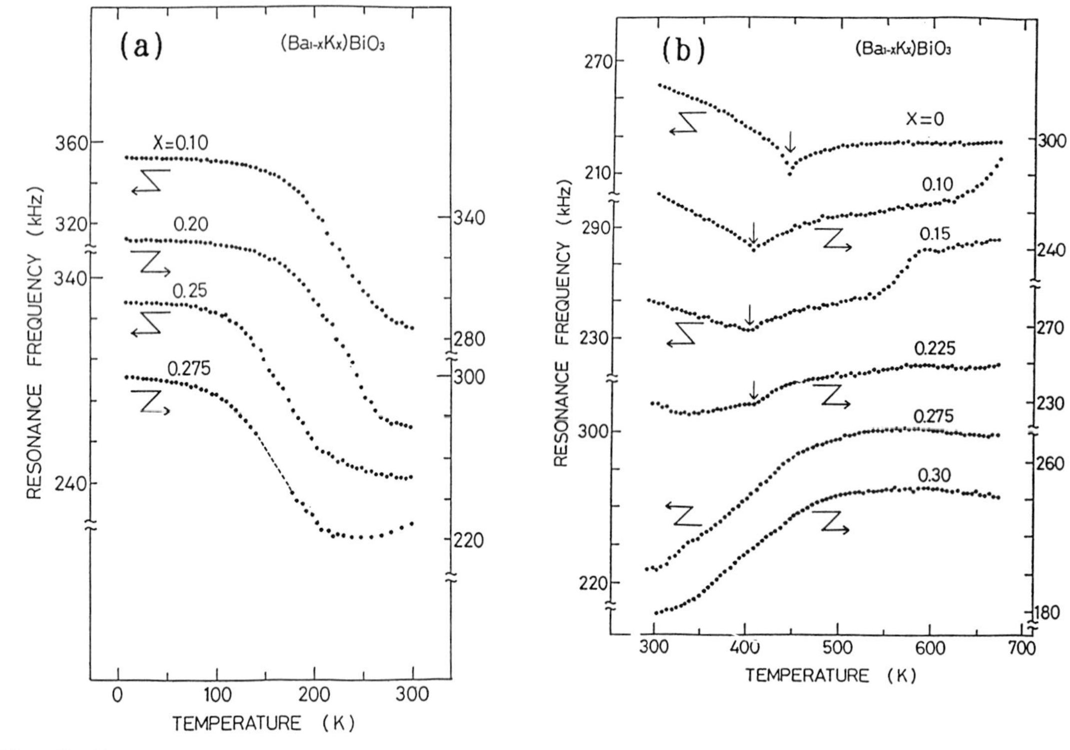

Fig. 5. Temperature dependence of the resonance frequency, (a)from 5K to room temperature and (b)from room temperature to 673K.

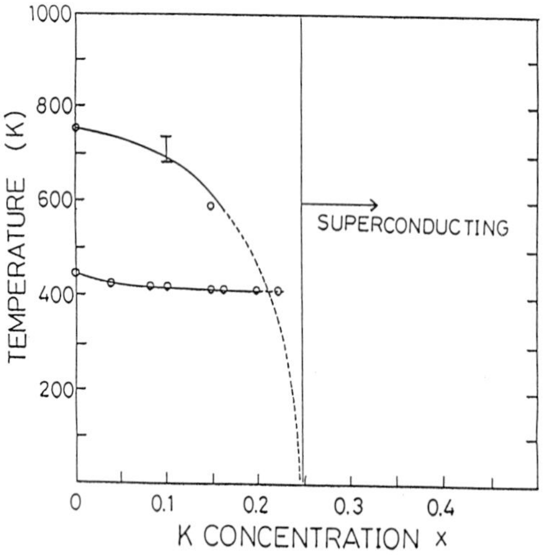

Fig. 6. The phase diagram determined by the temperature dependence of the elastic compliance in $Ba_{1-x}K_xBiO_3$ sintered in the sealed quartz tube.

REFERENCES

1. Cava RJ, Batiogg B, Krajewski JJ, Farrow R, Rupp Jr LW, White AE, Short K, Peak WF, Kometani T(1988) Nature 332: 814
2. Hinks DG, Dabrowski B, Jorgensen JK, Mitchell AW, Richards DR, Pei S, Shi D (1988) Nature 333: 836
3. Uwe H, Osada T, Iyo A, Murata K, Sakudo T (1989) Physica C 162-164: 743
4. Hinks DG, Richards DR, Dabrowski B, Mitchell AW, Jorgensen JD, Marx DT (1988) Physica C 156: 477
5. Yamaguchi H, Uwe H, Sakudo T, Sawaguchi E (1988) J. Phys. Soc. Jpn. 57: 147
6. Pei S, Jorgensen JD, Dabrowski B, Hinks DG, Richards DR, Mitchell AW (1990) Phys. Rev. B 41:4126

The Properties of Normal Carriers in $Bi_2Sr_2CaCu_2O_y$

TATSUYA HONMA and KAZUHIKO YAMAYA

Department of Nuclear Engineering, Hokkaido University, Sapporo, 060 Japan

ABSTRACT

Resistivity and Hall coefficient in the a-b plane of single-crystals $Bi_2Sr_2CaCu_2O_y$ (BSCCO) measured as a function of temperature have been analyzed by a simple two-band model involving hole and electron. It is found that the calculated values of the resistivity and Hall coefficient reproduce well the data, and the carrier concentration, mobility and mean-free-path obtained from the model are reasonable values in physics. It also is found that the model reproduces the data of resistivity and Hall coefficient of $YBa_2Cu_3O_y$ (YBCO) measured by Penney et al. and $Tl_2Ba_2CuO_y$ (TBCO) by Kubo et al.. These results indicate that the two-band model is very useful for investigation of transport properties in high-T_c oxides.

KEY WORDS: Resistivity, Hall coefficient, mobility, carrier concentration, two-band model

INTRODUCTION

Extensive studies on transport property, magnetism and specific heat etc. of high-T_c oxides have been carried out to investigate the mechanism of the superconductivity. Investigation on the properties of the normal carrier, such as carrier concentration and mobility, in high-T_c oxides is one of the clues to the mechanism of the superconductivity. In order to realize the properties of normal carrier, the transport properties are required to be interpreted consistently by some model. In such situation Eagles used a simple two-band model involving hole and electron to interpret consistently the temperature dependence of resistivity in the a-b plane ρ_{ab} and Hall coefficient R_H in a single-crystal $YBa_2Cu_3O_y$ (YBCO)[1]. From the linear temperature dependence of ρ_{ab}, he assumed that carrier concentrations do not vary with temperature T, and the temperature dependence of hole mobility and electron one are given by a formula of μ^{-1}=A(1+BT). The temperature dependence of ρ_{ab} and R_H in YBCO is reproduced well by his model. Therefore, it is very interesting to investigate whether the model can reproduce the transport properties in other systems.

It is known in $Bi_2Sr_2CaCu_2O_y$ (BSCCO) that the resistivity in the a-b plane is anisotropic and both resistivities along the a- (ρ_a) and b-axis (ρ_b) exhibit linear temperature dependence [2], and the Hall coefficient in the a-b plane with the magnetic parallel to the c-axis is positive and exhibits the temperature dependence[3,4]. Therefore BSCCO is a good sample to test the Eagles's model which is based on the linear temperature dependence of ρ. In the present paper, we analyze the data of resistivity and Hall coefficient of a single-crystal BSCCO by means of the model. Furthermore, data of YBCO measured by Penney et al.[5] and of TBCO by Kubo et al.[6] are analyzed by the same model. The model is discussed for its validity.

ANALYSIS

BSCCO is metallic, because of the T-linear resistivity. Then it is supposed that the carrier concentration n is independent on temperature. If the relation between ρ and μ is simply given by $\rho=(en\mu)^{-1}$, the temperature dependence of μ^{-1} is the same as that of ρ. Therefore, since the temperature dependence of the ρ is well expressed by a relation of $\rho=\alpha+\beta T$, that of the mobility is assumed to be represented by the formula of $\mu^{-1}=\alpha+\beta T$. This formula is the same as that proposed by Eagles. The fact that R_H depends on temperature suggests the multicarrier in BSCCO. From the positive R_H, one of multicarrier is hole carrier, at least. On the other hand, the thermoelectric power (TEP) in BSCCO[7] is observed to be negative, indicating conduction by electron. Positive R_H and negative

TEP are observed in YBCO as well as BSCCO[8]. Therefore it is reasonable to assume two carriers of the hole and electron in BSCCO as well as YBCO.

We adopt a two-band model involving hole and electron to analyze the temperature dependence of ρ_a, ρ_b and R_H in BSCCO. The hole concentration and electron one are p and n, respectively. Since the resistivity in the a-b plane is anisotropy, the mobility in the a-b plane also is anisotropic. Then, ρ_a ρ_b and R_H can be represented by using four kinds of mobility in the a-b plane (μ_{ha}, μ_{ea}, μ_{hb} and μ_{eb}) as following equations.

$$\rho_a = 1/\sigma_a = 1/e(p\mu_{ha} + n\mu_{ea}) \tag{1}$$
$$\rho_b = 1/\sigma_b = 1/e(p\mu_{hb} + n\mu_{eb}) \tag{2}$$
$$R_H = e(p\mu_{ha}\mu_{hb} - n\mu_{ea}\mu_{eb})/\sigma_a\sigma_b \tag{3}$$

where e is charge, p, μ_{ha} and μ_{hb} are carrier concentration, the mobility along the a- and b-axis for hole carrier, n, μ_{ea} and μ_{eb} are those for electron carrier, respectively. It is found from these eqs. that the temperature dependence of the resistivity depends on that of μ_h and μ_e, while that of R_H depends on the difference of temperature dependence between hole mobility and electron one. It is assumed that temperature dependence of all mobilities are represented by the formula of $(\alpha+\beta T)^{-1}$;

$$\mu_{ha}{}^{-1}=A(1+BT),\ \mu_{ea}{}^{-1}=C(1+DT),\ \mu_{hb}{}^{-1}=E(1+FT) \text{ and } \mu_{eb}{}^{-1}=G(1+HT). \tag{4}$$

From p, n and eqs.(1)-(4), ρ_a, ρ_b and R_H can be represented by ten parameters of p, n, and A - H. In order to determine the parameters, we use a non-linear least squares fitting routine. In the analysis, a datum consists of ρ_a, ρ_b and R_H measured at the same temperature. We used twenty data measured at temperatures above the temperature corresponding to a maximum of the R_H, because the origin of the maximum of R_H is unclear at the present stage.

RESULTS AND DISCUSSIONS

Figure 1 shows the temperature dependence of the calculated values and observed ones of ρ_a, ρ_b and R_H of BSCCO. We could obtain a fit to the data points to a root-mean-square accuracy of about 4%. It is found that the absolute values and the temperature dependence of ρ_a, ρ_b and R_H in BSCCO are reproduced well by the present model. As the result of this analysis, the concentration and hole mobility and electron one are determined simultaneously. A mean-free-path (m.f.p) can be estimated by assuming a simple free-electron model. The values of carrier concentrations, mobilities and m.f.p are shown in Table I. The values of ReO_3 and TiO_2 which are well known as a typical oxide conductor are also shown in Table I[9,10]. The magnitudes of carrier concentrations and mobilities in BSCCO are found to be almost equal to that of ReO_3 and TiO_2. The m.f.p along the a-axis of hole and electron in BSCCO are estimated to be 34A and 3.7A, respectively.

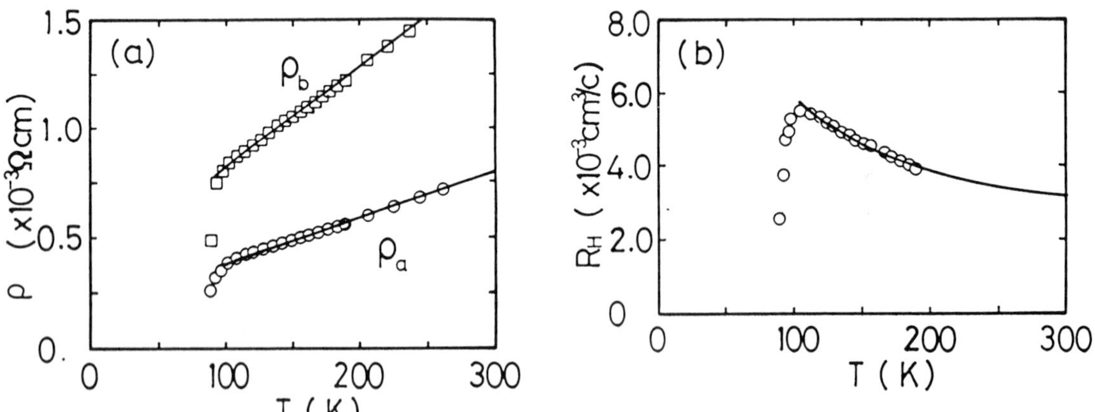

Fig. 1. Temperature dependence of ρ_a (a), ρ_b (b) and R_H (c) in $Bi_2Sr_2CaCu_2O_y$. Open circles and squares show the observed values, and solid lines show the calculated ones, respectively.

Table. I Carrier concentrations (p, n), mobilities (μ_h, μ_e) and mean-free-path (l_h, l_e) (300K) calculated by the present model.

	p (cm^{-3})	n (cm^{-3})	μ_h(cm^2/vs)	μ_e(cm^2/vs)	l_h (A)	l_e (A)
BSCCO (a-axis).	7.3 X 10^{19}	4.4 X 10^{21}	41.	1.1	34.	3.7
BSCCO (b-axis)			6.0	0.7	11.	2.4
YBCO	1.7 X 10^{21}	2.2 X 10^{21}	8.4	5.5	14.	10
TBCO	7.8 X 10^{20}	7.0 X 10^{20}	3.6	2.5	4.6	3.1
ReO_3	-	1.9 X 10^{22}	-	30	-	160
TiO_2..	-	1.4 X 10^{20}	-	0.4	-	0.5

These m.f.p are comparable to the unit-cell dimensions or larger than that. On the other hand, the m.f.p of ReO_3 and TiO_2 are estimated about 160A and 0.5A, respectively. The mechanism of the conduction for ReO_3 and TiO_2 is basically interpreted in terms of a band model in which the carrier of ReO_3 is a free-electron-like, while that of TiO_2 is a localized-electron-like[11]. The m.f.p of BSCCO are intermediate values between those of ReO_3 and TiO_2. These suggests that the electronic state of BSCCO is described with a band model. Thus it is found that the parameters estimated here are reasonable values in physics. Therefore, we conclude that the two-band model involving hole and electron is a very useful for analysis of the transport properties in BSCCO.

Measurement of angle-resolved photoemission on a single-crystal BSCCO by Takahashi et al. [12] showed two dispersive bands intersecting the Fermi level. They argue that the electronic structure of BSCCO can be described by a Fermi-liquid picture and there are at least two kinds of carrier at the Fermi level. Our results are not inconsistent with their argument.

In order to investigate the application of the present model, we try to analyze the data of the ρ and R_H measured in a single-crystal YBCO[5] and a sintered-sample TBCO[6] whose ρ's exhibit linear temperature dependence. Since the anisotropy of the resistivity in

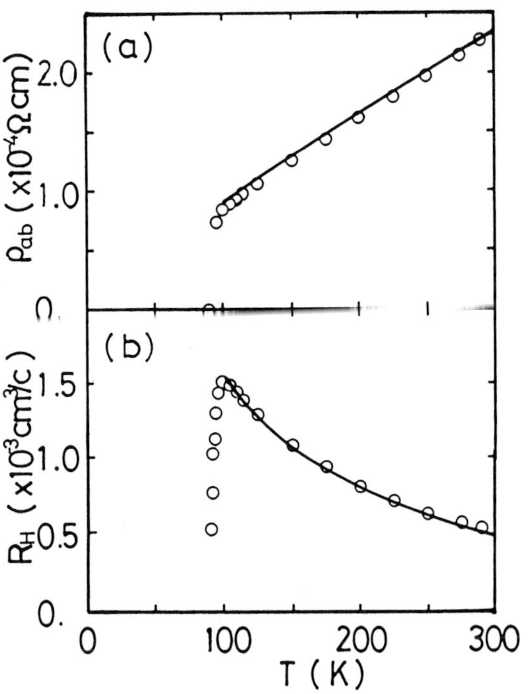

Fig. 2. Temperature dependence of ρ_{ab} (a) and R_H (b) in $YBa_2Cu_3O_y$. Open circles and solid lines show the observed values [5] and calculated ones, respectively.

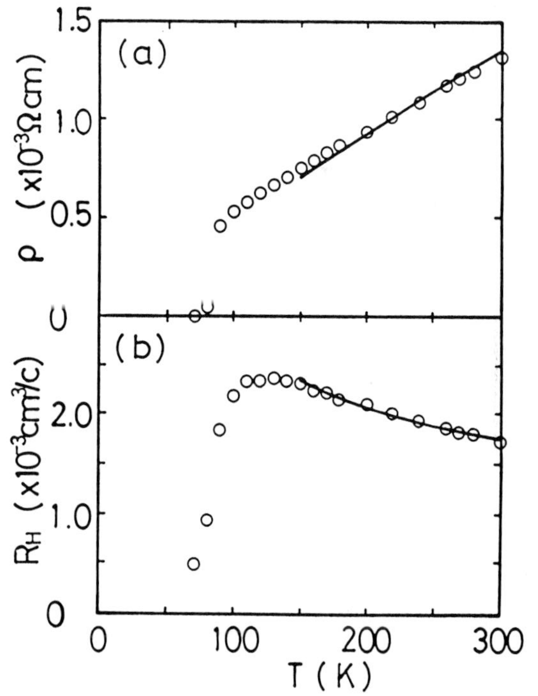

Fig. 3. Temperature dependence of ρ (a) and R_H (b) in $Tl_2Ba_2CuO_y$. Open circles and solid lines show the observed values [6] and calculated ones, respectively.

these samples is not known, the magnitude of the mobility is given by an average over the mobilities along each crystal axes. In eq.(4) we assume to be A=E, B=F, C=G and D=H. Figure 2 and 3 show the temperature dependence of ρ and R_H calculated for YBCO and TBCO, respectively. It is found that the temperature dependence of ρ and R_H are well reproduced. Table I shows the carrier concentrations, mobilities and m.f.p of YBCO and TBCO obtained from the analysis, which are found to be reasonable in physics. Therefore, it is concluded that the present model explains consistently the temperature dependence of the ρ and R_H for three systems of BSCCO, YBCO and TBCO.

It is very interesting to compare the carrier concentration and mobility between three systems of BSCCO, YBCO and TBCO. The p of BSCCO is about sixty times larger than the n. In spite of the positive R_H, the major carrier of BSCCO is electron, but not hole. In fact, a calculation shows that the electron conductivity σ_e is larger than the hole one σ_h. On the other hand, the μ_h's for both crystal axes are several times larger than the μ_e's. This leads to the positive R_H, as shown from eq.(3). In the case of YBCO, the p are comparable to the n, but the μ_h is about two times larger than the μ_e. As the result, the σ_h is larger than σ_e and the R_H is observed as a positive. In TBCO, both concentration and mobility of hole carrier are larger those of electron one. Therefore, the hole is dominant on the transport properties of TBCO. Thus we find that the hole mobility is larger than the electron one. This may be a feature in high-T_c cuprate oxides. To see the correlation between superconductivity and the properties of normal carrier, measurements and analyses of the transport properties on high-T_c oxides with various T_c are necessary, and are now in progress.

In summary, we have analyzed the temperature dependence of the resistivity and Hall coefficient in the a-b plane measured in single-crystals $Bi_2Sr_2CaCu_2O_y$ by a two-band model involving hole and electron, in which the mobility is anisotropic and its temperature dependence is represented by a relation of $\mu^{-1}=A(1+BT)$. The calculated values agreed with the data points with a root-mean-square accuracy of about 4%. It was confirmed that the magnitudes of the carrier concentration, the mobility and the mean-free-path obtained from the analysis are reasonable in physics. These results strongly support the two-band model in BSCCO. It also was found that the model explains well the temperature dependence of the resistivity and Hall coefficient in YBCO and TBCO. Therefore, we conclude that the present model is useful for analysis of the transport properties in high-T_c oxides which exhibits the T-linear dependence of ρ.

ACKNOWLEDGMENT

We thank Professor Y. Abe and Dr. Y. Okajima for fruitful discussions and Dr. S. Takekawa (National Institute for Research in Inorganic Materials) for supplying single crystals of $Bi_2Sr_2CaCu_2O_y$. This work was partly supported by Grant-in-Aid for Scientific Research on Priority Areas " Mechanism of Superconductivity " from the Ministry of Education, Science and Culture.

REFERENCES

1. Eagles DM (1989) Solid State Commun. 69: 229
2. Yamaya K, Haga T, Honma T, Abe Y, Minami F, Takekawa S, Tajima Y, Hidaka Y (1989) Physica C 162-164: 1009
3. Takagi H, Eisaki H, Uchida S, Maeda A, Tajima S, Uchinokura K, Tanaka S (1988) Nature. 332: 236
4. Honma T, Yamaya K, Minami F, Takekawa S, to be published.
5. Penney T, Molnar Sv, Kaiser D, Holtzberg F, Kleinsasser AW (1988) Phys. Rev. B 38: 2918
6. Kubo Y, Shimakawa Y, Manako T, Satoh T, Iijima S, Ichihashi T, Igarashi H (1989) Physica C 162-164: 991
7. Crommie MF, Liu AY, Cohen ML, Zettl A (1990) Phys. Rev. B 41: 2526
8. Ouseph PJ, RayO'Bryan M (1990) Phys. Rev. B 41: 4123
9. Pearsall TP, Lee CA (1974) Phys. Rev. B 10: 2190
10. Breckenridge RG, Hosler WR (1953) Phys. Rev. 91: 793
11. Tsuda N, Yanase A, Siratori K (1983) Electronic conduction in oxides. Syokabo Tokyo, pp150-178 [in Japanese]
12. Takahashi T, Matsuyama H, Katayama-Yoshida H, Okabe Y, Hosoya S, Seki K (1989) Phys. Rev. B 39: 6636

Specific Heat of $YBa_2Cu_4O_8$ near T_c

J.O. WILLIS, IZUMI TOMENO, TAKAYUKI MIYATAKE, TAKAO ITOH, KEISHI TAI, NAOKI KOSHIZUKA, and SHOJI TANAKA

Superconductivity Research Laboratory, International Superconductivity Technology Center, 10-13, Shinonome 1-chome, Koto-ku, Tokyo, 135 Japan

ABSTRACT

Specific heat measurements are reported for a polycrystalline, bulk sample of $YBa_2Cu_4O_8$ prepared at high oxygen pressure. The magnitude of the idealized specific heat jump $\Delta C_p/T$ at T_c (=80 ± 0.5K) is 15.5±2 mJ/K^2-mole. The transition width ΔT_c is about 3K. The size of the jump is greatly reduced compared to $YBa_2Cu_3O_7$. The normal state magnetic susceptibility is somewhat smaller than in $YBa_2Cu_3O_7$ and is temperature dependent. The results suggest a smaller electronic density of states for $YBa_2Cu_4O_8$.

KEY WORDS: $YBa_2Cu_4O_8$, specific heat, magnetic susceptibility, electronic density of states

INTRODUCTION

$YBa_2Cu_4O_8$ (Y124) was first seen as intergrowths in $YBa_2Cu_3O_7$ (Y123),[1] but can now be produced as high quality bulk single phase material with a transition temperature T_c of about 80 K.[2] The substitution of Ca for up to 10% of the Y has been shown by Miyatake and coworkers to raise T_c to as high as 88 K.[3] Unlike the Tl and Bi based superconductors, it is relatively easy to produce Y124 without other superconducting phases, and unlike Y123, there is no structural phase transition below the synthesis temperature and no, or possibly only a very small,[4] oxygen stoichiometry range, both of which may lead to nonideal behavior in Y123. These qualities may make Y124 useful technically and do make it a good candidate for studying the mechanism of superconductivity in the high T_c superconductors. Although many factors, such as the phonon spectra, magnetic correlations, etc., may be important for producing high T_c's in oxide superconductors, the electronic density of states (EDOS) must always be important. To investigate this property we have measured the specific heat and magnetic susceptibility of Y124 and compare these with measured and literature values for the same quantities for Y123 and Dy124 and discuss the implications for high temperature superconductivity.

EXPERIMENTAL

The polycrystalline sample of Y124 was prepared using an initial solid state reaction method and a final hot isostatic pressing (HIP) technique.[5] The X-ray powder diffraction measurements on this sample revealed that it was single phase Y124 with lattice parameters a = 3.84333 Å, b = 3.87296 Å, and c = 27.25448 Å, in good agreement with previous results.

The superconducting transition T_c occurs at 80 ± 0.5K as measured in a Quantum Design SQUID magnetometer in a field of 10 Oe. At 5K the zero field cooling susceptibility (shielding) and field cooling susceptibility (expulsion or Meissner effect) amount to 125% and 45% of the ideal volume susceptibility with no correction for sample demagnetization effects, which act to reduce the size of the measured value. The normal state susceptibility for Y124 in a magnetic field of 50 kOe is shown in Fig. 1. The 100 K susceptibility is about 240 μemu/mole and it increases slowly with temperature at a rate of 0.32 μemu/K. From this temperature dependence, we conclude that there is no evidence of magnetic impurity phases that order below 300 K.

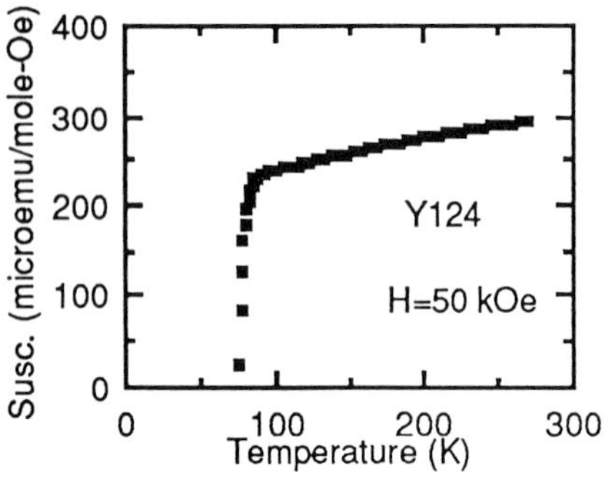

Fig. 1. The normal state molar susceptibility of Y124 in a field of 50 kOe plotted as a function of temperature.

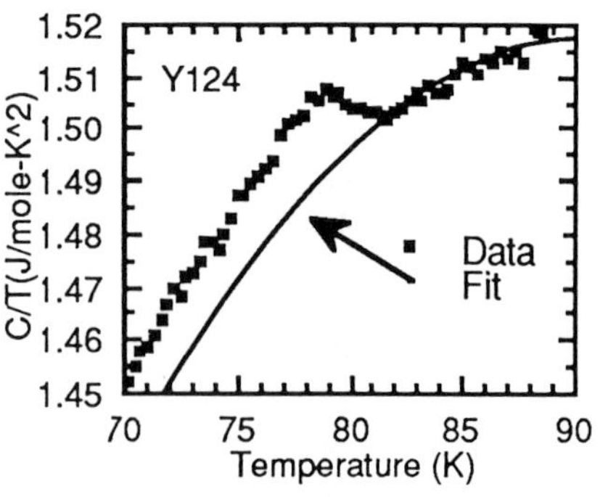

Fig. 2. The specific heat C_p of Y124 divided by temperature T plotted as a function of T. The solid line is an approximation to the specific heat above T_c.

The specific heat measurements were performed in an adiabatic calorimeter over the temperature range of about 60K to 300K. The approximately 19 g Y124 sample (the other samples were of similar mass) was coarsely crushed to ≤1 mm to improve filling of the sample holder and thermal equilibration. This sample and the others were measured several times with the reproducibility being 0.1% to 0.2%. The size of the specific heat jump $\Delta C_p/T_c$ at T_c was estimated by idealizing the shape of the superconducting transition by assuming it to be infinitely sharp while keeping the entropy constant. The normal state (mostly lattice) specific heat was approximated in the T_c region by fitting the data both near 60 K (well below the specific heat peak) and also from $T>T_c$ to T_c + 10K using a quadratic in C/T. This function was subtracted from the data and the entropy conserving construction was performed to extract $\Delta C_p/T_c$, T_c, and ΔT_c, the full width of the specific heat transition. The specific heat of Y124 plotted as C_p/T in the T_c region is shown in Fig. 2. A measurement of the specific heat of Y123 was made over the temperature range 30 to 180K for comparison purposes. The same data reduction techniques were employed as for the Y124 sample.

RESULTS AND DISCUSSION

The main specific heat results are summarized in Table I. The specific heat jump $\Delta C_p/T_c$ of Y124 is much smaller than that for Y123 and is in good agreement with the value reported for Dy124, which has a T_c of only 76 K.[6] The small size of $\Delta C_p/T_c$ has also been noted for Ho124 [7] ($\Delta C_p/T_c \approx 5$ mJ/K^2-mole) and for Y124 ($\Delta C_p/T_c \approx$ 12-15 mJ/K^2-mole) by other authors.[4,8] With the exception of the one measurement on Ho124, all the results for the 124 compounds have yielded values near the same value as reported here. The situation for the specific heat jump of Y123 is quite different. It is known to be very sample dependent, with reported values ranging from about 30 to as large as 75 mJ/K^2-mole.[9] In the Y123 system, there is a strong correlation between the size of $\Delta C_p/T_c$ and the contribution to a linear term in the specific heat at low temperature.[9,10] This is thought to be due to the presence of magnetic impurities primarily in the lattice, which can

Table I. Specific Heat Derived Quantities. T_c is the superconducting transition temperature, ΔT_c is the width of the specific heat transition, $\Delta C/T_c$ is the height of the jump at T_c, and, θ_{Debye} is the Debye temperature calculated at 100K.

Material	T_c (K)	ΔT_c (K)	$\Delta C/T_c$ (mJ/K^2-mole)	θ_{Debye} @ 100 K (K)
Y124	80±0.5	3±0.5	15.5±2	467±10
Dy124[a]	75.5±0.5	4±0.5	16±2	445±10
Y123	91±0.5	5±0.5	47±5	466±10

[a] Ref 6.

be observed in the normal state susceptibility, and that cause a degradation of the superconducting state presumably by pair-breaking interactions. The results for Y124 then are strikingly different, in that measurements on seven different samples all yield approximately the same (small) value, in spite of widely varying sample preparation techniques, sample quality, T_c values, and, likely, magnetic impurity content. This may be a reflection of the smaller possibility for disorder in the 124 system with its fully occupied double Cu-O chain sites, but a microscopic theory explaining this result is not yet available.

If we make the assumption that the specific heat jump for Y124 is BCS-like ($\Delta C_p/T_c = 1.43\gamma$), where γ is the electronic specific heat coefficient, which is proportional to the electronic density of states in the free electron model, then we are able to calculate a γ value of about 11 mJ/K^2-mole. The same calculation for Y123 from the present data would give a value three times larger (33 mJ/K^2-mole). The γ values then imply that the EDOS value for Y124 is much smaller than for Y123. This relation would also hold true if strong-coupling enhancements, which act to increase the value of 1.43 in the BCS expression given above, are equal for the two systems.

We have also calculated the Debye temperature from the data, correcting for the electronic term but neglecting the anharmonic contribution. The values at 100K are listed in Table I. The values for Y124 are in good agreement with those reported by Junod et al.[8] below about 200 K. We conclude that there is only a small difference between C_{ph} for the Y123 and Y124 systems suggesting that any contributions to strong coupling from the electron-phonon interaction should be similar in both cases.

The susceptibility results have been analyzed using the method of Lee and Johnston,[11] and are summarized in Table II. The total susceptibility χ_{tot} can be written as $\chi_{tot} = \chi_{core} + \chi_{vv} + \chi_{spin}$. The calculation for Y123 is similar to that reported in that paper and employs the Knight shift data of Takigawa et al.[12] χ_{core}, the contribution to the susceptibility from the ionic cores, was calculated from table values. χ_{vv}, the Van Vleck or orbital contribution to the susceptibility, was calculated from the anisotropic contribution for each Cu site as derived from Knight shift measurements on Y123. In the case of Y123 there is one Cu(1) (chain) site ion and two Cu(2) (plane) site ions, whereas for Y124 there are <u>two</u> chain and two plane site ions. This results in different values for the total Van Vleck contribution in Y123 and Y124. Thus using the measured value for χ_{tot} we can calculate χ_{spin} and find a value of 263 μemu/mol-Oe near 100K. Comparing this with the result for Y123 of 323 μemu/mol-Oe, we find that the spin susceptibility of Y124 is only about 80% of the Y123 value per mole formula unit. There is uncertainty in this result from two sources: Knight shift values for Y123 (rather than Y124) were used to calculate χ_{vv}; and the much stronger temperature dependence of the total susceptibility for Y124 introduces uncertainties into the correct value of χ_{tot} to be used in the calculation. With these caveats stated, we still conclude that the size of the spin susceptibility, which is proportional to the EDOS, appears to be smaller for Y124.

Table II. Calculation of Y123 and Y124 Spin Susceptibility. Contributions to the susceptibility of Y123 and Y124 from the two Cu sites are given. χ_{core}, χ_{vv}^x, $\langle\chi_{vv}\rangle$, χ_{spin} and χ_{tot}, are the ionic core term, the anisotropic Van Vleck or orbital electron term, its average value, the spin term, and the total (measured) susceptibility, respectively.

(μemu/ mol-Oe)	χ_{core}	χ_{vv}^c	χ_{vv}^b	χ_{vv}^a	$\langle\chi_{vv}\rangle$	χ_{tot}	χ_{spin}
[a]Cu (2)	-	98	17	17	43	-	-
[a]Cu (1)	-	22	30	84	45	-	-
[a]Y123	-175	214	64	118	132	280	323
Y124	-200	236	94	202	177	240	263

[a] Ref 11

CONCLUSIONS

A recent band calculation of the EDOS of Y124 has yielded the result[13] that, calculated by the same technique, the density of states for Y124 is only about one-half the value for Y123. This is consistent with both of the experimental results reported here in that the EDOS is reduced in Y124, but the actual values deduced from these three different sources are not in quantitative agreement. The very small values observed for γ as derived from the specific heat jump could be an underestimate because of a reduction in the size of $\Delta C_p/T_c$ in Y124 for the same reasons that the jump in Y123 may be reduced as discussed above. However, the much smaller variation in $\Delta C_p/T_c$ for Y124 suggests that this might not be the case. The relative size of the EDOS as deduced from the spin susceptibility is also subject to the uncertainties discussed in the previous paragraph. Nevertheless, the data strongly suggest and we are led to conclude that the EDOS is indeed smaller in the Y124 system. The fact that this decrease results in only a small (3-11%) change in T_c further suggests that the electronic density of states is not the only factor controlling T_c and the nature of superconductivity in the Y124 system.

ACKNOWLEDGEMENTS

This work is supported by NEDO under the management of the R & D of Basic Technology for Future Industries. JOW acknowledges partial support by the U. S. Department of Energy through the Exploratory Research and Development Center at Los Alamos National Laboratory.

REFERENCES

1. Zandbergen HW, Gronsky R, Wang K, Thomas G (1988) Nature 331: 596
2. Karpinski J, Kaldis, E, Jilek E, Rusiecki S, Bucher B (1988) Nature 336: 660
3. Miyatake T, Gotoh S, Koshizuka N, Tanaka S (1989) Nature 341: 41
4. Schilling A, Bernasconi A, Ott HR, Hulliger F (1990) Physica C 169: 237
5. Miyatake T, Yamaguchi K, Takata T, Gotoh S, Koshizuka N, Tanaka S (1989) Physica C 160: 541
6. Nichols TR (1990) unpublished data.
7. Bernasconi A, Schilling A, Guo JD, Ott HR (1990) Physica C 166: 393
8. Junod A, Eckert D, Graf T, Kaldis E, Karpinski J, Rusiecki S, Sanchez D, Triscone G, Muller J (1990) Physica C 168: 47
9. Phillips, Norman E, Fisher RA, Gordon JE, Kim S, Stacy AM, Crawford MK, McCarron, III EM (1990) Phys. Rev. Lett. 65: 357
10. Junod A, Eckert D, Graf T, Triscone G, Muller J (1989) Physica C 162-164: 1401
11. Lee WC, Johnston DC (1990) Phy. Rev. B 41: 1904
12. Takigawa M, Hammel PC, Heffner RH, Fisk Z, Smith JL, Schwarz RB (1989) Phys Rev B 39: 300
13. Freeman AJ (1990) private communication

Thermoelectric Power of T* Phase and $(L,Ce)_4(La,Ba,Sr)_4Cu_6O_{16+z}$

K. KUBO, S. IKEGAWA, K. MATSUURA, T. WADA, T. YAMASHITA, H. YAMAUCHI, and S. TANAKA

Superconductivity Research Laboratory, International Superconductivity Technology Center, 10-13, Shinonome 1-chome, Koto-ku, Tokyo, 135 Japan

ABSTRACT

Thermoelectric power (TEP) was measured for the ceramic samples of two sets of superconducting cuprates, in the temperature range between 50 K and 310 K. The first set was of the "4:4:6 compounds", i. e. $(L_{2/3}Ce_{1/3})_4(La_{1/3}Ba_{1/3}Sr_{1/3})_4Cu_6O_{16+z}$ where L were Eu (Tc=28. 5 K) and Y (Tc<2 K). The second set was "T* compounds" with $Ln_xLa_{1.8-x}Sr_{0.2}CuO_{4-d}$ where Ln was Sm (Tc=21K), Eu (Tc=12K) and Gd (Tc<2K). TEP of all the samples exhibited maxima with respect to temperature, being located at Tq=160-220 K.

This feature was not fit suffficiently to the RVB theory without a slight modification.

Key Words: Thermoelectric power, 4:4:6 compounds, T^*phase, RVB theory

INTRODUCTION

The Electrical transport properties in the normal state of superconducting cuprates have been so far studied because of their unusual behavior. The Seebeck effect is one of the important transport phenomena and gives complementary information to the Hall effect and electrical conductivity. In most of the superconducting cuprates , the general pattern is for the Seebeck coefficient to decrease with T at high temperatures.[1] This decrease is contrary to the expectation for the metallic diffusion thermopower due to uncorrelated electrons.[2] We have measured the temperature dependence of Hall coefficient, R_H, for the superconducting cuprates such as $(Nd, Ce, Sr)_2CuO_4$[3] (T^*phase) or $(L, Ce)_2(Ba, L)_2Cu_3O_{10-d}$ (L=Nd, Sm, Eu)[4] (4:4:6 phase), in order to know the reason why the Tc depends on ion substitution in these systems. The R_H-vs-T curves possessed a common feature that they all exhibited peaks at temperature Tp=120-150 K except for the magnitude, but which mechanisms remained unknown.

We measured thermoelectric power (TEP) for these samples and discuss the effect of atomic substitution and try to the fit to RVB model.

EXPERIMENTAL

Two sets of samples were investigated. The superconducting transition temperature, Tc, was defined at the onset of diamagnetic signal in dc magnetic susceptibility. The first set was of sintered samples of $(L_{2/3}Ce_{1/3})_4(La_{1/3}Ba_{1/3}Sr_{1/3})_4Cu_6O_Y$ (4:4:6 phase) where L=Eu and Y. The samples were the same as those used in the works reported in Ref. 3, which were annealed in O_2 gas of 3 atm. The magnitude of Tc for the sample with L=Eu was 28. 5 K. It was reported that Tc for this system decreased with decreasing the ionic radius of L^{3+}.[5] For the sample with L=Y, no superconducting transition was observed down to 2 K. The samples of the second set were $Ln_xLa_{1.8-x}Sr_{0.2}CuO_{4-d}$ (T^*phase) where Ln=Sm, Eu, and Gd, which were prepared by a standard solid state reaction method. The calcination was done at 850–1100 °C for 48 h in total with four times of intermediate regrinding. The samples were sintered at 1130 °C for 12h in air. And the samples were annealed at 600 °C for 5h and 400 °C for 5h in O_2 at 80 atm. The samples of T^* phase substituted by Eu and Gd were confirmed to be of single phase by powder X-ray diffraction. But Sm substituted sample could not be realized a single phase which was left T' phase for 8 to 10 percent. As the ion radius decreased, Tc increased monotonically as seen in 4:4:6 phase. We also measured TEP for $La_{1.85}Sr_{0.15}CuO_{4-d}$ (T phase with Tc=38.3K) for comparison.

The Seebeck coefficient was measured by a dc method using a 10 nV-resolution digital voltmeter at temperatures between 20 and 310 K. The sample dimension was 7X2.5X1 mm^3. The 1:2:4 high Tc cuprate was used as a reference material. Temperature was measured by means of a calibrated Si diode sensor. Temperature gradient in the sample was measured using two pairs of Cu-Constantan thermocouples. The two pairs of thermocouples were solderly contacted on hot and cold sides of the sample respectively. This sequence of the measurement was fully computer controlled.

RESULTS AND DISCUSSION

First of all, we show the measured Seebeck coefficient, S, for $La_{1.85}Sr_{0.15}CuO_{4-d}$ (T phase) which is plotted by open circles in Fig. 1. S-vs-T curve has a peak with respect to temperature around 90 K and decreases with increasing temperature above 90 K. Figure 1 also shows the temperature dependence of S and resistivity for the 4:4:6 phase. The S-vs-T curve for $(Eu, Ce)_4(La, Ba, Sr)_4Cu_6O_Y$ (crosses) has a peak at Tq of 170 K and is nearly identical with that for the T phase at temperatures above 200 K. For the sample of $(Y, Ce)_4(La, Ba, Sr)_4Cu_6O_Y$ (solid triangles in Fig. 1), S was larger in all the measured temperature range than the case of Eu. Tq was 190 K and slightly higher than that for the Eu derivative. Resistivity of the Y derivative was also larger than that for the Eu derivative, especially at temperatures below 100 K. The Hall coefficients of the two samples were nearly identical at temperatures above 150 K.[3] Below 150 K, R_H's for the Y derivative were larger than those for the Eu derivative.[3] Thus above 150 K, the replacement of Eu by Y caused an increase in S while R_H remained constant.

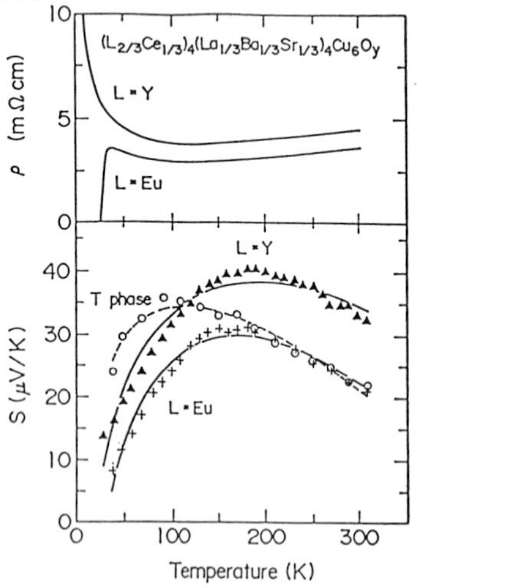

FIG. 1. Temperature dependence of Seebeck coefficients and resistivity for $(L_{2/3}Ce_{1/3})_4(La_{1/3}Ba_{1/3}Sr_{1/3})_4Cu_6O_Y$ with L=Eu (Tc=28.5 K) and Y (Tc<2 K). The solid curves in the lower part show fits to Eq. (1). Seebeck coefficients for $La_{1.85}Sr_{0.15}CuO_{4-d}$ are also shown by open circles. The dashed curve shows fit to Eq. (1).[8]

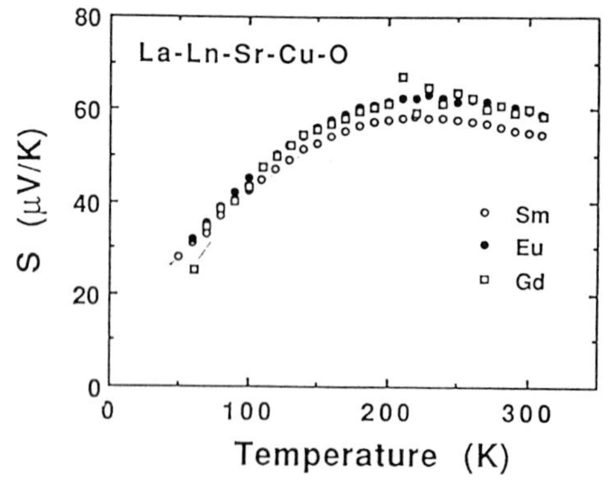

FIG. 2. Temperature dependence of Seebeck coefficients for $Ln_XLa_{1.8-X}Sr_{0.2}CuO_{4-d}$ (Ln=Sm, Eu and Gd)

Figures 2, 3 and 4 show the Seebeck coefficients, Hall coefficients and sucseptibility for $Ln_XLa_{1.8-X}Sr_{0.2}CuO_{4-d}$ (T^*phase) respectively, where Ln site was substituted with Sm, Gd and Eu. The S-vs-T curves for these three samples exhibited maxima around 220 K as common feature but the magnitude of S depended on the ion substitution, indicating that the smaller the ion is the larger is the Seebeck coefficient. The S-vs-T curves for the 4:4:6 and T^* phases investigated here were slightly different from that of $La_{1.85}Sr_{0.15}CuO_{4-d}$.

Recently, Nagaosa and Lee[6] proposed, based on a gauge-field theory in the uniform RVB state for the superconducting cuprates, that $S=S_F+S_B$ where $S_F \sim -(k_B/e)\,k_BT/E_F$ and $S_B \sim (k_B/e)\,[\,1 - \ln(2\pi x/mk_BT)\,]$. Therefore, we fitted the experimental data to the form :

$$S(T) = (k_B/e)\,[\,1 + F\ln(T/2\pi xG) - T/H\,] \tag{1}$$

using a least-squares method, where F, G and H were fitting parameters. The solid and dashed curves in the lower part of Figs. 1 and 2 are fitted curves in this way. The results for parameters F, G and H are summarised in Table 1. In the proposed model, F was unity. However, the experimental results do not reproduce unless non unity parameter F is introduced. This suggests that the thermal distribution function of bosons is slightly different from the proposed model. In the model,[6] parameters G ($\sim m^{-1}$) and H ($\sim E_F$) were expected to be of the order of J~1000 K. The values for G and H in Table 1 agree with this prediction.

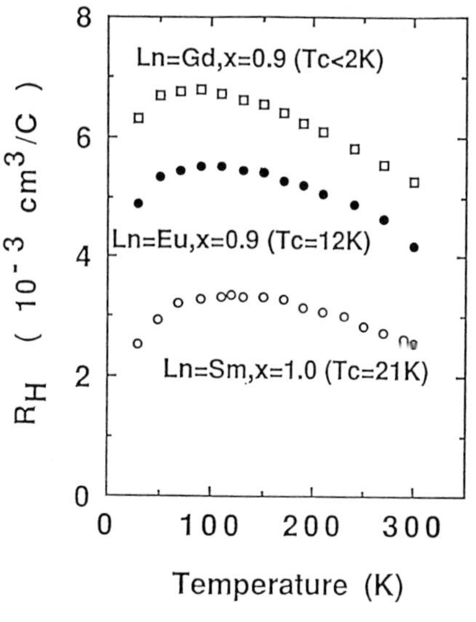

FIG. 3. Temperature dependence of Hall coefficients for $Ln_XLa_{1.8-X}Sr_{0.2}CuO_{4-d}$ (Ln=Sm, Eu and Gd)[9]

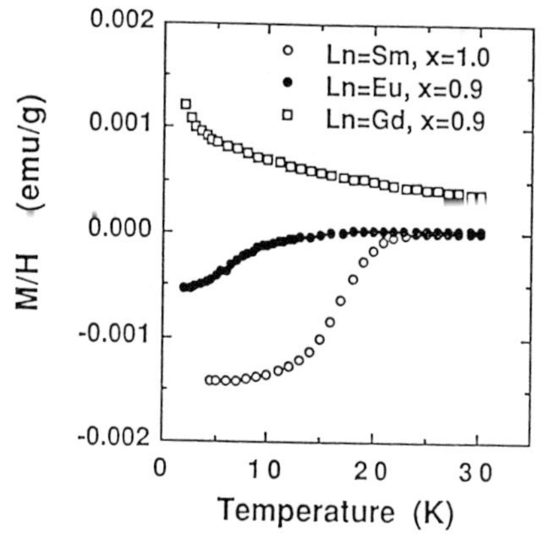

FIG. 4. Temperature dependence of susceptibility for $Ln_XLa_{1.8-X}Sr_{0.2}CuO_{4-d}$ (Ln=Sm, Eu and Gd)[9]

Table 1 Fitted values for parameters of S in Nagaosa-Lee theory. x is the experimental datum for the carrier density in the CuO_2 plane titrated by coulometry. (#, determined by Hall coefficients)

Sample	x	F	G(K)	H(K)
4:4:6-Eu	0.14#	0.408	352	414
4:4:6-Y	0.14#	0.333	426	575
T^*-Sm	0.133	0.469	224	530
T^*-Eu	0.106	0.562	228	443
T^*-Gd	0.076	0.662	282.9	362

SUMMARY

The Seebeck coefficient at temperatures between Tc and 310 K for the 4:4:6 and T^* phase exhibited a peak with respect to temperature located around 160-210 K. The model based on a gauge field theory in the uniform RVB state cannot be reproduced the present data of the S-vs-T curves, unless the thermal distribution function of bosons is modified. The changes in temperature dependences of S, by the Y substitution for Eu in the 4:4:6phase is similar to that of the Eu and Gd substitution for Sm in $Ln_xLa_{1.8-x}Sr_{0.2}CuO_{4-d}$ (T^* phase).

We would like to thank Dr. N. Nagaosa of the University of Tokyo for his helpfull discussions.

REFERENCES

1 For a review, A. B. Kaiser and C. Uher, in Studies of High Temperature Super conductors, Vol. 7, edited by A. V. Narlikar (Nova Science Publishers, New York, 1990).

2 D. K. C. MacDonald, Thermoelectricity: An Introduction to the Principles (Wiley, New York, 1962).

3 S. Ikegawa, T. Wada, A. Ichinose, T. Yamashita, T. Sakurai, Y. Yaegashi, T. Kaneko, M. Kosuge, H. Yamauchi and S. Tanaka, Phys. Rev. B 41, 11673 (1990).

4 S. Ikegawa, T. Yamashita, T. Sakurai, R. Itti, H. Yamauchi and S. Tanaka, Phys. Rev. B in press.

5 T. Wada, A. Ichinose, Y. Yaegashi, H. Yamauchi, and S. Tanaka, Jpn. J. Appl. Phys. 29, L266 (1990).

6 N. Nagaosa and P. A. Lee, Phys. Rev. Lett. 64, 2450 (1990).

7 M. F. Hundley, J. D. Thompson, S-W. Cheong, Z. Fisk, and R. B. Schwarz, Phys. Rev. B 40, 5251 (1989)7.

8 S. Ikegawa, T. Wada, T. Yamashita, A. Ichinose, T. Matsuura, K. Kubo, H. Yamauchi, and S. Tanaka, unpublished work (The Change in thermoelectricpower of $(L,Ce)_4(La,Ba,Sr)_4Cu_6Oy$ (L=Eu and Y) and zinc-doped T* phase, at ISTEC, SRL)

9 S. Ikegawa, T, Yamashita, T. Wada, H. Yamauchi and S. Tanaka unpublished work at SRL-ISTEC, Tokyo, 1990.

Transport Properties of Pb-Based Superconducting Copper Oxides

M. Kosuge, K. Sakuyama, T. Maeda, H. Yamauchi, and N. Koshizuka

Superconductivity Research Laboratory, International Superconductivity Technology Center, 10-13, Shinonome 1-chome, Koto-ku, Tokyo, 135 Japan

ABSTRACT

The Hall and Seebeck coefficient of $(Pb_{(1+x)/2}Cu_{(1-x)/2})Sr_2(Y_{1-x}Ca_x)Cu_2O_y$ were measured. Both Hall and Seebeck coefficient have positive values of about $6-10 \times 10^{-3}$ cm^3/C and 30-60 $\mu V/K$, respectively. The temperature dependence of the Hall coefficient and Seebeck coefficient were similar to other copper oxide superconductors such as $YBa_2Cu_3O_y$ and $(La,Sr)_2CuO_4$. Namely, the Hall coefficient increases monotonously with decreasing temperature and the Seebeck coefficient has a broad peak. The quenching caused the samples to become more metallic, and the Hall and Seebeck coefficients became smaller. These results are probably due to delocalization of holes trapped in the chain sites. Therefore in the "Pb-1212" system, removing the chain site oxygen appears to be necessary for superconductivity.

KEY WORDS: superconducting lead cuprates, "Pb-1212" phase, Hall effect, Seebeck effect, localization.

INTRODUCTION

Since Cava *et al.* discovered the first superconducting cuprate, $Pb_2Sr_2(Y,Ca)Cu_3O_8$ of the "2213" phase,[1] a variety of lead cuprates have been synthesized,[2-12] some of which exhibited superconductivity. In such lead cuprates "Pb-1212" phase has highest T_c (superconducting critical temperature).[7] This phase is isocrystallographic to "Tl-1212" phase; interestingly the Tl site is substituted not only by Pb but also by Cu.[2,5] Consequently, the crystal structure of the "Pb-1212" phase is quite similar to that of "123" ($YBa_2Cu_3O_7$). Namely, when an oxygen atom in the chain site of "123" is translated by (1/2 0 0), one obtains the "1212" structure.

The valence states of Pb and Cu in the (Pb,Cu)-O layer are considered to be close to 4 and 2, respectively, by Coulometric titration.[13] This is consistent with ionic radii calculations.[14] The insulating parent material of "1212" phase is $(Pb_{0.5}Cu_{0.5})Sr_2YCu_2O_y$. On the other hand, Ono and Uchida reported that $(Pb_{0.65}Cu_{0.35})Sr_2(Y_{0.7}Ca_{0.3})Cu_2O_y$ became superconducting when quenched into liquid nitrogen from high temperature.[15] It is strange that in this composition the positive charge introduced by Ca substitution for Y must be compensated by the increase in the Pb/Cu ratio, moreover superconductivity appears only when the sample is quenched. Recently, Maeda *et al.* successfully prepared single phase ceramics of $(Pb_{(1+x)/2}Cu_{(1-x)/2})Sr_2(Y_{1-x}Ca_x)Cu_2O_y$ and made them superconducting.[13] In the present paper we report transport properties (Hall and Seebeck effects) and discuss the carrier density of quenched and annealed samples.

EXPERIMENTAL

Samples were prepared by a solid state reaction using high purity (>99.9%) powders of PbO, $SrCO_3$, Y_2O_3, $CaCO_3$ and CuO. These powders were mixed into nominal composition of $(Pb_{(1+x)/2}Cu_{(1-x)/2})Sr_2(Y_{1-x}Ca_x)Cu_2O_y$ (x=0.3, 0.35). The mixed powders were calcined in air at 800 °C for 10 h, then pulverized and pressed into parallelepiped bars of 3x3x20 mm^3. The bars were fired in flowing O_2 gas at 1050 °C for 1 h followed by rapid cooling to 900 °C in a few minutes and slow cooling to room temperature at a rate of 1 deg/min. To make the electrical contacts, conductive gold paste was painted on the polished surface of sintered pellets, then annealed at 800 °C for 1 h in flowing O_2 gas followed either by slow cooling to room temperature at a rate of 1 deg/minute in the same atmosphere or by quenching to liquid nitrogen.

The Hall measurement was performed in a shielded room using a specially designed rotating sampleholder. The Hall coefficient was measured using current of 50 mA and a magnetic field of 8T. The sample dimension was about 2x0.15x6 mm^3. All the samples were polished at the same time so that sample thickness was the same within about 1 %. Consequently, a difference in Hall coefficient of more than 1 % between samples was considered meaningful. The Hall voltage was measured by a 10-nV resolution digital voltmeter (HP: 3458A) each time the sample was rotated by 180 degrees to change the effective direction of magnetic field. The measuring interval was about 1 minute. The temperature was swept at a rate of less than 1 deg/min. and monitored by a calibrated carbon glass resistance sensor or a platinum resistance sensor depending on the temperature range. To calculate the Hall coefficient we interpolate Hall voltage by using third ordered Lagrange polynomials. The correction for misalignment of the Hall contacts was also performed by a calculation.

The Seebeck coefficient of samples with dimension of about 2x2x10 mm^3 was performed by a dc method using the 3458A. High purity Cu wires of 20 μm diameter were used as reference material. Sample temperature was measured by a calibrated Si diode sensor and the temperature gradient in the sample was measured using two pairs of copper-constantan thermocouples.

RESULTS AND DISCUSSION

Figure 1 shows the temperature dependence of the resistivity for x=0.30 and 0.35 at a field of 8T, where "S" indicates a slowly cooled sample and "Q" a quenched one. The curves of x=0 and 0.1 are reproduced from Ref. 13. As shown in this figure, Ca doping makes the samples more conductive, and quenching makes the temperature dependence of resistivity metallic, and moreover superconductivity is realized in x=0.30 and 0.35. These results were very strange if we assume a hole type carrier, because in the nominal composition of $(Pb_{(1+x)/2}Cu_{(1-x)/2})Sr_2(Y_{1-x}Ca_x)Cu_2O_y$, carriers introduced by Ca substitution should be compensated by the increase in Pb content and quenching may introduce oxygen vacancies which also diminish hole density.

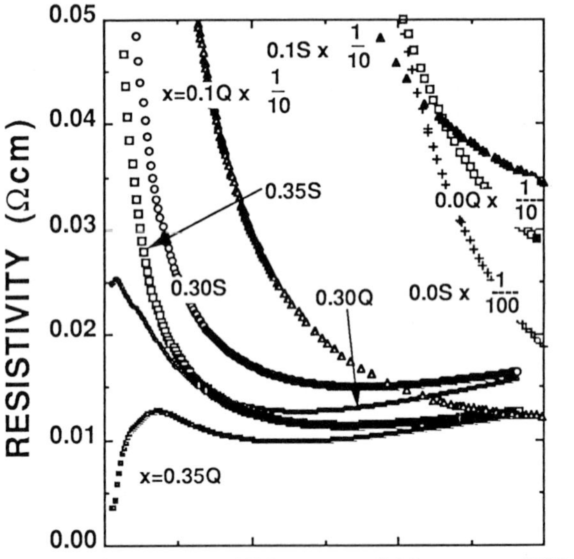

Fig.1 Temperature dependence of resistivity for $(Pb_{(1+x)/2}Cu_{(1-x)/2})Sr_2(Y_{1-x}Ca_x)Cu_2O_y$. Q and S indicate quenched and slowly cooled samples, respectively. The data of x=0.30 and 0.35 were measured at the field of 8T.

In order to clarify whether the carrier type is hole or electron, and whether the carrier density is changed or not by doping and quenching, we measured the Hall coefficient for x=0.30 and 0.35, and the Seebeck coefficient for x=0.35. The results are shown in Figs. 2 and 3. Although the temperature dependence of these coefficients is an unresolved problem so far, a drastic decrease in the Hall coefficient in the quenched sample is due to the superconducting transition. The difference in onset temperature between the resistivity curve

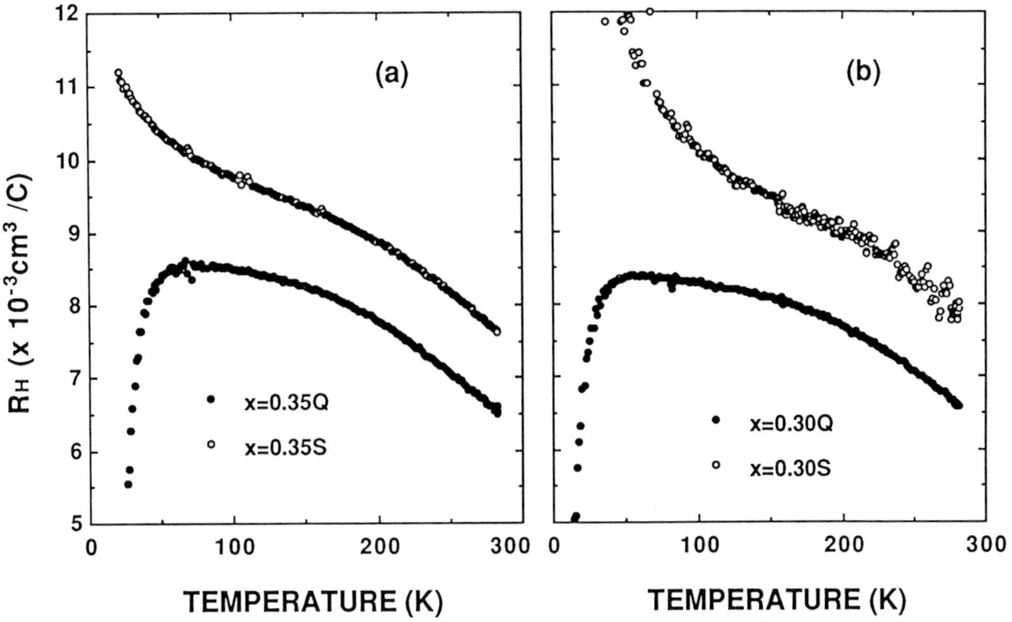

Fig.2 Temperature dependence of Hall coefficient of $(Pb_{(1+x)/2}Cu_{(1-x)/2})Sr_2(Y_{1-x}Ca_x)Cu_2O_y$. (a) x=0.35, (b) x=0.30. Q and S indicate quenched and slowly cooled samples, respectively.

and the Hall one at x=0.3 is probably because only small amounts of the high T_C phase. Measured Hall and Seebeck coefficients are all positive indicating hole type carriers. The Hall coefficients of x=0.35 are almost the same value as that of x=0.30 in the temperature range from 100 K to 270 K. Though this was not consistent with resistivity measurement at first, this result was consistent taking account of experimental error in the measurement of the absolute resistivity. However, the Hall coefficients of the quenched samples for both Ca contents are clearly smaller than that of the slowly cooled ones. This is consistent with the resistivity and Seebeck measurements. In the hole type copper oxide superconductor it is reported that the TEP (thermoelectric power) *vs.* temperature has a broad peak and the decrease in TEP with increasing temperature in the high temperature region is more rapid at higher carrier density.[16]

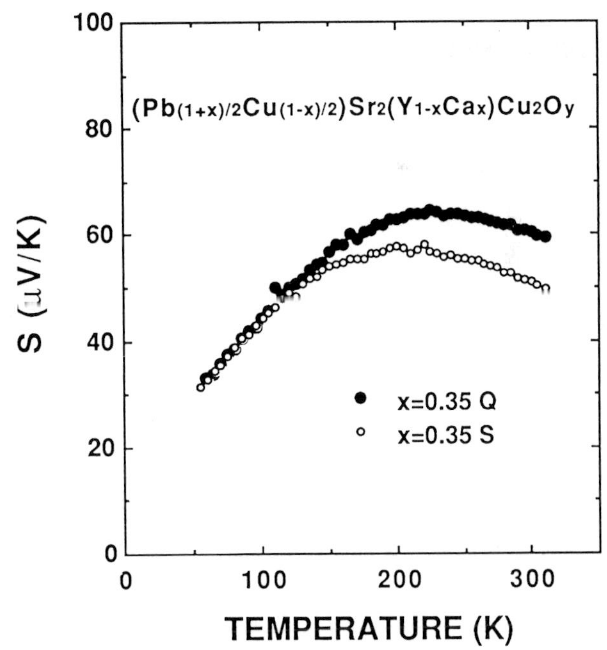

Fig.3 Temperature dependence of Seebeck coefficient: Q and S indicate quenched and slowly cooled samples, respectively.

Our preliminary oxygen analysis by coulometric titration [17,18] revealed that the slowly cooled samples with the compositions x=0, 0.1 and 0.3 had oxygen contents of 7.1, and the quenched one at x=0.3 had 7.0.[13,19] From the neutron structural analysis of the x=0.3 samples, excess amounts of oxygen located in the site corresponding to the chain site in the

"123" structure (abbreviated to chain site).[18] Therefore one of the plausible scenario explaining the substitution and quenching effects is as follows: Non-doped parent material of $(Pb_{0.5}Cu_{0.5})Sr_2YCu_2O_y$ prepared by slowly cooling has oxygen content y=7.1 and has a hole but it is probably localized to the excess oxygen in the chain site. If x in $(Pb_{(1+x)/2}Cu_{(1-x)/2})Sr_2(Y_{1-x}Ca_x)Cu_2O_y$ increases, this localized hole begins to be mobile. Therefore the resistivity and Hall coefficient become small but superconductivity does not yet appeared. (The delocalization mechanism, however, is not so clear.) When the chain site oxygen that traps the hole is removed by quenching, doped holes can move more freely and superconductivity appears.

SUMMARY

In the $(Pb_{(1+x)/2}Cu_{(1-x)/2})Sr_2(Y_{1-x}Ca_x)Cu_2O_y$ system, resistivity decreases with increasing x up to about 0.3. Quenching caused the samples to become more metallic, and positive Hall and Seebeck coefficient to decrease. These results are probably due to delocalization of holes trapped in the chain site. Therefore in the "Pb-1212" system, removing the chain site oxygen appears to be necessary for superconductivity.

ACKNOWLEDGMENT

The authors would like to thank K. Matsuura and K. Kubo of SRL-ISTEC for their kind use of the Seebeck measuring system. They also thank T. Machi of SRL-ISTEC for his help in operating the superconducting magnet and J. O. Willis of SRL-ISTEC for a critical reading of the manuscript.

REFERENCES

[1] Cava RJ, Battlog B, Krajewski JJ, Rupp LW, Schneemeyer LF, Siegrist T, van Dover RB, Marsh P, Peck Jr WF, Gallagher PK, Glaum SH, Marshall JH, Farrow RC, Waszczak JV, Hull R, Trevor P (1988) Nature **336**: 211

[2] Subramanian MA, Goparakrishnan J, Torardi CC, Gai PL, Boyes ED, Askew TR, Flippen RB, Farneth WE, Sleight AW (1989) Physica **C159**: 124

[3] Inoue O, Adachi S, Takahashi Y, Hirano H, Kawashima S (1989) Lpn J Appl Phys **28**: L60

[4] Rouillon T, Retoux R, Groult D, Michel C, Hervieu M, Provost J, Raveau B (1989) J solid state Chem **78**: 322

[5] Lee JY, Swinnea JS, Steinfink H, Mater J (1989) J Mater Res **4**: 763

[6] Zandbergen HW, Fu WT, van Ruitenbeek JM, de Jongh LJ, van Tendeloo G, Amelinckx S (1989) Physica **C159**: 81

[7] Rouillon T, Provost J, Hervieu M, Groult D, Michel C, Raveau B (1989) Physica **C159**: 201

[8] Tokiwa A, Oku T, Nagoshi M, Kikuchi M, Hiraga K, Syono Y (1989) Physica **C161**: 459

[9] Rouillon T, Groult D, Hervieu M, Michel C, Raveau B (1990) Physica **C167**: 107

[10] Cava RJ, Bordet P, Capponi JJ, Chaillout C, Chenavas, Fournier T, Hewat EA, Hodeau JL, Levy JP, Marezio M, Battlog B, Rupp Jr LW (1990) Physica **C167**: 67

[11] Adachi S, Inoue O, Kawashima S, Adachi H, Ichikawa Y, Setsune K, Wasa K (1991) Physica **C168**: 1

[12] Adachi S, Setsune K, Wasa K (1990) Jpn J Appl Phys **29**: L890

[13] Maeda T, Sakuyama K, Koriyama S, Yamauchi H, Tanaka S: submitted to PRB

[14] Shannon RD (1976) Acta Crysta **A32**: 751

[15] Ono A, Uchida Y (1990) Jpn J Appl Phys **29**: L586

[16] for exammple: Ando Y, Sera M, Yamagata S, Kondoh S, Onoda M, Sato M (1989) Solid State Commun **70**:303

[17] Kurusu K, Takami H, Shintomi K (1989) Analyst **114**: 1341

[18] Kosuge M, Kurusu K (1989) Jpn J Appli Phys **28**: L810

[19] Maeda M, Sakuyama K, Koriyama S, Yamauchi H, Tanaka S, Izumi F, Asano H: inpreparation

Tunneling Measurements on $YBa_2Cu_4O_8$ under Magnetic Fields

KOICHI NAKAO, SHIGETO FUJITA, MASANORI NEMOTO, KATSUMI SUZUKI, TAKAYUKI MIYATAKE, TADATAKA MORISHITA, NAOKI KOSHIZUKA, and SHOJI TANAKA

Superconductivity Research Laboratory, International Superconductivity Technology Center, 10-13, Shinonome 1-chome, Koto-ku, Tokyo, 135 Japan

ABSTRACT

Tunneling measurements on $YBa_2Cu_4O_8$ were performed systematically for the first time. Tunnel junctions were fabricated by depositing Au layers directly on the surface of sintered samples. Effects of magnetic fields ($\leq$ 6T) were studied. The overall feature of the tunneling spectrum is similar to that of $YBa_2Cu_3O_7$. The similarity includes slightly asymmetric background conductance, a gap-like structure at about $\pm$20mV, a subgap structure at about $\pm$7mV and a conductance peak which is observed in junctions with relatively large conductance. It is argued that the origin of the conductance peak is due to Andreev reflection.

KEY WORDS: $YBa_2Cu_4O_8$, tunneling spectrum, magnetic field

INTRODUCTION

Tunneling experiments have been an important probe for microscopic properties of superconductors. Since the discovery of high T_c oxide superconductors, many groups performed tunneling experiments using different techniques. But experimental results have not allowed a straightforward interpretation. Even the existence of an energy gap has not been verified unambiguously.

Difficulties are often attributed to the extraordinarily short coherence length, uncontrollable surface imperfections, two dimensional nature, and so on. In the case of $YBa_2Cu_3O_7$ (123), it was shown that rather reproducible data could be obtained for tunneling junctions fabricated by evaporating metals directly on the surface [1-3]. In such junctions, tunneling barriers are natural ones whose nature is not known exactly.

From an experimental point of view, the background conductance, which depends on temperature and voltage, masks the intrinsic structure of the tunneling spectrum. A conventional way to extract density of states from an experimental conductance curve is to normalize the curve below T_0 to that above T_c. But the fact that the background conductance is clearly temperature dependent makes this procedure less reliable. Another conventional technique for canceling the background conductance is to apply a magnetic field larger than H_{c2} to obtain normal state conductance at a temperature below T_c. But H_{c2}'s of high T_c compounds at low temperatures are usually too high to adopt this technique.

In the present work we applied magnetic fields ($\leq$ 6T) to junctions fabricated by depositing Au directly on the surface of sintered samples of $YBa_2Cu_4O_8$ (124). It has been made clear that the 124 compound is much more stable against oxygen loss from the surface than the 123 compound[4]. Therefore, we can expect the tunneling spectrum for the 124 compound to be related to the bulk properties more closely than for the 123 compound.

EXPERIMENTAL

$YBa_2Cu_4O_8$ was prepared with the oxygen HIP (hot isostatic pressing) treatment. The critical temperature was 80K. The details of the sample preparation were reported elsewhere[5]. A gold layer of 500A was deposited on the surface in a high vacuum evaporator. The area of the junctions was about 2x2mm^2. There was no special treatment for the surface before the deposition. The tunneling property varied greatly from junction to junction. In some cases, the tunneling resistance was so small that the tunneling spectrum could not be measured. Junctions with a considerable resistance were reasonably stable during a measurement procedure including the temperature sweep between 4.2K and a temperature much higher than T_c.

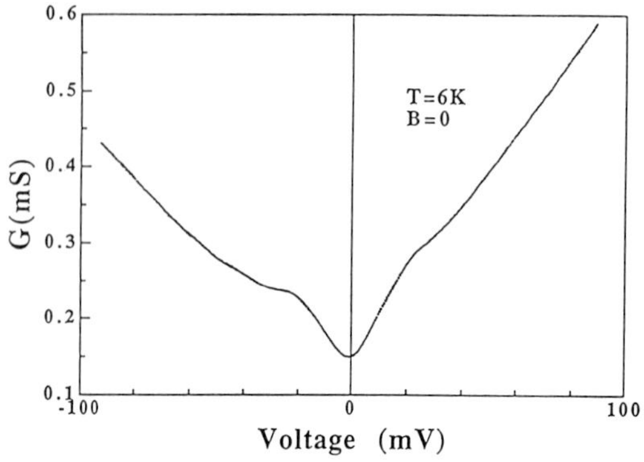

Fig.1. Tunneling conductance G of sample-1 at 6K and in zero magnetic field.

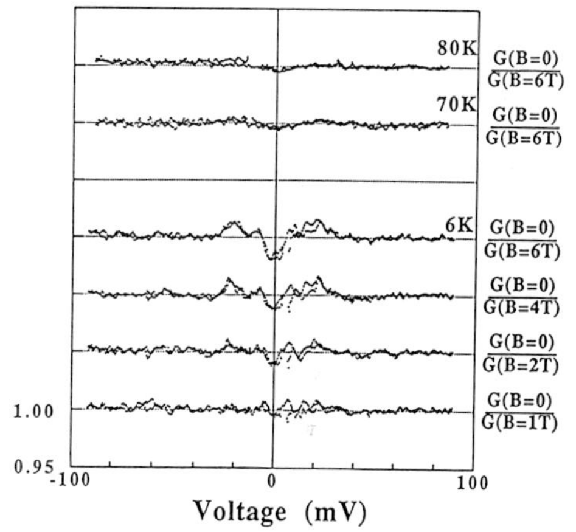

Fig.2. Ratio of tunneling conductance G of sample-1 in zero and finite magnetic fields. Curves except for the lowest one are shifted vertically for clarity.

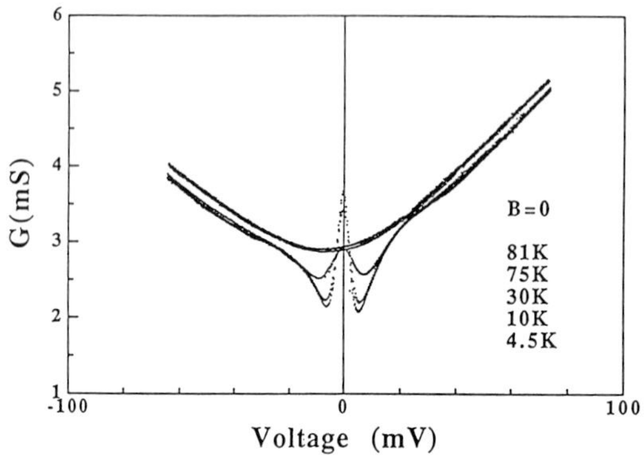

Fig.3. Tunneling conductance G of sample-2 at several temperatures in zero magnetic field.

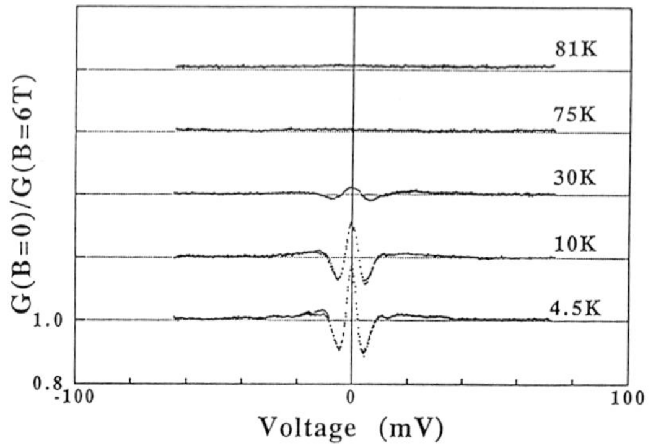

Fig.4. Ratio of tunneling conductance G of sample-2 in B=0 and in B=6T at several temperatures. Curves except for the lowest one are shifted vertically for clarity.

Magnetic fields were applied by a superconducting solenoid. The residual field of the superconducting solenoid was about 10Oe. When we measured the temperature and magnetic field dependence of the tunneling conductance, we varied the magnetic field at a fixed temperature and then changed the temperature to the value selected for the next measurement. Therefore in the "zero field" situation there may be a small number of vortices trapped within the sample. The tunneling conductance G=dI/dV was measured by the conventional lock-in technique.

We describe two typical types of tunneling spectra in the present paper. One (sample-1) has a small conductance and no conductance peak at zero bias, and the other (sample-2) has a larger conductance and a conductance peak at zero bias.

RESULTS AND DISCUSSION

Figure 1 shows the low temperature tunneling spectrum of sample-1. It is slightly asymmetric as is often observed in the case of the 123 compound[6]. It is a smooth curve with humps at $\pm$20mV(gap-like structure). In the 123 compound, similar gap-like structures have been often observed. Application of magnetic fields of 6T did not change the overall feature of the tunneling spectrum. But a small and systematic effect of the magnetic field can be seen when the tunneling conductance without magnetic field is normalized to that in a magnetic field. The results are shown in Fig.2. In the regions $|V|>50$mV, the conductance ratio is almost exactly unity and there is no structure. This fact strongly indicates that the magnetic field does not effect the background conductance. In the region $|V|<40mV$, there is a remarkable structure, although the effect of the magnetic field is less than 2% of the absolute conductance. This structure can be considered to correspond to the magnetic field dependence of the tunneling density of states, although it cannot be regarded as the density of states itself because the applied field was not intense enough to drive the material to the normal state. There are two pairs of peaks at $\pm$20mV and at $\pm$7mV. Peaks at $\pm$20mV probably correspond to humps on the original dI/dV curve (Fig.1), and can be interpreted that the humps are suppressed by the applied magnetic fields. This structure persists up to 70K and some trace can be seen even at 80K. Peaks at $\pm$7mV have no corresponding structure on the original dI/dV curve.

Figure 3 shows the tunneling conductance of sample-2. This sample has a larger conductance and the spectrum shows a distinct conductance peak at around V=0. Conductance peaks were observed in many other junctions with relatively large conductance. There are also humps at $\pm$ 20mV which are probably of the same origin as that shown in sample-1 (Fig.1). The conductance peak is strongly temperature dependent. While the humps at $\pm$ 20mV did not change considerably with a temperature increase from 4.3K to 30K, the conductance peak was diminished greatly. The conductance peak is also very sensitive to the magnetic field. Effects of a magnetic field at several temperatures are shown in Fig.4. Application of a magnetic field at T=4.3K decreased the conductance peak to half that at B=0. Effects of magnetic fields on the humps at $\pm$20mV are similar to the case of sample-1 shown in Fig.1 ($\leq$2%).

In an ideal BCS superconductor, an effect of applying a magnetic field, which is much larger than H_{c1} and much smaller than H_{c2}, is to reduce the gap energy. Our results in Figs.2 and 4, however, cannot be explained in such a simple way. On the contrary, it seems that the effect of a magnetic field is to reduce the amplitude of the gap-like structure, and not to alter the energy scale of the structure. This may have a relation to the fact that the gap-like structure in high T_c compounds usually does not change its energy scale as the temperature is increased[7]. In addition to this well-known temperature dependence, the present result of the magnetic field dependence makes it doubtful to interpret the gap-like structure at $\pm$20mV as the intrinsic energy gap.

Our tentative explanation for these experimental results is as follows. The fact that we sometimes have reasonable values of tunneling resistance in junctions fabricated by depositing Au directly on the 124 compound means that there is an insulating layer on the surface of such samples. In this situation, it is not unreasonable to suppose that there is a transition region between the insulating layer, which behaves as a tunneling barrier, and the bulk region. The transition region may be a normal metal or a superconductor with different properties from the bulk region. Then we have a configuration of the proximity effect. In such a situation, the pair potential $\Delta(x)$ depends on the position. A spatially varying pair potential modifies the tunneling density of states substantially. This situation was discussed by Yoshikawa et al.[8]. It has been known theoretically that a spatially varying pair potential can induce gaplessness in the density of states[9]. And it has also been known that in an extreme gapless region, the energy scale of the density of states is determined by a pair-breaking parameter and not by the pair potential Δ[10]. Published data on the 123 compound and the present results on the 124 compound remind us of these theoretical predictions. However, the theories were originally derived for superconductors in the dirty limit. Therefore it is not clear whether they are applicable to the 123 and 124 compounds, which are considered to be in the clean limit.

There have been many observations of subgap structures within the gap-like structure in the 123 compound[1-3]. Some authors attributed these to an anisotropy in gap values[2], and others to a second phase which exists only on the surface[3]. In the present work we did not find a subgap structure in the original dI/dV curve (Fig.1). But when we applied a magnetic field, an additional structure appeared at a smaller voltage (Fig.2). It is not clear yet what is the origin of the subgap structure. But the fact that it is seen in both the 123 and 124 compounds at similar voltages should give a clue to a successful interpretation.

Another characteristic feature which is observed in the tunneling experiments for the 123 compound is a conductance peak at zero bias. We also observed it in the 124 compound (Fig.3). The conductance peak is one of those which is often observed and has not been explained unambiguously yet. In general, it is observed when the overall tunneling conductance is large. An unlikely explanation is that it is due to the finite value of the supercurrent at grain boundaries within a sample. Because the present sample is a sintered one, the existence of such grain boundaries is quite possible. These grain boundaries can be considered to switch between the supercurrent and voltage states and may give rise to sharp and complicated structures in the conductance curve, as has been observed in many high T_c materials[11]. But the results shown in Fig.3 are quite different. The width of the conductance peak is rather large (~14mV) and nearly the same in all junctions. In addition, the conductance curve is very smooth and there is no sign of a switching effect. Furthermore, it is difficult to explain by the effect of grain boundaries within a sample why the conductance peak is observed only in junctions with large conductance.

A more likely explanation is that this is due to an Andreev-like reflection mechanism[12]. It is natural to consider that when the natural barrier is relatively weak, there are some spots whose area is small and the potential barrier is absent or very weak. At such spots, the Andreev reflection can take place and result in higher conductance below the gap energy[13]. This picture is consistent with the fact that the width of the conductance peak is constant and not sensitive to the magnetic field, although the height of the peak is very sensitive to the magnetic field. Furthermore, it should be noted that the width of the conductance peak (~14mV) is close to the distance between the pair of subgap structures observed for sample-1 in a magnetic field (Fig.2). In our preliminary experiment of point contact configuration, in which adjustment of the tunneling conductance is easier, we sometimes observed conductance peaks with a width of ~40mV together with peaks with a width of ~14mV at low temperatures. This is reasonable if the gap-like structures observed at ±~20mV and at ±~7mV really correspond to the gap values, because Andreev reflection should give higher conductance below the gap energy. But these gap-like structures cannot be interpreted easily as mentioned above, and more detailed study is needed before the relation between the gap-like structures and the conductance peaks is made clear.

In summary, we performed systematic tunneling measurements for $YBa_2Cu_4O_8$ for the first time. The results are similar to the case of the 123 compound. We carefully measured the effect of magnetic fields on the tunneling spectra. It was shown that the application of magnetic fields was useful in revealing the details of the tunneling spectrum even when the magnitude of the magnetic field was smaller than H_{c2}. The close relation of the experimental results to the gaplessness induced by a spatially varying pair potential was indicated. The conductance peak which was observed in relatively conductive junctions was suggested to be due to Andreev reflection.

REFERENCES

1. Geerk J, Xi XX, Linker G (1988) Z. Phys. B73: 329-336
2. Gurvitch M, Valles JM Jr., Cucolo AM, Dynes RC, Garno JP, Schneemeyer LF, Waszczak JV (1989) Phys. Rev. Lett. 63: 1008-1011
3. Fournel A, Sorbier JP (1990) Physica B 165&166: 1621-1622
4. Miyatake T, Gotoh S, Koshizuka N, Tanaka S (1989) Nature 341: 41-42
5. Miyatake T, Yamaguchi K, Takata T, Gotoh S, Koshizuka N, Tanaka S (1989) Physica C160: 541-544
6. see for example, Furuyama M, Iguchi I, Shirai K, Kusumori T, Ohtake H, Tomura S, Nasu M (1990) Physica B 165&166: 1603-1604
7. see for example, Ikuta H, Maeda A, Uchinokura K, Tanaka S (1988) Jpn. J. Appl. Phys. 27: L1038-L1040
8. Yoshikawa N, Murakami T, Sugahara M (1989) Proceedings of the 2nd International Symposium on Superconductivity (ISS'89): 1009-1012
9. Maki K (1969) Superconductivity. Marcel Dekker. New York, Vol.1 pp1035-1105
10. DeGennes PG (1969) Superconductivity of Metals and Alloys. Benjamin. New York, Chap.8
11. see for example, Ekino T, Akimitsu J (1989) Phys. Rev. B40: 6902-6911
12. Andreev AF (1964) Sov. Phys. JETP 19: 1228-1231
13. Klapwijk TM, Blonder GE, Tinkham M (1982) Physica 109&110B: 1657-1664

Resistive Transition in Magnetic Field for Single Crystals of Tl-Ba-Ca-Cu-O System

AKIHIKO ENDO and SHUNICHI NISHIKIDA

Research and Development Division, Sumitomo Metal Industries, Ltd.,
1-3, Nishinagasu-hondori, Amagasaki, 660 Japan

ABSTRACT

Single crystals of $Tl_1Ba_2Ca_2Cu_3O_y$ were grown with a melt method and the electrical resistivity was measured in several magnetic fields, which were applied parallel and perpendicular to the c-axis. The zero-resistance temperature (Tce) was 115K at a zero magnetic field. On the basis of the thermally activated flux creep model,the thermal activation energy Uo was estimated from the results. The values of Uo were larger than those of single crystals in Bi-system and the other Tl-system, such as $Tl_2Ba_2Ca_3Cu_4O_y$ and $Tl_1Ba_2Ca_1Cu_2O_y$. The dependence on the magnetic field was $Uo \propto H^{-1/2}$,differing from $Uo \propto H^{-1}$ proposed by Tinkham.

KEY WORDS: Tl-Ba-Ca-Cu-O system, resistivity, magnetic field, thermal activation energy

INTRODUCTION

Application of a magnetic field to a high temperature superconductor severely broadens the resistive transition. This phenomenon has been explained as thermally activated flux creep model in Bi-Sr-Ca-Cu-O system by Palstra et al[1]. In the case of Tl-Ba-Ca-Cu-O system, several studies have been reported about thermally activated processes for a polycrystalline of $Tl_2Ba_2Ca_2Cu_3O_y$ [2], for single crystals of $Tl_2Ba_2Ca_3Cu_4O_y$ [3] and $Tl_1Ba_2Ca_1Cu_2O_y$[4]. The activated energy Uo was proposed to depend on the temperature and the magnetic field as $Uo=A(1-T/Tc)^{2/3}/H$ by Tinkham [5]. In many cases of Y-Ba-Cu-O system[6,7], there is a good quantitative fit to Tinkham's theory, while in Bi-Sr-Ca-Cu-O and Tl-Ba-Ca-Cu-O system, the magnetic field dependence of Uo from the results doesn't seem to support the theory.

There is no report about the measurement of electrical resistivity on a single crystal of $T_1Ba_2Ca_2Cu_3O_y$ which we could synthesize recently, and the magnetic dependence of Uo in Tl-system has not been discussed in detail. In this study, we have characterized the single crystals in Tl-Ba-Ca-Cu-O system prepared with a melt method , and then measured the resistive transition in magnetic field . From these transition curves ,Uo was estimated on the basis of the flux creep model, and discussed with the results for the Bi-Sr-Ca-Cu-O system and the other phases of Tl-Ba-Ca-Cu-O system. Furthermore, we discussed the magnetic field dependence of Uo comparing with that of Tinkham model.

EXPERIMENTAL

Single crystals of $Tl_1Ba_2Ca_2Cu_3O_y$ were grown with a melt method similar to Ref.8. Starting materials were powders of Tl_2O_3,CaO,CuO and $BaCuO_2$ compounds prepared by reacting $BaCO_3$ and CuO powders. These were mixed with the nominal composition of $Tl_2Ba_2Ca_6Cu_6O_y$. Mixed powders were pressed into a pellet form. Several pellets were loaded into an alumina crucible. The total amount was about 60g at each charge. The firing profile was 1050°C for 1hr,slow cooling (5°C/hr) to 760°C,followed by furnace cooling. The entire growth run was performed under 1 atm. of pure oxygen. The typical sample size is about $1x1x0.1mm^3$ and its shape is of platelet. These crystals were annealed under 1 atm. of pure oxygen. The anneal was for 1hr at $400^{o}C$ with a 1hr cool down. The structure phase of the crystals were identified with X-ray diffraction pattern measurement(XRD) and energy-dispersive X-ray spectroscopy (EDX). The electrical resistivity was measured with a DC four-probe method. Electrical contacts were made with indium metal. Measurement current of 5mA was applied and its direction was in the ab-plane. Magnetic fields were applied parallel and perpendicular to the c-axis.

RESULTS AND DISCUSSION

Figure 1 shows the X-ray diffraction pattern of the sample. All peaks of this pattern could be indexed as 001 reflection. The lattice spacing was estimated to be 15.8A and almost consist with the measured value of c-axis of a sintered sample in $Tl_1Ba_2Ca_2Cu_3O_y$ [9]. The composition of grown crystal was Tl:Ba:Ca:Cu = 0.7:2.0:1.9:2.5 by EDX measurement. From these results, the crystal was guessed to be $Tl_1Ba_2Ca_2Cu_3O_y$ and the surface was ab-plane.

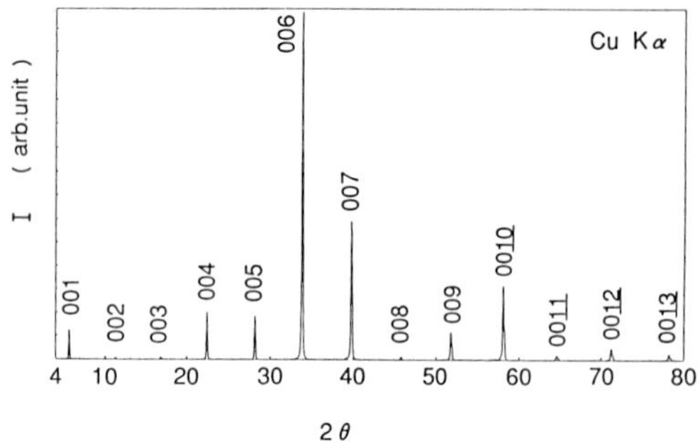

Fig.1 X-ray diffraction pattern of Tl-Ba-Ca-Cu-O single crystal.all peaks of this pattern could be indexed as 001 reflection.

In fig.2, we show the temperature dependence of resistivity for H⊥C-axis and H// C-axis with a magnetic field applied perpendicular to the transport current. The zero resistance temperature(Tce) was 115K at a zero magnetic field. Tce was defined as the temperature at which the voltage increased beyond 0.1μV. The temperature width of the superconducting transitions were more broadened in the field parallel to the c-axis than in perpendicular one. This behavior is similar to the other high Tc superconducting single crystals of Y-Ba-Cu-O [6], Bi-Sr-Ca-Cu-O [1] and Tl-Ba-Ca-Cu-O systems[3,4]. At 8T, the Tce decreased to 104K for H C-axis and to 69K for H c-axis. Figure 3 shows log R vs.T^{-1} plot. In the lower resistivity region below $5x10^{-3}\Omega$,log R changed almost linearly with T^{-1}. Assuming the temperature dependence of the resistivity resulted from the thermal activation, the activation energy Uo was calculated by following equation:

$$R=Ro \exp(-Uo/kT) \quad (1)$$

Figure 4 shows the activation energy Uo as a function of the magnetic field for both field directions. The value of Uo for H⊥c-axis is larger than for H//c-axis. We compared the obtained values of Uo with those of other samples in Tl and Bi-system, which were obtained in the same magnetic field with the same resistive method. For H// c-axis at 1T, Uo is 3000K. This value is about 1.5 times of those of Tl(2234) and Tl(1212) single crystals [3,4],and almost consistent with that of a Tl(2223) polycrystalline sample[2]. It is 6 times larger than that of Bi(2212) single crystal[1]. The difference of Uo among these systems seems to relate with the coherence length of these crystals. Actually, the coherence length was 5A for Tl(1212) [4], 1.6A for Bi(2212) [10] and 10A for this Tl(1223), which was estimated by the temperature dependence of Hc_2 defined by Tcmid. Figure 4 suggests a power-law dependence of $Uo(H)\propto H^{-\alpha}$, with exponent α . We find that α is from 0.12 to 0.6 ,which is similar to those of Tl(1212), Tl(2234) and Bi(2212) single crystals; these are from 0.17 to 0.4,from 0.13 to 0.68 and from 0.15 to 0.48 ,respectively. In the following part, we will discuss the magnetic field dependence of Uo for only H// c-axis because the Uo for H⊥c-axis is expected to be too sensitive to the small discrepancy of the direction between the applied field and ab-plane from the results of Iye et al [11].

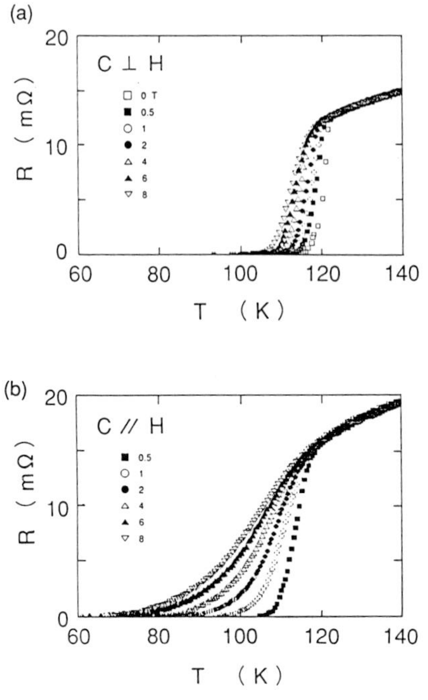

Fig.2 Temperature dependence of the electrical resistivity in various magnetic field for (a) H⊥c-axis and (b) H//c-axis.

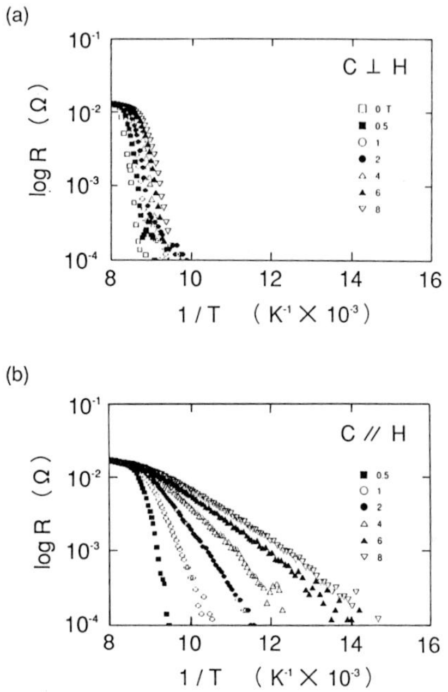

Fig.3 Log R vs. T^{-1} plot in various magnetic fields for (a)H⊥c-axis and (b)H//c-axis.

The activation energy Uo depends on the current density, the temperature and the magnetic field.[5] If the resistivity is measured at the high current density, the Uo is expected to decrease due to the increase of the Lorentz force on the magnetic flux. The current density ($5A/cm^2$) which we applied to measure the resistivity transition is assumed to be low enough not to influence on the Uo. The dependence on the temperature and the magnetic field is proposed by Tinkham as follows [5];

$$Uo = \beta Hc^2 \xi a_o^2$$
$$= A(1-t)^{2/3} H^{-1} \qquad (2)$$

where Hc is the thermodynamic critical field, ξ is the coherence length, a_o is the flux lattice spacing ,t is reduced temperature T/Tc, H is the magnetic field and β and A include all numerous factors. The equation (2) leads to the temperature width of transition (ΔTc) depending on the magnetic field as $\Delta Tc \propto H^{\gamma}$ (γ =2/3) at any level of R/Ro. In Fig.5, we show log(1-t) vs. H for H// c-axis , where 1-t is {Tce(0)-Tce(H)/Tce(0)}. It suggests a power-law dependence $(1-t) \propto H^{\gamma}$. We find γ =0.4 above 1 T. Combining the Uo dependence on 1-t based on Tinkham model and the power between (1-t) and H, we can estimate α' to be 0.6 in equation (3).

$$Uo = A'(1-t)^{3/2} H^{-\alpha'} \qquad (3)$$

The magnetic field dependence on Uo in eq.(3) is almost consistent with that of Uo in Fig.4. Although several data of resistivity in magnetic field support Tinkham model in Y-Ba-Cu-O system,that is α =1 [13], this study shows α =1/2. In Bi-Sr-Ca-Cu-O system, which has similar structure to Tl-Ba-Ca-Cu-O system, α was also reported to be from 1/3 to 1/2 [5]. From these results ,it might be caused by the difference of the structural anisotropy between Y-Ba-Cu-O system and Tl-Ba-Ca-Cu-O system. According to equation (2), the magnetic field dependence of Uo is determined by the power of a_0, that is the flux lattice spacing. Therefore this experimental result, $Uo \propto H^{-1/2}$, suggests that Uo for $H \propto$ c-axis is proportional to the flux lattice spacing above 1T.

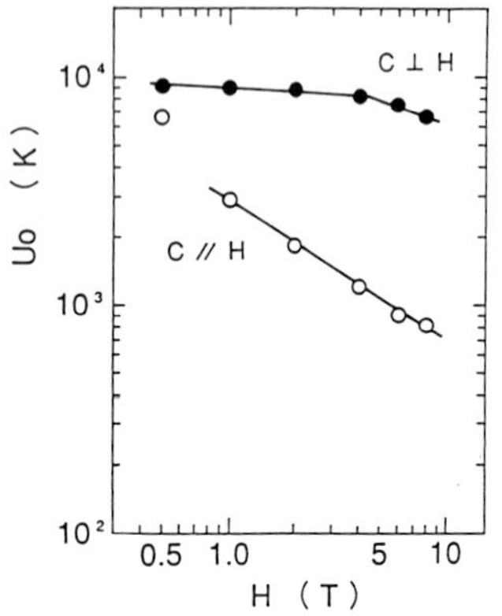

Fig.4 Magnetic field dependence of the activation energy Uo for H// c-axis (open circles) and for H⊥c-axis (filled circles).

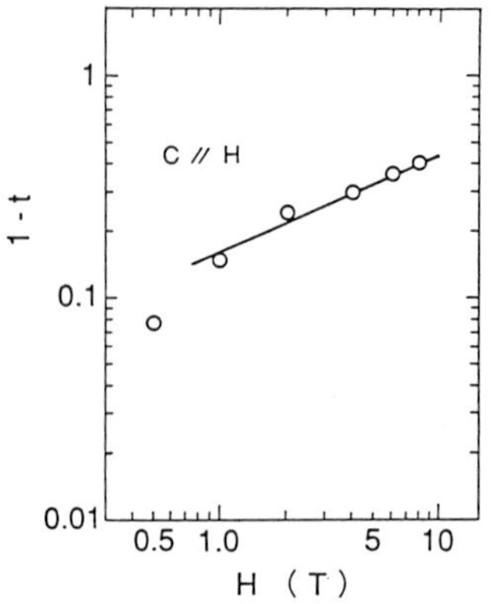

Fig.5 Magnetic field dependence of 1-t for H//c-axis, where 1-t is {Tce(0)-Tce(H)/Tce(0)}.

SUMMARY

We prepared the single crystal in Tl-Ba-Ca-Cu-O with a melt method. The material was characterized by XRD and EDX as $Tl_1Ba_2Ca_2Cu_3O_y$.The electrical resistivity was measured in the magnetic field up to 8T. They showed a significant anisotropy ,in accordance with the other reports of Tl-Ba-Ca-Cu-O single crystals. Thermally activated behavior was observed in low resistance part of the transition and the values of Uo were larger than those of single crystals of $Tl_1Ba_2Ca_1Cu_2O_y$,$Tl_2Ba_2Ca_3Cu_4O_y$ and $Bi_2Sr_2Ca_1Cu_2O_y$. The dependence of the activated energy on magnetic field was $Uo \propto H^{-1/2}$,differing from $Uo \propto H^{-1}$ proposed by in Tinkham.

REFERENCES

1. Palstla T.T.M, Batlogg B, Schneemeyer L.F and Waszczac J.V (1988) Phys.Rev.Lett. 64: 1662
2. Hettinger J.D, Swanson A.G, Brooks J.S, Hung Y.Z, Chen L.Q and Zhao Z.A (1989) Supercond.Sci.Technol. 1: 349
3. Togano K, Kumakura H, Mukaida H, Kawaguchi K and Nakao M (1989) Jpn.J.Appl.Phys. 28: 1907
4. Iwasaki H, Kobayashi N, Kikuchi M, Kajitani T,Syono Y and Muto Y (1989) Physica C 159: 301
5. Tinkham M (1988) Phys.Rev.Lett 61: 1658
6. Iye Y,Tamegai T,Takeya H and Takei H (1988) Jpn.J.Appl.Phys. Series 1: 46
7. Oh B, Chan K, Kent A.D, Naito M, Beasley M.R, Geballe T.H and Kapitulnik A (1988) Phys.Rev. B37: 7861
8. Kotani T, Kaneko T, Takei H and Tada K (1989) Jpn.J.Appl.Phys. 28: L1378
9. Nakajima S, Kikuchi M, Oku T, Hiraga K, Syouno Y and Muto Y (1989) Physca C 160: 476
10. Palstra T.T.M, Batlogg B, Schneemeyer L.F, VanDover R.B and Waszczak J.W (1988) Phys.Rev.B 38: 5102
11. Iye Y, Nakamura S and Tamegai T (1989) Physica C 159: 433
12. Zeldov E, Amer N.M, Koren G, Gupta A, McElfresh M.W and Gambino (1990) Appl.Phys.Lett. 56: 680
13. In a review by Kobayashi N (1990) Solid State Phys. 25: 755

Pressure Effect on the Superconducting Transition Temperature of Tl-Based High-T_c Superconductor (2212)

H. Yoshida[1], T. Kaneko[1], S. Abe[1], S. Miura[1], S. Nakajima[2], T. Atou[1], M. Kikuchi[1], and Y. Syono[1]

[1] Institute for Materials Research, Tohoku University, Sendai, 980 Japan
[2] Hachioji Research Center, CASIO Computer Co., Ltd., 2951-5 Ishikawa-cho, Hachioji, Tokyo, 192 Japan

ABSTRACT

The superconducting transition temperature T_c of Tl-based high-T_c superconductor(2212) under pressure was determined by measuring the electric resistance. It was found that there is a linear relationship between $J^{3/2}$ (J: electric current) and T_c and the precise T_c determination was carried out by investigating the current dependence of T_c. The value of dT_c/dP can be deduced from two ways; the values of dT_c/dP are obtained to be 0.07 and 0.1 K/kbar, respectively.

KEY WORDS: Tl SYSTEM SUPERCONDUCTOR, PRESSURE EFFECT, ELECTRIC RESISTANCE

INTRODUCTION

Since the existence of high-T_c superconductivity in the La-Ba-Cu-O oxide system was reported by Bednoz and Muller [1], very intensive investigations have been performed to find the higher superconducting transition temperature T_c and to clarify the mechanism of high-T_c. Many reports have been already published on the pressure dependence of T_c and lattice parameters for the La-Ba-Cu-O, Sr-Cu-O and Y-Ba-Cu-O oxide systems. Chu et al. [2] were the first to find that pressure remarkably enhances T_c of the La-Ba-Cu-O oxides. This T_c-increase under pressure provided a rough orientation for the higher T_c oxide development. Most of experimental results on the pressure dependence of T_c until now show that pressure enhances T_c of high-T_c oxides. This positive pressure derivative of T_c is considered at first to be an important characteristic of the high-T_c superconducting oxides in understanding the mechanism of high-T_c superconductivity, because most of the pure metal superconductors with low-T_c has a negative and small pressure derivative of T_c. On the other hand, a few experimental results, which the value of dT_c/dP (positive) decreases above 10 ~ 20 kbar [3,4] or dT_c/dP is negative [5], have been reported thereafter. It is not clear at present whether such contradictory results about the sign of dT_c/dP are originated from the intrinsic nature of each concerned sample or from the difference of experimental procedure for determination of T_c under pressure. Therefore, the establishment of the method for the exact determination of T_c is one of the important problems for high-T_c superconducting research. We have already proposed a method useful for the exact determination of T_c by taking account of the electric current J dependence of T_c which is varied with the current $J^{3/2}$, and applied it to the study of the pressure dependence of T_c of Bi system superconducting oxide [6]. Recently, Schirber et al. [7], Han Shun-hui et al. [8] and Morosin et al. [9] have studied the pressure effect on T_c of Tl(2212) and reported dT_c/dP to be 0.20 ~ 0.24 K/kbar. In this paper, we examine the pressure effect on T_c of Tl(2212), that is, $Tl_2Ba_2Ca_1Cu_2O_8$ and determine the pressure derivative of T_c using the method mentioned above.

EXPERIMENTAL

The polycrystalline sample of Tl system superconductor $Tl_2Ba_2Ca_1Cu_2O_8$ was prepared by a conventional powder method. Tl_2O, $BaCuO_2$ and CaO were mixed in nominal rate, pressed to be a pellet and covered with a sheet of gold foil. The pellet was placed in a furnace in which oxygen gas was flowed in the rate of 120 ml/min at 1 atmosphere and heated up to 910 °C. The pellet was held for 7 minutes at this temperature and cooled in the furnace.

The powder x-ray diffraction pattern for the obtained Tl(2212) sample was assighned with a tetragonal structure and the lattce parameters a and c were determined as 3.854 and 29.322 Å, respectively. Electric resistance measurements under pressure were carried out with a conventional 4 probe method utilizing a piston-cylinder type high pressure apparatus. The sample was ~ 1x1x5 mm^3 in size and was also used for the ac susceptibility measurement. The ceramic solder amalgamated with gold is used to have a good ohmic lead bonding. At the sample cell in our piston-cylinder type pressure apparatus, a small furnace in size of 10 mm x 10 ϕ made by BN was installed and the temperature was measured with the Alumel-Chromel thermojunction placed closely to the sample. Fluorinert (SUMITOMO 3M Co. JPN) was utilized as pressure transmitting medium.

RESULTS AND DISCUSSION

The temperature dependence of the electric resistance R for the various currents at 0 and 10.9 kbar is shown in Fig. 1 (A) and (B), respectively. In general, it is hard to define the intrinsic superconducting transition temperature T_c (energy gap → 0) from only these R-T curves. Instead of T_c, therefore, the onset(T_c^{on}) and offset temperature(T_c^{off}) are defined as shown in (A) of the figure. The transition temperatures become lower with increasing current J under a constant pressure and the transiting temperature region becomes wider. The Ohm's law is realized above ~ 125 K. In the temperature range between T_c and 125 K, however, the law is broken even

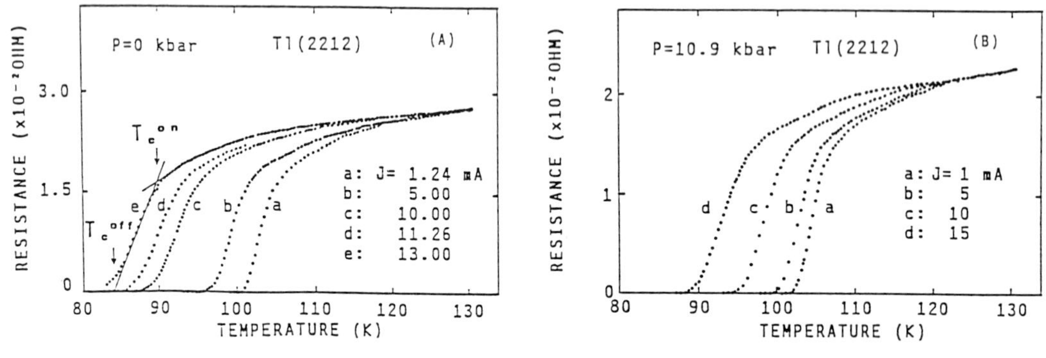

Fig. 1. Electric resistance vs. temperature curves of $Tl_2Ba_2Ca_1Cu_2O_8$ at various currents under pressures. (A) P = 0 kbar; (B) P = 10.9 kbar.

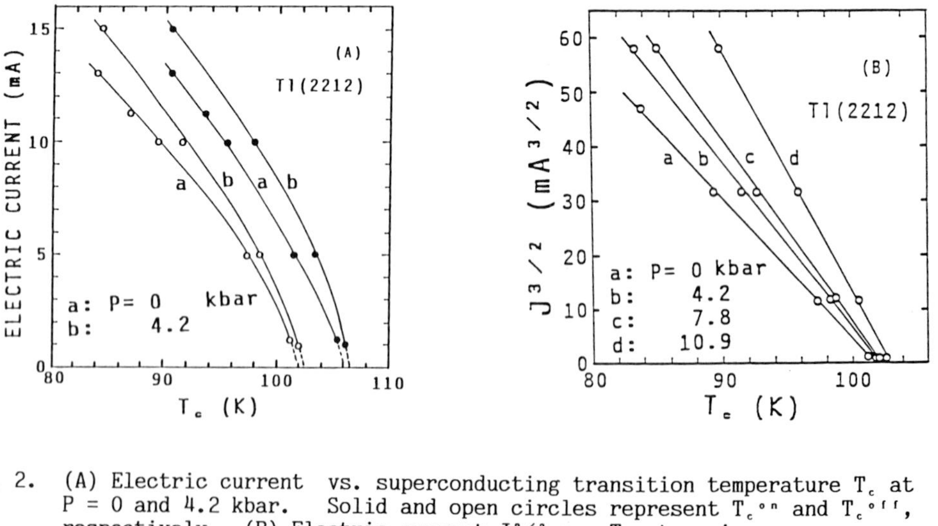

Fig. 2. (A) Electric current vs. superconducting transition temperature T_c at P = 0 and 4.2 kbar. Solid and open circles represent T_c^{on} and T_c^{off}, respectively. (B) Electric current $J^{2/3}$ vs. T_c at variuous pressures.

in nonsuperconducting state. It seems that the appearance of this region is due to the contamination of a small amount of sample having higher T_c or remaining the short range order till 125 K. It can be seen that the transition temperatures shift to higher temperatures by compression, by comparing the curves for J = 5 and 10 mA in (A) with those in (B).

As seen Fig. 1, the transition temperature depends on the value of current strongly. The current dependence of T_c^{on} and T_c^{off} at 0 and 4.2 kbar is shown in Fig. 2 (A). The shift of T_c by pressure at a constant J becomes large with increasing J. The relationship between T_c and J is nonlinear as seen in the figure. The linear relationship was obtained by plotting the transition temperatures against $J^{3/2}$ as well as for Bi system high-T_c oxide mentioned above. The relationship replotted for 0, 4.2, 7.8 and 10.9 kbar is shown in (B). We could determine the zero current transition temperatures by exrapolating the straight line to $J^{3/2} = 0$. This data arrangement is applicable to the sintered sample of Y and Bi system high-T_c superconductors [2]. The four straight lines in (B) appear to cross at ~ 105 K simultaneously. This temperature corresponds to a threshold on the ohmic breaking point.

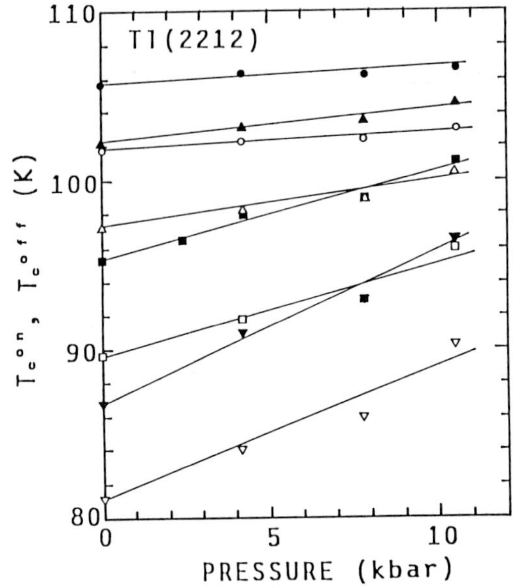

Fig. 3. Superconducting transition temeperature T_c vs. pressure at various currents. Circles, triangles, squarea and inverse triangles represent the values of T_c at J = 0, 5, 10 and 15 mA, respectively. Solid and open marks show T_c^{on} and T_c^{off}, repectively.

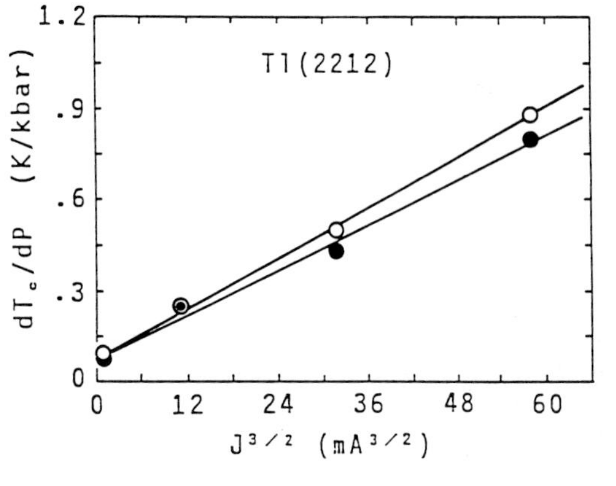

Fig. 4. Pressure derivative dT_c/dP vs. electric current $J^{3/2}$. Open and closed circles are the values of dT_c^{on}/dP, dT_c^{off}, respectively.

The offset temperatures determined as described above are 109 K for 0 kbar, 110.5 K for 4.7 kbar, 112 K for 7.8 kbar and 115 K for 10.8 kbar. The pressure effect on T_c^{on} and T_c^{off} for J = 0, 5, 10 and 15 mA is shown in Fig. 3. The T_c^{on} and T_c^{off} increase by compression and then the pressure derivative of the transition temperature has a positive sign and increases with increasing J. By extrapolating the experimental lines to higher pressures, the lines have a cross point at P_c ~ 20 kbar. However, the extrapolated curves for $J \neq 0$ cannot cross over beyond the curve for J = 0 at pressure above 20 kbar. Therefore, it is considered that the variation of the transition temperatures over 20 kbar traces on the line for J = 0. The values of dT_c^{on}/dP and dT_c^{off}/dP for J = 0 are obtained to be 0.1 ± 0.03 and 0.08 ± 0.03 K/kbar, respectively. It is interesting whether this enhancement of pressure effect of T_c on J is due to the intrinsic nature or the transport one of the boundary between crystalline grains.

The pressure effect on T_c can also be deduced from the current dependence of dT_c^{on}/dP and dT_c^{off}/dP. These derivatives is shown in Fig. 4 as a function of $J^{3/2}$. It is found that there is a linear relationship between these derivatives of the transition temperature and $J^{3/2}$. The values of the pressure derivative extrapolated to $J = 0$ are determined to be 0.07 ± 0.03 K/kbar for the onset and offset temperature.

It was payed attention at first that the La system superconductor oxides have the large pressure dependence of T_c. The pressure effect, however, for Y system and Bi system superconductors are so small that the experimental results with a positive sign of the pressure derivative of T_c and those with a negative one are reported. Therefore, it is important to determine T_c precisely. It was found that the precise determination of T_c can be done by measuring the current dependence of T_c in the resistance measurement even for the polycrystalline samples.

On the basis of the relation between T_c, J and P obtained in the present experiment, it was found that $T_c(J,P)$ can be well expressed as a function of J and P as follows:

$$T_c(J,P) = T_c(0,0)(1 + \alpha J^{3/2} + \beta P + \gamma J^{3/2}P),$$

where $T_c(0,0)$ is the transition temperature at $J = 0$ and $P = 0$, α, β and γ the coefficients. In the case of $T_c = T_c^{off}$, $T_c(0,0)$ and the coefficients are obtained as follows:
$T_c(0,0) = 101.6$ K, $\alpha = -3.46 \times 10^{-2}$ $1/(mA)^{3/2}$, $\beta = 9.84 \times 10^{-4}$ 1/kbar and $\gamma = 1.36 \times 10^{-4}$ $1/kbar(mA)^{3/2}$. The values of the pressure derivative of $T_c(0,0)$ and P_c are derived to be 0.1 K/kbar and 25 kbar from this formula, respectively. We have noted that the T_c vs. P curves for $J \neq 0$ should bend downward around $P_c \sim 20$ kbar on the basis of the results in Fig. 3. The results by Kurisu et al. [3] and Hor et al. [4] show that T_c increases with pressure and the rate of pressure change of T_c become small under pressure above $10 \sim 20$ kbar described earier. Such a tendency of pressure dependence of T_c is considered to appear, when the dT_c/dP was determined from the measurements of electric resistance in $J \neq 0$. $dT_c/dP \fallingdotseq 0.2$ K/kbar in ref. [7,8,9] is larger than $dT_c/dP|_{J=0}$ in this experiment and corresponds to that for $J = 3$ mA, which is obtained from the above formula.

REFERENCES

1. Bednorz JG, Muller KA (1986) Z. Phys. B-Condensed Matter 64: 189
2. Chu CW, Hor PH, Meng RL, Gao L, Huang ZJ, Wang YQ (1987) Phys. Rev. Lett. 58: 405
3. Kurisu M, Kumagai K, Maeno Y, Fujita T (1988) Physica C 152: 339
4. Hor PH, Gao L, Meng RL, Huang ZJ, Wang YQ, Forster K, Vassilious J, Chu CW, Wu MK, Ashburn JR, Torng CJ (1987) Phys. Rev. Lett. 58: 911
5. Driessen A, Griessen R, Koeman N, Salomons E, Brouwer R, de Groot DG, Heeck K, Hemmes H, Rector J (1987) Phys. Rev. B36: 5602
6. Yoshida H, Kaneko T, Fujimori H, Morita H, Noto K, Kamigaki K (1990) High Pressure Research 3: 114
7. Schirber JE, Morosin B, Ginley DS (1989) Physica C 157: 237
8. Han Shun-hui, Zhang Yu-lin, Han Han, Liu Yong (1988) Physica C 156: 1139
9. Morosin B, Ginley DS, Venturini EL, Hlava PF, Baughman RJ, Kwak JF, Schirber JE (1988) Physica C 152: 223

Anisotropic Resistive State and Dissipation in Magnetic Fields in Layered Superconductor $Bi_2Sr_2CaCu_2O_8$

KAZUO KADOWAKI[1], J.N. LI[2] and J.J.M. FRANSE[2]

[1] National Research Institute for Metals, Tsukuba Laboratories, 2-1, Sengen 1-chome, Tsukuba, Ibaraki, 305 Japan

[2] Natuurkundig Laboratorium der Universiteit van Amsterdam, Valckenierstraat 65, 1018-XE Amsterdam, The Netherlands

ABSTRACT

An anomalous field-induced resistive state in the superconducting state of high-quality single-crystalline $Bi_2Sr_2CaCu_2O_8$ near T_c has been studied in detail in various magnetic field and current configurations with respect to the crystallographic axes. The extremely sharp angular dependence of the resistivity observed in high fields and in the vicinity of the field orientation parallel to the ab-plane indicates unusually large anisotropic flux dynamics and can not be explained by the traditional flux dynamics such as flux-flow, which is essentially based on the Lorentz interaction. The fact that the anisotropy of the field-induced resistivity continuously grows through T_c in a wide temperature region suggests strong evidence of the gradual formation of the superconducting pairs well above T_c. The superconducting fluctuation model is considered as a possible origin due to the highly anisotropic superconducting order parameters in $Bi_2Sr_2CaCu_2O_8$.

key words: flux-flow, flux-creep, $Bi_2Sr_2CaCu_2O_8$ single crystal, flux dynamics, anisotropic resistive state

INTRODUCTION

An anomalous broadening behavior of the resistive transition in high T_c superconductors in magnetic fields has attracted much attention to experimentalists as well as theorists, because it is one of the peculiarities in these new class of superconductors. It has been recognized that the essence of the physics is partially associated with the weakly pinned flux-line-lattice dynamics in the highly anisotropic superconducting state in these compounds. The anisotropy of the coherence length in $Bi_2Sr_2CaCu_2O_8$ is indeed extremely large. In addition, ξ_c, the coherence length for the c-direction, is much shorter than the averaged CuO_2 layer distance. In such a system, it is expected that the superconducting order parameter is mainly confined in the superconducting CuO_2 layers and the fluxoid may not be well-defined between the CuO_2 layers. As a consequence, the flux dynamics will be quite different from the situations observed in conventional superconductors. After the early work of Palstra et al.[1] on the single crystals, a number of studies on this subject have been followed. The results in the most of the cases have so far been analyzed in the frame work of the traditional flux-creep and flux-flow based on the Lorentz force interaction.

The characteristic features found in these experiments are firstly that the field-induced resistivity at the onset region is of exponential type, suggesting that thermally activated processes are involved in the dissipation dynamics[2]. Secondly, strong relaxation phenomena in various properties are observed and become prominent as the temperature is increased. The decay of the superconducting diamagnetic magnetization is a typical phenomenon. The irreversibility rapidly disappears in a peculiar fashion in the B-T phase diagram[3]. This irreversibility line which distinguishes the reversible and irreversible regions has been studied by various techniques[3,4]. It seems that the relaxation phenomena point to the weak pinning effect at first glance but it must be emphasized that the phenomena can not be explained by the weakness of the pinning force only.

In order to elucidate the peculiar dissipation mechanism especially in the reversible state, we have chosen the $Bi_2Sr_2CaCu_2O_8$ system, partly because this system has a highly two-dimensional superconducting state. The preliminary results in single crystals[1] as well as in thin films[5] show much wider irreversible region in the B-T diagram than in $YBa_2Cu_3O_7$. This provide us a great advantage for the purpose of this study.

Here, we present experimental results of the field-induced resistivity on the issue of the flux dynamics and the dissipation mechanism. The anisotropy of the resistivity has precisely been measured with respect to the measuring current, the crystallographic axes and the magnetic field directions in a rotatable magnetic field up to 5 T on the well characterized single-crystalline $Bi_2Sr_2CaCu_2O_8$. An extremely sharp angular dependence of the field-induced resistivity is observed in the field direction near the ab-plane, irrespective of the relative configurations between the current and the magnetic field directions. This feature is seen in the whole temperature and field regions above the irreversibility line. The similar angular dependence is obtained even well above T_c. The fact that such an anisotropic behavior continues over the wide temperature range and even extends above T_c strongly suggests that a common mechanism should be responsible for the dissipation. Furthermore, the fact clearly indicates the existence of superconducting pairs even well above T_c. This casts a serious question concerning the existence of a clear-cut phase transition at T_c in magnetic fields. A model based on the superconducting fluctuations is applied in order to explain these experimental features consistently.

EXPERIMENTAL

Single-crystalline samples of $Bi_2Sr_2CaCu_2O_8$ have been grown by the travelling solvent floating zone method using the infrared-convergent type furnace[6]. The characterization of the samples have been done in detail by mean of various techniques. Here, we describe those briefly.

The elemental analysis has been performed by EPMA. The stoichiometry was determined to be 2.18 : 1.95 : 1.05 : 2.00 in a ratio of Bi : Sr : Ca : Cu in average. The oxygen content was not analyzed yet. Neither the second phase nor impurities has been detected.

The structure has been investigated by the x-ray Laue method and the neutron ω-scan technique around (2,0,0), (0,2,0) and (0,0,2) Bragg spots. The line shape was sharp with the half-width of 0.2 ° at most. The fundamental lattice constants were determined to be a = 5.402 Å, b = 5.420 Å and c = 30.840 Å with an orthorhombic symmetry.

The microscopic inclusions were studied by HREM (high resolution electron microscopy). It turns out that the number of inclusions such as the 2201 and 2223 phases are reduced by a factor of 10-100 in the present single crystals compared with those made by the flux method. The typical number of inclusions in a 10 μm sample thickness is about several layers mainly of the 2223 phase.

Superconductivity of the samples have been checked by low frequency ac-susceptibility measurements with varying the amplitude of the oscillating ac-field from 3 mOe to 10 Oe. With sufficiently low level of the amplitude, typically less than 100 mOe, the superconducting transition width becomes very sharp, approximately 50 - 100 mK. This feature does not change even in a magnetic field, although the onset of the transition is greatly reduced.

The electrical resistivity has been measured by the conventional four probe method with a low frequency ac-current density of a few A/cm^2. The contact resistance is about 0.1 Ω or less. The resistivity measurement is carried out in a rotatable Helmholtz magnet up to 5 T. The angular resolution is 0.2°. The temperature is controlled within 50 mK over 300 degree rotation.

RESULTS AND DISCUSSION

Angular dependence of the field-induced resistivity along the b-axis is shown in Figs 1 and 2 at various temperatures near T_c and in a 3 T magnetic field which is rotated in the bc-plane. The three curves from top were obtained above T_c, whereas the two curves were measured below T_c, at least in the field direction near the ab-plane. When the field is near the c-direction, this may not be so, especially at 83.1 K which is just below the zero field T_c = 84.56 K. Although the angular dependence of T_c is not clearly known yet, the fact that the angular dependence at 83.1 K is smooth and monotonous from b-direction to c-direction, suggests that there is no singularity at $B_{c2}(\theta)$ in the resistivity. At the field near the b-direction, the angular dependence is sharp as is seen in Fig. 2. The sharp feature is gradually diminishing as temperature is increased. Above about 140 K, the anisotropy of the field-induced resistivity becomes below 1 %. It is noted that the angular dependence is similar above and below T_c, and becomes sharper as temperature comes closer to T_c.

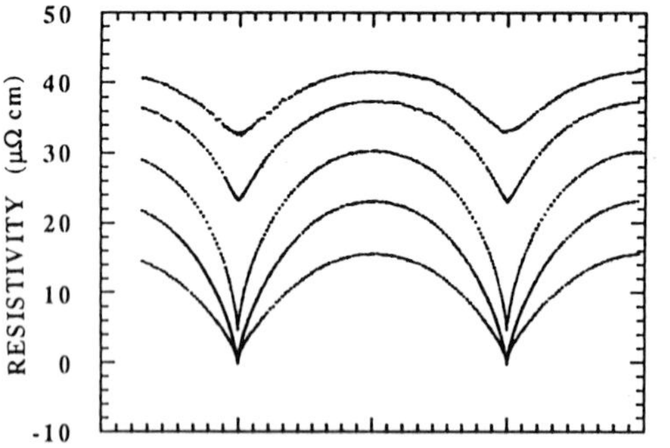

Fig. 1. The angular dependence of the resistivity curves in a field of 3 T rotated in the bc-plane at 87.8 K, 86.9 K, 85.1 K, 83.1 K and 79.5 K from top to bottom.

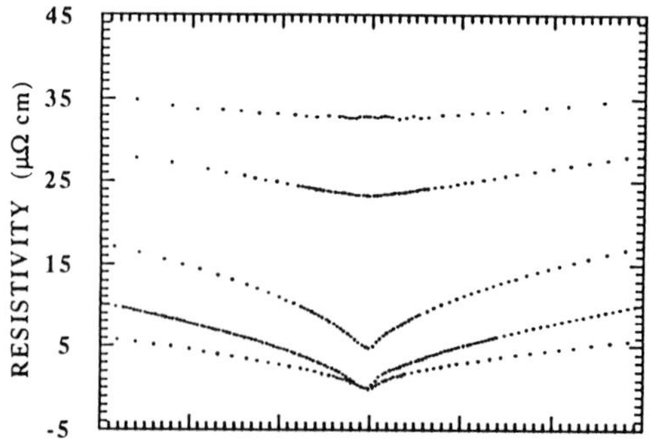

Fig. 2. The angular dependence of the resistivity curves in a field of 3 T rotated in the bc-plane at 87.8 K, 86.9 K, 85.1 K, 83.1 K and 79.5 K from top to bottom. The scale of the horizontal axis is expanded at the field direction near the b-axis.

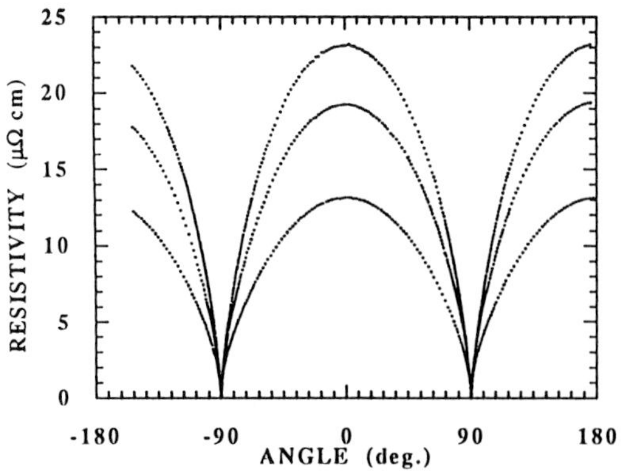

Fig. 3. The angular dependence of the resistivity curves at 83.2 K in fields of 1 T, 2 T and 3 T rotated in the bc-plane from bottom to top.

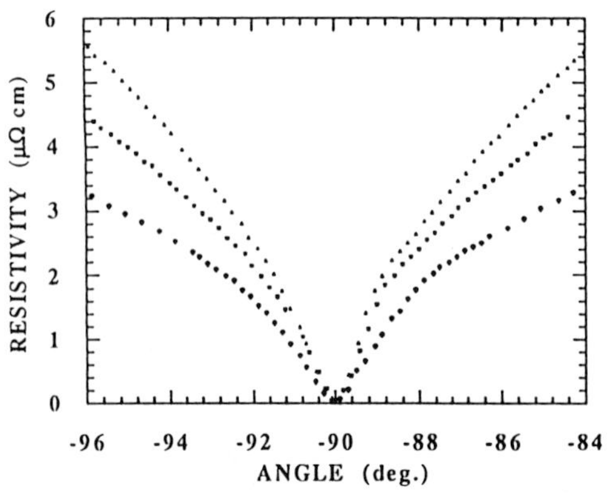

Fig. 4 The angular dependence of the resistivity at 83.2 K in fields of 1 T, 2 T and 3 T rotated in the bc-plane from bottom to top. The scale of the horizontal axis is expanded at the field direction near the b-axis.

Another interesting feature is found very near the field direction to the b-axis as seen in Fig. 2. Within 1° from the b-axis the angular profile has a kink and the resistivity drops sharper as the magnetic field direction approaches the b-axis. As the kink is also found in the angular dependence in the ac-plane, this kink is not a specific feature near the b-axis but it is a common feature near the ab-plane. This seems to appear only in the limited temperature region.

The magnetic field strength dependence of the field-induced resistivity is shown in Figs 3 and 4 at 83.2 K. The angular dependence becomes sharper as the field is increased. The kink seen in Fig. 2 is also observed in Fig. 4. The reason is not clear at present but may be related to the sudden penetration of the flux through the layers. In other words, within the angle 1° or so from the ab-plane, the situation that the magnetic flux is well confined between the CuO_2 layers may be realized. It is interesting to note that the similar anomaly is also found in the torque curve measured near the ab-plane[7].

The normalized angular dependence shown in Figs 1 and 2 is given in Figs 5 and 6 in addition to the |sine| curve. This angular dependence is expected by the two dimensional superconducting model proposed by Kes et al.[8]. As seen in Figs 5 and 6, the experimental results does not agree with the model in the whole angle. The disagreement becomes larger as the temperature comes closer to T_c and in higher magnetic fields.

The normalized angular dependence of the field-induced resistivity at 83.1 K in 3 T rotated in the bc-plane is compared with that rotated in the ac-plane as shown in Fig. 7 and 8. It is remarkable that the angular dependence is identical for both cases within the accuracy of the measurements. The angular scale near the ab-plane is extended in Fig. 8. Since the Lorentz forces acting on the flux are different, a different angular dependence is expected for the two cases. As a consequence, this discrepancy casts serious doubt in the flux-flow dynamics as an origin of the dissipation. This result also supports the previous conclusion that the field-induced resistivity depends on the relative configurations between the magnetic field and the crystallographic directions only. Therefore, it is of necessity to seek for the new dissipation mechanism.

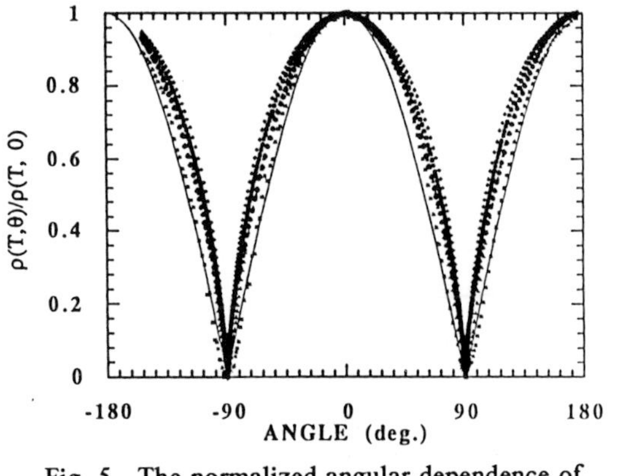

Fig. 5. The normalized angular dependence of the resistivity curves at 85.1 K, 83.1 K, 79.5 K, 86.9 K and 87.8 K from top to bottom. The solid line represents the |sine| curve for the comparison.

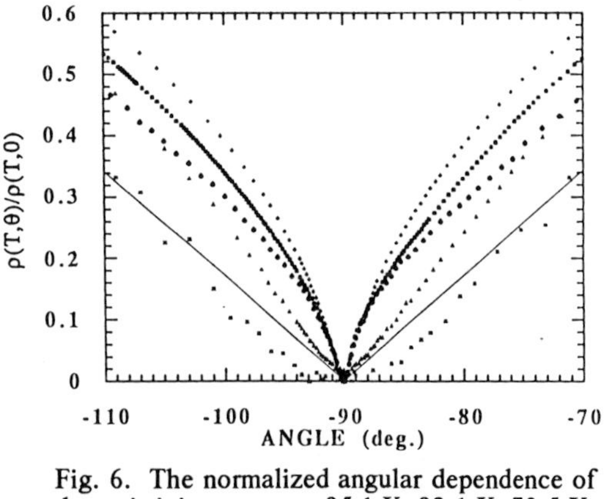

Fig. 6. The normalized angular dependence of the resistivity curves at 85.1 K, 83.1 K, 79.5 K, 86.9 K and 87.8 K from top to bottom. The solid line represents the |sine| curve for the comparison.

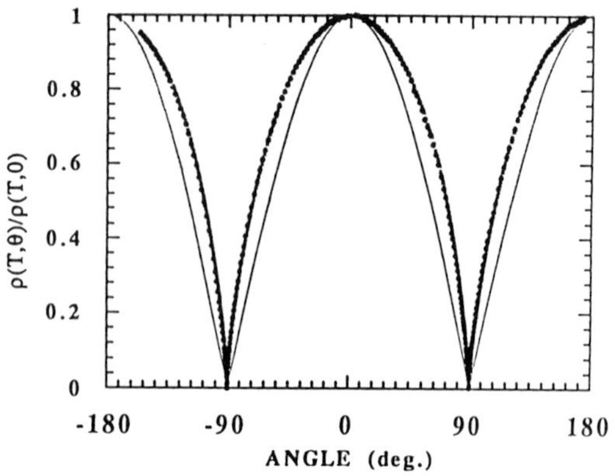

Fig. 7. The normalized angular dependence of the resistivity curves in a field of 3 T rotated in the ac-plane (•) and in the bc-plane (*).

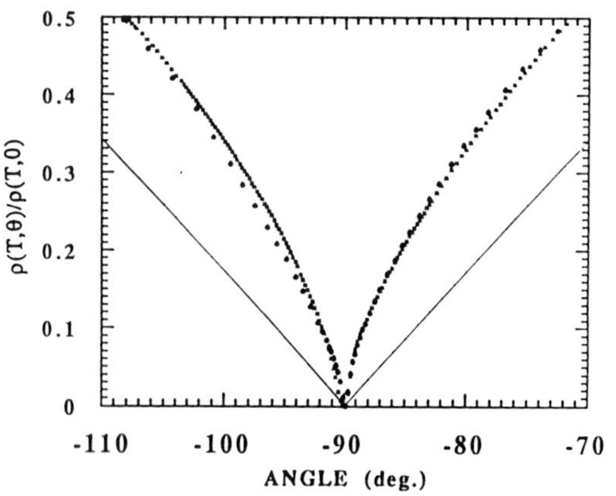

Fig. 8. The normalized angular dependence of the resistivity curves in a field of 3 T rotated in the ac-plane (•) and in the bc-plane (*). The scale of the horizontal axis is extended at the field direction near the ab-plane.

The nature of the superconducting state in magnetic fields may be quite different from the conventional one, because of the highly anisotropic electronic state in $Bi_2Sr_2CaCu_2O_8$. The main point here is related to the short coherence lengths and their large anisotropy. Assuming the values of $B_{c2}{}^{c}(0) \approx 60$ - 80 T[3] and $B_{c2}{}^{ab}(0) \geq 5.8 \times 10^{3}$ T based on the experimental result of $|dB_{c2}{}^{ab}/dT| \geq 100$ T/K obtained here, the coherence lengths for the ab- and c-directions can be estimated to be $\xi_{ab} \approx 20.3$ Å - 23.4 Å and $\xi_c \approx 0.3$ Å, respectively. As a consequence, a value for the anisotropic ratio $\gamma = \xi_{ab}(0)/\xi_c(0) \geq 72$ - 98 is obtained. This remarkable large anisotropy is in agreement with the result of the recent torque measurement[7, 9]. It is interesting to note that the anisotropic mass ratio of 10^5 in the normal state resistivity is also consistent with the value from the superconducting state. It is expected in such a case that the fluctuation effects become dominant, especially when the magnetic field is applied to the c-axis. Such a case has been treated by including the higher order term in the Ginzburg-Landau free energy[10]. The anisotropic features of the field-induced resistivity can reasonably be understood in this case as has been done in $YBa_2Cu_3O_7$ previously[11]. The scaling behavior of the resistivity broadening as is expected by the theory,

$$\Delta T = T/T_c - 1 \approx B^{2/3} \tag{1}$$

is also followed. However, as the temperature and the magnetic field approach the irreversibility line, it seems that the resistivity broadening is suppressed rather sharply, giving rise to significant deviation from eq. (1). This situation was found to be even stronger in $YBa_2Cu_3O_7$[2], than that in the present case. This indicates that the temperature region where the pinning effect becomes important can not be treated well by the present theory. The pinning effect on the superconducting fluctuations is to be solved in the future study.

In summary, the experimental results can better be explained by the fluctuation model. The crucial parameter here is the coherence length along the c-axis, which is much shorter than the distance between CuO_2 layers in which the superconducting order parameter is mainly confined. However, it is not clear at the present stage how to reconcile the fluctuation model to the flux-creep and flux-flow model near the irreversibility line. It seems that it is difficult to include the pinning problem into the theory, which becomes more important at low temperatures.

REFERENCES

1. T. T. M. Palstra, B. Batlogg, R. B. van Dover, L. F. Schneemeyer and J. V. Waszczak, Phys. Rev. 41 (1990) 6621.
2. K. Kadowaki, J. N. Li and J. J. M. Franse, Physica C 170 (1990) 298.
3. K. Kadowaki, J. N. Li and J. J. M. Franse, Proceedings of the satellite conference of LT19, Cambridge, 1990.
4. J. N. Li, K. Kadowaki, M. J. V. Menken, A. A. Menovsky and J. J. M. Franse, Physica B 165&166 (1990) 1429.
5. Y. Iye, S. Nakamura, T. Tamagi, Terashima, K. Yamamoto and Y. Kubo, Physica C166 (1990) 62.
6. K. Kadowaki and M. J. V. Menken, to be published in J. Cryst.
7. K. Okuda, private communication.
8. P. H. Kes, J. Aarts, V. M. Vinokur and C. J. van der Beek, Phys. Rev. Lett. 64 (1990) 1063.
9. D. E. Farrell, S. Bonham, J. Foster, Y. C. Chang, P. Z. Jiang, K. G. Vandervoort, D. J. Lam and V. G. Kogan, Phys. Rev. Lett. 63 (1989) 782.
10. R. Ikeda, T. Ohmi and T. Tsuneto, J. Phys. Soc. Jpn. 58 (1989) 1377.
11. R. Ikeda, T. Ohmi and T. Tsuneto, Physica B165&166 (1990) 1359.

2.4 Magnetic Resonance and Mössbauer Studies

Phase Stability, Superconductivity and ^{151}Eu Mössbauer Effect in T*-Phase $(La_{1-x}Eu_x)_{2-y}Sr_yCuO_4$

MASAFUMI TANIWAKI[1], HIDEAKI ITOH[2], and MIKIO KUSUHASHI[2]

[1] Faculty of Engineering, Hokkaido University, Sapporo, 060 Japan
[2] Muroran Research Laboratory, The Japan Steel Works Ltd., Muroran, 051 Japan

ABSTRACT

The phase stability and superconductivity in T*-phase $(La_{1-x}Eu_x)_{2-y}Sr_yCuO_4$ were studied and the microscopic state of Eu in T*-phase was examined by Mössbauer effect. Single and nearly single T*-phase samples were obtained at x = 0.5. T-phase and T'-phase appeared at a low Eu concentration and at a high Eu concentration, respectively. The samples annealed in 400 atm oxygen showed superconductivity(Tc = 17 - 29 K), while those annealed in 1 atm oxygen did not show it. The isomer shift of Mössbauer spectrum in single or nearly single T*-phase was 0.83-0.84 mm/s (showing Eu^{+3}), which is larger than that in T'-phase $(LaEu)_{2-y}Ce_yCuO_4$. The Debye temperature estimated from the line area was 180-200 K, which is larger than that in T'-phase. These reflect that Eu-O distance in T*-phase is small compared with that in T'-phase.

KEY WORD: T*-phase, LaEuSrCuO, ^{151}Eu Mössbauer effect, phase stability, isomer shift, Debye temperature

INTRODUCTION

In high temperature superconductors, the lanthanide atoms occupy the sites coordinated with eight oxygen atoms or the sites with nine oxygen atoms. Europium is the constituent element of '123', T'-phase and T*-phase superconductors. In all of these materials, Eu ions occupy the sites coordinated with eight oxygen atoms, i.e., T'-type sites [1-3]. These high Tc superconductors are different in Tc and carrier, which may be reflected on the electronic and lattice vibration states of Eu. Therefore, it is worthwhile to study ^{151}Eu Mössbauer effect in these high Tc superconductors and examine the correlation between the electronic and lattice vibration states and Tc or the carrier. We have studied ^{151}Eu Mössbauer effect in '123' and T'-phase superconductors, and the correlation between these states and Tc [4-8]. In the present work, we investigate the phase stability and superconductivity of T*-phase $(La_{1-x}Eu_x)_{2-y}Sr_yCuO_4$ by X-ray diffraction, resistivity measurement and Meissner effect, and the electronic and lattice vibration states of Eu by Mössbauer effect.

EXPERIMENTAL

The samples were prepared by solid-state reaction of mixtures of La_2O_3(99.99%), Eu_2O_3(99.99%), $SrCO_3$(99.99%) and CuO(99.9%) with nominal compositions of $(La_{1-x}Eu_x)_{2-y}Sr_yCuO_4$(x=0.33, 0.5 and 0.67, y=0.1 and 0.2). The mixed powders were calcined in flowing oxygen at 1273 K for 5 hours, ground, pressed into pellets and sintered in flowing oxygen at 1323K for 10 hours. Then some of the samples were cooled to room temperature at a cooling rate of 100 K/h. The other samples were quenched to room temperature from 1323 K, re-annealed at 1173 K in 400 atm oxygen and cooled at a cooling rate of 100 K/h. The crystal structures and the lattice constants in the samples were determined by X-ray powder diffraction. The resistivity and Meissner effect were measured by four-probe dc method and ac-inductance method, respectively. The Mössbauer spectrum was measured at room temperature using a 21.55 keV $^{151}SmF_3$ γ ray source. The velocity axis was standardized by EuF_3 and natural Fe at room temperature. In order to estimate the lattice vibration from the line area, Mössbauer samples with a constant σn(=3) were used, where σn is the product of the resonance cross section and the number of Mössbauer nuclei per unit area.

RESULTS AND DISCUSSION

X-ray diffraction pattern showed T- and T*-phases in the samples with a low Eu concentration(x=0.33), and T*- and T'-phases in the samples with a high Eu concentration(x=0.67). The T*-phase stability is sensitive to lanthanide and Sr concentration, and single T*-phase was obtained only in the sample of (x, y)=(0.5, 0.2). Figure 1 shows the observed phases and their abundances in the samples annealed in 1 atm oxygen. The annealing in high-pressure oxygen did not change this feature at all. This narrow concentration region for T*-phase stability in LaEuSrCuO samples heat-treated in 400 kbar oxygen accords with that of the samples heat-treated in 3 kbar oxygen [8].

The lattice constants of T*-phase in both samples annealed in 1 atm oxygen and in 400 atm oxygen are shown against Eu concentration in Fig. 2. We can see the tendency that the lattice constants, a and c, decrease with increasing Eu concentration. These decreases from x=0.33 to 0.67 are 0.4 % at most, which are very small compared with those in T'-phase LaEuCeCuO system, 1 % [7,8]. This is probably due to the difference of the real composition in T*-phase from the nominal one. T-phase, appearing at low Eu concentration, prefers the larger lanthanide ion, La, and T'-phase, appearing at high Eu concentration, prefers the smaller one, Eu. Because of it, T*-phase contains more Eu atoms than the nominal composition at a low Eu concentration, and less Eu atoms at a high Eu concentration. The dependences of the lattice constants on Sr concentration, y, is characteristic. The addition of larger Sr ions than La or Eu ions increased the lengths of c-axis and decreased b-axis at x=0.33 and 0.5. The heat treatment in 400 atm oxygen did not change the lattice constants systematically. In the samples of y=0.2, it increased the lengths of a- and c-axes, or changed little. It increased them in the sample of (x, y)=(0.5, 0.1), decreased in (x, y)=(0.67, 0.1), and increased the c-axis length and decreased the a-axis length in (x, y)=(0.33, 0.1).

The resistivity and Meissner effect of the samples of y=0.2 are shown in Fig. 3. All of the samples not annealed in 400 atm oxygen showed semiconductive behavior. By annealing in 400 atm oxygen, these all samples showed superconductivity and the resistivity in normal state (90 K) decreased by a factor of 10^2. These behaviors were similar in the samples of y=0.1. The superconducting volume fraction was about 10 % in the sample of (x, y)=(0.5, 0.1). The superconducting onset temperature,

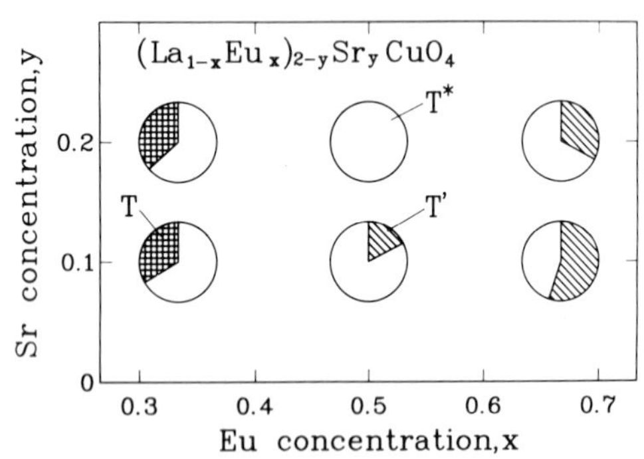

Fig. 1. Identified phases and their abundances in $(La_{1-x}Eu_x)_{2-y}Sr_yCuO_4$. T: La_2CuO_4 phase, T': Eu_2CuO_4 phase

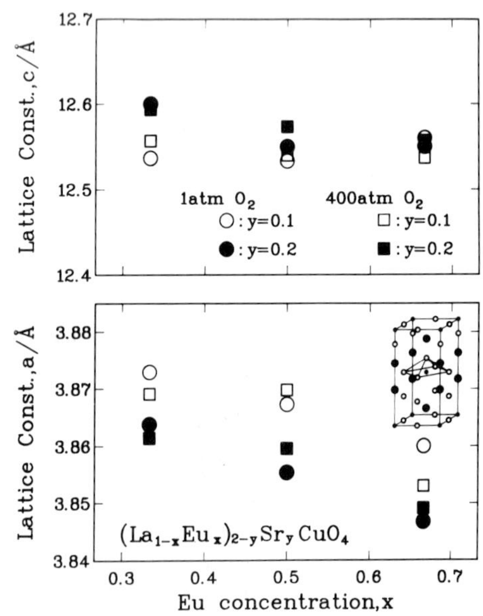

Fig. 2. Lattice constants in T*-phase are plotted against Eu concentration.

Tc, obtained from Meissner effect is shown against Eu concentration in Fig. 4. In both series(y=0.1 and 0.2), the samples of x=0.5 showed the highest Tc, 28.8 K and 21.8 K. At each x, Tc in the samples of y=0.1 was higher than that in ones of y=0.2, although the former samples contained more secondary phase than the latter.

All of the Mössbauer spectra obtained for $(La_{1-x}Eu_x)_{2-y}Sr_yCuO_4$ showed a single line as those observed in $Ba_2EuCu_3O_7$ or T'-phase $(La_{1-x}Eu_x)_{2-y}Ce_yCuO_4$ [4-8]. The spectra were analyzed by the nonlinear least square method using a Lorentzian. Figure 5 shows the isomer shift and the line area against Eu concentration. It must be kept in mind that all of the obtained ^{151}Eu Mössbauer spectra do not reveal the state of Eu in T*-phase, because some of Eu atoms will be dissolved in the secondary phases which appeared in quantity in the samples of x=0.33 and 0.67. It is considered that T'-phase appearing at a high Eu concentration contains many Eu atoms. However, Eu will be scarcely dissolved in T-phase since the lattice constants of T-phase (a=0.3784 nm and b=1.3122 nm) shows that this phase contains only La as lanthanide, not Eu [10]. Therefore, we can consider that the spectra of the samples of x=0.33 and 0.5 show the state of Eu in T*-phase. The isomer shift in the single lines at room temperature was 0.74-0.84 mm/s. This shows that the electronic state of Eu in T*-phase is essentially equal to those in T'-phase and '123'-superconductors, that is, Eu is tri-valent and not magnetic-ordered. However there is a little difference in isomer shift. The isomer shift in T*-phase was close to that in hole-doped type superconductor $Ba_2EuCu_3O_7$(0.77 mm/s), and larger than that in electron doped T'-phase(0.69-0.73 mm/s). This will be due to the difference in Eu-O distance as discussed in ref. 7. The ^{151}Eu isomer shift increases with pressure applied to the sample because of the increase of s-electron density, which means that the shortening in the Eu-O bond length increases the Eu isomer shift. The average Eu-O distance in T*-phase is about 0.24 nm, which is close to that in '123'-type and is smaller than that in T'-phase, 0.25 nm. The Debye temperature of Eu in T*-phase estimated from the line area was 180 - 210 K, which is lower than in '123'-phase and larger than T'-phase.

By increasing Eu concentration from 0.33 to 0.5, the isomer increased, which means the shortening of Eu-O bond. The line area of the samples of y=0.1, however, changed little, showing little change in the Eu lattice vibration. The line area of the sample of y=0.2 seems to be sensitive to the Eu concentration and the heat-treatment. In the 1 atm-oxygenerated samples of y=0.2, the line area

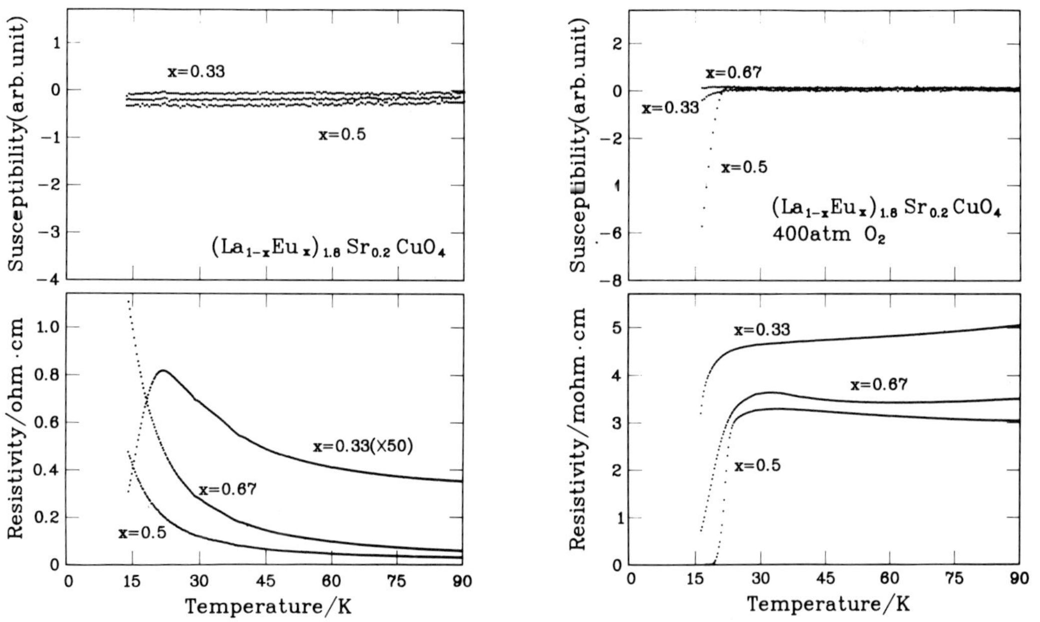

Fig. 3. Meissner effect and resistivity in $(La_{1-x}Eu_x)_{2-y}Sr_yCuO_4$ are plotted against Eu concentration. The left figure is for the samples annealed in 1 atm oxygen and the right for the samples annealed in 400 atm oxygen.

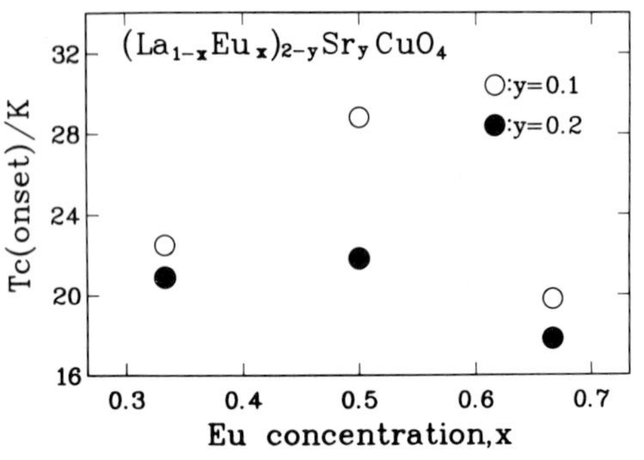

Fig. 4. Superconducting onset temperature, Tc, obtained from Meissner effect.

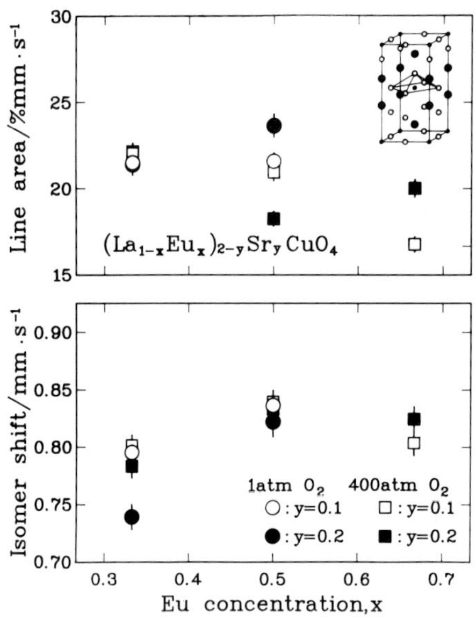

Fig. 5. Isomer shift and line area of [151]Eu Mössbauer effect in $(La_{1-x}Eu_x)_{2-y}Sr_yCuO_4$ are plotted against Eu concentration.

(Debye temperature) increased substantially by increasing Eu concentration, which means the shortening of Eu-O bond length as the changes of the lattice constants and the isomer shift do. In the 400 atm-oxygenerated samples of y=0.2, the line area decreased by increasing Eu concentration, which seems to contradict the data of the isomer shift and the lattice constants. The T*-phase lattice constants of the sample of (x, y)=(0.5, 0.2) was the most enlarged by 400 atm-oxygeneration. This means that oxygen was the most filled in this sample by this heat treatment. The oxygen filled at the apical sites in T-type part in crystal will pull Sr ion with the coordinated oxygen atoms. This may induce the increase of Eu-O bond in part and decrease the Debye temperature(line area) of Eu.

Although we tried to find some correlations between T*-phase superconductivity and Mössbauer parameters, this try is not necessarily successful at present. It seems that there is a correlation between Tc and the isomer shift, because of the similarity in the Eu concentration dependences. However we can not find the reason why Tc in the sample of (x, y)=(0.5, 0.1) is much higher than that in the sample of (x, y)=(0.5, 0.2), though these two samples have the nearly same isomer shift. Now we are proceeding to study [151]Eu Mössbauer effect using single T*-phase samples in order to obtain tight correlations between Mössbauer parameters and Tc.

REFERENCES

1. Izumi F, Asano H, Ishigaki T, Takayama-Muromachi E, Uchida Y and Watanabe N (1987) Jpn. J. Appl. Phys. **26**: L1193
2. Izumi F, Takayama-Muromachi E, Fujimori A, Kamiyama T, Asano H, Akimitsu J and Sawa H (1989) Pysica C **158**: 440
3. Izumi F, Matsui Y, Takagi H, Uchida S, Tokura Y and Asano H (1989) Physica C**158**: 439
4. Taniwaki M and Sasaki H (1988) Pysica C **135-155**: 1549
5. Taniwaki M, Muraki T and Shiramine K (1989) Physica C **162-164**: 568
6. Muraki T, Taniwaki M and Shiramine K (1989) In: Ishiguro, Kajiyama(eds) Proc. of 2nd Int. Sym. on Superconductivity(1989, Tsukuba), Springer-Verlag. Tokyo p563
7. Itoh H, Kusuhashi M and Taniwaki M (1990) Jpn J. Appl. Phys. **29**: L1604
8. Itoh H, Kusuhashi M and Taniwaki M (in this issue)
9. Fisk Z, Cheong S-W, Thompson JD, Hundley MF, Schwart RB, Kwei GH and Schirber JE (1989) Physica C **162-164**: 1681
10. Kishio K, Kitazawa K, Hasegawa T, Aoki M, Fueki K, Uchida S and Tanaka S (1987) Jpn. J. Appl. Phys. **26**:L391

NMR Study in $Tl_2Ba_2CuO_{6-\delta}$

Takayuki Goto[1], Takeshi Shinohara[1], Takashi Sato[1], Tetsuo Fukase[1], Satoshi Nakajima[2], Masae Kikuchi[1], and Yasuhiko Syono[1]

[1] IMR, Tohoku University, 1-1 Katahira 2-chome, Sendai, 980 Japan
[2] Hachioji Research Center CASIO, CASIO Computer Co., Ltd., Hachioji, 192 Japan

ABSTRACT

An intensive study for ^{205}Tl-NMR has been performed on the $Tl_2Ba_2CuO_{6-\delta}$ with T_c=82, 70, 25 and below 5K in the temperature range between 5 and 300K. Below T_c, the significant reduction in the resonance line shift was observed, and was revealed to be due mainly to the diamagnetic shielding rather than to the reduction in the Knight shift. This diamagnetization was found to be highly anisotropic, reflecting the H_{c2} anisotropy. The nuclear spin-lattice relaxation rate T_1^{-1} of all the samples showed the deviations from the Korringa's relation in the normal state, suggesting that the antiferromagnetic fluctuation of the Cu d-spins remains even in the overdoping region.

KEY WORDS: Tl system superconductor, NMR, H_{c2}, antiferromagnetic fluctuation

INTRODUCTION

The copper oxide superconductor $Tl_2Ba_2CuO_{6-\delta}$ (Tl2201) shows a wide range of T_c from 0 to 90K, depending on δ. Though the formal density of holes is zero on the assumption of the valences Tl^{3+}, Ba^{2+}, Cu^{2+} and O^{2-}, non superconducting phase of this compound is believed to be in an excessive hole concentration state [1-4]. The deletion of 0.1 atom oxygen per unit formula from the compound reduces the hole concentration and raises its T_c to 80K [1]. The origin of the carriers is not determined yet, although some ideas are proposed: excess oxygen atoms at the interstitial site between TlO_2 double layers [2], and the charge transfer from the CuO_2 planes to TlO_2 layers [3, 4]. Therefore, the TlO_2 layers seems to be deeply concerned with the origin of the carriers in Tl2201 system. In order to investigate the electronic state of the TlO_2 layers selectively, we have employed ^{205}Tl-NMR, which is a good probe for the electronic state, since ^{205}Tl nuclei (I=1/2) is free from quadrupolar interaction.

The anomalous temperature dependence of the nuclear spin-lattice relaxation rate T_1^{-1} is one of the unresolved problems for the normal state of the copper oxide superconductors. Deviation from the Korringa's relation, T_1T=const. above T_c is reported for the plane site Cu in YBCO, LSCO and other compounds [5]. The antiferromagnetic fluctuation of Cu d-spin is postulated for a possible origin [5, 6]. If this is the case, it may be meaningful to investigate the T_1^{-1} behavior for the overdoped compounds which are far from the antiferromagnetic region. In LSCO system, already, the Korringa type behavior of T_1^{-1} is reported for the overdoped compounds [7]. However, high concentration of Sr brings the system large inhomogeneity, which causes the distribution in T_1^{-1}, and hence the ambiguity. In Tl2201 system, on the other hand, the compounds in overdoped state can be obtained without any explicit doping, so that T_1^{-1} is expected to be determined with less ambiguity. The Korringa type behavior of T_1^{-1} reported for ^{89}Y and ^{17}O in YBCO [12, 13] is explained as the consequence of the cancellation of the fluctuation due to the structural symmetry [13]. Since the Tl2201 system has the double TlO_2 layers, the antiferromagnetic fluctuation of Cu is expected not to be canceled out even at Tl site.

In this paper, we report the results on the measurement of the shift, the Line width and T_1^{-1} for several specimens with different T_c.

EXPERIMENTAL

The polycrystalline samples were prepared by solid-state reaction of a mixture of the appropriate amount of Tl_2O_3, BaO_2 and CuO in the flowing oxygen. The oxygen defects are controlled by the treatments after sintering. Figure 1 schematically shows the sample synthesis method and the T_c, which were defined as the onset of the Meissner signal measured by the SQUID magnetometer. The DC susceptibility in the magnetic field of 1T was also measured. The X-ray diffraction patterns indicate that all the samples are tetragonal and are of single phase. Samples were ground into powder for the NMR measurements so as that the rf field penetrates into each grains. In order to measure the anisotropy in the shift, the powdered samples of about 0.5g were mixed with the epoxy (stycast 1266) and were cured in the magnetic field of 4T at room temperature. Partial alignment of the c-axis to the field direction was confirmed by the narrowing of the resonance lines.

The NMR data were taken in several magnetic fields from 0.8 to 4T, and in the temperature range of 5~300K. Spectra were obtained by integrating the spin-echo signal by a boxcar while sweeping the external magnetic field. The shift was determined as the center of the gravity of the spectra. In measuring T_1^{-1}, the nuclear spin system was first saturated completely by applying the pulse train, and the recovery of the magnetization was observed and fitted to the function $I_0(1-e^{-t/T_1})$, where I_0 is the magnetization in thermal equilibrium, and t, the time between the pulse train and the observation of the magnetization.

$Tl_2O_3 + BaO_2 + CuO$ Mixture
⇓ $890 \sim 900C, 5 \sim 30min, O_2 120m\ell/min$
Solid-Liquid Reaction
⇓
$Tl_2Ba_2CuO_6$ (a) ⟹ $T_c \simeq 25K$
⇓
(b) 500C O_2-anneal and slow cool-down ⟹ $T_c < 5K$
(c) 500C O_2-anneal and quenched in Liquid-N_2 ⟹ $T_c \simeq 70K$
(d) 600C Ar-anneal and quenched in R.T. ⟹ $T_c \simeq 82K$

Fig. 1 The method for sample synthesis and the T_c of the samples.

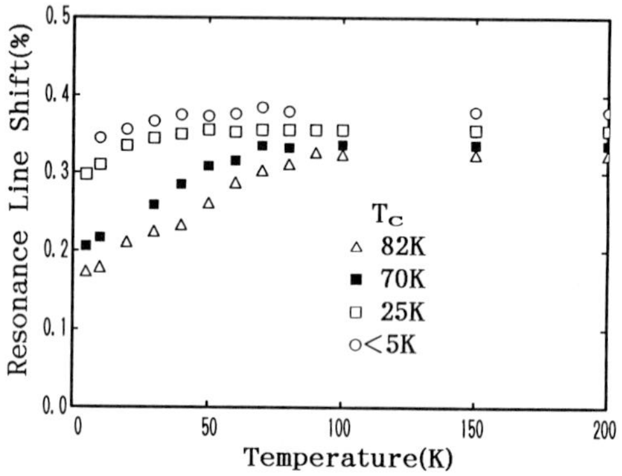

Fig. 2 The temperature dependence of the ^{205}Tl-NMR shift measured at 1.8108T, 44.6MHz.

RESULTS

The Shift and the Line Width

The ^{205}Tl-NMR spectra above T_c are of single peak with the tail at higher magnetic field. The asymmetry is attributed to the powder pattern of the axial anisotropy in the Knight shift or the chemical shift. Figure 2 shows that the temperature dependence of the shifts. All the samples show the positive shifts which are temperature independent above T_c, and are larger for the samples with lower T_c. Below T_c, the significant decrease was observed. As for the sample with T_c<5K, the shift stays constant. In Fig. 3, the difference between the shift at 5K and that above T_c is plotted against the inverse magnetic field. The higher the field is, the less the shifts decrease. In Fig. 4, the dependence of the shift below T_c on the orientation of the magnetic field is shown. The reduction in the shift becomes highly anisotropic below T_c.

The line widths in the low field about 1T are temperature independent above T_c, and start to increase below T_c. The sample with T_c<5K shows constant line width. The sample dependence of the line width above T_c is negligibly small. In high magnetic field of 4T, the temperature dependence of the line width drastically differs from that in 1T. The spectra of all the samples with different T_c similarly broaden with the lowering temperatures, suggesting that the two origins for the broadening exist.

T_1^{-1}

In the normal state, T_1^{-1}'s of all the samples show a sublinear temperature dependence rather than the Korringa type behavior as shown in Fig. 5. In Fig. 6, $(T_1T)^{-1}$ is plotted against temperature. The behavior of T_1^{-1} or $(T_1T)^{-1}$ in the normal state resembles what observed for the plane site Cu in YBCO and LSCO [5]. The smaller T_1^{-1} is observed for the samples with the lower T_c.

Below T_C, the significant reduction in T_1^{-1} is observed without any BCS like enhancements[9]. The temperature dependence below T_C is neither the power law nor the thermal activation type. The diverging behavior of $(T_1T)^{-1}$ at lowest temperatures for the samples with lower T_C suggests that a finite or temperature independent relaxation rate exists at the low temperatures. The T_1^{-1} for the sample with T_C about 80K is reported by the other authors[8] and agrees with our results.

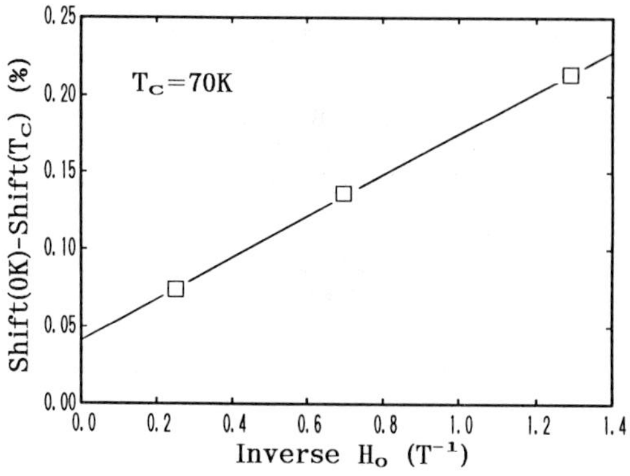

Fig.3 The difference between the shift at 5K and that at T_C is plotted against H_0^{-1}.

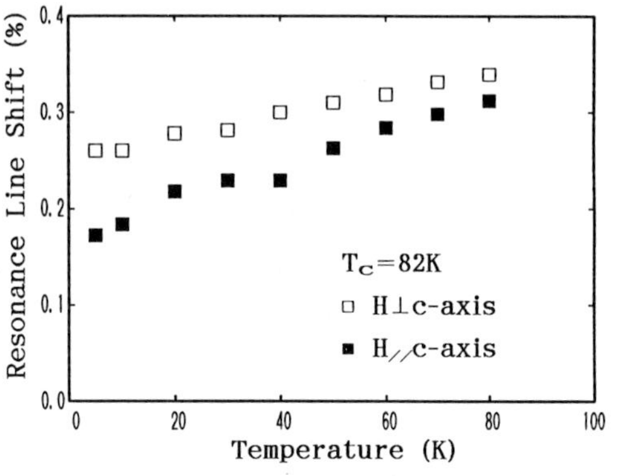

Fig.4 The reduction in the shift below T_C for the aligned powder sample with T_C=82K.

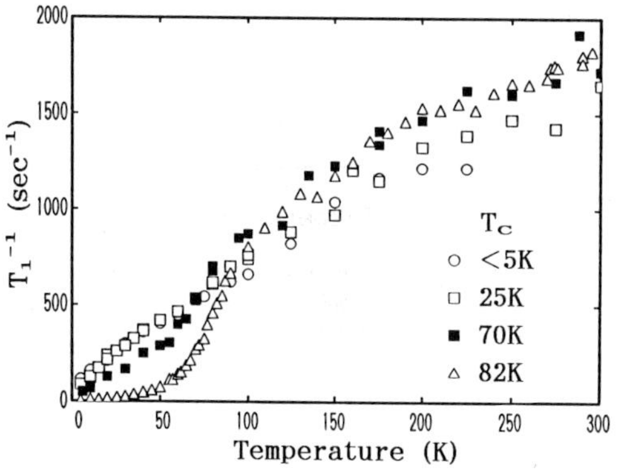

Fig.5 The temperature dependence of the T_1^{-1} measured at 1.3404T (3.961T for the sample with T_C=82K).

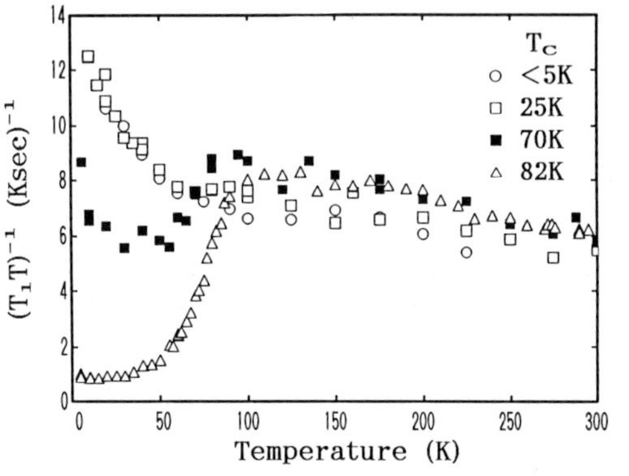

Fig. 6 The plot of $(T_1T)^{-1}$ against temperature (the data are the same as Fig. 5).

DISCUSSION

The Shift and the Line Width

There are two possible explanations for the interpretation of the decrease in the shifts observed below T_C. One is the reduction in the Pauli susceptibility χ_P due to the spin singlet paring of the electrons, while the other, the diamagnetization due to the Meissner effect. The two of them are tested by the dependence on the external magnetic field. If the former is the case, the field dependence must be none because the shift is proportional to χ_P, which is generally field independent. As for the latter, the change in the shift is proportional to $-M/H_0$, where M is the diamagnetization, and H_0, the applied magnetic field. The diamagnetization M can be regarded as almost field independent in the present case, because it depends on H_0 only logarithmically[9] in the region $H_{C1} \ll H_0 \ll H_{C2}$. Therefore, $-M/H_0$ is supposed to scale almost inversely with H_0. Figure 3 clearly indicates that the observed field dependence agrees with the latter interpretation. The reduction in the intrinsic Knight shift is estimated as small as 10% of the total shift. The strong spin-orbit interaction in Tl atoms may prevents the reduction in χ_P by mixing the electron levels below the gap with those far above[14]. For the spin-orbit coupling increases rapidly with the atomic number, the spin-orbit interaction of ^{205}Tl is supposed to be much larger than that of ^{63}Cu, for which the significant reduction in the Knight shift is reported for YBCO[10].

The large anisotropy in the reduction of the shift below T_c is explained as the consequence of the anisotropic diamagnetization M. Assuming that the thermodynamical critical field H_c is the same, M is the larger[9] for the smaller H_{c2} in the region $H_{c1} \ll H_0 \ll H_{c2}$. The observed anisotropy in M suggests that the $H_{c2//}$(H_0 parallel to c-axis) shall be larger than $H_{c2\perp}$(H_0 perpendicular to c-axis). In Tl2201 system, $H_{c2//}$ is reported to be ten times larger than $H_{c2\perp}$[11], of which results are consistent with our interpretation.

The field dependence of the broadening in the resonance line at low temperatures suggests that there exist the two origins. In the low field about 1T, the broadening which takes place below T_c is due to the field inhomogeneity of the vortex lattice[14]. In 4T, however, the resonance line of all the samples broaden similarly. Therefore the other origin such as the paramagnetic impurities which are distributed at random in the compounds must also exist. The large Curie-Weiss term in the DC susceptibility is observed actually for all the samples. The origin of the Curie-Weiss term is not clear, however. If it is assumed to be localized free spins with S=1/2, the amount of the spins is estimated to be 10% of mole-Cu.

The shifts in the normal state shows small difference for the samples with the different T_c, though the measurements of DC susceptibility suggest the significant change in the carrier concentration. Since both the Knight shift and the chemical shift differ for the samples with different T_c, the two of them may cancel each other. However, this is not the case, because T_1^{-1} in the normal state, too, shows small dependence on the samples. Therefore, the hyperfine coupling constant, which is a measure for the s-electrons at the nuclear site, is considered to be very small. A tentative estimation for the hyperfine coupling constant leads to the value which is about ten times smaller than that for Tl metal. This suggests that the valence of Tl in $Tl_2Ba_2CuO_6$ is +3 or +1 [3, 4].

T_1^{-1}

The deviation of T_1^{-1} from the Korringa's relation suggests that the Tl nuclei are affected by the antiferromagnetic fluctuation of Cu d-spin. The samples with lower T_c shows smaller T_1^{-1}, because they are far from the antiferromagnetic region, and are less affected by the fluctuation. The fact that the deviation takes place for the overdoped sample suggests the possibility of the remaining antiferromagnetic fluctuation even in the overdoped region.

In summary, the shift and the T_1^{-1} of Tl were studied for $Tl_2Ba_2CuO_6$ which are superconducting or overdoped. The reduction in the shift observed below T_c is found to be dominated by the diamagnetization, which is highly anisotropic due to the H_{c2} anisotropy. The deviation of T_1^{-1} from the Korringa's law observed for all the samples suggests the remaining of the antiferromagnetic fluctuation of Cu d-spins even in overdoping region.

ACKNOWLEDGEMENT

The author would like to thank Prof. N. Kobayashi for the DC susceptibility measurement by the SQUID magnetometer. This work was partly supported by the Grant-in-Aid for Scientific Research on Priority Areas "Mechanism of Superconductivity" given by the Ministry of Education, Science and Culture, Japan.

REFERENCES

1. Y. Shimakawa, Y. Kubo, T. Manako, T. Satoh, S. Iijima, T. Ichihashi and H. Igarashi: Physica C157 (1989) 279
2. Y. Shimakawa, Y. Kubo, H. Igarashi, F. Izumi, H. Asano: Phys. Rev. B (to be published)
3. S. Nakajima, M. Kikuchi, T. Oku, N. Kobayashi, N. Kobayashi, T. Suzuki, K. Nagase, K. Hiraga, Y. Muto and Y. Syono: Physica C160 (1989) 458
4. M. Kikuchi, S. Nakajima, Y. Syono, K. Nagase, R. Suzuki, T. Kajitani, N. Kobayashi, Y. Muto: Physica B (to be published)
5. T. Imai: J. Phys. Soc. Jpn 59 (1990) 2508, and references therein.
6. M. Horvatic, P. Segransan, C. Berthier, Y. Berthier and P. Butaud: Phys. Rev. B39 (1989) 7332
7. Y. Nakamura and K. Kumagai: Physica C161 (1089) 261
8. H. B. Brom, D. Reefman, J. C. Jol, D. M. de Leeuw and W. A. Groen: Phys. Rev. B41 (1990) 7261
9. M. Tinkham, "Introduction to Superconductivity" (McGraw-Hill, New York, 1975)
10. Barrett, D. J. Durand, C. H. Pennington, C. P. Slichter, T. A. Friedmann, J. P. Rice and D. M. Ginsberg: Phys. Rev. B41 (1990) 6283
11. H. Mukaida, K. Kawaguchi, M. Nakao, H. Kumakura, D. R. Dietderich and K. Togano: Phys. Rev. Lett. (to be published)
12. H. Alloul, T. Ohno and P. Mendels: Phys. Rev. Lett. 63 (1989) 1700
13. P. C. Hammel, M. Takigawa, R. H. Heffner, Z. fisk and K. C. Ott: Phys. Rev. Lett. 63 (1989) 1992
14. D. E. MacLaughlin: Solid State Physics 31 (1976) 1

Copper Nuclear Relaxation Rate and Knight Shift in $YBa_2Cu_4O_8$

TAKATO MACHI[1], IZUMI TOMENO[1], TAKAYUKI MIYATAKE[1], NAOKI KOSHIZUKA[1], TAKASHI IMAI[2], and HIROSHI YASUOKA[2]

[1] Superconductivity Research Laboratory, International Superconductivity Technology Center, 10-13, Shinonome 1-chome, Koto-ku, Tokyo, 135 Japan
[2] The Institute for Solid State Physics, The University of Tokyo, Roppongi, Minato-ku, Tokyo, 106 Japan

ABSTRACT

The temperature dependences of the ^{63}Cu nuclear spin-lattice relaxation rate $1/T_1$ for the Cu(1) chain sites and the planar Cu(2) sites in $YBa_2Cu_4O_8$ have been measured by nuclear quadrupole resonance(NQR). The in-plane component of the Knight shift for the Cu(2) sites, $K_{2\perp}$, and the anisotropies of $1/T_1$ for both sites have been also measured by nuclear magnetic resonance(NMR). In the normal state, $1/T_1$ for the Cu(2) sites increases rapidly from T_c up to about 250K and behaves approximately T-linearly above 250K. $1/T_1$ for the Cu(1) sites shows the superlinear temperature dependence. Knight shift $K_{2\perp}$ increases rapidly with temperature from T_c and saturates above 250K. The anisotropies of $1/T_1$ for the Cu(1) sites and the Cu(2) sites are ~0.6 and ~3.2, respectively.

KEY WORDS: NQR, NMR, $YBa_2Cu_4O_8$, spin-lattice relaxation, Knight shift

INTRODUCTION

The temperature dependence of the ^{63}Cu nuclear spin-lattice relaxation rate $^{63}(1/T_1)$ for the planar Cu sites have been measured in many oxide superconductors[1-11]. It is widely known that the behavior of $1/T_1$ in all high T_c compounds have several common features. In the superconducting state, neither the exponential variation nor the enhancement just below T_c, which are generally observed for the conventional s-wave superconductors, have been observed. In the normal state, $1/T_1$ increases rapidly with temperature from T_c up to about 150-200K[10]. Subsequently, $1/T_1$ increases approximately linearly with temperature.

$YBa_2Cu_4O_8$ with T_c = 80K has double Cu(1)-O chains and two Cu(2)-O_2 planes in the unit cell[12,13]. Because of the excellent thermal stability of the oxygen content[14], various physical properties can be measured in $YBa_2Cu_4O_8$ up to higher temperatures as compared with $YBa_2Cu_3O_{7-\delta}$. Zimmermann *et al.*[11] observed ^{63}Cu NQR and NMR in $YBa_2Cu_4O_8$ for the first time. According to their results, $1/T_1$ for the planar Cu(2) sites are constant between 300K and 500K and can be described by $1/T_1$=7.8T above 550K, from which they suggested the breakdown of the interaction between the quasiparticles at high temperature range. The temperature dependence of $^{63}(1/T_1)$ above room temperature has been measured only for $YBa_2Cu_3O_7$ except for $YBa_2Cu_4O_8$. According to the results of Imai *et al.*[10], $YBa_2Cu_3O_7$ shows an approximately linear temperature dependence of $^{63}(1/T_1)$ in a wide temperature range between 150K and 460K, where both ^{17}O and ^{89}Y relaxation rates show Korringa-like behavior[15,16,17]. Barret *et al.*[18] also confirmed that the linear behavior of $^{63}(1/T_1)$ persists up to 500K. Since the crystal structure of $YBa_2Cu_4O_8$ is quite similar to that of $YBa_2Cu_3O_7$, the anomalous $1/T_1$ behaviors observed by Zimmermann *et al.*[11] are quite surprising. Motivated by this, we performed a high precision measurement of $1/T_1$ for the planar Cu(2) sites up to 680K. Our results are similar to those of other high T_c superconductors, in contrast with the results of Zimmermann *et al.* In this paper, we also report on $1/T_1$ measurement for the chain Cu sites, anisotropies of $1/T_1$ and Knight shifts $K_{2\perp}$ for the planar Cu(2) sites in the case of the magnetic field applied perpendicular to the c-axis.

EXPERIMENTAL

$YBa_2Cu_4O_8$ was prepared by the solid state reaction method with the hot isostatic press(HIP) treatment in oxygen atmosphere as described in a previous paper[19]. A powder sample was used for NQR measurement. The NMR measurement was performed for the uniaxially oriented powders. All the NQR and NMR measurements were carried out in the similar manner as in Ref.[10] using the standard pulsed NMR spectrometers at the Institute for Solid State Physics(ISSP). Frequency spectra were observed in zero field at 1.3K. $1/T_1$ for Cu(1) and Cu(2) sites were measured by using NQR technique at 4.2K-280K and 4.2K-680K, respectively. To obtain high precision data of $1/T_1$ for the planar Cu(2) sites, the spin echo signal was averaged up to 10^5 times at the high temperature range. The NMR shift was measured for the planar Cu(2) sites for the external magnetic field applied perpendicular to the c-axis between T_c and 300K. NMR spectra were obtained at four fixed frequencies while sweeping the magnetic field. The results were analyzed by simulating the exact eigenvalue of the quadrupole perturbed nuclear Zeeman Hamiltonian[20]. In order to determine the anisotropy of $1/T_1$, $(1/T_1)_{ab}$ was measured by NMR technique. Here, the suffix ab stands for the rate measured under the external magnetic field applied perpendicular to the c-axis. We defined the anisotropy as $(1/T_1)_c/(1/T_1)_{ab}$. It should be noted that we defined $1/T_1$ as $2W$ in this paper, where W is the fundamental relaxation rate.

RESULTS

The intensity of the NQR spin-echo signal for the planar Cu(2) sites at 1.3K is shown in Fig.1 as a function of frequency. There is neither shoulder nor split for each peak. The ratio of the frequencies of the two peaks is equal to $^{63}\gamma/^{65}\gamma$, where $^{63}\gamma$ and $^{65}\gamma$ is the magnetogyric ratio of the two isotopes, ^{63}Cu and ^{65}Cu, respectively. The intensity ratio of the two peaks is nearly equal to the ratio of the natural abundances of the two copper isotopes, too. Therefore, there is no magnetic ordering in $YBa_2Cu_4O_8$. The recovery curves at several temperatures are shown in Fig.2. $1/T_1$ is determined from the slope of the recovery curve. The temperature dependence of $1/T_1$ for the planar Cu(2) sites is shown in Fig.3. Above 250K, $1/T_1$ exhibits approximately linear temperature dependence. Apparently no kink exists at 550K in contrast with the results of Zimmermann *et al*[11]. $1/T_1$ shows a rapid reduction with decreasing temperature from 250K to T_c. It is noteworthy that the spin contribution for $K_{2\perp}$, namely the remainder subtracted K_{orb} from the raw data, also shows a strong reduction with decreasing temperature from 250K to T_c as shown in Fig.4, where K_{orb} indicates the Van Vleck value obtained in $YBa_2Cu_3O_7$[18]. Our results of $K_{2\perp}$ are in good agreement with recent results independently obtained by Zimmermann *et al.*[21].

The anisotropy of $1/T_1$ obtained at 100K was ~3.2 for the planar Cu(2) sites. This is nearly equal to the value ~3.6 observed for $YBa_2Cu_3O_7$[5,9,22], implying that the hyperfine coupling constants for the Cu(2) sites are not so much different between $YBa_2Cu_3O_7$ and $YBa_2Cu_4O_8$.

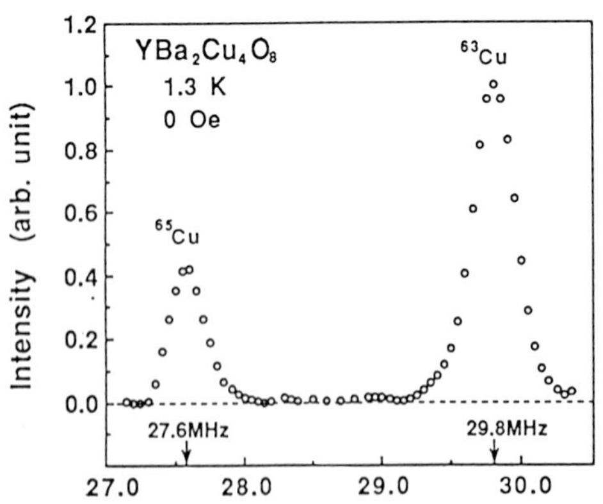

Figure 1. Frequency spectra for Cu(2) sites measured by using NQR technique at 1.3K.

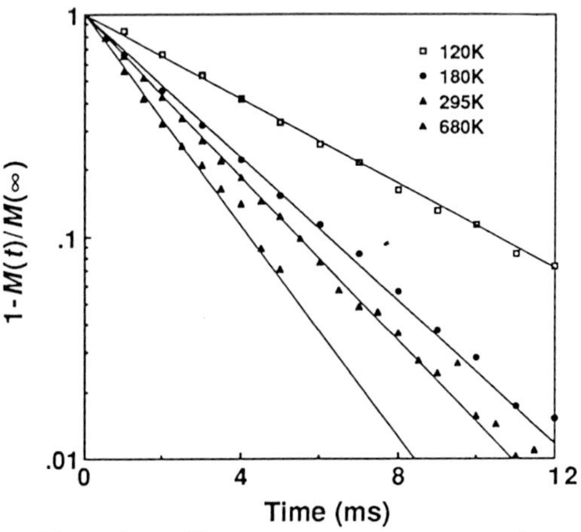

Figure 2. The recovery curves at several temperatures, where $M(t)$ indicates the magnetization after t(sec).

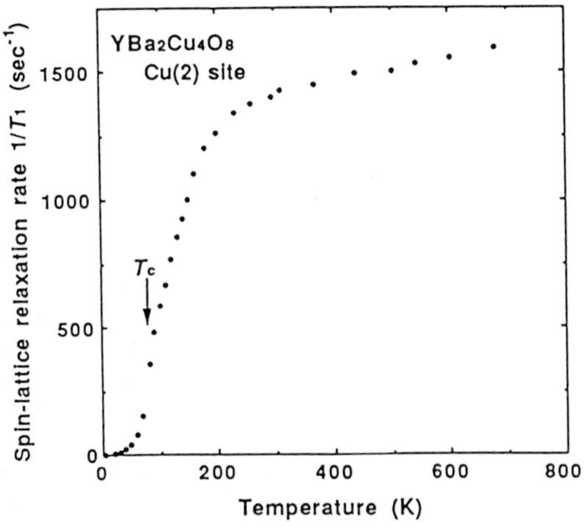

Figure 3. The temperature dependence of $1/T_1$ for the planar Cu(2) sites in $YBa_2Cu_4O_8$.

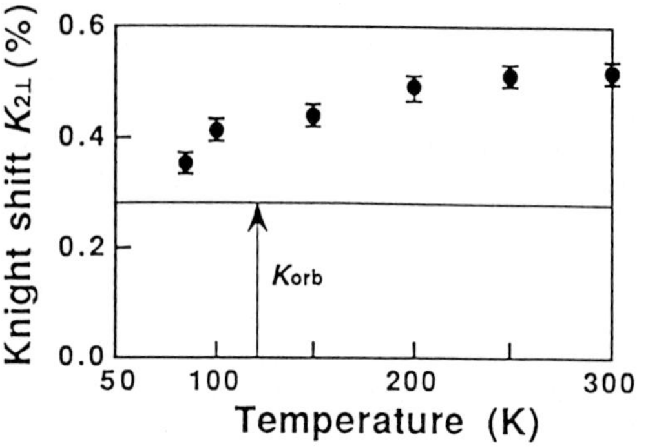

Figure 4. The temperature dependence of Knight shift $K_{2\perp}$ for the Cu(2) sites for the external magnetic field applied perpendicular to the c-axis. K_{orb} indicates the Van Vleck value of the Knight shift in $YBa_2Cu_3O_7$[18].

The temperature dependence of $1/T_1$ for the Cu(1) chain sites is shown in Fig.5. $1/T_1$ does not behave as Korringa-like T-linear dependence but as superlinear in the temperature. It is evident from Fig.4 that the observed temperature dependence of $1/T_1$ is quite similar to that reported for $YBa_2Cu_3O_7$[23]. The observed anisotropy, ~0.6, is also comparable with the case of $YBa_2Cu_3O_7$[5]. The absolute value of $1/T_1$ is, however, by a factor of 5-7 smaller compared with $YBa_2Cu_3O_7$. Although we do not have a firm idea to account for the observed behavior, the present observation implies that the superlinear behavior of $1/T_1$ is a characteristic of the chain structure.

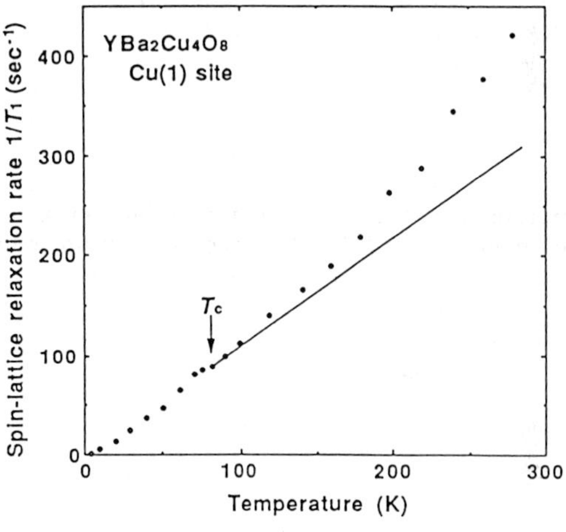

Figure 5. The temperature dependence of $1/T_1$ for the Cu(1) chain sites in $YBa_2Cu_4O_8$.

SUMMARY

We have reported the temperature dependences of $1/T_1$ for the planar Cu(2) sites in $YBa_2Cu_4O_8$ for a wide temperature range up to 680K. Our results indicated that $YBa_2Cu_4O_8$ shows quite similar behavior as other oxide high T_c superconductors. That is, $1/T_1$ follows an approximate T-linear behavior at high temperatures, where $K_{2\perp}$ is saturated. Both $K_{2\perp}$ and $1/T_1$ decreases rapidly below 250K. The behavior of $1/T_1$ for the chain Cu(1) sites is quite similar to that of $YBa_2Cu_3O_7$, suggesting that the superlinear temperature dependence is the characteristic of the one dimensional chain structure.

ACKNOWLEDGEMENT

We thank K. Shibutani of Superconducting & Cryogenic Technology Center, Kobe Steel, Ltd. for oxygen-HIP treatments.

REFERENCES

[1] M.Mali, D.Brinkmann, L.Pauli, J.Roos, H.Zimmermann and J.Hulliger; Phys.Lett.A 124(1987)112.
[2] W.W.Warren,Jr., R.E.Walstedt, G.F.Brennert, G.P.Espinosa and J.P.Remeika; Phys.Rev.Lett. 59(1987)1860.
[3] T.Imai, T.Shimizu, T.Tsuda, H.Yasuoka, T.Takabatake, Y.Nakazawa and M.Ishikawa; J.Phys.Soc.Jpn. 57(1988)1771.
[4] P.Butaud, M.Horvatic, Y.Berthier, P.Segransan, C.Berthier, P.Lejay and J.Y.Henry; Physica C 153-155(1988)741.
[5] C.H.Pennington, D.J.Durand, C.P.Slichter, J.P.Rice, E.D.Bukowski and D.M.Ginsberg; Phys.Rev.B 39(1989)2902.
[6] M.Horvatic, P.Segransan, C.Berthier, Y.Berthier, P.Butaud, J.Y.Henry, M.Couach and J.P.Chaminade; Phys.Rev.B 39(1989)7332.
[7] K.Ishida, Y.Kitaoka and K.Asayama; J.Phys.Soc.Jpn. 58(1989)36.
[8] T.Oashi, K.Kumagai, Y.Nakajima, T.Tomita and T.Fujita; Physica C 157 (1989)315.
[9] R.E.Walstedt, W.W.Warren,Jr., R.F.Bell and G.P.Espinosa; Phys.Rev.B 40(1989)2572.
[10] T.Imai, H.Yasuoka, T.Shimizu, Y.Ueda, K.Yoshimura and K.Kosuge; Physica C 162-164(1989)169.
[11] H.Zimmermann, M.Mali, D.Brinkmann, J.Karpinski, E.Kaldis and S.Rusiecki; PhysicaC 159(1989)681.
[12] P.Marsh, R.M.Fleming, M.L.Mandich, A.M.DeSantolo, J.Kwo, M.Hong and L.J.Martinez-Miranda; Nature 341(1989)41.
[13] P.Fisher, J.Karpinski, E.Kaldis, E.Jilek and S.Rusicki; Solid State Commun. 69(1989)531.
[14] J.Karpinski, E.Kaldis, E.Jilek, S.Rusicki and B.Bucher; Nature 336(1988)660.
[15] P.C.Hammel, M.Takigawa, R.H.Heffner, Z.Fisk amd K.C.Ott; Phys.Rev. Lett. 63(1989)1992.
[16] H.Alloul, P.Mendels, G.Collin and P.Monod; Phys.Rev.Lett. 61(1988)746.
[17] Y.Yoshinari, H.Yasuoka, Y.Ueda, K.Koga and K.Kosuge; J.Phys.Soc.Jpn. 59(1990)3698.
[18] S.E.Barrett, D.J.Durand, C.H.Pennington, C.P.Slichter, T.A.Friedmann, J.P.Rice and D.M.Ginsberg; Phys.Rev.B 41(1990)6283.
[19] T.Miyatake, K.Yamaguchi, T.Takata, S.Gotoh, N.Koshizuka and S.Tanaka; Physica C 160(1989)541.
[20] C.P.Slichter; Principles of Magnetic Resonance (Springer Verlag, 1980).
[21] H.Zimmermann, M.Mali, I.Mangelschots, J.Roos, L.Pauli, D.Brinkmann, S.Rusiecki and E.Kaldis; Proc. EMRS 1990 Spring Conference(to be published).
[22] T.Imai; J.Phys.Soc.Jpn. 59(1990)2508.
[23] T.Imai, T.Shimizu, H.Yasuoka, Y.Ueda and K.Kosuge; J.Phys.Soc.Jpn. 57(1988)2280.

Electron Spin Resonance Study of Paramagnetic Species Coexisted with the Fine Particles of the R-Ba-Cu-O (R=Y,Eu,Gd) Superconductors

MASAHIRO KAISE, HIROSHI KONDO, MASAGI MIZUNO, CHIZUKO NISHIHARA, HISAKAZU NOZOYE, and HITOSHI SHINDO

Basic Research Division, National Chemical Laboratory for Industry, Tsukuba, 305 Japan

ABSTRACT

An approach to the characterization of the high-Tc fine particles was presented using an electron spin resonance (ESR) method on powder mixtures of high-Tc cuprates and paramagnetic species.
R-Ba-Cu-O (R=Y,Eu,Gd) were synthesized by the sol-gel method and resistivity vs. temperature measurements were done by bulk oxides (Tc=91-93 K). ESR absorption was obtained with a powder mixture of a small amount of Cr^{3+} contained MgO and the ground high-Tc cuprates. Line broadening of Cr^{3+} has been observed at temperature lower than Tc of the bulk oxide. The results indicate that there was a pronounced decrease of Tc when the particle size was reduced.

KEY WORDS: R-Ba-Cu-O (R=Y,Eu,Gd) superconductors, electron spin resonance, fine particles

INTRODUCTION

Chemical and physical properties of a solid depend on its size, and fine particles exhibit unique physicochemical phenomena.[1] The electron spin resonance technique(ESR) has emerged as useful experimental probes in the effort to characterize the microscopic magnetic character of the high-Tc copper oxides.[2] The power of the method is derived largely from their high sensitivity and their highly local nature, which permits not only study of bulk compounds but also their fine particles. In this work we report the ESR observation of doped paramagnetic species (Cr^{3+}) in MgO which is magnetically affected by neighbouring super-oxides. An approach to the characterization of the high-Tc fine particles is presented with ESR measurements from 300 K to lower temperatures across the superconducting temperature Tc on powder mixtures of high-Tc cuprates and paramagnetic species.

EXPERIMENTAL

Samples with nominal composition of the YBCO ($YBa_2Cu_3O_{7-x}$), EBCO ($EuBa_2Cu_3O_{7-x}$), YEBCO ($Y_{0.9}Eu_{0.1}Ba_2Cu_3O_{7-x}$) and YGBCO ($Y_{0.9}Gd_{0.1}Ba_2Cu_3O_{7-x}$) were synthesized by the sol-gel method and once heat treated in air; calcination(940K, 12h) and annealing(470K, 6h). X-ray powder diffraction shows that the compounds are orthorhombic crystal structures and no impurity phases are contained. Resistivities were measured using a d.c. four-probe method with contact probes. Samples were semicircular disks with 12 mm in diameter and ca. 5 mm in thickness.

A small amount of Cr(3+) contained MgO powder were mixed into the ground fine powder of the high-Tc cuprates which were heat treated beforehand to eliminate any ESR active impurities.[5] Both powder size were adjusted between 200 and 325 mesh (74-44 micron). A Pyrex capillary (1 mm i. d.) was used as a ESR sample tube .

The ESR spectra were obtained using a X-band JEOL spectrometer (JES-RE3X) and magnetic field was calibrated with a proton gauss meter of ECHO Electronics (EFM-2000AX). The temperature was measured and controlled by a temperature indicator/controller of Scientific Instruments (Model 5500). The temperature accuracy was $\pm$ 0.5 K.

RESULTS AND DISCUSSION

Figure 1 gives electrical resistivity vs. temperature measurements of bulk tablets of the ceramic oxides. All materials show metallic temperature dependence of resistivity and sharp superconducting transition at Tc= 91-93 K except for EBCO. It reveals that the heat treatment condition is not optimum for the formation of EBCO superconductors. It is known that no depression of Tc was observed for full substitution of Y with other rare earth element, except La, Pr and Tb.[3] Obtained results here indicate that Tc dose not degrade by partial substitution of Y with Eu or Gd. However, the diagrams indicate a small difference of Δ Tc between YBCO and YEBCO,YGBCO. The same oxides were used for the following ESR measurements after grinding into powder.

As a paramagnetic probe, chromium doped MgO powder was used because the sample is stable and ESR signal exhibits narrow single line (g = 1.97995) of trivalent cation ^{52}Cr(I=0) and weak quartet from ^{53}Cr(I=3/2 ; natural abundance 9.50 %) Narrower ESR line width ($\Delta H_{m.s.l.}$ = 0.12 mT) is obtained copmared to that of solid DPPH (2,2-Diphenyl-1-picryl hydrazyl) if doping content of Cr is optimum. Generally used free radical DPPH has many hyper fines and is not stable at the suface of oxides especially for heating (m.p. 130 C). Fig. 2. gives relative line width (ΔH_R ; normalized with Cr(3+) spectra in genuine MgO powder) vs. temperature of each oxides. It is apparent from the figure that no appreciable line broadening is observed from 300 to 90 K for each oxides. Peak to peak intensities (Ip-p) were gradually decreased from about 85 K to lower temperature with increase of line width . We showed ESR spectra at 77 K in the insets of each graph. Ip-p decrease for EBCO is so slight that quartet peak of ^{53}Cr isotope can still been recognized at 77 K. Note that line broadening do not occur from Tc but from a lower temperature even for YBCO. Difference in line broadening phenomena among the oxides are distinct at 77 K. Introduction of different rare earth element like Eu and Gd into Y position causes distortion or strain in the YBCO crystal and degrades the quality of the superconductor without Tc decrease.

There are two type of line broadening in ESR spectra; homogeneous or inhomogeneous broadening.[5] In our experimental condition, main source of line broadening is considered to be inhomogeneous broadening ; the applied magnetic field inhomogeneities over the volume of the sample is caused by the magnetic flux change or fluctuation of inside as well as outside of oxides. If a powder oxide is pure single phase superconductor, line broadening should start at Tc and increase toward lower temperatures.[6]

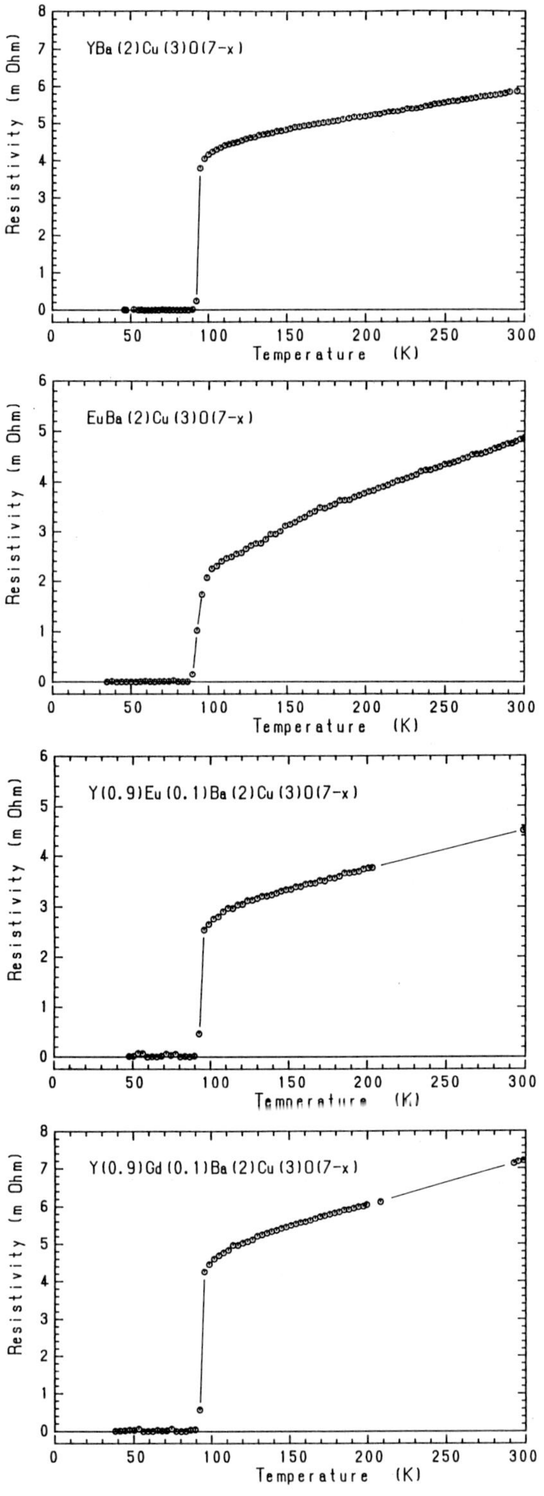

Fig. 1. Resistivity vs. temperature from 300 K to below Tc for the $RBa_2Cu_3O_{7-x}$ (R= Y,Eu,Gd) series.
Composite ratio of rare earth elements is shown in each diagram.

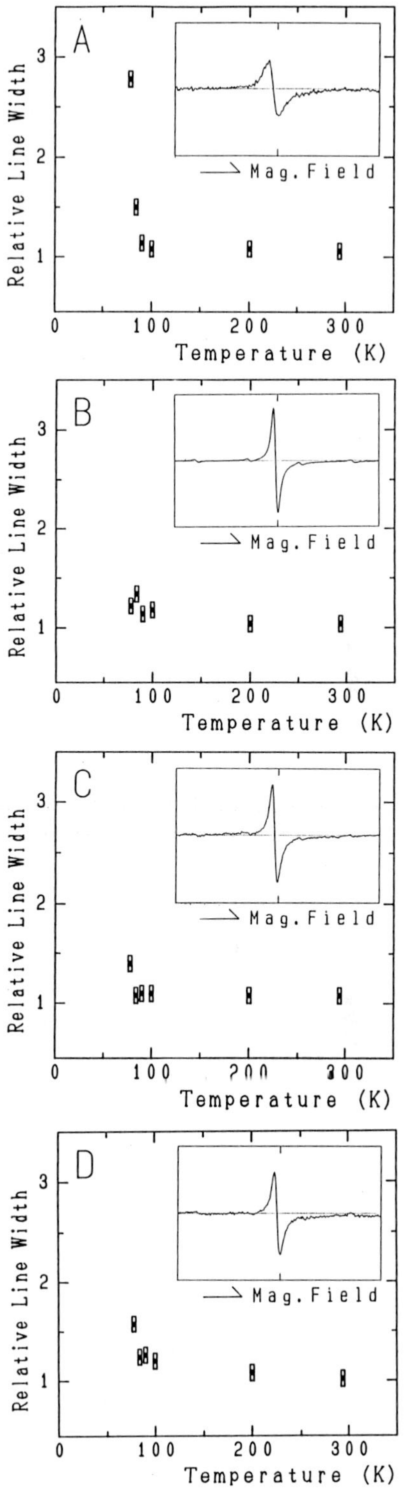

Fig. 2. The relative line width vs. temperature of the Cr^{3+} ESR in MgO coexisted with YBCO(A), EBCO(B), YEBCO(C), YGBCO(D) to MgO powder.
The insets are the ESR absorption at 77 K.
Range of magnetic field is 7 mT in full span.

However, starting temperature of line broadening in the prepared oxides is lower than Tc of each bulk oxides (compare Fig. 1 and Fig. 2). The facts indicate that Tc of the powder oxides are lower than those of corresponding bulk oxides. Increase of surface area, defects, voides, mismatch at grain boundaries and oxygen deficiency are attributed to the cause of Tc degradation. It is also apparent that ΔH_R(77 K) has close relation with Δ Tc of bulk oxides. The fact shows the quality characterization of bulk oxides are evaluated from the ΔH_R(77 K) measurement. A disadvantage of the method is the decrease of ESR absorption from Tc to lower temperater due to the shielding of microwave and modulation field which applied for phase sensitive detection besides the nonresonant microwave absorption [7]. No appreciable paramagnetic effect for line width and g shift by Eu^{3+} and Gd^{3+} [8] was observed in our measurements.

In conclusion, we showed that the characterization of the high-Tc fine particles as well as bulk oxides is possible with a paramagnetic-probe method of ESR on powder mixtures of the high-Tc cuprate and Cr^{3+} doped MgO. ESR measurements varying temperature for the each fine powder showed definite quality difference between YBCO and others. The phase transition (metallic-to superconductor) of the powder do not critically depend on temperature. Decrease of Tc was observed for the fine powder samples compared with their bulk state indicating quality degradation of the oxides ; reduction of superconducting phase.

REFERENCES

1. Henglein A (1989) Chem. Rev. 89: 1861-1873
2. Farach HA, Quagliata E, Mzoughi T, Mesa MA, Poole,Jr CP, Creswick R(1990) Phys. Rev. B 41: 2046-2048; Maniwa Y, Sato H, Mizoguchi K, Kume K (1990) Jap. J. Appl. Phys. 29: 268-269; Rakvin B, Mahl TA, Bhalla AS, Sheng ZZ, Dalal NS (1990) Phys. Rev. 769-771
3. Tarascon JM, McKinnon WR, Greene LH, Hull GW, Vogel EM (1987) Phys. Rev. 36: 226-234
4. Kaise M, Mizuno M, Nishihara C, Nozoye H, Shindo H (1990) Advances in Superconductivity II (Proceeding of 2nd ISS). Springer-Verlag, pp543-546
5. Poole,Jr CP (1967) Electron Spin Resonance. Interscience Publishers. John Wiley & Sons, pp699-705
6. Rakvin B, Pozek M, Dulcic A (1989) Solid State Commun. 72: 199-201
7. Stankowski J, Kahol PK, Dalal NS, Moodera JS (1987) Phys. Rev. B36: 7126-7128
8. Causa MT, Fainstein C, Nieva G, Sanchez R, Steren LB, Tover M, Zysler R. (1988) Phys. Rev. B38: 257-261

Microwave Absorption of Thin Films Formed by MOCVD

Y. YAMADA, H. HAYASHI, Y. SHIOHARA, and S. TANAKA

Superconductivity Research Laboratory, International Superconductivity Technology Center,
10-13, Shinonome 1-chome, Koto-ku, Tokyo, 135 Japan

ABSTRACT

Microwave absorption of thin films formed by MOCVD were measured using a conventional electron spin resonance(ESR) equipment. Three different types of microwave absorption signals were detected in a superconducting state. The one was giant absorption at a zero magnetic field just below Tc. The second appeared below 20K and its position was not at a field of zero. The third appeared at about 5K as a noisy-looking signal. These signals seem to be related to a motion of fluxoid and a coupling condition among grains.

KEY WORDS: high Tc superconductor, microwave absorption, thin film

INTRODUCTION

It is well known that high Tc superconductors have remarkable microwave absorption characteristics in a superconducting state at a low magnetic field[1,2]. There has been a lot of research groups studying the phenomena and most of them suggested that the phenomena are correlated with a weak link, however, the mechanism of the absorption has not been clear yet[3,4,5]. The appearance of absorption is very sensitive to the critical temperature of the specimen, suggesting that it depends on a condition of coupling between superconducting grains and the measurements of this behavior are expected to probe characteristics of the material. In this point of view, we have investigated microwave absorption. In this study, we found some unique behaviors which could not be explained well.

EXPERIMENTAL

The sample used in this study was Y-based thin films with thickness of 1000Å formed on a single crystal MgO substrate by MOCVD. This film was fabricated at 800°C in 20 Torr with flowing MO sources of metal β-diketonate, and it consisted of 123 phase and 124 phase as shown in figure 1. Its critical temperature of 82.5K was measured by a standard transport technique.

Microwave absorptions were detected by JEOL RE-3X ESR spectrometer operating at a frequency of 9.0 GHz (X-band).

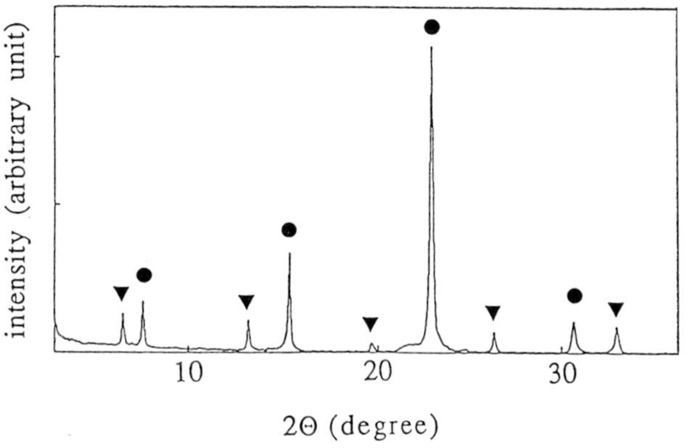

Fig.1 X-ray diffraction pattern of the thin film. The triangles and circles denote the line positions of the 124 phase and the 123 phase, respectively.

Temperature was controlled by helium gas flow cryostat from 4 to 300K. The static magnetic field Ho, which was modulated at 100kHz, was horizontal and microwave magnetic field H_1 was vertical at the sample position in the cavity. It is so important to measure the microwave absorption at about zero magnetic field that the electromagnet was improved to cancel the residual field by flowing electric current oppositely.

The sample was placed in the center of the cavity where the microwave magnetic component is the maximum. The film surface was parallel to the microwave magnetic field and normal to the static field. The sample dimension was $1x3x0.0001mm^3$.

RESULTS and DISCUSSION

Figure 2 shows the microwave absorption derivatives between -1000G to 1000G at various temperatures. The narrow signals at about 900G and -900G are originated from ESR of impurities of the MgO substrate. As the temperature decreased, giant microwave absorption occurs at several degrees below Tc, and one could notice that there is another weak signal which has an opposite slope at the zero field just below Tc. The occurrence of the oppositely sloped signal has already been reported on single crystals[4] and it was confirmed in thin films. It will be necessary to explain the cause of these two signals, not only by the giant absorption but also by the oppositely sloped signal, to reveal the mechanism of this type of absorption.

There are other interesting characteristics shown in figure 2. As the temperature decreased down to about 10 degrees below Tc, the giant microwave absorption disappears, however, another microwave absorption appears at a field of 950G below 20K.This type of signal was detected in another sample. This signal could be detected most of times, but not always, though the ESR signal was always detected. This indicates that it was not a ESR signal but a kind of microwave absorptions depending on the sample conditions which was not controlled in this measurement. A. Dulcic et. al. [6] have mentioned a signal whose position is not at the zero field. The signals detected in present measurements are similar to theirs, but it is not obvious that these are resulted from the same origin. If these signals are based on as the same mechanism as the giant absorption near Tc, i.e. the mechanism related with weak linked junctions, there might be junctions which have microwave absorbed at a peculiar magnetic field.

Another interesting behavior is appearance of a noisy-looking absorption signal at 5K. The amplitude of this signal increases as the strength of the static magnetic field increases. The signal remains as a noisy-looking oscillation when the magnetic field is not swept. Figure 3a shows the field

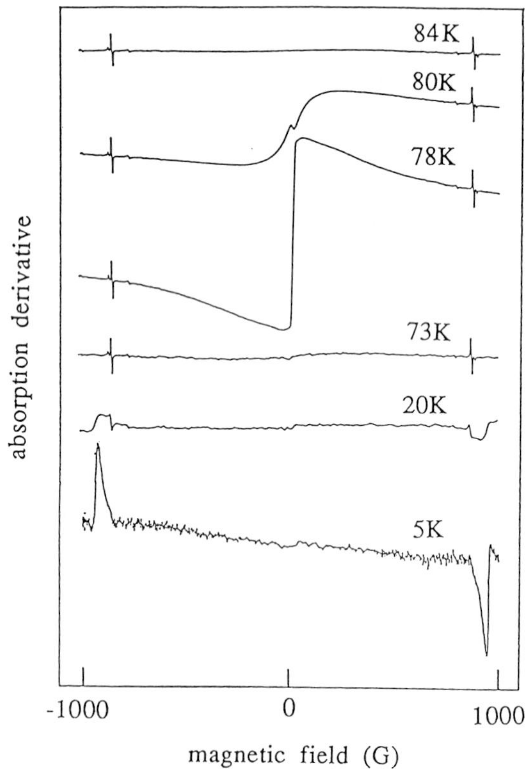

Fig.2 The magnetic field dependent microwave absorption. Modulation amplitude is 10G.

dependence of this signal up to 9000G. The signal is small at the zero field and its amplitude is the maximum at around 2000G. Above about 4000G, it reduces and remains in a very low level. The signal shows some patterns even the measured conditions were the same. Typical patterns are shown in figure 3. Figure 3b shows that the noisy-looking signal disappeared across to the zero field without changing conditions of the sample. Figures 3c and 3d show that signals oscillate in almost the same interval of 13G. These data in figure 3 indicate that this signal depends on the static magnetic field, but its appearance or shape are not directly related to the field. The oscillations shown in figures 3c and 3d resemble to resonating phenomena which happen to appear or disappear by trigger from outside of the system.

It seems that this behavior is also originated from the interaction of the static magnetic field with microwave through superconductor[7,8,9,10]. If it is true, a possible mechanism to explain this behavior could be correlated to the motion of fluxoids. When fluxoids vibrate around the pin potential and in this process they move in a weakly coupled junction among superconducting grains. Then the phases of the wavefunction of superconducting grains slip each other and the response of microwave absorption oscillate alternatively. To consider the similarity between this oscillation and a resonating phenomena, there is a possibility of the microwave resonance with the motion of flxoids.

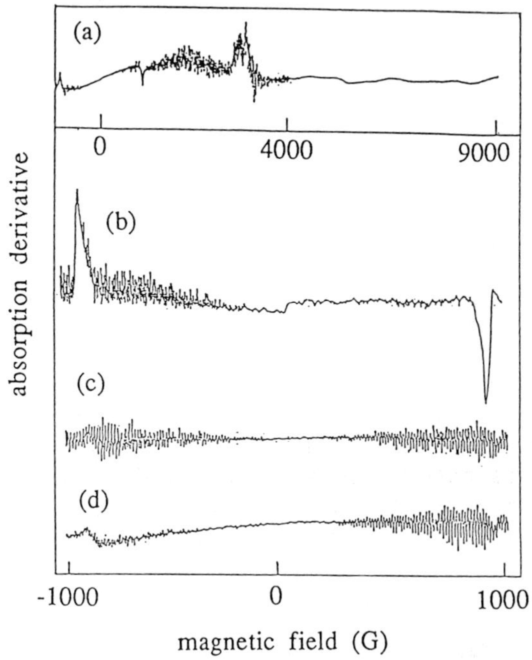

Fig.3 Typical patterns of a noisy-looking absorption. Modulation amplitude are 10G (a,b) and 2G (c,d).

In conclusion, we found three different types of microwave absorption signals. The first appears below near Tc and has a very large strength. The second appears below 20K at about 950G. The last one appears at about 5K and the signal oscillates like a noise. These three types seem to be originated from the interaction between magnetic field and microwave through the weak link in the superconductors.

ACKNOWLEDGMENT

A part of this work was supported by New Energy and Industrial Technology Development Organization under the management of the R&D Basic Technology for Future Industries.

REFERENCES

1. Glarum SH Marshall JH Schneemeyer LF (1988) Phys. Rev. B 37: 7491
2. Pozek M Dulcic A Rakavin B (1989) Solid State Communications 70: 889
3. Vendik OG, Gaidukov MM, Karpiuk A, Kolesov SG, Kowalewicz L, Kozyrev AB, Popov AYu (1990) Solid State Communications 74: 625

4. Dulcic A, Crepeau RH, Freed JH (1990) Phys. Rev. B 42: 2155
5. Pukulis EJ, Blum SE (1990) Appl. Phys. Lett. 57: 1464
6. Dulcic A, Crepeau RH, Freed JH (1988) Phys. Rev. B 38: 5002
7. Blazey KW, Portis AM, Muller KA, Bednorz JD, Holtzberg F (1988) Pysica C 153-155: 56
8. Vichery H, Beuneu F (1989) Pysica C 159: 823
9. Dulcic A, Crepeau RH, Freed JH (1989) Physica C 160: 223
10. Ting-kang Xia, Stroud D (1989) Phys. Rev. B 39: 4792

Correlation between T_c and Eu Lattice Vibration in $(La_{1-x}Eu_x)_{2-y}Ce_yCuO_{4-\delta}$ Observed by ^{151}Eu Mössbauer Effect

HIDEAKI ITOH[1], MIKIO KUSUHASHI[1], and MASAFUMI TANIWAKI[2]

[1] Muroran Research Laboratory, The Japan Steel Works Ltd., 4 Chatsu-machi, Muroran, 051 Japan
[2] Department of Electronic Engineering, Faculty of Engineering, Hokkaido University, Sapporo, 060 Japan

ABSTRACT

The electron doped T'-phase superconductor $(La_{1-x}Eu_x)_{2-y}Ce_yCuO_{4-\delta}$ were studied by X-ray powder diffraction and ^{151}Eu Mössbauer effect. The obtained isomer shift and the Debye temperature showed that Eu is trivalent but Θ_D was smaller than those in the hole-doped superconductors ($EuBa_2Cu_3O_{7-\delta}$ and T*-phase superconductor). With increasing Eu concentration, the lattice constants shortened and T_c increased. T_c enhancement can not be necessarily explained by the change of the lattice constants. A tight correlation between T_c and Θ_D was found. We try to explain it using McMillian's equation.

KEY WORDS: $(La_{1-x}Eu_x)_{2-y}Ce_yCuO_{4-\delta}$, ^{151}Eu Mössbauer effect, lattice constant, T_c enhancement, Debye temperature

INTRODUCTION

High-T_c superconductors consist of many elements, which causes difficulty in the understanding of its superconductivity. The direct observation of microscopic properties of the constituent elements is thus required. Several research groups[1-6], have studied the ^{151}Eu Mössbauer effect to examine the electronic, magnetic and lattice vibration states of the lanthanide in a hole-doped "1-2-3" superconductor, since the Mössbauer effect directly indicates the states of the Mössbauer atom. Through these studies, it has been established that Eu in the "1-2-3" superconductor is trivalent and not magnetically ordered. The Debye temperature of Eu obtained from the temperature dependence of the line area was 255-300K. Recently, Tokura et al. found an electron-doped superconductor, T'-phase $Ln_{2-x}Ce_xCuO_{4-\delta}$ (Ln=Pr,Nd and Sm)[7]. By Yoshimoto et al.[8], this was extended to $(La,Eu)_{2-x}Ce_xCuO_{4-\delta}$ where the weighted average ionic radius is nearly equal to that of Nd^{3+}. In this paper, the electronic and lattice vibration state of Eu in T'-phase $(La_{1-x}Eu_x)_{2-y}Ce_yCuO_{4-\delta}$ is studied by the ^{151}Eu Mössbauer effect and discussed with the lattice constant and the superconductivity. In particular, we aim to detect some correlations between the superconductivity and Mössbauer parameters by changing x and y.

EXPERIMENTAL

The samples were prepared by solid-state reaction from mixtures of La_2O_3(99.99%), Eu_2O_3(99.99%), CeO_2(99.9%) and CuO (99.9%) with the nominal compositions of $(La_{1-x}Eu_x)_{2-y}Ce_yCuO_{4-\delta}$ (x=0.33 and 0.5, y=0.1 and 0.2). The mixed powders were calcined in air at 1000°C for 5 hours, pulverized, pressed into pellets and sintered in a stream of Ar at 1050°C for 10 hours. The samples were cooled to room temperature at a cooling rate of 100°C/h in the same atmosphere.

The crystal structure in the sample and the lattice constants of the T'-phase were determined from an X-ray powder diffraction pattern. The superconducting transition onset temperature, T_c, was estimated by a conventional AC inductance method. The Mössbauer spectrum was measured at room temperature using a 21.55 keV $^{151}SmF_3$ γ-ray. In order to estimate the lattice vibration from the line area, Mössbauer samples with a constant σn(=3) were used, where σn is the product of the resonance cross section and the number of Mössbauer nuclei per unit area.

RESULTS AND DISCUSSION

The diffraction patterns showed only the T'-phase in the sample of (x,y)=(0.33,0.1), and the predominant T'-phase and a small amount of a secondary phase in the other three samples. Figure 1 shows the diffraction pattern of (x,y)=(0.2,0.33). The secondary phase is concluded to be $(La,Eu)_2O_3$ with a cubic Eu_2O_3 structure because of its lattice constant, a (ca. 11.04Å), is much larger than that of Eu_2O_3 (10.868Å). The volume fraction of these phase was estimated from the peak height of the most intensive line of the respective phase. These value and the lattice constants of the T'-phase are tabulated in Table I, with the superconducting-transition-onset temperature, T_c. The length of the a-axis did not depend on the Ce concentration, but the c-axis was shortened by the increase of Ce concentration. This behavior is the same as observed in $Nd_{2-x}Ce_xCuO_{4-\delta}$ [9,10]. The dependences of the lattice constants, a and c, on Eu concentration were almost the same in each y. The a- and c-axes were shortened by the increase of Eu concentration.

Figure 2 shows the ^{151}Eu Mössbauer spectrum of the single T'-phase ((x,y)=(0.33,0.1)). All Mössbauer spectra obtained in this investigation showed a single line as this spectrum. The spectra were analyzed by the non-linear least squares method using a Lorentzian. The line position (isomer shift) relative to EuF_3 at room temperature, the linewidth and the line area are tabulated in Table II. Comparing the line area with that of the sample with a known recoilless fraction, we obtained the recoilless fraction, f, and estimated the Debye temperature. The obtained parameters are listed in Table II together. The observed isomer shift shows that Eu in the electron-doped T'-phase superconductor is trivalent as the lanthanide in the hole-doped "1-2-3" superconductor[1-5] and T^*-phase superconductor[11]. The Eu isomer shift in the electron-doped superconductor is slightly less than those in hole-doped superconductors (0.80-0.86 mm/s). In T'-, T^*- and "1-2-3" phases, the local structure of Eu is similar; Eu is coordinated by eight oxygen atoms at the corners of a hexahedron. Therefore, it is considered that the difference in the isomer shift is attributed to the nature of Eu-O bonding. The ^{151}Eu isomer shift increases with pressure applied to the sample because of the increase in the electron density at the nucleus[12]. The isomer shift in three phases shows that the Eu-O distance in the T'-phase is larger than that in hole-doped superconductors, as is shown exactly by the neutron diffraction studies[13-15]. The Debye temperature of Eu in the T'-phase was less than those in T^*- and "1-2-3"-phase superconductors, which were above 200K[6,11]. This shows that the Eu lattice vibration in the T'-phase superconductor is soft. It coincide with consideration from isomer shift data.

Table I X-ray diffraction and superconductivity data on $(La_{1-x}Eu_x)_{2-y}Ce_yCuO_{4-\delta}$. T' and C denote tetragonal Nd_2CuO_4 phase and cubic $(La,Eu)_2O_3$ phase, respectively.

Sample x	Sample y	Identified phase	Lattice parameter of T'-phase a(Å)	c(Å)	T_c onset (K)
0.33	0.1	T'	3.978	12.259	13.8
0.33	0.2	T' + C(4%)	3.978	12.214	8.7
0.5	0.1	T' + C(1%)	3.962	12.182	17.0
0.5	0.2	T' + C(8%)	3.961	12.150	14.7

Table II ^{151}Eu Mössbauer spectrum parameters and lattice vibration parameters of $(La_{1-x}Eu_x)_{2-y}Ce_yCuO_{4-\delta}$.

Sample x	y	Isomer shift (mm/s)	Linewidth (mm/s)	Line area (%mm/s)	Recoilless fraction f	Debye temp. of Eu Θ_D(K)
0.33	0.1	0.733±0.013	2.45±0.05	16.9±0.5	0.315	172
0.33	0.2	0.689±0.008	2.37±0.03	20.1±0.4	0.396	192
0.5	0.1	0.707±0.015	2.38±0.06	14.8±0.5	0.265	160
0.5	0.2	0.703±0.013	2.37±0.05	16.2±0.5	0.298	168

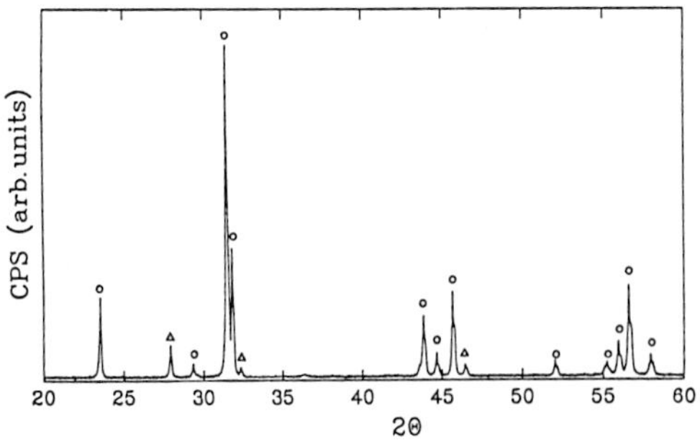

Fig.1 The X-ray diffraction pattern of $(La_{0.5}Eu_{0.5})_{1.8}Ce_{0.2}CuO_{4-\delta}$, ○:tetragonal Nd_2CuO_4 structure △:cubic $(La,Eu)_2O_3$ structure.

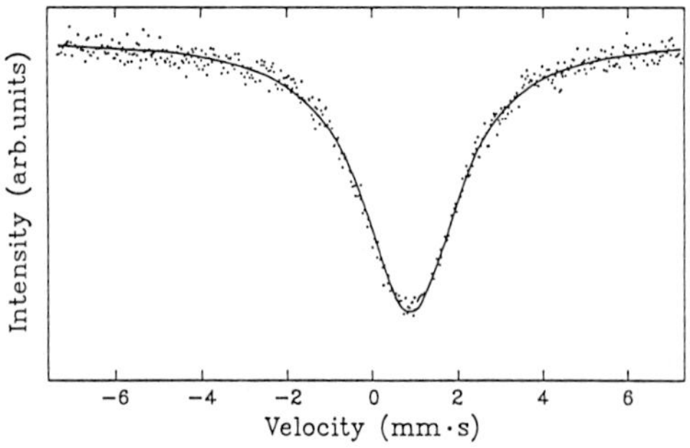

Fig.2 The ^{151}Eu Mössbauer spectrum of single T'-phase superconductor $(La_{0.67}Eu_{0.33})_{1.9}Ce_{0.1}CuO_{4-\delta}$ at room temperature.

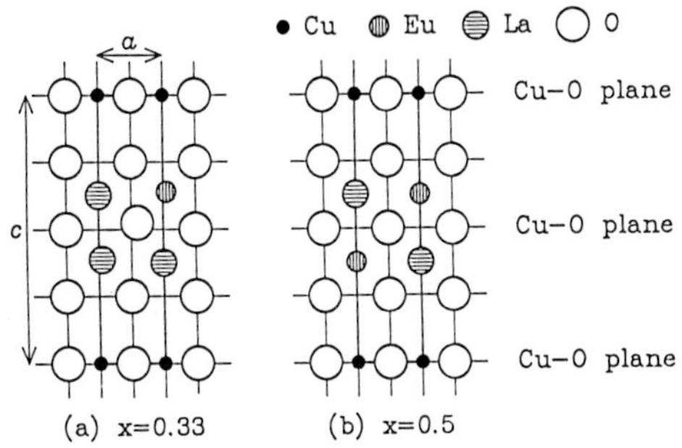

Fig.3 Schematic model of atomic arrangements in (100) plane of T'-phase $(La_{1-x}Eu_x)_{2-y}Ce_yCuO_{4-\delta}$,
(a) Eu concentration, x=0.33
(b) Eu concentration, x=0.5.

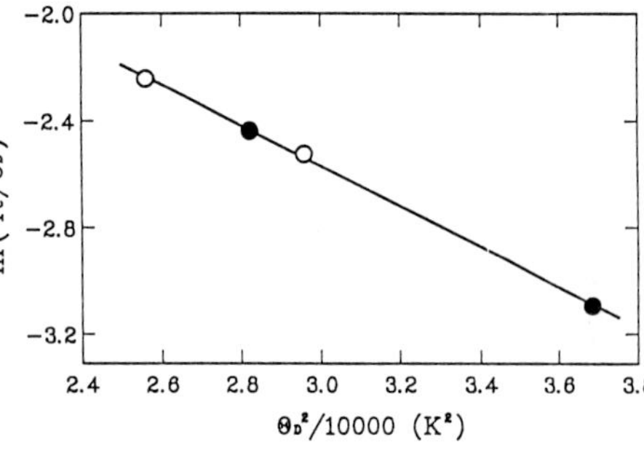

Fig.4 Plot of $\ln(T_c/\Theta_D)$ against $\Theta_D{}^2$ in $(La_{1-x}Eu_x)_{1.9}Ce_{0.1}CuO_{4-\delta}$(○) and $(La_{1-x}Eu_x)_{1.8}Ce_{0.2}CuO_{4-\delta}$(●).

In each y, the T_c was higher at x=0.5 than those at x=0.33. From the comparison of T_c with the lattice constant, we may derive the conclusion that the T'-phase superconductivity is enhanced with the decrease in lattice constant or Cu-O bond length, which is the same as reported for other T'-phase superconductors[16,17]. However, this conclusion cannot successfully explain the entire data. For example, nevertheless the sample of (x,y)=(0.33,0.2) has smaller lattice constants than the sample of (x,y)=(0.33,0.1), it shows the lowest T_c of the four samples. Furthermore, Θ_D increased with the decrease in lattice constant. Therefore, we should consider the change of the local structure around Eu. The increase in Eu concentration will change the Eu-O-La linear bonding in the (100) or (010) plane into Eu-O-Eu, as shown in Fig. 3. In the sample of x=0.33, Ln of Eu-O-Ln is almost La and the Eu-O distance is compressed due to the large ion radius of La^{3+}, 1.16Å, which is much larger than 1.07Å of Eu^{3+}[18]. To the contrary, in the sample of x=0.5, Ln of Eu-O-Ln linear bonding might be occupied by Eu. Unfortunately, we do not know the Eu-O distance in our samples; thus we estimated the Eu-O distance roughly. The Nd-O(1) distance in $Nd_{1.85}Ce_{0.15}CuO_{4-\delta}$ (a=3.9469Å and c=12.0776Å) is 2.659Å[15]. Assuming that the Ln-O(1) distance is proportional to $\sqrt{(a/2)^2+(c/8)^2}$, the average Ln-O(1) distance is 2.687Å and 2.674Å in the samples of (x,y)=(0.33,0.1) and (0.5,0.1), respectively. Although, the Eu-O distance in Eu-O-La will be smaller than 2.687Å due to the difference in ionic radius between La and Eu, 0.09Å. Assuming that the decrease of the Eu-O distance is half of 0.09Å, the Eu-O distance in Eu-O-La might be 2.642Å and smaller than that in Eu-O-Eu.

One notices a correlation between the Θ_D and T_c. The transition temperature, T_c, decreases with increasing Θ_D, which seems to indicate that the superconductivity in the T'-phase is enhanced by softening of the Eu lattice vibration. Plotting $\ln(T_c/\Theta_D)$ against Θ_D^2 in Fig. 4, we found a good linear relationship between $\ln(T_c/\Theta_D)$ and Θ_D^2, and the equation, $T_c=0.772\,\Theta_D\exp(-\Theta_D^2/13000)$ was obtained. In the following paragraph, we try to explain our data using McMillan's equation[19].

McMillan's equation is represented approximately as $T_c=(\hbar/k_B)\langle w\rangle\exp\{-(\lambda+1)/\lambda\}$, where $\langle w\rangle$ is the average phonon frequency and λ is an electron-phonon coupling constant which is $N(0)\langle g^2\rangle/M\langle w^2\rangle$ ($N(0)$: the electron density of states at the Fermi surface; $\langle g^2\rangle$: an average over the Fermi surface of the square of the electronic matrix element; $\langle w^2\rangle$: the average of the square of the phonon frequency). Using $\langle w^n\rangle = \int dw\, \alpha^2F(w)w^{n-1}/\int dw\, \alpha^2F(w)/w$ (α^2: an average of the electron-phonon interaction; $F(w)$: the phonon density of states), and assuming α^2=const. and $F(w)\propto w^2$, $T_c=0.245\,\Theta_D\exp(-k_B^2M\,\Theta_D^2/2\hbar^2N(0)\langle g^2\rangle)$ is derived as a theoretical equation which has the same form as the experimental one. Though $N(0)\langle g^2\rangle$ estimated from the experimental equation is 1.76 eV/Å^2, which is within the range of the values in many conventional superconductors, the pre-exponential factor, $0.772\,\Theta_D$, is larger than that in the theoretical equation by a factor of 3.15. Since this value is equal to $\sqrt{M/M'}$ (M,M': atomic masses of Eu and O), the optical phonon may contribute to the superconductivity. The above discussion suggests the possibility that the Cooper pair is formed through the interaction between electrons and the lattice vibration of the lanthanide or the oxygen bonding with it in the Cu-O plane.

In conclusion, a tight correlation between the lattice vibration and T'-phase superconductivity is clearly demonstrated in our experiment, although the contribution of the phonon to the high-temperature superconductivity has been considered to be unimportant. More effort should be made to clarify the lattice vibration in the high-T_c superconductors.

ACKNOWLEDGMENT

We are grateful to Prof. K. Kumagai of the Faculty of Science, Hokkaido University, who measured the AC susceptibility of one of our samples and advised us. We also thank Prof. Yoshiie of the Faculty of Engineering, Hokkaido University for his advice.

REFERENCES

1. Coey JM, Donnelly K (1987) Z. Phys. B67: 513
2. Wortmann G, Blumenroder S, Freimuth A (1988) Phys. Lett. A126: 434
3. Taniwaki M, Sasaki H (1988) Physica C153-155: 1549
4. Dalmas de Reotier P, Vulliet P, Yaounac A, Hartmann O, Chaudouet P, Garcon S, Senateur JP, Weiss F, Kalvius GM (1988) Physica C153-155: 1543
5. Taniwaki M, Muraki T, Shiramine K (1989) Physica C162-164: 568
6. Muraki T, Taniwaki M Shiramine K (1989) In: Ishiguro, Kajiyama (eds) Proc. of 2nd Int. Sym. on Superconductivity Springer-Verlag.Tokyo p 563
7. Tokura Y, Takagi H, Uchida S (1989) Nature 337: 345
8. Yoshimoto M, Hashimoto T, Takaya T, Koinuma H (1989) Jpn. J. Appl. Phys. 28: L1115
9. Uji S, Aoki H, Matsumoto T (1989) Jpn. J. Appl. Phys. 28: L563
10. Huang TC, Moran E, Nazzal AI, Torrance JB (1989) Physica C158: 148
11. Taniwaki M, Itoh H, Kusuhashi M (1990) 3rd Int. Sym. on Superconductivity, Sendai, PCP-12
12. Grandjean F, Long GJ (1990) Mössbauer spectroscopy applied to inorganic chemistry. Plenum Press. New York, pp 518-519
13. Izumi F, Asano H, Ishigaki T, Takayama-Muromachi E, Uchida Y, Watanabe N (1987) Jpn. J. Appl. Phys. 26: L1193
14. Izumi F, Takayama-Muromachi E, Fujimori A, Kamiyama T, Asano H, Akimitsu J, Sawa H (1989) Physica C158: 440
15. Izumi F, Matsui Y, Takagi H, Uchida S, Tokura Y, Asano H (1989) Physica C158: 439
16. Takayama-Muromachi E, Uchida Y, Kato K (1990) Physica C165: 147
17. Oh-Ishi K, Shono Y, Kikuchi M, Muto Y (1990) Solid State Commun. 73: 341
18. Shannon RD, Prewitt CT (1969) Acta Crystallogr. B25: 925
19. McMillan WL(1968) Phys. Rev. 167: 331

Lattice Vibration in Aligned $EuBa_2(^{57}Fe_{0.01}Cu_{0.99})_3O_7$ Observed by Mössbauer Spectroscopy

KEN-ICHI SHIRAMINE, TATSUYA MURAKI, and MASAFUMI TANIWAKI

Department of Electronic Engineering, Faculty of Engineering, Hokkaido University, Sapporo, 060 Japan

ABSTRACT

^{57}Fe Mössbauer spectra of *c* axis aligned $EuBa_2(^{57}Fe_{0.01}Cu_{0.99})_3O_7$ were measured at 80-330K. From the temprature dependence of the total line area of three doublets, the Debye temperature, Θ_D, was estimated. There was no significant difference between Θ_D in the lattice vibration of iron along the *c* axis and that in the basal plane. However this result does not mean the absence of the lattice vibration anisotropy of copper in $EuBa_2Cu_3O_7$.

KEY WORDS: $EuBa_2Cu_3O_7$, Fe substitution, ^{57}Fe Mössbauer spectroscopy, lattice vibration

INTRODUCTION

Since the discovery of high-Tc superconductor, enormous efforts were devoted to investigate its properties. Although it is established that it has anisotropic superconductivity, microscopic (site specific) property has not been well known.

In this work, we tried to clarify the lattice vibration in copper site of high-Tc superconductor $EuBa_2Cu_3O_7$ by Mössbauer spectroscopy. Mössbauer spectroscopy shows the electronic structure and the lattice vibration of particular site (Mössbauer nuclei). Since copper is not Mössbauer element, we substituted ^{57}Fe for copper of 1at.% in $EuBa_2Cu_3O_7$, for probe.

When iron is doped into $YBa_2Cu_3O_7$, Tc decrease monotonically and falls to 0K at 15 at.%[8,9]. At higher concentration, $YBa_2Cu_3O_7$ shows semiconductive behavior in the temperature dependence of the resistivity. Simultaneously, the lattice constant *a* and *b* become close, and the orthorhombic to tetragonal crystallographic phase transition occurs at 3 at.%[8,10]. It is considered that $EuBa_2Cu_3O_7$ shows a same behavior as $YBa_2Cu_3O_7$. At 1 at.% doping, Tc and crystal structure do not deviate largely from those of nondoped material.

It is reported that the *c* axis of $YBa_2Cu_3O_7$ powder is aligned parallel to applied magnetic field[1]; in this study, the samples were prepared by this method. It is possible to detect the lattice vibration anisotropy by Mössbauer spectroscopy, because its line area depends on the lattice vibraton parallel to the incident γ ray. Mössbauer spectra were measured as the incident γ ray was parallel to and perpendicular to the aligned *c* axis.

EXPERIMENTAL

Starting materials were weighed out so that compound becomes $EuBa_2(^{57}Fe_{0.01}Cu_{0.99})_3O_7$. Mixed powders of Eu_2O_3, $BaCO_3$, CuO and $^{57}Fe_2O_3$ were sintered at 1193K in air for 24h and was ground; this cycle was repeated three times. It was annealed at 823K in flowing oxygen for 24h. After that, $EuBa_2(^{57}Fe_{0.01}Cu_{0.99})_3O_7$ was ground to fine particles, mixed in epoxy, and healed in a magnetic field of 0.5-2T until epoxy solidified.

Mössbauer measurements were performed by the transmission method using aligned samples of σn=2 as absorbers (σ : Mössbauer cross section; *n*: surface density of ^{57}Fe). Velocity axis was standardized by natural iron. A gas proportional counter was used as a γ -ray detector. A 14.4keV γ -ray source was ^{57}Co-Rh at room temperature. The Mössbauer spectrum at temperatures in a range of 80-330K was recorded.

The observed spectrum was analyzed assuming three quadrupole doublets. In all analyses, line width was fixed to that of the nonaligned sample at room temperature(Table 1). Intensity asymmetry

Table 1 Line widths of nonaligned sample at liquid nitrogen temperature.

doublet	Line width [mm/s]
A	0.217
B	0.312
C	0.489

Table 2 Intensity asymmetries of quadrupole doublets (ratios of areas under positive to negative velocity resonances of doublets).

doublet	intensity asymmetry $\gamma \parallel c$	$\gamma \perp c$
A	0.999	0.660
B	1.625	2.155
C	1.271	0.927

Table 3 Mössbauer parameters at liquid nitrogen temperature. QS: quadrupole splitting, IS: isomer shift, AR: area ratio of doublet.

(a) γ ray $\parallel$ c

doublet	QS [mm/s]	IS [mm/s]	AR [%]
A	2.00	0.11	25.7
B	0.62	0.41	34.0
C	1.15	-0.06	40.3

(b) γ ray $\perp$ c

doublet	QS [mm/s]	IS [mm/s]	AR [%]
A	2.01	0.12	24.1
B	0.66	0.40	30.2
C	1.08	-0.08	45.7

of doublet in aligned sample, i.e., the ratio of area under positive to negative velocity resonance of doublet also was fixed to that at liquid nitrogen temperature(Table 2). The Debye model was assumed in lattice vibration analysis. The Debye temperature was estimated from the temperature dependence of line area, A, as follows[11]. If appropriate Debye temperature, Θ_D, is given, the recoil-free fraction, f, is a function of temperature, T:

$$f = \exp\left\{ -\frac{3E_R}{2k\Theta_D}\left[1+4\left(\frac{T}{\Theta_D}\right)^2 \int_0^{\Theta_D/T} \frac{u\,du}{\exp(u)-1}\right]\right\} , \tag{1}$$

where E_R is the recoil energy, and k is Boltzmann constant. Mössbauer thickness, T_A, is defined by

$$T_A = \sigma\, n\, f, \tag{2}$$

and line area, A, can be calculated from T_A,

$$A = \frac{\pi}{2}\Gamma\,\alpha\,[\,1-\exp(-T_A/2)\cdot J_0(iT_A/2)]. \tag{3}$$

where Γ is the line width, α is a constant, and $J_0(\cdot)$ is the zero order Bessel function. Line area, A, is a function of temperature, T, through Eqs.(1)-(3), and we can draw a $A(T)/A(T_0)$ curve (T_0: liquid nitrogen temperature). The Debye temperature was estimated from curve fitting of the A-T plot.

RESULTS AND DISCUSSION

The obtained spectra are shown in Fig. 1 and the typical parameters are tabulated in Table 3. The quadrupole splittings and line widths accord with other groups' results assuming three quadrupole

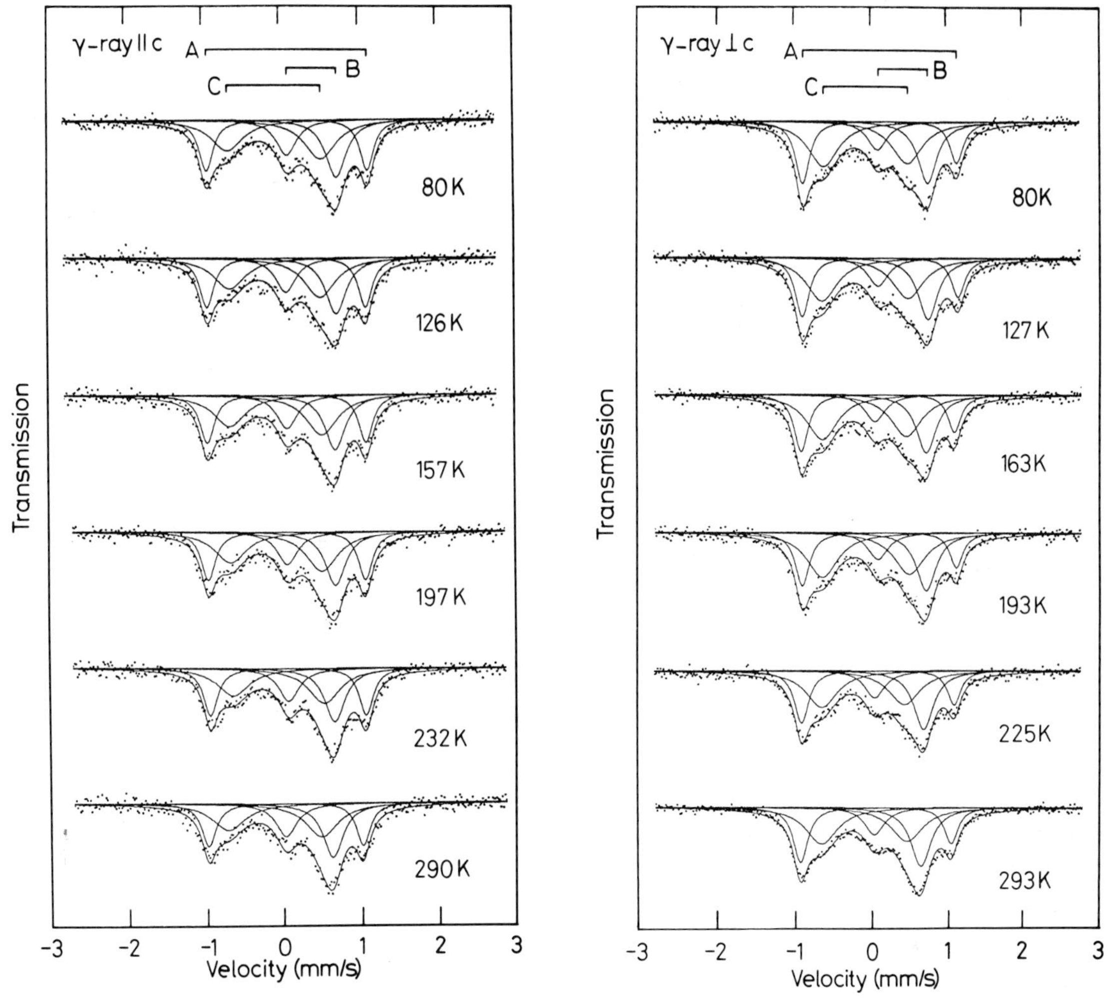

Fig. 1 Spectra of c axis aligned $EuBa_2(^{57}Fe_{0.01}Cu_{0.99})_3O_7$. (a) The γ ray is parallel to the c axis. (b) The γ ray is perpendicular to the c axis.

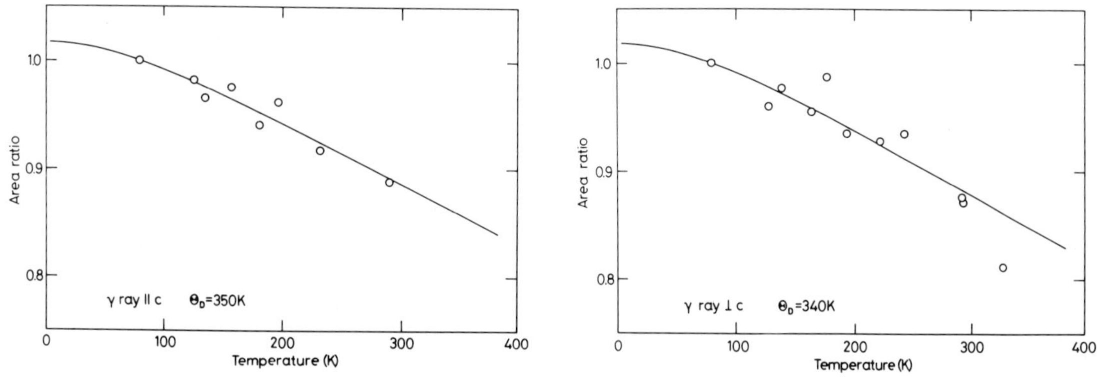

Fig. 2 Temperature dependence of line area normalized by that of liquid nitrogen temperature. (a) The γ ray is parallel to the c axis. (b) The γ ray is perpendicular to the c axis.

doublets. Mössbauer spectrum of ^{57}Fe is splitted to a doublet when charge distribution around ^{57}Fe has lower symmetry than cubic. Mössbauer spectrum of ^{57}Fe in $EuBa_2Cu_3O_7$ is composed of some doublets. Many reserch groups assume that this spectrum consists of three or four doublets and each doublet corresponds to particular oxygen ion coordination around the copper site substituted by iron atom. The line width of the doublet C is fairly larger than the natural line width of ^{57}Fe(~ 2mm/s). Several reserch groups consider that the doublet C consists of two doublets. The reported values of the line area ratio of the doublets are different from each other, because the site population of iron depends not only on iron concentration but also on heat treatment.

Figure 2 shows the plot of the total line area against temperature. The Debye temperature of the lattice vibration parallel to the *c* axis was 350K and that perpendicular to the *c* axis was 340K in copper site, which shows no significant difference in the lattice vibration of ^{57}Fe substituted for copper between these two directions.

$EuBa_2Cu_3O_7$ has two kinds of copper sites named Cu(1) and Cu(2) respectively. The Cu(1) site is in the Cu-O chain and has fourfold (square planar) coordination with oxygen ions. The Cu(2) site is in the Cu-O plane and has fivefold (pyramidal) coordination with oxygen ions. Iron atoms doped into $EuBa_2Cu_3O_7$ occupy mainly the Cu(1) sites at low concentration (for example, 1at.%), and occupy also the Cu(2) sites at high concentration[12,13].

The Cu(1) site has four oxygen ions as ligands; two on the *a* axis, two on the *c* axis, and nothing on the *b* axis. The lattice vibration displacement in the oxygen ion coordinated direction will be smaller than that in the no ion coordinated direction. Therefore the anisotropy of the lattice vibration is expected. Nevertheless, it was not clearly observed in our work. Two reasons for this are conceivable.

One is the difference between the oxygen coordination number around iron and that around copper. It is reported that the oxygen concentration in $YBa_2Cu_3O_7$ increases with increasing iron concentraion[10], which means that the oxygen coordination number around iron is larger than that around copper. The excess oxygen is expected to occupy the site on *b* axis. In this case, the anisotropy diminishes.

The other reason is the inadequacy of the analysis method. The Debye temperatures obtained above were estimated assuming a single value of recoil-free fraction in three components. So the anisotropy on each component was not estimated in the above discussion. However, the ^{57}Fe spectrum shows that there are three iron sites having different local environments from each other at least. It is reasonable to consider different recoil-free fractions for different sites. The line area of the doublet B decreases more slowly with increasing temperature than those of the doublets A and C. It shows that the lattice vibration displacement of iron site corresponding to the doublet B (site B) is smaller than those of the sites A and C. This means that the site B has more oxygen ions as ligands than the sites A and C have. The site B may be the Cu(2) site that has a fivefold coordination with oxygen ions[14].

REFERENCES

1. Farrell DE, Chandrasekhar BS, DeGuire MR, Fang MM, Kogan VG, Clem JR, Finnemore DK (1987) Phys. Rev. B36: 4025-4027
2. Pankhurst QA, Morrish AH, Zhou XZ (1988) Phys. Lett. A127: 231-234
3. Boolchand P, Blue C, Elgaid K, Zitkovsky I (1988) Phys. Rev. B38: 11313-11316
4. Bottyán L, Molnár B, Nagy DL, Szücs IS, Tóth J (1988) Phys. Rev. B38: 11373-11381
5. Baggio Saitovitch E, Scorzelli RB, Souza Azevedo I, Micklitz H (1990) Phys. Rev. B41: 2103-2108
6. Suharan S, Chadwick J, Hannon DB, Jones DH, Thomas MF (1989) Solid State Commum. 70: 817-821
7. Xia SK, Zhao ZB, Wang CR, Ma RZ, Cao GH, Ping JY (1989) Solid State Commun. 70: 141-144
8. Xu Y, Suenaga M, Tafto J, Sabatini RL, Moodenbaugh AR, Zolliker P (1989) Phys. Rev. B39: 6667-6680
9. Hennion M, Mirebeau I, Coddens G, Menelle A, Phillips TE, Moorjani K, Hervieu M (1989) Physica C159: 124-130
10. Dunlap BD, Jorgensen JD, Segre C, Dwight AE, Matykiewicz JL, Lee H, Peng W, Kimball CW (1989) Physica C158: 397-405
11. Frauenfelder H (1962) The Mössbauer effect. W. A. Benjamin, Inc.. New York
12. Bordet P, Hodeau JL, Strobel P, Marezio M, Santoro A (1988) Solid State Commun. 66: 435-439
13. Katano S, Matsumoto T, Matsushita A, Hatano T, Funahashi S (1990) Phys. Rev. B41: 2009-2016
14. Lines ME, Eibschutz M (1990) Physica C166: 235-247

Emission Mössbauer Spectra of Superconducting and Nonsuperconducting $YBa_2Cu_3O_y$ Doped with 57-Co

SERGEI S. YAKIMOV, VALERY M. CHEREPANOV, MICHAEL A. CHUEV, and VLADIMIR YA. GONCHAROV

I.V. Kurchatov Institute of Atomic Energy, Moscow, 123182 USSR

ABSTRACT

The temperature dependence of the parameters of 57-Co emission Mossbauer spectra reveals remarkable changes near 115 K, 155 K and 240 K. The largest anomaly is observed for the spectral linewidth. The broadening of spectral lines is suggested to be due to the quadrupolar relaxation process with characteristic fluctuation frequencies comparable with hyperfine Mossbauer ones ($\omega \approx 10^7-10^9\ s^{-1}$) in the temperature range of 90-130 K. The relation between the observed peculiarities and structural instabilities is discussed.

KEY WORDS: Y-123 system superconductor, emission Mössbauer spectroscopy, structural instabilities, linewidth anomaly, magnetic hyperfine structure

INTRODUCTION

Under the study of the high temperature superconductors (HTSC), a number of structural instabilities have been found in the normal state rising the question on their essential role in the superconducting mechanism. However, the nature of such structural anomalies have not been yet clarified. As to HTSC ceramics $YBa_2Cu_3O_y$ (Y-123 system), some peculiarities of structure-sensitive parameters were observed in the temperature range between ≈90 K and ≈250 K by X-ray and neutron diffraction, Raman scattering, positron annihilation, acoustic and other methods [1]. Meanwhile, the Mossbauer spectroscopy gives a powerful tool to search out local changes in electronic and crystal structures. In particular, the similar anomalies were observed in the temperature dependence of isomer shift δ, quadrupole splitting Δ, and Mossbauer effect probability f near about 100 K and 200 K for 57-Fe and 119-Sn microprobes substituting for Cu-sites in Y-123 [2]. In this case, one needs to introduce the iron impurity of about x=Fe/Cu≈1 at.% into the sample to get a spectrum with high statistical quality for the typical absorption Mossbauer spectroscopy (AMS) experiments. But, such 3d substituents as Fe or Co produce an orthorhombic tetragonal phase transition at x ≥ 2-3% and tend to increase the oxygen content, thus noticeably disturbing the crystal lattice and microstructure. Contrary to AMS, the emission Mossbauer spectroscopy (EMS) of 57-Fe daughter nuclei produced after 57-Co decay has an important advantage to study doped samples with extremely small 57-Fe impurity content x≈0.0001%. Such a low level of doping does not practically affect macroscopic properties and microstructure of matrix and allows to get a "pure" local information about a microprobe without clustering effects. However, only a few works was focused on study of the Y-123 system by EMS [3-7] in contrast with a large amount of AMS studies [2], which is due to more availability of the latter. In the present work which is an extended variant of Ref.[7], the results of the first detailed EMS study of the temperature dependence of 57-Fe Mossbauer spectra parameters for 57Co-doped Y-123 will be presented.

EXPERIMENTAL

The high quality single-phase polycrystalline pellets of "pure" Y-123 were prepared by a conventional solid phase method, simultaneously with the 57-Fe and 119-Sn substituted samples [8-10]. Then, a drop of carrier-free 57-Co in the form of $CoCl_2$ solution with about 0.1 mCi activity was deposited onto the surface of the pellets. After drying, the diffusion annealing was carried out

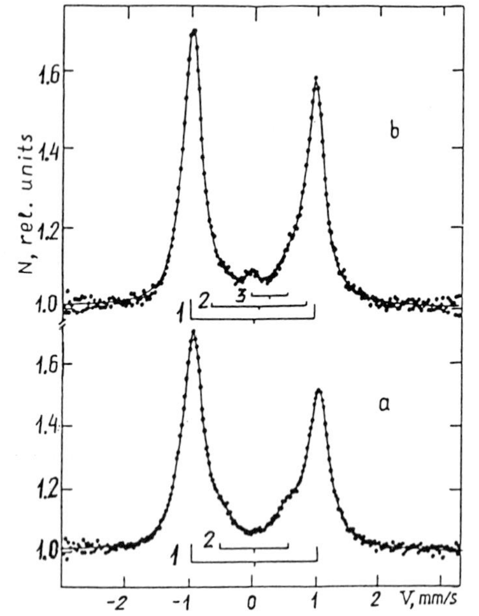

Fig. 1 Emission Mössbauer spectra of the daughter 57-Fe nuclei in the superconducting Y-123 at room temperature: (a) sample 1 with y≈6.95; (b) sample 2 with y≈6.6.

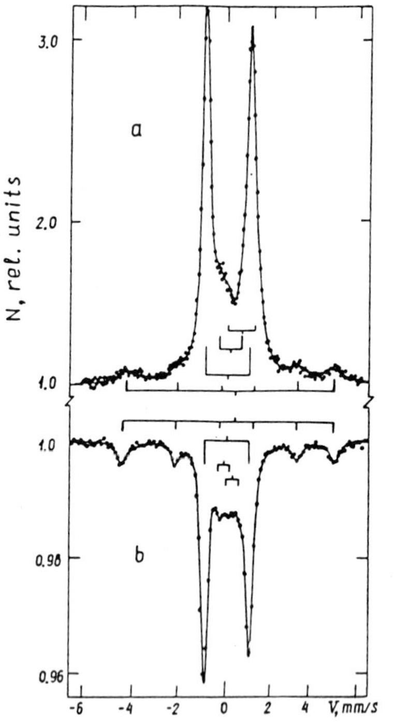

Fig. 2 Room temperature Mossbauer spectra of 57-Fe nuclei in the non-superconducting Y-123: (a) emission spectrum of the Co-doped sample; (b) absorption spectrum of the Fe-doped sample (Fe/Cu=1.7 at.%).

at 860 C for 8 h in air. Three types of the samples have been prepared: (1) slow-cooled superconducting orthorhombic-I (T_c=92(1) K; a=3.819 Å, b=3.887 Å, c=11.685 Å; y≈6.95(5)), (2) quenched to air superconducting orthorhombic-II (T_c=60(10) K; a=3.829 Å, b=3.833 Å, c=11.720 Å; y≈6.70(1)), and (3) quenched to liquid nitrogen nonsuperconducting tetragonal (a=3.859 Å, c=11.836 Å; y≈6.1(1)). The oxygen content y was estimated indirectly using the values of the lattice parameters and T_c's [11-12]. EMS measurements were performed with the NOKIA constant acceleration spectrometer and the resonant detector based on FeAl converter. The sample-source inside a plastic container was mounted on a cold finger of the cryostat inserted into a transport helium dewar. Its temperature could be varied in the range of 30-300 K and fixed within ΔT≈1 K.

RESULTS AND DISCUSSION

The emission Mossbauer spectra of 57-Fe daughter nuclei in the superconducting samples 1 and 2 are shown in Fig. 1. The shape of the spectrum of the oxygen-saturated sample 1 is similar to that observed in Refs.[4,5] for the samples synthesized in the same way, but it differs essentially from that in [3] where 57-Co was initially introduced into CuO followed by sintering. The analysis shows that the spectrum consists mainly of two doublets with the quadrupole splitting values of Δ1≈2.0 mm/s and Δ2≈1.1 mm/s and the relative intensities of I1≈82% and I2≈18%, respectively. Although the final identification of the spectral lines has not yet been established, the most of investigators attribute the doublets to Fe ions in the chain Cu(1) sites with four-fold (square) and fife-fold (pyramidal) oxygen coordinations. This is in agreement with the results reported on preferential substitution Co atoms for the Cu(1) sites [14]. The observed asymmetry in the doublet's lines intensities is due basically to texture effect in accordance with the spectrum taken at 55° orientation of the samples with respect to γ-beam. For compare, in AMS spectra of the Fe-doped Y-123 samples with $y \approx 7$, additional doublets with $\Delta 3 \approx 0.5$ mm/s and $\Delta 4 \approx 1.5$ mm/s have been observed [2]. The doublets were used to associate with Fe ions in the fife-fold (pyramidal) Cu(2) sites and six-fold (octahedral) Cu(1) sites, respectively.

On the other hand, quenching to air resulted in appearance of the doublet 3 ($I3 \approx 5\%$) and increase in the intensity I1 up to about 90% in the spectrum of the sample 2 (Fig. 1(b)). This is indicative of a transition of a small fraction of Co ions from Cu(1) to Cu(2) sites at about 900 C as well as of increase in the fraction of "square" Fe positions at the expense of the pyramidal ones in the Cu(1) sites by removing oxygen from O(5) sites with decreasing y.

Greater changes in the spectra are caused by quenching to liquid nitrogen (see Fig. 2 where the AMS spectrum of the Fe-doped Y-123 ceramic sample with x=1.7% is also shown). In that case, there arises a magnetic sextet ($I \approx 15\%$) with parameters $\delta \approx 0.3$ mm/s and $\Delta \approx 0.4$ mm/s close to those for the doublet 3 and characteristic of $Fe(3+)$ ion, and the value of the effective hyperfine field $Hn \approx 300$ kG. It was found that $Hn \approx 500$ kG at 77 K, and the sextet is not observed above 400 K, which is evidently related to a strong exchange coupling between $Fe(3+)$ ions and antiferromagnetically ordering Cu(2) spins with the Neel temperature of about 400 K for y=6.1 [14]. Thus, when quenching to liquid nitrogen, about 10% of Co atoms have time to remain in metastable state in the pyramidal Cu(2) sites, which is close to the corresponding magnitude for the case of Fe-doped samples. However, the observed spectrum of the semiconducting sample 3 differs strongly from that reported in [5] where deoxygenated samples were obtained by annealing at 700 C in nitrogen and revealed the sextet dominating ($I \approx 65\%$) in the spectra. In this connection, we have measured the emission spectra of the tetragonal 57Co-doped Y-123 samples obtained by removing oxygen at 700 C in vacuum, and there is no magnetic sextet observed in the spectra down to 77 K. Hence, there is no Co diffusion in the Cu(2) sites at 700 C in vacuum, but it goes intensively in nitrogen.

Then, the shape of the spectra of the samples 1 and 2 is found to be practically unchanged in the whole temperature range of 30-300 K. Due to the fact that the doublet 1 is dominant in the spectra of both the samples, its parameters have been determined with good accuracy. The largest anomaly (up to about 100% as compared to the value at room temperature) is observed in the temperature dependence of the linewidth $\Gamma 1$ whereas the variations of $\Delta 1$ do not exceed 2% and those of $\delta 1$ is not greater than 0.01 mm/s. As seen from Fig. 3 and 4, the peculiarities in $\Gamma 1(T)$ for the sample 1 are observed only at about 120 K (two more anomalies is revealed in $\Gamma 2(T)$ at about 220 K and 290 K) while the dependencies $\Gamma 1(T)$ and $\Gamma 2(T)$ for the oxygen-deficient sample 2 demonstrate an additional line broadening also near about 150 K and 240 K. It should be noted that an anomalous behavior of the structural parameters and sound velocity has been observed just in these temperature ranges [1], which is explained in terms of processes of redistribution and ordering in the oxygen-vacancy system. Moreover, in our recent AMS investigation on Fe-doped Y-123 with T_c=80 K, x=1.7%, and y=6.80±0.05, the anomalous line broadening was first found near 110 K, which has the Lorentz character, in other words, that is dynamical rather than static in nature [15]. The broadening is suggested to be due to the quadrupolar relaxation process with characteristic fluctuation frequencies comparable with hyperfine Mossbauer ones ($\omega \approx 10^7 - 10^9$ s^{-1}) in the temperature range of 90-130 K. Thus, the peak position in $\Gamma 1(T)$ at about 115 K does not depend essentially on y in the range of 6.6-6.95 whereas the other peaks in $\Gamma 1(T)$ displace with changing y. This can be associated with a complicated structure of the phase diagram in the range of lower temperatures where a number of structural transitions of I and II orders among the different phases of the oxygen ordering with the periods along the a-axis such as 1a, 2a, 3a etc. have been predicted [16].

In conclusion, it should be noted that (1) as compared with AMS, EMS gives a possibility to study the local properties of HTSC ceramics without disturbance of its microstructure by Mossbauer probe, and (2) EMS proved to be a good technique to detect structural instabilities in HTSC lattice equally with the other structure-sensitive methods.

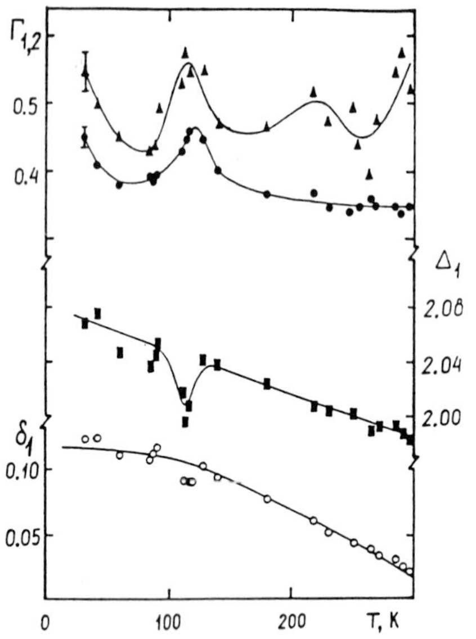

Fig.3 Temperature dependencies of the parameters of the emission spectra for the sample 1: isomer shift $\delta 1$ relatively to α-Fe (open circles); quadrupole splitting $\Delta 1$ (rectangles); linewidth $\Gamma 1$ (filled circles) and $\Gamma 2$ (triangles). All the values are given in mm/s.

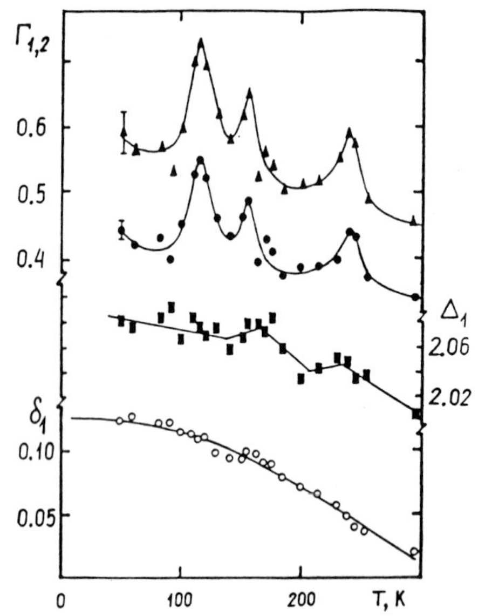

Fig.4 Temperature dependencies of the parameters of the emission spectra for the sample 2: isomer shift $\delta 1$ relatively to α-Fe (open circles); quadrupole splitting $\Delta 1$ (rectangles); linewidth $\Gamma 1$ (filled circles) and $\Gamma 2$ (triangles). All the values are given in mm/s.

REFERENCES

1. Gusakovskaya IG, Pirumova SI, Atovmyan LO (1990) Superconductivity: Physics, Chemistry, Technology 3: 1980 (in Russian)
2. Boolchand P., McDaniel D (1990) In: Narlikar AV (ed) Studies on High-temperature Superconductors. Nova Sci Publ, v 4, p 143
3. Bottiyn L, Molnar B, Nagy DL, Szucs IS, Toth J, Dengler J, Ritter G, Schober J (1988) Phys Rev B 38: 11373
4. Fontcuberta J, Obradors X, Vallet M, Gonzalez-Calbet JM (1988) Z Phys B 73: 143
5. Nath A, Nagy S, Baroum MW, Tyagi SD, Wei Y (1988) Solid State Commun 68: 181
6. Andrianov AV, Anisimova OL, Kozin MG, Pentin AYu, Semenov SI, Shpinel VS, Leonyuk LI, Moschalkov VV, Red'ko SV (1990) Physica C 166: 248
7. Goncharov VYa, Cherepanov VM, Chuev MA, Yakimov SS, (1988) Proceedings of All-union Meeting on Applied Mossbauer spectroscopy, Moscow, USSR, Abstracts, part II, p 134 (in Russian)
8. Cherepanov VM, Chuev MA, Yakimov SS, Goncharov VYa, Smirnov SA (1988) JETP Lett 47: 424
9. Cherepanov VM, Chuev MA, Yakimov SS, Goncharov VYa, Smirnov SA, Bush AA (1989) JETP Lett 49: 431
10. Cherepanov VM, Chuev MA, Yakimov SS, Goncharov VYa (1990) Physica B 163: 441
11. Jorgensen JD, Veal BW, Paulikas AP, Nowicki LJ, Crabtree GW, Claus H, Kwok WK (1990) Phys Rev B 41: 1863
12. Zhu Y, Moodenbaugh AR, Suenaga M, Tafto J (1990) Physica C 167: 363
13. Kajitani T, Kusaba K, Kikuchi M, Syono Y, Hirabayashi M (1987) Jap J Appl Phys 26: L1727
14. Nowik I, Kowitt M, Felner I, Bauminger ER (1988) Phys Rev B 38: 6677
15. Afanas'ev AM, Tsymbal EYu, Cherepanov VM, Chuev MA, Yakimov SS, Zinn W, Sauer Ch, Bush AA (in press) Solid State Commun
16. Mamsurova LG, Pigalskiy KS, Sakun VP, Shushin AI, Scherbakova LG (1990) Physica C 167: 11

2.5 Optical Properties

Electronic States of High-T_c Cuprates Investigated by Optical Measurements

SETSUKO TAJIMA

Superconductivity Research Laboratory, 10-13, Shinonome 1-chome, Koto-ku, Tokyo, 135 Japan

ABSTRACT

Based on the optical spectra, the electronic states of the high-Tc cuprates are discussed. In the high-Tc superconductivity composition, the charge transfer gap is almost completely destroyed, resulting in the appearance of a single band with a large Fermi surface, but the effect of the strong electron correlation remains. The unique feature in the electronic states of the high-Tc cuprates is that the charge transfer gap can be easily deformed or destroyed by a small amount of doping. The large mixing of the in-plane Cu- and O-orbital would be one of possible origins of such a doping effect. The strong hybridization of Cu- and O-orbital is evidenced by the fact that the in-plane Cu-O stretching mode frequency is very sensitive to the Cu-O bond distance.

KEY WORDS: Electronic States, Optical Spectrum, High-Tc Sperconductivity

INTRODUCTION

During these four years after the pioneering work by Bednorz and Muller[1], intensive studies have been accumulated on the electronic states of the high-T_c cuprates. On the experimental side, efforts have been made steadily to improve the sample quality and the measurement resolution. As a result, the rough outline of the electronic states for these high-T_c superconducting cuprates (HTSC) has been gradually revealed.
At present there is no doubt that the high-T_c superconductivity is induced by doping carriers into antiferromagnetic insulators with CuO_2-planes. However, the mechanism of high-T_c superconductivity seems to be still far away from elucidation. In this paper, the present understandings of the electronic states for HTSC are discussed, mainly based on optical studies from the following viewpoints: (i) Where is the high-T_c superconductivity region on the way from a Mott-insulator (large U-limit) to a band metal (small U-limit)? (ii) What is unique in the electronic states of HTSC? (iii) How can we evaluate this uniqueness?

COMPOSITIONAL DEPENDENCE OF THE OPTICAL REFLECTIVITY SPECTRUM

In order to discuss the first problem "Where?", it is necessary to see the compositional dependence of various physical properties. The most intensive study has been made on one of the typical HTSC'S, $La_{2-x}Sr_xCuO_4$, which has the advantages such as structural simplicity and the availability of specimens over wide compositional range from antiferromagnetic insulator to non-superconducting metal in the heavily doped region. [2,3] Figure 1 shows the in-plane(E c) reflectivity spectra for $La_{2-x}Sr_xCuO_4$ single crystals with various x. In the low doping region below x=0.06, the samples show a semiconductive temperature dependence of resistivity. The samples with x=0.10, 0.15 and 0.20 exhibit superconductivity with a maximum T_c of 23K at x=0.15, while the heavily doped samples with x=0.27 and 0.34 show no superconductivity but have lower resistivity than that of the superconducting samples. It should be noted in Fig.1 that in the parent material La_2CuO_4 the absorption across the energy gap is observed at 2eV which is smaller than the correlation energy U(~6-8eV).[4] This means that the system is not a simple Mott-insulator with the energy gap U but rather a charge-transfer (CT) insulator with the gap Δ_{CT} which is mainly determined by the energy difference between O-2p and Cu-3d level.

The doping effect on the reflectivity spectrum is clearly seen in Fig.1. In La_2CuO_4(x=0), three main phonon peaks dominate the far-infrared spectrum, while the excitation across the CT-gap appears at 2eV, followed by the higher energy interband excitations. With increasing Sr-content x, the CT-gap excitation decreases rapidly and instead the lower energy spectrum builds up, creating the reflectivity edge near 1eV. The energy spectrum below 1eV can not be simply described by the conventional Drude model, even by assuming a very large damping factor. The deviation from the Drude-curve can be more evidently seen in the optical conductivity spectra in Fig.2 which were calculated from the reflectivity spectra in Fig.1 by the Kramers-Kronig transformation. For the samples with low Sr-concentration, in addition to the far-infrared phonon peaks, a broad hump is observed in the mid-infrared (MIR) region of the conductivity spectrum. As x increases, this hump grows, moving its weight towards lower energy. Finally, the spectrum becomes a Drude-like curve without any hump or shoulder structure. This unique spectral change against composition is commonly observed among the other HTSC's, $YBa_2Cu_3O_y$ with T_{cmax}=90K[5], $Bi_2Sr_2CaCuO_z$ with T_{cmax}=80K[6] and the electron-doped system $(Nd,Ce)_2CuO_4$ with T_{cmax}=25K[7].

There may be two ways to describe the optical spectrum deviating from the Drude curve as mentioned above. One is to decompose the spectrum into the Drude part centered at ω =0 and the Lorentz oscillator part corresponding to the MIR-absorption. The other way is to describe the whole spectrum as an extended Drude curve in which the mass and the scattering time depend on energy. The first way seems to be applicable in the low-doping region. The MIR-absorption can be interpreted as the excitation from the O 2p

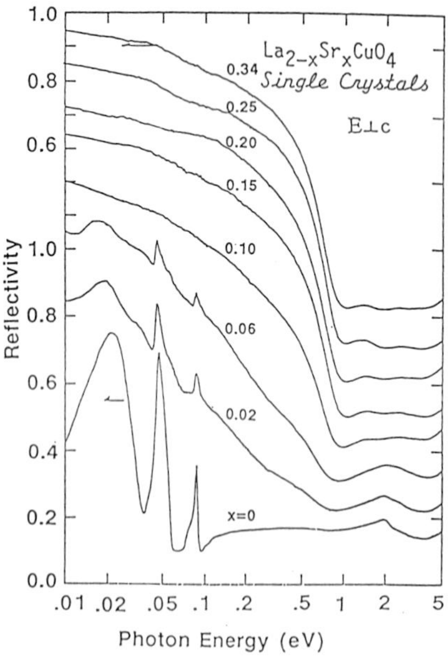

Fig.1 The room-temperature spectra of single crystals $La_{2-x}Sr_xCuO_4$ for various x with $E\perp c$.

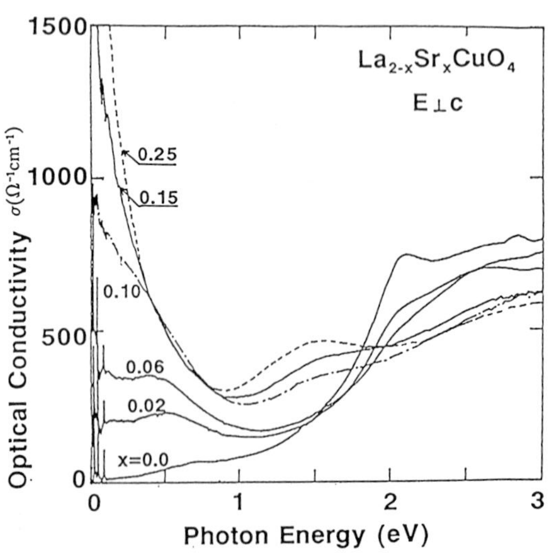

Fig.2 The optical conductivity spectra of $La_{2-x}Sr_xCuO_4$ calculated from the reflectivity spectra in Fig.1 by the Kramers-Kronig transformation.

band to the new states created by doping within the CT-gap.[6] However, in this regime is placed the constraint that the mid-gap states must grow so as to pin the reflectivity edge at around 1eV. In the heavily doped region, it becomes difficult to decompose the spectrum into two parts as above. The extended Drude description, which was successfully applied to the heavy fermion systems[8], seems to hold also for the spectra in the high doping region.[5,9] The effective mass shows ω-dependence even in the heavily doped region[3], which implies that the strong electron correlation is still surviving to some extent. However, this type of description reaches an impasse in the low-doping region, where the ω-dependence of the mass would turn out to be physically unreasonable.

Based on these experimental observations, the change of the electronic states can be depicted schematically as in Fig.3. When the holes (or electrons) are doped, new states are created within the CT-gap. The rapid weakening of the CT-excitation on doping as well as the accompanying saturation of the spectral change seems to indicate an essential change in the band structure accompanied with gap deformation (Fig.3(b)) rather than the behavior expected for the simple mid-gap state model, in which the CT-gap is essentially preserved.[6]. In the more complex scheme shown in Fig.3(b), a part of the density of states is transfered from the original conduction (Cu 3d) and valence (O 2p) band to the newly created states. Further doping considerably deforms the gap, and finally a single band with a large Fermi surface dominates the electronic states (Fig.3(c)), which may be regarded also as the gap destruction.

The high-T_c composition in question is situated between the two states in Fig.3(b) and (c). The optical spectrum can be described by either one of the above two ways. In this compositional region, neither the MIR-absorption nor the CT-gap absorption can be observed clearly. But they must remain because the spectrum shows a small but clear deviation from the Drude-curve. This, as well as the Raman scattering measurements and the neutron scattering measurements, strongly suggests that strong electron correlations and/or antiferromagnetic spin fluctuations remain in the HTSC region.

COMPARISON WITH RELATED NON-SUPERCONDUCTIVE MATERIALS

From the discussion above one may conclude that the low energy excitation termed as MIR-absorption is unique for the HTSC. In fact, many theorists attempt to explain this optical feature because they consider it as the most important and the most essential in order to understand the normal state of the HTSC.[9]

However, similar spectra are also observed in other materials. For example, Bi et al. have recently reported the reflectivity spectrum for a single crystal of $La_{1.8}Sr_{0.2}NiO_4$.[10] Although this material is not a superconductor, the spectrum shows the MIR-absorption

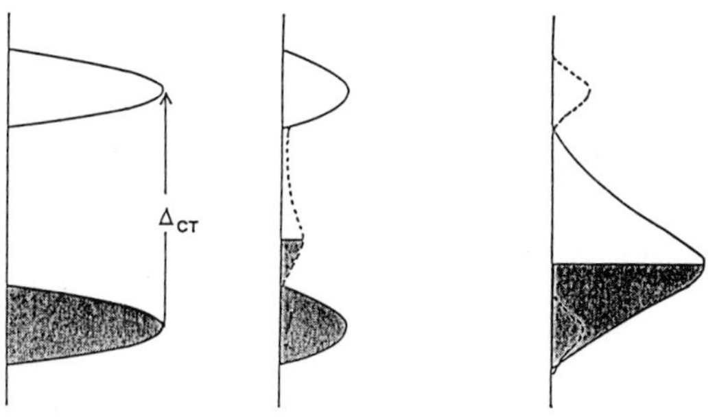

(a) (b) (c)

Fig.3 The schematical picture of the electronic states near the Fermi energy.

with the reflectivity edge at about 1eV. On the other hand, the reflectivity spectrum of non-superconducting (Nd,Sr)CoO_3 with cubic perovskite structure was systematically studied by Terasaki et al.[11] They found that the spectrum changes with doping similarly to the HTSC-system while the MIR-peak does not shift in energy. Watanabe et al. have grown a single crystals of non-superconductive $Bi_2M_3Co_2O_y$ (M=Ca,Ba and Sr) and found that their optical spectra also show the MIR-absorption.[12]

Therefore, the MIR-absorption itself is commonly observed in doped CT-insulators. What is really unique for the HTC is that the MIR-structure is rapidly smeared out, that is, the system becomes rapidly metallic even by a small amount of doping. Among the K_2NiF_4-type compounds, La_2MO_4 with M being the 3d-metal, only the cuprates becomes metallic by doping as little as 5%. In case of $(La,Sr)_2NiO_4$, as well as cubic (Nd,Sr)CoO_3, the transition to metal occurs at much higher level of doping of about 40%. For Co- and Fe-(214)-compounds, even 50% Sr-substitution cannot change the materials into metals. The far-infrared reflectivity spectra of $LaSrCoO_4$ and $LaSrFeO_4$ show clear phonon peaks as in the undoped La_2CuO_4 and La_2NiO_4.[13]

The next question is why is it only the cuprates that become metallic so easily. One may relate this unique feature of the cuprates to the strong hybridization of O 2p and Cu $3d_{x^2-y^2}$ orbitals the energies of which happen to be very close to each other. The covalent nature of Cu-O bonding is supported by the narrow band width of the optical phonon[15]. The exchange energy J is also a measure of hybridization, because it is related with the CT-gap energy Δ and the transfer energy t as $J \sim t^4/\Delta^2(\Delta+U)$. The large J-value estimated from the two-magnon energy in Raman spectrum indicates the strong orbital hybridization in the plane, i.e. large t and/or small Δ .[14] In this case, doped holes into the oxygen orbital could play an important role for reduction of the coupling between spins at Cu-sites. However, in the Ni- and/or Co-compounds, where the energy difference between oxygen and Ni (or Co) is larger than that of the cuprates, the spin ordering and/or the strong electron correlation may not so effectively be destroyed by doping, in spite of the fact that the correlation energy U is not so different from that of the cuprates.

EFFECTS OF STRONG ORBITAL HYBRIDIZATION

Since the hybridization of Cu and O-orbitals plays an important role in determining the electronic states of the HTSC, it is necessary to evaluate the degree of hybridization by some parameter. The parameter describing the hybridization should be sensitive to the Cu-O bond distance $\boldsymbol{d}$. One of the candidates for such parameters is the exchange energy J and the other is the charge transfer gap energy Δ_{CT}. In fact, the systematic $\boldsymbol{d}$-dependence of J and Δ_{CT} were observed among materials with the same type of oxygen coordination.[16,18] However, these parameters are determined not only by the Cu-O distance but also by other local structural factors such as the oxygen coordination around Cu.[19]

Another parameter that is likely to reflect the orbital hybridization is the phonon frequency for the Cu-O bond distance modulation mode. For the layered perovskite structures, there are two infrared active phonons in the CuO_2-plane. One is the stretching mode, which is the modulation of the in-plane Cu-O bond distance, and the other is the bending mode, which is the modulation of the Cu-O bond angle. In Fig.4 are plotted the TO-phonon frequencies of the two modes against the lattice parameter $\boldsymbol{a}$ for various layered cuprates.[13, 20] As is clearly seen in the figure, the $\boldsymbol{a}$-dependence of the stretching mode frequency is quite different from that of the bending mode frequency. The latter can be understood within conventional lattice dynamics, which predicts the relation $\omega \sim \boldsymbol{a}^{-3/2}$ as indicated by the dashed line in the figure, whereas the former shows anomalously strong ***a***-dependence. It is notable that the phonon frequency of the stretching mode is insensitive to the oxygen coordination around Cu.

In general, there are two origins of reduction of TO-phonon frequency. One is the dipole-dipole interaction characterized by local charge (static contribution), and the other is the electron-phonon coupling (dynamical contribution). In the case of HTSC, the former is not dominant because the local effective charge estimated from the phonon band width is relatively small and is not sensitive to ***a***. Therefore, it is concluded that the drastic decrease of the stretching mode frequency with ***a*** is caused by the strong coupling of this particular phonon mode with the electronic system. In a crude approximation, a change of TO-phonon frequency is described as $\Delta\omega^2=-\Xi^2/E_G$, where E_G is the average energy gap between the relevant bands and Ξ is the deformation potential corresponding to the phonon mode in question.[20] If Δ_{CT} is assumed to be the measure of E_G, the decrease of Δ_{CT} gives rise to the decrease of ω_{TO}. In fact, the remarkable decrease of Δ_{CT} with increasing ***a*** was observed in the (214)-compounds with T'-structure.[18] It suggests that the strong ***a***-dependence of Cu-O stretching mode frequency is dominantly attributed to the strong ***a***-dependence of Δ_{CT}, which indicates the direct coupling of Cu-O bond modulation phonon with the Cu-O charge transfer excitation. Thus, this remarkable a-dependence of phonon frequency just reflects the characteristic hybridization effect on the electronic system in the layered cuprates.

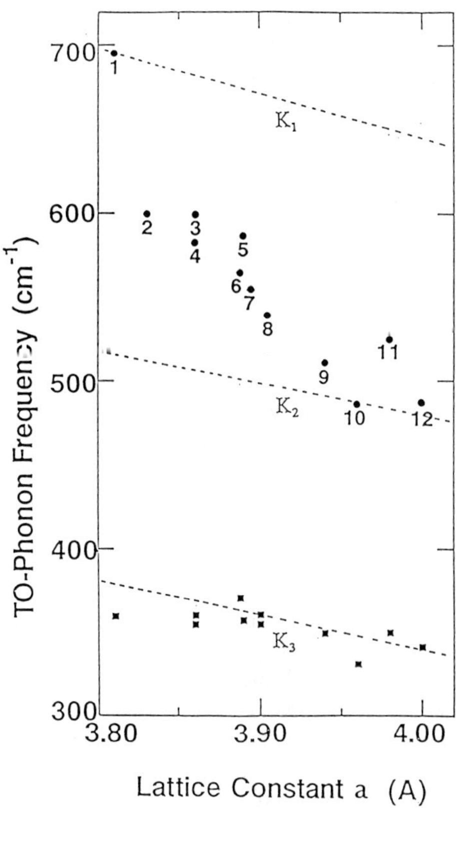

Fig.4 TO-phonon fequencies of the in-plane Cu-O stretching (●) and bending (■) mode against the lattice constant *a* for various layered cuprates. The data point numbers are as follows 1:La_2CuO_4 (T), 2:$Bi_2SrLaCuO_y$, 3:$(La,Dy)_2CuO_4$ (T*), 4:$(Ca,Sr)CuO_2$, 5:YBa_2CuO_6, 6-10:Ln_2CuO_4 (T') Ln=Gd(6),Eu(7),Sm(8),Nd(9) and Pr(10), 11:$Sr_2CuO_2Cl_2$ (T), 12:$Sr_2CuO_2Br_2$ (T). The dashed lines indicate the relation $\omega_{TO}=K(a/4.0)^{-2/3}$ with $K_1=645cm^{-1}$, $K_2=480cm^{-1}$ and $K_3=340cm^{-1}$.

SUMMARY

When holes (or electrons) are doped into a CT-insulator, the CT-gap is rapidly deformed and new states are created within the gap. Further doping almost completely destroys the gap, and a single band with a large Fermi surface manifests itself. In the high-Tc superconductivity compositions, the electronic states are situated in a critical region where the CT-gap is almost destroyed but the effect of strong electron correlation remains. The most noticeable feature in the electronic system of HTSC's is that these changes occur very rapidly with a small amount of doping. It is related more or less to the small energy difference between Cu and O level, i.e. the CT-energy. The large hybridization of Cu- and O-orbital is also evidenced by the fact that phonon frequency of the in-plane Cu-O stretching mode shows a strong dependence on the Cu-O distance, which indicates a strong coupling of this phonon with the CT-excitation.

ACKNOWLEDGEMENT

The author would like to thank Prof.S.Uchida at the University of Tokyo for stimulating discussions about the electronic states of HTSC.

REFERENCES

[1] Bednorz JG and Muller KA (1986) Z.Phys.B64: 189-.
[2] Takagi H,Ido T,Ishibashi S,Uota M,Uchida S and Tokura Y (1989) Phys.Rev.B40:2254-2261.
[3] Uchida S,Ido T,Takagi H,Arima T,Tokura Y and Tajima S: to appera in Phys.Rev.B.
[4] Fujimori A,Takayama-Muromachi E,Uchida Y and Okai B (1987) Phys.Rev.B35: 8814-8817. Frugel JC,Fink J and Nucker N (1988) Int.J.Mod.Phys.B2:1185-.
[5] Orenstein J,Thomas GA,Millis AJ,Cooper SL,Rapkine DH,Timusk T,Schneemeyer LF and Waszczak JV (1990) Phys.Rev.B42:6342-6362.
[6] Terasaki I,Nakahashi T,Takebayashi S Maeda A and Uchinokura K (1990) Physica C 165:152-160.
[7] Uchida S,Takagi H,Tokura Y,Koshihara N and Arima T (1989) "Strong Correlation and Superconductivity", Springer Sries in Solid State Sciences (Springer-Verlag) Vol.89, eds. Fukuyama H, Maekawa S and Malozemoff AP :194-203.
[8] Webb BC, Sievers AJ and Mihalisin T (1986) Phys.Rev.Lett.57:1951-1954.
[9] Schlesinger Z,Collins RT,Holtzberg F,Feild C,Koren G and Gupta A (1990) Phys.Rev.B 41:11237-11259.
[10] Bi XX,Eklund PC,McRae E,Zhang J-G,Metcalf P,Spalek J and Honig JM (1990) Phys.Rev.B 42: 4756-4759.
[11] Terasaki I,Nakahashi T,Maeda A and Uchinokura K (1990) to appear in Phys.Rev.B.
[12] Watanabe Y,Lyons SA,Tsui DC,Birmingham JT,Ong NP and Tarascon JM: to appear in Phys.Rev.B.
[13] Tajima S,Ido T,Ishibashi S,Itoh T,Eisaki H,Mizuo Y,Arima T,Takagi H and Uchida S, to appear in Phys.Rev.B.
[14] Sugai S (1990) "Mechanisms of High Temperature Superconductivity", Springer Series in Material Science 11, ed. Muller KA; 207-219.
[15] Tajima S,Uchida S,Ishibashi S,Ido T,Takagi H,Arima T and Tokura Y (1990)Physica C168: 117-122.
[16] Yoshida M,Koshizuka N and Tanaka S (1990): to appear in Phys.Rev.B 42.
[17] Tomeno I,Yoshida M,Ikeda K,Tai K,Takamuku K,Koshizuka N and Tanaka S, to appear in Phys.Rev.B.
[18] Arima T and Tokura Y, unpublished.
[19] Tajima S,Ishibashi S,Itoh T,Eisaki H,Takagi H,Arima T,Mizuo Y,Tokura Y and Uchida S (1990) Physica B 165&166: 1239-1240.
[20] Katayama S and Kawamura H (1977) Solid State Commun.21:521-524.

Far Infrared Spectra of High T_c Cuprate Superconductor

JIRO TANAKA, MICHIHIRO SHIMADA, and MASAAKI SHIMIZU

Department of Chemistry, Faculty of Science, Nagoya University, Chikusa, Nagoya, 464-01 Japan

ABSTRACT

Infrared and visible to uv spectra of $La_{2-x}Sr_xCuO_4$ are measured from 30 to 30000 cm^{-1} range with single crystals by using specially designed microspecstrophotometers. The spectra of insulating state shows all predicted lattice vibrations and origin of mid-infrared absorption band. The metallic crystal shows in addition the Drude type free carrier band.

KEY WORDS: Infrared spectra, lattice vibration, mid-infrared exciton band

INTRODUCTION

Infrared spectra of high Tc cuprate supeconductor consist of three parts, one is due to the free carrier of Drude type, the second is phonon structrue and the third is the so-called mid-infrared exciton band.
Far infrared spectra exhibit phonon structure and decay of free carrier occurring by the interaction with phonon and exciton. In order to find accurate dielectric function from optical data, precise measurement of reflectance in the far infrared region is essential for Kramers-Kronig transformation to be made with possibly wide spectral data. A serious difficulty exists in determining far infrared spectra with conventional spectrophotometer on small single crystals, since the light beam is not sharply collimated on the narrow crystal surface.
We overcome this difficulty by using a specially designed infrared lens of reflecting type. A Fourier type far-infrared spectrophotometer is constructed which is useful for 30 - 600 cm^{-1} range. The crystal size of 2mm square is sufficient to find far infrared spectra.

RESULTS AND DISCUSSIONS

We illustrate the reflection spectra of $La_{2-x}Sr_xCuO_4$ for x=0 and x=0.14 on their ab plane as shown in Fig.1. The crystals are kindly gifted from Prof.H.Kojima of Yamanashi University. The as grown crystal of x=0 was annealed in Ar atmosphere for 6 hrs, to remove extra oxygen. However, a small amound of oxygen residue still remains which exhibits a pristine origin of electronic transition in the mid-infrared region.

The Kramers-Kronig transformation of reflectivity gives optical conductivity spectra $\sigma(\omega)$ as shown in Fig.2. In the crystal of x=0, the peaks of $\sigma(\omega)$ show lattice vibration of TO mode, while peaks of $-Im(1/\varepsilon(\omega))$ correspond to LO phonon. The observed frequencies are compared with theoretical calculation of Prade et al.[1] as shown in Table 1; all predicted frequencies are found with reasonable agreement except the Cu-O stretching mode at 670 and 720 cm^{-1}. Several unclassified peaks are observed in $\sigma(\omega)$ at 500cm^{-1} and 620cm^{-1}, and two peaks in $-Im(1/\varepsilon(\omega))$ at 520 and 630cm^{-1}, respectively. The broad band extends from 500 cm^{-1} to 7000 cm^{-1}, which is an origin of mid-infrared exciton band. It is actually the interband transition disucssed in our previous paper on the energy band structure of the CuO_2 plane.[2] When a small amount of oxygen is incorporated into the crystal, the 500 cm^{-1} band appears together with broad mid-infrared band. Accordingly we assign the 500 cm^{-1} band to the band origin of the mid-infrared band. Our previous calculation showed the joint density of states between the filled and doped unoccupied band has a first peak at 500 cm^{-1} and broad band starts at 800 cm^{-1} extending to higher frequency. These features are found in the spectra of Fig.2. The shift of Cu-O stretching to higher frequency is another reason to assign the 500cm^{-1} band to the electronic origin.

By increasing the dopant concentration, the metallic state appears as Drude type contribution of free carrier and several vibrational peaks are found with increasing intensity. The 500cm^{-1} band shifts slighly to lower frequency and its intensity increases with doping. The maximum of braod mid-infrared band also shifts to lower frequency and its intensity increases steadly with doping. The full detail on the analysis of these bands and far infrared spectra of other cuprates will be discussed in separate papers.

ACKNOWLEDGEMENTS

The authors thank to Prof.H.Kojima and Dr.I.Tanaka for gift of crystals and to Profs. Y.Tsuda, S.Uchida and Dr.S.Sugai for helpful communications.

REFERENCES

1. Prade J, Kulkarni AD, de Wetter FW, Kress W, Cardona M, Reiger R and Schroder U (1987) Solid Stat. Comm. 64: 1267 .
2. Tanaka J, Kamiya K and Tanaka C (1989) Physica C 161: 451

Table 1 Calculated and Observed Infrared Active Modes in the ab plane.

	Calc.(cm^{-1})	Obs. (cm^{-1})
LO	617	720
TO	547	670
LO	483	460
TO	263	360
LO	256	280
TO	179	155
LO	135	150
TO	120	130

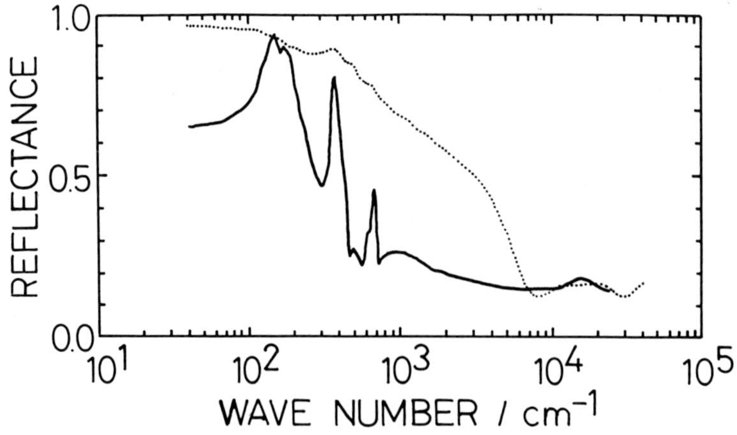

Fig.1 Reflection Spectra of Single Crystals of La_2CuO_4 (———) and $La_{1.86}Sr_{0.14}CuO_4$(··············).

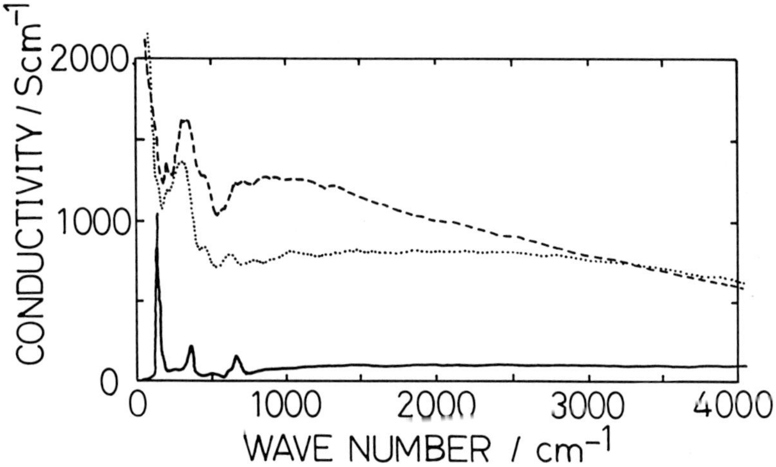

Fig.2 Optical Conductivity Spectra of La_2CuO_4 (———), $La_{1.86}Sr_{0.14}CuO_4$ (············) and $La_{1.81}Sr_{0.19}CuO_4$(-----).

Mid-*IR* Reflectivity of Single-Crystals $Bi_2Sr_2CuO_y$, $Bi_2M_{m+1}Co_mO_y$ (M=Ca,Sr,Ba; m=1,2), $Bi_2Sr_3Fe_2O_{9.2}$ and $Bi_2Sr_2MnO_{6.25}$

YUKIO WATANABE[1,2], N.P. ONG[3], D.C. TSUI[2], and J.M. TARASCON[4]

[1] Mitsubishi Kasei Research Center, Midori-ku, Yokohama, 227 Japan
[2] Department of Electrical Engineering, Princeton University, Princeton, NJ 08544, USA
[3] Department of Physics, Princeton University, Princeton, NJ 08544, USA
[4] Bell Communications Research, Red Bank, NJ 07701, USA

ABSTRACT

We report the room-temperature reflectivity of single crystals of $Bi_2Sr_2CuO_y$ and several quasi-two-dimensional oxides ($Bi_2M_{m+1}Co_mO_y$ (M= Ca, Sr, Ba; m=1,2), $Bi_2Sr_3Fe_2O_{9.2}$ and $Bi_2Sr_2MnO_{6.25}$), which are isomorphic to Bi-Cu based high *Tc* oxides. A broad absorption band is observed in the mid-*IR* range (near 0.5 eV) in all the compounds studied, including the insulators, and a new absorption band near 0.3 eV is observed in the insulating systems. The infrared spectra of $Bi_2M_3Co_2O_y$ system show striking similarities to those of the high-*Tc* cuprates; 1) the apparent plasma edge remains unshifted for all $Bi_2M_3Co_2O_y$ with different M, 2) the intensity of the mid-*IR* absorption approximately scales with that of the free carrier-like absorption.

KEY WORDS: Mid-IR absorption, Drude model, Bi-Cu superconductor, two-dimensional oxides

INTRODUCTION

The frequency dependence of the reflectivity in the mid-*IR* range, often referred to as the mid-*IR* absorption, is one of the puzzling normal state properties of the high-T_c cuprates. The *IR* reflectivity measurements on single crystals [1] and thin film samples of $YBa_2Cu_3O_{7-y}$ ("*123*") did not show the sharp feature near 0.5 eV which was observed in the early reports using the ceramic samples. We have reported studies of the *IR* reflectivity of the high-quality *123* superconductor (T_c>90K), and determined absolute values of the reflectivity spectra $R(\omega)$ which are very high and reproducible from sample to sample [2]. The reflectivity spectra $R(\omega)$ was featureless and the conductivity $\sigma(\omega)$ showed no sharp absorption peak, as reported in ref. [1]. However, both $R(\omega)$ and $\sigma(\omega)$ showed a clear deviation from a conventional Drude behavior up to the near-*IR* range, which was phenomenologically explained by an existence of a broad absorption band in the mid- and near-*IR* range. We have also reported evidence for the existence of a similar absorption band in the single-crystal $Bi_2Sr_2CuO_{6+\delta}$ [2] and in the electron type superconductor $Nd_{1.85}Ce_{0.15}CuO_{4-\delta}$ (T_c~20K) [3]. At present, there exist at least two viewpoints on the modeling of these reflectivity spectra : 1) absorption band model, 2) special scattering mechanism.

An important way to investigate this problem is to compare this absorption band with similar absorption bands in the non-cuprate layered oxide and in other simple oxides, and to understand this absorption band in the frame work of the optical properties of 3d-metal oxides. In this paper, we report the observation of similar absorption bands in quasi-two dimensional systems in which Cu in the "bismuthates" is completely replaced by one of the elements Co, Fe and Mn. As in the high-T_c Bi-Cu oxides, Bi-Co compounds undergo a metal-insulator transition upon doping. This study is intended to clarify a relationship between the relative position of the 3d electronic level to oxygen 2p electronic level and the mid-*IR* reflectivity (ref.[4]).

II. EXPERIMENTAL DETAILS

Crystals of the single- and double-plane Bi systems (in which Co, Fe or Mn replaces Cu entirely) were grown by self-flux method at Bellcore [5]. In addition to Co substitution for Cu, the alkaline earth metals Ba, Sr or Ca in $Bi_2M_3Co_2O_y$ (y~9) were used for M. The oxidation states of the 3d-metals, inferred from TGA, magnetic susceptivity and structural data, are ~3 in $Bi_2M_3Co_2O_y$ (M= Ca, Sr, Ba), ~2.5 in $Bi_2Sr_2CoO_{6.25}$, 3-δ in $Bi_2Sr_2MnO_{6.25}$ and 3+δ in $Bi_2Sr_3Fe_2O_{9.2}$ [6]. None of these systems are superconductors; the compounds $Bi_2Sr_3Fe_2O_{9.2}$, $Bi_2Sr_2MnO_{6.25}$, $Bi_2Sr_2CoO_{6.25}$ are insulators. The properties of the in-plane resistivity in $Bi_2M_3Co_2O_y$ are reported in refs. [5,7]. In general, the system becomes more conducting as the covalent radius of M in the formula $Bi_2M_3Co_2O_y$ becomes larger. Above 100 K, the in-plane resistivities of $Bi_2Ba_3Co_2O_y$ and $Bi_2Ca_3Co_2O_y$ have metallic and semiconductive temperature-dependence, respectively. The compound $Bi_2Sr_2CuO_{6+\delta}$ is metallic (T_c<4K, ~3 mΩ at 290 K). The reflectance between 0.08 eV and 1.4 eV was measured at near-normal incidence to the ab-plane of the crystals at room temperature. Instead of polishing the samples, gold was evaporated onto the crystals after measuring the crystals and the measurements were repeated.

RESULTS AND DISCUSSION

Figures 1 and 2 compare the reflectivity $R(\omega)$ and the frequency-dependent conductivity $\sigma(\omega)$ of the single 3d-O plane systems as derived by the KK analysis, respectively. Indicative of the strongly insulating nature of the compound $Bi_2Sr_2MnO_{6.25}$, a prominent unscreened phonon peak is observed near 600 cm^{-1}. In both $R(\omega)$ and $\sigma(\omega)$ of the insulating samples, a hump around 2000 cm^{-1} is evident. A corresponding structure is not seen in the insulating phases of the high-T_c cuprates (*123* [8], $La_{2-x}Sr_xCuO_4$[9], $Bi_2Sr_2(Ca,Nd)Cu_2O_{8+\delta}$[10]). As the number of d electrons in the 3d cation decreases, the compound appear more insulating in their reflectivity. This may be associated with an increase in the separation between the 3d levels and oxygen 2p levels. These trends agree qualitatively with the dc resistivity results. On the other hand, as in Fig.2, the intensity of the mid-*IR* absorption band (at ~0.5 eV) does not correlate with the d electron number nor with the magnitude of the low-frequency conductivity. The magnitude of the *IR* absorption band near 0.5 eV is almost same for both $Bi_2Sr_2CuO_{6+\delta}$ and $Bi_2Sr_2MnO_{6.25}$. Throughout, we reserve the term "mid-*IR* absorption " for the spectral range near 0.5 eV.

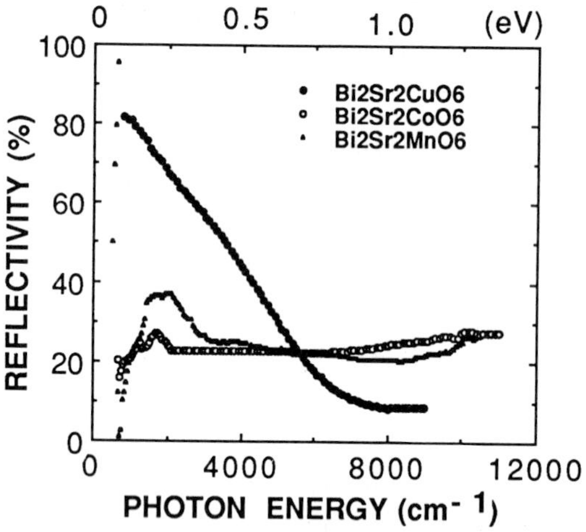

Fig.1 Reflectivity spectra $R(\omega)$ of single 3d-O plane bismuthates.

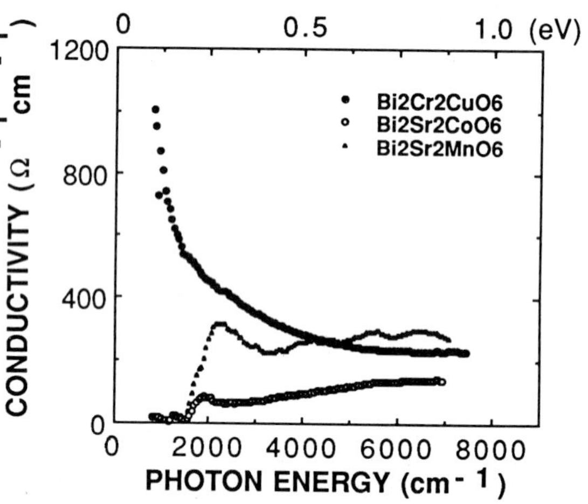

Fig.2 Dynamic conductivity $\sigma(\omega)$ of single 3d-O plane bismuthates.

Figures 3 and 4 compare the reflectivity $R(\omega)$ and the frequency-dependent conductivity $\sigma(\omega)$ of the double 3d-O plane systems as derived by KK analysis , respectively. The overall reflectivity $R(\omega)$ tends to be higher for the larger M's (Ba) in $Bi_2M_3Co_2O_y$. Compounds with large M are more conductive, as is apparent in $\sigma(\omega)$. However, the *apparent* plasma edge remains near 2500 cm^{-1}(0.3 eV). In addition, the measured $R(\omega)$-ω profile is more linear than that predicted from the Drude model. Similar behavior has been observed as well in the Cu-O based high-T_c systems. When La_2CuO_4 is doped with Sr, the Drude-like reflectivity increases but the apparent plasma edge always stays near 1 eV [9]. Similar observations in $YBa_2Cu_3O_{6+x}$ [8] and other high-T_c cuprate systems, and the non-cuprate system $BaPb_{1-x}Bi_xO_3$ [11] have been

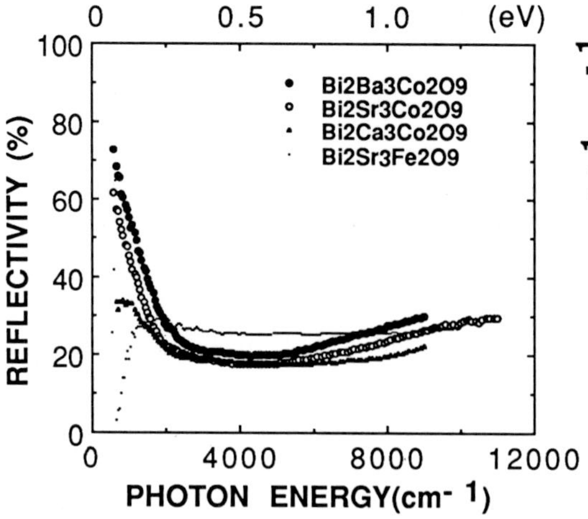

Fig.3 Reflectivity spectra $R(\omega)$ of double 3d-O plane bismuthates.

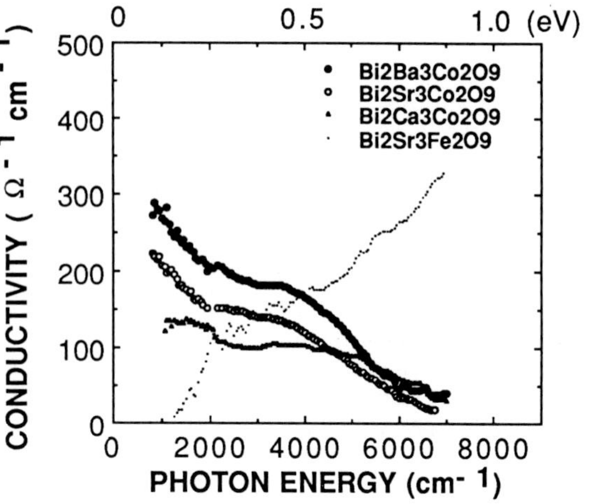

Fig.4 Dynamic conductivity $\sigma(\omega)$ of double 3d-O plane bismuthates.

reported, and frequently interpreted in terms of the mid-*IR* absorption [3,8]. We may also assume the existence of an absorption band below 0.3 eV in analyzing the spectra of the $Bi_2M_3Co_2O_y$ systems, or a frequency-dependence of the scattering rate $\tau^{-1}(\omega)$ ($= \tau_0^{-1} + \alpha\omega$) in the Drude model as in the high-T_c cuprates . Both models can fit the $R(\omega)$-ω profile, except the Drude- $\tau^{-1}(\omega)$ model fit to the $R(\omega)$-ω profile for $Bi_2Ca_3Co_2O_y$ [7].

In the $Bi_2M_3Co_2O_y$ compounds, the samples with a higher dc conductivity (as M changes) shows a more intense mid-*IR* absorption. The overall values of $\sigma(\omega)$ are much smaller than those of the high-T_c cuprates. Recent results on $YBa_2Cu_3O_{6+x}$ with different doping densities (x) show a similar dependence of the mid-*IR* absorption intensity on the free carrier absorption [8]. The hump structure around 0.3 eV observed in single 3d-O plane bismuthates appears also in the insulating sample of double 3d-O plane bismuthates ($Bi_2Ca_3Co_2O_y$ and $Bi_2Sr_3Fe_2O_{9.2}$). The $\sigma(\omega)$–ω profile of $Bi_2Sr_3Fe_2O_{9.2}$ seems qualitatively different from the other insulators studied here; $\sigma(\omega)$ of $Bi_2Sr_3Fe_2O_{9.2}$ increases steadily with ω. While $Bi_2Sr_2MnO_{6.25}$ and $Bi_2Sr_2CoO_{6.25}$ exhibit oxygen disorder or oxygen vacancy in the Bi-O plane, $Bi_2Sr_3Fe_2O_{9.2}$ does not [11]. This may account for the above difference in $\sigma(\omega)$. We could identify two absorption bands near 0.3 eV and 0.5 eV, probably because the apparent plasma edge is located at frequencies much lower than those in high-T_c cuprates. We speculate that this may be true as well in the Bi-Cu system, possibly with a much smaller absorption near 0.3 eV.

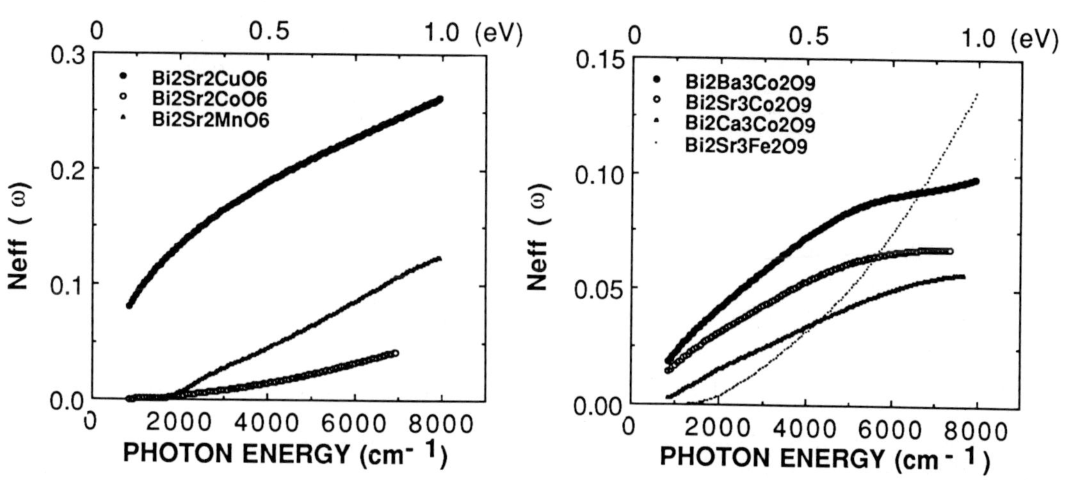

Fig. 5a (left), 5b (right). Effective carrier number per subunit-cell as function of energy (ω).

An absorption band similar to the mid-*IR* absorption is also observed in simple 3d-metal chalcogenides and oxides that show a metal-insulator transition [12]. In NiO, the intensity of a broad absorption between 0.2 eV and 1 eV is enhanced by doping [13], as in high-T_c cuprates. These observations suggest that the mid-*IR* absorption may be related to a general property of the metal-insulator transition.

It is useful to estimate the carrier number per subunit-cell volume containing a single 3d-O plane (V). Integrated spectral weight, or normalized effective carrier number is defined by,

$$N_{eff}(\omega) = (2Vm/\pi e^2)\int_0^{\omega} \sigma(\omega')\, d\omega',$$

where m is the free electron mass(Fig.5a, 5b). In all samples except $Bi_2Sr_2CuO_{6+\delta}$, the Neff values at 1 eV are similar (~0.1), and the major contribution to them comes from mid-*IR* range (assuming that the effective masses are almost same). In Fig. 5b, the increase of carrier density on doping is clear in the $Bi_2M_3Co_2O_y$ system.

CONCLUSION

A broad absorption in the mid-*IR* range was observed in all crystals of the quasi-two dimensional oxides studied, including $Bi_2Sr_2MnO_{6.25}$, a good insulator. This fact supports the absorption band model in the high-T_c cuprates rather than a special scattering mechanism. In comparison with the reflectivity in the cuprates, the much lower plasma edge in the former compounds allows the identification of two absorption bands below 1 eV. The non-superconductors $Bi_2M_3Co_2O_y$ show striking similarities to the high-T_c cuprates, such as a correlation of a Drude-like reflectivity with the intensity of the mid-*IR* absorption, the insensitivity of the apparent plasma edge to the doping level, and a quasi-linear $R(\omega)$-ω profile. These results suggest that the existence of a mid-*IR* absorption band is not sufficient for the occurrence of the high-T_c superconductivity. It appears to be present in several classes of 3d-metal oxides, especially those that undergo a metal-insulator transition.

ACKNOWLEDGMENTS

We thank Prof. S. Uchida, Prof. S.A. Lyons and Dr. G.A. Thomas for many helpful discussions.

REFERENCES

1. G.A. Thomas, J. Orenstein, D.H. Rapkine, and M. Capizzi, A.J. Millis, R.N. Bhatt, L.F. Schneemeyer, and J. V. Waszczak, Phys. Rev. Lett. **61**, 1313 (1988).
2. Y. Watanabe, Z.Z. Wang, S.A.Lyon, D.C. Tsui, N.P. Ong, J.M. Tarascon, P.Barboux, Phys. Rev. B **40** 6884 (1989).
3. Y. Watanabe, Z.Z. Wang, S.A. Lyon, D.C. Tsui, N.P. Ong, J.M. Tarascon, E.Wang, Solid State Commun. **74**, 757 (1990).
4. Y. Watanabe, Z.Z. Wang, S.A. Lyon, D.C. Tsui, N.P. Ong, J.M. Tarascon, P. Barboux and E.Wang, R. Kossowsky et al. ed., NATO ASI Physics and Materials Science of High Temperature Superconductors, 429, Kluwer Academic Publishers (1990).
5. J.M. Tarascon, R. Ramesh, P. Barboux, M.S. Hedge, G.W. Hull, L.H. Greene M. Giroud, Y. LePage, W.R. Mckinnon, J.V. Waszczak, and L.F. Schneemeyer, Solid State Commun. **71**, 663, (1989)
6. J.M. Tarascon, Y. LePage, W.R. Mckinnon, E. Tselepis, P. Barboux, B.G. Bagley and R. Ramesh, Proceeding of the MRS symposium, **156** (1989).
7. Y. Watanabe, S.A. Lyons, N.P. Ong, D.C. Tsui, and J.M. Tarascon, submitted to Phys. Rev.B.
8. J. Orenstein, G.A. Thomas, A.J. Millis, S.L. Cooper, D.H. Rapkine, T. Timusk, L.F. Schneemeyer, and J.V. Waszczak, (preprint)
9. M. Suzuki, Phys. Rev. B **39** 2312(1989).
10. I. Terasaki, T. Nakanishi, S. Takebayashi, A. Maeda and K. Uchinokura, Physica C **165**, 152 (1990) .
11. S. Tajima, S.Uchida, A. Masaki, H. Takagi, K . Kitazawa, S. Tanaka and A. Katsui, Phys. Rev. **32** , 6302 (1985).
12. I. Balberg and J.I. Pankove, Phys. Rev. Lett. **27** 596, (1971). A.S. Barker, Jr. and J. P. Remeika, Phys. Rev. B **10** , 987, (1974).
13. R. Newman and R.M. Chrenko, Phys. Rev. **114**, 1507 (1959).

Far-Infrared Absorption by Bulk $YBa_2Cu_3O_{6.89}$ Superconductor

ETSUO KAWATE[1], ATSUSHI ONAE[1], YASUHIRO NAKAZAWA[1], MASAYASU ISHIKAWA[2], YUKINOBU MIKI[2], and MASAHIRO OKAJI[1]

[1] National Research Laboratory of Metrology, 1-4, Umezono 1-chome, Tsukuba, Ibaraki, 305 Japan
[2] Institute for Solid State Physics, The University of Tokyo, 22-1, Roppongi 7-chome, Minato-ku, Tokyo, 106 Japan

ABSTRACT

We have developed a Far-Infrared-Laser Calorimetric method to investigate far-infrared absorption by high-Tc superconductors using an optically pumped CH_3OH laser. We obtained information on both transition temperature and superconducting energy gap. The sintered $YBa_2Cu_3O_{6.89}$ specimen absorbed photons below the gap energy even in the superconducting state. We found that this absorption would have an onset near 5 meV and consider that the existence of the onset will support a model of the low-energy excitation.

KEY WORDS: high-Tc superconductors, far-infrared absorption, an optically pumped CH_3OH laser, sintered $YBa_2Cu_3O_{6.89}$

INTRODUCTION

One of controversial problems in high-Tc superconductors is whether or not the far-infrared (FIR) reflectivity of these materials is 100 % for sufficiently low-energy photons in the superconducting state. The absolute FIR reflectivity (R) of high-quality $YBa_2Cu_3O_Y$ single crystals in the far-infrared region, where the superconducting gap may be observed (ε <70 meV), is very close to unity ($R \geqq 98\%$).[1,2] However, the detection limit of the traditional FIR reflectance measurements is usually about 1 %,[1,2] because it is restricted by detector noises and normalization errors. This gives a severe limitation to the detection of low-energy excitations of the superconducting state by reflectance measurements in the highly reflecting materials.

In this paper, we report direct measurements of the absorption of photons with energy between 7.6 and 17.6 meV by sintered $YBa_2Cu_3O_{6.89}$ specimens at the temperature range between 12 and 160 K.

EXPERIMENTAL

Single phase polycrystalline sample of $YBa_2Cu_3O_{6.89}$ was prepared by a gettered annealing technique, as previously reported.[3] The sample exhibited a Tc value of 93 K (onset) with a width of Δ Tc=0.5 K, measured by standard four-probe method. The optical arrangement and the schematic diagram of experimental apparatus for a Far-Infrared-Laser Calorimetric method (FILC) are shown in Fig.1 and Fig.2, respectively. Three types of excitation light sources were used. One was an optically pumped FIR CH_3OH laser[4] which have a choice of monochromatic photons with energies between 7.6 and 17.6 meV. Another was a CO_2 laser which gave photons of 124 meV. And the other was a tungsten lamp which was used as a white light source. Several powerful lines of the CH_3OH laser, used in the experiments, are summarized in Table 1. The

Table 1 Several powerful lines of a CH_3OH laser

Photon energy [meV]	2.6	4.9	7.6	10.4	12.8	17.6
Output Power [mW]	7.5	11	32	65	32	45

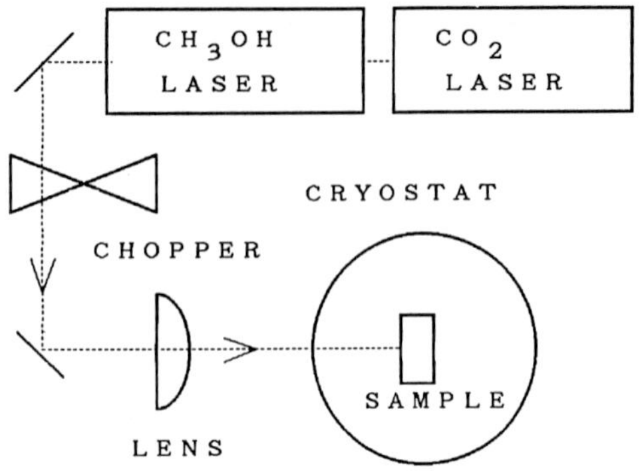

Fig.1 The optical arrangement.

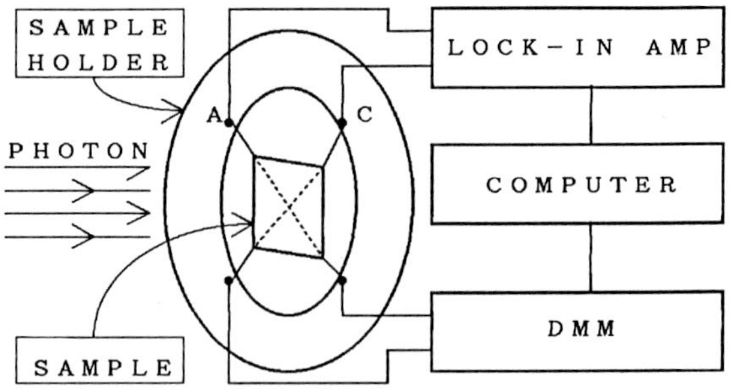

Fig.2 The schematic diagram of experimental apparatus. DMM; Digital MalutiMeter, A; Alumel, C; Chromel.

intensity of laser beam in the cryostat was less than a tenth of the output power of the laser due to losses caused by air, optical windows and so on. The light was chopped at 1.6 Hz and was focused with a lens onto the specimen surface of 2X2 mm^2 with a thickness of 0.1 mm. A thermocouple (Chromel / Alumel) was silver-pasted on the rear surface as temperature sensors. The amplitude of temperature oscillation (Δ Tac) induced by the chopped light was measured by a phase-sensitive lock-in amplifier.

RESULTS

The existence and magnitude of the superconducting energy gap strongly reflects the temperature and wavelength dependences of absorption coefficient, $\mu(\varepsilon,T)$, in the far-infrared region. A relation between the coefficient and absorption energy by the specimen, ΔQ, is $\Delta Q=\mu\, dSP$, where P,S and d are intensity of light, area of sample and penetration depth of light, respectively. The penetration depth of high-Tc superconductors is estimated to be about 200 nm[1] in the far-infrared region. In this case the relation[5] between the absorption energy and increment of sample temperature, ΔT_{ac}, was

$$\Delta Q = \omega C \Delta T_{ac}\left[1+\frac{1}{\omega^2\tau_e^2}+\omega^2\tau_i^2+\sqrt{40}\,\frac{\tau_i}{\tau_e}\right]^{1/2} \tag{1}$$

where f, C, τ_e and τ_i are a chopping frequency, a heat capacity, an external relaxation time and an internal relaxation time. In order to obtain the magnitude of the absorption coefficient from the experiment not from calculation, we coated the front surface of the same specimen with graphitic paint to keep absorptivity constant and irradiated its surface with a tungsten lamp. The white light is absorbed by the thin layer of paint, where the absorption coefficient of the paint, μ^{o}, can be considered constant, and heat is given to the specimen. The relation between the absorption energy by the paint, ΔQ^{o}, and the increment of sample temperature, ΔT_{ac}^{o}, is the same as eq.(1) The absorption coefficient $\mu(\varepsilon,T)$ in the FIR region is

$$\mu(\epsilon,T)=\frac{\Delta Q}{dSP}\propto\frac{\Delta Q}{\Delta Q^0}=\frac{\Delta T_{ac}}{\Delta T_{ac}^0} \tag{2}$$

Figure 3 exemplifies typical records of outputs of the lock-in amplifier. These were normalized at 105 K for convenience. When we turned off the laser at 12, 28 and 155 K, the outputs became zero as shown by closed circles. We can calculatn values of the amplitude of temperature oscillation (Δ Tac) from the outputs of the lock-in amplifier, considering the value of the thermoelectromotive force of our thermocouple at each temperature. Figure 4 represents these Δ Tac's. From the eq.(2), we can obtain the magnitude of $\mu(\varepsilon,T)$. Figure 5 represents the far-infrared absorption coefficient as a function of temperature. Δ Tac's and $\mu(\varepsilon,T)$'s were normalized by the intensity of laser beam. Figure 6 represents the $\mu(\varepsilon,T)$ as a function of photon energy. o's, x's, +'s and *'s represent the coefficient at temperature between 12 and 40, 50, 70 and 90 K, respectively.

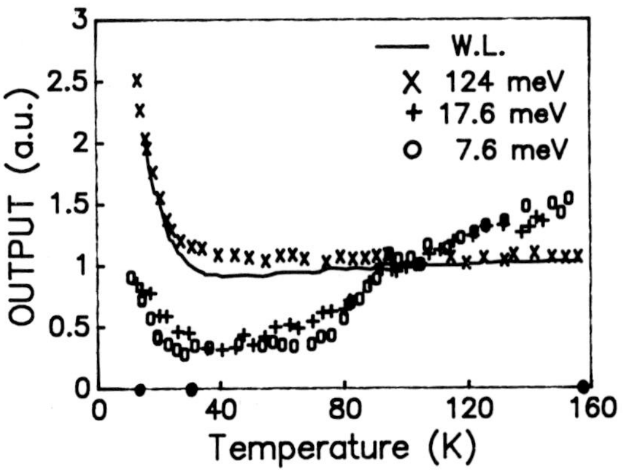

Fig.3 Output of lock-in amplifier vs. temperature. W.L.; white light.

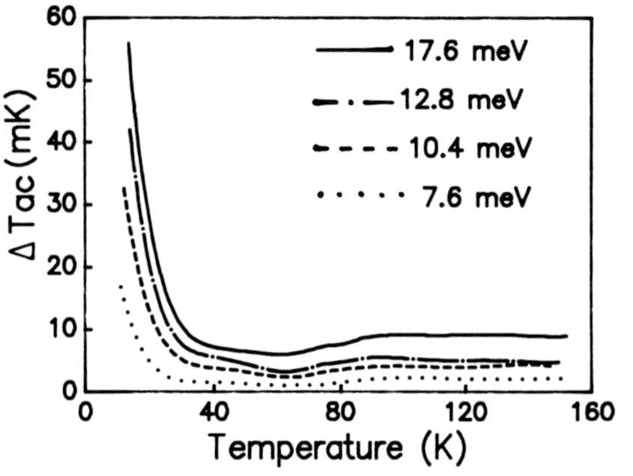

Fig.4 The amplitude of temperature oscillation, Δ Tac, vs. temperature.

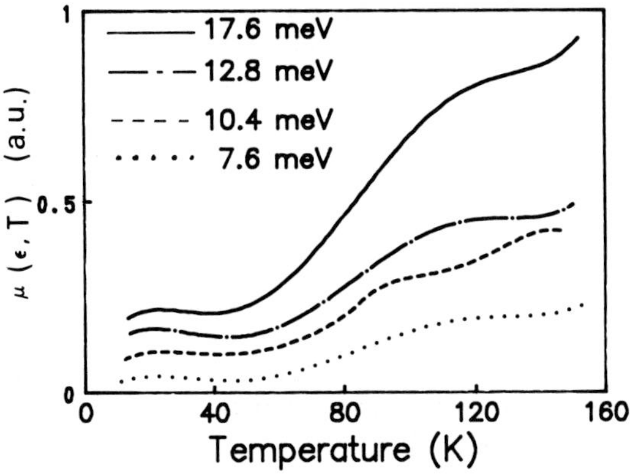

Fig.5 The absorption coefficient, $\mu(\varepsilon, T)$, vs. temperature.

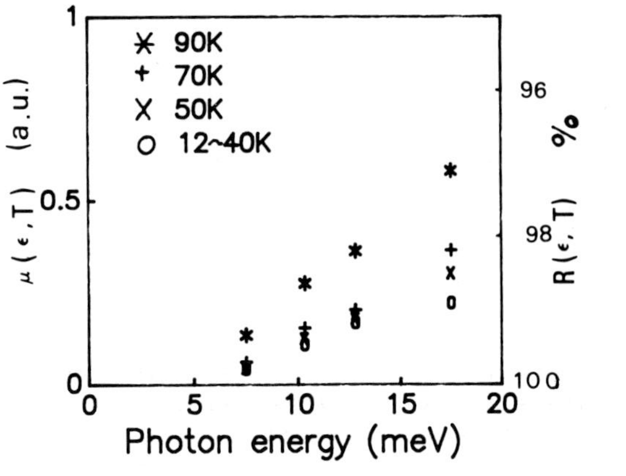

Fig.6 The absorption coefficient, $\mu(\varepsilon, T)$, and the reflectivity, $R(\varepsilon, T)$, vs. photon energy.

DISCUSSION

Considering both the relationship of R=1-A (R is reflectivity and A is absorptivity when transmissivity is zero) and the reflectivity of 98 %[1] for 100 μm at 100K, we can obtain the reflectivity from the absorption coefficient in Fig.6. As a result the absorption coefficient, $\mu(\varepsilon, T)$, and the reflectivity, $R(\varepsilon, T)$, are written alongside the left and the right ordinates as a function of photon energy, respectively.

The outputs became zero as shown in Fig.3 when we turned off the laser, so that the influence of a flux emitted by a black body at room temperature can be neglected in our FILC measurement and the signal to noise ratio (S/N) is larger than 25. Considering the 98%-reflectivity at 100 K, the detection limit of the FILC method is (1-0.98) / 25=0.08 %. Traditional FIR reflectance measurements the detection limit is about 1 %.[1,2] The detection limit of the FILC measurement is far superior to that of traditional FIR reflectance measurements. Thus the FILC method is fitted to the detection of changes in a very small absorption by these highly reflecting materials.

These superior properties are mainly due to three reasons discussed below. (1) The schematic diagram of Fig.2 is very similar to that of a well-known light irradiated ac calorimetric method. This method is highly sensitive due to ac measurement by using a lock-in amplifier. (2) The optical arrangement of Fig.1 is similar to that of a photoacoustic spectroscopy (PAS). In this method the stronger is the intensity of light, the bigger is the intensity of signal.

(3) There are processes of radiative and non-radiative transitions from an excited state to a lower state in solids. In the non-radiative decay the energy is released as phonons. The FILC method is a direct measurement of the temperature change closely related to phonons.

There were striking features in both o's and +'s compared with both x's and solid curve in Fig.3. The former dropped abruptly just below Tc, so that we could obtain a well-defined transition temperature. The decrement became steeper, as the photon energy decreased from 17.6 to 7.6 meV in Fig.3. x's were hardly affected by the superconducting phase transition. These results suggest that the energy gap exists in the region between 17.6 and 124 meV. This wide region is due to lack of available lines from the lasers.

We found, for the first time in the absorption measurement, "unexpected" absorption by the sintered $YBa_2Cu_3O_{6.89}$ specimen. This material absorbed photons below the gap energy even in the superconducting state. As a result, the reflectivity is not unity but 99.7 ± 0.08 % at sufficiently low temperature as shown in Fig.6. In order to estimate the photon energy at which the absorptivity becomes zero, we extrapolatre from the absorptivity curves at each temperature in Fig.6. The unexpected absorption had an onset near 5 meV which is independent of tempertaure. Three possibilities, as mentiond before,[1,2] can be completely ruled out because of the onset in the unexpected absorption : 1) the extrinsic origin by impurity phases or imperfections near the surface, 2) normal electrons that are thermally excited across the gap and 3) gapless superconductivity over a finite area of the Fermi surface. Two possibilities of optically active phonons and anisotropic energy gap can be neglected, because the phonon resonances in the $YBa_2Cu_3O_{6.89}$ have the maximum at higher photon energy than 17.6 meV and because the magnitudes of superconducting energy gaps are larger than 7.6 meV even if an anisotropic effect is considered. One possible model of the unexpected absorption is a low-energy excitation of spins on the CuO_2 planes.[6]

Conclusion

We have developed a Far-Infrared-Laser Calorimetric (FILC) method using an optically pumped CH_3OH laser. The method is a measurement of the increment of temperature induced by FIR absorption. The detection limit of the FILC method was 0.08 %, which is far superior to that of the traditional FIR reflectance methods. The FILC method is fitted to the detection of changes in a very small absorption by high-Tc superconductors. We could obtain information on both transition temperature and superconducting energy gap. The sintered $YBa_2Cu_3O_{6.89}$ specimen absorbed photons below the gap energy even in the superconducting state. With decreasing temperature or with decreasing photon energy, the absorption was monotonically decreasing and had an onset near 5 meV which was independent of tempertaure. Because of the existence of the onset, we consider that one possible model of this absorption is a low-energy excitation.

Acknowledgment
The authors wish to express their thanks to Dr. E. Sakuma and Dr. T. Kurosawa for their fruitful suggestions in the infrared measurements, and to Dr. H. Imai for his encouragement throughout this work.

References

1. Z.Schlesinger, R.T.Collins, F.Holtzberg, C.Feild, G.Koren, and A.Gupta (1990) Phys.Rev.B 41: 11237
2. G.A.Thomas, J.Orenstein, D.H.Rapkine, M.Capizzi, A.J.Millis, R.N.Bhatt, L.F.Schneemeyer, and J.V.Waszczak (1988) Phys.Rev.Lett. 61: 1313
3. Y.Nakazawa and M.Ishikawa (1989) Physica C 158: 381
4. T.Sakurai (1983) Appl.Phys.Lett. 42: 319
5. Paul F. Sullivan and G.Seidel (1968) Phys.Rev.173: 679
6. P.W.Anderson, G.Baskaran, Z.Zou, and T.Hsu (1987) Phys.Rev.Lett. 58: 2790

Optical Properties of Single-Crystal $YBa_2Cu_3O_y$ Thin Films

HIROSHI YASUOKA[1], HIROMASA MAZAKI[1], TAKAHITO TERASHIMA[2], and YOSHICHIKA BANDO[2]

[1] Department of Mathematics and Physics, The National Defense Academy, Yokosuka, 239 Japan
[2] Institute for Chemical Research, Kyoto University, Uji, 611 Japan

ABSTRACT

Optical transmittance spectra of single-crystal $YBa_2Cu_3O_y$ thin films (y = 6.0, 6.4, 6.8) were measured in a wavenumber range of 1500-50000 cm^{-1} at room temperature. Absorption spectrum for each oxygen content was obtained from the transmittance spectra. In $YBa_2Cu_3O_{6.0}$, an energy gap at 1.5 eV and an absorption peak at 4.1 eV were observed. When the oxygen content increases, the absorption in the energy gap increases and the 4.1-eV absorption peak becomes smaller. We found that the dependence of the 4.1-eV peak intensity on the oxygen content correlates with that of the absorption in the 1.5-eV gap. This correlation suggests that these behaviors come from the same origin.

KEY WORDS: Optical properties, Single-crystal thin film, $YBa_2Cu_3O_y$

INTRODUCTION

There have been many investigations on the electronic structure of copper-oxide high-T_c superconductors. Among these, optical properties in the normal state give us fundamental information on the electronic structure [1]. When the electric field of incident light is perpendicular to the c axis ($E \perp c$), information on the electronic state on the Cu-O planes is included in the observed spectra.

Studies of photoemission suggest that the perovskite-related copper oxides are based on charge-transfer-type semiconductors [2]. From optical studies of $La_{2-x}Sr_xCuO_4$ single-crystal thin films, it is known that semiconducting La_2CuO_4 makes a phase transition to the superconducting or metallic states by hole doping from the substitution of Sr for La [3]. In the case of $YBa_2Cu_3O_y$, it is reported that hole doping can be achieved by oxidizing $YBa_2Cu_3O_{6.0}$ [4]. However, details of the dependence of electronic states on the oxygen content are not yet clear. To get further information on this problem, we attempted to measure optical transmittance and absorption spectra for $E \perp c$ using single-crystal thin films of $YBa_2Cu_3O_y$.

For sintered and single-crystal bulk samples, Tajima et al. [1] measured optical reflectivity spectra and evaluated the optical conductivity spectra using a Kramers-Kronig analysis. For absorption properties, however, it is preferable to get absorption spectra directly from transmittance measurements, although for most metals, transmittance measurements are generally difficult. Fortunately, the reflectivity of copper-oxide superconductors is smaller than that of conventional metals. These features make it possible to carry out transmittance measurements for thin films of this kind of superconductors.

EXPERIMENTAL

The samples were prepared by the activated reactive evaporation method, of which the details were previously reported [5]. The films were epitaxially grown parallel to the <001> axis on the (100) surface of single-crystal MgO substrates. For three kinds of oxygen concentration, we prepared two samples of different thicknesses.

The samples were characterized by x-ray diffraction and complex susceptibility measurements. The oxygen content was estimated from the lattice parameter determined by x-ray diffraction using Ref. 6. Figure 1 is a typical result of complex susceptibility $\chi = \chi' - i\chi''$ for the y = 6.4 compound with 1200 A thickness, where the ac magnetic field was applied perpendicular to the film plane. The curves in Fig. 1 indicate that the superconducting transition temperature is about 50 K and the sample is a single phase superconductor.

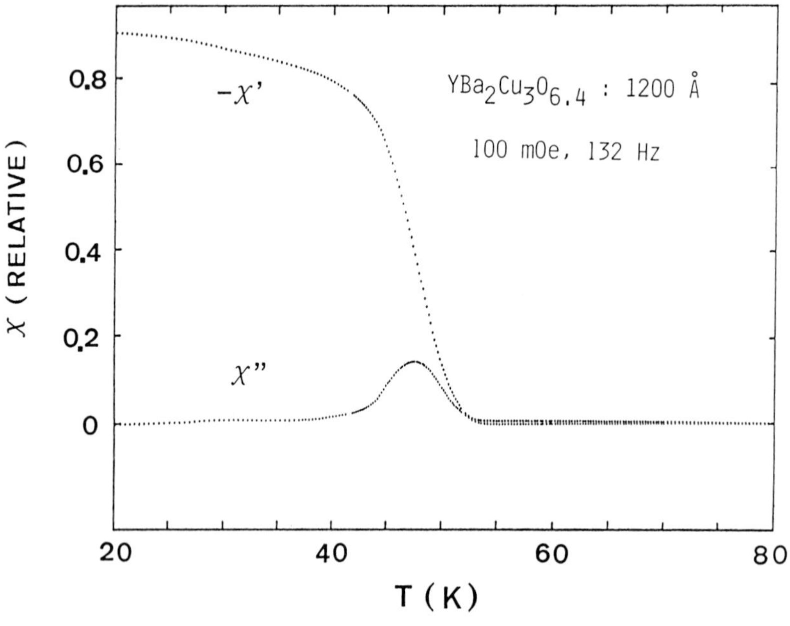

Fig. 1. Complex susceptibility for $YBa_2Cu_3O_{6.4}$ film (1200 A).

Transmittance measurements were carried out at room temperature. Measurements were made by using the Fourier-transform infrared spectrometer: Bruker Model IFS-113V for the infrared region, Cary Model-17 for the near infrared region and Shimazu Model MPS-2000 for the visible and ultraviolet region. The transmittance of a single-crystal MgO substrate of 0.5-mm thickness was more than 80% over a wavenumber range of 1500-38000 cm^{-1}.

RESULTS AND DISCUSSION

In Fig. 2, we show a typical transmittance spectra of 1200-A and 2400-A films of $YBa_2Cu_3O_{6.8}$, which had critical temperatures of about 85 K. The solid line below 10000 cm^{-1} was obtained using Bruker, the dotted and solid lines above 10000 cm^{-1} were from Cary and Shimazu, respectively. The continuity of the data obtained with the three spectrometers is satisfactory. The structure of these two spectra are similar, indicating that the structure does not arise from interference, but is due to the electronic state. The decrease in the transmittance below 10000 cm^{-1} comes from the absorption and reflectance of incident light by free carriers. Weak absorption bands are observed at around 12000 and 40000 cm^{-1}.

Optical absorption spectra were deduced from a pair of transmittance spectra by dividing the spectra of thicker films by those of the thinner ones. In Fig. 3, we present the absorption spectra thus obtained for three kinds of $YBa_2Cu_3O_y$ (y = 6.0, 6.4, 6.8). The results agree well with the results of sintered samples measured by a spectroscopic ellipsometer [7-9]. For y = 6.0, we observe an absorption edge at 1.5 eV and a sharp peak at 4.1 eV. When the oxygen content increases, the absorption in the energy gap increases and the 4.1-eV absorption peak becomes smaller, disappearing for y = 6.8. These features could be related to the electronic states of the Cu(1)-O chains and/or the Cu(2)-O planes, since the electric field of the incident light is perpendicular to the c axis.

As shown in Fig. 3, the spectra below 1.5 eV show that the band gap seen for

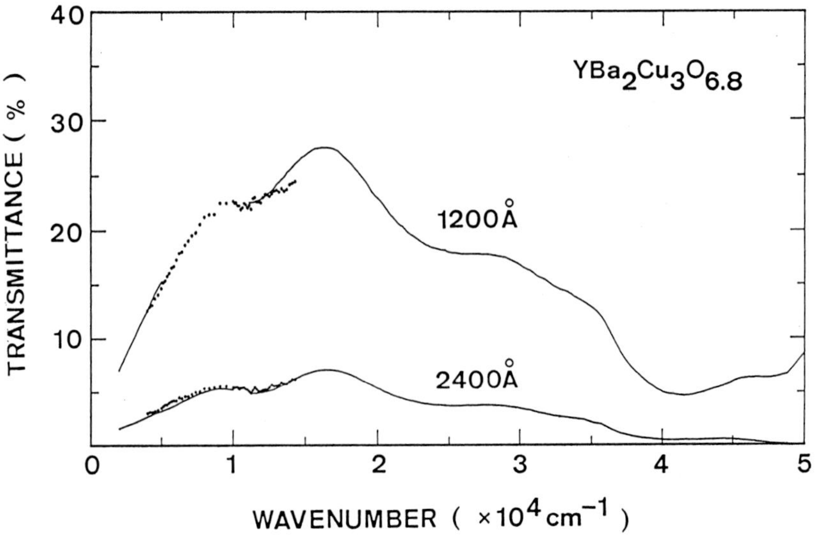

Fig. 2. Transmittance spectra of $YBa_2Cu_3O_{6.8}$ thin films.

$YBa_2Cu_3O_{6.0}$ disappears, as x increases. From this feature, the compound seems to become metallic from the semiconducting states upon oxidization. The disappearance of the energy gap in $La_{2-x}Sr_xCuO_4$ (2.0 eV) by carrier doping was also observed [3], where $La_{2-x}Sr_xCuO_4$ does not contain the Cu-O chains. From this fact, it is considered that the 1.5-eV gap found here comes from the electronic states of the Cu(2)-O planes. As to the source of this band gap, studies on the electronic states by photoemission and electron energy-loss spectroscopy suggest that, in semiconductor compounds, the top valence electrons of the Cu(2)-O planes have an oxygen 2p-like character, and that, in superconducting compounds, holes are introduced in the oxygen 2p band [2,10,11]. These works support the band picture that the oxygen 2p band is located between the separated Hubbard bands of the copper 3d electrons [12]. Therefore, it is plausible to consider that the observed 1.5-eV energy gap in $YBa_2Cu_3O_{6.0}$ corresponds to the charge-transfer gap between the oxygen 2p and the copper 3d bands. In addition, upon oxidization, the compound is expected to becomes metallic due to holes induced in the oxygen 2p band.

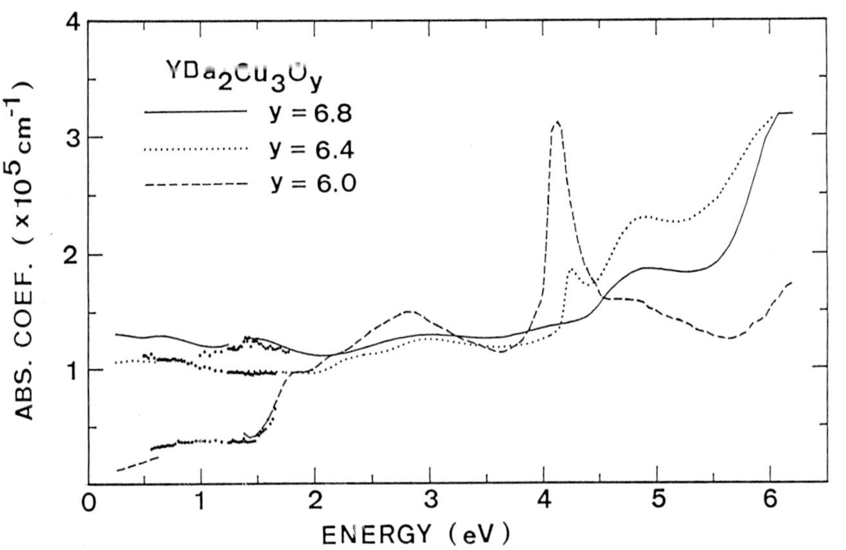

Fig. 3. Absorption spectra of three kinds of $YBa_2Cu_3O_y$ thin films.

The 4.1-eV absorption also depends drastically on the oxygen content. Meanwhile, no sharp absorption peak above the charge-transfer gap was observed in other semiconducting compounds like La_2CuO_4. Since La_2CuO_4 does not contain Cu-O chains, the 4.1-eV peak observed for $YBa_2Cu_3O_{6.0}$ is expected to come from the electronic state of the Cu(1)-O chains. From the fact that the largest absorption peak appears for y = 6.0 and disappears upon oxidization, we suppose, as pointed out in Ref. 8, that the 4.1-eV peak originates from the electronic transition in the Cu ions isolated from oxygen atoms.

In Fig. 3, we notice that the dependence of the 4.1-eV peak intensity on the oxygen content correlates with that of the absorption in the 1.5-eV gap. The decrease in the peak absorption at 4.1 eV on comparing y= 6.0 to 6.8 is 0.8 of the decrease in the absorption on comparing y = 6.0 to 6.8. Similar fraction increases in absorption are observed in the gap (e.g., at 1 eV). This suggests that both behaviors against the oxygen content have the same origin, indicating that the oxygen atoms deposited in the Cu(1)-O chains are essentially correlated to hole doping in the Cu(2)-O planes.

In summary, we measured optical transmittance spectra of $YBa_2Cu_3O_y$ (y - 6.0, 6.4, 6.8) single-crystal thin films at room temperature. We obtained absorption spectra from the transmittance spectra. In the y = 6.0 compound, we observed a charge-transfer gap at 1.5 eV and a sharp absorption peak at 4.1 eV. Upon oxidization, both the 1.5-eV gap and the 4.1-eV peak tend to disappear. From the correlation between the variations of the gap and the peak upon oxidization, we speculate that these behaviors have the same origin, i.e., deposition of oxygen atoms in the Cu(1)-O chains.

REFERENCES

1. Tajima S, Ishii H, Nakahashi T, Takagi T, Uchida S, Hidaka Y, Suzuki M, Murakami T, Oka K, Unoki H (1989) Opt. Soc. America 6: 475
2. Fujimori A, Takayama-Muromachi E, Uchida Y, Okai B (1987) Phys.Rev.B 35: 8814
3. Suzuki M, (1989) Phys. Rev. B 39: 2312
4. Tokura Y, Torrance JB, Huang TC, Nazzal AJ (1988) Phys. Rev. B 38: 7156
5. Terashima T, Iijima K, Yamamoto K, Bando Y, Mazaki H (1988) Jpn. J. Appl. Phys. 27: L91
6. Mazaki H, Ueda Y, Aihara Y, Kubozoe T, Kosuge K (1989) Jpn. J. Appl. Phys. 28: L368
7. Garriga M, Humlicek J, Cardona M, Schonherr E (1988) Solid State Commun. 66: 1231
8. Kelly MK, Barboux P, Tarascon J-M, Aspnes DE, Bonner WA, Morris PA (1988) Phys. Rev. B 38: 870
9. Kelly MK, Barboux P, Tarascon J-M, Aspnes DE (1989) Phys. Rev. B 40: 6797
10. Nucker N, Fink J, Fuggle JC, Durham PJ, Temmerman WM (1988) Phys. Rev. B 37: 5158
11. Nucker N, Romberg H, Xi XX, Fink J, Gegenheimer B, Zhao ZX (1989) Phys. Rev B 39: 6619
12. Emery VJ (1987) Phys. Rev. Lett. 58: 2794

Raman Scattering from Spin-Pair Excitations in $Pr_{2-x}Ce_xCuO_{4-y}$, $Nd_{2-x}Ce_xCuO_{4-y}$ and $Sm_{2-x}Ce_xCuO_{4-y}$

IZUMI TOMENO[1], MASASHI YOSHIDA[1], KAZUTO IKEDA[1], KEISHI TAI[1], KENSHI TAKAMUKU[1], NAOKI KOSHIZUKA[1], KUNIHIKO OKA[2], and HIROMI UNOKI[2]

[1] Superconductivity Research Laboratory, International Superconductivity Technology Center, 10-13, Shinonome, Koto-ku, Tokyo, 135 Japan
[2] Electrotechnical Laboratory, Umezono, Tsukuba, Ibaraki, 305 Japan

Abstract

Antiferromagnetic excitations in single crystals of $Pr_{2-x}Ce_xCuO_{4-y}$, $Nd_{2-x}Ce_xCuO_{4-y}$ and $Sm_{2-x}Ce_xCuO_{4-y}$ were investigated as a function of Ce concentration using Raman scattering at room temperature. The introduction of electrons by Ce doping significantly broadens the two-magnon peak in the B_{1g} mode, indicating that zone-boundary magnons become damped with increasing Ce concentration. The existence of broad spin-pair-excitation spectra in the B_{1g} and B_{2g} modes at x=0.15 suggests that short-range spin fluctuations persist in these electron-doped superconductors. The result for the spin-pair-excitation peak in $Pr_{2-x}Ce_xCuO_{4-y}$ up to x=0.2 demonstrates the absence of softening of the short-wavelength magnons.

KEY WORDS: electron-doped superconductor, Raman scattering, two-magnon, exchange constant

INTRODUCTION

The electron-doped superconductors $Ln_{2-x}Ce_xCuO_{4-y}$ (Ln=Pr, Nd, Sm) [1,2]. have the T' tetragonal crystal structure, where the Cu atoms are square-planar coordinated by oxygens. Nd_2CuO_4 and Pr_2CuO_4 are two dimensional spin-1/2 Heisenberg antiferromagnets with T_N=250-270 K [3-5]. Destruction of antiferromagnetic (AF) long-range order by doping is common to the occurrence of superconductivity in all layered Cu-O-based materials. We carried out Raman scattering measurements on single crystals of $Ln_{2-x}Ce_xCuO_{4-y}$ at room temperature. The present results strongly suggests that short-range spin fluctuations remain in the electron-doped superconductors.

EXPERIMENTAL PROCEDURE

Platelike crystals of $Ln_{2-x}Ce_xCuO_{4-y}$ were grown from a CuO flux. A single-crystal of $Pr_{1.85}Ce_{0.15}CuO_{4-y}$ was also prepared using the traveling-solvent-floating-zone (TSFZ) method.[6] The x=0.15 samples were annealed in an Ar atmosphere at 920 °C, and then quenched to room temperature. The Ce content was determined by inductively-coupled-plasma spectroscopy. Magnetization measurements were performed using a SQUID magnetometer. The T_c values for these crystals are presented in Figs. 2 and 3.

The Raman measurements were carried out in the backscattering configuration. Linearly polarized light of 50 mW at 4880 Å was focused on the (001) plane. Scattered light was analyzed with a Jobin Yvon double-monochromator. The geometries (y',x') and (y,x) yield the B_{1g} and B_{2g} Raman-spectra, respectively.[7]

RESULTS and DISCUSSION

Figure 1 shows the Raman spectra for the as-grown undoped crystals. Characteristic features of the B_{1g} spectra are summarized in Table I. Several points are notable for the B_{1g} results. (1) The energy $\hbar\omega_p$ for the peak-position is far larger than kT_N=190 cm^{-1}. (2) $h\omega_p$ increases with decreasing Cu-Cu distance a in the CuO_2 plane. (3) Each spectrum has an asymmetric lineshape whose center of gravity $\hbar\omega_c$ is higher than the $\hbar\omega_p$ value. The observed B_{1g} lineshape in Pr_2CuO_{4-y} is compared with the theoretical curve including the magnon-magnon interaction. The exchange constant J values derived from the relation $\hbar\omega_p$=2.7J.[8] are consistent with the values deduced from neutron scattering measurements.[9] However, the observed B_{1g} linewidth (FWHM) is much broader than the prediction. One possible explanation is that quantum spin fluctuation effects [10] give rise to the broadening of the B_{1g} spectrum. The observed relation $\hbar\omega_c > \hbar\omega_p$ is consistent with the quantum fluctuations.

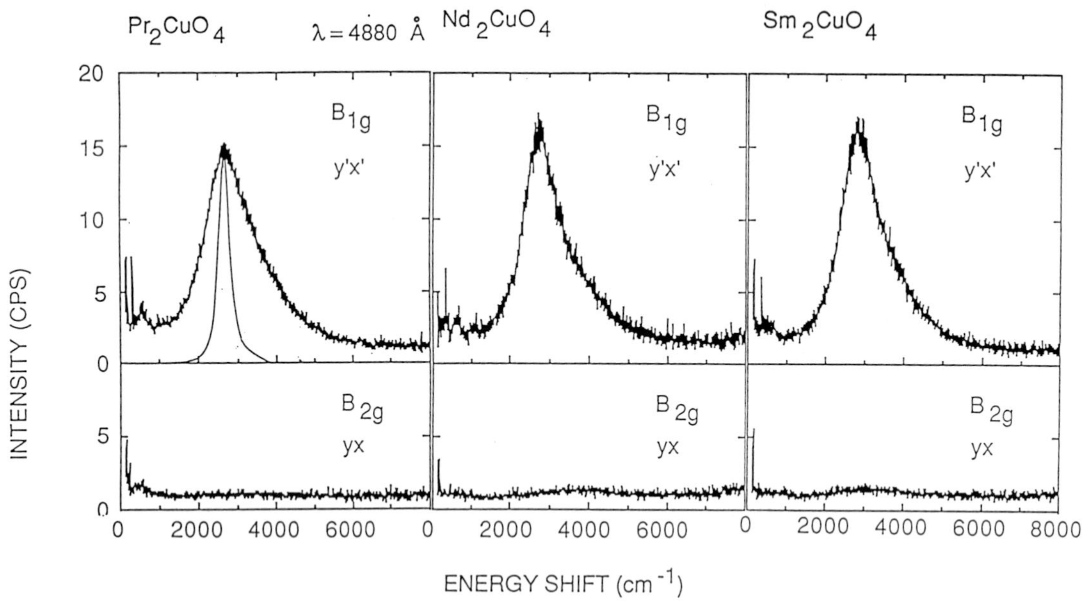

Figure 1. Raman spectra of Pr_2CuO_4, Nd_2CuO_4 and Sm_2CuO_4 at room temperature. Solid lines represent the theoretical B_{1g} and A_{1g} lineshapes based on the classical 2D Heisenberg model.

Table I. Raman scattering features and related physical parameters for crystals in the AF region. Energies $h\omega_p$ and $h\omega_c$ refer to the peak and the center of gravity for two-magnon B_{1g} spectra, respectively. Exchange constants J were obtained from the classical relation $h\omega_p$=2.7J.

	$h\omega_p$ (cm^{-1})	$h\omega_c$ (cm^{-1})	FWHM (cm^{-1})	J (cm^{-1})	T_N (K)	a (Å)
Pr_2CuO_4	2650	2990	1550	980	255[a]	3.948
$Pr_{1.91}Ce_{0.09}CuO_{4-y}$	2650	2940	2000	980	160[a]	3.950
$Pr_{1.87}Ce_{0.13}CuO_{4-y}$	2600	2880	2000	960	120[a]	3.953
Nd_2CuO_4	2750	3000	1200	1020	255[b]	3.934
Sm_2CuO_4	2800	3020	1200	1040	-	3.909
$Sm_{1.9}Ce_{0.1}CuO_{4-y}$	2800	2800	2400	1040	-	3.910

a.Ref.9. b. Ref.4.

The Ce concentration dependence of the Raman spectra is shown in Fig.2 for $Pr_{2-x}Ce_xCuO_{4-y}$. The flux-grown crystals span the AF, SC and metallic phases at 0 K. Doping in the AF region suppresses the B_{1g} peak intensity and significantly broadens its linewidth. The energy of the spin-pair-excitation peak in $Pr_{2-x}Ce_xCuO_{4-y}$ is not affected by doping up to x=0.2. This is similar in behavior to the $Sm_{2-x}Ce_xCuO_{4-y}$ system up to x=0.17. The existence of spin-pair excitations above T_N can be attributed to the fact that the intraplanar spin-spin correlation length ξ is very large even above T_N.

In view of the $KNi_{1-x}Mg_xF_3$ case [11], the B_{1g} results for $Ln_{2-x}Ce_xCuO_{4-y}$ are inconsistent with the localized picture that the nonmagnetic Cu^+ ions are randomly mixed with the S-1/2 Cu^{2+} ions. The Ce concentration may not be proportional to the decrease in average number <z> of nearest neighbor magnetic sites. The Ce doping broadens the B_{1g} lineshape significantly in $Ln_{2-x}Ce_xCuO_{4-y}$. The broadening is ascribed to significant magnon-damping near the zone boundary. Presumably the presence of nonmagnetic Cu^+ ions gives rise to the magnon-damping.

With increasing Ce concentration, an extremely broad B_{2g} peak appears near $\hbar\omega_p$. The observed behavior may be attributed to the change in magnetic symmetry due to the introduction of electrons. The fact that both (y',x') and (y,x) yield similar spectra in doped crystals supports the change in magnetic symmetry.

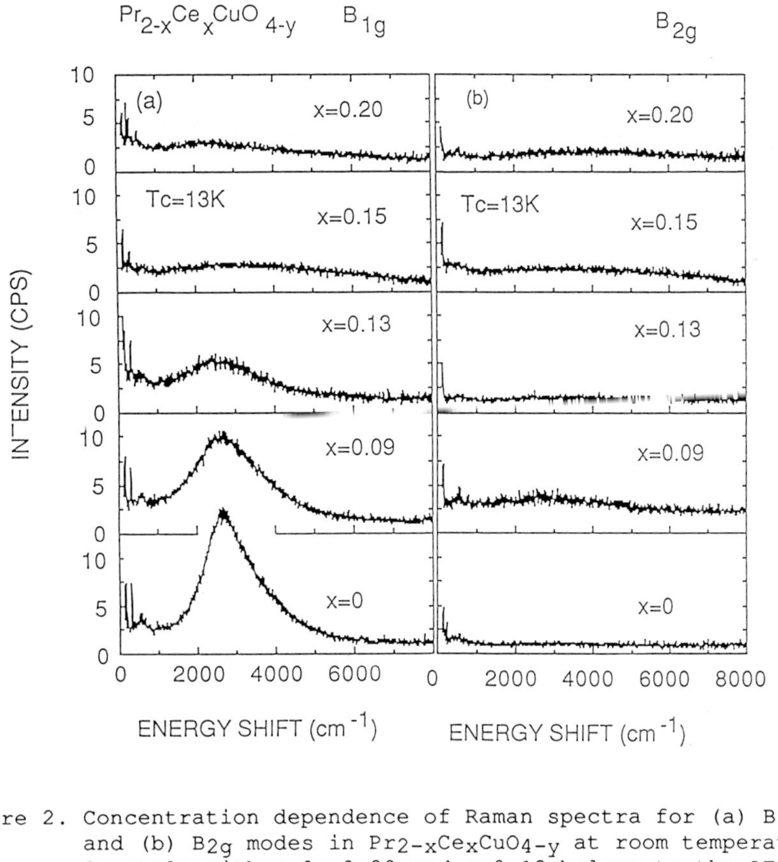

Figure 2. Concentration dependence of Raman spectra for (a) B_{1g} and (b) B_{2g} modes in $Pr_{2-x}Ce_xCuO_{4-y}$ at room temperature. Crystals with x=0, 0.09 and x=0.13 belong to the AF phase at 0 K. The x=0.15 crystal exhibits superconductivity at 13K, as shown in Fig.1. The x=0.2 crystal belongs to the metallic phase. All the spectra are taken using the flux-grown crystals.

Figure 3 shows the Raman spectra of $Ln_{1.85}Ce_{0.15}CuO_{4-y}$ single crystals. There is little discrepancy in the features between the flux and the TSFZ growth crystals, indicating the intrinsic spectra in the normal state of the electron-doped superconductor. In the SC and the metallic phases both B_{1g} and B_{2g} spectra exhibit extremely broad peaks whose positions are nearly equal to $h\omega_p$ in the AF phase. The extremely broad peaks in both B_{1g} and B_{2g} modes strongly suggest that the short-range spin fluctuations between Cu atoms persist in electron-doped superconductors. As shown in Fig.2, the magnetic excitations make no distinction between the SC and the metallic region. This means that the correlation length ξ decreases slowly with increasing Ce concentration x.

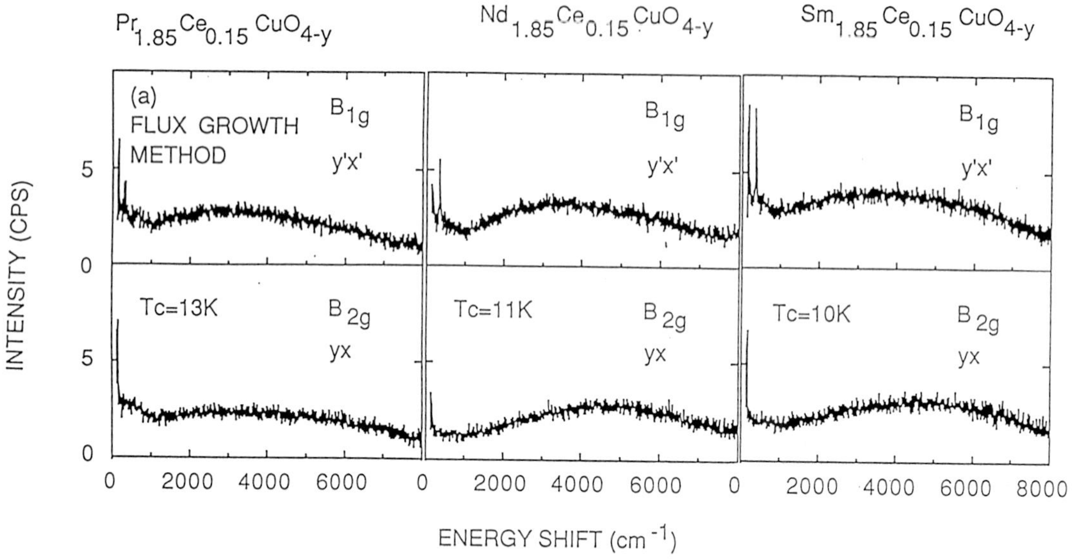

Figure 3. Raman spectra of $Pr_{1.85}Ce_{0.15}CuO_{4-y}$, $Nd_{1.85}Ce_{0.15}CuO_{4-y}$ and $Sm_{1.85}Ce_{0.15}CuO_{4-y}$ at room temperature.

REFERENCES

1. T.Tokura, H.Takagi, and S.Uchida (1989) Nature 337:345.
2. H.Takagi, S.Uchida, and Y.Tokura (1989) Phys.Rev.Lett.62: 1197.
3. J.Akimitsu, H.Sawa, T.Kobayashi, H.Fujiki, and Y.Yamada (1989) J.Phys.Soc.Jpn 58:2646.
4. Y.Endo, M.Matsuda, K.Yamada, K.Kakurai, Y.Hidaka, G.Shirane R.J.Birgeneau (1989) Phys.Rev.B40:7023.
5. D.E.Cox, A.I.Goldman, M.A.Subramanian, J.Gopalakrishnan, and A.W.Sleight (1989) Phys.Rev.B40:6998.
6. K.Oka and H.Unoki (1989) Jpn.J.Appl.Phys.28:L937.
7. K.B.Lyons and P.A.Fleury (1988) J.Appl.Phys.64: 6075.
8. J.B.Parkinson (1969) J.Phys.C 2: 2012.
9. T.R.Thurston, M.Matsuda, K.Kakurai, K.Yamada, Y.Endoh, R.J.Birgeneau, P.M.Gehring, Y.Hidaka, M.A.Kastner, T.Murakami and G.Shirane (1990) Phys.Rev.Lett.65: 263.
10. R.R.P.Singh, P.A.Fleury, K.B.Lyons, and P.E.Sulewski (1989) Phys.Rev.Lett.62: 2736.
11. P.A.Fleury, W.Hayes and H.J.Guggenheim (1975) J.Phys.C8: 2183.

Infrared Absorption in Single Crystals of Bi-Sr-(Ca)-Cu-O System

YOSHIKAZU HAYASHI, YOSHIYUKI HARADA, NOBUO FUKUOKA, and SHIGEMITSU NAKANISHI

Materials Science, University of Osaka Prefecture, 4-804, Mozuumemachi, Sakai, Osaka, 591 Japan

ABSTRACT

Lattice absorption measurements in the spectral region between 750 cm^{-1} and 400 cm^{-1} have been made in single crystals of semiconducting $Bi_2Sr_2CuO_6$ and superconducting $Bi_2Sr_2CaCu_2O_8$. The phenomenon known as the "infrared anomaly" has been investigated. One of the phonon mode at 629 cm^{-1} which is identified as the in-plane E_{u1} mode, cannot be observed in the superconducting compound although it has been clearly observed in the semiconducting crystal. In order to identify vibrational modes, measurements in powdered samples are also made.

KEY WORDS: High-Tc superconductor, Bi-Sr-Ca-Cu-O, Phonon spectra, Infrared anomaly

INTRODUCTION

It has been known that some of the infrared active lattice vibrational modes are not observable in high-Tc superconductors of La- and Y- based compounds although they have been observed in related insulators[1-5]. The phenomenon is called infrared anomaly. Recently, we have reported preliminary results of measurements in single crystals of Bi compounds[6]. In semiconducting crystals, having one sheet of Cu-O plane, we have observed an absorption at 629 cm^{-1}. On the other hand, absorption peak arround 600 cm^{-1} has not been detected in the superconducting crystals. However, these observations seem to contradict to the published results for sintered Bi compounds[7,8].

In order to resolve the discrepancy, we need to study whether the vibrational mode which have not been detectable in our previous experiment is detectable or not in the different experimental condition. Previously, the direction of the electric field vector of the infrared light was made parallel to the Cu-O plane. Therefore, it is necessary to make an experiment in which the direction of the electric field vector is perpendicular to the Cu-O plane. Absorption measurements in single crystals are preferable but are impossible because thin and large enough single crystals cut perpendicular to the Cu-O plane are not available. In this paper, results of two different measurements for searching extra absorption peaks arround 600 cm^{-1} are presented. Firstly, the range of measurements in single crystals is extended to 400 cm^{-1}. In this case, E_u modes is responsible for the absorption. Secondly, measurements in powdered samples are performed. If extra absorption peaks other than that found in single crystals are present, they must be due to the A_{2u} mode.

EXPERIMENTAL

Single crystals used in a present experiment are the same as those used in the previous experiment[6]. Infrared absorption measurements were made by making use of a Nicolet 60SXR FTIR(Fourier Transform Infrared) Spectrometer coupled with a microscope. The lower energy limit of the performance of this system is 400 cm^{-1}. Measurements were made also in powdered sample without microscope. The superconducting powder is obtained from the separately prepared sintered material since the small semiconducting crystals were found to precipitate in the single crystal of superconductor by the inspection with a microscope. About 1 mg of superconducting powder was mixed with about 400 mg of CsI and pressed to form a disk shaped sample with 2 cm diameter. All the measurements were performed at room temperature. Pressed pure CsI crystal is used as a standard for measurement. The CsI sample shows essentially no absorption above 400 cm^{-1}.

RESULTS AND DISCUSSION

Figure 1 shows the results of measurements of resistivities of single crystals. Curve (a) shows the temperature dependence of resistivity in semiconducting

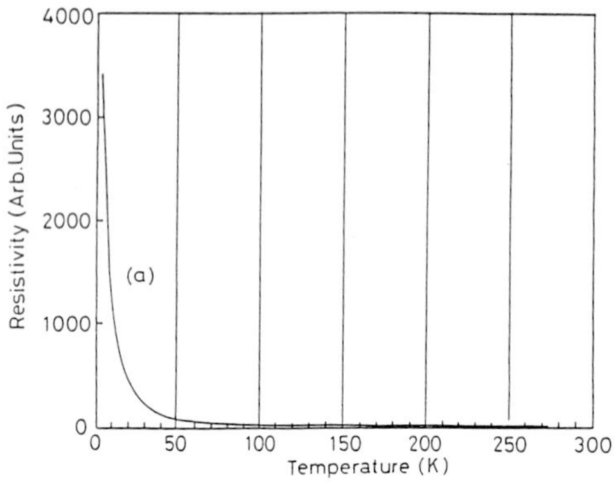

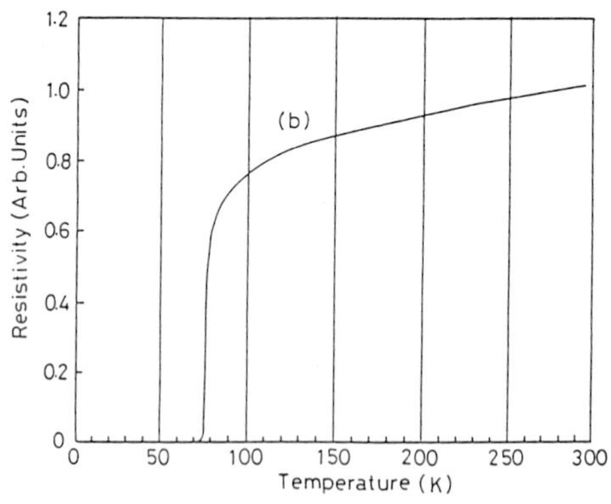

Fig.1 Temperature dependence of the resistivity for semiconducting (a) and superconducting (b) single crystals. The resistivities at room temperature are, 4Ω -cm for (a) and 4mΩ -cm for (b), respectively.

crystal. The resistivity at room temperature is about 4Ω -cm which increases to more than 2kΩ -cm at liq.He temperature. Curve (b) shows the resistivity change of the superconducting crystal. The resistivity of this crystal at room temperature is about 4mΩ-cm.

Typical examples of absorbance for single crystals are shown in Fig.2 Curves (a) and (b) show the absorption for $Bi_2Sr_2CuO_6$and $Bi_2Sr_2CaCu_2O_8$, namely, a

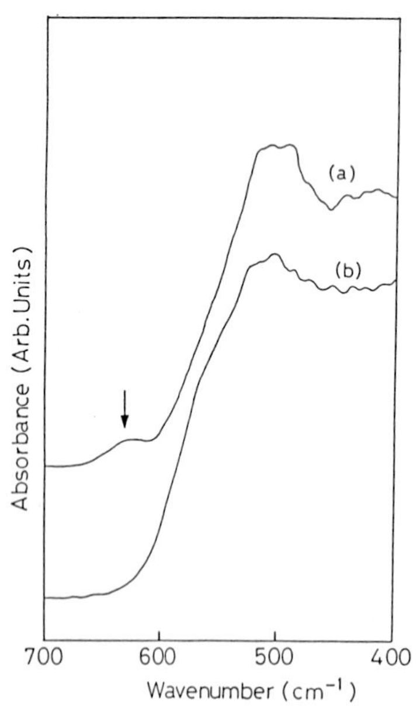

Fig.2 Infrared absorption of single crystals of Bi system. Curves (a) and (b) show the absorbance for semiconductor and superconductor respectively.

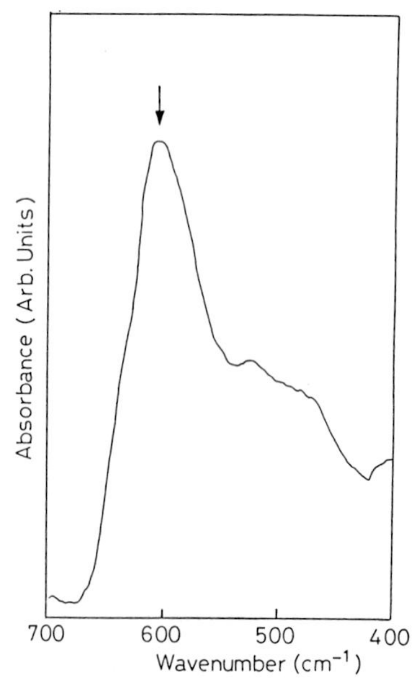

Fig.3 Infrared absoption for powdered superconducting sample. A new absorption peak indicated by an arrow appears.

semiconducting and a superconducting crystals, respectively. The absorption peak at 629 cm^{-1} is observed in curve (a), whereas, no absorption peak appears near 600 cm^{-1} in curve (b). The peak at 629 cm^{-1} has been reported before[6]. Although curves do not represent the precise absorbance at lower energy region because of the deterioration of the performance of the microscope, the absorption peak near 500 cm^{-1} can be recognized in both curves. The E_u vibrational mode is responsible for this absorption. In contrast to the 629 cm^{-1} peak, the absorption does not disappear in the superconducting crystal which may indicate that the screening effects due to doped holes are not so effective for this peak compared to that appearing at 629 cm^{-1}. Similar results have been reported for other materials[1-3].

Figure 3 shows a frequency dependence of the absorbance in powdered sample. A strong absorption peak near 600 cm^{-1} appears. The peak has not been obserbed before as shown in Fig.2 where the direction of the electric field vector of the infrared light is made parallel to the Cu-O plane. Hence, the strong peak near 600 cm^{-1} observed in powdered sample may be due to A_{2u} modes in which ions make a motion perpendicular to the Cu-O plane. The peak near 500 cm^{-1} is also observed but the intensity is weaker compared to that of the 600 cm^{-1} peak. The screening may be partially effective for this peak.

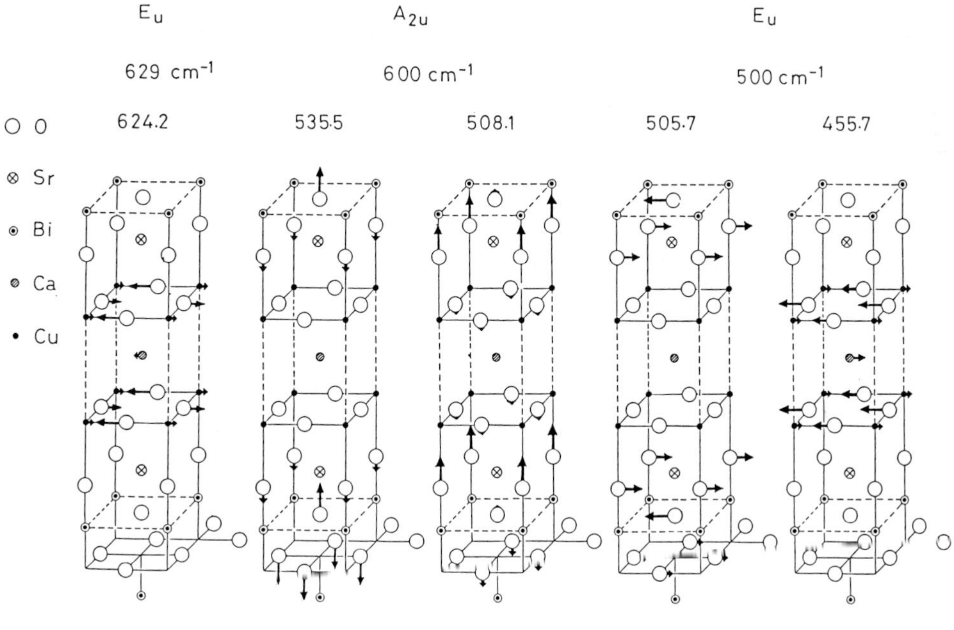

Fig.4 Schematic representations of five higher frequency vibrational modes in $Bi_2Sr_2CaCu_2O_8$ drawn after the theoretical results of reference [12]. Arrows represent polarization vectors. Numbers shown just above figures are the calculated results[12]. Observed modes and frequencies are shown at upper two lines.

In Fig.4, the comparison of observed absorption peaks with theoretical results by S.Mase and T.Yasuda[12] is given. At the first line, assignments of the types of modes infered from the directions of the polarization of the infrared light for present experiment are shown. The second and the third lines give observed and calculated phonon frequencies, respectively. Five calculated frequencies from the highest are presented with schematic drawings of corresponding modes drawn refering to Table VII of reference [12].

Qualitatively, the theoretical result for instance, the order of E_u and A_{2u} modes from the highest frequency E_u mode, is consistent with the present experiment. The appearance and nonappearance of the "Infrared Anomaly" can be also well understood qualitatively by reffering to Fig.4. In the highest frequency E_u mode, ions in Cu-O planes move out of phase each other. On the other hand, ions in Cu-O planes have no relation to the second E_u mode. Therefore, the infrared absorption peak due to this mode should be observable in superconducting compounds as has been confirmed in Fig.2 and Fig.3. According to the theoretical result, both A_{2u} and E_u modes should be splitted. However, these splittings are not clear in Fig.3. For the third E_u mode shown in Fig.4, "Infrared Anomaly" may possiblly be observable. Measurements at lower frequency region will be needed to cofirm the effect. Calculated values of vibrational frequencies of E_u modes agree considerablly well to experimental results. Values for A_{2u} modes seem to be smaller than observed frequencies.

ACKNOWLEDGEMENTS

We would like to express our thanks to Dr.M.Sato for his advice for taking IR spectra. Dr.H.Sasakura and other members of the Department of Materials Science of the University of Osaka Prefecture kindly supplied us well characterized superconducting samples. We also thank Professor S.Hashimoto and Dr.S.Sakuragi for giving us high quality single crystals of CsI used in the experiment.

REFERENCES

1. Ohbayashi K, Ogita N, Udagawa M, Aoki Y, Maeno Y and Fujita T (1987) Jpn.J. Appl.Phys. 26: L420
2. Schlesinger Z, Collins R T and Shafer M W (1987) Phys.Rev.B 35: 7233
3. Doll G L, Nicolls J T, Dresselhaus M S, Rao A M, Zhang J M, Lehman G W, Eklund P C, Dresselhaus G and Strauss A L (1988) Phys.Rev.B 38: 8850
4. Kuzmany H, Matus M, Faulques E, Pekker S, Hutiray Gy, Zsoldos E and Mihaly L (1988) Solid State Commun. 65: 1343
5. Crawford M K, Farneth W E, McCarron III E M and Bordia R K (1988) Phys. Rev.B 38: 11382
6. Hayashi Y, Fukuoka N, Harada Y and Nakanishi S (in press) J.Phys.Soc.Jpn.
7. Piro O E, Guida J A, Messa N E , Aymonono P J, Castellano E E, Basso H C, Gallo J N H and Martin A A (1989) Phys.Rev. B 39: 7275
8. Liu J, Yan Y, Zhou T, Che G, Chang D, Chen W, Wang Y and Zhao Z (1988) Modern Phys.Lett. 2: 707
9. Nakanishi S, Fukuoka N, Nakahigashi K, Kogachi M, Sasakura H, Minamigawa S and Yanase A (1989) Jpn.J.Appl.Phys. 28: L71
10. Popovic Z V, Kliche G, Cardona M and Liu R (1990) Phys.Rev.B 41: 3824
11. Prade J, Kulkarni A D, de Wette F W, Schroder U and Kress W (1989) Phys. Rev.B 39: 2771
12. Mase S and Yasuda T (1989) J.Phys.Soc.Jpn. 58: 658
13. Kulkarni A D, de Wette F W, Prade J, Schroder U and Kress W (1990) Phys. Rev.B 41: 6409

Phonons in $(Sm,Ce)_2(Ba,Sm)_2Cu_3O_x$

MASASHI YOSHIDA, SETSUKO TAJIMA, YURI MIZUO, TAKAHIRO WADA, YUJI YAEGASHI, ATARU ICHINOSE, and NAOKI KOSHIZUKA

Superconductivity Research Laboratory, International Superconductivity Technology Center, 10-13, Shinonome 1-chome, Koto-ku, Tokyo, 135 Japan

ABSTRACT

Polarized Raman spectra of $(Sm,Ce)_2(Ba,Sm)_2Cu_3O_x$ which has Cu-O_5 pyramid structure as well as $YBa_2Cu_3O_7$ have been measured at room temperature using single crystals. Peaks have been observed at 442 and 537 cm^{-1} in the A_{1g} symmetry configuration and at 200cm^{-1} in the B_{1g}. A peak due to a Raman-forbidden mode has been observed at 560cm^{-1}. The B_{1g} symmetry peak is assigned to the out-of-phase bending of oxygen in the Cu-O_2 plane. The energy of this mode is much lower than that of $YBa_2Cu_3O_7$ (330cm^{-1}). On the other hand, energies of the A_g-symmetry peaks are nearly the same as those in $YBa_2Cu_3O_7$ (440 and 500cm^{-1}, respectively), suggesting that these two peaks are associated with the vibrations of apical oxygen.

KEY WORDS: Raman scattering spectra, phonon, $(Sm,Ce)_2(Ba,Sm)_2Cu_3O_x$, $YBa_2Cu_3O_7$

INTRODUCTION

Since the discovery of superconductivity in $La_{2-x}Ba_xCuO_4$ by Bednorz and Müller,[1)] many cuprate-superconductors have been found. According to the oxygen configuration around Cu atom, they can be classified into three groups: the first is compounds with Cu-O_6 octahedra ($La_{2-x}Sr_xCuO_4$ etc.), the second compounds with Cu-O_5 pyramids ($YBa_2Cu_3O_7$ etc.) and the third compounds with Cu-O_4 planes ($Pr_{2-x}Ce_xCuO_4$ etc.).[2)] T_c depends critically on the Cu-O complexes and compounds with the Cu-O_5 pyramids have the highest Tc's.

Recently, Sawa et al found a new superconductor family $(Ln_{2/3}Ce_{1/3})_2(Ba_{2/3}Ln_{1/3})_2Cu_3O_9$ (Ln: Nd, Sm, Eu),[3)] which has the Cu-O_5 pyramid structure but the external environment of the pyramid is different from that of $YBa_2Cu_3O_7$. The T_c is around 30K,[3)] which is much lower than that of $YBa_2Cu_3O_7$ (92K).

Figure 1 shows primitive unit cells of $(Sm,Ce)_2(Ba,Sm)_2Cu_3O_9$ (hereafter abbreviated as Sm-223) and $YBa_2Cu_3O_7$ (Y-123). The space group of the former is I4/mmm while that of the latter is Cmca. Both of them contain two Cu-O_5 pyramids. They are very similar to each other except for the following points. (i) In Sm-223, there are two Sm(Ce) layers and one oxygen layer between them while there is only one Y layer in Y-123. (ii) In the Cu(1)-O plane of Sm-223, oxygen atoms populate randomly with occupancy of 0.5, while in the Cu(1)-O plane of Y-123, oxygen atoms align along the a-axis. In addition, there is a difference in the stacking of the primitive cell along the c-direction. In Sm-223 the primitive cells are stacked with the displacement (0.5,0.5) in the ab-plane while no such displacement occurs in Y-123.

In the present study, we measured Raman scattering spectra of Sm-223 using single crystals. Scattering peaks due to phonons associated with the Cu-O_5 pyramid were observed. The difference of the phonons between Sm-223 and $YBa_2Cu_3O_x$ (Y-123) is discussed.

EXPERIMENTAL

Single crystals of Sm-223 were grown with a small excess of copper oxide serving as the flux. Powders of Sm_2O_3, CeO_2, $BaCO_3$ and CuO with purity of 99.99% were mixed with cation ratio Sm:Ce:Ba:Cu=2:0.67:1.33:3.5. They were placed in alumina crucibles and were heated to 1400C in air, held for 10 hours, and cooled down to 1100C. After that, they were cooled gradually to 800C at the rate of 2 C/h and then furnace cooled to room temperature. The dimensions of the obtained crystals were about $1x1x0.1mm^3$. They were annealed at 500C in 400 atmospheres of oxygen for 5 hours to obtain superconducting samples.

With a single-crystal x-ray diffractometer, we found the crystals to be oriented with the c-axis perpendicular to the large surface of the platelet. The lattice parameters were a=b=3.855A and c=28.56A. The superconducting transition temperature of our sample was about 30K, which was nearly the same as that of sintered materials.[3)]

Raman spectra were observed at room temperature in the backscattering configuration using 5145A light from an Ar laser. The incident beam was focused on the sample surface with a diameter of about 0.1mm. The laser power was maintained below 50mW to avoid damaging samples. The scattered light was detected with a Jobin Yvon U-1000 double monochromator and the photon counting detection system with the photomultiplier Hamamatsu R-943.

RESULTS AND DISCUSSION

Figures 2 (a) and (b) show the polarized Raman spectra taken on the ab-plane and ac-plane, respectively, of Sm-223 samples, where X, Y and Z refer to the polarization of the incident or scattered electric fields along the a, b and c axes of the crystal. X' (Y') is the direction rotated 45 around the Z-axis from the X- (Y-)axis. In Fig. 2 (a) a peak is seen at around 200 cm^{-1} in the spectrum for the (X',Y') geometry but not for the (X',X'), exhibiting B_{1g} symmetry. Since there exists only one B_{1g} mode phonon in this crystal structure, we can identify it as the out-of-phase vibration of the O(3) atoms along the c-axis unambiguously. It should be noticed that the B_{1g} phonon energy of Sm-223 is far smaller than that of YBCO ($330cm^{-1}$)[4)]. On the other hand, a peak is seen at $560cm^{-1}$ both in the (X',Y') and (X',X') geometries. This peak is supposed to appear due to defects (no phonon with any symmetry is Raman-allowed in the (X',X') geometry).

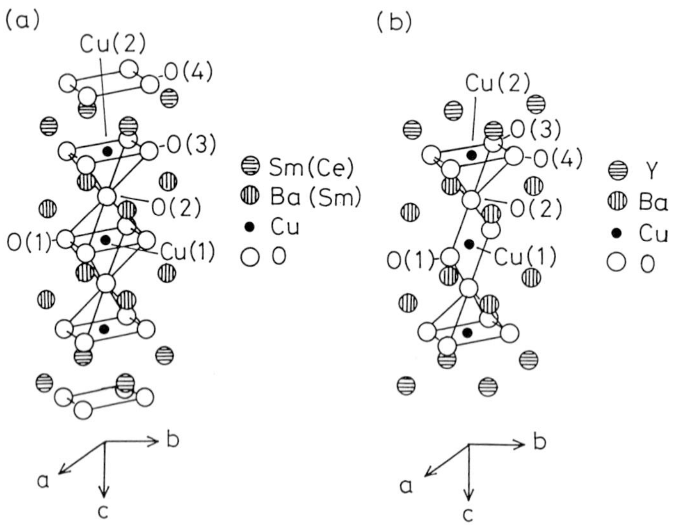

Figure 1 Primitive unit cells of (a) $(Sm,Ce)_2(Ba,Sm)_2Cu_3O_{10}$ and (b) $YBa_2Cu_3O_7$

In Fig. 2 (b) prominent peaks are seen at 537cm-1 and 442cm-1 in the (Z,Z) geometry . These peaks are not seen in the (Z,X) geometry exhibiting A_g symmetry. In the spectra of Y-123, prominent peaks with Ag symmetry are seen at $500cm^{-1}$ and $440cm^{-1}$ where the former is assigned to the vibration of O(2) along the c-axis and the latter to the in-phase vibration of O(3) and O(4) along the c-axis.[4)] The $537cm^{-1}$ peak in Sm-223 can be assigned to the vibration of O(2) along the c-axis. On the other hand, the $442cm^{-1}$ peak of Sm-223 cannot be assigned to the in-phase vibration of O(3) along the c-axis since it is unlikely that the energy of the out-of-phase vibration of O(3) of Sm-223, $200cm^{-1}$, is considerably lower than that of Y-123 while that of the in-phase vibration of O(3) is nearly the same. We consider the origin of the $442cm^{-1}$ peak below.

Atomic configurations around the $Cu-O_2$ plane in Sm-223 are different from those in Y-123, as described above. On the other hand, the configurations around the apical oxygen are nearly the same in both crystals. The energy of the B_{1g} phonon which is ascribed to the vibration of oxygen in the $Cu-O_2$ plane is considerably lower in Sm-223 than in Y-123, while the energy of the stretching vibration of the apical oxygen is nearly the same. These facts agree with the characteristics of the crystal structure of Sm-223. Since the energy of the $442cm^{-1}$ peak in Sm-223 is nearly the same as that of the $440cm^{-1}$ peak in Y-123, it is supposed that these peaks are associated with the vibration of apical oxygen.

There is only one A_g-symmetry mode associated with the apical oxygen (stretching mode) which is ascribed to the $537cm^{-1}$ peak. However, it often occurs, in ionic crystals, that infrared-active (Raman-forbidden) phonons appear in the Raman spectra due to the Fröhlich-type interaction.[5)] We suppose that the $442cm^{-1}$ peak in Sm-223 is due to bending vibration of the apical oxygen (E_u mode) which becomes Raman allowed through the Fröhlich type interaction.

The $442cm^{-1}$ peak has been observed using polarized light with polarization along the c-direction but not with polarization in the ab-plane. These polarization characteristics also support the assignment of the $442cm^{-1}$ peak to the vibration of apical oxygen, as follows.

The Raman scattering is the third-order perturbation process which consists of (i) virtual creation of an electron-hole pair by the absorption of light, (ii) scattering of the electron or hole by phonons, and (iii) radiative

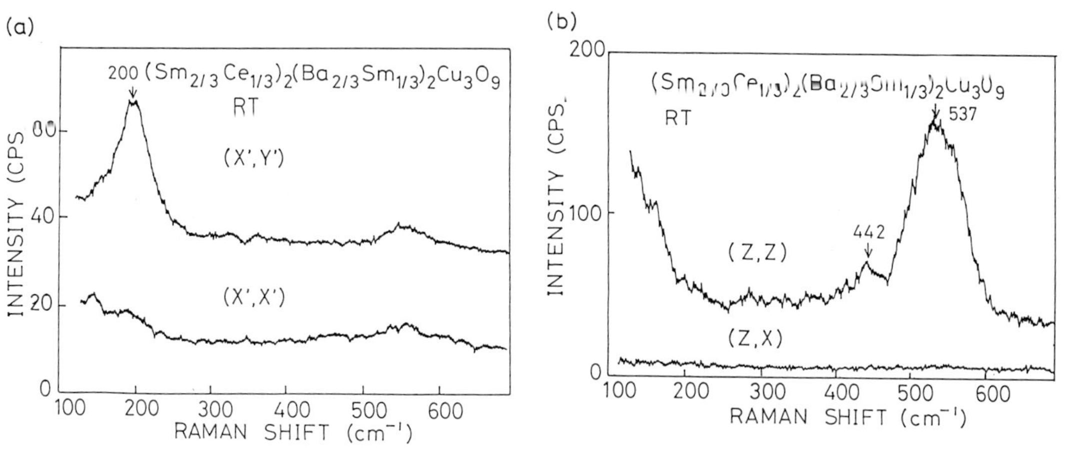

Figure 2 Polarized Raman spectra of $(Sm,Ce)_2(Ba,Sm)_2Cu_3O_x$ obtained on the ab-plane (a) and ac-plane (b) at room temperature.

annihilation of the electron-hole pair.6) The polarization of light has a crucial role in determining characteristics of the electron-hole pairs which take part in the Raman scattering.

The highest occupied states of Y-123 are supposed to be constructed by anti-bonding Cu-O bonds.[7)] On the other hand, the unoccupied state where electrons are excited by the absorption of light is supposed to be the upper Hubbard band of the Cu-3d states. A similar situation is expected to hold for Sm-223. Then, according to the selection rule of the dipole transition, the radiative transition occurs between the $2p_z$ state of O(2) and $3d_{x^2-z^2}$ state of Cu(1) for z-polarized light both in Y-123 and Sm-223. On the other hand, for x-polarized light the transition occurs between $2p_x$ of O(3) and $3d_{x^2-y^2}$ of Cu(2) or between $2p_x$ of O(1) and $3d_{x^2-z^2}$ of Cu(1). Thus, holes are created on the O(2) atoms by z-polarized light but on the O(1) or O(3) atoms by x-polarized light. Holes on the O(2) atoms (O(1) or O(3) atoms) are expected to couple strongly with the vibration of the O(2) atom (O(1) or O(3) atoms). The polarization dependence of the $442cm^{-1}$ peak agrees with that expected for the vibration of the apical oxygen. We suppose that the $440cm^{-1}$ peak in YBCO is also due to the bending mode of O(2) as was pointed out in Ref. 8.

SUMMARY

Raman scattering spectra of Sm-223 were measured. A B_{1g} symmetry peak was observed at $200cm^{-1}$ and ascribed to the out-of-phase bending of oxygen in the $Cu\text{-}O_2$ plane. The B_{1g} peak energy was considerably lower than that of Y-123. On the other hand, A_g-symmetry peak energies were found at $442cm^{-1}$ and $537cm^{-1}$, which were nearly the same as $440cm^{-1}$ and $500cm^{-1}$ peaks in Y-123. These peaks were ascribed to phonons associated with the apical oxygen (bending and stretching modes, respectively).

Acknowledgement

A part of this work was supported by the New Energy and Industrial Technology Developement Organization under the management of the R&D of Basic Technology for Future Industries.

REFERENCES

1. Bednorz J. G. and Müller K. A. (1986) Z. Phys. B64: 189
2. Yvon K. and Francois M. (1989) Z. Phys. B76: 413
3. Sawa H., Obara K., Akimitsu J., Matsui Y. and Horiuch S. (1989) J. Phys. Soc. Jpn. 58: 2252
4. Phonon scattering spectra of $YBa_2Cu_3O_x$ have been summarized in Feile R. (1989) Physica C 159: 1
5. Fröhlich H. (1954) Adv. Phys. 3: 325
6. Loudon R. (1964) Adv. Phys. 13: 423
7. Mattheiss L. F. and Hamann D. R. (1987) Solid State Commun. 63: 395
8. Yoshida Y., Gotoh S., Takata T., Koshizuka N. and Tanaka S. (1990) Phys. Rev. B 41: 11689

Contribution of Itinerant Electrons to Optical Anomalies in the Normal State of Oxide Superconductors

SHIGENORI TANAKA

Advanced Research Laboratory, Research and Development Center, Toshiba Corporation, 1, Komukai Toshiba-cho, Saiwai-ku, Kawasaki, 210 Japan

ABSTRACT

We have quantitatively investigated the contributions of itinerant electrons to optical conductivity and to Raman scattering intensity in the normal state of the high-T_c oxide superconductors on the basis of the Lindhard dielectric function in the relaxation-time approximation. Through inclusion of the frequency dependence of the relaxation time, our charged-Fermi-liquid model can appropriately account for the anomalous behaviors in optical conductivity in the mid-infrared region. As for Raman scattering intensity, however, inclusion of other contributions than those of the charged Fermi liquids is necessary to quantitatively explain the experimental results.

KEY WORDS: normal state, Fermi liquid, optical conductivity, Raman scattering

INTRODUCTION

The normal states of the high-T_c oxide superconductors are known to have a number of anomalous properties as compared with other metals. Examples include electrical resistivity, optical conductivity, Raman scattering intensity, tunneling conductance, nuclear relaxation rate, thermoelectric power, thermal conductivity and Hall coefficient. Among others, the anomalies observed in the Raman scattering experiments [1,2] and in the measurements of the optical conductivity [3-8] are especially significant and they seem to be intimately related to the occurrence of the high-T_c superconductivity.

In the Raman scattering experiments, nearly flat, large continuous spectra ranging from 0 to 4000 cm^{-1} were observed when the holes were doped in $La_{2-x}Sr_xCuO_4$, $YBa_2Cu_3O_{7-y}$ and $Bi_2Sr_2Ca_{1-x}Y_xCu_2O_{8+y}$ [1,2]. As for the optical conductivity, non-Drude peak in the mid-infrared region is remarkable and common to $YBa_2Cu_3O_{7-y}$ [3-5], $La_{2-x}Sr_xCuO_4$ [6], $Bi_2Sr_2Ca_{1-x}Nd_xCu_2O_{8+y}$ [7] and $Pr_{2-x}Ce_xCuO_{4-y}$ [8]. Several theoretical works [9-11] have been performed to account for these optical anomalies on the basis of the t-J model. The purpose of the present work is to provide another theoretical approach to this problem in the framework of the Fermi-liquid description [12]. We thus investigate the extent of contributions of charge fluctuations arising from itinerant carriers to the optical conductivity and to the Raman scattering intensity in the far- and mid-infrared regions.

MODEL

In order to simulate the actual situation occurring in the normal state of the high-T_c oxide superconductors, we consider a two-dimensional system consisting of a single species of charged fermions with effective electric charge e^*, effective mass m^*, number density n and temperature T embedded in a uniform background of neutralizing charges. We may regard such a system as a two-dimensional electron (or hole) liquid which may be characterized as a collection of strongly-correlated but itinerant, charged Fermi particles confined within the **ab**-plane of a copper oxide. We further consider that such two-dimensional planes are distributed uniformly with density λ along the **c**-axis and we neglect the interplane interaction. We assume that this 2+1-dimensional Fermi system constitutes a part of a whole superconductor system; the charged quasi-particles would then have a finite damping rate Σ through scattering by impurities, phonons, magnons, excitons and other excitations.

To describe the optical response of the system assumed above, we rely on the large-ω/k (frequency/wavenumber) limit of the Lindhard (longitudinal and transverse) dielectric function in the relaxation-time approximation [13-15]:

$$\varepsilon(k, \omega) = 1 - v(k)\chi_F(k, \omega). \tag{1}$$

Here,

$$v(k) = 2\pi e^{*2}/k \tag{2}$$

and

$$\chi_F(k, \omega) = nk^2/\{m^*\omega[\omega + i\Sigma(\omega)]\}. \tag{3}$$

In Eq. (3), $\Sigma(\omega)$ is the frequency-dependent damping function or the inverse of relaxation time. We can then calculate the dynamic structure factor $S(k, \omega)$ and the optical conductivity $\sigma(\omega)$ as

$$S(k, \omega) = -\hbar/[2\pi v(k)]\coth[\hbar\omega/(2k_B T)]\,\mathrm{Im}\{1/[\varepsilon(k, \omega)]\} \tag{4}$$

and

$$\sigma(\omega) = -\lim_{k\to 0}[(e^{*2}\omega/k^2)\,\mathrm{Im}\,\chi_F(k, \omega)]. \tag{5}$$

The frequency-dependent damping function of the quasi-particles should satisfy the requirements of causality. We therefore express the damping function in terms of a memory function as

$$\Sigma(\omega) = \Sigma(0) + [\Sigma(\infty) - \Sigma(0)][-i\omega\Gamma(\omega)]. \tag{6}$$

Here $\Sigma(0) = 1/\tau_0$ and $\Sigma(\infty) = 1/\tau_\infty$ are the low- and high-frequency limits of the damping function. The memory function $\Gamma(\omega)$ is expressed as

$$\Gamma(\omega) = \int_0^\infty dt\, \exp(i\omega t)\tilde{\Gamma}(t) \tag{7}$$

with $\tilde{\Gamma}(t=0) = 1$ and $\tilde{\Gamma}(t=\infty) = 0$. In the following we adopt the Gaussian approximation for the memory function as

$$\tilde{\Gamma}(t) = \exp(-t^2/\tau_c^2), \tag{8}$$

where τ_c is a characteristic relaxation time.

RESULTS

We have calculated the optical conductivity $\sigma(\omega)$ and the Raman scattering intensity $S(\omega/c, \omega)$ for some parametric combinations of n, λ, T, m^*, e^*, τ_0, τ_∞ and τ_c. We now show the results for two tentative choices of these parameters:
(a) Two-dimensional carrier density $n = 1.0\times10^{14}$ cm^{-2}, layer density $\lambda = 1.8\times10^{7}$ cm^{-1}, T = 100 K, $m^*/m = 1$, $e^{*2}/e^2 = 1$, $\hbar/\tau_0 = 300$ cm^{-1}, $\hbar/\tau_\infty = 200$ cm^{-1} and $\hbar/\tau_c = 1000$ cm^{-1}. The value of λ means that there exist planes in

which the carriers move at the rate of one layer per 5.7 Å on the average, which simulates the case of Cu-O planes in $YBa_2Cu_3O_7$. Physically, such a choice of the relaxation times represents those scattering processes well described by the simple Drude model. Since the time τ_0 or τ_∞ for which the carriers can move freely without being scattered is much greater than the relaxation time τ_c which characterizes the motion of the scatterers such as phonons, magnons or other electronic excitations, the scatterers can respond to the motion of the carriers quickly enough.
(b) The values for n, λ, T, m^*, e^* are the same as those in the case (a). The relaxation times are chosen as $\hbar/\tau_0$ = 5000 cm^{-1}, $\hbar/\tau_\infty$ = 3000 cm^{-1} and $\hbar/\tau_c$ = 1000 cm^{-1}, respectively. This case seems to correspond to those situations which may be well described in terms of the t-J model, where we presume that holes with kinetic energy ~ 0.5 eV move in the medium of localized spins which couple to the neighboring spins with interaction energy ~ 0.1 eV. The relaxation times τ_0 and τ_∞ characterize that process in which the holes transfer from site to site through hopping. This motion of the holes disturbs the antiferromagnetically ordered spin configurations localized on the Cu sites. The relaxation time τ_c refers to a characteristic time in which the spin configurations return to the initial state. Since $\tau_c > \tau_0, \tau_\infty$, this relaxation process cannot immediately follow the motion of the holes.

We show in Fig. 1 the results for the optical conductivity $\lambda\sigma(\omega)$ and in Fig. 2 the corresponding results for the normalized Raman scattering intensity $(Ry/\hbar n\alpha^2)S(\omega/c, \omega)$, where $Ry \equiv me^4/(2\hbar^2)$ is the Rydberg energy (~13.6 eV) and $\alpha \equiv e^2/(\hbar c)$ is the fine structure constant (~1/137).

As expected, the result (a) for optical conductivity shows a Drude-like behavior. Correspondingly, the result (a) for Raman scattering intensity gives a damped peak structure arising from the two-dimensional plasmon mode.

On the other hand, the result (b) for optical conductivity shows a non-Drude behavior that has a peak at ω ~ 2000 cm^{-1}. This anomalous behavior may be ascribed to the interrelationship between the relaxation times discussed above. We note that such a behavior occurs in our model when the condition that $1/\tau_0 - 1/\tau_\infty > (2/\sqrt{\pi})/\tau_c$ is satisfied. Physically, this anomaly may be regarded as a consequence of a marked effect of friction between the carriers and the scattering medium. The corresponding result (b) for Raman scattering amplitude gives a high intensity ranging over the mid-infrared region.

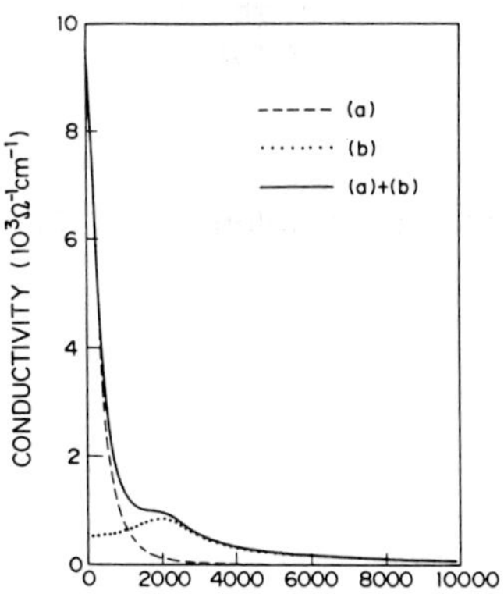

Fig. 1. Optical conductivity $\lambda\sigma(\omega)$. The dashed curve refers to the result for the case (a); the dotted curve, the case (b); the solid curve, the sum of contributions from (a) and (b).

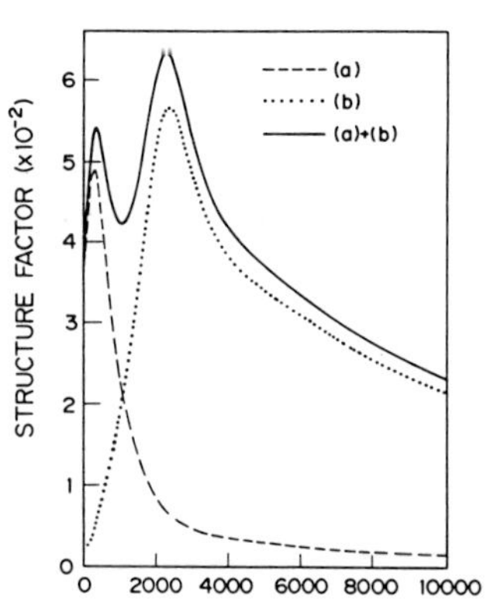

Fig. 2. Normalized Raman scattering intensity $(Ry/\hbar n\alpha^2)S(\omega/c, \omega)$. Otherwise the same as in Fig. 1.

Now let us sum up both contributions, (a) and (b). We thus regard the whole system as a mixture of two species of charged carriers. The result (a) + (b) for optical conductivity is analogous to the experimental one [3,4] obtained for $YBa_2Cu_3O_{7-y}$ at T = 100 K. The result (a) + (b) for Raman scattering intensity is, on the other hand, different from the experimental result [1,2] in that the intensity observed in the experiment does not show the peak structures found in Fig. 2; the fact that a high intensity remains over a wide range of the mid-infrared region, however, seems to be partially accounted for. In order to completely explain the overall shape of the experimental spectra, it is necessary to take into account the fluctuations that arise due to phonons, magnons, excitons and other excitations. In this connection we point out that the position of the plasmon peak should shift to higher-frequencies if we include the three-dimensionality of the charged Fermi liquids.

CONCLUDING REMARKS

To universally account for the anomalies observed in the normal state, Varma et al. [16,17] have proposed a phenomenological model called the marginal Fermi liquid (MFL). In this model, a two-carrier picture, which distinguishes between the Drude contribution and the non-Drude (direct-absorption) contribution, is adopted as a hypothesis. If the origin of the direct-absorption part is ascribed to such charge fluctuations as discussed in the present study, our charged-Fermi-liquid model may provide a microscopic (but still phenomenological) basis for the MFL model. The polarizations produced by the non-Drude carriers would then be responsible for the s-wave superconducting pairing force between the Drude carriers.

REFERENCES

1. Sugai S (1989) In: Mechanisms of High Temperature Superconductivity, Kamimura H, Oshiyama A (eds) Springer Series in Materials Science, Vol 11. Springer-Verlag Berlin, Heidelberg, pp 207-219
2. Sugai S, Enomoto Y, Murakami T (1989) Solid State Commun 72: 1193-1197
3. Thomas GA, Orenstein J, Rapkine DH, Capizzi M, Millis AJ, Bhatt RN, Schneemeyer LF, Waszczak JV (1988) Phys Rev Lett 61: 1313-1316
4. Orenstein J, Thomas GA, Rapkine DH, Millis AJ, Schneemeyer LF, Waszczak JV (1988) Physica C 153-155: 1740-1743
5. Schlesinger Z, Collins RT, Holtzberg F, Feild C, Blanton SH, Welp U, Crabtree GW, Fang Y, Liu JZ (1990) Phys Rev Lett 65: 801-804.
6. Tajima S, Tanaka S, Ido T, Uchida S (1990) In: Advances in Superconductivity, Ishiguro T, Kajimura K (eds) 2nd internatinal symposium on superconductivity. Springer-Verlag, Tokyo, pp 569-572
7. Terasaki I, Nakahashi T, Takebayashi S, Maeda A, Uchinokura K (1990) Physica C 165: 152-160
8. Cooper SL, Thomas GA, Orenstein J, Rapkine DH, Millis AJ, Cheong SW, Cooper AS, Fisk Z (1990) Phys Rev B 41: 11605-11608
9. Rice TM, Zhang FC (1989) Phys Rev B 39: 815-818
10. Kane CL, Lee PA, Read N (1989) Phys Rev B 39: 6880-6897
11. Inoue J, Maekawa S (1990) J Phys Soc Jpn 59: 2110-2123
12. Tanaka S (1990) Physica C 169: 271-274
13. Lindhard J (1954) K Dan Vidensk Selsk Mat-Fys Medd 28 No 8: 1-57
14. Kliewer KL, Fuchs R (1969) Phys Rev 181: 552-558
15. Mermin ND (1970) Phys Rev B 1: 2362-2363
16. Varma CM, Littlewood PB, Schmitt-Rink S, Abrahams E, Ruckenstein AE (1989) Phys Rev Lett 63: 1996-1999
17. Varma CM (1989) Int J Mod Phys B 3: 2083-2118

2.6 Magnetic Properties

Inelastic Neutron Scattering Study of Spin Dynamics in $YBa_2Cu_3O_{6+x}$

C. Vettier[1], P. Burlet[2], L.P. Regnault[2], J. Rossat Mignod[2], J.Y. Henry[2], and G. Lapertot[2]

[1] Institut Laue Langevin, 156X, 38042 Grenoble Cedex, France
[2] Centre d'Etudes Nucléaires, DRF/SPh-MDN, 85X, 38041 Grenoble Cedex, France

ABSTRACT

The recent results of extensive neutron scattering studies of the spin dynamics in $YBa_2Cu_3O_{6+x}$ single crystals are reviewed. In the antiferromagnetic region (x=0.15) the spin waves spectrum shows that $YBa_2Cu_3O_{6+x}$ is a quasi two-dimensional Heisenberg bilayer system with a weak XY anisotropy. Adding oxygen (x=0.37) creates holes in the CuO_2 layers which reduce correlation lengths. At low energy, magnetic excitations are strongly renormalized due to the presence of holes. In the superconducting region, (x=0.45 T_c=37K, x=0.51 T_c=47K, x=0.69 T_c=59K), dynamical magnetic correlations of the Cu(2) spins are still present and extend towards high energy. Below T_c, the magnetic response at low energy is suppressed, indicating the opening of a magnetic gap. When the oxygen content is increased towards O_7, the gap energy increases as does T_c, while the correlation length decreases. Above T_c, there is evidence for superconducting correlations.

KEY WORDS: 123 Superconductors, spin fluctuations, 2D systems

INTRODUCTION

The discovery of superconductivity in lamellar copper oxide materials has led to a huge amount of experimental and theoretical efforts in order to understand the mechanism of high T_c superconductivity [1,2]. If no agreement on the understanding of the origin of this phenomenon has been reached, various experimental methods have been used to investigate possible mechanisms. Neutron scattering techniques, which can probe lattice and spin modes, are quite appropriate to probe both spatial and temporal fluctuations which could induce couplings leading to superconductivity. Furthermore connections can be made with more local probes such as NMR. While several groups have embarked upon the investigation of lattice modes, we have concentrated on the magnetic properties of $YBa_2Cu_3O_{6+x}$ single crystals. It has been shown that these materials exhibit antiferromagnetic order with very high values of T_N. Actually, detailed investigations of the $(La,Sr,Ba)_2CuO_4$ system at Grenoble [3] and Brookhaven [4] have shown that defects play an extremely important role in the magnetic properties of the Cu spins and that spin fluctuations are present above T_N in these materials, as in conventional 2D systems.

At the Centre d'Etudes Nucléaires de Grenoble, interest has been focussed on the $YBa_2Cu_3O_{6+x}$ system because of the feasibility to grow large single crystals in which the oxygen content can be adjusted from O_6 to O_7. As a consequence, it has been possible to determine the magnetic phase diagram (fig.1) as a function of oxygen content, x, by means of diffraction experiments [5,6,7]. We will summarize briefly the most important results. At low x, up to x ≈ 0.20, a simple antiferromagnetic (AF) order exists at temperature below T_N which can be as high as 415K for x=0.0. Only the Cu sites in the Cu(2) show magnetic long range order with a low temperature moment $m_0=0.64\mu_B$. The Cu(1) sites are in a non magnetic state. The magnetic phase transition is characterized by a critical exponent $\beta=0.25\pm0.03$ typical of a quasi-2D XY system. For this oxygen concentration the magnetic Bragg peaks show no broadening, which indicates the absence of defects in the CuO_2 planes. In this range of x, the filling of oxygen in the Cu(1) planes induces electron transfers only within the Cu(1) planes, no holes are transferred to the Cu(2) planes. When increasing the oxygen content above x=0.20, a small amount of 2p-holes is created in the Cu(2) planes. Both T_N and m_0 decrease and no long range ordered magnetic moment is observed on the Cu(1) sites. The long range AF order disappears abruptly at $x_c \approx 0.41$. Furthermore, additional elastic magnetic scattering is observed along (1/2,1/2,l) rods indicating that p-holes induce some static disorder in the CuO_2 planes and in the stacking of the Cu(2) bilayers.

The in-plane correlation length ξ can be obtained from the Q-width of the magnetic rods. For small hole concentration (n_h), ξ can be taken as the distance between magnetic defects ($n_h=1/\xi^2$) provided the size of defects is much smaller than ξ. At x=0.37, it is found that ξ=7.5 a, where a is the in-plane lattice constant, corresponding to n_h=1.8%. Extrapolating to x_c shows that the 3D-AF order is destroyed by a hole concentration of the order of 2%, a result similar to that found in the $(La,Sr)_2CuO_4$ system [3]. Another important result is the observation of reentrant behavior at low temperature (fig.2) that we ascribe to hole localization. The Bragg peaks intensity is depressed and transferred to the magnetic rods indicating an additional reduction of the AF coupling between bilayer. At higher temperatures, the hole motion induces quantum fluctuations near the phase transition at T_N which lead to an increase of the critical exponent β and to an unusual behavior near T_N. For x larger than x_c, no static AF long range order is found to coexist with superconductivity.

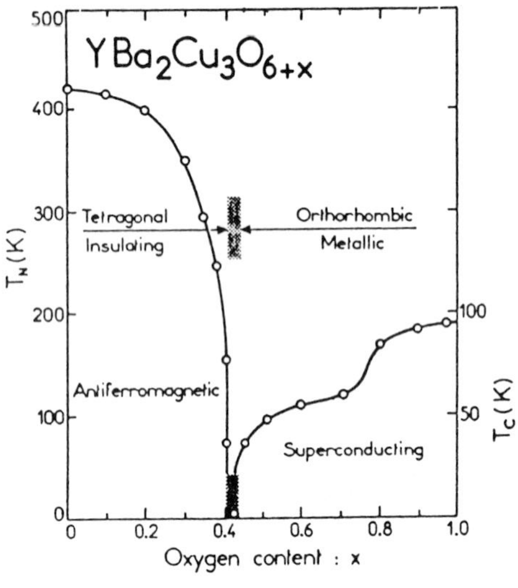

Fig. 1. Phase diagram of $YBa_2Cu_3O_{6+x}$ as a function of oxygen content.

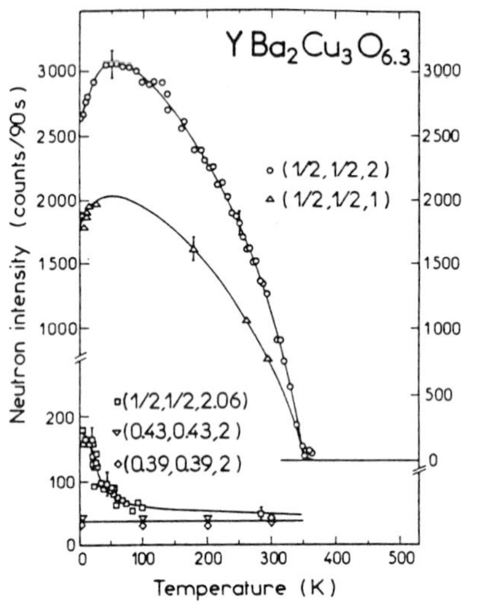

Fig. 2 Temperature dependence of magnetic Bragg peaks and diffuse scattering along the [1/2,1/2,l] rod.

However local spins do exist on the Cu^{2+} sites, and therefore spin dynamics should be observable in the superconducting region. Coupling to the carriers could be evidenced by measuring the dynamical magnetic response as a function of Q and ω. This is the purpose of this paper which is organized as follows. After a brief description of the experimental conditions, we present the spin dynamics in the AF state in order to observe the evolution of the magnetic response as a function of hole concentration. Then we report on the recent results obtained in the superconducting state which show the opening of a magnetic gap and the existence of superconducting fluctuations above T_c.

EXPERIMENTAL PROCEDURE

The dynamics of Cu(2) spins was investigated by means of inelastic neutron scattering. The experiments described here have been performed on three axis spectrometers at the High Flux Reactor of the Institut Laue Langevin. Most of the data reported here have been obtained on the IN8 spectrometer with Cu(111), Cu(220) or PG(002) crystals as monochromator and PG(002) or PG(004) as analyzer and 40'/40'/40' collimators.

The $YBa_2Cu_3O_{6+x}$ single crystals used in this work were grown at the Centre d'Etudes Nucléaires de Grenoble. Good quality single crystals (mosaic spread of 30' to 60') were grown with a special technique allowing an easy and homogeneous change of the oxygen content. The total sample volume was $\approx$0.40cm^3. All measurements were performed on the (hhl) scattering plane.

ANTIFERROMAGNETIC EXCITATIONS

The dynamics of the Cu(2) spins was first studied in the pure antiferromagnetic state [8]. As expected, well-defined spin-waves have been observed. From the steep dispersion curves of spin-waves and from the values of anisotropy gaps, it has been possible to determine the values of exchange interactions and the experimental results have established the two-dimensional character of the magnetic couplings. The very large in-plane spin-waves velocity (c_0=1.0±0.05 eVÅ) leads to a strong Cu-Cu superexchange interaction J_{Cu-Cu} =0.15 eV. The weak coupling between bilayers ($J'/J \approx 10^{-5}$) and within a bilayer ($J_b/J \approx 10^{-2}$-10^{-1}) make $YBa_2Cu_3O_{6+x}$ a very good approximation of a 2D-bilayer system with a weak in-plane anisotropy ($\Delta J/J \approx 10^{-4}$). The observed 3D ordering temperature (T_N=415K) is in agreement with the estimated reduction of the molecular field value T^{MF}=2SJ due to J'/J or ΔJ/J corresponding to a departure from the pure 2D Heisenberg model.
When p-holes are created in the CuO_2 planes, static disorder is induced and can be characterized by the finite correlation length ξ. Magnetic excitations with wave-vectors larger than $\Gamma_q=1/\xi$ are expected to propagate normally. Indeed, it is observed that in-plane spin excitations at low q are renormalized and overdamped (fig.3), whereas at large momentum transfer the bare spin-waves velocity is recovered [9]. The characteristic damping energy $\Gamma_\omega \approx 17$ meV is related to the renormalized spin-wave velocity c, $\Gamma_\omega = \left(\sqrt{2}\frac{2\pi}{a}\right) \times c \times \Gamma_q$. For x=0.37 we have found that c ≈0.4 eVÅ and that the anisotropy has been reduced by a factor of two. The presence of a small number of 2p-holes (less than 2%) in the planes affects the propagation of antiferromagnetic spin-waves as it modifies the static magnetic properties.

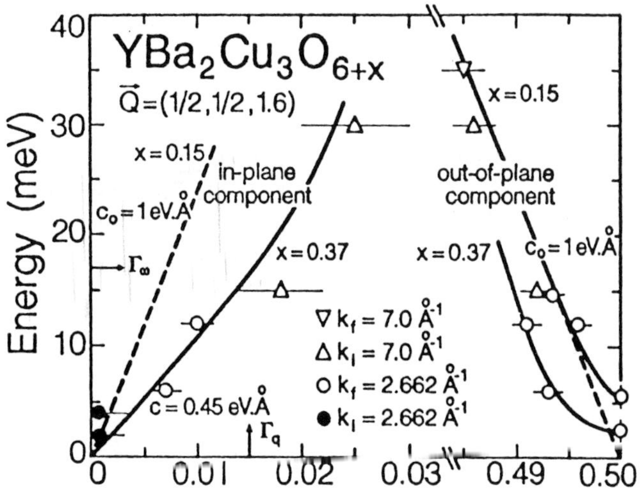

Fig. 3 Dispersion curves for the in-plane and out-of-plane spin excitations in the pure AF $YBa_2Cu_3O_{6.15}$ and doped AF state $YBa_2Cu_3O_{6.37}$. The dotted line corresponds to the spin-wave velocity c_0=1 evÅ in the undoped state.

SPIN DYNAMICS IN THE SUPERCONDUCTING STATE

A further increase of the oxygen content induces an insulating-metal transition due to a transfer of a large number of p-holes in the CuO_2 planes. So far, we have studied three different compositions in the superconducting state: x=0.45 (T_c=37K), x=0.51 (T_c=47K), x=0.69 (T_c=59K). As mentioned before, these holes destroy magnetic long range order down to zero temperature. But Q-scans performed at finite energy transfer (fig.4) clearly show the presence of magnetic scattering centered along the rods (1/2,1/2,l) indicating the existence of antiferromagnetic dynamical correlations between Cu(2) spins. It should be noted that the Cu(2) planes within a bilayer remain antiferromagnetically coupled in the superconducting phase, as demonstrated by the Q-dependence of the dynamical response. For x=0.51, this coupling persists up to T=150K. The spin fluctuations are characterized by three main features: a large Q-width, a maximum of the dynamical structure factor $S(Q,\omega)$ at large energy transfer Ω_M (fig.5), and the depression of the low energy part of the spectrum at temperatures below T_c(fig.6 and 7).The Q-width, which has been found to be independent of energy and temperature, is a measure of the hole density. For the two values of x=0.45 and x=0.51, which are very close, Γ_q=0.05±0.005 r.l.u. (ξ/a=2.2), whereas at x=0.69, Γ_q=0.07±0.01 r.l.u. (ξ/a=1.5), indicating an increase in the hole density. The dynamics of the Cu(2) spins is governed, in this energy range, by the motion of quasi-particles (hole+polaron). The energy scale of the fluctuations is related to the product of the velocity of these quasi-particles and their density. For x=0.45 and x=0.51, we find quite different values for Ω_M: 8 meV and 22 meV, respectively. Since the values of Γ_q are identical for these concentrations, we conclude that the quasi-particle velocity has increased from x=0.45 to x=0.51. The maximum at 23 meV in the scattered intensity at low temperature for x=0.69 does not correspond to the actual maximum Ω_M because of the transfer of intensity from the low energy side to the energy range just above the pseudo-gap to be discussed below. Ω_M can be determined at high temperature. The actual value (Ω_M=30 meV) is in agreement with the increase of hole density from x=0.51.

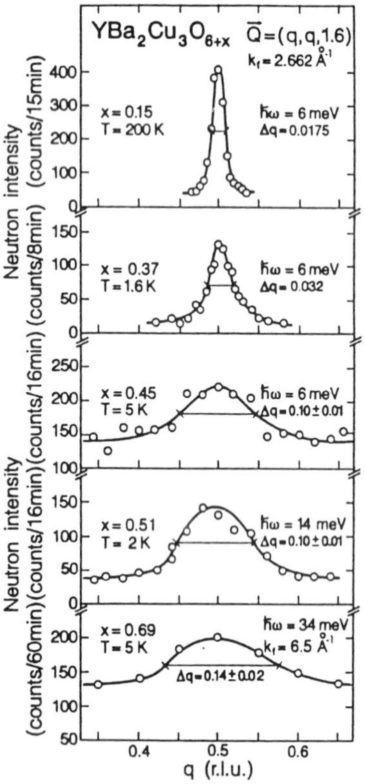

Fig. 4 Q-scans performed through the [1/2,1/2,l] rod at various oxygen content.

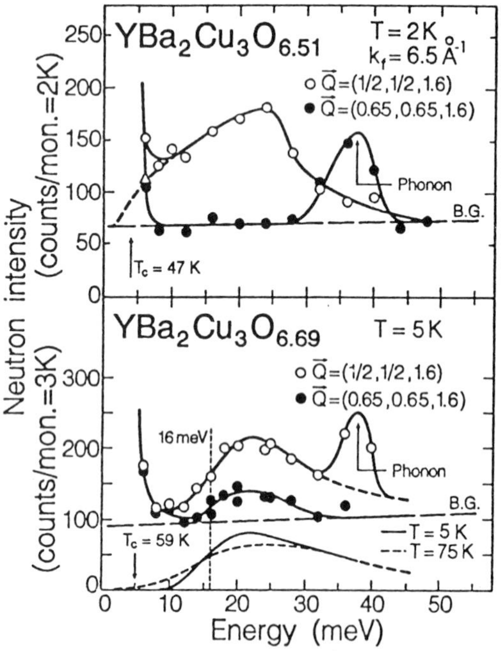

Fig. 5 Comparison between high energy transfer scans for $YBa_2Cu_3O_{6.51}$ and $YBa_2Cu_3O_{6.69}$. Background is obtained at Q=(0.65,0.65,1.6). The data at x=0.69 shows clearly a gap at 16 meV. With the experimental conditions, the gap at 4 meV for $YBa_2Cu_3O_{6.51}$ is not visible.

Below T_c the suppression of intensity at low energy in the dynamical response is evident on figure 6 and 7. This behavior has already been reported for x=0.45 [8,10,11], and the results for x=0.51 and x=0.69 confirm the effect. The magnetic signal is obtained by comparing measurements at Q=(1/2,1/2,1.6) on the 2D rod and at Q=(0.65,0.65,1.6) away from the rod.
Since all features in the magnetic response are broader than the instrumental resolution, the magnetic signal can be taken as proportional to the dynamical structure factor. Neglecting scale factors and

magnetic form factor, we have:

$$S(Q,\omega) = \frac{1}{\pi} \frac{1}{1 - \exp\left(-\frac{\hbar\omega}{kT}\right)} \, \mathrm{Im}\{\chi(q,\omega)\}$$

At low temperature, the magnetic scattering is directly proportional to $\mathrm{Im}\{\chi(q,\omega)\}$. The figures 6 and 7 show the energy dependence of $\mathrm{Im}\{\chi(q,\omega)\}$ at various temperatures. The slope of the curves at $\omega \to 0$ is related to the relaxation rate ($1/T_1T$) given by NMR measurements, which is also proportional to the limit of $\mathrm{Im}\{\chi(q,\omega)\}/\omega$ when $\omega \to 0$. The horizontal slopes (fig.6 and 7) at low temperature could be expected from the NMR experiments [12] which probe only the low energy part of the spectrum. At higher energies, an inflexion point is observed at $\hbar\omega_G$=3 meV (x=0.45), $\hbar\omega_G$=[illegible] meV (x=0.51) and $\hbar\omega_G$=16 meV (x=0.69). Well above T_c, $\mathrm{Im}\{\chi(q,\omega)\}$ recovers a linear dependence as a function of energy up to very large energy transfers (17 meV for x=0.51). Such a behavior is not consistent with a lorentzian lineshape with an energy scale of Γ_ω=22 meV. The lineshape at low temperature shows the effects of superconductivity: low excitations are suppressed and transferred to higher energies. This suppression is due to the onset of superconductivity. It should be pointed out that around $\hbar\omega_G$ at low T, $\mathrm{Im}\{\chi(q,\omega)\}$ has a finite slope, less steep than expected from the instrumental resolution which indicates that $\mathrm{Im}\{\chi(q,\omega)\}$ is a continuous function of ω. The values for the gap $\hbar\omega_G$ at low doping are of the order of kT_c, which is small compared to the BCS prediction. However, for x=0.69, $\hbar\omega_G$=3.2 kT_c, which is very close to the classical BCS result $2\Delta = 3.5kT_c$. Another important result is also shown on fig.7. At T=75K, above T_c, $\mathrm{Im}\{\chi(q,\omega)\}$ has not yet recovered a linear variation with ω. The reduction in $\mathrm{Im}\{\chi(q,\omega)\}$ can be attributed to superconducting fluctuations which start to develop below T_c^{MF} = 120K. Such a mean field value T_c^{MF} is twice as high as T_c in analogy to planar magnetic systems [13]. A similar behavior has been observed in NMR data [14].

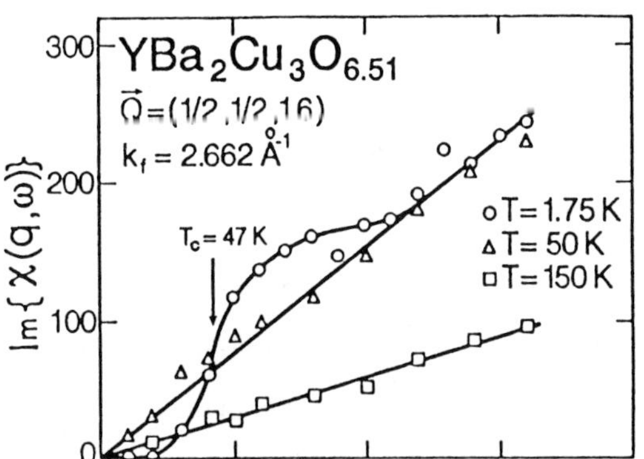

Fig. 6 $\mathrm{Im}\{\chi(q,\omega)\}$ (in arbitrary units) as a function of energy for $YBa_2Cu_3O_{6.51}$ at temperatures above and below T_c. An energy gap is clearly seen in the excitation spectrum.

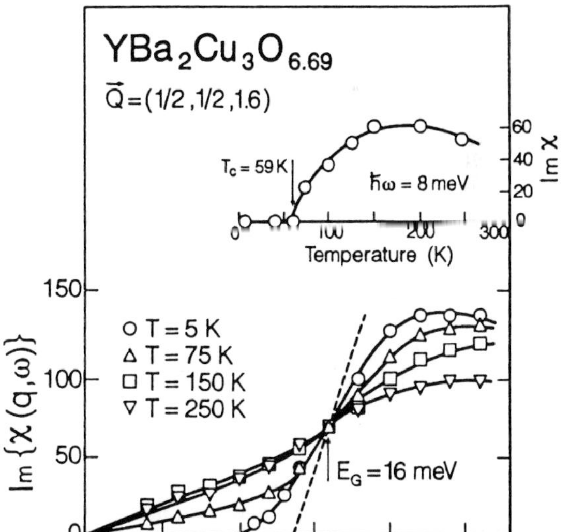

Fig. 7 $\mathrm{Im}\{\chi(q,\omega)\}$ (in arbitrary units) as a function of energy for $YBa_2Cu_3O_{6.69}$ at temperatures above and below T_c. The energy gap at 16 meV in the excitation spectrum persists above T_c.

CONCLUDING REMARKS

Our neutron scattering investigations have evidenced three major aspects of the properties of the $YBa_2Cu_3O_{6+x}$ materials, namely : the two-dimensional character of magnetic and superconducting properties, the extreme sensitivity of the Cu^{2+} magnetism to the p-holes and the coupling of the Cu spins to the superconducting carriers.

The 2D nature of the magnetic interactions in $YBa_2Cu_3O_{6+x}$ has been demonstrated by the observation of the spin-waves branches. Static magnetic properties are well accounted for by considering $YBa_2Cu_3O_{6+x}$ as a typical 2D Heisenberg system with a weak XY anisotropy.

Holes created in the CuO_2 planes induce magnetic disorder (magnetic polaron). These defects renormalize magnetic excitations. A small concentration of holes (≈ 2%) destroys magnetic long range order in the CuO_2 planes. At higher oxygen content, a large amount of holes is transferred to the CuO_2 planes yielding to a metallic state and superconductivity.

In the superconducting state, we have observed dynamic magnetic correlations near the antiferromagnetic wavevector. In this regime the Cu spins dynamics is dominated by the holes and the polarons. The energy scale is given by the density of holes and the mobility of polarons. Furthermore, we have observed the opening of a magnetic gap induced by superconductivity in the $YBa_2Cu_3O_{6+x}$ materials. Above T_c, magnetic fluctuations below the gap are reduced, which indicates the existence of 2D superconducting fluctuations. Obviously, the Cu spins are tighly coupled to superconducting carriers. Are they mediating the pairing interactions ?
More experiments on the $YBa_2Cu_3O_7$ system are now in progress. The increase of T_c from 60K to 90K could be due to larger coupling between layers, therefore to a more pronounced 3D character and to a reduction of the 2D superconducting fluctuations above T_c.

* Present address: ESRF BP220, 38043 Grenoble Cedex, France

REFERENCES.

1. Bednorz JG, Müller KA, (1986) Z Phys B 64: 189-193
2. Wu MK, Ashburn JR, Torng CJ, Horn PH, Meng RL, Gao L, Huang ZI, Wang YQ, Chu CW (1987) Phys.Rev.Lett.58: 908-910
3. Aeppli G, Hayden SM, Mook HA, Fisk Z, Cheong S-W, Rytz D, Remeika JP, Espinosa GP, Cooper AS (1989) Phys.Rev.Lett 62: 2052-2055
4. Vaknin D, Sinha SK, Moncton DE,Johnston DC, Newsam JM, Safinya CR, King HE, (1987) Phys.Rev.Lett 58: 2802-2806
5. Rossat-Mignod J,Burlet P,Jurgens MJ, Vettier C, Regnault LP, Henry JY, Ayache C, Forro L, Noel H, Potel M, Gougeon P, Levet JC (1988) J Physique C8:2119-2124
6. Rossat-Mignod J, Boucherle JX, Burlet P, Henry JY, Jurgens MJ, Lapertot G, Regnault LP, Schweizer J, Tasset F,Vettier C, (1989) in "Progress in High Temperature Superconductivity" Vol 21: 74-85
7. Tranquada JM, Cox DE, Konnmann W, Moudden AH, Shirane G, Suenage M, Zalliker P, Vaknin D, Sinha SK (1988) Phys.Rev.Lett. 60: 156-159
8. Rossat-Mignod J, Regnault LP, Jurgens MJ, Vettier C, Burlet P., Henry JY, Lapertot G (1990) Physica B163: 4-8
9. Vettier C, Burlet P, Henry JY, Jurgens MJ, Lapertot G, Regnault LP, Rossat-Mignod J (1989) Physica Scripta T29: 110-115
10. Rossat-Mignod J, Regnault LP, Jurgens MJ, Vettier C, Burlet P., Henry JY, Lapertot G (1990) in Dynamics of Magnetic Fluctuations in High Tc Materials, Eds.G.Reiter, P.Horsh G.Psaltakis Plenum Press
11 Rossat-Mignod J, Regnault LP, Vettier C, Burlet P., Henry JY, Lapertot G (1990) LT19 Conf. Physica B (to be published)
12. Horvatic M, Segransan P, Berthier Y, Butaud P, Kitaoka Y, Berthier C, Henry JY, Couach M (1989) Physica C162-164: 265-266
13. Regnault LP, Rossat-Mignod J (1990) in " Magnetic Properties of Layered Transition Metal Compounds" L.J.De Jongh (Ed.) Kluwer Academic Publishers: 271-321
14. Berthier C, Berthier Y, Butaud P, Horcatic M, Kitaoka Y, Segransan P (1990) in" Dynamics of Magnetic Fluctuations in High Tc Materials" eds. G.Reiter,P.Horsh and G.Psaltakis Plenum Press

de Haas-van Alphen Effect in YBCO

J.L. Smith, C.M. Fowler, B.L. Freeman, W.L. Hults, J.C. King, and F.M. Mueller

Los Alamos National Laboratory, Los Alamos, NM 87545, USA

ABSTRACT

Measurement of the de Haas-van Alphen effect in $YBa_2Cu_3O_{6.97}$ in pulsed, magnetic fields peaking at 100 T in powdered material with the field applied along the c-axis reveal two pieces of the Fermi surface. Their cross sections are 0.56 kT (0.054 $Å^{-2}$) and 0.78 kT (0.075 $Å^{-2}$) with effective masses of 2.8 and 4.4 respectively.

KEY WORDS: YBCO, high temperature superconductivity, de Haas-van Alphen effect, high magnetic fields

INTRODUCTION

The de Haas-van Alphen effect has continued to be important to understanding the nature of the ground state in many materials. Measurements on Fermi surfaces in high temperature superconductors have been delayed. This is not because theorists were not sure if the Fermi surfaces should exist[1] but rather because of the intrinsic difficulty of the experiments. The compound $YBa_2Cu_3O_7$ has probably been the most studied, where angle-resolved photoemission[2] and two-dimensional angular correlation of positron annihilation radiation[3] studies have led to reports of a Fermi surface. Measurements of the de Haas-van Alphen effect (and related magnetic oscillations) are perhaps the most difficult to perform because of the requirements of very high quality samples, high magnetic fields, and low temperatures[4]. Indeed as new materials are investigated the de Haas-van Alphen effect is usually one of the last experiments to be successful, as it was for heavy-Fermion superconductors[5]. For high temperature superconductors the difficulty of preparing high quality samples and the short electron mean free path have proved particularly daunting. Very high magnetic fields are a route to the de Haas-van Alphen measurements of the Fermi surface.

The application of a magnetic field to a metal can be viewed semi-classically as leading to the quantization of the momenta of electrons onto highly degenerate Laudau levels or nested tubes within the Fermi surface. As the field is varied, tubes will attach or detach themselves from extremal cross-sectional areas of the Fermi surface[4]. If the magnetization of the metal is monitored, an oscillatory component is observed due to this attaching or detaching that is called the de Haas-van Alphen effect which is periodic in the inverse of the magnetic induction (or for the case of this paper, the applied magnetic field). The period of this oscillation yields the cross-sectional area of the Fermi surface. Typically, orientational studies then allow the Fermi surface to be mapped. Temperature and field dependences, however, yield far more information than just the topology. Orbital averages of the electron effective masses, momentum values, and scattering rates can be deduced for each orbit that is observed as temperature and field are varied. Large effective masses and high scattering rates make some orbits difficult to observe. For materials with both of these properties, only high fields will provide the data. In this work we have taken the measurement of the de Haas-van Alphen effect to higher fields by more than a factor of two.

EXPERIMENTAL

We briefly discuss the novel 100 T magnetic flux compressor. We discharge a capacitor bank into a copper, wedge-shaped sandwich attached to a brass block with a 1.6 cm diameter hole in it for the sample. As the current reaches a maximum, explosives are set off that squeeze the magnetic flux into the hole in the brass block starting from the end like a tube of toothpaste[6]. The time dependence of the magnetic field is monitored with a pick-up coil that is in a slot on the side of the sample hole in the brass block.

Double-walled pyrex tubing (unsilvered, evacuated) conducts liquid helium from a two-liter liquid nitrogen dewar, through a sub-millimeter orifice, past the sample, and into a 3 meter long, 2 cm inside diameter rubber hose that then conducts the helium gas to a vacuum system. A carbon resistor, immediately downstream, monitors the sample temperature. This system has attained 2.3 K outdoors, where the experiment is performed, and 1.9 K in the laboratory.

The $YBa_2Cu_3O_{6.97}$ sample was prepared from oxides and barium carbonate by conventional techniques, with the addition of a brief excursion from 940°C to 890°C and back, every four hours during calcination and sintering[7]. The sample was then ball-milled to achieve an average particle size between 5 and 10 µm, with a few large pieces ranging up to 50 µm. This powder was mixed with epoxy and the c-axes were made parallel in a magnetic field as the epoxy hardened[8]. Pieces from the same sample batch have been used for the twenty YBCO experiments to date. The oxygen content was determined by titration and x-ray diffraction. The 2 degree alignment of the single-crystalline grains was confirmed by x-ray diffraction and NMR.

A piece of this sample was machined to a 1 mm diameter rod, 5 mm long, and a sample dummy of pure epoxy was machined with precision to match the sample size. These are both wound with about 100 turns of 30 µm copper wire, mounted parallel to each other, and matched inductively by adjusting the wire while monitoring the electrical parameters at an excitation frequency near 10 MHz. The difference signal is fed to a nearby passive filter that passes frequencies above 550 kHz.

Details of the data acquisition system and a longer description of the experiment have been published in a preliminary paper reporting the de Haas-val Alphen effect[9]. The duration of the experiment is about 100 µsec.

RESULTS AND DISCUSSION

Figure 1 shows the signal for a 2.8 K measurement. Because of the large time rate of change of the magnetic field (10^5T/sec produces a kV across the coil), we cannot put a second layer of turns on the coils, as in conventional phase sensitive detection systems. Although we achieve an overall system balance of less than one effective coil-turn, some dB/dt component shows up in the signal. This is the source of the large excursions between 90 and 92 µsec. We can check this by direct comparison to the field measurement coil. These large excursions do not have significant influence in the final Fourier transforms because their spectral content in the frequency range of interest is insignificant when compared to the full data set. These very reproducible excursions show up in every experiment and are due to non-linearities in the generator. The signal is not symmetric about the maximum field because during flux compression the field is retarded in its penetration into the superconductor[10]. We have put in pairs of dummy coils with the sample and its dummy in order to better record the noise. We think this system showed coupling between the four coils. The filtered signal from a second field measuring coil showed only noise that decreased as the reciprocal frequency, and thus did not resemble what we report here as the results. Our field calibration appears to be good to about 2%.

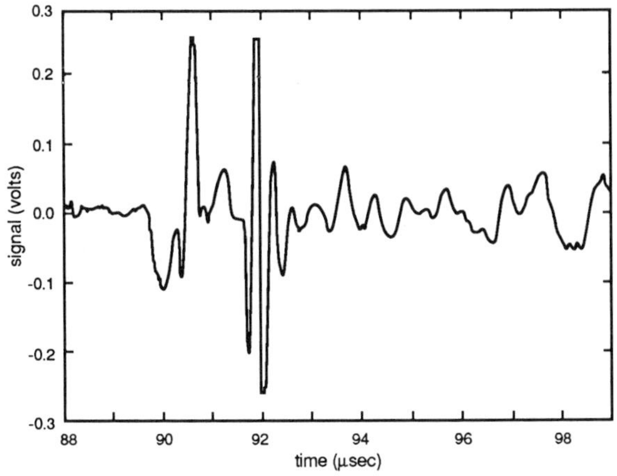

Fig. 1. de Haas-van Alphen signal from filter for YBCO at 2.8 K. The field is 84 T at 88 μsec, peaks at 100 T at 90.2 μsec, and is 65 T at 99 μsec.

Figures 2 and 3 show the Fourier transform[9] of the signal as a function of the inverse applied magnetic field for two temperatures. We shall only discuss the features labeled with letters, although it is possible that the other large features may be the de Haas-van Alphen effect in the copper pick up coil wire. A comparison of these figures suggests that feature A becomes A and B and that feature C becomes C and D at the lower temperature. Thus, the possibly barely resolved doublets in A and C in Fig. 2 have become resolved in the colder experiment of Fig. 3. These same features have been seen in every YBCO experiment. Thus, the spectral features of Fig. 3 are clearly stronger at the lower temperature and represent an enhanced signal to noise ratio. The unlabeled features at lower and higher frequencies are believed to be noise and do not reproduce between experiments. In our earlier work[9], we had identified C as a possible harmonic of A in Fig. 2. Now we can see that C and D in Fig. 3 do indeed occur at twice the frequency of A and B, within the accuracy of the field measurement. Thus features A and B do represent the two extremals of one sheet of the Fermi surface. Even with the limited data of Fig. 3, the presence of second harmonics permits a crude extraction of the unitless parameter, the electron effective mass[4]. These are shown in Table I for features A and B, where their Fermi surface areas are also shown. These masses are in good agreement with those of ref. 9, which were determined from the temperature dependence. The areas permit the cyclotron orbit radius to be calculated, which is also shown in Table I for H = 100 T. It is usually considered that the electrons on a particular piece of the Fermi surface must complete one cyclotron orbit before being scattered if the de Haas-van Alphen effect is to be observed. A reasonable lower bound for the electron mean free path in YBCO at T = 2 K is 400 Å[11]. As the circumference of the cyclotron orbits in Table I are about 300 Å, this condition has been met. This incidentially illustrates why very high magnetic fields will continue to be necessary until much better samples are available. Finally, we note that all of our measurements are above a reasonable estimate of the critical magnetic field at zero temperature for the field parallel to the c-axis of 40 T[12].

Table I. The Fermi surface areas A, effective masses m*, and cyclotron orbit radius r_c for peaks A and B in Fig. 3.

Peak	A (kilotesla)	A ($Å^{-2}$)	m*	r_c(Å)
A	0.56	0.054	2.8	48
B	0.78	0.075	4.4	57

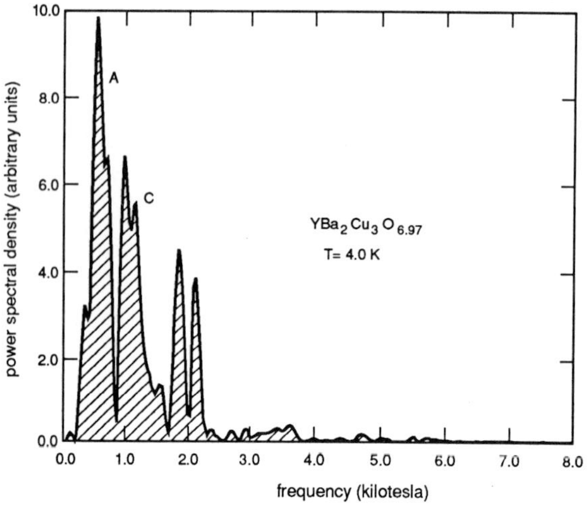

Fig. 2. Fourier transform of the signal versus the reciprocal applied magnetic field at T = 4.0 K.

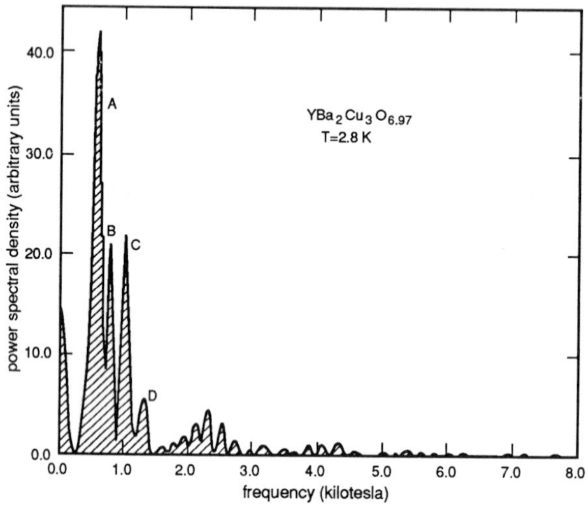

Fig. 3. Fourier transform of the signal versus the reciprocal applied magnetic field at T = 2.8 K.

In conclusion, we have found two small areas of the Fermi surface of YBCO. The masses are enhanced by a factor of 2 to 3 from the band masses[13]. So YBCO has displayed a de Haas-van Alphen effect that is evidence for a Fermi surface. Energy band models show larger pieces of the Fermi surface, which we are now seeking.

ACKNOWLEDGEMENTS

We are indebted to many people at Los Alamos for their assistance over the last two years. This work was performed under the auspices of the U. S. Dept. of Energy, Office of Basic Energy Sciences, Division of Materials Sciences. Part of this work was supported by the National Science Foundation under contract DMR 90-16241.

REFERENCES

1. Bedell KS, Coffey D, Meltzer DE, Pines D, Schrieffer JR (eds) (1990) High Temperature Superconductivity Proceedings. Addison-Wesley. Redwood City
2. Campuzano JC, Jennings G, Faiz M, Beaulaigue L, Veal BW, Liu JZ, Paulikas AP, Vandervoort K, Claus H, List RS, Arko AJ, Bartlett RJ (1990) Phys Rev Lett 64: 2308-2311
3. Hoffmann L, Manuel AA, Peter M, Walker E, Damento MA (1988) Europhys Lett 6: 61-66; Smedskjaer LC, Liu JZ, Benedek R, Legnini DG, Lam DJ, Stahulak MD, Claus H, Bansil A (1988) Physica 156C: 269-275
4. Shoenberg D (1984) Magnetic Oscillations in Metals. Cambridge University Press. Cambridge.
5. Taillefer L, Newbury R, Lonzarich GG, Fisk Z, Smith JL (1987) J Magn Magn Mater 63 & 64: 372-376
6. Fowler CM (1973) Science 180: 261-267; Fowler CM, Caird RS, Garn WB, Erikson DJ (1976) IEEE Trans on Magnetics MAG-12: 1018-1023
7. Smith JL, Hults WL, Clarke AP, Johnson KA (1989) Guidance and Control Information Center, Chicago, Report PR-89-02, pp. 11-16
8. Farrel DE, Chandrasekhar BS, DeGuire MR, Fang MM, Kogan VG, Clem JR, Finnemore DK (1987) Phys Rev B36: 4025-4027
9. Mueller FM, Fowler CM, Freeman BL, Hults WL, King JC, Smith JL (1990) Physica B, to appear
10. Mueller FM, Fowler CM, Freeman BL, King JC, Martinez AR, Montoya NN, Smith JL, Hults WL, Migliori A, Fisk Z, Arko AJ, List RS, Schwarz RB, Prueitt ML (1990) J Magn Magn Mater, to appear
11. Gurvitch M, private communication
12. Nakao K, Miura N, Tatsuhara K, Takeya H, Takei H (1989) Phys Rev Lett 63: 97-100
13. Massida S, Yu JJ, Freeman AJ, Koelling DD (1987) Phys Lett 122A: 198-201; Pickett WE (1989) Rev Mod Phys 61: 433-512

de Haas-van Alphen Measurement in $YBa_2Cu_3O_7$

G. Kido[1], K. Komorita[1], H. Katayama-Yoshida[2], T. Takahashi[2], Y. Kitaoka[3], K. Ishida[3], and T. Yoshitomi[3]

[1] Institute for Materials Research, Tohoku University, Katahira, Aoba-ku, Sendai, 980 Japan
[2] Department of Physics, Faculty of Science, Tohoku University, Aramaki, Aoba-ku, Sendai, 980 Japan
[3] Department of Material Physics, Faculty of Engineering Science, Osaka University, Toyonaka, Osaka, 560 Japan

ABSTRACT

Measurements of de Haas van Alphen effect have been carried out using modulation technique in steady high magnetic fields up to 27T at the temperatures range between 1.8 and 4.2K. A broad band centering at 540T, which corresponds to the extremal cross-section of $0.052\pm 0.003A^{-2}$, appeared in the Fourier component for H//c-axis. The effective mass was estimated to be $2.1^{+2}_{-0.5}\ m_e$ by the temperature dependence of the intensity of the band.

KEY WORDS: YBaCuO superconductor, high magnetic field, Fermi surface, de Haas van Alphen effect

INTRODUCTION

The measurement of the de Haas van Alphen effect has been interested by many physicists who concern the superconducting oxide, because the recent experiments, such as angle-resolved photoemission [1, 2], two-dimensional angular correlation of positron annihilation radiation [3], etc., strongly suggest the presence of Fermi surface.

However, it does not seem easy to measure the de Haas van Alphen effect in superconducting oxide because of its extremely high critical field and short scattering time. There are two ways to remove from these strictions. One is a usage of enormously high magnetic fields and the other is a measurement in mixed state [4] with a high quality sample. Mueller et al. first reported the oscillation in megagauss field range using a flux compression method [5]. On the other hand, we have measured the de Haas van Alphen effect in the steady high magnetic field up to 27T employing modulation technique and found characteristic Fourier spectrum [6]. Both of them measured with $YBa_2Cu_3O_7$ oriented powder, because this superconducting oxide is the stoichiometric and the scattering time may be longest among the various superconducting oxide.

EXPERIMENTAL

$YBa_2Cu_3O_7$ was prepared by sintering the mixture of $BaCO_3$, Y_2O_3 and CuO powders (purity of 99.99%), at 935℃ in air for 16 hours. The resultant powder was then annealed in flowing high purity O_2 gas at 350℃ for 20 hours, and slowly cooled down to the room temperature. The powder was confirmed to be in a single phase by an X-ray diffraction. The critical temperature was evaluated to be 93K by an ac susceptibility measurement. Since we measure the de Haas-van Alphen effect at a relatively high frequency, the powder was well ground into crystallites with diameters of smaller than 20μm. The crystallites were embedded in a non-conductive materials. In order to align the c-axis of crystallites, a mixture of crystallites and epoxy was exposed in a 5T field during the hardening process [7]. We have made 4 samples for the measurements.

Measurements of de Haas-van Alphen effect have been carried out using modulation technique in steady high magnetic fields up to 27 T produced by hybrid magnets, HM1 and HM2, installed at the High field laboratory, Tohoku University [8]. A block diagram and a coil system are schematically shown in Fig. 1. Since a major ripple component, which appears in the pick-up coil, is 600Hz in our case, the moderation frequency is chosen in the range of 1 to 2.5kHz. A modulation field up to $0.02T_{p-p}$ is generated using a high power amplifier (NF, TA-250), because no superconducting wire can be used in this field range. The field intensity is monitored by an H-coil.

When an ac field is generated in a strong magnetic field, very strong force works on the coil to vibrate. If the pick-up coil moves in a strong magnetic field, it induces a large voltage. We have to suppress the vibration of the modulation coil and movement of pick-up coil. Copper rings attached at the both side of the bobbin prevents the vibration of the modulation coil due to the electromagnetic reluctant forces. Pick-up coils are also supported by a copper block. This technique is very simple, but is quite effective in experiments using strong steady magnetic fields.

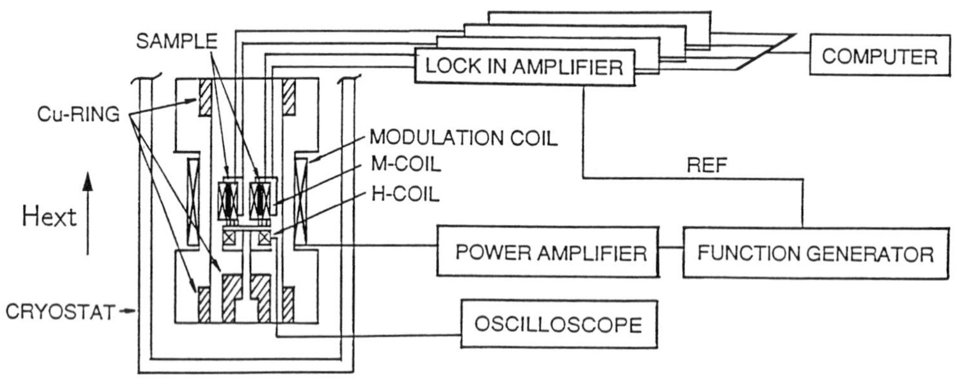

Fig. 1. Block diagram for the de Haas-van Alphen measurements.

For the efficient measurements, 4 samples are measured simultaneously in one sweep of the magnetic field by the use of 4 pick-up coils. Induced voltages in the pick-up coils are detected by 4 lock-in amplifiers (NF, 5610B) and are sent to an HP-310 computer. An intensity of the magnetic field is derived from the currents for a resistive magnet and a superconducting magnet which compose a hybrid magnet.

Temperature is changed by pumping liquid helium. Since the modulation coil consumes considerably large energy, the helium is continuously transferred during the sweep of the field in the higher temperature region than 2.5K. The magnetic fields are swept between 10T and the maximum field, 23 or 27T, which depends on the hybrid magnet, HM2 or HM1, respectively. The sweep velocity is in the range of 0.4 to 1 T/min.

RESULTS AND DISCUSSION

Figure 2 shows ac susceptibility of YBa_2Cu_3O as a function of 1/H at 1.8K. The signal level is approximately 1000 times larger than the noise level. The fine component in the spectrum is due to the motion of pined flux.

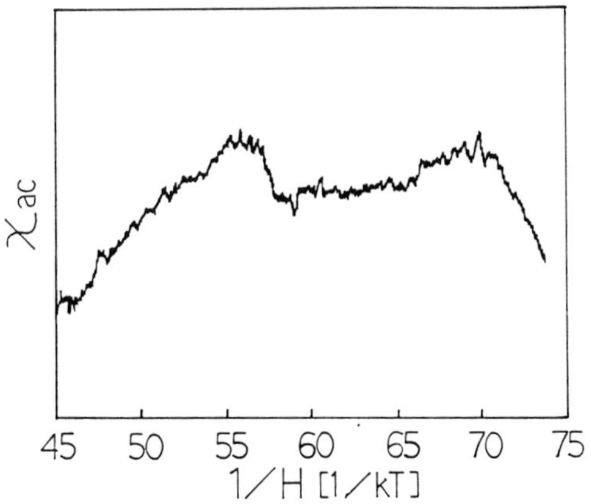

Fig. 2. ac susceptibility in $YBa_2Cu_3O_7$ at 1.8K.

Figure 3 shows Fourier spectra at different temperatures in sample #1. A broad band centering at 540± 30T is clearly seen in the measurement at 1.8K. The amplitude decreases with increasing temperature. This frequency corresponds to the extremal cross-section of 0.052± 0.003A^{-2}, which is almost equal to the theoretical value (0.042 A^{-2}) of the Fermi surface around S point due to one dimensional CuO chain [9].

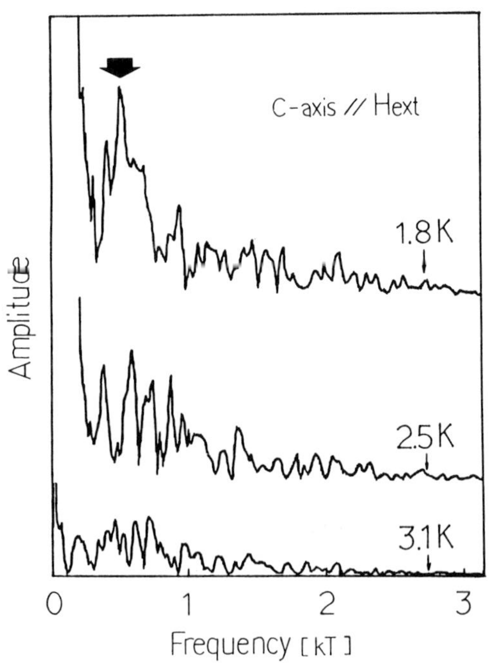

Fig. 3. Fourier transform of ac-susceptibility in $YBa_2Cu_3O_7$ at different temperatures. An arrow indicates the center of band.

The amplitude of oscillatory susceptibility, A, due to the de Haas-van Alphen effect is expressed as: $A = \alpha T \cdot \exp(-2\pi^2 kTm^*/e\hbar H)$, where m^* is the effective mass of carrier, k Boltzmann constant, α proportional constant. Reversely, the effective mass is derived from the slope of log(A/T) vs. T plot (see Fig.4). The effective mass was estimated to be $2.1^{+2}_{-0.5}\,m_e$, where m_e is the free electron mass. The observed values of the extremal cross-section and effective mass are in good agreement with those (600T and $2.9m_e$) of one of the two Fourier components obtained by F. M. Mueller et al. using megaguss fields [5].

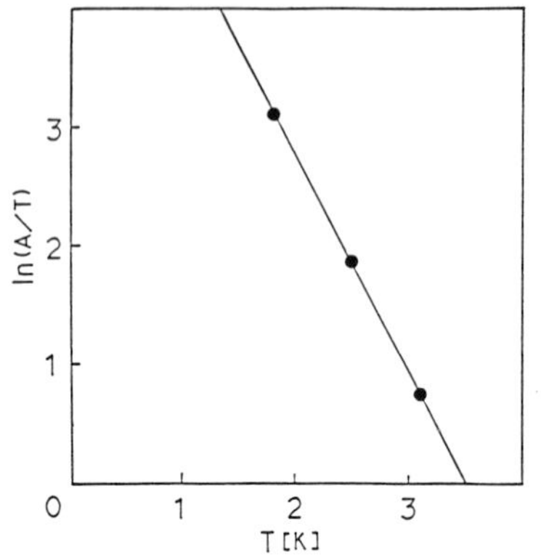

Fig. 4. log(A/T) vs. T plot.

We have measured 4 samples and found the band in three samples (#1-#3), though the signal intensity differs from sample to sample. When the modulation lowered, the characteristic Fourier spectrum has not been observed.

CONCLUSION

We have carried out the de Haas van Alphen measurements on $YBa_2Cu_3O_7$ and found a broad band centering at 540T ($0.052A^{-2}$), which possibly correspond to a small Fermi surface due to a one-dimensional Cu-O chain.

ACKNOWLEDGMENTS: We wish to thank Professor T. Suzuki who motivated us to make a de Haas van Alphen measuring system and offered a data processing software. We are also indebted to the members of High Field Laboratory for Superconducting Materials and Cryogenic Center for the operation of hybrid magnets. This work is partly supported by the Grant-in-aid for Scientific Research on Priority Areas from the Ministry of Education and Culture.

REFERENCES.

1. Takahashi T, Matsuyama H, Katayama-Yoshida H, Okabe Y, Hosoya S, Seki K, Fujimoto H, and Inokuchi H (1988) Nature 334: 691-692
2. Olson CG, Liu R, Yang AB, Lynch DW, Arko AJH, List RS, Veal BW, Chang YC, Jiang PZ, and Paulikas APY (1989) Science 245: 731-733
3. Smedskjaer LC, Liu JZ, Benedek R, Legnini DG, Lam DJ, Stahulak MD, Claus H, and Bansil A (1988) Physica 156C: 269-275
4. Markiewicz RS, Vagner LD, Wyder P, and Maniv T (1988) Solid State Commu 67: 43-45
5. Mueller FM. Fowler CM, Freeman BL, Hults WL, King JC, and Smith JL, Physica B, to appear
6. Kido G, Komorita K, Katayama-Yoshida H, Takahashi T, Kitaoka Y, Ishida K, and Yoshitomi T, Bull Jpn Phys Soc (Autum 1990) No.3, pp 93
7. Farrel DE, Chandrasekhar BS, DeGuire MR, Fang MM, Kogan VG, Clem JR, and Finnemore DK (1987) Phys Rev B 36: 4025-4027
8. Toukoku hybrid magnets are surbeyed in Sci Rep RITU A 33 (1986) pp 221-414
9. Massidda S, Yu JJ, Freeman AJ, and Koelling DD (1987) Phys Lett 122A: 198-202

Possibility of the Carrier Doping in a Ferromagnetic Copper Oxide: $La_4Ba_2Cu_2O_{10}$

HIROMU MASUDA, FUMIO MIZUNO, IZUMI HIRABAYASHI, and SHOJI TANAKA

Superconductivity Research Laboratory, Nagoya Division, International Superconductivity Technology Center, 4-1, Mutsuno 2-chome, Atsuta-ku, Nagoya, 456 Japan

ABSTRACT

The carrier-doping effects of an insulating copper oxide: $La_4Ba_2Cu_2O_{10}$ which shows a ferromagnetic transition at 5 K have been studied. By carrier-doping, two effects i. e., 1) possibility of the metal-insulator transition and 2) magnetic dilution of Cu^{2+} spin by non-magnetic Cu^{3+} ion are expected. By Ba substitution for La expecting hole-doping ($La_{4-x}Ba_{2+x}Cu_2O_{10}$), we could not observe the metal-insulator transition for the sample of which average valence number of the Cu ion is between 2 and 2.5. But the spin dilution effect on the effective magnetic moment and the ferromagnetic transition temperature was observed. On the other hand, By La substitution for Ba expecting electron-doping ($La_{4+x}Ba_{2-x}Cu_2O_{10}$) was not succeeded in this system. The spin dilution effect by Zn substitution for Cu ($La_4Ba_2(Cu_{1-x}Zn_x)_2O_{10}$ system) was also investigated.

KEY WORDS: ferromagnet, carrier-doping, magnetic dilution effect, metal- insulator transition, $La_4Ba_2Cu_2O_{10}$

INTRODUCTION

Recently we have found that an insulating copper oxide: $La_4Ba_2Cu_2O_{10}$ (La 422-phase) in the La-Ba-Cu-O family shows a ferromagnetic transition at around 5 K[1]. The effective magnetic moment of the La 422-phase from the Curie-Weiss plot was 1.54 μ_B/Cu, if the observed moment was assigned to every Cu atom. The saturation magnetic moment was 0.95 μ_B/Cu. These moment indicate that almost all Cu^{2+} 1/2 spins participate in the localized-type ferromagnetism of the La 422 compounds. The La 422-phase in the La-Ba-Cu-O system was first synthesized by Michel and Raveau in 1981 and its structure was analyzed with X-ray powder diffraction by them[2]. From their structural analysis and our neutron study[3], the La 422-phase has the square CuO_4 groups which are so isolated that one square planner CuO_4 is orthogonal to the nearest neighbor CuO_4, the each CuO_4 plane is parallel to the c-axis and perpendicular to the ab plane. The La 422-phase has the same stoichiometry as the green phase (the 211-phase) of the Y-Ba-Cu-O system[4], but its structure and magnetic properties are different. The 211-phase is orthorhombic with space group Pbnm(62) and shows antiferromagnetic transition at about 28 K[5]. The 422-phase is a tetragonal insulator with space group P4/mbm(127) and its color is not green but dark brown.

From our ESR study[6], the ground state of the Cu^{2+} d-hole in the 422-phase was described by the wave function $d_{x^2-y^2}$ which lies on CuO_4 plane. Therefore, the d-hole orbits of the ground state are orthogonal to the nearest copper one. In this case, only potential exchange interaction acts between the Cu ions, and there is no strong antiferromagnetic kinetic exchange due to the nonexistence of the oxygen between Cu ions. On the other hand, although the d-hole orbitals of the ground states between the nearest interlayer Cu ions are not orthogonal, the exchange interaction is also expected to be ferromagnetic. Because the distance between the interlayer Cu ions is large, the d-hole orbitals of the ground states have almost no overlap. Therefore the origin of the ferromagnetic of the La 422-phase is caused by that the whole exchange interaction which acts between the spins of the Cu^{2+} ion is ferromagnetic. It is similar to the case of the K_2CuF_4 which is famous as a Cu ferromagnet[7].

On the other hand, high-Tc superconductor has a feature that the superconductivity is arisen by carrier doping to CuO_2 plane. It is observed that the insulating phase is changed to high-Tc superconducting metallic phase and further non-superconducting metallic phase with increasing carrier density by doping in the $La_{2-x}Sr_xCuO_4$ (hole-like) system[8] and $Nd_{1-x}Ce_xCuO_4$ (electron-like) system[9].

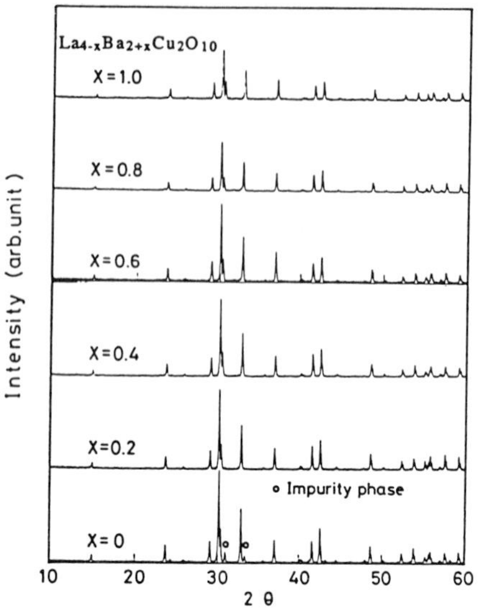

Fig. 1. X-ray diffraction patterns for $La_{4-x}Ba_{2+x}Cu_2O_{10}$.

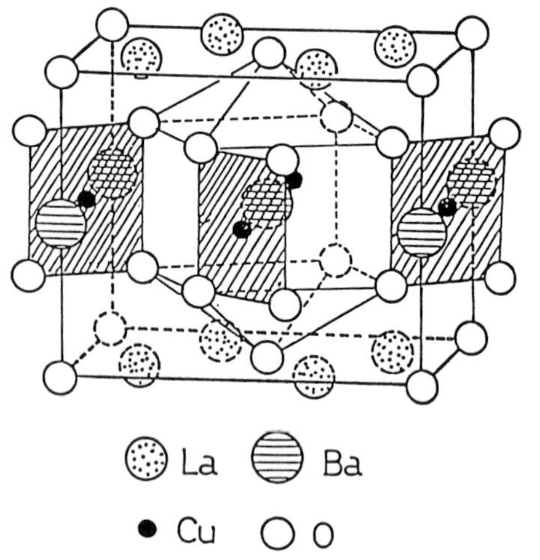

Fig. 2. Schematic view of the crystal structure of the $La_4Ba_2Cu_2O_{10}$.

We have tried the carrier-doping by Ba substitution for La: $La_{4-x}Ba_{2+x}Cu_2O_{10}$ (hole-doped type) and La substitution for Ba: $La_{4+x}Ba_{2-x}Cu_2O_{10}$ (electron-doped type) expecting possibility of the metal-insulator transition. In the $La_{4-x}Ba_{2+x}Cu_2O_{10}$ system, magnetic dilution effect of Cu^{2+} spin by non-magnetic Cu^{3+} is also expected. Then we have tried to realize magnetic dilution by Zn substitution for Cu: $La_4Ba_2(Cu_{1-x}Zn_x)_2O_{10}$.

EXPERIMENTAL

The sintered samples of the $La_{4-x}Ba_{2+x}Cu_2O_{10}$ ($0\leq x\leq 1$), $La_{4+x}Ba_{2-x}Cu_2O_{10}$ ($0\leq x\leq 1$) and $La_4Ba_2(Cu_{1-x}Zn_x)_2O_{10}$ ($0\leq x\leq 1$) were prepared by conventional solid state reaction from the mixture of high-purity La_2O_3, BaO_2, CuO and ZnO powders. The La_2O_3 powder was preheated at 1000 °C for 10 h in air, because La_2O_3 is highly hygroscopic. All the samples were prepared as follows. The mixed powders were calcined at 900 °C for 10 h in air. They were ground and pressed into disks and sintered at 1000 °C for 10 h in air. The structural phases were determined by X-ray diffraction using a monochromatized Cu-kα line. The magnetic susceptibility was measured by Quantum Design SQUID magnetometer.

RESULTS AND DISSUCUSSIONS

A. $La_{4-x}Ba_{2+x}Cu_2O_{10}$ and $La_{4+x}Ba_{2-x}Cu_2O_{10}$ ($0\leq x\leq 1$) system

The X-ray diffraction patterns for $La_{4-x}Ba_{2+x}Cu_2O_{10}$ ($0\leq x\leq 1$) are shown in Fig. 1. All the samples except x=0 (La 422-phase) are almost single phase, and the pure La 422-phase has a few impurity phases. The structure of La 422-phase from the Michel *et al.*'s analysis and our neutron diffraction study is shown in Fig. 2. According to the neutron diffraction study of the pure La 422-phase, La and Ba did not make solid solution but almost completely ordered. The result for $La_{3.6}Ba_{2.4}Cu_2O_{10}$ (x=0.2) indicated that La occupation at Ba site (4g site) and Ba occupation at La site (2b site) were 2.5 % and 11 %, respectively[10]. In this system, the lattice parameter a(=b) increases linearly with increasing x, but the lattice parameter c is constant. This is the result from which the ionic radius of Ba^{2+} is larger than that of La^{3+}. Therefore it seems that the doped Ba is preferentially accommodated in 2b site.

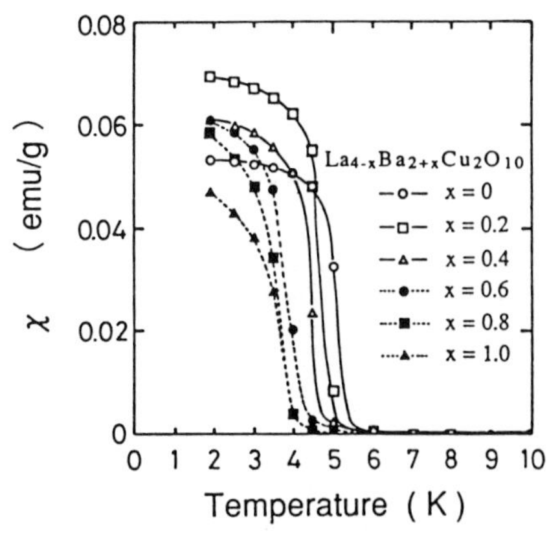

Fig. 3. The temperature dependence of the magnetic susceptibility for $La_{4-x}Ba_{2+x}Cu_2O_{10}$.

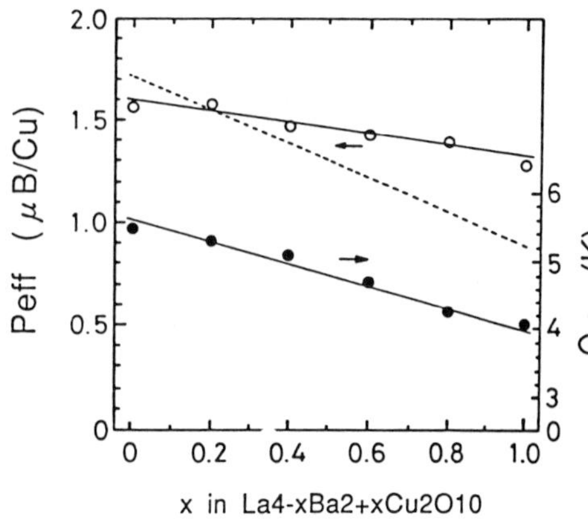

Fig. 4. The effective Bohr magneton P_{eff} and the ferromagnetic transition temperature Θ_f for x in $La_{4-x}Ba_{2+x}Cu_2O_{10}$. The broken line is the expected P_{eff} from the nominal composition.

The temperature dependence of the magnetic susceptibility χ for $La_{4-x}Ba_{2+x}Cu_2O_{10}$ is shown in Fig. 3. The magnitude of the positive divergence of χ at 2 K decreases with increasing x except x=0. Because the La-422 (x=0) sample has a few impurity phases, the magnitude of the magnetic susceptibility at 2 K is lower than most of the hole-doped samples. The effective Bohr magneton P_{eff} from the Curie-Weiss plot for x in $La_{4-x}Ba_{2+x}Cu_2O_{10}$ is shown in Fig. 4. The P_{eff} is decreased linearly with increasing x. This result indicates that the Cu^{2+} spin is diluted by non-magnetic Cu^{3+}. In the La-422 (x=0) samples, the P_{eff} is 1.54 μ_B/Cu and due to the existence of a few impurity phases, this value is smaller than the theoretical one ($P_{eff}=\sqrt{3}$) assuming the complete quenching of the orbital angular moment. On the other hand, the P_{eff} of all the sample in the range $0.4<x<1.0$ is larger than expected value from the nominal composition as shown by broken line in Fig. 4. For example in the case of the $La_3Ba_3Cu_2O_{10}$ (x=1) sample, the average valence number of Cu ion is 2.5 calculating from the nominal composition, the total number of Cu^{2+} ion and Cu^{3+} ion seems to be equal, so the value of the P_{eff} should be $\sqrt{3}/2$. The total number of Cu^{3+} ion is less than that expected by nominal composition. It may be caused by some structural defects or disorder which can not be observed by X-ray diffraction.

The ferromagnetic transition temperature Θ_f for x is shown in Fig. 4. This value decreases linearly with increasing x. The Θ_f is estimated from the formula by the molecular field approximation:

$$\Theta_f = 2zJS(S+1)/3k,$$

where z is coordination number, J is the magnitude of the exchange interaction, S is the quantum number of spin and k is the Boltzmann constant. When the Cu^{2+} spin is diluted by non-magnetic Cu^{3+}, z and Θ_f are decreased assuming J does not change so much by hole-doping.

On the other hand, in the electron doped La 422-phase ($La_{4+x}Ba_{2-x}Cu_2O_{10}$) system, Cu_2O and BaO_2 impurity phases are observed by the X-ray diffraction pattern even in the slightly doped sample and diffraction peak intensity of the La 422-phase becomes lower with increasing x. It is difficult to dope the electron to the La 422-phase. When the CuO_4 rectangles form strings of edge-sharing one, the valence number of Cu ion takes trivalent such as $NaCuO_2$ and $KCuO_2$[11]. It is not unusual that the Cu ion may take trivalent in the La 422-phase.

Another interest for the hole-doping is to investigate the possibility of metal-insulator transition. But, by the transport measurement all the samples of which average valence number of Cu ion is between 2 and 2.5 are insulator. It is because that the La 422-phase have not the continuous CuO plane as the high-Tc superconductors have but the CuO plane which is isolated each other.

B. $La_4Ba_2(Cu_{1-x}Zn_x)_2O_{10}$ ($0 \leq x \leq 1$) system

We have also tried to realize magnetic dilution by Zn substitution for Cu. The temperature dependence of the magnetic susceptibility χ and the inverse molar susceptibility χ_{mol}^{-1} for $La_4Ba_2(Cu_{1-x}Zn_x)_2O_{10}$ are shown in Fig. 5 (a) and (b), respectively. The magnitude of the positive divergence of χ decreases with increasing x. The slope of χ_{mol}^{-1} above 5 K is increased, in other words P_{eff} is decreased, with increasing x. But, the ferromagnetic transition temperature is almost constant in the whole x range, and is about 5 K. The reason of this results is that the phase separation between the $La_4Ba_2Cu_2O_{10}$ and $La_4Ba_2Zn_2O_{10}$ occurs in all composition range, because the crystal structure of the $La_4Ba_2Zn_2O_{10}$ is different from that of the $La_4Ba_2Cu_2O_{10}$[2]. In the case of the $La_4Ba_2Zn_2O_{10}$, oxygen ions around Zn are not quadrilateral as is shown in $La_4Ba_2Cu_2O_{10}$ but tetrahedral coordination.

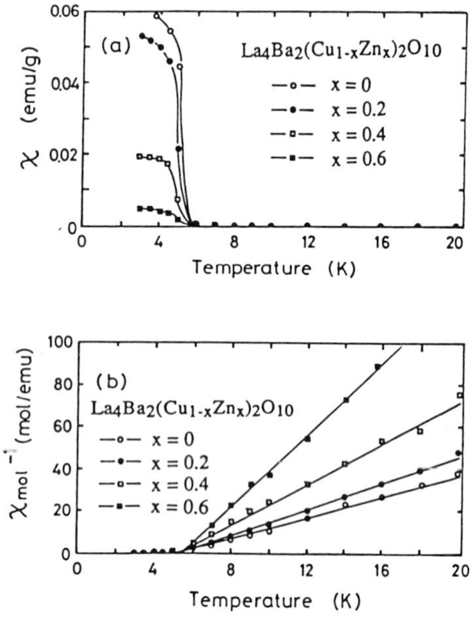

Fig. 5. Temperature dependence of (a) the magnetic susceptibility χ and (b) the inverse molar susceptibility χ_{mol}^{-1} for $La_4Ba_2(Cu_{1-x}Zn_x)_2O_{10}$.

In summary, for the hole-doped samples ($La_{4-x}Ba_{2+x}Cu_2O_{10}$) of which average valence number of Cu ion was between 2 and 2.5, we could not observe the metal-insulator transition. The spin dilution effect on the effective magnetic moment and the ferromagnetic transition temperature was observed in the same range of Cu average valence number. On the other hand, it was difficult to dope the electron to the La 422-phase. In the sample of Zn-doped La 422-phase (La4Ba2(Cu1-xZnx)2O10), the magnitude of the divergence of the magnetic susceptibility was decreased, but the transition temperature did not change from the value of the non-doped one, because the phase separation between the $La_4Ba_2Cu_2O_{10}$ and $La_4Ba_2Zn_2O_{10}$ occurred.

The authors acknowledge Drs. T. Mochiku, H. Asano, F. Izumi for their neutron diffraction study and helpful discussions.

REFERENCES

1. Mizuno F, Masuda H, Hirabayashi I, Tanaka S, Hasegawa M, Mizutani U (1990) Nature 345: 788
2. Michel C, Er-Rakho L, Raveau B (1981) J. Solid State Chem. 39: 161
3. Ogawa N, Mizuno F, Masuda H, Hirabayashi I, Tanaka S, Mochiku T, Asano H, Izumi F (1990) Physica B 165 & 166: 1687
4. Michel C, Raveau B (1982) J. Solid State Chem. 43: 73
5. Chattopadhyay T, Brown P.J, Kobler U, Wilhelm M (1989) Europhys. Lett. 8: 685
6. Masuda H, Mizuno F, Hirabayashi I, Tanaka S, submitted to Phys. Rev. B
7. Yamada I (1970) J. Phys. Soc. Japan 28: 1585
8. Torrance J. B, Tokura Y, Nazzal A, Parkin S. S. P (1988) Phys. Rev. Lett. 61: 1670
9. Takagi H, Uchida S, Tokura Y (1989) Phys. Rev. Lett. 62: 1197
10. Mochiku to be published in proc. 3rd ISS (1990) Sendai
11. Brese N. E, O'keeffe M, Von Dreele R. B, Young, Jr. V. G (1989) J. Solid State Chem. 83: 1

Magnetic Anisotropy of Single Crystalline $La_4Ba_2Cu_2O_{10}$ Ferromagnet

FUMIO MIZUNO, HIROMU MASUDA, and IZUMI HIRABAYASHI

Superconductivity Research Laboratory Nagoya Division, International Superconductivity Technology Center, 4-1, Mutsuno 2-chome, Atsuta-ku, Nagoya, 456 Japan

ABSTRACT

The magnetization of single crystalline $La_4Ba_2Cu_2O_{10}$ were measured as a function of magnetic field for both directions parallel and perpendicular to the c-axis. The direction of the easy axis is parallel to the c-axis. The measured magnetization and the estimated anisotropy energy suggest that the spin system is a Heisenberg system with Ising-like anisotropy. The characteristics of the spin system is more three-dimensional-like than that of the typical two-dimensional Cu ferromagnet K_2CuF_4. It is consistent with the result of the large magnetic entropy below T_c in the low-temperature specific heat which is well reproduced by a mean field approximation for three-dimensional S=1/2 system.

KEY WORDS: $La_4Ba_2Cu_2O_{10}$, ferromagnetism, magnetic anisotropy, specific heat

INTRODUCTION

We have found that $La_4Ba_2Cu_2O_{10}$ (the 422 phase) of the La-Ba-Cu-O family was an insulating ferromagnet with S=1/2, J/k=2.6K, and Tc=5.2K[1]. The 422 phase had been identified for the first time by Michel *et al.* during their studies on the La_2CuO_4+BaO system[2]. This phase is an insulator and dark brown in colour. The structure is tetragonal with space group P4/mbm and lattice parameters are a=6.8447(9) and c=5.8637(8) Å by the powder neutron diffraction analysis[3]. The structure selectively contains the edge-and face sharing LaO_8 and BaO_{10} polyhedra, and the square planar CuO_4 is orthogonal to the nearest neighbour CuO_4. All Cu-sites of the 422 phase exists in the same plane and a nearest neighbour distance between Cu ions is 4.84 Å. From the effective Bohr magneton (P_{eff}=1.66 μ_B/Cu) and saturation magnetic moment (μ_s=0.95 μ_B/Cu), this compound is a localized-type ferromagnet in which Cu^{2+} ions show a ferromagnetic ordering. The ferromagnetism relates to the orthogonal square-planar CuO_4 configuration. When the $Cu^{2+}d^9$-hole orbits of the ground state are orthogonal to the nearest neighbour copper one, only direct exchange interaction acts between Cu ions. Therefore, it is reasonable to expect that the whole exchange interaction which acts between spins of the Cu^{2+}ions is ferromagnetism[4].

In this report, we discuss the characteristics of the spin system of the 422 phase from both results of the magnetic anisotropy on single crystal and low-temperature spcific heat, comparing with those of the typical two-dimensional Cu ferromagnet K_2CuF_4[5].

EXPERIMENTAL

$La_4Ba_2Cu_2O_{10}$ samples were prepared by a conventional solid state reaction method[1]. Single crystals of $La_4Ba_2Cu_2O_{10}$ were grown from KCI flux, based on the phase diagram for the La_2O_3-BaO-CuO system[6]. To obtain the single crystal, the sintered samples, which were mixed in a concentration ratio of 1/2 La_2O_3: BaO: CuO=1-1.25: 1.8-2.3: 2.8-3.3 (molar ratio), were heated in alumina crucibles with KCI flux in N_2 atmosphere of 5 kg/cm^2 up to 1100 °C held at this temperature for 5 hrs, cooled to 700 °C at a rate of 3 °C/hr, and then furnace-cooled to room temperature. A number of needle-like single crystals grown along the c-axis with typical dimensions of 0.2x0.2x1.5 mm^3 were obtained on the surface of KCI flux. The structure of the single crystal is tetragonal with space group P4/mbm and lattice parameters are a=6.8443 and c=5.8684 Å. These results are consistent with the data by powder neutron diffraction analysis.

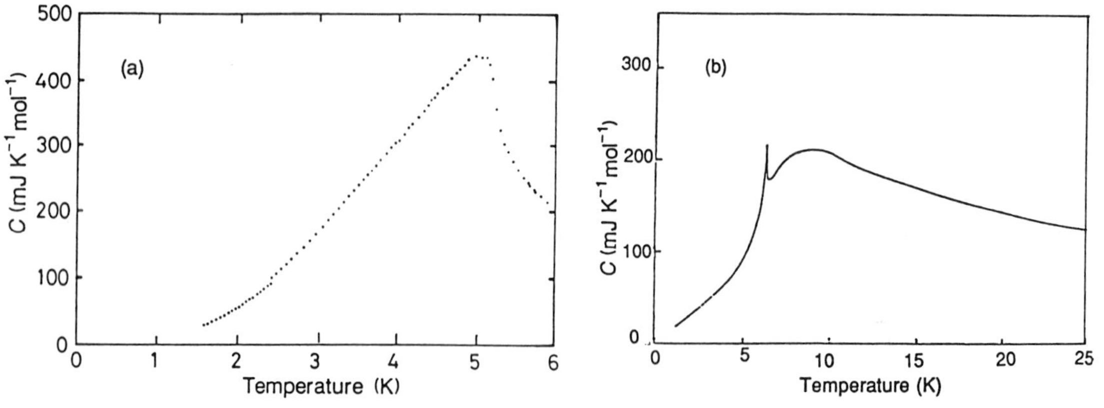

Fig 1 (a) The low-temperature specific heat for $La_4Ba_2Cu_2O_{10}$; (b) The magnetic specific heat C_m for K_2CuF_4 in ref. 8. The lattice contribution has been subtracted.

RESULTS AND DISCUSSION

LOW-TEMPERATURE SPECIFIC HEAT

The low-temperature specific heat was measured with an ordinary adiabatic DC method using a cylindrical sample in the temperature range a of1.5-6 K[7]. The specific heat of the ferromagnetic sample $La_4Ba_2Cu_2O_{10}$ is shown in Fig. 1(a). A very large magnetic specific heat is observed in the low-temperature region. The value of the specific heat has a peak at 5.2 K which corresponds to the ferromagnetic transition. The characteristic feature of the temperature dependence below T_c is consistent with that theoretically expected on the basis of the mean field approximation for a spin 1/2 three-dimensional ferromagnet. The temperature dependence of the magnetic specific heat curve due to the spin wave relates to the dimensionality of the spin system, which is expressed by $T^{-d/2}$, where d is a number of dimensionality. $T^{-d/2}$ dependence is not observed in this temperature region down to 2 K. The molar entropy for S=1/2 system is

$$\Delta S = R\ln(2S+1) = R\ln 2 = 5.763 \text{ J/K mol.} \tag{1}$$

The observed entropy change up to 5.2 K has been obtained to be 3.6 J/K mol, which corresponds to 63 % of the theoretical total magnetic entropy. The large Schottky type specific heat which relates to the short range correlation above T_c and the characteristic of low-dimensionality of the spin systems is not observed. The characteristic feature of the spin system is more three-dimensional-like. It is consistent with the results of the low-temperature specific heat which is well reproduced by a mean field approximation for three-dimensional S=1/2 system.

The magnetic specific heat of the typical two-dimensional Cu ferromagnet K_2CuF_4 which exhibits ferromagnetic transition at 6.25 K, is shown in the range 1.3-25 K in Fig. 1(b)[8]. The characteristic features of K_2CuF_4 indicate a typical low-dimensional property, namely, the λ shaped peak at 6.25 K, corresponding to T_c, and the Schottky type round peak at 9 K originating in the short range order. In the case of K_2CuF_4, the linear dependence of the magnetic specific heat on temperature is observed even in a few K. This result suggests that K_2CuF_4 belongs to the two-dimensional Heisenberg ferromagnet. The specific heat by the two-dimensional spin wave C_m is proportional to the temperature at T<<Tc with

$$C_m = (R/24S)(k/J)T, \tag{2}$$

where S is the quantum number of spin, and J is the magnitude of the exchange interaction within the layer, and k is the Boltzmann constant. The observed value corresponds to J/k=8.8 K. The magnetic entropy change below T_c is about 26 % of the theoretical total magnetic entropy and is 74 % above T_c. It shows that a great deal of short range order exists above T_c. From these results, it is reasonable to consider that K_2CuF_4 is a two-dimensional Heisenberg ferromagnet.

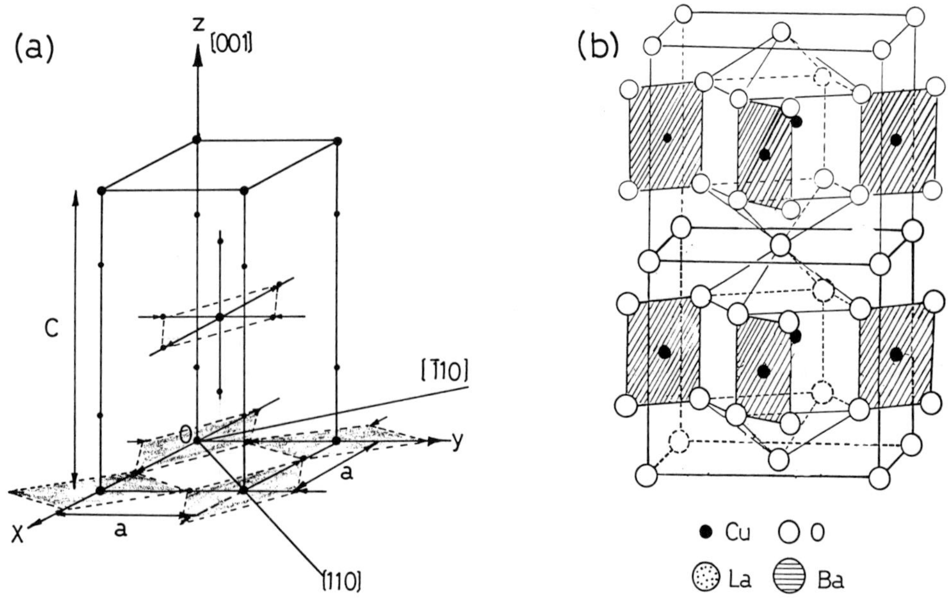

Fig. 2(a) The structure of K_2CuF_4. Only Cu^{2+} and F^- ions are represented. The arrows represent the displacements of the F^- ions from the middle of the a side, which are about 0.11Å; (b) The structure of the $La_4Ba_2Cu_2O_{10}$.

The characteristics of the spin system of $La_4Ba_2Cu_2O_{10}$ ferromagnet is more three-dimensional-like than that of typical two-dimensional Cu ferromagnet K_2CuF_4. It suggests that the interlayer interaction is comparably larger than that of K_2CuF_4 because both structural difference is reflected on dimensionality of the spin system. In the case of K_2CuF_4, Cu ion is not located at just above the Cu position in the under layer, but at the shifted position with 1/2 lattice periodicity toward each a- and b-direction as shown in Fig. 2(a). Therefore, the spin system in K_2CuF_4 tends to be two-dimensional-like. On the other hand, in $La_4Ba_2Cu_2O_{10}$, a Cu ion and another one in Cu layer just below belong to the same xy-coordinates as shown in Fig 2 (b). Therefore, interlayer Cu distance is smaller and the exchange interaction becomes larger.

MAGNETIC ANISOTROPY

It is important to investigate the magnetic anisotropy which offers easy magnetization axis (plane) for ferromagnet. One ion anisotropy does not exist because of such S=1/2 spin system as Cu^{2+} ion. Therefore, the spin system is fundamentally Heisenberg-like. Both the higher order anisotropic exchange interaction and the dipole-dipole interaction provide a small effective anisotropic energy in the vicinity of T_c and determine the direction of the ordered spins. In the case of the Cu^{2+} ferromagnetic system, both interactions contend with each other in the vicinity of T_c. The magnitude of the anisotropic exchange interaction energy can be estimated by $\lambda^2 J/\Delta^2$, where λ is a constant of the spin-orbit interaction, and Δ is a crystal field splitting energy. The isotropic exchange interaction energy, which is expressd by Eq. 3 with the value of T_c, is J/k=2.6 K.

$$T_c=2zJS(S+1)/3K, \quad (3)$$

where z is the coordination number of Cu^{2+}. The magnitude of the anisotropic exchange interaction energy is deduced with Eq. 3 and is as small as a few mK[4]. On the other hand, the magnitude of the dipole-dipole interaction energy can be estimated by Eq. 4

$$U=-(3M^2/r_{12}^3)(\cos^2\theta-1/3), \quad (4)$$

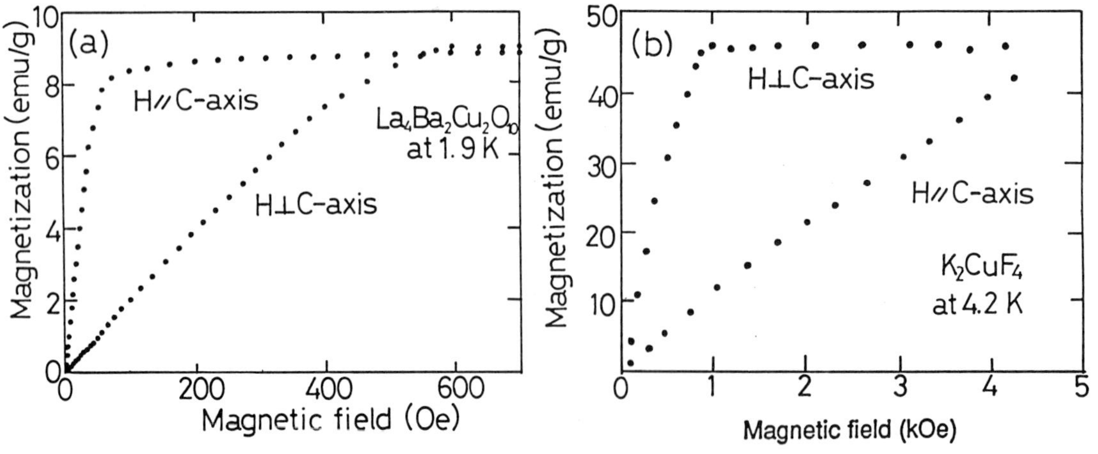

Fig. 3(a) Magnetization curves parallel and perpendicular to the c-axis at 1.9 K in $La_4Ba_2Cu_2O_{10}$; (b) Magnetizations curves parallel and perpendicular to the c-axis at 4.2 K in K_2CuF_4 from ref. 8.

where M is an effective Bohr magneton (1.66 μ_B/Cu), and r is a distance between Cu ions (4.84 Å) within a layer. The dipole-dipole interaction energy is order of 10 mK. The contribution of the dipole-dipole interaction is larger than that of the higher order anisotropic exchange interaction. In this case, if two-dimensionality of the spin system is strong enough, ab-plane would be an easy plane of the magnetization. Therefore, in K_2CuF_4 system XY-model holds at around T_c and Kosterlitz-Thouless transition has been observed[9].

To determine the magnetic anisotropy, the magnetizations parallel and perpendicular to the c-axis on the single crystal of the 422 phase were measured by SQUID magnetometer at 1.9 K. The direction of easy axis is parallel to the c-axis as shown in Fig. 3(a). The ferromagnetic saturation value of the magnetization reaches to 9.5 emu/g at 1.9 K. On the other hand, in the case of typical two-dimensional Cu ferromagnet K_2CuF_4, it seems obvious that the direction of easy axis is perpendicular to the c-axis as shown in Fig. 3(b). It is reasonable for each spin system, taking data of both low-temperature specific heat and magnetic anisotropy into consideration. In the 422 phase we focused on, the measured magnetizations and the estimated anisotropy energy with single crystal suggest that the spin system is a Heisenberg system with a small Ising-like anisotropy due to a comparably large interlayer coupling. The characteristics of the spin system is more three-dimensional-like than that of the typical two-dimenional Cu ferromagnet K_2CuF_4. It is consistent with the overall features of the low-temperature specific heat which is well reproduced by a mean field approximation for three-dimensional S=1/2 system.

REFERENCES

1. Mizuno F, Masuda H, Hirabayashi I, Tanaka S, Hasegawa M, Mizutani U (1990) Nature 345 : 788-789
2. Michel C, Er-rkho L, Raveau B (1981) J Solid State Chem 39: 161-167
3. Ogawa N, Mizuno F, Masuda H, Hirabayashi I, Tanaka S, Mochiku T, Asano H, Izumi F (1990) Physica B 165-166: 1687-1688
4. Masuda H, Mizuno F, Hirabayashi I, Tanaka S (1991) to be published in Phys. Rev. B
5. Yamada I (1970) J Phys Soc Japan 28:1585
6. Klibanow D, Sujata K, Mason TO (1988) J Am Ceram Soc 71:c-267-269
7. Mizuno F, Masuda H, Hirabayashi I, Tanaka S, Hasegawa M, Mizutani U, Mochiku T, Asano H, Izumi F (1991) to be published in Supercond Sci Technol
8. Yamada I (1972) J Phys Soc Japan 33:979-988
9. Hirakawa K (1982) J Appl Phys 53:1893-1898

Critical Fields of Anisotropic Superconductor Nb_3S_4

HIROYUKI OKAMOTO, YUTAKA ISHIHARA, and KITOMI TSUTSUMI

Faculty of Science, Kanazawa University, 1-1, Marunouchi, Kanazawa, Ishikawa, 920 Japan

ABSTRACT

On quasi-one dimensional Nb_3S_4 single crystal, we have measured the temperature dependence of the superconducting critical fields (H_{c1} and H_{c2}) close to T_c (3.8K). The result shows that Nb_3S_4 is an anisotropic type-II superconductor. Using an effective mass approach to the Ginzburg-Landau equations, we have obtained values for the coherence lengths and penetration depths. The ratio of the upper critical fields H_{c2} parallel and perpendicular to the c-axis is 3.86. The connection between these parameters and normal-state properties is discussed.

KEY WORDS: anisotropic superconductor, upper critical field, lower critical field

INTRODUCTION

The superconductivity of transition-metal chalcogenides with linear or layered structures has attracted considerable interest due to the marked anisotropy. Among them Nb_3S_4 is the first well-studied case of such a highly anisotropic superconductor. However, the measurement of the magnetization of Nb_3S_4 has not been made yet. Thus it is not clear whether Nb_3S_4 is a second kind superconductor. Nb_3S_4 was reported by Amberger et al. [1] as a superconducting compound of a quasi-one dimensional cluster. It's crystal symmetry is hexagonal [2]. The characteristic feature of the crystal is the existence of zig-zag Nb chains along the c-axis with intervening S atoms. The intrachain Nb-Nb distance is nearly equal to the corresponding value in Nb metal and the interchain Nb distance is larger than the corresponding value in Nb metal. As a consequence, quasi-one dimensional physical properties may be expected. However, the result of the anisotropy measurement carried out by Biebracher et al. [3] showed that Nb_3S_4 is an anisotropic but three dimensional superconductor.

In this paper we report the measurement of the magnetization of Nb_3S_4. Using the Ginzburg-Landau equations we have analyzed the anisotropic property in the superconductor. This approach is the only one that has received explicit theoretical attention in an effort to describe anisotropy. Discussions will be given with respect to the dimensionality of that compound.

EXPERIMENTAL

The crystals were prepared by the chemical vapor deposition method with the iodine as a transport agent following Amberger et al. [1]. The starting material is either elements of Nb or S in stoichiometric ratio or pressed pellets of sintered polycrystalline Nb_3S_4.

The starting material and the transport agent were sealed in a quartz ampoule 1.5cm in length after evacuation to 1×10^{-3} Torr. The ampoule was placed in a horizontal furnace. The temperature at the center of the furnace was kept at 1000℃ with temperature gradient of 3℃cm^{-1}. The location of the starting material was adjusted to come to the centre of the ampoule and at the same time to the centre of the furnace. Tiny needle-like single crystals up to several millimetres in length grew after a few weeks.

The crystal structure analysis has already been made by Ruysink et al. [2]. Nb_3S_4 belongs to the hexagonal crystal system. Based on their result the growth crystals were examined with the X-ray powder pattern analysis and identified to be Nb_3S_4. The needle-like crystals elongated always parallel to the crystallographic c-axis, which was confirmed with the X-ray oscillation photography method. The typical crystal habit is a hexagonal prism. The magnetization of Nb_3S_4 was measured by a Quantum Design MPMS SQUID magnetometer.

First we measured the field dependence of the magnetization in the temperature range close to T_c, where a Ginzburg-Landau analysis is most appropriate. The sample was mounted on the quartz tube sample-holder. The sample could be mounted either parallel or perpendicular to the applied field. The sample was first heated to a temperature well above T_c in order to remove the trapped flux. After cooling the sample to a given temperature below T_c, a magnetization curve was traced out. The lower critical field H_{c1} was determined by an onset of the deviation from perfect diamagnetization. The upper critical fields H_{c2} were determined as the intercept of the magnetization curves with the magnetic field axis.

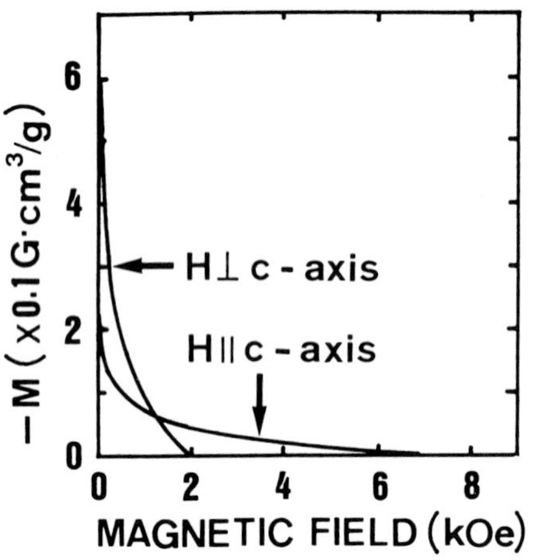

Fig.1. The magnetic field dependence of the magnetization M at 2.4K. The indices ⊥ and ∥ refer to the field direction perpendicular and parallel to the crystal c-axis, respectively.

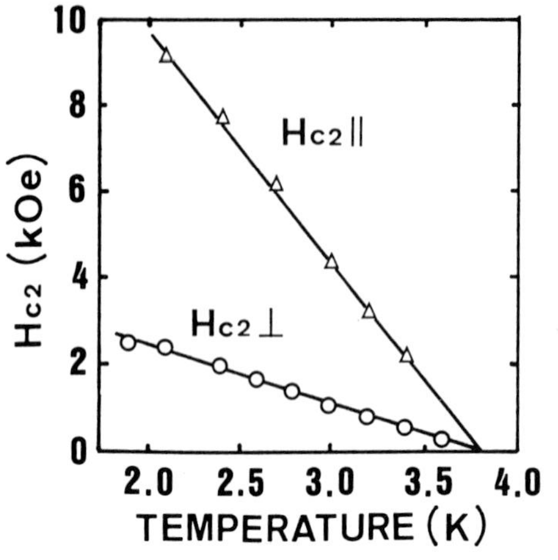

Fig.2. The temperature dependence of the upper critical fields $H_{c2\parallel}$ (△) and $H_{c2\perp}$ (○).

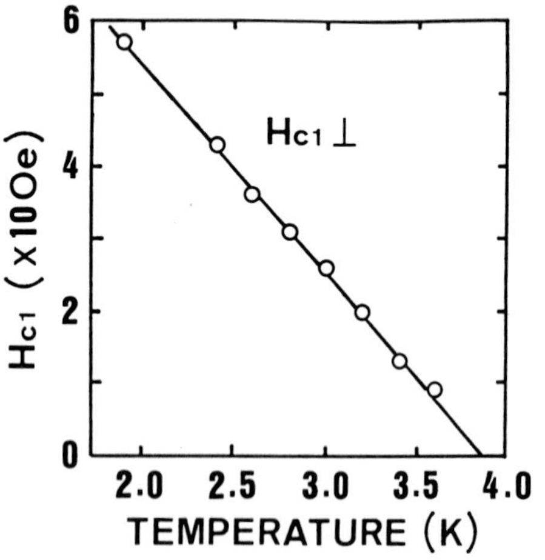

Fig.3. The temperature dependence of the lower critical field $H_{c1\perp}$.

TABLE I . List of the superconducting transition temperature T_c, anisotropy parameter ε, anisotropic Ginzburg- Landau parameters κ, coherence lengths ξ, and penetration depths λ for Nb_3S_4.

T_c	3.83 K
$\varepsilon = (m\perp/m\parallel)^{1/2}$	3.86
$\kappa\perp = 6.81$	$\kappa\parallel = 26.3$
$\xi\perp = 126$ Å	$\xi\parallel = 486$ Å
$\lambda\perp = 3310$ Å	$\lambda\parallel = 858$ Å

RESULTS

Figure 1 shows the magnetic field dependence of the magnetization of Nb_3S_4 at 2.4K (the indices ⊥ and ∥ refer to the field direction perpendicular and parallel to the crystal c-axis, respectively). The straight line of perfect diamagnetism is practically vertical in such a trace. Figure 2 shows the temperature dependence of $H_{c2\perp}$ and $H_{c2\parallel}$ of Nb_3S_4. Both the critical fields are almost proportional to $\Delta T=T_c-T$ in the range from 1.9K to 3.7K measured. The extrapolation to applied magnetic field H=0 yields T_c=3.80K. The anisotropy parameter $\varepsilon=H_{c2\parallel}/H_{c2\perp}$ is 3.86. Figure 3 shows the temperature dependence of $H_{c1\perp}$.

DISCUSSION

The experimental result shows that Nb_3S_4 is an anisotropic type-II superconductor. An effective-mass tensor in the Ginzburg-Landau equations has been used in order to describe anisotropy[4,5]. The effective-mass tensor can be written as

$$1/m_{ij}=1/(N2\pi^2h)\int v_iv_j\phi(k)^2dS/v_F \quad (1)$$

where N is the electron density, v_F is the Fermi velocity and the integral has to be taken over the Fermi surface. Without the gap function $\phi(k)$ this effective mass would simply be the effective mass appearing in the normal-state conductivity tensor σ_{ij}.

The anisotropy in the magnetic behavior of anisotropic superconductor is most conveniently described by an anisotropic Ginzburg-Landau parameter κ [6]. Experimentally, κ can be found by measuring the magnetic properties, in particular the critical fields. Only near T_c [5],

$$\kappa_\parallel=(1/\sqrt{2})(H_{c2\parallel}/H_c) \quad , \quad (2)$$

$$H_{c1\parallel}=(H_c/\sqrt{2})(\ln\kappa_\parallel/\kappa_\parallel) \quad \text{for } \kappa_\parallel>>1 \quad , \quad (3)$$

which can be written as

$$[(dH_{c2\parallel}/dT)/(dH_{c1\parallel}/dT)]_{Tc}=2\kappa_\parallel^2/\ln\kappa_\parallel \quad . \quad (4)$$

The value of $\kappa_\parallel$ obtained using Eq. (4) and $\varepsilon=H_{c1\perp}/H_{c1\parallel}$ is 26.3. The critical field $H_{c2\parallel}$ is related to the coherence length $\xi_\perp$ by

$$H_{c2\parallel}=\phi_o/(2\pi\xi_\perp^2) \quad , \quad (5)$$

where ϕ_o is the flux quantum. There we have the usual temperature dependence $\xi(T)=\xi(0)(1-T/T_c)^{-1/2}$, so that the coherence length $\xi_\perp(0)$ at 0K is related with the slope of $H_{c2\parallel}$ vs T using Eq. (5):

$$[dH_{c2\parallel}/dT]_{Tc}=-\phi_o/(2\pi\xi_\perp(0)^2T_c) \quad . \quad (6)$$

The coherence lengths $\xi_\perp(0)$ obtained from Eq. (6) is $\xi_\perp(0)$=486A. The penetration depth $\lambda_\perp$ can be written as

$$\lambda_\perp=\xi_\perp\kappa_\parallel \quad . \quad (7)$$

The anisotropy of these parameters is determined by ε :

$$\kappa_\perp=(1/\varepsilon)\kappa_\parallel \quad , \quad (8)$$

$$\xi_\parallel=\varepsilon\,\xi_\perp \quad , \quad (9)$$

$$\lambda_\parallel=(1/\varepsilon)\lambda_\perp \quad . \quad (10)$$

In Table I we have listed these parameters obtained. As can be seen from Table I the coherence length perpendicular to the chains extends over several chains, indicating good three-dimensional superconductivity.

In isotropic superconductors the microscopic theories able to link the superconductivity parameters with normal-state properties, such as density of states and Fermi velocity. It is clear from the Ginzburg-Landau equations that the situation is more complex. The anisotropy in the resistivity of Nb_3S_4 has been measured [7] and the result shows that the resistivity anisotropy $\rho\perp/\rho\parallel$ is about 15 over the temperature region from 50K to room temperature. The agreement between the superconductivity mass anisotropy $m\perp/m\parallel=\varepsilon^2=14.9$ and the resistivity anisotropy suggests that the inclusion of the $\phi(k)$ in Eq. (1) is not all that important. To ignore $\phi(k)$ completely and to obtain some estimate of averaged Fermi velocities in the two directions would be desirable and also possible if the Fermi surface were simple. But the Fermi surfaces proposed by Oshiyama shows three sheets, and it is not clear how they contribute to conductivity [8]. It is therefore not at all easy to extract Fermi velocity averages in each Fermi surface from superconductivity measurements.

ACKNOWLEDGMENTS

The authors thank T.Takesima for helpful discussions.

REFERENCES

1. E.Amberger, K.Polborn, P.Grimm, M.Dietrich & B.Obst (1978) Solid State Commun. **26**: 943

2. A.F.Ruysink, F.Kadijk, A.J.Wagner & F.Jellinek (1968) Acta Cryst. **B24**: 1614

3. W.Biberacher & H.Schwenk (1980) Solid State Commun. **33**: 385

4. R.C.Morris & R.V.Coleman (1973) Phys. Rev. **B7**: 991

5. P.de Trey & Suso Gygax & J.-P.Jan (1973) J. Low Temp. Phys. **11**: 421

6. D.R.Tilley (1965) Proc. Phys. Soc. (London) **85**: 1177

7. Y.Isihara & I.Nakada (1983) Solid State Commun. **45**: 129

8. A.Oshiyama (1982) Solid State Commun. **43**: 607

Magnetic and Transport Properties of the Ba(V,M)S3 [M+Ti,Nb] Systems

KIYOTAKA MATSUURA, TAKAHIRO WADA, TAKASHI NAKAMIZO, H. YAMAUCHI, and SHOJI TANAKA

Superconductivity Research Laboratory, International Superconductivity Technology Center, 10-13, Shinonome 1-chome, Koto-ku, Tokyo, 135 Japan

ABSTRACT

$BaVS_3$ shows a metal-to-semiconductor transition and a para-to-anti-ferromagnetic transition around 70 K. Upon substituting a small amount (5%) of Ti or Nb for V in $BaVS_3$, the electrical conduction above 70 K became semiconductive, and the magnetism became paramagnetic throughout the range 4.2 to 300 K. $BaNbS_3$ was found to show slight diamagnetism below 300 K and a metal-to-semiconductor transition around 620 K. These phenomena are thought to be caused by a charge density wave.

KEY WORDS: $BaVS_3$, $BaNbS_3$, transport property, magnetic property, charge density wave

INTRODUCTION

$BaVS_3$ is known to have a quasi-one-dimensional anisotropic crystal structure, characterized by linear chains of V^{4+} which has a $3d^1$ electron configuration running parallel to the c axis as shown Fig. 1 [1]. $BaVS_3$ has a structural phase transition from hexagonal to orthorhombic around 250 K, and shows a metal-to-semiconductor transition and a para-to-anti-ferromagnetic transition around 70 K [1-5]. Recently, we have observed that $BaVS_3$ has a metal(n-type)-to-semiconductor(p-type) transition around 70 K in the thermoelectric power measurement [6]. In this work, we study the effect of Ti(with $3d^0$) and Nb(with $4d^1$) substitution for V in the $BaVS_3$ compound on the magnetic and transport properties.

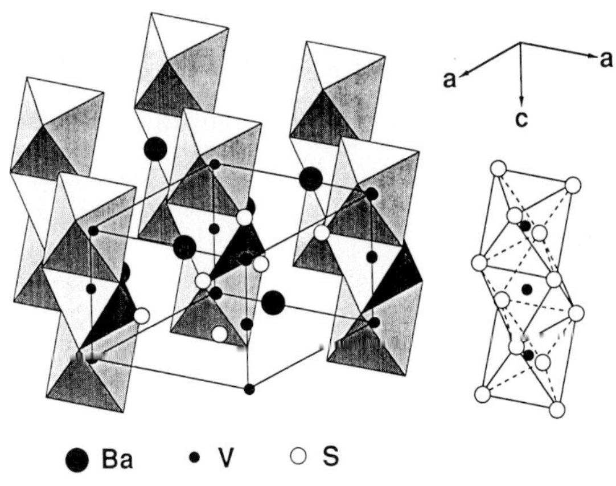

Fig. 1. Structure of $BaVS_3$

EXPERIMENTAL

$BaV_{1-x}Ti_xS_3(0.0\leq x\leq 0.2)$ and $BaV_{1-x}Nb_xS_3(0.0\leq x\leq 1.0)$ ceramics were synthesized by the following procedure. Highly pure powders of $BaCO_3$, V_2O_5 and TiO_2 or Nb_2O_5 were mixed in ethanol for the cation ratio of Ba : $(V_{1-x}M_x)$ = 1 : 1, where M=Ti and Nb. The mixed powder was calcined at 600 - 700 °C for 5 hours in H_2S-Ar mixture gas flow and then slowly cooled to room temperature. The resultant powder was ground, and the calcination was repeated three times. The calcined powder was pressed into pellets which were again calcined at 660 °C for 5 hours in H_2S-Ar mixture gas flow and then slowly cooled to room temperature. The pellets (except the $BaNbS_3$ sample) were then heated in evacuated quartz ampoules at 800 °C for 20 hours and then slowly cooled to room temperature. Electrical resistivity was measured using a standard four-probe technique. DC magnetic susceptibility was measured using a SQUID magnetometer (Quantum Design : Model MPMS). Thermoelectric power measurements were made in the temperature range of 40 K to room temperature by applying a temperature difference of ~ 0.3 K.

RESULTS AND DISCUSSION

The prepared compounds, i.e $BaV_{1-x}Ti_xS_3$ and $BaV_{1-x}Nb_xS_3$, had a quasi-one-dimensional anisotropic structure similar to $BaVS_3$. The temperature dependence of electrical resistivity of $BaV_{1-x}Ti_xS_3$ ($0.0 \leq x \leq 0.2$) and $BaV_{1-x}Nb_xS_3$ ($0.0 \leq x \leq 0.9$) are shown in Fig. 2. Even with a small amount (5%) of Ti or Nb substitution, the metallic portion disappeared and, for $x \geq 0.05$, $BaV_{1-x}Ti_xS_3$ exhibited semiconductive behavior in the temperature range of 4.2 K to 300 K. We suspect that this change was caused by an Anderson localization [7], as indicated by negative magnetoresistance.

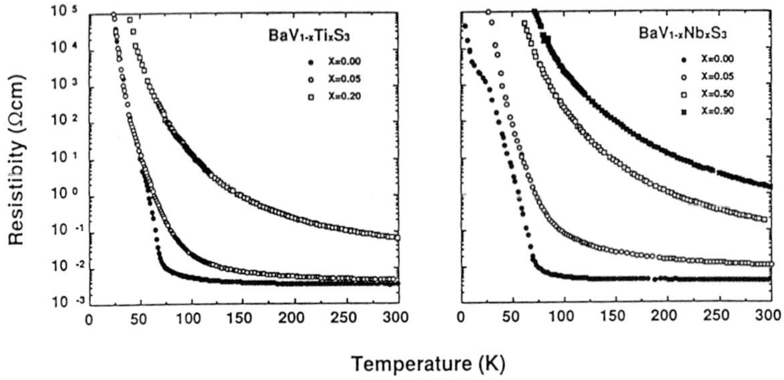

Fig. 2. Temperature dependence of electrical resistivity for $BaV_{1-x}Ti_xS_3$ and $BaV_{1-x}Nb_xS_3$

The temperature dependence of the Seebeck coefficient for $BaV_{0.8}Ti_{0.2}S_3$ is shown in Fig. 3. As temperature decreased, $BaV_{0.8}Ti_{0.2}S_3$ exhibited semiconductive behavior in the temperature ranges between 300 K and 150 K and also between 70 K and 40 K. If a single band model was employed, the carriers were electrons in the temperature range of 300 K to 150K, and holes in the temperature range of 70 K to 40 K. In the middle temperature range (i.e. between 150 K and 70 K), two kinds of carriers could coexist. The Seebeck data were thus not contradictory to the resistivity result shown in Fig. 2 that $BaV_{0.8}Ti_{0.2}S_3$ was semiconductive throughout the temperature range between 40 K and 300 K.

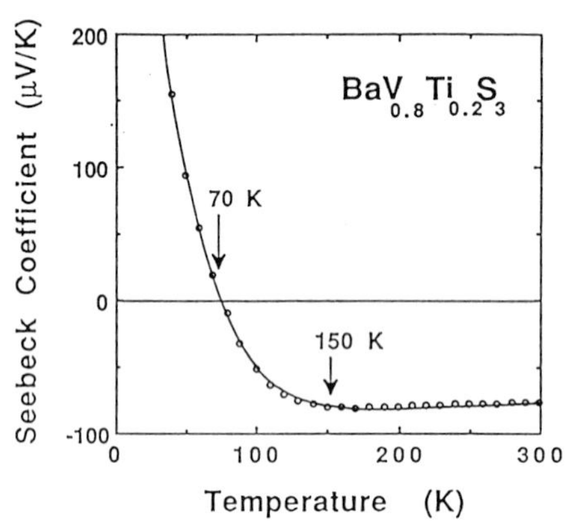

Fig. 3. Temperature dependence of the Seebeck coefficient for $BaV_{0.8}Ti_{0.2}S_3$

The temperature dependence of dc magnetic susceptibility of $BaV_{1-x}Ti_xS_3$ ($0 \leq x \leq 0.2$) and $BaV_{1-x}Nb_xS_3$ ($0.0 \leq x \leq 0.9$) is shown Fig. 4. The antiferromagnetism seems to be destroyed even by 5 % Ti or Nb substitution for V, and the samples with $x \geq 0.05$ became paramagnetic throughout the temperature range between 4.2 K and 300 K. By assuming that the susceptibility curve obeys the Curie-Weiss law in the high temperature range, the spin value for the V site was obtained in terms of the Ti and Nb content, x, as shown in Fig. 5. In this calculation, the value of 2.0 was employed for Lande's g-factor because there were no contributions from the orbital magnetic moment in the 3d transition metals. For the sample with x=0.0, the spin value was obtained to be 0.31. This value is 62 % of the spin when all the V^{4+} ions were assumed to be isolated, i.e. s=1/2. In Ti doped samples, as the Ti content increased, the spin value for the V site monotonically decreased and became nearly zero at x=0.4. On the other hand, in Nb doped samples, as the Nb content increased to about 0.4, the spin value was almost constant, and further increased the Nb content, the spin value monotonically decreased and became nearly zero at x=0.9.

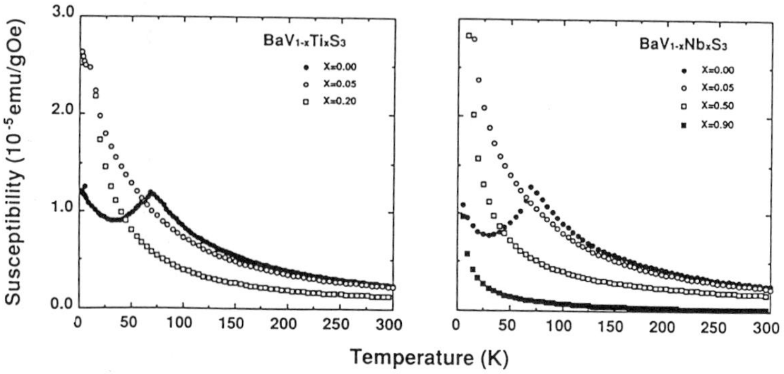

Fig. 4. Temperature dependence of magnetic susceptibility for $BaV_{1-x}Ti_xS_3$ and $Ba_{1-x}Nb_xS_3$

The dependence of the spin value on the Ti and Nb(x< 0.4) content may be explained as follows using the model proposed by O. Massenet et al [5], as schematically shown in Fig. 6. First, we understood that about 62 % of electrons are in the d_{xy} and $d_{x^2-y^2}$ band of the sample with x=0.0. By substituting progressively Ti^{4+} which have no d electrons, for the V site of $BaVS_3$, the number of d electrons decreased and the Fermi level went down. Consequently the spin value was decreased. As exceeds than 0.4, the spin value becomes zero when the Fermi level becomes located under the d_{xy} and $d_{x^2-y^2}$ band. On other hand, since Nb^{4+} has one d electron, the number of d electron for V site was constant and consequently the spin value was constant. However, it is impossible with this model to explain how the spin value of $BaV_{1-x}Nb_xS_3$ decreased as the Nb content, x, increased for x larger than 0.4. We suspect that the decrease of spin value in $BaV_{1-x}Nb_xS_3$ was induced locally by a charge density wave which is shown in the following analysis for $BaNbS_3$.

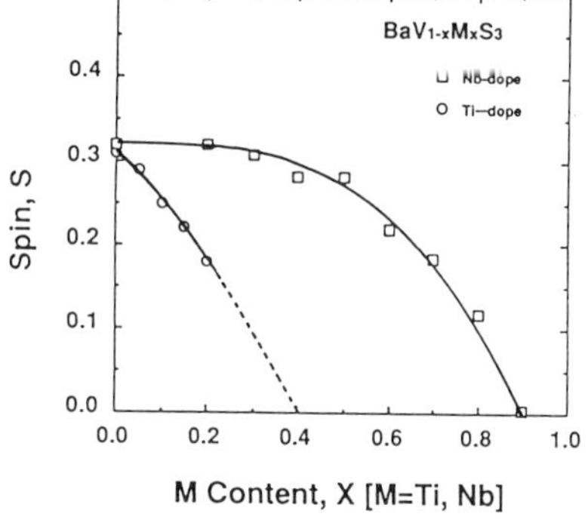

Fig. 5. Variation of spin at the V site in M-doped (M=Ti and Nb) $BaVS_3$ as a function of M content, x

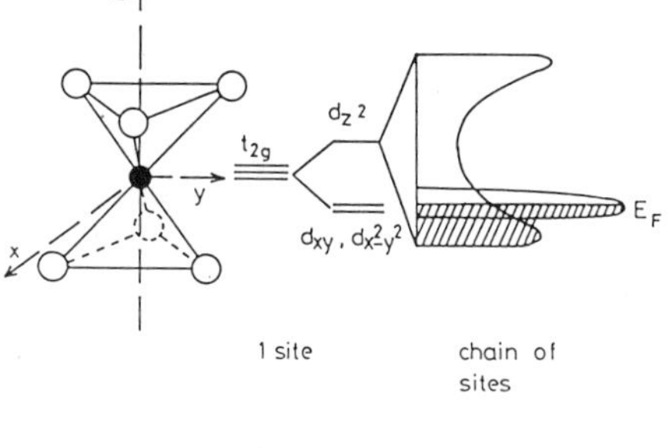

Fig. 6. Massenet et al's model (Ref. 5) for nearest neighbor sulfur environment at a vanadium site and corresponding electron level diagrams for one vanadium site as well as for a chain of vanadium sites: ●, Vanadium atom; ○, Sulfur atom

The temperature dependence of electrical resistivity of $BaNbS_3$ is shown in Fig. 7. Above 620 K, the temperature dependence was metallic, and below 620 K, it exhibited semiconductive behavior. Using an Arrhenius plot, the electrical conduction was of thermal activation type below 400 K, and the activation energy was estimated to be 220 meV. The thermoelectic power of $BaNbS_3$ exhibited n-type semiconductive behavior, and the Seebeck coefficient at 300 K was about -150 μ V/K . From X-ray Photoelectron spectroscopy for $BaNbS_3$, The Nb $3d_{5/2}$ is essentially equal to that of NbO_2, i.e. 205 eV. This result suggests that the Nb in $BaNbS_3$ is tetra valent. If a Nb ion in $BaNbS_3$ is tetra valent, we usually expect that $BaNbS_3$ has spin s=1/2 and is metallic in its transport property. Actually, $BaNbS_3$ showed a semiconductive behavior below 620 K, and the temperature dependence of dc magnetic susceptibility of $BaNbS_3$ indicated slightly diamagnetism below 300 K as shown in Fig. 8. This magnetic moment was about -1.6×10^{-7} emu/gOe, which was of the same order of the moment when all ions of $BaNbS_3$ were assumed to have closed shells. This indicates that a metal-to-semiconductor transition of $BaNbS_3$ might to be caused by a charge density transition.

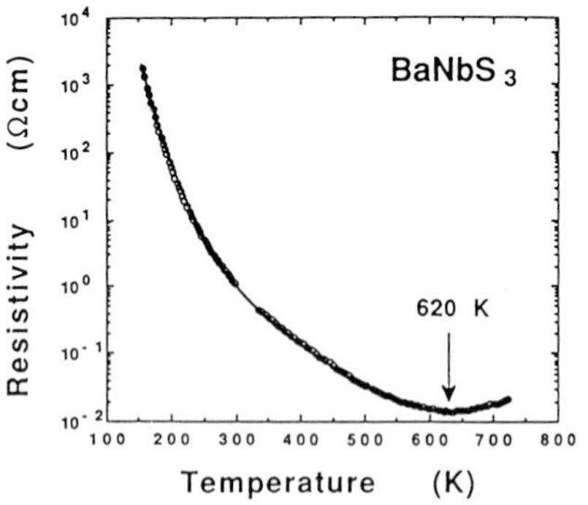

Fig. 7. Temperature dependence of electrical resistivity for $BaNbS_3$

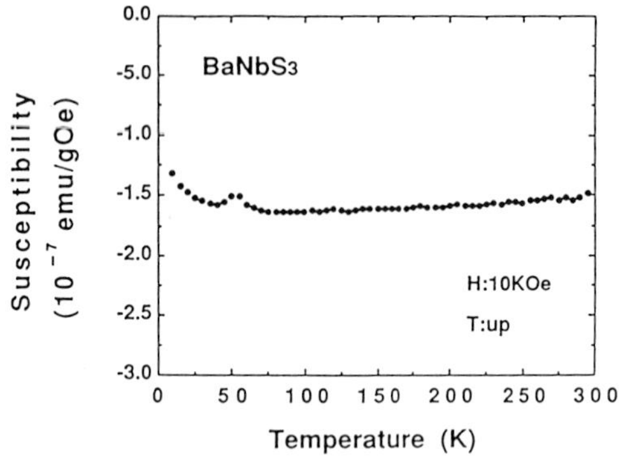

Fig. 8. Temperature dependence of magnetic susceptibility for $BaNbS_3$

In summary, the effect of Ti and Nb substitution for V in the $BaVS_3$ compound has been studied with respect to magnetic and transport properties. Upon substituting a small amount (5%) of Ti or Nb for V in $BaVS_3$, the electrical conduction above 70 K became semiconductive, and the magnetism became paramagnetic throughout the range 4.2 to 300 K. We observed a different substitution effect for V in $BaVS_3$ between $BaV_{1-x}Ti_xS_3$ and $BaV_{1-x}Nb_{1-x}S_3$ from susceptibility measurement. $BaNbS_3$ shows slight diamagnetism below 300 K and a metal-to-semiconductor transition around 620 K. We interpreted these phenomena as being caused by a charge density wave.

ACKNOWLEDGMENT

We are grateful to Prof. T. Fujiwara of the University of Tokyo, Dr. T. Ito and K. Kubo of SRL-ISTEC for their helpful discussions. We would also like to thank Prof. S. Uchida, Dr. H. Takagi and H. Eisaki of the University of Tokyo and T. Kawano of SRL-ISTEC for their help in the Seebeck measurement. We are indebted F. Munakata for his kind help in the high-temperature electrical resistivity measurement.

REFERENCES

1. Gardner R A, Vlasse M, Wold A (1969) Acta Cryst.B25: 781
2. Takano M, Kosugi H, Nakanishi N, Shimada M, Wada T, Koizuni M (1977) J.Phys.Soc.Japan.43: 1101
3. Massenet O, Buder R, Since J J, Sohlenker C, Mercier J, Kelber J, Stucky G D (1978) Mater.Res.Bull.13: 187
4. Wada T, Simada M, Koizumi M (1979) J.Solid State Chem.33: 357
5. Massenet O, Since J J, Mercier J, Avignon M, Buder R, Nguyen V D, Kelber J (1979) J.Phys.Chem.Solids 40: 573
6. Matsuura K, Wada T, Nakamizo T, Yamauchi H, Tanaka S (submitted) Phy.Rev.B
7. Anderson P W (1958) Phys.Rev.109: 1492

3 Chemistry

3.1 Material Synthesis

Superconducting Lead Cuprates Synthesized in Oxidizing Atmosphere

T. MAEDA, K. SAKUYAMA, H. YAMAUCHI, and S. TANAKA

Superconductivity Research Laboratory, 10-13, Shinonome 1-chome, Koto-ku, Tokyo, 135 Japan

ABSTRACT

Two kinds of superconducting lead cuprates, $(Pb,Cu)Sr_2(Y,Ca)Cu_2O_7$ (the "1212" phase) and $(Pb,Cu)(Sr,Eu)_2(Eu,Ce)_2Cu_2O_9$ (the "1222" phase), are synthesized and characterized. Although most of previously-known Pb-based cuprate superconductors form in a mildly reducing atmosphere, these two lead cuprates form in an oxidizing atmosphere. It is shown that single-phase ceramics of the "1212" lead cuprates are obtained for the chemical compositions of $(Pb_{(1+x)/2}Cu_{(1-x)/2})Sr_2(Y_{1-x}Ca_x)Cu_2O_z$ ($x = 0 - 0.35$ and $z = \sim 7$). The Pb-based "1222" compounds are the first "1222" superconductors and also the first superconducting lead cuprates containing "fluorite-type" block layers. The maximum values of the observed superconducting onset temperatures are 65 K and 28 K for the "1212" phase and for the "1222" phase, respectively.

KEY WORDS: high temperature superconductivity, superconducting lead cuprates, "1212" structure, "1222" structure, oxidizing atmosphere

INTRODUCTION

Since the first superconducting lead cuprate, $Pb_2Sr_2(Y,Ca)Cu_3O_8$ of the "2213" phase, was discovered by Cava et al.,[1] a variety of lead cuprates have been synthesized,[2-12] some of which exhibited superconductivity. These lead cuprates had crystallographic structures similar to those of Bi- and Tl-based superconducting cuprates and characteristically contained a block layer consisting of both Pb and O together with Cu. Crystallographically, such layers correspond to those consisting of Bi and O in the Bi-based superconducting cuprates or to those consisting of Tl and O in the Tl-based ones.

In most of those lead cuprates, typically in the "2213" phase which were reported to form in a mildly reducing atmosphere,[1] Pb and Cu constituting the block layer together with O were considered to be in the divalent state and in the monovalent state, respectively. Recently, we synthesized[13 15] superconducting lead cuprates, $(Pb,Cu)Sr_2(Y,Ca)Cu_2O_z$ ($z = \sim 7$, the "1212" phase) and $(Pb,Cu)(Sr,Eu)_2(Eu,Ce)_2Cu_2O_z$ ($z = \sim 9$, the "1222" phase) in an oxidizing atmosphere in contrast to the case of the "2213" phase. These two compounds contained (Pb,Cu)O monolayers in which Pb and Cu were thought to be respectively in the tetravalent state and in the divalent state. In this paper, synthesis and characterization of the two kinds of recently discovered lead cuprate superconductors are reported.

EXPERIMENTAL PROCEDURES

Samples were prepared by a solid state reaction using high-purity (> 99.9 %) powders of PbO, $SrCO_3$, Y_2O_3, $CaCO_3$ and CuO for the "1212" phase and PbO, $SrCO_3$, Eu_2O_3, CeO_2 and CuO for the "1222" phase. These powders were mixed into nominal compositions of $(Pb_{(1+x)/2}Cu_{(1-x)/2})Sr_2(Y_{1-x}Ca_x)Cu_2O_z$ ($x = 0 - 0.5$) or $(Pb_{0.5}Cu_{0.5})(Sr_{0.875}Eu_{0.125})_2(Eu_{0.75}Ce_{0.25})_2Cu_2O_z$. The mixed powders were calcined in air at 850 ℃ for 10 h and then pressed into parallelepiped bars of 3 x 3 x 20 mm^3. The bars were fired in O_2 gas flow at 1050 ℃ ("1212") or in air at 1080 ℃ ("1222") for 1 h, followed by rapid cooling to 900 ℃ and further slow cooling to room temperature at a rate of 60 deg/h. Then as-sintered "1212" samples were annealed at 800 ℃ for 1 h in air and then quenched into liquid nitrogen, while as-sintered "1222" samples were heat-treated in an O_2 atmosphere. The temperature-profile during the heat-treatment has been shown elsewhere.[15] The samples were characterized by an x-ray diffractometer equipped with a graphite monochrometer using CuKα radiation. Both electrical resistivity and dc-magnetic susceptibility were measured with respect to temperature by means of respectively a conventional four probe method and a SQUID magnetometry.

RESULTS AND DISCUSSIONS

(Pb,Cu)Sr_2(Y,Ca)Cu_2O_z (the "1212" phase)

Figure 1 schematically shows the crystallographic structure of the Pb-based "1212" phase, which has a tetragonal unit cell with the lattice constants, a = ~ 3.8 Å and c = ~ 11.8 Å. It is similar to but slightly different from that of $Pb_2Sr_2(Y,Ca)Cu_3O_8$[1] and also that of the "123" superconductor, $YBa_2Cu_3O_7$. X-ray powder diffraction (XRD) patterns for as-sintered samples with x = 0 - 0.4 are shown in Fig. 2. For x up to 0.35, single-phase samples were obtained. The XRD pattern for a quenched sample was practically the same as that of corresponding as-sintered sample except for slight shifts of diffraction lines due to the change in lattice constants induced by quenching.

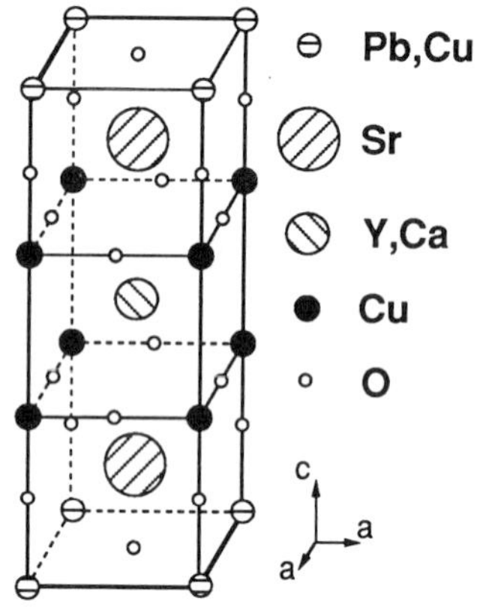

Fig. 1. Crystal structure of the "1212" phase.

Figures 3(a) and 3(b) show the temperature dependence of electrical resistivity of as-sintered and quenched samples with x = 0 - 0.4, respectively and Fig. 3(c) shows that of quenched sample with x = 0.5. For as-sintered samples, electrical resistivity tended to decrease as the Ca content, x, increased, but all samples did not exhibit superconductivity. Moreover, electrical resistivity was lowered by quenching. It also tended to decrease as x increased for quenched samples. Finally, superconductivity appeared only for quenched samples with x > 0.25. The superconducting onset temperatures (T_c^{on}) for the quenched samples with x = 0.3, 0.35, 0.4 and 0.5 were 25 K, 45 K, 52 K and 65 K, respectively.

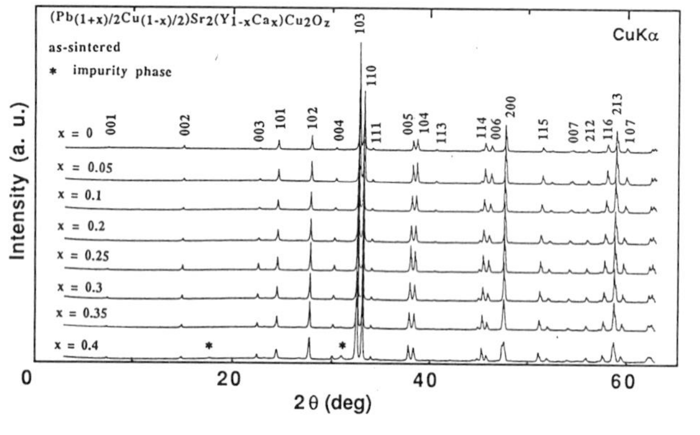

Fig. 2. XRD patterns for the samples with x = 0 - 0.4.

The dependences of the lattice constants, a and c/3, on the Ca content, x, for both as-sintered and quenched samples with x = 0 - 0.35 are shown in Fig. 4. The a-constant was nearly independent of x for both as-sintered and quenched samples. For all x, however, the a-constants were shrunken after quenching from ~ 3.818 Å to ~ 3.815 Å. In contrast, the c-constant elongated as x increased and remained unchanged after quenching. Using a coulometric titration technique,[17,18] the oxygen content, z, was determined for as-sintered (non-superconducting) samples and for quenched (superconducting) samples with x = 0.3 to be respectively 7.10(3) and 6.99(3). The latter value was practically equal to the stoichiometric value of 7.0. It is strongly suggested that the excess amount of oxygen contained in as-sintered samples was unfavorable for the appearance of superconductivity.

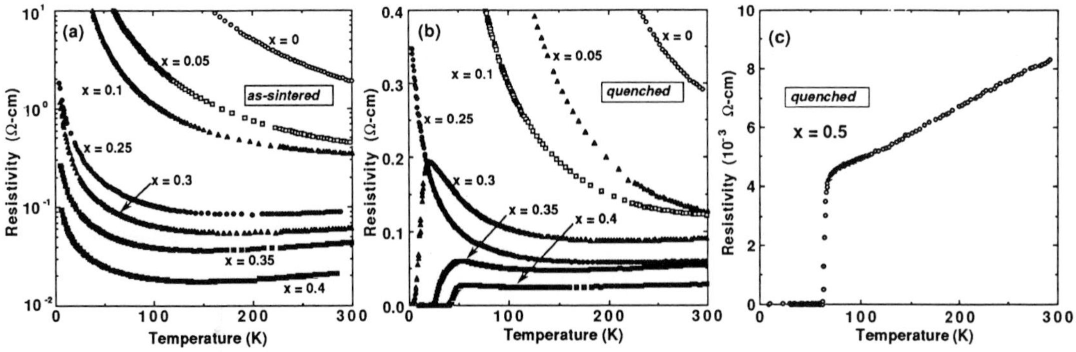

Fig. 3. Temperature dependence of electrical resistivity of (a) as-sintered samples, (b) quenched samples with x = 0 - 0.4 and (c) quenched samples with x = 0.5.

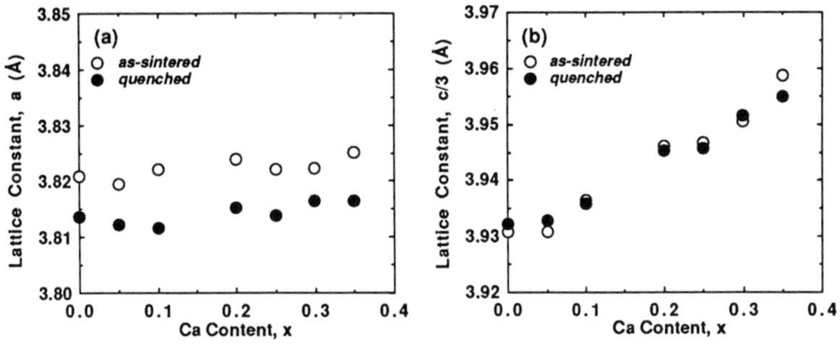

Fig. 4. Dependence on Ca content, x, of lattice constants, a and c/3.

(Pb,Cu)(Sr,Eu)$_2$(Eu,Ce)$_2$Cu$_2$O$_z$ (the "1222" phase)

Compounds having the "1222" structure were found for the first time in Tl-Sr-Nd-Ce-Cu-O system,[19] and later in Pb-Sr-Pr-Cu-O system.[11] However, these compounds were reported to be non-superconducting. We successfully synthesized superconducting Pb-based "1222" phase in Pb-Sr-Eu-Ce-Cu-O system,[15] which was also the first superconducting lead cuprate containing "fluorite-type" block layers. Figure 5 schematically shows the "1222" structure. This structure contains (Pb,Cu)O monolayers with "rock-salt-type" configuration instead of the BiO and TlO bilayers contained respectively in Bi- and Tl-based "2222" phases which were recently synthesized.[20] The XRD pattern of as-sintered sample with a nominal composition of (Pb$_{0.5}$Cu$_{0.5}$)(Sr$_{0.875}$Eu$_{0.125}$)$_2$(Eu$_{0.75}$Ce$_{0.25}$)$_2$Cu$_2$O$_z$ is shown in Fig. 6, which indicates that the sample was of nearly single "1222" phase. The x-ray Rietveld analysis suggested that the chemical formulae of the Pb-based "1222" phase should be written as (Pb,Cu)(Sr,Eu)$_2$(Eu,Ce)$_2$Cu$_2$O$_z$.[21] Figure 7 shows the temperature dependence of electrical resistivity for as-sintered and O$_2$-annealed samples. A superconductivity transition at ~ 25 K was observed only for the O$_2$-annealed sample. The temperature dependence of dc-magnetic susceptibility for the superconducting sample is also shown in Fig. 7, which comfirmed bulk-superconductivity below ~ 25 K. Considering that only the O$_2$-annealed sample exhibited superconductivity, the Pb-based "1222" compounds were thought to be in oxygen deficient state in contrast to the "1212" compounds which tended to contain excess amount of oxygen.

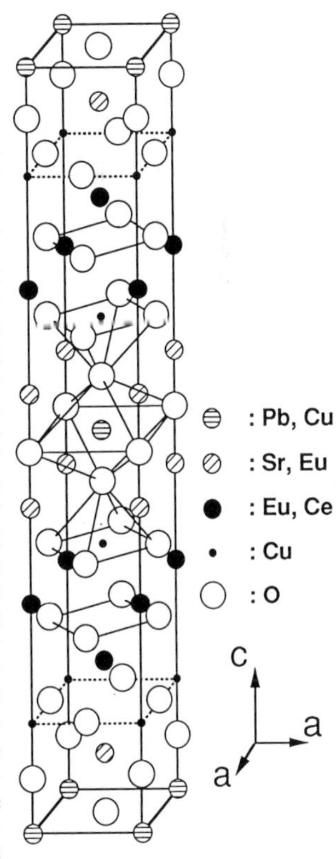

Fig. 5. Crystal structure of the "1222" phase.

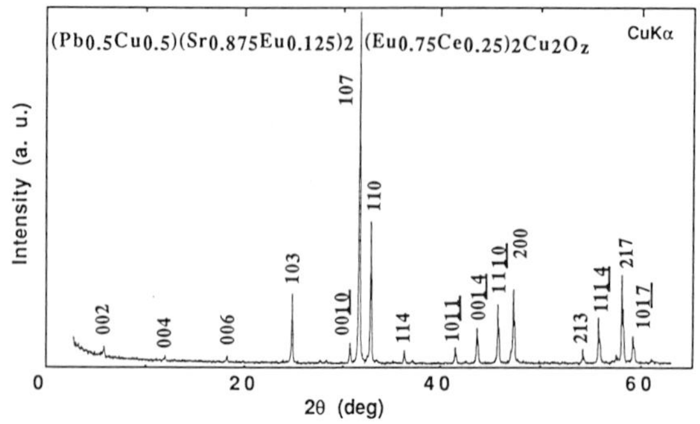

Fig. 6. XRD patterns of the Pb-based "1222" phase.

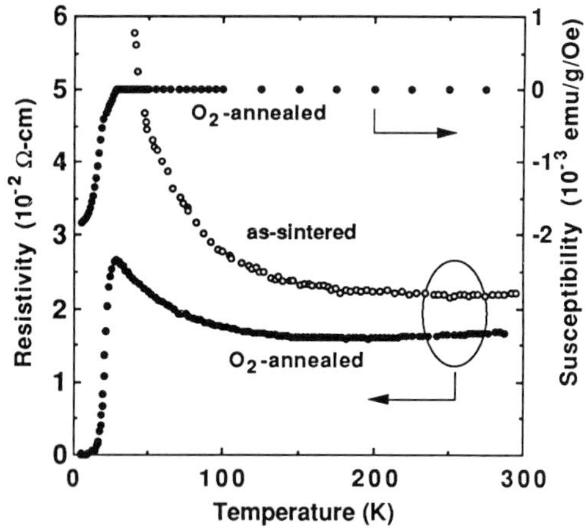

Fig. 7. ρ-T and χ-T curves.

SUMMARY

Superconducting lead cuprates, $(Pb_{(1+x)/2}Cu_{(1-x)/2})Sr_2(Y_{1-x}Ca_x)Cu_2O_z$ (the "1212" phase) and $(Pb_{0.5}Cu_{0.5})(Sr_{0.875}Eu_{0.125})_2(Eu_{0.75}Ce_{0.25})_2Cu_2O_z$ (the "1222" phase), were synthesized and characterized. For the "1212" phase, single-phase ceramics were obtained for x = 0 - 0.35 and superconductivity was observed for the samples containing appropriate amounts of Ca after removing excess amount of oxygen. T_c increased as x increased and the highest T_c^{on} was 65 K for multi-phase samples with x = 0.5. For the "1222" phase, superconductivity was observed at temperatures below 25 K for oxygen-annealed sample. It was suggested that the Pb-based "1222" phase was in an oxygen deficient state in contrast to the Pb-based "1212" phase.

ACKNOWLEDGMENTS

The authors would like to thank A. Ichinose of SRL-ISTEC for his helpful advises for sythesizing "1222" samples and his fruitful discussions. They also thank K. Kurusu of Furukawa Electric Co. Ltd. and S. Koriyama, T. Kawano and M. Kosuge of SRL-ISTEC for their kind help in coulometric titration.

REFERENCES

[1] R. J. Cava et al., Nature **336** (1988) 211.
[2] M. A. Subramanian et al., Physica **C159** (1989) 124.
[3] O. Inoue et al., Jpn. J. Appl. Phys. **28** (1989) L60.
[4] T. Rouillon et al., J. Solid State Chem. **78** (1989) 322.
[5] J. Y. Lee et al., J. Mater. Res. **4** (1989) 763.
[6] H. W. Zandbergen et al., Physica **C159** (1989) 81.
[7] T. Rouillon et al., Physica **C159** (1989) 201.
[8] A. Tokiwa et al., Physica **C161** (1989) 459.
[9] T. Rouillon et al., Physica **C167** (1990) 107.
[10] R. J. Cava et al., Physica **C167** (1990) 67.
[11] S. Adachi et al., Physica **C168** (1990) 1.
[12] S. Adachi et al., Jpn. J. Appl. Phys. **29** (1990) L890.
[13] T. Maeda et al., "Advances in Superconductivity II" (Proc. 2nd Int. Symp. Superconductivity, November 14 - 17, Tsukuba, ed. T. Ishiguro and K. Kajimura, Springer Verlag, Tokyo, 1990), pp. 91 - 94.
[14] S. Koriyama et al., Physica **C166** (1990) 413.
[15] T. Maeda et al., Physica **C169** (1990) 133.
[16] T. Maeda et al., submitted to Phys. Rev. **B**.
[17] K. Kurusu et al., Analyst **114** (1989) 1341.
[18] M. Kosuge and K. Kurusu, Jpn. J. Appl. Phys. **28** (1989) L810.
[19] C. Martin et al., Mod. Phys. Lett. **B3** (1989) 993.
[20] Y. Tokura et al., Nature **342** (1989) 890.
[21] T. Maeda et al., in preparation.

Uniaxial Stress Effect on the Superconductivity of Grain-Oriented Samples of High T_c Superconductors

YUICHI MOTOI, KATSUMI FUJIMOTO, HIROMOTO UWE, and TUNETARO SAKUDO

Institute of Applied Physics, University of Tsukuba, Tsukuba, Ibaraki, 305 Japan

ABSTRACT

Uniaxial stress effect on the superconductivity of grain-oriented samples of $La_{2-x}Sr_xCuO_4$ has been investigated. The T_c value monotonically decreases with increasing uniaxial stress along the c-axis up to 3.0 kbar. The result suggests the importance of the in-plane O-O or Cu-O bond length on the superconductivity.

KEY WORDS: uniaxial stress effect, $La_{2-x}Sr_xCuO_4$, grain-aligned composite magnetic field induced alignment

INTRODUCTION

Recent studies on hydrostatic-pressure effects of high T_c superconductors have attracted much attention. The T_c increases with pressure for most hole-doped superconductors, and the results were attributed to the decrease of distance between Cu and the apical O [1-6]. The experiment of uniaxial stress, which is complementary to that of hydrostatic-pressure effect, will reveal anisotropic nature of the relation between the bond length and the superconductivity.

EXPERIMENTAL

We prepared the grain-aligned samples of $La_{2-x}Sr_xCuO_4$ by the "magnetic-field induced alignment" method [7-9]. First, $La_{2-x}Sr_xCuO_4$ powder was synthesized by a solid state reaction method as follows: The stoichiometric mixtures of La_2O_3, $SrCO_3$ and CuO were calcined at 950°C for 15 hours. They were refired at 1050°C in O_2 for 20 hours, and were thoroughly ground. Next, the mixture of the $La_{2-x}Sr_xCuO_4$ powder (8g), methyl-ethyl ketone (2ml) and the monomers of epoxy resin (Comodino EP 007 2g) was casted into a Teflon tube. The mixture was suspended and polymerized in the magnetic field of 7T for 3 days. The rectangular bar with the size of typically $2.6\times 2.6\times 12.0$ mm^3 was cut from the polymerized precipitate for the measurement, the longest dimension being parallel to the applied magnetic field during the polymerization. The $La_{2-x}Sr_xCuO_4$ in the composite was evaluated to be about 50% in volume from the specific gravity. The degree of the alignment was estimated by an X-ray diffraction, in which the measurement surface was perpendicular to the applied magnetic field. The full-width of half maximum intensity in the rocking curve was about 10 degree, indicating the c-axes of most grains aligned along the same direction within 20 degree.

The ac magnetic susceptibilty of the specimen was measured from 4.2K to 50K with applying uniaxial stress along the aligned c-axis. Compressive force was applied to the specimen in a cryostat through a lever with lead weights, which produced a uniaxial stress inside the specimen up to 3.0 kbar. The inductance of the coil which surrounded the specimen was measured by an LCR meter.

RESULTS AND DISCUSSION

Figure 1 shows the T_c values against the applied stress along the aligned c-axis for the samples of $La_{2-x}Sr_xCuO_4$ with various composition x. The sign of stress coefficient $dT_c/d\sigma$

was found to be negative for all composition x, where σ means uniaxial stress along the aligned c-axis, and the maximum T_c was observed around x=0.16. The negative sign of $dT_c/d\sigma$ is in contrast to the posistve pressure coefficient dT_c/dp observed in the hydrostatic-pressure experiments over the whole range of x. Here we assume the compression of the c-axes and the expansion of the a-axes, on the average, for the grains embedded in the epoxy resin. Compared with the results of the hydrostatic-pressure experiment, the negative $dT_c/d\sigma$ observed here should be attributed to average increase of the bond length of Cu-O or O-O in the CuO_2-plane, because the compression of the c-axis may be commonly caused by uniaxial stress and hydrostatic pressure. We speculate that the change of the transfer probability of carriers from Cu site to O sites or among O sites in the CuO_2-plane may affect the superconductivity in this system.

In the above discussion, we neglected the effect of the internal displacement of apical oxygen induced by pressure, and the tilting of CuO_6 octahedra. Recently, non-linear behaviour of the apical Cu-O distance under hydrostatic pressure at room temperature was reported [10]. The measurement of the apical Cu-O distance under uniaxial stress will be required to clear the role of the apical oxygen on the superconductivity.

In conclusion, the effect of uniaxial stress along the aligned c-axis on the superconductivity of grain-aligned composites of $La_{2-x}Sr_xCuO_4$ has been investigated. The T_c decreases with increasing uniaxial stress along the aligned c-axis. The results suggests the importance of bond length in the CuO_2-plane on the superconductivity of $La_{2-x}Sr_xCuO_4$.

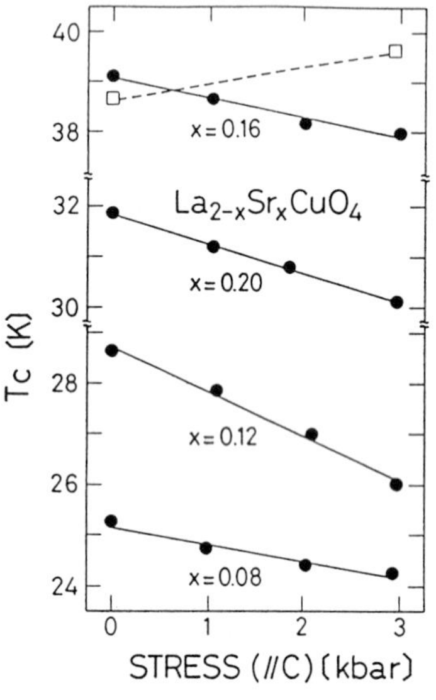

Fig. 1. Summary of the T_c value versus the applied uniaxial stress along the aligned c-axis for $La_{2-x}Sr_xCuO_4$ of various composition x. The open squares are the data of the hydrostatic-pressure experiment [1].

REFERENCES

1. Tanahashi N, Iye Y, Murayama C, Môri N, Yomo S, Okazaki N, Kitazawa K (1989) Jpn.J.Appl.Phys.28: L762
2. Griessen R (1987) Phys.Rev.B 36: 5284
3. Môri N, Takahashi H, Murayama C (1990) Proceedings of LT-19 satelite conference at Cambridge (to be published)
4. Murayama C, Môri N, Yomo S, Takagi H, Uchida S, Tokura Y (1989) Nature 339: 293
5. Ohta Y, Tohyama T, Maekawa S (1990) Physica C 166: 385
6. Matsukawa H, Fukuyama H (1990) J.Phys.Soc.Jpn. 59: 1693
7. Farrel DE, Chandrasekhar BS, Finnemore DK (1987) Phys.Rev.B36: 4025
8. Nakagawa Y, Yamasaki H, Obara H, Kimura Y (1989) Jpn.J.Appl.Phys. 28: L547
9. Yamaguchi Y, Tokumoto M, Waki S, Nakagawa Y, Kimura Y (1989) J.Phys.Soc.Jpn. 58: 2256
10. Nelmes RJ, Wilding NB, Hatton PD, Caignaert V, Raveau B (1990) Physica C 166: 329

Superconductivity of $Ln_{1.85-x}La_xCe_{0.15}CuO_4$ (Ln=Pr,Sm)

Akira Ito, Hideo Iwasaki, and Norio Kobayashi

Institute for Materials Research, Tohoku University, 1-1, Katahira 2-chome, Aoba-ku, Sendai, 980 Japan

ABSTRACT

Superconducting properties of the N-type superconductors $Ln_{1.85-x}La_xCe_{0.15}CuO_4$ (L=Pr and Sm) have been studied. Sintered samples were prepared by solid state reaction. It is possible to synthesize the samples with single phase of the Nd_2CuO_4 type structure in wide concentration ranges $0 \leq x \leq 1.2$ for Pr-based system and $0 \leq x \leq 0.5$ for Sm-based one. The lattice parameters a and c increase monotonically with increasing x for both systems. While it is expected that the substitution of Pr or Sm atoms by La does not change the carrier concentration of these systems because the oxygen content 4-y is kept almost constant. The superconducting transition temperature T_c decreases with increasing La-concentration x. The suppression of T_c is more rapid for Sm-based system than for Pr-based one. The x dependence of T_c for both systems could be scaled by the reduced lattice parameter $(a-a_0)/a_0$.

KEYWORDS : electron-doped superconductor, substitution effect, $Pr_{1.85}Ce_{0.15}CuO_4$, $Sm_{1.85}Ce_{0.15}CuO_4$

INTRODUCTION

Since the discovery[1] of the electron-doped system $Nd_{1-x}Ce_xCuO_{4-y}$, many investigations have been performed to find out another N-type superconductor with different crystal structure. It is, however, only known that $Ln_{1-x}M_xCuO_{4-y}$ (Ln=Pr, Nd, Sm or Eu[2] ; M=Ce or Th[2]) with Nd_2CuO_4-type (T'-type) structure exhibits the N-type superconductivity. In these compounds, trivalent rare-earth ions can be substituted by tetravalent Ce or Th in contrast with $La_{1-x}M_xCuO_{4-y}$ (M=Ca, Sr or Ba) hole-doped system with the K_2NiF_4-Type (T-type) structure. Furthermore, the compounds with T'-type structure must be reduced by heat treatment in a reducing condition, in order to obtain superconductivity. By this heat treatment, a small amount of oxygen is lost and superconductivity is realized.

Recently, Oh-ishi et al.[3] and Takayama-Muromachi et al.[4] reported that Nd sites in $Nd_{1.85}Ce_{0.15}CuO_{4-y}$ system could be substituted by La in wide concentration range and the transition temperature T_c was suppressed monotonically with increasing La concentration in spite of the fact that the La ions have the same trivalence as Nd ions.

It is very interesting to know what kind of mechanism is responsible for the T_c-suppression in the La-doped system. In this paper, we present experimental results on lattice parameters, oxygen deficiency and T_c in $Ln_{1.85-x}La_xCe_{0.15}CuO_{4-y}$ (Ln=Pr and Sm) system and discuss relations among them.

EXPERIMENTAL

Samples with compositions of $Ln_{1.85-x}La_xCe_{0.15}CuO_{4-y}$ (Ln=Pr and Sm) were prepared in the concentration ranges of $0 \leq x \leq 1.3$ for Ln=Pr and $0 \leq x \leq 0.5$ for Ln=Sm by a solid state reaction. Appropriate amounts of 4N-Pr_2O_3, Sm_2O_3, La_2O_3 CeO_2 and CuO powders were carefully mixed to improve homogeneity and pressed into pellets. The pellets were calcined at 950°C for 12h. Reacted materials were followed by grinding, pelletizing, reheating at 1150°C for 24h and then slowly cool down. These processes were done in air. A part of each sample was reduced in pure Ar atmosphere at 900°C for Pr system or 850°C for Sm system for 40 h and then cooled down in the furnace.

Powder x-ray diffraction of the samples were taken using Cu-Kα radiation. The diffraction patterns showed that any impurity phase was not observed up to x=1.2 for Ln=Pr and in the concentration range studied for Ln=Sm. Oxygen content 4-y of the $Pr_{1.85-x}La_xCe_{0.15}CuO_{4-y}$ samples was measured by the iodometric titration method with a TDA auto-titrator. The electrical resistance was measured by an usual four probe method. The temperature was measured by a carbon glass and a platinum resistance thermometers. DC susceptibility measurements were made by a SQUID magnetometer (Quantum Design).

RESULTS AND DISCUSSION

Figure 1 shows the concentration dependence of lattice parameters a and c. The lattice parameters for both Pr- and Sm-based systems increase almost linearly with increasing La concentration x. This behavior suggests that the T'-type structure is held in whole concentration range studied. On the other hand, the lattice parameters seem to become slightly longer by the annealing in Ar atmosphere. As seen in Fig. 2, the oxygen content for the Pr-based

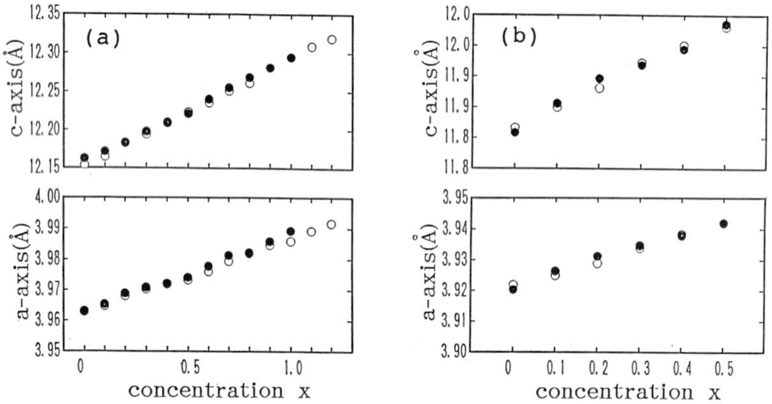

Fig.1 La-concentration dependence of lattice parameters a and c of as-grown (open circles) and reduced (closed circles) samples. (a) $Pr_{1.85-x}La_xCe_{0.15}CuO_{4-y}$, (b) $Sm_{1.85-x}La_xCe_{0.15}CuO_{4-y}$

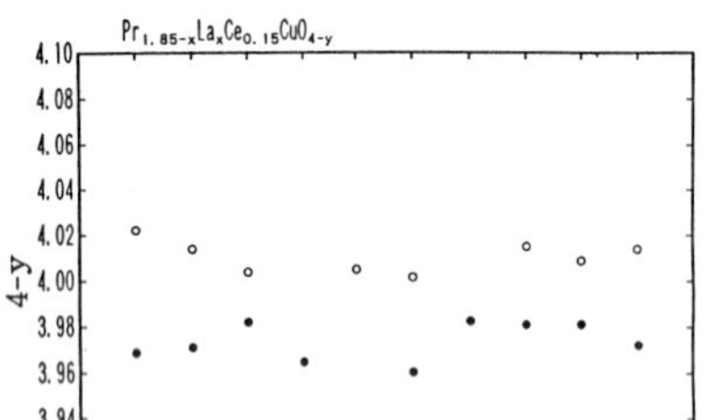

Fig.2 La-concentration dependence of the oxygen content in as-grown (open circles) and reduced (closed circles) samples of $Pr_{1.85-x}La_xCe_{0.15}CuO_{4-y}$

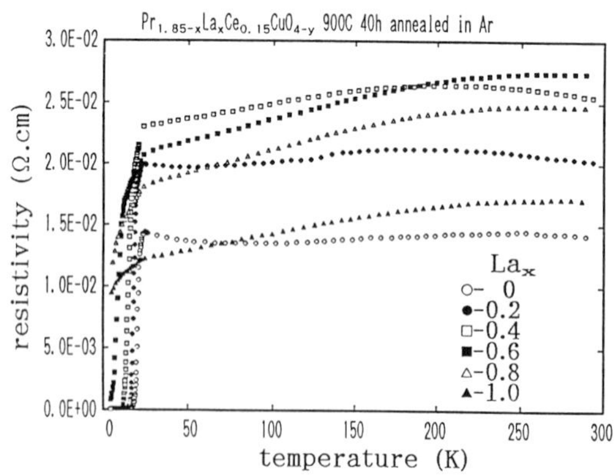

Fig.3 Temperature dependence of the electrical resistivity of $Pr_{1.85-x}La_xCe_{0.15}CuO_{4-y}$

system is independent of the La-concentration and decreases from 4.01±0.01 to 3.97±0.01 by the heat treatment under the reducing condition, though the obtained values are slightly distributed. This result is agreement with that for Nd-based system reported by Oh-ishi et al[3]. The slight change of the lattice parameters is probably due to this oxygen deficiency.

Figure. 3 shows the temperature dependence of the electrical resistivity of the reduced $Pr_{1.85-x}La_xCe_{0.15}CuO_{4-y}$ samples. The electrical resistivity of the normal state increases with x at low concentrations, and decreases at higher concentrations of x>0.6. However, the change in the resistivity value is not so large. The almost constant oxygen content and the small change of the electrical resistivity suggest that the carrier concentration can not be affected by the substitution of Pr and Sm by La as expected from the same chemical valence in these lanthanide ions.

The superconducting transition temperature T_c is determined by a midpoint of the extrapolated value of the normal resistivity. Concentration dependences of the T_c value are shown in Figs. 4(a) and (b) for the Pr- and Sm-based systems, respectively. The transition width defined by the temperature difference between 90% and 10% points of the extrapolated normal resistivity is also represented by a error bar in Figs. 4(a) and (b). The transition width is broadened with increasing La concentration. In particular, the superconducting onset temperature changes only a little. This result may be caused by compositional inhomogeneities of the La concentration and/or of the oxygen content. In order to determine more precisely the T_c value, magnetic measurements were performed on three samples of Pr-based system with x=0, 0.1 and 0.5 by using SQUID magnetometer. Temperature dependence of the magnetic susceptibility of the samples is shown in Fig. 5. The T_c value is determined by an extrapolation of the susceptibility to the normal state value. Since the magnetically determined T_c value is in good agreement with that defined by the midpoint of the normal state resistivity, in the following we use T_c determined by the midpoint value.

As shown in Figs. 4(a) and (b), the T_c value decreases monotonically with increasing La concentration. However, the suppression of T_c is more rapid for Sm-based system than that for Pr-based one. This tendency can not be explained by the change of the carrier concentration, because the carrier concentration is expected to be constant in the La-doped systems. On the other hand, Oh-ishi et al. pointed out [3] that the T_c value of $Nd_{1.85-x}La_xCe_{0.15}CuO_{4-y}$ is dominated by Cu-O bond length given by a half of the lattice parameter a. In our systems, changing rates of T_c against the Cu-O bond length are estimated to be $dT_c/dCu\text{-}O \sim -1000$ K/Å and -770 K/Å at low concentrations for Pr- and Sm-based systems, respectively. These values are two or three times as large as that (-380 K/Å) for Nd-based system [3].

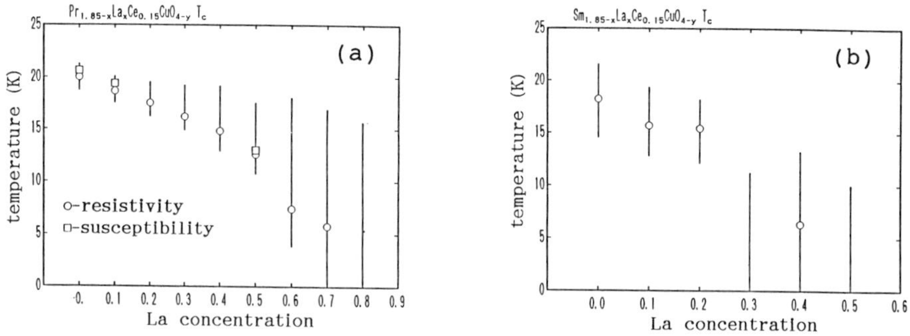

Fig.4 Dependence of the superconducting transition temperature T_c on the La-concentration in (a) $Pr_{1.85-x}La_xCe_{0.15}CuO_{4-y}$ and (b) $Sm_{1.85-x}La_{1-x}Ce_{0.15}CuO_{4-y}$. Squares are the T_c determined by the magnetic measurements.

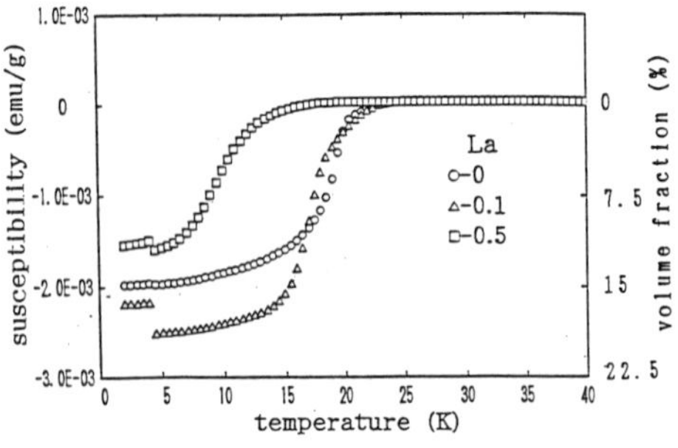

Fig.5 Temperature dependence of the Meissner signal measured by cooling the sample under the constant field of 20 Oe

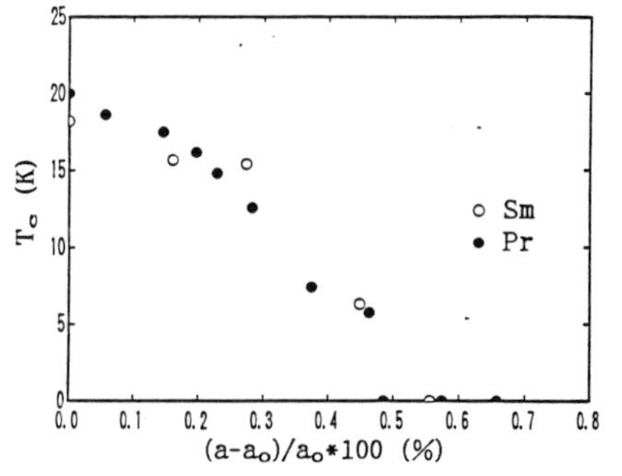

Fig.6 Transition temperature T_c as a function of reduced lattice parameter $(a-a_0)/a_0$, where a_0 denotes each the lattice parameter of $Pr_{1.85}Ce_{0.15}CuO_{4-y}$ or $Sm_{1.85}Ce_{0.15}CuO_{4-y}$.

In order to find out a relationship between T_c and the lattice parameters, Fig. 6 presents the change in T_c as a function of increasing rate of the lattice parameter $(a-a_0)/a_0$. where a_0 denotes the lattice parameter of the non-doped sample for each Pr- and Sm-based system. The T_c values of both systems fall on the almost same curve for the increase of $(a-a_0)/a_0$. However, similar relationship can not be seen for the elongation of the c-axis though the result is not shown. These results suggests that the change in the lattice parameter not along the c-axis but within the CuO_2 plane is mainly responsible for the T_c-suppression in the $Ln_{1.85-x}La_xCe_{0.15}CuO_{4-y}$ (Ln=Pr and Sm) systems. Finally, it should be noted that the T_c-suppression observed in $Nd_{1.85-x}La_xCe_{0.15}CuO_{4-y}$ system by Oh-ishi et al. is quantitatively in disagreement with our Pr- and Sm-based system. It may come from the different heat treatment at final stage. It is necessary to make clear the relationship between T_c and the number of the carrier simultaneously.

ACKNOWLEDGMENTS

The anthers would like to thank Mr. K. Oh-ishi for useful discussions. This work was supported by a Grant-In-Aid for Scientific Research on Priority Areas 'mechanism of Superconductivity' by Ministry of Education, Science and Culture, Japan.

REFERENCES

1. Tokura Y., Takagi H. and Uchida S. (1989) A superconducting copper oxide compound with electrons as the charge carriers. Nature 337: 345-347
2. Markert J. T., Early E. A., Bjørnholm T. , Ghamaty S. , Lee B. W., Neumeier J. J., Price R. D., Seaman C. L. and Maple M. B. (1989) Two new electron cuprate superconductors, $Pr_{1.85}Th_{0.15}CuO_{4-y}$ and $Eu_{1.85}Ce_{0.15}CuO_{4-y}$, and properties of $Nd_{2-x}Ce_xCuO_{4-y}$. Physica C 158: 178-182
3. Oh-ishi K., Syono Y., Kikuchi M., Kobayashi N., and Muto Y. (1990) Relation between T_c and Cu-O bond length within the basal plane of $Ce_{0.15}Nd_{1.85-x}M_xCuO_{4-y}$ (M=La nd Sm) and $Ce_{0.15}Pr_{1.85}CuO_{4-y}$. Solid State Commun. 73: 341-343
4. Takayama-Muromachi E., Uchida Y. and Kato K. (1990) Electron-doped system (La, Nd, Ce)$_2$CuO$_4$ and preparation of T'-type (La, Ln)CuO_4 (Ln=Ce, Y). Physica C165: 147-151

HIP-Processed Superconducting $La_{2-x}Sr_xCaCu_2O_6$

TAKESHI SAKURAI, TORU YAMASHITA, H. YAMAUCHI, and SHOJI TANAKA

Superconductivity Research Laboratory, International Superconductivity Technology Center, 10-13, Shinonome 1-chome, Koto-ku, Tokyo, 135 Japan

ABSTRACT

Superconducting 326 samples, $La_{2-x}Sr_xCaCu_2O_6$ with x=0.2 and x=0.4, which exhibited superconducting onsets at temperature(Tc^{on}'s) higher than 40K were successfully synthesized using hot-isostatic-pressing (HIP) techniques. Samples were sintered at various temperatures (T_s) for 72h in flowing O_2 gas, and were post-annealed at various temperatures (T_a) for 6h in a mixture of 80% Ar gas and 20% O_2 gas of total pressure at a 100MPa. Optimum combinations in terms of T_s and T_a were determined for obtaining superconducting samples. It was concluded that T_a was a crucially important factor for obtaining a superconducting 326 phase with T_c^{on}>40K. That is, T_a>1243K was required for the 6h HIP annealing.

KEY WORDS: $(La,Ca)_3Cu_2O_6$ structure, superconductivity, high pressure process

INTRODUCTION

Both systems of cuprates[1], $La_{2-x}Ca_{1+x}CuO_6$ and $La_{2-x}Sr_{1+x}CuO_6$, which are termed 326 compounds, possess crystallographic as well as chemical features common with conventional high T_c superconducting cuprates[2-4]. That is, the 326 compounds have double layers of CuO_5 pyramids and an effective valency per Cu ion which is higher than 2+ can be achieved by chemical doping to the La(or "A") sites. Among the perovskite-related cuprates with double layers of CuO_5 pyramids, many of which have been known to be superconducting, the 326 compound has the simplest crystallographic structure. A variety of attempts have been reported[5,6] for obtaining superconducting 326 compounds. Nontheless, no one had been successful up until Cava et al[7] recently reported that they had successfully synthesized $La_{1.6}Sr_{0.4}CaCu_2O_6$ which exhibited 60K-superconductivity.

In the present work, 326 samples of $La_{2-x}Sr_xCaCu_2O_6$ were prepared using powders of conventional carbonates and oxides as starting materials and sintering at a variety of temperatures(T_s) in flowing O_2 gas at 100KPa for 72h. Then, samples were processed by post-annealing at a variety of temperatures(T_a) in 80% Ar and 20% O_2 gas mixture at 100MPa for 6h. Conditions for obtaining superconductivity are sought in this work.

EXPERIMENTAL PROCEDURE

The starting materials were high purity(>99.9%) powders of La_2O_3, $SrCO_3$, $CaCO_3$ and CuO. The powders were weighted to nominal compositions of $La_{2-x}Sr_xCaCu_2O_6$ with x=0.2 and x=0.4 and mixed for 15h using a ball mill with ethanol. The mixed powders were formed into disc shaped pellets of 20mm diameter and 5mm thickness, applying a uniaxial pressure at 50MPa. The pellets were calcined at 1173K for 10h in air or O_2 gas flow, and then pulverized. This series of procedures of pellet-forming, calcination in either air or O_2 and pulverization was repeated four times. Then, the calcined powder was shaped into parallelepiped bars of $10x3x3mm^3$ under a uniaxial pressure at 100MPa. The bar shaped samples were sintered at a variety of temperatures(T_s) for 72h in O_2 gas flow. The sintered samples were post-annealed using an O_2-hot-isostatic press(HIP) at various temperatures(T_a) for an arbital period in the range of 6h to 100h. For the HIP processing, a mixture of 80% Ar and 20% O_2 of the total pressure at 100MPa (and O_2 partial presuure of 20MPa) was utilized. Each of the two kinds of samples, i.e. $La_{1.8}Sr_{0.2}CaCu_2O_6$(sample.1) and $La_{1.6}Sr_{0.4}CaCu_2O_6$(sample.2), was heat-treated at 16 different combinations of (T_s, T_a) as shown in Fig.1.

Crystallographic structures of the samples were analysed by x-ray powder diffraction. The lattice constants were determined by an extrapolation method using more than 15 diffraction peaks located in the 2θ range of 20° to 80°. Electrical resistivity was measured as a function of temperature using a conventional four-probe technique. The four electrodes were formed using a silver paste(Dupont4922) on the surface of the bar-shaped sample. Magnetic susceptibility was measured in a d.c. magnetic field of 10Oe using a SUQID magnetometer(Quantum Design, MPMS).

RESULTS AND DISCUSSION

In order to investigate the effect of post-annealing by HIP on the formation of superconducting 326 phases, two samples of the same composition, $La_{1.8}Sr_{0.2}CaCu_2O_6$, were processed differently after being sintered at 1198K: one was post-annealed at 1243K in flowing O_2 gas of 100KPa for 72h and the other was HIP-annealed at 1243K for 6h. The former exhibited superconductivity with T_c^{on}=25K, while the latter with T_c^{on}=62K. This demonstrated the HIP-annealing process was indispensable for obtaining superconducting 326 phase. It shoud be noted that, if T_c^{on}'s higher than 40K are observed and the 326 structure is confirmed by x-ray diffraction for samples of the La-Sr-Ca-Cu-O system, it can be safely said that the samples contain supercondiucting 326 phase.

The synthesis conditions and superconducting properties of samples of relationship between $La_{2-x}Sr_xCaCu_2O_6$ with x=0.2 and 0.4 are summarized in Fig.1. The samples were prepared with different combinations of sintering temperature, T_s, and HIP-annealing temperature, T_a. With in the frame work of the present experimental conditions, T_a higher than 1243K is required for obtaining superconducing 326 phase with T_c^{on}>40K for both compositions with x=0.2 and 0.4. When the combination of (T_s, T_a) is equal to (1273, 1243), T_c^{on} is far from the border value of 40K for the both compositions. This suggests that insufficient amounts of superconducting 326 phase were formed in these samples. Note that, for these samples, T_s was higher than T_a. This leads us to a supeculation that, among the (T_s, T_a) combinations which satisfy the condition,T_a>1243K, those in which T_s>T_a are inadequate for obtaining superconducting 326 phase. To form sufficient amounts of superconducting 326 phase in the superconducting samples, a HIP-annealing much longer than 6h will be required.

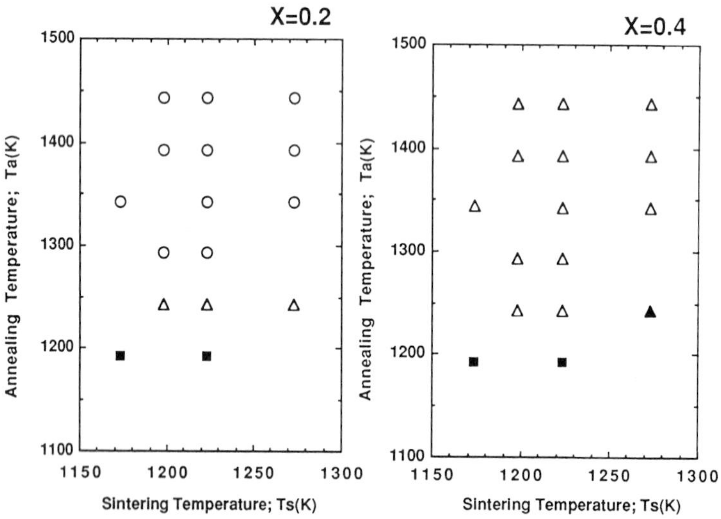

Figure.1 T_s -vs T_a diagrms for the composition of x=0.2 and 0.4. Open circles indicate samples which exhibit T_c^{on} higher than 40K and are of the 326 single phase. Open triangles indicate samples which exhibit T_c^{on} higher than 40K and consist of 326, $LaCuO_3$, $SrCu_2O_3$ and Cu_2SrO_2 phases. Closed triangles indicate samples which exhibit T_c^{on} lower than 40K and consist of 326, $LaCuO_3$, $SrCu_2O_3$ and Cu_2SrO_2 phases. Closed squares indicate samples which exhibit T_c^{on} lower than 40K and consist of 326, CaO and Cu_2SrO_2 phases.

The temperature dependence of electrical resistivity for samples with x=0.2 and sintered at T_s=1223K, is shown in Fig.2. As T_a increased, the normal-state resistivity decreased, while T_c^{on} remained constant around 60K, for all the samples except for one which was HIP-annealed at 1193K. Additionally, the higher T_a was, the more metallic was the normal-state electrical conduction and the narrower was the transition width between T_c^{on} and T_c^{zero}. Note that only samples HIP-annealed at 1293K, 1343K, 1393K and 1443K were of single phase.

Figure.3 shows magnetic suceptibility in terms of temperature for samples with the compositions, x=0.2 and 0.4, when the combination of (T_s, T_a) is equal to (1198, 1293). Note that sample with x=0.2 was of single phase and confirmed to have the oxygen content,z for $La_{1.8}Sre_{0.2}CaCu_2O_z$, equal to 6.01+0.01 according to coulometric titration[8]. Although both samples have diamagnetic onsets, T_c^{mag}, around 50-55K, the two curves are quite different. For sample with x=0.2, a kink appeared around 20K, indicating that there existed a secondary phase with T_c=20K. No such kinks were observed for sample with x=0.4. Although the sample with x=0.2 was of single-phase accrding to x-ray diffraction, certain secondary superconducting phases must be contained. One of the possibile choices is a 214 phase such as $(La,Sr)_2CuO_4$ or $(La,Sr,Ca)_2CuO_4$. Since the Sr content is higher in the sample with x=0.4 than the sample with x=0.2, the 214 phase in the sample with x=0.4 would be over-doped in terms of the hole concentration such that it might not be superconducting anylonger. This is a possible explanation for the sample with x=0.4 not to have such a kink around 20K in the susceptibility curve. The Meissner volume fraction of the 326 phase which exhibited T_c^{mag} around 55K was about 0.8% at 20K for the sample with x=0.2. This small value for the Meissner volume fraction would simply mean that the duration of HIP-annealing was short. In order to investigate the effect of the length of HIP-annealing on the Meissner volume fraction of superconducting 326 phase, the susceptibility curves for samples whose (T_s,T_a)is equal to (1223, 1293), with composition of x=0.2, are shown in Fig.4. It was found that T_c^{mag} was shifted from 39K to 59K and the Meissner volume fraction of the 326 phase increased at all temperatures below T_c^{mag}. Thus, a long period HIP-annealing is considered to be necessary for obtaining a significant amount of volume fraction of superconducting 326 phase.

In Fig.5, a lattic parameter diagram, i.e.a -vs- c, for the 326 phase, is given for samples studied in the present work. It is seen that the length of c-axis depends on the Sr content while the a-axis length is more-or-less independent of Sr content. In this figure the datum point reported by Cava, et. al. for their superconducting $La_{1.6}Sr_{0.4}CaCu_2O_6$[7] is also given, and it is observed that this datum point reasonably agrees with those of samples.2 of the same composition. In this figure, anthoer datum point for the as-sintered sample with T_c^{on}=25K which was annealed at 1198K in O_2 gas flow and the post-annealed at 1243K in flowing O_2 gas at 100KPa is also given. It is interesting to observe that even this datum point is in reasonable agreement those for samples which contain superconducting 326 phase. This may indicate that the formation of superconducting 326 phase with T_c^{on}=60K cannot be detected by measuring changes in the lattice constants. Such small crystallographic changes, if they exist, may include an ordering of Ca^{2+}, Sr^{2+} and La^{3+} ions which have rather similar ionic radii[9], 1.18A, 1.31A and 1.216A(for the cordination number of 9 at normal pressure) respectively. As suspected previously[5], Sr^{2+} ions which are mixed with Ca^{2+} ions located between the two CuO_2 planes may hinder the compound from being superconducting. The O_2-HIP annealling employed in the present work may have worked not only to supply sufficient amount of oxygen to the compound but also select Ca^{2+} ions from Sr^{2+} ions and/or La^{3+} ions and locate them exclusively between the two CuO_2 planes.

SUMMARY

Employing conventional carbonate and oxide powders as staring materials, superconducting $La_{2-x}Sr_xCaCu_2O_6$ samples, x=0.2 and 0.4, with T_c^{on}'s higher than 40K were successfully synthesized using HIP-annealing techniques. For the HIP-annealling process, a total pressure at 100MPa was employed. Optimum combinations of the sintering temperatire,T_s and HIP-annealling temperature,T_a, when sample preparation conditions other than T_s and T_a were fixed, were obtained for synthesizing superconducting 326 phase. Single-phase 326 samples were obtained only for the composition of $La_{1.8}Sr_{0.2}CaCu_2O_6$ when prepared at T_a>1243K under the condition that T_a>T_s. It was found that the HIP-annealling temperature was crucially important to obtain superconducting 326 phase: temperature higher than 1243K were required for HIP-annealing for 6h. In order to obtain a large volume fraction of superconducting 326 phase, it is suggested that the HIP-annealling duration much longer than 6h is neccesary.

ACKNOWLEDGMENTS

The authors are greatful to T. Maeda of SRL for his helpful discusions and to Toyo Engineering Corporation for their contibution to the sample preparation instrument. This work was supported by New Energy and Industrial Technology Development Organization as a part of the program for R&D of Basic Technology for Future Industries.

REFERENCES
1. Ngyen N, Er-Rakho L, Michel C, Choisnet J and Raveau B (1980) Mat.Res.Bull 15: 891
2. Torrance J B, Tokura Y, Nazzal A and Parkin S S (1988) Phy.Rev.Lett 60: 542
3. Izumi F, Takayama-Muromach E, Nakai Y and Asano H (1989) Physica C 157: 89
4. Caignaert V, Nguyen N and Raveau B (1990) Mat.Res.Bull 25: 199
5. Hiratani M, Sowa T, Takeda Y and Miyauchi K (1989) Solid.stat.comm 72: 541
6. Tamegai T and Iye Y (1989) Physica C 159: 181
7. Cava R J, Batlogg B, van Dover R B, Krajewski J J, Waszczak J R, Flemming R M, Peck Jr W F, Rupp Jr L W, Marsh P, James A C W P and Schneemeyer L F (1990) Nature 345: 602
8. Kurusu K, Takami H and Shintomi K (1989) Analyst 114: 1341
9. Shannon R D (1976) Acta Cryst A32: 751

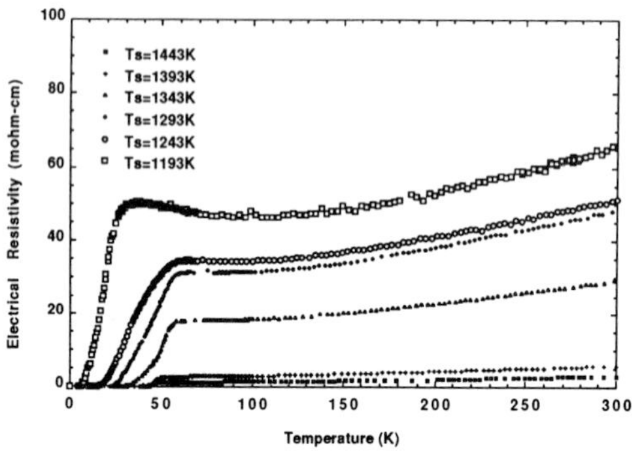

Figure.2 Electrical resistivity -vs- temperarute for samples anealed at various T_a, with T_s=1223K and X=0.2.

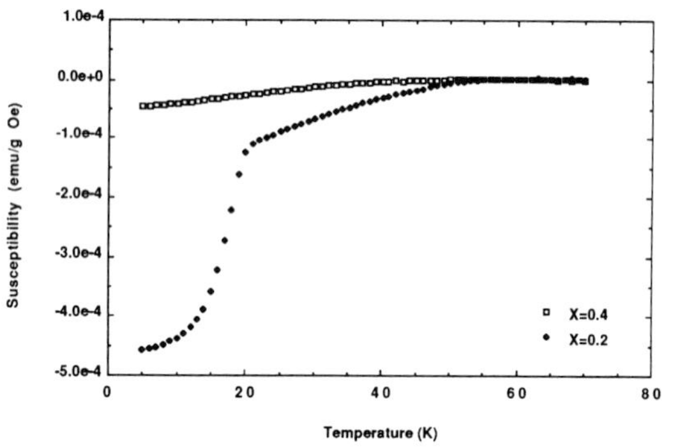

Figure.3 Magnetic susceptibility -vs- temperature for samples with the composition of X=0.2 and 0.4, when both samples have the same combination of (T_s, T_a)=(1198, 1293).

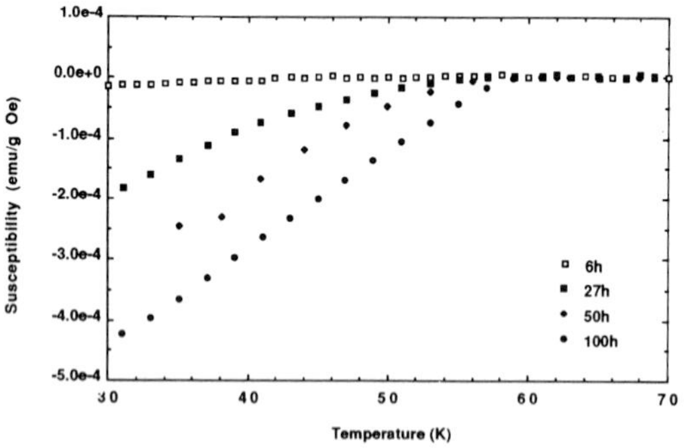

Figure.4 Magnetic suscptibility -vs- temperature for samples annealed for various period for 6, 27, 50h and 100h, with (T_s, T_a)=(1223, 1293) and x=0.2.

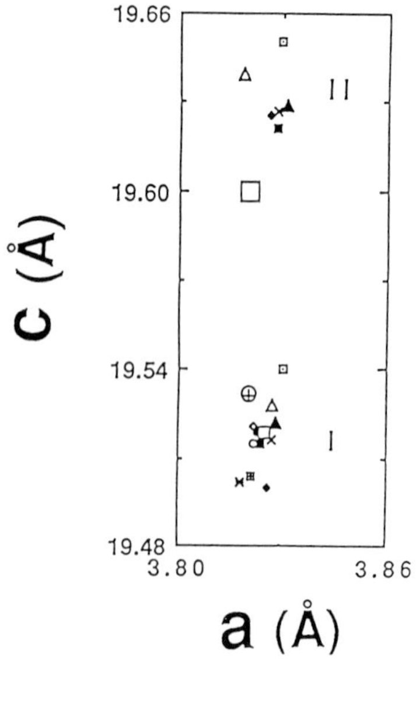

Figure.5 Lattice parameter diagram, a-axis -vs- c-avis. Group.1 is consist of the samples with x=0.2, and group.2 is consist of the samples with x=0.4.

Processing of $YBa_2Cu_3O_x$ Ceramic Materials: Effect of Additives

A. SRINIVASA RAO, L. APRIGLIANO, and OM P. ARORA

David Taylor Research Center, Annapolis, MD 21402, USA

ABSTRACT

The effect of the addition of Al_2O_3, TiO_2, PbO and Ag_2O, in the concentration range 0 - 30 wt. %, on crystal structure, morphology and superconductivity of $YBa_2Cu_3O_{6+x}$ was investigated. The results suggest that the addition of either Al_2O_3 or TiO_2 or $\geq$ 3wt.% PbO reduces the primary particle size of $YBa_2Cu_3O_{6+x}$. Additives, Al_2O_3, TiO_2 and PbO tend to stabilize the non superconducting tetragonal phase at the expense of superconducting orthorhombic phase. The oxygen from $YBa_2Cu_3O_{6+x}$ tends to migrate towards the additive and this process degrades the superconducting property of $YBa_2Cu_3O_{6+x}$ and increases the resistance. The addition of silver oxide, although, does not effect the particle size of $YBa_2Cu_3O_{6+x}$, it improves the electrical properties of the composite.

KEY WORDS : $YBa_2Cu_3O_{6+x}$, additive, superconductivity

INTRODUCTION

The new high temperature superconducting ceramic materials are limited in their application as bulk superconductors, because, of the inherent problems that are associated with the processing of brittle ceramic materials. Since, all the ceramic materials show significant plastic deformation at different high temperatures, it is possible that high temperature deformation may intelligently be used in forming these ceramic superconductors into useful shapes. For maximum superplastic deformation of ceramic materials, it was shown [1,2], that the grain size must be as small as possible (typically 100 - 500 nm). The present study was aimed to investigate methods to produce fine particles of $YBa_2Cu_3O_{6+x}$ by incorporating a few different additives prior to the sintering process without compromising the superconducting properties of the final sintered ceramic. In this paper, some of the results obtained for Al_2O_3, TiO_2, PbO and Ag_2O additions are presented.

EXPERIMENTAL

The basic superconductor, $YBa_2Cu_3O_{6+x}$, was prepared by solid state chemical reaction of commercial yttrium oxide, copper oxide and barium carbonate. The superconducting Y-Ba-Cu-O ceramic powder was ground into a fine powder. The fine powder was mixed thoroughly with predetermined amounts (in the range 0 - 30 wt.%) of any one of the four additives (Al_2O_3, Ag_2O, PbO and TiO_2) in a ball mill. The mixture was dry pressed into small discs which later were sintered. The general details of the processing of composites are given elsewhere [3].

It has to be pointed out that the selection of additives (Al_2O_3, TiO_2) was based on the fact that both these materials were reported to show superplastic deformation at temperatures above 1000°C [4,5]. The silver oxide was selected because, it decomposes on heating to the sintering temperatures and the free silver will be plastic during deformation and may assist the onset of superplasticity. The free oxygen will be taken up by the frequently oxygen deficient $YBa_2Cu_3O_{6+x}$ structure.

All sintered samples were analyzed using scanning electron microscope, microprobe, x-ray diffractometer and four point electrical resistance measurement apparatus.

RESULTS AND DISCUSSION

The morphology of sintered $YBa_2Cu_3O_{6+x}$ superconducting ceramic material suggests that the as-synthesized $YBa_2Cu_3O_{6+x}$ particles sinter into long elongated rods whose crystal structure is similar to that of orthorhombic $YBa_2Cu_3O_{6+x}$ materials that were reported in the literature.

A typical morphology of sintered $YBa_2Cu_3O_{6+x}$ ceramic materials containing 5 wt % additives (Al_2O_3, TiO_2, PbO and Ag_2O) is shown in Figure 1. The results suggest that the addition of Al_2O_3 and TiO_2 to the superconducting ceramic material significantly alter both the particle surface topography and the particle size. In addition, some

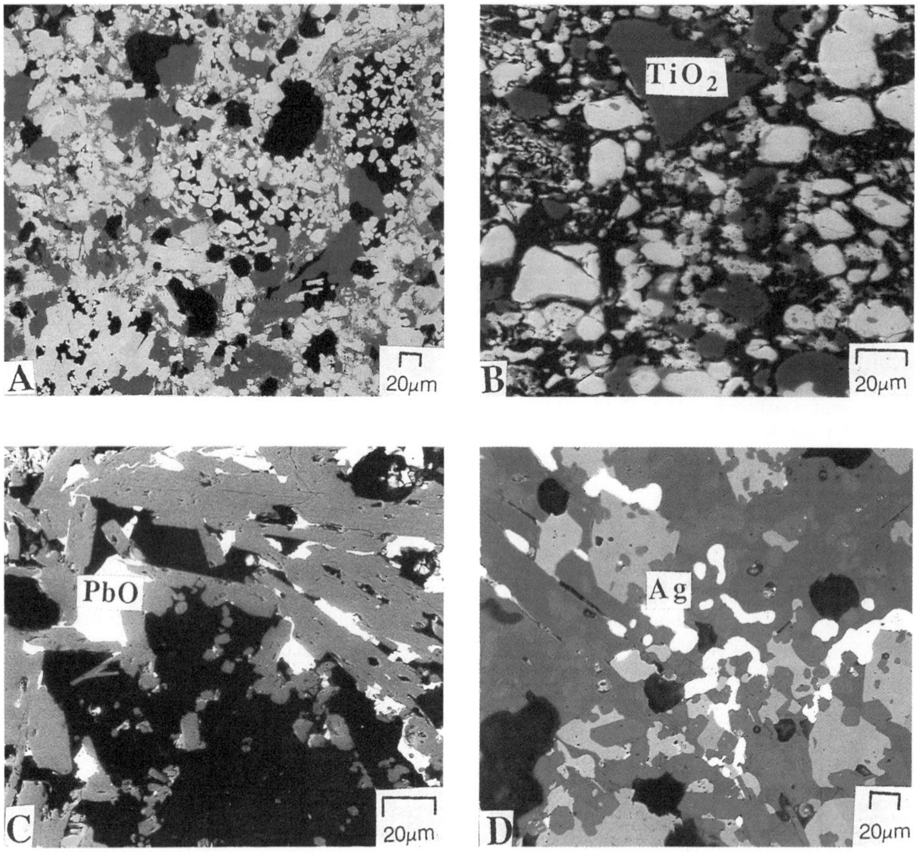

Figure 1. Morphology of sintered $YBa_2Cu_3O_{6+x}$ ceramic composite containing 5 wt. % (A) Al_2O_3, (B) TiO_2, (C) PbO and (D) Ag_2O.

diffusion of TiO_2 into $YBa_2Cu_3O_{6+x}$ grains was also noticed after sintering above 920°C. However, additives Ag_2O and PbO did not produce any physical changes in the superconductor. In order to quantify the effect of the additives on particle size, a number of both optical and scanning electron micrographs (representing different areas of the whole sample) were taken from polished samples as a function of additive concentration. From the micrographs, histograms were plotted in order to obtain the average particle size of the superconducting ceramic materials. Figure 2 shows the average particle size of $YBa_2Cu_3O_{6+x}$ in different composites. The results suggest that both additives Al_2O_3 and TiO_2 are very effective in reducing the particle size of $YBa_2Cu_3O_{6+x}$, and the degree of grain size reduction tends to increase with an increase in the additive concentration. However, no such particle size reduction trends are discernible for composites containing Ag_2O. The additive PbO shows a similar behavior as Ag_2O at lower concentrations (below 3 wt.%), but at higher concentrations (above 3 wt.%), the additive appears to reduce the particle size of $YBa_2Cu_3O_{6+x}$.

The interesting observation that was noticed in samples containing Ag_2O is that the particles of $YBa_2Cu_3O_{6+x}$ tend to line up and orient in the direction of pores. Figure 3 shows a typical scanning electron micrograph of 5 wt % Ag_2O / $YBa_2Cu_3O_{6+x}$ composite. Since, Ag_2O decomposes above 230°C to metallic silver, it is possible that during the calcination and sintering at 920°C (well above the decomposition temperature), the metallic silver (decomposition product of Ag_2O) flows towards the nearest pore and accumulates at the pores. Such a free flow of silver in liquid state would serve to orient the grains of $YBa_2Cu_3O_{6+x}$.

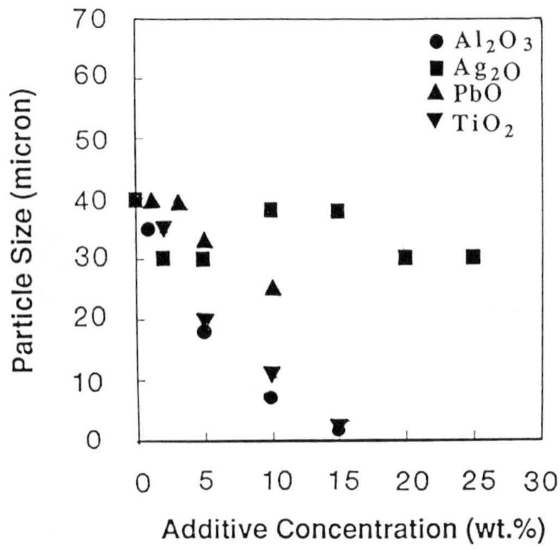

Figure 2. $YBa_2Cu_3O_{6+x}$ particle size distribution versus the additive concentration of sintered composite samples.

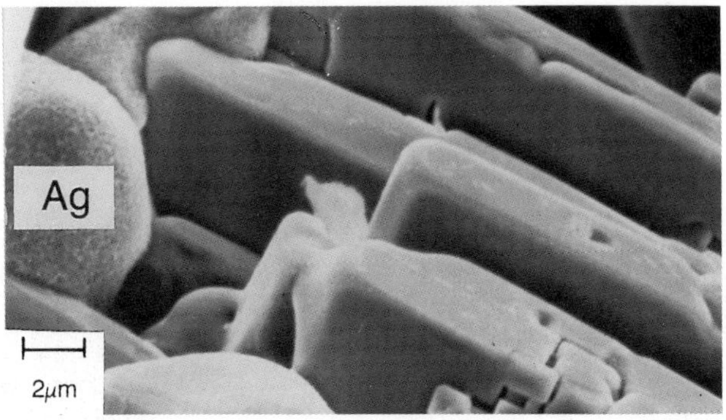

Figure 3. Scanning electron micrograph of aligned $YBa_2Cu_3O_{6+x}$ particles in sintered $YBa_2Cu_3O_{6+x}$ / 5 wt.% Ag_2O composites.

The structure of all $YBa_2Cu_3O_{6+x}$ samples containing the additives was determined as a function of superconducting (orthorhombic) or non superconducting (tetragonal) phase content of the superconductor and the detailed analysis is given elsewhere [3]. The results suggest that the superconducting orthorhombic phase of the $YBa_2Cu_3O_{6+x}$ material decreases with an increasing Al_2O_3 or TiO_2 or PbO content. However Ag_2O stabilizes the orthorhombic phase of $YBa_2Cu_3O_{6+x}$. In addition, it was also found that the oxygen liberated as the by product during the transformation of $YBa_2Cu_3O_{6+x}$ to Y_2BaCuO_y was taken up by the additives, and as a result, the additives tend to partially transform into a higher oxide.

From the electrical resistivity versus sample temperature measurement, the superconducting onset temperature (zero resistance temperature) for all samples was measured and the results are shown in Figure 4. The results suggest that the superconducting transition temperature (T_c) of samples containing Al_2O_3, PbO and TiO_2 decreases very sharply with an increase in the additive concentration. Once a critical T_c value (approximately 50 K for Al_2O_3 and 40 K for PbO) is reached, it appears that the effect of the addition of the additive (within the concentration range investigated) on T_c is not significant. However, the studies with TiO_2 as an additive (for the reduction of particle size) have suggested that the additive TiO_2 has a very adverse effect on the T_c of the composite. In addition, it was also noticed that the normal state resistance (measured at a given temperature) of all the above three composites increase with an increase in the additive concentration.

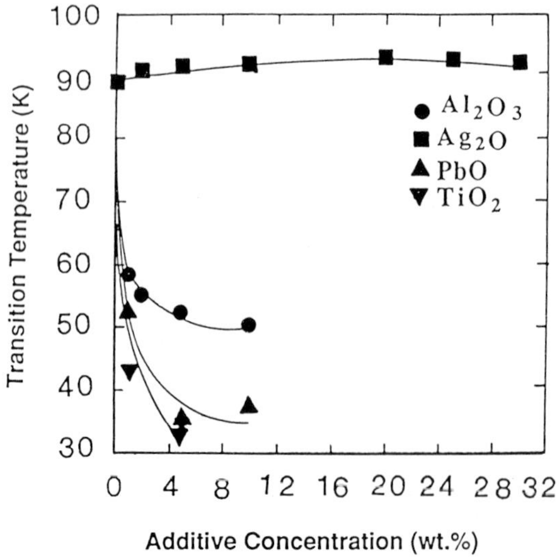

Figure 4. Superconducting transition temperature versus additive concentration of sintered $YBa_2Cu_3O_{6+x}$ composites.

$YBa_2Cu_3O_{6+x}$ / Ag_2O composites show completely opposite behavior to the above system. For example the superconducting transition temperature (zero resistance), T_c, as was measured by four point resistance method, of pure $YBa_2Cu_3O_{6+x}$ showed an increase from 88 - 90 K to 92 - 94 K with the addition of Ag_2O. The ac susceptibility measurements indicate that this increase can be as high as 97 K.

The control of the microstructure is essential for the development of both mechanical and electrical properties of high temperature superconductors. In general for superplastic flow, the important requirements are (1) a very small particle size, and (2) the retention of original crystal structure even after the deformation. Although, chemical coprecipitation methods often produce very fine particles, those methods are very tedious and time consuming. The present investigation clearly demonstrates that the addition of 15 wt. % Al_2O_3 or TiO_2 to $YBa_2Cu_3O_{6+x}$ prior to the sintering process will not only inhibit grain growth (Figure 3), but also decrease the primary particle size. Although, the present study failed to reveal an additive that will decrease the particle size without degrading the superconducting properties of $YBa_2Cu_3O_{6+x}$, it has given enough evidence to postulate a mechanism for the degeneration of the superconducting properties of $YBa_2Cu_3O_{6+x}$ in the composite. It is therefore possible that a proper selection of the additive on the basis of its chemical inertness towards $YBa_2Cu_3O_{6+x}$, may produce a preform consisting of fine particles of superconducting ceramic that is suitable for superplastic deformation.

In summary it can be suggested that the addition of Al_2O_3, TiO_2 and ($\geq$3 wt.%) PbO to fine $YBa_2Cu_3O_{6+x}$ powder prior to the sintering process produces composites with fine particle size. However, the superconducting properties of $YBa_2Cu_3O_{6+x}$ degrades with the increased concentration of the above three additives. The addition of silver oxide appears to show no effect on the particle size but it improves the superconducting properties of $YBa_2Cu_3O_{6+x}$.

ACKNOWLEDGMENT

The authors would like to thank Dr. L. F. Aprigliano for his discussion and comments, and acknowledge Ms. Lisa O'Connor and Dr. Syed Qadri for their help in obtaining the electrical resistivity measurement and x-ray diffraction analysis respectively.

REFERENCES

1. Harmer MP (1985) Adv. in Ceramics 10: 679
2. Edington J W, Melton R N and Cutler CP (1976) Prog. in Mater. Sci. 21: 61.
3. Rao AS, O'Connor LS, Aprigliano LF and Arora OP (1991) To be published in TMS Conf. Proc.
4. Crompton J and Escaig B (1980) J. Amer. Ceram. Soc. 63: 680 (1980).
5. Carry C and Mocellin A (1983) Proc. Brit. Ceram. Soc. 33: 101

Microstructure and AC-Susceptibility of (Y,Ca)$Ba_2Cu_4O_8$ Superconductors Prepared at Low (3 atm) and High (400 atm) Oxygen Pressure

NOBUO SUZUKI, TAKAHIRO WADA, ATARU ICHINOSE, YUJI YAEGASHI, and H. YAMAUCHI

Superconductivity Research Laboratory, International Superconductivity Technology Center, 10-13, Shinonome 1-chome, Koto-ku, Tokyo, 135 Japan

ABSTRACT

Superconductive $(Y_{1-x}Ca_x)Ba_2Cu_4O_8$ (x=0.0 and 0.05) ceramics were successfully prepared by firing for 160 h at 850 ℃ and at an oxygen partial pressure of 3 atm without using any catalysts. These samples were characterized by x-ray powder diffraction, scanning electron microscope (SEM) and electrical resistivity and ac magnetic susceptibility measurements. The x-ray powder diffraction pattern changed little after post-annealing for 6 h at 1070 ℃ and at oxygen partial pressure of 400 atm. The grains in the sample prepared at $P(O_2)$=3 atm were small and poorly sintered, while grains in the post-annealed sample at $P(O_2)$=400 atm were large and well sintered to be rectangular shaped. The temperature dependence of both the real and imaginary components, χ', and χ'', of the ac susceptibility was measured. The width of χ' at the superconducting transition temperature for the post-annealed samples was narrower than that for as-sintered samples.

INTRODUCTION

$YBa_2Cu_4O_8$ (1-2-4) is a superconducting material with T_c around 80 K [1]. This high-T_c superconductor has attracted much attention because of its high thermal stability in oxygen content up to about 800 ℃ [2]. Bulk 1-2-4 samples were first synthesized by means of a high oxygen pressure technique [3]. Recently, 1-2-4 samples were prepared even under normal atmospheric pressure by several independent groups [4-8]. Miyatake et al.[9] prepared calcium-substituted 1-2-4 sample by an O_2-HIP technique and discovered an increase of T_c to 90 K for $(Y_{0.9}Ca_{0.1})Ba_2Cu_4O_8$.

In this work, we synthesized $(Y_{1-x}Ca_x)Ba_2Cu_4O_8$ (x=0.0 and 0.05) samples having $(Y_{1-x}Ca_x)Ba_2Cu_3O_{7-y}$ precursors react with CuO. The samples were characterized with respect to the crystal structure, microstructure and superconducting properties. These results were compared with those of the post-annealed samples at 1070 ℃ for 6 h and at oxygen partial pressure of 400 atm.

EXPERIMENTAL PROCEDURE

Both $YBa_2Cu_4O_8$ and $(Y_{0.95}Ca_{0.05})Ba_2Cu_4O_8$ samples were prepared from high pure powders of Y_2O_3, $CaCO_3$, $BaCO_3$ and CuO. First, we prepared (Y,Ca)$Ba_2Cu_3O_{7-z}$ (1-2-3) powders and then the prepared 1-2-3 powders were mixed with appropriate amount of CuO powder to obtain stoichiometric 1-2-4. No catalysts such as Na_2CO_3 [4] and KNO_3 [6] were utilized in the present process. The mixed powder was uniaxially pressed and sintered at 850 ℃ for 160 h in oxygen gas of 3 atm. The presintered "1-2-3 + CuO" powder was also isostatically pressed by using a cold isostatic pressing (CIP) technique and sintered in a similar manner to the case of the uniaxially pressed samples. Some of the sintered ceramic bars were post-annealed for 6 h at 1070 ℃ in a mixed gas consisting of 80 % of Ar and 20 % of O_2 with a pressure of 2000 atm [i.e. $P(O_2)$=400 atm] employing a hot isostatic pressing (O_2-HIP) technique. The composition, pressing method, $P(O_2)$ for sintering and $P(O_2)$ for post-annealing for all the samples studied in this work are given in Table I .

The phases present were determined by x-ray powder diffraction using Cu-Kα radiation. Fracture surfaces of the ceramic samples were observed using a scanning electron microscope (SEM). Electrical resistivity was measured by a

Table I Compositions, pressing methods, oxygen partial pressures at sintering and post-annealing processes for $(Y_{1-x}Ca_x)Ba_2Cu_4O_8$ samples.

Sample n	Composition x	Pressing Method	$P(O_2)$ (atm) at sintering	at post-annealing
1	0.00	uniaxially	3	-
2	0.05	uniaxially	3	-
3	0.00	isostatically	3	-
4	0.00	uniaxially	3	400

conventional dc four-probe method. AC magnetic susceptibility was measured by an ac susceptometer (Lake Shore; Model 7000) as a function of temperature. The frequency of the applied field, f, was 133 Hz and the amplitude of the exciting field, H_m, ranged from 0.01 to 10 Oe. No attempts to shield the earth's magnetic field were made.

III . RESULTS AND DISCUSSION

A. X-RAY POWDER DIFFRACTION

X-ray diffraction patterns for as-sintered 1-2-4 samples with x=0.0 (sample 1) and 0.05 (sample 2) are shown in Figs. 1 (a) and 1 (b). It was clear that these samples were of 1-2-4 single phase and that no diffraction peaks for the starting 1-2-3 phase were observed. The post-annealed samples were also of 1-2-4 single phase and x-ray diffraction pattern for post-annealed sample 4 was shown in Fig. 1 (c) for reference. It was clear that the x-ray diffraction peaks for as-sintered samples were broader than those for the corresponding post-annealed samples. The broading of diffraction peaks was caused by 1-2-4 crystallites of poor quality [10, 11].

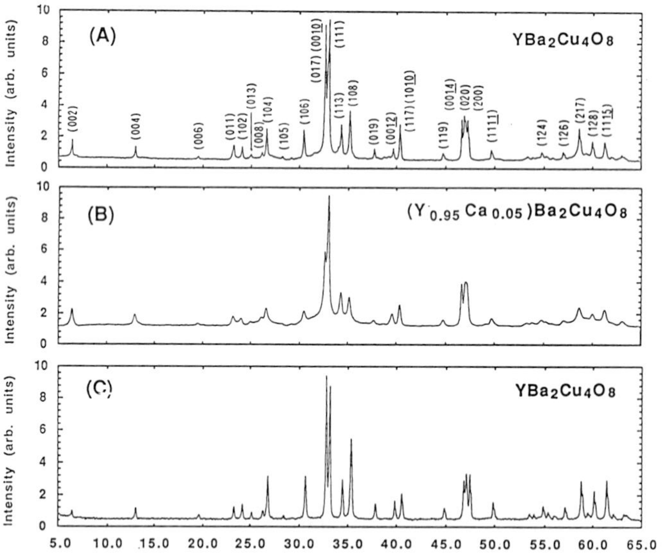

Fig.1 X-ray powder diffraction patterns for various 1-2-4 samples. (A): sample 1, (B):sample 2, (C):sample 4

B. OBSERVATION OF FRACTURED SURFACES

Scanning electron micrographs of the fractured surfaces of samples 1-4 are shown Fig. 2 (A)-(D). It is clearly observed that the grains in sample 1 are

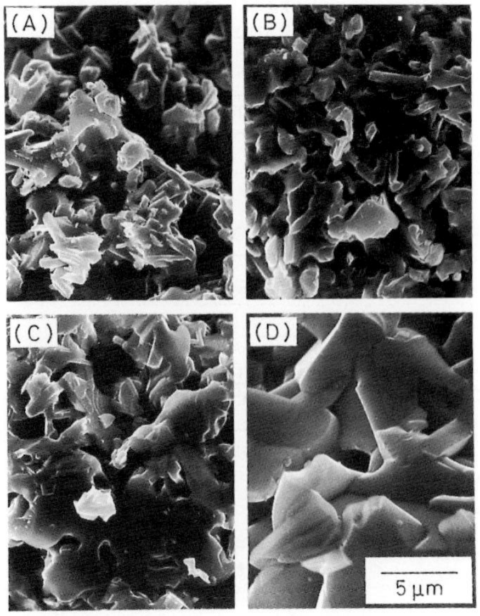

Fig.2
Microstructures for samples 1-4.
(A): sample 1, (B): sample 2,
(C): sample 3 and (D): sample 4.

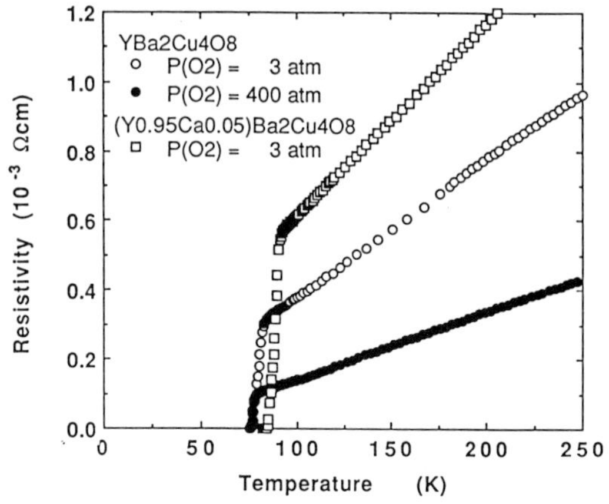

Fig.3
Temperature dependence of electrical resistivity for samples 1, 2 and 4.

small and poorly sintered, while the grains in sample 4 are large being well sintered to be rectangular-shaped. It was likely that the grown $YBa_2Cu_4O_8$ grains were of a rectangular shape and then sintered together during the post-annealing process. The grains in (Ca-doped) sample 2 and cold-isostatic-pressed) sample 3 were sintered better than those in the as-sintered sample.

C. ELECTRICAL RESISTIVITY

Figure 3 shows the temperature dependence of electrical resistivity for samples 1, 2 and 4. These samples showed metallic temperature dependence of resistivity and exhibited sharp superconducting transitions. The transition temperatures were: T_c^{on}=82 K and $T_c^{R=0}$=78 K for sample 1 and T_c^{on}=79 K and $T_c^{R=0}$=76 K for sample 4. The high resistivity for the as-sintered sample was probably due to poorly sintered grains contained in it. As sintered $(Y_{.95}Ca_{.05})Ba_2Cu_4O_8$ showed a sharp superconducting transition: T_c^{on}=91 K and $T_c^{R=0}$=85 K. The values for T_c's are higher than those for $YBa_2Cu_4O_8$ sample. The increase in T_c for Ca doped 1-2-4 may be explained by an increase in the hole concentration on the CuO_2 plane [7].

C. AC SUSCEPTIBILITY

Figure 4 (A) shows the temperature dependence of as susceptibility for sample (sample 1). All the values were normalized to the value at 5 K. The temperature dependence of both real and imaginary components, χ', and χ'', of the susceptibility indicated the presence of weak-links in the sample. For the small field amplitude, H_m of 0.1 Oe, a relatively sharp transition is observed at about 80 K. While for H_m as high as 1 or 10 Oe, the transition width significantly broadened. The widths of χ' at the superconducting transition temperature for the Ca doped sample (sample 2) and the cold-isostatically-pressed sample (sample 3) were narrower than that of the Ca-free 1-2-4 sample (sample 1). After the sample was post-annealed at 1050 ℃ and at $P(O_2)$=400 atm by employing a HIP technique, the width of χ'' became narrow. The sharp superconducting transition at high H_m for samples 4 was due to well sintered grains in these samples.

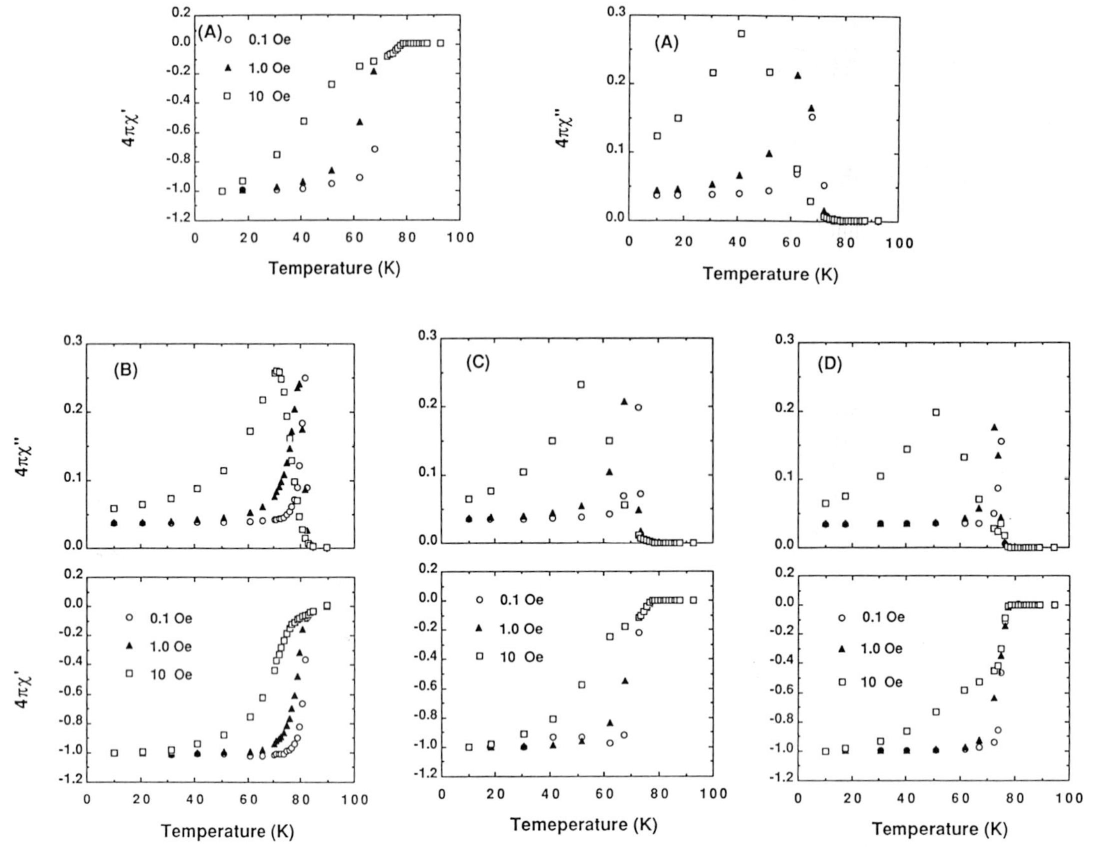

Fig.4 Temperature dependence of as susceptibility for samples 1-4. (A): sample 1, (B): sample 2, (C): sample 3 and (D): sample 4.

ACKOWLEDGMENT

We are grateful to T. Kawano for his help in ac susceptibility measurements and to K. Sakuyama for his help in SEM observations. This work was supported by New Energy and Industrial Technology Development Organization as a part of the program for R&D of Basic Technology for Future Industries.

REFERENCES

[1] Karpinski J Kaldis E Jilek E Rusiecki S Bucher B (1988) Nature **336**: 660.

[2] Wada T Suzuki N Ichinose A Yaegashi Y Yamauchi H Tanaka S (1990) Appl Phys Lett **57**: 81.

[3] Marshall AF Barton RW Char K Kapitulnik A Oh B Hammond RH Laderman SS (1988) Phys Rev **B37**: 9353.

[4] Cava RJ Krajewski JJ Peck Jr WF Batlogg B Rupp Jr LW Fleming RM Lames ACWP Marsh P (1989) Nature **338**: 328.

[5] Kourtakis K Robbins M Gallagher PK Tiefel (1989) J Mater Res **4**: 1289.

[6] Pooke DM Buckley RG Presland MR Tallon JL (1990) Phys Rev **B41**: 7220.

[7] Jin S O'bryan Gallagher PK Tiefel TH Cava RJ Fastnacht RA Kammlott GW (1990) Physica C **165**: 415.

[8] Murakami H Yaegashi S Nishino J Shiobara Y Tanaka S (1990) Jpn J Appl Phys **29**: L445.

[10] Miyatake T Yamaguchi K Takata T Goto S Koshizuka N Tanaka S (1989) Nature **341**: 41.

[11] Yamaguchi K Miyatake T Wada T Suzuki N Takata T Koshizuka N Yamauchi H Tanaka S (1990) Physica C **169**: 289.

[12] Wada T Suzuki N Yamaguchi K Ichinose A Yaegashi Y Yamauchi H Koshizuka N Tanaka S (1990) J Mater Res in publication.

[13] Miyatake T Yamaguchi K Wada T Suzuki N Willis JO Yamauchi H Koshizuka N Tanaka S (1990) J Appl Phys in publication.

Effects of Water and Organic Solvents on the Bi-Pb-Sr-Ca-Cu-O System

KATSUYOSHI HOTTA, HIDEKAZU MAGOME, YOSHIRO SUGIYAMA, TOSHIMASA SUZUKI, and HARUO HIROSE

Nippon Institute of Technology, 4-1, Gakuendai, Miyashiro, Minami-Saitama, Saitama, 345 Japan

ABSTRACT

We report the change in the superconductivity characteristics as a sample of the Bi-Pb-Sr-Ca-Cu-O system is soaked in the distilled water or in a certain organic solvent. The critical temperature shows a tendency to decrease with the water soaking time however both maximum magnetization and critical current density increase until about 18 hours after the soaking then rapidly decrease with the longer soaking time. For the effects of the organic solvents, each of such surface values as critical temperature, critical current density and maximum magnetization shows a monotonous decrease with the soaking time.

KEY WORDS: Bi system, distilled water, organic solvents

INTRODUCTION

A variety of high Tc oxide superconductors have been found so far and the applications have been also investigated intensively. Although the stabilization of the material characteristics is first required for the real applications, the related studies are still behind the other investigations. We have treated the Bi-Pb-Sr-Ca-Cu-O system for several years as it is known to be relatively stable against the liquid-water-caused deteriorations[1]. Therefore we noted the system Bi-Pb-Sr-Ca-Cu-O for the present research[2] where the samples with the system are soaked in the distilled water or in the organic solvents that are widely used in superconductor fabrication processes or with silver paste handled in laboratories. Then as a function of time during which the system is held in the solvent, we observe critical temperature, critical current density and magnetic properties as representative parameters of the superconductivity. The used solvents are methyl alcohol, ethyl alcohol, acetone and toluene[3-5].

EXPERIMENTAL

For the sample preparation, the solid phase reaction method was adopted by using the powders of Bi_2O_3, PbO, $SrCO_3$, $CaCO_3$ and CuO (purity: more than 99.9%). The powder mixture with such a weight ratio as Bi:Pb:Sr:Ca:Cu=0.7:0.3: 1.0:1.0:1.8 was stirred in a crucible for 3 hours, then pressed into a form of pellet (diameter 24 mm, thickness 7 mm). This pellet is semi-sintered in the atmosphere at the temperature 800 ℃ for 12 hours. Then the pellet is pulverized and filtered through 325 mesh to form a pellet (diameter 10 mm, thickness 2 mm) again by a press at the pressure 196 MPa. This pellet is sintered at the temperature 843 ℃ for 50 hours and cut into samples for various measurements. The adopted methods for observation and measurement are as follows:

Magnetic measurement: Vibrating Sample Magnetometer

Critical temperature Tc
and
Critical current density Jc: DC four terminal method
X ray refraction: Powder method
Surface observation: Scanning Electron Microscope.

RESULTS AND DISCUSSION

The relation between the magnetization σ_m at the liquid nitrogen temperature and the water soaking time is shown in Fig.1 where σ_m shows a rapid increase for 18 hours of the room temperature (23℃) water soaking and then decreases rapidly to a certain value for over 100 hours. This decreasing behavior of σ_m suggests that a kind of change in the superconducting phase might occur though the rising trend in the early time is considered to be attributable to the pinning effect. Some weak bonds of the structure might be broken and changed by the water invasion. The resultant critical temperature Tc and critical current density Jc were measured for such soaking periods of time as 0, 18, 240 and 860 hours. The Tc decreases monotonically with time, which is thought to be caused because the OH base or the crystal water invades the crystal lattices[6]. Figure 2 shows J-V characteristics at 77 K. The Jc after 18 hour water soaking shows a higher value than before. For the soaking time longer than 18 hours, the Jc decreases as well as σ_m. Such Jc behavior is also considered to be attributable to the same reason as mentioned for σ_m. It is noted that we can observe somewhat different behaviors with different sintering rates. Figure 3 shows an SEM image of the sample with 860 hour water soaking. With the figures including Fig.3, we can indicate that the grain boundaries are affected in connection with the water soaking time and the crystallites are broken at the weakened bonds. This observation may indicate that water will destroy the superconducting phase in the structure and the resultant damages are able to enhance the pinning effect or increase pinning sites at the certain stage of damage. The water after the experiment contains some white

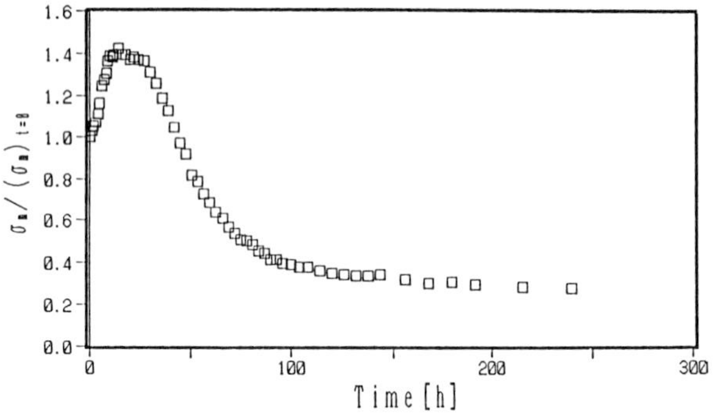

Fig.1 Water-affected behavior of the magnetization.

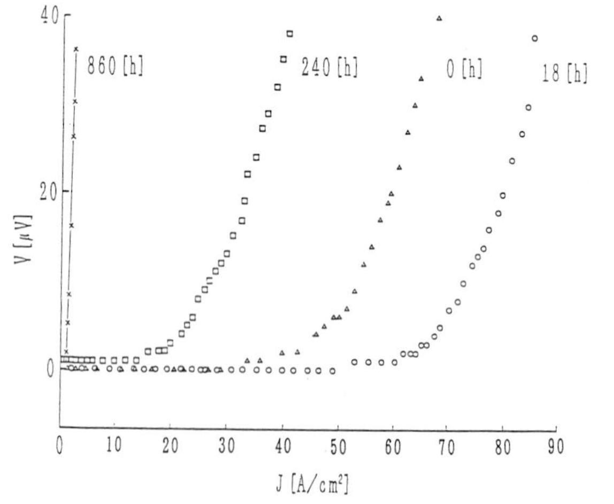

Fig.2 Voltage versus current density characteristics curves at 77 K with the soaking time as a parameter. H=0 Oe.

precipitate. The result of chemical analysis of the precipitate after 60 hour water soaking is tabulated in Table 1. We can recognize that Sr is most remarkable in weight and Ca is next. The amount of the other elements Bi, Pb and Cu are almost negligible since these elements are rarely dissolved in water in contrast with Sr and Ca. From these facts, we can consider that the changes in properties of Bi-Pb-Sr-Ca-Cu-O system with the water soaking are caused by the invasion of water, in the form of OH base or crystal water, into the crystal structure and by the dissolution of such constituent elements as Sr and Ca into water. The invasion of water will remind us of solvents such as methyl alcohol, ethyl alcohol, acetone and toluene. These will be used in the process of preparation of superconductors or in various treatments in laboratories. So that the similar experiments to the case of water are made using those solvents.

10 μ

Fig.3 SEM image of 860 hours water soaking sample.

Table 1. Result of chemical analysis of the sample after 60 hour water soaking.

Element	Bi	Pb	Sr	Ca	Cu	Total
Weight[mg]	0.175	0.053	4.030	1.377	0.102	5.737
$\times 10^{-6}$ mol	0.837	0.256	45.994	34.356	1.605	

Figure 4 shows the maximum magnetization σ_m at the liquid nitrogen temperature for the room temperature organic solvent soaking. The σ_ms show small decrease and in decreasing order are acetone, methyl alcohol, toluene and ethyl alcohol. Tc measurements as shown in Fig.5 decrease with the soaking time and X ray refraction pattern indicates decreasing of high Tc phase as well. Especially, the methyl alcohol sample was affected more severely than the others and the water absorption power of methyl alcohol is worth notice for the reason. However it is not confirmed yet which is more effective,

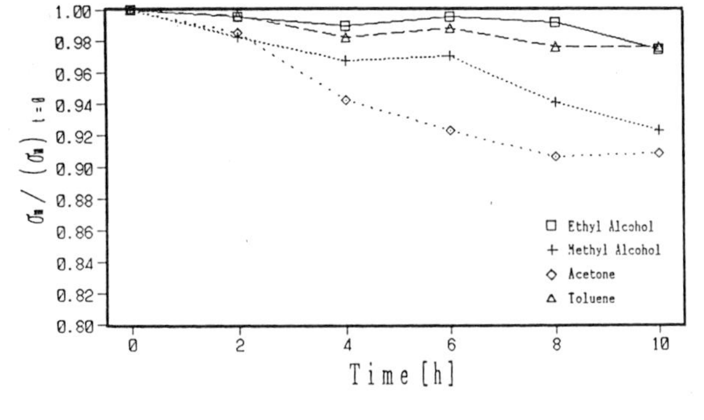

Fig.4 Organic solvent-affected behavior of the magnetization.

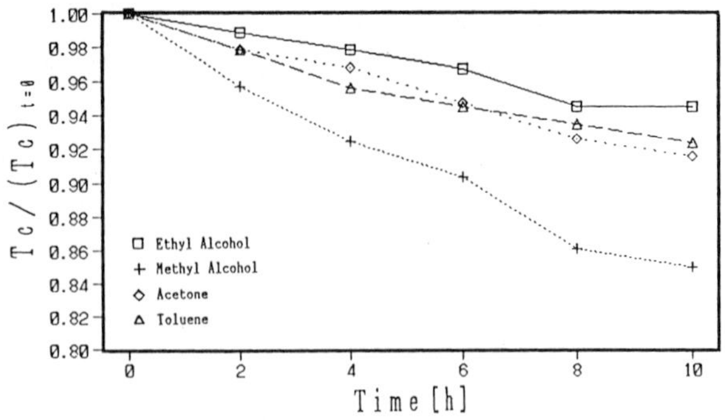

Fig.5 Critical temperature of the sample affected by organic solvent with the soaking time as a parameter.

Fig.7 SEM image of 5 hours acetone soaking sample.

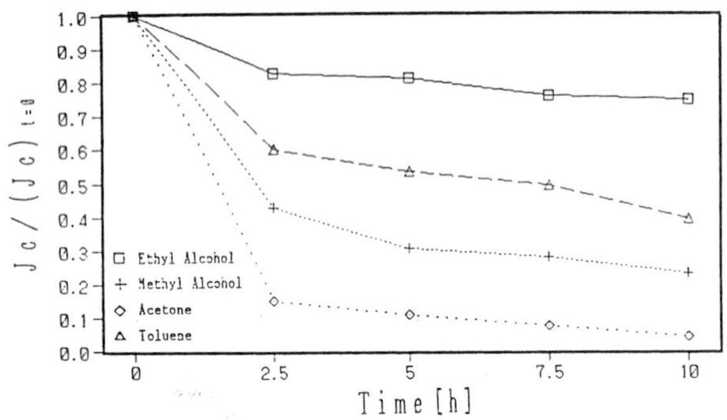

Fig.6 Critical current density of the sample affected by organic solvent with the soaking time as a parameter.

methyl alcohol itself or the water contained. In Fig.6, the critical current Jc defined as the current value at the voltage 1 μ V developed between the terminals is shown as a function of the soaking time at the temperature 77 K. The Jc decreases in the similar fashion to the σ_m and the order was acetone, methyl alcohol, toluene and ethyl alcohol. For the typical samples for 5 hour acetone soaking, an SEM image is given in Fig.7 where the sample surface after the soaking is seen influenced with the solvent. The inner part of the sample however is not affected. So that we can conclude that the organic solvents can affect at least the surface of the superconductor and should be paid attention in the process of the material. With regard to the yellowy precipitation remained after the solvent vapourization, the chemical analysis is not made yet.

REFERENCES

1. Maeda H, Tanaka Y, Fukutomi M, Asano T (1988) Jpn.J.Appi.Phys. 27: L209-210
2. Hotta K, Suzuki T, Koike Y, Hirose H (1988) J.Jpn.Soc.Powder & Powder Metallurgy 35: 970-973
3. Hotta K, Suzuki T, Koike Y, Hirose H (1989) J.Jpn.Soc.Powder & Powder Metallurgy 36: 523-526
4. Hotta K, Suzuki T, Hirose H (1990) J.Jpn.Soc.Powder & Powder Metallurgy 37: 41-45
5. Hotta K, Suzuki T, Magome H, Hirose H (1990) J.Jpn.Soc.Powder & Powder Metallurgy 37: 717-720
6. Yoshimura M, Inoue S, Ogasawara N, Ishikawa Y, Nakamura T, Takagi Y, Ryan Z Somiya S (1987) J.Jpn.Soc.Powder & Powder Metallurgy 34: 659-662

Effect of Particle Size in Raw Material CuO on Superconductivity of HTSC

SHUJI YOSHIZAWA, MANABU YANAGAWA, SHUETSU HASEYAMA, KOICHI KAWAZU, and YUICHI ITO

Central Research Laboratory, Dowa Mining Co., Ltd., Tobuki, Hachioji, Tokyo, 192 Japan

ABSTRACT

TlBaCaCuO (2223) ,ystem high-Tc superconductor was prepared, whereupon the particle size of raw material CuO as a starting material was changed from several μm to superfine powder 200A. The superconductivity represented by Tc of 123K and Jc of 530A/ cm^2 was highly improved by the use of the superfine CuO powder. In Bi system the same effect of the particle size in CuO was observed. It is suggested from the result of XRD patterns and SEM images that, as the specific surface area of CuO particle is increased, the solid-phase reaction is efficiently accelerated inducing the high-Tc phase (2223) of superconductors.

KEY WORDS: Tl system superconductor, raw material, CuO, particle size

INTRODUCTION

As synthesis and sintering of ceramics are classified as solid-state reaction, the characteristics of ceramics depend strongly on shape and size of raw materials as starting materials. There are few reports on the effect of raw materials on the superconductivity of high-Tc superconductor [1,2]. It was reported that in the case of Y system superconductors the superconductivity also depends on raw materials, particularly the form of CuO [1].

In this paper, the effect of particle size in the raw material CuO on the preparation and the superconductivity of TlBaCaCuO (2223) system high-Tc superconductor was studied as compared with Bi(Pb)SrCaCuO (2223) system.

EXPERIMENTAL

Superconducting powder was prepared by the solid-state reaction method. Figure 1 shows preparation flow of TlBaCaCuO (2223) calcined powder; powders of Tl2O3, BaCuO2, which was prepared from calcination of BaCO3 and CuO mixture, CaO, and CuO were mixed stoichiometrically at 4:2:2:3 ratio of Tl, Ba, Ca, and Cu, respectively. In order to compensate the volatile amount of Tl during calcination, excess amount of Tl was estimated by a preliminary experiment. The mixed powder was calcined in air at 830-870℃.

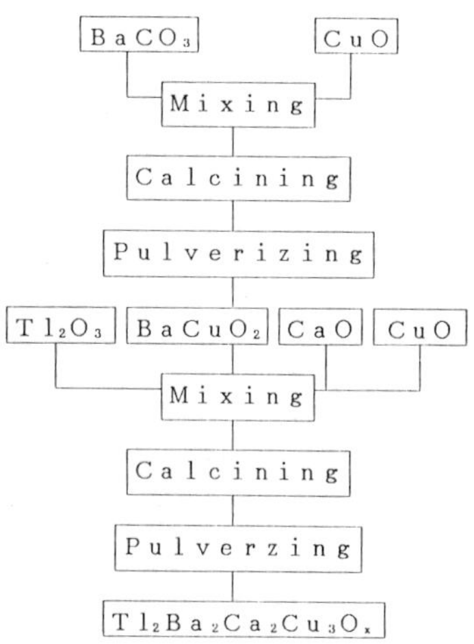

Fig.1 Preparation flow of TlBaCaCuO(2223) powder.

Raw material CuO used in this study was prepared by neutralization of copper nitrate. Particle size of CuO was dependent on the neutralization condition.
The shape and size of raw material CuO were studied by transmission electron micrography (TEM), as shown in Fig. 2. The particle size of specimens (a) and

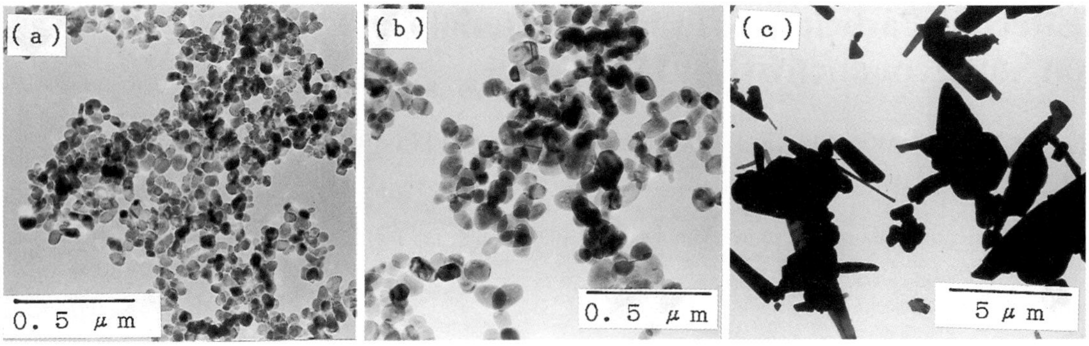

Fig.2 TEM photographs of (a) superfine CuO, S=60m²/g, (b) S=23m²/g, and (c) coarse CuO, S=0.9 m²/g.

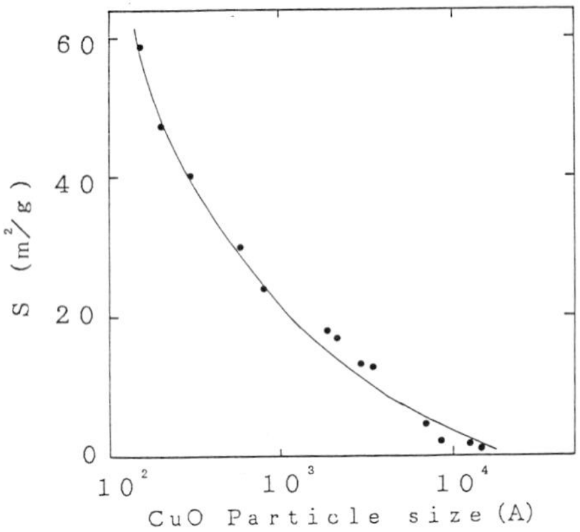

Fig.3 Relation between specific surface area (S) and particle size of CuO.

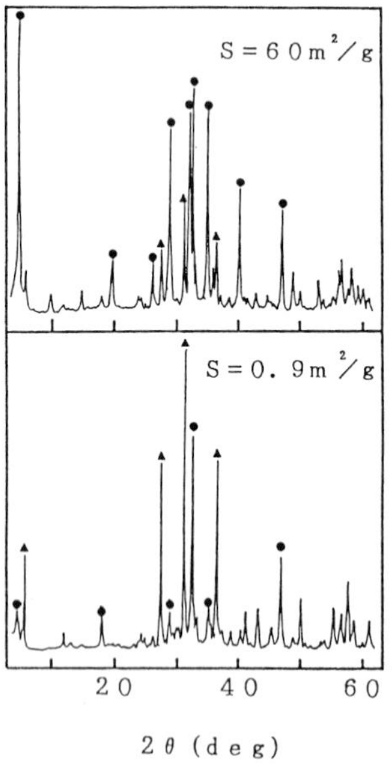

Fig.4 XRD patterns of calcined Tl system superconductive powder. ●:2223, ▲:2212.

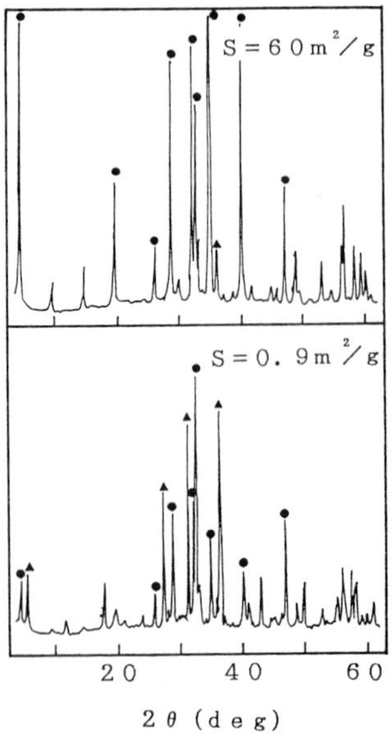

Fig.5 XRD patterns of sintered Tl system superconductive bulk. ●:2223, ▲:2212.

(b) were estimated as about 200A and 700A, respectively, whose shape were spherical. As the size increases, the shape changes to be needles and plate-like crystals shown in (c). We use specific surface area (S) determined by nitrogen absorption method in order to represent activity of CuO for reaction. Figure 3 shows the relation between specific surface area (S) and particle size of CuO. The specific surface area(S) of 200A CuO is observed to be about 60m^2/g, which is decreased with the increase of the particle size. The particle size of the other raw materials Tl_2O_3, $BaCuO_2$, and CaO was 3 μm >, 10μm > and 1μm >, respectively.

The superconductive powder was pressed into φ20 x 2mm thick pellets , and then, sintered in air at 870℃ for 10 hours.
The superconductivity was evaluated by electrical resistance and magnetic property of specimens.

RESULTS AND DISCUSSION

Reactivity in Calcining and Sintering

Reactivity of CuO with the other raw materials was studied by Xray diffraction patterns of calcined superconductive powder and sintered bulk. The diffraction patterns of Tl system which were prepared with superfine CuO and coarse CuO were shown in Figs. 4 and 5. High-Tc phase (2223) was well developed in samples prepared from superfine CuO (S=60 m^2/g), while in the specimen from CuO of 0.9m^2/g diffraction peaks based on low-Tc phase (2212) were observed.
In order to evaluate the relative proportion of high-Tc phase in the specimen, the intensity ratio of diffraction peaks (117) based on high-Tc phase and low-Tc phase. Figure 6 shows that the intensity ratio (H/H+L) of diffraction peaks of 2θ=28.78 for high-Tc phase and 2θ=31.53 for low-Tc phase increases with increasing the specific surface area (S) of CuO. This indicates the growth of high-Tc phase in the specimen. Figure 7 shows the photographs of scanning electron micrography (SEM) of calcined superconductive powder and sintered bulk. The grains are observed in Figs. 7 (a) and (b) to be well developed in both powder and bulk used S=60m^2/g CuO. In the sintered bulk (b) the plate-like crystals are developed as compared with (c) prepared from coarse CuO (0.9m^2/g).
These results suggest that using superfine CuO accelerates the solid-phase reation inducing high Tc phase and then, development of plate-like crystals, which is attributed to favorable contact of CuO and the other raw materials.

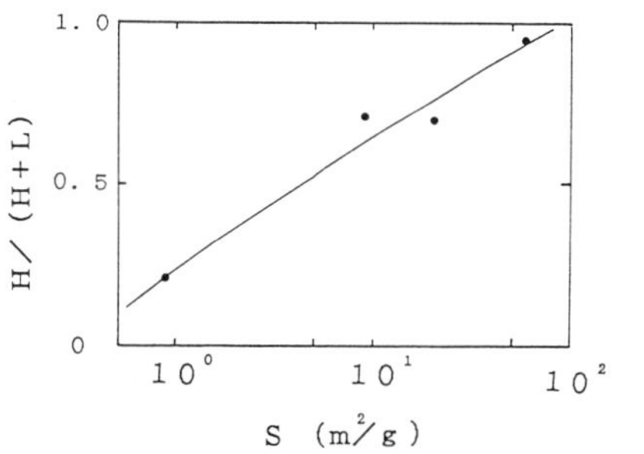

Fig.6 The growth of high-Tc phase (2223) (H) in sintered Tl system bulk with increasing specific surface area (S) of CuO.

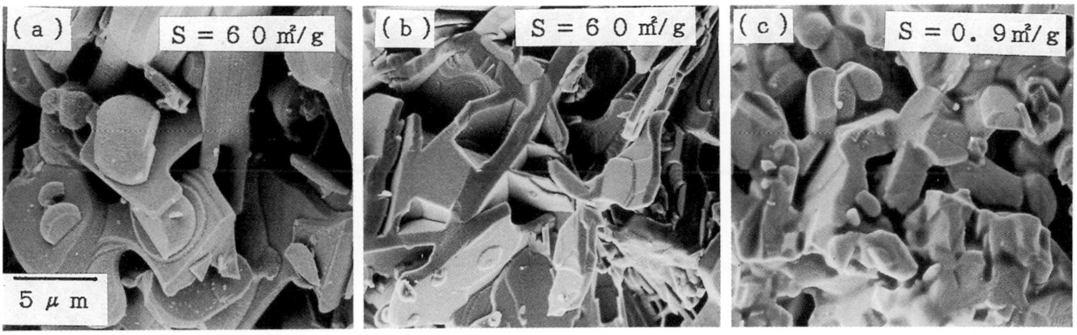

Fig.7 SEM photographs of (a) calcined Tl system superconductive powder, and (b) and (c) sintered Tl system superconductive bulk.

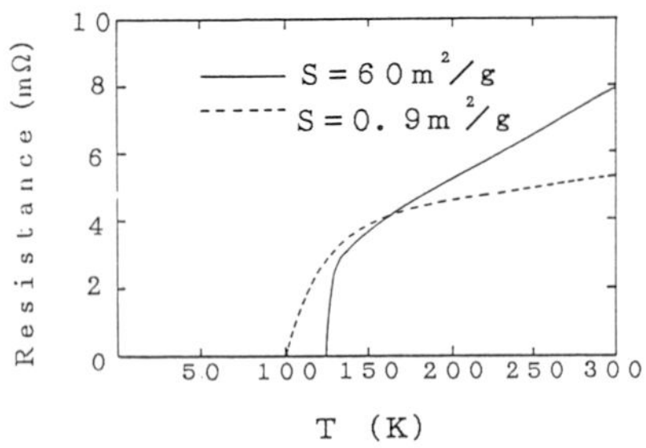

Fig.8 Temperature dependence of electirical resistenace for sintered Tl system bulk.

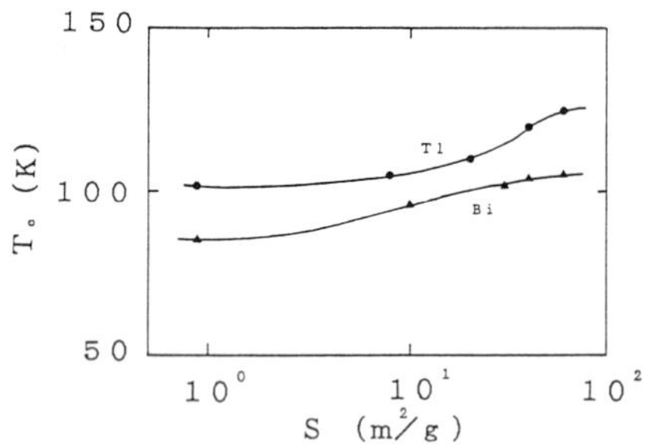

Fig.10 The behavior of Tc for Sintered Tl system and Bi system bulk with increasing specific surface area (S) of CuO.

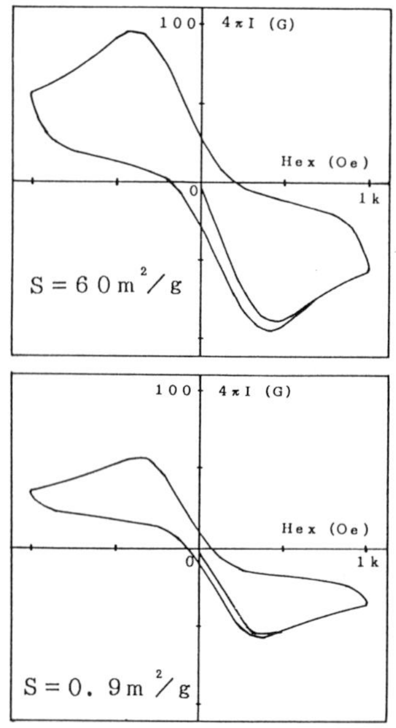

Fig.9 Magnetization curves at 77K for sintered Tl system bulk.

Superconductive Properties

The superconductivity of specimens were studied from ρ-T curve and magnetic hysteresis curve. From the temperature dependence of electrical resistance of sintered bulk in Fig. 8, Tc(zero) was estimated to be 123K and 100K for specimens by CuO of S=60m²/g and 0.9m²/g, respectively.
Figure 9 shows magnetization curves at 77K for sintered bulks. The width of hysteresis curve is dependent on the specific surface area (S), and as critical current density (Jc) calculated by Bean-London model, we obtained 530A/cm ^{2} for specimen by superfine CuO and 152A/cm ^{2} for specimen by coarse CuO. In the case of Bi(Pb)SrCaCuO (2223) system, the same results were obtained in the behavior of electric resistance and magnetic property.
The effect of the particle size of raw material CuO on the superconductivity of high-Tc Tl system and Bi system superconductors was summarized in Fig. 10. As the specific surface area increases up to about 40m²/g, Tc(zero) rises up to 120K for Tl system and 105K for Bi system.

CONCLUSION

We studied the effect of particle size of the raw material CuO on the preparation and the superconductivity of Tl (2223) system high-Tc superconductor as compared with Bi (2223) system.

(1) The Tc and Jc values in both Tl and Bi (2223) systems were affected by the particle size of raw material CuO.
(2) The Tc(zero) rises up to 120K for Tl system and 105K for Bi system by using superfine CuO as a raw material.
(3) Those improved superconductivity is due to the efficient inducing of the high-Tc phase (2223) in superconductor by the favorable contact of CuO with other the raw materials.

REFERENCES

1. Ohshima S, Okuyama K (1990) Jpn J Appl Phys 29: 1054-1058
2. Huang T.W, Huang M.P, Chin T.S, Yao P.S, Ong C.L, Hsu S.E (1989) Proc of Taiwan Inter Symp on Superconductivity, pp 319-330

Synthesis and Physical Properties of $(Tl_{1-x}Pb_x)Sr_2CuO_z$ ($z \sim 5$)

TETSUYUKI KANEKO, ATARU ICHINOSE, TAKAHIRO WADA, KAZUYUKI HAMADA, H. YAMAUCHI, and SHOJI TANAKA

Superconductivity Research Laboratory, International Superconductivity Technology Center, 10-13, Shinonome 1-chome, Koto-ku, Tokyo, 135 Japan

ABSTRACT

Single-phase $(Tl_{1-x}Pb_x)Sr_2CuO_z$ ($z\sim 5$) ("1201" phase) samples have been successfully synthesized for x = 0 - 0.5. The synthesized samples were orthorombic in the range of x = 0 - 0.12 and tetragonal in the range of x = 0.12 - 0.5. The orthorombic compounds have been obtained for the first time for family of layered oxide systems represented by the formula, $AB_2Ca_{n-1}Cu_nO_{2n+3}$ where A = Tl or (Pb, Cu) and B = Ba or Sr. All the samples for x = 0 - 0.5 showed metallic temperature dependence of resistivity. As Pb content increased, Temperature dependence of magnetic susceptibility became from Curie-type behavior to Pauli-paramagnetism and signal of thermoelectric power changed from positive to negative.

KEY WORDS: $TlSr_2CuO_5$ structure, orthorombic - tetragonal phase transition, magnetic susceptibility, thermoelectric power

INTRODUCTION

The oxide superconducting families containing Tl, Sr, Ca and Cu, $TlSr_2Ca_{n-1}Cu_nO_{2n+3}$ (n = 2 and 3), have been discovered previously [1]. Both families have superconductivity with a transition temperature 75 K and 100 K respectively. However, pure $TlSr_2CuO_5$ ("1201" phase) without other element was not successfully synthesized. The "1201" structure was realized by replacement at the Sr site by rare-earth element "RE", i. e. $Tl(Sr, RE)_2CuO_5$ [2-5]. In the compound represented by $TlSr_2Ca_{n-1}Cu_nO_{2n+3}$, the average Cu ion's valence, V(Cu), can be expressed by V(Cu) = 2 + 1/n. The V(Cu) strongly depends on the number of Cu-O layer, n. In the case of $TlSr_2CaCu_2O_7$ ("1212" phase) with n = 2 and $TlSr_2Ca_2Cu_3O_9$ ("1223" phase) with n = 3, the V(Cu)s are 2.5 and 2.33 respectively. While, in the case of $TlSr_2CuO_5$ with n = 1, the V(Cu) should be 3.0. This value is considerably higher than the valence of other copper oxide superconductor. Thus, it is difficult to synthesize $TlSr_2CuO_5$. Substitution of a rare earth element which reduces the copper ion's valence makes synthesis of "1201" phase easy. If $TlSr_2CuO_5$ can be realized, it is not expected to be a superconductor because it is heavily over-doped. However, it is very interesting material which has a very high copper oxidation state. The $(Tl_{0.5}Pb_{0.5})Sr_2CuO_z$ that did not contain a rare-earth element can be synthesized [6,7]. Therefore, we tried to obtain single-phase $(Tl_{1-x}Pb_x)Sr_2CuO_z$ with as little Pb content as possible.

EXPERIMENTAL PROCEDURE

Samples were prepared by a solid state reaction using highly pure powders of Tl_2O_3, PbO, SrO and CuO. The powders were mixed to nominal compositions of $(Tl_{1-x}Pb_x)Sr_2CuO_z$ (x = 0 - 1.0). The mixtures were pressed into pellets. The pellets were wrapped with an Au foil and sintered at 900 °C for 10 h in O_2 gas flow. The samples were analyzed by x-ray powder diffractometry (XRD) using Cu-Kα radiation. Electrical resistivity was measured as a function of temperature using a conventional four probe-method. DC magnetic susceptibility was measured using a SQUID magnetometer (Quantum Design: Model MPMS). Thermoelectric power was measured using a bar-shape sample of about 2 mm * 2mm * 10 mm. Temperature diffference between both electrodes was 0.3 K.

Synthesis and Physical Properties of $(Tl_{1-x}Pb_x)Sr_2CuO_z$ ($z \sim 5$)

TETSUYUKI KANEKO, ATARU ICHINOSE, TAKAHIRO WADA, KAZUYUKI HAMADA, H. YAMAUCHI, and SHOJI TANAKA

Superconductivity Research Laboratory, International Superconductivity Technology Center, 10-13, Shinonome 1-chome, Koto-ku, Tokyo, 135 Japan

ABSTRACT

T_c and oxygen content of $TlBa_2Ca_{n-1}Cu_nO_{y-v}$ (n= 2 and 3) have been investigated by the quenching experiment of the annealed specimen in oxygen atmosphere. The tetragonal unit cell dimension a increased and c decreased by introduction of oxygen loss. The oxygen loss, v, was determined by thermogravimetry in a nitrogen atmosphere using an oxygen-annealed specimen of $TlBa_2Ca_{n-1}Cu_nO_y$ and also by measurements of the weight differences before and after quenching. As-prepared specimen of $TlBa_2Ca_{n-1}Cu_nO_y$, n= 2 and 3, was found to be in an over-doping state and T_c increase to 110 K and 120 K respectively with increasing v up to v= 0.025-0.035.

KEY WORDS: Tl single layer system, over-doping state, quenching experiment

INTRODUCTION

High-temperature superconducting oxides of the Tl-Ba-Ca-Cu-O system with a layered structure showed a wide variety in both structure and composition [1]. Both double and single Tl layer structure have been found in this system and number of Cu layers in the perovskite units intervened by the Tl layers can be varied for 1-4 [2. 3] and 2-5 [4. 5], respectively. Non-superconductivity of oxygen annealed $Tl_2Ba_2CuO_6$ (2201) specimen has been reasonably explained in terms of over-doping due to charge transfer of Tl^{3-t}-$(Cu\text{-}O)^p$ and we have obtained evidence for the release of such over-doping by substituting Ba with La and the quenching experiment by liquid nitrogen [6. 7]. Shimakawa et. al [8] also reported that T_c variation was caused by change in oxygen content in $Tl_2Ba_2CaCu_2O_8$ and $Tl_2Ba_2Ca_2Cu_3O_{10}$.

T_c of Tl single layer system ($TlBa_2Ca_{n-1}Cu_nO_{2n+3}$) increases with increasing number of Cu layer up to n= 4, where it turns down [5]. We have already made a systematic study on superconductivity of $TlBa_2Ca_{1-x}RE_xCu_2O_7$, where RE is varied as Nd, Gd and Y, to observe the effect of amount of hole in the substituted system [9]. Starting from T_c= 80 K for x= 0, T_c increased with x to a maximum value of 100 K for x= 0.2-0.35. The average Cu valence of the stoichiometric $TlBa_2CaCu_2O_7$ and $TlBa_2Ca_2Cu_3O_9$ was calculated to be 2.50^+ and 2.33^+ respectively, suggesting an over-doping state which was more remarkable in $TlBa_2CaCu_2O_y$ than in $TlBa_2Ca_2Cu_3O_y$. Here we report the degree of over-doping of two specimens, n= 2 and 3, obtained by thermogravimetric analysis (TGA) of $TlBa_2Ca_{n-1}Cu_nO_y$ (n= 2 and 3) and the relationship between the oxygen content and superconductivity.

EXPERIMENTAL

The Tl single layer compounds can usually be synthesized by heat treatment for several hours [5]. $TlBa_2Ca_{n-1}Cu_nO_y$, n= 2 and 3, were synthesized by solid-state reaction of Tl_2O_3, BaO_2, CaO and CuO with the metal ratio of Tl:Ba:Ca:Cu= 2:2:1:2 or Tl:Ba:Ca:Cu= 2:2:2:3 respectively. The mixture was pressed into a pellet form of 10 mm diameter and 1 mm thickness. The pellet was wrapped with gold foil and fired at 885°C for 60-90 min in flowing oxygen gas of 120 ml/min. The prepared sample was checked by x-ray powder diffraction (XPD) using monochromated CuKα radiation and found to be single phase. Thermogravimetric analysis (TGA) was carried out by use of a thermobalance in flowing nitrogen gas. About 50 mg of powdered specimens was

RESULTS AND DISCUSSION

We tried to synthesize the samples with the nominal composition of $(Tl_{1-x}Pb_x)Sr_2CuO_z$ in range of x = 0 - 1.0. A nearly single-phase sample with "1201" structure from x = 0 to 0.5 colud be obtained. In the range of x greater than 0.5, peaks of impurity phase appeared in XRD patterns.

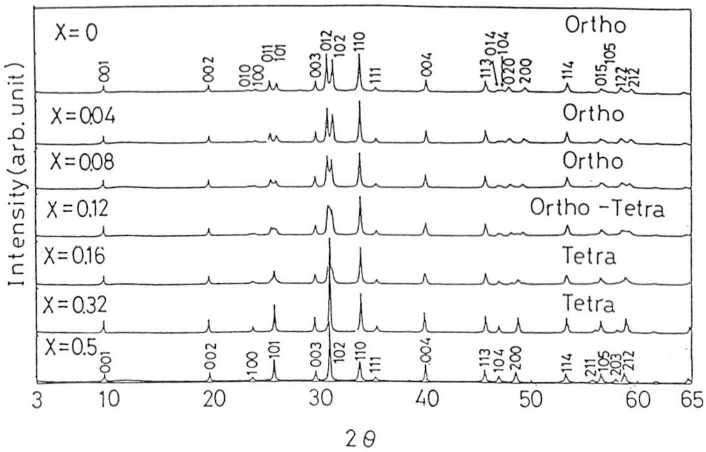

Fig. 1 Powder X-ray diffraction patterns for $(Tl_{1-x}Pb_x)Sr_2CuO_z$

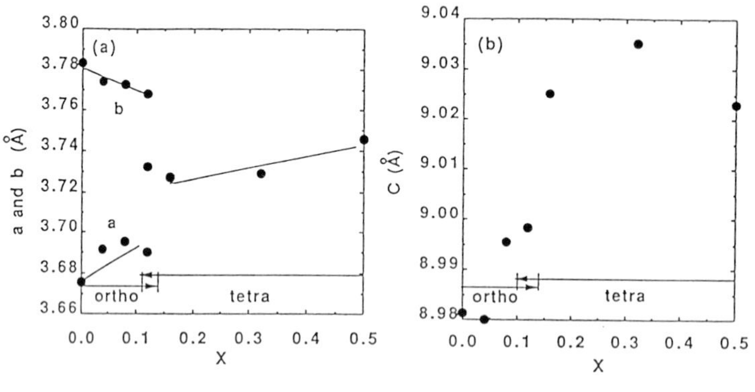

Fig. 2 Relationship between Pb content, X, and lattice constant (a) a and b, (b) c, for $(Tl_{1-x}Pb_x)Sr_2CuO_z$

Figure 1 shows XRD patterns for $(Tl_{1-x}Pb_x)Sr_2CuO_z$ samples with x = 0, 0.04, 0.08, 0.12, 0.16, 0.32 and 0.5 which are almost single-phase. The crystallographic structure of the samples with x = 0.16, 0.32 and 0.5 was tetragonal. In the samples with x $\leqq$ 0.12, the five peaks around 2θ = 23.7°, 26.7°, 31.0°, 48.6° and 49.6° corresponding to 100, 101, 102, 104 and 200 respectively in tetragonal phase spread as Pb content decreases. These peaks completely split in the samples with x less than 0.08. The XRD patterns with these splitting peaks can be indexed by a orthorhombic symmetric "1201" phase. The 101 and 102 peaks in the sample with x = 0.12 have three tops. These are due to the coexistence of the tetragonal and orthoromic phases. Thus, it is revealed that $TlSr_2CuO_z$ has an orthorombic symmetry and substitution of Pb for Tl site causes an orthorombic-tetragonal structural phase transition in $(Tl_{1-x}Pb_x)Sr_2CuO_z$. The lattice parameters, a, b and c, are shown in Fig. 2. In the orthorombic phase, the a and c axes extend and the b axis decreases with increasing Pb content. In tetragonal phase, both the a and c axes increase with increasing Pb content. The c axis has a maximum value at x = 0.32.

Figure 3 shows the temperature dependence of electrical resistivity for the samples with x = 0.0, 0.04, 0.12, 0.16, 0.32 and 0.5. All the samples show metallic behavior; the curve seem to be downward-convex rather than linear and the samples seem to have residual resistance. It is indicated that $(Tl_{1-x}Pb_x)Sr_2CuO_z$ are heavily carrier over-doped and have a Felmi-liquid like band structure. The trend seems to be that the magnitude of resistivity decreases as Pb content increases. We deduce that this is due to oxygen defects in the Cu-O plane which is shown by Reitvelt analysis [8]. That is, $TlSr_2CuO_z$ has oxygen defects in the Cu-O plane. This defect causes a decrease in the number of carriers and disturbs the passage of carriers. The substitution of Pb introduces oxygen and the Cu-O plane is fully occupied. Introduction of oxygen makes carriers and improves the passage of carriers.
Figure 4 shows the temperature dependence of dc magnetic susceptibility for the samples. This was measured with field cooling under a magnetic field of 2 T. $TlSr_2CuO_z$ has a large Curie term.

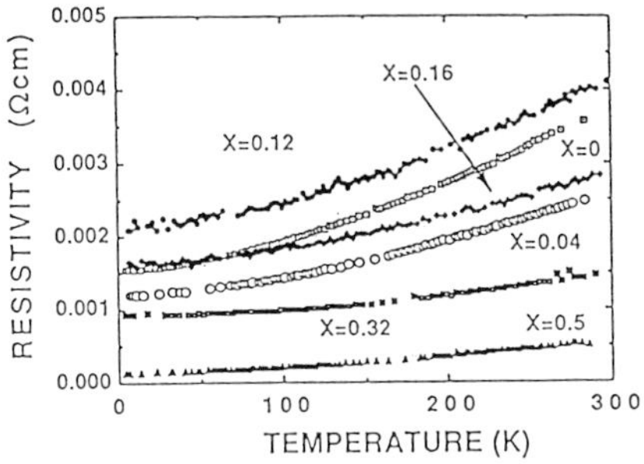

Fig. 3 Temperature dependence of resistivity for $(Tl_{1-x}Pb_x)Sr_2CuO_z$

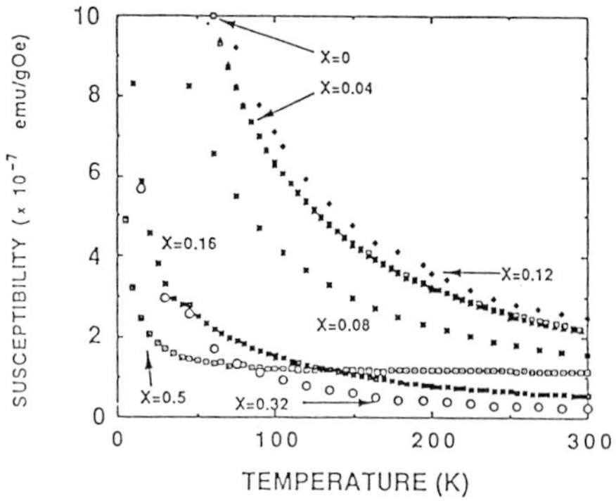

Fig. 4 Temperature dependence of dc magnetic susceptibility for $(Tl_{1-x}Pb_x)Sr_2CuO_z$

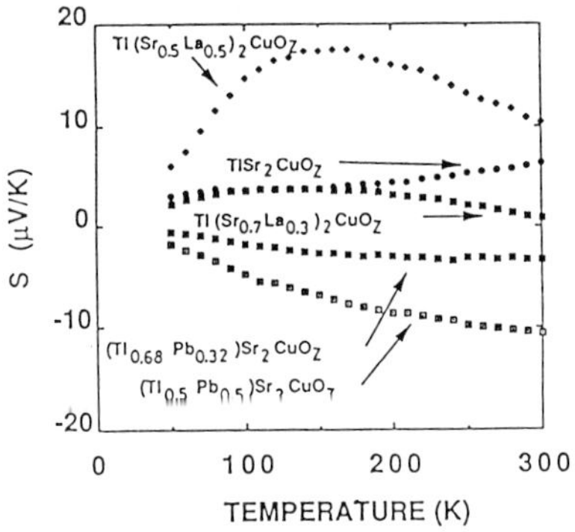

Fig. 5 Temperature dependence of thermoelectric power for "1201" phases.

The susceptibility of $TlSr_2CuO_z$ is fitted well by the Curie-Weiss law. As Pb content increases, the Curie term decreases and the susceptibility of $(Tl_{0.5}Pb_{0.5})Sr_2CuO_z$ shows a flat curve like Pauli-paramagnetism over 50 K. We assumed that the major contribution to the magnetic susceptibility is due to copper ions. Then, we calculated P_{eff} for one copper ion. The experimental value of P_{eff} for $TlSr_2CuO_z$ was derived to be 0.53. It seems that this value corresponds to the case that 30 % of copper ions have a theoretical value of 1.73 for a Cu^{2+} ion, caluculated by Hund's rule. We assume that this local moment is attributed to Cu ions that have oxygen defects. Substitution of Pb introduces oxygen and eliminates the local moment of the Cu ion. Thus, $(Tl_{0.5}Pb_{0.5})Sr_2CuO_z$ having nearly full oxygen content, z ~ 5, does not exhibit Curie-type behavior. It seems that the result for $(Tl_{0.5}Pb_{0.5})Sr_2CuO_z$, which shows Pauli-paramagnetic behavior, suggests that the "1201" phase with fully occupuied oxygen has a metallic band structure. This is consistent with the experimental observation of electrical resistivity.

Figure 5 shows the temperature dependence of thermoelectric power,"S", for some "1201" phase samples. The result for the $Tl(Sr_{0.5}La_{0.5})_2CuO_z$ sample, that is with Sr site replaced by La to reduce the valence of the copper ion, (V(Cu) ~ 2.0), and is superconducting with T_c of 40 K, shows a positive signal of S and a convex curve that has been observed in copper oxide superconductors [9]. $Tl(Sr_{0.7}La_{0.3})_2CuO_z$, which is expected to have a copper valence of 2.4 and is non-superconducting, exihibits the same behavior of positive S as one of $Tl(Sr_{0.5}La_{0.5})_2CuO_z$. However, the magnitude of the signal is much smaller than the magnitude for $Tl(Sr_{0.5}La_{0.5})_2CuO_z$. This result is consistent with the expected value for the copper valence. $TlSr_2CuO_z$ also shows positive S. However, the temperature dependence of this sample is monotonically increasing, like that of a metal. The results for the $(Tl_{0.68}Pb_{0.32})Sr_2CuO_z$ and $(Tl_{0.5}Pb_{0.5})Sr_2CuO_z$ show negative linear curve. This indicates that the main carriers are electrons in these two samples. This change of the thermoelectric power sign may be also due to the number of carriers. This phenomenon that the sign of the carriers changes from positive to negative in a Hall effect experiment for $(La_{1-x}Sr_x)_2CuO_4$ in the region of over-doping was observed as carrier concentration increases [10]. In the range of small Pb content, oxygen defects decrease the number of carriers. In the range of large Pb content, introduced oxygen produces more carriers.

Thus, the "1201" phase that has copper valence near 2+ shows a positive covex curve in the thermoelectric power as a function of temperature. As the copper valence increases, the magnitude of signal decreases, the temperature dependence becomes metallic and the sign of the signal changes from positive to negative.

SUMMARY

Single-phase $(Tl_{1-x}Pb_x)Sr_2CuO_z$ samples with x = 0 - 0.5 were successfully synthesized. All the samples having x = 0 - 0.5 did not exhibit superconductivity and showed metallic temperature dependence of resistivity, because they were over-doped in term of hole concentration. In the range of small Pb content (x = 0 - 0.12), the synthesized samples were orthorombic. In the range of large Pb content (x = 0.12 - 0.5), samples were tetragonal. The phase transition from tetragonal to orthorombic is probably caused by oxygen defects in the Cu-O plane as suggested by Reitvelt analysis. As Pb content increases, electrical conductivity improved, the temperature dependence of magnetic susceptibility changes from Curie-type behavior to Pauli-paramagnetic type, and the thermoelectric power turns from positive to negative. We have considered that the properties mentioned above are caused by oxygen content in the Cu-O layer.

ACKNOWLEDGEMENT

The authors would like to thank Mr. Matsuura and Mr. Kubo for their help in the measurement of thermoelectric power.

REFERENCES

1. Matsuda S, Takeuchi S, Soeta A, Suzuki T, Aihara K and Kamo T (1988) Jpn. J. Appl. Phys. 27: 2062
2. Adachi S, Inoue O, Hirano H, Takahashi Y and Kawashima S (1989) Jpn. J. Appl. Phys. 28: L775
3. Itoh T and Uchikawa H (1989) Jpn. J. Appl. Phys. 28: L1790
4. Mochiku T, Nagashima T, Watahiki M, Fukai Y and Asano H (1989) Jpn. J. Appl. Phys. 28: L1962
5. Subramanian M. A (1990) Mat. Res. Bull. 25: 191
6. Barry J. C, Iqbal Z, Ramakrishna B. L, Sharma R, Eckhardt H and Reidinger F (1989) J. Appl. Phys. 65: 5207
7. Kwei G. H, Shi J. B and Ku H. C: preprint
8. Ichinose A et. al.: proceeding of ISS'90.
9. Ando Y, Sera M, Fukuda Y, Kondoh S, Onoda M and Sato M (1989) Solid State Commun. 70: 303
10. Takagi K, Tokura Y and Uchida S (1989) Physica C 162-164: 1001

Synthesis and Characterization of $(Pb/Cu)SrLaCuO_y$

SEIJI ADACHI, KENTARO SETSUNE, and KIYOTAKA WASA

Central Research Laboratories, Matsushita Electric Industrial Co., Ltd.,
3-15, Yagumo-Nakamachi, Moriguchi, Osaka, 570 Japan

ABSTRACT

A new superconductor, $(Pb/Cu)SrLaCuO_y$, has been synthesized. The Rietveld refinement for powder X-ray diffraction data of a ceramic sample revealed that its structure was composed of a (Pb/Cu)-O single layer and a $Cu\text{-}O_6$ octahedron. This structure is recognized to be the simplest one among superconducting lead cuprates. Annealing treatments in nitrogen improved its superconducting properties. The single-phase sample annealed at 650°C for 24 hrs in nitrogen showed T_c(onset)=36K, T_c(zero resistivity)=34K and T_c(Meissner) of 36K.

KEY WORDS: oxide superconductor, high-T_c superconductor, lead cuprate, 1201 structure, powder X-ray diffraction, Rietveld analysis, annealing treatment

INTRODUCTION

Since the discovery of superconductivity in a La-Ba-Cu-O system by Bednorz and Müller,[1] various types of cuprate superconductors with layered structure and with a two-dimensional $Cu\text{-}O_2$ sheet have been found. Tokura pointed out that all the cuprate superconductors were composed of block layers in these crystal structures.[2] We classified p-type cuprate superconductors in terms of their structural patterns between two $Cu\text{-}O_5$ pyramids ordered along c-axis and discussed the strategy for discovery of a new superconductor.[3]

Pb-based cuprate superconductor, $Pb_2Sr_2(Ln,Ca)Cu_3O_y$, has been found by Cava et al.,[4] Subramanian et al.[5] and Inoue et al.[6] After that several variations of the Pb-based superconductor were discovered.[7-11] We designed a new material which consisted of a (Pb/Cu)-O single layer and a $Cu\text{-}O_6$ octahedron, namely "(Pb/Cu)-1201 structure", and have synthesized it successfully.[12] Moreover, superconductivity around 30K was found in the material. This "(Pb/Cu)-1201" superconductor has the simplest crystal structure in Pb-based cuprates. It is important for designing cuprates to characterize this material in comparison with other related materials and to investigate the role of block layers.

In this paper, the preparation, crystal structure [13] and annealing effects of (Pb/Cu)-1201 superconductor, $(Pb/Cu)SrLaCuO_y$, are reported.

EXPERIMENTAL

Ceramic samples were prepared by solid state reaction. PbO, $SrCO_3$, La_2O_3 and CuO powders as starting materials were mixed at nominal compositions of $(Pb_{0.6}Cu_{0.5})(Sr_{1-x}La_x)_2CuO_y$ (x=0.0-1.0 at 0.1 step). The mixture was calcined at 800°C for 5 hrs in air, pulverized and press-formed into rectangular bars. The bars were sintered at 1000°C for 2 hrs in flowing oxygen and then furnace-cooled. Temperature dependence of electrical resistance were measured by standard four-probe method and those of

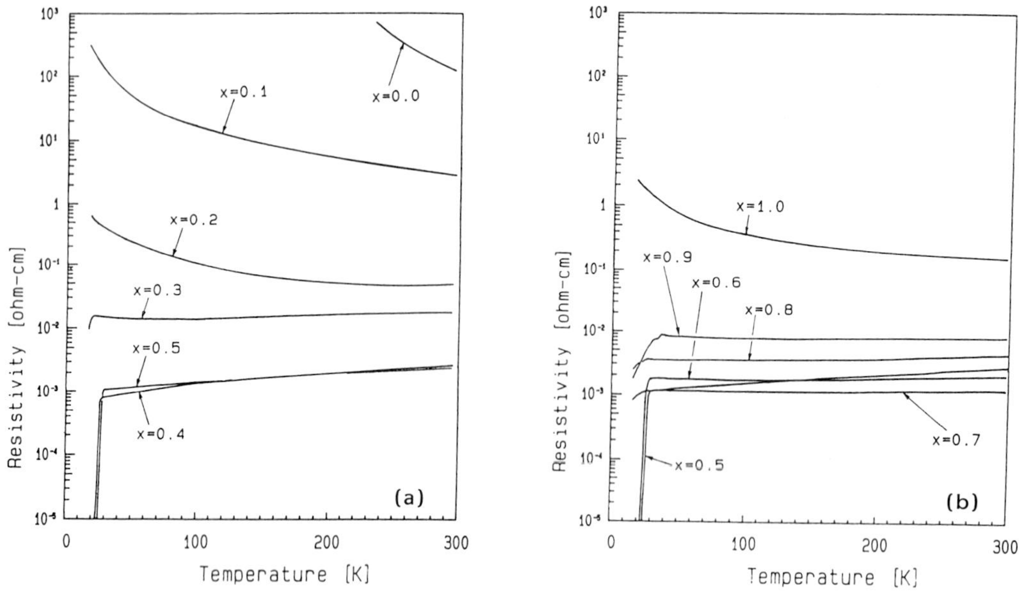

Fig.1. Temperature dependences of electrical resistivity for $(Pb_{0.6}Cu_{0.5})(Sr_{1-x}La_x)_2CuO_y$ ceramic. (a) x=0.0-0.5 (b) x=0.5-1.0

magnetization were measured under a "field cooing" condition of 10 Oe using a SQUID magnetometer (HOXAN HSM-2000). Structural properties of the samples were examined by powder X-ray diffraction (XRD; CuK_α radiation, graphite monochrometer). For a single-phase sample, the Rietveld analysis of the data collected at a step of 0.04° (in 2θ) for 4 s over a range of 15-115° was performed using a computer program "RIETAN".[14]

The samples slowly cooled at a rate of 100°C/h in flowing oxygen were also prepared. Thermogravimetric (TG) analysis (heating at a rate of 10°C/min. in dried argon) was performed, and then annealing effect in nitrogen were examined. Microstructure of fractured surfaces were observed using a scanning electron microscope (SEM).

RESULTS AND DISCUSSION

Figures 1(a) and 1(b) show the temperature dependences of electrical resistivity for the furnace-cooled samples with various values of x. Superconductivity at around 25K was observed for the samples at x=0.4-0.6. In case of x=0.5, T_c(onset) and T_c(zero resistivity) were 28K and 25K, respectively. Its temperature dependence of magnetization is shown in Fig.2. Volume fraction of the superconducting phase at liquid helium temperature was estimated to be approximately 20%. For the x=0.9 sample, T_c(onset) was observed at 38K. This behavior seemed to be caused by $(La,Sr)_2CuO_4$.

A single-phase sample was obtained at x=0.5. Figure 3 is its powder XRD pattern. Almost all peaks could be indexed by a tetragonal cell with a≅ 0.378 nm and c ≅ 0.866 nm. These cell parameters are quite reasonable for the (Pb/Cu)-1201 structure. It was considered that the (Pb/Cu)-1201 phase successfully synthesized. In case of x=0.9, peaks due to the K_2NiF_4-structure appeared.

The Rietveld analysis of powder XRD data for the x=0.5 sample was performed. An initial structure for the analysis was assumed as shown in Fig.4. Space group of this "(Pb/Cu)-1201" structure was tentatively assigned as P4/mmm (No.123) which is the same space group of "(Pb/Cu)-1212".[5] Peaks due to

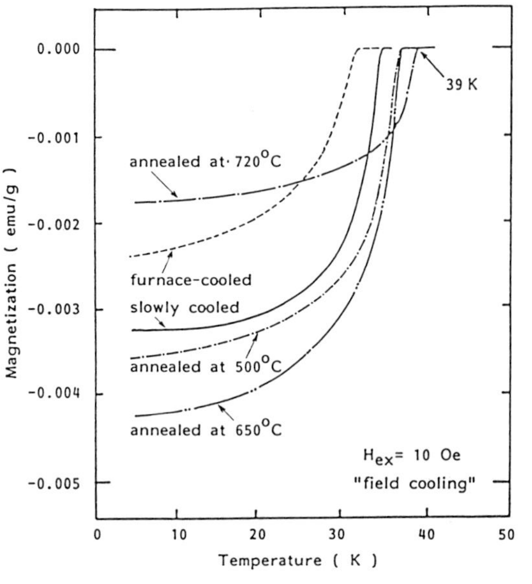

Fig.2. Temperature dependences of magnetization

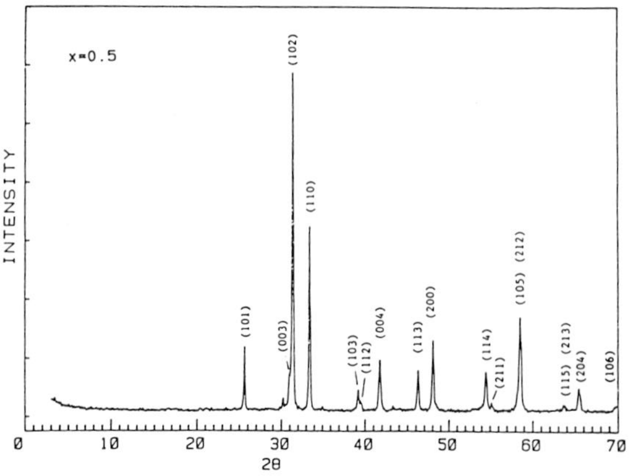

Fig.3. Powder XRD pattern of furnace-cooled sample (x=0.5)

Table I. Structural parameters of (Pb/Cu)SrLaCuO_y.

Atom	Site	x	y	z	g
Pb(1)/Cu(1)	1a	0	0	0	0.5/0.5
Sr(1)/La(1)	2h	0.5	0.5	0.2888(4)	0.5/0.5
Cu(2)	1b	0	0	0.5	1.0
O(1)	1c	0.5	0.5	0	1.0
O(2)	2e	0	0.5	0.5	1.0
O(3)	2g	0	0	0.230(3)	1.0

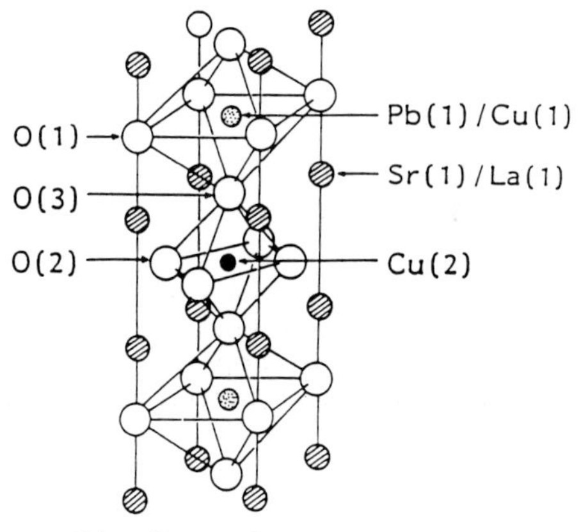

Fig.4. Schematic diagram of (Pb/Cu)SrLaCuO_y.

impurities over 2θ ranges of 21.0-21.8, 29.8-30.5 and 43.0-44.1 were eliminated for the refinement. The preferred orientation correction was made assuming that the cleavage plane was (001). The isotropic thermal parameters for cations and oxygen were fixed at 0.005nm^2 and 0.01nm^2, respectively. Occupancies for all sites were fixed at 1.00. The final structural parameters are listed in Table I. The refined lattice parameters were a=0.3787(2)nm and c=0.8652(1)nm. The reliability factors were R_{wp}=8.92%, R_p=6.47%, R_e=3.08%, R_I=9.53% and R_F=7.18%.

Annealing effects of the x=0.5 sample in nitrogen were investigated. The slowly cooled samples which absorbed sufficient oxygen under 1-atm oxygen were prepared. Its TG-curve in argon is shown Fig.5. No weight loss was observed below 500°C. Weight decreased slightly at 500-800°C and then decreased drastically. It is likely that oxygen atoms desorbed above 500°C and the decomposition with evaporation of lead occurred above 800°C. The samples were annealed at 500, 650 and 720°C. The XRD patterns for the slowly cooled sample (before annealing treatment) and the annealed samples (at 500°C and 650°C) indicated the 1201 single-phase with small impurity phase. In the XRD pattern for the sample annealed at 720°C, many peaks due to impurities were observed. Temperature dependences of magnetization are shown in Fig.2. Single-phase samples annealed at 500 and 650°C showed T_c(Meissner) of 36K. Regarding volume fraction of superconducting phase at liquid helium temperature, the sample annealed at 650°C had slightly larger fraction than that annealed at 500°C. For the sample annealed at 650°C,

T_c(onset) and T_c(zero resistivity) were 36K and 34K, respectively. Figure 6 shows an SEM photo of fractured surface for the sample annealed at 650°C. Fracture mode was intergranular . It seems that there were a lot of weak links in this ceramic sample.

SUMMARY

(Pb/Cu)SrLaCuO$_y$ whose crystal structure is recognized to be the simplest one among superconducting lead cuprates has been synthesized. The Rietveld refinement for powder X-ray diffraction data of a ceramic sample revealed that its structure was composed of a (Pb/Cu)-O single layer and a Cu-O$_6$ octahedron. Annealing treatments in nitrogen improved its superconducting properties. The single-phase sample annealed at 650°C for 24 hrs in nitrogen showed T_c(onset)=36K, T_c(zero resistivity)=34K and T_c(Meissner) of 36K.

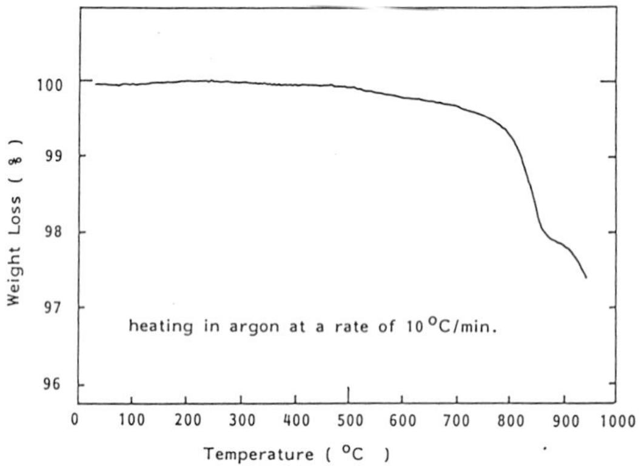

Fig.5. TG-curve for slowly cooled sample (x=0.5).

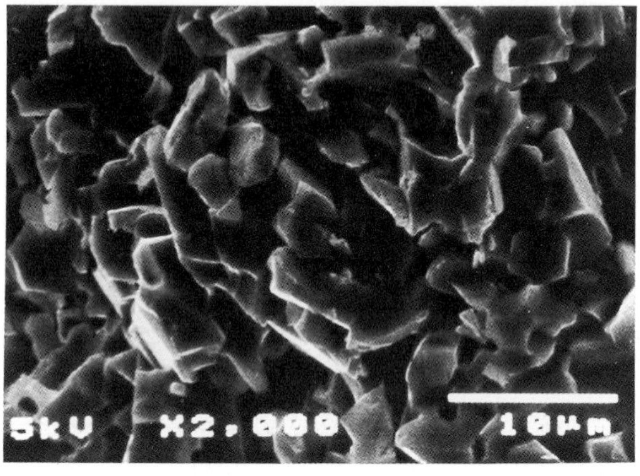

Fig.6. SEM photo of fractured surface for sample annealed at 650°C.

REFERENCES

[1] J.G. Bednorz and K.A. Müller: Z. Phys. B64(1986)189.
[2] Y. Tokura: Parity 5(1990)34 [in Japanese].
[3] S. Adachi, O. Inoue, S. Kawashima, H. Adachi, Y. Ichikawa, K. Setsune and K. Wasa: Physica C 168(1990)1.
[4] R.J. Cava, B. Batlogg, J.J. Krajewski, L.W. Rupp, L.F. Schneemeyer, T. Siegrist, R.B. van Dover, P. Marsh, W.F. Peck, Jr, P.K. Gallagher, S.H. Glarum, J.H. Marshall, R.C. Farrow, J.V. Waszczak, R. Hull and P. Trevor: Nature 336(1988)211.
[5] M.A. Subramanian, J. Gopalakrishnan, C.C. Torardi, P.L. Gai, E.D. Boyes, T.R. Askew, R.B. Flippen, W.E. Farneth and A.W. Sleight: Physica C 157(1989)124.
[6] O. Inoue, S. Adachi, Y. Takahashi, H. Hirano and S. Kawashima: Jpn. J. Appl. Phys. 28(1989)L60.
[7] H.W. Zandbergen, W.T. Fu, J.M. van Ruitenbeek, L.J. de Jongh, G. van Tendeloo and S. Amelinckx: Physica C 159(1989)81.
[8] T. Rouillon, J. Provost, M. Hervieu, D. Groult, C. Michel and B. Raveau: Physica C 159 (1989)201.
[9] A. Tokiwa, M. Nagoshi, T. Oku, N. Kobayashi, M. Kikuchi, K. Hiraga and Y. Syono: Physica C 168(1990)285.
[10] S. Koriyama, K. Sakuyama, T. Maeda, H. Yamauchi and S. Tanaka: Physica C 166(1990)413.
[11] T. Maeda, K. Sakuyama, S. Koriyama, A. Ichinose, H. Yamauchi and S. Tanaka: Physica C 169(1990)133.
[12] S. Adachi, K. Setsune and K. Wasa: Jpn. J. Appl. Phys. 29(1990)L890.
[13] S. Adachi, K. Setsune and K. Wasa: Jpn. J. Appl. Phys. 29(1990) No.10.
[14] F. Izumi: J. Crystallogr. Soc. Jpn. 27(1985)23 [in Japanese].

Syntheses and Characterization of Superconducting Lead Cuprates, $(Pb,Cu)(Ln,Ce)_2(Sr,Ln)_2Cu_2O_9$ (Ln: Rare-Earth Element)

KAZUHIRO SAKUYAMA, TOSHIHIKO MAEDA, and H. YAMAUCHI

Superconductivity Research Laboratory, International Superconductivity Technology Center,
10-13, Shinonome 1-chome, Koto-ku, Tokyo, 135 Japan

ABSTRACT

Nearly single-phase ceramics of Pb-based layered copper oxides, $(Pb_{0.5}Cu_{0.5})(Ln_{0.75}Ce_{0.25})_2(Sr_{0.875}Ln_{0.125})_2Cu_2O_z$ (z = ~ 9, Ln = Gd, Eu, Sm and Eu,Nd), were successfully synthesized in an oxidizing atmosphere. Superconductivity transitions with T_c's at 28 K, 17 K and 26 K were measured for the samples with Ln = Gd, Eu and Sm, respectively. The "1222" compounds, such as $(Pb_{0.5}Cu_{0.5})(Eu_{0.75}Ce_{0.25})_2(Sr_{0.875}Nd_{0.125})_2Cu_2O_z$ (z = ~ 9), containing two different rare-earth elements exhibited superconductivity at temperature of 32 K.

KEY WORDS : Pb-based layered copper oxides, Fluorite-type, Pb-Ln-Ce-Sr-Cu-O system, "1222" phase, Oxidizing atmosphere

INTRODUCTION

Since the superconducting lead cuprate, $Pb_2Sr_2(Y,Ca)Cu_3O_8$ of the "2213" phase with T_c near 70 K, was discovered by Cava et al. [1], a variety of Pb-based layered copper oxides have been synthesized [2] ~ [6]. Since they have the same or similar crystallographic structures to the Bi- and Tl-based superconducting cuprates [7] ~ [9], the Pb-based layered copper oxides are thought to be good candidates for superconductors having high onset temperature. Recently, after the discovery of $(Nd,Ce,Sr)_2CuO_4$ named the T* phase by Akimitsu et al. [10], several superconducting layered copper oxides, which characteristically contained "fluorite-type" block layers in the crystallographic structure, have been synthesized [11] ~ [13]. More recently, non-superconducting layered copper oxides having "fluorite-type" block layers were also synthesized [13] ~ [14]. Among the Pb-based layered cuprates, several copper oxides having "fluorite-type" structure, i.e. $Pb_2Sr_2(Ln,Ce)_2Cu_3O_{10}$ (the "2223" phase) and Pb-Pr-Sr-Cu-O system (the "1222" phase), were synthesized by Rouillon et al. [15] and Adachi et al. [16] respectively, but did not exhibit superconductivity. Previously, we reported that new superconducting Pb-based layered cuprates, $(Pb,Cu)(Eu,Ce)_2(Sr,Eu)_2Cu_2O_9$ (the "1222" phase), which had "fluorite-type" block layers and exhibited a superconductive transition at 25 K, were successfully synthesized as nearly single-phase samples after annealing in O_2 gas flow [17].

In the present study, we successfully synthesized several variations for the Pb-based "1222" phase superconductor, i.e. $(Pb_{0.5}Cu_{0.5})(Ln_{0.75}Ce_{0.25})_2(Sr_{0.875}Ln_{0.125})_2Cu_2O_z$ (z = ~ 9, Ln = Gd, Eu, Sm and Eu,Nd) in an oxidizing atmosphere. The highest superconducting onset temperature of 32 K was obtained for the sample with Ln = Eu,Nd.

EXPERIMENTAL

We prepared the samples by a conventional solid state reaction using high-purity (> 99.9%) powders of PbO, $SrCO_3$, CeO_2, Gd_2O_3, Eu_2O_3, Sm_2O_3, Nd_2O_3 and CuO. The powders were mixed into nominal compositions of $(Pb_{0.5}Cu_{0.5})(Ln_{0.75}Ce_{0.25})_2(Sr_{0.875}Ln_{0.125})_2Cu_2O_z$ (z = ~ 9, Ln = Gd, Eu, Sm and Eu,Nd). The mixture was calcined at 850°C for 10 h in air, and then pressed into parallelepiped bars of 3 x 3 x 20 mm^3. The bars were sintered at 1000 ~ 1120°C for 1 h in O_2 gas flow, then cooled to 900°C rapidly, followed by slow cooling to room temperature at a rate of 60°C/h. The phases were analyzed using an x-ray powder diffractometer equipped with a graphite monochrometer using CuKα radiation. The temperature dependence of electrcal resistivity in the range from room temperature down to 4.2 K was measured by a four-probe method. The measurements of dc magnetic susceptibility were carried out using a SQUID magnetometer.

RESULTS AND DISCUSSION

Figure 1 shows a schematic representation of the crystallographic structure of the Pb-based "1222" type layered cuprate, $(Pb,Cu)(Ln,Ce)_2(Sr,Ln)_2Cu_2O_9$ (z = ~ 9, Ln = Gd, Eu, Sm and Eu,Nd). The crystal system of this structure is body-centered tetragonal. The structure of the "1222" phase is similar to but slightly different from that of the "223" phase. The difference between the two structures is the position of the oxygen ions located in the middle plane of a unit cell. While the "223" phase has the oxygen deficient perovskite-type Cu-O layers between the apical oxygen ions of the $Cu\text{-}O_5$ pyramids, the "1222" phase has a "rock-salt" type (Pb,Cu)-O layers.

The x-ray powder diffraction (XRD) patterns measured in the range from 3° to 63° in 2θ for

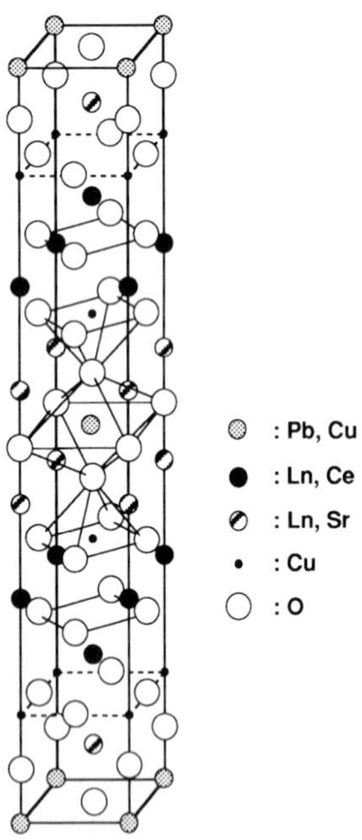

Fig. 1 Schimatic representation of a crystallographic structure of the Pb-based "1222" phase layered copper oxide, $(Pb,Cu)(Ln,Ce)_2(Sr,Ln)_2Cu_2O_9$ (z = ~ 9, Ln = Gd, Eu, Sm and Eu,Nd).

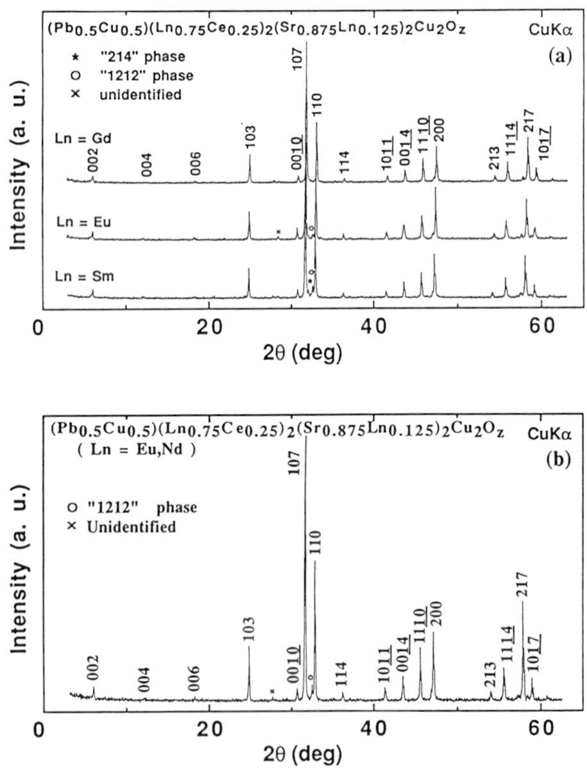

Fig. 2 Powder XRD patterns for the samples with (a) Gd, Eu and Sm as Ln and (b) Eu,Nd as Ln.

the samples of $(Pb_{0.5}Cu_{0.5})(Ln_{0.75}Ce_{0.25})_2(Sr_{0.875}Ln_{0.125})_2Cu_2O_z$ ($z = \sim 9$) with (a) Ln = Gd, Eu and Sm and (b) Ln = Eu,Nd are shown in figure 2 (a) and (b), respectively. It is shown that nearly single-phase samples of "1222" phase were obtained, although, in some samples, small amount of impurity phases such as "1212" or "214" compounds were contained.

Figure 3 shows the temperature dependence of electrical resistivity for the Pb-based "1222" phase samples with Gd, Eu and Sm as Ln. We had previously reported that, while "1222" type Pb-based cuprates with Ln = Eu sintered in air did not exhibit superconductivity, such samples became superconductive with $T_c \sim 25$ K after being annealed in O_2 gas flow. In the present work, it was confirmed that superconducting samples with Gd, Eu, Sm and Eu,Nd as Ln could be prepared by sintering in oxidizing atmosphere, followed by rapid cooling to 900°C and further cooling to room temperature at a rate of 60°C/h. The highest superconductivity onset temperature of 32 K was obtained for the sample containing Eu and Nd as Ln, and the temperature dependence of electrical resistivity for this sample was shown in figure 4.

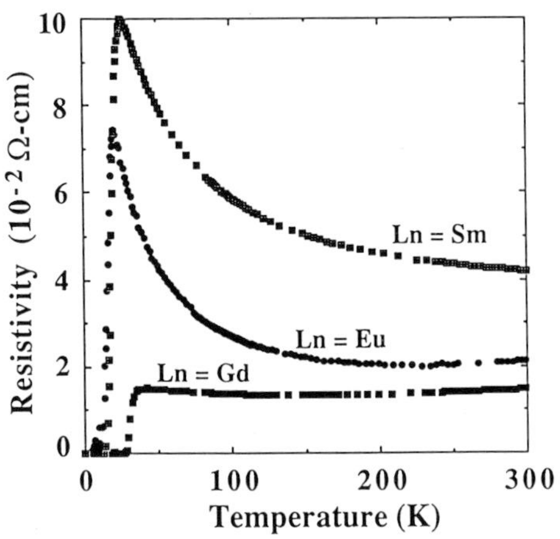

Fig. 3 Temperature dependence of electrical resistivity for the samples with Gd, Eu, Sm as Ln.

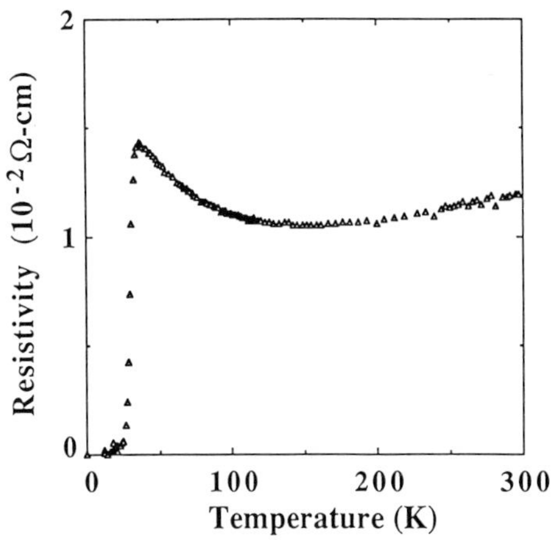

Fig. 4 Temperature dependence of electrical resistivity for the samples with Eu,Nd as Ln.

Figure 5 shows the dc magnetic susceptibility in a magnetic field of 10 Oe for the "1222" phase samples with Gd, Eu, Sm and Eu,Nd as Ln. Bulk superconductivity was observed for all the samples. Superconductivity transition temperatures 28 K, 17 K, 26 K and 32 K were measured for the samples with Ln = Gd, Eu, Sm and Eu,Nd, respectively.

In the $(Pb,Cu)(Ln,Ce)_2(Sr,Ln)_2Cu_2O_z$ ($z = \sim 9$) system, the highest T_c, 32 K, was obtained for the sample with two different rare-earth elements, Eu-Nd, as Ln while the "1222" phase with one element, Eu, as Ln exhibited the T_c at about 17 K. It seems that the substitution of Nd for Eu affects the superconductivity transition temperature. There is the possibility that the difference of ionic radii for Eu^{3+} and Nd^{3+} affects the

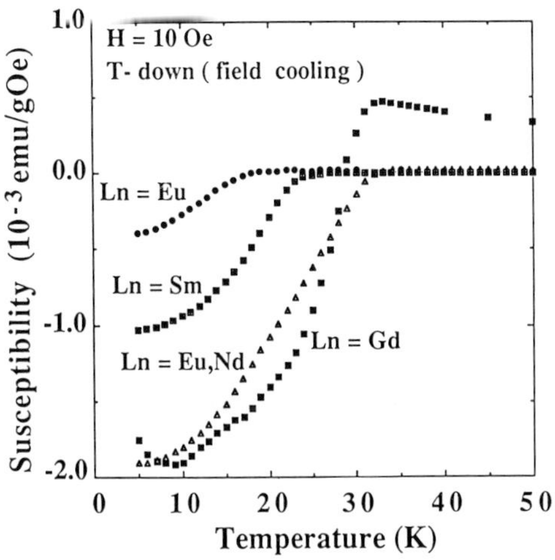

Fig. 5 Temperature dependence of dc magnetic susceptibility for the samples with Gd, Eu, Sm and En,Nd as Ln.

superconductive transition temperature. The Rietvelt analysis for the Pb-based "1222" phase cuprates is now in progress.

CONCLUSION

Nearly single-phase samples of "1222" type Pb-based layered copper oxides, $(Pb_{0.5}Cu_{0.5})(Ln_{0.75}Ce_{0.25})_2(Sr_{0.875}Ln_{0.125})_2Cu_2O_z$ (z = ~ 9, Ln = Gd, Eu, Sm and Eu,Nd), were synthesized in an oxidizing atmosphere. The samples containing two rare-earth elements such as Eu,Nd for Ln exhibited bulk superconductivity at temperature below 32 K, while the "1222" phase with Gd, Eu and Sm as Ln became superconducting at temperatures below 26 K, 25 K and 28 K, respectively.

ACKNOWLEDGEMENT

The authors would like to thank A. Ichinose of SRL-ISTEC for his helpful discussions.

REFERENCE

[1] R. J. Cava, B. Battlogg, J. J. Krajewski, L. W. Rupp, L. F. Schneemeyer, T. Siegrist, R. B. van Dover, P. Marsh, W. F. Peck Jr., P. K. Gallagher, S. H. Glarum, J. H. Marshall, R. C. Farrow, J. V. Waszczak, R. Hull and P. Trevor, *Nature* 336 (1988) 211.

[2] H. W. Zandbergen, W. T. Fu, J. M. van Ruitenbeek, L. J. de Jongh, G. van Tendeloo and Amelinckx, *Physica* C159 (1989) 81.

[3] T. Rouillon, J. Provost, M. Hervieu, D. Groult, C. Michel and B. Raveau, *Physica* C159 (1989) 201

[4] S. Koriyama, K. Sakuyama, T. Maeda, H. Yamauchi and S. Tanaka, *Physica* C166 (1990) 413.

[5] A. Ono and Y.Uchida, *Jpn. J. Appl. Phys.* 29 (1990) L586.

[6] A. Tokiwa, T. Oku, M. Nagoshi, M. Kikuchi, K. Hiraga and Y. Syono, *Physica* C161 (1989) 459.

[7] H. Maeda, Y. Tanaka, M. Fukutomi and T. Asano, *Jpn. J. Appl. Phys.* 27 (1988) L209.

[8] Z. Z. Sheng and A. M. Hermann, *Nature* 332 (1988) 55.

[9] S. S. P. Parkin, V. Y. Lee, E. M. Enger, A. I. Nazzal, T. C. Huang, G. Gorman, R. Savoy and B. Beyers, *Phys. Rev. Lett.* 60 (1988) 2539.

[10] J. Akimitsu, S. Suzuki, M. Watanabe and H. Sawa, *Jpn. J. Appl. Phys.* 27 (1988) L347.

[11] Y. Tokura, H. Takagi and S. Uchida, *Nature* 337 (1989) 345.

[12] H. Sawa, K. Obara, J. Akimitsu, Y. Matsui and S. Horiuchi, *J. Phys. Soc. Jpn.* 58 (1989) 2252.

[13] Y.Tokura, T.Arima, H. Takagi, S. Uchida, T. Ishigaki, H. Asano, R. Beyers, A. I. Nazzal, P. Lacorre and J. B. Torrance, *Nature* 342 (1989) 890.

[14] C. Martin, D. Bourgault, M. Hervieu, C. Michel, J.Provost and B. Raveau, *Mod. Phys. Lett.* B3 (1989) 993.

[15] T. Rouillon, D. Groult, M. Hervieu, C. Michel and B. Raveau, *Physica* C167 (1990) 107.

[16] S. Adachi, O. Inoue, S. Kawashima, H. Adachi, Y. Ichikawa, K. Setsune and K. Wasa, *Physica* C168 (1990) 1.

[17] T. Maeda, K. Sakuyama, S. Koriyama, A. Ichinose, H. Yamauchi and S. Tanaka, *Physica* C169 (1990) 133.

Preparation of Pb-Sr-Ca-Y-Cu-O Superconductors from Oxalate Precursors

T. IKEMACHI, T. KAWANO, S. KORIYAMA, and H. YAMAUCHI

Superconductivity Research Laboratory, International Superconductivity Technology Center, 10-13, Shinonome 1-chome, Koto-ku, Tokyo, 135 Japan

ABSTRACT

Pb-Sr-Ca-Y-Cu-O superconductors, $(Pb,Cu)Sr_2(Ca,Y)Cu_2O_7$("1212" phase) and $Pb_2Sr_2(Ca,Y)Cu_3O_8$("2213" phase), have been prepared by a coprecipitation method using metal acetates and a dimethyl oxalate. Precipitates having near starting-compositions for metal elements were obtained without a pH control. Superconductivity was observed for the samples sintered at temperatures lower than those for samples prepared by a solid state reaction method. The transition temperature onset(*Tc(onset)*) and the lattice parameters of the "1212" phase closely depended on the sintering temperature. *Tc(onset)* was 60K for samples sintered at 890°C (in O_2 gas flow) and decreased as the sintering temperature increased. In the case of the "2213" phase, *Tc(onset)* was nearly constant despite whatever sintering temperature was used. *Tc(onset)* and zero electrical resistance of samples sintered at 830°C (in Ar gas flow) were 83K and 68K, respectively.

KEY WORDS: Pb-Sr-Ca-Y-Cu-O, coprecipitation method, low temperature synthesis, "1212" phase, "2213" phase

INTRODUCTION

Since the discovery of the $Pb_2Sr_2(Ca,Y)Cu_3O_8$ superconductor reported by R. J. Cava et al.[1], many lead based cuprate superconductors have been synthesized. $Pb_2(Sr,La)_2Cu_2O_6$ and $(Pb,Cu)(Sr,La)_2CuO_5$ have one Cu-O_2 layer in each unit cell[2,3], so they are comparable to $Bi_2Sr_2CuO_6$, $Tl_2Ba_2CuO_6$, $TlBa_2CuO_5$ and so on. $(Pb,Cu)(Eu,Ce)_2(Sr,Eu)_2Cu_2O_9$ and $Pb_2Sr_2(Ln,Ce)_2Cu_3O_{10}$ (Ln=lanthanides) include a fluorite structure[4,5] that was observed in $Bi_2Sr_2(Ln,Ce)_2Cu_2O_{10}$ and $(Tl,Pb)(Sr,La)_2(Ln,Ce)_2Cu_2O_9$. The transition temperatures (*Tc*) of these Pb-cuprates are similar to those of the Bi-cuprates and Tl-cuprates, respectively, indeed $Pb_2Sr_2(Ln,Ce)_2Cu_3O_{10}$ and $(Tl,Pb)(Sr,La)_2(Ln,Ce)_2Cu_2O_9$ are not superconducting. However, in the superconductors which include two Cu-O_2 layers, the *Tc* values of Pb-cuprates, $Pb_2Sr_2(Ca,Y)Cu_3O_8$ and $(Pb,Cu)Sr_2(Ca,Y)Cu_2O_7$ are reported to be inferior to those of Bi-cuprates and Tl-cuprates[1,6~7]. Especially, the later *Tc*'s were reported to be 20~52K, are rather low values[6,7]. Also, there are several reports that the properties of these materials can be improved by changes in the preparation. For example, the $YBa_2Cu_4O_8$ system can be synthesized at a normal pressure by an alkoxide method[8], and *Tc(zero)* of $Bi_2S_2Ca_2Cu_3O_{10}$ system was improved to 117K by a coprecipitation method[9].

In this paper, we report on the preparation by a coprecipitation method and the properties of the Pb-Sr-Ca-Y-Cu-O superconductors, $(Pb_{1-x}Cu_x)Sr_2(Ca_{1-y}Y_y)Cu_2O_7$ and $Pb_2Sr_2(Ca_{1-x}Y_x)Cu_3O_8$. Compositions of "1212" phase having x=0.35 and y=0.2~0.7 were selected, according to report[6]. For "2213" phase, compositions of x=0.4~0.6 were selected, according to report[1].

EXPERIMENTAL

Acetates of $Pb(CH_3COO)_2$, $Sr(CH_3COO)_2{\cdot}9H_2O$, $Ca(CH_3COO)_2$, $Y(CH_3COO)_3$ and $Cu(CH_3COO)_2$ H_2O were used as metal precursors. These precursors were dissolved in 30~50% concentrations of acetic acid and mixed at nominal composition for metal elements. Next, a dimetyloxalate aqueous solution was added, and the resolved mixture was stirred for about one week. The original blue solution changed to a colorless liquid and a light blue coprecipitate. This coprecipitate of oxalates was filtered and dried at 100~120°C. It was then treated in a vacuum at 400~500°C to decompose

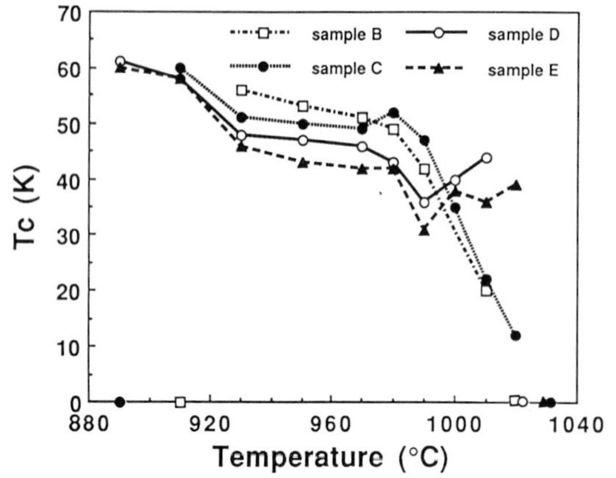

Fig.3. *Tc(onset)* vs. sintering temperature of $(Pb_{1-x}Cu_x)Sr_2(Ca_{1-y}Y_y)Cu_2O_7$. Plots on 0K line indicate that they are not superconducting.

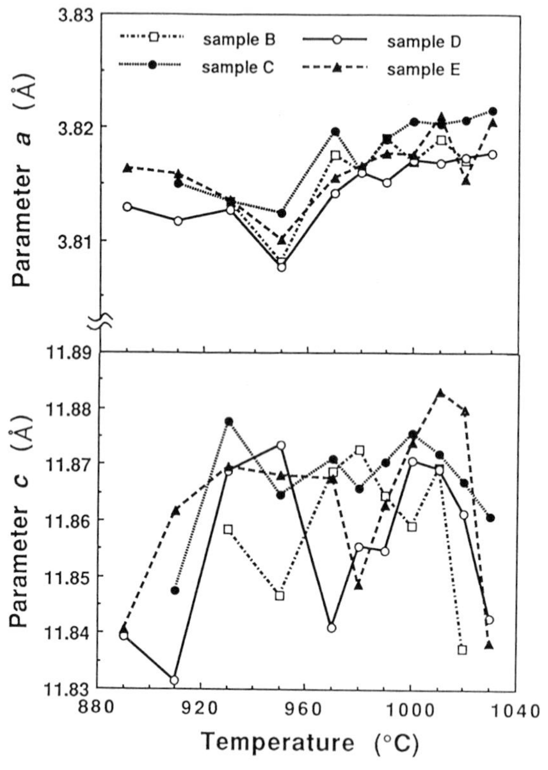

Fig.4. Lattice parameters(a, c) vs. sintering temperature of $(Pb_{1-x}Cu_x)Sr_2(Ca_{1-y}Y_y)Cu_2O_7$

seems to be no large difference in ratio of "1212" phase among these samples.

Figure 2 shows temperature dependences of resistivity for sample C. The resistivity increases and *Tc(onset)* decreases as the sintering temperature increases. In the case of samples sintered at lower temperatures the large resistivities and small transitions of resistivity are thought to be due to a lot of other phases. Similar tendencies are also recognized for samples B, D and E. The plots of *Tc(onset)* vs. sintering temperature for these samples are shown in figure 3. As the sintering temperature increases the *Tc* value decreases gradually until 970°C and drops steeply in the 970~1030°C range. Also, in relation to sample compositions in the 890~970°C range, the larger the content of Y, the higher the appearance temperature of the superconductor and *Tc* are. But in the 970~1030°C range the composition dependence seems to be different. And it is interesting that the samples sintered above 1020°C are nearly single "1212" phases, but have low *Tc* values or are not superconducting. Samples A and F, sintered at all temperature ranges, were not superconducting.

Figure 4 shows the lattice parameters(a, c) vs. sintering temperature. Parameter a, whose temperature dependence seems to be independent of composition, decreased to the sintering temperature of 950°C, where there seem to be gaps(about 0.007Å), and then increased. Parameter c seems to have two or three peaks with composition-dependent temperatures. For these matters, it is thought that large changes in "1212" phase occurred at 950~970°C and have influence on parameter a and *Tc*. And in the "1212" phase, Ca-Y in (Ca,Y) plane and Pb-Cu in (Pb,Cu)-O plane are solid-soluted, so this change of parameter a is believed that they are not related to metal element compositions, but are instead closely related to valence change ($Pb^{4+} \leftrightarrow Pb^{2+}$ or $Cu^{2+} \leftrightarrow Cu^{1+}$) in the (Pb,Cu)-O plane, oxygen content change and so on. The complex behavior of parameter c is thought to occur because of changes in oxygen concentration change, Ca-Y ordering, Pb-Cu ordering, changes in the valence of metal elements (Pb and Cu), and so on, but it can not be discussed specifically in this paper because of many unknown factors.

"2213" phase

Figure 5 shows x-ray diffraction patterns for sample H. The "2213" phase appeared below 700°C and became nearly single phase at 830°C. Above 850°C, partial melting occurred on samples. The temperature dependence of resistivity for sample H is shown in figure 6. It is recognized that there is not a large difference on *Tc(onset)*. Also, there was not a large difference among samples G~I.

oxalates. Nominal and measured compositions of samples are shown in Table I. Samples A~F were sintered at 890~1030°C for 1 hour in flowing O_2 and then dipped in liquid nitrogen for "1212" phase preparation. Samples G~I for "2213" phase preparation were sintered at 700~850°C for 1 hour before annealing at 300~500°C for 20 hours in a low oxygen pressure. Sintered samples 2 X2 X18 mm^3 in size were measured by a four probe method for electric properties. The compositions were observed by ICP(emission spectrochemical analysis) measurement. The product phases were characterized using x-ray powder diffraction analysis.

Table I. Nominal and measured compositions on metal elements

sample		composition (Pb : Sr : Ca : Y : Cu)	
type	No.	nominal	measured
1212	A	0.65 : 2.00 : 0.80 : 0.20 : 2.35	0.59 : 1.74 : 0.66 : 0.16 : 2.86
	B	0.65 : 2.00 : 0.70 : 0.30 : 2.35	0.67 : 2.11 : 0.65 : 0.28 : 2.29
	C	0.65 : 2.00 : 0.60 : 0.40 : 2.35	0.68 : 2.01 : 0.57 : 0.38 : 2.37
	D	0.65 : 2.00 : 0.50 : 0.50 : 2.35	0.67 : 2.15 : 0.46 : 0.46 : 2.26
	E	0.65 : 2.00 : 0.40 : 0.60 : 2.35	0.67 : 2.13 : 0.36 : 0.55 : 2.29
	F	0.65 : 2.00 : 0.30 : 0.70 : 2.35	0.65 : 2.20 : 0.27 : 0.64 : 2.24
2213	G	2.00 : 2.00 : 0.60 : 0.40 : 3.00	2.12 : 1.78 : 0.56 : 0.38 : 3.15
	H	2.00 : 2.00 : 0.50 : 0.50 : 3.00	2.05 : 1.95 : 0.52 : 0.47 : 3.02
	I	2.00 : 2.00 : 0.40 : 0.60 : 3.00	2.68 : 1.87 : 0.32 : 0.48 : 2.66

RESULTS and DISCUSSION

Nominal and measured compositions of samples are shown in Table I. During preparation by a coprecipitation method, a pH control should usually be done to verify composition[10]. But in our work, there is good agreement on these compositions without a pH control, so this method is regarded as useful for the preparation of these superconductors having several elements.

"1212" phase

The x-ray diffraction patterns for sample C are shown in Figure 1. The "1212" phase is observed from about 890°C, but some other phases that are thought to mainly consist of Sr_2PbO_4 or $Pb_2Sr_3CuO_x$[11] exist together. As the sintering temperature increases, the peaks of these other phases are linearly weakened in proportion to "1212" phase. In samples B, D and F this tendency was certainly observed. Also, there

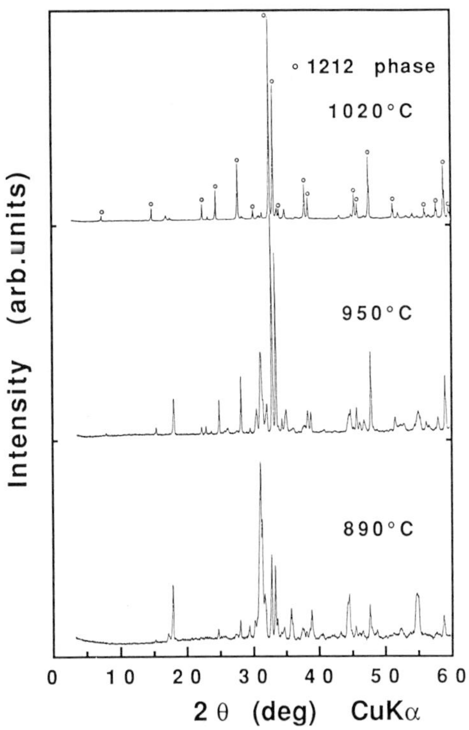

Fig.1. X-ray diffraction patterns for $(Pb_{1-x}Cu_x)Sr_2(Ca_{1-y}Y_y)Cu_2O_7$(sample C)

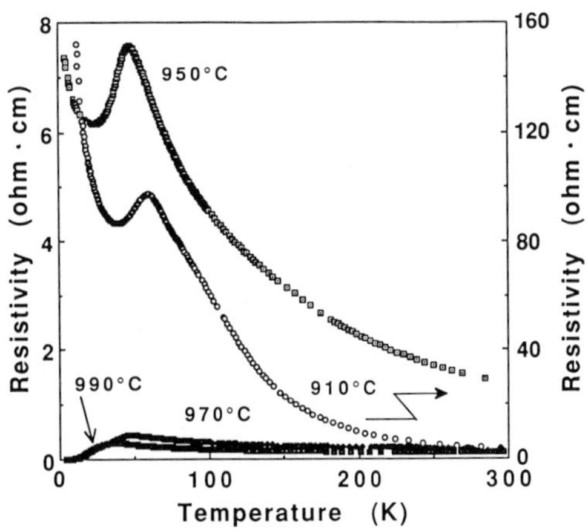

Fig.2. Temperature dependences of electrical resistivities for $(Pb_{1-x}Cu_x)Sr_2(Ca_{1-y}Y_y)Cu_2O_7$ (sample C)

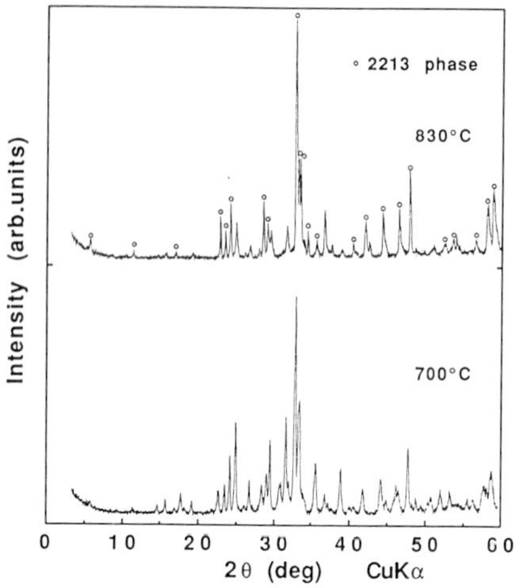

Fig.5. X-ray diffraction patterns for $Pb_2Sr_2(Ca_{1-x}Y_x)Cu_3O_8$(sample H)

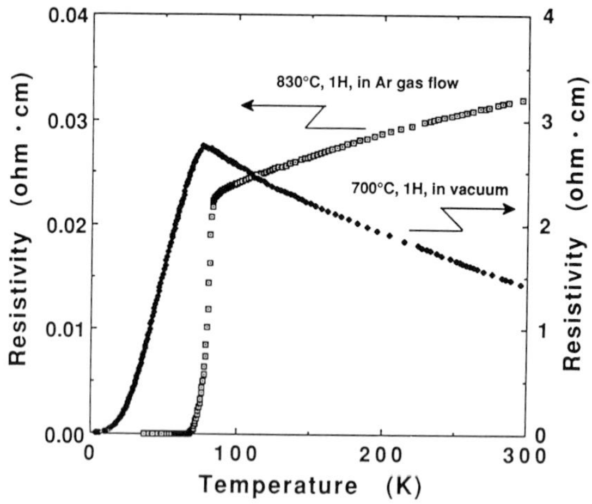

Fig.6. Temperature dependences of electrical resistivities for $Pb_2Sr_2(Ca_{1-x}Y_x)Cu_3O_8$(sample H)

CONCLUSION

(1) Pb-Sr-Ca-Y-Cu-O superconductors are successfully synthesized by a coprecipitation method. Without a pH control the metal composition was maintained at a nominal level. Single phases "1212" and "2213" were nearly obtained by sintering at temperatures lower than reported for samples prepared by the solid state reaction method.
(2) Superconductivity for the "1212" phase was observed for the samples sintered in the 890~1020°C range. *Tc* and lattice parameters were dramatically changed by the sintering temperature. This suggests that there are more than two regions on "1212" production. And the change of parameter *a* is believed to be caused by changes in valence of Pb or Cu in (Pb,Cu)-O plane, oxygen composition. In the case of "2213" phase, *Tc(onset)* was not largely dependent on sintering temperature.

ACKNOWLEDGEMENT

We would like to thank Mr. T. Maeda(SRL-ISTEC) for his useful suggestions on this work.
This work was supported by New Energy and Industrial Technology Development Organization under the management of the R & D of Basic Technology for Future Industries.

REFERENCES

1. R. J. Cava, B. Battlog, J. J. Krajewski, L. W. Rupp, L. F. Schneemeyer, T. Siegrist, R. B. van Dover, P. Marsh, W. F. Peck. Jr, P. K. Gallagher, S. H. Glaum, J. H. Marshall, R. C. Farrow, J. V. Waszczak, R. Hull and P.Trevor, Nature 336 (1988) 211.
2. H. W. Zandbergen, W. T. Fu, J. M. van Ruitenbeek, L. J. de Jongh, G. van Tendeloo and S. Amelinkx, Physica C 159 (1989) 81.
3. S. Adachi, K. Setsune and K. Wasa, Jpn. J. Appl. Phys. 29 (1990) L890.
4. T. Maeda, K. Sakuyama, S. Koriyama, A. Ichinose, H. Yamauchi and S. Tanaka, Physica C 169 (1990) 133.
5. T. Mochiku, M. Osawa and H. Asano, Jpn. J. Appl. Phys. 29 (1990) L1406.
6. T. Maeda, K. Sakuyama, S. Koriyama, H. Yamauchi and S. Tanaka, submitted to Phys. Rev. B.
7. A. Ono and Y. Uchida, Jpn. J. Appl. Phys. 29 (1990) L586.
8. H. Murakami, S. Yaegashi, J. Nishino, Y. Shiohara and S. Tanaka, Jpn. J. Appl. Phys. 29 (1990) L445.
9. J. Takada, H. Kitaguchi, T. Egi, K. Oda, Y. Miura, H. Mazaki, Y. Ikeda, Z. Hiroi, M. Takano and Y. Tomii, Physica C 170 (1990) 249.
10. Y. Zhang, Z. Fang, M. Muhammed, K. V. Rao, V. Skumryev, H. Medelius and J. L. Costa, Physica C 157 (1989) 108.
11. J. Takada, H. Kitaguchi, H. Kuniya, K. Sato, K. Oda, Y. Miura, Y. Ikeda, M, Takano, Y. Bando, R. Kanno and Y. Takeda, J. Jpn. Soc. Powder and Powder Metallurgy, 36 (1989) 533, in Japanese.

Crystal Structure and Superconductivity of New Pb Families with (Pb,Cu) Double Layer — $Pb(Ba,Sr)_2(Y,Ca)Cu_3O_y$ and $Pb(Ba,Sr)_2(Eu,Ce)_2Cu_3O_y$

A. Tokiwa[1], Y. Syono[1], T. Oku[1], M. Nagoshi[2], N. Kobayashi[1], M. Kikuchi[1], and K. Hiraga[1]

[1] Institute for Materials Research, Tohoku University, Katahira, Sendai, 980 Japan
[2] Steel Research Center, NKK Corporation, Kawasaki, 210 Japan

ABSTRACT

Crystal structure, superconductivity and phase transition of $Pb(Ba,Sr)_2(Y,Ca)Cu_3O_y$(Pb2212,y=7.0-8.4) and $Pb(Ba,Sr)_2(Eu,Ce)_2Cu_3O_y$(Pb2222,y=9.0-10.4) were investigated. Single phases of Pb2212 and Pb2222 were obtained in limited sintering and compositional conditions. $Pb(Ba,Sr)_2(Y,Ca)Cu_3O_y$ showed superconductivity with the maximum Tc of 60K. A reversible phase transition with oxygen uptake was observed in both Pb2212 and Pb2222. Lanthanoid substitutions of Pb2212 and Pb2222 were also discussed.

KEY WORDS: $Pb(Ba,Sr)_2(Y,Ca)Cu_3O_y$, $Pb(Ba,Sr)_2(Ln,Ce)_2Cu_3O_y$, oxygen uptake, (Pb,Cu) double layer, phase trasition

INTRODUCTION

Since the discovery of superconducting $Pb_2Sr_2(Y,Ca)Cu_3O_y$(Pb3212) by Cava et al.[1], several copper oxides with Pb containing layers were reported, such as $(Pb,Cu)(Ba,Sr)_2(Y,Ca)Cu_2O_y$(Pb1212)[2] with oxygen-defected lanthanoid layers, $Pb_2Sr_2NdCeCu_3O_y$(Pb3222)[3] and $(Pb,Cu)(Sr,Ln)_2(Ln,Ce)_2Cu_2O_y$(Pb1222)[4] with fluorite layers (refer to Fig. 1). Here, four digit numbers represent structure and composition, which are arranged in the order of the number of (Pb,Cu) block layers, alkaline earth (Ba,Sr) layers, oxygen-defected lanthanoid or fluorite layers, and Cu layers with CuO_2 sheets.

We also studied crystal structure and superconductivity of $Pb(Ba,Sr)_2(Y,Ca)Cu_3O_y$(Pb2212)[5-7] and $Pb(Ba,Sr)_2(Eu,Ce)_2Cu_3O_y$(Pb2222)[8]. Crystal structure of Pb2212 and Pb2222 were found to be a layered structure with (Pb,Cu) double layers, as shown in Fig. 1, by high-resolution electron microscopy[5,8]. Pb2212 showed superconductivity by Ca substitution for Y, same as Pb3212. However, no superconductivity was found in Pb2222 in spite of CuO_2 double sheets. In this paper, the effects of structure and composition on the superconductivity of Pb2212 and Pb2222 are discussed in comparison with other related compounds in the Pb system.

EXPERIMENTAL

The specimen was prepared from a mixture of PbO, BaO_2, Sr_2CuO_3, CaO, Y_2O_3, Eu_2O_3, CeO_2 and CuO. The pelletized material was sintered several times at 830 °C in a 1%O_2-N_2 mixed gas over 24 hours. The sintered pellet was subsequently quenched into liq. N_2 or slowly cooled in 1%O_2-N_2 from 830 °C, and a part of specimen was annealed in O_2 at 400-500 °C. The product was examined by X-ray powder diffraction (XPD) analysis with Cu Kα radiation. The unit cell dimensions were determined from measured d-spacings by least squares method.

Stoichiometry of cation and oxygen was examined by SEM-EDX analysis and iodide titration method, respectively. Thermogravimetic analysis (TGA) was carried out in 1%O_2-N_2. Change in lattice parameters with temperature was measured by high-temperature XPD analysis. The DC susceptibility measurement for the powder specimen was performed in a magnetic field of 20 Oe by SQUID.

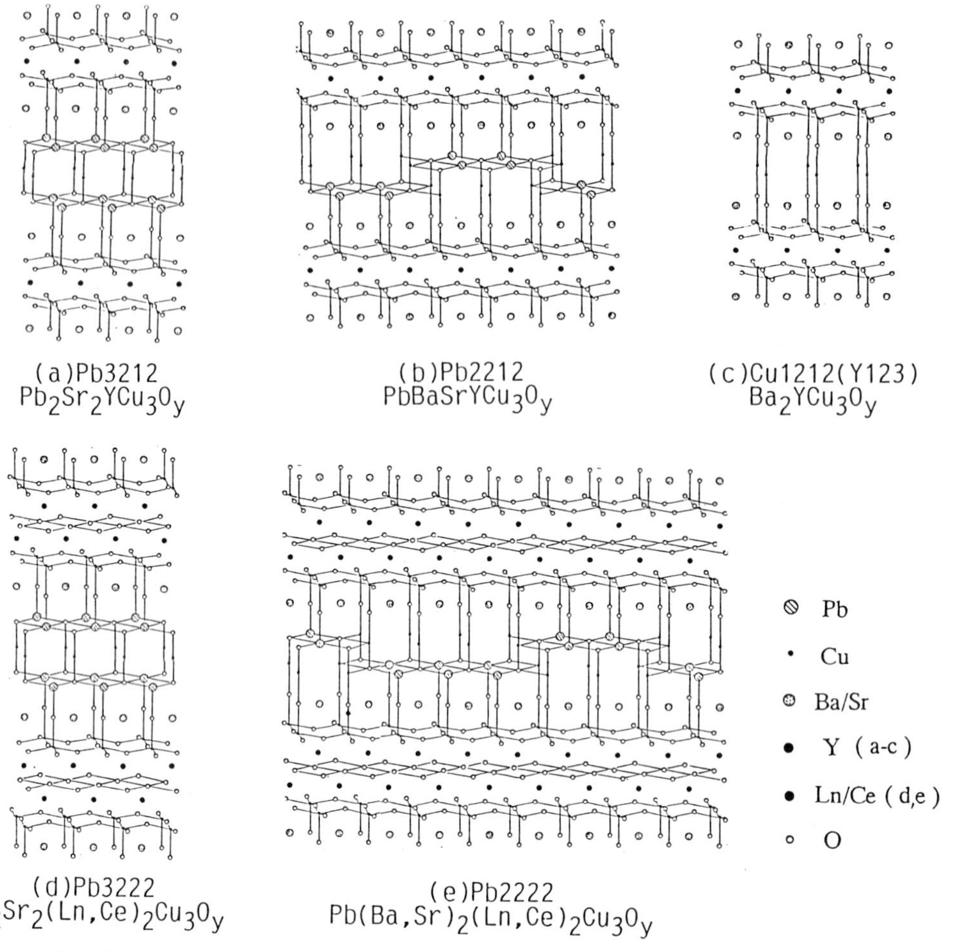

Fig. 1.Crystal structure models of superconducting-related lead-copper oxides

RESULTS AND DISCUSSION

Single phase specimen of $Pb(Ba,Sr)_2(Y,Ca)Cu_3Oy$(Pb2212) and $Pb(Ba,Sr)_2(Eu,Ce)_2$-Cu_3O_y(Pb2222) was obtained in limited sintering and compositional conditions. The proper oxygen pressure was found in about 1%O_2-N_2 atmosphere for both compounds, similar to Pb3212. Pb2212 and Pb2222 without impurities were obtained only for Sr/(Ba+Sr)=0.3-0.6 and Sr/(Ba+Sr)=0.65-0.75, respectively, although Pb3212 was obtained for Sr/(Ba+Sr)=0-1.0.

The lattice parameters, oxygen content and Tc of major compounds with (Pb,Cu) double layers are listed in Table 1. Both Pb2212 and Pb2222 have tetragonal symmetry in contrast to Pb3212 and Pb3222 with orthorhombic one. The c-axis difference between the quenched and oxygen annealed specimens are 0.18 A per (Pb,Cu)double layer in both compounds. Pb and Cu charges of the quenched specimens in (Pb,Cu) double layer are Pb^{2+} and Cu^{+}, and those of slowly cooled and oxygen annealed specimens are (Pb^{4+},Pb^{2+}) and Cu^{2+}. While Cu charge in CuO_2 sheets are always Cu^{2+} from consideration of charge neutrality and crystal structure.

Superconductivity appeared with hole introduction by Ca^{2+} substitution for Y^{3+} of Pb2212 because of constant oxygen content. The a-axis was shorter and c-axis was longer than those of the non-substituted. Ca could substitute Y up to 30% for Sr/(Ba+Sr)=0.5 and up to 40% for Sr/(Ba+Sr)=0.6 [5,6]. Tc increased with increasing Ca concentration. $PbBaSrY_{0.7}Ca_{0.3}Cu_3O_7$ and $PbBa_{0.8}Sr_{1.2}Y_{0.6}Ca_{0.4}Cu_3O_7$ showed Tc of 55K and 60K, respectively, with about 45% diamagnetic volume fraction, as shown in Fig. 2. It was noted higher Sr concentration lead to higher Ca concentration and increase in Tc.

Table 1. Lattice parameters, oxygen content, and Tc of Pb2212 and Pb2222. Every specimen was sintered in 1%O_2-N_2 at 830 °C before final heat treatment.

composition	final heat treatment	a(Å)	c(Å)	oxygen	Tc(K)
$PbBaSrYCu_3O_y$	quenched into liq. N_2	3.847(1)	27.48(1)	7.00	-
	cooled with 100 °C/h	3.852(1)	27.83(1)	8.21	-
	annealed in O_2 for 24h	3.857(1)	27.84(1)	8.40	-
$PbBaSrY_{.7}Ca_{.3}Cu_3O_y$	quenched into liq. N_2	3.837(1)	27.53(1)	7.02	55
$PbBa_{.8}Sr_{1.2}Y_{.6}Ca_{.4}Cu_3O_y$	quenched into liq. N_2	3.835(1)	27.50(1)	7.01	60
$PbBa_{.7}Sr_{1.3}EuCeCu_3O_y$	quenched into liq. N_2	3.854(1)	16.374(2)	9.00	-
	cooled with 100°C/h	3.858(1)	16.549(3)	10.15	-
	annealed in O_2 for 24h	3.859(1)	16.560(3)	10.43	-

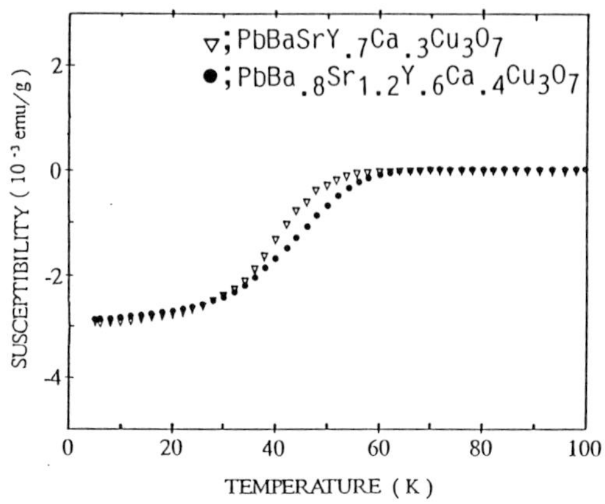

Fig. 2. Temperature dependence of DC susceptibility of Ca substituted Pb2212 with zero field coolig.

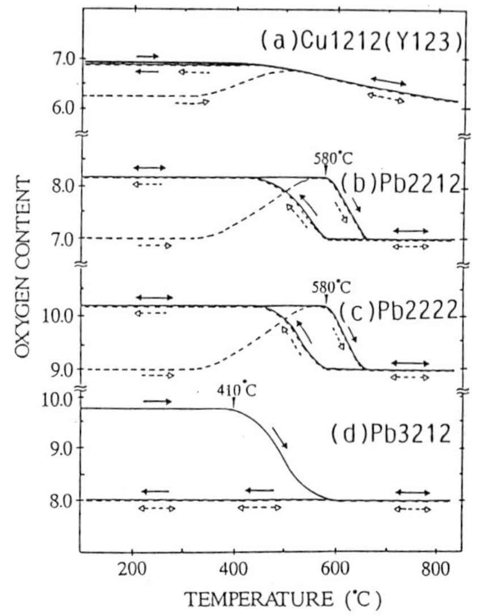

Fig. 3. Oxygen content change by thermogravimetric analysis (a)$Ba_2YCu_3O_y$, (b)$PbBaSrYCu_3O_y$, (c)$PbBa_{.7}Sr_{1.3}EuCeCu_3O_y$, (d)$Pb_2Sr_2YCu_3O_y$, ---- ;started from quenched specimen, ——;from slowly cooled specimen, heating rate;5 °C/min, and flowing gas; 1%O_2-N_2.

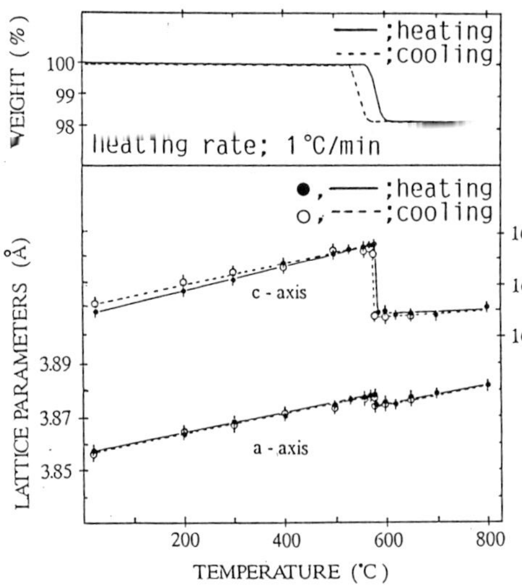

Fig. 4. Temperature dependence of weight (upper) and lattice parameters (lower) of $PbBa_{.7}Sr_{1.3}EuCeCu_3O_y$ (Pb2222) in flowing 1%O_2-N_2.

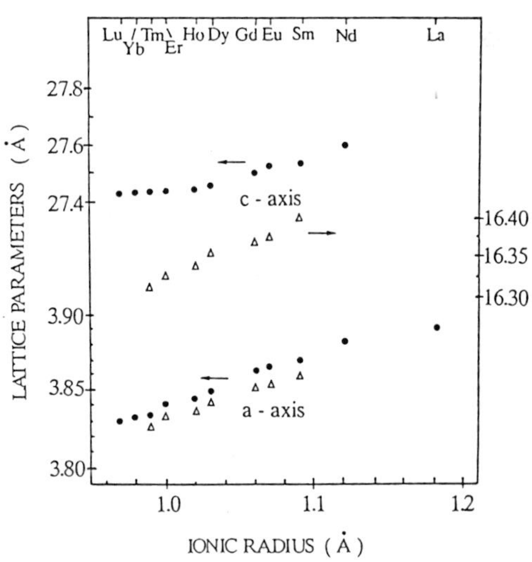

Fig. 5. Change of lattice parameters by lanthanoid(Ln) substitutions in $PbBaSrLnCu_3O_y$(Pb2212); ● and $PbBa_{.7}Sr_{1.3}LnCeCu_3O_y$(Pb2222);△.

Oxygen absorption and desorption of Cu1212(Y123), Pb2212, Pb2222 and Pb3212 with temperature is shown in Fig. 3. The slowly cooled specimens of Pb2212 and Pb2222 with oxygen content y=8.2 and y=10.2, respectively, lost 1.2 oxygen atoms above 580 °C with abrupt and reversible transition (Fig. 3(b),(c)). This reaction occurred in a stoichiometric fashion in contrast to well-known non-stoichiometric transition of Cu1212(Y123), as shown in Fig. 3(a). Pb3212 showed stoichiometric transition similar to Pb2212 and Pb2222, but the reaction was irreversible for oxygen uptake, as shown in Fig. 3(d)

Abrupt and reversible change in lattice parameters was also observed in high temperature X-ray powder diffraction analysis, as shown in Fig. 4. The starting temperature of oxygen desorption coincides with that in thermogravimetric analysis(TGA) (Fig. 4(a) and(b)). The transition interval in X-ray analysis was narrower than in TGA because of longer holding time. Both diffraction peaks of low and high temperature phases were observed during phase transition, indicating the stoichiometric phase transition.

Change in lattice parameters with lanthanoid(Ln) substitutions of Pb2212 and Pb2222 is shown in Fig. 5. Single phase of $PbBaSrLnCu_3Oy$(Pb2212) was obtained for all trivalent lanthanoid ions from La to Lu, but that of $PbBa_{0.7}Sr_{1.3}LnCe$-Cu_3O_y(Pb2222) was only from Sm to Tm. Both a- and c-axes increased with increasing ionic radii, except for c-axis of Pb2212 with small size lanthanoid substitutions, similar to Pb3212[9].

Pb2222 with CuO_2 sheets is expected to show superconductivity by introduction of holes. However, no superconductivity was found by cation substitutions due to instability of Pb2222 phase. High oxygen-pressure treatment may be effective to realize superconductivity in Pb2222, similar to $Bi_2Sr_2(Ln,Ce)_2Cu_2O_y$ (Bi2222)[10] with fluorite layers.

In summary, the lead-copper oxides with (Pb,Cu) double layers are characterized as follows. Pb2212 and Pb2222 are synthesized only in reducing atmosphere. Requirement for proper size of alkaline earth ions is severer than in Pb3212. $Pb(Ba,Sr)_2(Y,Ca)Cu_3O_y$ shows superconductivity with the maximum Tc of 60K. The reversible and stoichiometric phase transition occurs around 580 °C in $1\%O_2$-N_2, accompanied with oxygen absorption and desorption in (Pb,Cu) double layers. Almost single phases are obtained for all trivalent-lanthanoid substitutions of Pb2212 from La to Lu by sintering in $1\%O_2$-N_2, but those of Pb2222 are only from Sm to Tm with the same condition.

This work was partly supported by a Grant-in-Aid for Scientific Research by the Ministry of Education, Culture and Science, Japan.

REFERENCES

1. Cava RJ, Batlogg B, Krajewski JJ, Rupp LW, Schneemeyer LF, Siegrist T, van Dover RB, Marsh P, Peck WF Jr, Gallagher PK, Glarrum SH, Marshall JH, Farrow RC, Waszczak JB, Hull R ,Trevor P (1988) Nature 336: 211.
2. Koriyama S, Sakuyama K, Maeda T, Yamauchi H, Tanaka S (1990) Physica C 166: 413.
3. Rouillon T, Groult D, Hervieu M, Michel C, Raveau B (1990) Physica C 167: 107.
4. Maeda T, Sakuyama K, Koriyama S, Ichinose A, Yamauchi H, Tanaka S (1990) Physica C 169: 133.
5. Tokiwa A, Oku T, Nagoshi M, Kikuchi M, Hiraga K, Syono Y (1990) Physica C 161: 459.
6. Tokiwa A, Nagoshi M, Oku T, Kobayashi N, Kikuchi M, Hiraga K, Syono Y (1990) Physica C 168: 285.
7. Tokiwa A, Nagoshi M, Syono Y (1990) Physica C 170: 437.
8. Tokiwa A, Oku T, Nagoshi M, Shindo D, Kikuchi M, Oikawa T, Hiraga K Syono Y (1990) Physica C, in press.
9. Schneemeyer LF, Cava RJ, James ACWP, Marsh P, Siegrist T, Waszczak JV, Krajewski JJ, Peck WP Jr, Opila RL, Glarum SH, Marshall JH, Hull R, Bonar JM (1989) Chemistry of Materials.
10. Arima T, Tokura Y, Takagi H, Uchida S, Beyers R, Torrance JB (1990) Physica C 168: 79.

Crystal Structures and Physical Properties of Layered Perovskite Compounds $Sr_{n+1}V_nO_{3n+1-\delta}$ (n=1,2,3 and ∞)

NOBUAKI SUZUKI, TATSUO NORITAKE, NOBUYUKI YAMAMOTO, and TATSUMI HIOKI

Toyota Central Research and Development Laboratories Inc.,
Nagakute-cho, Aichi-gun, Aichi, 480-11 Japan

ABSTRACT

Single-phase polycrystalline materials of a series of layered perovskite compounds $Sr_2VO_{3.64}$ (n=1, δ=0.36), $Sr_3V_2O_{7.01}$ (n=2, δ=0), $Sr_4V_3O_{9.60}$ (n=3, δ=0.4), and $SrVO_{2.84}$ (n=∞, δ=0.16) were synthesized by the solid-state reaction in flowing hydrogen atmosphere. The crystal structures were determined by X-ray diffraction with Rietveld analysis and the electrical and magnetic properties were investigated. The n=1, 2, and 3 compounds have tetragonal structures (I4/mmm) and the a-axis lattice constant increases with increasing n. The n=1 compound is a semiconductor and the n=2, 3, and ∞ compounds exhibit metallic behaviors. The resistivity at room temperature decreases as n increases. The magnetic susceptibility for the n=2 and 3 compounds involves both Curie-Weiss and temperature-independent terms, while the n=∞ compound exhibits a temperature-independent susceptibility.

KEY WORDS: strontium vanadium oxide, structure, Rietveld analysis, transport, magnetic susceptibility

INTRODUCTION

The superconducting materials found thus far with transition temperatures above 40 kelvins have been confined to the layered-structure oxides that contain planes of copper and oxygen atoms. The structures of these high-Tc copper oxides are based on those of the layered perovskite oxides with general formula $AO(ABO_3)_n$. This structure can be built up by alternately layering the perovskite $(ABO_3)_n$ and rock-salt AO units. The layered structure and the existence of the spin 1/2 B-site Cu^{2+} ions are considered essential for these copper oxides to be superconducting at high temperatures.

A series of compounds $Sr_{n+1}V_nO_{3n+1-\delta}$ have the layered perovskite structure with the spin 1/2 B-site V^{4+} ions. Furthermore, these compounds are expected to exhibit a high electrical conductivity as reported for the cubic perovskite $SrVO_3$[1,2]. The investigation of these compounds, therefore, could give insight into the mechanism of the high-Tc superconductivity and lead to a new high-Tc superconductor that does not contain copper. For these compounds, however, only a few studies have been reported. The present authors[3] and Itoh et al.[4] have recently succeeded in synthesizing the n = 3 compound. For the n = 1 compound, Cyrot et al.[5,6] reported the physical properties and the structural determination by X-ray and neutron diffractions. For the n = 2 compound, the present authors[7] have reported the preparation of a single-phase material and the structural determination by X-ray diffraction with Rietveld analysis.

In the present paper, the synthesis of single-phase polycrystalline materials of the n=1, 2, 3, and ∞ compounds will be presented. The results of structural determination by X-ray diffraction with Rietveld analysis and of electrical and magnetic measurements will also be reported.

EXPERIMENTAL

All the samples were prepared from powders of $SrCO_3$ (99.9% pure) and V_2O_5 (99.99% pure) by the solid-state reaction. Mixtures of stoichiometric amounts of the powders, in which the atomic ratio V/Sr was 3/4 for $Sr_4V_3O_{10-\delta}$ for example, were thoroughly mixed for about one hour with an agate motor and pestle. The mixtures were then put into an alumina crucible and calcined in air for 6 hours at 600℃ and 6 hours at 650℃. After cooling down to room temperature, the calcined lumps were ground into powders and well mixed in an agate motar. Again the mixtures were fired in air for 6 hours at 670℃, for 6 hours at 700℃ and 6 hours at 900℃ successively. After these processes, a gray product composed of $Sr_3V_2O_8$ and a few impurity phases was obtained from the mixture of V/Sr = 3/4. The mixtures with V/Sr = 2/3, 1/2 and 1/1 yielded almost single-phase materials of $Sr_3V_2O_8$, $Sr_4V_2O_9$, and $Sr_2V_2O_7$, respectively. Powders of these products were pressed into pellets of 20 mm in diameter. The pellets were then placed on alumina boats and heated in flowing (~200 ml/min) hydrogen atmosphere. The pellets with V/Sr = 3/4, 2/3, 1/2, and 1/1 were fired for about 5 hours at 1300℃, 1110℃, 1100℃, and 1300℃, respectively. These procedures yielded single-phase samples as determined by X-ray diffraction.

X-ray powder diffraction was performed with Rigaku Denki RAD-B diffractometer. Step scanned intensity data were collected with CuKα radiation by accumulating counts for 8 sec at 0.02° (2θ) intervals over the 2θ range from 10° to 140°. These data were analyzed with the Rietveld analysis program RIETAN[8] to determine the crystal structure.

In order to determine the amount of oxygen, the samples were fired in an oxygen stream (200 ml/min) for 20 hours at 900 ℃. This process changed the black samples to white ones. The weight gain due to the oxydizing procedure was measured and used to calculate the oxygen content on assuming that this procedure completely changes V^{4+} or V^{3+} to V^{5+}.

Resistivity measurements were performed with a standard four-probe technique over the temperature range 2 K ≦ T ≦ 300 K. Magnetic susceptibility measurements were carried out over the temperature range 2 K ≦ T ≦ 280 K at a magnetic field of 1000 G with a Quantum Design Inc. SQUID magnetometer.

RESULTS AND DISCUSSION

The chemical compositions estimated from the oxygen analysis are $Sr_4V_3O_{9.60\pm0.01}$, $Sr_3V_2O_{7.01\pm0.01}$, $Sr_2VO_{3.64\pm0.02}$, and $SrVO_{2.84\pm0.01}$.

The X-ray diffraction data for the first three samples were analysed with the Rietveld method. The detailed results for $Sr_4V_3O_{9.60}$ and $Sr_3V_2O_{7.01}$ have been reported in ref.3 and ref.7. Here, the Rietveld analysis on the n=1 sample is described.

The Rietveld refinement pattern for the $Sr_2VO_{3.64}$ sample is shown in Fig.1. As a starting model, the structure parameters of Sr_2TiO_4 were used. In the refinement, the intensity data for the 2θ regions of 29.0° - 31.1° and 31.9° - 32.4° were excluded because in these regions, peaks of unknown phases were slightly observed. A preferred orientation correction[8] was applied for a (001) cleavage plane. The occupation factors of oxygen were varied as fitting parameters for all the oxygen sites. The final R factors[8] obtained for $Sr_2VO_{3.64}$ were R_{WP} = 19.25 %, R_P = 14.03 %, R_E = 6.27 %, R_I = 5.22 %, and R_F = 2.91 %. Table 1 lists the structural parameters determined by the present Rietveld analysis. The composition calculated from the oxygen occupancy is $Sr_2VO_{3.88}$. The value of δ therefore is somewhat different between the

Table 1. Structural parameters of $Sr_2VO_{4-\delta}$ (Lattice constants: a=3.8294(1) Å, c=12.6000(3) Å, Space group: I4/mmm). Numbers in the parentheses for Z and B are estimated standard deviations of the last significant digits, and those with out statistical uncertainties were fixed or constrained.

Atom	Site	X	Y	Z	B(Å ^{2})	g
Sr	4e	0.0	0.0	0.3545(5)	0.62(12)	1
V	2a	0.0	0.0	0.0	0.63(26)	1
O(1)	4c	0.0	0.5	0.0	0.70(97)	0.97(7)
O(2)	4e	0.0	0.0	0.1576(27)	0.67(93)	0.97(9)

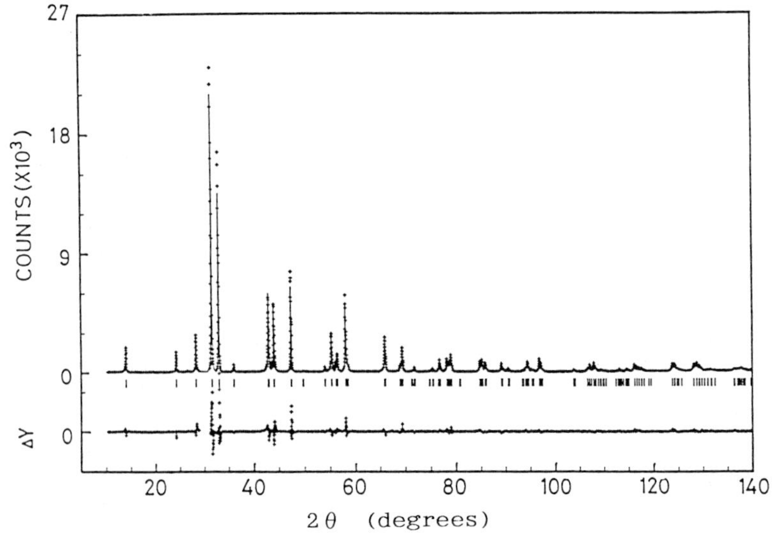

Fig. 1. Rietveld refinement pattern for $Sr_2VO_{3.64}$. + marks are the raw X-ray diffraction data, and the overlapped line is the calculated profile. Tick marks below the profile indicate the position of the allowed reflections for CuKα1 and CuKα2. The curve in the bottom, ΔY, is the difference between the observed and calculated intensities in the same scale.

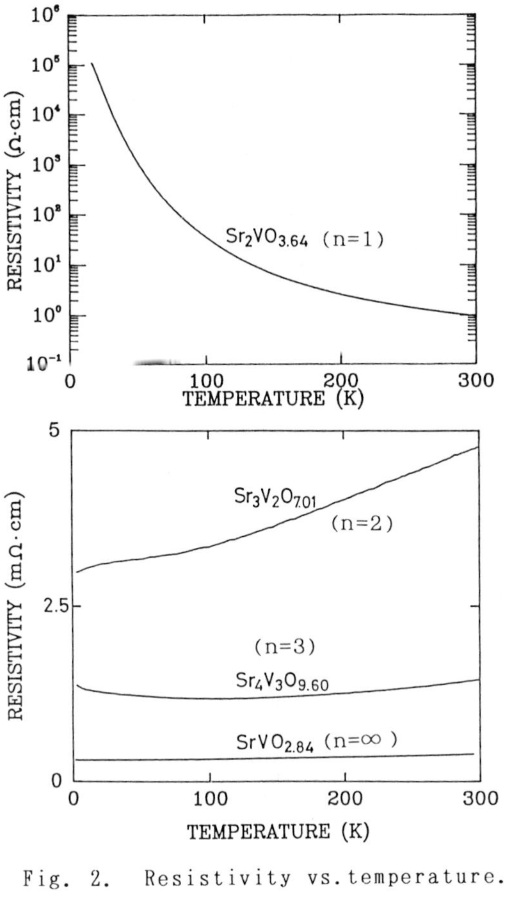

Fig. 2. Resistivity vs. temperature.

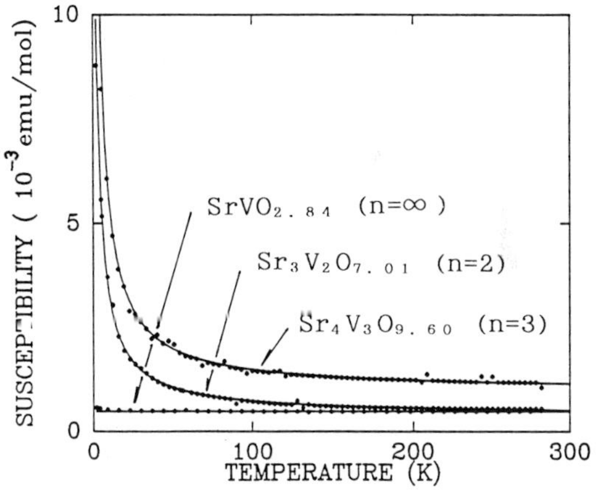

Fig. 3. Magnetic susceptibility vs. temperature. Points represent the raw data and the solid lines are the fitting curves calculated from Eq. (1). The curves for the n=∞, 2, 3 samples are

$\chi(T) = 0.48\times10^{-3}$,
$\chi(T) = 0.40\times10^{-3} + 34\times10^{-3}/(T+1.60)$,
$\chi(T) = 0.98\times10^{-3} + 52\times10^{-3}/(T+1.15)$,

respectively.

calculation and the experiment. This difference probably originates from the low sensitivity for X-rays to detect oxygen and the relatively high humid-sensitivity of the sample. The latter might lead to an experimental δ lower than the true value.

If we compare the lattice constants among the four samples, it is noted that the a-axis lattice constant increases from 3.829Å to 3.849Å [3] as n increases from 1 to ∞.

The dependences of electrical resistivity on temperature for the four samples are shown in Fig.2. The n = 1 compound exhibits a semiconductive behavior as reported by Cyrot et al.[5]. The resistivity for the n = 2 and ∞ samples increases almost linearly with increasing temperature in the temperature range studied. The resistivity of the n = 3 sample is 1.4 mΩ·cm at 300 K and decreases with decreasing temperature down to about 100 K followed by a slight increase with a further decrease of temperature. It is found that the resistivity at room temperature decreases with increasing number of the perovskite layers n.

The magnetic susceptibility as a function of temperature is shown in Fig.3 for the n = 2, 3, and ∞ samples. A Curie-Weis behavior is observed for the n = 2 and 3 samples, while the susceptibility of the n = ∞ sample is independent of temperature. These samples were synthesized from the same raw materials. Therefore, these Curie-Weiss behaviors are not due to magnetic impurities but the intrinsic properties of $Sr_3V_2O_{7.01}$ and $Sr_4V_3O_{9.60}$. As seen in Fig.3, the susceptibility $\chi(T)$ can be represented by,

$$\chi(T) = \chi_{TID} + \chi_C(T) \qquad (1)$$

with $\chi_C(T) = C/(T - \Theta)$ and $C = 3N_0(P_{eff}\mu_B{}^2/k_B)$, where χ_{TID} and χ_C are the temperature-independent and Curie-Weiss terms.

In summary, single-phase polycrystalline materials of a series of the layered perovskite compounds $Sr_{n+1}V_nO_{3n+1-\delta}$ (n = 1, 2, 3, ∞) were synthesized by the solid-state reaction under a flowing hydrogen aymosphere. The structures of the compounds were determined with Rietveld analysis of X-ray diffraction data. It was found that the n = 1, 2, and 3 compounds have tetragonal structures and the a-axis lattice constant increases as n increases from 1 to ∞. Electrical transport measurements revealed that the resistivity at room temperature decreases as n increases from 1 to ∞, and that the n = 2, 3 and ∞ compounds exhibit metallic behaviors whereas the n = 1 compound is a semiconductor. From the magnetic measurements, it was found that the susceptibility of the n = 2 and 3 compounds can be represented by a summation of temperature-independent and Curie-Weiss terms while the n = ∞ compound does not depend on temperature. Because $Sr_2VO_{4-\delta}$ is known to be antiferromagnetic[5], the above results indicate that the electronic structure of these compounds changes successively from an antiferromagnetic insulator to a temperature-independent paramagnetic metal as n changes from 1 to ∞. The localized $t_{2g}{}^*$ orbitals of the octahedral-site V^{4+} ions are considered to transform into collective narrow π^* bands as n increases from 1 to ∞.

The authors would like to thank S.Kobayashi, H.Hasegawa, M.Ohkubo, S.Noda, and A.Isogai for discussion and encouragement.

REFERENCES

1. Chamberland BL, Danielson PS (1971) J. Solid State Chem. 3: 243.
2. Dougier P, Fan JCC, Goodenough JB (1975) J. Solid State Chem. 14: 247.
3. Suzuki N, Noritake T, Yamamoto N, Hioki T (in press) Mater. Res. Bull.
4. Itoh M, Shikano M, Liang R, Kawaji H, Nakamura T (1990) J. Solid State Chem. 88: 597.
5. Cyrot M, Lambert-Anderson B, Soubeyroux JL, Rey MJ, Dehauht Ph, Cyrot-Lackmann F, Fourcaudot G, Beille J, Tholence JL (1990) J. Solid State Chem. 85: 321.
6. Rey MJ, Dehaudt Ph, Joubert JC, Lambert-Andron B, Cyrot M, Cyrot-Lackmann F (1990) J. Solid State Chem. 86: 101.
7. Suzuki N, Noritake T, Yamamoto N, Hioki T (Submitted) Mater. Res. Bull.
8. Izumi F (1985) J. Cryst. Soc. Japan 27: 23.

New Layered Perovskite Compounds, $Sr_{n+1}V_nO_{3n+1}$ (n=1,2,3 and ∞)

AYUMI NOZAKI, HIROSHI YOSHIKAWA, TAKAHIRO WADA, and H. YAMAUCHI

Superconductivity Research Laboratory, International Superconductivity Technology Center, 10-13, Shinonome 1-chome, Koto-ku, Tokyo, 135 Japan

ABSTRACT

New layered perovskite compounds, $Sr_{n+1}V_nO_{3n+1}$ (n=1, 2 and 3) and $SrVO_3$(n=∞), were synthesized and the physical properties of the compounds with n=1, 2 and ∞ were investigated. Sr_2VO_4 was semiconductive and antiferromagnetic, but a weak ferromagnetic contribution to the susceptibility was observed below 45 K. However, $Sr_3V_2O_7$ exhibited a metallic electrical conduction and a Pauli-paramagnetism. $SrLaVO_4$ was also successfully synthesized and was found to be semiconductive. The crystal structures and electrical and magnetic properties of $Sr_{n+1}V_nO_{3n+1}$ (n=1, 2, 3 and ∞) and $SrLaVO_4$ were compared.

KEY WORDS: strontium vanadium oxide, perovskite, K_2NiF_4 type, electrical resistivity, magnetic susceptibility, x-ray diffraction, antiferromagnetism

INTRODUCTION

Since the discovery of superconductivity in the La-Ba-Cu-O system[1], numerous experimental and theoretical studies have been reported for copper oxide superconductors. These copper oxide superconductors possess a common feature in the crystallographic structures. That is, there exists a two-dimensional array of CuO_6 (octahedron), CuO_5 (pyramid) or CuO_4 (square) clusters. For example, $(La, Ba)_2CuO_4$ has a K_2NiF_4-type structure which includes a two-dimensional array of CuO_6 octahedra[2]. It has been pointed out[3] that superconductivity in these copper oxides may be closely related to both the two-dimensional character in their crystal structures and the half spin (s=1/2) of a Cu^{2+} ion in the two-dimensional array.

In the present study, we are concerned with a new compound series, $Sr_{n+1}V_nO_{3n+1}$ (n=1, 2, 3 and ∞) and $SrLaVO_4$. $SrVO_3$ may be regarded as the compound with n=∞ in this series. These structures of this series are realized by piling up n-units of the perovskite-like layer along a single axis, such as the case of the compound series $Sr_{n+1}Ti_nO_{3n+1}$[4][5]. The valence of a V ion in this series is expected to be 4+ (=[2(3n+1)-2(n+1)]/n) and therefore V^{4+} ions are expected to have a half spin. Note that these features are also common to the known copper oxide superconductors. To date, only Sr_2VO_4[6] and $SrVO_3$[7][8] of the $Sr_{n+1}V_nO_{3n+1}$ series have been synthesized.

In the present work, a series of compounds, Sr_2VO_4, $Sr_3V_2O_7$, $Sr_4V_3O_{10}$ and $SrVO_3$, and substituted compound, $SrLaVO_4$ were synthesized and their electrical and magnetic properties were studied.

EXPERIMENTAL PROCEDURE

Samples were prepared by a solid-state reaction method using $SrCO_3$, La_2O_3 and V_2O_3 powders as starting materials. Powders with stoichiometric compositions were mixed and ground. The starting powders were ~ 99.9 % pure. The mixed powder was calcined at 1400℃ in an Ar atmosphere for 2 hours after being pressed into pellets. The calcined samples were re-ground and again pressed into rectangular bars of ~ 2 mm× 2 mm× 20 mm. The bars were encapsulated in evacuated silica tubes together with a reducing agent; TiO powder for Sr_2VO_4 ,Ti_2O_3 for $Sr_3V_2O_7$, $Sr_4V_3O_{10}$ and $SrVO_3$. The encapsulated samples were fired at 1050℃ for 100 hours. The final samples were obtained after several repetitions of firing with intermediate grinding. On the other hand, substituted sample $SrLaVO_4$ was obtained by firing twice at 1350℃ in atmosphere of 1% H_2 mixed Ar base for 10 hours and intermediate grinding , after calcination.

The crystal structures and the lattice constants were determined by an x-ray powder diffraction method using CuKα radiation. The oxygen contents of the samples were analyzed by an inert gas fusion nondispersive IR method(Horiba:EMGA-2800)[9]. Electrical resistivity was measured in the temperature range between 4.2 and 300 K using a standard dc four-probe technique and the magnetic properties were studied with a superconducting quantum interference device (SQUID) magnetometer(Quantum Design:MPM).

RESULTS

A. Sr_2VO_4 (n=1)

The x-ray powder diffraction pattern for Sr_2VO_4 is shown in Fig. 1(a). The diffraction peaks were successfully indexed for a tetragonal unit cell with the lattice constants, a=3.834(1) A and c=12.57(1) A. The measured oxygen content of the sample was 3.99± 0.05. The temperature dependence of electrical resistivity of Sr_2VO_4 is shown in Fig. 2. It is clear that Sr_2VO_4 showed a semiconductive behavior and the resistivity at room temperature was about 10 Ω cm. The temperature dependence of the magnetic susceptibility of Sr_2VO_4 is shown in Fig.3. The susceptibility, χ , of Sr_2VO_4 was sufficiently represented by the Curie-Weiss (C-W) law, i.e.

$$\chi = C/(T-\Theta_p),$$

at temperatures above 100K as demonstrated, but showed an unusual behavior at temperatures below 45 K which might suggest the existence of magnetic ordering at low temperatures. This magnetic order might be due to an antiferromagnetism because the resultant χ -vs-T curve angled sharply at 45 K. The Curie temperature, Θ_p were determined to be 47 K. The effective magnetic moment per V ion calculated from the value of Curie constant was about 1.6 times of the Bohr magneton(μ_B). This value agreed with the anticipated value for an isolated V^{4+} ion (1.73μ_B). The measured M-vs-H curve at 5 K was approximated by the superposition of a paramagnetism term and a ferromagnetism term which leveled off around 1kOe. The value of the saturated ferromagnetic magnetization was estimated by extrapolating the M-vs-H curve to zero magnetic field. The estimated ferromagnetic moment per V ion was about 10^{-4} μ_B. These

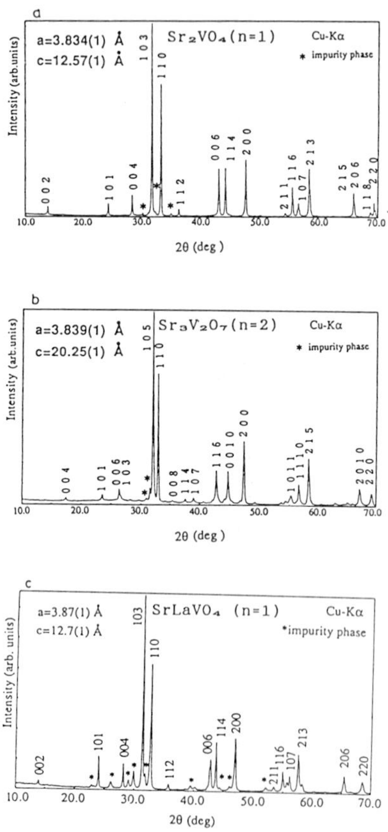

Fig.1. Powder X-ray diffraction patterns for (a) Sr_2VO_4(n=1), (b) $Sr_3V_2O_7$(n=2) and (c) $SrLaVO_4$ (n=1).

results suggest the existence of weak ferromagnetism in Sr_2VO_4 at low temperatures.

B. $Sr_3V_2O_7$ (n=2)
Figure 1(b) shows the x-ray powder diffraction pattern for the $Sr_3V_2O_7$ sample. The diffraction peaks were indexed for a tetragonal unit cell with the lattice constants, a=3.839(1) A and c=20.25(1) A. This diffraction pattern was parallel to that for $Sr_3Ti_2O_7$[5]. The oxygen content of the sample was 6.94± 0.05. The resistivity of $Sr_3V_2O_7$ was about $10^{-2}\Omega$ cm at room temperature and slightly decreased with decreasing temperature as shown in Fig.2. The electrical conduction of the compound was metallic.
The magnetic susceptibility of $Sr_3V_2O_7$ was approximated by a C-W curve, assuming the existence of a temperature independent part, χ_o, in the susceptibility at temperatures above 100K as shown in Fig.3. That is, χ was approximated by

$$\chi=\chi_o+C'/(T-\Theta_p').$$

The parameters, χ_o and Θ_p' were 1.4×10^{-6} emu/g and 24 K, respectively. The effective magnetic moment per V ion was $0.4\mu_B$. It is likely that an essential feature of the magnetic property of $Sr_3V_2O_7$ was Pauli-paramagnetism because $Sr_3V_2O_7$ was metallic in the electrical conduction.

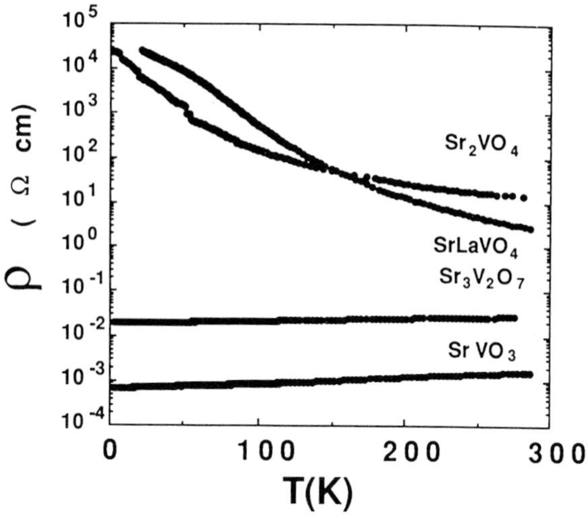

Fig.2. Temperature dependence of electrical resistivity of $Sr_{n+1}V_nO_{3n+1}$ (n=1, 2, ∞) and $SrLaVO_4$.

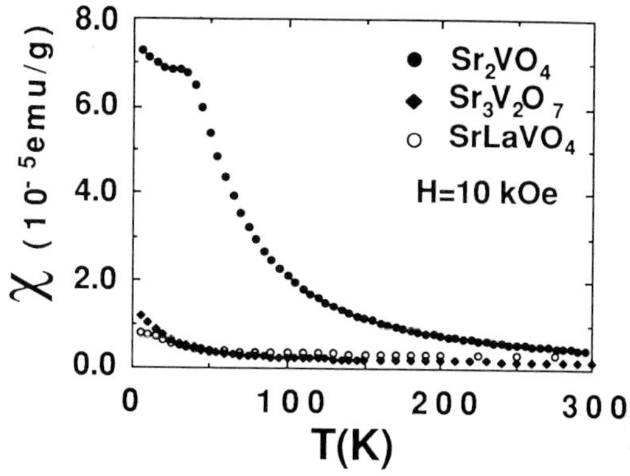

Fig.3. Temperature dependence of magnetic susceptibility of $Sr_{n+1}V_nO_{3n+1}$ (n=1, 2,) and $SrLaVO_4$. Measurements were carried out on cooling at a constant field of 10 kOe.

C. $Sr_4V_3O_{10}$ (n=3)
This sample contained $Sr_3V_2O_7$ as well as $SrVO_3$ as impurity phases. The lattice constants, a and c, were determined to be 3.84(1) A and 27.9(1) A, respectively.

D. $SrVO_3$ (n=∞)
Single-phase samples of the perovskite-type compound, $SrVO_3$, were synthesized. The crystal system was cubic and the lattice constant, a, was 3.842(3) A. The temperature dependence of electrical resistivity showed a metallic behavior at temperatures between 2.5 K and 300 K and the electrical resistivity at room temperature was about 10^{-3} Ω cm.

E. $SrLaVO_4$ (n=1)
$SrLaVO_4$ sample was synthesized but our sample included La_2O_3 as secondary impurity phase. The crystal system was tetragonal and lattice constants, a and c, were 3.87(1) A and 12.7(1) A, respectively. These lattice constants were good agreement with the previously reported values[10]. The temperature dependence of electrical resistivity was a semiconductive such as Sr_2VO_4 . However, the magnetic susceptibility was Pauli-paramagnetism-like such as $Sr_3V_2O_7$. This contradiction may be due to a heterogeneity in our sample.

DISCUSSION

The electrical and magnetic properties of the series of $Sr_{n+1}V_nO_{3n+1}$ compounds are summarized in Table 1. For n=1 i.e. Sr_2VO_4, the magnetic property was of the C-W -type at high temperature and the electrical property was semiconductive. Also, the antiferromagnetic order below 45 K for this compound suggests the existence of a ferromagnetic interaction between V^{4+} ions in same perovskite-like layer (P-layer) and an antiferromagnetic interaction between P-layer units, because Θp was a positive. With regard to weak ferromagnetism of this compound, similar behavior has been reported on β-$Na_{0.33}V_2O_5$[11]. However, for n≧ 2 i.e. $Sr_3V_2O_7$ or $SrVO_3$, the compounds showed Pauli-paramagnetism-like behavior in the magnetic susceptibility and a metallic behavior in the electrical conduction. This indicates that the $3d^1$ electrons are localized in Sr_2VO_4 (n=1) while they are delocalized in the compounds with n≧ 2.

Compound	n	Lattice constants (A) a	c	Magnetic property	Electrical property R_{300} (Ω cm)
Sr_2VO_4	1	3.834(1)	12.57(1)	C-W WF(T<45K)	Semi. 10
$Sr_3V_2O_7$	2	3.839(1)	20.25(1)	PP	Metallic 10^{-2}
$Sr_4V_3O_{10}$	3	3.84(1)	27.9(1)	PP (Ref.13)	Metallic 10^{-3} (Ref.13)
$SrVO_3$	∞	3.842(3)	——	PP (Ref.7)	Metallic 10^{-3}

WF: Weak ferro. PP: Pauli-para.-like

Table 1. Lattice constants and magnetic- and electrical-properties of $Sr_{n+1}V_nO_{3n+1}$ (n=1, 2, 3 and ∞).

Recently, properties of Sr_2VO_4[12] and $Sr_4V_3O_{10}$[13] were investigations by other groups. Our results are in agreement with their.

ACKNOWLEDGMENTS

We are grateful to Dr. K. Nakao for his helpful discussions and Dr. H. Yoshida of Tohoku University for his important suggestion.

This work was supported by New Energy and Industrial Technology Development Organization as a part of the program for R&D of Basic Technology for future Industries.

REFERENCES

[1] Bednorz JG, Muller KA (1986) Z. Phys. B64: 189
[2] Takagi H, Uchida S, Kitazawa K, Tanaka S (1987) Japan. J. Appl. Phys. 26: L123
[3] Anderson PW (1986) Science 235: 1196
[4] Ruddlesden SN, Popper P (1957) Acta Cryst. 10: 538
[5] Ruddlesden SN, Popper P (1958) Acta Cryst. 11: 54
[6] Feltz A, Schmalfuss S (1975) Z. Chem. 15: 289
[7] Chamberland BL, Danielson PS (1971) J. Solid State Chem. 3: 243
[8] Dougier P, Fan JCC, Goodenough BJ (1975) J. Solid State Chem. 14: 247
[9] Wada T, Suzuki N, Maeda A, Yabe T, Uchinokura K, Uchida S, Tanaka S (1989) Phys. Rev. B39: 9126
[10] Longo JM, Raccah PM (1973) J. Solid State Chem. 6: 526
[11] Schlenker C, Buder R, Nguyen VD, J. Dumas J (1979) J. Appl. Phys. 50: 1720
[12] Cyrot M, Lambert-Andron B, Soubeyroux JL, Rey MJ, Dehauht PH, Cyrot-Lackmann F, Fourcaudot G, Beille J, Tholence JL (1990) J. Solid State Chem. 85: 321
[13] Itoh M, Shikano M, Liang R, Kawaji H, Nakamura T (1990) J. Solid State Chem. 88: 597

Possible New Ternary Chalcogenide Superconductors with Composite Crystal Structure

YOSHINAO OOSAWA, YOSHITO GOTOH, and JUNJI AKIMOTO

National Chemical Laboratory for Industry, Tsukuba, Ibaraki, 305 Japan

ABSTRACT

New type of ternary sulfides with $(PbS)_n(TS_2)_2$(T = Ti, Nb, Ta; $n \simeq 1$) formula has been prepared and characterized by using X-ray diffraction and electric resistivity measurement. The sulfides have been prepared from elemens by heat treatment in evacuated silica tube and are black-grayish microcrystalline powder with luster. Their X-ray diffraction pattern resembles that of $(PbS)_nTS_2$(T = Ti, V, Nb, Ta; $n \simeq 1$) type of sulfides and only diffraction lines indexed to 0 0 ℓ are strong owing to extreme preferred orientation. Considering the difference of periodic distance in the stacking direction of the both types of sulfides, it has been suggested that the structure of the $(PbS)_n(TS_2)_2$ type of sulfides consists of two kinds of layers alternately stacked: one two-atom-thick layer of PbS and two three-atom-thick sandwiches of TS_2. Electric resistivity of $(PbS)_n(TaS_2)_2$ and $(PbS)_n(NbS_2)_2$ measured using d.c. four-probe method in the range of 1.7-300K showed metallic temperature dependence and its drastic decrease has been found around 3K which is probably due to superconducting transition.

KEY WORDS: chalcogenide superconductor, layered composite crystal structure, electric resistivity

INTRODUCTION

A few ternary chalcogenides with "MTX_3:(M = Pb, Sn; T = Nb, Ta; X = S, Se)" formula were reported as superconductor over fifteen years ago[1] but their structures were not reported for long time. Recently it has become apparent that these ternary chalcogenides belong to a new type of chalcogenide group whose structure consists of a two-atom-thick layer of MX and a three-atom-thick sandwich of TX_2 alternately stacked[2]. The chalcogenides have mutually incommensurate structure because the ratio of periodic lengths of b axes of MX and TX_2 sublattices is not simple rational and their ideal composition can be formulated as $(MX)_nTX_2$:$n \simeq 1$.

For these $(MX)_nTX_2$ type of chalcogenides, two series were known whose M are lanthanides, and IVb element(Pb or Sn). We found a new series whose M is Vb element(Bi or Sb)[3] and $(PbS)_{1.12}VS_2$[4]. In ISS'89 we reported preparation and characterization of $(BiX)_nTX_2$ and suggested through electric resistivity measurement that $(BiSe)_nNbSe_2$ and $(BiSe)_nTaSe_2$ are superconductors whose transition temperatures are around 1.5K[5].

In addition to these MTX_3 type of chalcogenides, a few chalcogenides with MT_2X_5 formula have been reported. Guemas et al reported $PbNb_2S_5$ and $SnTi_2S_5$[6]. We have preliminarily reported preparation and characterization of BiT_2X_5(T = Ti, Nb, Ta; X = S, Se)[7]. Considering the similarity between MTX_3 and MT_2X_5 in X-ray and electron diffraction pattern, it is suggested that the latter should be formulated as $(MX)_n(TX_2)_2$:$n \simeq 1$. In the present manuscript we report preparation and characterization of $(PbS)_n(TiS_2)_2$ and $(PbS)_n(TaS_2)_2$, laying stress on the latter. Electric resistivity measurement of $(PbS)_n(TaS_2)$ and $(PbS)_n(NbS_2)_2$ is also reported.

EXPERIMENTAL

$(PbS)_n(TiS_2)_2$ and $(PbS)_n(TaS_2)_2$ were prepared as follows. Starting materials(elements with purity above 99.9%) were mixed together and sealed in a silica tube under vacuum. The tube was placed in an electric furnace, heated at 500 ℃ then at 800 ℃ and quenched to room temperature in the air. Thus the sulfides were obtained as black-grayish microcrystalline powder with luster. The X-ray diffraction patterns were taken by the counter-diffractometer technique using Cu Kα radiation. In order to depress the preferred orientation, the sample was treated with cyanoacrylate type of adhesive reagent. The temperature dependence of electric resistivity was measured with pellet made of sample powder using a standard d. c. four-probe method in the range of 1.7-300K.

RESULTS AND DISCUSSION

The X-ray diffraction patterns of $(PbS)_n(TiS_2)_2$ and $(PbS)_n(TaS_2)_2$, before treatment with the organic adhesive reagent, consist of several strong peaks and a few medium and weak peaks. The strong peaks are attributable to a set of parallel planes (0 0 ℓ) as in the case of $(BiX)_n(TX_2)_2$. This fact suggest that the sulfides have some layered structure and take preferred orientation. Because the X-ray diffraction lines other than 0 0 ℓ are rather weak and broad probably owing to preferred orientation, we reported only the 0 0 ℓ diffraction data in a previous manuscript[8].

According to single crystal X-ray structure determination of $(PbS)_{1.12}VS_2$[4], $(PbS)_{1.14}NbS_2$[9], and $(PbS)_{1.13}TaS_2$[10], their structure in the layer-stacking direction consists of a two-atom-thick MS layer with distorted NaCl structure and a three-atom-thick TS_2 sandwich with T in octahedral or trigonal prismatic coordination, and the periodic length of the structure in the layer-stacking direction is around 12 Å. On the other hand, the periodic length of the present $(PbS)_n(TS_2)_2$ type of sulfides is in the range of 17.5-18 Å. The difference of periodic length between $(PbS)_n(TS_2)_2$ and $(PbS)_nTS_2$ is almost equal to the periodic length of layer stacking direction of corresponding disulfides. Therefore the $(PbS)_n(TS_2)_2$ type of sulfide seems to be represented by a model: a two-atom-thick layer of PbS and two adjacent three-atom-thick sandwiches of TS_2 are stacked alternately as in the case of $(BiX)_n(TS_2)_2$[8].

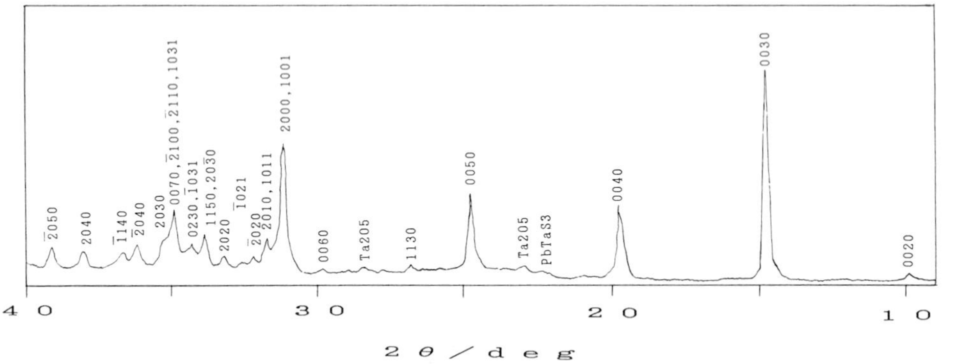

Fig. 1. Powder X-ray diffraction pattern of $(PbS)_n(TaS_2)_2$ after treatment with organic adhesive reagent. Indexing is carried out using 4 indices H, K, L, and M after four dimensional superspace group: H, common; K, K of PbS lattice; L, common; M, K of TaS_2 lattice.

When x was varied from 0.8 to 1.2 in the starting composition according to the formulation of $(PbS)_x(TaS_2)_2$, mixture of $(PbS)_n(TaS_2)_2$ and $(PbS)_{1.13}TaS_2$ or mixture of the former and TaS_2 was formed depending on x. And when x was 1.0, sample with neither $(PbS)_{1.13}TaS_2$ nor TaS_2 was obtained. Figure 1 shows the indexing of the X-ray diffraction pattern of the sample after treatment with the organic adhesive reagent. The pattern can be indexed as the superposition

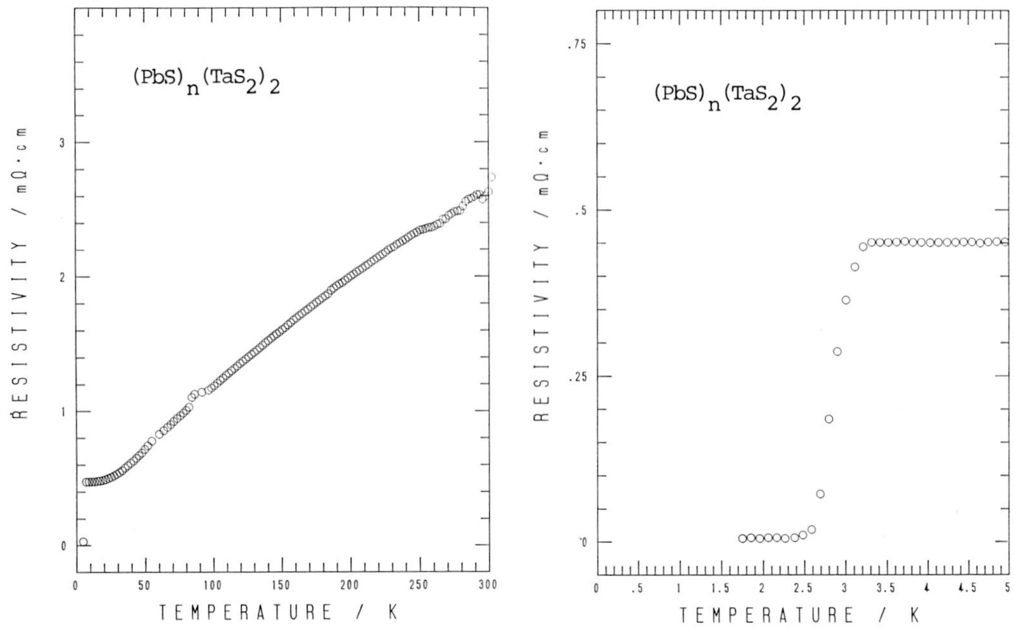

Fig. 2. Dependence of electric resistivity on temperature for $(PbS)_n(TaS_2)_2$

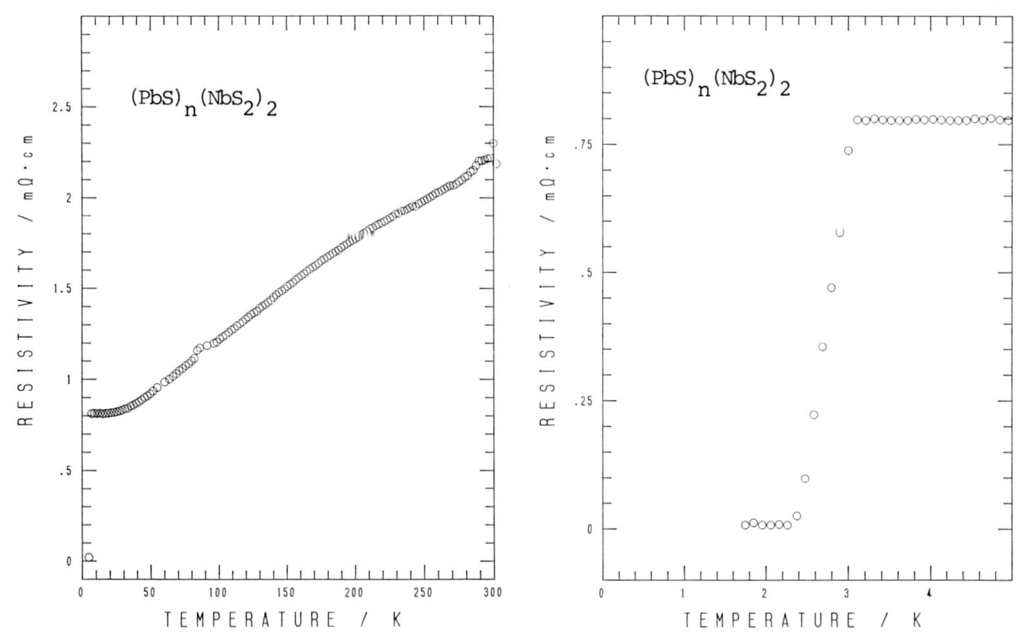

Fig. 3. Dependence of electric resistivity on temperature for $(PbS)_n(NbS_2)_2$

of PbS monoclinic sublattice(a = 5.754 A, b = 5.794 A, c = 18.016 A, β= 92 54') and TaS_2 monoclinic sublattice(a = 5.754 Å, b = 3.304 Å, c = 18.016 Å, β = 92°54'). The indexing is after four dimensional superspace group. The value of n calculated with the lattice constants of the both subcells is 1.14(3.304 x 2/5.794). There is comparatively large difference between the value and the composition expected from the starting stoichiometry: n = 1. The reason is not yet clear at the moment. The deviation of β from 90° may be explained on the assumption that two adjacent TaS_2 layers are stacked in 2Ha-type.

Figure 2 shows the temperature dependence of electric resistivity for $(PbS)_n(TaS_2)_2$:for the starting composition n was 1. The resistivity showed a metallic dependence from 300 to around 3K and its drastic decrease was observed at 3.2 K. In the present experimental scale, the resistivity was almost zero below 2.4 K. It is known that TaS_2(2Ha type) and $(PbS)_{1.13}TaS_2$ are superconductors and their transition temperatures are 1-2 K and 3.08 K[11], respectively. It seems that the superconducting property of $(PbS)_nTaS_2$ comes from TaS_2 layer and the rise in the transition temprature is caused by a kind of perturbation which originates in the presence of PbS layer. As has been explained above, it is suggested that the layer-stacking scheme of $(PbS)_n(TaS_2)_2$ is between TaS_2 and $(PbS)_{1.13}TaS_2$. Therefore it is most probable that the origin of the drastic decrease of resistivity observed in $(PbS)_n(TaS_2)_2$ is a superconducting transition. Figure 3 shows the temperature dependence of electric resistivity of $(PbS)_n(NbS_2)_2$. This sulfide is also probably a superconductor.

We have preliminarily reported that these $(PbS)_n(TS_2)_2$ type of sulfides have Van der Waals plane and intercalation of hydrazine into the plane occurs as in the case of $(BiX)_n(TX_2)_2$. It is known that state density at the Fermi level can be increased by the intercalation for dichalcogenides, and as a result the superconducting transition temperature can be raised in some case. For this type of chalcogenides formulated as $(MX)_n(TX_2)_2$, a similar effect may be expected. It will be a subject of further work.

REFERENCES

1. M. H. van Maaren, Phys. Lett., 40A, 353(1972).
2. G. A. Wiegers et al., Solid State Commun., 70, 409(1989). references therein.
3. Y. Oosawa, Y. Gotoh, and M. Onoda, Chem. Lett., 1989, 523.
4. (a) Y. Gotoh, M. Goto, K. Kawaguchi, Y. Oosawa, and M. Onoda, Mat. Res. Bull., 25, 307(1990). (b) M. Onoda, K. Kato, Y. Gotoh, and Y. Oosawa, Acta Cryst., B46, 487(1990).
5. Y. Oosawa, Y. Gotoh, and M. Onoda in T. Ishiguro and K. Kajimura(Eds.), Proc. 2nd Int. Symp. on Superconductivity, 1989, Tsukuba; Springer, Tokyo, pp193.
6. L. Guemas, P. Rabu, A. Meerschaut, and J. Rouxel, Mat. Res. Bull., 23, 1061(1988).
7. Y. Oosawa, Y. Gotoh, and M. Onoda, Chem. Lett., 1989, 1563.
8. Y. Gotoh, J. Akimoto, M. Sakurai, Y. Kiyozumi, K. Suzuki, and Y. Oosawa, 1990, 2057.
9. G. A. Wiegers, A. Meetsma, R. J. Haange, S. van Smaalen, and J. L. de Boer, Acta Cryst., B46, 324(1990).
10. J. Wulff, A. Meetsma, S. van Smaalen, R. J. Haange, J. L. de Boer, and G. A. Wiegers, J. Solid State Chem., 84, 118(1990).
11. D. Reefman, J. Baak, H. B. Brom, and G. A. Wiegers, Solid State Commun., 75, 47(1990).

A New Ambient-Pressure Organic Superconductor, κ-(BEDT-TTF)$_2$Ag(CN)$_2$H$_2$O

HATSUMI MORI[1], IZUMI HIRABAYASHI[1], SHOJI TANAKA[1], TAKEHIKO MORI[2], YUSEI MARUYAMA[2], and HIROO INOKUCHI[2]

[1] Superconductivity Reserach Laboratory Nagoya Division, International Superconductivity Technology Center, Mutsuno, Nagoya, 456 Japan

[2] Institute for Molecular Science, Okazaki, 444 Japan

ABSTRACT

A new ambient-pressure organic superconductor, κ-(BEDT-TTF)$_2$Ag(CN)$_2$H$_2$O (T_c=5.0 K) was found. The superconductivity is observed by zero electrical resistivity and the Meissner effect. The X-ray crystal structure analysis indicates that the packing mode of donors is κ-phase, while the anion constructs the unique two-dimensional network like ··NC-Ȧg-CN··H-O-H··NC-Ȧg-CN·· with hydrogen bonds. The Lorentzian ESR signal attributed to BEDT-TTF radical is observed. The g-value is independent of temperature, whereas the linewidth increases with lowering temperature although the electrical resistivity decreases to 5 K.

KEY WORDS: organic superconductor, BEDT-TTF, κ-phase

INTRODUCTION

Since the discovery of the first organic superconductor, (TMTSF)$_2$PF$_6$ with T_c=0.9 K under 12 kbar in 1980 (1), 36 kinds of organic superconductors have been found and the T_c reached 12.5 K recently (2). It is very interesting to study the mechanism of an organic superconductor since there is an experimental result which can not be explained based upon the BCS theory in κ-(BEDT-TTF)$_2$Cu(NCS)$_2$ (3) that first exceeds T_c=10 K (4). In order to investigate a new type of superconductor, more exploration of a new organic superconductor is necessary.

Recently we have found the 35th organic superconductor, κ-(BEDT-TTF)$_2$Ag(CN)$_2$H$_2$O (T_c=5.0 K) (5). In this paper, preparation, crystal and electronic structures, electrical resistivity, SQUID magnetization, ESR spectra, and thermoelectric power are presented.

EXPERIMENTAL

The single crystals were prepared by electrochemical oxidation of BEDT-TTF in the presence of KAg(CN)$_2$ and 18-crown-6 ether in 1,1,2-trichroloethane. After about a month, black rectangular needles, rhombohedral blocks, plates, etc. were harvested. Crystal data of a rectangular needle: monoclinic, space group P2$_1$, a=16.071(4), b=8.645(2), c=12.601(4) Å, β=109.29(2)°, V=1652.4(1) Å^3, and Z=2. The crystal structure is very similar to that found in (BEDT-TTF)$_2$Ag(CN)$_2$H$_2$O.(6) M. Kurmoo et al. mentioned that their sample was only metallic to 150 K.

The electrical resistivity was measured along the b-axis by a conventional four-probe method with using gold paint. The magnetic susceptibility was observed by utilizing d.c. SQUID (Quantum Design model MPMS). ESR measurement was carried out at X-band (9.2 GHz) in the temperature range of 289-4 K. The thermoelectric power was measured on a single crystal along the b- and c-axes.

RESULTS AND DISCUSSION

Electrical resistivity

The temperature dependence of electrical resistivity of κ-(BEDT-TTF)$_2$Ag(CN)$_2$H$_2$O was shown in Fig. 1a and b. The conductivity along the b-

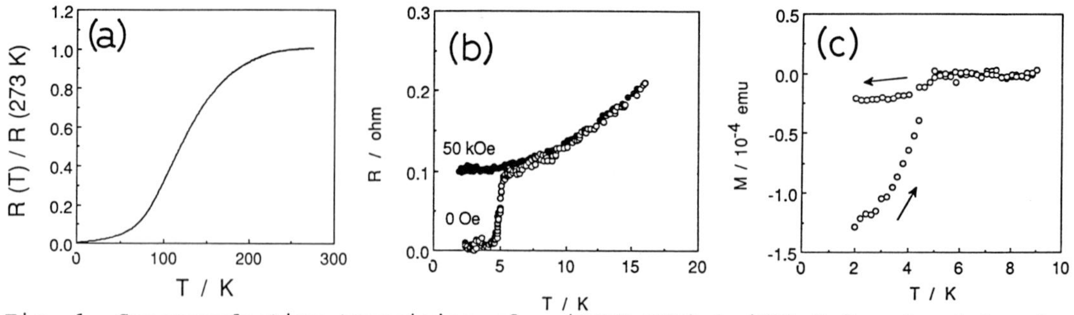

Fig. 1. Superconducting transition of κ-(BEDT-TTF)$_2$Ag(CN)$_2$H$_2$O under 1 bar by a), b)electrical resistivity, and c)SQUID magnetization measurements.

axis at room temperature is 27-37 Scm^{-1} which is in the same order as that of κ-(BEDT-TTF)$_2$Cu(NCS)$_2$, 14-37 Scm^{-1} (4). The temperature dependence, however, is totally different from each other; κ-(BEDT-TTF)$_2$Cu(NCS)$_2$ displays a broad resistivity maximum around 90-100 K (4,7) whereas the resistivity of κ-(BEDT-TTF)$_2$Ag(CN)$_2$H$_2$O decreases with lowering temperature without any resistivity peak. Around room temperature the resistivity shows a weak metallic behavior, its decrease gradually becomes steep, and $R(T) \propto T^2$ dependence was obtained below 100 K. The behavior seems to be closer to that of β-(BEDT-TTF)$_2$I$_3$ (8) rather than that of the isostructural salt, κ-(BEDT-TTF)$_2$Cu(NCS)$_2$.

Figure 1b depicts the electrical resistivity around superconducting transition; the onset of 5.7 K, the midpoint of 5.0 K, and completion at 4.2 K are displayed. The midpoint temperature was reproducible on another crystal in the same batch when a different apparatus with different thermometry was utilized. After applying the magnetic field of 50 kOe, the superconducting transition was suppressed as shown in Fig. 1b, indicating that the zero resistance without magnetic field is due to the real superconducting state.

D.c. magnetization

In order to confirm the superconductivity, the d.c. magnetization was measured with polycrystalline sample using a d.c. SQUID magnetometer. After cooling the sample from 9 K to 2 K in the zero applied field, the diamagnetic shielding signal was measured with increasing temperature in a constant magnetic field of 5 Oe (Fig. 1c). The volume diamagnetic signal at 2.0 K corresponds to ~26 % of the perfect diamagnetism. This exhibits that the superconductivity is a true volume effect in these crystals. The onset of the diamagnetic signal is 5.0 K in the magnetic field of 5 Oe, which agrees with the midpoint of electrical resistance drop. The magnetization above 5 K is almost constant, indicating the Pauli paramagnetism.

The magnetization under cooling from 9 K to 2 K in a magnetic field of 5 Oe was also shown in Fig. 1c. This magnetization traces that of the heating process down to 5.0 K and decreases more slowly than shielding signal down to

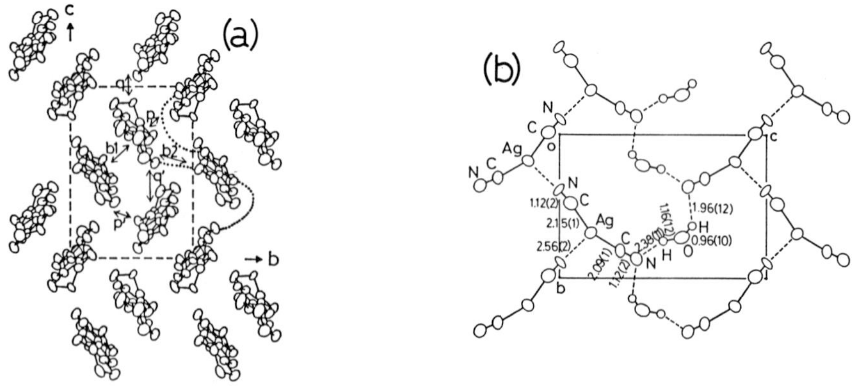

Fig. 2. a)Donor and b)anion arrangements of κ-(BEDT-TTF)$_2$Ag(CN)$_2$H$_2$O. The calculated overlap integrals are b1=23.4, b2=8.4, p=10.5, p'=10.4, q=-2.2, q'=-3.8 ($\times 10^{-3}$). The S···S contacts (<3.6 Å) are shown in dotted lines.

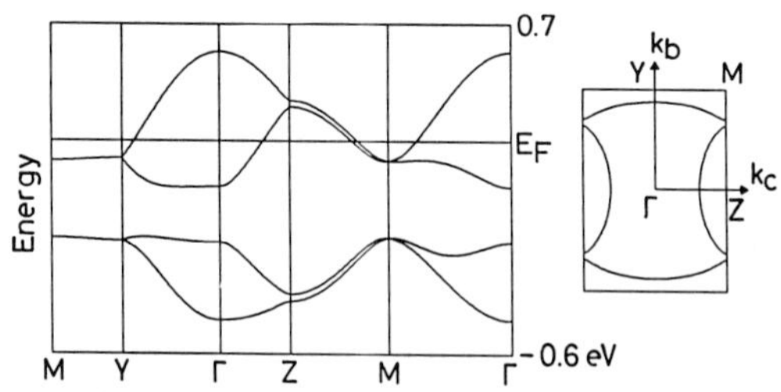

Fig. 3. Calculated Fermi surface and band structure of κ-(BEDT-TTF)$_2$Ag(CN)$_2$H$_2$O.

2 K due to the flux pinning effect in the sample.

Crystal and electronic structures

Figure 2 shows the crystal structure at room temperature. This salt has a layered-structure where two-dimensional donor layer and anion polymer sheet stack alternately along the a-axis. The packing pattern of donors is κ-phase which is similar to κ-(BEDT-TTF)$_2$Cu(NCS)$_2$ (9); two crystallographically independent donors form a pair and this pair is arranged perpendicular to neighbor pairs to construct two-dimensional network. The short S···S contacts (<3.6 Å) are drawn in dotted lines. (Fig. 2a) Only three kinds of intermolecular short contacts are observed at room temperature, while seven kinds of intra- and intermolecular contacts are obtained in κ-(BEDT-TTF)$_2$Cu(NCS)$_2$. The conformational disorder of terminal ethylene groups in donor molecules is smaller in κ-(BEDT-TTF)$_2$Ag(CN)H$_2$O ($2.9<B_{eq}(C_{terminal})<5.4$) than that of κ-(BEDT-TTF)$_2$Cu(NCS)$_2$ ($3.6<B_{eq}<12$).

As shown in Fig. 2b, the anion arrangement of κ-(BEDT-TTF)$_2$Ag(CN)$_2$H$_2$O is very unique; the terminal N atoms in a V-shape unit, NC-Ag-CN (∠C-Ag-C=155°), are linked to H atom in H$_2$O and neighbor Ag atom in Ag(CN)$_2$ to construct two-dimensional network in the bc plane like ··NC-Ag-CN···H-O-H···NC-Ag-CN··. The inclusion of H$_2$O to form anion polymer or anion cluster is also found in (BEDT-TTF)$_3$Br$_2$(H$_2$O)$_2$ (10a), (BEDT-TTF)$_3$Cl$_2$(H$_2$O)$_2$ (10b), and (BEDT-TTF)$_3$Li$_{0.5}$Hg(SCN)$_4$(H$_2$O)$_2$ (10c) as well as this salt.

Some specific cation-anion contacts were found in κ-(BEDT-TTF)$_2$Cu(NCS)$_2$ even at room temperature. However, no contact ((Ag···H)<2.36, (C···H)<2.90, (N···H)<2.75, and (H···H)<2.4 Å) is observed in κ-(BEDT-TTF)$_2$Ag(CN)$_2$H$_2$O.

In order to understand the electronic structure, the tight-binding energy band structure at room temperature was calculated on the basis of the extended Hückel method. The calculated overlap integrals and HOMO are shown in Figs. 2a and 3, respectively. The two-dimensional feature resembles to that of κ-(BEDT-TTF)$_2$Cu(NCS)$_2$ (11). Due to non-centrosymmetric space group, P2$_1$, the degeneracy on MZ is dissolved and both open and closed Fermi surfaces are obtained. The cross-section of the closed Fermi surface is 17 % of the first Brillouin zone which is close to 18 % of κ-(BEDT-TTF)$_2$Cu(NCS)$_2$. The investigation of the closed part by the measurement of Shubnikov-de Haas oscillation is under way.

ESR and thermoelectric power measurements

One broad Lorentzian ESR signal is obtained from 289 K to 5 K; the additional sharp signal that appeared in κ-(BEDT-TTF)$_2$Cu(NCS)$_2$ (12) was not recorded even at low temperatures in this salt. Figure 4a shows the angular dependence of

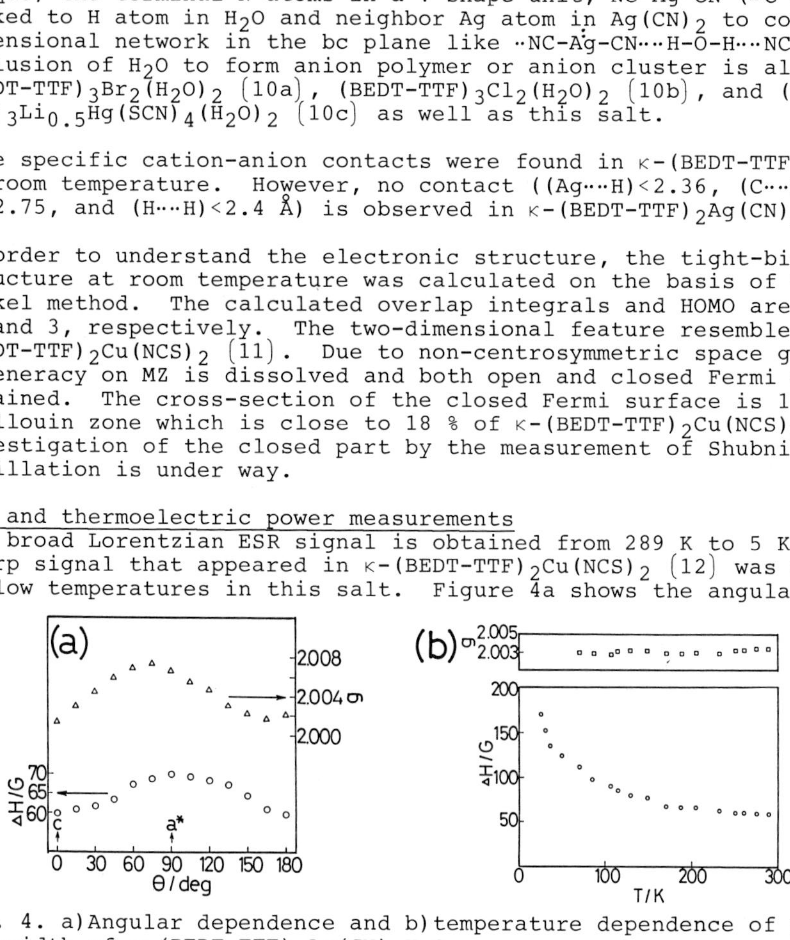

Fig. 4. a) Angular dependence and b) temperature dependence of g-value and linewidth of κ-(BEDT-TTF)$_2$Ag(CN)$_2$H$_2$O.

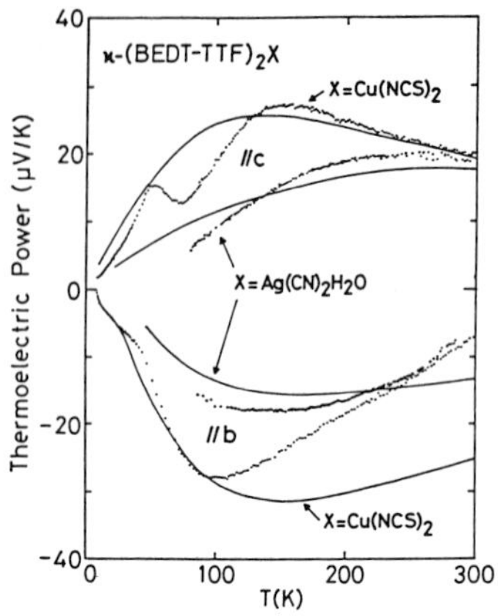

Fig. 5. Thermoelectric powers of κ-(BEDT-TTF)$_2$Cu(NCS)$_2$ and κ-(BEDT-TTF)$_2$Ag(CN)$_2$H$_2$O. The dotted from the experiment, and the solid curve from the theory (14).

g-value and linewidth of single crystal at room temperature. The maximum (2.0072) and the minimum (2.0018) of g-values were observed when the magnetic field was applied nearly parallel to the molecular long axis and perpendicular to the donor plane. The broad signal can be attributed to the BEDT-TTF cation radical (13). The temperature dependence of g-value and linewidth is plotted in Fig. 4b. The g-value is independent of temperature (2.0027-2.0034), whereas the linewidth increases from 57 G at 289 K to 170 G at 26 K although resistivity decreases to 5 K. The broadening of linewidth was also observed in κ-(BEDT-TTF)$_2$Cu(NCS)$_2$. The magnetic susceptibility is independent of temperature, indicating Pauli paramagnetism.

The thermoelectric powers along the b- and c-axes were measured as shown in Fig. 5. The absolute values are smaller than those of κ-(BEDT-TTF)$_2$Cu(NCS)$_2$, indicating the lower density of state of κ-(BEDT-TTF)Ag(CN)$_2$H$_2$O. It is consistent that T_c of κ-Ag(CN)$_2$H$_2$O salt, 5.0 K, is lower than that of κ-Cu(NCS)$_2$ salt, 10.4 K. The temperature dependence of thermoelectric powers down to 90 K is very similar to that of κ-Cu(NCS)$_2$ salt (12). The anisotropy of thermoelectric power ($S>0$ along the c-axis and $S<0$ along the b-axis) in the two-dimensional plane can be explained by the anisotropy of band structure.

In summary, we have found the new ambient pressure organic superconductivity with T_c=5.0 K in κ-(BEDT-TTF)$_2$Ag(CN)$_2$H$_2$O, exhibiting the zero electrical resistivity and Meissner effect. The H_{c2} and Shubnikov-de Haas oscillation measurements are under progress.

REFERENCES

1. Jerome D, Mazaud A, Ribault M, Bechgaard K, (1980) J. Phys. Lett. 41: L95.
2. Williams J M, Kini A M, Wang H H, Carlson K D, Geiser U, Montgomery L K, Pyrka G J, Watkins D M, Kommers J M, Boryschuk S J, Strieby Crouch A V, Kwok W K, Schirber J E, Overmyer D L, Jung D, Whangbo M -H, (1990) Inorg. Chem. 29: 3272.
3. Kanoda K, Takahashi T, Saito G, (1989) Physica C 162: 405.
4. Urayama H, Yamochi H, Saito G, Nozawa K, Sugano T, Kinoshita M, Sato S, Oshima K, Kawamoto A, Tanaka J, (1988) Chem. Lett. 55.
5. a)Mori H, Hirabayashi I, Tanaka S, Mori T and Inokuchi H, (1990) Solid State Commun. 76: 35; b)Mori H, Hirabayashi I, Tanaka S, Mori T, Maruyama Y, Inokuchi H, in the Proceeding of ICSM'90 (to be published in Synthetic Metals).
6. Kurmoo M, Talham D R, Patchard K L, Day P, Stringer A M, and Haward J A K, (1988) Synthetic Metals, 27: A177.
7. Oshima K, Urayama H, Yamochi H, Saito G, (1988) J. Phys. Soc. Jpn. 57: 730.
8. Ginodman V B, Gudenko A V, Zherikhina L N, (1985) JETP Lett. 41: 49.
9. Urayama H, Yamochi H, Saito G, Sato S, Kawamoto A, Tanaka J, Mori T, Maruyama Y, Inokuchi H, (1988) Chem. Lett. 463.
10. a)Urayama H, Saito G, Kawamoto A, Tanaka J, (1987) Chem. Lett. 1753; b)Mori T and Inokuchi H, (1987) Chem. Lett. 1657; c)Mori H, Tanaka S, Oshima M, Saito G, Oshima K, Mori T, Maruyama Y, Inokuchi H, in the Proceeding of ICSM'90 (to be published in Synthetic Metals).
11. Oshima K, Mori T, Inokuchi H, Urayama H, Yamochi H, Saito G, (1988) Phys. Rev. B38: 938.
12. Urayama H, Yamochi H, Saito G, Sugano T, Kinoshita M, Inabe T, Mori T, Maruyama Y, Inokuchi H, (1988) Chem. Lett. 1057.
13. Sugano T, Saito G, Kinoshita M, (1986) Phys. Rev. B34: 117.
14. Mori T and Inokuchi H, (1988) J. Phys. Soc. Jpn. 57: 3674.

New Layered Copper Oxides Containing Double-MO_2-Unit Fluorite Block

TAKAHIRO WADA, ATARU ICHINOSE, H. YAMAUCHI, and SHOJI TANAKA

Superconductivity Research Laboratory, International Superconductivity Technology Center, 10-13, Shinonome 1-chome, Koto-ku, Tokyo, 135 Japan

ABSTRACT

Two new families of layered oxides, $(Pb,Cu)Sr_2(Ho,Ce)_3Cu_2O_{11}$ [Pb-based "1232"] and $Sr_2(Ho,Ce)_3Cu_3O_{10+z}$ [Cu-based "1232"], have been synthesized. These compounds have tetragonal unit cells with the lattice constants, all being approximately equal to a=3.8 Å and c=17 Å. The Rietveld analysis using x-ray powder diffraction data shows that the crystal structure of $(Pb_{1/2}Cu_{1/2})Sr_2(Ho_{1/3}Ce_{2/3})_3Cu_2O_{11}$ consists of a new double-MO_2-unit fluorite block, $[(Ho,Ce)O_2]_2$, and a Pb-based "1212" block, $(Pb,Cu)Sr_2(Ho,Ce)Cu_2O_7$. The unit cell for $Sr_2(Ho,Ce)_3Cu_3O_{10+z}$ is considered to consist of the double-MO_2-unit fluorite block and an oxygen-deficient-triperovskite block, $(Ho,Ce)Sr_2Cu_3O_{6+z}$. The temperature dependence of resistivity for every one of these compounds was semiconductive but these compounds were expected to be parent materials for high T_c superconductors.

KEY WORDS: crystal structure, layered copper oxides, Rietveld analysis

INTRODUCTION

Since the discovery of superconductivity in $(La,Ba)_2CuO_4$ [1, 2], various types of copper oxide superconductors have been synthesized [3]. All the types are characterized as combinations of several different types of blocks such as perovskite block ($LaCuO_3$), rock salt block (SrO) and fluorite block (CeO_2). For example. the unit cell of $(La,Ba)_2CuO_4$ consists of a rock salt block, [(La,Ba)O], and a perovskite block [$LaCuO_3$] [4]. In this paper, we report syntheses and structural characterization of two new families of layered copper oxides, $(Pb,Cu)Sr_2(Ho,Ce)_3Cu_2O_{11}$ and $Sr_2(Ho,Ce)_3Cu_3O_{10+z}$. These compounds have crystal structures which contain a new type of fluorite block consisting of double-MO_2-units such as $[(Ho, Ce)O_2]_2$.

EXPERIMENTAL PROCEDURE

$(Pb_{1/2}Cu_{1/2})Sr_2(Ho_{1/3}Ce_{2/3})_3Cu_2O_{11}$, and $Sr_2(Ho_{1/3}Ce_{2/3})_3Cu_3O_{10+z}$ samples were prepared from high purity (≧ 99.9%) powders of Ho_2O_3, CeO_2, $SrCO_3$, CuO, and PbO. The starting powders were weighed to the compositions desired, and calcined three times at 1050 ℃ in O_2 gas flow, each time for 10 h with intermediate grinding. The mixed powder was pressed and sintered at 1060 ℃ for 20 h at an oxygen gas pressure of 3 atm. The phases present and the lattice constants were determined by x-ray powder diffraction using Cu-Kα radiation and electron probe microanalysis (EPMA). X-ray powder diffraction measurements for the Rietveld analysis were performed using an x-ray diffractometer (MAC Science: Model MPX18) equipped with a rotating-anode source and a curved graphite monochrometer. Diffraction data were collected with the step width of 0.04° (in 2θ) and the counting time of 4 s for each step over the 2θ range from 15 to 120° . The crystal structure was refined by the Rietveld analysis using a program, RIETAN, developed by Izumi [5].

RESULTS AND DISCUSSION

A. $(Pb,Cu)Sr_2(Ho,Ce)_3Cu_2O_{11}$

The x-ray powder diffraction pattern for $(Pb_{1/2}Cu_{1/2})Sr_2(Ho_{1/3}Ce_{2/3})_3Cu_2O_{11}$ is shown in Fig. 1 (a). The peaks of the pattern were successfully indexed for a tetragonal unit cell with the lattice constants, a=3.822 A and c=17.2 A. The diffraction pattern showed that this sample was of single phase. The crystallographic structure of $(Pb_{1/2}Cu_{1/2})Sr_2(Ho_{1/3}Ce_{2/3})_3Cu_2O_{11}$ was analyzed by the Rietveld analysis using x-ray powder diffraction data. Prior to the Rietveld refinement, the space group was determined to be P4/mmm (tetragonal) using a program, CELLO, developed by Takaki et al. [6]. Isotropic thermal parameters for all the rare earth elements were assumed to be the same and those of oxygen atoms which occupy four different sites in the crystal were also assumed to be the same. The structure parameters resulted from the Rietveld analysis for $(Pb_{1/2}Cu_{1/2})Sr_2(Ho_{1/3}Ce_{2/3})_3Cu_2O_{11}$ are listed in Table I. The refined lattice parameters were: a=3.82246(5)Å and c=17.2082(3) Å. The final values of the R factors were: R_{WP}=7.37 %, R_P=5.51 %, R_I=3.81 %, R_F=3.12 % and R_e=2.72 %. Figure 2 shows the final profile fit and difference pattern. A good agreement between the observed and calculated patterns was obtained. This strongly supports the following structural model. A schematic crystal structure of the model for $(Pb_{1/2}Cu_{1/2})Sr_2(Ho_{1/3}Ce_{2/3})_3Cu_2O_{11}$ is shown in Fig. 3 (c). The Pb-based "1212" and "1222" structures are also shown in Figs. 1 (a) and (b) respectively. The crystal structure of $(Pb_{1/2}Cu_{1/2})Sr_2(Ho_{1/3}Ce_{2/3})_3Cu_2O_{11}$ can be divided into two types of blocks; one is a new double-MO_2-unit fluorite block, $[(Ho,Ce)O_2]_2$, and the other is a Pb-based "1212" block, $(Pb,Cu)Sr_2(Ho,Ce)Cu_2O_7$ [7,8]. This structure resembles

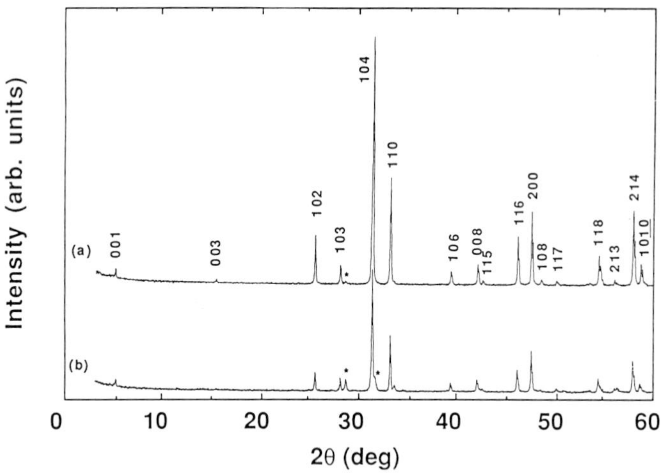

Fig. 1. X-ray powder diffraction patterns for $(Pb_{1/2}Cu_{1/2})Sr_2(Ho_{1/3}Ce_{2/3})_3Cu_2O_{11}$ (a) and $Sr_2(Ho_{1/3}Ce_{2/3})_3Cu_3O_{10+z}$ (b). The peaks for impurity phases are marked with asterisks.

Table I Structure parameters for $(Pb_{1/2}Cu_{1/2})Sr_2(Ho_{1/3}Ce_{2/3})_3Cu_2O_{11}$. Symbol g is the occupation factor and B is the isotropic thermal parameter. Figures in parentheses are estimated standard deviations.

atom	site	g	x	y	z	B(A^2)
Sr	2h	1	0.5	0.5	0.1393(5)	0.9(2)
Ho,Ce(1)	2h	1	0.5	0.5	0.3481(4)	0.2(1)
Ho,Ce(2)	1b	1	0.0	0.0	0.5	0.2(1)
Pb,Cu	4l	0.25	0.072(9)	0.0	0.0	1.3(5)
Cu	2g	1	0.0	0.0	0.2439(8)	0.3(3)
O(1)	4n	0.25	0.32(3)	0.5	0.0	1.4(5)
O(2)	2g	1	0.0	0.0	0.115(3)	1.4(5)
O(3)	4i	1	0.0	0.5	0.255(3)	1.4(5)
O(4)	4i	1	0.0	0.5	0.418(2)	1.4(5)

that of Pb-based "1222" structure, $(Pb,Cu)Pr_2(Pr,Sr)_2Cu_2O_9$ [9, 10], which consists of a single-MO_2-unit fluorite block, PrO_2 or $(Eu,Ce)O_2$, and a Pb-based "1212"block, $(Pb,Cu)(Sr,Pr)_2PrCu_2O_7$ or $(Pb,Cu)(Sr,Eu)_2(Eu,Ce)Cu_2O_7$. Therefore this new crystal structure may be termed as Pb-based "1232" structure.

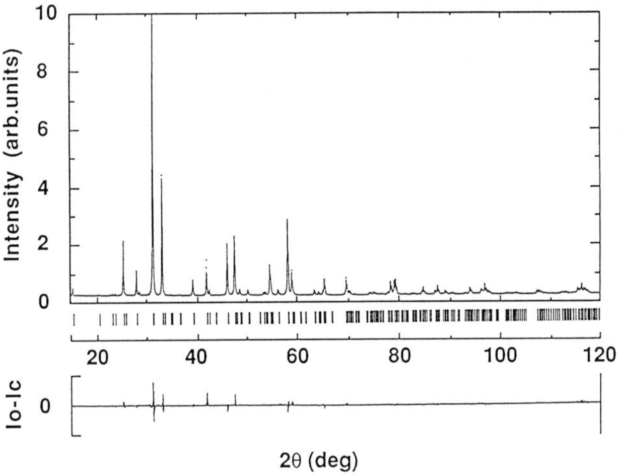

Fig. 2 X-ray Rietveld refinement pattern for $(Pb_{1/2}Cu_{1/2})Sr_2(Ho_{1/3}Ce_{2/3})_3Cu_2O_{11}$.

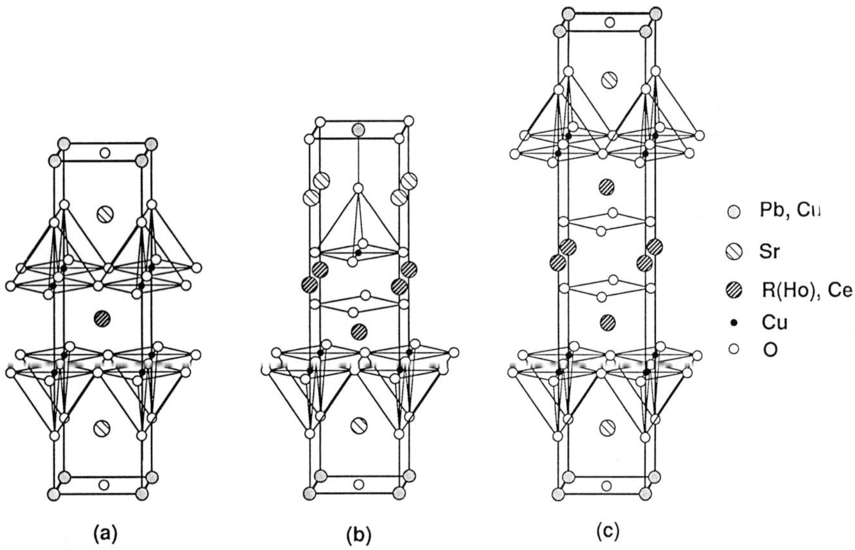

Fig. 3 Schematic crystal structures for $(Pb,Cu)(Sr,R)_2(R,Ce)_nCu_2O_{5+2n}$. (a):Pb-based "1212" [n=1], (b): Pb-based "1222" [n=2], (c):$(Pb_{1/2}Cu_{1/2})Sr_2(Ho_{1/3}Ce_{2/3})_3Cu_2O_{11}$ (Pb-based "1232") [n=3].

B. $Sr_2(Ho,Ce)_3Cu_3O_{10+z}$

The x-ray diffraction patterns for $Sr_2(Ho_{1/3}Ce_{2/3})_3Cu_3O_{10+z}$ is shown in Fig. 1 (b). The diffraction peaks were successfully indexed for a tetragonal unit cell. The peaks for impurity phases such as CeO_2 are marked with asterisks. The lattice constants of this compound are a=3.824 Å and 17.22 Å. The crystal structure of $Sr_2(Ho,Ce)_3Cu_3O_{10+z}$ is now supposed to consist of the new double-MO_2-unit fluorite block, $[(Ho,Ce)O_2]_2$, and an oxygen-deficient-triperovskite block, $(Ho,Ce)Sr_2Cu_3O_{6+z}$ [11]. A schematic structure for $Sr_2(Ho,Ce)_3Cu_3O_{10+z}$ is shown in Fig.4 (c). The oxygen-deficient-triperovskite ("123" or Cu-based "1212") structure and $(Eu,Ce)_2(Eu,Ba)_2Cu_3O_{8+z}$ (Cu-based

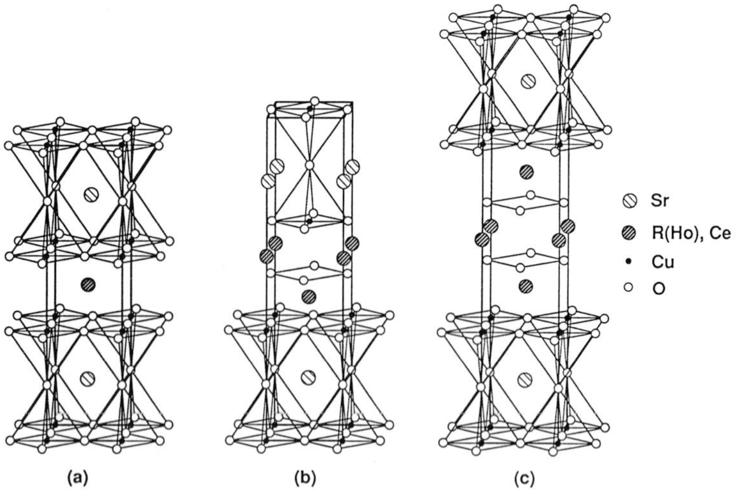

Fig. 4 Schematic crystal structures of $(Sr,R)_2(R,Ce)_nCu_3O_{4+2n}$ (a):Cu based "1212" or "123" [n=1], (b):"Cu based "1222" [n=2], (c):$Sr_2(Ho_{1/3}Ce_{2/3})_3Cu_3O_{10+z}$ (Cu-based "1232") [n=3].

"1222") structure, which consists of a single-MO_2-unit fluorite block and an oxygen-deficient-triperovskite block [12], were also shown in Figs. 4 (a) and (b) respectively. Therefore we could call this new crystal structure by Cu-based "1232" structure. The difference in the crystal structures given in Fig. 3 and Fig. 4 is concerned only with the positions of oxygen atom in the (Cu,Pb)O or CuO_z layer. The x-ray diffraction data are usually sufficient to determine the arrangements of metal ions only. Therefore, a precise crystallographic structure refinement based on neutron powder diffraction data is necessary to obtain reliable structure parameters for the oxygen atoms.

SUMMARY

Two new families of layered copper oxides, $(Pb,Cu)Sr_2(Ho,Ce)_3Cu_2O_{11}$ and $Sr_2(Ho,Ce)_3Cu_3O_{10+z}$ were discovered. The crystal structure of $(Pb,Cu)Sr_2(Ho,Ce)_3Cu_2O_{11}$ (Pb-based "1232" structure) consisted of a double-MO_2-unit fluorite block and a Pb-based "1212" block. The crystal of $Sr_2(Ho,Ce)_3Cu_3O_{10+z}$ (Cu-based "1232" structure) was considered to consist of the double-MO_2-unit fluorite block and an oxygen-deficient-triperovskite block. These compounds were semiconductors but were expected to be parent materials for high T_c superconductors.

ACKNOWLEDGMENTS

We are grateful to Mr. Tsutomu Takata for his help in electron probe microanalyses and Miss Akiko Nara for her help in sample preparation and x-ray powder diffraction.

REFERENCES

[1] Bednorz JG, Muller KA (1986) Z Phys B 64: 189.
[2] Takagi H,Uchida S,Kitazawa K,Tanaka S (1987) Jpn J Appl Phys 26: L123.
[3] Yvon K,Francois M (1989) Z Phys B 76: 413.
[4] Wada T: to be submitted.
[5] Izumi F (1985) J Crystallogr Soc Jpn 27: 23.
[6] Takaki Y,Taniguchi T,Yamaguchi,Ogura T (1987) J Ceram Soc Jpn 95: 610.
[7] Lee JY,Swinnea LS,Steinfink H (1989) J Mat Res 4: 763.
[8] Subramanian MA,Gopalakrishnan J,Torardi CC,Gai PL,Boyes ED,Askew TR, Flippen RB,Farneth WE, Sleight AW (1989) Physica C 157: 124.
[9] Adachi S,Inoue O,Kawashima S,Adachi H,Setsune K,Wasa K(1990)Physica C168:1
[10]Maeda T,Sakuyama K,Koriyama S,Ichinose A,Yamauchi H,Tanaka S (1990) Physica C 169: 133.
[11]Sunshine SA,Schneemeyer LF,Siegrist T, Douglass DC,Waszczack JV,Cava RJ, Gyorgy EM,Murphy DW (1989) Chemistry of Materials 1: 331.
[12]Sawa H,Obara K,Akimitsu J,Matsui Y,Horiuchi S (1989)J Phys Soc Jpn 58:2252

Superconductivity up to 127 K in Tl-Ba-Ca-Cu-O Ceramics

SEIJI ADACHI, KOICHI MIZUNO, KENTARO SETSUNE, and KIYOTAKA WASA

Central Research Laboratories, Matsushita Electric Industrial Co., Ltd.,
3-15, Yagumo-Nakamachi, Moriguchi, Osaka, 570 Japan

ABSTRACT

Suitable synthetic conditions to obtain Tl-Ba-Ca-Cu-O ceramics with T_c around 125 K have investigated. The sample sintered in quartz tubes at 760°C for longer periods than 240h at a nominal composition of $Tl_1Ba_1Ca_3Cu_3O_y$ showed T_c(shielding)=127.2K. Volume fraction of superconducting phase at 125K estimated from Meissner signals was approximately 10%. This synthetic condition could yield the sample showing Meissner effect over 125 K with good reproducibility.

KEY WORDS: Tl-Ba-Ca-Cu-O, synthesis, Meissner effect

INTRODUCTION

Since the first observation of superconductivity in a La-Ba-Cu-O system by Bednorz and Müller in 1986,[1] enormous efforts have been paid to find higher-T_c materials. Sheng et al. have discovered superconductivity above 100K in Tl-Ba-Ca-Cu-O in 1988.[2] Parkin et al. observed zero resistivity at 125K in $Tl_2Ba_2Ca_2Cu_3O_y$ ("2223" phase)[3] and suggested possible crystal structures in a Tl-Ba-Ca-Cu-O system.[4] Tl-Ba-Ca-Cu-O samples with T_c(zero resistivity) above 120K were generally prepared by heat treatments at around 900°C for short periods of shorter than several hours. Those superconductive properties must be remarkably affected by synthetic conditions. Although many researchers studied on this system, reproducibility of superconductivity at 125K was very poor. We attempted to look for more suitable synthetic conditions of superconductive Tl-Ba-Ca-Cu-O ceramics with high-T_c at around 125K by heat treatments at rather low temperatures for long periods.

EXPERIMENTAL

Samples were synthesized by a conventional solid state reaction. Precursor method using $BaCuO_2$ and Ca_2CuO_3 was adopted in order to eliminate the influence of carbonates. The former was made by firing the mixture of $BaCO_3$ and CuO powders (Ba:Cu=1:1) at 900°C for 10 h in oxygen, and the latter was by that of $CaCO_3$ and CuO (Ca:Cu=2:1) at 950°C. Tl_2O_3, CuO and two precursors were mixed at the ratio of Tl:Ba:Ca:Cu=1:1:3:3. This nominal composition yields the "2223" phase rather easily. At this nominal composition, Parkin et al. fabricated the sample with T_c(zero resistance) of 125K and Kotani et al. grew the "2223" single crystal.[5] Mixed powders were press-formed into rectangular bars. Then, they were wrapped in gold foil and sealed in evacuated quartz tubes, and sintered at 680-840°C for 8-720 hours.

Electrical resistances for the obtained samples were measured by the standard four-probe method. Diamagnetic shift (shielding effect) was checked by inductance measurement of a coil winding up the sample. Both measurements were performed with raising the temperature from liquid nitrogen temperature at a rate of 0.3°C/min. on the samples (about 0.3 g) mounted in 1 kg copper-sample holder. Temperature was monitored by a carbon glass resistance sensor (Lake Shore Cryotronics,Inc. CGR-1-1000, calibrated using the Chebychev polynomials) at the accuracy within 0.1°C. Magnetization measurement was performed under a "field cooling" condition at a cooling rate of 2°C/min in an external magnetic field of 10 Oe by using SQUID susceptometer (HOXAN HSM-2000). Crystal structure was examined using Powder X-ray diffraction (XRD; CuKα radiation).

RESULTS AND DISCUSSION

Figure 1 shows temperature dependences of electrical resistance for samples sintered at various temperatures for 24h. Superconductivity with T_c(onset) above 100K was observed for all samples sintered at 840-680°C. Above all, the sample sintered at 760°C showed the highest-T_c(zero resistance) of 123.6K.

Figure 2 shows temperature dependences of electrical resistance for samples sintered at 760°C for various periods. Improvement in the transition temperature was observed with prolongation of firing period. The resistance of the sample fired for 720h drastically decreased below about 133K and vanished at 124.9K after a little tailing. Extrapolation of the slope at around 127K suggested that T_c(zero resistance) might be about 126K if there were not the tailing.

Temperature dependences of inductance of a coil winding up the sample are shown in Fig.3. Start temperature of flux exclusion, T_c(shielding), was increased with prolongation of sintering period. In cases of 240h and 720h, T_c(shielding) was 127.2K.

Magnetizations for the samples are shown in Fig.4. Prolongation of sintering period brought about the increase of transition temperature of Meissner effect. For the sample sintered for longer period than 240h, volume fraction

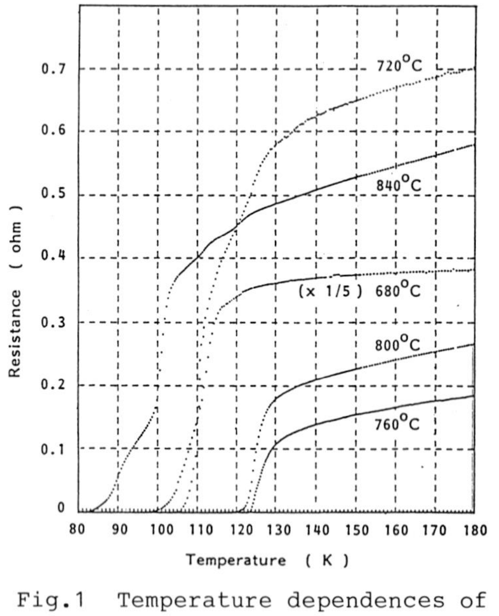

Fig.1 Temperature dependences of electrical resistance for samples sintered at various temperatures for 24h.

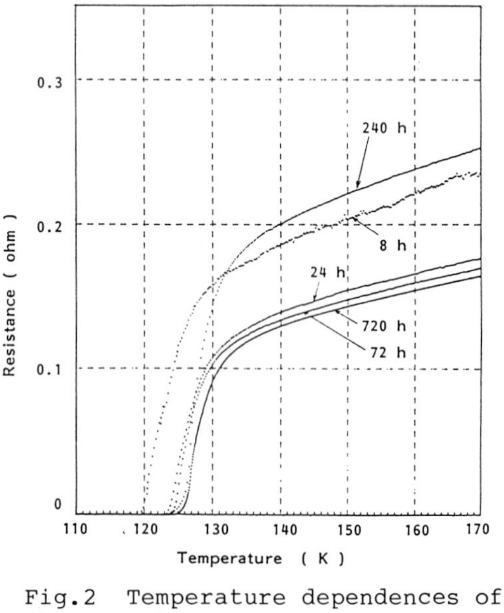

Fig.2 Temperature dependences of electrical resistance for samples sintered at 760°C for various periods.

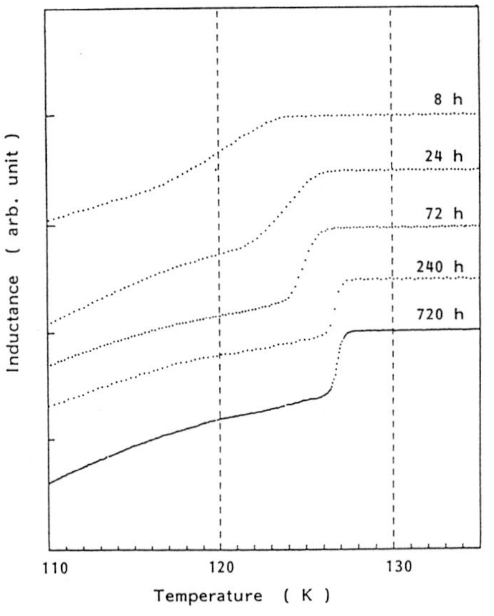

Fig.3 Temperature dependences of shielding effects for samples sintered at 760°C for various periods.

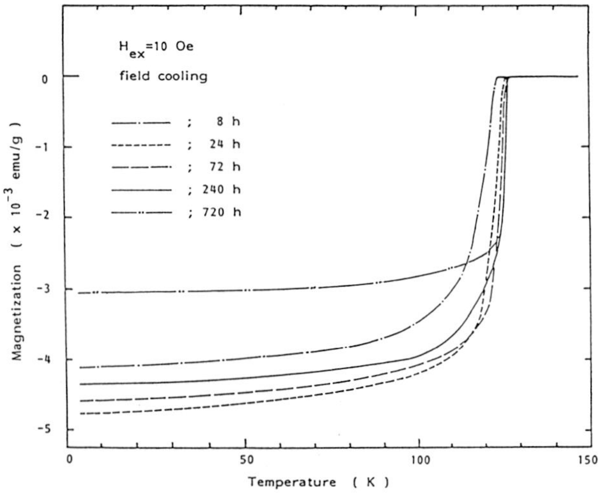

Fig.4 Temperature dependences of magnetization for samples sintered at 760°C for various periods.

of superconducting phase at 125.0K were roughly estimated to be approximately 10%. This result indicated that drastic decreases of electrical resistance at around 125K on these samples were not resulted from the superconductivity of filament-like phase but that of bulky one.

There were peaks due to the "2223" phase with impurity ones in the powder XRD patterns of obtained samples sintered for 240h. Observed superconductivity seemed to be caused by the "2223" phase because of its T_c value of around 125K. Figure 5 shows the (0 0 14) reflection peaks of the "2223" phase synthesized by two different methods. Peak (a) is the peak for the "2223" phase synthesized under 1-atm oxygen whose T_c(shielding) was 118K, peak (b) is that for the high-T_c sample (sintered at 760°C for 240hrs) obtained in this study. Dots are original plots and lines are data eliminated CuKα2 by calculation.[6] c-axis length for the sample synthesized under 1-atm oxygen was 3.5614 nm. Shimakawa et al.[7] and Kikuchi et al.[8] reported c-axis length of 3.55859(7) and 3.5572(2) nm for the "2223" phase synthesized under 1-atm oxygen. Their results were refined by Rietveld analysis using RIETAN.[9] On the other hand, our high-T_c sample (Fig.5(b)) had longer c-axis length of 3.5712 nm. Torardi et al. synthesized the "2223" phase with very high-T_c(onset) of 140K, and reported that its c-axis length was 3.59 nm.[10] It seems that long c-axis length corresponds high-T_c value for the "2223" phase.

Figure 6 shows as SEM photo of fractured surface for the sample sintered at 760°C for 240 hrs. Platy grains were observed. This morphology is different from that of Bi-based cuprate superconductors which is micaceous. It is considered that the manners of crystal growth for Tl-based and for Bi-based materials must be quite different.

SUMMARY

Tl-Ba-Ca-Cu-O ceramics fabricated by various sintering conditions at a nominal composition of $Tl_1Ba_1Ca_3Cu_3O_y$. The samples sintered in quartz tubes at rather low temperatures for long periods showed Meissner effect below 127.2K. Magnetization measurement revealed that superconductivity at 125K must reflected bulky characteristics. We emphasize that this synthetic condition could yield the sample showing Meissner effect over 125K with fairly good reproducibility.

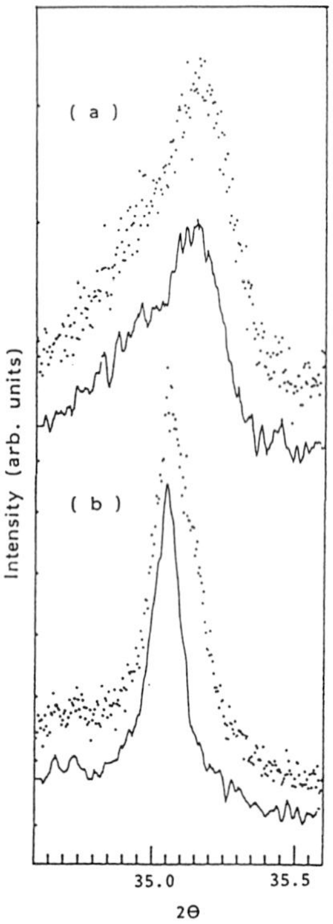

Fig.5 Powder XRD diffraction peaks of (0 0 14) reflection for "2223" phase.
(a) synthesized under 1-atm oxygen
(b) the high-T_c sample in this work (sintered at 760°C for 240 hrs)

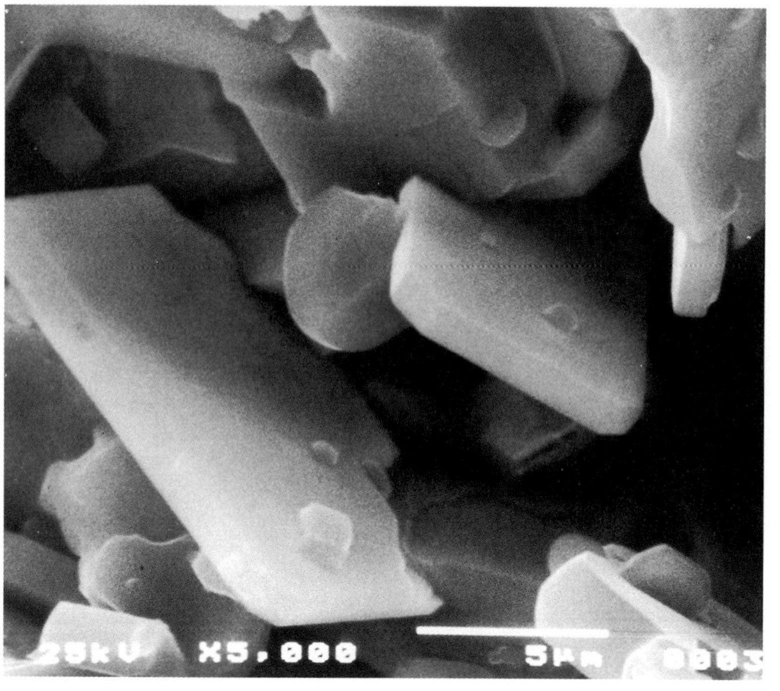

Fig.6 SEM photo of fractured surface for the sample sintered at 760°C for 240 hrs.

REFERENCES

[1] J.G. Bednorz and K.A. Müller: Z. Phys. B 64(1986)198.
[2] Z.Z. Sheng and A.M. Hermann: Nature 332(1988)138.
[3] S.S.P. Perkin, V.Y. Lee, E.M. Engler, A.I. Nazzal, T.C. Huang, G. Gorman, R. Savoy and R. Beyers: Phys. Rev. Lett. 60(1988)2539.
[4] S.S.P. Perkin, V.Y. Lee, A.I. Nazzal, R. Savoy and R. Beyers: Phys. Rev. Lett. 61(1988)750.
[5] T. Kotani, T. Kaneko, H. Takei and K. Tada: Jpn. J. Appl. Phys. 28(1989)L1378.
[6] D.T. Keating: Rev. Sci. Instr. 30(1959)725.
[7] Y. Shimakawa, Y. Kubo, T. Manako, Y. Kobayashi and H. Igarashi: Proc. 1st Int. Symp. on superconductivity (ISS'88) p.781, Aug. 28-31, 1988, Nagoya, Japan.
[8] M. Kikuchi, T. Kajitani, T. Suzuki, S. Nakajima, K. Hiraga, N. Kobayashi, H. Iwasaki and Y. Shono: Jpn. J. Appl. Phys. 28(1989)L382.
[9] F. Izumi: J. Crystallogr. Soc. Jpn. 27(1985)23 [in Japanese].
[10] C.C. Torardi, M.A. Subramanian, J.C. Calabrese, J. Gopalakrishnan, K.J. Morrissey, T.R. Askew, R.B. Flippen, U. Chowdhry and A.W. Sleight: Science 240(1988)631.

3.2 Structual Analysis

Phase Instability at the Maximum T_c in Oxide Superconductors: Phase Separation in La_2CuO_{4+x}, $Nd_{2-x}Ce_xCuO_4$, and $La_{2-x}Sr_xCuO_4$

J.D. Jorgensen[1], P. Lightfoot[2,3], Shiyou Pei[2], B. Dabrowski[1,4], D.R. Richards[1], and D.G. Hinks[1]

[1] Materials Science Division and [2] Science and Technology Center for Superconductivity, Argonne National Laboratory, Argonne, IL 60439, USA
[3] Chemical Crystallography Laboratory, University of Oxford, Oxford, UK
[4] Northern Illinois University, DeKalb, IL 60115, USA

ABSTRACT

Observation of similar behavior in three different superconducting oxides suggests that the maximum T_c is limited by phase instability. These conclusions are based on previously published neutron powder diffraction data for La_2CuO_{4+x} and $Nd_{2-x}Ce_xCuO_4$ and new data for $La_{2-x}Sr_xCuO_4$ which show a phase separated region for compositions adjacent to that which produces the highest T_c.

KEY WORDS: instability, neutron diffraction, miscibility gap, phase separation

INTRODUCTION

The concept that some form of instability limits the maximum T_c has been proposed for intermetallic superconductors and, more recently, for the oxide superconductors. In general terms, such arguments are based on the idea that the same electronic structure that results in superconductivity also gives rise to structural instability and can produce a transformation to a nonsuperconducting phase. For example, in the Chevrel phase superconductors, the highest T_c's are achieved for compositions adjacent to an electronically driven phase transition to a nonmetallic phase with triclinic symmetry.[1]

There are similar theories for the high-T_c oxide superconductors. Varma [2] has discussed the behavior of $Ba_{1-x}K_xBiO_3$, in which the highest T_c (30K) occurs adjacent to a transition to a charge density wave structure [3], in terms of a dielectric instability. He has proposed similar behavior for the copper-oxide superconductors. Markiewicz [4] proposes that a Van Hove singularity, characteristic of the two-dimensional nature of the copper-oxide superconductors, accounts for the high T_c's. As the composition varies away from the Van Hove singularity, an instability that manifests itself as a compositional phase separation involving superconducting and nonsuperconducting phases, or (if diffusion kinetics will not allow this) phase separation of the carrier themselves, occurs.

In this paper we review previously published data that show phase separation adjacent to the composition giving the highest T_c in La_2CuO_{4+x} [5] and $Nd_{2-x}Ce_xCuO_4$ [6] and report new data showing similar behavior for $La_{2-x}Sr_xCuO_4$. Taken together, these results suggest a consistent picture in which an instability that manifests itself as phase separation is associated with the maximum T_c in each compound. Our data for these three compounds imply a phase diagram of the general topology shown in Fig. 1. As composition is varied, T_c increases with increasing carrier concentration. Eventually the system becomes unstable and phase separates. The miscibility gap is a region where the sample consists of a superconducting phase of fixed composition with the maximum T_c mixed with a nonsuperconducting phase of a different fixed composition that may be insulating or metallic. Measurements of superconducting behavior (e.g., resistivity, magnetic susceptibility, etc.) for samples within the miscibility gap will, in general, be characterized by apparent decreases of T_c or the superconducting fraction, or both, depending on the measurement technique and the real-space distribution of the two-phase mixture.

THE La_2CuO_{4+x} SYSTEM

La_2CuO_{4+x} is the first oxide superconductor for which phase separation involving superconducting and nonsuperconducting phases was observed.[5] For bulk compositions within the miscibility gap, the superconducting phase has a composition near x=0.08, while the nonsuperconducting phase, which is an antiferromagnetic insulator, has a composition near x=0.00.[7] The superconducting phase exhibits a T_c of 35-40K, equal to the highest T_c obtained in the $La_{2-x}Sr_xCuO_4$ system. The two phases have almost identical orthorhombic structures. The excess oxygen occupies an interstitial site coordinated to four lanthanum atoms.[8]

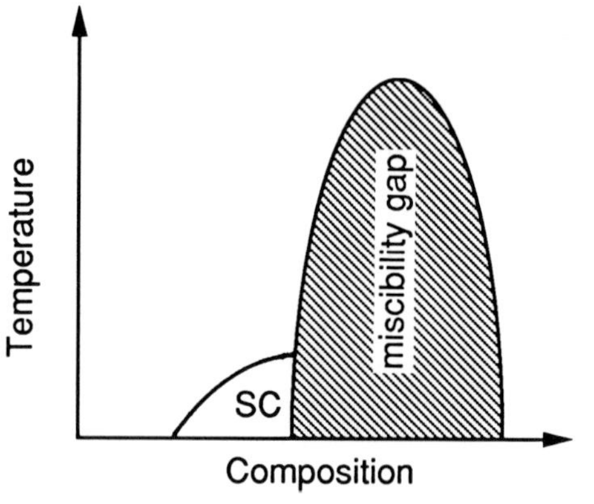

Fig. 1. Generalized phase diagram for oxide superconductors that exhibit phase instability at the composition corresponding to the maximum T_c.

The phase separation was first observed in high resolution neutron powder diffraction experiments. [5] A plot of the diffraction data and Rietveld refinement profile for a two-phase sample is shown in Fig. 2. The differences in the lattice constants of the two phases is so small that, even with the high resolution provided by the pulsed neutron source diffractometer, the minor phase (in this sample, the superconductor) appears only as shoulders on some of the Bragg peaks of the major phase. Such a subtle peak splitting is easy to overlook in routine diffraction measurements. Another remarkable feature of the phase separation in La_2CuO_{4+x} is that it occurs reversibly near room temperature.

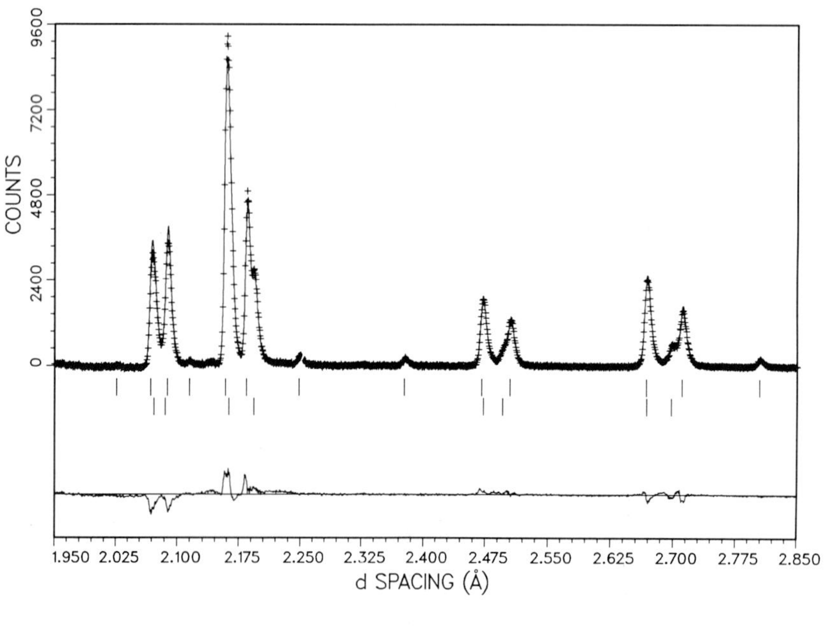

Fig. 2. Neutron powder diffraction data and Rietveld refinement profile for $La_2CuO_{4.03}$ showing overlapped patterns for two nearly identical orthorhombic phases. The tick marks below the data indicate the positions of allowed Bragg reflections for the two phases.

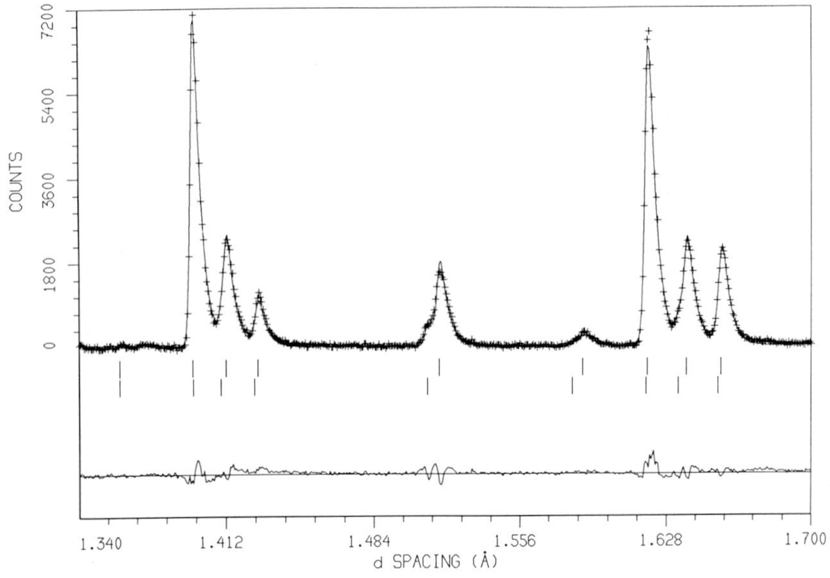

Fig. 3. Neutron powder diffraction data and Rietveld refinement profile for $Nd_{1.87}Ce_{0.13}CuO_4$ showing overlapped patterns for two nearly identical phases.

THE $Nd_{2-x}Ce_xCuO_4$ SYSTEM

We recently reported a similar phase separation phenomena in the $Nd_{2-x}Ce_xCuO_4$ system.[6] Again, the phase separation was first observed by high resolution neutron powder diffraction. The data and Rietveld refinement profile for a sample of composition x=0.13, which is in the two-phase region, are shown in Fig. 3. Just as for La_2CuO_{4+x} the two phases (which have essentially the same structure) are so similar in lattice constants that most of the Bragg peaks are overlapped. The second phase appears only as shoulders on some of the peaks. However, because the full composition range, $0 \leq x \leq 0.2$, can be more easily accessed than for La_2CuO_{4+x} it has been possible to correlate the superconducting behavior with the phase separation.

For the optimum superconducting composition, x≈0.17 for our samples, a single-phase diffraction pattern was obtained.[6] The important features of the phase diagram are shown schematically in Fig. 4. Because of the miscibility gap, there is a sharp onset for bulk superconductivity near x=0.15-0.17 (depending on synthesis technique), with T_c immediately exhibiting its highest value.[9] For further increases in x, T_c drops, but the range of bulk superconducting compositions is unusually narrow because of the solubility limit for cerium near x=0.2. The cerium-deficient phase, to the left of the miscibility gap in Fig. 4, is an antiferromagnetic insulator.[10]

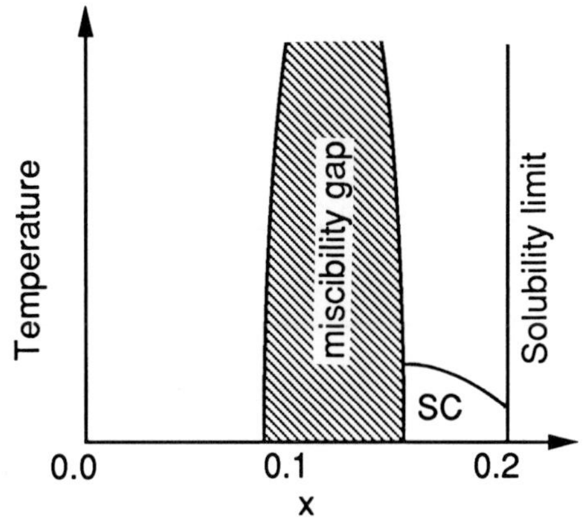

Fig. 4. Approximate phase diagram for $Nd_{2-x}Ce_xCuO_4$.

THE $La_{2-x}Sr_xCuO_4$ SYSTEM

In some of the earliest studies of $La_{2-x}Sr_xCuO_4$ it was noted that bulk superconductivity occurs only over a narrow composition range. Flemming [11] reported that the fraction of superconducting material increased smoothly with increasing x for x<0.2, peaked at the composition yielding the highest T_c, and then dropped sharply for x>0.2. The smooth increase of T_c and the superconducting fraction can be understood in terms of the increasing carrier concentration, and associated shortening of the magnetic penetration depth, resulting from Sr doping. However, the sharp decrease of the superconducting fraction for x>0.2 is not readily explained.

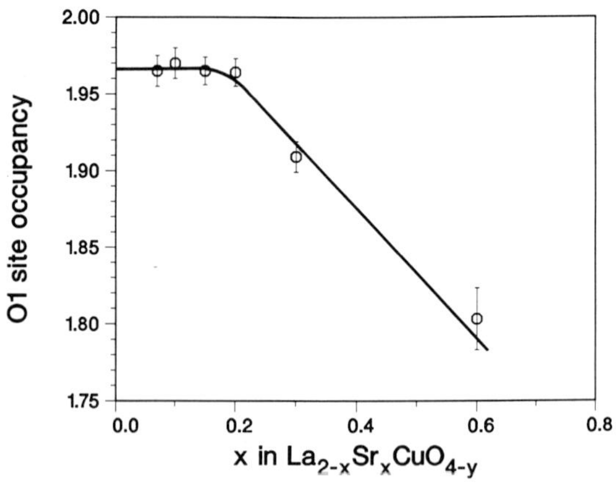

Fig. 5. Occupancy of the oxygen site in the CuO_2 plane (O1) of $La_{2-x}Sr_xCuO_4$ for samples quenched from high temperature. For x<0.2, the small (~0.03) offset from 2.00 is assumed to be a systematic error.

In previous work, we offered two explanations for this behavior.[12] For samples quenched from high temperature, oxygen vacancies were present on the O1 site (in the CuO_2 planes) for x>0.2 as shown in Fig. 5. Superconductivity is presumably destroyed by both the charge compensation effect of the vacancies and by the existence of defects in the critical CuO_2 planes. Both T_c and the superconducting fraction dropped sharply for x>0.2. For slowly cooled samples the oxygen vacancies were filled, but for x>0.2 we observed the formation of an $La_2SrCu_2O_6$ impurity phase in coexistence with $La_{1.8}Sr_{0.2}CuO_4$. T_c(onset) remained essentially constant, but the superconducting fraction dropped sharply for x>0.2.

After additional work, we now conclude that the formation of the $La_2SrCu_2O_6$ impurity phase can be minimized by processing the samples at a higher temperature. Such samples, when slowly cooled in oxygen, offer yet another way to investigate the behavior for x>0.2. Neutron powder diffraction studies of these samples show that phase separation into two phases with the $La_{2-x}Sr_xCuO_4$ structure, but with different compositions, occurs for x>0.2. The diffraction data and Rietveld refinement profile for a sample of bulk composition x=0.4 are shown in Fig. 6.

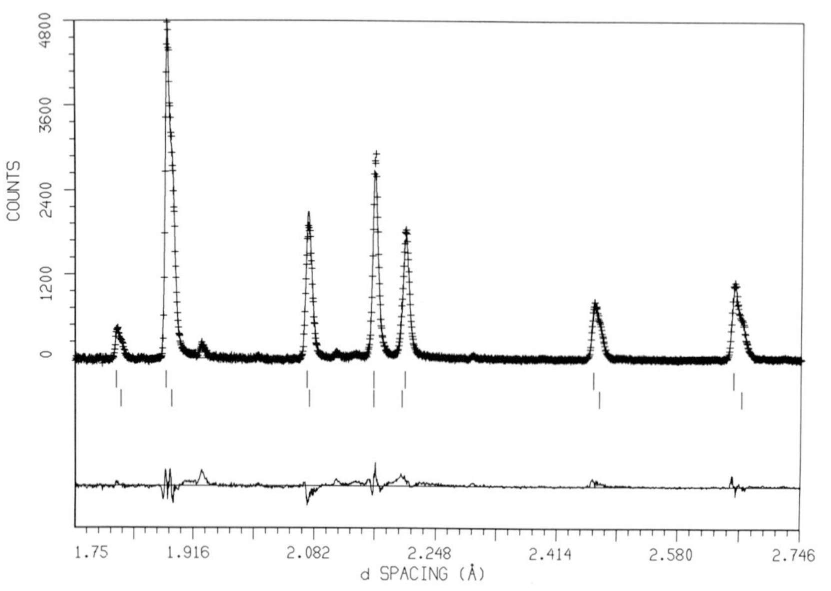

Fig. 6. Neutron powder diffraction data and Rietveld refinement profile for $La_{1.6}Sr_{0.4}CuO_4$ showing overlapped diffraction patterns for two nearly identical phases. The small peaks not fit are from a $La_2SrCu_2O_6$ impurity phase.

The appearance of two phases with lattice parameters differing by only a small amount is strikingly similar to that seen in La_2CuO_{4+x} and $Nd_{2-x}Ce_xCuO_4$. Rietveld refinement provides the lattice parameters of the two phases:

Major phase (70%):

a=b=3.7595(1) Å, c=13.2473(3) Å.

Minor phase (30%):

a=b=3.7747(1) Å, c=13.2227(1) Å.

From these lattice parameters and the volume fractions the compositions of the two phases are estimated to be x=0.5(2), and x=0.15(2), respectively. Since the superconducting fraction is decreasing rapidly while the T_c(onset) is remaining constant in this region, this result suggests that the minor phase is a superconductor with the optimum composition, and the major phase is a nonsuperconducting metal. The approximate phase diagram implied by these results is shown in Fig. 7. The highest T_c occurs on the Sr-deficient boundary of a miscibility gap that extends from x≈0.2 to x≈0.6.

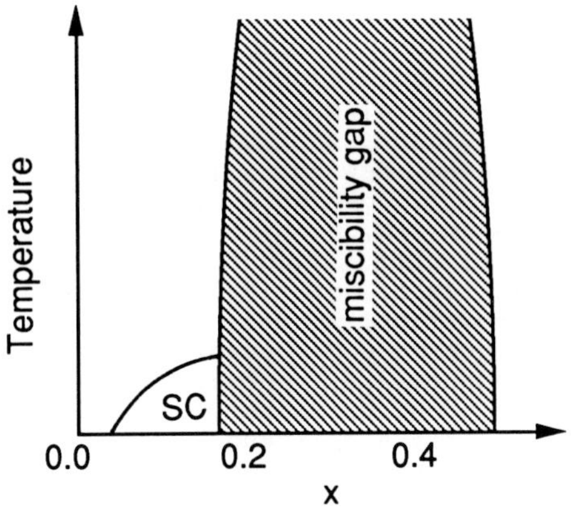

Fig. 7. Approximate phase diagram for $La_{2-x}Sr_xCuO_4$.

CONCLUSIONS

Several important conclusions can be drawn from these results. Since, in each of the three systems discussed, the miscibility gap is adjacent to the composition giving the highest T_c, it is logical to conclude that the phase separation is driven by an electronic instability that limits the maximum T_c.

For the $La_{2-x}Sr_xCuO_4$ system, one is led to reconsider the measurements of Torrance et al [13] who concluded that T_c decreases for increasing carrier concentration for x>0.2. Unless the samples are examined by diffraction methods with the same precision as in the present study, the existence of two-phase behavior cannot be ruled out. Similarly, one should reexamine the data for other systems in which it appears that T_c drops with increasing carrier concentration.

One might ask whether phase separation is a general phenomena that limits the maximum T_c in oxide superconductors. In this context, it should be noted that phase separation may be one of several phenomena associated with an instability that limits T_c. As we have shown, at least two different process (oxygen vacancy formation and phase separation) can limit T_c in $La_{2-x}Sr_xCuO_4$. Perhaps the formation of $YBa_2Cu_4O_8$ when one attempts to raise T_c above 92K by using high pressure to insert additional oxygen (above O_7) into $YBa_2Cu_3O_{7-x}$ is another example of phase instability. Additionally, it should be remembered that phase separation could occur in some systems over a short length scale, comparable to the superconducting coherence length, and would, in such cases, be invisible to the neutron powder diffraction techniques used in the present study. Clearly, the full investigation of these phenomena will require careful work.

ACKNOWLEDGEMENTS

This work is supported by the U. S. Department of Energy, Division of Basic Energy Science - Materials Sciences, under contract No. W-31-109-Eng-38 (JDJ, DGH, DRR) and the National Science Foundation, Science and Technology Center for Superconductivity, under grant No. DMR 88-09854 (SP, PL).

REFERENCES

1. Jorgensen J D, Hinks D G, and Felcher G P (1987) Phys. Rev. B 35: 5365.
2. Varma C M, in High Temperature Superconductors, Progress in High Temperature Superconductivity, Vol. 20, edited by Akachi T, Cogordan J A, and A A Valladares (World Scientific, 1989) p. 35; Varma C M, in Discussions on High Temperature Superconductivity, edited by Weber H W (Plenum, New York, 1988).
3. Pei S, Jorgensen J D, Dabrowski B, Hinks D G, Richards D R, Mitchell A W, Newsam J M, Sinha S K, Vaknin D, and Jacobson A J (1990) Phys. Rev. B 41: 4126.
4. Markiewicz R S (1990) Physica B 165&166: 1053; Markiewicz R S (1990) J. Phys. Cond. Matt. 2: 665.
5. Jorgensen J D, Dabrowski B, Pei S, Hinks D G, Soderholm L, Morosin B, Schirber J E, Venturini E L, and Ginley D S (1988) Phys. Rev. B 38: 11337.
6. Lightfoot P, Richards D R, Dabrowksi B, Hinks D G, Pei S, Marx D T, Mitchell A W, Zheng Y, and Jorgensen J D (1990) Physica C 168: 627.
7. Dabrowski B, Jorgensen J D, Hinks D G, Pei S, Richards D R, Vanfleet H B, and Decker D L (1989) Physica C 162-164: 99.
8. Jorgensen J D, Dabrowski B, Pei S, Richards D R, and Hinks D G (1989) Phys. Rev. B 40: 2187;Chaillout C, Cheong S W, Fisk Z, Lehmann M S, Marezio M, Morosin B, and Schirber J E (1989) Physica C 158: 183.
9. Takagi H, Uchida S, and Tokura Y (1989) Phys. Rev. Lett. 62: 1197; Liang G, Chen J, Croft M, Ramanujachary K V, Greenblatt M, and Hedge M (1989) Phys. Rev. B 40: 2646; Paulus E F, Yehia I, Fuess H, Rodriguez J, Vogt T, Strobel J, Klauda M, and Saemann-Ischenko G (1990) Solid State Commun. 73: 791.
10. Luke G M, Sternlieb B J, Uemura Y J, Brewer J H, Kadono R, Kiefl R F, Kreitzman S R, Riseman T M, Gopalakrishnan J, Sleight A W, Subramanian M A, Uchida S, Takagi H, and Tokura Y (1989) Nature 338: 49.
11. Flemming R M, Batlogg B, Cava R J, and Rietman E A (1987) Phys. Rev. B 35: 7191.
12. Hinks D G, Dabrowski B, Zhang K, Segre C U, Jorgensen J D, Soderholm L, and Beno M A (1988) Mat. Res. Soc. Proc., Vol. 99 (Materials Research Society, Pittsburgh) p. 9.
13. Torrance J B, Tokura Y, Nazzal A I, Bezinge A, Huang T C, and Parkin S S P (1988) Phys. Rev. Lett. 61: 1127.

High-Resolution Neutron Powder Diffraction of Superconducting $La_2CuO_{4+\delta}$

T. KAMIYAMA[1], H. ASANO[1], K. SHIBUTANI[2], R. OGAWA[2], Y. ENDOH[3], F. IZUMI[4], W.I.F. DAVID[5], R. IBBERSON[5], and Y. YAMAGUCHI[6]

[1] Institute of Materials Science, University of Tsukuba, Tsukuba, 305 Japan
[2] Kobe Steel, Ltd., Nishi-ku, Kobe, 673-02 Japan
[3] Department of Physics, Tohoku University, Sendai, 980 Japan
[4] National Institute for Research in Inorganic Materials, Tsukuba, 305 Japan
[5] Rutherford Appleton Laboratory, Chilton, Didcot, Oxfordshire, OX11 0QX, UK
[6] Institute for Materials Research, Tohoku University, Sendai, 980 Japan

ABSTRACT
High-resolution neutron powder diffraction measurements of $La_2CuO_{4+\delta}$ prepared by O_2-HIP confirmed the result of Jorgensen *et al.* that the sample consists of two orthorhombic phases with different oxygen contents below room temperature. The Meissner fraction is small compared with the volume fraction of the oxygen-rich phase. This indicates that the size of the superconducting domain is smaller than its magnetic penetration depth. The analysis of peak-broadening gives a domain size of the oxygen-rich phase to be less than 2000 Å, which is comparable to the magnetic penetration depth. This value is a little smaller than that obtained by Jorgensen *et al.*

KEY WORDS: neutron powder diffraction, $La_2CuO_{4+\delta}$, magnetic penetration depth, oxygen defect

INTRODUCTION

It is well known that $La_2CuO_{4+\delta}$ shows superconductivity near 35K. Jorgensen *et al.* [1] performed medium-resolution neutron powder diffraction on several $La_2CuO_{4+\delta}$ samples with Meissner fractions of 0.1 to 30% and showed that these samples consisted of two nearly-identical orthorhombic phases. The first is a non-superconducting phase with less oxygen while the second is an oxygen-rich superconducting phase. The amount of the second phase in the sample increases with the oxygen pressure at which the sample is annealed. For samples annealed in less than 0.1 kbar of oxygen, obtained Meissner fractions are very small compared with volume fractions of the oxygen-rich phase. They argued that the size of the superconducting domain estimated from broadening of Bragg peaks was comparable to the magnetic penetration depth (2000–2500 Å) [2] and Meissner-effect measurements yielded small values. On the other hand, the sample annealed in 3 kbar of oxygen gave almost the same Meissner fraction as the volume fraction of the oxygen-rich phase. In this case, no peak-broadening was observed and the crystallite size is at least 5000 Å. In order to obtain the crystallite size from small peak-broadening, however, it is necessary to undertake high-resolution diffraction studies, because the medium-resolution diffraction profile is subject to large instrumental broadening. In this work, we aimed at obtaining the microstructural evidence for the existence of small superconducting domains and performed a neutron powder diffraction measurement of $La_2CuO_{4+\delta}$ prepared by O_2-HIP using the highest-resolution diffractometer, HRPD, at the Rutherford Appleton Laboratory.

EXPERIMENTAL

An appropriate proportion of La_2O_3 and CuO powders was mixed according to the chemical formula La_2CuO_4. The powder was calcined at 900°C for 10 h, ground and calcined again under the same conditions, and then pressed into a pellet form. The pellet was annealed in 1.8 kbar of Ar+20%O_2 mixed gas at 500°C for 10 h with oxygen hot isostatic pressing (O_2-HIP) followed by 100 °C/h cooling. The final cooling rate was about 1-2°C/min.

The superconducting volume of the sample was estimated to be about 1 % by the initial slope of magnetization curve at 4.5 K after zero-field cooling. The resistivity data show superconductivity below 32 K.

Neutron diffraction data were taken at room temperature and 20 K on the medium-resolution time-of-flight (TOF) neutron powder diffractometer HRP ($\Delta d/d \approx 3\times10^{-3}$) at the KENS pulsed spallation neutron source at the National Laboratory for High Energy Physics [3], and at room temperature and 4.2 K on a high-resolution neutron powder diffractometer HRPD ($\Delta d/d \approx 5\times10^{-4}$) at the ISIS pulsed spallation neutron source at the Rutherford Appleton Laboratory [4].

RESULTS AND DISCUSSIONS

The neutron diffraction data taken on HRP were refined with a TOF neutron diffraction version of a RIETAN program [5]. Values of the coherent scattering length, b, used for the refinement were 8.24(La), 7.718(Cu) and 5.803(O) in the unit of fm (10^{-13} cm) [6]. The HRP data obtained at room temperature seemed to be a single phase. In a diffraction pattern taken on HRPD, however, extra shoulders are obviously visible at the positions where Bragg peaks from the second phase are expected. The first and the second phases have almost the same structures with similar lattice parameters and it is difficult to distinguish them by means of a medium-resolution diffractometer even if the Rietveld method is used. Since multi-phase capability for HRPD data is not available at the present stage, we analyzed the average structure by the Rietveld fitting of the HRP data. The good fit between observed and calculated patterns was obtained with the one-phase model. Table I lists final R factors [7], structural parameters and their standard deviations. Table II gives metal-oxygen interatomic distances calculated with ORFFE [8]. The analysis based on the average structure shows that both sites O(1) and O(2) are slightly deficient and the total amount of oxygen in the chemical formula is less than 4.

Table I. Structural parameters for $La_2CuO_{4+\delta}$ at room temperature obtained from the HRP data. Space group is Cmca. U_{ij}'s are anisotropic thermal parameters when the temperature factor is defined as $\exp[-2\pi^2(h^2a^{*2}U_{11}+k^2b^{*2}U_{22}+l^2c^{*2}U_{33}+2hka^*b^*U_{12}+2hla^*c^*U_{13}+2klb^*c^*U_{23})]$ (U_{12}=0 for Cmca). B_{eq} is the equivalent isotropic thermal parameter and g is the occupation factor. Numbers in parentheses following refined parameters represent the estimated standard deviations of the last significant digit(s), and parameters without any standard deviations were fixed. R_{wp}=4.17%, R_p=3.14%, R_e=3.62%, R_I=1.61%, R_F=1.62%. χ^2=1.33. a=5.36109(7) Å, b=13.1729(2) Å, c=5.39883(7) Å.

Atom	site	g	x	y	z	U_{11} (Å^2)	U_{22} (Å^2)	U_{33} (Å^2)	U_{13} (Å^2)	U_{23} (Å^2)	B_{eq} (Å^2)
La	8f	1	0	0.36135(6)	0.0060(3)	0.0052(5)	0.0053(4)	0.0064(5)	0	0.0019(8)	0.44
Cu	4a	1	0	0	0	0.0039(7)	0.0080(0)	0.0037(7)	0	-0.0016(13)	0.41
O(1)	8e	0.980(3)	1/4	0.0062(2)	1/4	0.0042(6)	0.0134(9)	0.0056(7)	-0.0018(4)	0	0.61
O(2)	8f	0.987(2)	0	0.1835(1)	-0.0280(4)	0.0211(8)	0.0046(7)	0.024(1)	0	0.000(1)	1.30

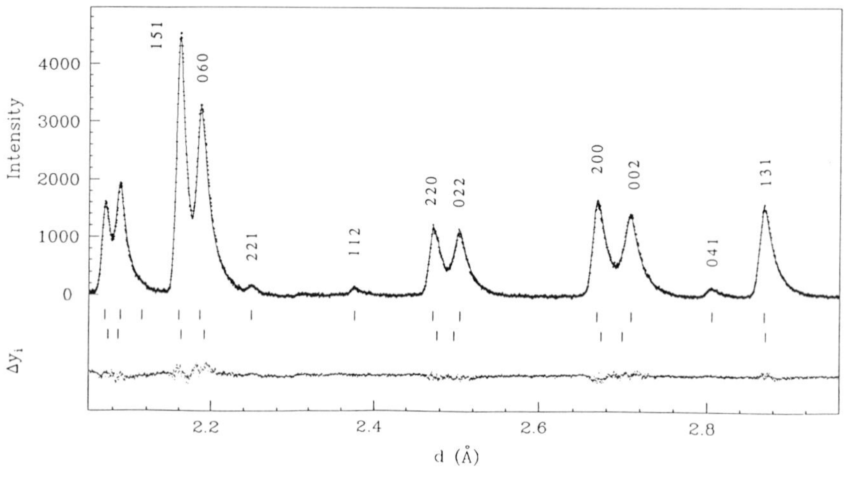

Fig. 1. Part of the Rietveld refinement pattern of the HRP data for $La_2CuO_{4+\delta}$ at 20 K. Background was fit as a part of the refinement but has been subtracted before plotting. Plus marks are the raw intensity data and the overlaid solid line is the calculated one. Δy_i in the bottom is the deference between them in the same scale. Upper and lower tick marks indicate the positions of allowed reflections from the first and the second phases, respectively. 10 reflections are indexed on the basis of Cmca space group.

On the other hand, the HRP data obtained at 20 K can be fitted with the two-phase model. We assume that an initial structural model is basically the same as that of Jorsensen *et al.*[1] and the space group of the second phase was assigned to be Fmmm [1] because the number of refinable parameters is much smaller in Fmmm than in Cmca [9]. Figure 1 illustrates final profile fit and difference patterns of the 20 K data. Table III lists final R

factors, structural parameters and their standard deviations. It should be noted that interstitial oxygen atoms exist at an O(3) site in the superconducting phase. The volume fraction of the second phase is 17 %, which is much larger than that of the Meissner fraction (several percent). The difference in lattice parameters between two short axes of the second phase was smaller than that for the first phase; at 20 K, the degree of orthorhombic distortion for the first phase 2x(c-a)/(c+a) is 1.50 % and that for the second phase 2x(b-a)/(b+a) is 0.93 %.

Table II. Selected metal-oxygen interatomic distances, l, in $La_2CuO_{4+\delta}$ at room temperature. Numbers of equivalent bonds follow "x" in parentheses.

Bond		l (Å)
La–O(1^{i})	(x2)	2.598(2)
La–O(1^{ii})	(x2)	2.678(2)
La–O(2^{i})	(x2)	2.7476(4)
La–O(2^{iii})	(x1)	2.585(3)
La–O(2^{iv})	(x1)	2.943(3)
Cu–O(1)	(x4)	1.9039(1)
Cu–O(2)	(x2)	2.423(2)

Symmetry codes:
i) 1/2-x, 1/2-y, -z;
ii) 1/2-x, 1/2+y, z;
iii) x, 1/2-y, 1/2+z;
iv) x, 1/2-y, -1/2+z.

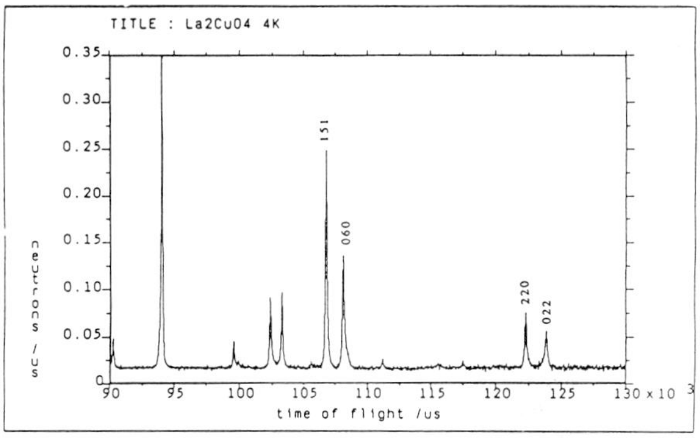

Fig. 2. Diffraction pattern of $La_2CuO_{4+\delta}$ by HRPD. The four reflections are indexed on the basis of Cmca space group.

Table III. Structural parameters for $La_2CuO_{4+\delta}$ at 20K obtained from the HRP data. B is the isotropic thermal parameter. R_{wp}=5.59%, R_p=4.19%, R_e=3.31%. χ^2=2.86.

Phase 1 (non-superconducting phase). Space group: Cmca. R_I=4.02%, R_F=2.42%. Volume fraction 83%. a=5.3368(2) Å, b=13.1200(4) Å, c=5.4174(2) Å.

Atom	site	g	x	y	z	B (Å^2)
La	8f	1	0	0.3620(2)	0.0083(5)	0.10(5)
Cu	4a	1	0	0	0	0.00(6)
O(1)	8e	1.00(1)	1/4	0.0090(3)	1/4	0.04(6)
O(2)	8f	1.01(2)	0	0.1828(3)	-0.0395(6)	0.39(7)

Phase 2 (superconducting phase). Space group: Fmmm. R_I=3.89%, R_F=2.24%. Volume fraction 17%. a=5.3433(6) Å, b=5.3934(5) Å, c=13.158(1) Å.

Atom	site	g	x	y	z	B(Å^2)
La	8i	1	0	0	0.358(1)	0.4(2)
Cu	4a	1	0	0	0	0.0(3)
O(1)	8e	0.99(5)	1/4	1/4	0	0.2(3)
O(2)	8i	0.95(7)	0	0	0.184(2)	0.8(4)
O(3)	8f	0.03(2)	1/4	1/4	1/4	0.8

Figure 2 and 3 shows a diffraction pattern taken on HRPD at 4.2 K. The second phase is clearly visible and the peak width of the second phase is much broader than that of the first phase. The domain size can be estimated from the width of the diffraction peaks. In the TOF method, peak-broadening effects by strains and the domain size can be differentiated because they have different wavelength dependences; the broadening by strain is proportional to d-spacing while the size broadening is proportional to the square of d-spacing. However, it can be reasonably assumed that interstitial oxygen atoms at the O(3) site distribute homogeneously in the second phase after the phase separation and the strain-broadening effect is negligible compared with size-broadening. We roughly estimated the domain size of the second phase by Gaussian fitting to several reflection peaks. Figure 3 shows a typical result of fitting. The obtained domain size is less than 2000Å for the second phase, which is comparable to the magnetic penetration depth. Therefore, the difference between the Meissner fraction and the volume fraction of the second phase is ascribable to the small superconducting domain size. Further analysis using a multi-phase Rietveld program for HRPD data is necessary to obtain detailed information.

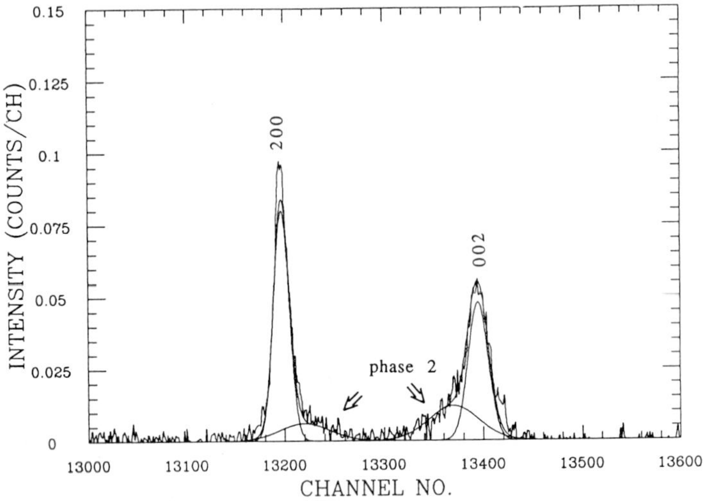

Fig. 3 Typical result of broadening analysis of HRPD data. The indices are based on the space group Cmca.

REFERENCES

1. J. D. Jorgensen, B. Dabrowski, Shiyou Pei, D. G. Hinks, L. Soderholm, B. Morosin, J. E. Schirber, E. L. Venturini, D. S. Ginley (1988) Phys. Rev. B 38: 11337
2. G. Aeppli, R. J. Cava, E. J. Ansaldo, J. H. Brewer, S. R. Kreitzman, G. M. Luk, D. R. Noakes, R. F. Kiefl (1987), Phys. Rev. B 35: 7129
3. N. Watanabe, H. Asano, H. Iwasa, S. Satoh, H. Murata, K. Karahashi, S. Tomiyoshi, F. Izumi, K. Inoue (1987) Jpn. J. Appl. Phys. 26: 1164
4. W. I. F. David, M. W. Johnson (1985) Rutherford Appleton Laboratory Report, RAL-85-112
5. F. Izumi, H. Asano, H. Murata, N. Watanabe (1987) J. Appl. Crystallogr. 20: 411
6. V. F. Sears (1986) In: Methods in Experimental Physics Vol.23, Neutron Scattering, Part A, eds. K. Skold, D. L. Price, Academic Press, New York, p.521.
7. R. A. Young, E. Prince, R. A. Sparks (1982) J. Appl. Crystallogr. 15: 357
8. W. R. Busing, K. O. Martin, H. A. Levy (1964) Report ORNL-TM-306, Oak Ridge National Laboratory, Tennessee
9. C. Chaillout, J. Chenavas, S. W. Cheong, Z. Fisk, M. Marezio, B. Morosin and J. E. Schirber (1990) Physica C 170: 87

Crystal Structure of the Ferromagnetic Copper Oxide $La_{4-x}Ba_{2+x}Cu_2O_{10}$

TAKASHI MOCHIKU[1], HAJIME ASANO[1], FUMIO MIZUNO[2], HIROMU MASUDA[2], IZUMI HIRABAYASHI[2], SHOJI TANAKA[2], and FUJIO IZUMI[3]

[1] Institute of Materials Science, University of Tsukuba, Tsukuba, 305 Japan
[2] Superconductivity Research Laboratory, Nagoya Division, International Superconductivity Technology Center, Nagoya, 456 Japan
[3] National Institute for Research in Inorganic Materials, Tsukuba, 305 Japan

ABSTRACT

The crystal structure of the ferromagnetic copper oxide $La_{4-x}Ba_{2+x}Cu_2O_{10}$ (x=0.4) has been refined by Rietveld analysis of TOF neutron powder diffraction data taken at room temperature. It is tetragonal with space group P4/mbm, and its lattice parameters are a=0.68652(1) nm and c=0.58698(1) nm. A La site is 11% occupied by excess Ba atoms, and 2.5% of a Ba site are substituted by La atoms. No deficiencies could be detected in copper and oxygen sites.

KEYWORDS: $La_{4-x}Ba_{2+x}Cu_2O_{10}$, neutron powder diffraction, Rietveld refinement

INTRODUCTION

$La_{4-x}Ba_{2+x}Cu_2O_{10}$ (x=0) has recently been found to be an insulator which exhibits ferromagnetic transition at about 5 K [1]. This oxide was first identified by Michel *et al.* [2] using X-ray powder diffraction. They reported that it contained MO_8 and MO_{10} coordinate polyhedra (M = La and Ba) and CuO_4 square planes and that it had deficiencies in copper and oxygen sites with a chemical formula of $La_{4-x}Ba_{2+x}Cu_{2-x/2}O_{10-x}$ ($0.3 \leq x \leq 0.5$). The structure parameters of oxygen and the distribution ratio of La and Ba atoms between the two metal sites could not accurately been determined by X-ray powder diffraction. A more detailed knowledge of the structure is desired to explore the mechanism of ferromagnetism. This study was undertaken to refine the crystal structure of $La_{4-x}Ba_{2+x}Cu_2O_{10}$ (x=0.4) by Rietveld analysis of neutron powder diffraction data.

EXPERIMENTAL

A sample of $La_{3.6}Ba_{2.4}Cu_2O_{10}$ was prepared by solid-state reaction of La_2O_3, BaO_2 and CuO powders of high purity. The La_2O_3 powder was preheated at 1000°C for 10 h in air, because La_2O_3 is highly hygroscopic. The mixture was calcined at 900°C for 10 h in air, ground, and then pressed into pellets. The pellets were sintered at 1000°C for 10 h in air.

Neutron powder diffraction data were taken on a time-of-flight (TOF) neutron powder diffractometer, HRP [3], at the KENS pulsed spallation neutron source at the National Laboratory for High Energy Physics. The sample was contained in a cylindrical vanadium cell 10 mm in diameter and 40 mm in height, and placed in aluminum vacuum chamber. Intensity data were collected at room temperature.

STRUCTURE REFINEMENT

The structure of $La_{3.6}Ba_{2.4}Cu_2O_{10}$ was refined using a Rietveld refinement program RIETAN [4] on the basis of the structure model given by Michel *et al.* [2]. Our previous refinement of the compound with a nominal composition of x=0 revealed that La^{3+} and Ba^{2+} ions were completely ordered to form LaO_8 and BaO_{10} polyhedra [5]. This result is reasonable because the larger Ba^{2+} ion prefer the coordination number 10 to 8. In the present nonstoichiometric compound $La_{3.6}Ba_{2.4}Cu_2O_{10}$, the occupation factors, g, of La(1), Ba(1), La(2) and Ba(2) sites (cf. Table I) have the linear constraints g(Ba(1))=1-g(La(1)), g(La(2))=0.9-0.5g(La(1)) and g(Ba(2))=0.1+0.5g(La(1)). Therefore, only g(La(1)) was refined under these constraints, and the other g values were calculated from g(La(1)). Consequently, g(La(2)) and g(Ba(1)) were determined to be 0.888(8) and 0.975(8), respectively. This result indicates that the ten-coordinate 2b site is occupied mostly by Ba atoms and that the excess Ba atoms locate in the 4g site. Thus, the random distribution of Ba and La suggested by Michel *et al.* has been denied by the present neutron diffraction study.

Next, we refined occupation factors of copper and oxygen sites. Michel *et al.* [2] claimed that deficiencies at the Cu and O(2) sites were inevitable as x was increased. However, we obtained neither occupacies for Cu, O(1) and O(2) sites deviating substantially from unity nor large thermal parameters indicative of deficiencies. In the final refinement, therefore, we fixed the g value of these site at 1 and assigned anisotropic thermal parameters to all the sites.

Table I lists final R factors [6], lattice and structure parameters and their standard deviations in parentheses. Table II gives metal-oxygen interatomic distances, l, calculated with ORFFE [7].

DISCUSSION

Michel *et al.* [2] reported that this compound had deficiencies in the copper and oxygen sites. However, the g values of the Cu, O(1) and O(2) sites obtained in the preliminary refinement were 0.994(5), 1.008(6) and 0.992(4), respectively. These g values give a chemical formula of $La_{3.6}Ba_{2.4}Cu_{1.998}O_{9.936}$ on the assumption g(O(1)) = 1. This formula is widely different from that expected from the model of Michel *et al.*: $La_{3.6}Ba_{2.4}Cu_{1.8}O_{9.6}$. The above g values of the three sites can be regarded as unity in view of the underestimation of standard deviations in Rietveld analysis. In addition, the B_{eq} values obtained by assuming $g(\mathrm{Cu}) = g(\mathrm{O(1)}) = g(\mathrm{O(2)}) = 1$ are quite normal, as shown in Table I. Thus, we can safely conclude that no deficiencies exist, at least, in our sample.

The effective copper valence, $2+p$, or the concentration, p, of doped holes per (Cu-O) unit can be determined from the chemical formula $La_{3.6}Ba_{2.4}Cu_2O_{10}$ under the condition of electrical neutrality in this compound. The value of p calculated in this way is 0.2, which indicates that an excess amount of Ba generates holes on the CuO_4 plane.

The average La,Ba(1)-O and La,Ba(2)-O distances are 0.298 nm and 0.256 nm, respectively, which agree with the ones evaluated from the weighted averages of effective ionic radii $r(^{X}La^{3+})$, $r(^{X}Ba^{2+})$, $r(^{VIII}La^{3+})$ and $r(^{VIII}Ba^{2+})$ [8]: 0.291 nm and 0.259 nm (Roman numerals denote coordination numbers).

The Cu-O(2) distance of 0.19214 nm is shorter than 0.197 nm and 0.200 nm calculated from $r(^{IV}Cu^{2+}) + r(^{VI}O^{2-})$ and $r(^{IV}Cu^{1+}) + r(^{VI}O^{2-})$, respectively. In $NaCuO_2$ and $KCuO_2$, copper ions with a formal charge of +3 are coordinated to four oxide ions [9]. The $^{IV}Cu^{3+}$-O^{2-} distances are 0.1839 nm in $NaCuO_2$ and 0.1829 nm in $KCuO_2$. The Cu-O(2) distance in the present compound lies between the Cu-O distance calculated from $r(^{IV}Cu^{2+})$ and the observed ones in $NaCuO_2$ and $KCuO_2$. This finding is consistent with the mixed valence state of Cu^{2+} and Cu^{3+} or $(Cu\text{-}O)^{+}$ holes on the CuO_4 plane.

The CuO_4 planes are completely isolated and arranged perpendicularly to each other (Fig. 1). We believe that complete CuO_4 planes with no copper and oxygen vacancies are essential for the appearance of ferromagnetism in the solid solution $La_{4-x}Ba_{2+x}Cu_2O_{10}$.

REFERENCES

1. Mizuno F, Masuda H, Hirabayashi I, Tanaka S, Hasegawa M, Mizutani U (1990) Nature (London) 345: 788
2. Michel C, Er-Rakho L, Raveau B (1981) J. Solid State Chem. 39: 161
3. Watanabe N, Asano H, Iwasa H, Satoh S, Murata H, Karahashi K, Tomiyoshi S, Izumi F, Inoue K (1987) Jpn. J. Appl. Phys. 26: 1164
4. Izumi F, Asano H, Murata H, Watanabe N (1987) J. Appl. Crystallogr. 20: 411
5. Ogawa N, Mizuno F, Masuda H, Hirabayashi I, Tanaka S, Mochiku T, Asano H, Izumi F (1990) Physica B 165-166: 1687
6. Young RA, Prince E, Sparks RA (1982) J. Appl. Crystallogr. 15: 357
7. Busing WR, Martin KO, Levy HA (1964) Report ORNL-TM-306, Oak ridge National Laboratory, Tennessee
8. Shannon RD (1976) Acta Crystallogr., Sect. A 32: 751
9. Brese NE, O'Keeffe M, von Dreele RB, Young VG Jr. (1989) J. Solid State Chem. 83: 1

Table I
Final results of the Rietveld refinement for $La_{3.6}Ba_{2.4}Cu_2O_{10}$ (tetragonal, P4/mbm, Z=1). B_{eq} is the equivalent isotropic thermal parameter.

R_{wp}	R_p	R_I	R_F	R_e	a (nm)	c (nm)
2.88%	2.28%	2.61%	1.60%	2.78%	0.68652(1)	0.58698(1)

Atom	Site	x	y	z	g	B_{eq}(nm^2)
La(1)	2b	0	0	1/2	0.025(8)	0.0073
Ba(1)	2b	0	0	1/2	0.975(8)	0.0073
La(2)	4g	0.17409(8)	=1/2+x	0	0.888(8)	0.0027
Ba(2)	4g	0.17409(8)	=1/2+x	0	0.112(8)	0.0027
Cu	2c	0	1/2	1/2	1	0.0035
O(1)	2a	0	0	0	1	0.0075
O(2)	8k	0.36187(8)	=1/2+x	0.2656(1)	1	0.0080

Table II
Selected interatomic distances in $La_{3.6}Ba_{2.4}Cu_2O_{10}$. N is the number of equivalent bonds.

Bond	l (nm)	N
Ba,La(1)-O(1)	0.29349(1)	2
Ba,La(1)-O(2^{i})	0.29941(5)	8
La,Ba(2)-O(1^{ii})	0.25366(2)	2
La,Ba(2)-O(2)	0.23988(11)	2
La,Ba(2)-O(2^{iii})	0.26618(9)	4
Cu-O(2^{iii})	0.19214(8)	4
Cu-Cu^{iii}	0.48544(1)	4
O(2)-O(2^{iv})	0.27520(17)	1
O(2)-O(2^{v})	0.26822(5)	1

Symmetry codes: i) $x,y-1,z$; ii) $x,y+1,z$; iii) $-y+1,x,z$; iv) $y-1,x,-z+1$; v) $-x+1,-y+2,z$.

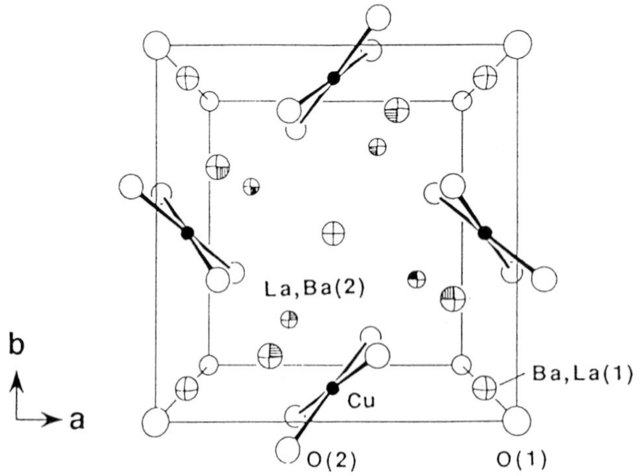

Fig. 1. Crystal structure of $La_{3.6}Ba_{2.4}Cu_2O_{10}$ viewd along the [001] direction. Oxygen atoms at (0,1/2,1/2) and (1,1/2,1/2) are not drawn for clarity.

Microstructure of Domain/Twin Intersections in Single Crystal $YBa_2Cu_3O_{7-d}$

K.N.R. Taylor[1], A.N. Khorasani[2], M.I. Darby[3], Y. Feng[4], I.R. Jones[4], M. Indro[1], and G.J. Russell[1]

[1] Advanced Electronic Materials Group, School of Physics, University of New South Wales, Kensington, NSW 2033, Australia
[2] Arrowcrest Group Pty. Ltd., Woodville North, S.A., 5012, Australia
[3] Department of Applied Physics, University of Salford, UK
[4] Department of Metallurgy and Materials, University of Birmingham, UK

ABSTRACT

In single crystal YBCO extensive domain and twin intersection can be observed in which the microstructural appearance of both the domains and the twin structure is unusual. Checkerboard structures are seen at both levels of magnification and domain width changes occur which are well correlated with position in the checkerboard. A model has been developed to account for these structures based on the orthorhombic strain energies and it is shown that twin boundary energy plays a significant role in the observed structures.

KEY WORDS: YBCO, Single Crystals, Microstructure, Bireflectance.

INTRODUCTION

Two of the best known features of the microstructure of YBCO are the optical domain structures visible in polarimetric microscopy, and the twinning observed in the electron microscope. Both these features are known to be associated with the oxygen ordering of the orthorhombic phase and result from the anisotropic property changes which occur as a result of the Cu-O chain formation in these materials. Several workers [e.g.1,2] have shown that the (001) plane of crystalline material is strongly bireflectant with the highest reflectivity lying parallel to the (100) direction.

While it is still generally assumed that there is a one-to-one correlation between the domain (optical) and twin (TEM) microstructures, we have recently shown quite unambiguously that under some conditions the situation is more complex than this [3]. In this study, it was observed that adjacent optical domains were twinned and untwinned respectively. The samples used in this work were incompletely oxygenated and this effect may be typical of such samples.

In fully oxygenated material, it is frequently reported that orthogonal sets of twins do not intersect, yet it is clear that at optical magnification levels extensive domain intersections occur in these crystals, in which the identity of individual domains is retained across the intersection as may be seen in Fig. 1. From an inspection of this figure, it is clear that in the intersection zone, the width of a single domain varies as it crosses successive orthogonal domains, suggesting that the lattice energy density resulting from the twinning mechanism may change from one region of the intersection checkerboard to the next. In order to attempt to understand the nature of these microstructural features we have recently extended our previous study to a combined optical and TEM investigation of domain/twin intersection regions in single crystals of YBCO. The results of this work, are described in the following.

EXPERIMENTAL

The techniques used in this work were the same as those of the previous study [3] and involved single crystals of YBCO grown by our alkali halide flux route [4]. These were optically polished on one side for polarimetric observations and ion beam thinned from the other until a central hole was developed which enabled electron beam microscopic techniques to be used through the thin tapered edge.

Fig. 2. shows a bright field TEM image obtained from an intersection region comparable to that of Fig. 1. Again, the checkerboard structure is clearly visible at this magnification along with the twinned/untwinned structure of the incident domains . It may also be seen, that where the intersection occurs the well defined checkerboard pattern consists of twinned and untwinned areas.

The former occur where one of the incident orthogonal domains is twinned and the other untwinned, while the latter are associated with the incident domains being either both twinned or both untwinned. It is also very noticeable, that as in the optical image the domain width varies across the intersection region, with the twinned sections of the checkerboard generally appearing to be narrower than the untwinned zones.This behaviour appears to be typical of the microstructure of these crystals, and many examples similar to Fig's 1 and 2 have been observed.

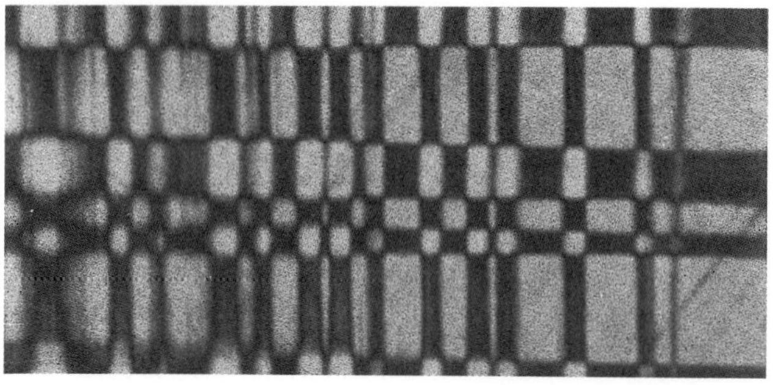

Fig. 1. An optical, polarimetric micrograph of a typical domain intersection region.

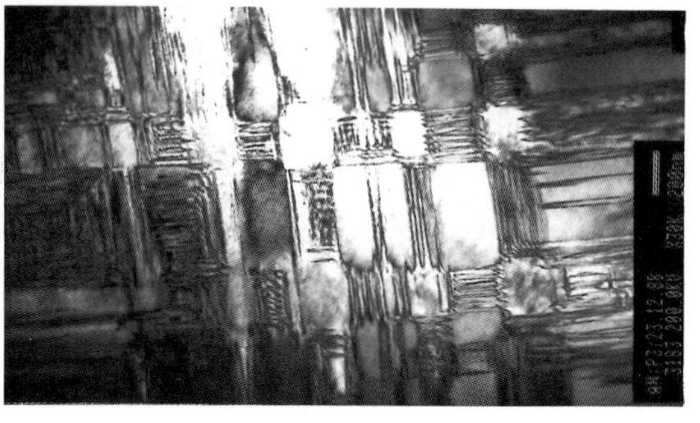

Fig. 2. A TEM image of a region of the crystal comparable to that shown in Fig. 1.

DISCUSSION

The general form of this microstructural feature can be readily modelled if we assume that at any point the local structural distortion is simply the resultant of the orthorhombic distortions of the incident orthogonal domains and twins. On this basis, the local structure may be derived [5] by the use of two simple rules, namely:

Rule 1. when two structural variants of like orthorhombicity intersect, the resultant retains the same orthorhombicity (Fig.3a).

Rule 2. when two variants with orthogonal strains intersect the resultant is an unstrained tetragonal zone (Fig. 3b).

These rules, of course, relate to the ordered oxygen chain structures and simply state that for Rule 1 the chain ordering in the intersection is the same as that in the individual twins while Rule 2 implies that the chain plane in the intersection region will be either fully oxygenated or totally deoxygenated.

Figure 3c shows the predicted microstructure obtained by the use of these rules for two sets of intersecting, orthogonal domains. The untwinned (1) and twinned (3) areas are similar in appearance to those of Fig. 2, however this model predicts a third type of microstructure, namely a mini-checkerboard structure (region 2 of Fig. 3c) which is not observed in the TEM image. Selected area diffraction (SAD) from this type of region in the TEM confirms that there is no reflection splitting due to twinning and consequently we assume that this mini-checkerboard structure (region 2) is unstable against transformation to a single structural variant. It is possible to interpret this behaviour in terms of the twin boundary energy, since the wall density, and hence the energy associated with predicted mini-

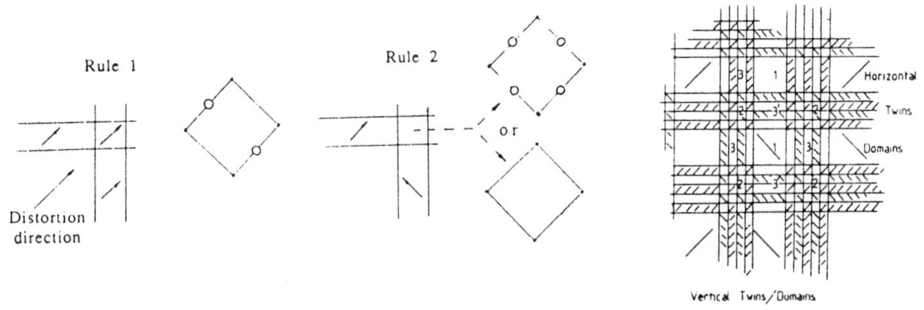

Fig. 3. a) Intersection of two 'like' domains
b) Intersection of domains with orthogonal strains
c) Predicted microstructure for a domain intersection region.

checkerboard structure will be considerably higher than that of the twinned domains and in turn higher than the untwinned domains.

It is relatively easily shown that the existence of twinned/untwinned domains, and the non-existence of a checkerboard twinned region 2 depends entirely on the relative dimensions of the two microstructural features and the boundary energy γ per unit area. Specifically this situation is satisfied by [6]:

$$\frac{2\gamma}{q} < (E_o - E_T) < \frac{8\gamma}{q'}$$

where q is a typical domain dimension (see Fig. 3) and q' the dimension of the mini divisions in region 2, i.e. a typical twin width. E_o and E_T are the energy per unit volume of the untwinned and twinned orthorhombic structures respectively. It is clear that the boundary energy per unit volume $8\gamma/q'$ will become large for small q' and that a stable checkerboard similar to that predicted for region 2 will only occur for q' > 4q!

The problem then remains as to the nature of the observed structure of region 2 which results from the intersection of the twinned and untwinned domains. Naively, it would seem likely that the lattice adopts a structure corresponding to the mean oxygen ordering in this region and consequently will correspond to an oxygen concentration close to the orthorhombic-tetragonal phase transition.

In order to attempt to establish the structure of region 2, and that of the other untwinned area in the checkerboard of Fig. 2 (i.e. region 1) we have used convergent beam techniques (CBED) to obtain HOLZ line patterns of both regions as shown in Fig. 4. It is clear from these, that the atomic structures of the two areas are different, and the best fits of lattice simulations are given in the figure caption.

A second source of information on the local microstructure in these twinned/untwinned domain structures has also been obtained by observation of the local bireflectance using a polarizing microscope and image digitization system [7]. These observations were understood in terms of a simple model in which the optical anisotropy of the untwinned domain is determined by the orthorhombic strain, while

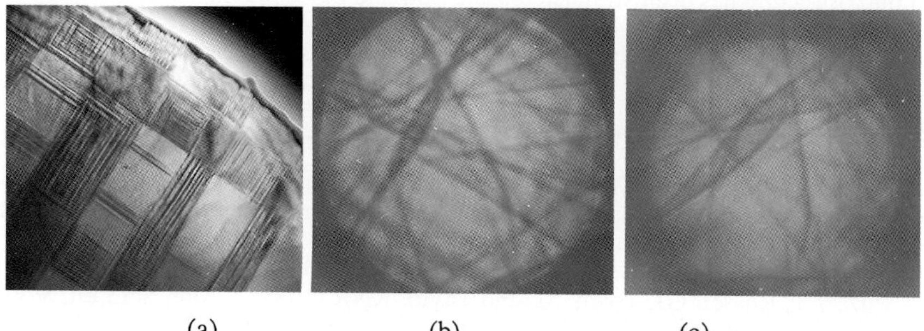

(a) (b) (c)

Fig. 4. (a) TEM micrograph of twin intersections (4mm = 200 nm). (b) and (c) CBED patterns from regions 1 and 2 respectively. Matching suggests region 2 is orthorhombic, with a = 0.383; b = 0.388; c = 1.169 nm, i.e. $O_{6.7-6.8}$. Region 1 appears close to tetragonal with a = b = 0.386 nm; c = 1.182 nm.

that of the twinned domain results from effects associated with the twin boundary conductivity. Measurements on a set of four units of the checkerboard structure shown in Fig. 2 (i.e. regions 1, 2 and 3) all show bireflectance, with the variation of reflected intensity with polarization angle for the two untwinned regions (1 and 2) being shown in Fig. 5.

The domains corresponding to this data are identified schematically in Fig. 4b, on which is shown the polarization angles, and the two relevant (100) and (110) crystallographic axis directions. Comparison with the data shows that domain 2, formed from the intersection of two twinned domains, has its major bireflectance axis parallel to either the (100) or (010) direction and we believe that this is consistent with the existence of an orthorhombic distortion in this area as predicted and as observed in the CBED observations. The reflectance variation of region 3 however, is more difficult to understand. If the bireflectance arises solely from local strains, then in this region of the intersection pattern involving two untwinned domains the single strain axis is parallel to either the (110) or (110) directions. This is, of course, the direction parallel to the twin boundaries in normal regions of the crystal and it is possible that in the transformation from the mini-checkerboard structure the oxygen plane only partially orders leaving a nett orthorhombic strain parallel to the (110) directions. Unfortunately at present we have insufficient evidence to develop this model further.

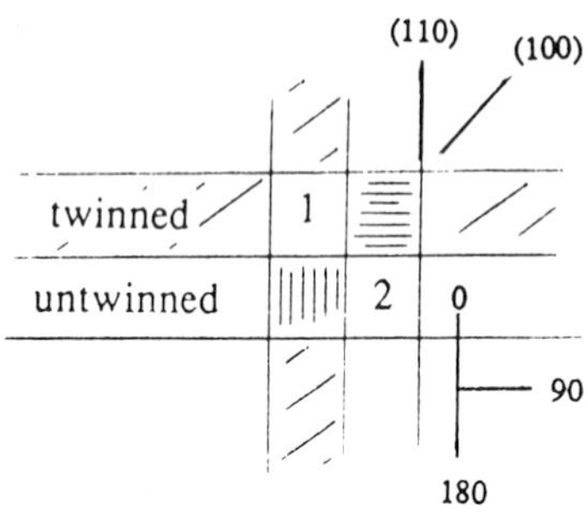

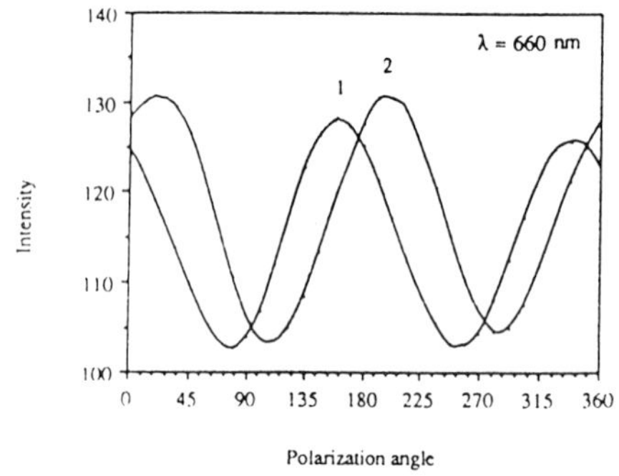

Fig. 5. The variation in reflected light intensity with polarization angle from two untwinned zones in the checkerboard.

CONCLUSIONS

In this continuing study of the correlation between optical domains and twins we have shown that the original findings [3] can be extended to the case of domain/twin intersections frequently observed in these crystals. The checkerboard microstructure has been shown to consist of three different structural zones, one of which is twinned and the others are untwinned. CBED observations have shown that differences exist between these two areas, including a variation in lattice parameter, and hence presumably the oxygen concentration. Similar variations were found in the previous work, and the present studies confirm that changes in the oxygen site occupancy can occur in these materials over very short distances.

The observed microstructure has been stimulated using a simple model based on the oxygen ordering state, and hence the orthorhombicity, of the incident twins. To obtain full agreement with the observations required that the twin boundary energy also be considered.

REFERENCES

1. Vlasko-Vlasov VK, Indenbom MV and Ossipyan YA (1988) Physics C 153-155: 1677.
2. Petrov MP, Grachev AI, Krasinkova MV, Nechitalov AA, Prokofiev VV, Poborchyv Shagin SI and Kartenko NF (1988) S.S. Comm. 67: 1197
3. Khorasani AN, Taylor KNR, Bosi S (1989) J. Cryst. Growth 98:461
4. Bosi S, Puzzer T, Russell GJ, Town SL and Taylor KNR (1989) J. Mat. Sci.Lett. 8: 497
5. Taylor KNR (1990) Studies of High Temperature Superconductors 6: 1
6. Taylor KNR and Darby MI. To be published.
7. Indro M, Taylor KNR and Russell GJ (1990). Submitted

TEM Observation of Microstructure in Sintered $YBa_2Cu_3O_{7-x}$

TAKASHI YOSHINO[1], YOSHITSUGU TOMOKIYO[1], and YOKO SUYAMA[2]

[1] Research Laboratory of High Voltage Electron Microscope, Kyushu University, Fukuoka, 812 Japan
[2] Japan Fine Ceramics Center, Nagoya, 456 Japan

ABSTRACT

Microstructures near a twin boundary of sintered $YBa_2Cu_3O_{7-x}$ have been investigated by means of transmission electron microscopy (TEM). The axial lengths of a- and b-axes shrink near a twin boundary in both sides and the angle between neighbouring c-planes across the boundary is measured to be about 0.8~1.0° suggesting the monoclinic distortion. A transformation from an orhtorohmbic phase to a tetragonal one occurs irreversibly during electron irradiation at 400 kV. The phase transformation proceeds in such a way that the orthorhombicity b_0/a_0 changes more fast near the twin boundary than in an inner part of the domains, the boundary fades away, and the tetragonal region of which front accompanies lattice strain propagates into the untransformed areas preferentially along a b-axis.

KEY WORDS: twin boundary, phase transformation, lattice distortion, orthorhombicity, microstructure

INTRODUCTION

The $YBa_2Cu_3O_{7-x}$ crystal exhibits a tetragonal phase at high temperature and an orthorhombic one at low temperature. The structural transition takes place when the oxygen content 7-x changes from 7 to 6 [1]. The orthorhombic phase easily transforms to the tetragonal one under electron beam irradiation in a vacuum in an electron microscope [2,3]. According to many reports on TEM observation, the phase transformation occurs either reversibly [4,5] or irreversibly [2,6] depending probably on oxygen pressure in the electron microscope. Although some mechanisms of the transformation have been proposed so far [7,8,9], the detailed mechanism remains controversial. One of the probable reasons for this is that the structural change caused by the transformation is very fast. A twin boundary seems to play a certain role in the transformation because the orthorhombic crystal is fragmented in fine coherent twin lamellae and an orthorhombic lattice is considered to be locally distorted at the twin boundary [10]. The twin boundary is also responsible for physical properties of superconductor such as Jc. In the present paper we will report some of the results on microstructures of twin boundaries. A specimen of $YBa_2Cu_3O_{7-x}$ was prepared by the standard sintering method. It showed x=0.136, a_0=0.3832, b_0=0.3879, c_0=1.1729 nm, Tc=93.8 K and Jc=403 A/cm^2. A thin specimen was prepared by ion-milling and observed by a JEM-4000EX microscope operated at 400 kV.

RESULTS AND DISCUSSION

1. Lattice distortion near twin boundary

Figure 1a shows a high resolution electron microscope (HREM) image of a twin boundary region observed along a [001] zone axis of orthorhombic crystal. Measurements of the inter-atomic spacing as a function of distance from the twin boundary indicate that the axial lengths of both a- and b-axes shrink within narrow region of about 10 nm in both sides of the boundary as shown in

Fig.1b. The lattice shrinkage is not brought about by a decrease in oxygen content but by the orthorhombicity, because the decrease in oxygen content should give rise to an increase in axial length a_0 and a decrease in b_0. There may be, of course, a slight and local difference of the occupation probabilities of oxygen atoms at lattice sites.

A low magnification image of twin lamellae viewed along [110] axis is shown in Fig.2 along with the corresponding electron diffraction pattern. The diffraction pattern shows no streaking to indicate stacking faults on (001) planes but splitting of the 00ℓ spot which implies a small orientation difference between different parts of the domains, as seen in the enlarged patterns in the insets of Fig.2b. As shown later, (001) planes of both sides on a twin boundary are not parallel but tilted by about 0.8~1.0° from each other, while (110) planes are parallel (see Fig.4). These results strongly suggest that there exists monoclinic lattice distortion in the vicinity of a twin boundary. We believe that the lattice strain caused by twinning is accomodated by shrinkage of the axial lengths a_0 and b_0 and inclination of c-plane near the twin boundary.

2. Phase transformation

High resolution electron microscopy provides valuable information on the phase transformation that proceeds under electron beam heating. A series of HREM images is shown in Fig.3a where the images are taken along [001] axis. We can see that the structure in Fig.3a-1 is orthorhombic, whereas that in Fig.3a-3, which was taken 4 min after Fig.3a-1, is tetragonal. In Fig.3a-1 lattices are slightly distorted near the twin boundary as indicated by a dark contrast along the boundary. The (110) lattice planes of both sides on the boundary are not parallel but tilted by 2θ, where $\tan\theta = a_0/b_0$. The twin boundary is initially seen clearly along a vertical line at a center of Fig.3a-1, as indicated by a black arrow. The twin boundary fades away and the tilting of (110) planes is reduced with the progress of the phase transformation as depicted in Fig.3b where the characteristic feature of Fig.3a is schematically drawn. It is noticeable in Fig.3a-2 that a dark line appears in parallel with the original twin boundary as shown by a horizontal arrow. This type of contrast has been interpreted as a new twin boundary produced during the transformation from the orhtorhombic phase to the tetragonal one. A detailed observation of HREM images, however, proved that there is no twin relation between the regions separated by the new dark line. As discussed later, the new dark line is explained as an interface between regions with different orthorhombicities.

Figure 4a shows a series of HREM images observed along [110] axis. We can recognize in Fig.4a-1 that (001) planes of both sides on the twin boundary are tilted by 0.8~1.0°, while (110) planes are parallel as already mentioned. We can see the process of the phase transformation by blurring of the twin boundary and extending of the region with lattice distortion toward the inner part of the domains. The structural change apparent from Fig.4a is schematically drawn in Fig.4b. In opposition to experiment by Muller and Gruehn [8] the phase transformation observed was irreversible and the length of c-axis increased by about 0.45 % upon the transformation.

From present observation we derived a mechanism of the transformation as schematically illustrated in Fig.5. The oxygen content is considered to decrease in whole area of the specimen but the change seems to be faster near the twin boundary as compared with an inner part of the domains. The interface between regions with different orthorhombicities, or between distorted and undistorted regions is not uniform and not well-defined. The front of distorted region, however, seems to propagate inward of the domain with a preferred orientation along a b-axis. The boundary between orthorhombic and tetragonal structures is also not well-defined as already suggested [6]. The structural changes of both sides on the twin boundary is not necessarily symmetrical against the boundary.

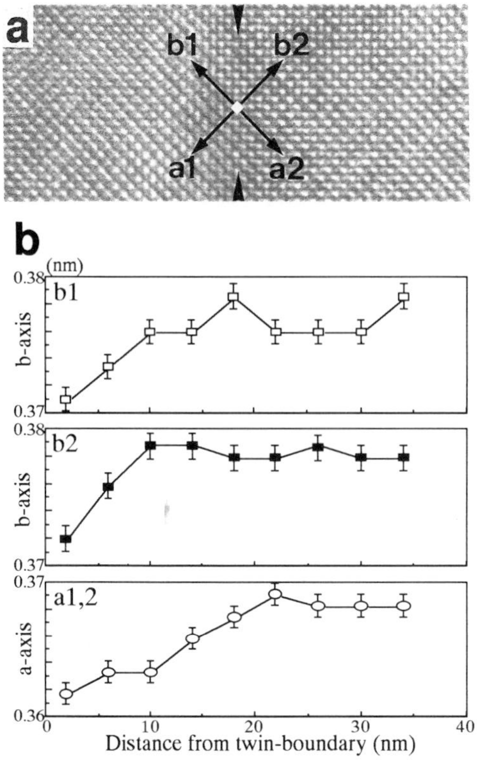

Fig.1 (a) HREM image of the of (110) twin boundary region observed along [001] axis. Changes in axial lengths a_0 and b_0 as a function of distance from the twin boundary in Fig.1a.

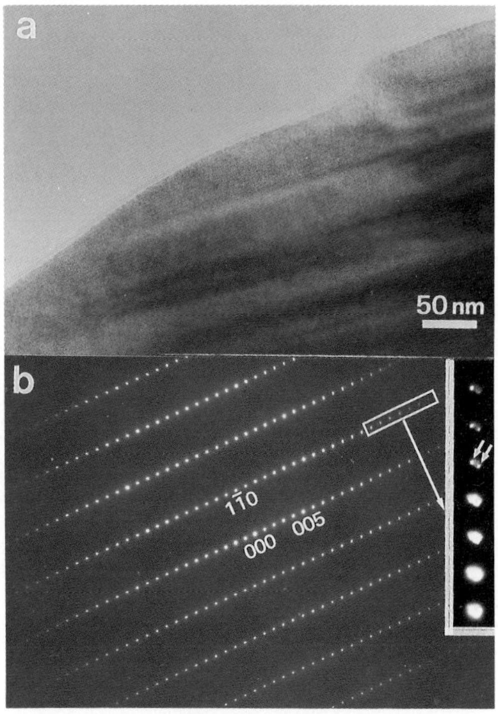

Fig.2 Low magnification image twin lamellae (a) and corresponding electron diffractio pattern (b) observed along [110] direction (b). Insets show the enlarged spots.

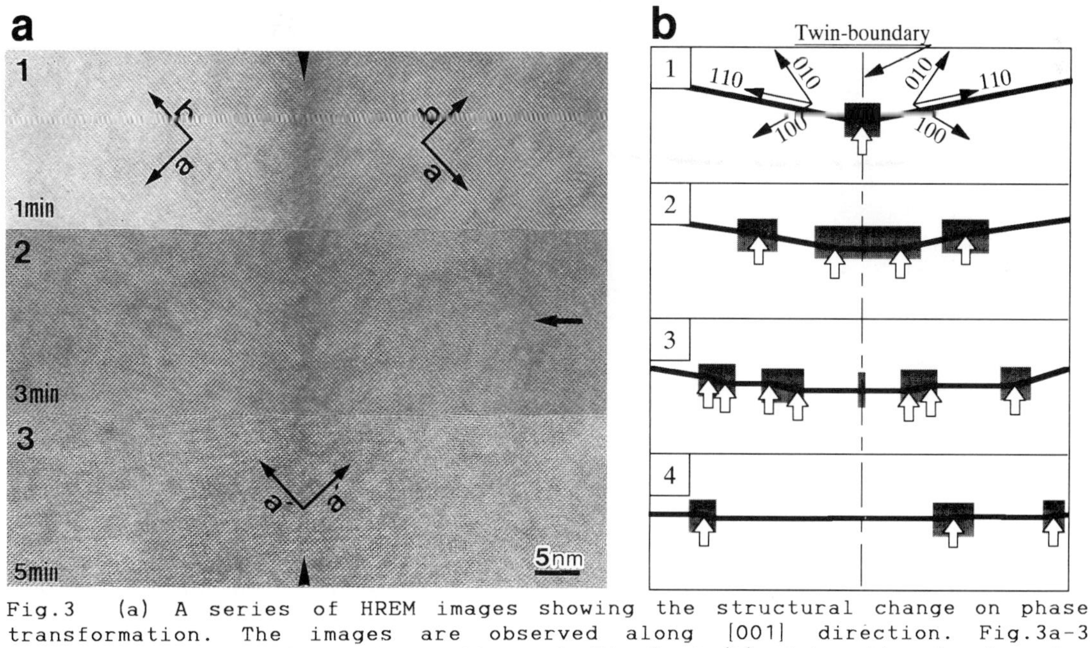

Fig.3 (a) A series of HREM images showing the structural change on phase transformation. The images are observed along [001] direction. Fig.3a-3 was taken 4 min after observation of Fig.3a-1. (b) Schematic drawing for the structural change in Fig.3a.

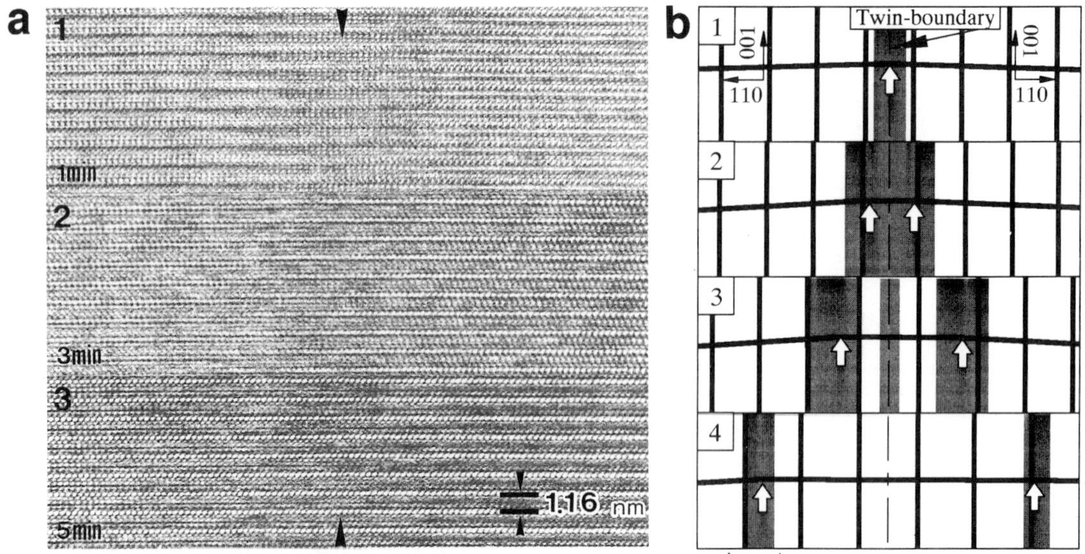

Fig.4 A series of HREM images observed along [110] direction, showing transformation evolution (a) and schematic explanation for the structural change upon phase transformation shown in Fig.4a (b).

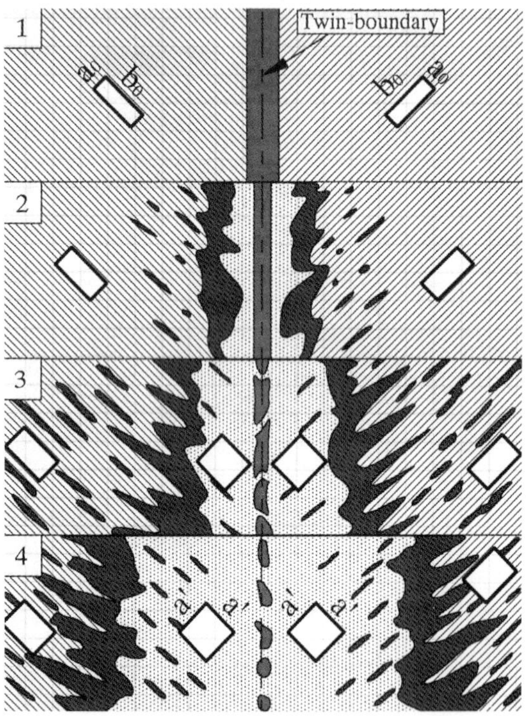

Fig.5 Schematic illustration for evolution of the transformation. : region with lattice distortion, ▭ :relative orthorhombocity exaggeratedly drawn :embryo of orthorhombic II or ordered phase originated from discrete depopulation of oxygen sites.

REFERENCES

1. Cava R.J., Hewat A.W., Hewat E.A, Batlog,B.,Marezio M., Rabe K.M., Krajewski J.J.,Peck Jr. W.F., and Rupp Jr.L.W. (1990) Pysica C 165: 419
2. Iijima S., Ichihashi T., Kubo Y., and Tabuchi J. (1987) Japanese J. Appl. Phys. 26: L1478
3. Zandbergen H.W., Tendeloo G. Van, Okabe T. and Amelinckx S. (1987) phys. stat. sol. (a) 103: 45
4. Sugiyama M., Suyama R., Inuzuka T., and Kubo H. (1987) Japanese J. Appl. Phys. 26: L1202
5. Tendeloo G.Van, Zandbergen H. W., and Amelinckx S. (1987) Solid State Com. 63: 389 and 603
6. Iijima S., Ichihashi T., Kubo Y., and Tabuchi J. (1987) Japanese J. Appl. Phys. 26: L1790
7. Ichihashi T., Iijima S., Kubo Y., Shimakawa Y., and Tabuchi J. (1987) Japanese J. Appl. Phys. 27: L594 and L1187
8. Müller J.H., and Gruehn R. (1989) Physica C 159: 527
9. Sasaki K., Kuroda K., and Saka H. (1989) Phil. Mag. Letters 59: 141
10. Hirotsu Y., Nakamura Y., Murata Y., Nagakura S., Nishihara T., and Takata M. (1987) Japanese J. Appl. Phys. 26: L1168

Microstructures in $(Y_{1-x}Ca_x)Ba_2Cu_4O_8$ Prepared at Low and High Oxygen Partial Pressure

K. YAMAGUCHI, T. MIYATAKE, T. WADA, N. SUZUKI, T. TAKATA, H. YAMAUCHI, and N. KOSHIZUKA

Superconductivity Research Laboratory, International Superconductivity Technology Center, 10-13, Shinonome 1-chome, Koto-ku, Tokyo, 135 Japan

ABSTRACT

Microstructures in $(Y_{1-x}Ca_x)Ba_2Cu_4O_8$ (x=0, 0.05 and 0.1) prepared at low and high oxygen partial pressures have been investigated by TEM. Crystal structures of the samples prepared at oxygen partial pressure of 400 atm show a small number of defects compared to the samples prepared at oxygen partial pressure of 3 atm. Such distortion of crystal structure may work as pinning center.

KEY WORDS : $YBa_2Cu_4O_8$, Oxygen partial pressure, Crystallographic defects, Transmission electron microscopy

INTRODUCTION

Superconducting $YBa_2Cu_4O_8$ ("124") was first found as a crystallographic defect in $YBa_2Cu_3O_7$ ("123") [1]. The superconducting transition temperature (T_c) of "124" is about 80 K, but recently, it has been reported that partial Ca substitution for Y in "124" causes an increasing of T_c up to a maximum of 91 K with increasing Ca content up to 10 % [2]. To synthesize bulk "124" compounds, the following two methods have been reported: (1) high oxygen pressure annealing methods [3,4]; and (2) ambient oxygen reaction methods [5,6]. Jin et al.[7] reported that the J_c of the bulk "124" prepared by the method (2) was nearly comparable to that of the "123", while the sample prepared by high oxygen pressure annealing exhibited lower and more sensitive J_c than the "123"[8]. These results suggest that the former sample has more flux pinning centers than the latter. Many candidates have been proposed as the flux pinning centers such as crystallographic defects or non-superconducting inclusions. In the above samples, the presence of non-superconducting phase was not reported. Thus the difference of J_c may be thought to be caused by the difference of the crystallographic defects. Therefore, a study of microstructures of $(Y_{1-x}Ca_x)Ba_2Cu_4O_8$ prepared at low and high oxygen partial pressure has been performed by transmission electron microscopy (TEM) to look for defects as possible pinning centers in these material.

EXPERIMENTAL

Two kinds of Ca substituted bulk "124" materials were prepared as follows [9]. First, we prepared $(Y_{1-x}Ca_x)Ba_2Cu_3O_{7-y}$ (x=0, 0.05 and 0.1) powder with no other phases by a conventional solid state reaction method. These "123" powders were mixed with appropriate amount of CuO powder and sintered at 850℃ for 160 h in oxygen gas of 3 atm. A part of each sample was post-annealed in 20 % O_2-Ar gas of 2000 atm ($P(O_2)$=400 atm) for 6 h at 1070 ℃ by oxygen-HIP (Hot Isostatic Press) technique, which is one of the high oxygen partial pressure annealing (HOP) methods. In this way, "124" materials with Ca content x=0, 0.05 and 0.1 were prepared at low oxygen pressure (LOP) of 3 atm and high oxygen partial pressure (HOP) of 400 atm (henceforth we name each sample as LOP-0, HOP-0.1 etc.). These six samples were examined by powder x-ray diffraction measurements, and confirmed to be of single phase except the LOP-0.1 sample.

Temperature dependence of the electrical resistivities for these five samples were measured by a conventional four-probe method and it is shown that T_c

depends on Ca content "x" but is independent of the oxygen partial pressure [9]. J_c of samples with x=0 were calculated from the results of the magnetization measurements and it was shown that the J_c of LOP-0 exhibited a few times larger than that of HOP-0 [10].

Fragments of five samples, which are confirmed to be of single phase, were crushed and dispersed on microgrids for observations of microstructure by the transmission electron microscope (TEM). All the observations here were performed by the microscope (JEM-4000EX) operated at 400 kV and having a point resolution of 0.17 nm.

RESULTS AND DISCUSSION

In the LOP samples, the crystal structure exhibited a large degree of distortion that varied from grain to grain. Figure 1-a and -b show micrographs of a slightly distorted and a severely distorted grains in the LOP-0 sample, respectively. In fig.1-a, sequence stackings along the c-axis are clearly seen and the electron diffraction pattern, inserted in the figure, shows a clear spot with streaks along the c*-axis, which indicates that the crystal of this region is not so distorted except there are stacking faults along the c-axis. On the other hand, in fig.1-b, the micrograph and the electron diffraction pattern show that the crystal structure of this region is severely distorted. In fig.2-a and -b, micrographs and diffraction patterns of the LOP-0.05 sample are shown. Figure 2-a shows that the crystal structure in this region is not so distorted, although it shows that the c-axis (ab-plane) tilts in this grain. In fig.2-b, a micrograph for a severely distorted grain, brighter spot lines are thought to be Cu-O chains and they show that the spacings and the directions of these lines are not uniform. The diffaction pattern shows no clear spot, which also means that the crystal structure of this grain is severely distorted.

Although these crystals given in figs.1-b and 2-b are so distorted as to determine the structure of defects, the distortions may be considered as follows. For the preparation method employs a reaction of "123" and CuO, "124" is synthesized by intercalation of CuO into "123". If the intercalation of CuO is not perfect, a part of "123" may remain unchanged and another type of stacking such as "125" may be formed partially. Therefore the spacings between the CuO chains vary from part to part and ab-planes tilt because the spacings of 001 plane differ depending on the stacking types.

On the other hand in HOP samples, the crystal structure does not vary from grain to grain in contrast to the case in LOP samples. Figure 3-a, -b and -c show typical micrographs of HOP-0, 0.05 and 0.1, respectively. In these micrographs, distorted structures as observed in LOP samples are not observed, although stacking faults are observed, which is thought to have a structure with single CuO chain [11,12].

For LOP samples, the distortion of crystal structures are so severe that the effect of Ca substitution for Y on the distortion of crystal can not be recognized. On the other hand, for HOP samples, it seems that the substitution causes the increase in the number of stacking faults, although the quantitative comparison can not be done only by these TEM observations.

Microstructural distortions causes stacking faults and they may introduce intergrowths of "123" or "247" material, or possibly non-superconducting materials. Furthermore, in LOP samples, other crystallographic defects as observed in figs.1 and 2 may exist, although the structures were not confirmed. All these defects may be expected to act as dispersed pinning centers in these "124" materials and to result in finite J_c values.

In summary, we found that high oxygen pressure is required to obtain high quality crystals, and low oxygen pressure oxygenated or Ca substituted "124" materials have more defects that may act as pinning centers and produce high J_c values. Control of the microstructure, as in the conventional superconductors, is important in realizing high J_c oxide superconductors too.

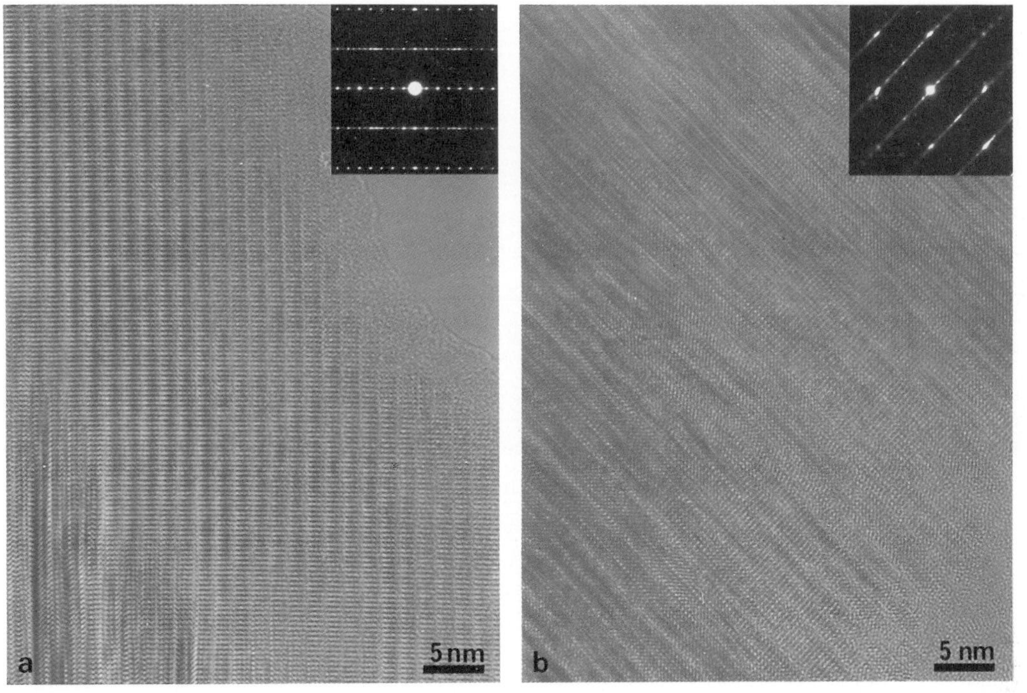

Fig.1 TEM images of a slightly distorted grain (a) and a severely distorted grain (b) in LOP-0 sample, together with corresponding diffraction patterns.

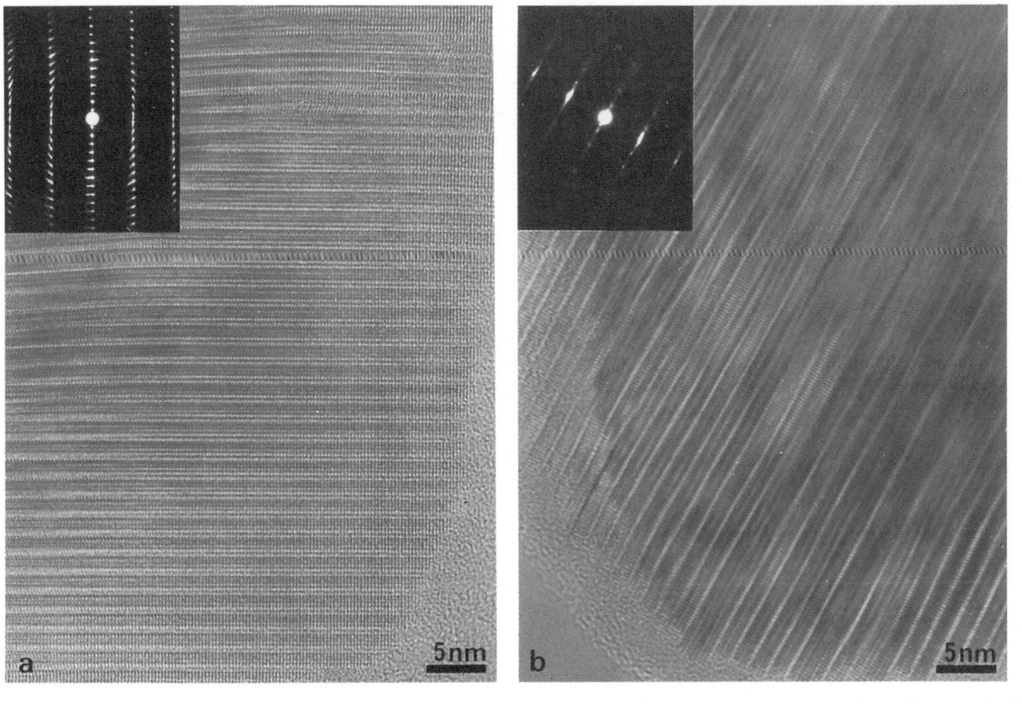

Fig.2 TEM images of a slightly distorted grain (a) and a severely distorted grain (b) in LOP-0.05 sample , together with corresponding diffraction patterns.

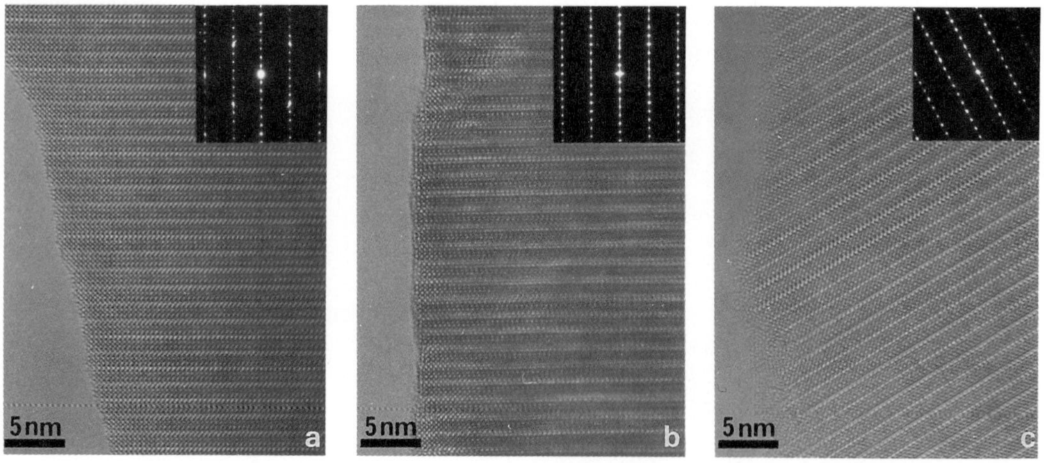

Fig.3 TEM images of HOP-0 sample (a), the HOP-0.05 sample (b) and the HOP-0.1 sample (c).

REFERENCES

1. Zandbergen HW , Gronsky R , Wang K , Thomas G (1988) Nature 331: 596
2. Miyatake T , Gotoh S , Koshizuka N , Tanaka S (1989) Nature 341: 41
3. Karpinski J , Kaldis E , Jilek E , Rusiecki S , Bucher B (1988) Nature 336: 660
4. Morris DE, Nickel JH, Wei JYT, Asmer NG, Scott JS, Scheven UM, Hultgren CT, Markelz AG, Post JE, Heaney PJ, Veblen DR, Hazen RM (1989) Phys.Rev.B 39: 7347
5. Cava RJ , Krajewski JJ , Peck Jr WF , Batlogg B , Rupp Jr LW , Fleming RM, Jemes ACWP, Marsh P (1989) Nature 338: 328
6. Cava RJ, Krajewski JJ, Peck Jr WF, Batlogg B, Rupp Jr LW (1989) Physica C 159: 372
7. Jin S , O'Bryan HM , Gallagher PK , Tiefel TH , Cava RJ , Fastnacht RA, Kammlott GW (1990) Physica C 159: 415
8. Miyatake T, Yamaguchi K, Takata T , Gotoh S , Koshizuka N , Tanaka S (1990) Jpn.J.Appl.Phys. 29: L1079
9. Wada T , Suzuki N , Yamaguchi K , Ichinose A , Yaegashi Y , Yamauchi H, Koshizuka N, Tanaka S (in press) J.Mater.Res.
10. Miyatake T , Yamaguchi K , Wada T , Suzuki N , Yamauchi H , Koshizuka N, Tanaka S (in press) J.Appl.Phys.
11. Yamaguchi K, Miyatake T, Takata T, Gotoh S, Koshizuka N, Tanaka S (1989) Jpn.J.Appl.Phys. 28: L1942
12. Yamaguchi K , Miyatake T , Wada T , Suzuki N , Takata T , Koshizuka N, Yamauchi H, Tanaka S (1990) Physica C 169: 289

X-ray Diffraction Study on 2201 Tl-Oxide Single-Crystals

Tsuyoshi Kajitani[1], Kenji Hiraga[1], Satoru Nakajima[2], Masae Kikuchi[1], and Yasuhiko Syono[1]

[1] Institute for Materials Research, Tohoku University, Sendai, 980 Japan
[2] CASIO Computer Co., Ltd., Ishikawa-cho, Hachioji, 192 Japan

ABSTRACT

Room temperature, X-ray single-crystal diffraction studies of heat-treated Tl-oxides are reported here. The structural parameters determined indicate that the apical-oxygen/copper distance of CuO_6 octahedrons is shorter, 2.645Å, in the no-Tc sample, but becomes longer, 2.666Å and 2.776Å in the samples with Tcs of 65K and 85K, respectively. No detectable deficiency or substitution of impurity ions at Tl-site was noticed.

KEY WORDS: Tl 2201, X-ray diffraction, single-crystal, crystal structure

INTRODUCTION

$Tl_2Ba_2CuO_{6\pm y}$ (Tl 2201 phase) is one of the most important and interesting superconducting cuprates, in which the superconducting critical temperature, Tc, can be easily changed from OK (non-superconducting) to 85K by the partial oxygen reduction. At present, there is some controversy whether the oxygen stoichiometry is more (6+y) or less (6-y) than six in this material [1-5].

Based on the powder neutron diffraction data, Parise et al. [5] argued that there is a partially occupied extra oxygen-site, namely the O(4)-site at 8f(1/4 1/4 1/4), in orthorhombic Tl 2201. However, Kikuchi et al. [2] assumed oxygen deficiency in Tl 2201, since the Tl-O double layers are too closely connected to each other, relative to Bi-O double layers, to store extra oxygens between them. Present authors [6] found quite a temperature-dependent oxygen desorption/absorption phenomenon coincidentally taking place at about 500K in the Tl 2201, Tl 2212 and Tl 2223 powders in air. The crystal structures of tetragonal and orthorhombic Tl 2201 crystals were studied previously by X-ray and neutron powder diffraction measurements [5, 7, 8] and by X-ray single-crystal diffraction measurement [9]. By growing tetragonal Tl 2201 single-crystals by the melt-growth method, we were able to do X-ray single-crystal diffraction measurements on an as-grown and two heat treated Tl 2201 samples with different Tcs ranging from OK(non-superconducting) to 85K. Our intention was to look for the Tc-dependent structural alteration. One important points, in which we were particularly interested, was the apical-oxygen/copper distance in the CuO_6 octahedron, because Ohta et al.[10] predicted that it became shorter in the high-Tc cuprates.

EXPERIMENTAL PROCEDURES

Tl 2201 single-crystals, of maximum size 0.3 x 0.3 x 0.1mm^3 , were grown by the melt-growth method. Tl_2O_3, BaO and CuO powders were mixed in a metal ratio of Tl/Ba/Cu=2/2/1, packed tightly in a gold tube of 5mmϕ x30mm and then sealed mechanically. The tube was annealed at 1193K for one hour in flowing oxygen and cooled slowly, at the rate of 1K/min. The resulting sample was an aggregate of small single-crystals, sintered powder and un-reacted raw materials. The aggregate was divided into small pieces and then two of them were annealed separately at 813K and 873K for 1-2 hours in 6N-Ar gas and 5N-N_2 gas, respectively. The sample annealed in Ar(#2 below) was quenched in at air at room temperature. The piece annealed at 873K in N_2 (#3) was quenched in liquid N_2. An as-grown single-crystal sample (#1), was also studied.
The superconducting critical temperatures of heat-treated single-crystals were estimated from the values determined on sintered powders heat-treated

identically with those of single-crystals. The Tcs determined by AC-susceptibility measurements were 65K and 85K and these were assumed to be those of the samples #2 and #3, respectively.

The single-crystal diffraction data were analyzed by the least-squares method, RADIEL [11].

RESULTS

Table 1 shows structural parameters obtained from the three samples, which are respectively as-grown(#1) and as-quenched from 813K(#2) and 873K(#3). The structural parameters of the tetragonal single crystals are obtained assuming the I4/mmm space-group. Since superlattice reflections of q=[0.08, 0.24, 1] type were observed by Parkin et al. [12] and Oku et al. [13] in the Tl 2201 phase, a real unit cell might be large, relative to that of the tetragonal lattice, and the real symmetry of the unit cell could not be tetragonal but monoclinic or orthorhombic. In this respect, obtained structural parameters show the atomic arrangement of the averaged lattice.

Table I. Structural parameters of Tl 2201 single-crystals, space group I4/mmm

Atom	As-grown Tc=0K sample #1	quenched from 813K, Tc=65K #2	quenched from 873K, Tc=85K #3
Tl 4e(0 0 z)	z=0.20261(3)	0.20269(2)	0.20259(2) /0.2302(4)
	$u_{11}=u_{22}$ =0.0256(1)Å^2	0.0230(1)	0.0250(1)
	u_{33}=0.0110(2)Å^2	0.0104(1)	0.0070(1)
	B=1.637	1.484	1.503
	Occ=1.00(1)	1.0	0.952/0.048(2)
Ba 4e(0 0 z)	z=0.41707(4)	0.41704(3)	0.41701(3)
	$u_{11}=u_{22}$ =0.0066(1)	0.0062(1)	0.0088(1)
	u_{33}=0.0123(3)	0.0113(2)	0.0121(2)
	B=0.671	0.624	0.782
Cu 2a(0 0 0)	$u_{11}=u_{22}$ =0.0113(5)	0.0087(3)	0.0111(4)
	u_{33}=0.0047(7)	0.0078(5)	0.0057(5)
	B=0.719	0.663	0.734
O(1) 4c(0 1/2 0)	u_{11}=0.005(1)	0.017(3)	0.007(3)
	u_{22}=0.000(3)	0.007(2)	0.004(3)
	u_{33}=0.014(4)	0.009(2)	0.014(3)
	B=0.49	0.87	0.66
O(2) 4e(0 0 z)	z=0.1140(6)	0.1148(4)	0.1195(4)
	$u_{11}=u_{22}$ =0.042(4)	0.018(2)	0.012(2)
	u_{33}=0.008(5)	0.008(3)	0.027(4)
	B=2.42	1.16	1.34
O(3) 4e(0 0 z)	z=0.2881(7)	0.2885(5)	0.2912(8)
	$u_{11}=u_{22}$ =0.135(11)	0.172(14)	0.358(27)
	u_{33}=0.004(6)	0.004(3)	0.017(7)
	B=7.21	9.16	19.3
Occ.	1.0	1.0	0.952(2)
a(Å)	3.8580(10)	3.8598(8)	3.8566(6)
c(Å)	23.206(9)	23.223(8)	23.229(9)
Reflections (unique)	1209 (293)	1552 (352)	1317 (362)
R_f	0.087	0.083	0.105
R_{wf}	0.066	0.061	0.075

For Tl and O(3), relatively large thermal factors, u_{11} and u_{22}, are noticed in the three crystals, and these of O(3) increase quite dramatically from sample

#1 to #3, which suggests that there is some correlation with the increase in Tc.

In sample #3, the difference (between the observed and calculated structure factors, Fob-Fcal) Fourier synthesis shows a relatively thick 'extra' electron cloud at a position 4e (0 0 z) with z~ 0.230. We assumed that a part of Tl atoms are attracted to O(2) due to an occasional loss of oxygen O(3) which may pull the Tl-atom back to Ba-atom side. With this assumption, the residuals R_f and R_{wf} decreased reasonably. The appropriate occupation probability at above extra-site was 5%, which might have corresponded to 5%-loss of the oxygen O(3), while 7.5%-loss of total oxygen in the as-quenched Tl 2201 was concluded by the chemical analysis[2].

The coordinate, z, of O(2), corresponding to the apical-oxygen/copper distance, increases from 0.1140(6) in sample #1 to 0.1148(4) and 0.1195(4) in samples #2 and 3, respectively.

Table II. Interatomic distances (Å)

Atom pair	As-grown Tc=0K sample #1	quenched from 813K Tc=65K #2	quenched from 873K Tc=85K #3
Cu-O(1)	1.929(1)	1.930(1)	1.928(1)
Cu-O(2)	2.645(14)	2.666(9)	2.776(9)
Jahn-Teller distortion $Cu-O_2/Cu-O_1$	1.36(2)	1.38(1)	1.44(1)
Tl-O(2)	2.056(15)	2.041(10)	1.930(9)
Tl-O(3)	1.984(28)	1.993(21)	2.058(19)
Tl-O(3) (ab-plane)	2.737(3)	2.737(2)	2.738(5)
Ba-O(3)	2.993(30)	2.985(21)	3.018(3)
Ba-O(2) (ab-plane)	2.822(6)	2.827(4)	2.868(4)

Table II shows calculated interatomic distances between the metal atoms and the first nearest neighbor oxygen atoms. The apical oxygen O(2)/copper distance increases from 2.645(14)Å in the sample #1 (non-superconducting) to 2.666(9)Å and 2.776(9)Å in the superconducting samples #2 and 3, respectively. The Tl-O(2) and Tl-O(3) distances also show systematic changes. It becomes obvious that the O(2) atom sits closer to the Tl atoms in the high-Tc samples relative to the non superconducting sample.

DISCUSSIONS AND CONCLUSIONS

Figures 1(a), (b) and (c) show the ORTEP-drawings of the as-grown tetragonal sample #1 and the as-quenched samples #2 and #3, respectively. Each thermal ellipsoid corresponds to the 50% existence region. Anomalously large thermal ellipsoids at the O(3)-site may be a characteritic of the crystal structure of the Tl 2201. The Tl- and O(3)-sites of the orthorhombic or the tetragonal unit cell are occasionally defined at 'split'-sites, e.g. the 16m-site in the orthorhombic unit cell by Parise et al. [5] or 16n-site in the tetragonal unit cell by Torardi et al. [9], in which one Tl- or O(3)-atom occupies four 16m- or 16n-sites statistically, i.e. the occupation probability is 25% in each.

In the present crystals, the Tl-site is quite normal, i.e. there is no indication of splitting. On the other hand, the O(3)-site could split into 16n-sites. But the estimated experimental error for the split exceeds the statistically meaningful level and then it is difficult to argue that the O(3)-site is a 'split'-site rather than the 'single'-site.

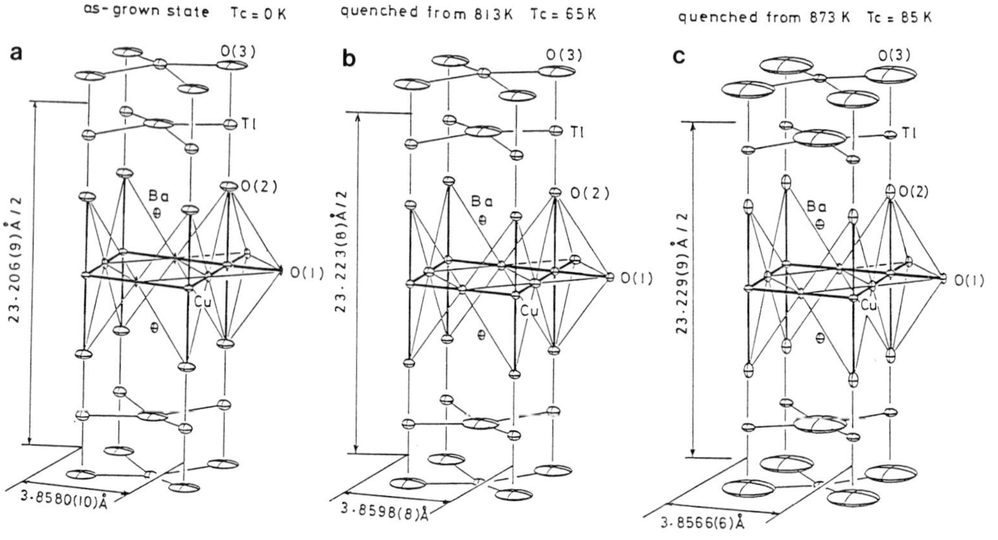

Fig. 1. ORTEP drawings of the 1/2 unit-cells of as-grown(a), as-quenched from 813K(b) and from 873K(c) samples.

The long period modulation, which occurs possibly in the Tl-O double layers [12,13], may have contributed to above uncertainty. The apical-oxygen, O(2)/copper distance is elongated in the as-quenched and high-Tc samples relative to the as-grown one, and this causes the O(2)-site to move closer to Tl-site in the as-quenched and high-Tc samples. This suggests that the electric charge of the Tl-O double layers becomes more-positive relative to the Cu-O layer. The observed pronounced structural and electric charge fluctuation shown in the Tl-O double layers in the high-Tc samples may be of benefit to the superconductivity.

ACKNOWLEDGMENT

This work was partly supported by a Grant-in-Aid for Scientific Research of Priority Area "Mechanism of Superconductivity" from the Ministry of Education, Science and Culture of Japan.

REFERENCES

1. Shimakawa Y, Kubo Y, Manako T, Igarashi H (1989) Phys. Rev. B 40: 11400
2. Kikuchi M, Nakajima S, Syono Y, Nagase K, Suzuki R, Kajitani T, Kobayashi N, Muto Y (1990) Physica C 166: 497
3. Ramanujachary K V, Li S, Greenblatt M (1990) Physica C 165: 377
4. Maignan A, Martin C, Hune M, Provost J, Hervieu M, Michel C, Raveau B (1990) Physica C 170: 350
5. Parise J B, Torardi C C, Subramanian M A, Gopalakrishnan J, Sleight A W (1989) Physica C 159: 239
6. Kajitani T, Hiraga K, Kikuchi M, Nakajima S, Nagase K, Syono Y (1990) Physica C 167: 212
7. Parise J B, Gopalakrishnan J, Subramanian M A, Sleight A W (1988) J. Solid State Chem. 76: 432
8. Huang T C, Lee V Y, Kariuri R, Beyers R, Parkin S S (1988) Mat. Res. Bull. 23: 1307
9. Torardi C C, Subramanian M A, Calabrese J C, Gopalakrishnan J, McCarron E M, Morrissey K J, Askew T R, Flippen R B, Chowdhry U, Sleight A W (1988) Phys. Rev. B 38: 225
10. Ohta Y, Tohyama T, Maekawa S (1990) Physica C 166: 385
11. Coppens P, Guru-Row T N, Leung P, Stevens E D, Becker P J, Yang Y W (1979) Acta Crystallogr. A 35: 192
12. Parkin S S, Lee V Y, Nazzal A I, Savoy R, Huang T C, Gorman G, Beyers R (1988) Phys. Rev. B 38: 6531
13. Oku T, Hiraga K,Shindo D, Kikuchi M, Nakajima S, Syono Y to be presented in ISS'90, Nov. 7-9, 1990, Sendai, Japan

High-Resolution Electron Microscopy of Tl-Ba-Ca-Cu-O Superconductors

TAKEO OKU[1], KENJI HIRAGA[1], DAISUKE SHINDO[1], MASAE KUKUCHI[1], SATORU NAKAJIMA[2], and YASUHIKO SYONO[1]

[1] Institute for Materials Research, Tohoku University, 1-1, Katahira 2-chome, Aoba-ku, Sendai, 980 Japan
[2] Hachioji Research Center, CASIO Computer Co., Ltd, Hachioji, 192 Japan

ABSTRACT

Crystal structures of high-Tc superconductors Tl-Ba-Ca-Cu-O ($Tl_2Ba_2CuO_6$, $Tl_2Ba_2CaCu_2O_8$ and $TlBa_2CaCu_2O_7$) have been investigated by high-resolution electron microscopy and electron diffraction, paying particular attention to the modulated structures. Diffuse scattering and satellite spots appeared at different positions in the observed diffraction patterns for the three crystals. High-resolution electron microscopic observations showed the displacements of metal atoms in Tl-O planes, and in their neighbourhood. It seems that the modulated structures are closely associated with ordering of the oxygen atoms.

KEY WORDS: Tl system superconductor, high-resolution electron microscopy, crystal structure

1. INTRODUCTION

Since the discovery of high-Tc Tl system superconductors, single Tl layer ($TlBa_2Ca_{n-1}Cu_nO_{2n+3}$ (n=1-5)) and double Tl layer ($Tl_2Ba_2Ca_{n-1}Cu_nO_{2n+4}$ (n=1-4)) structures have been investigated [1-3]. In addition to these basic layer structures, modulated structures accompanied with the satellite spots have also been reported. In $Tl_2Ba_2CuO_6$ (2201), some workers have observed the modulated structure [1,4,5], and Parkin et al. reported that the modulation wave vector is **q**=[$\pm$0.07,0.22,1], and that both orthorhombic and tetragonal phases exist in 2201 [1]. For $Tl_2Ba_2CaCu_2O_8$ (2212) [6,7] and $Tl_2Ba_2Ca_2Cu_3O_{10}$ (2223) [6], weak diffuse scattering also was observed, and the modulation wave vector is approximately **q**=[1/6,0,1] [7]. Again, in $TlBa_2CaCu_2O_7$ (1212), the modulation wave vector is **q**=[0.29,0,0.5] [1]. These modulated structures have been investigated mainly by electron diffraction. In the present work, we investigated the 2201, 2212 and 1212 materials by high-resolution electron microscopy and electron diffraction, paying particular attention to the modulated structures.

2. EXPERIMENTAL PROCEDURES

Bulk samples were prepared by reacting Tl_2O_3, CaO, BaO_2, CuO and $BaCuO_2$ as described previously [3,8,9]. The samples of 2201 and 1212 were reheated in the range 100-850℃ and then quenched in liquid nitrogen in order to control the oxygen content. 2201, 2212 and 1212 showed the superconductivity transition at 0-70K, 98K and 80-110K, respectively. Samples for high-resolution observation were prepared by dispersing crushed materials on holey carbon films. High-resolution images and electron diffraction patterns were observed on 400kV (JEM-4000EX) and 200kV (JEM-200CX) electron microscopes, with theoretical resolutions of 0.17nm and 0.23nm, respectively. A large-computer (ACOS-2020) was used for image calculations and image processing.

3. RESULTS AND DISCUSSION

In 2201, a modulated structure with a wave vector **q**=[$\pm$0.07-0.08,0.21-0.23,1] was observed in diffraction patterns. For this material, the indices are those of an orthorhombic unit cell (a=0.545nm, b=0.549nm, c=2.318nm). A high-resolution image of 2201 taken with [110] incidence is shown in Fig.1, together with the structural model (A) and the calculated image (B). The image shows directly the projected potential of metal atoms. The larger black dots in the zigzag array correspond to the Tl atoms. On both sides of the double Tl layers, there are layers of Ba atoms. In this image, Cu atoms, with a smaller atomic number, are represented as less dark

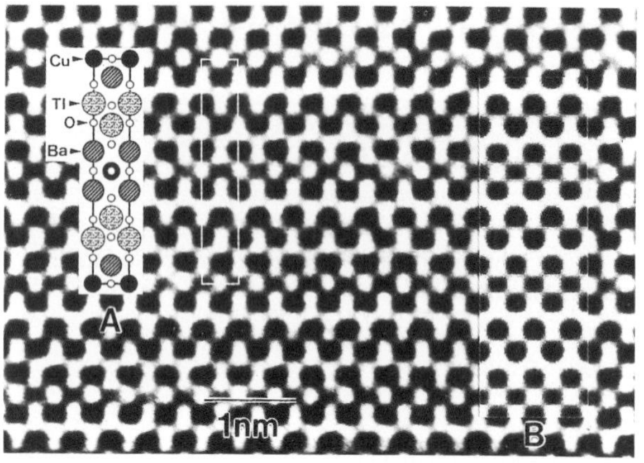

spots. The simulated image (Fig.1(B)), calculated by the multi-slice method at a defocus of 30nm and a crystal thickness of 1.9nm, agrees well with the observed image. Figure 2 shows an electron diffraction pattern (a) and a high-resolution image (b) of 2201 taken with [001] incidence. Sharp satellite spots are seen in Fig.2(a). In addition to the fundamental lattice fringes in Fig.2(b), dark and bright contrasts with a distance of 1.2nm and their twin relations can be seen. This image shows the direction of the modulated structure is near [130].

Fig.1. A high-resolution image of 2201 taken with [110] incidence, together with the structural model (A) and the calculated image (B).

Although twinning of the modulated structure appears on both {110} and near {100} planes, the twinning of fundamental lattice appears only on {110} planes, as indicated by the arrows in Fig.2. Figure 3 shows an electron diffraction pattern (a) and a high-resolution image (b) of 2201 taken with the [130] incidence. The modulated region (A) and the non-modulated region (B) are clearly distinguishable.

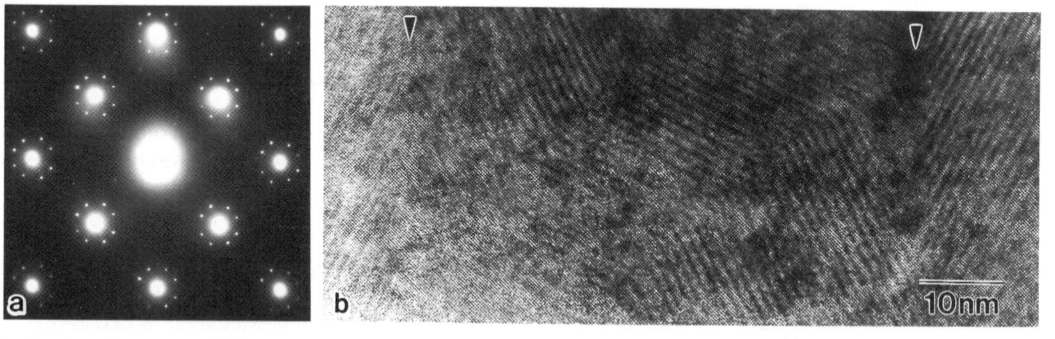

Fig.2. An electron diffraction pattern (a) and a high-resolution image (b) of 2201 taken with [001] incidence showing the modulated structure and its twinning.

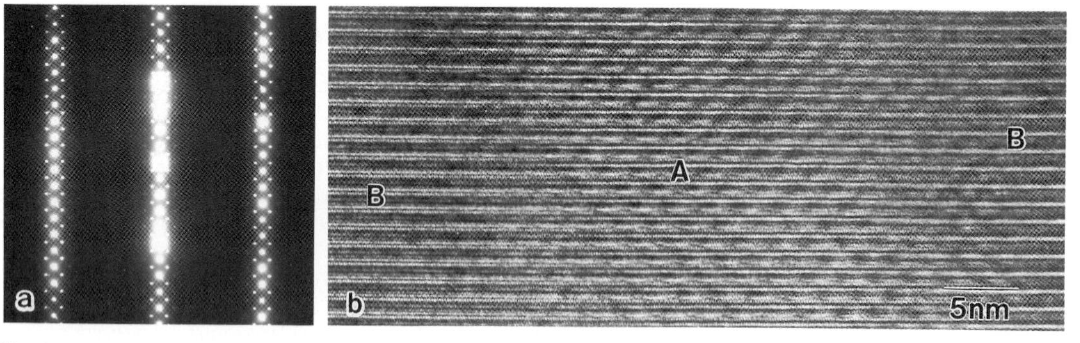

Fig.3. An electron diffraction pattern (a) and a high-resolution image (b) of 2201 taken with the [130] incidence showing clearly the modulated structure.

Figure 4 shows a processed high-resolution image of 2201 taken with [100] incidence, together with a Fourier diffractogram of the image. In order to enhance the contrast of the modulated structure, fundamental reflections in the diffractogram were reduced, so the satellite spots (indicated by small white arrows) can be seen. By careful observation of the modulation

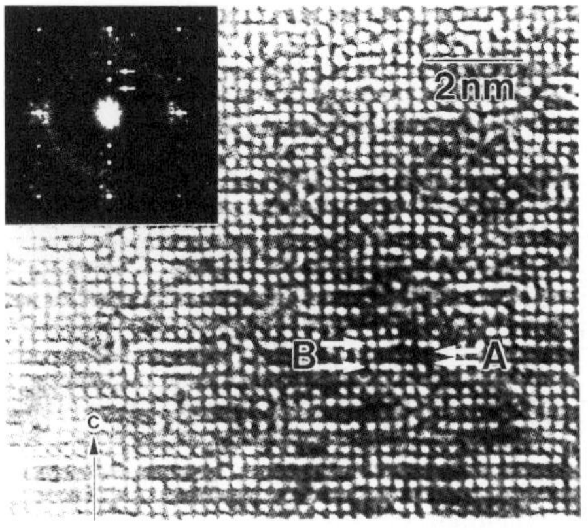

contrast, changes in both the darkness and position of Tl atoms can be seen, as indicated by arrows A. Moreover, with this image processing, the white dots between the Tl and Ba atoms are also emphasized, as indicated by arrows B. This shows that the origin of modulated structure exists on the Tl-O planes (at A) and between Tl and Ba atoms (at B). A theory which has been proposed to account for the origin of the modulated structure suggests that linear ordering of Tl and O vacancies occurs in the Tl-O planes [4]. As a small number of oxygen vacancies have been introduced in the samples observed in the present work [8], there is a possibility that periodic atomic displacements are caused by ordering of oxygen vacancies.

Fig.4 A processed high-resolution image of 2201 taken with [100] incidence, reconstructed by reducing the amplitudes of fundamental reflections, together with a Fourier diffractogram.

The correlation between the modulation and the superconductivity shows a tendency for lower Tc values in the orthorhombic phase (i.e. the phase showing strong modulation), but higher-Tc values (above 70K) have been reported also in the orthorhombic phases [11,12].

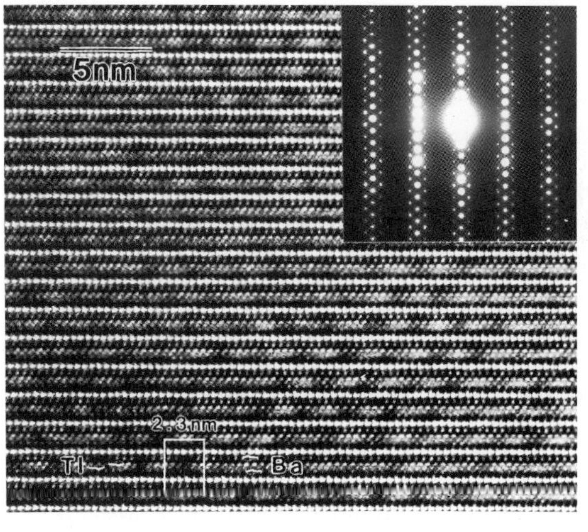

Diffuse scattering of the modulated structure is observed in diffraction patterns of 2212, but the modulation vector, **q**=[1/6,0,1], has a different direction from that of 2201. This modulation shows two-dimensional character, as observed previously [6], and the symmetry of the fundamental lattice remains tetragonal (a=0.385nm, c=2.92nm). A high-resolution image of 2212 taken with [100] incidence is shown in Fig.5, together with its diffraction pattern. The modulation contrast is observed clearly, and its periodicity in the [100] is approximately $6 \times a$ (2.3nm). Changes in both the darkness and the positions of Tl and Ba atoms show that the origin of the modulated structure exists in Tl-O planes and in their neighbourhood.

Fig.5 A high-resolution image of 2212 with [100] incidence, together with a diffraction pattern.

Some theories have been proposed to account for the origin of these modulated structures: short-range ordering due to displacements of Tl and O in the Tl-O planes [13], extra oxygen in the Tl-O planes [14], a partial substitution of Tl^{3+} by Tl^{+} [6,14], the mutual substitution of Tl and Ca atoms [7,15]. Again, there is a possibility of atomic displacements being caused by ordering of the oxygen vacancies, the ordering of the Ca atoms in the Tl-O planes or a charge density wave arising from the coexistence of Tl^{3+} and Tl^{+}.

Figure 6 (a) and (b) show electron diffraction patterns of 1212 taken with (a)[001] and (b)[100] incidence. The observed modulation wave vector is approximately **q**=[0.28,0,0.5], i.e. the modulation is incommensurate. As the modulation shows a two-dimensional character, the symmetry of the fundamental lattice remains tetragonal (a=0.385nm, c=2.92nm). The diffuse scattering becomes stronger as the oxygen loss is increased and also the Tc increases. A high-resolution image of 1212 with [001] incidence is shown in Fig.6(c). The image shows a very weak modulation contrast. As a small number of oxygen vacancies have been introduced in the samples observed in the present work [9], it is suggested that the modulation is caused by atomic displacements due to ordering of the oxygen vacancies.

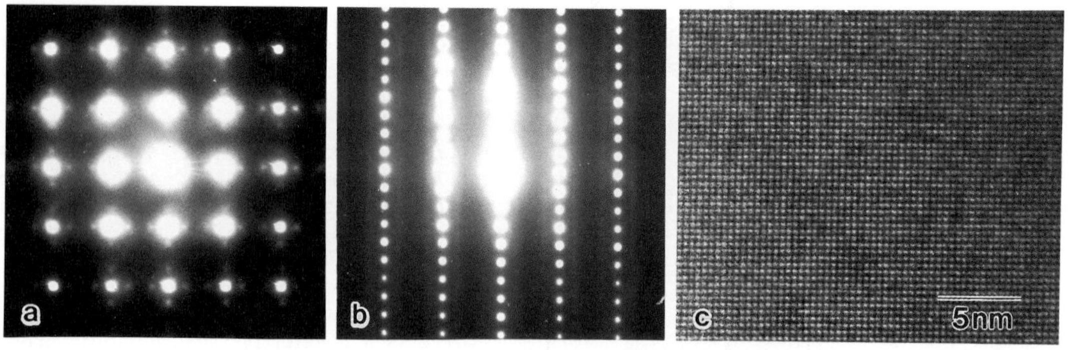

Fig.6 Electron diffraction patterns ((a),(b)) and a high-resolution image (c) of 1212 taken with (a)[001], (b)[100] and (c)[001] incidence, showing modulated structure.

4. CONCLUSION

High-Tc superconducting oxides were investigated by electron diffraction and high-resolution electron microscopy. Modulated structures were observed in each material. In 2201, 2212 and 1212, the modulation wave vector is **q**-[± 0.07-0.08,0.21-0.23,1], **q**-[1/6,0,1] and **q**-[0.28,0,0.5], respectively. It is suggested that the origin of the modulated structures is in the Tl-O planes, and further, that the modulated structures are closely associated with the ordering of oxygen atoms.

ACKNOWLEDGEMENT

The authors are grateful to F.J.Lincoln and T.Kajitani for critical reading of the manuscript and warm encouragement. One of the authors (TO) wishes to thank the Japan Society for the Promotion of Science for fellowship. This work was partly supported by the Grant-in-Aid for Scientific Research on Priority Areas "Mechanism of Superconductivity" provided by the Ministry of Education, Science and Culture, Japan.

REFERENCES

1. Parkin SSP, Lee VY, Nazzal AI, Savoy R, Huang TC, Gorman G, Beyers R (1988) Phys. Rev. B 38: 6531
2. Nakajima S, Kikuchi M, Syono Y, Oku T, Shindo D, Hiraga K, Kobayashi N, Iwasaki H, Muto Y (1989) Physica C 158: 471
3. Syono Y, Kikuchi M, Nakajima S, Suzuki T, Oku T, Hiraga K, Kobayashi N, Iwasaki H, Muto Y (1989) MRS Symp. Proc. 156: 229
4. Hewat EA, Bordet P, Capponi JJ, Chaillout C, Chenavas J, Godinho M, Hewat AW, Hodeau JL, Marezio M (1988) Physica C 156: 375
5. Iijima S (1989) in: Kamimura H, Oshiyama A (eds) in Mechanisms of High Temperature Superconductivity. Springer-Verlag, Berlin, pp 304-312
6. Zandbergen HW, Tendeloo GV, Landuyt JV, Amelinckx S (1988) Appl. Phys. A46: 233
7. Hewat AW, Hewat EA, Brynestad J, Mook HA, Specht ED (1988) Physica C 152: 438
8. Kikuchi M, Nakajima S, Syono Y, Nagase K, Suzuki R, Kajitani T, Kobayashi N, Muto Y (1990) Physica C 166: 497
9. Nakajima S, Kikuchi M, Syono Y, Nagase K, Oku T, Kobayashi N, Shindo D, Hiraga K (1990) Physica C 170: 443
10. Nakajima S, Kikuchi M, Oku T, Kobayashi N, Suzuki T, Nagase K, Hiraga K, Muto Y, Syono Y (1989) Physica C 160: 458
11. Parise JB, Gopalakrishnan J, Subramanian MA, Sleight AW (1988) J. Solid State Chem. 76: 432
12. Kikuchi M, Syono Y, Kobayashi N, Oku T, Aoyagi E, Hiraga K, Kusaba K, Atou T, Tokiwa A, Fukuoka K (1990) Appl. Phys. Lett. 57: 813
13. Dmowksi W, Toby BH, Egami T, Subramanian MA, Gopalakrishnan J, Sleight AW (1988) Phys. Rev. Lett. 61: 2608
14. Zandbergen HW, Groen WA, Mijlhoff FC, Tendeloo GV, Amelinckx S (1988) Physica C 156: 325
15. Hiraga K, Shindo D, Hirabayashi M, Kikuchi M, Kobayashi N, Syono Y (1988) Jpn. J. Appl. Phys. 27: L1848

X-Ray Rietveld Refinement of Orthorhombic and Tetragonal (Tl,Pb)Sr_2CuO_{5-z}

ATARU ICHINOSE, TETSUYUKI KANEKO, TAKAHIRO WADA, KAZUYUKI HAMADA, AKIKO NARA, and H. YAMAUCHI

Superconductivity Research Laboratory, International Superconductivity Technology Center, 10-13, Shinonome 1-chome, Koto-ku, Tokyo, 135 Japan

ABSTRACT

The crystal structures of orthorhombic ($Tl_{0.96}Pb_{0.04}$)Sr_2CuO_{5-z} and tetragonal ($Tl_{0.5}Pb_{0.5}$)Sr_2CuO_5 were analyzed by Rietveld refinement of x-ray powder diffraction data. The space group of the orthorhombic phase was Pmmm and that of the tetragonal one was P4/mmm. The determined lattice constants of the ($Tl_{0.96}Pb_{0.04}$)Sr_2CuO_{5-z} were a=3.6740(1), b=3.7686(1) and c=8.9690(3) Å, and those of the ($Tl_{0.5}Pb_{0.5}$)Sr_2CuO_5 were a=3.7405(1) and c=9.0097(2) Å. In both compounds, it is likely that the Tl-O octahedra in the crystal structures were displaced and/or rotated from ideal positions in the "1201" structure. It was also indicated for ($Tl_{0.96}Pb_{0.04}$)Sr_2CuO_{5-z} that the oxygen sites along the a axis in the Cu-O plane were not fully occupied.

KEYWORDS $TlSr_2CuO_5$, x-ray powder diffraction, Rietveld analysis, bond valence sum

INTRODUCTION

The synthesis of Tl based "1201" compound, $TlSr_2CuO_5$, is known to be difficult because of the high Cu valence (3+) in the compound [1]. The "1201" structure can be stabilized by substituting trivalent rare earth ions such as La^{3+} for Sr^{2+} and/or tetravalent ions such as Pb^{4+} for Tl^{3+}. Actually synthesis of superconductive Tl(La,Sr)$_2CuO_5$ [2] and (Tl,Pb)(R,Sr)$_2CuO_5$ (R=La, Pr, Nd and Sm) [3-5] and nonsuperconductive ($Tl_{0.5}Pb_{0.5}$)Sr_2CuO_5 [6] have been reported and all of these solid solutions have tetragonal unit cells. Recently, Pb- or rare earth element-free $TlSr_2CuO_{5-z}$ was synthesized and found to have an orthorhombic unit cell [7]. It was also found for the ($Tl_{1-x}Pb_x$)Sr_2CuO_{5-z} system that an orthorhombic-tetragonal structural phase transition occurred at $x\approx0.12$.

In this paper, the crystal structures of both orthorhombic and tetragonal (Tl,Pb)Sr_2CuO_{5-z} samples were refined by the Reitveld analysis of x-ray powder diffraction data. We discussed these structures on the crystallographic point of view.

EXPERIMENTAL PROCEDURES

Samples were prepared by a solid-state reaction method. High purity Tl_2O_3, PbO, SrO and CuO powders were weighed to the norminal compositions, ($Tl_{0.94}Pb_{0.06}$)Sr_2CuO_5 and ($Tl_{0.5}Pb_{0.5}$)Sr_2CuO_5. The mixed powder was pressed into parallelepiped bars, and then wrapped by gold foils and sintered at 900 °C in O_2 gas flow.

X-ray diffraction data were taken using a high intensity powder diffractometer with a roating anode source and a curved graphite monochrometer. Data were collected using CuKα radiation with the step-width of 0.04° (in 2θ) and a counting time of 2s over the 2θ range from 3° to 120°. The crystal structure was refined by the Rietveld analysis using the program "RIETAN" developed by Izumi [8].

RESULTS

I. Orthorhombic ($Tl_{0.96}Pb_{0.04}$)Sr_2CuO_{5-z}

The space group was determined to be P222 using the program "CELLO" developed by Takaki *et al.* [9]. Space group P222 includes two space groups, Pmm2 and Pmmm. In our refinement procedure, we selected more symmetrical space group, Pmmm. The data in the range of 3.0°- 19.0°, 27.0°- 28.0°, 32.0°- 33.5° and 41.0°- 42.0° were excluded owing to the appearance of minor impurity peaks. Preferred orientation was corrected on the same assumption that the cleavage plane is (001) using a Gaussian preferred-orientation function. The (Tl,Pb) site was occupied by the mixture of 96% Tl and 4% Pb, because the norminal composition of this sample was $(Tl_{0.96}Pb_{0.04})Sr_2CuO_{5-z}$

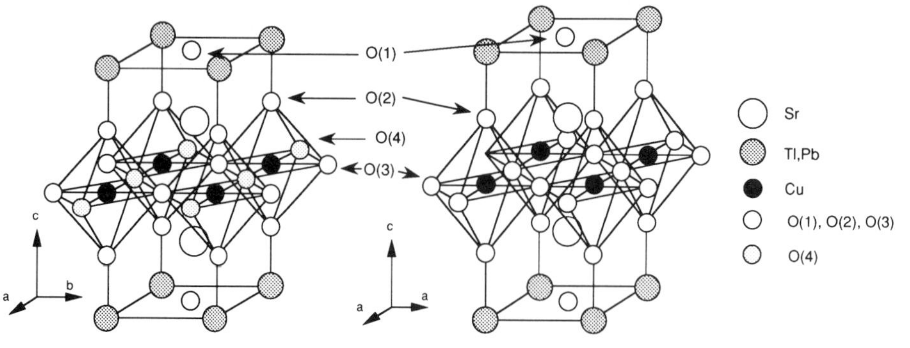

(a) $(Tl_{0.96}Pb_{0.04})Sr_2CuO_{5-z}$ (b) $(Tl_{0.5}Pb_{0.5})Sr_2CuO_5$

Fig.1 Crystal Structure

A schematic crystal structure of the orthorhombic "1201" phase is shown in Fig. 1(a). Preliminary refinements revealed that isotropic thermal parameters, *B*, for (Tl,Pb), O(1) and O(4) were larger than those for other sites. The crystal structure was refined on the basis of the vacancy model in each site and site split model. The isotropic thermal parameters in (Tl,Pb) and O(1) sites became smaller for the assumption of the (Tl,Pb) vacancy rather than the split of site. That is, the atomic scattering factor in the (Tl,Pb) site was small because of exsistence of vacancies and/or partial substitution of Cu. A similar result was obtained for the O(4) site. However, the obtained standard deviation was considerably large. In the final refinements, we assumed that the (Tl,Pb) site was fully occupied and part of the O(4) site was vacant. The determined occupancy of the O(4) site was 0.65 and accordingly *z* was 0.35. Therefore determined chemical formula of this compound was $(Tl_{0.96}Pb_{0.04})Sr_2CuO_{4.65}$. Table I lists the final *R* factors, lattice constants and structural parameters, and their standard deviations. Figure 2 shows the final profile fit and difference pattern.

Table I Final results of the Rietveld refinement for $(Tl_{0.96}Pb_{0.04})Sr_2CuO_{4.65}$ in space group Pmmm (orthorhombic). Symbol *g* is the occupation factor and *B* is the isotropic thermal parameter. R_{wp}=8.51%, R_p=6.58%, R_e=3.49%, R_I=5.63%, R_F=4.38%, *a*=3.6740(1) Å, *b*=3.7686(1) Å *c*=8.9690(3) Å, (*a*+*b*)/2=3.7213 Å, *V*=124.18 Å^3.

Atom	Site	*x*	*y*	*z*	*g*	*B*(Å^2)
Sr	2t	0.5	0.5	0.2912(7)	1	0.4(2)
Tl,Pb	1a	0	0	0	1	2.3(2)
Cu	1c	0	0	0.5	1	0.9(4)
O(1)	1f	0.5	0.5	0	1	3.8(25)
O(2)	2q	0	0	0.248(6)	1	0.7(10)
O(3)	1g	0	0.5	0.5	1	1.9(18)
O(4)	1d	0.5	0	0.5	0.65(12)	1.2(35)

II. Tetragonal $(Tl_{0.5}Pb_{0.5})Sr_2CuO_5$

The space group was assumed to be P4/mmm for the previously reported structural refinement of the tetragonal "1201" system [5]. The data in the ranges of 3.0°- 19.0°, 24.5°- 25.4° and 35.5°- 37.0° in 2θ were excluded from the analysis and the preferred orientation correction was also made. The (Tl,Pb) site was occupied by the mixture of 50% Tl and 50% Pb.

A schematic crystal structure of the tetragonal "1201" is shown in Fig. 1(b). Preliminary refinements revealed that the isotropic thermal parameters, *B*, for (Tl,Pb), O(1) and O(2) were large. Two reasons are considered for these results. One is that the occupation factor for the (Tl,Pb) site is smaller than 1.0. The other is that the sites of (Tl,Pb), O(1) and O(2) are displaced from the ideal positions. Since the determined occupation factor for the (Tl,Pb) site is 0.99(2), the abnormal isotropic thermal parameters does not seem to be caused by former reason. Therefore, we assumed *z*=0 and attempted to refine the displacements of the (Tl,Pb), O(1) and O(2) sites. The (Tl,Pb) site is displaced to 4l site (*x*, 0, 0) from 1a site (0, 0, 0). The O(1) site is displaced to 4n (*x*, 1/2, 0) or to 2h (1/2, 1/2, *z*) site from 1c site (1/2, 1/2, 0). The O(2) site is displaced to 8s site (*x*, 0, *z*) from 2g site (0, 0, *z*). However, after the refinement procedure, all the isotropic thermal parameters were little improved. (In this refinement, distortion of the (Tl,Pb)-O_6 octahedron was not taken into consideration.) Therefore, for the sake of simplicity only the (Tl,Pb) site was allowed to displace from the ideal position, and the (Tl,Pb), O(1) and O(2) sites did not displace and one fully occupied having their thermal parameters equal to 0.5. Table II lists the final *R* factors, lattice constants and structural parameters, and their standard deviations. Figure 3 shows the final profile fit and difference pattern. The determined chemical formula of this tetragonal compound was $(Tl_{0.5}Pb_{0.5})Sr_2CuO_5$.

Table II Final results of the Rietveld refinement for $(Tl_{0.5}Pb_{0.5})Sr_2CuO_5$ in space group P4/mmm (tetragonal). Symbol *g* is the occupation factor and *B* is the isotropic thermal parameter. R_{wp}=7.40%, R_p=5.63%, R_e=3.26%, R_I=5.44%, R_F=3.65%, *a*=3.7405(1) Å, *c*=9.0097(2) Å, *V*=126.06 Å^3.

Atom	Site	*x*	*y*	*z*	*g*	*B*(Å^2)
Sr	2h	0.5	0.5	0.2974(6)	1	0.7(2)
Tl,Pb	4l	0	0.069(3)	0	1	0.5
Cu	1b	0	0	0.5	1	0.8(4)
O(1)	1c	0.5	0.5	0	1	0.5
O(2)	2g	0	0	0.223(4)	1	0.5
O(3)	2e	0	0.5	0.5	1	0.9(10)

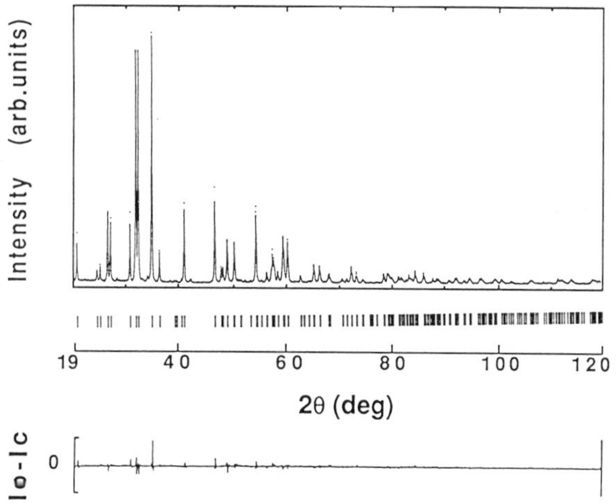

Fig. 2 Final profile fit and difference pattern for orthorhombic $(Tl,Pb)Sr_2CuO_{4.65}$

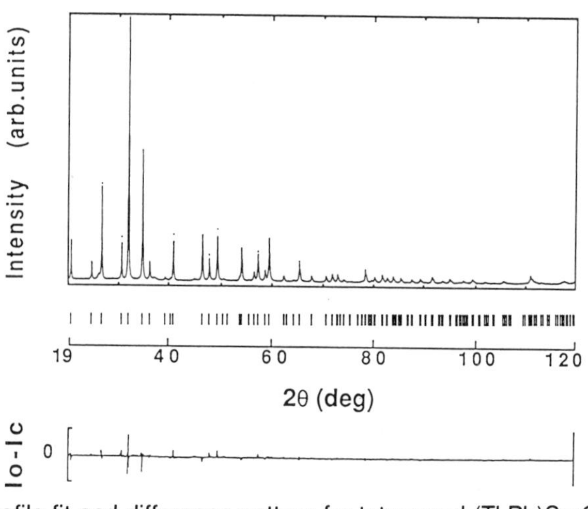

Fig. 3 Final profile fit and difference pattern for tetragonal $(Tl,Pb)Sr_2CuO_5$

DISCUSSION

The main differences in the crystal structure between orthorhombic and tetragonal (Tl,Pb)Sr_2CuO_{5-z} were (1) the distribution of oxygen around a Cu ion and (2) the positions of Sr and O(2) ions. In ($Tl_{0.5}Pb_{0.5}$)Sr_2CuO_5 the O(2), O(3) and O(4) sites were perfectly occupied and so the CuO_6 octahedra were perfect. In ($Tl_{0.96}Pb_{0.04}$)$Sr_2CuO_{4.65}$, a part of O(4) site was unoccupied. Therefore the CuO_6 octahedra were imperfect. Selected interatomic distances are given in Table III. The distances of Cu-O(2) and Cu-O(4) are shorter in ($Tl_{0.96}Pb_{0.04}$)Sr_2CuO_{5-z} than in ($Tl_{0.5}Pb_{0.5}$)Sr_2CuO_5. As increasing Pb content, the O(2) site approached to the (Tl,Pb) site because of the increase in the average valence at the (Tl,Pb) site. At this time, since the O(4) site was assumed to be fully occupied, the distances of Cu-O(3) and Cu-O(4) became the same value. That is, the crystal system of ($Tl_{0.5}Pb_{0.5}$)Sr_2CuO_5 was tetragonal.

Next, we consider the valences of Cu and Sr ions using a bond valence sum method. The bond valence sums were calculated using the method by Brown and Altermatt [10] on the basis of the bond length tabulated in Table III. In ($Tl_{0.96}Pb_{0.04}$)$Sr_2CuO_{4.65}$ the bond valence sums for Cu and Sr were 2.47 and 2.00, respectively. The bond valence sum for Sr atoms was equal to the formal valence of Sr atoms. If the valences of Sr, Tl and Pb ions are assumed to be respectively 2+, 3+ and 4+, the valence of Cu in ($Tl_{0.96}Pb_{0.04}$)$Sr_2CuO_{4.65}$ must be 2.26. The Cu valence obtained by a bond valence sum (2.47) is larger than one (2.26) calculated from the chemical formula. This result may mean that the effective Cu valence increases from chemically determined valence, while the effective (Tl,Pb) valence decreases. In ($Tl_{0.5}Pb_{0.5}$)Sr_2CuO_5 the bond valence sums for Cu atoms and Sr atoms were 2.61 and 2.04, respectively. The valence of Sr atoms is approximately equal to 2. When the valences of each cation except Cu ions are assumed to be equal to the values for the ($Tl_{0.96}Pb_{0.04}$)$Sr_2CuO_{4.65}$ the valence of Cu atom was calculated to be 2.5. Therefore, it is also plausible in ($Tl_{0.5}Pb_{0.5}$)Sr_2CuO_5 that the effective valence of Cu atoms increases slightly from chemically determined valence.

Table III Interatomic distance for ($Tl_{0.96}Pb_{0.04}$)$Sr_2CuO_{4.65}$ and ($Tl_{0.5}Pb_{0.5}$)Sr_2CuO_5

Bond	($Tl_{0.96}Pb_{0.04}$)$Sr_2CuO_{4.65}$	($Tl_{0.5}Pb_{0.5}$)Sr_2CuO_5
Tl,Pb-O(1)	2.632	2.645
Tl,Pb-O(2)	2.224	2.009
Cu-O(2)	2.260	2.496
Cu-O(3)	1.884	1.870
Cu-O(4)	1.837	(1.870)
Sr-O(1)	2.612	2.679
Sr-O(2)	2.660	2.729
Sr-O(3)	2.623	2.613
Sr-O(4)	2.657	(2.613)

REFERENCES

1. Wada T, Kaneko T, Yamauchi H (1990) ISTEC Journal 3(3): 21.
2. Itoh T, Uchikawa H (1989) Jpn J Appl Phys 28: L200.
3. Adachi S, Inoue O, Hirano H, Takahashi Y, Kawashima S (1989) Jpn J Appl Phys 28: L775.
4. Sabramanian MA (1989) Mat Res Bull 25:191.
5. Mochiku T, Nagashima T, Watahiki M, Fukai Y, Asano H (1989) Jpn J Appl Phys 28: L1926.
6. Kwei CH, Shi JB Ku HC: preprint
7. Kaneko T, Wada T, Hamada K, Yamauchi H :proceedings of ISS'90.
8. Izumi F (1985) J Crystallogr Soc Jpn 27: 23[in Japanese].
9. Takaki Y, Taniguchi T, Yamaguchi H, Ogura T (1987) J Ceram Soc Jpn Int Ed 95: 565.
10. Brown ID, Altermatt D (1985) Acta Crystallogr B 36: 244.

3.3 Substitution and Doping Effects

Effects of Disorder and Carrier Density on the Superconductivity in Oxide Systems

WEI YAO LIANG

IRC in Superconductivity, Madingley Road, Cambridge CB3 0HE, UK

ABSTRACT

Disorder due to slight oxygen deficiency in $YBa_2Cu_3O_{7-\partial}$ or a low substitution of Co for Cu in $YBa_2(Cu_{1-x}Co_x)_3O_7$ tends to destroy the coherence coupling across the layers. The materials become more anisotropic and exhibit considerable two dimensional fluctuations as observed in specific heat and resistivity measurements. It is proposed that in addition to charge carriers in the CuO_2 layers, the contributions from the charge reservoir layers should not be ignored, particularly in explaining the size of the Hall coefficients.

KEY WORDS: charge reservoir, site selective substitution, two dimensional fluctuations

INTRODUCTION

Since the first observation of superconductivity in $SrTiO_{3-\partial}$ ($T_c \sim 0.7$ K) in 1964, we have been aware of the fact that the characteristics of an oxide superconductor depend sensitively on the density of carriers of the material. This is because oxide superconductors are usually derived from an insulating parent and the carriers are introduced by cation substitution or changing oxygen stoichiometry. This distinguishes them from conventional superconductors which are intrinsically metallic at ordinary temperatures. Unfortunately, accompanying the carriers are defects and impurity scattering, which eventually degrade the superconductivity. The new cuprate superconductors have layered structure in which the active CuO_2 layers are often separated by many atomic distances. This gives rise to the possibility of selecting specific sites in which to place the doping atoms, similar to the process of modulation doping in semiconductor heterojunction multilayers. However unlike III-V semiconductors, charges in oxide superconductors can be localised at an atomic scale [1]. Consequently doping may not always lead to a direct increase in the number of carriers in the CuO_2 layers. This provides the experimentalist with a considerable freedom of choice, aswell as complicating interpretation of any doping experiment. This is highlighted, in particular, in the difficulty in understanding the relationship between the observed Hall number and the concentration of dopant [2], and in understanding the meaning of valency of some of the mixed valent metals present in the compound. This paper attempts to describe a variety of observations that may be better understood if the effects of carrier concentration and disorder are taken into account. We will discuss $YBa_2(Cu_{1-x}M_x)O_{7-\partial}$ where M = Co, Zn and ∂ either small or zero.

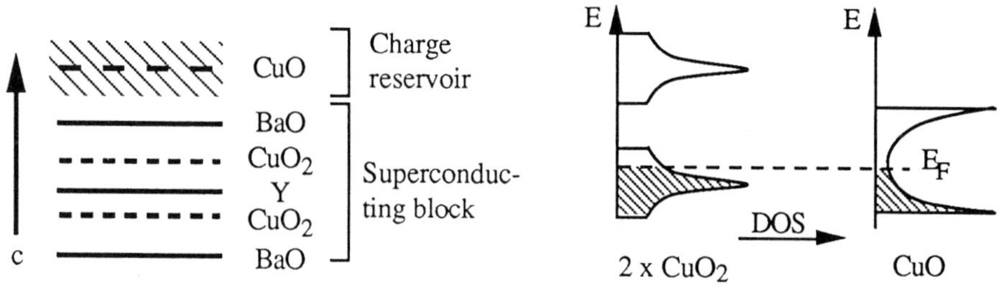

Fig. 1 Schematic diagram showing how $YBa_2Cu_3O_7$ may be viewed as being made up of a superconducting block $YBa_2Cu_2O_6$ and a charge reservoir block CuO. The corresponding density of states (DOS) curves are shown on the right.

The treatment of layered oxide superconductor as made up of a superconducting block and a charge reservoir block has been suggested by a number of authors [3,4]. As an example, fig.1 shows such a division for $YBa_2Cu_3O_7$ in which the charge reservoir is made up of CuO chain layer. The one and two dimensional density of states (DOS) curves are also shown. It must be noted that (1) the existence of a charge reservoir and the CuO_2 block implies that under ordinary circumstances, the Fermi level will intercept the energy bands derived from both regions, (2) when dopant concentration or oxygen stoichiometry are modified in either block, the energy of the effected block will also be modified relative to the other, in addition to the direct effect of the filling of the energy band, (3) a great deal of disorder has often been observed in the reservoir block, leading to a rapid localisation of the carriers in this block. For example, when oxygen is removed from $YBa_2Cu_3O_7$, the main loss of oxygen occurs in the reservoir block. This will not only result in an increased filling of the energy bands but also as the reservoir block becomes less negative, its band structure is lowered with respect to that of the CuO_2 block. Moreover the disorder this introduces into the block will quickly immobilise the carriers in their one dimensional motion. On the other hand, if Y^{3+} ions are partially substituted by Ca^{2+}, the CuO_2 block becomes less positive and its energy increases, together with the removal of electrons by Ca^{2+} the hole concentration in the CuO_2 block will increase rapidly. This is the reason for the superconductivity in $(Ca_{.25}Y_{.75})Ba_2Cu_3O_{6.1}$ [5] despite the apparant deficiency of hole carriers when compared with the minimum number of holes required in $YBa_2Cu_3O_{>6.4}$. It is clear that unequal amounts of hole carriers may be added or removed from the two blocks by selecting specific sites for doping.

PARTIAL SUBSTITUTION OF COBOLT AND ZINC FOR COPPER IN $YBa_2Cu_3O_7$

Low concentration ($x<0.05$) substitution of Co for Cu in $YBa_2(Cu_{1-x}Co_x)O_7$ is highly selective in that Co will only replace the Cu ions in the CuO chain layers. This is probably because Co will readily accept a four-fold rather than a five-fold ligand configuration. The immediate effects of the substitution are a disruption of the one dimensional order in the Cu-O chains thereby increasing the two dimensional properties of the parent compound. In addition, as Co ions enter into the Cu-O chain layers only, mismatch occurs between the Cu-O chain layer and the CuO_2 layer. This produces strained patterns which can easily be observed in an electron microscope [6]. The order parameter for the tetragonal-orthorhombic transition gradually weakens with increasing Co concentration until at $x \sim 0.028$, the orthorhombic phase collapses and it is believed that there is a high degree of two dimensional disorder in the Cu-O chain layer. By contrast Zn ions appear to enter the Cu lattice at random and will substitute Cu ions in both the chain and the plane layers. Lattice strain and orthorhombic-tetragonal transition are not observed in this case. These structural studies correlate very well with the measurements of normal resistivity, ac susceptibility, and specific heat.

Normal State Resistivityand Matthiessen's Rule

As reported by Cooper et al [7], Zn substitution has resulted in a direct lowering of the superconducting transition temperature but without noticeable broadening and there is no change in the resistivity-temperature slope for a series of ceramic samples with $x =< 0.07$. To obtain such a result, Cooper [7] normalised the density of all the samples against some ideal density; in this case the maximum X-ray density has been used. A series of resistivity curves, after density normalisation, are reproduced in fig.2a. It can be seen that Matthiessen's Rule is found to hold and that the linear temperature dependent carrier scattering mechanism is not affected by the presence of Zn as extra scattering centres. The residual resistivity, defined as the intercept at $T = 0$ axis, however, is found to be proportional to the Zn concentration. The rapid fall in T_c is believed to be due to the presence of Zn in the CuO_2 planes frustrating the antiferromagnetic interactions.

The substitution by Co has produced qualitatively different results. The normal state resistivity as a function of temperature is plotted in fig.2b. Matthiessen's Rule is believed to still apply but now major differences appear as compared with the Zn substituted data. (1) evidence of fluctuations increases with Co concentration, (2) despite an increase in the scattering due to the impurity, the superconducting transition temperature first increases slightly (up to $x \sim 0.02$) before dropping at higher Co concentrations. The increased fluctuation is believed to be due to the destruction of the Cu-O chain order, confining superconductivity to the quasi-two dimensional CuO_2 block. At small Co concentrations, apart from the loss of coherence across the layers, the pair interaction within the CuO_2 planes probably remains relatively unaffected. This effect is similar to that produced by a small deviation of oxygen stoichiometry from 7 as will be shown later.

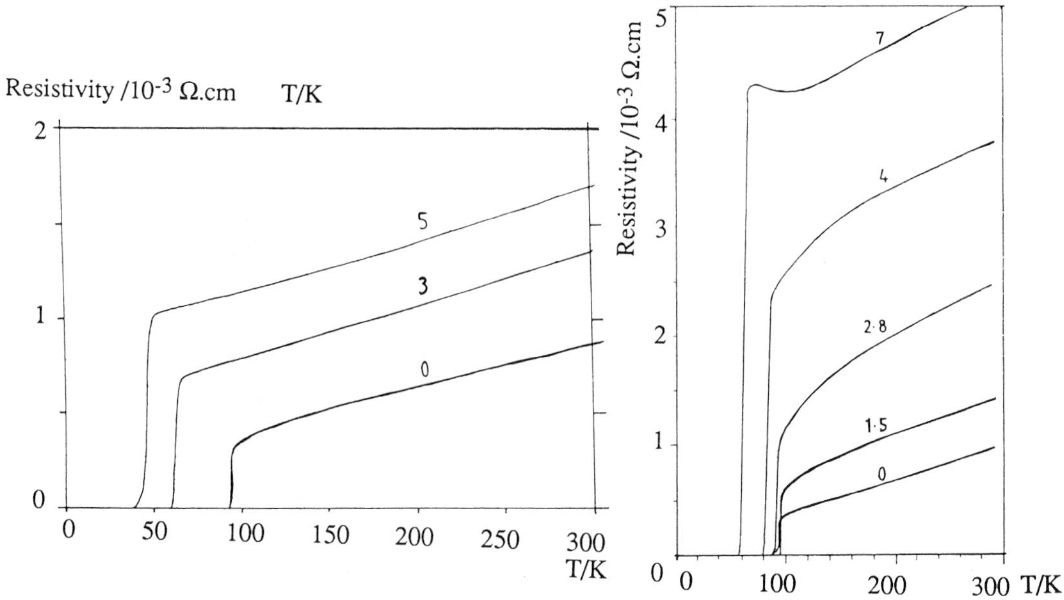

Fig. 2 Normal state resistivity as a function of temperature for a series of (a) Zn and (b) Co substituted $YBa_2(Cu_{1-x}M_x)O_7$ where M = Zn and Co. The numbers next to the curves are 100 x.

AC Susceptibility

The results of ac susceptibility [8] correlate very well with the above data. As shown in fig.3 for the ac susceptibility of magnetically aligned powdered samples, the anisotropy for H // c and H $\perp$ c in susceptibility remains unaffected in the 0.05 Zn (T_c = 42 K) sample compared with that of pure YBCO (T_c = 93 K) but clearly increases in the 0.04 Co (T_c = 82 K) sample. The magnetic penetration length λ deduced from these results confirms that the main effect of Co substitution is to change λ parallel to c but Zn changes λ in both perpendicular and parallel directions. The values of λ at 5 K are given in table 1.

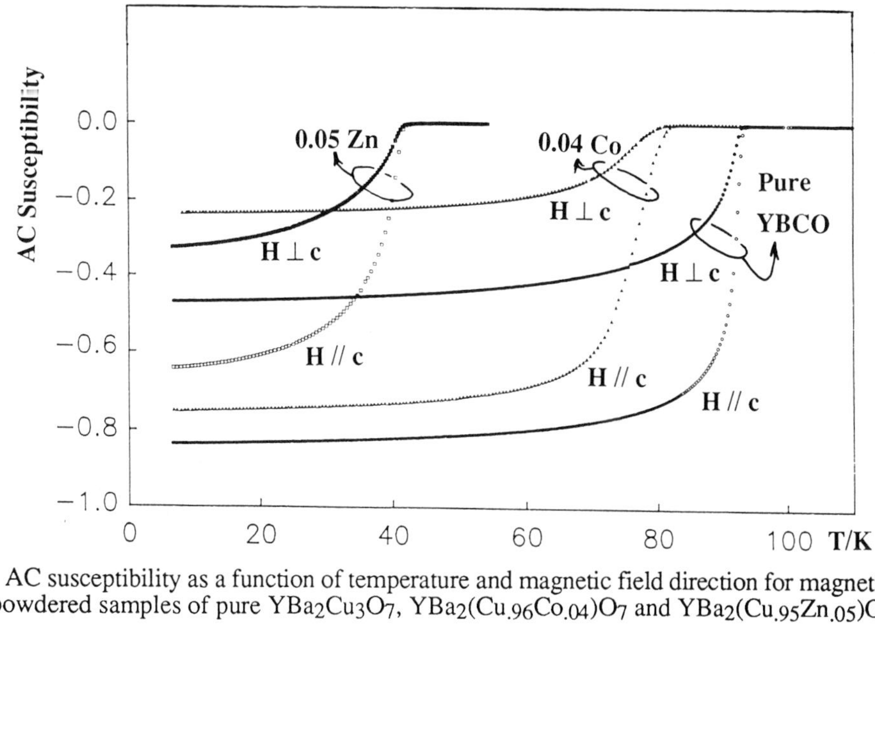

Fig. 3 AC susceptibility as a function of temperature and magnetic field direction for magnetically aligned powdered samples of pure $YBa_2Cu_3O_7$, $YBa_2(Cu_{.96}Co_{.04})O_7$ and $YBa_2(Cu_{.95}Zn_{.05})O_7$.

Table 1 The magnetic penetration lengths λ for pure and substituted YBCO at 5 K deduced from ac susceptibility data of magnetically aligned powdered samples.

	Pure YBCO	0.015 Co	0.04 Co	0.05 Zn
λ_{ab} /μm	0.14	0.19	0.19	0.31
λ_c /μm	0.89	1.0	1.7	1.3

EFFECTS OF DISORDER IN THE Cu-O CHAIN LAYER

When a small fraction of oxygen ($\partial < 0.1$) is removed from an otherwise fully oxygenated $YBa_2Cu_3O_{7-\partial}$, the main loss of oxygen appears to come from the Cu-O chain layer. This results in a disruption and considerable disorder in the Cu-O chains. It follows that the coherent properties across the layers are broken down and instead of a three dimensional superconductor with highly anisotropic properties, the material now behaves like a quasi-two dimensional solid. The same has also been observed with a small fraction of Co substitution for Cu.

Specific Heat Measurements

The electronic specific heat of whole series of Co and Zn substituted samples have been reported by Loram and Mirza [9]. Their measurements are being continued with slightly oxygen deficient samples of $YBa_2Cu_3O_{7-\partial}$ [10], and the results are shown in fig. 4 which confirm that disorder, whether produced by a slight (6.9) oxygen deficiency or small (0.015) Co substitution, removes the complex structure observed in the specific heat anomaly of fully oxygenated samples. The order parameter associated with the sharp feature at the low temperature side of the transition is therefore related with the one dimensional chain structure of $YBa_2Cu_3O_{7-\partial}$. More detailed discussions concerning the split transition can be found in reference [10].

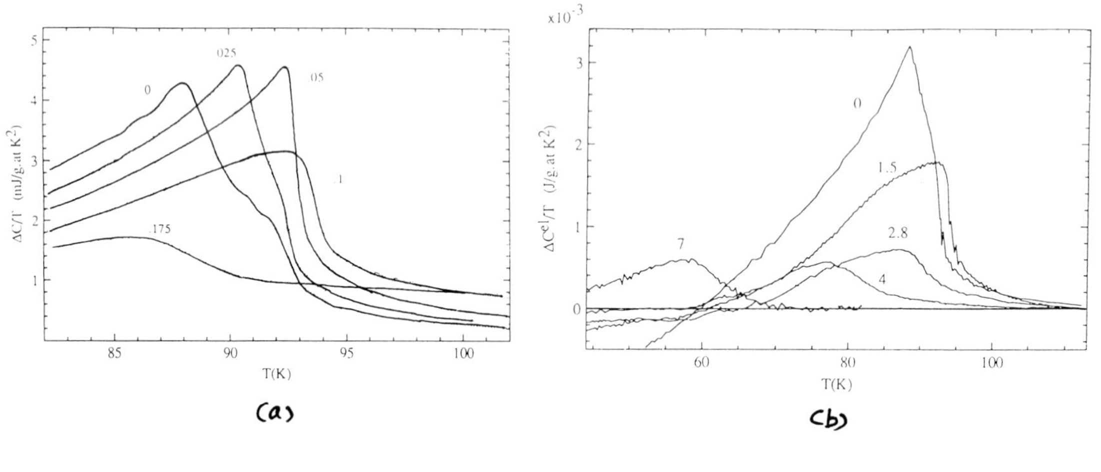

Fig. 4 Specific heat anomalies near the superconducting transition for (a) slightly oxygen deficient samples of $YBa_2Cu_3O_{7-\partial}$, and (b) low Co concentration samples of $YBa_2(Cu_{1-x}Co_x)O_7$; values of ∂ and x are given next to their respective curves.

Two Dimensional Fluctuations

One of the consequences of the disorder in the chain layer is to localise the carriers in this layer. This gives rise to a quasi-two dimensional behaviour which manifests itself in both specific heat as well as resistivity measurements. As the anomaly in specific heat ΔC is given by the expression

$$\Delta C \sim k_B / \Omega(1-T/T_c^*)^n \quad (1)$$

where Ω is the coherence volume, T_c^* is a fitted critical temperature whose value should be close to the superconducting transition temperature and n has the value 1 or 1/2 depending whether the system is two or three dimensional respectively. Data in the region above T_c can therefore be tested for two or three dimensional fluctuations. Fig. 5a plots the reciprocals of $\Delta C/T$ versus T for a number of oxygen deficient samples. The results indicate that $n = 1$ and the existence of two dimensional fluctuations with a coherence volume decreasing from ~ 1600 Å^3 for $\partial = 0$ to ~ 400 Å^3 for $\partial = 0.1$ [10].

Resistivity for the same set of ceramic samples also shows similarly strong fluctuations with increasing ∂ and this has been plotted in fig. 5b for the reciprocal of $\Delta\sigma$ versus temperature, where σ is conductivity and $\Delta\sigma$ is a measure of the conductivity differential when the resistivity deviates from the linear temperature behaviour. The linearity of the plot again demonstrates the existence of two dimensional fluctuations.

In the fully oxygenated samples, oxygen atoms in the chain layer are ordered and the carriers in this layer should become superconducting along with those in the CuO_2 planes via the proximity effect. This will depress the transition temperature of the CuO_2 plane somewhat. However, when carriers in the chain layer become localised, the planes alone are superconducting giving rise to a slightly higher temperature of transition. This is believed to be the reason why the O(6.9) sample has the highest T_c.

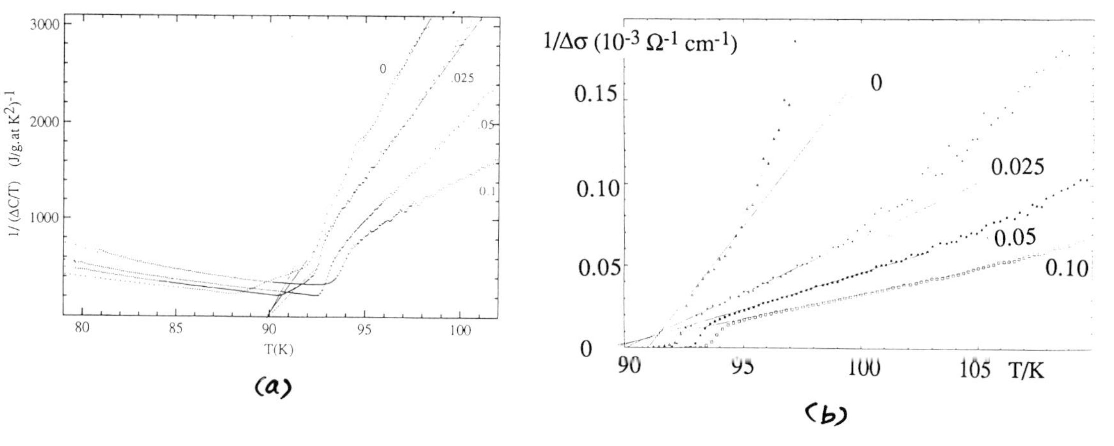

Fig. 5 (a) A plot of the reciprocal of $\Delta C/T$ with temperature showing evidence of two dimensional fluctuations, (b) a plot of reciprocal paraconductivity further confirming two dimensional fluctuations.

Hall Measurements

As pointed out by Ong [2], the carrier density of oxide superconductors deduced from Hall coefficients follows the concentration of the dopant only for dilute systems, and in general the numbers predicted by Hall measurements are well in excess of what might be reasonably expected by a factor of 3 to 6, particularly the values at room temperature and for the fully oxygenated YBCO. This is perhaps not surprising if the DOS curves of fig. 1 hold, where it is suggested that the Fermi level generally cuts across at least two bands, and the measured Hall coefficient will be smaller than for one band only. On the other hand if the carriers in the chain layer are localised due to the disorder discussed above, they will not make a significant contribution to the Hall voltage, and the measured Hall density might be more representative of the carriers in the CuO_2 plane. This appears to be qualitatively borne out by experiments as shown in fig. 6 [11]. The Hall density shown in fig. 6 as a function of oxygen deficiency ∂ appears in two parts. In the region $\partial <$ 0.1, the density is well in excess of 1 carrier per formula unit even when considering the data taken at 103

K. This is because the carriers in the chain make a significant contribution to the measured Hall coefficients, thereby artificially enhancing the measured carrier density. Although the apparant density is still large at room temperature, the numbers for $\partial > 0.1$ are more in line with the expected density of carriers.

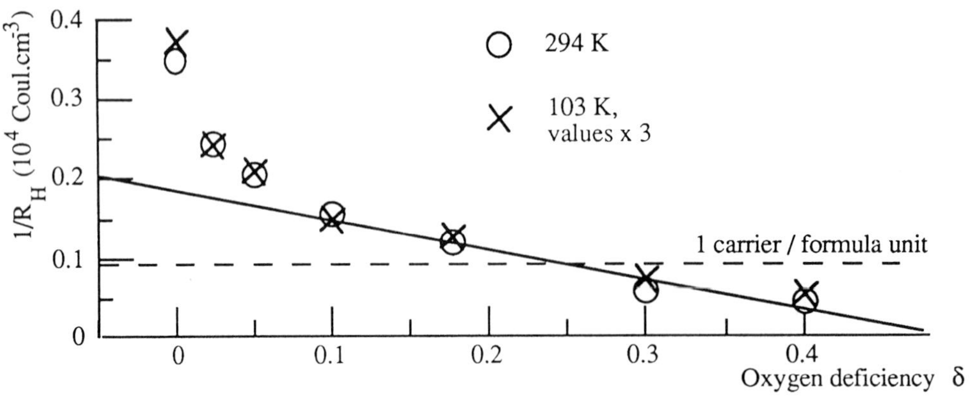

Fig. 6 Carrier density for $YBa_2Cu_3O_{7-\partial}$ from Hall coefficients for $\partial < 0.4$.The value 0.1 x 10^4 Coul.cm^{-3} corresponds to a density of 1.1 carriers per formula unit [11].

CONCLUSION

Ceramic $YBa_2Cu_3O_{7-\partial}$ and $YBa_2(Cu_{1-x}M_x)O_7$ where M = Co or Zn have been investigated close to the range of highest T_c. The occurance of two dimensional fluctuations is linked to the disorder in the Cu-O chain layers. This disorder is caused by the loss of oxygen or by the site selective substitution of Co for the Cu ions in the chain layers.

ACKNOWLEDGMENTS

I like to thank JR Cooper, SD Obertelli, DN Zheng for allowing me to use their data prior to publication. The work is part of an on going project and detailed results will be published later. Together with RS Liu, JW Loram, NF Mott, EKH Salje, they have given generously their time and attention in numerous helpful discus-sions with me.

REFERENCES

1. Cava RJ (1990) Science 247: 656
2. Ong NP (1990) in: DM Ginsberg (ed) in Physical Properties of High Temperature Superconductors II. World Scientific, Singapore, pp 459
3. Liang WY (1989) Physica C 162-164: 197-198
4. Jorgensen JD, BW Veal, AP Paulikas, LJ Nowicki, GW Crabtree, H Claus and WK Kwok (1990) Phys Rev B41: 1863
5. McCarron III EM, MK Crawford and JB Parise (1989) J Sol St Chem 78: 192-196; Liu RS, JR Cooper, JW Loram, W Zhou, W Lo, PP Edwards and WY Liang (1990) Solid St Comm 76: 679-683
6. Schmahl WW, A Putnis, E. Salje, P Freeman, A Graeme-Barber, R Jones, KK Singh, J Blunt, PP Edwards, J Loram and K Mirza (1989) Phil Mag 60: 241L
7. Cooper JR, SD Obertelli, PA Freeman, DN Zheng, JW Loram and WY Liang (1990) in: Proc LT-19 Satellite Conf. on High Temperature Superconductivity, August 1990. To be publ. in:Superconductor Science and Technology, 1991
8. Cooper JR and DN Zheng (1990), to be published
9. Loram JW and KA Mirza (1990) in: Electronic Properties of High T_c Superconductors and Related Compounds. Springer-Verlag, to be published
10. Loram JW, JR Cooper and KA Mirza (1990) in: Proc LT-19 Satellite Conf. on High Temperature Superconductivity, August 1990. To be publ. in: Superconductor Science and Technology, 1991
11. Cooper JR and SD Obertelli (1990), to be published

Anion Substitution Effects in Nd_2CuO_4 and $La_{1.7}Sr_{0.3}CuO_4$

JUN SUGIYAMA, YOSHIHUMI OJIMA, and H. YAMAUCHI

Superconductivity Research Laboratory, International Superconductivity Technology Center, 10-13, Shinonome 1-chome, Koto-ku, Tokyo, 135 Japan

ABSTRACT

We report the effects of substitution of fluorine for oxygen in Nd_2CuO_4 and $La_{1.7}Sr_{0.3}CuO_4$. For the samples of $Nd_2CuO_{4-x}F_x$, x-ray powder diffraction analysis shows that the fluorine ions have successfully doped into the oxygen sites of tetragonal Nd_2CuO_4 lattice and the maximum fluorine solubility limit is $x_{cr.}$=0.3. Although the lattice constant **a** increases linearly with increasing x up to 0.3, the lattice constant **c** decreases linearly. For the samples with x≥0.2 sintered at 1050 and 1100 ^{o}C, superconducting transition around 27 K is observed by both resistivity and magnetic susceptibility measurements. On the other hand, superconducting transition around 30 K is observed even for the sample of $La_{1.7}Sr_{0.3}CuO_4$. Moreover, it is found that the substitution of fluorine for oxygen in $La_{1.7}Sr_{0.3}CuO_4$ gives both a rise of T_c up to about 40 K and a large increase in the intensity of diamagnetic signal. These results indicate that the carrier density has been controlled not only in the n-type superconductor but also in the p-type superconductor by the substitution of fluorine for oxygen.

KEY WORDS : fluorine substitution, carrier density, 214 system, n-type superconductor, p-type superconductor

1. INTRODUCTION

For the high temperature superconducting cuprates, it is widely believed that the magnitude of T_c strongly depends on the charge carrier density of the CuO_2 planes[1-4]. The magnitude of T_c exhibits a maximum in terms of the carrier density. Usually, the carrier density is controlled by the substitution of cations and by oxidation or reduction process. For the p-type superconductors such as those of the La-Sr-Cu-O system, electrons are removed by the substitution of Sr^{2+} for La^{3+} as well as by the oxidation during the annealing process in O_2 gas flow[1]. On the other hand, for the n-type superconductors such as those of the Nd-Ce-Cu-O system, electrons are added by the substitution of Ce^{4+} for Nd^{3+} and by the reduction during the annealing process in N_2 gas flow[5].

The recent discovery of superconductivity in $Nd_2CuO_{4-x}F_x$ [6] indicated that the effects of anion substitution worked parallel to those of cation substitution in terms of the control of the carrier density. Therefore, for p-type superconductors, there must exist the possibility that the carrier density is controlled by anion substitution.

In this paper, we report the synthesis and physical properties of superconducting $Nd_2CuO_{4-x}F_x$ and the effects of the substitution of fluorine for oxygen in an over-doped p-type material, $La_{1.7}Sr_{0.3}CuO_4$.

2. EXPERIMENTAL

Samples of $Nd_2CuO_{4-x}F_x$ with x=0-0.5 were prepared by a solid state reaction from reagent-grade Nd_2O_3, NdF_3 and CuO powders. Stoichiometric mixtures of starting materials were thoroughly mixed by a planetary ball mill using ethanol as solvent. After drying, the mixtures were calcined at 860 ^{o}C for 10 hours in 20 % O_2-Ar gas mixture flow. After re-grinding, the calcined powder was pressed into rectangular bars and sintered at T_S=950-1150 ^{o}C for 14 hours in 1 % O_2-Ar gas mixture flow. Then, the samples were annealed at 910 ^{o}C for 14 hours in nitrogen gas flow and furnace-cooled to room temperature with a rate of about 4 ^{o}C/min.

Samples of $La_{1.7}Sr_{0.3}CuO_{4-x}F_x$ with x=0-1.0 were prepared in a similar way to the case of the samples of $Nd_2CuO_{4-x}F_x$. That is, stoichiometric mixtures of La_2O_3, LaF_3, $SrCO_3$ and CuO powders were calcined at 900 ^{o}C for 5 hours in 20 % O_2-Ar gas mixture flow. After re-grinding, the calcined powder was pressed into rectangular bars and sintered at 1050 ^{o}C for 50 hours in oxygen gas flow. Then, the samples were furnace-cooled to room temperature with a rate of about 1 ^{o}C/min.

The compositions of the samples were determined by electron-probe microanalysis (EPMA). The crystal structure of the samples was characterized by x-ray powder diffraction (XRD). The micro-structure of the sample was observed by scanning electron-microscope (SEM). Electrical resistivity of the samples was measured by a dc four-probe method. Magnetic susceptibility measurements were made using a superconducting-quantum-interference-device magnetometer (SQUID).

3. RESULTS AND DISCUSSION

3-1. $Nd_2CuO_{4-x}F_x$

The XRD analysis indicated that the samples of $Nd_2CuO_{4-x}F_x$ with x≤0.2 were of nearly single phase of the Nd_2CuO_4-type tetragonal structure. On the other hand, the samples with x≥0.3 were mixtures of $Nd_2CuO_{4-x}F_x$, NdOF and CuO phases. These results were in good agreement with EPMA observations. Figure 1 shows the lattice constants, **a** and **c**, with respect to the fluorine content x. These samples were sintered at T_S=1100 ^{o}C and then annealed at 910 ^{o}C. The value of **a** linearly increases as x increases up to 0.2 and then starts to level off to a constant value when x exceeds 0.3. On the contrary, the value of **c** decreases linearly with increasing x and then levels off to a constant value. Since the ionic radius of a F^- ion (r=0.133 nm)[7] is smaller than that of an O^{2-} ion (r=0.140 nm)[7], the increase in **a** due to the fluorine substitution may suggest that the fluorine ions does not substitute into the oxygen sites in the CuO_2 planes but into the oxygen sites around the Nd^{3+} ions.

Figures 2(a) and 2(b) show the SEM pictures of the $Nd_2CuO_{3.8}F_{0.2}$ samples sintered at T_S=950 and 1100 ^{o}C, respectively. The sample sintered at 950 ^{o}C seems to be a collection of small grains and exhibits high degree of porosity at the grain boundaries. The average dimension of each grain is about 1 μm. On the other hand, the sample sintered at 1100 ^{o}C consists of large grains of the average size of about 5 μm and intragrain pores.

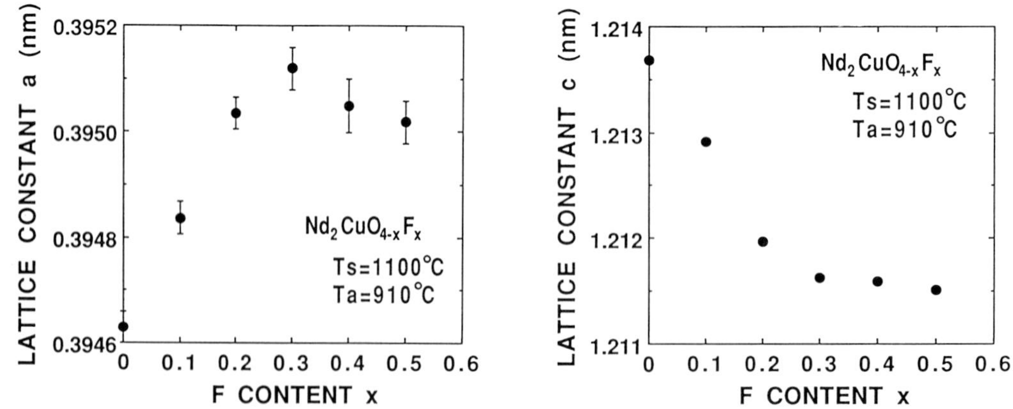

Fig.1 The dependences of the lattice constants, **a** and **c**, on the fluorine content, x, for $Nd_2CuO_{4-x}F_x$ samples. The samples were sintered at 1100 ^{o}C and then annealed at 910 ^{o}C.

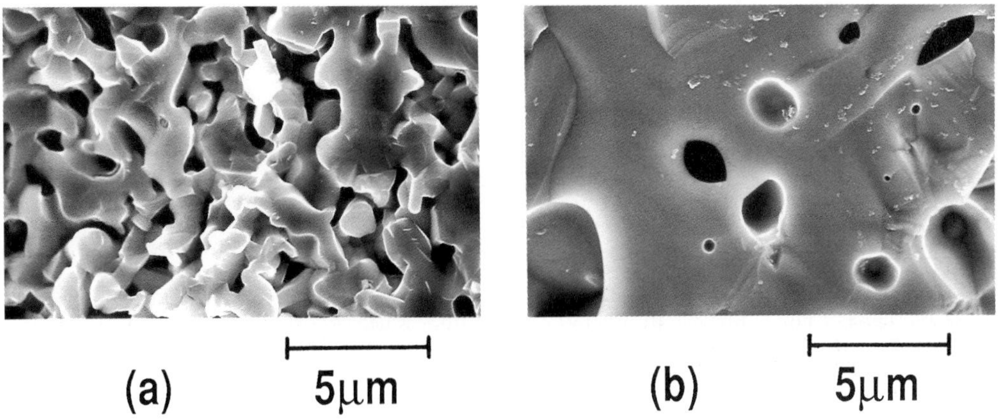

Fig.2 SEM pictures for the $Nd_2CuO_{3.8}F_{0.2}$ samples after the annealing at 910°C, (a) sintered at 950 °C, (b) sintered at 1100 °C.

The temperature dependence of magnetic susceptibility, χ, is shown in Fig. 3 for the $Nd_2CuO_{3.8}F_{0.2}$ samples sintered at T_s=950-1150 °C. The data were measured in the field-cool mode with a magnetic field of H=10 Oe. For each sample, χ decreases suddenly at 26 K due to the superconducting transition. In addition, the intensity of the diamagnetic signal seems to be nearly independent of T_s, except for the sample sintered at T_s=1150 °C. Thus, for each of the samples sintered at T_s=950-1100 °C, bulk superconductivity was observed below 26 K. Since XRD analysis showed no evidences of decomposition of the samples in the temperature region of T_s=950-1150 °C, the sharp decrease in the intensity of diamagnetic signal from the sample sintered at T_s=1150 °C may indicate a possibility of the vaporization of fluorine around 1150 °C.

Figure 4 shows the temperature dependence of resistivity of the $Nd_2CuO_{3.8}F_{0.2}$ samples sintered at T_s=950-1150 °C. The magnitude of resistivity of the samples in the normal state decreases with increasing T_s, except for the sample sintered at T_s=1150 °C. Although the values of the diamagnetic signal in the superconducting state are approximately constant for the samples sintered at T_s=950-1100 °C, sharp drops in resistivity are observed at ~27 K only for the samples sintered at T_s=1050 and 1100 °C. This means that the electrical conduction pass does not connect in the samples sintered at T_s=950 and 1000 °C, because the samples consist of small grains and high density of pores, as shown in Fig. 2(a).

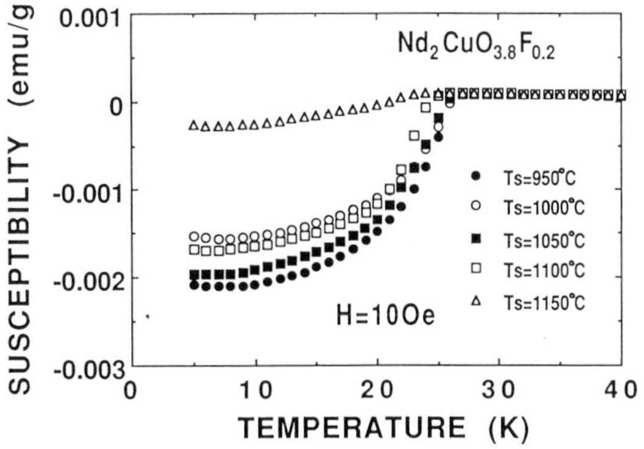

Fig.3 The temperature dependence of magnetic susceptibility of the $Nd_2CuO_{3.8}F_{0.2}$. The samples were sintered at 950-1150 °C and then annealed at 910 °C.

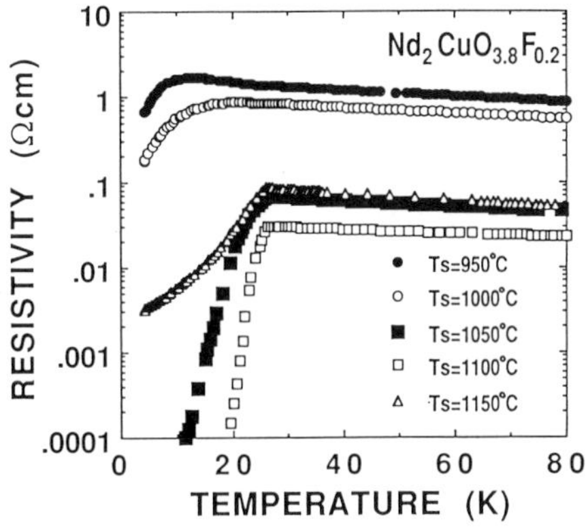

Fig.4 The temperature dependence of resistivity of the $Nd_2CuO_{3.8}F_{0.2}$ samples. The samples were sintered at 950-1150 °C and then annealed at 910 °C.

Furthermore, for the samples with x=0.3-0.5, similar results to the case of the sample with x=0.2 were obtained in resistivity and magnetic susceptibility measurements.

3-2. $La_{1.7}Sr_{0.3}CuO_{4-x}F_x$

The temperature dependence of resistivity is shown in Fig. 5 for the samples of $La_{1.7}Sr_{0.3}CuO_{4-x}F_x$ with x=0-1.0. The resistivity of each of the sample with x=0 and 0.1 shows steps around 30 and 15 K, respectively. This means that even $La_{1.7}Sr_{0.3}CuO_4$ without fluorine dope exhibits the superconductivity. On the contrary, the resistivity of each of the sample with x≥0.2 shows a sharp drop around 40 K. Moreover, the resistivity in the normal state of the samples increases with increasing x. These results indicate that the hole density in the sample decreases due to the substitution of fluorine for oxygen.

Figure 6 shows the temperature dependence of magnetic susceptibility, χ, for the samples of $La_{1.7}Sr_{0.3}CuO_{4-x}F_x$ with x=0-1.0. The data were measured in the field-cool mode with a magnetic field of H=10 Oe. For the sample of $La_{1.7}Sr_{0.3}CuO_4$, χ decreases slightly around 10 K as temperature lowers. For each of the sample with x≥0.1, a few factors larger decrease in the intensity of the diamagnetic signal is observed. Moreover, for each of the sample with x≥0.2, χ decreases around 35 K with decreasing temperature. Thus, the fluorine substitution for oxygen in $La_{1.7}Sr_{0.3}CuO_4$ brought about both the rise of the magnitude of T_c and the increase in the volume of the superconducting phase in the sample.

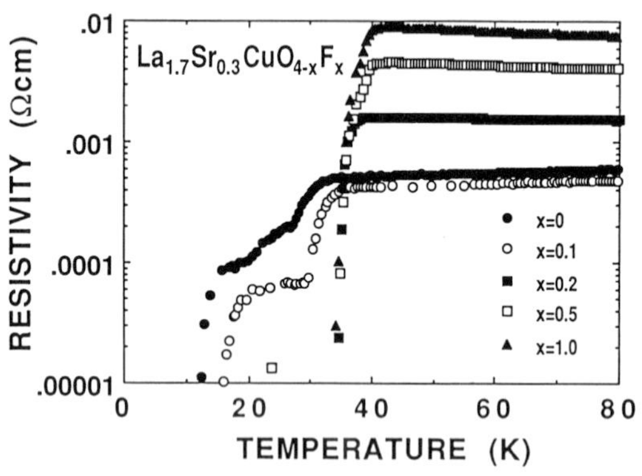

Fig.5 The temperature dependence of resistivity of the $La_{1.7}Sr_{0.3}Cu_{4-x}F_x$ samples with x=0-1.0

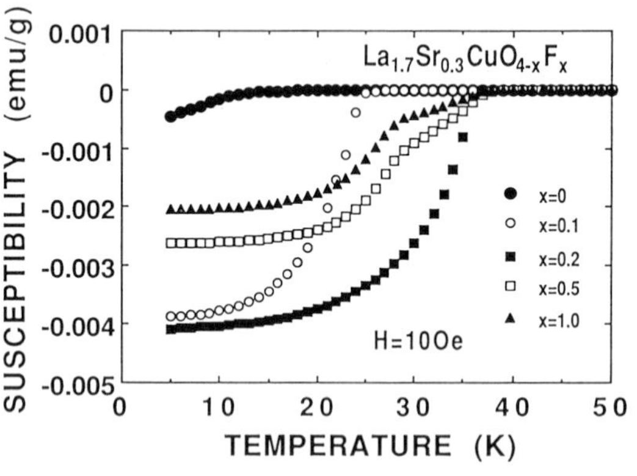

Fig.6 The temperature dependence of magnetic susceptibility of the $La_{1.7}Sr_{0.3}CuO_{4-x}F_x$ samples with x=0-1.0.

ACKNOWLEDGMENT

We would like to thank T.Takata and K.Sakuyama their kind assistance in EPMA measurements and SEM observation. We would also like to thank Dr.H.Hidaka for his helpful discussions.

REFERENCES

[1] Takagi H, Ido T, Ishibashi S, Uota M, Uchida S (1989) Phys.Rev. **B40**: 2254
[2] Takagi H, Uchida S, Tokura Y (1989) Phys.Rev.Lett. **62**: 1197
[3] Ong N.P (1990) In: Ginsberg D.M(eds) in Physical Properties of High Temperature Superconductors II. World Scientific, Singapore, pp459-507
[4] Whangbo M.H, Torardi C.C (1990) Science **249**: 1143
[5] Tokura Y, Takagi H, Uchida S (1989) Nature **337**: 345
[6] James A.C.W.P, Zahurak S.M, Murphy D.W (1989) Nature **338**: 240
[7] Shannon R.D (1976) Acta Cryst.**A32**: 751

Possibility of Substitution of Sn for Cu in High-T_c Superconductors — Solubility of Sn in CuO

TATSUYA MURAKI, KEN-ICHI SHIRAMINE, MASAFUMI TANIWAKI, and KOICHI MUKASA

Department of Electronic Engineering, Faculty of Engineering, Hokkaido University, Sapporo, 060 Japan

ABSTRACT

Possibility of substitution of Sn for Cu oxide is investigated by X-ray diffraction and TEM. By firing of mixtures of CuO and SnO_2 below 1080℃, it was clarified that few Sn atoms were substituted for Cu in CuO by conventional method (firing at 900 ℃). Above 1080 ℃, Sn-solubility in CuO suddenly increased and Sn atoms were substituted for Cu of 0.45 % in Cu oxide at 1090℃.

KEY WORDS: Cu-Sn oxide, Cu-O, Sn-solubility, high-Tc oxide

INTRODUCTION

Mössbauer spectroscopy (MS) has provided many available informations and been established as a superior method for investigating structural, magnetic and lattice dynamics properties related to the novel superconducting state. All of the atomic components of the high-Tc oxides are not applicable to MS. Therefore the sample for MS must contain Eu, Gd, Er, Dy etc at rare-earth element sites[1-3], or ^{57}Fe, ^{57}Co and ^{119}Sn etc at copper sites[4-13].
We are going to the subject to investigate the lattice vibration and electronic state at Cu sites by ^{119}Sn Mössbauer effect. Many ^{57}Fe MS studies of high-Tc oxides have been already reported[4-7] and it has been established that Fe impurities are almost subsutituted for Cu in Cu-O chain. On the other hand, the results of ^{119}Sn MS studies are confusing. For example, reported ^{119}Sn spectra show different shapes. For Sn doped 1-2-3 system, the spectrum consisted of a strong singlet and a weak doublet [9,11], and the spectrum consisted of only one doublet[12] were reported. A research group reported that the spectrum was a singlet at low Sn concentration but a doublet-like at high concentration, and that the shape of the spectrum changed as a function of time[13]. From such a confusion, one conceives that Sn atoms were not substituted for Cu sites in these samples. In fact, there are few studies on the substitution of Sn for Cu in high-Tc oxides so that it is not established. P.Pasternak and R.D.Taylor considered that Sn was not substituted for Cu by usual method (sintering at 900℃), from the change of the Debye temperature of ^{119}Sn in Cu-Sn oxide with annealing at various temperatures[10,14]. ^{119}Sn MS is convenient to investigate superconducting properties, because Sn is not magnetic atom. In addition, it may be suitable for the study of Cu-O plane, because Sn is considered to occupy Cu sites in Cu-O plane[9-14]. For ^{119}Sn Mössbauer experiment, it is necessary to prepare the high-Tc oxide samples in which ^{119}Sn atoms are perfectly substituted for Cu sites.
This study is to examine the structural changes of Cu-Sn oxide with the various heat treatments and to obtain the condition of solution of Sn into the CuO matrix in order to present fundamental data for synthesis of the high-Tc oxides containing tin.

EXPERIMENTAL

The samples were prepared by the solid state reaction from the mixtures of CuO and SnO_2 with nominal compositions of 1, 2, 3, and 5 mole % SnO_2. The thoroughly ground mixture of Cu- and Sn-oxides were fired for 12 ~ 24 hours at 920-1100℃, and quenched to room temperature. The structure of obtained oxides was estimated by powder X ray diffraction(Mo K α line) and TEM. We estimated the solubility of Sn in CuO from the ratio of the 110 reflection line intensity of SnO_2 to 110 line intensity of CuO. If Sn atoms are perfectly soluted in CuO, the reflection line of SnO_2 cannot be detected. The samples in which SnO or Cu_2O appeared were re-annealed at 900℃ in a O_2 flow for 8 hours for re-oxidization, and then cooled slowly to room temperature. After this treatment X ray diffraction pattern was measured.

RESULS AND DISCUSSION

The X ray diffraction pattern showed two phases, monoclinic CuO-phase(a=4.653Å, b=3.410Å,

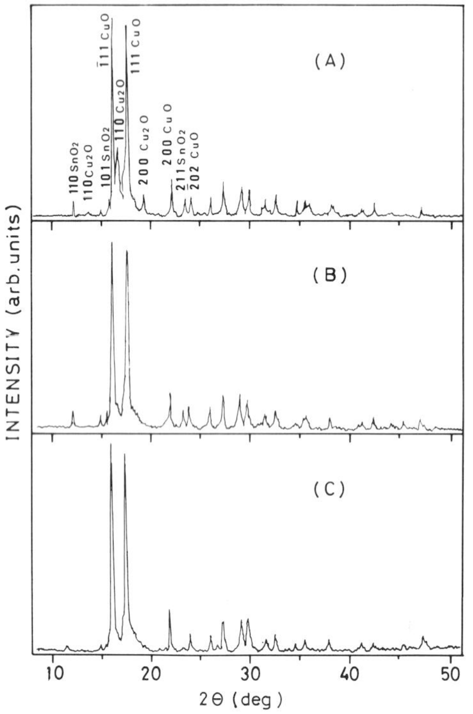

Fig.1 Comparison of the powder X ray diffraction patterns of Sn 2at.% with firing condition
(A) fired at 1070℃ in air for 24 hours
(B) following annealed in a O_2 flow at 900℃ for 8 hours
(C) then annealed in a O_2 flow at 900℃ for 8 hours after fired at 1100℃ for 24 hours.

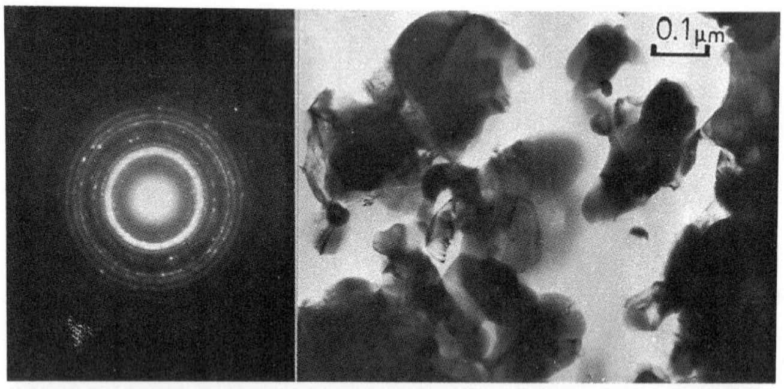

Fig.2 Selected area electron diffraction pattern and bright field micrograph of the CuO-5% SnO_2 powder sample fired at 1000℃ for 24 hours.

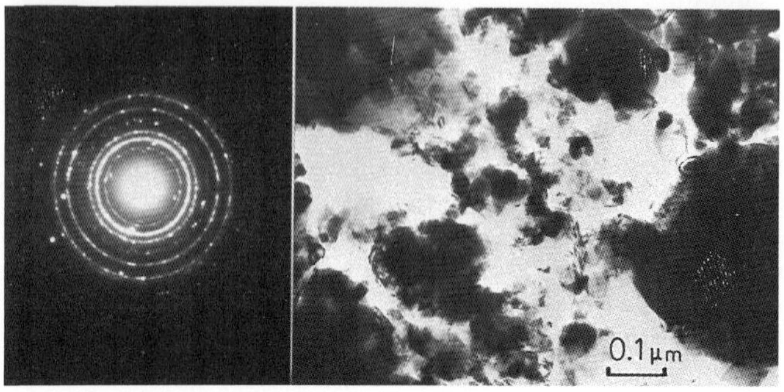

Fig.3 Selected area electron diffraction pattern and bright field micrograph of the CuO-5% SnO_2 powder sample fired at 1090℃ for 24 hours.

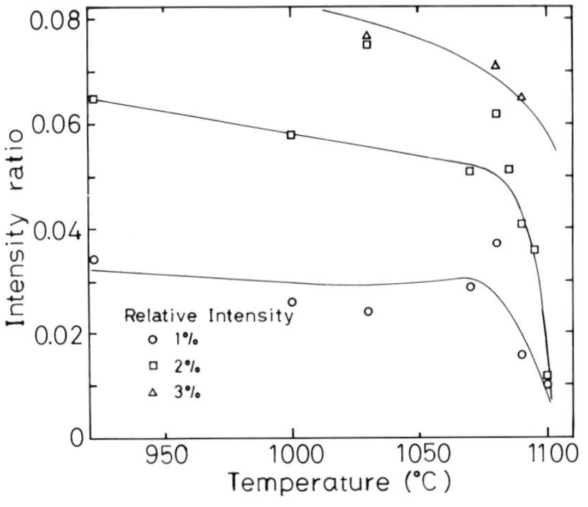

Fig.4 Temperture dependence of the ratio of 110 $_{SnO2}$ line intensity to 110$_{CuO}$.

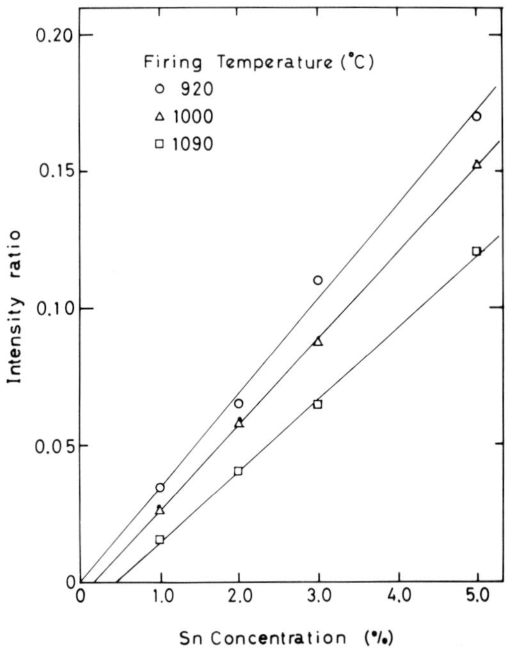

Fig.5 The line intensity ratio of 110 $_{SnO2}$ to 110$_{CuO}$ is shown against Sn concentration. The Sn solubilityateach temperatures is obtained by extraporation of line intensity ratio to 0.

c=5.108Å, β=99° 27′) and tetragonal SnO_2-phase(a=4.737Å, c=3.186Å) in all samples annealed below 1030 ℃. In the samples annealed above 1030 ℃, cubic Cu_2O-phase(a=4.270Å) and tetragonal SnO-phase(a=3.802Å, c=4.836Å) appeared. The X ray diffraction patterns of the sample of CuO-1 % SnO_2 are shown in Fig.1. By the high temperature firing in the air, Cu_2O phase (110, 200 peaks) and SnO_2 phase (110, 101, 211 peaks) appeared as shown in Fig.1-(A). By annealing at 900 ℃ in a O_2 flow, Cu_2O-phase disappeared (in Fig.1-(B)). Increasing the firing temperature to 1100 ℃, the sample melted. In the sample annealed at 900 ℃ in a O_2 flow after melting, SnO_2 phase was scarcely observed (C). The diffraction pattern showed single CuO phase. Figures 2 and 3 show the selected area electron diffraction patterns and blight field micrographs of the powder sample of CuO-5 % SnO_2 annealed at 1000 ℃ and those annealed at 1090 ℃. In the sample annealed at 1000℃, CuO-phase was dominant. SnO_2 particles were observed in this sample although they are not shown in this micrograph. In the sample annealed at 1090 ℃, Cu_2O-phase was dominant and CuO-phase was scarcely observed. SnO_2 particles were observed in this sample, also.

In all samples containing 1-5 % SnO_2 and fired at 900 ℃-1100 ℃, SnO_2 or SnO phase was observed in both of X ray diffraction and TEM data. This shows that heat treatment at 900 ℃-1000 ℃ can not dissolve 1 % Sn in Cu-oxide. This Sn concentration is the typical concentration in ^{119}Sn MS samples which have ever been performed by many researchers. Here, we will examine the two factors of heat treatment, 1) firing temperature and 2) impurity concentration. The relative line intensities of SnO_2 for CuO obtained from the X ray powder diffraction pattern were plotted against firing temperature, in Fig.4 using Sn concentration as a parameter. Here, the line intensity ratio of SnO_2 110 line to the CuO 111 is adopted. For the oxide containing a low concentration of Sn, the line intensity ratio is nearly constant up to about 1080 ℃, which shows that the replacement of Sn for Cu didn't occur so much. At T>1080 ℃, the ratio decreased suddenly. In the sample containing 2 % SnO_2, Sn seems to be substituted for Cu substantially near 1100 ℃. Nearly half of tin atoms are soluted in 1 % SnO_2 sample at 1100 ℃. On 5 % SnO_2 sample, little change was observed. In X ray diffraction pattern of this sample, SnO_2 and SnO phases were revealed clearly after firing at 1200 ℃ for 24 hours, where the sample has melted. Therefore it is considered that Sn atoms can not be substituted for Cu of 5 % even by firing at 1200℃.

Figure 5 shows the SnO_2 X ray line intensity against SnO_2 concentration for several firing temperature (920, 1000 and 1090℃). From the extrapolation to zero line intensity, the soluble Sn concentration in Cu oxide is obtained under each heat treatment condition. In the sample fired at 920℃, Sn is scarcely substituted for Cu as shown in Fig.5. Tin atoms can be substituted for the Cu of 0.18 and 0.45 at.% in Cu oxide fired at 1000 and 1090℃, respectively.

CONCLUSION

The substitution of Sn for Cu in copper oxide was investigated by X ray diffraction and TEM. The firing at 900℃ scarcely substitutes Sn for Cu. The firing at 1100℃ substitute about 0.45 % Sn for Cu. The conventional sintering method(firing at 900℃) is unsuitable for the preparation of ^{119}Sn Mössbauer samples.

REFERENCES

1. Coey JMD, Donnell K (1987) Z. Phys. B-Condensed Matter **67**:513
2. Bornemann HJ, Czjzek G, Ewert D, Meyer C, Renken B (1987) J. Phys. F:Met. Phys. **17**:L337
3. Hodges JA, Imbert P, Marimon da chunha JB, Sanchez JP (1989) Phsica C **160**:49
4. Nasu S, Kitagawa H, Oda Y, Kohara T, Shinjo T, Asayama K, Fujita FE (1987) Physica B **148**:484
5. Blue C, Elgaid K, Zitkovsky I, Boolchand P, McDanield, Joiner WCH, Oos988) tens J, Huff W (1988) Phys. Rev. **B37**:5905
6. Simopoulos A, Niarchos D (1988) Phys. Rev. **B38**:8931
7. Baggio Saitovitch E, Scorzelli RB, Souza Azevedo I, Micklitz H (1990) Phys. Rev. **B41**:2103.
8. Nath A, Homonnay Z (1989) Physica C **161**:205
9. Boolchand P, Enzweiler RN, Zitkovsky I, Wells J, Bresser W, MacDniel D, Meng RL, Hor PH, Chu CW, Huang CY (1988) Phys. Rev. **B37**:3766
10. Pasternak M, Taylor RD, (1989) Solid State Commin. **69**:1157
11. Nishida T, Katada M, Matsumoto Y (1990) Jpn. J. Appl. Phys. **29**:259
12. Yuen T, Lin CL, Myer GN, Salomon RE, Schlottmann P, Bykovetz N, Herman WN (1988) Phys. Rev. **B37**:3770
13. Kuzmann E, Homonnay Z, Vértes A, Gál M, Torkos K, Csákvári B, Solymos GK, Horváth G, Bánkuti J, Kirschner, Korecz L (1989) Phys. Rev. **B39**:328
14. Pasternak MP, Taylor RD (1990) Solid State Commn. **73**:33

Effect of Oxygen Deficiency on the Superconducting Properties of YBCO

YOKO SUYAMA, MAMORU MATSUMOTO, SACHIKO KAGEYAMA, and ISAO SATO

Japan Fine Ceramics Center, 4-1, Mutsuno 2-chome, Atsuta-ku, Nagoya, 465 Japan

ABSTRACT

Sintered $YBa_2Cu_3O_x$ samples with different oxygen-deficiency were prepared and the chemical composition and impurities of the samples were determined precisely. The effects of the oxygen deficiency on Tc, Jc or lattice constants were examined. The critical temperature decreases continuously with increasing oxygen deficiency. No flat region on the Tc vs. oxygen content curve was observed. The critical current density increases with decreasing oxygen oxygen deficiency. The relationship between lattice constants and oxygen deficiency of YBCO was elucidated.

KEY WORDS : $YBa_2Cu_3O_x$ Superconductor, oxygen deficiency, lattice constants, critical temperature, critical current density

INTRODUCTION

The oxygen content of $YBa_2Cu_3O_x$ superconductors varies from 7 to 6 in molar ratio, and it is well known that the properties of $YBa_2Cu_3O_x$ are strongly dependent on the oxygen deficiency. However, the effects of the oxygen deficiency on the properties have not been elucidated. In this study, the effects of oxygen deficiency on various properties such as Tc, Jc, and lattice constants of YBCO were examined by using sintered $YBa_2Cu_3O_x$ samples with various oxygen content.

EXPERIMENTAL

Seven kinds of $YBa_2Cu_3O_x$ samples with various oxygen contents were prepared as follows: $YBa_2Cu_3O_x$ powder was synthesized by a solid state reaction of a mixture of high purity(4N) Y_2O_3, $BaCO_3$, and CuO powders at 930 ℃ for 12 hours in air. The YBCO powder was ground, pressed into 3mmX2mmX30mm plates, and sintered at 950 C for 24 hours in O_2. The sintered bodies were annealed at various temperatures between 500°～920 °C for 12 hours to adjust the oxygen content, followed by quenching.

The chemical composition, especially the oxygen content, of the prepared samples were analyzed precisely by using oxygen analyzer, ICP-AES, wet chemical analysis techniques, etc. as described before[1].

The crystal structure of the samples was examined by X-ray diffraction of CuKα radiation monochromatized with curved graphite crystal. Intensity data collected at 0.02° step width for 9s over a 2θ range from 18° to 100° were analyzed by the RIETVELD method by using a RIETAN program made by Izumi[2].

The superconducting properties, e.g.,Tc and Jc were measured by a conventional for-point probe method.

RESULTS AND DISCUSSION

Chemical Composition

The chemical composition and impurities of the YBCO samples prepared are shown in table 1. The chemical composition was determined precisely with four significant figures. The molar ratio of oxygen of the samples is 6.864~6.151. The oxygen content decreases in the order of samples A(6.868)>H(6.619)>H(6.537)>B(6.520)>I(6.307)>C(6,220)>G(6.151). A small amount of Fe, Ca, Sr, Si, C, and H_2O was detected as impurities.

Superconductig Properties

The critical temperature(onset) and the critical current density of the samples are shown in Table 2. Figure 1 shows the variation of Tc with oxygen content, in comparison with the results of previous studies [3,4,5]. The critical temperature decreases continuously with increasing oxygen deficiency. The earlier papers mention that Tc takes a constant value at a region of 6.7~6.5 in oxygen content as shown in Fig.1 and there exist a different superconducting phase at this region . However, any flat region on the Tc vs. oxygen content curve was not observed in this study. No evidence of the existence of a different superconducting phase was also obtained form the results of X-ray diffraction and observation with TEM.

Crystal Structure

The crystal structure of the samples was analyzed by the Rietveld profile analysis of X-ray diffraction data. No phase besides $YBa_2Cu_3O_x$ was detected in every sample. The lattice constants and occupations of oxygen atoms are calculated. The observed and calculated profiles showed a satisfactory agreement for every sample. The lattice constants, occupation factor, and R factors are shown in Table 3. The orthorhombic structure of YBCO approaches to the tetragonal structure with increasing oxygen deficiency. From X-ray diffraction patterns, it was difficult to determine the crystal structure, i.e., orthorhombic or tetragonal,when the oxygen content became around 6.2. According to the Rietveld analysis, the crystal structure of YBCO sample C with 6.220 in oxygen content has orthorhombic structure , whereas YBCO sample G with 6.151 in oxygen content has tetragonal structure. It is considered that the crystal structure transforms from orthorhombic to tetragonal at around 6.15~6.20 in oxygen content.

Figure 2 shows the relationship between lattice constants and oxygen content. Several results reported by other researchers [6,7,8,9] are also shown in Fig.2. Although there are many papers presented the change of lattice constants with oxygen content, the results are different among the researchers, especially the difference is large in a low oxygen content region, thus the relationship between the lattice constants and oxygen content has not been established yet. This indicates that a precise analyses of oxygen content and lattice constants are required for the establishment of the relationship. A very good relationship between lattice constants and oxygen content is considered to be obtained in the present work, because of the precise determination of oxygen content and the refinement of lattice constants.

CONCLUSIONS

The effects of oxygen deficiency on various properties such as Tc, Jc, and lattice constants of $YBa_2Cu_3O_x$ were investigated by using sintered YBCO with various oxygen contents determined precisely. The critical temperature and the critical current density increase with decreasing oxygen deficiency. The relationship between lattice constants and oxygen deficiency of YBCO was elucidated.

REFERENCES

1.Suyama Y, Matsumoto M, Hayakawa T,(1990) In:T.Ishiguro,K.KAjimura(eds) Advances in Superconductivity II, Springer-Verlag, Tokyo,pp.103-106
2.Izumi F, (1985) J Cryst Soc Japan 27: 23-31
3.Cava RJ, Batlogg B, Chen CH, Rietman EA, Zahurak SM, Werder D (1987) Phys Rev B 36:5719-5722
4.Kubo.K,Ichihasi T,Manako T,Baba K,Tabuchi J,Igarashi H (1988): Phys Rev

B 37:7858-7860
5.Takayama-Muromachi E, Uchida Y, Fujimori A (1988) Jpn J Appl Phys 27:L233-L22
Yanase A (1988) Jpn J Appl Phys 27:L1228-L1230
7.Nakanishi S,Kogachi M,Sasakura H, Fukuoka N,Minamigawa S, Nakahigashi K,Yanase A ibid 27:L329-L332
8.Mckinnon WR, Post ML, Selwyn LS, Pleizier G (1988) Phys Rev B 38:6543-6551
9.Kosuge K, Ueda H (1988) Funtai Oyobi Funmatsuyakin 35:821-829

Table 1 Chemical Composition and Impurities Of YBCO Samples

Sample	chemical composition, wt%				Impurities, ppm			
	Y	Ba	Cu	O	Fe	Ca	Sr	Si
A	13.26 (1.000)	41.45 (2.024)	28.57 (3.014)	16.38 (6.864)	13	10	63	14
E	13.44 (1.000)	41.73 (2.010)	28.78 (2.996)	15.81 (6.537)	14	10	55	21
B	13.38 (1.000)	41.83 (2.024)	28.78 (3.009)	15.70 (6.520)	13	8	62	10
C	13.50 (1.000)	42.16 (2.022)	29.05 (3.011)	15.11 (6.220)	15	18	62	26
G	13.57 (1.000)	42.12 (2.009)	29.07 (2.997)	15.02 (6.151)	16	9	56	24

( ):molar ratio based on 〔Y〕

Table 2 Tc and Jc of YBCO Samples

Sample	Oxygen content (molar ratio)	Tc* (K)	Jc (A/c㎡)
A	6.864	93.8	243 (77K)
H	6.619	71.5	520 (32K)
E	6.537	59.0	403 (38K)
B	6.520	60.6	21 (39k)
I	6.307	20.9	
C	6.220	<14	
G	6.151	<14	

*Value of onest

Table 3 Results of Rietveld Analysis of $YBa_2Cu_3O_x$ Samples

Sample	Oxygen Content (molar ratio)	lattice constants (Å)			Occupation factor		R factor (%)				
		a_0	b_0	c_0	$(\frac{1}{2}00)$	$(0\frac{1}{2}0)$	Rw	Rp	Re	Ri	Rf
A	6.864	3.820	3.883	11.680	0.029	0.84	7.3	5.5	4.0	2.5	1.7
H	6.619	3.827	3.883	11.712	0.14	0.47	8.6	6.3	3.7	4.1	2.7
E	6.537	3.832	3.879	11.729	0.16	0.38	6.9	5.2	3.8	1.9	1.2
B	6.520	3.834	3.881	11.725	0.034	0.49	7.5	5.6	3.7	3.0	1.9
I	6.307	3.852	3.869	11.772	0.043	0.26	8.1	5.8	3.8	2.7	1.5
C	6.220	3.858	3.865	11.806	0.019	0.20	7.4	5.5	3.5	3.1	1.8
G	6.151	3.865		11.805	0.08	0.058	10.3	7.5	5.0	3.6	2.1

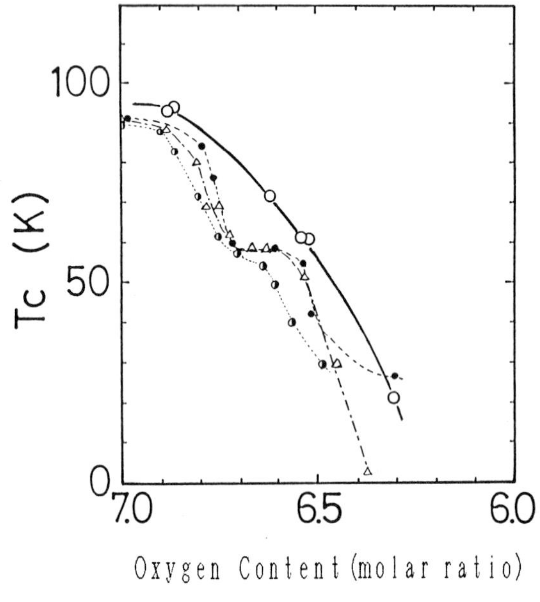

Fig.1 Variation of Tc with oxygen content
solid line : this work
dotted lines : after references [3,4,5]

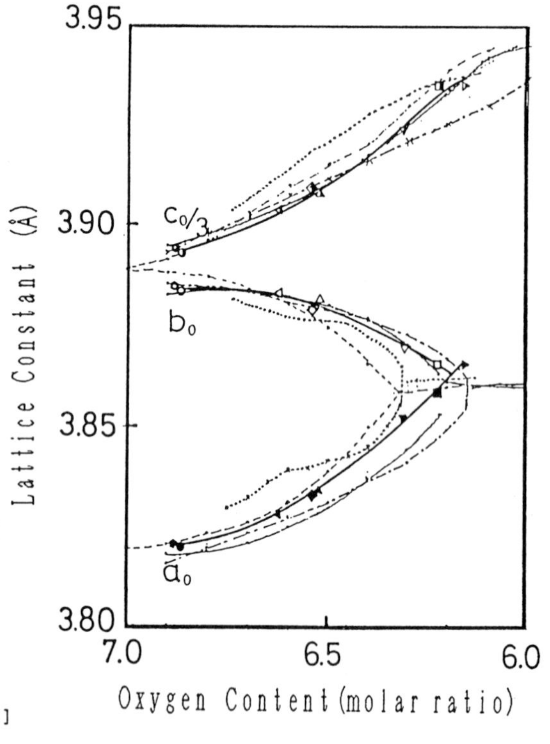

Fig.2 Relationship between lattice constants and oxygen content
solid lines : This work
dotted lines: after references[6,7,8,9]

ACKNOWLEDGMENT

This work was performed under the management of JFCC via ISTEC as a part of the R & D of Basic Technology for Future Industries sponsored by NEDO.

1-2-3 Superconductors with High Hole Density

Y. YAEGASHI, T. WADA, N. SUZUKI, A. ICHINOSE, and H. YAMAUCHI

Superconductivity Research Laboratory, International Superconductivity Technology Center, 10-13, Shinonome 1-chome, Koto-ku, Tokyo, 135 Japan

ABSTRACT

$(Yb_{1-x}Ca_x)(Ba_{0.8}Sr_{0.2})_2Cu_3O_z$ ($0\leq X\leq 0.4$) solid solutions of the 1-2-3 (oxygen deficient triperovskite) structure have been successfully synthesized. These samples have been characterized with respect to oxygen nonstoichiometry, crystal structure and superconducting properties. The fully oxygenated samples with $Z\sim 6.8$ had an orthorhombic unit cell and exhibited bulk superconductivity as shown by magnetic susceptibility measurements. Their superconducting transition temperatures decreased with increasing Ca content, X. The $(Yb_{0.9}Ca_{0.1})(Ba_{0.8}Sr_{0.2})_2Cu_3O_z$ and $(Yb_{0.7}Ca_{0.3})(Ba_{0.8}Sr_{0.2})_2Cu_3O_z$ samples with various oxygen content, Z ($6.36\leq Z\leq 6.89$ for X=0.1 and $6.1\leq Z\leq 6.78$ for X=0.3) were prepared by the reduction of fully oxygenated samples. One of the samples with X=0.3 and Z=6.56 had a tetragonal unit cell and exhibited a sharp superconducting transition at 82 K. The superconducting transition temperatures (T_c) are discussed on the basis of the hole concentration in the CuO_2 sheet.

KEY WORDS: crystal structure, hole concentration, superconducting transition temperature, oxygen nonstoichiometry, Ca doping

INTRODUCTION

The importance of the hole concentration in copper-oxide superconductors is widely recognized. Many studies concerning the relationship between the hole concentration (p) and the superconducting transition temperature (T_c) have been conducted. In the $La_{2-x}Sr_xCuO_4$ system, as Sr content (X), and p are increased, T_c first increases, and then peaks, becoming nonsuperconducting for X ≥ 0.26 [1-3]. $YBa_2Cu_3O_{6+Y}$ (the 1-2-3 system) has two kinds of oxygen deficient perovskite blocks, the Y-CuO_2-BaO block having fixed oxygen content and the BaO-CuO_Y-BaO block having variable oxygen content [4]. The T_c is proportional to the concentration of mobile holes in the CuO_2 plane in the Y-CuO_2-BaO block [5]. However, detailed studies concerning the high hole density region in the 1-2-3 system have not been reported, because $(Y_{1-x}Ca_x)Ba_2Cu_3O_{6+Y}$ samples with high Ca content could not be synthesized.

In this study, a sample with high-hole density (average copper valence was 2.3) was successfully synthesized. We used the $(Yb_{1-x}Ca_x)(Ba_{0.8}Sr_{0.2})_2Cu_3O_z$ ($0\leq X\leq 0.4$) system as a model compound with the 1-2-3 structure. Samples were characterized with respect to oxygen nonstoichiometry, crystal structure and superconducting properties.

EXPERIMENTAL

Samples were prepared by a conventional solid-state reaction method. High purity powders of Yb_2O_3, $CaCO_3$, $BaCO_3$, $SrCO_3$ and CuO were used as starting materials. Appropriate amounts of powders for a nominal composition of

$(Yb_{1-x}Ca_x)(Ba_{0.8}Sr_{0.2})_2Cu_3O_z$ were mixed, calcined and compacted. The bars were sintered at 850 ℃ for 20 h in 99%N_2 + 1%O_2 gas flow and held successively at 400 ℃ for 20 h and at 300 ℃ for 20 h in O_2 gas flow. In order to control oxygen content, the samples with X=0.1 and 0.3 were re-heat treated at 300 ℃ or 500 ℃ for 3 h in air and then quickly cooled in liquid nitrogen. Some samples quenched from 500 ℃ were reduced at about 500 ℃ using an appropriate amount of Ti_2O_3 powder in an evacuated silica tube.

The lattice constants were determined by powder x-ray diffraction using Cu-Kα radiation. The oxygen content was analyzed by coulometric titration technique [6]. Electrical resistivity measurements were carried out by a conventional dc four-probe method, and dc magnetic susceptibility measurements were performed using a SQUID magnetometer.

RESULTS

1. FULLY OXYGENATED $(Yb_{1-x}Ca_x)(Ba_{0.8}Sr_{0.2})_2Cu_3O_z$ SAMPLES WITH Z~ 6.8

Figure 1 shows x-ray powder diffraction patterns of the $(Yb_{1-x}Ca_x)(Ba_{0.8}Sr_{0.2})_2Cu_3O_z$ (0≦ X≦ 0.4) samples. These samples were of single-phase or near single-phase. The crystal structures were all orthorhombic and of the 1-2-3 type. The lattice constants, a and c increased a little with increasing Ca content. Figure 2 shows the oxygen content (Z) of the fully oxygenated $(Yb_{1-x}Ca_x)(Ba_{0.8}Sr_{0.2})_2Cu_3O_z$ (0≦ X≦ 0.4) samples. Z decreased with increasing Ca content from 6.875 for X=0.0 to 6.712 for X=0.4. The maximum average Cu valence was 2.30 for the sample with X=0.2 and Z=6.85.

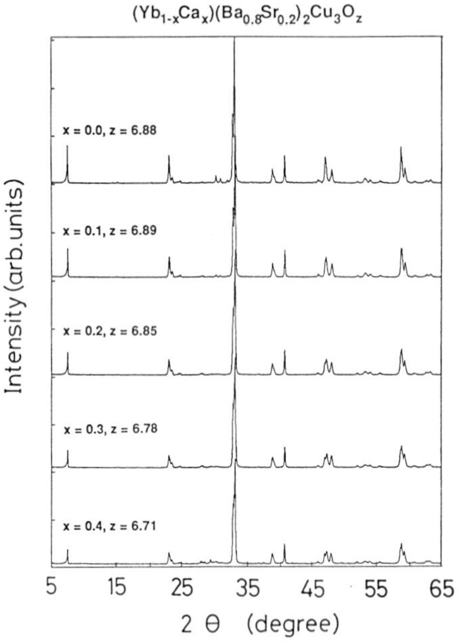

Fig.1 X-ray powder diffraction patterns of fully oxygenated $(Yb_{1-x}Ca_x)(Ba_{0.8}Sr_{0.2})_2Cu_3O_z$ samples with Z~ 6.8.

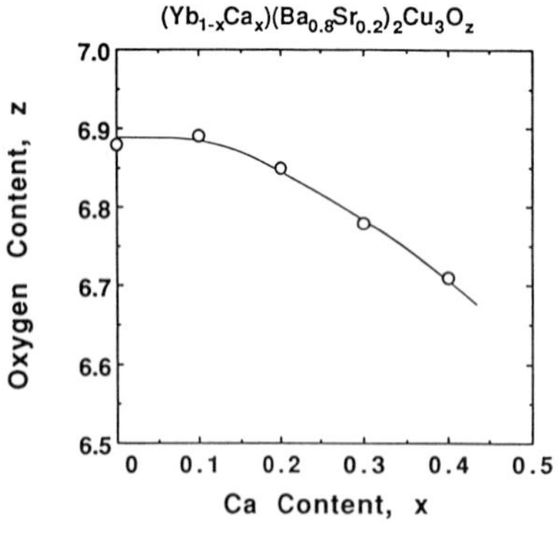

Fig.2 Oxygen contents (Z) of the fully oxygenated $(Yb_{1-x}Ca_x)(Ba_{0.8}Sr_{0.2})_2Cu_3O_z$ samples.

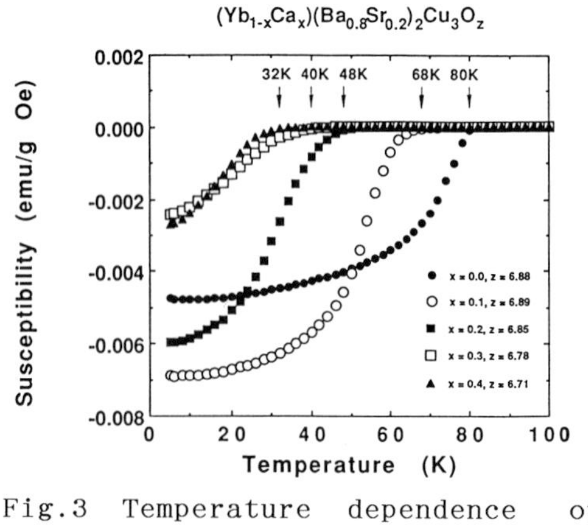

Fig.3 Temperature dependence of dc-magnetic susceptibility for the fully oxygenated $(Yb_{1-x}Ca_x)(Ba_{0.8}Sr_{0.2})_2Cu_3O_z$ samples.

The fully oxygenated $(Yb_{1-x}Ca_x)(Ba_{0.8}Sr_{0.2})_2Cu_3O_z$ samples exhibited sharp superconductivity with T_c value ranging from 85 to 45 K from electrical resistivity measurements. Figure 3 shows the temperature dependence of the dc-magnetic susceptibility. The samples exhibited bulk superconductivity. The magnetically determined T_c's decreased almost linearly with increasing X, from 80 K for X=0.0 to 32 K for X=0.4.

2. REDUCED $(Yb_{0.7}Ca_{0.3})(Ba_{0.8}Sr_{0.2})_2Cu_3O_z$ SAMPLES WITH $6.79 \geq Z \geq 6.10$

The x-ray powder diffraction patterns of $(Yb_{0.7}Ca_{0.3})(Ba_{0.8}Sr_{0.2})_2Cu_3O_z$ samples with various oxygen content $(6.79 \geq Z \geq 6.10)$ are shown in Fig. 4. The samples with $6.79 \geq Z > 6.6$ were orthorhombic and the samples with $6.6 > Z \geq 6.10$ were tetragonal. Figure 5 shows the relationship between the oxygen content, Z, and the lattice constants, a, b and c/3, in this system. The c-axis linearly increased with decreasing oxygen content. The orthorhombic and tetragonal phase transition occurred at about Z=6.6.

The temperature dependence of dc-magnetic susceptibility is shown in Fig.6. The samples with $6.79 \geq Z \geq 6.10$ exhibited bulk superconductivity. These samples also exhibited sharp superconducting transitions in electrical resistivity measurements.

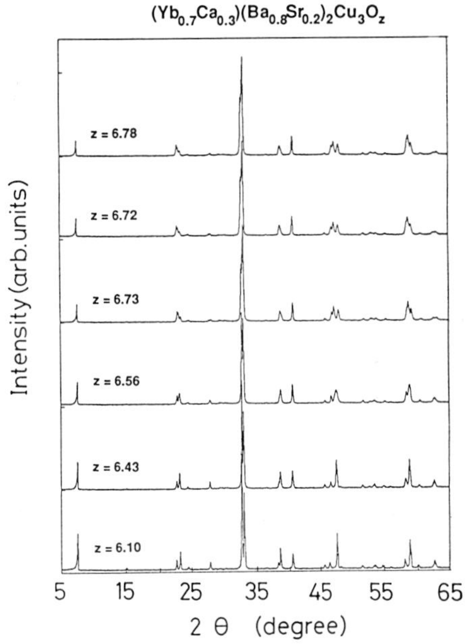

Fig.4 X-ray powder diffraction patterns of $(Yb_{0.7}Ca_{0.3})(Ba_{0.8}Sr_{0.2})_2Cu_3O_z$ samples with various oxygen content.

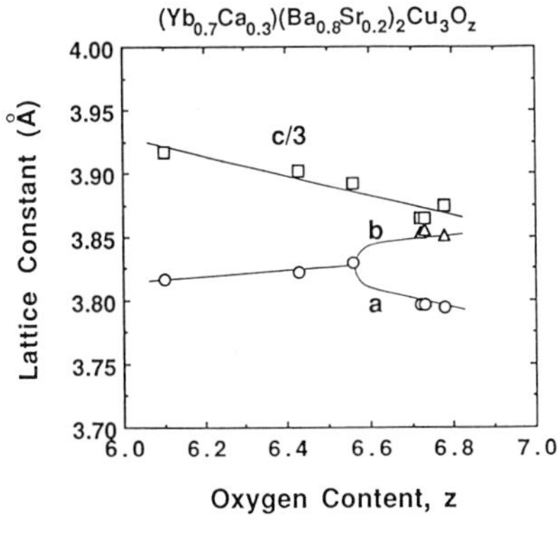

Fig.5 Relationship between the oxygen content, Z, and the lattice constants, a, b and c/3, in $(Yb_{0.7}Ca_{0.3})(Ba_{0.8}Sr_{0.2})_2Cu_3O_z$.

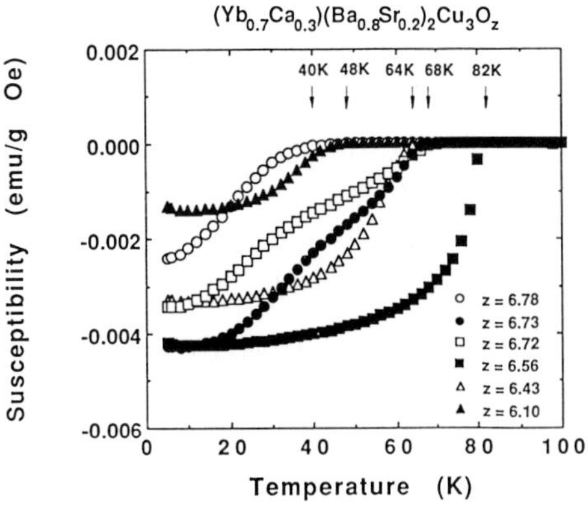

Fig.6 Temperature dependence of dc-magnetic susceptibility for $(Yb_{0.7}Ca_{0.3})(Ba_{0.8}Sr_{0.2})_2Cu_3O_z$.

The relationship between oxygen content and magnetically determined T_c is shown in Fig.7. In the present system, as oxygen content, Z, and hole concentration, p, decreased, the T_c first increased, then peaked and decreased to 50 K for Z=6.1. The maximum T_c was achieved for the tetragonal sample with Z=6.56. Similar results were obtained for the $(Yb_{0.9}Ca_{0.1})(Ba_{0.8}Sr_{0.2})_2Cu_3O_z$ system.

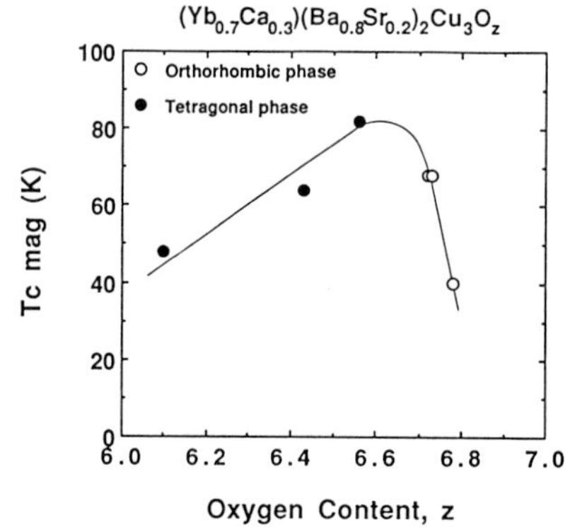

Fig.7 Relationship between oxygen content and magnetically determined T_c for $(Yb_{0.7}Ca_{0.3})(Ba_{0.8}Sr_{0.2})_2Cu_3O_z$ system.

DISCUSSION

The obtained results showed that the T_c of this system depended on calcium content and oxygen content. We know that the T_c of the 1-2-3 system depends on the concentration of mobile holes in the CuO_2 sheet [5]. The mobile hole concentrations in CuO_2 sheet $[p_{sh}]$ were calculated by the method proposed by Tokura et. al. First, the average Cu valence, V(Cu), was calculated using the Ca and oxygen contents. Next, average hole concentration per Cu atom, p, was calculated as follows:

$$p = V(Cu)-2,$$

and the critical hole concentration per Cu atom at chain site, p_c, was obtained from the following expansion:[5]

$$p_c = -2.21+0.34Z.$$

Consequently, p_{sh} was:

$$p_{sh} = 3(p-p_c)/2.$$

The T_c's of the samples with $p_{sh} > 0$ were plotted against p_{sh} (Fig. 8). As the p_{sh} was increased, T_c first increased, then peaked at 0.18 and decreased to 32 K at p_{sh}=0.304 for the sample with X=0.4 and Z=6.714. This curve exhibited similar shape as the one obtained in the $La_{2-x}Sr_xCuO_4$ system.

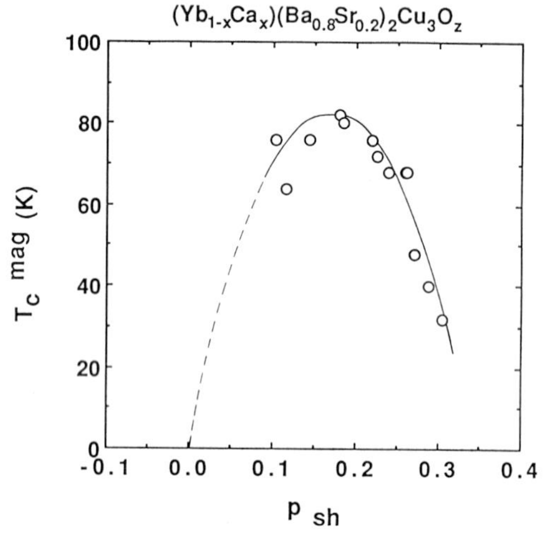

Fig.8 Relationship between T_c and p_{sh} in the $(Yb,Ca)(Ba_{0.8}Sr_{0.2})_2Cu_3O_z$ system.

REFERENCES

1. Shafer MW, Penney T, Olson BL (1987) Phys.Rev.B.36: 4047.
2. Torrance JB, Bezinge A, Nazzal AI, Huang TC, Parkin SSP, Keane DT, LaPlaca SJ, Horn PM, Held GA (1989) Phys.Rev.B.40: 8872-8877.
3. Takagi H, Ido T, Ishibashi S, Uota M, Uchida S, Tokura T (1989) Phys.Rev.B.40: 2254.
4. Izumi F, Asano H, Ishigaki T, Takayama-Muromachi E, Uchida Y, Watanabe N, Nishikawa T (1987) Jpn.J.Appl.Phys.26: L649.
5. Tokura Y, Torrance JB, Huang TC, Nazzal AI (1988) Phys.Rev.B.38: 7156.
6. Kurusu K, Takami H, Shintomi K (1989) ANALYST.114: 1341-1343.

Superconductivity in Ni Substituted $YBa_2Cu_4O_8$

YASUHARU KODAMA[1], YUH YAMADA[2], NORIMITSU MURAYAMA[1], MASANOBU AWANO[1], and TAKAHIKO MATSUMOTO[2]

[1] Government Industrial Research Institute, Nagoya, 1-1, Hirate-cho, Kita-ku, Nagoya, 462 Japan
[2] National Research Institute for Metals, 3-12, Nakameguro 2-chome, Meguro-ku, Tokyo, 153 Japan

ABSTRACT

The effects of the substitution of Ni for Cu in $YBa_2Cu_4O_8$ were investigated. The $YBa_2Cu_4O_8$ structure was successfully synthesized up to 10% substitution of Ni at 1010°C under a 40 MPa oxygen partial pressure. With increasing Ni content the critical temperature decreased continuously and the superconductivity disappeared at 10% substitution.

KEY WORDS: $YBa_2Cu_4O_8$, Ni substitution, high oxygen pressure synthesis, depression of T_c

INTRODUCTION

$YBa_2Cu_4O_8$ (124phase) has the characteristic feature which other high-T_c oxide superconductors do not possess, that is, it is stable up to 850°C without oxygen desorption in air [1]. The large pressure dependence of T_c is another feature of the 124 phase [2,3]. The crystal structure of the 124 phase is similar to that of $YBa_2Cu_3O_{7-y}$ (123phase) except for the structure of Cu-O chain [4,5]. The Cu-O chain in the 124 phase is double one which consists of edge-shared CuO_4 square, whereas that in the 123 phase is single one which consists of corner-shared CuO_4 square. Therefore, it seems that the features of the 124 phase mentioned above originates in the double Cu-O chain.

Though the substitution for Y and Ba in the 124 phase was reported by several authors with various elements [6-12], the substitution for Cu, which is important for superconductivity and crystal structure, was only reported with Fe [13,14]. As the first step of considering the contribution of the double Cu-O chain to the property of the 124 phase, we attempted to substitute Ni, which occupied both Cu(1) (in the CuO chain) and Cu(2) (in the CuO_2 plane) site in the case of the 123 phase [15,16], by high oxygen partial pressure synthesis method. Then we investigated the influence on the crystal structure and the superconductivity and compared the result with that of the substitution of Ni in the 123 phase [17,18].

EXPERIMENT

Samples of $YBa_2(Cu_{1-x}Ni_x)_4O_8$ (x=0.0-0.1) were synthesized under high oxygen pressure. Appropriate amounts of Y_2O_3, $BaCO_3$, CuO, and NiO were mixed thoroughly and calcined three times at 900°C in flowing oxygen for 12 hrs with intermediate grinding. The calcined powders were confirmed to be mixture of the 123 phase and CuO by powder X-ray diffractometry (XRD). Small amounts of NiO peaks were also observed in higher Ni content ($x \geq 0.05$). They were pressed into pellets and reacted at 980°C and 1010°C for 6 hrs in the mixture gas of 80% Ar and 20% O_2 at a pressure of 200MPa, which corresponds to an oxygen partial pressure of 40MPa, by using O_2-HIP apparatus made by Kobe Steel Co. Ltd. The resulting powders were characterized by XRD with Cu Kα radiation. The lattice parameters were determined by a least-squares method using Ge as an internal standard. Resistivities were obtained by the standard four-probe method. Superconducting transitions were determined by both resistivity and magnetization, the latter was obtained with a superconducting quantum interference device (SQUID) magnetometer while cooling in a field of 10 Oe. Normal state susceptibilities were measured with a magneticgravimeter in a field of 7kOe.

RESULTS AND DISCUSSIONS

From XRD measurement the samples reacted at 980°C were the 124 phase, but CuO and NiO were also detected for all x values. Though NiO was detected for $x \geq 0.05$ in the samples reacted at 1010°C, they were the 124 phase with good crystallinity without CuO for all x values as shown in Fig. 1. In the pure (non-substituted) 124 phase, reaction at 950°C was enough to synthesize the CuO free 124 phase. The stability region of the 124 phase on the oxygen partial pressure-temperature diagram became narrower by elementary substitution. (e.g. substitution of Ca [8] and rare-earth elements [19] for Y, substitution of La [12] for Ba) Substitution of Ni may also cause the narrowing of the stability region, but it is less drastic than that of Fe and Co in the 124 phase [14,20]. Figure 2 shows the variation of the lattice parameters a, b, and c. These parameters also list in Table I. The a and b lattice parameters scarcely change. In contrast, the c lattice parameters decrease with increasing Ni concentration. This behavior is consistent with the results of the 123 phase [18].

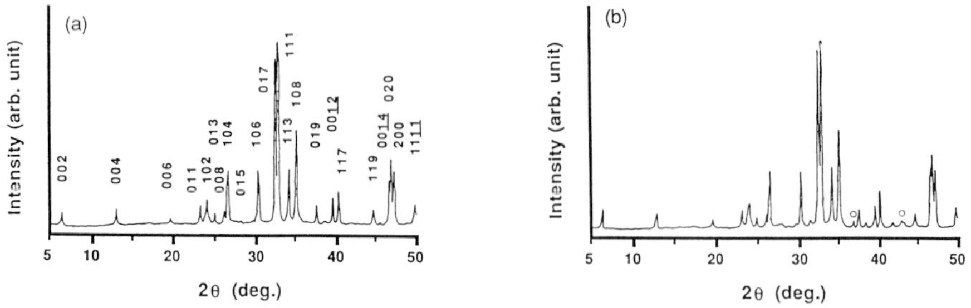

Fig.1 X-ray diffraction patterns for $YBa_2(Cu_{1-x}Ni_x)_4O_8$ synthesized at 1010°C under a 40 MPa oxygen partial pressure. (a) x=0.02, (b) x=0.08. Open circles represent the peaks from NiO.

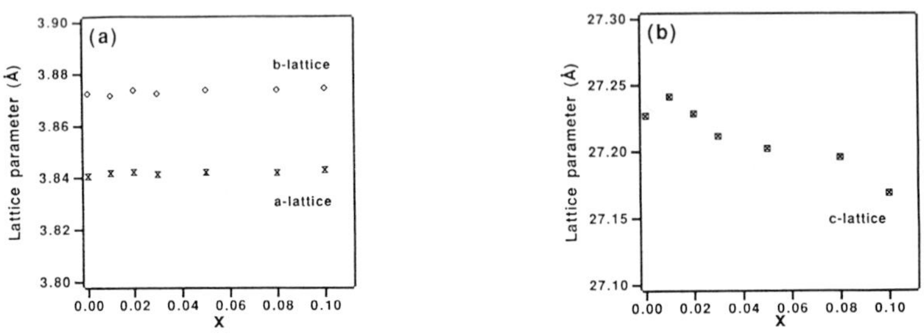

Fig. 2 Lattice parameters, (a) a and b, (b) c, as a function of x.

Table I Various parameters of $YBa_2(Cu_{1-x}Ni_x)_4O_8$. x is starting composition. a, b, and c are the lattice parameters. $T_{c\ onset}$ and $T_{c\ zero}$ are determined from resistivity. ΔT_c is width (10%-90%) of resistivity transition. T_c (M) is determined from a diamagnetic onset. χ_0, θ, and P_{eff} are the parameters obtained from a fit to a Curie-Weiss formula; χ_0 is the temperature independent susceptibility; θ is the Weiss temperature; P_{eff} is the effective Bohr magneton number.

x	a (Å)	b (Å)	c (Å)	$T_{c\ onset}$ (K)	$T_{c\ zero}$ (K)	ΔT_c (K)	T_c (M) (K)	χ_0 (10^{-6} emu/g)	θ (K)	P_{eff} (μ_B)
0.0	3.841	3.872	27.227	76.5	72.5	3.5	80	0.36	-34	0.08
0.01	3.842	3.872	27.241	75.5	71.5	3.5	···	·····	····	·····
0.02	3.842	3.874	27.229	69.5	63	5	66	0.15	6	2.88
0.03	3.841	3.872	27.212	65	51.5	8.5	···	0.04	2	2.32
0.05	3.842	3.874	27.202	44.5	30.5	12	42	0.20	7	1.01
0.08	3.842	3.874	27.196	28	8.5	17	28	0.31	6	1.43
0.10	3.843	3.874	27.169	·····	·····	···	···	·····	····	·····

The superconducting critical temperatures (T_c) determined both resistively and magnetically are listed in Table I. The transition widths, ΔT_c, which are determined from the 10% to 90% points on the resistive transition, are also listed in Table I. The temperature dependence of the relative resistivity is shown in Fig.3. The relative resistivities were normalized by the resistivities at 300K. With increasing the Ni concentration, T_c decrease continuously and normal state resistivity shows semiconducting behavior. The superconducting transition is disappeared at x=0.1. With increasing x the transition width is broader and the transition is tailed in a high measuring current. This means an inhomogeneity and a weak connection between grains in the sample of large x value. The temperature dependence of the magnetization is shown in Fig. 4. The diamagnetic onset and the resistivity drop have good agreement. This means that the depression of T_c is caused by the substitution of Ni for Cu in the 124 phase.

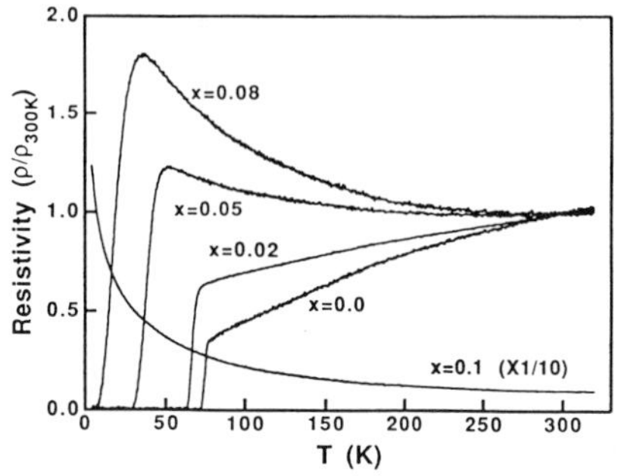

Fig. 3 Temperature dependence of the reduced resistivity for various x. The reduced resistivity was normalized by the resistivity at 300K.

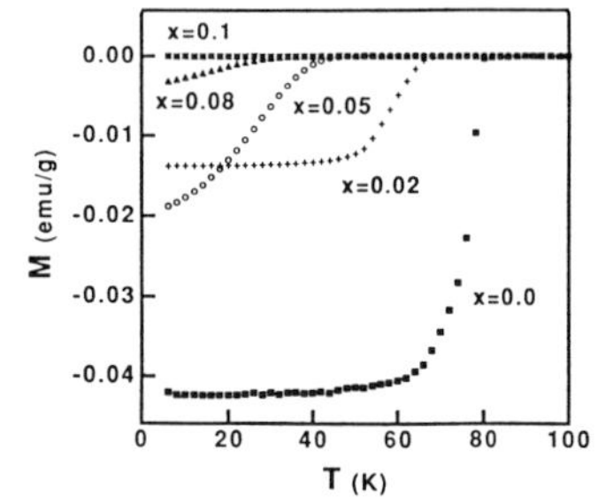

Fig. 4 Temperature dependence of the magnetization in a field of 10 Oe for various x.

The variation of T_c determined by the magnetization as a function of nominal x is shown in Fig. 5. That variation for the 123 phase [17,18] is also shown as a reference. In the case of the 123 phase, several different results of this variation were reported. One type is reported by Wu et al. [18] that the depression of T_c is less drastic than the substitution of Fe and Co, and another type is reported by Tarascon et al. [17] that it is as drastic as the substitution of Fe and Co. Our result of the 124 phase is the same degree as the result of Tarascon et al. But the depression of T_c of the samples substituted by Fe and Co is more drastic than Ni in the 124 phase [13,14,20]. The depression of T_c in the 123 phase become more drastic in order of the substitution of Ni, Fe and Co, and Zn, which occupies in both Cu(1) and the Cu(2) sites, the Cu(1) site, and the Cu(2) site, respectively. Although this reason has not been clarified yet, it is suggested that the substitution of Zn introduce a local structural disorder into the CuO_2 plane and that the substitution of Fe and Co cause a hole redistribution between the chain and the plane [21]. The effect on T_c of Ni substitution is considered as follow: divalent Ni occupied in the Cu(1) site influences the hole distribution less than trivalent Fe and Co, and Ni in the Cu(2) site mainly contributes the depression of T_c. The difference of the depression of T_c by the Ni substitution may originate in the difference of the sample quality, e.g. oxygen deficiency, small amounts of impurity phases etc. In the 124 phase Ni does not occupy in the double chain as easily as in the single chain of the 123 phase, because the temperature at which the 124 structure was successfully synthesized was increased by the Ni substitution. Then we consider that the preferential substitution of Ni in the plane site makes the depression of T_c more drastic. But more precise information about the occupation sites is needed to solve this problem.

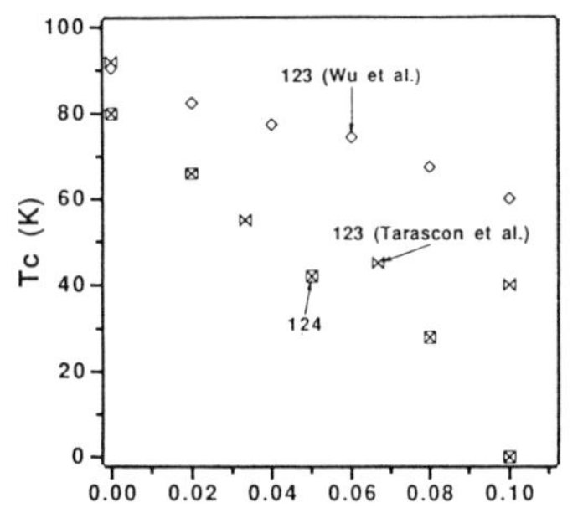

Fig. 5 Variation of T_c determined from the magnetization as a function of x. The data of the 123 phase cited from Wu et al. [18] and Tarascon et al. [17] are also plotted.

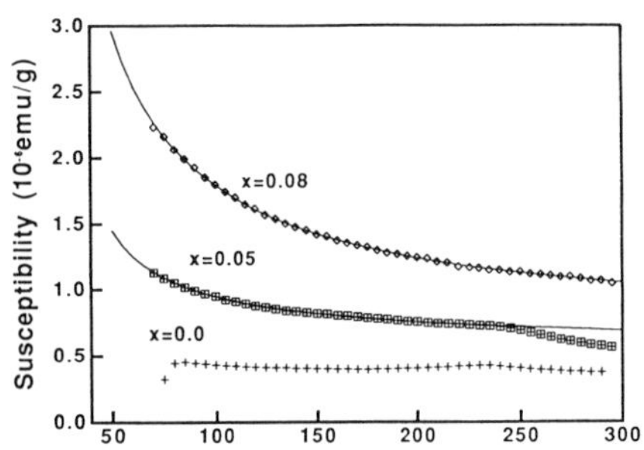

Fig. 6 Magnetic susceptibility as a function of temperature for x=0.0, 0.05, and 0.08. The solid lines show a fit to a Curie-Weiss formula.

The magnetic susceptibility data are shown in Fig. 6. The data were successfully fitted to a Curie-Weiss formula

$$\chi=\frac{C}{T+\theta}+\chi_0 \quad (1) \qquad C=\frac{NP_{eff}^2\mu_B^2}{3k_B} \quad (2)$$

where C is the Curie constant, θ is the Weiss temperature, χ_0 is the temperature independent susceptibility, N is the number of magnetic atoms per gram, P_{eff} is the effective Bohr magneton number, μ_B is the Bohr magneton, and k_B is the Boltzmann constant. In the Ni substituted samples, a fit was performed after subtracting the susceptibility of the non-substituted sample (χ_{Cu}) from the total susceptibility ($\chi_{eff}=\chi-\chi_{Cu}$) in order to obtain the contribution from Ni. The Weiss temperature, the temperature independent susceptibility, and the Bohr magneton number are listed in Table I. The fitting results are also shown in Fig. 6 as a solid line. The reason why those parameters do not show a systematic behavior is considered that the ratio of the Ni atoms inside to outside the 124 phase varies by the synthesis procedure. To discuss the magnetic properties of the Ni atoms in the 124 phase, a sample of good quality is required.

SUMMARY

$YBa_2(Cu_{1-x}Ni_x)_4O_8$ up to x=0.1 was successfully synthesized at 1010°C under a 40 MPa oxygen partial pressure. But at a large x value NiO was also detected. The a and b lattice parameters scarcely changed whereas the c lattice parameter decreased with increasing x. The depression of T_c by the substitution was as drastic as the result in the 123 phase reported by Trascon et al. [17] and the superconductivity disappeared at x=0.1. The magnetic susceptibility data were successfully fitted to a Curie-Weiss formula although the magnetic parameters did not show a systematic behavior because of the Ni atoms outside the 124 phase. For precise investigations, e.g. the decision of the occupation site and the magnetic behavior in the 124 phase, we need to synthesis a sample of good quality.

REFERENCES

[1] Karpinski J., Kaldis E., Jilek E., Rusiecki S., and Bucher B. (1988) NATURE 336: 660-662

[2] Bucher B., Karpinski J., Kaldis E., and Wachter P. (1989) Physica C 157: 478-482

[3] Yamada Y., Matsumoto T., Kaieda Y., and Môri N. (1990) Jpn. J. Appl. Phys. 29: L250-253

[4] Marsh P., Fleming R.M., Mandich M.L., DeSantolo A.M., Kwo J., Hong M., and Martinez-Miranda L.J. (1988) NATURE 334: 141-143

[5] Hazen R.M., Finger L.W., and Morris D.E. (1989) Appl. Phys. Lett. 54: 1057-1059

[6] Morris D.E., Nickel J.H., Wei J.Y.T., Asmar N.G., Scott J.S., Scheven U.M., Hultgren C.T., Markelz A.G., Post J.E., Heaney P.J., Veblen D.R., and Hazen R.M. (1989) Phys. Rev. B 39: 7347-7350

[7] Miyatake T., Gotoh S., Koshizuka N., and Tanaka S. (1989) NATURE 341: 41-42

[8] Morris D.E., Narwankar P., Sinha A.P.B., Takano T., Fayn B., and Shum V.T. (1990) Phys. Rev. B 41: 4118-4122

[9] Buckley R.G., Tallon J.L., Pooke D.M., and Presland M.R. (1990) Physica C 165: 391-396

[10] Shibutani K., Hayashi S., Fukumoto Y., Ogawa R., and Kawate Y. (1990) Physica C 167: 102-106

[11] Wada T., Sakurai T., Suzuki N., Koriyama S., Yamauchi H., and Tanaka S. (1990) Phys. Rev. B 41: 11209-11213

[12] Morris D.E., Narwankar P.K., Sinha A.P.B., Takano K., and Shum V.T. (1990) Appl. Phys. Lett 57: 715-717

[13] Pradhan S., McDaniel D., Kiling A., Huff W., Boolchand P., and Farrell D.E. (1990) Bull. Am. Phys. Soc. 35: 208

[14] Morris D.E., Marathe A.P., and Sinha A.P.B. (1990) Physica C 169: 386-390

[15] Bridges F., Boyce J.B., Claeson T., Geballe T.H., and Tarascon J.M. (1990) Phys. Rev. B 42: 2137-2142

[16] Yang C.Y., Moodenbaugh A.R., Wang Y.L., Xu W., Heald S.M., Welch D.O., Suenaga M., Fischer D.A., and Penner-Hahn J.E. (1990) Phys. Rev. B 42: 2231-2241

[17] Tarascon J.M., Barboux P., Miceli P.F., Greene L.H., Hull G.W., Eibschutz M., and Sunshine A. (1988) Phys. Rev. B 37: 7458-7469

[18] Xu Y., Sabatini R.L., Moodenbaugh A.R., Zhu Y., Shyu S.G., Suenaga M., Dennis K.W., and McCallum R.W. (1990) Physica C 169: 205-216

[19] Morris D.E., Asmar N.G., Wei J.Y.T., Nickel J.H., Sid R.L., Scott J.S., and Post J.E. (1989) Phys. Rev. B 40: 11406-11409

[20] Kodama Y. et al. (unpublished)

[21] Kakihana M., Börjesson L., Eriksson S., Svedlindh P., and Norling P. (1989) Phys. Rev. B 40: 6787-6796

Superconductive Property of the Bi 2201 Phase

SHIRO KAMBE[1], YASUHIRO MURAKOSHI[2], MAKI KAWAI[3], and TOMOJI KAWAI[4]

[1] FRP, The Institute of Physical and Chemical Research, 2-1, Hirosawa, Wako, Saitama, 351-01 Japan
[2] Tokyo Gakugei University, 1-1, Nukuikita-machi 4-chome, Koganei, Tokyo, 184 Japan
[3] RLEM, Tokyo Institute of Technology, 4259, Nagatsuta, Midori-ku, Yokohama, 227 Japan
[4] ISIR, Osaka University, Mihogaoka, Ibaraki, Osaka, 567 Japan

ABSTRACT

Superconductive property of the Bi 2201 phase is studied. The observed changes in the value of Tc is discussed in connection with the total valence of the system (the total valence), Bi and Cu valence, and its lattice structure. Nitrogen annealing leads to the increase of Tc from 22.5-23.0K to 32.5-39.6K. It was revealed that the value of Tc should not be discussed only as a function of the total valence. Transformation of the position of oxygen atoms may cause the change in the distribution of the total valence, which leads to the increase of Tc.

KEY WORDS: Bi system superconductor, oxygen content, total valence, Bi and Cu valence, lattice structure

INTRODUCTION

While the $Bi_2Sr_2CuO_{6+z}$ (2201 phase) superconductor has an octahedral Cu-O plane in a unit cell similar to the $Tl_2Ba_2CuO_{6+z}$ (2201 phase) and the $(La,Sr)_2CuO_4$ (214 phase) superconductors [1-3], Tc of the $Bi_2Sr_2CuO_y$ so far reported is much lower than those of $Tl_2Ba_2CuO_{6+z}$ and $(La,Sr)_2CuO_4$. Tc of the 2201 phase in the Bi system is 8K [4]. For the La substituted 2201 phase of the Bi system, a maximum Tc of 25K has been reported [5-9]. Comparing the Tc value of a cuprate superconductor with one CuO_2 layer, this Tc value is smaller than 80K of the 2201 phase in the Tl 2201 phase [8,10] and 39K of the 214 phase [11].
We have reported previously that a maximum Tc of 98.5K is attainable for the 2212 phase of the Pb-substituted Bi system after annealing in N_2 or Ar and 39.5K for the 2201 phase [12-15] These Tc values are 15-20K higher than those prepared in ambient oxygen and are closer to those for the Tl system [12,15].
In order to understand the changes in the Tc value, the relation between Tc, lattice structure and Bi, Cu valences is discussed.

EXPERIMENTAL

Sample preparation and the measurement of superconductivity are carried out in the same way as reported previously [12]. $Bi_{2-x}Pb_xSr_{2-y}M_yCuO_{6+z}$ with x=0-0.4 and y=0-0.4 (M=La, Nd) in were prepared. The peaks observed by XRD were assigned according to those indexed by A. Maeda et al. [9] and Y. Ikeda et al. [4]. The total valence was determined by ordinary iodometric titration method [9]. The valence of Bi was determined by the permanganate titration method [16]. The valence of Cu was estimated by subtracting the Bi valence from the total valence [16].

RESULTS AND DISCUSSION

In Fig.1, temperature dependences of magnetic ac susceptibility are displayed. The result indicates that N_2 annealing leads to the increase of Tc from 22.5-23.0K to 32.5-39.6K, consistent with the temperature dependences of the resistance. A maximum Tc of 39.6K is observed at x=0.4 after annealing in ambient N_2 and its volume fraction is about 10% at 10K.
All the peaks that appeared in the X-ray diffraction are indexed as those in the 2201 phase (Fig.2). Neither 2212 nor $La_{2-x}Sr_xCuO_4$ (the 214 phase) is present.

23.0K to 32.5-39.6K, consistent with the temperature dependences of the resistance. A maximum Tc of 39.6K is observed at x=0.4 after annealing in ambient N_2 and its volume fraction is about 10% at 10K.
The X-ray diffraction patterns show that all the peaks are originated from the 2201 phase and not from the 2212 phase as shown in Fig.2. The diffraction patterns also show that no $La_{2-x}Sr_xCuO_4$ (the 214 phase) is formed. Furthermore, Tc for $Bi_{2-x}Pb_xSr_{1.6}Nd_{0.4}CuO_{6+z}$ increased from 19.5-26.0K to 29.5-37.0K after N_2 annealing. $Nd_{2-x}Sr_xCuO_4$ is a non-superconductor, suggesting that the extremely high Tc after N_2 annealing is not originated from the 214 phase, but from the 2201 phase itself.
The Tc values of the 2201 phase are plotted against the total charge (Fig.3). The total charge was changed by the substitution of La for Sr and Pb for Bi.For samples prepared in ambient oxygen (represented by □), the plot is approximately consistent with the results reported by Maeda et al. [9]. The optimum value of the total charge gave the highest Tc of around 25K ranges between 0.10 and 0.15.
In contrast to these results, the Tc value of the 2201 phase obtained by N_2 treatment is clearly different (as represented by ○) and the plots show anomalous features. The total charge, δ, is between 0.05 and 0.07, indicating that after reducing the amount of oxygen, the value of Tc should not be treated only as a function of the total charge. In the 2212 phase, reduction treatment raised Tc from 73.5K to 98.5K and decreased the total charge to 2.01 [10], showing that the high Tc of 98.5K can not be explained only by the relation between the Tc and the total charge.
In this study, very similar results are obtained for the 2201 phase.

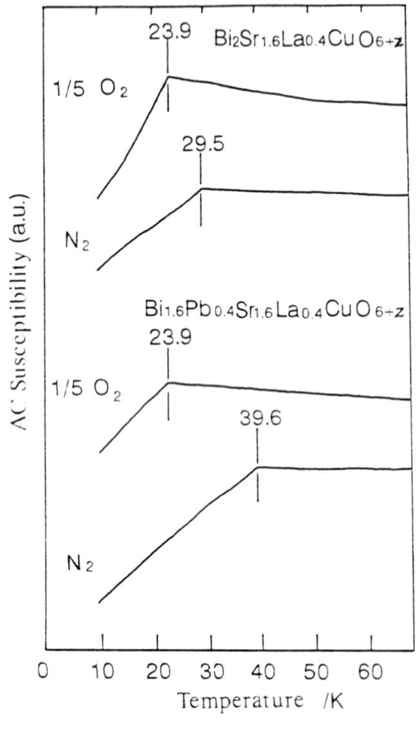

Fig.1. AC magnetic susceptibility vs. temperature in $Bi_{2-x}Pb_xSr_{1.6}La_{0.4}CuO_{6+z}$ in ambient O_2 or N_2. Annealing in N_2 leads to the increase of Tc value.

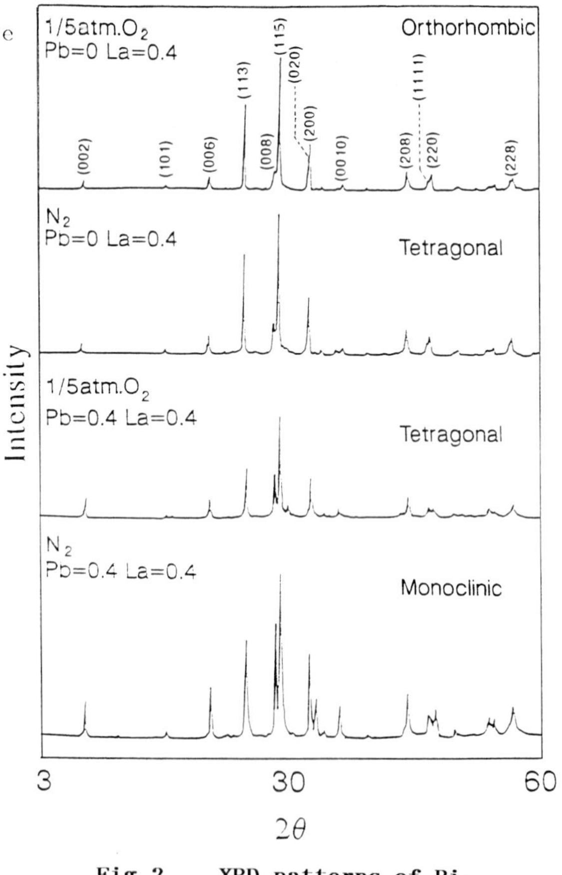

Fig.2. XRD patterns of $Bi_{2-x}Pb_xSr_{1.6}La_{0.4}CuO_{6+z}$. Neither 2212 nor 214 phase is detected. N_2 annealing causes the change in the lattice structure.

For Pb-free samples, Bi and Cu valences were separated. Oxygen content was changed by substituting La for Sr. As shown in Fig.4, the Cu and Bi valences were dramatically changed by N_2 annealing. For the samples annealed in 1/5 O_2, the total charges were divided both into Bi and Cu valences. The increase of oxygen content lead to the decrease of Bi valence and the increase of Cu valence. For the samples annealed in ambient N_2, Bi and Cu valences were fixed at 3.02 and 2.03-2.06, respectively. This result shows that the distribution of charge between Cu and Bi is dramatically changed by annealing in N_2, suggesting that N_2 annealing causes the structural change.

For the 2212 phase, the reduction of oxygen amount caused the structural change from a pseudotetragonal to a monoclinic cell, suggesting that such an improvement of Tc must be closely correlated with the changes in the crystal structure. As shown in Fig.2, the XRD patterns of the 2201 phases were changed by N_2 annealing. By annealing in N_2, orthorhombic structure changed into tetragonal for $Bi_2Sr_{1.6}La_{0.4}CuO_{6+z}$ and tetragonal into monoclinic for $Bi_{1.6}Pb_{0.4}Sr_{1.6}La_{0.4}CuO_{6+z}$. However, it is still difficult to explain the increase of a Tc value systematically only by the change in the lattice structures. The XRD diffraction pattern strongly depends on the positions of large cationic atoms particularly on Bi and Pb, and is insensitive to the change of the oxygen position in a unit cell. From the fact that the charge distribution between Cu and Bi is dramatically changed by N_2 annealing, the high Tc value may be correlated with a structural change due to oxygen positions, particularly the apical one to Cu-O plane or the one incorporated in the Bi-O layer.

While the 2201 phase in the Bi system has an octahedral Cu-O plane in a unit cell similar to $Tl_2Ba_2CuO_{6+z}$ and $(La,Sr)_2CuO_4$, the highest Tc value so far reported was 25K for the La-substituted 2201 phase in the Bi system [34,35]. This value was much lower than 80K for $Tl_2Ba_2CuO_{6+z}$ and 39K for $(La,Sr)_2CuO_4$. Now we have found that the compound with octahedral Cu-O plane have a potential for high Tc of 39K. Among these three systems having octahedral Cu-O plane, Tc of 80K in $Tl_2Ba_2CuO_6$ is exclusively high, implying that $Tl_2Ba_2CuO_6$ has a unique mechanism for high Tc over 80K.

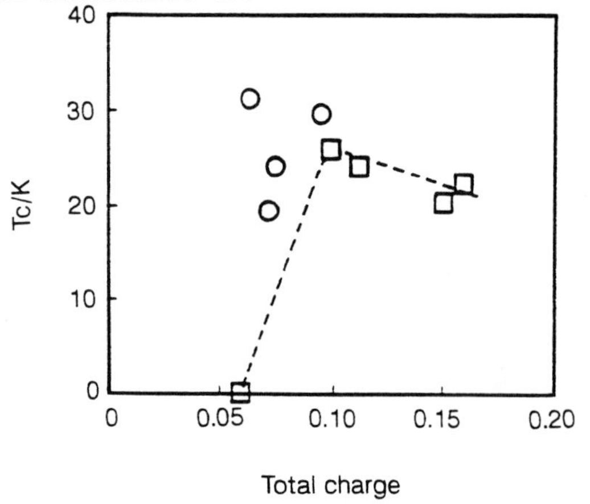

Fig.3. Tc dependence of the 2201 phase on the total valence. The 2201 phase annealed in N_2 (○) and that without annealing (□) are shown. The trend of Tc dependence for the samples prepared in N_2 annealing is clearly different from that of the samples prepared in ambient oxygen (···).

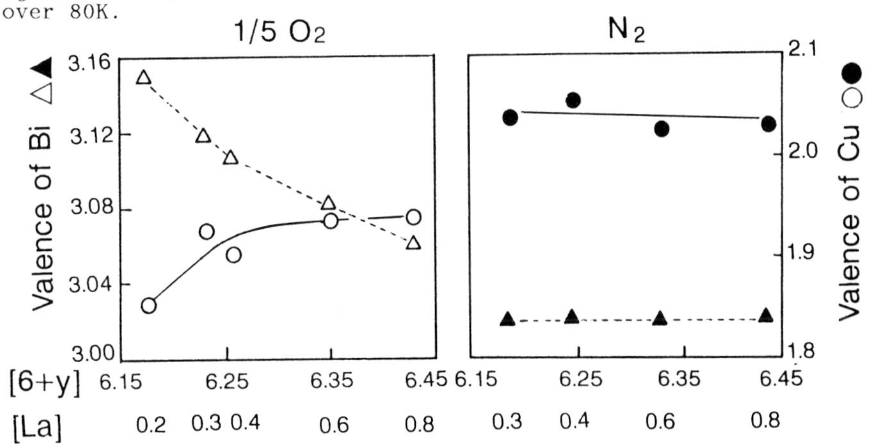

Fig.4. Bi and Cu valences against oxygen (La) content for $Bi_2(Sr,La)_2CuO_{6+z}$. The Cu and Bi valences are dramatically changed by N_2 annealing.

Conclusion

The $Bi_{1.6}Pb_{0.4}Sr_{1.6}La_{0.4}CuO_{6+z}$ superconductor (2201 phase), prepared by annealing in the absence of oxygen shows a high Tc value of 39.6K. As well as the 2212 phase, the formation of the 2201 phase with the high Tc was difficult to explain simply by consideration of the optimum total charge. Distribution of the charge between Cu and Bi planes dramatically changes after annealing in N_2, suggesting that N_2 annealing causes the structural change including the oxygen positions so as to affect the distribution of the total valence. It must be closely correlated with the high Tc for the 2201 phase. As a result, the superconductors with an octahedral Cu-O plane seems to have a potential for high Tc over 39K.

Acknowledgement

We are grateful to Prof. Y. Hirokawa of Tohoku University and Prof. K. Fueki of Science University of Tokyo for useful discussions about the permanganate titration.

References

[1] Tarascon JM, Mckinnon WR, Barboux P, Hwang DM, Bagley BG, Greene LH, Hull GW, Page YL and Giroud M (1988) Phys Rev.B 38: 8885

[2] Torardi CC, Subramanian MA, Calabrese JC, Gopalakrishnan J, McCarron EM, Morrissey KJ, Askew TR, Flippen RB, Chowdhry U and Sleight AW (1988) Phys.Rev.B 38: 225

[3] Pichett WE (1989) Rev.Mod.Phys. 61: 433

[4] Ikeda Y, Ito H, Shimomura S, Oue Y, Inaba K, Hiro Z and Takano M (1989) Physica C 159: 93

[5] Den T, Yamazaki A and Akimitsu J (1988) Jpn.J.Appl.Phys. 27: L1620

[6] Kijima T, Tanaka J and Bando Y (1988) Jpn.J.Appl.Phys., 27: L1620

[7] Bauhofer W, Mattausch H, Kremer RK, Murugaraj P and Simon A (1989) Phys.Rev.B 39: 7244

[8] Groen WA, Leeuw DM and Stollman GM (1989) Solid State Commun. 72: 697

[9] Maeda A, Hase M, Tsukada I, Noda K, Takebayashi S and Uchinokura K (1990) Phys.Rev.B 41: 6418

[10]Shimakawa Y, Kubo Y, Manako T and Igarashi H (1989) Phys.Rev.B 40: 11400

[11]Torrance JB, Tokura Y, Nazzal AI, Bezinge A, Huang TC, Parkin SSP (1988) Phys.Rev.Lett. 61: 1127

[12]Kambe S, Matsuoka T, Takahasi M, Kawai M and Kawai T (1990) Phys.Rev.B 42: 2669

[13]Kambe S, Matsuoka T, Kawai M, Kawai T and Takahasi M (1990) In: Ishiguro T and Kajimura K (eds) in Advances in Superconductivity. Springer-Verlag, Tokyo, pp 215-218

[14]Kambe S, Matsuoka T, Kawai M, Kawai T and Takahasi M (1990) Mol. Cryst. and Liq. Cryst. 184: 135

[15]Kambe S, Kawai M and Kawai T (1990) Solid State Comm. 75: 435

[16]Idemoto Y and Fueki K (1990) Physica C 168: 167

Oxygen Loss from $Bi_2Sr_2CaCu_2O_y$ by Annealing in Vacuum

Masayasu Nagoshi[1], Yasuo Fukuda[2], Ayako Tokiwa[3], Junichi Murata[2], Noriaki Sanada[2], Teruo Suzuki[1], Yasuhiko Syono[3], and Masashi Tachiki[3]

[1] Steel Research Center, NKK Corporation, Kawasaki, 210 Japan
[2] Research Institute of Electronics, Shizuoka University, Hamamatsu, 432 Japan
[3] Institute for Materials Research, Tohoku University, Sendai, 980 Japan

ABSTRACT

The amount of oxygen loss, Δy, from $Bi_2Sr_2CaCu_2O_y$ by annealing in vacuum ($\sim 10^{-8}$ Pa) was estimated by the integration of oxygen partial pressure measured by a quadruple mass spectrometer. The Δy of $Bi_2Sr_2CaCu_2O_y$ was in the order of 10^{-3} for the annealed sample at 500 - 600 °C, leading to the increase of Tc up to about 90 K. The superconductivity of $Bi_2Sr_2CaCu_2O_y$ is very sensitive to the amount of oxygen loss. The other gases such as H_2O and CO_2 were also evolved from the superconductors by annealing. The total amount of these gases evolved was several times more than that of the evolved oxygen in weight, which implies that the amount of oxygen loss obtained by thermogravimetric method is overestimated.

KEY WORDS: $Bi_2Sr_2CaCu_2O_y$, oxygen loss, annealing in vacuum, mass spectrometry

INTRODUCTION

Since high - Tc superconductivity was discovered in Cu-O_2 based oxides [1], many studies on the relationship between the oxygen content and the Tc have been carried out for various systems [2 - 5]. It has been noted that the Tc of oxide systems is strongly affected by the change in oxygen content. Clarifying the relationship between the oxygen content and superconductivity in the oxides is important in order to understand the mechanism of high - Tc superconductivity and also to improve Tc for practical use with these materials.

In the case of the $Bi_2Sr_2CaCu_2O_y$ system, it has been reported that the Tc is sensitive to change in oxygen content [5 - 10]. That is to say, the Tc of materials synthesized in air increases as the oxygen content decreases. However, the amount of oxygen loss Δy from $Bi_2Sr_2CaCu_2O_y$ oxides, which has been generally estimated by the thermogravimetric methods, has a large scatter ranging from ~ 0 [7] to 0.4 [5, 6] and is still controversial. Therefore, the effect of oxygen loss on the superconductivity is still unclear in this system.

In this paper, we attempt to estimate the accurate value of oxygen loss from $Bi_2Sr_2CaCu_2O_y$ by the integration of oxygen partial pressure measured using mass spectrometry, because the thermogravimetric method generally used measures all gases evolved from samples as weight loss. The Δy was obtained by using a calibration curve between integrated oxygen partial pressure and the oxygen loss measured by an iodometric titration for $YBa_2Cu_3O_{7-\delta}$ system. We will discuss the relationship between the amount of oxygen loss and the Tc in the $Bi_2Sr_2CaCu_2O_y$ system.

EXPERIMENTAL

The polycrystalline $Bi_2Sr_2CaCu_2O_y$ and $YBa_2Cu_3O_{7-\delta}$ synthesized by the solid state reaction method were used as the starting materials. The mixtures of powder with proper cation ratios were fired and reground three times. Then they were pelletized and fired at 880 °C for 12 hours in air for the Bi system and at 930 °C for 12 hours in an oxygen stream (1 atm) for the Y system. These starting materials were confirmed to have almost entirely a single phase by x-ray powder diffraction.

In order to estimate the amount of oxygen evolution from the superconductors, the oxygen partial pressure, P_O, was measured as a mass number 32 (O_2) by a quadrupole mass spectrometer (VG Scientific SX-200) as a function of heating time t. Before each measurement, the pressure was confirmed to be under 10^{-8} Pa. The samples were heated up with the constant heating ratio of 5 °C/min, and then the temperature was kept at between 450 and 600 °C until the P_O decreased down to the pressure of at least 2 orders lower than the maximum P_O. The amount of oxygen evolution, q_O, is defined by the following equation,

$$q_O = S \times \int_{t=0}^{t=t'} P(t)dt, \tag{1}$$

where S is the exhaust velocity of a vacuum pump (an oil diffusion pump with a liquid nitrogen trap was used in this study) and t' is the heating time when the measurement is finished.

The oxygen loss of $YBa_2Cu_3O_{7-\delta}$ annealed in vacuum was also measured by means of an iodometric titration. If the S is constant in the pressure range of 10^{-9} - 10^{-6} Pa, Δy of $Bi_2Sr_2CaCu_2O_y$ can be estimated from a relationship between the amount of oxygen loss measured by the iodometric titration and the integrated oxygen partial pressure, q_O/S, for the $YBa_2Cu_3O_{7-\delta}$ systems by using the following equation,

$$\Delta y = K \times q_O/S. \qquad (2)$$

Where K is a slope of the amount of oxygen loss versus q_O/S for $YBa_2Cu_3O_{7-\delta}$.

The partial pressures of other gases evolved from the oxides were also measured in the range of mass numbers 1 - 50.

RESULT AND DISCUSSION

Figure 1 shows the relationship between the amount of evolved oxygen measured by the iodometric titration and q_O/S for $YBa_2Cu_3O_{7-\delta}$. The calibration curve, shown as a solid line, was drawn by the assumption that the S is constant. It was found that the oxygen loss can be estimated, even if it is very small, by using this calibration curve. The Δy of $Bi_2Sr_2CaCu_2O_y$ estimated by using the equation (2) are 3.2 ($\pm$ 0.8) $\times$ 10^{-3} and 3.9 ($\pm$ 1.0) $\times$ 10^{-3} , assuming y = 8.2 in the starting material, for the samples annealed at 500 °C and 600 °C in vacuum, respectively. These values are 1 - 2 orders smaller than that (0.05 - 0.4) measured previously by the weight loss [5, 6, 8 - 10]. On the other hand, the Tc's of these samples were increased from 65 K and 78 K up to 90 K and 87 K by annealing at 500 °C and 600 °C, respectively. This implies that the Tc of $Bi_2Sr_2CaCu_2O_y$ is very sensitive to the amount of oxygen loss. These results are shown in Table 1.

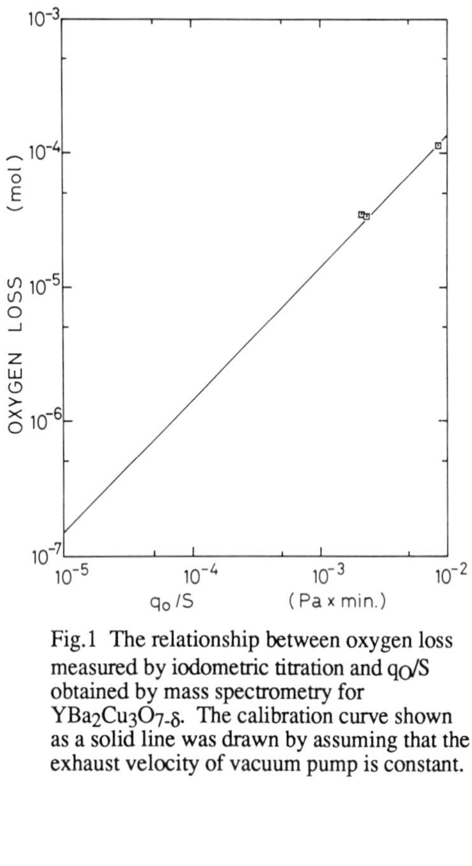

Fig.1 The relationship between oxygen loss measured by iodometric titration and q_O/S obtained by mass spectrometry for $YBa_2Cu_3O_{7-\delta}$. The calibration curve shown as a solid line was drawn by assuming that the exhaust velocity of vacuum pump is constant.

The relationship between the hole concentration and the Tc in the high - Tc superconducting oxides is widely known. Tc increases with increasing hole concentrations n_H for $n_H < n_H^{opt}$ (n_H^{opt} denotes optimum vale for maximum in Tc) and decreases with increasing n_H for $n_H > n_H^{opt}$. $Bi_2Sr_2CaCu_2O_y$ synthesized in air or in a high oxygen stream is thought to be in the over - doping state; $n_H > n_H^{opt}$. The decrease in n_H which causes the increase in Tc from the starting value to the maximum value were estimated to be in 10^{-1} order per Cu - O_2 unit from Hall measurements for $Bi_2Sr_2CaCu_2O_y$, in which n_H was controlled by oxygen loss [11] and atomic substitutions [12, 13]. These values are in the same order as those of other superconducting oxides. For $YBa_2Cu_3O_{7-\delta}$, it was reported that the change in n_H almost agrees with that expected from oxygen nonstoichiometry for $\delta > 0.5$ [14].

Table 1. Integrated oxygen partial pressure q_O /S, weight, Δy and Tc of $Bi_2Sr_2CaCu_2O_y$ annealed at 500 and 600 °C in vacuum.

annealing temperature (°C)	q_O /S (Pa $\times$ min.)	weight (g)	Δy	Tc (K) starting material	Tc (K) annealed
500	1.36×10^{-5}	0.0522	3.2 ($\pm$ 0.8) $\times 10^{-3}$	78	87
600	1.76×10^{-4}	0.5710	3.9 ($\pm$ 1.0) $\times 10^{-3}$	65	90

For the $Bi_2Sr_2CaCu_2O_y$ system, however, the change in n_H expected from the oxygen loss obtained here is about two orders smaller than that estimated from the Hall measurement [11] if one oxygen defect removes two holes. This suggests that the mechanism for the change of hole concentrations in the Bi system differs from that in the Y system, although the changes are caused by oxygen loss in both cases. Therefore, the other mechanisms such as a charge redistribution between hole donors and Cu - O_2 planes [7, 8, 11] should be taken into account for explaining the change in n_H caused by the oxygen loss in the $Bi_2Sr_2CaCu_2O_y$ system.

Figure 2 (a) shows the change of the oxygen partial pressure (measured by mass number 32) and the sample temperature as a function of heating time for $Bi_2Sr_2CaCu_2O_y$. The temperature was kept at 500 °C after it was increased up to 500 °C with a heating rate of 5 °C /min. Oxygen starts to evolve at about 280 °C. The oxygen partial pressure, p_O, has a peak of about 400 °C and then it decreases quickly. After the sample is annealed for about 180 min. the P_O decreases to as low as 10^{-8} Pa.

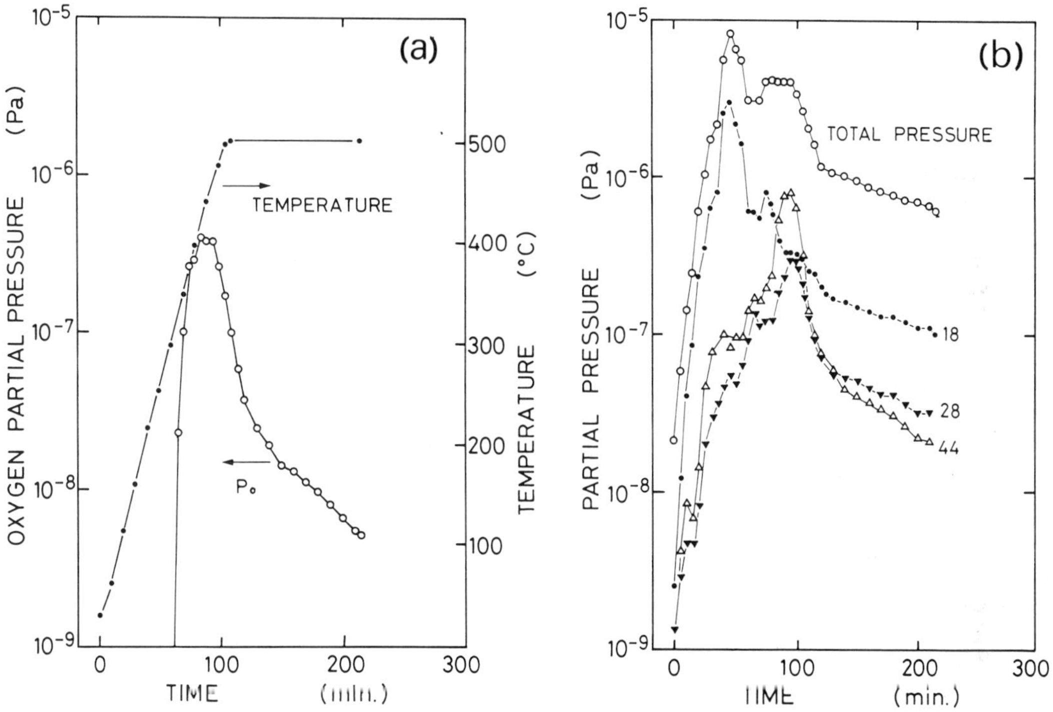

Fig.2 Changes of oxygen partial pressure, measured by mass number 32, and sample temperature: (a), and partial pressures of mass numbers 18, 28, and 44 along with total pressure: (b) as a function of heating time for $Bi_2Sr_2CaCu_2O_y$ superconductor.

Table 2. Integrated partial pressures of evolved gases from $Bi_2Sr_2CaCu_2O_y$ superconductor and the sample holder by annealing at 500 °C in vacuum. These results were calculated from Fig. 2.

	integrated partial pressures, q /S (10^{-5} Pa × min.)				
mass number	2 (H_2)	18 (H_2O)	28 (CO)	32 (O_2)	44 (CO_2)
sample + sample holder	20.20	9.77	1.62	1.36	2.76
sample holder	10.83	5.53	1.16	0	1.42
sample	9.37	4.24	0.46	1.36	1.34
(weight ratio: %)	(8.9)	(36.3)	(6.1)	(20.7)	(28.0)

Figure 2 (b) shows the partial pressures of mass numbers 18 (H_2O), 28 (CO and / or N_2), and 44 (CO_2) along with total pressure as a function of heating time for the same experiment as shown in Fig.2 (a). The partial pressures of mass number 18, 28 and 44 peak at about 200 °C, 450 °C, and 450 °C, respectively. The partial pressure of 28 has a similar behaviour to that of 44, which shows that the mass number 28 consists of mostly CO which seems to be the fragment of CO_2. The integrated partial pressures, q/S, of these gases are shown in Table 2, where the relative ionization cross sections of these gases to oxygen are assumed to be 1.

In order to estimate the amount of gases evolved from the sample holder, the same experiments were repeated without a sample. The q/S of these gases are also shown in Table 2 in which the result for gases evolved only from the sample are obtained. This result clearly shows the evolution of other gases such as H_2O, CO, and CO_2 from the $Bi_2Sr_2CaCu_2O_y$ oxide. It should be noted that the percentage of oxygen compared to the total amount of evolved gases is only about 20 wt.%. We emphasis here that the amount of the oxygen loss obtained previously by thermogravimetric methods has been overestimated thus leading to the large values of Δy up to 0.4.

SUMMARY

The amount of oxygen loss Δy from $Bi_2Sr_2CaCu_2O_y$ caused by annealing in vacuum was estimated by measuring the partial pressure of evolved oxygen with the mass spectrometry. It was confirmed that the Δy can be measured, even if it is very small, by this method. Although Δy is very small (in of the order of 10^{-3}), the oxygen loss strongly affects the Tc of this system. The other gases such as H_2O and CO_2, which are about 80 wt. % of the whole evolved gases, were also evolved from the samples by annealing. This shows that the amount of oxygen loss obtained by the thermogravimetric methods is overestimated.

ACKNOWLEDGMENT

The authors thank Mr. Kyoya, KKS Corporation for his assistance in the oxygen loss measurement. Their thanks also go to Dr. K. Sato for his useful advice. This work was partly supported by the Grant in Aid for Scientific Research on "Chemistry of New Superconductors" (Y.F.) and "Mechanism of Superconductivity" (Y.S. and M.T.) from the Ministry of Education, Science, and Culture of Japan.

* present address, Electronic Research Center, NKK Corporation, Toyama 934, Japan

REFERENCES

1. Bednorz JG, Muller KA (1986) Z.Phys. B 64: 189
2. Tarascon JM, Greene LH, Mckinnon WR, Hull GW, Geballe TH (1987) Science 235: 1373
3. Tokura Y, Torrance JB, Huang TC, Nazzal AI (1988) Phys. Rev. B 38: 7156
4. Shimakawa Y, Kubo Y, Manako T, Satoh T, Iijima S, Ichihashi T, Igarashi H (1989) Physica C 157: 279
5. Buckley RG, Tallon JL, Brown IWM, Presland MR, Flower NE, Glberd PW, Bowden M, Milestone NB (1988) Physica C156: 629
6. Morris DE, Hulgtren CT, Markelz AM, Wei JY, Asmar NG, Nickel JH (1989) Phys.Rev. B 39: 6612
7. Groen WA, De Leeuw DM (1989) Physica C 159: 417
8. Zhao J, Seehara MS (1989) Physica C 159: 639
9. Imai K, Matsuba H (1989) In: Ishiguro T, Kajimura K (eds) Advances in Superconductivity II, Proceedings of the 2nd Internal Simposium on Superconductivity (ISS'89). Springer - Verlag, Tokyo, pp 149 - 152
10. Motoi Y, Ikeda Y, Uwe H, Sakudo T (1989) Physica C 162 - 164: 929
11. Nagoshi M, Suzuki T, Fukuda Y, Terashima K, Nakanishi Y, Ogita M, Tokiwa A, Syono Y, Tachiki M (submitted) Phys.Rev. B
12. Clayhold J, Hagen SJ, Ong NP, Tarascon JM, Barboux P (1989) Phys.Rev. B 39: 7320
13. Koike Y, Iwabuchi Y, Hosoya S, Kobayashi N, Fukase T (1989) Physica C 159: 105
14. Wang ZZ, Clayhold J, Ong NP, Tarascon JM, Greene LH, Mckinnon WR, Hull GW (1987) Phys.Rev. B 36: 7222

The Effect of Oxygen Content on the Properties and Structure of the Low T_c Phase in Bi-Sr-Ca-Cu-O Superconductors

SHIGEKI KOBAYASHI

Toyota Central Research & Development Laboratories, Inc., Nagakute, Aichi, 480-11 Japan

ABSTRACTS

The oxygen content of the low Tc phase in Bi-Sr-Ca-Cu-O superconductors has been varied by annealing at 500-750℃ and its effect on the properties and structure has been examined. The oxygen stoichiometry affected greatly the properties of low Tc phase in BSCCO. Annealing in N_2 increased the Tc of low Tc phase up to 100K and expanded the length of c-axis by 0.01nm. The change in the stoichiometric parameter δ in $Bi_2Sr_2CaCu_2O_{8+\delta}$ was as small as 0.1-0.2.

KEY WORDS:oxygen stoichiometry, annealing, low Tc phase, BSCCO

INTRODUCTION

The critical temperature of $YBa_2Cu_3O_y$(YBCO) superconductor is dependent on the concentration of oxygen in the lattice and could be increased by annealing in oxygen[1,2]. Similar annealing treatment in Bi-Sr-Ca-Cu-O (BSCCO) was not necessarily successful and it has been believed that oxygen stoichiometry does not play an important role on the superconducting properties in BSCCO system [3,4]. However, as shown in Fig.1, BSCCO superconductors do absorb or desorp oxygen and their Tc's are strongly dependent on the content of oxygen, especially in the low Tc phase[5]. In this study, we focus on the low Tc phase and investigate the variation of properties and structure with oxygen content.

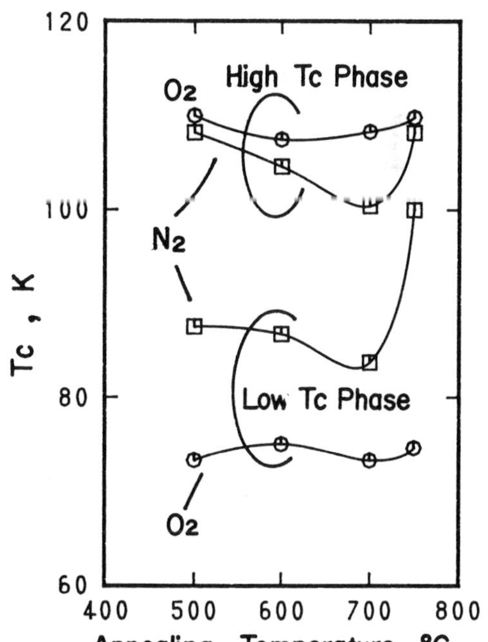

Fig. 1. The variation of Tc's with annealing temperature and atmosphere.

EXPERIMENTAL

The powders were prepared by a solid-state reaction of Bi_2O_3, $CaCO_3$, $SrCO_3$ and CuO. The compositions were chosen to be $BiSrCaCu_2O_x$. The sintered bulk samples were obtained by a similar process described in the previous paper[6]. Bulk samples for AC susceptibility measurement were sintered in air at 870℃ and they contained both the high Tc (Tc=110K,c=3.7nm) and low Tc phases (Tc=80K, c=3.0nm). Samples sintered at 860℃ contained mainly the low Tc phase and used for determining the variation of oxygen content during annealing. Powder samples for structural analysis were prepared by heating in air at 860℃ to form only low Tc phase. Samples were annealed in oxygen and nitrogen at 500-750℃. After annealing for 2h, specimens were furnace-cooled with an approximate rate of 100 ℃/min. The variation of properties was evaluated by measuring AC susceptibility at 1kHz. The measurement was performed during cooling from room temperature to 4.2K.

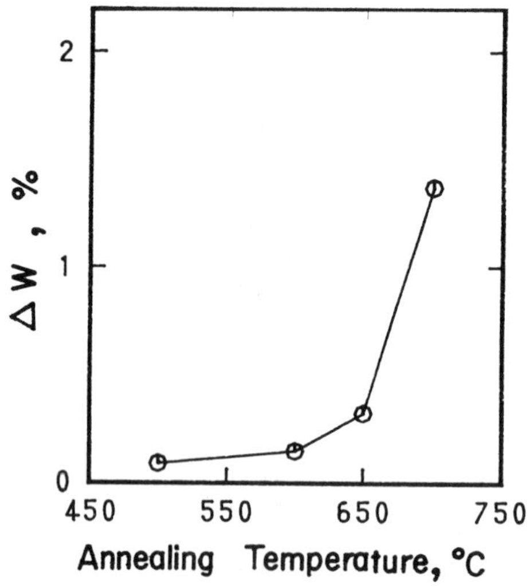

Fig. 2. The variation of weight change (Δ W), which is the weight difference between samples annealed in O_2 and N_2.

RESULTS AND DISCUSSION

The variation of Tc's with annealing atmosphere and annealing temperature is shown in Fig. 1. The behavior was basically same for all the annealing temperatures; annealing in nitrogen caused a decrease in the Tc of high Tc phase and an increase in the Tc of low Tc phase. The critical temperature Tc did not depend largely on the annealing temperature except for at 750 ℃. Since slow cooling was employed in this experiment, the change of oxygen content during cooling could make the temperature dependence obscure. The effect of annealing is more evident in the low Tc phase. The difference in the Tc of the low Tc phase between samples annealed in nitrogen and in oxygen is as high as 25K at 750℃ and the Tc of low Tc phase annealed at 750℃ in nitrogen reached up to 100K.

As shown above, annealing treatment clearly changes the superconducting properties in BSCCO system. The phenomenon was consistent with the recent reports on annealing[7-12]. The variation of Tc is considered to be associated with oxygen content in the lattice. This was confirmed by the reversibility of properties and the corresponding weight change. The weight change of bulk sample was determined using the sintered sample containing mainly the low Tc phase. The sample was annealed in O_2 and N_2 at 500, 600 and 700℃ successively. The weight change was defined as the weight difference between annealing in O_2 and N_2 (Fig. 2). The change was small (0.1-0.3%) but reversible at T $\leqq$ 650 ℃. At higher temperatures, the reduction and decomposition of superconducting phases took place and caused much larger weight change. The level of weight change is comparable to the reported values for BSCCO system [3,13-15]. The change in the stoichiometric parameter δ in $Bi_2Sr_2CaCu_2O_{8+\delta}$ was calculated to be 0.1-0.2, which is much smaller than that for YBCO system (about 0.5). The variation of Tc in the low Tc phase was 20-30 K (Fig. 1), which is comparable to that for YBCO system (about 30K).

Using powder samples containing only the low Tc phase, structural change during annealing was examined by X-ray diffraction (XRD). As shown in Fig. 3, no drastic change of the structure was detected but the length of c-axis changed with the oxygen content(Fig. 4). The lattice constants were estimated from the position of major (h00) and (00l) peaks. The absolute values of c-axis were slightly different from those obtained by the Rietveld analysis of X-ray diffraction but the difference between samples annealed in O_2 and N_2 was almost same. The c-axis is longer for the samples annealed in N_2 but the difference is only 0.01 nm. As for the a-axis, the samples annealed in N_2 always showed longer a-axis, but the difference is of the order of only 0.001 nm, which is in the same order of experimental error.

As clearly seen from the above results, annealing treatment greatly affected the properties of low Tc phase in BSCCO system. Annealing in N_2 caused oxygen to desorp from lattice and increased Tc of low Tc phase. This behavior can be contrasted with that in YBCO system where Tc is increased by annealing in O_2. Koike et al.[16] measured the hole concentration for the low Tc phase of various cation-substituted BSCCO systems. They showed that there is a optimum hole concentration where Tc takes a maximum and that non-substituted BSCCO is in the over-doped state, i.e., its hole concentration is higher than the optimum hole concentration. From this picture, the results obtained in this study can be easily understood; i.e., reducing the oxygen content decreases the hole concentration and then increases Tc until the optimum concentration is attained. The problem to be solved is to determine the lattice site where the oxygen atoms desorped are located. The Rietveld analysis of X-ray diffraction was tried but it turned out unsuccessful because of the existence of the impurity phases such as CuO and also the preferred orientation. High resolution TEM is very useful to analyse the detailed structure such as modulated structure. A example of HRTEM micrograph is shown in Fig.5. Pb-substitution is known to fade out the modulation[17] and the same effect can be expected for oxygen removal. At this moment, the number of observation is limited and more work is needed to draw any conclusion.

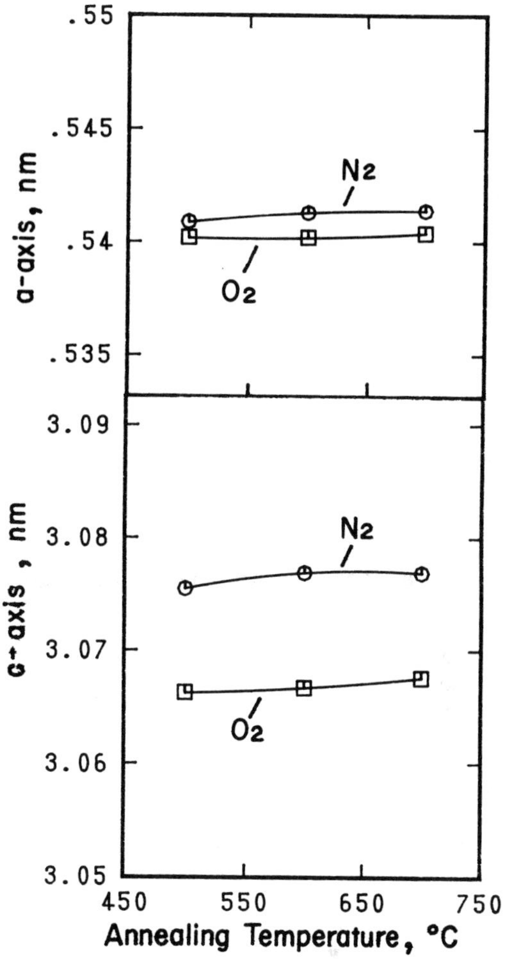

Fig. 4. Effect of annealing on the lattice constants of the low Tc phase.

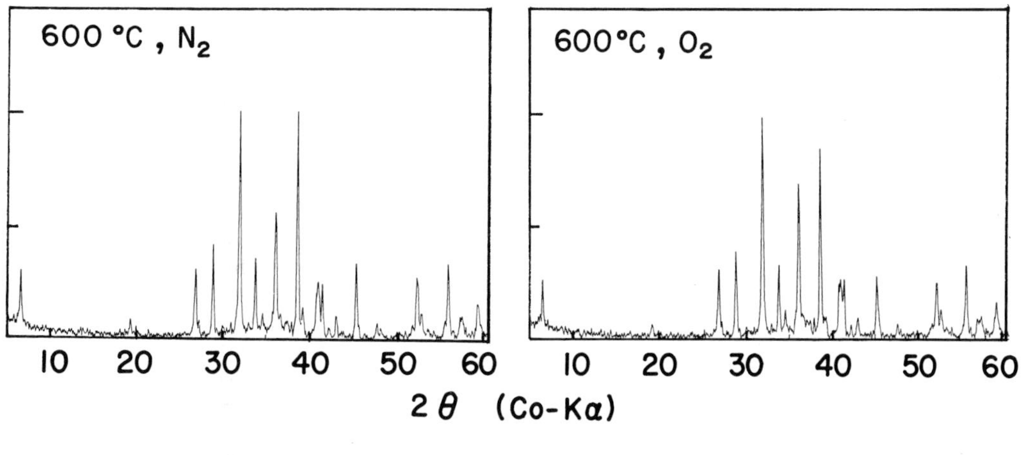

Fig. 3. X-ray diffraction patterns for samples annealed at 600 ℃.

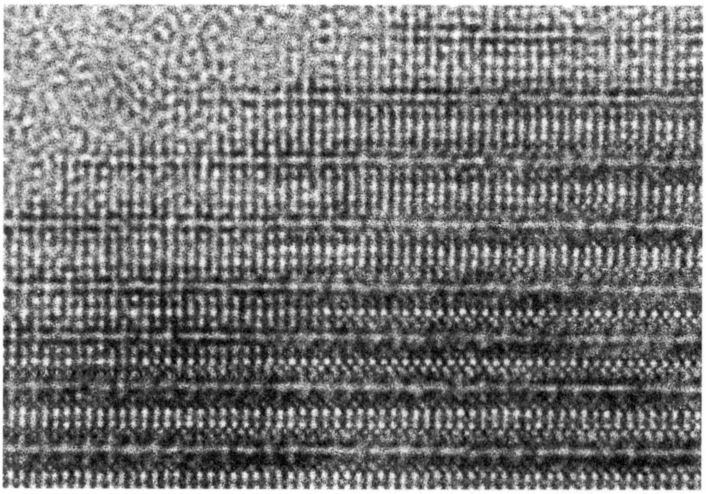

Fig. 5. A high resolution TEM micrograph taken from the sample annealed in N_2 at 700℃ with the incident beam parallel to b-axis.

ACKNOWLEDGEMENTS

The author wishes to acknowledge the members of superconductor research project, especially Dr. Tatsumi Hioki, Mr. Nobuaki Suzuki and Mr. Masataka Ohkubo for their interest and helpful discussions. He also appreciates Mr. Noritomo Suzuki for high resolution TEM observation and Mr. Tatsuo Noritake for Rietveld analysis of X-ray diffraction.

REFERENCES

1. Takayama-Muromachi E, Uchida Y, Ishii M, Tanaka T, Kato K (1987) Jpn. J. Appl. Phys. 26: L1156.
2. Cava RJ, Batlogg B, Chen CH, Rietman EA, Zahurak SM, Werder D (1987) Nature 329: 423.
3. Tarascon JT, Page YL, Barboux P, Bagley BG, Greene LH, McKinnon WR, Hull GW, Giroud M, Hwang DM (1988) Phys. Rev. B 37: 9382.
4. Narumi S, Ohtsu H, Iguchi I, Yoshizaki R (1989) Jpn. J. Appl. Phys. 28: L27.
5. Kobayashi S, Wada S (in press) J. Mater. Sci. Lett.
6. Kobayashi S, Saito Y, Wada S (1989) Jpn. J. Appl. Phys. 28: L772.
7. Terada H, Ido T, Muto S (1988) Jpn. J. Appl. Phys. 27: L1219.
8. Miura N, Suzuta H, Deshimaru Y, Shimizu Y, Sakashita H, Yamazoe N (1989) Jpn. J. Appl. Phys. 28: L1112.
9. Pena O, Perrin A, Sergent M (1988) Physica C 156: 489.
10. Buckley RG, Tallon JL, Brown IWM, Presland MR, Flower NE, Gilberd PW, Bowden M, Milestone NB (1988) Physica C 156: 629.
11. Tallon JL, Buckley RG, Gilberd PW, Presland MR (1989) Physica C 158: 247.
12. Groen WA, Leeuw DM (1989) Physica C 159: 417.
13. Ono A (1988) Jpn. J. Appl. Phys. 27: L2276.
14. Ono A (1988) Jpn. J. Appl. Phys. 27: L1213.
15. Niu H, Fukushima N, Ando K (1988) Jpn. J. Appl. Phys. 27: L1442.
16. Koike Y, Iwabuchi Y, Hosoya S, Kobayashi N, Fukase T (1989) Physica C 159: 105.
17. Ikeda Y, Hiroi Z, Ito H, Shimomura S, Takano M, Bando Y (1989) Physica C 165: 189.

Li Substitution of $Bi_2Sr_2Ca_{n-1}(Cu_{1-x}Li_x)_nO_{2n+4}$ (n=1,2)

T. Horiuchi[1], T. Kawai[1], S. Kawai[1], and K. Ogura[2]

[1] The Institute of Scientific and Industrial Research, Osaka University, Mihogaoka, Ibaraki, Osaka, 567 Japan

[2] Faculty of Science and Technology, Kinki University, Kowakae, Higashi-Osaka, Osaka, 577 Japan

ABSTRACT

Substituted Li ions occupy the Cu site of 2201-phase and 2212-phase. The amount of oxygen and the length of c-axis of 2201-phase do not change for Li substitution , and resistance-temperature (R-T) curve indicates semiconductive behavior yet. In the case of 2212-phase, the amount of oxygen decreases and the length of the c-axis becomes longer with the Li substitution. T_c values raise at the same time.

KEY WORD: Li substitution, Bi 2212-phase, Bi 2201-phase

INTRODUCTION

The $Bi_2Sr_2Ca_1Cu_2O_8$ superconductor (2212-phase or BSCCO) has crystal structure with double CuO_2 sheets and square-pyramidal coordinations of Cu ions, and has T_c value of about 80 K for samples prepared in air. The experiment of substitution of 3-d metals (Fe,Co,Ni and Zn ions) for Cu ions show that the T_c value decrease with increasing of concentration of these metals [1]. Recently, we found that T_c value of the 2212-phase was raised by the incorporation of Li ions into BSCCO lattice [2]. For the nominal composition of Bi:Sr:Ca:(Cu:Li) = 2.2: 1.8: 1.0:(1.2: 0.8), $T_{c,onset}$ of 92K was observed with Meissner effect. Matsubara et al. showed that the doping of Li is effective to raise T_c for both the 2212-phase and 2223-phase [3].

$Bi_2Sr_2Cu_1O_y$ (2201-phase) has single CuO_2 plane in the unit cell while 2212-phase has double CuO_2 planes, and moreover, they have the different coordinations of Cu ions, octahedral and square-pyramidal. Consequently, it seems very interesting for us to study Li-doped 2201-phase, comparing with the Li-doped 2212-phase.

We report here the results of chemical analysis, SEM observations, X-ray diffractions and the measurements of T_c for the Li doped-samples in the 2212- and 2201-phases, and the sites of the doped Li ions are discussed.

EXPERIMENTAL

Samples were prepared from starting materials with the following compositions,
1) Bi:Sr:Ca:(Cu:Li)= 2.2:1.8:1.05:(2.15-x:x) (x=0,0.2,0.4,0.7),
2) Bi:Sr:(Cu:Li)=2.2:1.8:(1-x:x) (x=0,0.1,0.2,0.33).
Samples were precalcined at 550 ^oC to 700 ^oC for 5h to 20h before sintering. They were sintered at 700 ^oC to 840 ^oC and quenched to room temperature. The morphology of the samples was observed by a scanning electron microscopy (SEM), and the crystal structure was identified by X-ray diffraction patterns. Lattice constants were precisely determined using Si as a standard. The compositions were determined with an analytical electron microscope for microanalysis and by an atomic absorption spectrochemical analysis for total analysis, together with quantitative analysis of oxygen content. The electrical resistance of the samples were measured by a DC four probe method, and magnetic susceptibility measurements were made using a superconducting quantum interference device magnetometer (SQUID).

RESULTS AND DISCUSSION

1.2212-phase

1-1. Effects of Li substitution on superconductivity

Alkaline metal (Li,Na,K,Rb,Cs) addition to BSCCO lowered the formation temperature of the 2212-phase drastically. Moreover, the value of $T_{c,onset}$ in the Meissner measurement became higher than that of standard BSCCO. In particular, the increase in T_c was largest in the case of Li-doped BSCCO by about 10 K. [2]. Based on above findings, we examined the dependency of the formation of the Li-doped 2212 single phase on the starting composition, in order to clarify whether Li ions occupy the Cu sites or not. The standard 2212-phase was prepared from the starting material with the composition Bi:Sr:Ca:Cu = 2.2:1.8:1.0: 2.15 (Fig.1(a)). Samples added with Li ions to the standard sample , its composition: Bi:Sr:Ca:Cu:Li = 2.2:1.8:1.0:2.15:0.7, showed two phases, that is, 2212- and CuO phase (Fig.1(b)). Its $T_{c,zero}$ was about 85K. This result suggests that the added Li ions incorporate into the Cu sites of the lattice and the excess Cu ions precipitate as CuO. When we, however, used the starting materials with composition: Bi:Sr:Ca:Cu = 2.2:1.8:1.0:1.2, which corresponded to removal of only Li ions from the nominal composition: Bi:Sr:Ca:(Cu:Li) = 2.2:1.8:1.0:(1.2:0.8), 2212- and 2201-phases appeared. From this result that Cu ions seem to be deficient for preparing the single 2212-phase (Fig.1(c)). $T_{c,zero}$ value was only 56K. By adding Li to such the composition, we obtained X-ray pattern corresponding to the single 2212-phase (Fig.1(d)) because deficient Cu ions for the formation of the single phase was compensated by adding Li ions. High $T_{c,zero}$ of 91K and Meissner signal at 92 K (Fig.2(d)) were obtained. These behaviors indicate that the Li ions incorporate into the lattice and occupy the Cu sites.

1-2.Compositions and morphology.

It has been ascertained from an atomic absorption spectrochemical analysis that Li ions do in fact remain in the sample (Table 1). Measurements of the oxygen content revealed that the amount of oxygen decreases, while the valency of Cu keeps constant at a value of 2.2-2.3. The samples of Bi:Sr:Ca:(Cu:Li) = 2.2 :1.8 :1.05 :(2.15-x : x) for x=0 and x=0.7, consist mainly of plate-like grains. Quantitative analysis of the composition of the grains showed that the sample with x=0 has a nearly 2212-composition (Table 2). The amount of Li in the grains are randomly distributed among the plate-like grains of BSCCO, that is about half of the grains of the samples is deficient by 20% in Cu concentration, the rest is nearly 2212 composition (Table 2).

1-3.Influences of Li substitution on the lattice constant.

The length of a- or c-axis of the tetragonal unit cell were determined for samples with x = 0, 0.2, 0.4 and 0.7 by measurements of the X-ray diffraction peaks using Si as a standard. The results showed that the lattice constant c with x > 0 is larger than that of the sample with x = 0 (Fig.3). On the other hand, the constant a does not change appreciably. T_c value of 2212-phase is known to be increased by reduction treatments[4,5] because this procedure may reduce the excess amount of holes in the 2212-phase. The length of c-axis becomes longer by the reduction treatments [6].

In the case of Li addition, the length of c-axis becomes longer presumably because the amount of oxygen is decreased by the introduction of Li. The formal charge of Cu remains constant at around 2.3 (2.29 for x=0 and 2.32 for x=0.7) as estimated from the results of the composition analysis (Table 1). Thus, the rise of T_c by Li incorporation cannot be explained only by the reduction of hole concentration.

2. 2201-phase

We examined the dependence of the formation of the 2201-phase on the starting composition so as to show that Li ions occupy the Cu sites. The standard 2201-phase was prepared from the starting materials mixed with composition,

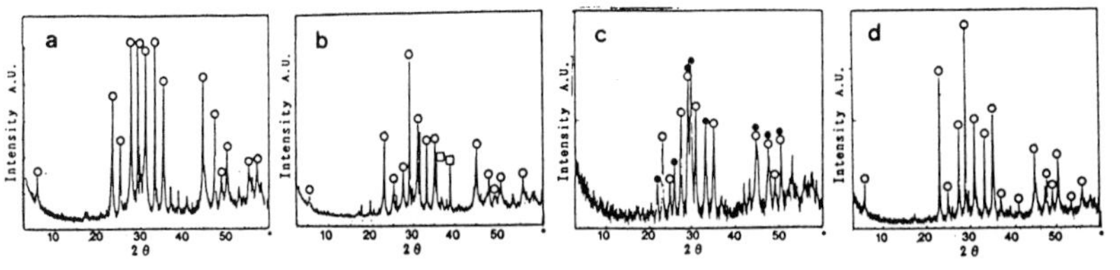

Fig.1. X-ray diffraction patterns of samples prepared from starting materials with Bi:Sr:Ca:(Cu:Li) compositions of (a) 2.2:1.8:1.05:2.15:0 (b) 2.2:1.8:1.0:2.15:0.7 (c) 2.2:1.8:1.0:1.2:0 and (d) 2.2:1.8:1.45:0.7. (o) 2212-phase, (□) CuO and (•) 2201-phase.

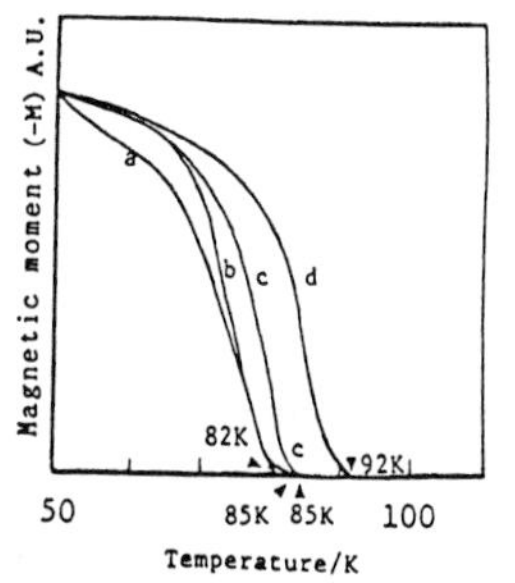

Fig.2. Meissner curves of the samples corresponding to (a)-(d) in Fig.1.

x	Bi	Sr	Ca	Cu	Li	O
0	2.16	1.79	0.94	2.15	0	8.43
0.7	2.10	1.75	0.98	1.45	0.48	7.81

Table 1 Total elemental analysis of the sintered sample with the nominal compositions of Bi:Sr:Ca:(Cu:Li) = 2.2:1.8:1.05:(2.15-x:x). Normalized by Cu=2.15 for x=0 and Cu = 1.45 for x =0.7.

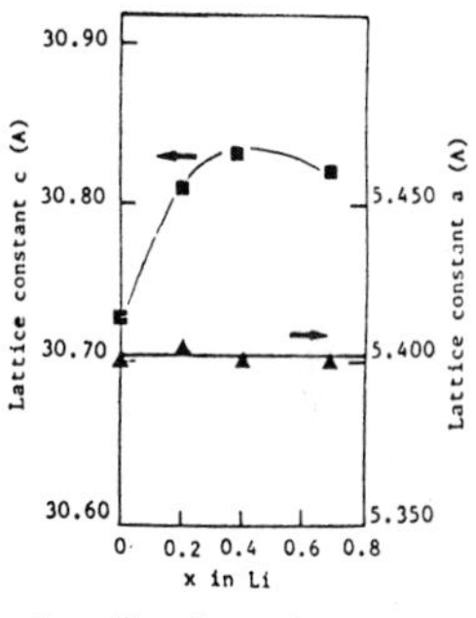

Fig.3. The dependence of the lattice constants a and c on the Li content of the samples with Bi:Sr: Ca:(Cu:Li) = 2.2:1.8:1.05 (2.15-x:x).(■) c-axis,(▲) a-axis

x	Bi	Sr	Ca	Cu
0	2.0±0.1	1.6±0.2	1.1±0.2	2.1±0.1
0.7	2.0±0.1	1.7±0.1	0.8±0.1	1.6±0.1

Table 2. Compositions of grains in the sintered sample with the nominal compositions of Bi:Sr:Ca:(Cu:Li) = 2.2:1.8:1.05:(2.15-x:x) where x=0, 0.7. The compositions were normalized by taking Bi=2.0. Cu is deficient in the Li substituted particles.

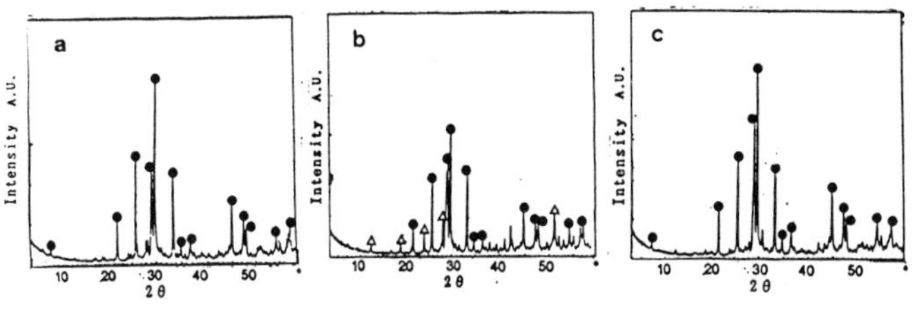

Fig.4. X-ray diffraction patterns of the samples with nominal Bi:Sr:Cu:Li compositions of (a) 2.2:1.8:1 (standard), (b) 2.2:1.8:0.67 (Cu subtraction) and (c) 2.2:1.8:0.67:0.33 (Cu subtraction and Li addition). (•) 2201-phase, (△) other phase.

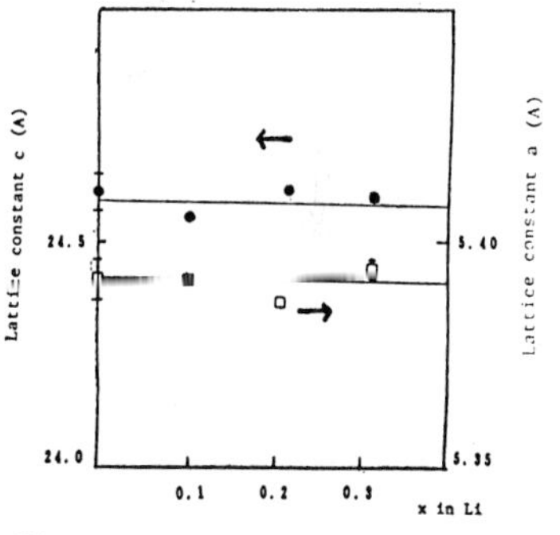

Fig.5 The dependence of the lattice constants a and c on the Li content of the samples with Bi:Sr:(Cu:Li)=2.2:1.8:(1-x:x). (•) c-axis, (□) a-axis.

x	Bi	Sr	Cu	Li	O
0	2.2	1.8	1.0	0	6.3
0.33	2.2	1.8	0.62	0.35	6.3

Table 3. Total elemental analysis of the sintered samples with the nominal compositions of Bi:Sr:(Cu:Li) = 2.2:1.8:(1-x:x). Normalized by Bi= 2.2.

x	Bi	Sr	Cu
0	2.20±0.06	1.53±0.05	0.97±0.08
0.33	2.20±0.02	1.55±0.03	0.74±0.04

Table 4. Compositions of particles in the standard sample with the nominal compositions of Bi:Sr:(Cu:Li) = 2.2:1.8:(1-x:x) where x = 0, 0.33. The compositions were normalized by taking Bi= 2.2. Cu is deficient in the Li substituted particles.

Bi:Sr:Cu = 2.2:1.8: 1 (Fig.4(a)). But, when we used the starting materials with composition: Bi:Sr:Cu = 2.2:1.8:0.67, 2201-phase and other phase appear(Fig.4(b)). This result suggests that Cu ions were deficient for preparing single 2201-phase. By adding Li to such the composition, we obtained X-ray pattern corresponding to the single 2201-phase (Fig.4(c)) because deficient Cu ions for the formation of the single phase, was compensated by adding Li ions. These behavior indicated that the Li ions got into the lattice and occupied the Cu site.

It has been confirmed from atomic absorption spectrochemical analysis that Li ions do in fact remain in the sample (Table 3). Measurements of the oxygen content revealed that the amount of oxygen remains constant, while the valency of Cu increases presumably because 2201-phase have octahedral coordination for Cu ion and compounds including Li ions tend to have this coordination for Li ions. The samples from the starting materials with composition Bi:Sr:(Cu:Li)= 2.2:1.8:(1-x:x) for x=0 and x=0.33, consist mainly of plate-like grains. Quantitative analysis of the composition of the grains showed that the sample with x=0 has a nearly 2201-composition (Table 4). The amount of Li randomly distributed among the plate-like grains of 2201-phase, that is about half of the grains of the samples is deficient by 20% to 30% in Cu concentration, the rest is near 2201 composition (Table 4).

The lattice constant a and c of the tetragonal unit cell were determined for samples prepared from materials with composition Bi:Sr:(Cu:Li) = 2.2:1.8:(2.15-x:x) (x=0, 0.1,0.2 and 0.33) by measurements of the X-ray diffraction peaks using Si as a standard. The results show that the lattice constants are constant for substitution of Li for Cu(Fig.5) because the amount of oxygen remains constant.

R-T curve of standard 2201-sample used starting materials with composition Bi:Sr:Cu = 2.2:1.8:1, indicated semiconductive behavior. R-T curves of Li-added samples prepared from that mixed with composition: Bi:Sr:(Cu:Li) = 2.2:1.8:(0.67:0.33), also indicated semiconductive behavior.

COMPARISON BETWEEN 2212- AND 2201-PHASE, AND CONCLUSION

We conclude from the results of X-ray diffraction, composition analysis and measurement of the length of a- and c-axis that Li ions substitute for the Cu sites of 2201- and 2212-phase.

The results of R-T measurement and measurement of magnetic susceptibility showed a great difference between an effect of the substitution on raising T_c of 2201-phase and that of 2212-phase. R-T curves of Li-doped 2201-phase are semiconductive, and Li-doped 2212-phase has high $T_{c,zero}$ of 91K and Meissner signal at 92K.

The results of composition analysis indicates that there is no difference between the amount of the substituted Li ions per one CuO_2 plane in 2201- and 2212-phase. It seems to be important that the amount of oxygen in 2201-phase remained constant, but in 2212-phase, it decreases for the substitution of Li ions. As a result, the amount of holes increases and the length of c-axis remained constant in 2201-phase, but in 2212-phase, the amount of holes remains constant and length of c-axis becomes longer. These results show that a change of c-axis is caused by the decrease of the amount of oxygen and it may affect the value of T_c of 2201-and 2212-phase.

ACKNOWLEDGMENTS

We would like to thank Dr. H.Takenaka, Dr. M.Motone and Dr. I. Matsubara of the Government Industrial Research Institute of Osaka for the analytical electron microscope measurements. We acknowledge Dr. K.kitahama for the precise measurement of the Meissner effect.

REFERENCES

[1] A.Maeda, T.Yabe, S.Takebayashi, M.Hase and K.Uchinokura (1989) submitted to Phys. Rev.B.
[2] T.Kawai, T.Horiuchi, K.Mitsui, K.Ogura, S.Takagi and S.Kawai (1989) Physica C 161: 561-566
[3] I.Matsubara, H.Tanigawa, T.Ogura, H.Yamashita, M.Kinoshita and T.Kawai (1990) Appl.Phys. Lett. 56, 2141-2143.
[4] S.Kambe, T.Matsuoka, M.Kawai, T.Kawai and M.Takahashi (1989) Proc. ISS 89.
[5] N.Fukushima, H.Niu, S.Nakamura, S.Takeno, M.Hayashi and K.Ando (1989) Physica C 159: 777.
[6] M.Kawasaki, S.Nagata, K.Takeuchi and K.Koimuma (1988) Jpn.J.Appl.Phys. 27 L2227

Comparative Study of the Superconductivity of $Tl_2Ba_2CuO_w$ and $Bi_2Sr_2CuO_y$

MASAE KIKUCHI, HIROFUMI NAMEKI, SATORU NAKAJIMA*, AYAKO TOKIWA, YASUHIKO SYONO, and NORIO KOBAYASHI

Institute for Materials Research, Tohoku University, Katahira, Aoba-ku, Sendai, 980 Japan
* Hachioji Research Center, CASIO Computer Co., Ltd., 2951-5 Ishikawa-cho, Hachioji, 192 Japan

ABSTRACT

Oxygen contents of Bi-2201, Pb substituted Bi-2201 and Tl-2201 were determined by iodometric titration and the relationship between T_c and oxygen content was examined. Oxygen number of Bi-2201 series increases with La substitution for Sr and reaches 6.52 with Lanthanum content of 1.0 in air annealed specimen. Lattice parameter, a, changes with La and Pb substitution, indicating its dependence on Cu charge. La and Pb atoms act as phase stabilizer of 2201 through charge control and structural stabilization respectively.

KEY WORDS: $Bi_2Sr_2CuO_y$, $Tl_2Ba_2CuO_w$, La substitution, over-doping, excess oxygen

INTRODUCTION

The superconductivity of $Tl_2Ba_2CuO_w$(Tl-2201) has been known strongly dependent on synthesis condition such as temperature, atomosphere and cooling rate [1,2]. We have already shown that the enhancement of Tc in $Tl_2Ba_2CuO_w$ was due to release of an over-doping state in oxygen annealed specimens by a small amount of oxygen loss [1] or La substitution for Ba [2]. On the other hand, the isostructural $Bi_2Sr_2CuO_y$(Bi-2201) also shows the enhancement of Tc by La substitution for Sr [3,4]. In $Bi_2Sr_2CuO_y$ superconductor, an excess oxygen in Bi-double layer is may play an important role for superconductivity [5,6,7]. These two isostructural materials of $Tl_2Ba_2CuO_w$ and $Bi_2Sr_2CuO_y$ are prototype of the layered superconductors. The comparative studies to examine the simirarity and difference in both compounds, $Tl_2Ba_2CuO_w$ and $Bi_2Sr_2CuO_y$ will be usefull to elucidate the superconductivity of the double layered structure system.

The difficulty of estimation of Cu valence of Tl-series and Bi-series, prevented from understanding of the superconductivity of the double layered structure system. Fortunately, oxygen contents of $Bi_2Sr_2CuO_y$ which is soluble with hydrochloric acid easily in contrast to $Bi_2Sr_2CaCu_2O_y$(Bi-2212) can be determined accurately with potentiometric titration of iodine. Here we report the relation between the superconducting transition temperature, Tc, lattic parameter, and oxygen content for the La substituted materials for alkali earth elements and discuss about the enhancement of Tc in both series.

EXPERIMENTAL

Solid solution of $Tl_2Ba_{2-x}La_xCuO_y$(x=0-0.3) was synthesized by solid state reaction of Tl_2O_3, BaO_2, La_2O_3 and CuO with stoichiometric metal ratio. The mixture was pressed into a pellet of 10 mm diameter and 1 mm thickness, wrapped with a gold foil, and then fired at around 900-910 C for 5-10 min in oxygen flow at a rate of 120 ml/min. The specimen was slowly cooled in the furnace to assure the maximum oxygen content close to 6. The prepared sample was examined by X-ray powder diffraction(XPD) using monochromated CuK radiation and found to be nearly single phase up to x=0.3.
A mother pellet of $Bi_{2-z}Pb_zSr_{2-x}La_xCuO_y$(z=0 and 0.4, x=0-1.0) was synthesized by solid state reaction of stoichiometric mixture of Bi_2O_3, PbO_2, $SrCO_3$, La_2O_3 and CuO. The specimen for z=0 series was first fired at 800 C for 24 hrs and then at 875 C for 24 hrs in air, while the specimen with z=0.4 was fired at 800 C for 10 hrs and then at 875 C for 20 hrs in air. A part of the mother pellet was pelletized into a disk of 10 mm diameter and of 1 mm thickness and fired at 600 C for 12 hrs in air, or N_2/Ar, or quenched in liquid nitrogen for

the oxygen number attains 6.52 for Bi-2201(z=0) specimen at [x]=1.0. Judging from the distance between Bi-Bi double layer, the Bi-double layer is most probable position in which excess oxygen content enters. The oxygen content of Bi-2201 prepared in air is estimated to be 6.15 by the extrapolation to [x]=0. Oxygen number exceeding 6 is almost the same as that in the Bi double

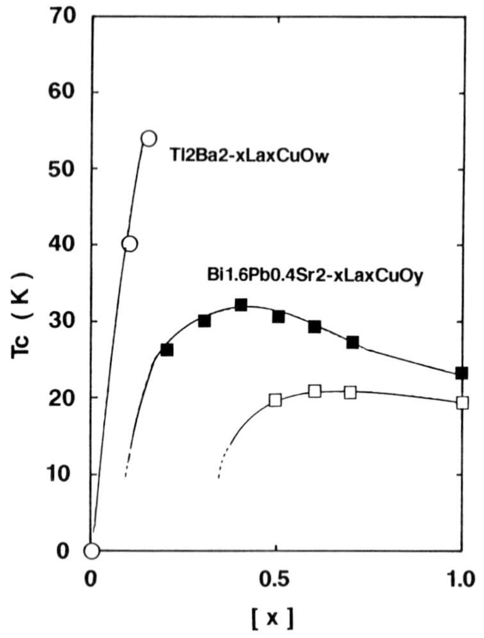

Fig. 2
Variation of T_c vs. La content, x.
(○): O_2 annealed
(□): air annealed
(■): N_2 annealed

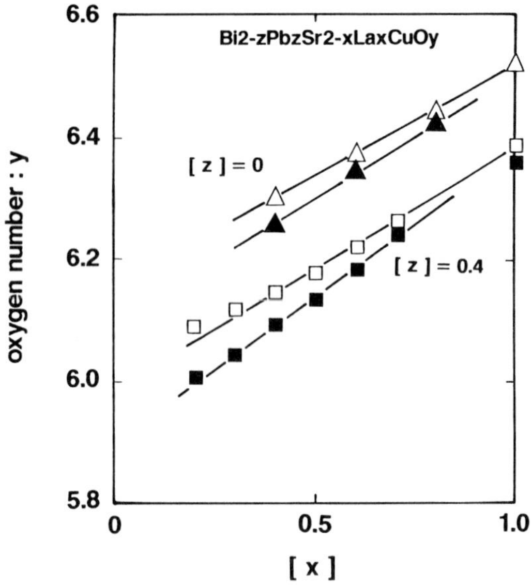

Fig. 3
Oxygen number, y, vs. La content, x.
(△): air annealed
(▲): N_2 annealed
(□): air annealed
(■): N_2 annealed

layer compound in the previous paper [6,7]. The oxygen number of the specimen annealed specimen in N_2 atmosphere is lower than that of the specimen synthesized in air.

Figure 4 shows the variation of Cu valence and lattice parameters of a and b with La content in Pb substituted Bi-2201. For the specimen annealed in air, the shorter axis, b, expanded with La substitution, but the longer one, a, showed little change. However, for annealed specimen in N_2 atmosphere, both a and b axes did not show remakable change. These observations correspond to larger change of Cu valence in air annealed specimen than in N_2 annealed one, as shown in fig 4-a. The increase in basal axes, i. e. bond length within Cu-O plane, is explained by rapid increase of ionic radii of Cu atom with decrease in Cu valence by La substitution.

Figure 5 shows the variation of Cu valence and of a and b parameters with Pb substitution for [x]=0.4 in the Bi series. The lattice parameter, a, did not show remarkable change but b decreased with increase of Pb content, corresponding to increasing Cu valence as shown in fig. 5-a. Thus Pb substitution tended to increase hole concentration. The orthorhombicity increased with increasing Pb concentration and was larger for the specimen annealed in reducing atmopsphere, N_2 flow, than for the specimen annealed in air. The phase stabilization of the Bi 2201 by Pb substitution seems to be due to structural origin rather than charge balance, while that by La substitution is performed by charge control of Cu atom.
The enhancement of T_c by La substitution in Pb substituted Bi system due to release of the over-doping state is small because of increase in oxygen number and suppressing effect of Pb atom, while a sharp increase in T_c in Tl-2201

z=0, and at 760 C for 10 hrs in N_2 for z=0.4. The prepared specimen was found to be nearly single phase by XPD.

The unit cell dimensions were determined from observed d-spacing by the least squares method. The oxygen content was determind by potentiometric titration method of iodine with an auto-titrator TOWA DENPA AUT-301 and the formal charge of Cu was calculated by assuming that the formal charges of Bi and Pb are trivalent and divalent respectively. The accuracy of idometric titration for Bi-series is very high and the experimental error mainly depends on sample handling. The superconducting properties were determined by measuring the susceptibility using superconductivity test systems(Sumitomo Heavy Industry and Chino Type III) and Tc was estimated from the observed temperature variation of the diamagnetic signals. The lower limit of temperature is 12 K in these system.

RESULTS AND DISCUSSION

The X-ray powder diffraction patterns of $Tl_2Ba_{2-x}La_xCuO_y$ showed almost single phase only for x<0.3, indicating that the solubility range is very narrow for La substitution for Ba in $Tl_2Ba_2CuO_w$(Tl-2201). The unit cell dimension was calculated by indexing tetragonal indices to the diffraction peaks. $Bi_2Sr_{2-x}La_xCuO_y$(Bi-2201) was found to be nearly single phase for a wide solubility range of 0.1<x<1.0 and XPD patterns were indexed with a pseudo-tetragonal cell, although synthesis of the specimen for x<0.2 was very difficult, suggesting that the decrease in Cu charge by La substitution for Sr contributes to the phase stability of the Bi-2201 phase. As observed in Bi-2223 [5], Pb substitution for Bi was another method for phase stabilization of the Bi-2201 phase. $Bi_{2-z}Pb_zSr_{2-x}La_xCuO_y$ became nearly single phase for [z]=0.2 and 0.4 even for [x]=0, resulting in a solid solution ranging from [x]=0 to 1.0. XPD patterns of Pb substituted specimen were indexed with orthorhombic indices and the orthorhombicity was found to depend on Pb concentration.

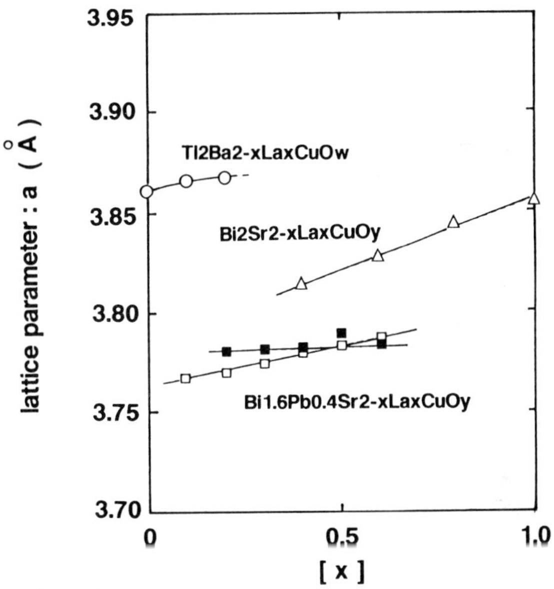

Fig. 1
Lattice parameter, a, indexed with tetragonal indices vs. La content, x.
(○): O_2 annealed
(△): air annealed
(□): air annealed
(■): N_2 annealed

The variation of the lattice parameter, a, with La concentration in Tl-2201, Bi-2201 and Pb substituted Bi-2201 is shown in fig.1. For easier comparison, the lattice parameter of tetragonal or pseudo-tetragonal cell in three series are shown. The variation of a is very remarkable for Bi-2201 annealed in air as well as for Pb-substituted Bi-2201 annealed in air. There are no remarkable change of a-axis in Tl-2201 annealed in O_2 and Bi-2201 annealed in N_2 atmosphere.

The relation between T_c and La concentration of Tl-2201 and Pb substituted Bi-2201(z=0.4) is shown in fig.2. T_c of Tl-2201 was strongly enhanced with the substitution of a small amount of La, but T_c of Pb-substituted 2201 annealed in N_2 atmosphere gradually increased from below 12 K for x=0 to 32 K for x=0.4 and then decreased to 23 K at x=1.0. The oxygen content of Tl-2201 as annealed in oxygen flow was close to 6.0 and increase in oxygen content with La substitution was not found in a preliminaly experiment.

The oxygen content of Bi-2201(z=0) and Pb substituted Bi-2201(z=0.4) determined with iodometric titration is shown in fig. 3. The increase in oxygen number with La content is nearly linear over a wide La concentration range and

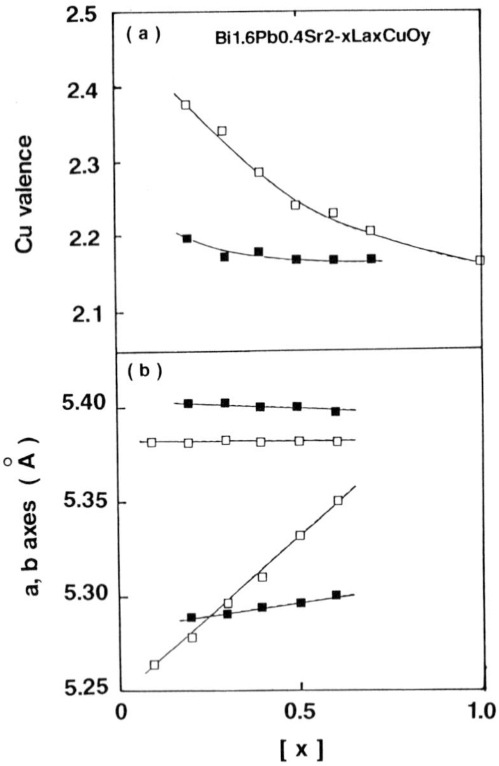

Fig. 4
(a) Cu valence vs. La content, x.
(b) Lattice parameter, a and b, vs. La content, x.
(□): air annealed
(■): N_2 annealed

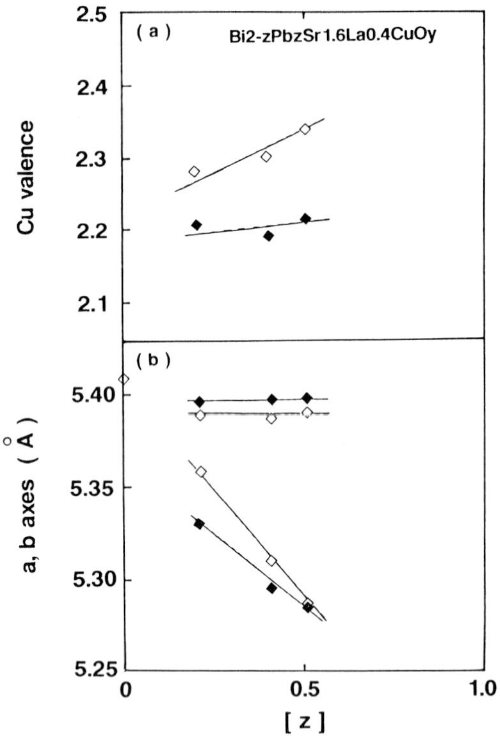

Fig. 5
(a) Cu valence vs. Pb content, z.
(b) Lattice parameter, a and b, vs. Pb content, z.
(□): air annealed
(■): N_2 annealed

indicates that the release of the over-doping state is achieved without change of oxygen number.

ACKNOWLEDGMENTS

The authors are grateful to Prof. K. Nagase and Mrs. R. Suzuki, College of General Education, Tohoku University and Prof. K. Watanabe and Mr. S. Awaji, Institute for Materials Research, Tohoku University for kind advice and technical support. This work was partly supported by the Grant-in-Aid for Scientific Research on Priority Area "Mechanism of Superconductivity" given by the Ministry of Education, Science and Culture, Japan.

REFERENCES

1. Kikuchi M, Nakajima S, Syono Y, Nagase K, Suzuki R, Kajitani T, Kobayashi N and Muto Y (1990) Physica C 166: 497
2. Nakajima S, Kikuchi M, Oku T, Kobayashi N, Suzuki T, Nagase K, Hiraga K, Muto Y and Syono Y (1989) Physica C 160: 458
3. Hwang GH, Shi JB and Ku HC (1989) Chinese J. Phys. 27: 273
4. Kanbe S, Kawai M and Kawai T (1990) Solid State Commun. 75: 435
5. Takano M, Takada J, Okada K, Kitaguchi H, Miura Y, Ikeda Y, Tomii Y and Mazaki H (1989) Jpn. J. Appl. Phys. 28: L1112
6. Le Page Y, McKinnon WR, Tarascon JM and Barboux P (1989) Phys. Rev. B 40: 6810
7. Ren J, Jung D, Whangbo MH, Tarascon JM, Le Page Y, McKinnon WR, and Torardi CC (1989) Physica C 158: 501

Superconductivity of $TlBa_2Ca_{n-1}Cu_nO_y$ System in Over-Doping State

SATORU NAKAJIMA[1], MASAE KIKUCHI[2], YASUHIKO SYONO[2], KENZO NAGASE[3], and NORIO KOBAYASHI[2]

[1] Hachioji Research Center, CASIO Computer Co., Ltd., 2951-5 Ishikawa-cho, Hachioji, 192 Japan

[2] Institute for Materials Research, Tohoku University, 1-1, Katahira 2-chome, Aoba-ku, Sendai, 980 Japan

[3] College of Education, Tohoku University, Kawauchi, Aoba-ku, Sendai, 980 Japan

ABSTRACT

T_c and oxygen content of $TlBa_2Ca_{n-1}Cu_nO_{y-v}$ (n= 2 and 3) have been investigated by the quenching experiment of the annealed specimen in oxygen atmosphere. The tetragonal unit cell dimension a increased and c decreased by introduction of oxygen loss. The oxygen loss, v, was determined by thermogravimetry in a nitrogen atmosphere using an oxygen-annealed specimen of $TlBa_2Ca_{n-1}Cu_nO_y$ and also by measurements of the weight differences before and after quenching. As-prepared specimen of $TlBa_2Ca_{n-1}Cu_nO_y$, n= 2 and 3, was found to be in an over-doping state and T_c increase to 110 K and 120 K respectively with increasing v up to v= 0.025-0.035.

KEY WORDS: Tl single layer system, over-doping state, quenching experiment

INTRODUCTION

High-temperature superconducting oxides of the Tl-Ba-Ca-Cu-O system with a layered structure showed a wide variety in both structure and composition [1]. Both double and single Tl layer structure have been found in this system and number of Cu layers in the perovskite units intervened by the Tl layers can be varied for 1-4 [2. 3] and 2-5 [4. 5], respectively. Non-superconductivity of oxygen annealed $Tl_2Ba_2CuO_6$ (2201) specimen has been reasonably explained in terms of over-doping due to charge transfer of Tl^{3-t}-$(Cu-O)^p$ and we have obtained evidence for the release of such over-doping by substituting Ba with La and the quenching experiment by liquid nitrogen [6. 7]. Shimakawa et. al [8] also reported that T_c variation was caused by change in oxygen content in $Tl_2Ba_2CaCu_2O_8$ and $Tl_2Ba_2Ca_2Cu_3O_{10}$.

T_c of Tl single layer system ($TlBa_2Ca_{n-1}Cu_nO_{2n+3}$) increases with increasing number of Cu layer up to n= 4, where it turns down [5]. We have already made a systematic study on superconductivity of $TlBa_2Ca_{1-x}RE_xCu_2O_7$, where RE is varied as Nd, Gd and Y, to observe the effect of amount of hole in the substituted system [9]. Starting from T_c= 80 K for x= 0, T_c increased with x to a maximum value of 100 K for x= 0.2-0.35. The average Cu valence of the stoichiometric $TlBa_2CaCu_2O_7$ and $TlBa_2Ca_2Cu_3O_9$ was calculated to be 2.50^+ and 2.33^+ respectively, suggesting an over-doping state which was more remarkable in $TlBa_2CaCu_2O_y$ than in $TlBa_2Ca_2Cu_3O_y$. Here we report the degree of over-doping of two specimens, n= 2 and 3, obtained by thermogravimetric analysis (TGA) of $TlBa_2Ca_{n-1}Cu_nO_y$ (n= 2 and 3) and the relationship between the oxygen content and superconductivity.

EXPERIMENTAL

The Tl single layer compounds can usually be synthesized by heat treatment for several hours [5]. $TlBa_2Ca_{n-1}Cu_nO_y$, n= 2 and 3, were synthesized by solid-state reaction of Tl_2O_3, BaO_2, CaO and CuO with the metal ratio of Tl:Ba:Ca:Cu= 2:2:1:2 or Tl:Ba:Ca:Cu= 2:2:2:3 respectively. The mixture was pressed into a pellet form of 10 mm diameter and 1 mm thickness. The pellet was wrapped with gold foil and fired at 885°C for 60-90 min in flowing oxygen gas of 120 ml/min. The prepared sample was checked by x-ray powder diffraction (XPD) using monochromated CuKα radiation and found to be single phase. Thermogravimetric analysis (TGA) was carried out by use of a thermobalance in flowing nitrogen gas. About 50 mg of powdered specimens was

Superconductivity of $TlBa_2Ca_{n-1}Cu_nO_y$ System in Over-Doping State

Satoru Nakajima[1], Masae Kikuchi[2], Yasuhiko Syono[2], Kenzo Nagase[3], and Norio Kobayashi[2]

[1] Hachioji Research Center, CASIO Computer Co., Ltd., 2951-5 Ishikawa-cho, Hachioji, 192 Japan
[2] Institute for Materials Research, Tohoku University, 1-1, Katahira 2-chome, Aoba-ku, Sendai, 980 Japan
[3] College of Education, Tohoku University, Kawauchi, Aoba-ku, Sendai, 980 Japan

ABSTRACT

Single-phase $(Tl_{1-x}Pb_x)Sr_2CuO_z$ ($z\sim 5$) ("1201" phase) samples have been successfully synthesized for x = 0 - 0.5. The synthesized samples were orthorombic in the range of x = 0 - 0.12 and tetragonal in the range of x = 0.12 - 0.5. The orthorombic compounds have been obtained for the first time for family of layered oxide systems represented by the formula, $AB_2Ca_{n-1}Cu_nO_{2n+3}$ where A = Tl or (Pb, Cu) and B = Ba or Sr. All the samples for x = 0 - 0.5 showed metallic temperature dependence of resistivity. As Pb content increased, Temperature dependence of magnetic susceptibility became from Curie-type behavior to Pauli-paramagnetism and signal of thermoelectric power changed from positive to negative.

KEY WORDS: $TlSr_2CuO_5$ structure, orthorombic - tetragonal phase transition, magnetic susceptibility, thermoelectric power

INTRODUCTION

The oxide superconducting families containing Tl, Sr, Ca and Cu, $TlSr_2Ca_{n-1}Cu_nO_{2n+3}$ (n = 2 and 3), have been discovered previously [1]. Both families have superconductivity with a transition temperature 75 K and 100 K respectively. However, pure $TlSr_2CuO_5$ ("1201" phase) without other element was not successfully synthesized. The "1201" structure was realized by replacement at the Sr site by rare-earth element "RE", i. e. $Tl(Sr, RE)_2CuO_5$ [2-5]. In the compound represented by $TlSr_2Ca_{n-1}Cu_nO_{2n+3}$, the average Cu ion's valence, V(Cu), can be expressed by V(Cu) = 2 + 1/n. The V(Cu) strongly depends on the number of Cu-O layer, n. In the case of $TlSr_2CaCu_2O_7$ ("1212" phase) with n = 2 and $TlSr_2Ca_2Cu_3O_9$ ("1223" phase) with n = 3, the V(Cu)s are 2.5 and 2.33 respectively. While, in the case of $TlSr_2CuO_5$ with n = 1, the V(Cu) should be 3.0. This value is considerably higher than the valence of other copper oxide superconductor. Thus, it is difficult to synthesize $TlSr_2CuO_5$. Substitution of a rare earth element which reduces the copper ion's valence makes synthesis of "1201" phase easy. If $TlSr_2CuO_5$ can be realized, it is not expected to be a superconductor because it is heavily over-doped. However, it is very interesting material which has a very high copper oxidation state. The $(Tl_{0.5}Pb_{0.5})Sr_2CuO_z$ that did not contain a rare-earth element can be synthesized [6,7]. Therefore, we tried to obtain single-phase $(Tl_{1-x}Pb_x)Sr_2CuO_z$ with as little Pb content as possible.

EXPERIMENTAL PROCEDURE

Samples were prepared by a solid state reaction using highly pure powders of Tl_2O_3, PbO, SrO and CuO. The powders were mixed to nominal compositions of $(Tl_{1-x}Pb_x)Sr_2CuO_z$ (x = 0 - 1.0). The mixtures were pressed into pellets. The pellets were wrapped with an Au foil and sintered at 900 ℃ for 10 h in O_2 gas flow. The samples were analyzed by x-ray powder diffractometry (XRD) using Cu-Kα radiation. Electrical resistivity was measured as a function of temperature using a conventional four probe-method. DC magnetic susceptibility was measured using a SQUID magnetometer (Quantum Design: Model MPMS). Thermoelectric power was measured using a bar-shape sample of about 2 mm * 2mm * 10 mm. Temperature diffference between both electrodes was 0.3 K.

used for each run. Heating and cooling rates were 5 °C/min. In the quenching experiments, the pellet samples of $TlBa_2Ca_{n-1}Cu_nO_y$ were heated to temperatures in the range of 100-700°C with the same heating rate of TGA run, kept at a constant temperature for 30-60 min and then cooled rapidly by dropping into liquid nitrogen. T_c was determined from the temperature dependence of resistivity and DC susceptibility. Resistivity was measured with a current of 1 mA. DC susceptibility was measured by a SQUID magnetometer in a field of 10 Oe.

RESULTS AND DISCUSSION

X-ray powder diffraction patterns (XPD) of $TlBa_2Ca_{n-1}Cu_nO_{2n+3}$ (n= 2 and 3) measured with CuKα radiation are shown in fig. 1. XPD showed no evidence of impurity phases. XPD patterns were indexed with the tetragonal indices and unit cell dimensions were determined from measured d-spacings by least squares method. Temperature dependences of the resistivity and DC susceptibility of $TlBa_2CaCu_2O_y$ are shown in fig. 2 and fig. 3 respectively. Low resistivity (10^{-3} ohm·cm) and large Meissner fraction (20-40%) were observed for the quenched specimen, indicating bulk superconductivity. From these data, the relation between T_c and the quenching temperature (T_q) is shown in fig. 4. T_c of $TlBa_2Ca_{n-1}Cu_nO_{y-v}$ (n= 2 and 3) increases with increasing quenching temperature and reaches a maximum value at around T_q= 500° C in both samples.

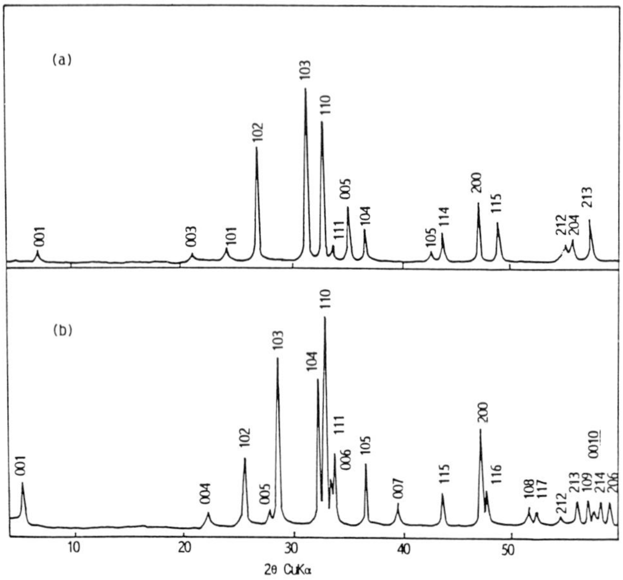

Fig.1 X-ray powder diffraction patterns of Tl single layer compounds; (a)$TlBa_2CaCu_2O_y$, (b)$TlBa_2Ca_2Cu_3O_y$.

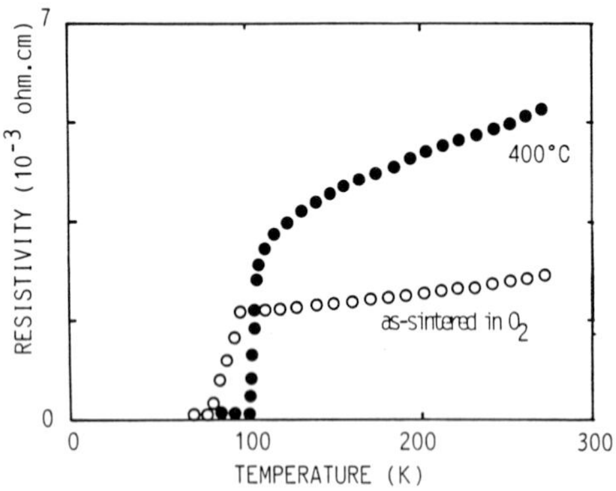

Fig.2 Temperature dependence of resistivity for $TlBa_2CaCu_2O_{y-v}$ quenched in a nitrogen atmosphere at T_q= 400° C and as-sintered in O_2.

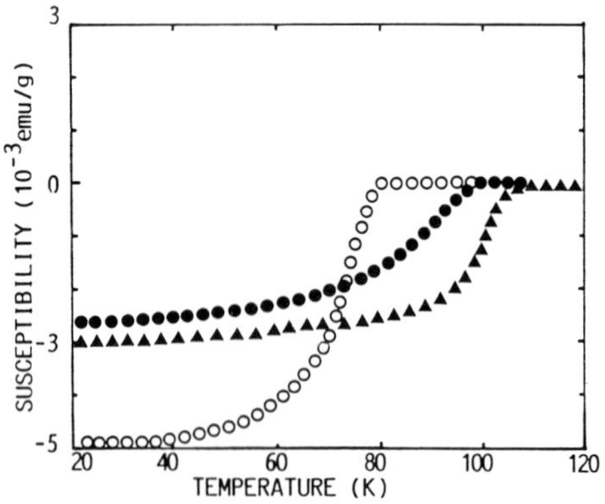

Fig.3 Temperature dependence of DC susceptibility obtained from the Meissner effect for $TlBa_2CaCu_2O_{y-v}$ quenched in a nitrogen atmosphere at various temperatures; (○)as-sintered in O_2, (●)T_q= 400° C, (▲)T_q= 500° C.

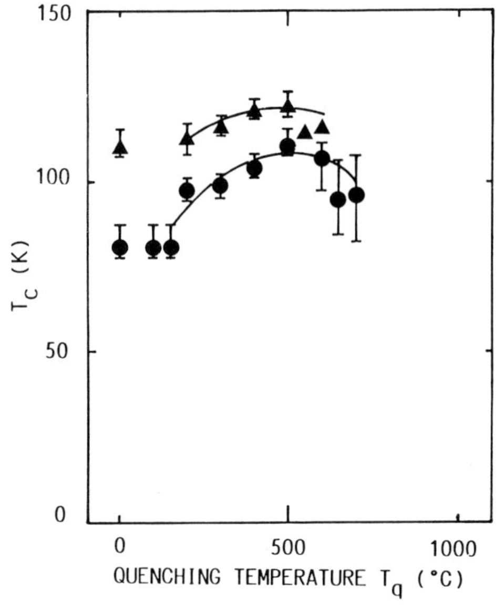

Fig.4 Variation of T_c with quenching temperature(T_q); (●)1212, (▲)1223.

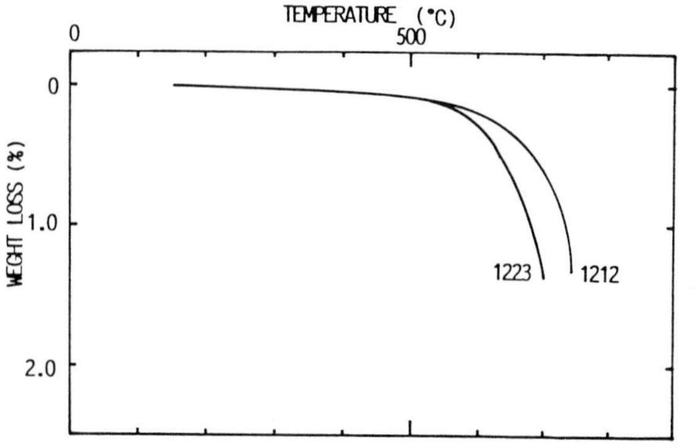

Fig.5 Weight change of $TlBa_2CaCu_2O_y$ (1212) and $TlBa_2Ca_2Cu_3O_y$ (1223) measured with heating rate of 5 °C/min.

After the quenching experiments using liquid nitrogen the specimens were examined by XPD and quenched specimens of T_q<500°C were found to be single phase. Thermogravimetry of $TlBa_2Ca_{n-1}Cu_nO_y$ in nitrogen gas, as shown in fig. 5, showed a small decrease in weight at temperatures ranging from 150° C to about 500°C and an obvious weight loss at temperatures of about 600-700° C. The weight loss in $TlBa_2Ca_2Cu_3O_y$ started at lower temperatures than in $TlBa_2CaCu_2O_y$. The weight loss of specimens quenched from various temperatures were also estimated from the weight difference before and after quenching. These weight loss data estimated by two methods are in good agreement.

Figure 6 shows the relation of tetragonal unit cell dimensions, a and c, versus oxygen loss, v, of $TlBa_2Ca_{n-1}Cu_nO_{y-v}$ (n= 2 and 3). The unit cell dimension, c, of $TlBa_2CaCu_2O_{y-v}$ decreased with increasing v up to v= 0.02 and a increased monotonously up to v= 0.05, while the unit cell dimension, a and c, of $TlBa_2Ca_2Cu_3O_{y-v}$ leveled off with small oxygen loss of about v= 0.01. The relation between T_c and oxygen loss for $TlBa_2CaCu_2O_{y-v}$ and $TlBa_2Ca_2Cu_3O_{y-v}$ are shown in fig. 7. T_c of $TlBa_2CaCu_2O_{y-v}$ and $TlBa_2Ca_2Cu_3O_{y-v}$ increases with increasing oxygen loss. T_c of $TlBa_2CaCu_2O_{y-v}$ reaches its maximum value of 110 K at about v= 0.035 and $TlBa_2Ca_2Cu_3O_{y-v}$ also reaches 120 K at about v= 0.025, which is considered to give the optimum doping level in these compound. These results suggested that the over-doping state of $TlBa_2Ca_{n-1}Cu_nO_y$ system is relieved with increasing number of Cu layer. This trend is in agreement with the Tl double layer system [8].

The increase of T_c was achieved by an appropriate reduction of hole number, which is accompanied by expansion of the a-axis. T_c of $TlBa_2Ca_{n-1}Cu_nO_y$ increases with increasing number of Cu layers up to n= 4, where it turns down [5]. T_c of these Tl single layer systems are inversely correlated with the variation of unit cell dimension, a, which is approximately twice the Cu-O distance within the c-plane. The T_c of two specimens, $TlBa_2CaCu_2O_y$ and $TlBa_2Ca_2Cu_3O_y$, obtained by quenching experiments showed different maximum T_c values, and with different unit cell dimensions, a, suggesting the importance of the Cu-O distance within the c-plane for superconductivity. We have already reported that the over-doping state of $TlBa_2CaCu_2O_y$ was relieved by substitution of Ca^{2+} with RE^{3+} (where RE is Nd^{3+}, Gd^{3+} and Y^{3+}) [9] and also an as-sintered specimen of $TlBa_2CaCu_2O_y$ with Tc= 80 K was found to be in an over-doping state. From stoichiometry the Cu valence of $TlBa_2Ca_{n-1}Cu_nO_y$,

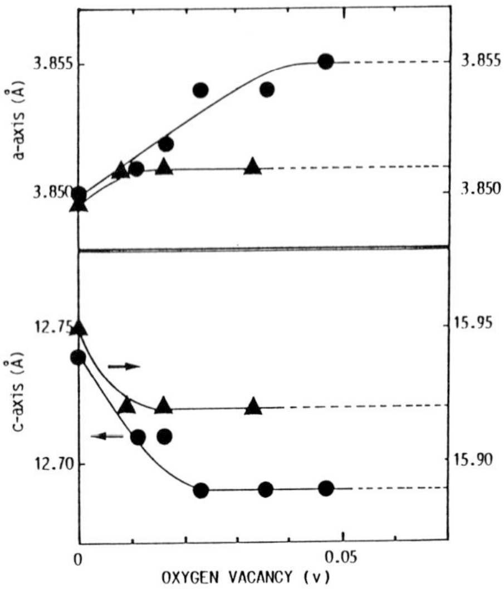

Fig.6 Unit cell dimensions a and c versus oxygen loss measured bu TGA; (●)1212, (▲)1223.

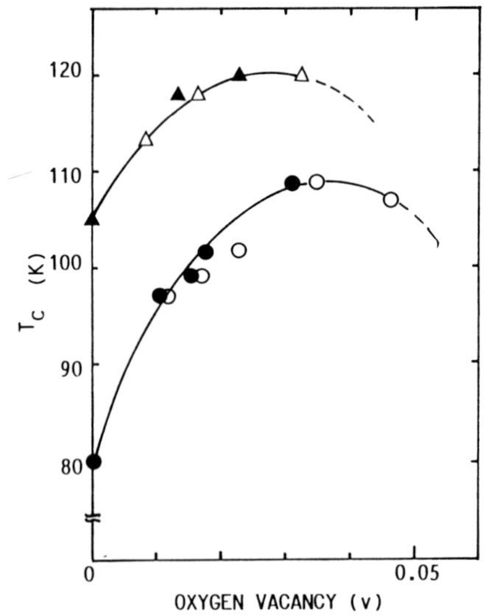

Fig.7 Dependence of T_c on the oxygen loss, v, determined by TGA (open symbols) and calculated from the weight difference before and after quenching (solid symbols); (○,●)1212, (△,▲)1223.

n= 2,3,4 and 5, system are calculated to be 2.5^+, 2.33^+, 2.25^+ and 2.2^+ respectivery. From the present results the over-doping state in oxygen-annealed specimen with n= 2 and 3 has been demonstrated.
Experiments on the specimens with n= 4 and 5 are in progress. Improvement in superconductivity of $TlBa_2Ca_{n-1}Cu_nO_y$ (n= 2 and 3) system by oxygen loss suggested the importance of small amount of hole.

ACKNOWLEDGEMENT

The authors are grateful to Mrs. R. Suzuki, College of General Education, Tohoku University, for warm encouragement and technical support. This work was partly supported by a Grant-in-Aid for Scientific Research on Priority Areas "Mechanism of Superconductivity", given by the Ministry of Education, Science and Culture, Japan.

REFERENCES

1. Sheng ZZ and Hermann AM (1988) Nature 332: 138
2. Kikuchi M, Kajitani T, Suzuki T, Nakajima S, Hiraga K, Kobayashi N, Iwasaki H, Syono Y and Muto Y (1989) Jpn. J. Appl. Phys. 28: L382
3. Kikuchi M, Nakajima S, Syono Y, Hiraga K, Oku T, Shindo D, Kobayashi N, Iwasaki H and Muto Y (1989) Physica C 158: 79
4. Parkin SSP, Lee VY, Nazzal AI, Savoy R, Beyers R and La Placa SJ (1988) Phys. Rev. Lett. 61: 750
5. Nakajima S, Kikuchi M, Syono Y, Oku T, Shindo D, Kobayashi N, Iwasaki H and Muto Y (1989) Physica C 158: 471
6. Nakajima S, Kikuchi M, Oku T, Kobayashi N, Suzuki T, Nagase K, Hiraga K, Muto Y and Syono Y (1989) Physica C 160: 458
7. Kikuchi M, Nakajima S, Syono Y, Nagase K, Suzuki R, Kajitani T, Kobayashi N and Muto Y (1990) Physica C 166: 497
8. Shimakawa Y, Kubo Y, Manako T and Igarashi H (1989) Phys. Rev B 40: 11400
9. Nakajima S, Kikuchi M, Syono Y, Kobayashi N and Muto Y (1990) Physica C 168: 57

Structure and Superconducting Properties of $Bi_{2-x}Pb_xSr_2Ca_{1-y}Y_yCu_2O_{8+d}$

NOBURU FUKUSHIMA, HIROMI NIU, and KEN ANDO

Advanced Research Laboratory, Research and Development Center, Toshiba Corporation, Saiwai-ku, Kawasaki, 210 Japan

ABSTRACT

The relation between the carrier concentration and the transition temperature was examined in the modulation-free Bi 2212 high Tc superconductor. Modulation-free $Bi_{2-x}Pb_xSr_2Ca_{1-y}Y_yCu_2O_{8+d}$ samples were prepared with appropriate values of x and y, which gave no extra oxygen in the Bi-O layer. Tc showed the initial increase and decreased with increasing Y concentration. At the same time, the volume fraction of the superconductivity decreased rapidly. These changes in the superconducting properties are in the same manner as in $Bi_2Sr_2Ca_{1-y}Y_yCu_2O_{8+d}$ samples, but the temperature range in which the Tc changes occurred was higher in the modulation free, Pb-substituted samples. Thus, the carrier concentration change by Ca/Y replacement and the Tc increase derived by Bi/Pb replacement are considered to occur independently. These results indicate that the increase in Tc by Bi/Pb substitution is not attributed to the change in the carrier concentration, but to the relaxation of the structure modulation itself.

KEY WORDS: Bi-Sr-Ca-Cu-O superconductor, structural modulation, carrier concentration, phase separation

INTRODUCTION

The Bi-Sr-Ca-Cu-O superconductors have incommensurate structural modulation along the b axis[1,2]. This type of modulation has been intensively studied, and several models[3] for the origin of the modulation have been proposed. Among them, the additional oxygen model[3,4], in which one in ten additional oxygen ions was intercalated into the Bi-O plane of $Bi_2Sr_2CaCu_2O_{8.2}$, was in good agreement with the structural data, derived from X-ray diffraction, neutron diffraction and electron microscopy. Oliver, et al.[5], explained the change in the structural modulation in $Bi_2Sr_2Ca_{1-z}Y_zCu_2O_{8+d}$ and other systems, where some elements were replaced by rare-earth elements, by means of an increase in the additional oxygen content in the Bi-O layer.

We have investigated the change in the structural modulation in $Bi_{2-x}Pb_xSr_2CaCu_2O_{8+d}$ and $Bi_{2-y}Pb_ySr_2YCu_2O_{8+d}$, which revealed that the b-axis decreased and the periodicity of modulation along the b-axis increased with increasing Pb content, and the modulation disappeared in a high Pb concentration region[6]. These results supported the additional oxygen model for the structural modulation. In fact, the precise structural refinement by means of the neutron diffraction confirmed the additional oxygen in the Bi-O layer[7].

Furthermore, a narrower transition width and an increase in Tc had been observed in our study for $Bi_{2-x}Pb_xSr_2CaCu_2O_{8+d}$, as x increased. The change in the superconducting properties was attributed to the relaxation of the structural modulation. The same tendencies were reported by other groups for Bi2212[8], Bi2201[9] and even Bi2222[10] systems. The origin of the changes in the superconductivity, though, has not yet been clarified.

In this study, the simultaneous substitution of Bi/Pb and Ca/Y was carried out in Bi 2212 system. The modulation-free samples, which have various carrier concentrations, can be expected to be obtained by selecting appropriate values for x and y in $Bi_{2-x}Pb_xSr_2Ca_{1-y}Y_yCu_2O_{8+d}$. Thus, the changes in the superconducting properties can be discussed in both aspects of the carrier concentration and the structural modulation. After summarizing the previous results in $Bi_{2-x}Pb_xSr_2CaCu_2O_{8+d}$, the structural features and the superconducting properties in $Bi_{2-x}Pb_xSr_2Ca_{1-y}Y_yCu_2O_{8+d}$ will be discussed.

EXPERIMENTAL

$Bi_{2-x}Pb_xSr_2CaCu_2O_{8+d}$ ($0 \le x \le 1$), $Bi_{2-y}Pb_ySr_2YCu_2O_{8+d}$ ($0 \le y \le 1$) and $Bi_{2-x}Pb_xSr_2Ca_{1-y}Y_yCu_2O_{8+d}$ ($0 \le x \le 0.6$, $0 \le y \le 0.4$) samples were synthesized by the ordinary solid state reaction method. Here, the values of x and y in $Bi_{2-x}Pb_xSr_2Ca_{1-y}Y_yCu_2O_{8+d}$ were selected so as to maintain the excess oxygen value d=0, considering the changes in oxygen content which accompany the cationic substitutions. These oxygen quantity changes will be described later in this issue. After calcination, $Bi_{2-x}Pb_xSr_2CaCu_2O_{8+d}$ and $Bi_{2-x}Pb_xSr_2Ca_{1-y}Y_yCu_2O_{8+d}$ samples were sintered at 750 °C for 24 hours in nitrogen flow and $Bi_{2-y}Pb_ySr_2YCu_2O_{8+d}$ was sintered at 840 °C in air. The samples were analyzed using an X-ray diffractometer (Philips PW-1700) and an electron transmission microscope (JEOL 4000FX 400 KV). Magnetization was measured by a SQUID magnetometer (HOXAN HSM-2000).

RESULTS AND DISCUSSION

1. STRUCTURAL MODULATION

Only single phase was obtained for $Bi_{2-y}Pb_ySr_2YCu_2O_{8+d}$ ($0 \leq y \leq 1$). The lattice parameters and modulation periodicity were calculated by a four dimensional description for the crystal structure of these systems[11]. The lattice parameters (a, b, c and the modulation periodicity s) were plotted versus the Pb content in Fig.2. In Fig.1, the lattice parameters for the $Bi_2Sr_2Ca_{1-z}Y_zCu_2O_{8+d}$ system, reported elsewhere[12], are displayed for comparison. A continuous decrease in the b-axis and an increase in the modulation periodicity, s, with increasing Pb content were observed in the $Bi_{2-y}Pb_ySr_2YCu_2O_{8+d}$ system, contrary to the case of the $Bi_2Sr_2Ca_{1-z}Y_zCu_2O_{8+d}$ system. Furthermore, the intensity of the (0211) peak in the XRD patterns became lower with increasing Pb content, which indicated a relaxation of the structural modulation and an increase in the s value in this system.

Single phase samples were also obtained for $Bi_{2-x}Pb_xSr_2CaCu_2O_{8+d}$ for x up to 0.6. A slight amount of secondary phase was observed, for x larger than 0.8, but the higher Tc phase, $Bi_2Sr_2Ca_2Cu_3O_{10+d}$, was not seen. Figure 3 shows the lattice parameters and modulation periodicity for the $Bi_{2-x}Pb_xSr_2CaCu_2O_{8+d}$ system. The relaxation of the structural modulation was also observed by the electron diffraction along the c-axis for both systems.

For the case of the Bi/Pb replacement, the oxygen content decreases with increasing Pb content, assuming that the Pb ions introduced into the system are divalent, to result in less additional oxygen in the Bi-O layer. The less additional oxygen in the Bi-O layer yields a longer periodicity of the modulation and a shorter b-axis. Furthermore, no additional oxygen exists beyond a certain limiting value of the Pb content, and the ordinary structural modulation (Bi-type modulation)[13], which is observed in $Bi_2Sr_2CaCu_2O_{8+d}$ disappears. It has been pointed out that the Bi/Pb replacement yields a new modulation mode (Pb-type modulation)[13] which can not be detected by an electron diffraction along the c-axis. This new type of modulation was caused by the Bi-Pb ordering and is not assumed to be accompanied with the additional oxygen.

To examine the relationship between the oxygen content and the structural modulation, quantitative analyses of the oxygen content in $Bi_2Sr_2Ca_{1-z}Y_zCu_2O_{8+d}$ and $Bi_{2-y}Pb_ySr_2YCu_2O_{8+d}$ were carried out by a titration method. The results, shown in Figs.4 and 5, demonstrate a full agreement with the assumption described above, namely the composition which shows the longer modulation periodicity has less additional oxygen. This supports the validity of the additional oxygen model for the structural modulation.

The additional oxygen model will also give a solution to the question concerning the origin of the charge carriers in these compounds. That is, holes doped into the Bi-Sr-Ca-Cu-O superconductors are attributed to the additional oxygen in the Bi-O layers.

As mentioned above, the origin of the ordinary "Bi-type" modulation can be considered to be the additional oxygen in the Bi-O layer. The results of the oxygen quantity analyses denote that the charge difference of Bi^{3+}/Pb^{2+} substitution was fully compensated for by the decrease in oxygen content, while the Ca^{2+}/Y^{3+} replacement was only partially compensated for. Thus, it can be expected to synthesize the modulation-free (the additional oxygen value d=0) samples with various carrier concentration values, by selecting the appropriate values for x and y in $Bi_{2-x}Pb_xSr_2Ca_{1-y}Y_yCu_2O_{8+d}$. We synthesized the samples with x=0.4+0.4y and x=0.4+0.5y ($0 \leq y \leq 0.4$) and obtained the modulation-free samples with less carrier concentrations. Figure 6 shows the zone axis pattern for $Bi_{1.48}Pb_{0.52}Sr_2Ca_{0.7}Y_{0.3}Cu_2O_8$. The satellite spots, caused by the structural modulation, were not observed in these electron diffraction patterns.

2. Superconducting Properties

Figure 7 shows the results of magnetization measurements for the $Bi_{2-x}Pb_xSr_2CaCu_2O_{8+d}$ system. The measurements were carried out at 1 mT after zero-field cooling. The transition temperature width became narrower for samples with high Pb contents, namely the Pb introduction into this system defined the superconducting transition more sharply. Furthermore, the transition shifted toward the higher temperature side, up to 95 K for x=0.4, and then Tc decreased to the lower temperature side for higher values of x.

Since no evidence for the existence of the $Bi_2Sr_2Ca_2Cu_3O_{10+d}$ phase was observed in the samples, according to the XRD results, and the volume fraction of superconductivity above 85-90 K was fairly large for samples which showed enhanced transition temperatures, the change in superconducting properties is considered to occur within the 2212 phase. There are several possible reasons for these changes, such as:

(1) Altered electronic states, caused by the relaxation of structural

modulation and/or a shorter ionic distance between Cu and O along the b-axis.

(2) A change in hole concentration, caused by the Bi/Pb replacement, which was not fully compensated for by the change in oxygen content.

The relationship between the hole concentration and the transition temperature in high-Tc oxide superconductors has been clarified. In these systems, the transition temperature shows an initial increase in proportion to the carrier concentration, and, after displaying the maximum, Tc decreases with the increasing carrier concentration. The enhanced Tc in Bi/Pb-replaced samples could be considered as the result of the decrease in the carrier concentration. The oxygen quantitative analyses data, though, do not support the change in the hole concentration.

The results of the magnetization measurements for $Bi_{2-x}Pb_xSr_2Ca_{1-y}Y_yCu_2O_{8+d}$ were displayed in Fig.8. The initial increase in the onset temperature and the drastic decrease in the superconductive volume fraction were observed, which indicate the same tendency as the $Bi_2Sr_2Ca_{1-y}Y_yCu_2O_{8+d}$ system. However, the temperature range was higher in the $Bi_{2-x}Pb_xSr_2Ca_{1-y}Y_yCu_2O_{8+d}$. The superconducting properties in $Bi_{2-x}Pb_xSr_2Ca_{1-y}Y_yCu_2O_{8+d}$ show the same tendency as seen in $Bi_2Sr_2Ca_{1-y}Y_yCu_2O_{8+d}$, namely the decrease in the carrier concentration, which implies that Bi/Pb-replaced samples are still staying in the high-carrier concentration region and the effects of Bi/Pb and Ca/Y replacements occur independently. Thus, we consider that the enhanced Tc and the sharp transition in Bi/Pb replaced samples are attributed to the relaxation of the structural modulation.

In addition, we would like to add comments about the hole concentration dependence of the transition temperatures and the superconductive volume fractions. Although the clear "dome" shape has been reported for the Tc versus p concentrations[14], the results of the presently reported magnetization experiments did not reproduce the same situation, even in the simple $Bi_2Sr_2Ca_{1-y}Y_yCu_2O_{8+d}$ system. This problem may come from the definition of Tc, but it is considered that the phase separation[15-17] will be more likely, because of the drastic decrease in the superconductive volume fraction.

In any case, the phase separation is also considered to occur, accompanied with the decrease in the carrier concentration, independent occurrence of the relaxation of the modulation by Bi/Pb replacement and the change in p concentration by Ca/Y replacement is still valid to explain the results in the $Bi_{2-x}Pb_xSr_2Ca_{1-y}Y_yCu_2O_{8+d}$ system.

CONCLUSION

The change in the structural modulation was investigated in $Bi_{2-x}Pb_xSr_2CaCu_2O_{8+d}$ and $Bi_{2-y}Pb_ySr_2YCu_2O_{8+d}$. The b-axis decreased and the periodicity of modulation along the b-axis increased with increasing Pb content. The modulation disappeared in a high Pb concentration region. The results were well explained, employing the additional oxygen model for the structural modulation, where the additional oxygen content decreased with increasing Pb content. A narrower transition width and an increase in Tc were observed in $Bi_{2-x}Pb_xSr_2CaCu_2O_{8+d}$, as x increased. The change in the superconducting properties was attributed to the relaxation of the structural modulation, considering the superconducting properties of the $Bi_{2-x}Pb_xSr_2Ca_{1-y}Y_yCu_2O_{8+d}$ system. The occurrence of the phase separation is suspected also in these systems.

Acknowledgments-The authors would like to thank S.Takeno, S.Nakamura and M.Hayashi for sample characterization and valuable discussions.

REFERENCES

1) Y.Matsui,H.Maeda,Y.Tanaka and S.Horiuchi:Jpn.J.Appl.Phys.27(1988)L361
2) T.M.Shaw,S.A.Shivashanker,S.J.LaPlaca,J.J.Cuomo,T.M.McGuire,R.A.Roy, K.H.Kelleher and D.S.Yee:Phys.Rev.B37(1988)9856
3) H.W.Zandbergen,W.A.Groen and F.C.Mijlhoff:Physica C 156(1988)325
4) E.A.Hewat,J.J.Capponi and M.Marezio:Physica C 157(1989)502
7) S.Oliver,W.A.Groen,C.van der Beek and H.W.Zandbergen:PhysicaC157(1989)531
6) N.Fukushima,H.Niu,S.Nakamura,S.Takeno,M.Hayashi and K.Ando: Physica C 159(1989)777
7) A.Yamamoto,M.Onoda,E.Takayama-Muromachi,F.Izumi,T.Ishigaki and H.Asano: Phys.Rev.B42(1990)4228
8) S.Kambe,T.Matsuda,M.Takahashi,M.Kawai and T.Kawai:Phys.Rev.B42(1990)2669
9) S.Kambe,M.Kawai and T.Kawai:Solid State Commun.75(1990)435
10) K.Remschnig,J.M.Tarascon,R.Ramesh,G.W.Hull and P.Rogl:PhysicaC170(1990)284
11)M.Onoda,A.Yamamoto,E.Takayama-Muromachi and S.Takekawa: Jpn.J.Appl.Phys.27(1988)L833
12)H.Niu,N.Fukushima and K.Ando:Jpn.J.Appl.Phys.28(1989)L784
13)H.W.Zandbergen,W.A.Groen,A.Smit and G.van Tendeloo:PhysicaC168(1990)426
14)W.A.Groen,D.M.de Leeuw and L.F.Feiner:physicaC165(1990)55
15)J.D.Gorgensen,B.Dabrowski,S.Pei,D.G.Hinks and L.Soderholm: Phys.Rev.B38(1988)11337
16)P.Lightfoot,D.R.Richards,B.Dabrowski,D.G.Hinks,S.Pei,D.T.Marx,A.W.Mitchel Y.Zheng and J.D.Jorgensen:PhysicaC168(1990)627
17)K.Yoshimura,Y.Nishizawa,Y.Ueda and K.Kosuge:J.Phys.Soc.Jpn,59(1990)3073

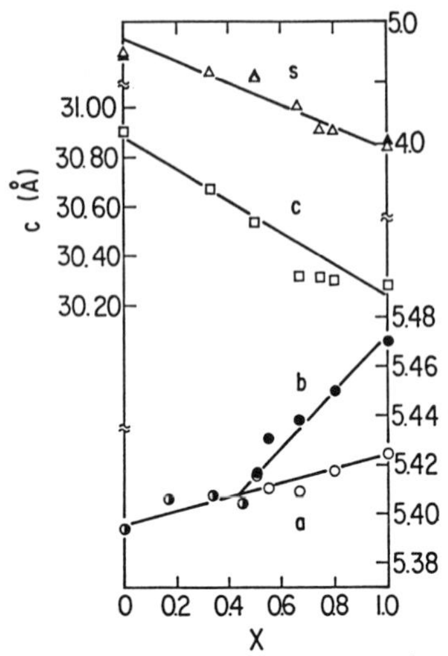

Fig.1 Lattice parameters for $Bi_2Sr_2Ca_{1-x}Y_xCu_2O_{8+d}$

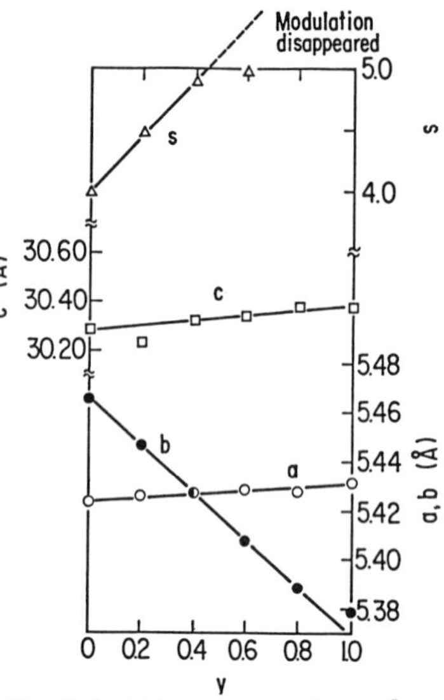

Fig.2 Lattice parameters for $Bi_{2-x}Pb_xSr_2YCu_2O_{8+d}$

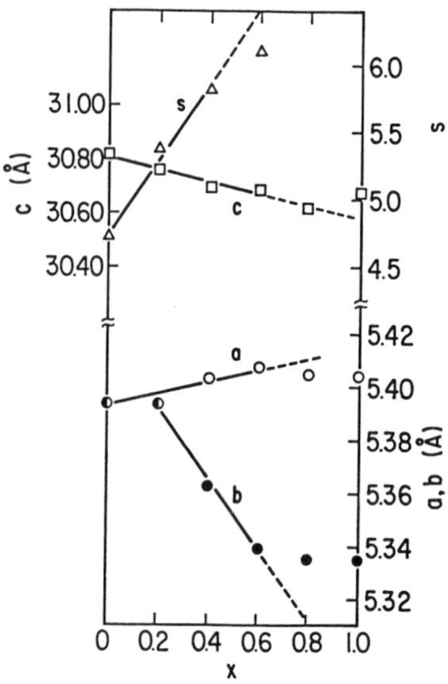

Fig.3 Lattice parameters for $Bi_{2-x}Pb_xSr_2CaCu_2O_{8+d}$

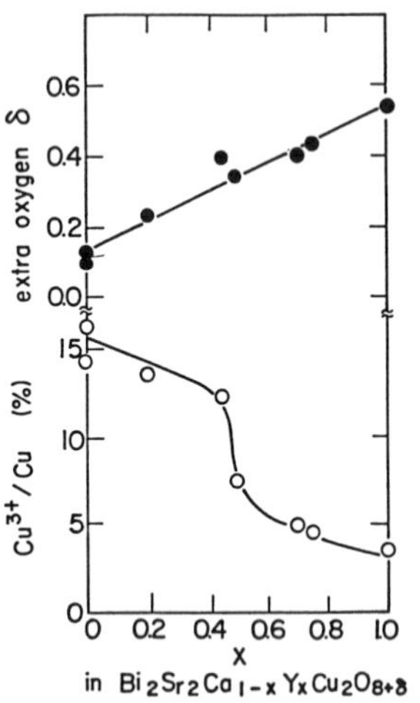

Fig.4 Additional oxygen values for $Bi_2Sr_2Ca_{1-x}Y_xCu_2O_{8+d}$

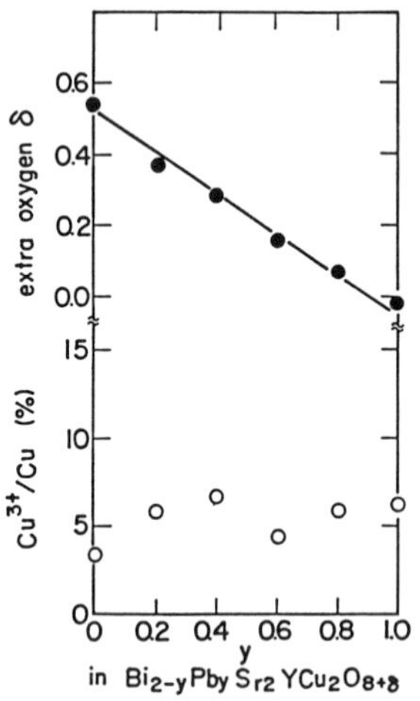

Fig.5 Additional oxygen values for $Bi_{2-x}Pb_xSr_2YCu_2O_{8+d}$

Fig.6 Zone axis electron diffraction pattern of $Bi_{1.48}Pb_{0.52}Sr_2Ca_{0.7}Y_{0.3}Cu_2O_{8+d}$ with incident beam along the c-axis

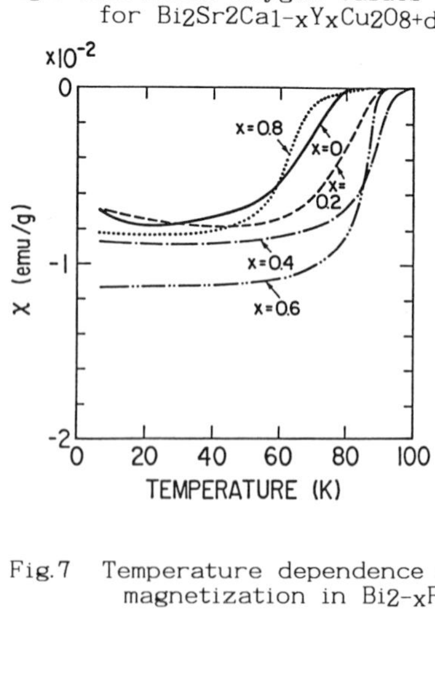

Fig.7 Temperature dependence of magnetization in $Bi_{2-x}Pb_xSr_2CaCu_2O_{8+d}$

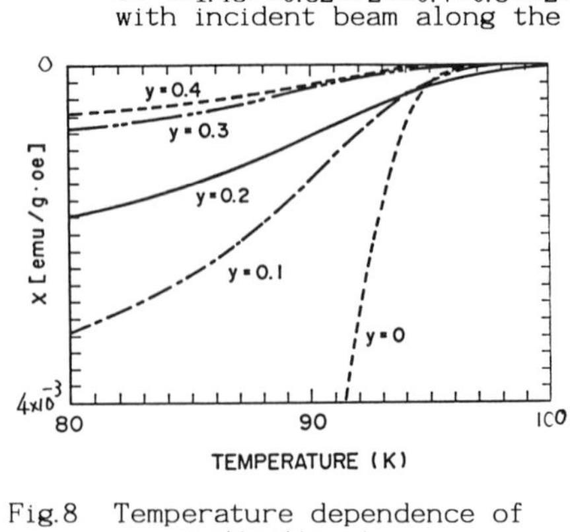

Fig.8 Temperature dependence of magnetization in $Bi_{2-x}Pb_xSr_2Ca_{1-y}Y_yCu_2O_{8+d}$ $(x=0.4+0.4y, 0 \le y \le 0.4)$

Superconducting Properties in Alkali Element Doped $Bi_2Sr_2CaCu_2O_8$

S.X. Dou[1], H.K. Liu[1], W.M. Wu[1], W.X. Wang[1], C.C. Sorrell[1], R. Winn[2], and N. Savvides[2]

[1] School of Materials Science and Engineering, University of New South Wales, Sydney, NSW 2033, Australia

[2] CSIRO Division of Applied Physics, Sydney, NSW 2070, Australia

ABSTRACT

The effect of alkali element doping on the superconducting properties of the Bi-Sr-Ca-Cu-O system has been investigated. It was found that a T_c of 94 K and T_o of 90 K in alkali element doped $Bi_2Sr_2CaCu_2O_8$ were achieved through the use of a melt-processing technique. Alkali elements assist the melting process, promote grain growth and grain alignment, and produce dense dislocation lines. A possible mechanism for the T_c enhancement is proposed in the bulk materials.

KEY WORDS: Transition temperature, Alkali doping, Melting

INTRODUCTION

It has been established that there are three superconducting phases in the Bi-Sr-Ca-Cu-O system: $Bi_2Sr_2CuO_6$ (2201) with a T_c up to 20 K [1], $Bi_2Sr_2CaCu_2O_8$ (2212) with a T_c near 85 K [2-4], and $Bi_2Sr_2Ca_2Cu_3O_{10}$ (2223) with a T_c near 110 K [2,4-6]. Single-phase 2212 has been prepared with the T_c varying from 70 to 92 K depending on the processing conditions [7-9]. The T_c of 2212 can be raised to 92 K by quenching from high temperature or treating in a low oxygen partial pressure [9]. Recently, an improvement in the T_c of 2212 by doping with Li has been reported by Kawai and co-workers [8]. They found that Li raised the T_c of 2212 while K depressed the T_c significantly, and Na did not affect the T_c. In the present work, similar results for Li doping were obtained. However, the results for K and Na doping are considerably different in that K and Na were found to have a beneficial effect on the T_c of 2212.

EXPERIMENTAL PROCEDURE

Mixtures of Bi_2O_3 (99.9%), $SrCO_3$ (99.5%), A_2CO_3 (99%) (where A = Li, Na, or K), $CaCO_3$ (99%), and CuO (99.9%) in appropriate proportions were ground with a mortar and pestle and calcined at 720°C for 12 h and at 780°C for 10 h. The calcined powders were then pressed into pellets and sintered or partially melted in Ag boats at 830 to 890°C for 5 min to 15 h.

The electrical resistivity was measured by the standard four-probe d.c. technique. The a.c. magnetic susceptibility was measured by a mutual inductance method at a frequency of 83 Hz. Microstructural and compositional analyses were performed with a JEOL JSM-840 scanning electron microscope (SEM) and JEOL FX 2000 transmission electron microscope (TEM), both equipped with a Link Systems AN10000 energy dispersive spectrometer (EDS). X-ray powder diffraction (XRD) patterns were obtained with a Philips type PW1140/00 powder diffractometer using CuKα radiation and scanning rate of 2θ = 1°/min. Chemical analyses were carried out with a Labtam International Plasmalab inductively coupled plasma atomic emission spectrometer.

RESULTS

Figure 1 shows resistivity-temperature curves for samples with the cation ratio Bi:Sr:Ca:Cu:A = 2.2:1.8:1.05:(2.15-x):x, where A = Li, x = 0.7; A = Na, x = 0.6; and A = K, x = 0.7. The heat treatment conditions for these samples were as follows. The Li-doped sample was sintered at 830°C for 1 h in nitrogen and cooled at a rate of 10°C/h to 790°C and then at 60°C/h to room temperature. The Na-doped sample was treated at 885°C for 5 min in air and cooled at a rate of 60°C/h to room temperature. The K-doped sample was treated at 850°C for 10 h and cooled at a rate of 60°C/h to room temperature. It can be seen that Na raised the T_c to 94 K and T_o to 90 K, while K and Li raised the T_c the same degree, where the T_c = 92 K and T_o = 88 K. The undoped standard sample, which was partially melted at 890°C for 1 h in air and cooled at 60°C/h to room temperature, showed a T_c = 84 K and T_o = 80 K. The a.c. magnetic susceptibility measurements confirmed the enhancement in the T_c, as shown in Figure 2.

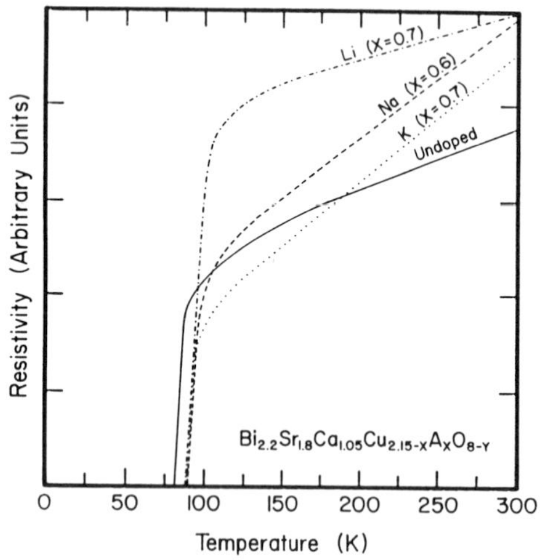

Fig. 1. Effect on the transition of alkali doping

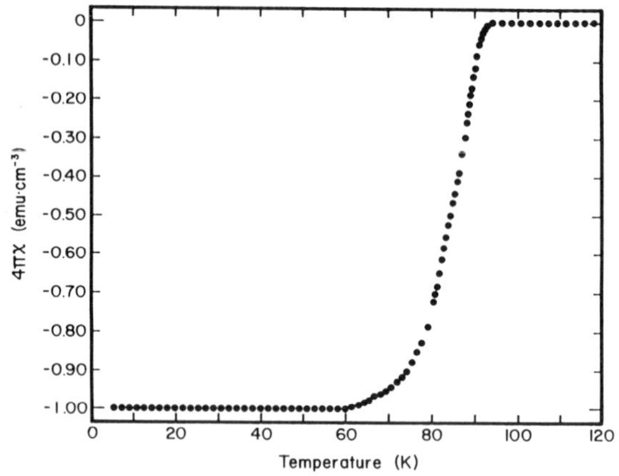

Fig. 2. Effect on the a.c. magnetic susceptibility of the Na-doped sample (0.01G)

The alkali element doped samples contained a high density of dislocations, as shown in Figure 3. The dislocation line density in the sintered sample is typically 10^7 lines/cm^2 though it can vary depending on the detailed sample preparation procedure [10]. The dislocation line density in K-doped samples [Figure 3] is estimated at 2 x 10^{11} lines/cm^2, which is much higher than that of normally sintered samples and is comparable with the values observed in melt-texture-growth samples [11]. The density of dislocation lines in the alkali element doped samples is comparable with the magnetic flux lines at a moderate field. For example in a field of 2 T, the flux lines in the high T_c superconductor are established at a spacing of 30 nm [12], while dislocation lines in the alkali element doped sample have a spacing 20 nm. Thus, it is expected that the dislocation lines interact with the magnetic flux lines, acting as effective pinning centres.

DISCUSSION

The enhancement of the T_c through alkali element doping may be attributed to three possible factors: the presence of the high T_c phase (2223), the fluxing action of the alkali elements, and the substitution of the alkali elements in the superconducting phases. In the following, the three possible mechanisms for the T_c enhancement are discussed.

1. Absence of the 2223 phase in alkali doped samples

Since Bi-based materials form a series of superconducting phases, which are often found in an intergrowth pattern, the enhancement of the T_c may be attributed to the presence of small amounts of the 2223 phase. Figure 4 shows X-ray diffraction patterns for alkali element doped samples. The XRD data indicate that all samples consisted of 2212 as the major phase. The minor phases were 2201 and some impurity phases including $SrCaCu_2O_4$, CuO, and (Li,Na,K)-Bi-Sr-Ca-O. However, no 2223 phase was detected. It is believed that the enhancement in the T_c for the alkali element doped samples cannot be attributed to the presence of the high-T_c phase (2223).

2. Effect of alkali element substitution

It has been reported that a T_c of 92 K for the 2212 phase has been achieved by annealing in low oxygen partial pressures or by quenching [9]. The substitution of monovalent ions for divalent ions should have a reducing effect similar to that of low oxygen partial pressure treatment. In the present work, EDS analyses of single crystals revealed that Na was incorporated in the superconducting phase. The results listed in Table 2 are average values of 8 individual analyses on a TEM powder sample. K^+ was also detected in the superconducting phase in the samples sintered for 10-30 h at 850°C, as listed in Table 1. It is interesting to note that the K content in the samples decreased with increasing sintering times, vanishing in the sample sintered for the longest period of time, although the T_c remained nearly unchanged at 90 K. Since the T_c was unaffected while K volatilised out of the samples, it is unlikely that the incorporation of K in the superconductor structure was responsible for the enhancement of the T_c.

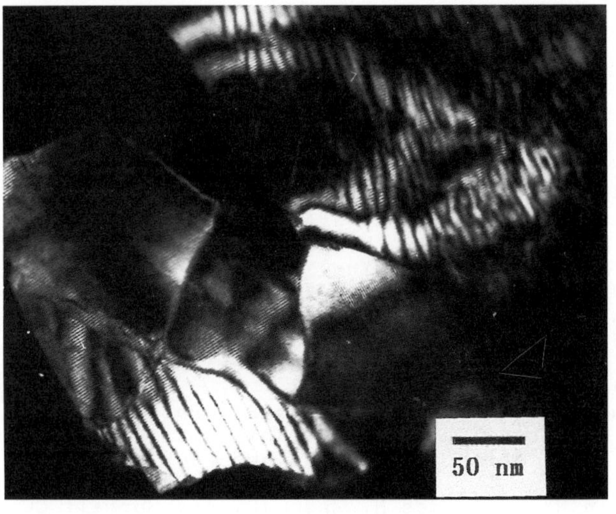

Fig. 3. TEM image showing dislocation lines and modulation in the K-doped sample

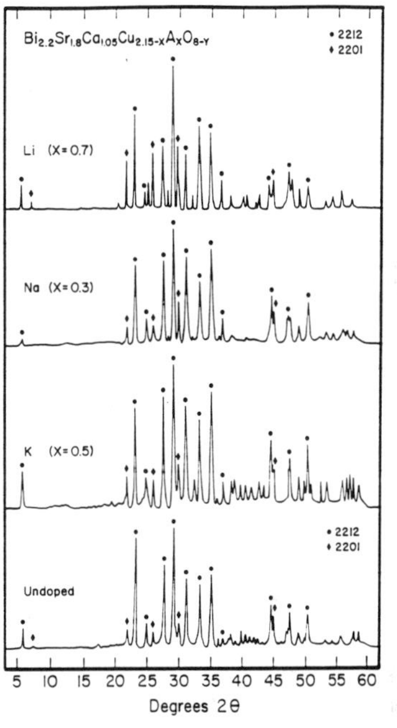

Fig. 4. X-ray diffraction patterns for alkali element doped samples

3. Effect of fluxing action by alkali element doping

At this stage, it can be argued that the enhancement of the T_c by alkali element doping may be attributed to liquid-phase sintering. Li_2O and Bi_2O_3 form a eutectic at a temperature as low as 700°C at Li_2O/Bi_2O_3 = 0.1 [13]. This explains why low sintering temperatures could be used with Li additions. The liquids provide fast diffusion paths for Ca, Sr, and Cu, which accelerates the reaction to form the superconductor. This is evident from the fact that exaggerated grain growth and alignment were achieved through liquid-phase sintering within a short period of time. From Table 1, it is evident that the major phase (2212) in alkali element doped samples has a relatively high Ca/Sr ratio, indicating that Ca is enriched during melting due to assisted Ca diffusion through the liquid phase. In undoped samples, Ca is often depleted, resulting in low T_c values [7]. The fact that the T_c was enhanced to the same degree by K, Na, and Li doping strongly suggests that the enhancement of the T_c may be attributed mainly to the liquid-phase sintering resulting from alkali metal additions.

In summary, the enhancement of the T_c in alkali element doped Bi-Sr-Ca-Cu-O through melt-processing is reported. Values of T_c of 94 K and T_o of 90 K have been achieved in K-, Na-, and Li-doped samples. It is tentatively concluded that the mechanism for the enhancement of the T_c is by a fluxing action that provides a fast diffusion path for Ca and Cu to the superconducting phase, leading to the desirable 2212 stoichiometry. On the other hand, the substitution of alkali elements in the superconducting phase may give rise to a reducing effect of the oxygen content, leading to an enhancement of the T_c.

Table 1. EDS analyses for alkali element doped Bi-Sr-Ca-Cu-O materials

Bi :	Sr :	Ca :	Cu :	Na	
2.20	1.62	0.97	2.00	0.31	(sintered at 885°C for 5 min)
Bi :	Sr :	Ca :	Cu :	K	
2.11	1.73	0.98	1.68	0.88	(sintered at 850°C for 10 h)
2.26	2.16	0.66	1.67	0.24	(sintered at 850°C for 20 h)
2.26	1.94	0.95	1.93	0.00	(sintered at 850°C for 35 h)

ACKNOWLEDGEMENTS

The authors are grateful to Metal Manufactures Ltd. and to the Commonwealtl Department of Industry, Technology and Commerce for financial support.

REFERENCES

1. Michael L, Hervieu M, Bordet MM, Grandi A, Deslandes F, Provost J, and Raveau B, (1987) Zeit.f.Phys. B68:421.
2. Maeda H, Tanaka Y, Fukolami M, and Asano T, (1988) J.Appl.Phys. 27:L209.
3. Hazen RM, Prewitt CT, Angel RJ, Ross NL, Finger LW, Hadidiacos CG, Veblen DR, Heaney PJ, Hor PH, Meng RL, Sun YY, Wang YQ, Xue YY, Huang ZJ, Gao L, Bechtold J, and Chu CW, (1988) Phys.Rev.Lett. 60:1174.
4. Subramanian MA, Torardi CC, Calabrese JC, Gopalakrishnan J, Morrissey KJ, Askew TR, Flippen RB, Chowdhry U, and Sleight AW, (1988) Science 239:1015.
5. Tarascon JM, Le Page Y, Barboux P, Bagley BG, Greene LH, Mckinnon WR, Hull GW, Giroud M, and Hwang DM, (1988) Phys.Rev. B 37:9382.
6. Sunshine SA, Siegrist T, Schneemeyer LF, Murphy DW, Cava RJ, Batlogg B, Van Dover RB, Fleming RM, Glarum SH, Nakahara S, Farrow R, Krajewski JJ, Zahurak SM, Waszczak JV, Marshall JH, Marsh P, Rupp LW, and Peck WF, (1988) Phys.Rev. B 38:893.
7. Dou SX, Liu HK, Bourdillon AJ, Tan NX, Savvides N, Zhou JP, and Sorrell CC, (1988) Supercond.Sci.Technol. 1:178.
8. Kawai T, Horiuchi T, Mitsui K, Ogura K, Takagi S, and Kawai S, (1989) Physica C, 161: 561.
9. Tallen JL, Buckley RG, Gilberd PW, Presland MR, Brown IWM, Bowden ME, Christian LA, and Goguel R, (1988) Nature 333:153.
10. Chen CH, (1990) In: Ginsberg DM(ed), in Physical Properties of High Temperature Superconductors II World Scientific Singapore, p 199.
11. Nakahara S, Jin S, Sherwood RC, and Tiefel TH, (1989) Appl.Phys.Lett. 54: 1926.
12. Kitagawa K, (1989) Amer.Ceram.Soc.Bull. 68:880.
13. Levin EM and Roth RS, (1964) J.Res.Nat.Bur.Stand. 68A:198.

3.4 Phase Diagrams

Phase Diagram of Y-Ba-Cu-O System

WEI ZHANG and KOZO OSAMURA

Department of Metallurgy, Kyoto University, Sakyo-ku, Kyoto, 606 Japan

ABSTRACT

The phase diagrams of Y-Ba-Cu-O four component system and the related ternary systems were investigated. The isothermal sections of Y-Ba-Cu-O quaternary system at 1263, 1223 and 1173K were determined under artificial air. The sequence diagram of univariant reactions in this four component system was proposed. The existence of two invariant reactions were also suggested.

KEY WORDS: phase diagram, Y-Ba-Cu-O system, isothermal section

INTRODUCTION

Since the discovery of superconductivity in the system Y-Ba-Cu-O, the phase diagram of this system has been investigated extensively[1-3]. In order to understand this system completely, we investigated this quaternary system and the related ternary systems under various oxygen partial pressures[3]. The pollution of $CO_3{}^{2-}$ to these systems were also investigated.

EXPERIMENTAL

After given amounts of Y_2O_3, $BaCO_3$ (BaO_2) and CuO powder were weighed, they were mixed and ground for 4 h. The mixture was calcined at 1173K for 24 h and pressed with 1 MPa to form a pellet. The specimens with a melting point below 1173K were calcined at 1123K for 24 h. The pellets were sintered at various temperature and then quenched. As it has been known that CO_2 impurity in air affects seriously the phase relation of oxide system, the atmosphere was controlled. When the experiments were carried out under natural air without any manipulation, the atmosphere is called "natural air", in which oxygen with 0.21 atm partial pressure and impurities like CO_2 and H_2O are included. When the mixture of oxygen and argon gas were used, it is called "artificial air", where the partial pressure of oxygen is 0.21 atm. The content of carbon in the oxide phases was analyzed by means of oxygen combustion infrared absorption method. In order to investigate the temperature and oxygen dependence of phase relation, thermogravimetric and differential thermal analysis(TG-DTA) has been performed under a flow of mixed gas with argon and oxygen. The heating and cooling rate was 5K/min. Al_2O_3 powder was used as a standard reference. A Pt crucible was used in the present experiment. The extrapolated onset temperature on the DTA curve was used. The microstructure was examined by scanning electron microscope(SEM) and identification of the phases was performed by the X-ray diffraction(XRD) analysis and electron probe microanalysis(EPMA). In this study, some phases were represented by symbols for easy representing the phase relations in the figure.

RESULTS AND DISCUSSION

For the Y_2O_3-BaO quasibinary system, the phase diagram has been firstly reported by Roth[1], where four intermediate phases were insisted to exist. But two of them, $Y_2Ba_2O_5$ (Y_2B_2) and $Y_2Ba_4O_7$ (Y_2B_4) were pointed out to be oxycarbonates by Abbattista[2]. In present study, we confirmed these two oxycarbonates by carbon content analysis. On the other hand, we examined the samples which were annealed under artificial air and only the Y_2BaO_4 (Y_2B) and $Y_4Ba_3O_9$ (Y_4B_3) phases were detected. Therefore, only the Y_2B and Y_4B_3 phases are confirmed to be rigid oxide.

The isobaric section of BaO-CuO quasibinary system was investigated[4] in order to make clear the four component system. In this system, the

Ba_2CuO_x (B_2C) phase is formed by the peritectic reaction at 1193K by the reaction L+BaO ⇆ B_2C. At 1083K, its high temperature phase with tetragonal symmetry transforms to the low temperature one with orthorhombic symmetry. The $BaCuO_2$ (BC) phase is formed by the synthetic reaction at 1289K. Two eutectic reactions as expressed by the reaction L ⇆ BC+CuO and L ⇆ B_2C+BC occurs at 1163 and 1173K, respectively. As the coexistence of three condensed phases in equilibrium corresponds to univariant system, the reaction temperature may shift depending on oxygen partial pressure.

The isobaric section of $YO_{1.5}$-CuO_x quasibinary system in natural air[5] and under various oxygen partial pressure[6] were also investigated. In the natural air, an intermediate phase $Y_2Cu_2O_5$ (Y_2C_2) is formed by the peritectic reaction L+Y_2O_3⇆Y_2C_2 at 1393K. The liquid phase equilibrates with Y_2C_2 and Cu_2O at 1368K by the eutectic reaction L⇆Y_2C_2+Cu_2O. Cu_2O(C) phase transforms to CuO(C) at 1299K by the reaction 2CuO⇆Cu_2O+1/2O_2.

Informations on the isothermal section under constant pressure are primarily necessary to study the phase diagram. Fig.1 shows the change of the relative X-ray intensity of each phase at different temperature as a function of composition. For instance, at 1263K (Fig.1(a)), the 123 phase exists in the composition region less than 50mol%CuO and the Y_2C_2 phase appears in the region larger than 55mol%CuO. Combining with the EPMA analysis, the phases existing under equilibrium in this part at 1263K were judged as listed in the figure. Fig.2 shows the isothermal section determined under artificial air. Four phases $YBa_2Cu_3O_x$ (123), Y_2BaCuO_5 (211) and $YBa_6Cu_3O_x$ (163) and $YBa_4Cu_3O_x$ (143) are present inside the triangle. The whole section is separated in many segments. In each of the segments, two phase equilibrium or three phase one is realized except the range of liquid phases. The isothermal section under natural air was also investigated in this study. Comparing the two sections, it is found that the phase relation in the BaO rich part is different between the two conditions. Instead of the 163 and 143 phase, a $YBa_8Cu_4O_x$ (184) phase was detected to exist with a solid solution[6]. The carbon content measurement of the specimens in the BaO rich part showed that the oxycarbonate is easy to be formed in this part under the existence of CO_2. The 184 phase is confirmed to be another oxycarbonate phase except of Y_2B_2 and YB_2 phases, which do not appear under purified air excluding CO_2 or under artificial air.

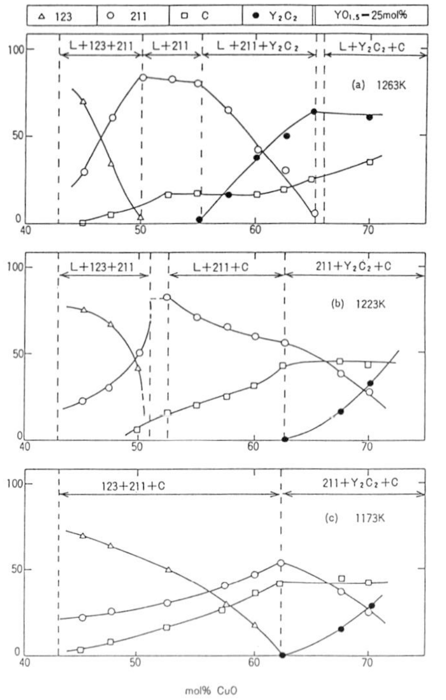

Fig.1 X-ray intensities of phases as a function of CuO mol%.

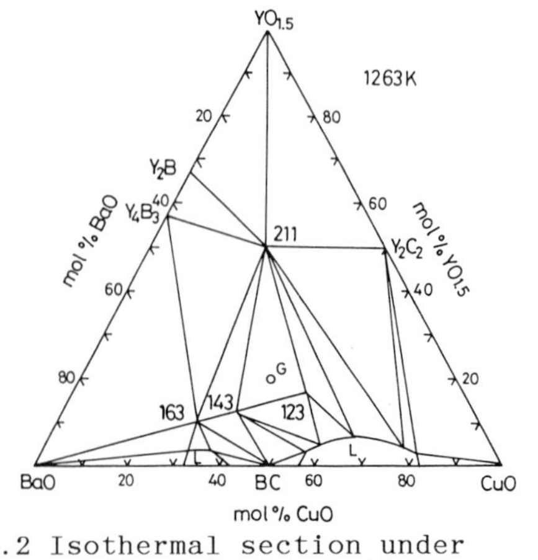

Fig.2 Isothermal section under artificial air at 1263K.

The isothermal section at 1223K under artificial air is shown in Fig.3. The phase relation changed mainly in two regions comparing with that in Fig.2. In the region limited by 211-Y_2C_2-C-L, two three-phase triangles, Y_2C_2+211+C and 211+L+C at 1223K are present, while another set of triangles, Y_2C_2+211+L and Y_2C_2+L+C existed at 1263K, as shown in Fig.1(a) and (b). These phase change can be expressed by the reaction U_5: L+Y_2C_2 ⇆ 211+C. From the thermal analysis, the reaction temperature was determined to be 1245K. Similarly, in the 123-143-BC-L region, another transition reaction(U_4) occurred as given by the equation U_4 : L+143 ⇆ 123+BC.

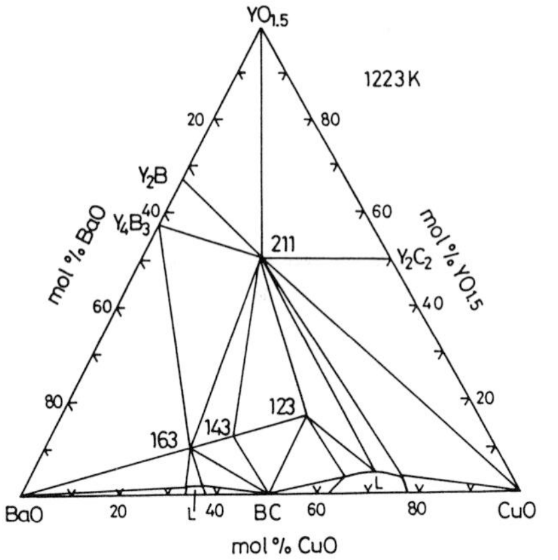

Fig.3 Isothermal section under artificial air at 1223K

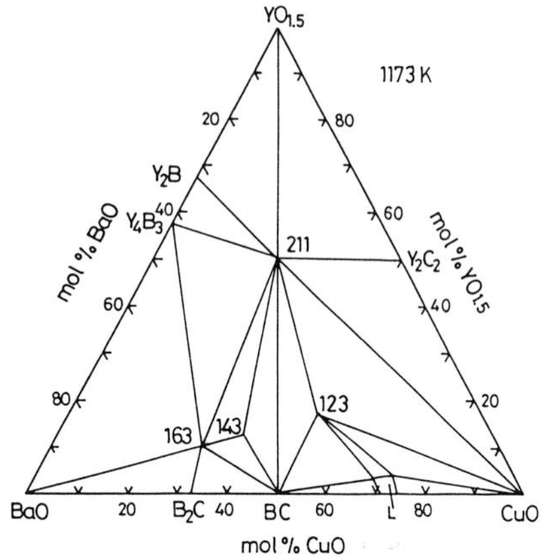

Fig.4 Isothermal section under artificial air at 1173K

Figure 4 shows the isothermal section at 1173K under artificial air. Still the liquid phase remains in the region limited by 123-BC-C. Two univariant reactions are estimated to occur at temperatures between 1223 and 1173K, by comparing two diagrams shown in Fig.3 and 4. From the phase relation change as shown in Fig.1(b) and (c), a transition reaction related with the liquid phase was suggested to occur in the region limited by 123-211-L-C and was expressed by U_6 : L+211 ⇆ 123+C. The reaction temperature was determined to be 1208K from the change in TG-DTA curves. Another transition reaction in the region surrounded by 211-143-BC-123 was confirmed to occur by means of XRD and expressed by U_{S3} : 123+143 ⇆ 211+BC. The reaction temperature was determined to be nearly 1183K from similar XRD experiment at different temperature.

The liquid phase remains still at 1173K as shown in Fig.4. As in the quasibinary system BaO-CuO, the eutectic reaction L ⇆ BC+C occurs at 1163K with composition of BaO-71.5mol%CuO. Near there, the ternary eutectic reaction E_1 was found to exist as expressed by E_1 : L ⇆ 123+BC+C. From the DTA analysis, a common reaction was observed to occurred at 1157K for all the specimens examined. The eutectic composition was determined to be 3%$YO_{1.5}$-30%BaO-67%CuO.

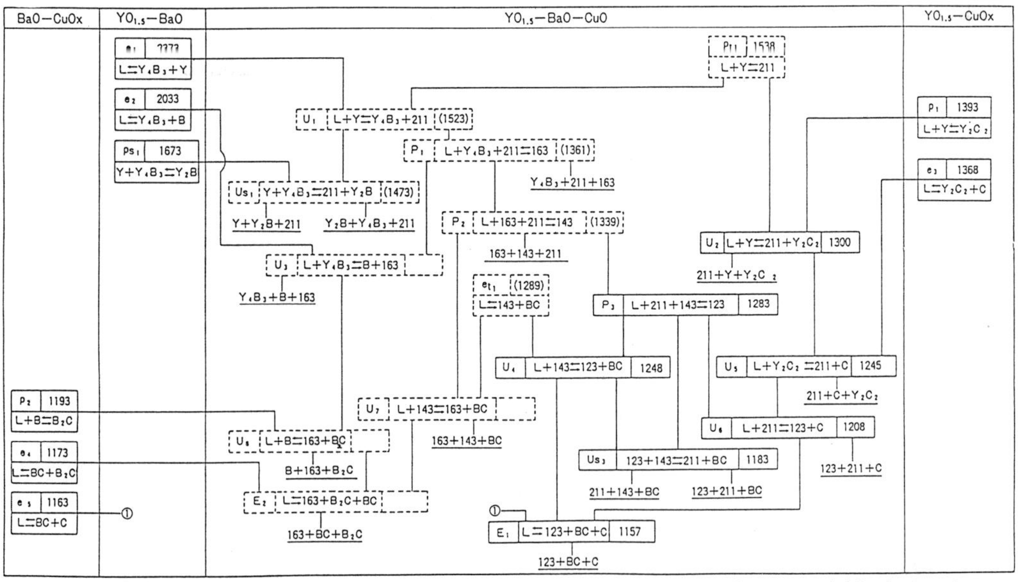

Fig.5 Sequence diagram of univariant reactions under artificial air

The 123 phase is well known to melt incongruently. In this experiment, a large thermal change occurs at 1283K under 0.21atmO_2. When the specimen G as shown in Fig.3 with composition 20%$YO_{1.5}$-39%BaO- 41%CuO was annealed at 1263 or 1303K and quenched to room temperature, three phases 123, 143 and 211 were detected in equilibrium at 1263K as shown by Fig.3. On the other hand, the main phases at 1303K were detected to be 211, 143 and BC phases. As the specimen were melted in 1303K and the melting temperature of BC phase is lower than 1303K, the BC phase detected in specimen G at 1303K were considered to be formed from liquid. Therefore, a ternary peritectic reaction was suggested to be given by the equation P_3: L+211+143 $\leftrightarrows$ 123. This reaction indicates that the three phase triangle L+211+143 exists above the reaction temperature and the solid three phase triangle 123+211+143 should exist below it. For the reaction is an univariant one, the melting temperature of 123 phase would change according to oxygen partial pressure.

It is possible to connect sequentially the above mentioned univariant reactions. A sequence diagram for univariant reactions in the present Y-Ba-Cu-O four component system is proposed as shown in Fig.5. The diagram is restricted under artificial air, where the oxygen partial pressure is constant of 0.21 atm. The univariant reactions in terminal three component systems as in first, second and the fourth columns have been mentioned in the ternary systems. These reactions are extended to four component system. As mentioned above, some of the univariant reactions have been confirmed by experiment, but almost all univariant reactions located in the Y_2O_3 rich and BaO rich sides are only hypothetical and expressed by dotted lines. This sequence diagram is important to know the framework of phase relationship in the present four component system under a constant oxygen pressure.

The further study about the oxygen dependence of each univariant reaction is undergoing now. Some results showed the existence of invariant reaction in the Y-Ba-Cu-O four component system. Fig.6 shows the result of the oxygen dependence of the ternary eutectic reaction E_1 and the transition reaction U_6 together with the decomposition reaction of 2CuO $\leftrightarrows$ Cu_2O+1/2O_2. The temperature of each reaction decreases with the decreasing of the oxygen partial pressure. Both of the curves of reaction E_1 and U_6 cross the decomposition curve of CuO at A and B, respectively. It is possible to consider that the four solid phases coexist with the liquid phase at each of the point, as expressed below, respectively. For point A, a reaction by L$\leftrightarrows$123+BC+CuO+Cu_2O occurs at almost 1103K under 0.0032 atmO_2, and for B, a reaction expressed by L+211 $\leftrightarrows$ 123+CuO+Cu_2O occurs at 1143K under 0.0085 atmO_2.

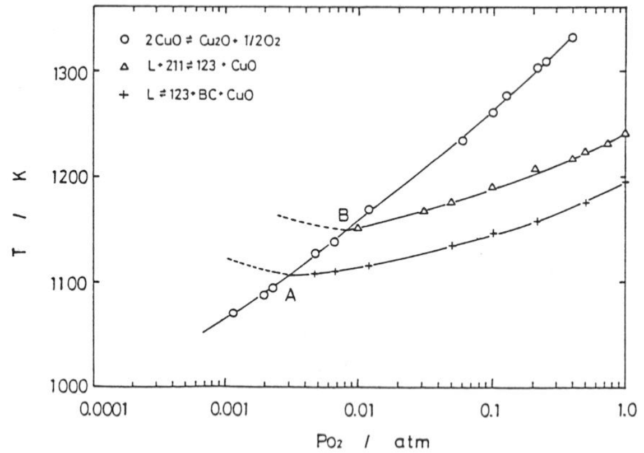

Fig.6 Oxygen dependence of univariant reactions

CONCLUSION

Some phases which are formed in natural air are not rigid oxide, but oxycarbonate. The pollution of CO_3^{2-} affects the phase diagram seriously. The phase diagrams of four component system and the three terminal ternary systems were made clear. The sequence diagram of univariant reactions has been proposed based on the experiment data. The existence of invariant reactions was also confirmed.

REFERENCES

[1] R.S.Roth, K.L.Davis and J.R.Dennis, Adv.Ceramic Mater., 2(1987) 303.
[2] F.Abbattista, M.Vallino and D.Mazza, Mat.Chem. and Phys., 21(1989) 521.
[3] K.Osamura and W.Zhang, Z.Metallkde., submitted(1990).
[4] W.Zhang and K.Osamura and S.Ochiai, J.Am.Ceram.Soc., 73(1990) 1958.
[5] W.Zhang and K.Osamura, Metall.Trans., 21A(1990) 2243.
[6] W.Zhang and K.Osamura, Z.Metallkde., 81(1990) 196.
[7] K.Osamura and W.Zhang, Jpn.J.Appl.Phys., 26(1987) L2094.

Phase Diagram of Bi-based Superconductors (2)

ARIO NAKAMURA, TERUO IZUMI, and YUH SHIOHARA

Superconductivity Research Laboratory, International Superconductivity Technology Center,
10-13, Shinonome 1-chome, Koto-ku, Tokyo, 135 Japan

Abstract

Phase equilibrium between liquid and solid was studied in the Bi-Sr-Ca-Cu-O system using a quench method together with microscopic observation and composition analysis by EPMA. Equilbrium phases at different temperatures were investigated and plotted on a pseudo-binary temperature versus composition diagram. Such phase diagrams were to predict a pseudo-ternary phase diagram. The ratio of Sr : Ca was fixed at 2 : 1 in this study to examine the pseudo-ternary phase diagram in which Sr : Ca is 2 : 1. From the results of quenches, 2201 and CuO were observed as the primary phase in the experimental compositions. The series of the primary phases were then investigated in the pseudo-ternary phase diagram in which Sr : Ca is 2 : 1. In these results, the phase including 2212 was not observed.

Key word : BSCCO, Phase diagram

Introduction

The discovery of superconductivity in Bi-Sr-Ca-Cu-O system has generated great interest in the field of condensed matter material science. Since the first report was submitted by Michel [1] and Maeda et al.[2] , research has been concentrated in the areas of crystal structures, phase diagrams, superconducting properties, and synthesis of Bi-Sr-Ca-Cu-O. We have been undertaking flux method, Czochralski method, etc. to obtain the Bi-based superconductors (2212 type) as grown. When we employ these techniques, the phase diagram is helpful, since knowledge from the diagram allows us to select starting liquid compositions from which the superconducting phases crystallize. We have reported that CaO was a primary phase of the compositions near 2212 and 2223 in the previous paper[3]. In the present paper, we have studied the pseudo-ternary phase diagrams in which Sr : Ca is 2 : 1 to find the phase having coexistence of liquid and 2212.

Experimental

We selected the plane in which Sr : Ca is 2 : 1 in the pseudo-tetra phase diagram, and studied this plane as the pseudo-ternary phase diagram as shown in figure 1. This plane would be called the (Sr_2,Ca) plane. Reagent grade starting materials of Bi_2O_3, $SrCO_3$, $CaCO_3$, CuO were used. The nominal compositions of the samples are shown in table 1 and figure 2. The batches were weighed and mixed, heated to the reaction temperature of 790°C at a rate of 400°C/h, and then held for 12 hours in the air. The reacted powders were pulverized and melted at 1400°C, quenched, and pulverized again to reduce segregation. The powders were confirmed to be amorphous by X-ray diffraction. These powders were placed in platinum crucibles (5mm in diameter x 5mm in height) and heated and held at different temperatures until equilibrium was supposed to be reached, then quenched in water. The phases existed in these samples were investigated by EPMA. These phases at equilibrium was plotted on pseudo-binary phase diagrams.

	Bi	Sr	Ca	Cu
1	30.0	25.3	12.7	32.0
2	32.0	25.3	12.7	30.0
3	33.0	25.3	12.7	29.0
4	34.0	25.3	12.7	28.0
5	38.0	25.3	12.7	24.0
6	40.0	25.3	12.7	22.0
7	42.0	25.3	12.7	20.0
8	35.0	24.0	12.0	29.0

	Bi	Sr	Ca	Cu
9	36.0	22.6	11.4	30.0
10	37.0	21.3	10.7	31.0
11	34.0	24.0	12.0	30.0
12	35.0	22.6	11.4	31.0
13	36.0	21.3	10.7	32.0
14	33.2	26.6	13.4	26.8
15	33.2	27.2	13.6	26.0

Table 1. Nominal compositions of the samples. (mol.%)

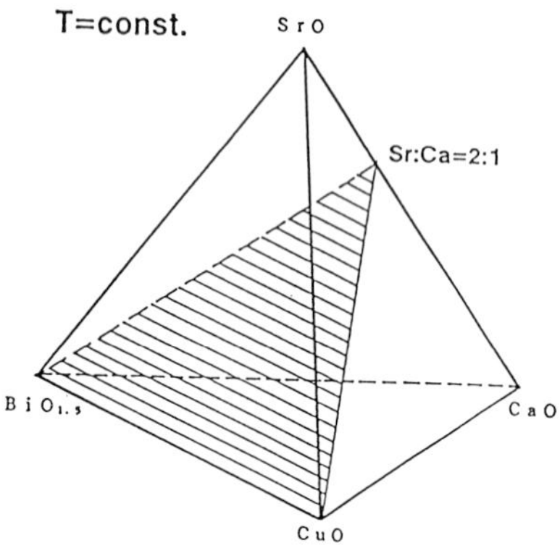

Figure 1. The (Sr_2,Ca) plane in pseudo-tetra phase diagram.

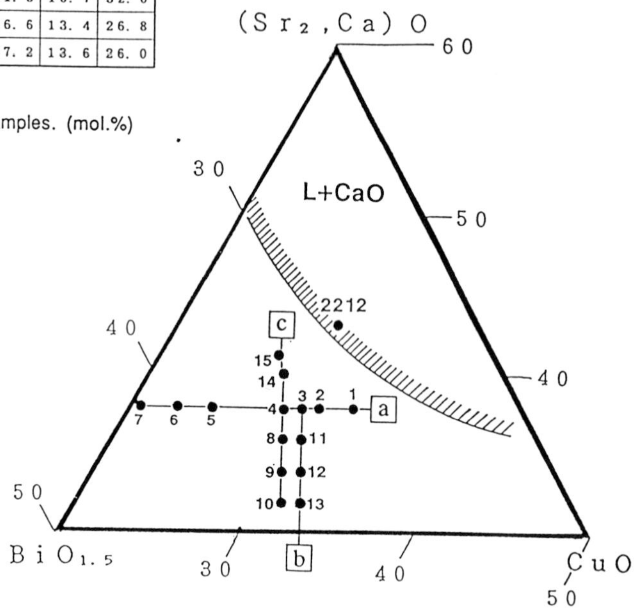

Figure 2. Nominal compositions and positions of pseudo-binary phase diagrams a, b, and c in (Sr_2,Ca) plane.

Results and Discussion

Results of these quenches were plotted as pseudo-binary phase diagrams. In figure 2, a,b, and c show the positions of the pseudo-binary phase diagrams in the (Sr_2,Ca) plane. The region of L+CaO in figure 2 was referred from the previous paper[3]. The pseudo-binary phase diagram a is shown in figure 3. Complete melting in sample 2 was observed at 865°C. A primary crystal of $Bi_2Sr_2CuO_x$(2201 phase) was observed in most of the samples in the Cu-poor region. In the Cu-rich region, $(Sr,Ca)_{14}Cu_{24}O_{41}$ was observed as a primary crystal at 880°C in sample 1. In the intermediate region, the coexistence of 2201, CuO, and liquid phases was observed between 845°C and 860°C. There was no primary phase region of CuO in this pseudo-binary diagram, but that was found in b as shown in figure 4. This coexisting phase consists of the primary phases of 2201 and CuO.

In figure 4 (the pseudo-binary phase diagram along b), a primary crystal of 2201 was observed in sample 3 between 860°C and 865°C, and the primary phase 2201 was seen in the Sr,Ca-rich region. The primary crystal CuO was observed in sample 13 at 855°C, and the primary phase CuO was supposed in the Sr,Ca-poor region. The coexistence of 2201, CuO, and liquid phases, which are the same as in figure 3, was observed at 850°C. The structure is a result of a eutectic reaction with which 2201 and CuO precipitated. The line of the eutectic was supposed to go through sample 12 at 860°C.

The pseudo-binary phase diagram along c was shown in figure 5. This phase diagram is almost the same as figure 4. The primary crystal 2201 was observed in sample 4 at 855°C and 860°C, and 2201 was also seen as the primary crystal in Sr,Ca-rich region. CuO was not observed as the primary crystal in the range of our experiments. However, extrapolation from the figure 4 suggests the CuO primary crystal region in Sr,Ca-poor composition. The coexistence of 2201, CuO, and liquid phases was observed between 845°C and 855°C, the same as in figure 3.

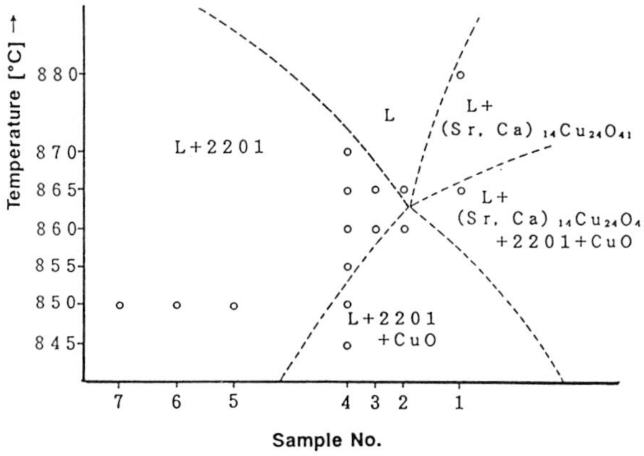

Figure 3. Pseudo-binary phase diagram a.

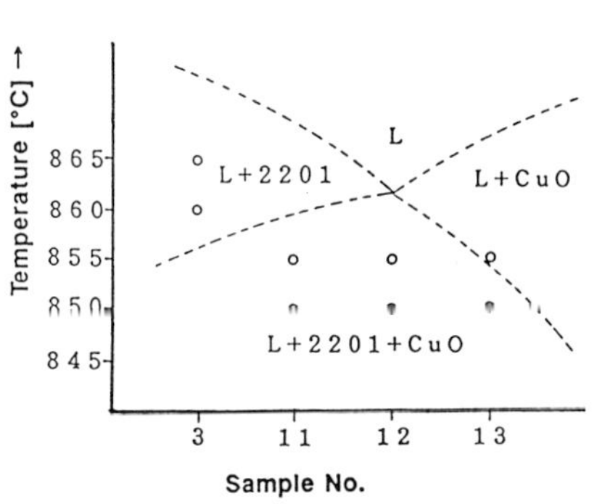

Figure 4. Pseudo-binary phase diagram b.

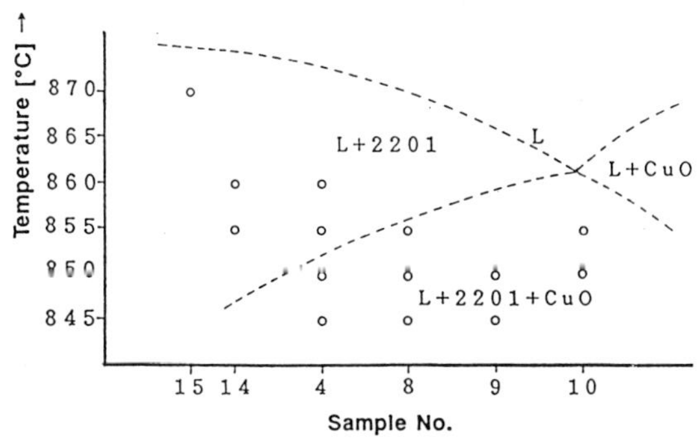

Figure 5. Pseudo-binary phase diagram c.

In these results, the coexistence of 2201, CuO, and liquid phases was observed. The line of the eutectic was supposed to go through samples 10 and 12 at 860°C. The pseudo-ternary phase diagram in this system was investigated by showing coexistence of 2201, CuO, and liquid phases into isothermal phase diagrams at 850, 855, and 860°C, as shown in figure 6. This coexisting phase region was found to become narrow for increasing temperature. Above 860°C, this coexisting phase disappeared; the whole region of experimental composition was covered with the primary phases of 2201 and CuO. Added to these phases, the primary phases of CaO and $(Sr,Ca)_2CuO_3$ in previous paper [3] were extended to the (Sr_2,Ca) plane as shown in figure 7. Figure 7 shows the primary phases on (Sr_2,Ca)plane. In this study, coexisting of 2212 and liquid phases was not observed. The shaded area was not investigated for this paper, and is the subject of future studies.

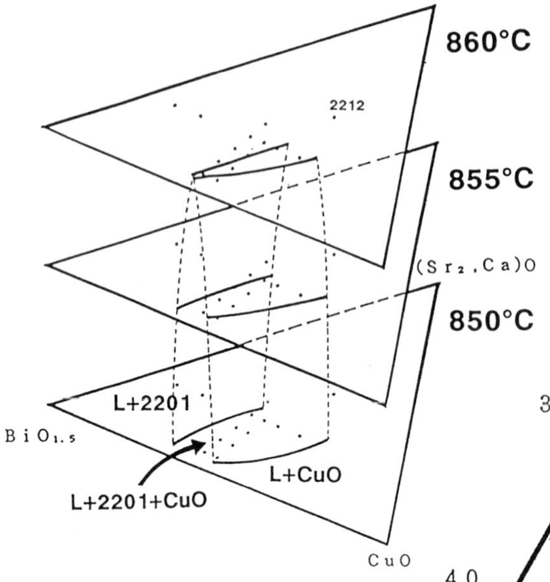

Figure 6. Isothermal phase diagrams at 850, 855, and 860°C.

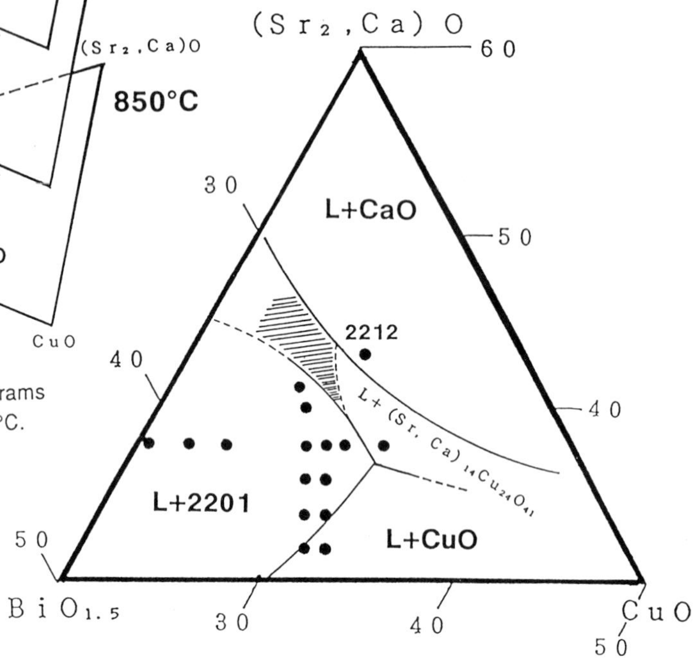

Figure 7. Primary phases on (Sr_2,Ca) plane.

These results show several features;

(1) These experimental compositions were covered with the primary phases of 2201 and CuO above 860°C.

(2) The coexisting of 2212 and liquid phases was not observed in this study.

Acknowledgement

This work has been supported by the R&D Basic Technology for Future Industries through the New Energy and Industrial Technology Development Organization(NEDO).

Reference

1. C. Michel, M. Hervieu, M. M.Borel, A. Grandin, F. Deslandes, A. Provost and B. Raveau, Z. Physik B68 (1987)
2. H. Maeda, Y. Tanaka, M. Fukutomi, and T. Asao, Jpn. J. Appl. Phys. 27 (1988) L209
3. T. Noda, T. Izumi, A. Nakamura, and Y. Shiohara, Proc. of 2nd Int. Symp. on Superconductivity (1989) 239

A Study of the Temperature Dependence of the Heat Capacity in a High T_c Superconductor $HoBa_2Cu_3O_y$ Single Crystal

K. TSUTSUMI[1], H. OKAMOTO[1], Y. ISHIHARA[1], T. TAMURA[2], and K. KITAZAWA[2]

[1] Department of Physics, Kanazawa University, Kanazawa, 920 Japan
[2] Department of Industrial Chemistry, The University of Tokyo, Tokyo, 113 Japan

ABSTRACT

We have measured the temperature dependence of the heat capacity C_p on a large rectangular single crystal of a high T_c superconductor $HoBa_2Cu_3O_y$. An anomaly in the temperature dependence of the heat capacity was observed around 27 K which is much lower than the superconductivity transition temperature T_c. We speculate that this temperature might correspond to that at which the lower critical field H_{c1} begins to increase very rapidly when we measure H_{c1} with decreasing temperature.

KEY WORDS : heat capacity, lower critical field, single crystal

INTRODUCTION

Since the discovery of the high-T_c oxide superconductor by Bednorz and Muller [1] the very intensive studies have been performed and being continuing both experimentally and theoretically in order to reveal the mechanism of the high-T_c oxide superconductivity [2]. Experimentally there are many works on structural, electrical-transport and magnetic properties. However those on the thermal-transport property seem to be a little few compared with works mentioned just above. One reason might be as follows; because the T_c of the oxide superconductor is much higher than that of the ordinary superconductor, which have also been studying intensively, the contribution of the phonon system to the thermal-transport properties is very large. Therefore such as the measurement of the heat capacity anomaly at T_c and the electron system contribution to the thermal conductivity are very difficult in the high-T_c oxide superconductor.

EXPERIMENTAL

We succeeded in synthesizing the large single crystals of a high-T_c oxide superconductor $HoBa_2Cu_3O_y$ (HBCO).The size and form are 2.0 mm x 2.0 mm x 0.5 mm and a rectangular one respectively. So we tried to measure the temperature dependence of the heat capacity C_p by means of an AC calorimetry method after thinning of the crystal up to 0.2 mm. We used two pairs of chromel-alumel thermocouple wires whose diameters are both 25.0 μm. We found a very faint anomaly near 77.5 K in the temperature dependence of C_p/T where T is temperature. We investigated the faint anomaly precisely. The estimated transiton temperature from the C_p/T versus temperature curve is 77.2 K. We show the Cp versus temperature curve in Fig. 1. Ordinarily the transition

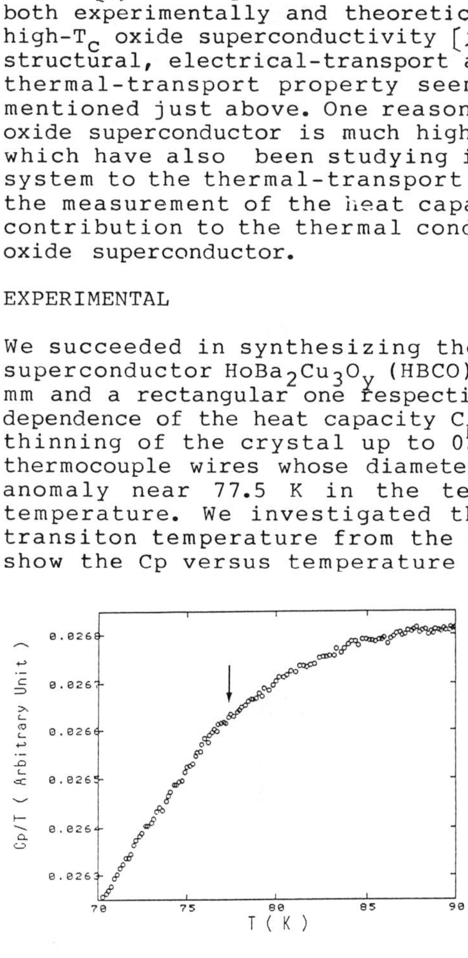

Fig. 1. Temperature dependence of Cp/T of a large single crystal in an oxide high-T_c superconductor $HoBa_2Cu_3O_y$ near the superconducting transition temperature T_c.

temperature T_c of HBCO is around 90 K. The transition temperature of our HBCO single crystal is a little lower and the anomaly of Cp at T_c is also very faint. We consider that the origin of this lower T_c and the faint anomaly are due to the insufficiency of the annealing in the oxygen atmosphere.

We also measured the temperature dependence of the heat capacity Cp below 77 K using the same sample and same method, namely, using the chromel-alumel thermocouple. The measurements were made three times. Each measurement was performed only with increasing temperature. The results of the first measurement are shown in Figs. 2-4. Figure 2 shows the temperature dependence of Cp below 40 K. As is clear in Fig. 2, we observed a very clear kink in the vicinity of 27 K. We plotted these data in the form where the vertical axis is Cp/T and the horizontal axis the square of temperature. The plotted results are shown in Fig. 3. A jump of Cp/T is seen clearly. We also calculated the temperature-derivative of Cp/T. The [-d(Cp/T)/dT] versus temperature curve of the first measurement is shown in Fig. 4.

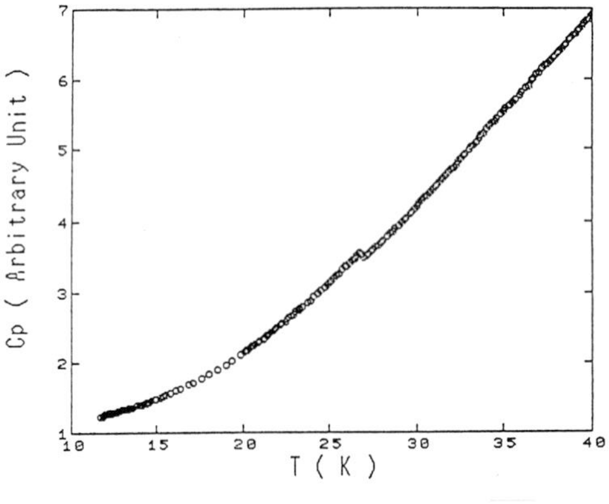

Fig. 2. Temperature dependence of the heat capacity Cp down below T_c. A clear anomaly is observed near 27 K.

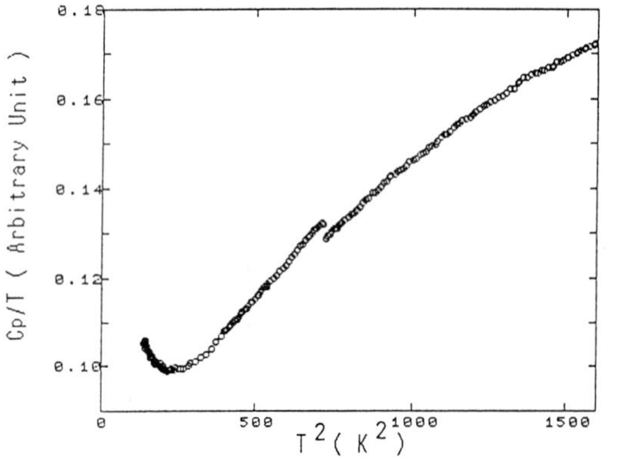

Fig. 3. The Cp/T versus the square of temperature curve. A clear jump of Cp/T is seen.

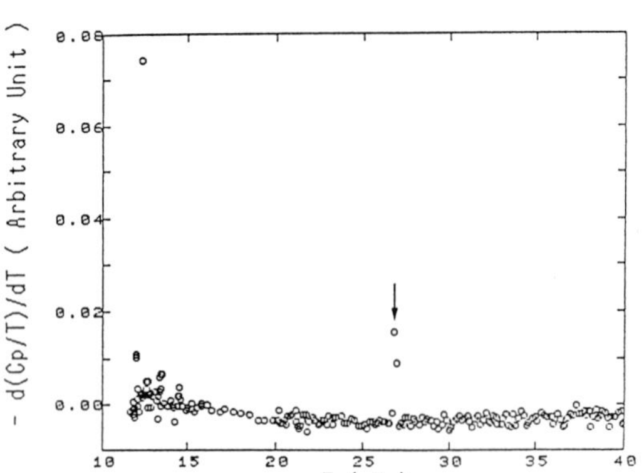

Fig. 4. The temperature derivative of Cp/T versus temperature curve. A very sharp peak is observed near 27 K.

Figure 5 shows the temperature dependence of Cp obtained in the second measurement. The same plotting and temperature-derivative calculation as the data obtained in the first measurement were performed. The results are shown in Figs. 6 and 7. We can certainly find a peak although it is not so sharp as that in Fig. 4. Figure 8 shows the temperature-derivative versus temperature curve in the third measurement. We can find a broad peak around 27 K too.

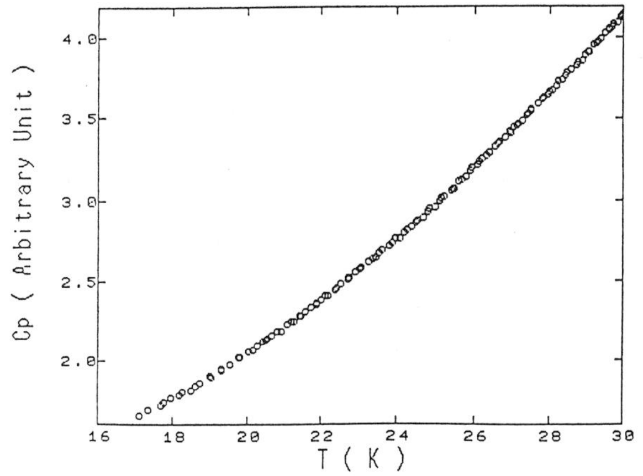

Fig. 5. Temperature dependence of Cp obtained in the second measurement.

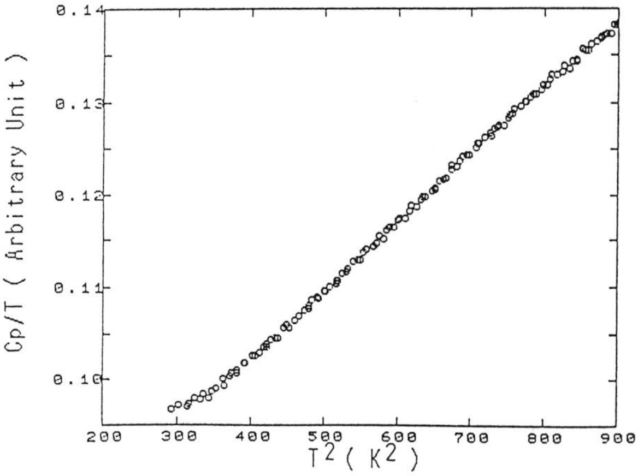

Fig. 6. The Cp/T versus the square of temperature curve of the data obtained in the second measurement. No distinct change is observed.

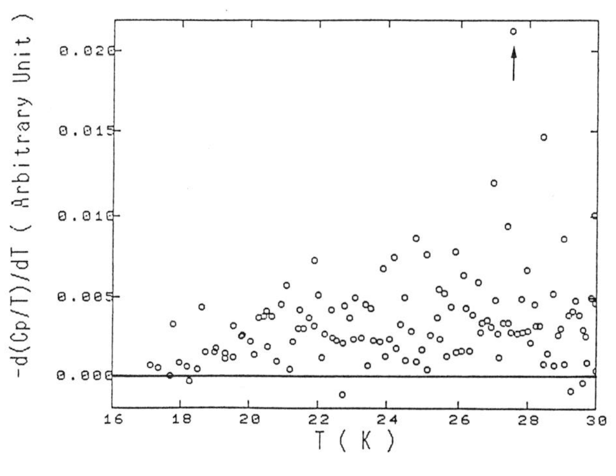

Fig. 7. The temperature derivative of Cp/T versus temperature curve on the data obtained in the second measurement. A broad peak is seen around 27 K.

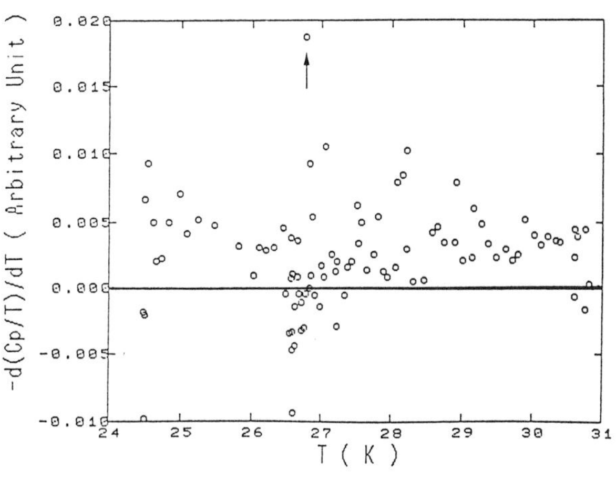

Fig. 8. The temperature derivative of Cp/T versus temperature curve of the data obtained in the third measurement. A broad peak is seen too.

Furthermore we also measured the temperature dependence of the lower critical field H_{c1} on another single crystal. We show the results in Fig. 9. As is clear in Fig. 9 , H_{c1} begins to increase very rapidly at the temperature which is a little lower than 30 K. So we insist that the temperature at which the heat capacity Cp shows an anomaly might correspond to that at which H_{c1} begins to increase very rapidly.

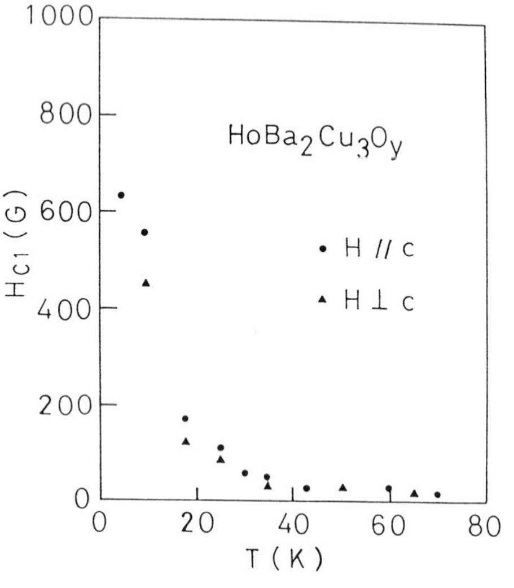

Fig. 9 . Temperature dependence of the lower critical field H_{c1}. The closed circles present the data obtained when the magnetic field is parallel to the c-axis and the closed triangles present when the magnetic field is perpendicular to the c-axis.

DISCUSSION

We discuss on the inconsistency of the data obtained during the three measurements mentioned above. The first measurement strongly supports our insistence on the correlation between the anomalous behavior of both H_{c1} and Cp in the temperature range down below T_c. However those obtained in the second and third measurements do not support our above insistence so strongly. Why does only the first measurement show the anomaly clear ? As the reason we might consider the followings. The circumstance of the sample, which is set on the thermocouple, is in a soft vacuum. Therefore we consider the deoxygenation effect of the sample. Recently Collocott et al.[3] have reported that when we describe the heat capacity in the following formula ;$Cp = AT + BT^3$, the value of A decreases below 10 K in the high-T_c oxide superconductor on deoxygenation. If we compare Fig. 3 with Fig. 6 precisely, we can find that below 25 K the value of A in Fig. 3 is larger than those in Fig. 6. So we consider that in the second and third measurements we obtained the poor data on the temperature dependence of the heat capacity Cp, because we performed the measurement on the more deoxygenated sample than that in the first measurement. So we again insist that the temperature at which the heat capacity Cp shows an anomaly might correspond to that at which H_{c1} begins to increase very rapidly.

REFERENCES

1 J. G. Bednorz and K. A. Muller, Z. Phys. B 64 (1986) 189.
2 Studies of High Temperature Superconductors, ed. A.V. Narlikar, (Nova Science Pub., New York ,1989).
3 S. J. Collocott , R. Driver and E.R. Vance,Phys.Rev.B41 (1990) 6329.

Study of the Structural Phase Diagram, Oxygen Bulk In-Diffusion, and Equilibrium Partial Pressure of $YBa_2Cu_3O_{6+x}$

N.H. Andersen, B. Lebech, and H.F. Poulsen

Physics Department, Risø National Laboratory, DK-4000 Roskilde, Denmark

ABSTRACT

An experimental technique based on powder neutron diffraction and gas volumetry has been used for on-line preparation of $YBa_2Cu_3O_{6+x}$ samples under equilibrium conditions and experimental studies in an extended part of (x,T)-space. The phase line separating the tetragonal and the orthorhombic structures, and the twin-domain size in two different orthorhombic phases have been determined from the neutron powder diffraction data. The oxygen equilibrium pressure proves the thermodynamic stability of the orthorhombic double cell structure (the OII-phase) and the oxygen chain formation in the tetragonal phase. The behaviour of the bulk oxygen in-diffusion data indicates that microcracks and the twin-domain boundaries may contribute significantly to the oxygen diffusion.

KEY WORDS: '123' high-T_c superconductor, O-T phase transition, twin-domain size, oxygen equilibrium pressure, bulk oxygen in-diffusion

INTRODUCTION

Recent studies have revealed that the superconducting properties of $YBa_2Cu_3O_{6+x}$ (YBCO) are strongly influenced by the oxygen stoichiometry x ($0<x<1$) [1] and the oxygen defect ordering properties [2]. There are strong indications that the basal plane oxygen defect ordering is essentially two-dimensional (2D) [3], and may be accounted for theoretically by a three parameter 2D Ising model with one nearest neighbour and anisotropic next-nearest neighbour interactions, the socalled ASYNNNI model [4]. However, the experimental data for determination of the interaction parameters and testing the model predictions have not been available. Such data are reported in this paper where we give a short presentation of an extensive study of the structural phase transition line separating the tetragonal disordered and the orthorhombic ordered phases, determination of twin-domain sizes, oxygen equilibrium pressures, and bulk oxygen in-diffusion kinetics in an extended part of (x,T)-space ($0.15<x<0.92$ and 25 °C$<T<$725 °C). An experimental technique based on neutron powder diffraction and gas volumetry for on-line sample preparation under equilibrium conditions and combined measurements of structural and oxygen pressure data has been developed and used for the study. A more comprehensive account for our experimental technique and data will appear in Refs. [5,6].

EXPERIMENTAL

A cylindrical ceramic sample with diameter 11.3 mm, height 28.8 mm, and 81% density was prepared by standard solid state reaction technique based on oxides. The particle size was determined by laser light scattering technique and showed a flat logarithmic volume size distribution with diameters ranging from 2 µm to 100 µm. Samples with variable oxygen stoichiometries were prepared in a gas volumetric cell from an intially reduced state with $x_{red}=0.15(2)$ at 725 °C by equilibration with the oxygen content of a closed volume reservoir. By step cooling to desired temperatures the sample absorbs oxygen and points in (x,T)-space may be mapped out by use of different starting pressures and the ideal gas law [5]. The decay rate in the oxygen pressure during absorbtion supplies information about bulk oxygen in-diffusion kinetics, and the final pressure about oxygen equilibrium pressures, and thereby the oxygen chemical potential. Sixteen traces with a total of 350 points in (x,T)-space were mapped out (see Fig. 1). Absolute values of oxygen stoichiometries were determined from iodiometric titration and refinements of neutron powder diffraction data in combination with weight measurements and gas volumetry [5]. For fully oxidized YBCO we obtain $x_{ox}=0.92(2)$, $T_c=93$ K and $\Delta T_c=1$ K.

The neutron powder diffraction measurements were carried out on the multi-detector powder neutron diffractometer at the DR3-reactor, Risø National Laboratory using a neutron wavelength of $\lambda=2.32$ Å. For $Q<3.0$ Å^{-1} the resolution width (FWHM) is 0.013 Å^{-1}. The diffraction data were analyzed by a program that refines the lattice parameters (a,b,c), the zero angle, a constant

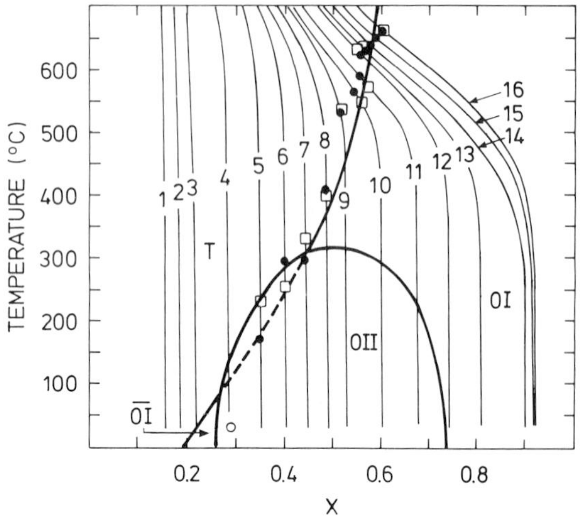

Fig. 1 The structural phase diagram of $YBa_2Cu_3O_{6+x}$. Thin curves are the trajectories of measurements in (x,T)-space,• (and o) are the experimental transition temperatures separating the tetragonal and the orthorhombic phases. Thick curves are theoretical model calculations [4] of the phase transition lines between tetragonal T-phase and orthorhombic O-phases (see text). □ are the temperatures below which in-plane diffraction peaks are broadened (from [6])

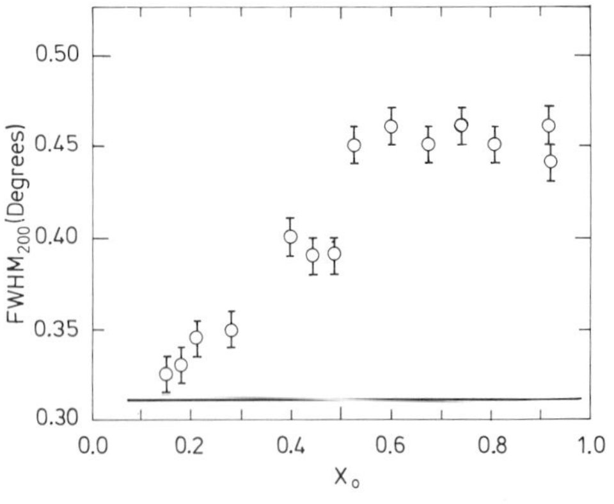

Fig. 2 Neutron powder diffraction peak broadening (FWHM) of the (200) reflections measured at room temperature. The horizontal line is the experimental resolution $FWHM_{res} = 0.312°$. The twin-domain sizes determined in orthorhombic phases from the peak broadenings are 350 Å (0.35<x<0.5) and 250 Å (0.5<x) (from [6])

background, the peak profile parameters (u,v,w), and independent Bragg peak intensities. All refinements were carried out assuming orthorhombic symmetry. Tetragonal structure was assumed when the difference Δ(a,b) between the refined values of a and b became smaller than the experimental resolution.

RESULTS AND DISCUSSION

The strucutral phase diagram of YBCO has been established from analysis of 350 data set in (x,T)-space. All the data sets were well accounted for by single phase refinements. The transition temperatures separating the tetragonal T-phase from the orthorhombic O-phases are shown in Fig. 1 as solid circles (the open circle is the extrapolated x-value where the orthorhombic distortion disappears at room temperature). For scan traces nos. 1-4 no orthorhombic splitting was observed. None of the spectra contained superlattice peaks which could be used to distinguish different O-phases. In Fig. 1 our experimental results are compared with the quite recent model calculations of Ceder et al.[4] using the cluster variational method (CVM) on the ASYNNNI model (bold curves). The interaction parameters in the calculations are the nearest neighbour interaction V_1, and the two next-nearest neigbour interactions, V_2 along the b axis being mediated by hybridization via the Cu atoms, and V_3 the direct interaction along the a axis, determined by Sterne and Wille [7] from first principles electron band calculations. The orthorhombic OI-, $\overline{OI}$-, and OII-phases correspond to basal plane structures where the oxygen sites along the b axis are either all 'full' (CuO chains) or all 'empty' (Cu chains), and have ordering sequences along the a axis with all 'full', all 'empty', and alternating 'full' and 'empty' chains (the double cell structure), respectively. A remarkable agreement is observed between experiment and the model calculation without adjustable parameters. However, in a quite recent paper Aukrust et al. [8] have carried out transfer matrix and Monte Carlo studies on the ASYNNNI model and found that the CVM results of the transition temperatures deviate up to 21% from their more accurate values, and they do not find the $\overline{OI}$-phase.

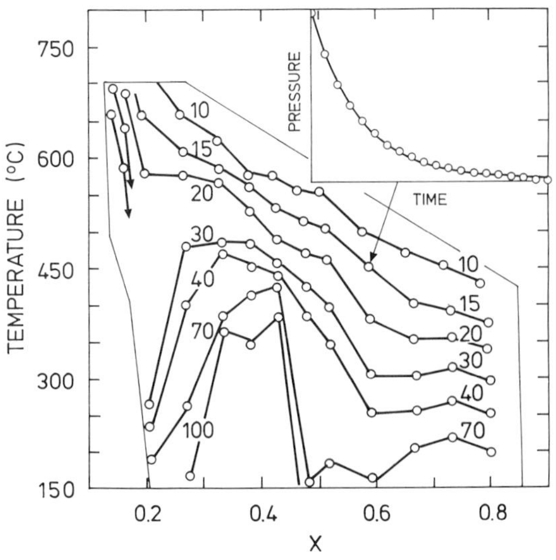

Fig. 3 Isochrones in minutes of bulk oxygen in-diffusion relaxation times τ determined from decay rates of gas volumetric pressure data. As exemplified in the inset, a simple exponential decay expression (curve) gave satisfactory agreement with all the data (o) (from [6])

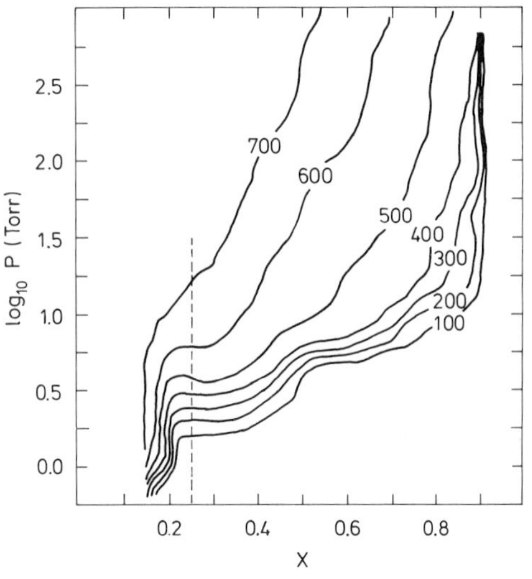

Fig. 4 Isotherms in °C of oxygen equilibrium pressure of $YBa_2Cu_3O_{6+x}$. The isotherms for 25 °C and 100 °C are almost identical. The dashed line is the ASYNNNI model calculation of the disorder line by Rikvold et al. [9] (from [5])

Analyses of the neutron powder diffraction profiles, based on fits to individual peaks and the full profile (u,v,w)-data show that the in-plane (h,k,0) reflections broaden below temperatures (open squares in Fig. 1) which are in close agreement with the structural phase transitions. This is consistent with the assumption that the peak broadening results from the formation of twin-domains. In the orthorhombic phase the peak widths are temperature independent. The room temperature peak widths (FWHM) determined from the (200) reflections are shown in Fig.2. Three regions of different peak broadenings, which we suggest are related to the tetragonal phase ($x_0<0.3$) with essentially no broadening, and the two orthorhombic phases, OII ($0.3<x_0<0.5$) and OI ($0.5<x_0$), are observed. By this suggestion we are in conflict with the theoretical phase diagram, shown in Fig. 1, which suggest that $x_0=0.73$ separates the OII- and OI-phases at room temperature. This will be discussed further below. The twin-domain sizes, determined by deconvoluting the (200) peak profile with the experimental resolution function, are 250 Å (OI-phase) and 350 Å (OII-phase), respectively.

Despite the rather broad particle size distribution of our sample all the oxygen pressure data and thereby the oxygen in-diffusion flow have a simple exponential decay as exemplified in the inset of Fig. 3. This indicates that micro-cracks or twin-domain boundaries rather than the grain size determine the diffusion length. From analysis of the pressure decay as a result of a temperature change we obtain the relaxation time τ and the total pressure difference ΔP. In Fig. 3 are shown the results of isochronal τ-contours determined by interpolations in the data points in the part of (x,T)-space where realistic values could be obtained. The contours have a regular behaviour at higher temperatures, whereas strong anomalies, with a significant increase in the relaxation times for $0.25<x<0.50$ and a decrease for $0.20<x<0.25$, are observed at low temperatures. The increase in the relaxation time for $x>0.25$ is not a direct consequence of the structural phase transition (cf. Fig. 1). It is rather related to the strong tendency towards chain formation, which develops in this part of phase space (see below) and reduces the diffusivity. The reduction in the relaxation times for $x>0.50$ we interprete to result from the difference in twin-domain size, and a suggestion that the twin-domain boundaries may be effective oxygen diffusion pathways. An order of magnitude calculation based on the experimental data for τ and twin-domain sizes (cf. Fig. 2) support this suggestion [6]. At high temperatures we observe activated behaviour with activation energy 0.55 eV for $x<0.6$ (T-phase), and 0.25 eV for $x>0.6$ (OI-phase). The oxygen flow, which is proportional to $\Delta P/\tau$, has a similar anomalous behaviour as τ, but the activation energies are significantly larger (between 1.5 eV and 1.0 eV).

Isotherms of the oxygen equilibrium pressures are presented in Fig. 4. From these data the oxygen chemical potential μ may be calculated. Three significant features should be commented on. Firstly, the close lying contours at stoichiometries close to $x=0.9$ and to some extend close to $x=0.15$ indicate a stability range of $0.15<x<0.92$ for oxygen pressures 1 Torr $<P<$ 760 Torr. Secondly, the maxima or kinks in the isotherms around $x=0.22$ is in close agreement with recent ASYNNNI model calculation by Rikvold et al. [9] of the disorder line (dashed line), defined as the peak position in the non-ordering susceptibility $X_{\Theta}=2(\partial x/\partial)_T$. X_{Θ} may be calculated directly from our pressure data and is positioned at the maxima or kinks in the isotherms shown in Fig. 4. A striking coincidence with the theoretical disorder line is observed. Thirdly, we mention that the jump in the low temperature equilibrium pressures at $x=0.5$ proves the thermodynamic stability of the orthorhombic OII-phase. Actually, the size of the jump supplies important information about the interaction parameter in the ASYNNNI model. From initial analyses using Monte Carlo simulations and the pressure data we estimate that the V_3 parameters determined by Sterne and Wille [7] should be reduced, and in consequence that the theoretical phase line separating the OII- and the OI-phase at low temperature should be shifted towards smaller x-values. Results of our Monte Carlo simulations will be published elsewhere.

To conclude, in this work we have established the structural phase diagram in an extended part of phase space for the first time, and supplied new information about twin-domain sizes, bulk oxygen in-diffusion processes and oxygen equilibrium pressures, and thereby the chemical potential of the basal plane oxygens. Comparison between recent model calculations and our data gives strong support to the validity of the ASYNNNI model description of the oxygen basal plane ordering and to some extend to the first principles calculations of the interaction parameters. Our data may be used in coombination with further model calculation and Monte Carlo simulations to increase the understanding of the oxygen disorder properties, and thereby the influence of oxygen ordering on superconductivity.

REFERENCES

1. Cava R.J., Hewat A.W., Batlogg B., Marezio M., Rabe K.M., Krajewski J.J., Peck Jr. W.F., Rupp Jr. L.W. (1990) Physica C 165: 419.
2. Jorgensen J.D., Pei S., Lightfoot P., Shi H., Paulikas A.P., Veal B.W. (1990) Physica C 167: 571.
3. Beyers R., Ahn B.T., Gorman G., Lee V.Y., Parkin S.S.P., Ramirez M.L., Roche K.P., Vazques J.E., Gür, T.M., Huggins R.A. (1989) Nature 340: 619.
4. Ceder G., Asta M., Carter W.C., Kraitchman M., deFontaine D., Mann M.E., Sluiter M. (1990) Phys.Rev.B 41: 8698.
5. Andersen N.H., Lebech B., Poulsen H.F. (1990) Physica C: in press.
6. Poulsen H.F., Andersen N.H., Lebech B. (1990) submitted.
7. Sterne P.A., Wille L.T. (1989) Physica C 162-164: 223.
8. Aukrust T., Novotny M.A., Rikvold P.A., Landau D.P. (1990) Phys.Rev.B 41: 8772.
9. Rikvold P.A., Novotny M.A., Aukrust T. (1990) to be published.

3.5 Crystal Growth

Growth and Superconductive Properties of La_2CuO_4 Single Crystals

ISAO TANAKA, HIDEHIRO TAKAHASHI, and HIRONAO KOJIMA

Institute of Inorganic Synthesis, Faculty of Engineering, Yamanashi University, Miyamae 7, Kofu, 400 Japan

ABSTRACT

Single crystals of La_2CuO_4 5 mm in diameter and 50 mm in length were grown at an oxygen pressure of 0.2 MPa by the traveling solvent floating zone method (TSFZ method). The as-grown crystals of La_2CuO_4 were superconducting, but were a mixture of a superconducting phase and an antiferromagnetic phase. The single crystals of La_2CuO_4 were changed reversibly from superconductors to nonsuperconductors by annealing. In the nonsuperconducting state, the single crystals of La_2CuO_4 were semiconducting and had a highly anisotropic resistivity, where the resistivity ratio $\rho_{\perp}/\rho_{\parallel}$ was about 28,000 at RT.

KEY WORDS: lanthanum cupric oxide, single crystals, TSFZ method, anisotropic resistivity, annealing effect

INTRODUCTION

La_2CuO_4 has been recognized as an antiferromagnetic semiconductor, and it has been discovered that substitution of Sr^{2+} or Ba^{2+} for La^{3+} in La_2CuO_4 induces superconductivity [1-6]. Several laboratories have recently reported on the superconducting behavior of La_2CuO_4 [7-13]. Superconductivity in La_2CuO_4 was attributed to the presence of La vacancies or excess oxygen. Superconducting La_2CuO_4 has been synthesized by the sintering method, and the fraction of the superconducting phase in La_2CuO_4 depends on the synthesis conditions, where annealing in high-pressure oxygen at low temperature increases the fraction. The superconductivity of La_2CuO_4 has been investigated using polycrystalline La_2CuO_4. However, the K_2NiF_4-type structure of La_2CuO_4 suggests that the magnetic and electrical properties are anisotropic, so the measurement of these properties using single crystals of superconducting La_2CuO_4 would give us further information on the superconducting mechanism of La_2CuO_4. Therefore, it would be desirable to obtain good-quality and sizable single crystals of superconducting La_2CuO_4.

Single crystals of undoped La_2CuO_4 were previously grown by the flux method and the top seeded solution method, but the grown crystals were nonsuperconducting [14-18]. Good-quality and sizable single crystals of $La_{2-x}Sr_xCuO_4$ have been grown by the traveling solvent floating zone method (TSFZ method), and the anisotropic resistivities of the $La_{2-x}Sr_xCuO_4$ single crystals have been reported [19-21]. In this study, the single crystals of superconducting La_2CuO_4 were grown by the TSFZ method. The Meissner fraction and anisotropic resistivity of the La_2CuO_4 single crystals were measured, and then the annealing effect on the superconductivity of the La_2CuO_4 single crystals was discussed.

EXPERIMENTAL

Stoichiometric amounts of La_2O_3 and CuO were mixed in ethanol, dried, and calcined at 1123 K for 12 h in air. The preheated powder was formed into a cylindrical shape 6 mm in diameter by 60 mm in length, sintered at 1473 K for 12 h in oxygen, and then used as the feed rod. The apparatus for crystal growth was an infrared heating furnace of the double ellipsoidal type with two 1.5 kW halogen lamps as the heat source. The growth conditions were as follows: the solvent composition was 80 mol%CuO, the growth rate was 1.0 mm/h, and the growth atmosphere was oxygen at the oxygen pressure of

0.2 MPa for preventing the vaporization of CuO from the melt. Preparation of the feed and crystal growth by the TSFZ method were described in detail elsewhere [19].

For investigation of the anisotropic resistivity and the effect of annealing on the superconductivity of La_2CuO_4 single crystals, not only the as-grown crystals but also the crystals annealed in argon or oxygen were used as samples to be measured for superconductive properties. The Meissner signal was observed by the ac magnetization method between 15 K and room temperature. The electrical resistivity was measured by a standard four-probe method. The single crystals of La_2CuO_4 were cut to rectangular bars measuring 4x2x1 mm^3 with the long axis perpendicular to the Cu-O planes or parallel to the Cu-O planes, respectively. The samples were attached to gold leads with silver paste and heated in air at 673 K for 20 min. The resistivity measurements were performed between 15 K and room temperature.

RESULTS AND DISCUSSION

Figure 1 shows an as-grown crystal of La_2CuO_4. The as-grown crystal was sizable (5 mm in diameter and 50 mm in length) and was black with metallic luster. The grown crystals were determined to be La_2CuO_4 by the powder XRD method and quantitative analysis using an electron probe microanalyzer (EPMA). The grown crystals were further identified as single crystals by back-reflection Laue X-ray photography and optical microscopy.

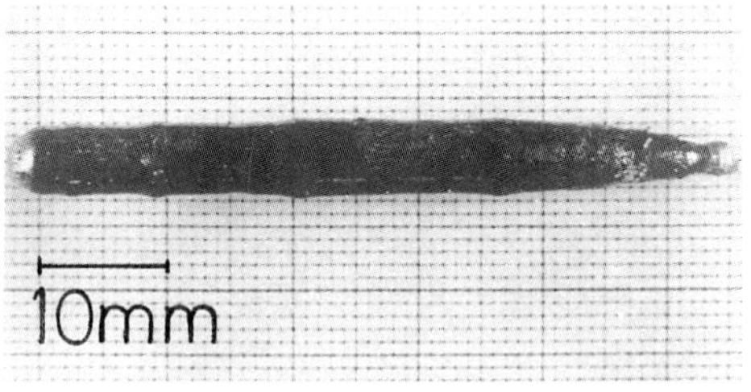

Fig. 1. As-grown crystal of La_2CuO_4.

Figure 2 shows the temperature dependence of the ac magnetic susceptibility of the as-grown and the annealed crystals. The Meissner signal was observed below 32 K in the as-grown crystals, and the Meissner fraction of the as-grown crystals was determined to be about 60 % that of the $La_{1.85}Sr_{0.15}CuO_4$ single crystals grown under the same condition by the TSFZ method. It is noted that the single crystals of La_2CuO_4 grown at low oxygen pressure of 0.2 MPa were superconducting, although superconducting La_2CuO_4 had ever been synthesized at very high oxygen pressure above 1 GPa. When the as-grown crystals of La_2CuO_4 were annealed in argon flow at 1273 K for 24 h, the

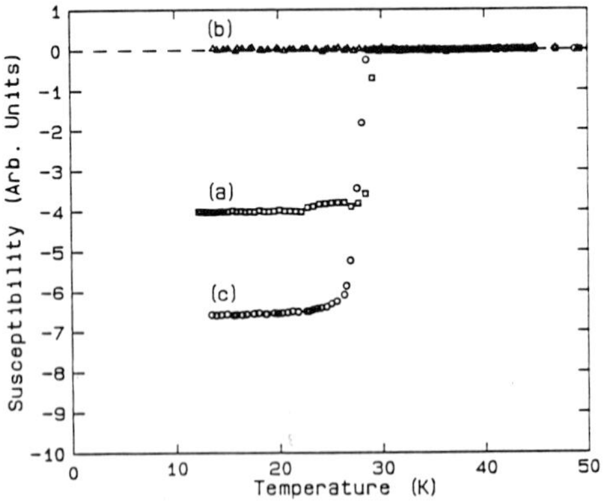

Fig. 2. Temperature dependence of the ac magnetic susceptibility of La_2CuO_4 single crystals: (a) as-grown crystals; (b) crystals annealed in argon at 1273 K for 24 h; (c) crystals annealed in oxygen at 1273 K for 48 h.

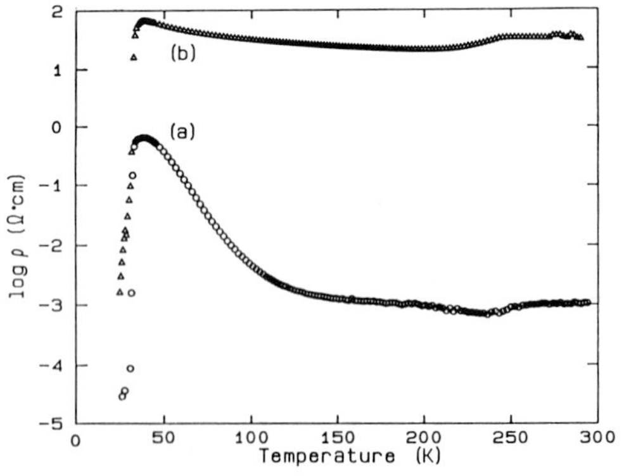

Fig. 3. Temperature dependence of the electrical resistivity of La_2CuO_4 as-grown crystals: (a) $\rho_{//}$; (b) $\rho_{\perp}$.

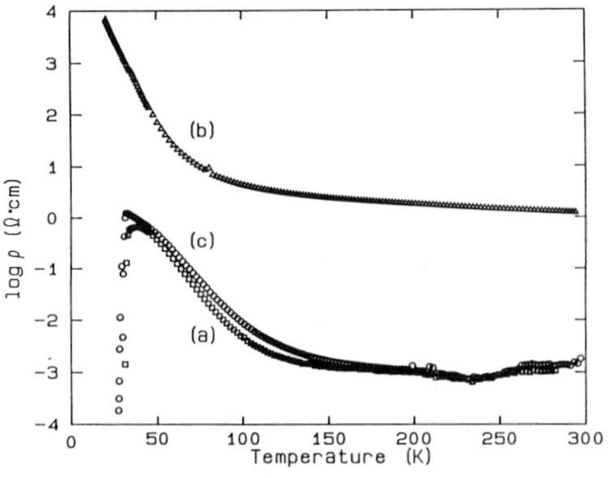

Fig. 4. Temperature dependence of the resistivity parallel to the Cu-O planes of La_2CuO_4 single crystals ($\rho_{//}$): (a) as-grown crystals; (b) crystals annealed in argon at 1273 K for 24 h; (c) crystals annealed in oxygen at 1273 K for 48 h.

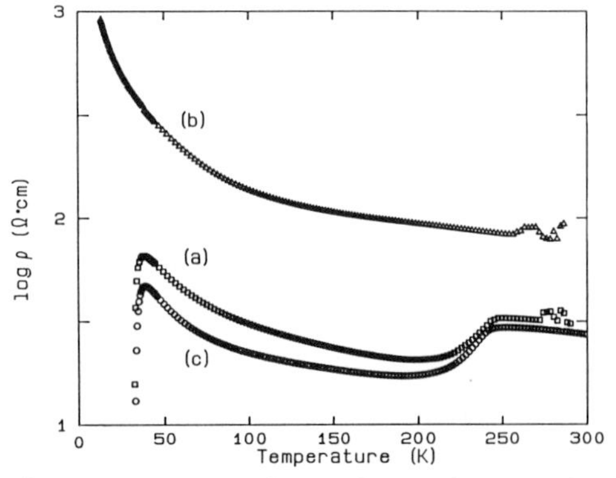

Fig. 5 Temperature dependence of the resistivity perpendicular to the Cu-O planes of La_2CuO_4 single crystals ($\rho_\perp$): (a) as-grown crystals; (b) crystals annealed in argon at 1273 K for 24 h; (c) crystals annealed in oxygen at 1273 K for 48 h.

Meissner signal disappeared. When the annealed crystals were further annealed in oxygen flow at 1273 K for 48 h, the Meissner signal appeared again, and the Meissner fraction of the annealed crystals became higher than that of the as-grown crystals. This result suggests that the La_2CuO_4 single crystals are changed reversibly between superconductors and nonsuperconductors by annealing, and the Meissner fraction of La_2CuO_4 depends on the oxygen concentration in La_2CuO_4.

Figure 3 shows the temperature dependence of the electrical resistivity of the as-grown crystals of La_2CuO_4. The resistivities parallel to the Cu-O planes ($\rho_{//}$) and perpendicular to the Cu-O planes ($\rho_\perp$) of the as-grown crystals indicated a superconducting behavior below about 40 K. The superconducting transition of the as-grown crystals was shape, and the transition mid-temperature (T_c) and the transition width (ΔT_c) were 33 K and 3 K, respectively. Both $\rho_{//}$ and $\rho_\perp$ were changed anomalously at about 250 K owing to a paramagnetic-antiferromagnetic phase transition. It was found that the antiferromagnetic phase coexisted with the superconducting phase in the as-grown crystals of La_2CuO_4. In the normal state, both $\rho_{//}$ and $\rho_\perp$ increased as temperature decreased, indicating semiconducting behavior. The value of $\rho_{//}$ especially increased from 1.07 mΩ·cm to 371 mΩ·cm with decreasing temperature from room temperature (RT) to 50 K, and $\rho_{//}$ at 50 K became about 350 times higher than that at RT. However, $\rho_{//}$ and $\rho_\perp$ of $La_{2-x}Sr_xCuO_4$ single crystals indicated a metallic and a semiconducting behavior, respectively [19]. Dabrowski *et al.* reported that $La_2CuO_{4+\delta}$ with δ =0.08 was metallic in the normal state and consisted of only the superconducting phase, but $La_2CuO_{4+\delta}$ with $0<\delta<0.08$ was semiconducting in the normal state and consisted of the superconducting phase and the antiferromagnetic phase [13]. Therefore, the La_2CuO_4 as-grown crystals were a mixture of the superconducting phase and the antiferromagnetic phase, so that the behavior of the normal-state resistivity of the La_2CuO_4 single crystals was different from that of the $La_{2-x}Sr_xCuO_4$ single crystals. As compared to the anisotropy of the resistivity at room temperature, the resistivity ratio $\rho_\perp/\rho_{//}$ of the La_2CuO_4 as-grown crystals was about 28,000 at RT, and was much higher than that of the $La_{2-x}Sr_xCuO_4$ single crystals, about 1,000. Thus, the resistivity of the La_2CuO_4 single crystals was highly anisotropic.

Figures 4 and 5 show the annealing effect on the resistivity of the La_2CuO_4 single crystals. When the as-grown crystals were annealed in argon flow at 1273 K for 24 h, both $\rho_{//}$ and $\rho_\perp$ increased as the temperature decreased from RT to 15 K, and indicated a semiconducting behavior. The values of $\rho_{//}$ and $\rho_\perp$ of the annealed crystals became about 1000 times and 10 times as high as those of the as-grown crystals, respectively. When the as-grown crystals were annealed in oxygen flow at 1273 K for 48 h, the resistivity of the annealed crystals was close to that of the as-grown crystals. This result shows that the oxygen-annealed crystals were superconducting, but the antiferromagnetic phase was contained in them. Therefore, annealing at the oxygen pressure of 0.1 MPa had no effect on the resistivity of the La_2CuO_4 single crystals, but annealing at higher oxygen pressure may have an effect on the resistivity.

In summary, single crystals of La_2CuO_4 5 mm in diameter and 50 mm in length were grown at an oxygen pressure of 0.2 MPa by the TSFZ method. The as-grown crystals were superconducting below about 40 K and semiconducting in the normal state, and had a highly anisotropic resistivity in the normal state, $\rho_{\perp}/\rho_{\parallel}$ of about 28,000 at RT. However, the as-grown crystals included an antiferromagnetic phase as well as the superconducting phase. The single crystals were changed from superconductors to nonsuperconductors by annealing in argon flow, and became superconducting by annealing in oxygen flow. Thus it was found that the as-grown crystals of La_2CuO_4 were due to an excess of oxygen, and that the superconductivity of the La_2CuO_4 single crystals depended on the excess oxygen.

ACKNOWLEDGMENTS

This work was partially supported by a Grant-in-Aid for Scientific Research on Priority Areas, "Chemistry of New Superconductors", from the Ministry of Education, Science and Culture.

REFERENCES

1. Bednorz JG, Müller KA (1986) Z. Physik B64: 189
2. Uchida S, Takagi H, Kitazawa K, Tanaka S (1987) Japan. J. Appl. Phys. 26: L1
3. Takagi H, Uchida S, Kitazawa K, Tanaka S (1987) Japan. J. Appl. Phys. 26: L231
4. Chu CW, Hor PH, Meng RL, Huang ZJ, Wang YQ (1987) Phys. Rev. Letters 58: 405
5. Cava RJ, van Dover RB, Batlogg B, Rietman EA (1987) Phys. Rev. Letters 58: 408
6. Vaknin D, Sinha SK, Moncton DE, Johnston DC, Newsam J, Safinya CR, King H (1987) Phys. Rev. Letters 58: 2802
7. Grant PM, Parkin SSP, Lee VY, Engler EM, Remirez ML, Vazquez JE, Lim G, Jacowitz RD, Greene RL (1987) Phys. Rev. Letters 58: 2482
8. Fine SM, Greenblatt M, Simizu S, Friedberg SA (1987) Phys. Rev. B36: 5716
9. Shaheen SA, Jisrawi N, Lee YH, Zhang YZ, Croft M, McLean WL, Zhen H, Rebelsky L, Horn S (1987) Phys. Rev. B36: 7214
10. Hsu SW, Tsaur SY, Ku HC (1988) Phys. Rev. B38: 856
11. Demazeau G, Tresse F, Plante Th, Chevalier B, Etourneau J, Michel C, Hervieu M, Raveau B, Lejay P, Sulpice A, Tournier (1988) Physica C 153-155: 824
12. Jorgensen JD, Dabrowski B, Pei S, Hinks DG, Soderholm L, Morosin B, Schirber JE, Venturini EL, Ginley DS (1988) Phys. Rev. B38: 11337
13. Dabrowski B, Jorgensen JD, Hinks DG, Pei S, Richards DR, Vanfleet HB, Decker DL (1989) Physica C 162-164: 99
14. Birgeneau RJ, Chen CY, Gabbe DR, Jenssen HP, Kastner MA, Peters CJ, Picone PJ, Thio T, Thurston TR, Tuller HL, Axe JD, Böni P, Shirane G (1987) Phys. Rev. Letters 59: 1329
15. Picone PJ, Jenssen HP, Gabbe DR (1987) J. Crystal Growth 85: 576
16. Cheong SW, Fisk Z, Kwok RS, Remeika JP, Thompson JD, Gruner G (1988) Phys. Rev. B37: 5916
17. Bykov AB, Demianets LN, Zibrov IP, Kanunnikov GV,Melnikov OK, Stishov SM (1988) J. Crystal Growth 91: 302
18. Picone PJ, Jenssen HP, Gabbe DR (1988) J. Crystal Growth 91: 463
19. Tanaka I, Yamane K, Kojima H (1989) J. Crystal Growth 96: 711
20. Kambe S, Kitazawa K, Naito M, Fukuoka A, Tanaka I, Kojima H (1989) Physica C 160: 35
21. Kitazawa K, Kambe S, Fukuoka A, Naito M, Tanaka I, Kojima H (1989) Mat. Res. Soc. Symp. Proc. 156: 153

Growth and Structure of (Tl,Bi,Pb)$Sr_2CaCu_2O_x$ Single Crystal

Toshihiro Kotani, Yoshinobu Ueba, Hiromi Takei, Koji Tada, and Maumi Kawashima

Basic High-Technology Laboratories, Sumitomo Electric Industries, Ltd., 1-3, Shimaya 1-chome, Konohana-ku, Osaka, 554 Japan

ABSTRACT

Crystal growth and structural analysis of (Tl,Bi,Pb)$Sr_2CaCu_2O_x$ single crystals were performed. A single crystal with superconducting onset transition temperature of 85K was grown in a sealed gold tube by the slow-cooling method. The crystal structure was determined to be tetragonal(P4/mmm) with lattice constants of a=3.80Å and c=12.09Å. The refinement of the crystal structure was carried out based on the single-crystal X-ray diffraction data.

KEY WORDS: Tl-Bi-Pb-Sr-Ca-Cu-O system, crystal growth, crystal structure, X-ray diffraction

INTRODUCTION

Soon after the Tl-based high-Tc superconducting oxides were found, their superconductivity and structures were intensively studied, mainly with sintered materials. However, a single crystal is indispensable for precise analysis of their structures. Therefore, many efforts have been devoted to growing a single crystal[1,2], though these oxides show an incongruent melting and vaporization of Tl-oxide, and there is a lack of data on the phase relations. Recently, new Tl-based compounds were prepared by partial substitution of Bi and Pb for Tl in the Tl-Sr-Ca-Cu-O system[3,4]. Much interest is focused on their superconductivity and crystal structures. However, these compounds have not been examined in single crystals.

In this work, we investigated the crystal growth of (Tl,Bi,Pb)$Sr_2CaCu_2O_x$(Tl-Bi-Pb 1212-phase) according to a study on the phase relations in this system, and then examined the crystal structure by means of single-crystal X-ray diffraction measurement.

EXPERIMENTAL

Crystal growth was carried out by the slow-cooling method based on a preliminary study of the phase relations in this system. Raw materials were prepared by thoroughly mixing oxide powders of Tl_2O_3, Bi_2O_3, PbO, SrO, CaO and CuO in the ratios of Tl:Bi:Pb:Sr:Ca:Cu=1:0.5:0.5:2:(n-1):n (n=2-7). The materials were charged in sealed gold tubes, and then the tubes were put into a platinum crucible with a tight lid. The crucible was inserted into a furnace which had been heated to 990°C, kept for 1 hour at this temperature, and then slowly cooled to room temperature at a rate of 5-10°C/hour. Crystals were mechanically separated from the heat-treated materials described above.

Grown crystals were identified by X-ray diffraction measurements, energy-dispersive X-ray spectrometry(EDX) and DC magnetic susceptibility measurement. Single-crystal X-ray diffraction measurement was made on an automated diffractometer with a four-circle goniometer(Rigaku RASA-5R) using MoKα radiation operated at 50kV and 150mA at room temperature. A black plate-like crystal with dimensions of 0.1mm x 0.1mm x 0.01mm was selected for data collection by an examination of the oscillating photograph. The structure was analyzed by the direct method(MITHRIL) and refined by full-matrix least squares analysis. Details of the analysis are given in Table 1.

Table 1. Summary of crystallographic information for $(Tl,Bi,Pb)Sr_2CaCu_2O_7$. Estimated standard deviations in the least significant figure are given in parentheses.

Dimensions (mm)	0.1 x 0.1 x 0.01
Radiation	MoKα
Monochromator	Graphite
Crystal composition	$Tl_{0.7}Bi_{0.1}Pb_{0.3}Sr_2Ca_{0.9}Cu_2O_7$
Crystal system	Tetragonal
Space group	P4/mmm (No. 123)
Cell constants (Å)	a=3.796(2), c=12.091(2)
Temperature	Ambient
Calculated density	$6.49 gcm^{-3}$ (Z=1)
Scan mode	$2\theta-\omega$
2θ range	5 to 55°
Absorption coefficient	$448.0 cm^{-1}$
Absorption correction	DIFABS
Transmission factors	0.52 to 1.38
Total reflections	906
Unique reflection	788
Observed reflection ($I>3\sigma$)	571
Data/parameter	27.2
R_f	0.084
R_w	0.105

RESULTS AND DISCUSSION

Although all the heat-treated materials prepared had a Tl-Bi-Pb 1212-phase, an almost single phase was obtained from the starting composition of n=2. A grown crystal separated from this material was thin, plate-like, and had a mirror-like surface, as shown in Fig. 1. Typical dimensions of a crystal were 0.1-0.5mm x 0.1-0.5mm x 0.02mm. The as-grown crystal showed sharp superconducting transition at 85K, as shown in Fig. 2. This transition temperature is the same as that for a sintered sample of $Tl_{0.5}Pb_{0.5}Sr_2CaCu_2O_7$ (Tl-Pb 1212-phase) [5] and higher than the 75K for a sintered sample of $TlSr_2CaCu_2O_7$ (Tl 1212-phase)[6]. EDX analysis revealed a crystal composition of Tl:Bi:Pb:Sr:Ca:Cu=0.7:0.1:0.3:2:0.9:2. It has been reported that the Tl/Pb ratio in the Tl-Pb 1212-phase is close to one[5]. However, a Tl/(Bi,Pb) ratio of 1.75 was obtained in the Tl-Bi-Pb 1212-phase. In addition, this analysis demonstrated that Bi-content was less than Pb-content in this crystal.

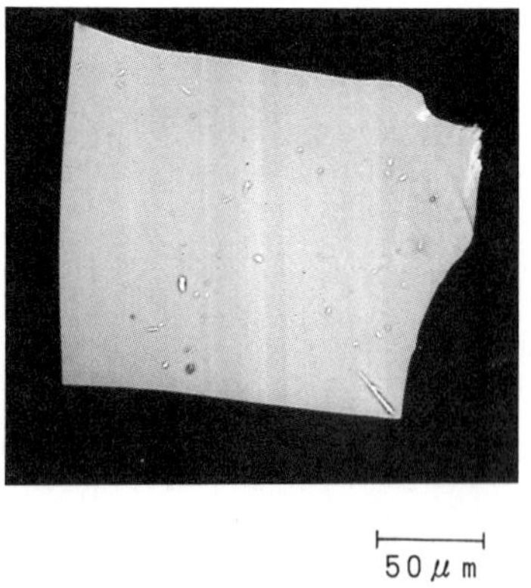

Fig. 1. SEM micrograph of a grown crystal.

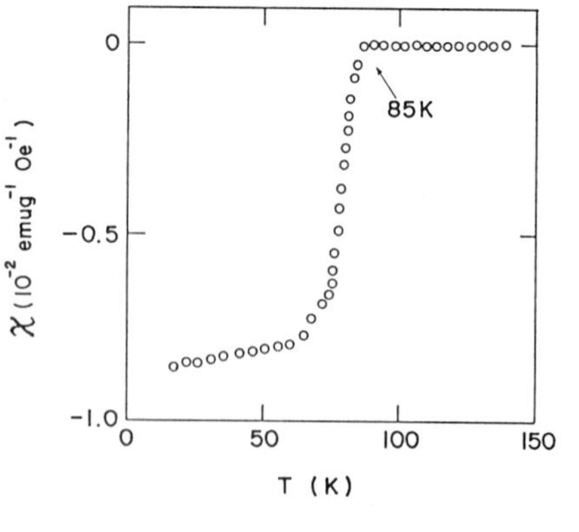

Fig. 2. Temperature dependence of DC magnetic susceptibility for $(Tl,Bi,Pb)Sr_2CaCu_2O_7$ single crystal. The data were obtained while cooling in a 10 Oe applied field parallel to the c-axis of a crystal.

Table 2. Crystallographic data of $(Tl,Bi,Pb)Sr_2CaCu_2O_7$. B_{eq} is the equivalent isotropic temperature factor. Estimated standard deviations in the least significant figure are given in parentheses.

Atom	site	x	y	z	B_{eq} (Å^2)	g
Tl(Bi,Pb)	1a	0	0	0	2.70(5)	1
Sr	2h	1/2	1/2	0.2166(2)	0.84(6)	1
Ca	1d	1/2	1/2	1/2	0.4(1)	0.91
Tl(Bi,Pb)	1d	1/2	1/2	1/2	0.5	0.09
Cu	2g	0	0	0.3641(3)	0.56(7)	1
O(1)	1c	1/2	1/2	0	6(1)	1
O(2)	2g	0	0	0.166(2)	1.2(5)	1
O(3)	4i	0	1/2	0.369(1)	1.0(6)	1

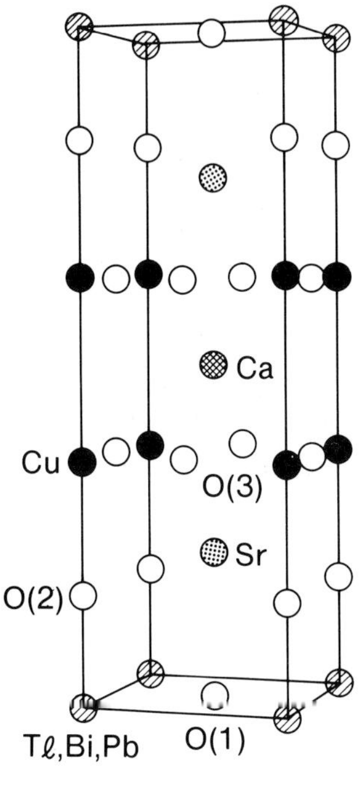

Fig. 3. Crystal structure of $(Tl,Bi,Pb)Sr_2CaCu_2O_7$.

Table 3. Anisotropic temperature factors(Å^2) Factors are given by the expression: $\exp[-2\pi^2(h^2U_{11}a^{*2}+\cdots)]$; $U_{ij}=0$ for $i\neq j$.

Atom	U_{11}	U_{22}	U_{33}
Tl(Bi,Pb)	0.0491(9)	0.049	0.004(1)
Sr	0.0110(8)	0.011	0.010(1)
Ca	0.006(2)	0.006	0.002(3)
Cu	0.005(1)	0.005	0.010(2)
O(1)	0.11(3)	0.11	0.01(2)
O(2)	0.018(7)	0.018	0.01(1)
O(3)	0.012(7)	0.012(7)	0.015(8)

Since the X-ray scattering factors of Tl, Bi and Pb are very similar, these atoms cannot be distinguished from each other. For the final refinement based on the diffraction data, we chose the space group P4/mmm, where the occupancy factors g of all atoms except for Tl(Bi,Pb) and Ca are fixed at unity and an anisotropic temperature parameter can be refined. The structure of the Tl-Bi-Pb 1212-phase, as shown in Fig. 3, is the so-called 1212-type reported for the Tl-based compounds[7]. The refined atomic positions are listed in Table 2 and thermal parameters in Table 3. As reported in the other Tl-based compounds, thermal parameters of Tl(Bi,Pb) and O(1) atoms in the Tl(Bi,Pb)-O layer have large values, which suggests an apparent displacement of their ideal positions in an averaged tetragonal structure[8]. The refinement of the occupancy factors of Tl(Bi,Pb) and Ca reveals that about 9% of Ca-sites were substituted by Tl(Bi,Pb) atoms. It seems that EDX analysis showing a higher (Tl,Bi,Pb)-content and a lower Ca-content than the ideal value, 1, supports this result. Also, this feature is observed in the other Tl-based compounds[6-8].

Table 4. Interatomic distances(Å^2) and lattice constants(Å^2) for (Tl,Bi,Pb)$Sr_2CaCu_2O_7$(Tl-Bi-Pb 1212), $Tl_{0.5}Pb_{0.5}Sr_2CaCu_2O_7$(Tl-Pb 1212)[5] and $TlSr_2CaCu_2O_7$(Tl 1212)[6]. Estimated standard deviations in the least significant figure are given in parentheses.

	Tl-Bi-Pb 1212	Tl-Pb 1212	Tl 1212
a	3.796(2)	3.8023(3)	3.7859(1)
c	12.091(2)	12.107(1)	12.104(1)
Tl(Bi,Pb)-O(1)	2.684(1)	2.35-3.05(4)	2.68
Tl(Bi,Pb)-O(2)	2.01(2)	1.996(7)	2.05
Sr-O(1)	2.619(3)	2.764(2)	2.58
Sr-O(2)	2.752(5)	2.656(6)	2.73
Sr-O(3)	2.65(1)	2.648(4)	2.67
Ca-O(3)	2.470(9)	2.470(2)	2.47
Cu-O(2)	2.39(2)	2.443(7)	2.31
Cu-O(3)	1.8988(9)	1.9027(2)	1.90

The interatomic distances are listed in Table 4, as compared with those of the Tl-Pb 1212-phase[5] and Tl 1212-phase[6] which were determined using a powder X-ray diffraction measurement. In the Tl-Bi-Pb and Tl-Pb 1212-phases, O(2) atoms bridging Tl(Bi,Pb) and Cu atoms are more strongly drawn towards the Tl(Bi,Pb) atom than in the case of Tl 1212-phase. A partial substitution of Bi and/or Pb, which have higher oxidation states of Bi^{5+} and Pb^{4+}, for Tl seems to cause a shift of the O(2) atom to the Tl(Bi,Pb) atom from the Cu atom, keeping the Cu-O(3) distance almost constant. Since this change leads to a lower oxidation state of Cu (hence, a hole concentration closer to an optimum value for maximum transition temperature), transition temperature might vary from 75K for the Tl 1212-phase to 85K for the Tl-Bi-Pb and Tl-Pb 1212-phases.

In conclusion, we have grown a (Tl,Bi,Pb)$Sr_2CaCu_2O_x$ single crystal by the slow-cooling method and have investigated the crystal structure by means of single-crystal X-ray diffraction. The structure analysis suggests that Ca-sites are partially substituted by Tl(Bi,Pb) atoms, and that Cu-O(2) distance is stretched with an introduction of these atoms to the Tl-O layer. These features may be closely related to their superconducting properties.

REFERENCES

1. Ginley DS, Morosin B, Banghman RJ, Venturini EL, Schirber JE, Kwak JF (1988) J. Cryst. Growth 91: 456
2. Kotani T, Kaneko T, Takei H, Tada K (1989) Jpn. J. Appl. Phys. 28: L1378
3. Torii Y, Takei H, Tada K (1989) Jpn. J. Appl. Phys. 28: L2192
4. Kaneko T, Wada T, Yamauchi H, Tanaka S (1990) Appl. Phys. Lett. 56:1281
5. Parise JB, Gai PL, Subramanian MA, Gopalakrishnan J, Sleight AW (1989) Physica C 159: 245
6. Doi T, Usami K, Kamo T (1990) Jpn. J. Appl. Phys. 29: L57
7. Parkin SSP, Lee VY, Nazzal AI, Savoy R, Beyers R, LaPlaca SJ (1988) Phys. Rev. Lett. 61:750
8. Subramanian MA, Torardi CC, Gopalakrishnan J, Gai PL, Cala brese JC, Askew TR, Flippen RB, Sleight AW (1988) Science 242: 249

4 Flux Pinning

4.1 Flux Pinning Properties

Pinning by Oxygen Point Defects in Proton-Irradiated $Ba_2YCu_3O_{7-\delta}$

R.B. van Dover, E.M. Gyorgy, A.E. White, L.F. Schneemeyer, R.M. Fleming, R.J. Felder, and J.V. Waszczak

AT&T Bell Labs, Murray Hill, NJ 07974, USA

ABSTRACT

Proton irradiation of single-crystal $Ba_2YCu_3O_{7-\delta}$ reproducibly increases the critical current tenfold, but does not alter the activation energy for flux creep. This can be accounted for in terms of a conventional flux pinning model. Based on this interpretation, as well as circumstantial structural evidence from TEM and X-ray studies, we propose that the chief effect of low-fluence proton irradiation is to create a higher density of oxygen point defects (which exist even in undamaged crystals). This conclusion is supported by observations of the effect of irradiation on electronic parameters.

INTRODUCTION

Radiation-induced damage provides a reproducible means to alter the properties of materials, and specifically allows us to examine the influence of defects and disorder on superconductivity in a controlled manner. The usefulness of this technique for understanding the high-T_c superconductors has been apparent since early 1987 and has been vigorously investigated since then[1].

Both neutron and proton irradiation have been shown[2–5] to create defects which can lead to a large critical current at 77 K in $Ba_2YCu_3O_{7-\delta}$. Neutron irradiation has the advantage of a large depth of penetration, so that bulk samples can be modified. Alternatively, protons with energies in the range 2-3.5 MeV are readily obtained, do not induce radioactivity in the sample, and penetrate $Ba_2YCu_3O_{7-\delta}$ up to 60 µm, an ample depth for the study of thin single crystals. In any case it is desirable to study single crystals, since they simplify the interpretation of most measurements. Even single crystals exhibit a range of behavior[3,6], so it is important to distinguish behavior that is characteristic of the material from behavior which is idiosyncratic to individual crystals. In the work leading to this paper we have examined crystals of varying quality; here we report only the behavior characteristic of the highest quality crystals. The crystals were grown from a CuO flux[7] in a ZrO_2 crucible, and had no detectable impurities at the 0.1 at% level. The crystals were annealed at 400 °C for six month in O_2. Two measures of the quality of the crystals are first, that the critical current at 77 K and 1 T is only 5×10^3 A/cm^2 (recall that a defect-free crystal would presumably have $J_c = 0$, at least for $\mathbf{H}\perp c$), and second, that the normal-state resistivity is only 120 µΩ-cm at 300 K. Typical crystal dimensions are $1\times 1\times 0.04$ mm^3.

EXPERIMENTAL

Proton irradiation was performed in a Van de Graaf accelerator on a liquid- nitrogen-cooled stage at an energy of 2-3.5 MeV and a flux ϕ= .1-10 µA/cm^2, for a fluence ϕt= 1E15-2E17 cm^{-2}. The samples were warmed to room temperature before measurement. Critical currents were inferred from magnetization measurements using the critical state model.[8] Resistivity measurements were performed on other samples which were contacted with evaporated Ag pads. Note that the properties of these high-quality crystals are quite reproducible (for example, the value of $J_c(T= 77\text{ K}, H= 1\text{ T})$ varies by ~ 20% from crystal to crystal) so that comparisons from sample to sample are reliable. Other experimental details have been published previously.[3,4]

RESULTS AND DISCUSSION

The effect of low-fluence proton irradiation on the critical current of $Ba_2YCu_3O_{7-\delta}$ is shown in Figures 1 and 2. Figure 1 shows the critical current at a field of 1 T as a function of temperature for an unirradiated crystal and a crystal irradiated at the fluence which results in the highest J_c at low temperatures. In this figure $J_c^{1/2}$ is plotted versus T, so the straight-line fits imply $J_c(T, H= 1\ T) \propto (T_c^*- T)^2$. This square-law dependence is unusual in superconductors but is consistently observed in our crystals. Thin film studies usually report $J_c(T, H= 0) \propto (T_c^*- T)$ but we note that films with a particularly high degree of structural perfection have recently been found[9] to obey $J_c(T, H= 1\ T) \propto (T_c^*- T)^2$. Note in Figure 1 that T_c^* (the extrapolated temperature at which $J_c \rightarrow 0$) is lower than T_c (indicated by the arrow) and does not change over the range $0< \phi t< 3E16\ cm^{-2}$. This is probably related to the fact that the irreversibility line does not change upon proton irradiation in this range.[5]

The temperature dependence of $J_c^*(T= 0,H= 1T)$ on fluence is shown in Figure 2, where $J_c^*(0, 1\ T)$ is the zero temperature critical current as extrapolated in Figure 1, and is not necessarily the actual critical current that would be measured at T= 0. Each point represents a different crystal. Note the reproducibility of the results and that a broad peak near $\phi t= 3E16\ cm^{-2}$ is found. Using a Monte-Carlo simulation[10] and making the plausible assumptions that it takes 20 eV to displace an atom and that the energy to break a bond is 1 eV, we infer a displacement density of $1.4\times 10^{20}\ cm^{-3}$ at the peak, which corresponds to an average defect spacing of ~ 20 Å. Note that the critical current at the maximum is roughly $2\times 10^7\ A/cm^{-2}$, a value which approaches 10% of the depairing limit. As pointed out by Hylton and Beasley[11] in the context of thin films, this is an extraordinarily high critical current, and certainly justifies an interpretation in terms of strong pinning (many pinning sites per vortex and negligible vortex correlation effects at $B<< B_{c2}$).

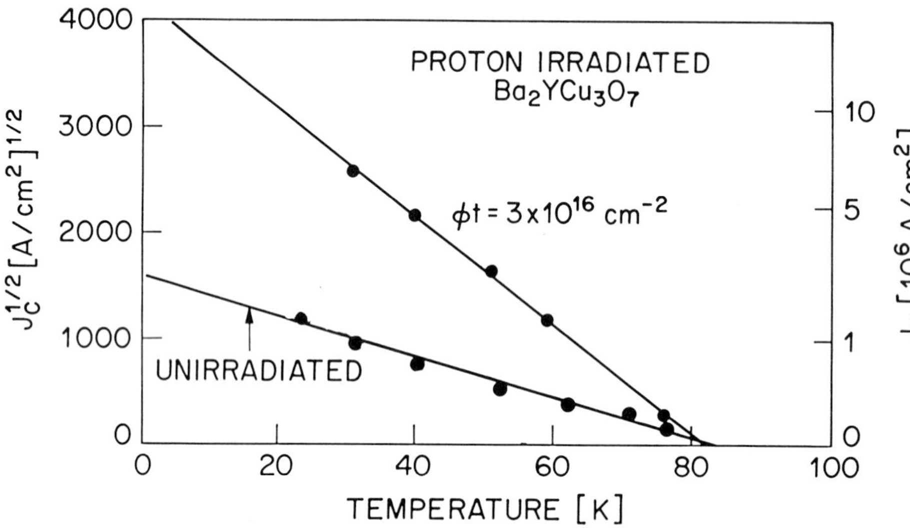

Figure 1. Temperature dependence of the critical current at a field of 1 Tesla, for two $Ba_2YCu_3O_{7-\delta}$ crystals. The results are typical of many samples. The critical currents were inferred from magnetization measurements.

Figure 3 shows the effect of irradiation on the resistive transition of a typical high-quality crystal. We have found that the detailed shape of the resistive transition in a magnetic field after irradiation can be a sensitive test of the quality of a crystal: inferior crystals show a non-monotonic increase of ρ within the resistive transition. Again, in this paper only the highest-quality crystals are discussed. The broadening of the transitions in a magnetic field in Figure 3 is attributed partly to the upper critical field, $H_{c2}(T)$ of the crystal, and partly to ill-understood dissipation in the superconducting state. By measuring the magnetic moment of the samples in the ergodic region near T_c, we can fairly reliably infer dH_{c2}/dT, which changes from 1.74 T/K to 2.5 T/K for the sample irradiated with a fluence of $3E16\ cm^{-2}$. This corresponds to change in coherence length $\xi_{ab}(0)$ from 17.7 to 14.9 Å,

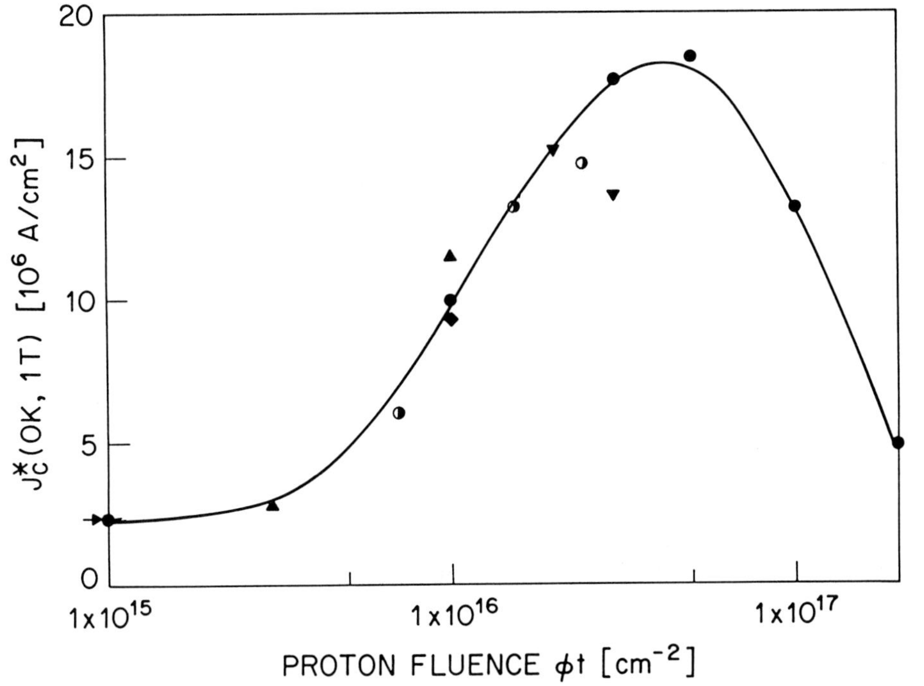

Figure 2. Fluence dependence of the low-temperature critical current for proton-irradiated $Ba_2YCu_3O_{7-\delta}$. $J_c^*(0,1\ T)$ is the value at T= 0 extrapolated from plots as inFigure 1.

and correlates approximately with the broadening in $\rho(T)$ at the empirical criterion $\rho = 0.9\rho_n$. The change in ξ_{ab} suggests that one effect of the irradiation is to decrease the mean free path in the **ab** plane. The associated change in the normal-state resistivity, $\rho_n(T)$, is consistent with this picture: for low fluences we find that the zero temperature extrapolation $\rho_n^*(0)$ increases, while the linear slope of $\rho_n(T)$ changes very little (it decreases slightly, in fact). These observations lend credence to the interpretation that low-fluence proton irradiation generates local disorder without creating even moderate-sized insulating regions (cf. Ne irradiation[12].)

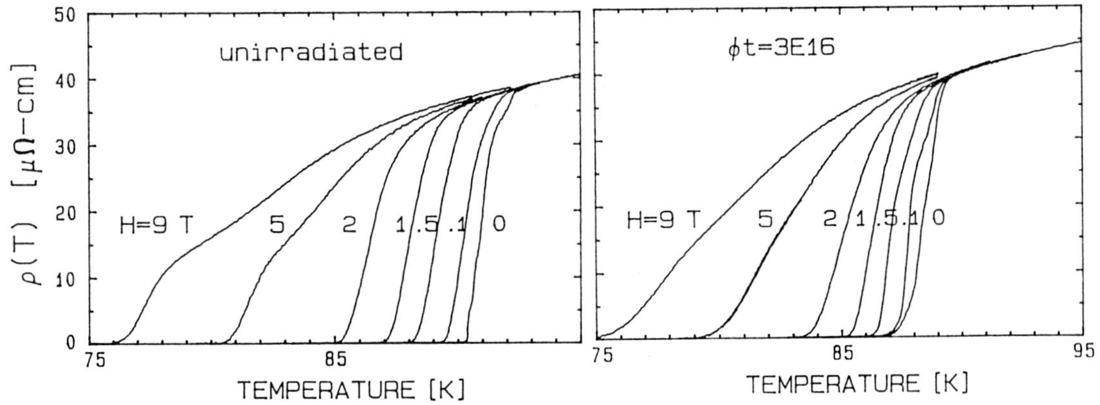

Figure 3. The resistive transition for a $Ba_2YCu_3O_{7-\delta}$ crystal. The current density for the measurement was 27 A/cm^2. Lower current densities did not alter the data within experimental resolution.

The low-resistance tail of the resistive transition can be accounted for in terms of conventional flux creep. In this picture we expect the resistivity to decrease exponentially with an activation energy U(T,H):

$$\rho = \rho_0 \exp(-U/kT)$$

Here ρ_0 is the resistivity that would be measured in the absence of pinning. The physics underlying this parameter are not well-understood; for the present purposes a sufficient approximation is $\rho_0 \sim \rho_n(T)$, the normal-state resistance. This allows us to infer U(T) from:

$$U(T) = kT\ell n(\rho_n(T)/\rho(T))$$

as suggested by Palstra, et al.[13] The result is plotted in Figure 4 for the same crystal as in Figure 3, for three fluences. Notice that, except for a slight shift in T_c after the initial irradiation, the curves are virtually unchanged. This result is comparable to the behavior of $Bi_{2.2}SrCa_{0.8}CuO_4$ upon irradiation,[13] and is consistent with the observation that the "irreversibility line" in $Ba_2YCu_3O_{7-\delta}$ is essentially unchanged after proton irradiation.[5]

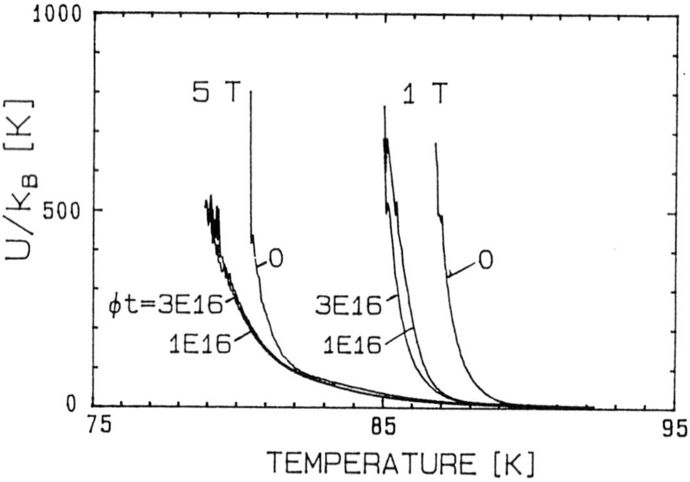

Figure 4. The pinning energy for a proton-irradiated $Ba_2YCu_3O_{7-\delta}$ crystal inferred from the resistive transitions of Figure 3 as described in the text. The fluences ϕt are as indicated.

That J_c reliably increases 10-20 fold, yet U(T) is unchanged after proton irradiation at a fluence of 3E16 cm^{-2} may at first seem unexpected. But it can can be readily understood in terms of the strong pinning model, assuming that the effect of irradiation is merely to increase the density of the same defects which extensively pervade the unirradiated crystal. Then U(T), the energy required to move part of a vortex away from the nearest pinning site, should remain unchanged. Note that here we picture a somewhat different model from that of Hylton and Beasley, who assume an initially linear vortex, which after depinning is bent, increasing the line tension of the vortex. We assume that the vortex is already kinked on the scale of the random defects, and that, on average, the line tension is unchanged after depinning. Then the density of defects does not, to first order, affect the pinning energy. The critical current is not determined by this energy. Instead, it is determined by a force density, namely the pinning force per unit length of a vortex. Increasing the defect density increases the the number of pinning centers per unit length, thus it increases the critical current. By this argument, we would expect to find that the critical current increases roughly as the linear density of vortices, *i. e.*, as $(\phi t)^{1/3}$. In fact, $J_c^*(0, 1\ T)$ does vary with fluence in approximately this manner, which can be verified by replotting the data of Figure 2.

We therefore suggest that the dominant effect of low-fluence proton irradiation on $Ba_2YCu_3O_{7-\delta}$ is to simply increase the number of oxygen point defects present in the crystal. Structural studies provide supporting, albeit indirect, evidence for this contention. TEM study, for example, does not reveal any systematic increase in the number of observable defects in the irradiated crystals.[14] No doubt some large defects due to collision cascades could be found given a sufficiently extensive TEM search, but these defects are clearly not present with high enough density to account for the critical current of the irradiated samples. But we note that TEM studies also do not reveal the oxygen defects that are known to be present in bulk $Ba_2YCu_3O_{7-\delta}$ where $\delta > 0$. A null observation on the irradiated crystals suggests that here we may well have the same type of defects, only with a higher density to account for the higher critical current.

Extensive x-ray studies on the irradiated and unirradiated crystals have been performed using a monochromatic high intensity synchrotron light source, which allows us to examine both the linewidth and background scattering in detail. We have concentrated on the (110), (020) and (006) lines. No increases in the linewidths are observed in the proton-irradiated sample, allowing us to conclude that the crystal lattice maintains coherence over a length scale in excess of 1 μm. More significantly, the diffuse scattering between the peaks also does not increase, to a resolution of approximately 1 part in 10^6. This implies a lack of scattering due to point defects. Since we know that even unirradiated $Ba_2YCu_3O_{7-\delta}$ has extensive oxygen point defects, it must be that oxygen is such a weak x-ray scatterer that it does not cause measurable diffuse scattering. (This is not surprising since the scattering power of each element varies as Z^2). Again, the null measurement supports the interpretation that irradiation merely generates a higher density of oxygen point defects. We also do not observe any change in the lattice constants of the crystals after irradiation, within the experimental error of ± 0.01 Å. This puts an upper limit on any change in the oxygen stoichiometry or in the reduction of orthorhombicity due to chain-oxygen site disorder.

Point defects are generally unimportant sources of pinning in superconductors, but it can be shown that oxygen point defects can qualitatively account for the observed pinning energies and critical currents. Tinkham has argued[15] that the pinning energy due to oxygen point defects should be roughly 400 K. We have repeated his calculation using our measured values for H_{c2} and κ and obtain an expected pinning energy of 500 K—a satisfying but probably fortuitous level of agreement. Both values are comparable to 400 K, the value inferred by Civale, et al.,[5] from low temperature flux creep (using the conventional interpretation), as well as the values inferred at high temperature from the activated resistivity (Figure 4). These values are sufficient to account for a critical current of 10^6 A/cm^2 at low temperatures, assuming that the pinning force is roughly $f_p \sim U_p/\xi_{ab}$. A more quantitative analysis would not be appropriate at this time considering the ambiguity about the detailed nature of the oxygen defects in both unirradiated and proton-irradiated $Ba_2YCu_3O_{7-\delta}$. The important concept is that the evidence suggests that low-fluence proton irradiation does little other than simply increase the density of oxygen defects in $Ba_2YCu_3O_{7-\delta}$. To obtain a higher pinning energy we argue that either cation defects or extended defects will have to be produced on a fine scale within the crystalline matrix.

REFERENCES

1. For an extensive and continually updated bibliography see Ozawa, K. "A Bibliography of Irradiation Effect on High Temperature Superconductors" (Ozawa, Hitachi ERL, Ibaraki Japan) (unpublished).
2. van Dover RB, Gyorgy EM, Schneemeyer LF, Mitchell JW, Rao KV, Puzniak R, and Waszczak JV (1989) Nature 342: 55.
3. van Dover RB, Gyorgy EM, Schneemeyer LF, White AE, Glarum S, Felder RJ, and Waszczak JV (1990) Materials Res. Soc. Proceedings 169: 911.
4. van Dover RB, Gyorgy EM, White AE, Schneemeyer LF, Felder RJ, and Waszczak JV, (1990) Appl. Phys. Lett. 56: 2681.
5. Civale L, Marwick AD, McElfresh MW, Worthington TW, Malozemoff AP, Holtzberg FH, Thompson JR, and Kirk MA (1990) Phys. Rev. Lett. 65: 1164
6. Daeumling M, Seuntjens JM, and Larbalestier DC (1990) Nature 346: 332.
7. Schneemeyer LF, Waszczak JV, Siegrist T, vanDover RB, Rupp LW, Batlogg B, Cava RJ, and Murphy DW, (1987) Nature 328: 601.
8. Gyorgy EM, van Dover RB, Jackson KA, Schneemeyer LF, and Waszczak JV, (1989) Appl. Phys. Lett. 55: 283.
9. Siegal MP, Phillips JM, Hebard AF, van Dover RB, Farrow RC, Tiefel TH, and Marshall, JH (unpublished).
10. Ziegler FJ and Biersack JP *The Stopping and Range of Ions in Solids*, (Pergamon, New York, 1985).
11. Hylton TL and Beasley MR (1990) Phys. Rev. B. (in press).
12. White AE, Short KT, Dynes RC, Levi AFJ, Anzlowar M, Baldwin KW, Polakos PA, Fulton TA, and Dunkelberger LN (1988) Appl. Phys. Lett. 53: 1010; Valles JM, White AE, Short KT, Dynes RC, Garno JP, Levi AFJ, Anzlowar M, and Baldwin K (1989) Phys Rev. B 39: 11599.
13. Palstra TTM, Batlogg, B, Schneemeyer LF, Waszczak JV (unpublished).
14. Chen CH (private communication).
15. Tinkham, M (1988) Helv. Phys. Acta. 61: 443.

Flux Pinning in CVD Processed YBaCuO Film

K. WATANABE[1], S. AWAJI[1], N. KOBAYASHI[1], H. YAMANE[1], T. HIRAI[1], Y. MUTO[1], and T. YAMASHITA[2]

[1] Institute for Materials Research, Tohoku University, 1-1, Katahira 2-chome, Aoba-ku, Sendai, 980 Japan
[2] Department of Electronics, Nagaoka University of Technology, 1603-1, Kamitomioka-cho, Nagaoka, 940-21 Japan

ABSTRACT

The anisotropy of J_c and B_{c2} of high-J_c CVD-$Y_1Ba_2Cu_3O_{7-\delta}$ films was resistively measured. The small anisotropic character of J_c near zero magnetic field at 77.3 K was explained by the model of the flux pinning centers of thin plate-like precipitates. The angular dependence of B_{c2}*, whose value is affected by the flux creep, contributed largely to the anisotropy of J_c.

KEYWORDS : $Y_1Ba_2Cu_3O_{7-\delta}$ film, CVD, anisotropy, flux pinning

INTRODUCTION

Investigations for practical use of high-T_c superconducting oxides have many subjects to be solved. Most important issue for application is to hold large critical currents up to a high magnetic field. In addition, major problems such as the flux lattice melting[1)] and flux creep[2)] and the anisotropic behaviors[3)] must be overcome. Thin films of $Y_1Ba_2Cu_3O_{7-\delta}$ prepared by CVD[4)] have already demonstrated the splendid J_c properties more than 10^4 A/cm^2 in high fields up to 30 T at 77.3 K so far. These thin films were provided with the single crystalline morphology. It is not possible yet for poly-crystalline superconducting oxides to apply the high magnetic field at 77.3 K, while highly textured $Bi_2Sr_2CaCu_2O_{8+x}$ tapes[5)] indicated relatively large J_c in fields at 4.2 K. In the Bi-system wires and tapes, the very large history effect[6)] of J_c appeared at 4.2 K. The weak-links at grain boundaries in poly-crystalline superconducting oxides are considered to be responsible for the poor J_c properties at 77.3 K and the history effect at 4.2 K. For realization of practical use, the key points are how the weak-links can be got rid of as well as how the grain boundaries contributing to the flux pinning can be introduced.

The elucidation of the flux pinning mechanism in high-T_c superconducting oxides is needed. At the present stage, the common understanding of the flux pinning centers in high-T_c superconducting oxides is insufficiently established yet. We reported that fine disk-shaped precipitates observed by transmission electron microscopy for CVD-YBaCuO films are conceivable as a candidate of the flux pinning centers[7)]. Moreover, it was also reported that defects produced by neutron irradiation in $Y_1Ba_2Cu_3O_{7-\delta}$ single crystal can act as strong pinning sites[8)]. Therefore, the flux pinning mechanism seems to be related to many finely dispersed precipitates, some defects and so on because of the short coherence length. If we succeed in introducing strong pinning centers adequate to superconducting oxides, the phenomena peculiar to the flux lattice melting and flux creep may not affect so seriously for application.

Another problem of the high-T_c superconducting oxides is that the large anisotropic behaviors due to the layered structure are observed in the superconducting characteristics such as the upper critical field and the critical current density. This anisotropy may be associated with the flux pinning mechanism. The investigation on the anisotropy will offer an important information to understand the flux pinning.

In this study, the anisotropic critical current densities in CVD-$Y_1Ba_2Cu_3O_{7-\delta}$ films are presented. The anisotropy is described on the basis of the flux pinning mechanism, and also the angular dependence of J_c is discussed in terms of the flux scaling law.

EXPERIMENTAL

(1) Sample preparation and morphology

Source gases of metal β-diketonates of $Y(DPM)_3$, $Ba(DPM)_2$ and $Cu(DPM)_2$ were introduced with a carrier gas of Ar into the CVD reactor with a hot-wall. Substrates used were $SrTiO_3(100)$ single crystals. The relationship between the deposition temperature of 850 °C and the oxygen partial pressure of 3.3 Torr corresponded to the condition of so-called Bormann's line[9], whose condition was found in the investigation of $Y_1Ba_2Cu_3O_{7-\delta}$ superconducting oxides with high quality. The CVD-$Y_1Ba_2Cu_3O_{7-\delta}$ films prepared on Bormann's line also indicated high-T_c and high-J_c characteristics.

CVD-$Y_1Ba_2Cu_3O_{7-\delta}$ films indicated preferred c-axis orientation perpendicular to $SrTiO_3(100)$ substrates, and many precipitates were observed in a dense matrix of $Y_1Ba_2Cu_3O_{7-\delta}$ by scanning electron microscope. There existed two kinds of precipitates. Large precipitates with sizes of a few microns grew from the surface of the $SrTiO_3$ substrate, and the energy dispersive X-ray spectra and the X-ray diffraction patterns indicated that the large precipitates[10] are considered as CuO. On the other hand, very fine precipitates[7] with sizes of a few hundred angstroms were observed dispersively around large precipitates of CuO. These fine precipitates may be also regarded as CuO, because CVD-YBaCuO films with high J_c character were made in a Cu-rich compositional region.

Moreover, CVD-YBaCuO films included many precipitates as well as a small amount of grains with the a-axis orientation perpendicular to substrates[10]. The quantity of the a-axis oriented grains in the preferred c-axis oriented matrix was controllable. For the flow of the transport current, however, CVD-YBaCuO films looked like a single crystal without being separated by grain boundaries. This single-crystalline morphology was also the reason why Noge et al.[11] could not observed the Josephson tunneling using native grain boundaries of CVD-YBaCuO films deposited onto $SrTiO_3(100)$ substrates.

(2) Electrical measurements

Measurements of T_c, B_{c2} and J_c were performed by a four-probe resistive method. External magnetic fields were applied by use of a hybrid magnet[12] (HM-1b, 28 T, ϕ52 mm room temperature bore, 0.1%/24 mm DSV homogeneity), a water-cooled resistive magnet (WM-2, 16 T, ϕ52 mm room temperature bore, 0.4%/20 mm DSV homogeneity) or a superconducting magnet (13 T, ϕ50 mm bore, 0.2%/20 mm DSV homogeneity).

The sample holder in an adiabatic vacuum can is rotatable for measuring the angular and temperature dependence of B_{c2}, and enables us to measure the angular dependence in various temperatures above 4.2 K[13]. The flow direction of the transport current is always normal to the applied magnetic field as shown in Fig.1. The field direction was varied in the xz plane in this figure by rotating the sample. The angle θ was defined as that between the magnetic field and the c-axis (z-axis), and a relative accuracy was 0.5°. The angular dependence of J_c at 77.3 K was measured by use of the similar rotatable cryostat with a relative accuracy of 0.2°. This sample holder of J_c was directly immersed in liquid nitrogen, so that a flexible current lead had a capacity of about 50 A.

The sample for J_c measurements was patterned by excimer laser (KrF, λ=2480 Å). A narrow bridge with dimensions of 30 μm wide, 300 μm long and 1 μm thick was made by irradiation with energy density of 0.3 - 1 J/cm^2. A mask of NbN deposited onto a quartz substrate was used, in order to make a optically reduced printing of a bridge on a scale of about 1/22.

RESULTS AND DISCUSSION

Anisotropic behaviors have been measured to search the flux pinning mechanism of CVD-YBaCuO films[14]. A typical example of anisotropic J_c at 77.3 K is shown in Fig.2. The anisotropy was very large in a high field region but became

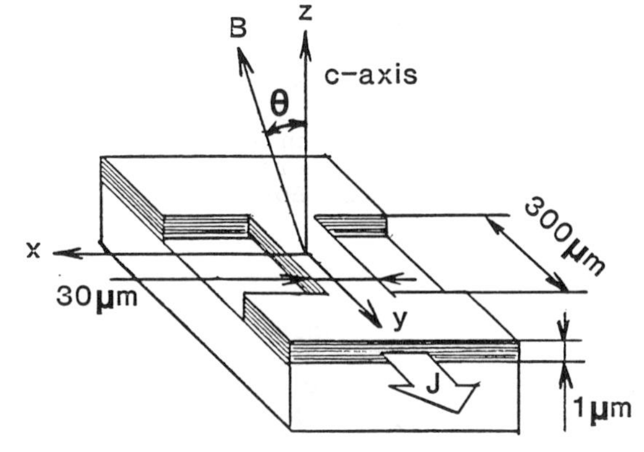

Fig.1 The x and y axes are parallel to the film surface (ab-plane) and the z axis is parallel to the c-axis. The magnetic field B is applied with angle θ in the xz plane.

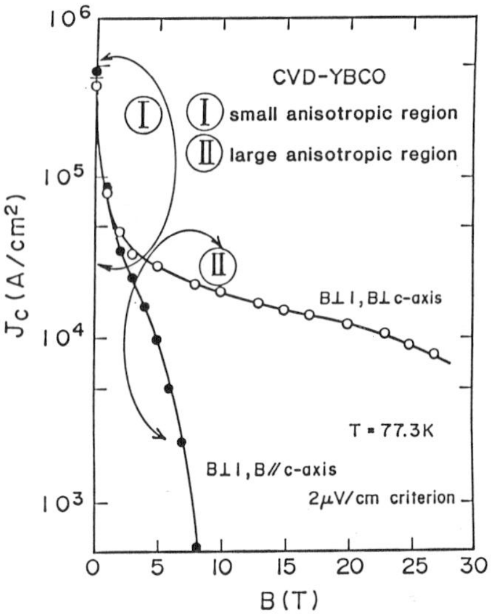

Fig.2 Anisotropy of the transport critical current densities in fields up to 27 T at 77.3 K for a CVD-$Y_1Ba_2Cu_3O_{7-\delta}$ film.

smaller in a low field region. In particular, one can notice that almost the same J_c values in both fields perpendicular and parallel to the c-axis (B⊥c-axis and B//c-axis) were obtained in low fields below 1 T and that the anisotropic behavior became indistinct in the vicinity of zero magnetic field. It is very interesting to know how this small anisotropic behavior close to zero magnetic field is related to the flux pinning mechanism. We have already reported the calculated results of the global pinning force in B⊥c-axis, where fine disk-shaped precipitates were conceived as strong pinning centers[7]. These results supported that thin plate-like precipitates parallel to the ab-plane of CVD-YBaCuO films can be regarded as pinning centers in fields perpendicular to the c-axis at 77.3 K.

Here we think the flux pinning in the vicinity of B=0 for both directions of B⊥ and B//c-axis. Figure 3 represents the simple model of such a thin plate-like precipitate. As already reported in ref.7, the global pinning force for B⊥c-axis ($F_p^{\perp}$) was described as follows,

$$F_p^{\perp}=N_p f_p^{\perp} P^{\perp}=N_p \frac{B_c^2}{\mu_0}\frac{\xi_{ab}}{\xi_c}(Lt)^2\frac{B}{\phi_0}\left(1-\frac{B}{B_{c2}^{\perp}}\right), \tag{1}$$

where N_p is the pin density, $f_p^{\perp}$ elementary pinning force for B⊥c-axis and $P^{\perp}$ the probability in which the flux lines come across a pinning center. In the case of B//c-axis, we can write the elementary pinning force $f_p^{/\!/}$ in Fig.3 as

$$f_p^{/\!/}=\frac{U_p}{\xi_{ab}}=\frac{B_c^2}{2\mu_0}\left(1-\frac{B}{B_{c2}^{/\!/}}\right)\pi\xi_{ab}t. \tag{2}$$

The pinning for B//c-axis is effective only on the border of a plate-like precipitate due to the short coherence length. Therefore, the probability $P^{/\!/}$ for B//c-axis is given by

$$P^{/\!/}=\frac{4(2\xi_{ab}L)}{d_p^2}\left(\frac{d_p}{a_f}\right)^2=\frac{8\xi_{ab}LB}{\phi_0}, \tag{3}$$

where the average spacing of the pinning center is $d_p=N_p^{-1/3}$ and the flux line spacing $a_f=(\phi_0/B)^{0.5}$. Thus, the global pinning force $F_p^{//}$ for B//c-axis is

$$F_p^{//}=N_p f_p^{//} p^{//}$$

$$=N_p \frac{B_c^2}{2\mu_0}(1-\frac{B}{B_{c2}^{//}})8\pi\xi_{ab}^2 t\frac{LB}{\phi_0}. \quad (4)$$

We get the ratio of $F_p^{\perp}/F_p^{//}$ under the condition of $B \ll B_{c2}$ as follows,

$$\frac{F_p^{\perp}}{F_p^{//}}=\frac{1}{4\pi}\frac{Lt}{\xi_{ab}\xi_c}. \quad (5)$$

Therefore, we have the ratio of $F_p^{\perp}/F_p^{//}\approx 0.7$-$1.6$ in low fields near B=0 for ξ_{ab}(77.3 K)≈40 Å, ξ_c(77.3 K)≈7 Å[15], L≈250-300 Å and t≈10-20 Å. The values of the coherence lengths used were comparable to those derived from the resistively measured B_{c2} using the midpoint definition for CVD-YBaCuO films. This result indicates that the value of $J_c^{\perp}$ almost agrees with that of $J_c^{//}$ in fields close to B=0 at 77.3 K, but that the anisotropy in the vicinity of B=0 will become clear again with decreasing temperature or with varying the size of the precipitate.

On the other hand, it is considered that B_{c2} will predominantly affect the anisotropic behavior of J_c in high fields. The anisotropic B_{c2} is originated from the layered structure of the CuO_2 planes in high-T_c superconducting oxides. We anticipate that the $J_c^{//}$ will rapidly decrease with applying magnetic field, because of the small $B_{c2}^{//}$ in comparison with the large $B_{c2}^{\perp}$. We measured the temperature dependencies of B_{c2} at various angles. Although the resistively determined values of B_{c2} are ambiguous due to phenomena of a broad resistive transition, the values of B_{c2} determined by zero resistance are important quantities. It is because the value of B_{c2}(R=0) corresponds to the reversible point in magnetization measurements, and its reversible point means that J_c becomes zero at that point. The derived angular dependencies of $-dB_{c2}/dT$ were compared with the anisotropic Ginzburg-Landau theory[14]. Figure 4 shows the obtained data and the theoretical curve together with those by the midpoint definition for reference. The angular dependence of $-dB_{c2}/dT$ roughly fitted the GL theory, and in particular the data points by the midpoint definition gave good agreement with the GL prediction. As the result, we

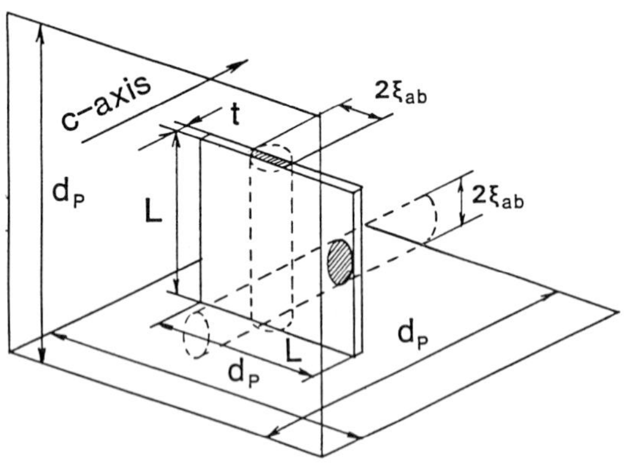

Fig.3 Model of the flux pinning in a plate-like precipitate. Both directions of B⊥c-axis and B//c-axis for flux lines are indicated.

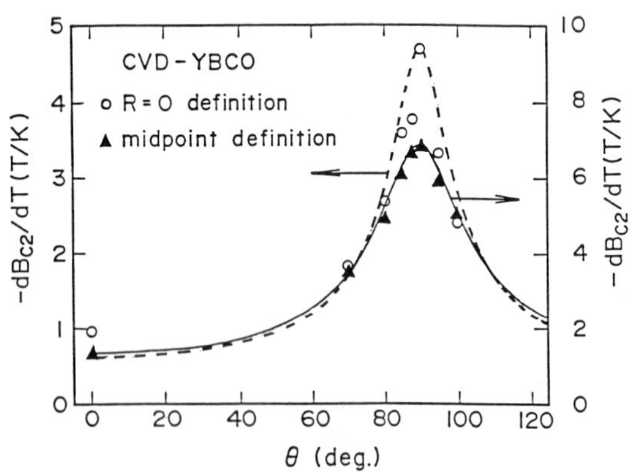

Fig.4 Angular dependence of $-dB_{c2}/dT$. The data points obtained by R=0 or midpoint were fitted in the GL expression of $-dB_{c2}/dT=0.62(0.13^2\sin^2\theta+\cos^2\theta)^{-1/2}$ or $-dB_{c2}/dT=1.3(0.20^2\sin^2\theta+\cos^2\theta)^{-1/2}$, respectively.

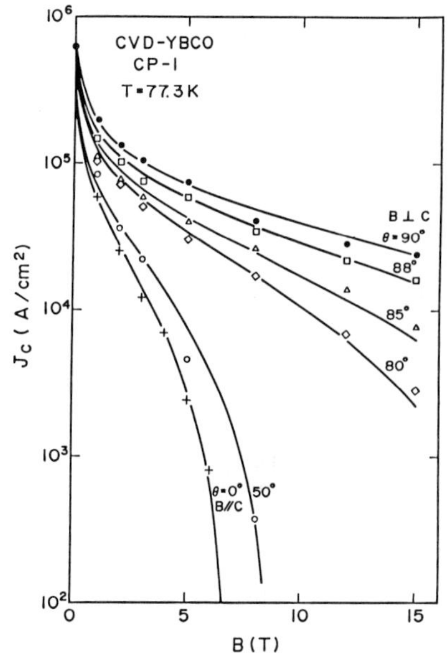

Fig.5 J_c vs B curves in various angles at 77.3 K for a CVD-$Y_1Ba_2Cu_3O_{7-\delta}$ film. The values of J_c were determined by a 10 μV/cm criterion.

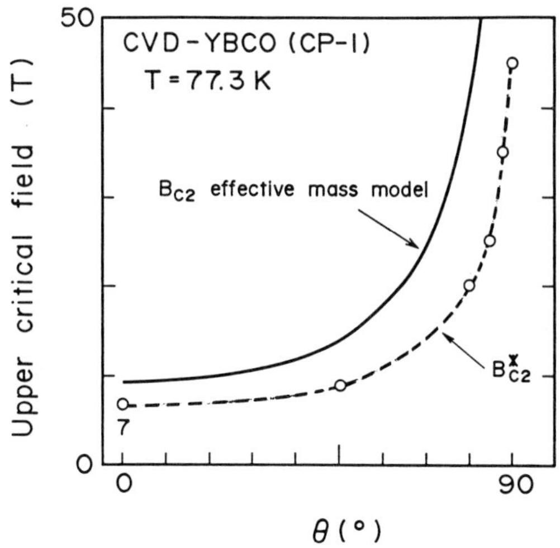

Fig.6 Comparison between B_{c2} determined by R=0 and B_{c2}* used in the flux pinning scaling law for a CVD-$Y_1Ba_2Cu_3O_{7-\delta}$ film. The typical angular dependence of B_{c2} described well by the GL theory is demonstrated here for comparison.

could almost understand the anisotropic behaviors of J_c for CVD-$Y_1Ba_2Cu_3O_{7-\delta}$ films by introducing the angular dependence of B_{c2}[14]. However, the influence to J_c due to the large flux creep at 77.3 K in high-T_c superconducting oxides must be examined. We should adopt the more appropriate value of B_{c2} to explain the J_c vs B curves at various angles in detail.

Figure 5 shows the measurements of the magnetic field dependence of J_c at various angles at 77.3 K. We estimate the J_c vs B curves at various angles using the flux pinning scaling law [16] given by

$$J_c = K[B_{c2}]^{m-p}B^{p-1}(1-\frac{B}{B_{c2}})^q, \tag{6}$$

where m, p and q are the scaling parameters, and K is the fitting parameter. Scaling parameters of m=1, p=0.5 and q=2 were used, although these parameters depended on samples. The relationship between the scaling parameters and microstructures is under investigation. One faces some difficult problems, that is, major problem lies in the magnetic field dependence in the flux pinning scaling law different from that in the model of plate-like precipitates. This field dependence remained to be solved out theoretically. Furthermore, the flux creep behavior will directly affect the value of B_{c2}. In this respect, we intended that the values of B_{c2}* smaller than B_{c2}(R=0) should be adopted as indicated in Fig.6[17]. The values of B_{c2}* were determined by fitting the data of J_c vs B measured at various angles. The solid lines in Fig.5 were fitted well to the measured data of J_c vs B at various angles from B⊥c-axis to B//c-axis.

In conclusion, anisotropic behaviors of J_c and B_{c2} were investigated for high-J_c CVD-$Y_1Ba_2Cu_3O_{7-\delta}$ films. Nearly the same values of J_c in both B⊥c-axis and B//c-axis close to B=0 at 77.3 K were explained by the model of the flux pinning centers of thin plate-like precipitates. It was found that the angular dependence of J_c in fields except near B=0 is fitted well by introducing the flux scaling law and the angular dependence of B_{c2}*.

ACKNOWLEDGMENTS

We would like to thank Profs. K. Hiraga and T. Matsushita for their nice discussion. We also thank to H. Kurosawa (RIKEN Co.) for his contribution in sample preparations. This work is supported by Grant-in-Aid for Scientific Research on Priority Area from the Ministry of Education, Science and Culture, Japan.

REFERENCES

1. Gammel PL, Schneemeyer LF, Waszczak JV, Bishop DJ (1988) Evidence from mechanical measurements for flux-lattice melting in single-crystal $YBa_2Cu_3O_7$ and $Bi_{2.2}Sr_2Ca_{0.8}Cu_2O_8$. Phys Rev Lett 61 : 1666-1669
2. Yeshurun Y, Malozemoff AP (1988) Giant flux creep and irreversibility in an Y-Ba-Cu-O crystal : an alternative to the superconducting-glass model. Phys Rev Lett 60 : 2202-2205
3. Roas B, Schultz L, Saemann-Ischenko G (1990) Anisotropy of the critical current density in epitaxial $YBa_2Cu_3O_x$ films. Phys Rev Lett 64 : 479-482
4. Watanabe K, Yamane H, Kobayashi N, Hirai T, Muto Y (1990) CVD route for high J_c superconductors. In : Narlikar A (Ed) Studies of high temperature superconductors (in press)
5. Enomoto N, Kikuchi H, Uno N, Kumakura H, Togano K, Watanabe K (1990) The high field magnetic dependence of critical current density at 4.2 K for Ag-sheathed $Bi_2Sr_2CaCuO_y$ superconducting tape. Jpn J Appl Phys 29 : L447-L449
6. Watanabe K, Noto K, Morita H, Fujimori H, Mizuno K, Aomine T, Ni B, Matsushita T, Yamafuji K, Muto Y (1989) Anisotropy and hysteresis of transport critical currents in high temperature Ln-Y-Ba-Cu-O superconductors. Cryogenics 29 : 263-267
7. Watanabe K, Matsushita T, Kobayashi N, Kawabe H, Aoyagi E, Hiraga K, Yamane H, Kurosawa H, Hirai T, Muto Y (1990) Strong flux pinning centers in Y-Ba-Cu-O films prepared by chemical vapor deposition. Appl Phys Lett 56 : 1490-1492
8. van Dover RB, Gyorgy EM, Schneemayer LF, Mitchell JW, Rao KV, Puzniak R, Waszczak JV (1989) Critical current near 10^6 Acm^{-2} at 77 K in neutron-irradiated single-crystal $YBa_2Cu_3O_7$. Nature 342 : 55-57
9. Bormann R, Nolting J (1989) Stability limit of the perovskite structure in the Y-Ba-Cu-O system. Appl Phys Lett 54 : 2148-2150
10. Yamane H, Hirai T, Kurosawa H, Suhara A, Watanabe K, Kobayashi N, Iwasaki H, Aoyagi E, Hiraga K, Muto Y (1990) High-T_c Superconducting oxide films prepared by CVD. In : Ishiguro T, Kajimura K (Eds) Advances in Superconductivity II : 767-772
11. Noge S, Yamashita T, Warg Z, Matsui T, Kurosawa H, Yamane H, Hirai T (1989) Bridge type Josephson junction in MO-CVD thin films. Jpn J Appl Phys 28 : L1581-L1584
12. Muto Y, Nakagawa Y, Noto K, Miura S, Hoshi A, Watanabe K, Kido G, Ichikawa H, Fujioka T, Sato Y, Osaki O, Takano H (1986) General survey of Tohoku hybrid magnet system. Sci Rep RITU A33 : 221-237
13. Noto K, Lee ZH, Toyota N (1983) A temperature-variabele sample rotating cryostat in high magnetic fields. Cryogenics 23 : 201-202
14. Watanabe K, Awaji S, Kobayashi N, Yamane H, Hirai T, Muto Y (1990) Angular dependence of the upper critical field and the critical current density for $Y_1Ba_2Cu_3O_{7-\delta}$ films. submitted to J Appl Phys
15. Welp U, Kwok WK, Crabtree GW, Vandervoot KG, Liu JZ (1989) Magnetic measurements of the upper critical field of $YBa_2Cu_3O_{7-\delta}$ single crystals. Phys Rev Lett 62 : 1908-1911
16. Campbell AM, Evetts JE (1972) Flux vortices and transport currents in type II superconductors. Adv Phys 21 : 199-428
17. Watanabe K, Kobayashi N. Yamane H. Hirai T, Muto Y, Kawabe H, Kurosawa H (1990) Critical current and flux pinning in $Y_1Ba_2Cu_3O_{7-\delta}$ films prepared by chemical vapor deposition. In : Proceedings of International Cryogenic Materials Conference (ICMC'90). 9-11 May 1990, Garmisch-Partenkirchen (FRG)

Irreversibility Temperatures in Oxide and Metallic Superconductors

M. Suenaga, A.K. Ghosh, Youwen Xu, and D.O. Welch

Brookhaven National Laboratory, Upton, NY 11973, USA

ABSTRACT

We measured the irreversibility temperatures, $T_r(H)$ for both oxides [pure and alloyed Y(123) and Bi(2212, 2223)] and metallic [Nb, NbTi, and Nb_3Sn]superconductors. These results are compared and discussed in terms of $T_r(H)$ being the depinning line and/or the melting line OF the flux line (crystalline or disordered) lattice in the H-T plane.

KEY WORDS: Nb_3Sn, NbTi, Y(123), Bi(2212, 2223), flux lattice melting

INTRODUCTION

Nearly all of the applications for superconductors require high critical current densities at operating magnetic fields and temperatures. One of the major disappointments associated with the high T_c superconductors was their inability to have sufficiently high critical current densities J_c at high temperatures. Although the values of J_c in thin films have improved to the point where they are useful, the intragranular critical current densities in single crystal and ceramic oxide superconductors are still significantly lower than the desired values. It is believed that a part of this difficulty is associated with very low values of the so-called "irreversibility temperature, $T_r(H)$" (the lowest temperature at which the flux lines can be moved reversibly). Also, these low vales of $T_r(H)$ are related to very weak coupling between the conducting layers (CuO_2) in these highly anisotropic layered superconductors [1]. However, what this irreversibility line in the H-T plane represents, whether a thermodynamic phase transition in the state of the flux line lattice such as melting or a kinetic cross-over such as a depinning line, is not clear at this moment and this question is still one of the highly discussed subjects in superconductor studies. In the conventional superconductors, it is generally believed that since they are isotropic, the temperature region of the reversibility is so small that it is very difficult to measure and is not of practical importance. However, recently we have measured and reported [2] surprisingly large temperature regions where the flux lines are reversible in Nb_3Sn and NbTi multifilamentary wires. Here, we will compare these results with similarly measured values of $T_r(H)$ [3] for the oxide superconductors, Y(123) and Bi(2212, 2223), and discuss these in terms of various models for $T_r(H)$: crystalline [4] or disordered [5] flux lattice melting and thermally-assisted depinning [6].

EXPERIMENTAL PROCEDURE

For the purpose of measuring the magnetic-field-dependent irreversibility temperature $T_r(H)$ and the critical temperature $T_c(H)$, bulk sintered specimens of the desired compositions were prepared [7,8]. The specimens were powdered (~10 μm) and aligned along the c-axis. For the conventional wires, state-of-the-art multifilamentary Nb_3Sn and NbTi wires were selected. In addition, in order to study the effect of variations in the value of the Ginzburg-Landau constant κ on the irreversibility temperatures, a fine filamentary (~3 μm) cold-drawn Nb wire was also studied. These wires were cut in a length of ~6 mm segments and several lengths were placed horizontally (perpendicular to applied fields) in a superconducting quantum interference device (SQUID) magnetometer (Quantum Design, Inc.) while the aligned oxide powders were placed in the magnetometer such that the c-axis of the oxides are parallel to the applied field. As previously described [3], a standard procedure for measuring $T_c(H)$ was used to determined $T_r(H)$.

The results of such measurements of the magnetic moments for a Nb-Ti-alloy and a Nb wire are illustrated in Fig. 1. Also shown in the figures are the criteria which were used for determining the irreversibility temperature $T_r(H)$ and the critical temperature $T_c(H)$ [or the mean-field critical magnetic field $H_{c2}(H)$]. Here we define $T_r(H)$ at the temperature which an observable deviation is noted in the moments after the temperature cycle. It was found that by this criteria the difference in the moments for the warming and the cooling cycle was consistently ~0.1% at $T_r(H)$.

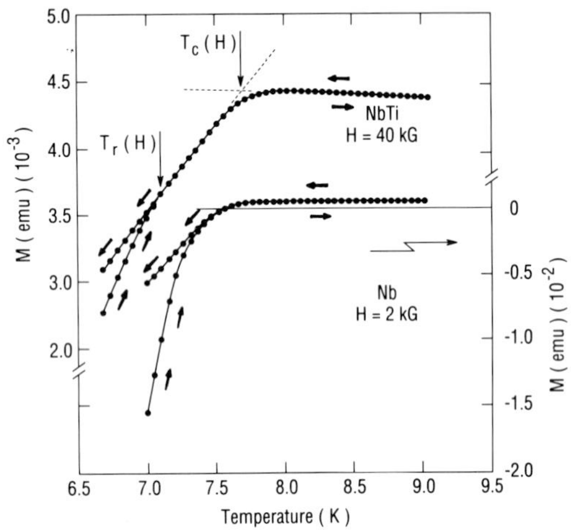

Fig. 1. The reversible temperature regions for a Nb-Ti alloy and a pure Nb wire are illustrated.

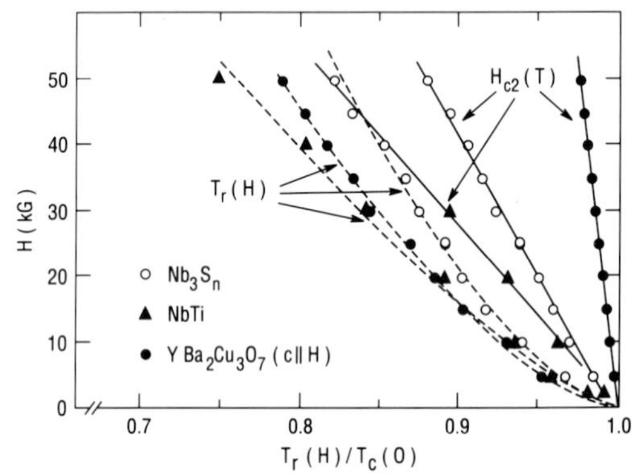

Fig. 2. A comparison of the temperature region of the reversible magnetic flux line motion in Nb_3Sn, NbTi and Y(123).

RESULTS AND DISCUSSION

The results of the magnetic field dependence of the values of $T^r(H)$ for Nb_3Sn, Nb-Ti; and Y(123) and for the Bi(2212) and Bi(2223) specimens are summarized in Figs. 2, 3 and 4. As shown in Fig. 3, the data for all of the pure and alloyed Y(123) specimens follow a relationship:

$$H = a[1 - T_r(H)/T_c(0)]n \quad (1)$$

with n ~ 1.5 and "a" is a constant of proportionality. We measured the magnetic hysteresis width ΔM [at $T/T_c(0) \cong 0.1$ and 0.5 and H = 10 kG] for each specimen. We find that "a" decreases nearly proportionally with decreasing ΔM. Thus, it is clear that the pinning strength of a specimen has a strong correlation with the irreversibility temperature in the case of alloyed $YBa_2Cu_3O_7$. Also, it was found that the data above do not fit the predicted dependence of $T_r(H)$ on H predicted on the basis of flux-lattice melting by Houghton et al. [4]. In the case of the Bi-oxides, the relationship in Eq. (1) does not fit at all as shown in Fig. 4. However, the relationship between the strength of the flux pinning and the width of the reversible temperature region is similar to that of Y(123), i.e. the stronger the pinning, the higher the values of $T_r(H)$. Thus, these results indicate that the $T_r(H)$ lines in these oxides are not likely to be a thermodynamic melting temperature of the crystalline flux line lattice unless the melting temperature depends upon the state of flux pinning.

Since both the flux creep model and Fisher's glass-to-liquid transition model [5] for $T_r(H)$ depend on the strength of the pinning, we have to study other

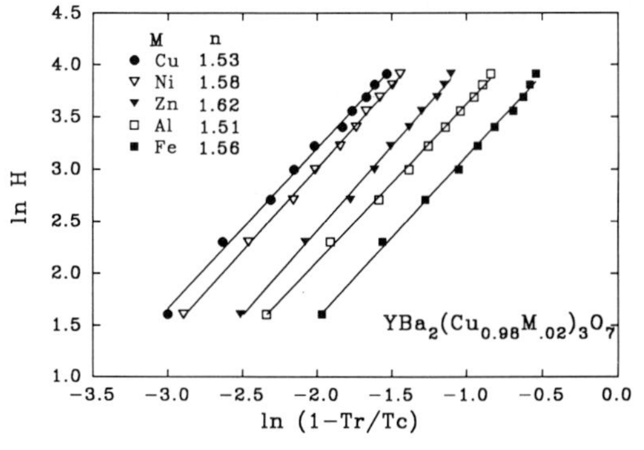

Fig. 3. The irreversibility temperatures for a pure and alloyed Y(123) as a function of applied fields.

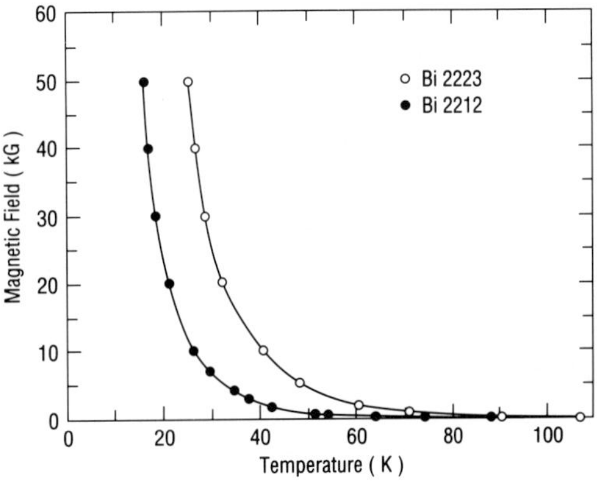

Fig. 4. The irreversibility temperatures for a Bi(2212) and a Bi(2223) as a function of applied field.

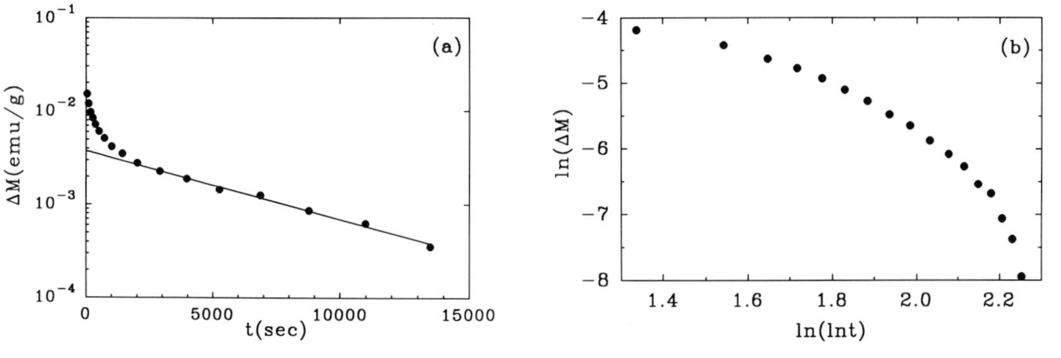

Fig. 5. The temporal dependence of the hysteresis width (the critical current) for a Bi(2223).

properties of these materials in order to distinguish these models. In order to do this, we note that only in the limit of $t \to \infty$, the predictions of the temporal dependence of the flux creep are distinguishable i.e. Fisher's model predicts,

$$\Delta M \sim (\ell n\ t)^{-1/\mu} \quad \text{where } \mu < 1 \tag{2}$$

while the creep model [6] predicts

$$\Delta M \sim e^{-\alpha t} \text{ where } \alpha \text{ is a constant.} \tag{3}$$

If we assume that the regime $\Delta M \to o$ is equivalent to $t \to \infty$, we can achieve this condition in the Bi-oxides at low fields (1-3 kG) and at temperatures very close to T_r. The result for the Bi(2223) specimen is shown in Fig. 5. This clearly shows that the creep in this specimen follows an exponential temporal dependence for $t \to \infty$ [Fig. 5(a)] and does not follow the $(\ln t)^{-1/\mu}$ dependence [Fig. 5(b)]. Thus, we conclude that in these materials, $T_r(H)$ is more likely to be a depinning line in the H-T plane rather than a glass-to-liquid phase transition line.

The present interest in studies of the magnetic state of the oxide superconductors als prompted reexamination of the conventional superconductors to seek clarification of some of the observed phenomena in the oxides. As shown in Fig. 1, state-of-the-art Nb-Ti wires exhibited rather large reversibility regions. Also, in Fig. 2 a comparison of the sizes of the reversibility temperature regions is illustrated for Nb_3Sn, Nb-Ti, and pure Y(123) on a reduced temperature scale, $T_r(H)/T_c(O)$. The regions, $[T_c(H) - T_r(H)]$, for Nb_3Sn and Nb-Ti are nearly 1/3 as large as that for Y(123). This is quite surprising, but what is more surprising is the fact that the field dependence of $T_r(H)$ for both the Nb_3Sn and Nb-Ti wires fits the predicted dependence of the melting temperature calculated with the theory of Houghton et al. [2]. In addition, the predicted κ dependence was also founded to follow the theory as illustrated in Fig. 1, i.e. smaller the value of κ, the smaller the reversibility region and κ for Nb is an order of magnitude smaller than Nb-Ti. Furthermore, a large reduction in J_c (~5 times) or in the size of filaments in Nb-Ti wires did not result in observable difference in the size of $[T_c(H) - T_r(H)]$. Thus, all these results point toward flux lattice melting in the high-κ and high-Jc conventional soperconductors. The only experimental observation which raises a question about this interpretation, is the fact that when the irreversibility temperature is measured by the cycling magnetic field at a constant temperature, this temperature $H_r(T)$ is significantly closer to $H_{c2}(T)$ than $T_r(H)$. At present, the reason for this discrepancy is not clear and further study is required to confirm the possible observation of flux-lattice melting in Nb_3Sn and Nb-Ti.

In summary, it appears that the irreversibility lines are flux-line depinning lines in the oxides and are possibly flux-line-lattice melting lines in the conventional superconductors. Particularly, the latter is very unexpected and further study is needed to understand the nature of flux line motion in these materials.

ACKNOWLEDGMENT

This work was supported by the U.S. Department of Energy, Division of Materials Sciences, Office of Basic Energy Sciences under Contract No. DE-AC02-76CH00016.

REFERENCES

1. For review, see for example, Tinkham M, (1991) IEEE Trans. Magn. Mag-27: (to be published).
2. Suenaga M, Ghosh AK, Xu Youwen, and Welch DO Phys. Rev. Lett.(submitted).
3. Xu Youwen, Suenaga M, Gao Y, Crow JE, and Spencer ND Phys. Rev. B (to be published).
4. Houghton A, Pelcovits RA , and Sudo S (1989) Phys. Rev. 40: 6763.
5. Fisher MPA, (1989) Phys. Rev. Lett. 62: 1416.
6. Griessen R, Flipse CFJ, Hagen CW, and Lensink J (1989) J. Less-Common Metals 151: 39.
7. Xu Youwen, Suenaga M, Tafto J, Sabatini RL, Moodenbaugh AR, and Zolliker P (1989) Phys. Rev. B 39: 6667.
8. Xu Youwen, Sabatini RL, Moodenbaugh AR, Zhu Yimei, Shyu Shin-guang, Suenaga M, Dennis KW, and McCallum RW (1990) Physica C 169: 205.

Properties of Single Phase 2223-Bi Cuprates: Substitution Effect of Ln for Ca

RYOZO YOSHIZAKI[1], HIROSHI IKEDA[1], MASAYA WATAHIKI[1], KOZO YOSHIKAWA[2], and NOBUYOSHI TOMITA[3]

[1] Institute of Applied Physics and Cryogenics Center, University of Tsukuba, Tsukuba, 305 Japan
[2] Takasago Reserach and Development Center, Mitsubishi Heavy Industries, Ltd., Takasago, Hyogo, 676 Japan
[3] Kobe Shipyard and Machinery Works, Mitsubishi Heavy Industries, Ltd., Kobe, 652 Japan

ABSTRACT

We have synthesized single phase 2223-Bi compounds and studied the substitution effect of Ln(=La or Gd) for Ca on magnetic flux-creeping properties. We found from the analysis of the distribution of pinning potentials that the substitution of Ln resulted in increase of high energy pinning potentials as well as low energy ones.

KEY WORDS: 2223-phase Bi system, Ln substitution for Ca, substitution effect on pinning potentials, distribution of pinning potentials

INTRODUCTION

In the field of power application of high-Tc superconductors, it is desired to fabricate them with higher Jc properties. In order to improve Jc-properties, one must increase pinning energies of magnetic flux in superconductors [1]. As a first step, we have to clarify the origin of pinning mechanism in the high-Tc superconductors. For this purpose we introduced point defects into the 2223-phase Bi cuprates by substituting Ln for Ca and studied their effect on the distribution of the pinning potentials and on the magnetic Jc profiles.

EXPERIMENTAL

We synthesized high-quality, highly oriented and dense ceramics of single phase Bi compounds by a hot-press method [2]. The replacement of Ca by Ln was attainable up to about 2% with maintaining the single phase of 2223, which was confirmed from a Rietveld analysis of the X-ray powder diffraction data. As a Ln atom we chose La or Gd, since it is a typical nonmagnetic and magnetic element, respectively. Magnetic measurement in the present experiment was performed by a SQUID magnetometer for the H⊥c-axis configuration. Pinning potentials were estimated from the decay profiles of the remanent magnetic moment.

RESULTS AND DISCUSSION

Figure 1 shows the temperature dependence of magnetic susceptibilities for non-doped and doped samples measured in field-cooling (FC) process at H=100 Oe. The onset temperature of the superconductivity did not change, but the magnitude of the diamagnetic susceptibility at low temperature was decreased by the doping of Ln which was substituted for Ca. This fact suggests that the amount of the flux-expulsion (Meissner effect) becomes small for the La or Gd substituted compounds. Since the small flux-expulsion is correlated with the strong pinning force [3], we expect, as the doping effect, that the pinning potentials would increase and the Jc-profile must be improved.

Figure 2 shows the magnetic hysteresis loops measured at 77.3 K after zero-field cooling. Since the magnitude of the hysteresis is proportional to the magnetic Jc in the Bean's critical-state model [1], Jc profile was decreased with the replacement of Ca by La or Gd, as seen from the figure. The magnetic field where the hysteresis loop merges into reversible curve is known as a measure of the critical field of the flux flow-region. This critical field was also decreased by the substitution of La or Gd, suggesting weakened pinning potentials in the doped samples. This result contradicts the expectation from the Meissner signals mentioned above.

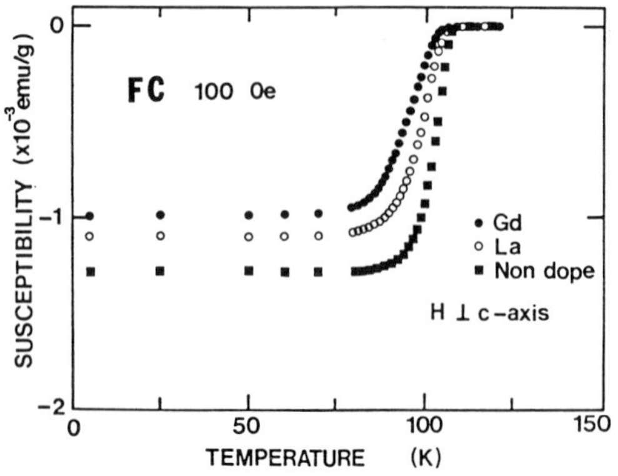

Fig. 1 Temperature dependence of the magnetic susceptibilities measured in field cooling for the non-doped, La-doped and Gd-doped 2223 Bi-cuprates, which are denoted by solid squares, open circles and solid circles, respectively.

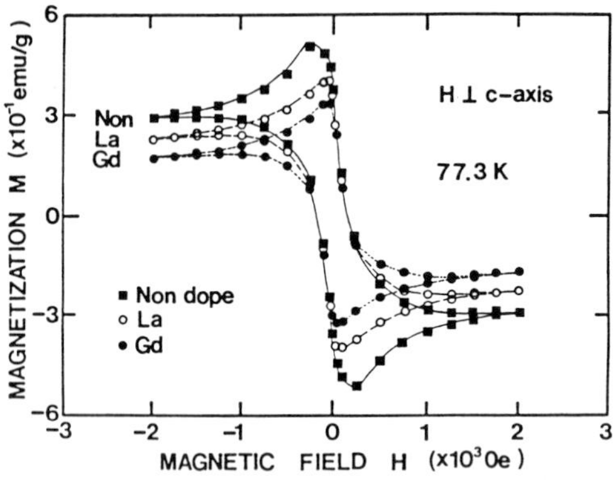

Fig. 2 Magnetic hysteresis loop of the non-doped, La-doped and Gd-doped 2223 Bi-cuprates, which are denoted by solid squares, open circles and solid circles, respectively. The lines are the guides for eyes.

In order to clarify the difference between the two experimental results, we studied the distribution of pinning potentials of the non-doped and doped samples from the the measurement of relaxation rate of the remanent magnetic moment [4]. The relaxation rate of the remanent moment (M), -dM/dlnt, is plotted as a function of temperature in Fig. 3. The peak of the relaxation rate is around 10 K for these 2223-phase Bi cuprates. The greater value of -dM/dlnt, in general, corresponds to the sooner decay of the pinned flux, and suggests the weaker pinning potentials which are effective at that measuring temperature. From this qualitative point of view, the substitution effect appears remarkably at about 10 K in Fig. 3; The peak value of the relaxation rate for the non-doped sample becomes to decrease by the substitution. Thus, the shallow pinning potentials, which is effective only at low temperature, seem to be improved by the substitution. The difference between the magnetic and nonmagnetic atoms for the replacement is not so large.

In the measurement of the Meissner effect, the magnetic flux starts to be pinned below the flux-flow region with decreasing temperature from Tc. Hence, the magnitude of the Meissner diamagnetism is determined dominantly by the strength of pinning potentials at the higher temperature near Tc. In the present samples the flux-flow region is considered to be in the range from Tc to 90-100 K for H=100 Oe. Then the distribution of pinning potentials at that temperature would affect to the magnitude of the Meissner diamagnetism critically. The high temperature parts of the relaxation rates in Fig. 3 are enlarged in Fig. 4. We must note that the simple equation of -dM/dlnt=kT/U does not hold in this temperature region because of non-negligible probability of backward flux-scattering, where U is a pinning potential [5]. Nevertheless, the lower relaxation rate for the doped samples than the one for the non-doped sample suggests the increase of deep pinning potentials in the doped samples, and it must be attributed to the substitution effect.

The relaxation rates of the three samples change their magnitude's order at 77 K compared to their order at the higher temperatures as seen in Fig. 4, although the difference is quite small. This reversed order is able to explain the order of the magnetic hysteresis in Fig. 2; The small hysteresis loops for the doped samples are probably due to the weak pinning potentials at this temperature. According to this qualitative interpretation, the hysteresis loops at 10 K should be larger for the doped samples compared with the non-doped one. The result was, however, not consistent with this simple expectation, but the hysteresis loop for the non-doped one was the biggest.

The quantitative analysis of the distribution of the pinning potentials was carried out based on the Hargen-Griessen's analysis [4]. The calculated re-

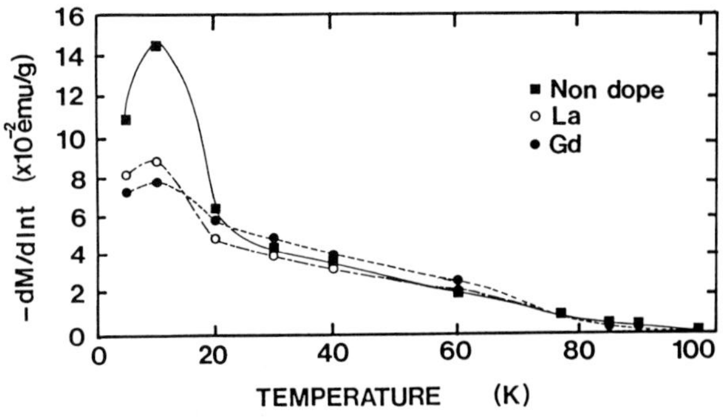

Fig. 3 Relaxation rate of the remanent moment (-dM/dlnt) plotted as a function of temperature for the non-doped, La-doped and Gd-doped 2223 Bi-cuprates. The lines are the guides for eyes.

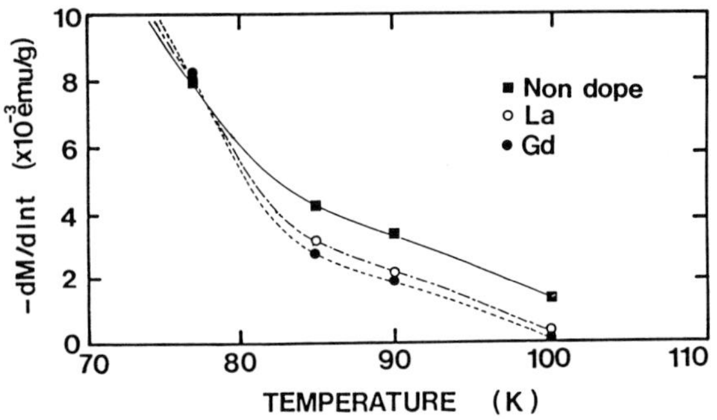

Fig. 4 High-temperature part of the relaxation rate of the remanent moment. The lines are the guides for eyes.

sults of the distribution function ($m(E^*)$) of the activation energies (E^*) are shown in Fig. 5. The results indicate that the distribution peak of the activation energies locates around 30 meV for the non-doped sample. This fact is similar to the results for the 2212-phase Bi cuprates [6]. The substitution of Ln resulted in the increase of the low activation energies below 20 meV as well as the increase of the higher energy parts. The difference of the distribution functions between the La and Gd substituted samples is observed at low activation energies. This is due to the difference of the ionic radius and/or the spin states of the doped atoms. Referring to the the distribution functions, the experimental results of the magnetic measurement will be interpreted as follows: The magnitude of the Meissner signal is correlated to the distribution of the high-energy pinning potentials. Moreover, the magnitude of

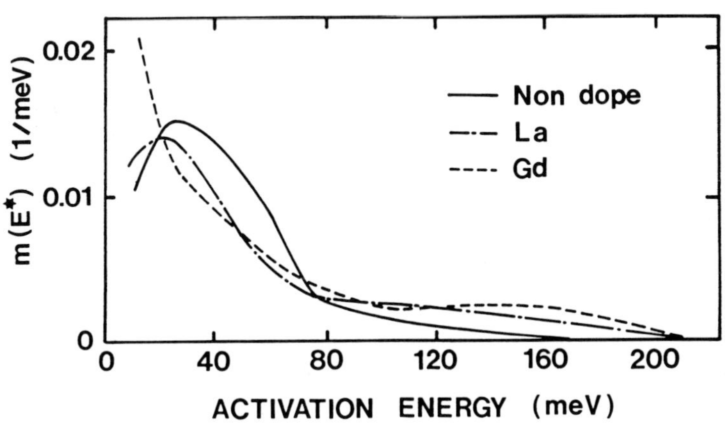

Fig. 5 Distribution function (m) of the activation energies for the non-doped, La-doped and Gd-doped 2223 Bi-cuprates. The lines are the guides for eyes.

the magnetic hysteresis loop is not directly correlated to the pinning potentials which is effective at the measuring temperature, but seems to be related to the integrated value of the potential distribution in the range of the higher energies than the observed temperature.

In summary, we have investigated the substitution effect of Ln for Ca on the distribution of the pinning potentials in the 2223 single-phase Bi cuprate system. We observed the suppression of the Meissner signals for the doped samples, suggesting the enhancement of the flux pinning. On the other hand, the magnetic hysteresis loops at 77.3 K became small by the doping, suggesting the weakened pinning forces. The change of the distribution functions of the activation energies were obtained from the relaxation measurement of the remanent moments of those compounds; the distribution of the middle energies decreased and both the lower energies and the higher energies increased by the doping.

ACKNOWLEDGEMENTS

The authors are grateful to K. Kawaguchi for helpful discussion on the calculation of the pinning energies. They also thank H. Asano and T. Ishigaki for the X-ray diffraction measurement and the Rietveld analysis of the data. This work is partly supported by Grand-in-Aid for Scientific Research on Priority Areas, "Mechanism of Superconductivity" from the Ministry of Education, Science and Culture.

REFERENCES

[1] Bean CP (1962) Phys. Rev. Lett. **8:** 250
[2] Ikeda H, Yoshizaki R, Yoshikawa K and Tomita N (1990) Jpn. J. Appl. Phys. **29:** L430
[3] Kitazawa K, Nakamura O, Matsushita T, Tomioka Y, Motohira N, Murakami M and Takei H (1990) Adv. in Superconductivity II: 609
[4] Hargen CW and Griessen R (1989) Phys. Rev. Lett. **62:** 2857
[5] Malozemoff AP, Worthington TK, Yandrofski RM and Yeshurun Y (1988) Intern. J. Modern Phys. B **1:** 1293
[6] Kobayashi N, Iwasaki H, Kawabe H, Watanabe K, Yamane H, Kurosawa H, Masumoto H, Hirai T and Muto Y (1989) Physica C **159:** 295

Heat Treating Condition Dependence of J_c-B Characteristics for Silver Sheathed $Y_{1-x}Ca_xBa_2Cu_4O_8$ Superconductors

KAZUYUKI SHIBUTANI, SEIJI HAYASHI, RIKUO OGAWA, and YOSHIO KAWATE

Superconducting and Cryogenic Technology Center, Kobe Steel Ltd., 5-5, 1-chome, Takatsukadai, Nishi-ku, Kobe, 651-22 Japan

ABSTRACT

The decomposition temperature of the Ca-doped Y124 materials in the silver sheathed wires rose from 775 °C to 845 °C with increasing oxygen partial pressure (Po_2) from Po_2=0.2bar (in air) to Po_2=1.0 bar (in pure oxygen). The critical transport current density (Jct) of 2.01×10^4 A/cm^2 was obtained for silver sheathed Y(Ca)124 wire at 4.2K and 0T by selecting the oxygen partial pressure and the annealing temperature. The grain boundary coupling of this specimen was found to be insufficient by AC susceptibility measurement.

KEY WORDS: $Y_{1-x}Ca_xBa_2Cu_4O_8$, O_2-HIP technique, silver sheathed wires, oxygen partial pressure, critical current density

INTRODUCTION

After the discovery of high critical temperature oxide superconductors, the many attempts to fabricate the oxide superconducting wires have been carried out. Y124 material of which the critical temperature is around 80K has been reported to be thermally stable up to 800 °C [1]. Miyatake et al. [2] found a remarkable effect of Ca substitution on the Tc of Y124 phase. By partial substitution of Ca for Y, Tc increased from 80K to 90K. Moreover, We could synthesize the single phase bulk Y(Ca)124 material in short reacting time by using O_2-HIP technique with the oxides only as the starting materials [3]. Jin et al.[4] showed the possibility of introducing a pinning site in Y124 system by partial decomposition technique. They reported the intragranular critical current density (Jcm) was raised up to 1.0×10^5 A/cm^2 at 77K in 0.9T by short heat treatment at 920 °C for 1.5 minutes. Recently Wada et al. [5] reported the phase stability and decomposition of Y(Ca)124 superconductors. Although there is a report showing the possibility of fabrication of Y124 thick film on silver tape by spray pyrolysis of organic acid salts at atmospherical pressure [6], there has been no report for fabrication of the Y(Ca)124 silver sheathed superconductor.
Recently Kase et al. [7] has reported that Bi2212 textured tapes with very high Jc were synthesized by doctor-blade casting and partial melt growth process on silver tape (open air system). Dietderich et al.[8] also reported similar results.
The purpose of this work are (1) to clarify the relationship between decomposition temperature of Y(Ca)124 material and oxygen partial pressure, and to optimize the sintering condition for fabrication of silver sheathed Y(Ca)124 superconductors. (2) To make the guide line for improving the Jct-B characteristics of silver sheathed Y(Ca)124 superconductors, through making comparison between Jct-B characteristics and Jcm-B characteristics for both Y(Ca)124 and Bi2212 systems.

EXPERIMENTAL

Samples with nominal compositions of $YBa_2Cu_4O_8$ (x=0) and $Y_{0.9}Ca_{0.1}Ba_2Cu_4O_8$ (x=0.1) were prepared by a solid state reaction method using a high oxygen partial pressure technique (O_2-HIP). The O_2-HIP treatment was carried out directly without pre-sintering, using only oxides as the starting materials. The starting materials, Y_2O_3, CaO, BaO and CuO were mixed in various combinations, ground and compacted into pellets. These pellets were treated directly with O_2-HIP technique using gas mixture of 20% O_2 and 80% Ar at 1050 C for 6 hours under Po_2=360 bar. The magnetic susceptibility and the X-ray diffraction patterns were measured to identify the crystalline phase and to determine the superconducting volume fraction value R(%). In result, these powders were found to be almost single phase (R>95%)[3].
$Bi_2Sr_2CaCu_2O_8$ powder was also prepared by a solid state reaction. The starting materials, Bi_2O_3, $SrCO_3$, $CaCO_3$ and CuO were mixed in Bi:Sr:Ca:Cu=2:2:1:2, ground and calcined at 820°C for 36 hours in air.
These powders were packed into the silver pipe (4/6 mm in inner/outer diameter and 150 mm long). These silver pipes were drawn to wires of 0.3 mm in diameter using the roller dices. Both silver sheathed Y(Ca)124 and Bi2212 wires were pressed into the tape forms. In order to

determine the decomposition temperature, the thermogravimetry (TG) and differential thermal analysis (DTA) were measured for the reacted Y(Ca)124 powders, the reacted powders mixed with a silver powder and the silver sheathed Y(Ca)124 wires (with closed both ends). The temperature was increased in air and pure oxygen at a rate of 5°C/min. The silver sheathed Y(Ca)124 tapes were sintered in various conditions.
The sintering conditions were as follows:

T=770 and 790 °C in air (Po_2=0.2 bar) for 100 hours,
T=830, 840 and 850 °C in pure oxygen (Po_2=1.0 bar) for 100 hours,
T=850, 870 and 890 °C in Po_2=10.0 bar (using O_2-HIP equipment) for 6 hours.

The silver sheathed Bi2212 tapes were heat treated in air using the following 4 schedules (see fig.1 (A)-(D)). These samples were heated to T_1=860-883 °C and cooled slowly to T_2=840-850°C at a rate of 3-6°C/hour to get the aligned Bi2212 grains, and kept at 840-850°C for 12 hours. And all samples were slowly cooled to room temperature at a rate of 100 °C/hour.
The fractured surfaces of these silver sheathed tapes (Y(Ca)124 and Bi2212) were observed using high resolution scanning electron microscope. The magnetic field dependence of Jcm and Jct at 4.2K for these samples were measured using M-H hysteresis curves and ordinary four probe method. The criterion for Jct determination was defined as 0.5 uV/cm. AC magnetic susceptibility measurement was carried out to evaluate the grain boundary coupling.

RESULTS AND DISCUSSION

Table 1 shows the results of DTA and TG measurements for the powder samples of $Y_{1-x}Ca_xBa_2Cu_4O_8$(x=0 and 0.1). These measurements were carried out for the several samples , the powder mixed with a silver powder (Y:Ag = 1:0.2) and the silver sheathed wires both end of which were closed in order to decide the decomposition temperature under various oxygen partial pressure. For both samples of x=0 and x=0.1, the decomposition temperature increases with increasing the oxygen partial pressure, and the de-oxidation temperature of x=0 sample is higher than that of x=0.1 sample. This result shows that Ca doping reduces the thermal stability in Y124 system. There exists good agreement between our results and Wada's results [5] for the powder samples. The decomposition temperature of the powder mixed with a silver powder are lowered in pure oxygen, but almost the same in air and slightly higher in pure nitrogen. This means that the silver powders become an oxygen absorber and vary the local partial pressure. In the case of silver sheathed wires, the decomposition temperatures are about 50 °C lower than those of the powder samples. However, the dependence of oxygen partial pressure is still seen in both x=0 and x=0.1 samples.
The transport critical current density Jct was determined at 4.2K by 4 probe method and the critical temperature Tc was determined by SQUID magnetometer for silver sheathed wires sintered in various conditions are summarized in Table 2. The Jct values of x=0.1 samples are found to be higher than those of x=0 samples. Ca doping has an effect to increase the critical current density as well as increasing the critical temperature. In the case of sintering at Po_2=0.2 and 1.0, the samples were sintered in 100 hours, but at Po_2=10.0 the samples were sintered in 6 hours by using O_2-HIP(Hot Isostatic Press). O_2-HIP treatment is effective to increase the de-oxidation temperature and to shorten the sintering time, but as the melting point of silver decreases remarkably with increase in oxygen partial pressure, upper sintering temperature is limited.
The relations between the decomposition temperature and the oxygen partial pressure in Table 1 and 2 are shown in Fig.2. The dotted line in Fig.2 shows the melting point of silver. In this figure, it is found that the optimum condition for sintering the silver sheathed wire is around 900 C at Po_2=4-5 bar (see * mark in Fig.2).
SEM micrographs of the fractured surfaces of silver sheathed Y(Ca)124 wires are shown in Fig.3. Fig.3 (A) and (B) show the surfaces of the samples sintered in 100 h at 820 °C and 840 °C (Po_2=1.0) respectively. Fig.3 (C) and (D) show the surfaces sintered in 6 hours at 850 °C and 870 C by O_2-HIP. Typical grain growth is observed in (C) and (D). Especially, plate-like grain growth which is supposedly effective to control the preferred orientation is seen in the sample sintered at 870 °C.
Fig. 4 shows the SEM micrographs of the fractured surfaces of silver sheathed Bi2212 wires heat-treated in the patterns shown in Fig.1. The sample slow-cooled from 860 °C (Fig.4 (A)) shows a trace of melting but no oriented grains. In the samples slow-cooled from the temperatures above 870 °C, well-oriented grains are visible. Especially, slow-cooling from 883°C at the rate of 3°C/h, large plate-like grains well-aligned parallel to the silver wall were obtained. This result suggests that the partially melting and growth technique is available to get well-oriented grains for silver sheathed Bi2212 wires.
The field dependences of the intragranular critical current density Jcm and the transport critical current densities Jct are shown in Fig.5. Jcm was obtained by using Bean's critical state model.

$$Jcm(A/cm^2)=20xM(emu/cm^3)/D(cm)$$

where, M is hysteresis of magnetization and D is a average diameter of grains.
In result Jcm at 4.2K and 1T for Y(Ca=0.1)124, Y124 and Bi2212 were $6.0x10^4$ A/cm^2, $3.9x10^4$ A/cm^2 and $3.4x10^4$ A/cm^2. Jct were measured by conventional 4 probe method for silver sheathed

Y(Ca)124, Y124 and Bi2212 wires. The magnetic field direction was parallel to the tapes flat surface and the perpendicular to the current direction. Jct values at 4.2K and 0T for silver sheathed Y(Ca)124, Y124, Bi2212(880 ℃) and Bi2212(870 ℃) wires were 1.6x10^4 A/cm^2, 3.4x10^3 A/cm^2, 1.7x10^4 A/cm^2 and 1.1x10^4 A/cm^2. Although Jct values at 0T are almost the same, the field dependence of Y(Ca)124 and Y124 wires are larger than those of Bi2212 wires. This means that the samples of Y(Ca)124 and Y124 wires have much week links of grain boundaries compared with Bi2212 wires.

In order to evaluate the grain boundary weak links, AC susceptibility measurement was carried out. The temperature dependence of X' and X" for silver sheathed Y(Ca)124 and Bi2212 wires and Bi2212 single crystal are shown in Fig.6. The Bi2212 single crystal was grown by TSFZ (traveling solvent floating zone) method. Jcm at 4.2K in 3.5T of this single crystal was 1.1x10^6 A/cm^2. The detailed result for this single crystal will be presented elsewhere [9]. The measurements were done at 0.1mT and 100Hz applying the field perpendicular to the tape plane of wires and a-b plane of the single crystal. X" peak relating to the coupling loss of the sample become broader with the sequence of single crystal, Bi2212 and Y(Ca)124. This means that the grain boundary coupling of Y(Ca)124 wires is not enough compared with that of Bi2212 wires and thereafter coupling loss is observed in the wide temperature range because of the weak link which disappeared at low temperatures. In order to improve the grain boundary coupling, higher temperature sintering is required as shown in Fig.2.

REFERENCES

1. Karpinski J, Kaldis E, Jilek E, Rusiecki S and Bucher B (1988) Nature 336:660
2. Miyatake T, Gotoh S, Koshizuka N and Tanaka S (1989) Nature 341: 41
3. Shibutani K, Hayashi S, Fukumoto Y, Ogawa R and Kawate Y (1990) Physica C 167: 102-106
4. Jin S, Tiefel TH, Nakahara S, Graebner JE, O'Bryan HM, Fastnacht RA and Kammlott GW (1990) Appl. Phys. Lett. 56: 1287
5. Wada T, Suzuki N, Ichinose A, Yaegashi Y, Yamauchi H and Tanaka S (1990) Jpn. J. Appl. Phys.29: L1079-1082
6. Amemiya I, Kobayashi H, Yamaguchi K and Hasegawa T (1990) J. Jpn. Soc. Powder and Powder Metallurgy 37: 780-783
7. Kase J, Togano K, Kumakura H, Dietderich DR, Irisawa N, Morimoto T and Maeda H (1990) Jpn. J. Appl. Phys. 29: L1096-1099
8. Dietderich DR, Ullmann B, Freyhardt HC, Kase J, Kumakura H, Togano K and Maeda H (1990) Jpn. J. Appl. Phys. 29: L1100-1103
9. Hayashi S, Shibutani K, Shigaki I, Ogawa R, Kawate Y, Maret V, Kitahama K and Kawai S (1990) Physica C (to be published)

Table 1. De-oxidation temperature in $Y_{1-x}Ca_xBa_2Cu_4O_8$ (℃)

Ca consentration x	x=0			x=0.1		
Po_2 condition	in O_2 Po_2=1.0	in Air Po_2=0.2	in N_2 Po_2=0	in O_2 Po_2=1.0	in Air Po_2=0.2	in N_2 Po_2=0
without sheath	895	838	700	880	828	700
with silver powder	870	837	728	864	826	705
with silver sheath	845	775	-	845	775	-

Table 2. Jct and Tc onset after various heat treatments.

T(℃) / Po_2(bar)	770	790	830	840	850	870	890
			Ca consentration x=0.1				
0.2	12.0 85.2	5.1 78.1					
1.0			20.1 86.6	15.5 85.2	4.5 86.2		
10.0					10.2 85.7	14.5 86.0	Ag melt 86.9
			Ca consentration x=0				
0.2	3.0 78.9	1.1 70.2					
1.0			6.2 79.8	7.5 79.2	2.3 71.0		
10.0					8.6 79.4	12.3 80.2	Ag melt 80.0

Up column: Jct in x10^3 A/cm^2, Down column: Tc onset in K.

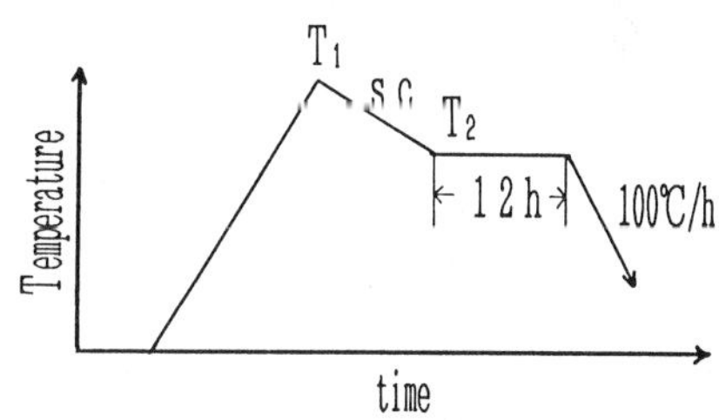

	T_1	sc	T_2
(A)	860℃	6℃/h	840℃
(B)	870℃	6℃/h	840℃
(C)	880℃	6℃/h	840℃
(D)	883℃	3℃/h	850℃

Fig.1 Schematic explanation and heat treatment schedules for partial melting and growth technique.

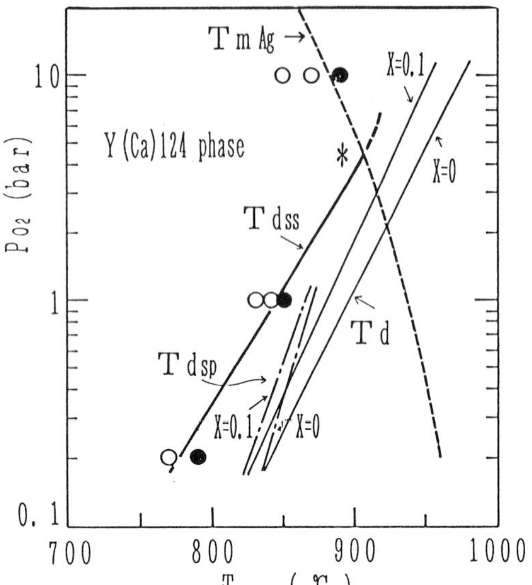

Tdss: with silver sheath TmAg: melting temperature of silver
Tdsp: with silver powder Td : without silver sheath
○: Y(Ca)124 was stable.
●: Y(Ca)124 was de-oxidized or silver melting.

Fig.2 P-T diagram for sintering the silver sheathed Y(Ca)124 wires.

(A) T=820℃, Po2=1.0bar, t=100hours (B) T=840℃, Po2=1.0bar, t=100hours

(C) T=850℃, Po2=10.0bar, t=6hours (D) T=870℃, Po2=10.0bar, t=6hours

Fig.3 The fractured surfaces of the cross section of silver sheathed Y(Ca)124 wires.

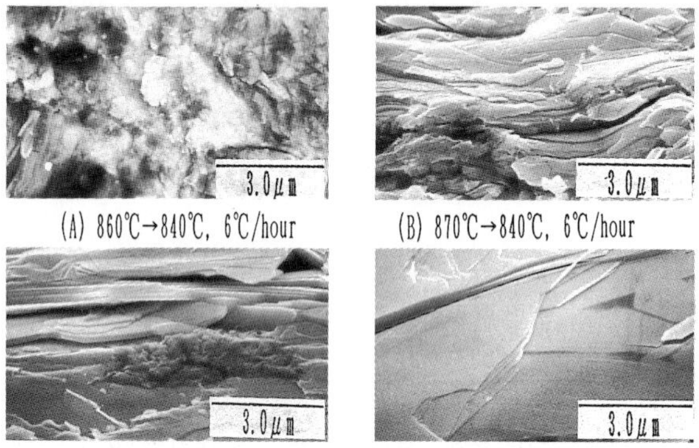

(A) 860℃→840℃, 6℃/hour (B) 870℃→840℃, 6℃/hour

(C) 880℃→840℃, 6℃/hour (D) 883℃→850℃, 3℃/hour

Fig.4 The fractured surfaces of the cross section of silver sheathed Bi2212 wires.

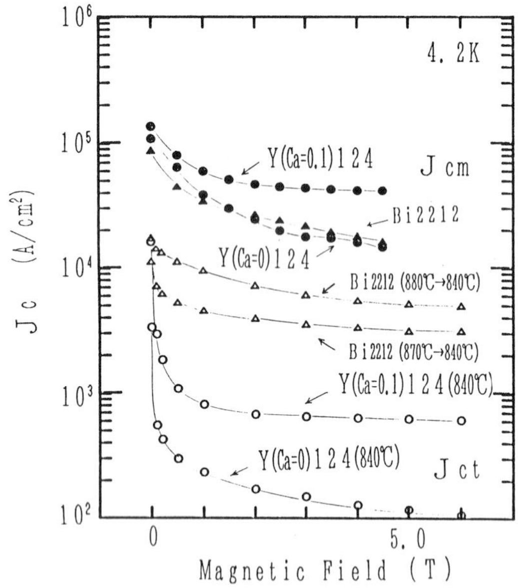

●: Jcm of Y(Ca)124 ▲: Jcm of Bi2212
○: Jct of Y(Ca)124 △: Jct of Bi2212

Fig.5 Jcm-B and Jct-B characteristics for Y(Ca)124, Y124 and Bi2212 powders and silver sheathed wires. Jcm was determined from the measurement of M-H curves for the powder and Bi2212 single crystal. Jct was measured by 4 prove method for silver sheathed wires.

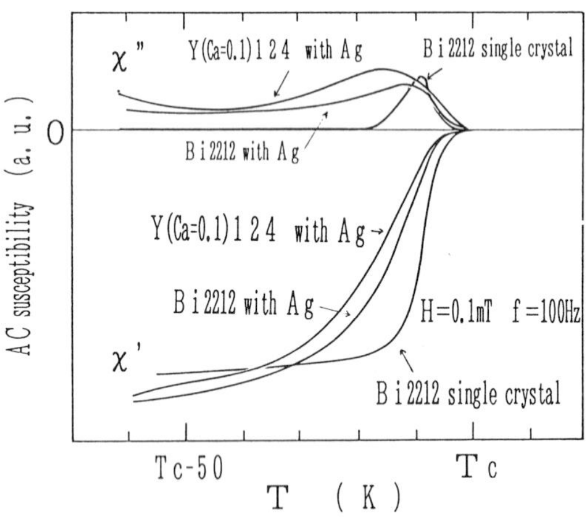

Fig.6 Temperature dependence of the real part X' and imaginary part X" of complex susceptivility obtained at 100Hz with AC field amplitude of 0.1mT for silver sheat hed Y(Ca)124 and Bi2212 and single crystal of Bi2212.

Distribution of Pinning Energies in Oxide Superconducting Thin Films and Single-Crystalline Powders

KENICHI KAWAGUCHI, IWAO TAGO, and MASAO NAKAO

SANYO Tsukuba Research Center, 2-1, Koyadai, Tsukuba, 305 Japan

ABSTRACT

A role of oxygen vacancies in Cu-O chains as pinning centers has been investigated for single-crystalline $YBa_2Cu_3O_{7-\delta}$ powders, and pinning features in an $ErBa_2Cu_3O_{7-\delta}$ thin film have been compared with those in the powders. Distribution functions of pinning energies determined from magnetic relaxation data indicate that there exist three kinds of pinning centers in the powders: one is independent of the oxygen deficient structure, and the others are associated with the oxygen vacancies. For the thin film, the flux pinning is mainly dominated by extended defects which show a wide range of pinning energies.

KEY WORDS: rare earth system superconductor, flux pinning, oxygen vacancy

INTRODUCTION

Ideal crystals of type II superconductors including few defects can carry only a small current with zero resistivity in the presence of magnetic fluxons. However, high-T_c superconductors exhibit large critical current densities even if they are pure crystals which have no precipitates. It was reported that twin boundaries effectively act as pinning centers at 76 K, whose pinning strength is strongly anisotropic [1]. In contrast, back-ground isotropic pinning dominates the pining behavior in the lower temperature region. This suggests that some other inherent defects mainly act as pinning centers in this range of temperature. $YBa_2Cu_3O_{7-\delta}$ has a large number of oxygen vacancies in Cu-O chains. A spatial change in oxygen content induces a spatial fluctuation of superconducting order parameters, which can pin magnetic fluxons [2]. In addition, an order-disorder transition of the oxygen vacancies in the Cu-O chains was reported [3]. Such ordering produces an additional fluctuation of superconducting order parameters.

In this paper, we focus our attention on a role of the oxygen vacancies in the Cu-O chains as pinning centers. We present relaxation data of the magnetization in $YBa_2Cu_3O_{7-\delta}$ single-crystalline powders with different oxygen deficiencies and with different ordering schemes of oxygen. Furthermore, we describe a comparison of the results of magnetic measurements in an $ErBa_2Cu_3O_{7-\delta}$ thin film with those in the $YBa_2Cu_3O_{7-\delta}$ powders. Within the framework of a flux creep model by Anderson and Kim [4-5], distribution functions of pinning energies are deduced from the magnetic relaxation data using an inversion scheme recently proposed by Hagen and Griessen [6]. Finally, the contribution of the oxygen vacancies in the Cu-O chains to flux pinning is discussed.

EXPERIMENT

Master batch of $YBa_2Cu_3O_{7-\delta}$ was prepared from Y_2O_3(4N), $BaCO_3$(4N), and CuO(4N) by firing at 850-920 ℃ in air for 57 h with intermediate grindings. The powder was then pressed into pellets of weight 0.5 g and fired at 920 ℃ in air for 1 h. The resulting pellets were then annealed at 450-730 ℃ in air for 85-10 h. The annealing was followed by a rapid quench into liquid nitrogen. After quenching, the pellets were ground and mixed with pure Al_2O_3 powder for the purpose of electric isolation between the superconducting particles. The $ErBa_2Cu_3O_{7-\delta}$ thin film was prepared by rf magnetron sputtering. The film was deposited on a (100) MgO substrate with a 77% O_2/23% Ar gas flow mixture at 26 mTorr total pressure at a substrate temperature of 650 ℃. As deposited, the film exhibits a superconducting transition at 58 K, and the c axis of the film is predominantly oriented perpendicular to the substrate plane.

Cell dimensions of the $YBa_2Cu_3O_{7-\delta}$ powders were determined by powder x-ray diffraction using a Rigaku diffractometer RINT 1400 with Cu Kα radiation. Magnetic measurements were performed on a HOXAN superconducting quantum interferene device (SQUID) magnetometer HMS-2000. Superconducting transition temperatures were defined as onsets in temperature dependences of the magnetization in a field of 100 Oe.

In a high magnetic induction, fluxoids move co-operatively with neighboring fluxoids. Thus, observation on a behavior of isolated fluxoids requires measurements in a low magnetic induction. However, critical current densities are more sensitive to a change in the magnetic induction for a low magnetic induction than for a high magnetic induction. It was reported that virgin magnetization curves (M-H curves) of oxide superconductors at different temperatures can be scaled by the minimum magnetizations (M_m) in the M-H curves at each temperature and fields (H_m) where the magnetizations reache the M_m's [7]. This suggests that critical states in equivalent fields of the scaled M-H curves at each temperature are similar to each other. Therefore, we take applied fields in relaxation measurements as the equivalent fields of M-H curves at each temperature for the purpose of the measurements in low fields. The applied fields were twice the fields of H_m for powders and as three times for the thin film at each temperature.

The procedure for acquiring the magnetic relaxation data consists of cooling in zero field (ZFC) and, after attaining a stable temperature, application of the filed. The first magnetization point was taken after 200 s required to latch the field. Subsequent magnetization points were taken every 25 s for the duration of the experiment. Between runs the field was removed and the sample heated to above its T_c in order to expel all remaining flux.

RESULTS AND DISCUSSION

Table 1 summarizes annaeling temperatures, critical temperatures, and cell dimensions of the powders used in the magnetic measurements. Oxygen content in $YBa_2Cu_3O_{7-\delta}$ increases linearly with decreasing cell dimension of the c axis, c_0, over the range of c_0 in this study [3]. Thus, c_0 can be a measure of the oxygen content. It was reported that T_c of $YBa_2Cu_3O_{7-\delta}$ as a function of oxygen content has a double plateau structure at 90 K and 60 K for oxygen contents of $0 \leqq \delta \leqq 0.2$ and $0.4 \leqq \delta \leqq 0.6$, respectively [3]. The oxygen content in the sample with a T_c of 85 K is located on the lower-oxygen-content edge of the 90-K plateau. Note that the samples annealed at 620 ℃ and 680 ℃ show ruoghly the same T_c irrespective of their different contents of oxygen. This indicates that the degree of oxygen ordering in the sample annealed at 680 ℃ is higher than that in the sample annealed at 620 ℃.

Figure 1 illustrates typical temperature dependences of logarithmic relaxation rates R(T) for the $ErBa_2Cu_3O_{7-\delta}$ thin film and the $YBa_2Cu_3O_{7-\delta}$ powder with a T_c of 85 K. The relaxation rate for the thin film reaches a peak around 7 K, and monotonously decreases with increasing temperature. On the other hand, the R(T)-T curve for the $YBa_2Cu_3O_{7-\delta}$ powder has a complicated structure. The curve has a small peak around 10 K and a shoulder around 23 K. It also has another peak below 5 K, since R(T) should be zero at 0 K. The crystalline axis of the powder sample is randomly oriented. However, the obtained relaxation behavior mainly ascribed to the magnetic moment induced by currents in the a-b plane, because critical current densities in the a-b plane for pure $YBa_2Cu_3O_{7-\delta}$ single crystals are extremely larger than those along the c axis andan

Table 1 Annealing temperature, critical temperature, and cell dimension for $YBa_2Cu_3O_{7-\delta}$ powders used in magnetic measurements.

Annealing temperature (℃)	T_c (K)	Cell dimensions c_0 (Å)	a_0 (Å)	b_0 (Å)
500	85	11.692	3.886	3.823
620	61	11.716	3.884	3.832
680	57	11.738	3.881	3.836

anisotropy of pinning energies is small in the temperature range of this study [8]. Consequently, the difference between the R(T)-T curve of the film and that of the powder is attributed to a difference in pinning mechanisms.

The distribution function of the pinning energy $\rho(U_0)$ is determined from the relaxation data using the inversion scheme on the basis of the Bean critical state model [9] and the flux creep model of the Anderson-Kim theory. Figure 2 shows the $\rho(U_0)$ in the $ErBa_2Cu_3O_{7-\delta}$ thin film. The $\rho(U_0)$ for the thin film has a peak around 25 meV and a high-energy tail. On the other hand, the $\rho(U_0)$ for the $YBa_2Cu_3O_{7-\delta}$ powder with a T_c of 85 K, shown in fig. 3(a), has distinct two peaks around 25 meV and 50 meV. It should also have a peak below 20 meV, since a integration of $\rho(U_0)$ over a whole range of U_0 is unity. These three peaks suggest that there exist three kinds of defects which act as pinning centers in the powder sample.

Figure 3(b) illustrates the $\rho(U_0)$ for the powder with a T_c of 61 K. The 50-meV peak is shifted to a lower energy with a lower T_c. The energy of this peak remains constant as long as the T_c does not change, as shown in fig. 3(c), however the oxygen content changes. Anderson and Kim have suggested that U_0 should scale as $p(H_c^2/8\pi)V$, where p is a numerical factor which describes the fraction of the condensation energy available as pinning energy, H_c is a thermodynamic critical field, and V is an activation volume [4,5]. Accordingly, the shift of 50-meV peak is due to the decrease in H_c with decreasing T_c, and the nature of the defect related to this peak does not change by decreasing content of oxygen. Thus the defect is not associated with the oxygen vacancies in the Cu-O chains.

The 25-meV peak is shifted little with decreasing T_c. Moreover, the peak is

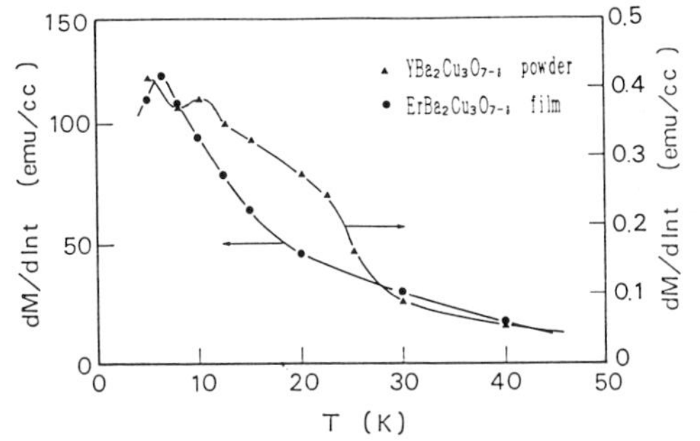

Fig. 1 Temperature dependence of relaxation rates in an $YBa_2Cu_3O_{7-\delta}$ powder with T_c of 85 K (▲) and an $ErBa_2Cu_3O_{1-\delta}$ thin film with T_c of 58 K (●).

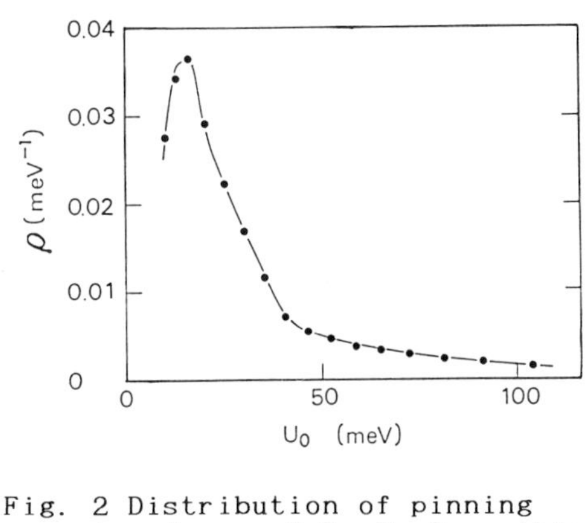

Fig. 2 Distribution of pinning energies for an $ErBa_2Cu_3O_{7-\delta}$ thin film with T_c of 58 K.

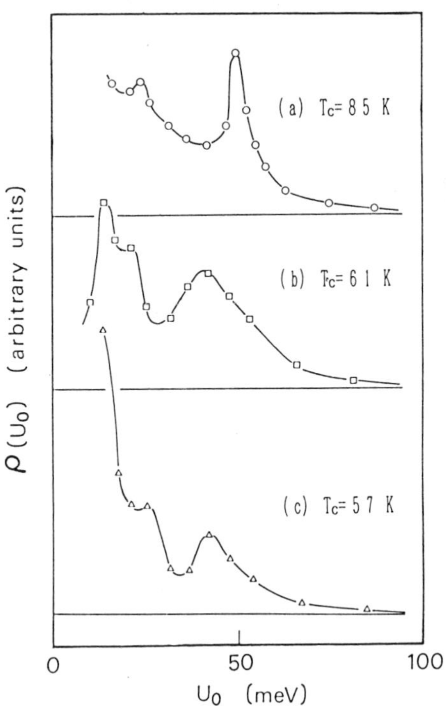

Fig. 3 Distribution of pinning energies for $YBa_2Cu_3O_{7-\delta}$ powders with T_c's of (a) 85 K, (b) 61 K, and (c) 57 K.

shifted to a higher energy with decreasing content of oxygen, comparing the $\rho(U_0)$-U_0 curve of fig. 3(b) with that of fig. 3(c). The ratios of the energies for the 25-meV peak to those for the 50-meV peak are 0.45, 0.5, and 0.59 for the powders with T_c's of 85 K, 61 K, and 57 K, respectively. The change in this ratio corresponds to a deviation from a thermodynamic change of the pinning energy induced by the decrease of H_c. The ratio increases monotonously with decreasing oxygen content. This indicates that the chracteristic of the defect related to the 25-meV peak changes systematically with decreasing content of oxygen. This change can be explained if we assume that the defects are fine domains in which oxygen sites in the Cu-O chains are vacant. The average size of the domains increases with increasing oxygen vacancies, and the size is associated with the activation volume V. Thus the change in the energy for the 25-meV peak is thought to be caused by a change in V. The reason why the 25-meV peak shifted little with decreasing T_c from 85 K to 61 K is that the decrease in U_0 induced by the degradation of H_c was cancelled by the increase in U_0 due to the increase in V.

The peak below 20 meV for the powder with a T_c of 57 K is remarkably higher than that for the powder with a T_c of 61 K. This is originated in the increase in a fractional amount of the pinning centers with U_0 below 20 meV. The ordering of oxygen ions in the Cu-O chains is achieved in the powder with lower oxygen content. This suggests that pinning centers corresponding to this energy region are associated with the oxygen ordering in the Cu-O chains. The oxygen ions in the Cu-O chains affect superconductivity in Cu-O planes indirectly. Moreover, the period of the oxygen ordering structure is rather smaller than the superconducting coherence length in the a-b plane of $YBa_2Cu_3O_{7-\delta}$. Consequently, the full condensation energy is not available for the pinning energy, and the numerical factor p is smaller than unity. This is consistent with the small pinning energy.

In a range of U_0 above 60 meV, the $\rho(U_0)$'s for the powders are smaller than that for the thin film. In addition, the $\rho(U_0)$ for the thin film has a simple structure with only one peak. This suggests that one kind of an extended defect dominates the pinning behavior in the thin film. The thin film was fabricated in a non-equilibrium condition, and the surface of the film was damaged by activated sputtering gasses over the duration of the growth. Accordingly, lots of extended defects like stacking faults were induced during the growth, and a fractional amount of the extended defects in the thin film is larger than that of the inherent defects mentioned above. Moreover, such extended defects are thought to have a wide range of pinning energies. This gives rise to the presence of the high-energy tail.

In conclusion, we find three kinds of pinning centers in $YBa_2Cu_3O_{7-\delta}$ single-crystalline powders in a range of U_0 below 100 meV. The first one has U_0 around 50 meV. The nature of the defects related to this pinning center is independent of oxygen deficiency. The second one has U_0 around 25 meV. The U_0 of this pinning center changes systematically with decreasing content of oxygen. This suggests that the defects corresponding to this kind of pinning center are fine domains where oxygen sites in the Cu-O chains are vacant. The third one has U_0 below 20 meV. The fractional amount of this pinning center is strongly associated with the oxygen ordering in the Cu-O chains. For the $ErBa_2Cu_3O_{7-\delta}$ thin film, extended defects like stacking faults mainly act as pinning centers.

REFERENCES

1. Gyorgy EM, van Dover RB, Schneemeyer LF, White AE, O'Bryan HM, Felder RJ, Waszczak JV, Rhodes WW (1990) Appl. Phys. Lett. 56: 2465-2467
2. Nakao M, Furukawa H, Kawaguchi K, Mukaida H, Shikichi K (1990) Proc. 2nd ISTEC Workshop on Superconductivity, Kagoshima, 1990: 68-71
3. Cava RJ, Hewat AW, Hewat EA, Batlogg B, Marezio M, Rabe KM, Krajewski JJ Peck Jr WF, Rupp Jr LW (1990) Physica C 165: 419-433
4. Anderson PW (1962) Phys. Rev. Lett. 9: 309-311
5. Anderson PW, Kim YB (1964) Rev. Mod. Phys. 36: 39-43
6. Hagen CW, Griessen R (1989) Phys. Rev. Lett. 62: 2857-2860
7. Wolfus Y, Yeshurun Y, Felner I, Sompolinski H (1989) Phys. Rev. B 40: 2701-2703
8. Kawaguchi K, Nakao M (1990) Read at the 45th Annual Meeting of the Phys. Soc. Jpn., Osaka, March, 1990, 31p-PS-83. (Advance abstract vol.3, p.267)
9. Bean CP (1962) Phys. Rev. Lett. 8: 250-253

Temperature Dependence of the J_c-B Characteristics of $Bi_{2.2}Sr_{1.8}Ca_{1.0}Cu_{2.0}O_{8.2}$ Single Crystal

SEIJI HAYASHI[1], KAZUYUKI SHIBUTANI[1], ICHIRO SHIGAKI[1], YOSHITO FUKUMOTO[1], HIROSHI KJIKAWA[1], RIKUO OGAWA[1], YOSHIO KAWATE[1], VINCENT MARET[2], KATSUKI KITAHAMA[3], and SHICHIO KAWAI[3]

[1] Superconducting and Cryogenic Technology Center, Kobe Steel Ltd., 5-5, Takatsukadai 1-chome, Nishi-ku, Kobe, 651-22 Japan

[2] School of Industrial Physics and Chemstry of Paris, France

[3] The Institute of Scientific and Industrial Research, Osaka University, Ibaraki, Osaka, 567 Japan

ABSTRACT

Measurements of the magnetization vs. magnetic field hysteresis for $Bi_{2.2}Sr_{1.8}Ca_{1.0}Cu_{2.0}O_{8.2}$ single crystal were performed in the temperature range from 4.2K to 25K. Very large critical current density in a-b plane (Jc_{ab}) of $1.8x10^6$ A/cm^2 was observed at 4.2K. The falloff of Jc_{ab} with applied magnetic field was small at 4.2K, while Jc_{ab} decreased rapidly with increasing temperature. As for the anisotropy of this single crystal, we measured M-H hysteresis curve as a function of the angle between magnetic field and a-b plane. The results support very well the Kes's model that Bi-based high-Tc superconductors are composed of 2-dimensional superconducting layers separated by normal conducting insertion layers.

KEY WORDS: critical current density, pinning potential, $Bi_{2.2}Sr_{1.8}Ca_{1.0}Cu_{2.0}O_{8.2}$ single crystal, anisotropy

INTRODUCTION

The discovery of high-temperature superconducting (HTSC) materials that have critical temperature higher than 77K has created the possibility of widespread applications[1]. The possibility of these applications depends on how large critical current density can be achieved. In traditional superconductors, it is now of common knowledge that it is necessary to introduce artificial pinning center to prevent flux movement and to accomplish high critical current density (Jc). On the other hand, in HTSC materials it seems that higher Jc-value is achieved for the material with better quality. For example, the highest Jc-value in bulk form have been obtained for single crystals as Dinger et al for YBCO [2] and Dover et al [3] for $Bi_{2.2}Sr_2Ca_{0.8}Cu_2O_{8+x}$. In these cases, very large Jc values larger than $10^6 A/cm^2$ were observed at 4.2K. There should be some kind of pinning center in the single crystal.

The prominent feature of HTSC is its large anisotropy. Recently, Kes [4] suggested that Bi-2212 actually behaves as if the CuO_2 planes were decoupled, that is, the field parallel to the CuO_2 planes penetrates completely. The 2-dimensional layer structure is one of the main causes for low Jc. Almost all of the superconducting current flows along the a-b plane and little current flows along the c-axis. Therefore, it is important to investigate crystal anisotropy in these HTSC materials.

Dover et al [3] and Martin et al [5] have already reported the measurement of Jc for Bi-based single crystal. However Dover's sample showed a two-step transition and therefore it was not a true single-phase crystal. Martin et al made a measurement of transport-Jc and also of its anisotropy, but their measurements were done only near 77K and magnetic measurements were not conducted entirely. Therefore, we measured the magnetic characteristics for $Bi_{2.2}Sr_{1.8}Ca_{1.0}Cu_{2.0}O_{8.2}$ single crystal in order to investigate pinning mechanism and anisotropy in this material.

EXPERIMENTAL PROCEDURES

Single crystal of $Bi_{2.2}Sr_{1.8}Ca_{1.0}Cu_{2.0}O_{8.2}$ was produced by the travelling solvent floating zone method [6]. Two typical types of samples were prepared: one was a plane with thickness about 30 μm along c-axis and area about 2mmx4mm parallel to the a-b plane that was torn off from the bulk crystal, the other was a rectangular bar with thickness 0.25mm along c-axis and length 1.7mm width 0.9mm parallel to a-b plane. Magnetization was measured using vibrating sample magnetometer (VSM). In this apparatus, we can change the angle between crystal surface and magnetic field by rotating the vibrating rod with accuracy within 1 degree. Thereby we can detect anisotropy in magnetic characteristics of single crystal. In case of anisotropy measurements, only the bar-shaped sample was used. Magnetic field was applied up to 5×10^4Oe, and sample temperature was changed from 4.2K to about 25K with accuracy of 0.3K. Two types of measurements were performed; magnetization vs magnetic field (M-H) hysteresis curve, and decay of the magnetization (M-t). The M-t measurements were performed applying magnetic field perpendicular to the a-b plane. As for the M-H hysteresis curves, measurements were performed as a function of temperature and also as a function of angle between magnetic field and a-b plane of the crystal.

RESULTS AND DISCUSSION

Figure 1 shows M-H hysteresis curves measured at several temperatures. In this case, magnetic field was applied perpendicular to the a-b plane. According to the Bean's critical state model, we can calculate Jc_{ab} by the following equation; $Jc_{ab}=(20 \times \Delta M \times \rho)/d$; ($Jc_{ab}$ in A/cm^2, ΔM in emu/g, ρ in g/cm^3, d in cm), where ΔM is the magnetization difference for increasing and decreasing fields and ρ is the density and d is the width of the crystal. Calculated Jc_{ab} vs H curves are shown in Fig. 2. As shown in this figure, very large Jc_{ab} of 1.8×10^6A/cm^2(H=0) was obtained at 4.2K. And decreasing rate of Jc_{ab} by the application of magnetic field is very small compared to the sintered bulk materials. However, Jc_{ab} decreases rapidly with increasing temperature, and becomes strongly dependent on magnetic fields. Strong temperature dependence of Jc_{ab} can be understood by the small pinning potential as described below.

Figure 3 shows a decay of normalized magnetization for several magnetic fields in case of magnetic field perpendicular to the a-b plane. In this case measurements were performed as follows; sample was cooled to 4.2K in zero-magnetic field, then magnetic field was raised up to desired value, and held at that value. From the time when the magnetic field reached some constant value, The measurement of the decay of magnetization was started. As shown in this figure, magnetization decays almost linearly to the logarithm of time. From the equation[7]; $d[M(t)/M_0]=-(kT/U)d\ln(t/t_0)$; we can calculate pinning potential for each magnetic field and the results are shown in Fig. 3. In these cases, magnetic field was confined to the range that it perfectly penetrated into the inner-most of the sample. Pinning potential was almost the same value of 8meV irrespective of magnetic fields above 2T and same level as reported earlier for Bi-based materials [9,10]. When there is some kind of normal conducting pinning site whose size is smaller than ξ, the energy required to remove one flux line from the pinning center is given as; $Up=(Bc^2/2\cdot\mu_0) \times \pi \xi_{ab}^2 \times 2\xi_c$. When we let ξ_{ab}=40A, ξ_c=2.1A [8], and Bc=0.6T [3,11], we obtain 19meV for Up. At present, we cannot explain the reasons why the magnitude of U is the same level as single flux pinning potential Up, and why U does not depend on magnetic fields. These facts suggest that the increase of Jc in single crystal is mainly due to the elimination of weak links, and it will be necessary to introduce artificial pinning centers lying across several $Cu\text{-}O_2$ layers for 77K operation.

Figure 4 shows M-H hysteresis curves as a function of the angle between magnetic field and a-b plane. θ corresponds to the angle that is shown in the inset of Fig.4. Two prominent features are seen in Fig.4. The first is the fact that almost reversible straight lines are observed for small θ angle, the slope of the line increases with increasing θ. The second is the fact that hysteresis of the loop increases with increasing θ. At first, we did this measurement with the aim to calculate critical current density along c-axis, Jc_c. However, it is impossible to consider the data for small θ to be the reflection of shielding current flowing along the c-direction for the reasons described below. Almost straight line may be interpreted as due to strong pinning and thereby extremely high magnetic field necessary to saturate the magnetization. However, this is not probable, because if so, there should be more larger hysteresis. And first of all, initial slope of the magnetization curves should coincide with $M/H = -1/4\pi$ for $H \leq Hc_1$ irrespective of the angle.

In order to explain our data, we let three assumptions according to the Kes's model:

1: Bi-2212 crystal is composed of CuO_2 superconducting layers and normal conducting insertion layers.

2: Magnetic flux can penetrate freely along the normal conducting layer.

3: Magnetization is induced only by the magnetic field components perpendicular to the a-b plane ($H_\perp = H\sin\theta$).

The nearly straight line and angle dependence of its slope can be explained easily by this model. When θ is small, the situation $H_\perp = H\sin\theta \leq Hc_1$ is realized, and we obtain $M/H = -\sin\theta/4\pi$. This equation explains the fact that the slope of the M-H hysteresis curve increases with θ. Furthermore, the hysteresis will be small because of no flux pinning in Meissner state. To confirm this more quantitatively, we analyzed the results as shown in Fig.5. Figure 5 shows $-4\pi M$ as a function of $H\sin\theta$. Magnetization M was obtained by extrapolating the initial magnetization curve to H=0.1T. As shown in Fig.5, data points coincide very well with the straight line of perfect diamagnetism. Therefore, our assumption explains very well the data for small θ-region. As for the large θ-region, we analyzed the saturation field H_0. Nominal saturation field in M-H hysteresis curve will become large with decreasing θ from 90 degrees, because saturation will take place at $H\sin\theta = H_0$. As shown in Fig.4, determination of H_0 from M-H hysteresis curve contains large uncertainty. Therefore, we tried to estimate it as follows. We let H_{max} as maximum applied field, and H^* as the field at which magnetization becomes 0. Then difference ΔH is expressed within the framework of Bean's model as; $\Delta H = H_{max} - H^* = \sqrt{2}H_0/(\sqrt{2}+1)$. When we change angle θ, nominal ΔH is written as; $\Delta H\sin\theta = \sqrt{2}H_0/(\sqrt{2}+1)$. We can understand qualitatively from this equation the reason why nominal ΔH increases with decreasing θ from 90 degrees. Therefore, the data for large magnetic field region also supports our assumption well.

In conclusion, very large critical current density Jc_{ab} of $1.8\times10^6 A/cm^2$ was obtained at 4.2K. Jc_{ab} decreased rapidly with increasing temperature because of its small pinning potential. As for the anisotropy of single crystal, experimental results can be explained very well by the Kes's model.

REFERENCES

1) M.K.Wu, J.R.Ashburn, C.W.Torng, P.H.Hor, R.L.Meng, L.Gao, Z.J.Huang, Y.Q.Wang and C.W.Chu: Phys. Rev. Lett. 58(1987)908.
2) T.R.Dinger, T.K.Worthington, W.J.Gallagher and R.L.Sandstrom: Phys.Rev. Lett. 58(1987)2687.
3) R.B. van Dover, L.F.Schneemeyer, E.M.Gyorgy and J.V.Waszczak: Appl. Phys. Lett. 52(1988)1910.
4) P.H.Kes, J.Arts, V.M.Vinokur and C.J. van der Beek: Phys. Rev. 64(1990)1063.
5) S.Martin, A.T.Fiory, R.M.Fleming, G.P.Espinosa and A.S.Cooper: Appl. Phys. Lett. 54(1989)72.
6) I.Shigaki, Y.Kawate, R.Ogawa, K.Shibutani, K.Kitahama, T.Kawai, S.Kawai and J.Shirafuji: Cryogenic Engineering 25(1990).
7) A.M.Campbell and J.E.Evetts: Adv. Phys. 21(1972)299.
8) Y.Koike, T.Nakano and T.Fukase: Jpn. J. Appl. Phys. 27(1988)841.
9) S.Hayashi, K.Shibutani, Y.Fukumoto, R.Ogawa and Y.Kawate: Proc. of ISS'89 pp.363-366.

10) M.Mimura, N.Enomoto, N.Uno and M.Nakajima, H.Kumakura and K.Togano: Proc. of ICMC'90 Topical Conference.
11) T.T.M.Palstra, B.Batlogg, L.F.Schneemeyer and R.J.Cava: unpublished

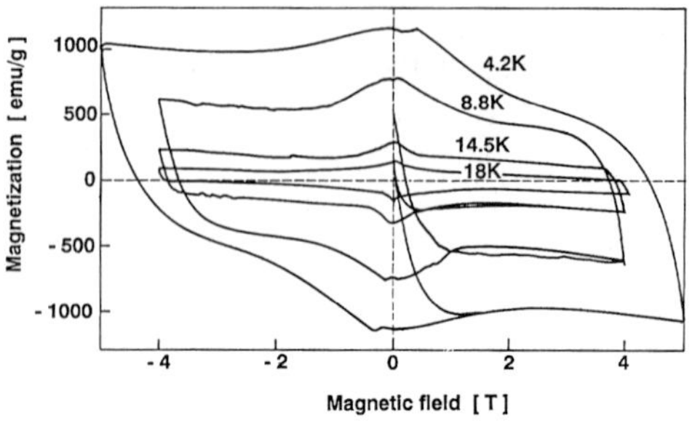

Fig.1 M-H hysteresis curves as a function of temperature. Magnetic field was applied perpendicular to the a-b plane of the crystal.

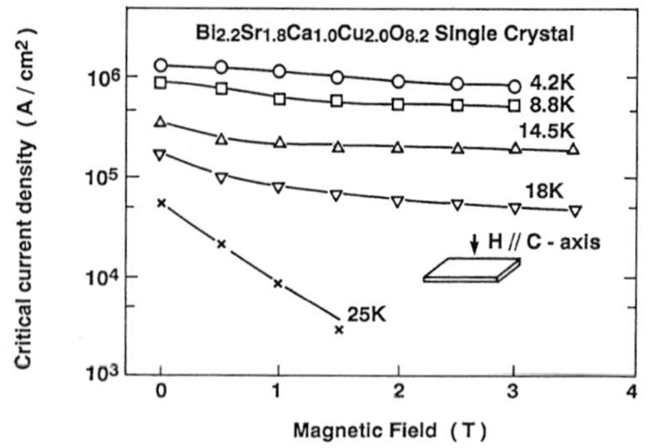

Fig.2 Critical current density in the a-b plane Jc_{ab} as a function of magnetic field and temperature. Very large Jc_{ab} was observed at 4.2K.

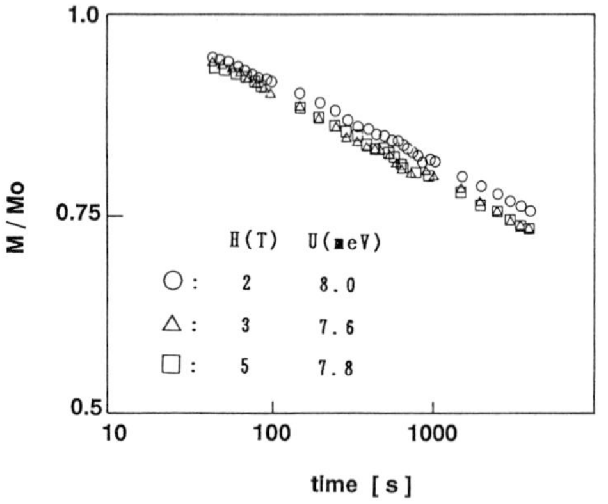

Fig.3 Decay of the normalized magnetization in case of magnetic field perpendicular to the a-b plane. Pinning potential is almost constant for the range of magnetic field from 2T to 5T.

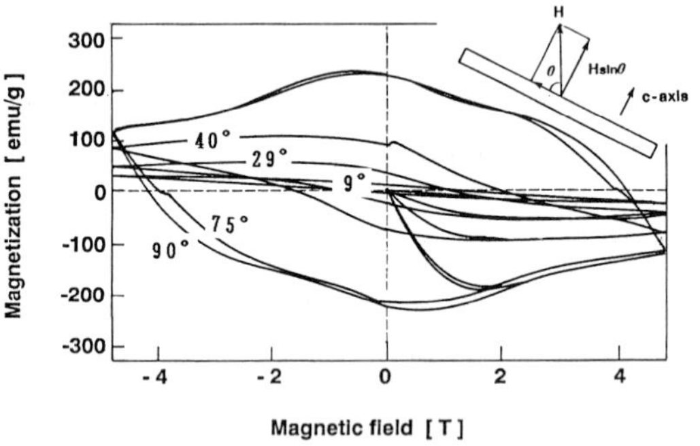

Fig.4 M-H hysteresis curves as a function of angle between magnetic field and a-b plane. Almost straight lines are observed for small θ-angle, their slopes increaseing with increasing the angle θ.

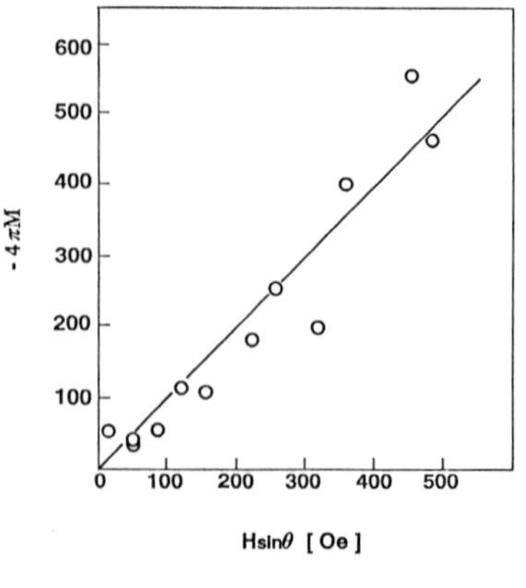

Fig.5 $-4\pi M$ as a function of $H\sin\theta$. Straight line corresponds to the perfect diamagnetism. This result can be explained by Kes's model very well.

Critical Current Density $j_c(H,T)$ and Current Voltage Characteristic in Sintered and Melt Processed Ceramics of Y-, Bi, and Tl- Based Superconductors

W. PAUL, TH. BAUMANN, H. DERSCH, F. STUCKI, J. RHYNER, and J. BLATTER

ABB Corporate Research, CH-5405 Baden, Switzerland

ABSTRACT

The field and temperature dependence of j_c the and the current voltage characteristic in Y-, Bi-, and Tl- based HTSC ceramics, sintered and melt processed, was determined inductively by using ring shaped samples. In all our sintered materials except Bi2212 $j_c(H)$ exhibits a double-step characteristic typical for weak link coupling with $j_c(H=0)$ being self-field limited. Between 77 K and 4.2 K $j_c(H=0)$ increases almost linearly from about 700 A/cm^2 to 2000 A/cm^2. Melt processing of Tl2223 reduces the weak-link behaviour and increases $j_c(H=0)$ by a factor of about 4. In our melt processed Bi2212 bulk material $j_c(H)$ exhibits only one plateau and $j_c(H=0)$ is not self-field limited. Between 77 K and 4.2 K $j_c(H=0)$ increases from 1700 A/cm^2 to 35000 A/cm^2 (field criterion $E=1\mu V/cm$). J_c was found to depend strongly on the field criterion in Bi2212. We checked this by measuring E vs j and found a power law dependence $E = cj^{\alpha}$ over 10 decades in E and 4 decades in j.

KEY WORDS: Ceramic processing, Inter- and intragranular current transport, Ring currens, I- V characteristic

1. MATERIAL PREPARATION

1.1. Conventionally sintered material

YBCO ($YBa_2Cu_3O_{7-\delta}$)
In a comparison between different preparation procedures it was found that the ceramic as well as the superconducting quality of YBCO are optimum when a fine grained precursor is used. Best results were achieved with precursors prepared by thermal decomposition of coprecipitated oxalates at 700 C (15 h in air followed by 24 h under O_2 flow). The precursor powder consists of primary particles with diameters below 0.1 µm showing homogeneous distribution of the elements. From this precursor material pellets were pressed (standard size 13 mm diam, 1mm height). Reactive sintering was performed at 925 C for 24 h under oxygen followed by a oxygen anneal step at 450 C for another 24 h. Temperature ramps of 60 C/h were used both in the heating and the cooling cycle. The result was a ceramic with <5% porosity consisting of particles of 5-10 µm diameter. Phase purity was estimated from XRD to be >90% with $BaCuO_2$, Y_2BaCuO_5, and CuO as the main contaminates.

Bi 2212 ($Bi_2Sr_2Ca\,Cu_2O_{8+\delta}$)
Bi2212 was prepared starting from a mixture 3N purity powders of Bi_2O_3, CuO, and $Sr(NO_3)_2$ together with freshly prepared CaO. Phase pure Bi2212 material was obtained after 3 heat treatments at 800 C- 850 C under air, 24 h each, with intermittent milling steps. The grain size after final milling and sieving was below 32 mm. For the sintering process again pellets were pressed which were sintered at 800 C to 880 C in air for 24 h or more and removed quickly from the oven. Noticeable sintering as indicated by crystallite growth set in above 830 C. In contrast to YBCO the ceramic pellets of Bi2212 did not shrink, i.e. the relative density equals the density of the green body of 45-55%.

Bi(Pb)2223 ($Bi_{1.84}Pb_{0.34}Sr_{1.91}Ca_{2.05}Cu_{3.06}O_{10+\delta}$)
Following a receipt given by Koyama et al /1/. our starting mixture composed of carbonates (Pb, Sr, and Ca) and oxides (Bi and Cu) had a cation stoichiometry of Bi: Pb: Sr: Ca: Cu= 1.84: 0.34: 1.91: 2.03: 3.06. This mixture was dissolved in 50% HNO_3 and transformed to a nitrate-oxide mixture by dissicating. An pure- oxide precursor powder was achieved by reacting at ca 750 C for 10 h, which was finally milled and sieved. Reactive sintering of the pelletized powder was performed at 830 C in an air- nitrogen atmosphere containing ca 7% oxygen for 100 h followed by a pure oxygen treatment during the cooling cycle below 750 C. After a repetition of this program the material was phase pure according to XRD. Attempts to melt process the 3 layer compound did not lead to phase pure material even after anneal times of more than 300 h.

Tl 2212 and Tl2223 ($Tl_2Sr_2CaCu_2O_{8+\delta}$ and $Tl_2Sr_2Ca_2Cu_3O_{10+\delta}$)
Beside toxicity, the main problem in the preparation of Tl based superconductors is to get control over the Tl stoichiometry which may change under sinter conditions due to the high volatility of Tl_2O_3. To prevent evaporation we used as containments small envelopes of Pt foil which were closed by tight folding and clamping. No reaction between pellet and container material and no weight loss even for sintering times of more than 100 h at 850 C were observed. Whilst the 2212 formation from stoichimetric precursors is easily accomplished between 770 C and 870 C, satisfactory yields of 2223 were only obtained in a fairly narrow temperature regime. Optimum conditions for the formation of Tl2223 (>95% Tl2223, the rest mainly Tl2212) are found in air at 855 C for reaction times of at least 4 h. Deviations from this temperature lead to an increasing amount of Tl2212. At T< 760 C and T> 870 C only Tl2212 was obtained despite of a starting composition according to 2223.

1.2. Meltprocessing

Bi2212
Bi2212 melts incongruently with an onset of melting at 890 C in air (900 C in O_2) and a melting range of >50 C, according to DTA. In its generic form melt processing of Bi2212 is done well above these onset temperatures /2/, where the melt is assured to be homogeneous. However we found best values for j_c when Bi2212 was only partially melted (typically 905- 915 C in O_2 for 20 min). Typical furnace cooling rates (ca 100 C/h) result in the formation of Bi2201 + (Sr, Ca)- cuprates. Transformation to 2212 is accomplished in a predominantly solid state reaction at 800 C- 860 C in air or O_2. Installation of optimum superconducting properties require a long time exposure to O_2 (>48 h) around 800 C followed by cooling in inert gas atmosphere.

Tl2223
Prereacted powder was isostatically pressed to cylinders of 8-9 mm diameter and 50 mm length. These cylinders were placed into Pt ampoules fitted with a stopper. Melting was performed at 1020 C- 1100 C for 10- 30 min in a preheated furnace. After room temperature cooling and dismantling of the foil no reaction between the melted body and Pt was observed. For further processing 1 mm slices were cut from the chunk which were tightly packed in Pt-envelopes for the following annealing procedure (ca 100 h at 855 C to transform the sample from predominantly Tl2212 into Tl2223). In a short time anneal (1h) of the uncovered samples in flowing oxygen at 750 C- 800 C superconducting properties were optimized by the uptake of at least 1 formula unit oxygen.

2. INDUCTIVE MEASUREMENT METHOD

In order to determine the intergrain critical current density $j_c(H)$, $j_c(T)$ or the current- voltage characteristic, E(j), we analyzed the hysteresis loop (magnetic moment M versus magnetic field H) of ring shaped samples. The rings were carved out of pellets using a dentist´s drilling machine. Typical dimensions of the rings were 4 mm for the inner radius, 5 mm for the outer radius, and the height of 1 mm. (The cross section of the rings was almost rectangular.)
The measurements were either performed in an AC Magnetometer (ACM) or in a Vibrating Sample Magnetometer (VSM). The measurements using the ACM were performed in a bath of LN_2 or LHe, respectively. The VSM (PAR 4500) is combined with an Variable Temperature Cryostat (Janis 153) which allows measurements between T = 2 K and room temperature. The maximum field available was about 2000 Oe for the ACM and about 7000 Oe for the VSM. The modulation frequency of the external field can be varied from 0.5 mHz to 0.1 Hz in the case of the VSM and from 0.1 Hz to 1KHz in the case of the ACM.
The intergranular critical current density j_c is determined from the flux change that can be shielded by the ring current when the external field is changed with defined sweep rate dB/dt. The applied voltage is given by $U_{ind} = \pi r^2 dB/dt$. Details of our method will be described elsewhere.

3. RESULTS AND DISCUSSION

3.1 Conventionally Sintered YBCO, Tl2212, Tl2223,Bi2223

The critical current density of the conventionally sintered YBCO, Tl2212, Tl2223 and Bi2223 samples exhibits almost the same field dependence. As an example, fig 1 shows such a measurement for Tl 2223 at 77 K and 4.2 K. The $j_c(H)$ curve can be divided into three sections:
1. In the low field regime j_c is independent of external fields. $j_c(H)$ starts to decrease at the intersection point with the self-field line H= bj/4 which represents the averaged field produced by a current carrying conductor of cross section b^2 at its surface. So j_c is obviously limited by its self-field and $j_c(H=0)$ depends on the cross section of the conductor which in our case is a ring. The temperature dependence of this zero field current is almost linear (fig.2).

2. In the following region j_c decays obeying the power law dependence $j_c= a (H/1\ Oe)^{-\beta}$ with $\beta \approx 3/2$ for all sintered samples. This decay of j_c is interpreted as the decoupling of Josephson junctions. The purely experimentally determined exponent $\beta= 3/2$ agrees well with a result of a model calculation considering the current transport through grains connected by junctions in form of circular discs /3/. In contrast to j(H=0) the values of a and β (see table 1) are true material parameters.
3. At high fields j_c levels out into a plateau regime with an almost field independent critical current, j_p. In this regime j_c is only about 10 A/cm^2 at 77 K. Nevertheless we find a pronounced hysteresis in the M(H) loops which unambiguously can be attributed to the intragrain currents. Assuming an averaged grain diameter of about 10 µm and using Beans model this intergranular critical current density can be estimated to be some 100´000 A/cm^2.

Table 1. Comparison of the high field values of j_c and the material parameters a, β which describe the decoupling of the weak links in the conventionally sintered materials. The values are given for T = 77 K / 4.2 K.

Material	YBCO	Bi2223	Tl2223	Tl2212
a (A/cm^2)	18 000 / 500 000	12 000 / 580 000	11 000 / 110 000	33 000 / 270 000
β	1.46 / 1.51	1.48 /1.39	1.49 / 1.48	1.32 /1.20
j_p(A/cm^2)	17 / 50	7 / 110	15 / 70	9 /150

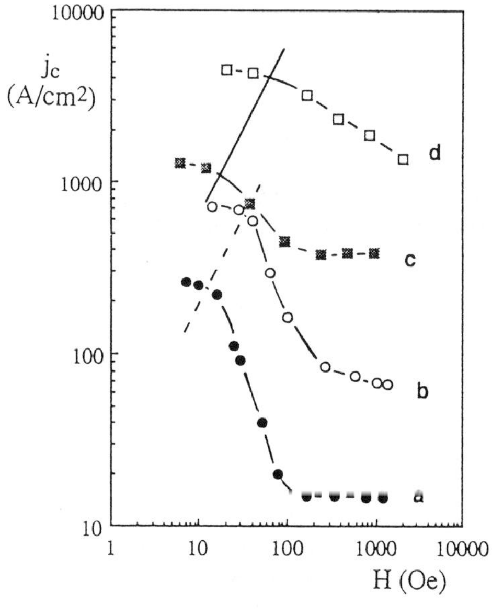

Fig. 1 j_c(H) of Tl2212 (1µV/cm). Curves (a) and (b) are for sintered material at 77 K and 4.2 K, curves (c) and (d) for melt processed material at 77K and 4.2 K. The straigt broken and solid lines represent the self-field lines calculated from the geometry of the sintered and melted ring, respectively.

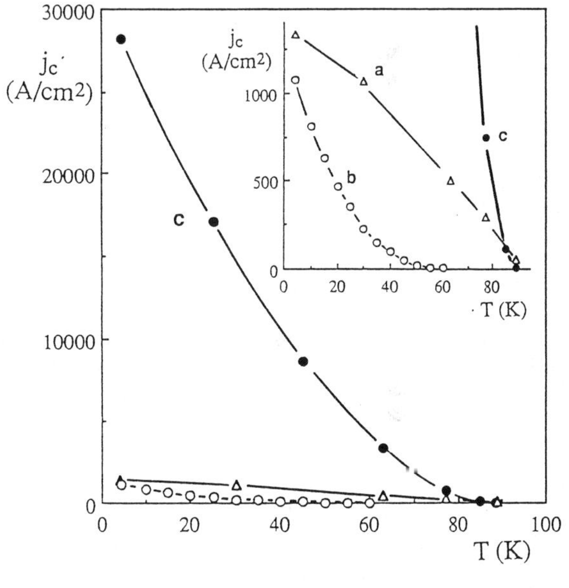

Fig. 2. Temperature dependence of jc (10nV/cm) of sintered YBCO (a), sintered Bi2212 (b), and melt-processed Bi2212 (c).

3.2 Melt processed Tl2223

In Tl based superconductors melt processing increases j_c significantly. Values of j_c(H=0) > 1700 A/cm^2 are frequently found. We argue that the reason for this improvement is not only the better ceramic density of melt processed material compared to its sintered counterpart but that melt processing improves the ratio of strong links/ weak links. This is clearly demonstrated in fig 1 which shows a plateau value of the critical current of j_p= 400 A/cm^2 in Tl2223 at 77 K. This is an improvement of x 27 compared to j_p in the sintered material whereas the values at H_{ext}= 0 differ only by the factor of 4.

3. 3 Bi2212 (sintered and melt processed)

Although T_c in sintered Bi2212 was found to be 95 K the material shows completely reversible M(H) loops at 77 K indicating that not only the intergranular but also the intragranular current density is zero. The irreversibility temperature, T_{irr}, marking the onset of flux pinning and supercurrents was found to be ca. 60 K. J_c increases in an overlinear manner to about 1100 A/cm^2 at 4.2 K (inset in fig 2, curve b). J_c(H) does not exhibit the typical weak link behaviour (i. e. a steep decay in low fields, see curve a and b in fig.1.) but is almost field independent in the available field region (fig 3, curve a).

Melt processing was found to increase T_{irr} in Bi2212 up to 90 K (Tc =95 K). With decreasing temperature j_c increases rapidly up to 30'000 A/cm^2 (fig. 2) At 77 K,.j_c = 800 A/cm^2 for H< 100 Oe (fig 3) was found without being self-field limited. Above 100 Oe j_c decreases rapidly to zero. There is no high field plateau as in the other materials. We do not find any weak link behaviour. The M(H) loops exhibit no hysteresis at fields where j_c= 0. We observe that the decay of both inter- and intragrain currents follow the same field dependence and that there is no decoupling between the grains. A disadvantage of the material is its weak pinning potential which becomes obvious in the rapid decrease of j_c with increasing temperature. Possibly also a consequence of weak pinning is the very peculiar current voltage characteristic. Instead of a well defined transition from R= 0 to the flux flow regime, we do not find any true transition behaviour at high temperatures (65 K-85 K): The power law dependence $E = cj^{\alpha}$ which was found to describe the E vs j behaviour over several decades does not allow to define any critical current density $jc \neq o$.

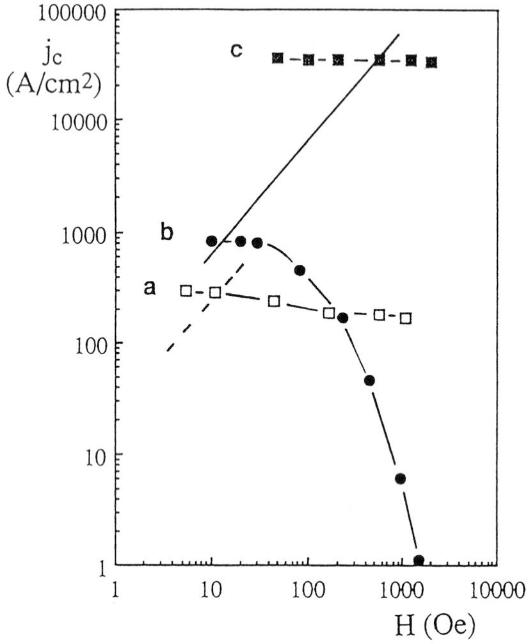

Fig 3. jc(H) (1μV/cm) of sintered Bi2212 at 4.2 K (a) and of meltprocessed Bi2212 at 77 K (b) and 4.2 K (c) The broken and solid lines represent the self-field lines calculated from the geometry of the sintered and melted ring, respectively.

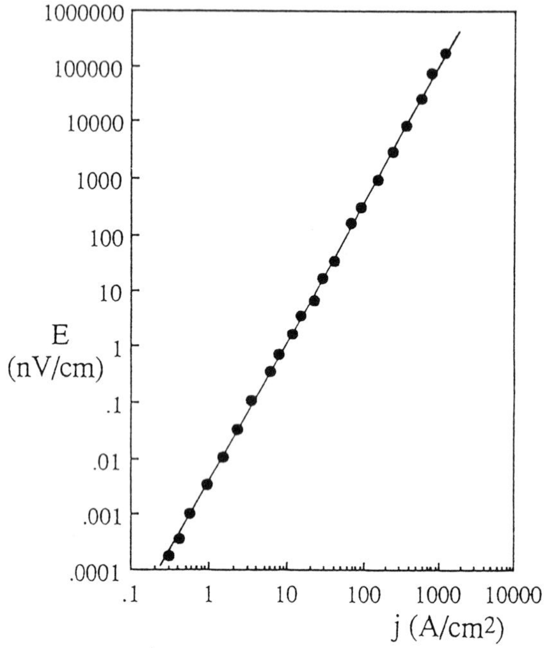

Fig 4. Current voltage characteristic of melt processed Bi2212 at T= 77 K and H = 200 Oe.

REFERENCES

1 S.Koyama, U. Endo, T. Kawai ,Jap. J. appl. Phys, 27, L1861- L1836 (1988)
2 J. Bock, E. Preissler, Sol State Comm. 72, 453 (1989)
3 R.L. Peterson, J.W. Ekin, Phys. Rev. B 37, 9848 (1988)

Pinning Characteristics and Their Anisotropy in a MPMG-Prepared High-T_c Oxide Superconductor

BAORONG NI[1], KAORU YAMAFUJI[1], NOBUYUKI MORI[2], KEISAKU ŌGI[2], TERUO MATSUSHITA[3], and KAZUO FUNAKI[4]

[1] Department of Electronics, [2] Department of Metallurgy, and [4] Research Institute for Superconducting Magnets, Kyushu University, 10-1, Hakozaki 6-chome, Higashi-ku, Fukuoka, 812 Japan
[3] Department of Computer Science and Electronics, Kyushu Institute of Technology, 680-4, Kawazu, Iizuka, 820 Japan

ABSTRACT

The critical current density, J_c, of bulk Y-Ba-Cu-O samples prepared with MPMG (Melt-Powder-Melt-Growth) method was investigated by an ac inductive measurement. A significant anisotropy was observed between J_c's parallel and perpendicular to the c-axis under a magnetic field parallel to the a-b plane. The temperature dependence of J_c was also measured. The results were discussed in terms of pinning mechanism by nonsuperconducting 211 particles and background pinning centers such as point defects.

KEY WORDS: MPMG process, critical current density, anisotropy, pinning mechanism

INTRODUCTION

It is well known that a pronounced anisotropy of critical current density, J_c, not only with respect to the direction of the magnetic field, but also with the direction of current exists in high T_c superconducting oxides, both for Y- and Bi-based compounds[1, 2]. However, for granular bulk specimens prepared, e.g., by solid-state sintered method, it is impossible to evaluate the anisotropic characteristics accurately, because of random orientations and fatal weak links between grains.

From the prospect of applications of these materials, another important problem which has to be solved immediately is an identification of the pinning mechanism and discovery of the method for improvement of J_c and pinning force.

Fortunately, fabrication techniques of bulk Y-Ba-Cu-O specimen have recently been progressed greatly, such as Melt Textured Growth (MTG) process[3], Quench and Melt Growth (QMG) process[4] and Melt-Powder-Melt-Growth (MPMG) process[5]. Bulk specimens with fairly well crystalline orientation and prominently enhanced critical current characteristics can be prepared by these methods. It makes the studies on anisotropy, pinning mechanism and other characteristcs of this material possible. We have reported that 211 phase may play a certain role as pinning center[6] in YBaCuO.

In this contribution, we report the results on anisotropies and temperature dependences of J_c in specimens prepared by MPMG process under various dc magnetic field. An ac inductive method was employed to measure the critical current density. Discussion was made on results, in terms of pinning mechanism by nonsuperconducting 211 particles and background pinning centers such as point defects. For background pinning, we referd to the experimental results by Ref.7.

EXPERIMENTS

YBaCuO bulk specimen was fabricated by MPMG technique. Precursor powder ($YBa_2Cu_3O_{7-\delta}$ with Y_2O_3-rich composition) in a Pt crucible was melt by heating in a furnace preheated to 1450℃ for 2 minutes, and then quenched using two copper plates. The quenched material was pulverized, well-mixed, and formed into a pellet by pressing. This process provides a uniform distribution of Y_2O_3 in solidified liquid phase. The pellet was again heated to 1090℃ for partially melting. After preserving for 30 minutes, it was cooled down gradually ranging from 1060℃ to 960℃, with a temperature gradient of 1.5℃/h in order to get a uniaxial grain growth. Then it was cooled down slowly to room temperature. An annealling in oxygen was employed at 600℃ for 10 hours later. The critical temperature T_c obtained from ac susceptibility measurement was 90K.

The Laue Pattern was used for determination of c-axis orientation of the pellet. Two slabs were cut out of the pellet, with different c-axis directions. The orientation of the c-axis is parallel and perpendicular to the thickness direction in specimens (I) and (II), respectively, and perpendicular to long axis. Figure 1 shows an optical micrograph of specimen. The white particles are nonsuperconducting 211 phase.

Critical current densities were measured by using an ac inductive method[8]. Dc and superposed ac magnetic fields were applied along the long axis of the specimens, i.e., parallel to the a-b plane. Because of a significant anisotropy of J_c, penetrating rates of ac flux into the slab from each side are different. Since the current density estimated directly from the ac inductive measurement is an average value of two components along the directions parallel ($J_{c\parallel}$) and perpendicular ($J_{c\perp}$) to the a-b plane, the separation of the two components can be realized by measuring two slabs with different geometries, supposing they have the same electromagnetic characteristics.

Fig.1 Optical micrograph of the structure in MPMG specimen. White particles denote the 211 phase.

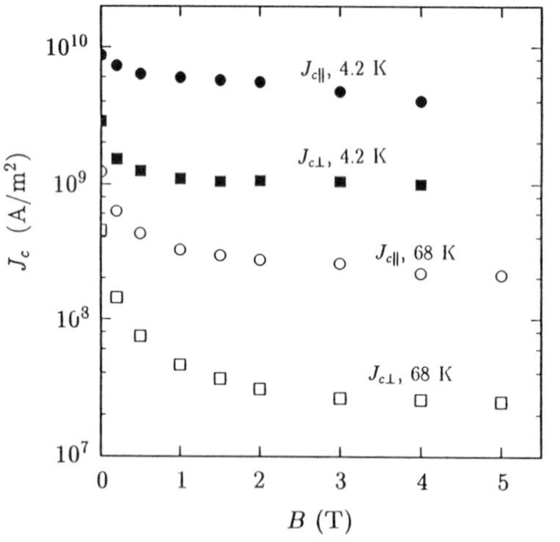

Fig.2 Dc magnetic field dependence of critical current densities.

RESULTS AND DISCUSSION

The obtained critical current densities parallel ($J_{c\parallel}$) and perpendicular ($J_{c\perp}$) to the a-b plane are shown in Fig.2. The anisotropy ratio, $J_{c\parallel}/J_{c\perp}$, was about 5 at 4.2K and 9 at 68K. Considering an error of determination of the c-axis, the real values seem to be larger somewhat.

Although the reason for this significant anisotrpy has not yet been exactly understood, following candidates can be given:
(1) The intrinsic anisotropy of critical current density associated with crystal structure. CuO plane is expected to contribute to a better transport current.
(2) The anisotropy from the flux pinning due to the anisotropy in the coherence length. That is, $J_{c\|}$ is considered to be larger by a factor of ξ_{ab}/ξ_c than $J_{c\perp}$, as will be shown later.
(3) Nonsuperconducting layers grown uniaxially and microcracks which restrict the current along the c-axis.
Each candidate explains larger J_c in the a-b plane and it can not be declared which is the real reason. It is also possible that all of them contribute to the J_c's anisotropy.

Supposing a direct summation of elementary pinning forces of large normal particles[9] for estimation of the pinning force density, we obtain the critical current density as

$$J_c = \frac{\pi {B_c}^2 S\xi}{4\mu_0 B a_f}\left(1 - \frac{B}{B_{c2}}\right), \tag{1}$$

where a_f is a fluxoid spacing, S a total effective surface area of normal particles in a unit volume and ξ the coherence length in a corresponding direction. If we assume the volume fraction of particles as to be f, S can be obtained as $S = f\langle D^2\rangle/\langle D^3\rangle$, where D is the size of a particle and $\langle\ \rangle$ represents an average value.

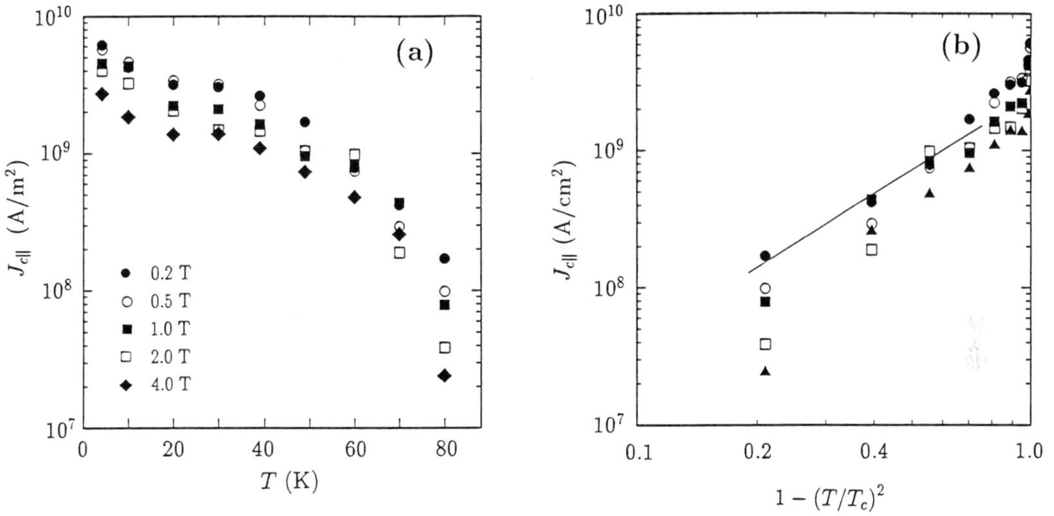

Fig.3 (a) Temperature dependence of critical current densities and (b) replots of (a) as a function of logarithmic $1 - (T/T_c)^2$.

At $B = 1.0$T ($a_f = 48.9$nm) and $T = 68$K, the critical current density along the a-b plane is estimated as $J_{c\|} \doteq 1.32 \times 10^9$ A/m^2, where we used $\xi_{ab} = 2.36$nm, $S = 9.76 \times 10^4$m^{-1} and $B_c = 0.67$T. This value is larger by a factor of 4 than the experimental value shown in Fig.2. The difference indicates that some other factors, such as flux creep and weak links at domain boundarys, reduce the critical current characteristic sufficiently.

If B is constant and sufficiently smaller than B_{c2}, the temperature dependence of J_c is given by

$$J_c \propto {B_c}^2\xi. \tag{2}$$

Noticing $B_{c2} = \phi_0/2\pi\xi^2$ and using an empirical temperature dependence of $B_{c2}(T) \propto 1 - (T/T_c)^2$, Eq.(2) can be expressed as

$$J_c \propto B_{c2}^{3/2}(T) \propto \left[1 - \left(\frac{T}{T_c}\right)^2\right]^{3/2}. \tag{3}$$

That is to say, the critical current density is proportional to 3/2th power of $1 - (T/T_c)^2$.

Figures 3(a) and 3(b) show the temperature dependence of J_c and replots of J_c versus logarithm of $1 - (T/T_c)^2$, respectively. In the region of high temperatures (ranging from 40K to 80K), the solid line in Fig.3(b) shows an exponent of 1.8 which coincides approximately with the expectation given in Eq.(3). This suggests that 211 particles act as dominent pinning centers at high temperatures. The reason why the observed exponent is larger somewhat than theoretical one seems to be an effect of flux creep, especially at higher temperatures.

On the contrary, the exponent observed in low temperature region deviates from 3/2 greatly, meaning that 211 particles are not dominant in this region. In fact, as shown by Gyorgy *et al.*[7], the background pinning centers become superior when temperature falls sufficiently. Point defects such as oxygen deficiencies or stacking faults are expected to contribute to the background pinning. In addition, the weak links are considered to be degraded with rising temperature. A rapid decrease in the critical current density at very low temperature region shown in Fig.3(b) seems to be caused by weak links. This expectation agrees with the above-mentioned fact that observed values of J_c are much smaller then theoretical ones.

In conclusion, the bulk specimen prepared by MPMG process shows a remarkble anisotropy in critical current density perpendicular and parallel to the c-axis. The sufficient high value of J_c in high temperature region seems to be ascribed to the pinning interaction of nonsuperconducting 211 particles. Nevertheless, at lower temperatures, background pinning with a strong temperature dependence becomes superior. In addition, the degradation of weak links at domain boundarys seems to occur while temperature goes up in the low region.

REFERENCES

[1] C. Keller, H. Küpfer, R. Meier-Hirmer, U. Wiech, V. Selvamanickam and K. Salama: Cryogenics **30** (1990) 401

[2] T. Umemura, K. Egawa, M. Wakata, K. Yoshizaki, B. Ni and T. Matsushita: Physica **B 165 & 166** (1990) 1411

[3] S. Jin, T. H. Tiefel, R. C. Sherwood, R. B. vanDover, M. E. Davis, G. W. Kammlott and R. A. Fastnacht: Phys. Rev. B **37** (1988) 7850

[4] M. Murakami, M. Morita, K. Doi, K. Miyamoto and H. Hamada: Jpn. J. Appl. Phys. **28** (1989) 399

[5] H. Fujimoto, M. Murakami, T. Oyama, S. Gotoh, K. Yamaguchi, T. Tanaka, Y. Shiohara, N. Koshizuka and S. Tanaka: Cryog. Eng. **25** (1990) 77 [in Japanese]

[6] T. Matsushita, B. Ni and K. Yamafuji: Adv. Cryog. Eng. (Mater.) **36** (Plenum Press, New York, 1990) 403

[7] E. M. Gyorgy, R. B. van Dover, L. F. Schneemeyer, A. E. White, H. M. O'Bryan, R. J. Felder, J. V. Waszczak and W. W. Rhodes: Appl. Phys. Lett. **56** (1990) 2465

[8] B. Ni, T. Munakata, T. Matsushita, M. Iwakuma, K. Funaki, M. Takeo and K. Yamafuji: Jpn. J. Appl. Phys. **27** (1988) 1658

[9] A. M. Campbell, J. E. Evetts and D. Dew-Hughes: Philos. Mag. **18** (1968) 313

Pinning Characteristics in Superconducting Bi-Pb-Sr-Ca-Cu-O

Toshio Umemura[1], Kunihiko Egawa[1], Mitsunobu Wakata[1], Shin Utsunomiya[1], Baorong Ni[2], Masakatsu Takeo[3], Kaoru Yamafuji[2], Edmund Soji Otabe[4], and Teruo Matsushita[4]

[1] Materials & Electronic Devices Laboratory, Mitsubishi Electric Corporation, 1-57, Miyashimo 1-chome, Sagamihara, 229 Japan
[2] Department of Electronics, and [3] Research Institute for Superconducting Magnets, Kyushu University, 10-1, Hakozaki 6-chome, Higashi-ku, Fukuoka, 812 Japan
[4] Department of Computer Science and Electronics, Kyushu Institute of Technology, 680-4, Kawazu, Iizuka, 820 Japan

ABSTRACT

Critical current densities, J_c, in bulk Bi-Pb-Sr-Ca-Cu-O have been measured ac inductively under various magnetic fields and temperatures. The J_c drops abruptly at high fields and high temperatures and such a significant reduction seems to be attributed to thermally activated flux creep. The irreversibility field at which J_c goes to zero is investigated and compared with a theoretical prediction using observed macroscopic pinning force density. The mechanism of flux pinning in Bi-based superconductors is also argued.

KEY WORDS: Bi-based superconductors, flux pinning, critical current density, flux creep, irreversibility line

INTRODUCTION

Oxide superconductors have large anisotropy such as axial anisotropies of critical current density, J_c, and coherence length. From a viewpoint of material, it is easier in Bi-based compounds to obtain a textured structure by an intermediate pressing and so on than in Y-based superconductors. In addition, Bi-based superconductors are less influenced by a weak link problem at grain boundaries than in Y-based ones. Therefore, the former materials may be superior for application to a superconducting wire to the latter. While the J_c's ensure the practical use at 4.2 K, however, they drop seriously at high temperatures, T, and high magnetic fields, B.

In a previous study, we have reported the magnetic field dependence and the axial anisotropy of J_c measured inductively at various temperatures, and also pointed out its abrupt degradation at high temperatures and high magnetic fields[1]. The abrupt degradation seems to be attributed to an influence of significant flux creep due to a large thermal energy and a weak pinning energy. It is desired, therefore, to improve the flux pinning strength. However, the pinning centers in Bi-based superconductors have not yet been clarified.

In order to investigate the flux pinning characteristics under the influence of thermal activation, we have measured the temperature dependent J_c under various magnetic fields, and drove the irreversibility line in B-T plane on which J_c is reduced to zero. The irreversibility line is one of the characteristic issue from the flux creep and depends strongly on the flux pinning strength. Its relation with the pinning strength was argued by Matsushita *et al.*[2]. In this paper, the observed irreversibility line is compared with their theoretical prediction.

EXPERIMENTS

Specimens were prepared by solid-state reaction by using coprecipitated powder with a nominal composition of Bi:Pb:Sr:Ca:Cu=1.6:0.4:2:2:3. The powder was calcined at 750℃ for 12 hours in air. After powdering by ball milling with alcohol, the powder was pressed into slabs of 3 $mm^{w}\times 1$ $mm^{t}\times 20$ mm^{l} and sintered at 845℃ in N_2-50%O_2. Sintering was repeated three times with the intermediate pressing at several tons per square centimeter. The c-axis of each grain of the sintered specimens was preferencially oriented toward the press direction. The specimens used for the experiment were mechanically cut from the slab. Their sizes are typically 1 $mm^{w}\times 0.2$ $mm^{t}\times 20$ mm^{l} and the flat surface is parallel to the c-axis.

The transition temperature, T_c, measured resistively is 106 K. The specific weight is 5.7 g/cm^3. The critical current density was measured ac inductively at various temperatures in a deware with a temperature control facility. The temperature was observed using a Au-Fe 0.07% vs Ag thermo-couple and its stability was within $\pm$0.3 K during the measurement. In the ac inductive measurement, the dc and superposed small ac magnetic fields were applied parallel to the flat surface and the long axis of the specimen and hence, normal to the c-axis. The frequency of the superposed ac magnetic field was 35.0 Hz. Critical current densities at each dc magnetic field have been estimated from the magnetic flux going in and out the specimen measured as a function of the ac magnetic field amplitude. The irreversibility magnetic field, B_i, was determined as a function of temperature under the criterion of the critical current density being 10^5 A/m^2.

RESULTS AND DISCUSSION

Observed critical current densities of specimen 1 in an increasing field process at various temperatures are shown in Fig. 1. Although J_c keeps a relatively high value up to 5 T at temperatures lower than 25 K, it decreases abruptly around 1 T at 77 K. Figure 2 represents the temperature dependence of the irreversibility field in specimen 2. The temperature dependence is usually expressed as

$$B_i(T) \propto \left[1 - \left(\frac{T}{T_c}\right)^2\right]^n , \qquad (1)$$

where $n = 1.5$ has been reported for Y-based superconductor[3] and $n = 3.2$ for Bi-based one[4]. The solid line in Fig. 2 shows the dependence of $n = 4.1$.

According to Matsushita *et al.*[2], the irreversibility field is theoretically expressed as

$$B_i(T) = \left(\frac{K}{T_c}\right)^{4/(3-2\gamma)} \left[1 - \left(\frac{T}{T_c}\right)^2\right]^{2m/(3-2\gamma)} , \qquad (2)$$

where m and γ are the parameters representing temperature and magnetic field dependences of ideal pinning force density in the absence of flux creep as

$$F_p(B,T) = AB_{c2}^m(T)B^\gamma \left[1 - \frac{B}{B_{c2}(T)}\right]^\delta . \qquad (3)$$

A constant, K, in Eq. (2) is given by

$$K = \frac{0.835 g^2 [AB_{c2}^m(0)]^{1/2}}{\zeta^{3/2} \ln(Ba_f\Omega/E_c)} , \qquad (4)$$

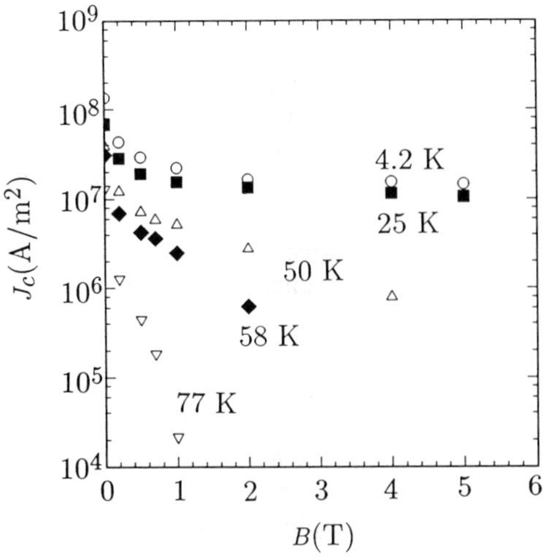

Fig. 1. Critical current density vs magnetic field in specimen 1 at various temperatures.

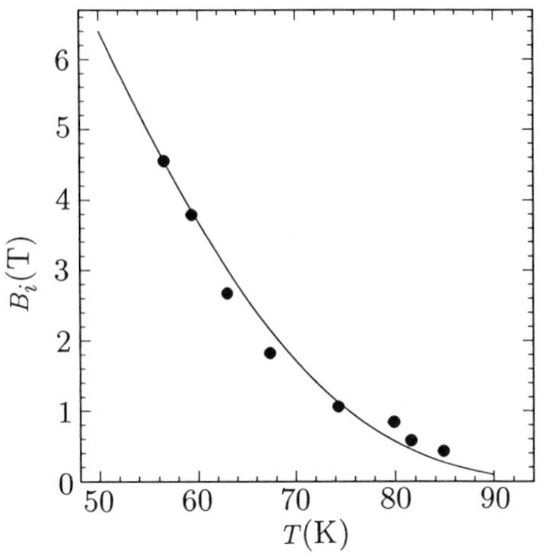

Fig. 2. Temperature dependence of irreversibility field in specimen 2. Solid line represents the characteristic given by Eq.(1) with $n = 4.1$.

where ζ is a constant of about 2π, a_f a fluxoid spacing, Ω an oscillation frequency of flux bundle, E_c an electric field criterion in resistive measurements. Substitution of typical values into the logarithmic term leads to $\ln(Ba_f\Omega/E_c) \sim 14$. A factor g^2 represents a number of fluxoids in the bundle. In the elastic limit of the fluxoid lattice, we have

$$g = \frac{\ell_{66}}{a_f} = \frac{1}{a_f}\left(\frac{C_{66}}{\alpha_L}\right)^{1/2}, \tag{5}$$

where ℓ_{66} is a transverse elastic correlation length of the fluxoid lattice, C_{66} is the shear modulus and α_L is the Labusch parameter representing a sharpness of the pinning potential. If g is independent of temperature, as in strongly pinned materials, we have

$$n = \frac{2m}{3 - 2\gamma}. \tag{6}$$

In order to compare the present experimental results with the theoretical prediction, we need the parameters, m and γ, representing the ideal pinning characteristic free from the thermal activation. These parameters seem to be approximately estimated from the results observed at sufficiently low temperatures. Figure 3 is a replot of the results in the low temperature region shown in Fig. 1. At very low temperatures, i.e., around 1 in the abscissa, J_c drops abruptly. This decrease is considered to be caused by degradation of weak link nature at grain boundaries[5]. Hence, the characteristic in this region is not useful for the present estimate. The straight line in the figure represents $m = 4.4$. Figure 4 shows magnetic field dependence of J_c replotted from Fig. 1. The straight line gives $\gamma = 0.66$. Substitution of these parameters into Eq. (6) yields $n = 5.2$, which is fairly higher than the experimental result, $n = 4.1$. However, the tendency of larger exponent than in Y-based superconductors, which is caused by a larger m value[2], agrees qualitatively with experiments.

Secondly we shall quantitatively compare the observed irreversibility field with the theoretical result. Although the real pinning force density is not clear due to the weak link nature as above mentioned, we use the observed values. If the results $B_i = 4.55$ T at $T = 56.6$ K and $T_c = 106$ K are substituted into Eq. (2), we have $K = 4.19 \times 10^2$. At 4.2 K and 1 T, the pinning force density is estimated as $F_p = 2.22 \times 10^7$ N/m^3. Hence, we have $AB_{c2}^m(0) \simeq AB_{c2}^m(4.2) = 2.22 \times 10^7$ N/m^3 from Eq. (3). This leads to $g \simeq 4.7$, suggesting that about 20 fluxoids consist a bundle. On the other hand, at 4.2 K and 1 T ($a_f = 4.9 \times 10^{-8}$ m) α_L is estimated as 2.7×10^{15} N/m^4 with the

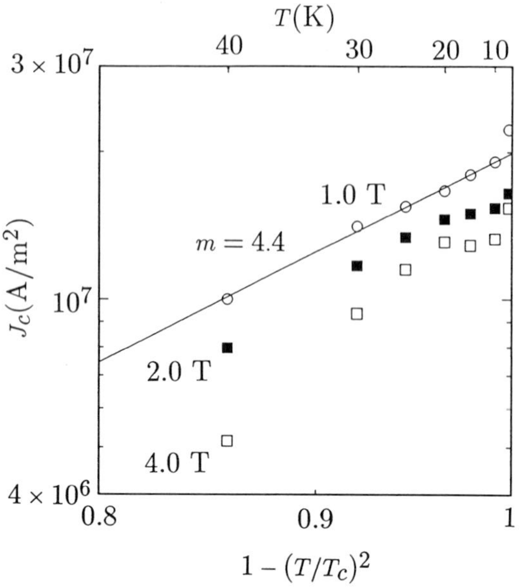

Fig. 3. Critical current density vs log$[1-(T/T_c)^2]$ in specimen 1.

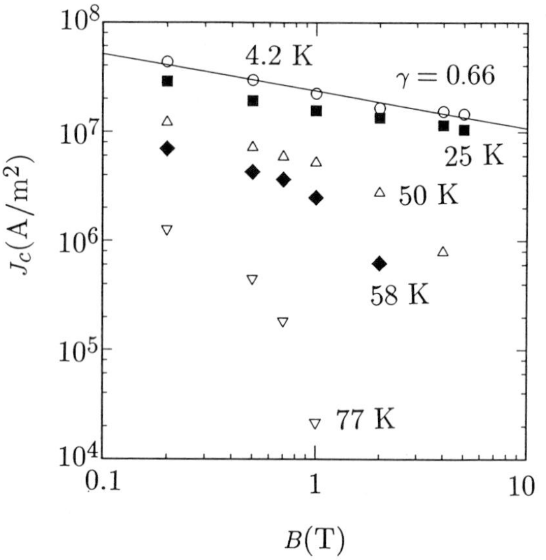

Fig. 4. Critical cureent density vs log B in specimen 1.

assumption of $\zeta = 2\pi$. If we assume $B_c \simeq 1$ T, we have $C_{66} \simeq 0.2{B_c}^2/\mu_0 \simeq 1.6 \times 10^5$ N/m^2. From Eq. (5), these values yield $g \simeq 157$, which is much larger than the above estimate from the direct comparison of the irreversibility field.

Here we shall discuss the cause of difference between the experimental and theoretical results. As above mentioned, the observed critical current density is considered to be influenced by connections at grain boundaries. That is, the real pinning strength is expected to be much larger than the present observation. This yields an underestimate of α_L and an overestimate of g. Another reason for large difference in g may be found in an occurrence of plastic deformation around defects in the fluxoid lattice. That is, the local plastic deformations lead to a reduction of g. The difference in the exponent, n, seems to be attributed to a neglect of temperature dependence of g. If we take account of an increase of g due to a steeper decrease of α_L with elevating temperature, g increases with temperature and qualitatively agrees with the observed result.

The temperature dependence of the pinning force density in the present Bi-based superconductors is characterized by $m \simeq 4.4$. This is close to that of background pinning in a single crystal Y-Ba-Cu-O ($m \simeq 3.8$ is obtained in [6]) and seems to be attributed to a collective pinning by point defects[5]. Since any defects which seem to dominate the pinning in Bi-based superconductors have not yet been detected, the above expectation that point defects are effective is reasonable. However, the kind of point defects, such as oxygen deficiencies or lattice defects, is not known.

References

1 Umemura T, Egawa K, Wakata M, Yoshizaki K, Ni B, Matsushita T (1990) Physica B 165&166: 1411-1412.
2 Matsushita T, Fujiyoshi T, Toko K, Yamafuji K (1990) Appl. Phys. Lett. 56: 2039-2041.
3 Yeshurun Y, Malozemoff AP (1988) Phys. Rev. Lett. 60: 2202-2205.
4 Green SM, Lobb CJ, Greene RL (1990) to be published in Proc. Appl. Superconductivity Conf.
5 Matsushita T. Otabe ES, Ni B, Kimura K, Morita M, Tanaka M, Kimura M, Miyamoto K, Sawano K, submitted to Jpn. J. Appl. Phys.
6 Gyorgy EM, van Dover RB, Schneemeyer LF, White AE, O'Bryan HM, Felder RJ, Waszczak JV and Rhodes WW (1990) Appl. Phys. Lett. 56: 2465-2467.

History Effect of Critical Current Density in Superconducting Bi-Pb-Sr-Ca-Cu-O Wires

TAKESHI HIKATA[1], KEN-ICHI SATO[1], BAORONG NI[2], EDMUND SOJI OTABE[3], TERUO MATSUSHITA[3], MASAKATSU TAKEO[4], and KAORU YAMAFUJI[2]

[1] Osaka Research Laboratories, Sumitomo Electric Industries, Ltd., 1-3, Shimaya 1-chome, Konohana-ku, Osaka, 554 Japan

[2] Department of Electronics and [4] Research Institute for Superconducting Magnets, Kyushu University, 10-1, Hakozaki 6-chome, Higashi-ku, Fukuoka, 812 Japan

[3] Department of Computer Science and Electronics, Kyushu Institute of Technology, 680-4, Kawazu, Iizuka, 820 Japan

ABSTRACT

Magnetic field dependence of J_C and its history effect were measured at various temperatures by ac inductive method for silver sheathed BiPbSrCaCuO high$-T_C$ superconducting wires. It was revealed that the result was consistent with the 4 probe dc method. It was found that the J_C was influenced by weak links at very low temperature region, where the history effect was observed. This suggests that the history effect is caused by weak links.

KEY WORDS: BiPbSrCaCuO, superconducting wire, critical current density, history effect, weak link

INTRODUCTION

Since the critical temperature (T_C) of oxide superconductors exceeded nitrogen temperature (77.3K), the transport current properties were investigated intensively in order to apply them in liquid nitrogen[1–5]. It is important for the realization of high$-T_C$ superconducting wires for superconducting cables and magnets, to improve the critical current density (J_C). In a polycrystalline wire, J_C was limited by the properties of grain boundaries with weak links. It is possible to increase J_C by improvement of weak links at the grain boundaries because J_C inside the grains are much larger than the J_C through grain boundaries. It has been known that the weak link's properties were reflected in the magnetic field dependence of J_C and the history effect[6]. In this paper, we report temperature dependence of J_C–B properties and the history effect which reflect the weak link behavior measured by using ac inductive method.

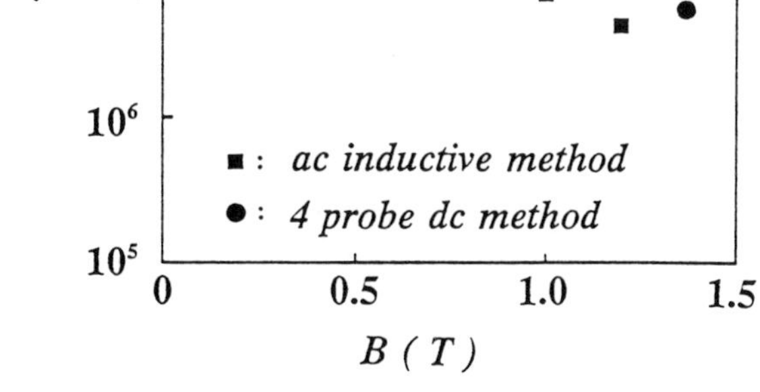

Fig.1 Magnetic field dependence of J_C for BiPbSrCaCuO Ag–sheathed wire with 1.7x 10^8 A/m^2 by 4 probe dc method and ac inductive method.

EXPERIMENTAL

Appropriate amount of Bi_2O_3, PbO, $SrCO_3$, $CaCO_3$ and CuO powders with 3N to 4N purity were mixed, calcined and ground. The ground powder was put into silver tubes. The composites were drawn to round wires, deformed to taped wires, and then sintered. Those deformation and sintering processes were repeated twice[3]. The typical sample size was 2.5–4.5mm in width and 0.1–0.2 mm in thickness. T_C was 106K. The J_C–B characteristics were measured by 2 kinds of methods *i.e.*, 4 probe dc method and ac inductive method which could be done without transport current[6]. The ac inductive method was carried out by superposition of dc and ac magnetic fields which were applied parallel to the long axis of the wire. The frequency of the ac field was 35.0 Hz. The dc field was applied up to 5 Tesla. The temperature was controlled in the range from 4.2K to 77K. The magnetic flux Φ going in and going out of the sample was measured as a function of the ac field amplitude, $\boldsymbol{b}$, and the penetration depth of the ac field was derived as follows: $\lambda'=(1/2w)\partial\Phi/\partial \boldsymbol{b}$ (where w shows the effective width of the superconducting section in the tapes). The slope of the λ' vs $\boldsymbol{b}$ gives $1/\mu_0 J_C$ (μ_0 denoting the permeability of the vacuum). λ' saturated to a constant value when the magnetic field reached to the center of the sample[7].

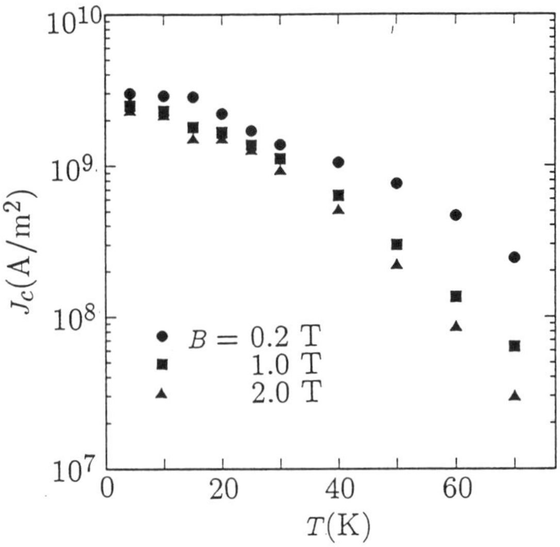

Fig.2 Temperature dependence of J_C of sample #1 in 0.2T, 1.0T and 2.0T by ac inductive method.

RESULTS AND DISCUSSION

Figure 1 shows the J_C–B properties measured at 77.3 K for silver–sheathed wires by ac inductive method and 4 probe dc method. J_C of these wires was 1.7 x 10^8 A/m^2 in a zero magnetic field measured by the dc method. This result shows that the magnetic field dependence of J_C can be described by ac inductive method without inducing transport current as well as 4 probe dc method. This means that it is possible to easily measure the temperature dependence of J_C–B properties (J_C–B–T) of the wires with high critical current without the influence of heating at the transport current terminals in flowing gas Helium. The transport critical current of 233 A (J_C=2.1x10^9 A/m^2) at 4.2 K in a wire with the dimension of 4.2mm^w x 0.15mm^t was achieved, so we need a large heat flux to coolant or extremely low contact resistant joints in the case of transport current measurements [8]. There is a little difference between dc and ac method because J_C by dc method was dominated by the local weak links, while J_C by ac method gives an average of inter– and intragrain currents. Figure 2 shows the temperature dependence of J_C in 0.2, 1.0 and 2.0T for sample #1 (J_C(77.3K,0T)=2.5 x 10^8 A/m^2) by ac method. It was found that the J_C decreased large in the temperature region over 30K with the increase of magnetic field from 0.2T to 2T. Figure 3 shows the magnetic field dependence of J_C up to 5T and 60K. At 4.2K and 25K, the magnetic field dependence of J_C is small and especially small around 5T, while at 60K J_C has large field dependence. J_C can be represented under a fixed magnetic field as follows:

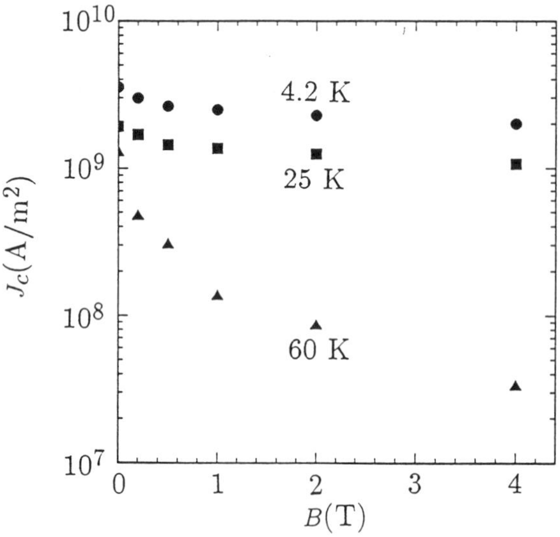

Fig. 3 Magnetic field dependence of J_C in 4.2K, 25K and 60K.

$$J_C \propto B_{C2}^m(T) \propto [1-(T/T_C)^2]^m, \quad (1)$$

using an empirical temperature dependence of $B_{C2}(T) \propto 1-(T/T_C)^2$ [9]. m value is a parameter determined by the sort of pinning sites. Figures 4 and 5 show the double logarithmic plots of J_C vs $1-(T/T_C)^2$ for sample #1 and sample #2 (J_C(77.3K,0T) = 6.3 x 10^7 A/m^2) For sample #1, m is 3.43 at high temperature region and 10.05 at low temperature region. For sample #2 we have m = 2.75 and m=16.3 at high and

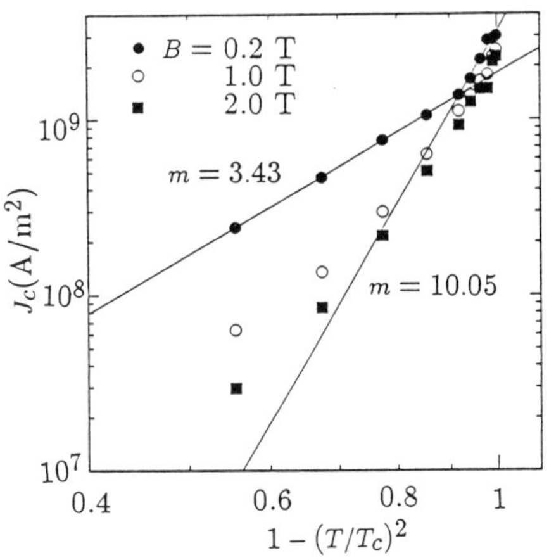

Fig. 4 Logarithmic plots of J_C vs 1-$(T/T_C)^2$ at 0.2T, 1T and 2.0T for sample #1.

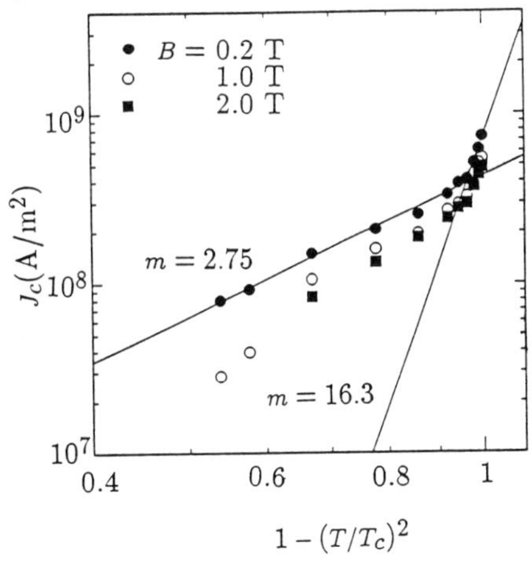

Fig. 5 Logarithmic plots of J_C vs 1-$(T/T_C)^2$ at 0.2T, 1T and 2.0T for sample #2.

low temperature regions, respectively. These data exhibit that the J_C–B properties were dominated by more than 2 kinds of pinning mechanism at least. Usually m ~ 1.5 is obtained for planar pinning sites and m ~ 4 for point pinning sites. However, these are not fitted in this case. Polycrystalline Bi-based high T_C superconductor has large temperature dependence due to flux creep and the irreversibility line increases with the improvement of J_C [10]. Therefore, effects of another factors, such as weak links and flux creep, should be taken in account for more comprehensive understanding of pinning. Especially in low temperature region, the influence by weak links is large because m value was improved much from 16.3 to 10.5 with the increase of J_C (77.3K) from 6.3 x 10^7 A/m^2 to 2.5 x 10^8 A/m^2.

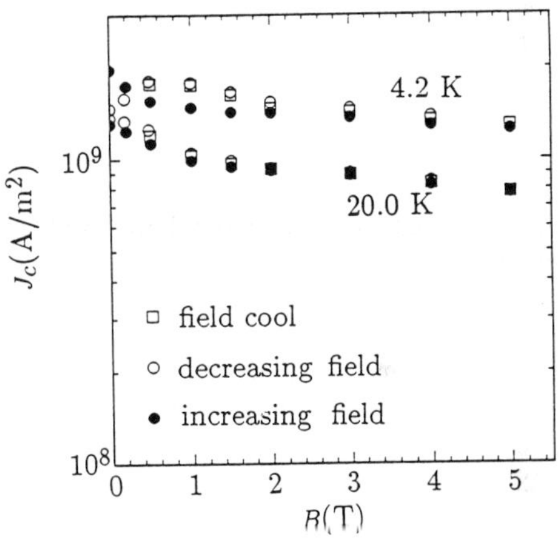

Fig. 6 History effects of sample #3 in 4.2K and 20K.

Figures 6 and 7 show the history effects of J_C–B properties for sample #3 (J_C(77.3K,0T)=3.0 x 10^8 A/m^2) and sample #2 at 4.2K and 20K. Solid circles indicate the increasing process of magnetic field and open circles indicate the decreasing one in the case of a zero field cooling. The obtained critical current density takes a smaller value in the increasing field process than the decreasing one. This data is also consistent with one of 4 probe dc method [11]. Open squares indicate J_C–B properties in the field cooling process. In this case, the intragrain shielding currents were not induced, and so J_C was not influenced by the weak

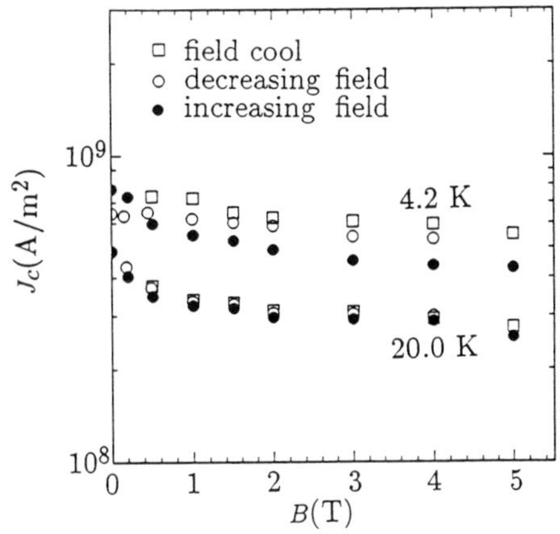

Fig. 7 History effects of sample #2 in 4.2K and 20K.

links at the grain boundaries. Therefore J_c in the field cooling case must take an intermediate value of those in increasing and decreasing (major) lines [6]. In sample #3 with high J_C, the field cooling line is in approximately intermediate region, while in sample #2 with low J_C, the field cooling line is larger than the decreasing field line in a zero field cooling process. It is considered that when the history effect exists up to much high magnetic field than maximum in the measurement, the decreasing field line cannot return completely to the major decreasing line. That is, the history effect shown in Fig. 7 is smaller than true one. Here the degree of history effect is represented by the defference in J_C between major increasing line and field cooling line, [$\Delta J_C=J_C(FC)-J_C(increasing)$]. This difference, ΔJ_C, in low J_C sample is larger than those in the higher samples. However this figure shows that the history effect almost simultaneously disappears near 20K for both samples, which can be seen in Fig. 8. This data is consistent with temperature dependence of transport J_C–B by using Liq.He and Liq.H_2 [8]. This suggests that 20K is a characteristic temperature at which tunneling of superconducting currents appears throgh through weakly couple grains.

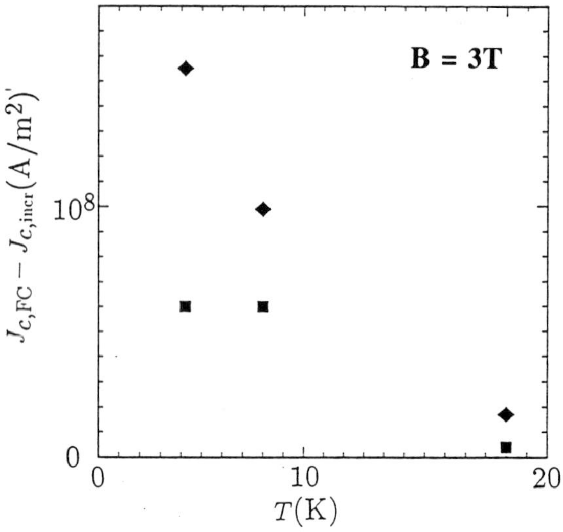

Fig. 8 Temperature dependence of history effect for sample #2 (◆:$J_C(77.3K)=6.3 \times 10^7$ A/cm^2) and #3 (■:$J_C(77.3K)=3.0 \times 10^8$ A/cm^2).

CONCLUSION

The temperature dependence of J_C–B properties was measured effectively in wide temperature range for BiPbSrCaCuO silver sheathed wires with high critical current and critical current density. In low temperature region, it was found that J_C was influenced by weak links and the history effect was observed clearly. It was shown that the history effect was improved with the increase of J_C and disappeared at around 20K for the samples with J_C of different levels. This suggests that much higher J_C can be achieved by improvement of weak links between grains.

REFERENCES

1. Maeda H, Tanaka Y, Fukutomi M and Asano T (1988) Jpn. J. Appl. Phys. 27: L209.
2. Hikata T, Sato K and Hitotsuyanagi H (1989) Jpn. J. Appl. Phys. 28: L82.
3. Hikata T, Nishikawa T, Mukai H, Sato K and Hitotsuyanagi H (1989) Jpn. J. Appl. Phy. 28: L1204.
4. Sato K, Hikata T, Mukai H, Masuda T, Ueyama M, Hitotsuyanagi H, Mitsui T and Kawashima M (Springer–Verlag,1990) Adv. Superconductivity: p.335.
5 Sato K, Hikata T, Mukai H, Ueyama M, Shibuta N, Kato T, Masuda T, Nagata M, Iwata K and Mitsui T (1990) Applied Superconductivity Conference, Sept. 24–28, Snowmass.
6. Matsushita T, Ni B, Yamafuji K, Watanabe K, Noto K, Morita H, Fujimori H and Muto Y (Springer–Verlag, Tokyo, 1989) Adv. Superconductivity: P.393.
7. Ni B, Munakata T, Matsushita T, Iwakuma M, Funaki K, Takeo M and Yamafuji K (1989) Jpn. J. Appl. Phys. 27: L1658.
8. Sato K, Hikata T and Iwasa Y (1990) Appl. Phys. Lett. 57: 1928.
9. Matsushita T, Otabe ES, Ni B, Kimura K, Morita M, Tanaka M, Kimura M, Miyamoto K and Sawano K (1990) US–Japan Workshop on Superconductors, Sept. 20–21, Buffalo.
10. Hikata T, Sato K and Iwasa Y (private communications).
11. Hikata T, Ueyama M, Mukai H and Sato K (in press) Cryogenics.

J_c and J_c-B Enhancement with Micro Structure Improvement of Bi-Based High-T_c Superconductor

TAKESHI HIKATA, MUNETSUGU UEYAMA, HIDEHITO MUKAI, NOBUHIRO SHIBUTA, TAKESHI KATO, and KEN-ICHI SATO

Osaka Research Laboratories, Sumitomo Electric Industries, Ltd., 1-3, Shimaya 1-chome, Konohana-ku, Osaka, 554 Japan

ABSTRACT

BiPbSrCaCuO silver sheathed wires were produced by solid reaction method. Maximum J_C at 77.3K were 4.7x10^4 A/cm^2 in a zero magnetic filed and 1.1x10^4 A/cm^2 at 1 Tesla. It was found that the improvement of magnetic filed dependence of J_C had strong relations with the size of non–superconducting phases and the c–axis component of transport current across the grain boundaries. At 4.2 K, it was revealed that maximum J_C was 1.03x10^5 A/cm^2 at 23Tesla and the anisotropy was small even at 23T.

KEY WORD: Bi–based superconductor, silver–sheathed wire, critical current density, weak links irreversibility line

INTRODUCTION

High–T_C superconductor, BiPbSrCaCuO [1–3] has some excellent features for applications of superconducting wires. Critical current temperature (T_C) is high (110K) and the superconducting devices can be operated with large temperature margin from 77.3K. Superconducting grains are aligned easily along c–axis and bonded strongly by a combination of plastic deformation and heat treatment. An improvement of critical current density (J_C) in the magnetic field is important for actual applications. Superconducting cables are used mainly around 0.1Tesla, and superconducting magnets are used at higher magnetic field. Therefore it is required that J_C enhanced in each magnetic field. In this paper, we report the relation between the enhancement of J_C and the micro structures of BiPbSrCaCuO high–T_C superconductor in the silver sheathed wires.

EXPERIMENTAL

Appropriate amount of Bi_2O_3, PbO, $SrCO_3$, $CaCO_3$ and CuO powders with 3N to 4N purity were mixed, calcined and ground. The ground powder was put into silver tubes. The composites were drawn to round wires and rolled or pressed to tape wires. The typical demensions of the tape wires were 0.1–0.2 mm in thickness and 2.5–4.5mm in width. Typical T_C was 106 K [3–10]. Those deformation and sintering process were repeated twice. The first sintering is for the homogenization of high–T_C phase (110K) and the second sintering is for good contact at the grain boundaries [5]. The critical current was measured by 4 probe dc method. The magnetic field was applied parallel and perpendicular to the tape surface and perpendicular to the transport current in temperature ranges from 4.2K to 77K. The hybrid magnet was used to apply the magnetic field up to 23 Tesla at HFLSM, Tohoku University. The micro structure of high–T_C superconductors of the polished cross section of the wires were observed through SEM and EDX. Superconducting and non–superconducting phases were distinguished by using EDX.

RESULTS AND DISCUSSION

Through the improvement of materials and fabricating process, maximum J_C achieved was 4.7 x 10^4 A/cm^2 at 77.3K and a zero magnetic field. This wire exhibited J_C=1.1x10^4 A/cm^2 at 1 Tesla as shown in Fig. 1.

It is shown that the J_C of BiPbSrCaCuO silver-sheathed wire is close for the useful level to the cables and magnets. For actual applications, J_C was required to be over 10^4 A/cm^2 at least in each operating magnetic field. Figure 2 shows the magnetic field dependence of pinning force (J_C x B) for wires having each levels of J_C ,(a) 47,000A/cm^2, (b) 18,000 A/cm^2 and (c) 5,500 A/cm^2. Pinning force was improved in whole range up to 3 T with the increase of J_C in a zero magnetic field. The magnetic field at which the pinning force is maximum value, does not change. Figure 3 shows the pinning force normalized with maximum pinning force value. The magnetic field dependence of the pinning force was improved especially large in the high magnetic field region. This means that the improvement of the pinning force was caused with the improvement of irreversibility line with the improvement of J_C in a zero magnetic field. Figure 4 shows the SEM photographs of superconducting section of wires having J_C of (a) 5,500 A/cm^2, (b) 10,600 A/cm^2 and (c) 47,000 A/cm^2. In sample (a) with low J_C, three kinds of non-superconducting phases, such as phase I (SrCaCuO), phase II ($(Ca,Sr)_2PbO_4$) and phase III (Ca_2CuO_3), were found with large granular shape. In sample (b), phase I disappeared where phase II became small and thin shape. However, phase III remained with the large size. In sample (c) with high J_C, all non-superconducting phases were dispersed in fine and thin shape. The volume ratio of non-superconducting phases was estimated by EDX to be 30% for sample(a), 20% for sample(b) and 8% for sample(c) versus the volume of high-T_C phase. The improvement of J_C-B properties of the wires in Fig.2 have strong relation with these improvement of morphology as follows. In low J_C wire such as sample (a), the transport current flows percolatively and has the large component along c-axis around large sized non superconducting phases as shown in Fig. 4(a). When the magnetic field is applied parallel to the tape plane, the transport current along c-axis and the magnetic field were acted reciprocally generating Lorentz force which works along the grain boundaries perpendicular to c-axis. Figure 5 shows the temperature dependence of an irreversiblity line which was determined by the criterion of temperature with zero resistivity when J = 90 A/cm^2. The temperature dependence was non-linear and dH_C/dT increased with the decrease of temperature. This exhibits that the H_C was influenced in the weak links at grain boundaries. The pinning force has anisotropy and weak for the direction perpendicular to c-axis, that is parallel to grain boundaries. When the Lorentz force is acted

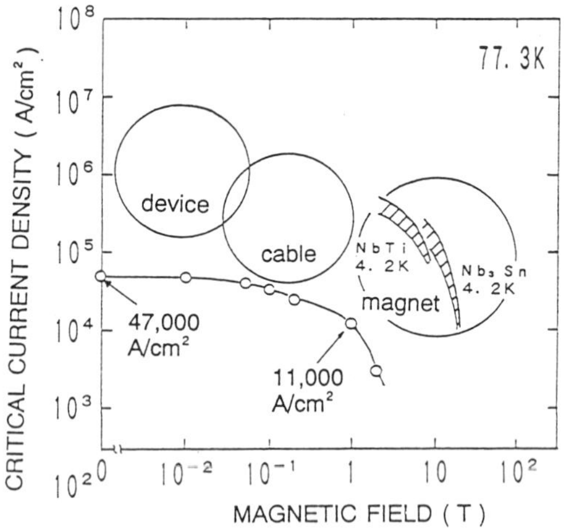

Fig. 1 Magnetic field dependence of J_C of BiPbSrCaCuO Ag-sheathed wire.

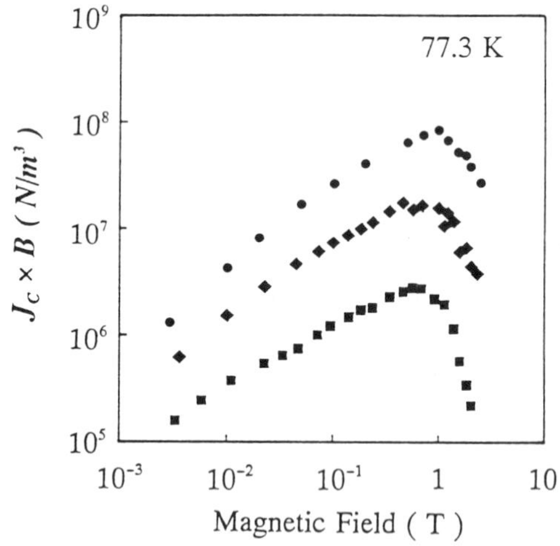

Fig. 2 Magnetic field dependence of pinning force for sample (a), (b) and (c).

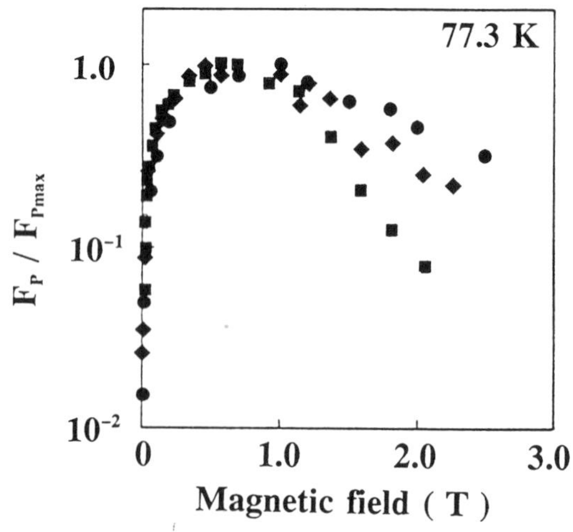

Fig. 3 Magnetic field dependence of normalized pinning force for sample (a), (b) and (c).

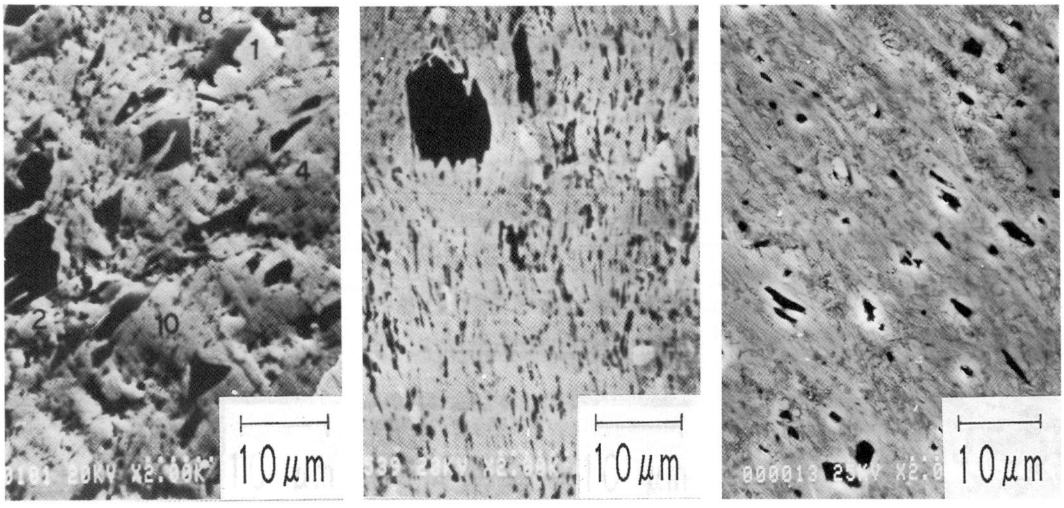

Fig. 4 SEM photographs of polished surface for sample (a) 5,500 A/cm^2, (b) 10,600 A/cm^2 and (c) 47,000 A/cm^2.

locally for the current crossing grain boundaries, the flux is slipped along those at 77.3 K. It is considered that these phenomena is related with the component of transport current density along c–axis and the area size. Therefore by reducing the size of non–superconducting phases, the transport current density along c–axis becomes relatively small and theLorentz force becomes small.

Figure 6 shows the magnetic field dependence of J_C at 4.2 K for tape wires having J_C of $1.4x10^4$ to $4.4x10^4$ A/cm^2 in a zero magnetic field and at 77.3K. The magnetic filed was applied parallel to the tape plane up to 23 Tesla. J_C changed little in the magnetic field from 5 to 23 Tesla. The J_C at 4.2K was also increased with the improvement of J_C at 77.3 K. The wire with J_C(77.3K)=$4.4x10^4$ A/cm^2 has a J_C of $1.03x10^5$ A/cm^2 at 23 Tesla. Figure 7 shows the magnetic field direction dependence of the wires with J_C(77.3K)=$2.8x10^4$ A/cm^2. When the magnetic field was applied parallel to the tape plane, J_C(4.2K,23T) was $4.6x10^4$ A/cm^2. When the magnetic field was applied perpendicular to the plane, J_C(4.2K,23T) was $3.5x10^4$ A/cm^2. The difference was about only 30%. These properties are much more excellent than the

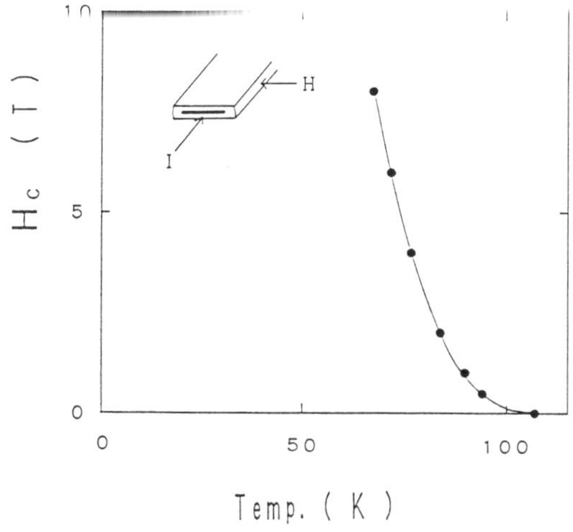

Fig. 5 Temperature dependence of an irreversility line when J=90 A/cm^2 for the sample with J_C(77.3K) =30,000 A/cm^2.

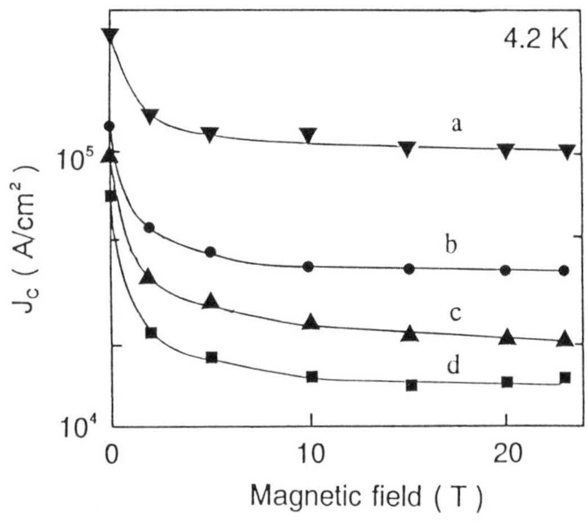

Fig. 6 Magnetic field dependence of J_C at 4.2K for samples a (J_C(77.3K)=47,000A/cm^2), b (25,000 A/cm^2), c (18,000 A/cm^2) and d (14,000 A/cm^2).

metallic superconductors such as Nb_3Sn because J_C of the metallic superconductor becomes smaller around 20 Tesla. Therefore it was found that BiPbSrCaCuO sheathed wires had a sufficient high potential for the application in super high magnetic field over 20 Tesla at 4.2K.

CONCLUSION

High J_C BiPbSrCaCuO silver sheathed superconducting wires were produced through the solid reaction method. The maximum J_C achieved $4.7x10^4$ A/cm^2 at 77.3K and a zero magnetic field. The magnetic field dependence of J_C was improved with the increase of J_C in a zero magnetic field with the fine dispersion of non-superconducting phases. At 1 Tesla J_C=$1.1x10^4$ A/cm^2 was obtained. The improvement of J_C–B properties were explained through the suppression of local Lorentz force perpendicular to c-axis between transport current component along c-axis and the magnetic field when the field was applied parallel to the tape plane. At 4.2K high J_C over 10^5 A/cm^2 was obtained at 23 Tesla. The anisotropy for applied magnetic fields was small.

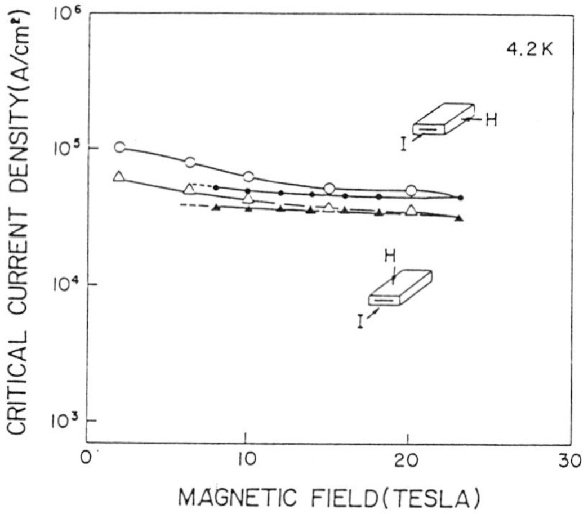

Fig. 7 J_C–B properties at 4.2K for magnetic field applied parallel and perpendicular to the tape plane.

ACKNOWLEDGMENTS

We express the appreciation to Professor K. Watanabe and his co-workers of HFLSM at Tohoku University for their assistance and advice in critical current measurements up to 23Tesla. We also express the appreciation to Professor Y. Iwasa, Dr. L. Rubin and their co-workers of FBNML at MIT for their collaborative research on stability of high-T_C superconducting wires.

REFERENCES

1. Maeda H, Tanaka Y, Fukutomi M and Asano T (1988) Jpn. J. Appl. Phys. 27: L209.
2. Takano M, Takada J, Oda K, Kitaguchi H, Miura Y, Ikeda Y, Tomii and Y, Mazaki H (1988) Jpn. J. Appl. Phys. 27: L1041.
3. Hikata T, Sato K, Hitotsuyanagi H (1989) Jpn. J. Appl. Phy. 28: L82.
4. Sato K, Hikata T, Mukai and M, Hitotsuyanagi H (1989) Proc. of ISTEC Workshop on Superconductivity, Oiso (ISTEC), February 1–3: 119
5. Hikata T, Nishikawa T, Mukai M and Hitotsuyanagi H (1989) Jpn. J. Appl. Phys. 28: L1024.
6. Sato K, Mukai H and Hikata T (1989) submitted to Proc. of Int. Atomic Energy Agency Technical Committee Meeting on Superconducting Materials and Magnets, Tokyo, September 4–6.
7. Sato K, Hikata T, Mukai H, Masuda T, Ueyama M, Hitotsuyanagi H, Mitsui and T, Kawashima M (1990) Proc. 2nd Int. Symposium on Superconductivity (ISS'89), Tsukuba(Springer-Verlag): 335–340.
8. Hikata T, Ueyama M, Mukai and H, Sato K: Cryogenics in Press.
9. Sato K, Hikata T and Iwasa Y (1990) Appl. Phys. Lett. 57: 1928.
10. Sato K, Hikata T, Mukai H, Ueyama M, Shibuta N, Kato T, Masuda T, Nagata M, Iwata K and Mitsui T (1990) Applied Superconductivity Conference, Sep. 24–28, Snowmass.

Magnetization Properties of YBCO Prepared by MPMG Process

HIROYUKI FUJIMOTO, MASATO MURAKAMI, SATOSHI GOTOH, and NAOKI KOSHIZUKA

Superconductivity Research Laboratory, International Superconductivity Technology Center,
10-13, Shinonome 1-chome, Koto-ku, Tokyo, 135 Japan

ABSTRACT

Magnetization properties of YBaCuO samples prepared by melt-powder-melt-growth (MPMG) process are different from those of the sintered samples, because the MPMG processed sample does not have weak-links and the shielding current flows around the whole sample. Anisotropy of Jc was estimated from M-H hysteresis curves, using Bean's model. As a result, the MPMG processed YBaCuO sample has no weak-links, a large pinning force, and a small anisotropy of close to 3:1.

KEY WORDS: $YBa_2Cu_3O_x$, Y_2BaCuO_5, magnetization, melt process, critical current density, anisotropy.

INTRODUCTION

Since the Y-Ba-Cu-O superconductor with a critical temperature above the boiling point of liquid nitrogen was discovered in 1987 [1], considerable efforts have been made to increase the critical current density, Jc, which is important for applications, and Jc has been increased remarkably. Samples prepared by melt processes such as MTG [2,3], QMG [4~7] and MPMG [8] have large critical current densities exceeding 10^4 A/cm^2 at 77 K due to grain alignment. However, the introduction of additional pinning centers is needed to improve Jc.

In the MPMG process, it is possible to introduce fine 211 phase particles into the 123 superconducting matrix. The average size of the 211 particles is around 1 µm. We believe that the 211 particles can contribute to flux pinning, which is supported by the fact that the larger the volume fraction of the 211 phase, the higher the critical current density. A Jc of 3×10^4 A/cm^2 is achieved for the sample with 211 precipitates, and 4×10^3 A/cm^2 for that without 211 precipitates at 77K and 1T, which are calculated from magnetization hysteresis curves based on Bean's model [9].

Large anisotropy of Jc due to crystal structure, coherence length and penetration depth has been reported [10], but the anisotropy is different between samples because of the structural sensitivity of Jc. In this paper, the anisotropy of Jc for the MPMG processed samples is investigated using magnetization measurements.

EXPERIMENTAL

Samples were prepared by the melt-powder-melt-growth (MPMG) process [8]. Appropriate amounts of mixtures of Y_2O_3, $BaCO_3$ and CuO were calcined at 900 °C; then the powders were melted at over 1200 °C and rapidly cooled down to room temperature. After the melted oxides were pulverized and pressed into pellets, the samples were reheated to 1100 °C and cooled slowly to room temperature. At the final stage, the samples were annealed at 600 °C and cooled down to room temperature in flowing oxygen. Starting compositions were $Y_{1.0}Ba_{2.0}Cu_{3.0}O_{7.0-x}$, $Y_{1.2}Ba_{2.1}Cu_{3.1}O_{7.0-x}$, $Y_{1.3}Ba_{2.15}Cu_{3.15}O_{7.0-x}$ and $Y_{1.5}Ba_{2.25}Cu_{3.25}O_{7.0-x}$.

Magnetization properties were measured using a vibrating sample magnetometer (VSM) in magnetic fields between -15000 and 15000 Oe at 77K.

RESULTS AND DISCUSSION

Figure 1 shows schematic illustrations of the superconducting shielding currents. In the case of Fig. 1 (a), the angle between the c-axis and the direction of the applied field is varied from 0 to 90 degree. In the case of (A) of Fig. 1 (a), the magnetic field is applied parallel to the c-axis, and for the case (B) of Fig. 1 (a), the applied field is perpendicular to the c-axis. Magnetic flux penetrates into the sample from all the edges of the cross section of the sample based on Bean's model. For the first case, Jc within the ab plane is estimated at the largest penetrated region: the direction of critical current is along the a or b axis. For the latter case, Jc within the plane perpendicular to the ab plane is estimated at the largest penetrated region: the direction of critical current is along the c-axis. In the case of Fig. 1 (b), the direction of the applied field is always perpendicular to the c-axis. For both the cases shown in Fig. 1 (b) (A) and (B), the applied field is perpendicular to the c-axis and Jc within the ab plane is estimated at the largest penetrated region: the direction of critical current is along the a or b axis.

The results of M-H loop measurements for the sample $Y_{1.5}Ba_{2.25}Cu_{3.25}O_{7.0-x}$ are shown in Fig. 2. The sample shape and the direction of the applied field is the same as that shown in Fig. 1 (a). The M-H loop shown in Fig. 2 (a) is for the applied field parallel to the c-axis, and Fig. 2 (b) for the applied field perpendicular to the c-axis. The ratio of anisotropy of Jc

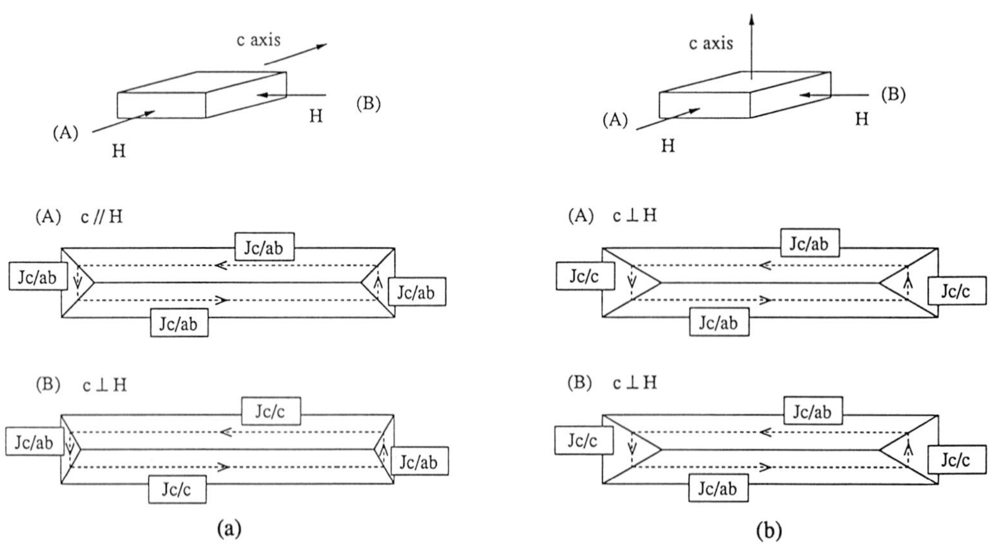

Fig. 1 Schematic illustrations of shielding currents. (a) The external field is perpendicular and parallel to the c-axis, (b) the external field is perpendicular to the c-axis.

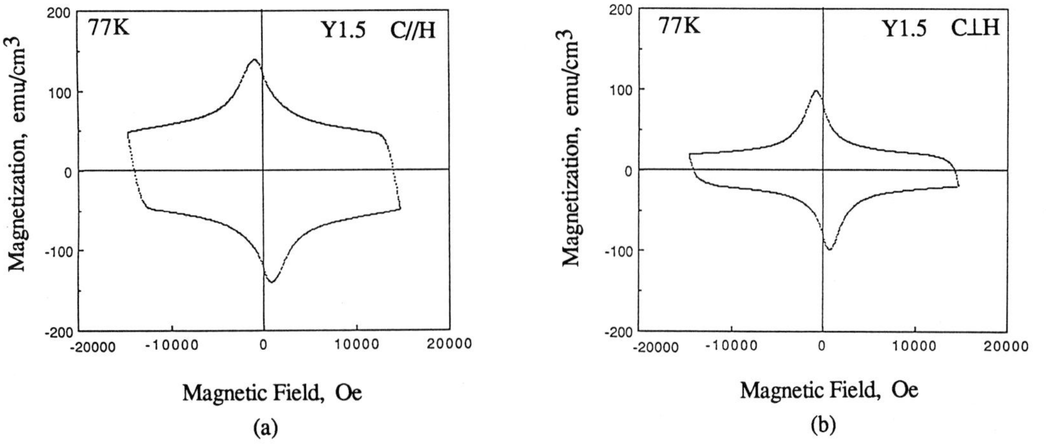

Fig. 2 M-H loop for $Y_{1.5}$ sample with (a) the c-axis parallel to the magnetic field H and (b) the c-axis perpendicular to the magnetic field H.

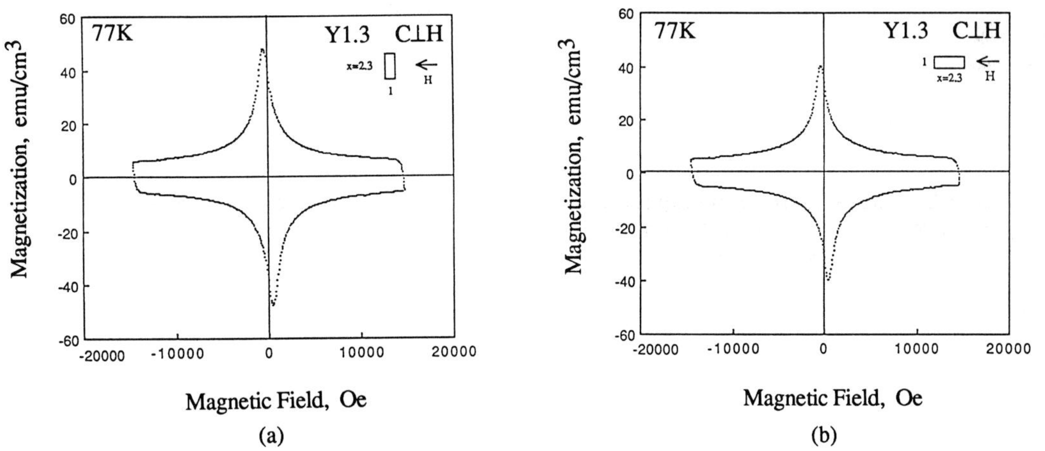

Fig. 3 M-H loop for $Y_{1.3}$ sample with the c-axis perpendicular to the magnetic field H. The area of the cross sections of the sample are different between (a) and (b).

for the applied field parallel to the c-axis, to that for the applied field perpendicular to the c-axis was close to 3:1 at 77K. For the case of Fig. 1(a)(A) and Fig. 2(a), when the c-axis is parallel to the applied field, Jc within the ab plane is estimated, and for the case of Fig. 1(a)(B) and Fig. 2(b), when the c-axis is perpendicular to the applied field, Jc within the plane perpendicular to the ab plane is estimated. This anisotropy of close to 3:1 is equivalent to that of Jc in the direction of the a or b axis and the c-axis. Based on the direct summation model, Jc is considered to be proportional to the elementary pinning force f_p which is given by [11],

$$f_p = U_p/2\xi = (1/2\mu_0) B_c^2 \pi \xi^2 d/2\xi,$$

where U_p is the elementary pinning energy, ξ is the Ginzburg-Landau coherence length, B_c is the thermodynamical critical magnetic induction, and d is the size of non-superconducting phase (211 inclusion). When the applied field is parallel to the c-axis,

$$f_p = U_p/2\xi_{ab} = (1/2\mu_0)B_c^2\pi\xi_{ab}^2 d/2\xi_{ab} = (1/4\mu_0)B_c^2\pi d\xi_{ab},$$

and when the applied field is perpendicular to the c-axis,

$$f_p = U_p/2\xi_{ab} = (1/2\mu_0)B_c^2\pi\xi_{ab}\xi_c d/2\xi_{ab} = (1/4\mu_0)B_c^2\pi d\xi_c.$$

Therefore, Jc(the a or b axis direction):Jc(the c-axis direction) is equal to $\xi_{ab}:\xi_c$. At 77K $\xi_{ab}:\xi_c$ = 3.7:1 has been reported [12], so the anisotropy of Jc can be attributed to the anisotropy of these parameters.

When the sample geometry and the direction of the applied field is the same as that shown in Fig. 1 (b), the results of the M-H loop measurements are shown in Fig. 3. The sample was $Y_{1.3}Ba_{2.15}Cu_{3.15}O_{7.0-x}$. The applied field was perpendicular to the c-axis in both Fig. 3(a) and 3(b). The aspect ratio of length to width was 2.3. The shapes and amplitudes of the M-H loop for both cases (a) and (b) were about the same at 77K : the Jc values are therefore also about the same. The illustration of Fig. 1 (b) explains the results, because Jc within the ab plane is estimated for both the cases shown in Fig. 3(a) and 3(b).

CONCLUSIONS

The effect of sample geometry, the direction of the sample axis, and the direction of the applied field on the magnetization hysteresis were investigated. There are two cases that Jc in the direction of both the a or b axis and the c-axis are measured and that Jc in the direction of only the a or b axis are measured. The anisotropy of Jc along either the a or b direction and along the c direction was close to 3:1 for the MPMG processed YBaCuO, as calculated from the M-H loops using Bean's model.

ACKNOWLEDGEMENTS

The authors would like to thank Mr.T.Oyama (SRL,ISTEC) for sample preparation.

REFERENCES

1) M.K.Wu, J.R.Ashburn, C.J.Torng, P.H.Hor, R.L.Meng, L.Gao, Z.J.Huang, Y.Q.Wang and C.W.Chu : Phys.Rev.Lett., **58** (1987) 908.
2) S.Jin, T.H.Tiefel, R.C.Sherwood, R.B.van Dover, M.E.Davis, G.W.Kammlott and R.A.Fastnacht : Phys.Rev.B, **37** (1988) 7850.
3) S.Jin, T.H.Tiefel, R.C.Sherwood, M.E.Davis, R.B.van Dover, G.W.Kammlott, R.A.Fastnacht and H.D.Keith : Appl.Phys.Lett., **52** (1988) 2074.
4) M.Murakami, M.Morita, K.Doi, K.Miyamoto and H.Hamada : Jpn.J.Appl.Phys., **28** (1989) L399.
5) M.Murakami, M.Morita, K.Doi and K.Miyamoto : Jpn.J.Appl.Phys., **28** (1989) 1189.
6) M.Murakami, M.Morita and K.Miyamoto : Proc. Osaka Univ. Int. Symp. on New Development in Applied Superconductivity, Suita, 1988 (World Scientific, Singapore, 1989) p.95.
7) M.Murakami and M.Morita : Proc. Mater. Res. Soc. Meeting, 1988 (MRS, Tokyo, 1989) p.641.
8) H.Fujimoto, M.Murakami, S.Gotoh, N.Koshizuka, T.Oyama, Y.Shiohara and S.Tanaka : Advances in Superconductivity II (Springer-Verlag, Tokyo, 1990) p.285.
9) C.P.Bean : Phys.Rev.Lett., **8** (1962) 250.
10) T.R.Dinger, T.K.Worthington, W.J.Gallagher and R.L.Sandstrom : Phys.Rev.Lett., **58** (1987) 2687.
11) M.Murakami, S.Gotoh, N.Koshizuka, S.Tanaka, T.Matsushita, S.Kambe and K.Kitazawa : Cryogenics **30** (1990) 390.
12) Y.Iye, T.Tamegai, H.Takeya and H.Takei : Jpn.J.Appl.Phys., **26** (1987) L1507.

Enhancement of Critical Current Density of $Ba_2YCu_3O_y$ by 28 MeV Electron Irradiation

TAKAYUKI TERAI[1], TAMAKI MASEGI[1], YOICHI TAKAHASHI[1], KOHJI KISHIO[1], KOICHI NAKATANI[1], YOUICHI ENOMOTO[2], and SHUGO KUBO[2]

[1] Faculty of Engineering, The University of Tokyo, Hongo, Bunkyo-ku, Tokyo, 113 Japan
[2] NTT Applied Electronics Laboratories, Tokai, Ibaraki, 319-11 Japan

ABSTRACT

Effect of 28 MeV electron irradiation was investigated on the critical current density (J_c) of high-T_c superconductor $Ba_2YCu_3O_y$. For the thin film sample irradiated with a fluence of 4.7×10^{16} electrons/cm^2, transport J_c determined from I-V characteristics was enhanced by about 8% at temperatures from 75 K to 85 K without magnetic field. For the single-phased poly-crystalline sample irradiated with a fluence of 8.8×10^{17}electrons/cm^2, intragranular J_c determined from magnetization measurement was also enhanced by 1.4-4.5 times at temperature from 4.2 K to 77 K in magnetic field up to 9 T. The difference in the enhancement ratio of J_{cM} and J_{cT} can be attributed to the effect of grain boundary. On the contrary, the critical temperature (T_c) was unchanged in the fluence range investigated.

KEYWORDS: $Ba_2YCu_3O_y$, electron irradiation, critical current density, pinning center

INTRODUCTION

Since the discovery of superconductivity above 90 K in the Y-Ba-Cu-O system, it has been pointed out that one of the most critical problems for the application of this new material is its low critical current density (J_c). Thus, the enhancement of J_c of this compound is an important subject to be studied. For this purpose, the introduction of intragranular microprecipitates and twin boundaries to crystal grains has been examined as to whether or not they can act as pinning centers against flux motion. Radiation defects generated by irradiation are also considered to act as pinning centers. From this point of view, irradiation effects of $Ba_2YCu_3O_y$ (BYCO) have been studied by many groups. These studies showed that in single crystal or polycrystal BYCO, J_c values were considerably enhanced due to irradiation of neutron and charged particles[1-9]. In these studies,most of the J_c values were deduced from the magnetization hysteresis curves. On the other hand, in the case of sintered samples, the transport J_c (J_{cT}) values directly measured were reported to degrade in spite of the enhancement of the magnetization J_c (J_{cM}) value[10]. The reason for the difference in J_c behavior between these J_c values was considered to be attributable to the defects at intergrain regions generated by irradiation. Thus, it has been important to distinguish the radiation effect on crystal grains from that on grain boundaries. We have been studying the effect of irradiations of charged particles on c-axis-oriented BYCO thin films, in which the effect of grain boundary is much smaller than that in a polycrystalline sample with no grain orientation, and reported previously the enhancement of transport J_c values [11,12]. In this paper, we describe experimental results on the changes of T_c and J_c values by the irradiation of 28 MeV electrons on both c-axis-oriented BYCO thin films and a BYCO sintered sample to compare with each other.

EXPERIMENTAL

The BYCO thin films were grown on $SrTiO_3$ (100) substrates by the activated reactive evaporation method from metals of Y, Ba and Cu under the following conditions : substrate temperature;650 °C, O_2 flow rate ; 20 cm^3 /min, rf power ; 120 W , and typical total deposition rate ; 0.4nm/s. The thicknesses of these films were 300 - 700 nm. These films consisted of many grains of 0.1 mm order in diameter, and their c-axes were oriented perpendicular to $SrTiO_3$ (100) planes. Their c-axis lattice constant was 1.172 nm , and the resistivities were about $4\times10^{-4}\Omega$cm at 300 K. The J_c values were in the order of $10^5 A/cm^2$ at 77 K, and the transition widths of the critical temperature were about 5 K without magnetic field. A single phased poly-crystalline sample was prepared by the conventional sintering method. The sintered pellet was cut off to the rectangular size of $2.8 \times 3.1 \times 1.6$ mm^3.

The irradiation of 28 MeV electrons was performed with a linear accelerator (LINAC) at the University of Tokyo at room temperature in air. Bunched electron pulses were irradiated on to the samples frequently through a beam slit with a hole of 3-5 mm in diameter. The width and the frequency of the electron pulse were 2 nsec and 50 Hz, respectively. The averaged electron-beam current was about 1.6 mA/cm^2, and the electron fluence ranged from 1.2×10^{15}-8.8×10^{17} $electrons/cm^2$. Because the penetration depth of the irradiated electrons is much larger than the thickness of the samples, radiation-induced defects distribute almost entirely over the films. No significant temperature increase of the sample films was observed during irradiation.

Before and after the electron irradiation, temperature-dependent *I-V* characteristics of these films were measured at temperatures from 75 to 85 K. In the *I-V* measurements, the conventional four-probe method was used with 100-200 Hz sine-wave current. Prior to the measurement, no significant influence of frequency was confirmed. The films were etched chemically to form a conduction path of 240 μm long and 30 μm wide in order to reduce the total current. The temperature of the film was sensed using a calibrated Au/Fe-chromel thermocouple. The values of J_c were defined as the current at the voltage of 10 μV/cm, and those of T_c (which means the end-point temperature) were defined as the temperature at which the critical current became less than 10^{-5} A. No significant differences in the values of J_c and T_c between warming and cooling the sample were observed within the accuracy of $\pm$ 2% and $\pm$ 0.5 K, respectively. Magnetization measurement was performed on the sintered sample with a vibrating sample magnetometer (VSM) at temperatures from 4.2 K to 77 K before and after the electron irradiation. Maximum applied magnetic field was 9 T and the sweep rate was 8 mT/s or 20 mT/s.

RESULTS AND DISCUSSION

Figure 1 shows the experimentally determined J_c values of the film irradiated with a fluence of 4.7×10^{16} $electrons/cm^2$. As shown in this figure, at temperatures below 85 K, the J_c values after the irradiation were enhanced by about 8%. On the other hand, T_c value measured in this experiment (which is equivalent to the value of T_c(end)) was found to be not affected within the accuracy of 0.5 K. From the fact that T_c(end) is generally more sensitive to change by irradiation than T_c(onset) and T_c(mid point), neither of these T_c values were thought to be changed by irradiation in this fluence. The J_c -vs-T curve after the irradiation rose steeper with decreasing temperature than that before the

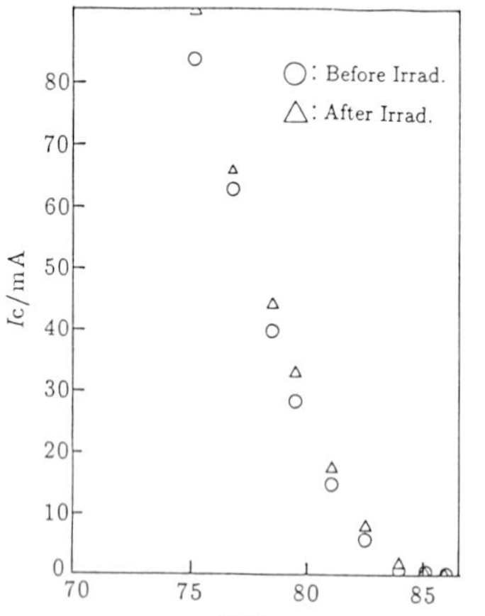

Fig.1 Effect of 28 MeV electron irradiation with a fluence of $4.7 \times 10^{16} e^-/cm^2$ on the transport I_c of BYCO thin film[13].

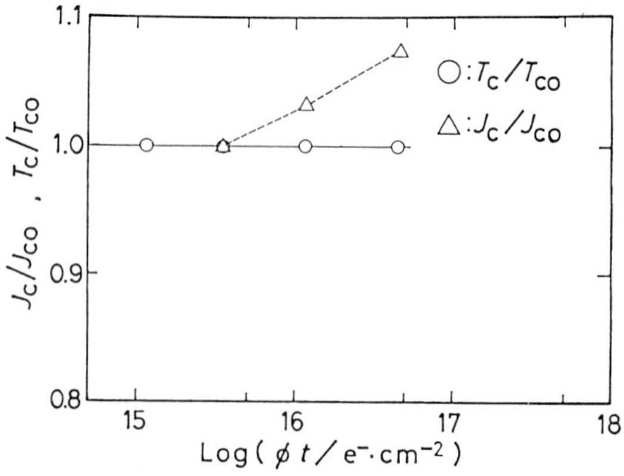

Fig.2 T_c/T_{c0} and transport J_c/J_{c0} vs. 28 MeV electron fluence[13].

irradiation. For all the samples investigated, the J_c values were represented as $J_c = A(1-T/T_c)^2$, where A is a parameter. This temperature dependence was not affected by irradiation, but the value of A changed due to irradiation. The similar irradiation effect was observed in the case of O^+ ions impinged to BYCO thin films, as reported in our previous paper[11].

The changes of T_c and J_c due to the irradiation of 28 MeV electrons are summarized in Fig. 2, where T_{c0} and T_c represent the critical temperature before and after the irradiation, respectively, and J_c and J_{c0} values are the critical current densities at the temperature of $0.9T_{c0}$. In this fluence range investigated, T_c did not change and J_c was enhanced at fluences above 1.2×10^{16} electrons/cm^2. Considering that atmospheric degradation occurs easily in BYCO thin films and that radiation damage of the grain boundary might degrade transport supercurrents, the observed enhancement of the directly measured J_c values in this experiment strongly suggests that defects formed by electron irradiation act as pinning centers.

J_{cT} seems to increase with increasing fluence above 1.2×10^{16} electrons/cm^2. However, the value of J_c/J_{c0} should approach zero in the region of very high fluence, as observed in the case of O^+ irradiation[11,12]. Shiraishi et al. also reported that on 3 MeV electron irradiation to BYCO sintered pellets, J_c obtained from magnetization curve has the maximum value at 1.5×10^{18} electrons/cm^2[7]. Therefore, it is natural that J_c/J_{c0} should have the maximum value at a higher fluence than that examined in this study. To decide the most effective fluence for J_c increase is an important subject to be studied.

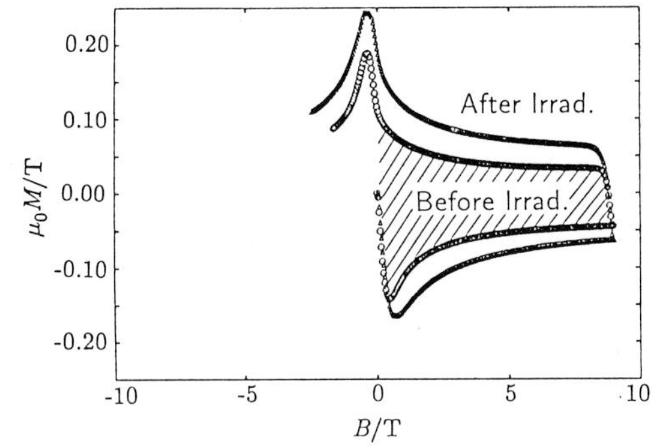

Fig.3 Magnetization curves at 4.2 K before and after the electron irradiation with a fluence of 8.8×10^{17}e$^-$/cm^2.

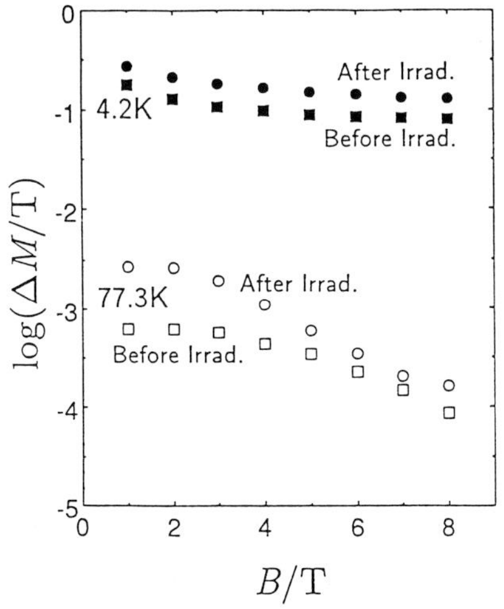

Fig.4 Magnetic field dependence of ΔM before and after the 28 MeV electron irradiation with a fluence of 8.8×10^{17}e$^-$/cm^2.

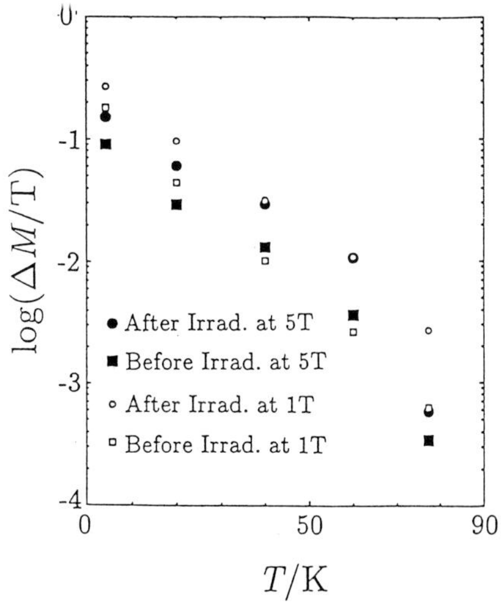

Fig.5 Temperature dependence of ΔM before and after the 28 MeV electron irradiation with a fluence of 8.8×10^{17} e$^-$/cm^2.

Magnetization curves before and after irradiation at 4.2 K are shown in Fig. 3. The magnetization hysteresis (ΔM) increased by the electron irradiation, which suggests that irradiation-induced defects act as pinning centers. From the Bean-model, the critical current density (J_{cM}) is proportional to ΔM, as follows:

$$J_{cM}\ (\mathrm{A/m^2}) = 2\times\Delta M(\mathrm{T})/(\mu_0\ (\mathrm{H/m})\cdot d(\mathrm{m})),$$

where d represents the thickness of the sample perpendicular to the applied magnetic field. In this study, since the irradiated sample was poly-crystalline, the loop of the critical magnetization current might be shielded by its grain boundaries. Thus, estimation of J_{cM} has some difficulties, and the irradiation effects was analyzed only on magnetization hysteresis in this paper.

Figure 4 shows the dependence of ΔM on applied magnetic field at 4.2 K and 77 K before and after the electron irradiation. As shown in this figure, magnetization hysteresis increased at each applied magnetic field up to 9 T. The slope became slower with magnetic field at 4.2 K, while it becomes steeper with magnetic field at 77 K. Figure 5 shows the temperature dependence of ΔM at 1 T and 5 T before and after the electron irradiation. As shown in this figure, magnetization hysteresis increased at each temperature, and the enhancement ratio was larger at higher temperature. This result suggests that pinning-centers induced by irradiation may be more effective than other defects. In other words, irradiation-induced defects might have a pin-potential deeper than that of other pinning-centers. It is reported that in high-T_c superconductor, critical current density decreases drastically at high temperatures. This result suggests that the modification of critical current density by irradiation has a bright prospect to enhance a critical current density at relatively high temperatures.

The enhancement ratio of J_{cM} due to the irradiation was 1.4-4.5 times at 77 K, which was much larger than that of J_{cT}. The fluence was somewhat larger in the sintered pellet than that in the thin films, and a larger enhancement ratio of J_{cT} would be expected at the same fluence as the sintered sample. Nevertheless, the difference between the enhancement ratios in J_{cT} and J_{cM} is too large to be attributed to the difference in fluence. The reason of the difference is considered to be the existence of grain boundaries in the thin film samples. Transport critical current is controlled by some weaklinks at grain boundaries as well as those in crystal grain. Irradiation decreases the critical current density at grain boundaries and increase the one in crystal grain at the same time, and as a result of these two effects the apparent transport J_{cT} was observed to increase by 8 %. This result is somewhat different from the fact that J_{cT} decreased by irradiation , reported in the literature[10]. This shows that the structure of the grain boundaries is much better in these thin films than that in the literature.

REFERENCES

1) Umezawa A, Crabtree GW, Liu JZ, Weber HW, Kwok WK, Nunez LH, Moran TJ, Sowers CH, Claus CH (1987) Phys Rev B36: 7151

2) Cost JR, Willis JO, Thompson JD, Peterson DE (1988) Phys Rev B37: 1563

3) Clark GJ, Marwick AD, Koch RH, Laibowitz RB (1987) Appl Phys Lett 51: 139

4) Wisniewski A, Baran M, Przyslupski P, Szymczak H, Pajaczkowska A, Pytel B, Pytel K (1988) Solid State Commun 65: 577

5) van Dover RB, Gyorgy EM, Schneemeyer LF, Mitchell JW, Rao KV, Puzniak R, Waszczak JV (1989) Nature 342: 55

6) Kato T, Shiraishi K, Kuniya J (1989) Jpn J Appl Phys 28: L766

7) Shiraishi K, Kato T, Kuniya J (1989) Jpn J Appl Phys 28: L807

8) Müller P, Gerstenberg H, Fisher M, Schindler W, Strobel J, Saemann-Ischenko G, Kammermeier H (1989) Solid State Commun 65: 223

9) Küpfer H, Apfelstedt I, Schuer W,Flugiger R, Meier-Hirmer R, Wuhl H, Scheurer H (1987) Z Phys B69: 167

10) Przyslupski P, Wisniewski A, Kolesnik S, Dobrowolski W, Pajaczkowska A, Pytel K, Pytel B (1988) Physica C 153-155: 345

11) Masegi T, Terai T, Takahashi Y, Enomoto Y, Kubo S (1989) Jpn J Appl Phys 28: L1521

12) Terai T, Masegi T, Takahashi Y, Enomoto Y, Kubo S (1990) Mol Cryst Liq Cryst 184: 165

13) Terai T, Masegi T, Takahashi Y, Enomoto Y, Kubo S (1990) Jpn J Appl Phys 29: in press.

Changes of Superconductivity and Crystal Structure on $Ba_2YCu_3O_y$ by Ion Irradiation

TAKAYUKI TERAI[1], TAI FURUTA[1], TAMAKI MASEGI[1], KAZUYUKI KUSAGAYA[1], YOICHI TAKAHASHI[1], YOUICHI ENOMOTO[2], and SHUGO KUBO[2]

[1] Department of Nuclear Engineering, Faculty of Engineering, University of Tokyo, Hongo, Bunkyo-ku, Tokyo, 113 Japan
[2] NTT Applied Electronics Laboratories, Tokai, Ibaraki, 319-11 Japan

ABSTRACT

Effect of 1 MeV Ar^+ion irradiation as well as 0.5 MeV He^+ ion irradiation was investigated on thin films of high-T_c oxide superconductor $Ba_2YCu_3O_y$ (BYCO) . Increase of transport critical current density (J_{cT}) by irradiation was observed with a fluence below 1.3×10^{11} Ar^+/cm^2, while the critical temperature (T_c) was unchanged in this fluence region. With a fluence above 3.4×10^{11} Ar^+/cm^2 or 5.0×10^{13} He^+/cm^2, both T_c and J_c decreased with increasing fluence. Regarding the change of the crystal structure, expansion of the c-axis lattice constant and increase of FWHM (Full Width at Half Maximum) value of (00l) peak were observed by the X-ray diffraction above the fluence of 1.0×10^{13} Ar^+/cm^2. In the irradiation of three kinds of ions (He^+, O^+ and Ar^+), a simple relationship was observed between critical fluence for decreasing T_c and mass of ions.

KEYWORDS: superconductor, $Ba_2YCu_3O_y$, thin film, Ar^+ irradiation, He^+ irradiation, critical current density, critical temperature

INTRODUCTION

It has been reported that some properties of high-T_c oxide superconductor are affected by the irradiation of high energy particles such as neutrons[1-2], electrons[3-4], ions[5-8] and photons (γ-ray[9] and X-ray[10]). In most of these studies, T_c decreases and normal-state resistivity increases by heavy irradiation, while, with low fluence, J_c increases without change of T_c. From the viewpoint of J_c enhancement, high energy particles are considered to introduce radiation defects effective for pinning centers. The size and the density of the radiation defects can be controlled by the irradiation condition such as the kind, the energy and the fluence of incidental particles. In order to find the optimum condition, we have been studying the effect of irradiations of charged particles to c-axis-oriented BYCO thin films up to the present[12-14]. In this paper, we report some results about Ar^+ and He^+ irradiations to show that J_c is enhanced with low fluence and that both J_c and T_c decrease and the crystal structure is damaged with high fluence.

EXPERIMENTAL

BYCO thin film samples were grown on $SrTiO_3$ (100) substrates by the activated reactive evaporation method, in which Y,Ba and Cu were used as metal sources[14]. The thicknesses of these films were about 300 nm, and their c-axes were oriented perpendicular to $SrTiO_3$ (100) substrates. Their J_c values were in the order of $10^5 A/cm^2$ at 77 K.

Before and after ion irradiation , the temperature-dependent I-V characteristics of these films were measured by the conventional four-probe method with 200 Hz sine-wave current. The films were etched chemically to form a conduction path of 240μm long $\times$ 30μm wide in

Before and after ion irradiation , the temperature-dependent I-V characteristics of these films were measured by the conventional four-probe method with 200 Hz sine-wave current. The films were etched chemically to form a conduction path of 240μm long $\times$ 30μm wide in order to reduce the total current in the I_c measurements. The temperature of the film was sensed by a calibrated Au/Fe-Chromel thermocouple. The values of J_c were defined as the current at voltage of 10μV/cm, and those of T_c were defined as the temperature at which the critical current values were less than 10^{-5} A. No significant difference in the values of J_c and T_c between warming and cooling the sample was observed within the accuracy of $\pm$ 2% and $\pm$ 0.5K, respectively. For the samples irradiated with high fluence, X-ray diffraction (XRD) and electrical resistance measurements were carried out.

The BYCO thin film samples were irradiated by 1.0 MeV Ar^+ ion or 0.5 MeV He^+ ion at room temperature in a vacuum of 3 $\times$ 10^{-5} Pa with a Van de Graaff accelerator at the University of Tokyo. These ions have penetration depths which are larger than the thickness of the BYCO thin films used in this experiment, which means radiation-induced defects distribute almost entirely over the films.

RESULTS AND DISCUSSION

Figure 1 shows temperature-dependent I_c curves of the film subjected to a fluence of 1.7×10^{10} Ar^+/cm^2. As shown in this figure , I_c values after irradiation are enhanced by about 30% around 80K without change of the T_c . This I_c -vs- T_c curve is fitted in a form of $I_c = A(1-T/T_c)^2$ as pointed out in the previous papers[12-14]. With fluences above 3.4×10^{11} Ar^+/cm^2 , however, I_c and T_c decreased and the values of "A" also decreased.

Figure 2 visualizes the Ar^+ irradiation effects on J_c and T_c values with increase of fluence. In this figure , J_c/J_{c0} and T_c/T_{c0} vs ion fluence are plotted, where J_{c0} and J_c represent the J_c values before and after the irradiation at T=0.9T_c , and T_{c0} and T_c represent T_c values before and after the irradiation , respectively. In this case, J_c/J_{c0} is equivalent to I_c/I_{c0} because the cross section of the current path for each sample did not change by the irradiation. The observed enhancement of directly measured J_c values below a fluence of 1.3 $\times$ 10^{11} ions/cm^2 strongly suggests that defects formed by Ar^+ ion irradiation act as pinning centers, considering that atmospheric degradation occurs easily in BYCO thin films and that radiation damage particular to heavy ions might degrade transport supercurrents through grain boundaries. On the other hand, above 3.4 $\times$ 10^{11} Ar^+/cm^2, J_c values are lowered due to radiation damage.

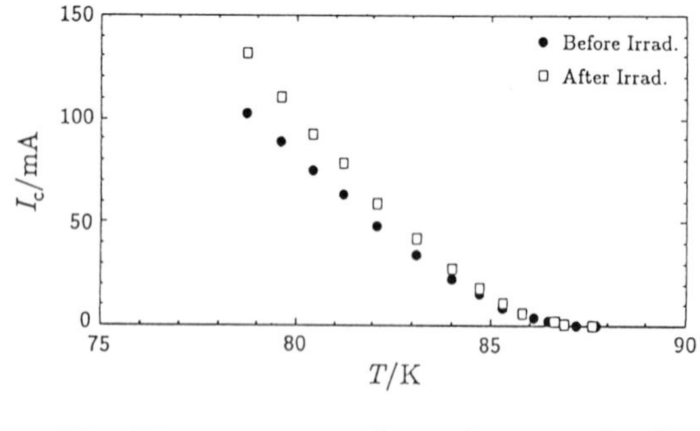

Fig.1 Temperature dependence of I_c before and after 1MeV Ar^+ irradiation with a fluence of 1.7 $\times$ 10^{10} ions/cm^2.

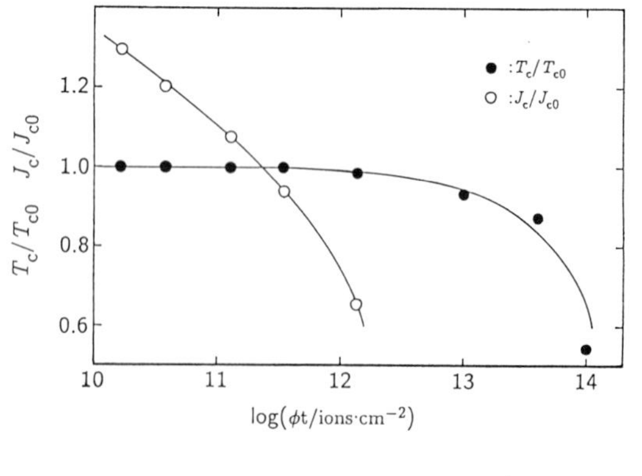

Fig.2 Irradiation effect of 1MeV Ar^+ irradiation on J_c and T_c.

Under the irradiation with high fluence, the normal-resistivity increased gradually with fluence. The transition temperature from normal state to superconducting state was also affected strongly by the irradiation. Figure 3 shows the changes of T_c^{onset}, T_c^{mid}, T_c^{end}, with fluence. These three T_c values decreased, at least, above 1.0 × 10^{13} Ar^+/cm^2. In particular, T_c^{end} was most strongly affected, and the transition width of the critical temperature became broader with fluence. The similar results have been observed as an usual irradiation effect for high-T_c superconductor also by other research groups. Also in the case of 0.5 MeV He^+ irradiation, the similar tendency was observed, as shown in Fig.4. Though plots are somewhat scattered due to chemical degradation by humidity, clear decreases of T_c's can be seen with a fluence above 7.0 × 10^{13} He^+/cm^2.

In Fig.5, the critical fluences $(\phi t)_c$ giving decreasing T_c in the irradiation of three kinds of ions (0.5MeV He^+, 0.5MeV O^+, 1MeV Ar^+) are plotted against the mass number M of the ion. From this figure, we can find the following simple relationship between them: $(\phi t)_c$ = B M^{-2}, where B is a constant. The origin of this simple relationship may be induced from the computational calculation for radiation damage.

Figure 6 shows the change of the d-spacing calculated by the diffraction angle for the (007) peak position with increase of the fluence. 'Error Bars' in this figure indicate the d-spacing range corresponding to FWHM (Full Width at Half Maximum) value of each (007) peak.

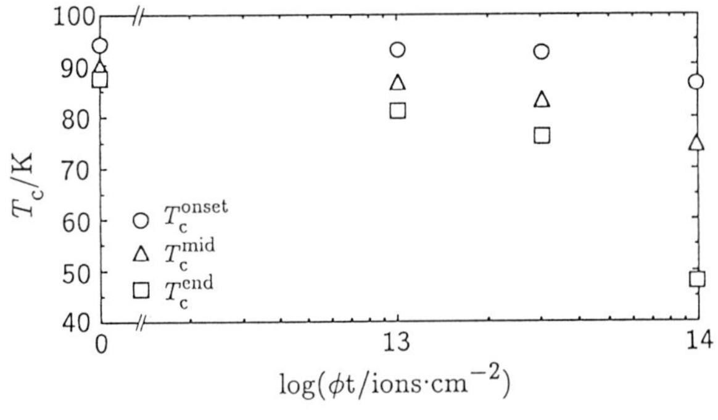

Fig.3 Changes of T_c^{onset}, T_c^{mid} and T_c^{end} by 1MeV Ar^+ irradiation.

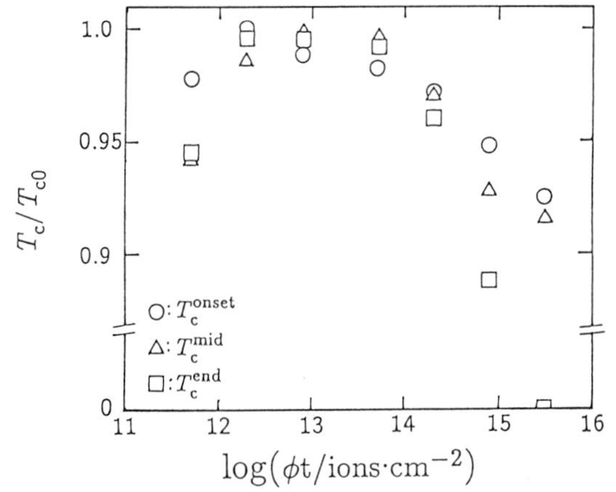

Fig.4 Changes of T_c^{onset}, T_c^{mid} and T_c^{end} by 0.5MeV He^+ irradiation.

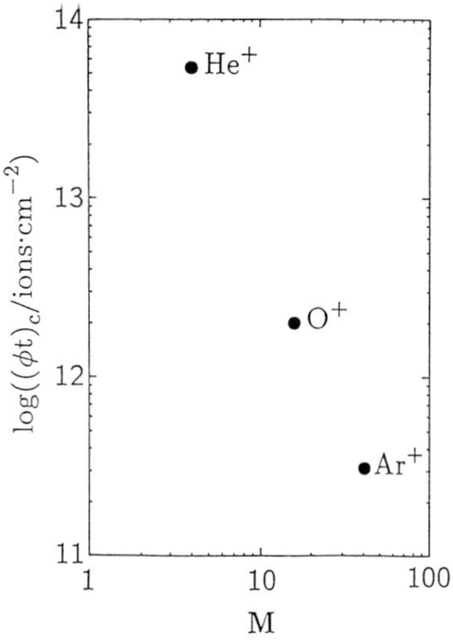

Fig.5 Mass number of ion vs critical fluence.

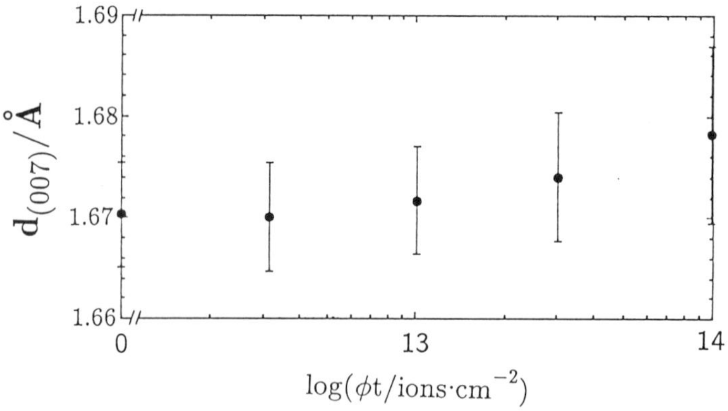

Fig.6 Change of d-spacing by 1MeV Ar^+ irradiation.

The value of d-spacing has a threshold at 10^{13} ions/cm^2, and above this limiting fluence it increases gradually. Similar results were also obtained for other (00*l*) peaks. Two interpretations were considered to explain this phenomenon: First, the interval between ab-planes were expanded by the knocked-on atoms derived from cascade damages as interstitial atoms. Second, the decrease of the oxygen stoichiometry of the samples by irradiation caused the expansion of the c-axis lattice constant. The values of T_c^{onset} and T_c^{end} observed in this experiment were compared with those estimated using the relationship among the values of T_c's, oxygen stoichiometry in BYCO and the c-axis lattice constant, where as the observed values, the c-axis lattice constant were used[15]. It was concluded that the origin of the elongation of the c-axis cannot be attributed only to the oxygen non-stoichiometry. Changes of a/b-axis length could not be discussed here because the samples used in this experiment were the c-axis highly-oriented thin films and no peaks except (00*l*) peak could be detected by XRD.

REFERENCES

1) Umezawa A, Crabtree GW, Liu JZ, Weber HW, Kwok WK, Nunez LH, Moran TJ, Sowers CH, Claus CH (1987) Phys Rev B36: 7151

2) Cost JR, Willis JO, Thompson JD, Peterson DE (1988) Phys Rev B37: 1563

3) Hofmann A, Kronmüller H, Moser N, Reisser R, Schüle P, Dworschak F. (1988) Physica C 156: 528

4) Kato T, Shiraishi K, Kuniya J (1989) Jpn J Appl Phys 28: L766

5) Roas B, Hensel B, Saemann-Ischenko G, Schultz L (1988) Appl Phys Lett (1987) 54: 1051

6) Summers GP, Burke EA, Chrisey DB, Nastasi M, Tesmer JR (1989) Appl Phys Lett 55: 1469

7) Xiong GC, Li HC, Linker G, Meyer O (1988) Phys Rev B38: 240

8) Bourgault D, Bouffard S, Toulemonde M, Groult D, Provost J, Studer F, Nguyen N, Raveau B (1988) Phys Rev B39: 6549

9) Marwick AD, Clark GJ, Yee DS, Laibowitz RB, Coloman G, Cuomo JJ (1988) Phys Rev B39: 9061

10) Vasek P, Smrcka L, Dominec J, Pesek M, Smrckova O, Sykorova D (1988) Solid State Comm 69: 23

11) Izumi T, Iwasaki K, Kurihama T, Nakajima T (1990) Jpn J Appl Phys 29: L591

12) Masegi T, Terai T, Takahashi Y, Enomoto Y, Kubo S (1989) Jpn J Appl Phys 28: L1521

13) Terai T, Masegi T, Takahashi Y, Enomoto Y, Kubo S (1990) Mol Cryst Liq Cryst 184: 165

14) Terai T, Masegi T, Takahashi Y, Enomoto Y, Kubo S (1990) Jpn J Appl Phys 29: in press.

15) Terai T, Furuta T, Masegi T, Kusagaya K, Takahashi Y, Jpn J Appl Phys :submitted

Critical Current Density and Microstructure of $YBa_2Cu_4O_8$

T. MIYATAKE, K. YAMAGUCHI, T. WADA, N. SUZUKI, H. YAMAUCHI, and N. KOSHIZUKA

Superconductivity Research Laboratory, International Superconductivity Technology Center, 10-13, Shinonome 1-chome, Koto-ku, Tokyo, 135 Japan

ABSTRACT

We investigated the intragranular critical current density Jc in YBa2Cu4O8 prepared by two routes with different oxygen pressures. A relationship between the Jc's and the microstructures observed by transmission electron microscopy is discussed. The magnitude of Jc for the low oxygen processed sample is higher by a factor of 5 than that for the high oxygen pressure sample. Transmission electron microscopy reveals that the sample prepared at high oxygen pressure shows an almost perfect stacking sequence, but the low oxygen pressure prepared one has a severely distorted structure. We conclude that the difference in Jc(H) for the samples must be related to the different microstructures.

KEY WORDS: YBa2Cu4O8, critical current density, microstructure

INTRODUCTION

One of the primary difficulties for practical applications of the well-known 90K superconductor $YBa_2Cu_3O_7$ (1-2-3) is the instability of the oxygen content, causing a depression of the superconductivity. On the other hand, $YBa_2Cu_4O_8$ (1-2-4) with T_c=80K, containing double Cu-O chains running along the b-axis, has a stable oxygen stoichiometry at high temperatures compared with the 1-2-3 [1]. Recently, we reported that partial substitution of Ca for Y in the 1-2-4 compound increased T_c to 90K[2]. Thus, the Ca-doped 1-2-4 compound alleviates the difficulty of the unstable oxygen content associated with the 1-2-3 compound. Jin et al.[3] have reported that J_c(H) in the 1-2-4 prepared at ambient pressure is about the same as that in the 1-2-3. However, in our previous paper[4], we have demonstrated that J_c(H) values in the 1-2-4 prepared by high oxygen pressure annealing are lower and more sensitive to the applied magnetic fields than those in the 1-2-3. In general, J_c(H) in bulk superconductors strongly depend on microstructures, such as imperfections in the crystalline lattice. Therefore, we anticipate that the difference of J_c(H) between these 1-2-4 samples may arise from their microstructures.

In this article, we describe intragranular critical current densities in $YBa_2Cu_4O_8$ prepared by two routes with different oxygen pressures. In addition, their microstructures are observed by using transmission electron microscopy. We consider a correlation between the intragranular critical current density and the microstructure in $YBa_2Cu_4O_8$.

EXPERIMENTAL

Samples were prepared by the following two methods: (1) high oxygen pressure annealing and (2) low oxygen pressure annealing. Preparation of the 1-2-4 sample by high oxygen pressure annealing ($P(O_2)$=20MPa) was conducted using hot isostatic pressing treatments in a mixed gas environment of argon with 20% oxygen (O_2-HIP), as reported in ref.5. Another 1-2-4 sample was synthesized by a heat treatment at 850℃ for 160h under low oxygen pressure ($P(O_2)$=0.3MPa). Details of the sample preparation at low oxygen pressure have been described elsewhere [6]. High and low oxygen pressure processed samples are hereinafter referred to as the O_2-HIP and LOP, respectively. Values of T_c (82K) obtained from magnetization measurements were the same for both samples. To evaluate the intragranular J_c, magnetic hysteresis measurements up to a maximum field of 22kOe were performed on powdered samples using a SQUID magnetometer with the scan length set at 25mm. Hysteresis was measured at various temperatures. The samples were examined by transmission electron microscopy (TEM) employing the technique of crushed powders on a microgrid.

RESULTS AND DISCUSSION

Typical magnetic hysteresis loops at 10K up to the maximum field of 22kOe for O_2-HIP and LOP processed $YBa_2Cu_4O_8$ samples are shown in Fig.1. The loop for the O_2-HIP sample exhibits a significant decrease at lower magnetic fields compared with that for the LOP sample. We used the difference $\Delta M(H)$ between the field increasing and decreasing magnetizations at an applied magnetic field H to estimate the critical current density $J_c(H)$ using the Bean critical state model[7]. Assuming that samples have a circular cross section, we can calculate the $J_c(H)$ by the following relation:

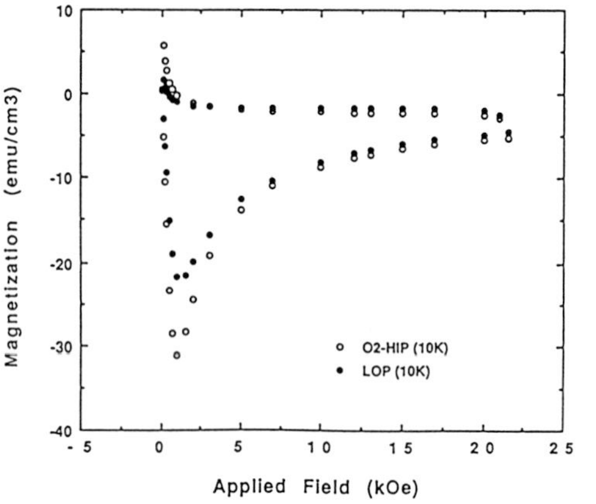

Fig.1 Typical magnetic hysteresis loops at 10K for O_2-HIP and LOP processed $YBa_2Cu_4O_8$ samples

$$J_c(H)=30\Delta M(H)/D, \qquad (1)$$

where $J_c(H)$ is in A/cm^2, $\Delta M(H)$ is in emu/cm^3 and D is the sample diameter in cm. By scanning electron microsopy, the average grain sizes were evaluated to be 6.0mm and 1.5mm for O_2-HIP and LOP samples, respectively. In general, the critical state model is not established for sintered samples with weak-link networks at grain boundaries, i.e. for inhomogeneous superconductors. However, if the sample dimension is comparable to the grain size, $\Delta M(H)$ is directly proportional to that dimension[8]. Therefore we ground sintered samples to powder comparable to their grain sizes. Then we used the grain sizes as the powder diameters to calculate the intragranular $J_c(H)$ values. Figure 2 shows the field dependence of the intragranular critical current densities at various temperatures for the O_2-HIP (a) and LOP (b) processed $YBa_2Cu_4O_8$ samples. At low temperature, we observe little dependence of $J_c(H)$ on the applied magnetic field, while at higher temperature, $J_c(H)$ decreases significantly with increasing magnetic field. The $J_c(H)$ data for the LOP sample are higher by a factor of 2~5 than those for the O_2-HIP sample. This difference in $J_c(H)$ increases with field.

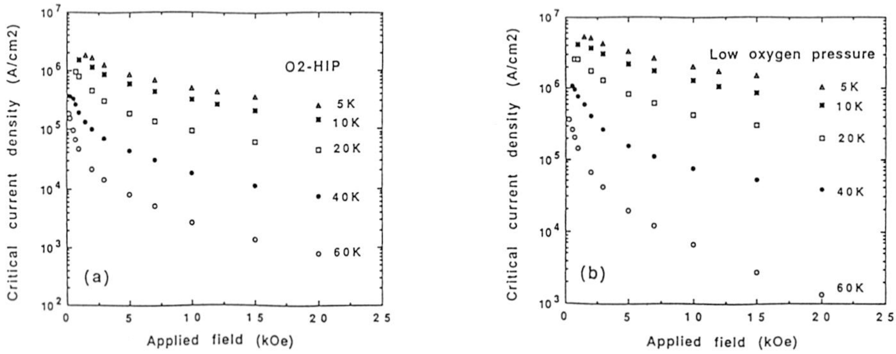

Fig.2 Field dependence of the intragranular critical current densities at various temperatures for the O_2-HIP (a) and LOP (b) processed $YBa_2Cu_4O_8$ samples.

Intragrain J_c values at 77K and 10kOe in the sintered 1-2-3 material by using the AC inductive method is reported to be $3.1 \times 10^4 A/cm^2$ [9]. Intragranular J_c data at 60K and 10kOe in the 1-2-4 compounds yield about $2.7 \times 10^3 A/cm^2$ and $6.7 \times 10^3 A/cm^2$ for O_2-HIP and LOP samples, respectively. 77K for the 1-2-3 materials and 60K for the 1-2-4 materials correspond to nearly the same reduced temperature t ($=T/T_c$). Thus, J_c in $YBa_2Cu_3O_7$ is higher by about an order of magnitude than in $YBa_2Cu_4O_8$ at a same reduced temperature.

In order to clarify the results of different J_c's in the 1-2-4 system, we performed transmission electron microscopy observation on the O_2-HIP and LOP processed $YBa_2Cu_4O_8$ samples. Figure 3 shows a typical TEM image for the O_2-HIP (a) and LOP (b) samples. The TEM image for the O_2-HIP sample was taken with the incident electron beam along the [100] direction and that for the LOP sample was with the beam along the [110] direction. In the TEM image for the O_2-HIP sample (Fig.3-(a)), the stacking sequence appears to be almost perfect, while the lattice image for the LOP sample shows a severely distored structure (Fig.3-(b)). The brighter lines, which we believe are double Cu-O chains, also show distortion for the LOP processed sample.

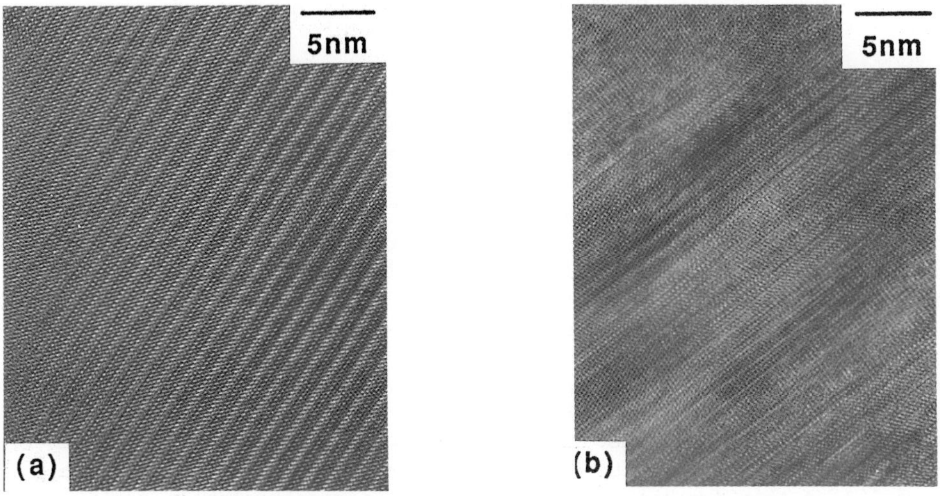

Fig.3 Typical TEM image for the O_2-HIP (a) and LOP (b) samples.

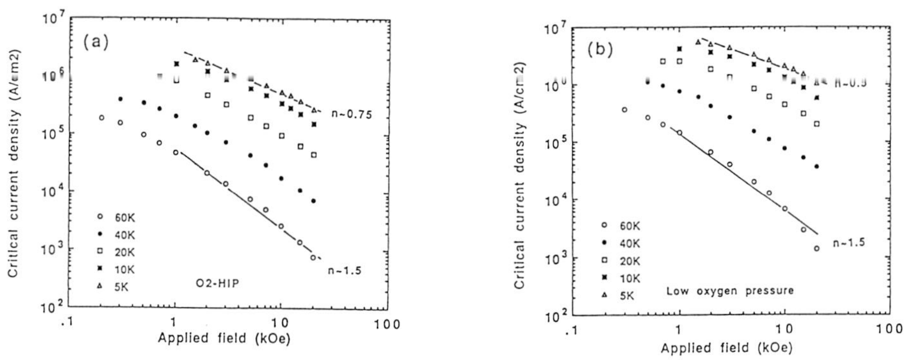

Fig.4 Plot of log J_c versus log H obtained from data of fig.2.

In the $YBa_2Cu_3O_7$ system, there are three main candidates for flux-pinning centers: (1) normal-state precipitates, (2) twinning planes and (3) inhomogeneous oxygen distributions. However, the $YBa_2Cu_4O_8$ system contains no twinning planes and has a fixed oxygen stoichiometry. Thus the last two candidates may be not applicable for the 1-2-4 compound. Murakami et al.[10] have proposed an expression for $J_c(H)$ in the melt processed 1-2-3 materials by assuming that normal precipitates such as Y_2BaCuO_5 (2-1-1) can act as dominant pinning centers. In this case, J_c is expressed by a relation

$$J_c(H)=J_{c0}H^{-1/2}. \qquad (2)$$

Thus, according to this model, J_c should vary with $H^{-1/2}$ at a given temperature. Figure 4 shows the plot of log J_c versus log H obtained from Fig.2. The field dependence for both samples is $J_c(H) \propto H^{-n}$ with $0.5 \le n \le 1.5$ at magnetic fields above 1kOe. The value of n for the O_2-HIP sample varies between 0.75 and 1.5, and that for the LOP sample between 0.5 and 1.5, in both cases increasing at higher temperatures. The n of 0.5 which corresponds to eq.(2) is realized only for the LOP sample at 5K. Values of n for the LOP sample are somewhat larger than that for O_2-HIP sample at the same temperature and magnetic field. Consequently, the intragranular critical current density in $YBa_2Cu_4O_8$ can not be determined by this simple flux-pinning model of only normal-state precipitates. However, the 1-2-4 samples with different microstructures seem to obey a similar flux-pinning temperature and field dependence. As a result, the difference in the magnitude of J_c for the samples is a result of the difference in the number of pinning centers.

A TEM study of the 1-2-4 compound [11] has shown that 1-2-3 sub-units are observed in the 1-2-4 host compound as planar defects. These units, forming $Y_2Ba_4Cu_7O_{15}$ (2-4-7) [12] with a typical T_c of 40K have been considered to act as pinning centers above the T_c of the 2-4-7 compound. The present results do not show any enhancement of flux-pinning in either temperature or field dependences of J_c within our experimental conditions. Therfore, it may be that this type of defect does not cause the same type of anomalous $J_c(H)$ behavior reported by Larbalestier et al.[13] However the n value, implying the field dependence of J_c, for O_2-HIP and LOP processed $YBa_2Cu_4O_8$ is slightly larger in the latter case. The disordered structures, including many defects, for the LOP processed $YBa_2Cu_4O_8$ can give rise to the enhancement of flux-pinning.

In summary, we measured magnetic hysteresis loops up to a maximum field of 22kOe at various temperatures for high and low oxygen pressure processed $YBa_2Cu_4O_8$ with different microstructures, and estimated the critical current density J_c using the Bean critical-state model. The magnitude of J_c for the low oxygen processed sample was higher by a factor of 5 than that for the high oxygen pressure sample. The transmission electron microscopy observation indicated that the sample prepared at a high oxygen pressure showed an almost perfect stacking sequence but the sample at a low oxygen pressure showed a severely distorted structure. Therefore it is thought that the difference in $J_c(H)$ for the two samples must come from the different microstructures.

REFERENCES

1. Karpinski J, Kaldis E, Jilek E, Rusiecki S, Bucher B (1988) Nature 336: 660.
2. Miyatake T, Gotoh S, Koshizuka N, Tanaka S (1989) Nature 341: 41.
3. Jin S, O'Bryan HM, Gallagher PK, Tiefel TH, Cava RJ, Fastnacht RA, Kammlott GW (1990) Physica C 165: 415.
4. Miyatake T, Yamaguchi K, Takata T, Gotoh S, Koshizuka N, Tanaka S (1990) Jpn.J.Appl.Phys. 29: L1079.
5. Miyatake T, Yamaguchi K, Takata T, Gotoh S, Koshizuka N, Tanaka S (1990) Physica C 160: 541.
6. Wada T, Suzuki N, Yamaguchi K, Ichinose A, Yaegashi A, Yamauchi H, Koshizuka N, Tanaka S (in press) J.Mater.Res.
7. Bean CP (1962) Phys.Rev.Lett. 8: 250.
8. Shimizu E, Itoh D (1989) in: (eds) Kitazawa K, Ishiguro T in *Advances in Superconductivity*. Springer-Verlag, Tokyo, pp.451.
9. Ni B, Munakata T, Matsushita T, Iwakuma M, Funaki K, Takeo M, Yamafuji K (1987) Jpn.J.Appl.Phys. 28: 1658.
10. Murakami M, Gotoh S, Koshizuka N, Tanaka S, Matsushita T, Kambe S, Kitazawa K (1990) Cryogenics 30: 390.
11. Yamaguchi K, Miyatake T, Takata T, Gotoh S, Koshizuka N, Tanaka S (1989) Jpn.J.Appl.Phys. 28: L1942.
12. Bordet B, Chaillout C, Chenavas J, Hodeau JL, Marezio M, Karpinski J, Kaldis E (1988) Nature 336: 660.
13. Larbalestier DC, Daeumling M submitted to Phys.Rev.Lett.

Microstructure and Flux Creep Behavior of Ag-Sheathed Bi-Pb-Sr-Ca-Cu-O Tapes

KAZUTAKA YAMAMOTO, SATORU MURASE, SHIROU TAKENO, and SHIN-ICHI NAKAMURA

Toshiba R&D Center, 4-1, Ukishima-cho, Kawasaki-ku, Kawasaki, 210 Japan

ABSTRACT

TEM observations and flux creep measurements have been carried out for the Bi-Pb-Sr-Ca-Cu-O (BPSCCO) tapes, with and without the intermediate press process. A lot of dislocations were found in the grains of the pressed tape, which had 20,000 A/cm^2 transport critical current density at 77 K, 0 T. Field and temperature dependence of the pinning potential U_0 was summarized in ternary diagrams. It could be clearly recognized that the U_0 for the BPSCCO tape was enhanced by the press. As a result, the intra-grain critical current density proved to be increased, and dislocations seemed to play a role in a pinning center.

KEY WORDS: Bi-Pb-Sr-Ca-Cu-O tape, Dislocation, Magnetization, Flux creep, Pinning potential

INTRODUCTION

It has been reported that an uni-axial intermediate press process is effective to improve the transport critical current density (Jct) of oxide superconductors, especially of the bismuth based ones [1,2]. The press has been thought to help to reduce the porosity and to align the grains of the oxide superconductor. However, the authors considered that the press affected not only the alignment of the grains, but also the superconducting properties of each grain itself. The U_0 of the BPSCCO tape is much smaller than that of conventional superconductors [3]. With the enhancement of the intra-grain critical current density (intra-Jc) or the pinning potential U_0 , the Jct will be also improved. In this work, flux creep measurements were carried out in order to investigate whether or not the intra-Jc of the Bi-Pb-Sr-Ca-Cu-O tapes was enhanced by the press.

EXPERIMENTAL

Ag-sheathed BPSCCO tapes were fabricated by the powder-in-tube method [4]. Co-precipitated oxalate powder, having a $Bi_{1.84}Pb_{0.34}Sr_2Ca_2Cu_3O_x$ composition, was calcined at 800 ℃ for 40 hours in air. A silver tube with the calcined powder inside it, was cold-worked by drawing and rolling, to be formed into a tape with 0.15 or 0.1 mm thickness. Some parts of it were cut into short specimens, 15 mm in length. All of the specimens were heat-treated in 7.7%O_2 +Ar(bal.). Some of them were pressed once or twice with 5 to 20 t/cm^2 pressure.

Jct for the tapes was measured by the standard four probe technique in the liquid nitrogen, with 1 μV/cm criterion. Microstructure was observed by TEM (Transmission Electron Microscopy). Magnetization was measured by the D.C.SQUID (Quantum Design Model MPMS) to simultaneously investigate both the magnetization hysteresis (Δ M) and U_0 for the tape [5]. The specimens for the magnetization measurement were about 2 × 4 × 0.08 mm and 1.9 mg. Silver sheath was removed to avoid the effect of normal metal on the measurements. The magnetic field was applied perpendicular to the rolled or pressed surface of the tape.

RESULTS AND DISCUSSION

Transport critical current density of the tape

Jct changed, depending on both the tape thickness and the pressure applied, as can be seen in Fig.1. For the 0.1 mm tape, the maximum Jct value of 20,000 A/cm^2 was obtained for the sample pressed twice at a 15 t/cm^2 pressure. On the other hand, tape without being pressed showed 4,000 A/cm^2 at most. It is quite obvious that the press improves the Jct value for the BPSCCO tape.

Dislocations observed in the BPSCCO crystal

TEM observation for the pressed tape revealed that a lot of dislocations were formed in some of the crystal grains in the core, as shown in Fig.2. In some cases, it was also observed that the dislocations made up a network. With the diffraction contrast analysis technique, the Burgers vectors of the dislocations were always perpendicular to the dislocation line, so that it was possible to determine them as the edge dislocation. Every dislocation causes a strain field around the dislocation line. It was possible to estimate the strain field radius from the dislocation line width, observed in the TEM photograph, to be about 20 Å, which is approximately equal to half of the coherent length, ξ_{ab} = 40 Å [6]. It is good for those dislocations to interact with the fluxoid. Though their density in the pressed tape, which was 10^9 cm^{-2}, is not sufficient to explain the pinning in the grain completely, they could be partly responsible for the pinning. Most of dislocations observed in this study seemed to be produced in the intermediate press process, because their density in the pressed tape had increased to a value several times larger than that in the non pressed tape. That means it is possible to control the dislocation density or some parts of pinning centers with an appropriate press and heat treatment.

Magnetization hysteresis ΔM and pinning potential U_0

Hysteresis loop and relaxation of the magnetization were measured by D.C.SQUID. Figure 3 shows the magnetization hysteresis curves at 5, 20, 40, 60 and 77.3 K, (a): for non pressed tape with 4,000 A/cm^2, and (b): pressed

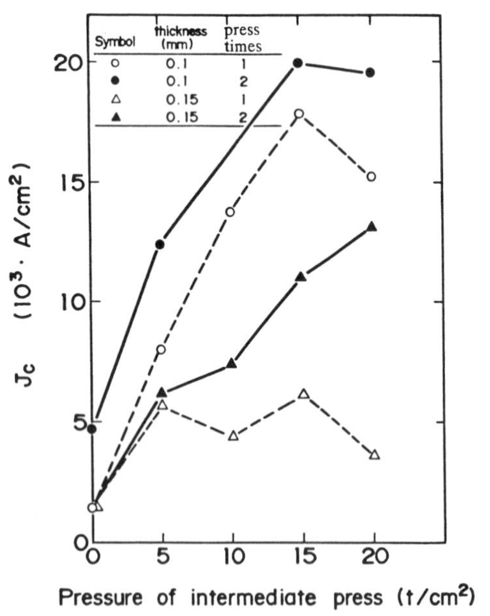

Fig.1 Transport critical current density of BPSCCO tapes.

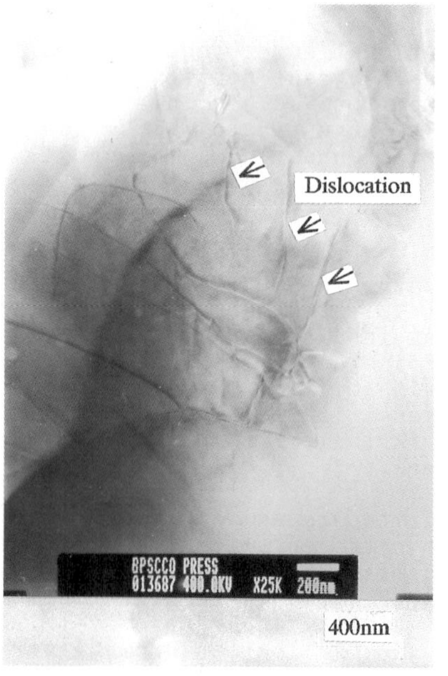

Fig.2 TEM photograph of pressed BPSCCO tape with an electron beam parallel to the c-axis. Dislocations can be observed in the grain.

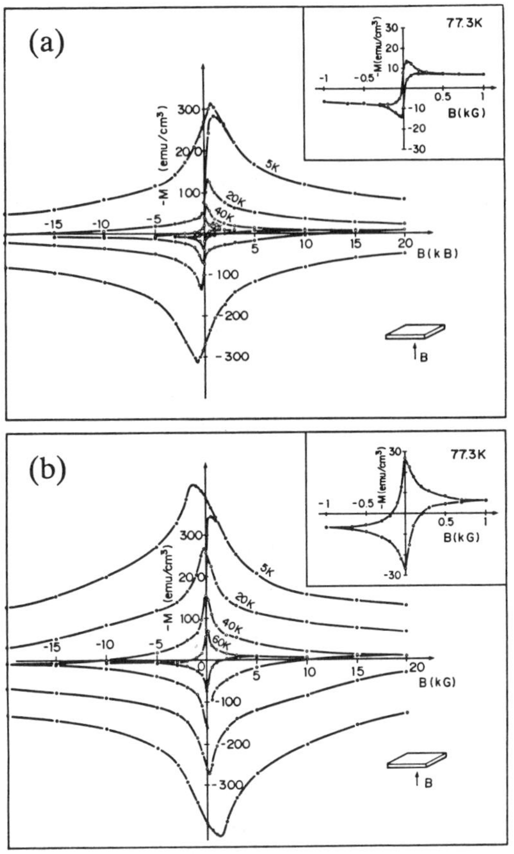

Fig.3 Magnetization hysteresis curve at 5, 20, 40, 60 and 77,3 K. (a): Non pressed tape. (b): Pressed tape. Applied field was perpendicular to the rolled surface of the tapes.

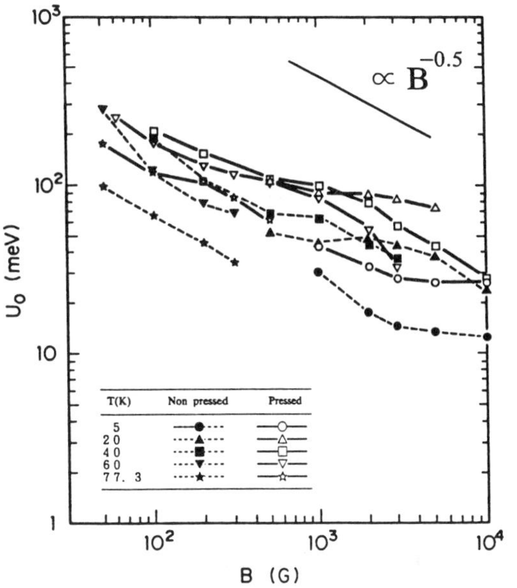

Fig.4 Log-log plot for the pinning potential U_0 vs. field B, indicating that the U_0 for the pressed tape decreases proportionally to $B^{-0.5}$.

tape with 20,000 A/cm^2. Δ M of (b) was larger than that of (a) in all the temperature range. Moreover, the ratio of Δ M for (b) to that for (a) became larger, as the temperature increased. For example, the ratio reached four times at 77.3 K, 0.01 T, though the ratio at 5 K, 1 T was twice. This means the effect of the intermediate press on Δ M is more prominent in the higher temperature range.

On the other hand, the U_0 values for the tapes were calculated according to the well-known equation, $dM/d(\ln t) = (\Delta M/2) \cdot (kT/U_0)$. A plot of $\ln(U_0)$ vs. $\ln(B)$ is shown in Fig.4. The solid lines stand for the pressed tape and the dotted lines stand for the non pressed tape. It was found that the U_0 values decreased as the field increased and the gradient for each line was approximately - 0.5. It is coincident fairly good with the $B^{-0.5}$ dependence, which is derived on condition that the lattice constant for the fluxoid, a_f, is much larger than the correlation length for each fluxoid, l_{66} [7].

A ternary diagram of U_0-B-T seems to be useful in order to recognize the U_0 in a given field distribution and temperature easily. Figure 5 shows those diagrams, (a):for the non pressed tape, and (b):for the pressed tape, respectively. It is obvious that the U_0 for the pressed tape is enhanced in the whole range of fields and temperature. For example, the U_0 value for the pressed tape was 122 meV, whereas it was 65 meV for the non pressed tape at 77.3 K, 0.01 T. The U_0 had a peak value at the middle temperature range, between T = 0 and T = Tc, which moved toward the lower end as the field increased. Furthermore, in Fig.5 (a) and (b), the points where each extrapolated iso-field line intersects the plane of U_0 = 0, give the irreversibility line. It is also possible to evaluate the line, from a s eries of fields where the Δ M becomes zero in the magnetization hysteresis curve at a given temperature. The fact that the irreversibility line of the specimen (b) goes through at a twice higher field than that for specimen (a), implies that the irreversibility line enhancement is also possible, corresponding to that of the pinning potential U_0.

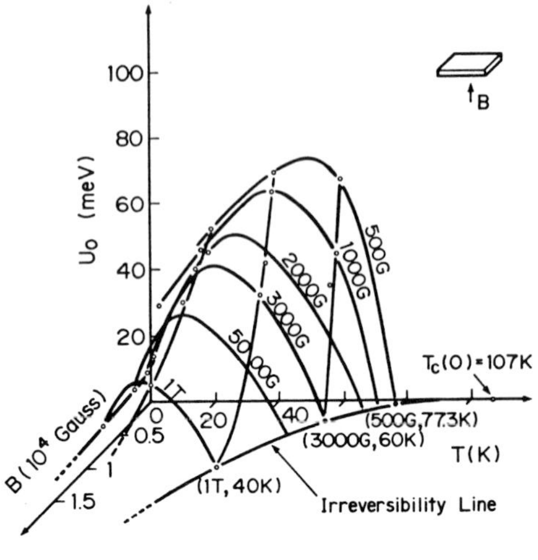

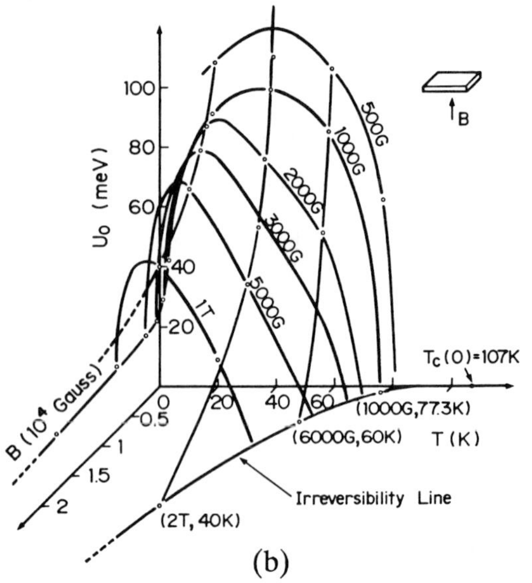

Fig.5 Ternary diagrams, indicating the relation of pinning potential U_0 to the field and temperature. The points where an iso-field line of the U_0 cross the $U_0=0$ plane, makes an irreversibility line. (a): Non pressed tape. (b): Pressed tape.

CONCLUSIONS

Transport critical current density (Jct), microstructure, and flux creep behavior for the BPSCCO tapes were investigated in this work. The intermediate press process improved the Jct value for the pressed tapes, and also enhanced the magnetization hysteresis and pinning potential U_0. Furthermore, it increased the density of dislocations existing in the grains of the tape. It can be concluded that the intermediate press process is effective for the Ag-sheathed BPSCCO tapes, not only to increase the intergrain Jc, but also to increase the intragrain Jc. Dislocations introduced by the press seem to form one of the artificial pinning centers. Ag-sheathed BPSCCO tapes will be able to have better superconducting properties with the further introduction of microdefects, such as dislocations.

ACKNOWLEDGMENTS

The authors are grateful to Y. Kitamura and T. Hasegawa of the Showa Electric Wire and Cable Co. Ltd., for their help in fabricating the Ag-sheathed tape. They are also grateful to Dr. Y. Yamada and Dr. O. Horigami for their encouragement.

REFERENCES

1. Asano T, Tanaka Y, Fukutomi M, Jikihara K, Machida J and Maeda M (1988) Jpn. J. Appl. Phys. 27 : L1652
2. Yamada Y, Jikihara K, Hasebe T, Yanagiya T, Yasuhara S, Ishihara M, Asano T and Tanaka Y (1989) Jpn. J. Appl. Phys. 29 : L456
3. Kumakura H, Togano K, Yanagisawa and Maeda H (1989) Appl. Phys. Lett. 55: 185
4. Yamada Y, Murase S, Yamamoto K and Kamisada M (1990) Cryogenics 30: 581
5. Xu Y and Suenaga M (1990) submitted to Phys. Rev. B
6. Koike Y, Nakanomyo T and Fukase T. (1988) Jpn. J. Appl. Phys. 27 : L841
7. Yamafuji K (1990) Cryogenic Engineering 25: 2

Texture Study of Pressed $YBa_2Cu_3O_{7-x}$ Polycrystals by Electron Channeling Patterns

WHANG JIN KANG[1], SHUJI HANADA[2], KYOSUKE YOSHIMI[2], SAKAE SAITO[2], YONOSUKE MURAYAMA[2], SHIGEYUKI HAYASHI[2], and AKIHIKO NAGATA[3]

[1] Graduate student, Tohoku University, Sendai, 980 Japan
[2] Institute for Material Research, Tohoku University, Sendai, 980 Japan
[3] Department of Metallurgy, Mining College, Akita University, Akita, 010 Japan

ABSTRACT

(001) pole figures of $YBa_2Cu_3O_{7-x}$ tapes and bulk specimens with various values of critical current density (Jc) were drawn by analyzing electron channeling patterns (ECPs) taken from individual grains of the specimens. It was found that preferred orientation of (001) formed in the tape with decreasing thickness of the tape in agreement with the X-ray result. It is concluded that the ECP technique is very useful for the analysis of orientation distribution in small sized $YBa_2Cu_3O_{7-x}$ specimens such as tapes or thin films.

KEY WORDS: ECP, pole figure, orientation distribution, critical current density

INTRODUCTION

Many of the applications depend on the ability of the oxide superconductors to transport sufficiently high current densities (Jc), i.e., more than $10^5 A/cm^2$ at 77K in high magnetic fields. Reported values of Jc are, however, too low for practical applications. There are several factors controlling the critical current density of the oxide superconductors; i.e., weaklink at grain boundaries, flux pinning, flux creep and strong anisotropy in superconducting properties. In this work preferential texture along the c-axis is investigated in relation to critical current density.

The pole figure technique with X-ray[1] or neutron diffraction[2] is most commonly used to demonstrate texture. Spread in texture is sometimes estimated by rocking curves[3]. However, these techniques show crystallographic orientation on the statistical average and require a large sized sample. On the other hand, the twin trace technique, as Verhoeven et al[4-5] pointed out, is applicable to a small sized sample but it is less accurate than the ECP technique.

In this work the texture of small sized $YBa_2Cu_3O_{7-x}$ specimens is analyzed by the ECPs and is compared to pole figures by X-ray technique. Measuring the pole figures of the specimens in which Jc is evaluated, the relation between texture and Jc is discussed.

EXPERIMENTAL TECHNIQUE

Powders of $YBa_2Cu_3O_{7-x}$ composition was prepared by a typical solid state reaction technique. Pellets were sintered for 72 hr at 1273K in flowing oxygen and then slowly cooled at 1K/min down to room temperature. The processing produced a large grained material of grain size $10\sim 100\mu$ m. The surface of specimen was mechanically polished and then electropolished in a solution of 10% perchloric acid in ethylene glycol monobutyl ether at $10\sim 30$V for $1\sim 5$min. ECPs were taken at an accelerating voltage of 25kV and a working distance of 7mm on a JEOL scanning electron microscope (JSM-840A) equipped with selected area channeling pattern (SACP) accessories.

The calcined powders were placed in a silver tube with 9mm i.d. and 13mm o.d. The composite was grooved rolled to 1.0mm in diameter and then fabricated into tapes by pressing in the range between 40 and 200μ m thickness. The tapes were sintered at 1203K for 12hr in flowing oxygen. Critical transport current was measured with the four probe dc method at liquid nitrogen temperature in zero magnetic field.

After measuring Jc the tape specimens were immeresed in a nitric acid solution to remove the sheathed silver and then electropolished. The analysis of orientation distribution for $YBa_2Cu_3O_{7-x}$ was performed with ECP method, which was developed quite recently by the present authors[6]. These results are compared to pole figures by measuring of X-ray diffraction peak. X-ray pole figure data were collected for the (006)peak by the Schulz reflection method.

RESULTS AND DISCUSSION

Figure 1 is the standard channeling map of $YBa_2Cu_3O_{7-x}$ for [001-$\bar{1}$10-$\bar{1}\bar{1}$0] poles sweeping out an arc whose radius corresponds to 90° . Most of ECPs were taken from sintered pellets. In addition, a single crystal was used to make the map within 25° from [001] pole. Poles and lines in Fig.1 are indexed as shown in Fig.2. The angular scale factor for this map have been reported in Ref.[6].

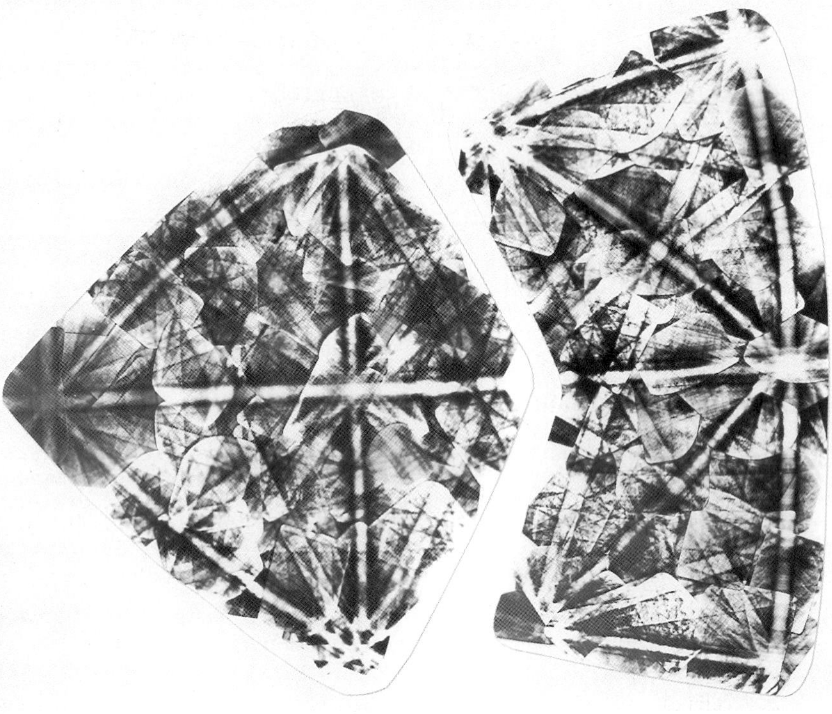

Fig.1 Standard electron channeling map of $YBa_2Cu_3O_{7-x}$ for quadrant [001-$\bar{1}$10-$\bar{1}\bar{1}$0] poles.

Figure 3 shows the Jc of the specimens prepared by grooved rolling and pressing as a function of pressed tape thickness and areal reduction ratio of the initial to the final cross section area (S_i/S). One can see a tendency that Jc depends on thickness and areal reduction ratio of tape. Osamura et al[7]. have reported that the increment of Jc in pressed tape is correlated with preferred orientation, densification and macroscopic homogeneity. Although the obtained Jc in this study is lower than that of Osamura et al., the tendency for the thickness dependence of Jc is similiar. The main object of this study is not improvement of Jc but investigation of the crystallographic orientation distribution using the ECP technique. Therefore, the experiment was performed for the 60μ m thick tape with a Jc of 1300 A/cm^2 at 77K, the 200μ m thick tape with a Jc of 200 A/cm^2 at 77K and as-sintered pellets.

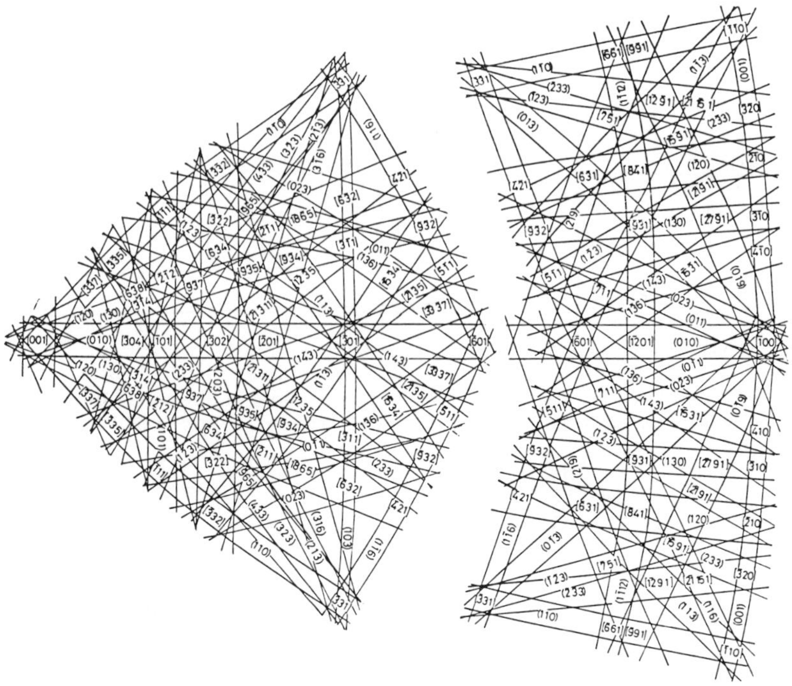

Fig.2 Indexing of Fig.1.

Figure 4 shows (001) pole figures for the pressed tapes and as-sintered pellet of $YBa_2Cu_3O_{7-x}$ which are accomplished by the analysis of ECPs. These pole figures are illustrated in correspondence to crystallographic orientation for electropolished surface of specimens. Figure 4(a) and 4(b) indicate that the (001) plane of each grain in the 60μ m specimen is concentrated strongly on the center but that in the 200μ m specimen is not. The inclined angle of the (001) plane in each grain to the specimen surface is within 30° and 50° for the 60μ m and 200μ m thick tapes, respectively. Figure 4(c) shows that the (001) plane of each grain in the as-sintered pellet is randomly oriented to the specimen surface. Pole figures by X-ray technique for the 60μ m and 200μ m thick tapes are shown in Fig.5(a) and 5(b). It was found that c-axis of grains is inclined within 30° to the surface normal in Fig.5(a). In the case of 200 μ m thick tape as shown in Fig.5(b), inclined angle is more disperse to the surface normal. These results are in excellent agreement with those from analysis of ECP as shown in Fig.4(a) and 4(b).

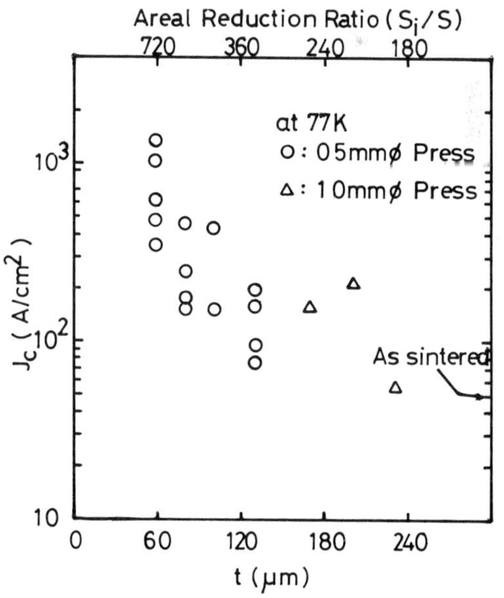

Fig.3 Jc as a function of pressed tape thickness and areal reduction ratio.

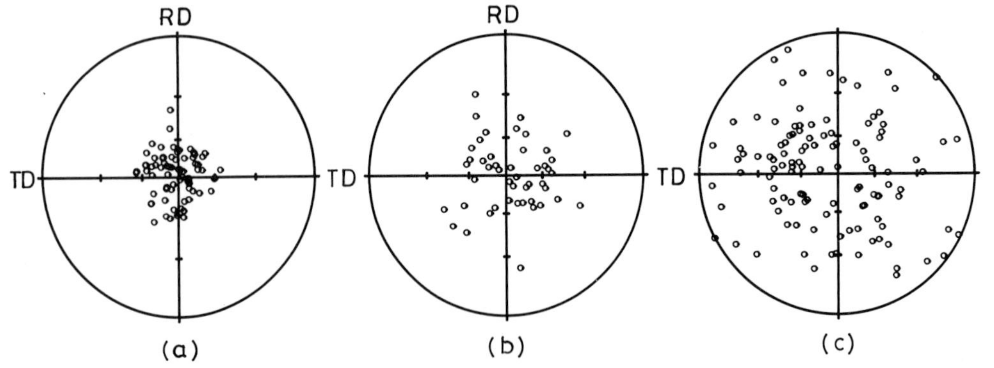

Fig.4 (001) pole figures for the pressed tapes of 60μ m (a), 200μ m (b) in thickness and as-sintered pellet (c).

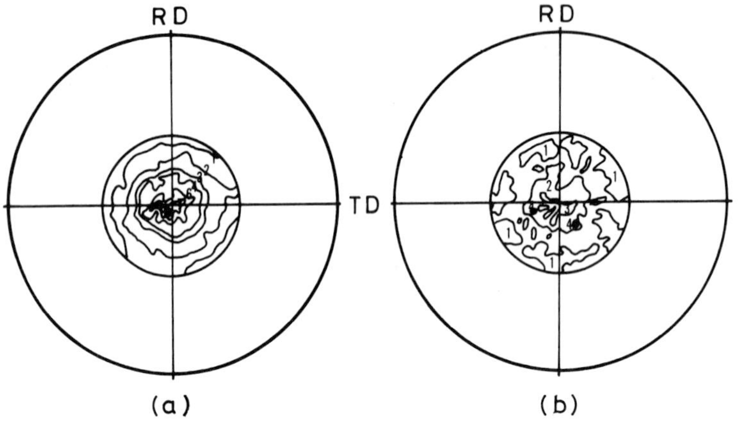

Fig.5 (006) pole figures for the pressed tapes of (a)60μ m, (b)200μ m

CONCLUSION

The standard channeling map for $YBa_2Cu_3O_{7-x}$ superconductors in the range of a stereographic quadrant was produced. Analysis of orientation distribution of pressed tapes in which Jc was measured was performed by taking ECPs for individual grains. In the 60μ m thick tape, (001) texture is considerably strong and (001) planes of most grains are inclined within 30° to the specimen surface. In the 200μ m thick tape, (001) texture is weak in comparison to the 60 μ m tape. No apparent preferred orientation is observed in the as-sintered pellet. High Jc in pressed $YBa_2Cu_3O_{7-x}$ polycrystals is ascribed to (001) texture. Pole figures by X-ray technique are in excellent agreement with results from analysis of ECP. It is concluded that the ECP technique is very useful for the analysis of orientation distribution in small sized specimens such as tapes or thin films.

REFERENCE

1. Knorr DB, Livingston JD (1989) Supercond.Sci.Technol. 1: 302
2. Choi CS, Prask HJ, Ostertag CP (1989) J.Appl.Cryst. 22: 465
3. Arendt RH, Gaddipati AR, Garbauskas MF, Hall EL, Hart HR, Lay JR.,KW, Livingston JD, Luborsky FE, Schilling LL (1988) Mat.Res.Soc.Proc.Vol. 99: 203
4. Verhoeven JD, Gibson ED (1988) Appl.Phys.Lett.52: 1190
5. Verhoeven JD, Gibson ED, Chumbley LS, McCallum RW, Baker (1988) J.Appl. Phys. 64: 761
6. Kang WJ, Hanada S, Yoshimi K, Saito S, Murayama Y, Hayashi S,Nagata A (1990) J.Appl.Phys. to be published
7. Osamura K, Takayama T, Ochiai S (1989) Supercond.Sci.Technol. 1: 302

3000 A cm^{-2} Thick Film YBCO: Preparation and Flux Pinning Characteristics

K.N.R. Taylor, A. Bailey, D.N. Matthews, and G.J. Russell

Advanced Electronic Materials Group, School of Physics, The University of New South Wales, Kensington NSW 2033, Australia

ABSTRACT

By optimizing all aspects of the production method for the melt processing of thick YBCO films on YSZ substrates it has been possible to produce current carrying capacities which are high enough for device fabrication. The details of these processing techniques will be given, and the effects of variations in the different parameters discussed. The origin of the big increase in critical current density appears to be the result of high texturing in the film and improvement of the intergranular contact resistances.Measurements of the variation of $j_c(H)$ show a better performance than has been obtained for films prepared by normal sintering techniques. The flux pinning energies of these films have been measured and shown to be comparable to those of bulk YBCO material.

KEY WORDS: YBCO, Thick Films, Flux Pinning Energies

INTRODUCTION

Since the discovery of the high temperature superconducting YBCO material, many groups have attempted to produce high quality superconducting thick (5-250 μm) films on various substrates [1-2]. Unfortunately substrate/film interactions and incompatible thermal expansion coefficients have led most workers to use rapid, low temperature (~930°C) processing cycles which result in weakly interconnected granular films that are characterised by low critical current densities. It has been found recently [3,4] that the use of yttria stabilized zirconia substrates allows for high temperature film processing above the peritectic decomposition temperature without significant film contamination. Scanning electron microscopy and associated x-ray electron dispersive spectroscopy show that an inert $BaZrO_3$ barrier layer with dispersed CuO needles and very small (211) YBCO crystals forms at the film/substrate interface, a result previously found [5], which inhibits atomic diffusion and film poisoning.

However, there are many other processing parameters that lead to the optimized route for the production of high current carrying capacity (~3000 A cm^{-2}) thick YBCO films. Details of these parameters are given below, along with the dependence of the superconducting characteristics on the processing conditions.

EXPERIMENTAL

Thick film formation is a complex route to obtaining surface coatings of thickness <5 μm to >200 μm. While it consists essentially of spreading a suitably prepared ink over the surface, and subsequently thermally processing the layers, the final products depend critically upon a number of parameters. Those include the precursor powder for the inks, the nature of the carrier liquids and the presence of additives to increase grain connectivity and possible flux pinning energies. It also includes the processing temperatures and times once the surface has been coated, the layer thickness and the substrate material.

The production of YBCO thick films is no exception to this complexity and in the following we describe the optimization of a number of these parameters for films formed on YSZ substrates. Specifically, we have studied the effects of the reactivity of the precursor powder on the final film performance, the effects of processing temperature, the enhancement of superconducting parameters which can be obtained by silver addition and the use of impurity phases in an attempt to improve the flux pinning energies.

RESULTS AND DISCUSSION

(I) PRECURSOR POWDER

Precursor powder reactivity was controlled by varying the sintering times for the precursor YBCO material used to prepare the final inks. A double processing cycle (reaction and sintering) was used for the production of YBCO pellets, with equal times for both cycles at 900°C. The sintered pellets were then

ground to ~ 1 μm particle size, suitable for the inks, mixed ultrasonically in triethanolamine and the ink applied to the polished surfaces of YSZ substrates by the doctor blade method.
For optimization of the precursor powder, ad hoc film processing conditions were used which involved heating to 1015°C at 5°C/min, firing at 1015°C for 5 minutes, and cooling to room temperature at 2°C/min. These conditions had previously been shown to give a reasonable film quality.

Figure 1 shows the temperature variation of resistivity for the final films prepared in this manner. The transition improved markedly to the 2x24 firing time, above which the normal state resistivity increases slowly. Table 1 tabulates the transition temperature T_c, the critical current density, j_c, and the c-axis texturing figure of merit, $F= I_{006}/I_{103}$, for each film type. A range is given for each j_c value listed and represents the variation in this quantity found for a number of films prepared from the same starting powder processed under identical conditions.

Table 1. Film characteristics for Various Processing Conditions

Powder batch	T_c (K)	j_c at 77K ($A\ cm^{-2}$)	Barrier layer thickness (μm)	YBCO film thickness (μm)	c-axis texturing $\frac{I_{006}}{I_{103}}$
2x6 hr	87	25-35	6.3	3.4	0.37
2x12 hr	90	80-100	8.5	11.8	0.67
2x24 hr	91.5	750-1200	9.6	15.0	1.05
2x48 hr	91.5	200-300	≤0.2	14.0	1.52

The well defined peak in j_c at 2x24 hour firing time we believe to be due to optimization of both the reactivity of the powder and its conversion to the YBCO compound. This improved reactivity enables the formation of the $BaZrO_3$ barrier layer and the smaller j_c values at shorter firing times is almost certainly due to inadequate reaction of the starting compounds.

The high value of critical current density obtained after this initial optimization step is already close to an order of magnitude improvement in performance over previous work.

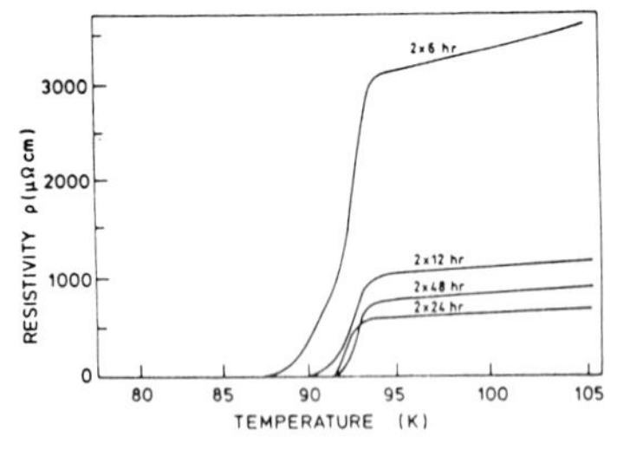

Fig. 1: Temperature variation of resistivity, near the transition temperature, for type different precursor powders.

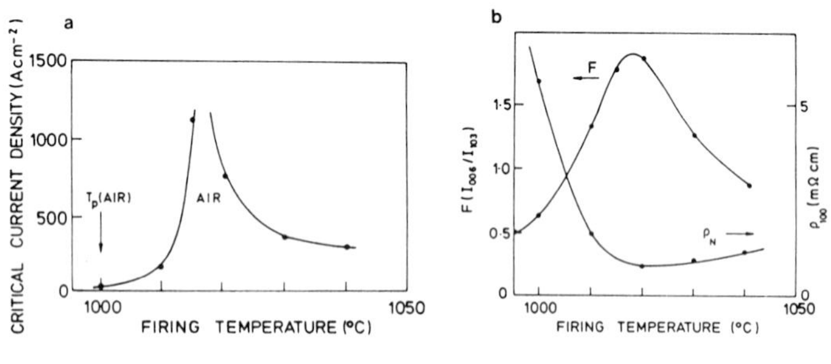

Fig. 2: (a) Optimization of the critical current density with film firing temperature.
(b) Variation of the c-axis texturing figure of merit and the normal state resistivity corresponding to (a).

(II) FILM FIRING TEMPERATURE

Using ink prepared from the optimized (2x24 hour) precursor powder, an investigation of the effect of the firing temperature, over the range 1000-1040°C in air, on the film properties again showed a well developed maximum in the critical current density as shown in Fig. 2(a). Figure 2(b) shows the corresponding variation of the c-axis texturing figure of merit and the normal state resistivity.

(III) SILVER-ADDITION

In an attempt to enhance grain growth and improve the intergranular connectivity , increasing amounts of very fine silver powder were added to the optimized (123) ink before film fabrication. The silver-addition was examined over the concentration range 0-15 wt % Ag using the above optimized processing temperature of 1015°C, and once the optimum silver concentration was determined, the film firing temperatures were again varied to determine if the optimum temperature had changed.

Figure 3 shows the dramatic influence silver-addition has on both the degree of c-axis texturing and the critical current density, the latter reaching a peak value of 2857 A cm^{-2} at 77K for 10 wt %Ag, approximately an order of magnitude improvement over films prepared by more conventional methods [1,6] and a factor of three improvements over the pure YBCO melt processed films.

Figure 4 shows the effect of varying the film firing temperature for the optimum silver concentration of 10 wt % Ag. Again, a temperature of 1015°C gives the optimum film properties. Despite the high j_c and c-axis texturing values, the current capacity of the films is still strongly dependent on external magnetic fields and as Fig. 5 shows for magnetic fields in the range 500 to 3500 Oe and either parallel or perpendicular to the film surface, the j_c value for the optimum film is approximately constant, and has the value of 85 A cm^{-2}.

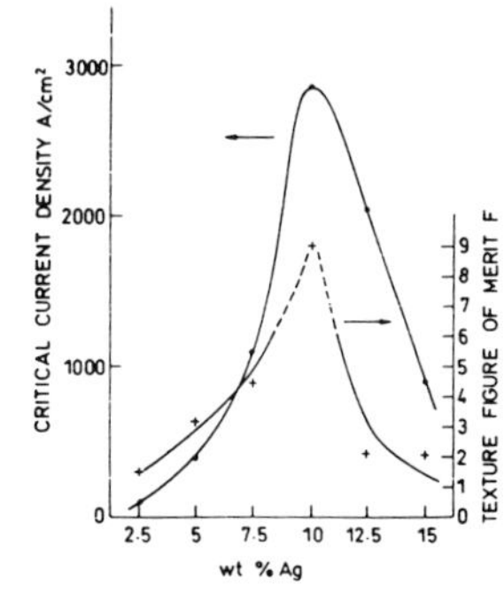

Fig. 3: Variation of c-axis texturing figure of merit and critical current density with silver concentration.

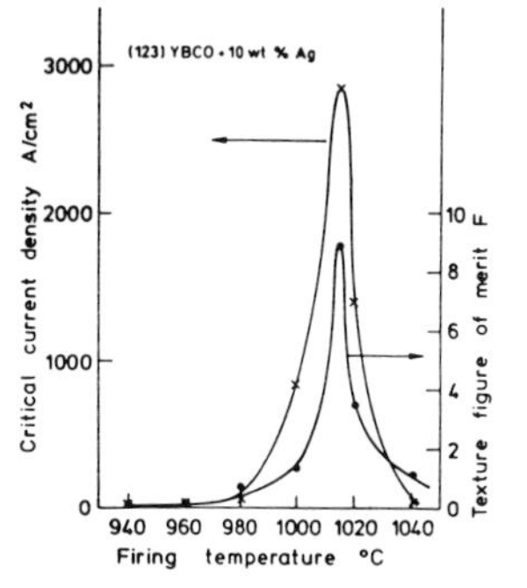

Fig. 4: Variation of critical current density and c-axis texturing figure of merit with film firing temperature.

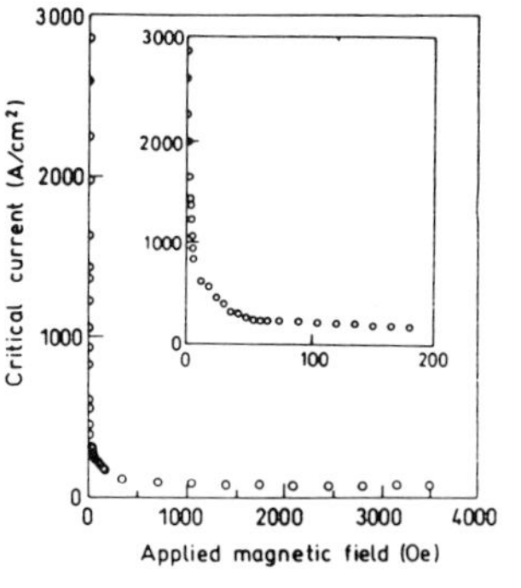

Fig. 5: Variation of j_c, for an optimum thick film, as a function of applied magnetic field which is perpendicular to the film surface.

(IV) FLUX PINNING ENERGIES

For the resistive transition in the presence of magnetic fields the exponentially decreasing resistivity on the low-temperature side of the transition is attributed to flux creep, and from flux creep theory, the resistivity ρ due to flux creep is given by [7] $\rho = \rho_o \exp(-U/kT)$ where the activation energy for flux motion, U, is temperature and magnetic field depend. In the following analysis we assume that U is independent of T as only a narrow range of low temperatures has been used.

Figure 6 shows the resistive transition for the (123)+10 wt % Ag film at a field of 2100 Oe and various current densities. Expansion of the flux creep region indicates that U is almost independent of current density and to first order it will be assumed that U is field dependent only.

Figure 7 shows the resistive transition for a high quality (123) + 10 wt % Ag film at a fixed current density of 143 A cm^{-2}, and various magnetic fields. Assuming that the curves in the flux creep region may be approximated by straight lines, then the field dependent pinning energies shown in Table 2 are found.

Table 2: Pinning Energies for Flux Motion in Various Thick Films

Applied Magnetic Field (Oe)	U (meV) (123)	123+10wt%Ag	1%Zr	3% Zr	4% Zr
350	195	166	221	173	186
700	161	149	189	134	113
1050	147	125	160	113	115
1400	138	129	149	110	
2100	138	125	120	96	107
2800	128	121	120	96	99

The flux pinning energies for pure (123) and (123) + 10 wt % Ag films are essentially the same and comparable to those of bulk YBCO material [8]. These results suggest that the improvements introduced by the presence of silver arising dominantly from improvements in the microstructure of the films (grain size, connectivity, texturing, etc.) rather than from changes in the flux pinning mechanisms.

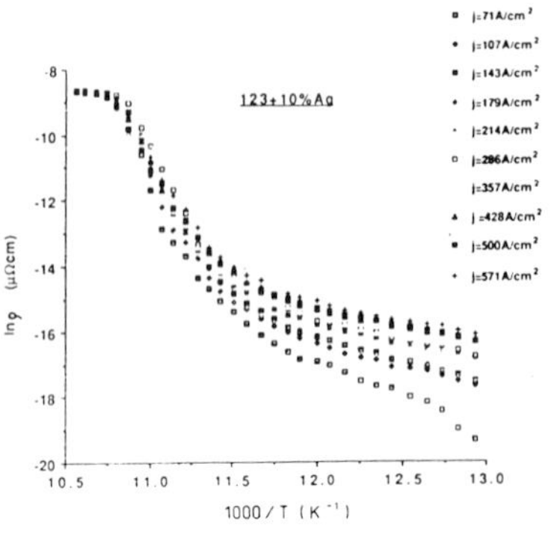

Fig. 6: Arrhenius plot of the resistivity for the (123) + 10 wt % Ag film at a magnetic field of 2100 Oe and various current densities.

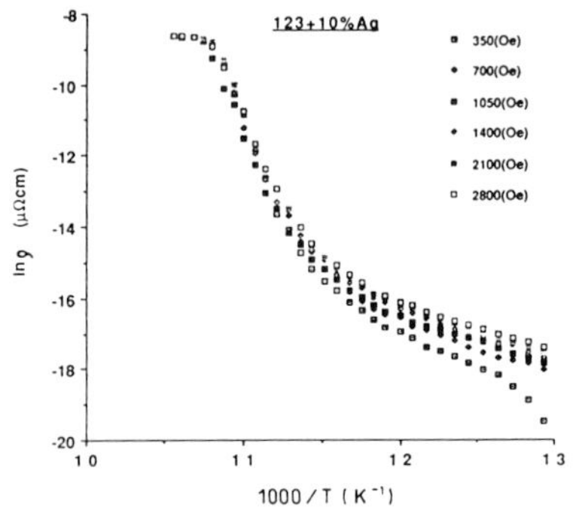

Fig. 7: Arrhenius plot of the resistivity for a (123) + 10 wt % Ag film at a constant current density and various magnetic fields.

(V) Zr AND (211) ADDITIONS

In an attempt to improve the above pinning energies small additions of Zr and (211) powders were added to the high quality pure (123) ink before film fabrication. Using the standard optimized film processing route detailed above the superconducting properties of these films are given in Table 3, while a selection of activation energies for the (123) + Zr films is given in Table 2.

Table 3: Film Characteristics for Various Zr and (211) Additions

(123)+wt%	Addition	T_c (K)	j_c A cm^{-2}	Film Thickness (μm)
	0	91.5	~1000	12
Zr	1	91.5	485	10
	3	90.5	109	40
	4	91.0	62	50
	1	91.5	383	15
(211)	2	91.0	253	16
	3	90.5	186	15
	4	90.5	120	18

The Zr and (211) additions significantly reduce the critical current density of the pure (123) film, but have little effect on T_c (R=0). However, the additions do change the flux energetics for the thick film system and further work is being undertaken to illucidate these changes.

CONCLUSIONS

It has been shown, that by careful and systematic optimization of the processing parameters it is possible to improve the current carrying capacity of YBCO thick films by at least an order of magnitude. The origin of the increase in j_c appears to be the result of high c-axis texturing in the film and improvement of the intergranular contact resistances. Measured activation energies for pure (123) and (123) + 10 wt % Ag films are essentially the same and comparable to those of bulk YBCO material. These results suggest that the improvements introduced by the presence of silver arises dominantly from improvements in the microstructure of the films. Attempts to improve the flux pinning energies by the use of Zr and (211) additions to the (123) ink have not been successful, however, the flux energetics in the thick film system was changed.

REFERENCES

1. Aponte JM, Octavio M (1989) J Appl Phys 66: 1480.
2. Lakshimi MVS, Ramkumar K, Satyam M (1989) J Phys D: Appl Phys 22: 373.
3. Bailey A, Town SL, Alvarez G, Russell GJ, Taylor KNR (1989) Physica C 161: 347.
4. Shields TC, Wellhofer F, Abell JS, Taylor KNR, Holland D (1989) Physica C 162-164: 1265.
5. Tabushi J, Utsumi K (1988) Appl Phys Lett 53: 606.
6. Sacchi M, Soritti F, Morten B, Prudenziati M (1988) Appl Phys Lett 53: 1110.
7. Anderson PW (1962) Phys Rev Lett 9: 309.
8. Matthews DN, Russell GJ, Taylor KNR (1990) Physica C, to be published.

Limiting Factors of Critical Current for Bi-Pb-Sr-Ca-Cu-O Oxide Superconductor

TSUNEYUKI KANAI, TOMOICHI KAMO, and SHIN-PEI MATSUDA

Hitachi Research Laboratory, Hitachi Ltd., 4026, Kuji-cho, Hitachi, Ibaraki, 319-12 Japan

ABSTRACT

Critical current limiting factors for Bi-Pb-Sr-Ca-Cu-O oxide superconductor were investigated. A mixture of the oxide powders with a composition of $Bi_{1.84}Pb_{0.34}Sr_{1.9}Ca_{2.0}Cu_{3.1}O_x$ was calcined and the powders were placed in silver tubes. The tubes were processed by drawing and rolling to give resulted in 0.3mm thick tapes. The tapes were annealed at 825-843°C for 300h with intermediate pressing. Different annealing temperatures gave three types of samples with different microstructures. The critical current density Jc of the samples depended on the kind of foreign phases such as the Cu free (Bi,Pb)-(Sr,Ca)-O phase or 2201 phase as well as the a-b plane orientation. The weak link structure of the samples was an S-N-S junction regardless of the annealing conditions.

KEY WORDS: Bi-Pb-Sr-Ca-Cu-O superconductor, Ag-sheathed tape, Critical current, Preferred a-b plane orientation, Foreign phase

INTRODUCTION

Since a Bi-Sr-Ca-Cu-O oxide was discovered to have a high transition temperature Tc [1], its superconducting properties have been extensively investigated. The system contains three of superconducting phases, mainly characterized by their composition: the $Bi_2Sr_2Ca_2Cu_3O_x$ (2223) high-Tc phase, the $Bi_2Sr_2CaCu_2O_x$ (2212) low-Tc phase and the $Bi_2Sr_2CuO_x$ (2201) phase. The partial substitution of bismuth with Pb stabilizes the high-Tc phase [2].

Many efforts to put the material into practical use have been carried out. One of the most promising candidates is a superconducting magnet made of a metal-sheathed tape. For this application, critical current density is the most significant property. In order to obtain high critical densities, there are many factors to be controlled. One of the main factors is crystal orientation of the constituent grains. Due to the crystallographic anisotropy of the superconducting phases, a large transport current flows only when the a-b plane of each grain aligns along the direction of current flow. The preferred orientation of the grains has been reported to be achieved by repeated cold-working and heat treatment in the Ag-sheathed Bi(Pb)-Sr-Ca-Cu-O superconductor [3]. However, other main factors have not yet been examined, such as different superconducting phases, foreign phases and the combination of these phases and crystal orientation.

The purpose of this study was to investigate the effect of these factors on the Jc.

EXPERIMENTAL

The preparation of the tape specimens was essentially based on a powder-in-tube method. The mixture of oxide powders with a composition of $Bi_{1.84}Pb_{0.34}Sr_{1.9}Ca_{2.0}Cu_{3.1}O_x$ was calcined twice at 800°C for 10h. Silver tubes, filled with the powder, were processed by drawing and rolling to give a 0.3mm thick tape. The tape was annealed at three different temperatures, i.e., 825, 834 and 843°C for 100h. After this annealing, pressing at $50kg/mm^2$ and annealing at the same temperature were repeated twice.

Crystal phases of the samples were examined by X-ray diffraction analysis using Cu-Kα radiation. The crystallographic orientation of the grains was investigated by an X-ray pole figure technique. The cross-sectional microstructure was observed by SEM. The transport critical current measurements were performed in the temperature range of 77-4.2K with a 1μV/cm criterion. The magnetic field was applied parallel or perpendicular to the tape surface and perpendicular to the current flow up to 1T. The critical current density was defined as the current divided by the cross-sectional area of the oxide phase.

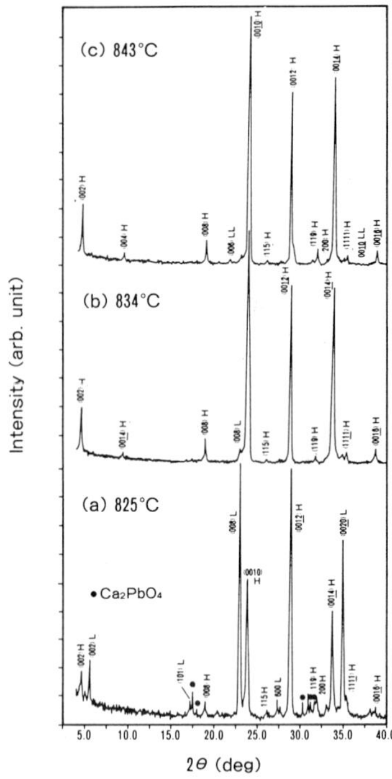

Fig.1. X-ray diffraction patterns of the samples annealed at three different temperatures; i.e., 825, 834 and 843°C.

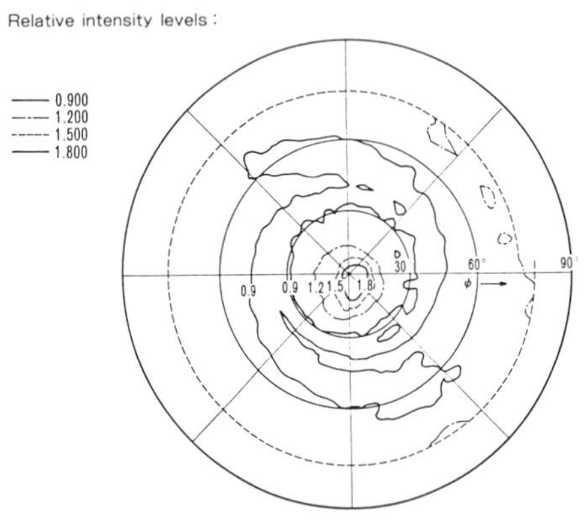

Fig.2. Pole figure of the (00<u>10</u>) crystallographic plane of the high-Tc phase from the surface of the tape annealed at 834°C.

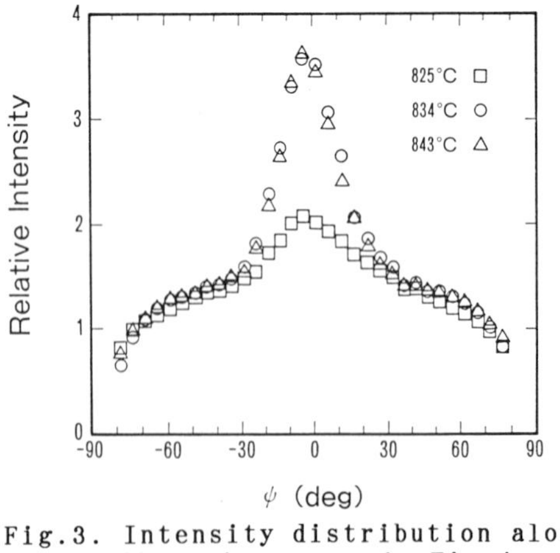

Fig.3. Intensity distribution along the ϕ direction shown in Fig.4.

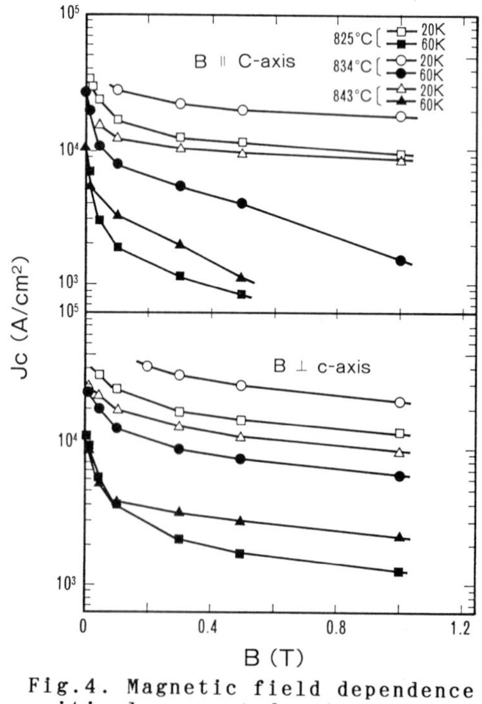

Fig.4. Magnetic field dependence on critical current density at 60K and 20K.

RESULTS AND DISCUSSION

X-ray diffraction patterns from the tape surface prepared at three different temperatures are shown in Fig.1. The sample prepared at 825°C consisted of a mixture of the high-Tc phase, low-Tc phase and small amounts of Ca_2PbO_4 phase. On the other hand, the samples prepared at 834°C and 843°C were almost single phased high-Tc phase containing very small peaks of other superconducting phases. The minor phases of the samples prepared at 834°C and 843°C were the low-Tc phase and the mixed low-Tc phase and 2201 phase, respectively. Fig.2 shows a pole figure of the (00<u>10</u>) crystallographic plane of the high-Tc phase from the surface of the tape sample prepared at 834°C. The intensity increased towards the center and an anisotropic intensity profile was observed in the angular dependence. This suggested that the c-axis of the high-Tc phase was directed, on average, perpendicular to the tape surface. The degree of the preferred orientation was evaluated from the intensity distribution along the transverse direction ϕ in Fig.2. Fig.3. shows the intensity distribution of the samples prepared at 825, 834 and 843°C. All samples had a preferred a-b plane orientation and the degree of the orientation was almost the same for the samples prepared at 834 and 843°C, and smaller for 825°C. These crystallographic analysis revealed that the samples prepared at three different temperatures consisted of different crystal structures and different degrees of a-b plane orientation.

The different crystal structures of the three samples are expected to affect the transport properties in different ways. Fig.4 shows the magnetic field dependence on the critical current density Jc at 60K and 20K. In general, the Jc values showed a large difference between the magnetic field directions to the tape surface regardless of the heat treatment conditions and the magnitude of the magnetic field; i.e., Jc was higher when the magnetic field was applied parallel to the tape surface than the perpendicular direction. This resulted from the preferred a-b plane orientation of the samples. The tendency decreased with decreasing temperature. However, particular attention should be called to the annealing conditions, the order of the Jc values did not correspond to the order of the degrees of the a-b plane orientation shown in Fig.3. That is to say, the Jc of the sample prepared at 843°C, which had the preferred a-b plane orientation equivalent to the 834°C sample, showed almost the same Jc as the sample prepared at 825°C with lesser a-b plane orientation. This suggested that the Jc strongly depended on different superconducting phases or foreign phases as well as the a-b plane orientation.

The distribution of these phases in the matrix was examined by SEM. Fig.5 shows cross-sectional SEM photographs of the samples prepared at the three different temperatures. The sample annealed at 834°C consisted of the high-Tc phase with small amounts of CuO phase. No foreign phases such as the 2212 phase were observed in the matrix. In the sample annealed at 825°C, the net-like Cu free (Bi,Pb)-(Sr,Ca)-O phase was distributed in the matrix, while the 2201 phase was precipitated around the grain boundaries in the sample prepared at 843°C. This meant that the kind of grain boundary between the high-Tc phase grains was significantly different in each sample. In order to examine the grain boundary properties of the samples, it was important to evaluate the temperature dependence of the critical current. Typical examples of the temperature dependence of the critical current density under the magnetic field of 0.1T are shown in Fig.6. The Jc dependence on the temperature seemed to follow the behavior of the S-N-S proximity junction, which satisfied eq. (1) [5].

$$Jc(T) \sim (1-(T/Tc))^2 \exp(-2d_n/\xi_N) \qquad (1)$$

where d_n is thickness of a normal layer and ξ_N is length of an electron pair diffusion into the normal phase. Fig.7 shows the dependence of the $Jc^{1/2}$ on the temperature for the samples. The Jc dependence on the temperature of the three samples was linear. This suggested that the weak link structure of the samples was an S-N-S junction regardless of the annealing conditions. From eq. (1), the slope became smaller for the larger d_n/ξ_N, which meant that the larger d_n and/or the shorter ξ_N gave the smaller slope. The slope of the lines in Fig.7, an index of proximity between superconductors, was in the descending order of the sample prepared at 834, 825 and 843°C.

In summary, critical current limiting factors for Bi-Pb-Sr-Ca-Cu-O oxide superconductor were investigated by changing annealing conditions. The critical current density strongly depended on foreign phases such as the Cu free (Bi,Pb)-(Sr,Ca)-O phase or 2201 phase as well as the a-b plane orientation. In the sample prepared at 843°C, the 2201 phase controlled Jc more than the a-b plane orientation. The weak link structure of the samples was an S-N-S junction, regardless of the annealing conditions.

REFERENCES

1. Maeda H,Tanaka Y, Fukutomi M, Asano T (1988) Jpn. J. Appl. Phys. 27: L209
2. Tanaka M, Takada J, Okada K, Kitaguchi H, Miura Y, Ikeda Y, Tomii Y, Mazaki H (1988) Jpn. J. Appl. Phys. 27: L1041
3. Asano T, Tanaka Y, Fukutomi M, Jikihara K, Machida J, Maeda H (1988) Jpn. J. Appl. Phys. 27: L1652
4. Uno N, Enomoto N, Kikuchi H, Matsumoto K, Miura M, Nakajima M (1989) 2'nd ISS Nov.14-17
5. Gennes P. G. (1964) Rev. Mod. Phys. 36: 226

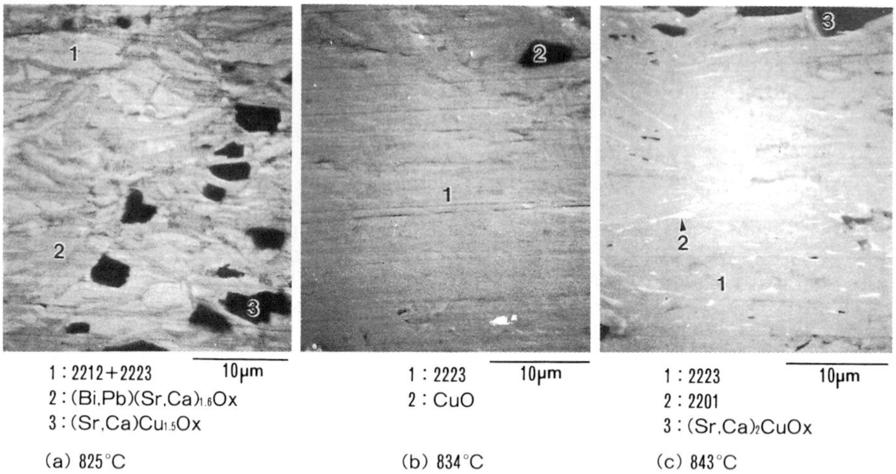

Fig.5. Cross-sectional SEM photos of the samples annealed at three different temperatures; i.e., 825, 834 and 843°C.

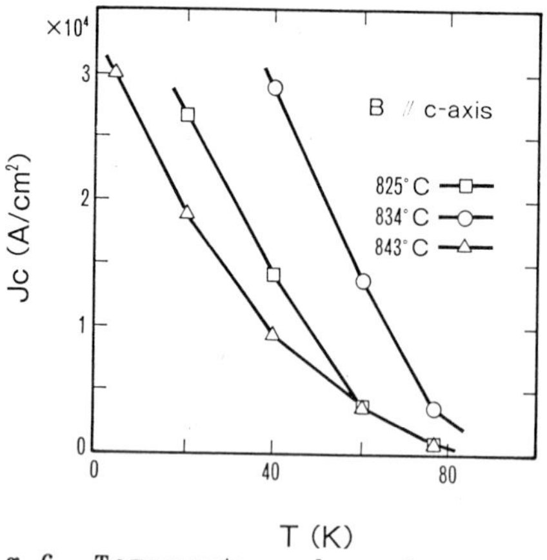

Fig.6. Temperature dependence of Jc for the samples annealed at three different temperatures; 825, 834 and 843°C.

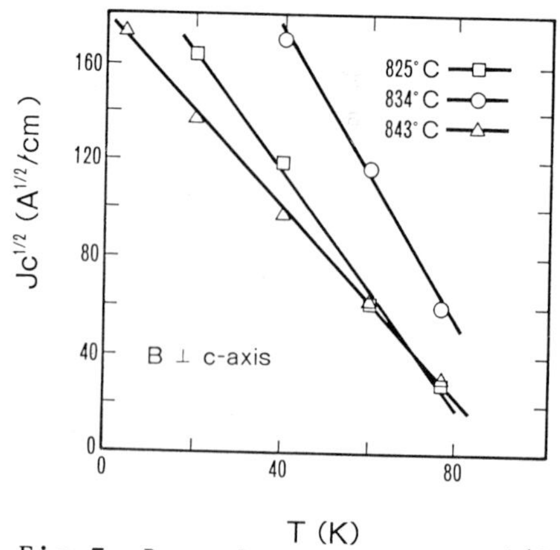

Fig.7. Dependence of the $Jc^{1/2}$ on temperature for the sample shown in Fig.6.

Anisotropy of the Critical Current in Layered Superconductor

TAICHI YAMAGUCHI[1], SHIN'YA AOKI[1], KAZUNORI ONABE[1], NAOHIRO FUTAKI[1], NOBUYUKI SADAKATA[1], OSAMU KOHNO[1], SHIGERU NAKANO[2], and YOSHINORI KUWASAWA[2]

[1] Fujikura Ltd., 5-1, Kiba 1-chome, Koto-ku, Tokyo, 135 Japan
[2] Department of Physics, Faculty of Science, Chiba University, 1-33, Yayoi-cho, Chiba, 260 Japan

ABSTRACT

The critical current density $J_c(B,T,\theta)$ of c-axis and ab-axis oriented $YBa_2Cu_3O_x$ films was measured at 77K as a function of angle between applied magnetic field and film surface. Furthermore, we made the same measurement for $Nb/Nb_{0.5}Zr_{0.5}$ multilayer. The c-axis oriented $YBa_2Cu_3O_x$ film and the metallic multilayer showed analogous characteristics in the anisotropy of J_c caused by the layer structure.

KEYWORDS: anisotropy of J_c, layer structure, flux pinning

INTRODUCTION

In an oxide high-T_c superconductor (HT_cS), impurities, defects, precipitates and grain boundaries will not act as strong pinning centers because of the very long penetration depth and very short coherent length, whereas experimental data showed that the c-axis oriented high quality films fabricated by chemical vapor deposition and laser deposition had a J_c value more than $10^6 A/cm^2$ at 77K, zero tesla.[1-3] These results require a strong pinning force acts on the vortices in the oxide HT_cS. Recently, theoretical approach has been performed by Tachiki et al.[4,5] They proposed that the layer structure itself in the oxide HT_cS may work as strong pinning centers for the vortices that axis are parallel to the layers. According to this pinning model, the J_c of c-axis oriented HT_cS films may be strongly dependent on the direction of the applied magnetic field. In actuality Roas et al.[6] observed the anisotropy of critical current of c-axis oriented YBCO film prepared by laser deposition. In this paper, we investigate the change of J_c for ab-axis and c-axis oriented YBCO films as a function of angle *theta* between applied magnetic field and film surface. Further, we tried to make same measurements for Nb/NbZr multilayer which has composed of two superconducting materials with equal T_c and different electronic diffuse constants. We try to make a discussion that the c-axis oriented HT_cS and the metallic multilayer show the similar characteristics in the anisotropy of critical current density.

EXPERIMENTAL

The YBCO film was prepared by chemical vapor deposition (CVD) process on a single crystalline $SrTiO_3(100)$ substrate. The process in detail has been reported in previous paper.[7,8] Two different crystallographic orientation of YBCO film were designated. One was that the (nn0) plane of the crystal was parallel to the substrate surface (ab-axis orientation). The other was that the (00n) plane was parallel to the substrate (c-axis orientation). Figure 1 shows the X-ray diffraction pattern by a *theta*-2*theta* method using Cu-K_{alpha} radiation, where the X-ray was turned off near 23° and 46° to avoid the strong peaks from substrate. The ab-axis oriented film had a T_c=89K, and c-axis oriented film had a T_c=92K. The J_c of YBCO film in liquid N_2 was measured by DC four probe method. The angle between applied magnetic field and film surface was variable from -5° to 95°. The Nb/NbZr multilayer was prepared by magnetron sputtering from separate Nb and NbZr targets.[9] This film has a period d=54nm (d_{Nb}=24nm, d_{NbZr}=30nm) for emulating the oxide superconductor.[10]

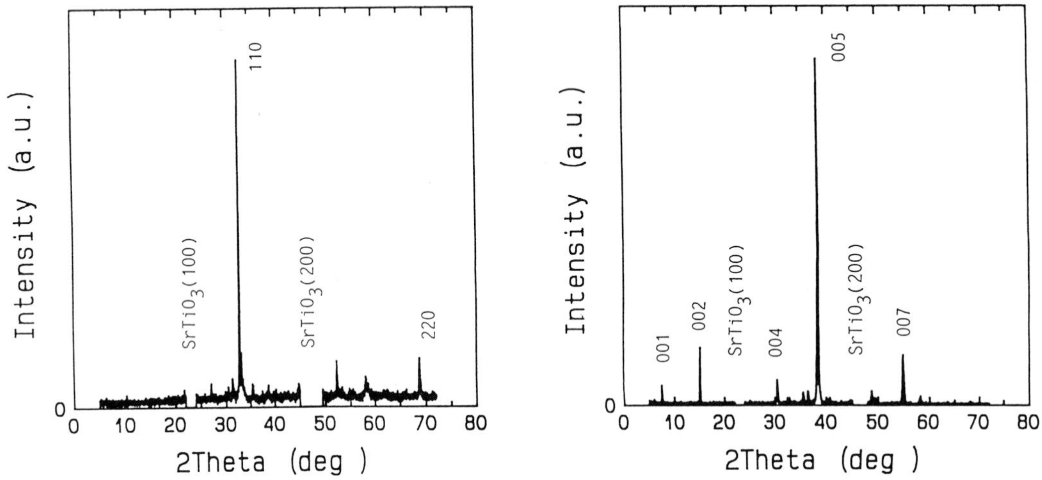

Figure 1. X-ray diffraction pattern of the YBCO film (CVD) deposited on a $SrTiO_3$(100) substrate. (Left:ab-axis orientation, Right:c-axis orientation)

RESULTS AND DISCUSSION

The anisotropy of volume pinning force (F_p) in YBCO films was shown in figure 2. The value of F_p is fairly small when the magnetic field applied perpendicular to the film surface unrelated to the crystallographic orientation of films. However, both the value of $F_p(Z_{//})/F_p(Z_\perp)$ and $b(Z_{//})/b(Z_\perp)$, b means the value of B where $(dF_p/dB)=0$, was small in case of the c-axis oriented film. This phenomenon suggests that the c-axis oriented film had some strong pinning centers for the vortices whose axis are perpendicular to the c-axis.

The angular dependence of J_c is displayed in figure 3. There is a peak near *theta*=0 for both ab-axis and c-axis oriented films. But the J_c(*theta*) function is more anisotropic in case of c-axis oriented film. Additionally, the peaks near *theta*=0 become narrower with increasing applied magnetic field. The shape of function J_c(*theta*) in magnetic field above 3[T] fitted by the expression obtained due to Tachiki et al.[5] by assuming the vortices pinned in extrinsic pinning centers such as twin-boundaries and the trapped vortices in the intrinsic pinning layers. It is given by

$$J_c = \frac{J_{c2}}{|\cos(90-theta)|^{1/2}} \tag{1}$$

where, J_{c2} is the value of critical current density when the magnetic field applied perpendicular to the film surface.

The Nb/NbZr multilayer that superconducting order parameter localized in Nb layers showed the similar characteristics in the anisotropy of J_c with the c-axis oriented YBCO films in high magnetic field.[10] Figure 4 displays the reduced J_c as a function of tilted angle for the Nb/NbZr multilayer, The shape of J_c(*theta*) also fitted to Eq.(1).

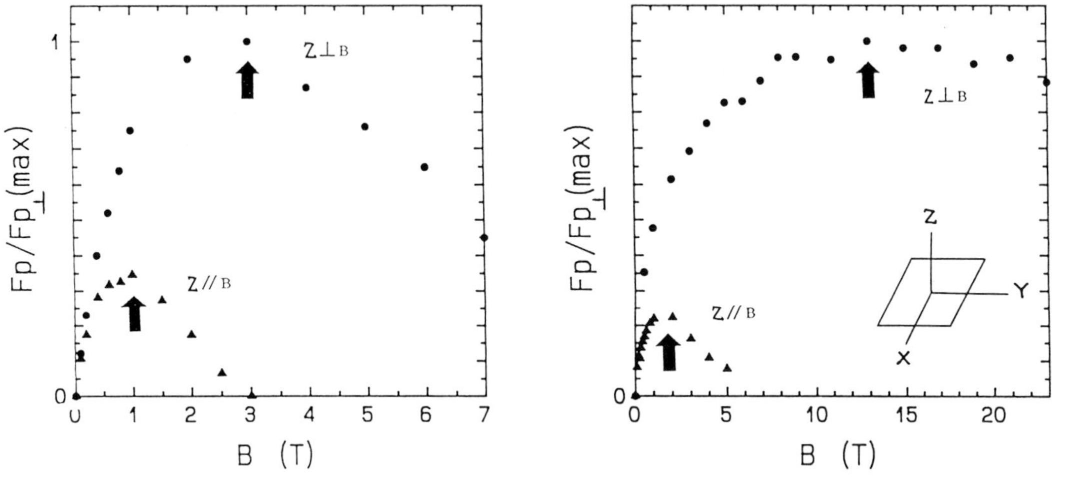

Figure 2. Anisotropy of the volume pinning force in YBCO films.(L:ab-axis, R:c-axis) The maximum F_p value of ab-axis and c-axis oriented film is $5.5x10^5$ (3[T]), $3.25x10^7$[dyn/cm^3] (13[T]) respectively.

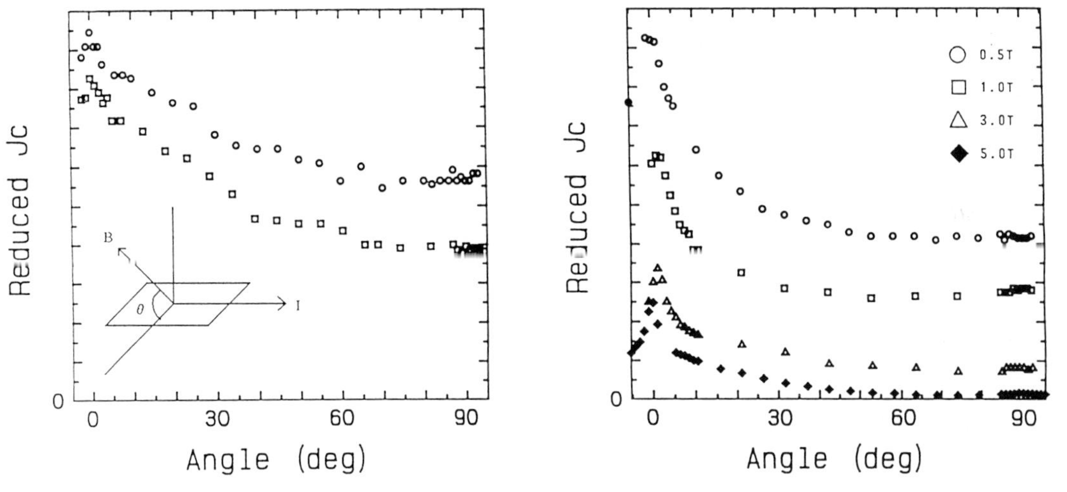

Figure 3. Angular dependence of the J_c in YBCO films as a function of tilted angle *theta*. (L:ab-axis, R:c-axis) The J_c value at *theta*=0 of ab-axis and c-axis oriented film in B=1[T] is $1.4x10^3$, $2.2x10^4$[A/cm^2] respectively.

CONCLUSION

We have demonstrated the angular dependence of critical current density in magnetic field of the ab-axis and c-axis oriented YBCO films and Nb/NbZr multilayer composed of two superconducting materials. We found that c-axis oriented YBCO film and metallic multilayer showed the similar characteristics in the anisotropy of critical current density caused by the layer structure. The behaviors of our c-axis oriented YBCO film in high magnetic field at 77K fitted in Tachiki's model. The intrinsic and extrinsic pinning centers determine the critical current density in layered superconductor for particular direction of the magnetic field. The pinning effect caused by the layer structure may depend on the strength of magnetic field.

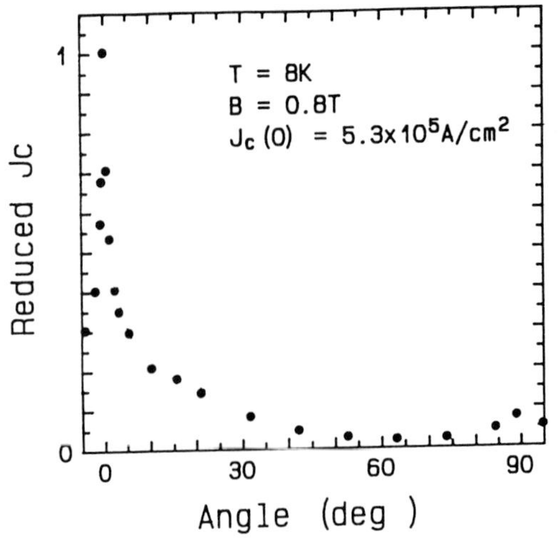

Figure 4. Angular dependence of the J_c in Nb/NbZr multilayer as a function of tilted angle *theta*.

Acknowledgment ---- The measurements of J_c at high magnetic field were performed at High Field Laboratory for Superconducting Materials, Institute of Materials Research, Tohoku university by use of 23[T] hybrid magnet HM-2. The authors are grateful to Prof. K.Watanabe and the stuffs of IMR. Preparation of YBCO thin films by chemical vapor deposition method has been collaborated with Chubu Electric Power Co.

REFERENCES

[1] K.Watanabe, H.Yamane, H.Kurosawa, T.Hirai, N.Kobayashi, H.Iwasaki, K.Noto, Y.Muto, Appl. Phys. Lett. 54 (6) 6.February 1989

[2] S.Matsuno, F.Uchikawa, K.Yoshizaki, Jpn. J. Appl. Phys. Vol.29, No.6, June 1990, pp.L947-L948

[3] B.Roas, L.Schultz, G.Endres, Journal of the Less-Common Metals, 151 (1989) 413-418

[4] M.Tachiki, S.Takahashi, Solid State Communications, Vol.70, No.3, pp.291-295, 1989

[5] M.Tachiki, S.Takahashi, Solid State Communications, Vol.72, No.11, pp.1083-1086 1989

[6] B.Roas, L.Schultz, G.Saemann-Ischenko, Phys. Rev. Lett. volume 64, number 4, 479 (1990)

[7] T.Yamaguchi, S.Aoki, N.Sadakata, O.Kohno, H.Osanai, Appl. Phys. Lett. 55 (15), 9 October 1989

[8] S.Aoki, T.Yamaguchi, N.Sadakata, A.Kagawa, O.Kohno, Proceedings of 2nd International Symposium on Superconductivity (ISS89) pp.439 (1989)

[9] Y.Kuwasawa, U.Hayano, T.Tosaka, S.Nakano, S.Matsuda, Physica C 165 (1990) 173

[10] Y.Kuwasawa, T.Yamaguchi, T.Tosaka, S.Aoki, S.Nakano, Physica C 169 (1990) 39

Angular Dependence of Magnetic Field on Critical Current Density in $Tl_2Ba_2Ca_2Cu_3O_x$ Thin Film

TOSHIHIDE NABATAME, YUKIO SAITO, KATSUZO AIHARA, TOMOICHI KAMO, and SHIN-PEI MATSUDA

Hitachi Research Laboratory, Hitachi Ltd., 4026, Kuji-cho, Hitachi, Ibaraki, 317 Japan

ABSTRACT

The angular dependence of magnetic field on the critical current density(Jc) has been investigated for a polycrystalline $Tl_2Ba_2Ca_2Cu_3O_x$(2223) thin film which was prepared on a MgO(100) substrate using excimer laser ablation and post-annealing. The film was composed entirely of the 2223 phase and had a highly c-axis oriented structure. The film showed Jc of $1.5X10^5A/cm^2$, which correspond to 9.8A, in zero magnetic field at 77K. A drastic decrease in Jc was observed under magnetic fields with increasing angle between the film plane and the magnetic field. When this angle was equal to 0 and 90 degrees with B⊥I, the Jc was $1.3X10^5A/cm^2$(1T) and $1.3X10^3A/cm^2$(0.7T), respectively. Magnetic field dependence on Jc was formulated as a function of perpendicular component of the magnetic field to the a-b plane.

KEYWORD: $Tl_2Ba_2Ca_2Cu_3O_x$ thin film, $1.5x10^5A/cm^2$ at 77K, Angular dependence of B on Jc

INTRODUCTION

Since the discovery of the $Tl_2Ba_2Ca_2Cu_3O_x$(Tl-2223) superconductor with the highest critical temperature(Tc) of 122K[1], many studies have been done to fabricate a single phase film of this system. Recently, Liou et al.[2] prepared a $Tl_2Ba_2Ca_2Cu_3O_x$ superconducting thin film with Tc of 110K and Jc of $1X10^4A/cm^2$ at 77K, 0T using a Nd:YAG laser evaporation. Itozaki et al.[3] have achieved a Tc of 115K and Jc of $3.2X10^6A/cm^2$ at 77K, 0T. They observed a very large degradation of the Jc by the magnetic field. In a previous paper[4], we fabricated Tl-2223 thin films with a single 2223 phase and highly c-axis oriented structure.

In this letter, we report their superconducting properties, especially angular dependence of Jc on magnetic field.

EXPERIMENTAL

The details of the films preparation by laser ablation were reported in the previous paper[4]. Briefly, an excimer laser with 308nm wavelength, 20ns pulse duration, 10Hz repetition rate and 250mJ pulse energy was used to deposit an amorphus film from a $Tl_{1.6}Ba_2Ca_2Cu_3O_x$ target. The film was deposited on single-crystal MgO(100) at 100°C for 20min. The as-deposited film was sandwiched between two $Tl_2Ba_2Ca_2Cu_3O_x$ pellets and calcined in a sealed alumina crucible at 840°C for 20h. X-ray analysis confirmed a highly c-axis oriented structure and a single 2223 phase. The sample dimension was about $5X6mm^2$ and 1.3μm thick. Au-Pd stripes were deposited by rf-sputtering on four separated parts of the film and Ag wires were bonded to each part using Ag paste.

Critical current densities were measured in magnetic fields up to 1T generated by a splitting magnet. For the angular dependence measurements, the angle(θ) between the the applied field and plane of the film was varied from 0 to 90 degree with a resolution of 0.1 degree by a computer controlled stepping motor. The current direction in the measurements was held perpendicular to the field.

The cross section of the Tl-2223 thin film was also studied by transmission electron microscopy(TEM).

RESULTS AND DISCUSSION

Fig. 1 is a low magnification TEM image of the Tl-2223 film, together with the corresponding electron diffraction pattern. Most of the grain boundaries were parallel to the substrate surface. The incident beam is parallel to (110) of the two grains, respectively. The diffraction patterns of upper and lower grains are interpreted as having their c-axes perpendicular to the substrate surface. This data indicates that this is a highly c-axis textured polycrystalline film.

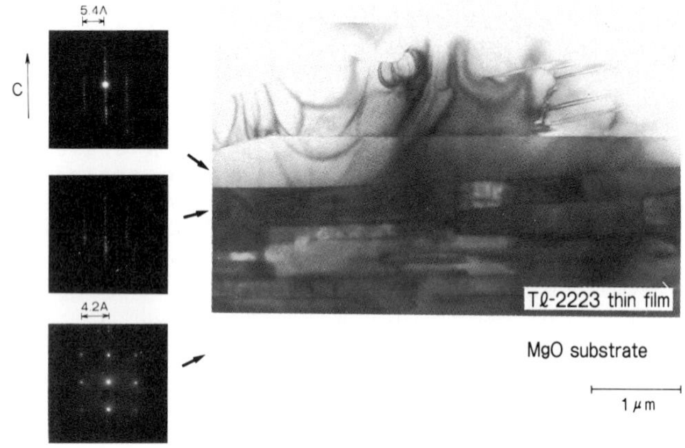

Fig.1 TEM image and electron diffraction patterns of c-axis oriented $T\ell_2Ba_2Ca_2Cu_3Ox$ thin film

The magnetic field dependence of Jc for the film at 77K using the 2μV/cm criterion is shown in Fig. 2. A critical current(Ic) of 9.8A and a Jc of $1.5X10^5A/cm^2$ at 77K in zero magnetic field were obtained. In the case of B ⊥I, B//a-b plane, Jc is as high as $1.3X10^5A/cm^2$ even at 1T. This degradation of Jc by magnetic field is much smaller than reported previously[3]. Enhancement of Jc was realized by purification of grain boundaries and highly c-axis oriented structure in the Tl-2223 film[4]. But when the magnetic field was applied perpendicularly to the current direction and the a-b plane, Jc decreased drastically to $1.3X10^3A/cm^2$ at 0.7T. This anisotropy of Jc indicates the existence of strong pinning centers parallel to the a-b plane.

Effect of changing the incident angle(θ) of the applied magnetic field to the a-b plane on Jc is shown Fig. 3. When θ was equal to 0 degree, Jc was almost the same in both magnetic fields of 0.5 and 1T. When the direction of the magnetic field is slightly tilted from the a-b plane, then the Jc decreases drastically.

In this measurement, the incident angle θ and perpendicular component of magnetic field Bp obeyed the relation $Bp = B\sin\theta$, where B is the applied magnetic field. Jc vs Bp characteristic for Tl-2223 film is shown in Fig. 4. The solid curve is taken from Fig. 2 and the triangles and circles are from Fig. 3. The good agreement is apparent. This data indicates that the properties of Jc on magnetic field are dominated by the component of the magnetic field perpendicular to the a-b plane. These results are due to the difference of pinning force between perpendicular and parallel to the a-b plane. This result can be explained using the intrinsic pinning model discussed by Tachiki et al.[5].

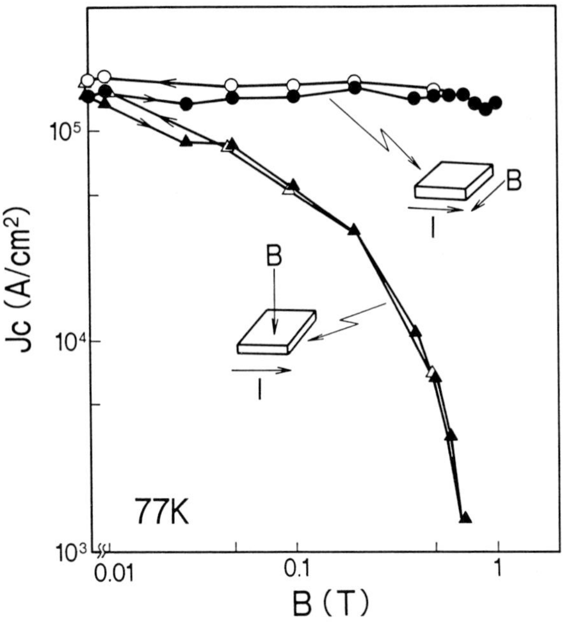

Fig. 2 Magnetic field dependency of Jc of Tl-2223 thin film.

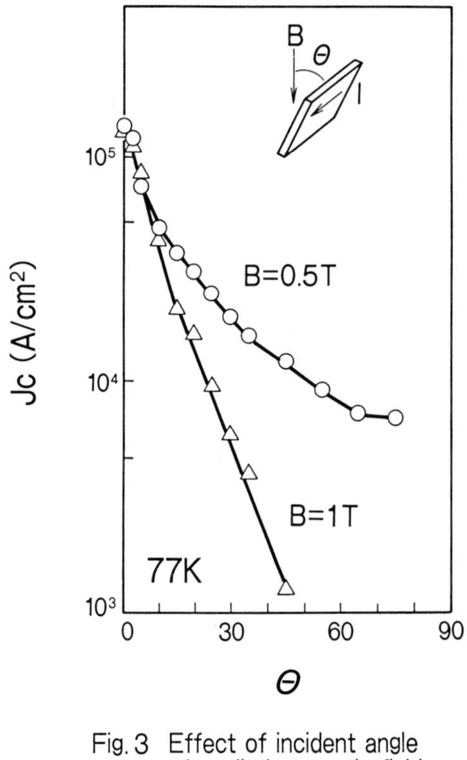

Fig. 3 Effect of incident angle of applied magnetic field to a-b plane on Jc.

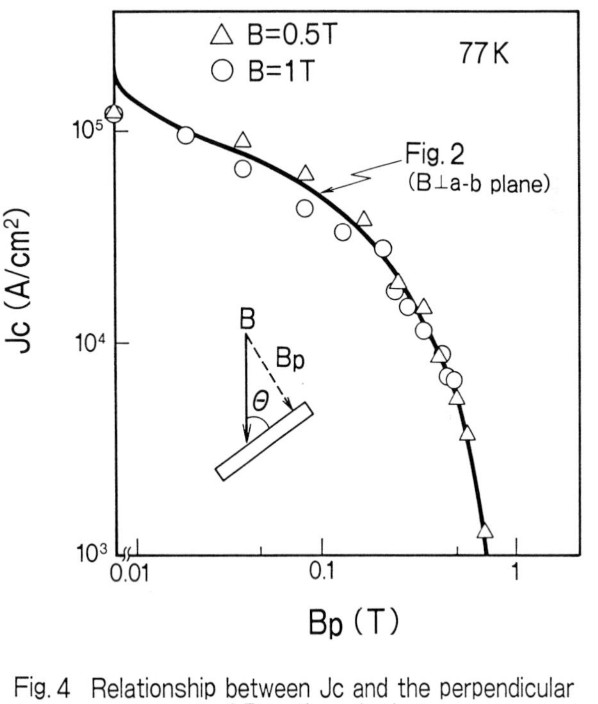

Fig. 4 Relationship between Jc and the perpendicular component of B to the a-b plane.

In summary, angular dependence of magnetic field on Jc investigated for the polycrystalline Tl-2223 thin film. When the incident angle(θ) was changed 0 and 90 degree under $B \perp I$, the Jc was $1.3X10^5 A/cm^2$(1T) and $1.3X10^3 A/cm^2$(0.7T),respectively. As θ increased, Jc decreased drastically. Degradation of Jc by the magnetic field was dependent on the component of B perpendicular to the a-b plane. This behavior is explained by an intrinsic pinning model.

REFERENCES

1. A. Soeta, T. Suzuki, T. Kamo, K. Usami and S. Matsuda:Jpn.J.Appl. Phys. 28(1989) L1186
2. S.H. Liou, K.D. Aylesworth, N.J. Iwano, B. John, D. Thompson, J.A. Woollam and C. Barry:Appl.Phys.Lett. 54(1989) 760
3. H. Itozaki, S. Tanaka, K. Higaki, K. Harada, S. Yazu and K. tada:Proc. 1st Istec Workshop on Superconductivity, Oiso, Japan, February, 1989,P63
4. T. Nabatame, Y. Saito, K. Aihara, T. Kamo and S. Matsuda(in press) Jpn.J.Appl.Phys.
5. M. Tachiki and S. Takahashi:Solid State Commu. 72(1989) 1083

Critical Currents of (103)- and (001)-Oriented Y-Ba-Cu-O Thin Films

Shugo Kubo, Yoichi Enomoto, and Tsunekazu Iwata

NTT Applied Electronics Laboratories, Tokai, Ibaraki, 319-11 Japan

ABSTRACT

Critical currents and flux pinning were examined in (103)- and (001)-oriented $YBa_2Cu_3O_{7-x}$ thin films at liquid nitrogen temperature (77.3K) under magnetic fields applied perpendicular to the film surface. Measurements were made in the [010] and [301] transport current directions for the (103)-oriented films. The field-dependent terms of the pinning force density F_p were found to be described by the single Kramer scaling law irrespective of crystallographic orientation and transport current direction: $F_p/F_p^{max} \propto b^{1/2}(1-b)^2$, where F_p^{max} is the maximum pinning force density and b is the reduced field defined as $b \equiv B/B_i$ (B_i: irreversibility field).

KEY WORDS: $YBa_2Cu_3O_{7-x}$ film, critical current, Kramer scaling law

INTRODUCTION

The current carrying capacity of high-T_c oxide superconductors is an important area for study for both high-magnetic field and electronic device applications. There have been several reports on the critical currents of oxide superconductor thin films in the transport current direction parallel to the Cu-O planes[1-3]. On the other hand, there has been little work on critical current behavior when the current travels across the Cu-O planes.

In this paper, we examine the crystallographic direction dependence of critical currents at liquid nitrogen temperature (77.3K) and flux pinning behavior in $YBa_2Cu_3O_{7-x}$ thin films having two kinds of crystallographic orientation: One is a usual (001)-oriented film, and the other has a (103) orientation. In the latter orientation, the transport current along the [301] direction travels across the Cu-O planes.

EXPERIMENTAL

$YBa_2Cu_3O_{7-x}$ thin films were prepared by a reactive evaporation technique. Ytterium was evaporated by an electron-beam gun and barium and copper were evaporated by effusion cells. Oxygen gas was introduced into the deposition chamber through nozzles attached near the substrate. Low power RF (13.56MHz) plasma was generated to assist oxygen activation. The oxygen gas pressure in the deposition chamber was about 1.0×10^{-4}Torr during the evaporation process. $SrTiO_3$ single crystals with (100) and (110) planes were used as substrates. The substrate temperature was maintained at 650°C during deposition and then slowly decreased (−5°C/min) in the oxygen plasma after deposition. The deposition rate was about 2Å/s and the prepared films were all about 2000Å thick.

The prepared films were characterized by X-ray diffraction (XRD) and high-resolution SEM. Films on the (110) substrates were examined by four-axis X-ray diffraction in order to distinguish (103) from (110) orientation[4].

Critical currents were measured by the dc method using patterned samples at 77.3K. The measured samples were 100μm wide and had a 4mm potential-terminal distance. These samples were patterned by a photolithographic technique and Ar ion beam etching. Samples with two current directions at right angles were fabricated on one substrate. Critical currents were determined by the 1μV/cm criterion. The samples were immersed in liquid nitrogen and measured under magnetic fields of up to 8T applied perpendicular to the film surface.

RESULTS AND DISCUSSION

Films on (100) substrates are mainly (001) orientation, with a small amount of the (100)-oriented phase. This is consistent with SEM observations, in which (100)-oriented rectangular precipitates are scattered on the film surface. Films on (110) substrates were proved to have a completely (103) orientation through an examination using a four-axis XRD technique[4]. The (103) film surface is uniform and free from precipitates. The crystallographic directions of (103)-oriented $YBa_2Cu_3O_{7-x}$ are shown in Fig. 1.

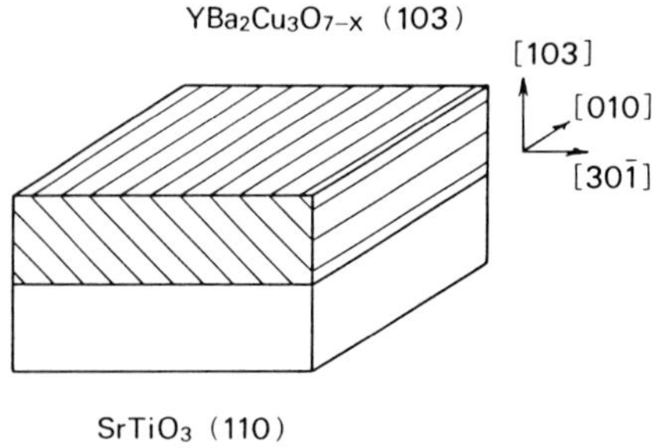

Fig. 1 Crystallographic directions in a (103)-oriented $YBa_2Cu_3O_{7-x}$ film. In this figure inclined planes represent the Cu-O planes.

Zero-resistance critical temperatures T_{c0} of the measured samples were 86∼87K. Typical temperature dependences of electrical resistivity for the [010] and [301] directions in a (103)-oriented film are shown in Fig. 2. Here the [010] direction is parallel to the Cu-O planes, and the [301] direction is across the Cu-O planes, as shown in Fig. 1. The anisotropy factor of these two directions in terms of electrical resistivity at room temperature is about 5. The resistivity behavior of (001)-oriented films is nearly the same as that of the [010] direction in (103)-oriented films.

An example of the magnetic field B dependence of critical current density J_c for a [010] direction sample in a (103)-oriented film is shown in Fig. 3. In this figure, results are also shown for larger critical current determination criteria (10μV/cm and 100μV/cm). The field dependences for the different criteria behave in basically the same manner. The $J_c(B)$ curves have tails under fields above 5T. In this tailing region, the current I versus voltage V relationship becomes proportional, and consequently $J_c = 0$ according to the offset criterion for critical current determination[5]. Vortices are completely in a flow state in this region. The magnetic field value where the $I - V$ relationship starts to be proportional in the present measurement corresponds almost exactly to the irreversibility field B_i in the flux creep theory[6,7].

The $J_c(B)$ curves of [301] direction samples in (103)-oriented films and of (001)-oriented film samples have a similar shape to those of [010] direction samples in (103)-oriented films. In general, however, zero-field critical current densities $J_c(0)$s for (001)-oriented films tend to be higher than those for the [010] direction in (103)-oriented films. On the other hand, $J_c(B)$s for the [301] direction are about one order of magnitude smaller than those for the [010] direction in the same film over the whole range of applied magnetic fields.

Magnetic field B dependences of pinning force density F_p, where $F_p = J_c \times B$, are shown in Fig. 4. For a conventional type-II superconductor, the pinning force obeys the Kramer scaling law[8]:

$$F_p(B,T) = [B_{c2}(T)]^m f_p(b), \tag{1}$$

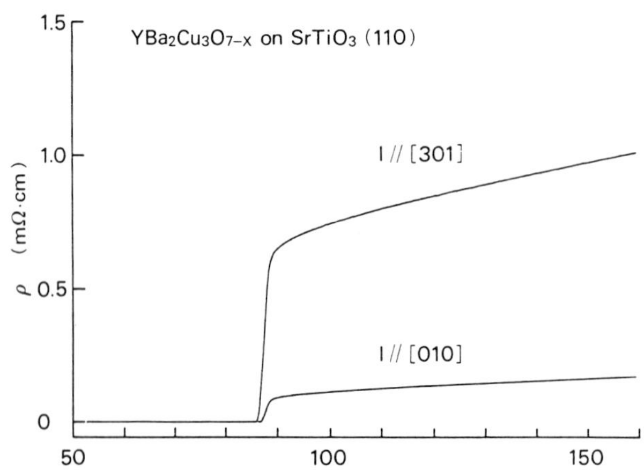

Fig. 2 Typical temperature T dependences of electrical resistivity ρ for the [010] and [301] directions of a (103)-oriented $YBa_2Cu_3O_{7-x}$ film. The former direction is parallel to the Cu-O planes, while the latter is across the Cu-O planes.

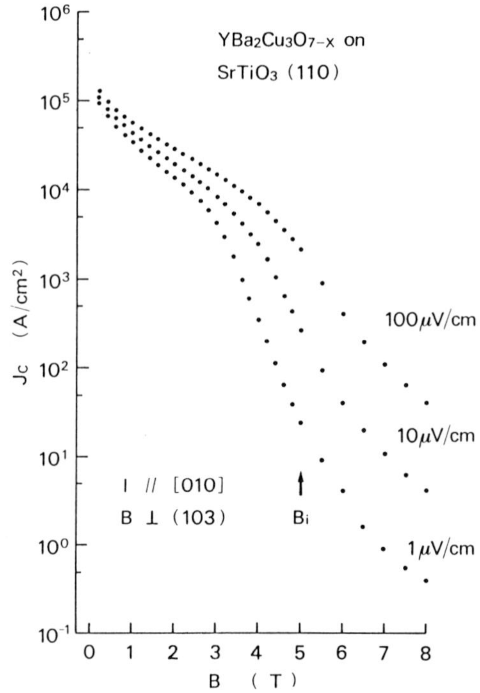

Fig. 3 An example of applied magnetic field B dependences of critical current density J_c for a [010] direction sample of a (103)-oriented film. Results are shown corresponding to three different critical current determination criteria; 1μV/cm, 10μV/cm and 100μV/cm.

where B_{c2} is the upper critical field. The second term $f_p(b)$ is a function of only the reduced field $b(\equiv B/B_{c2})$ and is given in the form:

$$f_p(b) = A \cdot b^p(1-b)^q, \tag{2}$$

where A is a numerical parameter. In the present experiment, we did not measure temperature dependences of critical current density, and thus we consider only the field scaling of the pinning force density in the following.

The reduced field b dependences of the normalized pinning force density $f_p(\equiv F_p/F_p^{\max})$, where $F_p^{\max}$ is the maximum pinning force density, are shown in Fig. 5. Here the irreversibility field B_i replaces the upper critical field B_{c2} and the reduced field is defined as $b \equiv B/B_i$. The full curves in Fig. 5 represent Eq.(2) where $p = 1/2$, $q = 2$ and $A = 3.5$. The f_p versus b data coincide quite well with the standard Kramer scaling law,

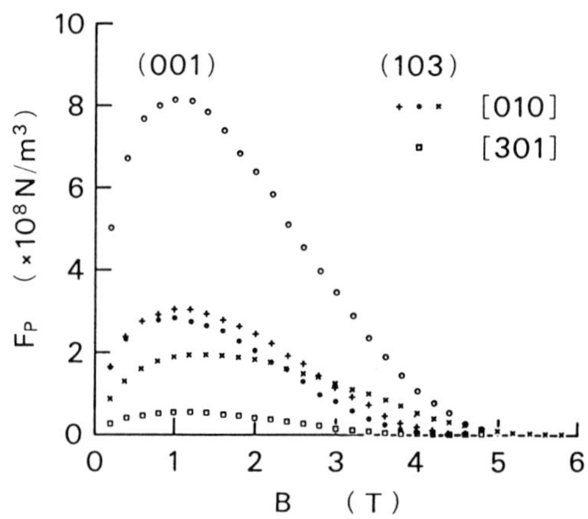

Fig. 4 Pinning force densities F_ps of (103)- and (001)-oriented $YBa_2Cu_3O_{7-x}$ films at 77.3K.

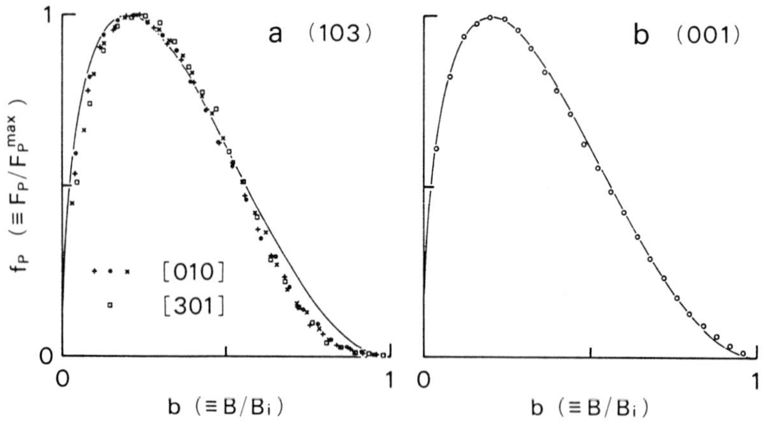

Fig. 5 Replot of Fig. 4 in terms of the normalized pinning force $f_p(\equiv F_p/F_p^{max})$ and the reduced field $b(\equiv B/B_i)$. Results for (**a**)(103)-oriented films, and for (**b**) a (001)-oriented film. The full curves in both the figures represent Eq.(2) where $p = 1/2$, $q = 2$ and $A = 3.5$.

irrespective of crystallographic orientation and transport current direction, as seen in Fig. 5.

The physical meaning of the present result is not entirely clear, but the result indicates that the basic concept of the critical shear strength of the flux-line lattice[8] can also be applied to the model describing the pinning force of $YBa_2Cu_3O_{7-x}$ at high temperatures. On the other hand, a different scaling ($p = 2$ and $q = 4$) has been reported in the measurement of bulk single-crystalline $YBa_2Cu_3O_{7-x}$ samples[9].

Pinning centers in the present samples have not been determined yet. The twin planes, which are produced perpendicular to the Cu-O planes, however, do not seem to be the pinning centers of these samples because they contribute in a quite different manner to pinning when the field direction does not coincide with the twin planes. Non-anisotropic defects are thought to be the pinning centers in $YBa_2Cu_3O_{7-x}$ thin films. Precipitates and/or stacking-disorder domains are plausible candidates for the pinning centers in the present samples.

The authors would like to express their sincere thanks to Dr. M. Suzuki for his helpful discussions and to Dr. A. Yamaji for his continuous support during this study.

REFERENCES

1. K. Watanabe, H. Yamane, H. Kurosawa, T. Hirai, N. Kobayashi, H. Iwasaki, K. Noto and Y. Muto, Appl. Phys. Lett. **54**, 575 (1989).
2. J. D. Hettinger, A. G. Swanson, W. J. Skocpol, J. S. Brooks J. M. Graybeal, P. M. Mankiewich, R. E. Howard, B. L. Straughn and E. G. Burkhardt, Phys. Rev. Lett. **62**, 2044 (1989).
3. B. Roas, L. Schuitz and G. Saemann-Ischenko, Phys. Rev. Lett. **64**, 479 (1990).
4. T. Iwata, Y. Enomoto, S. Kubo, K. Moriwaki and A. Yamaji, in *Proceedings of the Second International Symposium on Superconductivity, Tsukuba, 1989* (Springer-Verlag, Tokyo, 1990), p.773.
5. J. W. Ekin, Appl. Phys. Lett. **55**, 905 (1989).
6. Y. Yeshurun and A. P. Malozemoff, Phys. Rev. Lett. **60**, 2202 (1988).
7. T. Matsushita, T. Fujiyoshi, K. Toko and K. Yamafuji, Appl. Phys. Lett. **56**, 2039 (1990).
8. E. J. Kramer, J. Appl. Phys. **44**, 1360 (1974).
9. J. N. Li, F. R. De Boer, L. W. Roeland, M. J. V. Menken, K. Kadowaki, A. A. Menovsky, J. J. M. Franse and P. H. Kes, Physca C **169**, 81 (1990).

Anisotropy in Magnetization and Critical Current Density of La-Sr-Cu-O Single Crystal

KOHJI KISHIO[1], KOICHI NAKATANI[1], KOICHI KITAZAWA[1], ISAO TANAKA[2], HIRONAO KOJIMA[2], and KAORU YAMAFUJI[3]

[1] Department of Industrial Chemistry, The University of Tokyo, Tokyo, 113 Japan
[2] Institute of Inorganic Synthesis, Yamanashi University, Kofu, 400 Japan
[3] Department of Electronics, Kyushu University, Fukuoka, 812 Japan

ABSTRACT

Magnetization curves of $(La_{1-x}Sr_x)_2CuO_4$ (x=0.065-0.07) single crystals were measured at temperatures from 4.2 to 35K under magnetic field of up to 9T. The field was applied along the a- or the c-axis of the two types of the crystals having different macroscopic geometry. The analysis of the hysteresis data revealed strong anisotropy in flux pinning and critical current density. At 4.2 K and 5 T, for example, $J_c{}^{//ab\text{-plane}}$(H//c-axis), $J_c{}^c$(H//ab) were determined to be 2.1x10^5, and 3.0x10^4 A/cm^2, respectively and $J_c{}^{ab}$(H//ab) was estimated to be 4.2x10^5 A/cm^2. It was also found that the field and temperature dependence was quite appreciable in $J_c{}^{//ab}$(H//c), as expected from the generally observed behavior in cuprate-based HTSC oxides.

KEY WORDS: anisotropy, magnetization curve, critical current density, flux pinning

INTRODUCTION

Magnetization measurements on single crystals under various orientations of applied field, provide basic information on anisotropic J_c and flux pinning behavior, and a number of studies [1-5] have been reported. However, most of the studied systems have been limited to $Ba_2YCu_3O_7$. In addition, the single crystals utilized in these studies were generally thin rectangular platelets having very short dimension (the order of several tens of μm) in the direction of the crystal c-axis. When the magnetic field is applied parallel to the short edge of such crystals, the necessary corrections for the demagnetization effects are extremely difficult and determined parameters such as B_{c1} and J_c for this orientation have been rather unreliable. On the other hand, this problem is avoided if single crystals having long dimension parallel to the c-axis can be used. By utilization of such crystals, Naito et al.[5] have successfully determined the anisotropic B_{c1} values of $(La_{1-x}Sr_x)_2CuO_4$ and also estimated the field-independent J_c near B_{c1}. The purpose of the present paper is to perform magnetization measurements at high fields up to 9 T with use of similar crystals, and to determine the anisotropic J_c values as functions of field and temperature, in order to provide complementary data to the previous work [5].

EXPERIMENTAL

Single crystals of $(La_{1-x}Sr_x)_2CuO_4$ were prepared by a TSFZ (Traveling Solvent Floating Zone) technique as described [6] previously. Slab specimens having different crystal orientations were cut from two separate crystal boules. One specimen had a dimension of 4.8x1.7x0.45 mm^3 in which the direction of the slab thickness was parallel to the tetragonal c-axis. Since LSCO transforms to the orthorhombic structure at low temperatures, we call this specimen, the "ab" crystal, hereafter. Another specimen had dimension of 3.2x2.5x0.45 mm^3 with the 2.5 mm edge parallel to the c-axis and the other long and the shortest edges parallel to the a-axis. This specimen having long dimension along the c-axis is called, the "ac" crystal.

Zero-field-cool (ZFC) and field-cool (FC) susceptibility measurement were performed on these crystals using a SQUID magnetometer. Magnetic field of 10.3 Oe was applied parallel to the a-axis in both cases. ZFC diamagnetic signals for the "ab" and "ac" crystals were 1.39 and 1.36x10^{-2} emu/g, respectively, and corresponded to about 110% in the superconducting volume fraction. The slight excess over 100% is probably due to a small contribution of the demagnetization effect, and this guarantees that both crystals are of good quality. The FC (Meissner) signals were 3.4 and 11.5 % for the "ab" and "ac" crystals, which clearly indicated the flux pinning and the specimen size dependence along the direction of the flux expulsion [7]. T_c determined from the onset of the FC signals were 33 and 35 K for the "ab" and "ac" crystals, respectively. These T_c's were reasonable from the analyzed Sr composition, x=0.065 and 0.07$\pm$0.005, by X-ray micro-probe analysis.

Magnetization curves were measured at temperatures from 4.2 to 35 K using a parallel-field vibrating sample magnetometer (Princeton Applied Research, Model 4500 equipped with a 12 T magnet, Janis Research). In this study, magnetic field was scanned at a rate of 20mT/s up to ±9 T. Specimens were always set and vibrated so that the magnetic field was parallel to the slab in order to suppress the demagnetization effect. Since the crystals were rectangular with two edges on the slab faces, the magnetic field was applied in two directions, parallel to either these long or the short edge.

RESULTS AND DISCUSSIONS

Figure 1 shows the observed magnetization curves of the "ab" crystal at 4.2 K for two orientations with respect to the direction of the applied magnetic field. The curve (a) was measured when the short edge of the crystal was set parallel to the field and the curve (b) was obtained with the long edge parallel to the field. It is clear in Fig. 1 that the curve (a) exhibits much larger hysteresis than (b). If we consider that the flux density gradient exists, as generally expected in isotropic materials, in the direction of the slab thickness of the "ab" crystal (the c-axis direction), the two curves must coincide with each other. One may argue that this difference results from the demagnetization effect, because the slopes of the initial (virgin) magnetization curves are different. However, it must be stressed that the slopes are strictly -1 up to B_{c1} which has been determined to be less than 3×10^{-2} T for this orientation [5], and is approximately -1 only when the flux pinning is sufficiently strong over the B_{c1} [5].

It is seen in Fig. 1 that the penetration magnetic field, B_p, when the penetrated flux reaches the center of the specimen and where the virgin curve reaches the cyclic curve, is larger in the curve (a) than in (b). The difference of B_p between the two orientations in Fig. 1 therefore suggests that the large flux gradient exists parallel to the wide surface of the slab. This situation is understandable from the anisotropy of the flux pinning and J_c; J_c along the c-axis (denoted as J_c^c) is expected to be much smaller than J_c within the ab-plane (J_c^{ab}). If we assume that the flux penetrates only in the direction parallel to the slab as an extreme case and the standard Bean model is applied, the magnetization hysteresis ΔM should be expressed as

$$J_c\ [A/m^2] = 2 \times \Delta M\ [A/m] \ /\ D\ [m] \qquad (1)$$

where units are given in brackets.

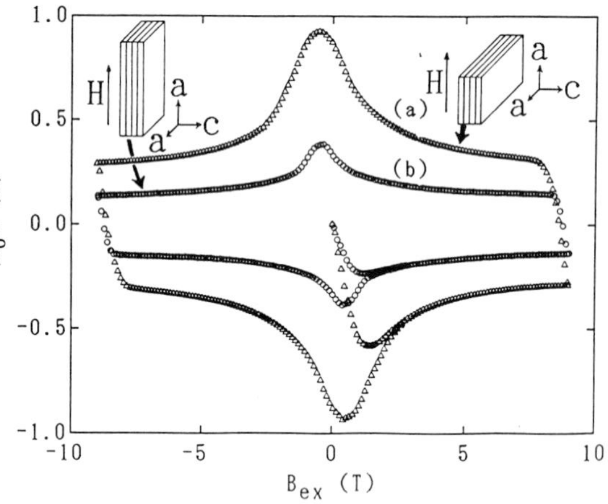

Fig. 1 Magnetization curves of "ab" crystal at 4.2 K with applied field parallel to the short (a) and the long edge (b).

Eq. (1) must be interpreted as $J_c = J_c^c$ and D being the edge length perpendicular to the field, instead of the slab thickness. Then, it is expected that the ratio of the ΔM for the two orientations in Fig. 1, must be equal to the edge length ratio of the "ab" crystal face; 4.8/1.7=2.8. The observed ΔM ratio is about 2.6 to 2.2 depending on the field and slightly smaller than the expected value. This suggests that Eq. (1) is not strictly satisfied, and the flux penetration and expulsion occurs two dimensionally in an anisotropic manner.

In analyzing the similar magnetization data of $Ba_2YCu_3O_7$ single crystals, Gyorgy et al. [2] proposed calculation of J_c based on an extended Bean model. Depending on the sample geometry, they considered two cases and expressed ΔM in terms of two anisotropic J_{c1}, J_{c2} and L (//J_{c1}), t(//J_{c2}) which are the dimensions of the specimen perpendicular to the magnetic field.

$$\Delta M = J_{c1}\, t \ /\ 2 \times (1 - t\, J_{c1} \ /\ 3L\, J_{c2}) \qquad \text{when} \quad J_{c1}/J_{c2} < L/t \qquad (2)$$

$$\Delta M = J_{c2}\, L \ /\ 2 \times (1 - L\, J_{c2} \ /\ 3t\, J_{c1}) \qquad \text{when} \quad J_{c1}/J_{c2} > L/t \qquad (3)$$

We have tentatively applied Eq.(2) for the curve (a) shown in Fig. 1 and Eq.(3) for the curve (b). Selecting 5 T as a representative point, the calculation gave $J_{c1} = J_c^{ab} = 4.2 \times 10^5$ A/cm^2 and $J_{c2} = J_c^c = 3.4 \times 10^4$ A/cm^2; anisotropy ratio of roughly 12. Although the condition of Eq.(2) is not strictly satisfied, it is considered that Eq.(2) can still give rough estimate of J_{c1} ($=J_c^{ab}$). Moreover, J_c^c value which is more reliable than J_c^{ab} in this analysis, is consistent with that estimated from the result of the "ac" crystal measurement, described next.

It should be kept in mind, however, the extended Bean model used above assumed the field-independent J_c. When the dimension of the sample is large in the direction of the flux penetration, like in our present case, large variation of the local field gradient should be established within the sample crystal under a given external field. In such a case, the assumption of B_{ex}-independent Jc may lose its validity. Reexamination of the experimental data with a more precise critical state model adopting the field dependence of J_c such as the Irie-Yamafuji model [8], is then desirable.

Next, we would like to turn our discussion on the "ac" crystal measurement which is the central result of the present work. Figures 2 and 3 show the observed magnetization curves of the "ac" crystals at various temperatures for the magnetic field applied parallel to the a-axis and the c-axis, respectively. In order to avoid jamming of the curves, the data curves obtained at 4.2, 20 and 20 K are only shown. It is noticed that observed field dependence in Figs. 2 and 3 is quite different from each other. When H is applied along the a-axis, the magnetization has relatively weak H-dependence, except at low fields below at 0.5 T. In the curves at 10 and 20 K, even larger ΔM can be observed by increasing B_{ex} from less than 1 to a few or several Tesla, which is the similar behavior observed in BYCO systems [4,9]. On the other hand, in Fig. 3, it is clear that the field dependence is much stronger than in Fig. 2. In addition, it must be noticed that ΔM for H//c near zero field is approximately 10 times larger compared to that for H//a.

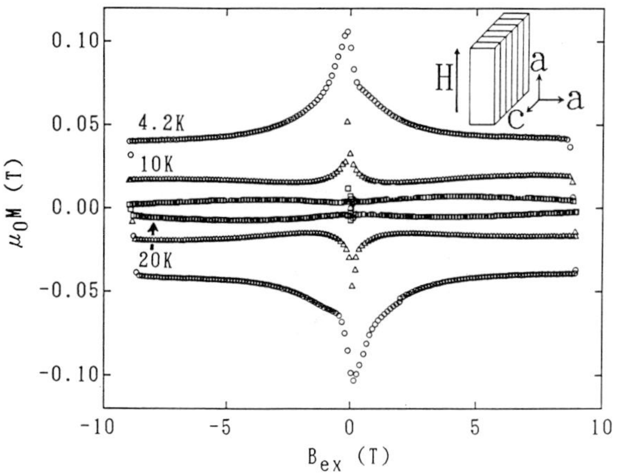

Fig. 2 Magnetization curves of "ac" crystal at 4.2, 10 and 20 K with applied field parallel to the a-axis

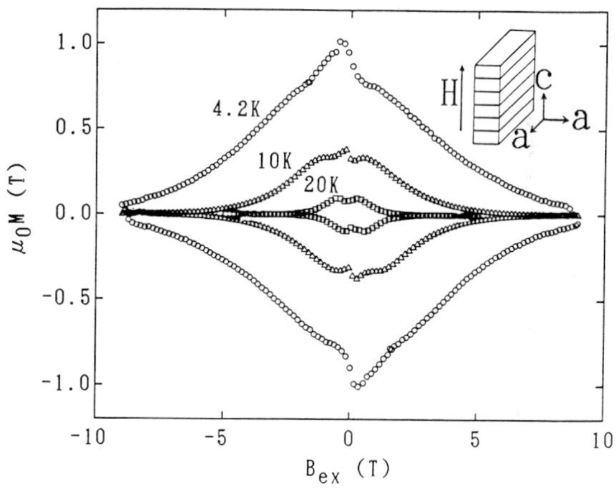

Fig. 3 Magnetization curves of "ac" crystal at 4.2, 10 and 20 K with applied field parallel to the c-axis.

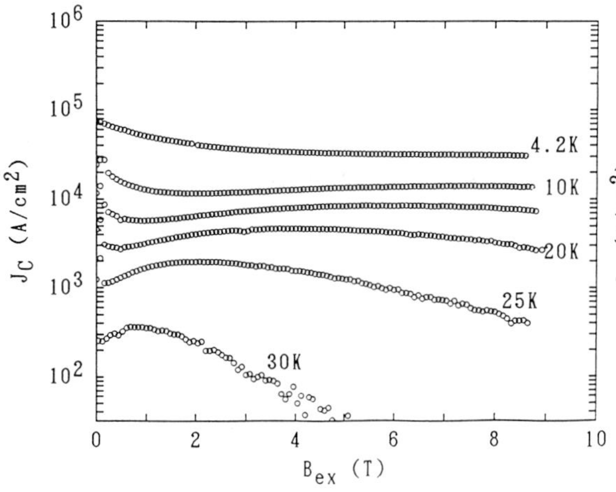

Fig. 4 Magnetic field dependence of critical current density J_c^c (H//a) for various temperatures.

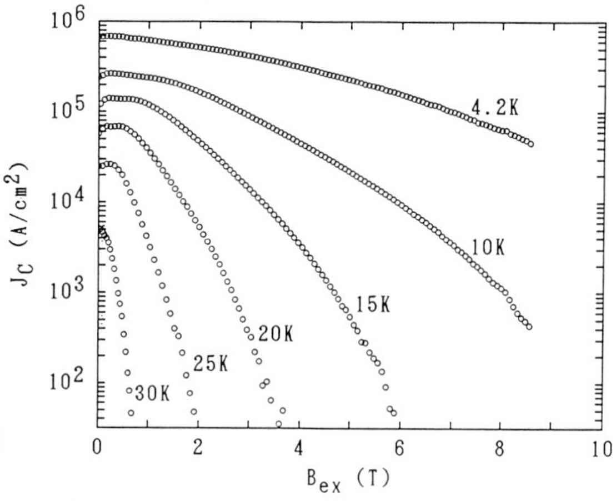

Fig. 5 Magnetic field dependence of critical current density J_c^{ab} (H//c) for various temperatures.

The analysis of the magnetization data of Figs. 2 and 3 are more straightforward compared to the case of Fig. 1. When the "ac" crystal is measured with H//a (Fig. 2), the flux can easily penetrate perpendicular to the slab, since it is the shorter direction and J_c^c(//ab)>J_c^{ab}(//ab). In the case of H//c (Fig. 3), the flux penetration occurs quasi- two-dimensionally, but it can be approximated as one dimensional penetration perpendicular to the slab. The correction by the contribution of the flux penetration parallel to the slab is, as in Eq. (3), t/3L which is about 6%. Therefore, both magnetization curves can be safely analyzed through Eq.(1) with D taken as the slab thickness. Shown in Figs. 4 and 5 are the determined values of J_c^c(H//a) and J_c^{ab}(H//c) as a function of applied field at various temperatures. Figure 6 further shows the temperature dependence of these values at selected points of the applied field.

It is noticed in Fig. 6 that J_c^{ab}(H//c) exhibits its nearly exponential temperature dependence at low temperatures but deviates appreciably downward at high temperatures. It is also clear that this crossover to high temperature regime occurs at lower temperatures when the applied field is larger. Such a transition of temperature dependence from the low temperature to the high temperature regime, has been also observed in the studies of $Ba_2YCu_3O_7$ single crystals [10] and even in sintered Bi(Pb)-Sr-Ca-Cu-O studied by the present authors [11], and therefore considered as a universal phenomenon in cuprate superconductors. Nevertheless, the functional form of temperature dependence of J_c in the high temperature regime is still unknown. It is also pointed out that H//c is the orientation by which widely-known resistivity broadening of most of the cuprates is observed, and these phenomena might be correlated with each other.

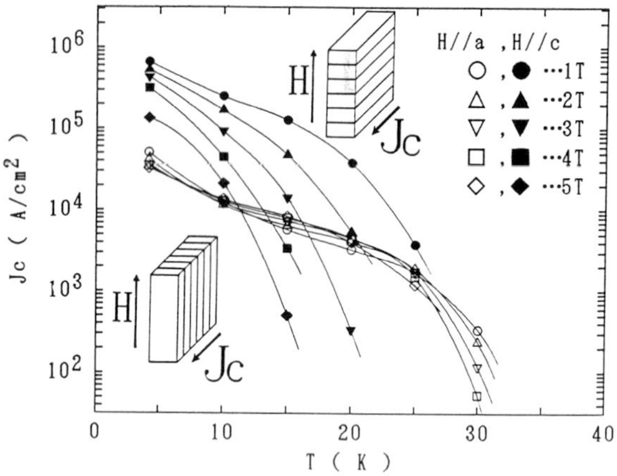

Fig. 6 Temperature dependence of J_c^{ab} (H//c) and J_c^c (H//a) under various applied fields.

Temperature and field dependence of J_c^c(H//a) is quite contrasting to the J_c^{ab}(H//c). Temperature dependence is much weaker, and even crossover; J_c^c>J_c^{ab} occurs at high temperatures. Increase of the field even increases the J_c^c at the mid-temperature ranges which clearly suggest different types of flux pinning mechanism [9] is operative in this region, in addition to the basic pinning mechanism at low temperatures.

Finally, it is pointed out that J_c^{ab}(H//ab) could not be calculated unambiguously in this study but was only estimated based on the extended Bean model [2]. This was because the dimensional ratio t/L of our crystal was not small enough to overcome the large anisotropy ratio of J_c^{ab}/J_c^c(both H//ab). Apparently, J_c^{ab}(H//ab) determined by Naito et al [5] has suffered the same difficulty, and it is necessary to reexamine this configuration by using various sizes of the single crystals.

ACKNOWLEDGEMENTS

The authors would like to thank Messrs. Y. Tomioka, N. Motohira T. Noda, S. Komiya and H. Nakashima for experimental assistance and discussions. This work has been supported by a Grant-in-Aid for Scientific Research from the Ministry of Education, Science and Culture and by Special Coordination Funds from the Science and Technology Agency of Japan.

REFERENCES

1. Dinger TR, Worthington TK, Gallagher WJ, Sandstrom RL (1988) Phys Rev Lett 58: 2687-2690
2. Gyorgy EM, van Dover RB, Jackson KA, Schneemeyer LF, Waszczak JV (1989) Appl Phys Lett 55: 283-285
3. Swartzendruber LJ, Roitburd A, Kaiser DL, Gayle FW and Bennett LH (1990) Phys Rev Lett 64 483-486
4. Franse JJM, Li JN, Kadowaki K (1989) Proc Int Conf on High T_c Thin Films and Single Crystals, Ustron (Poland), in press
5. Naito M, Matsuda A, Kitazawa K, Kambe S, Tanaka I, Kojima H (1990) Phys Rev B41: 4823-4826
6. Tanaka I, Kojima H (1988) Nature 337: 21-22
7. Kitazawa K, Matsushita T, Nakamura O, Tomioka Y, Motohira N, Tamura T, Hasegawa T, Kishio K, Tanaka I, Kojima H (1990) Proc Int Conf Superconductivity, Bangalore (India), in press
8. Irie F, Yamafuji K (1957) J Phys Soc Jpn 23: 255-268
9. Seuntjens JM, Daeuming M, Larbalestier DC, preprint
10. Sennoussi S, Oussena M, Collin G, Campbell I A (1988) Phys Rev B37: 9792-9795
11. Kishio K, Green M L, Nakatani K, Kitazawa K (1990) Mol Cryst Liq Cryst 184: 369-376

4.2 Flux Line Lattice and Flux Distribution

Dynamical Magnetic Flux Behavior in High-T_c Oxide Superconductors

SATOSHI GOTOH, NAOKI KOSHIZUKA, MASASHI YOSHIDA, HIROYUKI FUJIMOTO, and MASATO MURAKAMI

Superconductivity Research Laboratory, International Superconductivity Technology Center, 10-13, Shinonome 1-chome, Koto-ku, Tokyo, 135 Japan

ABSTRACT

The dynamical magnetic flux behavior in MPMG-processed YBaCuO was observed using the motion of stripe magnetic domains on an iron garnet film placed on the specimen. Optical observation of the domain structure in the iron garnet film clearly reveals the magnetic flux expulsion, penetration, and trapping and the magnetic field distribution. It was found that the flux motion reflected the uniform distribution of Y_2BaCuO_5 inclusions introduced in the MPMG-process and was in good agreement with that predicted by the critical state model.

KEY WORDS: $YBa_2Cu_3O_x$, Y_2BaCuO_5, iron garnet, stripe domain, flux gradient

INTRODUCTION

Magnetization measurements for superconductors generally show the averaged values for the total sample volume. The magnetic flux distribution in type II superconductors is not directly observable by magnetization measurements, so that the macroscopic flux density gradient or distribution, which corresponds to the critical current density, Jc, is usually analyzed by the critical state model[1]. A more direct method using the decoration technique[2] has also been applied to observe the flux line lattice in the mixed state of high-Tc oxide superconductors[3]. However, it was difficult to observe the behavior of flux line lattice over the total volume of the sample, and dynamic observations while varying the temperature and magnetic field were not possible. Alers[4] first observed the flux motion of type I superconductors by means of the Faraday effect. This method has been improved with an evaporated magneto-optical thin film technique by Kirchner[5] and was then applied to type II superconductors by Habermeier et al.[6] and Klein and Kronmüller et al[7]. Then, Moser et al.[8] and Koblischka et al.[9] observed the flux penetration into both single crystal and sintered $YBa_2Cu_3O_x$ samples at 10 K using the Faraday effect of an evaporated magneto-optical EuS/EuF_2 film.

Recently, a new simple observation method for the flux behavior in high-Tc superconductors has been introduced by Indenbom et al.[10] and Gotoh et al.[11,12] using the Faraday effect of iron garnet films. We have named this method HICOM (high-contrast observation method for magnetic flux density) since a high-contrast image of the flux distribution is obtained. In this paper, we report on the observations of flux penetration and trapping in the Melt-Powder-Melt-Growth(MPMG)-processed YBaCuO crystals by this technique.

EXPERIMENTAL

Samples of $Y_{1.2}Ba_{2.1}Cu_{3.1}O_x$ were prepared by the Melt-Powder-Melt-Growth (MPMG) process[13]. Y_2O_3, $BaCO_3$ and CuO powders were calcined at 900°C and then heated to 1400°C followed by quenching by copper hammer plates. The quenched sample was crushed into powder and mixed well. The powder was then pressed into pellet and reheated to 1100°C for 20 min and cooled down to 1000°C at the rate of 100°C/h followed by slow cooling. The above process was done in air. Finally the samples were annealed at 600°C for 1h and slowly cooled in flowing oxygen. Rectangular specimens with dimensions approximately 1.5x3x3mm^3 were cut from the MPMG-processed plates so that the C axis was parallel to the surface (3x3mm^2) of the plate. Mechanical polishing was carried out on the plates to obtain a flat surface.

The magneto-optical thin film of $(BiY)_3(GaFe)_5O_{12}$ for the optical detection was prepared on a transparent gadolinium gallium garnet (GGG) substrate[14]. The thickness of the (BiY)IG film and the GGG substrate were 4μm and 0.4mm, respectively. The optical detection system with an electromagnet, an optical cryostat and a polarization microscope was used. The YBaCuO plate was fixed on a cooling holder inside the cryostat and then the film/substrate was placed onto the sample surface film-side down and in good thermal contact with the sample. The temperature is variable from 7.7 K to 300 K by controlling both the flow of cooled helium gas and the power of a heater inside the copper sample holder. The polarization microscope (OLYMPUS, BHT-P) has a resolution of about 1.3μm and a working distance of up to 8mm. The electromagnet was placed under the cryostat to apply a magnetic field perpendicular to the sample surface. Magnetization properties were measured using a SQUID magnetometer. The applied field was perpendicular both to the c axis and the surface. A large demagnetization effect was brought about by this configuration and the demagnetization factor was about 0.5 in the perfect diamagnetism state.

RESULTS AND DISCUSSION

The changes in the stripe domain patterns of the iron garnet film on the MPMG-processed YBaCuO sample zero-field cooled to 10 K and measured at 10 K with increasing applied fields are shown in Fig. 1. The principle of HICOM is to observe the image of the stripe domains in the iron garnet film, which reflect the flux density inside the sample. The stripe domains are caused by the perpendicular magnetization in the iron garnet film. The domain width with the magnetization parallel to the applied field (white region in the figure) expands, and the domain width with the opposite magnetization (black region in the figure) becomes narrower, when a magnetic field is applied perpendicular to the surface. The value of the magnetic flux density in the sample can therefore be estimated from changes in the domain width.

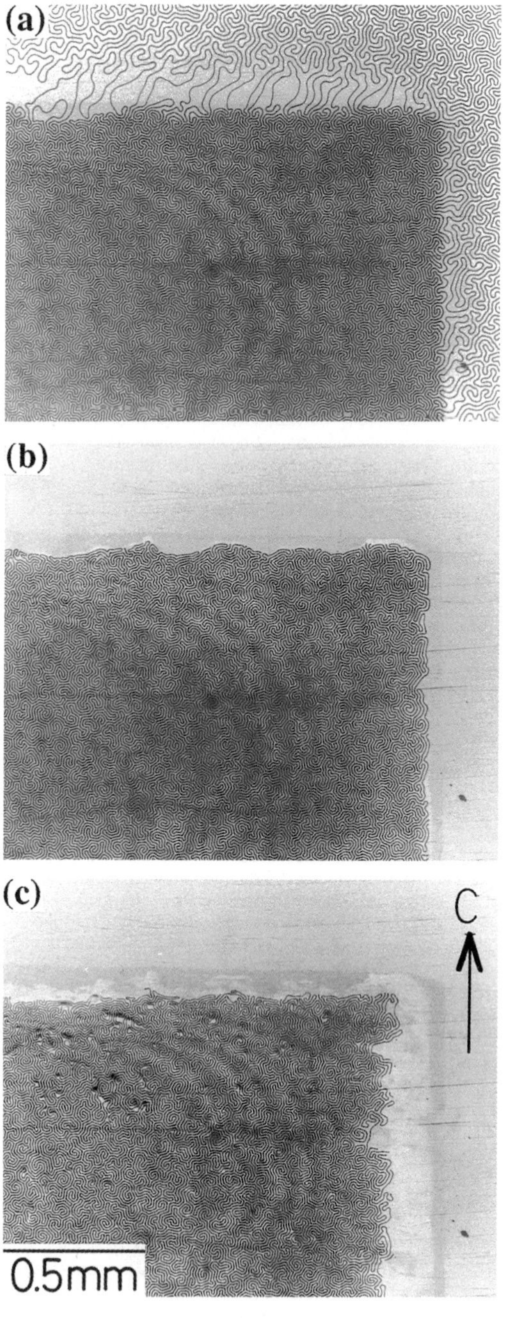

Fig. 1. Stripe domain structures of iron garnet film on MPMG-processed YBaCuO for increasing fields at 10 K after zero-field cooling. (a)H=84 Oe; (b)H=209 Oe; (c)H=753 Oe.

The white stripe domain reflects the homogeneity of the applied field except for the superconducting sample itself and its vicinity, as shown in Fig. 1(a). A narrow domain structure remains in the interior of the sample because the applied field does not penetrate into the sample due to the Meissner effect. However, in the region surrounding the superconducting sample, the white domain width increases as flux is excluded from the sample and concentrated in a narrow region. The stripe domain structure surrounding the sample disappears and a single white domain appears as the applied field exceeds a temperature-dependent threshold value. The applied flux penetrates from the edges of the sample when the applied field exceeds the lower critical field, H_{c1}, as shown in Figs. 1(b) and 1(c).

Fig. 1(b) shows the domain structure whose front reflects the sample shape, just after the field penetrates. The field beginning to penetrate into the sample is about 200 Oe without correcting for the demagnetization factor. When the applied field increases further, the flux penetration starts smoothly with a rectangular flux front, which is in contrast with the penetration in the QMG-processed[15] material with a nonuniform distribution of Y_2BaCuO_5(211) inclusions[10,11], indicating that fine 211 inclusions are homogeneously distributed in the MPMG-processed sample.

It is also found that the difference of the flux penetration between parallel and perpendicular to the c axis indicates the anisotropy in Jc, as shown in Fig.1(c). The depth of the flux penetration is inversely proportional to Jc. The observed anisotropy in Jc, Jc(I//ab)/Jc(I//c), is about 3 which may reflect the anisotropy in the coherence length. This value of anisotropy is also in good agreement with that obtained by the macroscopic magnetization measurements[16].

Figure 2 shows the initial magnetization curve at 10 K. As the applied field increases, the curve gradually deviates from the straight line of perfect diamagnetism. Since magnetization can be considered to be nearly perfect diamagnetism in this curve, a demagnetization correction was made for the applied field assuming a constant demagnetization factor, N=0.54. From the definition, H_{c1} can be determined to be the field at which the curve deviates from a straight line, and is 200 Oe for the applied field and corresponding to 437 Oe for the effective field as indicated by the arrow in the figure. This value was determined by the change of differential value($\Delta M/\Delta Ha$) at each field. The value of H_{c1} is in good agreement with the result in HICOM so that it is reasonable to define H_{c1} by this method.

Figure 3(a) shows the domain pattern at 77 K after field-cooling in a magnetic field of 142 Oe. There is no change of the domain structure both inside and outside the sample. This indicates that almost all applied field is trapped in the sample below Tc due to the large pinning force. DC susceptibility measurements yield a very small Meissner signal upon field-cooling in 142 Oe, which also indicates strong pinning, as shown in Fig. 4.

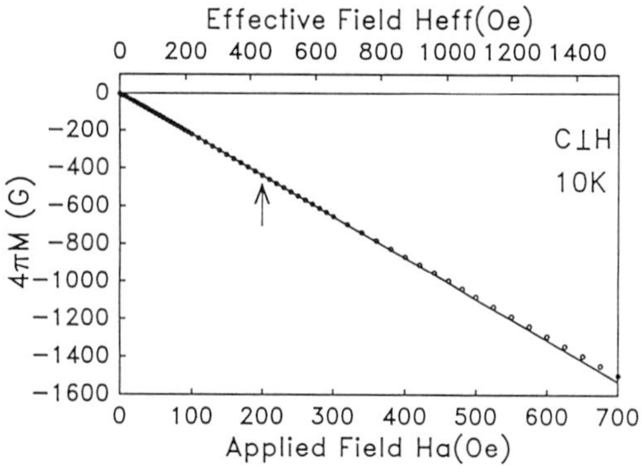

Fig. 2. Initial magnetization curve at 10 K after zero-field cooling.

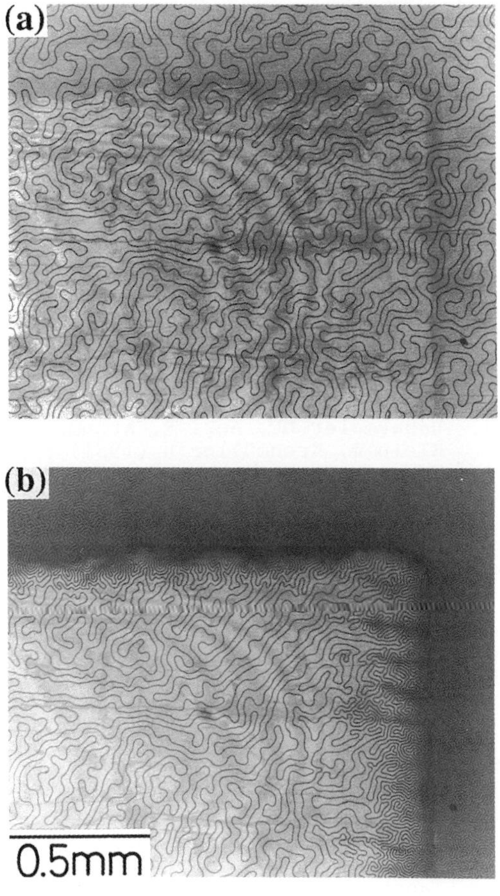

Fig. 3. Stripe domain structures at 77 K after field-cooling in a magnetic field of 142 Oe. (a)H=142 Oe; (b)H=0 Oe.

When the applied field was removed, the trapped flux appears as positive remanent magnetization as shown in Fig. 3(b). Although large trapped flux remains in the center of the sample, the flux abruptly decreases at the end of the sample with large flux gradients. This flux profile is in good agreement with dH/dx=4π Jc/10 which is predicted by the critical state model.

CONCLUSIONS

Dynamic flux motion was directly observed in the MPMG-processed YBaCuO with a rectangular shape using the Faraday effect of an iron garnet film with stripe domains. When the applied field exceeds H_{c1}, the flux gradually penetrates into the sample with a rectangular flux front since the fine 211 inclusions, which can work as pinning centers, are homogeneously distributed in the sample. According to the observation in HICOM, the value of H_{c1} can be determined by the field at which the initial magnetization curve deviates from the straight line of perfect diamagnetism. The anisotropy in Jc, Jc(I//ab)/Jc(I//c), is estimated to be about 3 by the depth of the flux penetration.

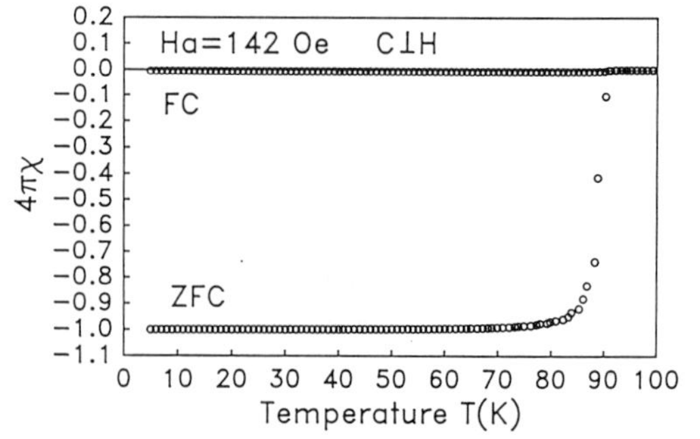

Fig. 4. Temperature dependence of DC susceptibility in both zero-field cooling and field-cooling processes.

ACKNOWLEDGMENTS

We would like to thank Dr.Y.Yokoyama of Electrotechnical Laboratory for the preparation of the iron garnet films used. This work was supported by New Energy and Industrial Technology Development Organization under the management of the R&D of Basic Technology for Future Industries.

REFERENCES

1. Bean CP (1962) Phys. Rev. Lett. 8: 250
2. Träuble H, Essmann U (1968) Phys. Status Solidi 25: 373
3. Gammel PL, Bishop DJ, Dolan GJ, Kwo JR, Murray CA, Schneemeyer LF, Wasczak JV (1987) Phys. Rev. Lett. 59: 2592
4. Alers PB (1957) Phys. Rev. 105: 104
5. Kirchner H (1971) Phys. Status Solidi A4:531
6. Habermeier HU, Aoki R, Kronmüller H (1978) Phys. Status Solid A60: 187
7. Klein W, Kronmüller H (1981) Phys. Status Solidi A67: 109
8. Moser N, Koblischka MR, Kronmüller H, Gegenheimer B, Theuss H (1989) Physica C 158: 117
9. Koblischka MR, Moser N, Gegenheimer B, Kronmüller H (1990) Physica C 166: 36
10. Indenbom MV, Kolesnikov NN, Kulakov MP, Naumenko IG,Nikitenko I, Polyanskii AA, Vershinin NF and Vlasko-Vlasov VK (1990) Physica C 166: 486
11. Gotoh S, Koshizuka N, Yoshida M, Murakami M, Fujimoto H, Tanaka S (1990) Jpn. J. Appl. Phys. 29: L1083
12. Gotoh S, Koshizuka N, Yoshida M, Murakami M, Fujimoto H, Tanaka S (1990) to be published in Superconductor Science and Technology
13. Fujimoto H, Murakami M, Gotoh S, Oyama T, Shiohara Y,Koshizuka N, Tanaka S (1990) In: Advances in Superconductivity II, Ishiguro, Kajimura (eds) the 2nd International Symposium on Superconductivity(ISS'89). Springer-Verlag, Tokyo, pp 285-288
14. Yokoyama Y, Koshizuka N, Takeda N (1985) IEEE Trans. Magn. MAG-21: 1666
15. Murakami M, Morita M, Doi K, Miyamoto K(1989) Jpn. J. Appl. Phys. 28: 1189
16. Murakami M, Gotoh S, Fujimoto H, Yamaguchi K, Koshizuka N, Tanaka S (1990) to be published in Superconductor Science and Technology

ESR of DPPH as a Probe of Magnetic Flux Fluctuations on the Surface of YBaCuO Made by the MPMG Method

KAZUSHI SUGAWARA, TERUTSUGU OYAMA, MASATO MURAKAMI, KAZUHITO YASUIKE, YUH SHIOHARA, and SHOJI TANAKA

Superconductivity Research Laboratory, International Superconductivity Technology Center, 10-13, Shinonome 1-chome, Koto-ku, Tokyo, 135 Japan

ABSTRACT

Magnetic flux fluctuations on the surface of Y-Ba-Cu-O samples (powder and bulk), which were made by the MPMG method, were studied by means of ESR of DPPH coated on these samples. The temperature dependence of the ESR linewidth, ΔHpp, and the effect of applied magnetic field on the ΔHpp were studied. ΔHpp gradually increases with decreasing temperature below Tc for all the samples. Analyses revealed that the excess ESR linewidth, $\delta(\Delta Hpp)=\Delta Hpp(T)-\Delta Hpp(T\simeq Tc)$, is nearly proportional to $(1 - T / Tc)^{a}$ with $a\simeq 1.4$ at temperatures between about 80K and 89K ($\simeq$Tc). It is also found that ΔHpp is broadened by applied magnetic field.

KEY WORDS : ESR, DPPH, magnetic flux, YBaCuO superconductor, MPMG

INTRODUCTION

When a superconductor is placed in a static magnetic field, the magnetic field on the surface of the superconductor may fluctuate with time and in space because of thermal fluctuation of fluxons trapped at the pinning centers, Meissner effect, etc.. The physical origin of the magnetic fluctuation is quite complicated. However, it may be directly correlated with fluxon dynamics. Thus, it is of interest to obtain information about the magnetic fluctuation, for example, its magnitude and temperature dependence. To obtain such information, in the present work we have studied the ESR of DPPH(1,1-diphenyl-2-picryl-hydrazyl) coated on the surface of high Tc superconductor, Y Ba Cu O, which was made by the MPMG (Melt-Powder-Melt-Growth) method. This idea is similar to the detection of magnetic field distribution and/or fluctuation on the surface of electronic materials and devices [1]. In this work ESR experiments were performed using a conventional ESR spectrometer operating at 9GHz. Powder and bulk samples were used for the experiments. Particular attention was paid to the temperature dependence of the ESR linewidth, ΔHpp, below Tc($\simeq$89K). The effect of applied magnetic field on ΔHpp was also studied. The temperature dependence of the excess ESR linewidth, $\delta(\Delta Hpp)=\Delta Hpp(T)-\Delta Hpp(T\simeq Tc)$, was analyzed, and was discussed in conjunction with the evaporation of vortice.

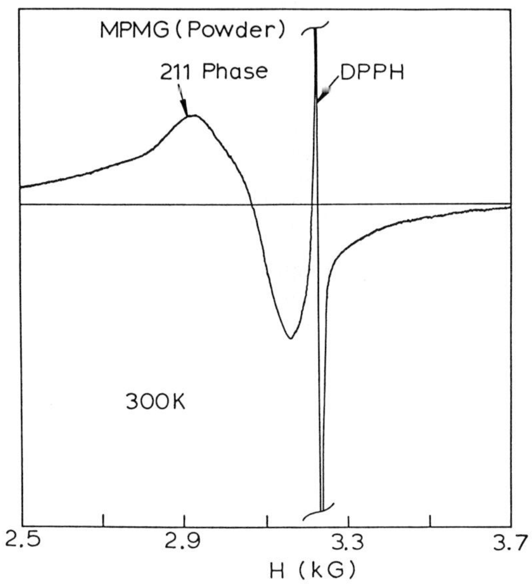

Fig. 1. Typical ESR of MPMG sample at room temperature.

EXPERIMENTAL

Sample Preparation

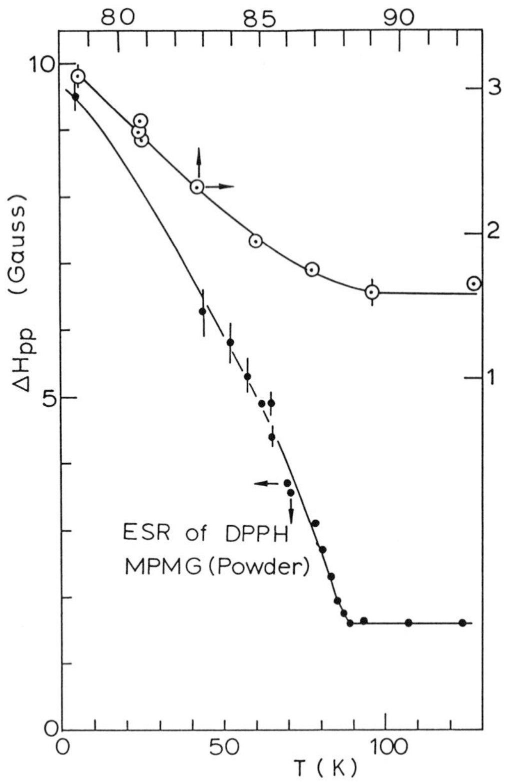

Fig. 2. Temperature dependence of ESR linewidth of DPPH coated on MPMG sample (powder).

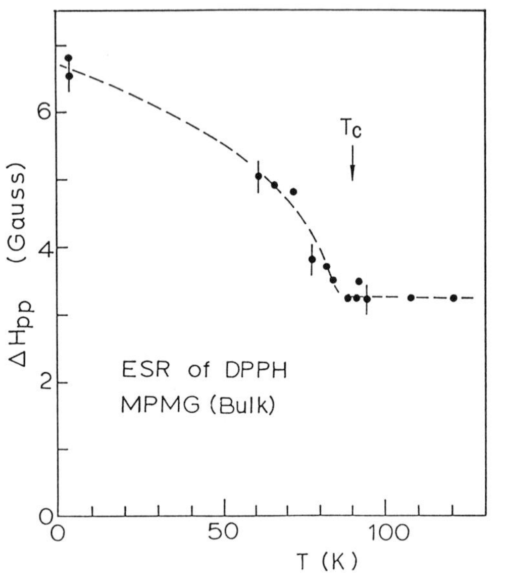

Fig. 3. Temperature dependence of ESR linewidth of DPPH coated on MPMG sample (bulk).

The high-Tc superconductor, Y-Ba-Cu-O, used in this study was made by the MPMG method [2]. The nominal composition of the starting mixed powder was $Y_{1.5}Ba_{2.25}Cu_{3.25}O_y$, and the final composition was $Y_1Ba_2Cu_3O_y$ with $Y_2Ba_1Cu_1O_5$ inclusion, in which the atomic fraction of the 211 phase is considered to be about 25 %. The zero-resistance temperature, Tc, of the Y-Ba-Cu-O crystal was about 89K. A typical critical current density, Jc, of MPMG sample is 3.5×10^4 A/cm^2 at 77K and 1T, in which Jc was obtained from transport and magnetization measurements [2]. By breaking the mother MPMG crystal, two samples (powder and bulk) were prepared for ESR measurements. The weight of each sample was about 23 mg. These samples were immersed in a solution of DPPH in acetone and then dried in air. From weight measurements of the samples, the thickness of the DPPH coating was estimated to be about 10 μm.

ESR Measurements

A conventional ESR spectrometer (JEOL RE3X) operating at about 9GHz was used for the ESR measurements. A cylindrical cavity with TE_{011} mode was used. The microwave absorption signal was detected by modulating the externally applied magnetic field at 100kHz. The modulation amplitude was 0.1 Gauss. Incident microwave power to the sample was about 1mW. The Y-Ba-Cu-O samples were put into standard ESR quartz sample tubes (with ID of about 3mm). In order to prevent the motion of sample by the Meissner effect below Tc, the samples were wrapped in vinyl sheets and pressed to the bottom of ESR sample tubes.

RESULTS AND DISCUSSION

A typical ESR of a powder specimen obtained at 300K is shown in Fig. 1. The broad signal in Fig. 1 originates from $Y_2Ba_1Cu_1O_5$ in the sample. The temperature dependence of ΔHpp are shown in Fig. 2 and Fig. 3 for powder and bulk samples, respectively. The direction of experimental run is from higher temperature to lower temperature, i.e. the samples were cooled below Tc (~89K) in magnetic fields of about 3.2KGauss. As can be seen in Figs. 2 and 3, ΔHpp substantially increases with decreasing temperature. We have also found that ΔHpp strongly depends on the magnetic field, which was applied to the samples just before ESR measurement. This behavior was found for both powder and bulk specimens as shown in Fig. 4 and Fig. 5, respectively. The broadening of ΔHpp by the applied field may be due to trapped magnetic fluxoids in the samples. The ΔHpp, however, sharply decreases with increasing temperature as shown in Fig. 6 for a bulk specimen. Similar behavior was also found for powder

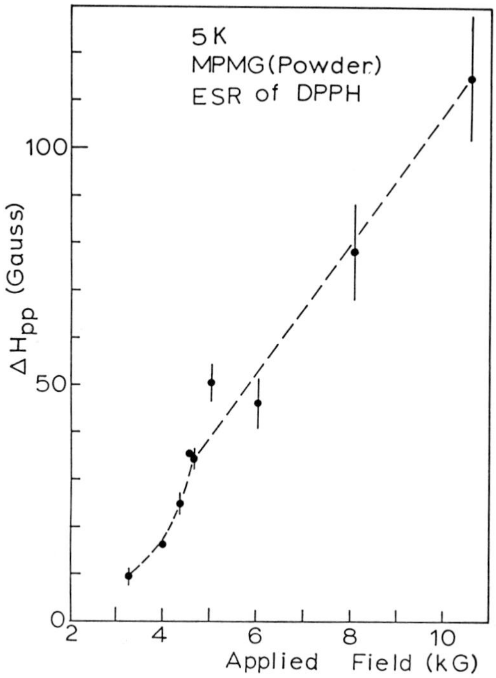

Fig. 4. Applied magnetic field effect on ΔHpp of DPPH coated on MPMG sample (powder) at 5K.

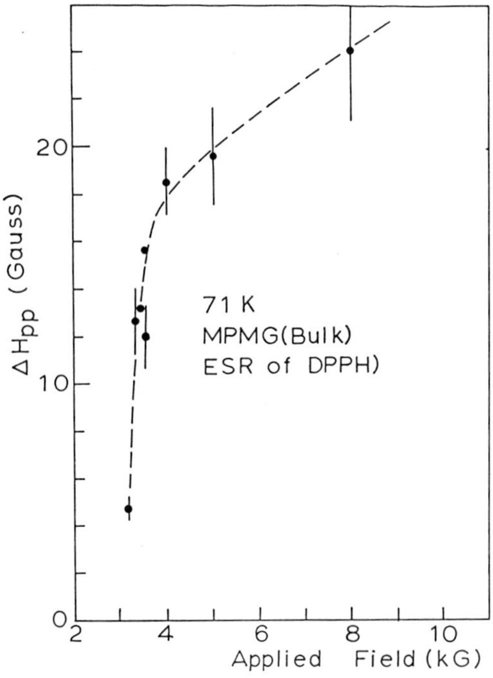

Fig. 5. Applied magnetic field effect on ΔHpp of DPPH coated on MPMG sample (bulk) at 71K.

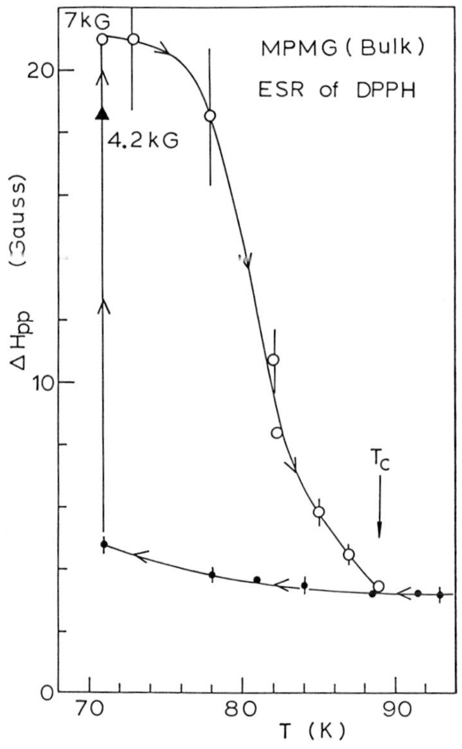

Fig. 6. Temperature dependence of ΔHpp of DPPH coated on MPMG sample (bulk).

specimen. The sharp drop in ΔHpp may be due to the evaporation of vortices. In the following, the temperature dependence of the excess ESR linewidth, $\delta(\Delta Hpp) = \Delta Hpp(T) - \Delta Hpp(Tc{\sim}89K)$, will be discussed. The physical origin of $\delta(\Delta Hpp)$ may be due to the dynamic behavior of vortices on the surface of samples. ΔHpp(89K) are about 1.59 Gauss and 3.2 Gauss for powder and bulk samples, respectively. By using Figs. 2 and 3, $\delta(\Delta Hpp)$ were obtained. $\delta(\Delta Hpp)$ are plotted against $(1 - T/Tc)$ (Tc~89K) for powder and bulk specimens in Figs. 7 and 8, respectively. As can be seen in these figures, $\delta(\Delta Hpp)$ is approximately proportional to $(1 - T/Tc)^a$ with $a{\simeq}1.4$ for both samples at temperatures between about 80K and 89K. (In Figs. 7 and 8, C_0 and C are numerical constants.) According to Tinkham [3], the unperturbed pinning potential is proportional to $(1 - T/Tc)^{1.5}$. The temperature dependence of $\delta(\Delta Hpp)$ obtained in our study is close to that of theoretical pinning potential. From this it is inferred that the ESR linewidth of DPPH may be associated with magnetic fluxoids in the sample.

CONCLUSION

In this study we have found that ΔHpp of DPPH increases with decreasing temperature below Tc, and

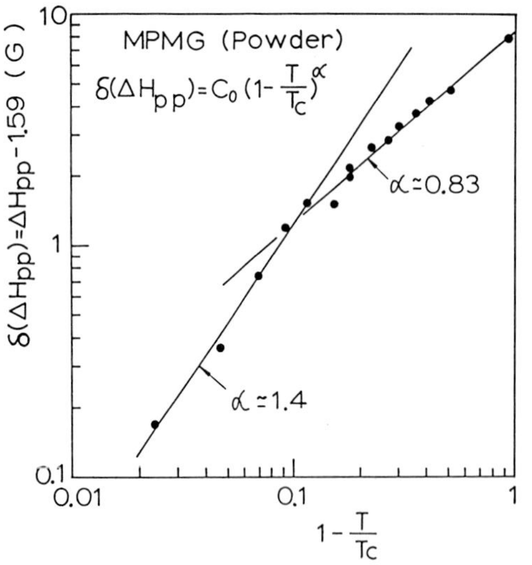

Fig. 7. Excess ESR linewidth $\delta(\Delta Hpp)$ of DPPH coated on powder MPMG sample vs 1 - T/Tc, where Tc=89K.

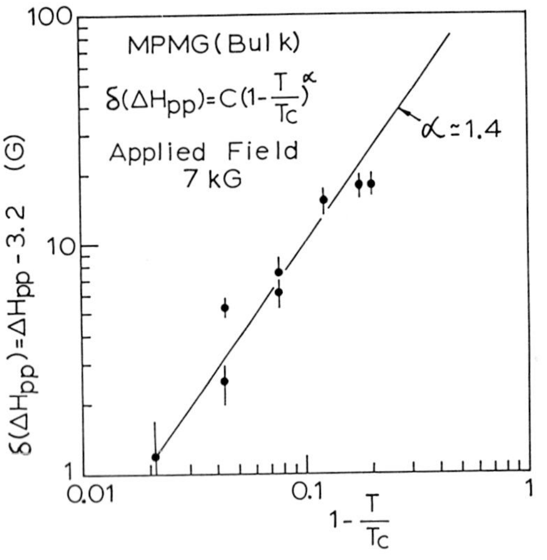

Fig. 8. Excess ESR linewidth $\delta(\Delta Hpp)$ of DPPH coated on bulk MPMG sample (bulk) vs 1 - T/Tc. A magnetic field of 7KG was applied to the sample prior to ESR data acquisition.

at temperatures between about 80K and Tc($\simeq$89K) the excess ESR linewidth $\delta(\Delta Hpp)$ is nearly proportional to $(1 - T/Tc)^a$ with a$\simeq$1.4. This temperature dependence may be associated with the fluxon motion on the surface of high-Tc superconductors. We have also found that the ΔHpp is sensitive to the applied magnetic field. In this study, we have performed ESR experiments for two different kinds of samples; powder and bulk. ΔHpp values for powder specimens are substantially larger than those of bulk samples. This indicates that the magnitude of the DPPH ESR linewidth may depend on sample configuration and particle size. In order to find any correlation between DPPH ESR and critical current density, further experiments are needed.

ACKNOWLEDGEMENT

This work was supported by New Energy and Industrial Technology Development Organization under the management of the R & D of Basic Technology for Future Industries.

REFERENCES

1. Sugawara K : Patent document 60-73136, April 1985, Japan.
2. Murakami M, Gotoh S, Koshizuka N, Tanaka S, Matsushita T, Kambe S and Kitazawa K (1990) Cryogenics 30: 390.
3. Tinkham M (1988) Phys. Rev. Lett 61: 1658.

Size Effects in the Magnetization of Epitaxial $YBa_2Cu_3O_x$ Films

SHIN KOSAKA[1], HARUHIKO OBARA[1], MASAICHI UMEDA[1], YOICHI KIMURA[1], HISANORI YAMANE[2], TOSHIO HIRAI[2], and HIDEYUKI KUROSAWA[3]

[1] Electrotechnical Laboratory, 1-4, Umezono 1-chome, Tsukuba, Ibaraki, 305 Japan
[2] Institute for Materials Research, Tohoku University, 1-1, Katahira 2-chome, Aoba-ku, Sendai, 980 Japan
[3] RIKEN Co., 810 Kumagaya, Kumagaya, 360 Japan

ABSTRACT

Systematic measurements of magnetization dependence on sample size have been performed on epitaxial YBCO films in order to investigate granularity characteristics of oxide superconductor thin films. YBCO films were deposited by using a molecular beam epitaxy(MBE) technique and chemically etched into a disks of diameters between 10μm to 1 mm. Magnetization of these films have exhibited a remarkable linear dependence of the magnetization on the diameter, which shows Bean-critical-state model has been fully established and no granularity has existed in these epitaxial YBCO films. Results have been compared with YBCO films deposited by chemical vapor deposition(CVD) technique.

INTRODUCTION

Magnetization measurements are useful to observe the critical current properties of superconductors[1]. In oxide superconductors, however, it is the issue whether Bean-critical-state model is fully established in the whole scale of the sample or not. It has been reported that powdered samples of sintered YBCO show granular characteristics and magnetization of these samples does not scale to the sample size[2], which means that the sample is subdivided into a smaller scale by existence of weak links due to grain boundaries so that the critical state is not established in the whole scale of the sample size. However, even in single crystals, it has been reported that magnetization does not scale to the sample size and these properties are attributed to existence of intra-grain weak links[3-6]. On the other hand, in melt processed YBCO bulk samples, magnetization has been reported to be linearly proportional to the sample size[7]. We also have observed a linear magnetization dependence on sample size in a series of magnetization measurements for an epitaxial YBCO film by reducing the sample size by cutting in a rectangular shape[8]. As far as the existence of the intra-grain weak link concerns, its depends strongly on the micro structure of the sample. Up to now, there is no general understanding about in what case magnetization scales to the sample size or otherwise granularity appears. To clarify the mechanism for the granular characteristics is very important for investigating the pinning mechanism in oxide superconductors. It is a straightforward way to measure the size scaling characteristics of the magnetization for investigating the granular nature of the sample. The purpose of this paper is to systematically study the magnetization characteristics of YBCO thin films as a function of the sample size. Two kinds of YBCO films have been studied. One is deposited by a molecular beam epitaxy(MBE) method and the other is deposited by a metal organic chemical vapor deposition(MOCVD) method.

EXPERIMENTAL

We used a thin film of YBCO deposited by MBE and two films deposited by MOCVD. Details are given in Table 1. The MBE film(#10626) was prepared by coevaporation in an ultra high vacuum system using three effusion cells for Y, Ba and Cu metal sources and electron cyclotron resonance(ECR) ion source for oxygen source, and deposited on an MgO substrate with (100) plane at the substrate temperature of 650 C. The detail of film preparation and their superconductive properties have been described elsewhere[8,9]. The film was epitaxially grown with their <001> axis normal to the (100) surface of the MgO substrate. The MOCVD films(P27, P41) were deposited on a MgO(100) substrate and $SrTiO_3(100)$ substrate, respectively, at the substrate temperature of 850 C using ß-diketone metal chelates. Detail of the preparation process and critical current properties of the films have been described in reference 10.

Table 1 Details of samples

Sample	Source	Technique	Substrate	Thickness	Comments
#10626	ETL*	MBE	MgO(100)	0.18 μm	Ref. 9
P41	IMR**	MOCVD	MgO(100)	0.5 μm	Ref. 10
P23	IMR**	MOCVD	$SrTiO_3(100)$	0.5 μm	Ref. 10

*Electrotechnical Laboratory
**Institute for Materials Research, Tohoku University

In order to study the size effect systematically, films were patterned in disk shape having various diameters. For varying the diameter accurately in a wide range, photoresist mask(Microposit 1400-31) and etylenediaminetetraacetic acid(EDTA)[11] were used for etching. Four groups of disks having diameter of 1 mm, 100 μm, 50 μm and 25 μm were patterned on a single film(area is about 6 mm by 6 mm), and each group were mechanically cut from the substrate after the etching was completed. So as to avoid the effect of decrease in magnetic moment caused by reducing the sample size, total volume of the superconductors of each sample is designed to be equal to that of a single 1 mm diameter disk. Actual diameter of the disks are undersized from the designed value due to side etching effect. The diameter values measured by using an optical microscope were used for analyzing the magnetization data but nominal values were used to identify each data sets.

The magnetic moment measurements were made on a Quantum Design SQUID magnetometer. The magnetic field was applied perpendicular to the film surface throughout this experiments. Special attention was paid so as to achieve very low field conditions(~0.1 Oe) by using the superconducting magnet at field lower than Hc1 of the superconducting wire in the magnet. High field measurements were done in the completely separate experimental runs and the low field measurements were performed after heating the whole superconducting magnet up to the room temperature to eliminate the effect of the residual magnetic field due to trapped magnetic flux.

RESULTS AND DISCUSSIONS

Figure 1 shows the temperature dependence of zero-field-cooled(ZFC) moment for MBE deposited samples(#10626) of various diameters. In this measurements, the sample was initially cooled in zero field to 10K and subsequently a filed of 10 Oe was applied. The sample was warmed in this field and the moment was measured by the SQUID magnetometer. As all the samples have nearly the same volume of YBCO, neglecting the effect of undersizing, the magnetization of each sample is roughly proportional to its moment data. As shown in the figure, the moments of the samples were systematically decrease with the sample size. If we calculate the magnetization and plot it as a function of the diameter of the sample, we can get a remarkable linear dependence of the magnetization on the diameter. This effect can be interpreted by assuming the perfect diamagnetism of the film and taking into account the demagnetizing factor of the disk shaped sample[8]. We don't go into this point further in this paper, but note that the shape of the sharp drop in the diamagnetic moment at the transition temperature was not changed by decreasing the diameter of the sample. We also have compared the transition with a sample cut from the same film without using chemical etching process, and observed no distinct difference in transition between these etched and unetched samples. So it can be said that the chemical etching process used in this experiments does not deteriorate the superconducting properties of the film at least by measuring the magnetization properties.

Figure 2 shows the remanent moment of the MBE deposited samples(#10626) obtained by turning off the field after field-cooling and measuring as a function of increasing temperature. A field of 200 Oe was used for field-cooling because above this field the value of remanent moment saturates and become field independent. Using this data and measuring the actual diameter of the disk, the remanent magnetization was calculated and plotted as a function of the diameter for various temperatures in Fig. 3. As shown in the figure, linear diameter dependence of magnetization was confirmed at the diameter range from 20μm to 1mm and all the temperature range measured. Linear extrapolation of the data at zero diameter were less than 20 emu/cc. We can say that the magnetization of the MBE deposited epitaxial film is linearly proportional to sample diameter and its extrapolation to zero diameter goes through the origin.

Figure 4 and 5 shows the remanent magnetization as a function of the sample diameter at various temperatures for MOCVD film deposited on $SrTiO_3$(P23) and MgO(P41) substrates, respectively. As shown in Fig. 4, the MOCVD

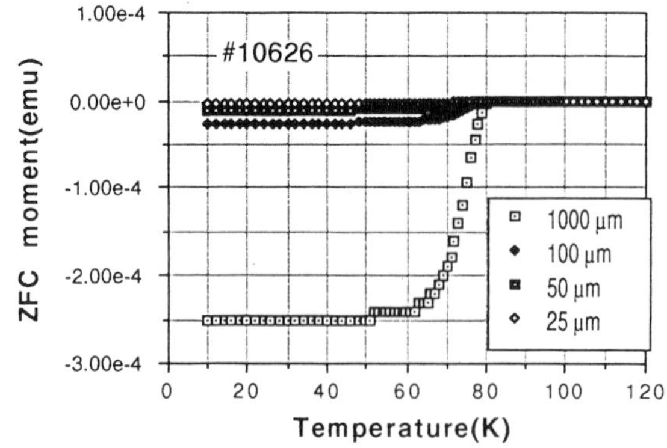

Fig. 1. Zero-field-cooled(ZFC) moment vs temperature for the MBE films of various diameters. Applied field=10 Oe.

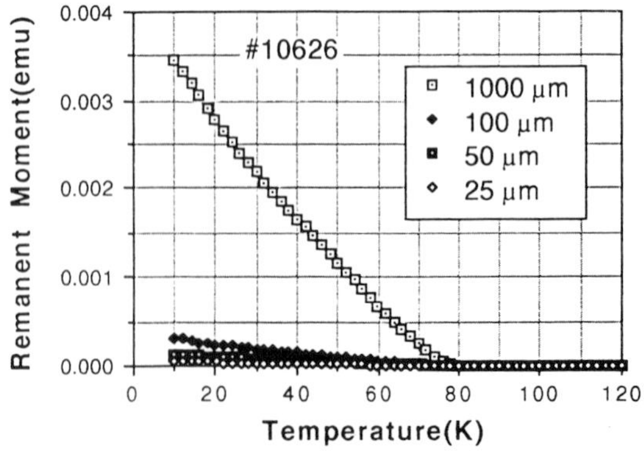

Fig. 2. Remanent moment vs temperature for the MBE films of various diameters. Applied field=200 Oe.

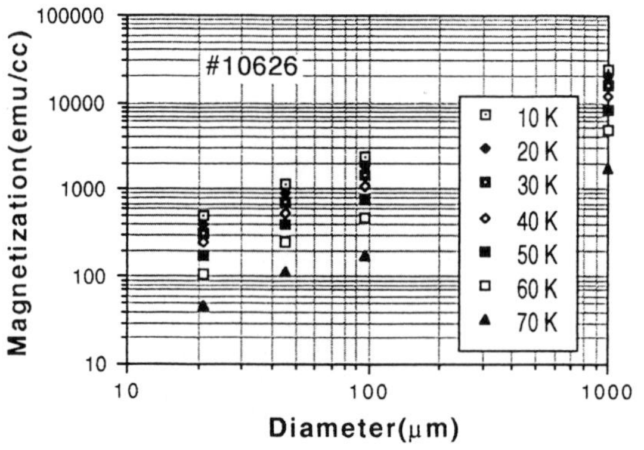

Fig. 3. Remanent magnetization vs diameter for the MBE film.

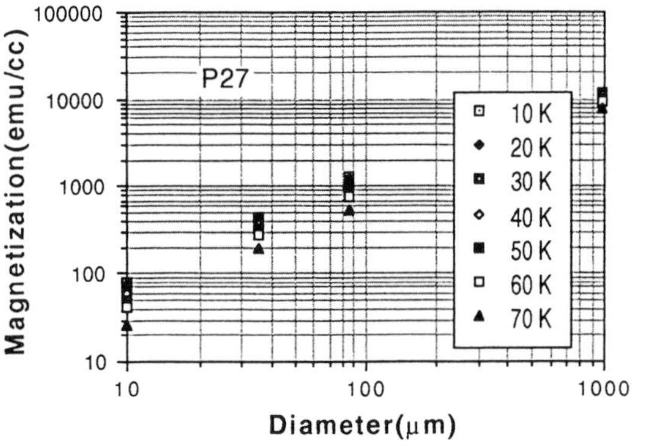

Fig. 4. Remanent magnetization vs diameter for the MOCVD film on $SrTiO_3$ substrate.

film deposited on $SrTiO_3$ substrate also shows the linear and through-origin magnetization dependence on diameter, similar to the MBE film. The value of the magnetization is also close to each other. On the other hand, the magnetization dependence on the diameter of the MOCVD film deposited on MgO substrate is very different from the former two films, as shown in Fig. 5. The magnetization of this film shows almost no diameter dependence and the absolute value of the magnetization is more than 100 times smaller than that of the former two films.

Hysteresis loop measurements of magnetization versus field up to 10 kOe have been carried out for the MBE film(#10626) at the temperature of 5 K. In Fig. 6, width of the magnetization hysteresis loop at various field were plotted as a function of the sample diameter. Again it shows a linear and through-origin diameter dependence up to the field of 10 kOe. At fields less than 1 kOe, magnetization hysteresis were almost constant in each sample. At fields higher than 10 kOe, magnetization dropped off rapidly and became the same level as the background magnetization of the sample holder of the SQUID magnetometer at about 20 kOe.

Both in the MBE film(#10626) and the MOCVD film on the $SrTiO_3$ substrate(P23), the remanent magnetization has shown linear and through-origin size-scaling results at the temperature range of 10 K to 70K. These results strongly indicate that no granularity exists in these films and shielding current flows in the whole of sample, in other word, Bean-critical-state is fully established in the whole scale of the sample in these films. This no granular nature of the film is also verified in the magnetic field range up to 10 kOe for the MBE film. As the magnetization of these films are well described by the Bean-critical-state model, we can estimate the critical current density(Jc) of the films from the magnetization data using an expression Jc=30M/R, where M is the magnetization in unit of emu/cc, R is the radius of the sample in cm, and Jc is the critical current density in A/cm^2. Jc's of #10626 and P23 are calculated to be $1.5x10^7$ A/cm^2 and $7.3x10^6$ A/cm^2 at 10 K, respectively. On the contrary, the magnetization of the MOCVD film deposited on the MgO(P41) does not scale to the size and Jc cannot be directly estimated from the magnetization data.

Recently a transmission electron microscopy(TEM) study has been reported on MOCVD deposited YBCO films[12]. The MOCVD deposited YBCO film on MgO is composed of almost-<001> oriented polycrystalline grains with the

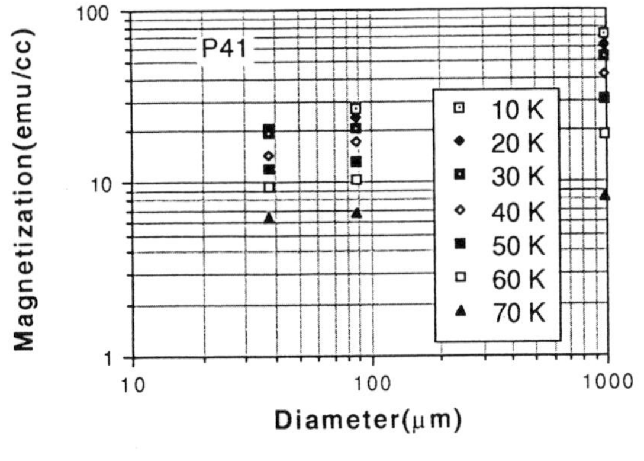

Fig. 5. Remanent magnetization vs diameter for the MOCVD film on MgO substrate.

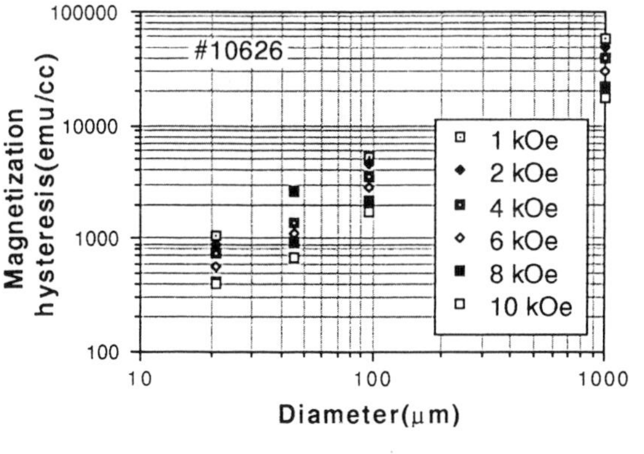

Fig. 6. Magnetization hysteresis vs diameter for the MBE film at 5 K.

mean grain size of about 0.5 μm and with small-angle grain boundaries of about 5°. The granularity of the film observed in the present experiment can be understood to be due to the weak links caused by this small-angle grain structure of the film. It also has been reported that the MOCVD deposited YBCO film on $SrTiO_3$(100) has no clear grain boundary structure[12]. This is also consistent with the present result that the MOCVD film on $SrTiO_3$ didn't show any granularity in the magnetization measurements. The epitaxial MBE film and the MOCVD film on $SrTiO_3$ are seemed to have no such explicit grain boundaries as observed in the MOCVD film on MgO.

The critical current density of the MOCVD film on MgO has been reported to be two order of magnitude smaller than that of the film on $SrTiO_3$[10]. This difference in Jc is consistent with the difference in the granularity nature of the films observed in the present study. The decrease in Jc on MgO seems to be also caused by the weak links observed in the magnetization measurements. In order to relate the magnetization data to the critical current density, we need the information about the scaling length where the shielding current is continuously flowing. In the present experiment, the minimum patterning size used was about 10 μm which is more than one order of magnitude larger than the average grain size of the film on MgO observed by TEM[12]. Though experiments using finer pattern are needed to clarify the real scaling length of the MOCVD film on MgO, thinking the grain size of the film as the scaling length my be a good assumption in the present stage.

In this experiments, no granularity was observed in the epitaxial MBE film and in the MOCVD film on $SrTiO_3$ substrate. This result is contrast to the existence of the intra-grain granularity reported for some bulk single crystals[3-6]. It seems that the intra-grain granularity is not an intrinsic characteristics for the oxide superconductors but is dependent on the microstructure of the samples. More precise study for relating the microstructure with magnetization characteristics is required to clarify the granular characteristics of the oxide superconductors.

CONCLUSIONS

The size effects of the magnetization characteristics of YBCO thin films has been studied. The epitaxial film deposited by MBE on MgO substrate and the film deposited by MOCVD on $SrTiO_3$ substrate have shown a remarkable linear and through-origin magnetization characteristics as a function of sample diameter under wide range of temperature and magnetic field, which shows that Bean-critical-state is fully established in the whole scale of the sample in these films. On the other hand, the MOCVD film deposited on MgO has shown a clear granular characteristics whose magnetization exhibit almost no sample size dependence. The difference in the granular characteristics in these films have been strongly related to the microstructure of the films. It has been experimentally clarified that there is no granularity or weak link in the high quality YBCO films.

ACKNOWLEDGEMENT

The authors would like to express their thanks to Drs. K. Kajimura and M. Ono for their encouragement throughout this research and Dr. J. S. Edmends for helpful discussions and helps in magnetization measurements.

REFERENCES

1. Chaudhari P, Koch RH, Laibowitz RB, McGuir TR, Gambino RJ (1987) Phys Rev Lett 58: 2684-2686
2. Shimizu E, Ito D (1989) Phys Rev B39: 2921-2923
3. Daeumling M, Seuntjens JM, Larbalestier DC (1988) Appl Phys Lett 52: 590-591
4. Kupfer H, Apfelstedt I, Flukiger R, Keller C, Meier-Hirmer R, Runtsch B, Turowski A, Wiech U, Wolf T (1989) Cryogenics 29: 268-280
5. Hibbs AD, Campbell AM (1989) IEEE Trans Magn 25: 2142-2145
6. Daeumling M, Seuntjens JM, Larbalestier DC (1990) Nature 346: 332-335
7. Murakami M, Morita M, Koyama N (1989) Japan J Appl Phys 28: L1125-L1127
8. Obara H, Kosaka S, Umeda M, Kimura Y (1990) Physica B 166: 1413-1414
9. Obara H, Kosaka S, Umeda M, Kimura Y (1990) In: Ishiguro T, Kajimura K (eds) Advances in Superconductivity 2. Springer-Verlag, Tokyo pp 801-804
10. Yamane H, Hirai T, Kurosawa H, Suhara A, Watanabe K, Kobayashi N, Iwasaki H, Aoyagi E, Hiraga K, Muto Y (1990) In: Ishiguro T, Kajimura K (eds) Advances in Superconductivity 2. Springer-Verlag, Tokyo pp 767-772
11. Shokoohi FK, Schiavone LM, Rogers CT, Inam A, Wu XD, Nazar L, Venkatesan T (1989) Appl Phys Lett 55: 2661-2663
12. Suzuki H, Kurosawa H, Miyagawa K, Hirotsu Y, Era M, Yamashita T, Yamane H, Hirai T (1990) Japan J Appl Phys 29: L1648-L1651

4.3 Electromagnetic Properties

Ac Susceptibility of High-Temperature Ceramic Superconductors: Critical State and Creep at Grain Boundaries and in Grains

K.-H. MÜLLER[1], M. NIKOLO[2], N. SAVVIDES[1], and R. DRIVER[1]

[1] CSIRO Division of Applied Physics, Sydney, 2070 Australia
[2] Department of Physics, University of Colorado, Boulder, CO 80303, USA

ABSTRACT

The ac susceptibilities of different ceramic high-temperature superconductors were measured as a function of temperature, ac magnetic field amplitude and frequency. The susceptibility data are well described employing a critical state model for a granular superconductor where intergranular Josephson vortices and intragranular Abrikosov-like vortices sweep in and out of the superconductor during each ac cycle. By incorporating the Anderson flux creep model, the frequency dependence of the ac susceptibility can be predicted.

KEY WORDS: ac susceptibility, critical state, flux creep, pinning potential

INTRODUCTION

The magnetic properties of a superconductor can be assessed by measuring its ac susceptibility. The dependence of $\chi'(T)$ and $\chi''(T)$ on ac field amplitude, H_m, and frequency, f, can be utilized to obtain information about flux pinning and creep.
In this paper we show that ac susceptibility data of ceramic high-temperature superconductors, where χ' and χ'' are measured as a function of T, H_m and f, can be described in detail by employing a critical state model for a granular superconductor. The model considers inter- and intragranular magnetic field profiles and takes account of profile changes due to flux creep during the ac cycle. Values can be obtained for the inter- and intragranular pinning force densities and pinning potentials.

MODEL

The real and imaginary parts, χ' and χ'', of the ac susceptibility are determined by the lowest Fourier components of $< B > (t)$, where $< B >$ is the flux density which is spatially averaged over the cross-sectional area of the sample [1]. By definition,

$$\chi' = -1 + \frac{2f}{\mu_o H_m} \int_0^{1/f} < B > (t) \cos(2\pi f t)\, dt \tag{1}$$

and

$$\chi'' = \frac{2f}{\mu_o H_m} \int_0^{1/f} < B > (t) \sin(2\pi f t)\, dt \quad , \tag{2}$$

where H_m is the amplitude of the applied ac field and f the frequency. In granular materials like the sintered ceramic high-temperature superconductors, where grains are weakly Josephson-coupled, the spatially averaged flux density $< B >$ consists of two components, $< B > = < B_J > + < B_g >$, where $< B_J >$ is the spatially averaged *intergranular* flux density and $< B_g >$ the *intragranular* one.

For a sample of long cylindrical shape of radius R, where the applied ac field is parallel to the cylinder axis, the spatially averaged intergranular flux density $< B_J >$ is given by

$$< B_J > = \mu_{eff} \frac{2\mu_o}{R^2} \int_0^R H_J(r)\, r\, dr \quad , \tag{3}$$

where $H_J(r)$ is the intergranular magnetic field distribution and μ_{eff} the effective permeability, representing the grain diamagnetism. If the grains are in the Meissner state, μ_{eff} depends on the ratio $R_g/\lambda_L(T)$ where R_g is the average grain radius and λ_L is the London penetration depth for grains (neglecting anisotropy) [2].

In order to avoid complicated demagnetizing effects, one can approximate the grains by long cylinders of radius R_g, equal to the average grain radius. The spatially average intragranular flux density $< B_g >$ is then given by

$$< B_g > = (1 - \mu_{eff}) \frac{4\mu_o}{R^2 R_g^2} \int_0^R \int_0^{R_g} H_g(r, r')\, r\, r'\, dr'\, dr \quad , \tag{4}$$

where r' is the radial coordinate inside a grain which is located at r inside the sample and $H_g(r,r')$ is the intragranular magnetic field distribution.

The intergranular field $H_J(r)$ in Eq. (3) forms when the applied magnetic field penetrates the granular superconductor along the grain boundary weak links and Josephson vortices are created. These vortices are pinned at grain boundary intersection points or inhomogeneities along grain boundaries. If $H_J(r)$ is greater than the lower critical field H_{c1g} of grains (neglecting anisotropy), Abrikosov-like vortices form inside grains and magnetic flux penetrates beyond the London penetration depth, λ_L, into each grain, producing the intragranular magnetic field $H_g(r,r')$. The vortices inside grains are pinned at regions of low order parameter e.g. between copper-oxygen sheets or planar defects.

Because of the existence of inter- and intragranular flux pinning, the magnetic field profiles $H_J(r)$ and $H_g(r,r')$ are solutions of the critical state equations

$$\frac{dH(r)}{dr} = \pm\frac{F_{PJ}(B_J)}{\mid B_J(r)\mid} \quad \text{and} \quad \frac{dH_g(r,r')}{dr'} = \pm\frac{F_{Pg}(B_g)}{\mid B_g(r,r')\mid} \quad , \tag{5}$$

where the boundary conditions for an ac susceptibility measurement are $H_m\cos(2\pi ft) = H_J(R)$ and $H_J(r) = H_g(r,R_g)$. The quantities F_{PJ} and F_{Pg} are the inter- and intragranular pinning force densities. The intergranular pinning force density, F_{PJ}, is derived from the Anderson flux creep model [3], assuming that the pinning sites for Josephson vortices are on average $2R_g$ (average grain size) apart and the average intergranular pinning potential, which is of depth U_{oJ}, is $2R_g$ wide [4]. One obtains

$$F_{PJ} = \frac{kT\mid B_J\mid}{2R_g\Phi_o}\,\text{arcsinh}[\frac{\nu_h}{2\nu_o}\exp(\frac{U_{oJ}(0)(1-T/T_c)^\alpha}{kT}\frac{H_o}{\mid H_J\mid + H_o})] \quad , \tag{6}$$

where ν_o is the attempt frequency for Josephson vortex hopping which can be related to the flux line volume around a pinning site and the velocity of sound. The exponent α has a value between 1 and 2. The field H_o is defined as

$$H_o = \frac{\Phi_o}{4\mu_o\lambda_L R_g} \quad . \tag{7}$$

In the case of an applied ac magnetic field, the "minimal observable" hopping rate, ν_h, in Eq.(6) depends on the ac field amplitude H_m and the frequency f and one finds the approximate expression

$$\nu_h \simeq \frac{8\Phi_o H_m R_g^2}{RkT}f \quad . \tag{8}$$

In a similar way the intragranular pinning force density can be obtained. For simplicity, we neglect here the effect of creep on the intragranular critical state profile $H_g(r,r')$ and employ the "zero-temperature" Bean approach where for $H_{c1g} < H_g << H_{c2g}$ one uses $F_{Pg}/\mid B_g\mid = J_{cg}(T)$ [5].

Figure 1 shows schematically $H_J(r)$ and $H_g(r,r')$ inside a granular superconductor. The form of the critical state profile depends on the applied field, H_a, and changes during the ac cycle as discussed in Refs. [1] and [6].

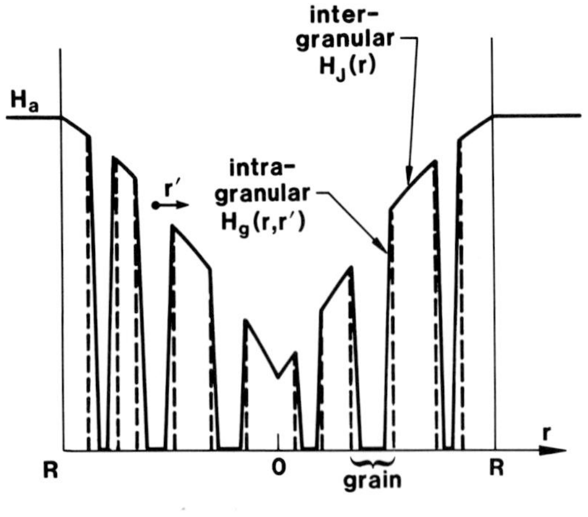

Fig. 1. Schematic drawing of the inter- and intragranular magnetic fields, H_J and H_g, inside a granular superconductor. The sample is of cylindrical shape with radius R, where R is only about 10 R_g. H_a is the applied magnetic field, $H_a = H_m\cos(2\pi ft)$.

RESULTS AND DISCUSSION

Figure 2 shows the calculated ac susceptibility which is typical for a ceramic YBCO sample employing the above model. Inter- and intragranular contributions are indicated. The material parameters used were $R = 1.5$ mm, $R_g = 10$ μm, $\lambda_L(0) = 0.4$ μm, $(\lambda_L(T) = \lambda_L(0)(1-(T/T_c)^4)^{-1/2})$, $\mu_{eff}(0) = 0.2$, $U_{oJ}(80\text{ K}) = 5$ eV, $\alpha = 2$, and $J_{cg}(80\text{ K}) = 10^7$ Acm^{-2} $(J_{cg}(T) = J_{cg}(0)(1 - T/T_c))$. The broad peak of χ'' at lower temperatures represents the intergranular loss. The maximum of χ'' appears at a temperature T_{PJ} where the intergranular pinning force density is such that the intergranular field, H_J, just penetrates the centre of the sample. A narrow intragranular loss peak occurs near T_c at a temperature T_{Pg} where intragranular flux just penetrates the centre of each grain. The intragrain χ'' peak temperature, T_{Pg}, shifts less strongly to lower temperature with increasing amplitude, H_m, than T_{PJ}. This is because the pinning force density for the intragrain Abrikosov-like vortices is much greater than the intergranular Josephson-vortex pinning force density [1]. The weak Josephson-vortex pinning is responsible for the low transport critical current density found in ceramic high-temperature superconductors [6]. Though the intergranular pinning potential depth, U_{oJ}, is found to be quite large, the pinning force density is small. This is due to the large average range of the intergranular pinning potential which is equal to the average grain radius, assuming homogeneous Josephson junctions at grain boundaries. The situation is different for conventional Abrikosov vortex pinning, where the potential depth and range are smaller.

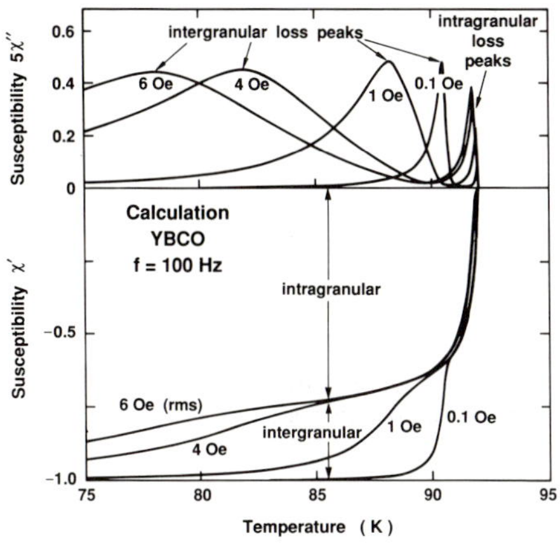

Fig. 2. Calculated ac susceptibility as a function of temperature at different magnetic field amplitudes which agrees well with measurements done on ceramic YBCO samples [1].

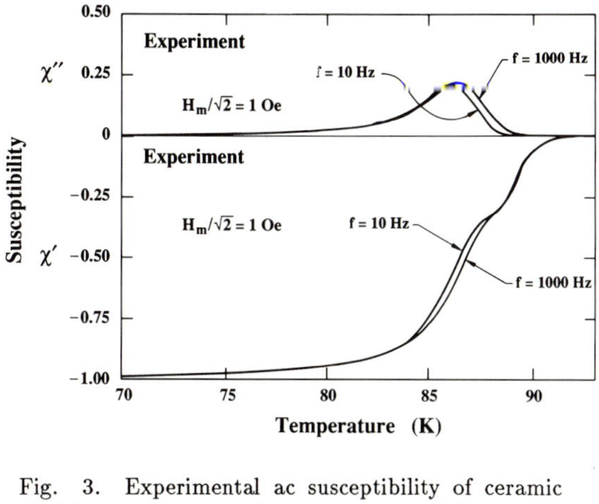

Fig. 3. Experimental ac susceptibility of ceramic YBCO as a function of temperature at different frequencies for a magnetic field amplitude of 1 Oe (rms).

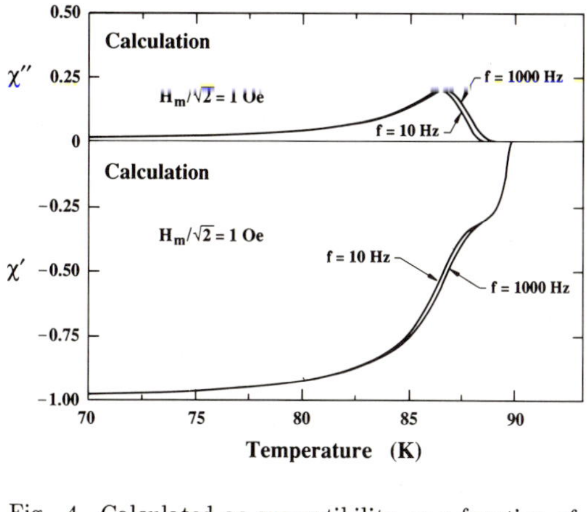

Fig. 4. Calculated ac susceptibility as a function of temperature at different frequencies for a magnetic field amplitude of 1 Oe (rms).

By varying the frequency, f, the effect of inter- and intragranular flux creep on the ac susceptibility, occurring on a millisecond time scale, can be investigated [7,8]. Figure 3 shows the experimental data of χ' and χ'' as a function of temperature for an amplitude of 1 Oe rms at frequencies $f = 10$ Hz and 1000 Hz for a ceramic YBCO sample. The χ' and χ'' curves shift slightly to higher temperature with increasing frequency. Figure 4 shows the result of the calculation which agrees well with the experimental data of Fig. 3. The shift of T_{PJ} with frequency can be understood in terms of the effect of flux creep on the intergranular magnetic field profile H_J. At a higher frequency the critical state profile, H_J, has less time to relax according to the flux creep model and therefore the "effective" pinning force density increases with increasing frequency causing T_{PJ} to shift to higher temperature. A reversible region can be seen in Fig. 4 between 88 K and 90 K where the intergranular pinning force density is practically zero.

It has been found experimentally [7] that T_{PJ} depends on f like

$$(kT_{PJ})^{-1} = -(E_{aJ})^{-1}\ln(f/f_o) \quad , \tag{9}$$

where f_o is a constant and E_{aJ}, defined by Eq.(9), is an energy which is only a function of the ac field amplitude H_m. Equation (9) can also be derived directly from the above model and E_{aJ} can be calculated [8,9].

We have measured T_{PJ} at different frequencies (10 - 1000 Hz) and rms amplitudes (0.01 - 10 Oe), and have determined the energy E_{aJ} defined by Eq. (9) as a function of H_m. Figure 5 shows E_{aJ} as a function of H_m for a $YBa_2Cu_3O_{7-\delta}$ and a $Bi_{1.84}Pb_{0.34}Sr_{1.91}Ca_{2.03}Cu_{3.06}O_{10}$ ceramic sample. The curves in Fig. 5 represent the calculated results which agree well with the experimental data. The material parameters U_{oJ} and α were obtained by fitting the $\chi'(T)$ and $\chi''(T)$ data at one frequency ($U_{oJ}(80\text{ K}) = 4.6$ eV, $\alpha = 1$ for YBCO and $U_{og}(80\text{ K}) = 2.2$ eV, $\alpha = 1.5$ for BPSCCO). The average grain size, R_g, in the case of the BPSCCO sample, was estimated using polarized light microscopy. Measurements are underway to determine the dependence of the intragranular χ'' peak-temperature, T_{Pg}, on frequency and ac amplitude. Preliminary results show that the intragranular E_{ag} (defined by an equation like Eq.(9)) is smaller than E_{aJ}, indicating that the intragranular pinning potential depth is smaller than the intergranular one.

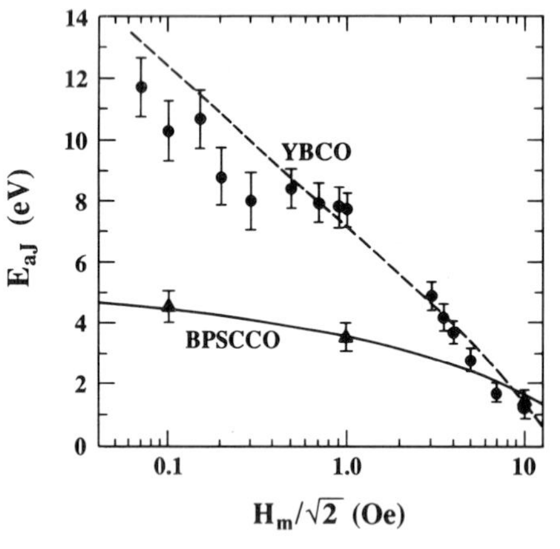

Fig. 5. Energy E_{aJ} as a fuction of the magnetic field amplitude H_m for a YBCO and a BPSCCO ceramic sample.

In summary, the dependence of the ac susceptibility of ceramic high-temperature superconductors on ac magnetic field and frequency can be described quantitatively by employing a "finite-temperature" critical state model for a granular superconductor. The observed frequency dependence is caused by flux creep. The model allows us to assess the inter- and intragranular pinning force densities, pinning potentials and their dependence on temperature and local magnetic field. In addition, the effective permeability (or Meissner-fraction) can be determined.

Acknowledgement:

The authors would like to thank R.B. Goldfarb from the National Institute of Standards and Technology in Boulder, Colorado and T. Ciszek and R. Schwerdtfeger from the Solar Energy and Research Institute in Golden, Colorado for providing access to their ac susceptometers and for their kind assistance.

REFERENCES

1. Müller K-H (1989) Physica C 159: 717-726
2. Clem J R (1988) Physica C 153-155: 50-55
3. Anderson P W (1962) Phys.Rev.Lett. 9: 309-311
4. Lobb C J, Abraham D W, Tinkham M (1983) Phys.Rev.B 27: 150-157
5. Bean C P (1962) Phys.Rev.Lett. 8: 250-253
6. Müller K-H, Macfarlane J C, Driver R (1989) Physica C 158: 69-75
7. Nikolo M, Goldfarb R B (1989) Phys.Rev.B 39: 6615-6618
8. Müller K-H (1990) Physica C 168: 585-590
9. Müller K-H, Nikolo M, Driver R (1991) Phys.Rev.B: (in press)

Flux Pinning Studies for Plasma-Sprayed $YBa_2Cu_3O_{7-x}$ Superconductors

T.R. FINLAYSON, Y.C. FAYMAN, D.B. LOWE, Z. PRZELOZNY, and T.F. SMITH

Monash University, Clayton, 3168 Australia

ABSTRACT

Plasma sprayed $YBa_2Cu_3O_{7-x}$ has been studied using magnetization measurements. Differences in magnetic hysteresis at very low fields following different heat treatments indicate a variation in the intergranular coupling strength. The influence of twin plane density on activation energy is discussed within the constraints of Ginzburg-Landau theory. A model based on pinning by twin boundaries accounts for the behaviour at higher fields.

KEY WORDS: $YBa_2Cu_3O_{7-x}$ superconductor, magnetization, activation, twins

INTRODUCTION

Plasma spraying is a relatively inexpensive method for preparing $YBa_2Cu_3O_{7-x}$ superconductors. It could have future application for the production of superconducting components. As with all polycrystalline ceramic samples of $YBa_2Cu_3O_{7-x}$, the high-current applications are severely limited by intrinsically low critical current densities resulting from intergranular weak links and giant flux creep at high temperatures. It is important to understand these phenomena if such materials are to be exploited.

Our objective is the production of a prototype superconducting magnetic energy storage device by plasma spraying. In a recent paper [1] reported data indicate the applicability of the Bean model to flux penetration into the material. More recently we have observed that the magnetic behaviour is dependent on the thermal history of the sample in that heat treatment, which is essential to the production of a superconducting microstructure, has a marked effect on properties. In this paper we compare the effects of two different heat treatments. Microstructurely, the most noticeable difference between the two materials is the density of the twins which are formed on account of the transformation from the tetragonal to the orthorhombic phase during cooling [2].

The elementary pinning force of a planar defect is mainly caused by electron scattering. From the Ginzburg-Landau (GL) theory for dirty superconductors, the local variation in the mean free path, l, leads to $\delta\kappa$ pinning [3]. Activation energies derived from creep data for samples of different twin densities are consistent with the behaviour expected from a derived pinning model.

EXPERIMENTS

The materials under study are heat-treated, plasma-sprayed deposits. Their preparation and characterisation were discussed elsewhere [1]. In this paper we compare the electromagnetic behaviours of materials heat treated in flowing oxygen at 875°C and 950°C for 20h and furnace-cooled (designated $875/20/O_2$ and $950/20/O_2$, respectively). Magnetization versus field and temperature data have been recorded in a vibrating-sample magnetometer, described elsewhere [4]. The former have been used to investigate appropriate models for the critical state

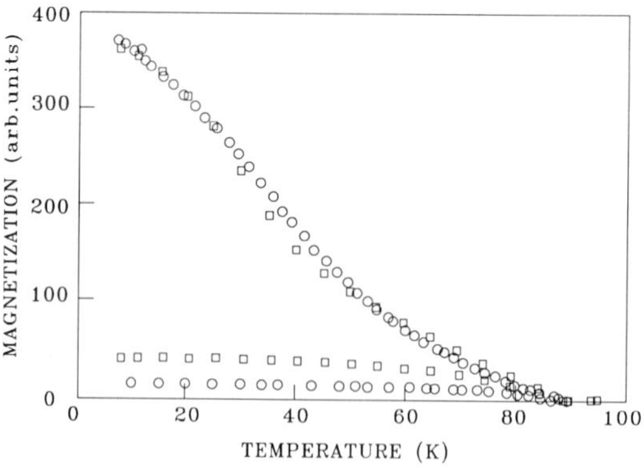

Fig. 1. ZFC and FC magnetization data for a 1kOe field, for samples heat treated at 875°C (□) and 950°C (o).

and to measure magnetization losses in the materials. From the latter the temperature dependences for zero-field-cooled (ZFC) and field-cooled (FC) magnetizations at specific applied fields, have been measured. Flux creep measurements have been made by recording the decay of magnetization with time.

RESULTS

The low-field magnetization versus field characteristics for these two samples are quite different. For 875/20/O_2, the usual low-field hysteresis attributed intergranular superconductivity was observed [1] but for 950/20/O_2 this was absent. From an a.c. susceptibilty, ($\chi(T)$), experiment on the 950/20/O_2 material, carried out several months prior to the present experiments, a weak intergranular peak in $\chi''(T)$ had been observed. However a further measurement of $\chi''(T)$ at the time of the present experiments, verified that any intergranular superconductivity had disappeared. Comparable changes in the 875/20/O_2 sample with time, have not been found.

ZFC and FC magnetizations at 1kOe are shown in Fig. 1. The low-field behaviour for 950/20/O_2 prevents appropriate calibration of the magnetization. Thus the comparison in Fig. 1. involves only arbitrary units for magnetization data. From transmission electron microscopy performed on pieces from these two batches of material, we note that each sample is heavily twinned, with the most noticeable difference between the two being the twin size. For 875/20/O_2 there are approximately 17 twin planes per μm and about 3.5/μm in 950/20/O_2.

PINNING FORCES ARISING FROM TWIN PLANES

The GL functional in non-reduced units has the form [5]

$$F = \int dr\ \{-A|\Delta|^2 + \tfrac{1}{2}B|\Delta|^4 + C|(-i\nabla - (2e/\hbar c)\mathbf{A})\Delta|^2 + \frac{H^2}{8\pi}\} \tag{1}$$

where $A \sim N(0)(1-t)$, $B \sim N(0)/(kT_c)^2$ and $C \sim N(0)\hbar v_f l/(kT_c)$, for $l < \xi_0$. N(0) is the density of states at the Fermi level, $t = T/T_c$ with T_c the critical temperature, v_f the Fermi velocity, and ξ_0 is the BCS coherence length. The local variation of l, produces a first order perturbation in the third term only. Then the change in the free energy becomes

$$\delta F_1 = \int dr\ C|(-i\nabla - (2e/\hbar c)\mathbf{A})\Delta|^2\ \frac{\delta l}{l} \tag{2}$$

Introducing the notation [5] $\psi(\mathbf{r}) = (2mC/\hbar)^{1/2}\Delta(\mathbf{r})$, $\alpha = (\hbar^2/2m)A/C$, and $\beta = (\hbar^2/2m)^2 B/C^2$ one has $\alpha^2/(2\beta) = H_c^2/(8\pi)$, and the order parameter in zero field is $\psi_o^2 = -\alpha/\beta$. After the transformations $f = \psi/\psi_o$, $h = H\xi\lambda/\phi_o$, and $H_c = 1/\sqrt{2}$, the dimensionless GL equations take the well known form

$$(i\nabla/\kappa + \mathbf{A}_o)^2 f_o = f_o - f_o^3 \quad \text{and} \quad -\text{curl } \mathbf{h} = f_o^2 \mathbf{A}_o \tag{3}$$

where $f_o = f\exp[-i\phi]$, $\mathbf{A}_o = \mathbf{A} - \nabla\phi/\kappa$. Since the high-$T_c$ superconductors are characterized by a high κ, from (3) one has

$$\{(\text{curl } \mathbf{h})/f_o^2\}^2 f_o = f_o - f_o^3 \quad \text{and} \quad A_o^2 f_o^2 = f_o^2(1-f_o^2). \tag{4}$$

The first-order perturbation from (2) is given by

$$\delta F = \frac{H_c^2}{8\pi}\int dr\, f_o^2(1-f_o^2)\frac{\delta l}{l} \quad \text{or} \quad \delta F \cong \frac{H_c^2}{8\pi}\int dr\, f_o^2 \frac{\delta l}{l} B/H_{c2}. \tag{5}$$

This is the starting point for our computation for planar defects in the yz plane, located at points x_i. We assume $\delta l/l \approx p_{tr}\,\delta\{(x - x_i)/l\}$ where p_{tr} is an effective scattering probability for the twin planes. Because the orientation of the flux line lattice (FLL) perpendicular to this plane is unstable [3], we consider mirror twins of high and low densities.

For a high density of twin planes, or for intervortex spacing a_o, greater than twin spacing, L,

$$j_c\phi_o s_o = \frac{H_c^2}{8\pi}\int dr\, f_o^2 \frac{\delta l}{l} B/H_{c2} \tag{6}$$

where the integration extends over an area a_o^2. j_c denotes a critical current density and s_o is the maximum displacement of the FLL in equilibrium. Including the above assumption, one obtains

$$j_c\phi_o s_o = \frac{H_c^2}{8\pi}\, l\, a_o \frac{p_{tr} f_o^2}{H_{c2}} \sum_{i=1}^{a_o/L} B(x_i) \tag{7}$$

and because $a_o >> L$, one has $j_c = H_c^2/(8\pi)\,l\, p_{tr}\, f_o^2/(H_{c2}Ls_o)$. Assuming $s_o \approx \xi$, as for the dense pinning centre distribution, one derives

$$j_c = \frac{H_c^2\xi}{8\pi\phi_o}\, l\, p_{tr} f_o^2/L \tag{8}$$

and the total pinning force takes the form

$$F_p = \frac{H_{co}^2\xi_o}{8\pi\phi_o}\, l\, p_{tr} B\,(1-\frac{H}{H_{c2}})\,(1-t)^{\frac{3}{2}}/L\,. \tag{9}$$

For a low density of twins, or for magnetic fields high enough so that $a_o < L$, the elementary pinning energy (from (5)) is $f_p (= H_c^2 l\,\xi/(8\pi\kappa) p_{tr} f_o^2)$. Because many vortices interact with the same planar boundary, they are displaced by s_o as a flux bundle of volume $V = X_c^2 L_c$. We derive X_c and L_c from its total GL energy

$$\delta F = -\,f_p X_c L_c (B_b/\phi_o)^{\frac{1}{2}} + \frac{1}{2}c_{44}(s_o^2/L_c)X_c^2 + \frac{1}{2}c_{66}\, s_o^2 L_c\,. \tag{10}$$

In (10) we have assumed the local form of the elastic energy [6]. From the condition of minimum free energy one obtains $X_c = s_o^2 c_{66}/\{f_p(B_b/\phi_o)^{1/2}\}$, $L_c = X_c\sqrt{(c_{44}/c_{66})}$. The critical state is now determined by the equation $j_c B_b s_o X_c = f_p(B_b/\phi_o)^{1/2}$ and one derives $j_c = (f_p c_{66})^{1/2}/\phi_o$, $B_b = (BL)^2/\phi_o$. For fields close to H_{c2}, $c_{66} = 0.1(H_{c2}/\kappa)^2(1-b)^2$ and $j_c \approx (1-t)^{3/2}$. The resulting displacement is $s_o = (f_p/c_{66})^{1/2}$, or, in terms of the elementary pinning force, p_m $(= f_p/s_o)$ one obtains $s_o = p_m/c_{66}$, which is consistent with the local Labush theory [6].

DISCUSSION AND CONCLUSIONS

The low-field magnetization for these samples is well described by the Bean model [1]. This is indicated by the creep data in Fig. 2 where we observe rates proportional to H^2 for $0 \leq H \leq 4$kOe for each sample, and a constant creep rate for higher fields. The Bean critical state model is appropriate up to 12kOe for $875/20/O_2$ and to 6kOe for $950/20/O_2$.

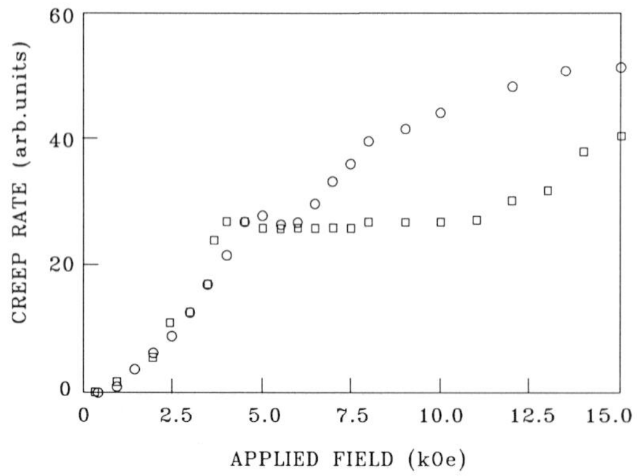

Fig. 2. Flux creep rate versus field at 5K for $875/20/O_2$ (□) and for $950/20/O_2$ (o).

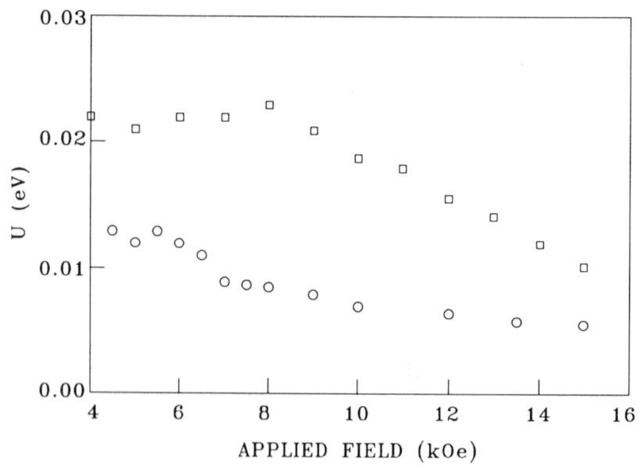

Fig. 3. Activation energies estimated from creep data in Fig. 2 for $875/20/O_2$ (□) and $950/20/O_2$.(o).

To compare the experimental results with the discussed pinning forces, we consider activation energies. Generally, the activation energy is described by $U = j_cBVX_p$, where X_p is an effective width of a pinning barrier which is of order a_o. For high twin density, or for low magnetic fields, $V = a_o^2L_c$. One obtains L_c from the expression $2c_{44}s_o/L_c^2 = j_cB$ and hence $U = \phi_o(2j_c\xi\phi_o)^{1/2}$. Thus

$$U = \phi_o \{ H_c^2 \phi_o\, l\, p_{tr}(1-H/H_{c2})/(4\pi L H_{c2})\}^{\frac{1}{2}}. \qquad (11)$$

For low twin density, or for high magnetic fields, $V = X_c^2L_c$ which leads to the following activation energy

$$U = \phi_o^2 \{ H_c^2 \xi\, l\, p_{tr}(1-H/H_{c2})/(8\pi\kappa)\}^{\frac{1}{2}}/(BL^2) \ . \qquad (12)$$

It follows from (11) that for low magnetic fields the activation energy is almost field independent, in agreement with the data in Fig. 3. For fields higher than 6kOe (corresponding to $a_o \approx 0.1$ µm), the U for sample $950/20/O_2$ (with $L \approx 0.3$ µm) decreases with increasing magnetic field, as described by (12). The sample $875/20/O_2$, with higher twin density (giving $L \approx 0.06$ µm), follows the same behaviour, but with the decrease in U with field commencing at a field much higher than 6kOe, in accordance with the predictions of our model.

ACKNOWLEDGEMENTS

This work is supported by an I R & D Board GIRD Grant under Project No. 15011. One of us (TRF) acknowledges financial support from the State Electricity Commission of Victoria and Olex Cables (A Division of Pacific Dunlop) in order to attend this International Symposium.

REFERENCES

1. Przelozny Z, Finlayson TR, Garvie DA, Lowe DB and Smith TF (1990) Superconductor Sci & Technol (in press)
2. Van Tendeloo G, Zandbergen HW and Amelinckx S (1987) Solid State Commun 63: 389-393
3. Kes PH, Pruymboom A,van den Berg J and Mydosh JA (1989) Cryogenics 29: 228-231
4 Finlayson TR, Garvie DA, Lowe DB, Przelozny Z and Smith TF (1990) Cryogenics 30: 467-473
5. De Gennes P (1966) Superconductivity of Metals and Alloys. W.A. Benjamin, Inc. New York p 175
6. Campbell AC and Evetts JE (1972) Adv Phys 21: 36-37

Magnetic Shielding Effect of Bi Oxide Superconducting Vessels with a SQUID

ATSUSHI KOIKE[1], KAZUTOMO HOSHINO[1], HIROSHI OHTA[2], EICHI SUDOH[1], KOICHI NUMATA[1], KAZUHIKO KATO[1], SHIGERU YAMAZAKI[1], HIDEFUSA TAKAHARA[1], and MASAKAZU AONO[2]

[1] Corporate R & D Center, Mitsui Mining & Smelting Co., Ltd., 1333-2 Haraichi, Ageo, Saitama, 362 Japan
[2] RIKEN (The Institute of Physical and Chemical Research), Wako, Saitama, 351 Japan

ABSTRACT

We have developed a relatively large high-Tc Bi-Pb-Sr-Ca-Cu-O superconducting magnetic shield vessel of 150mm in inner diameter, 320mm in depth and 10mm in thickness. Magnetic fields inside the vessel were measured by using an RF SQUID magnetometer. The shielding factors, defined as the ratio of the magnetic field at 290K to that at 77K, were 1×10^5 in the very low frequency range through 0.2 to 100Hz, which was measured near the bottom position of the vessel. Our experimental data are most important because the magnetic fields from human brains are as weak as 10^{-12}T with low frequencies below 10Hz. Such a high attenuation ratio in the very low frequency region indicates that magnetic shielding of a high-Tc superconductor is promising for neuromagnetic measurement with SQUIDs.

KEY WORDS: oxide superconductor, Bi-Pb-Sr-Ca-Cu-O, magnetic shield, SQUID

INTRODUCTION

Superconducting magnetic shields may be one of the most important applications of high-Tc oxide superconductors, as they have advantages over other magnetic shields [1]. Magnetic fields from human brains, whose measurement is known as magnetoencephalography (MEG) [2], are as weak as 10^2-10^3fT at low frequencies such as 10Hz. Recently developed SQUID magnetometers with second-order gradiometers can have a sensitivity of $20\mathrm{fT}/\sqrt{\mathrm{Hz}}$. Since the magnetic field of the earth is 3×10^{-4}T, heavy magnetic shielding is essential for MEG. However, shielding of low-frequency magnetic fields below 10Hz is quite difficult in a conventional magnetic shield using a Permalloy [3] Measurements of MEG require carefully designed shielding room, which consists of six or seven shells with permalloy and aluminum layers. Such a shielding room is usually very expensive. The minimum magnetic fields inside a permalloy shield is limited by residual magnetic fields due to mechanical stress on the shielding vessel. In addition, the shield of a metal superconductor must be cooled down to a liquid helium temperature.

We have successfully fabricated large magnetic shield vessels for high-Tc oxide superconductors of the Bi-Pb-Sr-Ca-Cu-O system, which are applicable to biomagnetic measurements with SQUID magnetometers [4]. We have experimentally confirmed that the strength of the magnetic field inside the vessel is more than one thousand times weaker at liquid nitrogen temperature than at room temperature [4]. In the experiments, we used two different methods: "the induction method" and the measurement using SQUID. In these methods, we confirmed a good attenuation ratio sufficient for MEG. However, the frequency range of magnetic fields in the induction method was limited to above 10-100Hz, and in the both method the external magnetic fields which were applied to the superconducting vessel were inhomogeneous. It is therefore indispensable to measure the magnetic shielding effect of high-Tc superconducting vessels in a weak ,homogeneous, low-frequency magnetic field.

In this paper, we reported the magnetic shielding effect of our superconducting vessel when a homogeneous magnetic field was applied. The magnetic field inside the vessel in a superconducting state at 77K was compared with that at 290K by using an RF

SQUID magnetometer. The shielding factors thus measured are promising as magnetic shielding for neuromagnetic measurements with SQUIDs.

EXPERIMENTAL

We pressed Bi-Pb-Sr-Ca-Cu-O powders with a high-Tc (110K) phase into a hollow cylindrical shape using cold isostatic pressing ,and heated it in air [4] . The critical temperature Tc and the critical current density Jc of pellet samples prepared by the same process as was the vessel were about 103K and 100-200A/cm^2,respectively.
We estimated the magnetic shielding effect using the SQUID method [4] . Figure 2 shows a schematic arrangement for the SQUID measurements. A Helmholtz coil (diameter: 940mm) generating homogeneous external magnetic fields (B_{ex}) was located outside of the liquid nitrogen vessel. The RF SQUID of an all-niobium quasi-planar Josephson junction (Shimadzu MB240) was used, where the second-order gradiometers were detached and the magnetometer was directly coupled with the SQUID. The magnetic field sensitivity of the SQUID was selected in this experiment to be 1nT/V or 10nT/V, where the flux sensitivity corresponds to 0.2V/ϕ or 0.02V/ϕ. The noise level of the SQUID was less than 20fT/$\sqrt{\text{Hz}}$. The SQUID magnetometer was inserted into the vessel and placed at the position of L/D=1 or L/D=1.7,where D and L are the diameter of the vessel and the distance from the open end of the vessel, respectively. Alternating currents at various frequencies from 0.2 through 200Hz were applied to the exciting coil to generate magnetic fields.

Fig.1. Superconducting vessels (at the center; inner diameter: 150mm, depth: 320mm, thickness:10mm).

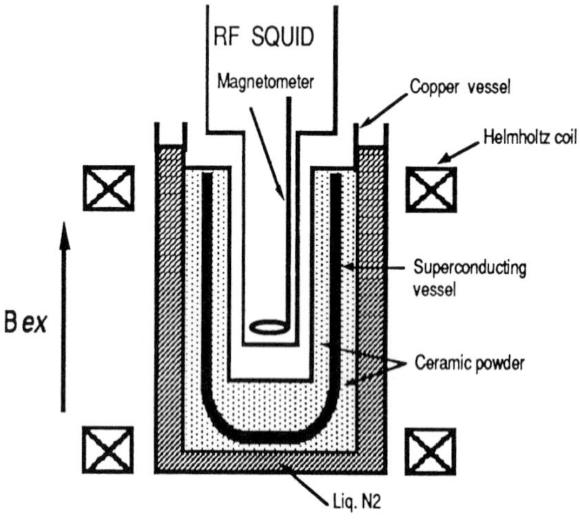

Fig.2. A schematic arrangement of the SQUID magnetometer, coils and superconducting vessel

The shielding factor S is usually defined as the ratio of the external field H_e to the field inside the shield H_i:

$$S = H_e / H_i. \qquad (1)$$

In this experiment, we assume that S is the ratio of the magnetic field inside the vessel in a normal state at room temperature to that in a superconducting state at 77K. The magnetic field at room temperature was evaluated based on the electromagnetic theory using the value of the current applied to the exciting coil and the number of turns of the coil. The magnetic field at 77K was estimated from the output voltage and the field sensitivity of the SQUID. This experiment was carried out in the electromagnetic shielding room.

RESULTS AND DISCUSSION

The shielding behavior in our relatively large vessel of the high-Tc Bi-Pb-Sr-Ca-Cu-O superconductor at 77K is highly promising for the application to MEG. Figure 3 shows the frequency dependence of S estimated from the SQUID measurements. Shielding behavior obtained from the μ-metal shielding vessel is also shown in this figure. The vessel is 180mm in diameter and 500mm in depth with double-layer shells of 1mm each in thickness. First of all, we wish to emphasize that the SQUID magnetometer can operate in our superconducting vessel without a second-order gradiometer, which is usually used for reduction of the effect of the magnetic field of the earth. The typical strength of Bex was around 1×10^{-5}T. The magnitude of magnetic fields detected by SQUID was typically 10^{-10}-10^{-12}T. In the superconducting magnetic shield vessel, we obtained very good shielding factors of 10^5 even in the low frequency range of 0.2-10Hz. This is most important for biomagnetic measurements because the frequencies of the magnetic fields from human brains are as low as 10Hz. On the other hand, the shielding factors of the μ-metal vessel were very poor in that frequency range, where S is on the order of 10^2.

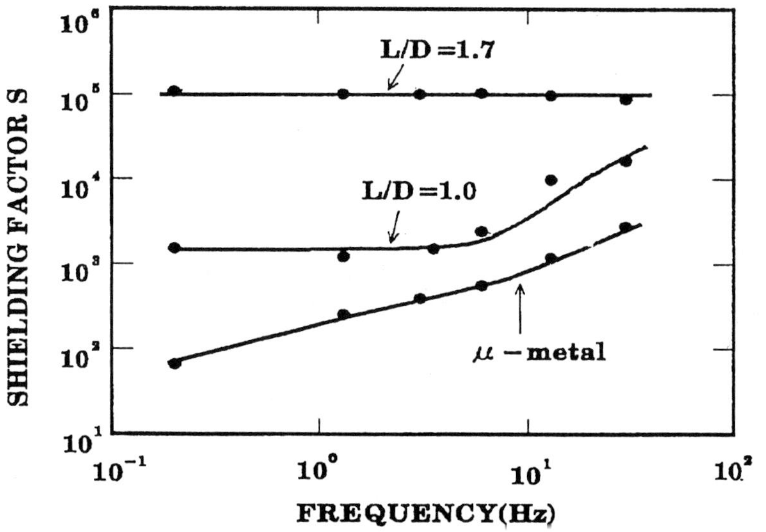

Fig.3 The frequency dependence of the shielding factor S estimated from SQUID measurements.

The measured factor S at low frequencies below 10Hz has no frequency dependence. Since the 1/f noise of our SQUID is less than 0.05Hz [5] , the influence of the 1/f noise is insignificant even at such low frequencies as in this experiment. Moreover, the fluctuations of the trapped magnetic flux inside the vessel when it was cooled down in the magnetic field of the earth were not important in this experiment. The detailed measurements are now being performed. According to the theoretical calculation for a perfect semi-infinite superconducting tube when a homogeneous external magnetic field is applied, the interior field of the tube is expressed as [6]

$$H_{/\!/} = H_0 \exp(-7.6\ L/D) \tag{2}$$

and

$$H_{\perp} = H_0 \exp(-3.6\ L/D), \tag{3}$$

where $H_{/\!/}$ and $H_{\perp}$ represent the axial and transverse interior magnetic fields, respectively. H_0 is the external magnetic field. These equations assume that the tube has an infinite current density. From eq.(2), the attenuation of magnetic fields inside the tube can be calculated to be 5×10^{-4} at L/D=1 and 2.5×10^{-7} at L/D=2. These attenuation ratios are better than our experimental data by a factor of 10-100.

Magnetic fields from human brains are as weak as 10^{-12}T with very low frequencies below 10Hz, so that we should attenuate the ambient magnetic field to at least less than 10^{-6} for those measurements because fluctuations of the magnetic field of the earth are in the order of 10^{-6}T. Our superconducting vessel has such a good attenuation even in a very low frequency range. The influence of trapped magnetic flux inside the superconducting vessel on the shielding effect is still an open question. It may be eliminate to cancel the magnetic field of the earth with a large Helmholtz coil just before cooling the vessel. Moreover,it is recognized that flux creep phenomena may be one of the sever problems in high-power applications of high-Tc oxide superconductors. Flux creep, however, may not be serious in a very weak magnetic field because of its high activation energy rapidly increasing with the decrease of the magnetic field strength [7] .

SUMMARY

We have successfully fabricated relatively large Bi-Pb-Sr-Ca-Cu-O superconducting magnetic shield vessels operating at 77K,and experimentally confirmed, from the measurement with a SQUID, that the shielding factors of the vessel are 10^5 at the position of L/D=1.7 for frequencies below 10Hz and weak magnetic fields such as 10^{-4}T.

ACKNOWLEDGEMENT

We would like to thank Mr.M.Takahata of Shimadzu Corporation for his helpful assistance with the SQUID measurements.

REFERENCES

1. W.O Hamilton: Rev. Phys. Appl., 5, 41-48, 1970.
2. D.Cohen: Science, 175, 664-665, 1972.
3. S.N.Erne, H.-D.Hahlbom, H.Scheer and Z.Trontelj: *Biomagnetism*, eds.,S.N.Erne, H.-D.Hahlbom and H.Lubbig (Walter de Gruter & Co.,Berlin, 1981) p.78.
4. K.Hoshino,H.Ohta,E.Sudoh, K.Katoh, S.Yamazaki, H.Takahara and M.Aono: Jpn. J. Appl. Phys., 29, L1435-L1438, 1990.
5. H.Ohta, M.Takahata, Y.Takahashi, K.Shinada, Y.Yamada, T.Hanasaka, Y.Uchikawa, M.Kotani, T.Matsui and M.Komiyama: IEEE Trans. Magnetics, 25, 1018-1021, 1989.
6. B.Cabrera: Stanford Univ.,Ph. D. Thesis, 1975.
7. M.Foldeaki, M.E.McHenry and R.C.O'Handley: Phys. Rev. B, 39, 4784-4787, 1989.

Theory of AC Loss of High-T_c Superconductors in Flux-Flow State

YOICHI ANDO and SHIRABE AKITA

Central Research Institute of Electric Power Industry, 11-1, Iwato-kita 2-chome, Komae, Tokyo, 201 Japan

ABSTRACT

The AC loss of a high-T_c superconductor in the absence of flux pinning is theoretically examined to determine if such a conductor can be practical for AC applications. The AC loss is calculated for external-field and self-field conditions, with the simplest Bardeen-Stephen model being assumed for the flux-flow resistivity ρ_f. In the self-field condition, it is found that the nonlinearity occurring from the magnetic field dependence of ρ_f plays an important role in the reduction of the loss to a sufficiently low level.

KEY WORDS: flux-flow, anisotropy, AC loss, superconducting wire, superconducting cable

INTRODUCTION

If we consider the application of high-T_c superconductors to the AC apparatus used in electric power systems, the most appealing point of a high-T_c superconductor is that it can be used at liquid nitrogen temperature. A practical 100 W-class refrigerator requires only 18 W to remove heat generation of the amount of 1 W at 77 K, although it requires as much as 740 W if the heat is generated at 4.2 K [1]. This means that the restraint of the AC loss on high-T_c superconductors for the realization of AC application is much less than on conventional low-T_c superconductors. In addition to this, the phenomenon of flux creep causes no problem in AC application because a superconductor shows resistivity if alternating current flows in it. Therefore, one can conclude that high-T_c superconductors are more suitable than low-T_c superconductors for AC application.

Practical superconductors are always considered to be used in the critical state [2]. If an AC field is applied to a superconductor in the critical state, hysteresis loss, which is the essential part of the AC loss of a superconductor, is generated. Although the hysteresis loss can be decreased by reducing the diameter of a superconductor, it would be very difficult to apply the same method to high-T_c superconductors which have little plasticity. In this paper, the AC loss of a high-T_c superconductor in flux-flow state [3] is calculated to determine whether such loss can be made smaller than the hysteresis loss in the critical state, and it is found that the flux-flow loss will indeed become smaller under some conditions. Since the flux-flow loss does not depend on the diameter of a superconductor, a "flux-flow superconductor" provides an alternative possibility for a low-loss AC superconductor.

FLUX-FLOW RESISTIVITY OF A HIGH-T_c SUPERCONDUCTOR

Since high-T_c superconductors have strong anisotropy, one must take account of the anisotropy in considering the flux-flow resistivity. We start with the simplest Bardeen-Stephen model [4], which expresses the flux-flow resistivity ρ_f in the form

$$\rho_f = \rho_n \frac{B}{\mu_0 H_{c2}(0)} . \tag{1}$$

Here, ρ_n is the normal-state resistivity at the measuring temperature, and $H_{c2}(0)$ is the upper critical field at zero temperature. $H_{c2}(0)$ appears in the formula because the flux-flow resistivity is proportional to the area of the normal core of a vortex; therefore, for a high-T_c superconductor, we should simply use the $H_{c2}(0)$ in the direction of the applied field. The ρ_n in eq. (1) represents the energy dissipation when the transport current crosses the normal core on the vortex motion; therefore, for an anisotropic superconductor, one should take the normal-state resistivity in the direction of the transport current.

The $H_{c2}(0)$ of a high-T_c superconductor is quite large when the applied field is perpendicular to the c-axis. This means that the flux-flow resistivity can be made small if one makes good use of the anisotropy. It is minimized when the field is applied perpendicular to the c-axis and the transport current flows in the ab-plane. This desirable situation is shown schematically in Fig. 1.

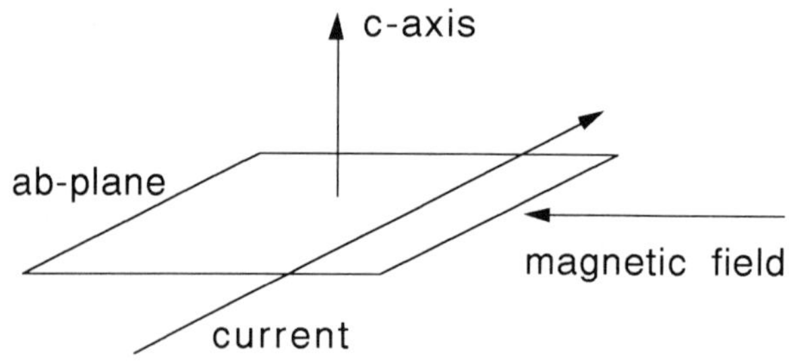

Fig. 1 Direction of the current and the magnetic field relative to the crystal orientation, which realizes the minimum flux-flow resistivity.

In the calculations below, we take the Bi-Sr-Ca-Cu-O (2212) phase as an example. This oxide has a ρ_n of 1×10^{-6} Ωm [5] and $\mu_0 H_{c2}(0)$ of 500 T [6] as typical values. Substituting these values into eq. (1), ρ_f is calculated to be 0.2×10^{-8} Ωm at 1 T, which is of comparable order to the resistivity of copper or silver at 77 K.

AC LOSSES IN FLUX-FLOW STATE

In superconductors, the relationship between B and H is nonlinear. To be precise, this nonlinearity should be included in eq. (1), but we neglect it for high-T_c superconductors in this paper because H_{c1} is very small for high-T_c superconductors at 77 K (it is of the order of 10 Oe) and therefore, the difference between B and H is very small. Hereafter, we consider two cases: a high-T_c superconductor in an external AC field and in a self-exciting AC field. The former corresponds to the winding of a coil, and the latter corresponds to a power transmission cable.

External AC Field

When the effect of the self-field and the skin effect are neglected, the flux-flow loss w_f of a high-T_c superconducting wire in the absence of pinning is calculated as [7]

$$w_f = \frac{\rho_n H_0}{\pi H_{c2}(0)} J_0^2 \left(1 + \frac{1}{3} \cos 2\delta \right) \qquad (\mathrm{W/m^3}), \qquad (2)$$

where H_0 is the amplitude of the AC field, J_0 is the amplitude of the AC current, and δ is the phase difference between the external field and the AC current. In order to minimize the flux-flow loss, the wire should have the crystal orientation shown in Fig. 2. We have compared this loss to the hysteresis loss of a wire used in the critical state and a copper wire at 77 K, which has the resistivity of 0.3×10^{-8} Ωm. The result is shown in Table I. Here we assumed that the amplitude of the external AC field is 0.5 T, frequency is 60 Hz, $J_0=1\times10^8$ A/m^2, and $\delta=0$. For a critical-state wire, for simplicity, it is assumed that the superconducting filament has a square cross section, of width D, and the critical current density is 10^9 A/m^2. Note that the flux-flow loss does not depend on the size of the superconductor, and it becomes the same level as the hysteresis loss of a multifilamentary critical-state wire whose filament is as thin as 0.1 mm.

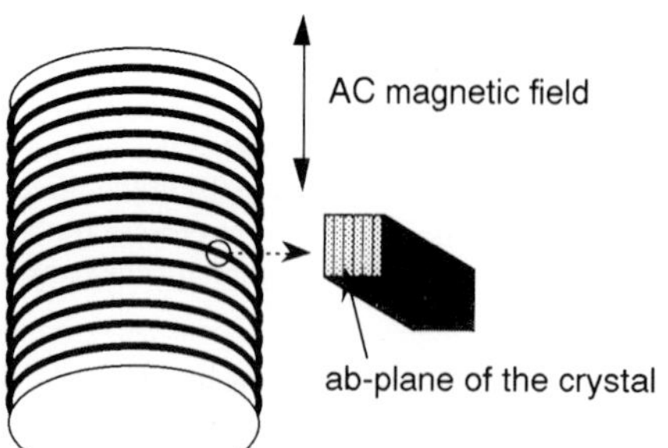

Fig. 2 Schematic diagram of a flux-flow superconducting wire and a coil wound with it.

Table I. The formulae for the losses and the evaluated losses of the wires considered in the text. We assumed that $\rho_n=1\times10^{-6}$ Ωm and $\mu_0H_{c2}(0)$=500 T. The losses are estimated under the condition that the peak value of the AC current density, J_0, is 1×10^8 A/m², δ=0, μ_0H_0=0.5 T, $J_c=10^9$ A/m², and the frequency is 60 Hz. The hysteresis loss is evaluated under two conditions: D=0.1 mm [*] and D=1 mm [**].

	Formula (W/m³)	Evaluated loss (10^6W/m³)
Flux-flow wire	$w_f = \frac{\rho_n H_0}{\pi H_{c2}(0)} J_0^2 (1 + \frac{1}{3}\cos 2\delta)$	4.24
Critical state wire	$w_h \cong \mu_0 J_c D H_0 f$	3.00*, 30.0**
Normal metal wire	$w_J = \frac{1}{2}\rho J_0^2$	15.0

Self-exciting AC Field

A superconductor is affected by an alternating concentric circular field when a large alternating current flows in a superconducting power transmission cable. In this case, the flux-flow superconductor should be a solid cylinder, and the crystal orientation inside it should have the structure shown in Fig. 3 in order to reduce the flux-flow loss to its minimum.

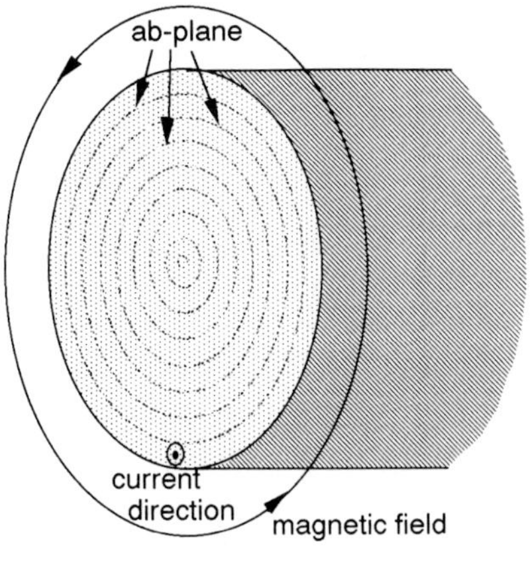

Fig. 3 Schematic diagram of a flux-flow superconducting cable.

The current distribution inside the cable is essential for loss calculation, because the magnetic field inside it varies, and the flux-flow resistivity also varies according to the magnetic field. If the skin effect is neglected, the current distribution is solved as [7]

$$J_0(r) = \frac{3 I_0}{4\pi R^{3/2} r^{1/2}} , \qquad (3)$$

where J_0 is the peak value of the current distribution, I_0 is the peak value of the total current, and R is the radius of the cable. Using this current distribution, the flux-flow loss W_f per unit length of a cable in the absence of flux pinning is calculated as

$$W_f = \frac{1}{2\pi^3} \frac{\rho_n}{H_{c2}(0)} \frac{I_0^3}{R^3} \quad \text{(W/m)} . \qquad (4)$$

We have also compared this loss to the hysteresis loss of a superconducting cable used in the critical state and a copper cable at 77 K. It is assumed that I_0=25 kA, R=4 cm, and the frequency is 60 Hz. The loss of a critical state cable is calculated for two values of J_0: 10^8 and 10^9 A/m². The result is shown in Table II. It is clear that the flux-flow superconducting cable will have advantages over other cables. This is because the current density shows a $1/\sqrt{r}$-dependence on the radial coordinate for a superconducting cylinder in the flux-flow state, in which the current density is larger in the central part of the cable, where the magnetic field is small, than in the outer part.

Table II. The formulae for the losses and the evaluated losses of the cables considered in the text. The losses are estimated under the condition that the peak value of the AC current I_0 is 25 kA, R=4 cm, and the frequency is 60 Hz. The hysteresis loss is evaluated under two conditions: $J_c=10^8$ A/m² [*] and $J_c=10^9$ A/m² [**].

	Formula (W/m)	Evaluated loss (W/m)
Flux-flow cable	$W_f = \frac{1}{2\pi^3} \frac{\rho_n}{H_{c2}(0)} \frac{I_0^3}{R^3}$	10.0
Critical state cable	$W_h \cong \frac{\mu_0 I_0^3 f}{6\pi^2 J_c R^2}$	126^*, 12.6^{**}
Normal metal cable	$W_J = \frac{\rho I_0^2}{2\pi R^2}$	188

DISCUSSION

Before the discovery of high-T_c superconductors, Clem *et al.* studied the electromagnetic behavior and the AC loss of a superconductor in the flux-flow state [8]. They found that the flux-flow skin effect is significant in an ideal type-II superconductor in an AC field, and the AC loss is similar to the eddy-current loss of a normal conductor. However, his study was confined to the case where large DC magnetic field persists and the small-amplitude-AC field can be treated as a perturbation. In that case, equations are linearized and become similar to the equations for a normal conductor. In our calculation, the magnetic field dependence of ρ_f is essential. As a result of this complexity, the skin effect in the self-field flux-flow superconductor becomes highly nonlinear. In order to analyze the skin effect in this case, one must solve the equation for the current density:

$$\frac{\partial}{\partial r}\left(\frac{J(r,t)}{r}\right)\int_0^r s\,J(s,t)\,ds + J^2(r,t) = \frac{\mu_0 H_{c2}(0)}{\rho_n\, r}\frac{\partial}{\partial t}\int_0^r s\,J(s,t)\,ds\ . \tag{5}$$

This equation cannot be solved analytically. However, it is clear from the structure of this equation that the current distribution consists of many higher-order harmonics. Further study of the skin effect is necessary to ascertain the influence on AC loss.

CONCLUSIONS

The flux-flow loss of a high-T_c superconductor in the absence of flux pinning is calculated. If the anisotropy of a high-T_c superconductor is taken care of, the flux-flow loss becomes sufficiently small, and a flux-flow-state operation would provide an alternative approach to realizing a high-T_c superconductor with low AC loss.

REFERENCES

[1] Brechna H (1973): *Superconducting Magnet Systems*. Springer-Verlag, Berlin, p 481
[2] Carr WJ Jr (1983): *AC Loss and Macroscopic Theory of Superconductors*. Gordon and Breach, New York
[3] Kim YB, Stephen MJ (1969) In: Parks RD (ed) *Superconductivity*. Marcel Dekker, New York, Vol 2, Chap 19, p 1116
[4] Bardeen J, Stephen MJ (1965) Phys Rev 140: A1197
[5] Martin S, Fiory AT, Fleming RM, Espinosa GP, Cooper AS (1989) Appl Phys Lett 54: 72
[6] Palstra TTM, Batlogg B, Schneemeyer LF, van Dover RB, Waszczak JV (1988) Phys Rev B38: 5102
[7] Ando Y, Akita S (in press) J Appl Phys
[8] Clem JR, Kerchner HR, Sekula ST (1976) Phys Rev B14: 1893

Current Limiting Resistance Characteristics of High-T_c Superconductor

YASUKAZU ITO[1], KOICHI NAKAMURA[1], YOSHIHIRO ABE[1], and MOTOTAKA SONE[2]

[1] Nagoya Institute of Technology, Gokiso, Shouwa-ku, Nagoya, 466 Japan
[2] Musashi Institute of Technology, Tamatsutsumi, Setagaya-ku, Tokyo, 158 Japan

ABSTRACT

Resistance appears in the superconductor element when the superconductive condition is broken down. (1) The resistance when the current through the element is greater than that of the critical current, is called "flow resistance", and (2) the resistance when the element temperature becomes higher than that of the critical temperature, is called "normal conductive resistance".
A current limiting element using superconductor is proposed to limit the short circuit current in the power systems. In this paper the resistance characteristics of two kinds of High-Tc superconductor materials were studied and compared. A rough size estimation was tried with a current limiting element for the practical use.

KEY WORDS: Y-type and Bi-type High-Tc Superconductor, Current and Temperature Dependence of Resistance, Flow Resistance, Normal resistance, Current Limiting Element

INTRODUCTION

In this research High-Tc superconductor is studied to apply to a current limiting element in the electric power fields, where the resistance changing from the super state to the normal state is utilized. Figure 1 shows the resistivity vs. temperature characteristics of Bi-type superconductor (Bi-Sr-Ca-Cu-O). The critical offset temperature Tc is 88 K. The resistivity in the normal state indicates the order of milliohm·cm. This value is comparatively larger than that of the usual metal. For example, the resistivities of tungsten and nichrome metals are $3.2*10^{-2}$ milliohm·cm and 0.1 milliohm·cm at 77 K, respectively. When the temperature of the superconductor sample is lower than Tc, and at the same time the current density is below the critical current density Jc, the resistance is zero. If the current through the sample is greater than the critical value, some voltage appears, which means resistance.
Figures 2 (a) and 2 (b) show the characteristics of the voltage E and the current density J of the Y-type and Bi-type superconductor samples under the various cryogenic temperatures, respectively. The voltage E is given with the unit length, V/cm. In Fig. 2 (a), the relation of E and J is linear in the temperature above 90 K, and non-linear below the critical offset temperature of 89 K. The resistance in this non-linear area is the flow resistance, which attributes to the movement of the fluxon. The flow resistivity is obtained from the slope of the characteristics E vs. J. In Fig. 2 (b), the non-linearity appears below 88 K.
The value of E/J simply implies the element resistivity and corresponds to the current limiting resistance. The characteristics of the resistivity vs. current density of the Bi-type sample in the different temperature between 75 K and 89 K are shown in Fig. 3. The more the temperature of the element is close to the critical temperature, and the larger the current density through the element is, the larger the element resistance becomes. In Fig. 3 the maximum value of the resistance is around 0.2 milliohm·cm.
When the element temperature increases higher than Tc due to the Joule heating, the superconductor becomes the normal conductor. Thus there are two kinds of resistance phenomena in the superconductor.
Which resistance is available to apply practically to the current limiting device? The flow resistance is comparatively smaller but the response is instant. The normal resistance is larger but the response is delayed. Then if an appropriate resistance for the current limiting element is given, the size of the superconductor can be roughly estimated.

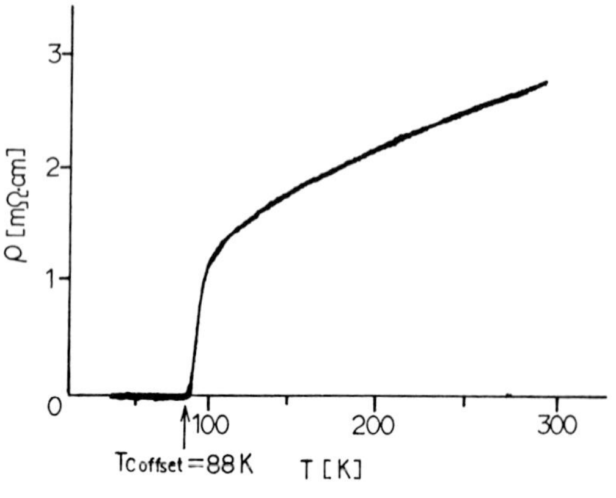

Fig. 1 Resistivity vs. temperature characteristics

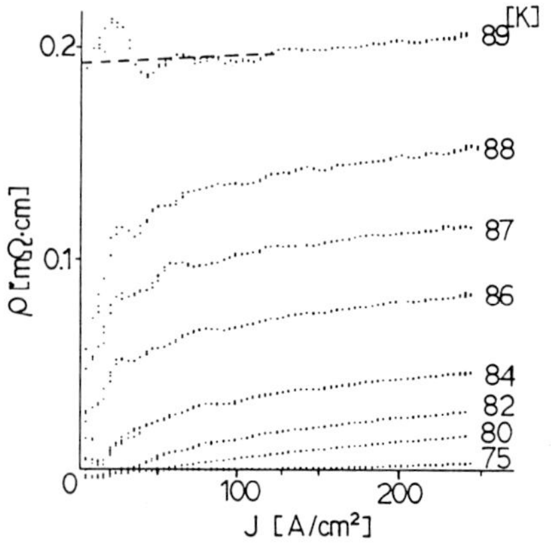

Fig. 3 Resistivity vs. current density characteristics

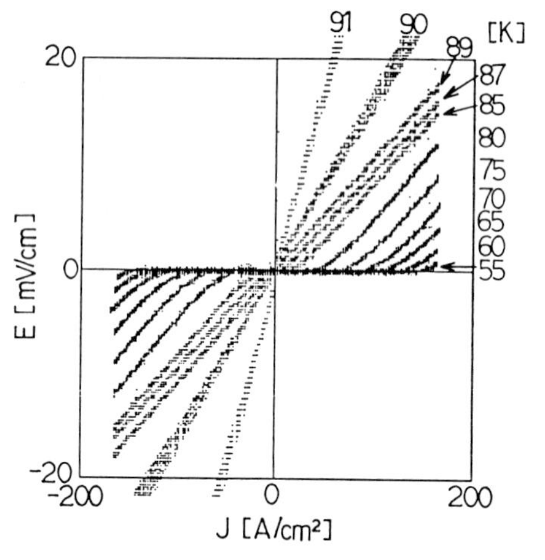

Fig. 2 (a) Voltage E vs. current density J characteristics of the Y-type element

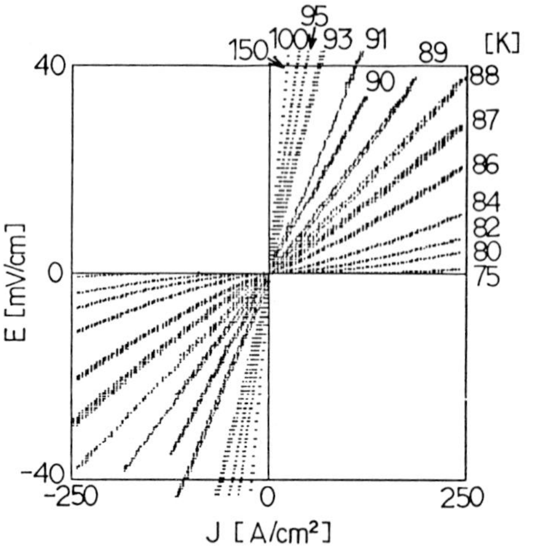

Fig. 2 (b) E vs. J characteristics of the Bi-type element

EXPERIMENTAL

In order to study the basic characteristics of current limiting some small superconductor samples were used. They were a Y-type and a Bi-type superconductor. Atomic ratio of the Y-type was Y:Ba:Cu=1:2:3 and produced by the processes of calcining, pressing, sintering and annealing. Atomic ratio of the Bi-type was Bi:Sr:Ca:Cu=2:2:1:2. This element is a glass-ceramic[2]. The size of samples are 5 mm * 2 mm * 45 mm (plate) of Y-type, and 2 mm * 60 mm (bar) of Bi-type. Tc and Jc are 89 K and 15 A/cm^2 for the Y-type and 88 K and 40 A/cm^2 for the Bi-type, respectively. The measurement circuit is shown in Fig. 4. The current is 60 Hz. Here the liquid nitrogen was used for cooling.

RESULTS AND DISCUSSIONS

The current and the voltage wave forms with the Bi-type are shown in Fig. 5. The peak value of the current density in the first half cycle of (a) is 1200 A/cm^2 which is 30 times greater, and of (b) is 2100 A/cm^2 which is 50 times greater than that of the critical current density, respectively. The change of resistivity are shown in Fig. 6. In the figure the characteristics of (a) show the resistivity periodically changing between zero and 0.02 milliohm·cm. This resistivity depends only on the flow resistance. The characteristics of (b) indicate the resistivity which periodically change in the earlier

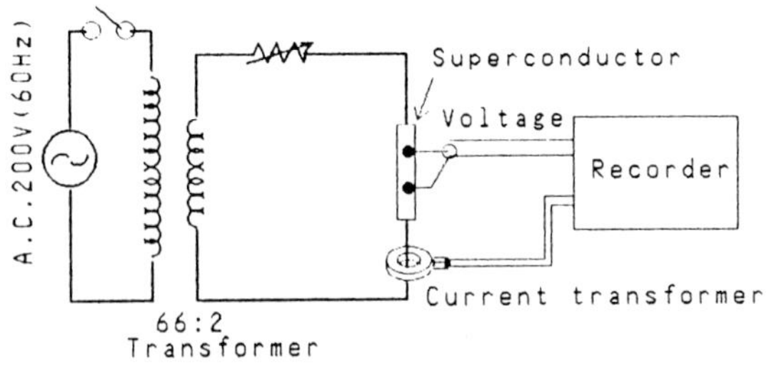

Fig. 4 The measurement circuit

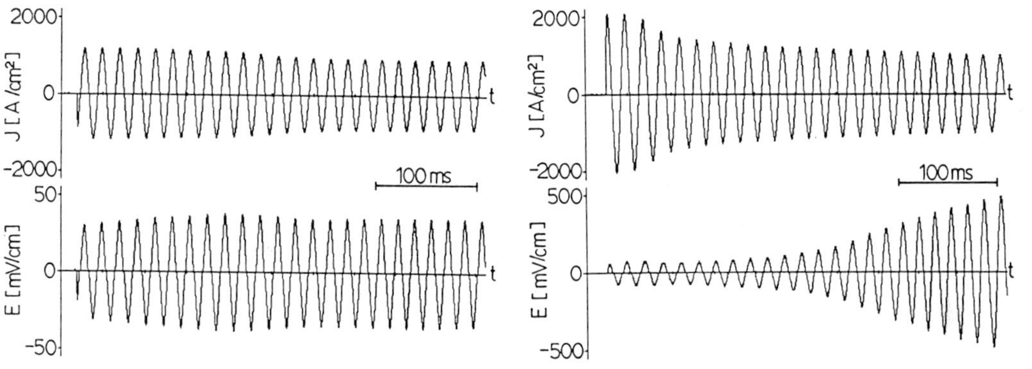

(a) 30 times greater than Jc (b) 50 times greater than Jc

Fig. 5 Current and voltage waveforms

period and gradually increase after the lapse of around 200 ms. This resistivity includes the both of the flow resistance and the normal resistance.

The resistivity of the Y-type sample is shown in Fig. 7. Where the peak value of the current density in the first half cycle for the characteristics (a) was 750 A/cm^2 which is 50 times greater, and for the characteristics (b) was 800 A/cm^2 which is 55 times greater than the critical current density, respectively. Those characteristics (a) and (b) include the both of the flow and the normal resistance. The normal component appears around after the lapse of 100 ms. The greater the current is, the earlier normal resistance tends to appear.

Fig. 8 shows the relation of the resistivity vs. accumulated electric power par unit volume which was calculated by the time integral of the voltage and the current as shown in Fig. 5 for the Y-type and the Bi-type samples, respectively. The remarkable changes of slope are seen near the value of 10 J/cm^3 in the characteristics (a), (b) for the Y-type and (b) for the Bi-type samples. This means the flow resistance continues until the energy consumed in the sample increases to 10 J/cm^3, in spite of the different materials of the two kinds of superconductor, and of the different processing. The normal resistance appeared above the energy of 10 J/cm^3.

We will try to estimate the size of current limiting element using the superconductor as mentioned above. Here, the following assumptions are given. (1) The necessary resistance R of the current limiting element is 2 ohms and the short circuit reactance of the power system is 2 ohms. (2) The normal load current I_0 and the short circuit current are 2 kA and 80 kA, respectively. (3) The critical current Ic is equal to the normal load current. From the above assumptions, and from the critical current density Jc and the resistivity ρ of element applied, the necessary cross section s and the length l of the element are evaluated in the following equations.

$$s = \frac{\sqrt{2} \cdot I_0}{Jc} \qquad l = \frac{\sqrt{2} \cdot R \cdot I_0}{\rho \cdot Jc}$$

For the Bi-type material, s=70 cm^2 and l= 6 * 10^6 cm and for the Y-type material, s=180 cm^2 and l=7.5 * 10^5 cm were resultantly obtained, respectively. Where Jc=40 A/cm^2, ρ=0.025

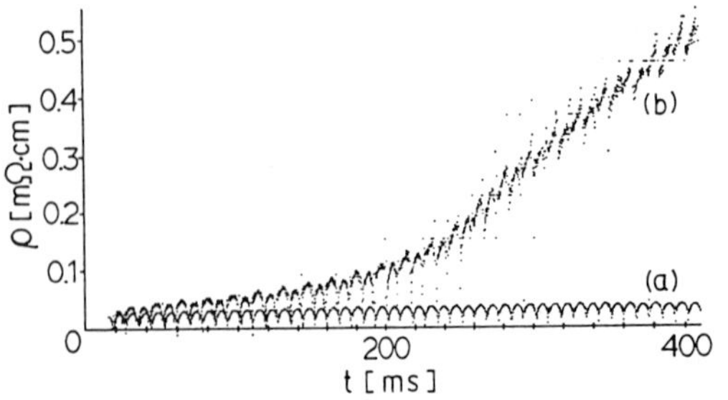

Fig. 6 Resistivity changing of the Bi-type element

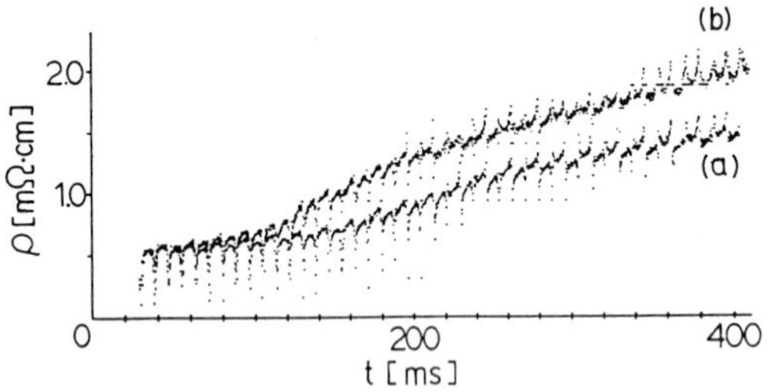

Fig. 7 Resistivity changing of the Y-type element

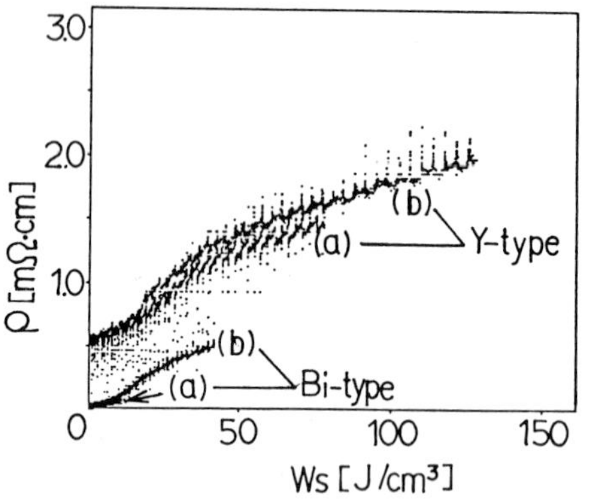

Fig. 8 Resistivity vs. accumulated power

milliohm·cm, for the Bi-type, and Jc=15 A/cm^2, ρ=0.5 milliohm·cm for the Y-type superconductor samples.

CONCLUSIONS

The resistance appeared in the samples of superconductor, when the current through the sample was increased to 50 times greater than the critical current, was measured and discussed. The following results were obtained for a Y-type and a Bi-type superconductors.

(1) The maximum resistivity obtained from the flow resistance appeared at 88 K near the critical temperature were around 0.2 milliohm·cm for the Bi-type, and 0.5 milliohm·cm for the Y-type samples used, respectively.

(2) When the temperature of the element rose above Tc, the normal conductive resistance appeared and the resistivity increased to the order of 1 $\sim$ 2 milliohm·cm .

(3) The total accumulated electric power until the appearance of the normal conductive resistance was about 10 J/cm^3 and the value was almost equal in the both of Y-type and Bi-type samples.

A rough size estimation with a current limiting element was tried by considering these data mentioned above for the practical use in the power system.

ACKNOWLEDGMENT

The authors would like to thank Chubu Electric Power Co. Inc., and Aichi Denki Co. Ltd. for the help of equipment, and also the Ministry of Education, Science and Culture of Japan for the financial support.

REFERENCES

1. Y.Ito, K.Nakamura, T.Kawashima(1990) Proc. of the 1st Power and Energy Conf.,PES,I.E.E. of Japan:505-510
2. Y.Abe, H.Hosono, M.Hosoe, J.Iwase,Y.Kubo(1988) Appl.Phys.Lett.:53(14)1341

5 Tape and Bulk

5.1 Wires and Tapes

Properties of Ag/Bi-Based Superconducting Long Wire and Test Coil

HIDEHITO MUKAI, NOBUHIRO SHIBUTA, TAKATO MASUDA, TAKESHI HIKATA, MUNETSUGU UEYAMA, TAKESHI KATO, and KEN-ICHI SATO

Osaka Research Laboratories, Sumitomo Electric Industries, Ltd., 1-3, Shimaya 1-chome, Konohana-ku, Osaka, 554 Japan

ABSTRACT

Fundamental problems in the application of high-Tc superconductors, such as fabrication of long wire, strain-resistant conductors and large current conductors, were investigated. A 60 m-long wire showed a uniform transport property of 10.5 A critical currents defined as 10^{-13} $\Omega \cdot m$. 1,296-core multifilamentary wires sustained 100% of initial transport property even after a strain of 0.33% repeated 20 times. A composite superconductor 50 cm in length showed a critical current of 573 A. Test coil made by the wind & react process generated 500 Gauss at 77.3 K. A react & wind-processed coil using 4.3 m-long 1,296-core wire could maintain the initial transport current after being covered with polyvinyl formal and wound on 15 mm-diameter steel core (13 turns x 7 layers). 1,500 A class current leads made of high-Tc superconducting wires suppressed the liquid helium consumption of 0.93 W/kA, which is lower than with commercially available copper current leads. A flexible cable conductor of 75 cm in length was fabricated using 22 m-long 1,296-core wire. This conductor was proved to have a critical current of 80 A after winding spirally on a 20 mm-diameter spacer.

KEY WORDS: high-Tc superconducting wires, current lead, multifilamentary wire, bundle conductor, react & wind coil, wind & react coil, flexible cable conductor

INTRODUCTION

The powder-in-tube method is one of the most convenient processes for making a high Tc superconductor into wire, not only because the conventional metal working process is suitable for long wire fabrication, but also because the wire can be covered with low resistivity material as stabilizer. Many efforts have been made to improve the critical current density (Jc) which is the most important parameter for actual usage of high-Tc superconductor[1-6]. Until now, Jc values of 4.7×10^4 A/cm^2 were attained at 77.3 K, 0 T and 1.1×10^4 A/cm^2 at 77.3 K, 1.0 T in Bi-based superconducting wire[7]. These values are high enough for starting the investigation of actual application.

In this paper, fundamental problems in the usage of high-Tc superconducting wires were investigated. After confirming the satisfactory results, prototypes, such as small coils, current leads and cable conductors, were made and examined in detail.

EXPERIMENTAL PROCEDURE

Samples with a nominal composition of $Bi_{1.8}Pb_{0.4}Sr_{2.0}Ca_{2.2}Cu_{3.0}O_y$ were prepared from powders of Bi_2O_3, PbO, $SrCO_3$, $CaCO_3$ and CuO with 3 to 4 N purity. The powders were mixed, reacted and ground several times. These powders were put into silver tubes and drawn to round wires. The round wires were shaped into taped wires using a cold-rolling machine. Repeated procedures of cold-rolling and sintering improved the Jc with good alignment and strong bonding of grains at grain boundaries.

Multifilamentary taped wires were fabricated using the double-stacking technique, up to 1,296 multifilamentary wires. The critical current (Ic) measurement of all samples was done using a four-probe DC method.

RESULTS AND DISCUSSION

LONG WIRE PROPERTIES

In order to ascertain the uniformity of the transport current property, a 60 m-long wire was sintered. An Ic of 10.5 A (Jc=2,450 A/cm^2) was attained over the full length with a resistivity of 10^{-13} Ω · m. Figure 1 shows the distribution of the Ic every 4.3 meters. The deviation from the mean value of 10.6 A was only 3.8%.

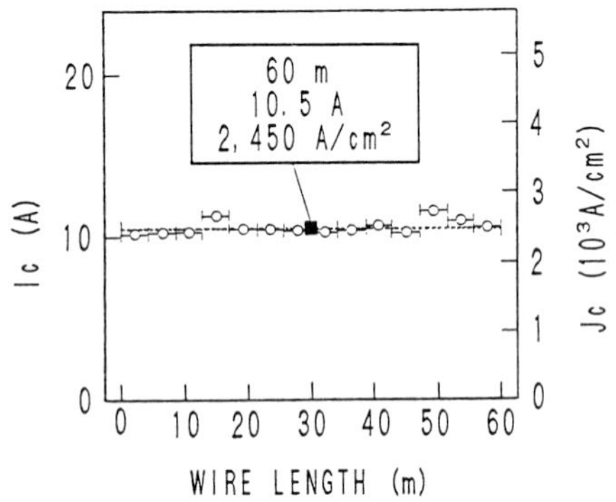

Fig. 1. Ic distribution of a 60 m-long wire.

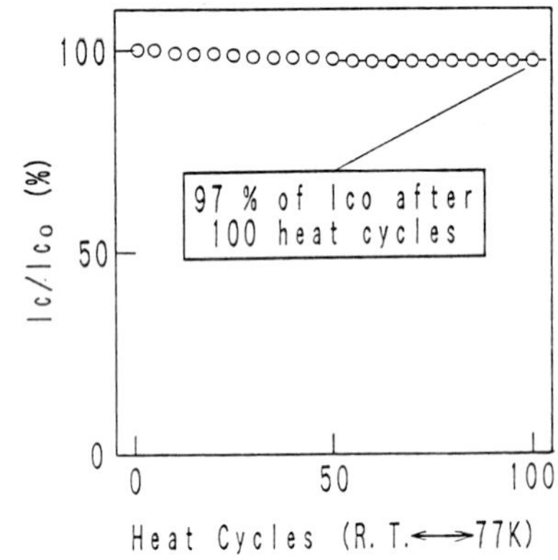

Fig. 2. Heat cycles-Ic property of a composite conductor.

BUNDLE CONDUCTOR AND HEAT CYCLE vs Ic PROPERTIES

For actual usage of high-Tc superconductors, not only the improvement of Jc but also the development of a large current conductor is needed. The Jc is improved by increasing the reduction ratio of the taped wire thickness, generally. In fabricating a large current conductor, the Jc-B characteristics are taken into account because of the relatively larger self-field of the conductor. The Jc-B characteristics of Bi-based high-Tc superconductors are excellent compared to other high-Tc superconductors. Furthermore, improvements of the Jc in a zero magnetic field caused great improvements of the Jc in a magnetic field in the Bi-based high-Tc superconductors. Therefore, high Jc and large Ic conductors can be obtained by collecting the thin taped wires. A "bundle conductor" is composed of wires which are stacked and diffusion-bonded to each other during sintering.

Larger current conductors are fabricated by collecting the bundle conductors around the substrate. In composing the large current conductors, mechanical properties, especially the thermal expansion properties of each element, are taken into account. Figure 2 shows the heat cycle vs Ic property of a composite which was made of the bundle conductor and fiber reinforced plastics (FRP). The composite maintained 97 % of the initial critical current after 100 heat cycles between liquid nitrogen temperature and room temperature.

MULTIFILAMENTARY WIRES

The transport property of sintered high-Tc superconductors suddenly drops with increasing bending strain. Cracks initiated in the high-Tc superconductors rapidly propagate with the concentration of stress and then the superconductors fracture. The crack propagation is thought to be prevented through the stacking structure of the silver and the divided superconductors in the multifilamentary taped wires.

Single-core and multifilamentary wires were fabricated and wound on various-sized mandrels after sintering. Figure 3 shows the Jc-strain characteristics of single-core, 36-core, 762-core and 1,296-core wires. 0.7% strain (30 mm diameter mandrel) does not reduce Jc of 1,296-core multifilamentary wires. Though Fig. 3 shows only the bending property when bent once, that is important when we consider making magnets. Cyclic bending properties are required to make cables. When making cables, the wires are wound on a spacer, wound on a drum and rewound to be placed in the conduit. The wires must experience bending strain many times between the sintering and the energizing. Figure 4 shows the cyclic bending characteristics of 1296-core multifilamentary wires. Initial Ic was maintained even after a strain of 0.33% repeated 20 times.

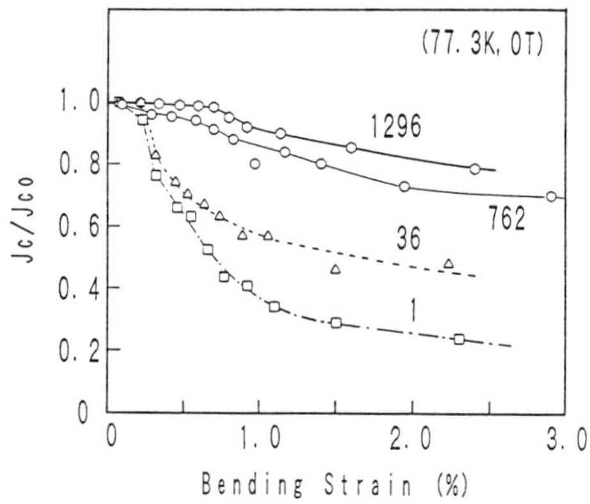

Fig. 3. Jc-strain characteristics of single-core, 36-core, 762-core and 1,296-core wires.

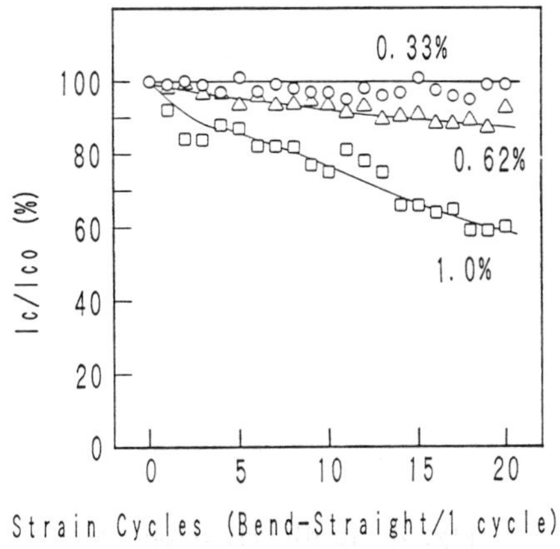

Fig. 4. Cyclic bending strain characteristics of 1,296-core wires.

WIND & REACT COIL

Using a 1 m-long bundle conductor (total wire length of 20 m), a wind & react coil was fabricated as shown in Fig. 5. This coil was wound from 20 wires 1 m-long with 1 layer of 10 turns. The inner diameter of the coil is 27.5 mm and the height is 44 mm. Figures 6 and 7 show the load lines of the coil at 77.3 K and 4.2 K, respectively. At 77.3 K, the coil generates a maximum self-field (Bm) of 142 Gauss with a critical current of 54.2 A, and was studied with a back-up field of up to 990 Gauss, showing 22.7 A of Ic. At 4.2 K, a critical current of 355 A is achieved with a Bm of 867 Gauss. With a 3.7 Tesla back-up field, the Ic was 137 A generating 335 Gauss. These results show that we can obtain over 0.1 T at 77.3 K and a super-high field at 4.2 K with high-Tc superconducting coils.

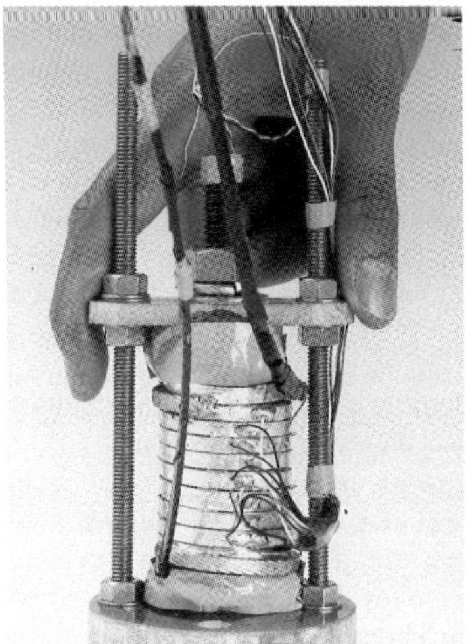

Fig. 5. Overview of wind & react-processed coil.

Using a 6 m-long bundle conductor (total wire length of 30 m), a single pancake coil was fabricated as shown in Fig. 8. An electrical insulator composed of silica, alumina and magnesia was used. The inner diameter of the coil was 16 mm, and the outer diameter was 90 mm. Transport current of 37.1 A was achieved showing a Bm of 500 Gauss at 77.3 K.

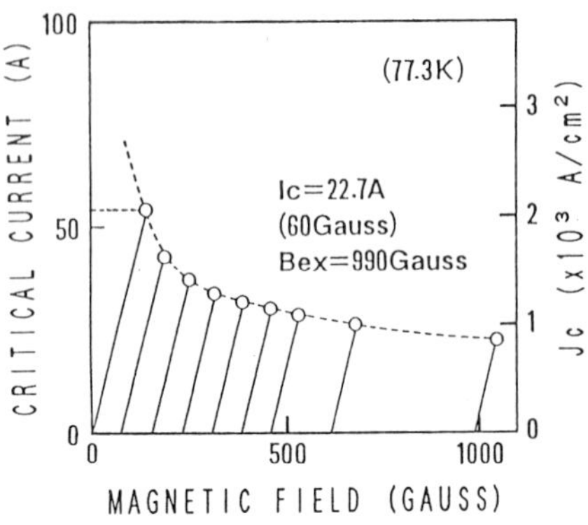

Fig. 6. Load line of a coil at 77.3 K.

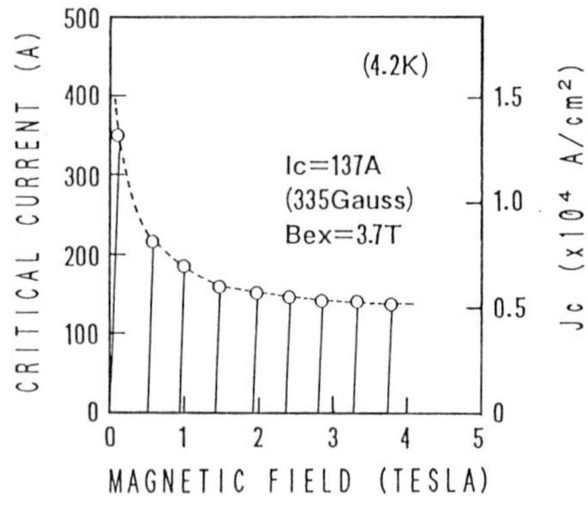

Fig. 7. Load line of a coil at 4.2 K.

Fig. 8. Overview of a single pancake coil.

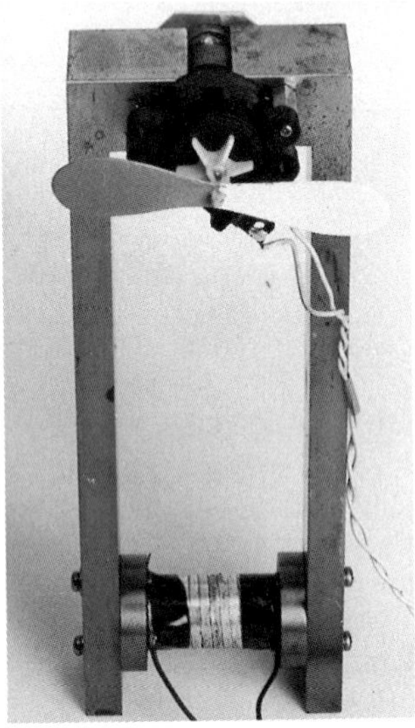

Fig. 9. Overview of a react & wind-processed coil.

REACT & WIND COIL

As mentioned before, the strain-resistant properties of multifilamentary wires are essential for application of cables and react & wind-processed magnets. Figure 9 shows the overview of a DC motor composed of a copper rotor and a layer-wound coil made of 1,296-core multifilamentary wire. The 4.3 m long high-Tc wire was rolled, sintered, covered with polyvinyl formal (PVF) and then wound on a 15 mm-diameter steel core (13 turns x 7 layers). This coil has a critical current density of 7,800 A/cm^2 (Ic=7.4 A) at 77.3 K and can generate 150 Gauss within 22 mm-gap of the steel core when operated at 6 A.

CURRENT LEAD AND LARGE CURRENT CONDUCTOR

Current leads provide the current to the magnets which are operated at liquid helium temperature. Low liquid helium consumption is required, so the cross-sectional area of the current lead must be designed for the balance of thermal penetrations and joule heating. Usage of high-Tc superconductors as current leads has advantages of non-joule heating and low thermal conductivity.

In the case of silver-sheathed superconducting wires, the thermal conducting properties of silver are dominant over the efficiency of high-Tc superconducting current leads. Therefore, the wires should be designed to maximize the operating current per unit cross-sectional area of the silver sheath. Figure 10 shows an overview of 1,500 A class current leads. Using a 20 m-long wire, four bundle conductors 50 cm in length were fabricated. Two bundle conductors were embodied in "H"-shaped frames of FRP. Transport current was 242 A at 77.3 K. Helium consumption was measured by soaking the lower end of the conductor in liquid helium. A net helium consumption of 0.93 W/kA through the high-Tc superconducting current leads is attained when operating at a transport current of 1,500 A. This should result in lower helium consumption than that of commercially available current leads made from copper (1.0-1.2 W/kA). Figure 11 shows an overview of 600 A class conductor operated at 77.3 K. Using a 50 m-long wire, 10 bundle conductors 50 cm in length were fixed on to the FRP pipe. Transport current properties were measured in liquid nitrogen and proved to carry 573 A with a criterion of 10^{-13} Ω · m. This conductor was thought to be 4,000 A class current leads. Larger current conductor will be attained by increasing the number of the bundle conductors and the strands.

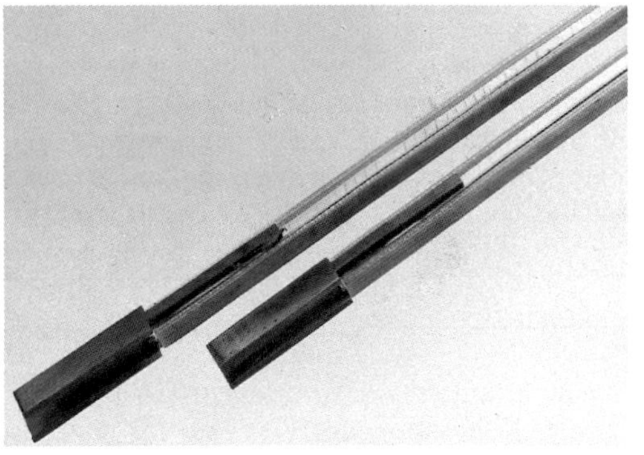

Fig. 10. Overview of 1,500 A class current leads.

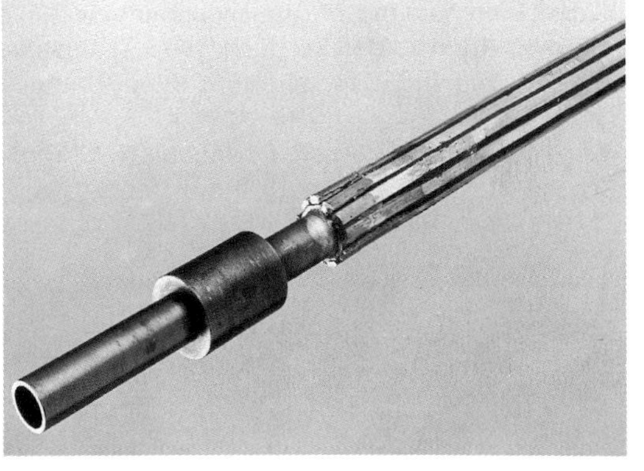

Fig. 11. Overview of a 600 A class composite conductor.

FLEXIBLE CABLE CONDUCTORS

We also tested the strain-resistant multifilamentary wires for a flexible cable conductor. The 1,296-core wires (1 m x 4 strands) were wound spirally (60 mm pitch) on a Teflon rod (20 mm-diameter). This composite was bent with a radius of 10 cm as shown in Fig. 12. After bending, the wires could sustain an initial transport current of 4.8 A.

A flexible cable conductor of 75 cm in length was fabricated using 22 m-long 1,296-core wires. 22 wires of 1 m-length were sintered and wound spirally on 20 mm-diameter copper spacer. This conductor was proved to have a critical current of 80 A with a resistivty of 10^{-13} Ω·m at 77.3 K. Thus, multifilamentary wires can be adapted for flexible cable conductors.

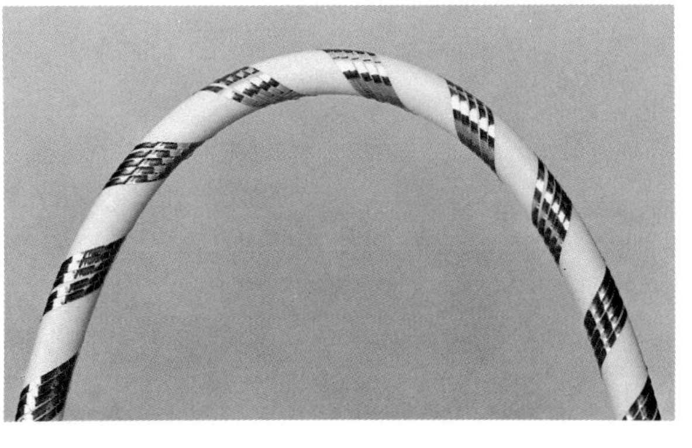

Fig. 12. Overview of a flexible cable conductor. (4 strands and one dummy strand)

CONCLUSION

We have studied the basic problems in the actual usage of silver-sheathed high-Tc superconducting wires and satisfactory results were obtained. Prototypes, such as small coils, current leads and cable conductors, were fabricated and examined in detail. High-Tc superconducting current leads revealed excellent properties from the point of liquid helium consumption compared to those of commercially available current leads.

REFERENCES

1. Mukai H, Shibuta N, Hikata T, Sato K, Nagata M and Hitotsuyanagi H: New Developments in Applied Superconductivity, ed. by Murakami, World Scientific, Singapore (1989) 101
2. Hikata T, Sato K and Hitotsuyanagi H: Jpn. J. Appl. Phys. 28(1989)L82
3. Sato K, Hikata T, Mukai H and Hitotsuyanagi H:Proc. of ISTEC Workshop on Superconductivity, Oiso(ISTEC), February 1-3, 1989, pp 119-122
4. Hikata T, Nishikawa T, Mukai H, Sato K and Hitotsuyanagi H: Jpn. J. Appl. Phys. 28(1989)L1024
5. Sato K, Mukai H and Hikata T: submitted to Proc. of Int. Atomic Energy Agency Technical Committee Meeting on Superconducting Materials and Magnets, Tokyo, September 4-6,1989.
6. Sato K, Hikata T, Mukai H, Masuda T, Ueyama M, Hitotsuyanagi H, Mitsui T and Kawashima M : Proc. 2nd Int. Symposium on Superconductivity (ISS'89), Tsukuba (Springer-Verlag),1990, pp 335-340
7. Sato K, Hikata T, Mukai H, Ueyama M, Shibuta N, Kato T, Masuda T, Nagata M, Iwata K and Mitsui T: submitted to 1990 Applied Superconductivity Conference, Colorado, September 24-28, 1990, MM-1

Recent Developments in BiSrCaCuO Based Wires and Tapes

H. KRAUTH[1], K. HEINE[1], J. TENBRINK[1], M. WILHELM[2], and H.-W. NEUMÜLLER[2]

[1] Vacuumschmelze GmbH, Gruener Weg 37, D-W 6450 Hanau, Germany
[2] Siemens AG, Corporate Research, Post Box 3220, D-W 8520 Erlangen, Germany

ABSTRACT

BiSrCaCuO/Ag based wires and tapes were produced with the powder-in-tube technique by swaging, drawing, rolling and pressing with intermediate and final heat treatments. At 77 K j_c values close to $3 \cdot 10^4$ A/cm^2 in zero field were achieved in highly textured Bi(Pb)-2223 tapes. In Bi-2212 tapes and wires the zero field j_c is about an order of magnitude smaller. At 4.2 K j_c values between 1 and $5 \cdot 10^4$ A/cm^2 were achieved in very high field up to 26 T in textured Bi(Pb)-2223 and Bi-2212 tapes as well as in untextured wires. In addition, j_c (B) and anisotropy factors were measured as a function of temperature. Formal pinning forces were calculated indicating similar pinning mechanisms in Bi(Pb)-2223 and Bi-2212. The inception of flux creep as a function of temperature depends strongly on magnetic field strength and orientation and occurs at a significantly higher temperature in 2223 as compared with 2212.

KEY WORDS: Bi system superconductors, wires, tapes, critical current density

INTRODUCTION

Several preparation routes are presently under investigation to explore the capabilities of polycrystalline High-T_c-Superconductor materials (HTSC) for use as Technical Superconductors in energy and magnet technology. A simple and at present one of the most promising approach is the powder-in-tube method with Ag as the tube material. It became clear relatively soon that the critical current density in $Y_1Ba_2Cu_3O_7$/Ag wires (YBaCuO) is limited by the pronounced weak link behaviour of polycrystalline $Y_1Ba_2Cu_3O_7$. As for the BiSrCaCuO system it could be shown, first for the $Bi_2Sr_2Ca_1Cu_2O_8$ (Bi 2212) phase, that the weak link limitation is much less important and that relatively high critical current densities can be achieved, at least at 4.2 K up to very high fields beyond 20 T [1]. Whereas this is true for Bi-2212 even in wires with untextured HTSC material, $(Bi,Pb)_2Sr_2Ca_2Cu_3O_{10}$ (Bi(Pb)-2223) must be highly textured to allow similar performance at low temperature [2]. At 77 K the critical current density in Bi-2212 is limited to rather low values (of about 10^3 A/cm^2 at 0T) due to the rather low T_c (85-90 K) and large flux creep effects. In Bi(Pb)-2223 the current density can be made much higher but depends strongly on the degree of texture and on the orientation of the magnetic field, due to the large anisotropy at this temperature. A variety of wires and tapes of both, Bi-2212 and Bi(Pb)-2223 using different swaging, drawing, rolling, pressing and heat treatment schedules were produced and analyzed with respect to microstructure, texture and current carrying capacity. Critical current measurements were performed as a function of temperature and magnetic field strength and orientation with respect to the conductor, in order to better understand flux pinning and flux creep effects as well as the influence of anisotropy and texture.

EXPERIMENTAL

HTSC powders were prepared by conventional ceramic techniques using high purity oxides and carbonates in order to prevent deleterious effects from

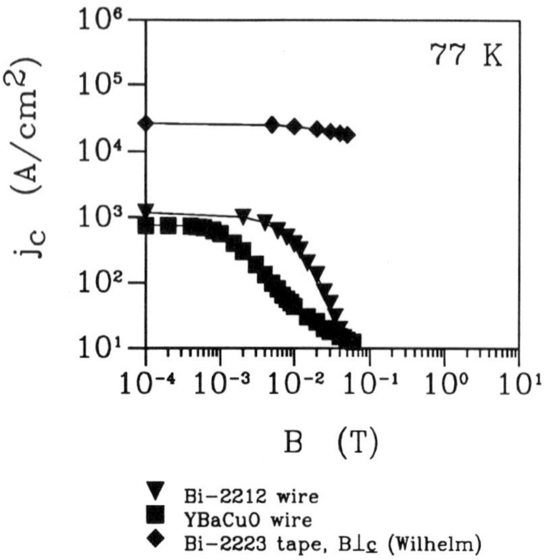

Fig. 1 Comparison of the $j_c(B)$ characteristics at 77 K of untextured YBaCuO and Bi-2212 wires and Bi-2223 tapes

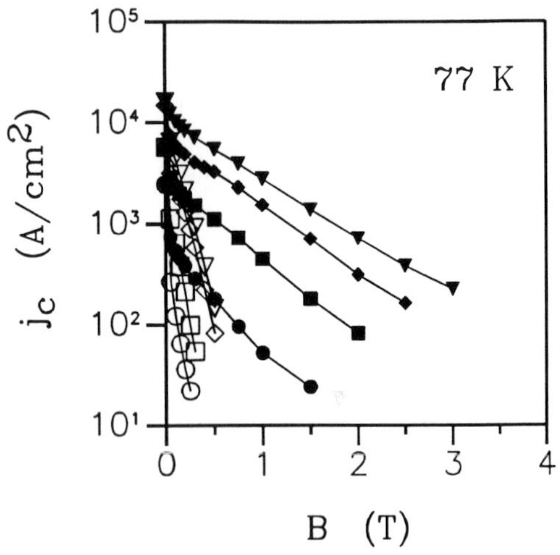

Fig. 2 Comparison of the $j_c(B)$ curves at 77 K of Bi-2223 tapes with different quality with B parallel (a,b)-plane (dark symbols) and B perpendicular (a,b)-plane (open symbols)

impurity elements. Appropriate amounts of Bi_2O_3, $SrCO_3$, CaO, CuO and PbO (in case of 2223 material) were mixed and calcined at temperatures between 800°C and 850°C with intermediate grinding in order to improve the homogeneity and to reduce the content of carbon [2].

Powder-in-tube wires were produced by filling the calcined powder into Ag tubes with 8 mm outer and 6 mm inner diameter, typically, and subsequent swaging and drawing. Tapes were produced by additional rolling or pressing using thicker Ag tubes. The superconducting properties were optimized by final heat treatment at different temperatures leading to sintering or even partial melting of the HTSC material followed by an annealing heat treatment [2,3]. Typical final wire diameters are 1 mm with a HTSC: Ag ratio of about 1. Typical outer dimensions of tapes are 0.1 mm · 2.5 mm with a HTSC core thickness of 25-30 μm. This low value turned out to be required in order to achieve a high degree of favourable texture in the superconductor material. Consequently the HTSC: Ag ratio is about 1 : 5. This ratios have to be taken into account when comparing the current carrying capacity and overall critical current density of wires and tapes with respect to future application. Usually short straight samples of 40 to 50 mm were taken to measure the critical current density as a function of magnetic field strength and orientation and of temperature. Different test facilities were used depending on temperature and field level. If not otherwise stated, critical current densities were calculated for a sensitivity criterion of 1 μV/cm. In addition the samples were characterized by metallographic and x-ray analysis, e. g. to determine the phase purity and the degree of texture.

RESULTS AND DISCUSSION

Critical current density at 77 K

Typical critical current densities at 77 K of completely untextured Bi-2212/Ag circular wires and of highly textured Bi(Pb)-2223/Ag tapes are compared in Fig. 1 as a function of magnetic field. The data for $Y_1Ba_2Cu_3O_7$/Ag wires are given for comparison. In partially melted Bi-2212 wires j_c values of up to $1.2 \cdot 10^3$ A/cm^2 were achieved. In only sintered material very poor current carrying capacity was observed. In any case the $j_c(B)$ dependence is very steep most probably due to flux creep effects, whereas the even steeper behaviour for YBaCuO wires is attributed to weak

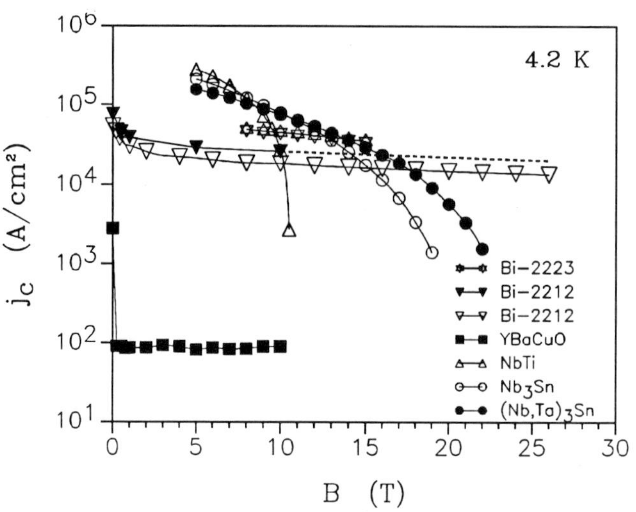

Fig. 3 Comparison of the $j_c(B)$ curves at 4.2 K for YBaCuO wires, Bi-2212 wires (two curves) and Bi-2223 tapes. Bi-2212 tapes (not shown) exhibit j_c values similar to Bi-2223 tapes. The performance of conventional superconductors is shown to demonstrate the superiority of HTSC in high fields.

links. In Bi(Pb)-2223 tapes critical current densities of up to $2.6 \cdot 10^4$ A/cm^2 at 0T were achieved in a reproducible way. The $j_c(B)$ dependence for these tapes was relatively flat (see Fig. 1). Measurements at samples of different quality are shown in Fig. 2 for two orientations of the external field. All samples had a texture of $F > 95$ % according to the definition in [9]. No correlation between F and the $j_c(B)$ characteristics could be seen. Very pronounced is the difference in the slope of the $j_c(B)$ curve, especially for the favourable orientation of B parallel to the tape surface i. e. parallel to the (a, b)-plane. In case of B perpendicular to the (a,b)-plane the improvement is less impressive such that the j_c anisotropy with respect to field orientation is very high at 77 K even at relatively low fields. Nevertheless, the improvement of j_c with B parallel to the tape surface is very encouraging and further enhancement seems to be possible by optimizing the fabrication process [4]. Anisotropy as a function of temperature will be discussed in a subsequent section.

Critical current density at 4.2 K

At 4.2 K the $j_c(B)$ characteristics are very different from that at 77 K as shown in Fig. 3. Whereas in YBaCuO the week link limitation is still dominant, the $j_c(B)$ curves are very flat for Bi-2212 wires and tapes as well as for Bi(Pb)-2223 tapes. It is important to note that the normalized field dependence $j_c(B)/j_c(0T)$ is very similar for Bi-2212 and Bi(Pb)-2223 independent on the fabrication process. Even sintered and untextured Bi-2212 wires with very poor performance at 77 K exhibit essentially the same normalized dependence as the best textured tapes [3, 4, 5, 6, 7]. This supports the conclusion that weak links play a minor role in the limitation of j_c in Bi based material. Nevertheless there are observed hysteresis effects in the $j_c(B)$ curves at 4.2 K for both, Bi-2212 and Bi(Pb)-2223 (Fig. 4). Similar behaviour at 77 K for $Y_1Ba_2Cu_3O_7$ can be satisfactoryly explained by weak links and related flux trapping. The width of the hysteresis (several Tesla) for the Bi based materials seems to exclude such a simple explanation. It is supposed that the observed behaviour is connected to the very high anisotropy of the current transport and flux pinning mechanisms for the current direction parallel and perpendicular to the (a,b)-plane, respectively. In a polycrystalline material, even with a high degree of texture (i. e. orientation of the grains), the current has to flow perpendicular to the (a,b)-plane in order to pass from one grain to the other. This assumption is supported by the fact that the hysteresis can be very large especially in highly textured material (Fig. 4a). Another observation in highly textured tapes is the missing of any Lorentz force effect on j_c [7]. The authors observe no effect of the angle between j_c and B for the case that B orientation is varied in (a, b)-plane (i. e. tape plane) from 0 to 90°C. This may be interpreted by the assumption that the transport current limitation is given by the current transfer from one grain to the other perpendicular to the (a, b)-plane. If above assumption is correct further improvement of the j_c values can be achieved by optimizing grain size and morphology.

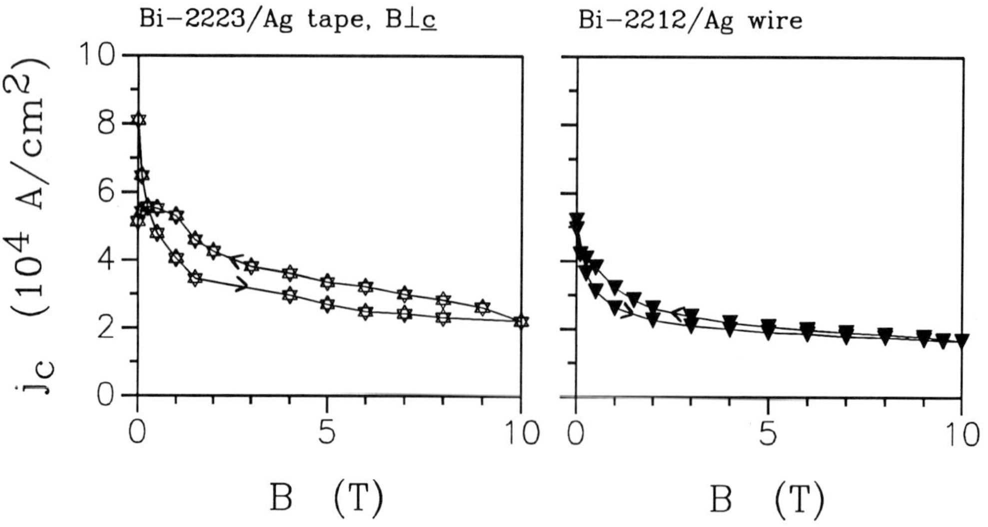

Fig. 4 Hysteresis of the $j_c(B)$ curves at 4.2 K for highly textured Bi-2223 tapes (Fig. 4a) and untextured Bi-2212 wires (Fig. 4b)

Temperature dependent measurements

Temperature dependent $j_c(B)$ measurements were performed to investigate flux pinning and flux creep effects and to find out the temperature and magnetic field regions exhibiting j_c values of technical relevance. It is helpful to note that for a given compound there usually is a qualitatively fair correlation of j_c values at 77 K, 0T and 4.2 K, 10 T, respectively, such that the selection of samples to be investigated may be made on the basis of j_c at 77 K (Fig. 5).

To ease interpretation $j_c(B,T)$ data were transformed into a formal volume pinning force $j_c \cdot B$ [2,8]. Fig. 6 shows F_p (B,T) data for Bi-2212 wires (Fig. 6a) and for Bi(Pb)-2223 tapes with B parallel to the (a,b)-plane (Fig. 6b) and B perpendicular to the (a,b)-plane (Fig. 6c), with B always perpendicular to the transport current direction.

At low temperature $F_p \propto B^m$ with m = 0.75 for both, Bi-2212 wires and Bi(Pb)-2223 tapes, as expected from conventional scaling laws for flux pinning, though this exponent can not be explained by one of the known elementary pinning mechanisms [2,8]. At higher temperatures a new mechanism starts to limit the formal volume pinning force and the critical current density. This is supposed to be connected to flux creep effects.

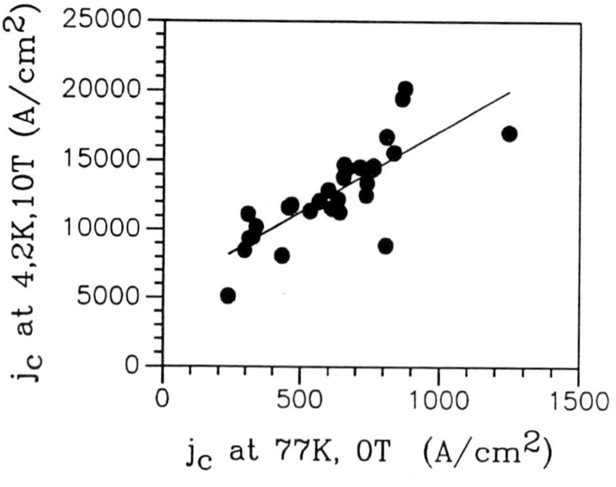

Fig. 5 Correlation of j_c (77 K, 0T) and j(4.2 K, 10 T) for Bi-2212 wires

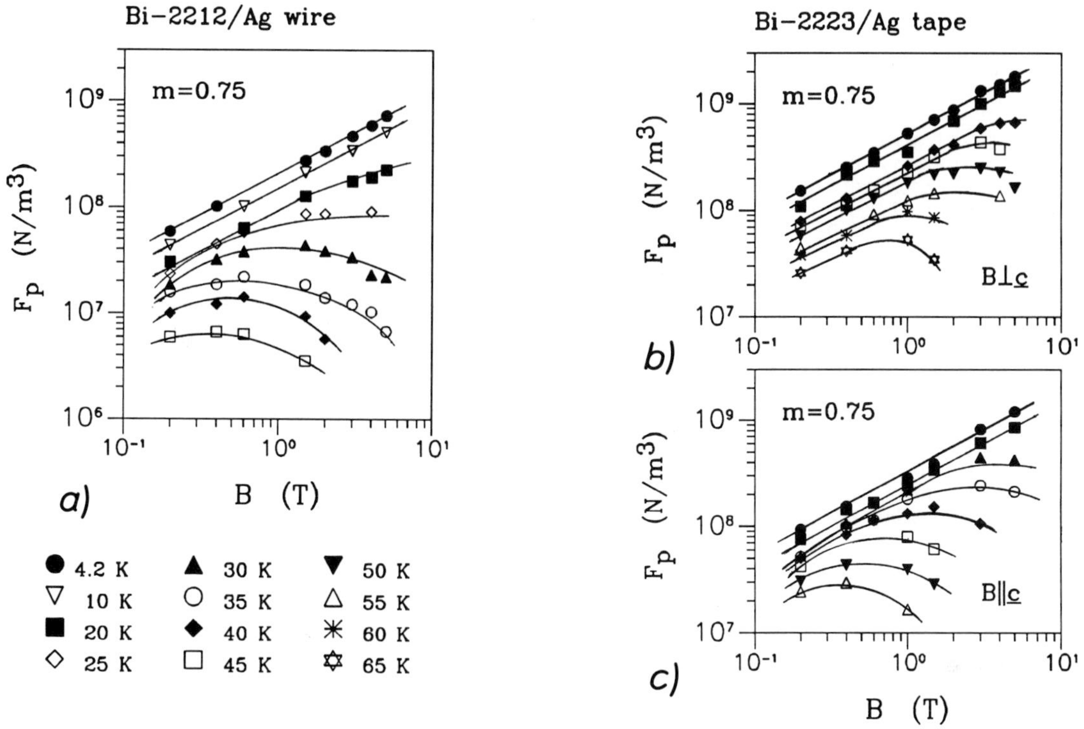

Fig. 6 Formal pinning force as a function of magnetic field for different temperatures from 4.2 K to 65 K for Bi-2212 wires (Fig. 6a) and Bi-2223 tapes with B parallel (a,b)-plane (Fig. 6b) and B perpendicular to (a,b)-plane (Fig. 6c).

The inception of this limitation depends on the magnetic field. For intermediate fields of 3 T - 5 T the flux creep inception is at a temperature of about 20 K for Bi-2212 wires. For the Bi(Pb)-2223 tapes this temperature is shifted to about 40 K for the favourable direction of B parallel (a,b)-plane. For the direction of B perpendicular to (a,b)-plane, the inception temperature is lowered and the effect is much more dramatic. This leads to a strongly temperature dependent anisotropy factor in these tapes (Fig. 7).

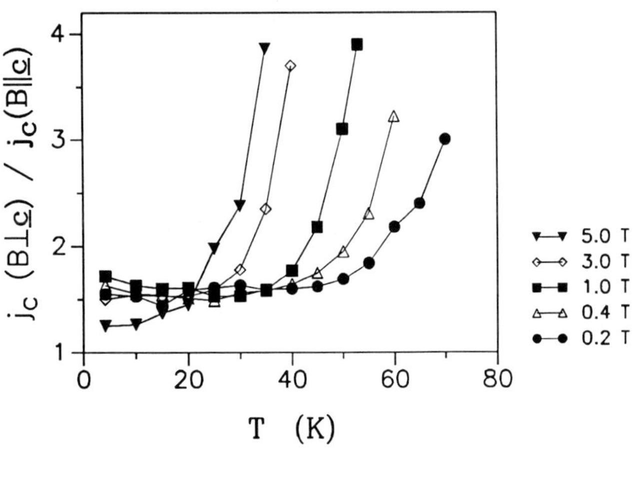

Fig. 7 Anisotropy factor with respect to magnetic field orientation of Bi-2223 tapes as a function of temperature for different magnetic fields

At low temperatures the anisotropy factor is about 1.5. At a certain temperature the anisotropy factor increases steeply, depending on the magnetic field. The corresponding values may be interpreted as constituting an "irreversibility line"for the situation present in a textured polycrystalline material with the transport current parallel to the (a,b)-plane of the grains and the field perpendicular to this plane. This line may be shifted to higher temperatures by further improvement of the microstructure.

CONCLUSION

Bi-2212 wires and tapes and Bi(Pb)-2223 tapes were produced by the powder-in-tube technique. At 77 K the critical current density approaches technically interesting values in the low field region (< 1 T) for highly textured Bi-2223 tapes with the field parallel to the tape surface. At 4.2 K a very flat $j_c(B)$ dependence is observed for Bi-2212 wires and tapes as well as for Bi(Pb)-2223 tapes with critical current densities well above 10^4 A/cm² even at field levels of 26 T. It is still not clear whether Bi-2212 wires or Bi(Pb)-2223 tapes represent the better option for ultra high field magnets at 4.2 K, despite of the fact that the tapes exhibit the higher j_c values. The 2212 phase is much easier to obtain than 2223, the wire fabrication process is much simpler than the tape fabrication and the Ag area can be kept small, whereas in tapes the Ag:HTSC ratio is presently very high. With the present performance both compounds may also already be suitable for medium field applications (up to about 5 T) for operational temperatures up to 20 K (Bi-2212) or 40 K (Bi(Pb)-2223). Future development work has to concentrate on further improvement of flux pinning and creep properties, on fabrication of long lengths [2] and on mechanical properties of the conductors.

REFERENCES

[1] Heine K., Tenbrink J., Thöner M., Appl. Phys. Lett. 55 (1989) 2441
[2] Tenbrink J., Wilhelm M., Heine K., Krauth H., Paper MM-2, Appl. Supercond. Conf., Snowmass, USA, Sept. 24-28, 1990
[3] Krauth H., Heine K., Tenbrink J., Proc. ICMC Topical Conf. HTSC Materials Aspects, Garmisch-Partenkirchen, May 9-11, 1990
[4] Sato K., Hikata T., Mukai H., Keyama M., Shibuta N., Kato T., Masuda T., Nagata M., Iwata K., Mitsui T., Paper MM-1, loc. cit. [2]
[5] Krauth H., Physics and Materials Science of High Temp. Superconductors, (Kossowsky R. Ed.), Kluwer Academic Press, Dordrecht, 1990
[6] Tenbrink J., Heine K., Krauth H., Cryogenics 30 (1990) 422
[7] Mimura M., Enomoto N., Uno N., Nakajima M., loc. cit. [3]
[8] Heine K., Tenbrink J., Krauth H., loc. cit. [3]
[9] Uno N., Enomoto N., Kikuchi H., Matsumoto K., Mimura M., Nakajima M., Advances in Superconductivity (Ishiguro T., Ed.), Springer Verlag Tokyo, 1990, p. 341

Processing and Properties of BiSrCaCuO Superconducting Tapes

K. TOGANO[1], H. KUMAKURA[1], D.R. DIETDERICH[1], H. MAEDA[1], J. KASE[2], and K. NOMURA[3]

[1] National Research Institute for Metals, Tsukuba, 305 Japan
[2] Asahi Glass Co., Ltd., Research Center, Kanagawa-ku, Yokohama, 221 Japan
[3] Hitachi Cable Ltd., 3550, Kidamari-cho, Tsuchiura, Ibaraki, 300 Japan

ABSTRACT

Highly textured $Bi_2Sr_2CaCu_2O_x$ tapes were prepared by applying partial melting and slow cooling process for BiSrCaCuO/Ag composite prepared by doctor blade process and for Ag-sheathed BiSrCaCuO composite prepared by powder-in-tube method. Jc at 77K was greater than $10^4 A/cm^2$ in zero magnetic field, however, it is very sensitive to an applied magnetic field and shows a large anisotropy with resepct to field direction. These sensitivity and anisotropy are significantly reduced at lower temperatures. Jc's at 4.2K reach practical level($\sim 10^5 A/cm^2$) even in high magnetic fields above 20T and shows a small anisotropy factor ($\sim$1.5), which should not limit practical application.

KEY WORDS : BiSrCaCuO superconductor, doctor blade process, Ag-sheathed tape, texturing, critical current density

INTRODUCTION

High-Tc oxide superconductors have had a great impact not only on basic physics but also on the field of superconductor application. Especially, the constructions of superconducting magnet using the new high-Tc oxides are expected in various technological fields. However, the utilization of a superconductor material in a magnet requires the winding of tape or wire conductors, with sufficient flexibility and high superconducting current capacity, into a solenoid. Since the oxide superconductors are intrinsically brittle, as in the case for A15 intermetallic compounds, special techniques must be developed for making tape or wire conductors. Another problem of the high-Tc oxide superconductor is its low critical current density Jc. One of the causes for this low Jc is a two dimensional layer structure. Almost all of the superconducting current flows in the a-b plane and little current flows along the c-axis. Furthermore, the current transfer across the grain boundary is strongly limited by the misorientation between the grains, reducing significantly the transport Jc in bulk polycrystalline materials. Those problems have led researchers to try to produce large single crystals or polycrystalline materials with highly oriented grain structure(texturing) in order to have sufficiently high transport Jc.

As reported by many investigators[1], $(Bi,Pb)_2Sr_2Ca_2Cu_3O_x$(2223) phase is more amenable to a mechanical grain alignment process such as cold pressing and cold rolling. On the other hand, we recently found that the $Bi_2Sr_2CaCu_2O_x$ (2212) phase suitable for a melt process because of easy crystal growth from molten oxide, in contrust with the difficult growth of 2223 crystals[2]. In this paper, we report the results of applying this melt growth process for both 2212/Ag composite tape prepared by the doctor blade process and silver-sheathed 2212 tape prepared through the powder-in-tube method.

EXPERIMENTALS

The detail of the preparation of $Bi_2Sr_2CaCu_2O_x$(2212)/Ag composite tape by doctor blade process (DBP) was already described in our previous papers[2-4]. In brief, a slurry prepared by mixing the 2212 powder with various organic materials was cast under a doctor-blade into green sheet. Tape samples cut from the green sheet were put on a 50μm-thick Ag foil and then subjected to various heat treatments. The most typical heat treatment program is as follows : first the sample is heated at 500°C for 2h to remove the organic formulation. Then the temperature is increased to 890°C, slowly cooled to 870°C at a rate of 2-20°C/h to grow the aligned 2212 grains, and kept at 870°C for 1-24h. And lastly, the samples are either quenched from 870°C or slowly cooled to room temperature at a rate of 30-100°C/h.

Ag-sheathed $Bi_2Sr_2CaCu_2O_x$/Ag composite was prepared by the so-called powder-in-tube method. The 2212 powder was put into a Ag tube of 20mm outer diameter and 12mm inner diameter. The composite was cold worked into a round wire of 0.5-1mm diameter by swaging and drawing, and then rolled into a tape of 2-3mm in width and 40-200μm in thickness. Short samples cut from the tapes were then subjected to the heat treatment of partial melting and slow cooling. Typical heat treatment program is composed of initial heating to 875-890°C, keeping for 1-6h, slow cooling at a rate of 5-10°C/h to 860°C, keeping for 6-12h and final slow cooling to room temperature at a rate of 100°C/h.

RESULTS AND DISCUSSIONS

(1) Microstructures

After the partial melting and slow cooling process, the oxide layers of both the DBP-2212/Ag and Ag-sheathed 2212 composite tapes show a well-aligned microstructure of plate-like 2212 phase grain, which can be easily observed on the fractured cross section of the composite tapes by scanning electron microscopy (SEM). X-ray diffraction (XRD) pattern shows that the material has a strong preferred orientation of c-axis alignment perpendicular to the tape surface.

In order to study the grain alignment mechanism, three heat treatment schedules were studied. The first schedule (a) proceeds as follows : the sample is melted at 890°C for 10min, quenched, and reheated at 870°C for 3h. The second schedule (b) is similar to the first but held for longer time(5h) at 890°C. The third schedule (c) is the standard one, including a slow cooling process, that is, melted at 890°C for 10min, slowly cooled to 870°C at 10°C/h, and kept at 870°C for 3h. Just after the quenching from 890°C in the

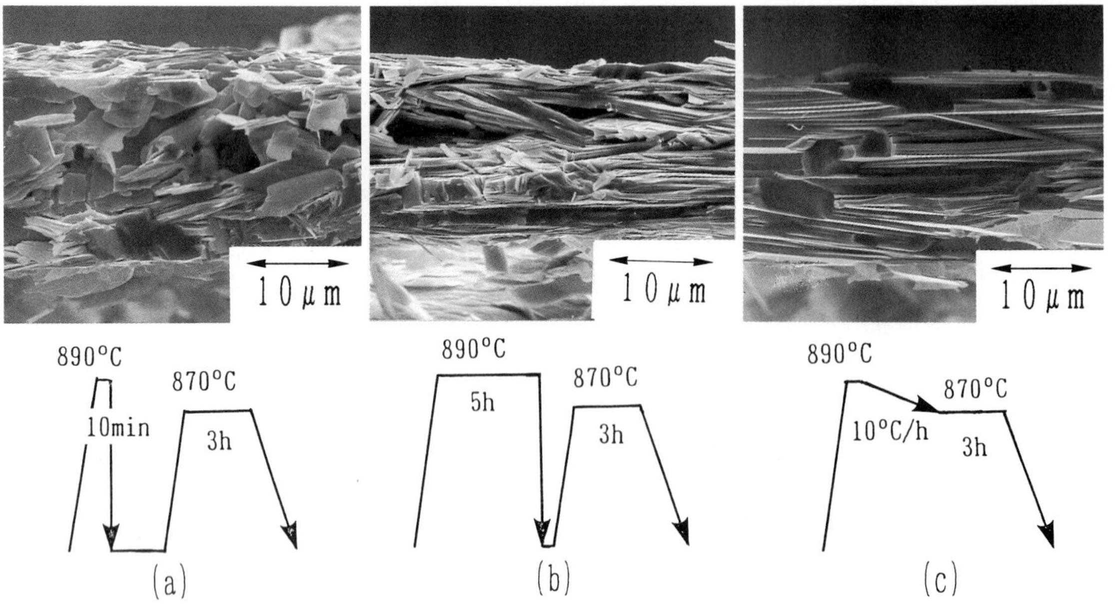

Fig. 1 SEM micrographs on the fractured cross section for the DBP-2212 /Ag composite tapes processed by different schedule (a), (b) and (c).

schedules (a) and (b), the samples are composed of the 2201 phase matrix with a composition of near $Bi_2(Sr,Ca)_2Cu_1O_x$ and precipitates of Bi-free Cu-free phases. We also observed a CuO-rich regions in the portion of the 2201 matrix, which is thought to be the quenched liquid. However, after the second heat treatment at 870°C, the matrix composition and structure change to the 2212 phase and volume fraction of the particles decreases. These results indicate that the phase transition from 2201 to 2212 has proceeded at 870°C. Figure 1 shows scanning electron micrographs (SEM) of the fractured surface of these three samples. The sample treated by the schedule (a), that is, treated with a short time in the molten state and quenched, has a dense 2212 matrix but no alignment, while the one held for 5h (schedule (b)) has developed a partially aligned structure. Judging from these results, the alignment should be produced by grain growth of the 2201 phase, not 2212 phase, during the partially molten state at 890°C. In contrast with the melt and quenched processed samples, a slow cooled sample (schedule (c)) has a highly oriented structure. Most of the x-ray diffraction peaks from the surface can be indexed as (001) reflections of 2212 phase. This result indicates a strong orientation with the c-axis aligned perpendicular to the tape surface. It can be concluded that slow cooling between 890°C and 870°C promotes the aligned grain growth of the 2201 phase and subsequently the 2201 phase transforms to the 2212 phase.

For a more detailed study of the grain alignment mechanism, green tapes with various thicknesses from 20 to 200μm were layed on a 50μm-thick silver foil and given the standard heat treatment. Figure 2 shows the fractured cross section of the tapes with various oxide thickness. The thickness of the final oxide layer is from 1/3 to 1/4 of the original green tape thickness. A highly aligned grain structure is observed for the tapes whose final oxide layer is thinner than 20μm. When the oxide layer is thicker than 30μm, an oriented structure is only observed at the surface of the oxide and near the oxide/silver interface, but with less regularlity. These results suggest that the texturing proceeds from the surface and possibly from the oxide/silver interface inward.

(2) Superconducting Properties

The superconducting transition temperature of the tape is strongly dependent on the cooling condition from 870°C. The quenched sample shows a higher Tc of 89K(zero-resistance), while the slowly cooled sample has a Tc below 77K.

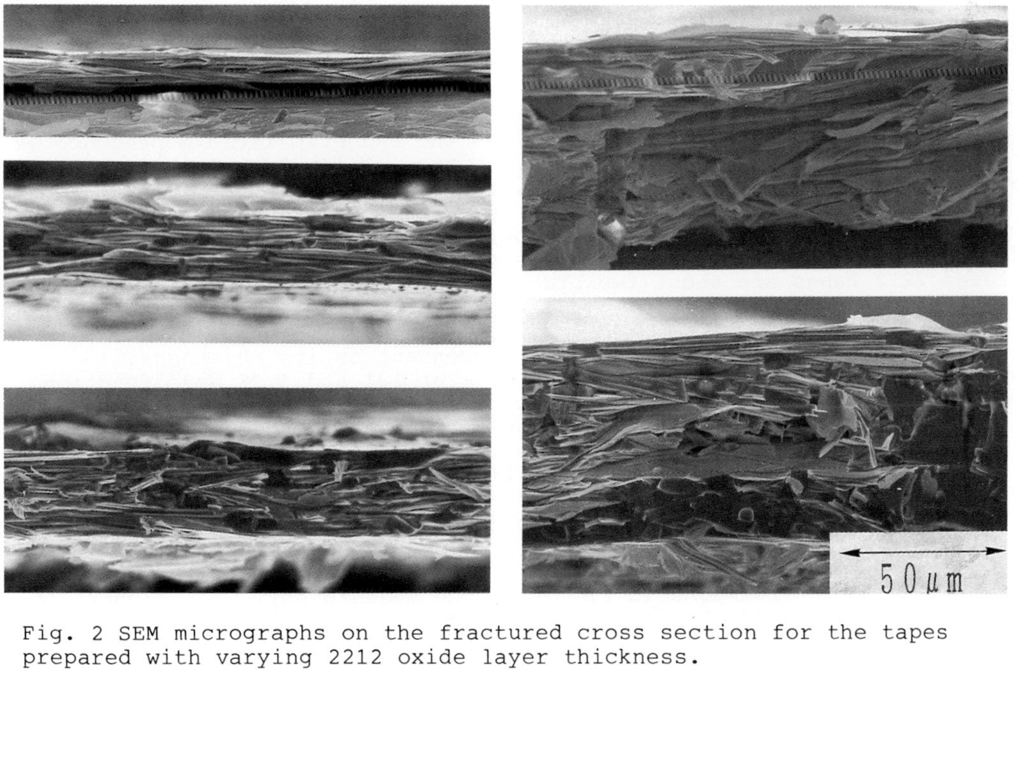

Fig. 2 SEM micrographs on the fractured cross section for the tapes prepared with varying 2212 oxide layer thickness.

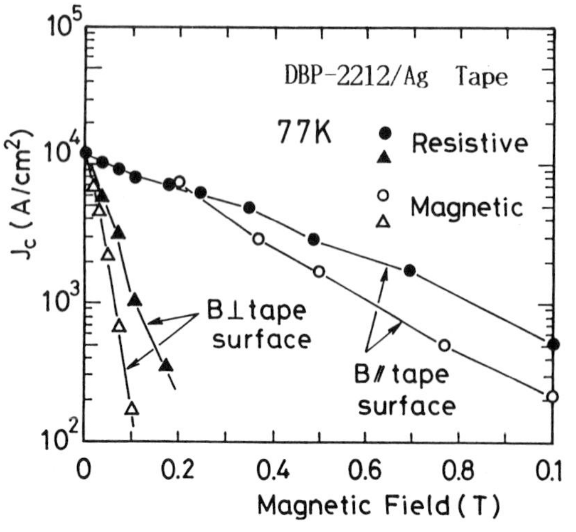

Fig. 3 Jc-B curves at 77K in parallel and perpendicular fields for DBP-2212/Ag tape. Jc values were obtained by both the resistive and magnetic method.

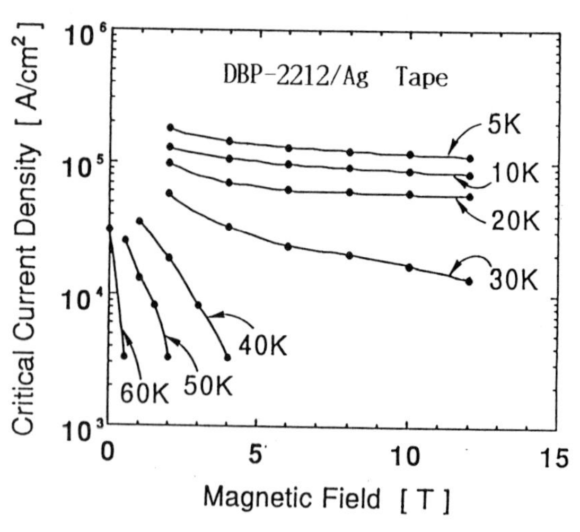

Fig. 4 Temperature dependence of the Jc-B curves obtained by resistive method for DBP-2212/Ag composite tape. Magnetic field was applied parallel to the tape surface.

Figure 3 shows the magnetic field dependence of Jc at 77k in parallel and perpendicular magnetic fields for a textured DBP-2212/Ag composite with Tc of 89K. J_c values obtained by both the resistive method and magnetization method are shown in figure. The tape has a large anisotropy in Jc with respect to the field direction. At low magnetic fields Jc values deduced from magnetization is nearly equal to the resistive Jc values. This indicates that the shielding superconducting currents are not restricted to each grain but flow uniformly throughout the sample due to the strong coupling of grains(5). However, Jc deduced from magnetization decreases more rapidly with increasing magnetic field than the resistive Jc for both magnetic field directions.

Although the rapid decrease of Jc in magnetic fields at 77K discourages us in applying this material for practical use in liquid nitrogen, the Jc-B charactreristic is significantly improved by lowering the temperature. Figure 4 shows the temperature dependence of the transport Jc in magnetic fields parallel to the tape surface for the DBP-2212/Ag tape. Jc increases rapidly as the temperature decreases. Especially, the magnetic field dependence of J_c is significantly reduced at lower temperatures. Below 30K, temperature dependence of Jc-B curve becomes small, and Jc's greater than 10^4A/cm^2 can be obtained at 12T. This result suggests that textured Bi-based oxides may not be restricted to liquid helium operation. Another coolant such as liquid hydrogen, for example, may be effectively used. These Jc values for the textured 2212 tapes are about two orders higher than those of non-textured 2212 tapes.

The Jc's of the oxide layer is strongly dependent on its thickness. Figure 5 shows critical currents(Ic) and Jc at 10T and 4.2K as a function of oxide thickness. The Ic decreases with decreasing thickness, while the Jc increases with decreasing thickness. The Jc curve appears to have two regions one above and one below 20μm. This Jc behavior corresponds with the oxide structure, which is shown in Fig. 2. When the oxide layer thickness exceeds 20μm, more non-textured regions appear in the oxide layer. Similar Jc dependence on oxide layer thickness was also obtained for the Ag-sheathed 2212 tape as shown in Fig. 6. The average thicknesses of oxide core in the composite are shown in the figure. The J_c increases with the decrease of thickness in entire magnetic fields in the measurement. The highest Jc at 4.2K and 12T is 1.3x10^{5}A/cm^2.

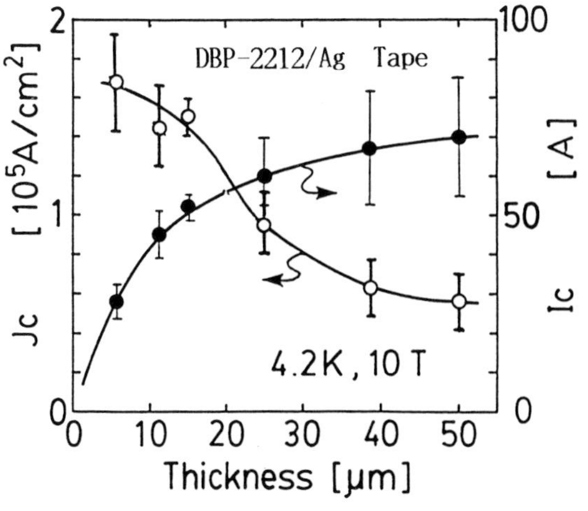

Fir. 5 Ic and Jc at 4.2K and 10T as a function of oxide layer thickness of DBP-2212/Ag composite tape. (See Fig. 2)

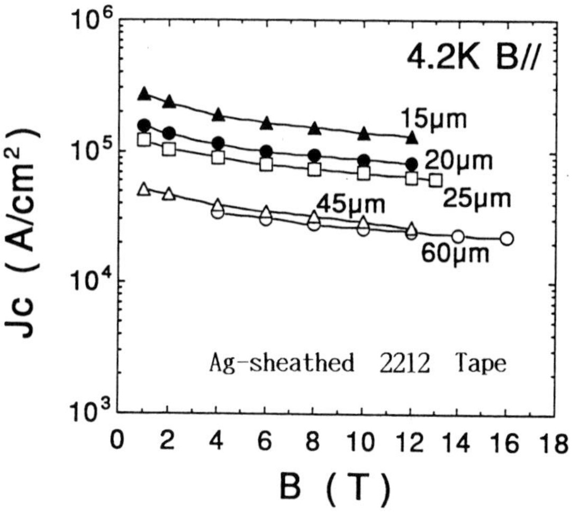

Fig. 6 Jc-B curves at 4.2K for Ag-sheathed 2212 tapes with various oxide core thicknesses.

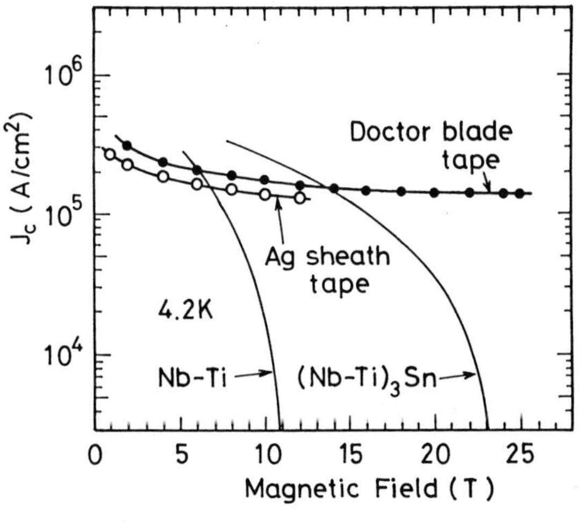

Fig. 7 Typical Jc-B curves at 4.2K for DBP-2212/Ag and Ag-sheathed 2212 composite tapes. Curves for the conventional metallic superconductors are also shown for comparison. High field measurements were done at Tohoku University.

Figure 7 shows typical transport Jc values at 4.2K in magnetic fields parallel to the tape surface for both the DBP-2212/Ag and Ag-sheathed 2212 tapes. For comparison, Jc values of conventional metallic superconductors Nb-Ti and $(Nb,Ti)_3Sn$ are also shown in the figure. The DBP/Ag tape has excellent Jc values, higher than $10^5 A/cm^2$ in magnetic fields up to 25T. A slightly smaller Jc values for the Ag-sheathed tape is probably due to less oriented microstructure. Due to the extraordinally high Bc2 of the 2212 oxide at 4.2K, a smaller magnetic field dependence of Jc is observed than that of metallic superconductors such as $(Nb,Ti)_3Sn$ and Nb-Ti, such that, the oxide tapes have excellent Jc's above 20T. These results indicate that textured 2212 tapes are very promising for high magnetic field application.

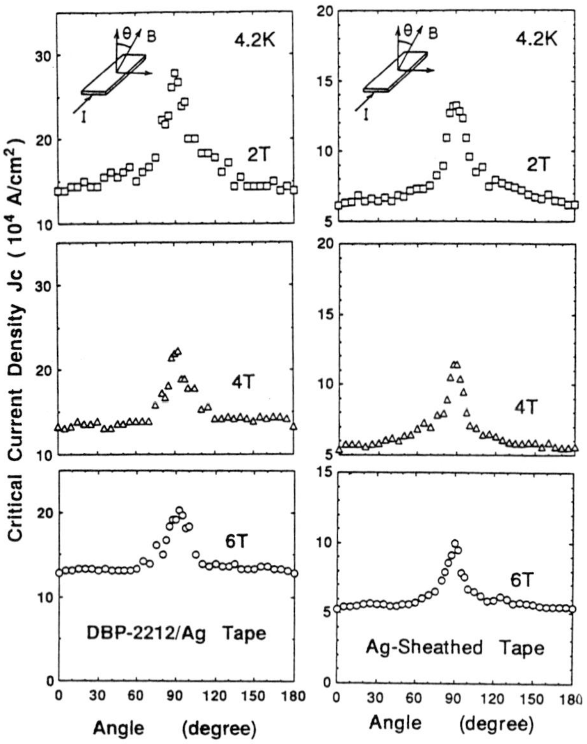

Fig. 8 Transport Jc at 4.2K versus magnetic field orientation for fields of 2,4 and 6T for DBP-2212/Ag and Ag-sheathed 2212 tapes.

Figure 8 shows the angular dependence of transport Jc at 4.2K in magnetic field of 2, 4 and 6T for the textured DBP-2212/Ag composite tape and the Ag-sheathed 2212 composite tape. The angular dependence of Jc decreases with increasing magnetic field in the field range of these measurements. Jc in perpendicular magnetic field ($\Theta=0°$) is about 50-60% of the Jc in parallel magnetic field ($\Theta=90°$). This anisotropy is small enough for a practical application. It has been reported that highly textured Bi-based thin films have a very sharp increase at $\Theta=90$, which is in contrast with the present results[6]. In our tapes, grain orientation is not perfect, such that, a peaks at $\Theta=90°$ is smeared out.

ACKNOWLEDGEMENTTS

We would like to thank Prof. Y. Muto and Dr. K. Watanabe of Tohoku University for their giving us the opportunity for using their high magnetic field facilities.

REFERENCES

1. For example, Asano T, Tanaka Y, Fukutomi M, Jikihara K, Machida J, Maeda H (1988) Jpn. J. Appl. Phys. 27: L1652.
2. Kase J, Togano K, Kumakura H, Deitderich D.R, Irisawa N, Morimoto T, Maeda H (1990) Jpn. J. Appl. Phys. 29 : L1096.
3. Togano K, Kumakura H, Dietderich D.R, Maeda H, Kase J, Morimoto T, Ullmann B, Freyhardt H.C (in press) Proc. ICMC'90 990).
4. Kase J, Morimoto T, Togano K, Kumakura H, Dietderich D.R, Maeda H (1991) Proc. 1990 ASC, IEEE Trans. on Magn. (to be published).
5. Kumakura H, Togano K, Dietderich D.R, Maeda H, Kase J, Morimoto T (1991) Proc. 1990 ASC, IEEE Trans. on Magn. (to be published).
6. Raffy H, Labdi S, Laborde O, Monceau P (1990) Physica B 165&166 : 1423.

The Transport Critical Current Properties of BSCCO Superconducting Wires

NORITSUGU ENOMOTO

Yokohama R&D Laboratories, The Furukawa Electric Co., Ltd.,
4-3, Okano 2-chome, Nishi-ku, Yokohama, 220 Japan

Abstract

The critical transport current density (Jc) for BSCCO wires were measured at 77K and 4.2K in magnetic field up to 30T. The Jc of wire specimen was significantly improved by this grain orientation of low-Tc phase. Although the maximum Jc at 77K and 0T were enhanced up to 3.5×10^4 A/cm^2 consequently, the magnetic field dependence of Jc is not sufficient for a magnet coil and showed large anisotoropy with respect to the magnetic field for pursuing crystal orientation up to over 95% F value. While the Jc at 4.2K and 10T was 2.1×10^5 A/cm^2, which is almost kept constant up to 30T parallel to the tape surface. The Jc anisotropy was dependent on the angle between the magnetic field direction and the tape surface and the Jc reduction at 4.2K in the field normal to the surface remained in about one-half up to 10T. By using this transport property at 4.2K, traial coils of BSCCO superconductor were prepared and tested. Magnetic field of 1220 gauss and 370 gauss were generated by BSCCO coil in the external magnetic field of 0T and 10T, respectively. This indicates the applicability of BSCCO wire for high field superconducting apparatuses.

KEY WORDS : transport critical current, F value, magnetic field dependence, grain alignment, high-Tc superconducting wire, coil

Introduction

Since the discovery of the high Tc oxide superconductors, the metal sheathed technique by powder-in-tube has been widely applied for wire conductors. The critical current density (Jc) of the wires have been improved by some approaches of materials processes and conductivity mechanism.

Bi-based materials have strong structual and superconductive anisotropy, therefore improvement of alignment of crystal grains resulted reduction of weak link in grain boundaries[1],[2],[4]. We have reported the relationship between Jc at 77K and c-axis orientation for low Tc tape in cooperation with Tokyo Electric Power Company[3].

Although the Jc-B property of BSCCO wire was improved remarkably at low magnetic field in comparison with YBCO wires, this high Jc degradated over 0.1T field, which is supposedly due to both the flux creep Uo and the weakness of intragrain pinning rather than intergrain coupling. Recently it was reported that BSCCO wire with a few cm length was less sensitive to magnetic field over 20T at 4.2K than the conventional superconducting wire [2],[3],[4],[18]. However it is necessary for electric power applications to produce long wires with uniform superconducting properties.

In this letter, We report some features of short BSCCO tapes for a magnet application and trial small coils using long BSCCO wires. The coil was tested at 4.2K in external magnetic field up to 10T as a magnet and as a long wire using V-I curve analysis[7].

Experimental

The calcined powder in a nominal composition (Bi:Sr:Ca:Cu=2:2:1:2) was put into silver tube and the composite was cold worked into a long tape wire with the thickness 0.1- 0.5 mm. The wires were sintered at 800-900 C for up to 50h in oxygen atmosphere [3]. X-ray diffraction analysis were performed on the wide surface of the wire after sheathes were stripped off and the degree of c-axis orientation (F) was estimated from X-ray diffraction peaks [5]. The measurements of Jc were carried out at 77K and 4.2K in magnetic fields by four prove method. The Jc measurements at high fields up to 30T were carried out at Tohoku University. The dependence of the Jc on the field direction was studied at 77K, 200 gauss and 4.2K up to 10T as shown in Fig.1. The DC magnetization was measured by a SQUID. The activation energy for the flux creep Uo of the wire was estimated from the magnetic relaxation[6]. The Jc distribution of a composite wire were evaluated by using following equation[7].

$$V(I)=A\int_0^\infty (I-I)f(I)dI \cdots (1) \qquad \langle Ic \rangle = \int_0^\infty I f(I)dI \cdots (3)$$

$$d^2V/dI^2=Af(I) \cdots (2) \qquad \int_0^\infty f(I)dI=1 \cdots (4)$$

where, f(I) the distribution of critical currents in the sample

By using this transport property at 4.2K, traial coils of BSCCO superconductor were prepared. The type of coils were single or double pancake. The coils obtained were tested in the applied magnetic field up to 10T by dc operation and the generated magnetic fields were monitored by GaAs Hall device located on the center of the coil.

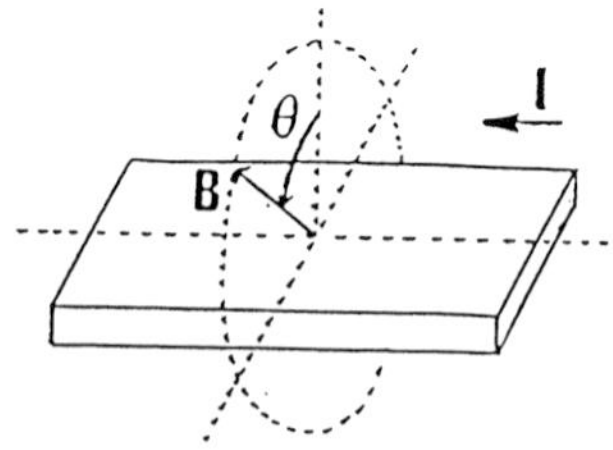

Fig.1 The configuration of transport current and applied magnetic field direction for Jc angular measurement.

Result and Discussion

The influence of the degree of c-axis orientation (F) to the critical current density in the various magnetic field for Bi-based Ag-sheathed wires at 77K is shown in Fig.2[1]. The Jc in the various magnetic fields was significantly dependent on F value over the wide region and increased steeply with F value. Although the maximum Jc was improved up to 3.5×10^4 A/cm^2 at 77K and 0T around 99% F, the Jc decreased steeply in the magnetic field over 0.1T. And the Jc was apparently depended on the magnetic field direction. The Jc anisotropy with respect to the applied magnetic field direction showed large reduction as shown in Fig 3. The $Jc(\theta=0)/Jc(\theta=90)$ was 0.29 at 77K and 200G and this result indicated that the magnetic field dependence is not sufficient for the magnet application at 77K.

The Jc in the magetic field at 4.2K increased with F values, too. The Jc of the wire with F=99% as shown in Fig.4 was 5.5×10^5 A/cm^2 at 1T and 2.1×10^5

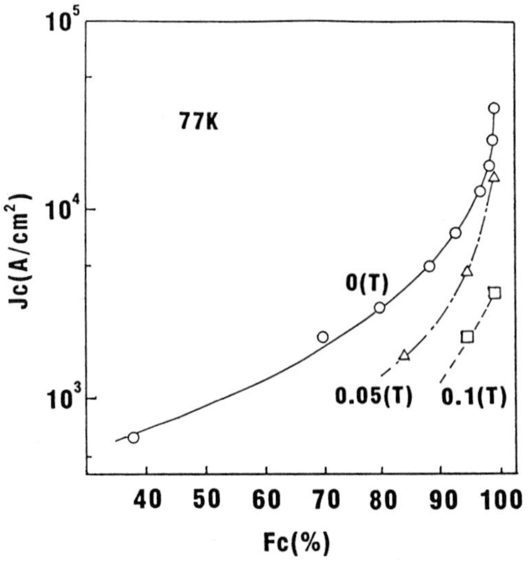

Fig. 2 The influence of the degree of c-axis orientation (F) to the critical current density (Jc) for Ag-sheathed BSCCO tapes at 77K.

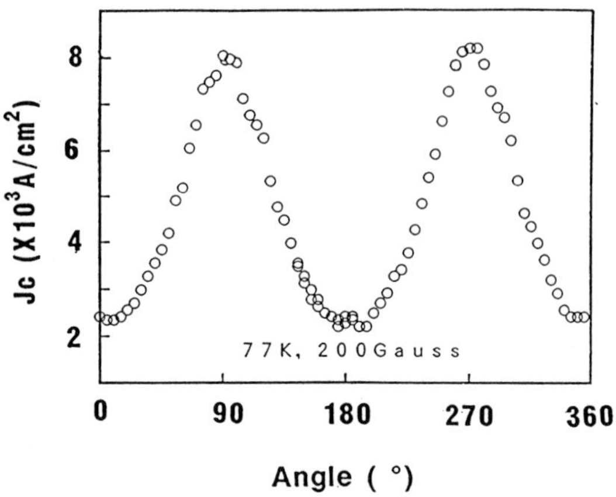

Fig.3 The angular dependence of the Jc as a function of the tilt angle for the Ag-sheathed BSCCO tape at 77K and 200G.

A/cm^2 at 10T, which is almost kept constant up to 30T. This Jc is much less sensitive to magnetic field than that of the conventional superconductors such as Nb_3(Al,Ge) and Chevrel phases due to the extraordinaly high Bc_2 values of handreds Tesla. Recently, many workers have tested the temperature dependence of Jc in magnetic field parallel to the surface for textured BSCCO 2212 thick film, tape and BSCCO 2223 tape[9],[10],[11]. The magnetic field dependence of Jc is significantly reduced at lower temperatures. Its dependece of Jc-B curve becomes smaller below about 20K, and the Jc greatrer than 10^4 A/cm^2 can be obtained over 10T.

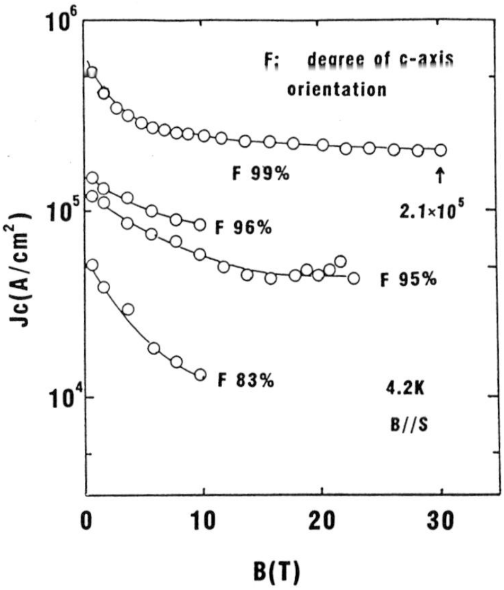

Fig. 4 Magnetic field dependence of Jc at 4.2K for low-Tc tapes with various F values.

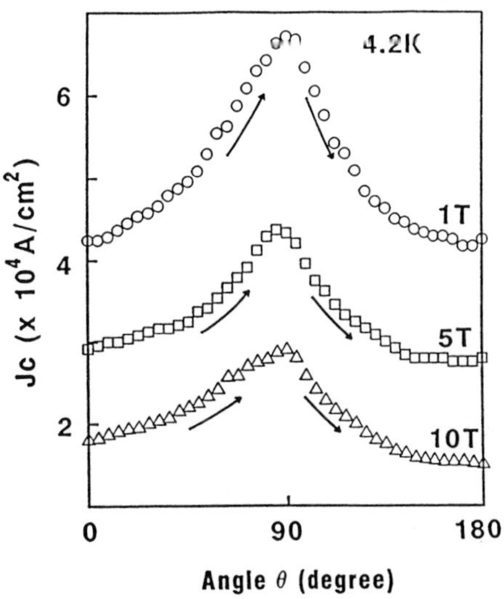

Fig.5 Angular dependence of Jc in magnetic fields 1T,5T and 10T.

Especially, the tendency for Bi(2212) phase was more prominent at lower temperature in comparison with Bi(2223) phase.

The activation energy Uo was calculated from the magnetic relaxation at 4.2K and 0.5T for the wires with differrent F values[8]. The Uo(F=99%) is almost equalto the Uo(F<75%) in the both field dirctions of normal and parallel to wire surface although the Jc (F=99%) was about 20 times as large as the Jc (F<75%). These results indicate that the improvement of Jc-B property is mainly due to the improvement of the coupling in the grains by texuring for crystal alignment.The flux creep have no influence for a magnet application at 4.2K.

Figure 5 shows the angular dependence of the Jc for BSCCO wire with F=96% in various magnetic fields up to 10T. In this case, the Jc was dependent on the angle between the field direction and tape surface. As magnetic fields increased from 1T to 10T, the Jc(θ=0)/Jc(θ=90) were about 0.60, which were hardly changed, while the Jc(θ=0)/Jc(θ=180) slightly increased. These results suggest that Jc anisotropy is not influenced by the magnitude of magnetic field and the history effect of Jc anisotropy increases with magnetic field. As the F value for BSCCO tape wire increased, the influnce of the angular dependence on Jc increased[8]. These results indicate that the pinning force is stronger in the magnetic field parallel to the ab-plane but is very weak in the magnetic field normal to the ab-plane. In this case, the possible pinning sites which can explain this anisotropy are boundaries induced by intergrowthes, twist boundaries, an intrinsic pin[12],[13] and grain boundaries. The transmission electron microscopy (TEM) observations apparently show the existences of a intergrowth and a twist boundary[5],[14]. Similar results have been reported by several investigators [15],[16],[17].

By using these transport properties at 4.2K, single pancake coils were manufactured for trial. The specification of the coil is listed in Table 1. Photograph of BSCCO single pancake coil and measurement system for coil are shown in Fig.6 and 7. With respect to preparation of oxide superconducting coil, the homogeneous of Ic distribution for the long wire is the most important problem because the performance of coil is governed by the property of the wire.
The Jc distribution of BSCCO wire for pancake coil under external magnetic field from 0T to 10T were evaluated using the equation (2) as shown in

Fig.6 Photograph of single pancake coils.

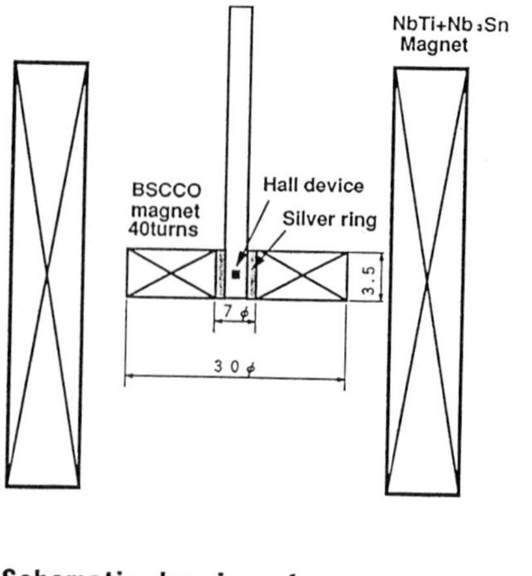

Fig. 7 Schematic drawing of measurement of BSCCO single pancake coil.

Fig.8. The Jc ditributed widely and the apparent shape of the Jc ditribution change with external magnetic field. But the Jc distribution hardly change with fields intrinsically. This indicates that BSCCO long wire at 4.2K have no weak-link which disturbs transport current in higher magnetic field. The result of pancake coil at 4.2K in the extenal magnetic field from 0T to 10T are listed in Table 2 where Ic was determined with criterion of 10^{-12} Ω·m. The relationship between the maximum self-magnetized field of coil and Jc with various criterion are shown in Fig.9. The maximumself-magetized fields of coil (Bo) in the external fields of 0T and 10T at 4.2K were obtained up to 1220 gauss and 370 gauss, respectively. The coil current dencity was smaller than that the short sample. In order to enhancethe maximum self-magnetized fields of coil, it is essential to develop the homogeous long wire with not only high Ic but also excellent magnetic fieldproperty.

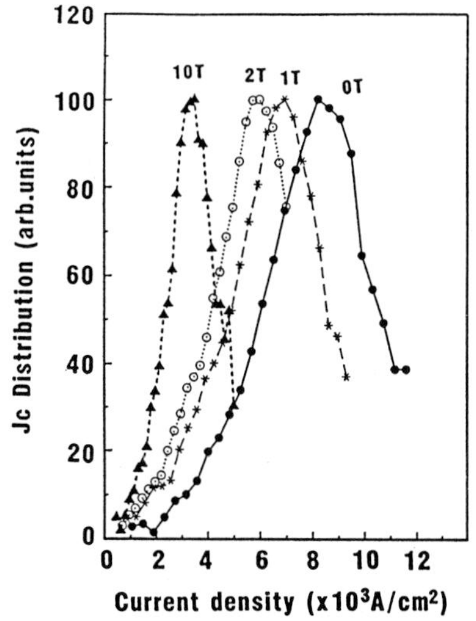

Fig.8 Jc distribution calculated by d^2V/dI^2.

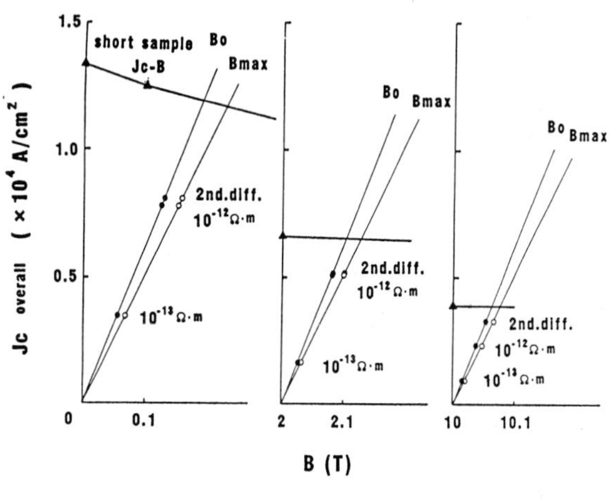

Fig.9 Magnetic field generated by BSCCO magnet in external field by conventional magnet at 4.2K.

Table 1. Specification of single pancake coil

Wire	Size	$0.2^t \times 3.5^w \times 2300^l$ (mm)
	Materials	BSCCO (2212) / Ag
Coil	Type	Single pancake
	Inner diameter	φ 7 (mm)
	Outer diameter	φ 30 (mm)
	Number of turns	40turns

Table 2. Test results of BSCCO coil

Bex (T)	0	1	2	10
Ic (A)	43.6	37.1	28.5	13.2
Jc overall (A/cm²)	7790	6620	5080	2350
Bo (G)	1220	1040	800	370
Bmax (G)	1520	1290	990	460

Conclusion

The Jc of wire specimen was significantly improved by this grain orientation of low-Tc phase ($Bi_2Sr_2CaCu_2O_x$). The Jc at 4.2K and 10T was $2.1x10^5$ A/cm^2, which is almost kept constant up to 30T parallel to the tape surface. The Jc reduction at 4.2K in the field normal to the surface remained in about one-half up to 10T. By usingthis transport property at 4.2K, traial coils of BSCCO superconductor were prepared and tested. The BSCCO coil successfully generated magnetic field (Bo) of 1220 gauss and 370 gauss in the external magnetic field of 0T and 10T, respectively. These results indicate the applicability of BSCCO wire for high field superconducting apparatuses.

References

[1] S.Shiga:Japan Science and Technology,31(1990)55,Japan Science Foundation, Science Museum

[2] N.Enomoto, H.Kikuchi, N.Uno, M.Ikeda, H.Kumakura, K.Togano, K.Watanabe: "11th International Conference on Magnet Technology, Tsukuba, Japan, 28 Aug.-1 sep. 1989"

[3] N.Enomoto, H.Kikuchi, M.Mimura, M.Nakajima, N.Uno, T.Hara, K.Okaniwa and T.Yamamoto: "2nd International Symposium on Superconductivity, Tsukuba, Japan Nov.14-17,1989"

[4] M.Mimura, H.Kumakura, K.Togano, H.Maeda:Appl.Phys.54(1989)1582

[5] N.Enomoto, H.Kikuchi, N.Uno, H.Kumakura, K.Togano and K.Watanabe:Jpn.J. Appl. Phys.29(1990)L447

[6] N.Uno, N.Enomoto, H.Kikuchi, M.Mimura, M.Nakajima: "2nd International Symposium on Superconductivity, Tsukuba, Japan Nov.14-17,1989"

[7] W.H.warnes and D.C.Larbalestier:Cryogenics (1986) Vol26 P643-653

[8] M.Mimura, N.Enomoto, N.Uno and M.Nakazima:ICMC.May 9-11(1990)

[9] E.Narumi, L.W.Song, S.Hwa, J.Ye, F.Yang, Y.H.Kao, Patel, and D.Shaw: Applied Superconductivity Conference,Sep.24-28,1990,Snowmass,USA

[10] H.Kumakura, K.Togano, D.R.Dietderich,J.Kase, T.Morimoto and H.Maeda: Applied Superconductivity Conference,Sep.24-28,1990,Snowmass,USA

[11] H.Kumakura, K.Togano, E.Yanagisawa, J.Kase and H.Maeda:Jpn.J.Appl.Phys. 29(1990)L1652

[12] M.Tachiki and S.Takahashi:Solid state Communication.70(1989)

[13] M.Tachiki and S.Takahashi:Solid state Communication.72(1989)

[14] H.Kumakura, K.Togano, E.Yanagisawa, K.Takahashi, M.Nakao and H.Maeda: Jpn.J.Appl.Phys.28(1989)L24

[15] H.Kumakura, K.Togano, M.Uehara, H.Maeda, K.Takahashi, M.Nakao:IEEE. Trans. Magn.25(1989)

[16] Y.Iye, S.Nakamura, T.Tamegashi:Physica C159(1990)433

[17] Y.Iye, S.Nakamura, T.Tamegashi:Physica C166(1990)62

[18] J.Tembrink, M.Wilhelm, K.Heine, H.Krauth: Applied Superconductivity Conference,Sep.24-28,1990,Snowmass,USA

Preparation of Bi-Sr-Ca-Cu-O Superconducting Tape by Laser Zone Melting Method

K. TOMOMATSU[1], A. KUME[1], N. SADAKATA[1], O. KOHNO[1], T. HARA[2], K. OKANIWA[2], and T. YAMAMOTO[2]

[1] Materials Research Laboratory, Fujikura Ltd., Koto-ku, Tokyo, 135 Japan
[2] Engineering Research Center, Tokyo Electric Power Compnay, Chofu, Tokyo, 182 Japan

ABSTRACT

Superconducting tapes of 50μm and 90μm in thickness on the MgO substrate were prepared by the laser zone melting method. Those tapes of 50um in thickness were grown at the different growth rate and temperature gradient at the growth interface. The superconductors grown at 5mm/h and 10mm/h consist of the single 80K-phase, and have well aligned grains parallel to the longitudinal axis of the tape. The critical current was measured by the direct current measurement as high as 0.5A (740A/cm^2,77K,0T) for the tape grown at 5mm/h and followed by heat treatment. The tape about 90um in thickness grown at 10mm/h also had the highly oriented structure.

KEY WORDS: Bi-Sr-Ca-Cu-O system, 80K-phase, melting process, orientation, critical current

INTRODUCTION

A large anisotropy in resistivity[1-3] and critical current[4] has been reported for Bi-Sr-Ca-Cu-O superconducting material. The superconducting materials by a zone melting process is expected to have high critical current density because of its dense and highly oriented grain structure. The fibers with highly oriented 80K-phase grain structure were prepared by the floating zone method.[5,6] The structure and composition of this material grown by the zone melting process were remarkably influenced by the growth rate.[5,7] It was reported that $Bi_2(Sr,Ca)_2CuO_y$ and $(Sr,Ca)CuO_y$ are precipitated at the higher growth rate.[5,7] Those facts are considered to be caused by that the composition of the 80K-phase is incongruently melting.[7] On the other hand, the substrate is necessary to keep enough strength of the long superconducting cable. In this study, Bi-Sr-Ca-Cu-O superconducting tapes, which have thick superconductor grown from the melt on the substrate, have been prepared by laser zone melting method. The effects of temperature gradient at the growth interface, as well as the growth rate, are discussed on the fraction of the 80K-phase and the grain structure.

EXPERIMENTAL

Figure 1. shows the schematic diagram of the laser zone melting method used in this study. A CO_2 laser beam was focused in order to melt the source powder on the substrate. In order to maintain the appropriate composition of the melt, the beam was focused in rectangular shape by the cylindrical lens and form the melting zone as narrow as 0.5mm. Another CO_2 laser beam was focused on the back side of the substrate. The external electric furnace heats the source powder and the substrate, and vary the temperature gradient at the growth interface. The composition of the source powder was $Bi_2Sr_2Ca_1Cu_2O_y$. MgO (100) single crystal (10x2x0.5mm^3) was

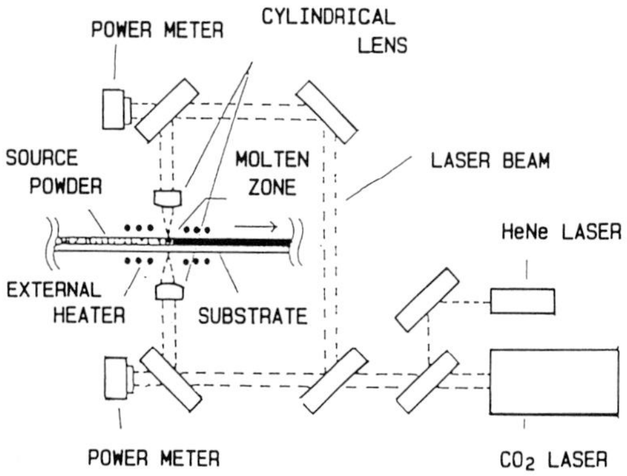

Fig.1. Schematic diagram of the laser zone melting method.

used for the substrate. Tapes were grown from 5mm/h to 50mm/h in the atmosphere heated at 600°C and 750°C by the electric furnace. Temperature gradient is measured by a thermoviewer JEOL TG-4100. Powder X-ray diffraction was carried out in order to evaluate the fraction of the 80K phase in the as-grown tapes. The grain structure was observed by scanning electron microscope. The tapes about 50μm in thickness were prepared and used for the powder X-ray diffraction and the observation of the structures.

RESULTS AND DISCUSSION

Figure 2 shows the volume fraction of the 80K-phase in the as-grown tape as a function of the growth rate. The peaks of 80K-phase, $Bi_2(Sr,Ca)_2CuO_y$ and slight amount of $(Sr,Ca)CuO_y$ were found in the powder X-ray patterns. The volume fraction of the 80K-phase was estimated by the relative intensity of the (008) peak of the 80K-phase to the (006) peak of the $Bi_2(Sr,Ca)_2CuO_y$. In the tapes heated at 750°C, the volume fraction of the 80K-phase is close to 100% at the growth rate of 5mm/h and 10mm/h, and becomes lower with the increase of the growth rate. The volume fraction of the each tape heated at 600°C is lower than that of the tape heated at 750°C. It is observed that the both tapes heated at 600°C and 750°C have the threshold growth rate between 10mm/h and 20mm/h where the volume fraction of the 80K-phase decreases abruptly. It was reported that the amorphous state was obtained by quenching in Bi-Sr-Ca-Cu-O system, then $Bi_2(Sr,Ca)_2CuO_y$ was crystallized by annealing the amorphous sample between 450°C and 750°C.[8] This suggests that $Bi_2(Sr,Ca)_2CuO_y$ is stable under 750°C. $Bi_2(Sr,Ca)_2CuO_y$ in those tapes was considered to be crystallized under supercooling condition. Figure 4 shows the 80K-phase fraction as a function of the cooling rate at the crystallization. In Fig. 3, the cooling rate determined by a product of the temperature gradient at the growth interface and the growth rate. Temperature gradients at the growth interface heated at 600°C and 750°C were measured as high as 1000°C/cm and 700°C/cm respectively. If the temperature gradient at the growth interface is the only factor which is different between the two heating condition, all tapes must be on the same curve in Fig. 3. In the floating zone fibers grown by Gazit et al., fibers grown slower than 10mm/h were single phase and those grown faster than 20mm/h were multiphase and this growth rate threshold was between 10mm/h and 20mm/h[7]. Temperature gradient at the growth interface was reported higher than 1000°C/cm in this floating zone method.[5] Hence the tapes heated at 600°C also have the growth rate threshold between 10mm/h and 20mm/h, it is considered that a slight difference of the fraction of the 80K-phase was caused by some reasons except the cooling rate.

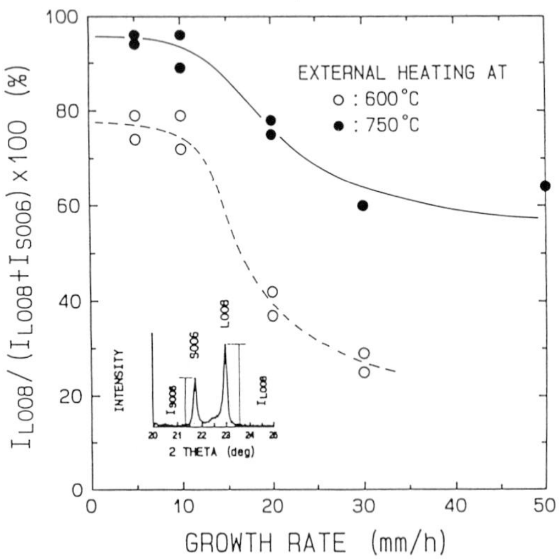

Fig.2. Relationship between the fraction of 80K-phase and the growth rate.

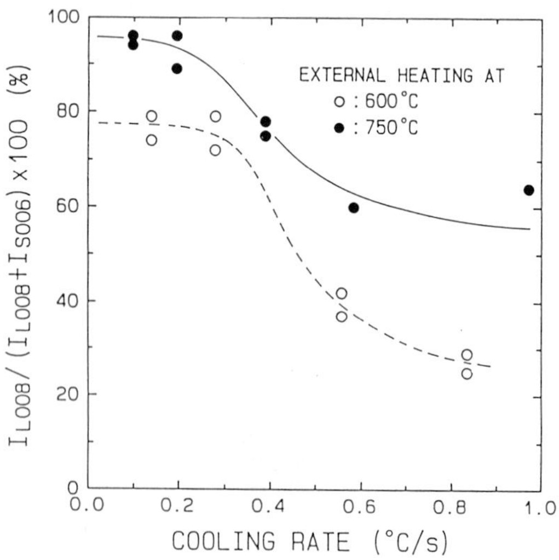

Fig.3. Relationship between the fraction of 80K-phase and the cooling rate.

Figure 4 shows the scanning electron micrographs of cleaved plane parallel to the longitudinal axis of the tapes grown with the heat at 750°C. The tapes grown at 5mm/h and 10mm/h have well aligned long grains parallel to the longitudinal axis of the tape. It is easily observed that the grains of the tape grown at 5mm/h are aligned as their a-b plane is lying parallel to the flat plane of the substrate. In Fig. 4(b), the tape grown at 10mm/h does not exhibit the simple aligned structure as shown in Fig. 4(a), however, we did not find any grain whose c-axis parallel to its longitudinal direction. This indicates that the structure shown in Fig. 4(b) is enough oriented to obtain the high critical current density. In the tape grown at 20mm/h, long flake like grains were found near the surface and the bulky grains are observed in the other part of the tape.
Figure 5 shows the the scanning electron micrographs of the cleaved plane of the tapes heated at 600°C. The grain structures of the tape grown at 5mm/h and 10mm/h are similar to the structure shown in Fig. 4(a) and Fig. 4(b), however, the alignment was not uniform in comparison with the grain structure shown in Fig. 4(a) and Fig. 4(b). No flake like grain was not found in the tape grown at 20mm/h.
Figure 6 shows the grain structure of the tape as thin as 10 μm grown at 5mm/h heated at 600°C was shown in Fig. 6. The well aligned grains were obtained in the thin tape, even in the tape grown at the high temperature gradient. Furthermore, in the tape grown at the low temperature gradient (10mm/h with heat at 750°C), the highly oriented structure was observed even in the tape as thick as 90μm. This structure is shown in Fig. 7. Those results indicate that the thickness of the tape also can be the factor having the highly oriented structure. It is supposed that the difference of the temperature between the surface and substrate can become smaller with being thinner tape, or with increasing the temperature of the substrate heated by the external electric furnace. The thicker and more highly oriented tape would be grown by controlling the both vertical and horizontal temperature gradients in the molten zone.

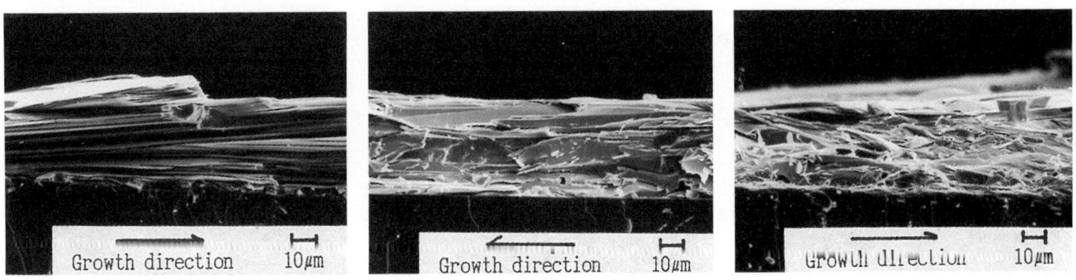

(a)5mm/h (b)10mm/h (c)20mm/h

Fig.4. Scanning electron micrographs of the cleaved plane of the tapes grown with the external heating at 750°C

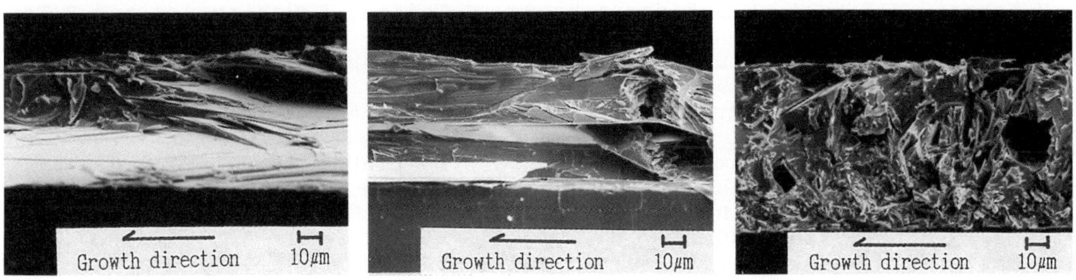

(a)5mm/h (b)10mm/h (c)20mm/h

Fig.5. Scanning electron micrographs of the cleaved plane of the tapes grown with the external heating at 600°C

The critical temperature and the critical current were measured by the four-probe direct current method for the tapes grown at 5mm/h and 10mm/h with the heating at 750°C and followed by heat treatment at 840°C for 100h. The results are summarized in Table 1. Both tapes revealed the similar critical current densities. It was reported that the critical temperature of the 80K-phase could be increased to near 90K by having the appropriate composition and heat treatment.[9] Increasing the critical temperature is considered to be an effective means of increasing the critical current density.

In summary, superconducting tapes about 50μm and 90μm in thickness on the MgO substrate were prepared by the laser zone melting method. The superconductors grown at 5mm/h and 10mm/h consist of the single 80K-phase, and have well aligned grains parallel to the longitudinal axis of the tape. The critical current was measured by the direct current measurement as high as 0.5A (740A/cm^2,77K,0T) for the tape grown at 5mm/h and followed by heat treatment. The tape about 90μm in thickness grown at 10mm/h also had the highly oriented structure. We are pursuing preparation of thicker and more highly oriented tape, which would be grown by controlling the both vertical and horizontal temperature gradients in the molten zone.

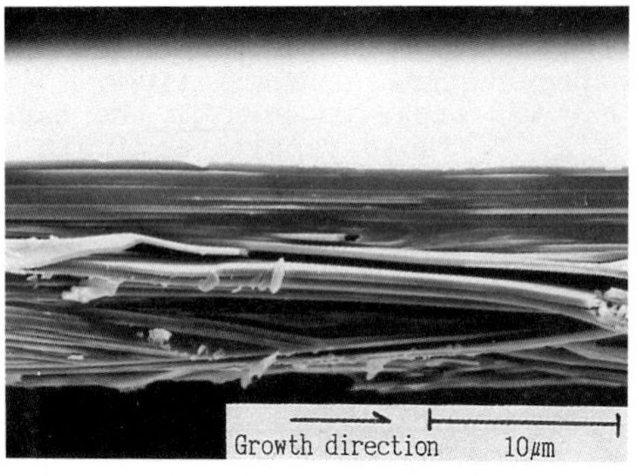

Fig.6 SEM images of the tape 10um in thickness grown at 5mm/h with external heating at 600°C

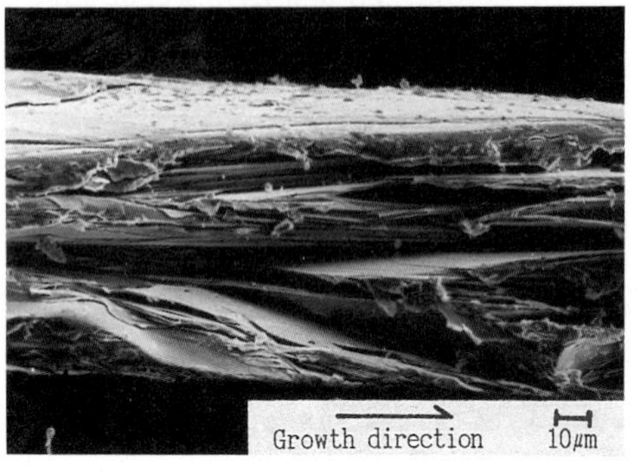

Fig.7 SEM images of the tape 90um in thickness grown at 10mm/h with external heating at 750°C

Table 1. The critical temperature and the critical current

Tape	Size Width (mm)	Thickness (um)	External heating (°C)	Growth rate (mm/h)	Annealing condition	T_c (K)	I_c (A)	J_c (A/cm^2)
A	1.3	52	750	5	840°Cx100h	82	0.50	740
B	1.4	54	750	10	840°Cx100h	85	0.55	730

REFERENCES

1. H.Takagi, H.Eisaki, S.Uchida, A.Maeda, S.Tajima, K.Uchinomura and S.Tanaka (1988) Nature **333**: 17
2. T.Hidaka, M.Oda, M.Suzuki, Y.Maeda, Y.Enomoto and T.Murakami (1989) Jpn.J.Appl.Phys. **27**: L538
3. S.Martin, A.T.Fiory, R.M.Fleming, G.P.Espinosa and A.S.Cooper (1989) Appl.Phys.Lett. **54**: 71
4. S.Nomura,Y.Yamada,T.Yamashita and H.Yoshino (1990) J.Appl.Phys **67**: 547
5. D.Gazit and R.S.Feigelson (1988) J.Cryst.Growth **91**: 318
6. A.Kurosaka, M.Aoyagi, H.Tominaga, O.Fukuda and H.Osanai (1989) Appl.Phys.Lett. **55**: 390
7. D.Gazit, P.N.Peszkin, L.V.Moulton and R.S.Feigelson (1989) J.Cryst.Growth **98**: 545
8. T.Kanai, T.Kumagai, A.Soeta, T.Suzuki, K.Aihara, T.Kamo and S.Matsuda (1988) Jpn.J.Appl.Phys **27**: L1435
9. A.Ono (1989) J.Jpn.Appl.Phys **28**: L1372

Tl-2212 Films Prepared on Flexible YSZ Tape by Ion Beam Sputtering

YUKIO SAITO[1], RYOU NISHIWAKI[1], TOSHIMI MATSUMOTO[1], TOMOICHI KAMO[1], and RYOICHI SHIOBARA[2]

[1] Hitachi Research Laboratory, Hitachi Ltd., 4026 Kuji-cho, Hitachi, Ibaraki, 319-12 Japan
[2] Engineering Research Association for Superconductive Generation Equipment and Materials, Nishi-tenma, Kita-ku, Osaka, 530 Japan

ABSTRACT

$Tl_2Ba_2CaCu_2Ox$(Tl-2212) films were prepared on polycrystalline flexible yttria-stabilized zirconia(YSZ) substrates by high deposition rate ion beam sputtering of Ba_2CaCu_2Ox followed by annealing in Tl vapor. A Tc of 104 K and a Jc of 1.1×10^4 A/cm^2 at 77K in zeromagnetic field were measured. X-ray diffraction showed highly c-axis oriented film structure. The decrese in Jc in a magnetic field was rather small as compared to Ag-sheathed Tl wire. Almost constant Jc was observed up to 7T at 4.2 K.

KEY WORDS: Tl-2212 superconducting film, flexible YSZ substrate, wire

INTRODUCTION

Many of the envisioned applications of high-Tc superconductors are based on the availability of high critical current superconducting wires. Some of the methods to produce such wires have included Ag-sheathed Bi-Sr-Ca-Cu-O[1] and Tl-Ba-Ca-Cu-O[2] superconducting wires, doctor blade tapes[3] and the quench melt growth of bulk materials[4]. We are attempting to produce high-Tc superconducting wires using film fabrication methods because of the higher critical currents obtainable in oriented films. It is also desirable that these "thin-film wires" be flexible. Thin, flexible yttria-stabilized zirconia tape is an ideal substrate for this purpose. Two reports related to the making of superconducting films on flexible YSZ tape substrates have been reported[5,6], but they have been restricted to the Y-Ba-Cu-O system and have required the substrate to be heated during the film deposition.
In this paper we report the fabrication of Tl-2212 thin films on flexible YSZ by high deposition rate ion beam sputtering and post annealing. Tl-2212 was chosen because of it's high critical temperature(110K) and easier producibility than the Tl-2223 phase. The morphorogy, structure and superconducting properties are reported.

EXPERIMENTAL

Tl-2212 films were prepared on polycrystal flexible YSZ substrate by high rate ion beam sputtering followed by annealing in Tl vapor. The sputtering target was a complex oxide of Ba_2CaCu_2Ox which was made by calicining a mixture of BaO CaO and CuO. Flexible YSZ substrates 10mm wide, 20mm long and 0.1mm thick were used. Experimental conditions are shown in Table 1. The sputtering gas was a mixture of Ar and O_2(<10%). The substrate temperature remained below 200℃ during the sputtering. As-deposited films did not contain Tl and were amorphous. In order to form the desired Tl-2212 structure , the films were annealed in Tl vapor and air. The as-deposited films were put into a closed alumina crucible with Tl_2O_3 pellets. The best films were annealed at 830℃ for one hour. Superconducting properties were measured by four-probe method. Composition and crystal structure of the films were investigated by inductively coupled plasma photometry(ICP), energy-dispersive X-ray analysis(EDX), scanning electron microscopy(SEM) and x-ray diffraction(XRD).

RESULTS AND DISCUSSION

ICP and EDX analysis of the as-deposited films showed a stoichiometry close to Ba_2CaCu_2. The morphology of the annealed films was studied by SEM. Superficial and sectional SEM images of an annealed film are shown in Fig.1. Particle-like or platelet-like grains parallel to the substrate can be seen in these images. The grains are 2~10μm in diameter and somewhat poorly connected. The film thickness is about 1 μm so that a deposition rate of 2.4μm/h was achieved which is faster than conventional sputtering. The X-ray diffraction pattern of this film is shown in Fig.2. Strong and sharp periodic peaks can be observed in the pattern. These peaks can be assigned to diffraction from (00l) planes of Tl-2212 phase. This indicates that the film is highly c-axis orieted.
Resistance versus temperature curves for an annealed film is shown in Fig.3. The superconducting transition starts around 110K and reaches zero resistance at 104K. This Tc is comparable to or a little higher than that of the Tl-2212 films prepared on single crystal MgO substrate by sputtering[7] or laser ablation[8].

Table 1. Experimental conditions

Target	:	$Ba_2Ca_1Cu_2Ox$
Substrate	:	flexible polycrystal YSZ
Accelerator voltage	:	800~1200V
Beam current	:	150~200mA
Pressure	:	$1\sim2\times10^{4}$ Torr.
Substrate temperature	:	~100℃

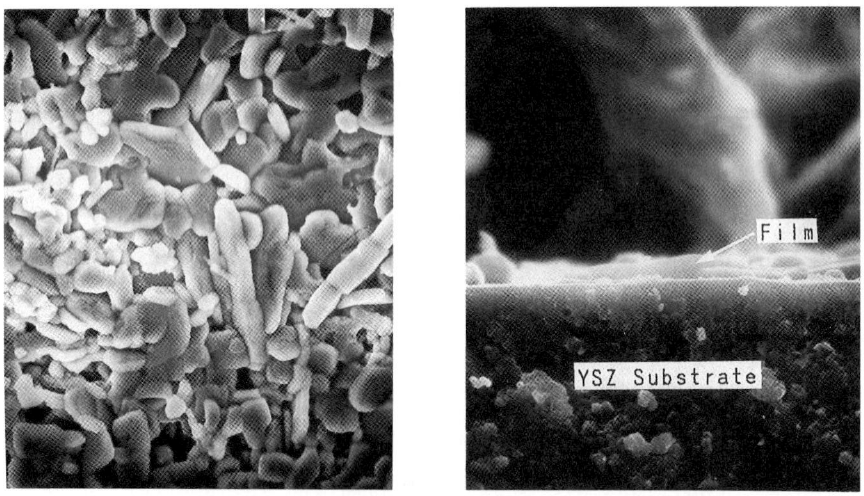

(A) (B)

Fig.1 SEM images of an annealed Tl-2212 film
A : Surface B : Section

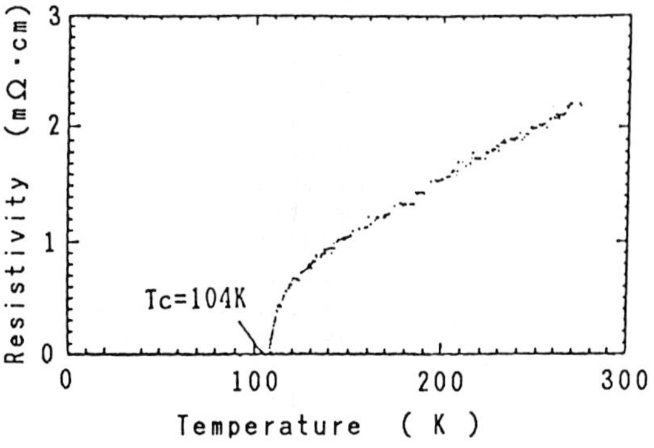

Fig.3 Resistivity vs temperature curve of the annealed film.

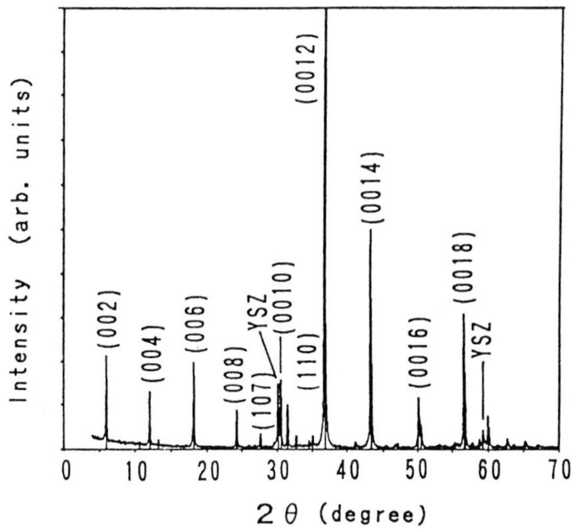

Fig.2 X-ray diffraction patterns of an annealed film showing exclusively the (00ℓ) c-axis oriented peaks.

The transport critical current density(Jc) of this film was 1.1×10^{4}A/cm^{2} at 77K in zero magnetic field. The magnetic field dependence of the Jc at 77K and 4.2K are shown in Fig.4. The magnetic field was applied perpendicular to the current flow and parallel to the a-b plane. A rapid decrease of Jc with increasing B was observed at 77K. But the rate of Jc decrese in magnetic field was less than that for not so highly oriented Ag-sheathed Tl wire. This is due to the high c-axis oriented of the film. Some hysterisis of Jc is also observed. This depends on flux trapping within the grains. Little decrease of Jc in magnetic fild up 7T was observed at 4.2K. At low temperature, weak links in the film are coupled strongly and the super current flow at high magnetic field.

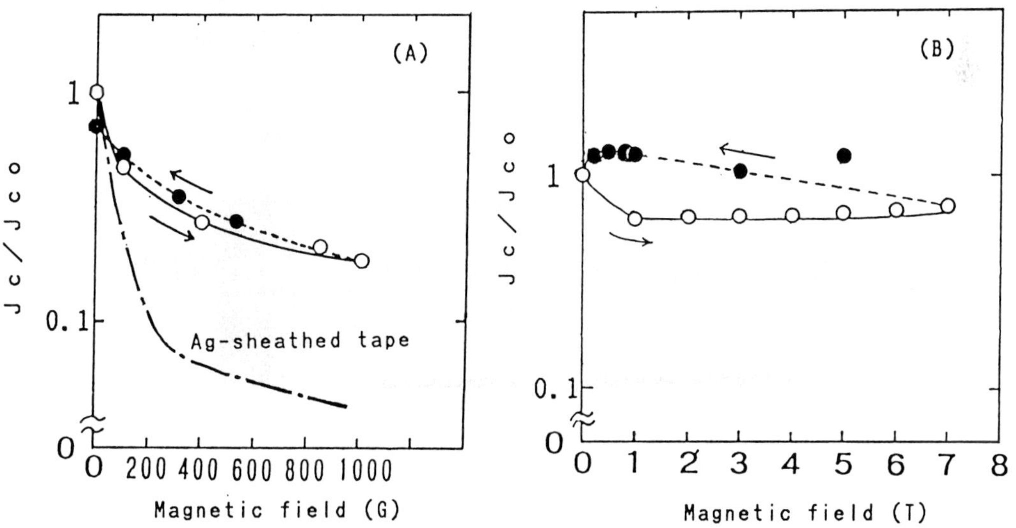

Fig.4 Magnetic field dependence on Jc at 77K(A) and 4.2K(B).

In summary, we have shown that well-oriented c-axis perpendicular Tl-2212 superconducting thin films can be grown by ion beam sputtering and post annealing in Tl vapor on polycrystal flexible YSZ substrates with Tc=104K, Jc=$1.1\times10^4 A/cm^2$ at 77K. Degradation of Jc in a magnetic field is less than for Ag-sheathed Tl superconducting wires at 77K. At 4.2K, Jc is very insensitive to magnetic fields. These results support the concept of making superconducting wires by film fabrication processes.

ACKNOWLEDGEMENT

This work was performed as a part of "R&D on Superconductivity Technology for Electric Power Apparatus" under the Moonlight Project of AIST of MITI, being consigned by NEDO.

REFERENCES

1. M.Mimura, H.Kumakura, K.Togano and H.Maeda, Appl. Phys. Lett., 54(1989)158.
2. M.Okada, R.Nishiwaki, T.Kamo, T.Matsumoto, K.Aihara, S.Matsuda and M.Seido, Jpn. J. Appl.Phys., 27(1988)2345.
3. J.Kase, N.Irisawa, T.Morimoto, K.Togano, H.Kumakura, D.R.Dietderich and H. Maeda, Appl. Phy. Lett., 56(1990)970.
4. M.Murakami, M.Morita, K.Doi, K.Miyamoto and H.Hanada, Jpn. J. Appl. Phys., 28(1989)399.
5. D.P.Norton, D.H.Lowndes, J.B.Budai and D.K.Christen, Appl. Phys. Lett., 57(1990)1164.
6. S.Takano, N.Yoshida, S.Okuda and M.Nagata, Proc.2nd Istec Workshop on superconductivity, Kagoshima, Japan, May,1990, p17.
7. Y.Ichikawa, H.Adachi, K.Setsune, S.Hatta, K.Hirochi, K.Wasa, Appl. Phys. Lett., 53(1988)919.
8. B.Johs, D.Thompson, N.J.Lanno and J.A.Woollam, Appl. Phys. Lett., 54(1989) 1810.
9. M.Okada et al, 1989 MRS Fall Meeting.

Properties of TL-2223 Superconducting Thick Films Formed by Plasma Spraying

Y. YOSHIDA[1], T. KANAI[1], T. KAMO[1], T. MATSUMOTO[2], and R. SHIOBARA[3]

[1] Hitachi Research Laboratory, Hitachi Ltd., Kuji-cho, Hitachi, Ibaraki, 319-12 Japan
[2] Corporate Planning & Development Office, Hitachi Ltd., Kanda-Sugadaira, Chiyoda-ku, Tokyo, 101 Japan
[3] Engineering Research Association for Superconductive Generation Equipment and Materials, Nishi-tenma, Kita-ku, Osaka, 530 Japan

ABSTRACT

We report on the fabrication and properties of superconducting thick films of $Tl_2(Ba_{1.6}Sr_{0.4})Ca_2Cu_3O_x$(Tl-2223)[1] produced by the plasma spray method on Ag substrates. Various post-annealing conditions were investigated. Scanning electron microscope (SEM) and energy dispersive X-ray analysis (EDX) showed the annealed films to be composed primarily of the Tl-2223 phase with a small amount of Ca-Cu-O. The Jc of these films reached 3800A/cm^2 at 77K and 0T.

KEY WORDS ; high-Tc superdonductor , Tl-(Ba/Sr)-Ca-Cu-O , plasma spray , thick films.

INTRODUCTION

Plasma spraying , a mass production processes , is a useful method of forming superconducting thick and large area films.[2]~[4] We previously reported the Tc and Jc values (91K and 1120A/cm^2 (77K,0T) respectively) for $Y_1Ba_2Cu_3O_x$ films produced by plasma spraying.[5] These films were deposited on the Ni-based substrate (Hasterroi X) with a yttria-stabilized zirconia (YSZ) buffer layer. In this study , we tried to make Tl-based superconducting thick films by plasma spray method on Ag substrate which is inactive to Tl.

EXPERIMENTAL

The Tl-2223 superconducting film fabrication process by the plasma spray method is shown in Fig.1. Tl is a toxic material , so Tl-free films which have the nominal composition method $Ba_{1.6}Sr_{0.4}Ca_{2.0}Cu_{3.0}O_x$ were formed on the Ag substrate first. Powders for spraying were prepared by grinding a sintered bulk mixture of high-purity BaO , SrO , CaO and CuO powders at 800 ℃ in air for 10h. The powders were sieved into groups of different particle sizes. The average particle size of 84 μ m was used as feed powder. The powders were then sprayed onto Ag substrates at 300 ℃. The resulting films were about 100 μ m thick and amorphous. These films were then annealed at 830 ℃ in Tl-vapor for 5h-100h to make then superconducting. Several films were prepared using different annealing conditions.

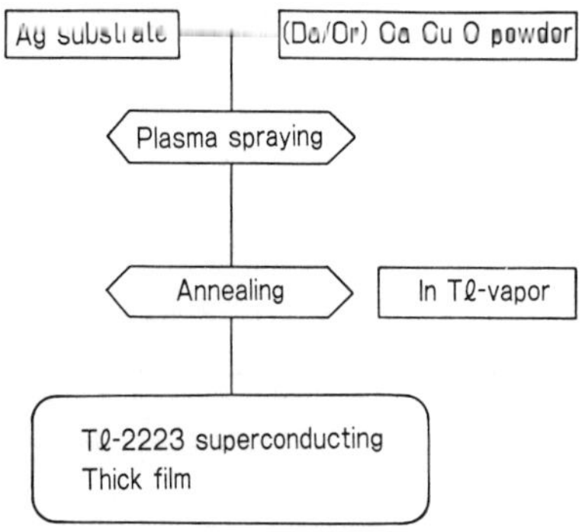

Fig. 1 Tl-2223 thick film formation process by prasma spraying.

The superconducting properties of these films were measured by a standard dc four probe method. The critical current density (Jc) was defined by the 1 μ V/cm criteria . The structure and composition of the films were analyzed by X-ray diffraction and electron dispersive X-ray analysis (EDX). Their microstructure was observed by scanning electron microscope (SEM) and optical microscope.

Results and Discussion

Fig.2 shows the X-ray diffraction patterns of plasma sprayed films annealed at 830 ℃ for 3h(a),10h(b) and 50h(c). Diffraction peaks assigned to the low-Tc Tl-2212 phase , BaO, $BaCO_3$, CaO and CuO were observed in the film annealed for 3h, and no preffred orientation was observed. The high-Tc Tl-2223 phase appeared in the film annealed for 10h. Increasing the annealing time to 50h decreased the Tl-2223 phase and the low-Tc Tl-1223 phase , $BaCuO_2$ and Ca_2CuO_3 appeared. This was due to a loss of Tl from the film by long annealing.

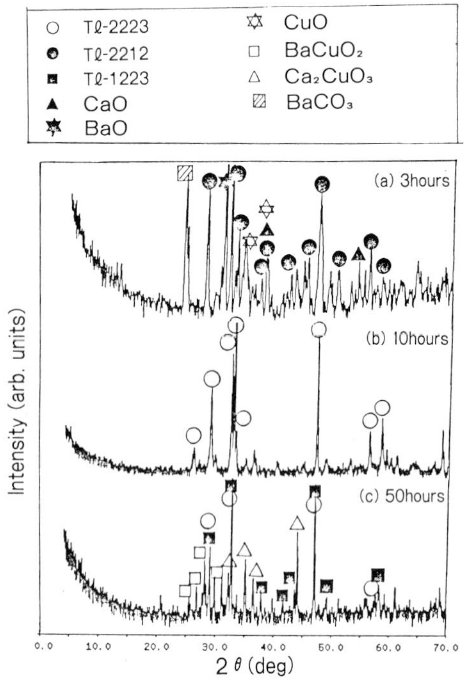

Fig. 2 X-ray diffraction pattern of the films anneaaling at 830℃ for 3h(a), 10h(b) and 50h(c).

Fig.3 shows the SEM micrographs of the three films after polishing. In the film annealed for 3h , the Tl-2212 phase (spot A), (Ca/Tl)-Cu-O (B) and Ca-Cu-O (C) were detected. (Ca/Tl)-Cu-O and Ca-Cu-O phases were not detected by XRD analysis. These nonsuperconducting phases may be amorphous phases. We conclude that 3h is too short a time to crystalize the as-deposited thick films. The film annealed for 10h is composed of Tl-2223 (D) mainly with a small amount of Ca-Cu-O. The Tl-2223 phase decomposed into Ba-Cu-O (a) and Ca-Cu-O (B) after the 50h.

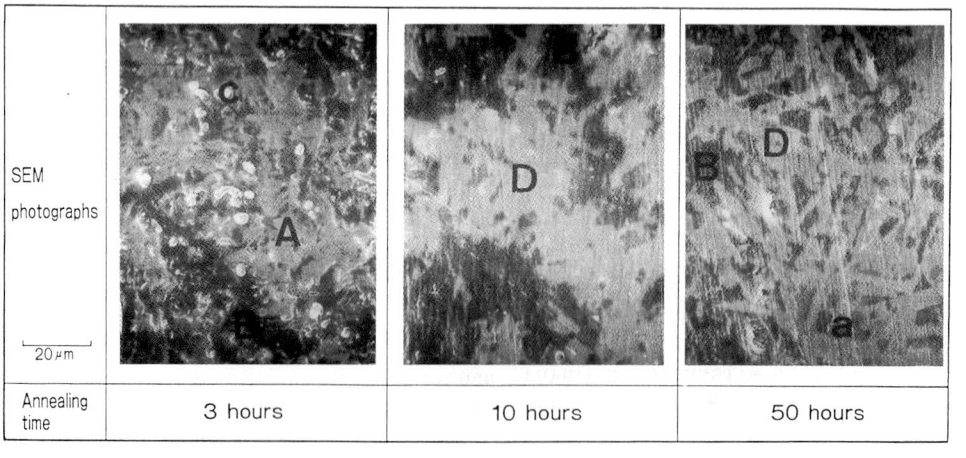

Fig. 3 Scanning electron of the films at 830℃ for 3 to 50h.

Fig.4 shows the annealing time on the critical current density , Jc. The Jc increased dramatically to 3800A/cm^2 by increasing the annealing time from 2h to 10h. This Jc value is the highest reported for a film prepared by the plasma spray method. The Jc decreased gradually with annealing times greater than 10h. This gradual decrease in Jc is caused by decomposition of the Tl-2223 phase as described above. An annealing time of 10h is near optimum for making the Tl-2223 phase.

The magnetic field dependence on Jc at 77K and 4.2K is shown in Fig.5. The magnetic field was applied perpendicular to the current flow . A rapid decrease in Jc is observed up to 200Gauss. Above 200Gauss an almost constant Jc was measured. Hysterisis of Jc was also observed . These are due to weaklink within the grains(6). A Jc of 6700A/cm^2 was measured at 4.2K in zero magnetic field and a rapid decrease in Jc with increasing magnetic field was not observed. A Jc of 5800A/cm^2 was obtained in a 10000Gauss magnetic field. Thus these films show promise for use at temperatures well below 77K.

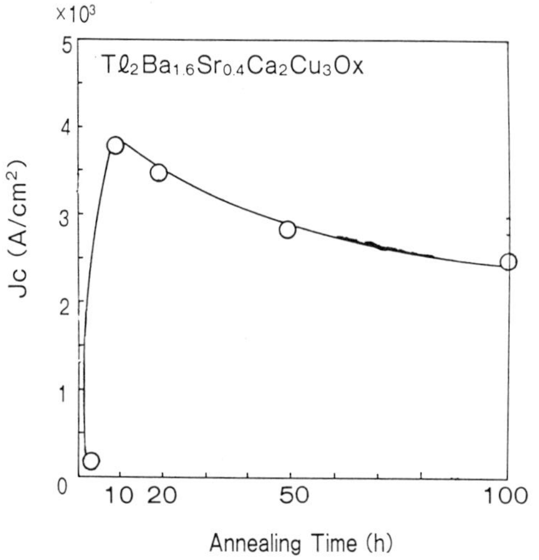

Fig. 4 Annealing time dependence on Jc (77K, 0T)

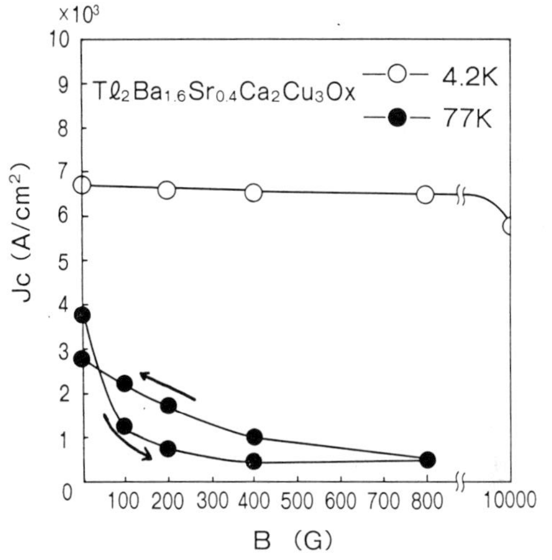

Fig. 5 Jc in magnetic field at 77K and 4.2K. Films were sintered at 830℃ for 10h.

CONCLUSIONS

We prepared thick Tl-2223 films with a Jc of 3800A/cm^2 at 77K and 6700A/cm^2 at 4.2K. The Jc value at 77K is the highest of any film made by the plasma spray method. A large decreased in Jc was observed in a magnetic field at 77K but not at 4.2K. This is due to weak link within the grains. Purification and orientation of the grains should improve the Jc in a magnetic field.

This work was performed as a part of " R & D on Superconductivity Technology for Electric Power Apparatus " under the Moonlight Project of AIST of MITI , being consigned by NEDO.

REFERENCES

(1) A.SOETA , T.SUZUKI , S.TAKEUCHI , T.KAMO , K.USAMI , and S.MATSUDA Jap.J.Appl.Phys. 26,7,(1989)

(2) I.SANAKAWA , T.KONKA , T.MATSUURA , and K.ISHIHARA Jap.J.Appl.Phys. 27,6,(1988)

(3) Y.YOSHIDA , S.KITAHARA , K.TOGANO , M.YUYAMA , K.INOUE , H.MAEDA , and M.FUNAKI Jap.J.Appl.phys. 28,4,(1989)

(4) N.MORI , Y.ITOI , and M.OKUYAMA Jap.J.Appl.phys. 28,2,(1989)

(5) Y.WADAYAMA , T.MATSUMOTO , K.AIHARA , and S.MATSUDA Advances in Superconductivity 2

(6) D.P.NORTON , D.H.LOWNDES , J.B BUDAI , and D.K.CHRISTEN Appl.Phys.Lett ,57 (1990) 1164

Preparation of Superconducting Thick Films of Bi-Pb-Sr-Ca-Cu-O by Gas Deposition of Fine Powder

S. KASHU[1], M. KAITOU[1], K. HATANAKA[2], and C. HAYASHI[2]

[1] Vacuum Metallurigical Co., Ltd., Yokota, Sanbu-machi, Chiba, 289-12 Japan
[2] ULVAC JAPAN Ltd., 2500 Hagisono, Chigasaki, 253 Japan

ABSTRACT

The fine particles with the composition of Bi-Pb-Sr-Cu-O(BPSCCO) were fast sprayed in the form of aerosol upon MgO substrates in air atmosphere through a mozzle, and the resultant fine-particle on the substrates were heat-treated by using forcusing laser beams. The substrates moved at a rate of 1 mm/min to prepare high-temperature superconducting thick films continuously.
Since the fine particles were transferred by air at room temperature,it was heated before the fine particles are sprayed. As to the transition temperatures of BPSCCO thick films deposited and heat-treated on the substrates, Tc(on) and Tc(end) were 112 and 71 K, respectively.

KEY WORDS: gas deposition,fine powder,spot heating

INTRODUCTION

With regard to the preparation of high-temperature superconducting thick films,the authors,in our previous reports, prepared as-deposited films by using ① ultra-fine Y-Ba-Cu-O particles resulting from vaporization of each component element in gas atmosphere and subsequent mixing[1] and ② composition-adjusted BPSCCO fine powder[2][3],and then heat-treated the as-depsited films in an electric furnace to prepare high-temperature superconducting thik films.
Lately, as-grown high- temperature superconducting films have been tentatively prepared in the field of high-temperature superconducting thin films, so the authors tried to prepare a thick films showing high-temperature superconducting characteristics by means of gas deposition. In this process, fine particles with a superconducting composition are deposited on the substrate by the gas deposition method to form a film, followed by spot heating of the deposited film by laser beams. Spot heating is also effective for heat treatment.
The aims of the gas deposition method are ① preparation of as grown high-temperature superconducting thick films, ② local heat treatment by short-time heating, and ③ possible teratment in air atmosphere.

EXPERIMENT

A. Preparation of thick films

Figure 1 shows the experimental method for depositing the BPSCCO particles and the subsequent heat treatment of the resultant films in the gas deposition method. The composition of the BPSCCO fine powder is adjusted in advance to $Bi_{1.4}$-$Pb_{0.6}$-Sr_2-Ca_2-Cu_3-O_x.

The BPSCCO fine powder of 20g is fed to the aerosol chamber, and the dry air is introduced into the chamber at a rate of 0.4 l/min. The fine powder is fluidized to form aerosol with the dry air, and pressurized to 0.24 MPa.

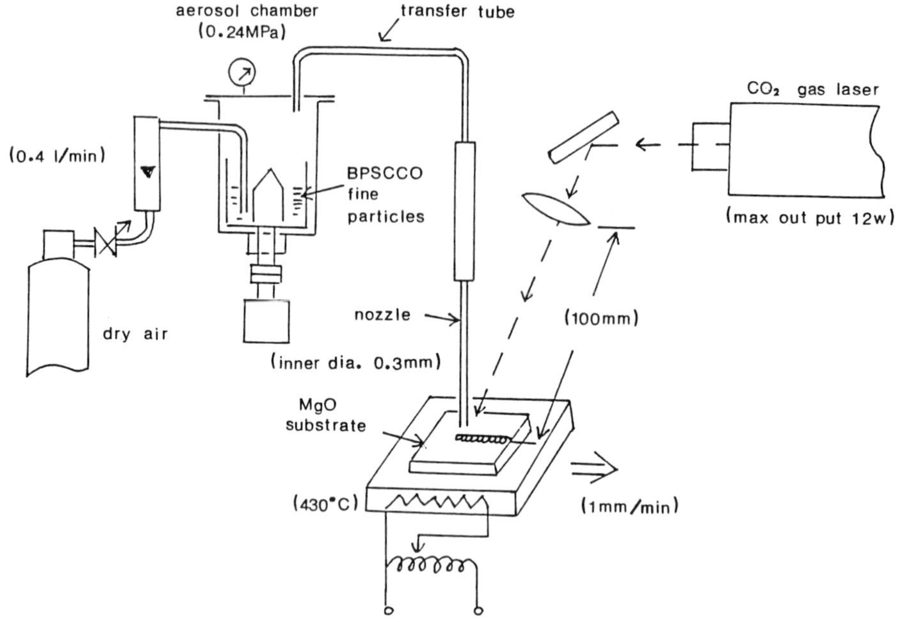

Fig.1 Preparation method of BPSCCO superconducting thick film.

Since deposition onto the substrate takes place under atmospheric pressure (0.1 MPa), the dry air and fine powder are transferred in the state of mixture by a pressure difference of 0.14MPa between the aerosol chamber and on the surface of the substrate, pass through the nozzle tube of 0.3mm in diameter at a rate of 90m/s, and are sprayed upon the MgO substrate, resulting in the deposition of the fine powder. The substrate is heated up to 430° C from underneath, and moves at a rate of 1mm/min toward the arrow-marked direction. Carbon dioxide gas laser is used for spot heating. The length from the focusing lens of the optical system to the heating surface is 100mm with the focusing spot being about 1mm. Because the method shown in Fig.1 sprays room-temperature air upon the to-be-heated zone, the method indicated in Fig.2 was employed for heating air, air flowing through the tube at high speed is directly heated by the electric current flowing through the stainless steel nozzle for heat exchange.

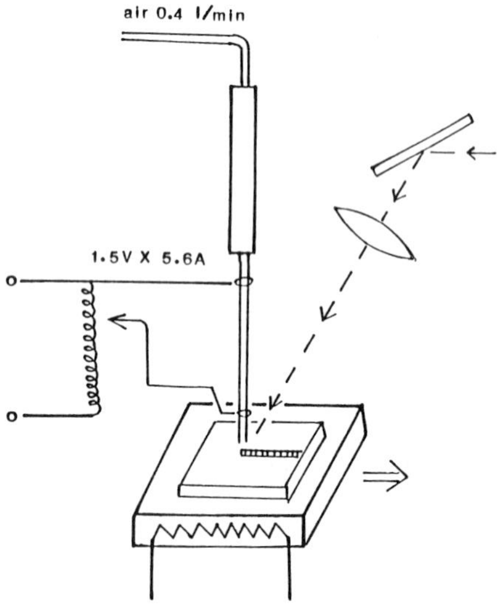

Fig.2 Spray gas heating method

The quantity of heat required for raising 0.4 l/min air from room temperature to 800° C is 1.8 cal/sec (7.5W),so the nozzle was directly heated electrically at 1.5V×5.6A, which give the Joule heat corresponding to the wattage shown above.
The output of laser was 12W at a maximum, and the optimum output was obtained by measuring Tc(end) after heat teratment.

B. Evaluation of thick films

The measurement of resistivity-temperature characteristics and X-ray diffraction were carried out on the superconducting films prepared in accordance with the method described in the preceding section. Also, the grain growth in the film surfaces were observed by SEM.

RESULTS AND DISCUSSION

A. Transition temperatures

In the resistivity-temperature characteristics given in Fig.3, the X axis shows temperatures, whereas the Y axis does ratios of resistivity of the superconducting film at each temperature to that at 300k.The superconducting-initiation temperature Tc(on) is 112K, irrespective of heating theto-be-speayed air. On the other hand, Tc(end)s, which shows complete superconductivity, are 71K for hot air spraying and 58K for roomtemperature air spraying,indicating the effectiveness of heating the to-be-sprayed air.

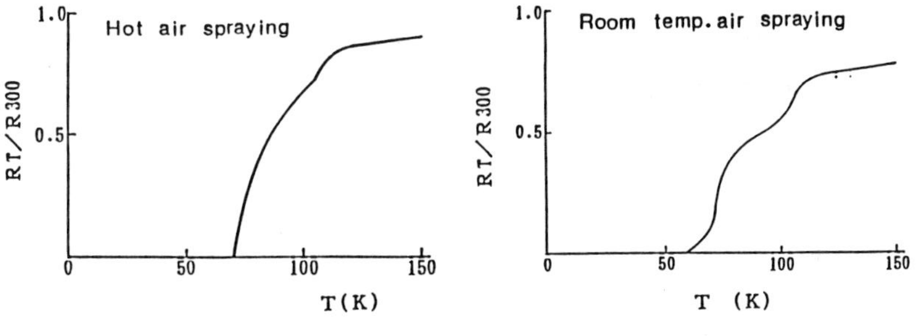

Fig.3 Resistance-temperature characteristic of BPSCCO thick films.

B. X-ray diffraction

Fig.4 gives the X-ray diffraction patterns, showing the minor high-temperature phase peaks and several low-temperature phase peaks.No outstanding impurity phase is observed.

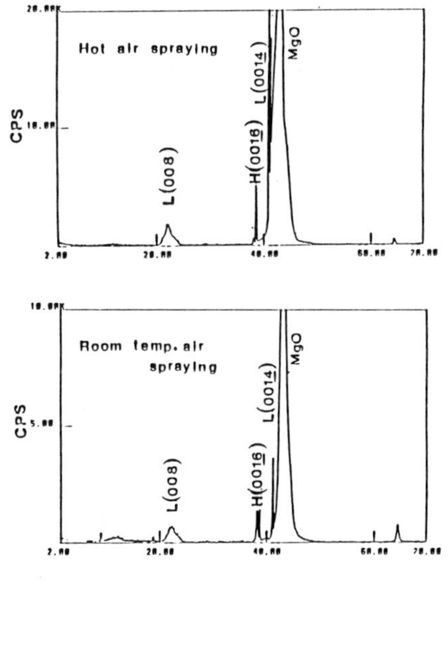

Fig.4 X-ray diffraction pattern of BPSCCO thick films.

C. Microstructures of films

The surfaces of the think films were observed by SEM, and the resultant photographs are shown in Fig.5. The particles in both films grow into neadle crystals, but the particles in the hot-air sprayed film are greater than those in the room-temperature sprayed one, showing the higher effect of heat treatment.

Hot air spraying　　　　Room-temp.air spraying

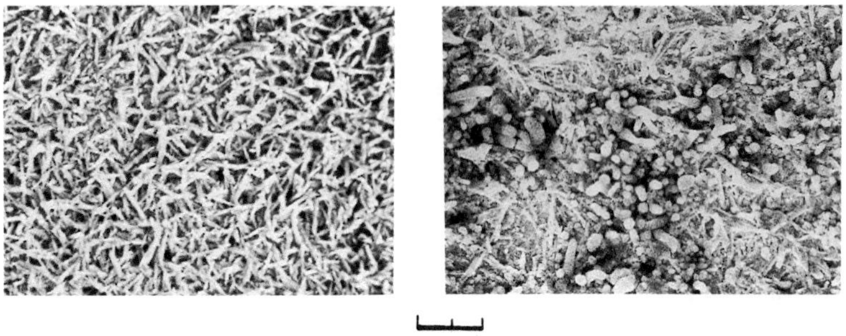

20 μm

Fig.5 SEM photogragh of the surface of BPSCCO thick films.

CONCLUSIONS

1) In the gas deposition method, depositon of the BPSCCO fine powder and spot heating by CO_2 laser were applied to form and heat treat the films simultaneously with the to-be-sprayed air heated. At present, the value of Tc(end) is 71k.
2) The films were formed in air stmosphere, despite conventionally under reduced pressures. The merits are ① no composition change in the BPSCCO as deposited film due to heating and ② simplicity of the method and equipment.

ACKNOWLEDGEMENT

This research is forwarded by the support of High Technology Consortium which is prometed by Research Development Corporation of Japan.

REFERENCES

1) K.Hatanaka, M.Kaitou, M.Umehara, S. Kashu and C.Hayashi, Proc.Ist Intern. Symp. on Superconductivity, 341(1988)
2) S. Kashu, Y.Matsuzaki, M.Kaitou, M.Toyokawa, K.Hatanaka and C.Hayashi, Proc. 2nd Intern. Symp. on Superconductivity, 413(1989)
3) K.Hatanaka, M.Kaitou, Y.Matsuzaki, M.Toyokawa, S. Kashu and C.Hayashi, Proc. 2nd Intern. Symp. on Superconductivity, 1005(1989)

Superconducting YBCO Film on Ag Tape Using ICB Deposition

HISASHI YOSHINO[1], MUTUKI YAMAZAKI[1], HIROYUKI FUKE[1], T.D. THANH[1], KEN ANDO[1], and YOSHIO KOBAYASHI[2]

[1] Advanced Research Laboratory, Research and Development Center, Toshiba Corporation, 1, Komukai, Toshiba-cho, Saiwai-ku, Kawasaki, 210 Japan

[2] Super-GM, 14-10, Nishi Tenman 5-chome, Osaka, 530 Japan

ABSTRACT

A high Jc superconducting tape with Tc of 84K and Jc of 1.5×10^4 A/cm^2 at 77K,0 T was obtained using the ICB deposition method. This tape had no buffer layer and had a high c-axis orientation of YBCO to the Ag tape surface. Increasing the acceleration voltage for the Cu cluster was found to be effective to improve the surface smoothness of the YBCO film. The cross-sectional view of a high-resolution transmission micrograph showed no amorphous layer or precipitates in the boundary between the YBCO film and the Ag substrate.

KEY WORDS: high Jc superconducting tape, Ag substrate, ICB deposition method,

INTRODUCTION

Since the discovery of high-Tc oxide superconductors, many efforts to fabricate high-Tc superconducting wires and tapes have been made for applications to the power electric field[1-4]. A more than 10^5 A/cm^2 Jc will be needed when we consider the practical use of the wire. Moreover, the stability of the superconducting wire against quenching is required to protect the superconducting system. Although Jc values over 10^6 A/cm^2 at 77K have been obtained in thin films of YBCO deposited on oxide single crystals such as $SrTiO_3$[5], the Jc's for wires and tapes are still low.
The authors have been trying to obtain a high Jc superconducting tape directly deposited on a Ag tape without a buffer layer using the ICB (ionized cluster beam) deposition method. It has been clarified that the acceleration for the cluster is effective for the improvement of the c-axis orientation of a YBCO film deposited on a polycrystal Ag substrate[6].
In this paper, the authors investigated the parameters in ICB deposition method to improve Tc and Jc of the superconducting tape. The characterization of the superconducting tape has also been carried out.

EXPERIMENTAL

The details of the method of ICB deposition are described elsewhere [6]. The source materials of Y, Ba, and Cu are vaporized through each nozzle of three refractory crucibles into a vacuum chamber, and clusters are made by adiabatic expansion. Each cluster is ionized positively by electron bombardment and accelerated toward the substrate by an electrode. Activated oxygen plasma generated by rf glow discharge was supplied through a nozzle made of quartz. The estimated pressure near the substrate was the order of 10^{-3} Torr. The temperature of the substrate made of Ag tape (1mm x 0.3mm x 70 mm) was kept about 650°C during deposition by lamp heaters. The deposition rate was about 0.1 nm/sec. After the deposition of 200 nm thick films, they were cooled to 400°C in an oxygen plasma atmosphere at the rate of about 500°C/h. After holding 1 h at 400°C in the same atmosphere, they were cooled to room temperature.
The critical temperature and the critical current density of the obtained tapes were measured by the four-probe method. The distance between the voltage measurement electrodes was about 40 mm. The surface morphology and the cross-section micrograph of the superconducting tapes were observed by SEM and TEM, respectively.

RESULTS AND DISCUSSION

It is considerd that improvements in the surface flatness and the c-axis orientation of the YBCO film are important to increase the Jc of superconducting tapes. So, the authors first studied the effect of the acceleration voltage for the Cu cluster Va(Cu) on the surface morphology of the YBCO films. Va(Cu) was changed between 0 V and 0.5 kV with keeping the acceleration voltage Va(Y) and Va(Ba) at 0.2kV. An improvement in the surface flatness withincreasing Va(Cu) was observed. Typical surface morphologies of YBCO are shown in Fig.1 for the conditions of Va(Cu)=0 and 0.5 kV, respectively. It seems that the enhancement of the migration of atoms on the substrate by increasing the kinetic energy of the Cu cluster is effective in improving the flatness and the density of YBCO films. The substrate temperature was changed between 620℃ and 660℃. The superconducting tapes with Tc's over 80K were obtained at 650℃ and 660℃. But the Tc's decreased when the substrate temperatures were below 630℃.

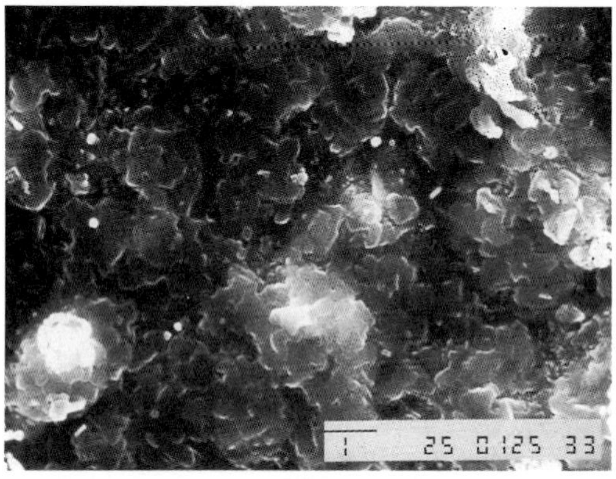

(a)

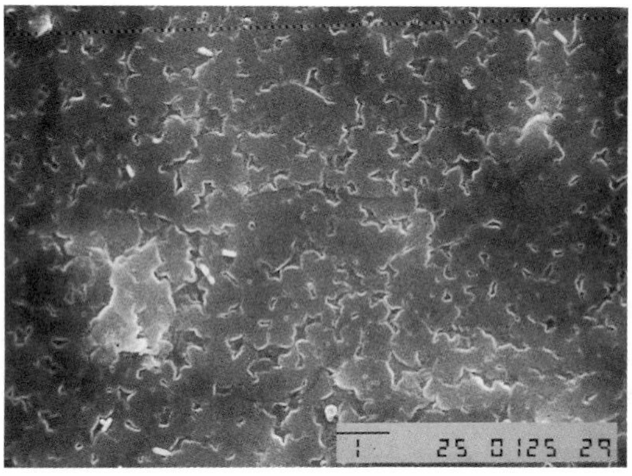

(b) 1μm

Fig.1 Surface morphology of YBCO films deposited on Ag tape by ICB. a) is for Va(Cu)=0 kV and b) is for Va(Cu)=0.5 kV. Va(Y) and Va(Ba) were 0.2 kV for both cases.

The cross-sectional view was observed by a transmission electron microscope (TEM)to clarify the boundary structure between the YBCO film and the Ag substrate. It is difficult to prepare a thin sample for TEM observation because of the brittleness of YBCO and the softness of Ag. The authors prepared the thin films as follows. First, a superconducting tape was fixed between Si wafers using an epoxy adhesive. After slicing with a micro cutter, the samples were dimpled using a dimpler without water. Finally they were thinned slowly by an Ar ion milling apparatus. Figure 2 shows a TEM micrographof the boundary between the YBCO film and the Ag substrate. The Cu-O plane is parallel to the surface of the Ag substrate and the boundary is very sharp. We can not observe any precipitate or an amorphous phase in the YBCO film and in the region of the boundary.

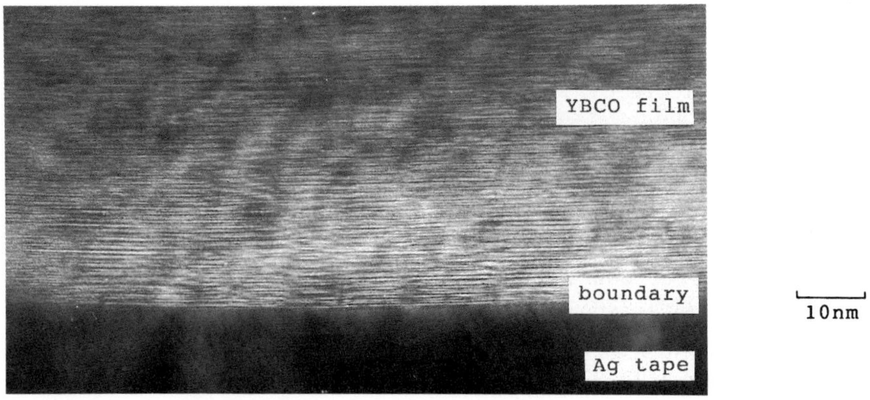

Fig.2 Cross-sectional micrograph observed by TEM for boundary between YBCO film and Ag tape.

It is supposed that the reaction between YBCO and Ag is very small. To clarify the reaction between them further, RBS depth profile analysis was carried out for the samples after annealing at 500-800℃ x 1 h in an oxygen atmosphere. The profile is shown in Fig.3. The profile broadened slightly by increasing the annealing temperature. The calculated diffusion coefficient was small of the order of 10^{-14} cm^2/sec at 800℃.

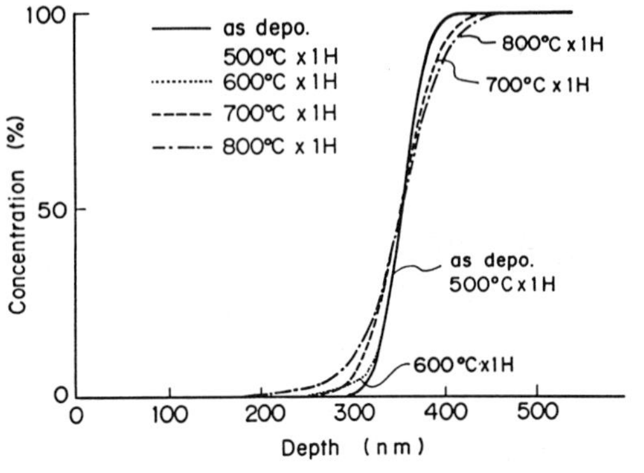

Fig.3 Depth profile of Ag measured by RBS for superconducting tape after annealing at various temperatures.

Figures 4 and 5 show the temperature dependence of resistivity and the current-voltage curve at 77K for the obtained superconducting tape, respectively. The resistivity dropped sharply and the zero resistance Tc was 84.5K. The Jc under the criterion of 0.1 µV/cm was fairly high of $1.5x10^4$ A/cm^2 at 77K and 0 T.

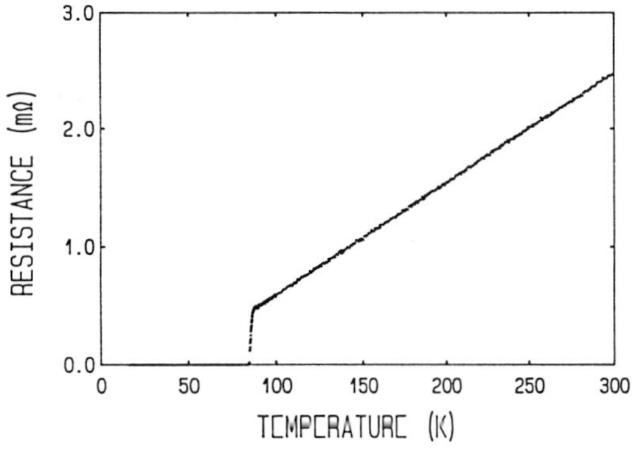

Fig.4 Temperature dependence of resistance for the superconducting tape.

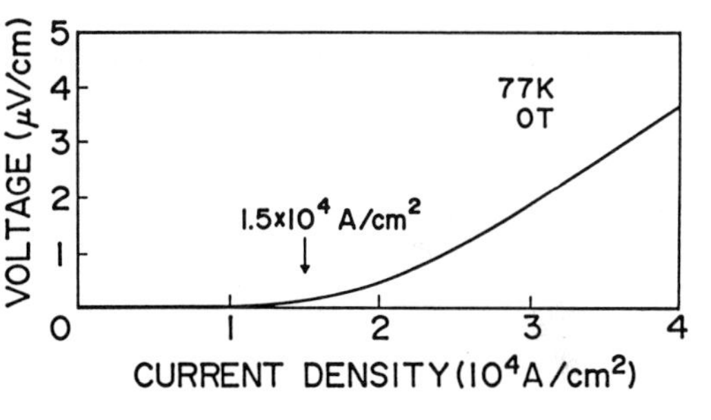

Fig.5 Critical current density for superconducting tape. YBCO film was directry deposited on Ag tape.

The magnetic field dependence of Jc for this superconducting tape was also measured at 77K. The Jc value decreased rapidly with increasing magnetic field, as shown in Fig.6. The degree of decrease in Jc was small when the applied magnetic field was parallel to the tape surface. However, the absolute value of Jc was small compared with that of a YBCO film deposited on $SrTiO_3$. The authors think that the reason is due to the difference in the surface roughness of the Ag tape and the $SrTiO_3$ substrate. Namely, the Ag substrate recrystallized easily during the deposition at 650℃. And a deep grain boundary(the depth was about 1 µm) appeared at the surface of the Ag substrate. Consequently, the superconducting tape had grooves, as shown in Fig.7. When the superconducting tape has many grooves, the value of Jc is suppressed lower. Therefore, the elimination of grooves on the Ag substrate is required in order to obtain a more higher Jc superconducting tape.

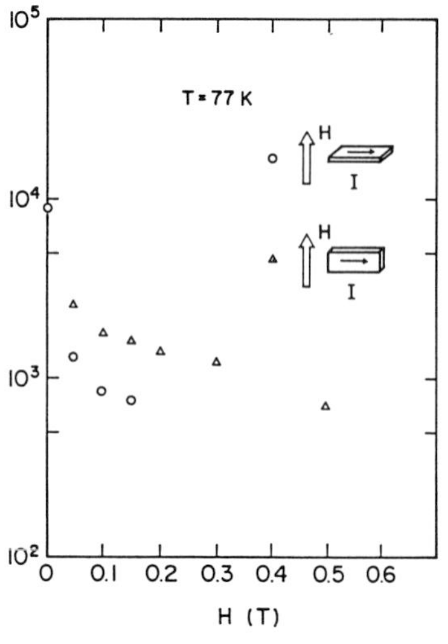

Fig.6 Magnetic field dependence of Jc for superconducting tape.

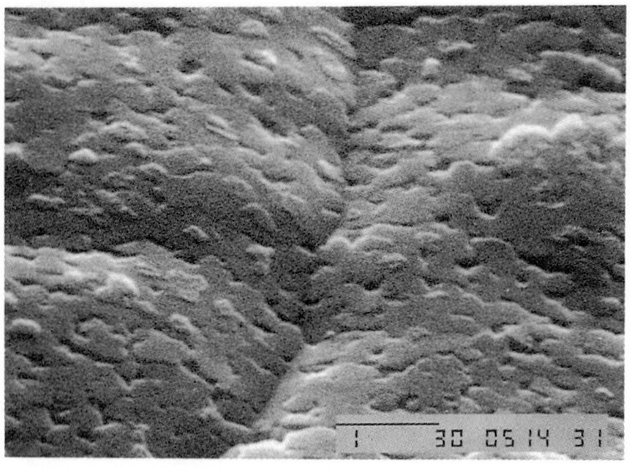

Fig.7 Surface morphology of YBCO film near grain boundary of Ag substrate.

In conclusion, the authors succeeded in preparing high Jc superconducting tapes using the ICB deposition method having Tc of 84K and Jc of $1.5x10^4 A/cm^2$ at 77K, 0 T. The increase in the acceleration voltage for the Cu cluster has been found to be effective in improving the surface smoothness of the YBCO film. The cross-sectionl micrograph observed by TEM showed a high c-axis orientation in spite of the rough surface of the Ag tape. And no amorphous layer nor precipitate were observed in the boundary between the YBCO film and the Ag tape. The grain boundaries on the Ag tape which are induced by recrystallization during deposition is considered to be one of the obstructive factors in obtaining a higher Jc superconducting tape.

This work was performed as a part of R&D Superconducting Technology for Electric Power Apparatuses under the Moonlight Project of Agency of Industrial Science and Technology, MITI, being consigned by New Energy and Industrial Technology Development Organization (NEDO).

REFERENCES

1. Yamada Y, Fukushima N, Nakayama S, Yoshino H, Murase S (1987) Jap.J.Appl. Phys. 26 203
2. Hikata T, Nishikawa T, Mukai H, Sato K, Hitotuyanagi H (1989) Jap.J.Appl. Phys. 28 L1204
3. Saitoh J, Fukutomi M, Tanaka Y, Asano T, Maeda H, Takahara H (1990) Jap.J. Appl. Phys. 29 L1117
4. Kase J, Togano K, Kumakura H, Dietderich D.R, Irisawa N, Morimoto T, Maeda H, (1990) Jap.J.Appl.Phys. 29 L1096
5. Terashima T, Bando Y, Iijima K, Yamamoto K, Hirata K, (1988) Appl.Phys.Lett. 53 (22) 2232
6. Yoshino H, Yamazaki M, Thanh T.D, Yamshita T, Ando K, (1989) Proceedings of ISS'89, Tsukuba

Composition Dependence of Critical Current Density in Ag Sheathed Bi-Pb-Sr-Ca-Cu-O Tapes

SANG-SOO OH, KOZO OSAMURA, and SHOJIRO OCHIAI

Department of Metallurgy, Kyoto University, Sakyo-ku, Kyoto, 606 Japan

ABSTRACT

In order to investigate more specifically about the microstructure change due to the difference of nominal composition in Bi-Pb-Sr-Ca-Cu-O system and its influences on critical transport current density (Jc), we prepared various Ag sheathed tapes by changing the parameter, y in the nominal composition of 1.8 (or 1.6) : 0.4 : (1.6-0.1y) : (2+0.04y) : (2.8+0.0.6y) for Bi:Pb:Sr:Ca:Cu. Jc showed maximum at the composition parameter, y = 0 and decreased with increasing amount of Sr or Ca and Cu. By XRD analysis and SEM observation, it was found that Jc depends sensitively on the amount of impurity phases such as $(Sr,Ca)_3Cu_5O_x$, 11905 phase(2.4 nm with c-axis) and the volume fraction of 2223 phase.

KEY WORDS : Bi-Pb-Sr-Ca-Cu-O superconducting tape, critical current density, composition dependence, impurity phase

INTRODUCTION

It has been known that the " powder-in-tube method " adopted TMT process is a powerful technique for preparing ceramic supeconductors with high critical current density (Jc) in Bi-Pb-Sr-Ca-Cu-O (BPSCCO) system.[1,2] By using this method, microstructure control of Bi-Pb-Sr-Ca-Cu-O superconducting tapes can be achieved by optimizing the TMT condition and combination of cold working processes. Recently, we investigated and reported [1] about optimum condition of preparation of Ag sheathed Bi-Pb-Sr-Ca-Cu-O tapes. On the other hand, it is recognized that microstructure of BPSCCO sensitively depends on the nominal composition for the case of multi-component system like Bi-Pb-Sr-Ca-Cu-O. [3] Not only volume fraction of 2223 phase but the morphology of impurity phase is related with the nominal composition, nevertheless it has not been studied enough for the case of Ag sheathed BPSCCO tapes comparing with the case of BPSCCO bulk.

In this study, we investigated the microstructure change for BPSCCO tapes prepared by changing mole ratio of Sr,Ca,Cu in the nominal composition and its influence on the superconducting properties, mainly critical transport current density at 77.3 K without magnetic field.

EXPERIMENTAL

High purity Bi_2O_3, PbO, $SrCO_3$, $CaCO_3$ and CuO powders were weighed by changing the composition parameter x,y in the nominal composition of $Bi_{(1.6+0.2x)}Pb_{0.4}Sr_{(1.6-0.1y)}Ca_{(2.0+0.04y)}Cu_{(2.8+0.06y)}O_z$. In this paper, we indicated them of x = 0 and 1 as #BI1.6 and #BI1.8 samples under consideration of the mole ratio of Bi. After mixing enough, calcination was performed at 1023 K for 12 h and the oxide powders consisting of 2212 phase and impurity phase were prepared by repeating calcination and pulverization. This powders were filled into Ag pipe and reduced the diameter by swaging and deformed to tape by uniaxial cold press. This tapes were heat-treated firstly at 1108 K for 100 h or 1113 K for 100 h for samples #BI1.6 and #BI1.8, respectively and repressed and heat-treated at same temperature with 1st heat-treatment for 50 h. Critical transport current and transition temperature were measured by 4 terminal method at 77.3 K. Microstructure was observed by SEM and identification of existed phase was performed by powder XRD (Cu kα) analysis and EPMA.

RESULTS AND DISCUSSION

Fig.1 illustrates the composition of samples as a function of mole ratio y for Sr,Ca and Cu elements. As shown in composition map, parameter, y was changed

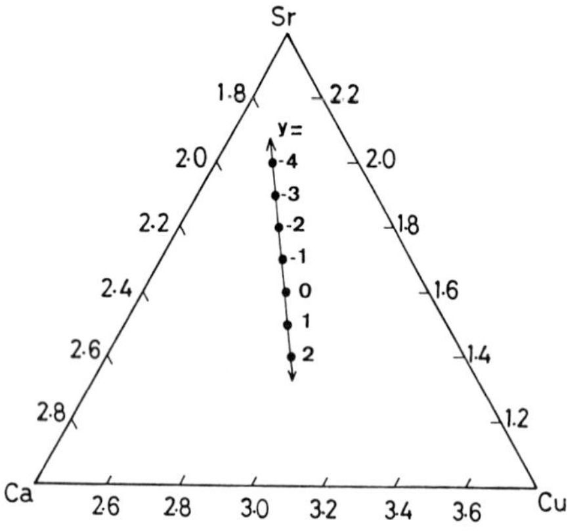

Fig.1 Composition map indicating the mole ratio y for Sr,Ca and Cu elements.

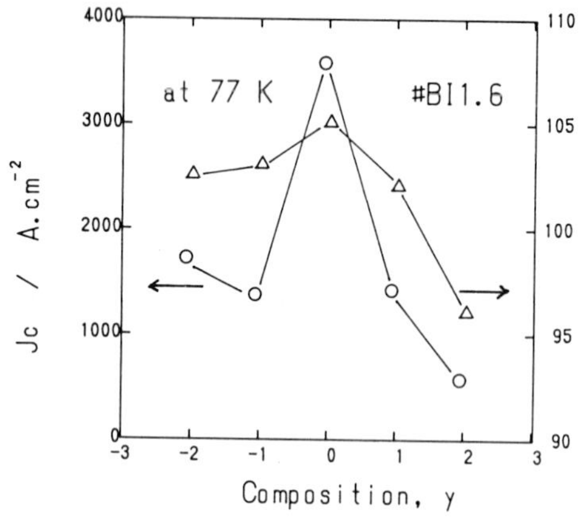

Fig.2 Composition dependence of Jc and Tc for #BI1.6 tapes

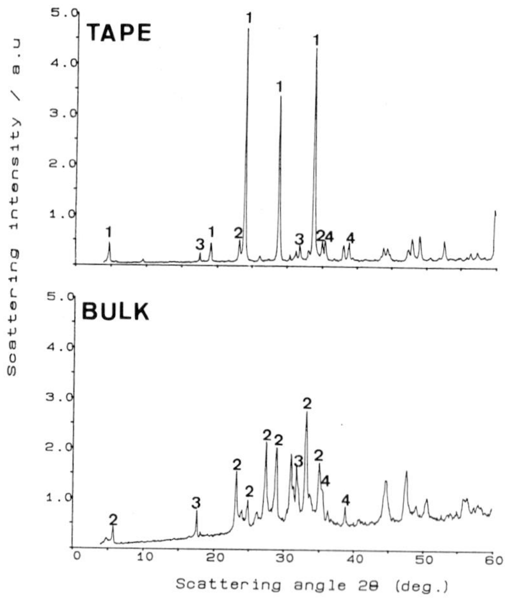

Fig.3 Powder XRD patterns for tape and bulk of #BI1.6(y=0),where 1: 2223, 2:2212, 3:Ca_2PbO_4, 4: CuO

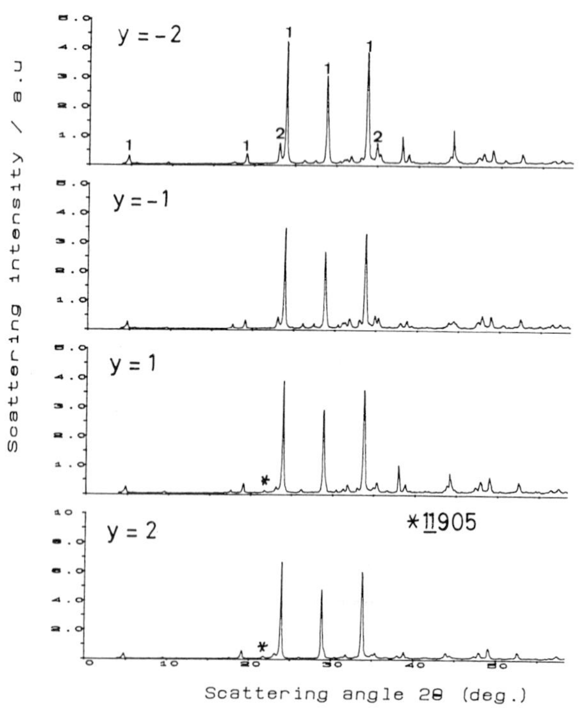

Fig.4 Powder XRD patterns for various tapes of #BI1.6 (y = -2,-1,1,2)

in two directions from y = 0, where Sr increases with decreasing Ca and Cu, and Sr decreases with increasing Ca and Cu. Fig.2 shows the composition dependence of critical transport current density and critical temperature for #BI1.6

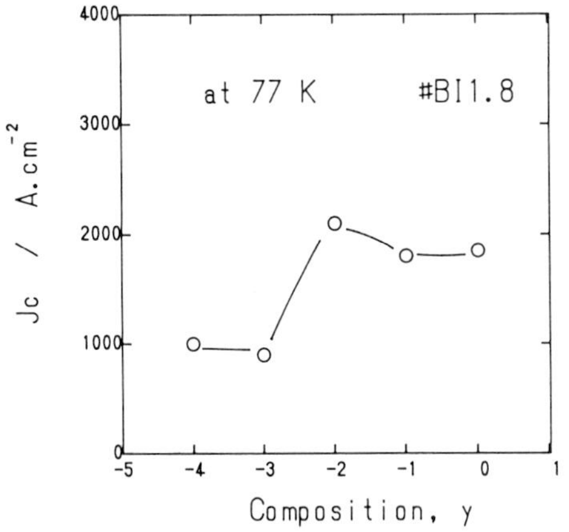

Fig .5 Composition dependence of Jc for the #BI1.8 tapes

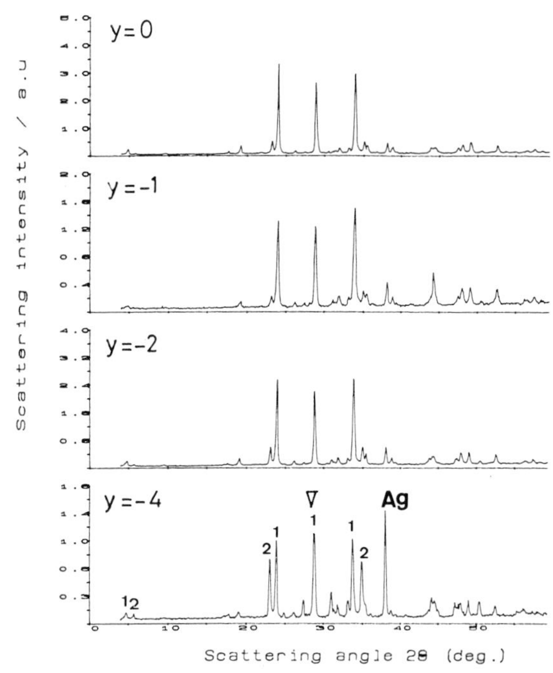

Fig.6 Powder XRD patterns of various tapes for #BI1.8(y=0,-1,-2,-4).

tapes. It was found both Jc and Tc shows maximum at y = 0. They decreased with increasing the mole ratio of Sr or Ca+Cu. It can be known that the degradation of superconducting properties occurred remarkably at compositions where both Ca and Cu are rich.

Fig.3 shows the XRD patterns for #BI1.6 tape and bulk with composition, y = 0. Nevertheless the heat-treatment was same for the two cases, different microstructure was presented due to the difference of production method. It can be confirmed that the 2223 phase with (00L) orientation appeared as a major one for the Ag sheathed BPSCCO tape. In the bulk sample, it was found that the 2212 phase became major one and the volume fraction of impurity phase is larger than the case of Ag sheathed tape. This difference of microstructure is attributed to the difference in degree of cold working or the influence of Ag sheath.

Fig.4 shows the XRD patterns for the various #BI1.6 tapes with different composition parameter, y. In all case, the 2223 phase appeared as a major one. The X-ray intensity of 2212 phase increased with decreasing the the value of y.

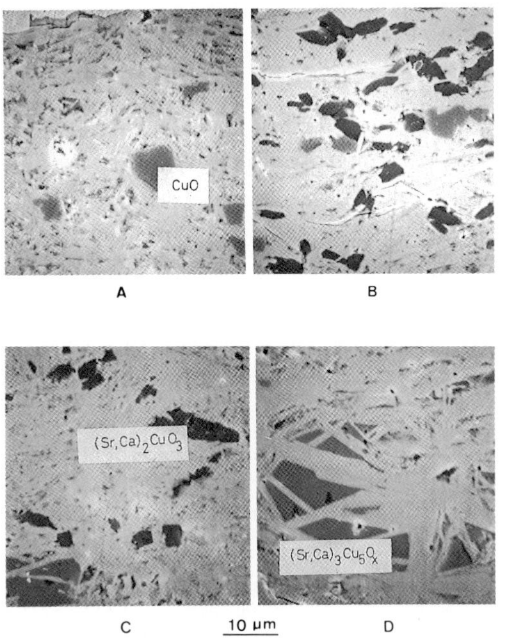

Fig.7 Microstructure for the tapes ,A:#BI1.6 and y=0, B:#BI1.8 and y=0, C:#BI1.8 and y=-2, D:#BI1.8 and y=-4

This means that the volume fraction of 2212 phase increased with increasing amount of Sr. Therefore, the degradation of Jc shown in Fig.3 is thought to be mainly resulted from the decrease of volume fraction in 2223 phase. On the other hand, 11905 phase (c = 2.4 nm), which was found by J.S.Saggio et al. [4]

was detected with a small amount for tapes with y = 1 or 2. This phase was thought to be formed by decomposition of 2223 phase accompanying with partial melting. When the decomposition occurred at grain boundary of 2223 phase, it is suggested that both Jc and Tc decrease by precipitation of 11905 phase in the grain boundary.
Fig.5 shows the composition dependence of critical transport current density for the #BI1.8 tapes. Jc increased little at y = -2 and decreased remarkably at -3 and -4 of y. It was confirmed that the Jc of #BI1.8 tapes with y = 0 was lower than that of the case for the #BI1.6 tape. According to this, It is considered that the Jc of Ag sheathed tape sensitively depends on the Bi mole ratio with respect to the Pb mole ratio like the case of BPSCCO bulk [3].
Fig.6 shows the XRD patterns for the #BI1.8 tapes with various composition parameters. The X-ray intensity of 2212 phase increased continuously with decreasing value of y. The fact that the increasing volume fraction of 2212 phase with increasing mole ratio of Sr is consistent with the result from Fig.3. The peak of 2223 phase indicated by is overlapped with the strongest peak of $(Sr,Ca)_3Cu_5O_x$ and increases relatively for the tape with y = -4.
Fig.7 shows the microstructure of BPSCCO tapes heat treated at 1113 K. It was observed that $(Sr,Ca)_2CuO_x$ phase, which has sharp interface with matrix distributed with large amount for the #BI1.8 (y = 0). Critical transport current density of the #BI1.8 (y = 0) tape is thought to be lower than that of #BI1.6 (y = 0) tape because of an existence of $(Sr,Ca)_2CuO_x$ phase. It was found that the amount of $(Sr,Ca)_3Cu_5O_x$ increased and CuO decreased with increasing the amount of Sr as shown in photographs C and D. Especially, coarse $(Sr,Ca)_3Cu_5O_x$ and superconducting phases were observed in photograph, D ,where the 2223 and 2212 phase could not be distinguished each other.

CONCLUSION

Jc of Ag sheathed BPSCCO tapes shows a maximum at nominal composition of $Bi_{1.6}Pb_{0.4}Sr_{1.6}Ca_{2.0}Cu_{2.8}$. It was found that Jc decreases with increasing mole ratio of Sr or Ca+Cu from this composition. It is attributed to the existence of impurity phases such as $(Sr,Ca)_3Cu_5O_x$ and 11905 phase, as well as the decreasing volume fraction of superconducting phases.

ACKNOWLEDGMENTS

The authors wish to express their gratitude to Messers. I.Nakagawa and T.Unesaki at Kyoto University for their help in the SEM and EPMA studies. This study was made possible by a Scientific Research Grant-in-Aid (Project No. 02226220 from the Ministry of Education, Science and Culture, Japan)

REFERENCES

1. S.-S.Oh, K.Osamura and S.Ochiai : Proc. of 2nd Int. Symp. on Superconductivity (ISS'89) ed. by T.Ishiguro and K.Kajimura (Springer Velag, 1990).
2. K. Sato, T.Mikata, H.Mukai, T.Matsuda, M.Uneyama, H.Hitotsuyanagi, T.Mitsui and M.Kawasima : Proc. of 2nd Int. Symp. on Superconductivity (ISS'89) ed. by T.Ishiguro and K.Kajimura (Springer Velag, 1990).
3. U.Endo, S.Koyama and T.Kawai : Jpn.J.Appl.Phys. 28(1989)L190.
4. J.S.Saggio, K.Sujata, J.Harn, S.J.Hwu, K.R.Poeppelmeier and T.O Maison ; J.Am.Cerami.Soc.72[5] (1989) 849.

Properties of Silver Sheathed Superconducting Wires

HIROYUKI KIKUCHI[1], NAOKI UNO[1], YASUZO TANAKA[1], TSUKUSHI HARA[1], KIYOSHI OKANIWA[2], and TAKAHIKO YAMAMOTO[2]

[1] Yokohama R&D Laboratories, The Furukawa Electric Co., Ltd., 4-3, Okano 2-chome, Nishi-ku, Yokohama, 220 Japan
[2] Engineering Research Center, Tokyo Electric Power Co., 4-1, Nishi-tsutsujigaoka 2-chome, Chofu, Tokyo, 182 Japan

ABSTRACT

$Bi_2Sr_2CaCu_2Ox$ silver sheathed superconducting wires were prepared by the powder in tube technique, and transport critical current density (Jc) of 10^4 (A/cm^2) at 77(K), 0(T) has been obtained with 10 cm length. The long wires have the small variations in Jc, cross sectional areas of the oxide and the degree of crystal orientation (F-value) as to the longitudinal direction. This homogeneous Jc is discussed by the Jc-distribution estimated from the current-voltage curves in each wire.

KEY WORDS: $Bi_2Sr_2CaCU_2Ox$, silver sheathed wire, critical current density, super-normal transition

INTRODUCTION

Transport critical current density (Jc), in a series of silver sheathed BiSrCaCuO superconducting wires, was improved more than 10^4 (A/cm^2) as a result of refinements for grain coupling, alignment and crystal orientation. The maximum Jc of $3.5x10^4$ (A/cm^2) in $Bi_2Sr_2CaCu_2Ox$ at 77 (K) 0(T), was obtained [1]. In this case, Jc measurements have been carried out for a specimen with a few cm long. However, it was necessary for the wires to be lengthened and to be uniform in properties to the longitudinal direction for applications such as cables, the other conductors.

For the conventional multi-filament type long wires, Jc-distributions obtained by d^2V/dI^2 from current-voltage curves were found to be a dominant factor to determine the quality of wires[2]. In this paper, we reported some properties for long wires, moreover, and showed that the d^2V/dI^2 analysis was also an available method for the determination of the quality of the oxide superconducting long wires by introducing the Jc-distributions.

EXPERIMENTAL

Superconducting wires were prepared by the conventional powder in tube method as shown in Figure 1. Bi_2O_3, $SrCO_3$, $CaCO_3$ and CuO were used as starting pow-

Fig.1 Sample preparation

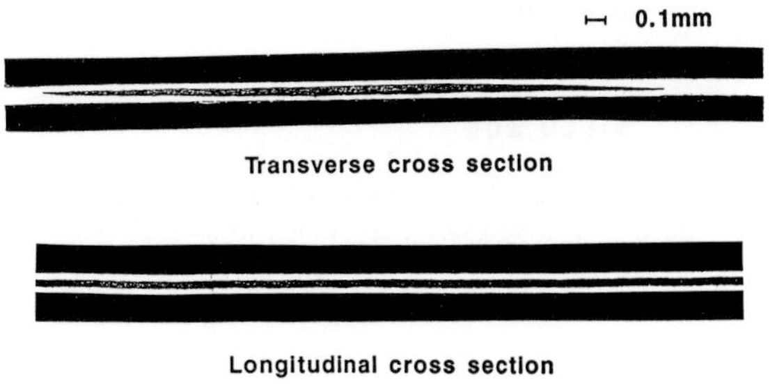

Fig.2 Cross section of a Ag-sheathed BSCCO superconducting wire

ders. The powders were mixed and calcined at 800-850 C in oxygen atmosphere to form the $Bi_2Sr_2CaCu_2Ox$ phase. The $Bi_2Sr_2CaCu_2Ox$ powder was pressurized to a higher dense cylinder by cold isostatic pressing (CIP) at 300MPa. The cylinders were packed into silver tubes of 7(mm) outer and 4(mm) inner diameter, and cold worked into tape wires with width of 3-4(mm) and thickness of 0.08-0.15(mm). The wires with 10(cm) length were cut from the long wire, and heat treated at 800-900(°C) up to 50(H). The transport critical current Ic at 77(K), 0(T) was measured by a standard four point resistive measurement technique, using the criterions, which were electric field criterion of 1(μV/cm) and resistive criterions of 10^{-10} (ohm.cm) and 10^{-12} (ohm.cm). Jc was calculated by Ic and cross sectional area of the superconducting wire. The distance between two adjacent voltage contacts was 5-20(mm), which were placed at equal interval the tape wire. The microstructure was evaluated by scanning electron microscope (SEM), and the degree of crystal orientation (F-value) was obtained by X-ray diffraction analysis as follows,

$$F=(Po-Poo)/(1-Poo)$$
$$P= I(001)/ I(hkl)$$

where, Poo : X-ray intensity ratio for random sample
Po : X-ray intensity ratio for oriented sample

Jc - distribution {f(J)}of the long wire was evaluated by the current-voltage (I-V) analysis as follows,

$$V(J)=A\int(J-J')\cdot F(J')\cdot dJ'$$
$$d^2V/dI^2=Af(J)$$

where, J is the current density, J' is the local critical current density, F(J') is the distribution of critical current in the wire, A is a factor describing the flux flow.

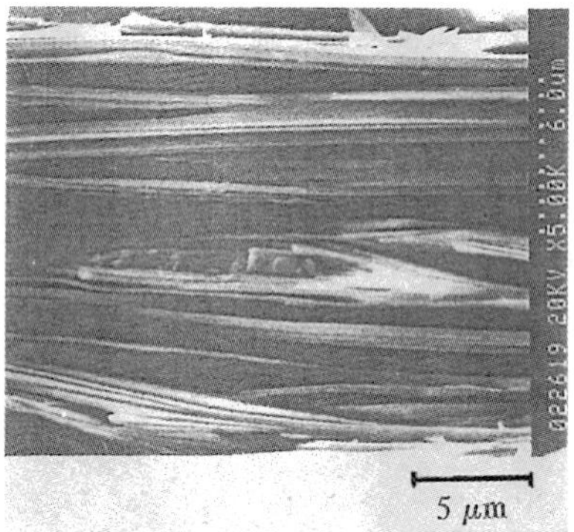

Fig.3 SEM photograph of the fractured cross section inside Ag sheathed BSCCO wire

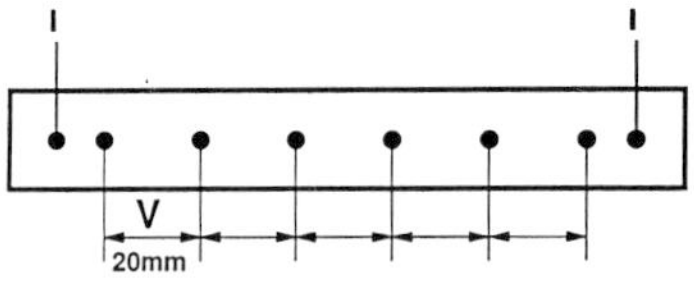

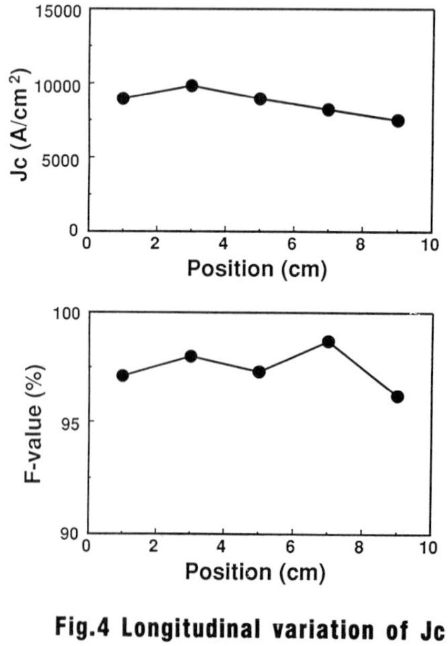

Fig.4 Longitudinal variation of Jc and F-value for BSCCO wire

RESULT AND DISCUSSION

Figure 2 shows, that the changes in the thickness of BSCCO oxide was found along the transverse direction, but that was very small along the longitudinal direction. The scanning electron microscope (SEM) photograph as shown in Figure 3 indicated the fractured cross section of highly oriented grain. The specimen was composed of laminar grains aligned in the same direction. Figure 4 shows variations of Jc and the degree of crystal orientation (F-value) along the longitudinal direction of the tape wire with 10(cm) length. These variation were very small, which of F-value were higher than 96(%), Jc were about 10^4 (A/cm^2) were obtained in the same position. Figure 5 shows the Jc as a function of the distance between the voltage contacts and Jc with different criterions also. Jc decreased from 8.5×10^3 (A/cm^2) to 7.3×10^3 (A/cm^2) with the distance, but Jc was determined by the lowest Jc value in the distance. Jc with a resistivity criterion of 10^{-10} (ohm.cm) was similar to that with a criterion of 1(μV/cm). But Jc in case of a resistivity

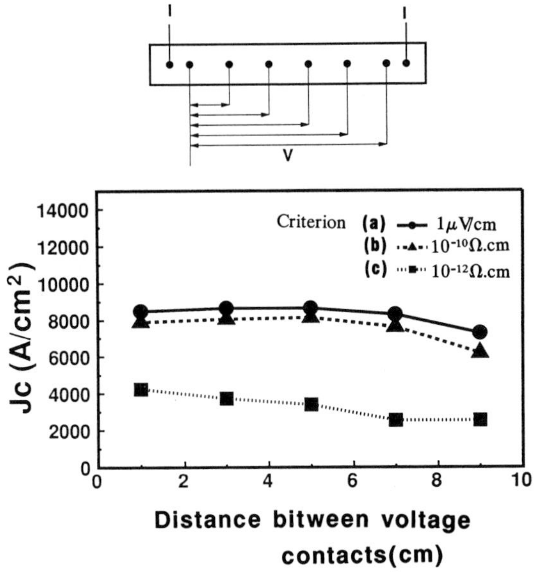

Fig.5 Jc as a function of the distance between voltage contacts

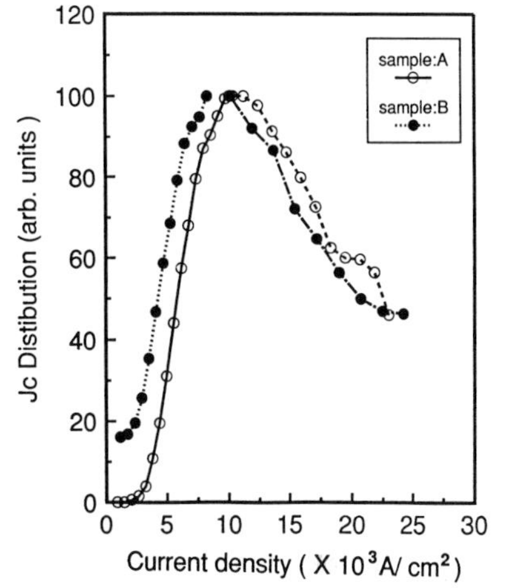

Fig.6

The Jc distribution for BSCCO wire

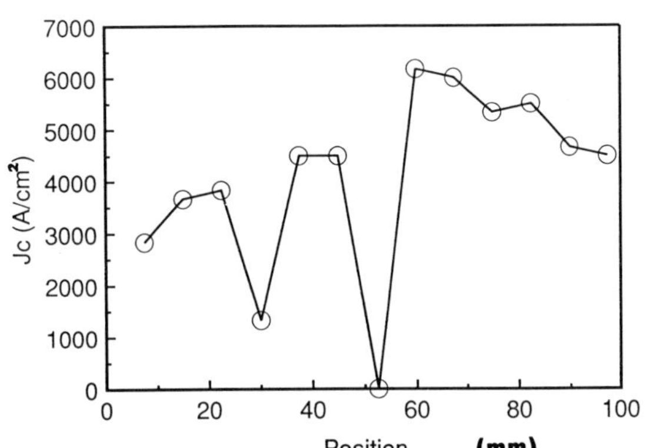

Fig.7

Longitudinal variation of Jc for BSCCO wire (Sample B)

criterion of 10^{-12} (ohm.cm) was decreasing to $3x10^3$-$4x10^3$ (A/cm^2). It is caused by small n-value, which has been previously reported by the other paper. [3]

Figure 6 shows the Jc-distributions calculated by the d^2V/dI^2 applied to sample A and B of different types. Jc in the tape wire A does not change greatly along the longitudinal direction as shown in Figure 4, Jc in the tape wire B changes greatly as shown in Figure 7. The tape wire B has some cracks near the center in a wire with 10cm length. In the case of B the Jc-distribution shifts to the direction in low Jc. Thus, the analysis of d^2V/dI^2 can reveal the Jc-distribution in long wire.

CONCLUSIONS

$Bi_2Sr_2CaCu_2Ox$ silver sheathed superconducting wires were prepared by powder in tube technique, and Jc of about 10^4 (A/cm^2) at 77(K), 0 (T) has been obtained for 10 (cm) long wires. The long wires have small variations in Jc, cross sectional area of the HTSC oxide and the degree of c-axis orientation (F-value) along the longitudinal direction. The analysis of d^2V/dI^2 can reveal the Jc-distribution using two voltage contacts in long wire..

REFERENCES

1) Enomoto,N., Kikuchi,H., Mimura,M., Nakajima,M., Uno,N., Hara,T., Okaniwa,K, and Yamamoto,T. Proceeding of the 2nd International Symposium on Superconductivity (ISS,89) November 14-17, 1989, Tsukuba

2) W.H.Warnes and D.C.Larbalestier Cryogenics 1986, Vol 26, December

3) Uno,N., Enomoto,N., Kikuchi,H. and Matsumoto,K,. Presented at Symposium on Superconductivity, November 14-17, 1989 Tsukuba, Ibaraki Japan.

Electromagnetic Properties and Detailed Preparation Methods of Ag-sheathed Bi-Pb-Sr-Ca-Cu-O Superconducting Tapes

KEISUKE YAMAMOTO, JUNICHI KAI, and MAKOTO HIRAOKA

Central Research Laboratory, Mitsubishi Cable Industries, Ltd., 8, Nishino-cho, Higashimukaijima, Amagasaki, Hyogo, 660 Japan

ABSTRACT

We conducted researches on Bi-Pb-Sr-Ca-Cu-O powders, for preparing Ag-sheathed superconducting tapes with high critical current density (Jc) at 77.3 K. Three different powder compositions ($Bi_8Pb_2Sr_{10}Ca_{10}Cu_{15}O_x$, $Bi_9Pb_2Sr_{10}Ca_{11}Cu_{15}O_x$, and $Bi_{10}Pb_2Sr_{10}Ca_{10}Cu_{15}O_x$) were investigated. With these powders, superconducting tapes about 0.2 mm in thickness were prepared by powder-in-tube method and the process combining sintering and intermediate uniaxial pressing. The highest Jc was 19,500 A/cm^2(77.3 K,0 T) on the specimen using $Bi_9Pb_2Sr_{10}Ca_{11}Cu_{15}O_x$. Influence of powder compositions on electromagnetic properties such as Jc and A.C. susceptibility in various magnetic fields are also reported.

KEY WORDS: Bi-Pb-Sr-Ca-Cu-O, Ag-sheathed superconducting tape, Jc, A.C. susceptibility

INTRODUCTION

In fabricating long wires of oxide superconductors with high transition temperatures (Tc), winding the superconductors into coils and stabilizing the superconductivity , it is advantageous to make the superconductors in the form of metal-clad tapes, films and multifilaments. In addition, for practical applications such as superconducting cables operated in liquid nitrogen, it is necessary to improve the Jc at 77.3 K. For these purposes, we are developing the processing of Ag-sheathed $Bi_2Sr_2Ca_2Cu_3O_x$ (2223 phase)[1] superconducting tapes. In this paper, the experimental results of influence of Bi-Pb-Sr-Ca-Cu-0 powder compositions[2] on electromagnetic properties are reported and the problem of weak links such as Josephson junction of intergrain is discussed.

EXPERIMENTAL

Bi_2O_3,PbO,$SrCO_3$ and CuO powders were mixed in atomic proportion of Bi:Pb:Sr:Ca:Cu=A, B and C (A=8:2:10:10:15, B=9:2:10:11:15, C=10:2:10:10:15) and calcined at 780°C for 30 hours in air. The calcined powders were packed into silicone-rubber molds and pressed by a CIP (Cold Isostatic Pressing) equipment. By the pressing, green rods 4.9 mm in diameter and 100 mm in length were obtained. The rods were then inserted into silver tubes 7 mm in outer diameter, 5 mm in inner diameter and 330 mm in length. Both ends of the tubes were filled by Pb-Sn solder. The composite tubes were cold-drawn and cold-rolled, and tapes about 10 m in length, 0.2 mm in thickness and 2.6 mm in width were obtained. These tapes were cut in 4 cm-long, and subjected to the combined processes[3][4] of sintering (S) at 830 $^{\circ}$C for 40-160 hours in air and intermediate uniaxial pressing (P) at 1.7×10^3 MPa.

The transport critical current (Ic) at 77.3 K in external magnetic fields varying from 0 to 4.5×10^{-2} T was measured by standard D.C. four-probe method and determined with 1μV/cm criterion. Jc was calculated on the cross-sectional area of the tapes excluding silver sheath. The A.C. susceptibility was measured at four different values of applied A.C. field (H') amplitude between 0.1 Oe and 12.5 Oe for a frequency of 313 Hz. Microstructures of the specimens were investigated by a scanning electron microscope (SEM) and an energy dispersive spectroscope (EDS). Identification of the superconducting phase was carried out using the X-ray diffractometer (XRD).

RESULTS AND DISCUSSION

Figure 1 shows the effect of intermediate uniaxial pressing on the Jc(77.3 K,0 T) of the specimens for three different powder compositions. It is shown that Jc increased as increasing number of the pressing. While this effect was already reported by K.Osamura et al.[5], it is found further that the specimens using powder B more clearly show the effect of the pressing than those using powders A and C. Therefore the difference in powder compositions has influence on the critical current density. Then we investigated the reason why high critical current density was obtained by powder B.

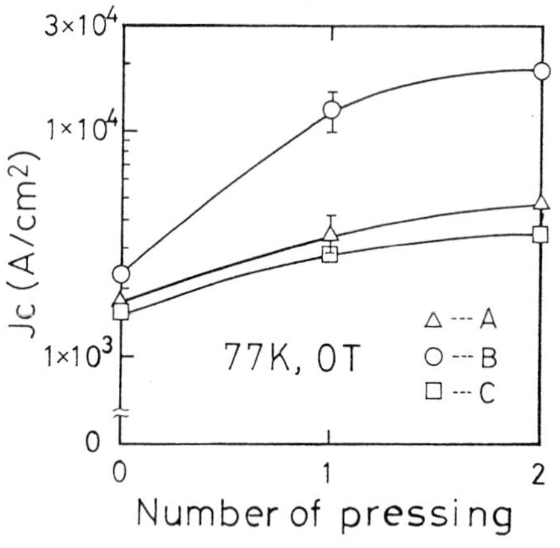

Fig.1 Effect of intermediate uniaxial pressing on Jc(77.3 K, 0 T)

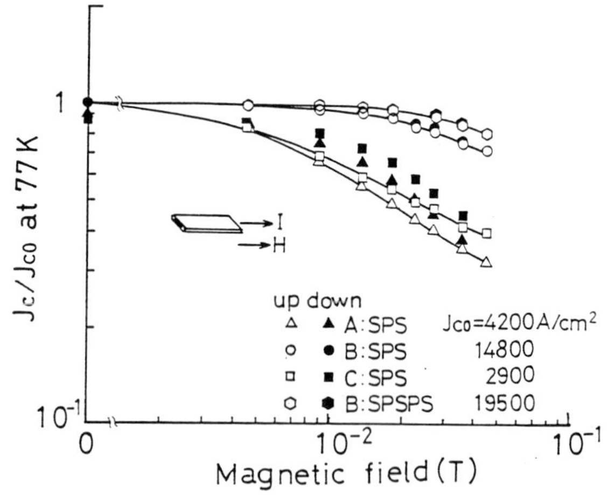

Fig.2 Magnetic fields dependence of Jc at 77.3 K

Figure 2 shows low magnetic field dependence of Jc at 77.3 K for the specimens using three different powders, obtained in D.C. magnetic fields up to 4.5×10^{-2} T were applied parallel to the tape surface and the current flow, where SPS denotes one intermediate pressing between sintering, and SPSPS denotes twice intermediate pressing. In the SPS condition, it was found that hysteresis of Jc-B curves for the specimens using powder B was smaller than these for the specimens using powders A and C. Osamura et al.[5] have reported that in the case of Ag-sheathed Bi-Pb-Sr-Ca-Cu-O tapes, it might be suggested that a much higher Jc can be achieved and the hysteresis greatly reduced by eliminating weak-link parts and improving weak pinning sites. Therefore it is suggested that the weak-link parts were almost eliminated in the specimens prepared by using powder B and intermediate pressing.

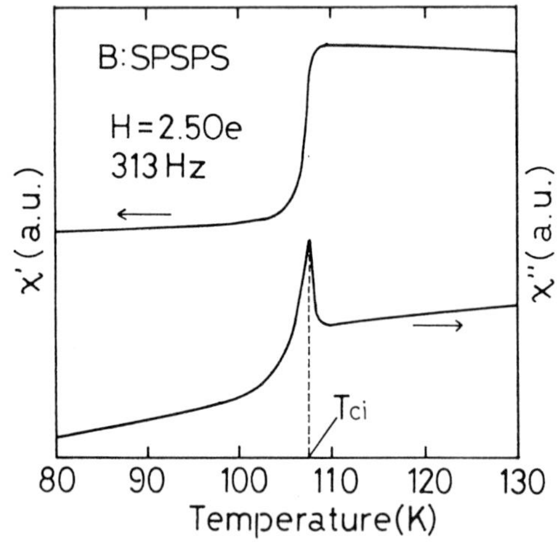

Fig. 3 Temperature dependence of A.C susceptibility, X' and X" denote real parts and imaginary parts, respectively

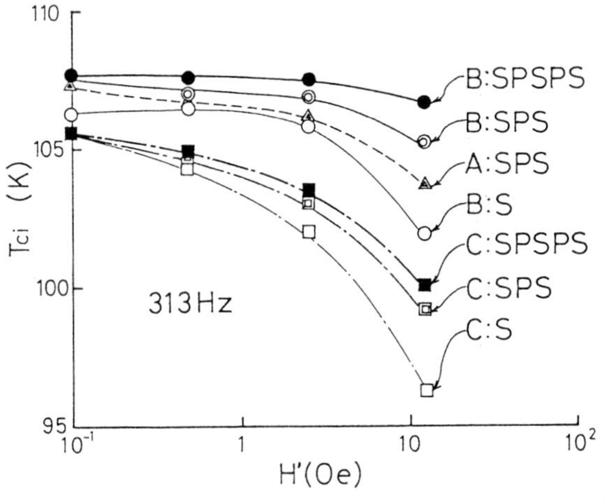

Fig. 4 Applied fields (H') dependence of peak temperature (Tci) of X"

Figure 3 shows a typical result of A.C. susceptibility measurement, i.e., the peak temperature (Tci) of imaginary parts of complex magnetic susceptibility. A.C. applied fields (H') dependence of Tci for the specimens prepared by various conditions are shown in figure 4. It is found that Tci shifts to lower temperatures as increasing H', but the shift in the specimens using powder B is smaller. This behavior suggests that improving weak pinnig sites were improved by eliminating weak-link parts.

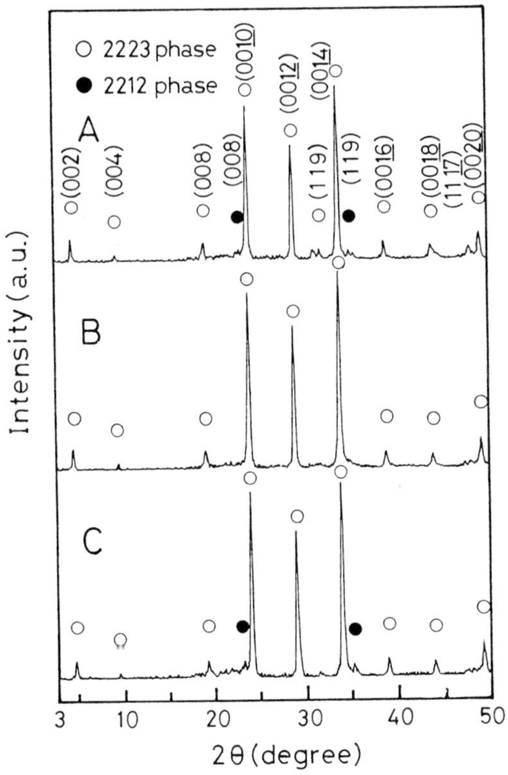

Fig.5 XRD patterns of surface of tapes after the partial stripping of silver sheath

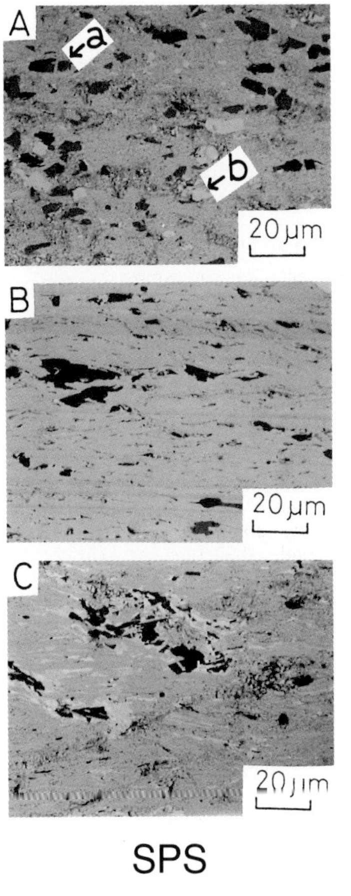

Fig.6 BEI for the polished cross-section of tapes by SEM, where mark a and mark b denote Sr-Ca-Cu-O and Ca-Pb-O non-superconducting grains, respectively

Figure 5 shows the XRD patterns of surface of tapes, prepared by using three different powders and the condition SPS, after the partial stripping of silver sheath. These patterns indicate that c-axis orientation of 2223 phase is perpendicular to the tape surface, and a little $Bi_2Sr_2CaCu_2O_y$ (2212 phase) peaks are observed on the patterns of A and C.

Figure 6 are the back-scattered electron image (BEI) photographs of the polished cross-section of the tapes by SEM. As the results of analysis using EDS, it was found that the specimen using powder B contains less non-superconducting grains such as Sr-Ca-Ca-O and Ca-Pb-O than A and C.

In summary, we suppose that the reason why high critical current density was obtained on powder B is the weak-link parts were almost eliminated and weak pinning sites were improved by decreasing of such non-superconducting grains as Sr-Ca-Cu-O and Ca-Pb-O .

Since these Jc(77.3K) obtained in this experiments were much lower than Jc(4.2K) of Nb-Ti and Nb_3Sn superconducting wires, it is necessary to increase Jc remarkably by improvement in fabrication processes with creation of effective pinning sites.

REFERENCES

1. H.Maeda, Y.Tanaka, M.Fukutomi, and T.Asano, Jpn.J.Appl.Phys.,27,L209 (1988)

2. M.Takano, J.Takada, K.Oda, H.Kitaguchi, Y.Miura, Y.Ikeda, Y.Tomii, and H.Mazaki, Jpn.J.Appl.Phys.,27,L1041 (1988)

3. T.Hikata, T.Nishikawa, H.Mukai, K.Sato, and H.Hitotsuyanagi, Jpn.J.Appl.Phys.,28,L1204 (1989)

4. K.Yamamoto, Y.Nishikawa, and M.Hiraoka, Cryogenic Engineering,25,No.2,p.103 (1990), (in Japanese)

5. K.Osamura, S.S.Oh and S.Ochiai, Supercond.Sci.Technol.,3,p.143 (1990)

Superconducting Properties of Hot-Pressed Bi-Pb-Sr-Ca-Cu-O Thick Films

NORIMITSU MURAYAMA[1], YASUJI HIRAMATSU[2], and YASUYOSHI TORII[1]

[1] Government Industrial Research Institute-Nagoya, 1-1, Hirate-cho, Kita-ku, Nagoya, 462 Japan
[2] IBIDEN Co., Ltd., 3-200, Gama-cho, Ohgaki, Gifu, 503 Japan

ABSTRACT

Superconducting $Bi_{1.84}Pb_{0.34}Sr_{2.03}Ca_{1.9}Cu_{3.06}O_y$ thick films were prepared on MgO substrates by screen-printing. The thick film sandwiched between the two MgO substrates was then hot-pressed in air at 810-860 °C under a constant pressure of 500 kg/cm^2 for 270 min. The thickness of the film was 20-40 μm. The degree of grain orientation increased with increasing hot-pressing temperature. When hot-pressed at 855 °C, the critical current density had a maximum value of 9500 A/cm^2 at 77 K, 0 Oe, and the critical temperature was 110 K. The relationship between the J_c(H) dependence and the hot-pressing temperature was examined.

KEY WORDS: Bi system superconductor, thick film, hot-pressing, J_c-H curve

INTRODUCTION

There are two superconducting phases (The 80 K and 110 K phases) in Bi-(Pb)-Sr-Ca-Cu-O system with critical temperatures (T_c) higher than the liquid nitrogen temperature.The screen-printing is a simple process to prepare thick films. Homogeneous 110 K-phase thick films could be prepared by this process when the powder used for screen-printing is single 110 K-phase. In this case, the subjects are densification and grain orientation in solid state sintering. It is difficult to prepare a dense Bi-based superconducting ceramics by normal sintering because the shape of the grain is platelet. Hot-pressing is an effective technique for preparing dense and grain-oriented ceramics. We have already succeeded in the hot-pressing of screen-printed 110 K-phase thick films[1]. It was found that the critical current density (J_c) was increased by the hot-pressing. In this study, the relationship among superconducting properties, microstructures and hot-pressing temperatures is systematically investigated in the 110 K-phase thick films, and the magnetic field dependence of the J_c is discussed.

EXPERIMENTAL PROCEDURE

Superconducting powder with the starting composition of $Bi_{1.84}Pb_{0.34}Sr_{2.03}Ca_{1.9}Cu_{3.06}O_y$ was prepared by a solid-state reaction. The paste was prepared by mixing ethylcellulose, terpineol and the powder. Afterwards, it was printed on a single-crystal MgO substrate through a 165-mesh screen. The paste was then dried at 120 °C for 30 min. The dried thick films were heat treated at 500 °C in air for 1 h. Figure 1 shows the procedure for the hot-pressed thick film preparation. The two substrates were piled up so as to bring thick films into contact with each other, and a uniaxial pressure was further applied perpendicular to the surface of the thick film at 500 kg/cm^2 at temperatures from 810 °C to 860 °C in air for 270 min. The thick film sandwiched between two substrates was prepared in the form of a mechanically strong body. The hot-pressed samples were cut into bars. The size of the thick films were 0.65 mm wide, 20-40 μm thick and 15 mm long. The values of T_c and J_c were measured by a four-probe dc method. The critical current (I_c) was measured in liquid nitrogen and at magnetic fields up to 5 kOe. The

I_c was defined as the current which induced 1 μV across a 10 mm length of the thick film , and J_c was calculated by dividing I_c by the cross-sectional area. The microstructures of the hot-pressed thick films were examined by scanning electron microscopy (SEM) and energy dispersive X-ray spectroscopy (EDX).

RESULTS AND DISCUSSION

Apart from the 110 K phase, there was no appreciable second phase in the X-ray diffraction pattern for the powder used for screen-printing. The SEM studies revealed that the powder was in the form of a plate with a thickness of about 0.5 μm. In the differential thermal analysis (DTA) curve for the powder,.the endothermic peak, which was due to the melting of the 110 K phase, appeared at 894 °C. The endothermic peak, which was observed in the calcined powder or the 80 K-phase powder, did not appeared[2].

The EDX analysis on the hot-pressed sample revealed the presence of $SrCaCu_3O_x$ phase and Bi,Pb-rich phase as minor phases. Figure 2 shows the cross sections of the samples with the different hot-pressing temperatures which were chemically etched in 0.1 % HNO_3 water solution for 10 min. It was difficult for each grain boundary to be etched because grains were thin; the degree of grain orientation could be, however, estimated because the Bi,Pb-rich phase, which was located along the ab-plane of the 110 K phase, was etched. The grains were preferentially oriented so that the c-axis of the grain was parallel to the hot-pressing direction. The image analysis on the SEM micrographs for the chemically etched cross sections was conducted as shown in Fig.3 in order to evaluate the degree of grain orientation quantitatively. The average angle θ_m between the chemically etched portion and the plane perpendicular to the hot-pressing direction was then calculated. The value of θ_m decreases with increasing degree of grain orientation. As shown in Fig.4, the degree of grain orientation increased with increasing hot-pressing temperature.

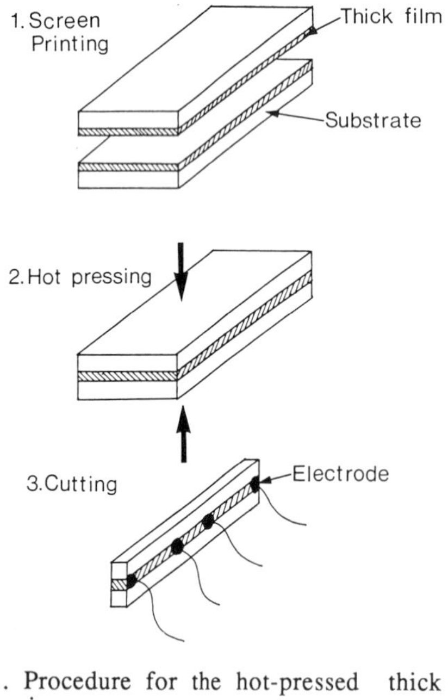

Fig.1. Procedure for the hot-pressed thick film preparation.

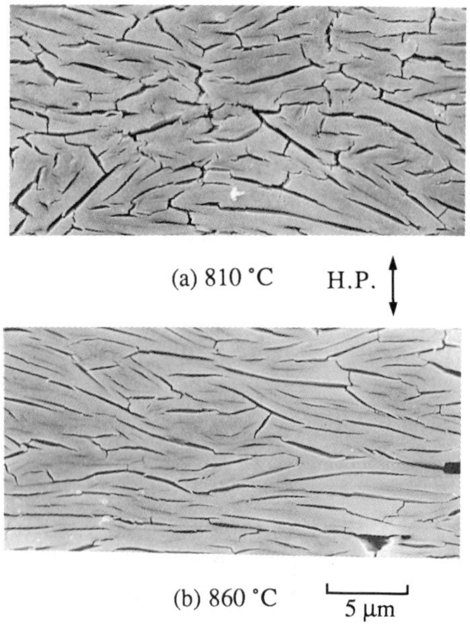

Fig.2. SEM micrographs for the chemically etched cross section of the screen-printed thick films hot-pressed at 500 kg/cm^2 at 810 °C (a) and at 860 °C (b) in air for 270 min. The cross sections are parallel to the hot-pressing direction.

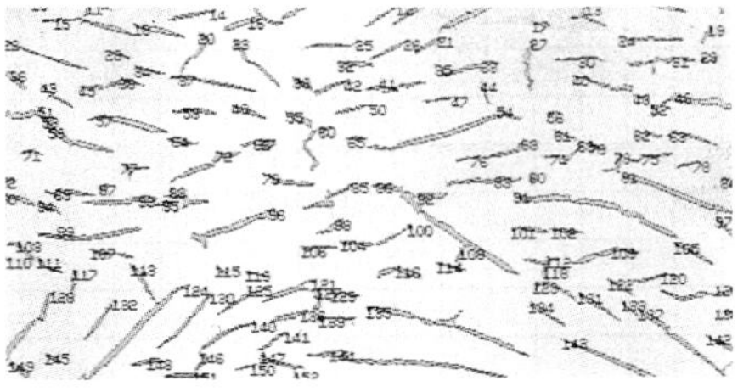

Fig.3. Image analysis on Fig.2 (a).

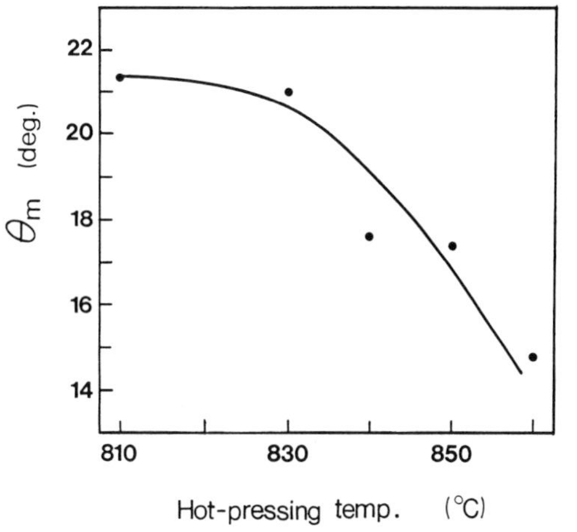

Fig.4. Relationship between the degree of grain orientation and hot-pressing temperature. θ_m represents the average angle between the chemically etched portion in the cross section and the plane perpendicular to the hot-pressing direction.

Figure 5 shows the relationship between the T_c, J_c at zero magnetic field ($J_c(0)$) and the hot-pressing temperature. The T_c increased with increasing hot-pressing temperature and reached a constant value of 110 K above 850 °C. The $J_c(0)$ increased with increasing hot-pressing temperature below 855 °C and had a maximum value of 9500 A/cm^2 at 855 °C. The changes in the T_c and $J_c(0)$ were small below 830 °C.

Figure 6 shows the magnetic field dependence of the J_c for samples with different hot-pressing temperatures. The magnetic field was applied parallel to the hot-pressing direction. The J_c was constant at very low magnetic fields. The maximum value among magnetic fields, where the J_c was almost constant, increased with increasing hot-pressing temperature. Followed by this regime, the J_c varied as $H^{-0.7}$ up to about 200 Oe and decreased rapidly at about 2 kOe. When the hot-pressing temperature was high, the $J_c(H)$ dependence had a tendency of a double-step character; the slope of the $\log(J_c)$-$\log(H)$ curve decreased prior to the rapid decrease in the J_c.

The J_c-H curve is strongly dependent on the coupling strength between grains. When J_c is governed by Josephson weak-link decoupling, it is related to the magnetic field as follows[3],

$$J_c \propto H^{-\gamma} \qquad (1 \leq \gamma \leq 1.5) \qquad (1)$$

When Jc is governed by flux pinning, it is related to the magnetic field as follows[4],

$$J_c \propto H^{-0.5}(1-H/H_{c2})^{\delta} \qquad (1 \leq \delta \leq 2) \qquad (2)$$

where H_{c2} is the upper critical field. When flux creep occures, J_c decreases rapidly at the magnetic field H_i which is less than H_{c2}. When there are partially percolation conduction paths, where the pinning force is stronger in a sample, a plateau region appears in the J_c-H curve[5].

For the sample hot-pressed at 810 °C, the Jc-H curve was fitted by the eq. (2) with the parameters $H_i \risingdotseq 3$ kOe and $\delta \risingdotseq 1.5$, and no appreciable plateau region was observed in the J_c-H curve. This result suggests that the J_c is governed by flux pinning and that the pinning force is spatially uniform in this sample. In addition, it seems that the pinning force increases at some places in the sample with increasing hot-pressing temperature.

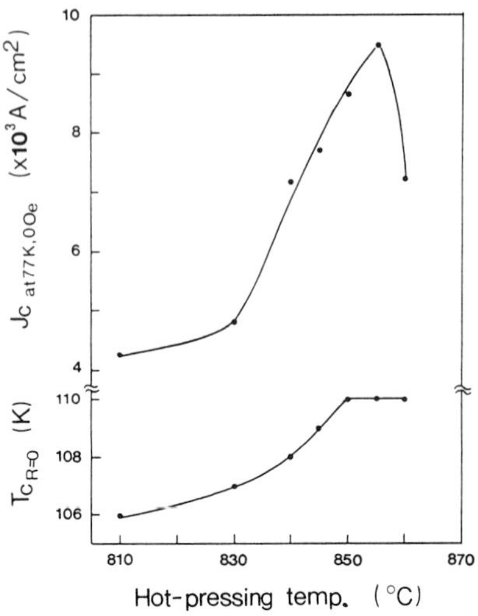

Fig.5. Hot-pressing temperature dependence of T_c and J_c at 77 K at zero magnetic field for the screen-printed thick films hot-pressed at 500 kg/cm^2 in air for 270 min.

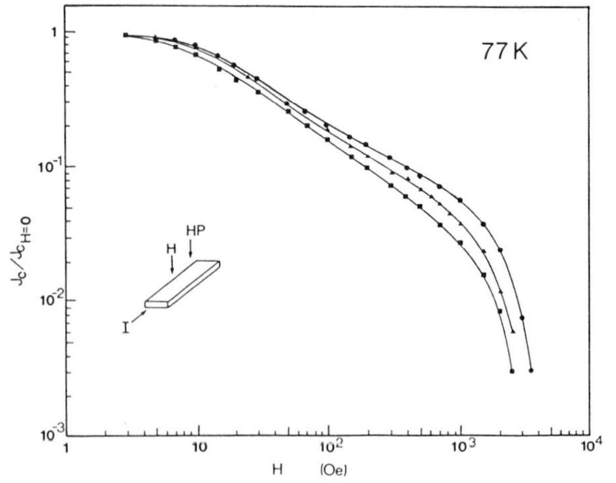

Fig. 6. Magnetic field dependence of J_c for the hot-pressed thick films. Magnetic field is applied parallel to the hot-pressing direction.
810 °C (■), 840 °C (▲), 860 °C (●)

CONCLUSION

The screen-printed $Bi_{1.84}Pb_{0.34}Sr_{2.03}Ca_{1.9}Cu_{3.06}O_y$ thick film was successfully hot-pressed. The degree of grain orientation was qualitatively evaluated by means of image analysis on the SEM micorgraphs for the chemically etched cross sections of the samples. The degree of grain orientation increased with increasing hot-pressing temperature. It was found that a high-J_c polycrystalline Bi-Pb-Sr-Ca-Cu-O sample could be prepared from the 110 K-phase powder by solid state sintering for a very short duration under the condition that the temperature was controlled within a narrow range just below the melting point of the 110 K phase. From the J_c-H curve for the sample, it was suggested that the J_c is governed in the sample not by Josephson weak-link decoupling but, rather, by flux pinning.

ACKNOWLEDGMENTS

The authors would like to acknowledge T.Matsusita for helpful comments and O.Abe for the image analysis.

REFERENCES

1. Murayama N, Hiramatsu Y and Torii Y (1990) Jpn.J.Appl.Phys. 29: L875-L877
2. Aota K, Hattori H, Hatano T, Nakamura and Ogawa K (1989) Jpn.J.Appl.Phys. 28: L2196-L2199
3. Peterson RL and Ekin JW (1989) Physica C 157: 325-333
4. Yasukochi K, Ogawawara T and Ushino N (1964) J.Phys.Soc.Jpn. 19: 1649-1661
5. Ekin JW, Larson TM, Hermann AM, Sheng ZZ, Togano K and Kumakura H (1989) Physica C 160: 489-496 .

Preparation of High J_c Y-Ba-Cu-O Thick Films by a Paint-on Method

MASAMICHI MIYAJIMA, SHIGEO NAGAYA, IZUMI HIRABAYASHI, and SHOJI TANAKA

Superconducting Research Laboratory, Nagoya Division, International Superconductivity Technology Center, 4-1, Mutsuno 2-chome, Atsuta-ku, Nagoya, 456 Japan

ABSTRACT

The structure and Jc of Y-Ba-Cu-O thick films prepared using three different kinds of starting powders were reported. The Jc value of the film made from melt quenched powders has been improved by introducing effective pinning centers. Hysteresis loop at 77 K became larger than those of films made from other starting powders. Fine particles of the 211 phase with a submicron order in diameter were homogeneously distributed in the 123 grains. The Jc value exhibited 2.5 x 10^4 A/cm^2 at 77 K and 0.1 T by the magnetization measurement. The c-axis orientation was promoted by the addition of Ag_2O to melt quenched powders.

KEY WORDS: high-Tc superconductor, paint-on method, thick film, Jc, pinning center

INTRODUCTION

Since the discovery of high-Tc superconductors, there has been widespread interest in producing both thin and thick films of these materials. Because it is necessary to fabricate films for several applications, such as circuits, devices, tapes, and shieldings, many different techniques have been utilized, including sputtering, evaporation, molecular beam epitaxy, as well as thermal decomposition of organometallics. These methods generally involve complicated and expensive vacuum apparatus by which the deposition rates are very low, and a critical current density (Jc) is higher than that by other methods. On the other hand, thick-film methods [1, 2] such as coating and printing are simple, less costly, and easy to apply to a long and large area, but Jc has not been yet high enough. It is necessary for the application of thick films to improve Jc.

We have already reported that highly oriented films of Y-Ba-Cu-O systems were obtained on polycrystalline YSZ (yttria stabilized zirconia) substrates by a paint-on method, and discussed the relationship between the crystal orientation and process conditions [3]. The grain alignment is one of the significant factors to improve Jc, since superconductors with a layered type structure have the large anisotropy on electrical conduction paths.

In type II superconductors, the introduction of effective pinning centers is also required in order to obtain high Jc. Large Jc values have been achieved only in melt processed bulk materials, such as QMG and MPMG processes [4, 5, 6], except for epitaxially grown thin films. The origin of high Jc in these materials seems to be the suppression of the week-links by the grain alignment and the introduction of effective pinning centers.

Recently, we have improved the Jc value of thick films by introducing effective pinning centers, and found that the silver addition to pastes promoted the c-axis orientation. In this letter, we report that the use of melt quenched powders as starting materials and the addition of Ag_2O are key elemental processes to fabricate thick films with high Jc.

EXPERIMENTAL

Three different kinds of pastes made from superconducting particles (SP), precursors (PC), and melt quenched powders (MQP) were prepared in order to investigate the effect of starting powder characteristics on the grain alignment and Jc as well. SP were prepared by an ordinary solid state reaction method. Appropriate amount of mixtures of Y_2O_3, $BaCuO_2$, and CuO were adopted as PC. MQP were prepared as follows. The calcined $YBa_2Cu_3O_{7-y}$ powders were rapidly heated to 1400 °C and held for 3 min. in platinum crucibles, then subsequently quenched by copper hammers. The quenched platelet samples were pulverized and ground using alumina ball milling. Two types of melt quenched powders, with and without 10 wt% Ag_2O, were prepared. Subsequently, these powders were mixed with an appropriate amount of organic vehicles using a three-roll mill device to form pastes. The pastes were then painted on flat ceramic substrates using a metal mask and a squeeze. Flexible polycrystalline YSZ (10 mm x 10 mm x 0.1 mm) was used as substrates. After dried at 120 °C for 30 min., films were sintered at 1000-1050 °C for 12 min. in flowing oxygen atmosphere.

Crystal structure and surface morphology of the films were characterized by X-ray diffraction (XRD) and scanning electron microscopy (SEM), respectively. From SEM, the cross section of films was observed and the thickness was determined. The composition of the films was examined by energy dispersive X-ray microanalysis (EDX). A critical temperature was determined by the electrical resistivity and ac susceptibility. A critical current density and its temperature dependence were measured by a standard dc/four-prove method with indium electrodes. Jc was defined by the current density where the generated voltage reached 1 μV from the I-V characteristics of the films. The magnetization was measured by a SQUID magnetometer.

RESULTS AND DISCUSSION

Figure 1 shows X-ray diffraction patterns of the films annealed at 1040 °C for 12 min. The diffraction patterns of each film consist of the peaks of the 123 orthorombic phase and YSZ substrates. In the case of the films made from SP and PC, intensities of *(0 0 l)* peaks are larger than those of other peaks. On the other hand, the film made from MQP shows only slight orientation and the peak intensities are lower than those of the films made from other starting powders. Three kinds of films showed a sharp transition around 90 K by the resistivity and ac susceptibility measurement. Y_2BaCuO_5 and $BaZrO_3$ were detected as impurities.

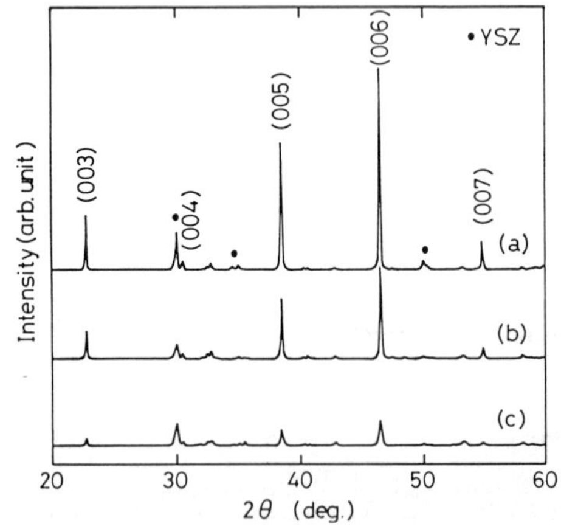

Fig. 1 X-ray diffraction patterns of the films made from (a) superconducting particles, (b) precursors, and (c) melt quenched powders, annealed at 1040 °C for 12 min.

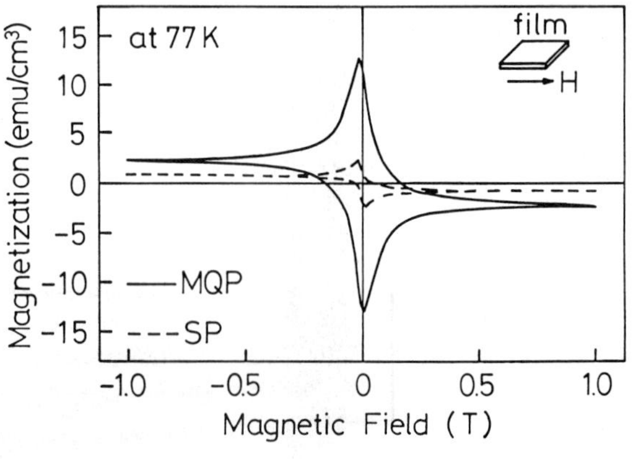

Fig. 2 Magnetization hysteresis curves at 77 K of the films made from superconducting particles (a broken line) and melt quenched powders (a solid line) with the applied magnetic field parallel to the film surface.

Figure 2 shows magnetization hysteresis curves of the films at 77 K with the applied magnetic field parallel to the film surface. The hysteresis loop of the film made from MQP was larger than those from other starting powders in spite of the lower crystal orientation. The surface and the cross sectional SEM micrographs of the film made from MQP are shown in Fig. 3. Fine particles with a submicron order in diameter are observed in the 123 grains, although the film has not a few pores, grain boundaries, or cracks. These particles were identified as the 211 phase by the EDX analysis. In the case of the films made from SP and PC, the dense textured feature was observed, but the 211 particles were not observed in the 123 grains. Cracks and pores were fewer than those of the film made from MQP. For comparison, those of SP and PC were shown in Fig. 4.

The Jc value exhibits 2.5×10^4 A/cm^2 at 77 K and 0.1 T by the magnetization measurement using eq. (1).

$$Jc = 20\ \Delta M/d \qquad (1)$$

Fig. 3 SEM micrographs of the film made from melt quenched powders: (a) surface, (b)(c) cross section.

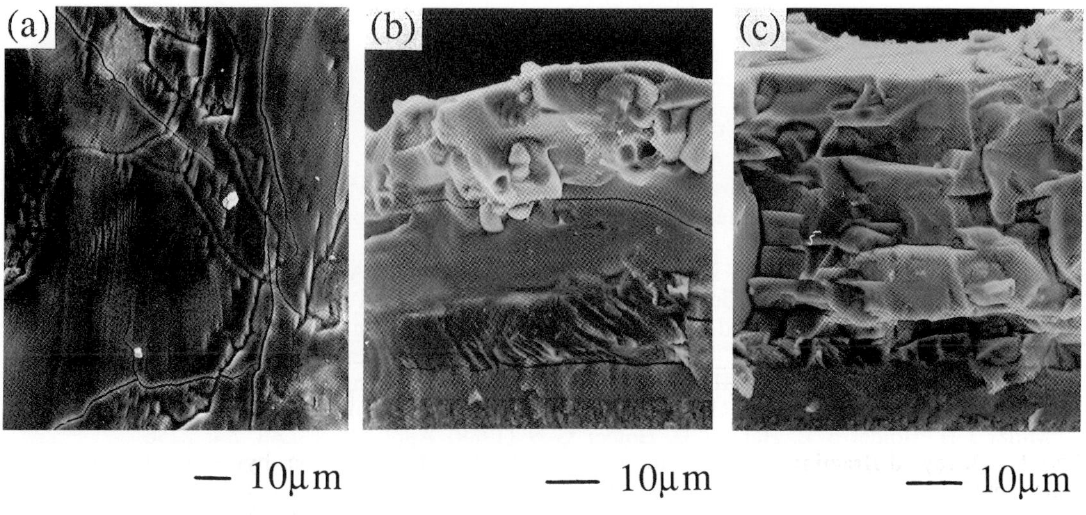

Fig. 4 SEM micrographs of the films made from superconducting particles: (a) surface, (b) cross section and precursors: (c) cross section.

where Jc is in A/cm^2, magnetization difference (ΔM) is in emu/cm^3 and sample thickness (d) is in cm. We regard the film thickness as d in order to avoid the overestimation. This value is five times larger than those of the films made from other starting powders. On the other hand, transport properties were not consistent with this result. The transport Jc values of the films made from SP, PC, and MQP were 1100, 2400, and 1100 A/cm^2 at 77 K and 0 T by the four-probe method, respectively. It seems that not a few pores or cracks weaken electric conduction paths in the film made from MQP.

Figure 4 shows X-ray diffraction patterns of the films made from MQP with and without Ag_2O. By the addition of Ag_2O, intensities of *(0 0 l)* peaks are enhanced, and the optimum annealing temperature decreased from 1040 °C to 1000 °C. We expect that the Ag addition prevents from the formation of some pores and/or cracks and improves grain boundary characteristics, as reported by others [7, 8].

In summary, we have improved Jc of Y-Ba-Cu-O thick films using melt quenched powders as starting materials of pastes, and successfully introduced effective pinning centers in thick-film methods. The film fabricated by the MQP method consists of the 123 grains with the fine dispersion of the 211 phase. By the Ag addition, the optimum annealing temperature decreases and the c-axis orientation of the films is improved. Further study for the optimization of process conditions will lead to thick films with higher critical current densities. We expect that this simple and easy method enables us to apply to a long and large area such as tapes and shieldings.

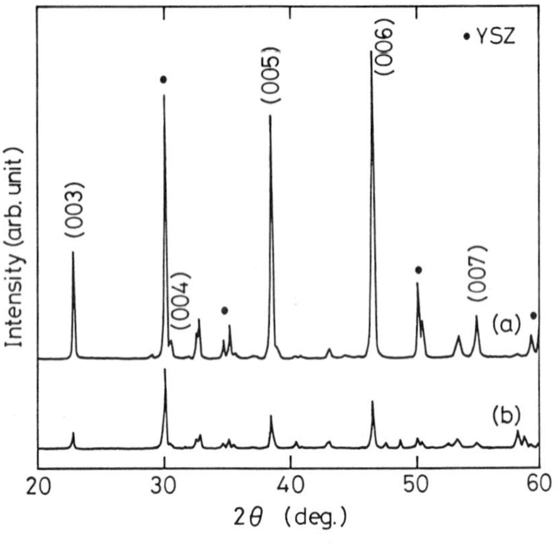

Fig. 5 X-ray diffraction patterns of the films made from melt quenched powders (a) with and (b) without 10 wt% Ag_2O.

ACKNOWLEDGMENTS

The authors would like to thank Mr. Ogawa (Dowa Mining Co., LTD.) for the pastes preparation, and Dr. Murakami, Dr. Shiohara (SRL, ISTEC) and Mr. Sawada (JGC CORPORATION) for helpful discussion.

REFERENCES

1. Shin I, Qiu CX (1988) Appl. Phys. Lett. 51: 1277
2. Tabuchi J, Utsumi K (1988) Appl. Phys. Lett. 53: 606
3. Miyajima M, Nagaya S, Hirabayashi I, Ogawa Y, Mitsune Y, Ishikawa Y, Yoshizawa S (1989) Proceedings of the Second International Symposium on Superconductivity. Tsukuba, pp427-430
4. Murakami M, Morita M, Doi K, Miyamoto K (1989) Jpn. J. Appl. Phys. 28: 1189
5. Fujimoto H, Murakami M, Gotoh S, Koshizuka N, Oyama T, Shiohara Y, Tanaka S (1989) Proceedings of the Second International Symposium on Superconductivity. Tsukuba, pp285-288
6. Nagaya S, Miyajima M, Hirabayashi I, Shiohara Y, Tanaka S (1990) Proceedings of Applied Superconductivity Conference. Colorado (to be published)
7. Miller J H, Holder S L, Hunn J D, Holder G N (1989) Appl. Phys. Lett. 54: 2256
8. Murakami M, Oyama T, Fujimoto H, Taguchi T, Gotoh S, Shiohara Y, Koshizuka N, Tanaka S (1990) Jpn. J. Appl. Phys. (to be published)

Preparation of Bi-Sr-Ca-Cu-O Superconducting Fibers by Laser Pedestal Growth Method

K. HAYASHI[1], H. NONOYAMA[1], M. NAGATA[1], and K. TAKAHASHI[2]

[1] Osaka Research Laboratories, Sumitomo Electric Industries, Ltd., 1-3, Shimaya 1-chome, Konohana-ku, Osaka, 554 Japan

[2] Super-GM, 14-10, Nishitenma 5-chome, Kita-ku, Osaka, 530 Japan

ABSTRACT

2212 BiSrCaCuO superconducting fibers were prepared by the laser pedestal growth method. The influence of fiber diameter and growing speed on micro-structures and superconducting properties was investigated. Temperature gradient and growing speed were important factors in obtaining highly aligned micro-structures. The temperature gradient increases as the fiber diameter decreases. So highly aligned structures were obtained when the fiber diameter was small and the growing speed was low. When the nominal diameter of grown fiber was 0.5mm and the growing speed was 30mm/h, Jc of 17,500A/cm^2(77.3K,0T) was obtained. In the case of small fiber diameter, aligned crystal structures and relatively high Jc's were maintained in spite of the high growing speed.

KEY WORDS : Bi system superconductor, laser pedestal growth, Jc

INTRODUCTION

Since the discovery of high-Tc oxide superconductors[1], many efforts have been made to prepare high Jc wires for application in electric power apparatuses. As high-Tc oxide superconductors have strong anisotropy of crystal structures and of superconducting properties, alignment of crystal micro-structures is needed for obtaining high Jc in poly-crystalline samples. It is well-known that melt processed oxide superconductors have high density and highly aligned structures with high Jc.[2,3] As for the YBaCuO system, a completely melted state was difficult to obtain and the viscosity of melt is low. In contrast, for the BiSrCaCuO system it was easy to obtain a completely melted state and superconducting phase. The laser pedestal growth method is known as the manufacturing process of single crystalline oxide fibers.[4] Recently, R.S.Feigelson et.al. applied the laser pedestal growth method to the BiSrCaCuO system and obtained well-aligned superconducting fiber.[5] A similar process, such as the halogen lamp focused floating zone melting method[6], was also applied to the BiSrCaCuO system. However, high Jc's were obtained only in cases of lower growing speed. For the purpose of industrially manufacturing fibers by the laser pedestal growth method, higher growing speed with high Jc is needed. So, in order to study the possibility of obtaining high Jc at higher growing speed, the relationships between superconducting properties and crystal growth conditions, such as diameter of grown fiber and growing speed, were investigated.

EXPERIMENTAL PROCEDURE

Bi_2O_3, $SrCO_3$, $CaCO_3$ and CuO were mixed in a ratio of 2:2:1:2(Bi:Sr:Ca:Cu). A sintered rod about 4mm in diameter was used for the feed rod. Figure 1 is a

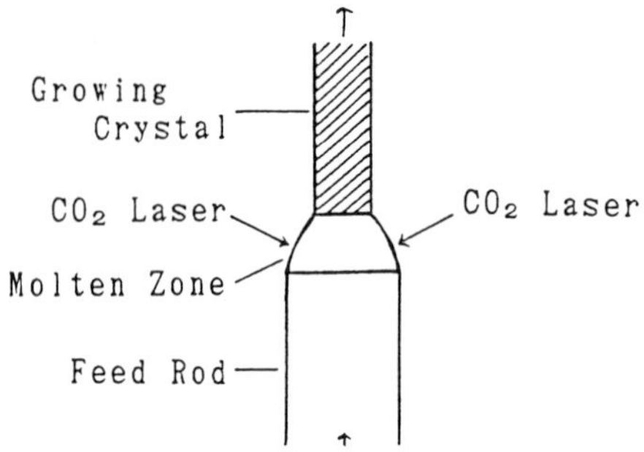

Fig.1 Schematic illustration of the laser pedestal growth method.

schematic illustration of the laser pedestal growth method used in this study.
The two-beam method was used. The top of the feed rod was melted by CO_2 laser and the fibers were grown from the molten zone. The seed crystal was the same sintered rod as the feed rod. Growing speed and feed speed were controlled individually. In many cases, the feed rod and/or grown fiber was rotated. The diameter of grown fibers was controlled by the diameter of the feed rod, and the ratio of growing speed and feed speed. All the samples after crystal growth were annealed in air.
The temperature gradient at the liquid-solid interface was measured by infrared thermal video system. Superconducting properties were measured by the conventional four-terminal method. The current flow direction was parallel to the growing direction. The sensitivity of the critical current was 1μ V/cm. The magnetic field dependence of Jc was measured at 77.3K and 4.2K. The micro-structure was observed by optical microscope and SEM-EDX analysis.

RESULTS AND DISCUSSION

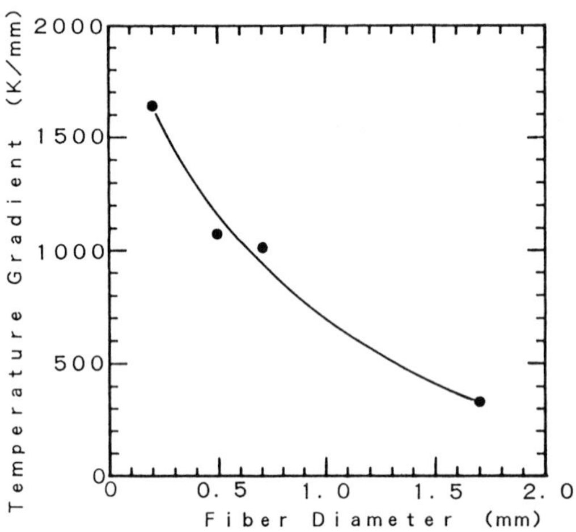

Fig.2 Relationship between temperature gradient at liquid-solid interface and fiber diameter.

In order to obtain a highly-aligned crystal structure by a directional solidification method, the temperature gradient at the liquid-solid interface (G) and the growing speed (R) are important factors, and a large G/R is needed. If an aligned structure at a high growing speed is considered, a much larger temperature gradient is needed. The laser pedestal growth method is known as a high temperature gradient process, because laser beams are easy to focus on the molten zone by ZnSe lenses. The laser pedestal growth method is one of the floating zone melting methods, so the molten zone is maintained by its own surface tension. When the fiber diameter is small, a small volume molten zone is needed to grow successfully. For these reasons, the fiber diameter dependence of temperature gradient at the liquid-solid interface was measured. Results are shown in figure 2 in the case of 30mm/h growing speed.
The temperature gradient increases as the fiber diameter decreases. 1,640K/mm at 0.2mm of fiber diameter is a larger value as compared with other directional solidification methods, such as the Bridgman method using an electrical resistance heater, by two orders of magnitude.
In view of the application of the laser pedestal growth method to wiring processes, a small diameter sample which simultaneously achieves flexibility, higher growing speed and higher Jc is needed. In the case of small fiber diameter, temperature gradient G is very large and crystal alignment is expected to be superior to the samples grown in large fiber diameter, so experi-

ments of small diameter were done. Figure 3 shows the relationship of Jc (77.3K,0T), growing speed and fiber diameter. Samples grown in 0.5mmϕ fiber diameter showed higher Jc in all growing speed conditions. As-grown fibers grown at 30mm/h mainly consisted of 2201 phase and 0112 phase, and these phases were well-aligned. During annealing, these phases reacted with each other and as a result, a 2212 superconducting phase was produced aligned along the alignment of 0112 phase. Samples grown in 0.5mmϕ diameter had an aligned 0112 phase even in higher growing speed conditions, such as 120mm/h, and matrix phases were randomly produced in as-grown condition. But an aligned 2212 phase was produced along the 0112 phase after post annealing. Therefore, in the case of 0.5mmϕ nominal fiber diameter, Jc of 17,500A/cm^2 and 5,060A/cm^2 were obtained for 30mm/h and 120mm/h growing speeds, respectively. In the case of 1.7mmϕ fiber diameter, more than 100A of critical current, which is enough for practical use, was obtained. A longitudinal cross-section of a 17,500A/cm^2 sample is shown in Photo 1 and the sample has a well-aligned structure. This photograph shows that the sample is not single phase. The sample consists of unreacted 2201 phase, 0112 phase and 2212 phase. It was found by EDX analysis that the 2212 phase of this sample did not have ideal composition but had compositional fluctuation between 2201 and 2212.

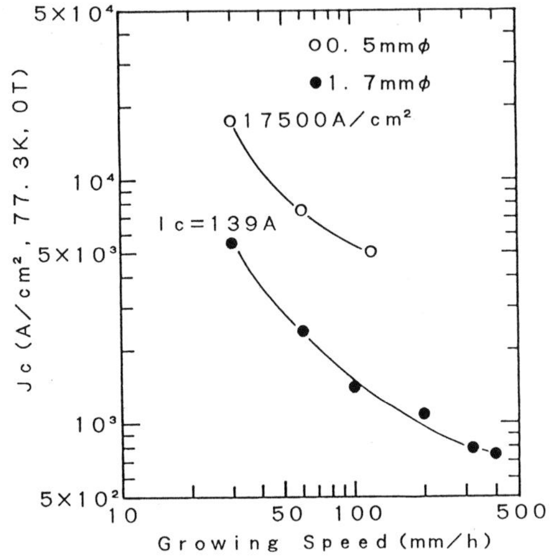

Fig.3 Relationship between Jc(77,3K ,0T) and growing speed for some fiber diameters.

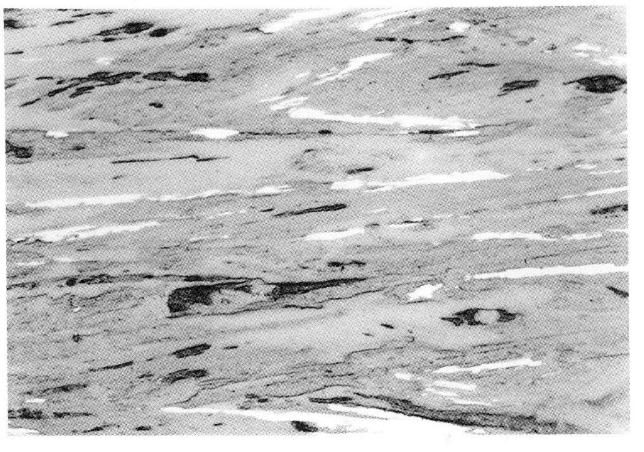

Photo 1 Longitudinal cross-section of 17,500A/cm^2 (77,3K,0T) sample.

Figure 4 shows the magnetic field dependence of Jc at 77.3K for some samples. Jc-B characteristics of the samples which consist of an as-grown 2212 phase for 1mm/h growing speed are inferior to that of the samples which consist of a 2212 phase produced by post annealing. It is believed that the difference in these Jc-B characteristics is caused by the perfection of superconducting phases. However, all these samples had low Jc values above 800gauss. In contrast, the 2212 phase has a large temperature dependence of Jc. Jc's were much higher at 4.2K than 77.3K. Figure 5 shows the magnetic field dependence of Jc at 4.2K. More than 10^5A/cm^2 of Jc was achieved at 4.2K and 0T for the sample which had 5,390A/cm^2 of Jc at 77.3K and 0T. Compared with 77.3K, temperature dependence of Jc was much improved. More than 10^4A/cm^2 of Jc in a magnetic field of around a few tesla was also achieved for the samples which had around 3,000A/cm^2 of Jc at 77.3K. If the Jc at 77.3K is improved, 10^5A/cm^2 in high magnetic field at 4.2K, which is comparable or superior to conventional superconductors such as NbTi alloy, will be achieved. This study is now underway.

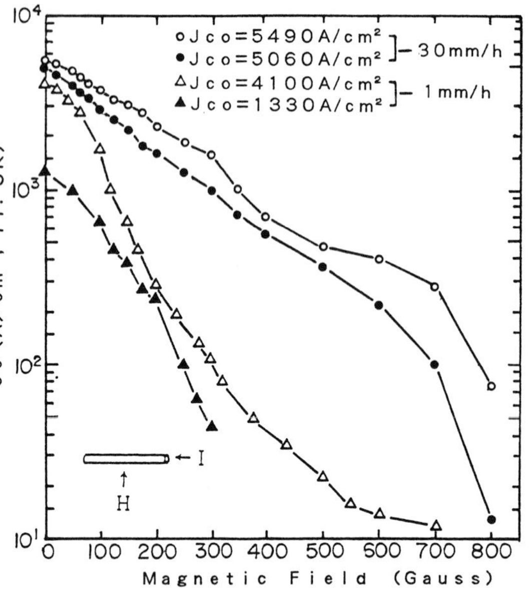

Fig.4 Jc-B characteristics at 77.3K for several samples.

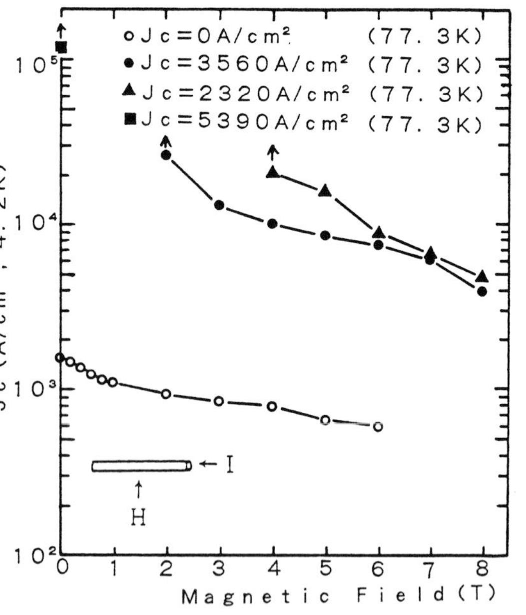

Fig.5 Jc-B characteristics at 4.2K for several samples.

In conclusion, relatively high growing speed and high Jc were achieved simultaneously for small fiber diameter samples. The reason for this result is believed to be as follows. ① higher temperature gradient at liquid-solid interface with smaller fiber diameter ② highly aligned 0112 phase in high growing speed condition and fine micro-structure in as-grown condition ③ aligned 2212 phase produced along aligned 0112 phase after post annealing.
The aligned 2212 phase produced by post annealing had good Jc-B characteristics compared with the aligned 2212 phase produced by lower growing speed conditions.

ACKNOWLEDGMENT

This work was performed as a part of "R & D on Superconducting Technology for Electric Power Apparatuses" as a subject of Super-GM under the Moonlight Project of Agency of Industrial Science and Technology, MITI, being consigned by New Energy and Industrial Technology Development Organazation (NEDO).

REFERENCES

1.J.G.Bednorz and K.A.Muller, Z.Phys. (1986)B4:189
2.K.Hayashi, H.Nonoyama, M.Nagata and H.Hitotsuyanagi,Supercond. Sci. Technol. (1990)3:81
3.K.Hayashi,H.Nonoyama,M.Nagata, H.Hitotsuyanagi and M.Kawashima:T.Ishiguro, K.Kajimura(Eds.) Advances in Superconductivity II, Springer-Verlag Tokyo (1990):309
4.R.S.Feigelson: E.Kaldis(Eds.) Crystal Growth of Electronic Materials, Elsevier Science Publishers (1985):127
5.R.S.Feigelson, D.Gazit, D.K.Fork and T.H.Geball, SCIENCE (1988)240:1642
6.S.Takekawa, H.Nozaki, A.Umezono, K.Kosuda and M.Kobayashi, J. Crystal Growth (1988)92:687

10M Long Silver-Sheathed Tl-Ba/Sr-Ca-Cu-O Superconducting Tapes

FUMIKAZU HOSONO[1], AKIRA NOMOTO[1], TADASHI UMEZAWA[1], MASAHIRO SEIDO[1], TOSHIMI MATSUMOTO[2], and KATSUZO AIHARA[2]

[1] Metal Research Laboratory of Hitachi Cable Ltd., Tsuchiura, Ibaraki, 300 Japan
[2] Hitachi Research Laboratory of Hitachi Ltd., Hitachi, Ibaraki, 319-12 Japan

ABSTRACT

Silver-sheathed Tl-Ba/Sr-Ca-Cu-O tapes, 10m long, were fabricated by a newly developed continuous pressing technique instead of the conventional rolling or short sample pressing technique. This continuous pressing technique provides highly densified polycrystalline oxide cores in the 10m superconducting tapes, similar to those in the pressed high-quality short samples. Critical current densities at several sections of the 10m long continuous pressed tape reached 10^4 A/cm^2 at 77K without external magnetic fields, and deviations of the Jc values were within +2% throughout the whole tape length . Furthermore, the 10m long tape was wound into a pancake type coil of 30mm bore diameter. This coil generated a maximum magnetic field of 103 gauss and also a magnetic field of 47 gauss at the center of the coil.

INTRODUCTION

Since oxide superconductors with a high critical temperature above 77K were discovered [1]-[3], many studies looked at the oxide superconducting wires, which could be used in the future for superconducting coils and cables operated in liquid nitrogen. These applications require long wires with a high critical current density.

Among the many fabrication processes for the oxide superconducting wires, a drawing-rolling method [4] was particularly studied. Critical current densities (J_c values) of silver-sheathed oxide superconducting tape shaped wires have been enhanced by the drawing-rolling method with a pressing technique [5]. It is, however, difficult to fabricate a long and uniform tape with high J_c value by the short sample pressing method. We developed a continuous pressing method as a new fabrication process for a long silver-sheathed oxide superconducting tape. Ten-meter-long Tl-Ba/Sr-Ca-Cu-O(TBSCCO) tapes were fabricated by the continuous pressing method. Furthermore, the 10m tape was wound into a pancake type coil. This paper presents the tape properties and the coil performance.

EXPERIMENTAL

$Tl_2Ba_{1.6}Sr_{0.4}Ca_2Cu_3O_{10}$ [6] powder was prepared by the following method. Stoichiometric amounts of BaO(99.9%), SrO(99.9%), CaO(99.99%), CuO(99.99%) powders were mixed, ground and calcined in air at the temperature between 1173 and 1223 K for 10 hours. The Ba/Sr-Ca-Cu-O precursors were ground and mixed with an appropriate amount of Tl_2O_3(99.99%), and the resultant mixtures were pressed into cylindrical pellets (30 mm in diameter) under a pressure of about 1000kg/mm^2. The pellets were sintered in air at the temperature between 1073 and 1173 K for 3 to 10 hours. Then, the Tl-Ba/Sr-Ca-Cu-O pellets were reground into powder.

Silver-sheathed TBSCCO superconducting tapes were fabricated by the newly developed continuous pressing method shown in Fig.1. Superconducting powder was put into a silver tube. The diameter of the tubing was reduced using a drawing or a swaging. Next, the drawn tubing was continuously pressed into a tape at high pressure by a newly developed pressing machine, and was sintered in a flowing oxygen atmosphere at the temperature between 1073 and 1173 K for 2 to 100 hours.

There is a deformation pattern difference between continuous pressing and conventional rolling. Table 1 shows longitudinal strains of the deformed tapes during the continuous pressing and rolling process. The strain of the continuous pressed tape was smaller than that of the rolled tape. The tape was elongated mainly in the longitudinal direction by the rolling. On the other hand, for the continuous pressing process, the tape was deformed very little in the longitudinal direction, rather main deformation was in the thickness direction of the tape and the transverse wide direction, as shown in Fig.2.

The 10m long continuous pressed tape was wound into a pancake type coil on a silver bobbin of 30mm diameter, and was sintered after winding. The layer between the tapes was insulated by a ceramic tape.

J_c values were measured by the four probe method. The criterions for the critical current (I_c value) were $1\mu V/cm$ for a short tape and $0.01\mu V/cm$ for the 10m long tape or coil. Furthermore, n-value was calculated from V-I characteristics between $0.2\mu V/cm$ and $1\mu V/cm$. Here the n-value was defined by V I^n. The microstructures of the TBSCCO core were observed using a polarized optical microscope and a SEM.

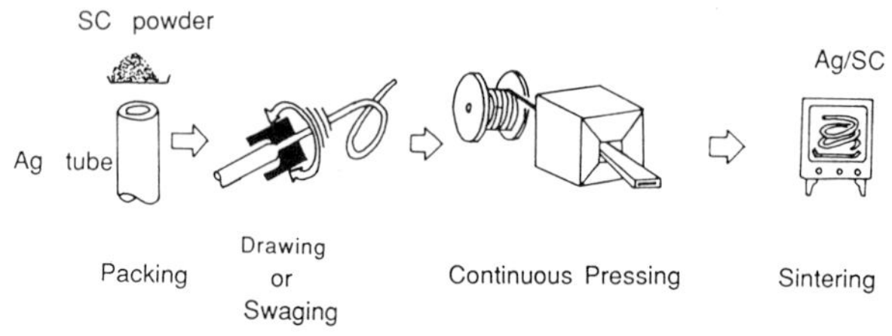

Fig.1 Fabrication process of continuous pressing method.

Table.1 Comparison of strains between continuous pressing and rolling.

Process	ϵ_t (%)	ϵ_l (%)	ϵ_w (%)
Rolling	-1 55	13.1	1 00
Continuous pressing	-1 55	3.9	1 33

*ϵ_t, ϵ_l, ϵ_w are logalithmic strains in the thickness, longitudinal and transeverse direction.

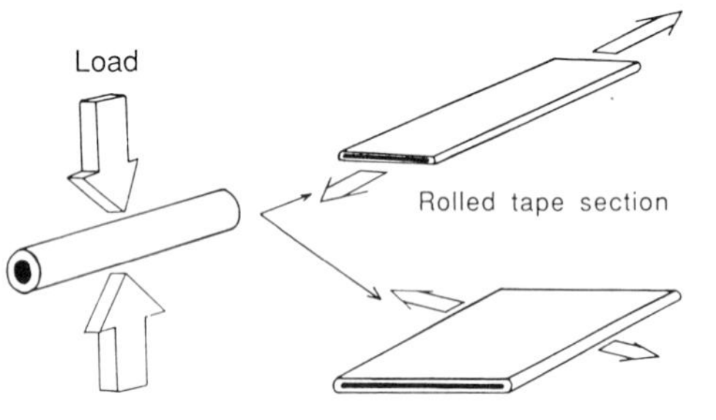

Fig.2 Deformation patterns of continuous pressing and rolling.

RESULTS AND DISCUSSION

PROPERTIES OF 10M LONG TAPE

Figure 3 shows the J_c values of the 10m continuous pressed tape at 77K without external magnetic fields. The tape was wound in a solenoid type coil on a bobbin of 60mm diameter and was sintered. The J_c values were measured at 1m intervals throughout the whole 10m length using 10mm voltage terminals. J_c values reached 10^4 A/cm^2 at each section, and deviations were within +2% throughout the whole length of the tape. The J_c also reached 10^4 A/cm^2 when it was measured at a voltage terminal section of 10m. The voltage rose rapidly, and n-value of the 10m long tape was about 30 which was equal to that of a short tape of 30mm length.

Figure 4 compares the J_c values of the continuous pressed tape, short sample pressed tape and rolled tape [7] at 77K in an external magnetic field. It was clearly found that the J_c of the tape in the external magnetic field was much enhanced by continuous pressing, and that the B-J_c property of the continuous pressed tape corresponded to that of the short sample pressed tape.

Figure 5 shows optical micrographs of the longitudinal section of the continuous pressed tape and the rolled tape. Microstructural examinations revealed that the superconducting oxide core of the continuous pressed tape was uniform and highly densified polycrystal as compared with that of the rolled tape. The microstructures were induced by the difference of the deformation pattern between two working methods. As a result, deviations of the J_c values of the 10m long continuous pressed tape were very small.

Figure 6 shows SEM images of the fractured core surface in the rolled tape and the continuous pressed tape. The contact between grains in the core was improved by continuous pressing. Therefore, enhancement of J_c was thought to be caused by improvement of the grain boundary weak link.

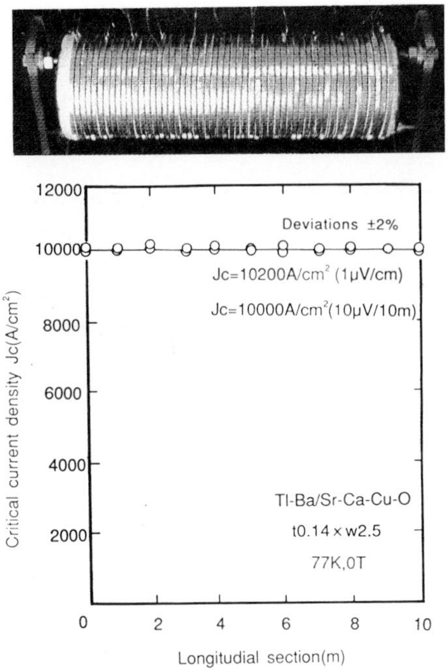

Fig.3 Distribution of the Jcs in a 10m long silver sheathed TBSCCO tape

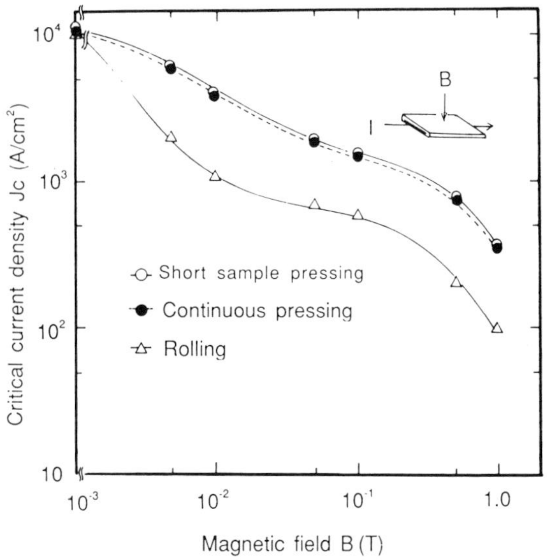

Fig.4 Jcs of a continuous pressed tape in an external magnetic field compared with that of a rolled tape and a short sample pressed tape.

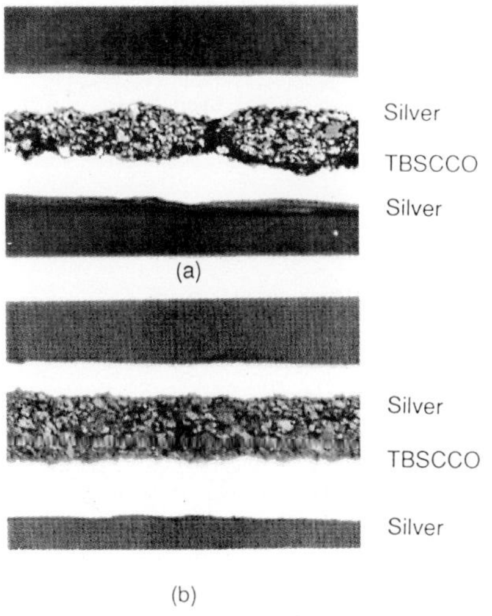

Fig.5 Optical micrographs of longitudial view of silver sheathed TBSCCO tape:
(a) a rolled tape (b) a continuous pressed tape

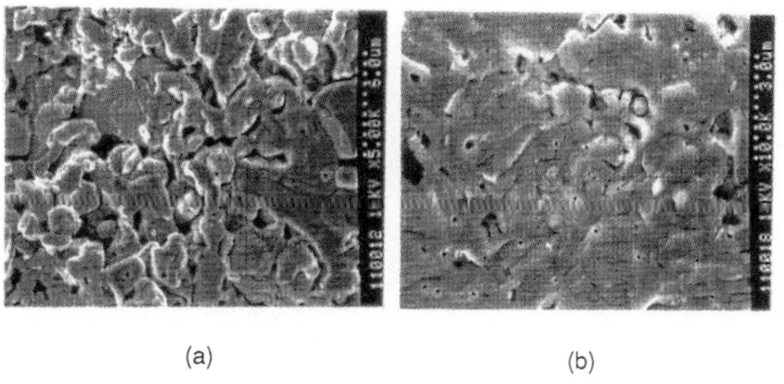

Fig.6 SEM images of the fractured core surface in silver sheathed TBSCCO tape :
(a) a rolled tape (b) a continuous pressed tape

PERFORMANCES OF A SMALL PANCAKE TYPE COIL

Specifications and a photograph of a small pancake type coil were shown in Table 2 and Fig.7, respectively. The sizes of the coil were inner diameter of 30mm and outer diameter of 65mm. Figure 8 shows V-I characteristics of the coil at 77K. The I_c of the coil was 3A in the criterion of 10 μV, which was equivalent with the criterion of 0.01μV/cm. The voltage, which generated less than 1μV between 0A and 3A, was thought to be the coil inductance generated by the current sweeping and a noise. An n-value of the coil was 10 which was equal to that of a short tape.

Figure 9 shows the J_c value of a short straight sample cut from a long tape at 77K in an external magnetic field and load lines of the coil. The magnetic field was applied perpendicularly to the short tape. This coil demonstrated the generation of maximum magnetic field of 103 gauss and also the magnetic field of 47 gauss at the center of the coil. The I_c of this coil was restricted by the I_c of the tape part where maximum self-magnetic field of the coil was induced.

Table 2 Coil specifications

Innerdiameter	mm	30
Outerdiameter	mm	65
No. of turns		67
Coil Ic	A	3.0
Magnetic field (Bo)	G	47
(Bm)	G	103
Wire length	m	10
Superconductor Materials		TBSCCO/Ag
Size	mm mm	t0.14 w2.5
Short sample Ic	A	7.0

Fig.7 Photograph of a pankake type coil

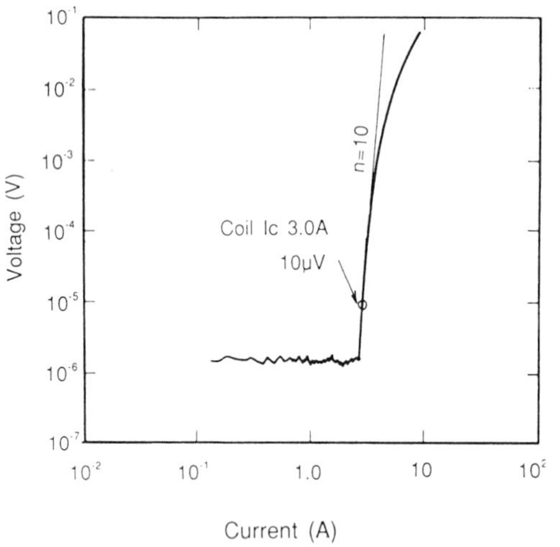

Fig.8 V-I characteristics of pancake type coil in liquid nitrogen.

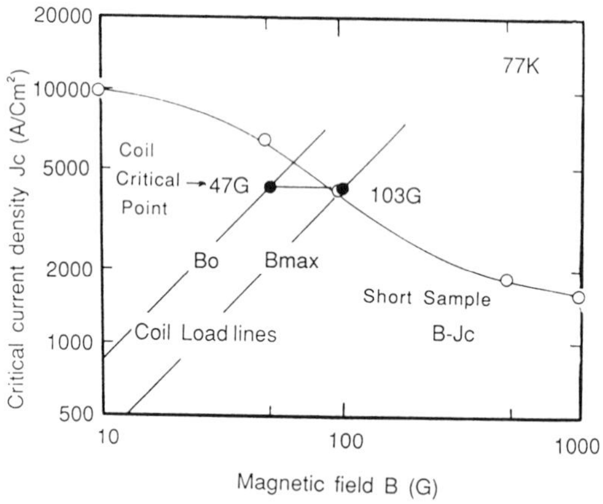

Fig.9 Comparison of short sample and pancake type coil in liquid nitrogen.

CONCLUSION

10m long silver-sheathed TBSCCO superconducting tapes were fabricated by a newly developed continuous pressing method. The J_c values of the tape reached 10^4 A/cm^2 at each section, and deviations of the Jc values were very small. Furthermore, a coil fabricated from the 10m long tape demonstrated the generation of maximum magnetic field of 103 gauss and also the magnetic field of 47 gauss at the center of the coil.

REFERENCES

[1] M. K. Wu et al., Phys. Rev. Lett., 58 (1987) 908.
[2] H. Maeda et al., Jpn. Appl. Phys., 27 (1988) L209.
[3] Z. Z. Sheng et al., Phys. Rev. Lett., 60 (1988) 937.
[4] M. Okada et al., Jpan. J. Appl. Phys., 27 (1988) L185.
[5] M. Seido et al., Proc. 2nd. Symp. on Superconductivity, (1988).
[6] A. Soeta et al., Jpn. J. Appl. Phys., 28 (1989) L1185.
[7] T. Matsumota et al., ISTEC Workshop on Superconductivity, (1989) p111.

AC Power Loss Characteristics of Ag-Sheathed High-T_c Superconducting Tapes

T. SASAOKA[1], M. KANAOKA[1], Y. SEKII[1], S. HOSONO[2], and M. SEIDO[2]

[1] Cable Research Laboratory and [2] Metal Research Laboratory, Hitachi Cable Ltd., Hitachi, Ibaraki, 319-14 Japan

ABSTRACT

In adopting High-Tc superconductors for AC power systems, the evaluation of the AC power loss characteristics of those superconductors is indispensable. Using two kinds of AC loss measuring methods, we evaluated the AC power loss of Ag-sheathed Tl-series superconducting tape at 4.2K and 77K. It was confirmed that the values of the AC power losses under the alternating transport current were almost equal to the values estimated from the DC magnetic method, and that the AC power loss is proportional to the frequency. These results suggest that the AC power loss of our specimen consists mainly of the hysteresis loss.

KEY WORDS: AC loss, Tl-series Ag-sheathed superconducting tape, power application

INTRODUCTION

We have been studying the feasibility of the application of High-Tc superconductors for power equipment including power transmission lines[1]. The AC power loss of the superconductor is one of the crucial factors for power systems using superconductors. Though there are many reports on electrical properties of High-Tc superconducting tapes under the DC transport current, only some results of experimental studys for the evaluation of the AC power loss characteristics are reported[2][3]. From this point of view, we developed a method of evaluating the AC power loss in High-Tc superconducting tapes. It is often said when discussing High-Tc superconductors that the critical current density Jc measured under the DC transport current is quite different from the value of Jc evaluated from the measurement of magnetization[4]. These phenomena have been explained by some thoughts about a " weak link " connected with the grain boundary in the High-Tc superconductor[5] or explained by a sausaging effect, which means that the cross section of the core of superconductors changes along with the direction of the current pass. On the other hand, the hysteresis loss, which is one of the AC power losses, is strongly related to the value of Jc[6] Considering a superconducting power transmission line, it is important to evaluate the AC power loss not only under the alternating magnetic field corresponding to the induced current, but also under the alternating transport current. In this report we adopt two experimental methods and compare the measured values of the total loss of AC power dissipation under the alternating transport current with the loss value estimated from the DC magnetic hysteresis. The measurements in the Ag-sheathed Tl-series superconducting tape were conducted at 4.2K and 77K.

TABLE 1. Characteristics of the tested sample

sample construction	Ag-sheathed tape single core Tl-Ba/Sr-Ca-Cu-O 2223 phase
Ag-thickness(mm)	0.3
tape width (mm)	4
tape length (mm)	100
Jc (A/cm^2)	3400 at 77K, 0 Tesla
Jc (A/cm^2)	2400 at 77K,10mTesla
Jc (A/cm^2)	800 at 77K, 1 Tesla
Jc (A/cm^2)	21000<< at 4.2K, 0 Tesla

EXPERIMENT

The characteristics of the tested superconducting specimen are shown in Table 1[7]. The values of critical current density Jc in the table 1 were measured under the DC transport current. Figure 1 shows a block diagram of an AC power loss measuring system under the alternating transport current. The voltage appearing along the longitudinal direction in the specimen was measured using the low-noise coaxial cable, whose copper leads were soldered directly with indium metal onto the Ag-sheathed specimen. The tiny loss component can be obtained by compensating the large inductive voltage, the phase of which is 90° different from the phase of the transport current. This can be done by using a properly adjusted variable mutual inductance to minimize the input voltage. The loss signal voltage was measured using a digital storage oscilloscope and a lock-in amplifier. We confirmed that the error of the measured voltage, corresponding to the loss signal, was less than 0.1 micro-volt, through the evaluation of the AC resistance of the copper. The DC resistance of the copper was 0.577 micro-ohm as measured by a DC four-probe method at 4.2K. We also confirmed that the deviation from right angle of the phase difference between the primary and the secondary coil of the variable mutual inductance, that was developed less than 0.01°. Thewaveforms of loss signal voltage at the current density below Jc were almost sinusoidal. On the other side, the waveforms of the voltage beyond Jc contains orther high-harmonic peaks. These phenomena are similar to that obserevedin Nb-Ti alloy superconducting wires 8).

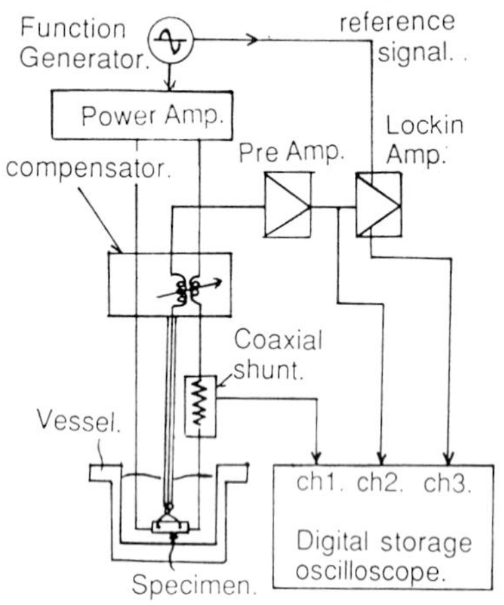

Fig.1 Block diagram of the measurement system.

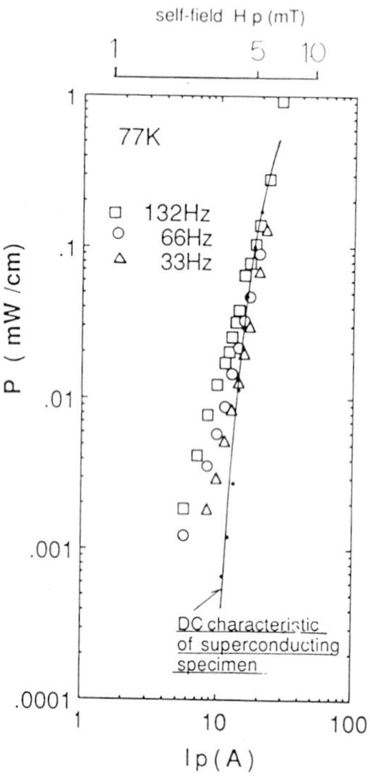

Fig.2 Current dependence of AC power loss of Ag-sheathed Tl-series tape at 77K.

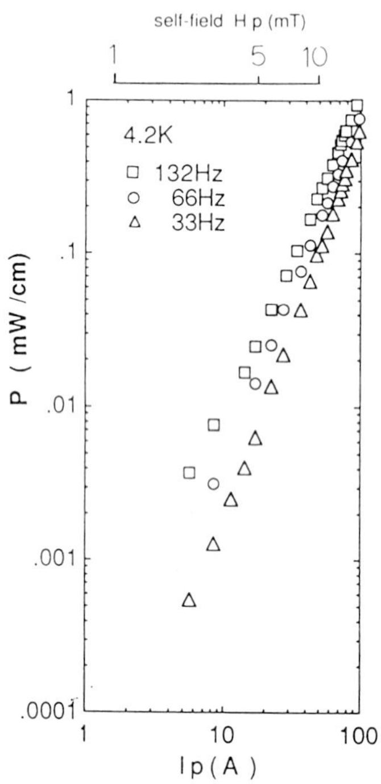

Fig.3 Current-dependence of AC power loss of Ag-sheathed Tl-series tape at 4.2K.

OBSERVATIONS

Figure 2 and figure 3 show the alternating current dependence of the AC power loss, derived from measurements at 77K and 4.2K for Ag-sheathed Tl-series superconducting tape, respectively. The value of the horizontal axis Ip in figure 2 and 3, denotes peak values of the AC current. In figure 2 the DC power dissipation of the specimen is also shown. Finite losses, which are nearly proportional to $I^{3.5}$ at 77K, and are proportional to $I^{2.5}$ at 4.2K, were observed when the AC transport current Ip is smaller than the DC critical current Ic. When Ip is larger than Ic, the AC power loss characteristics become independent on the frequency.

Figure 4 shows the frequency dependence of the AC power loss under the alternating transport current below Ic (Ip < Ic). This figure shows that the AC power total losses depend almost linearly on the frequency both at 77K and 4.2K. These results suggest that the hysteresis loss is the dominant mechanism in the AC power total loss of Ag-sheathed high-Tc super-conducting tapes.

In addition to the measurement of the AC power total loss by the electrical method mentioned above, the magnetic hysteresis of the same sample was measured to estimate the hysteresis loss Ph caused by the magnetic-field corresponding to the alternating transport current given in figure 2 and figure 3. Figure 5 shows the magnetic hysteresis curves obtained with 3 different external magnetic fields, being perpendicular to the surface of the Ag-sheathed superconducting specimen at 77K.

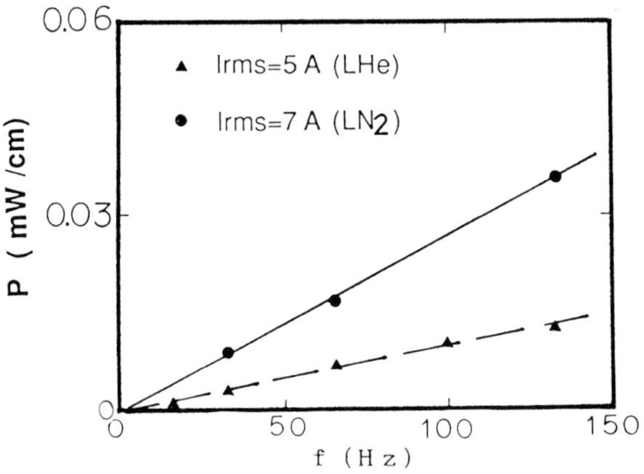

Fig.4 Frequency-dependence of AC power loss

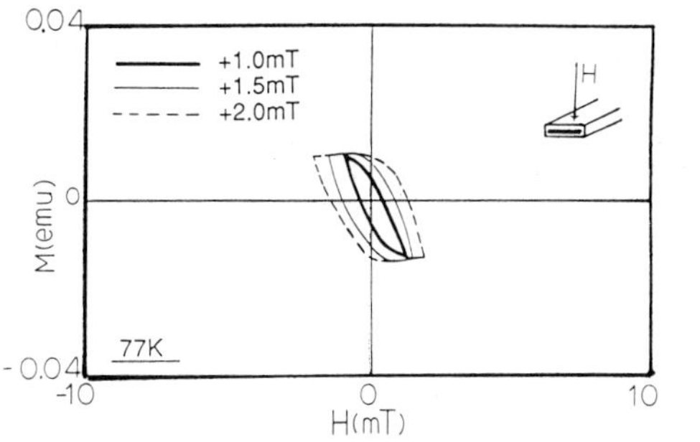

Fig.5 Magnetic-Hysteresis curve of Ag-sheathed Tl-series superconducting tape at 77K.

DISCUSSION

In Figure 6, the current dependence of the AC power loss of the Ag-sheathed superconducting tape at 77K are given. The open circles show the values of the AC loss measured under the alternative transport current and the solid circles show the values estimated from the DC magnetization curve in figure 5.

Total AC power loss of a superconducting tape is supposed to be composed nearly of the hysteresis loss Ph which is proportional to the frequency f ; of the eddy current loss Pe, which is proportional to f^2 ; and of the coupling loss Pc , which is proportional to f^3 . Thus, the AC power total loss P is given in the following equation(1),

$$P = Ph + Pe + Pc = C1 f + C2 f^2 + C3 f^3 \quad \ldots\ldots (1)$$

where C1 and C2 and C3 are the coefficients.

The AC power total loss measured (open circles les in Fig.6) was almost equal to the estimated Ph (closed circles). Also, as shown in figure 4, the AC power total loss P measured under the alternating transport current is proprtional to the frequency f . It was confirmed that these characteristics of the specimen are also found at 4.2K, as shown in figure 7.

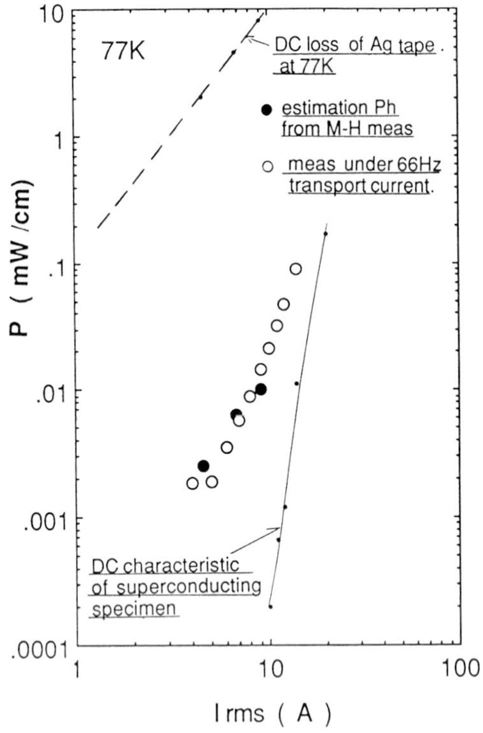

Fig.6 AC power loss of Ag-sheathed Tl-series superconducting tape at 77K. (For comparision the DC loss of Ag tape and theDC power loss of the superconducting specimen at 77K are given)

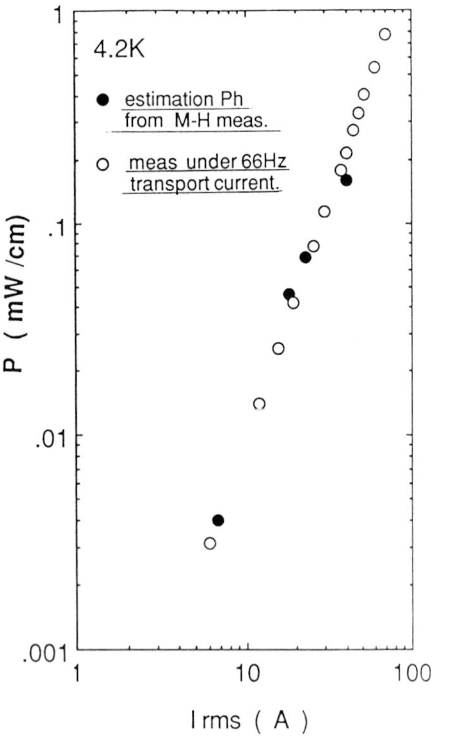

Fig.7 AC power loss of Ag-sheathed Tl-series superconducting tape at 4.2K.

In conclusion our experimental results suggest that the total loss of AC power dissipation of the Ag-sheathed Tl-series superconducting tape is dominated mainly by the hysteresis loss. In other words, the characterestics,which suggest eddy current loss Pe or coupling loss Pc, were not observedin the region of the frequency up to 130Hz below Ic. It was also found that the evaluated AC power loss under the alternating transport current is almost the same as the value determined from the DC magnetization measurement in the case of such comparably weak magnetic fields as up to 10mT.

For comparison, both the DC power dissipation of the superconducting specimen and the DC power loss at 77K of the silver tape, whose cross section is equal to that of the silver cladding of the superconducting specimen, are also given in Figure 6 . The equivalent AC resistivity of the Ag-sheathed Tl-series superconducting tape is less than 1 percent of the DC resistivity of silver tape at 77K. This fact demonstrates the posibility of utilizing our Ag-sheathed Tl-series superconducting tape for power application on liquid nitrogen cooling systems.

REFERENCES

1) T.Sasaoka, M.Kanaoka, Y.Sekii, I.E.E of Japan No 864 in Japanese.
2) T.Hara, K.Okaniwa, T.Yamamoto et al, 1st Power and Energy conference . I.E.E of Japan No217 in Japanese.
3) Yunhui Xu, Weigen Guan, and K.Zeibeig, Appl.Phys.Lett.54(17),24 April 1989.
4) S.Zannella, J.Tenbrink, K.Heine, V.Ottoboni, A.Ricca,and G.Ripamonti ,Appl. Phys.Lett. 57(2),9,July 1990.
5) H.Kupfer, I.Apfelstedt et al, Cryogenics 1988 Vol 28 October.
6) C.P.Bean, Rev. Mod.Phys.36,31(1964).
7) M.Seido, F.Hosono, T.Umezawa, A.Nomoto, K.Nomura,2nd ISS89 PWB-17.
8) N.Yamada and M.Kogawa, Japan.J.Appl.Phys.7(1968)291.

Superconductor/Silver Composites and Tapes Based on Bi-Pb-Sr-Ca-Cu-O

N. SAVVIDES[1], S.X. DOU[2], K.-H SONG[2], H.K. LIU[2], and C.C. SORRELL[2]

[1] CSIRO Division of Applied Physics, Sydney, 2070 Australia
[2] School of Materials Science and Engineering, University of New South Wales, Sydney, 2033 Australia

ABSTRACT

High-J_c bulk composites, containing 10–80 vol%Ag, and silver-sheathed tapes, 100 μm thick, were fabricated using the $(Bi_{1.6}Pb_{0.4})Sr_2Ca_2Cu_3O_{10-y}$ compound. A process of pressing and rolling with intermediate sintering was used to promote grain alignment (up to 90% texture) and improve flux pinning. AC susceptibility measurements on the composites show improvement of flux pinning with 20–30 vol%Ag, giving maximum J_c values of 1600 A cm^{-2} at 77 K. The resistivity of composites decreased in the range 10^{-3} – 10^{-6} ohm cm with increasing silver content. J_c in tapes depended on phase purity and the degree of c-axis texture. The best tapes had $J_c = 6.5 \times 10^3$ A cm^{-2} (77 K) in zero field, decreasing with field to $\sim 10^3$ A cm^{-2} for H$\perp$c-axis and to $\sim$30 A cm^{-2} for H$\|$c-axis at 0.4 T. The temperature dependence of J_c was fitted by the flux creep model using pinning energy $U(0) \simeq 90$ meV.

KEY WORDS: critical current density, flux pinning, flux creep, superconducting tape, superconductor composite.

INTRODUCTION

For the commercial exploitation of high-T_c superconductors in high-temperature or high-field large scale applications, such as magnet coils and power transmission lines, it is essential to develop metal-sheathed wire or tape conductors which have high critical current density $J_c = 10^4 - 10^5$ A cm^{-2} at 77 K [1,2]. Low resistivity superconductor/metal composites are also required to form reliable low ohmic contacts to normal metals. Silver appears to be the metal of choice due to its low reactivity with the superconducting oxides, its permeability to oxygen for the reoxygenation process, and its ductility [3,4]. Because of favourable cryophysical properties of the high-T_c superconductors and that of silver at 77 K it is predicted that for operation at liquid nitrogen temperatures cryostability and stability against flux jumping are achievable by using metal/superconductor ratios of 1 to 2 and superconductor filament diameters of about 100 μm [5].

The 110 K phase superconductor $(Bi_{1.6}Pb_{0.4})Sr_2Ca_2Cu_3O_{10-y}$ was used to prepare bulk silver/superconductor composites containing 10–80 vol%Ag, and silver-sheathed tapes. In previous papers [6,7] we reported on the interaction of silver with this superconductor system, and outlined a procedure of heat treatment under an oxygen partial pressure which prevented the degradation of superconducting properties. In this paper we report magnetic and transport properties of composites and tapes including measurements of J_c as a function of silver content, temperature and magnetic field. J_c of the tapes is found to increase with increasing degree of c-axis texture, and the temperature dependence of J_c is flux creep controlled with pinning energy of about 90 meV. The AC susceptibility of composites provides evidence that flux pinning can be enhanced with the addition of silver.

EXPERIMENTAL TECHNIQUES

Homogeneous single phase powder of the 110 K superconductor $(Bi_{1.6}Pb_{0.4})Sr_2Ca_2Cu_3O_{10-y}$ was produced and used to fabricate both the bulk silver/superconductor composites and the silver-sheathed tapes. The bulk composites were prepared by mixing powder of the superconductor with silver metal powder ($\leq 10\mu$m particle size), pressing into pellets under uniaxial pressure of 200 MPa, sintering at 800 – 840 °C for 50 h, then repeating the process of pressing and sintering to produce specimens about 1 mm thick. Specimens containing $\geq$ 50 vol%Ag became flexible and could be reduced to thin sections, about 0.15 mm thick. The silver-sheathed tapes were fabricated by the powder-in-tube method. A process of pressing and rolling with intermediate sintering followed by lengthy annealing at about 840 °C was used to promote grain growth and favourable texturing. The high temperature heat treatment of all materials was carried out in an oxygen-nitrogen gas mixture with the oxygen partial pressure controlled close to 0.06 atm to preserve the superconductor in the single, high-T_c phase [7].

The specimen resistivity and critical current density were determined from standard four-probe measurements. The bulk specimens were necked down at the centre to about 1 mm in width. J_c for the tapes was calculated using their superconducting core (2 mm width, 100 μm thickness). The critical current criterion was 1 μV/cm. The AC susceptibility was measured using a mutual inductance susceptometer with the field parallel to the long axis of the specimens [8].

RESULTS AND DISCUSSION

Silver/Superconductor Composites

Figure 1 shows normalized resistivity as a function of temperature for bulk specimens of silver/superconductor composites. Above the transition the specimens have metallic conductivity, and the resistivity ratio ρ_{300}/ρ_0 increases with increasing silver content. Examination of polished sections of composites show that the silver remains as isolated islands up to 30 vol%Ag content. Above this level island-to-island contact is evident and matrix inversion takes place gradually, but is never complete even at the maximum silver content. The low resistivity of the composites and the fact that superconductivity can be maintained up to high silver content suggest applications as graded cable terminations allowing low resistivity contact between the superconductor and normal metals. It is also of interest to note that thin sections of composites containing more than 50 vol% Ag become progressively more flexible with silver addition and can be cold worked into various shapes before the final heat treatment.

The critical current density and the resistivity of composites depend on the silver content as shown in Fig. 2. The resistivity results are indicative of percolative type conduction with a percolation threshold at about 35 vol%Ag. J_c is sensitive to the silver content, and it increases with the initial addition of silver, reaching a maximum of 1600 A cm^{-2} at 20~30 vol%Ag. While cryostabilization provided by the silver would contribute to raising J_c, a more valid reason is increased flux pinning due to silver. Since some chemical reaction can occur between silver and the oxides of copper and lead, the silver may provide effective pinning sites at grain boundaries and within grains, thereby raising J_c. Increasing the silver content beyond 20~30 vol%, however, leads first to the formation of S-N-S junctions and then to discontinuities in the superconductor matrix which combine to reduce J_c.

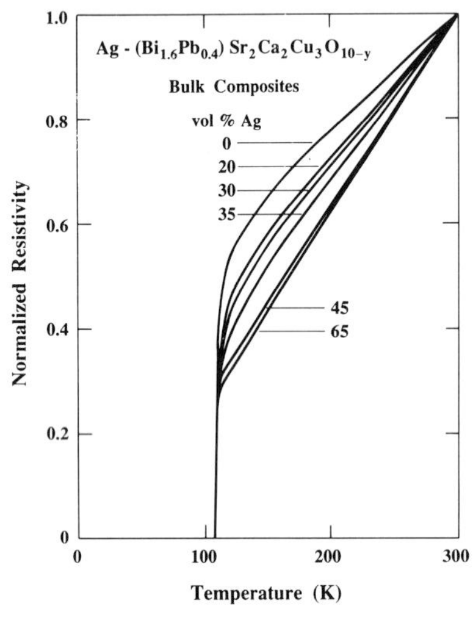

Fig.1. Resistivity vs. temperature for composites.

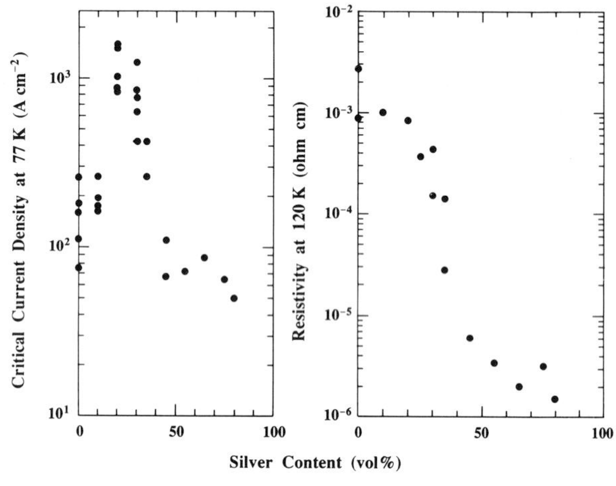

Fig.2. J_c and ρ_{120K} vs. silver content for composites.

Measurements of the complex AC susceptibility, $\chi = \chi' + i\chi''$, can be used to obtain qualitative information regarding flux pinning. In a typical sintered or granular high-T_c superconductor the susceptibility has a sharp decrease in χ' near T_c, a consequence of diamagnetic shielding, and two peaks in χ'', representing hysteresis loss. Müller [9] developed a model for granular superconductors in which superconducting grains are coupled via weak (Josephson) links. In such materials, χ has both coupling or intergrain and intragrain components, and the intergrain component is very sensitive to the applied AC field. Measurements of the field and temperature dependences of the χ'' intergrain and intragrain loss peaks can be used to determine the pinning force density. In Fig. 3 the susceptibility χ'' of composites containing 30 and 75 vol%Ag is compared with that of a bulk specimen that does not contain silver. Two loss peaks are observed, at sufficiently high applied field, one due to the motion of intergranular (Josephson) vortices, the other due to movement of Abrikosov vortices inside the superconducting grains. As the temperature increases toward T_c from below, flux penetrates first the intergrain material, which has a low pinning force, then the intragrain material, which has a higher pinning force. At full flux penetration of the intergrain material the losses reach a maximum (peak in χ'') at a temperature T_m^J, then drop off to zero at T_c.

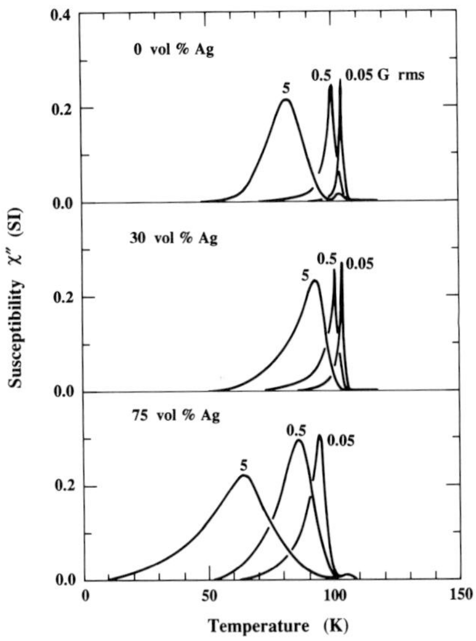

Fig.3. AC susceptibility χ'' for composites.

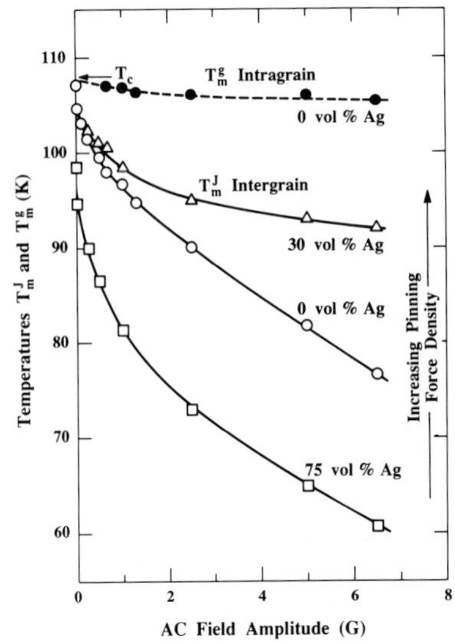

Fig.4. χ''-peak temperatures, T_m^J and T_m^g, as a function of AC field amplitude for composites.

Flux penetration of the grains also leads to losses, with the intragrain loss peak occuring at a higher temperature T_m^g due to the stronger flux pinning within the grains. Above the intergrain peak the superconducting grains are effectively decoupled. Figure 4 shows the variation of the temperatures T_m^J and T_m^g with the AC field amplitude. Increasing pinning force density raises the temperatures T_m^J and T_m^g toward T_c. The non-linear dependence of T_m^J on field [9] is believed to be due to flux creep. It is seen that the 30 vol%Ag specimen has the highest pinning force density. This result and the fact that the intragrain loss peaks are absent in this specimen indicate flux pinning can be enhanced by some silver addition. Excessive amounts of silver, however, lead to weak links (S-N-S) and consequently the pinning force density decreases. These observations correlate with the critical current density results shown in Fig. 2.

Silver-sheathed Tapes

Recent studies of the transport and magnetic properties of high-T_c superconductors focusing on the dynamics of flux movement and dissipation [9-12] have shown that J_c can be severely limited by the weak links. From a practical standpoint of obtaining high critical current densities, weak-link minimization must be achieved as well as pinning-force enhancement within grains.

Figure 5 shows the temperature dependence of J_c for silver-sheathed tapes. The data points represent the results for single-phase tapes having different degree of texture (c-axis normal to the tape sheat) [4]. These tapes were heat treated under low oxygen partial pressure to produce the (2223) phase, $(Bi_{1.6}Pb_{0.4})Sr_2Ca_2Cu_3O_{10-y}$. The experimental data are well represented by the Anderson-Kim flux creep model discussed by Savvides [10], i.e. J_c is controlled by flux creep. The movement of flux is thermally activated, with a zero-temperature pinning energy $U(0)$, and leads to decreasing J_c with increasing temperature. The solid curves are theoretical fits to the data using $U(0) \simeq 90$ meV. Also shown in Fig. 5 by the dashed curves are data for tapes which were heat treated in air to form both the (2223) phase and the (2212) low-T$_c$ phase (T_c = 80 K). Consequently J_c has a weak-link behaviour that can be identified as resulting from S-N-S junctions [10], with the low-T_c phase acting as a normal metal for temperatures above about 80 K.

Figure 6 compares the magnetic field dependence of J_c at 77 K of a single-phase tape and a bulk sintered specimen. J_c in the bulk materials is very sensitive to the applied field, and the steep fall of J_c with increasing field is a consequence of intergranular weak links. The tape has a much higher J_c value (in zero field) due to the special microstructure produced by the process of rolling and pressing, i.e. a high degree of crystalline alignment (90% c-axis texture), and a high pinning force density. In the presence of a DC field J_c of the tape is anisotropic. For H$\|$c-axis, J_c decreases rapidly with increasing field reaching a value of about 30 A cm^{-2} at 0.4 T, while for H$\perp$c-axis, J_c remains high ($\sim 10^3$ A cm^{-2}) after an initial drop. It is clear from these results that, although special processing techniques can lead to improvements in J_c through texture, further improvements of flux pinning are needed in order to take advantage of the high intrinsic critical parameters of the high-T_c superconductors for large scale applications.

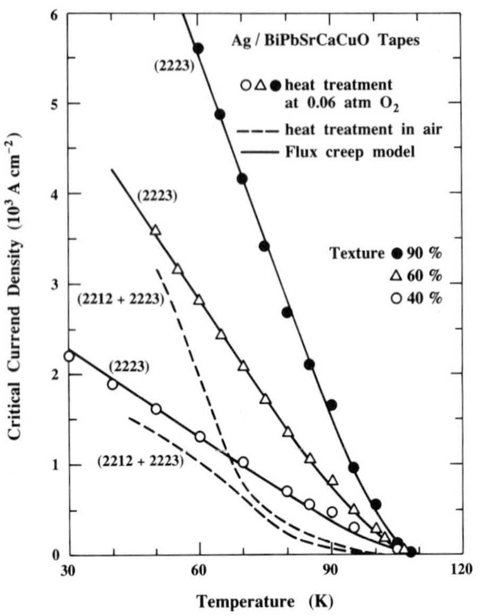

Fig.5. J_c vs. temperature for tapes. The solid curves are fits to the data by the flux creep model [10].

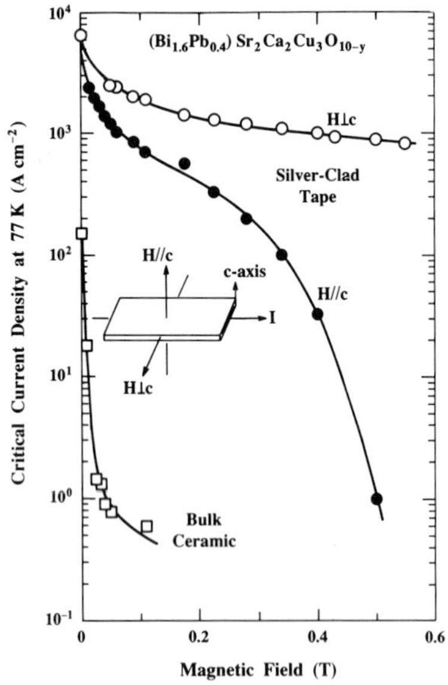

Fig.6. Magnetic field dependence of J_c.

CONCLUSION

Silver/superconductor bulk composites, containing up to 80 vol%Ag, and silver-sheathed tapes of the compound $(Bi_{1.6}Pb_{0.4})Sr_2Ca_2Cu_3O_{10-y}$ were prepared by special microstructure and processing techniques. The resistivity of the composites decreased by several orders with increasing silver content while their J_c at 77 K reached a maximum value of 1600 A cm^{-2} at about 20 vol%Ag. Such composites should find application as graded superconductor-to-metal low resistivity terminals. The critical current density of tapes depended on the phase purity and the degree of texturing. The best tapes (superconducting core 2 mm x 100 μm, with 90% c-axis crystalline alignment) had $J_c = 6.5 \times 10^3$ A cm^{-2} at 77 K in zero field, decreasing to 10^3 A cm^{-2} for H⊥c-axis and to 30 A cm^{-2} for H||c-axis when the field increased to 0.4 T. The temperature dependence of J_c was thermally activated and could be fitted to the flux creep model using a zero-temperature pinning energy $U(0) \simeq 90$ meV.

Acknowledgement: This work was supperted by Metal Manufactures Ltd and Department of Industry Technology and Commerce under GIRD grant No.15017 .

REFERENCES

1. D. Dew-Hughes, *Proc. Instn. Mech. Engrs.* **202** (1988) 147.
2. T. H. Geballe and J. K. Hulm, *Science* **239** (1988) 367.
3. Y. Yamada, K. Jikihara, T. Hasebe, T. Yanagiya, S. Yasuhara, M. Ishihara, T. Asano and Y. Tanaka, *Jpn. J. Appl. Phys.* **29** (1990) L456-L458.
4. N. Uno, N. Enomoto, H. Kikuchi, K. Matsumoto, M. Mimura and M Nakajima, *Advances in Superconductivity II*, ed. T. Ishiguro and K. Kajimura (Springer-Verlag, Tokyo 1990), p.341.
5. E. W. Collings, *Advances in Superconductivity II*, ed. T. Ishiguro and K. Kajimura (Springer-Verlag, Tokyo 1990), p.327.
6. S. X. Dou, K.-H. Song, H. K. Liu, C. C. Sorrell, M. H. Apperley, A. J. Gouch, N. Savvides, and D. W. Hensley, *Physica C* **160** (1989) 533.
7. S. X. Dou, K.-H. Song, H. K. Liu, C. C. Sorrell, M. H. Apperley, and N. Savvides, *Appl. Phys. Lett.* **56** (1990) 493.
8. N. Savvides, S.J. Collocott, C. Andrikidis, K.-H. Müller and E.R. Vance, *Physica C* **171** (1990).
9. K.-H. Müller, *Physica C* **159** (1989) 717.
10. N. Savvides, *Physica C* **165** (1990) 371; *Advances in Superconductivity II*, ed. T. Ishiguro and K. Kajimura (Springer-Verlag, Tokyo 1990), p.669.
11. J.W. Ekin, T.M. Larson, A.M. Hermann, Z.Z. Sheng, K. Togano and H. Kumakura, *Physica C* **160** (1989) 489.
12. D. Dimos, P. Chaudhari and J. Mannhart, *Phys. Rev. B* **41** (1990) 4030.

Preparation of Bi-Pb-Sr-Ca-Cu-O Wires/Tapes by Partial-Melt Mediated Process

D.Y. JEONG, D.W. HA, M.H. SOHN, K.S. RYU, M.S. YUN, S.B. BYON, and U. AHN

Applied Superconductivity Laboratory, Korea Electrotechnology Research Institute, Changwon, 641-120 Korea

ABSTRACT

This study was focused on finding a proper process using the benefits of partial melting in the preparation of 2223 phase wires/tapes by the powder-in-tube method. The wires/tapes were prepared by the combination process of partial melting, cold-working and sintering. The preparation parameters were the temperature and the stage of the partial melting, the sintering atmosphere and the cold-worked state. The results are discussed in connection with Tc, Jc, crystallography and microstructure. The results suggest various preparation processes of the 2223 tapes with high Jc using the benefits of the partial melting method.

KEY WORDS : BiPbSrCaCuO, Superconducting wire, partial melting, powder-in-tube method.

INTRODUCTION

It is well known that the melt growth method [1, 2, 3] and the repetitive process [4, 5, 6] of pressing and sintering enhance the densification and the grain alignment in $Bi_2Sr_2CaCu_2O_x$ (2212) and $(Bi,Pb)_2Sr_2Ca_2Cu_3O_x$(2223), respectively, resulting in the Jc's of 10^4 A/cm² and the weak dependence of the Jc's on magnetic field intensity [2, 5]. There were reports [7, 8] that the partial melting method in the 2223 decreased electrical Tc due to easy formation of nonsuperconducting phases although it enhanced the formation and growth of the 2223 phase. In this study, it was attempted to find a proper process using a partially melted state as an intermediate state for the preparation of 2223 phase wires/tapes with high Jc. The process-dependence of Bi-Pb-Sr-Ca-Cu-O (BPSCCO) wires/tapes prepared by partial-melt mediated process is presented in relation to mainly their electrical properties and microstructures.

EXPERIMENTS

The BPSCCO wires/tapes were prepared by combination processes of partial melting, cold-working and sintering, in addition to so-called powder-in-tube technique. The powder mixture of Bi_2O_3:PbO:$SrCO_3$:$CaCO_3$:CuO = 0.92:0.34:1.97:2.03:3.06 was calcined at 800°C for 48 hr in air. The calcined mixture was ground and packed into silver tubes with 6 mm in outer diameter and 3.5 mm in inner diameter. The tubes were drawn to wires with 1.24~2.24 mm in outer diameter. The wires were cut into 3cm-long pieces and heat-treated or / and rolled according to the different preparation processes as shown in Fig.1. The preparation parameters are temperature and stage of the partial melting, cold-worked state, and sintering atmosphere. The partial melting temperature varied from 860°C to 900°C with an interval of 10°C. The partial melting was conducted after calcination at 800°C for 48 hr in air, after sintering at 840°C for 48 hr in air, and after full growth of the 2223 phase. The wires/tapes were quenched to room temperature by taking them out of the furnace. The wires in the process Ⅳ experienced rolling after the calcination and drawing. In the processes Ⅴ, the wires were rolled after the 48 hr sintering. The wires/tapes were sintered at 860°C in air (Process Ⅰ) or at 840°C in an atmosphere of Ar:O_2

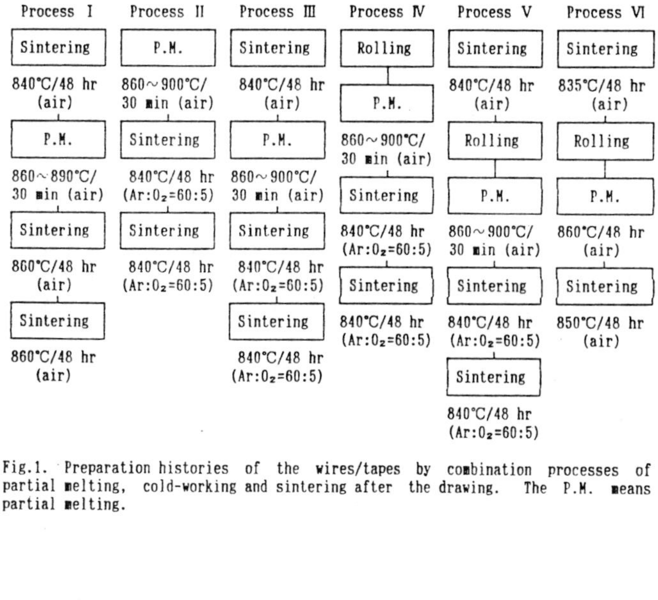

Fig.1. Preparation histories of the wires/tapes by combination processes of partial melting, cold-working and sintering after the drawing. The P.M. means partial melting.

= 60:5 (mℓ/min) (Processes II ~ V) upto 96 hr after the partial melting. In the process VI, the tapes were sintered upto 96 hr in air at a partially melted state of 850 ~860°C. To examine the reproducibility of the high-Tc in the wires prepared by the process II and the effects of partial melting and rolling after full growth of the 2223 phase, a series of experiments (processes VII~IX) were conducted as shown in Fig.2, with 890°C of partial melting temperature.

The Tc and Jc of the wires/tapes were measured by the standard 4-probe dc method. The test current for R vs. T curves was 30 mA. 1 μV/cm potential drop was used as a criterion for Jc. The microstructures were observed by optical and scanning electron microscopes. The X-ray diffraction patterns were obtained to identify the phases present in the wires/tapes.

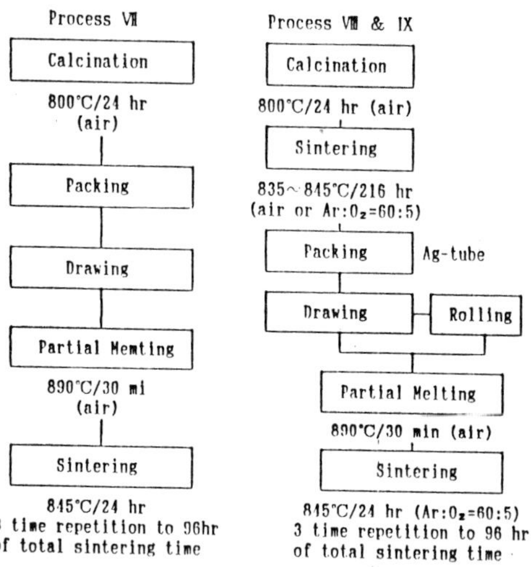

Fig.2. The preparation histories of the wires/tapes to study the reproducibility of the high-Tc in the wires prepared by the process II and the effects of partial melting and rolling after full growth of the 2223 phase.

RESULTS AND DISCUSSION

The Tc's in the wires/tapes prepared by the processes I ~ VI are listed in Table 1 with the partial melting temperatures used. Jc's measured at 77 K and 0 T in the wires/tapes showing relatively high Tc are listed in Table 2. In general, the 2223 wires/tapes could be prepared by sintering them in an atmosphere of $Ar:O_2$ = 60:5(mℓ/min) not in air after the partial melting. In the wires/tapes prepared via sintering in an atmosphere of $Ar:O_2$=60:5(mℓ/min), the sintering at 840°C for 48 hr prior to the partial melting resulted in high Tc regardless of their cold-worked state. In that case, the Tc's and Jc's are higher in the tapes than in the wires, indicating that the rolling process after the sintering enhanced densification and growth of the 2223 phase. On the other hand, the wires/tapes prepared via partial melting after calcination, i.e., without a sintering process prior to the partial melting, showed quite different behavior compared to the previous case. The high Tc could be observed only in the wires prepared with partial melting at 870~890°C without a rolling process. The rolling process after the calcination seems to hinder the formation and growth of the 2223 phase during both processes of partial melting and sintering after the partial melting.

Table 1. Tc's (K, R = 0) in the BiPbSrCaCuO wires/tapes prepared by the processes I ~ VI including partial melting. (Test current = 30 mA)

	Partial melting temperature (°C)				
Process No.	860	870	880	890	900
I	70 *	35	67	62	
	70 **	59	55	54	
II	— ***	52	86	72	—
	—	92	—	99	—
III	—	63	83	90	63
	87	95	93	96	83
IV	53	—	68	69	50
	58	60	69	71	72
V	93	90	75	93	92
	100	96	97	88 ****	99
VI	95				
	100				

*, ** The Tc's in the first and the second rank of each process are the values measured in the wires/tapes sintered for 48 hr and 96 hr after the partial melting in the process, respectively.

*** The mark "—" means that the superconducting transition was not observed down to 40~50 K at which our system can distinguish the potential drops due to the superconducting transition and the resistance of the silver sheath.

**** The Tc was taken in the tape bent when the tape was dealt for the measurement.

Table 2. Jc's (A/cm²) measured at 77 K and 0 T in the wires/tapes with high Tc. The Jc was obtained in the wires/tapes sintered for 96 hr after the partial melting.

	Partial melting temperature (°C)				
Process No.	860	870	880	890	900
II				149	
III	74	112	238	157	112
V	318	145	264	114	391
VI	240 *				
	640 **				

* The Jc was obtained in the tape sintered for 48 hr in a artially melted state at 860 °C

** The Jc was obtained in the tape sintered at 850°C for 48 hr after the above process.

The relatively high Tc observed in the wires/tapes prepared with the sintering prior to partial melting is attributed to the formation of 2223 phase during the prior sintering process, as verified by the step-like resistence drop near 100 K in R vs. T curve taken in a bulk specimen sintered at 835°C for 48 hr in air as shown in Fig.3. The 2223 phase is stable at 860~900°C used for the partial melting and acts as a trigger for further formation of the 2223 phase in the final sintering process [9]. The observed higher Tc and Jc in the tapes than in the wires are attributed that a rolling only after the 2223 phase formation is effective in the growth and the alignment of the 2223 phase grains. It is noticeable in the wires that the partial melting at 870~890°C resulted in high Tc regardless of the prior formation of the 2223 phase, implying that the 2223 phase can form in the temperature region in air in a relatively short period of time.

When the wires/tapes were prepared by partial melting at 890°C after full growth of the 2223 phase followed by sintering at 845°C upto 96 hr, the wire and the tape sintered for 72 hr showed the highest Tc's of 88 K and 97 K, respectively, as shown in Table 3, indicating that the rolling accelerated the recovery of 2223 phase during the sintering process. Among the wires/tapes prepared by the processes I ~ IX, the highest Jc of 640 A/cm² was obtained in the tape prepared by the process VI as shown in Table 2. However, 100 K of Tc and 480 A/cm² of Jc were obtained in the wire prepared by sintering for 48 hr in the process VII. It is expected, therefore, that the tapes with high Jc can be prepared in a relatively short period of time by incorporating the process II or VII into the repetition process of rolling and sintering.

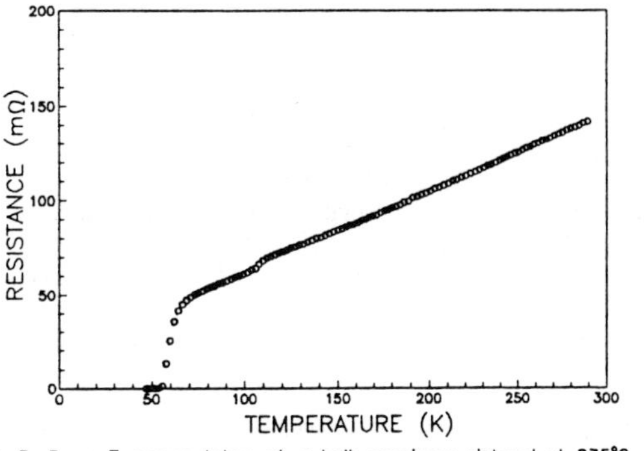

Fig.3. R vs. T curves taken at a bulk specimen sintered at 835°C for 48 hr in air after calcination at 800°C for 24 hr.

Table 3. Tc's (K, R = 0) in the BiPbSrCaCuO wires/tapes prepared by the processes VI, VII and IX including partial melting at 890°C. (Test current = 30 mA)

Process No.	Total sintering time (hr) 24	48	72	96
VI	40	100	92	96
VII	65	75	88	72
IX	73	92	97	95

Fig. 4 shows an XRD pattern taken in the tape sintered for 96 hr after partial melting at 900°C in the process V. The volume fraction of (001) of the 2223 phase estimated by using % = (I_{hkl} / ΣI_{hkl}) x 100 was 73.4%. All the tapes prepared by the process V showed more than 50% of (001). Unfortunately we could not identify at present that the melt-mediated process added to the rolling enhanced the directional growth of 2223 phase further.

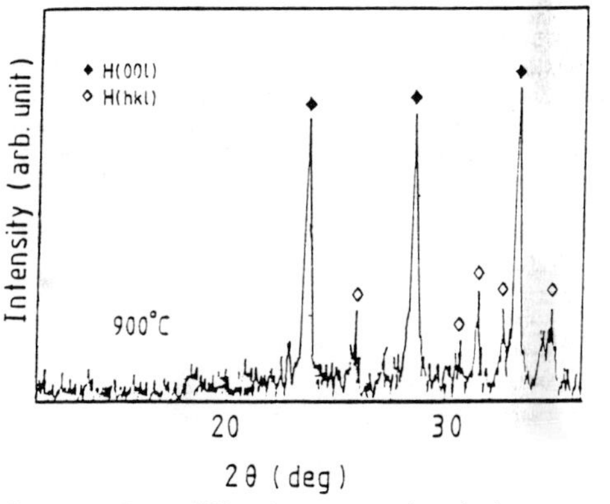

Fig.4. An X-ray diffraction pattern taken in the tape prepared by the process V with partial melting at 900°C. (001) peaks of the 2223 phase are prominent.

Optical micrographs taken at the polished surfaces of the wires/tapes showed quite distinct features depending on the degree of the 2223 phase growth and the cold-worked state prior to the partial melting process. All the wires/tapes except the tape prepared by the process IV showed microstruc-tures with needle-like or extended needle-like grains. An example of the features are shown in Fig.5. (a), (b), (c), (d) and (e) in Fig.5 are optical micrographs taken in the wires/tapes prepared by the processes I, II, III, IV and V, respectively, with partial melting at 890°C. Fig. 5(f) was taken in the tape sintered at 860°C for 48 hr in air in the process VI. The degree of needle-like feature increased in the order of process II, I or III, VIII, V, IX and VI. In the wires, the needle-like feature decreased with increasing the distance from the interface between the Ag sheath and the BPSCCO material, implying that the pressure effect due to the drawing in the processes decreased in that direction. In the tapes prepared by the processes V, VI and IX, the microstructures were even textured. The grain-alignment or texturing became more prominent by sintering at a partially melted state after rolling. The tapes prepared by the process IV, however, showed microstructures with little needle-like grains. Fig. 6 (a) and (b) shows SEM photographs taken at polished surfaces of the tapes prepared by the process V with partial melting at 860 and 900°C, respectively. The higher temperature used for the partial melting seems to result in the wider grains in c-direction.

In summary, the 2223-phase wires/tapes were prepared by various combination processes of partial melting, cold-working and sintering. The highest textured structure was obtained by sintering at a partially melted state after rolling. The fast growth of the 2223 phase during the sintering process after partial melting at 870~890°C may imply that the partial-melt mediated process can be used to prepare high-Jc 2223 tapes in a relatively short period of time by incorporating it into the repetition process of rolling and sintering.

(a) (b) (c) (d) (e) (f)

Fig.5. Optical micrographs taken on the longitudinal or thin longitudinal surfaces of the wires/tapes prepared by the processes I ~ VI. (a), (b), (c), (d) and (e) were taken in the wires/tapes prepared by the processes I, II, III, IV and V, respectively, with partial melting at 890℃. (f) was taken in the tape sintered for 48 hr at a partially melted state of 860℃ for 48 hr in air in the process VI.

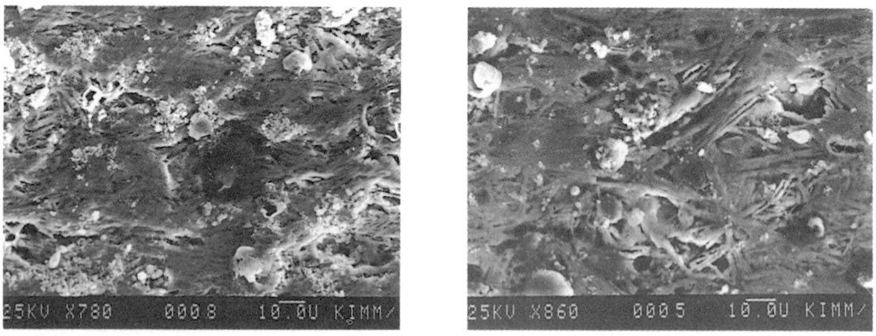

(a) (b)

Fig.6. Scanning electron micrographs taken on the thin longitudinal surfaces of the tapes prepared by the process V with partial melting at (a) 860℃ and (b) 900℃.

REFERENCES

1. Kase J, Irisawa N, Morimoto T, Togano K, Kumakura H, Dietderich DR, Maeda H (1990) Appl. Phys. Lett. 56: 970
2. Asano T, Tanaka Y, Fukutomi M, Maeda H (1990) Jpn. J. Appl. Phys. 29: L1066
3. Kase J, Togano K, Kumakura H, Dietderich DR, Irisawa N, Morimoto T, Maeda H (1990) Jpn. J. Appl. Phys. 29: L1096
4. Mimura M, Kumakura H, Togano K, Maeda H (1989) Appl. Phys. Lett. 54: 1582
5. Kumakura H, Togano K, Maeda H, Mimura M (1990) J. Appl. Phys. 67: 3443
6. Sato K, Hikata T, Mukai H, Masuda T, Ueyama M, Hitotsuyanagi H, Mitsui T, Kawashima M (1989) Proc. 2nd ISS, (Nov. 14-17, Tsukaba): 335
7. Hatano T, Aota K, Ikeda S, Nakamura K, Ogawa K (1988) Jpn. J. Appl. Phys. 27: L2055
8. Oota A, Ohba K, Ishida A, Kirihigashi A, Iwasaki K, Kuwajima H (1989) Jpn. J. Appl. Phys. 28: L1174
9. Horiuchi S, Shoda K, Wu X, Nozaki H, Tsutsumi M (1990) Physica C 168: 205

Possibility of Manufacturing Superconductive Film by Laser Treatment of Plasma Sprayed Oxide Ceramic Layer

HAJIME MIYAZAWA, KATSUYOSHI HOTTA, SHUICHI WATANABE, SHOJIRO MIYAKE, HARUO HIROSE, and MASAO MURAKAWA

Nippon Institute of Technology, 4-1, Gakuen-dai, Miyashiro-machi, Minamisaitma-gun, Saitama, 345 Japan

ABSTRACT

A basic study on the preparation of superconductive film through a laser treatment of a plasma sprayed oxide ceramic layer was conducted. As the first step of this study, an experiment was carried out to recover the superconductive crystal structure which was damaged during the plasma spraying operation. A superconductive ceramic powder of a Y-Ba-Cu-O system was first plasma sprayed onto a stainless steel substrate. A CO_2 laser beam was then irradiated onto the sprayed surface film together with the oxygen gas flow. Results showed that an appropriate combination of laser parameters produce a sprayed film having the superconductive crystal structure.

KEY WORDS: Y-Ba-Cu-O system superconductor, plasma spray, CO_2 laser, superconductive crystal structure

INTRODUCTION

It has been reported that a superconducting oxide ceramic film can be obtained by using the plasma spraying technique[1,2]. However, a subsequent heat treatment for as long a period as several hours in oxygen or air atmosphere is required to reproduce or produce satisfactory superconductive characteristics such as transition temperatures and magnetic properties (Here, the roll of the heat treatment is to supply oxygen into the sprayed film which is dissociated during the plasma spraying operation). The above mentioned necessity of heat treatment in the furnace spoils the significant advantage of using the plasma spraying technique i.e., the advantage being applicable to a wide area.

In an attempt to produce a superconductive film having a wide area, a basic study on the preparation of superconductive film using a laser treatment of a plasma sprayed oxide ceramic film was conducted.

A typical superconductive ceramic powder of Y-Ba-Cu-O system was selected as a target and the experiment was carried out to recover the superconductive crystal structure of the original powder which was damaged during the plasma spraying operation when transforming the powder into a film.

EXPERIMENTAL

It is known that significant losses of oxygen and copper happen during the plasma spraying operation. Then, in the preparatory step to arrange the spraying powder, as much as 40 at.% copper element against YBa_2Cu_3Ox system (to be refered to as "YBCO" hereinafter) was added to the previously calcined YBCO powder compensating the loss of the copper. The additional copper element was supplied through CuO powder. These two kinds of powders were put in a small capsule and shaken about one hour to mix completely. An austenitic stainless steel plate (7 x 14 x 1 mm) was used as a substrate.

A procedure of producing a laser treated specimen is as follows.

1) Apply a grid blast treatment of the substrate surface using aluminum powder: This treatment is conducted for the purposes of cleaning the surface of the substrate and to improve the bonding characteristics of the sprayed film onto the substrate.

2) Conduct plasma spraying of NiCrAlY onto the substrate: A NiCrAlY layer as thin as 100 μm is prepared onto the substrate to prevent the cracking in the YBCO film being prepared afterwards.

3) Apply plasma spraying of YBCO powder: The thickness of the prepared film is about 100 μm. The characteristics of the plasma sprayed film differs not only by the spraying instruments used but also by spraying conditions.

In the preliminary experiments, it was found that the color of the sprayed film changes drastically into green, blue, brown, gray, and black in accordance with the spraying conditions. We eventually selected the spraying condition which yields a film as black as the calcined YBCO powder in this experiment. A PLASMADYNE SG-100 spray gun was used to prepare the film.

4) Apply a laser treatment to the sprayed YBCO layer: It is known that the melting of the YBCO material spoils its superconductive characteristics. Accordingly, The laser condition which does not melt the sprayed surface but can heat up the film high enough to turn the surface red at the laser irradiated spot (about 5 mm in dia.) was adopted. Then the coated material was moved linearly at an adequate speed so that the whole surface of the film could be laser treated. Oxygen assist gas was blown onto the heated red spot co-axially with the laser beam axis to promote the oxidization of the film. A Mitsubishi ML-1000P carbon dioxide laser was used for the treatment of the sprayed layer.

Fig.1 shows the the schematic illustration of the plasma sprayed specimen and Fig.2 shows the set up of the laser treatment.

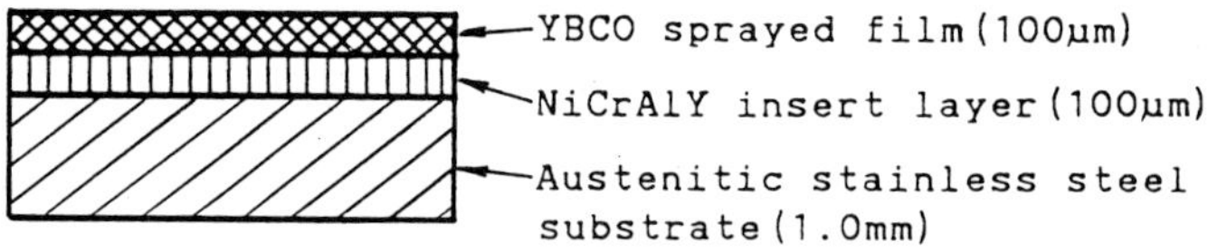

Fig.1 Schematic illustration of the plasma sprayed specimen

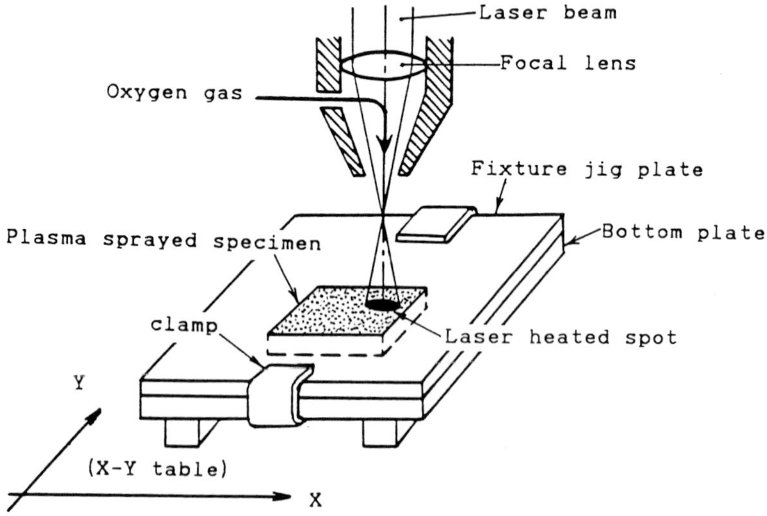

Fig.2 Schematic illustration of the set up of the laser treatment

To be more specific, the plasma sprayed specimen was placed on the bottom plate and fixed tightly in the rectangular groove in the austenitic stainless steel fixture jig plate to 1) minimize the difference of temperature distribution especially at the edges and the outer line portion of the specimen and to 2) avoid the specimen blown away by the above mentioned oxygen assist gas pressure. The experiment was conducted on the X-Y table and the table was moved linearly changing its directions X and Y alternatively so that the whole surface of the specimen could then be laser treated.
Table 1 summarizes the experimental conditions.

Table 1 Experimental conditions

spray powder : calcined YBa_2Cu_3Ox + CuO(40at.%)		
Plasma spraying conditions		
Arc gas (Argon)	:	44 SL/min
Powder gas (Argon)	:	10 SL/min
Powder hopper setting	:	3.5 rpm
Operating mode	:	subsonic
Laser treatment conditions		
Laser	:	Continuous wave CO_2 laser
Output power	:	100 W
Defocus distance	:	60 mm
Assist gas(Oxygen)	:	0.5 kgf/cm^2
Beam scan speed	:	160 mm/min
treating number	:	up to 4 times

(SL:standard litters)

The crystal structure of the film was observed by X-ray diffractometer. The film surface was characterized by SEM (scanning electron microscope) observation and by EDX (energy dispersive X-ray) analysis.

RESULTS AND DISCUSSION

Fig.3 shows the X-ray diffraction patterns of the as-sprayed film (a) and those laser treated films (b), (c) and (d) treated at different times. Although not shown in the figure, the diffraction pattern of laser film treated ten times revealed almost the same results as the pattern of Fig.3(d). For comparison, the diffraction pattern of the superconductive calcined YBCO powder is also shown in Fig.3(e).
Fig.3(a) shows that the superconductive crystal structure is destroyed thoroughly during plasma spraying operation. It was then observed that laser treated one time is not sufficient at all to bring back YBCO crystal structure film. It was found that multi treatment is required to restore the YBCO crystal structure.
Fig.3(d) shows that a laser treated four times can provide almost the same crystal structure with original calcined YBCO powder. Although the quantative analysis has not yet been done, we surmise that the oxygen which dissociated during plasma spraying operation would be restored through the laser treatment in the oxygen gas flow.

Fig.4 shows the SEM photos and corresponding EDX mappings for yttrium, barium, and copper of the films respectively (a) for the as-sprayed and (b) for the laser treated (The EDX mappings for Ba and Cu of as-sprayed film are not shown in the figure because there is not such an observable difference in the distribution patterns from those of the laser treated ones).

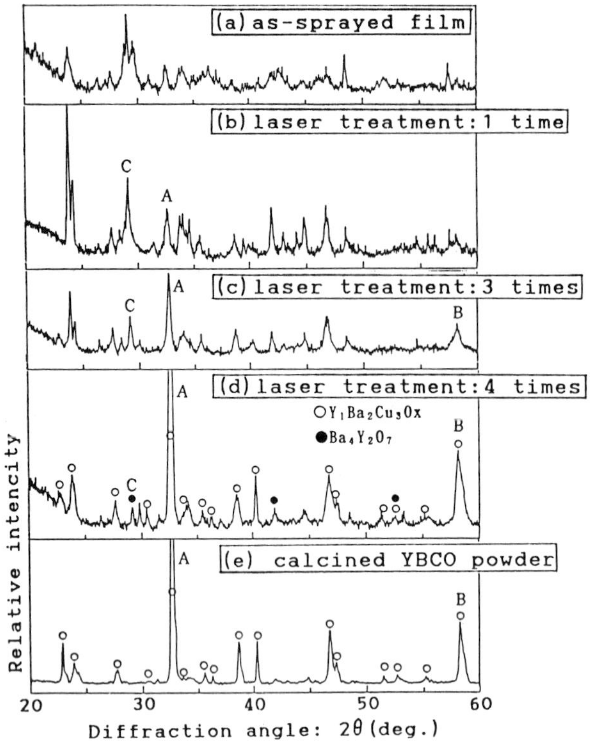

Fig.3 X-ray diffraction patterns of the films and the calcinated YBCO powder

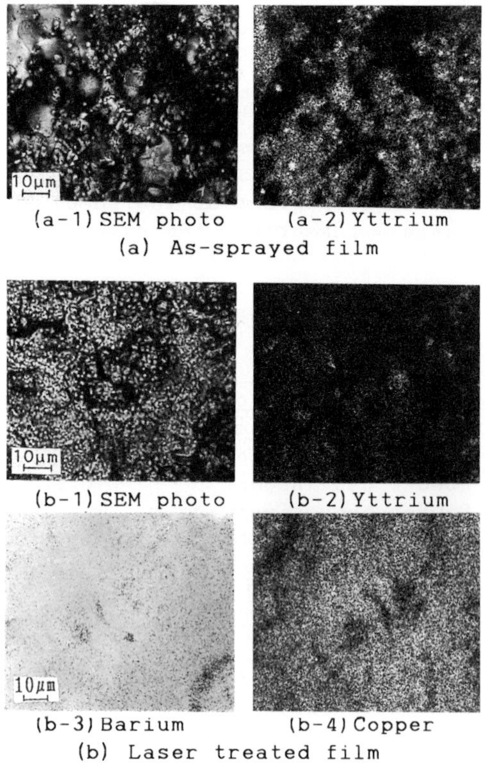

Fig.4 SEM photos and EDX mappings of the films

It should be noticed that the laser treatment can improve the distribution condition of an element on the sprayed film i.e., the segregated yttrium in the as-sprayed film (concentrated white dots in Fig.4(a-2)) can be dispersed by laser treatment. The distribution of the other elements (barium and copper) is observed to be in good condition.

CONCLUSION

Using a mixed powder of calcined YBa_2Cu_3Ox and CuO powders, a basic study on the preparation of superconductive film using a laser treatment of a plasma sprayed oxide ceramic layer was conducted. The following results were obtained.

(1) The appropriate combination of laser parameters (including treating number) can give a sprayed film a superconductive crystal structure.

(2) Laser treatment can disperse the segregated yttrium element in the as-sprayed film.

REFERENCES

1. Neiser RA, Kirkland JP, Elam WT, Skelton EF, Herman H, Bancke GA, Sampath S, Gansert D, Wang HG (1987) Proceedings of the National Thermal Spray Conference, Orlando, Florida. pp195-201
2. Suryanarayanan R, Brun G, Bouanani AM, Meguellati F, Rateau M, Pankowska H, Gorochov O, Bhandage GT (1989) Proceedings of the 12th International Conference on THERMAL SPRAYING, London, UK. pp(47-1)-(47-5)

Preparation of Superconducting Oxide Films by Sol-Gel Method

AKIYO OKUHARA, MAKOTO HIRAOKA, and YOSHINORI TAKADA

Central Research Laboratory, Mitsubishi Cable Industries, Ltd., 8, Nishinocho, Higashimukaijima, Amagasaki, 660 Japan

ABSTRACT

For preparing superconducting wires and tapes, the sol-gel method is one of the most useful techniques. We prepared highly-oriented Y-Ba-Cu-O and Bi-Sr-Ca-Cu-O superconducting films using a solution of acetate and propionic acid. The Y-Ba-Cu-O films on Ag substrate were dense and homogeneous and achieved T_c of 88K and J_c of 7,500A/cm^2(77K, 0T). The Bi-Sr-Ca-Cu-O films were prepared on MgO substrate. The film including some high-T_c phase and T_c=87K was obtained with the starting composition of 2223.

KEY WORD: Y-Ba-Cu-O system, Bi-Sr-Ca-Cu-O system, sol-gel method

INTRODUCTION

Since the discovery of the oxide superconductors with a high critical temperature many studies have been made for the application of these materials. Preparing long wires and tapes has great impact on the application, such as development of superconducting magnets so that various techniques have been investigated up to date. Among many processes, we noticed that sol-gel method had some advantages for preparing long wires. The method allows to mix each component at molecular level because it starts from the liquid state. Therefore homogeneous and high purity starting materials can be obtained. In addition, it allows to prepare superconductors in various shapes easily. Actually, many reports have presented the preparation of superconductor by the sol-gel method[1-5].

As the first step, we searched the best combination of solute and solvent to prepare homogeneous precursor. In general alkoxide is used for solute but it is expensive and its stability is low. Besides, its solubility to solvent is low so that it is difficult to obtain enough concentration[1,3,4]. In view of avoiding precipitation of solute during preparation of solution and film coating, the optimum combination was found to be acetate as the solute and propionic acid as the solvent. This combination is stable at room temperature. In addition to the ceramic substrates, flexible metal substrates were investigated.

EXPERIMENTAL

1. Y-Ba-Cu-O System

Firstly, Y, Ba and Cu acetates of reagent grade were dissolved into propionic acid with the molar ratio of Y:Ba:Cu=1:2:3. Then the solution was stirred and heated at 80°C. This starting solution was dark green-blue and homogeneous without precipitation at room temperature. This coating solution was spin-coated onto MgO poly-crystal and Ag substrate and then dried in vacuum and heated to decompose. Decomposition temperature was determined by thermogravimetric analysis (TGA) and differential thermal analysis (DTA) measurements. Figure 1 shows the result of TG-DTA. The weight rapidly decreased as the temperature rised from 200°C to 400°C and a few exothermal peaks were remarkably observed. These correspond to the oxidation reaction of metal components and also decomposition of organic components. The decomposition temperature was determined to 480°C because decomposition of these organic components ended. Film thicknesses were controlled by repeating the process from the coating to the decomposition for several times. Finally, the films were sintered under various conditions in flowing oxygen.

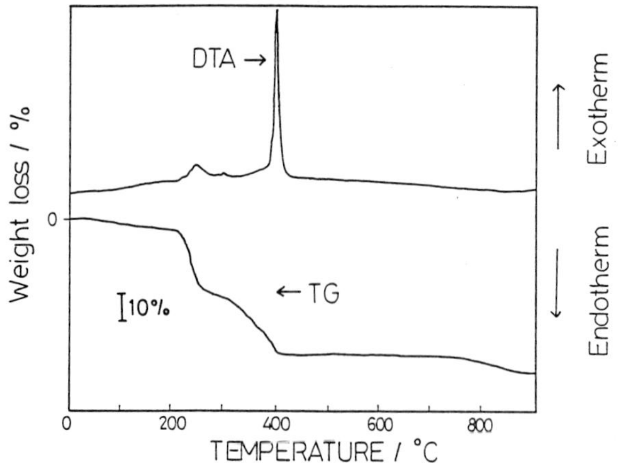

Fig.1 Thermogravimetry(TG) and differential thermal analysis(DTA) curves of Y-Ba-Cu-O precursor.

2. Bi-Sr-Ca-Cu-O System

Bi, Sr, Ca and Cu acetates of reagent grade were dissolved into propionic acid with the molar ratio of Bi:Sr:Ca:Cu=2:2:2:3. TG-DTA was conducted in the same procedure as that for Y-Ba-Cu-O and the decomposition temperature was determined to 700°C in Bi-Sr-Ca-Cu-O. Other conditions were the same as those for Y-Ba-Cu-O system. As the substrate, MgO was investigated. Finally the Bi-Sr-Ca-Cu-O film was sintered around the melting point in the air.

Crystal structure of the films was determined by X-ray diffractometry (XRD). Surface and cross-sectional morphology were observed by a scanning electron microscope (SEM). T_c and J_c(77K, 0T) were measured by standard DC four-probe method. Criterion of the critical current(I_c) was 1μV/cm , and J_c was calculated on the cross-sectional area of the film.

RESULTS AND DISCUSSION

1.Y-Ba-Cu-O System

Figure 2 shows the XRD patterns of the films on Ag substrate sintered at 900°C and 920°C for 6h. The (00l) peaks of the $YBa_2Cu_3O_{7-y}$ phase are dominant, indicating that the films are oriented with the (00l) plane of the $YBa_2Cu_3O_{7-y}$ parallel to the substrate surface.

Figure 3 shows the film thickness dependence of T_c, measured on the films sintered at various temperatures for 2h. According to this figure, T_c of the films on the MgO substrates less than 1μm-thick was below 60K. As increasing the thickness, T_c increased rapidly, and saturated around the thickness of 4-5μm. The maximum value of T_c was 82K obtained on a 10μm-thick film. The J_c of this film was 32A/cm^2. On the other hand, the films on Ag substrate indicated relatively higher T_c but the T_c also depends on thickness and saturated at 4-5μm. We consider the reason for this as below. In the case of thin films less than 1μm in thickness, superconducting path seems difficult to be connected. In addition the reaction between film and substrate occurs on the films on MgO substrate so that Mg in the substrate contaminated films. And it is known that a trace of Mg_2 reduces the transition temperature largely[6]. The maximum value of T_c, 88K and J_c, 7,500A/cm^2 were obtained on the 2.5μm-thick films on the Ag substrate sintered at 920°C.

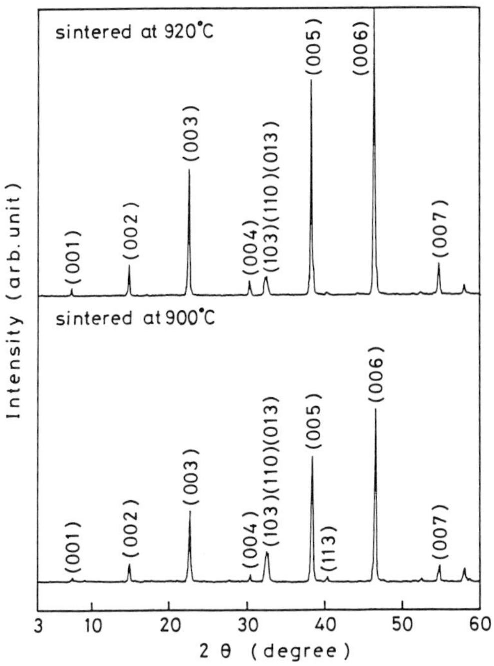

Fig.2 X-ray diffraction patterns of Y-Ba-Cu-O films on Ag substrate.

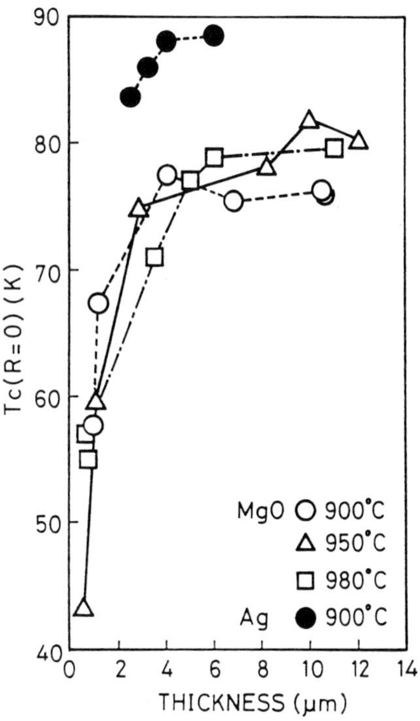

Fig.3 Film thickness dependence of the critical temperature (T_c) on MgO and Ag substrates.
○ : 900°C △ : 950°C □ : 980°C (on MgO)
● : 900°C (on Ag)

Figure 4 shows a SEM photograph of $YBa_2Cu_3O_{7-y}$ film on Ag substrate sintered at 900°C for 2h. It is observed that the film has high density and that the grains grow parallel to the substrate. However, pores and voids are partially observed so that it is necessary to enhance density, orientation and homogeneity of the grains further.

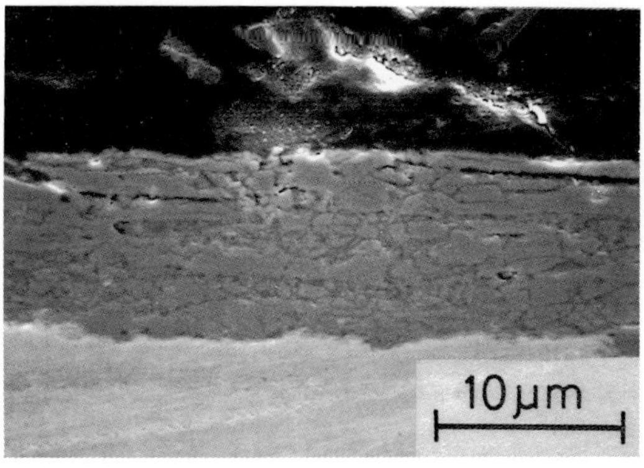

Epoxy resin

$YBa_2Cu_3O_{7-y}$ film

Ag substrate

Fig.4 Cross section photograph of the film sintered at 900°C for 2h.

2.Bi-Sr-Ca-Cu-O System

Figure 5 shows the XRD patterns of the films sintered at 840°C-860°C for 1h from nominal composition of 2223. It is observed that the (00l) peaks of the low-T_c phase are dominant in whole, indicating that the films are oriented with the plane parallel to the substrate surface. The formation of the high-Tc phase is most remarkable at 850°C. Therefore, to enhance the proportion of the high-T_c phase, the films were sintered at 850°C for various hours,10min, 1h, 6h, 24h. As a result, the (00l) peaks of the high-T_c phase appeared at the film sintered for 1h, though this is a considerably short time. The proportion of the high-T_c phase decreased as the sintering time was prolonged. The formation of the 2201 phase was recognized for 6h sintering, and remarkably for 24h sintering. Figure 6 shows the temperature dependence of the resistance of these films. As the sintering time increases from 10min, 1h, 6h, up to 24h, the values of Tc varied from 81K,87K,69K, to 30K. A little resistance drop at 110K was observed on the film sintered for 10 min. For 1h sintering, the resistance drop increased and maximum value of T_c was obtained. As the sintering time is prolonged from 1h, the resistance drop gradually decreased. These results correspond to the results of XRD analysis. Low values of T_c of the films sintered for a long time depend on the amount of the 2201 phase. The values of J_c of the films sintered at 850°C for 10min and 1h were 500A/cm^2 and 250A/cm^2, respectively.

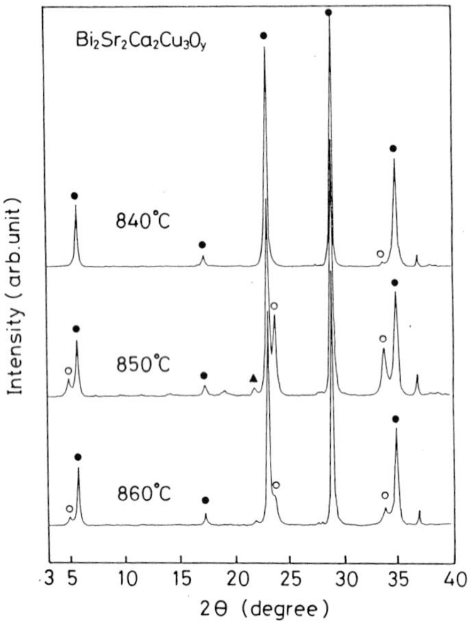

Fig.5 X-ray diffraction patterns of Bi-Sr-Ca-Cu-O films on MgO substrate sintered at 840°C to 860°C for 1h.
○ : high-T_c phase ● : low-T_c phase
▲ : 2201 phase

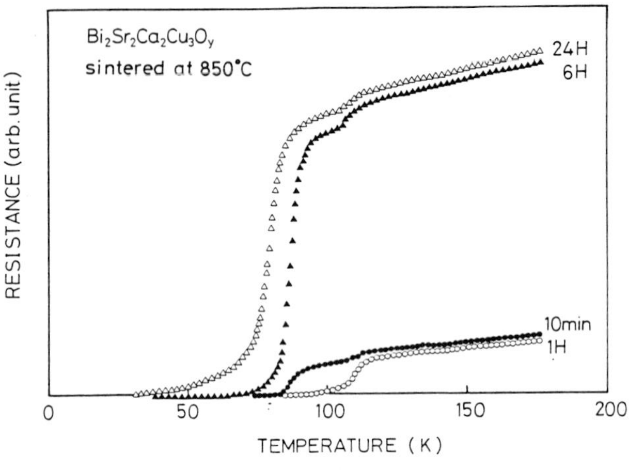

Fig.6 Temperature dependence of the resistance for Bi-Sr-Ca-Cu-O films sintered at 850°C for various hours.
● : 10min ○ : 1h ▲ : 6h △ : 24h

CONCLUSION

Superconducting Y-Ba-Cu-O and Bi-Sr-Ca-Cu-O films were prepared by the sol-gel method using the acetate and propionic acid as the starting materials. By use of a substrate with low reaction between film and substrate, the $YBa_2Cu_3O_{7-y}$ film with excellent properties, T_c=88K, J_c=7,500A/cm^2, was obtained on Ag substrate and the possibility of preparing long tapes was indicated.

In the Bi-Sr-Ca-Cu-O system, the $Bi_2Sr_2Ca_2Cu_3O_y$ film with T_c=87K was obtained. The XRD analysis showed that the films have dominant (00l) peaks of the low-T_c phase but high- T_c phase was also remarkable.

REFERENCES

1. Kramer S, Wu K, Kordas G (1988) Mat.Res.Soc.Symp.Proc. 99, 323-326
2. Tanaka K, Nozue A, Kamiya K (1989) Jap.J.Appl.Phys. 28: L934-936
3. Matsumura K, Nobumasa H, Shimizu K, Arima T, Kitano Y, Tanaka M, Susida K (1989) Jap.J.Appl.Phys. 28: L1797-1799
4. Hirano S, Hayashi T, Tomonaga H (1990) Jap.J.Appl.Phys. 29: L40-42
5. Gupta A, Jagannathan R, Cooper EI, Giess EA, Landman JI, Hussey BW (1988) Appl.Phys.Lett. 52: 2077-2079
6. Komatsu T, Meguro H, Sato R, Tanaka O, Matusita K (1988) Jap.J.Appl.Phys. 27: L2063-2066

Atmosphere Control During Preparation of $YBa_2Cu_3O_{7-x}$ Magnet Windings

R.B. Poeppel, S.E. Dorris, J.J. Picciolo, U. Balachandran, M.T. Langan, C.Z. Zhang, K. Merkle, Y. Gao, and J.T. Dusek

Materials and Components Technology Division, Materials Science Division, Argonne National Laboratory, 9700 South Cass Avenue, Argonne, IL 60439, USA

ABSTRACT

Large coils of $YBa_2Cu_3O_{7-x}$ can be fired successfully if the furnace atmosphere is carefully controlled. Organics added during processing produce CO_2 during the initial portions of the firing schedule. Transmission electron microscopy of material fired in atmospheres containing various levels of CO_2 clearly shows the extent of grain boundary degradation caused by CO_2. Coils with acceptable critical current density can be produced if the rate of CO_2 removal is adequate.

KEY WORDS: $YBa_2Cu_3O_{7-x}$, CO_2, grain boundary degradation

INTRODUCTION

Since the discovery of high-T_c superconductors, people have envisioned many possible applications, ranging from small-scale items such as electronic components to large-scale devices for generating, storing, and distributing energy. For most of these applications, the required critical current density, J_c, is in the range of 10^4–10^6 A/cm^2. Thin-film techniques can produce superconductors whose performances meet these requirements, with J_c ranging up to $\approx 10^6$–10^7 A/cm^2 [1]. However, these techniques are limited to production of small-scale superconductors, and bulk fabrication processes will be required to produce the long lengths of continuous superconductor necessary for large-scale applications.

In this paper, we describe bulk processes by which multilayer superconductor coils can be fabricated with a $YBa_2Cu_3O_{7-x}$ (YBCO)/silver composite as the superconductor and Y_2BaCuO_{25} (211) as an insulator. In particular, we discuss the importance of carefully controlling the furnace atmosphere during the firing of coils, and demonstrate that even small concentrations of CO_2 lead to degradation of grain boundaries. Firing extruded coils at reduced total pressure allows for close control of the atmosphere, and prevents decomposition of YBCO, whereas firing coils at ambient pressure leads to decomposition of YBCO during removal of organic constituents. With the low-pressure process, large, multilayer coils with acceptable J_c can be produced.

EXPERIMENTAL

Details of the coil fabrication process are given elsewhere [2–4]. To examine the reaction between YBCO and CO_2, pellets were fired at 900-1000°C for ≈5 h in various O_2/CO_2 gas mixtures (0–5% CO_2). J_c was measured by the four-probe technique with a criterion of 1 μV/cm. Other details are given in Refs. 5 and 6.

RESULTS AND DISCUSSION

Table 1 shows the firing conditions for coils made by extrusion, their J_c's, and the magnetic fields generated when they operated just below their critical currents. Although the J_c's are well below those necessary for many large-scale applications, an important accomplishment is that the length of continuous superconductor in these coils ranges up to ≈12 m. Moreover, the measurements were made in magnetic fields as high as 73 Gauss, and fringing effects at the coil ends probably increase the field even further at the extreme end turns. Considering that just a few years ago it was not possible to consistently obtain such performance on even short lengths of superconductor in zero field, these results represent significant improvement in the fabrication of bulk superconductors. In the green state, superconducting coils contain ≈10 wt.% organics, which must be completely removed in a way that does not damage the superconductor. Incomplete removal of the organics can cause decomposition of the superconductor or leave carbon-rich material at the grain boundaries, either of which will degrade superconducting properties. Organics can be easily removed by thermal decomposition at 240-350°C, but if the decomposition proceeds too rapidly, the

Table 1. Description and Properties of Three Superconductor Coils

Coil Description	Firing Conditions	J_c (A/cm^2)	B^a (Gauss)	B(Fe Core) (Gauss)
1 Uncoated 21 Turns	2 mm Hg 875°C	225	36	---
2 Coated 211 2 Layers 42 Turns	10 mm Hg 875°C	150	42	160 @ 77 K
3 Coated 211 5 Layers 75 Turns	10 mm Hg 875°C	150	73	330 @ 77 K 420 @ 73 K

[a]Magnetic field.

coils can become severely bloated and, in some cases, even explode. Also, decomposition of the organics produces significant concentrations of CO_2 and H_2O, two species that are known to react with YBCO [7,8]; therefore, the rate at which organics are removed must be carefully controlled.

When superconducting coils are fired at a reduced total pressure, CO_2 and H_2O are removed as they are produced, thereby minimizing their concentrations and preventing decomposition of YBCO. But when coils are fired at ambient pressure, the harmful gaseous products accumulate and lead to the decomposition of YBCO. To demonstrate this, mixtures were made from YBCO powder and the same organics that are used in extrusion: they were fired in flowing oxygen at either ambient or reduced pressure (≈2 mm Hg). Two different powders were used: powder produced by solid-state reaction at reduced pressure, and powder produced by a liquid mix technique. Samples were taken from the mixtures at 240, 300, and 350°C, and their X-ray patterns were obtained. Figures 1 and 2 are schematic illustrations of the major peaks in these patterns. Figure 1 shows that, when the YBCO/organic mixtures were fired at ambient pressure, both samples of YBCO decomposed, the liquid-mix powder at 240°C, the solid-state powder at 300°C. Figure 2 shows, however, that gross decomposition of YBCO did not occur when the mixtures were fired at reduced total pressure, even though the relative intensities of peaks varied as a result of changing oxygen content.

Figure 2 shows that gross decomposition of YBCO is avoided by firing coils at reduced total pressure; however, small concentrations of impurities (<5 wt.%) would not be apparent in X-ray diffraction results. It was possible, therefore, that coils fired at reduced total pressure could still be degraded by small, residual concentrations of CO_2 in the furnace atmosphere. Therefore, it was important to investigate the effect of small concentrations of CO_2 on the properties of YBCO.

It has been shown [5] that, as the CO_2 partial pressure in the firing atmosphere increased, J_c (zero field/77 K) of sintered pellets decreased and finally became zero. Resistivity measurements showed that the materials with $J_c = 0$ were semiconductive. On the basis of such measurements, the stability region for superconducting YBCO was derived with respect to the partial pressure of CO_2 at the four firing temperatures, 910, 940, 970, and 1000°C. (Fig. 3). Magnetization measurements indicated that major portions of the semiconductive samples were still superconducting, and that the onset temperature of superconductivity (≈90 K) was almost the same for all of the samples. These results suggest that CO_2 reacts with YBCO to form a thin layer of nonsuperconducting second phase at grain boundaries, but does not degrade the superconducting grain interior. When enough of the grain boundaries become coated with second phases, as in the semiconductive samples, the passage of superconducting current is effectively blocked: however, because of the superconducting-grain interiors, a sharp change in magnetization is still evident.

These arguments are supported by TEM observations [6], which show the presence of secondary phases at some grain boundaries. An example of one such grain boundary is shown in Fig. 4 for a sample fired at 970°C in a 0.5% CO_2/O_2 gas mixture. The second phase was determined to be $BaCuO_2$ and $BaCO_3$ by X-ray energy-dispersive spectroscopy. The width of this grain boundary phase is much larger than the coherence length in YBCO, so that the phase can completely obstruct the superconducting current and reduce the overall J_c. However, this type of grain boundary accounts for only ≈10% of the observed grain boundaries; the majority appear quite sharp, with no obvious evidence of a second phase. Because of the multitude of possible percolation paths, the value of J_c would not become zero if only 10% of the grain boundaries are coated with a second phase. For the superconducting current to be completely blocked, as in the case of

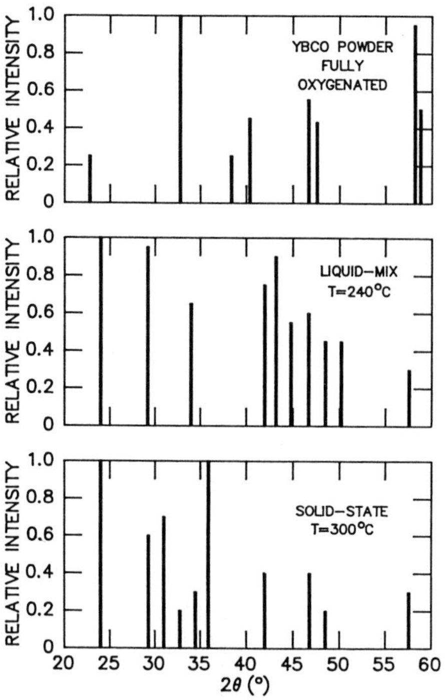

Fig. 1. Schematic X-ray patterns of fully oxygenated YBCO powder (as reference) and two YBCO powders that were first mixed with organics used in extrusion, then fired at ambient pressure. The patterns show that both powders completely decomposed during firing.

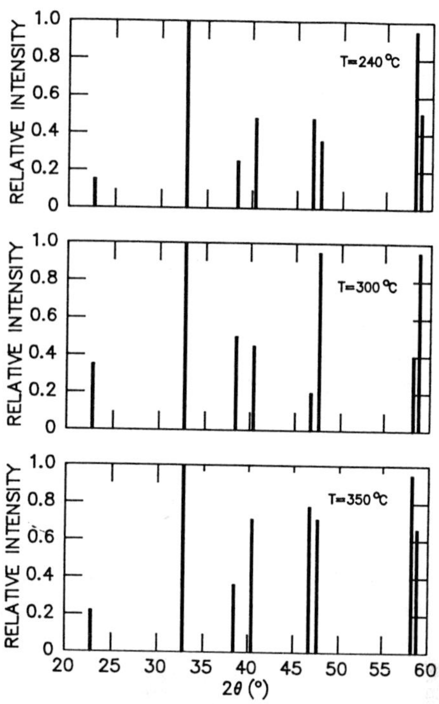

Fig. 2. Schematic X-ray patterns of YBCO powder as a function of temperature. The powder was mixed with organics used in extrusion, then fired at reduced total pressure. The patterns show no signs of decomposition of YBCO.

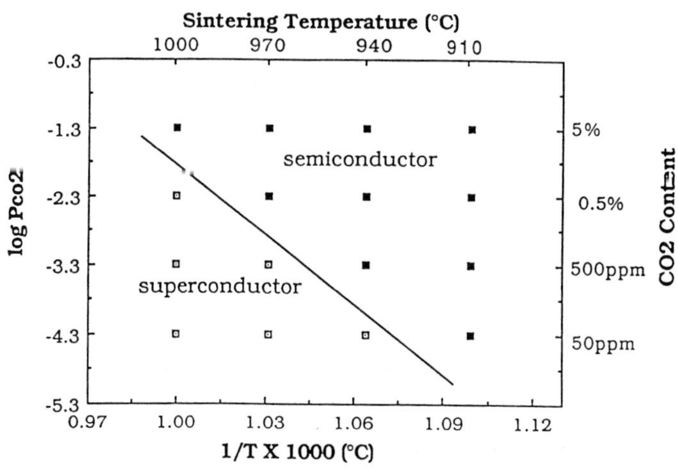

Fig. 3. Plot showing stability of YBCO with respect to partial pressure of CO_2 at various temperatures.

semiconductive samples, a majority of the grain boundaries must be nonsuperconducting. High-resolution electron microscopy (HREM) images of the grain boundaries suggested that, in fact, the majority of grain boundaries probably are nonsuperconducting.

Careful study of the HREM images showed that the structure near the sharp grain boundaries was not orthorhombic, but some other phase, possibly tetragonal YBCO. By careful measurement of the interplanar spacing, it was found that the spacing, c, is ≈1.19 nm at the region near the grain boundary, while c ≈1.17 nm in the region far from the grain boundary. Neutron diffraction data [9] show that tetragonal YBCO has c = 1.19 nm. Another indication of tetragonal material is the termination of twinning near the grain boundaries, which can be taken as the demarcation line between orthorhombic and tetragonal structures,

Fig. 4. TEM micrograph of YBCO sample sintered at 970°C in 0.5% CO_2/O_2 atmosphere. The thick grain boundary layer was identified as $BaCuO_2$.

because the tetragonal structure has no twins. A possible explanation for the phase transformation from orthorhombic to tetragonal structure is the incorporation of carbon into the lattice, due to the presence of CO_2 in the firing atmosphere [5,6]. Segregation of carbon at grain boundaries or in regions near grain boundaries has been confirmed in a previous study [6] by secondary ion mass spectroscopy. Carbon can diffuse into the lattice, and expel oxygen from the orthorhombic structure, thereby forming a nonsuperconducting tetragonal structure that can block the superconducting current.

Large differences in size and geometry make it difficult to compare the J_c results of pellets and coils; nevertheless, comparison suggests that CO_2 had minimal impact on the superconducting properties of coils that were fired at reduced total pressure. Keeping in mind that the J_c of bulk materials drops dramatically with magnetic field, and that the coils were measured in fields up to at least 73 Gauss, it is suggested that the coil results agree most closely with the results for pellets fired in 0% CO_2/100% O_2 (J_c of such a pellet was ≈300 A/cm^2 in zero field). This further suggests that large, multilayer, superconducting coils can be successfully fabricated by firing at reduced total pressure, and that the superconducting properties of these coils are representative of bulk materials.

CONCLUSIONS

It has been shown that very small concentrations of CO_2 in the firing atmosphere can react with YBCO, leaving superconducting grain interiors encased in nonsuperconducting grain boundary phases. In the firing of large, multilayer coils, decomposition of YBCO can be avoided and the concentration of CO_2 can be minimized by firing at reduced total pressure. As a result, superconducting coils can be fabricated that produce magnetic fields up to 73 Gauss with an air core, and 330 Gauss with an iron core.

ACKNOWLEDGMENTS

Work was supported by the U.S. Department of Energy, Conservation and Renewable Energy, as part of a program to develop electric power technology, and Basic Energy Sciences-Materials Science, under Contract W-31-109-Eng-38, and by the National Science Foundation through the Science and Technology Centers for Superconductivity, under Contract DMR88-09854.

REFERENCES

1. T Nakahara (1988) In: World Congress on Superconductivity, C G Burnham and R D Kane (eds.) World Scientific, Singapore, p. 88 1988.
2. U Balachandran, R B Poeppel, J E Emerson, S A Johnson, M T Lanagan, C A Youngdahl, Shi, K C Goretta, N G Eror (1989) Mater. Lett. **8**(11,12): 454.
3. J P Singh, H J Leu, R B Poeppel, E Van Voorhees, G T Goudey, K Winsley, D Shi (1989) J. Appl. Phys. **66**(7): 3154.
4. S E Dorris, J T Dusek, J J Picciolo, R Russell, J P Singh, and R B Poeppel (1990) In: Proceedings of the International Conference on Electrical Machines (ICEM), Cambridge, MA, August, 1990, in press.
5. Y Gao, Y Li, K L Merkle, J N Mundy, C Zhang, U Balachandran, and R B Poeppel (1990) Mater. Lett. 9: 347.
6. Y Gao, K Merkle, C Zhang, U Balachandran, and R B Poeppel (1990) J Mater. Res. 5: 1363.
7. M S Jahan, D W Cooke, H Sheinberg, J L Smith, and D P Lianos, (1989) J Mater. Res. 4: 759.
8. P K Gallagher, G S Grader, and H M O'Bryan (1988) Mat. Res. Bull. 23: 1491.
9. J D Jorgensen, M A Beno, D G Hinks, L Soderholm, K J Volin, R L Hitterman, J D Grace, I K Schuller, C U Segre, K Zhang, and M S Kleefisch (1987) Phys. Rev. B36: 3608.

Suspension-Spun High-T_c Oxide Superconducting Filament with High J_c

TOMOKO GOTO, CHIAKI YAMAOKA, YUZURU KINO, SHUICHI IWATA, and KAZUMICHI KOJIMA

Department of Materials Science & Engineering, Nagoya Institute of Technology, Gokiso-cho, Showa-ku, Nagoya, 466 Japan

ABSTRACT

High Tc oxide supercdonducting filaments with high Jc were successfully prepared using a technique in which suspension spun and subjected to heating treatment. High Jc at 77 K, 0 T of more than 10^4 A/cm² was attained for Tl-(Ba,Sr)-Ca-Cu-O filaments by phase decomposition from 2223 phase to 1223 or 1234 phase, for Bi-Sr-Ca-Cu-O filaments by zone melting and for Y-Ba-Cu-O filament by Quench and Melt Growth process, respectively.

KEY WORDS: suspension spinning, high critical current density, Tl-(Ba,Sr)-Ca-Cu-O filament, Bi-Sr-Ca-Cu-O filament, Y-Ba-Cu-O filament.

INTRODUCTION

High Tc oxide superconducting wires have great potential for applications involving superconducting magnet and power cables etc.. However the applications have been hindered by the difficulty in fabrication of the brittle ceramic materials into tapes or wire and by the weak link problem in the sintered materials.

We have developed a technique to fabricate oxide superconducting filaments by the suspension spinning method, which is a simple manufacturing process for wire-making with good processability. However, the Jc values of the filament produced by this method were as low as 2000 A/cm² at 77 K, 0 T. This paper deals with the preparation of suspension-spun high Tc oxide filament with high Jc of more than 10^4 A/cm².

EXPERIMENTAL

Phase decomposition process, zone melting and Quench and Melt Growth process were used to enhance the Jc of the suspension spun Tl, Bi and Y system filaments, respectively.

$Tl_2Ba_{1.6}Sr_{0.4}Ca_2Cu_3O_x$ samples were prepared from a mixture of Ba-Sr-Cu-O, Tl_2O_3 and CaO. Appropriate amounts of Tl_2O_3, CaO and $Ba_{1.6}Sr_{0.4}Cu_3O_4$ were completely mixed and ground with pestle using ethanol and pressed into a pellet. Then the pellet was calcined at 860 ℃ for 300 s, sintered at 910 ℃ for 600 s in air and pulverized. The powder was a single high Tc 2223 phase with Tc of 113 K.

The resultant powder was suspended in a mixed poly vinyl alcohol solution of dimethyl sulfoxide and hexamethyl phosphoric triamide. The viscous suspension was extruded as a filament into a precipitating medium of methyl alcohol and coiled on a winding drum. The as-drawn filament of 250 μm in diameter was pressed and pyrolyzed in an alumina box at 500 ℃ for 1800 s to remove volatile component. The filament was then cold-pressed and was set in a gold vessel. The vessel was put into an alumina crucible with a cup on it, which was kept at various sintering temperatures. After heating in air, the crucible was rapidly cooled in the room temperature.

The powder of the low-Tc 2212 phase of Bi system was prepared by the conventional solid-state reaction and long filaments were made by suspension spinning method. The as-drawn filaments were pyrolyzed, sintered at 820 ℃ for 14 h and zone melted The zone melting apparatus consists of a vertical Pt tube furnace heated by an induction heating coil, in which the filament is supported. The filament was automatically fed at a rate ranging from 2 mm/h to 30 mm/h. The maximum temperature of hot zone was between 900 ℃ and 1000 ℃. The solidification rate was controlled by varying the sample speed through the hot zone and the associated cooling gradient. The gradient was more than 100 ℃/mm. The zone melting was performed in air and the filament was annealed in a separate furnace at 840 ℃ for 40 h.

Appropriate amounts of Y_2O_3, $BaCO_3$ and CuO were mixed into the composition of $Y_{1.2}Ba_2Cu_3Ox$, ground and calcined at 900 ℃ for 24 h. The powder were melted in a platinum crucible at 1400 ℃ for 900 s. The molten oxides were rapid-quenched by soaking in liquid nitrogen. Then the oxides were pulverized. The resultant powder was suspension spun and the as-drawn filaments were reheated under various conditions in flowing oxygen.

The electrical resistivity (ρ) of the filaments obtained by these methods was measured by a standard four-probe method. Silver paint was used to connect gold sputtered parts of the filament with Ag electrodes 50 μm in diameter. The specimen temperature was measured using a calibrated chromel-gold + 0.07 % iron thermocouple. The transport Jc measurement was performed at 77 K, 0 T with criterion of 1 μV/cm.

RESULTS AND DISCUSSION

Tl-(Ba,Sr)-Ca-Cu-O filaments

The pyrolyzed filament was sintered at various heating conditions after cold press and the Jc of the filament obtained was measured. The high Jc of 12000 A/cm^2 was observed for the filament sintered at 840 ℃ for 1200 s. The filament exhibited superconducting at Tc of 102 K. A diamagnetic response due to the Meissner effect below 102 K, consistently with the resistivity data was found. The normal resistivity value such as 3 mΩ・cm was relatively high as compared with a value of 0.5 mΩ・cm for a Tl system filament, which was suspension spun and subjected to be sintered under oxygen flow and exhibited Jc of 2300 A/cm^2 at 77 K 0 T.

An X-ray diffraction patterns of the pyrolyzed filaments and the filaments sintered at 840 ℃ for 1200 s using Cu Kα radiation are shown in Fig.1. Although some impurities were observed for these filaments, the main peaks for the pyrolyzed filament are indexed at that of the high Tc 2223 phase having a tetragonal unit cell with a = b = 0.385 nm and c = 3.561 nm. On the other hand, the sintered filament consists of a mixed Tl monolayer phases of 1223 and 1234. The 1223 phase in the filament decreased with increasing sintering time at 840 ℃. It is considered that the high Tc 2223 phase is instable at the sintering temperature of 840 ℃ and resulted in the Tl monolayer phases such as 1223 and 1234 by vaporizing the Tl oxide due to the large surface area of the filament.

We have investigated the chemical compositions of the filaments with the highest Jc by means of an electron probe X-ray microanalyzer. The filaments approximately consisted of $Tl_{10}Sr_8Ba_{15}Ca_{22}Cu_{45}Ox$. The microstructure obtained by backscattered electron image from the polished and etched surfaces on the cross section of the filament with the highest Jc is shown in Fig.2. The chemical composition of the brighter needles (A) in Fig.2 was approximately $Tl_{16}Sr_9Ba_{12}Ca_{20}Cu_{43}Ox$, while the darker one (B) was CuO and/or CaCuO and black one (C) was CaO. It should be noted that the 1223 and 1234 phases coexist and are formed as a mixture. It was reported on the preparation of Ag-sheathed Tl-Ca-Ba-Cu-O superconducting wire that the Jc of the $(Tl,Pb)_1Ca_3Ba_2Cu_4O_{11}$ wire was superior to that of the $Tl_2Ca_2Ba_2Cu_3O_{10}$ wire. Fine non superconductive impurity phases and the interface of intergrown supercoducting phase were thought to be candidates of pinning sites [1]. It was also investigated that significantly increased critical current density were obtained in the $Y_1Ba_2Cu_3Ox$ superconductor prepared by decompositive process starting from the $Y_1Ba_2Cu_4Ox$ compound precursor. The cause of the Jc improvement was attributed to the enhanced flux pinning by the presence of fine scale transformation-induced defects [2]. For the present filament the decomposition from the 2223 phase to 1223 or 1234 phase is considered to result in flux-pinning enhancement.

Bi-Sr-Ca-Cu-O filaments

The suspension spun $Bi_2Sr_2Ca_1Cu_2Ox$ filaments were zone melted under various feeding speeds and the electrical resistivity of the filament obtained was

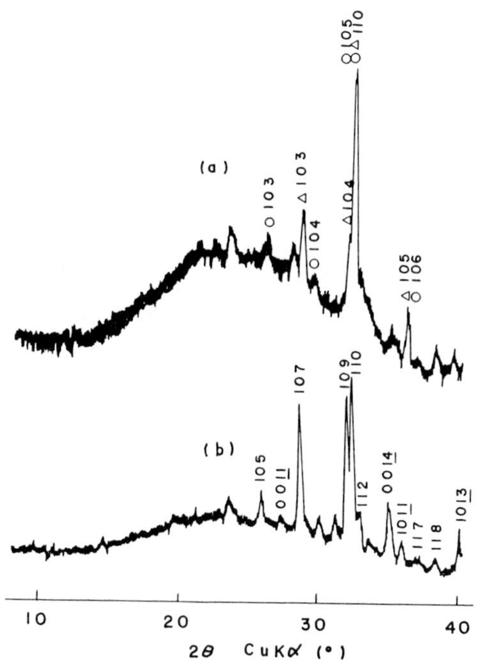

Fig.1 X-ray diffraction pattern of the Tl system filaments using Cu Kα radiation. (a) filament at 840℃ for 1200 s ○-1234 phase △-1223 phase (b) pyrolyzed filament

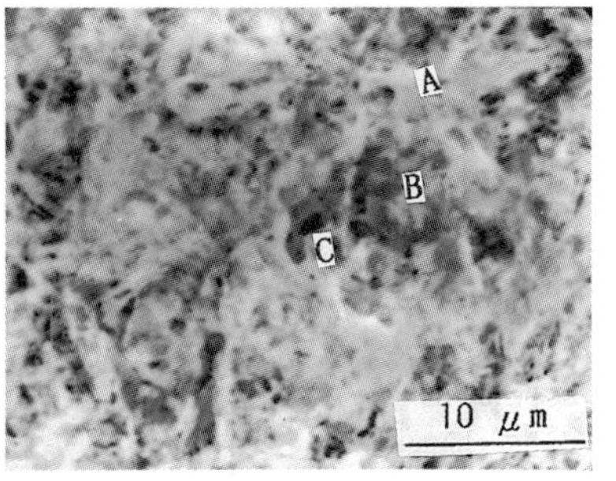

Fig.2. Backscattered electron image from the polished and etched surfaces on the longitudinal cross-section of the Tl system filament

measured. The filament, which was spun and solidified at a feeding speed of 22 mm/h, was superconducting with a Tc of 83 K and a Jc of 210 A/cm² at 77 K, 0 T. The as-solidified filament was annealed at 840 ℃ for 40 h to improve the Jc value. The feeding speed of the filament played an important role in the high Jc value. The Jc for the sintered filament is higher than that for the filament without sintering. A maximum current density, which is exceeded the measurement capacity (maximum current capacity of 22 A and Jc of 2 x10⁴ A/cm²) was attained for the filament zone melted by a feeding speed of 22 mm/h with and without sintering. It is noticeable that a high Jc more than 10⁴ A/cm² was attained for the filament without sintering, which makes handling of the zone-melting process easy. The zone melted filaments with high Jc' s were characterized by SEM. Figure 3 (A) shows a scanning electron microscopy image of the polished and etched of a filament cross section. The sample shows a very dense and well aligned structure consisting of plate-like grains. The near parallel $Bi_2Sr_2Ca_1Cu_2Ox$ plates were sometimes separated by a particle of nonsuperconducting phase, identified as $(Ca,Sr)_1Cu_1O_2$ and/or $(Ca,Sr)_2Cu_1O_3$.

An X-ray transmission Laue pattern of the zone melted filament with a high Jc was taken by a flat cassete camera to examine the orientation of the crystals. A single [100] fiber texture was deduced for the zone melted filament. The fiber axis appears to coincide with the direction of the tetragonal phase in the superconducting layer. Such crystallographic alignment should be beneficial for the flow of supercurrent in this severely anisotropic layer compound.

Y-Ba-Cu-O filaments

The suspension-spun $Y_{1.2}Ba_2Cu_3Ox$ filaments were partial-melted at various temperatures ranged from 1050 ℃ to 1150 ℃ for 20 min and the temperature was reduced by 100 ℃. After 30 min, the sample were cooled at the rate of 10 ℃/h to 900 ℃ followed by furnace cooling. The relationship between Jc of the filament and heated temperature ranging from 1050 ℃ to 1150 ℃ was examined, and the maximum Jc of 1200 A/cm² was observed at a heating temperature of 1100 ℃. The effect of the heating time from 5 min to 30 min at 1100 ℃ on the Jc of the filament was also examined. However no remarkable enhancement of the Jc was observed. The filament heated at 1100 ℃ and cooled down above condition consists of needle like grains and some pores were found. Therefore the partially melted filament was held at 1000 ℃ for various times subjecting the sample to the slow cooling. The Jc of the filament was dependent on the holding time at 1000 ℃. The maximum Jc of the current density exceeded the measurement capacity (maximum current capacity of 22 A) and Jc of more than 45000 A/cm^2 was attained for the filament held at 1000 ℃ for 60 min. It is noticeable on the resistivity measurement that some samples of the filament held at 1000 ℃ for 60 min were burnt out by employing a current of more than 1 A. The cooling rate from 5℃/h to 20 ℃/h after holding the sample at 1000 ℃ for 60 min was also examined. The maximum Jc was observed at a cooling rate of 10 ℃/h. The filament with the highest Jc exhibited superconductivity at a Tc of 86 K, and a diamagnetic responce due to the Meissner effect below 86 K, consistent with the resistivity data was observed. The normal resistivity value such as 4 mΩ・cm was relatively high as compared with the value of 0.29 mΩ・cm for a sintered $Y_1Ba_2Cu_3Ox$ filament.

The filaments with the highest Jc were characterized by scanning electron microscopy. Figure 3 (B) shows the polished and etched surface on the longitudinal cross-section of the filament. The sample shows a very dense structure and has a well-aligned texture and the matrix is sometime cut off by the bulk grain with different orientation. The burn out of the sample on the resistivity measurement is considered due to the grain boundary of the kinked bulk grains. An X-ray diffraction powder pattern of the filament with high Jc using CuKα radiation was examined. The filament was a mixed phase of $Y_1Ba_2Cu_3Ox$, $Y_2Ba_1Cu_1O_5$, CuO and $BaCuO_2$.

Some kinds of starting powders were also investigated and it was found that quenching from the melt is most important in this method and pulverization of the oxides after quenching enables homogeneous distribution of Y_2O_3 ; consequently, fine dispersion of the 211 phase in the 123 superconducting phase can be realized, leading the high Jc [3].

ACKNOWLEDGEMENT

This work was partly supported by a Grant-in-Aid for Scientific Research on Chemistry of New Superconductors for the Ministry of Education, Science and Culture (No. 0222712).

REFERENCES

1. Takei H, Torii Y, Hikata H, Sato K, Hitotsuyanagi H, Tada,K (1989) in: Ishiguro T, Kajimura K (eds) in Addvances in Superconductivity Ⅱ.Springs-Verlag,Tokyo, p.475
2. Jin S, Tiefel HT, Nakahara S, Graebner JE, Bryan HMO, Fastnacht RA, Kammlott GW (1990) Appl.Phys.Lett.56:1287
3. Fujimoto H, Murakami M, Gotoh S, Koshizuka N, Oyama T, Shiohara S, Tanaka S (1989) in: Ishiguro T, Kajimura K (eds) in Addvances in Superconductivity Ⅱ. Springs-Verlag,Tokyo, p-285

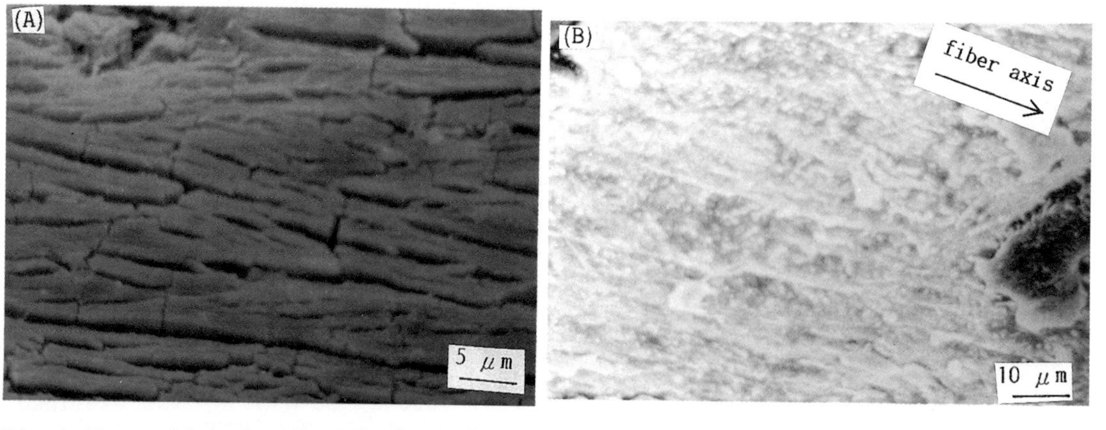

Fig.3 The polished and etched surface on the longitudinal cross-section of the filament. (A) Bi system filament (B) Y system filament.

5.2 Bulk Materials

Critical Current Characterization of Melt-Processed $YBa_2Cu_3O_{7-x}$

H. KÜPFER[1], C. KELLER[1], A. GUREVICH[1], K. SALAMA[2], and V. SELVAMANICKAM[2]

[1] Kernforschungszentrum Karlsruhe, Institut für Technische Physik und Universität Karlsruhe, W 7500 Karlsruhe, Germany

[2] Department of Mechanical Engineering, Texas Center for Superconductivity at University of Houston, Houston, TX 77004, USA

ABSTRACT

Upper critical field, irreversibility line, weak superconducting coupling and thermal relaxation are investigated in melt processed $YBa_2Cu_3O_{7-x}$ in the geometry magnetic field parallel and perpendicular to the a,b plane. Granularity within the highly c-axis oriented grains is excluded from comparison between transport and shielding current if the current flows within the a,b plane. The current along the c-axis is not degraded at the grain boundaries but becomes nonuniform with increasing field and temperature. This information was obtained from a more detailed investigation of magnetization measurements. Oxygen deficiency and inhomogeneous oxygen distribution are proposed as a major reason for this behaviour. Flux creep phenomena were studied by magnetic relaxation and inductive flux profile measurements. Temperature and field dependent activation energies were determined. The increase of the activation energy with temperature is discussed and compared with different explanations. Sweep rate dependent dc and ac measurements which allow access to relaxation in the millisecond regime exclude in the low temperature region a short time decay of the magnetic moment before the relaxation measurements starts.

KEY WORDS: melt processed YBaCuO, critical currents, granularity, relaxation

INTRODUCTION

High T_c oxide superconductors are characterized by an extremely short coherence length and a large anisotropy of the upper critical field. Further they are considered to be used in applications at 77 K. These three parameters, coherence length, anisotropy and operation temperature are by one order of magnitude or more different in low T_c conventional superconductors. They are therefore responsible for the key problems which have to be solved in order to achieve transport currents at 77 K for high field applications for instance critical current densities of 10^5 A/cm^2 at fields of 10 T with a specific resistivity of less than 10^{-8} μΩcm. The small coherence length causes low pinning energy and weak superconducting coupling at planar defects. Weak coupling or granularity at the grain boundaries or within a grain or single crystal results in transport currents much lower than the shielding currents flowing inside the decoupled regions. Pronounced thermal relaxation is initialized by the small ratio activation energy divided by the higher operation temperature. In oxide superconductors at 77 K these values of U_0/k_BT are between one and two orders of magnitude smaller than in low T_c superconductors at 4 K. At the irreversibility line $B_{irr}(T)$ the activiation energy becomes comparable to the thermal energy k_BT resulting in zero current above. This "reversible region" between $B_{irr}(T)$ and the upper critical field B_{c2} is large, for instance in $YBa_2Cu_3O_{7-x}$ at 77 K, B_{irr}/B_{c2} is between 0.1 and 0.2. Whether this borderline is inherent to the oxide superconductors by flux melting or it results from low pinning energy that means thermal relaxation, is an essential question because the irreversibility field is only around 5 T at 77 K if the field is applied parallel to the c-axis of the crystal.

These problems: low pinning energy, thermal relaxation, granularity and the extended reversible region above the irreversibility line are present in all high T_c oxides in bulk material as well as in thin films. Considerable progress has been made towards a solution of all these obstacles. For instance strong grain alignment in conjunction with a better quality of grain boundaries [1,2,3] leads to higher transport currents and avoids decoupling, or the refinement of the 211 second phase precipitates shifts the irreversibility line to considerably higher values disproving flux melting as the origin of this problem [4].

In this contribution we discuss these problems in oriented grained $YBa_2Cu_3O_{7-x}$ prepared by a liquid phase processing method [3]. The highly c-axis oriented grains have a platelike shape with typical dimensions of several mm in the a,b plane and 10 to 20 μm along the c-axis. Transport and shielding currents, granularity, and magnetic relaxation were investigated in the geometry field parallel and perpendicular to the a,b plane of the crystal. The samples were cut in a quadratic shape typically 3 x 3 mm with a thickness of 0.5 mm. The c-axis is parallel to one direction of the 3 mm side of the square resulting in the same demagnetization for both geometries and in the same aspect ratio of the areas in which the shielding currents flow. The anisotropic Bean model gives minor corrections for obtaining the current along the c-axis. For the estimation of the current density we always used the macroscopic scaling length of the specimen but never the grain size.

UPPER CRITICAL AND IRREVERSIBILITY FIELD

Reversible properties were obtained by measuring the magnetization M as a function of temperature T at a constant applied field B [5]. Between the transition into the superconducting state and the irreversibility line M(T) results in the same values at increasing and decreasing temperature. This reversible behaviour is caused by a relaxation much faster than the characteristic experimental time scale and perhaps by a melting of the flux line lattice in the vicinity of the upper critical field. The value of B_{c2} is estimated from the intersection of the linear extrapolation of M(T) in the normal and superconducting state. The irreversibility field at which M(T) in increasing and decreasing field becomes different from each other corresponds to a voltage criterion for the critical current density j_c of about 1 μV/cm. Figure 1 shows B_{c2} and B_{irr} as a function of temperature for both geometries B parallel and perpendicular to the a,b plane. From these measurements we obtain dB_{c2}/dT of -6.1 T/K and -2.4 T/K and the Ginzburg-Landau parameter κ of 293 and 92 in B‖a,b and B⊥a,b respectively. The relation $B_{irr} = \beta\,(1-T/T_c)^{\alpha}$ fits the measurement with β of 230 T and 51 T and with α of 1.65 and 1.33 in B‖a,b and B⊥a,b respectively. The alignment between B and crystal orientation could be achieved only within 10°. For this reason the difference between the values in both directions are higher. Especially in B‖a,b, B_{c2} is larger due to the sharp maximum of B_{c2} versus angle in this geometry. At 77 K the ratio B_{irr}/B_{c2} is about 0.13 in both geometries.

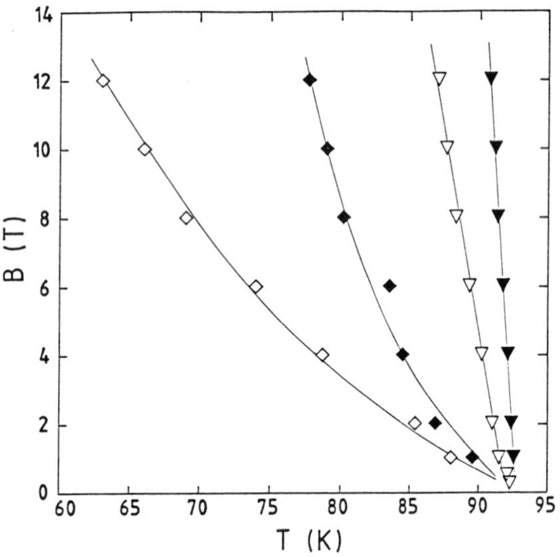

Fig. 1.
Upper critical fields B_{c2} (triangles) from reversible magnetization measurement and irreversibility fields B_{irr} (squares) measured with a voltage criterion of about 1 μV/cm versus temperature T. The field was directed parallel (filled symbols) and perpendicular to the a,b plane (open symbols) respectively.

Recently Ekin et al. [6] achieved transport current measurements with current and field perpendicular to each other but both within the a,b plane up to very high fields in oriented grained $YBa_2Cu_3O_{7-x}$. A critical current density of 10^3 A/cm² was obtained with 10 μV/cm at 77 K, 30 T which points to a much larger B_{irr} (77 K) in B‖a,b than the value extrapolated from Fig. 1. The reason for this discrepancy might be the different voltage criterion, a different misalignment between field and a,b plane or the different direction of the Lorentzforce. The last possibility is correlated with a different pinning interaction and would therefore disprove an inherent origin of the irreversibility field.

CRITICAL CURRENT AND GRANULARITY

Transport critical currents were determined using a resistive pulse technique [7]. Figure 2 shows j_c flowing along the a,b plane of the crystal versus magnetic field oriented perpendicular to the a,b plane at 77 K. The transport current in zero applied field, 77 K, reaches about $7.5 \cdot 10^4$ A/cm² [3]. Figure 2 shows in addition dc transport current measurements from Ref. 8 performed in the same geometry with 10 μV/cm criterion at fields above 5 T. The results of dc and pulse current technique are in satisfying agreement. Shielding currents obtained from flux profile or magnetization measurements show j_c values of similar size. Field and temperature dependences are the same as from transport current measurements and quite different from a weak link current in a ceramic $YBa_2Cu_3O_{7-x}$ or in an intragrain decoupled single crystal [9]. This difference in $j_c(B,T)$ and the satisfying agreement between shielding and transport currents excludes a granular behaviour of the current flowing along the a,b plane in the mixed state. These similar values of transport and shielding current demonstrate that the transport current is a homogeneous bulk current not degraded by partial decoupling or percolation.

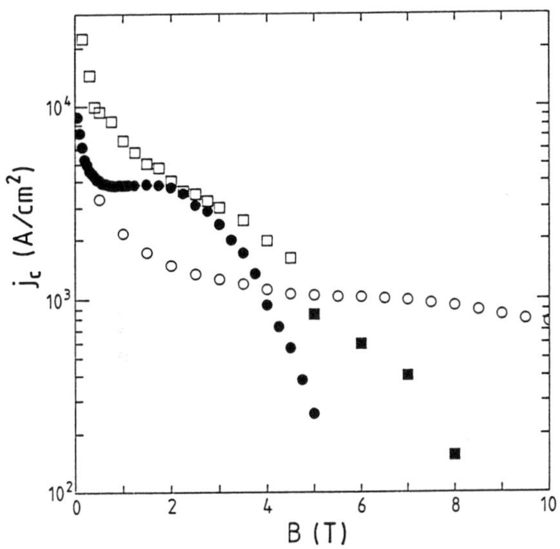

Fig. 2.
Transport current density (squares) and shielding current density (circles) at 77 K versus field B. The transport current and the shielding current (filled circles) are measured in the geometry field perpendicular and current parallel to the a,b plane. Values shown by open squares are obtained from pulsed transport and filled squares from dc transport measurement from Ref. 8. The open circles represent the shielding current in the geometry field within the a,b plane.

In the geometry B||a,b the current is directed partly along the c-direction in which the grain thickness is between 10 and 20 μm. The question whether these grain boundaries behave as weak links was investigated by three different ac methods. First the imaginary part of the ac susceptibility measurement did not show two maxima which reflect the intragrain transition at T_c and the transition of the junction network into a coherent state respectively. Second increasing the amplitude of the ac field or a superimposed dc field did not shift the maximum to lower temperatures to such an extent as observed in polycrystalline materials [7,10]. The ac susceptibility measurement shows only the transition from the reversible into the irreversible state at the irreversibility line but no indication of a phase locking temperature below. Third inductive measurements of the flux profile do not show two current systems [11] as in granular materials. This nongranular behaviour is further supported by a comparison of the shielding currents obtained in both geometries as shown in Fig. 2 using the macroscopic sample size as scaling length for calculating j_c. Assuming a decoupling at the boundaries would require to use an about 20 times smaller scaling length in this geometry which results then in j_c values about 5 times larger than for the current flowing within the a,b plane. To sum up the obtained results show no features of an intergrain decoupling at the grain boundaries for the current flowing along the c-axis.

A more difficult question is the possibility of intragrain decoupling within the grains along this direction. The measurements mentioned above are only reliable for a sufficiently low coupling energy, i.e. a sufficiently large difference between the two current systems. A direct measurement of transport and shielding current along the c-direction [12] results in 3000 A/cm^2 and 30 000 A/cm^2 respectively at 77 K, zero applied field. But this comparison which points to intragrain decoupling suffers mainly from the conditions of zero field, further the self field in both measurements as well as the exact geometries are different.

The field dependence of the shielding current shows quite often a pronounced minimum at low fields followed by a broad maximum. This pecularity called "fishtail effect" is correlated with intragrain decoupling [13,14]. It was shown in Ref. [9] in a low quality $YBa_2Cu_3O_{7-x}$ single crystal that the transport current becomes considerably smaller than the shielding current at fields above its minimum, whereas in Ref.[14] in a high quality single crystal the decoupling starts at fields above the maximum of the shielding current. In our oriented grained specimen no fishtail effect was observed as shown in Fig. 2, but indications for granularity are present in B||a,b.

Recently we used another method in order to detect granularity or deviations from a uniform bulk current flow [15]. From magnetization measurements the shielding current density is obtained from the Bean model as $j_c = 3\ \Delta M/(2\ R)$ where ΔM is the difference of the magnetization obtained in increasing and decreasing field. On the other hand, j_c can also be estimated from the field interval ΔB necessary to fully invert the orientation of the critical state, resulting in $j_c = \Delta B/(\mu_0\ 2\ R)$. The scaling length 2 R cancels in the ratio of the two current values: $\Delta B/\ (3\ \mu_0\ \Delta M) = 1$ and obviously no information about granularity is obtained. But the measurements result in much larger ratios than 1 if the current flows in a multiply connected specimen caused by partial decoupling, inhomogeneities or percolation. Uncertainties in the determination of ΔB and ΔM result in a scatter of the ratio between 0.5 and 1.5 for a uniform bulk current or in the case of a complete decoupling for zero transport current. In the intermediate case between a homogeneous current flow and a complete decoupling we regard values of this ratio larger than about 2 as an indication for a nonuniform current or a precursor of granularity. This method is not applicable if the sample dimension or the size of the decoupled regions are of the order of the reversible penetration depth of the pinned flux lattice [16]. Similar considerations about a nonuniform bulk current flow are made in Refs. 17 and 18.

If the shielding current flows within the a,b plane in B⊥a,b the ratio is about 1 in the temperature region up to 80 K and at fields close up to the irreversibility line in full agreement with the results discussed above. In the same geometry insufficient oxygen loading results in a value of 4 at 77 K, 2 T which decreases with further oxygen treatment. This supports the correlation between granularity and oxygen content [14]. In the other geometry in which the current flows along the c-direction, the ratio deviates considerably from 1 and increases with B and T as shown in Fig. 3. This indicates the tendency towards intragrain decoupling with increasing field and temperature. Especially at 77 K the ratio is of the order of 10 at low fields. The nonuniform current flow may be caused by planar defects or oxygen deficient regions. Low fluence neutron irradiation decreases the ratio considerably [15] and supports that the inhomogeneous current flow is caused by overlapping of oxygen deficient regions which become more homogeneous due to irradiation enhanced oxygen diffusion.

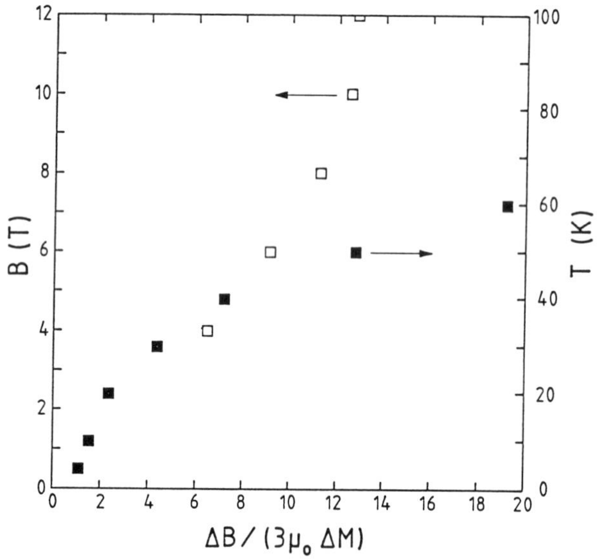

Fig. 3.
Ratio of the critical current density obtained from ΔB and ΔM versus temperature at 12 T (filled squares) and versus field at 50 K (open squares) in the geometry field within the a,b plane. The value $\Delta B/(3\,\mu_0\,\Delta M) \approx 1$ indicates a uniform current flow. The ratio at low fields cannot be determined by this simple equation because j_c is not constant within ΔB.

THERMAL RELAXATION

The activation energy as a function of field and temperature was obtained from the measurement of the magnetic moment M versus time by means of a vibrating sample magnetometer. In order to establish the full critical state at B and T the sample was cooled to the desired temperature T then a field B_s was applied with $(B_s - B) > 2\,\mu_0\,R\,j_c\,(B)$ and after the field was decreased to the value B. The first measurement of M was made at a time t_i of 15 s after the field was lowered to B. The decay of M(t) was found to be logarithmic in time within our experimental time window up to $5{\cdot}10^4$ s. Deviations from $M \sim \ln t$ are observed in the vicinity of the irreversibility line.

Within the framework of conventional flux creep theory [19,20] the ratio U_0/k_BT is obtained from the slope of dM/d lnt and the initial value $M(t_i)$:

$$U_0/k_BT = [1 - (k_BT/U_0)\ln(1 + t_i/\tau_0)]^{-1}\,M(t_i)\,/\,(dM/d\ln t) \tag{1}$$

We demonstrated that τ_0 is not the inverse attempt frequency Ω_0 but a time constant [21] given in terms of the unrelaxed critical current density j_{co}, the electric field $E(j_{co})$ and the sample radius R:

$$\tau_0 = \mu_0\,R^2\,k_BT\,j_{co}\,/\,(6\,U_0\,E) \tag{2}$$

Using typical values as $U_0/k_BT = 10$, $j_{co} = 10^4$ A/cm^2, E = 1 μV/cm and R = 0.1 cm equation (2) results in values of τ_0 within the order of seconds, much larger than $1/\Omega_0$ which is between 10^{-6} and 10^{-12} s. Therefore t_i/τ_0 is of the order of 1 and the correction in equation (1) can be neglected for $U_0/k_BT > 10$ resulting then in the well known equation:

$$U_0/k_BT = M(t_i)\,/\,(dM/d\ln t) \tag{3}$$

The initial value $M(t_i)$ was corrected for paramagnetic and reversible diamagnetic contributions obtained from the measurement of M(B). From equation (3) we calculated U_0 as shown in Fig. 4 versus temperature at B = 4 T in B||a,b. The increase of U_0 with T extends up to a maximum value U_{om} at a temperature T_m. Both values U_{om} and T_m decrease with B for instance from about 0.3 eV, 85 K, 1 T to 0.1 eV, 50 K, 12 T. In the other geometry the increase of $U_0(T)$ is more linear with T and the values of T_m are shifted considerably to lower values. Below $T_m(B)$ the values U_0/k_BT are between 30 and 80 comparable with low T_c superconductors. In this temperature region U_0 shows a modest dependence on B up to 12 T, whereas above U_0 drops rapidly towards the irreversibility line in connection with a pronounced decrease with field.

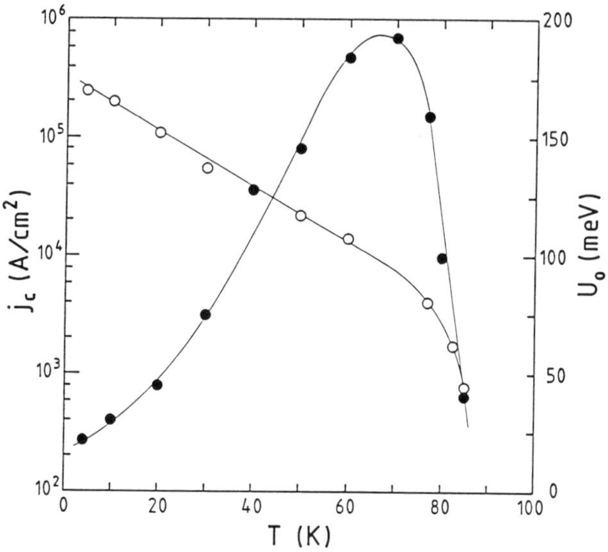

Fig. 4.
Activation energy U_0 (filled circles) and critical current density j_c (open circles) versus temperature at 4 T in the geometry field parallel to the a,b plane. The sample was irradiated with a fast neutrons fluence of 10^{17} cm^{-2} as described in Ref. 15.

Explanations given for the increase of U_0 with T suggest either that it results from an interplay between experimental uncertainties and assumptions made in order to obtain equation (3) or they are based on an inhomogeneous superconducting structure which mask the expected temperature dependence of U_0. Hagen and Griessen [20] regarded a distribution of activation energies, the measured value corresponds to the mean value of all energies above the experimental detection threshold. Assuming a suitable temperature dependence of the distribution and the threshold the remaining high energy tail which corresponds to the measured U_0 may increase with T. Such a model is based on the statistical independence of the thermally activated depinning events and requires therefore a special local arrangement of the various U_0's of the distribution. Then the measured time dependent decay of the magnetic moment is the sum over all contributions, each of them decaying with its own logarithmic slope corresponding to its U_0, the observable time dependence is nonlogarithmic.

In equation (3) a linear dependence between potential and driving force is assumed. A nonlinear dependence however - first discussed in Ref. 19 - results in an increasing observable activation energy if the current decreases from the critical one [22,23,24]. The assumption that the decay from the critical current to the first measured value at t_i becomes larger with temperature gives a natural explanation for the increase of U_0 with T. Also this model results in a nonlogarithmic time dependent decay of M in contradiction to the logarithmic relation observed in our experimental time window. Both explanations discussed above predict a change of the measured U_0 if the experimental time scale for instance t_i varies. Comparisons with measuremts of U_0 in $YBa_2Cu_3O_{7-x}$ made with squid magnetometers with a t_i larger by one order of magnitude does not show any remarkable difference. An increase of $M(t_i) \sim j$ with shorter t_i as required in both explanations is further not observed comparing the initical value $M(t_i)$ with values obtained from magnetization measurements with sweep rates between 0.1 and 12 mT/s. The values are the same in the temperature region below the maximum of U_0. Critical current measurements with the flux profile technique offer 1 T/s and a delay time of ms between generation and measurement of the profile. In the temperature, field region below 60 K, 10 T this measurement shows j_c values independent of db/dt [25] disproving a decay before the start of the magnetic relaxation measurement. This demonstrates further that the about exponential temperature dependence (Fig. 4) of j_c below the maximum of U_0 does not result from an increasing relaxation. The origin of this unexpected j_c (T) dependence at low temperatures at which the superconducting properties are constant is uncertain, one possibility are thermal fluctuations of vortex lines [26]. In the temperature region around the maximum and above variations of the time scale or the sweep rate are of essential influence of M and the observable U_0.

Assuming flux bundle activation, an increase of U_0 with T can be related to an increase of the flux bundle volume [27]. This explanation may also hold in the low temperature region because $j_c(T)$ is not constant there. But an estimation of the maximum attainable increase of the flux bundle volume accounts to a factor of 5, much lower than the observed U_0 variation [28]. A distribution of activation energies and a relaxation assumed not to be statistically independent was recently investigated [29]. Thermally activated depinning occurs at the percolation threshold energy at which flux can move through the specimen. The observable activation energy is predicted to consist of the percolation energy which is constant at low T and a term governed by the distribution. If the percolation threshold is on the high energy side of the distribution the second term increases with T.

A further approach is based on a change of the defect structure with temperature [28] by normal driven oxygen deficient regions similar to their change with field [14]. The activation energy increases with T due to these additional defects and the restrictive flux movement from the enhanced inhomogeneous structure. Both explanations are in qualitative agreement with experimental observations obtained from subsequent oxygen treatment and low fluence irradiation [15]. In both cases a more homogeneous oxygen structure is expected which reduces the percolation threshold and the restricted flux movement. Both experiments show a pronounced decrease of U_0 in the region of its maximum.

SUMMARY

Oriented $YBa_2Cu_3O_{7-x}$ shows high transport current densities at 77 K and no granularity for the current flowing along the a,b plane. A highly nonuniform current flow is observed along the c-direction of the material. Reason for this behaviour are not intergrain weak links at the grain boundaries but a tendency for intragrain decoupling within the grains which increases with field and temperature. Different oxygen treatment and low fluence neutron irradiation point to oxygen deficiency and an inhomogeneous oxygen structure as the main reason for this observation. The increase of the activation energy with temperature is not influenced or caused by the experimental time scale of the measurement. This observation rejects explanations based on a short-time decay of the current before the measurement of the magnetic relaxation starts. We favour a model based on a percolative movement of flux lines through an inhomogeneous distribution of activation energies. The maximum of U_0 becomes smaller and the increase with T less pronounced with improved homogeneity of the superconducting properties.

REFERENCES

1. Jin S, et al. (1988) Appl. Phys. Lett. 52: 2074 - 2076
2. Murakami M, et al. (Oct. 1988) Proc. Osaka University Int. Symp. Osaka, Japan
3. Salama K, et al. (1989) Appl. Phys. Lett. 54: 2352 - 2354
4. Murakami M, et al. (1991) to be published in IEEE Trans. on Magnetics
5. Welp U, et al.(1989) Phys. Rev. Lett. 62: 1908 - 1911
6 Ekin JW, Salama K, Selvamanickam V, preprint
7. Keller C, et al. (1990) Cryogenics 30: 401-409
8. Francavilla TL, et al. (1990) Cryogenics 30: 606 - 609
9. Küpfer H, et al. (1989) Progress in High Temp. Superconductivity 18: 172-182
10. Küpfer H, et al. (1989) Appl. Phys. Lett. 55: 1903 - 1905
11. Küpfer H, et al. (1988) Cryogenics 28: 650 - 660
12. Selvamanickam V, Salama K (1990) Appl. Phys. Lett. 57: 1575 - 1577
13. Küpfer H, et al. (1989) Cryogenics 29: 268 - 279
14. Däumling M, Seuntjens JM, Larbalestier DC (1990) Nature 346: 332 - 335
15. Küpfer H, et al. (1991) to be published in IEEE Trans. on Magnetics
16. Campell AM, (1971) J. Phys. C 4: 3186 - 3198
17. Matsushita T, (1989) Jpn. J. Appl. Phys.
18. Cave JR, et al. (1991) to be published in IEEE Trans. on Magnetics
19. Beasley MR, Labusch R, Webb WW (1969) Phys. Rev. 181: 682 - 700
20. Keller C, et al. (1990) Cryogenics 30: 410 - 416
21. Hagen CW, Griessen R (1989) Phys. Rev. Lett. 62: 2857 - 2860
22. Xu Y, Suenaga M, Moodenbaugh AR., Welch DO (1989) Phys. Rev. B 40: 10882 - 10890
23. Maley MP, et al. (1990) Phys. Rev. B 42: 2639 - 2642
24. Welch DO (1991) to be published in IEEE Trans. on Magnetics
25. Küpfer H, et al. (1990) Phys. Rev. B 41: 838 - 841
26. Feigel'man MV, Vinokur VM (1990) Phys. Rev. B 41: 8986 - 8990
27. Fujiyoshi T, et al. (1989) Jpn. J. Appl. Phys. 28: L1906
28. Keller C, et al. (Oct. 1990) to be published in J. Appl. Phys.
29. Gurevich A, Küpfer H, Keller C, submitted to Phys. Rev. Lett.

Preparation and Properties of Large Bulk YBCO Superconductor Quench and Melt Growth (QMG) Process

KIYOSHI SAWANO, MITSURU MORITA, MASAMOTO TANAKA, KEIICHI KIMURA, SEIKI TAKEBAYASHI, MASAO KIMURA, and KATSUYOSHI MIYAMOTO

R&D Labs-I, Nippon Steel Corporation, 1618, Ida, Nakahara-ku, Kawasaki, 211 Japan

ABSTRACT

A large bulk YBCO superconductor was prepared by the modified Quench and Melt Growth (QMG) process. The process included the enlargement precursor and unidirectional growth. Single-grained bulk as large as 45mm in diameter and 15mm thick was obtained by this technique. The single grained bulk material exhibited a strong repulsion force in relation to a permanent magnet and a strong attraction force after field cooling. These phenomena were explained by the resistance to the magnetic flux movement in the superconductor. Avoiding grain boundary proved to be effective in preparing the high performance bulk superconductor. The critical current density (Jc) was improved by optimizing the QMG process. The starting composition was effective and Jc increased to 3.7 x 10^4 A/cm^2 at 1T and 77K when 50 mol% of excess Y_2BaCuO_5 was added.

KEY WORDS: Superconductor, Melt process, Critical current density, Magnetic levitation, Magnetic flux trapping

INTRODUCTION

Since the discovery of the superconductors in which the transition temperature exceeds liquid nitrogen temperature[1], much effort have been made to improve their practical properties such as critical current density (Jc). Melt processes are among the best processes to produce high Jc bulk materials[2]. We have reported that the Quench and Melt Growth (QMG) Process can prepare high Jc YBaCuO bulk material whose Jc exceeds 10^4 A/cm^2 at 1 T[3]. This high Jc is primarily achieved by the absence of grain boundary which acts as a weak link and reduces the Jc. On the other hand, applications for the bulk superconductors have been proposed. Bulk magnets and magnetic bearings are among these applications[4]. For these applications, the bulk has to be sufficiently large and retain the properties of a small-sized sample. Attempts to increase the size of QMG material have been reported[5]. However, mere size increase is not adequate because it often results in including grain boundaries and reducing the properties of the small-sized sample. In this paper, we report on the modified QMG process for large bulk preparation and its properties. Large bulk was evaluated by several methods without reducing its size. These include repulsion force or attraction force in relation to a permanent magnet and trapped magnetic flux density. Dominant factors in superconducting properties were discussed in terms of microstructure. Improvement of Jc was also attempted by optimizing the process.

EXPERIMENTAL

Samples were prepared by the QMG process. The process consisted of four steps. The first step was a melt quench. A rapidly melted appropriate composition of Y-Ba-Cu-O was quenched to room temperature. The second step was partial melting of the quenched material. The third step was slow cooling of the material around the peritectic reaction to form $YBa_2Cu_3O_x$ (123) from Y_2BaCuO_5 (211) and liquid. The forth step was cooling to room temperature at a sufficiently slow rate for transformation to the orthorhombic superconducting phase. Subsequently, the samples were annealed in oxygen. Details of the process are described elsewhere[3].

First of all, the QMG process was optimized in terms of composition. The 211 inclusions are usually observed as a residue of the peritectic reaction in the QMG material and considered to act as dispersoids in a composite to prevent cleavage cracking in the 123 matrix. Therefore, excess 211 is effective to reinforce the QMG material and increase Jc. In addition, 211 may act as a pinning site. However, too much 211 should reduce current path and decrease Jc. There should exist an optimum composition. For these reasons, the composition was varied. The various mixtures of 123 and 211 were used as the starting composition.

Secondly, the QMG process was modified to obtain large bulk materials. Modification included two additional processes. One was enlargement of the precursor for the subsequent partial melting and growth step. The other was unidirectional growth and control of the crystal direction. Two methods were employed to enlarge the size of the precursor. One was to pile layers of quenched plate by repetitive quenching up to the required thickness. The other was to mix fine Y_2O_3 powder with quenched and ground Ba-Cu-O powder and to form it into the required shape. By these techniques, the large precursor was obtained[6].

Unidirectional growth and control of crystal direction was achieved by temperature or compositional gradient and seeding. The compositional gradient method is based on the fact that 123 formation temperature is dependent on the kinds of rare earth element in the Y site of 123. The formation temperature decreases with an increasing atomic number in the lanthanide series. For instance, while Sm-123 formed at about 1060°C, Yb-123 started to form around 900°C[7]. A mixture of these different rare earth 123 produces an intermediate formation temperature. Therefore, the gradient in a rare earth composition can act as a temperature gradient. Seeding was performed between the partial melting step and the growth step. Details of the process are described in ref.[6].

The microstructure of the large bulk was evaluated mainly by a polarized-light optical microscope because of its large size. SEM or TEM were also used. Modified X-ray diffraction techniques were used to evaluate the inner structure of the single grain. These technique includes modified Schultz method and microfocused X-ray method[8].

Superconducting properties were mainly measured by a vibrating sample magnetometer. Magnetic Jc was calculated by a Bean model using d as a sample thickness. As a bulk superconductor, the repulsion force of a zero-field-cooled sample and the attraction force of field-cooled sample in relation to a permanent magnet and the trapped magnetic flux density of field-cooled sample were measured. Details of these measurements are described in ref. 9.

RESULTS AND DISCUSSION

Optimization of Process for Jc

The starting composition significantly affected the microstructure and Jc. Figure 1 shows the relation between composition and magnetic Jc. From the stoichiometric 123, Jc increased with an increasing amount of 211 up to 50 mol% excess. At a 50 mol% excess, Jc reached as high as 3.7×10^4 A/cm^2 at 1 T and 77K. This value was about three times larger than that of the stoichiometric composition. Above a 50 mol% excess, Jc decreased to the level of the stoichiometric.

Figure 2 shows three optical micrographs of the final microstructure from different compositions. In each micrograph, matrices are 123 and dark inclusions are 211. With an increasing amount of 211, the number of inclusions increased. In addition, the amount of cracking decreased. Since 211 is a residue of the peritectic reaction which forms 123 and is needle-shaped before the reaction, 211 can act as a dispersoid and reinforce the 123 matrix. One reason for the Jc increase was attributed to the reduction of cracks which can disturb current path. Contribution of the 211 inclusions as a pinning center was not clear but is expected to increase pinning force. The reason why Jc dropped above a 50 mol% 211 excess is also not clearly understood. Too much 211 may reduce the current path.

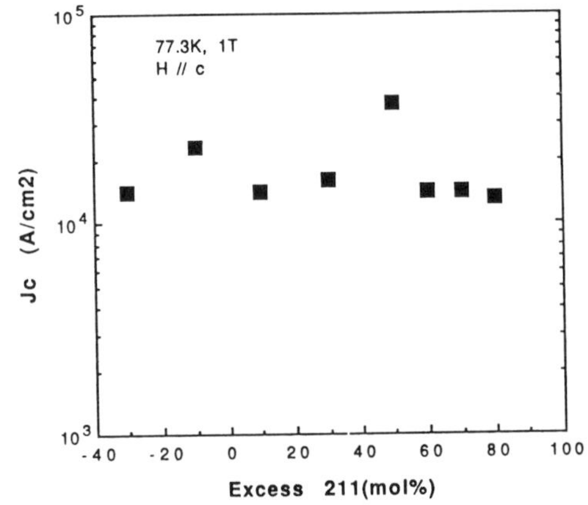

Fig. 1 Effect of starting composition on Jc of QMG processed materials.

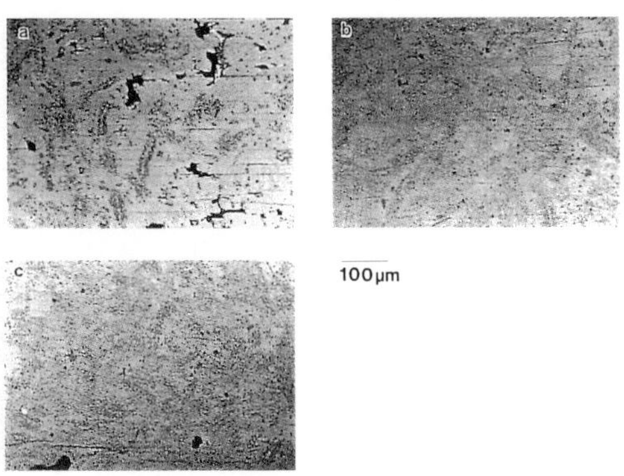

Fig. 2 Optical micrographs of QMG processed materials from different starting compositions.
(a) 10 mol% excess 211; Jc = 1.4 x 10^4 A/cm^2
(b) 30 mol% excess 211; Jc = 1.6 x 10^4 A/cm^2
(3) 50 mol% excess 211; Jc = 3.7 x 10^4 A/cm^2
All Jc was calculated from magnetization at 77K & 1T

Large Bulk Preparation

Figure 3(a) shows an optical micrograph of the enlarged sample by the modified QMG process. It can be clearly seen that the whole specimen consists of a single grain. The size of the grain exceeded 40 mm. As not seen in the figure, the thickness of the sample wasabout 15 mm. From its microstructure, the c-axis was perpendicular to the plane of the sample. Domain structure was observed in the grain. Each domain possessed a common c-axis and rotated around the c-axis. Small crystallites existed in domains. However, the whole single grain could be considered as a twinned large crystal in the macroscopic point of view[7].

From the viewpoint of grain size, unidirectional growth and seeding were key technologies in the process. When the sample was neither seeded nor unidirectionally grown, the bulk became polygrained as shown in Fig. 3(b). Randomly oriented mm-sized grains composed the bulk. From these results, the modification was proved to strongly contribute to improving the microstructure of the QMG bulk material.

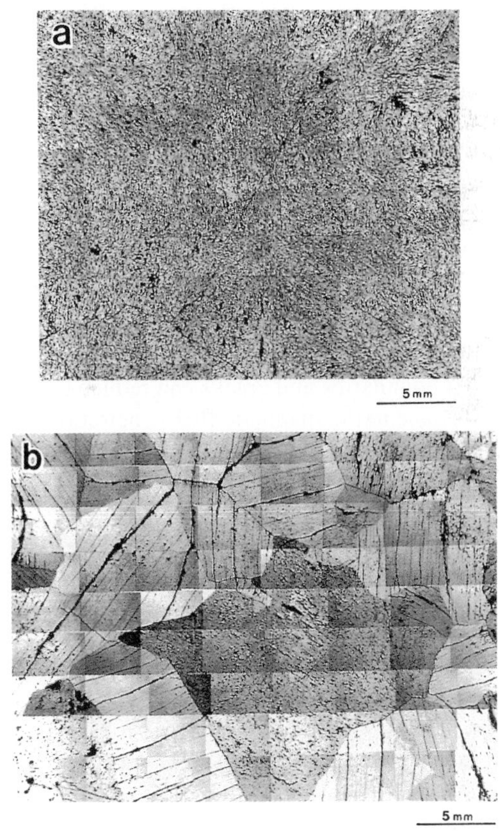

Fig. 3 Optical micrographs of enlarged QMG samples.
(a) seeded, (b) non-seeded. Note that no grain boundary is observed in seeded sample.

Interaction of Bulk Materials with Permanent Magnets

Repulsion and attraction force and trapping of the magnetic flux were measured for samples of approximately the same size. Figure 4 shows two different curves of repulsion force during the approach of a Sm-Co magnet and the subsequent force on the magnet during removal. One was for a seeded and unidirectionally grown sample while the other was for a neither seeded nor directionally grown sample. The seeded sample obviously showed much stronger repulsion force than the non-seeded sample. The maximum repulsion force was 5.6 kgf. When a Nd-Fe-B magnet was used, the maximum repulsion force increased to 10.2 kgf (Fig. 5). During removal of the magnet, the non-seeded sample showed strong hysteresis of force which often became negative (attraction), while the seeded sample showed only small hysteresis. These differences were attributed to the difficulty of magnetic flux penetration. Since intra-grain Jc is very high on the order of 10^4 A/cm^2, the magnetic flux could hardly penetrate the grain. However, if there are grain boundaries and/or cracks, the flux can easily penetrate the bulk through them. Therefore, the existence of grain boundaries and cracks can reduce the resistance to the magnetic flux penetration which results in lower repulsion force. This penetration through the boundary also contributes to increasing the apparent surface of the bulk in relation to the magnetic field. And since the crystal direction of the seeded sample is controlled and no grain boundary exists, the whole sample can act as a single grain. Consequently, only small amount of the flux was trapped in the bulk by pushing a magnet against it.

However, those samples showed different types of behavior when field-cooled. The attraction force as a function of distance of two field-cooled samples were shown in Fig. 6. As expected from the repulsion results, the seeded sample exhibited a higher attraction force than the non-seeded sample. The difference again can be explained by the difference in microstructure. In the sample without a weak region such as cracks or grain boundaries, it was hard for the trapped flux to escape. This resulted in a higher attraction force. The seeded sample exhibited 0.32 T of surface magnetic flux density when field-cooled in 1 T of a magnetic field.

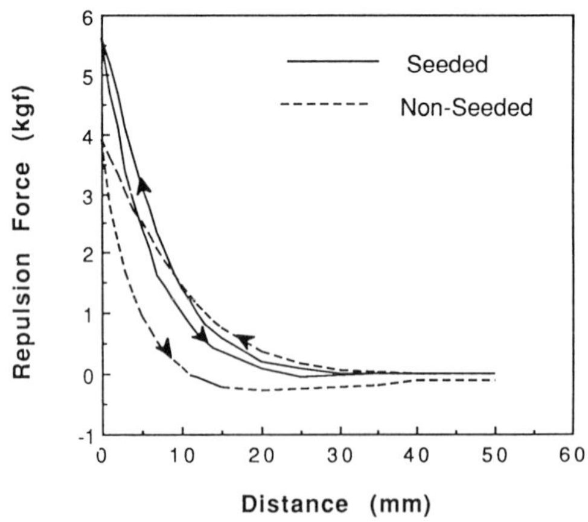

Fig. 4 Repulsion forces to a permanent magnet by two zero-field-cooled QMG samples. Note that a seeded sample exhibits higher repulsion and smaller hysteresis than a non-seeded one.

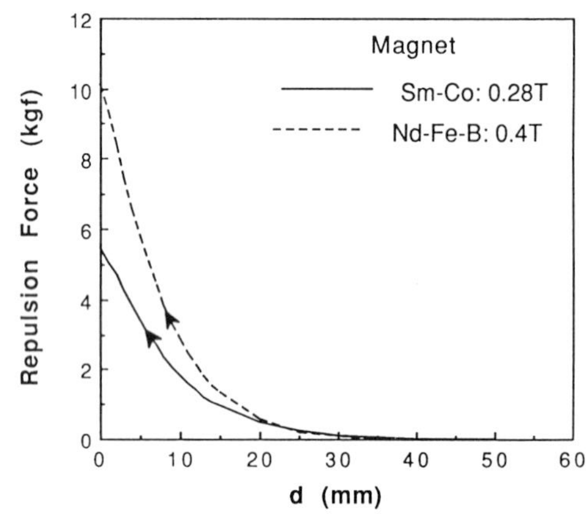

Fig. 5 Effect of types of permanent magnets on repulsion force of a seeded QMG sample.

A stronger magnet (Nd-Fe-B) clearly receives more repulsion force from the sample.

From these results, the seeded sample was proved to be better than the non-seeded sample. This indicates that the modified QMG process which was able to produce large bulk without a grain boundary and control crystal direction is significantly effective in improving the properties.

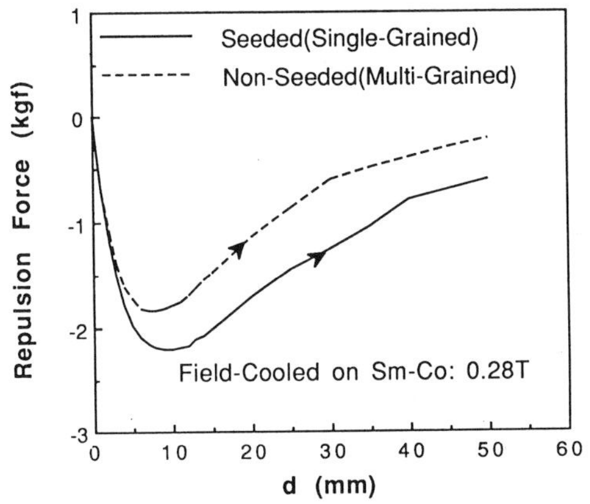

Fig. 6 Attraction forces to a Sm-Co magnet by field-cooled two QMG processed samples.

CONCLUSIONS

1. By optimization of the starting composition, the critical current density of the QMG material was improved to 3.7 x 10^4 A/cm^2 at 1 T and 77K. Increase of the 211 amount in the final microstructure was effective in reducing the number of cracks due to reinforcement by 211 inclusions.

2. By modification of the QMG process through seeding and unidirectional growth, the bulk which consisted of a single grain as large as 45 mm in diameter and 15 mm thick was obtained. The seeding also allowed to control of the crystal direction of the grain. The domain structure was observed in the grain.

3. The seeded sample exhibited 10.2 kgf of repulsion force in relation to a 0.4 T Nd-Fe-B permanent magnet. This value was the highest in a single grained sample. Grain boundaries reduced the repulsion force.

4. Attraction force when field-cooled was also higher in the seeded sample. The attraction force of 2.2 kgf was obtained by the seeded sample using a 0.28T Sm-Co magnet. Trapped surface magnetic flux density was as high as 0.32 T when field-cooled in 1 T.

5. The modification of the QMG process was proved to be effective in producing a bulk superconductor with superior properties.

ACKNOWLEDGEMENT

Authors gratefully acknowledge Professor Matsushita of Kyushu Institute of Technology for fruitful discussion.

REFERENCES

1. M.K. Wu et al, Phys. Rev. Lett. 58 908 (1987)

2.S. Jin et al, "Melt-Textured Growth of Polycrystalline $YBa_2Cu_3O_{7-d}$ with high transport Jc at 77K," Phys. Rev. B 37 [13] 7850-3 (1988)

3. M. Morita et al, "Processing and Superconducting Properties of High Jc Bulk YBaCuO Prepared by Melt Process," to be published in Physica C

4. F.C. Moon et al, "High Speed Rotation of Magnets on High Tc Superconducting Bearings," Appl. Phys. Lett. 56 [4] 397-9 (1990)

5. H. Fujimoto et al, "Melt Processing of YBaCuO Oxide Superconductors," pp285-8, Adv. in Superconductivity II, Springer-Verlag,Tokyo (1990)

6. M. Morita et al, "Quench and Melt Growth (QMG) Process for Large Bulk Superconductor Fabrication," submitted to Proc. ISS'90

7. K. Sawano et al,"Effect of Rare Earth Substitution on $REBa_2Cu_3O_x$ Superconductors by Quench and Melt Growth Process," pp277-80, Adv. in Superconductivity II, Springer-Verlag,Tokyo (1990)

8. M. Kimura et al,"Microstructural Characterization of $YBa_2Cu_3O_{7-x}$ Prepared by the Quench and Melt Growth (QMG) Process, " pp313-6, Adv. in Superconductivity II, Springer-Verlag,Tokyo (1990)

9. K. Miyamoto et al,"Quench and Melt Growth (QMG) Processed Superconducting Materials for Levitation and Magnetic Flux Trapping," submitted to Proc. ISS'90

QMG Processed Materials with High J_c and Their Mechanism

MASAMOTO TANAKA, MITSURU MORITA, SEIKI TAKEBAYASHI, MASAO KIMURA, KEIICHI KIMURA, MASAAKI SUGIYAMA, KATSUYOSHI MIYAMOTO, and KIYOSHI SAWANO

R&D Labs-I, Nippon Steel Corporation, 1618, Ida, Nakahara-ku, Kawasaki, 211 Japan

Abstract

The bulk materials of $YBa_2Cu_3O_{7-x}$ prepared by the Quench and Melt Growth (QMG) process were characterized from the microstructural points of view and the contribution of the 211 particles. The relationships between the magnetization difference and sample thickness for the QMG processed sample were measured. The critical current density exceeded 7 x 10^8 A/m^2 at 77 K and 1 T when the sample thickness was reduced to 0.05 mm.The contributions of the 211 particles turned out to be clear as well as the meaning of the effective percent of the increasing critical current density.

KEY WORDS: superconductor, melt process, critical current density, microstructure, magnetic property

1. Introduction

One of the most important key parameters for the practical application of an oxide high critical temperature superconductor is critical current density (Jc), which determines its potential. For major applications of superconductors, Jc on the order of 10^9 A/m^2 is required in the presence of a large magnetic field. Large critical current densities have been obtained in thin films[1)] in which the reported values of 1 x 10^{11} A/m^2 proved the high potentiality of these materials. On the other hand, the critical current density in bulk materials such as sintered sample and wires have so far been below the required value. Such a characteristic in critical current density was caused by the weak linking nature of the grain boundaries[2)]. Therefore, it is necessary to eliminate weak links from bulk type materials. Since a melt process such as Melt Textured Growth (MTG)[3)] or Quench and Melt Growth (QMG)[4-6)] is able to avoid weak links, a specimen obtained by the QMG process exhibits high Jc value exceeding 3.7 x 10^8 A/m^2 at 77 K and 1 T in the recent report [7)]. In spite of such a high Jc value, this Jc value is smaller than that of the thin film value. Therefore, it may be considered that there are some weak couplings besides the grain boundaries.

On the other hand, it is known that the presence of effective pinning centers is able to achieve high Jc values. In the $YBa_2Cu_3O_{7-x}$ superconductor, twin plane, oxygen defect[8)] and Y_2BaCuO_5 (211) inclusions have been proposed to serve as pinning centers. However, it is still unclear what constitutes an effective pinning center. In the conventional superconductor, it is required that the size of the most effective pinning center be on the same order of magnitude in size as in coherence length. The grain sizes of the 211 particles in a QMG material are 0.5 to 10 μm and the average is about 1.5 μm. These sizes are much larger than the coherence length and greater than the flux line lattice spacing in a large magnetic field.

The present paper discusses two possible features to determine Jc; one is the existence of other weak links that decrease the Jc which is a grain boundary, and the other is how the 211 particles contribute to an increase of Jc. For these purposes, we carried out some experiments; a sample thickness dependence of magnetization, a magnetic properties measurement, microarea X- ray diffraction and an optical microscope and TEM observation. Through the above experiments, the relation between microstructure and superconducting properties and the contribution of the 211 particles were investigated.

2. Experimental

Samples were prepared by the Quench and Melt Growth (QMG) process; the details of the process are described elsewhere[4-6]. In this process, the 123 grain contained the 211 inclusions which are residues of the peritectic reaction that was shown in Fig. 1. In an addition of this nature, it is possible to control the amount of the 211 inclusions through the starting composition.

The samples for the thickness dependence of magnetization were originally about 3 x 5 x 1.2 mm^3. The sample thickness was reduced from 1.2 to 0.05 mm. For a certain interval of the sample thickness, the magnetization properties were measured using a vibrating sample magnetometer (VSM) at 77 K and up to 1.0 T. The microstructures were studied by using an improved technique of the apparatus for microarea X-ray diffraction and transmission electron microscopy (TEM). The magnetization properties were also measured using VSM for each different excess value of the 211 particles, and magnetic Jc was calculated from those magnetization differences using the Bean model.

Fig. 1 Polarized optical micrograph of the microstructure of YBaCuO fabricated by the QMG process. Dark spots in the figure are the 211 particles.

3. Results and Discussion

3.1 Thickness Dependence and Domain Structure

Figure 2 shows the typical data for dM and Jc as a function of the sample thickness of QMG materials, where dM is the magnetization hysteresis between increasing and decreasing magnetic fields.

The Jc increased remarkably under a thickness of 0.1 mm and exceeded 7 x 10^8 A/m^2 at a thickness of 0.05 mm. It was found there were three regions for dM as shown by the dotted line. One region was above the thickness of 0.8 mm in which dM was saturated. The second was between a thickness of 0.2 and 0.8 that had a moderate gradient of dM. The last one was under a thickness of 0.2 mm and that had a sharp gradient. We think this moderate gradient corresponds to the bulk critical current density and the sharp gradient corresponds to the local current density which was pointed out by Matsushita et al[9]. According to the Bean model, the magnetization (dM) has a constant gradient in the grain. Therefore, the strongly linked area can be estimated by measuring the magnetization by changing the thickness. In considering the more general implications of our results, the above result suggests the existence of some weak links. That is, there is a weak link between 0.2 and 0.3 mm in thickness because Jc increases sharply between them.

In order to understand the above experimental results, we carried out microarea X-ray diffraction for the microscopic analysis. Figure 3 shows the results of the intensity mapping of 2010 and 0210 along a certain line on the single QMG sample with an interval of 60 µm. The change with a periodicity of 150 - 250 µm showed the existence of crystallite of about 200 µm in size. Though the intensities showed a tendency to decrease from the starting point to the final point, the crystallographic relations were maintained in the whole measured area where five or six crystallites existed. We have called this structure the domain structure. More detailed discussions are described elsewhere[10]. These experimental results showed good agreement with the results of the thickness dependence experiment. That is, both the experimental results suggest the existence of the domain structure. From these results, this domain boundary disturbs an increasing Jc. Microscopic analysis of the domain boundary has not been completed. However, we consider that these domain boundaries have the same nature as oxygen defects or precipitations.

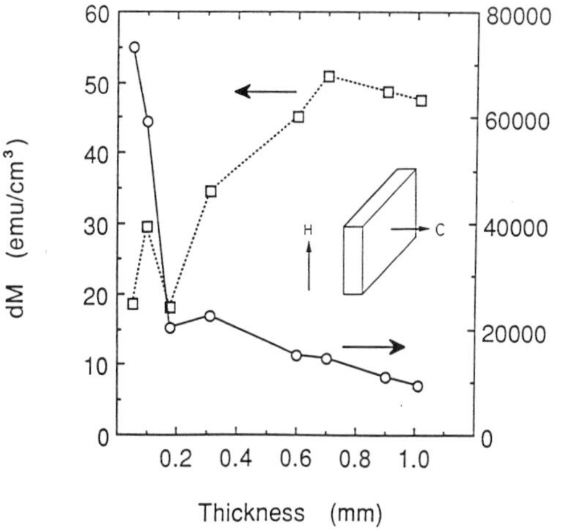

Fig. 2 The thickness dependence of the magnetization differences (dM) and critical current densities (Jc) with the magnetic field perpendicular to the c axis at 77K and 1T.

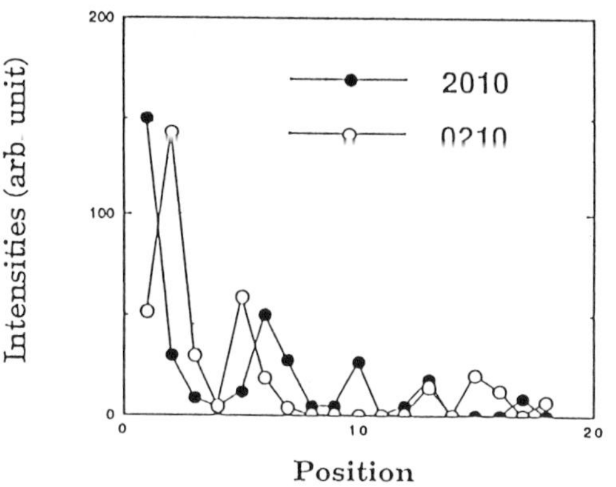

Fig. 3 The results of the intensity mapping of 2010 and 0210 reflection along the surface of QMG processed crystal.

3.2 Contribution of Y_2BaCuO_5

The contributions of the 211 particles were investigated by veiws from microscopic to macroscopic. The typical TEM image of the QMG processed specimen is shown in Fig. 4. In this micrograph, twins can be observed in the 123 matrix running up to the interface. A characteristic feature can also be observed around the 211 particles where twins have kinks. Since twins appeared when the phase transition proceeded from an orthorhombic to a tetragonal, the kinks in the twins were attributed to the strain field that we thought the kinks were introduced by the 211 particle. Stacking faults can also be observed in the 123 matrix that is shown in Fig. 5. Since the size of the stacking faults have an atomic scale and are on the same order of coherence length, they are also considered to serve as a pinning center.

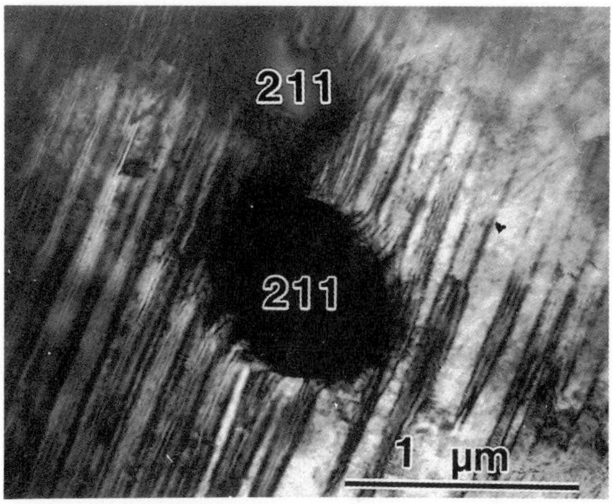

Fig. 4 TEM image of the microstructure of YBaCuO superconductor fabricated by the QMG process.

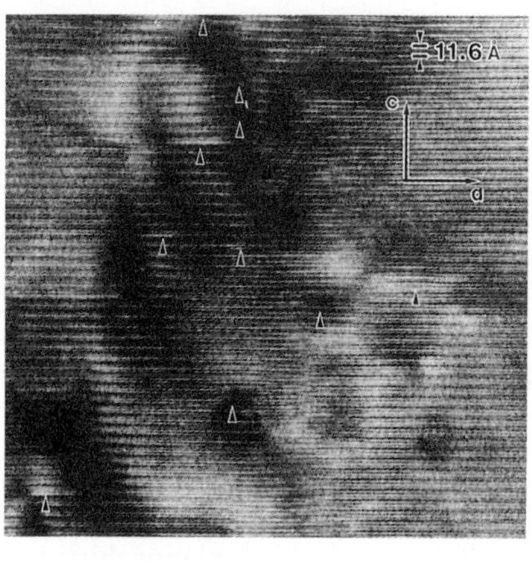

Fig. 5 The high resolution TEM image of the 123 matrix. The stacking faults are marked by arrow.

Figure 6 shows the contribution of the 211 particles in increasing Jc as a function of an excess volume of the 211 particles. The maximum Jc was obtained with a 50 mol% excess of the 211 particles. The relationship between the excess volume ratio and Jc does not show linearity or curve. It seems that the Jc is saturated above a 60 mol% excess. In addition this behavior, the amount of cracking decreased with the increasing amount of the 211 particles. We think that an increasing Jc can be considerably attributed to the reduction in cracks which can disturb the current path. Too many 211 particles may reduce the current path. The 211 particles can be regarded as working for the Jc increase, but the contribution of 211 particles in increasing pinning force is not clear from this experiment.

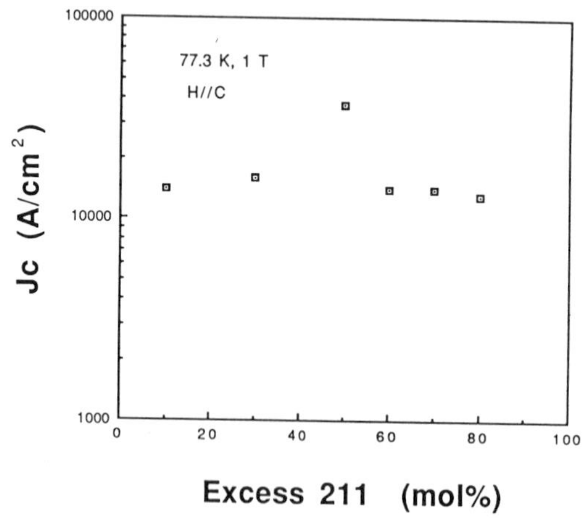

Fig. 6 Logarithmic plot of Jc vs linear plot of the excess mol% of the 211 particles at 77K and 1T.

Furthermore, the contribution of the 211 in increasing Jc was discussed in other experimental results. If we assume a direct summation of individual interaction force of large normal particles and use an empirical temperature dependence of $Bc_2(T)=1-(T/Tc)^2$, the logarithmic Jc is proportional to logarithmic $[1 - (T/Tc)^2]$. It is expected that the coefficient of this relationship is 1.5, which was introduced from a theoretical projection[11, 12]. Figure 6 represents double logarithmic plots of Jc versus the functional form of Bc_2 for three samples. Sample A and B were processed by the QMG method, which were different in the excess volume of 211 particles; A was 5 vol% and B was 30 vol%. Since the sintered sample did not include the 211 particles, the sintered sample C is also described on the same figure for comparison.

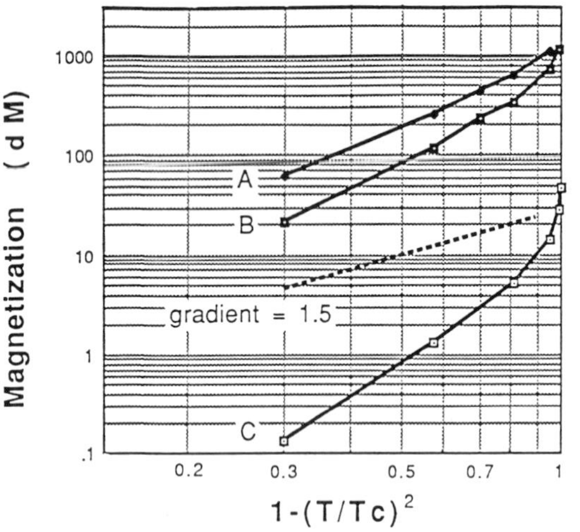

Fig. 7 Logarithmic plots of Jc vs $1-(T/Tc)^2$ at B = 4T and 77K for different volume ratios of the 211 particles (A:5%, B:30%) and C:sintered sample.

These results showed that the temperature dependence was a little different from the prediction even in the high temperature region, and it became even more different at a low temperature. Furthermore, samples with and without the 211 particles had nearly the same gradient. That is, the contribution of increasing Jc with the 211 particles serving as pinning centers may not be so large according to the present results.

4. Conclusions

To summarize, QMG processed materials with high Jc have been investigated. We have found the domain structure experimentally, there are two kinds of current density in QMG materials, bulk and local current densities. The domain boundaries disturb an increasing Jc, however this disturbance may be very small in comparison with the grain boundaries. The contribution of increasing Jc with the 211 particles serving as pinning center may not be so grate in comparison with the sintered specimen which does not include the 211 particles. Furthermore, it was found that 211 phase particles stopped the growth of cracks. Namely, that 211 particles can be regarded as decreasing the defects such as cracks. On the other hand, there were many stacking faults and strain fields around 211 phase particles by TEM observations. Such stacking faults and strain fields may work as effective pinning centers. An investigation of effective pinning centers at high temperature will also be necessary.

Acknowledgments

We would like to thank Professor T. Matsushita for the worth-while discussion results. We are also grateful to the members of the superconductor group of Nippon Steel Corporation for their collaboration.

References

1)S. Tanaka and H. Itozaki: Jap. J. Appl. Phys. 27(1988)L622.
2)D. Dimos, P. Chaudhari, J. Mannhart and F. K. LeGoues: Phys. Rev. Lett. 61(1988)219.
3)S. Jin, T. Tiefel, R. Sherwood, R. van Dover, M. Davis, G. Kammlott and R. Fastnacht: Phys. Rev. B 37(1988)7850.
4)M. Morita, K. Miyamoto, K. Doi, M. Murakami, K. Sawano and S. Matsuda: Physica C, (to be published)
5)M. Morita, M. Murakami, K. Miyamoto, K. Sawano and S. Matsuda: Pyhsica C, 162-164(1989)1217.
6) M. Murakami, M. Morita, K. Doi and K. Miyamoto: Jap. J. Appl. Phys. 28(1989)1189.
7)S. Sawano, M. Tanaka, M. Morita, K. Kimura, S. Takebayashi and K. Miyamoto: ISS'90, Sendai, 1990.
8)P. Kes, A. Pruymboom, J. van den Berg and A. Mydosh: Cryogenics 29(1989)228.
9)T. Matsushita, B. Ni, M. Murakami, M. Morita, K. Miyamoto, M. Saga, S. Matsuda and M. Tanino: Jap. J. Appl. Phys. 28(1989)L1545.
10)M. Kimura, M. Tanaka and K. Sawano: Physca C, (submitted).
11)A. Campbell, J. Evetts and D. Dew-Hughes: Philos. Mag. 18(1968)313.
12)T. Matsushita, E. S. Otabe, B. Ni, K. Kimura, M. Morita, M. Tanaka, M. Kimura, M. Kimura, K. Miyamoto and K. Sawano: Jap. J. Appl. Phys. (submitted).

Quench and Melt Growth (QMG) Processed Superconducting Materials for Levitation and Magnetic Flux Trapping

KATSUYOSHI MIYAMOTO, MITSURU MORITA, MASAMOTO TANAKA, SEIKI TAKEBAYASHI, KEIICHI KIMURA, MASAO KIMURA, and KIYOSHI SAWANO

R&D Labs-1, Nippon Steel Corporation, 1618, Ida, Nakahara-ku, Kawasaki, 211 Japan

ABSTRACT

Interaction between permanent magnets and YBCO bulk superconductors prepared by the Quench and Melt Growth (QMG) process was investigated. An apparatus to measure forces to a permanent magnet from bulk superconductors was developed by modification of a tension/compression testing machine. Strong repulsion force as high as 10.2 kgf was observed. This high value is attributed to the resistance to magnetic flux penetration into the bulk. The sample without grain boundary showed the highest repulsion force. This resulted from the avoidance of flux penetration through the grain boundary. The attraction force and trapped magnetic flux density after field cooling was also larger in the sample without grain boundary. The modified QMG process which enabled us to obtain large YBCO bulk proved to be effective for high a performance bulk superconductor.

KEY WORDS: Superconductor, YBaCuO, Melt process, Magnetic levitation

INTRODUCTION

YBCO bulk superconductor prepared by the Quench and Melt Growth (QMG) process is known to show high critical current density (Jc) even in a high magnetic field[1].

Recently, the Jc was improved to 3.7 x 10^4 A/cm^2 at 1 T by optimizing its composition and other processing factors[2]. The size of the QMG processed material increased by a modification of the process larger than 40 mm in diameter and 15mm in thickness without reducing the superconducting properties[3]. This development in the process enabled us to utilize this material for applications such as magnetic levitation or magnetic bearings[4]. For these applications, a quantitative evaluation of the materials without changing their large size was required. Measurement of the levitation force against a permanent magnet is one of these methods[5]. In this study, the levitation and attraction force against a permanent magnet is evaluated using tension/compression testing machine at the liquid nitrogen temperature (77K). Contributing factors such as the microstructure of the material are discussed.

EXPERIMENTAL

Samples were prepared by the modified QMG process. The details of the process are described elsewhere [3]. The key techniques of the modified QMG process are the enlargement of the melt -quenched precursor and seeding before the growth step. By this process, a sample which was comprised of a single grain could be obtained in size of a 45 mm in diameter and 15 mm thick. The crystal direction of the sample could be controlled. In most samples, c-axis was controlled to become perpendicular to the plane of a disc-shaped sample. The size of the samples used was changed to understand the size effect. For comparison, non-seeded-processed samples were used.

The tension/compression testing machine was revised for the measurement of levitation and attraction. The apparatus is schematically shown in Fig. 1.

A Samarium-Cobalt (Sm-Co: 50 mmϕ x 10 mm)or Neodymium-Iron-Boron (Nd-Fe-B: 45 x 60 x 10 or 20 mm) permanent magnet was attached on the end of a load-cell. The surface magnetic flux density of the Sm-Co and the Nd-Fe-B magnets were 0.28 T and 0.4 T, respectively. Levitation (repulsion) force was measured by slowly moving the magnet toward the zero-field-cooled sample in the liquid nitrogen. Subsequently, the magnet was removed from the specimen and the force was also measured. The attraction force was measured by removing the magnet from the field-cooled sample. The trapped magnetic flux density was measured by a Hall-probe after field-cooling under various magnetic fields.

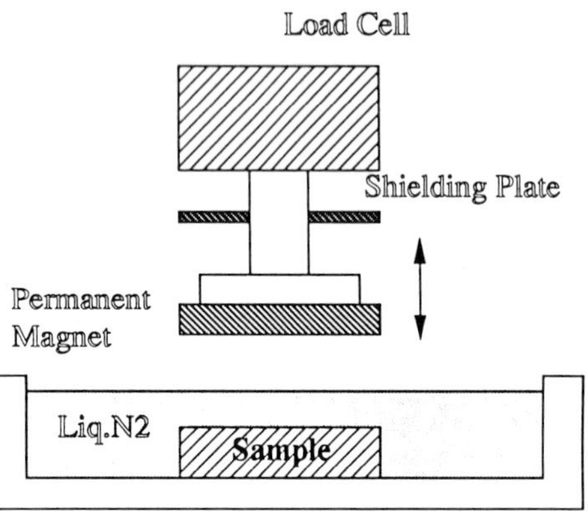

Fig. 1 Apparatus for levitation and attraction force measurement

RESULTS AND DISCUSSION

Figure 2 shows typical curves of the force as a function of the distance, arrows indicate the direction of the magnet's movement. The repulsion force sharply increased when the magnet moved closer than approximately 10 mm. A hysteresis was observed between moving and removing the magnet. The force often changed to attraction from repulsion during removal of the magnet. The amount of hysteresis became maximum at about 10 mm in distance. The difference in the repulsion force is mainly discussed in terms of the microstructure and the size of the sample.

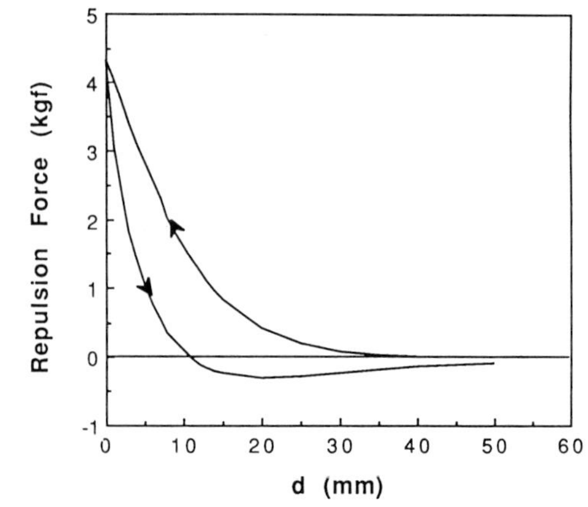

Fig. 2 Typical repulsion force during approach of a permanent magnet. Arrows indicate the direction of magnet movement.

Effect of Sample Size

Since the diameter of the samples was the same or smaller than the magnet, the repulsion force was mostly dependent of the size of the sample. The effect of the thickness was clearly observed in the case of thin samples. Figure 3 shows the force versus distance curves in samples with various thicknesses. The samples had a similar microstructure. It is easily seen that thinner sample showed lower repulsion force. The force seemed to saturate at an area very close to the specimen. This was attributed to the magnetic flux which run around the sample.

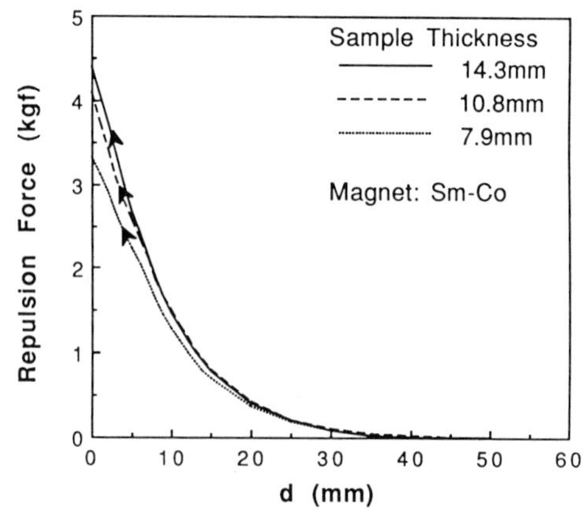

Fig. 3 Repulsion forces of the QMG samples with various thickness.

Effect of Microstructure

The effect of microstructure on the repulsion force was examined using two samples of approximately the same size (42 mmϕ x 15 mm). One was processed with seeding and the other without seeding. The seeded sample was single-grained and the non-seeded sample was comprised of more than ten grains. Figure 4 shows the microstructure of the two samples. Figure 5 shows curves of the force as a function of distance in both directions when a Sm-Co magnet was used. The arrows indicate the direction of the magnet's movement. The repulsion force of the seeded sample is obviously higher than that of the non-seeded. The maximum repulsion force of the seeded sample was as high as 5.6 kgf for a Sm-Co magnet. The same seeded-sample exhibited 10.2 kgf of the maximum repulsion force in the case of a Nd-Fe-B magnet (Fig. 6).

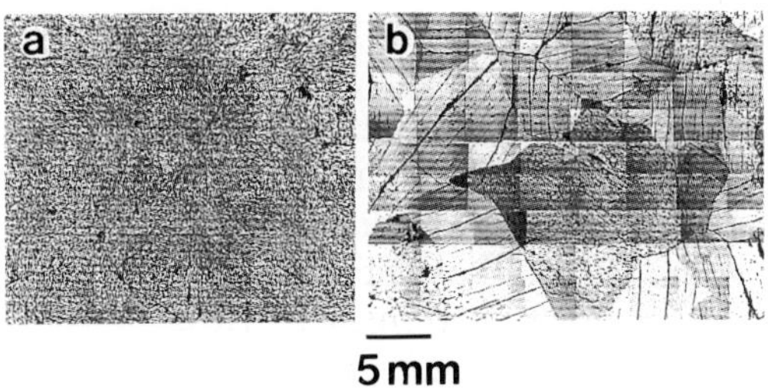

Fig. 4 Optical micrographs of (a) seeded and (b) non-seeded QMG processed samples.

This large repulsion force was not explained by the Meissner effect. This was caused by the resistance to the penetration of magnetic flux into the bulk. As indicated by Moon et al., the force may be expressed as the product of the magnetic moment of the magnet and the gradient of the field outside the superconductor produced by magnetization alone[6].

The difference in force in the two samples can be explained by the existence of grain boundaries. Since the grain boundary acts as a weak link, magnetic flux can penetrate into the bulk more easily through the grain boundary. As a result, less force is necessary for magnetic flux penetration in a sample with grain boundaries. If a weak region does not exist, as in the case of a single-grained sample, flux pinning should be dominant. As described above, since the QMG samples exhibit high Jc, this implies a strong pinning force and higher resistance to the flux penetration.

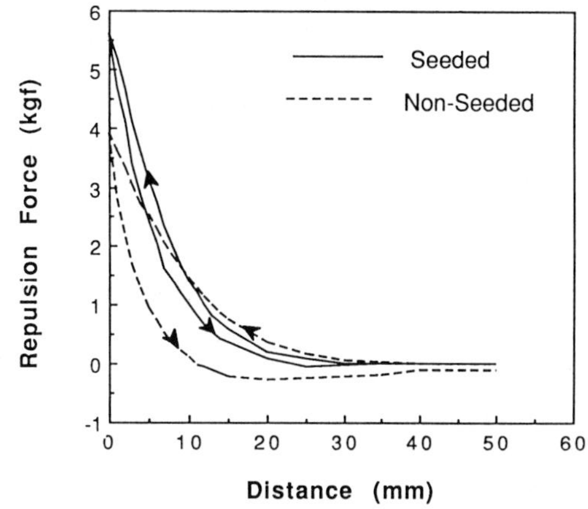

Fig. 5 Repulsion force curves of two types of QMG processed materials. Note that a seeded sample exhibits higher repulsion and less hysteresis than a non-seeded sample.

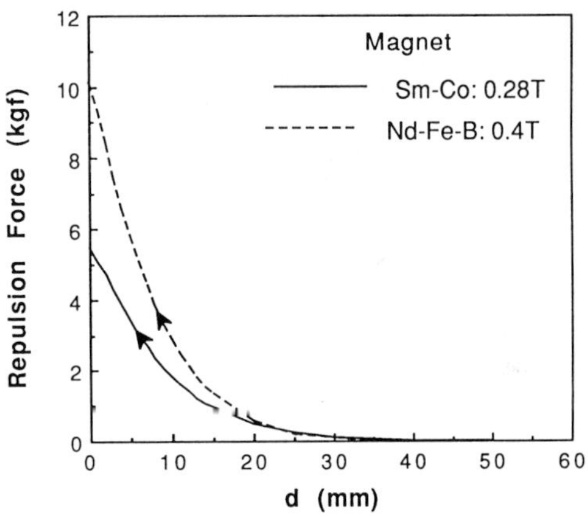

Fig. 6 Repulsion force toward different types of magnets by the seeded QMG sample.

Another difference in both curves was the degree of hysteresis. The seeded sample showed much less hysteresis than the non-seeded sample which exhibited attractive force during the removal of the magnet. This can be understood by the amount of magnetic flux trapped in the bulk. When the magnet approaches the sample, some amount of magnetic flux penetrates the sample. Since the amount of the penetrating flux increases by increasing number of grain boundaries in the bulk, more flux can be trapped in the sample with grain boundaries. This seems to conflict with the fact that polycrystalline sintered materials do not exhibit strong hysteresis. This is because the amount of trapped flux is dependent on the size of the grain and the strength of the magnetic field. The size which gives the maximum flux trapping should exist in a given magnetic field. The grain size of the sintered material is much smaller than this size. The flux penetration and trapping of single-grained and multigrained samples are schematically shown in Fig. 7. If we assume a critical state model and constant Jc in the magnetic field, the grain size for maximum flux trapping can be calculated by assuming that the zero flux density position is at the center of the grain. Using Bean's critical state model, size d can be calculated as $d = \alpha J c B_S$ where B_S is surface flux density, α is constant depending on the unit. When the external field of 0.4 T and 10^4 A/cm^2 were used, the grain size was calculated to be 15 mm. The size was increased by increasing the external field or decreasing Jc. The trapped flux from the approaching magnet could be maximized at this size. The grain size of the seeded single-grained sample was more than 40 mm and was too large for this field with a permanent magnet.

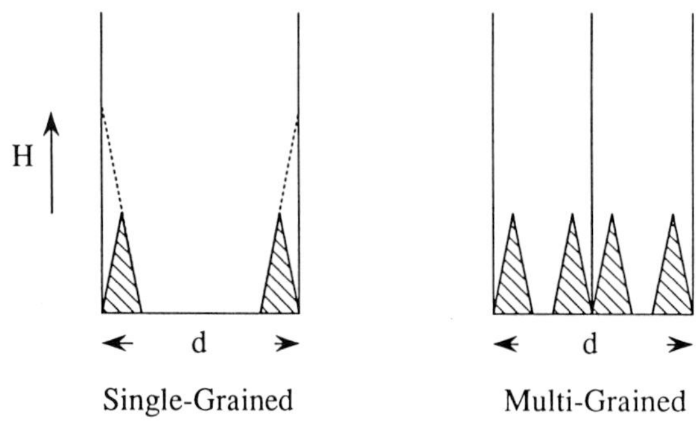

Fig. 7 Schematic drawings of the magnetic flux trapped in single-grained and multi-grained bulk.

Because of the small hysteresis, the seeded sample did not show stable suspension by simply pushing the magnet toward the sample. Instead, field cooling exhibited excellent attractive force as a magnet. Figure 8 shows the attraction force of the seeded sample and non-seeded sample after field-cooling using a Sm-Co magnet. A maximum attraction force of 2.2 kgf was exhibited by the seeded sample, while the non-seeded sample exhibited 1.8 kgf. This indicates the strong pinning force in this seeded sample.

When the sample was cooled in a magnetic field, the sample trapped the magnetic flux and acted like a permanent magnet. The magnetic field of 1 T was applied for field cooling down to 77K for the 42 mmϕ x 15 mm seeded sample. After removing the magnetic ficld, the sample still kept 0.32 T of the surface magnetic flux density. No flux jump was observed during measurement. For a 20 mmϕ x 5 mm sample, field cooling using a 0.28 T Sm-Co permanent magnet was performed. After removing the magnet, 0.17 T of the flux density was observed. An analysis of the trapped flux is now in progress.

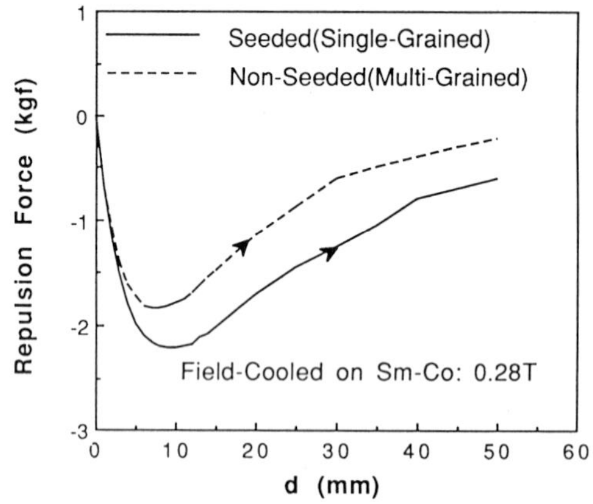

Fig. 8 Attraction forces to a magnet by field-cooled seeded and non-seeded samples.

CONCLUSIONS

1. The repulsion force of a permanent magnet from QMG samples was measured at liquid nitrogen temperature. As much as 10.2 kgf of the repulsion force was observed for a seeded QMG processed sample. A seeded single-grained sample exhibited higher repulsion force than a non-seeded multi-grained sample.

2. Such a high repulsion force is not explained by the Meissner effect. The resistance to

magnetic flux penetration due to the mixed effect of pinning and weak links are the probable mechanism.

3. Hysteresis of the force was observed between the approaching and removal of the magnet. The origin of the hysteresis is considered to be the trapped magnetic flux during the approach of the magnet. When the sample included grain boundaries, more hysteresis was observed. Single-grained samples showed much less hysteresis because of the penetration only from the surface, while the sample with grain boundaries could trap more flux which penetrated through the boundaries. The size for maximum trapping can be calculated by assuming a critical state model and weak-link.

4. Surface magnetic flux density of the single-grained sample was measured after field cooling. 0.32 T of the magnetic flux density was observed when cooled in 1 T field. 0.17T was observed after being field-cooled on a 0.28 T Sm-Co permanent magnet.

REFERENCES

1. M. Morita, K. Miyamoto et al, "Processing and Superconducting Properties of High Jc Bulk YBaCuO Prepared by Melt Process," accepted to Physica C

2. K. Sawano et al, "Preparation and Properties of Large Bulk YBCO Superconductor by Quench and Melt Growth Process," submitted to Proc. ISS'90

3. M. Morita et al, "Quench and Melt Growth (QMG) Process for Large Bulk Superconductor Fabrication," submitted to Proc. ISS'90

4. T. Higuchi, "Application of Superconductors in Machine Components," J. Jpn. Soc. Mech. Engineers, 91 [835] 567-70 (1988)

5. F.C. Moon et al, "Hysteretic Levitation Forces in Superconducting Ceramics," Appl. Phys. Lett., 52 [18] 1534-36 (1988)

6. F.C. Moon et al, "Levitation Forces, Relaxation and Magnetic Stiffness of Melt-Quenched $YBa_2Cu_3O_x$," J. Jpn. Appl. Phys., 29 [7] 1257-8 (1990)

Quench and Melt Growth (QMG) Process for Large Bulk Superconductor Fabrication

M. MORITA, S. TAKEBAYASHI, M. TANAKA, K. KIMURA, K. MIYAMOTO, and K. SAWANO

R&D Labs-I, Nippon Steel Corporation, 1618, Ida, Nakahara-ku, Kawasaki, 211 Japan

Abstract

Enlargement of QMG crystals was performed by seeding and compositional gradient techniques. First, the precursors were enlarged by the repetitive quench method. Secondly, nucleation and crystal direction were controlled by seed crystals which have higher formation temperature in the $REBa_2Cu_3O_{7-x}$ phase. Thirdly unidirectional growth was performed by compositional gradient of RE elements in a uniform temperature. Using these techniques, we obtained large crystals of up to about 50 cm^3.

KEY WORDS: $REBa_2Cu_3O_{7-x}$, melt process, seeding, compositional gradient

1.Introduction

It is known that the QMG process can produce a bulk $YBa_2Cu_3O_{7-x}$ superconductor with high critical current density(Jc) even at 77K and a high magnetic field[1,2]. The QMG processed sample is a pseudo crystal which has finely dispersed Y_2BaCuO_5 phases in the $YBa_2Cu_3O_{7-x}$ matrix. The QMG process can be regarded as one of the crystal growth methods using two peritectic reactions. One is between Y_2O_3 and the liquid phase into acicular Y_2BaCuO_5, another is between Y_2BaCuO_5 and the liquid phase into $YBa_2Cu_3O_{7-x}$.

Because of excellent Jc properties, QMG materials have the possibility of being used for a bulk superconducting magnet, superconducting bearing, current lead, and so on. Enlargement of QMG crystals is an important technology on applying this material to such practical uses. Enlargement of QMG crystals by using unidirectional growth in a temperature gradient was reported previously[3]. In the melt process, a promising method is desired to control nucleation and crystal direction. In this paper, The authors report enlargement of the precursor by repetitive quenching and enlargement of the QMG crystal by using the difference of formation temperatures in the $REBa_2Cu_3O_{7-x}$(123) phases.

2. Enlargement of quenched precursor.

To make large QMG crystals, large and dense precursors without voids are required. To obtain a large precursor by quenching, a repetitive quench method was tried. Fig.1 shows the process of the repetitive quench method schematically. First, a precursor with a thickness of about 2 mm was made by quenching from the melt by using two copper plates. The precursor was dense as long as the thickness was less than 2mm. Secondly melt was poured on the precursor and pressed by a copper plate. A precursor whose thickness was about 4mm was obtained. By repeating quenching in a similar manner, a thicker precursor was obtained.

Adhesion of layers was strong when quenching was done without long intervals. On the contrary, the layers exfoliated easily when the surface of the precursor was exposed to the atmosphere for about 2h. Fig.2 shows the micro structure of a cross section of layered precursor. From the results of EDX analysis, it was found that there is a thin CuO layer on the surface of the precursor and the CuO layer strongly binds the precursor and solidified melt and that Y_2O_3 particles were almost uniformly dispersed in BaCu oxide. A thick precursor prepared by the repetitive quenching method does not break by rapid reheating up to around 1200°C Large and dense precursors were obtained by the repetitive quench method.

As described above, the nature of the QMG process is that acicular 211 phases were finely dispersed in the liquid phase by peritectic reaction between Y_2O_3 and liquid phases before crystallization of the 123 phase. From this point of view, it is not necessary to use quenched material for the precursor. The precursor has to be a mixture of Y_2O_3 and Ba and Cu compound. We tried to use this by forming a mixture of Y_2O_3 powder and quenched and ground BaCu-oxide powder for the precursor. Pseudo crystals which have finely dispersed 211 phases in the 123 matrix were obtained by using the formed mixture. It was found that the formed mixture worked as well as melt-quenched material for precursor. The advantage of using this powder mixed precursor is ease of shape forming. The disadvantage of these techniques is that QMG crystals occasionally contain many voids.

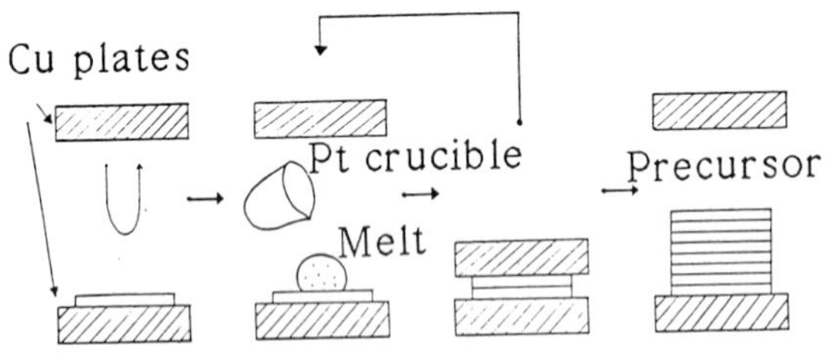

Fig.1 Schematic explanation of repetitive quench method.

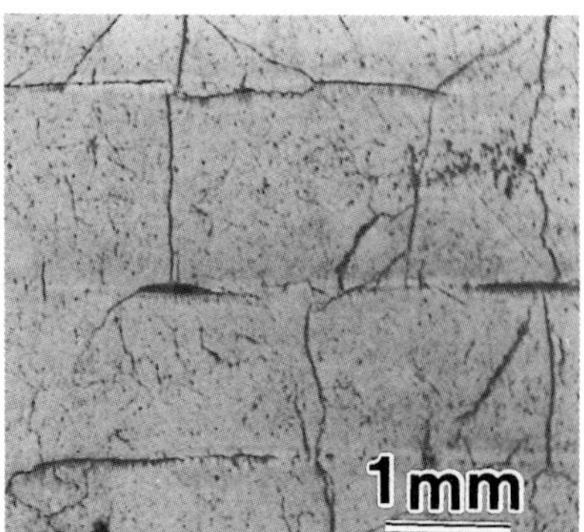

Fig.2 Optical micrograph of repetitive quenched precursor.

3. Enlargement of QMG crystal

a) Formation temperature of $REBa_2Cu_3O_{7-x}$ phase[4].

123 phases have a complicated crystal structure and the crystallization of the 123 phase is accompanied by a large entropy change. Therefore, the 123 phase grows with a stable facet[5]. Since nucleation of 123 crystals does not occur quickly and spontaneously at around the formation temperature (Tf), long time holding of a constant temperature is necessary to determine the Tf. In the QMG process, 123 crystals are block-shaped. It seems that the growth of 123 crystals is not so anisotropic, in spite of the two dimensional crystal structure. To estimate Tf of RE systems, a temperature gradient was employed. First, several layered precursors were prepared by the repetitive quench method in each system. Secondly, the precursors were reheated up to about 1200°C and hold for 20 minutes by using a Pt holder. Thirdly, a temperature gradient of 20°C/cm was applied in the furnace, and the samples were kept for several hours to grow 123 crystals. Finally, the samples were quenched to room temperature, and Tf was determined by the position of the growth front. The results of experiments are shown in tab.1. The growth front had random facets, Therefore we consider that each value of Tf has an error of about +10°C.

Tab.1 Formation temperature of $REBa_2Cu_3O_{7-x}$ phases in the air for QMG process.

Elements	Y	Sm	Eu	Gd
Tf (°C)	1000	1060	1050	1030
	Dy	Ho	Er	Yb
	1010	990	970	900

On the basis of this result, with an increase of in the ion radius of the RE element, Tf increases. However, in the La system, a 123 phase was not obtained. In the Nd system, $Nd_{1-x}Ba_{2+x}Cu_3O_{7-y}$ crystals which did not completely transit superconductive at 92K were obtained.

Moreover, diffusion of RE elements was investigated in a plural RE system. Three layered precursors which contained Ho, Y and Dy in each layer were prepared by the repetitive quenching method. The precursor was reheated to 1200°C and hold for 20 minutes. The 123 phase grew with slow cooling from 1020°C to 900°C at a rate of 2°C/h. To observe the distribution of each RE element, mapping of RE element was measured by EDX analysis. From the results, it was found that almost each RE element remained in each layer. however, in boundary region of the Dy and Y layer, 123 facets which contained Dy elements were slightly observed in the Y layer.

Furthermore, variations in Tf were investigated for the compositional ratio of RE elements in the Yb-Y system. Y and Yb elements can easily be mixed into $Y_xYb_{1-x}Ba_2Cu_3O_{7-y}$ and $Y_{2x}Yb_{2-2x}BaCuO_5$ powder by calcination. By using such powders, an eleven layered precursor in which the compositional ratios Yb to Y in each layer were changed from 0% to 100% every 10% were prepared by the repetitive quench method. The precursor was reheated to 1200°C for 20 minutes. and kept in a temperature gradient of 20°C/cm for 8h to grow the 123 phase. The sample was quenched to room temperature. The position of the growth fronts of each layer was observed. From the results of the observation, it seems that Tf in the plural mixed RE system is given a molar mean value of Tf in each singular RE system. It should be noted that Tf in the 123 phase is controllable by changing the RE composition. The authors applied these properties to enlarge the 123 crystals.

b) Control of growth direction using seed crystal with higher Tf.

QMG crystals which consist of a Y element were made by using $SmBa_2Cu_3O_{7-x}$ seed crystals. The Tf of Y and Sm systems are about 1000°C and 1060°C, respectively. Therefore, seeding was done at around 1030°C. Seed crystals were prepared by the conventional QMG process in advance. A cleaved surface of seed crystals was put on the precursor before the growth step in which the precursor consisted of Y_2BaCuO_5 and a liquid phase. And the seeded sample was slowly cooled to 985°C at the rate of 1°C/h, kept for 20h and slowly cooled to room temperature. Fig. 3 shows the Sm seed and seeded Y-QMG sample. On the surface of the seeded Y-QMG sample, a characteristic pattern was observed. A schematic explanation of the pattern is shown in Fig. 4. The hatched area in the center is the region which contacted the seed crystal. The pattern may be understood as a result of the growth process that first in occurred crystal growth in this region. Secondaly,it proceeded at the kink, then a rectangular faceted region was formed and the crystal grew with these facets. In fact, a rectangular crystal was obtained, when the sample was quenched to room temperature to stop growth. The growth direction was maintained as the direction of the seed crystal. These results mean that crystal direction and nucleation can be controlled by seeding and heat treatment at a uniform temperature.

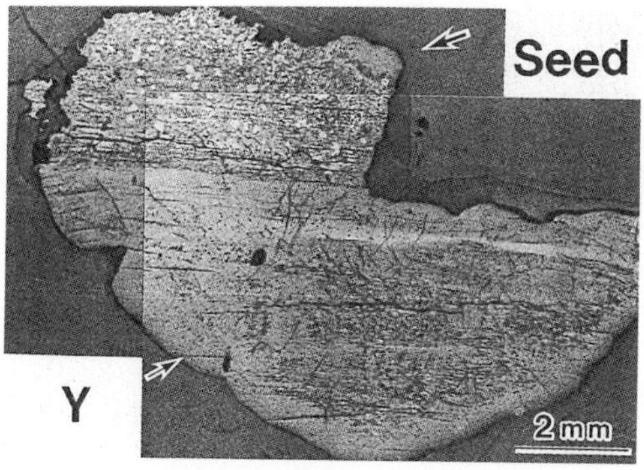

Fig.3 Optical micrograph of a seed crystal consistingof a Sm system and a seeded crystal consisting of a Y system.

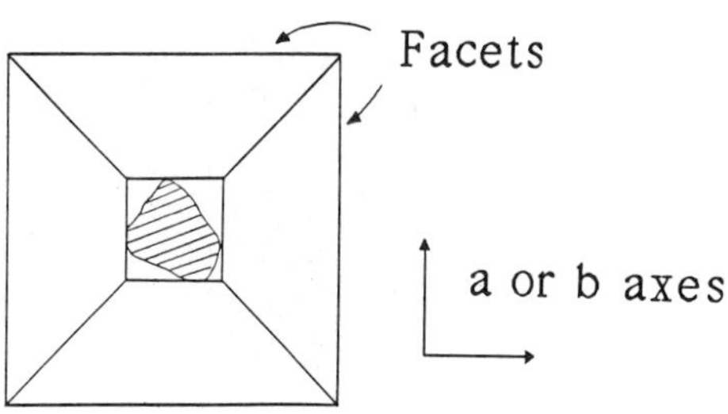

Fig.4 Shematic explanation ofpattern on the surface of a seeded crystal.

c) Unidirectional growth by compositional gradient in uniform temperature

Seeding and control of supercooling are not enough for the enlargement of a single QMG crystal, because long time maintenance in a supercooling state leads to nucleation which often prevent preferred crystal growth. To reduce the supercooling area, a compositional gradient was employed in a precursor. A fifteen layered precursor whose compositional ratio of Yb to Y elements was changed from 0% to 50% with every 5% was prepared by repetitive quench. The precursor was reheated to 1200°C for 20 minutes and cooled to 1030°C. Sm seed was put on 100% of the Y side of the precursor. The sample was slowly cooled to 930°C at the rate of 1°C/h. Fig.5 shows the compositional mapping of the Yb element. To ensure the crystal direction, twin patterns were observed at the opposite side of seeding. The twin patterns were perpendicular each other. This twin patterns suggest that the bottom of the sample has the same crystal direction as the top of the sample. Therefore, the sample is a single QMG crystal. It can be concluded that a large QMG crystal can be made by seeding and compositional gradient. Since these heat treatments can be performed in uniform temperature, These techniques are regarded as practical fabrication method of large bulk superconductor.

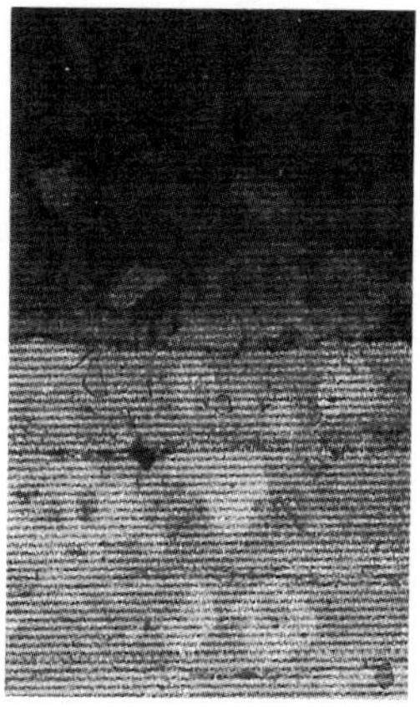

Fig.5 Compositional mapping of Yb element in the sample which was nidirectionally grown by a compositional gradient.

4.CONCLUSIONS

1) A large and dense precursor can be prepared by the repetitive quench method.
2) The formation temperature of $REBa_2Cu_3O_{7-x}$ was controllable by changing the RE components.
3) Nucleation and crystal direction can be controlledby using $REBa_2Cu_3O_{7-x}$ seeds which have higher a formation temperature.
4) Unidirectional growth by compositional gradient can be produced in a uniform temperature.

QMG crystal size of about 50 cm^3 was obtained using these techniques.

REFERENCES

1 M. Morita and S. Matsuda; NSMF NEWS (1988)No,10 p15

2 M. Morita, K. Miyamoto, K. Doi, M. Murakami, K. Sawano, and S.Matsuda; Physica C, (to be published)

3 M. Morita, M. Murakami, K. Miyamoto, K.Sawano and S. Matsuda; PhyscaC 162-164 (1989)1217

4 K.Sawano, M. Morita, K. Kimura, K. Doi and K Miyamoto; Proceedings of Second International Symposium on Superconductivity, Nov.14-17, (1989) in Tsukuba, JAPAN

5 W. D. Kingery, H. K. Bowen and D. R. Uhmann(1960) Introduction to Ceramics. A Wiley-Publication pp336-346

Improvement of Critical Current Density by Addition of Other Elements in $YBa_2Cu_3O_{6+x}$ Superconductor

KOZO OSAMURA, YUKIHIDE KUSUMOTO, and SHOJIRO OCHIAI

Department of Metallurgy, Kyoto University, Sakyo-ku, Kyoto, 606 Japan

ABSTRACT

The effect of addition of other elements to superconducting properties has been investigated. Various types of oxide complex are formed from additive element by a reaction with the superconducting YBaCuO phase. Nine oxide phases of $BaZrO_3$, Ba_2YNbO_6, $BaHfO_3$, $Ba(Y_{0.5}Ta_{0.5})O_3$, $Al_2Y_4O_9$, Ba_2SiO_4, $Ga_2Ba_4O_7$, Ba_2GeO_4 and $BaSnO_3$ have been confirmed to be formed. Especially the addition of Sn is very effective for improving critical current densities as well as for refining the microstructure. Oxide complex, $BaSnO_3$ distributes finely in the grains and at grain boundaries. The AC magnetization measurements indicated that Sn addition causes intragrain critical current density to increase at 77 K in the magnetic fields up to 0.12 T. It exceeded 10^6 A/cm^2 at 0.04 T for the 3-mol%-SnO_2-added specimen. When the silver sheathed tapes were prepared by using the Sn-added powder, the transport critical current density is found to be improved by factor two comparing with non-additive tapes.

KEY WORDS: high-Tc superconductor, $YBa_2Cu_3O_{6+x}$, critical current density, flux pinning, $BaSnO_3$ precipitates

INTRODUCTION

Recently the mechanism of flux pinning in high Tc oxide superconductors has attracted much attention. Various kinds of crystal imperfections can act as flux pinning. Structural imperfections like twin boundaries, irradiation induced defects and stacking faults, and impurity phases such as CuO and Y_2BaCuO_3 associated with non stoichiometric composition have been reported to improve flux pinning. The addition of other elements such as Ag, Zr /1/, Si/2/, Ca/3/ and Sn/4/, are also very effective as flux pinning. Those elements are classified into three categories; substituting the lattice site of 123 structure, forming the oxide complex reacted with 123 phase, and being non reactive with 123 phase. They affect characteristically the microstructure after sintering and the superconducting properties. We investigated systematically the effect of the addition of other 12 elements to the microstructure and superconducting properties.

EXPERIMENTAL PROCEDURE

Oxide powders of Y, Ba and Cu were mixed with a composition of the 123 phase. The mixture was calcined at 1173 K for 86.4 ks and then reground. After this process was repeated two times, a small amount of additive oxide powder was added by 0.5 to 5 mol% to the 123 powder. The oxides of Ti, V, Zr, Nb, Hf, Ta, Al, Si, Ga, Ge, In and Sn have been examined in the present study. Again, grinding and calcining were performed. This calcined powder was compacted into pellets of 10 mm diameter and 2 mm thick. The pellet was sintered at 1223 K for long times and then cooled gradually.

The microstructures were studied by using the scanning electron microscope (SEM), electron probe microanalyzer(EPMA) and X-ray diffraction(XRD) in order to determine the grain size and precipitate morphology. The density was measured according to Archimedes' principle. The transport critical current density was resistively measured with a criterion of 1 μV/cm at 77 K and 0 T, and the transition temperature was also resistively determined. The

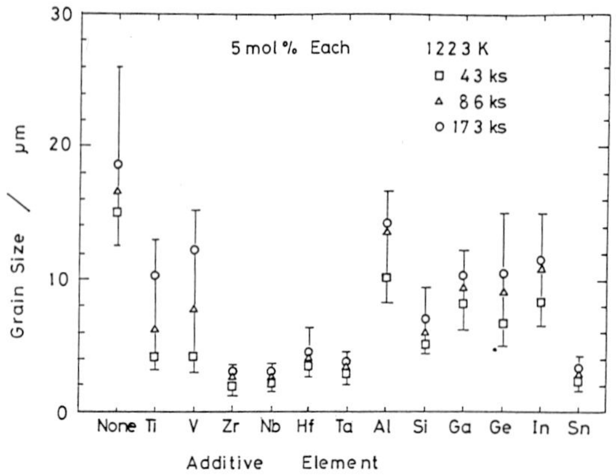

Fig. 1 Change of grain size for the specimens sintered at 1223 K after addition of another 5 mol % oxide.

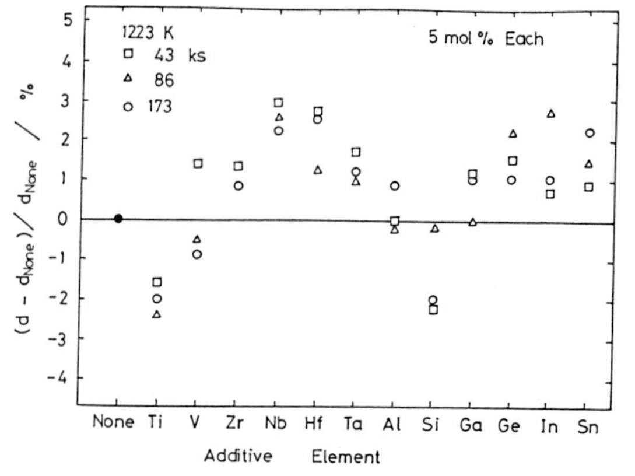

Fig. 2 Relative change of density as a function of additive element.

AC magnetization measurements (the Campbell method) were performed. From the flux profile, the intragrain critical current density (J_c^m) was estimated. The details have been reported elsewhere/4/.

RESULTS AND DISCUSSION

Fig. 1 shows the change of grain size in the sintered specimens when it was sintered at 1223 K for various times after addition of another oxide by 5 mol%. Comparing with the non-additive case, it became small for all specimens including additive elements and especially for Zr, Nb, Hf, Ta and Sn oxides. The relative change of density was measured as shown in Fig. 2, where d_{None} is the value of density for the non-additive specimen. Depending on additive element, the density changes in a range of few per cent. It is found that the densification is remarkablly enhanced by addition of Zr, Nb, Hf, Ge, In and Sn oxides. However, Ti and Si oxides prevent densification. As shown in Fig. 3, the transport critical current density degraded in general by addition of other oxides except in the case of Sn oxide. From these preliminary examinations, it was made clear that Sn addition is much effective. However for other elements, a further investigation should be done to get exact conclusion.

When the specimen was sintered at the higher temperature, the density became greater. As shown in Fig. 4, the density became 95% of the theoretical value for the non-additive specimen sintered at 1223 K for 260 ks. By addition of

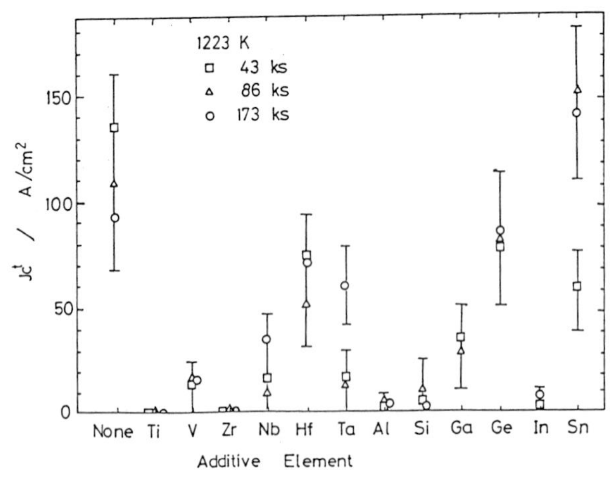

Fig. 3 Influence of additive element on the transport critical current density of the sintered polycrystalline specimens.

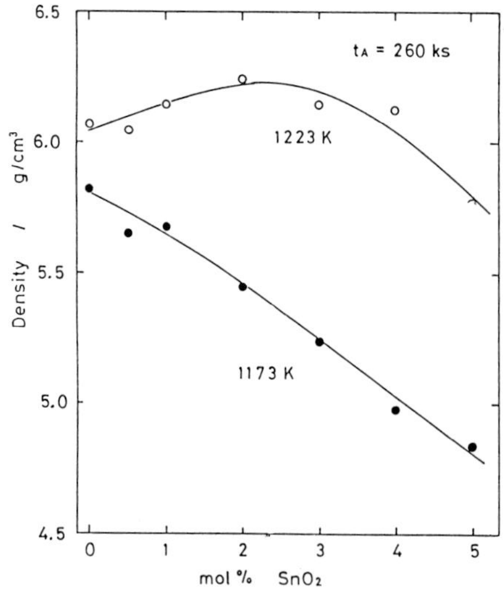

Fig. 4 Change of density as a function of amount of SnO_2 for the sintered specimens.

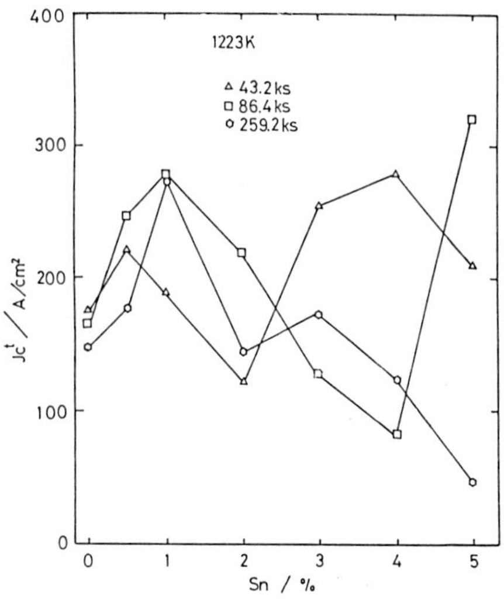

Fig. 5 Change of transport critical current density as a function of amount of SnO_2.

Sn oxide, the density became greater and reached 6.25 g/cm^3 for the 2-mol%-SnO_2-added specimen, which corresponds to 99 % of the theoretical density. On the other hand, the density decreased with increasing Sn content for the specimens sintered at 1173 K. As shown in Fig. 1, the average grain size was smaller for the specimen with the higher Sn oxide and became extremely small for the specimen with more than 3 mol% SnO_2.

In the absorption electron image of SEM, the spherical precipitates distributed finely in size of submicron in the grains and at grain boundaries. They are so small in size that the quantitative analysis of composition by EPMA was difficult, but the enrichment of Sn and the deficiency of Cu in these precipitates were confirmed. In the matrix, no Sn exceeding the detectable limit larger than 0.1 mol% Sn oxide was observed. Using a specimen with 20 mol% Sn oxide sintered at the same condition, the phase analysis was performed by XRD. The coexistence of the four phases of 123, CuO, Y_2BaCuO_5 and $BaSnO_3$ was determined. Accordingly, the following chemical reaction is proposed to occur at the sintering temperature of 1223 K: 2 $YBa_2Cu_3O_{6+x}$ + 3 SnO_2 = 5 CuO + Y_2BaCuO_5 + 3 $BaSnO_3$. Therefore, the Sn-rich precipitates correspond to $BaSnO_3$. In the present study, however, the other phases mentioned above could not be detected in the specimens with less than 5mol% Sn oxide. From the similar microstructure analysis, various types of oxide complex were detected in each oxide system as follows; $BaZrO_3$, Ba_2YNbO_6, $BaHfO_3$, $Ba(Y_{0.5}Ta_{0.5})O_3$, $Al_2Y_4O_9$, Ba_2SiO_4, $Ga_2Ba_4O_7$ and Ba_2GeO_4.

Figure 5 shows the Sn concentration dependence of transport critical current density for the specimens sintered at 1223 K for various times. By the first 0.5 mol% addition, the critical current density is ensured to increase. However it decreased irregularly by further Sn addition. Perhaps this is attributed to the presence of microcracks in the so high densified specimens. As shown in Fig. 6, the on-set value of critical temperature determined from the resistive measurement became nearly constant regardless of the change of Sn content, but the transition width tended to increase slightly for the specimen with the higher Sn oxide.

From the AC magnetization measurements, the critical current density inside the grains has been determined in the magnetic field region between 0.02 and 0.12 T. Figure 7 shows the intragrain critical current density as a function

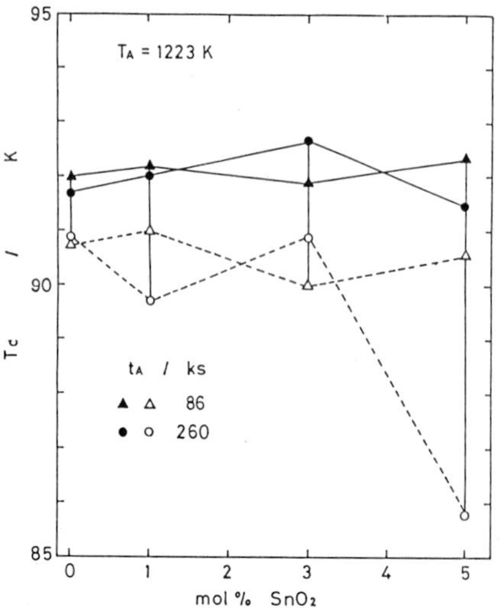

Fig. 6 Change of critical temperature as a function of Sn oxide.

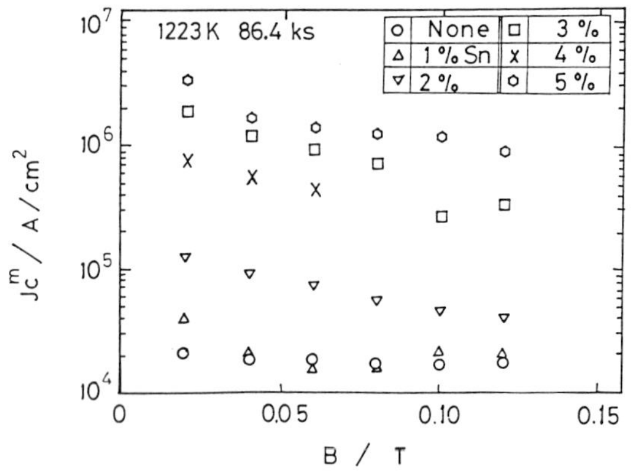

Fig. 7 Intragrain critical current density as a function of magnetic field.

of magnetic field for the specimens sintered at 1223 K for 86.4 ks. The current density decreases gradually with increasing magnetic field, but it increases remarkably with increasing Sn oxide content and exceeds 10^6 A/cm^2. As discussed previously /4/, its increase by the addition of Sn oxide suggested that the Sn-rich precipitates act as pinning centers. However the contribution of other products yielded by the reaction above mentioned to flux pinning cannot be neglected.

Ag sheathed tapes have been prepared by the powder-in-tube method reported elsewhere /5/. When the grain size is compared for both specimens with and without Sn addition, it became very samll for the Sn added specimen, even though the tapes were prepared by the same condition. Table 1 indicates the transport critical current density for the Ag sheathed tapes with or without Sn oxide heat-treated at 1203 K, where 4 mol% Sn oxide was added for each specimen. For these tapes, the thickness of oxide layer were about 30 μm. Comparing with the transport critical current density for both tapes with or without Sn oxide, the observed values for Sn added specimens is by the factor two larger than those for non-added specimens.

Table 1 Transport critcal current density for Ag sheathed tapes

SnO_2 added			not added		
Sample No.	t_o /μm	J_c^t /A/cm²	Sample No.	t_o /μm	J_c^t /A/cm²
SG1731	34	1700	NS1731	35	1300
SG3532	33	1980	NS3532	30	920
SG5033	32	1990	NS5033	24	1010
SG6034	30	1820	NS6034	32	670
SG0035	29	1580	NS0035	43	1010

Acknowledgment:This study was made possible by a Scientific Research Grant in Aid (Project No. 02226220) from the Ministry of Education of Japan.

REFERENCES

1. K.C.Goretta, M.Kullberg, O.D.Lacy, U.Balachandran, R.B. Poeppel, D.Shi, M.Xu and J.L.Routbort: TMS Meeting Anaheim (1990).
2. S.H.Whang, Z.X.Li, D.X.Pang, M.Suenaga, D.O.Welch, S.Jin and T.H.Tiefel: Physica C (1990).
3. H.Masuda, F.Mizuno and I.Hirabayashi: Jap. J. Appl. Phys. 28 (1988) L1226.
4. K.Osamura, N.Matsukura, Y.Kusumoto, S.Ochiai, B.Ni and T.Matsushita; Jap. J. Appl. Phys. 29 (1990) L1621.
5. K.Osamura, T.Takayama and S.Ochiai; Cryogenics 29 (1990) 430.

Preparation of Bi-Based Superconducting Rods by Melt-Quenching and Floating Zone Method

KAZUHIKO SAWADA[1], MAKOTO HIRAOKA[1], TAKAYUKI KOMATSU[2], and KAZUMASA MATUSITA[2]

[1] Central Research Laboratory, Mitsubishi Cable Industries, Ltd., 8, Nishinocho, Higashimukaijima, Amagasaki, 660 Japan

[2] Department of Chemistry, Nagaoka University of Technology, 1603-1, Kamitomioka, Nagaoka, 940-21 Japan

ABSTRACT

Bi-based superconducting rods have been grown by means of the melt-quenching and floating zone method. The influences of growth rates and annealing conditions on structure and superconducting properties are investigated. The T_c and J_c values of these rods are largely improved by post-annealing. The temperature dependence of ac susceptibility at various ac fields, however, reveals that the superconducting coupling at grain boundaries is still weak.

KEY WORDS : high-Tc superconductor, Bi-Sr-Ca-Cu-O system, melt-quenching method, floating zone method

INTRODUCTION

Since the discovery of new high-T_c superconducting Bi-Sr-Ca-Cu-O ceramics by Maeda et al. [1], many researchers have expended much effort to prepare high-performance superconducting Bi-Sr-Ca-Cu-O ceramics. Komatsu et al. [2] succeeded in preparing superconducting Bi(Pb)-Sr-Ca-Cu-O ceramics with high transition temperature T_c using the melt-quenching method. However, since critical current densities J_c in these samples were still low because of porous microstructure, random arrangement of superconducting grains with strong anisotropy and Josephson-like weak links at boundaries between superconducting grains, various techniques have been applied to improve superconducting properties. Recently, Sawada et al. [3] reported that the Bi-based superconducting rods with dense microstructures were prepared and the orientation of superconducting grains were improved by using the melt-quenching and floating zone method. The low-T_c phase was dominant in these samples irrespective of nominal composition, and influences of weak links on the superconducting properties were not clear. Further work is therefore necessary in order to obtain samples with good superconducting properties. In this paper, we report the effects of growth rates and post-annealing conditions on the superconducting properties for Bi-Sr-Ca-Cu-O rods and discuss briefly the influence of weak links.

EXPERIMENTAL

The feed rods and the seed for floating zone experiments were prepared by the melt-quenching method. The appropriate amounts of commercial powders of guaranteed reagent Bi_2O_3, $SrCO_3$, $CaCO_3$ and CuO were mixed together and calcined at 800°C for 12h in air. The nominal composition examined in this study is $Bi_2Sr_2CaCu_2O_y$. The calcined powders were melted in a platinum crucible at 1150°C for 20min in an electric furnace. The melts were pumped up into silica glass tubes with 4.5mm diameter. After the outer glass tube were removed, the obtained rods were used as the feed rods and seeds. The floating zone apparatus with an infrared convergence-type image furnace with two halogen lamps as a radiation source was used for present experiments. A feed rod and a seed were put coaxially in the center of a silica glass tube, and were rotated in opposite directions at 25rpm. The bottom of the feed rod was melted with the infrared radiation focused with the ellipsoidal reflection mirror. The feed rod was moved downward to contact between the molten zone and the seed, and the molten zone was traversed along the feed rod. The growth rates were 1, 2, 5 and 10mm/h. After floating zone experiments, some samples were annealed at several conditions including the condition at 840°C for 3h in air.

Measurements of transition temperature T_c and J_c (77 K, zero magnetic field) were made by using a four-point probe method. As a measure of the diamagnetic susceptibility, mutual inductance were measured using an ac Hartshorn-type bridge. Crystal phases present in the samples were examined by X-ray diffraction (XRD) analyses.

RESULTS AND DISCUSSION

Figure 1 shows the temperature dependence of resistivity for the as-grown $Bi_2Sr_2CaCu_2O_y$ rods prepared at the growth rates of 1, 2, 5 and 10mm/h. The value of T_c was increased with decreasing growth rate, and the highest T_c value in these samples was 85K for the sample grown at 1mm/h. The normal-state resistivity for these samples tends to decrease with slower growth rate. These results indicate that the superconducting properties on these samples prepared at slower growth rate are better than that at faster growth rate. Figure 2 shows the XRD patterns for the plane parallel to the growth direction in the as-grown rods prepared at the growth rate of 1, 2, 5 and 10mm/h. In all samples, the (00n) peaks of low-T_c phase are dominant, indicating that the samples are orientated with the (00n) plane of the low-T_c phase parallel to the growth direction. The (00n) peaks of a sample prepared at slower growth rate were sharper than that prepared at faster growth rate. From SEM observations and EDX analyses, it was found that the amount of non-superconducting phase decreased and the size of superconducting phase increased with decreasing growth rate.

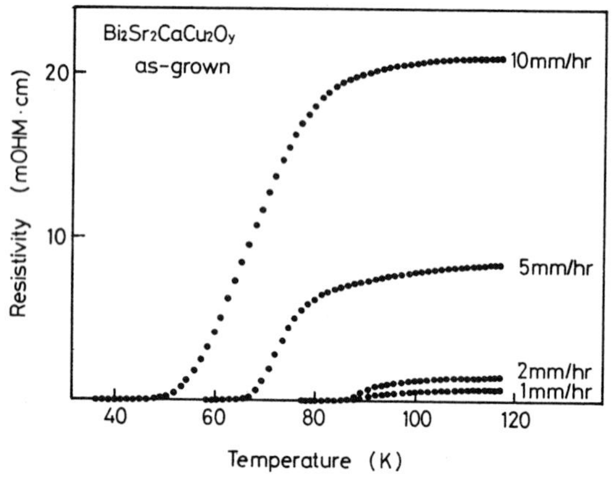

Fig. 1 Temperature dependence of resistivity for the superconducting $Bi_2Sr_2CaCu_2O_y$ rods (as-grown) prepared at the growth rates of 1, 2, 5 and 10mm/h.

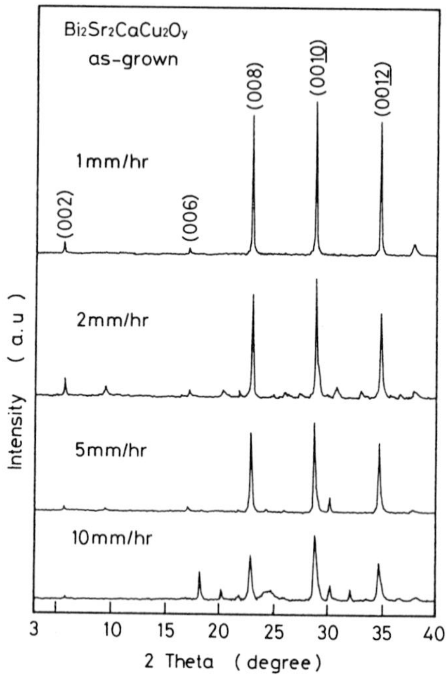

Fig. 2 XRD patterns for the plane parallel to the axial direction in the superconducting $Bi_2Sr_2CaCu_2O_y$ rods prepared at various growth rates.

In order to obtain more high-performance superconducting materials, we also carried out the post-annealing for the as-grown samples. Figure 3 shows the T_c of the samples annealed at 840°C for 3h in air, together with the T_c values of as-grown samples. In the as-grown samples (open circles), the highest T_c value was 85K obtained at 1mm/h. The T_c value was decreased steeply increasing growth rate and gone down 39K at 10mm/h. After post-annealing (solid circles), the T_c values of all samples were increased (i.e. 90K and 85K grown at 1mm/h and 10mm/h, respectively). Similar results were obtained for J_c measurements at 77K in zero magnetic field on these samples, as shown in figure 4. In the as-grown samples (open circles), the highest J_c value was 310A/cm^2 for the sample obtained at growth rate of 1mm/h. After annealing (solid circles), the J_c values of all samples were increased, and the highest value was 598A/cm^2 at growth rate of 1mm/h. This J_c value was almost twice as high as the J_c value of as-grown sample grown at 1mm/h. From this figure, it is clear that the J_c value has been improved by the post-annealing. For both as-grown and annealed samples, the J_c values were increased with decreasing growth rate.

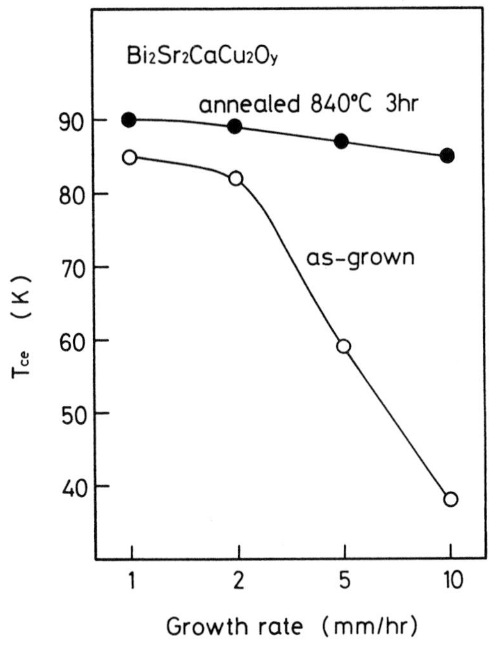

Fig. 3 T_c values of as-grown and post-annealed samples prepared at various growth rates.

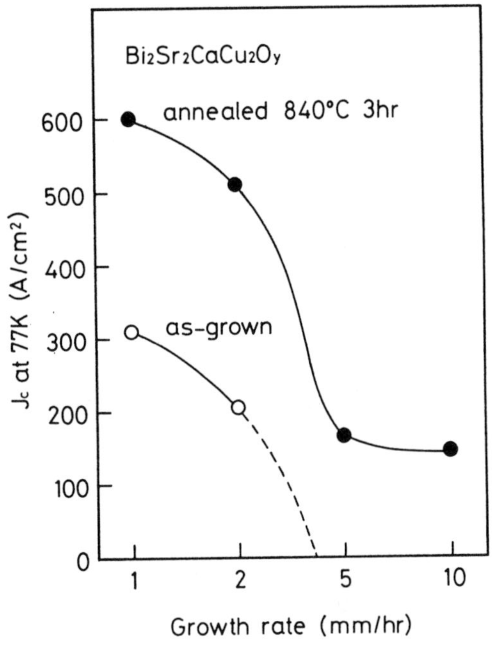

Fig. 4 J_c values of as-grown and post-annealed samples prepared at various growth rates.

Figure 5 shows the XRD patterns for both as-grown and post-annealed samples prepared at the growth rate of 2mm/h, in which the incident X-rays are perpendicular to the rods, i.e. to the growth direction. Although the T_c and J_c values were improved by post-annealing, similar XRD patterns were obtained for both samples. SEM observation and EDX analyses also revealed that the microstructure of post-annealed samples was almost the same as that of as-grown samples. These results suggest that the improvements in T_c and J_c through the post-annealing are not due to the change in crystal structure or orientation.

In order to study the influence of weak links on the superconducting properties, we measured the temperature dependence of imaginary part of ac susceptibility x". Figure 6 shows the ac field dependence of x" for both as-grown and post-annealed samples prepared at 1mm/h. In the as-grown sample, a sharp peak was observed at the transition temperature T_c at the applied ac magnetic field of 0.1Oe. The peak shifts largely to lower temperature and the width of the peak becomes broader with increasing ac magnetic field, indicating that the superconducting coupling at grain boundaries in the as-grown sample is very weak. Similar results were obtained for post-annealed sample, however, the degree of the shift to lower temperature of the peak is small compared with the as-grown sample. These results indicate that the superconducting weak links at grain boundaries have been improved through the post-annealing. It is expected that more high-performance superconducting materials would be obtained by optimized preparing and annealing condition.

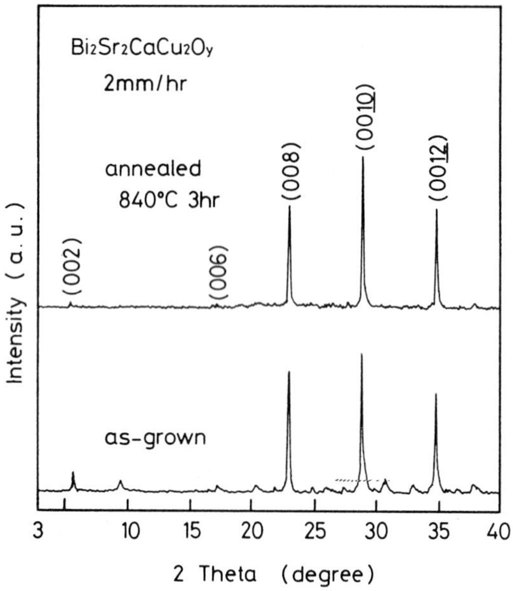

Fig. 5 XRD patterns for the as-grown and post-annealed samples prepared at growth rate of 2mm/h.

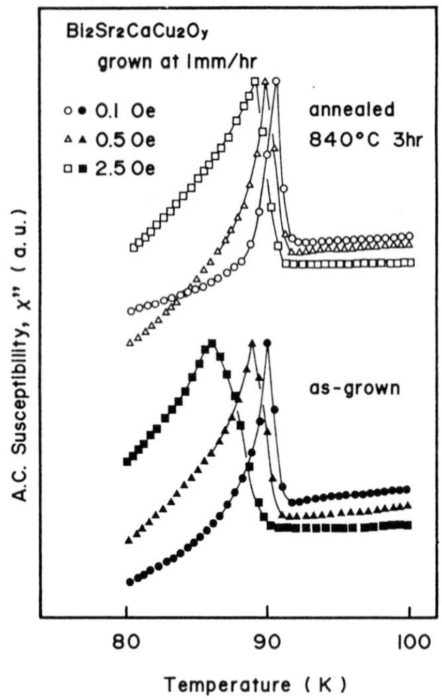

Fig. 6 Temperature dependence of ac susceptibility x" at various magnetic fields for the as-grown and post-annealed samples.

CONCLUSIONS

Superconducting $Bi_2Sr_2CaCu_2O_y$ rods were prepared by the melt-quenching and floating zone method. The low-T_c phase was dominant and orientated unidirectionally in these samples. The T_c and J_c values were improved by the post-annealing. From the measurement of ac susceptibility x" at various magnetic fields, it was found that the weak links at boundaries between superconducting grains existed in these samples and improved by post-annealing.

REFERENCES

1. H. Maeda, T. Tanaka, M. Fukutomi and T. Asano, Jpn. J. Appl. Phys., 27, L209-L210, (1988)

2. T. Komatsu, R. Sato, C. Hirose, K. Matusita and T. Yamashita, Jpn. J. Appl. Phys., 27, L2293-L2295, (1988)

3. K. Sawada, M. Hiraoka, T. Shintani, T. Komatsu and K. Matusita, Advances in Superconductivity 2, Proceedings of the 2nd International Symposium on Superconductivity (ISS'89), 301-304, (1990)

A Successful Control Processing for Composite Particles of Oxide Superconductor

YASUMASA TAKAO, MASANOBU AWANO, and HIROYOSHI TAKAGI

Government Industrial Research Institute, Nagoya, 1-1, Hirate-cho, Kita-ku, Nagoya, 462 Japan

ABSTRACT

A successful processing for the preparation of the composite oxide superconductor was developed, and the properties of that composite were evaluated. The purposes of this processing are the control of dispersions of fine pinning-centers in the matrix phase and the design of structures of the sintered composite body. The composite powder of the fine particles were fabricated by Mist-Pyrorysis method, and the sintered composite bodies were made by using these powder. In this paper, the sinterability and superconductive properties for the sintered composite bodies of oxide superconductor dispersed Ag particles in Y-Ba-Cu-O matrix were evaluated.

KEY WORDS: composite, control of dispersions of fine pinning-centers, Y-Ba-Cu-O system, Ag, Mist-Pyrorysis method

INTRODUCTION

High density and the dispersions of fine pinning centers are important points for the improvement of Jc on the oxide superconductor [1]. As one approach to that purpose, the sintered composite bodies dispersed fine pinning-centers in the matrix phase were prepared by using the composite powder of the fine particles. High-purified chemically homogeneous powder are necessary for the fabrication of superconducting wires, tapes, and magnetic shields [2]. Furthermore, the fine particles are favorable for the control of mixing the composite elements, and are necessary to improve the deformation processing properties of the sintered body [3]. In this paper, the composite powder of the fine particles of Y-Ba-Cu-O and Ag fabricated by Mist-Pyrorysis method and the sintered composite body dispersed Ag particles in Y-Ba-Cu-O matrix were described. It's reported that Mist-Pyrorysis method is favorable for the preparation of high-purified homogeneous fine particles [3,4], and Ag enhances the sinterability and magnetic pinning properties [2,5,6]. Therefore, the composite with Ag particles was selected as the beginning of the study for a successful processing of composite oxide superconductor.

EXPERIMENT

Figure 1 shows the concept of the preparation for the composite in this paper. Composite powder are prepared by each fine particle mixed uniformly, and the sintered composite body dispersed fine pinning-centers in the matrix phase are made by using this composite powder. Figure 2 shows the experimental apparatus. The system consists of ultrasonic nebulizers, reaction furnaces, a filter, and a rotary pump. The composite particles were synthesized by the following process; each starting solution contained the matrix componets and secondary components was atomized by the ultrasonic nebulizer. Produced droplets separated by each size were carried by carrier gas into reaction zone. Passing through the reaction zone, droplets were dried, decomposed, and crystallized as the matrix phase and secondary phase.

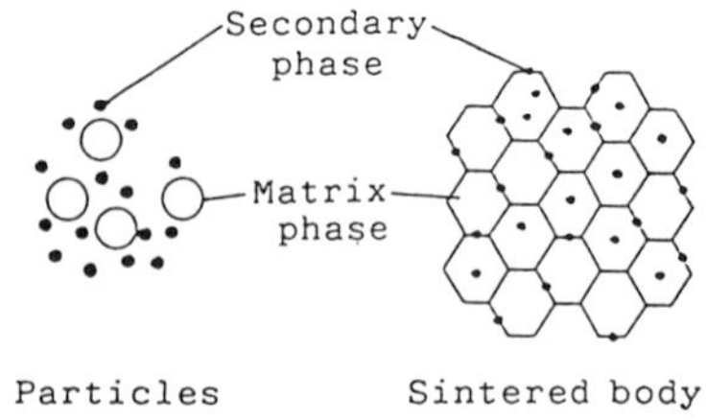

Fig.1: The concept of the preparation for the composite powder and sintered body.

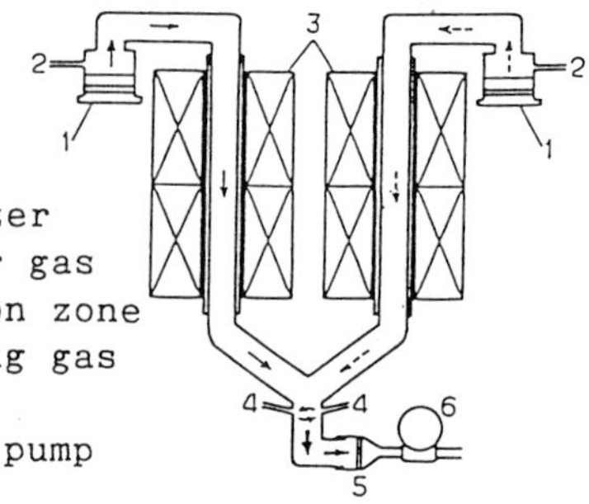

Fig.2: Schematic diagram of the experimental system.

Two phases were uniformly mixed by stirring gas, and the prepared composite powder were trapped to a filter.
The size of reactor tubes were 50mm inner diameter and 500mm in length. Ultrasonic nebulizers were operated at 1.65MHz. Reaction temperature of each zone is 950° C at middle part of the reactor tube. Starting solutions were prepared as follows; Y-Ba-Cu-O nitrate solution consisted of Y, Ba, and Cu in a 1:2:3 molecular ratio, and the concentration was 0.03 mol/l. Ag solution were with 0.0005 mol/l and 0.001 mol/l. Consequently, three types of the powder were prepared by the kinds of starting solutions, (1);the composite of Y-Ba-Cu-O(0.03 mol/l) and Ag(0.0005 mol/l) and (2);the composite of Y-Ba-Cu-O(0.03 mol/l) and Ag(0.001 mol/l), and (3);the superconductive powder of Y-Ba-Cu-O(0.03 mol/l). The carrier gas was O_2 . The flow rate of the carrier gas for Y-Ba-Cu-O droplets was 0.5 l/min. And the Ag droplets was carried at two kinds of rates by the varieties of starting solutions, (1);9.0 l/min to the composite of Y-Ba-Cu-O(0.03 mol/l) and Ag(0.0005 mol/l) and (2);4.5 l/min to the composite of Y-Ba-Cu-O(0.03 mol/l) and Ag(0.001 mol/l). The contents of Ag particles to Y-Ba-Cu-O matrix was 5wt.% for each composite powder. Stirring gas was Ar, and the flow rate of the stirring gas was 0.4 l/min. As forming treatment, CIP was operated at 100 kg/cm^2 . The formed samples were fired at 905° C for one hour. Annealing was operated at 500° C for 48 hour in O_2 atmosphere.
The properties of the composite powder and sintered bodies were evaluated by X-Ray diffraction, SEM, and TEM with Energy Dispersive Spectroscopy. Superconductive properties were evaluated by the measurement for the relationship between resistivity and temperature of four prove method.

RESULTS AND DISCUSSION

Figure 3 shows the secondary electron image for the composite powder at the concentration of the starting solutions with Y-Ba-Cu-O(0.03 mol/l) and

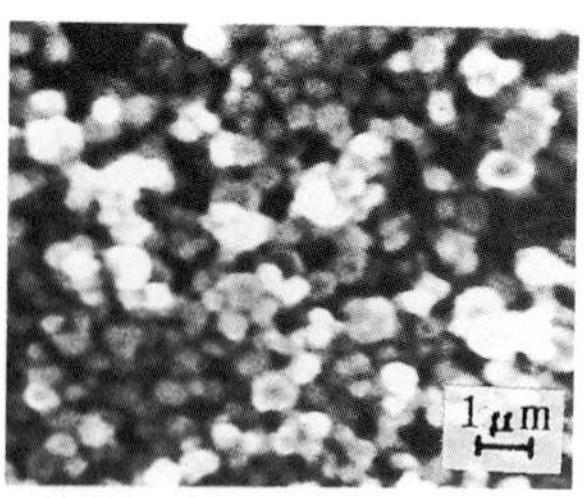

Fig.3: SEM image for the composite powder.

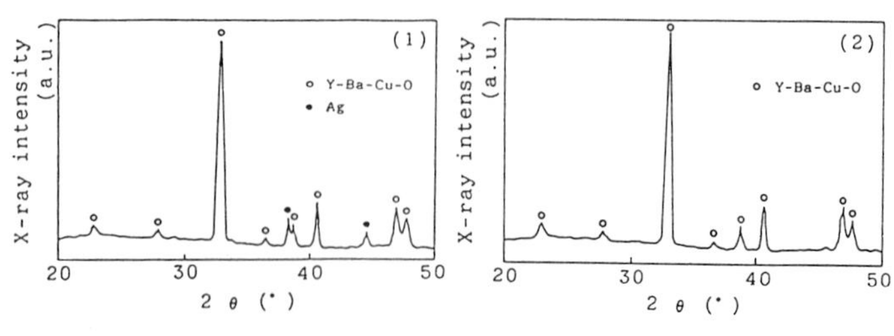

Fig.4: X-Ray diffraction patterns for
(1); Y-Ba-Cu-O powder containing Ag
(2); Y-Ba-Cu-O powder.

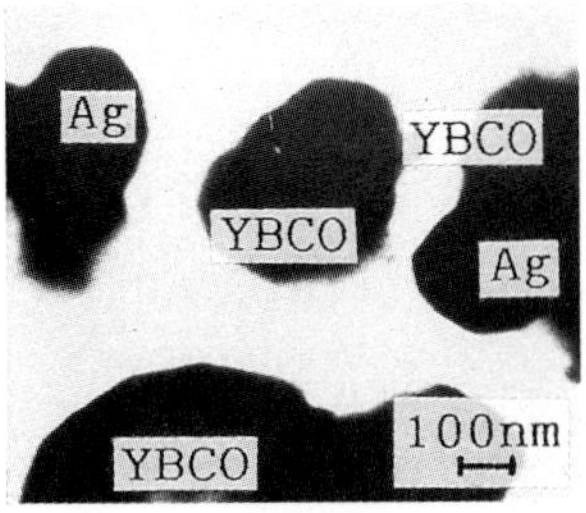

(1);TEM image

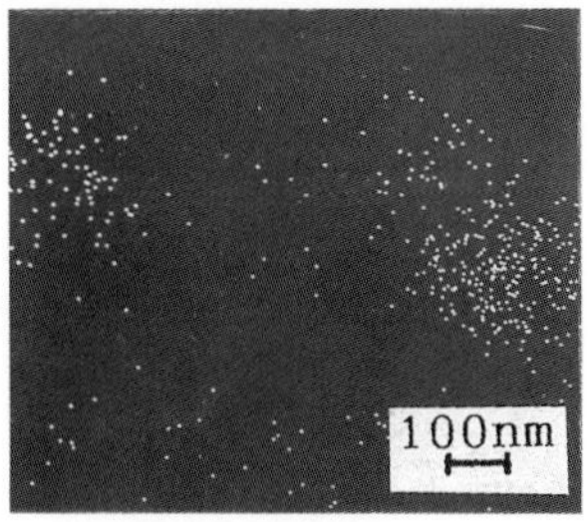

(2);X-Ray image of Ag-Kα

Fig.5: TEM image with X-Ray image of Ag-Kα for the composite powder at the concentration of the starting solutions with Y-Ba-Cu-O(0.03 mol/l) and Ag(0.001 mol/l).

Ag(0.001 mol/l). The average size of these particles was sub-micron size. Figure 4 are X-Ray diffraction patterns for (1);the composite powder at the concentration of the starting solutions with Y-Ba-Cu-O(0.03 mol/l) and Ag(0.001 mol/l) and (2);the superconductive powder of Y-Ba-Cu-O(0.03 mol/l). Figure 4 shows that Ag in composite powder is reduced to the metallic silver at 950°C reaction temperature. Figure 5 shows TEM image with X-Ray image of Ag-Kα for the composite powder at the concentration of the starting solutions with Y-Ba-Cu-O(0.03 mol/l) and Ag(0.001 mol/l). According to figure 5, Ag particles are dispersed uniformly among Y-Ba-Cu-O particles.

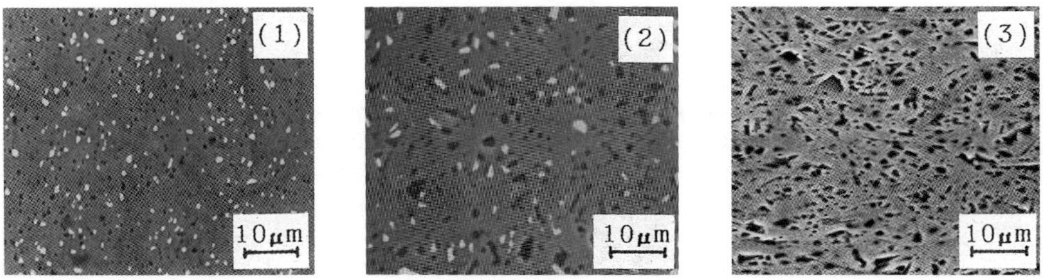

Fig.6: SEM images for the composite and superconductive sintered body,
(1); the sintered composite body at the concentration of the starting solutions with Y-Ba-Cu-O(0.03 mol/l) and Ag(0.0005 mol/l)
(2); Y-Ba-Cu-O(0.03 mol/l) and Ag(0.001 mol/l), in the same way
(3); the sintered body of Y-Ba-Cu-O(0.03 mol/l).

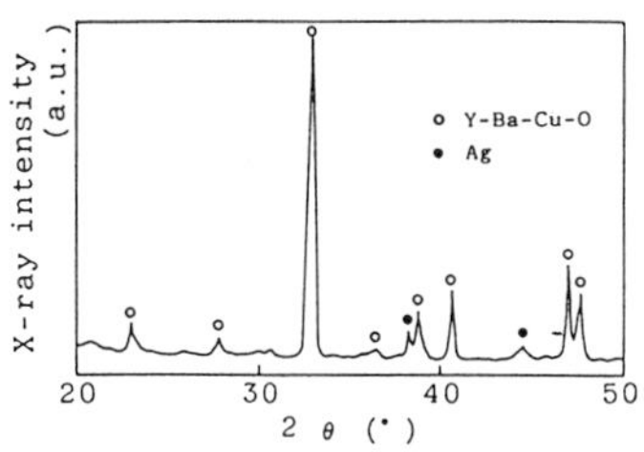

Fig.7: X-Ray diffraction pattern for the sintered composite body.

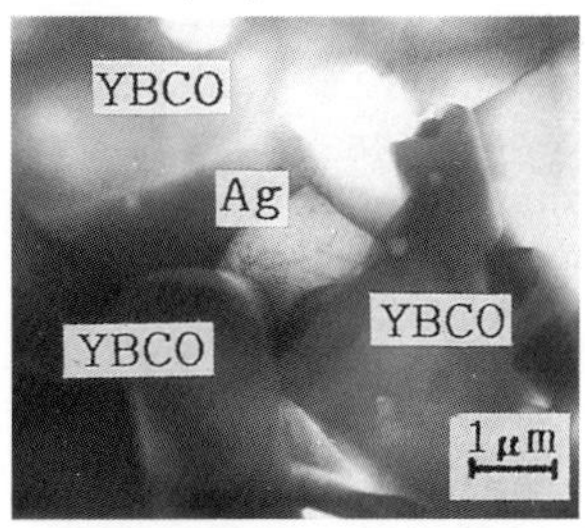

Fig.8: TEM image of the grain boundary for the composite sintered body.

Figure 6 shows SEM images for (1);the sintered composite body at the concentration of the starting solutions with Y-Ba-Cu-O(0.03 mol/l) and Ag(0.0005 mol/l) and (2);the sintered composite body of Y-Ba-Cu-O(0.03 mol/l) and Ag(0.001 mol/l), and (3);the sintered body of Y-Ba-Cu-O(0.03 mol/l). Our successful control processing method could arbitrarily disperse Ag particles in Y-Ba-Cu-O matrix ranging from sub-micron size to several micron size

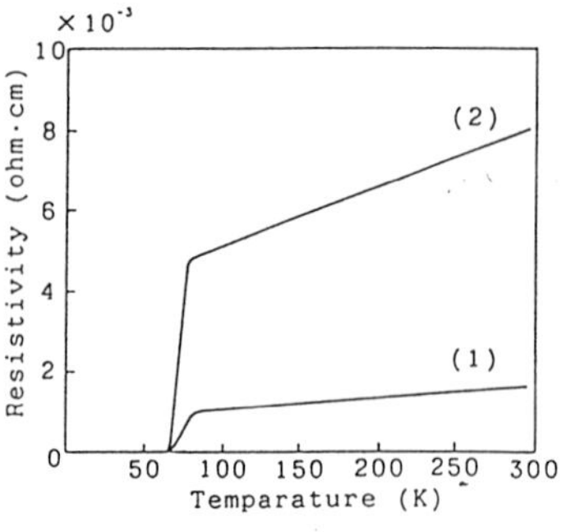

Fig.9: The relationship between resistivity and temperature of four prove method for
(1); Y-Ba-Cu-O sintered composite body containing Ag
(2); Y-Ba-Cu-O sintered body.

by the control for the concentration of the starting solution of Ag. Furthermore, there is the possibility of preparation for more kinds of the composite superconductor by changing the reaction temperatures , the flow rates of droplets, and the varieties of the secondary phase etc. According to figure 6, the hole contents of the sintered composite body (figure 6-(1) and 6-(2)) gradually reduce than the contents of Y-Ba-Cu-O (figure 6-(3)). These results suggest the accelerating effects of Ag particles on the sinterability among the grains of Y-Ba-Cu-O. Figure 7 shows X-Ray diffraction pattern for the sintered composite body at the concentration of the starting solutions with Y-Ba-Cu-O(0.03 mol/l) and Ag(0.001 mol/l), and figure 8 shows TEM image of the grain boundary for the same sintered composite body. According to figure 7 and 8, metallic Ag particles are distributed among the grain boundaries of Y-Ba-Cu-O. The sintered composite body for the concept of preparation at figure 1 could be synthesized.

Figure 9 shows the relationship between resistivity and temperature of four prove method for (1);the sintered composite body at the concentration of the starting solutions with Y-Ba-Cu-O(0.03 mol/l) and Ag(0.001 mol/l) and (2); the sintered superconductive body of Y-Ba-Cu-O(0.03 mol/l). Tc_{onset} and Tc_{end} are the same values, respectively. The resistivity of the composite superconductor reduces than that of Y-Ba-Cu-O superconductor. These results show the positivity of the accelerating effects of Ag particles on sintering of the weak-linked grains of Y-Ba-Cu-O.

CONCLUSION

A successful processing method for the preparation of the composite oxide superconductor dispersed the fine pinning-centers was reported. Our method could arbitrarily disperse the fine metallic Ag particles in the sintered body of Y-Ba-Cu-O matrix ranging from sub-micron size to several micron size. The hole contents of sintered body and the resistivity of R-T curve for the composite superconductor were reduced compared with the superconductor without Ag particles.

REFERENCES

1.Murakami M (1990) Mod.Phys.Lett. B 4-3: 163-179
2.Nishio K, Sakai T, Ogawa N, Hirabayashi I (1989) Proceedings of the 2nd International Symposium on Superconductivity, Tsukuba: 873-876
3.Awano M, Takagi H (1989) Proceedings of the 2nd International Symposium on Superconductivity, Tsukuba: 393-396
4.Tohge N, Tatsumisago M, Minami T, Okuyama K, Adachi M, Kousaka Y (1988) Jap.J.Appl.Phys. 28-6: L1086-L1088
5.Matsumoto Y, Hombo J, Yamaguchi Y, Nishida M, Chiba A (1990) Appl.Phys. Lett. 56(16): 1585-1587
6.Dwir B, Affronte M, Pavuna D (1989) Appl.Phys.Lett. 55(4): 399-401

Preparation of Fine Particles Composed of Y System Oxide by Mist Pyrolysis Method

MUNEO AYABE, RYOUZO TATSUGAE, HIROAKI NISHIKAWA, and CHIE FUKUOKA

Research Institute, Ishikawajima-Harima Heavy Industries Co., Ltd., 1-15, Toyosu 3-chome, Koto-ku, Tokyo, 135 Japan

ABSTRACT

Fine particles composed of Y system oxide have been successfully prepared by using of the mist pyrolysis method.
An aqueous solution of Y,Ba,and Cu nitrate was atomized by the nebulizer with a small orifice (ca.50 μm). The generated mist controlled less than 0.5 μm in diameter by using of an impaction collector, they were carried by O_2 gas into a diffusion dryer for evaporation to dryness, and finally into thermal reactor kept at 1223K for pyrolysis. Produced fine particles were deposited on cold substrate (MgO) by thermal precipitation effect.
The mean diameter of the particles was mainly varied as a function of the solution concentration. For example, the solution concentration of 0.003 and 0.03 mol, respectively yielded the particles of 0.025 and 0.050 μm in diameter.
The superconducting thick film with zero-resistance of 84K was formed on the MgO substrate with post-annealing process.

KEY WORDS: ultra-fine powder, Y system oxide, mist pyrolysis, thick film, gas jet atomizer

INTRODUCTION

Since the discovery of high-Tc surperconductors in the mixed oxide systems, a considerable research effort has been carried out. From the viewpoint of material processing, the superconductive oxides have several problems to be solved out for practical use. In order to obtain the dense and homogeneous materials with high-Jc by solid-state syntheses,fine-homogeneous superconductive powder have been required as starting materials.
There have been enormous efforts to prepare the fine-homogeneous particles by using various methods. One of the useful method to prepare the fine particles is mist pyrolysis method with a ultrasonic nebulizer[1-4].
However it is difficult to obtain ultra-fine particles in the ultrasonic-nebulizing systems; the mist size produced is comparatively large.
In this work,ultra-fine particles composed of Y system oxide were successfully prepared by a mist pyrolysis method aided by a gas jet atomizer and a diffusion dryer with an aqueous solution of Y,Ba,and Cu nitrates.

EXPERIMENTAL

A schematic diagram of the apparatus used for the mist pyrolysis and the measurement system of particle size is shown in Fig.1. An aqueous solution of Y, Ba, and Cu nitrates with a molar ratio of 1:2:3 was atomized with the gas jet atomizer (shown in Fig.2) to form a mist. The droplets in the mist were controlled less than 0.5 μm in diameter by inertial impaction on a collection wall. The mist was passed through a diffusion dryer with an absorber reagent to evaporate the water from the droplets, and then carried by O_2 gas into an electric furnace with 4 heating zones controlled independently. The dryer was consisted of two cylinders,with a smaller and screen cylinder inside. The void between the cylinders was filled with granular silica-gel and the mist was passed through inner cylinder. Passing through the electric furnace, mists were decomposed to oxide and turned out to be ultra-fine particles as a superconductor. The particle-size distribution in process stream, was measured using the method described previously.[5] The particles deposited on substrate were annealed with flowing O_2 to form a thick film.
The compositions and the structures of samples were examined by inductively coupled plasma emission spectroscopy (ICP) and X-ray diffraction (XRD).
The surface morphology were characterized by scanning electron microscopy (SEM), and transmission electron microscopy (TEM). The temperature dependence of resistivity was measured by a conventional four probe method.

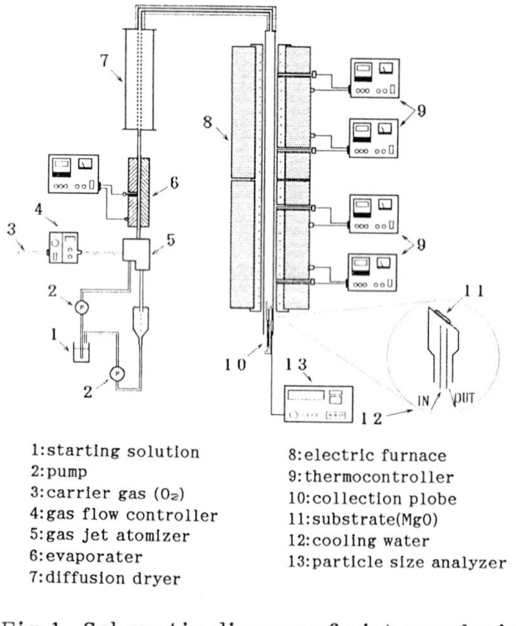

Fig.1. Schematic diagram of mist pyrolysis and particle size measurement

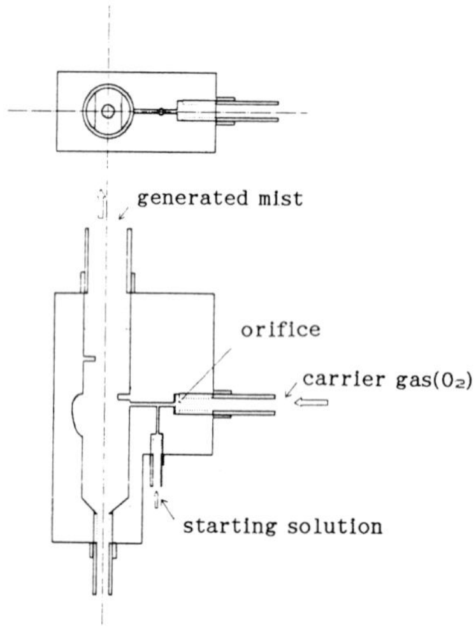

Fig.2. Gas jet atomizer diagram

RESULT AND DISCUSSION

Figure 3 shows the particle-size distribution patterns in process stream obtained at different concentrations of starting solution. The mean diameter of the particles produced by the mist pyrolysis method is 1 order of magnitude

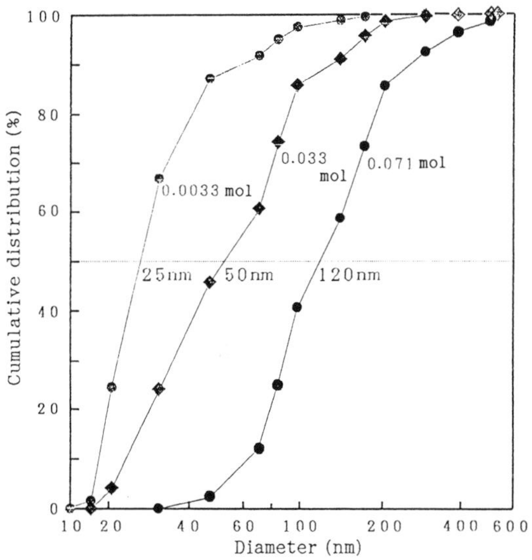

Fig.3. Particle-size distribution obtained from various liquid concentrations

Table 1. Chemical compositions analyzed by ICP.(refered to Y=1.0)

	composition ratio (Y=1.0)		
	Y	B a	C u
starting solution	1.00	2.08	2.97
powder deposited	1.00	2.07	2.96
film annealed	1.00	2.02	2.91

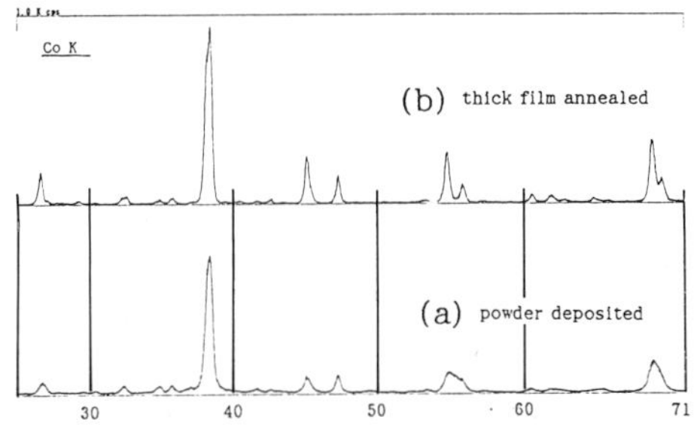

Fig.4. X-ray diffraction patterns of the powder deposited and film annealed.

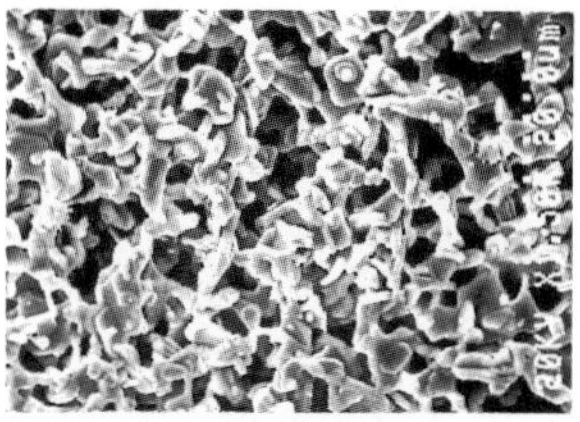

surface(1223K)

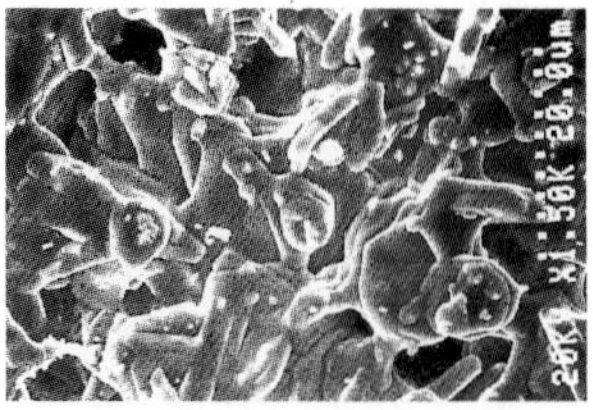

surface(1243K)

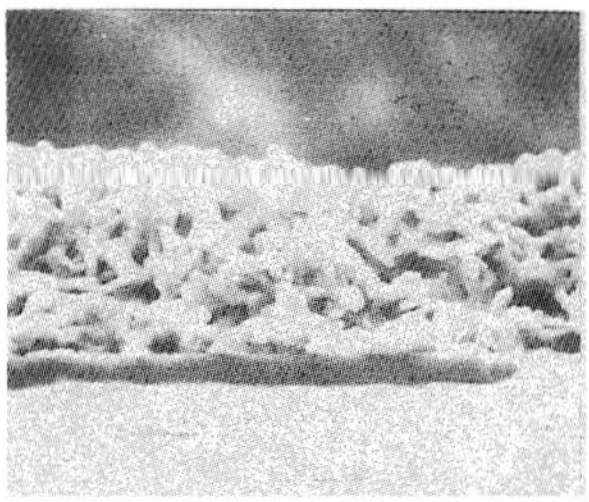

film

MgO

cross-section(1223K)

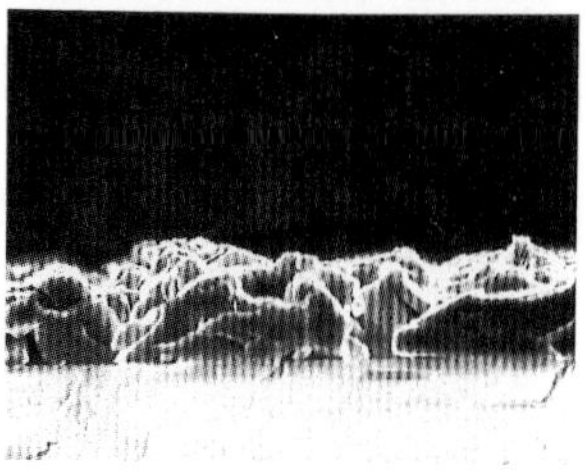

film

MgO

cross-section(1243K)

Fig.5. SEM photographs of the surface and cross-sectional morphorogies of films sintered at 1223K and 1243k for 1 hour.

smaller than the data reported previously, and is mainly varied as a function of the solution concentration. It has been recognized that the particle size will be controlled by the size of droplets, and so it is effective to use the gas jet atomizer which can easily produce the droplets less than 0.5 μm in diameter. Table 1 shows the chemical composition analyzed by ICP of the starting solution, the powder deposited, and the thickfilm annealed. The ICP analysis reveals the stoichiometry of the particles and the film composition.

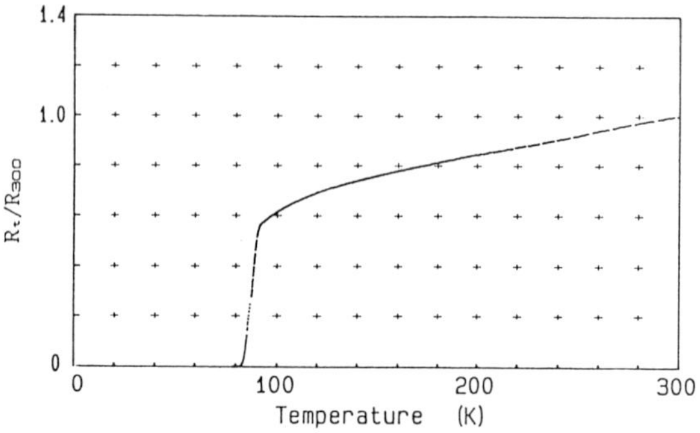

Fig.6. Temperature dependence of the resistivity for the film: the film was produced at 1223K, sintered at 1243K for 1 hour and annealed at 823K for 5 hours in flowing O_2.

Fig.4(a) shows the result of XRD (Co Kα) analysis of the powder obtained at 1223K. Superconductive particles are synthesized directly as orthorhombic phase. These results indicate that the ultra-fine and homogeneous powder composed of Y system oxide, even in an orthorhombic phase, can be prepared directly by the mist pyrolysis method.

Figure 5 shows SEM micrographs of the surface and cross-sectional morphologies of films sintered at 1223K and 1243K for 1 hour in flowing O_2. The degree of the union of grains increased as the sintering temperature increased from 1223K to 1243K. Fig.4(b) shows the result of XRD analysis of the film sintered at 1243K for 1hour and annealed at 823K for 5 hours, suggesting the crystalline growth with the heat treatment in flowing O_2. Fig.6 shows the temperature dependence of resistivity of the film sintered at 1243K for 1 hour and annealed at 823K for 5 hours in flowing O_2.

The superconducting transition of the film was observed at Tc(onset)=93K, and Tc(zero resistance)=84K.

CONCLUSIONS

The ultra-fine and homogeneous powder composed of Y system oxide,even in an orthorhombic phase,could be prepared by the mist pyrolysis method aided by the gas jet atomizer.

The superconducting thick film with zero-resistance of 84K was formed on the MgO substrate with post-annealing process.

REFERENCES

[1] K. Okuyama, Y. Kousaka, N. Tohge, and M. Adachi: J. Mater. Sci.Lett.,6,1466 (1987).

[2] N. Tohge, M. Tatsumisago, T. Minami, K. Okuyama, M. Adachi, and Y. Kousaka: Jpn. J. Appl. Phys.,27,L1086(1988).

[3] T. Ogawa et al: Bull. Chem. Soc. Jpn.,62,1844(1989).

[4] M. Awano and H. Takagi:Proc. of ISS'89,Nov.,393(1989).

[5] J. Mogi, K. Suzuki, T. Kondou,and M. Ayabe:Ishikawajima-Harima gihou,25(6), 1(1985).

Large Levitation Force Due to Flux Pinning in MPMG Processed YBaCuO Superconductors with Ag Doping

M. MURAKAMI, T. OYAMA, H. FUJIMOTO, T. TAGUCHI, S. GOTOH, Y. SHIOHARA, N. KOSHIZUKA, and S. TANAKA

Superconductivity Research Laboratory, International Superconductivity Technology Center, 10-13, Shinonome 1-chome, Koto-ku, Tokyo, 135 Japan

ABSTRACT

An extremely large levitation force of 90N and a suspension force of 25N were achieved in Ag-doped YBaCuO fabricated by the Melt-Powder-Melt-Growth process using a repulsive and an attractive force between the superconductor and a 0.4T rare-earth magnet at 77K. The combination of a large J_c value and large shielding current loop is the source of such a large suspension or levitation force.

KEYWORDS: levitation force, flux pinning, YBaCuO, Ag doping, MPMG

INTRODUCTION

While one of the most attractive properties of the superconductor is resistanceless current, the response of the superconductor to magnetic field is also attractive for some applications[1]. In magnetic fields below H_{c1}, type II superconductors can prevent magnetic flux penetration absolutely, which is known as the Meissner effect. Under such circumstances, a magnet can levitate above the superconductor and vice versa. Superconducting rotating devices such as non-contacting bearings operate by utilizing the Meissner effect. However, the repulsive force due to the Meissner effect is small, which limits the application of such devices. On the other hand, a repulsive force can also be obtained above H_{c1}, when the superconductor exhibits pinning force[2-4]. Although the external field enters the superconductor, the internal field has a gradient corresponding to $dH/dx = J_c$, where H is the magnetic field in A/m and J_c is the critical current density in A/m^2. The physical origin of the flux-density gradient is the pinning of the magnetic flux lines (Abrikosov vortices) by structural defects in the material. Therefore, when J_c is high, a large portion of the superconductor can be shielded by flux pinning, suggesting that a large repulsive force would be obtainable if bulk superconductors with large J_c values could be produced. In this paper, we show that the levitation force as high as 90N can be obtained at 77K.

It should also be noted that a large magnetic field can be trapped in the superconductor exhibiting high J_c values. The suspension of the superconductor below a magnet is explained by the attractive force due to the trapped field. Therfore, the suspension force can be increased with increasing flux pinning force. We then show that a mass of 2.5 kg can be suspended below a magnet using the attractive force between the superconductor and the magnet.

EXPERIMENTAL

The YBaCuO samples were prepared by the MPMG process, the details of which are described elsewhere[5,6]. In this process, the Y-Ba-Cu-O sample is first melted at temperatures above 1200°C and cooled down to room temperature. At this stage, the sample consists of Y_2O_3 and a solidified liquid phase. Here the sample is crushed into powder and then pressed into the desired shape. It is reheated up to the temperature region of 1000 - 1200°C and rapidly cooled down to the temperature at which the superconducting phase starts to nucleate. This is followed by slow cooling. During the remelting, Y_2O_3 reacts with the liquid to produce Y_2BaCuO_5, and during the final slow cooling Y_2BaCuO_5 reacts with liquid to produce $YBa_2Cu_3O_x$ phase. When the starting composition is changed from the stoichiometry of Y:Ba:Cu=1:2:3 toward the 211-rich region, fine 211 particles about one micron in diameter are dispersed in the matrix of the 123 phase. Fine dispersion of these 211 particles contributes to the increase in J_c[7)], and a J_c value exceeding 10^9 A/m^2 has been achieved at 77K and zero field[7)]. In order to reduce the amount of cracking, AgO particles were added during the solid state reaction.

The repulsive and attractive forces were measured as a function of distance between the superconductor and the 0.4 T rare-earth magnet 2 cm in diameter.

RESULTS AND DISCUSSION

The force between the superconductor and the magnet can be described as

$$F = m\, dH/dx,$$

where m is the magnetic moment of the superconductor and dH/dx is a field gradient produced by a magnet. m is equal to Mv where M is the magnetization per unit volume and v is the volume. dH/dx is a constant provided we use the same magnet, while M is dependent on J_c and the size of the shielding current loop according to the following relation,

$$M = A\, J_c\, r,$$

where A is a constant which depends on sample geometry and r is the radius of the shielding current loop[8)]. This size dependence of M is a characteristic of the superconductor and is the key to large levitation force.

In order to increase F, we must increase either J_c or r. In most high T_c samples, r is not proportional to sample size, and the shielding current is localized in individual grains, since most grain boundaries act as weak-links. It is possible to fabricate Y-Ba-Cu-O samples consisting of grains with diameter greater than 1 cm using the MPMG process, which can contribute to an increase in r.

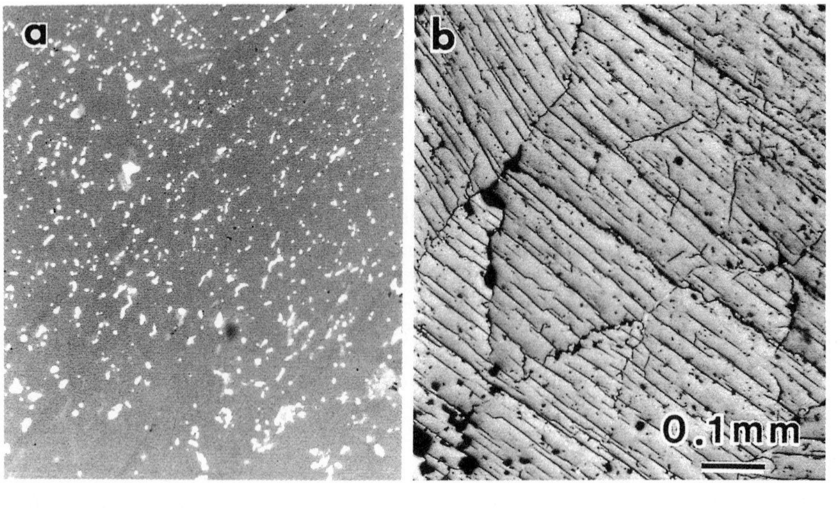

Fig. 1. Microstructure of MPMG-processed Y-Ba-Cu-O: (a) with 10 mass% Ag and (b) without Ag. Note that the amount of cracking can be reduced by Ag dispersion.

However, we have found that cracking is likely to take place in large grains, resulting in the reduction of r. The cracking is ascribed to the strain due to oxygen absorption associated with tetragonal to orthorhombic transformation. Despite large J_c values in the MPMG-processed samples, the levitation force is limited by cracking. Fine dispersion of 211 particles is effective in preventing cracks from propagating, and the levitation force can be increased up to 10N in samples with the composition $Y_{1.8}Ba_{2.4}Cu_{3.4}O_x$, but cracking is not completely avoidable.

In order to reduce the amount of cracking, we tried to disperse silver into the matrix[9], since silver particles may absorb strain and do not deteriorate the superconducting property of the matrix. Using the MPMG process, it is easy to incorporate other elements into the superconductor, which can be performed by solid-state mixing after the quenched Y-Ba-Cu-O sample has been crushed into powder. Figure 1 shows the microstructure of the MPMG-processed Y-Ba-Cu-O with and without Ag doping. It is notable that silver particles are dispersed in the matrix and the number of cracks decreased significantly. Figure 2 shows levitation force versus distance for the 10 mass%Ag-doped $Y_{1.8}Ba_{2.4}Cu_{3.4}O_x$. The data were taken as a 0.4T magnet 2 cm in diameter was moved toward and away from the superconductor 3 cm in diameter and 2 cm in thickness. The levitation force at 1mm distance exceeded 50N, which indicates that an object which weighs 5kg can be levitated over the superconductor at this height. We also found that the repulsive force reached 90N at a distance of about 0.1mm. The source of such a large levitation force is the combination of high J_c and large shielding current radius in the MPMG processed Ag-doped Y-Ba-Cu-O.

It is also notable that an attractive force can be observed when we removed the magnet from the superconductor as shown in Fig. 2. The attractive force is attributable to the trapped magnetic field. In such case, the attractive force will be increased if the trapped magnetic field is increased. In order to confirm this, a magnetic field was trapped in the field-cooled process, where the superconductor was cooled with a

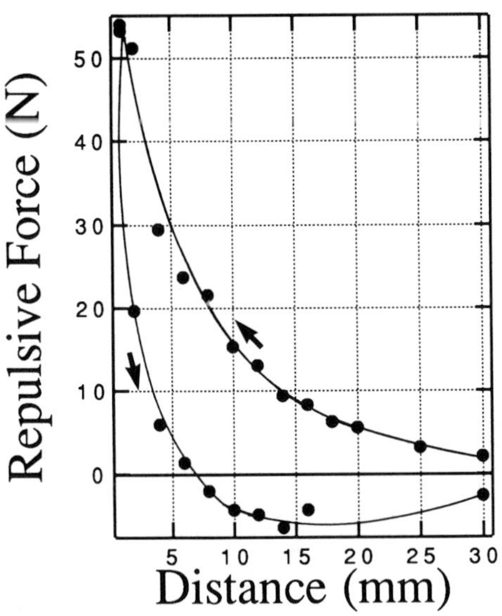

Fig. 2. Levitation force versus distance. The data were taken as a 0.4T magnet was moved toward and away from the MPMG-processed Ag-doped Y-Ba-Cu-O.

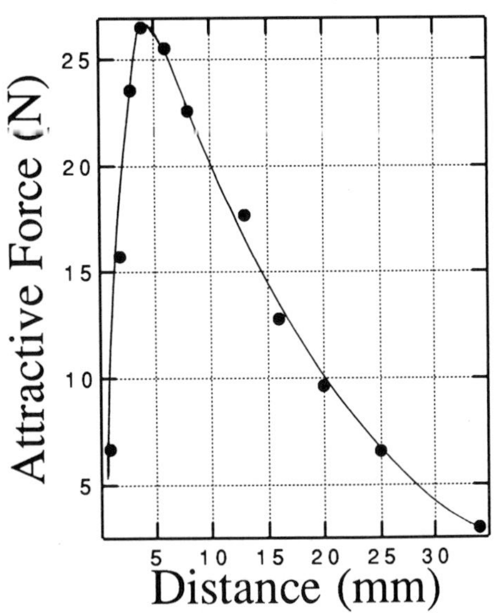

Fig. 3. Force versus distance in the field-cooled process. Note that the attractive force can be increased significantly by the field-cooled process.

Fig. 4. MPMG processed Y-Ba-Cu-O with Ag doping can be suspended below a magnet by the attractive force due to trapped flux. Iron clips are attracted by the superconductor, indicating flux penetration.

magnet placed on the superconductor. Figure 3 shows the force versus magnet distance in the field-cooled process. It is clear that a much larger attractive force can be obtained in this experiment, supporting the idea that the attractive force is attributed to the trapped magnetic field. At the peak, the force value reached 27N, indicating that a material which weighs 2.7kg can be suspended below the superconductor.

Figure 4 shows the suspension of the superconductor below the magnet. The superconductor is attracting iron clips, indicating that the flux is surely penetrating the superconductor.

CONCLUSIONS

An extremely large levitation force and attractive force can be obtained using the force between high J_c superconductor and the magnet. The source of such large forces is strong flux pinning.

REFERENCES

1) D. Wolfstein , T. E. Seidel, D. W. Johnson, Jr. and W. W.Rhodes: *J. Superconductivity* **2** (1989) 211.
2) F. C. Moon, M. M. Yanoviak and R. Ware : *Appl. Phys. Lett.* **52** (1988) 1534.
3) E. H. Brandt: *Appl. Phys. Lett.* **53** (1988) 1554.
4) W. G. Harter, A. M. Hermann and Z. Z. Sheng: *Appl. Phys.Lett.* **53** (1988) 1119.
5) H. Fujimoto, M. Murakami, S. Gotoh, T. Oyama, Y. Shiohara, N. Koshizuka and S. Tanaka: *Advances in Superconductivity II* (Springer-Verlag, Tokyo, 1990) p. 285.
6) M. Murakami: *Mod. Phys. Lett.* **4** (1990) 163.
7) M. Murakami, S. Gotoh, N. Koshizuka, S. Tanaka, T. Matsushita, S. Kambe and K. Kitazawa: *Cryogenics* **30** (1990) 390.
8) M. Murakami, M. Morita, K. Doi and K. Miyamoto: *Jpn. J. Appl. Phys.* **28** (1989) L1125.
9) P. A. Curreri, P. N. Peters, R. C. Sisk, M. K. Wu and C. Y. Huang: *Met. Trans.* **21A** (1990) 257.

New Melt Process of High-T_c Oxide Superconductor by a Rapid Solidification Using a Nd-YAG Laser

SHIGEO NAGAYA, MASAMICHI MIYAJIMA, IZUMI HIRABAYASHI, YUH SHIOHARA, and SHOJI TANAKA

Superconductivity Research Laboratory, Nagoya Division, International Superconductivity Technology Center, 4-1, Mutsuno 2-chome, Atsuta-ku, Nagoya, 456 Japan

ABSTRACT

A nonequilibrium phase transformation is realized by a rapid solidification using the laser zone melting method which has a high temperature gradient at a solid-liquid interface and a high growth rate. The 123 orthorhombic superconducting phase crystallizes directly from the melt of the 123 composition by introducing a high interfacial undercooling. The obtained samples show superconducting transition at 94 K by the magnetic measurement in as-grown. Moreover, the structure with finely dispersed Y_2O_3 was obtained at the lower growth rate. It is possible to use this structure as a long and narrow precursor for the QMG process, and by annealing at 1000 °c the critical current density of the sample was $1.5x10^4$ A/cm^2 at 77K, 0T from the magnetization measurement.

KEY WORD: rapid solidification, nonequilibrium phase transformation, temperature gradient, growth rate, laser zone melting

INTRODUCTION

Since the discovery of high-Tc superconductors, tremendous efforts have been made to synthesize the materials with the high critical current density (Jc). To obtain the high Jc superconductors, we should resolve three problems. The first is controlling the crystal orientation, the second is improvement of the grain boundary characteristics, and the last problem is introducing the pinning center. In particular, the melting-solidification process has been investigated as a promising process to control the crystal orientation and to improve the grain boundary characteristics. However, in Y-Ba-Cu-O systems [1], the 123 phase can not be obtained directly from the liquid of the 123 composition. Because it melts incongruently and decomposes into the 211 phase and the liquid under the equilibrium phase transformation [2]. Therefore it has been generally produced through the peritectic reaction of Y_2BaCuO_5 and the liquid. The successful melt processes, such as Melt-Texture-Growth (MTG) process by Jin et. al. [3], Quench and Melt-Growth (QMG) process by Murakami et. al. [4], and Melt-Powdering-Melt-Growth (MPMG) process by Fujimoto et. al. [5] utilized the following peritectic reaction.

$Y_2O_3 + L_1 \rightarrow Y_2BaCuO_5 + L_2 \rightarrow YBa_2Cu_3O_7$, (1)

where L_1, L_2 means liquid phases which are different compositions.

But, these processes of using the peritectic reaction have several problems, for example, the long reaction time for the crystallization, the strong reactions with the crucible materials, and the inhomogeneity in the structures. The most difficult problem among these problems is the less controllability of the crystal orientation by the peritectic reaction.

Then, if the direct crystallization from the liquid phase realizes, we could control the crystal orientation. That is, if we are able to suppress the crystallization of Y_2O_3 and the 211 phase, in other words, able to give a high undercooling to them, the 123 phase may crystallize from the liquid directly, and then, the direction of the crystal growth could be controllable [6].

Figure 1 shows the schematic of the pseudo-binary nonequilibrium phase diagram for the Y-Ba-Cu-O system with high undercooling. Considering this nonequilibrium phase transformation by the addition of this phase diagram, when the undercooling is realized, the liquidus lines of both Y_2O_3 and the 211 phase disappear, and then, the liquidus line of the 123 phase can be extended as the solid line in Fig. 1. We consider that the crystal growth under the metastable state, that is, the direct crystallization of the 123 phase from the liquid will be realized by a high interfacial undercooling, and this undercooling can be attended by a rapid solidification with a very high growth rate. In the rapid solidification process, the cooling rate is given by the product of the temperature gradient (G) and the growth rate (R) [7]. Therefore, we investigated this process using the laser zone melting method. Because this method enable us to realize both the high G and the high R.

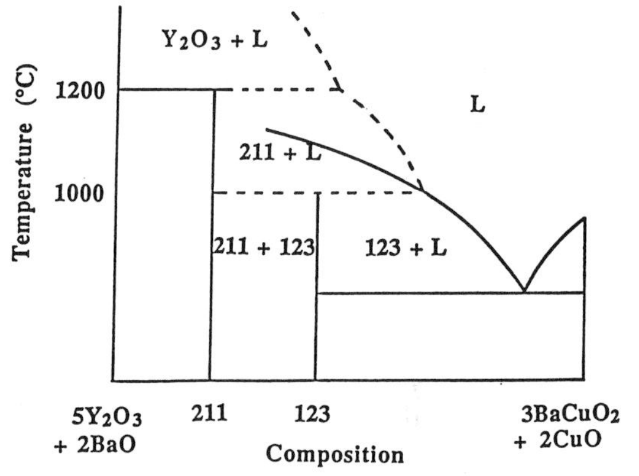

Fig.1 A schematic pseudo-binary nonequilibrium phase diagram for the Y-Ba-Cu-O system under the undercooling.

EXPERIMENTAL

The laser using this work is a multi mode Nd-YAG laser, and its maximum output power is 150 W. The irradiated sample was a conventional solid-phase sintered pellet, and its size was 15 millimeters in diameter and one millimeter in thickness. The pellet was placed on the high speed traveling stage, and traversed under the focused laser beam. As the region processed by the laser was very small, we identified the crystalline phase of the solidified regions and measured the degree of the orientation by the micro-X-ray diffractometer equipped with the curved position sensitive proportional counter. The diameter of the size of the X-ray spot was 50 μm or 100 μm depending of the size of solidified regions. The superconductivity was measured by a DC SQUID magnetometer.

RESULTS AND DISCUSSION

Figure 2 shows SEM micrograph of the irradiated sample in this process. The laser irradiated power was 50 W and the laser scanning velocity was 10 cm/sec. A dendritic structure was observed on the surface and in the cross section, and it was different from the microstructure of the unirradiated one. These results suggest that the rapid solidification resulting from the high growth rate with relatively high undercooling at the solid-liquid interface can be achieved.

Figure 3 shows the micro-X-ray diffraction data for the polished cross section of the several irradiated conditions changed the scanning velocity at 50 W. All diffraction patterns consist of peaks of the 123 phase, Y_2O_3, and an unknown phase. The unknown phase disappeared by the annealing at 600 °c and the peaks of $BaCuO_2$ appeared. This result is considered that the unknown phase is the nonequilibrium phase originated in the Ba-Cu-O. It should be noticed that the 211 phase does not exist in these patterns. *Why is the 211 phase not observed in these patterns?* Two reasons are considered, one is that the 211 phase has been reacted to the 123 phase completely by the peritectic reaction (1), and the other is that the 211 phase is not formed by this process from the beginning. Since a peritectic reaction is, in general, a very slow reaction including the diffusion in the solid, it is difficult to consider that all 211 phase crystals react into the 123 phase during this process.
Therefore, we consider two other possible reaction;

$L_1 \rightarrow Y_2O_3 + L_2 \rightarrow YBa_2Cu_3O_7$, (2)

and $L \rightarrow YBa_2Cu_3O_7$. (3)

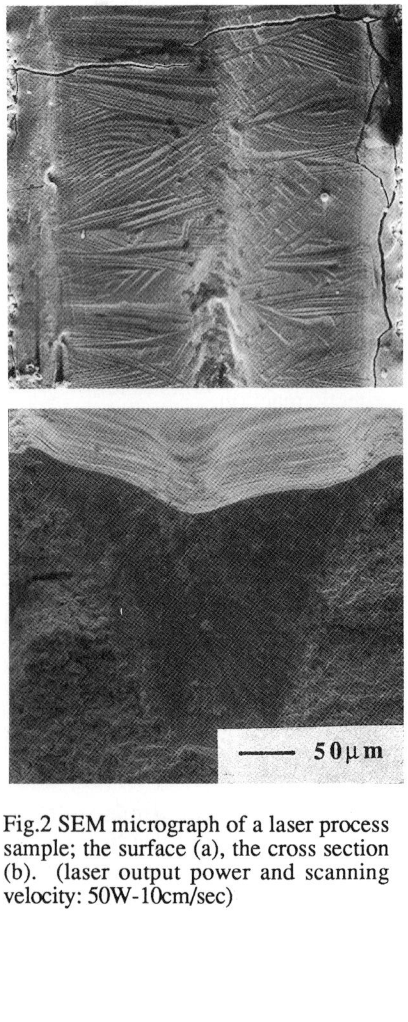

Fig.2 SEM micrograph of a laser process sample; the surface (a), the cross section (b). (laser output power and scanning velocity: 50W-10cm/sec)

The first reaction (2) is the single peritectic reaction which skip over the 211 phase. In this case the 123 phase crystallized by the reaction with Y_2O_3 and the liquid.
The next possible reaction (3) is the direct crystallization from the liquid.

The dependences of the laser scanning velocity on the relative intensity for the 123 phase and yttria are shown in Fig. 4. Increasing the laser scanning velocity, the amount of the 123 phase increases, while that of Y_2O_3 decreases. If the 123 phase are formed through the single peritectic reaction with Y_2O_3 and the liquid, the amount of Y_2O_3 is expected decrease with decreasing the velocity. Because Y_2O_3 has a tendency to react with the surrounding liquid and transform into the 123 phase. But this result is not the case, Y_2O_3 increased in the low scanning velocity, and the 123 phase decreased. It suggests that the reaction in this process is not the peritectic reaction (2) but the direct crystallization from the liquid (3).

Is the direct crystallization realized in this process? In order to answer the question, we measured the pole figure of the sample and checked the crystal orientation, since the crystal orientation might be improved in the sample, if the 123 phase crystallized from the liquid. Figure 5 shows the pole figure of the (006) peak for the 123 phase. In the figure, the laser scanning direction is taken to the horizontal axis. The (006) peak distributes perpendicularly to the horizontal axis. It shows that the c-axis of the 123 phase orients perpendicularly to the laser scanning direction, that is, the ab planes orient along the laser scanning direction. It indicates that the direction of the crystal growth, the ab planes in YBCuO, is parallel to the temperature gradient. This result suggests that the direct crystallization of the 123 phase is realized by a rapid solidification using the laser zone melting and it makes it possible to control the crystal orientation which is necessary to obtain the high Jc superconductor.

Moreover, the peak splits which mean the existence of the orthorhombic superconducting phase were observed in the diffraction patterns. Superconducting phase transitions due to the ortho-1 and ortho-2 were observed in the magnetization measurement. The temperature dependence of magnetization of the samples is shown in Fig. 6. Increasing the scanning velocity, their signal intensities of super-conductivity increase and it is consistent with the X-ray diffraction data. Interestingly, the rapid solidification samples show the superconducting phase transition at 94 K. This Tc is a little higher than the usual one. The non-equilibrium solidification might give some factors which rose up Tc, such as the anomalous distribution of oxygen and the lattice distortion.

Figure 4 shows that Y_2O_3 becomes the main crystalline phase in the melt-solidified region under the lower R condition. By the measurement of atomic distribution using EPMA, the coalescence of Y_2O_3 was observed and its primary size was considered to be 2 to 3 μm.

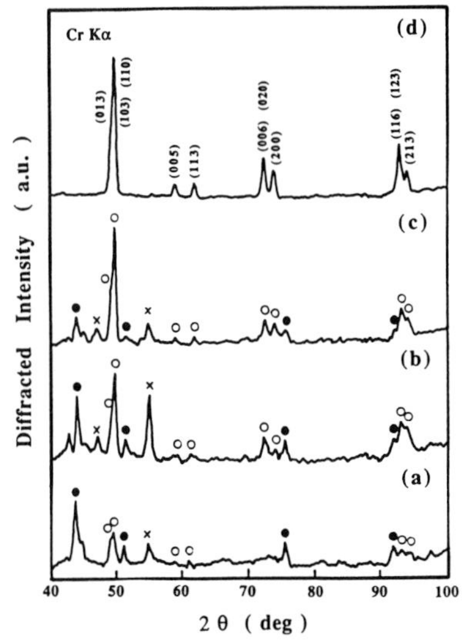

Fig.3 Micro-X-ray diffraction patterns and phase identification showing the 123 phase (o), Y_2O_3 (•), and unknown Ba-Cu-O phase (x). The laser irradiated conditions are 50W-5cm/sec (a), 50W-10cm/sec (b), 50W-15cm/sec (c), and unirradiated original sample (d).

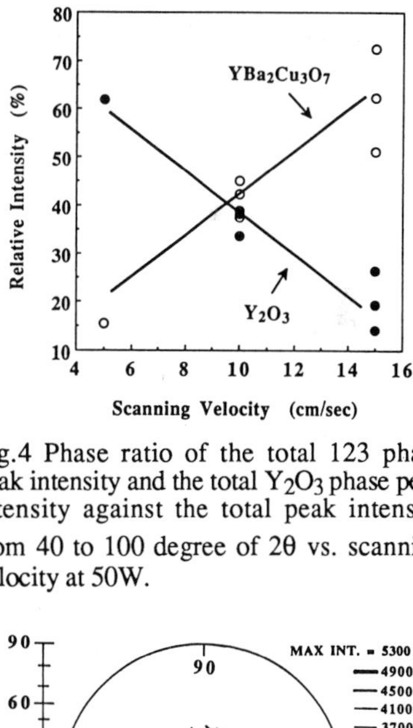

Fig.4 Phase ratio of the total 123 phase peak intensity and the total Y_2O_3 phase peak intensity against the total peak intensity from 40 to 100 degree of 2θ vs. scanning velocity at 50W.

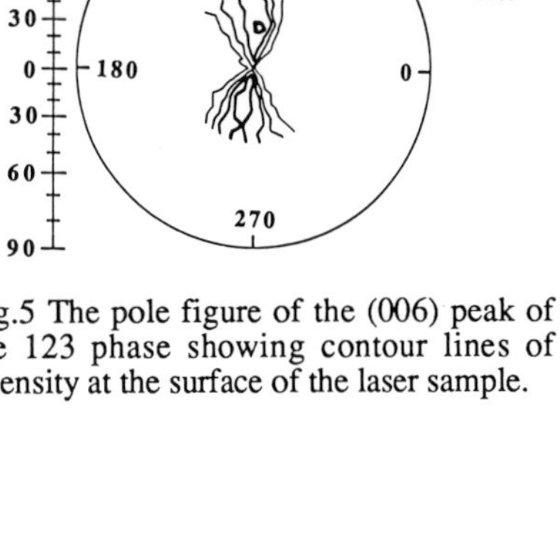

Fig.5 The pole figure of the (006) peak of the 123 phase showing contour lines of intensity at the surface of the laser sample.

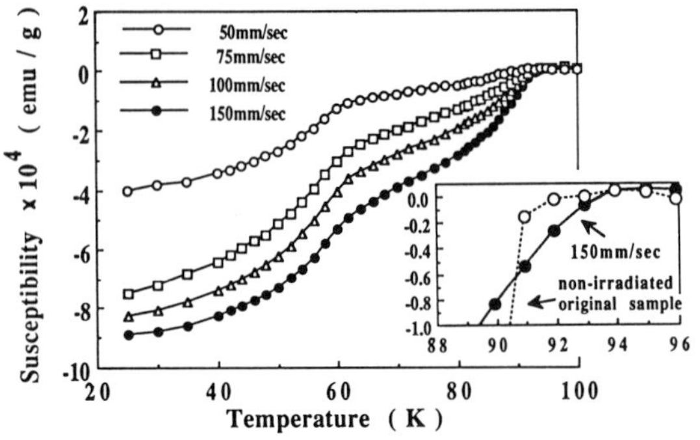

Fig.6 Temperature dependence of the field cooled susceptibility (50 Oe) of laser process samples with 50 W.

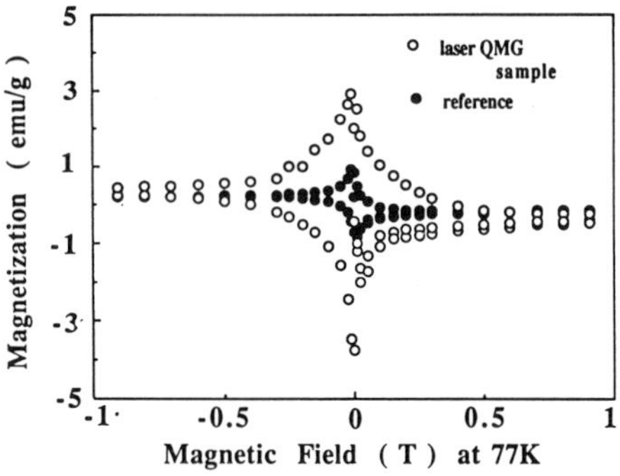

Fig.7 Magnetization loops for the laser QMG process sample and the unirradiated sample with the same annealing.

The solidified structure with Y_2O_3 dispersion can be utilized for a suitable precursor of the QMG process. As described before, the QMG process utilized the peritectic reaction (1) which used the reaction of the 211 phase and the liquid, and the reaction of Y_2O_3 and the liquid. The key point of the QMG process is controlling the dispersion of Y_2O_3, because the continuous growth of the 123 phase requires the finely dispersed 211 phase and this fine dispersion of the 211 phase is originated from the fine dispersion of Y_2O_3. But in the QMG process it is difficult to obtain the uniform and fine dispersion of Y_2O_3 due to the existence of the segregation of Y_2O_3 and the reaction with the crucible materials. In the laser method, the macroscopic segregation of Y_2O_3 is reduced by the rapid-melting/rapid-solidification and the contamination with the crucible can be avoided. Moreover, the most different point from the QMG process is a possibility to obtain an arbitrary long precursor.

By annealing at 1000 °c, the structure with the fine dispersion of the 211 phase was obtained. This sample showed the perfect demagnetization in the zero-field cooling process with the sharp transition by a DC SQUID measurement. Figure 7 shows magnetization loops for the samples at 77 K. The improvement of Jc is depicted comparing with the unirradiated sample with the same annealing.

CONCLUSION

We developed two new melt processes by a laser zone melting method. One is the direct crystallization of the 123 phase from the liquid phase. This crystallization is realized by the high undercooling achieved by the high temperature gradient and the high growth rate. This process enable us to control the crystal orientation. The other process is the laser QMG, and its precursor is made by the laser melting. The precursor had a structure in which Y_2O_3 dispersed finely, and by the partial melt annealing the high Jc sample was obtained. We expect that these two processes enable us to produce the high Jc long superconductor like a tape or wire.

REFERENCES

1. Wu M, Ashburn J, Torng C, Hor P, Meng R, Gao L, Huang Z, Wang Q, Chu C (1987) Phys.Rev.Lett. 58: 908
2. Oka K, Nakano K, Ito M, Saito M, Unoki H (1988) Jap.J.Appl.Phys. 27: L1065
3. Jin S, Tiefel T, Shevwood R, Davis M, Van Dover R, Kammlott G, Fastnacht R, Keith H (1988) Appl.Phys.Lett. 52: 2074
4. Murakami M, Morita M, Doi K, Miyamoto K (1989) Jap.J.Appl.Phys. 28: 1189
5. Fujimoto H. Murakami M, Gotoh S, Koshizuka N, Oyama T, Shiohara Y, Tanaka S (1989) Proceedings of 2nd Int. Symp. Superconductivity. Tsukuba, p285
6. Nagaya S, Miyajima M, Hirabayashi I, Shiohara Y, Tanaka S (1990) Proceedings of Applied Superconductivity Conference, Snowmass-Village Colorado. IEEE Magnetics (to be published)
7. Shiohara Y, Nakagawa M, Suga T, Ishige K, Oyama T, Izumi T, Nagaya S, Miyajima M, Hirabayashi I, Tanaka S (1989) Proceedings of 2nd Int. Symp. Superconductivity. Tsukuba, p263

Preparation of High J_c Y-123 by Addition of Reactive Fine Powders

TAKENOBU SAKAI, KOZI NISHIO, NAOYUKI OGAWA, and IZUMI HIRABAYASHI

Superconductivity Research Laboratory, International Superconductivity Technology Center, 4-1, Mutsuno 2-chome, Atsuta-ku, Nagoya, 456 Japan

ABSTRACT

We report a new method for introducing pinning centers by a partial melting processing using reactive fine powders, such as SiC, SiO_2, Al_2O_3, and ZrO_2. Since this method lowers processing temperature, it may become a useful technique for the production of high Jc metal-sheathed tapes and wires. YBCO powders were mixed with 25mol% SiC powder of 0.3 μm, sintered at 980°C for 15min and cooled at 3°C/hr under the flowing oxygen gas. The obtained samples show prominent grain growth and the reacted phases (Y_2BaCuO_5, Ba_2SiO_4) which are submicron size and uniformly dispersed in the YBCO matrix. Magnetic measurements show the improvement of the critical currents (over 7500A/cm^2 at 77K 0.04T) and the sample shows the so-called fishing behavior at 77K. It seems that the reacted phases (Y_2BaCuO_5, Ba_2SiO_4) work as the effective pinning centers.

KEY WORDS: high-Tc superconductors, critical current density, flux pinning

INTRODUCTION

Since the discovery of high-Tc oxide superconductors (1), many efforts have been concentrated on searching for the processing which yields high critical current density (Jc). In the previous work, Jin et al. presented the "melt-textured growth" (MTG) method to obtain highly oriented structures in YBCO materials (2). Murakami et al. developed the "quench and melt growth" (QMG) method to obtain single crystal-like YBCO bulk materials which contain fine particles of Y_2BaCuO_5 (3).

Though these methods enable us to obtain high Jc bulk materials, they have common problems that a complete melting temperature above 1200°C is indispensable. Therefore, it is very difficult to apply these methods for the production of high Jc metal-sheathed tapes and wires because any metals for sheath have lower melting temperatures than 1200°C.

In the papers (4)(5), we have reported an effective method introducing pinning centers by a partial melting processing with reactive fine powders, such as SiC, SiO_2, Al_2O_3, and ZrO_2. This method lowers processing temperature.

In this paper, we report on the reaction between microstructure and superconductivity for the YBCO by addition of fine SiC powder.

EXPERIMENTAL

YBCO powders were prepared from Y_2O_3, $BaCO_3$, and CuO powders of 99.9% purity by the conventional solid-state reaction (4). The sintered pellets were crushed, and ground to grains of about 10 μm in diameter. Powder X-ray diffraction showed that the prepared powders were the orthorhombic YBCO (123) phase. The YBCO powder and the fine SiC powder were mixed by ball-mills in ethyl-alcohol using nylon balls and pots, as a mixing rate of 2.5-25 mol%. After drying, the mixed powders were pressed into pellets, 15 mm in diameter and 2 mm thick, at a pressure of 3000kgf/cm^2. Samples sintered at 950-1030°C for 15min, cooling down to 50°C below the maximum temperature at 3°C/hr, and cooling at 100°C/hr to room temperature under the flowing oxygen gas at 1 atm.

Characterization of the samples was carried out by powder X-ray diffraction, electron-probe micro-analysis (EPMA), scanning electron microscopy (SEM), optical microscopy, electrical resistivity, and magnetic susceptibility.

RESULTS AND DISCUSSION

Thermal Analysis and Sintering Density

Thermogravimetric (TG) and differential thermal analysis (DTA) were performed on the mixed powders in the stream of O_2-N_2mixed gas with the oxygen partial pressure of 1 to 0 atm. The melting temperature of the mixed powders determined by the TG-DTA curves are shown in Fig. 1. The melting temperature are lowered by the low oxygen partial pressure and by addition of the SiC powder.

Fig. 2 shows the dependence of the sintering density on the sintering temperature of the various SiC added samples. The densities of the samples were measured from their weights and dimensions. We find the increase of the sintering density for the undoped YBCO sample as increasing sintering temperature. The same tendency is observed for the SiC 13-25 mol% added samples except that the temperature of the maximum of the density lowers as decreasing of melting temperature. Small SiC added samples show the two maxima of the density. It seems that the SiC addition works two different effects on the sintering process of the YBCO/SiC system. Then, we investigated the microstructures of these samples to confirm its relation to the sintering density.

Microstructure

On examination of the powder X-ray diffraction of every SiC added sample sintered at 950-1030°C, the reaction products are identified as Y_2BaCuO_5, Ba_2SiO_4, and CuO. In spite of the same products, there is much difference on the microstructure of these samples. Fig. 3 shows the optical micrographs of the 5.5 mol% SiC added samples sintered at various temperatures. The sample sintered at 950°C shows the smaller grain size than that of undoped YBCO which observed grain growth of the average size of 20 μm. According to EPMA spot analysis, the reaction products accumulate at the grain boundary. By the lower temperature sintering, the reaction products inhibit the grain growth. The sample sintered at 1000°C shows the grain growth morphology due to the liquid-phase-reaction on the grain surface and grain size reaches to 20-30 μm. Further, the sample sintered at 1025°C shows prominent grain growth, typically 0.2 mm wide and 1 mm long, due to the partial melting. According to EPMA analysis, submicron-size Y_2BaCuO_5 and Ba_2SiO_4 fine particles uniformly disperse in the YBCO matrix. The sample sintered at 1030°C has considerable amount of voids due to the intense decomposition with the outbreak of O_2 gas, as shown below in eq (2).

Reaction Equations

On the basis of these results, we conclude that the samples undergo a partial melting growth near the melting temperature, and the reaction between YBCO and SiC can be written as follows,
(a) oxidation of SiC under the existence of YBCO which acts as a catalyst and a supplier of oxygen (250-500°C):

$$SiC + 2O_2 \xrightarrow{\text{Ba-compound}} SiO_2 + CO_2, \text{ and} \qquad (1)$$

(b) decomposition of YBCO by silica (900-1050°C):

$$4YBa_2Cu_3O_7 + 3SiO_2 \rightarrow 2Y_2BaCuO_5 + 3Ba_2SiO_4 + 10CuO + O_2. \qquad (2)$$

Superconductive Properties

Fig. 4 shows the SiC content dependence of the critical temperatures (Tc) for the reacted samples sintered at 950-1030°C. We find the decrease of Tc for the sample sintered at 950-970°C as increasing SiC content. In this case, the reaction products form transport barriers at the grain boundaries. Such results have already been reported in the previous studies as the problem of the reaction between YBCO and the substrates in the field of thin film fabrications (6). On the other hand, higher temperature sintering leads to keep the high Tc values, in spite of increasing SiC content. On the sample sintered over the melting temperature, the Tc shows rapid drop due to the intense decomposition.

The Jc was estimated from the magnetization hysteresis loops using Bean's critical state model for the thin slab configuration (7). Fig. 5 shows the field dependence of Jc at 60K for the YBCO samples with 25 mol% SiC content and without SiC sintered at 980°C. The Jc values of every samples show relatively small drop than that of undoped YBCO sample. The magnetic Jc value at 77K and 0.04T for the partial melting sample is estimated to be 7500A/cm^2 and the sample shows the so-called fishing behavior at 77K. This value is about three times higher than that for the conventional sintered undoped YBCO sample. It seems that the reacted phases and decomposition products which are well dispersed in the YBCO matrix work as the effective pinning centers.

Transport Jc measurements were performed by a four-probe technique using a pulse current in order to suppress the heat generation from electric contacts. The transport Jc value of 25 mol% SiC content sample was 250A/cm^2 at 77K and 0T. This is a half of that of the conventionally sintered undoped YBCO sample. The sample shows prominent grain growth, but still has the weak coupling between the grains. If we are able to grow the grains enough to observe intragrain Jc by transport, the Jc value will be much higher. From the result of the temperature dependence of magnetic susceptibility for the partial melting processed sample in the zero-field-cooling (ZFC) and the field-cooling (FC) , the superconducting volume fraction is calculated to be 40vol% at 77K. It seems that these two factors; weak link and small volume fraction, lower the transport Jc value.

CONCLUSIONS

The chemical interactions between the YBCO superconductor and the SiC powder have studied at various temperatures. We found;
(1) The sample reacted at 950°C shows the small grain due to inhibition of grain growth by reacted products which accumulate at grain boundaries. The reacted products form transport barriers at the grain boundaries.
(2) The sample reacted just below the melting temperature shows the grain growth due to the liquid-phase-reaction on the grain surface.
(3) According to sintering at the melting temperature, the samples show the prominent grain growth and the reacted phases (Y_2BaCuO_5 and Ba_2SiO_4) are submicron size and uniformly dispersed in the YBCO matrix. Magnetic measurements show the strong pinning effect and the fishing behavior. The best Jc value at 77K and 0.04T is 7500A/cm^2.

REFERENCES

1. J.G.Bedonorz and K.A.Muller; Z.Phys.B 64, (1986), 189 .
2. S.Jin, T.H.Tiefel, R.C.Sherwood, R.B.van Dover, M.E.Davis, G.W.Kammlott and R.A.Fastnacht; Phys.Rev.B 37 (1988) 7850.
3. M.Murakami; Modern Physics Letters B, Vol. 4, No. 3 (1990) 163-179.
4. T. Sakai, K. Nishio, N. Ogawa, I. Hirabayashi, S. Tanaka; Proc. of 7th CIMTEC Satellite Symposium 4 (Italy, July 2-5, 1990).
5. T. Sakai, K. Nishio, N. Ogawa, I. Hirabayashi, S. Tanaka; Submitted to Jap.J.Appl.Phys.
6. C.T.Cheung and E.Ruckenstein; J.Mater.Res., Vol.4, No1, Jan/Feb 1989, 1.
7. C.P.Bean; Phys.Rev.Lett.; 8 (1962) 250.

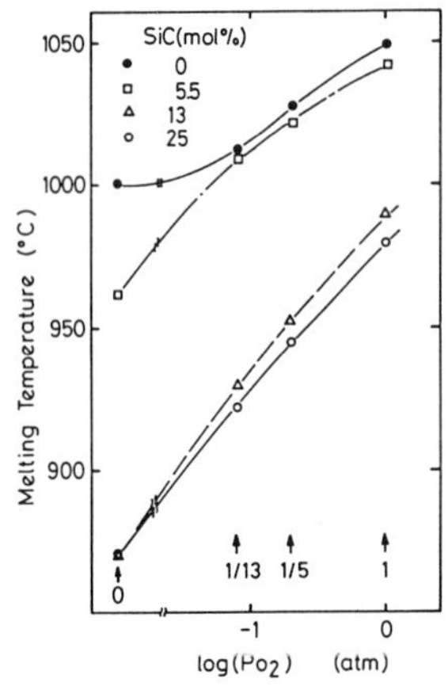

Fig. 1 The oxygen partial pressure dependence of the melting temperature for the various SiC content samples.

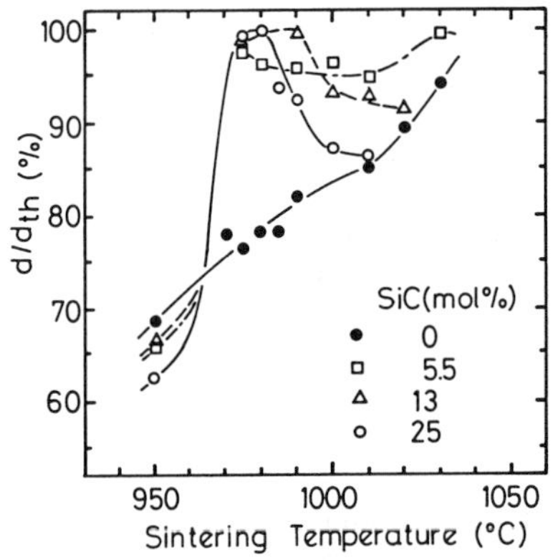

Fig. 2 The sintering temperature dependence of the bulk density.

(a) (b) (c) (d) 20 μm

Fig. 3 Optical microphotographs of 5.5mol% SiC content samples sintered at (a)950°C, (b)1000°C, (c)1025°C, (d)1030°C.

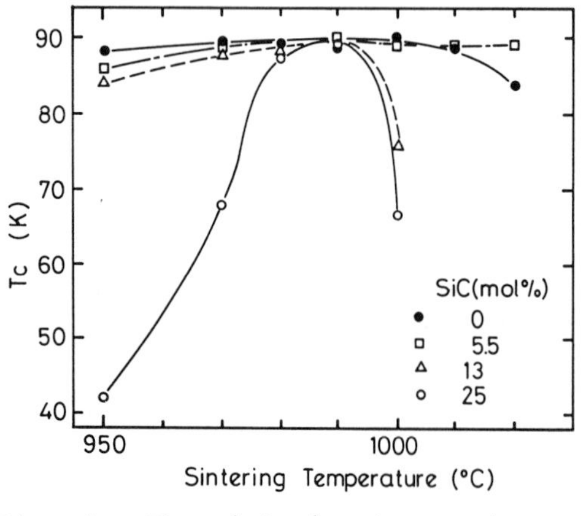

Fig. 4 The sintering temperature dependence of the Tc for the various SiC content samples.

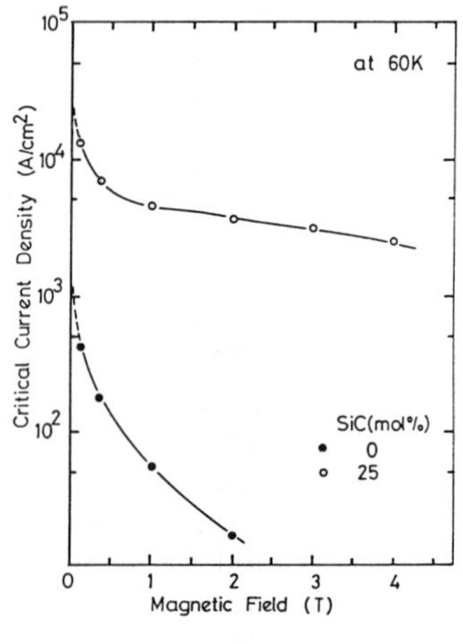

Fig. 5 Field dependence of Jc for various samples.

Highly Orientated Growth of YBCO Superconductors by Partial Melt Process

NAOYUKI OGAWA, KOZI NISHIO, TAKENOBU SAKAI, and IZUMI HIRABAYASHI

Superconductivity Research Laboratory, Nagoya Division, International Superconductivity Technology Center, 4-1, Mustuno 2-chome, Astuta-ku, Nagoya, 456 Japan

ABSTRACT

The highly orientated bulk YBCO superconductors were prepared by partial melt and slow cooling. The orientated $YBa_2Cu_3O_y$ crystals contaning small particles of Y_2BaCuO_5 have been grown. We measured anisotropy of magnetic properties applying the field in perpendicular to ab-plane or parallel to ab-plane. If we apply low sweep field up to 100 gauss, anisotropy was caused by the microstructure which was evident by optical micrograph. On the other hand, if we apply the field with sweep range from 5T to -5T, the response is dominated by the intragranular properties and its anisotropy which agree with those of single crystal. The grown YBCO samples are so large that oxygen hardly diffuses completely. The magnetic hysteresis shows a fish tail profile which has been reported as anomalous pinning effect caused by small oxygen defects.

KEYWORDS High-Tc superconductor, melt process, magnetic property, microstructure, critical current density

INTRODUCTION

Since the discovery of high Tc superconductors, extensive efforts have been made to obtain the high critical current density. However, in bulk superconductors the weak coupling between grains and absence of pinning centers are to be prevented to improve critical current density especially in magnetic field [1,2]. To attain high Jc, in these materials, it is very important to control microstructure properly. It is well known that melting process is very effective to solve these problems [3,4]. Using these techniques, the samples which are single crystal like YBCO bulk materials containing fine particles of Y_2BaCuO_5 were obtained. We have recently made highly oriented YBCO superconductors by partial melt and slow cooling process. We have measured anisotropic magnetic properties as well as microstructure of the samples. In this paper we also discuss the anisotropy of critical currents and flux behavior comparing with microstructure of the sample.

EXPERIMENTAL

The raw materials of YBaCuO oxide superconductors were prepared by mixture of 3N purity Y_2O_3, $BaCO_3$ and CuO. The nominal composition of the samples was Y:Ba:Cu:=1.5:2:2.8. After calcination, the powder was well mixed by using wet mill for 10 hours in polyethylene jar containing methanol and ZrO_2 balls. The wet milled slurry was dried and pressed into pellets. The pressed samples placed on the MgO single crystalline substrates were heated in a few hours and kept at 1100°C for 1 hour. In the temperature range 1000 to 930°C, the samples are cooled slowly at various cooling rate. Then, the samples were annealed with oxygen atmosphere at 600°C at least more than 24 hours. The samples were polished using diamond paste. Microstructure was observed with an optical microscope. Superconducting properties were measured using dc-SQUID magnetometer.

RESULT AND DISCISSION

The orientated YBaCuO samples have been grown under the cooling rate of 1°C/hr in the temperature range from 1000 to 930°C. Crystal orientation was examined by X-Ray diffraction. The X-ray pattern shows that all reflections from the cleaved surface were indexed by (00l). Therefore, the cleaved plane was ab-plane. In this plane, there were twin boundaries crossing perpendicularly each other. There are no grain boundaries. However, some cracks were observed (Fig.1(a)). On the other hand, in the plane perpendicular to ab-plane, there were many grain boundaries parallel to ab-plane which is the growth direction as well as twin boundaries. However, there were no cracks along the grain boundaries (Fig.1(b)). Small white particles in the grown YBaCuO crystal were Y_2BaCuO_5 precipitates of which size was relatively large and distribution was inhomogeneous.

Figure 2 shows the magnetization as a function of temperature at 100 gauss. Discrepancy between field cooling (FC) signal and zero-field cooling (ZFC) signal indicates existing strong pinning centers in these samples. However, two types of ZFC signal are observed by difference of the external field direction. If we apply the field in the direction of perpendicular to ab-plane, ZFC signal has a sharp transition at 90K, but, in the direction of parallel to ab-plane, ZFC signal shows a broad transition. Therefore, in spite of the close susceptibility values of two ZFC signals below 60K, susceptibility obtained applying field in perpendicular to ab-plane is larger than that in parallel to ab-plane in the temperature range from 60 to 90K.

In the case of low-field hysteresis loop which is associated with intergranular effects, there is a difference on the field direction. If the field is applied in perpendicular to ab-plane, hysteresis loop was scarcely observed because of perfect exclusion of flux (Fig.3(a)). But, we can see clear hysteresis, when the field is applied to parallel to ab-plane (Fig.3(b)).

These low field anisotropies seem to relate closely to the microstracture. We consider flux penetrates into the sample through the grain boundaries even at less than 100 gauss if we apply the field parallel to ab-plane. In the case of zero-field cooling, as temperature approaches to Tc, flux penetrates into the sample. Therefore ZFC signal decreases gradually. Some flux penetrates into the sample even at 77K and the flux trapped at grain boundaries causes low-field hysteresis loop [5].

(a) (b)

Fig.1 Optical microstracture of polishing surface showing (a) parallel to ab-plane and (b) perpendicular to ab-plane.

When the sweep amplitude range of the external field is 5T, we obtain magnetization curves at variable temperatures as shown figure 4. For the field perpendicular to ab-plane, the hysteresis at 77K has a hump at 0T and fades away blow 5T decrease sharply with increasing field and it shows so-called "fish tail" profile. According to Daeumling et al, this anomalous magnetization characteristics caused by small oxygen defects [6]. In our sample, Y_2BaCuO_5 particles were larger than that prepared by MPMG or QMG and were not distributed enough homogeneous [7]. So, oxygen can hardly diffuse in our samples. Therefore, small oxygen defects may remain in our samples even after annealing above 24 hours in oxygen. On the other hand, for the field parallel to ab-plane the irreversible magnetization persists up to maximum field 5T or -5T until at 85K. The magnetization width ΔM, however, is a few times smaller than that for perpendicular to ab-plane at 0T. Anisotropy of these hysteresis loops are similar to that measured using single crystal by J.N.Li [8]. We consider that the anisotropy of our sample is due to the common mechanism to the crystalline cases.

Figure 5 shows time dependence of remanent magnetization on logarithmic time scale in the magnetic field of 1 and 10 kOe at 77K. It shows that relaxation rates are very small in our sample as same as QMG- or MPMG-processed samples [9]. We estimated the magnetic relaxation based on the thermallyactivated model proposed by Campbell-Evetts [10],

$$(1/M_0)dM/d\ln t = kT/U_0$$

where M_0 is the initial magnetization and t is time. We obtain U_0 values of 0.9eV and 0.8eV for the field of 1kOe and 10kOe for the field perpendicular to ab-plane and 0.6eV (1kOe) and 0.4eV (10kOe) for the field parallel to ab-plane. This anisotropy is also similar to that of single crystal reported by Y.Yeshurun and A.P.Malozemoff [2]. We consider that the difference of U_0 is suggestive of the role of twin boundaries which would act as pinning centers most effectively for the field perpendicular to ab-plane as they mentioned. The U_0 values in our samples are larger than that obtained by single crystal because of existing effective pinning centers.

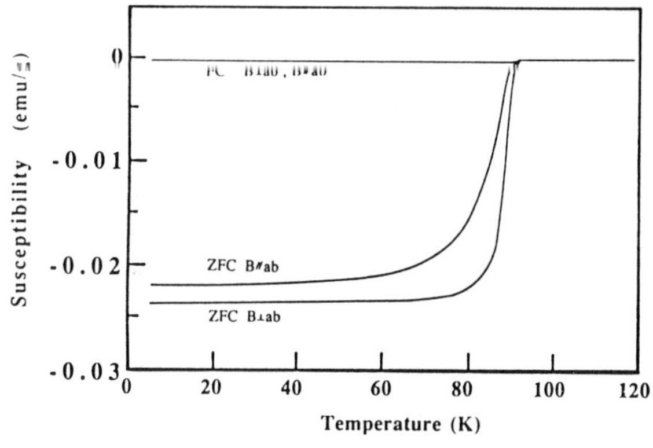

Fig.2 Temperature dependence of susceptibility with an external magnetic field 100 Oe.

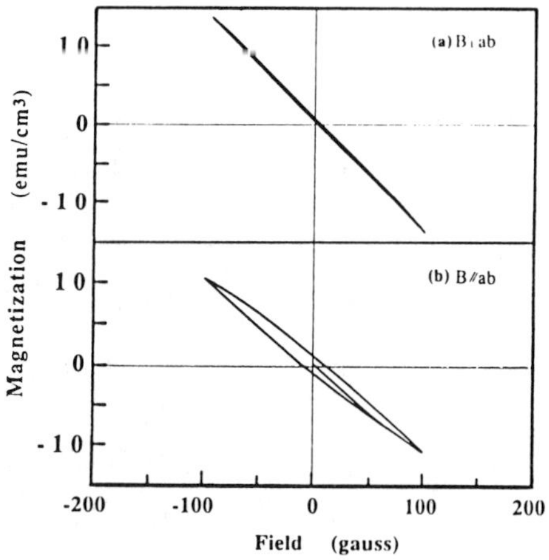

Fig.3 Magnetization hysteresis loops (a) for the field perpendicular to ab-plane and (b) for the field parallel to ab-plane at 77K.

CONCLUSION

The highly orientated YBaCuO oxide superconductors were prepared by partial melt and slow cooling. We have measured magnetic properties. We recognized that the anisotropy of magnetization hysteresis in the low-field sweep is owing to microstructuresuch as grain boundaries and that in the high-field sweep is similar to that of single crystal. We have achieved to obtain the large magnetic hysteresis loop by controlling intra-grain granularity and crystal orientation. The largest Jc values estimated by using Bean's critical state model are 10000A/cm^2 at 1T [11].

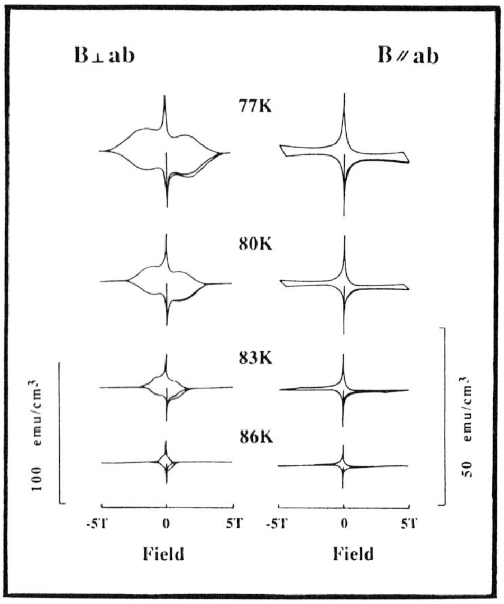

Fig.4 Magnetization hysteresis loops for the field perpendicular to ab-plane and for the field parallel to ab-plane.

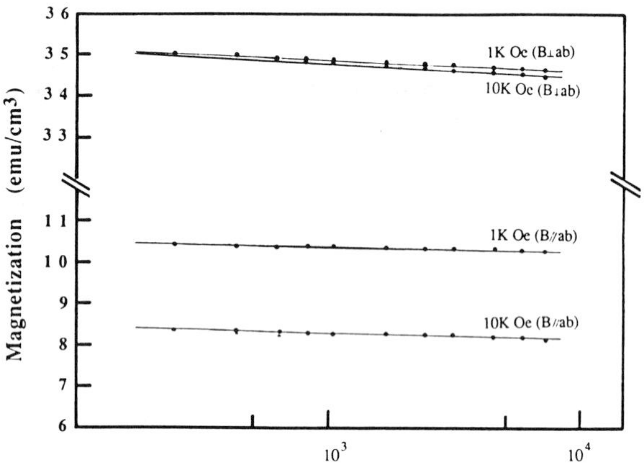

Fig.5 Magnetic relaxation for the field perpendicular to ab-plane and for the field parallel to ab-plane at 77K.

REFERENCES

1. Ekin JW (1987) Adv. Ceram. Mat. 2:586.
2. Yeshurun Y and Malozemoff AP (1988) Phys. Rev. Lett. 60: 2202.
3. Jin S, Tiefel TH, Sherwood RC, van Dover RB, Davis ME,Kammlott GW, Fsatnacht RA (1988) Phys. Rev. B 37:7850
4. Murakami M, Morita M, Miyamoto K (1989): Progress in High Temperature Superconductivity 15 (World Scientific, Singapore, 1989) p.95.
5. Mizuno F, Masuda H, Hirabayashi I, Tanaka S (1990) Phys. Rev. B 41: 7270
6. Daeumling M, Seuntjens JM, Larbalestier DC (1990) Nature 346: 332
7. Murakami M, Morita M, Doi K, Miyamoto K (1989) Jpn. J.Appl. Phys. 28: L1189
8. Li JN, De Boer FR, Roeland LW, Menken MJV, Kadowaki K, Menovsky AA, Franse JJM, Kes PH (1990) Physica C 169: 81
9. Murakami M, Morita M, Koyama N (1989) Jpn. J.Appl. Phys. 28: L1754
10. Campbell AM, Evetts JE (1972) J. Appl. Phys. 21: 199
11. Bean CP (1962) Phys. Rev. Lett.8: 250

Laser Zone Melting of $Bi_2Sr_2CaCu_2Li_{0.5}Al_yO_x$ Superconductors

K. Ishige[1], T. Suga[1,2], and Y. Shiohara[1]

[1] Superconductivity Research Laboratory, International Superconductivity Technology Center, 10-13, Shinonome 1-chome, Koto-ku, Tokyo, 135 Japan
[2] The Kansai Electric Power Co., Inc.

ABSTRACT

Quenched $Bi_2Sr_2CaCu_2Li_{0.5}Al_YO_X$(Y=0~0.35) ceramic precursors were remelted and directionally solidified by a laser surface zone melting method. The X-ray diffraction pattern of the quenched precursor sample shows a glassy characteristic. Samples traversed at the speed of 0.5 mm/s under continuous-wave laser beams which were focused through a set of lens on their surfaces. After the zone melting, the treated surfaces were characterized by an x-ray diffraction technique. It is shown that the $Bi_2Sr_2CaCu_2O_X$ phase was primarily observed in the samples containing over 0.2 aluminum.

KEY WORDS : high-Tc superconductor, laser, zone melting, Bi-Sr-Ca-Cu-O, doping

INTRODUCTION

Since the discovery of high-Tc superconductors[1,2], many researchers have tried to produce superconductors with high critical current densities by melting and solidification method[3-6]. In a laser zone melting method, a focused beam is scanned across a material to raise the temperatures of the surfaces locally above the melting point. Two potential advantages of this method are as follows. First, the high rate of energy supply enables very high rates of temperature increase to be attained which is followed by a similar rate of cooling due to self-quenching by heat conduction into the bulk material[7-10]. Second, extremely high thermal gradients can be attained because the heating effect is concentrated in a very thin and small region at the material surface.

Superconductive phases have been produced at very slow solidification rates less than several tens of mm/h, and then the solidification of those methods were restricted to occur under a local equilibrium condition. The rapid solidification process of laser zone melting is a method which enables achievement of controllable crystallization under the metastable state[11]. Though materials which have a complex crystal structure like $Bi_2Sr_2CaCu_2O_X$ require extensive local diffusional rearrangements to form the structure during the rapid solidification, it is possible to produce superconductors at high speed and also to prevent segregations which cause the weak links between the grains. Furthermore, in spite of rather rapid solidification, it is possible to form textured microstructures along the scanning direction because of the steep thermal gradient.

We reported that Bi 2212 phases were produced by the laser zone melting and solidification process at the range of cooling rate between 5 and 20×10^3K/s[12]. Within the condition of this cooling rate limitation, we chose high thermal gradient at the solid-liquid interface(G) and small solidification rate(R) so as to get high G/R value. We can expect textured microstructure because of the unidirectional solidification under high G/R condition. It was reported that aluminum[13] or lithium[14] doping improved the superconducting properties of the bulk materials. In this paper, we report the effects of aluminum and lithium doping on the laser processed samples.

EXPERIMENTAL PROCEDURE

Glassy $Bi_2Sr_2CaCu_2Li_{0.5}Al_YO_X$ precursors for the laser melting process with thickness of about 1mm were prepared by conventional splat quenching. Commercial powders of Bi_2O_3, $SrCO_3$,

$CaCO_3$, CuO, $LiCO_3$ and Al_2O_3 were mixed and calcined at 1100K for 12hrs. Subsequently, the calcined powders were melted in platinum crucibles at about 1450K for 5 mins in an electric furnace. The melts were poured onto a copper plate and pressed quickly. After the quenching samples were annealed at 623K for 2 hours in air to prevent the samples from cracking during further processes such as polishing and laser melting. Figure 1 shows the X-ray diffraction patterns of as annealed samples.There is no difference between the patterns of Y=0 and Y=0.35 samples and both patterns show a glassy characteristic.

Figure 2 shows a schematic illustration of the laser zone melting apparatus. The samples were placed on an enclosed hot stage in air. And the stage traversed under the beam of a YAG laser (CW TEM_{00} mode, wave length is 1.06μm) so as to move a molten zone. The beam was focused on the sample surface with a set of lens (f=100mm) to supply an approximate point heat source. Laser scanning velocity(scanning speed of samples, v), laser input power(Q) and sample temperature(Tsub.), i.e. hot stage temperature, were controlled as experimental parameters. In this experiment, v, Q and Tsub. were fixed at 0.5mm/s, 0.7W and 573K respectively to clear the effects of aluminum and lithium doping. Calculated G and cooling rate are 8.3K/μm and 4.2×10^3K/s at this condition[12]. Figure 3 shows effects of G, R and cooling rates on the phases appeared in the laser processed samples with nominal composition of $Bi_2Sr_2CaCu_2O_X$[12]. In the figure an asterisk indicates the condition of this experiment.

Laser processed samples were characterized by optical and scanning electron microscopy and electron probe microanalyzer (EPMA). Moreover, laser beams were scanned across all the surface of samples with constant intervals so as to get the sufficient intensity of X-ray diffraction for characterization.

RESULTS AND DISCUSSION

Figure 4 shows a typical optical micrograph of a laser processed surface. The width of melted zone is about 650μm. Figure 5 shows the X-ray diffraction patterns of laser processed samples with different aluminum contents. Bi 2212 or 2201 phase should appear in the sample with no aluminum (Y=0) according to figure 3, if lithium does not affect the phase

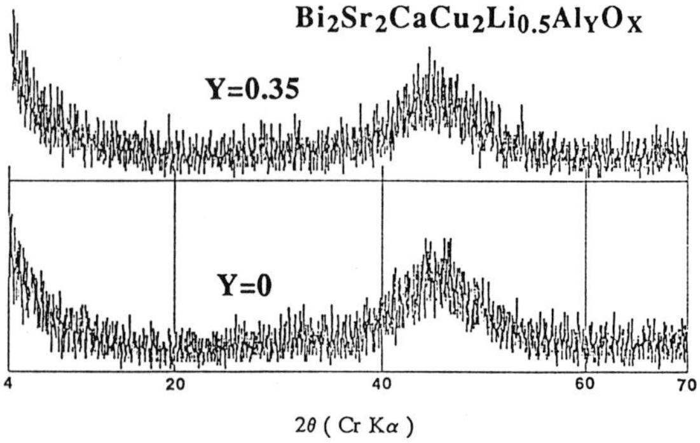

Fig. 1. The X-ray diffraction patterns of as annealed precursors with different aluminum contents.

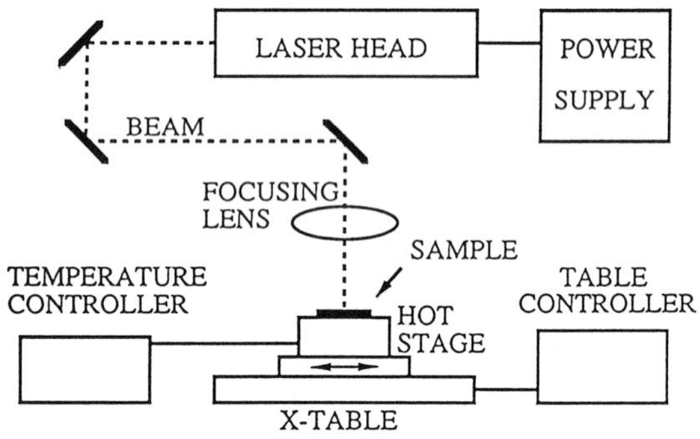

Fig. 2. Schematic of laser zone melting apparatus.

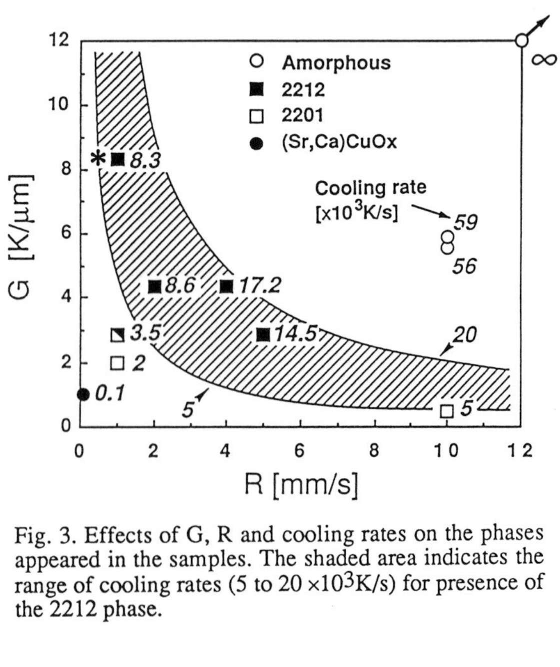

Fig. 3. Effects of G, R and cooling rates on the phases appeared in the samples. The shaded area indicates the range of cooling rates (5 to 20 $\times10^3$K/s) for presence of the 2212 phase.

stability. Thus it is clear that lithium stabilizes in a glassy state, because only a little crystallization occurs in the sample without aluminum. On the contrary, very sharp peaks of Bi 2212 phase are observed in the sample containing 0.35 aluminum. Thus, it is thought that aluminum stabilizes the 2212 phase in this laser process condition with the presence of 0.5 lithium. In addition, the intensity enhancement of the *(00l)* peak shows that c-axis is oriented normal to the surface.

Figure 6 shows the temperature dependence of resistivities for the laser processed $Bi_2Sr_2CaCu_2Li_{0.5}Al_{0.35}O_X$ sample. In the case of high measuring current (I=1 mA), the sample shows a semiconductive characteristic. In the case of low measuring current (I=0.01 mA), Tc-onset is observed at 60K. However, Tc-zero is not observed and semiconductive characteristic is still observed below the Tc-onset. This figure suggests that the linkage between superconductive phases is poor.

Fig. 4. Optical micrograph of laser processed surface (v=0.5mm/s, Q=0.7W, Tsub.=573K).

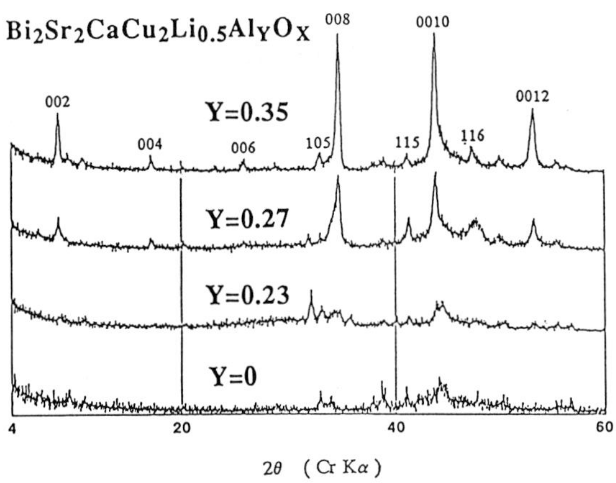

Fig. 5. The X-ray diffraction patterns of laser processed samples with different aluminum contents. (v=0.5mm/s, Q=0.7W, Tsub.=573K).

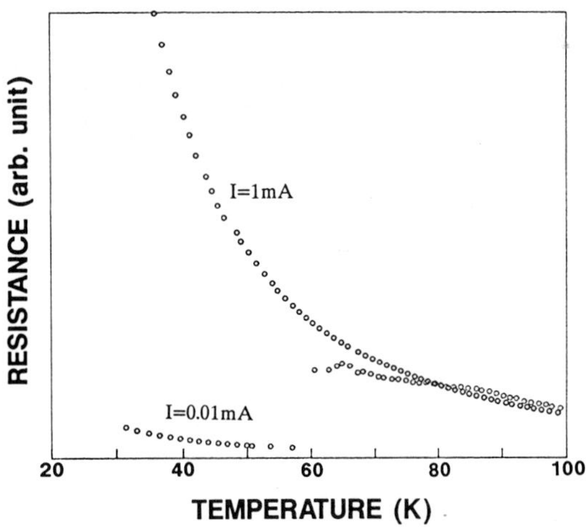

Fig. 6. Temperature dependence of resistivities with different measuring current for the laser processed $Bi_2Sr_2CaCu_2Li_{0.5}Al_{0.35}O_X$ sample.

Figure 7 (A) shows the topographic image of a laser processed sample produced by EPMA. The surface is relatively flat. However, it is shown that the distribution of chemical composition is quite heterogeneous from the reflected electron image of figure 7 (B). Figures 7 (C), (D), (E) and (F) show chemical distributions of Bi, Sr, Ca and Cu along the white line. At the dark regions which look like grain boundaries, strontium (Sr) is rich and copper (Cu) is poor compared to the bright region. Because of these segregations in spite of rather rapid solidification, it is thought that the linkage between superconductive phases is poor.

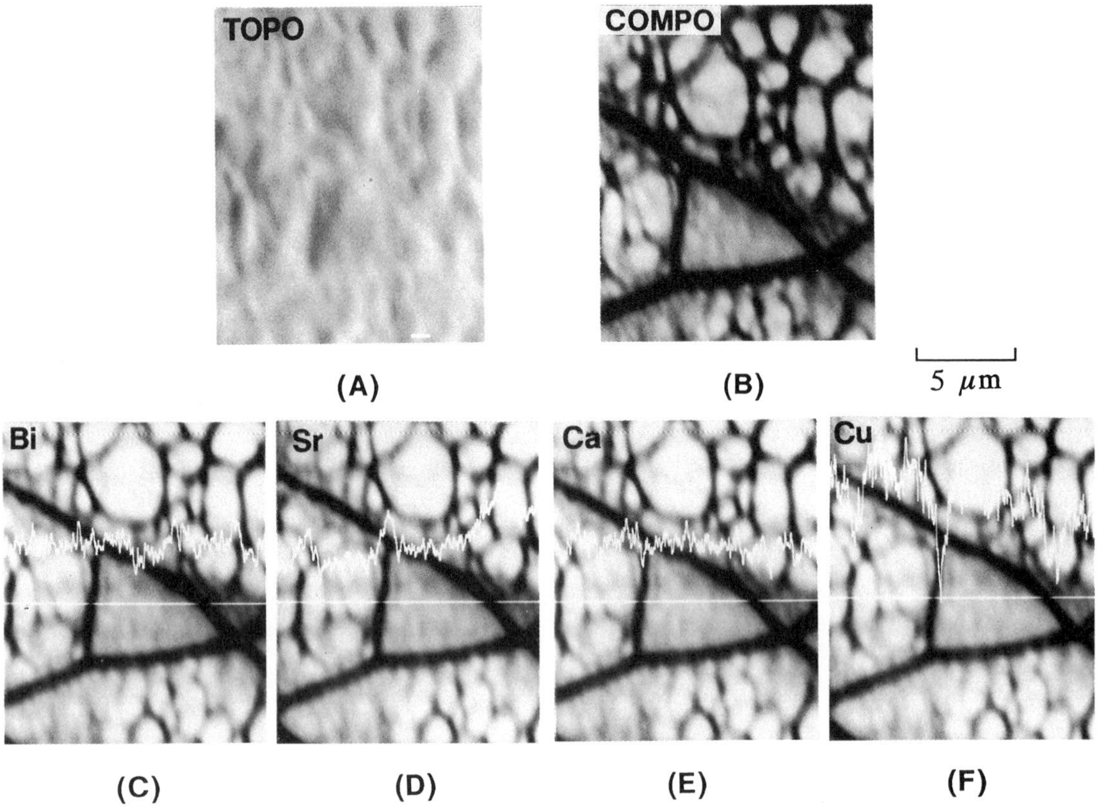

Figs. 7. EPMA images of the laser processed surface.((A) topographic image, (B) reflected electron image, (C),(D),(E) and (F) chemical distributions of Bi, Sr, Ca and Cu)

CONCLUSION

1. Effects of lithium or aluminum doping on the appearance of Bi 2212 phase in laser processed samples were investigated.
2. Bi 2212 phase was observed in $Bi_2Sr_2CaCu_2Li_{0.5}Al_{0.35}O_X$ sample by X-ray diffraction but Tc-zero was not observed because of poor linkage at grain boundary where segregation occurred.

REFERENCES

1. Bednorz JG, Muller KA (1986) Z. Phys. B64:189
2. Maeda H, Tanaka Y, Fukutomi M (1988) Jpn. J. Appl. Phys.Lett., 27:209
3. Jin S, Tiefel TH, Sherwood RC, Davis ME, van Dover RB, Kammlott GW, Fastnacht RA, Keiyh HD (1988) Appl. Phys. Lett. 52:2074
4. Feigelson RS, Gazit D, Fork DK (1988) J. Crstal Growth, 91:318
5. Kurosaka A, Aoyagi M, Tominaga H, Fukuda O, Osarai H (1989) Appl. Phys. Lett., 55(4), 24:390
6. Levinson M, Shah SSP,Wang DY (1989) Appl. Phys. Lett. 55(16): 1683
7. Anthony TR, Cline HE (1977) J. Appl. Phys., 48:3888
8. Cline HE, Anthony TR (1977) J. Appl. Phys., 48:3895
9. Sepold G (1968) Grundlagen fur das Schweißen mit strahlen hoher Intensitat, Dissertation TU Hannover
10. Bergmann HW, Hunger G, Fritsch H.U (1981) J. of Mat. Science 16:1935
11. Mordike BL, Bergmann HW (1984) Mat. Res. Soc. Symp. Proc., 28:45
12. Ishige K, Suga T, Shiohara Y (1989) Proc. of the 2nd Int. Symp. on Superconductivity : 321-324
13. Komatsu T, Sato R, Imai K, Matusita K,Yamashita T (1988) Jpn. J. Appl. Phys.Lett., 27:550
14. Kawai T, Horiuchi T, Mitsui K, Ogura K, Takagi S, Kawai S (1989) Physica C 161:561

Crystal Growth of Oxide Superconductors Prepared by the Bridgman Method

TERUO IZUMI and YUH SHIOHARA

Superconductivity Research Laboratory, International Superconductivity Technology Center,
10-13, Shinonome 1-chome, Koto-ku, Tokyo, 135 Japan

ABSTRACT

The effects of temperature gradient (G) and growth rate (R) on the microstructure and characteristics of Y-based superconductors grown from the partially molten state by the horizontal Bridgman method were studied. The structure showed highly aligned $YBa_2Cu_3O_y$ crystals containing non-aligned finely dispersed Y_2BaCuO_5 crystals for high G/R ratio. Coarsening of the $YBa_2Cu_3O_y$ crystals was found to occur as GR was decreased. A sample grown under conditions of high G/R ratio (482 Kh/cm^2) and low GR (5.8 K/h) had a critical current density of 19000 A/cm^2 at 77K in a magnetic field of 1T.

KEY WORD; crystal growth, Bridgman method, YBCO-system, melting process, critical current density

INTRODUCTION

Following the discovery of high-Tc superconducting oxides[1)2)3)], a number of studies have been carried out aimed at obtaining higher critical current density for various applications. Murakami et al.[4)] reported the highest critical current density (30000 A/cm^2 at 77K in 1T) in bulk materials by using a melt processing method (MPMG method). In order to expand development in applications, however, it will be necessary to align fibers continuously. In the Bi-based system, Hayashi[5)] reported critical current density as high as 3070 A/cm^2 (77K,0T) using the laser pedestal growth method. In a previous paper[6)], the effects of G/R and GR were reported on Bi-system crystals prepared by the vertical Bridgman method. Later Cima et al.[7)] reported the similar results using the laser-heated floating zone method. In the Y-based system, Kase[8)] studied applications of the unidirectional melting method and R.L.Meng[9)] further investigated the effect of cooling rates. For more advances, detailed investigations of the mechanism of crystal growth in directional solidification is required.

In this paper , the effects of processing parameters (G and R) on solidification from partial melting in the YBCO-system were investigated by the horizontal Bridgman method, which allows G and R to be varied independently.

EXPERIMENTAL

The experimental apparatus is shown schematically in Figure 1 ; further details of construction were given in the previous paper[6)]. The mixed powders, in a composition ratio of Y:Ba:Cu=1.2:2.1:3.1, were calcined at 900°C for 12 hours and then melted at 1400°C for 10 minutes followed by quenching. The quenched lumps were ground and pressed into rods (5mm in diameter and 100mm in length) by cold isostatic pressing at 2000 atm. Processing was begun by loading a rod into a crucible (Y_2O_3-stabilized ZrO_2) and then heating to 1050°C for partial melting. The crucible was then pulled out of the furnace under predetermined conditions of temperature gradient (G) and growth rate (R). The various conditions used in this work are listed in Table 1. The microstructures were studied by optical microscopy and electron probe microanalysis (EPMA). Superconducting characteristics were investigated by vibrating sample magnetometry (VSM) at a temperature of 77K.

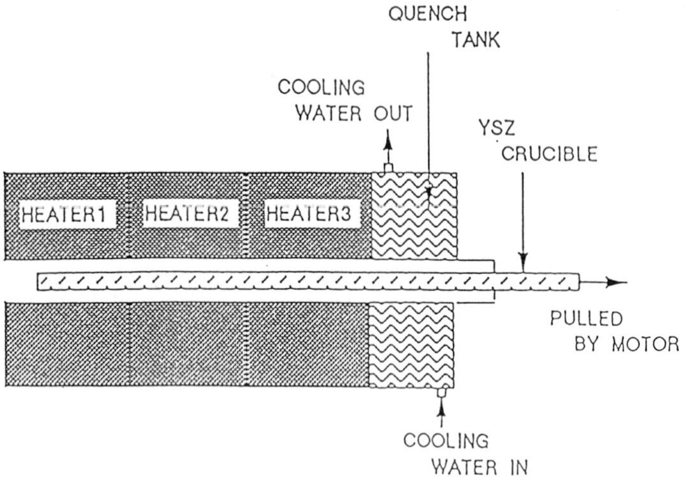

Figure 1 A schematic of the growth apparatus

Table 1 The used conditions of crystal growth

Sample	G (K/cm)	R (cm/h)	G/R	GR
A-1	53	0.11	482	5.8
A-5	53	0.50	106	26.5
A-10	53	1.02	52	54.1
B-2	26	0.23	113	6.0
B-10	26	1.02	25	26.5
B-20	26	2.05	13	53.3

RESULTS AND DISCUSSION

1. THE EFFECT OF G/R

In general, unidirectionality of the crystals depends strongly on the G/R ratio. At higher values of G/R, more highly aligned grain structures were obtained. Because a constitutionally supercooled region hardly existed in front of planar interface at higher G/R, solidification could proceeds only with a planar liquid-solid interface.

Figure 2 shows the microstructures of two samples grown at different G/R, but with the same GR i.e. with the same cooling rate. The higher G/R sample , A-1 , shows strongly aligned grains along the growth direction. Figure 3 shows that the grains consist of a $YBa_2Cu_3O_y$ matrix containing fine Y_2BaCuO_5 crystals, as determined by EPMA.

On the other hand, the grains of the lower G/R sample, B-2 , were not continuous at the liquid-solid interface. This means that the G/R value was too low to maintain unidirectional solodification with a planar interface.

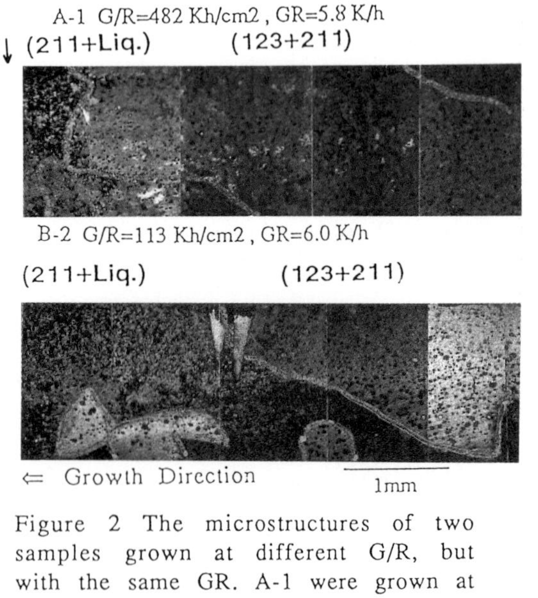

Figure 2 The microstructures of two samples grown at different G/R, but with the same GR. A-1 were grown at G/R equal to 482 Kh/cm^2 and B-2 were at 106 Kh/cm^2. Observed grain boundaries are marked by solid lines (white).

In order to further clarify the differences in grain alignment in these two samples, magnetic properties were investigated by VSM where the directions of the applied magnetic fields were changed as shown in Figure 4. Because the grains within the samples are aligned, it is expected that the magnetization values will be different for the two orientations. As the degree of alignment of the grains increases, the difference in the magnetizations will be increased. The largest and the smallest magnetization curves of for the sample A-1 are shown in Figure 5. The ratio of the magnetization was 4.0 at an applied field of 1T. On the other hand, sample B-2, grown under the lower G/R condition, showed a magnetization ratio of 2.5 at 1T, indicating a lesser degree of grain alignments. The magnetization curves for this sample are shown in Figure 5. These results indicate that the degree of grain alignment depends on the G/R ratio and that a microstructure with more aligned grains was obtained with the higher G/R ratio.

The Bean model was used to calculate Jc values, taking the sample thickness for "d". The A-1 sample had a Jc of 19000 A/cm^2 at 1T(77K) when measured with the sample orientation that produced the highest value of magnetization. The corresponding value for the B-2 sample (lower G/R) was 9380 A/cm^2.

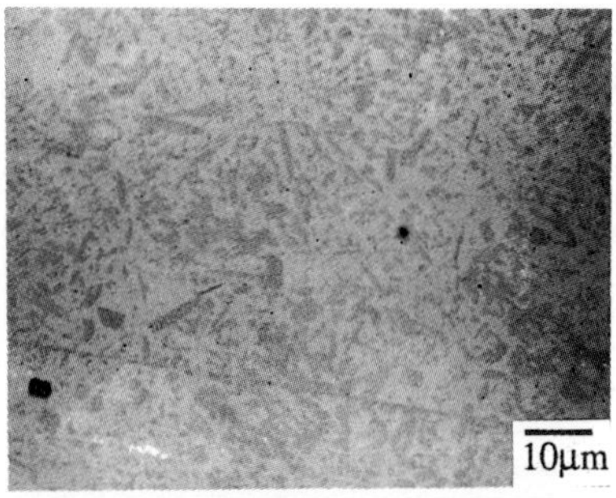

Figure 3 The grains consist of a $YBa_2Cu_3O_y$ matrix containing fine Y_2BaCuO_5 crystals.

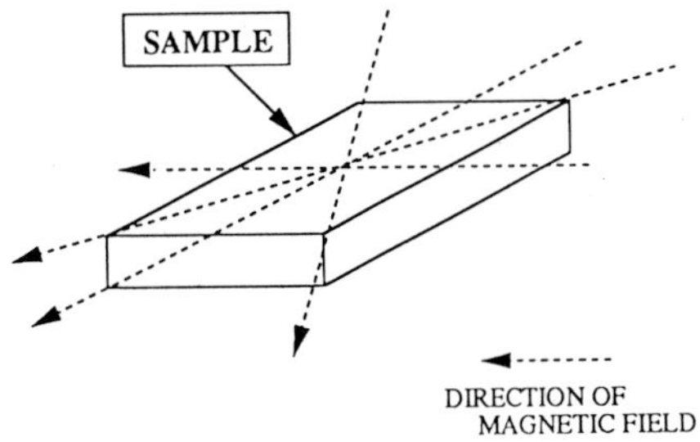

Figure 4 The directions of magnetic field in the measurements of VSM

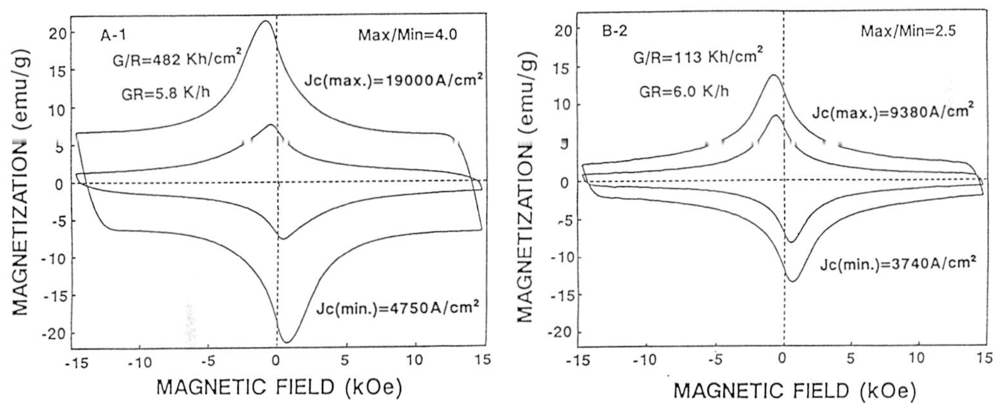

Figure 5 Magnetization hysterisis loops of two samples grown at different G/R. The curves pertaining to max. and min. magnetization loops are as andicated.

2. THE EFFECT OF GR

In directional solidification processing, the GR value gives the local cooling rate at the interface, since G and R correspond to dT/dx and dx/dt, respectively.

Figure 6 shows the microstructures of the A-5 sample grown with GR equal to 26.5 K/h comparing with the B-2 sample grown with GR equal to 6.0 in Figure 2. For both samples, the G/R ratios were the same, 106 Kh/cm^2. The grains of A-5 were smaller than those of B-2 and so Jcmax of A-5 was smaller than that of B-2 as shown in Figure 7. This means that coarsening of the microstructure occurred for low cooling rate.

A-5 G/R=106 Kh/cm^2, GR=26.5 K/h

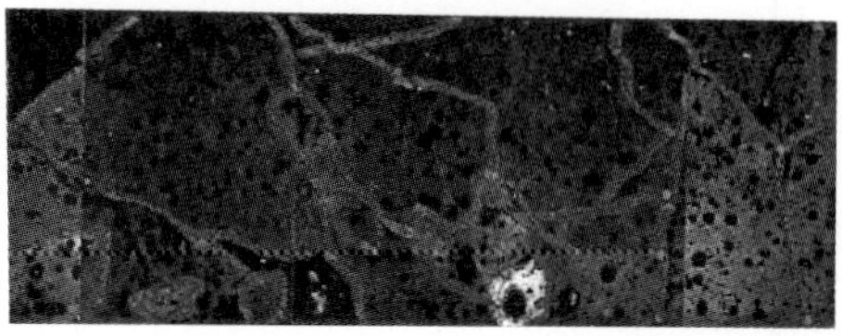

⇐ Growth Direction 1mm

Figure 6 The microstructure of A-5 grown at high cooling rate

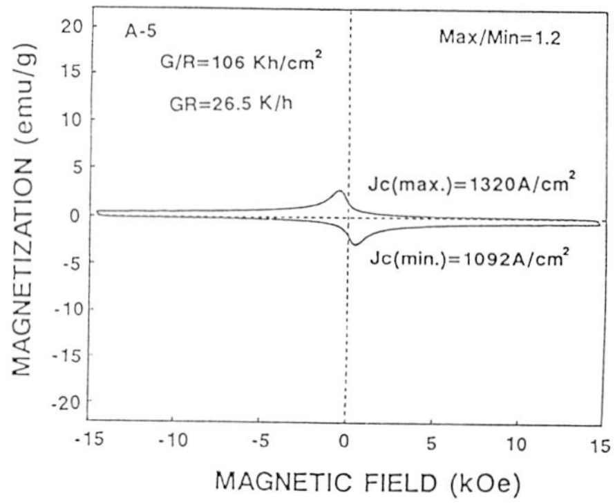

Figure 7 Magnetization hysterisis loop of A-5

CONCLUSION

Highly textured, large $YBa_2Cu_3O_y$ grains , containing fine Y_2BaCuO_5 within the $YBa_2Cu_3O_y$ grains were grown by the horizontal Bridgman method. Under the conditions of G/R of 482 Kh/cm^2 and GR of 5.8 K/h, a sample with Jc of 19000 A/cm^2 (1T,77K) was produced. The orientation of $YBa_2Cu_3O_y$ crystals was found to depend strongly on the G/R value, showing an increase in alignment for higher G/R value. The limited G/R value for continuous solidification was inferred to be between 106 and 482 Kh/cm^2. Coarsening of the microstructures was found to occur for low cooling rate.

In summary, higher G/R and lower GR were important for obtaining high-Jc textured samples.

This work was supported by New Energy and Industrial Technology Development Organization under the management of the R&D of Basic Technology for Future Industries.

REFERENCES

1)J.G.Bednorz and K.A.Muller, Z.Physik B64 (1986) 189

2)C.W.Chu, P.H.Hor, R.L.Meng, L.Gao, Z.J.Haung, and Y.Q.Wang, Phys.Rev.Lett. 58 (1987) 405

3) H.Maeda, Y.Yanaka, M.Fukutomi, and T.Asano, Jpn.J.Appl.Phys. 27 (1988) 209

4)M.Murakami, T.Oyama, H.Fujimoto, S.Gotoh, K.Yamaguchi, Y.Shiohara, N.Koshizuka, and S.Tanaka, to be published on IEEE Trans. MAG

5)K.Hayashi, H.Nonoyama, N.Nagata, H.Hitotsuyanagi, and M.Kawashima, Proc. of 2nd Int. Symp. on Superconductivity (1989) 309

6)T.Izumi, T.Oyama, and Y.Shiohara, Proc. of 2nd Int. Symp. on Superconductivity (1989) 289

7)M.J.Cima, X.P.Jiang, H.M.Chow, J.S.Haggerty, and M.C.Flemings, J.Mater.Res. 5 (1990) 1834

8)J.Kase, J.Shimoyama, E.Yanagisawa, S.Kondoh, T.Mastubara, T.Morimoto, and M.Suzuki, Jpn.J.Appl.Phys. 29 (1990) 277

9)R.L.Meng, C.Kinalidis, Y.Y.Sun, L.Gao, Y.K.Tao, P.H.Hor, and C.W.Chu, Nature 345 (1990) 326

Superconducting Textured Crystal Produced by a Partial Melt Process

TAMOTSU NODA and YUH SHIOHARA

Superconductivity Research Laboratory, International Superconductivity Technology Center, 10-13, Shinonome 1-chome, Koto-ku, Tokyo, 135 Japan

ABSTRACT

A partial melt process is one of the processes useful for obtaining textured crystals. The sample was pressed at above the solidus temperature (where solid and liquid phases coexist). When the solid phase is a superconductive one, superconducting textured crystals can be obtained by squeezing out the liquid phase through the partial melt process. The texture, phase, critical temperature (Tc) and critical current (Jc) of the specimens were investigated. As an example, a sample pressed at 850°C with nominal composition of Bi:Sr:Ca:Cu:O=2:1.6:0.8:1.9:y is in a highly textured manner. Three phases, Bi2212 (low Tc Phase), Bi2201, and CuO were observed in the sample. The values of Tc and Jc of this sample were 60K and 740A/cm^2 at 30K, respectively.

KEY WORDS:textured crystal, partial melt , Bi-Sr-Ca-Cu-O, Jc

INTRODUCTION

Since high Tc oxides were discovered, many efforts have been made to improve the Jc . The Jc of the bulk oxide superconductor is lower than that of thin film or single crystal. In bulk superconductors , it is considered that the agreement of non-textured crystals reduces the critical current density. Therefore, the production of textured crystals is one of the most important factors for producing bulk material with superior properties . Experimental results using several different processes for obtaining textured crystals have been reported , including the doctor-blade method ,the hot-pressing method (prepared by solid state reaction) ,and the Ag-sheathed method. We applied the partial melt process to obtain the textured crystals.

EXPERIMENTAL

Table-1 shows the procedure of a partial melt process. In this work, we prepared two types powder. One sample with starting nominal composition was Bi:Sr:Ca:Cu=2:1.6:0.8:1.9 (labelled type A powder.). Powders of Bi2O3, SrCO3, CaCO3 and CuO were mixed,and then melted at 1400°C in air for 10minites. The melt was quenched by copper plate, and then ground. Type B powder cosisted of silver (10wt%) plus type A powder. These two types of powder (types A and B) were pressed to form pellets (diameter=20mm,hight is about 3mm) by applying 1.6ton/cm2 pressure. We used a hot-press furnace as shown schematically in Fig-1. The sample was put between the upper and lowew MgO plates. The sample was surrounded by a cylinder with a piston in it. Table-2 shows experimental parameters, reaction temperatures, nominal composition, and pressing or non-pressing. Fig-2 shows the temperature control program and pressing procedure. The samples were heated to each reaction temperature (840, 850, 860, 870) and held for 5 hours. After this reaction process, these samples were pressed to one half of the original thicness. The press speed was 2Kgf/sec.The Tc values of all the samples were measured by a conventional D.C. four-probe method as a function of temperature. Jc values of some of the samples were measured by pulse D.C. four-probe method under zero magnetic field. The surface of the samples were analysed by EPMA. The structures of the samples were analysed by XRD.

EXPERIMENTAL RESULTS

Fig-3 shows REF(Refracted electron image) micrograph of cross section of the hot-pressed "850A" sample. The crystals were textured and separated into three phases;2212,2201 and CuO. Composition of phases which were analysed by EPMA. X-ray diffraction analysis was carried out for all samples by using Cr K radiation. The results for "850A" and "850C" are shown in Fig-4. The (00l) reflection peaks of "850A" sample are strong,and thus the crystal was textured. In type A samples, the higher the temperature during reaction, the better was the texture. Howevere in type B samples, the lower the temperature during reaction, the better was the texture. The reason was that the silver in the type B sample melted at the reaction temperature. We think that too much liquid phase changed uni-axial pressure to isostatic pressure.The Tc of the "850A" sample was 71K, but Tc's of the other samples were below 21K. (shown in Table-3) In case of the "850A" sample, contacted to the phase was made within the partial melt stage. The resistivity-temperature curve for the "850A" sample and "850C" sample are shown in Fig-5. The resistivity of "850A" behaved a 2212 phase, but "850C" has two peaks. the higher one is caused by 2212 phase. The Tc of the "850C" sample was 18K, which was the result of 2201 phase. The Jc of the "850A" sample is 740A/cm^2 at 30K.

CONCLUSION

Superconducting textured crystal of the Bi-system was produced by a partial melt process.A Jc value of 740A/cm^2 at 30K was obtained. Jc was lower than the other bulk processing methods, possibly because there was too much Bi2201 and CuO precipitate in the superconducting phase.

Table-1:Experimental procedure

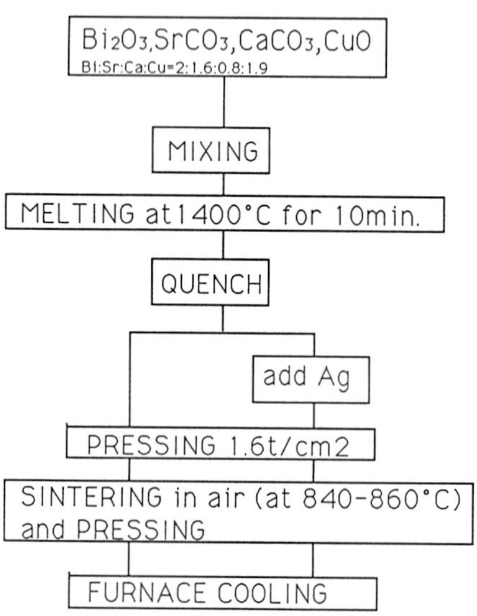

Table 2:Experimental paramater

Temp. °C \ type	A	B	C
840	840A	840B	
845		845B	
850	850A	850B	850C
855	855A		
860	860A		

A:type A samples+hot pressing
B:type B samples+hot pressing
C:type A samples

Table 3 Tc of samples.

Temp·C \ type	A	B	C
840	21K	NS	
845		NS	
850	71K	NS	18K
855	17K		
860	NS		

A:type A samples+hot-pressing
B:type B samples+hot-pressing
C:type A sample
NS:Tc is nothing or below 16K.

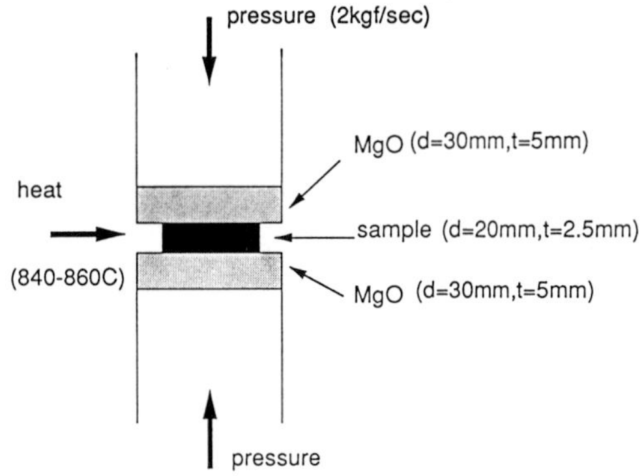

Fig.1 Schematic drawing of a hot-press furnace.

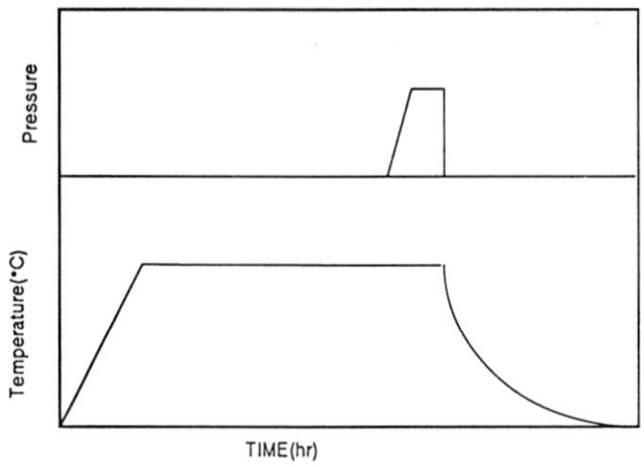

Fig.2 Aheating and pressing program

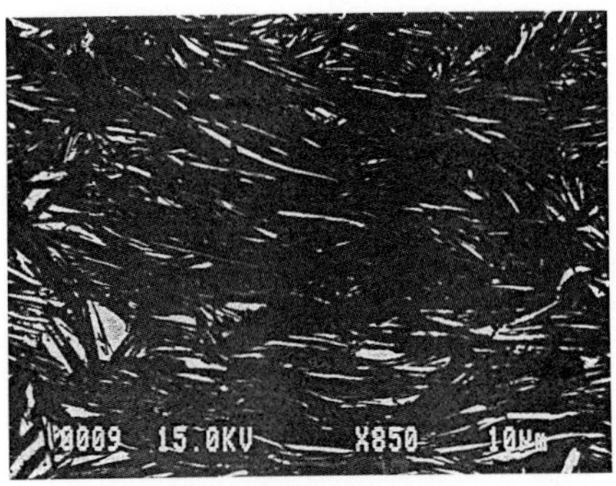

Fig.3 Refracted electron image of cross section of the hot-pressed "850A" sample.

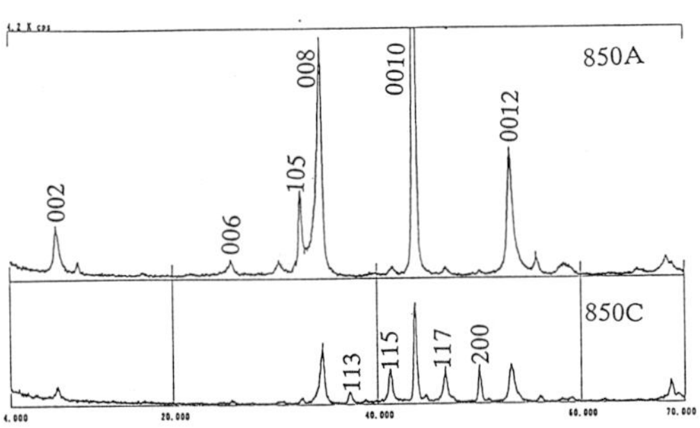

Fig.4 XRD patterns of "850A" and "850C" samples.

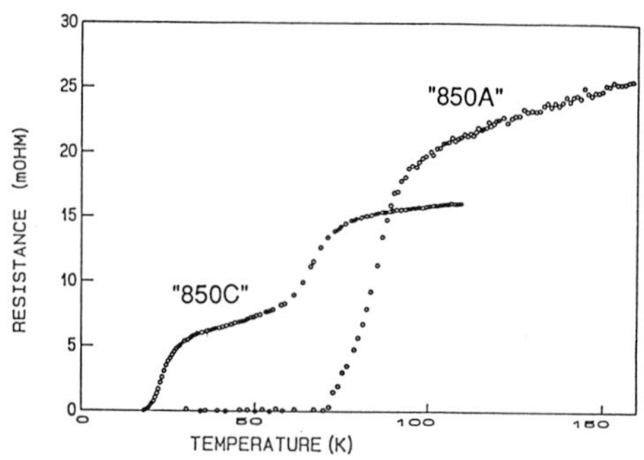

Fig.5 Temperature dependence of resistivity for "850A" and "850C" samples.

Preparation Conditions of 80K-Phase Bi (Pb)-Sr-Ca-Cu-O Single Crystals

WATARU FUTO, SATORU KISHIDA, HEIZO TOKUTAKA, MAKOTO CHIHAYA, KATSUMI NISHIMORI, and NAGANORI ISHIHARA

Department of Electrical and Electronic Engineering, Tottori University, Koyama, Tottori, 680 Japan

ABSTRACT

The 80K-phase Bi(Pb)-Sr-Ca-Cu-O single crystals were prepared by the self-flux method, and the optimum composition of the starting material, which was (Bi,Pb):(Sr,Ca):Cu=1:1:1, Ca/Sr=1 and Pb/(Bi+Pb)=0.15, was clarified. The prepared single crystals were plate-like and their typical sizes were 7×7 mm^2. The critical temperature (Tc) of the specimen was ~74K.

KEY WORDS : Bi-system superconductors, single crystal, preparation condition

INTRODUCTION

Bi-Sr-Ca-Cu-O (Bi system) has three superconducting phases of $Bi_2Sr_2Cu_1O_y$(7K-phase), $Bi_2Sr_2Ca_1Cu_2O_y$(80K-phase) and $Bi_2Sr_2Ca_2Cu_3O_y$(110K-phase) [1]. Therefore, Bi-system superconductor is a very interesting material from a physical and applicative point of view. The ceramics and thin films, which were nonoriented and polycrystalline, have been used in the studies of the physical property of the Bi-system superconductor. Since the ceramics and films often contain the impurity phase, the properties of Bi-system superconductors can not exactly be determined. Therefore, it is nescessary to prepare large and high quality Bi-system single crystals in order to investigate the physical properties.

In this study, we attempted to grow the Bi-system 80K-phase single crystals using the self-flux method and clarified the optimum composition of the starting materials.

EXPERIMENTAL

80K-phase Bi-system single crystals were grown by the self-flux method. The starting materials with various metal ratios were mixed using Bi_2O_3, PbO, $SrCO_3$, $CaCO_3$ and CuO powders. They were heated in the furnace at 1000℃ for 36h, then, cooled to 800℃ at rate 3℃/h, followed by cooling to 600℃ at rate of 40℃/h, finally to room temperature in the furnace. The single crystals were taken out from the alumina crucible by destroying it mechanically.

The resistance-temperature (R-T) characteristics were measured by conventional four-probe method. The Au electrodes were formed by the resistive heating method on the crystal surfaces and then, Cu wires were fixed using Ag paste on the Au electrodes. The X-ray diffraction (XRD) measurements were carried out on the single crystals cleaved using the tape.

RESULTS AND DISCUSSION

Figure 1 shows (Bi,Pb)-(Sr,Ca)-Cu compositional diagram. The numerical values in fig.1 show the compositions of starting materials. In fig.1, the large single crystals were not

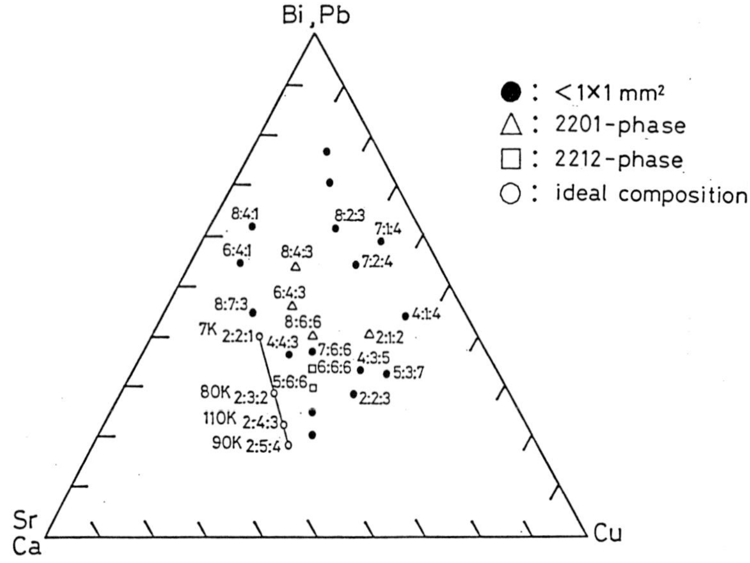

Fig. 1 (Bi,Pb):(Sr,Ca):Cu compositional diagram showing the composition of the starting materials.

obtained at the starting material compositions of closed circles (●) and their dimensions were always less than 1×1 mm^2. Triangles (△) and squares (□) indicate the presences of 7K-phase and 80K-phase single crystals with the dimensions (>1×1 mm^2), respectively, and open circles (○) show the ideal compositions of Bi-system superconductor. The large 7K-phase single crystals were obtained in the wide region of the composition of the starting materials. However, the large 80K-phase single crystals were obtained in the narrow region of the composition, as shown in fig.1. From the results, the optimum composition of the starting materials, where the large 80K-phase single crystals were obtained, were (Bi,Pb):(Sr,Ca):Cu = 6:6:6 and 5:6:6, and different from the ideal composition of the 80K-phase single crystals, (Bi,Pb):(Sr,Ca):Cu=2:3:2. This indicates that excess Bi and Cu are needed to prepare the large 80K-phase single crystals as flux material. When the 80K-phase single crystals were prepared using (Bi,Pb):(Sr,Ca):Cu=6:6:6, their crystal size was 5×5 mm^2, while in the case of (Bi,Pb):(Sr,Ca):Cu=5:6:6, the

Table 1 The superconducting phase and size of the Bi-system single crystals prepared as a function of the Ca/Sr ratios in the starting material of $Bi_{1.75}Pb_{0.25}Sr_xCa_yCu_2O_z$.

composition Bi:Pb:Sr:Ca:Cu	Ca/Sr ratio	superconducting phase	size of single crystal
1.75:0.25:2.0 :0.0 :2	0.00	—	—
1.75:0.25:1.8 :0.2 :2	0.11	7 K	8×3 mm^2 plate-like
1.75:0.25:1.6 :0.4 :2	0.25	7 K	7×6 mm^2 plate-like
1.75:0.25:1.4 :0.6 :2	0.43	7 K	7×6 mm^2 plate-like
1.75:0.25:1.2 :0.8 :2	0.67	80 K	3×3 mm^2 plate-like
1.75:0.25:1.0 :1.0 :2	1.00	80 K	7×7 mm^2 plate-like
1.75:0.25:0.95:1.05:2	1.11	80 K	5×5 mm^2 plate-like
1.75:0.25:0.9 :1.1 :2	1.22	80 K	3×3 mm^2 plate-like
1.75:0.25:0.8 :1.2 :2	1.50	80 K	1×1 mm^2 plate-like
1.75:0.25:0.6 :1.4 :2	2.33	—	—

crystal size was 3× 3 mm^2. From the results, we concluded that the optimum composition of the starting materials should be (Bi, Pb):(Sr, Ca):Cu=6:6:6, that is, 1:1:1.

When the starting material composition was (Bi, Pb):(Sr, Ca):Cu=1:1:1, the optimum Ca/Sr ratio was clarified. Table 1 shows the results prepared as a function of the Ca/Sr ratio in the starting material of $Bi_{1.75}Pb_{0.25}Sr_xCa_yCu_2O_z$. When the Ca/Sr ratio was less than 0.43, 7K-phase crystals were obtained. On the other hand, when the Ca/Sr ratio were more than 1, the 80K-phase single crystals were obtained and their sizes became small with increasing the Ca/Sr ratio. When the Ca/Sr ratios were larger than 1.5, the dimensions of the prepared single crystals were always less than 1× 1 mm^2. The results indicate that the Sr/Ca ratios considerably influence the superconducting phases as well as dimensions of the prepared single crystals. When the single crystals were prepared using the starting materials with the Sr/Ca ratio of 1.0, the largest 80K-phase plate-like single crystal with the dimension of 7× 7 mm^2 was obtained.

Table 2 The superconducting phase and size of the Bi-system single crystals prepared as a function of the Pb substitution ratios in the starting material of $Bi_{2-x}Pb_xSr_1Ca_1Cu_2O_y$.

composition Bi:Pb:Sr:Ca:Cu	Pb substitution ratio*	superconducting phase	size of single crystal
2.0 :0.0 :1:1:2	0.0	80 K	5× 5 mm^2 plate-like
1.9 :0.1 :1:1:2	0.05	80 K	3× 6 mm^2 plate-like
1.8 :0.2 :1:1:2	0.1	80 K	3× 4 mm^2 plate-like
1.75:0.25:1:1:2	0.125	80 K	7× 7 mm^2 plate-like
1.7 :0.3 :1:1:2	0.15	80 K	7× 7 mm^2 plate-like
1.6 :0.4 :1:1:2	0.20	—	2× 2 mm^2 plate-like
1.2 :0.8 :1:1:2	0.40	80 K	2× 2 mm^2 plate-like
0.8 :1.2 :1:1:2	0.60	—	—

*:Pb substitution ratio is defined as Pb/(Bi+Pb).

Next the optimum Pb substitution ratio, which was defined as R=Pb/(Bi+Pb), was investigated. Table 2 shows the results prepared as a function of the subsutitution ratio R in the starting material of $Bi_{2-x}Pb_xSr_1Ca_1Cu_2O_y$. When the subsutitution ratio R was more than 0.2, the demensions of the prepared single crystals became vary small, and in the case of R=0.60, large size single crystals could not be obtained. When R was 0.125 and 0.15, the large plate-like 80K-phase single crystals were obtained and their typical sizes were 7× 7 mm^2. Although the size of crystals prepared using the substitution ratio of R=0.125 was the same as that using the ratio of R=0.15, the volume fraction of the single crystals occupied in the crucible due to the latter was more than that due to the former. From the results, we concluded that the optimum Pb subsutitution ratio was 0.15. Therefore, the optimum composition of the starting materials for obtaining the large 80K-phase Bi-system single crystals was Bi:Pb:Sr:Ca:Cu=1.7:0.3:1:1:2.

Figure 2 shows the XRD pattern of the 80K-phase single crystal prepared using this starting material. Figure 3 shows the resistance-temperature characteristics of the single crystal. From the result of the XRD pattern, the cleaved surface was a-b plane and the single crystal contained almost no 7K-phase. The prepared single crystals showed the metallic temperature dependance of the resistance in normal state and the zero-resistance at 74K.

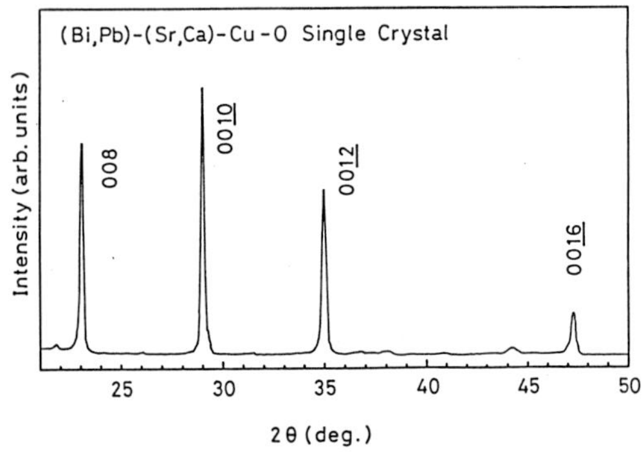

Fig. 2 X-ray diffraction pattern of the 80K-phase Bi-system single crystal prepared using the starting material of Bi:Pb:Sr:Ca:Cu=1.7:0.3:1:1:2.

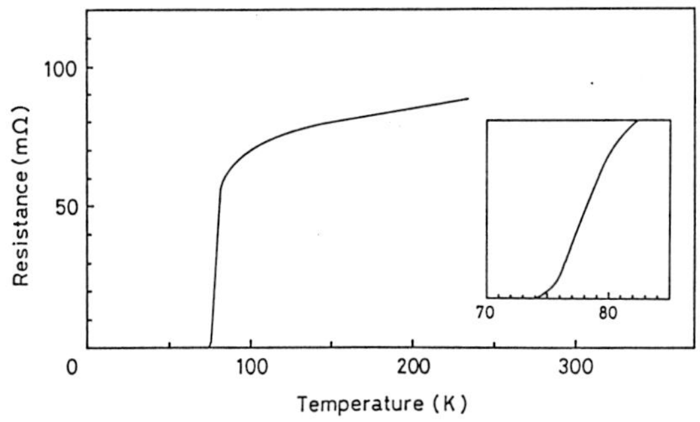

Fig. 3 The resistance-temperature characteristics of the 80K-phase Bi-system single crystal prepared using the starting material of Bi:Pb:Sr:Ca:Cu=1.7:0.3:1:1:2.

In summary, 80K-phase single crystals were prepared by the self-flux method. The optimum composition of the starting materials was (Bi,Pb):(Sr,Ca):Cu=1:1:1, Sr/Ca=1 and Pb/(Bi+Pb)=0.15, that is, Bi:Pb:Sr:Ca:Cu=1.7:0.3:1:1:2. The 80K-phase single crystals prepared using this composition were plate-like and their typical sizes were 7×7 mm^2.

REFARENCES

1. Maeda H, Tanaka Y, Fukutomi M, Asano T (1988) Jpn.J.Appl.Phy.27: L209

Preparation and Properties of the High-T_c $Bi_2Sr_2Ca_2Cu_3O_{10}$ Whiskers and Crystals

ICHIRO MATSUBARA[1], HIDEO TANIGAWA[1], TORU OGURA[1], HIROSHI YAMASHITA[1], MAKOTO KINOSHITA[1], and TOMOJI KAWAI[2]

[1] Government Industrial Research Institute Osaka, Ikeda, Osaka, 563 Japan
[2] The Institute of Scientific and Industrial Research, Osaka University, Ibaraki, Osaka, 567 Japan

ABSTRACT

Whiskers of the high-Tc $Bi_2Sr_2Ca_2Cu_3O_{10}$ (2223) phase have been prepared by annealing $Bi_2Sr_2CaCu_2O_8$ (2212) whiskers in a $Bi_2Sr_2Ca_4Cu_6Pb_{0.5}O_x$ powder. By the same method, single crystals of the 2212 phase have also been converted to the 2223 phase, resulting in 2223 sheet crystals as large as approximately $1mm^2$. In the case of the $Bi_2Sr_2CuO_x$ (2201) single crystals as a starting material, the 2223 phase has been partially formed after annealing in the powder, because the rate of the phase conversion is slower than that for the 2212 phase. A zero-resistance temperature of 107K and a maximum Jc value of $7.3x10^4A/cm^2$ at 77K in a zero magnetic field were obtained for the 2223 whisker. The 2223 whisker shows Jc values above $1x10^4A/cm^2$ up to 1T at 77K.

KEY WORDS: oxide superconductor, $Bi_2Sr_2Ca_2Cu_3O_{10}$ phase, whisker, critical current density,

INTRODUCTION

For the Bi-system, three types of superconducting phases are known; $Bi_2Sr_2CuO_x$ (2201 phase)[1], $Bi_2Sr_2CaCu_2O_x$ (2212 phase)[2], and $Bi_2Sr_2Ca_2Cu_3O_x$ (2223 phase)[2], with superconducting transition temperatures (Tc) around <20K, 80K and 110K, respectively. Powders of each of these phases have been synthesized by controlling the initial composition and the heating conditions. Single crystals of the 2201 phase and the 2212 phase have been obtained by a self-flux method[3], by using alkali-halide fluxes[4], and by the floating zone method[5]. On the other hand, large crystals of the 2223 phase have not been previously obtained in spite of their great usefulness in revealing the fundamental properties of this phase. We have previously reported the formation of flexible Bi-Sr-Ca-Cu-O superconducting whiskers which are composed of several 2212 platelike single crystals stacked in a layered structure[6,7].

In this paper, we report on the formation of the 2223 whiskers which are obtained by annealing these 2212 whiskers in a calcined $Bi_2Sr_2Ca_4Cu_6Pb_{0.5}O_x$ powder. For the 2223 whisker, a Tc_{zero} value of 107K and a maximum Jc value of $7.3x10^4A/cm^2$ at 77K in a zero magnetic field were obtained . We have also applied this method to 2212 and 2201 single crystals grown by the self-flux method. In the case of the 2212 single crystals, we have obtained 2223 sheet crystals as large as 1mmx1mm in area while retaining the outline of the form in the original crystals. However, fully converted 2223 sheet crystals have not been obtained from the 2201 single crystals even in the same annealing condition. In this case, X-ray diffraction pattern of the annealed samples shows the mixed phase of the 2223 phase and the original 2201 phase.

EXPERIMENTAL

Three types of samples, 2212 whiskers, 2212 single crystals, and 2201 single crystals were used as starting material. The 2212 whiskers with the dimensions of 1-5um thick, 30-500um wide, and 1-15mm long were prepared by heating glassy melt-quenched plates in a stream of O_2 gas, as previously described [6,7]. Single crystals of the 2212 phase and the 2201 phase were prepared by the flux method. Bi:Sr:Ca:Cu=2:1:1:2 for the 2212 phase or Bi:Sr:Cu=2:1:1 for the 2201 phase oxide or carbonate powders were heated in an alumina crucible at 1000°C, held at this temperature for 10h, cooled to 800°C at a rate of 5°C/h, cooled to 600°C at 40°C/h and finally cooled to a room temperature in the furnace. The whiskers and single crystals of the 2212 or the 2201 phase were placed in a powder with the composition of $Bi_2Sr_2Ca_4Cu_6Pb_{0.5}O_x$ which had previously been calcined at 820°C for 20h. The samples containing the whiskers and the single crystals were heated at 840°C for 60h-180h in air. After cooling to room temperature, the samples were removed mechanically. Characterization of the annealed samples were made by electron microprobe analysis, X-ray diffraction, and susceptibility measurement. The critical current density (Jc) and the temperature dependence of the resistance were measured by a standard four-probe method. The susceptibility was measured with a superconducting quantum interference device (SQUID) in a magnetic field of 1.5Oe from 120K to 5K (field cooling).

RESULTS AND DISCUSSION

Figure 1 shows the X-ray diffraction pattern of the flat surfaces of the annealed whiskers (840°C, 180h), with that of the original 2212 samples for comparison. For the annealed whiskers, several lines corresponding to the (00n) indices of the 2223 phase can be observed, indicating that the 2212 phase was converted to the 2223 phase during the annealing. Sheet crystals obtained by annealing the 2212 single crystals show the same XRD pattern. Diffraction patterns of annealed samples crushed into powder also consist solely of lines of the 2223 phase, which indicates that the whiskers and sheet crystals were fully converted into the 2223 phase as far as X-ray diffraction can detect.

Figure 2 shows the X-ray diffraction patterns of the annealed (840°C, 180h) and unannealed 2201 single crystals. Diffraction lines of the 2223 phase and the 2201 phase are observed for the annealed samples. The 2212 samples are fully converted to the 2223 phase by annealing at the same temperature and the same time, as mentioned above. This result indicates that the 2212 phase and the 2201 phase have the different rate of the phase conversion to the 2223 phase. The 2212 phase is easily converted to the 2223 phase as compared with the 2201 phase. This difference would be due to the crystal structure; the 2201 phase has one octahedral CuO_6 unit forming CuO_2 plane between Bi_2O_2 rock salt layers, the 2212 phase has two pyramidal CuO_5 units forming CuO_2 planes between Bi_2O_2 layers. For the 2201 phase, the process of the phase conversion involves Cu-O (apical) bond cleavage. On the other hand, the 2212 phase can be converted to the 2223 phase by intercalating Ca and one square planar CuO_4 unit between the pyramidal CuO_5 units without Cu-O bond cleavage. It seems that the bond cleavage process reduces the rate of the phase conversion.

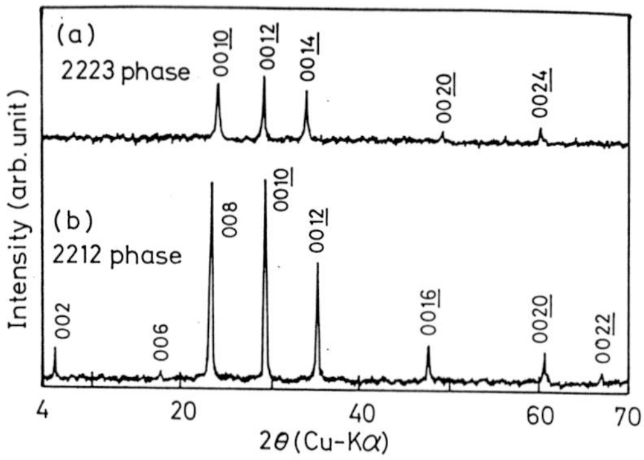

Fig. 1. X-ray diffraction patterns for (a) whisker annealed in a Bi2Sr2 Ca4Cu6Pb0.5Ox powder and (b) unannealed whiskers.

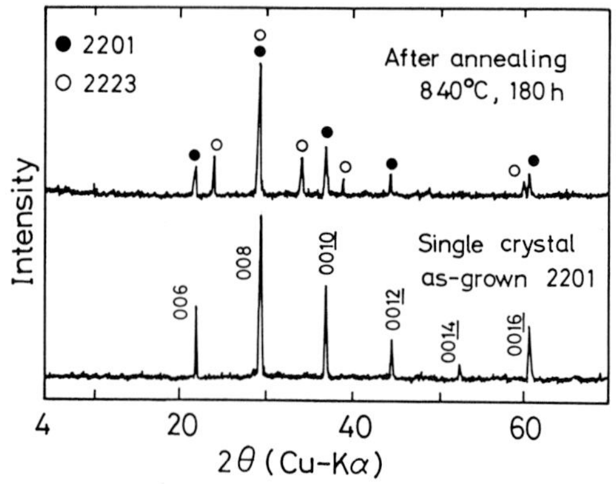

Fig. 2. X-ray diffraction patterns for unannealed 2201 single crystals and annealed samples.

Another interesting point is the absence of the 2212 phase in the process of the conversion from the 2201 to the 2223 phase. For samples annealed in short period, 60h and 120h, neither the 2212 nor the 2223 phase was observed in the X-ray diffraction patterns. It seems that the 2201 phase changes directly to the 2223 phase without passing through the 2212 phase as an intermediate. Formation mechanisms of the 2223 phase have been proposed by several groups. Shi et al.[8] have reported that the 2223 phase grows at the interfaces between the 2212 phase and Cu_2CuO_3 by diffusing the calcium and copper into the 2212 phase. The occurrence of partial melting promotes the diffusion of the elements required for the formation of the 2223 phase [9]. Other inner-grain reaction mechanisms such as intergrowth [10] or disproportionation [11] have also been reported. In these proposed mechanisms, the 2212 phase is regard as an intermediate. However, the present result indicates that the 2223 phase can be formed directly from the 2201 phase by diffusing the calcium and copper, while the rate of the phase conversion is smaller than that for the 2212 phase.

In order to examine the existence of Pb in the annealed samples, we made analytical electron microscopic study. Figure 3 shows energy-dispersive X-ray spectra for the unannealed whisker and whisker annealed in a $Bi_2Sr_2Ca_4Cu_6Pb_{0.5}O_x$ powder. Since the atomic numbers of Bi and Pb are close, the characteristic X-ray of these elements overlap at low energy region. A small peak at 10.5KeV corresponding to the Pb L -peak is observed for the annealed whisker. This confirms that Pb exists in the 2223 phas . Pb is diffused from the powder to the whiskers together with Ca and Cu during the annealing. Atomic ratio of Bi:Pb:Sr:Ca:Cu=2:0.35:1.71:1.91:3.40 is calculated from the X-ray spectra for the annealed whisker.

The original surface of the 2212 whisker or single crystal is flat and smooth, while the surface is rough with many steps after annealing in the calcined powder. The edge becomes round, suggesting the existence of a liquid phase around the whiskers during the annealing. The annealing temperature is in the partial melting region of the calcined $Bi_2Sr_2Ca_4Cu_6Pb_{0.5}O_x$ powder, as determined by DTA. The phase conversion could be due to diffusion of Ca, Cu and Pb into the 2212 phase through the liquid phase. The 2223 samples that have been obtained previously by the solid state reaction method are polycrystals with grain size smaller than 100um. In contrast, we have obtained large 2223 whiskers and crystals by applying the present method.

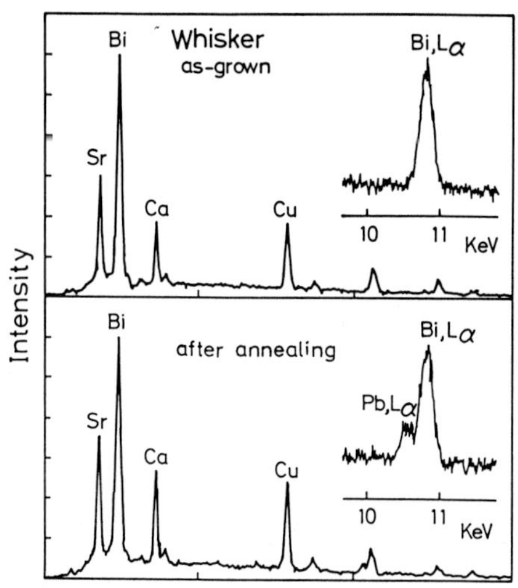

Fig. 3. Energy-dispersive X-ray spectra for unannealed whisker and whisker annealed in a Bi2Sr2Ca4Cu6 Pb0.5Ox powder.

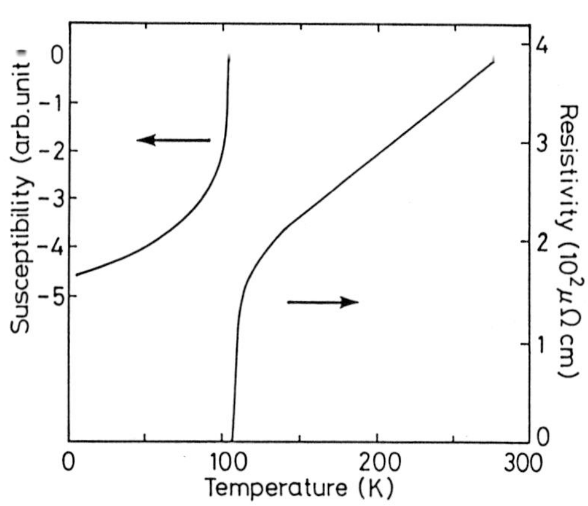

Fig. 4. Temperature dependence of resistivity and susceptivility for the 2223 whisker.

Temperature dependence of susceptibility and resistivity of the annealed 2223 whisker are shown in Fig. 4. A sharp superconducting transition at 105K with a large Meissner signal is obvious from Fig. 4. The resistivity decreases sharply at 110K, and the zero-resistance state is achieved at 107K. The resistivity of the whisker is 380u cm at room temperature and decreases linealy against temperature down to 130K.

The magnetic field dependence of Jc for the 2223 whisker and the 2212 whisker at 77K are shown in Fig. 5. The 2212 whisker contains Li as a dopant which is effective for increasing Tc, and has a Tc_{zero} value of 81K [12,13]. A maximum Jc value is usually obtained when the magnetic field is aligned along the ab plane. For the 2223 whisker, the Jc value gradually decreases with increasing the magnetic field, nevertheless the Jc is as high as $1x10^4 A/cm^2$ at B=1T in the case of B c. On the other hand, the Jc value decreases steeply in magnetic fields higher than 0.2T for the 2212 whisker. The Jc value of the 2223 whisker is hundred times larger than that of the 2212 whisker ($1x10^2 A/cm^2$) at 1T. This would be attributed to a difference in the temperature margin between the zero-resistance temperature and 77K. The whiskers and the sheet crystals would be good samples to study the basic properties of the 2223 phase. For practical applications, it is desirable to connect the whiskers in a shape of wire or tape. If the bonded portion has a Jc value comparable to that of the 2223 whisker, the wire will be useful as a superconductor operating at liquid nitrogen temperature.

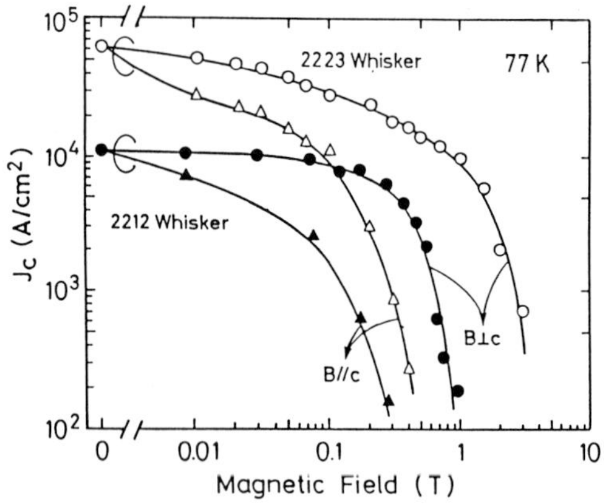

Fig. 5. Magnetic field dependence of Jc for the 2223 whisker and the 2212 whisker at 77K.

REFERENCES

1. Michel C, Hervieu H, Bolel MM, Grandia A, Deslandes F, Provost J, Raveau M (1987) Z.Phys. B68: 421.
2. Tallon JL, Buckley RG, Gilberd PW, Preland MR, Brown IWM, Bowden ME, Christian LA, Goguel R (1988) Nature 333: 153.
3. Lin JJ, Benitez EL, Poon SJ, Sabramanian MA, Gopalakrishnan J, Sleight AW (1988) Phys.Rev. B38: 5095.
4. Schneemeyer LF, van Dover RV, Clarum SH, Sunshine SA, Fleming RM, Batlogg B Siegrist T, Marshall JH, Waszczak JV, Rupp LW (1988) Nature 332: 423.
5. Takekawa S, Nozaki H, Umezono A, Kusuda K, Kobayashi H (1988) J.Cryst.Growth 92: 687.
6. Matsubara I, Kageyama H. Tanigawa H, Ogura T, Yamashita H, Kawai T (1989) Jpn.J.Appl.Phys. 28: L1121.
7. Matsubara I, Tanigawa H, Ogura T, Yamashita H, Kinoshita M, Kawai T (1989) Jpn.J.Appl.Phys. 28: L1358.
8. Shi D, Boley MS, Chen JG, Xu M, Vandervoort K, Liao YX, Zangvil A (1989) Appl.Phys.Lett. 55: 699.
9. Hatano T, Aota K, Ikeda S, Nakanura K, Ogawa K (1988) Jpn.J.Appl.Phys. 27: L2055.
10. Ikeda S, Ichinose H, Kimura T, Matsumoto T, Maeda H, Ishida Y, Ogawa K (1988) Jpn.J.Appl.Phys. 27: L999.
11. Nobumasa H, Shimizu K, Kitano Y, Kawai T (1988) Jpn.J.Appl.Phys 27: L846.
12. Matsubara I, Tanigaea H, Ogura T, Yamashita H, Kinoshita M, Kawai T (1990) Appl.Phys.Lett. 56: 2141.
13. Matsubara I, Tanigaea H, Ogura T, Yamashita H, Kinoshita M, Kawai T (1990) Physica C167: 503.

Preparation of High Magnetization YBaCuO Bulk Superconductor by Quasi-melt Process

H.G. Lee, S.C. Kwon, H.J. Lee, J.C. Lee, I.S. Chang, C.S. Rim, C.J. Kim, and D.Y. Won

Superconductivity Research Department, Korea Atomic Energy Research Institute, Daeduk Science Town, Daejun, 305-353 Korea

ABSTRACT

The development of microstructure and its effects on magnetization characteristics were investigated for the YBaCuO bulk superconductor prepared by quasi-melt process(QMP). Samples with high magnetization have been prepared successfully using the materials with the nominal compositions of $Y_1Ba_2Cu_3O_x$, $Y_1Ba_{1.8}Cu_3O_y$ and $Y_{1.2}Ba_{2.1}Cu_{3.1}O_z$ as a starting material. Anomalous double-peaks in M-H curve were observed for the samples with the nominal compositions of $Y_1Ba_{1.8}Cu_3O_y$ and $Y_{1.2}Ba_{2.1}Cu_{3.1}O_z$ as a starting materials. It was observed that the nucleation of $YBa_2Cu_3O_{7-x}$ occurred around 980 °C and grain growth was nearly completed around 950 °C.

KEYWORDS: YBaCuO bulk superconductor, high magnetization, anomalous double-peaks, quasi-melt process

INTRODUCTION

Since the discovery of the high Tc YBaCuO superconductor[1], the enhancement of the critical current density(Jc) in this system has been of great interest to scientists. It has been known that the high Jc in the range of 10^5 - 10^6A/cm^2 is obtained for single crystals[2] and epitaxially grown thin film of $Y_1Ba_2Cu_3O_x$[3,4] whereas not for bulk samples.

Recently, several techniques to fabricate the bulk YBaCuO superconductor with high Jc have been reported. Jin et al[5] firstly reported that directional solidification method was very useful for obtaining high Jc YBaCuO bulk sample(17,000 A/cm^2 at 77K , H = 0). Salama et al[6] also developed slow cooling method and obtained the Jc of 18,500 A/cm^2 at 77K, H = 0. More recently, Murakami et al[7] controlled composition and the heat treatment cycle to introduce fine Y_2BaCuO particles as second phase and obtained the high Jc = 15,000 A/cm^2 at 77K, H= 1 T from the M-H curve using Bean's critical state method. Kase et al[8] synthesized high Jc bulk superconductor with Jc of 12,000 A/cm^2 at 77 K, H = 0.25 T and Jc = 4,600 A/cm^2 at 77 K, H = 1 T by utilizing directional solidification method with the mixture of $Y_1Ba_2Cu_3O_x$ and $Y_2Ba_1Cu_1O_y$ powders. Recently, we have been successfully prepared the high magnetization bulk superconductor using quasi-melt process for various compositions in YBaCuO system.

In this study, the development of microstructure and its effect on magnetization characteristics were investigated for the YBaCuO bulk superconductos prepared by quasi-melt process. The effect of starting composition was also investigated for the materials with the nominal compositions of $Y_1Ba_2Cu_3O_x$, $Y_1Ba_{1.8}Cu_3O_y$ and $Y_{1.2}Ba_{2.1}Cu_{3.1}O_z$.

EXPERIMENTAL

YBaCuO bulk superconudctors used in this study were made by quasi-melt process(QMP) using Y_2O_3, $BaCO_3$, and CuO powders as starting materials. Powders with compositions of $YBa_2Cu_3O_x$(123), Y_2BaCuO_y(211) ,and $YBa_{1.8}Cu_3O_z$ were prepared by calcining at 900 °C for 24h in air. Calcined powders of $YBa_2Cu_3O_x$(Specimen A) , $YBa_{1.8}Cu_3O_y$(Specimen B) ,and the mixtures of 123 phase and 10 % 211 phase(Specimen C) were pelletized and heat treated in air; held for 10min at 1050 °C, cooled to 1010 °C with a rate of 40 °C/h and then slowly cooled to 900 °C with a rate of 3 °C/h. Samples were extracted at intermediate stage of reaction for the observation of microstructure develpoment. Oxygenation was done for 24h at 450 °C in air. The formation of 123 phase during QMP was examined by X-ray diffraction(XRD) and microstructure. Field -dependent magnetization(M-H) curve was taken at 77 K by using a sueprconducting-quantum-interference-device(SQUID) magnetometer.

RESULTS AND DISCUSSION

Fig. 1 shows the XRD data for the specimen C. Samples were extracted at intermediate stages during heat treatment. It can be seen that the 123 phase is completely decomposed into YBaO, CuO and 211 phase by holding for 10 min at 1050 °C. XRD peaks from 123 phase are detected for the specimen extracted at 980°C but much amount of liquid phase present at this temperature. Formation of 123 phase via peritectic reaction was nearly completed at 950°C but the small amount of liquid phase still remained at this temperature. It is very interesting that the XRD peaks from 211 phase were weakened for the specimen extracted at 950 °C and strengthened for the specimen extracted at 900 °C.

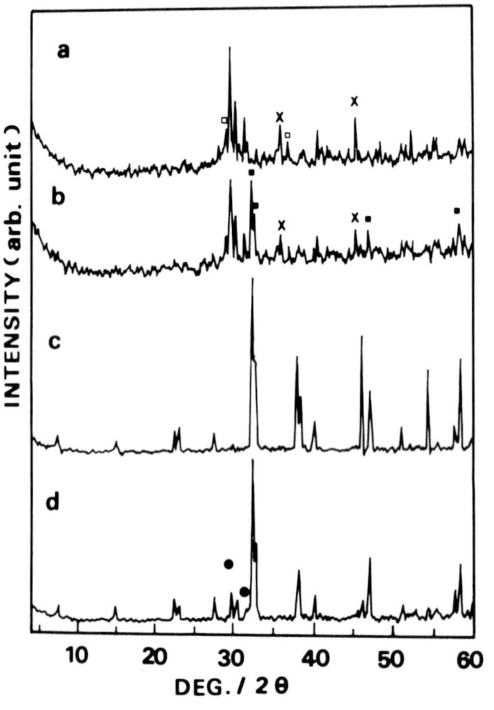

Fig.1 XRD patterns of specimen C. Samples were extracted at 1050 °C , (b) 980 °C, (c) 950 °C, and (d) 900 °C. ●, ■ , x ,and □ denote 211 phase, 123 phase, CuO, and YBaO, respectively.

Fig. 2 reveals the microstructures for the specimen C. Samples were extracted at various stage of heat treatment. It can be seen that the 123 phase appears at

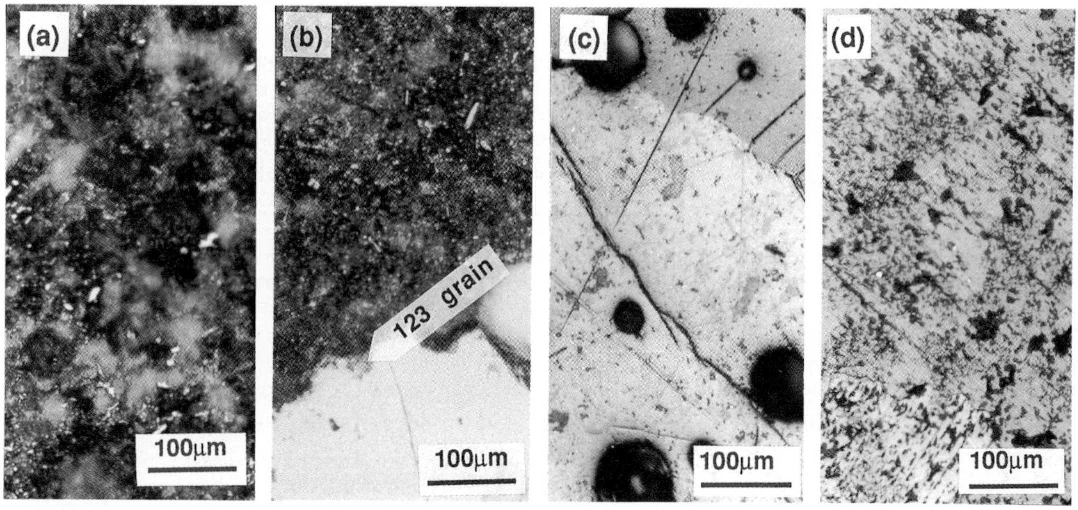

Fig.2 Micrographs with optically polarized light. Samples were extracted at 1050 °C, (b) 980 °C, (c) 950 °C, and (d) 900 °C.

980°C and covers nearly whole the specimen at 950°C. These results are well coincided with those obtained from XRD data. In Fig.2(d), It can be also seen that the 211 phase with a size of 5 - 10 μm are trapped in large grains of 123 phase. It is considered that the $Y_2Ba_1Cu_1O_5$ particles might not take a role of flux pinning center due to its large size[9].

Fig.3 shows the M-H curve for the specimen A, specimen B, and specimen C. It can be seen that the shape of the curve appears differently depending on the nominal composition of starting material. The magnetization is larger for the specimens B and C than for specimen A. It means that the starting compostion is very important factor in introducing the flux pinning center for the bulk YBaCuO superconductor with high magnetization. SEM observations showed that $Y_2Ba_1Cu_1O_5$ particles had a wide range of distribution from sub-micron to 10 micron[10]. Therefore, it is considered that the $Y_2Ba_1Cu_1O_5$ particles with a size of

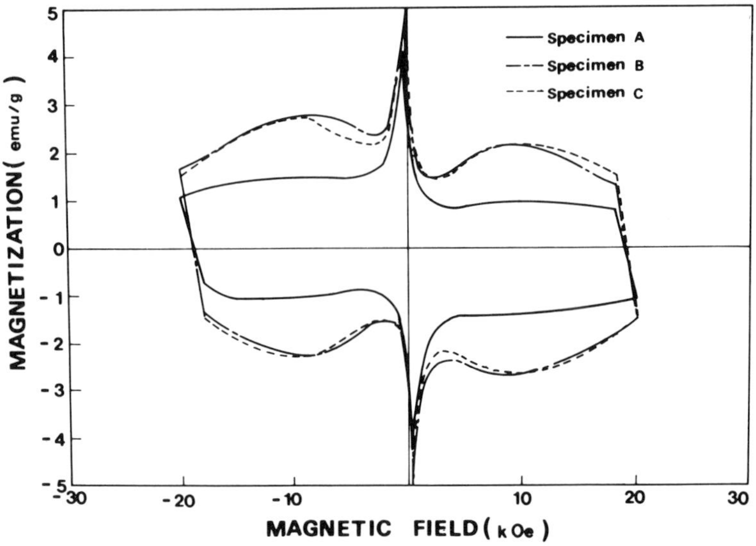

Fig.3 Field-dependent magnetization(M-H) curve for the specimens A, B, and C.

sub-micron acted as pinning center in the specimens prepared by QMP. It is also noticeable that the anomalous double peaks appears for the specimen B and specimen C. Kase et al's results[8] showed that the anomalous peak observed in this study(specimen B) did not appeared for the directionally solidified specimens using the mixed powders of 123 phase and 211 phase as a starting materials. Seuntjens et al[9] reported that the presence of the anomalous double peaks dissapeared for the specimens prepared by melt process as oxygenation proceeded. Therefore, it is considered that insufficient oxygenation heat treatment for specimens B and C resulted in the anomalous double-peaks in M-H curve.

Summarizing, the development of microstructure and its effects on magnetization characteristics were investigated for the YBaCuO bulk superconductor prepared by quasi-melt process(QMP). Samples with high magnetization have been prepared successfully using the materials with the nominal compositions of $Y_1Ba_2Cu_3O_x$, $Y_1Ba_{1.8}Cu_3O_y$ and $Y_{1.2}Ba_{2.1}Cu_{3.1}O_z$ as a starting materials. Anomalous double-peaks in M-H curve were observed for the samples with the nominal compositions of $Y_1Ba_{1.8}Cu_3O_y$ and $Y_{1.2}Ba_{2.1}Cu_{3.1}O_z$ as a starting materials. It was observed that the nucleation of $YBa_2Cu_3O_{7-x}$ occurred around 980 ^oC and grain growth was nearly completed around 950 ^oC. Clearly further work is needed to find the optimum composition for obtaining higher magnetization in YBaCuO bulk superconductor.

ACKNOWLEDGEMENT

The authors are grateful to Lee K.W., Korea Standards Research Institute for magnetic measurements. This work was financially supported by Ministry of Science and Technology.

REFERENCES

1. Wu M.K, Ashburn J.R, Torng C.J, Hor P.H, Meng R.L, Gao L, Huang Z.G, Wang Q, Chu C.W(1987) Phys. Rev. Lett. 58: 908
2. Schneemeyer L.F., Gyorgy E.M, Waszczak(1987) Phys. Rev. B36: 8804
3. Laibowith R.B, Koch R.H, Chaudari P, Gambino R.J(1987) Phys. Rev. B 35: 8821
4. Watanabe K, Yamane H, Kurosawa H, Hirai T, Kobayashi N, Iwasaki H, Noto K, Muto Y(1989) Appl. Phys. Lett. 54: 575
5. Jin S, Tiefel T.H, Sherwood R.C, Davis M.E, Dover R.B, Kammlott G.W, Fastnacht R.A,Keith H.D(1988) Appl. Phys. Lett. 52: 2074
6. Salama K, Selvamanickam V, Gao L, Sun K(1989) Appl. Phys. Lett. 54: 2352
7. Murakami M, Morita M, Koyama N(1989) Jpn J. Appl. Phys. 28: L1189
8. Kase J, Shimoyama J, Yanagisawa E, Kondoh S, Matsubara T, Morimoto T, Suzuki M(1990) Jpn J. Appl. Phys. 29: L277
9. Seuntjens J.M, Daeumling M, Larbalestier D.C,(1990) preprint
10. Kim C.J, Lee H.J, Rim C.S, Chang I.S, Lee H.G, Hong G.W, Won D.Y(1990) this volume

Preparation of YBaCuO Thick Film on (001) MgO Single Crystal by the Peritectic Reaction

C.J. Kim, H.J. Lee, C.S. Rim, I.S. Chang, H.G. Lee, G.W. Hong, and D.Y. Won

Korea Atomic Energy Research Institute, P.O. Box 7 Daeduk Danji, 305-353 Korea

ABSTRACT

YBaCuO thick films were prepared by the peritectic reaction of liquid phase + 211 phase -> 123 phase. polycrystalline 123 grains were grown on (001) MgO single crystal substrates with a c-orientation. The particle size and distribution of the 211 phase trapping in the 123 matrix could be controlled by crushing and uniform mixing of starting materials. The trapped amount of the 211 phase was also controlled by changing the Y content in $Y_{1+x}Ba_2Cu_3O_y$. The formation of the 123 phase on the (001) MgO substrate was discussed in this report.

KEY WORDS: YBaCuO thick film, c-orientation, Y_2BaCuO_5 particle, trapping

INTRODUCTION

Since the discovery of the YBaCuO superconductor with transition above liquid nitrogen temperature of 77 K, many studies have been concentrated on increasing the critical current density(Jc) of this system for practical application. The bulk YBaCuO prepared by the conventional solid state reaction method using powders, has a low Jc of 10 - 1000 A/cm^2 at 77 K and 0 T. In order to apply the oxide superconductor to high current system such as transmission cables, a superconducting magnet and the superconducting energy storage system, high Jc of 10^5 - 10^6 A/cm^2 is required even in a high magnetic field. The low Jc of the oxide superconductor has been found to be originated from the weak link properties of grainboundary, poor connectivity among grains, phase inhomogeneity, microcracks caused by the tetragonal-to-orthorhombic phase transition and anisotropic current flow in a crystal. Thus, it is thought that proper control of its microstructure may solve the Jc problem in YBaCuO system.

Recently, several improved fabrication techniques such as MTG(melt-texture-growth)[1], liquid phase processing method[2], and MPMG(melt-powder-melt-growth) method[3] have been successively developed. All the methods control their microstructures by the liquid phase sintering with thermal gradient or variation of cooling rate through the peritectic temperature region. The obtained microstructure by these processes is characterized by the growth of the large $YBa_2Cu_3O_y$ grains of a few mm size where small $YBa_2Cu_3O_y$(123 phase) grains were directionally aligned and an introduction of non-superconducting Y_2BaCuO_5(211 phase) particles which were uniformly distributed in the 123 matrix. These samples exhibit high Jc value exceeding 10^4 A/cm^2 at 77 K and 1 T. Controlling the particle size of the 211 phase may contribute to the Jc value, because the 211 phase is considered to be the pinning center of YBaCuO system.

It is well known that the 211 phase can be formed through the two peritectic reactions of Y_2O_3 + liquid phase(1) -> 211 phase at 1200 - 1500°C and liquid phase(2) + 211 phase -> 123 phase at 1100 - 980°C. By the second reaction, the 211 phase and liquid phase react to form the 123 phase. Though the reaction was completed, the 211 phase was remained as a trapped form in 123 phase because of fast growth rate of the 123 phase. The trapped 211 phase is an object for Jc research when consider the current carrying capacity in this system.

In this study, we report a preparation condition of the YBaCuO thick film on (001) MgO single crystal by the peritectic reaction of liquid phase + 211 phase -> 123 phase. Effects of particle size of initial materials and variation of Y content on the formation of the 211 phase and 123 phase were investigated in $Y_{1+x}Ba_2Cu_3O_y$(x=0, 0.2, 0.4, 0.6) systems.

EXPERIMENTAL

Powder precusors were prepared by the conventional solid state reaction method using powders of CuO, $BaCO_3$ and Y_2O_3 with 99.9% purity. The content of the 211 phase in the precusors was varied from $Y_{1.0}Ba_2Cu_3O_y$ to $Y_{1.6}Ba_2Cu_3O_y$ by changing the Y content by 0.2 mole. The powder mixtures were well mixed by grinding in an agate mortar, calcined in an aluminar crucible at 910 C for 48 h in air and then cooled to room temperature. The calcined powders were ball-milled with ethanol of 99.9% purity for 4 h and were dried in an oven at 200°C for 5 h. In order to remove the carbon pick-up from the ethanol, the powders were heat-treated at 700°C for 20 h in air and cooled in air. Figure 1 shows the X-ray powder diffraction(XRD) patterns for the calcined powders. The peaks of the 211 phase were denoted by closed circle. As illustrated in Fig. 1(a), the XRD pattern of Y=1.0 reveaved that the well crystallized single orthorhomic 123 phase with a=0.382 nm, b=0.389 nm anc c=1.169 nm was formed. However, in the sample of Y-rich composition, the 211 phase was formed with the 123 phase. The amount of the formed 211 phase increases when Y content increased. The powder precusors were mixed with suspension media of (ethylengrycol + ethanol) and were spread on (001) MgO substrates. The prepared films were dried at 200°C for 4 h to remove the suspension media. These samples were heated to 1000 - 1100°C, maintained for 10 - 60 mins, cooled down to 900 - 950°C with cooling rate of 1 - 3°C/h and cooled again to 450°C with cooling rate of 20°C/h. The microstructure was investigated by an scanning electron microscope(SEM) and an optical polarized microscope. The formed phases were identified by the XRD analysis using Cu(Kα) target.

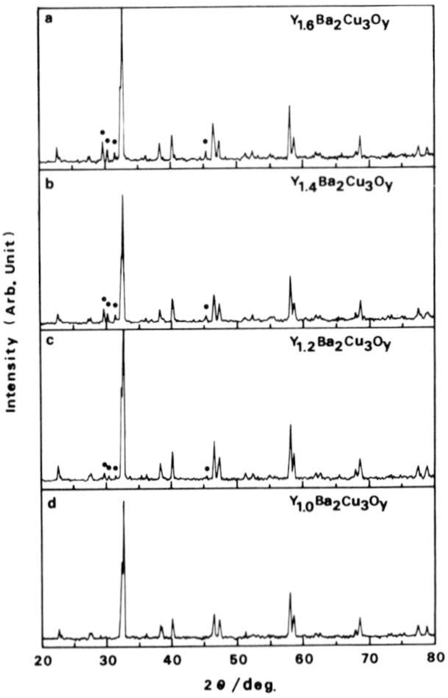

Fig. 1. The XRD patterns for the calcined $Y_{1+x}Ba_2Cu_3O_y$(x=0, 0.2, 0.4, 0.6) powders.

RESULTS AND DISCUSSION

Figure 2 shows the XRD pattern for the surface of $Y_{1.0}Ba_2Cu_3O_y$ thick film grown on a (001) MgO substrate. The film thickness of this sample is about 5 microns. The observed peaks can be indexed to the orthorhombic 123 phase and MgO. It should be noticed that the peaks with high intensity are corresponding to the (001) peak of 123 phase. This c-oriented film growth is always observed irrespect to the Y content in $Y_{1+x}Ba_2Cu_3O_7$ systems. In the samples with Y-rich compositions, the peaks of the 211 phase are observed together with the c-oriented 123 peaks. These results implies that the 123 phase was grown with specific directionality to a (001) plane of the MgO substrate. A similar result was reported in the sample prepared by the peritectic reaction from a mixed powder of 123 phase and 211 phase[4].

We have observed the relation between film thickness and the c-oriented growth of 123 grain. The XRD pattern for the surface of $Y_{1.6}Ba_2Cu_3O_y$ thick film with a thickness of 10 - 20 microns, is illustrated in Fig. 3. In this sample, the strong (001) peaks are also observed but the intensity of the peaks was considerably reduced in comparison to the sample shown in Fig. 2. With the strong (001) peaks of orthorhombic 123 phase, other 123 peaks such as (013), (110)/(103), (026) peak and the peaks of the 211 phase are observed. It means

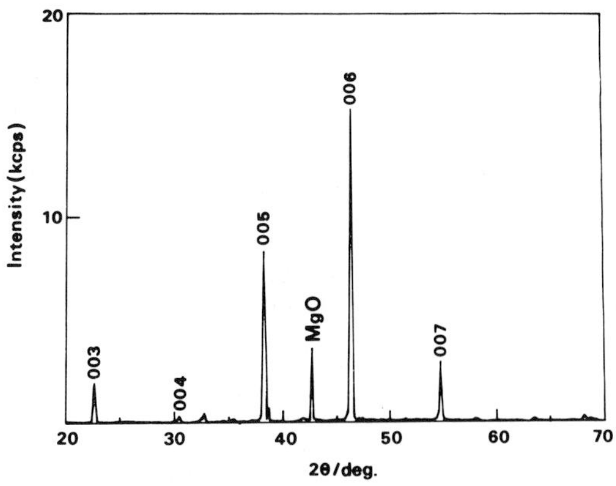

Fig. 2. XRD pattern for $Y_{1.0}Ba_2Cu_3O_y$ thick film showing the c-oriented tex-ture.

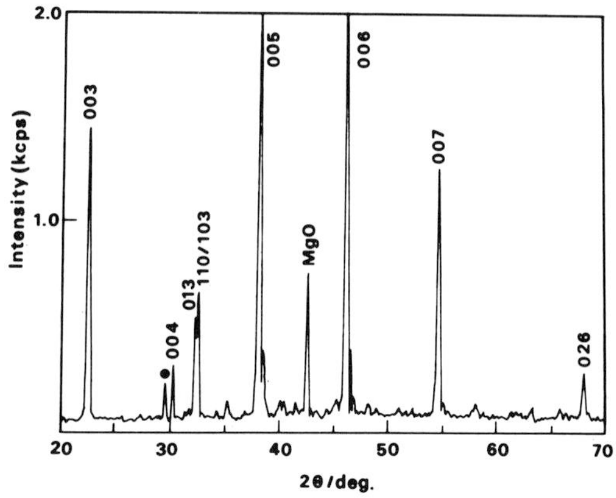

Fig. 3. XRD pattern for $Y_{1.6}Ba_2Cu_3O_y$ thick film with a thickness of 10 - 20 microns.

that the 123 phase could be formed with random orientations as well as the c-orientation in the sample with a thickness above 10 micron.

Figure. 4 shows (a) an SEM micrograph and (b) an optical polarized micrograph for the surface of $Y_{1.0}Ba_2Cu_3O_y$ thick film. It can be seen that the large 123 grains were grown with a round surface. Small pores and second phases are also found to be trapped in the large 123 grains. From the twin morphology in Fig. 4(b), the growth direction of the 123 grains could be determined. Perpendicular crossing of twin direction indicates the the growth direction of the film is the a-b plane of the unit cell. This observation is well agreed with the XRD result of Fig. 2. However, these results does not mean that the 123 phase was epitaxially grown but means that was grown as a polycrystalline form with a specific orientation to the MgO substrate. It is thought that the poly-crystalline form might due to the lattice mismatch between 123 phase and the (001) plane of the MgO single crystal.

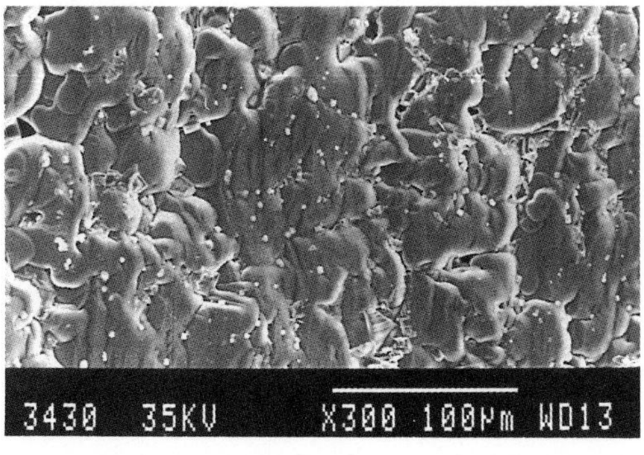

Fig. 4(a). An SEM micrograph for the surface of $Y_{1.0}Ba_2Cu_3O_y$ thick film.

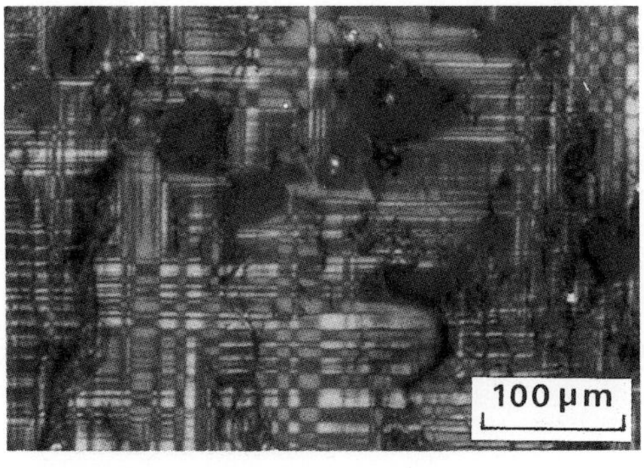

Fig. 4(b). An optical polarized micrograph for Fig. 4(a) showing optical twins.

In our process, the content of the 211 phase trapping in a 123 matrix and its particle size and distribution could be controlled by changing the Y content in $Y_{1+x}Ba_2Cu_3O_y$ and by ball milling of the powder precusor. The results of our processing are presented in Fig. 5. (a) and (b) are the optical polarized micrograph for a polished surface and and SEM micrograph for the film surface of $Y_{1.6}Ba_2Cu_3O_y$. As seen in Fig. 5(a), the 211 phase was uniformly distributed in the 123 matrix. The trapped 211 phase has two different particle shapes. The 211 phase with small particle size below 3 microns has a spherical shape but the phase above 5 microns has a plate-like shape. The origin of the particle shape is thought to be related to a strain energy between the 211 phase and 123 phase and further study is now in progress. Average particle size of the 211 phase in this sample is about 2 - 3 microns. This value is compared to that prepared by the MPMG method 3 . High magnification of Fig. 5(a) shows the very fine 211 particles in 123 matrix. An average size of the particle is 0.1 micron. These small particle is considered to act as an effective pinning center in this system.

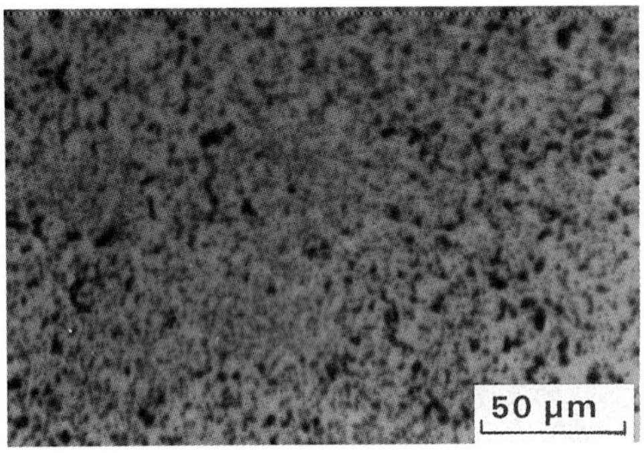

Fig. 5(a). An optical poralized micro -graph of $Y_{1.6}Ba_2Cu_3O_y$ showing the 211 phase trapped in a 123 matrix.

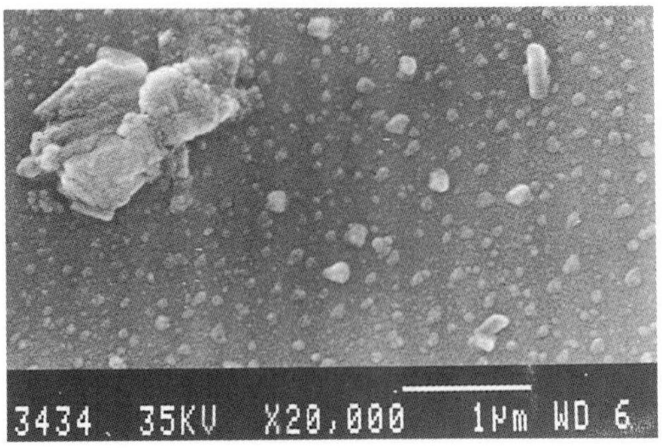

Fig. 5(b). An SEM high magnification of Fig. 5(a) showing submicron sized 211 particles.

In summary, the YBaCuO thick film with small 211 particles was prepared by the peritectic reaction of liquid + 211 phase → 123 phase. The 123 film grown on (001) MgO substrate contains mostly c-oriented grains. The particle size and its distribution could be controlled by crushing and uniform mixing of a starting materials. And the formed amount of the 211 phase is also controlled by changing the Y content in $Y_{1+x}Ba_2Cu_3O_y$ systems. The two type of the 211 phase with different particle shapes were observed in 123 matrix. The shape of the 211 phase is thought to be related to the equilibrium shape of the 211 phase in liquid phase and growth kinetics of the 123 phase during the peritectic reaction.

ACKNOWLEDGEMENT

The authors wish to thank Ki B. Kim for his help in the experiments. This work was financially supported by Ministry of Science and Technology of Korea.

REFERENCES

1. S. Jin, T.H. Tiefel, R.C. Sherwood, M.E. Davis, R.B. van Dover, G.W. Kammlott, R.A. Fastnacht, and H.D. Keith(1988) Appl. Phys. Lett. 52: 2074
2. K. Salama, V. Selvamanickam, L. Gao, and K. Sun(1989) Appl. Phys. Lett. 54: 2352
3. T. Matsushita, B. Ni, M. Murakami, M. Morita, K. Miyamoto, M. Saga, S. Matsuda, and M. Tanino(1989) Jpn. J. Appl. Phys. Lett:28: 1545
4. P. Fournier, R. Gagnon, and M. Aubin(1990) J. Appl. Phys. 67: 3776

The Impact of Oxygen on Melt Processing of Bi-HT Superconductors

J. BOCK, S. ELSCHNER, and E. PREISLER

Hoechst AG, D-5030 Hürth, D-6230 Frankfurt am Main 80, Germany

ABSTRACT

In the Bi-Sr-Ca-Cu-O system single phase 2-layer material with enhanced properties: Tc = 90 K (R=0), jc(5K) = 35 000 A/cm^2 (0T) and 13 000 A/cm^2 (5T) is yielded with melt processing in contrast to usual sintering procedures. The jc-values are due an intergrain current. Partial melting during the formation of the 2-layer phase supports the rapid and complete progress of the multistep reaction. The oxygen partial pressure plays an important role. The oxygen content of the pure 2-layer phase has a significant impact on Tc. Melt casting (instead of melt quenching) and annealing offers promising shaping possibilities for rigid conductors.

KEY-WORDS

$Bi_2Sr_2CaCu_2O_x$ melt casting, oxygen impact on annealing and on Tc, enhanced jc and Tc, shaping possibilities

INTROCUCTION

Preparation methods involving a melt are straightforward for synthesis of superconducting compounds in the Bi-Sr-Ca-Cu-O system, especially for the so called 2-layer phase. Melt quenching results in small sized particles in the glassy state which can be crystallized to superconducting material within one annealing step [1-3]. Melt casting leads to polycrystalline material because of comparatively slow cooling rates. Even those multiphase samples may be transfered in homogeneous superconducting material with enhanced properties (high jc-values <20 K) if the impact of oxygen during the annealing step is carefully taken into account [4].

We prefer casting instead of quenching because it offers the opportunity of obtaining bulk material in a wide variety of shapes. The impact of oxygen on the formation of the superconducting phase and the dependence of Tc on oxygen content were studied in detail.

EXPERIMENTAL

Binary oxides were melted in the stoichiometrical ratio 2:2:1:2. Homogeneous melts were obtained after heating to 1000-1100 °C for 10 to 30 minutes. The melt was cast into copper moulds were slow cooling to room temperature occured. Phase analysis (XRD, EMPA) revealed four different phases in the heterogeneous microstructure (fig.1) of the melt cast material: 1-layer phase and $(Sr,Ca)CuO_x$, embedded in a eutectic of Cu_2O and $(Sr,Ca)_3Bi_2O_6$.

If the casting was ground to powder and annealed at 815 °C for 24 h, the material obtained was a mixture of the 1-layer- and 2-layer compounds in addition to other phases, exhibiting a Tc of 60-70 K.

If the annealing was performed under the same conditions using bulk material, e.g. in form of ingots, it resulted in the complete transformation of the heterogeneous mass into the 2-layer phase in a state of high density (fig.2), exhibiting Tc = 90 K (R=0). These experiments underline the influence of oxygen for the formation of the Bi-containing superconductors via the melt.

Fig.1: Micrograph of a BSCCO 2212 melt in the "as cast" state

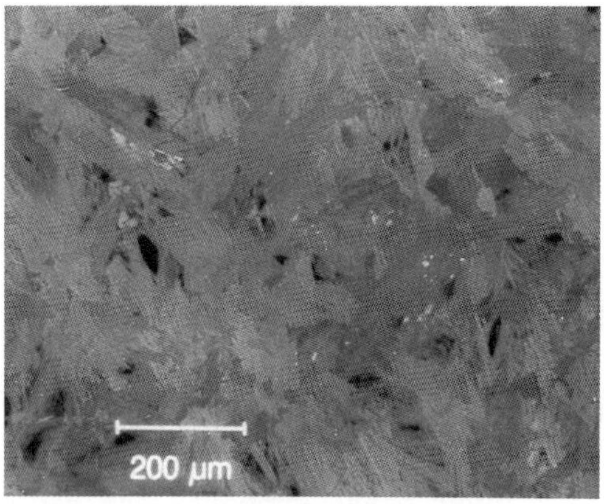

Fig.2: Micrograph of a cast BSCCO 2212 melt after annealing

OXYGEN UPTAKE OF CAST MELT

A common feature for the different compounds in the Bi-Sr-Ca-Cu-O system is the strong loss of oxygen in the course of melting. The oxygen is not regained during cooling of the melt. Investigations of the oxygen content in the cast melt of $Bi_2Sr_2CaCu_2O_x$ revealed x-values close to 7.5, indicating that half of the copper is monovalent.

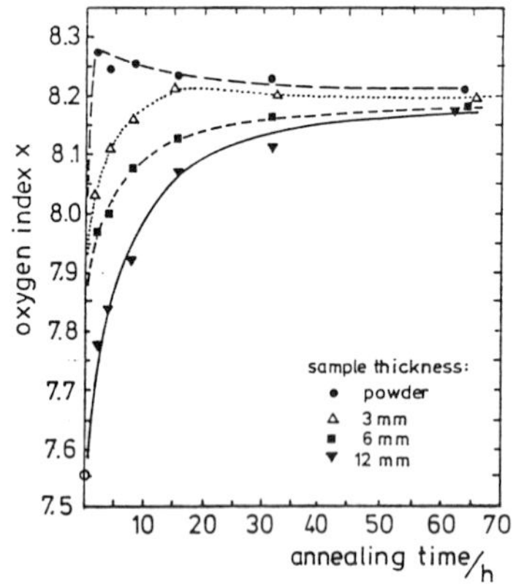

Fig.3: Oxygen uptake of a cast BSCCO 2212 melt (powder and block material) during annealing at 815 °C in air

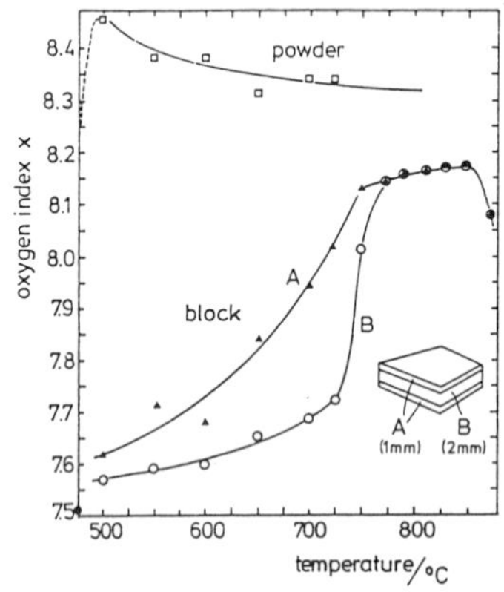

Fig.4: Oxygen uptake of a cast BSCCO 2212 melt (powder and block material) at different temperatures for 24 h in air

The oxygen uptake was investigated as a function of material thickness and annealing time at the given temperature 815 °C and as a function of temperature for a distinct material thickness at a given annealing time.

Annealing of the powder samples proceeds in a rapid oxidation (fig.3) to high values, even at 500 °C (fig.4). Oxidation of the compact ingots is kinetically determined by the material thickness.

The oxidation of the compact material takes place only at the surface of the blocks below 700 °C. At about 750 °C a steep increase in oxygen uptake and a vanishing difference between the inner and the outer part of the blocks was observed (fig.4). This indicates a sudden oxygen mobility and is due to melting of a eutectic in the heterogeneous samples. Partial melting in the compact material was confirmed by thermal analysis (DSC/TG) under different conditions [5].

THE EFFECT OF OXYGEN CONTENT

The superconductive properties of the 2-layer phase are strongly affected by the oxygen content in the material. Systematic studies were undertaken for $Bi_2Sr_2CaCu_2O_x$ in the range $8.00 < x < 8.30$. This was achieved by annealing of highly oxidized single phase powders under inert atmosphere in order to remove oxygen in small steps. Heating was performed at temperatures between 300 and 750 °C for 24 h in order to equilibrate the oxygen content in the samples.

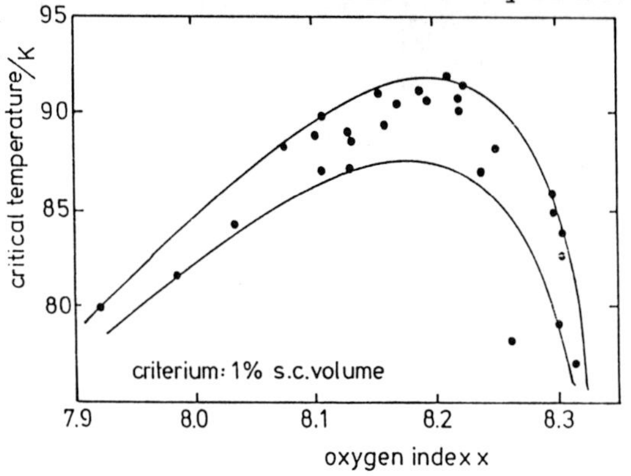

Fig.5: Influence of oxygen index on Tc in pure $Bi_2Sr_2CaCu_2O_x$

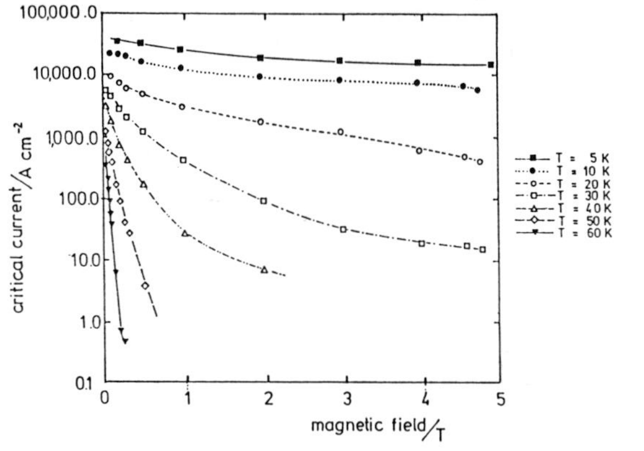

Fig.6: Jc-values in melt processed $Bi_2Sr_2CaCu_2O_x$ (x=8,17) as a function of temperature and magnetic field

The relation between the oxygen index x and Tc was taken from DC-susceptibility measurements (fig.5). While in $YBa_2Cu_3O_{7-x}$ high values of x (close to 7.0) stand for high Tc-values, we observed a dramatic decrease of Tc in connection with high oxygen contents in $Bi_2Sr_2CaCu_2O_x$.

CRITICAL CURRENT DENSITIES

A striking feature of the melt-processed bulk materials are high critical current densities jc at low temperatures, even in high magnetic fields. Magnetic measurements were performed at little rods (radius: 0,5 mm, length: 8 mm). Evaluation of jc according to the "Bean model" (jc = 1.5 M/r with r = radius of the rodlet) yielded 35 000 A/cm^2 in the zero field and 13 000 A/cm^2 at 5 T (fig. 6). Earlier investigations [6] of jc by transport measurements exhibited 13 000 A/cm^2 (4K, OT) and 6000 A/cm^2 (4K, 8T). This suggests that the values obtained from the magnetic measurements correspond to an intergrain current over the total rodlet. The promising behaviour in magnetic fields is due to the absence of weak links (clear grain boundaries) in the melt processed bulk material. The sharp decrease of jc with increasing temperature to about 400 A/cm^2 at 77 K (0 T) indicates the lack of pinning forces.

SHAPING

An important advantage of the melt casting process is the opportunity to obtain bulk material in a great variety of shapes e.g.

- plates
- rods (with length up to 500 mm)

- rings (with diameters up to 200 mm)
- tubes
- hollow cylinders (with diameters of up to 100 mm)

Fig.7: Tube (22x15x60 mm^3) of BSCCO 2212 obtained with melt processing

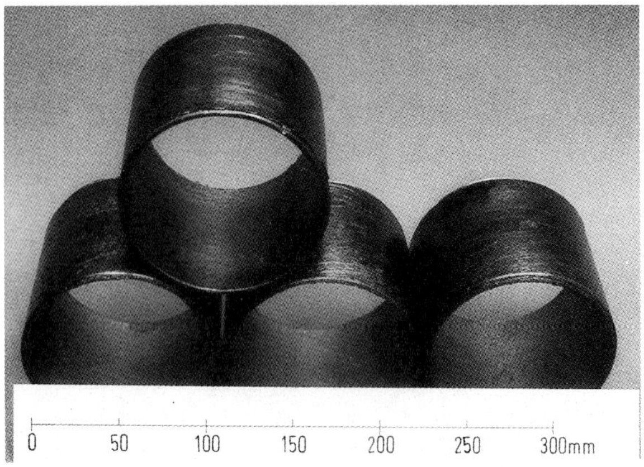

Fig.8: Hollow cylinders (110x100x4 mm^3) of BSCCO 2212 obtained with melt processing

These bulk materials may be used as rigid conductors for current leads, magnetic shielding, motors and generators etc.

RESULTS AND DISCUSSION

When the precursor oxides in the ratio 2:2:1:2 are heated to above 1000 °C a homogeneous melt is formed which releases oxygen during melting. The reduced state is kept on solidification and causes a heterogeneous mass of different phases including an eutectic. These reactive phases together with their intimate contact in the cast are essential for the complete transformation into single phase 2-layer material. Partial melting supports the rapid proceeding of this reaction and is probably responsible for clear grain boundaries and therefore the absence of weak links. In order to realize the partial melting it is necessary to perform the annealing under low access of oxygen, at least in the first period of the reaction. The lack of pinning forces in the material still remains a serious problem. Yet, melt processing offers promising shaping possiblities for this HTc-material.

LITERATURE

[1] Haixing Zheng, Ren Xu and J.D.Mackenzie, J.Mater. Res.4, 911-915 (1989)
[2] N.P.Bansal and M.R.DeGuire , J.Am.Ceram.Soc., in press
[3] T.Komatsu, T.Ohki, C.Hirose and K.Matusita, J.Non-Crystal. Sol. 113 (1989) 274-281
[4] J.Bock and E.Preisler, Solid State Comm. 72 (1989) 453-458
[5] J.Bock, S.Elschner and E. Preisler, Proc. ICMAS-90, Grenoble, October 15-16, 1990
[6] M.Peuckert, W.Becker, J.Bock, B.Hettich, H.-W.Neumüller and M.Schwarz, Physica C 162-164 (1989) 893-894

Performance of Current Lead of YBaCuO Bulk in Solid Conduction Cooling System

HIROSHI YAMAMOTO[1], MASAHIKO SANO[1], SUSUMU OZAWA[1], MASAICHI TANAKA[1], YOUICHI MATSUBARA[2], and TAKESHI OGASAWARA[2]

[1] Nihon University, College of Science and Technology, Narashinodai, Funabashi, 274 Japan
[2] Nihon University, Atomic Energy Research Institute, Kanda Surugadai, Chiyoda-ku, Tokyo, 101 Japan

ABSTRACT

A $YBa_2Cu_3O_7$ sample with Tc of above 80 K and Jc of about 3000 Acm^{-2} at 50 K was prepared by a melt powdering melt growth (MPMG) technique. Thermal conductivity of the sample was about 10^{-2} $Wcm^{-1}K^{-1}$ at 15 K. The maximum quench current attained up to about 80 A at 26 K in the sample with the thickness of 2.3 mm and the current junction of 6.9x8.7 mm^2. Obtained results reveal that the MPMG bulk superconductor is a hopeful material for superconducting current leads.

KEY WORDS: MPMG YBCO, current lead, thermal conductivity, quench current

INTRODUCTION

The discovery of high Tc oxide superconductors inevitably enables a new class of a superconducting application which is used in a comparatively high temperature region. Various applications have been investigated. One of them is a current lead which is adopted for a power supplying line in a superconducting magnet system. Demanded factors for a current lead are typically a high critical current, a low thermal conductivity and a high strength. Already a superconducting current lead of a high critical current above 1000 A has been developed with BiSrCaCuO tape conductors.[1] On the other hand, recently a group of Murakami[2] newly developed a Melt Powdering Melt Growth(MPMG) process for getting a high Jc $YBa_2Cu_3O_7$(YBCO) superconductor. A feasibility of the MPMG bulk has been studied from various points of view. A low thermal conductivity is remarked because the high Jc MPMG bulk includes many fine Y_2BaCuO_5 particles as pin centers.

The purposes of this work are to study fundamental superconductivity and thermal conductivity of the MPMG YBCO and to investigate its feasibility for a small size of current lead. Experimental results of current flow are discussed based on the measured fundamental properties. In particular the MPMG YBCO is evaluated as a material for superconducting current leads which will be used in a superconducting magnet system[3] shown in Fig. 1. Since in the system the magnet is cooled by solid conduction, thermal flow through the current leads must be suppressed low as possible.

EXPERIMENTAL

The sample was prepared by the MPMG process as following. Mixed powders of Y_2O_3, $BaCO_3$ and CuO with the mole ratio of 1:4:6 were calcined at 900 °C for

24 h. The calcined powder was heated and melted in a Pt crucible at 1400°C for 15 min. The melt sample was cooled down to room temperature on a copper plate. Then the solid sample was crushed into powders and pressed into a disk. After the disk was again heated up to 1100 °C and held for 20 min, it was cooled down rapidly to 1000 °C with a rate of 2 °Cmin^{-1} and slowly to 500 °C with a rate of 0.16 °Cmin^{-1}. Finally thermal anneal at 450 °C for 24 h was done in the atmosphere of oxygen.

The x-ray diffraction indicated that the sample included a small amount of Y_2BaCuO_5 or Y_2O_3 in $YBa_2Cu_3O_7$ matrix. The YBCO sample showed Tc above 80 K. The resistivity in the vicinity of Tc was few～few tens $m\Omega$ cm. A temperature dependence of Jc of the sample was measured with transport current induced by pulse magnetic field[4]. Thermal conductivity λ was measured by a steady heat flow method of which details was appeared in elsewhere. The sample was cut into a rectangular of which the size was typically about 6x5x4 mm^3. After polishing the sample by a sand paper, the surface of the sample was coated by evaporated indium film. Almost same two samples were strongly sandwiched by three copper blocks on which carbon-glass(CGR) thermometers were embedded. Heat flux flew through the two samples from the heater which was sandwiched by the samples. System temperature was controlled by another heater. The observed λ is the value including the term from the thin indium plates which were used for thermal contacts.

An experimental system of current flow is shown in Fig. 2. The MPMG YBCO bulk was formed into a rectangular shape with sides of few mm. An electric contact surface was coated by an evaporated silver film and silver paste. An anneal at about 450°C in the atmosphere of oxygen was effective to decrease a contact resistance between the sample and electrodes. The two electrodes were embedded on a cooling head of a cryorefrigerator. Temperature of the sample was monitored at a copper block by a CGR thermometer. After cooling the sample to a cryogenic temperature, V-I characteristics were plotted by changing a current sweep time, simultaneously measuring temperature of the sample. The maximum current was 100 A.

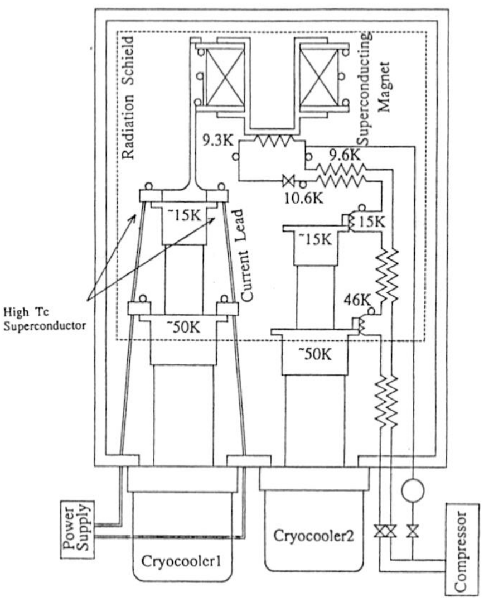

Fig.1 Superconducting magnet system cooled by solid conduction.

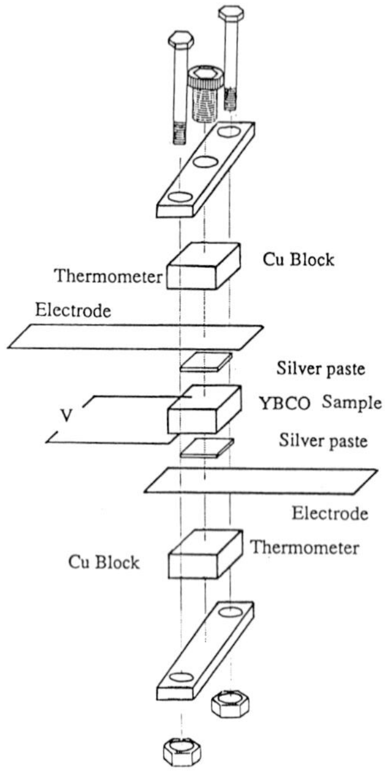

Fig.2 Measuring system adopted for experiments of passing current.

RESULTS AND DISCUSSION

Typical Jc(T) and thermal conductivity λ (T) of the MPMG YBCO bulk are shown in Fig. 3. The value of Jc was about 3000 Acm^{-2} at 50 K. In the range of temperature from 10 K to about 50 K λ decreased as decreasing temperature. The little dependence of λ on temperature suggests that the sample includes many defects. The value of λ was an order of $10^{-2}Wcm^{-1}K^{-1}$ in the range of measurement temperature. A value of λ of a single crystal YBCO is much larger than that of a sintered one. Then the variation of λ is attributed to differences in crystal structure and/or porosity. The value of λ of the MPMG sample is an intermediate value between that of the single crystal[5] and the sintered one[6].

Hull[7] discussed about high temperature superconducting current leads and concluded that a heat flow into a heat sink is characterized by the quantity of $\rho\lambda$. The value of $\rho\lambda$ in the vicinity of Tc is estimated about 10^{-4} W K^{-1} of which the value is nearly 1～2 order of magnitude larger than that of a pure metal. The result reveals a high feasibility of the sample for an application of superconducting current leads.

Figure 4 shows results of V-I characteristics in the sample with the area of the cross section of 6.9x8.7 mm^2. In the region of low current flow, voltage appeared from a contact resistance and increased linearly as increasing temperature. The contact resistance of the current junction was about 0.5 mΩ. A rapid increase of voltage meant a quench of the sample. The value of the quench current decreased as increasing temperature. The result is shown in Fig. 5 as a function of temperature. Data were obtained with good reproducibility, exception of the one point of the first run. The maximum quench current was about 80 A at 26 K in the sample.

The observed quench current Ic(T) was fairly small comparing with the expected value from Jc(T). However, since the temperature dependence of Ic was qualitatively similar to that of Jc, the suppression of Ic may be caused by an effective decrease of the area of the current junction rather than the degradation of the surface of the superconductor. In other words, the current sharing took place at the current junction. The area of the current junction should be wider to get the higher quench current. There are still many technical subjects to be solved in the interface region of the current junction. One of them is a problem about a material stability of the surface of the superconductor. As shown in Fig. 6, the superconductivity of the sample became worse and the normal resistivity increased slightly after passing a lot of current. Further investigations about materials and/or a preparation process are needed to fablicate the better current junction.

In summary the properties of the MPMG YBCO bulk were investigated about its feasibility for applications to a superconducting current lead. Thermal conductivity of the high Jc sample is fairly small, about 10^{-2}W$cm^{-1}K^{-1}$ at 15 K. The maximum current attained up to about 80 A at 26 K in the sample with the thickness of about 2.3 mm and the area of the current junction of 6.9 x8.7 mm^2. Obtained results reveal that the MPMG YBCO bulk is a very hopeful material for superconducting current leads. Some considerations are, however, needed to get a stable metallic surface of the superconductor. Furthermore the shape of the current junction should be optimized from the point of view of the current sharing.

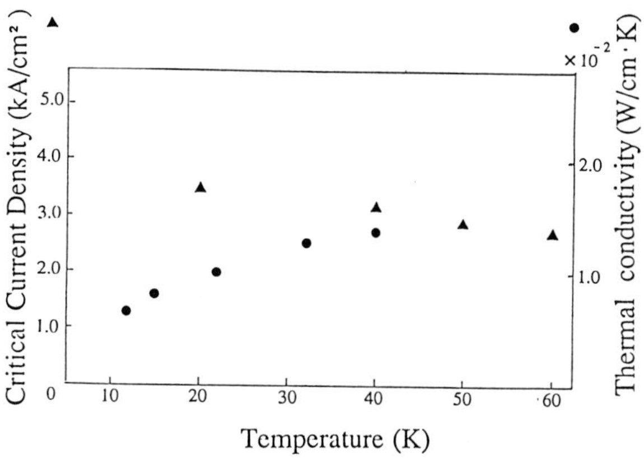

Fig.3 Thermal conductivity and critical current density Jc vs. temperature of the MPMG sample.

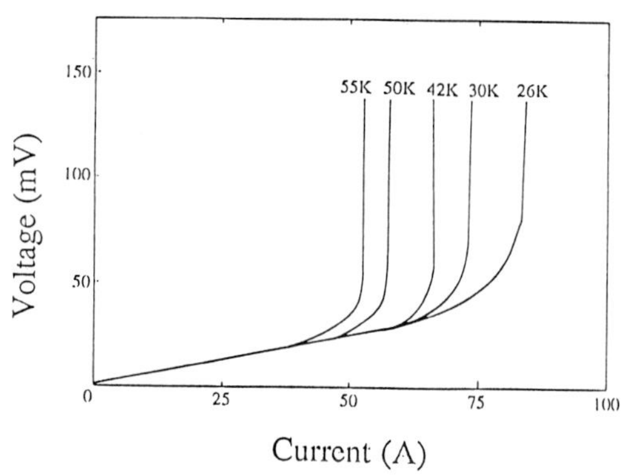

Fig.4 Voltage-current characteristics observed in the MPMG sample.

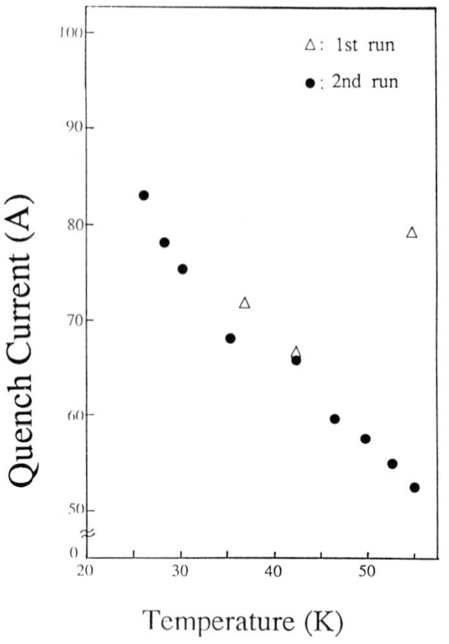

Fig.5 Quench current vs. temperature.

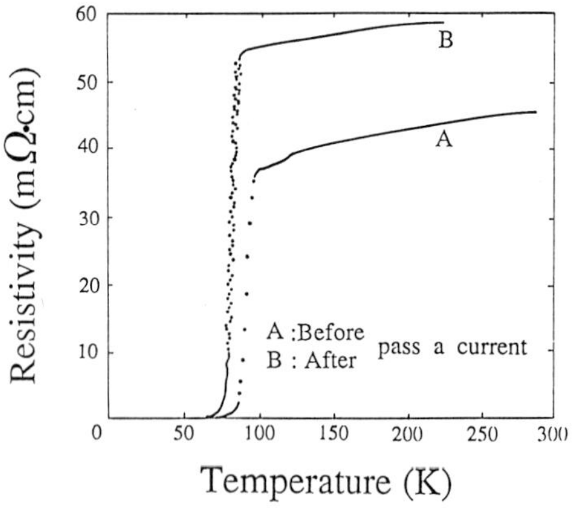

Fig.6 Change of resistivity of the MPMG sample before and after passing current.

ACKNOWLEDGMENTS

The authors would like to thank Mr. A. Kawashima and Mr. M. Seki for their assistances in the preparation of samples.

REFERENCES

1. Nagata M (1991) Extended Abstracts (The 51th Autumn Meeting, 1991); The Japan Society of Applied Physics Vol.0: 1217
2. Murakami M (1990) Mod.Phys.Lett. 4: 163
3. Furuyama M, Yamamoto H, Tanaka M, Kaneko M, Matsubara Y and Ogasawara T (1990) Advances in Cryogenic Engineering 35: 625
4. Yamamoto H, Ozawa S, Sano M, Kouma N, Tanaka M, Takizawa T, Kanbara Kand Katoh K(1990) Proc. 65th Topical Symposium of the Magnetics Society of Japan MSJ65-12: 73(in Japanese)
5. Kirk W P, Kobiela P S, Tsumura R N and Pandey R K (1989) Ferroelectric 92 :151
6. Noto K, Watanabe K, Morita H, Mori K, Sasakawa M, Isikawa Y, Sato K, Isihara T, Inukai E,Fujimori H and Muto Y (1989) Cryogenics 29: 648
7. Hull J R (1989) Cryogenics 29: 1116

High Temperature Deformation and Textures in Oxide Superconductors

KENJI HIGASHIDA and NOBUTAKA NARITA

Department of Metallurgy, Kyoto University, Yoshida-honmachi, Sakyo-ku, Kyoto, 606 Japan

ABSTRACT

High temperature deformation and its role on texture formation were investigated using $YBa_2Cu_3O_y$ (YBCO) and $Bi_{0.7}Pb_{0.3}SrCaCu_{1.8}O_y$ (BPSCCO) ceramics. The YBCO and BPSCCO ceramics are fairly ductile to deform over 50% in compressive strain at temperatures above 840°C and 720°C, respectively. In compression tests of YBCO and BPSCCO, their yield stresses are markedly decreased with increasing temperature and also with decreasing strain rate. After the tests, preferred orientations in the loading direction were examined by means of pole figures and Debye-Sherrer patterns. The results showed that the intensity of (00n) reflections was increased by high temperature deformation in both YBCO and BPSCCO specimens. This indicates that the YBCO and BPSCCO can deform by slip at high temperatures so as to develop deformation textures.

KEY WORDS: oxide superconductor, Y-Ba-Cu-O, Bi-Pb-Sr-Ca-Cu-O, high temperature deformation, texture

INTRODUCTION

Oxide ceramics are usually brittle so that their plastic forming such as wire-drawing is very limited. The study of plastic deformability in superconducting oxides must be an important subject for their practical application. Recently, Suzuki and Takeuchi predicted the possibility of slip deformation in $YBa_2Cu_3O_y$ ceramics at temperatures above 800°C from the theoretical consideration about the intrinsic strength of ceramics having a perovskite structure [1]. If the superconducting oxides can be easily deformed by slip on (001), it may contribute not only to plastic working but also to the improvement of superconducting properties by developing the (001)-preferred orientation. The purpose of the present study is to make clear the deformation characteristics of $YBa_2Cu_3O_y$ (YBCO) and $Bi_{0.7}Pb_{0.3}SrCaCu_{1.8}O_y$ (BPSCCO) ceramics at high temperatures with particular attention to the texture formation.

EXPERIMENTAL

YBCO and BPSCCO specimens (13mmϕ × 5mm) were prepared by a conventional sintering method described previously [2-4]. Compression tests for YBCO and BPSCCO specimens were made in air at high temperatures with strain rates 3×10^{-5} and 3×10^{-4}/sec. After the compression tests, specimens were slowly cooled down with the rate of 1.7°C/min. Structural characterization was made by X-ray measurements using Cu-Kα radiation. Preferred orientations in specimens subjected to the high temperature deformation were examined using the Schulz's reflection method.

RESULTS AND DISCUSSION

Figure 1 shows stress-strain curves of YBCO specimens deformed in compression at temperatures between 840°C and 900°C with a strain rate 3×10^{-5}/sec. As is seen from this figure, the YBCO specimens are deformed over 50% in strain at high temperatures and their yield stress is markedly lowered with increasing temperature. In the tests at 840°C and 860°C, the flow stress is gradually decreased as the deformation proceeds after yielding. This behavior seems to be a sort of yield-drop phenomena but is quite large compared with that in

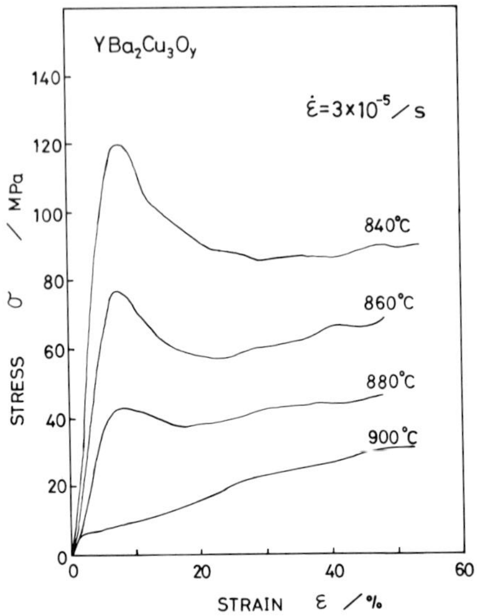

Fig.1 True stress-true strain curves for YBCO specimens deformed in compression.

Fig.2 Change in the shape of a YBCO specimen by a compression test at 900° C (ε = 52%).

metals. Such large yield-drop might be rationalized in case when the density of mobile dislocations on the basal plane is originally low and is gradually increased during deformation. When deformation temperature is raised, the peculiar yield-drop becomes obscure, while work-hardening becomes obvious instead. This may suggest that secondary dislocations operate to interact with basal dislocations at higher temperatures. Figure 2 shows a change in the shape of a YBCO specimen by a compression test at 900° C. The upper and lower photographs correspond to the top and side views of the specimen, respectively. After the deformation of 52% in strain, the specimen is fairly thinned without any severe cracking. YBCO specimens were also tested at temperatures below 840° C, but were fractured before macroscopic yielding. From these results, it is evident that high temperature deformation at 840° C or above is required for the plastic forming of YBCO ceramics at the strain rates employed in the present study. Figure 3 represents the effect of strain rates on the yield stress of YBCO specimens deformed at various temperatures. It is seen that the yield stress is rather sensitive to the strain rate as well as the deformation temperature. Deformation behavior of BPSCCO specimens was also investigated; the stress-strain curves are shown in Fig.4. The BPSCCO specimens are fairly ductile at high temperatures, but their stress-strain curves are

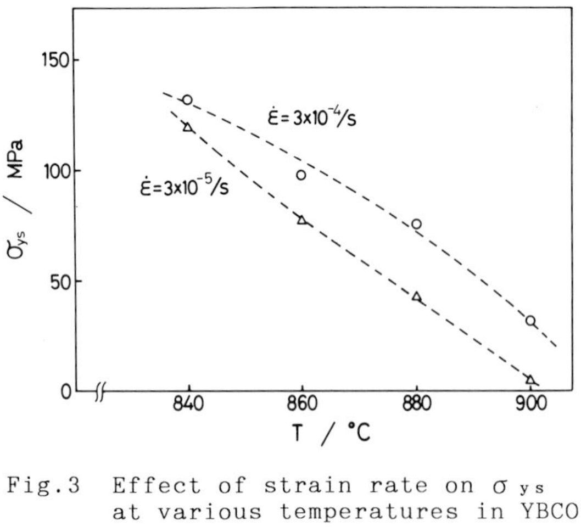

Fig.3 Effect of strain rate on σ_{ys} at various temperatures in YBCO.

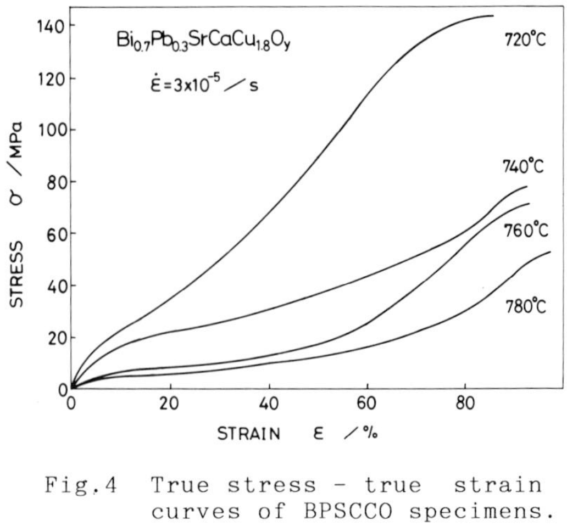

Fig.4 True stress - true strain curves of BPSCCO specimens.

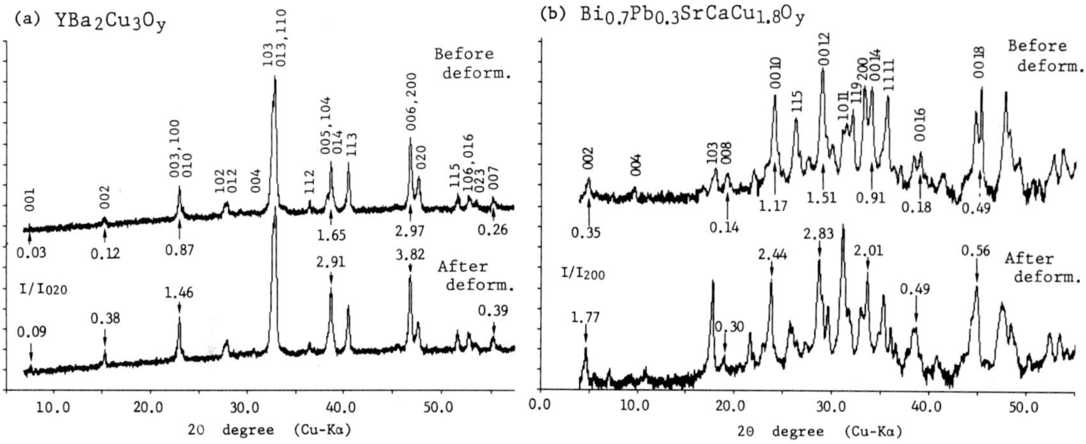

Fig.5 Change of X-ray diffraction patterns by compression tests at high temperatures. (a) YBCO deformed at 840°C, (b) BPSCCO deformed at 720°C.

quite different from those in YBCO. They have very low yield stresses and do not exhibit any yield-drop. The lack of the yield-drop may suggest that a number of mobile dislocations exist on the basal plane and their multiplication causing a yield-drop is not much required in the early stage of deformation. As the deformation proceeds after yielding, the work-hardening rate is gradually increased. Such work-hardening may take place from the same reason as that described above for YBCO.

In order to examine the development of preferred orientations at the interior of specimens by high temperature deformation, the surface layers were removed from the top surfaces (loading plane) and subsequently X-ray diffraction analyses were made. Figures 5(a) and (b) show the results for YBCO and BPSCCO specimens, respectively. The upper and lower patterns in these figures correspond to those before and after the compressive deformation, respectively. In Fig.5(a), it is shown that the intensity ratios of (00n) peaks (n=1,2,3,5,6,7) to a (020) peak, $I/I_{(020)}$, are increased by the deformation in the YBCO. In case of the BPSCCO, it is also seen from Fig.5(b) that the ratios of (00n) peaks (n=2,8,10,12,14,16,18) to a (200) peak, $I/I_{(200)}$, are increased by the deformation. Such changes in the X-ray patterns suggest that the (001)-preferred orientation is developed by the compressive deformation at high temperatures in both YBCO and BPSCCO specimens. Figures 6(a) and (b) show the (001) pole figures corresponding to Figs.5(a) and (b), respectively. In these figures, the dark regions at the center of the pole figures are more evolved by the deformation, indicating that the (001)-preferred orientation is developed by the high temperature deformation in both YBCO and BPSCCO specimens. It is to be noticed that such preferred orientations somewhat exist even before the deformation since the calcined powders being of the aggregates of thin particles are pressed before sintering so as to be preferentially oriented.

Figure 7 shows the slip systems for YBCO crystals. In this lattice, the interplanar spacing of (001) is large and the shortest vector of primitive translations lies along [010] or [100]. Therefore, the most operative slip system must be <010>(001). In YBCO ceramics, dislocations of this slip system are actually introduced by deformation [5]. In order to deform polycrystals, it is necessary to operate five independent slip systems, as pointed out by von Mises [6]. Since the slip system of <010>(001) has only two independent slip systems, secondary slip systems must operate to satisfy the von Mises condition. As suggested by Suzuki and Takeuchi [1], a possible secondary system is of the $\langle 0\bar{3}1\rangle\{113\}$ shown in Fig.7. In $BaTiO_3$ having a perovskite structure, dislocations of this type actually exist and split into two half dislocations [7]. The split reaction in the case of YBCO can be expressed as

$$(1/3)[0\bar{3}1] = (1/9)[\bar{3}\bar{3}2] + (1/9)[3\bar{6}1] \quad . \qquad (1)$$

Since the systems of <010>(001) and $\langle 0\bar{3}1\rangle\{113\}$ provide five independent slip

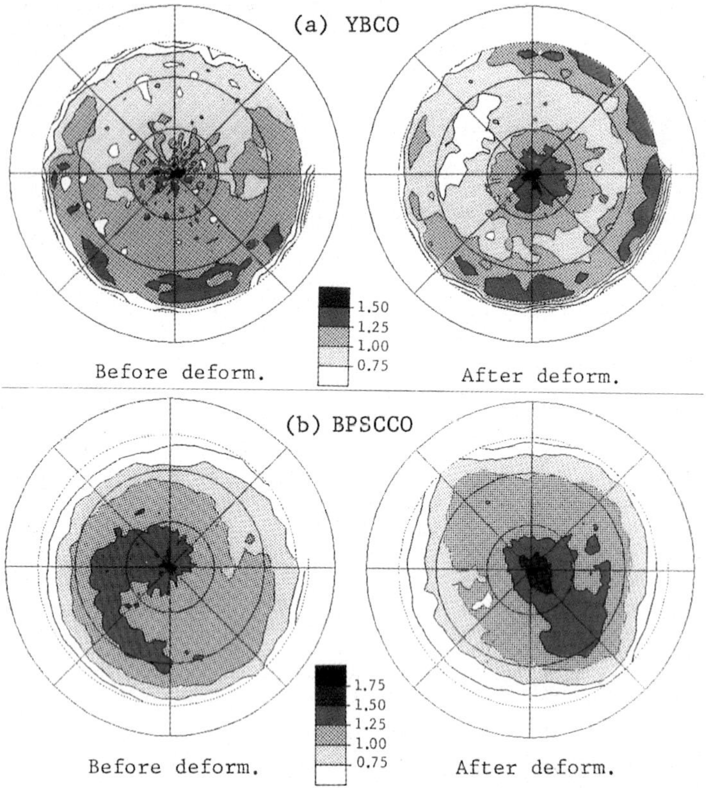

Fig.6 (001) pole figures of YBCO and BPSCCO specimens before and after deformation.

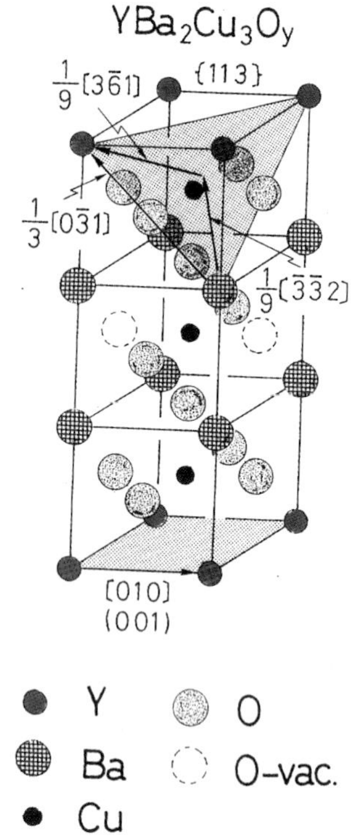

Fig.7 Slip systems for YBCO crystals

systems, their operation gives rise to the plastic deformability of YBCO ceramics. Our experimertal results agree with the theoretical prediction that both slip systems in YBCO can operate in the deformation at 800°C or above (Suzuki and Takeuchi [1]). This suggests that the observed deformation is mainly caused by slip. However, since the crystal rotation due to secondary slip is different from that of the (001) slip, abundant operation of the secondary slip may suppress the development of the (001)-preferred orientation. Such suppression of texture formation was actually observed in YBCO specimens deformed at 900°C. It is therefore concluded that suitable deformation temperatures should be employed in order to develop (001) textures.

To sum up, YBCO and BPSCCO ceramics are fairly ductile to deform over 50% in compressive strain at high temperatures. The (001)-preferred orientation is developed by the compressive deformation in both YBCO and BPSCCO specimens.

This work has been supported in part by Grant-in-Aid for Scientific Research from the Ministry of Education and Culture.

REFERENCES

1. Suzuki T, Takeuchi S (1989) In: JJAP Series 2, Lattice Defects in Ceramics. Publication Office, JJAP, Tokyo, pp 9-15
2. Leng S, Narita N, Higashida K (1987) Jpn J Appl Phys 26: L1394-L1397
3. Narita N, Leng S, Higashida K (1989) In: Proc MRS Int Mtg on Adv Mat Vol 6. MRS, Pittsburgh, pp 403-408
4. Narita N, Higashida K, Mishina S (1989) In: Proc ISS'89. Springer-Verlag, Tokyo, pp 109-112
5. Saka H, Kuroda K (1990) Bull Jpn Inst Metals 29: 693-697
6. von Mises R (1928) Z angew Math Mech 8: 161-185
7. Takeuchi S, Suzuki K, Ichihara M, Suzuki T (1989) In: JJAP Series 2, Lattice Defects in Ceramics. Publication Office, JJAP, Tokyo, pp 17-24

Effects of Melting Temperature on the Properties of High-T_c Superconducting $Bi_2Sr_2CaCu_2O_x$ Glass-Ceramics

WON-HYUK LEE, HIDEO HOSONO, and YOSHIHIRO ABE

Department of Materials Science and Engineering, Nagoya Institute of Technology, Gokiso-cho, Showa-ku, Nagoya, 466 Japan

ABSTRACT

$Bi_2Sr_2CaCu_2O_x$ superconductors can be fabricated by controlled crystallization of the corresponding glasses. In this study we examined the effects of melting temperatures(T_m) of the glasses on the T_c of the resulting glass-ceramics. The Cu(II)/ΣCu ratio and CO_3^{2-} concentration vary to a considerable extent with T_m. The T_c of the glass-ceramics obtained by the same heat-treatment increases with T_m of the mother glasses, where the CO_3^{2-} concentrations are not important. The increase of T_c of the glass-ceramics with the increase of T_m is considered to be due to the increase of the mean copper valence in glass-ceramics.

KEY WORDS: $Bi_2Sr_2CaCu_2O_x$ glass-ceramics, melting temperature, Cu(II)/ΣCu ratio, CO_3^{2-} concentration, T_c

INTRODUCTION

Bi-Sr-Ca-Cu-O high T_c superconductors have been fabricated using glass-ceramic process[1-4]. Precursor glasses in this system are usually prepared from the melts above 1000°C using calcined oxides and carbonates as raw materials. It is expected that melting temperature(T_m) of the glasses should have an effect on both the valence of copper ion and CO_3^{2-} concentrations dissolved in glasses.

With increasing temperatures, CuO should dissociate into Cu_2O and O_2[5]. In the Bi-Sr-Ca-Cu-O system, Cu(II)/ΣCu ratio can be controlled by T_m so that both Cu(I) and Cu(II) ions coexist in these glasses. Therefore, the glasses with different melting history are expected to have different ratio of Cu(II)/ΣCu. The mean copper valence in superconducting $Bi_2Sr_2CaCu_2O_x$ phase is reported to be 2.07[6], while the mean copper valence in Bi-Sr-Ca-Cu-O glasses is much lower than that in the superconducting $Bi_2Sr_2CaCu_2O_x$ phase. During the crystallization process, the valence change from Cu(I) to Cu(II) should take place for the glass-ceramics to exhibit superconductivity. Superconducting properties of resultant glass-ceramics are expected to depend on Cu(II)/ΣCu ratio in the mother glasses.

On the other hand, the concentration of CO_3^{2-} dissolved in the melt decreases with increasing T_m. It may be considered that CO_3^{2-} residuals in glasses are distributed in grain boundaries between superconducting phases during crystallization process, resulting in deleterious effects on the superconducting properties.

Zheng et.al.[7] investigated the influence of glass preparation temperature on the Cu(I)/ΣCu ratio in Bi-Sr-Ca-Cu-O glasses and effects of Cu(I)/ΣCu ratio on the glass properties and crystallization behavior.

In this paper, we report effects of melting temperatures on the properties of high T_c superconducting $Bi_2Sr_2CaCu_2O_x$ glass-ceramics.

EXPERIMENTAL

Three kinds of powder mixture with different preparation process were used as starting materials as follows;

(a)A-mixture: A given mixture of premixed reagent powders corresponding to nominal composition of $Bi_2Sr_2CaCu_2O_x$ which were made by coprecipitation from oxalates of each metal and subsequent calcination(Furuuchi Chemical Co.).
(b)B-mixture: Non-calcined mixture of conventional oxide and carbonate powders corresponding to nominal composition of $Bi_{2.2}Sr_2CaCu_2O_x$ which are composed of reagent Bi_2O_3, $SrCO_3$, $CaCO_3$, and CuO.
(c)C-mitxure: Calcined B-mixture at 800°C for 20 h.
A- and B-mixture were melted in a platinum crucible at 1100-1400°C for 5 min and then quenched by iron pressing. To obtain high Cu(II)/ΣCu ratio of the melt-quenched glass, O_2 bubbling into the melt of A-mixture at 1100°C was specially carried out. C-mixture was also melted at 1100°C to obtain a smaller amount of CO_3^{2-} in the glass than that prepared from B-mixture. The melt-quenched samples were heat-treated at 800-880°C for 50 h in air. After heat-treatment, the samples were removed out of the furnace and cooled rapidly in air.

Electrical resistivities are measured using four-probe method. The quantitative determination for Cu(I) ion and total copper ion was conducted for the melt-quenched materials by cerate titration[8] and inductively coupled plasma spectrometry(ICPS), respectively. The Cu(II) ion content was obtained by subtracting Cu(I) ion content from total copper ion content. The CO_3^{2-} concentrations of glass samples were determined by infrared spectroscopy. The weight gains of the glasses during heating at a rate of 5°C/min were measured using the Thermogravimetric Analyzer(Seiko Instrument Inc.) or were also obtained by direct measurement before and after the heat-treatment.

RESULTS AND DISCUSSION

Figure 1 shows that the higher the glass melting temperature is, the lower is Cu(II)/ΣCu ratio. When the glass melting temperature varied from 1100°C to 1400°C, the ratio of Cu(II)/ΣCu shifted from 0.36 to 0.2. CuO dissociates into Cu_2O and O_2 at higher temperatures[5] as following reaction

$$4CuO_{(s)} = 2Cu_2O_{(s)} + O_2 \quad (1)$$

The same dissociation should take place during the melting in the Bi-Sr-Ca-Cu-O system.

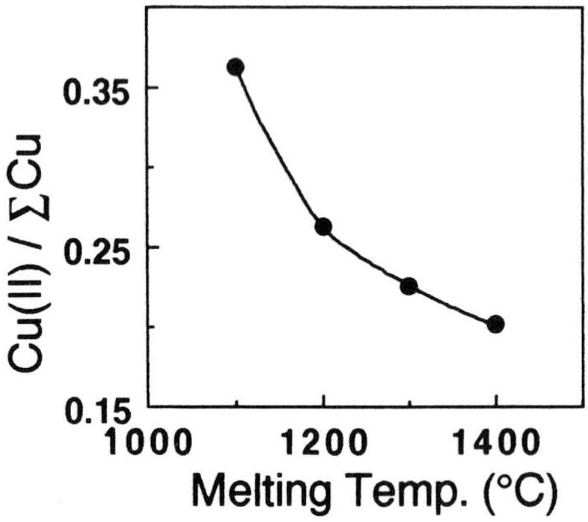

Fig.1 Dependence of the Cu(II)/ΣCu ratio of $Bi_2Sr_2CaCu_2O_x$ glasses on the melting temperature.

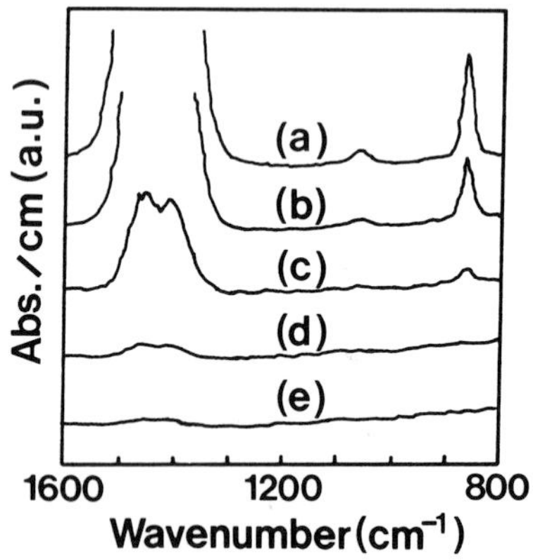

Fig.2 Infrared absorption spectra of $Bi_{2.2}Sr_2CaCu_2O_x$ glasses. Melting temperature:(a)1100°C, (b)1100°C(calcined powder), (c)1200°C,(d)1300°C,(e)1400°C.

The variation of standard free energy change with temperature for the reaction(1) is expressed by following equation[9].

$$\Delta G° = 69,900 + 12.2T\ln T - 88.6T \quad (2)$$

The highest Cu(II)/ΣCu ratio(0.48) among the melt-quenched glasses was achieved by the bubbling of oxygen gas in the melt.

Figure 2 shows infrared absorption spectra for the melt-quenched glasses prepared from B-mixture. Three absorption bands were observed in IR spectra; the first is broad band split into two bands at 1460-1410cm^{-1} and the second is 1060cm^{-1} and third is 860cm^{-1}. These bands were assigned to the vibrational spectra of the carbonate ions. It is noted that the peak intensities gradually decreases as the glass melting temperature increases.

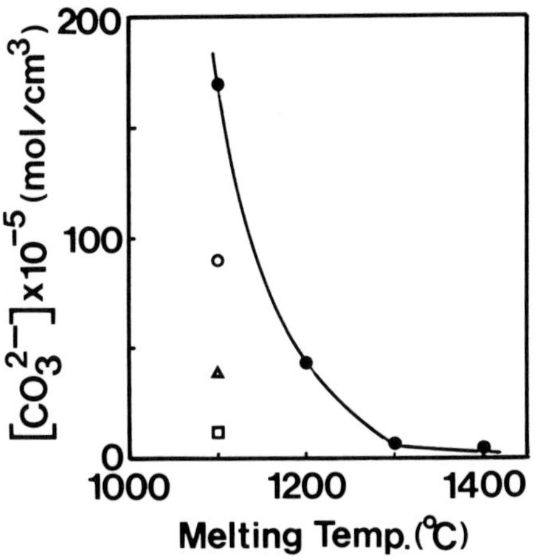

Fig.3 Dependence of CO_3^{2-} concentration of $Bi_{2.2}Sr_2CaCu_2O_x$ glasses on the melting temperature. Melting conditions: ●○ each temperature x 5min (○ calcined powder), △1100°C x 20min, □1100°C x 60min.

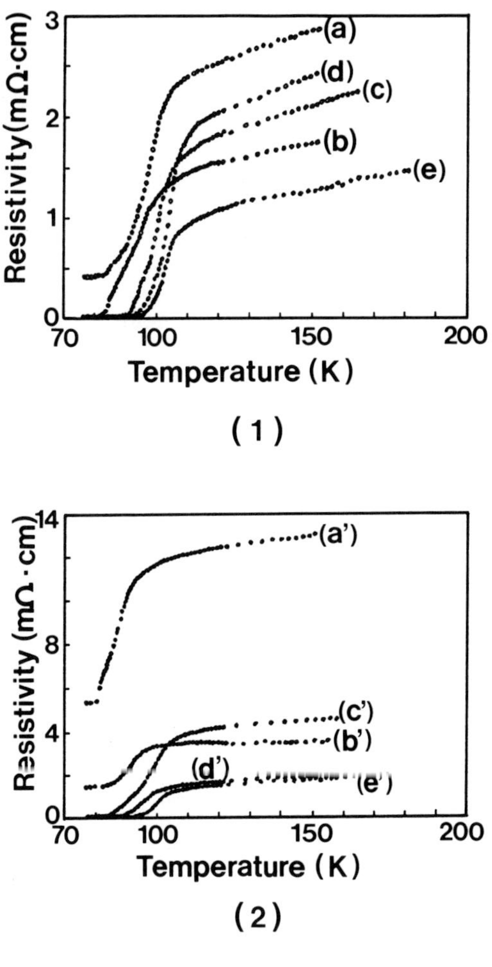

Fig.4 Temperature dependence of the resistivities of (1)$Bi_2Sr_2CaCu_2O_x$ (820°C x 50h) and (2)$Bi_{2.2}Sr_2CaCu_2O_x$(850°C x 50h) glass-ceramics. Melting conditions: (a)1100°C(O_2 bubbling), (a')1100°C(calcined powder), (b) and (b')1100°C,(c) and (c')1200°C, (d) and (d')1300°C, (e) and (e') 1400°C.

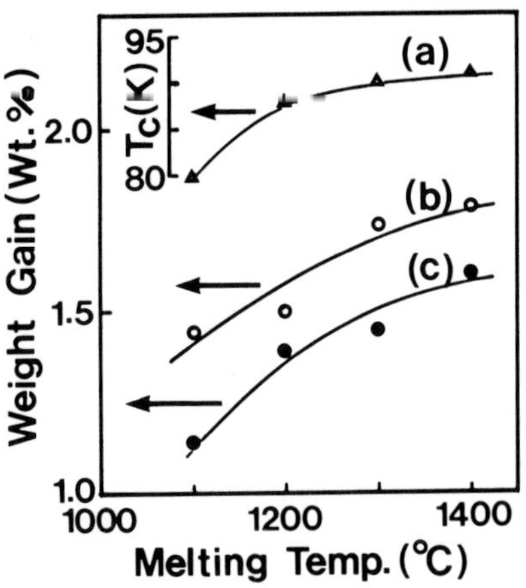

Fig.5 Melting temperature dependence of T_c and weight gains: (a)T_c of $Bi_2Sr_2CaCu_2O_x$ glass-ceramics heat-treated at 820°C for 50h, (b)weight gains by heating at 820°C for 50h, (c)maximum weight gain in TGA curve(heating rate = 5°C/min, up to 900°C).

The CO_3^{2-} concentrations in the glass were determined from IR spectra as shown in Fig.3. As the melting temperature increases, the concentrations of CO_3^{2-} dissolved in glass samples decrease. When the glass melting temperature varies from 1100°C to 1400°C, the CO_3^{2-} concentrations vary from $170x10^{-5}$ to $3.2x10^{-5}$ mol/cm^3. Above 1300°C a very small amount of CO_3^{2-} remains in the melt. At the same melting temperature of 1100°C, as the melting time varies from 5 min to 60 min, the CO_3^{2-} concentrations vary from $170x10^{-5}$ to $12x10^{-5}$ mol/cm^3. The CO_3^{2-} concentration in the sample prepared from C-mixture is about half of that in the sample prepared from B-mixtyre under the same melting conditions(1100°C for 5 min).

Figure 4 shows the temperature dependence of resistivities for the heat-treated samples (1) at 820°C for50 h, which were prepared from A-mixture and (2) at 850°C for 50 h, which were prepared from B- or C-mixture. The T_c of the glass-ceramics obtained by the same heat-treatment increases with increasing melting temperature of the mother glasses. The glasses with the highest Cu(II)/ΣCu ratio (0.48) which were obtained by the bubbling of O_2 in the melt at 1100°C were crystallized by the same heat-treatment; the T_c was the lowest among all the specimens in Fig.4(1). Although the concentration of CO_3^{2-} in the specimens prepared from B-mixture, the T_c of the former is lower than that of the latter as shown in Fig.4(2). Therefore, it is concluded that a determining factor to affect on superconducting properties is the Cu(II)/ΣCu ratio but not the CO_3^{2-} concentrations.

Figure 5 shows the melting temperatre dependence of T_c of the glass-ceramics and weight gains during heating of mother glass specimens; the curve (a) is for T_c and the curve (b) and (c) are for weight gains, respectively. The weight gain increases with increasing T_m; this is because the content of Cu(I) to be oxidized during the heat-treatment is higher in the glass with higher T_m. The mean copper valence in glass-ceramics, which was calculated on the assumption that weight gains are entirely due to the oxidization of Cu(I) ions, also increases with T_m. The mean copper valence in $Bi_2Sr_2CaCu_2O_x$ phase is reported to be 2.07[6].

In this study, as the melting temperature varies from 2.06 to 2.19 whose values may deviate from a true value but relatively right. The increase of T_c with increasing T_m is considered to be caused by the increase of mean copper valence or hole concentrations in the glass-ceramics.

In summary, both Cu(II)/ΣCu ratio and CO_3^{2-} concentrations of the mother glasses vary to a considerable extent, depending on T_m. When the melting temperatures were varied from 1100°C to 1400°C, the ratio of Cu(II)/ΣCu shifts from 0.36 to 0.2 and CO_3^{2-} concentrations vary $170x10^{-5}$ to $3.2x10^{-5}$ mol/cm^3. The T_c of the glass-ceramics obtained by the same heat-treatment of the mother glasses increases with T_m of the resulting glasses of which mechanism is not known yet; this may be explained by the increase of mean copper valence or hole concentrations in the glass-ceramics, while the CO_3^{2-} ions do not affect perceptibly the superconducting properties of the glass-ceramics.

ACKNOWLEGEMENT: One of the authors(Abe) thanks support by a Grand-in-Aid (No.0255160) from the Japanese Ministry of Education, Science and Culture.

REFERENCES

1. Abe Y, Hosono H, Hosoe M, Iwase J, Kubo Y (1988) Appl Phys Lett 53: 1341
2. Abe Y, Arakawa H, Hosoe M, Hikichi Y, Iwase J, Hosono H, Kubo Y (1989) Jap J Appl Phys 28: L1929
3. Zheng H, Mackenzie JD (1988) Phys Rev B 38: 7166
4. DeGuire MR, Bansal NP, Kim CJ (1990) J Am Ceram Soc 73: 1165
5. Gadalla AMM, Ford WF, White J (1963) Trans Brit Ceram Soc 62: 45
6. Endo H, Tsuchiya J, Kijima N, Sumiyama A, Mizuno M, Oguri Y (1988) Jap J Appl Phys 27: L1906
7. Zheng H, Colby MW, Mackenzie JD, private communication
8. Riebling EF (1973) Inorg Chem 12: 2213
9. Kubaschewski O, Alcock CB (1979) Metallurgical thermochemistry. Pergamon Press. London, pp379

5.3 Others

Evaluation of AC Loss in High-T_c Superconductors for Power Cable Application

MASAHIKO NAKADE[1], TSUKUSHI HARA[1], TAKAHIKO YAMAMOTO[1], HARUHIKO HOSHINO[2], SATORU TANAKA[3], NAOTAKA ICHIYANAGI[3], and ZENSUKE IWATA[3]

[1] Electric Power Department, Engineering Research Center, Tokyo Electric Power Co., Inc., 4-1, Nishi-Tsutsujigaoka 2-chome, Chofu, Tokyo, 182 Japan
[2] Power and Substation Department, Kandenko, 8-33, Shibaura 4-chome, Minato-ku, Tokyo, 108 Japan
[3] Chiba Research Laboratory, The Furukawa Electric Co., Ltd., 6 Yawata-Kaigandori, Ichihara, Chiba, 290 Japan

ABSTRACT

An important parameter which should be evaluated in detail for the design of a compact high-Tc superconducting cable is the AC loss associated with the alternative transport current. We have commenced the quantitative evaluation of the AC loss in high-Tc sintered bulks and Ag-sheathed wires. In order to clarify the differences in AC losses which are dependent on the frequency of applied magnetic fields, two methods have been used. One method is to measure magnetization curves by applying 50 Hz AC magnetic fields (AC method). The other is to measure magnetization curves by means of a SQUID susceptometer (DC method).

KEY WORDS: AC loss, magnetization curve, flux creep, Bean model

INTRODUCTION

With the rapidly mounting demand for electric power in Japan's large cities, construction costs for underground transmission lines are also rising. The authors conducted comparative analysis of transmission costs for future main underground transmission lines, the results of which suggest that the replacement installation of compact, large-capacity AC high-Tc superconducting cables in existing ducts and tunnels would be more cost-efficient than the expansion of conventional cable lines that would be accompanied by further construction of tunnels and ducts [1,2]. When applied to electric power cable, however, high-Tc superconductors pose a problem in terms of the lower critical field, which is one order of magnitude smaller than that of metal superconductors. The surface current on the conductor decreases in proportion to a decrease in the lower critical field. For this reason, the transport current must be conducted not only on the superconducting layer surface but also within it in the case of a compact AC high-Tc superconducting cable. The AC loss in the superconducting layer of this cable is considered to be substantially greater than that of its metal superconductor counterparts [3]. The present paper describes comparative studies of AC losses in Y-based and Bi-based high-Tc superconducting bulks and Ag-sheathed tapes measured by the AC method and DC method with reference to the theory of AC loss.

EXPERIMENTAL

There are several conventional methods for measuring AC losses with relatively high sensitivity. For example, one method is to apply static magnetic fields or quasi-static magnetic fields to superconducting wires to determine their degree of magnetization (DC method). Another method is to conduct an AC current in the wires to determine the voltage induced in them. The former method provides superior sensitivity, but fails to determine eddy current losses developed in the metal for stabilization used in wires, so that AC losses in the entire wire cannot be measured. On the other hand, the latter method also cannot be sufficiently depended upon for practical use, because the critical current density in the wires is still small so that the magnetic field induced by AC conduction is considerably smaller than the field applied to the wire during cable operation. In order to solve these problems, the authors have developed a new system with the capability of measuring an AC loss actually induced by high-Tc superconductors in the AC magnetic field of commercial frequency. This system permits measurements to be performed by the AC method [4].

Figure 1 shows an overview of the AC loss measurement system based on the AC method, developed by the authors. This system is designed to apply AC magnetic fields of up to 0.22 T in the range of 10 to 100 Hz to short samples about 30 mm long, and employs pickup coils to detect magnetization developed in the samples. Field and magnetization signals are con-

Fig. 1. AC Loss Measurement System Based on AC Method

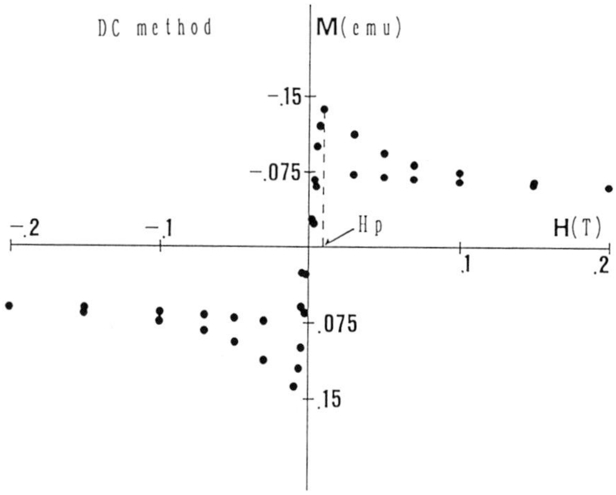

Fig. 2. Magnetization Curve for Y-Based Bulk Obtained by DC Method (at 77 K)

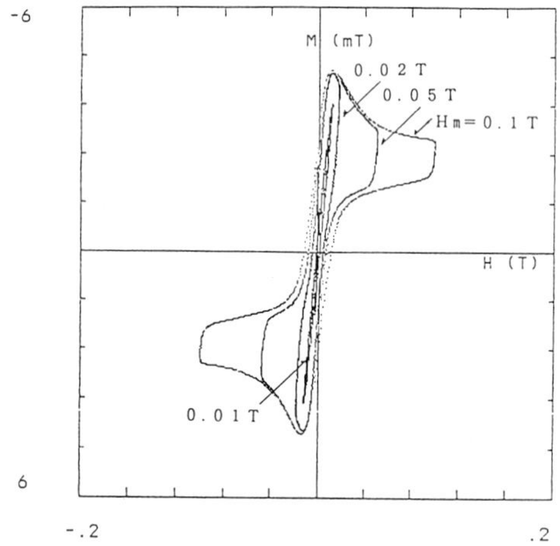

Fig. 3. Magnetization Curves for Y-Based Bulk Obtained by AC Method (at 77 K)

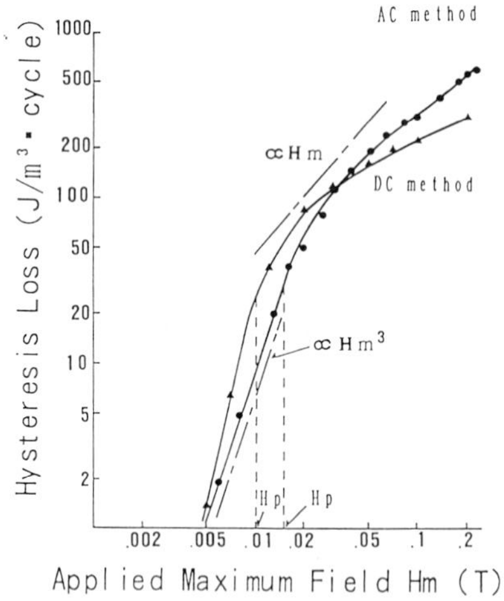

Fig. 4. AC Losses in Y-Based Bulk (at 77 K)

verted to digital form with the aid of a personal computer to yield magnetization curves. The magnitude of AC loss can then be determined by calculating the area surrounded by the resulting magnetization curves. The DC method relies on a SQUID susceptometer to measure AC losses. In the DC method, the magnitude of magnetization relative to a certain magnetic field must be determined 10 to 30 min after field application in order to ensure that the magnetic field applied was stabilized. In contrast hand, the AC method requires only 20 msec to permit one magnetic curve to be derived at 50 Hz.

RESULTS AND DISCUSSION

Figures 2 and 3 show magnetization curves obtained by the DC method and AC method, respectively, for Y-based bulk samples (2 mm thick, 5 mm wide, 25 mm long). Assuming that the magnitude of applied magnetic field where the absolute value of magnetization reaches a maximum is Hp, the values of Hp obtained by the DC method and AC method were 0.010 T and 0.015 T, respectively. Hp represents the magnitude of applied magnetic field observed when the flux penetrates to the center of the superconducting layer. The magnetization curves derived in this way represent the hysteresis of magnetization due to flux pinning as in the case of metal superconductors, thus suggesting that hysteresis corresponding to the area surrounded by the magnetization curves took place in the superconductor. Figure 4 shows AC losses derived from these magnetization curves. The results obtained by the AC method indicate that AC losses were proportional to the third power of magnetic fields below Hp, and proportional to the first power above Hp. This finding shows a tendency similar to the Bean model [5]. In the Bean model, the hysteresis loss W (J/m^3·cycle) in a superconductor is given by:

$$Hm \leqq Hp \qquad W = \frac{2\mu 0}{3} \cdot \frac{Hm^3}{Jc \cdot d}$$

$$Hm \geqq Hp \qquad W = 2d\mu 0JcHm\left(1 - \frac{2}{3} \cdot \frac{Hp}{Hm}\right)$$

where Hp=Jc·d, Hm is the applied magnetic field, $\mu 0$ is the space permeability, Jc is the critical current density, and d is half the thickness of the superconductor.

It was found that values of Hp obtained by the DC method were smaller than those obtained by the AC method. The AC loss was greater when determined by the DC method below 0.03 T than when determined by the AC method, and vice versa when it was determined above 0.03 T. Such differences between the results obtained by the two methods may be explained by flux creep and the Bean model. More specifically, flux creep is more likely to occur for measurements obtained by the DC method than by the AC method, with an attendant decrease in the critical current density Jc. The Bean model suggests that Hp is proportional to Jc, and that AC losses are in inverse proportion to Jc in magnetic fields below Hp while they are proportional to Jc above Hp. This explains why values of Hp determined by the DC method were smaller than those determined by the AC method. Accordingly, measurements of AC losses by the DC method were greater than those by the AC method in magnetic fields below Hp, and vice versa in magnetic fields above Hp.

Figure 5 shows the results of AC loss measurements by the AC method for an Y-based Ag-sheathed tape (0.02 mm thick, 5 mm wide, 33 mm long). As in the case of the Y-based bulk sample, AC losses were proportional to the third power of magnetic fields below Hp, while they were proportional to the first power above Hp. Because AC losses in the Ag-sheathed tape consist of hysteresis losses in the superconductor and eddy current losses in silver, these two components of AC loss must be separated from each other. Figure 6 shows the results of eddy current loss measurements by the AC method for a silver sheet (1 mm thick, 10 mm wide, 45 mm long). Because the eddy current loss is proportional to the square of the thickness of a sample, eddy current losses in silver sheath 50 μm thick were estimated by calculations based on the results of the above measurements. The resulting estimated values were found to be about one four-hundredth the AC losses obtained in Figure 5, suggesting that eddy current losses account for a negligibly small percentage of the total AC loss. Figure 7 shows magnetization curves obtained by the AC method for a Bi-based Ag-sheathed tape (0.17 mm thick, 3 mm wide, 30 mm long). The results indicate that the hysteresis was smaller than that observed in the Y-based bulk sample, suggesting that the pinning force was weak. Figure 8 shows AC losses in the Bi-based Ag-sheathed tape. For the Bi-based Ag-sheathed tape, AC losses measured by the AC method were also proportional to the third power of magnetic fields below Hp and to the first power above Hp. It should be noted that values of Hp were too small in the measurements by the DC method to yield any measurement data

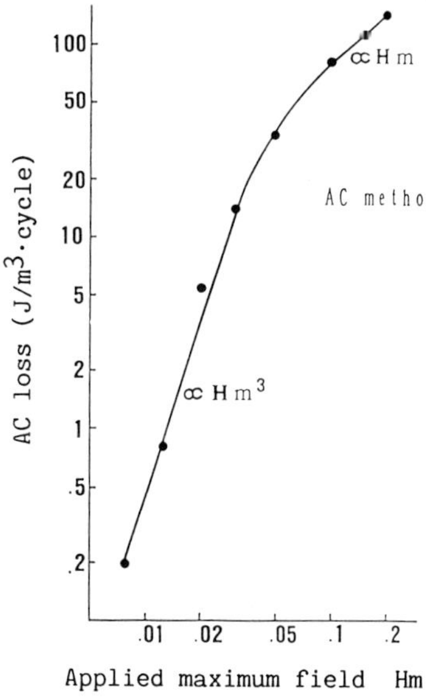

Fig. 5. AC Losses in Y-Based Ag-sheathed Tape Obtained by AC Method (at 77 K)

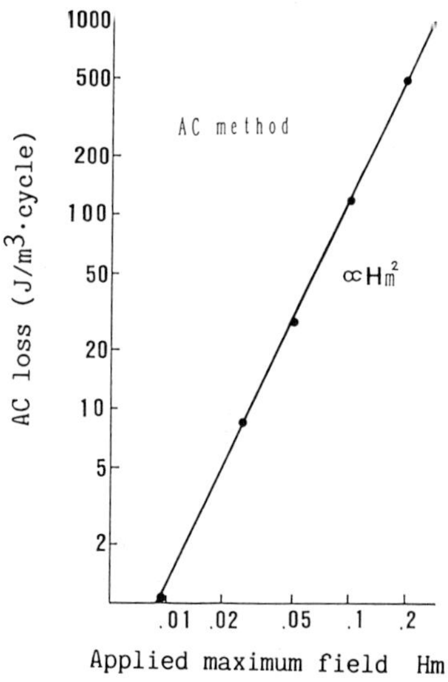

Fig. 6. AC Losses in Ag Sheet Obtained by AC Method (at 77 K)

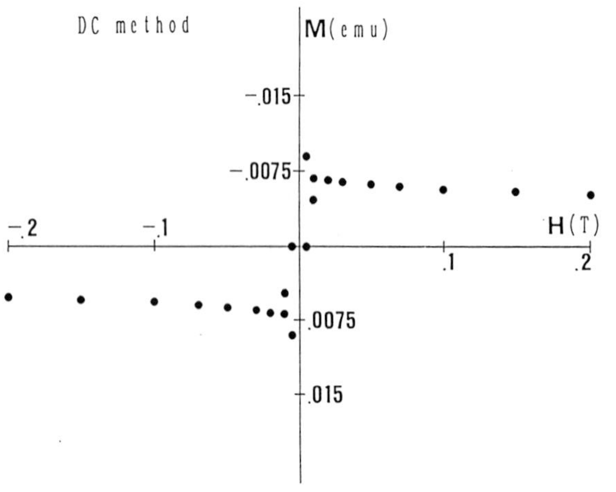

Fig. 7. Magnetization Curve for Bi-Based Ag-Sheathed Tape Obtained by DC Method (at 77 K)

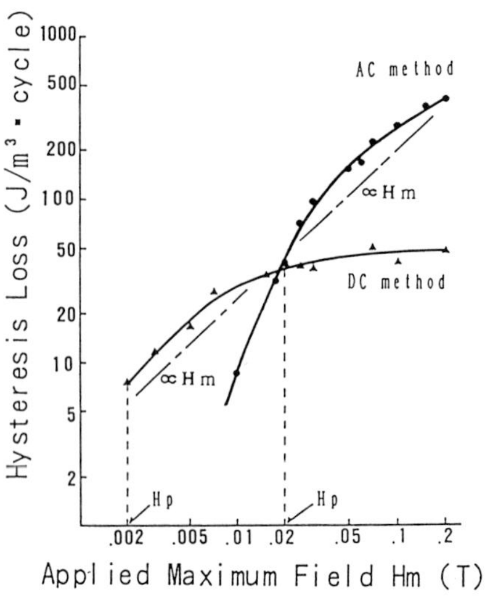

Fig. 8. AC Losses in Bi-Based Ag-Sheathed Tape (at 77 K)

below Hp. There was also a greater discrepancy between values of Hp obtained by each method for the Bi-based samples than for the Y-based samples. Further, differences between values of AC loss obtained by each method and the tendency toward deviation from the line representing the first power of magnetic fields above Hp obtained by the DC method were also more substantial for the Bi-based samples than for the Y-based samples. These findings may be explained by the supposition that more flux creep occurs in Bi-based samples than in Y-based samples, which, in turn, de-creases the critical current density substantially.

SUMMARY

AC losses in high-Tc superconducting wires were measured by the AC method and DC method. The following findings were obtained:

(1) For both Y-based and Bi-based samples, AC losses measured by the AC method were proportional to the third power of magnetic fields below Hp and to the first power above Hp. This result shows a tendency similar to that suggested by the Bean model.

(2) For both Y-based and Bi-based samples, values of Hp obtained by the DC method were smaller than those obtained by the AC method. Values of AC loss obtained by the DC method were greater than those obtained by the AC method in relatively low magnetic fields, and vice versa in relatively high magnetic fields. Such differences between the data obtained may be explained by flux creep and the Bean model.

(3) Values of Hp in Bi-based samples obtained by the DC method were considerably smaller than those in Y-based samples, indicating that differences between values of AC loss obtained by the two methods are more pronounced for the Bi-based samples than for the Y-based samples. Such differences between the Bi-based and Y-based samples are due to a greater degree of flux creep developed in the B-based samples than in the Y-based samples.

(4) With regard to the Ag-sheathed wires used in the present study, eddy current losses in the metal for stabilization were substantially smaller than AC losses in high-Tc superconductors.

REFERENCES

1. T. Hara , K. Okaniwa, S. Tanaka, N. Ichiyanagi (1990) 1990 IEEE Winter Meeting 075-2PWRD
2. T. Hara, M. Nakamura (1990) IEE of Japan 110-B, 58-66
3. N. Ichiyanagi, S. Tanaka, F. Enokubo, Y. Furuto (1990) Cryogenic Engineering 25, 12-17
4. M.P. Maley, J.D. Thompson, L.R. Newkirk (1981) Cryogenics, 675-678
5. C.P. Bean (1962) Physical Review Letters 8, 250

Influence of Contact Resistance on Ic Measurement of Bi Based Oxide Superconductor

KENICHI TAKAHASHI[1], YOSHIMASA KAMISADA[1], TETSUJI OGAWA[1], KAZUO MICHISHITA[2], YUICHI IKUHARA[2], TSUTOMU TSUNOOKA[2], and YUKIO KUBO[2]

[1] Super-GM (Engineering Research Association for Superconductive Generation Equipment and Materials), 14-10, Nishitenma 5-chome, Kita-ku, Osaka, 530 Japan
[2] Japan Fine Ceramics Center, 4-1, Mutsuno 2-chome, Atsuta-ku, Nagoya, 456 Japan

ABSTRACT

In a measurement of critical current (critical current density) using 4-probe method, a number of factors, such as contact resistance, influence the measurement results according to Ic increasing. In this study, using Bi based superconductor prepared by FZ method, influences of temperature and magnetic field on electrode contact resistance and influence of pulsating current conditions on critical current measurement were studied. Using the $10^{-5}\Omega\ cm^2$ sample we obtained almost same Ic in the conditions of pulsating current in the range of 50～ 2000msec. On the other hand, samples with high contact resistivities were melt in the conditions of long pulsating current. Moreover we observed the microstructure with the immediate vicinity of contact by EPMA analysis.

KEY WORDS: Bi based superconductor, Critical Current, Contact Resistance, 4-probe method

INTRODUCTION

In a research and development of wire fabrications for the purpose of applying oxide superconductor to electric power apparatuses, it is a large subject how to increase the current capacity as well as the improvement of critical current density. A number of factors, such as contact resistance, influence the results of critical current measurement according to increasing transport current. In this study, using Bi based superconductor prepared by FZ method, we examined the influences of Ag sputtering treatment and annealing on the contact resistivities for the electrode of 4-probe method, the dependences of temperature and magnetic field on contact resistivities, and the influence of pulsating current condition on Ic measurements.

Moreover, we observed the microstructure with immediate vicinity of contact by EPMA and studied the correlation between microstructure and contact resistivity.

EXPERIMENT

$Bi_2Sr_2CaCu_2O_x$ powders coprecipitated with oxalic salt were sintered twice in 1123K and crystal samples were grown by the floating zone (FZ) method at the rate of 2～ 5mm/hr. Samples were cut in size of about 1 × 0.5 × 15～ 20mm using non-water soluble fluid and annealed adequately. Bi based superconductor contained 10 wt% Ag for reducing contact resistivities were also prepared[1].

Fig.1 shows the copper block for supporting a sample and arrangements of voltage terminals for measuring Ic and contact resistance by 4-probe method. In the forming of electrode, 99.99% Ag underlayer in the

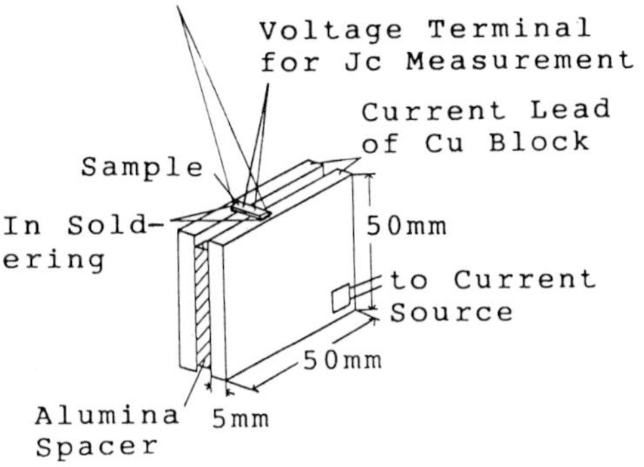

Fig.1 Schematic diagram of Cu block and arrangement of voltage terminals.

thickness of about 1.5 μ m was prepared by RF sputtering. Some samples were annealed at 773K in oxygen flow.
Electrodes were formed by ultrasonic soldering using 99.99%In(1 mmϕ wire) and fluorinate resin coated copper wire was used as the lead wire.
The dependence of temperature on contact resistance was measured in the cryostat which can control the temperature from about 10 to 300K. Ic and contact resistivities were measured in liquid nitrogen(77K).
Ic was measured with the criterion of 1 μ V/cm at 0T in the conditions of 50~2000 msec interval of rectangle pulsating current per 5sec (the current increases by 0.1A). The spacing of voltage terminals was 3.5mm constantly.

EXPERIMENTAL RESULTS AND DISCUSSIONS

(1) Contact resistivities

V-I characteristics of In soldered sample was measured at 77K in order to study the influence of Ag sputtering treatment on contact resistivities using Bi based superconductor and 10 wt% Ag doped Bi based superconductor and also in order to study the influence of annealing on contact resistivities using 10 wt% Ag doped Bi based superconductor with Ag sputtering. Voltage terminals for measuring the contact resistance were taken at the copper block and contact resistance Rc(Ω) was defined as voltage drop at a current of 1 A and contact resistivities were defined as products of contact resistance and contact area. It is favorable to enlarge the area of contact for radiating heat which generates at Ic measurement and therefore Ic was measured under constant area of $0.06cm^2$.
Fig.2 shows a linear relationship of V-I characteristics of 10 wt% Ag doped Bi based superconductor with Ag sputtering and reveals ohmic characteristics of contact resistance. On the other hand, the increase of contact resistance of undoped Bi based superconductor without Ag sputtering is depressed with increase of current and V-I characteristics shown in Fig. 3 reveals non-ohomic characteristics. Contact resistivities of undoped Bi based superconductor shows $10^{-2}\sim 10^{-3}\Omega\ cm^2$. Ag doped Bi based superconductor shows the magnitude of $10^{-5}\Omega\ cm^2$, which is lower in vary of 2~3 order of magnitude, and reveals $2.1\times 10^{-5}\Omega\ cm^2$ at the lowest value.

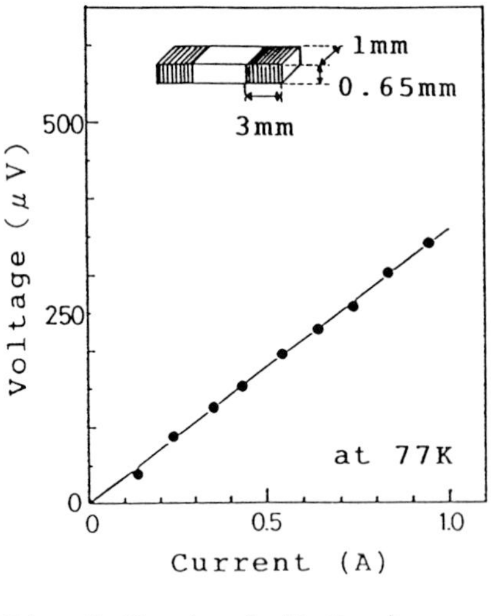

Fig.2 Typical V-I characteristics of Ag doped and sputtered sample.

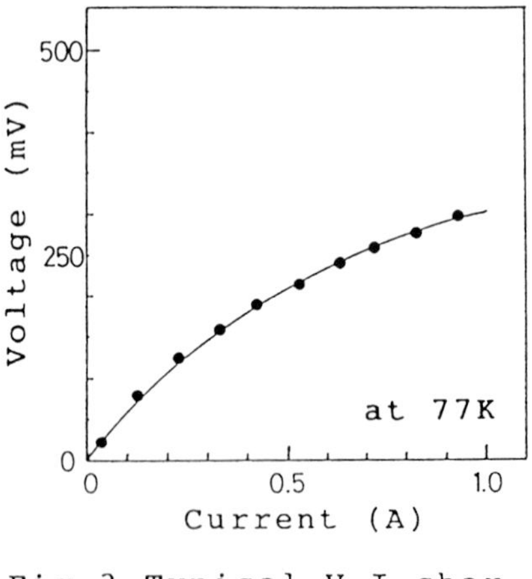

Fig.3 Typical V-I characteristics of undoped sample.

(2) Dependences of temperature and magnetic filed on contact resistivities

The dependence of temperature on contact resistivities was measured at the decreasing rate of 2K/min. by 4-probe method using 10 mA current.

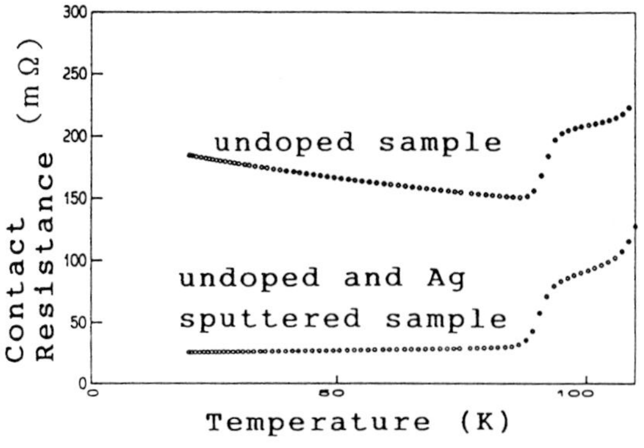

Fig.4 Temperature dependence of contact resistances.

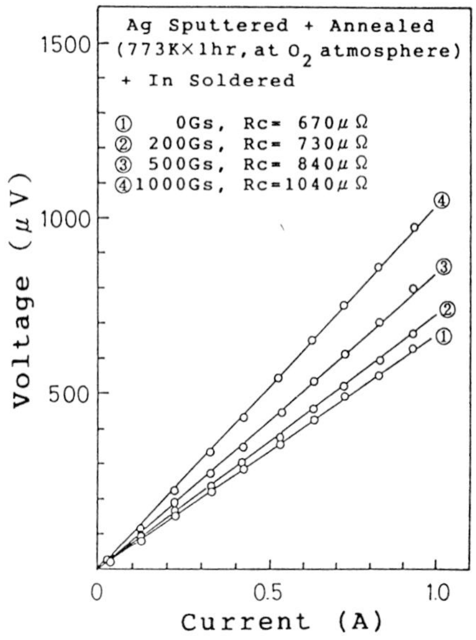

Fig.5 I-V curves for a Ag-doped sample at 77K as a function of applied field strength.

Fig.4 shows the dependence of temperature on contact resistance of undoped Bi based superconductor. The resistivities of undoped Bi based superconductor without Ag sputtering decrease according to the temperature increasing and reveals characteristics of semiconductor. But the dependence of temperature on resistivities of undoped Bi based superconductor with Ag sputtering reveals metallic in character.
The dependence of magnetic filed on contact resistivities perpendicular to the longitudinal direction of the sample was measured to 1000 Gauss as the maximum.
Fig.5 shows the V-I characteristics of Ag doped Bi based superconductor in various magnetic fields and as the magnetic field is increased, the contact resistivities increase. Miyake etal.[2] reported that contact resistivities of Y based superconductor increased according to increasing of magnetic field and it is necessary to further study the influence of magnetic field on contact resistivities.

(3) Dependence of pulsating current conditions on critical current

Fig.6 shows the results of Ic measurement using Ag doped Bi based superconductor, which contact resistivities is $2.1\times 10^{-5}\Omega\,cm^2$, in the range

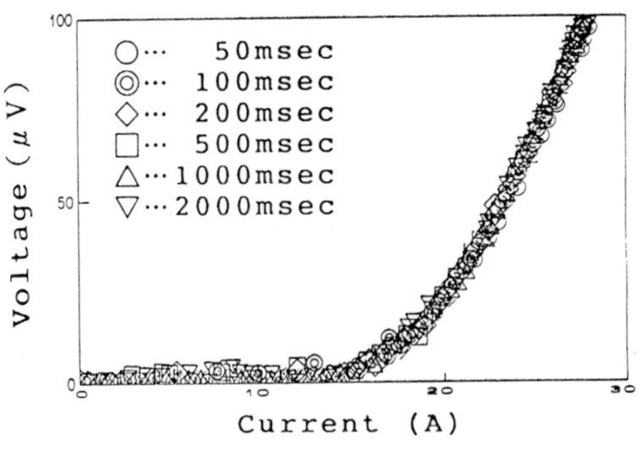

Fig.6 Results of Ic measurement for a Ag-doped sample as a function of current sweep rate. Contact resistivity is $2.1\times 10^{-5}\Omega\,cm^2$.

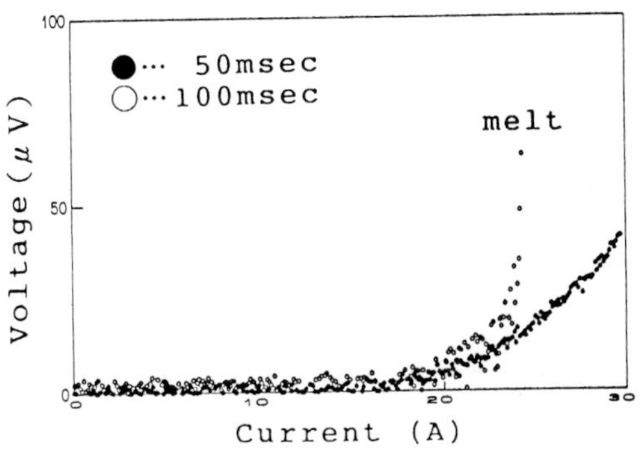

Fig.7 Results of Ic measurement for an undoped sample as a function of current sweep rate. Contact resistivity is $2.4\times 10^{-3}\Omega\,cm^2$.

of 50~ 2000 msec pulsating current. In $10^{-5}\Omega\ cm^2$ magnitude, Ic are almost same at these current sweep rate, using samples which can transmit 2~ 13 A . On the other hand in $10^{-3}\Omega\ cm^2$ magnitude, undoped Bi based superconductor is melt in the conditions of 10~ 17 A and 100 msec pulsating current shown in Fig. 7. For accurate Ic measurement, it is necessary to shorten the pulsating current length less than 50 msec or to reduce the contact resistivities.

(4) EPMA analysis of contacts

The correlation with microstructure near contact and contact resistivities was partly studied by EPMA analysis. There are no differences of the distributions of Bi, Sr, Ca and In between Ag doped Bi based superconductor and undoped Bi based superconductor. In the undoped Bi based superconductor, Cu and O are dispersed in the In solder. On the other hand in the doped Bi based superconductor, Ag is dispersed in the In solder and the tendency is observed with depressing the diffusion of Cu. Kusaka etal.[3] analyzed that the high resistivities contacts of Y based superconductor was contained tetragonal crystal phase and oxygen defect phase by ESCA analysis and Ekin etal.[4] reported that Ag layer between Y based superconductor and In did not contain O and In layer contained O by the AES analysis.
Cu in In solder may be diffused from Bi based superconductor and In_2O_3 may be considered to form at the interface of In and Bi based superconductor.
In case of forming In_2O_3 at the interface of In and oxide superconductor it may be to show the non-ohmic characteristics in Fig. 4 because In_2O_3 is oxide semiconductor at 77 K[5].

CONCLUSION

We investigated the dependences of temperature and magnetic field on contact resistivities and the dependence of pulsating current conditions on Ic using Ag doped and undoped Bi based superconductor with Ag sputtering and annealing treatment. In the case of Ag sputtering treatment the contact resistivities of Ag doped Bi based superconductor was $2.1\times 10^{-5}\ \Omega\ cm^2$ at the lowest value.
In order to reduce the contact resistivities of contacts, it is necessary to optimize the annealing and Ag sputtering treatment conditions furthermore by considering the results of interface analysis and to investigate the conditions for increasing transport current capacity by considering the heat generation at contacts.

ACKNOWLEDGEMENT

This research has been carried out to support the R&D of oxide superconducting wires under the Moonlight Project. The authors would like to thank the staffs concerned in this research.

REFERENCES

1.Shimizu N, Michishita K,Higashida Y,Yokoyama H,Hayami Y, Kubo Y,Inukai E, Saji A,Kuroda N, Yoshida H (1989) Jpn.J.Appl.Phys.28:L1955.
2.Miyake S, Aoyama T, Suzuki Y, Kusaka T, Yotsuya T, Ogawa S (1989) Proc.Japan.Sympo.Appl.Phys. 3a-A-11;131
3.Kusaka T, Suzuki Y, Aoki A, Aoyama T, Yotsuya, Ogawa S (1988) Progress in High Temperature Superconductivity vol.15:309
4.Ekin J W, Larson T M, Bergen N F, Nelson A J, Swaltzlander A B, Kazmerski L L, Panson A J, Blankenship B A (1988) 52(21),23:1819
5.Muller H (1968) Phys. Status Solidi.27:723

Low Surface Resistance of Ag-Added $YBa_2Cu_3O_{7-x}$ Ceramics

Osamu Ishii[1], Masakatu Senda[1], Tuneo Konaka[2], and Makoto Sato[2]

[1] NTT Applied Electronics Laboratories, Tokai, Ibaraki, 319-11 Japan
[2] NTT Transmission Systems Laboratories, Tokai, Ibaraki, 319-11 Japan

ABSTRACT

The surface resistance (Rs) of screen-printed $YBa_2Cu_3O_{7-x}$ (YBCO) thick films is investigated using cavities of 17 and 50 GHz in a temperature range of 30 K to 300K. Marked decreases in Rs are realized by using Ag-alloy-coated zirconia substrates. The optimum YBCO film thickness is also determined to be about 7 μm, where the lowest Rs values at 77 K (8 mΩ at 17 GHz and 48 mΩ at 50 GHz) are achieved. This improvement in Rs is attributed to the grain growth and c-axis orientation of the YBCO film.

KEY WORDS: superconductor, surface resistance, $YBa_2Cu_3O_{7-x}$

INTRODUCTION

Among the high Tc oxide superconductors, $YBa_2Cu_3O_{7-x}$ (YBCO) was the first material found to have a Tc higher than liquid nitrogen temperature (~77 K) giving a large potential for the device application[1]. From the many possible applications of YBCO, microwave devices (i.e. antennas and resonators) seem to be the most promising candidates[2,3]. To this end, it is very important to reduce the surface resistance (Rs) of YBCO. Although epitaxially grown YBCO thin films exhibit very low Rs[4], their fabrication processes are too complicated to enable the preparation of large-size devices. On the other hand, the screen-printing method[5] can offer large-scale YBCO thick films. We have already reported that screen-printed YBCO thick films exhibit a low Rs when Ag-coated substrates are used[6,7]. In this paper, we investigate the crystal grain morphology and c-axis orientation of screen-printed YBCO films in detail to lower their Rs even further.

EXPERIMENTAL

YBCO thick films (26.5 mm in diameter and 5 μm thick) were fabricated on Y_2O_3-stabilized ZrO_2 (YSZ) substrates and films (26.5 mm in diameter and 5~50 μm thick) were fabricated on Ag-alloy-coated ZrO_2 (Ag/YSZ) subtrates by the screen-printing method[8]. The thickness of the Ag-coating was 5~10 μm. The printed powders were annealed in an oxygen flow at 1025~1050 ℃ for 10 min. The fabrication processes were the same as those in ref.5.

The Rs of the films was measured by the cavity method from 30 to 300 K. Both TE_{011} and TE_{013} resonating cavities of 17 and 50 GHz were used to measure the Q values. Other measuring conditions and the Rs calculation method from the Q value have been reported in ref.9. The Rs of oxygen-free Cu was also measured for comparison.

A scanning electron microscope (SEM) was used to observe the crystal grain morphology. Auger electron spectroscope (AES) observation was carried out to analyze the Ag distribution. X-ray diffractometry was also used to evaluate the orientation of the YBCO grains. The orientation factor (P) was calculated from (002) and (013)(103)(110) X-ray diffraction intensities[10].

RESULTS AND DISCUSSION

Figure 1 shows SEM photographs of 30-μm-thick YBCO films formed on (a) a YSZ substrate, YBCO(30)/YSZ, and (b) an Ag/YSZ substrate, YBCO(30)/Ag/YSZ. The crystal grains of YBCO(30)/YSZ are much smaller than those of YBCO(30)/Ag/YSZ, indicating that the Ag-coating accelerated the grain growth. The X-ray diffraction patterns of both films are shown in Fig.2. In YBCO(30)/YSZ, both the YBCO phase (123-phase) and the Y_2BaCuO_5 phase (211-phase) are detected. On the other hand, only the 123-phase is detected in the YBCO(30)/Ag/YSZ. These results prove that the Ag-coating acts as a buffer layer which isolates the YBCO grains from YSZ and prevents 211-phase formation.

Figure 3 shows SEM photographs of (a) 50-, (b) 15-, (c) 7- and (d) 5-μm thick YBCO films formed on Ag/YSZ substrates. From these pictures, it can be seen that the grain growth was enhanced as the YBCO film thickness decreases from 50 to 7 μm and that the 5-μm-thick YBCO film was not formed uniformly because of the agglomeration of the YBCO powders. This result is consistent with the report by Greuter et al. that the growth of YBCO grains starts at the bottom of YBCO films screen-printed on Ag substrates[11].

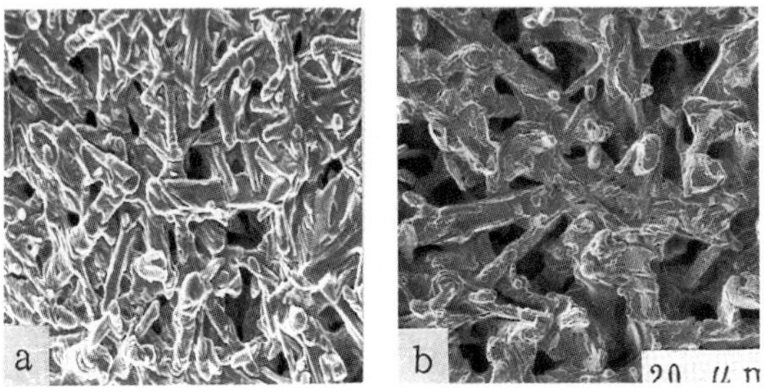

Fig.1 SEM photographs of (a) YBCO(30)/YSZ and (b) YBCO(30)/Ag/YSZ

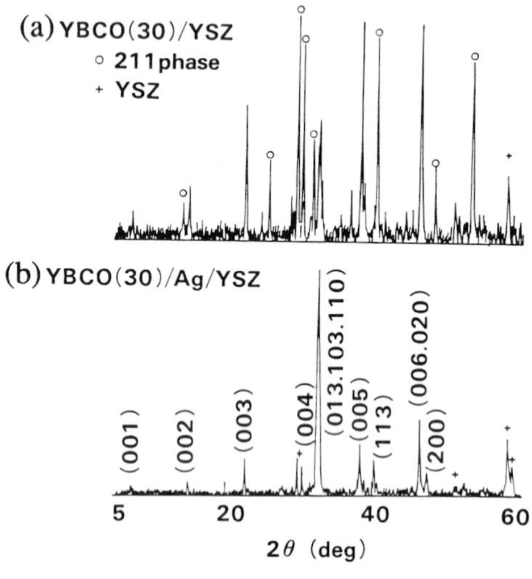

Fig.2 X-ray diffraction patterns of (a) YBCO(30)/YSZ and (b) YBCO(30)/Ag/YSZ

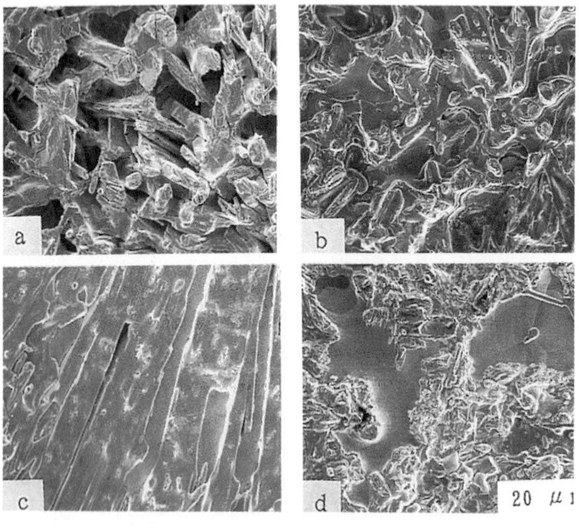

Fig.3 SEM photographs of (a) 50- (b) 15- (c) 7- and (d) 5-μm thick YBCO films formed on Ag/YSZ

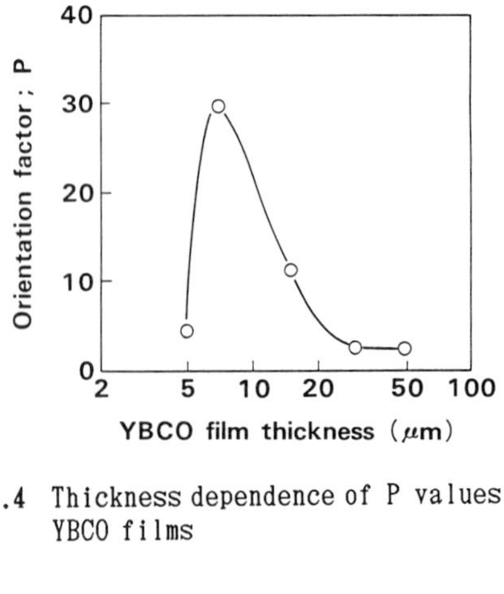

Fig.4 Thickness dependence of P values for YBCO films

Figure 4 shows the YBCO film thickness dependence of the P values. The P value shows a sharp maximum peak at a thickness of 7 μm, which indicates that the c-axis orientation of the YBCO grains improved in a similar manner to the YBCO grain growth.

Figure 5 shows cross sections of (a) SEM-, (b) Ag- (c) O- and (d) Cu- images for the 7-μm-thick YBCO film on an Ag/YSZ, YBCO(7)/Ag/YSZ. No Ag was detected on the surface, which indicates that the sample was entirely covered with coarse plate-like YBCO grains. The Ag, O and Cu images also proved that the Ag was not oxidized but included in the YBCO crystals. Since the sintering temperature (1025~1050 ℃) was higher than the melting point of Ag (960.5 ℃) and the density of Ag (10.5 g/cc) is higher than that of YBCO (6.3 g/cc), the YBCO grains should be suspended in the liquid phase Ag during the sintering process. Such highly mobile YBCO powders are considered to enhance the grain growth and c-axis orientation.

Figure 6 shows the thickness dependence of the Rs at 50 GHz for YBCO/YSZ (open square: 30 K, filled square: 77 K) and YBCO/Ag/YSZ (open circles: 30K, filled circles: 77 K). YBCO/YSZ exhibits an Rs several times higher than that of YBCO/Ag/YSZ, showing that the 211-phase and the fine YBCO crystal grains caused considerable Rs deterioration. For YBCO/Ag/YSZ, the Rs values at 77 K decreased with decreasing film thickness from 220 mΩ for 50 μm thickness to 48 mΩ for 7 μm thickness followed by an abrupt increase to 621 mΩ for 5 μm thickness. The Rs value of 48 mΩ is approximately equal to that of oxygen-free Cu (~50 mΩ). The higher Rs of YBCO(5)/Ag/YSZ is attributed to the discontinous film structure. The Rs value of YBCO/Ag/YSZ at 30 K also reaches the minimum at 7 μm film thickness. It should be noted that the Rs difference between 77 K and 30 K is smallest at 7 μm film thickness, indicating a steep Rs drop around the transition temperatures (~91K). These phenomena agree well with the report by Sato et al. that as the YBCO grains in sintered pellets become coarser, the Rs drop around Tc becomes sharper[12]. These results prove that the grain growth and c-axis orientation of the YBCO film markedly improve the Rs.

Figure 7 shows the Rs frequency dependences at 77 K for YBCO/YSZ (filled circles) and YBCO/Ag/YSZ (open circles) together with our previously reported data[6]. The measuring accuracy at 17 GHz was of several milliohm order. The Rs values of YBCO/Ag/YSZ are lower than those of Ag-doped and Ag-nondoped YBCO pellets and those of YBCO/YSZ. In par-

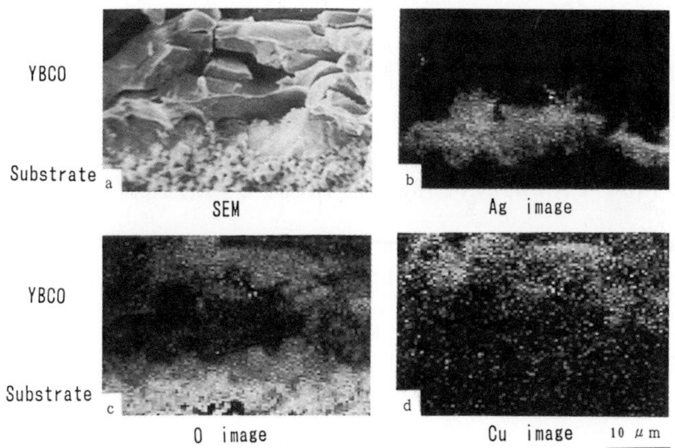

FIG.5 AES cross sections of (a) SEM- (b) Ag- (c) O- and (d) Cu- images for YBCO(7)/Ag/YSZ.

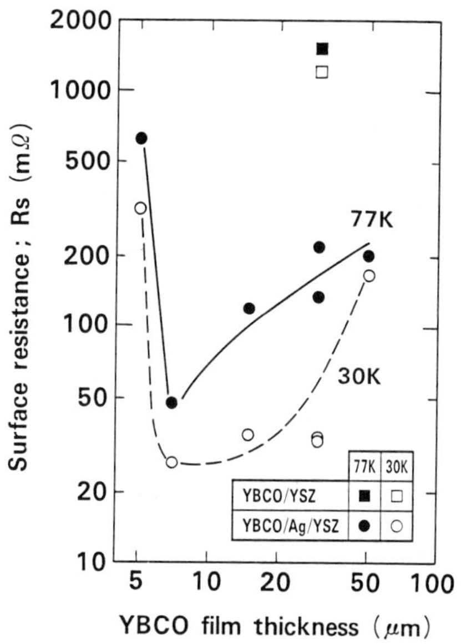

Fig.6 Rs thickness dependences for YBCO films at 50 GHz
Filled circles: YBCO/Ag/YSZ at 77 K
Open circles: YBCO/Ag/YSZ at 30 K
Filled squares: YBCO/YSZ at 77 K
Open squares: YBCO/YSZ at 30 K

ticular, YBCO(7)/Ag/YSZ shows the lowest Rs of all. This film exhibited an extremely low Rs at 17 GHz (8 mΩ), which is almost one third that of oxygen-free Cu (~20 mΩ). Since the Rs of metal is proportional to the square root of frequency[13], the almost squared frequency dependence of the Rs for YBCO/Ag/YSZ ($Rs \propto f^{1.5} \sim f^{1.8}$) suggested that a much lower Rs than that of Cu is attainable in the frequency range below 17 GHz.

CONCLUSION

It is proved that the Rs of screen-printed YBCO films is much improved by using Ag-coated substrates which enhances YBCO grain growth without forming the Y_2BaCuO_5 phase. It is also revealed that both the grain growth and c-axis orientation of YBCO crystals are enhanced when thin and uniform YBCO films are formed on the Ag-coated substrates. These improvements in YBCO film structure result in an extremely low Rs which is lower than that of pure Cu in the frequency range below 50 GHz at 77 K. Consequently, screen-printed YBCO films are confirmed to have great potential for microwave application.

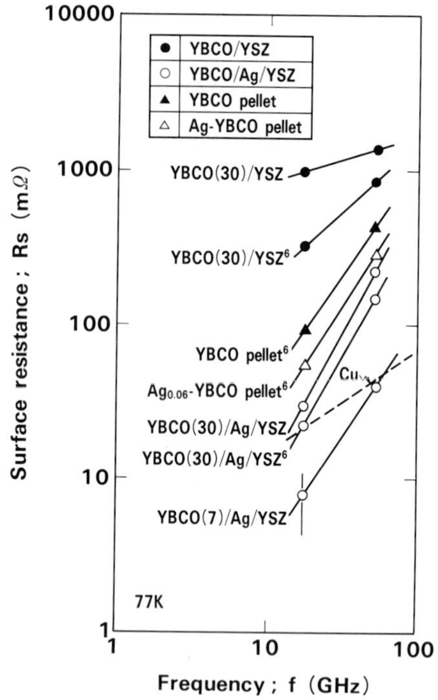

Fig.7 Rs frequency dependences at 77 K for YBCO samples
Filled circles: YBCO/YSZ
Open circles : YBCO/Ag/YSZ
Filled triangles: YBCO pellet[6]
Open triangles : Ag-doped YBCO pellet[6]

Acknowledgments

The authors would like to thank Mr. Yasuhiro Koshimoto for his helpful advice, The authors also thank Mr.Hideji Kuwashima and Mr.Shozo Yamana at Hitachi Chemical Co.,Ltd. for assistance in the fabrication of the YBCO thick films.

References

1 M.K.Wu, J.R.Ashburn, C.J.Torng, P.H.Hor, R.L.Meng, L.Lao, Z.J.Huang, Y.Q. Wang and C.W.Chu (1987) Phys.Rev.Lett.58: 908.
2 G.E.Peterson, R.P.Stawicki and N.McN.Alford (1989) Appl.Phys.Lett.55: 1798.
3 S.K.Khamas, M.J.Meher. T.S.M.Maclean, G.E.Gough, N.McN.Alford and M.A.Harmer (1988) Elect.Lett.24: 460.
4 H.Piel, M.Hein, N.Klein, U.Klein, A.Michalke, G.Muller and L.Ponto (1988) Physica C 153-155: 1604.
5 H.Kuwashima, S.Yamana, T.Ashizawa and T.Kosugi(1989) Hitachi Hyouron,71: 101 [in Japanese].
6 O.Ishii, T.Konaka, M.Sato and Y.Koshimoto (1990) Jpn.J.Appl.Phys.,29: L1075.
7 O.Ishii, T.Konaka, M.Sato and Y.Koshimoto: to be published on Jpn.J.Appl. Phys.
8 S.Yamana, Abstract of MRS Spring meeting (1990, San Francisco)D7.10.
9 M.Sato, T.Konaka and I.Sankawa (1989) Proc.ISEC'89,PH-36: 247.
10 S.Yamana, H.Kuwajima, T.Ashizawa, S.Shimoda, K.Sumiya, T.Kosugi and T.Kamo: (1989) Proc. ISS'88: 305.
11 F.Greuter, C.Schuler,P.Kluge-Weiss and W.Paul(1990) Proc.ISS'89:377.
12 M.Sato, T.Konaka and K.Ishihara(1990) Proc. ISS'89:499.
13 J.M.Victor and W.H.Hartwig(1968) J.Appl.Phys.39:2539.

Contact Resistance in BiPbSrCaCuO Superconducting Rods

Koichi Numata, Kazutomo Hoshino, and Hidefusa Takahara

Corporate R&D Center, Mitsui Mining & Smelting Co., Ltd., 1333-2, Ageo, Saitama, 362 Japan

ABSTRACT

Low contact surface resistivity (product of contact resistance and area) between an oxide superconductor and silver was achieved by pressing silver foil isostatically onto oxide superconducting rods and sintering them. We applied a current of over 100 A to the rods at 77K, and no contact resistance was observed in the sensitivity of 1μV. The temperature dependence of the contact resistance was observed to be metallic. SEM observations revealed that silver foil was attached very well on the oxide surface. Contact surface resistivity was estimated to be less than 10 nΩ•cm^2 below 126 A.

KEY WORDS: Bi system superconductor, contact resistance, critical current

INTRODUCTION

It is essential for practical usage of oxide superconductors to reduce contact resistance between metal and the oxide superconductor. When we apply the oxide superconductors as power leads using a conventional refrigerator, a current of over 1000 A flows in the power leads. Then the contact resistance is required to be less than 1μΩ because heat generation of 1 W ($I^2R = 1000^2 \times 10^{-6} = 1$) can be removed by a conventional refrigerator at 4.2 K.

Many reports on contact resistance have already been published. In the reports, the contacts are prepared by sputter deposition of gold or silver[1)], plasma-arc spraying of silver[2)], and melting silver or gold pads onto the oxide superconductors[3)]. The lowest contact surface resistivity (power of contact resistance and area) of 10^{-12} Ω•cm^2 is reported by Jin et al. by embedding silver wire or powder in the oxide superconductors[4)]. Although their contact surface resistivities are low, their samples are relatively small and the current for measuring the contact resistance is also small, a few amperes.

In this report, we describe the fabrication of low contact surface resistivity on the relatively large oxide superconducting rods (12 ~ 20 mm in diameter and 100 ~ 150 mm in length) and the estimation of contact surface resistivity (ρ_c) and critical current (I_c).

EXPERIMENTAL

Oxide superconducting rods (12~20 mm in diameter, 100~150 mm in length) were fabricated using a cold isostatic press of fine powders of the Bi system high T_c phase (nominal composition; Bi:Pb:Sr:Ca:Cu = 0.8:0.2:0.8:1.0:1.4, Tc ~ 103 K). Preparation of the fine powders is described in another paper[5]. After sintering the rods, silver foil (20μm in thickness) as contacts was attached onto the rods and pressed isostatically. Then the rods were sintered again. Figure 1 shows the rod.

Measurements of contact resistance and critical current were carried out by the four-probe and three-probe techniques, respectively. Figure 2 shows the system of measurement. The voltage generation was measured by an X-Y recorder (YOKOGAWA 3036 X-Y RECORDER; 5μV/cm) or digital multimeter (KEITHLEY 181 NANOVOLTMETER). Figure 2 shows the diagram of the circuit of the measurement.

A scanning electron microscope was used to observe the interface between the silver and the oxide superconductor.

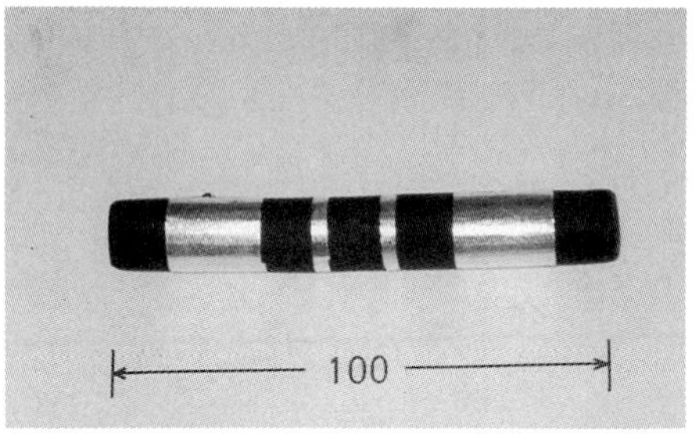

Fig. 1 A photograph of oxide superconducting rod.

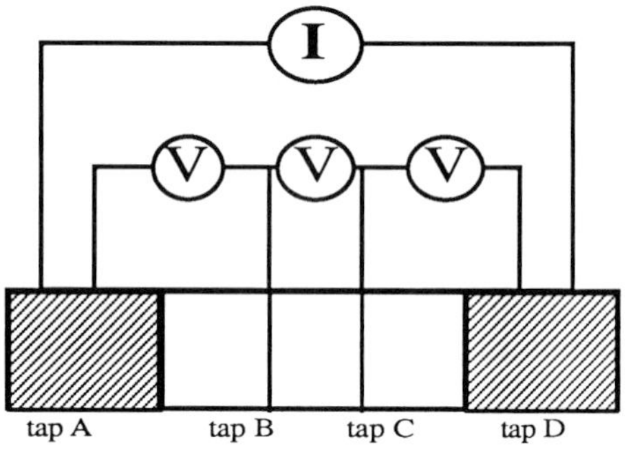

Fig. 2 A diagram of circuit of the measurement.

RESULTS AND DISCUSSION

Figure 3 shows the superconducting transition curves of a sample which had somewhat worse contact measured by the (a) four-probe and (b) three-probe techniques. Figure 3 (a) shows that the critical temperature of this sample is about 103 K, and Figure 3 (b) shows that there is remnant resistance below 103 K caused by contact resistance. The contact surface resistance of this sample is calculated to be about 10^{-6} Ω•cm^2, and the dependence of the contact resistance on temperature is metallic. However, remnant resistance usually could not be observed in the samples with better contact within the sensitivity.

Figure 4 shows the I-V curves measured by the four-probe and three-probe techniques for a sample with diameter of 12 mm. The critical current of this sample is 140 A (criterion : 1μV/cm), and then the contact resistance can not be observed until the applied current reaches to 126 A in the sensitivity of 0.1μV. The contact surface resistivity is estimated to be less than 10^{-8} Ω•cm^2 below 126 A. The critical current in the I-V curves of A-B and C-D shown in Fig.4 (at which the voltage of 0.1μV was detected) is nearly equal to or smaller than that of B-C. Thus the difference in the critical currents measured in the taps of A-B, B-C and C-D seems to come from the inhomogeneity of the current characteristic if the sample itself or the heat generation at the taps of A and D.
The highest critical current achieved so far is 480 A for a sample with diameter of 20 mm.

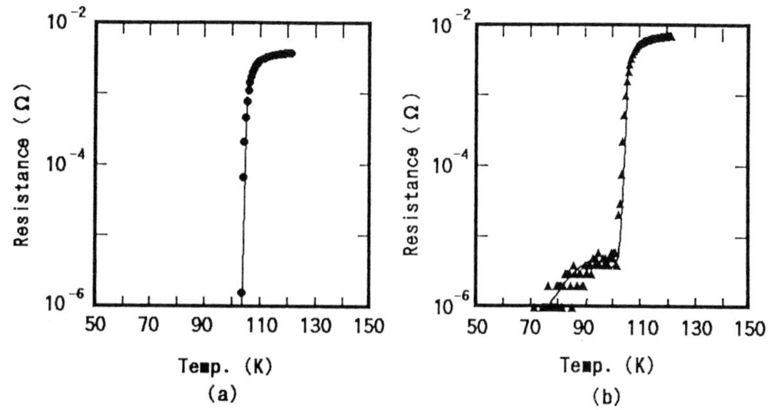

Fig.3 Transition curves of the rod measured by the (a) four-probe and (b) three-probe techniques

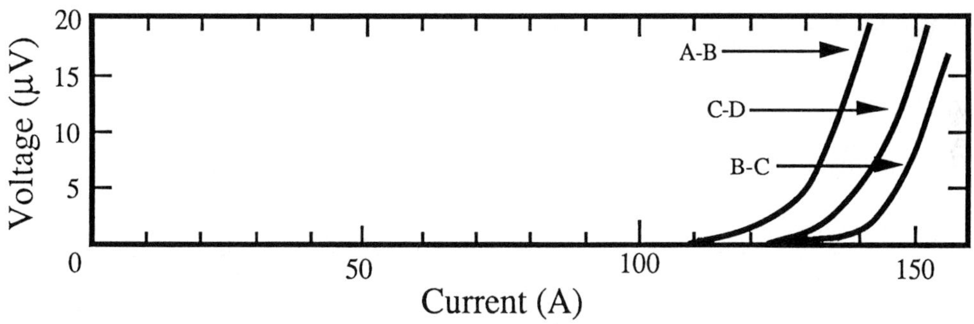

Fig.4 I-V characteristics of the rod. A,B,C,D indicate the tap illustrated in Fig.2

Figure 5 shows a SEM photograph of the interface between silver and oxide superconductor with (a) secondary electron image and (b) back electron image (compositional image). In the oxide region, voids and some second phases observed in Fig.5 correspond to black and gray regions, respectively. The second phase was identified by energy dispersive X-ray analysis to be copper oxide. At the interface, no other phases than seen in the inner oxide region can be observed.

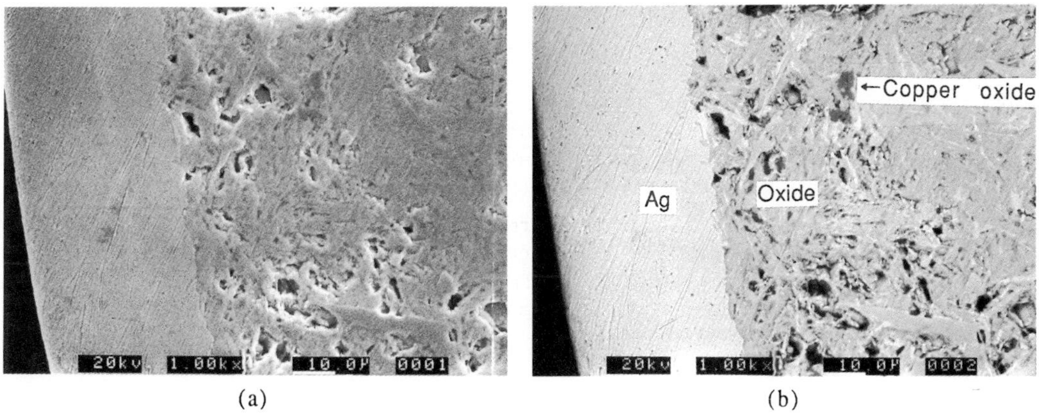

Fig.5 SEM photographs of the interface between silver and oxide superconductor. (a) secondary electron image (b) back electron image (compositional image)

The silver foil is firmly attached onto the surface of oxide superconductor, and no intermediates can be observed at the interface. However if we tear off the silver foil from the oxide superconductor, many oxide grains stick to the silver foil. Thus, there seems to be a very narrow reaction region at the interface, which is caused by the interdiffusion between silver and oxide elements during high temperature annealing.

In summary, we successfully fabricated a low resistive contact to the Bi oxide superconductor by pressing silver foil isostatically onto the oxide superconductor and sintering. The surface contact resistivity was estimated to be less than 10^{-8} $\Omega \cdot cm^2$ below 126 A. The highest critical current at 77 K achieved so far was 480 A for the sample with diameter of 20 mm. Although we could not measure the temperature dependence of the contact resistance in the usual samples because of its low resistivity, even in the sample with worse contact the temperature dependence of it was metallic. The difference in the critical current behavior among the taps of A-B, B-C and C-D seems to be due to the inhomogeneity of the critical current property in the sample itself or the heat generation at the taps of A and D.

ACKNOWLEDGMENT

The authors wish to thank Drs. D. Itoh, H. Ogiwara and Mr. K. Tasaki of Research and Development Center of Toshiba Corporation for measurements of critical current and helpful discussions.

REFERENCES

1) J.W. Ekin, T.M. Larson, N.F. Bergren, A.J. Nelson, A.B. Swartzlander, L.L. Kazmerski, A.J. Panson, and B.A. Blankenship (1988) Appl. Phys. Lett. 52:1819
2) J.D. Katz, J.O. Willis, M.P. Maley and R.G. Castro (1989) J. Appl. Phys. 65:1792
3) R. Selim, R. Caton, A.M. Buoncristiani, C.E. Byvik, R.A. Edahl and S. Wise (1990) J. Appl. Phys. 67: 376
4) S. Jin, M.E. Davis, T.H. Tiefel, R.B. van Dover, R.C. Sherwood, H.M. O`Bryan, G.W. Kammlott and R.A. Fastnacht (1989) Appl. Phys. Lett. 54:2605
5) K. Hoshino, H. Ohta, E. Sudoh, K. Katoh, S. Yamazaki, H. Takahara and M. Aono (1990) Jpn. J. Appl. Phys. 29:L1435

Magnetic Separation of High-T_c Superconducting Particles

EIJI OBA[1], HIDEKI MIYAJIMA[1], YUUICHI ISHIKAWA[2], and SHUJI YOSHIZAWA[2]

[1] Department of Physics, Keio University, Hiyoshi, Kohoku, Yokohama, 223 Japan
[2] Central Research Laboratory, Dowa Mining Co., Ltd., Hachioji, Tokyo, 192 Japan

ABSTRACT

A method to separate superconducting particles magnetically from high-T_c oxide powder is developed. The powder cooled at about 80 K drops slowly from a pin hole and passes through a magnetic field gradient generated by two permanent magnets. The superconducting particles having a large diamagnetic susceptibility are pushed to the direction along the field gradient, while no force acts on the non-superconducting particles whose susceptibility is small. The magnetic separation is useful to select and to condense superconducting particles from a sintered specimen including non-superconducting components.

KEY WORDS : Y-Ba-Cu-O superconductor, fine particles, magnetic separation, extraction of superconducting component

INTRODUCTION

It is well known that a superconductor levitates above a permanent magnet (magnetic levitation effect) and that it can be hung down with a magnet (fishing effect [1]). These phenomena are related deeply to the diamagnetism or the flux pinning in the superconductor. Such forces acting on superconductors can be applied to a magnetic separation or condensation of high T_c superconductors which usually consist not only of many kinds of superconducting phases but also of non-superconducting substances. In the YBaCuO system prepared by a sintering method, for example, there are $Y_1Ba_2Cu_3O_7$ superconducting phase and non-superconducting phases, that is, $Y_2Ba_1Cu_1O_y$ and so on. The superconducting characters such as the critical current and the sharpness of the critical temperature will be improved when pure superconducting phases are condensed.
In this paper, a method which can separate superconducting particles from a sintered oxide superconductor is proposed, and the experimental results for YBaCuO system are presented.

EXPERIMENTAL PROCEDURE

(1) Sample Preparation of Y-Ba-Cu-O Powder
The pellet samples of $Y_1Ba_2Cu_3O_y$ (y is undetermined) were prepared by a usual sintering method. The calcined pellet was ground in an agate mortar for appropriate time so as to obtain particles of a few micrometers in size. The size of the particles is adjusted to be the same order of single grains. In this experiment, the size was desirable to be same and fine as far as possible. After that, the powder was put into a powder vessel described below, and the magnetic separation was carried out.
(2) Magnetic Separation Method
The schematic illustration of the instrument used in this experiment is shown in Fig. 1. It comprises three components: an ultrasonic vibrator, two permanent magnets and a collector of particles.
The most important matter in the magnetic separation is how to isolate the fine particles from one another. Ultrasonic isolation was useful to break the agglomeration of the particles. Three kinds of equipment were designed and put to the test. Figure 2 shows the ultrasonic vibrators consisting of a magnetostrictive transducer and a powder vessel. In the type (a), the ultrasonic vibration of the transducer excited at a frequency of 50 kHz is amplified through a horn made of aluminum and vibrates longitudinally the powder vessel which contains the powder cooled at about 80 K so as to be superconductive. The particles drop through a pipe (3mm in inner diameter and about 10 mm in length) located at the center of the bottom, and then the beam of the particles are restricted not to diffuse with an aperture (3mm in diameter) positioned under the powder vessel. This

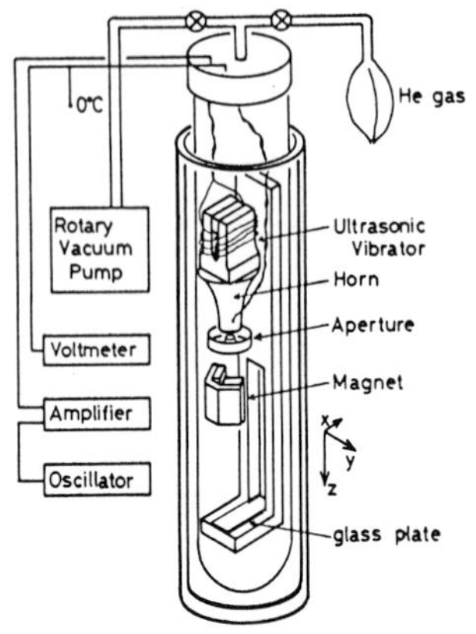

Fig. 1: Schematic illustration of the magnetic separation

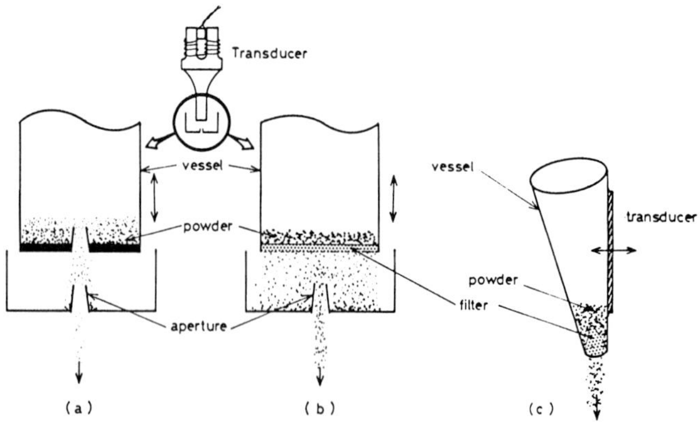

Fig. 2: Configuration of powder vessel and ultrasonic vibration

method, however, was useful to get a single isolated particle but not so good for the magnetic separation, since the vibration may arise the temperature of the particles up above the critical temperature. Instead of using the pipe, the type (b) makes use of a paper filter fitted at the bottom of the vessel. The pore size is about 10 μm. Only the particles smaller than about 10 μm in size can pass through the filter. In this type, the thermal energy introduced by the ultrasonic vibration was removed when going through the filter cooled at 80 K, but the efficiency of the separation was rather low. Finally improved one is shown in Fig. 2(c). The powder vessel is comprised with a paper filter of pore size 10 μm and a circular cone of which open angle is about 20 degree. The vessel is vibrated parallel to the plane of the corn at a frequency of 10 kHz. During passing through the filter and a pin hole (2mm in diameter) at the top of the corn, the particles are cooled down below the critical temperature. High efficiency and good separation of the particles were realized by using this method, nevertheless there were some agglomerations of the particles.

These isolated particles fall freely and then enter into an air gap in a magnetic circuit consisting of two permanent magnets ($SmCo_5$) connected by an iron yoke. As shown in Fig. 3, two magnets are opposite to each other and inclined by an angle about 30 degrees so as to obtain a strong magnetic field gradient. When the particles pass through the gap, their trajectories are bent by the magnetic force which is given by the product of the magnetization and the field gradient, since the superconducting particles have a large diamagnetic susceptibility but the non-superconducting particles have a small paramagnetic one. The magnetic force acting on the particles and the trajectory are described

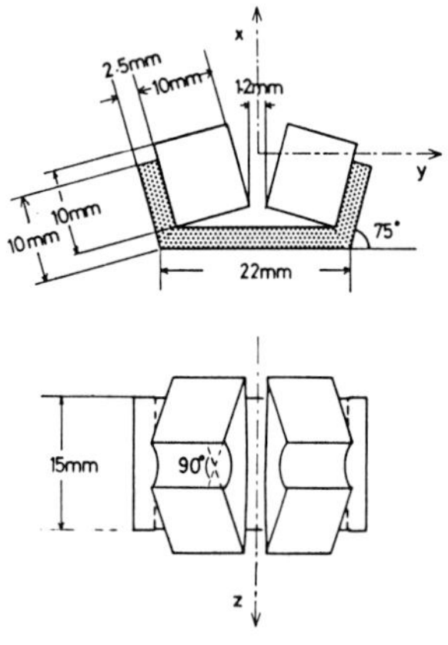

Fig. 3: Magnetic circuit containing two magnets and a yoke.

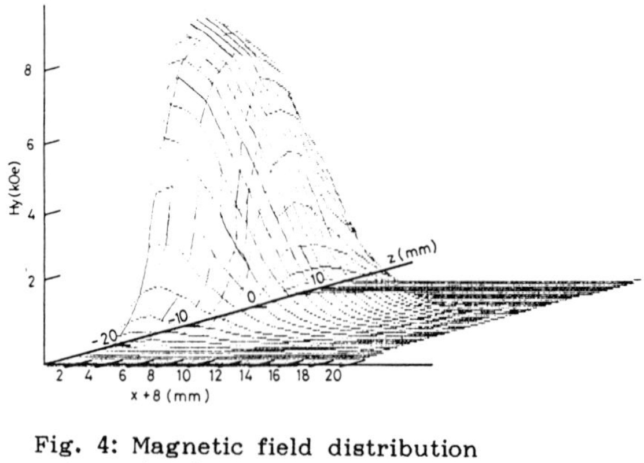

Fig. 4: Magnetic field distribution in the x-z plane.

below. After that, the particles drop on a glass plate (collector) placed at a distance of about 200 mm under the magnets. The particles are separated and distributed on the collector; the superconducting particles come together at a distant position, while the non-superconducting ones gather just under the aperture.
The equipment is installed into a pirex glass chamber and in a liquid nitrogen dewar. After the inside of the chamber is evacuated, a small amount of heat exchange gas of He is injected into the inside so as to be at a pressure about 10 mmTorr, and then it is cooled down to about 80 K with liquid nitrogen. If the pressure inside the chamber is high, the convection of gas occurs and disturbs the free falling of the fine particles. The temperature in the powder vessel is monitored with a thermocouple.

(3) Magnetic Force
In this section, the magnetic force acting on the particles is described. Here, let **i**, **j**, **k**, be the unit vectors of a Cartesian coordinate system (*x*, *y*, *z*) whose origin is at the position of the outlet of the powder vessel, and let the *x*, *y* and *z* axes be as shown in Fig. 3. The force, $\boldsymbol{F}$, acting on a particle of mass m is given by

$$\boldsymbol{F} = mg\,\boldsymbol{k} + m\nabla(\boldsymbol{M}\cdot\boldsymbol{H}) \tag{1}$$

$$= m\left(M_y\frac{\partial H_y}{\partial x}+H_y\frac{\partial M_y}{\partial x}\right)\boldsymbol{i}+m\left(g+M_y\frac{\partial H_y}{\partial z}+H_y\frac{\partial M_y}{\partial z}\right)\boldsymbol{k} \tag{2}$$

where g is the gravitational acceleration that is parallel to the positive z direction, $\boldsymbol{M}$ the magnetization per unit mass, and $\boldsymbol{H}$ the magnetic field at a position of (x, y, z). It is assumed that $H_x=H_z=0$ and $\partial H_x/\partial y=\partial H_z/\partial y=0$. The second term in Eq. (1) represents the magnetic force caused by the field gradient. In order to get good separation, the field gradient along the y direction needs to be large as far as possible, so the gap of the magnets widens along the x direction. By considering the magnetic field distribution and the field dependence of the magnetization, the trajectory of a particle is calculated numerically by solving Eq. (2). The distribution of the magnetic field was measured with a Hall probe. The distribution in the x-z plane is shown in Fig. 4. The field gradient took a maximum of 1.1 kOe/mm. In a simple case that the particle is in the Meissner state of which magnetic susceptibility is given by $-1/4\pi$ in cgs units, the trajectory of a particle is bent ahead of the magnets by the repulsion force. On the contrary, the magnetic force acting on non-superconducting particles is very small, since their susceptibility is of order of 10^{-5}. Therefore, these particles fall freely regardless of the magnetic fields. In reality, however, as the magnetization of the superconducting particles depends not only on the magnetic field but also on the hysteresis, the magnetic hysteresis must be considered. The hysteresis curve of the bulk YBaCuO superconductor before the pulverization was measured and is shown in the inset of the Fig. 5. Considering the above mentioned situations, the trajectory of the YBaCuO superconducting particles is calculated by using Eq. (2). The results are shown in Fig. 5 as a function of the distance, x, from the origin. The trajectory of a perfect diamagnetic particle is illustrated by a broken line. In this case, the separation distance on the collector is about 5.8 times larger than that in the practical case in which the magnetic hysteresis is considered.

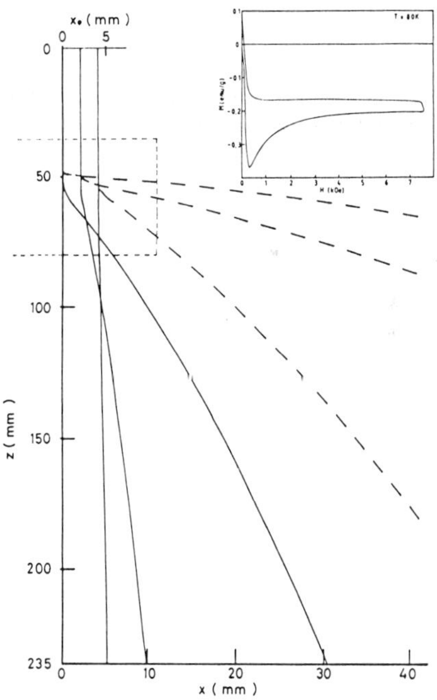

Fig. 5: Trajectory of a particle in the Meissner state (broken line) and in the real case (solid line) considered with the magnetization curve (inset). The trajectory varies with the initial position of falling.
The rectangle drwaing with broken line indicates the position of the magnet.

EXPERIMENTAL RESULTS AND DISCUSSION

Figure 6 is a photograph that shows the separation of pure $Y_1Ba_2Cu_3O_y$ powder. The small spot seen in the left side is the particles which are dropped at room temperature. As no magnetic force acts on these particles, they fall down straightly from the powder vessel. The right side spot is obtained by dropping the particles cooled at about 80 K. The separation distance from the former spot is about 3 cm. On the other hand, the YBaCuO particles including some non-superconducting phases is distributed narrowly along the y axis but long along the x axis, as shown in Fig. 7. The distance of the separation was a few centimeters from the origin (just under the aperture). The maximum distance, however, is only a half time of the distance evaluated by the above mentioned calculation. It is considered that the convection of the exchange gas disturbs the free falling and, in addition, that the particles may be hard to be isolated perfectly by the ultrasonic vibration only. Therefore, a lump of several particles including non-superconducting ones falls from the powder vessel.

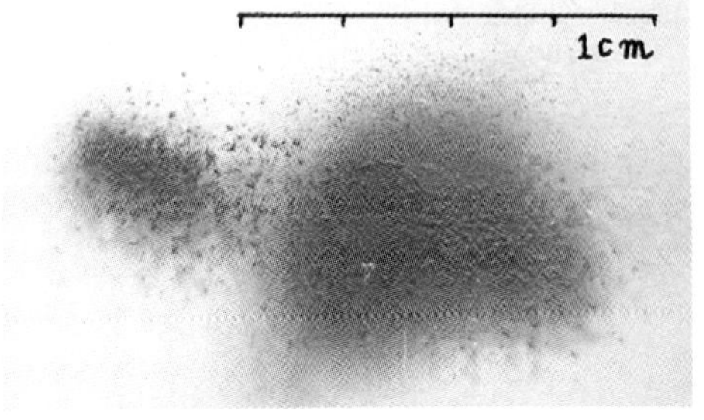

Fig. 6: Photograph of the $Y_1Ba_2Cu_3O_y$ particles

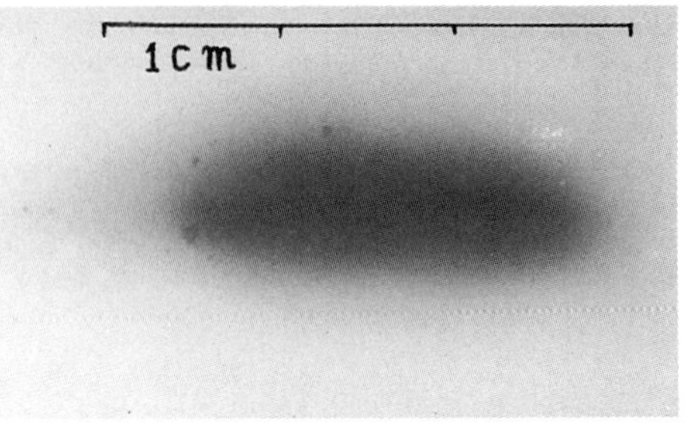

Fig. 7: Photograph of the as sintered Y-Ba-Cu-O particles

X-ray diffraction measurements using Co target were carried out on the powder collected from the plate. Figure 8 shows the X-ray diffraction patterns of the mixed powder of $Y_1Ba_2Cu_3O_y$ particles and the Y_2O_3 particles. The powder is picked up from at x=0 mm (a) and at x=20 mm (b). It should be noted that the particles are preferred to orient their c-axis perpendicular to the collector plane. Since the magnetic flux trapped in the YBaCuO crystal is usually directed along the c-axis [2], the magnetized particles interact with one another when they drop on the collector and are oriented its direction.

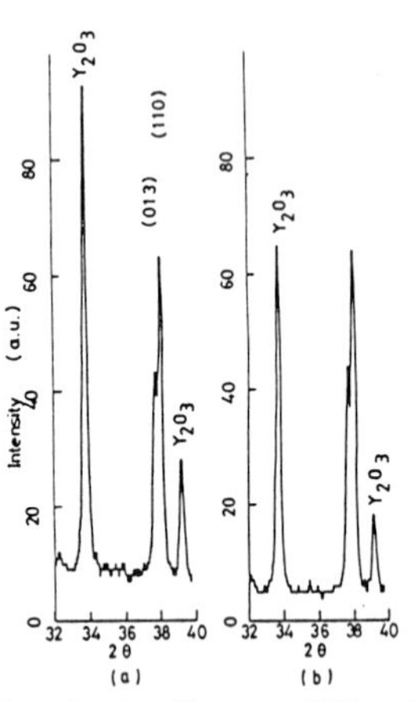

Fig. 8: Co X-ray diffraction patterns of the mixed powder of as sintered Y-Ba-Cu-O particles and Y_2O_3 powder
(a): particles near x= 0 mm
(b): particles near x=20 mm

The efficiency to separate the particles was not always high in this experiment. It is attributed to the distribution of size of the particles and also to the agglomeration of the particles. The experiments are now carried out on the Bi and Tl system superconductors. The magnetic separation will be useful in these systems, as they comprise many kinds of superconducting phase exhibiting different T_c according to the number of the layer. Furthermore, their higher T_c by about 20 K or 30 K than that of the Y system works effectively in the separation.

In conclusions, a conventional instrument to condense superconducting particles with a magnetic force is designed and constructed. It is favorable to separate the superconducting phase from high T_c superconductors.

The authors would like express their thanks to Prof. S. Chikazumi for his valuable advises to the experiment.

REFERENCES

[1] Peters P. N.,Sick R. C., Urban E. W., Huang C. Y., Wu M. K. (1988) Appl. Phys. Lett. 52 : 2066

[2] Iye Y., Tamegai T., Takeya H., and Takei H. (1987) Physica B+C 148 : 224

Pole Figure Analysis of Aligned High-T_c Superconductors Using Low-Indexed X-Ray Diffraction

MIKIO NAKAGAWA, YUH SHIOHARA, and SHOJI TANAKA

Superconductivity Research Laboratory, International Superconductivity Technology Center, 10-13, Shinonome 1-chome, Koto-ku, Tokyo, 135 Japan

ABSTRACT

A micro-region pole figure measurement apparatus was developed using the Schulz reflection method. Choice of low-indexed diffraction peaks, such as (001) and (002) enabled an independent analysis of orientation of a specific superconductivity phase in a multi-phase specimen. Unsymmetry in distribution and/or discrepant maximum c-axis density from the expected orientation, e.g. normal to film surface, was often found out in Y-based and Bi-based film specimens though they displayed c-axis alignment in θ-2θ scan diffractometry. This fact indicates that a pole figure analysis is indispensable for evaluation of the crystal alignment in high Tc superconductors.

KEY WORDS : pole figure analysis, crystal alignment, thin film, X-ray diffraction

INTRODUCTION

Many efforts have been made to enhance critical current density(Jc) in the high temperature oxide superconductors. Single crystal would be a solution for the problem since one can avoid the effect of weak link at grain boundaries, however it should be very difficult or impractical to fabricate a single crystal conductor for a power use. It was pointed out that Jc was reduced with discrepancy of crystal orientation. The fact indicates that control of the crystal alignment is indispensable to attain high Jc in polycrystalline materials. Thus, the quantitative evaluation of the alignment is important for developing processes to fabricate materials with the high quality. However, only qualitative investigation by a θ–2θ scan x-ray diffraction method has been applied in most cases. In this paper the authors will show that investigation by a θ–2θ scan or a rocking curve method is insufficient and a pole figure measurement is indispensable for its evaluation.

EXPERIMENTAL

Due to complex crystal structures, diffraction patterns of the materials are complicated. (00ℓ) peaks are neighbored with other peaks, such as (006) and (200), (020) in Y-based 123 phase. Interference with such unexpected diffraction peaks will introduce errors in measurements. In addition, specimens could have secondary phases and diffraction from these phases would make errors. As the secondary phases, e.g. superconducting 124, 247 or non-superconducting 211 phase, could act as pinning centers and improve Jc in the materials, interference from the phases must be eliminated in the measurements. These interference can be avoided by choosing a low angle diffraction peak from (001) or (002) plane of a superconducting phase.

The pole figure measurement apparatus used in this study was developed modifying the micro-region x-ray diffraction system, RIGAKU RINT-1500. Schematic drawings of the apparatus is shown in figure 1. X-ray generated with a Cr target is introduced onto a specimen surface through a collimator. Diameter of the collimator used in the experiments was 100μm. When diameter of x-ray beam is 100μm and incident angle is 5°, irradiated area is about 0.1×1 mm^2. Diffracted x-ray quanta were counted by a position sensitive proportional counter(PSPC). To enable detection of low angle diffraction($2\theta<10°$), the range of incident angle(ω) was determined as $0.5<\omega<60°$. The x-ray diffraction pattern in the range of $1<2\theta<150°$ can be also measured with this apparatus.

Pole figure analyses were carried out by the Shulz reflection method. The incident angle was determined and fixed to be a half of the diffraction angle. The diffraction peak detected by PSPC was integrated with a detection channel over a range and subtracted the background. The integrated value was plotted as a function of α and β with contour lines on a (001) pole figure. Range of α was limited with a mechanism of the apparatus; $0<\alpha<45°$. Change of diffraction intensities with α due to absorption in the specimens was not compensated in this experiments, but the absorption is negligible because the specimens were thin and range of α is not large.

Specimens used in this paper were films on MgO(100) substrates prepared by MOCVD method[1, 2]. Typical thickness of the films was 0.2μm. These films contained multiple superconducting phases; 123 and 124 in Y-based films or 2201, 2212, and 2223 in Bi-based films.

RESULTS AND DISCUSSION

Overlap of the diffraction peaks from superconducting phases were not seen in either Y-based or Bi-based specimens. Thus, we could analyze the crystal alignment of each phase independently.

Figures 2 a), b) show the alignment of 123 and 124 phases in the Y-based film. Though an x-ray diffraction pattern of the specimen by θ-2θ scan method suggested a good c-axis alignment of the two phases, the pole figure analysis gave a different result. The maximum of c-axis densities was discrepant with the expected, e.g. normal to the film surface, and the c-axes distributed in a range as wide as 15°.

Figures 3 a), b) also show the alignment of 2212 and 2223 phases in the Bi-based film. Distribution

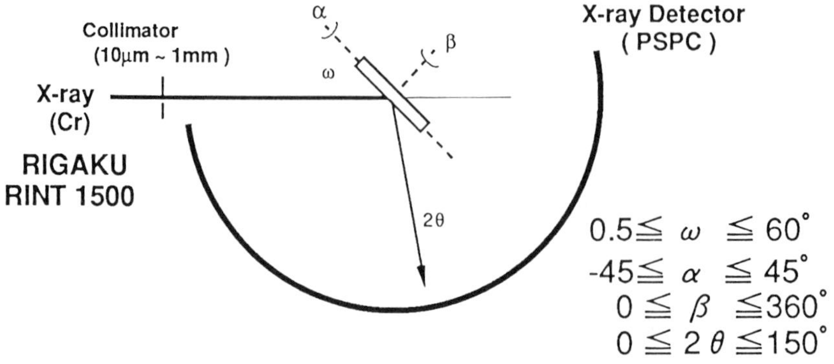

Fig. 1. Schematic Drawing of Pole Figure Measurement Apparatus.

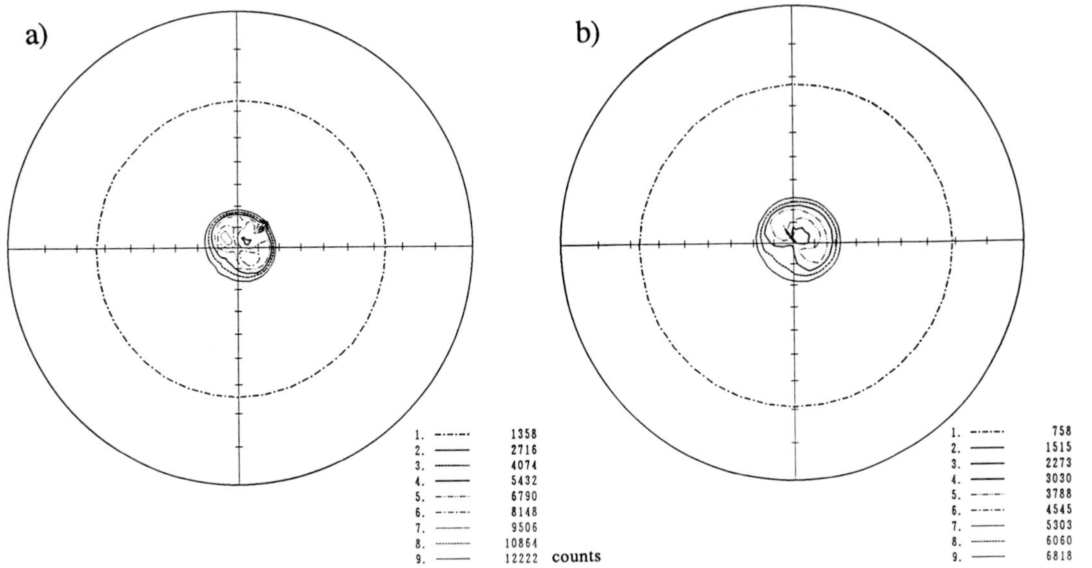

Fig. 2 Results of pole figure analysis for a Y-based film.
a)123 phase
b) 124 phase

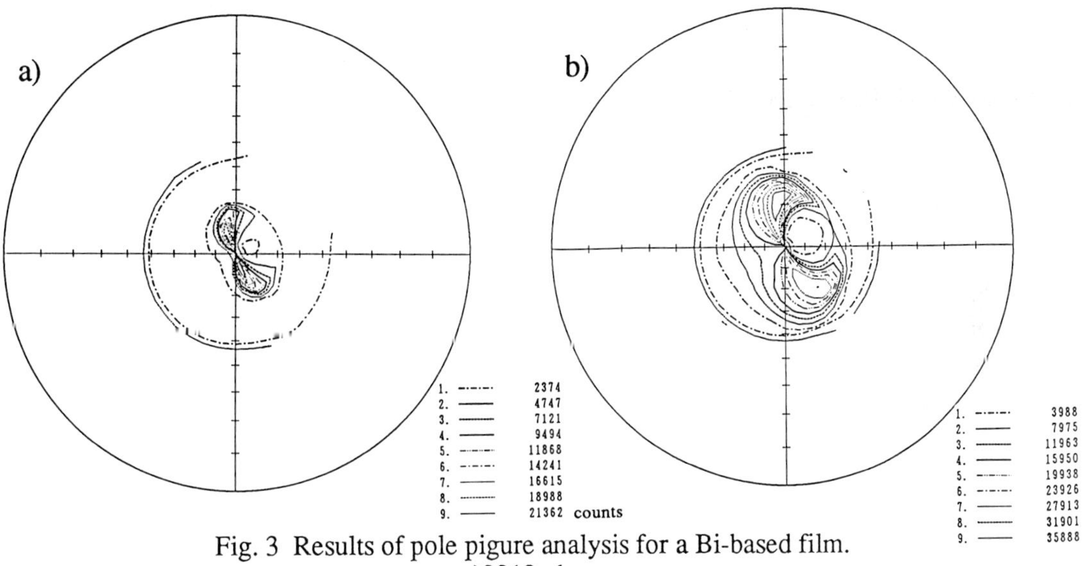

Fig. 3 Results of pole pigure analysis for a Bi-based film.
a)2212 phase
b) 2223 phase

of the c-axis had two peaks. Distribution of c-axes of 2212 was sharper than that of 2223, but they had the same tendency. Relation to the orientation of the MgO substrate was not clear at present. It is difficult to attribute two fold symmetry in the distribution to the effect of MgO substrate which has four fold symmetry. Origin of such distribution should be argued with respect to growth mechanism of the crystals.

The rocking curve taken by the θ-scan diffraction method is equivalent to a central section of a pole figure. The rocking curve measurement is effective enough to evaluate crystal alignments provided

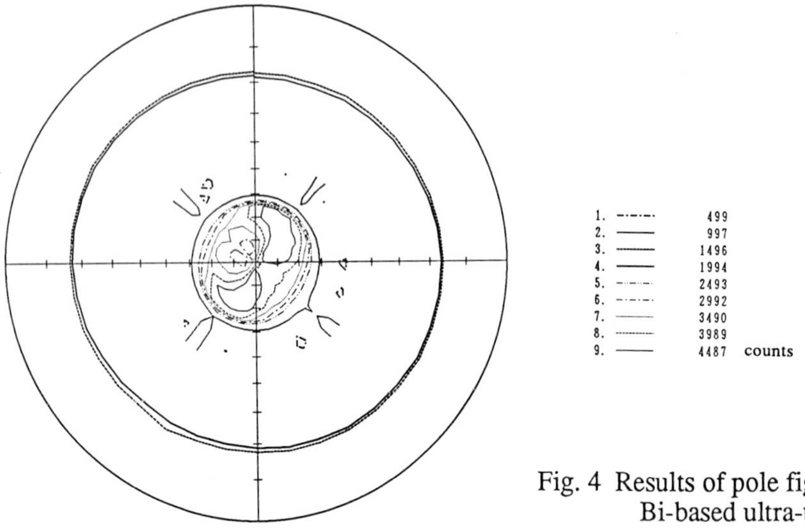

Fig. 4 Results of pole figure analysis for a Bi-based ultra-thin film.

that the distribution is symmetrical with respect to the normal. However, the curves would have different widths, and even a different shapes depending on the sections measured, if the distribution is unsymmetrical. It is clear that a rocking curve measurement will not give a sufficient information about the crystal alignment.

Figure 4 shows alignment of a Bi-based ultra-thin film[2]. Thciknes of the film was about 50Å. Only diffraction from 2212 phase was observed with the specimen. Intensity of diffracted x-ray is lowered with decreasing film thickness in general, however, sufficient strength of (002) peak was attained in this analysis thanks to low incident angle.

CONCLUSION

Authors have developed a pole figure measurement appratus and analyzed crystal alignments in Y-based and Bi-based superconductor films. Alignments of a specific phase could be analyzed independently by choosing (001) or (002) diffraction peak. The alignment often showed an unsymetrical distribution and/or the discrepant maximum from the expected, normal to the surface. These facts indicate that the pole figure analysis is indispensable for evaluation of crystal alignment in high Tc superconductors.

Acknowledgment

We thank T. Sugimoto, H. Hayashi, and Y. Yamada at our laboratory for providing film samples.

REFERENCES

[1]. H. Hayashi, K. Sugawara, and Y. Shiohara: submitted to *Proc. 2nd Int. Symp. on Superconductivity (ISS'90)*

[2]. T. Sugimoto, A. Kondo, M. Yoshida, K. Sugawara, and Y. Shiohara: submitted to *Proc. 2nd Int. Symp. on Superconductivity (ISS'90)*

Angular Dependence of Critical Current Density of High-T_c BiPbSrCaCuO Superconductor

K. ISAWA[1], C. TAKAHASHI[1], S. OGOTA[1], M. NAGANO[1], H. TAKAHASHI[1], K. HAMADA[2], S. KAMADA[2], K. SUGAWARA[2], and A. NAGATA[2]

[1] Applied Technology R&D Center, Tohoku Electric Power Co., Inc.,
2-1, Nakayama 7-chome, Aoba-ku, Sendai, 981 Japan
[2] Department of Metallic Materials for Engineering, Mining College, Akita University,
1-1, Tegatagakuenmachi, Akita, 010 Japan

ABSTRACT

We report measurements on the magnetic field dependence of transport critical current density(Jc) and the angular dependence of Jc in bulk ceramic samples of high-Tc superconductors in the magnetic field(B) up to 8 T. The raw superconducting materials of BiPbSrCaCuO were prepared by conventional solid state reaction, and their Tc were around 110 K. Samples of oriented Ag-sheathed tapes were fabricated by rolling. The measurements of Jc at 4.2 K and 77.3 K were carried out by a four probe dc method. The magnetic field dependence of Jc was observed; the Jc reduced with increasing B. And the angular dependence of Jc was also observed; it was more remarkable at 77.3 K than at 4.2 K.

INTRODUCTION

Since the discovery of high-Tc oxide superconductors, extensive studies on the superconducting properties of the oxides and on the preparation of the oxides wires have been made [1-3]. Bi based high Tc superconductors are expected to be used as magnet wires, since Ag-sheathed BiPbSrCaCuO wires can carry the Jc of the order of $10^4 A/cm^2$ at 4.2 K even in high magnetic field of 23 T [4]. Anisotropic behaviors of Jc in Ag-sheathed BiPbSrCaCuO wires have been reported[5], and the experimental results suggest that the Jc may be sensitive to the direction of applied magnetic field. Therefore, it is very important from both viewpoints of fundermental research and practical application to investigate the anisotropic properties in Jc.

In this paper, we report the masurements of the magnetic field dependence of Jc and the angular dependence of Jc on Ag-sheathed BiPbSrCaCuO oxide superconducting tapes.

EXPERIMENTAL

Samples measured were Ag-sheathed BiPbSrCaCuO superconducting tapes, and its fabricating process is shown in Fig.1. In the first procedure, appropriate amounts of Bi_2O_3, PbO, $SrCO_3$, $CaCO_3$ and CuO powders in the nominal composition (Bi:Pb:Sr:Ca:Cu= 2.05 : 0.41 : 2.12 : 3.24 : 4.18) were calcined at 800 °C for 15 hours. After grinding, the powders were wrapped with Ag sheets and then fabricated into tapes by rolling. The tapes were sintered at 825 °C for 96 hours in air. The samples, about 3 mm wide and 15 mm long, were uniaxially pressed at 0.5 GPa then sintered at 825 °C for 12 hours. The pressing and sintering processes were repeated twice.

The measurements of transport critical current were carried out by a four probe dc method, and the current were always passed perpendicular to the magnetic field direction. The criterion for Jc determination was defined as the transport current of 1 $\mu V/cm$. The angle(θ), between the direction of applied magnetic field and the oriented plane of sample(pressing surface), was determined by the micro meter connected to the θ-rotation unit in the sample holder. The angle $\theta = 0°$ corresponds to $B \parallel$ tape surface and $\theta = 90°$ corresponds to $B \perp$ tape surface.

RESULTS AND DISCUSSION

The SEM micrographs, taken on (a) the surface of the sample after removing Ag sheath and (b) the fracture surface of a transverse cross section, are shown in Fig. 2. Thin platelike grains with the two dimensional size difference can be seen in Fig.2(a), and fine microcrystals mainly due to the re-pressing process are also seen in Fig.2(b).

The typical magnetic field dependence of Jc at 4.2 K for $B \parallel$ tape surface and $B \perp$ tape surface are shown in Fig.3. The magnetic field dependence of Jc is small at 4.2 K. When the field direction is perpendicular to the surface($B\perp$), for example, the Jc initially tends to reduce with increasing B up to about 2 T and then keeps almost constant up to 8 T. The reduction of Jc is about 1/5 : $Jc \simeq 8.0\times10^3 A/cm^2$ at 0 T, $Jc \simeq 1.5\times10^3 A/cm^2$ at 2 T. A rather large hysterisis is observed on the Jc vs B curve between the increasing and decreasing B. This hysterisis can be explained by the weak coupling between each grain[6].

Figure 4 shows the magnetic field dependence of Jc at 77.3 K for the same sample measured at 4.2 K. In contrast to the results at 4.2 K, the magnetic field dependence of Jc at 77.3 K is remarkable, and two significant features are observed; one is the rapid drop in Jc with increasing B, and the other is the extremely small hysterisis on the Jc vs B curve. In addition, the anisotropic Jc vs B behavior is also observed; the Jc depends on the direction of the magnetic field, which may be due to the anisotropy and/or alignment of grains in the sample.

Figures 5 and 6 show the angular dependence of Jc at 4.2 K and 77.3 K in various magnetic fields, respectively. At 4.2 K, the ratio of $Jc/Jc,max$ does not change so much, but reduces gradually with increasing θ up to θ=90° in any magnetic field. The lowering of $Jc/Jc,max$ becomes clear with increasing B. At 77.3 K, as seen in Fig.6, the ratio of $Jc/Jc,max$ is very sensitive to both B and θ, and reduces rapidly with increasing θ up to about θ=60°.

Therefore , from Figs. 5 and 6, the angular dependence of Jc can be said to be more remarkable at 77.3 K than at 4.2 K. This angular dependence may depend on the value of upper critical field, and it shoud be explored in detail.

In summary, we measured the magnetic field dependence of Jc and the angular dependence of Jc on Ag-sheathed BiPbSrCaCuO oxide superconducting tapes at 4.2 K and 77.3 K. The angular dependence of Jc was clearly observed; it was more remarkable at 77.3 K than at 4.2 K. The magnetic field dependence of Jc was also observed; the Jc reduced with increasing B.

REFERENCE

1. The proceedings published in Jpn.J.Appl.Phys.26, Supplement 26-3(1987).
2. The proceedings of ISS '88, Nagoya(1988).
3. The proceedings of ISS '89, Tsukuba(1989).
4. K.Watanabe, K.Noto, H.Morita, H.Fujimori, K.Mizuno, T.Aomine, B.Ni, T.Matsushita, K.Yamafuji and Y.Muto: Cryogenics,29(1989),263.
5. K.Sato, T.Hikata, H.Mukai, T.Masuda, M.Ueyama, H.Hitotsuyanagi, T.Mitsui and M.Kawashima: The proceedings of ISS '89, Tsukuba(1989),335.
6. T.Matsushita, B.Ni, K.Yamafuji, K.Watanabe, K.Noto, H.Morita, H.Fujimori and Y.Muto: The Proceedings of ISS '88, Nagoya(1988),393

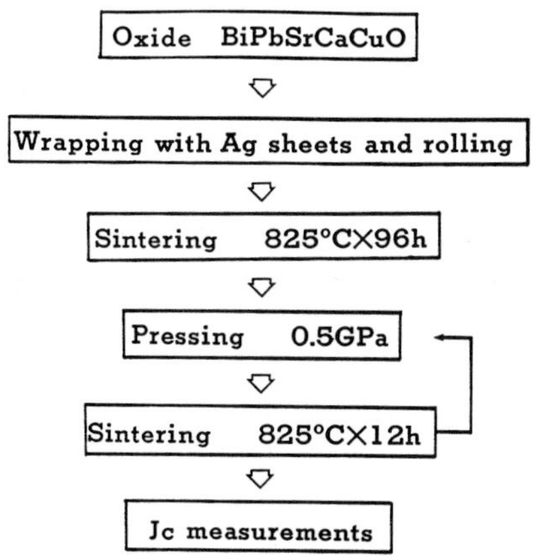

Fig.1 A basic process for fabricating Ag sheathed BiPbSrCaCuO superconducting tapes .

Fig.2 SEM micrographs of tape;
(a) The surface of BiPbSrCaCuO superconductor after removing Ag sheath.
(b) The fractured transverse cross section.

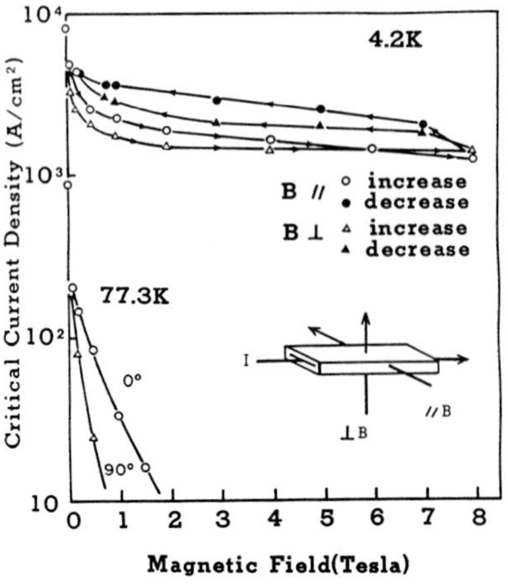

Fig.3 Magnetic field dependence of Jc at 4.2 K for Ag sheathed BiPbSrCaCuO superconducting tapes

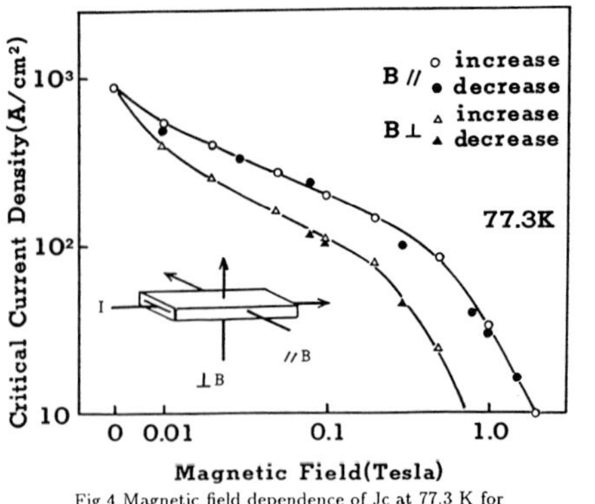

Fig.4 Magnetic field dependence of Jc at 77.3 K for Ag sheathed BiPbSrCaCuO superconducting tapes

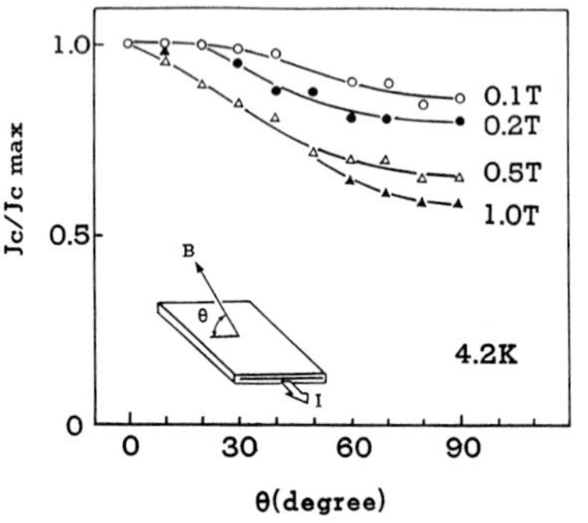

Fig.5 Angular dependence of Jc at 4.2 K in various magnetic fields.

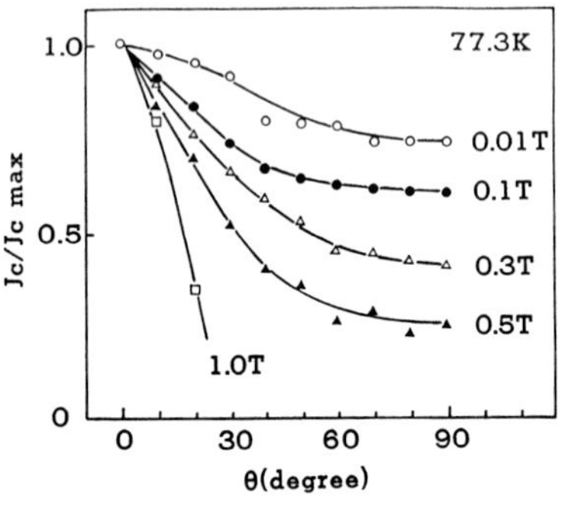

Fig.6 Angular dependence of Jc at 77.3 K in various magnetic fieids.

Superconducting Current Lead Development of BiPbSrCaCuO

KENJI TASAKI[1], ERIKO S. YONEDA[1], DAISUKE ITO[1], HIROYASU OGIWARA[1], KOJI NUMATA[2], and KAZUTOMO HOSHINO[2]

[1] Toshiba R&D Center, 4-1, Ukishima-cho, Kawasaki-ku, Kawasaki, 210 Japan
[2] Mitsui Mining & Smelting Co., Ltd., 1333-2, Haraichi, Ageo, Saitama, 362 Japan

ABSTRACT

Cylindrical superconducting rods composed of Pb doped Bi-Sr-Ca-Cu-O, 10∼15 cm long and 10∼20 mm in diameter, were fabricated by a cold isostatic pressing (CIP) method. To evaluate their electric properties as current leads, critical current (Ic) and contact resistance to the ceramic superconductor were investigated. As a result of improvements, transport Ic value of 480 A at 77 K, that of more than 1000 A at 4.2 K, and contact resistance of less than 1.7×10^{-9} Ω (contact resistivity of less than 1.3×10^{-8} Ωcm^2) were achieved.

INTRODUCTION

To design a superconducting magnet system, heat leak from current leads into a liquid helium pool in a cryostat must be paid attention to, since the refrigeration power required to recondense vaporized helium back to the liquid state is more than five hundred times as much as heat leak power, i.e. 1W heat leak power is equivalent to spending 500W refrigeration power. The quantity of heat leak depends mainly on Joule's heating and heat conduction at the current lead. Therefore, low resistance metal such as copper has been used as the material of a current lead, to reduce the Joule's heating. However, such a material has at the same time high thermal conductivity, which makes it difficult to reduce the heat leak.

On the other hand, a ceramic superconductor has not only low thermal conductivity but also no Joule's heating when it is used in the superconducting state. Therefore, a low heat leak current lead will be obtained, using ceramic superconductors. However, such a ceramic current lead also has heat generation due to contact resistance to a superconductor at electrodes.

For heat conduction, the heat leak increases with increasing cross sectional area of the rod. Therefore, it is important to raise Jc of the rod, with regard to fabricating low heat leak current leads. For contact resistance, the techniques to decrease it have already been reported(1)-(5) and constant resistivity of $10^{-9}\sim10^{-12}$ Ωcm^2 range has been obtained. However, these reports don't refer to whether techniques introduced there can be applied to large current capacity devices, though those techniques will be effective for small devices . The authors aimed to make large current capacity superconductors for current leads, fabricated large superconducting rods and investigated their electrical characteristics.

SAMPLE PREPARATION AND EXPERIMENTAL

Pb doped Bi-Sr-Ca-Cu-O powder, prepared by a coprecipitation method, was fed into a rubber tube to make a cylindrical sample. The tube was put into the CIP chamber for compaction at 1 ton/cm^2 hydrostatic pressure. Then the sample was removed from the tube and sintered at 845 ℃ for 24 hours in air. After cooling the sample, two pieces of silver foil, which were used as electrodes, were rolled at both sides of the sample. Afterwards, the CIP and sintering process was repeated once again. The sample preparation procedure is shown in Fig.1.

Samples, fabricated as described above, have 80% density and have following approximate sizes as shown in Fig.2; 10∼15cm length, 10∼20mm outer diameter, 4∼9cm distance between silver electrodes and a 2cm wide silver electrode.

To connect the sample with a power supply, copper wires were attached on the silver electrodes with solder. Then, for measurement of Ic and contact

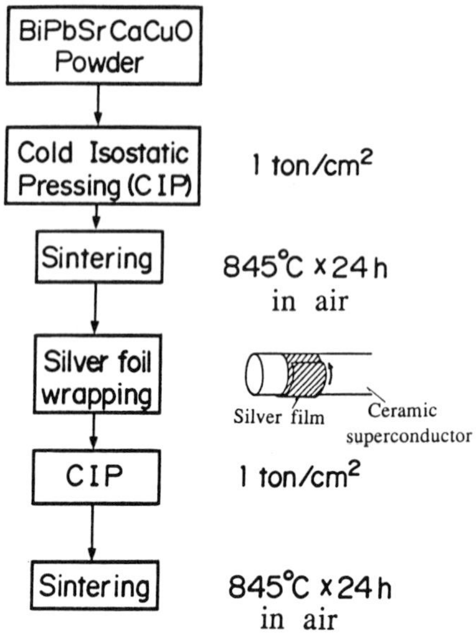

Fig.1 Sample preparation procedure.

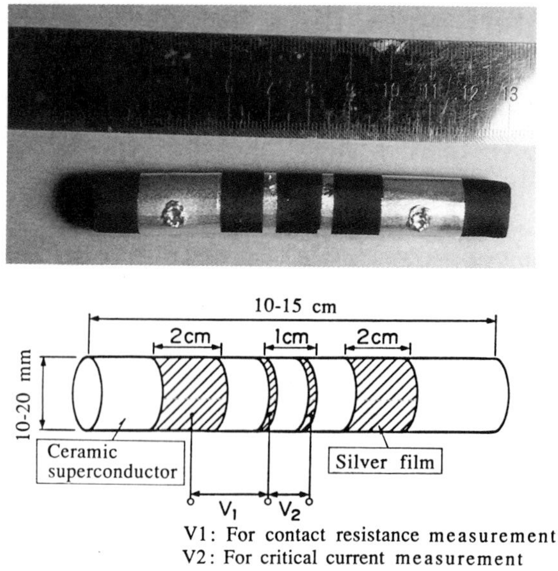

Fig.2 Sample shape.

resistance at the silver electrodes, three potential leads were attached to the sample as shown in Fig.2. Afterwards, the sample was fixed mechanically, to prevent the sample from moving. The sample was immersed in liquid nitrogen or liquid helium. DC current was supplied to it and its value was measured by a shunt resistor. Transport current value (I_t) and voltages at V_1 and V_2 in Fig.2 were recorded by an X-Y recorder. Critical current was determined from a V_2-I_t curve, using a 1 μV/cm voltage criterion in a standard four-terminal method. In the same way, contact resistance was determined from the V_1-I_t curve.

RESULTS AND DISCUSSION

Contact resistance

Figure 3 shows a typical voltage-current (V_1-I_t) curve. Voltages exceeding 0.2 μV were not observed, up to 120 A transport current. Besides, the voltage increase is probably not due to contact resistance but due to normal zone developing somewhere between voltage taps. Therefore, contact resistance at 77 K turns out to be less than 1.7×10^{-9} Ω (contact resistivity is less than 1.3×10^{-8} Ωcm^2, using a silver electrode area of 7.5 cm^2). This is the lowest value that have ever been reported for large current capacity (more than 100

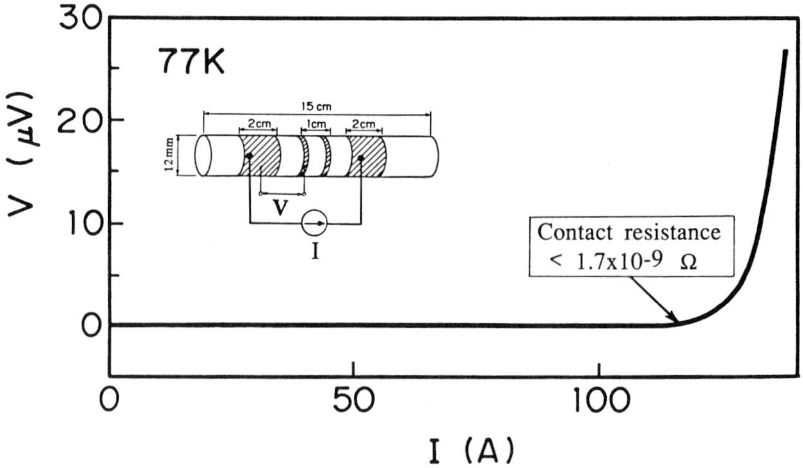

Fig.3 V-I Curve for contact resistance measurement.

Fig.4 SEM micrograph of interface between ceramic superconductor and Ag foil.

A class) ceramic superconductors.

A SEM micrograph of the interface between the ceramic superconductor and the silver electrode is shown in Fig.4. These two materials have good contact, i.e. the rough ceramic surface was covered tightly with silver foil. This condition was obtained during the CIP and the sintering process and probably contributed to make contact resistance to a low value.

The 10^{-9} Ω order resistance is sufficiently small for a kA-class current lead, because the several mW class Joule's heating at 1 kA can be easily cooled by a small refrigerator.

Critical current

Figure 5 shows a typical V_2-I_t curve. The maximum Ic value in zero magnetic field at 77 K results in 480 A (Jc value was 153 A/cm^2), using 1 μV/cm voltage criterion. This is the highest Ic obtained for bulk high Tc superconductors at 77 K ever reported.

Current was further increased up to 500 A as shown in Fig.5, over critical current, i.e. in flux flow state, and held at the state for 5 minutes. During the period, no voltage instability, i.e. voltage increasing or decreasing, was observed, which means that the bulk ceramic superconductor without any stabilizer has good thermal stability, in spite a ceramic superconductor has low thermal conductivity. This result suggests that a practical current lead can be designed as operating at high current load factor.

Ic value at 4.2 K was also measured and resulted in more than 1000 A (Jc

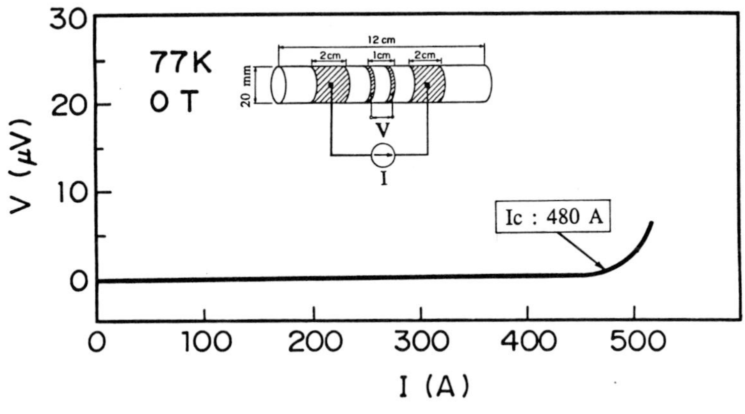

Fig.5 V-I Curve for critical current measurement.

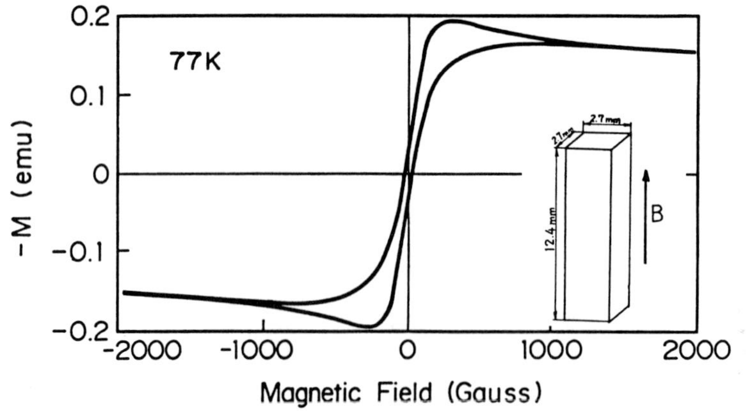

Fig.6 Field dependence of magnitization at 77K.

was more than 885 A/cm^2), with another low Jc sample whose Ic at 77 K was 142 A. External copper wires in our measurement system had not so high current capacity exceeding 1000A, therefore, the critical current value for the sample at 4.2 K could not be measured.

Magnetization

Magnetization of two different size samples (2.7×2.7×12.4 mm and 1.3×1.3× 12.4mm) were measured by SQUID (field dependence of magnetization in the former size sample is shown in Fig.6). However, sample size dependence of magnetization per unit volume could not be observed. The result suggests that the magnetization in Fig.6 is mainly due to shielding current in grains. Therefore, 80% density is not enough to improve the weak link between grains.

CONCLUSION

In a cylindrical bulk superconducting rod, composed of Pb-BiSrCaCuO, Ic value of 480 A at 77 K, that of more than 1000 A at 4.2 K, and contact resistance of less than 1.7×10^{-9} Ω (contact resistivity of less than 1.3×10^{-8} Ωcm^2) were achieved, respectively. Each of these is the best value in the present. Contact resistance of 10^{-9} Ω order is small enough to design the kA-class current lead. Critical current is expected to be raised by improving weak link between grains further.

REFERENCES

1. J.W.Ekin, A.J.Panson and B.A.Blankenship: Appl.Phys.Lett.52(4),25 January 1988
2. J.W.Ekin, et al: Appl.Phys.Lett.52(21),23 May 1988
3. Y.Suzuki, et al: Appl.Phys.Lett.54(7),13 February1989
4. S.Jin, et al: Appl.Phys.Lett.54(25),19 June 1989
5. John Talvacchio: IEEE Transactions on C.H.M.T. vol.12, No.1,March 1989

6 Preparation and Properties of Thin Films

6.1 YBCO and Related Films

Rheed Oscillations During Growth of High-T_c Superconducting Oxides

T. TERASHIMA[1], K. SHIMURA[1], Y. BANDO[1], K. IIJIMA[2], K. YAMAMOTO[2], K. HIRATA[2], K. HAYASHI[2], K. KAMIGAKI[3], and H. TERAUCHI[3]

[1] Institute for Chemical Research, Kyoto University, Uji, 611 Japan
[2] Research Institute for Production Development, Kyoto, 606 Japan
[3] Department of Physics, Kansei-Gakuin University, Nishinomiya, 662 Japan

ABSTRACT

We have found strong intensity oscillations in RHEED during epitaxial growth of $BaTiO_3$, La_2CuO_4, and $YBa_2Cu_3O_{7-x}$ with perovskite-type structures. The oscillation period corresponds to the height of the minimum unit of each oxide that satisfies the chemical composition and the electrical neutrality. The discovery of the RHEED oscillations during the growth of the oxides provides a means of the precise control over the crystal growth on the atomic scale.

KEY WORDS: RHEED oscillations, reactive evaporation, $YBa_2Cu_3O_{7-x}$, 2D nuclei, growth unit

INTRODUCTION

Reflection high-energy electron diffraction (RHEED) specular intensity oscillations are well known in the molecular beam epitaxy (MBE) growth of semiconductors and metals.[1] The RHEED oscillations have played a significant role in the understanding of the microscopic processes involved in the epitaxial growth[2,3]. And it has aided in the development of extremely precise control over the crystal growth of these materials on the atomic scale.[4] The RHEED oscillations are interpreted as a consequence of the nucleation of 2D islands and their growth into a flat terrace, in a cyclic process.

Since the discovery of the high-Tc superconducting oxides, much attention has been attracted to the epitaxial film growth of these oxides for the study of basic physics as well as for wide applications. Most preparation methods including reactive thermal evaporation, reactive sputtering, and laser ablation are essentially co-deposition of the elements. The major point of interest concerning the growth process is to find what the growth unit is : All the element atoms coexist on a substrate surface. The growth unit for single-element metals and semiconductors has been confirmed to be one atomic layer from the observation of RHEED oscillations. However, in ionic oxide systems, the growth unit may be subject to charge neutrality and, therefore, may not be an atomic single layer.

In this paper, we report on RHEED oscillations observed during the epitaxial growth of $BaTiO_3$, La_2CuO_4, and $YBa_2Cu_3O_{7-x}$ by reactive evaporation.[5]

EXPERIMENTAL

The deposition system is schematically shown in Fig.1. Essentially, this growth method is a co-evaporation of the metal elements under an oxygen atmosphere. In previous work, we have reported on *in situ* RHEED observations during the growth of $YBa_2Cu_3O_{7-x}$(001) on $SrTiO_3$(100) and on MgO(100).[6] The experimental system used in the present study was the same as in Ref.6, except for the method of oxygen activation. In this study oxygen gas was partially converted into ozone by a commercially available ozonizer (O_3 content $\leq 5\%$). The typical deposition conditions were the following: The substrate temperature was 680 °C and the deposition rate was about 0.5Å/s. The substrate was the $SrTiO_3$ (100) surface which was etched by an Ar^+ ion beam (600V) at 650 °C before the deposition. The acceleration voltage of the incident electron beam for the RHEED observation was 20 keV. The intensity of the RHEED pattern was measured via an optical fiber using a photo-multiplier.

RESULTS AND DISCUSSION

Figure 2 (a) shows RHEED oscillations observed during the epitaxial growth of $BaTiO_3$(001) on $SrTiO_3$

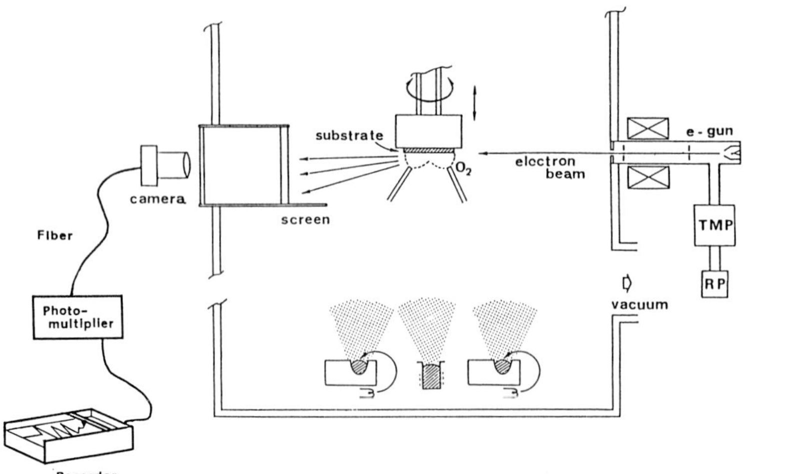

Fig. 1 Schematical illustration of the deposition system.

(100), where the direction of the incident beam is [100]. The intensity I decreased rapidly as soon as the growth was initiated and then continuous oscillations were observed whose amplitude decreased gradually. If the growth was interrupted, the intensity increased back to initial levels and showed definite oscillations again.

In order to identify the period of the RHEED oscillations, we must compare the number of the oscillations with the film thickness on an atomic scale. One way to realize this is by the measurement of X-ray diffraction from the ultrathin films.

In general, the intensity of X-ray diffraction is described by:

$$I = I_e F^2 G^2 \qquad (1)$$

where $I_e = (c/8\pi) < E_0^2 > (e^2/mc^2R)^2(1+\cos^2 2\theta)/2$, F is a structure factor, and the Laue function G^2 for a one dimensional system is given as:

$$G^2 = \frac{\sin^2 \pi N x}{\sin^2 \pi x} \qquad (2)$$

where x is the Miller index and N is the number of the unit cells in a crystal. In the plane of the crystal N is large and G^2 has finite values only when x is integer. However, along the direction of the thickness where N is small, G^2 has (N-2) sub-maximum points of finite intensity between the peaks of (00ℓ) and $(00\ell+1)$. And also (N-1) minimum points appear when $x = \ell + 1/N, \ell + 2/N, \cdots, \ell + (N-1)/N$ (ℓ is integer). The thickness L of the ultrathin film can be determined from the angles of the minimum points of G^2 by:

$$L = \frac{\lambda}{2 \times (\sin\theta_n - \sin\theta_{n-1})}, \qquad (3)$$

where θ_n and θ_{n-1} are the angles of the nth and $(n-1)th$ minimum points, respectively.

Figure 3(a) shows an X-ray diffraction pattern scanned along the [001] direction for the $BaTiO_3(001)$ film the growth of which was finished at the $16th$ peak position of the RHEED oscillations. The sub-maxima peaks of the Laue function are seen around the fundamental peaks of (001) and (002). The thickness is determined from Eqs.(3) to be 67.0 Å. The thickness can also be determined from Scherrer's formula using the full width of the half maximum (FWHM) value of the (002) peak ($\Delta 2\theta = 1.44°$) to be 66.0Å, neglecting an instrumentation broadening. These values are in good agreement with sixteen times the lattice spacing c_0 which is determined from the (002) peak position to be 4.14Å. The RHEED oscillation period thus precisely corresponds to the height of one unit cell of $BaTiO_3$, *i.e.* the thickness of two atomic layers-one each of (BaO) and (TiO_2).

The RHEED oscillations observed during the epitaxial growth of GaAs and Si etc. under the layer-by-layer growth condition are interpreted by the single scattering theory [7] to be a consequence of the repetition of the formation of 2D nuclei and growth to a flat terrace. One significant experimental fact which supports this growth model is that the intensity of the oscillations is different when observed

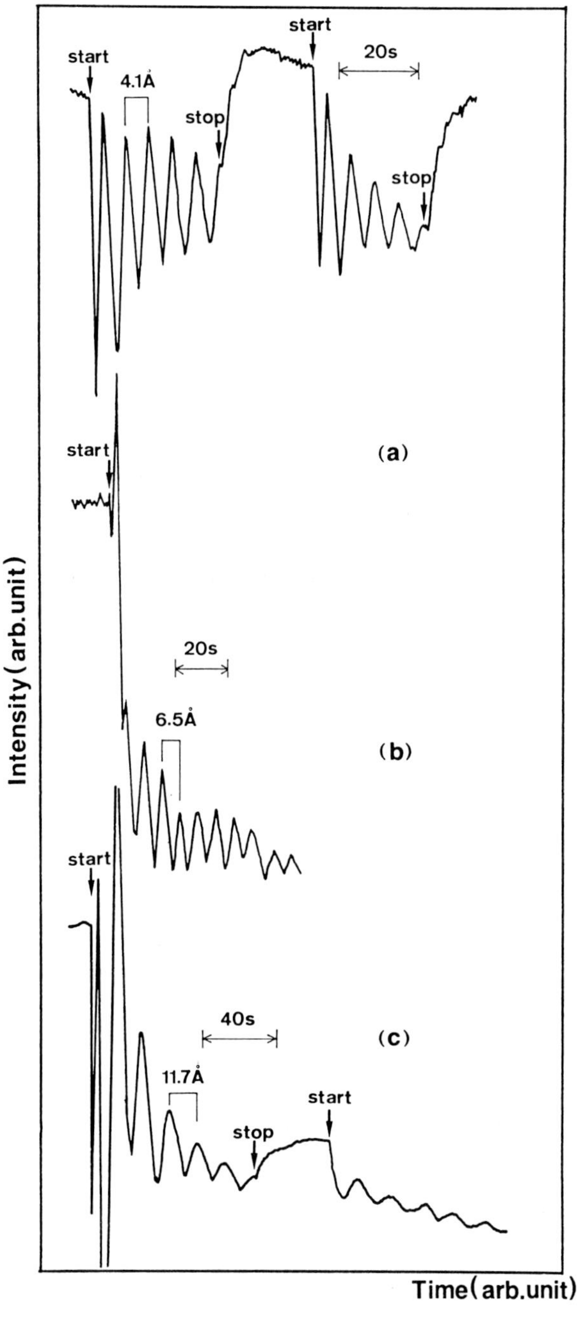

Fig. 2 RHEED intensity oscillations of the specular beam during the epitaxial growth for (a) $BaTiO_3(001)$; (b) $La_2CuO_4(001)$; and (c) $YBa_2Cu_3O_{7-x}(001)$ on $SrTiO_3(100)$. The acceleration voltage of the incident beam is 20 keV and the incident angle is 1.9° (off-Bragg condition); [100] azimuth. The starting and ending points of the growth are indicated by the arrows. T_s=680 °C. Inserts are the RHEED patterns observed during the growth of each oxide.

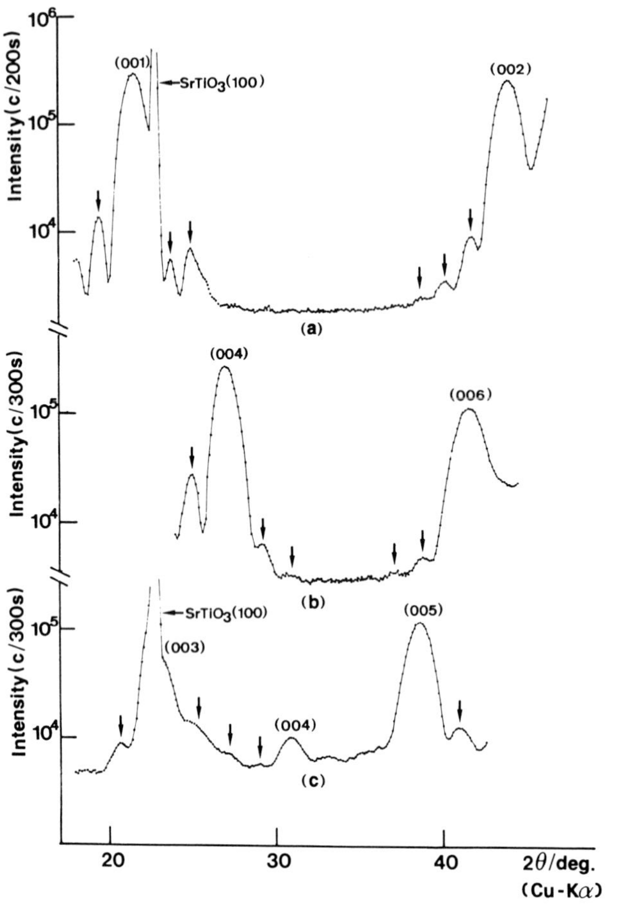

Fig. 3 X-ray diffraction patterns scanned along the [001] direction for (a) $BaTiO_3$ (001) film; (b) La_2CuO_4(001) film; and (c) $YBa_2Cu_3O_{7-x}$ film of which growth was finished at the 16*th*, 10*th* and 5*th* peak positions of the RHEED oscillations, respectively. The peaks indicated by the arrows are those of the Laue function.

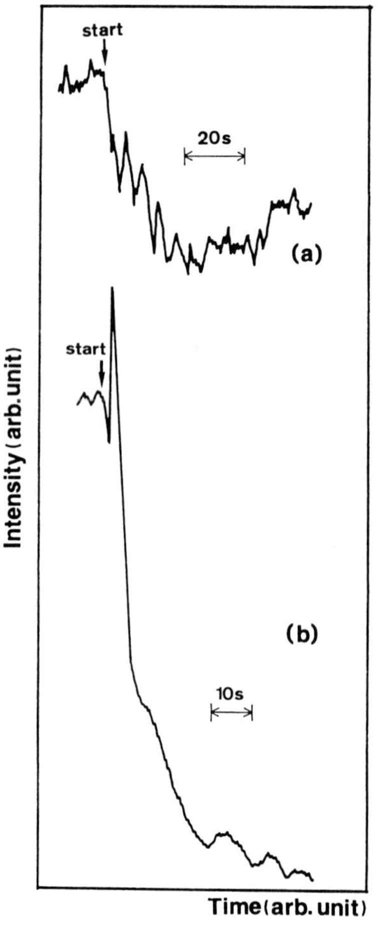

Fig. 4 RHEED specular beam intensity during the epitaxial growth for (a) $BaTiO_3$(001); and (b) $YBa_2Cu_3O_{7-x}$ (001) on $SrTiO_3$(100) in the in-phase (Bragg) condition.

under in-phase (Bragg) and out-of-phase conditions. In an in-phase condition, where the electrons are insensitive to surface steps, intensity oscillations are much weaker. We have confirmed that the intensity oscillations during the growth of $BaTiO_3$ also become weaker when the incident beam angle is near the in-phase Bragg condition, as shown in Fig.4(a). This result supports the conclusion that the oscillations of Fig.2(a) is attributed to the two dimensional layer-by-layer growth mode.

As mentioned already, the intensity recovered and the oscillation amplitude increased after the growth was interrupted. This recovery of the intensity is commonly observed in the growth of GaAs by MBE. It is interpreted to be due to an increase in the mean terrace width of the surface and, hence, a reduction in the surface step density by the migration of surface adatoms to the flat terrace edge. From the remarkable recovery of the RHEED intensity, it is assumed that the migration of surface adatoms easily occurs in the $BaTiO_3$ system. Figure 5(a) shows a schematical illustration of a deposition-and-growth process model for $BaTiO_3$.

Figure 2(b) and (c) show RHEED oscillations observed during the growth of La_2CuO_4(001) on $SrTiO_3$ (100) and $YBa_2Cu_3O_{7-x}$(001) on $SrTiO_3$(100). In the first place, we must note that an abrupt increase of the RHEED intensity is observed in both cases as soon as the growth is initiated. An initial transient increase of the RHEED intensity is well-known in the MBE growth of GaAs.[8,9] It was revealed that the transient behavior strongly depended on the diffraction conditions, *i.e.* the angle and the azimuth of the

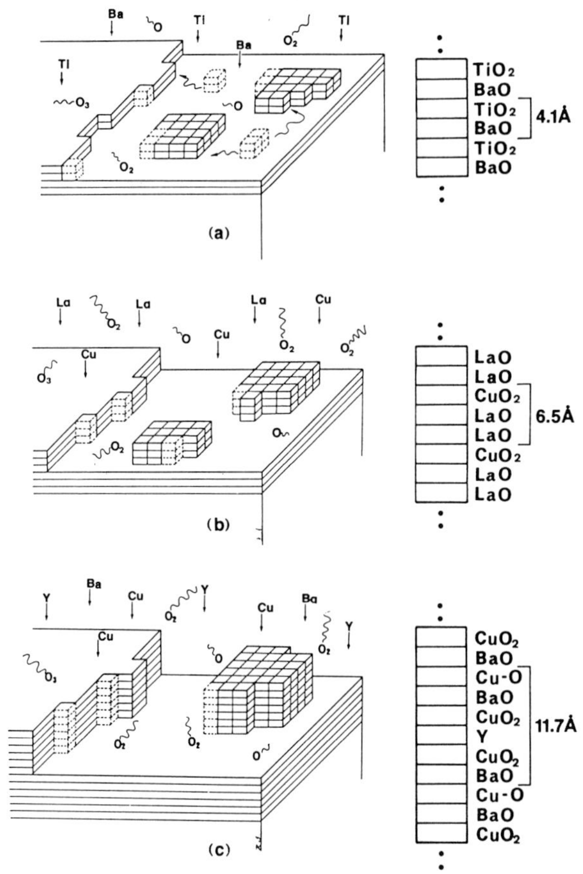

Fig. 5 Schematical illustrations for the deposition and growth process of (a) $BaTiO_3$ (001); (b) La_2CuO_4(001); and (c) $YBa_2Cu_3O_{7-x}$ (001) on $SrTiO_3$(100). The inserts are the stacking sequences of the atomic layers in each crystal, and the units of the growth are indicated. Note that the stacking sequences in the growth units shown in the figures have no specific meanings, the top layer can not be specified.

incident beam. The initial transient behavior has been explained to be caused by an initial smoothing of steps already present on the substrate surface [8] or a reconstruction of the substrate surface.[9] Anyway, as the period to the first peak from the growth initiation is too short as compared with those of the subsequent regular oscillations, we can distinguish between the first peak and others.

Figure 3(b) shows an X-ray diffraction pattern for a La_2CuO_4(001) film whose growth was finished at the $10th$ peak position of the RHEED oscillations. In the numbering of the oscillation peaks, we neglected the initial transient peak. For this film, the thickness was determined from the Laue function and Scherrer's formula ($\Delta 2\theta(004) = 1.40°$) to be 66.0 Å and 65.0Å, respectively. From this result, one period of the RHEED oscillations was determined to be 6.5 Å. For La_2CuO_4, the chemical repetition unit consists of three atomic layers (LaO), (LaO) and (CuO_2), as shown in the inset of Fig.5(b), and this unit is stacked on itself with a relation ($a/2+b/2$) shifted in the (001) plane. The lattice spacing of this unit along the [001] direction is determined to be 6.50Å from the (006) peak position, and this value is in good agreement with one period of the RHEED oscillations.

Finally, Fig.3(c) shows an X-ray diffraction pattern for the film of $YBa_2Cu_3O_{7-x}$ where the growth was finished at the $5th$ peak position. In the numbering of the RHEED oscillation peaks, we also neglected the initial transient peak. As shown in the inset of Fig.5(c), a unit cell of $YBa_2Cu_3O_{7-x}$ consists of six atomic layers along the c-axis, the length of which is determined to be 11.7 Å from the (005) peak position. The thickness of this film was determined to be 58.9 Å and 59.0 Å using the Laue function and Scherrer's formula ($\Delta 2\theta(005) = 1.57°$). This value is in good agreement with five times the lattice spacing c_0. This result revealed that one period of the RHEED oscillations observed during the growth of $YBa_2Cu_3O_{7-x}$ (001) corresponded to the height of the unit cell. Such a long period of the RHEED oscillations was first observed in this system. We have confirmed the intensity oscillations during the growth of $YBa_2Cu_3O_{7-x}$ become weaker in the in-phase condition, as shown in Fig.4 (b). This indicates that the growth of $YBa_2Cu_3O_{7-x}$ also occurs in the two-dimensional growth mode.

The present results have significance for an understanding of the epitaxial growth of ionic oxides. The observed period in the RHEED oscillations indicates that the 2D nuclei of $YBa_2Cu_3O_{7-x}$, La_2CuO_4, and $BaTiO_3$ all have the height of one minimum unit needed to satisfy the chemical composition and the electrical neutrality. That is, the chemical and neutral unit is the unit of crystal growth of ionic materials. The deposition and growth process models for La_2CuO_4(001) and $YBa_2Cu_3O_{7-x}$(001)are schematically shown in Fig.5(b) and 5(c).

A significant problem is derived from this result. There should exist a definite stacking sequence of atomic layers in the growth unit, and therefore, the top atomic layer in the growth unit should be always the same. In the present stage, we can not specify the metallic element of the top atomic layers in the growth unit, and it still remains as an extremely important problem for the understanding of epitaxial growth of ionic oxides.

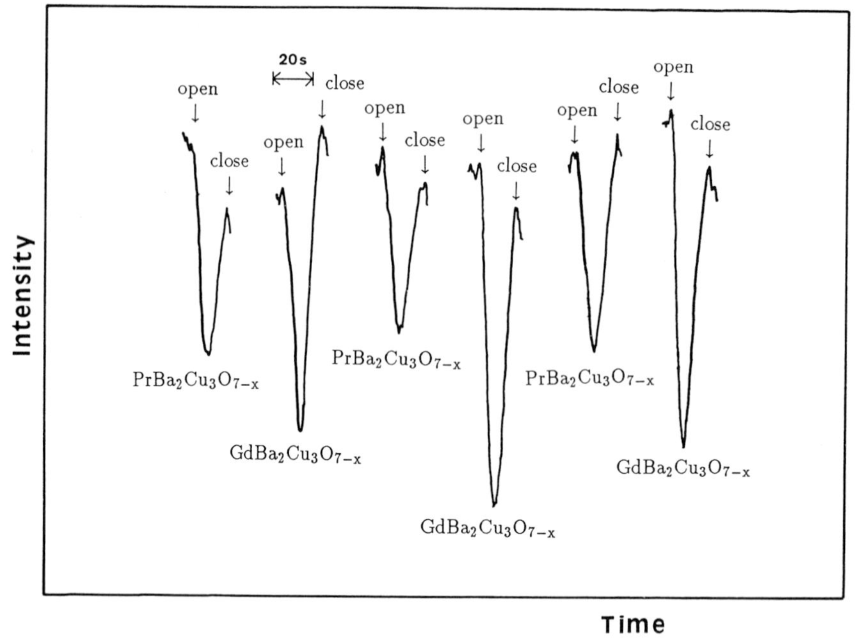

Fig. 6 RHEED intensity oscillations during the growth of the superlattice : $PrBa_2Cu_3O_{7-x}$(1 unit cell)/$GdBa_2Cu_3O_{7-x}$(1 unit cell).

The discovery of the RHEED oscillations during the growth of oxides provides a means for the precise control over the crystal growth on the atomic scale. Especially, the RHEED oscillations are of great advantage to fabricate superlattices. Recently, superlattices composed of a few unit cells of $YBa_2Cu_3O_{7-x}$ and those of other 1:2:3 compounds, e.g. $PrBa_2Cu_3O_{7-x}$,[10] have been fabricated in order to study the two dimensional nature of superconductivity. In Fig.6, we demonstrate the growth of a superlattice composed of one unit cell $GdBa_2Cu_3O_{7-x}$ and one unit cell $PrBa_2Cu_3O_{7-x}$ by monitoring the RHEED oscillations. The well defined super-structure can be made by this technique.

In conclusion, we have observed for the first time RHEED oscillations during the epitaxial growth of oxides caused by the layer-by-layer (unit-by-unit) growth and have found that one period of the RHEED oscillations correspond to the height of the minimum unit which satisfies the chemical composition and the electrical neutrality of the oxide. The data provides a basic understanding of the epitaxial growth mechanism of oxides. Finally, we believe that the formation of superlattices composed of different kinds of oxides by monitoring the RHEED oscillations bring a hopeful new field of study.

REFERENCES

1. Harris J, Joyce B, Dobson P (1981) Surf. Sci. Lett. 103: L90
2. Kaneko T, Imafuku M, Kokubu C, Yamamoto R, Doyama M (1986) J. Phys. Soc. Jpn. 55: 2903
3. Sakamoto T, Kawamura T, Nago S, Hashiguchi G, Sakamoto K, Kuniyoshi K (1987) J. Cryst. Growth 81: 59
4. Sakamoto T, Funabashi H, Ohta K, Nakagawa T, Kawai N, Kojima, Bando Y (1985) Superlattices & Microstruct. 1: 347
5. Terashima T, Iijima K, Yamamoto K, Bando Y, Mazaki H (1988) Jpn. J. Appl. Phys. 27: L91
6. Terashima T, Iijima K, Yamamoto K, Hirata K, Bando Y, Takada T (1989) Jpn. J. Appl. Phys. 28: L987
7. Lent C, Cohen P (1984) Surf. Sci. 139: 121
8. Briones F, Golmayo D, Gonzalez L Miguel J (1985) Jpn. J. Appl. Phys. 24: L478
9. Dobson P, Joyce B, Neave J, Zhang J (1987) J. Cryst. Growth 81: 1
10. Triscone J, Fischer Ø, Brunner O, Antognazza L, Kent A, Karkut M (1990) Phys. Rev. Lett. 64: 804

Low Temperature In-Situ Growth of YBCO High-T_c Superconducting Thin Films Using MOCVD and Plasma-Enhanced MOCVD (PE-MOCVD)

J. ZHAO[1], Y.Q. LI[2], C.S. CHERN[3], W. HUANG[3], P. NORRIS[1], B. GALLOIS[2], B. KEAR[3], P. LU[3], and F. COSANDEY[3]

[1] EMCORE Corporation, 35 Elizabeth Ave., Somerset, NJ 08873, USA
[2] Stevens Institute of Technology, Hoboken, NJ 07030, USA
[3] Rutgers, The State University, New Brunswick, NJ 08854, USA

ABSTRACT

YBCO thin films with a critical current density of 2.3 x 10^6 A/cm^2 and 1x10^6 A/cm^2 at 77.7 K and 0 T were prepared, in-situ, by metalorganic chemical vapor deposition (MOCVD) and PE-MOCVD. The films were formed in-situ on $LaAlO_3$ at a substrate temperature of 570-730°C in 2-20 torr partial pressure of N_2O. Resistivity and magnetic susceptibility measurements show a sharp superconducting transition (less than 1K wide) at 88-90 K . X-ray diffraction and high resolution electron microscopy measurements indicate that films grew epitaxially with the c-axis perpendicular to the surface of the substrate.

INTRODUCTION

The development of a low temperature, in-situ process for the formation of high temperature superconducting (HTSC) thin films is crucial for the integration of HTSC technology with economically important active device technologies. A low temperature process should permit the deposition of high quality, high density, HTSC films with smooth suface morphology, on a variety of substrates with a minimal degree of film-substrate interaction. Metalorganic Chemical Vapor Deposition (MOCVD) process technology offers the advantages of a high degree of compositional control, adaptability for large scale production and the potential for low temperature fabrication. The capability of operating at high oxygen partial pressure is particularily suitable for in-situ formation of HTSC films. Additionally, the use of a plasma-enhanced (PE) process substitutes activated species for conventional thermal equilibrium ones and enhances compound formation through active oxygen generated from the dissociation of oxidizer precursors. This increase in chemical activity allows deposition to proceed at lower substrate temperatures. The PE-MOCVD process also offers the advantage of easy transfer from R&D to low cost, large scale fabrication.

The successful growth of HTSC thin films by MOCVD was reported by several groups in 1988[1,2,3] . Although early results showed low values of T_c, less than one year later, a high J_c of 2x10^6 A/cm^2 (77.3 K,0 T, 5μV/cm criterion) and T_c=91.5 K on (100) $SrTiO_3$ was reported[4]. However these results required a deposition temperature of 850°C, which is considered too high for many technologically important substrate materials. Tsuruoka et al[5] used the greater reactivity of N_2O in order to achieve significant reductions in growth temperature to as low as 650°C. At this low growth temperature, T_c had decreased to 79 K and no J_c values were reported. Further progress was made by Kanehori et al[6] who used magneto-microwave PE-MOCVD with O_2 in order to further reduce substrate temperatures for the growth of YBCO to 515-580°C. For a deposition temperature of 515°C, on (100) $SrTiO_3$, with 200 w. of microwave power, T_c was 60 K. At 580°C, however, T_c increased to 84 K with 200 w. of power and J_c values of 10^5 A/cm^2 at 77 K were measured. When the excitation power was reduced to 100 w. , T_c decreased to 30 K. This strong dependance of T_c on plasma excitation level indicates the important role of the active oxidizer species in YBCO growth at low substrate temperatures. In this paper, we report the results of two sets of experiments; one with MOCVD using N_2O and the other using both N_2O and PE-MOCVD in an attempt to maximize YBCO HTSC properties at low growth temperatures. In both cases the results indicated that excellent YBCO films, with T_c in the range 88-90 K and J_c values in excess of 10^6 A/cm^2 at 77 K could be obtained.

MOCVD and PE-MOCVD

MOCVD, due to its capability of operating at high oxygen partial pressures, is particularly suitable for in-situ formation of HTSC films. Fig.1 is a phase diagram for YBCO, where oxygen partial pressure is plotted vs. temperature indicating the location of the phase transitions of $YBa_2Cu_3O_y$ together with experimental data points from the literature[7] for several PVD in-situ growth techniques. The critical phase stability boundary is at an oxygen content, y, equal to 6.0. Below y=6.0, YBCO is unstable and disproportions into Y_2BaCuO_5, $BaCuO_2$ and Cu_2O. At y=6.5 a phase transition from tetragonal to an orthorhombic structure occurs, resulting in superconducting YBCO with a T_c of about 50K. At y = 6.9, T_c increases to about 90K. The filled circle in Fig.1 represents our present PE-MOCVD growth conditions. It is significant to note that PE-MOCVD appears to be capable of directly forming the orthorhombic superconducting phase in the as-deposited state which may lead to the fabrication of untwinned YBCO film. This potential for low temperature deposition makes the PE-MOCVD process very desirable for the in-situ formation of 90K YBCO superconducting films (y =6.9).

GROWTH of In-situ HTSC FILMS of YBCO by MOCVD and PLASMA-ENHANCED MOCVD

A prototype EMCORE System 5000 with a remote plasma coupling system was developed for the PE-MOCVD of HTSC films, as described in previous publications[8]. For PE-MOCVD studies, a N_2O plasma was generated by a 2.45 GHz modified microwave source with a variable power output of 10 - 100 Watt.

Among the possible volatile organometallic precursors for high T_c MOCVD, metal ß-diketonates are attractive since they are easily synthesized, purified, and handled in air. ß-diketonate complexes $Y(dpm)_3$, $Cu(dpm)_2$, and $Ba(dpm)_2$ (dpm = dipivaloymethanate), were used as yttrium, copper, and barium precursors, and maintained at 110,120 and 220°C, respectively. Deposition was carried out at a system pressure of 2-20 Torr and substrate temperature of 570°C - 730°C with a rotation speed of 1100 rpm. Film deposition rate was controlled at 0.5 - 1µm/hour and the films were usually 0.1 - 1µm thick.

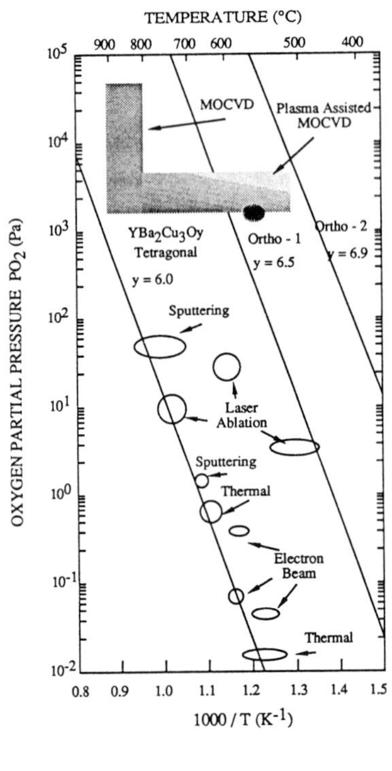

Fig 1. Phase Diagram of Oxygen Partial Pressure vs. 1/T for YBCO

CHARACTERIZATION

The superconducting properties were characterized by both four-probe dc resistivity and ac magnetic susceptibility measurements. The resistivity vs. temperature measurements, as shown in Fig.2, of MOCVD YBCO films grown at 730°C on (100) $LaAlO_3$ were performed with a dc current of 0.5 mA and reveal that the films have a sharp zero resistance transition temperature (T_c) of 89K. The films exhibit a clearly metallic characteristic in the normal state with room temperature resistivity of 220μΩ-cm and extrapolation to zero resistance at T=0 K. The corresponding magnetic susceptibility characteristic is given in Fig.3, as measured by an ac mutual inductance method at a frequency of 2 KHz and a magnetic field of 0.03 Oe. These data also show a sharp superconducting transition less than 1K wide. The agreement between the onset of the magnetic response and the zero resistance transition temperature of 89K indicates highly homogeneous thin films. The critical current density (J_c) was measured on a 120μm x 40μm bridge patterned by a conventional wet photolithographic techniques. J_c was determined from plots of current vs. voltage with a criterion of 1μV/cm. I-V curves, measured at 77.7K, showed a sharp increase in voltage which indicated a J_c of $2.3x10^6$ A/cm^2. This is one of the highest values reported for YBCO films on $LaAlO_3$. Fig.4 shows the temperature dependence of the critical current density which is very similar to that reported for reactive co-evaporation[9]. X-ray diffraction (XRD) measurements showed the films to be highly oriented with the c-axis perpendicular to the surface of the substrate. The films were also found to have grown epitaxially on the $LaAlO_3$ substrates as indicated by the sharp x-ray rocking curves with a FWHM of 0.32° and HREM studies. SEM examination of the thin films showed a smooth surface morphology with some a-b axis oriented crystallites and precipitates.

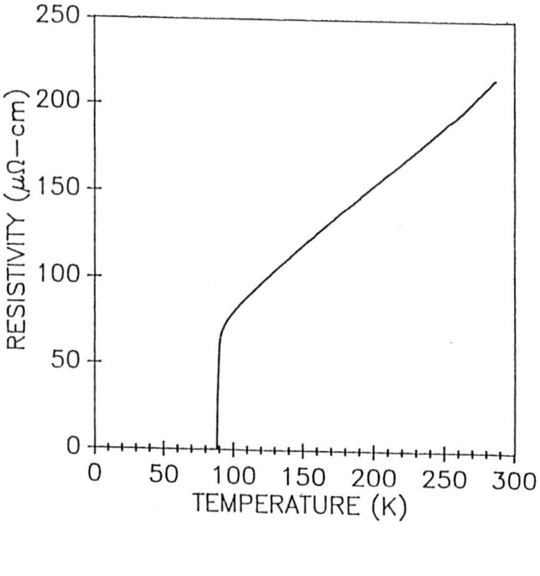

Fig. 2. Resistivity vs. Temperature Characteristic for YBCO Deposited on $LaAlO_3$ at 730°C by MOCVD using N_2O

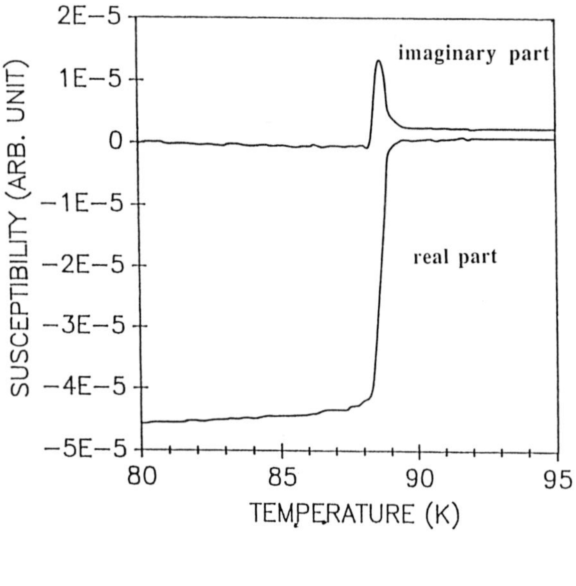

Fig. 3 AC Susceptibility vs. Temperature Characteristic for YBCO Deposited on $LaAlO_3$ at 730°C by MOCVD using N_2O

XRD data on PE-MOCVD films grown at 670°C on (100) $LaAlO_3$ showed the films to be single 1-2-3 phase which was highly oriented c-axis perpendicular to the substrate surface. Resistivity vs. temperature results on these films are given in Fig.5 and also show a sharp superconducting transition with T_c=89.8 K. The films exhibit metallic characteristics above Tc with a normal state resistance extrapolation to zero at 0 K. In addition, the magnetic susceptibility data of Fig.6 also shows a sharp superconducting transition. J_c was measured at 77.7K and found to be 1×10^6 A/cm^2, which is comparable to films on $LaAlO_3$ prepared by coevaporation[10]. This indicates that, with high quality, high J_c YBCO films can be grown at substrate temperatures at least 60°C lower through the use of PE-MOCVD.

The effects of further reduction in growth temperature were also investigated. XRD data for an in-situ film grown at 570°C on a (100) YSZ substrate followed by cooling in one atmosphere of pure O_2 showed that the $Y_1B_2C_3O_{7-x}$ phase is the dominant phase with a high degree of preferential c-axis orientation. Moreover, a mirror-like surface smoothness and a highly dense morphology are observed for the HTSC films deposited using PE-MOCVD. These films exhibited a greatly increased scratch resistance compared to normal MOCVD-grown YBCO. Resistivity vs. temperature results reveal that the superconducting transition onset occurs at ≈80 K and reach zero resistance at 72K. YBCO films deposited on (100) $SrTiO_3$ substrates show a lower normal state metallic resistance at temperatures above 90K than the films deposited on YSZ, however they show identical T_c values. Other samples having similar transition temperatures have been reproducibility grown under the same conditions. The relatively low transition temperature of the film may be due to a slight deviations in film stoichiometry, as indicated by EDX, or to an oxygen deficiency.

Carbon is a common impurity in films prepared by MOCVD processes, especially for films deposited at low temperatures. Auger electron spectroscopy measurements on the films prepared at 610°C by PE-MOCVD revealed that the carbon content in the films is below the limit of detection, or less than one atomic percent.

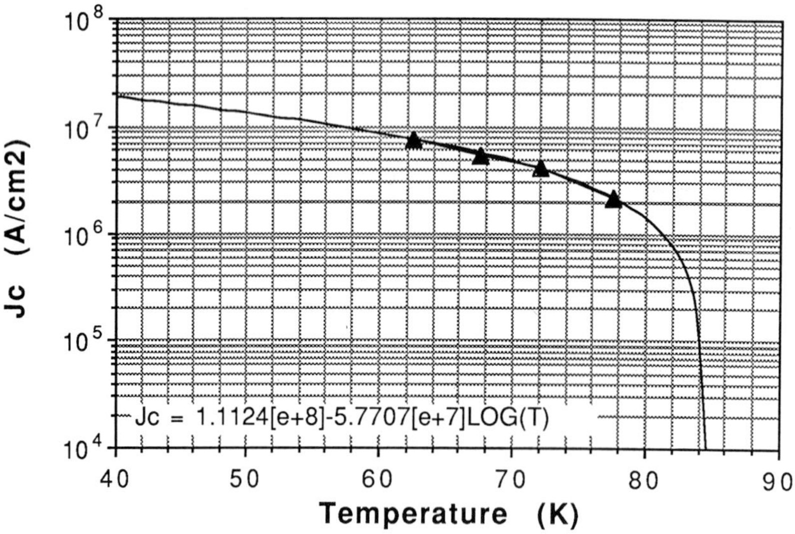

Fig. 4 Critical Current vs. Temperature for YBCO Deposited on $LaAlO_3$ at 730°C by MOCVD using N_2O (4 probe microbridge, 1μV/cm criterion)

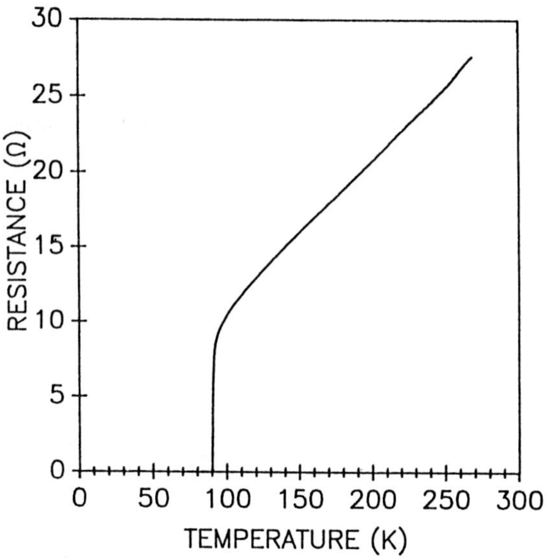

Fig. 5 Resistivity vs. Temperature Characteristic for YBCO Deposited on $LaAlO_3$ at 670°C by PE-MOCVD using N_2O

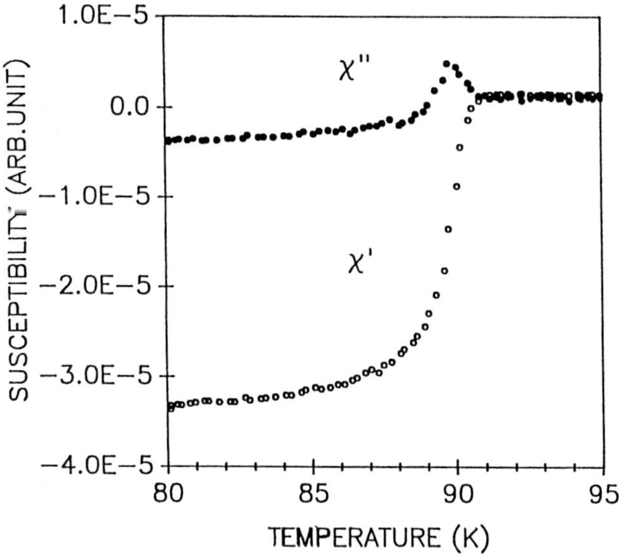

Fig. 6 AC Susceptibility vs. Temperature Characteristic for YBCO Deposited on $LaAlO_3$ at 670°C by PE-MOCVD using N_2O

It is interesting to compare the data above with the O_2 PE-MOCVD results cited earlier. At nearly equal substrate temperatures,and for equivalent microwave power levels the T_c in the case of N_2O plasma was 72 K, as compared with 30 K in the case of the O_2 plasma. This indicates clearly the beneficial effects of both N_2O and PE-MOCVD. These results suggest that with additional development of the PE-MOCVD process, fabrication of HTSC thin films at even lower growth temperatures may be possible. This development, in turn will increase the probability of sucessfully integrating HTSC and active device technologies.

REFERENCES

[1] A.D. Berry, D.K. Gaskill, R.T. Holm, E.J. Cukauskas, R. Kaplan, R.L. Henry, Appl. Phys. Lett. **52**, 1743 (1988)

[2] H. Yamane, H. Masumoto, T. Hirai, H. Iwasaki, K. Watanabe, N. Kobayashi and Y. Muto, AppL. Phys. Lett. **53**, (1988) 1548

[3]T.Nakamori, H.Abe, T.Kanamori and S.Shibata, Jpn. J.Appl.Phys.27(1988)L1473

[4] H. Yamane, H. Kurosawa, T. Hirai,K. Watanabe, H. Iwasaki, N. Kobayashi and Y. Muto, Supercond. Sci. Technol. **2** (1989) 115-117

[5] T. Tsuruoka, R. Kawasaki, and H. Abe, Jpn. J. Appl. Phys. 28, L607, 1989.

[6] K. Kanehori, Fall MRS Meeting 1989, ISTEC Workshop 1990.

[7] R.H. Hammond and R.Bormann, Physica C162-164, 703 (1989).

[8]J. Zhao, D.W. Noh, C. S. Chern, Y.Q. Li, P. Norris, B. Gallois, and B. Kear, Appl. Phys. Lett. 56, 2342

[9] V. Marijasevic, P.Rosenthal, K.Shinohara, A.F.Marshall, R.H.Hammond and M.R.Beasley (to be published in J. Mater. Res.)

[10] M. C. Nuss, P.M. Mankiewich, R.E. Howard, B.L. Straughn, T.E. Harvey, C.D. Brandle, G.W. Berkstresser, K.W. Goossen, and P.R. Smith, Appl. Phys. Lett. 54, 2265 (1989).

In-Situ Preparation of High-T_c Thin Films by Co-Evaporation Using Ozone Vapor Oxidation

D.D. BERKLEY

School of Physics and Astronomy, 116 Church St. S.E., University of Minnesota, Minneapolis, MN 55455, USA

ABSTRACT

Distilled ozone vapor has been used in a co-evaporation process to produce superconducting thin films of $YBa_2Cu_3O_{7-x}$ *in-situ* at background evaporation pressures on the order of 10^{-7} Torr. Ozone is distilled at 77 K from a dilute O_3/O_2 mixture to form solid ozone with a purity greater than 98%. When prepared in this manner the ozone can be warmed to provide a well characterized source of reactive oxygen for production of the superconductor and oxidation of subsequent overlayers. In addition thin overlayers of Y metal applied *in-situ* to the surface of $YBa_2Cu_3O_{7-x}$ films are passivated in the presence of the ozone vapor. This is seen in an XPS analysis of the oxidation state of Cu at the surface of the superconductor.

KEY WORDS: Ozone, high T_c, co-evaporation

INTRODUCTION

A number of methods have been developed in the last years for preparing thin films of the high T_c superconductors which hope to take advantage of the inherent flexibility of a Molecular Beam Epitaxy (MBE) or co-evaporation process.[1] Precise rate control and an ability to conveniently create multilayer structures are factors which motivate the use of the co-evaporation technique to produce films of these materials. More importantly, however, an MBE method affords an opportunity to study the growth dynamics of the superconductor as it is grown using, for example, Refection High Energy Electron Diffraction (RHEED). Electron surface probes can be used only if the vacuum pressure during evaporation can be maintained low enough to eliminate debilitating scattering effects of the electron probe. This is done by employing some form of activated oxygen during the evaporation which can efficiently form the superconducting phase at the lowest possible flow rate and pressure.

A number of techniques have been developed to prepare activated species of oxygen for use in high T_c thin film making. Radio frequency and microwave excitation have been employed in various configurations to convert diatomic oxygen to a mixture of neutral and ionized atomic and molecular oxygen.[2] The composition of this product is determined, if not completely known, by the specific frequency and power of the generator and the vacuum environment of the resulting oxygen radicals on the way to the substrate surface. In additon Electron Cyclotron Resonance (ECR) is a technique for producing oxygen radicals which can maximize the efficiency of O_2 dissociation by tightly confining the ionized plasma in a magnetic field.[3]

Recently a number of investigators have developed *in-situ* co-evaporation processes which use distilled ozone as a highly reactive oxidizer at low gas flows and system pressures. Ozone is the triatomic azeotrope of oxygen which is unstable with respect to diatomic oxygen by 148 J/molecule. This is roughly the amount of energy released per molecule in the oxidation of hydrogen and is an indication of the power of the pure substance in the oxidation of metals. The technology of reducing a dilute ozone mixture to 98% pure solid ozone was borrowed from a group at the University of Minnesota[4] which is principally involved in the study of the reaction kinetics of gases found in the upper atmosphere.

BACKGROUND AND MOTIVATION

The utility of ozone as an oxidant for use in thin film processes has been demonstrated in a number of diverse ways. In one example the effect of ozone annealing on the properties of tantalum oxide thin films has been investigated by Isobe and Saitoh[5] and found to dramatically improve the dielectric properties of Ta_2O_5 capacitor barriers annealed at low O_3 concentrations.

In another example the reactivity of ozone with metals and contaminants has been demonstrated and exploited in the task of cleaning the surface of substrates used in surface analysis and film deposition techniques. GaAs[6] and Si[7] single crystal substrates have been cleaned using ozone generated by the dissociation of O_2 subjected to UV radiation at the substrate surface. A dramatic decrease in the amount of carbon on the surface of the ozone treated substrate is seen in the Auger and X-ray Photoelectron Spectroscopy (XPS) analysis.

A demonstration of the efficiency of oxidation of pure ozone with respect to diatomic oxygen was carried out by observing the Cu 2p doublet of the XPS spectrum of the surface of a Cu film. In Fig. 1 the spectrum of the pure film is shown in (a) and the spectrum of the same film after exposure to 5 minutes of pure ozone gas at a flow rate of approximately 10^{17} molecules/s cm^2 is shown in (b). During the exposure the pressure in the vacuum chamber remained below 10^{-5} Torr and the temperature of the film substrate was maintained at approximately 0°C. The surface of the film is oxidized completely by the ozone gas as evidenced by the growth of satellite peaks characteristic of the CuO spectrum.

Though it is possible to create O_3 by exposing O_2 to UV radiation at any local application where it is required, it is quite difficult to determine the actual concentration of O_3 produced for the desired reaction. In order to work with a well characterized material of the highest purity it is desirable to use distilled O_3. This can be prepared with a purity of greater than 98% and delivered to the desired application at any desired flow rate and concentration.

The distilled ozone used in these studies is prepared first by generating a low concentration of O_3 in O_2 using a silent discharge capacitor cell operating at about 10 kV at 60 Hz. The discharge cell is approximately an order of magnitude more efficient at producing ozone than UV cells now commonly available at the same price. The O_3/O_2

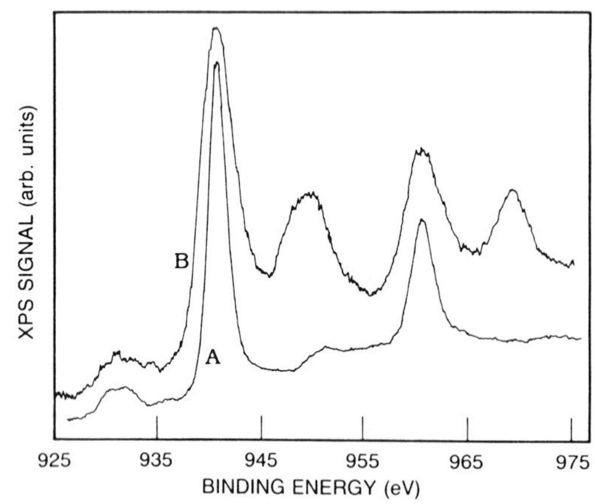

Fig. 1 Cu 2p XPS signal for (a) a 500 Å thick Cu film and (b) the same film exposed to O_3 at a flow rate of 10^{17}/s cm^2 for 5 minutes.

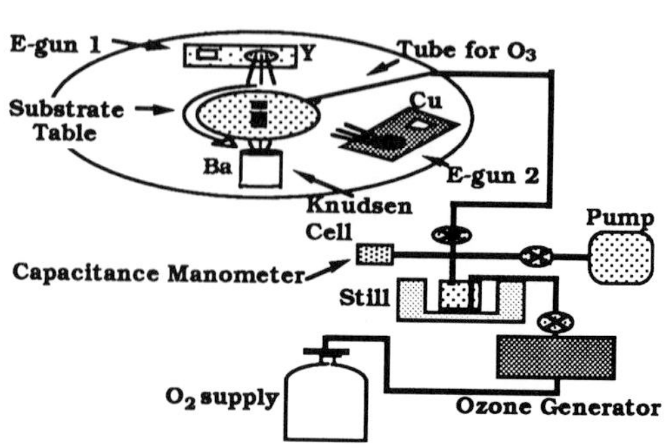

Fig. 2 Shematic diagram of UHV co-evaporation system and ozone distillation apparatus.

mixture is distilled at liquid nitrogen temperatures by maintaining a pressure of 10-20 Torr above the still as the mixture is applied to a calibrated leak valve at about 5 psi.[8] The O_3/O_2 phase diagram at 77 K indicates the resulting product will be greater than 98% solid O_3.[9] Figure 2 is a schematic diagram of the distillation apparatus and evaporation system used to produce $YBa_2Cu_3O_{7-x}$ thin films.

APPLICATIONS OF OZONE OXIDATION TO HIGH T_C FILM PROCESSES

The original demonstration of the ozone coevaporation process used distilled ozone and electron beam and Knudsen cell sources to produce superconducting $YBa_2Cu_3O_{7-x}$ thin films at a background pressure during evaporation of $2x10^{-7}$Torr and a total process pressure which was less than $1x10^{-5}$ Torr.[10] This was done using a small prototype Ultra High Vacuum (UHV) system by evaporating Y and Cu from electron beam sources and Cu from a Knudsen cell. Films with transition temperatures of 85 K which exhibited primarily c-axis orientation were routinely produced in this system when prepared at a substrate temperature of 700°C. In Fig. 3 (a) and (b) the resistance as a function of temperature is shown for films produced at 700°C and 590°C respectively. The film prepared at the colder temperature has a T_c of about 50 K and an expanded c-axis lattice parameter. This expansion is due probably to a defect structure which can be stabilized at lower temperatures as reported in the study by Eom, et al.[11]

In an effort to further advance the state of this prototype system toward the ideal of a MBE environment, the electron beam guns were replaced completely with Knudsen cells and the Dy-Ba-Cu-O compound became the subject of study. In the absence of stray electrons and fields coming from electron beam sources this arrangement would be able to support electron diffraction surface probes that could be installed at a later date. This system produced high quality DyBaCuO films with transition temperatures as high as 88 K.[12] In addition superconducting films with thicknesses of only 30 Å were prepared. The author noted a remarkable insensitivity of T_c to film composition.

Important contributions to the understanding of the growth kinetics of the Bi-Sr-Ca-Cu-O family of superconductors have been made by Eckstein, et al.,[13] using Knudsen cell sources and distilled ozone oxidation. Low background evaporation pressures allowed the use of RHEED surface analysis during growth and the relationship between phase formation and various shuttering schemes to be investigated. This MBE process has produced films of the 2223 compound with T_c's of 86 K.

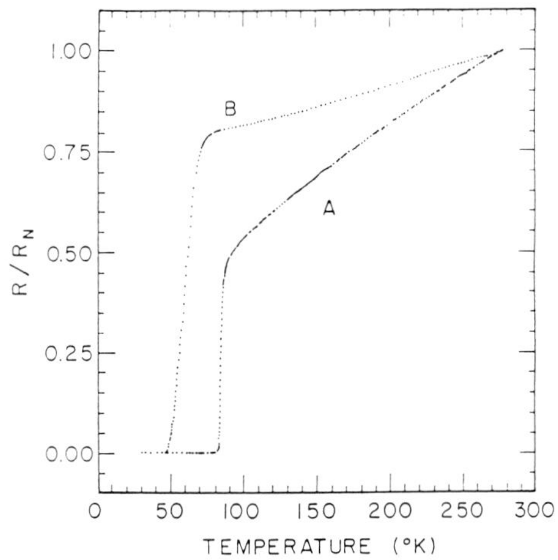

Fig. 3 Resistance vs. temperature of two $YBa_2Cu_3O_{7-x}$ thin films prepared using a substrate temperature of (a) 700°C and (b) 590°C.

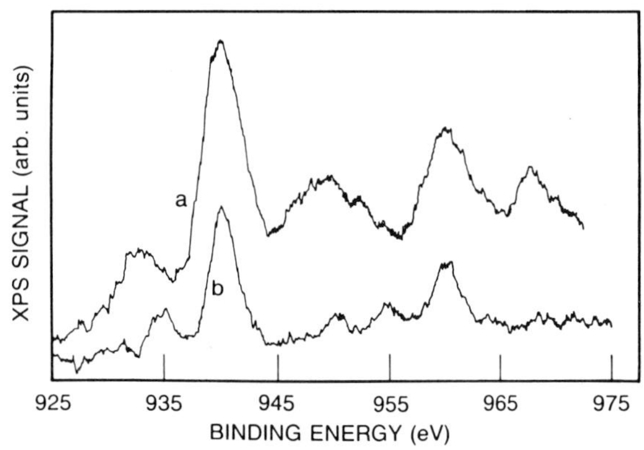

Fig. 4 Cu 2p XPS signal for (a) an $YBa_2Cu_3O_{7-x}$ film and(b) the same film with a 10 Å Y overlayer. Oxygen is removed from the surface Cu by the Y, reducing the intensity of the Cu 2p satellite peaks.

PASSIVATION OF AN $YBa_2Cu_3O_{7-x}$ THIN FILM SURFACE USING AN OZONE VAPOR

The utility of a convenient source of reactive oxygen is not limited to the production of the superconductor itself. In the application of insulating overlayers to the film surface the ozone gas can be quite useful in the preparation of an interface which is minimally perturbed from the bulk state. The ozone jet can maintain the oxygen stoichiometry of the superconductor even as a thin overlayer of potentially disruptive material is applied to the surface.

Yttrium was used as an overlayer material with and without ozone at "cold" and "hot" substrate temperatures to test the ability of the O_3 vapor to maintain the oxidation state of the surface copper.

The effect of depositing clean Y metal onto the surface of an $YBa_2Cu_3O_{7-x}$ film is investigated by observing changes in the XPS Cu doublet due to possible Y disruption of the Cu oxidation state. For this measurement 10 Å of Y metal is evaporated at a system pressure of about 5 x 10^{-9}Torr onto the just prepared film at a substrate temperature of about 20 ^{0}C. Figure 4 is a comparison of the Cu XPS doublet of an $YBa_2Cu_3O_{7-x}$ film immediately after preparation (a) and the same film after the Y evaporation (b). The smaller satellite peaks observed in Fig. 4(a) are typical of fully oxidized copper and are not present in the spectrum of pure Cu metal. As seen in Fig. 4(b) Y metal severely diminishes the intensity of these peaks as the surface copper loses oxygen to the Y overlayer. This situation is unacceptable in the process of preparing tunneling junctions since the superconducting order parameter will be quenched at the surface.

The procedure described above was modified to include an ozone flux directed at the substrate simultaneous to the application of a 10 Å Y metal overlayer. The flux rate is maintained at the value employed in the preparation of the superconductor. The substrate temperature was about 20 ^{0}C. The plot of the Cu XPS doublet of the bare superconductor and the metal/ozone overlayer can be seen in Figs. 5(a) and 5(b) respectively. The ratio of the main and the satellite peak heights is unchanged by the application of the overlayer. The ozone serves to oxidize the Y metal and passivate any competition of the Y and Cu for oxygen at the surface.

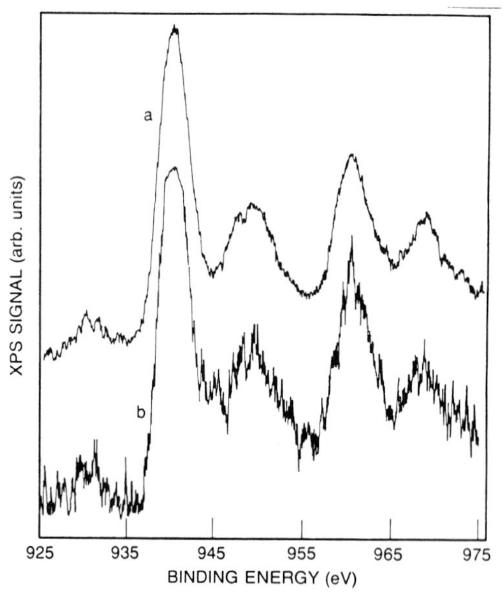

Fig. 5 Cu 2p XPS signal for (a) an $YBa_2Cu_3O_{7-x}$ film and (b) the same film with a 10 Å overlayer of Y_2O_3. The Y_2O_3 does not disturb the relationship between the main and satellite peak intensities.

In another test of the effect of Y on the superconductor surface a thin layer of Y was applied immediately after the deposition of the $YBa_2Cu_3O_{7-x}$ film by leaving the Y source shutter open to deposit another 10 Å after the superconductor was deposited. This was done while the ozone was flowing and the substrate temperature remained at 650°C, the value used during film growth. There was no effect on the relative height or shape of the Cu XPS satellite peaks with respect to the main doublet except for the expected reduction in height of all of the peaks. The importance of this study is to assess our ability to apply a crystalline barrier to the high T_c surface with excellent electrical and mechanical properties. A high temperature, high mobility deposition may produce a smoother barrier with fewer pinholes than a "quench condensed" film.

CONCLUSIONS

Distilled ozone provides a convenient and well characterized form of activated oxygen for use in the preparation of the rare earth and bismuth class of superconducting thin films. Because background evaporation pressures can be maintained at low levels during film growth, in-situ surface characterization of the superconductor is possible as the film is deposited. Yttrium metal applied to the film surface in the presence of ozone is not disruptive to the oxidation state of the surface Cu as seen by XPS. This has implications for the *in-situ* preparation of tunneling barriers which do not disturb the superconducting order parameter at the film surface.

ACKNOWLEDGEMENTS

I would like to thank P.R. Broussard and V. Cestone for invaluable help with work conducted at the NRL. I gratefully acknowledge the contributions of A.M. Goldman, B.R. Johnson, T-F. Wang, K.M. Beauchamp, and J. Morton at the University of Minnesota.

*Currently a National Research Council post-doctoral fellow tenured at the Naval Research Laboratory, Washington, D.C.

REFERENCES

1. See Naito M, Hammond RH, Oh B, Hahn M, Hsu JW, Rosenthal P, Marshall A, Beasley MR, Geballe TH, and Kapitulnik A, J. Mat. Res. **2**, 713 (1987) and references therein.

2. Kwo J, Hong M, Trevor DJ, Fleming RM, White AE, Mannaerts JP, Farrow RC, Kortan AR, and Short KT, Physica C **162-164**, 703 (1989) and Missert N, Hammond R, Matisjasevic V, Rosenthal P, Geballe TH, Kapitulnik A, Beasley MR, Laderman SS, Lu C, Garwin E, Barton R, IEEE Trans. Mag. **25**, 2418 (1989).

3. Yamamoto K, Lairson BM, Eom CB, Hammond RH, Bravman JC, and Geballe TH, Appl. Phys. Lett. **57**, 1936 (1990).

4. Many thanks to the people of Prof. K. Mauersberger's lab for help with the development of the ozone distillation apparatus.

5. Isobe C and Saitoh M, Appl. Phys. Lett. **56**, 907 (1990).

6. Solomon JS and Smith SR, Thin Films-Interfaces and Phenomena, Part of the Fall 1985 Meeting of the Materials Research Society. 449-54, (1986).

7. Baunack S, Zehe A, Staus Solidi A, **115**, 223 (1989).

8. Berkley DD, Goldman AM, Johnson BR, Morton J, and Wang TF, Rev. Sci. Instr. **60**, 3769 (1989).

9. Horvath M, Bilitzky L, Huttner J, Ozone (Elsevier, New York, 1985).

10. Berkley DD, Johnson BR, Anand N, Beauchamp KM, Conroy L, Goldman AM, Maps J, Mauersberger K, Mecartney M, J. Morton J, Tuominen M, and Zhang YJ, Appl. Phys. Lett. **53**, 1973 (1988).

11. Eom CB, Sun JZ, Yamamoto K, Marshall A, Luther KE, Geballe TH, Laderman SS, Appl. Phys. Lett. **55**, 595, (1989).

12. Johnson BR, Beauchamp KM, Wang TF, McGreer K, Wan JC, Tuominen M, Zhang YJ, Mecartney M, and Goldman AM, Appl. Phys. Lett. **56**, 1911 (1990).

13. Eckstein JN, Bozovic I, von Dessonneck KE, Schlom DG, Harris JS, Jr., Baumann SM, Appl. Phys. Letts. **57**, 931, (1990).

Low Temperature Growth of YBCO Films by Plasma Assisted Vapor Deposition

KEIICHI KANEHORI, AKIRA TSUKAMOTO, and NOBUYUKI SUGII

Central Research Laboratory, Hitachi, Ltd., Kokubunji, Tokyo, 185 Japan

ABSTRACT

The effect of plasma excitation on reactive evaporation and chemical vapor deposition of superconducting Y (Ho) BCO thin films are studied. It is revealed that superconducting thin films, whose zero-resistivity temperatures are 89K and 84K, can be grown by plasma assisted reactive evaporation and plasma-enhanced metalorganic chemical vapor deposition at 580 ℃. Both films grown by the plasma assisted reactive evaporation at 510 ℃ and grown by the plasma-enhanced metalorganic chemical vapor deposition at 580℃ had a critical current density of 10^5 A/cm^2 at 77K. Spectroscopic studies revealed that activated metal and oxygen species, Ho, Ba, Ba^+, Cu, O_2^+, and O exist during plasma assisted reactive evaporation and also revealed that activated metalorganic species exist during plasma-enhanced MOCVD.

KEY WORDS : Oxide Superconductor, Thin Film, Plasma Excitation, Reactive Evaporation, Metalorganic Chemical Vapor Deposition

INTRODUCTION

Low-temperature growth of high Tc oxide superconductors is one of the most important subjects concerning superconductor applications to electronic devices, since it makes it possible to fabricate multi-layer structures with clean interfaces. In addition, the low temperature growth process seems to be a key technology for studying new superconducting properties since it enables the preparation of metastable structures, superlattices and artificial lattices.

In the early stages of this field, a high temperature (above 700℃) was required to grow superconducting YBCO thin films by conventional physical vapor deposition or conventional chemical vapor deposition.

Recently, however, successful results have been reported on low temperature growth of oxide superconducting films. Witanachchi prepared superconducting YBCO films at 400℃ by plasma assisted laser evaporation [1]. Terashima succeeded in growing superconducting YBCO films at 600℃ by RF-plasma assisted reactive evaporation [2]. Similarly, Moriwaki grew superconducting YBCO films by ECR plasma assisted MBE[3].

By employing microwave plasma excitation, the authors found that superconducting Y(Ho)BCO films could be grown at low temperature by reactive evaporation or chemical vapor deposition [4,5]. However, the effects of plasma excitation has not yet been clarified.

This paper reports on electrical and crystallographic properties of Y(Ho)BCO films grown by microwave plasma assisted reactive evaporation and microwave plasma-enhanced metalorganic chemical vapor deposition (MOCVD), and also reports spectroscopic studies on the effects of plasma excitation.

EXPERIMENTAL

The plasma assisted reactive evaporation system used in this report consists of two differentially pumped chambers: an evaporation chamber and a deposition chamber. The microwave source consists of a magnetron and three magnets is attached to the deposition chamber. The microwave frequency is 2.45GHz, and the magnetic field around the discharge tube is approximately 875G. Evaporators using EB guns and Knudsen cells are contained in the evaporation chamber. The deposition rates of Ho, Ba, and Cu are individually measured with quartz thickness monitors. The typical growth conditions are as follows: substrate is MgO(100), microwave power is 120 W, oxygen pressure in the deposition chamber is 1 X 10^{-4} Torr, deposition rate of HoBCO is 1.2nm/min. The deposition system configuration is illustrated in the previous paper[4].

The plasma-enhanced MOCVD apparatus consists of an evaporation system of source materials and a deposition chamber. The configuration of the deposition chamber is analogous to that of the plasma assisted reactive evaporation system. The tetramethyl-heptadione (THD) complexes of Y, Ba, and Cu were employed as metal sources. They were sublimated and transferred to the growth chamber with Ar carrier gas. Oxygen was used as an oxidant. The typical growth conditions are as follows: substrate is $SrTiO_3$(100), microwave power is 200 W, total pressure in the deposition chamber is 0.3 Torr, oxygen partial pressure is 0.1 Torr, deposition rate is around 1nm/min. The exact configuration of the deposition system and the deposition procedure are reported in the previous paper[5].

Emission spectra of the substrate region were recorded through a quartz viewing port during both deposition processes. The detected wavelength range was 300nm to 700nm.

RESULTS AND DISCUSSION

The temperature dependences of electrical resistivity for films grown by reactive evaporation and MOCVD at 580℃ are shown in Fig. 1. The film grown by reactive evaporation showed a sharp superconducting transition. The zero-resistivity temperature was 89K. The zero-resistivity temperature subsequently lowered to 80K and 23K with decreasing growth temperature to 510 ℃ and 450 ℃. However, the superconducting properties of the films were improved when $SrTiO_3$(110) was used as the substrate, that is, films grown on $SrTiO_3$(110) at 510℃ and 450℃ indicated zero-resisivity at 87K and 83K. The critical current density of the films grown at 510℃ was 10^5 A/cm^2 at 77K.

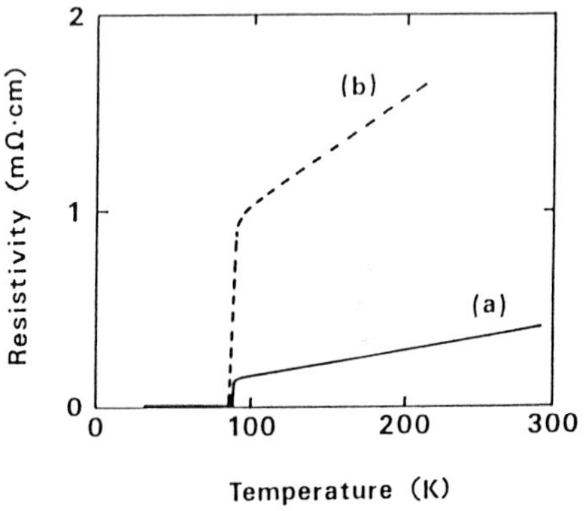

Fig. 1 Resistivity of films grown at 580℃,
(a) HoBCO film grown by plasma assisted reactive evaporation,
(b) YBCO film grown by plasma-enhanced CVD.

The films grown by the plasma-enhanced MOCVD at 580℃ indicated zero-resistivity at 84K. The critical current density was 10^5 A/cm^2 at 77K. The zero-resistivity temperature lowered to around 60K and 50K with decreasing growth temperature to 515 ℃ and 500 ℃. After a comparison with the electrical properties of films grown by conventional MOCVD, it was found that the deposition temperature for obtaining superconducting films decreased over 100 ℃ by plasma-enhancement.

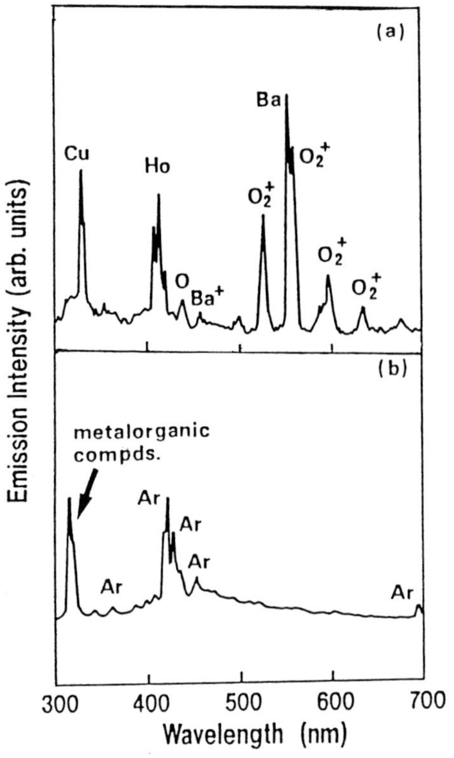

Fig. 2 Emission spectra around substrate region during deposition, (a) reactive evaporation, (b) Chemical vapor deposition

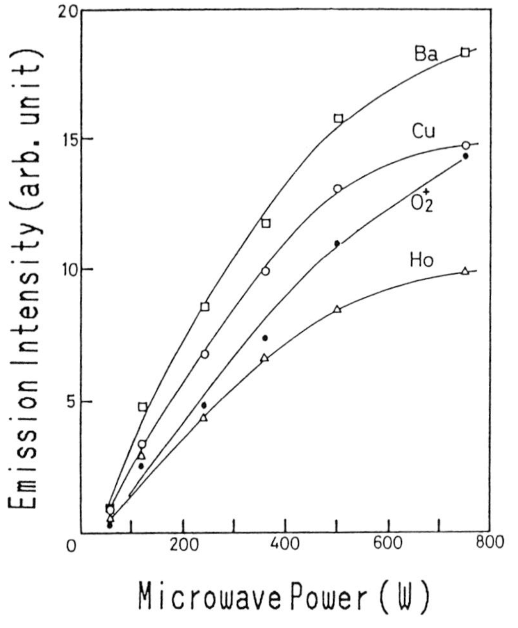

Fig. 3 Effect of microwave power on emissions from activated species during reactive evaporation

An X-ray diffraction study on the film grown by reactive evaporation at 580℃ indicated that the film had the orthorhombic structure and good crystallinity. In addition, the film was found to have an orientation where the c-axis was perpendicular to the substrate. This orientation deteriorated with decreasing deposition temperature. When $SrTiO_3$ (110) was used as the substrate, grown film had an orientation where the c-axis was parallel to the substrate.

The film grown by the plasma-enhanced MOCVD at 580℃ was found to have an orientation where the c-axis was perpendicular to the substrate. This orientation deteriorated with decreasing deposition temperature. In addition, the full width at half maximum of the (007) peak decreased with increasing microwave power.

Emission spectra of the substrate region are shown in Fig. 2. In the case of reactive evaporation, emissions from Ho, Ba, Ba^+, Cu, O_2^+, and O are detected. This result indicates that plasma is effective for exciting not only oxygen but also metals. The intensity of emissions from activated metals and oxygen increased with increasing microwave power as shown in Fig. 3. This indicates that plasma excites a part of evaporated metals and introduced oxygen gas at the typical growth condition: 120 W.

In the case of MOCVD, an emission from the metalorganic compounds was detected at 316nm in addition to emissions from Ar as shown in Fig. 2. On the other hand, no emisson from oxygen was detected since it co-existed with Ar. In addition, it was found that the intensity of the emission from the metalorganic compounds increased with increasing microwave power[6]. This result indicates that the concentration of activated species increased with increasing microwave power. As shown previously, full width at half maximum of the (007) peak decreased with increasing microwave power. From these results, an improvement in crystallinity is achieved by the excitation of metalorganic compounds.

Conclusion

The effects of plasma excitation on reactive evaporation and metalorganic chemical vapor deposition of superconducting Y(Ho)BCO films have been studied.It has been revealed that the deposition temperature for growing superconducting films can be decreased to around 500℃ by plasma excitation. Spectroscopic studies suggested that this effect is due to the excitation of metals and oxygen in the case of reactive evaporation, and the excitation of metalorganic compounds in the case of metalorganic chemical vapor deposition.

References

1) Witanachchi S, Kwok HS, Wang XW, Shaw DT (1988) Appl. Phys. Lett., 53: 234
2) Terashima T, Iijima K, Yamanoto K, Bando Y, Mazaki H (1988) Jpn. J. Appl. Phys., 27: L91
3) Moriwaki K, Enomoto Y, Kubo S, Murakami T (1988) Jpn. J. Appl. Phys., 27: L2075
4) Aida T, Tsukamoto A, Imagawa K, Fukazawa T, Saito S, Shindo K, K.Takagi Miyauchi K (1989) Jpn. J. Appl. Phys., 28: L635
5) Kanehori K, Sugii N, Miyauchi K (1990) MRS symp. Proc. vol.169 585
6) Kanehori K, Sugii N, (1990) SPIE Proc. from Conference 1394 (to be published)

Uniform, Large Area Deposition of Superconducting Y-Ba-Cu-O Thin Films by Single Cathode Magnetron Sputtering

KEIJI ISHIBASHI[1], KAZUO HIRATA[1], NAOKICHI HOSOKAWA[1], YASUO SHIRAKAWA[2], and YASUHIRO YAMAMOTO[2]

[1] ANELVA Corporation, 8-1, Yotsuya 5-chome, Fuchu, Tokyo, 183 Japan
[2] Hosei University, 7-2, Kajino-cho 3-chome, Koganei, Tokyo, 184 Japan

ABSTRACT

In order to improve superconducting film properties by suppressing the degradation induced from energetic negative ions in a sputtering system, influences of the magnetic field intensity and the shape of erosion area in magnetron sputtering have been studied. The composition and the critical temperature (Tce) of Y-Ba-Cu-O thin films have been investigated using three kinds of permanent magnet assembly sets within a 100mm diameter planer magnetron cathode. With increasing the magnetic field intensity from 200G up to 1000G, the target bias voltage was decreased from -150V to -65V at an input discharge power of 150W. Accordingly the energy of oxygen ions bombarding the growing film was reduced, however, the uniformity of the film composition could not be improved within the whole substrate area. On the other hand as the diameter of erosion area center at the target surface was expanded, the deviation in the film composition from that of the target was much suppressed in larger area around the substrate central region. When a film was prepared on (100)MgO substrate at 700°C with the target of $Y_1Ba_2Cu_3O_x$, the Tce of 62 ± 4K was obtained within an area of 50mm in diameter.

KEY WORDS: Y-Ba-Cu-O, magnetron sputtering, uniform film composition

INTRODUCTION

Since the discovery of high temperature superconductive oxides including Y-Ba-Cu-O system[1,2], extensive research efforts have been directed to the preparation of superconducting thin films for microelectronic device applications. Because one can usually obtain films of relatively complicated composition without sophisticated techniques as required for other thin film deposition methods, single cathode magnetron sputtering has been most widely used among various film deposition methods[3,4]. However, in the synthesis of Y-Ba-Cu-O thin films using parallel substrate and target system, it has been reported that the distribution of the film composition over the substrate was not uniform; the film composition was very different from that of the target especially in the substrate region projected normally from the erosion area of the target[5,6]. Therefore, stoichiometric and superconducting films could be obtained only within a very small region above the center of the target. This results in poor reproducibility of obtained film properties. The cause of compositional deviation has been considered to be resputtering by bombardment of energetic oxygen atoms emitted from the target as negative ions and accelerated in the cathode fall[7,8]. In some cases resputtering resulted in the etching of the substrate[8]. This resputtering effect can be suppressed by two methods. (1)When sputtering is carried out at higher gas pressure[9,10], the kinetic energy of the energetic oxygen ions will be reduced by collisions with gas atoms. (2)When the substrate is mounted at places not directly facing the target[11,12], it will receive no intense ion bombardment. Stoichiometric films could be actually obtained by both methods. Nevertheless there were disadvantages of low deposition rates and small effective area.

In this paper, the trial to obtain uniform and large area deposition of superconducting Y-Ba-Cu-O thin films by a single cathode planer magnetron sputtering is described. Special interest is taken in the influence of the magnetic field intensity and the shape of the erosion area of the target on the distribution of the film composition.

EXPERIMENT

A magnetron sputtering system was employed with a 100mm diameter single cathode. Three different magnet assemblies were used. They will be referred to as M1, M2, and M3. With these magnet assemblies each diameter of the center of the target annulus erosion was 50mm for M1 and M2 and 75mm for M3. Magnetic filed intensities parallel to the target surface at the same positions were about 200G, 1000G, and 550G for M1, M2, and M3, respectively. Two kinds of sintered targets, $Y_1Ba_{2.22}Cu_{4.02}O_x$ and $Y_1Ba_2Cu_3O_x$, were used. They were 100mm in diameter and 5mm in thickness. Y-Ba-Cu-O thin films were deposited on Si wafers at an ambient temperature to study the distribution of the film composition and on (100)MgO substrates at 700°C to evaluate superconducting film properties. Substrates were placed at a distance of 30mm from the target surface. Sputtering was carried out in an Ar+50%O_2 atmosphere with the pressure of 3.33Pa at the input discharge power of 150W rf or dc.

Contents of Y, Ba, and Cu in films were measured with an energy dispersive X-ray micro analyzer (EDX: HORIBA EMAX-3700 combined with HITACHI SEM S-2500CX). Tce and crystal structure of films were evaluated using a dc four-probe technique and an X-ray diffraction measurement, respectively.

RESULTS AND DISCUSSION

Distributions of the film composition for the two different magnet assemblies, M1 and M2, are shown in Fig.1. These films were deposited by rf magnetron sputtering using the $Y_1Ba_{2.22}Cu_{4.02}O_x$ target. The target composition is also shown in this figure. The target bias voltage (Vt) was -150V for M1 and -65V for M2. In both distributions, the largest compositional deviation of the film from the target composition can be recognized at 25mm from the center. This position corresponded to the normal projection of the most eroded area of the target. With increasing the magnetic field intensity from 200G up to 1000G, the Vt was decreased from -150V to -65V at the input discharge power of 150W. The deposition rates were 300A/min for M1 and 200A/min for M2 at the substrate center, respectively. It can be found that the deviation was suppressed at the substrate center because the energy of oxygen ions bombarding the growing film was reduced. However, noticeable improvement in uniformity was not recognized. It is considered that oxygen ions had sufficient energy to resputter the growing film even in the case of Vt=-65V. Therefore, in order to eliminate the influence of the resputtering, the Vt must be reduced still lower. However, in lower voltages, the deposition rate will be decreased, and at worst sputtering will not take place. Accordingly it is considered to be unpractical to eliminate the resputtering with reducing the Vt.

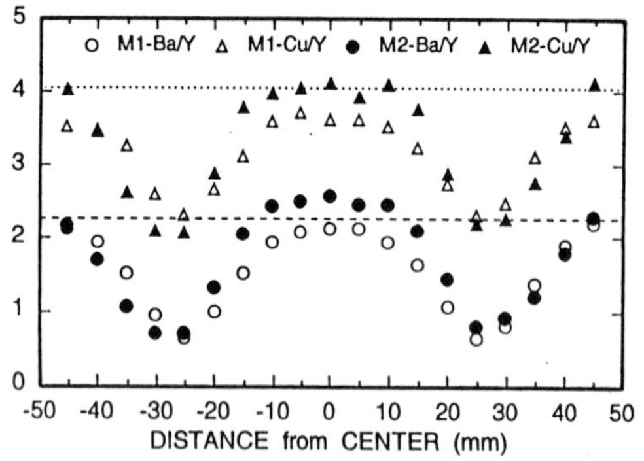

Fig.1. Film composition for M1 and M2 of different magnetic field intensity. Si substrate (ambient temperature), $Y_1Ba_{2.22}Cu_{4.02}O_x$ target, rf discharge. Dashed lines indicate atomic ratios of Cu/Y (lower) and Ba/Y (upper) in the target.

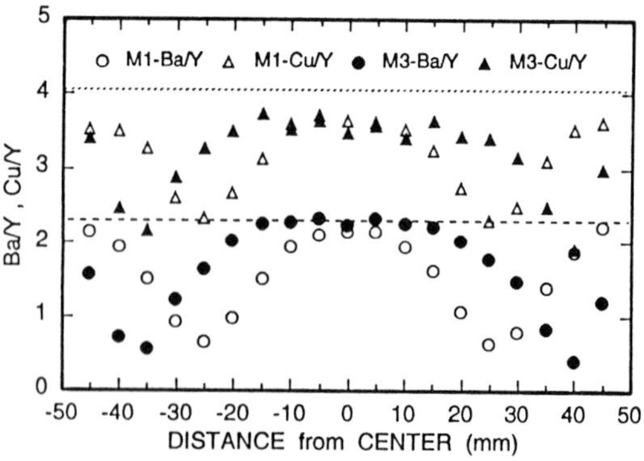

Fig.2. Film composition for M1 and M3 of different target erosion. Si substrate (ambient temperature), $Y_1Ba_{2.22}Cu_{4.02}O_x$ target, rf discharge.

It was reported by Migliuolo et al. that a uniform distribution of the film composition was obtained within a whole substrate area above the target by magnetron sputtering using an intense magnetic field similar to this experiment[6]. Important differences between their experiment and ours are target-substrate distance and sputtering pressure. They took the distance of 64mm and a little larger sputtering gas pressure of 6.65Pa resulting in the increase of the collision probability of oxygen ions with gas atoms.

Distributions of the film composition for the two different magnet assemblies, M1 and M3, are shown in Fig.2. Films were deposited under the same sputtering conditions adopted in the previous experiment. The largest compositional deviations were recognized at the positions projected normally from the most eroded area of the target: 25mm from the center for M1 and 37.5mm from the center for M3. In the case of M3 larger uniform area could be obtained compared with M1. As the diameter of erosion area center at the target surface was expanded, the more uniform film composition could be obtained in the larger area around the substrate central region. In the EDX measurement of the film composition, Cu content was estimated a little smaller than the actual value, because oxygen content in the film was not taken into account. Therefore, the film composition around the substrate central region in Fig.2 was probably almost the same as that of the target.

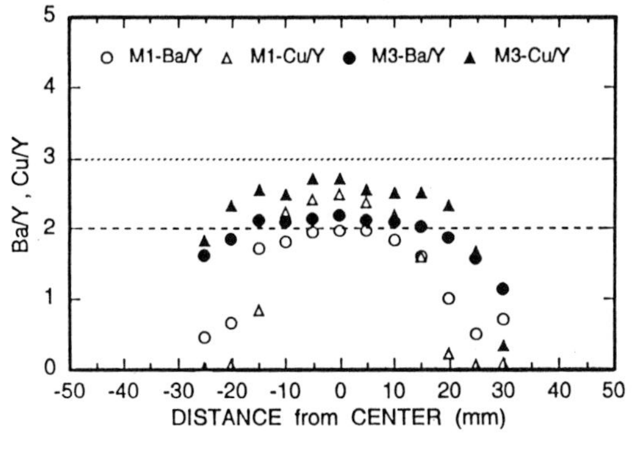

Fig.3. Film composition for M1 and M3. MgO substrate (700°C), Y1Ba2Cu3Ox target, dc discharge.

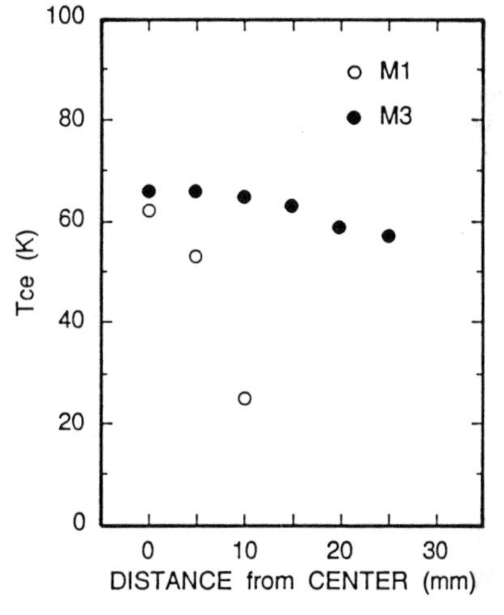

Fig.4. Distribution of the Tce for M1 and M3. MgO substrate (700°C), Y1Ba2Cu3Ox target, dc discharge.

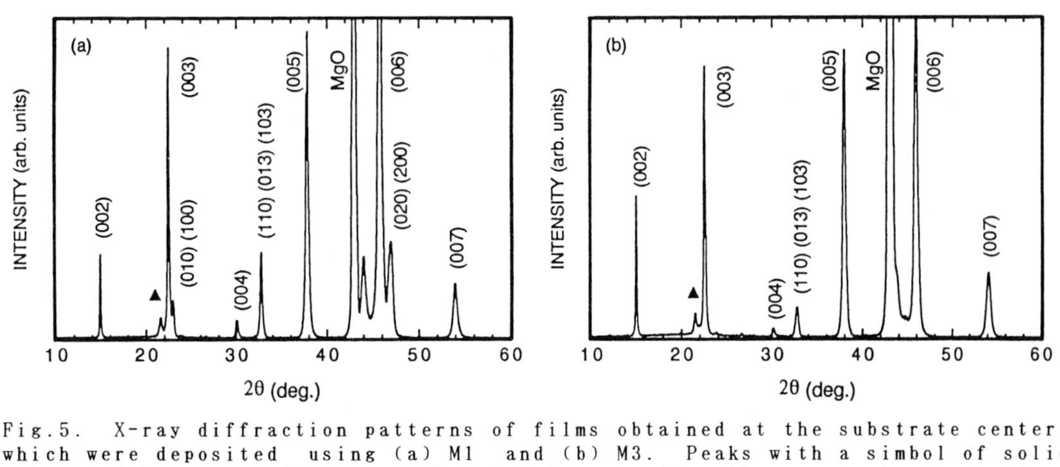

Fig.5. X-ray diffraction patterns of films obtained at the substrate center, which were deposited using (a) M1 and (b) M3. Peaks with a simbol of solid triangle(▲) are attributed to the clay for mounting the sample to the holder.

Figure 3 indicates distributions of the film composition for the different two magnet assemblies of M1 and M3, which were deposited on MgO substrates at the substrate temperature of 700°C with the $Y_1Ba_2Cu_3O_x$ target by dc magnetron sputtering. Also under these conditions the result similar to that of the previous experiment (Fig.2) was obtained. Figure 4 shows distributions of the Tce of these two films. In the film sputtered with M1 the Tce was 62K at the substrate center and drastically dropped with increasing the distance from the center, whereas in the film sputtered with M3 the Tce was relatively constant and 62 ± 4K within an area of 50mm in diameter. X-ray diffraction patterns obtained at the center of these two films are indicated in Fig.5 (a) for M1 and (b) for M3. Similar patterns as shown in Fig.5 were obtained within an area of 20mm in diameter for M1 and within an area of 50mm in diameter for M3. These area corresponded to the superconducting area as shown in Fig.4. In every position within the superconducting area the lattice constant of c-axis calculated from the position of the peak of (002) was 11.75-11.79A, which was longer than that of the stoichiometric crystal (11.677A)[13,14]. This result indicates that oxygen content x in the film was about 6.7, which was smaller by 0.3 than the value of the stoichiometric composition.

Further experiments using M3 is now in progress in order to prepare the single crystal film.

CONCLUSIONS

The effect of the magnetic field intensity and the shape of erosion area on film properties in the magnetron sputtering have been studied. Although the target bias voltage was decreased from -150V to -65V at an input power of 150W with increasing the magnetic field intensity from 200G up to 1000G, the uniformity of the film composition could not be improved within the whole substrate area. However, the deviation in the film composition from that of the target could be suppressed in larger area around the substrate central region by expanding the diameter of erosion area center of the target surface with appropriate designing of the magnet assembly. When a film was prepared on a (100)MgO substrate at 700°C with the target of $Y_1Ba_2Cu_3O_x$, the Tce of 62 ± 4K was obtained uniformly within an area of 50mm in diameter for a 100mm diameter planer magnetron cathode. This will contribute to the reproducible- and practical use of a single cathode magnetron sputtering for the deposition of thin films of high Tc superconductor.

REFERENCES

1. Bednorz JG, Müller KA (1986) Z. Phys. B 64: 189-193
2. Wu MK, Ashburn JR, Torng CJ, Hor PH, Meng RL, Gao L, Huang ZJ, Wang YQ, Chu CW (1987) Phys. Rev. Lett. 58: 908-910
3. Eltoukhy AH (1986) J. Vac. Sci. Technol. A 4: 539-542
4. Sreenivas K, Sayer M, Garrett P (1989) Thin Solid Films 172: 251-267
5. Ishibashi K, Hirata K, Hosokawa N, Asamaki T, Kawakita K, Inada T (1989) Extended Abstracts of 6th International Workshop on Future Electron Devices. Research & Development Association for Future Electron Devices. Tokyo, pp39-44
6. Migliuolo M, Belan RM, Brewer JA (1990) Appl. Phys. Lett. 56: 2572-2574
7. Rossnagel SM, Cuomo JJ (1988) A. I. P. Conference Proceedings. American Institute of Physics. New York, pp106-113
8. Shah SH, Carcia PF (1987) Appl. Phys. Lett. 51: 2146-2148
9. Migliuolo M, Stamper AK, Greve DW, Schlesinger TE (1989) Appl. Phys. Lett. 54: 859-861
10. Muroi M, Matsui T, Koinuma Y, Okamura Y, Tsuda K, Nagano M, Mukae K (1989) J. Mater. Res. 4: 781-786
11. Sandstrom RL, Gallagher WJ, Dinger TR, Koch RH, Laibowitz RB, Kleinsasser AW, Gambino RJ, Bumble B, Chisholm MF (1988) Appl. Phys. Lett. 53: 444-446
12. Kageyama Y, Taga Y (1989) Appl. Phys. Lett. 55: 1035-1037
13. Takayama-Muromachi E, Uchida Y, Ishii M, Tanaka T, Kato K (198) Jpn. J. Appl. Phys. 26: ppL1156-L1158
14. Cava RJ, Batlogg B, Chen CH, Rietman EA, Zahurak SM, Werder D (1987) Phys. Rev. B 36: 5719-5722

Preparation and Characterization of $(Y, Gd)_1Ba_2Cu_3O_x$ Thin Films

SHUICHI FUJINO, TADASHI SUGIHARA, and TAKUO TAKESHITA

Central Research Institute, Mitsubishi Metal Corporation, 297, Kitabukuro-cho 1-chome, Omiya, 330 Japan

ABSTRACT

Thin films of the oxide superconductor (Y,Gd)BaCuO system on MgO(100) substrates were prepared by Facing Targets Sputtering (FTS) deposition method. These thin films possess the following properties ;Y-Ba-Cu-O(YBCO) thin film has Tc(0)=72[K], Jc(at 4.2[K])=6.2×10^6[A/cm^2] ; Gd-Ba-Cu-O(GBCO) thin film has Tc(0)=58[K], Jc(at 4.2[K])=2.2×10^5[A/cm^2] ; (Y,Gd)-Ba-Cu-O ((Y,G)BCO) thin fim has Tc(0)=69[K], Jc(at 4.2[K])=3.2×10^5[A/cm^2]. Current-voltage characteristics of these thin films are as follows ; YBCO thin film has define hysteresis, (Y,G)BCO thin film's critical current(Ic) drops in I-V curve with micro-wave irradiation.

key words: (Y,Gd)BaCuO , thin film, I-V characteristic

INTRODUCTION

Since the high-Tc superconducting oxide was discovered[1], many studies have been made on high-Tc superconductors in bulk form or in thin film form. In paticurar, thin film preparation of high-Tc superconducting oxides for application of microelectronic devices, sensores, and so on were studied by many methods such as magnetron sputtering[2], electron beam evapolation[3], chemical vapor deposition[4], and so on. Preparation of Josephson junction was attempted with various methods of micro processing technique. There were many cases of preparation of bridge type Josephson junction, and most bridgetype junctions[5] were grain boundary junctions. Their reproducibility is not high.

In this paper, we prepared films having very small grain (the grain size was under 0.05μm) or ones having few grains. A bridge type Josephson junction with 5μm width bridge was prepared on this YBCO thin film. Its I-V characteristics was measuerd and was found that this bridge does not show good Josephson junction behavior because of its high critical current Ic.

In this study, we prepared (Y,G)BCO thin films which have lower Ic but about the same Tc and film morphorogy as YBCO thin film. A bridge type Josephson junction with 5μm width bridge was prepared and its Jc, Tc, and I-V characteristics were measured..

EXPERIMENTAL

1.Thin films preparation

(Y,Gd)BaCuO films on MgO(100) substrates were deposited by Facing Targets Sputtering (FTS).

Table-1 shows targets composition used in FTS, and compositions of films prepared are about 1-2-3. Typical deposition conditions are shown in Table-2. Substrate temperature used was 750℃ with total pressure of 1mTorr, and Ar/O_2 ratio 1/1. RF power was 200W, and sputtering deposition rate was about 16Å/min. The thickness of the

Table-1. Target composition of prepared films

	Target①	Target②
YBaCuO	$Y_1Ba_2Cu_3O_x$	$Y_1Ba_2Cu_{3.5}O_x$
GdBaCuO	$Gd_1Ba_2Cu_3O_x$	$Gd_1Ba_{2.2}Cu_{4.2}O_x$
(Y,Gd)BaCuO	$Y_1Ba_2Cu_{3.5}O_x$	$Gd_1Ba_2Cu_3O_x$

Table-2. Typical sputtering condition of films

substrates	MgO(100) 15×15×0.5 mm
substrates Temp.(Tsub.)	750 ℃
Total pressure (Ptotal)	1 mTorr
Ar:O_2 ratio	1:1
Deposition rate	~16 Å/min
Deposition time	2 hrs
Cooling condition	Natural cooling in 10 Torr O_2

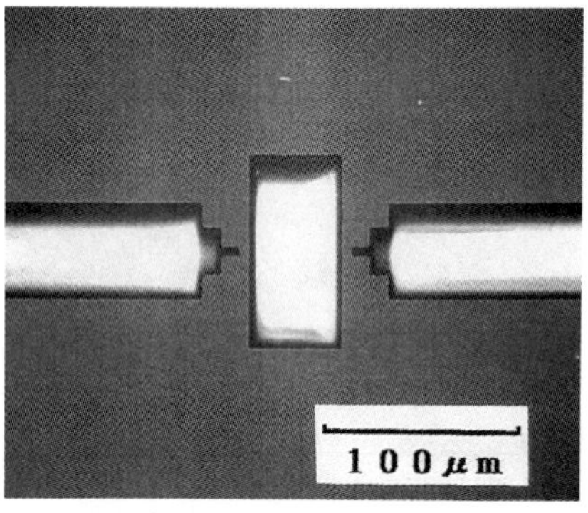

Fig.1 Bridge type DC-SQUID pattern of oxide superconductor film

Table-3. Typical film composition of films

YBaCuO	Y:Ba:Cu = 1:2.11:3.14
GdBaCuO	Gd:Ba:Cu = 1:2.18:3.18
(Y,Gd)BaCuO	(Y:Gd):Ba:Cu=(0.51:0.49):2.33:3.22

films prepared are 1000~2000A. After deposition, films were cooled in the chamber in 10Torr O_2 atmosphere. Superconducting properties and micro structures of the films were evaluated by X-ray diffraction, SEM, EPMA, and conventional DC four probe resistivity measurement.

2.Patterning of bridge type junction

5μm line width bridge type DC-SQUID circuits were prepared on films by photolithography technique and ion-beam sputtering etching as shown in Figuer 1. Superconducting properties of this junction, Tc(0), Jc, and I-V characteristic with or without micro-wave irradiation, were measured .

RESULTS AND DISCUSSION

Table-3 shows the film compositions, and Figuer 2 shows X-ray diffraction patterns of YBCO, GBCO, and (Y,G)BCO films. The composition of each film is about 1-2-3. X-ray diffraction peaks of each film could be assigned mainly to (001) planes, thus they have good c-axis orienta-

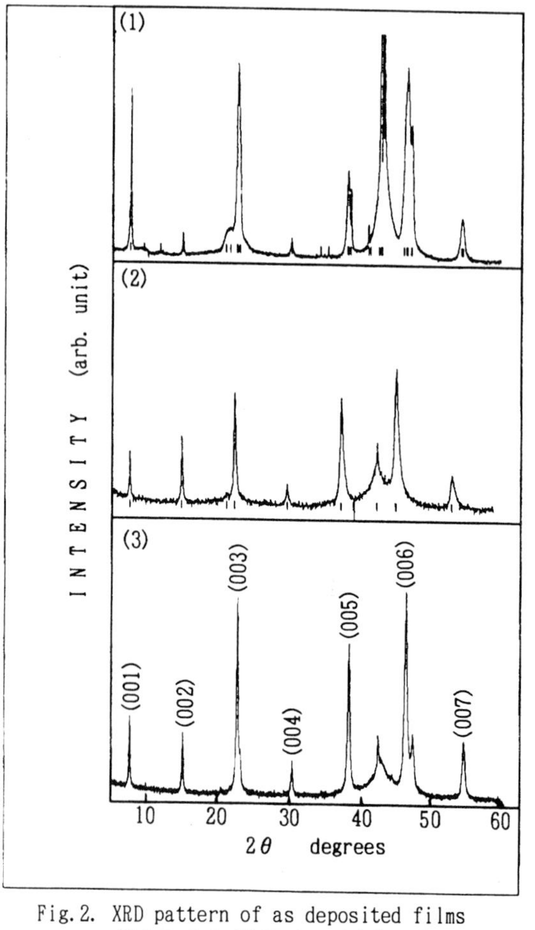

Fig.2. XRD pattern of as deposited films (1)GdBaCuO (2)YBaCuO (3)(Y,Gd)BaCuO

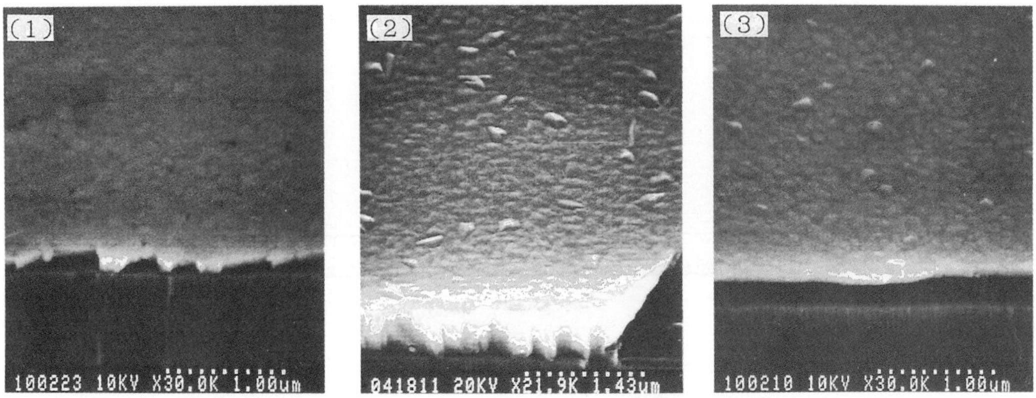

Fig. 3 SEM photographs of as deposited films
(1) GdBaCuO (2) YBaCuO (3) (Y, Gd) BaCuO

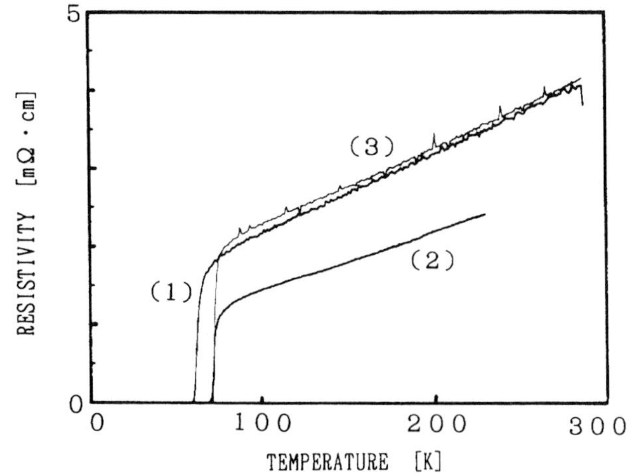

Fig. 4. Temperature dependence of resistivity of films
(1) GdBaCuO (2) YBaCuO (3) (Y, Gd) BaCuO

Table-4. Critical current density of films at 4.2K

YBaCuO	6.2×10^{6} [A/cm^2]
GdBaCuO	2.2×10^{5} [A/cm^2]
(Y, Gd) BaCuO	3.2×10^{5} [A/cm^2]

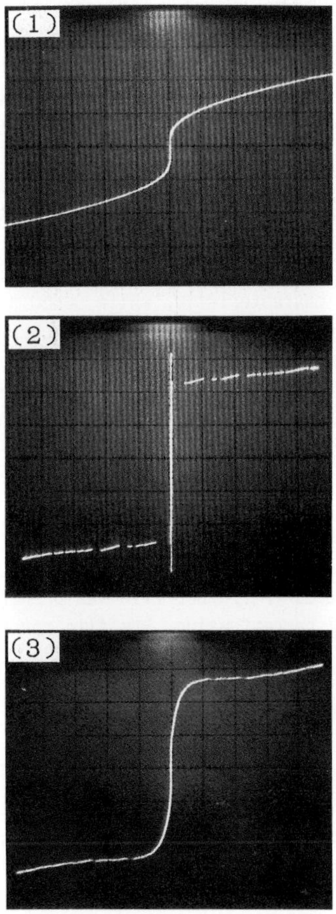

X:200[mV/div] ; Y:20[mA/div]

Fig. 5. I-V characteristic of films at 4.2K
(1) GdBaCuO (2) YBaCuO (3) (Y, Gd) BaCuO

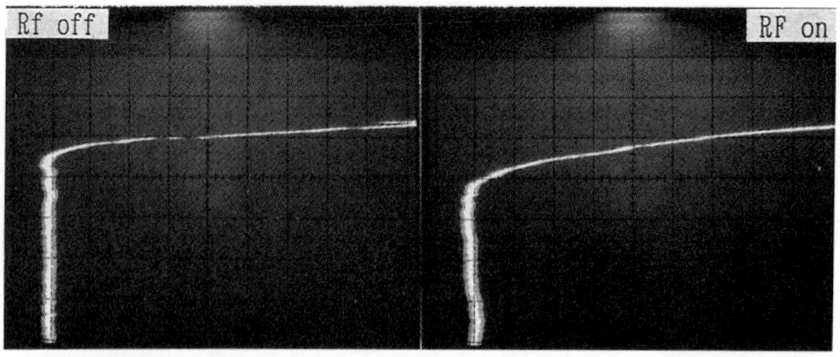

X:20[μV/div] ; Y:1[mA/div]

Fig.6. I-V characteristic of (Y,Gd)BaCuO film with or without 12GHz microwave irradiation

tion perpendicular to the substrate surface plane. However comparison of YBCO film's pattern with GBCO film's pattern, shows that the intensity of low angle peak is very different. And (Y,G)BCO film's pattern is a mixture of YBCO and GBCO films, indicating that in (Y,G)BCO films, Y and Gd were mixed homogeneously. This is also seen by SEM photographs as shown in Figuer 3. The YBCO film has scarcely any grains, and consists of densely columner crystals. The GBCO film consists of small grains, the size of which is about 0.05μm. (Y,G)BCO film is made up of small grains, the size of which is under 0.05μm, and the morphology of this film is nearly the same as that of YBCO film These results are consistent with that of XRD.

Figuar 4 and Table-4 show temperature dependence of resistivity and the critical current density of each film respectively. These were measured by using 5μm line width bridge type DC-SQUID pattern shown in Figuer 2. We determined the value of critical current as the one which induced 1μV between voltage electrodes. The critical current density of (Y,G)BCO film was lowered by one order of magnitude of that of YBCO film.

Figuer 5 shows the I-V characteristic of each film. Only YBCO film shows clear hysteresis in its I-V characteristic, while GBCO and (Y,G)BCO films do not show hysteresis. (Y,G)BCO film having low Jc was prepared, and its I-V characteristic with 12GHz microwave irradiation is shown in Figuer 6. The value of critical current was slightly lowered but the Shapiro step due to microwave irradiation was not observed. The reason of this may be that the value of critical current is still too large for the bridge size, and we believe that we need to lower the value of Ic by one more order.

CONCLUSION

In summary, (Y,G)BCO film was prepared. It has similar Tc(0) to YBCO film, but with one order lower critical current value. Its film morphology is not as good as YBCO films. The I-V characteristic of 5μm line width bridge of (Y,G)BCO film did not show the Shapiro step.

It will be interesting to examine the addition of Pr, Lu, and other rare earths elements which have different ionic radii than that of Y.

REFERENCES

[1] J.G.Bednorz, and K.A.Muller : Z.Phys. B Condenced Matter. 64 (1986) 189
[2] S.Tanaka, and H.Itozaki : Jpn. J. Appl. Phys. 27 (1988) L622
[3] T.Terashima, K,Iijima, K.Yamamoto, Y.Bando, H.Mazaki : Jpn. J. Appl. Phys. 27 (1988) L91
[4] H.Hamane, H.Kurusawa, H.Iwasaki, T.Hirai, N.Kobayasi, and Y.Muto : Jpn J. Appl. Phys. 28 (1989) L127
[5] H.Tanabe, M.Tomouchi, T.Kobayashi et. al. : Jpn. J. Appl. Phys. 26 (1987) L1961

Superconducting YBCO Thin Films with (013) and (103) Orientation

KOICHI TSUDA, YUKO OKAMURA, TAKESHI SUZUKI, TOSHIYUKI MATSUI, MEGUMI NAGANO, and KAZUO MUKAE

Fuji Electric Corporate Research and Development, Ltd., 2-1, Nagasaka 2-chome, Yokosuka, 240-01 Japan

ABSTRACT

We studied the YBCO thin films without c-axis orientation. (013) oriented films were obtained by a sputtering at a low gas pressure. On the other hand, the film deposited by activated reactive deposition showed (103) orientation. The surface of sputtered film was smooth enough to fabricate a Josephson junction, but stripe-like cracks were often found in the films with thickness of more than 1500 A. We assumed these cracks were caused by the large change in a-axis lattice constant associated with the phase transition during cooling.

KEY WORDS: YBCO, thin film, orientation, intrinsic stress

INTRODUCTION

The discovery of oxide superconductors with high critical temperatures(Tc), including $YBa_2Cu_3O_{7-x}$ (YBCO)[1], has progressed research activity in this field. Much study has been focused on the fabrication of high quality thin films both scientific and technological aims. Among the various deposition systems, sputtering with a single oxide target is one of the simplest methods of depositing the films such as YBCO. There was, however, a major problem in sputtering system, namely the resputtering of the growing film by negative oxygen ions. We successfully prevented the resputtering during the deposition even at a low working gas pressure of 0.5 Pa by arranging a substrate right above the center of a target[2], instead of sputtering at a relatively high gas pressure[3].
By the way, it has been reported to easily fabricate c-axis oriented films of YBCO by some deposition methods. It is difficult, however, to make Josephson junctions using c-axis oriented films owing to the shorter coherence length of these surfaces. Consequently, we have attempted to obtain the surfaces with the longer coherence length by control of the orientation.
In this paper, we report the properties of (013) and (103) oriented YBCO films obtained by the sputtering at a low working gas pressure and activated reactive evaporation (ARE), respectively.

EXPERIMENTAL PROCEDURE

Thin films of YBCO have been mainly prepared by rf-magnetron sputtering using a single oxide target at a low working gas pressure of 0.5 Pa. The sputter was performed under the same conditions reported in a previous paper[2] using $SrTiO_3$(110) substrates instead of MgO. The substrate was kept at a temperature between 550°C and 650°C during the sputtering, then oxygen gas of 1 atom was introduced during cooling. We have also deposited the films using ARE to study difference in preferred orientation of the films between the sputter and the ARE. The ARE was carried out with metal sources and an oxygen gas pressure of 0.017 Pa at a substrate temperature of 600°C using $SrTiO_3$(110) substrates. Rf-plasma was introduced between the evaporating sources and the substrates to accelerate oxidation of the sources during the deposition and cooling[4]. Deposition rate were 2.0 A/s and 3.0 A/s for the sputtering and the ARE, respectively.
Structural information was given by X-ray diffraction (XRD) and four-axis XRD using CuKα radiation. Surface morphology of the films was observed by scanning electron microscopy (SEM). Resistivity-temperature characteristics (R-T

characteristics) were measured by a conventional dc four-probe method. In order to study surface properties of the films, we have fabricated a YBCO(013)/Au/Pb junction with a junction area of 0.5 x 0.5 mm^2. Au and Pb were evaporated by resistive heating.

RESULTS AND DISCUSSION

We first show about some difference between the sputtered and the ARE films. Figure 1 and 2 show XRD patterns of the sputtered and the ARE films respectively. These films were determined as (013) orientation for the sputtered film and (103) for the ARE film by four-axis XRD since we were not able to distinguish their reflections only by XRD due to almost the same Bragg angles[4]. The orientation of the sputtered films did not change at the substrate temperature between 550°C and 650°C. Figure 3 shows the SEM photograph of the sputtered film's surface deposited at 650°C. It has a considerably flat and mirror-like surface. On the other hand, the surface morphology of the ARE film was somewhat rough.

Figure 4 shows R-T characteristics of the sputtered film including its anisotropic behavior of <013> and <100> directions. Tc of <013> direction is slightly lower than that of <100>, and resistivity at the onset temperature of <013> direction is about ten times larger than that of <100>. The aniso-

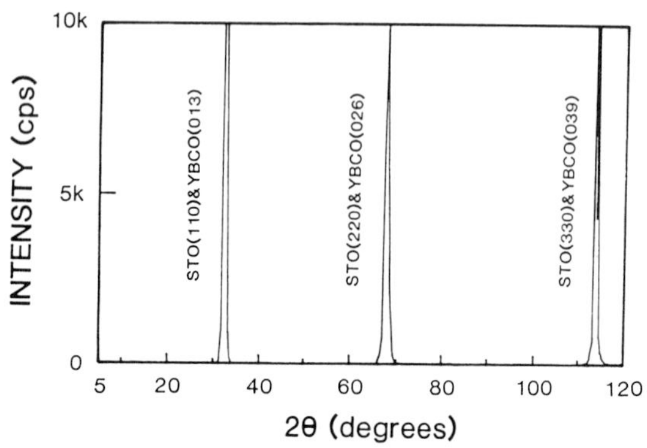

Fig.1 XRD pattern of the sputtered film

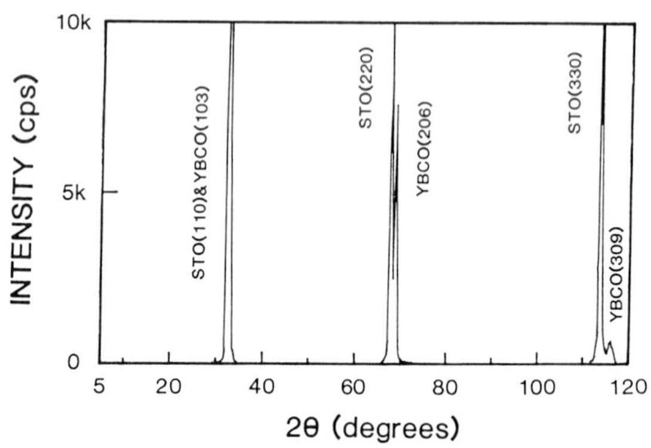

Fig.2 XRD pattern of the ARE film

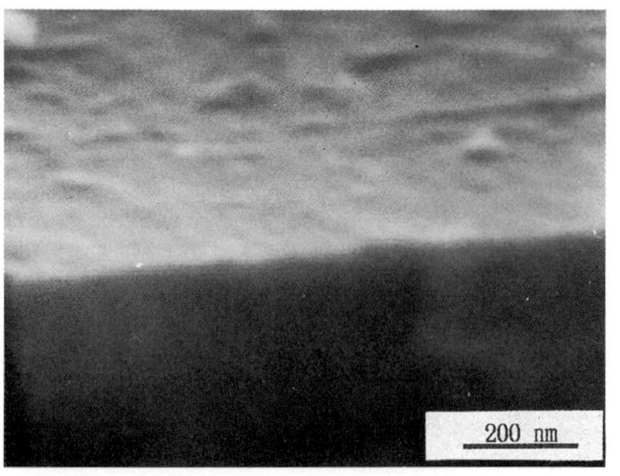

Fig.3 SEM photograph of the surface of the film deposited at 650 °C by the sputter

tropic R-T characteristics of the ARE film slightly weak in comparison with the sputtered film, as illustrated in Fig. 5, but it clearly shows that current flows less along the <103> direction than <010>.

Figure 6 shows a V-I characteristic of a YBCO(013)/Au/Pb junction with the Josephson current. We also found the Fraunhofer pattern depending on a magnetic field. This fact means that the surface of the sputtered film is smooth enough to make a Josephson junction.

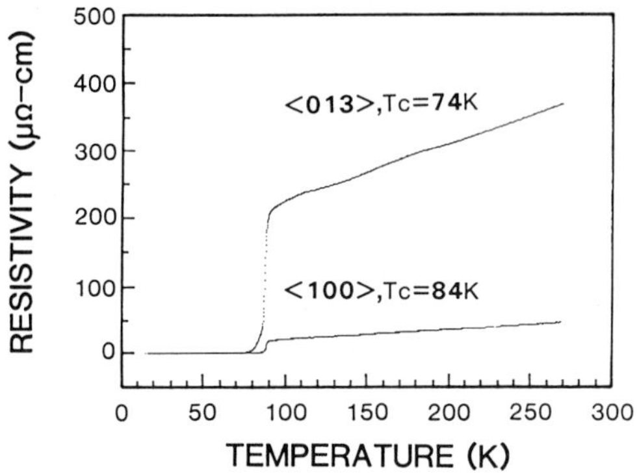

Fig.4 R-T characteristic (the sputter)

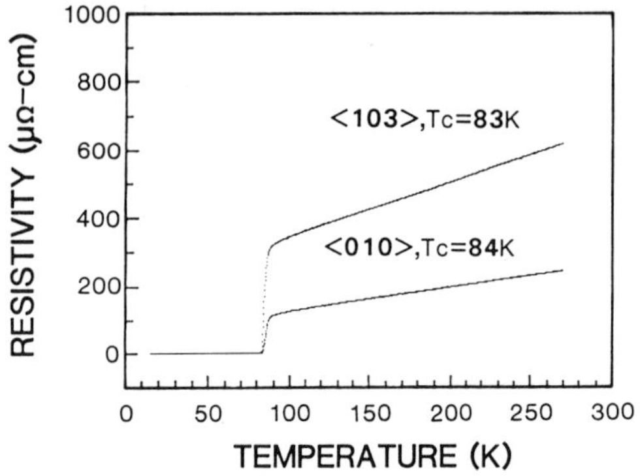

Fig.5 R-T characteristic (ARE)

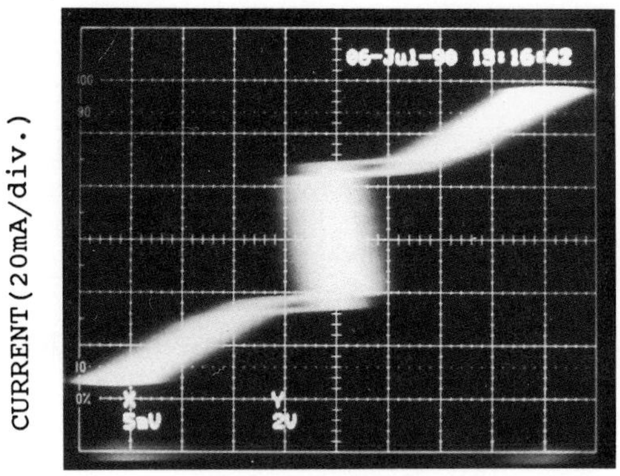

Fig.6 V-I characteristic of a YBCO(013)/Au/Pb junction (at 4.5K)

Though the sputtered film had excellently smooth surface, we often found that stripe-like cracks ran parallel with <013> direction at intervals of a few μm in the film with the thickness of more than 1500 A, as shown in Fig. 7. In contrast, no crack was found in the ARE film even with the thickness of more than 3000 A.

Here, we discuss why our sputtered film shows (013) orientation instead of (103) reported by T.Terashima et al.[5] and what causes the cracks in the film. We considered the preferred orientation in view of an intrinsic stress of the film during the deposition. Let us think of a change in a cross section per unit cell of the film keeping in a contact with the substrate as shown in Figure 8. The area painted out is given by

$\sqrt{b_o^2 + (c_o/3)^2} \times a_o$ for the (013) oriented film, and $\sqrt{a_o^2 + (c_o/3)^2} \times b_o$ for (103), where a_o, b_o and c_o are lattice parameters of a-, b- and c-axis respectively. The change of the (013) oriented film is larger than that of (103)[6].

In general, thin films are in a state of stress which is composed of a ther-

mal stress and an intrinsic stress. A sputtered film deposited at low gas pressures is commonly in a compressive stress. In contrast, the intrinsic stresses are typically tensile for an evaporated film and a sputtered film deposited at high working gas pressures[7].
It is assumed that the growing film by the sputtering at such a low gas pressure as 0.5 Pa is in the compressive stress. Consequently, it could be natural for the film to become (013) orientation at the phase transition from tetragonal to orthoronbic because of releasing the compressive stress. it is also reasonable for the film deposited by the ARE to select (103) orientation in order to free from the tensile stress, assuming it to be true.
Once the orientation of the sputtered film becomes (013) in the early stage of the cooling process, the orientation is kept (013) during cooling. We suppose that the cracks in the sputtered film were caused by the large change in a-axis lattice constant associated with the phase transition from tetragonal to orthoronbic during cooling[6].
It is expected that the sputtered films without cracks will be obtained by reducing the compressive stress by means of slight increase of the gas pressures or change of the gas species[7].

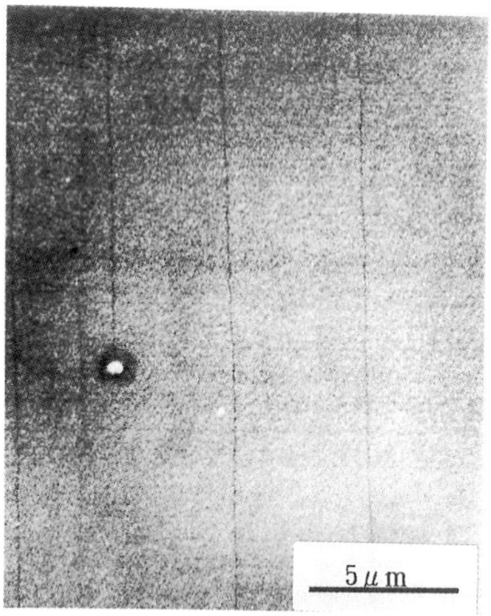

Fig.7 SEM photograph of cracks on the sputtered film

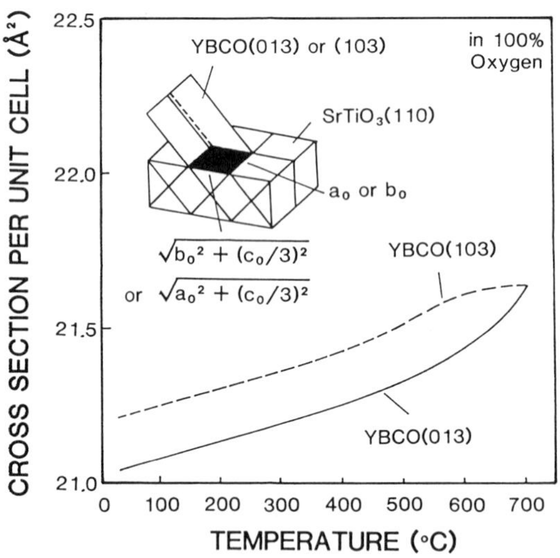

Fig.8 changes in cross section per unit cell keeping contact with the substrate

CONCLUSION

The surface of YBCO films sputtered at a low gas pressure was smooth enough to fabricate a Josephson junction, but stripe-like cracks were often found in the films with thickness of more than 1500 A, which were caused by phase transition. It is expected that the films without cracks will be obtained by reducing the compressive stress during the deposition.

REFERENCES

1.M.K.Wu,J.R.Ashburn,C.J.Torng,P.H.Hor,R.L.Meng,L.Gao,Z.J.Huang,Y.Q.Wang and C.W.Chu(1987) Phys.Rev.Lett.58 :908
2.M.Muroi,Y.Okamura,T.Suzuki,K.Tsuda,M.Nagano and K.Mukae(1990) Jpn.J.Appl. Phys.29 :69
3.G.Linker,X.X.Xi,O.Meyer,Q.Li and J.Geerk(1989) Solid State Commun.69 :249
4.T.Terashima,K.Iijima,K.Yamamoto,Y.Bando and H.Mazaki(1988) Jpn.J.Appl.Phys. 27 :L91
5.T.Terashima,Y.Bando,K.Iijima,K.Yamamoto and K.Hirata(1988) Appl.Phys.Lett. 53 :2232
6.J.D.Jorgensen,M.A.Beno,D.G.Hinks,L.Soderholm,K.J.Volin,R.L.Hitterman, J.D.Grace and Ivan K.Schuller(1987) Phys.Rev.B36 :3608
7.J.A.Thornton(1982) In: Rointan F.Bunshah(eds) Deposition Technologies for Films and Coatings. Noyes Publications. Park Ridge ,pp.221-225

Low Temperature Synthesis of Cu, Y-Cu, Ba-Cu Oxide Film by Reactive Sputtering

TOSHIYUKI NAKAMURA, YOSHINORI MORIKAWA, and HIROSHI YAMAOTO

College of Science & Technology, Nihon University, 24-1, Narashinodai 7-chome, Funabashi, 274 Japan

ABSTRACT

Crystal growth of Cu-O, Y-Cu-O and Ba-Cu-O sputtered film were investigated in order to get informations with respects to a low temperature synthesis of a superconducting YBaCuO thin film. A triclinic (111) plane of a CuO phase was epitaxially grown on a MgO (100) plane at about 450 °C by a reactive dc sputtering. The Y-Cu-O and Ba-Cu-O films were prepared by a reactive rf sputtering in an atmosphere of mixed gas of Ar, O_2 and N_2O. In particular an irradiation of ultraviolet ray was effective to lower the crystallization temperature of a single $Y_2Cu_2O_5$ phase by about 100 °C.

KEY WORDS: Y-Ba-Cu-O, reactive N_2O gas, UV irradiation, low temperature growth

INTRODUCTION

Since a YBaCuO superconductor with a 90 K class of Tc was discovered[1], most of thin film processes[2-4] have been tried and achieved to synthesize the superiór YBaCuO film. A sputtering deposition technique has been most popularly adopted because an apparatus is comparatively simple even for synthesis of a complex oxide with many kinds of constituent metals. Many technical subjects have been investigated in the sputtering deposition of YBaCuO films in order to control a serious change of a film composition from a target one[5,6], to promote oxidation reactions of the film, or to suppress crystal growth of nonsuperconducting phases. Nowadays it is required to achieve a low temperature process[7,8] in order to apply it for fabrications of superconducting electronic devices.

The final goal of this research is to realize a low temperature sputtering synthesis of a YBaCuO thin film. For the purpose, in this work the film growing process is investigated with respects to elementary oxides: Cu-O, Y-Cu-O and Ba-Cu-O. Sputtering circumstances are studied to activate a crystal growth or an oxidation reaction, in particular focusing effects of N_2O gas and an irradiation of an ultraviolet ray.

EXPERIMENTAL

The Cu-O film was deposited from a copper metal target by a reactive dc sputtering. A sputtering atmosphere was mixed gas of Ar and O_2, and the respective partial gas pressure was controlled by a mass flow meter. Oxygen gas was sprayed toward the substrate through the nozzle. Single crystal MgO with the (100) surface was used as substrate. The substrate temperature was changed from ambient to about 700 °C. The typical sputtering conditions were as following : an applied voltage of 600 ~ 700 V, a sputtering current density of about 6 mA/cm^2, a total sputtering gas pressure of about 16 Pa, a deposition rate of about 2.0 nm/min.

The Y-Cu-O and Ba-Cu-O films were prepared by a reactive rf sputtering. Two kinds of targets were investigated. One was an alloy target and the other was sintered oxide one. The metal composition of the targets were 1:1 in the mole ratio. Sputtering gas used were a mixture gas of Ar, O_2 and/or N_2O. The substrate temperature was changed from ambient to above 700 °C. The substrate used was a (100) MgO. The typical sputtering conditions were as following: a sputtering power density of 1.3 W/cm^2, a total gas pressure of 7.8 Pa, a distance between the target and substrate of 2.5 cm, and a deposition rate of about 0.2 nm/min. Here ultraviolet ray (UV) of wavelength of 160 nm from a heavy hydrogen lamp was irradiated into the film growing space through a quartz window. The power of the lamp was 144W.

The film composition was measured by ICP. A crystal structure of the film was analyzed by XRD or RHEED. Chemical binding energy spectra of elementary atoms were measured by XPS.

RESULTS AND DISCUSSION

Figure 1 shows the phase diagram observed in the preparation of the Cu-O sputtered film. The obtained film was considerably influenced a substrate temperature, Ts and a partial pressure of O_2, P_{O2}. At Ts's from 350 °C to 500 °C and in P_{O2} from 8 Pa to 2 Pa, a crystalline CuO film was obtained. At Ts's under 250 °C, the Cu-O film was not crystallized. The film was not deposited at Ts's above 500 °C.

Figure 2 shows XRD patterns of the typical CuO films. The CuO film with a strongly oriented structure grew at Ts of about 450 °C in P_{O2} of about 8 Pa. At Ts of about 350 °C in P_{O2} of about 6 Pa, the polycrystal film was deposited which revealed a strong $(\bar{1}11)$ peak and small $(\bar{1}11)$ and $(\bar{2}02)$ peaks. Even at Ts of 450 °C, less P_{O2} of about 6 Pa resulted in a polycrystal CuO film. A crystalline film was not obtained at Ts of 250 °C in P_{O2} of 5 Pa. The RHEED

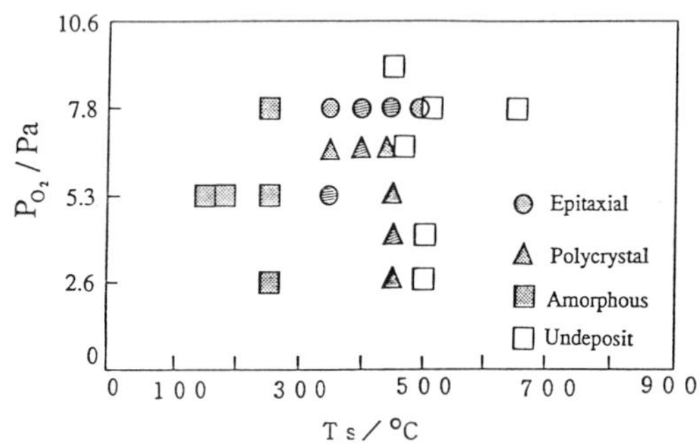

Fig.1 Phase diagram of Cu-O reactively sputtered film.

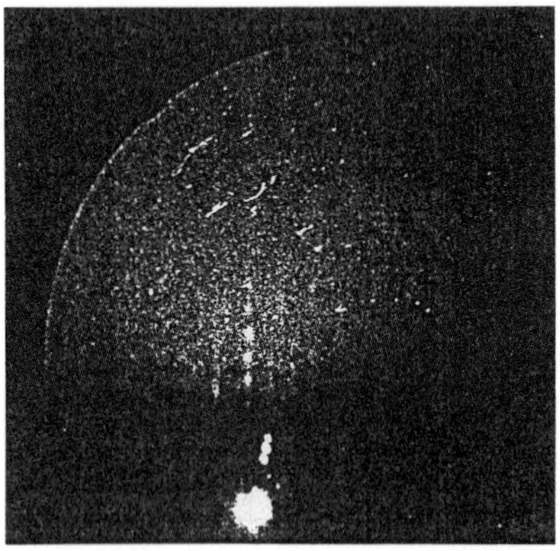

Fig.3 RHEED pattern of epitaxial CuO sputtered film.

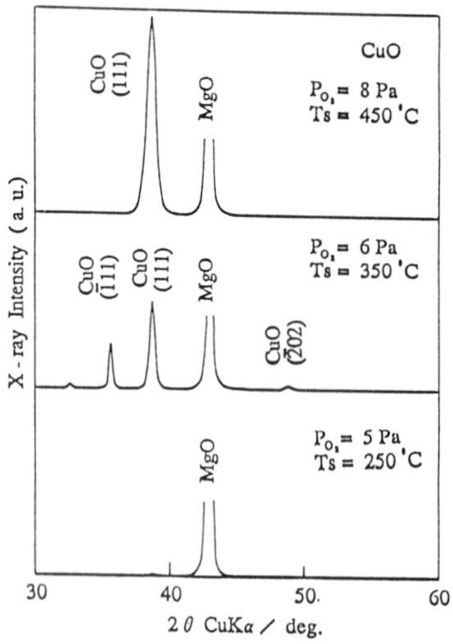

Fig.2 Typical XRD patterns of sputtered CuO films.

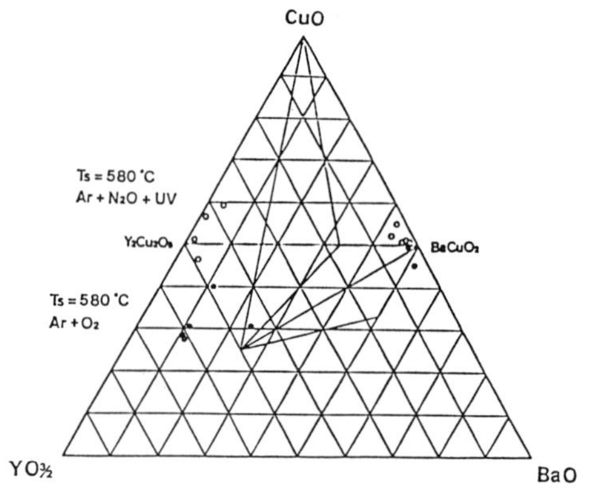

Fig.4 Composition of Y-Cu-O and Ba-Cu-O sputtered film.

pattern of the strongly oriented CuO film is shown in Fig. 3. The epitaxial growth of the CuO phase was confirmed from the diffraction pattern which revealed definite spots from the CuO (111) planes and no clear diffraction ring from any polycrystal phase.

It was found from a XPS spectrum of $Cu_{2p3/2}$ of the epitaxial film that a valence of constituent copper ions in the film was almost two and the film was thoroughly oxidized.

Figure 4 shows a distribution of composition of Y-Cu-O and Ba-Cu-O films in the tricomponent system, when alloy targets were sputtered. The composition of Y-Cu-O films sputtered in $Ar+O_2$ gas significantly shifted toward a Y-rich side. When films were deposited under a mixture gas of N_2O and the UV irradiation, the film composition converged to the stoichiometry. Since the

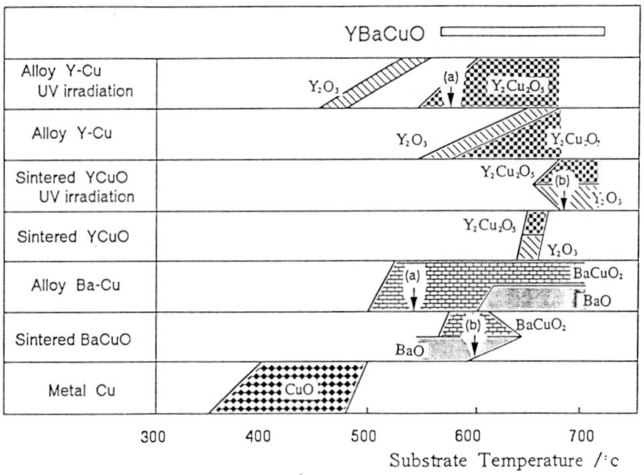

Fig.5 Phase diagram of elementary oxide films prepared under various sputtering conditions.

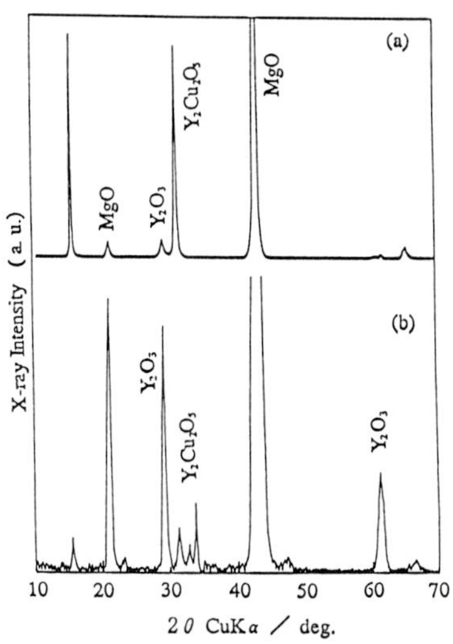

Fig.6 XRD patterns of Y-Cu-O films sputtered under UV irradiation at Ts of 580 °C from an alloy target (a), at Ts of 680 °C from a sintered target (b).

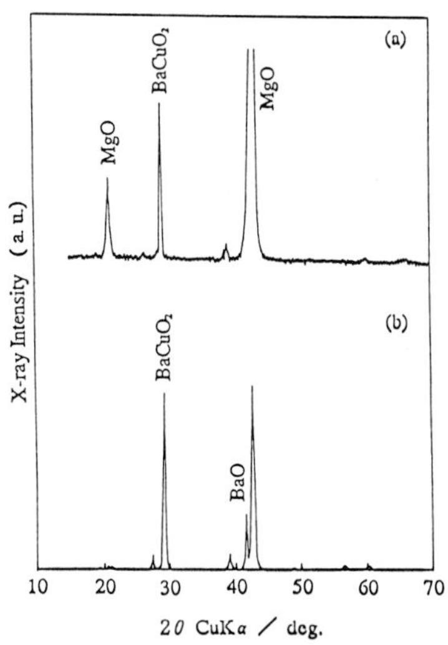

Fig.7 XRD patterns of Ba-Cu-O films sputtered without UV at Ts of 540 °C from an alloy target (a), at Ts of 600 °C from a sintered target (b).

CuO was decomposed on a hot substrate above 500 ℃ as shown in the case of CuO film preparation, a more stable Y_2O_3 phase selectively remained in the film. On the other hand, radical oxygen atoms which are generated by dissociation of N_2O molecules proceed oxidation reactions. As a result $Y_2Cu_2O_5$ phase may be stabilized even on the hot substrate. Such the mechanism wasn't, however, so available for the case of the Ba-Cu-O film.

When the sintered targets were sputtered, observed results were quietly different. As a general tendency the sintered oxide targets promote oxidation reactions far more than the alloy targets. The phase diagram of related oxide thin films obtained from different sputtering conditions was summarized in Fig. 5. Only the CuO phase is well crystallized even at a low Ts. The two component oxides are crystallized at the region of Ts from 500 ℃ to 700 ℃. At the same region of Ts as-grown synthesis of YBaCuO thin films are also realized as reported by many groups. In the atmosphere without the UV irradiation, the growth of the single $Y_2Cu_2O_5$ and $BaCuO_2$ phase takes place only at high Ts of about 700 ℃ from the sintered targets, and the selective growth of the single phase does not occur from any alloy targets. On the contrary, it was possible by the UV irradiation to obtain the single phase at comparatively low Ts of about 600 ℃ only from alloy targets. In other words, the UV irradiation in the atmosphere of N_2O gas brings a remarkable effect for a low temperature process only under the insufficiently oxidizing condition.

Figure 6 and 7 shows typical XRD patterns of Y-Cu-O and Ba-Cu-O films prepared in the sputtering conditions remarked in Fig. 5. It is confirmed that the selective growth of the $Y_2Cu_2O_5$ and/or $BaCuO_2$ phase takes place at a comparatively low Ts under the adequate conditions, though they are very restricted ones.

In summary crystal growth of Cu-O, Y-Cu-O and Ba-Cu-O sputtered film were studied. A triclinic (111) plane of a CuO phase was epitaxially grown on a MgO (100) plane at about 450℃ by a reactive sputtering in the atmosphere of $Ar+O_2$. For the preparation of films with the single Y-Cu-O and/or Ba-Cu-O phase, the UV irradiation in the atmosphere of mixed gas of Ar, O_2 and N_2O was effective to lower the crystallization temperature, when alloy targets were reactively sputtered. These informations about a low temperature process will be made use of for the purpose of a low temperature synthesis of the YBaCuO thin film.

ACKNOWLEDGEMENTS

This work was partially supported by Grant-in-Aid for Scientific Research on Priority Areas from The Ministry of Education, Science and Culture. The authors would like to thank Prof. M. Tanaka for his comments and suggestions. The authors would like to thank also Mr. N. Kobayashi and Mr. T. Yajima for their assistances in the preparation of samples.

REFERENCES

1. Wu MK, Ashburn JR, Torng CJ, Hor PH, Meng RI, Gao L, Huang ZJ, Wang YQ and Chu W (1987) Phy.Rev. Lett. 58: 908
2. Terashima T, Iijima K, Yamamoto K, Bando Y and Mazaki H (1988) Jap.J.Appl.Phys. 27: L91-L93
3. Abe H, Tsuruoka T and Nakamori T (1988) Jap.J.Appl.Phys. 27: L1473-L1475
4. Goto T, Masumoto H and Hirai T (1989) Jap.J.Appl.Phys. 28: L88-L90
5. Terada N, Ihara H, Jo M, Hirabayashi M, Kimura Y, Matsuhiro K, Hirata K, Ohno E, Sugise R and Kawashima F (1988) Jap.J.Appl.Phys. 27: L639-L642
6. Enami H, Shinohara T, Kawahara N, Kawabata S, Hoshizaki H and Imura T (1990) Jap.J.Appl.Phys. 29: L728-L784
7. Miura T, Terashima Y, Sagoi M and Kubo K (1988) Jap.J.Appl.Phys. 27 : L1260-L1261
8. Michikami O, Asahi M and Asano H (1989) Jap.J.Appl.Phys. 28: L448-L451

Angular Dependence of Critical Current Density of YBaCuO Thin Film

S. OGOTA[1], C. TAKAHASHI[1], K. ISAWA[1], M. NAGANO[1], H. TAKAHASHI[1], H. MORITA[2], and M. SUZUKI[2]

[1] Applied Technology R&D Center, Tohoku Electric Power Co., Inc., 2-1, Nakayama 7-chome, Aoba-ku, Sendai, 981 Japan
[2] Department of Electrical Engineering, Tohoku University, Sendai, 980 Japan

ABSTRACT

Experimental studies on the angular dependences of resistive transition and transport critical current density (Jc) under magnetic fields have been carried out on thin film of YBaCuO. Investigated samples are c-axis oriented polycrystalline films prepared by dc magnetron sputtering from a single sintered target on MgO(100) substrates, and their transition temperature(Tc) ranges from 82 to 89 K. The Tc of samples varies depending on the magnetic field strength(B) and field angle(θ) relative to the film plane. As B increases up to 7 T, the Tc is generally lowered and a large reduction of Tc takes place at $\theta > 30°$. Values of $Jc(\theta = 90°)/Jc(\theta = 0°)$ obtained in every field appreciably decreases with increasing B up to 7 T.

INTRODUCTION

High Tc oxide superconductors are expected for use in future electronics devices and magnets. In order to realize the applications of oxide superconductors in the electronics field, thin films of high quality must be produced. Recently, YBaCuO thin films with $Tc \simeq 90\ K$ and $Jc > 10^6 A/cm^2$ at 77.3 K and 0 T have been prepared by various techniques such as chemical vapor deposition[1], sputtering [2] and laser ablation [3]. These films indicates strong anisotropies in the resistivity[4], the upper critical field (Hc_2)[5] and the critical current density[6]. However, a few data have been reported because of difficulties in preparing high quality films and in measuring anisotropic properties. Further detailed studies are required for a better understanding of anisotropies in Jc and Hc_2 which are important properties from the viewpoint of practical application.

In this paper, we report the magnetic field dependences of Tc and Jc, and the angular dependences of Tc and Jc in YBaCuO thin films prepared by sputtering.

EXPERIMENTAL

Samples measured were YBaCuO thin films deposited by dc magnetron sputtering on MgO(100) substrates with the c-axis perpendicular to the film plane. Typical film thickness was about 165 nm. The sputtering target 50 mm in diameter was prepared by sintering and its composition was $Y_1Ba_{1.7}Cu_3O_y$.

Sputtering was carried out under the following conditions: substrate temperature 720 $^{\circ}C$, mixture gas pressure 0.3 $Torr$ ($Ar : O_2 = 30 : 1$), deposition rate 0.05 nm/sec.

Resistive transition of the films were measured by a four probe method. The measurements of Jc were carried out by a transport current method at 77.3 K ,and Jc was determined with 1 $\mu V/cm$ criterion. Currents were passed perpendicular to c-axis under any angle magnetic field applied. The angle of the a-b plane of film and the field direction was defined θ. Both resistivity and Jc were measured under magnetic field up to 7 T, changing the applied angle from perpendicular($\theta = 0^{\circ}$) to pararell($\theta = -90^{\circ}, 90^{\circ}$) to the c-axis.

RESULTS AND DISCUSSION

Resistive transition curves under various magnetic fields up to 7 T at $\theta = 0^{\circ}$ and 90° are shown in Fig.1. At the first glance, two significant features are observed; one is the significant anisotropy of the resistive transition curves, and the other is the broadening of the resistive transition by the applied fields. The similar trend was observed on the thin films of other high Tc superconductors .

Figure 2 shows the magnetic field dependence of Tc, mid and $Tc, zero$. Tc, mid and $Tc, zero$ simply decrease with increasing B. We estimated the values of $-dHc_2/d$T and $Hc_2(0)$ at various angles using Tc, mid or $Tc, zero$ as shown in Table 1. As the superconducting transition becomes broad under magnetic field, the value of $Hc_2(\theta = 0^{\circ})/Hc_2(\theta = 90^{\circ})$ may indicate the degree of anisotropy. However, the value estimated from Tc, mid does not agree with that from $Tc, zero$.

Figure 3 shows the angular dependence of $Tc, zero$ under magnetic field. $Tc, zero$ doesn't change so much, but the lowering of $Tc, zero$ is clearly observed under high magnetic field. The changing of $Tc, zero$ can be clearly seen in the range of 0° to 30° under the high magnetic field of 5 to 7 T.

The anisotropic Jc vs B behaviors at 77.3 K at various angles are shown in Fig.4. Critical current density drops rapidly with increasing B up to 1 T.

Figure 5 shows the angular dependence of Jc at 77.3 K under various magnetic fields. Only one Jc peak can be seen at $\theta = 0^{\circ}$ in the range from -90° *to* 90° and the angular dependence of Jc becomes clear with increasing B.

Roas et al. [7] observed a Jc peak at $\theta = 90^{\circ}$ for the YBaCuO film prepared by laser deposition. They explained the peak caused by the pinning force of twin boundaries at the a-b plane. However, in our measurements, Jc peak can not be seen at $\theta = 90^{\circ}$.

In summary, we measured the magnetic field dependences of Tc and Jc, and also measured the angular dependence of Tc and Jc of YBaCuO thin films prepared by sputtering. The angular dependence of Jc was clearly observed; the angular dependence of Jc became clear with increasing B, and only one peak of Jc existed at $\theta = 0^{\circ}$. The angular dependence of Tc was also clearly observed up to $\theta = 30^{\circ}$. The Jc dropped rapidly with increasing B, however, the Tc simply decreased with increasing B.

REFERENCES

1. H.Yamane, H.Kurosawa, T.Hirai, K.Watanabe, H.Iwasaki, N.Kobayashi, and Y.Muto; Supercond.Sci.Technol.2,115(1989).
2. X.X.Xi, G.Linker, O.Meyer, E.Nold, B.Obst, F.Ratzel, R.Smithey, B.Strechlan, F.Weschenfelder, and J.Geerk; Z.Phys. B 74,13(1989).
3. G.Koren, A.Gupta, E.A.Giess, A.Segumuller, and R.B.Laibowitz; Appl.Phys.Lett.54,1054(1989).
4. Y.Iye; Studies of High Temperature Superconductors, edited by A.Narlikar, Nova Science Pub, New York, vol.2,199(1989).
5. Y.Muto, K.Watanabe, N.Kobayashi, H.Kawabe, H.Yamane, H.Kurokawa, and T.Hirai; Physica C 162 – 164,105(1989).
6. Y.Enomoto, T.Murakami, M.Suzuki, and K.Moriwaki; Jpn.J.Appl.Phys.26,1248(1987).
7. B.Roas, L.Schultz, and G.Saemann-Ischenko; Phys.Rev.Lett.64,479(1990).

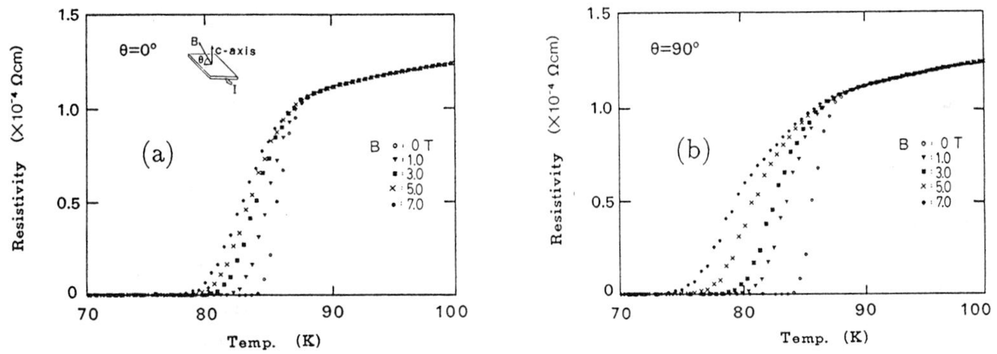

Fig.1 Temperature dependence of the resistivity transition for various magnetic fields
(a) applied perpendicular to the c-axis ($\theta = 0°$).
(b) applied pararell ($\theta = 90°$).

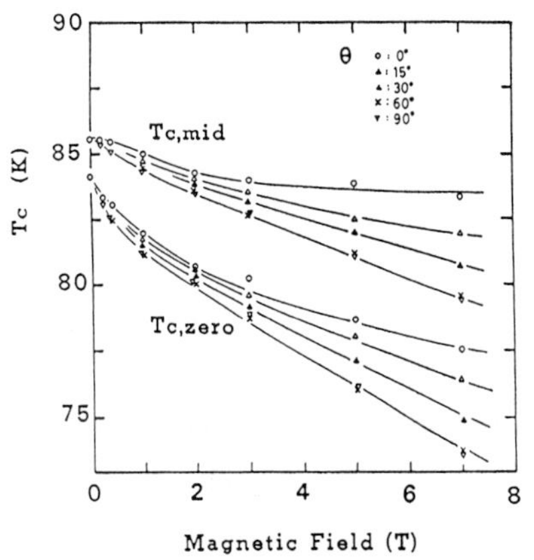

Fig.2 Magnetic field dependence of $T_{c,mid}$ and $T_{c,zero}$.

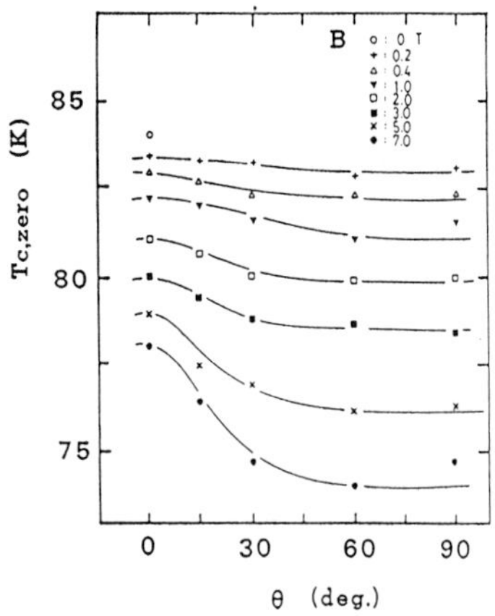

Fig.3 Angular dependence of $T_{c,zero}$ under various magnetic fields.

Table I $-dH_{c2}/dT$ and H_{c2} estimated from $T_{c,mid}$ or $T_{c,zero}$

θ	-dHc₂/dT estimated from Tc,mid	-dHc₂/dT estimated from Tc,zero	Hc₂(0) Tc,mid	Hc₂(0) Tc,zero
0°	5.26	1.44	446	118
15°	2.14	1.19	182	97.7
30°	1.61	0.96	137	78.6
60°	1.19	0.79	101	65.3
90°	1.19	0.79	101	65.3

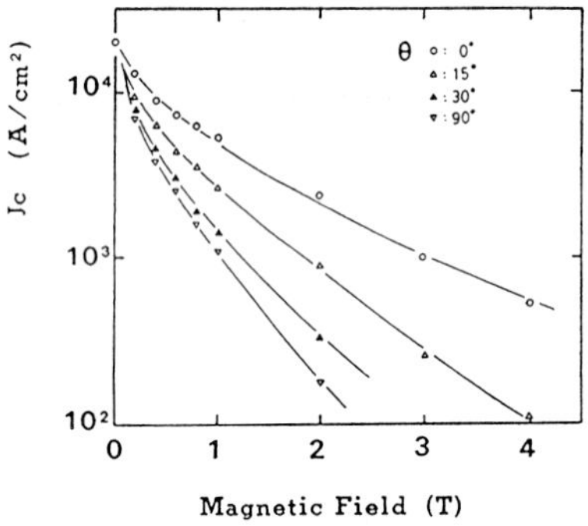

Fig.4 Magnetic field dependence of Jc for YBaCuO thin film at $77.3K$.

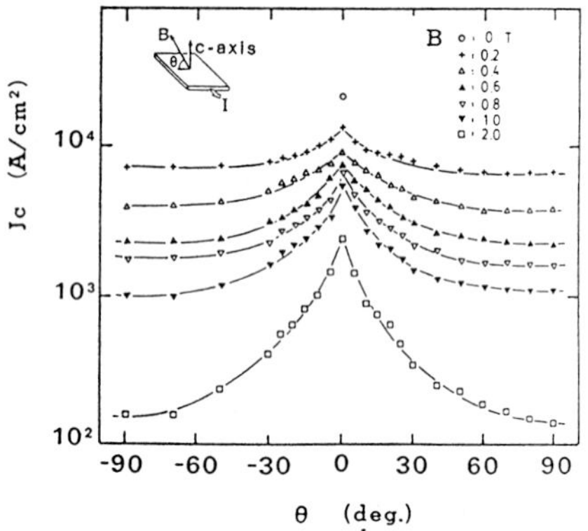

Fig.5 Angular dependence of J_c at $77.3K$ under various magnetic fields .

Epitaxial Growth of Superconducting $YBa_2Cu_3O_{7-x}$ Films by the RF Sputtering Method with the Oylindrical Substrate Holder

H. KAJIKAWA, Y. FUKUMOTO, K. SHIBUTANI, S. HAYASHI, R. OGAWA, and Y. KAWATE

Superconducting & Cryogenic Technology Center, Kobe Steel Ltd., 5-5, Takatsukadai 1-chome, Nishi-ku, Kobe, 673-02 Japan

ABSTRACT

Epitaxial $YBa_2Cu_3O_{7-y}$ films were formed by the sputtering system using a cylindrical substrate holder, on which the substrates were set vertically to the target surface. In this system, the re-sputtering phenomenon could be reduced so that it could become easier to investigate each effect of the deposition parameters on the film properties separately. The films obtained by optimizing the substrate temperature and the oxygen partial pressure showed the sharp transition and the high critical current density of $1.6x10^6A/cm^2$ at 77K.

KEY WORDS: $YBa_2Cu_3O_{7-y}$ films, re-sputtering, critical current density, ac-susceptibility

INTRODUCTION

In the conventional co-planar sputtering system using oxide $YBa_2Cu_3O_{7-y}$target, the re-sputtering caused by high energy negative ions taken out of the target have made it difficult to analyze the effect of the deposition parameters on the film properties. In order to avoid the rush of the negative ions upon the film, several researchers have improved the sputtering method by using the off-axis arrangement of the substrates [1], using cylindrical substrate holder [2], using the inverse cylindrical target [3], or setting the shield plate in front of the substrate [4]. We also introduced the cylindrical substrate holder (temple-bell type substrate holder) on which the substrates are put vertically to the target surface in the RF-diode sputtering system and it became very easy to obtain epitaxial $YBa_2Cu_3O_{7-y}$ films with a critical temperature Tc of >80K and a critical current density Jc of $>10^5A/cm^2$ [5]. It was then expected that the film properties would be improved by higher density plasma of the magnetron target. In this work, the properties of the films deposited by the RF magnetron sputtering system with the temple-bell type substrate holder were investigated and the sputtering deposition process was compared with other deposition methods.

EXPERIMENTAL

The magnetron sputtering system with the temple-bell type substrate holder is shown in Fig.1. The dimensions of the system were described in the reference [5]. The sputtering conditions were shown in Table 1. Both the powder target and the sintered bulk target with the stoichiometric composition of $YBa_2Cu_3O_x$ were used, and they were placed on the copper target holder. The substrate temperature (Ts) was measured by the sheathed thermo-couple attached on the surface of the dummy substrate. The post heat treatment was made for some films at 880℃ for 30 minutes and at 500℃ for 2 hours in oxygen gas. Tc and

Jc of the films were estimated by the dc four point probe method. The crystallinity was estimated by the X-ray diffraction (XRD). The ac-susceptibility was also measured at 1G and at 100Hz to observe the homogeneity of the films.

RESULTS AND DISCUSSION

The crystal orientations of the films deposited at RF power (Po) of 50W on the $SrTiO_3$ (100) substrates are plotted on the phase diagram for YBaCuO in Fig.2 [6]. The notations of "a", "c" and "ac" show a-axis oriented film, c-axis oriented film and mixed a-axis and c-axis oriented film, respectively. The typical XRD patterns for the films deposited at Ts=638°C and 713°C on the $SrTiO_3$ (100) substrates are shown in Fig.3. It seemed that these crystal orientations were related to the lattice matching between the substrate and the initial phase of $YBa_2Cu_3O_{7-y}$ crystals just after the deposition [5], [6], [7]. The tetragonal phase was predominantly formed at a high Ts of >685°C with oxygen partial pressure (Po_2) of 6-95Pa and might have the better lattice matching with the substrate when both a- and b-axes lay along the [010] or [001] of the single-crystalline substrate. On the contrary, at a low Ts of <635°C with Po_2 of 95Pa the orthorhombic phase (Ortho-I phase) was predominantly formed, which might have the better lattice matching when both a- and c-axes lay on the substrate. Noting that in the segregation region the films could be formed, the above results suggest that the two borderlines (one between the segregation region and the tetragonal region, and the other between the tetragonal region and the ortho-I region) on the phase diagram in Fig.2 may be shifted to the left, when film surfaces were exposed to activated oxygen atoms or molecules [8]. The crystal orientations of the films on the MgO (100) substrates had the similar tendency to those on the $SrTiO_3$ (100) substrates, but the c-axis orientation was predominant because of the large lattice mismatch and the large growth rate of the $YBa_2Cu_3O_{7-y}$ crystals in the basal plane [5].

The superconducting properties (Tc and Jc) were the best in the c-axis oriented films deposited at Ts=713°C. The as-deposited films showed a Tc of 84-86K and a Jc of $1\text{-}3\times10^4 A/cm^2$ at 77K. Tc increased as Po_2 decreased from 95 to 6Pa as shown in Fig.4, which is contrary to the results obtained by the RF diode sputtering method. It may be due to that the oxygen gas was much more excited in the magnetron sputtering method than in the diode sputtering method. Thus in the magnetron sputtering method the crystallinity might be degraded by the in-situ production of some amount of the orthorhombic phase, which caused the lattice distortion by neighboring the tetragonal phase, or by the production of stable metal oxides in the vapor phase, as the Po_2 increased. After the post heat treatment, Jc was very much improved to attain to $3\text{-}6\times10^5 A/cm^2$ at 77K, while Tc were almost unchanged. The highest Jc in this work was obtained in the film deposited at Ts=713°C and Po_2=6Pa and was $1.6\times10^6 A/cm^2$ at 77K as shown in Fig.5. The ac-susceptibility curves of the film are shown in Fig.6-(a). The sharp transition in the real part x' and the rapidly vanishing tail end of the imaginary part x" show that the film has a high homogeneity. This contrasts with the ac-susceptibility obtained from the film deposited by the RF-diode sputtering method shown in Fig.6-(b).

Our results compared with those reported by Xi et al [3], Kwo et al [10] and Terashima et al [11] may indicate that high quality films with a large Jc of $>10^6 A/cm^2$ at 77K will be obtainable when films are deposited near the boundary between the segregation region and the tetragonal region. It may be mainly because that in this region, only the tetragonal phase can be well formed, avoiding the in-situ formation of extra orthorhombic phase. The complete tetragonal crystals without lattice distortion will be easily oxidized during the cooling or the low temperature heat treatment, and will transform into the orthorhombic crystals. From this point of view, the depositions have been made with the locally increased Po_2 near substrates using excited species of oxygen in the MBE method [10] or in the activated reactive evaporation method [11], where the background Po_2 is usually too low. On the contrary, in the RF magnetron sputtering system Po_2 should be rather lowered as possible using effectively the high density plasma above the target.

SUMMARY

By the RF magnetron sputtering method using a temple-bell type substrate holder, the superconducting $YBa_2Cu_3O_{7-y}$ films with Jc of > $10^6 A/cm^2$ at 77K were formed, putting the deposition parameters close to the boundary between the segregation and the tetragonal phase regions.

REFERENCES

1. Terada, H.Ihara, M.Jo, M.Hirabayashi, Y.Kimura, K.Matsutani, K.Hirata, E.Ohno, R.Sugise, and F.Kawashima (1988) Jpn.J.Appl.Phys.,27:L639
2. W.Y.Lee, J.Salem, V.Lee, C.T.Lettner, G.Lim, R.Savoy and V.Deline (1988) AIP Conf. Proc. No.165:95
3. X.X.Xi, G.Linker, O.Meyer and J.Geerk (1989) J.Less-Common Met. 151:295
4. P.H.Ballentine, A.M.Kadin and W.R.Donaldson (1990) Proceedings for The SPIE Conference Superconductivity Applications for Infrared and Microwave Devices (to be published)
5. H.Kajikawa, Y.Fukumoto, S.Hayashi, K.Shibutani, R.Ogawa, and Y.Kawate (1990) Proceedings for 1990 Applied Superconductivity Conference (to be published)
6. D. Talor (1985) Trans. & J.Br.Ceram.Soc,84:181
7. E.D.Specht, C.J.Sparks, A.G.Dhere, J.Brynestad, O.B.Cavin and D.M.Kroeger (1988) Phys.Rev.,B37:7426
8. Y.Terashima, M.Sagoi, K.Kubo, Y.Mizutani, T.Miura, J.Yoshida and K.Mizushima (1989) Jpn.J.Appl.Phys. 28:L653
9. R.H.Hammond and R.Bormann, Appl.Phys.Lett. (to be published)
10. J.Kwo, M.Hong, T.A.Hulton, P.L.Gammel and J.P.Mannearts (1989) Proceedings of Films for High Tc Superconducting Electronics.SPIE,1187:57
11. T.Terashima, K.Iijima, K.Yamamoto, Y.Bando and H.Mazaki (1988) Jpn.J.Appl.Phys.,27:L91

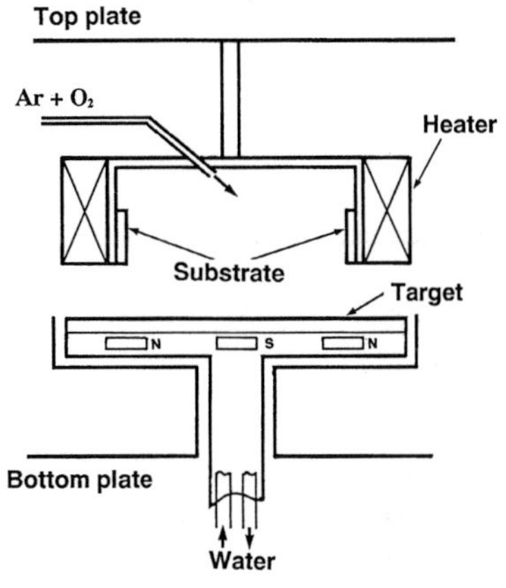

Table 1 Sputtering conditions.

Target	$Y_1Ba_2Cu_3O_x$ (Stoichiometric)
Substrate	MgO (100) STO (100)
Sputtering Gas	Ar (4Pa) + O_2 (6-95Pa)
RF Power	50W
Growth Temperature	585-713℃
Film Thickness	100-300nm

Fig.1 Schematic diagram of the sputtering chamber.

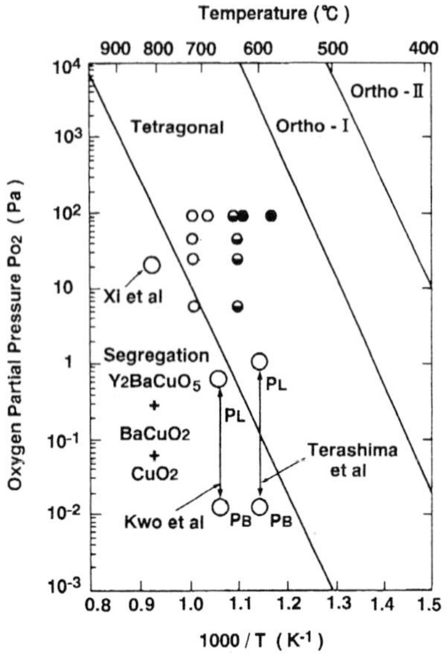

PL: Local pressure, PB: Background pressure

Fig.2 Deposition parameters vs crystal orientations of the films shown on the phase diagram.

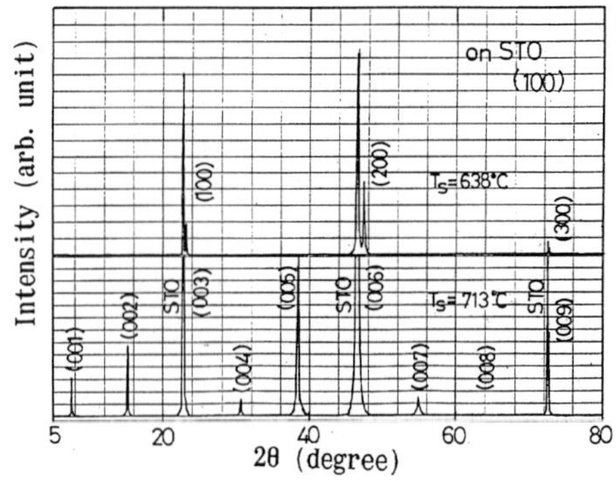

Fig.3 X-ray diffraction patterns for the films deposited (a)at 638℃ and (b)at 713℃, with Po_2=95Pa.

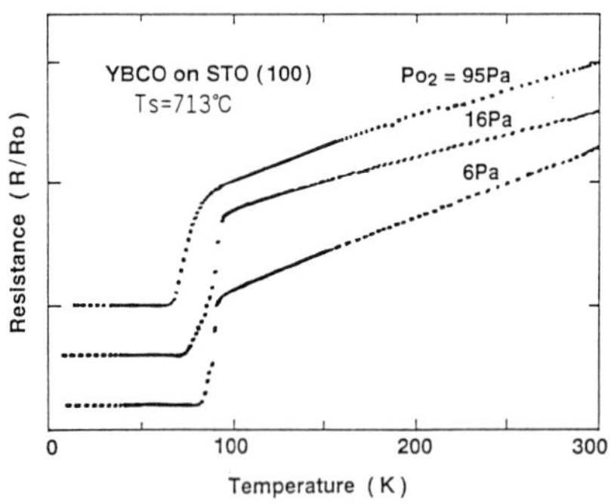

Fig.4 R-T curves with Po_2 as a parameter.

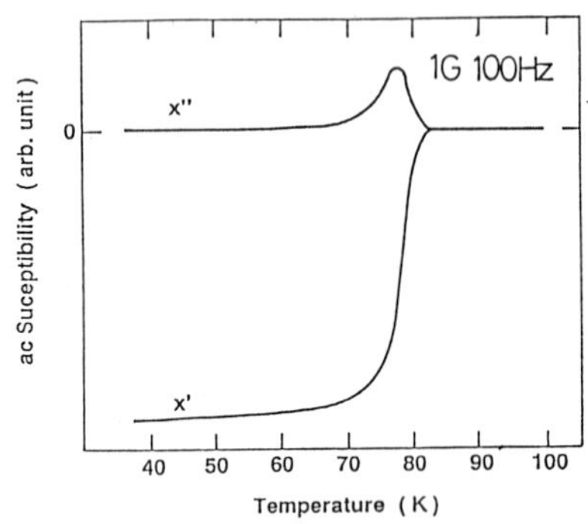

(a)

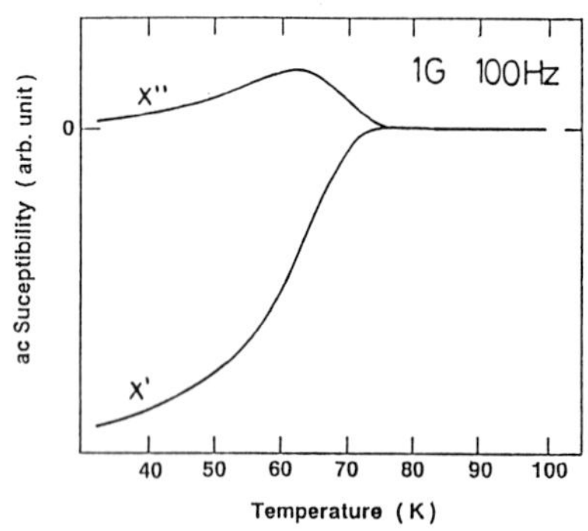

(b)

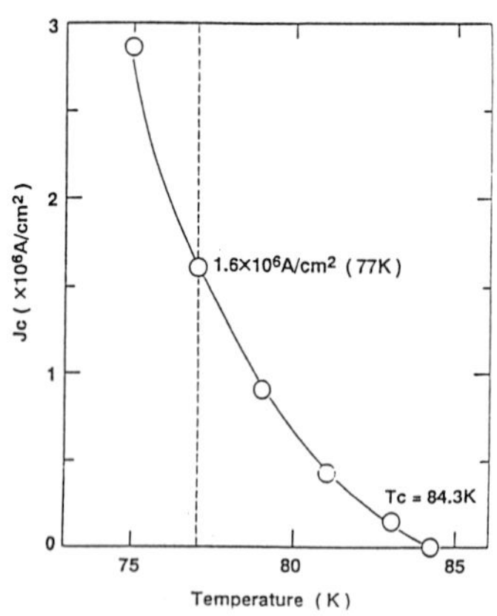

Fig.5 Temperature dependence of Jc of the film deposited at 713℃ and Po_2 =6Pa, and post-annealed at 880℃ for 30 min. and 500℃ for 2h in oxygen gas.

Fig.6 Ac-susceptibilities of the films (a)deposited by the RF magnetron sputtering method and (b)deposited by the RF diode sputtering method.

Preparation of Y-Ba-Cu-O Films by ECR Plasma-Assisted Sputtering

TAKAMICHI FUJII[1], TOSHIHIKO ANNO[2], KENJI FURUSHO[1], TSUYOSHI KOYANAGI[1], KOJI MOROFUJI[1], and KAKUEI MATSUBARA[1]

[1] Department of Electronics, Faculty of Engineering, Yamaguchi University, 2557, Tokiwadai, Ube, 755 Japan

[2] Ube Laboratory, Ube Industries Ltd., 1978-5, Kogushi, Ube, 755 Japan

ABSTRACT

Using ECR plasma assisted-sputtering, epitaxial growth of $YBa_2Cu_3O_{7-y}$ films with c-axis orientation has been accomplished on a (100) $SrTiO_3$ substrate. The $YBa_2Cu_3O_{7-y}$ films prepared by the ECR plasma-assisted sputtering exhibited a shorter lattice constant c and a higher T_c than those prepared by only the magnetron sputtering. To investigate the relationship between the superconducting properties of the films and the plasma species in the ECR plasma, measurements of an optical emission spectroscopy in the ECR plasma have been made. The superconducting properties of the films were closely related to the activated particles of oxygen in the ECR plasma.

KEYWORDS:ECR plasma-assisted sputtering, $YBa_2Cu_3O_{7-y}$ films, optical emission spectroscopy analysis

INTRODUCTION

Recently, ECR microwave plasma has been attracted much attention for use in thin film deposition and etching [1], [2]. The attractive features of ECR plasma are absence of heated filament, high degree of ionization, and processes at low gas pressures. In several techniques using ECR plasma for thin film deposition, ECR sputtering is successful in preparation of high-quality thin films at low substrate temperatures with high deposition rates [3-5]. In the case of reactive sputtering, a fabrication process of thin films using magnetron sputtering combined with ECR plasma has many advantages, since the reaction of metal atoms sputtered from targets with reactive gas is important in synthesizing films, such as oxides, nitrides and carbides films.

We have fabricated the magnetron sputtering apparatus combined with an ECR plasma source, hereafter called as ECR plasma-assisted sputtering, and have prepared Cu-O films using this system as a basic research for thin film formation of high-T_c oxide superconductors [6]. From this study, it was found that the oxygen content of the Cu-O films could be controlled by changing the microwave power of the ECR plasma source.

In this paper, the preparation of $YBa_2Cu_3O_{7-y}$ films using the ECR plasma-assisted sputtering and their superconducting properties are reported. The properties of $YBa_2Cu_3O_{7-y}$ are discussed in connection with several results of the plasma diagnostics by the optical emission spectroscopy analysis.

EXPERIMENTAL SYSTEM AND PROCEDURES

The schematic diagram of the ECR plasma-assisted sputtering system is shown in Fig.1. In this system, the ECR plasma source is equipped facing to the magnetron sputtering target, and the substrate is placed on the side of the target. O_2 gas is introduced directly into the ECR plasma source, and Ar gas is used as the sputtering gas, which is introduced near the target. Oxygen plasma is generated by supplying a microwave of 2.45 GHz to the resonance cavity in the ECR plasma source. A magnetic field of 875 $\sim$ 900 Gauss is applied to the resonance cavity by a magnetic coil surrounding the cavity. The oxygen plasma is flowed toward the target and the substrate along a divergent magnetic field formed by the magnetic coil.

In the magnetron sputtering, a magnetic field on the target surface is formed by permanent magnets behind and beside the target. A DC electric field is applied to the target, crossing the magnetic field on the target surface.

A magnetic field generated by the magnetic coil in the ECR plasma source forms a mirror and a cusp field by the permanents magnets beside and behind the target, respectively. These fields control the plasma stream near the target. Figures 2 (a) and (b) show the views of the plasma for only the magnetron sputtering and the ECR plasma-assisted sputtering, respectively. The plasma in Fig.2 (a) is generated only near the target surface, while the plasma stream along the magnetic flux is observed in Fig.2 (b). A magnetic flux formed by the magnetic coil in the ECR plasma source is compressed to the target surface by the cusp field, leading to enhancement of the deposition rate due to confinement of plasma in the region near the target surface. On the other hand, the plasma between the target and the ECR plasma source is confined by the mirror field. As a result, the substrate is in the plasma-free region, and the radiation damages from ion bombardment to the film reduce satisfactorily.

The sputtering conditions are listed in Table 1. The $YBa_2Cu_3O_{7-y}$ films were deposited on a (100) $SrTiO_3$ substrate at substrate temperatures of 750 ℃, using a sintered planer target (Y:Ba:Cu)=(1:2:3) of 75mm in diameter. The total gas pressure (P_g) was changed from 1 to 100 mTorr ($Ar+50\%O_2$). The microwave power (P_m) was changed from 0 to 340 W at the constant sputtering power of 60 W. The crystal orientations and phases of deposited $YBa_2Cu_3O_{7-y}$ films were checked by the x-ray diffraction (XRD) method using CuKα radiation. Electrical resistivity measurements were made using a standard four-probe DC method.

In order to investigate the relationship between the plasma species generated in ECR plasma and the properties of the obtained $YBa_2Cu_3O_{7-y}$ films, the optical emission spectroscopy analysis of the ECR oxygen plasma was made in the wavelength region from 300 to 800 nm [7], [8].

EXPERIMENTAL RESULTS AND DISCUSSIONS

Figure 3 shows the typical optical emission spectrum from the ECR oxygen plasma, generated at P_m = 400 W and P_g = 5 mTorr ($O_2$100%). In this spectrum, the optical emissions due to the transitions in O_2^+ molecular ions (525,

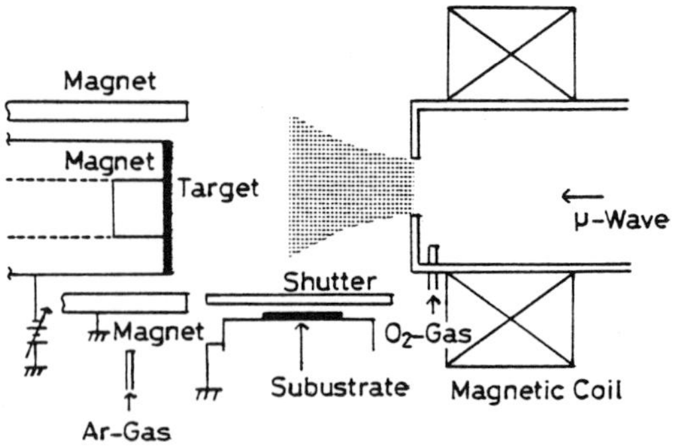

Fig.1 Schematic diagram of the ECR plasma-assisted sputtering system.

(a)

(b)

Fig.2 Views of the plasma for (a) only the magnetron sputtering. (b) ECR plasma-assisted sputtering

Table 1. Sputtering conditions.

Target	Y:Ba:Cu=1:2:3
Sputtering Gas	$Ar+50\%O_2$
Gas Pressure	1∿100mTorr
Sputtering Power	60W
Microwave Power	0∿340W (P_m)
Substrate	$SrTiO_3$(100)
Sub.Temperature	750℃

559, 597, 635 nm) and O^* atomic radicals (616, 777 nm) are observed. This result indicates that ECR oxygen plasma contains a lot of the highly activated particles of oxygen, which are useful for the formation of $YBa_2Cu_3O_{7-y}$ films.

Figure 4 shows the microwave power dependences of the emission intensities from O_2^+ molecular ions (559 nm) and O^* atomic radicals (777 nm) at P_g = 5 mTorr. The emission intensities from these activated particles increased monotonously with increasing the microwave power. From the result, it was found that the activated particles of oxygen in the ECR plasma increased with increasing the microwave power.

Figure 5 shows the typical x-ray diffraction pattern of a $YBa_2Cu_3O_{7-y}$ film on a (100) $SrTiO_3$ substrate prepared by the ECR plasma-assisted sputtering. The film exhibited (001) orientation of a $YBa_2Cu_3O_{7-y}$ orthorhombic phase perpendicular to the substrate surface, and no other phase was included.

Structural deformation derived from oxygen vacancies, which is strongly influenced on the superconducting properties of $YBa_2Cu_3O_{7-y}$, can be evaluated by the length of the c-axis. Figure 6 shows the dependences of the lattice constants c of $YBa_2Cu_3O_{7-y}$ films on the total gas pressure when the films were prepared by only the magnetron sputtering and the ECR plasma-assisted sputtering. The lattice constant c decreased monotonously with increasing the total gas pressure. For the $YBa_2Cu_3O_{7-y}$ films prepared by only the magnetron sputtering (P_m = 0 W), the lattice constants $c \geqq 11.735$ Å, indicating that the

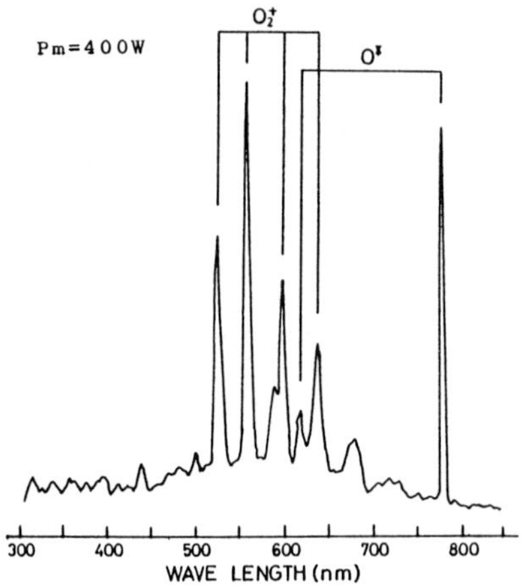

Fig.3 Typical optical emission spectrum from O_2 ECR plasma.

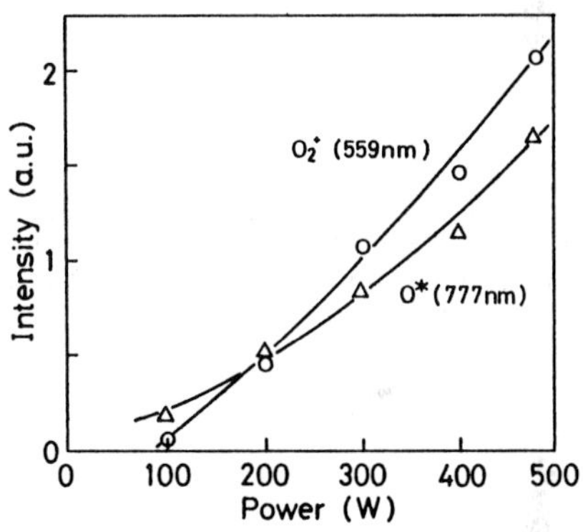

Fig.4 Microwave power dependences of the emission intensities from O_2^+ (559nm) ions, and O^* (777nm) radicals.

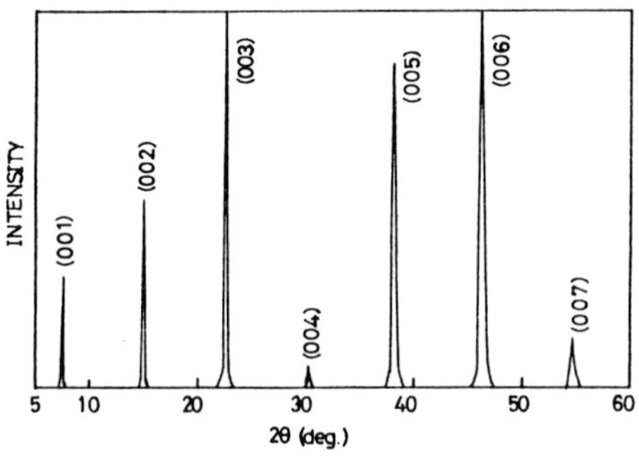

Fig.5 Typical x-ray diffraction pattern of a $YBa_2Cu_3O_{7-y}$ film.

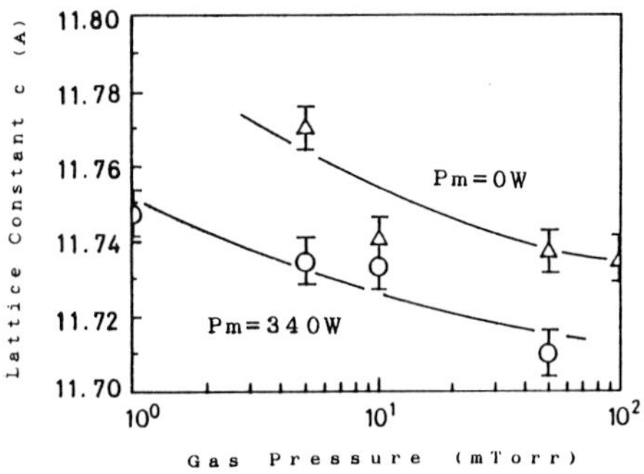

Fig.6 Dependences of the lattice constants c of the $YBa_2Cu_3O_{7-y}$ films on the total gas pressure with and without using the ECR plasma source.

oxygen content in $YBa_2Cu_3O_{7-y}$ films was poor. On the other hand, when the microwave power was P_m = 340 W using the ECR plasma source, the lattice constants *c* were shorter than that of the films deposited without using the ECR plasma source. The length of the *c*-axis of the films, which is related to the oxygen content in the films, depends on the amount of activated particles of oxygen such as O_2^+ molecular ions and O^* atomic radicals in the ECR plasma during deposition of the films, as shown in Fig.3. Consequently, these activated particles in ECR plasma act as an important role in oxidation of $YBa_2Cu_3O_{7-y}$ films.

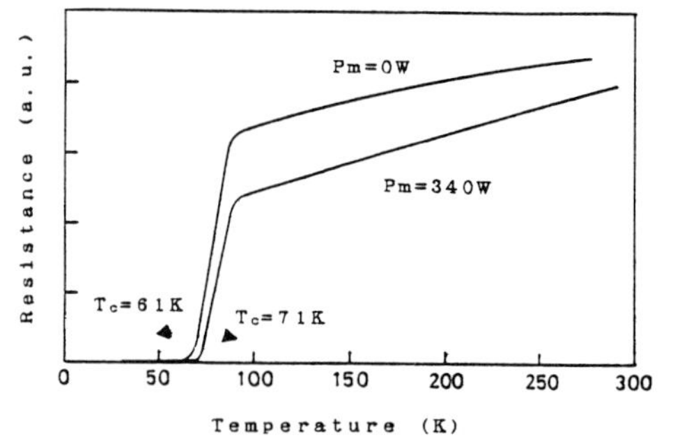

Fig.7 Temperature dependences of the resistance of the $YBa_2Cu_3O_{7-y}$ films deposited with and without using the ECR plasma source.

Figure 7 shows the resistance temperature (R-T) characteristics of the films deposited by only the magnetron sputtering (P_m = 0 W) and the ECR plasma-assisted sputtering (P_m = 340 W) at P_g = 50 mTorr. The film prepared by only the magnetron sputtering exhibited a superconducting transition at T_{cend} = 61 K. On the other hand, a superconducting transition temperature of the film prepared by the ECR plasma-assisted sputtering was T_{cend} = 71 K, which is higher than the transition temperature for the film prepared by only the magnetron sputtering. This result corresponds to the shorter length of *c*-axis of the films, and demonstrates that the ECR plasma-assisted sputtering is more suitable to obtain high-T_c superconducting $YBa_2Cu_3O_{7-y}$ films.

CONCLUSION

We have prepared the $YBa_2Cu_3O_{7-y}$ films using the ECR plasma-assisted sputtering deposition technique. The films prepared using the ECR plasma-assisted sputtering exhibited a shorter lattice constant and a higher superconducting temperature T_c than those prepared using only the magnetron sputtering. The relationship between the properties of the films and the plasma species in the ECR oxygen plasma was investigated by the optical emission spectroscopy analysis. From this result, it was found that the O_2^+ molecular ions and O^* atomic radicals play an important role in oxidation of $YBa_2Cu_3O_{7-y}$ films, which leads to improvement of the superconducting properties of $YBa_2Cu_3O_{7-y}$ films.

REFERENCES

[1] Matsuo S, Adachi Y (1982) Jpn.J.Appl.Phys. 21: L4

[2] Kitagawa M, Setune K, Manabe Y, Hirai T (1988) Jpn.J.Appl.Phys. 27: 2026

[3] Matsuoka M, Ono K (1988) J.Vac.Sci.Technol. A 6: 25

[4] Matsuoka M, Ono K (1989) J.Vac.Sci.Technol. A 7: 2975

[5] Goto T, Masumoto H, Hirai T (1988) Jpn.J.Appl.Phys. 28: L88

[6] Fujii T, Anno T, Koyanagi T, Hirai H, Matsubara K Proc.Symp. on ISIAT'90. (Tokyo): 241

[7] Miyake K, Kimura S, Warabisako T, Sunami H, Tokuyama T (1984) J.Vac.Sci.Technol. A 2: 496

[8] Ying QY, Shaw DT, Kwok HS (1988) Appl.Phys.Lett. 53: 1762

Y-Ba-Cu-O Films Grown on Flexible Polycrystalline Substrates by Excimer Laser Ablation

NORIYUKI YOSHIDA[1], MARI KUBOTA[1], SHIGERU OKUDA[1], SATOSHI TAKANO[1], MASAYUKI NAGATA[1], TSUKUSHI HARA[2], KIYOSHI OKANIWA[2], and TAKAHIKO YAMAMOTO[2]

[1] Osaka Research Laboratories, Sumitomo Electric Industries, 1-3, Shimaya 1-chome, Konohana-ku, Osaka, 554 Japan

[2] Engineering Research Center, Tokyo Electric Power Company, 4-1, Nishi-Tsutsujigaoka 2-chome, Chofu, Tokyo, 182 Japan

ABSTRACT

We report the results of a study on as-grown Y-Ba-Cu-O films deposited on flexible and random oriented polycrystalline yittria-stabilized zirconia substrates using excimer laser ablation technique. Zero resistance Tc's of 88-90 K are reproduciblly obtained. The best Jc(77.3K,0Tesla) is $2.9x10^4$ A/cm^2, which is one of the highest reported for superconducting films on any polycrystalline substrate. Films are c-axis oriented, while the in-plane orientation has been found to be random. This random orientation in the a-b plane is thought to be related to the existence of large angle grain boundaries which block the super-current. We have also examined the flexibility of our samples. Jc value is never degraded up to the compressive strain of 0.2%.

KEY WORDS: Y-Ba-Cu-O film, laser ablation, polycrystalline substrate, bending strain

INTRODUCTION

Recently laser ablation technique has been extensively used to fabricate superconducting films of cuprous oxide ceramics. Many researchers reported high current density of films on single crystalline substrates grown in-situ by this process[1],[2],[3]. We recently have shown that as-grown Y-Ba-Cu-O films with Jc of several million amperes per square centimeter can be routinely obtained on single crystalline MgO[4], even at the rate of $3.7\mu m$/min[5] by excimer laser ablation. This process seems suitable for fabricating a high current superconductor which will be constructed as a superconducting film deposited on a long flexible tape. Some groups reported the growth of Y-Ba-Cu-O films with Jc at 77.3K of around $1x10^4 A/cm^2$ on randomly oriented polycrystalline substrates[6],[7]. In this paper we report the results of a study on as-grown Y-Ba-Cu-O films deposited on flexible and random polycrystalline yittria-stabilized zirconia(poly-YSZ) substrates(0.1-0.2mm thick).

EXPERIMENTAL

The experimental arrangement have been previously described[4]. Briefly, a rotating target was irradiated by a focused KrF excimer laser with 248nm output operated at several to several tens pps. During the laser deposition a mirror-polished poly-YSZ substrate(10x20mm) was heated at 700-750℃ in a few hundred mTorr oxygen ambient. After the deposition the chamber was backfilled with oxygen up to 1 atm and the sample was then slowly cooled down. Typical deposition rate was approximately 1Å/pulse. The film thickness was between $0.5\text{-}1.5\mu m$.

Structural features of the film were analysed using scanning electron microscopy(SEM), transmission electron microscopy(TEM) and X-ray diffraction measurement. Especially X-ray pole figure measurements were employed for determining the in-plane orientation of the film. For comparison the film on single crystalline MgO[MgO(100)] was also characterized in the same methods. The conventional four probe technique was used for determining the zero resistance critical temperature(Tcend) and the critical current density(Jc) at 77.3K. The Jc was defined by the onset voltage of $1\mu V$ across a bridge(0.1x0.2mm).

The effects of bending strain on the superconducting properties[8] have also been determined. These measurements were made by the same four probe technique under condition such as the Y-Ba-Cu-O film was bent convexly (tensile strain) and concavely (compressive strain) on a round copper block. Several blocks with different radius R were used for changing the amount of strain ε applied to the Y-Ba-Cu-O film. ε is approximated as $t_s/2R$, where t_s is a thickness of the poly-YSZ substrate.

RESULTS AND DISCUSSION

X-ray diffraction pattern of a Y-Ba-Cu-O film on poly-YSZ substrate is shown in Fig.1. This denotes that the film is highly oriented with its c-axis perpendicular to the substrate. This high orientation has been found to be possible only in the case that mirror-polished poly-YSZ substrates are used. This indicates that each Y-Ba-Cu-O grains can be nucleated so that its a-b plane parallels the facet of the substrate surface. Such a nucleation originates possibly in anisotropy of the growth rate and the surface free energy for the grain. The Bragg peaks corresponding to the (00l)[l=1-13] are as sharp as those of a film on MgO(100), indicating good crystallinity of the grains of Y-Ba-Cu-O film. The c-axis lattice parameter calculated from the Bragg angle of (0010) reflection is 11.68Å which is expected for the fully oxygenated Y-Ba-Cu-O.

Fig.2 shows the magnetic field dependences of Jc at 77.3K for a film on poly-YSZ as well as on MgO(100) substrate. The best Jc(77.3K,0Tesla) on poly-YSZ is $2.9\times10^4 A/cm^2$ which is among the highest Jc values reported for high-Tc film on any polycrystalline substrate. Tcend's of 88-90K are reproducibly obtained. In the figure, stronger dependence of Jc on low magnetic field(<0.1Tesla) than that of the film on MgO(100) is observed, indicative of weak-link behavior of grain boundaries[9],[10]. Fig.3 is a comparative result of surface morphology of the films with two diferent thicknesses. The thicker film[Fig.3(a)] with higher Jc has been found to have a densely packed morphology. Narrowing intergrain structures, which are visible for the thinner film in Fig.3(b), seems to reduce an area of the super-current path section, resulting in the lower Jc.

Although the densely packing morphology is considered one of the important factors to realize a high current density yet the obtained Jc for the film on poly-YSZ is still two orders magnitude lower than that on MgO(100). This substantial difference in Jc's seems related to some intergrain structures of these two kinds of films. SEM photographs in Fig.4 demonstrate

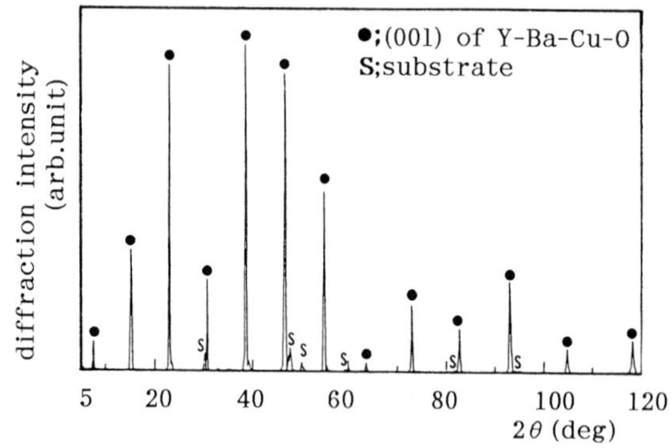

Fig.1 X-ray diffraction pattern of a Y-Ba-Cu-O film on poly YSZ substrate.

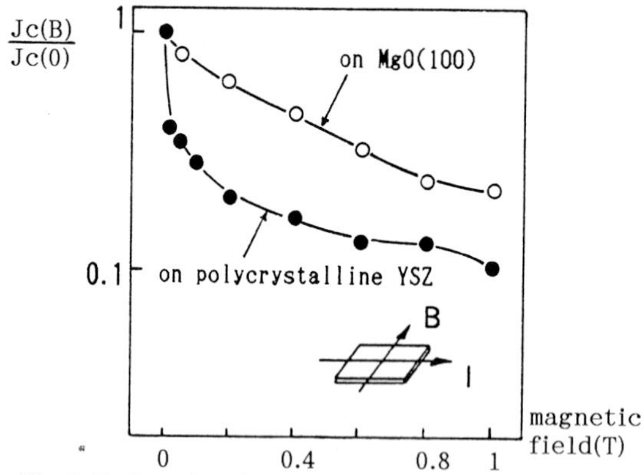

Fig.2 Reduced critical current density vs magnetic field for Y-Ba-Cu-O films grown on poly-YSZ and on MgO(100).

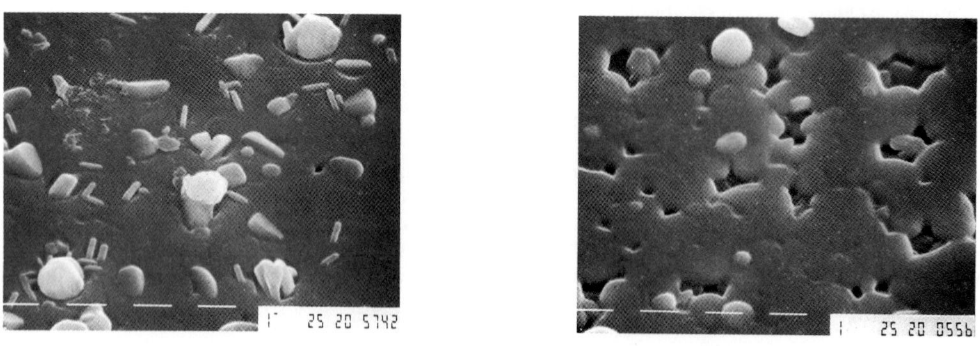

(a) 1μm (b) 1μm

Fig.3 SEM micrographs of the surface of Y-Ba-Cu-O film grown on poly-YSZ.
(a): thickness;1.2μm, $Jc=2.4\times10^4 A/cm^2$ (b): thickness;0.7μm, $Jc=1.6\times10^3 A/cm^2$

the initial nucleation behavior of the film growth on poly-YSZ and on MgO(100). Thicknesses are both around 500Å, correcponding to 480 laser pulses. The grains for the film on poly-YSZ in Fig.4(a) composed of sub-micron crystallites seem oriented randomly. In contrast, all the grains roughly square-shaped on MgO(100)[Fig.4(b)] appear to be aligned in the a-b plane. A quantitative evaluation of the in-plane alignment can be done using X-ray pole figure analysis. Figs.5(a) and 5(b) show the azimuthal scans of (103) poles of Y-Ba-Cu-O films on poly-YSZ and on MgO(100), respectively. The Schulz geometry was employed in this measurement. The azimuthal angle β was defined as a rotating angle of the sample in its own plane around its normal i.e. c-axis of the film. The sample was tilted so that (103) plane of Y-Ba-Cu-O reflects the X-ray in the diffractometer plane. θ-2θ resolution in this technique did not allow to separate the contribution of (103) and (013) poles into signal. Azimuthal distibution of the pole intensity can be interpreted as a statistic of in-plane orientation for grains composing the film. The plot in fig.5(a) for the film on poly-YSZ has no regularity and denotes that the grains are randomly oriented in the a-b plane, whereas the four peaks located at every 90° of β in Fig.5(b) mean the almost perfect in-plane alignment for the film on MgO(100). The full width at half maximum for one of these peaks is estimated to be 5°. A TEM observation reported elsewhere[11] demonstrated that the in-plane structure of the film on MgO(100) has entirely orthogonal twin boundaries, consisting with the result of Fig.5(b).

It is concluded from the result of the pole figure analysis that grain boundaries exisiting in the film on poly-YSZ are formed by adjacent grains with various in-plane orientations. Figs.6(a) and 6(b) are TEM photographs showing contrastive two types of grain boundaries of the film on poly-YSZ. A grain boundaries whose adjacent grains meet with a large angle[approximately 45° in Fig.6(a)] has found to be relatively thick(several to several tens nm) and continuous. Fig.6(b) shows a small angle(less than a few degree) grain boundary. It looks like an array of descrete defects. The short coherence length for Y-Ba-Cu-O, which is reported to be 1.6 nm along the a,b axis[12], prevents super-current from going across the large angle grain boundary. Openings between the descrete defects in Fig.6(b) have a possibility to be a free path for super-current.

Lastly we will refer to the flexibility of our sample. As well known, sintered zirconia sheet with a thickness of around 0.1-0.2mm can be easily bent elastically. The bending strain dependences of Jc at 77.3K are plotted in Fig.7. Another abscissa represents the bending radius for the thickness of 0.2 mm. In the case of tensile bending, the Jc of 1.9×10^4 A/cm^2 at 77.3K is gradually degraded to around 4×10^3 A/cm^2 as the strain was increased up to 0.4%. On the other hand, no degradation in Jc is observed up to the compressive strain of 0.2%. The strain

0.3 μm

Fig.4 SEM micrographs of the Y-Ba-Cu-O films at the initial stage of film formation. (a):grown on poly-YSZ. (b):grown on MgO(100).

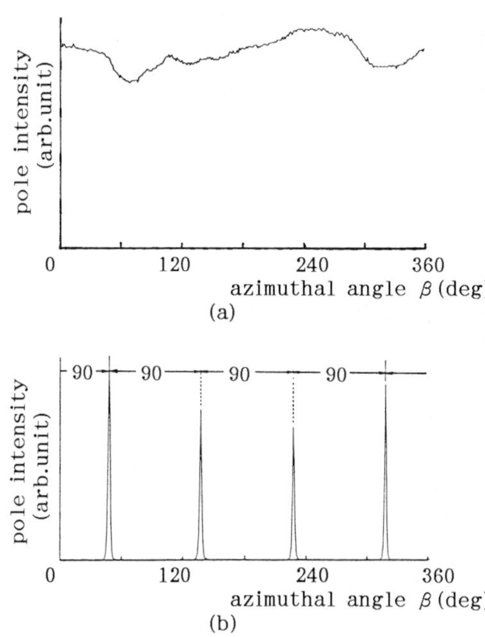

Fig.5 The azimuthal scans of (103) pole of Y-Ba-Cu-O films. (a):grown on poly-YSZ. (b):grown on MgO(100).

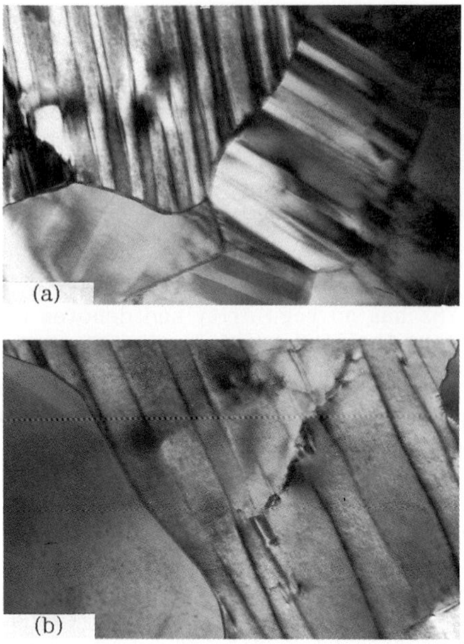
0.1 μm

Fig.6 TEM photographs showing two types of grain boundaries of the films on poly-YSZ.(a):large angle grain boundary.(b):small angle grain boundary

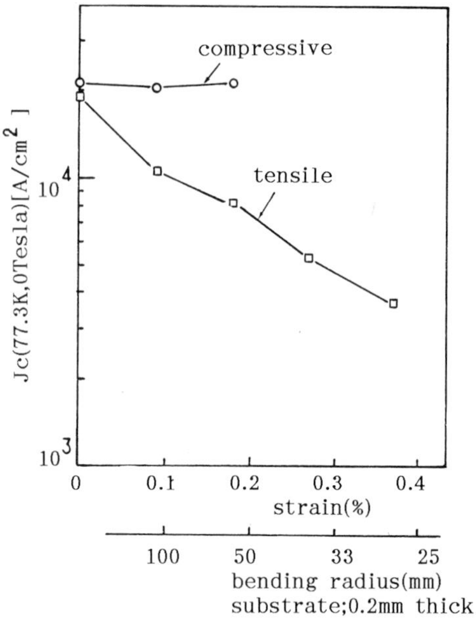

Fig.7 The effects of bending strain on the Jc for film on poly-YSZ.

of 0.2% for compressive bending is equivalent to the bending radius of 50mm for our 0.2mm thick sample. This shows a possibility of flexible Y-Ba-Cu-O superconducting tape for practical use.

In summary, we have successfully deposited Y-Ba-Cu-O films on randomly oriented polycrystalline YSZ substrates using excimer laser ablation technique. Jc at 77.3K of 2.9x10^4 A/cm^2 is obtained. This value for Jc is one of the highest reported for superconducting films on any polycrystalline substrate. Two orders magnitude difference of the Jc's between of the film on poly-YSZ and of that on MgO(100) may possibly be related to the in-plane alignment and the structure of grain boundaries of the films. To improve further the Jc value of the film on poly-YSZ, some locking mechanism to align the grains each other may be necessary. We have also examined the effects of bending strain on the superconducting properties. In the case of compressive bending, the Jc is never degraded up to the strain of 0.2%, corresponding to the bending radius of 50mm for our 0.2mm thick sample.

REFERENCES

[1] R.Roas,L.Schulz and G.Endres: Appl. Phys. Lett.,**53**, (1988) p1557.
[2] T.Venkatesan,X.D.Wu,B.Dutta,A.Inam,M.S.Hedge,D.M.Hwang,C.C.Chang,L.Nazar and B.Wilkens: Appl. Phys. Lett.,**54**, (1989) p581.
[3] R.K.Singh,J.Narayan,A.k.Singh and J.Krishnaswamy: Appl. Phys. Lett.,**52**, (1989) p2171.
[4] N.Yoshida,M.Kubota,S.Takano,K.Sato,H.Hitotsuyanagi,M.Kawashima,T.Hara,K.Okaniwa and T.Yamamoto: in Proc. the 2nd Int. Symp. on Superconductivity(ISS'89), Tsukuba, 1989, (Springer-Verlag, Tokyo, 1990) p853.
[5] S.Takano,N.Yoshida,S.Okuda,M.Nagata,T.Hara,K.Okaniwa and T.Yamamoto: in Proc. the 2nd ISTEC Workshop on Superconductivity, Kagoshima, 1990, p17.
[6] D.P.Norton,D.H.Lowndes,J.D.Budai,D.K.Christen,E.C.Jones,K.W,Lay and J.E.Tkaczyk: Appl. Phys. Lett.,**57**, (1990) p1164.
[7] E.Narumi,L.W.Song,F.Yang,S.Patel,Y.H.Kao and D.T.Shaw: Appl. Phys. Lett.,**56**, (1990) p2684.
[8] K.Rokudo, J.Yamaguchi,H.Miyazaki,T.Mitsui,H.Hitotsuyanagi and N.Yoshida: in Proc. MRS Int. Meeting on Advanced Materials, **vol.6**, Tokyo,1988, (Mat.Res.Soc.,Pittsburg,1988) p891.
[9] P.Chaudhari,J.Mannhart,D.Dimos,C.C.Tsei,J.Chi,M.M.Oprysko and M.Scheuermann: Phys. Rev. Lett.,**60**, (1988) p1653.
[10] J.Manhart,P.Chaudhari,D.Dimos,C.C.Tsei and T.R.McGuire: Phys. Rev. Lett.,**61**, (1988) p2476.
[11] S.Okuda,N.Yoshida,M.Kubota,S.Takano,K.Sato,H.Hitotsuyanagi,M.Kawashima,T.Hara,K.Okaniwa and T.Yamamoto: in Proc. ICMC Topical Conf.,Garmisch-Partenkirchen,1990, (in press).
[12] U.Welp,W.K.Kwok,G.W.Crabtree,K.G.Vandervoort and J.Z.Liu: Phys. Rev. Lett.,**62**, (1989) p1908.

Superconducting Properties of YBCO Prepared by Laser Ablation

K. ONABE[1], Y. IIJIMA[1], N. FUTAKI[1], N. Sadakata[1], O. KOHNO[1], and Y. IKENO[2]

[1] Materials Research Laboratory, Fujikura Ltd., 5-1, Kiba 1-chome, Koto-ku, Tokyo, 135 Japan
[2] Super-GM, 14-10, Nishitenma 5-chome, Kita-ku, Osaka, 530 Japan

ABSTRACT

We prepared YBCO superconducting compound with J_c=4.4x10^{6}A/cm^2(T_c=89.5K) at 77K, 0T on $SrTiO_3$(100) substrate by Ar-F excimer laser ablation method. We investigated the magnetic field(B) dependence of J_c and the anisotropy of J_c as a function of the angle theta(θ) between B and the sample normal for these high J_c samples. Also in the case of depositing on metallic substrate, hastelloy, with YSZ thin buffer layer, YBCO with J_c=1.4x10^{4}A/cm^2(T_c=89.5K) at 77K, 0T was obtained, and it showed strong c-axis grain alignment. We recognized that the width of XRD rocking curve (FWHM, full width at half maximum) of the YBCO(005) diffraction peak on hastelloy was 2.3 degrees and that on $SrTiO_3$ was 0.60 degrees, which represented highly oriented YBCO compound on metallic substrate.

KEY WORDS: YBCO, high J_c, magnetic field, anisotropy, hastelloy

INTRODUCTION

Considering the long wire fabrication of oxide superconducting compound, it is necessary to deposit on the substrate that is easy to process and flexible like metallic substrate, hastelloy, stainless steel, etc. Laser ablation technique offers several desirable characteristics for the conductor fabrication, including rapid deposition rate and simple target requirement. Up to now, several reports using this technique have been published about the preparation on single-crystal and metallic substrate[1-4] to investigate the magnetic field(B) dependence of J_c and the angular dependence of J_c as a function of the angle θ between B and sample surface[5].

We prepared YBCO superconducting compound on $SrTiO_3$(100) by laser ablation method to investigate the J_c properties, such as the magnetic field dependence of J_c and the anisotropy of J_c as a function of the angle θ between B and sample normal. And also we prepared on metallic substrate, hastelloy, in the same condition on $SrTiO_3$ single-crystalline substrate[6].

EXPERIMENTAL

193nm Ar-F excimer laser was employed to deposit the superconducting compound. Laser light was controlled at 80-100mJ of the pulse energy and at 5Hz of the repetition rate. It was introduced into the vacuum chamber by the two reflection mirrors and through a condensing lens and a quartz glass window, and was focused at the surface of target at the angle of 45 degrees from the surface. Substrate was placed parallel to the target surface and depositing condition was as follows, target-to-substrate distance; 35-45mm, substrate surface temperature; 700-760°C, oxygen pressure in deposition; 200mtorr. The target was prepared by standard sintering techniques and the composition was $Y_{1.0}Ba_{2.0}Cu_{3.0}O_x$.

Crystal structure of the samples were investigated by XRD and critical currents were measured by the standard four-probe method using 1 μV/cm criterion after etched the pattern of 0.4-0.5μm in thickness, 150-250μm in width and 500-1000μm in length. An anisotropy of J_c was measured by tilting the sample from θ=0 degree(c // B ⊥ i) to θ=90 degree(c ⊥ B ⊥ i) at the applied magnetic field of 1T and 3T.

RESULTS AND DISCUSSION

Prepared YBCO samples showed constantly T_c=89-90K and J_c>10^6A/cm^2(77K,0T). The deposition rate of the samples was approximately 2 μm per hour. Resistivities of these samples were few hundreds $\mu\Omega\cdot$cm at 100K, and it showed high quality superconducting compounds have been obtained. Figure 1 shows the typical T_c curve of the YBCO. Figure 2 shows XRD pattern(theta-2theta) of the sample(a) which had the J_c value of 4.4x10^6A/cm^2(77K,0T). Only YBCO(00L) diffraction peaks were recognized and other phases were not observed. Width of the XRD rocking curve(FWHM, full width at half maximum) of YBCO(005) diffraction peak of this sample was 0.6 degrees, so that it showed well c-axis grain alignment.
Figure 3 shows the result of the magnetic field dependence of J_c(J_c-B) of the samples(b),(c) with 2x10^6A/cm^2(77K,0T), and figure 4 shows the result calculated from the data of the magnetic field dependence of pinning force(F_p-B), where F_p was calculated as J_cxB. From figure 3 it was recognized the decrease of J_c of sample(c) was less than of sample(b) with increasing magnetic field. And as shown in the figure 4, if the magnetic field was applied normal to the sample surface(c // B ⊥ i), the maximum value of the F_p was near 1T for sample(b) comparing with 2T for sample(c). Figure 5 shows the anisotropy of J_c as a function of the angle θ between B(1T,3T) and the sample normal(J_c-θ) for the samples. It was recognized an anisotropy for the direction of magnetic field each samples, too. In this case, J_{cn}/J_{cp} values of these samples were 0.5 at 1T and 0.2 at 3T, where J_{cp} and J_{cn} represent the Jc at θ =90 degrees and θ =0 degree, respectively. The sample(b) showed the another small peak near θ=20 degrees, that was different from the results presented by Tachiki[7] and Roas[5] and others.

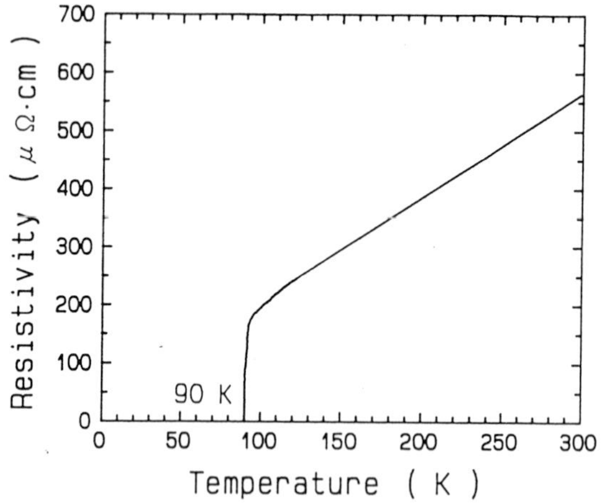

Fig. 1 ρ-T curve of the YBCO prepared by laser ablation.

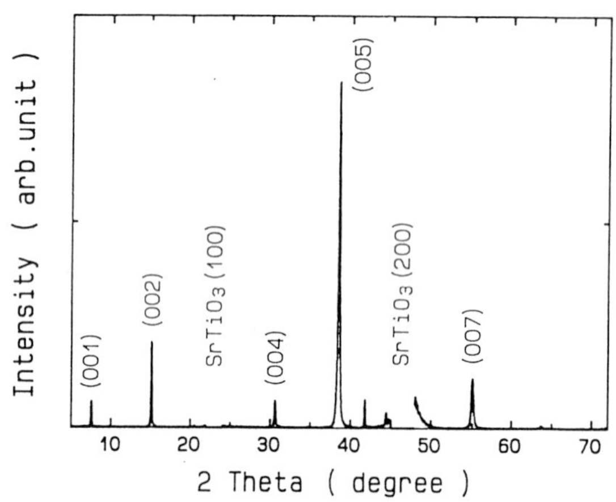

Fig. 2 Typical XRD pattern of the YBCO deposited on $SrTiO_3$(100).

In case of the YBCO prepared on metallic substrate, hastelloy, the reaction between YBCO and hastelloy at the interface degrades the superconductivity. To prevent this reaction, YSZ thin buffer layer with 0.5μm in thickness on polished hastelloy prepared by RF-magnetron sputtering method was formed before depositing YBCO. The sample(d) which showed T_c=89.5K, J_c=1.4x10^4cm^2(77K,0T) was obtained on the metallic substrate described above. Figure 6 shows XRD pattern of the sample(d). Similar to the sample(a) prepared on $SrTiO_3$(100), only YBCO(00L) peaks and YSZ peaks were observed, and it showed the strong c-axis grain alignment. FWHM of the (005)peak of this sample was 2.3 degrees, and it was slightly broader than that deposited on

$SrTiO_3$. It is considered that the condition of the crystal and the smoothness of the surface of YSZ thin buffer layer was different from $SrTiO_3$ single crystalline substrate.

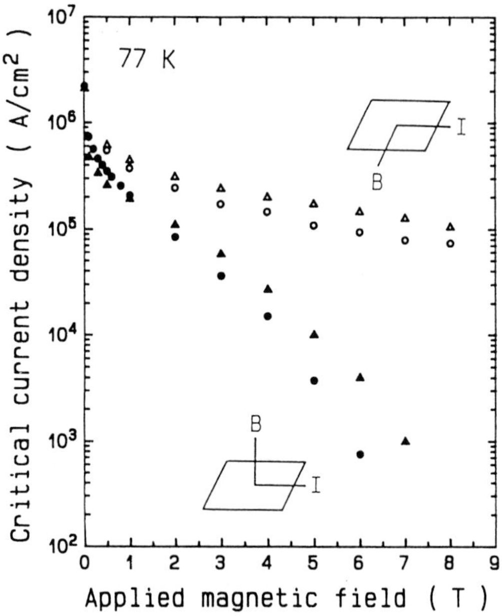

Fig. 3 Jc-B properties of the sample(b)-○● and (c)- △▲ at 77K.

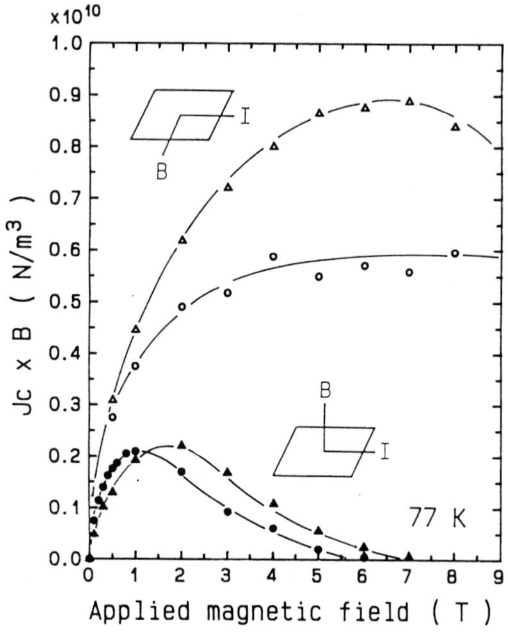

Fig. 4 Fp-B properties of the sample(b)-○● and (c)- △▲ at 77K.

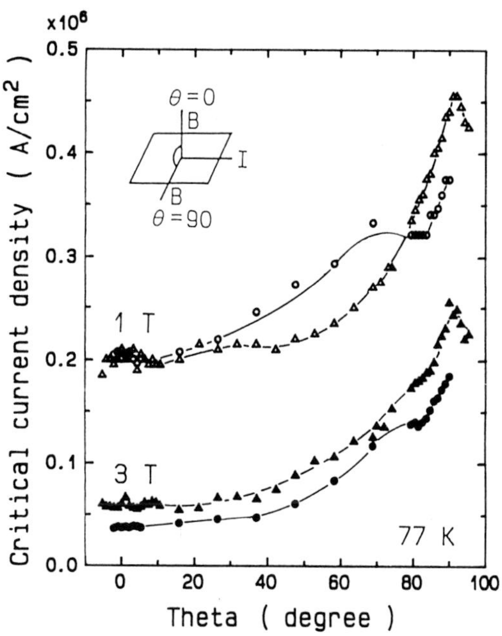

Fig. 5 Jc- θ properties of the sample(b)-○● and (c)- △▲ at 77K.

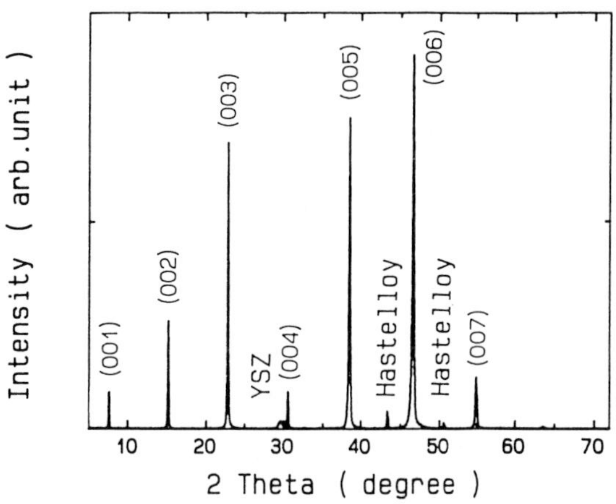

Fig. 6 XRD pattern of the YBCO deposited on hastelloy with YSZ thin buffer layer by laser ablation.

CONCLUSION

We prepared YBCO superconducting compound by laser ablation technique. Sample with T_c=89.5K and J_c=4.4x10^{6}A/cm^2(77K,0T) on $SrTiO_3$(100) single crystalline substrate was obtained. J_c-B and J_c-θ properties of the high J_c samples were investigated. Also in the case of hastelloy substrate, the YBCO with T_c=89.5K, J_c=1.4x10^{4}A/cm^2(77K,0T) was obtained. FWHM of the (005)peak of this sample was 2.3 degrees, and it was slightly broader than that deposited on $SrTiO_3$ single crystal. By improving width of the FWHM of the YBCO on the metallic substrate, the surface morphology of the YSZ buffer layer and that of metallic substrate, higher J_c can be achieved on flexible metallic substrate.

ACKNOWLEDGMENT

This work was performed as a part of " R & D on Superconducting Technology for Electric Power Apparatuses " under the Moonlight Project of Agency of Industrial Science and Technology, MITI, being consigned by New Energy and Industrial Technology Development Organization (NEDO).

REFERENCES

1.Roas B, Schultz L, and Endres G (1989) J. Less-Common Metals **151**: 413.

2.Narumi E, Song LW, Yang F, Patel S, Kao YH, and Shaw DT (1990) Appl. Phys. Lett. **56**: 2684.

3.Russo RE, Reade RP, McMillan JM, and Olsen BL (1990) J. Appl. Phys. **68**: 1354.

4.Saitoh J, Fukutomi M, Tanaka Y, Asano T, Maeda H, and Takahara H (1990) Jpn. J. Appl. Phys. **29**: L1117.

5.Roas B, Schultz L, and Saemann-Ischenko G (1990) Phys. Rev. Lett. **64**: 479.

6.Onabe K, Sadakata N, and Kohno O (1989) In :Ishiguro and Kajimura (eds) in Advances in Superconductivity 2th. Springer-Verlag, Tokyo, pp849-852.

7.Tachiki M, and Takahashi S (1989) Solid State Communications **72**: 1083.

Hetero-Epitaxial Growth of MgO/YBCO Thin Films by Excimer Laser Deposition

MASAKAZU MATSUI[1], KIYOSHI YAMAMOTO[1], MINORU NAKAJIMA[1], KUNIO DOI[2], and TAKESHI KOBAYASHI[3]

[1] Yokohama R&D Laboratories, The Furukawa Electric Co., Ltd., Okano, Nishi-ku, Yokohama, 220 Japan
[2] Super-GM, 14-10, Nishitenma 5-chome, Kita-ku, Osaka, 530 Japan
[3] Department of Electrical Engineering, Faculty of Engineering Science, Osaka University, Toyonaka, Osaka, 560 Japan

ABSTRACT

Hetero-epitaxial thin films of superconducting YBCO and insulating MgO were successfully grown on an MgO(100) substrate by an excimer laser deposition. The growth orientations of YBCO and MgO films on MgO(100) substrates were investigated. The results of cross-sectional transmission electron microscopy(TEM) observations indicated that there was no disruption at the interface of the MgO/YBCO single hetero-epitaxial system with the 2nm-thick MgO layer, but that the interfaces of the YBCO/MgO/YBCO double hetero-epitaxial system were disrupted. At the deposition of the top YBCO layer, the smooth surface of the MgO film was deformed to rugged one, and then the intermediate MgO layer became five times as thick as the original one.

KEYWORDS : $Y_1Ba_2Cu_3O_{7-X}$, MgO, pulsed laser deposition, epitaxial growth, TEM

INTRODUCTION

Many processes have been described [1] for the deposition of a single layer of thin films of the high-temperature superconductor $YBa_2Cu_3O_x$(YBCO). The technology to control the orientation of a YBCO thin film has been developed[2,3]. The next step is to deposit multilayer structures for electronics applications. YBCO/MgO/YBCO[2], YBCO/Y_2O_3/YBCO[4], YBCO/$SrTiO_3$/YBCO[5], YBCO/PrBCO/YBCO[6,7], YBCO/Bi-Sr-Cu-O/YBCO[8], and $Bi_2Sr_2Ca_1Cu_2O_y$/$Bi_2Sr_2Cu_1O_y$/$Bi_2Sr_2Ca_1Cu_2O_y$[9] structures have been proposed and investigated. Recently, these multilayer systems have another task, the study on the flux pinning. Gross et al[10,11] have reported a high critical current density($1.1x10^7$ A/cm^2) of strained multilayer thin films of YBCO/$Nd_{1.83}Ce_{0.17}CuO_x$ grown heteroepitaxially by a laser deposition. Since YBCO has short coherence length, a heteroepitaxial growth of layered structures is required for both electronics applications and the study on the flux pinning. Kobayashi and Iyori[12] reported a successful MgO epitaxial growth on both the (001) and (100) YBCO layers by rf magnetron sputtering. They also reported the YBCO(001) epitaxial growth on the YBCO(001) layer via the MgO intermediate layer by the same method.

In this paper, we describe the process we have developed for producing heteroepitaxial structures of YBCO/MgO/YBCO on MgO(100) substrate by a pulsed laser deposition which, we believe, will be suitable for a variety of applications in superconducting materials at 77K. The purposes of this work are: (1) to clarify the deposition conditions for the growth of (001) oriented YBCO films and (100) oriented MgO films on MgO(100) single crystalline substrates, which are constituent layers of the multilayer system; (2) to present how YBCO/MgO/YBCO structure is grown heteroepitaxially on MgO(100) substrate; (3) to study its microstructure.

EXPERIMENTAL

A $Y_1Ba_2Cu_3O_{7-X}$ thin film was grown on a (100) MgO substrate by a pulsed laser deposition technique. We use a Lambda Physik LPX210F KrF excimer laser with 248nm wavelength, 30ns pulse, a repetition rate of 5Hz and a pulse energy of 200mJ. The laser beam was focused onto a sintered $Y_1Ba_2Cu_3O_x$ pellet in a vacuum chamber. The laser energy density was chosen to obtain the desired 1-2-3 stoichiometric composition in the films. The material, which was evaporated perpendicularly to the target surface, was then deposited on the substrate at a distance of 35-55mm. The substrate heater could be heated from 650C to 850C, controlled by a thermocoupler. Oxygen gas of 20sccm was introduced into the chamber during deposition and the gas pressure was maintained at 50-500mTorr. After the deposition, oxygen was also introduced into the chamber. The films were then cooled down to the room temperature in 1hr in 500Torr oxygen. As substrate materials, we used polished single-crystalline MgO with (100) orientation. The deposition took 15min, corresponding to 4500 laser pulses, after 4min pre-sputtering. With a typical deposition rate of about 0.2A/pulse, the film thicknesses were 1000A. The transport properties were measured using a standard four-probe arrangement.

An MgO thin film was also grown on a (100) MgO substrate by a pulsed laser deposition technique. We used a KrF excimer laser with 248nm wavelength or a CO_2 laser with 10.6um wavelength, and compared an excimer laser deposition with a CO_2 laser deposition. Table 1 listed the deposition parameters of MgO films using excimer and CO_2 lasers. The major difference between deposition parameters of YBCO films and ones of MgO films was to introduce no oxygen into the chamber during deposition of MgO films. Because MgO was sublimationally evaporated from MgO targets.

Table 1. Deposition parameters of MgO films

	Excimer laser depo.	CO_2 laser depo.
Pulse duration (FWHM)	30nsec	9.5msec
Laser energy density	$50 J/cm^2$	$50 J/cm^2$
Pulse energy	0.4J	6.7J
Repetition Rate	5Hz	5Hz
Block Temperature	200-500C	200-500C
O_2 pressure	$< 10^{-5}$ Torr	$< 10^{-5}$ Torr
Target-substrate distance	48mm	48mm
Target	sintered MgO pellet	sintered MgO pellet

MgO/YBCO and YBCO/MgO/YBCO multilayer systems were produced by the ex-situ process. Of course, in-situ process was desired for clean interface of two layers. But our laser deposition system had only one target, so it was necessary to exchange targets after deposition of one layer. No post-annealing was performed. Reflection high energy electron diffraction(RHEED) observations using 20kV electrons were used for crystallographic analysis. Scanning electron microscopy(SEM) was used for morphology observations. Cross-sectional transmission electron microscopy(TEM) with an acceleration voltage of 300 kV was used for microstructure observations of a hetero-epitaxial growth.

RESULTS AND DISCUSSION

YBCO Thin Films

The analysis of the X-ray diffraction spectra of YBCO films grown on MgO(100) substrates showed that the YBCO films grew highly textured in the applied block heater temperature range of 650-850C and in the oxygen pressure range of 50-500mTorr during deposition. It was found that at higher temperatures the films grew with the c-axis perpendicular to the substrate surface and that at lower temperatures the c-axis of the film was in the film plane and the a-axis perpendicular to the substrate. Between the two extreme cases, mixed orientations of the two directions, showing only (001) and (h00) lines, observed. We showed the growth behavior of the films in Fig.1-(a). In the cases of using several deposition techniques, such as electron beam co-deposition and sputtering, only the substrate temperature(Ts) during deposition was the crucial parameter determining the growth orientation of the films[3,4,12]. But in the case of using a pulsed laser deposition, the growth orientation of the films was strongly dependent on the oxygen gas pressures(PO_2) during deposition, as well as Ts. What is the reason for this difference between a pulsed laser deposition and other deposition methods?

The dependence of growth orientations of the films on the target-substrate distances(Lt-s) indicated this answer. Figure 1-(b) showed the growth orientation v.s. Lt-s at 300mTorr oxygen pressures and at 850 C block-heater temperatures during deposition. At the larger Lt-s, the growth orientations changed from pure c-axis direction to a-axis and c-axis mixed directions. From the direct observation of the plasma plume from the targets produced by laser irradiations, it seemed that the dependence of the growth orientation of the films on PO_2 and Lt-s was caused by the relative position of the substrate to the plasma plume. At the lower PO_2, the plasma plume from the target was greater and reached to the substrate. To the contrary, at the high PO_2 it was smaller and didn't reach to the substrate. So at the high PO_2, the growth orientation was similar to the one at the case where Lt-s was large and vice versa. From these discussions, in the case of using a pulsed laser deposition, the growth orientation depended on the PO_2 and Lt-s, as well as the Ts. The transport properties of the YBCO film v.s. Lt-s were also shown in Fig.1-(b). The maximum Tc and Jc values were got at the optimal deposition parameter, Lt-s=45mm, where the growth orientation of the film just changed from pure c-axis direction to a-axis and c-axis mixed directions. The deposition parameter, Lt-s of a pulsed laser deposition was crucial one for crystallinity and superconducting properties of the films.

For the epitaxial growth with the c-axis perpendicular to the substrate, we used the YBCO films produced at such deposition parameters as 300mTorr, 850C, and Lt-s=45mm shown in Fig.1, which had perfectly (001) oriented YBCO films with good superconducting properties.

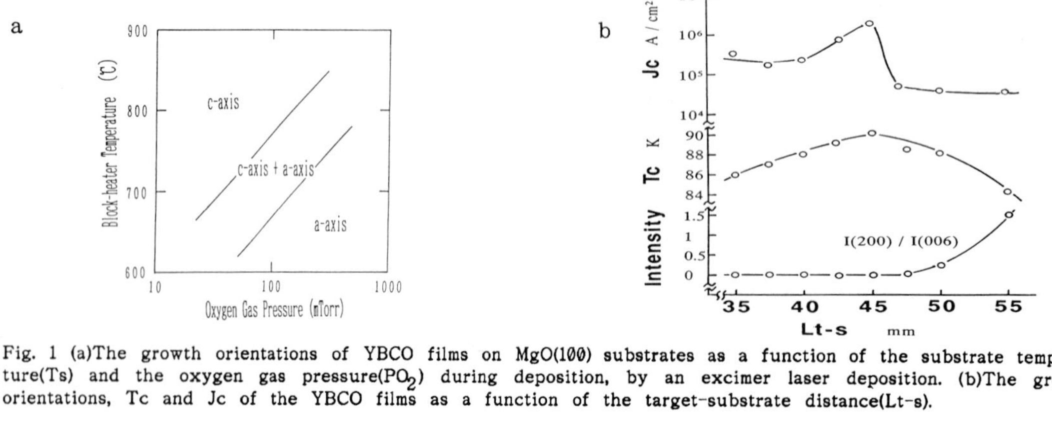

Fig. 1 (a)The growth orientations of YBCO films on MgO(100) substrates as a function of the substrate temperature(Ts) and the oxygen gas pressure(PO_2) during deposition, by an excimer laser deposition. (b)The growth orientations, Tc and Jc of the YBCO films as a function of the target-substrate distance(Lt-s).

MgO Thin Films

In order to produce MgO/YBCO and YBCO/MgO/YBCO structures, we studied to control the growth orientation of MgO films for the (100) epitaxial growth. Here, we investigated temperature dependence of the morphology and crystallinity of MgO films on MgO(100) single crystalline substrates, i.e. deposition parameters of the homo-epitaxial growth of MgO films. The MgO films with 100nm thickness were deposited on MgO(100) substrates at several substrate temperatures using a CO_2 laser and an excimer laser. Figure 2 showed RHEED patterns and SEM photographs of MgO films using a CO_2 laser, and Fig.3 using an excimer laser. The direction of the incident electron beam was [010] relative to the substrate. In the case of using a CO_2 laser, and at the lower substrate temperaturès, there were tetrahedron-shape structures on the surface, and the RHEED patterns were weak. There were also Debye-Scherrer rings diffracted from poly-crystaline structures. In view of the fact that the tetrahedron shape was the facet of MgO, it was found that MgO films grew mainly with the (100) orientation and contained a small part of the (111) growth of poly-crystalline MgO. On the other hand, in the case of using an excimer laser, the surface morphology of the films was very smooth. And at the higher substrate temperature, the RHEED pattern of MgO films was the same as one of the MgO single crystal substrate. Then the range of substrate temperatures of the homo-epitaxial growth of MgO films using an excimer laser was lower than those using a CO_2 laser. For the epitaxial growth with the (100) orientation, we used the MgO films produced at such a deposition parameter as Ts=400C, using an excimer laser, which had perfectly (100) oriented MgO films with smooth surface.

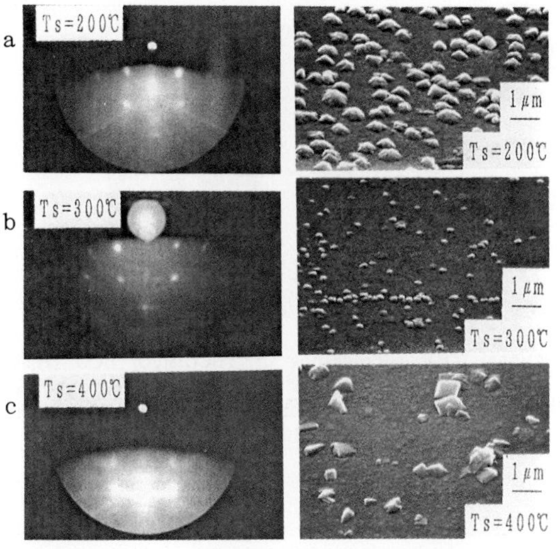

Fig. 2 RHEED patterns and SEM photographs of 100nm MgO films, grown homo-epitaxially at different substrate temperatures, using a CO_2 laser : (a)200C (b)300C (c)400C.

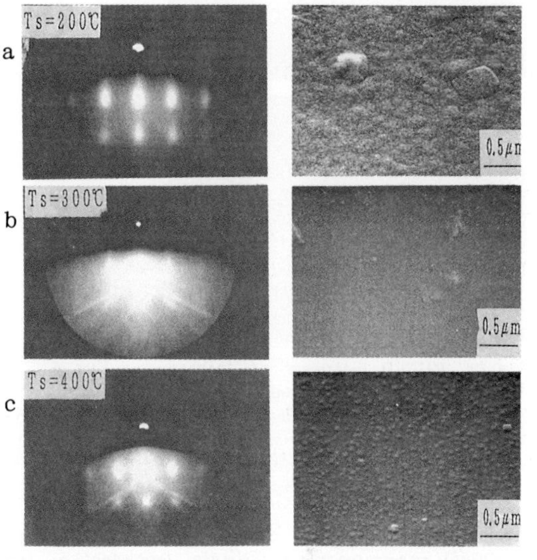

Fig. 3 RHEED patterns and SEM photographs of 100nm MgO films, grown homo-epitaxially at different substrate temperatures, using an excimer laser : (a)200C (b)300C (c)400C.

MgO/YBCO and YBCO/MgO/YBCO Multilayers on MgO(100) Substrates

It was presented that each YBCO and MgO film could be epitaxially grown with the (001) and (100) orientation, respectively. At the next step, we produced MgO/YBCO and YBCO/MgO/YBCO multilayers on MgO(100) substrates. Using RHEED, the crystal structures of grown layers were investigated. Figure 4 showed the typical RHEED patterns. Figures 4-(a),(b) and (c) were the patterns of the base YBCO layer with 100nm thickness, the intermediate MgO layer with 2nm, and the top YBCO layer with 100nm, respectively. The sharp streaks of each layer suggested that the film surfaces were smooth, and highly ordered structures. So, hetero-epitaxial thin films of superconducting YBCO and insulating MgO were successfully grown on an MgO(100) substrate by an excimer laser deposition. To investigate the microstructures of two multilayers, cross-sectional TEM observations were performed. Typical results of MgO/YBCO and YBCO/MgO/YBCO structures were shown in Fig.5 and Fig.6, respectively. Electron diffraction patterns of both cases revealed that the films grew epitaxially with the desired orientations, shown in Fig.5-(d) and Fig.6-(c). The reflection spots of YBCO and MgO were superimposed in the <100> MgO zone axis. High-resolution cross-sectional TEM images indicated that there was no disruption at the interface of the MgO/YBCO single hetero-epitaxial system with the 2nm-thin MgO layer, shown in Fig.5-(a),(b)and(c), but that the interfaces of the YBCO/MgO/YBCO double hetero-epitaxial system were disrupted, shown in Fig.6-(a)and(b). At the deposition of the top YBCO layer, the smooth surface of the MgO film was deformed to rugged one, and then the intermediate MgO layer became five times as thick as the original one. It seemed that this disruption of the interface was caused by the interchange diffusion between an MgO layer and two YBCO layers, at the deposition of the top YBCO layer, when a 2nm-thin MgO layer was sandwiched between the base and top YBCO layer under the high temperature. Actually, after the heat treatment of MgO/YBCO single hètero-epitaxial system under the same condition of the deposition of the top YBCO layer, no disruption was observed at the MgO/YBCO interface. This disruption would affect the properties of electronic devices using hetero-epitaxial structures, such as a Josephson junction. Thus this problem should be solved in order to develop superconducting devices.

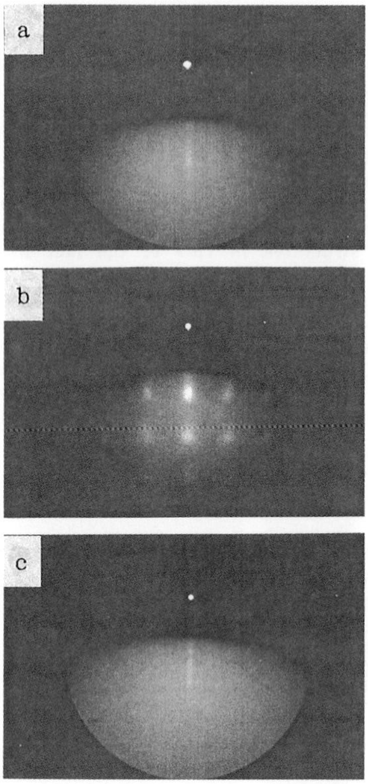

Fig. 4 RHEED patterns of (a)base YBCO layer, (b)intermediate MgO layer and (c)top YBCO layer, in the <100> MgO azimuth.

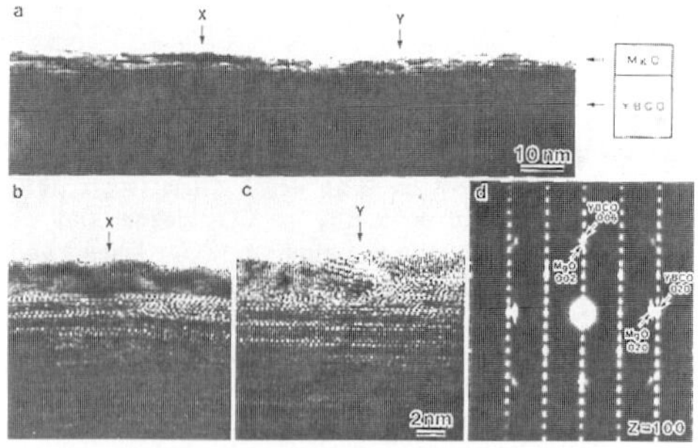

Fig. 5 Cross-sectional TEM micrographs of MgO/YBCO double-layer system ; (a), (b)and(c) high-resolution images obtained at the interface. (d)selected area diffraction pattern(SAD) from the interface. The incident electron beam direction was parallel to $<100>_{YBCO}$.

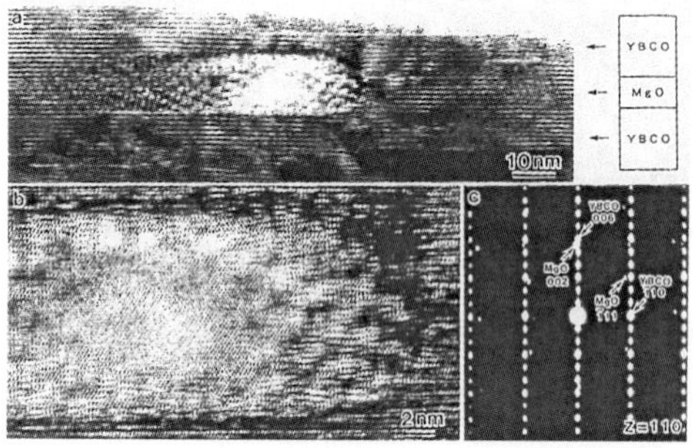

Fig. 6 Cross-sectional TEM micrographs of YBCO/MgO/YBCO trilayer system ; (a)and(b) high-resolution images obtained at the interface. (c)selected area diffraction pattern(SAD) from the interface. The incident electron beam direction was parallel to $<110>_{YBCO}$.

CONCLUSIONS

In summary : (1)In the case of using an excimer laser deposition, the growth orientations of YBCO films on MgO(100) substrates were strongly dependent on the oxygen gas pressures during deposition and target-substrate distances, as well as substrate temperatures. ; (2)The crystallinity of a homo-epitaxial growth of MgO films on MgO(100) substrates using an excimer laser was studied, compared with films using a CO_2 laser. The range of substrate temperatures of the homo-epitaxial growth of MgO films using an excimer laser was lower than that using a CO_2 laser. ; (3)Hetero-epitaxial thin films of superconducting YBCO and insulating MgO were successfully grown on an MgO(100) substrate by an excimer laser deposition. ; (4)Cross-sectional transmission electron microscopy(TEM) images indicated that there was no disruption at the interface of the MgO/YBCO single hetero-epitaxial system with the 2nm-thick MgO layer, but that the interfaces of the YBCO/MgO/YBCO double hetero-epitaxial system were disrupted. At the deposition of the top YBCO layer, the smooth surface of the MgO film was deformed to rugged one, and then the intermediate MgO layer became five times as thick as the original one.

This work was performed as a part of "R&D on Superconducting Technology for Electric Power Apparatuses" under the Moonlight Project of Agency of Industrial Science and Technology(AIST), MITI, being consigned by New Energy and Industrial Technology Development Organization(NEDO).

REFERENCES

1. For a review, see the Proceedings of the Appl. SC Conf.(1988); IEEE Trans. Magn. MAG-25, 719(1989)
2. Tanaka S, Nakanishi H, Higaki K, Itozaki H(1990) Jpn. J. Appl. Phys. 29(6): L1059-1065
3. Linker G, Xi XX, Meyer O, Li Q, Geerk J(1989) Solid State Communications 69(3): 249-253
4. Hirata K, Yamamoto K, Iijima K, Takada J, Terashima T, Bando Y, Mazaki H(1990) Appl.Phys.Lett. 56(7): 683
5. Kingston JJ, Wellstood FC, Lerch P, Miklich AH, Clarke J(1990) Appl. Phys. Lett. 56(2): 189
6. Rogers CT, Inam A, Hegde MS, Dutta B, Wu XD, Venkatesan T(1989) Appl. Phys. Lett. 55(19): 2032
7. Venkatesan T, Inam A, Dutta B, Ramesh R, Hegde MS, Wu XD Nazar L. Chang CC, Barner JB, Hwang DM, Rogers CT(1990) Appl. Phys. Lett. 56(4): 391
8. Harada K, Tanaka S, Itozaki H, Yazu S(1990) Jpn. J. Appl. Phys. 29(7): L1114-1116
9. Mizuno K, Higashino K, Setsune K, Wasa K(1990) Appl. Phys. Lett. 56(15): 1469
10. Gross R, Gupta A, Olsson E, Segmuller A, Koren G(1990) Appl. Phys. Lett. 57(2): 203
11. Gupta A, Gross R, Olsson E, Segmuller A, Koren G, Tsuei CC(1990) Phys. Rev. Lett. 64(26): 3191
12. Kobayashi T, Iyori M(1989) Extended Abstracts of 2nd Workshop on High-Temperature Superconducting Electron Devices, Hokkaido(1989), p415

Preparation of YBCO Thin Films by Plasma Assisted Pulsed Laser Deposition

KATSUMI OHATA, TAKASHI SAWADA, HIROHIKO IZUMI, KATSUMI SUZUKI, TADATAKA MORISHITA, and SHOJI TANAKA

Superconductivity Research Laboratory, International Superconductivity Technology Center, 10-13, Shinonome 1-chome, Koto-ku, Tokyo, 135 Japan

ABSTRACT

YBCO films were deposited in a RF discharge oxygen plasma by pulsed laser deposition. Critical temperatures (Tc) decreases with an increasing RF power. Since Tc increases up to almost the same value by annealing at 500 °C in oxygen followed by slow cooling, films deposited in an oxygen plasma were deficient in oxygen. Scanning electron microscopic observations revealed that the crystal grain size in films deposited in an oxygen plasma was larger than that of the film without plasma, which corresponds to a smaller oxygen diffusivity into films prepared in oxygen plasma

KEY WORDS: YBCO, pulsed laser deposition, oxygen plasma,

INTRODUCTION

Pulsed laser deposition is one of the most promising techniques for *in-situ* preparation of superconducting films. Preparation of high quality films with a high critical temperature and high critical current density [1] and fabrication of the heterostructural multilayer [2] by pulsed laser deposition have been reported. Various excitation techniques such as DC discharge oxygen plasma [3], DC biasing to the ring electrode between the substrate and the target [1], N_2O atmosphere [4], secondary light irradiation to the substrate surface [5] and irradiation of the second laser to N_2O gas in order to promote decomposition of N_2O molecules [6] are utilized. We have reported that substrate biasing technique is also effective for improving the superconducting properties and crystallinities at lower growth temperatures, which is probably due to the new light emission in addition to an ordinary light emission from the plume [7][8]. The new emission is resulted from particles evaporated with lower velocity than that of an ordinary plume and the oxygen plasma with a short life time.

In this work, we have prepared YBCO films in RF discharge oxygen plasma generated between the substrate and the target in order to investigate effects of an oxygen plasma on the film properties.

EXPERIMENTAL

Figure 1 shows a schematic diagram around the substrate and target of the experimental system. A rotating target is irradiated by ArF excimer laser (193 nm) and ablated particles are deposited on to a substrate supported 50 mm above a target. To locally generate a high density oxygen plasma between the substrate and the target , a quartz tube (30 mm in diameter) is located. RF power (13.56 MHz) is supplied to the ring shaped electrodes around the tube to generate the plasma inside and near it. Materials ablated at the target by laser pulses reached the substrate after passing through the plasma. Oxygen gas is introduced into the tube through the thin pipe at the side of the tube. The reason why we used RF excitation is that it is easy to generate a high density plasma and that there is less influence on acceleration of ionic species from the target compared to DC excitation. Since the inside wall of the quartz tube had been coated by ablated target materials, contamination from sputtered quartz was ignored. The increase in substrate temperature by being exposed in a plasma is measured less than 20 °C at a RF power of 100 W.

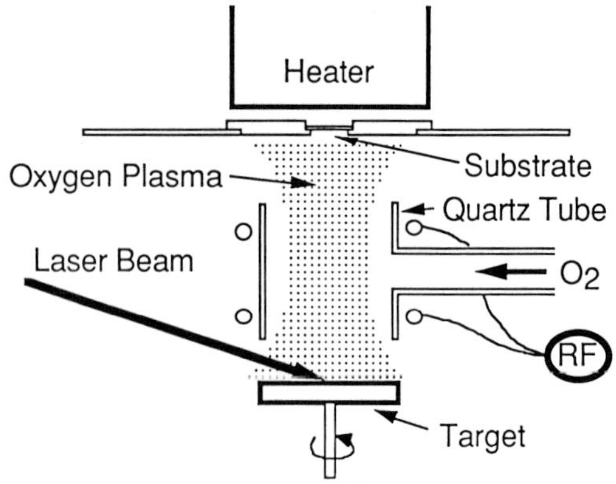

Fig.1 Schematic diagram of the experimental system

$YBa_2Cu_3O_{7-x}$ superconducting thin films were prepared in an oxygen plasma using this system. Growth conditions are as follows. A sinterd $YBa_2Cu_3O_{7-x}$ target was irradiated by laser at a fluence of about 2 J/cm^2 and a repetition rate of 5 Hz in 100 mTorr oxygen. Single crystal MgO(100) was used as a substrate and kept at 700 °C during deposition. RF power was supplied up to 100 W to generate oxygen plasma during deposition. After deposition, the oxygen pressure increased to 300 Torr and cooled down to room temperature at a rate of 15 °C/min.

RESULTS AND DISCUSSION

Zero-resistance temperatures (Tc) of resulting films prepared at various RF power were measured by 4-probe method, and are shown by open circles in Fig. 2. Tc gradually decreased with an increasing RF power until 76 K at 100 W, whereas the film prepared without the plasma shows 82 K. However, X-ray diffraction patterns indicate that all of these films are oriented with their c-axes perpendicular to the substrate surface, and that the peak intensity and c-axis lattice parameter are almost same for them. Therefore, superconducting properties of these films are strongly depend on cooling process incorporating oxygen to cause the structural transition from a nonsuperconducting tetragonal to a superconducting orthorhombic phase.

To make clear the effect of an oxygen plasma during deposition, films deposited with a plasma (50 W) and without a plasma were rapidly cooled in vacuum (10^{-5}Torr) after deposition. Both films show tetragonal crystal structure and c-axis orientation. However, the film prepared in a plasma is stable in the air whereas the film prepared without a plasma was decomposed in a few hours. We were not able to distinguish a difference in the crystallinity between samples prepared with and without a plasma by X-ray diffraction measurements in both cases of cooling in vacuum and cooling in oxygen. Nevertheless, chemical stability of film is quite different.

The samples shown in Fig. 2 were cooled in 300 Torr oxygen at a rate of 15 °C/min. Probably, this condition was not enough for the films deposited in a plasma to incorporate oxygen. After annealing at 500 °C for 2 hours in flowing oxygen gas followed by slow cooling at a rate of 1.2 °C/min, the films increased Tc up to about 83 K as shown by closed circles in Fig. 2. It means that it is not easy to incorporate oxygen in the films deposited in a plasma compared to the films without a plasma.

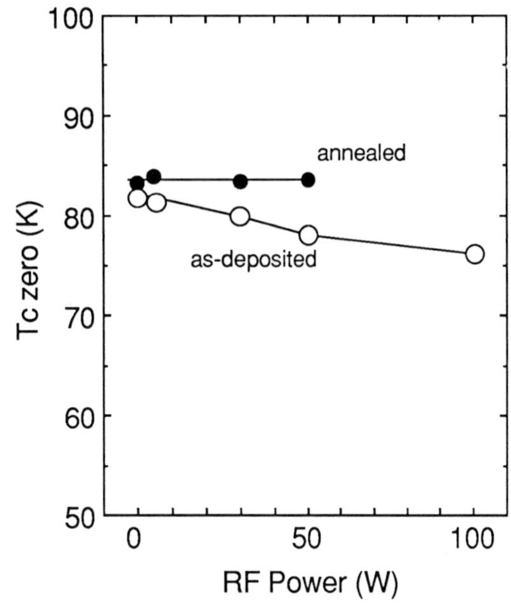

Fig.2 Dependence of Tc on RF power. The open circles indicate as-deposited films and the closed circles indicate the films annealed at 500°C for 2 hours.

Surface morphology of films were observed by a scanning electron microscope as shown in Fig. 3. The grain size is larger in the film deposited in a plasma than that without plasma, and the difference in a grain size causes different diffusivity of oxygen. The size of grains may depend on the condition of nucleation and growth of the film, but it is now under consideration how it concerns in the oxygen plasma.

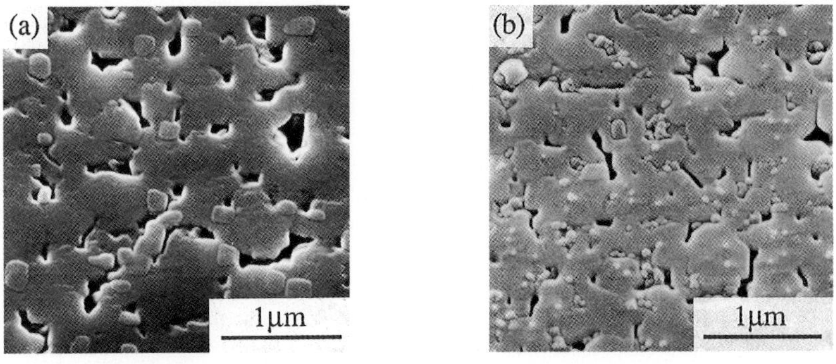

Fig.3 Scanning electron micrographs for the film deposited (a) in an oxygen plasma at a RF power of 50 W and (b) without plasma.

CONCLUSION

YBCO films have been prepared in an oxygen plasma. By introducing an oxygen plasma, the growth of grains has improved, resulting in less incorporation of oxygen in post-deposition treatment.

ACKNOWLEDGEMENT

This work was supported by New Energy and Industrial Technology Development Organization under the management of R&D of Basic Technology for Future Industries.

REFERENCES

1. R.K.Singh, J.Narayan, A.K.Singh and J.Krishnaswamy, Appl. Phys. Lett. 54, 2271 (1989).
2. X.D.Wu, X.X.Xi, Q.Li, A.Inam, B.Butta, L.DiDomenico, J.A.Martinez, B.J.Wilkens, S.A.Schwarz, J.N.Barner, C.C.Chang, L.Nazar and T.Venkatesan, Appl. Phys. Lett. 56, 400 (1990).
3. S.Witanachchi, H.S.Kwok, X.D.Wang and D.T.Shaw, Appl. Phys. Lett. 53, 234 (1988).
4. M.Kanai, T.Kawai and S.Kawai, Appl. Phys. Lett. 54, 1802 (1989).
5. S.Otsubo, T.Minamikawa, Y.Yonezawa, T.Maeda, A.Morimoto and T.Shimizu, Jpn. J. Appl. Phys. 28, 2211 (1989)
6. G.Koren, A.Gupta and R.J.Baseman, Appl. Phys. Lett. 54, 1920 (1989)
7. K.Ohata, H.Izumi, T.Hase, K.Suzuki, T.Morishita and S.Tanaka, to be published in Proc. Mater. Res. Soc. Spring Meeting, April 1990, San Francisco, U.S.A.
8. H.Izumi, K.Ohata, T.Hase, K.Suzuki, T.Morishita and S.Tanaka, to be published in J. Appl. Phys.

Superconducting Properties of $YBa_2Cu_3O_x$ Films on Y_2O_3 Single Crystal

Akira Oishi, Hidekazu Teshima, Hirohiko Izumi, Katsumi Ohata, Shinji Kawamoto, Tadataka Morishita, and Shoji Tanaka

Superconductivity Research Laboratory, International Superconductivity Technology Center, 10-13, Shinonome 1-chome, Koto-ku, Tokyo, 135 Japan

ABSTRACT

The c-axis oriented $YBa_2Cu_3O_X$(YBCO) film has epitaxially grown on Y_2O_3(100) substrates by laser deposition for the first time. The orientation of YBCO<110> is parallel to Y_2O_3<001>. The critical temperature at zero resistance is 86.2K with a transition width of 1.2K. The critical current densities at 5K was estimated to be 1.4 $\times 10^7$, and 2.0 $\times 10^6$A/cm^2 without and with a magnetic field of 5Tesla, respectively.

KEYWORDS; Y_2O_3(100) substrate, $YBa_2Cu_3O_X$ film, epitaxial growth

INTRODUCTION

Substrate candidate materials must meet several criteria including reasonable compatibility with respect to lattice matching and thermal expansion, little or no chemical reactivity with the superconducting material for epitaxial growth, and the lower dielectric constant for the sake of device making. Most common materials for the substrate are MgO and $SrTiO_3$. Recently many studies using $LaAlO_3$[1], $YAlO_3$[2], and $NdGaO_3$[3] as a substrate have reported. They are superior to MgO and $SrTiO_3$ with regard to lattice matching.

The lattice constant, a of this cubic Y_2O_3 is 10.604Å ($a'=a/2\sqrt{2}=3.749$Å, a' is the atomic spacing in the [011] direction of Y_2O_3). As compared with a, b, and c/3 of YBCO, the lattice mismatch is 1.9, 3.6, and 3.9%, respectively. The thermal expansion coefficient ($\alpha=9.3 \times 10^{-6}/°C$) is smaller than that of YBCO ($\alpha=1.3 \times 10^{-5}/°C$). And Y_2O_3 consists of only Y and O, which are components of YBCO. The dielectric constant of Y_2O_3 ($\varepsilon=14$) is comparable to that of MgO ($\varepsilon=10\sim17$), and smaller than $SrTiO_3$ ($\varepsilon=310$)[4]. In spite of these merits , there is no study of YBCO films using the Y_2O_3 substrate, probably because of difficulty in growing a large single crystal.

In this work, YBCO films have successfully deposited on Y_2O_3 single crystal by means of in-situ processing. Then, we have estimated availability of Y_2O_3 single crystal as a substrate from the view point of crystalline and superconducting properties of a film deposited, and the stability of the interfaces between the film and the substrate.

EXPERIMENTAL

The substrates were cut from Y_2O_3 single crystal grown by the Verneuil method. The YBCO films were deposited by the pulsed laser ablation method onto Y_2O_3 under the conditions of 100mTorr oxygen partial pressurė, and ~700°C substrate temperature. The film which thickness was about 3000Å. Films and substartes were characterized by an X-ray diffractometer with a CuKα source and a scanning electron microscopy. The interfacial stability between the film and the substrate was estimated by depth the profile of Auger electron spectroscopy. As for superconducting properties, we determined T_C^0, and ΔT from the resistivity versus temperature curves, and the current density J_C from the magnetization hysteresis loop.

RESULTS AND DISCUSSION

The substrates used a single crystal with some defects shown in Fig.1. The Y_2O_3(400) peaks by ω scan indicates there exists the deviation a-axis orientation and distribution of unit cell length. And it also indicates that the lattice parameter corresponding with the main peak is 10.628Å, and that the substrate has the texture of grains and/or subgrains whose a-axis is distributed within ±2deg. and whose lattice parameter ranges are 10.628 to 10.677Å. The lattice parameter $a'(=a/2\sqrt{2}=3.758\text{~}3.775\text{Å})$ is a little smaller than a, b, and c/3 of YBCO.

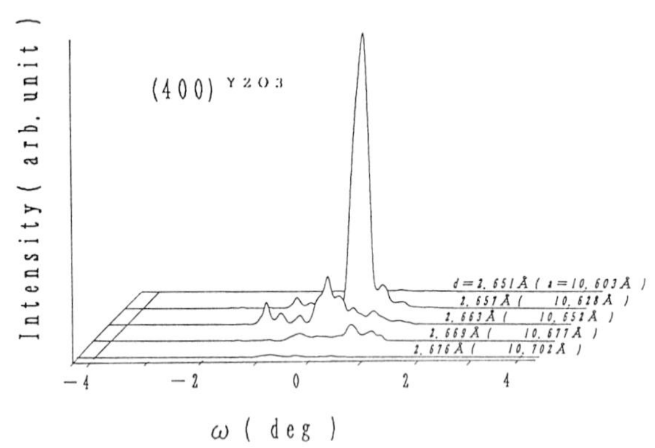

Fig.1 Rocking curves of Y_2O_3 (400) peak

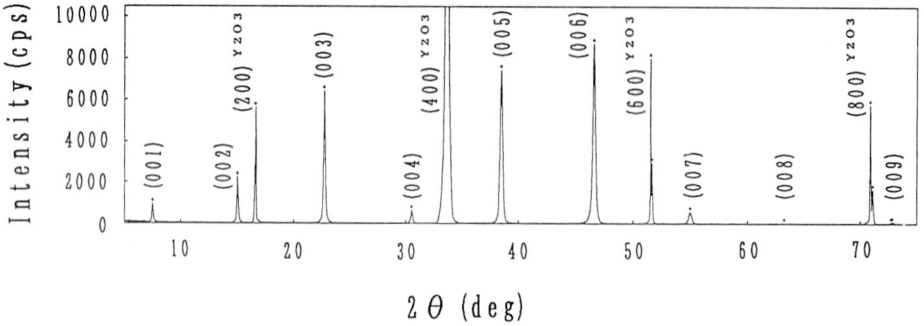

Fig.2 X-ray diffraction pattern of YBCO film on Y_2O_3(100)

Fig.2 shows X-ray diffraction pattern for YBCO films on $Y_2O_3(100)$. A 2θ-θ scan shows only the {00ℓ} of the film, which indicates that the film is oriented with the c-axis perpendicular to the substrate ($<001>^{YBCO}//<100>^{Y2O3}$) The degree of c-axis orientation in the films can be examined by observing the rocking curve for the (005) peak. A full width of half maximum of 0.87deg. is neary equal to that of our case of the MgO substrate (0.85deg.). Fig.3 shows the ω scan peaks of $Y_2O_3(311)$, YBCO(018), and YBCO (108) in the [011] direction of Y_2O_3. The splitting of the $Y_2O_3(311)$ peak reflects on the YBCO(018) and YBCO(108) peaks. Therefore YBCO is epitaxially grown on $Y_2O_3(100)$.

Fig.4 shows the resistivity versus temperature plot for YBCO film. The critical temperature (zero resistance) T_C^0 is 86.2K with a transition width ΔT of 1.2K. The resistivity curve is extrapolated, crossing the positive temperature axis. A similar phenomenon have been reported in the case of c-axis oriented films with compressive stress in the a-b plane by substrates[5] or the interlayers[6]. In our case, the ratio a/a_0=99.7%, b/b_0=98.2%, and c/c_0=100.1%, where a_0, b_0, and c_0 are lattice constants of bulk YBCO, indicate the existence of compressive stress in film.

Fig.5 shows the critical current densities at 5K, J_C^{5K} calculated from the magnetization hysteresis loop using the Bean critical state model[7]. J_C^{5K} is 1.4 x10^7 and 2.0 x10^6A/cm^2 without and with the magnetic field of 5Tesla, respectively.

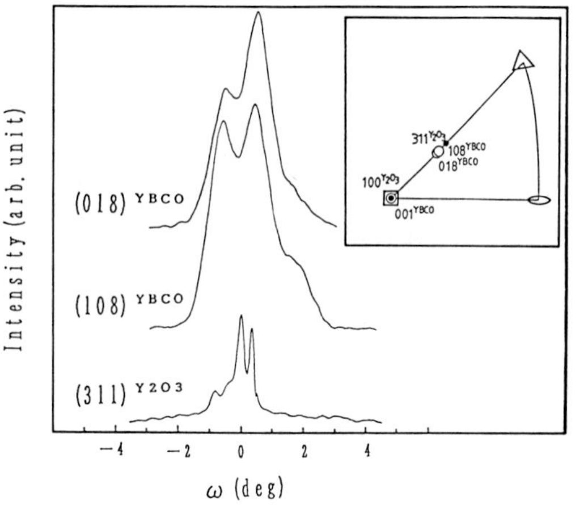

Fig.3 Rocking curves of $Y_2O_3(311)$, YBCO(018), and YBCO(108) peaks in the [011] direction of Y_2O_3

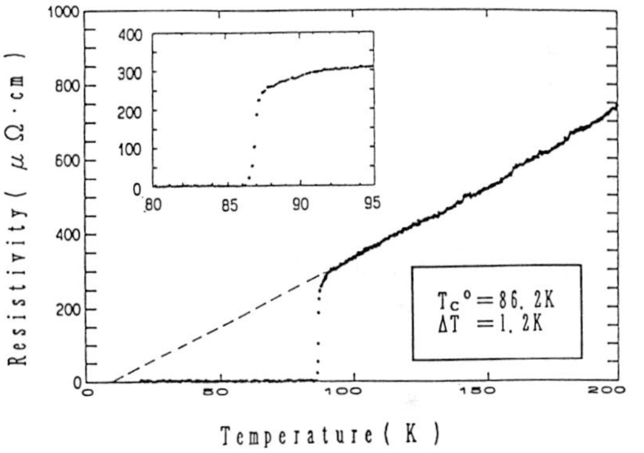

Fig.4 Resistivity versus temperature plot for YBCO film

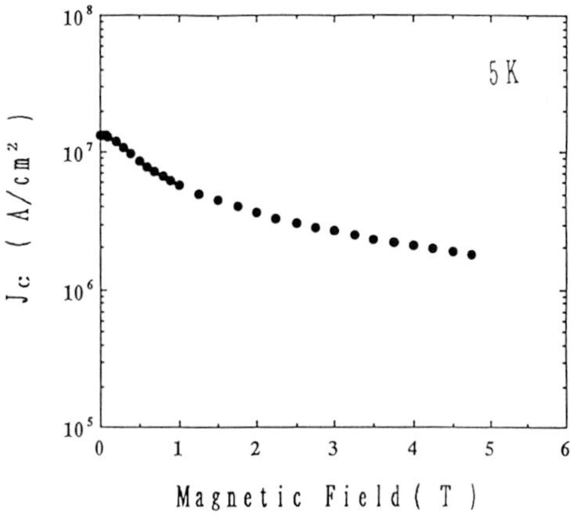

Fig.5 Magnetic field dependence of critical current density of YBCO film on $Y_2O_3(100)$

SUMMARY

Epitaxial films of YBCO have been grown on Y_2O_3 (100) by laser ablation. The relative orientation between YBCO and Y_2O_3 is $<001>^{YBCO}$ // $<100>^{Y2O3}$ and $<110>^{YBCO}$ // $<001>^{Y2O3}$. The unit cell of YBCO film is smaller than that of bulk YBCO, resulting in the compressive stress in films. The film is superconducting by T_C^0=86.2K and exhibits at 5K J_C=1.4 x10^7A/cm^2 without a magnetic field and 2.0 x10^6A/cm^2 in 5Tesla.The Y_2O_3 single crystal well excepted substrate for superconducting films.

ACKNOWLEDGMENTS

A part of this work was supported by New Energy and Industrial Technology Development Organization under the management of R&D of Basic Technology for Future Industries.

REFERENCES

[1] H.C.Li, H.R.Yi, R.L.Wang, Y.Chen, B.Yin, X.S.Ron, and L.Li; Appl.Phys. Lett. **56**,2454(1990)

[2] H.Asano, S.Kubo, O.Michikami, T.Tokairin, K.Shiroki; Extended Abstracts of 37th Spring Meeting of the Japan Society of Applied Physics and on the Relation Society , Saitama,April, 1990,

[3] M.Mukaida, S.Miyazawa, M.Sasaura; Extended Abstracts of 51st Autumn Meeting of the Japan Society of Applied Physics and , Iwate,September, 1990, 27p-Z-5

[4] G.A.Samara; J.Appl.Phys.**68**.4214(1990)

[5] M.Sasaura, S.Miyazawa, M.Mukaida; Extended Abstracts of 51st Autumn Meeting of the Japan Society of Applied Physics and , Iwate,September, 1990, 27-Z-6.

[6] D.K.Fork, D.B.Fenner, R.W.Bartin, Julia M.Phillips, G.A.N.Connell, J.B.Boyce, T.H.Geballe; Appl. Phys.Lett. **57**, 1161(1990)

[7] C.P.Bean; Phys. Rev. Lett. **8**, 250(1962)

Structure of High-T_c Superconducting Thin Film by Laser Ablation

TATSUOKI NAGAISHI, HIDEO ITOZAKI, TAKASHI MATSUURA, NOBUHIRO OTA, NAOJI FUJIMORI, and SHUJI YAZU

Itami Research Laboratories, Sumitomo Electric Industries, Ltd., 1-1, Koya-kita 1-chome, Itami, 664 Japan

ABSTRACT

High quality $YBa_2Cu_3O_{7-x}$ thin films were prepared on (100)$SrTiO_3$ substrate by laser ablation. One of the best samples showed the critical current density (Jc) of $8.0x10^6 A/cm^2$ at 77.3K under a zero magnetic field. The characteristics of this sample were investigated in comparison with the sample which has Jc of $2.0x10^6 A/cm^2$. The characteristics of the film, mis-oriented crystallites and stacking faults, were revealed by transmission electron microscopy (TEM). Furthermore, the preferential orientations of the film with various preparation conditions by laser ablation were also studied.

KEYWORDS: $YBa_2Cu_3O_{7-x}$ thin film, excimer laser, laser ablation, high critical current density, pinning site

INTRODUCTION

High Tc superconducting thin films with Jc of more than $1.0x10^6$ A/cm^2 were reported by several preparation methods in recent years.[1-6] Among them, laser ablation can be one of the most promising preparation methods from the point of the controllability of the deposition conditions and the simplicity of the deposition system.

Recently, Gross et al. succeeded in making a high Tc superconducting thin film with Jc of $1.1x10^7$ A/cm^2 by fabricating the strained multilayer of $YBa_2Cu_3O_{7-x}/Nd_{1.83}Ce_{0.17}CuO_x$ by a laser ablation. This makes use of the idea that strain field and additional defects at the interface relate to the flux pinning. However, the mechanism of the high Jc process is still a controversial issue in this field. Ramesh et al.[7,8] observed the stacking defects corresponding to the cationic stoichiometry of the 248, 247 and 224 compounds and edge dislocations and antiphase boundaries in high Jc laser ablation thin films. They discussed the possibility of the high Jc [illegible]g caused by these defects which come from the nonequilibrium process. Roas et al.[9] ex[illegible]mentally showed evidence for intrinsic pinning between the CuO layers and pinning at twin boundaries or stacking faults with the laser ablation films.

In the present study, in addition to these defects as pinning sites, other possible pinning centers were found with the high Jc laser ablation film. Furthermore, the preferential orientations of the film by the preparation conditions were investigated with the laser ablation method.

EXPERIMENTAL

A 193nm ArF excimer laser (Lambda Physik LPX315i) was used for film preparation. The laser pulse duration was 20 nsec and the rate of pulse shot was 1 Hz. The laser beam was focused to about 1.5mm x 5mm on the target, which was a 1.5cm diameter sintered pellet of $YBa_2Cu_3O_{7-x}$ superconductor. The laser power density was about $0.7J/cm^2$ at the surface of the target. 10mm x 10mm (100)MgO and (100)$SrTiO_3$ were used as substrates. The pressure of the deposition chamber was maintained from 1Pa to 100Pa with oxygen gas during the deposition. The substrate temperature (Ts) was varied from 550°C to 750°C, measured by a thermocouple clipped on the surface of the substrate. The distance between the target and the substrate was fixed as 40mm. The deposition was performed for 15 minutes. The film thickness was 200nm. After the deposition, Ts was decreased to 200°C within 4 minutes with the gas pressure in the chamber reaching 1 atm.

The characteristics of the two different films with the Jc of $2.0x10^6A/cm^2$ (Tc=85.5K) and $8.0x10^6A/cm^2$ (Tc=90.5K) were compared with various techniques. (Hereafter, the films with Jc of $2.0x10^6A/cm^2$ and $8.0x10^6A/cm^2$ will be named as sample (I) and sample (II), respectively.) Sample (I) was deposited on the (100)MgO substrate with the Ts of 710°C. Sample (II) was deposited on the (100)$SrTiO_3$ substrate with the Ts of 700°C. Electric properties were measured by the four-probe transport method. The films were patterned by a photo-lithographic technique to produce a 20μm-wide and 20μm-long microbridge.

RESULTS AND DISCUSSION

X-ray diffraction (XRD) patterns of the films show sharp (00l) (l=1,2,3...) peaks, indicating that the films are highly oriented with the c-axis perpendicular to the substrate surface. The full width at half maximum of for (005) peak is about 0.24° for sample (I) and about 0.21° for sample (II). These values are considered to designate the fairly good crystallinity of the films. Surface observation of sample (I) by SEM seems to show some depressions near the interface between a few hundred nm to 1μm-sized domains. (Fig.1(a)) Surface observation of sample (II) shows small voids as shown in Fig.1(b). The voids in Fig.1(b) are usually revealed in the condition of high Ts or high oxygen pressure from the experiments. The films which showed these voids usually had high Tc and high Jc.

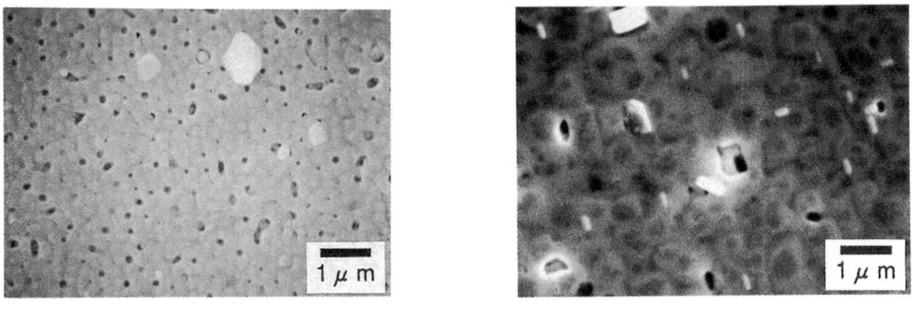

(a) (b)

Fig.1. SEM micrographs of the $YBa_2Cu_3O_{7-x}$ film: (a) sample (I) and (b) sample (II)

TEM observation was performed to investigate the microstructure of the films. The bright field image of sample (I) in Fig.2(a) is considered to reveal twin boundaries which cross each other. In Fig.2(b), in addition to the twin boundaries such as the ones shown in Fig.2(a), some rectangular fragments which show fringes are observed. Since the distance between each fringe in the rectangular fragments is about 1.17nm, these fringes were identified as CuO layers. In short, the c-axes of these fragments (hereafter referred to as a, b crystallites) are parallel to the substrate. Another characteristic of sample (II) from the TEM micrograph is moire fringe contrast in Fig.2(b). The moire fringe contrasts are considered to be formed by stacking faults or thin precipitates. The a, b crystallites and the moire fringe contrast were dispersed in about 5% of the film from the area about $0.59cm^2$ in the planar [001] section. As for sample (II), the a, b crystallites and the moire fringe contrast were not clearly observed. Therefore, a, b crystallites or the defects revealed as the moire fringe contrasts may be responsible for pinning to achieve the extra high Jc of $8.0x10^6A/cm^2$. However, the authors think that too many a, b crystallites rather degrade the Jc characteristics of the sample because the current volume is reduced at the fragments where the CuO plane is perpendicular to the current direction. The voids observed by SEM are also considered to be strong pinning centers.

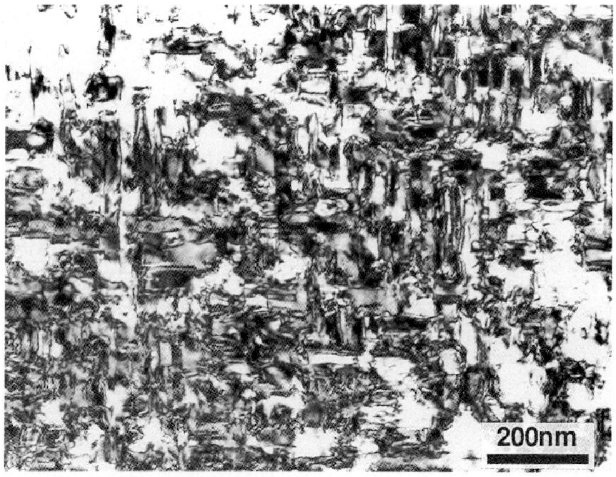

(a)

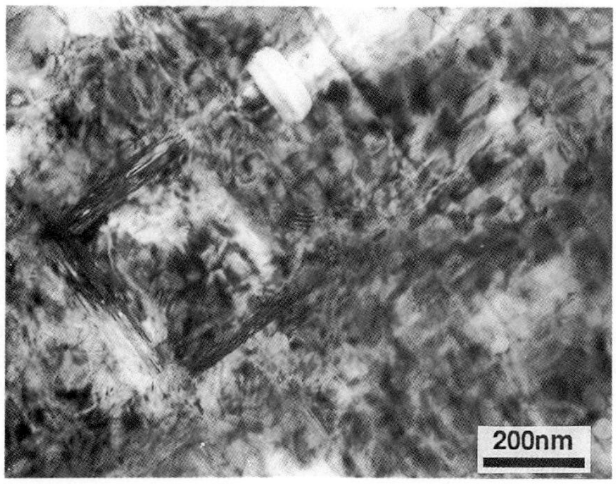

(b)

Fig.2. TEM micrographs of the [001] projection of: (a) the sample (I) and (b) the sample (II).

Next, the preferential orientation of the film regarding the deposition temperature was investigated. Figures 3(a), (b) and (c) show the typical XRD patterns of the laser ablation film. The deposition was performed under the oxygen gas pressure of 50 Pa and other preparation conditions were the same as the ones described in the experimental section. As shown in Fig.3, at the Ts of 650°C the film has c-axis orientation. At the Ts above 650°C up to 710°C, c-axis orientation of the film was observed by XRD results. At the Ts of 600°C a-axis oriented fragments are growing because the mixture of (003) and (100) peaks near the theta of 23° became highr relative to (005) peak compared to the result at the Ts of 650°C. At the Ts of 580°C the film has mainly a-axis orientation and only the small (005) peak from the c-axis orientation is shown. This result is consistent with that of Asano et al.[10] Recently Eom et al.[11] and Arikawa et al.[12] reported that the a-axis oriented film was formed by reducing the Ts about 80°C

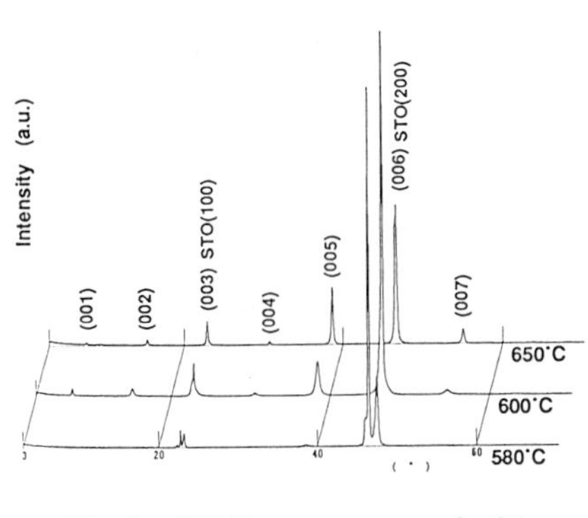

Fig.3. XRD patterns of film prepared at Ts of: (a) 650°C, (b) 600°C and (c) 580°C.

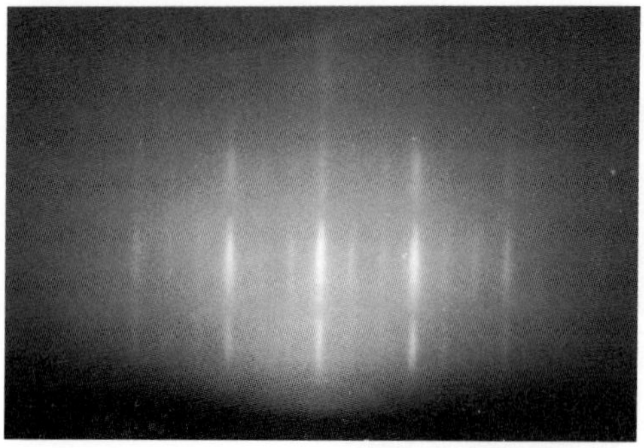

Fig.4. RHEED pattern of the a-axis oriented film.

from the c-axis orientation condition. To get good a-axis oriented film, the experiment was performed with a variety of oxygen gas pressures at Ts of 600°C. The preparation conditions of the laser pulse repetition rate and the laser power density were changed to 5Hz and 1.2J/cm2, respectively. From the XRD patterns, each film showed the a-axis orientation and some had Tc. In the previous experiments with various Ts, non-superconducting behavior was given at the Ts of 600°C. At oxygen pressure of 10Pa, the films show maximum Tc of 70K and RHEED reveals a long-range ordering pattern of the c-axis clearly (Fig.4). Since the long-range ordering patterns were observed from the two directions which cross each other, the film is composed of fragments with two different c-axis directions (c-axes crossed).

CONCLUSION

In summary, superconducting thin films have been successfully prepared by the laser ablation technique. With the extra high Jc film of c-axis orientation, tiny a, b crystallites and moire fringe contrast were observed by TEM while the sample with lower Jc did not show any. The voids on the surface of the film are also revealed by SEM as a characteristic of the high Jc films. The preferential orientation of the film also was investigated and an a-axis oriented film was successfully prepared.

These works were performed under the Research and Development Program on "Advanced Material Processing and Machining Systems," conducted under a program of the New Energy and Industrial Technology Development Organization.

REFERENCES

1) Y. Enomoto, T. Murakami, M. Suzuki: Jpn. J. Appl. Phys. 26(1987)L1248.
2) S. Tanaka, H. Itozaki: Jpn. J. Appl. Phys. 27(1988)L622.
3) T. Terashima, Y. Bando, K. Iijima, K. Yamamoto, and H. Hirata: Appl. Phys. Lett. 53(1988)2232.
4) J. Geerk, Q. Li, G. Linker, O. Meyer, and X. X. Xi: Program and abstracts of the ISTEC Workshop on Superconductivity, Oiso, February , 1989, p.67.
5) R. Gross, A. Gupta, E. Olsson, A. Segmuller, and G. Koren: Appl. Phys. Lett. 57(1990)203.
6) A. Inam, M. S. Hedge, X. D. Wu, T. Venkatesan, P. England, P. F. Miceli, E. W. Chase, C. C. Chang, J. M. Tarascon, and J. B. Wachtman: Appl. Phys. Lett. 53(1988)908.
7) B. Roas, L. Schultz, and G. Endres: Appl. Phys. Lett. 53(1988)1557.
8) T. Venkatesan, X. D. Wu, B. Dutta, A. Inam, M. S. Hedge, D. M. Hwang, C. C. Chang, L. Nazer, and B. Wilkens: Appl. Phys. Lett. 54(1988)581.
9) B. Roas, B. Hensel, G. Saemann-Ischenko, and L. Schultz: Appl. Phys. Lett. 54(1989)1051.
10) G. Koren, A. Gupta, and R. J. Baseman: Appl. Phys. Lett. 54(1989)1920.
11) G. Koren, A. Gupta, R. J. Baseman, M. I. Lutwyche, and R. B. Laibowitz: Appl. Phys. Lett. 55(1989)2450.
12) R. K. Singh, L. Ganapathi, P. Tiwari, and J. Narayan: Appl. Phys. Lett. 55(1989)2351.
13) R. Ramesh, D. M. Hwang, T. Venkatesan, T. S. Ravi, L. Nazar, A. Inam, X. D. Wu, B. Dutta, G. Thomas, A. F. Marshall, and T. H. Geballe: Science 247(1990)57
14) R. Ramesh, D. M. Hwang, J. B. Barner, L. Nazar, T. S. Ravi, A. Inam, B. Dutta, X. D. Wu, and T. Venkatesan: J. Mater. Res. 5(1990)704
15) B. Roas, L. Schultz, and G. Saemann-Ischenko: Phys. Rev. Lett. 64(1990)479

Ultrathin $YBa_2Cu_3O_x$ Films on Lattice Matched $NdGaO_3$ Substrates

MASASHI MUKAIDA, SHINTARO MIYAZAWA, and MASAHIRO SASAURA

NTT LSI Laboratories, 3-1, Morinosato Wakamiya, Atsugi, 243-01 Japan

ABSTRACT

We report ultrathin superconducting $YBa_2Cu_3O_x$ films grown on lattice-matched $NdGaO_3$ substrates by a laser ablation deposition technique. The 40 Å-thick ultrathin film has a T_C as high as 63 K. Scanning electron micrographs of films on (110) just-cut and about 2 %-off-cut $NdGaO_3$ substrates show that the films grow epitaxially.

KEY WORDS: YBCO superconductor, ultrathin films, $NdGaO_3$ substrate, laser ablation

INTRODUCTION

There is a lot of interest applying recent discovered oxide superconductors [1-2] to microelectronics. In such microelectronics device applications, these superconductors will probably be in the form of ultrathin films. Of further interest in the area of ultrathin films are the basic principles governing the occurrence of superconductivity in two-dimensional CuO_2 sheet system and the least number of layers necessary for superconductivity to occur. These are just two of the reasons, it is necessary to prepare films as thin as possible. However, many difficulties remains in the preparation of ultrathin films.

Until now, many studies focused on the preparation of ultrathin films on $SrTiO_3$ and MgO substrates [3-5]. An oriented film as thin as 100 Å with T_C of 82 K by reactive evaporation has been demonstrated on $SrTiO_3$ substrate by Bando *et al.* [3]. Venkatesan *et al.* [4] reported 100 Å-thick $YBa_2Cu_3O_x$ films with T_C of 84 K on $SrTiO_3$ substrates by a KrF laser ablation deposition technique. However their 50 Å-thick films did not exhibit zero-resistance. Michikami *et al.* [5], on the other hand, prepared $EuBa_2Cu_3O_x$ ultrathin films with a protection layer on MgO substrates by magnetron sputtering.

When films are very thin, the influence of the substrate becomes an extremely important factor. To date, $SrTiO_3$ and MgO have been widely used as the substrates for oxide superconducting thin films. It is often said that for compound semiconductors a lattice mismatch within 0.5 % is strictly demanded for epitaxial growth. However, $SrTiO_3$ substrate and MgO have a lattice mismatch with $YBa_2Cu_3O_x$ in excess of 1.1 % and 8 %, respectively.

More recently, we have successfully grown $NdGaO_3$ single crystals [6] which have a lattice mismatch of 0.36 % to $YBa_2Cu_3O_x$ at a deposition temperature of about 700°C by a conventional Czochralski (CZ) growth method. In addition, we have grown high crystalline quality $YBa_2Cu_3O_x$ thin films by laser ablation on lattice matched $NdGaO_3$ substrates [7]. These films exhibited the preferred *c*-axis orientation and had a full width at half maximum as narrow as 0.12 degrees for an X-ray rocking curve of (005) peak. This value is a half of that reported for the films on conventional $SrTiO_3$ substrates.

This paper reports high crystalline quality ultrathin films epitaxially grown on lattice matched $NdGaO_3$ substrates.

EXPERIMENTAL

Ultrathin films were prepared by an ArF excimer laser (Lambda Physik 20 ns pulse duration, 1 Hz, 1 J/pulse) ablation technique. The incident beam had an angle of about 30° with respect to the surface plane of the rotating target. The target was a 40 mm diameter and 10 mm thick disk of superconducting $YBa_2Cu_3O_x$. The substrates were optically polished just-cut and about 2 %-off-cut $NdGaO_3$ (110) of about 12 mm square. They were affixed with silver paste to a resistively heated up substrate holder, which is located about 5 cm away from the target. Before deposition, the substrate was heated to 800 °C in vacuum. The background pressure of the chamber was $1x10^{-6}$ Torr. Typical substrate surface temperature and atmosphere were about 700 °C and 0.2 Torr of oxygen. The film thickness was controlled by growth time. Typical deposition rate was about 25 Å/min. After deposition, the heater was turned off and immediately ventilated by oxygen. The films exhibited superconductivity in the as-grown state.

The film thickness was measured by using a mechanical stylus. Resistivity versus temperature curves of the as-grown ultrathin films were measured by the conventional four probe technique. Crystallinity such as *c*-axis length is estimated by X-ray diffractometer using Cu - K_α radiation. Surface morphology of the films was observed by scanning electron microscope (SEM; HITACHI S-900) with 5 kV acceleration voltages. The composition of the films was estimated by X-ray fluorescence spectroscopy analysis (XFA) calibrated by using inductively coupled plasma spectroscopy. There was no remarkable difference in composition between the film and the target.

RESULTS AND DISCUSSION

Figure 1 shows the temperature dependence of normalized electrical resistance ($R(T)/R(300)$) for as - grown films with different thicknesses. As the film thickness decreases, T_C decreases. For films more than 80 Å-thick, the R - T curves showed a metallic behavior; however, at 40 Å, the film's resistance behavior deviated from linearity. Even the 40 Å-thick film has a T_C of 63 K. *c*-axis lengths of these films as a function of film thickness are shown in Fig. 2. In the figure, the solid line is the measured *c*-axis length of the grown films calculated by the least square method for $(00L)$ (L = integer) peaks up to (0.0.10). The deviation of the calculated length is about 0.006 Å. *c*-axis length gradually increases as film thickness deceases. At 40 Å, the *c*-axis lengthens rapidly. The broken line is a calculated *c*-axis length from their T_C values by using Cava's Data [8]. The measured *c*-axis of 40 Å-thick film is about 11.75 Å, however, *c*-axis length of an oxygen deficient $YBa_2Cu_3O_x$ bulk with a T_C of 63 K is 11.72 Å. *c*-axis length is always longer than the calculated one. Then, the increased *c*-axis length of the film may be caused by a lattice distortion and / or the existence of inhomogeneous oxygen deficient layer.

The relationship between T_C's and the film thickness for our films, the ultrathin films on $SrTiO_3$ from Venkatesan *et al.* [4] and those on MgO from Michikami *et al.* [5] are provided in Fig. 3 for comparison. Ultrathin film on $NdGaO_3$ substrate as thin as 40 Å has a T_C of 63 K. On the contrary, for the 50 Å-thick films on $SrTiO_3$ was non superconducting and that on MgO with protect layer had a T_C of 50 K. The effect of the different substrate is quite apparent in ultrathin films thinner than 50 Å. Then, it can be said that the lattice matching resulted in good film quality.

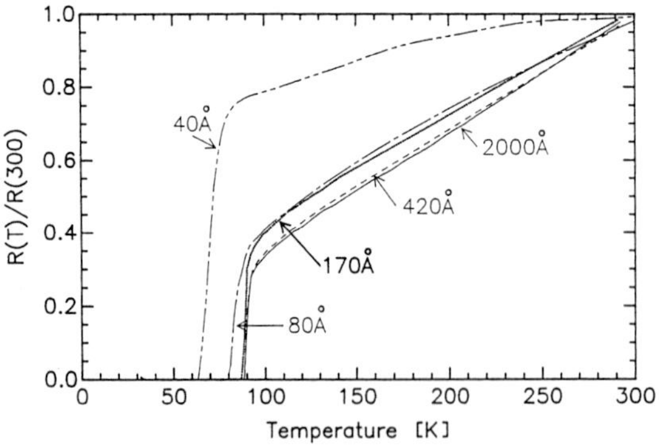

Fig. 1. Temperature dependence of normalized electrical resistance ($R(T)/R(300)$) for as-grown films with different thicknesses. Even 40 Å-thick film has T_C of 63 K.

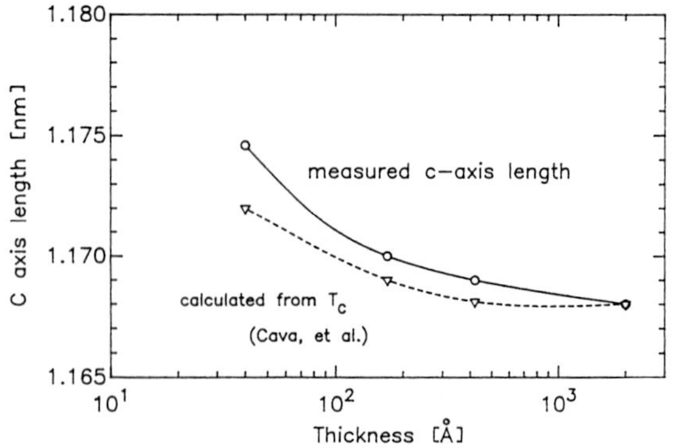

Fig. 2. The thickness dependence of *c*-axis length. The solid line is measured by the least square method for $(00L)$ (L = integer) peaks up to (0.0.10) and the broken line is a calculated from their T_C values using Cava's Data [8].

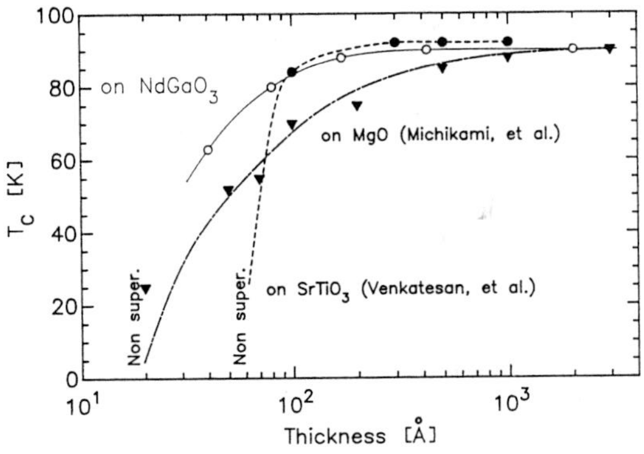

Fig. 3. Comparison of T_C's and the film thickness for ultrathin films on $NdGaO_3$, $SrTiO_3$ and MgO substrates.

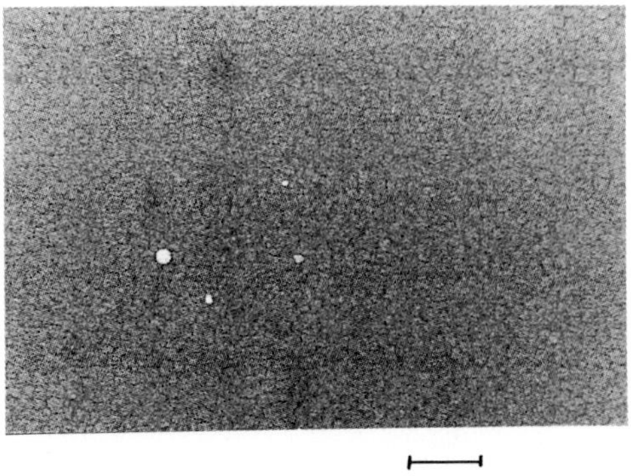

Fig. 4. Surface morphology of the films on $SrTiO_3$ substrates. Mosaic textures are observed in the figure.

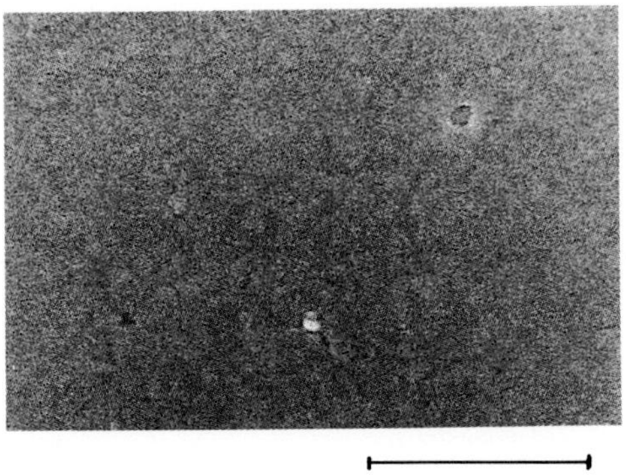

Fig. 5. Surface morphology of the films on $NdGaO_3$ substrates. No apparent texture are observed.

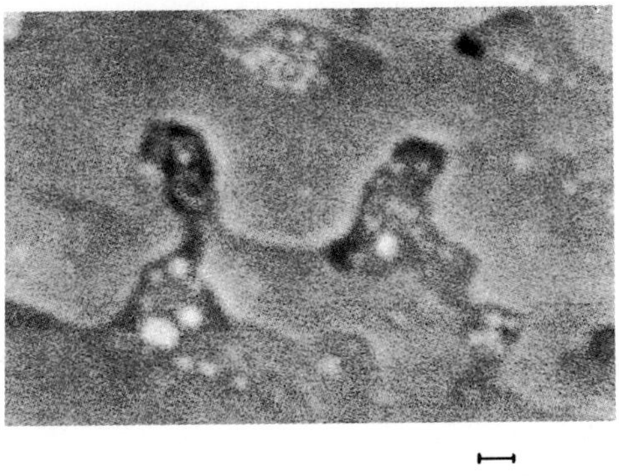

Fig. 6. Scanning electron micrograph of the film on about 2 %-off-cut $NdGaO_3$ substrates. There can be seen many terraces with a little exclusion on the edges of the terraces.

To compare the surface morphology of the films, $YBa_2Cu_3O_x$ films on $SrTiO_3$ substrates were also prepared by the same procedure as the films on $NdGaO_3$ substrates. Both films have a T_C of 90 K. These films on both substrates had a smooth surface morphology. No apparent structure can be observed by common SEM. By using high resolution and low-accele ration-voltage SEM, the surface morphology of the films on $SrTiO_3$ substrates and those on $NdGaO_3$ are shown in Fig. 4, and Fig. 5, respectively. In Fig. 4, there is little dust, however, the film has a mosaic texture. This texture, it must be noted, is only observable using high resolution and low-acceleration-voltage SEM. In contrast, the film in Fig. 5 has no apparent mosaic textures. We would like to stress that these mosaic textures are due to the lattice mismatch and / or difference in thermal expansion coefficient between the films and the substrates. In addition, $YBa_2Cu_3O_x$ films are thought to be epitaxially grown on $NdGaO_3$ substrates.

To certify the epitaxy, we grew thin films on about 2 %-off-cut $NdGaO_3$ substrates. Figure 6 shows a scanning electron micrograph of the film. Many terraces are revealed with a little exclusion on the edge of the terrace. These are bunching effects, which are well known in epitaxial growth of compound semiconductors. This aspect gives an evidence that $YBa_2Cu_3O_x$ films grows two dimensionally. Then, we can say that the films grew epitaxially on lattice matched $NdGaO_3$ substrates.

In conclusion, ultrathin $YBa_2Cu_3O_x$ films were successfully grown on lattice-matched $NdGaO_3$ substrates. Films only 40 Å-thick had a T_C as high as 63 K. Judging from scanning electron micrographs of films on just-cut and about 2 %-off-cut $NdGaO_3$ substrates, the films grew epitaxially. These results show that it is absolutely essential to use lattice-matched substrates for obtaining a high T_C in ultrathin oxide superconducting films with good crystallinity.

ACKNOWLEDGEMENTS

The authors would like to acknowledge Drs. Y. Tazoh, T. Ito, and Mr. J. Kobayashi for their useful comments and encouragement. They also thank to Ms. Ogawa for helping with X-ray measurements.

REFERENCE

1. Bednorz JG, Müller KA (1986) Z. Phys. **B64**: 189.
2. Wu MK, Ashburn JR, Torng CJ, Hor PH, Meng RL, Gao L, Huang ZJ, Wang YQ, Chu CW (1987) Phys. Rev. Lett. **58**: 908.
3. Bando Y, Terashima T, Iijima, K, Yamamoto K, Mazaki H (1988) Physica C, **153-155**: 810.
4. Venkatesan T, Wu XD, Dutta B, Inam A, Hegde MS, Hwang DM, Chang CC, Nazar L, Wilkens B (1989) Appl. Phys. Lett. **54**: 581.
5. Michikami O, Asahi M, Asano H (1990) Jpn. J. Appl. Phys. **29**: L298.
6. Sasaura M, Miyazawa S, Mukaida M (1990) J. Appl. Phys. **68**: 3643.
7. Mukaida M, Miyazawa S, Sasaura M, Kuroda K (1990) Jpn. J. Appl. Phys. **29**: L936.
8. Cava RJ, Batlogg B, Chen CH, Rietman EA, Zahurak SM, Werder DJ (1987) Phys. Rev. **b36**: 5719.

New Substrate $PrGaO_3$ for High-T_c Superconducting $YBa_2Cu_3O_x$ Epitaxial Film

MASAHIRO SASAURA, MASASHI MUKAIDA, and SHINTARO MIYAZAWA

NTT LSI Laboratories, 3-1, Morinosato Wakamiya, Atsugi, 243-01 Japan

ABSTRACT

In this letter, a new substrate material, $PrGaO_3$ crystal, for epitaxial films of superconducting $YBa_2Cu_3O_x$ is reported. The lattice mismatch between $PrGaO_3$ and $YBa_2Cu_3O_x$ is 0.02 % at deposition temperature of ~700 °C, and the thermal expansion coefficient of $PrGaO_3$ is close to that of $YBa_2Cu_3O_x$. The $YBa_2Cu_3O_x$ film deposited by the laser ablation method grows epitaxially on $PrGaO_3$ substrate, and surface morphology is much better than that on $SrTiO_3$. Pr inter-diffusion does not provide any crucial problems, and zero-resistance temperature of the as-deposited 500 Å thick film is 90 K.

KEY WORDS: $YBa_2Cu_3O_x$, substrate, $PrGaO_3$, epitaxial growth

1. INTRODUCTION

Among the high-T_C oxide superconductors investigated so far, $YBa_2Cu_3O_x$ oxide is one of the promising. To date, there have been many attempts at preparing thin films necessary for applications to electronic devices. One of the crucial problems with these films has been the lack of extremely high quality, i.e., no *epitaxial* single-crystalline film. The films deposited on avaliable substrates, e.g., $SrTiO_3$ [1] and MgO [2], did not overcome this problem because of lattice mismatch and/or difference in their thermal expansion coefficients between substrates and films deposited on them.

Generally speaking, following three requirments are essentially requested for epitaxial growth;

1. Lattice matching between substrates and films at growth temperature must be less than 0.5 %. In the case of $YBa_2Cu_3O_x$ films, lattice matching at 700 °C is essential because the usual deposition temperature is ~700 °C.
2. Thermal expansion coefficients of the substrates and the films must be close to each other. During cooling process, the difference between these values produces dislocations and/or the amount of residual thermal stress in the deposited films.
3. Crystal structure of the substrates must be similar to that of the films. Such similarities determine how to stack at initial stage of the epitaxial growth. Substrate materials for the $YBa_2Cu_3O_x$ film must have a Perovskite-like structure consisting of oxygen octahedra.

Our research to realize such "real" *epitaxial* $YBa_2Cu_3O_x$ thin films has led us back to basic study on the substrate material. We have been investigating various lanthanum-series gallates [3,4,5,6], and present a new highly suitable substrate material, $PrGaO_3$ in this letter.

2. EXPERIMENTAL RESULTS AND DISCUSSION

2.1 Characterization of $PrGaO_3$ crystal

$PrGaO_3$ single crystals were grown by the conventional Czochralski method. X-ray diffraction patterns with Si powder as a reference revealed that the crystal had a $GdFeO_3$ structure (distorted perovskite) and that lattice constants were a=5.462, b=5.486, and c=7.731 Å in an orthorhombic system at room temperature. The thermal expansion coefficients along the a, b, and c axes were estimated by high-temperature X-ray diffraction data to be 8.4×10^{-6}, 7.3×10^{-6}, and 7.0×10^{-6} /°C, respectively.

From these measures, the lattice mismatch with $YBa_2Cu_3O_x$ at the deposition temperature of ~700 °C was examined. Because of the structural similarity between $PrGaO_3$ and $YBa_2Cu_3O_x$, it is more likely that gallium octahedra on the surface of the $PrGaO_3$ substrate match preferentially copper octahedra of the $YBa_2Cu_3O_x$ film. The lengths of the Ga-O-Ga cell edges at 700 °C in the < 100 >, < 010 >, < 001 >, and < 110 > directions were caliculated to be 5.493, 5.513, 3.884, and 3.893_7 Å, respectively. In contrast, the length of the tetragonal Cu-O-Cu cell edge at 700 °C was 3.893_4 Å, reported by Jorgensen *et al.* [7]. Figure 1 shows the temperature dependence of the Ga-O-Ga cell edges in the < 001 > and < 110 > directions. The length of the Ga-O-Ga cell edge in the < 110 > direction is the closest to that of the Cu-O-Cu cell edge, and the lattice mismatch is only 0.02 %. The thermal expansion coefficient along the < 110 > direction was caliculated to be 9.0×10^{-6} /°C, which is also close to that of $YBa_2Cu_3O_x$ (9.4×10^{-6} /°C). So, it will be the best to use (001)-oriented wafers for depositing $YBa_2Cu_3O_x$.

Figure 2(a) shows the (001)-oriented wafers prepared from the grown boule. Twins could not be eliminated yet completely, and misoriented subgrains could be observed in twin-free regions under a transmission polarizing microscope. So, the full width at half maximum (FWHM) measured by an X-ray double-crystal diffracrometry was 220 seconds, as shown in Fig. 2(b). This value is comparable with that of not-qualified $SrTiO_3$ substrate.

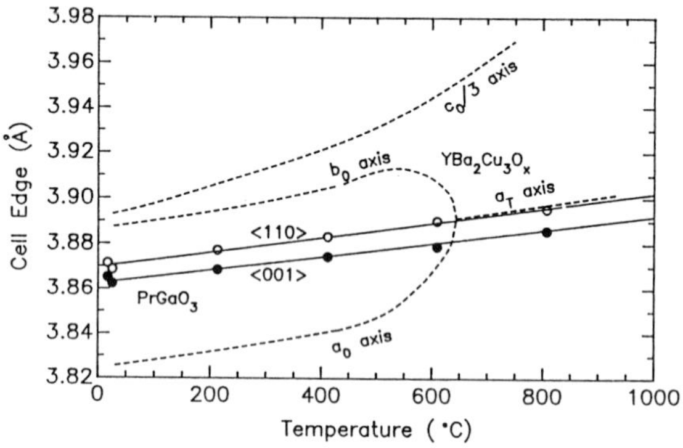

Fig.1 Temperature dependence of Ga-O-Ga cell edges along < 001 > and < 110 > directions of $PrGaO_3$(solid lines). Broken lines are Cu-O-Cu cell edges of $YBa_2Cu_3O_x$ in 2 % O_2 atomosphere (see Ref.7).

2.2 Properties of YBCO film on $PrGaO_3$ substrate

To ascertain the suitability of $PrGaO_3$ as a substrate for $YBa_2Cu_3O_x$ films. The films were successfully grown by the laser ablation method at deposition temperature of 700 °C. Deposition procedure has already been reported [8]. Film quality was evaluated, as compared with the films on conventional $SrTiO_3$ substrates.

(a)

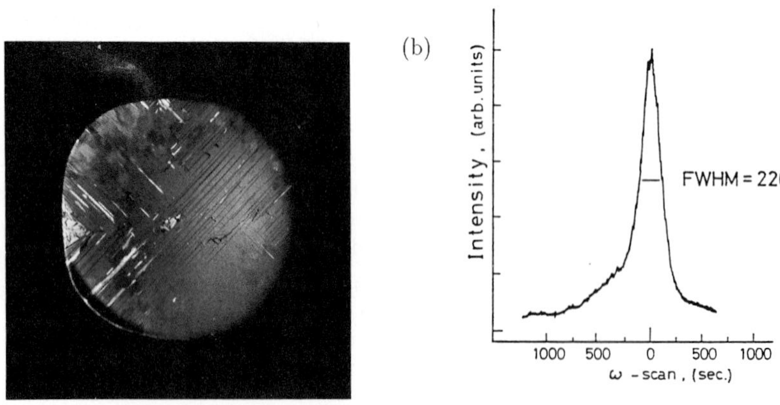

Fig.2 (a) (001) wafer cut from $PrGaO_3$ crystal. The bar is 10 mm. (b) X-ray double-crystal diffracrometry curve of twin-free region in the (001) $PrGaO_3$ substrate.

The temperature dependence of resistivity for the as-deposited $YBa_2Cu_3O_x$ film with a thickness of 500 Å on the $PrGaO_3$ substrates exhibited quite metallic; the zero-resistance temperatures were 90 K, and resistivity at 100 K was ~90 $\mu\Omega$·cm. The film did not show any remarkable degradation of superconductivity by Pr inter-diffusion, which will be discussed later.

X-ray diffraction pattern of the $YBa_2Cu_3O_x$ film revealed only ($00l$) peakes, indicating a c-axis orientaion normal to the (001) substrate. The FWHM of the (005) peak on a $2\theta/\theta$-scan was only 0.07 degrees, which is less than the value for the film grown on $SrTiO_3$ (0.12 degrees). This fact indicates that the cell edge length of $YBa_2Cu_3O_x$ in the c-direction is much more uniform for $PrGaO_3$ than for $SrTiO_3$ as will be shown later by a Transmission Electron Microscope (TEM) observation. On the other hand, the FWHM of the (005) peak on an ω-scan was 0.24 degrees on $PrGaO_3$ and 0.23 degrees on $SrTiO_3$. This poor quality on $PrGaO_3$ must be due to the subgrains existed in the $PrGaO_3$ substrate. Rutherford Back-scattering Spectroscopy (RBS) measurements yielded that $YBa_2Cu_3O_x$ peaks did not completely disappear in the aligned spectrum and the substrate intensity of the aligned spectrum was only ~50 % less than that of the random intensity. This implies that the poor crystallinity of the film is due to substrate quality.

To investigate surface and interface morphology, a high-resolution Scanning Electron Microscope (Å-SEM) and a TEM were used. Figure 3(a) shows the surface morphology of $YBa_2Cu_3O_x$ on $SrTiO_3$ by the Å-SEM. It exhibited sharp and mosaic-like irregular grain boundaries probably caused by the large difference in thermal expansion coefficients, although observation by a conventional SEM could not detect such morphology. On the other hand, these irregular boundaries could not recognized on the film grown on $PrGaO_3$, as shown in Fig.3(b). Lattice image shown in Fig. 4 demonstrates that the tri-layers of the $YBa_2Cu_3O_x$ film is very uniform, as described above, and an interface seems quite sharp. It gives evidently that the $YBa_2Cu_3O_x$ film grew epitaxially on the $PrGaO_3$ substrate with very little distortion of the c-planes, as can be seen quite parallel.

(a) (b)

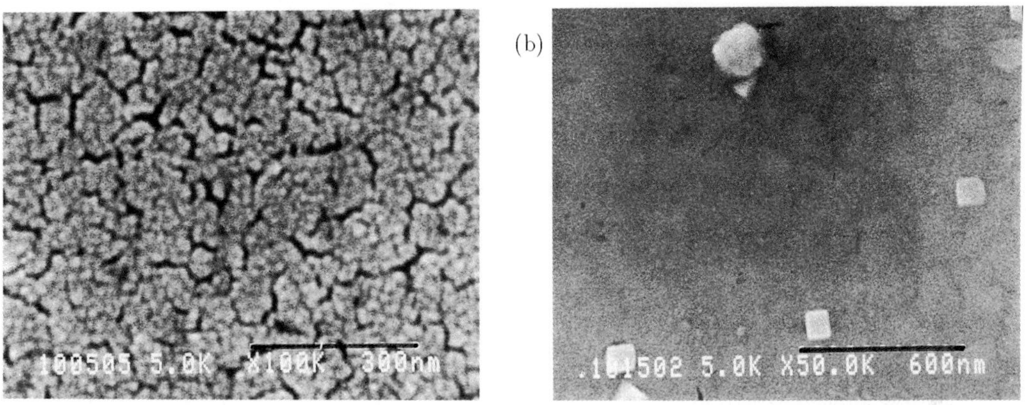

Fig.3 (a) Surface morphorogies of a $YBa_2Cu_3O_x$ film on a $SrTiO_3$ with a thickness of 2000 Å under a high-resolution scanning electron microscope. The bar is 300 nm. (b) Surface morphorogies of a $YBa_2Cu_3O_x$ film on a $PrGaO_3$ substrate with a thickness of 500 Å. The bar is 600 nm.

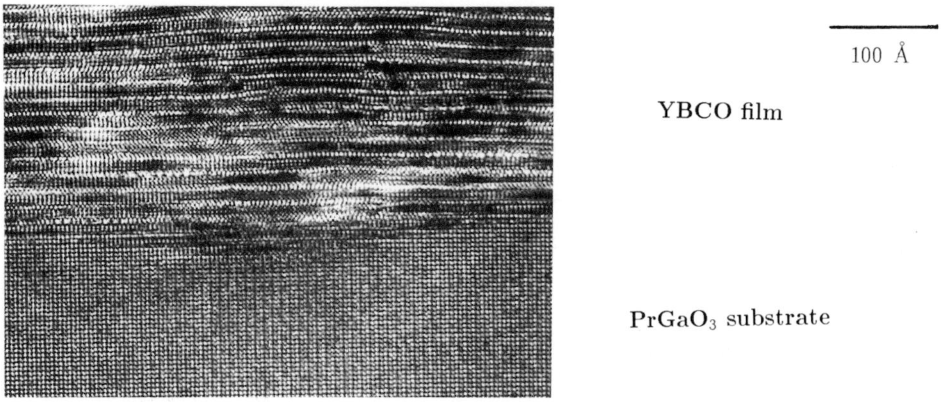

Fig.4 Sectional lattice image of a 500 Å thick $YBa_2Cu_3O_x$ film on $PrGaO_3$.

It has been well known that Pr ions degrade superconductivity of $YBa_2Cu_3O_x$ [9]. An Auger Electoron Spectroscopy (AES) was carried out to know inter-diffusion of Pr into the grown $YBa_2Cu_3O_x$ film. Figure 5 shows depth profiles of each element. Inter-diffusion layer into the film was roughly estimated to be less than 40 Å in this case. Therefore, it can be said that Pr inter-diffusion dose not provide any crucial problems, as mentioned before.

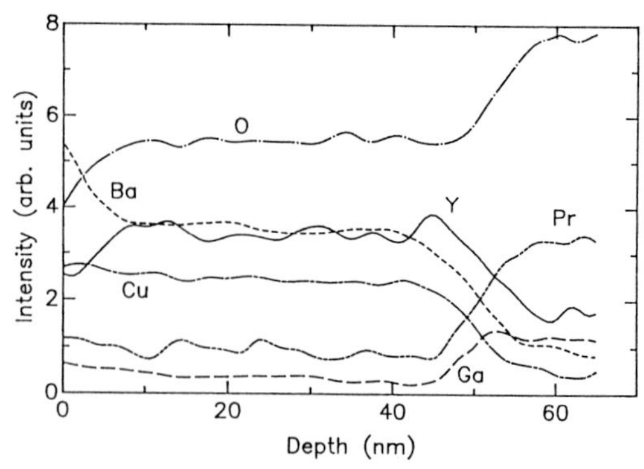

Fig.5 Auger Electron Spectroscopy along the film depth direction in a 500 Å thick $YBa_2Cu_3O_x$ film on $PrGaO_3$.

3. SUMMARY

We have investigated a new substrate material, $PrGaO_3$ crystal, for growing epitaxial thin films of $YBa_2Cu_3O_x$. Crystallographic properties are well-suited to the growth of $YBa_2Cu_3O_x$ films: the lattice mismatch between $PrGaO_3$ and $YBa_2Cu_3O_x$ was only 0.02 % at the deposition temperature ~700 °C. The thermal expansion coefficient of an (001) wafer was close to that of $YBa_2Cu_3O_x$. The superconductig $YBa_2Cu_3O_x$ film on the $PrGaO_3$ substrate included no sharp and irregular grain boundaries, and actually grew *epitaxially*. Though the film was deposited on twin-free regions of $PrGaO_3$ substrate, the existence of subgrains in the substrates seriously degraded the film quality. It can be virtually concluded that twin-free and subgrain free $PrGaO_3$ crystal must be an excellent substrate for the *epitaxial* growth of $YBa_2Cu_3O_x$ film.

REFERENCES

1. Narayan J, Sharan S, Singh RK, Jagannadham K (1989) Mater.Sci.Eng.B 2: 333
2. Ramesh R, Hwang D, Ravi TS, Inam A, Barner JB, Nazar L, Chan SW, Chen CY, Dutta B, Venkatesan T, Wu XD (1990) Appl.Phys.Lett. 56: 2243
3. Miyazawa S (1989) Appl.Phys.Lett. 55: 2230
4. Miyazawa S, Mukaida M, Sasaura M, Tazoh Y (1990) In: The Electrochemical Society, Symposium of high T_c superconductor technologies. Montrial, Canada, No.953HTS
5. Sasaura M, Mukaida M, Miyazawa S (1990) J.Appl.Phys. 68: 2643
6. Sasaura M, Mukaida M, Miyazawa S (1990) Appl.Phys.Lett. 57: (in press)
7. Jorgensen JD, Beno MA, Hinks DG, Soderholm L, Volin KJ, Hitterman RL, Grace JD, Schuller IK, Segre CU, Zhang K, Kleefisch MS (1987) Phys.Rev.B 36: 3608
8. Mukaida M, Miyazawa S, Sasaura M (1990) Jpn.J.Appl.Phys. 29: L936
9. Matsuda A, Kinoshita K, Ishii T, Shibata H, Watanabe T, Yamada T (1988) Phys.Rev.B 38: 2910

Development of a Four-Source EB Gun for High-T_c Superconductor Film Deposition

HISASHI YAMAMOTO, TOSHIO KIKUCHI, and NOBUHARU OHSAKO

ANELVA Corporation, 8532-28, Narusawa, Minamitsuru, Yamanashi, 401-03 Japan

ABSTRACT

We developed a four-source EB gun which incorporated four crucibles into a unit for deposition of oxide superconductor thin film by means of simultaneous evaporation. The major characteristics of this gun are that the size of the evaporation system chamber can be reduced greatly in comparison with those of four separate EB guns of a similar capacity, high- frequency electron beam sweeping is possible, the gun is bakable and can be used in ultrahigh vacuum systems, and few electrons, which may reach the substrate and deteriorate the quality of the evaporated film, are scattered.

KEY WORDS: Oxide superconductor film, EB gun, simultaneous evaporation, electron beam sweeping, circular magnetic circuit

INTRODUCTION

Recently, research and development on oxide superconductors have become very active at various research institutions. There are several choices of methods of deposition of superconductor thin films; deposition by simultaneous evaporation is considered to be most appropriate for R&D because stoichiometrical control of the film is possible and the evaporation materials are relatively easily obtainable. Of the various means of evaporation, evaporation by EB gun is widely preferred due to its high deposition rate and controllability.

However, for this method of deposition, three to five electron guns must be arranged in an evaporation chamber, and the following can be listed as disadvantages in using the method.

1) Close arrangement of EB guns is not possible. This problem arises from the fact that the magnetic fields applied to EB guns for deflection or sweeping of electron beams interfere with neighboring EB gun magnetic fields.
2) Because of point 1), the evaporation sources of each material are separated; thus, a difference in the distribution of film thickness at the substrate from each evaporation source is caused, resulting in deterioration of the composition ratio of the film.
3) EB guns must be placed a certain distance from each other on a single plane; thus, a large chamber is required.

In order to solve the above problems, the authors developed an EB gun in which four independent evaporation sources are integrated into a single body avoiding mutual interference of magnetic fields. The report explains the structure and performance of the EB gun unit called the "TETRA GUN."

STRUCTURE AND PERFORMANCE OF THE TETRA GUN

Magnetic Circuit for Electron Beam Deflection

Figures 1 and 2 show the external view of the TETRA GUN. Four crucibles arranged on a circumference of 60mm in diameter are installed on a single copper body. An electron emitter is attached to the outside of each crucible. To deflect electron beams to the crucibles and irradiate them, four magnets are arranged between the crucibles as shown in Figure 2 to produce magnetic field, and the circular magnetic circuit (as shown in Fig. 3) is established. This configuration of magnets produces the required magnetic field (approx. 200G) for electron beam deflection on each crucible.

With the use of a common magnetic circuit to produce magnetic fields for four evaporation sources, instead of creating an independent magnetic circuit for each evaporation source, it is possible to arrange the four sources on a 60mm-diameter circumference without mutual interference of magnetic fields.

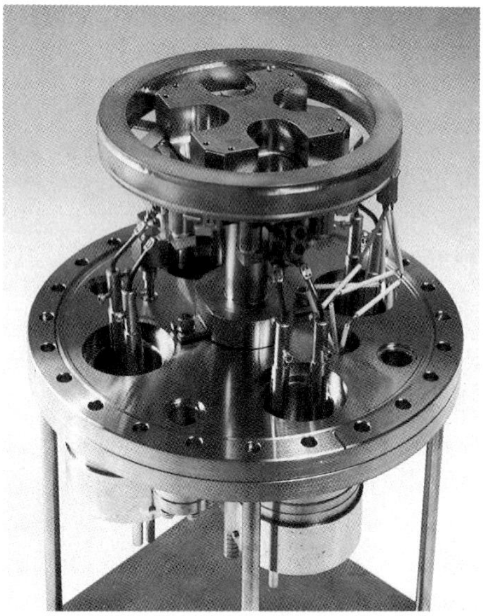

Fig. 1 Photograph of TETRA GUN

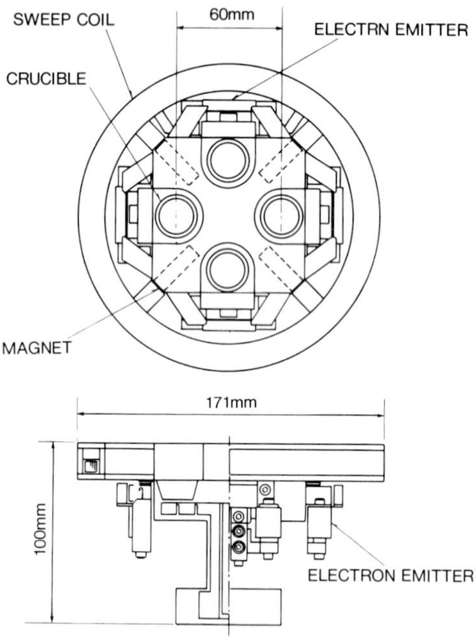

Fig. 2 External drawing of TETRA GUN

Conventionally, in EB guns with similar capacities, it was not possible to arrange crucibles on a circumference below 180mm in diameter because of mutual magnetic field interference. As shown in Fig. 4, assuming the distribution of evaporation rate by EB guns to be $Cos^{2.5}\theta$, in order to maintain the difference of film thickness from two sources at the edge of the 8-inch substrate which is placed above the midpoint of the diagonally placed two EB guns within ±5%, 50cm vertical spacing is required for two crucibles placed 60mm away from each other; however, 85cm is required for 180mm-separated crucibles. We can say that increasing the vertical spacing from 50cm to 85cm reduces the deposition rate on the substrate to one-third, as well as increases the size of the chamber.

Furthermore, the advantage of the circular magnetic circuit is that the magnetic field gets stronger toward the center of the EB guns; the magnetic field strength increases toward the side of the crucible opposite to the electron emitter. Therefore, the secondary electrons emitted from the surface of the evaporation materials due to the irradiation of the primary electrons, and reflected electrons at the evaporation surface are trapped by the magnetic field, and they are only slightly scattered toward the substrate. Figure 5 shows the results of measurements of the amount of scattered electrons. Only about 1% of the primary emission current of electrons was observed.

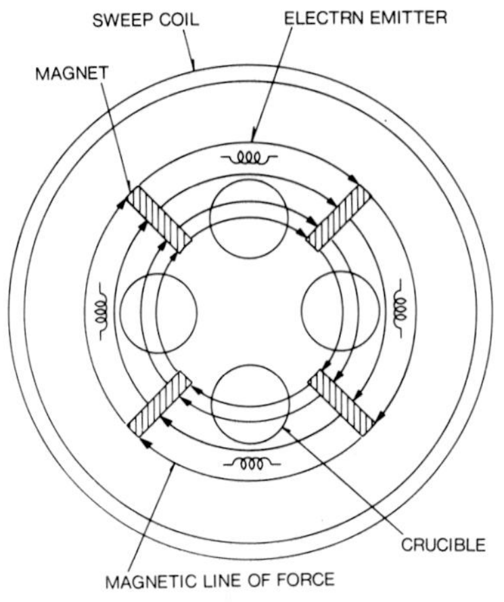

Fig. 3 Schematic diagram of circular magnetic circuit

The sections with hatched lines show the locations of four permanent magnets. The concentrically positioned arrows indicate the magnetic lines of force produced by magnets.

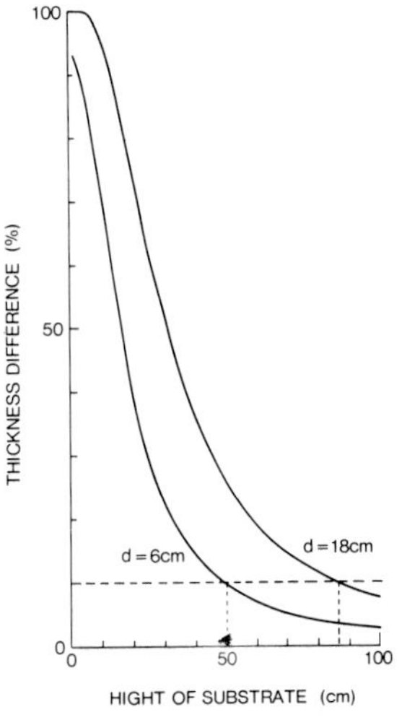

Fig. 4 Calculated thickness difference

The horizontal axis indicates the height of the 8-inch substrate located immediately above the midpoint between the two crucibles separated by distance d. The vertical axis indicates in percent the maximum values of the difference in the thickness from the two crucibles at the edge of the substrate.

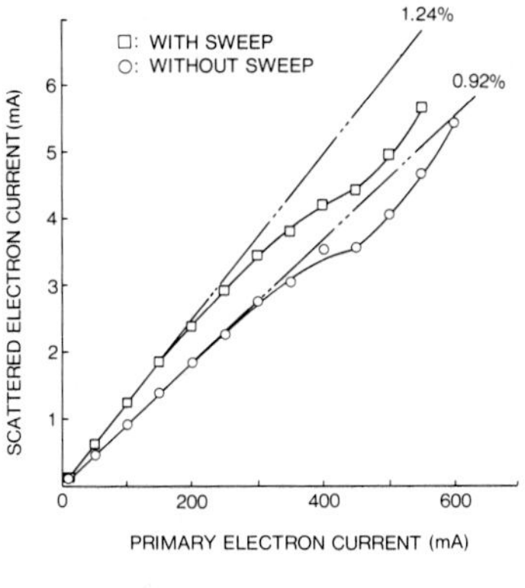

Fig. 5 Scattered electron current

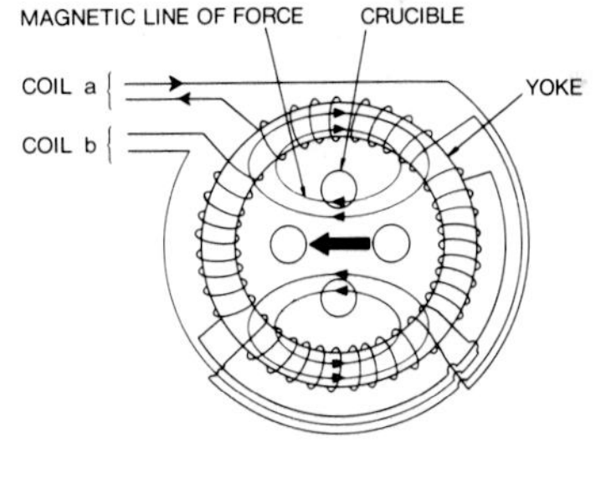

Fig. 6 Schematic diagram of sweep coil

When an illustrated current is applied to COIL a, illustrated magnetic lines of force are produced.

Coils for Electron Beam Sweeping

Figure 6 shows the structure of the coil unit for electron beam sweeping and explains the resulting magnetic field. The coil unit consists of four separated 200-turn coils wound around an annular iron core. The diagonal coil pairs (two coils) are connected in series; each pair is provided with a current from an independent power supply. The coil pair, a, in the figure produces a magnetic field with a horizontal vector in the figure. Conversely, the other pair, b, produces a magnetic field with a vertical vector in the figure. Thus, by providing appropriate currents for coil pairs a and b, a magnetic field with the desired vector can be created in the circle surrounded by the coils. This enables the sweeping of the electron beams of four evaporation sources at the same time by a single coil unit. In both pair a and pair b, sweeping across the whole crucible area was possible with an excitation current of 2.5A.

Also, high-frequency electron beam sweeping is required in the evaporation of insulator materials. In this coil, avoiding heat loss due to eddy current by using a laminated steel core enables high-frequency sweeping at 400Hz.

SUMMARY

By using a common magnetic circuit and sweeping coil for electron beam deflection, four-source EB guns which unify four evaporation sources have been developed. The outermost diameter of the gun unit is 171mm and its height is 100mm. Hence, it can be housed in a relatively small evaporation chamber. To enable deposition under ultrahigh-vacuum conditions, the unit was designed to tolerate baking at temperatures of up to 200°C. Figure 7 shows an example of data obtained by the four-source simultaneous evaporation. Table 1 shows the specifications of the TETRA GUN.

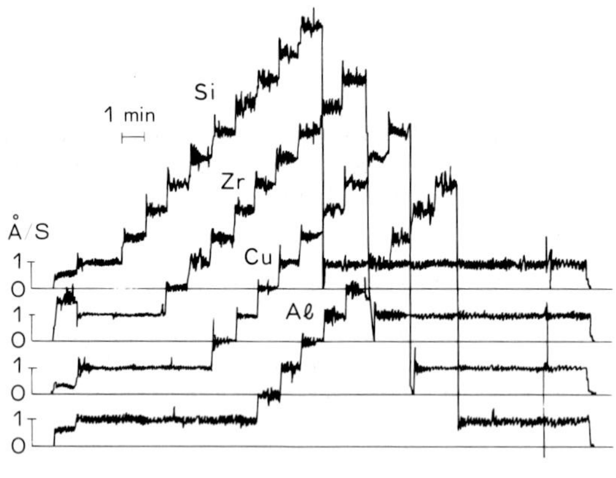

Fig. 7 Four-source simultaneous evaporation

Each evaporation rate was controlled and measured independently by a crystal thickness monitor.

Table 1 Specifications

Electron acceleration voltage		5kV
Electron emission current		0-600mA
Sweep coil current		3A(Max.)
Crucibles	Number	4
	Size	ϕ24mm × ϕ20mm × t10
	Contents	3.8cc
Deflection angle		225°
Required water flow		14L/min
Baking temperature		200°C
Weight		10kg

Oxygen Content Dependencies of Superconducting and Normal State Properties in Epitaxially Grown $YBa_2Cu_3O_x$ (X=6−7) Thin Films

MASATAKA OHKUBO, TATSUMI HIOKI, and TOHRU KACHI

Toyota Central Research and Development Laboratories, Nagakute-cho, Aichi-gun, Aichi, 480-11 Japan

ABSTRACT

Epitaxially grown thin films of $YBa_2Cu_3O_x$ with x=6−7 were prepared by rapid cooling following laser deposition. The thin films have uncommon properties. For example, the thin films have no two plateaus at 60 K and 90 K in the Tc−x plot. The highest T_c(=90.5 K) is at x=6.85, and T_c decreases in x>6.85. We have simulated the rapid cooling with Monte Carlo calculation based on an asymmetric Ising model for the Cu(1) plane. From oxygen configurations by the calculation, it has been expected that there is a linear relation between hole concentration and oxygen content. This explains some properties of the thin films.

KEY WORDS: $YBa_2Cu_3O_x$ thin films, Laser deposition, Ising model, Monte Carlo calculation

INTRODUCTION

It is well known that oxygen content affects the crystal structure and the electrical properties of $YBa_2Cu_3O_x$. As x decreases from 7, the structure is transformed from a superconducting orthorhombic phase into an antiferromagnetic tetragonal phase. The orthorhombicity is caused by an oxygen ordering. The oxygen ordering has been treated as superstructures with the full chains (O-Cu-O) and the empty chains (V-Cu-V, where V is a vacant O site) along the b axis in the Cu(1) planes. In polycrystalline samples of $YBa_2Cu_3O_x$, there are many reports on changes of superconducting or normal state properties with x. It has been observed that polycrystalline $YBa_2Cu_3O_x$ samples annealed well exhibit superconductivity in x≥6.2−6.4, and there are two plateaus at 60 K and 90 K in the T_c−x plot.[1,2,3,4] When x is reduced from 7.0, ρ at room temperature changes with a nonmonotonic unusual manner that ρ has a minimum at x≈6.7.[1] In addition, there appears semiconductive behavior between 60 K- and 90 K-phases.[1] On the other hand, properties of epitaxially grown thin films with x=6−7 prepared by us are different from the above properties.[5] In order to understand the properties of the thin films, we have investigated oxygen in-diffusion into $YBa_2Cu_3O_6$ by Monte Carlo calculation based on an asymmetric Ising model.[6,7,8] The calculation allows us to observe nonequilibrium oxygen configurations in the Cu(1) plane.

EXPERIMENTAL

The epitaxially grown $YBa_2Cu_3O_x$ thin films with x=6−7 were synthesized on $SrTiO_3$(001) at 730°C by an ArF excimer laser. The oxygen content was adjusted by oxygen pressure during rapid cooling following the deposition. Cooling time from 730 to 400°C was approximately 2 min. The mechanism of the oxygen variation has been explained by oxygen in-diffusion into $YBa_2Cu_3O_6$ during the cooling.[9,10] Resistivity in the ab plane (ρ_{ab}) was measured by a dc four-probe method down to 12 K. Oxygen content was evaluated from a relation between x and c lattice parameter.[11]

Model and Monte Carlo calculation

We used a model on the Ising square lattice consisting of fixed copper atoms and mobile oxygen atoms. Essential features of the model are three interactions between oxygen atoms: which are V_1, repulsive nearest-neighbor interaction; V_2, attractive next-nearest-neighbor interaction with a Cu atom; V_3, repulsive next-nearest-neighbor interaction without a Cu atom.[6] The Hamiltonian is given by

$$H = -V_1 \sum_{ij} n_i n_j - V_2 \sum_{ij} n_i n_j - V_3 \sum_{ij} n_i n_j - \mu \sum_{i} n_i \qquad (1)$$

where n_i=0,1 is an occupation variable and μ is oxygen chemical potential simulating oxygen pressure Po_2. We used the interaction strengths adapted in ref.12: V_1=−6.9mRy, V_2=2.4mRy, and V_3=−1.1mRy by Sterne and Wille. We have investigated nonequilibrium state due to the oxygen in-diffusion into $YBa_2Cu_3O_6$ by the Monte Carlo technique. Since the thin films were prepared by the rapid cooling at various P_{O_2}, the system was cooled at various μ values. Figure 1 shows cooling schedule in the Monte Carlo calculation. The time is expressed by Monte Carlo steps per site. There is no simple relation between the MCS/s time and the real time. At 0 MCS/s, temperature of the system is 0.46(1000 K) in kT/V_1. The temperature decreases exponentially, and is 0.05(110 K) at 5000 MCS/s. The absolute temperatures in parentheses are in the rough estimation, because they vary greatly with V_1. The time of 5000 MCS/s is insufficient in order for the system to reach equilibrium, so that the system shows such nonequilibrium state as inhomogeneous distribution of the oxygen atoms. The cooling schedule can thus simulate the real cooling in the experiment. The initial condition is of c=0 at 0 MCS/s. The oxygen concentration in the Cu(1) plane (c) is related to x as x=6.0+2c. Oxygen atoms are incorporated from edges of the square lattice including 40 x 40 oxygen sites. The incorporated oxygen atoms diffuse inside the lattice. Such nonequilibrium properties have been first investigated by Andersen, et al.[13]

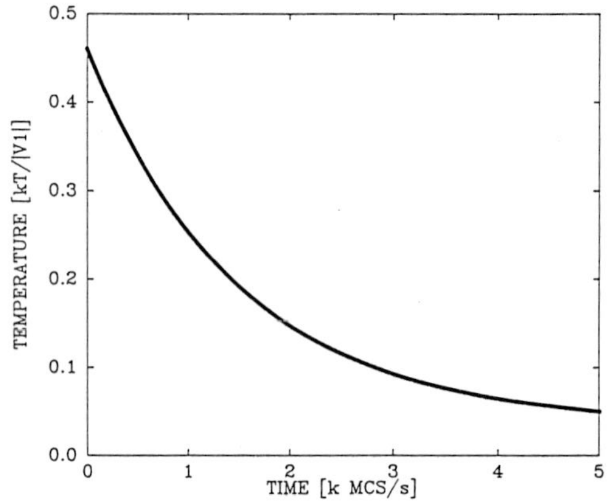

Fig. 1. Cooling schedule for the calculation.

RESULTS AND DISCUSSION

Figure 2 and 3 show ρ_{ab}-T curves and ρ_{ab}(280) and T_c as a function of x.[5] The thin film with x=5.86 is semiconducting in whole temperature range. At x=6.19, ρ_{ab}(280 K) becomes one-fortieth of that at x=5.86, and the ρ_{ab}-T curve shows metallic behavior above 100 K with the rapid increase below 50 K. This suggests that the hole doping into the $Cu(2)O_2$ planes occurs even in x slightly above 6.0. As x increases, the gradient of the ρ_{ab}-T curve in normal state decreases, while ρ (280 K) increases. Both changes are monotonic, and are different from the properties of the $YBa_2Cu_3O_x$ polycrystalline samples.[1]

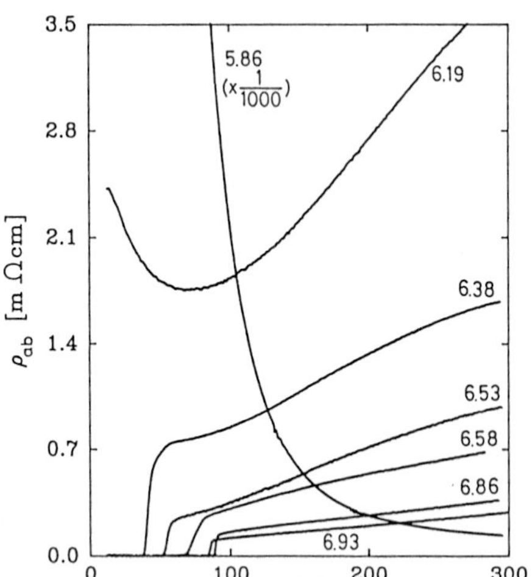

Fig. 2. ρ_{ab}-T curve of the thin films. The number for each curve is oxygen content x.

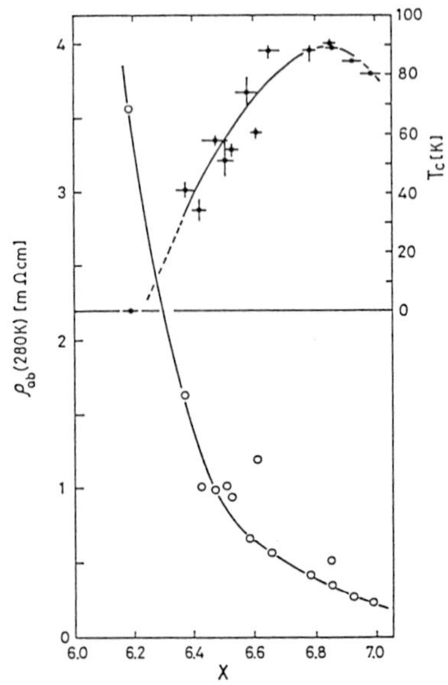

Fig. 3. ρ_{ab} at 280 K and T_c as a function of x. The open circles are ρ_{ab}(280 K). The deviant ρ_{ab} at x=6.61 may be caused by a crack. The closed circles are T_c at half transition.

The lowest value of ρ_{ab}(280 K) is 0.24 mΩcm at x=6.99, while the highest T_c at half transition is 90.5 K at x=6.85. As x increases, T_c increases in x<6.85, while T_c decreases in x>6.85. The T_c−x plot is similar to the T_c−y plot in $La_{2-y}Sr_yCuO_4$.[14] This suggests that the phases with x>6.85 are in over hole doping, and that the manner of the hole doping by oxygen in the $YBa_2Cu_3O_x$ thin films is similar to that by Sr in $La_{2-y}Sr_yCuO_4$. The above arguments postulate that in the thin films hole concentration in the Cu(2)O_2 plane is proportional to c.

The above properties may be affected by oxygen ordering. Figure 4 shows an oxygen configuration constructed by the Monte Carlo calculation at μ=0.6 and 5000 MCS/s. In this calculation, parameters of the system are c=0.336(x=6.67), N_1(the number of the oxygen pairs interacting with V_1)=0, N_2(with V_2)=435, N_3(with V_3)=270. It is observed that the ortho-II phase core is surrounded by the orth-I phase near the edges. This inhomogeneous structure is the same as that observed by Andersen et al.[13] In μ<0.2, c is less than 0.22, and oxygen atoms cannot diffuse to the center of the lattice. Even in this low c range, most oxygen atoms form the Cu-O chains. This can be a reason that the thin films made by the rapid cooling show metallic behavior in the low x range. It has been reported that the evolution of the Cu-O chains from a disordered phase at high temperature is very slow, because the chains quickly collide.[15] On the other hand, since the system in this paper has an initial condition of c=0, the chains rarely collide, so that they can extend easily. In μ>0.6, the core of the ortho-II phase shrinks and turns into the ortho-I. However, the time of 5000 MCS/s is insufficient to fill the whole lattice with orth-I.

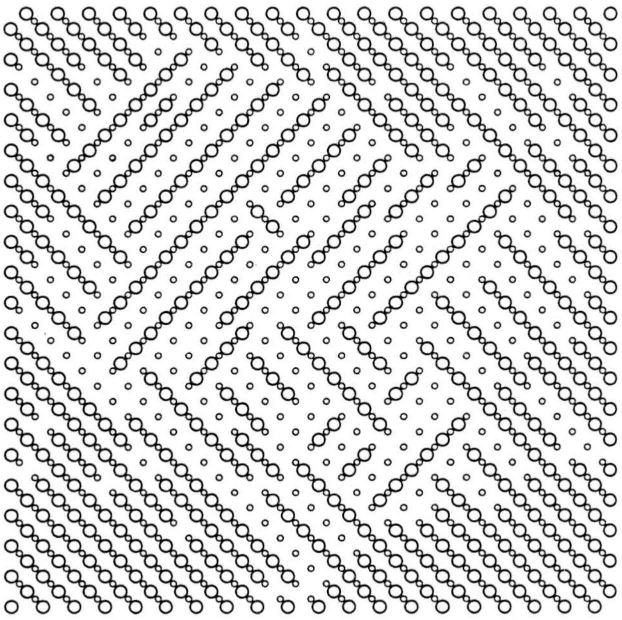

Fig. 4. Oxygen configuration at μ=0.6 and 5000 MCS/s. The small circles are copper atoms, and the large circles are oxygen atoms.

Figure 5 shows the N_2−c plot. It is apparent that N_2 increases linearly with increasing c in c<0.4. In c>0.4, N_2 deviates upward from the linear relation. This is caused by a fact that the number of the fragmented Cu-O chains decreases rapidly in this region, that is, continuous chains almost cover the whole lattice. When we postulate that one oxygen pair in the Cu-O chain supplies one hole into the Cu(2)O_2 plane, hole concentration per CuO_2 equals N_2/800. As a result, we can conclude that hole concentration is proportional to c in the thin films made by the rapid cooling.

It has been reported that a long time annealing at low temperature results in phase separation into phases with different oxygen orderings which differ slightly in x. These structures have been observed in electron diffraction experiments.[16] This phase separation is believed to cause the plateaus in the Tc−x plot.[17] On the other hand, no plateaus in the T_c−x plot have been observed in quenched samples.[11] Therefore, it is reasonable that the short cooling time in this paper prohibit the phase separation, so that the thin films have no plateaus.

CONCLUSIONS

The epitaxially grown $YBa_2Cu_3O_x$ thin films by the rapid cooling following the laser deposition have uncommon properties. The thin films have no two plateaus at 60 K and 90 K in the $Tc-x$ plot. The metallic behavior has been observed even at x=6.19. The highest T_c(=90.5 K) is at x=6.85, and T_c decreases in x>6.85. The properties may be affected by the oxygen configurations in the Cu(1) plane, which are made by the oxygen in-diffusion into $YBa_2Cu_3O_6$ during the rapid cooling. The in-diffusion has been simulated by the Monte Carlo calculation based on the asymmetric Ising model. The in-diffusion makes the formation of the Cu-O chains easy even in the low x range. It has been revealed that N_2, the number of the oxygen pairs interacting with V_2, is proportional to c, oxygen concentration. This corresponds to the linear hole doping by oxygen in the Cu(1) plane in x<0.4. Additionally, the short cooling time can prohibit the phase separation, which causes the plateaus in the T_c-x plot. In x>6.85, the $YBa_2Cu_3O_x$ thin films have shown a sign of over hole doping, as observed in $La_{2-y}Sr_yCuO_4$. This is consistent with a report that the system of $(Y_{1-x-y}Ca_yPr_x)Ba_2Cu_3O_{7-\delta}$ indicates that the highest T_c is obtained by reducing hole concentration in the $Cu(2)O_2$ plane from that of $YBa_2Cu_3O_7$.[18] As a result, the T_c-x plot in the thin films is similar to the T_c-y plot in $La_{2-y}Sr_yCuO_4$. One possible reason for 90 K-T_c in polycrystalline $YBa_2Cu_3O_7$ samples is that there are boundary phases with x lower than that of the inside of grains, even though measurements of oxygen content show x≈7.

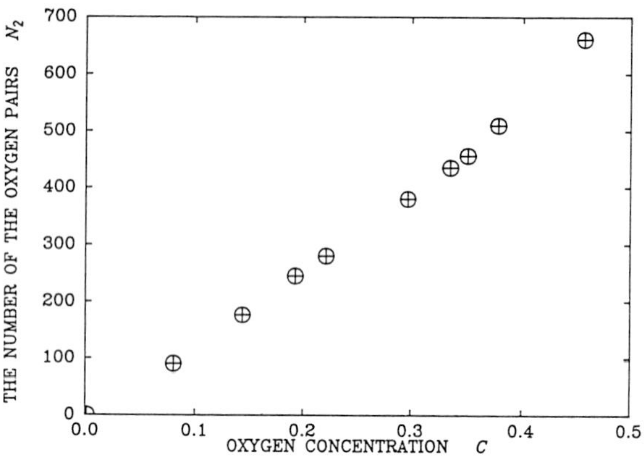

Fig. 5. N_2-c plot in the Monte Carlo calculation. N_2 is the number of the oxygen pairs in the Cu-O chains, and c is oxygen concentration in the Cu(1) plane.

1. Cava RJ, Batlogg B, Chen CH, Rietman EA, Zahurak SM, Werder D (1987) Phys Rev B36: 5719; Cava RJ, Hewat AW, Hewat EA, Batlogg B, Marezio M, Rabe KM, Krajewski JJ, Peck Jr WF, Rupp Jr LW (1990) Physica C165: 419
2. Jacobson AJ, Newsam JM, Johnston DC, Goshorn DP, Lewandowski JT, Alvarez MS (1989) Phys Rev B39: 254
3. Nakazawa Y, and Ishikawa M (1989) Physica C158: 381
4. Mazaki H, Ueda Y, Aihara Y, Kubozoe T, Kosuge K (1989) Jpn J Appl Phys 28: L368
5. Ohkubo M (1990) Solid State Commun 74: 785
6. Will LT, Berera A, de Fontaine D (1988) Phys Rev Lett 60: 1065
7. Kikuchi R, Choi J-S (1989) Physica C160: 347
8. Aukrust T, Novotny MA, Rikvold PA, Landau DP (1990) Phys Rev B41: 8772
9. Ohkubo M, Kachi T, Hioki T, Kawamoto J (1989) Appl Phys Lett 55: 899
10. Ohkubo M, Kachi T, Hioki T (1990) J Appl Phys 68: 1782
11. Tranquada JM, Heald SM, Moodenbaugh AR, Xu Y (1988) Phys Rev B38: 8893
12. de Fontaine D, Ceder G, Asta M (1990) Nature 343: 544
13. Andersen JV, Bohr H, Mouritsen OG (1990) Phys Rev B42: 283
14. Torrance JB, Tokura Y, Nazzal AI, Bezinge A, Huang TC, Parkin SSP (1988) Phys Rev Lett 61: 1127
15. Burmester CP, Wille LT (1989) Phys Rev B40: 8795
16. Beyers R, Ahn BT, Gorman G, Lee VY, Parkin SSP, Ramirez ML, Roche KP, Vazquez JE, Gür TM, Huggins RA (1989) Nature 340: 619
17. Beyers R, Shaw TM (1989) Solid State Physics 42: 135
18. Neumeier JJ, Bjørnholm T, Maple MB, Schuller IK (1989) Phys Rev Lett 63: 2516

Preparation and Superconductive Properties of $YBa_2Cu_3O_y/PrBa_2Cu_3O_y$ Layered Structure

HARUHIKO OBARA, SHIN KOSAKA, and YOICHI KIMURA

Electrotechnical Laboratory, 1-4, Umezono 1-chome, Tsukuba, 305 Japan

ABSTRACT

Using molecular beam epitaxy (MBE), we have successfully prepared $PrBa_2Cu_3O_y$ (PBCO) films and PBCO/$YBa_2Cu_3O_y$ (YBCO) layered structures. Epitaxial growth of these films were confirmed by *in-situ* reflection high-energy electron diffraction (RHEED) and coverage of $YBa_2Cu_3O_y$ by $PrBa_2Cu_3O_y$ was estimated using *in-situ* x-ray photoelectron spectroscopy (XPS). The temperature dependence of resistivity in the PBCO film showed semiconducting behavior. The lattice parameters of $Pr_{1+x}Ba_{2-x}Cu_3O_y$ films decreased when x increased and almost coincided with that of well-oxidized $YBa_2Cu_3O_y$ films ($y\sim 7$) when $x \sim 0.3$. Finally the superconductive properties of these layered structure have been measured by magnetization measurements using a SQUID magnetometer.

KEY WORDS: oxide superconductor, $YBa_2Cu_3O_y$ thin film, $PrBa_2Cu_3O_y$, layered structure, magnetization

INTRODUCTION

Since the discovery of high-T_c oxide superconductivity, thin film processes have been extensively studied. One of the purposes of this research is to apply the high-T_c superconductors to electronic devices and another is to observe the intrinsic properties of these systems made possible by high quality thin films. For electronic devices, layered structures of superconducting and non-superconducting materials are needed. The complicated structure of the oxide superconductors, however, makes it difficult to layer different materials.

The PBCO system, which is one of the non-superconducting phases in the $RBa_2Cu_3O_y$ system, is the ideal material for a normal layer because it has the same structure as the YBCO system. Several groups have reported the preparation of PBCO/YBCO layered structures using sputtering [1,2] and laser deposition [3].

The MBE technique, which we used in the present work, is an ideal method for thin film preparation because high quality samples can be prepared, and this system has the possibility of the epitaxial growth of films. So we have tried to prepare PBCO films and PBCO/YBCO layered structures using MBE and observed the epitaxial growth of these materials.

EXPERIMENTAL

In the present work, samples were prepared by coevaporation in an ultra- high vacuum (UHV) MBE system. The background pressure was 1×10^{-10} Torr and the pressure during deposition was less than 1×10^{-4} Torr. The evaporation sources were four effusion cells (the maximum temperature is 1800°C) and Y, Ba, Cu and Pr metal elements were individually evaporated. Ba and Cu were continuously evaporated and the layered structure was controlled by shutters of Y and Pr sources. The deposition rate was typically ~ 0.5 Å/s and the substrate temperature was about 650°C. For oxidation of the samples, we used an electron cyclotron resonance (ECR) ion source and oxygen gas was introduced to the ECR ion source during deposition. The distance between the discharge tube of the ECR ion source and the sample was approximately 10 cm. The ion current density at the sample was typically 0.1 mA/cm^2, which depends on the pressure and voltage of the ECR ion source. The details of sample preparation have been described elsewhere [4].

DC magnetization measurements were performed using a SQUID magnetometer. Samples were cut in a rectangular shape and magnetic field was applied perpendicular to films.

RESULTS AND DISCUSSION

A typical X-ray diffraction pattern of a PBCO film is shown in Fig. 1. The PBCO films were deposited on MgO substrates with (100) plane and the thicknesses of the films were about 1000 Å. All peaks can be indexed to $(00l)$ reflections of PBCO. The Films are highly c-axis oriented perpendicular to the substrates. The conditions of growth of the PBCO films, (i.e., the temperature of substrate, oxygen pressure and deposition rate), are the same as the YBCO films. The common condition of growth of YBCO and PBCO films indicates that the layering of these system is easy and that good quality layered structure can be achieved.

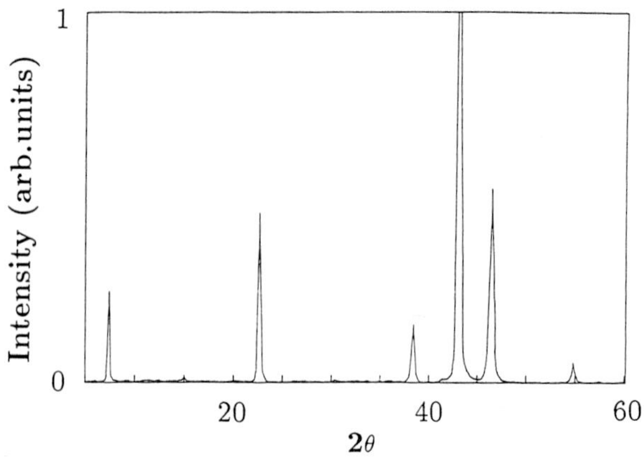

Fig. 1 X-ray diffraction pattern of a PBCO film on a MgO (100) substrate.

Figure 2 shows the temperature dependence of the resistivity of a PBCO film. The resistance measurement was performed using a four terminal technique. We have clearly observed semiconducting behavior of resistivity. The rare-earth substitution for Y in $RBa_2Cu_3O_y$ does not affect superconductivity except for R = Ce, Tb, Pm, and Pr. The non-superconductivity of PBCO is due to the Pr^{4+} valence of the Pr [5]. The reduction of the hole concentration and the magnetic ordering of the Pr ion is important for the transport properties of PBCO. The extensive study of transport in epitaxial PBCO films will give information regarding the magnetism and superconductivity in high-T_c oxide superconductors.

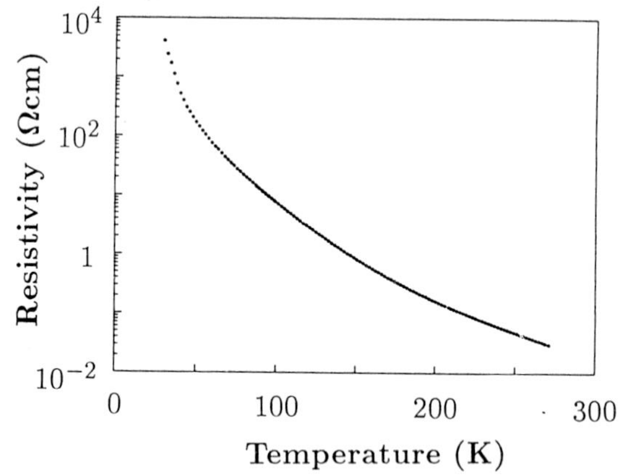

Fig. 2 Temperature dependence of resistivity of PBCO film.

We have also succeeded in the epitaxial growth of YBCO films using MBE. Details of the superconducting properties of these YBCO films have been described elsewhere [4]. In the present work, we have tried to layer YBCO and PBCO films using MBE technique and succeeded to layer these films.

Figure 3 shows the superconductive properties of PBCO/YBCO/PBCO layered structures which were measured using a SQUID magnetometer. The thickness of bottom PBCO layer was 200 Å, which was deposited on MgO substrate as the bufferlayer, and the thickness of top PBCO layer was 80 Å, which prevented the degradation of YBCO films. In Fig. 3, T_c was determined as the onset temperature of magnetization and the magnetization at $T_c/5$ was also plotted as a function of thickness of YBCO layer. The rapid decrease of magnetization was observed in very thin films of YBCO (< 40 Å). The magnetization measurements of thin films is sensitive for homogeneity of samples [6]. Large magnetization which was observed in relatively thick film (> 80 Å) indicates that YBCO films was enough homogeneous to have high critical current density ($> 10^6 A/cm^2$). On the other hands, very thin films of less than 40 Å showed inhomogeneous properties, which is due to the island structure or step structure of these layered films.

In Fig. 4, we show the x-ray photoelectron spectroscopy (XPS) spectra of PBCO films on YBCO films of various PBCO thicknesses. Above 60 Å, the peak of Y 3d disappears and complete coverage of the YBCO film was observed. The small intensity of the Y 3d peak of the sample with the PBCO 30 Å upper layer indicates that the films were not smooth in the scale of 30 Å , because of large unit cell of these systems. These XPS data were consistent with the superconducting properties of PBCO/YBCO/PBCO system.

Moreover, we have prepared Pr rich and Ba poor samples, i.e., $Pr_{1+x}Ba_{2-x}Cu_3O_y$. Figure 5 shows the lattice parameter of a YBCO/PBCO bilayer. The thickness of each film is 300 Å and the lattice parameters were calculated from YBCO (004) peaks and PBCO (003) peaks of the x-ray diffraction pattern. In such a relatively thick layered films (~ 300 Å), lattice parameters of each layer were same as single PBCO and YBCO films. The lattice parameter of $Pr_{1+x}Ba_{2-x}Cu_3O_y$ decreases as x increases and tends to coincide with that of YBCO when $x \sim 0.3$. The similarity in the lattice parameters of the YBCO and PBCO films is advantageous for layering of these materials.

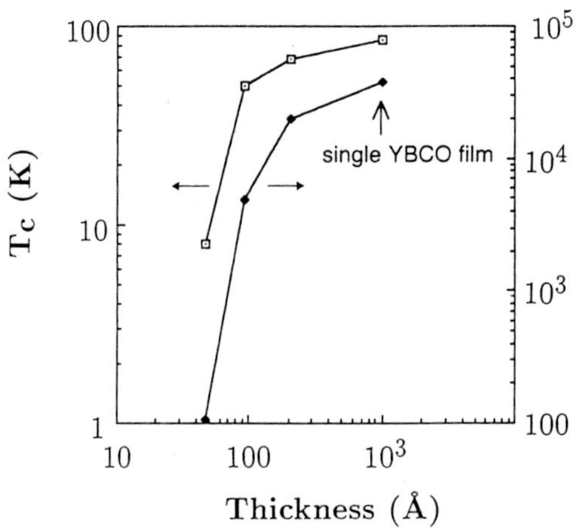

Fig. 3 YBCO thickness dependence of T_c (the onset temperature of magnetization) and remanent magnetization at $T_c/5$ of PBCO/ YBCO/ PBCO layered structure.

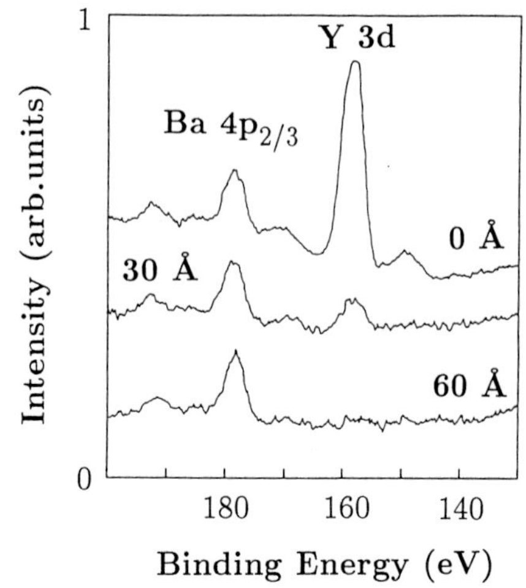

Fig. 4 XPS spectrum of YBCO, PBCO (30 Å) on YBCO (300 Å) and PBCO (60 Å) on YBCO (300 Å).

The substitution of Pr for Ba is easy because the ionic radii of Pr and Ba are similar and Pr rich / Ba poor sample has a shorter c-axis lattice parameter than the stoichiometric PBCO system [7,8]. Of course, lattice matching in the a-b plane is needed in the film, which is c-axis oriented perpendicular to the substrate. For low x ($x < 0.4$), $Pr_{1+x}Ba_{2-x}Cu_3O_y$ has an orthorhombic structure and the lattice parameters a and b decrease as c decreases. So we can estimate the lattice matching of YBCO/PBCO system from the measurements of lattice parameter c in ordinary x-ray diffraction patterns.

Figure 6 show the RHEED patterns of a YBCO/ $Pr_{1.1}Ba_{1.9}Cu_3O_y$/ YBCO film. The substrate is also MgO (100), the top layer is YBCO and the thickness of each layer was 300 Å. Clear streak patterns were observed and confirm the epitaxial growth of the YBCO/$Pr_{1.1}Ba_{1.9}Cu_3O_y$/YBCO films on MgO. The lattice matching of YBCO and $Pr_{1.1}Ba_{1.9}Cu_3O_y$ is better than that of YBCO and PBCO, which has stoichiometric composition. The controllability of the lattice matching in PBCO/YBCO thin films will make it possible to change the properties of the YBCO film between the PBCO films. For example, artificial control of the defects at the interface between YBCO and PBCO films is one of the attractive prospects in regard to technological aspects of the oxide superconductors.

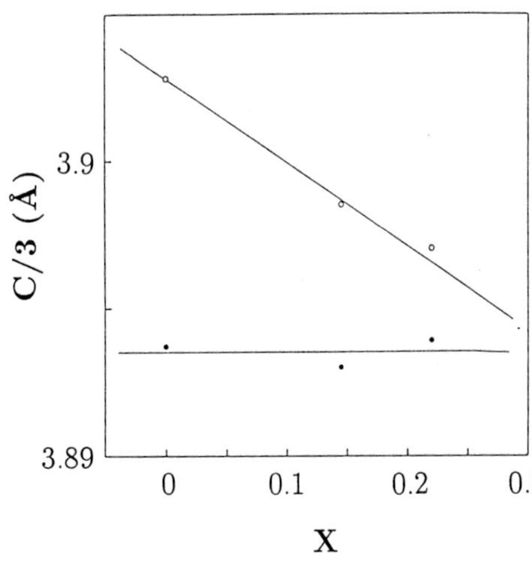

Fig. 5 Lattice constant $c/3$ of YBCO/ $Pr_{1+x}Ba_{2-x}Cu_3O_y$ bilayer. $c/3$ of YBCO (•) is calculated from (004) peaks and that of $Pr_{1+x}Ba_{2-x}Cu_3O_y$ (∘) is calculated from (003) peaks of the x-ray diffraction pattern.

CONCLUSION

In summary, using MBE techniques, we have successfully prepared epitaxial films of PBCO that have a semiconducting temperature dependence of resistivity. The growth condition of PBCO film is same as YBCO films. Large magnetization which was observed in PBCO (200 Å)/YBCO (80 Å)/PBCO (80 Å) layered films which indicates that 80 Å YBCO films was enough homogeneous to have high critical current density ($> 10^6 A/cm^2$). Moreover, the lattice parameter, c, in Pr rich and Ba poor sample decreased as Pr content increased. The RHEED pattern of the YBCO/ PBCO/ YBCO system shows clear evidence of epitaxial growth on MgO substrates. These experimental results indicate that this system is very suitable for layered structure of oxide superconductors.

Fig. 6 RHEED pattern of YBCO / $Pr_{1.1}Ba_{1.9}Cu_3O_y$ / YBCO film of < 100 > MgO azimuth.

ACKNOWLEDGMENT

The authors would like to express their thanks to Dr. Kajimura for his encouragement throughout this research.

REFERENCES

1. J.-M, Triscone, ϕ. Fischer, O. Brunner, L. Antognazza and A.D. Kent, Phys. Rev. Lett. **64**,804 (1990).

2. C.L. Jia, B. Kabius, H. Soltner, U. Poppe, K. Urban, J. Schubert and Ch. Buchal, Physica C **167**, 463 (1990).

3. T. Venkatesan, I. Inam, B. Dutta, R. Ramesh, M.S. Hegde, X.D. Wu, L.Nazar, C.C. Chang, J.B. Barner, D.M. Hwang, and C.T. Rogers, Apple. Phys. Lett. **56**,391 (1990).

4. H. Obara, S. Kosaka, M. Umeda and Y. Kimura, *Advances in Superconductivity* II (Springer, Tokyo, 1990), p.801.

5. D.W. Cooke, R.S. Kwok, E.L.Lichati, T.R Adams C. Boekema, W.K. Dawson, A. Kebede, J. Schwegier, J.E. Crow and T.Mihalisin: Phys. Rev. **B41**, 4801 (1990).

6. H. Obara, S. Kosaka, M. Umeda and Y. Kimura: Physica B **166** 1413 (1990).

7. A. Suzuki, Sampathkumaran, K. Khon, T. Shibuya, A. Tohdake and M. Ishikawa: Jpn. J. Appl. Phys. **27**, L792 (1988).

8. M. Izumi, T. Yabe, T. Wada, A. Maeda, K. Uchinokura and S. Tanaka: Phys. Rev. **B40**, 6771 (1989).

Application of NO_2 Supersonic Molecular Beam to Oxide Superconductor Fabrication in MBE

HIDEHIKO NONAKA, TAKASHI SHIMIZU, and KAZUO ARAI

Electrotechnical Laboratory, 1-4, Umezono 1-chome, Tsukuba, 305 Japan

ABSTRACT

We have developed a new type of supersonic molecular beam source to supply a high-density oxidizing agent locally to the substrate in MBE system without raising the ambient pressure. With NO_2 beam, oxide superconductor films were grown in proper MBE conditions under in-situ monitoring of the growing surface by RHEED.

KEY WORDS: NO_2, supersonic molecular beam, RHEED

INTRODUCTION

Molecular beam epitaxy (MBE) is a challenging and promising method for fabrication of oxide superconductors with an atomic level control. The method will give the possibility to investigate the mechanism of superconductivity by making and studying model compounds and even to design new materials. The MBE is a preparation method where crystal films grow from fluxes of constituent elements supplied only in the form of molecular beam (without any disturbance from the surroundings because of ultra-high vacuum conditions). In the case of oxides, as gaseous oxidizing agents are usually introduced through an effusive nozzle, the deposition pressure tends to become too high for the proper MBE conditions due to the scattered gases from the nozzle out of the direction to the substrate.

We have adopted the supersonic molecular beam (SMB) technique for supplying a large flux of oxidizing agent locally to the substrate without increasing the ambient pressure. For this purpose, we have developed a new type of SMB source with a liquid nitrogen trap, which is small enough to be installed in our MBE system. Supplying the oxidizing agent in the form of SMB enables not only the film growth in high vacuum, but also the use of various means for monitoring the growing surface of the films including reflection high energy electron diffraction (RHEED) because there is neither mechanical nor electrical hindrance around the substrate such as nozzles or electrical plasma. With an NO_2 beam, which is the best oxidizing agent for the reasons described later, as-deposited superconducting $ReBa_2Cu_3O_{7-\delta}$ (Re=Nd, Dy, Y) films were grown under monitoring of the growing surface with the RHEED.

SUPERSONIC MOLECULAR BEAM

In the last two decades, the technique for generating supersonic molecular beams has been established especially in the field of physical chemistry [1]. Figure 1 shows the basic idea to generate the SMB. When a high pressure gas is blown out from a nozzle with a pin-hole aperture into a low-pressure ambient background, it expands adiabaticaly having a region of supersonic flow along the beam centerline. The centerline beam is then extracted with a skimmer, a cone-shaped filter, to obtain a well-collimated supersonic beam. The beam is often further narrowed by another cone-shaped collimater as shown in the figure.

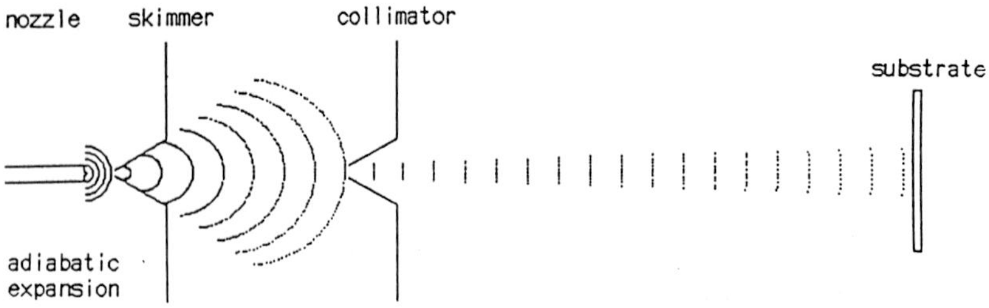

Fig. 1. Basic idea to generate supersonic molecular source.

However, in order to extract successfully the supersonic jet from the nozzle, the gas molecules scattered by the skimmer must be fully pumped out of the region. The situation is the same at the collimator. Therefore, usual SMB sources need differential pumping system of a large pumping speed to extract the beam efficiently and their sizes are often too large to use in the conventional MBE system. We have developed a new type of SMB source with a cryogenic pumping, which is small enough to be installed easily in our MBE chamber through an ICF-152 flange.

Figure 2 shows a schematic diagram of our SMB source. The source has a liquid N_2 trap (A, 43 mmϕ × 200 mm) for cryogenic pumping of the gas instead of a large pumping system. The pumping speed by the trap is estimated to be more than 2000 ls^{-1} for gases which solidify and have negligible vapor pressures at liquid N_2 temperature. The source has a nozzle with a 0.1mm-aperture (B), followed by a 1-mm skimmer (C) and a 3-mm collimator (D). The nozzle can be moved along the beam axis to optimize beam conditions and has a tungsten heater around it to prevent the gases from condensing at the aperture.

In table 1 the physical properties [2] and chemical reactivities of the candidates for oxidizing gases which could be used in our SMB source are listed up. Oxygen, O_2, and ozone, O_3, are not suitable because their vapor pressures at 77 K are too high to pump out by the liq. N_2 trap though the reactivity of O_3 is very high. Among nitrous oxide, N_2O, nitrogen dioxide, NO_2, and hydrogen peroxide, H_2O_2, NO_2 is the best oxidizing agent because N_2O has a similar reactivity to O_2 and handling of pure H_2O_2 seems extremely difficult at high stagnation pressure while NO_2 is rather stable at room temperature even at high pressure and is expected to react vigorously with metals at high temperature. Therefore, we chose NO_2 for the SMB source gas [3].

Table 1. Properties of various oxidizing gases

Gases	melting point (K)	Vapor pressure at 77K (Pa)	Chemical reactivity
O_2	54.8	>100	poor
O_3	80.5	4	good
N_2O	182.4	4×10^{-5}	poor
NO_2	262.0	$<1\times10^{-6}$	good
H_2O_2	272.4	$<1\times10^{-6}$	?

Figure 3 shows the NO_2 beam profile measured with a conventional pressure gauge (B-A type) at a setup shown in fig. 2. The gauge was modified to have a hole of 3.2 mm in diameter [E] through which the beam goes into the bulb of the gauge. The beam intensity was measured as an equilibrium pressure between going in and out of the gas molecules through the hole. The flux of the beam is thus estimated to be more than 10^{16} molecules $cm^{-2}s^{-1}$ with the observed full width at half maximum of 7 mm when the stagnation pressure is 1 atmospheric pressure, i.e., room temperature, and the distance between the source head and the gauge hole was 70 mm.

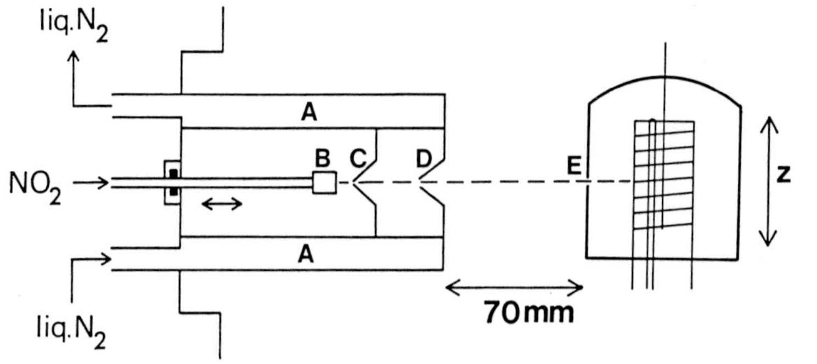

Fig. 2. Schematic Diagram of the supersonic molecular beam source (**A**. Liquid N_2 trap. **B**. Nozzle with a 0.1mm aperture. **C**. 1-mm skimmer. **D**. 3-mm collimator. **E**. B-A gauge with 3.2 mmϕ hole).

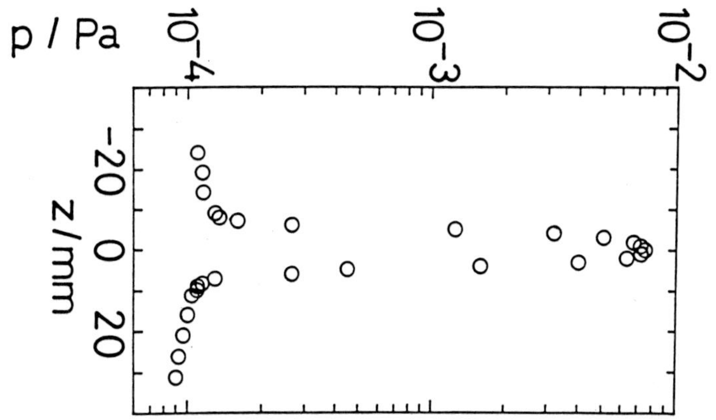

Fig. 3. NO_2 beam profile measured as a function of the position from the beam center and the equilibrium pressure therein at a stagnation pressure of 1 atmospheric pressure over a distance of 70 mm from the source head.

PREPARATION OF OXIDE SUPERCONDUCTOR FILMS BY MBE WITH NO_2 BEAM

The $ReBa_2Cu_3O_{7-\delta}$ (Re=Nd, Dy, Y) films were deposited on either MgO (100) or $SrTiO_3$ (100) substrates by coevaporation of pure metals with the NO_2 beam, using a conventional MBE system [4]. Elemental Re and Cu were evaporated using individual electron beam guns and Ba from a Knudsen cell. The SMB source was directed vertically to the substrate surface and the distance between the source head and the substrate was 110 mm. Typical substrate temperature, was 700°C-750 ℃. The overall oxide growth rate was about 0.1 nm·s^{-1}. Although the vapor pressure of NO_2 at 77 K is less than 1×10^{-6} Pa, ambient pressure during deposition was $\sim 1\times 10^{-5}$ Pa, probably because of NO, a product of thermally decomposed NO_2, which cannot be trapped by the liquid N_2 shrouds. When the deposition was finished, the films were cooled down to about 250 ℃ with the NO_2 beam incident on the sample.

The NO_2 beam during deposition was so well collimated that the film was locally oxidized as recognized by a black circle with a definite periphery. The diameter of the circle was around 13 mm as expected from the beam intensity distribution at the distance of 110 mm. According to the in situ measurement of X-ray photoelectron spectroscopy, the film does not contain any detectable amount of nitrogen. The X-ray diffraction pattern of the films showed that the films were highly oriented along c-axis. The as-grown $DyBa_2Cu_3O_{7-\delta}$ film showed zero electrical resistances at about 30 K while it came up to 88 K after annealing at 450 ℃ in O_2.

During deposition, we observed the growing surface of the films by RHEED. In the in-situ RHEED patterns, clear streak lines were observed indicating a smooth surface of epitaxial film growth. As shown in fig. 4, the RHEED intensity oscillation for the specular beam was observed for the early stage of the deposition of $YBa_2Cu_3O_{7-\delta}$ on $SrTiO_3$. The observed period of 15.2 s may correspond to the time for the growth of one unit cell if the actual growth rate is 0.08 $nm \cdot s^{-1}$.

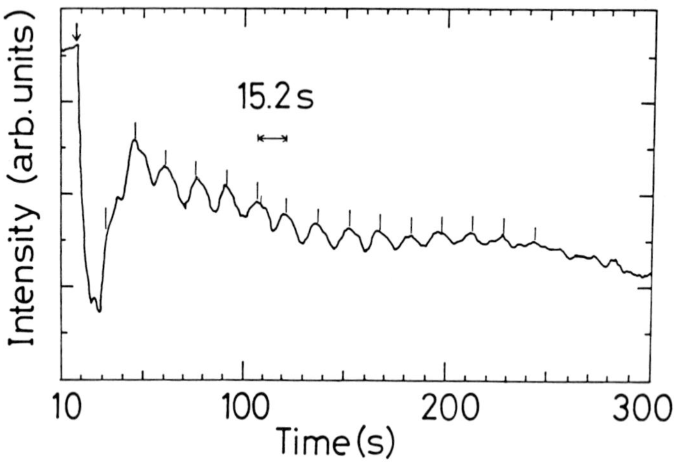

Fig. 4. RHEED intensity oscillation for the specular beam observed during the deposition of $YBa_2Cu_3O_{7-\delta}$ on $SrTiO_3$.

CONCLUSION

The efficient oxidizing gas beam source we have developed has made it possible to use a conventional MBE system for preparation of oxide superconductor without any modification. During deposition the epitaxial growth of the films was observed in situ by RHEED. The method can be applied to preparation of any kinds of oxide films, for example, SiO_2 crystal, in UHV by MBE method. As an extended study we are now comparing the total effectiveness of NO_2 and O_3 [5] as an oxidizing agent in MBE system.

We would like to thank Dr. Takashi Nagata of the University of Tokyo for his valuable advice and suggestion on designing the supersonic molecular beam source. We wish to thank the members of Liaison Committee on Superconductivity Research in our Laboratory for their efforts to support and promote our research program.

REFERENCES

1. Basic techniques and applications for supersonic molecular beams are explained in detail in Atomic and Molecular Beam Methods, Scoles G (1988) Oxford University Press. New York Oxford and references therein.
2. Weast RC (1989) Handbook of Chemistry and Physics. CRC Press, Boca Raton.
3. Nonaka H, Shimizu T, Arai K, Okada K, Nakamichi I, Tsuboi A, and Nakashita T, Spring Meeting of Appl. Phys. Soc. Jpn. (Mar. 1990) 31a-V-4.
4. Nonaka H, Shimizu T, Arai K, Appl. Phys. Lett. in press.
5. Ichimura S, Hosokawa S, Nonaka H, Shimizu T, and Arai K, this volume.

Preparation of Y-Ba-Cu-O Films by Activated Reactive Evaporation

YUJI AOKI, TOSHIO USUI, RAINER FROMKNECHT, KATSUMI SUZUKI, TADATAKA MORISHITA, and SHOJI TANAKA

Superconductivity Research Laboratory, International Superconductivity Technology Center, 10-13, Shinonome 1-chome, Koto-ku, Tokyo, 135 Japan

ABSTRACT

Y-Ba-Cu-O superconducting thin films were prepaered on MgO(100) substrate at 550°C by activated reactive co-evaporation with in-situ annealing. The crystallinity and surface morphology of the films have been investigated using XRD, SEM, and TEM. In-situ annealing at 550°C in oxygen plasma after deposition was quite effective to improve crystallinity and surface morphology of the films.

KEYWORDS: YBaCuO, activated reactive evaporation, in-situ annealing, TEM

INTRODUCTION

Several methods have been applied to prepare the superconducting thin films, such as sputtering[1], chemical vapor deposition[2], laser ablation[3], and reactive co-evaporation[4]. From the viewpoint of fabrication of superconducting electronic devices, it is necessary to get high-Tc films with homogeneity in electronic properties and smooth surface.

In this paper, we report the preparation of Y-Ba-Cu-O thin films with flat and smooth surfaces by activated reactive co-evaporation with in-situ annealing.

EXPERIMENTAL

The films of $YBa_2Cu_3O_{7-x}$ were deposited on MgO(100) substrate at 550°C in oxygen plasma by co-evaporation of Y, Ba, and Cu. The substrate temperature was measured with a pyrometer. Oxygen gas was introduced into a chamber through a stainless ring nozzle, below and 10cm apart from the substrate, to oxidize the metal sources. Moreover, oxygen plasma was generated by a coil electrode with an RF power of 100W in order to accelerate the oxidation. Oxygen pressure during deposition was measured as 2.0×10^{-4} Torr by an ionic gauge, being 20cm apart from the substrate. After deposition, the films were in-situ annealed at 550°C in oxygen plasma for 0~30 min, and then cooled down to room temperature in oxygen plasma. The film thickness was about 100nm and the growth rate was 0.02nm/sec. Crystallinity and surface morphology of the resulting films were investigated using X-ray diffraction (XRD), scanning electron microscopy (SEM) and transmission electron microscopy (TEM).

RESULTS AND DISCUSSION

Crystalinity of the Y-Ba-Cu-O films was investigated as a function of the in-situ annealing time after deposition. Figure 1 shows XRD patterns of the films after in-situ annealing for (a) 0 min, (b) 15 min, and (c) 30 min. All the films were c-axis oriented normal to the substrate; the c-axis orientation was strengthened with increasing in-situ annealing time. In addition, the full width at half maximun (FWHM) of the rocking curve of the (006) peak for corresponding samples was determined as (a) 2.00°, (b) 1.27°, and (c) 0.61°, respectively. These results indicate that the in-situ annealing was quite effective to improve crystallinity of the film at 550°C.

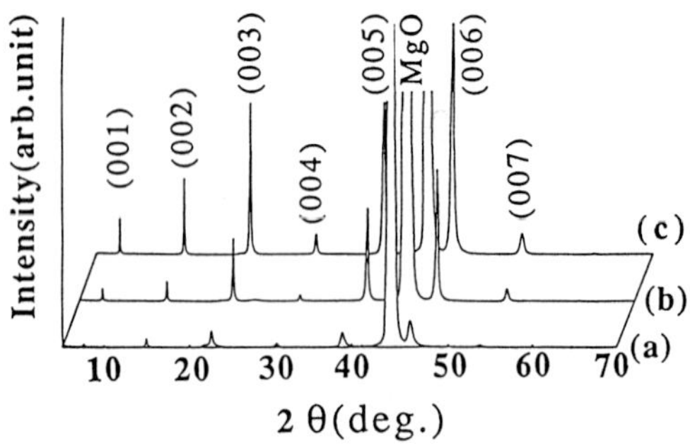

Fig.1. X-ray diffraction patterns of YBCO films on MgO substrate with in-situ annealing for (a) 0 min, (b) 15 min, and (c) 30 min.

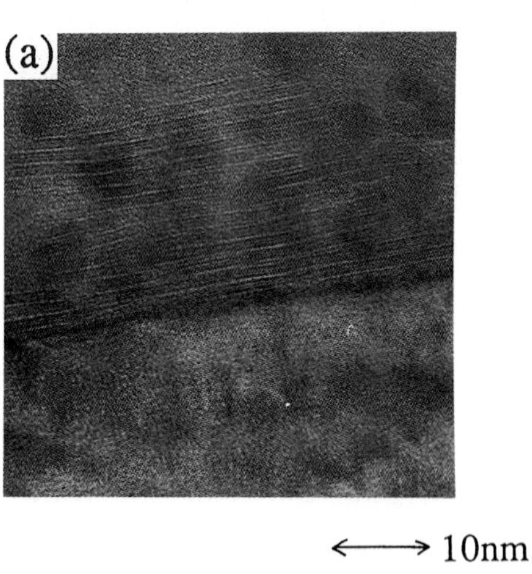

Fig.2(a). TEM micrograph of YBCO film without in-situ annealing.

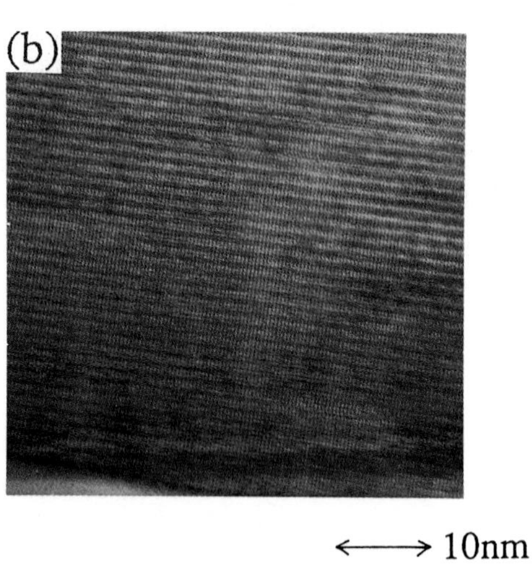

Fig.2(b). TEM micrograph of the YBCO film with in-situ annealing for 30 min.

The difference in crystal microstructure of the films without and with in-situ annealing for 30 min was investigated using TEM. Corresponding micrographs are shown in Figs.2(a) and 2(b), respectively. Figure 2(a) shows that the a-b planes were wavy due to frequent falts in the stacking sequence.

On the other hand, Fig.2(b) shows that the film with annealing for 30 min clearly consists of stacking structure of 123 phase with c-axis orientation. The wedge of a- or b-axis oriented domain sticking in the c-axis oriented layer is found in a low magnification TEM micrograph (Fig.3) of Fig.2(b). This is quite interesting for undrestanding growth mechanism of Y-Ba-Cu-O thin film. The same phenomenon has also been observed in the films deposited on $LaAlO_3(100)$ substrates by Ramesh et al.[6~7]

Surface morphology of the films without and with in-situ annealing for 30 min are shown in Figs.4(a) and (b), respectively. After in-situ annealing for 30 min, surface morphology of the film became smooth.

The temperature dependence of the electorical resistance for films with in-situ annealing for 0, 15, and30 min are shown30 min are

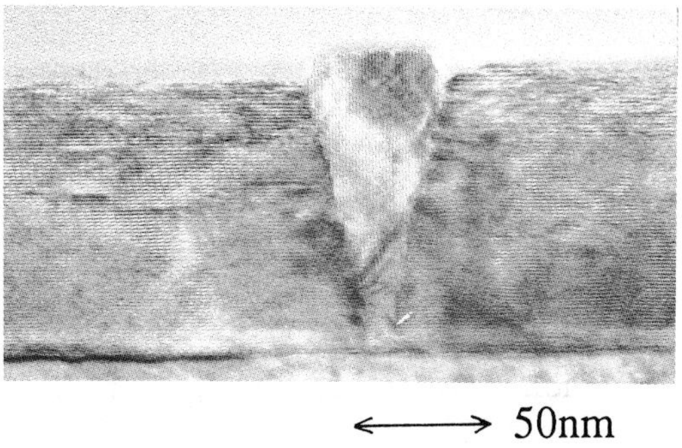

Fig.3. Low magnification TEM micrograph of Fig.2(b).

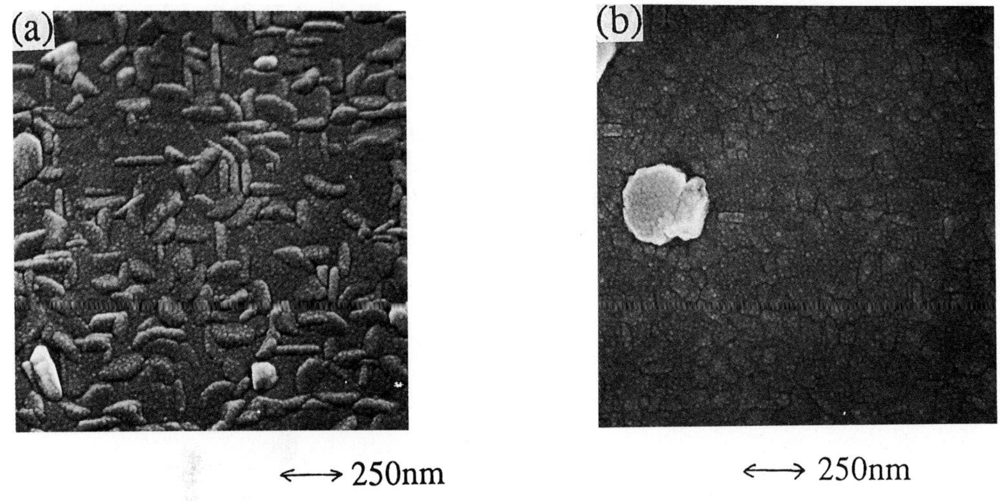

Fig.4. SEM micrographs for YBCO films (a) without and (b) with in-situ annealing for 30 min.

shown in Figs.5(a), (b), and (c), respectively. The film without annealing shows a semiconducing behavior, but others shows a superconducting behavior with the onset at 78K and zero resistance at 60K.

As discussed above, in-situ annealing in oxygen plasma is effective to improve the crystallinity and surface morphology of the films, but superconducting properties of the films are still not sufficiently satisfied.

CONCLUSIONS

$YBa_2Cu_3O_{7-x}$ thin films have been prepared on a MgO(100) substrate at 550°C by activated reactive co-evaporation following in-situ annealing in oxygen plasma. The annealing in oxygen plasma is effective on the crystallization and the improvement of surface morphology of the films.

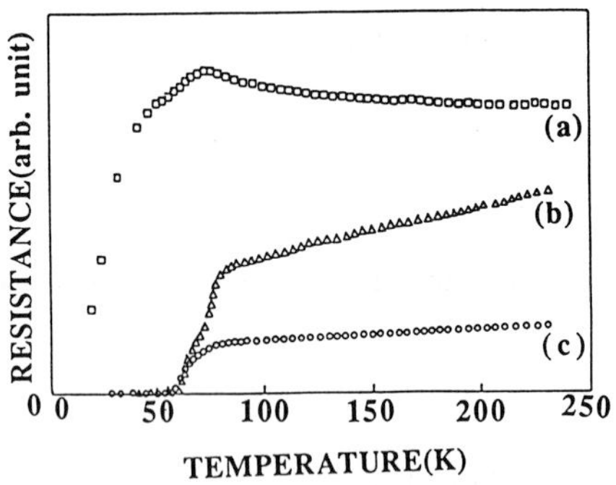

Fig.5. Temperature dependence of a resistance in a 100 nm-thick YBCO films with in-situ annealing for (a) 0 min, (b) 15 min, and (c) 30 min.

REFERENCES

1. O. Michikami, M. Asahi and H. Asano, Jpn. J. Appl. Phys. Lett., **28**, L91 (1989).
2. A. D. Barry, D. K. Gaskiil, R. T. Holon, E. J. Cukauskas, R. Kaplan and R. L. Henry,Appl. Phys. Lett., **52**, 1743 (1988).
3. B. Koren, A. Gupta, E. A. Giess, A. Segmuller and R. B. Laibowits, Appl. Phys., Lett., **54**, 1054 , (1989).
4. T. Terashima, K. Iijima, K. Yamamoto, Y. Bando and H. Mazaki, Jpn. J. Appl. Phys. Lett., **27**, L91 (1988).
5. R. Ramesh, D. M. Hwang, J. B. Barner, L. Nazar, T. S. Ravi, A. Inam, B. Dutta, X. D. Wu and T. Venkatesan, J. Mater. Res., **5**, 704 (1989).
6. R. Ramesh, C. C. Chang, T. S. Ravi, D. M. Hwang, A. Inam, X. X. Xi, Q. Li., X. D. Wu and T. Venkaesan, Appl. Phys. Lett., **57**, 1064 (1990).

Design and Performance of Ozone Generator for Preparation of Superconducting Oxide Thin Films

SHINGO ICHIMURA, SHUNSUKE HOSOKAWA, HIDEHIKO NONAKA, TAKASHI SHIMIZU, and KAZUO ARAI

Electrotechnical Laboratory, 1-4, Umezono 1-chome, Tsukuba, 305 Japan

ABSTRACT

A new ozone generator which enables a stable supply of ozone flux (stability within ±2%) was constructed. Special care was taken for the safety by adoption of an ozone vessel made of stainless steel and by attachment of a rapture disk to it. Ozone vapor was produced by evaporation of liquid ozone which was accumulated by liquefaction of ozone/oxygen mixture gas from an ozonizer, under precise temperature control (accuracy within 0.1 K) of the liquid ozone with a cryocooler and a heater. It was confirmed that the ozone vapor in the vessel had a life time of 2 hours at 95 K, and that the ozone concentration of the ozone vapor was higher than 70%. The successful performance of the ozone generator was confirmed through the formation of superconducting oxide thin films by MBE method.

KEY WORDS: ozone generator, MBE, high temperature superconductor

INTRODUCTION

The formation of superconducting oxide thin films by MBE (molecular beam epitaxy) method requires the supply of highly concentrated oxidizing agent to a deposition substrate. The pressure of the reagent in a vacuum chamber should be low enough, e.g., less than 10^{-4} Pa, to control molecular beam flux precisely. From this view point, ozone is now considered as one of the ideal oxidizing agents because higher activity of ozone than molecular oxygen makes it possible to lower the pressure in a vacuum chamber. Some applications of ozone in the field of the preparation of superconducting oxide thin films have been already reported [1-3].

The simplest way of supplying ozone during the film formation is to use ozone/oxygen mixture gas from an ozonizer. However, due to the limitation of the ozone production ability of the ozonizer and the safe handling of highly concentrated ozone gas, the concentration of ozone in the mixture gas is usually limited to about 10 %. At an atmospheric pressure the mixture gas with high ozone concentration (above ~10 vol.%) may possibly cause an explosion [4]. Thus, for the safety, high purity ozone has been obtained by evaporation of a liquid or solid ozone condensed from the mixture gas within the low vapor pressure region [1-3]. For the liquefaction and evaporation of ozone, the combination of an ozone vessel made of quartz glass and liquid nitrogen as a coolant has been applied. Since liquid ozone also has a high potential of ozone explosion, special care and skill are often required when using the combination of quartz vessel and liquid nitrogen.

In order to improve the reliability and controllability of ozone handling, we have constructed a new ozone generator. The generator has a ozone vessel made of stainless steel instead of quartz glass. To provide against ozone explosion in case of an accidents, a rupture disk is attached just above the ozone vessel. The vessel temperature was precisely controlled, within an accuracy of 0.1 K, with a cryocooler and a heater. This resulted in the stability of ozone flux within ±2 % as is described elsewhere [4]. Precise control of ozone flux will surely be necessary in the formation of oxide thin films by MBE method since oxygen is also a constituent element.

The aim of the present paper is to describe the design and performance of the ozone generator from the viewpoint of its usefulness as an oxidation source.

EXPERIMENTAL

Figure 1 shows a schematic diagram of the ozone generator system. The procedure to get high purity ozone jet with the system was as follows: First, ozone/oxygen mixture gas with ozone concentration of 5 vol.% was produced by passing oxygen gas through the ozonizer. Then the mixture gas was flowed through the ozone vessel and was evacuated with a rotary pump (pumping speed; 260 l/m), keeping the valve 2 open while the valve 1 close (see Fig.1). During the gas flow the vessel temperature was controlled at a certain value (temperature of liquefaction). Due to higher vapor pressure of oxygen than ozone, e.g., about 3 orders of magnitude higher at 100 K and further at lower temperature region, high purity liquid ozone can be accumulated. After a certain amount of liquid ozone was condensed in the ozone vessel, the gas flow was stopped and the vessel was kept at a desired temperature (temperature of evaporation). During film deposition high purity ozone evaporated from the liquid was introduced into a deposition chamber by opening the valve 1, being the valve 2 closed. After the film deposition by MBE method the remaining ozone in the vessel was pumped out by the rotary pump through the ozone killer, a stainless pipe heated at 400 °C to decompose ozone to oxygen. The procedure is described in detail elsewhere [5].

Following cares have been taken to improve the reliability of the system: 1) a foreline oil trap cooled with liquid nitrogen was placed between the ozone vessel and the rotary pump in order to avoid back-diffusion of the pump-oil. The oil in the vessel, if it should exist, would often trigger the ozone explosion. 2) the ozone vessel is made of stainless steel instead of quartz glass. Since steel has higher strength and thermal conductivity than quartz, it enables a uniform temperature control of the liquid ozone even for an abrupt heat invasion from outside. 3) a rupture disk is installed immediately above the ozone vessel to release the sudden pressure rise in case of the explosion. The released ozone is ventilated through the draft to avoid leakage of highly concentrated ozone gas to the working environment.

In order to estimate the concentration of ozone in the vapor, an experimental chamber equipped with a quadrupole mass analyzer (ANELVA: AGA-360) was used. The mass analyzer was set at the opposite side (about 30 cm in distance) of a pipe through which the ozone vapor in the ozone vessel was supplied to the chamber.

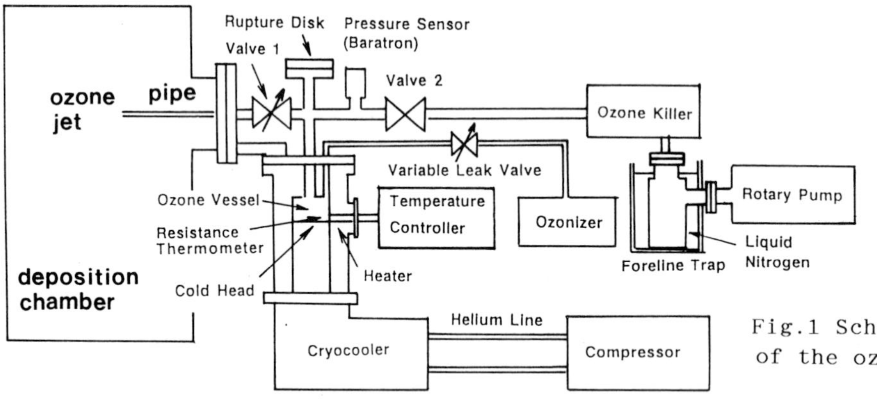

Fig.1 Schematic diagram of the ozone generator

RESULTS AND DISCUSSION

To estimate the quality of ozone vapor obtained with the generator, the ozone-oxygen mixture gas from the ozonizer was first liquefied at various vessel temperatures between 75 K and 95 K. Figure 2 shows vapor pressure of liquid ozone which was accumulated at 82 K and 90 K, respectively, as a function of evaporation temperature. A capacitance manometer (Baratron: 2 Torr head) was used for the pressure measurement. Open marks represent the results obtained at the first stage of the generator operation, while filled marks represents the results measured after a long (2 months) use. Two things are clear from the figure. First, the equilibrium vapor pressure of liquid ozone accumulated at 82 K tends to deviates from that of pure liquid ozone (solid line) at low evaporation temperature. This is partly due to the mixing of oxygen into the liquid ozone, and partly to the decrease of effective pumping speed at this pressure region as is discussed elsewhere [5]. Second, the liquid accumulated at 90 K after the long use showed the same vapor pressure as that of pure liquid ozone [4]. This is probably due to that the surface of the vessel was stabilized by liquid ozone during the long use.

Figure 3 shows mass spectra for the ozone vapor evaporated at 85 K and 90 K from the liquid ozone which was accumulated at 82K (Fig.3(a)) and 90 K (Fig.3(b)), respectively, at the initial stage of the operation (correspond to open marks in Fig.2). Ozone peak (m/e=48) is higher when the liquid was accumulated at 90 K. The signal of diatomic oxygen (m/e=32) is attributable mainly to oxygen molecules evaporated from the liquid oxygen which existed as a small amount of impurity in the liquid ozone [6]. Taking as a reference the mass intensities measured directly for the ozone/oxygen mixture gas from the ozonizer, the ozone concentration was estimated for the vapor evaporated at 85 K and 90 K. The results indicated that the liquid ozone accumulated at temperatures higher than 90 K has the ozone concentration of 70 % [6]. The concentration higher than 70% will be obtained for the vapor evaporated from the liquid which was accumulated after long use (see Fig.2).

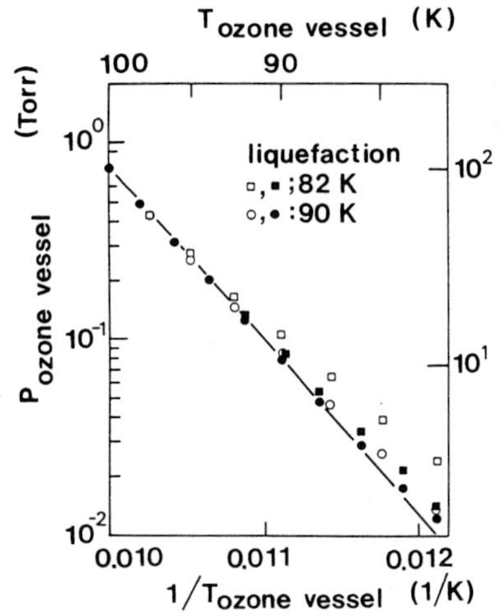

Fig.2 Vapor pressure of liquid ozone accumulated at vessel temperatures of 82 K and 90 K.

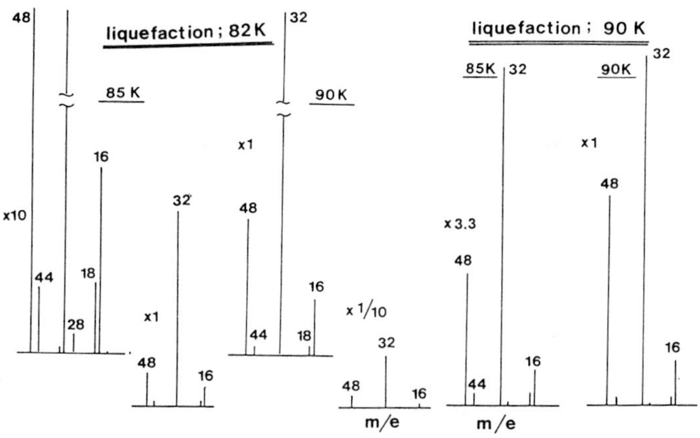

Fig.3 Mass spectra of ozone vapor evaporated at 85 K and 90 K from liquid ozone accumulated at 82 K and 90 K.

Figure 4 shows the pressure of the ozone vessel measured with a capacitance manometer during exposure of ozone vapor to the deposition chamber. Before the exposure the vessel temperature was kept at 83 K, which was increased to 95 K during the exposure. Then the temperature was kept for about 40 min. The figure shows that the pressure is very stable (stability ; +2%) during the exposure. The stability coincides well with the pressure change due to the variation of vessel temperature (+0.1 k). At a time indicated by an arrow in the figure, both the valve 1 and 2 were closed. Then the pressure increased linearly as is shown in the figure. If we assume that the pressure rise is due to the decomposition of ozone molecule into oxygen molecule, and that the diminished ozone molecule is spontaneously supplied from the liquid ozone to keep the equilibrium ozone vapor pressure, we can expect the pressure change in the vessel obeys the following equation [6].

$$P(t) = Ps\left(1+\frac{3\,t}{2\tau}\right)$$

Here, τ denotes the life time of the ozone molecule in the ozone vessel and Ps and P(t) are the equilibrium vapor pressure of the ozone and the time-dependent pressure in the vessel at a constant temperature. From the equation and the vapor pressure of ozone at 95 K, we can estimate the life time of ozone in the vessel to be about 2 hours. This result also supports the high purity of ozone in the jet.

Using the ozone jet produced with the ozone generator, a YBCO film was fabricated with an MBE equipment. Ozone vapor evaporated at 96 K from the liquid accumulated at 82 K was used as an oxidizing reagent. The pressure in the deposition chamber during the film formation was about 4 x 10^{-4} Pa. Figure 5 shows an Cu photoelectron peaks measured for the film. In addition to the two photoelectron peaks of $2p_{1/2}$ and $2p_{2/3}$, which appear at binding energy of 933 eV and 953 eV, respectively, strong satellite peaks are observed. The satellite peaks are characteristic for the oxide layer of CuO. The deposited YBCO film showed, at as-deposited condition, zero resistance at temperatures as high as 55 K. The substrate was $SrTiO_3$, and the deposition rate was about 0.8 A/s. The results and the results of the formation of BSCCO film [7] showed the usefulness of the ozone generator especially for the formation of superconducting oxide by MBE method.

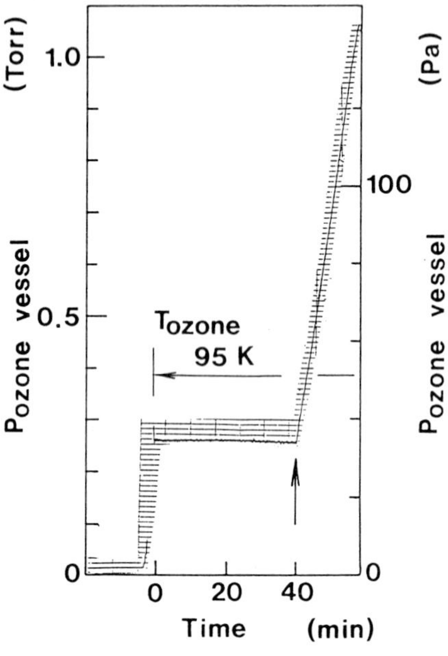

Fig.4 Pressure in the ozone vessel during the evaporation at 95 K. At the time indicated by the arrow the valves were closed.

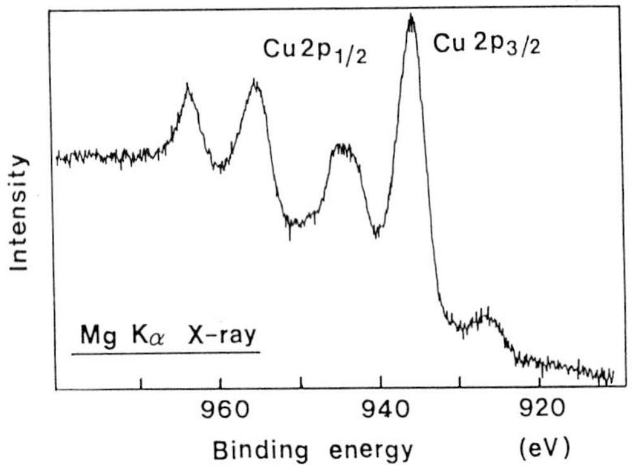

Fig.5 An XPS spectra of Cu measured for a YBCO film prepared with the ozone generator.

CONCLUSIONS

The ozone vapor produced with a new ozone generator was evaluated. The precise controllability of the generator has clarified the following facts.

1)Liquid ozone accumulated at 90 K had higher purity than the liquid accumulated at 82 K. The vapor pressure of the liquid ozone accumulated at 90 K after long use showed the same vapor pressure with pure liquid ozone.

2)Ozone vapor from liquid ozone had a higher ozone concentration when it was evaporated at higher temperature. By evaporation at 90 K, an ozone vapor with ozone concentration of ~70 % was obtained.

3)Ozone vapor in the vessel had a lifetime of 2 hours at the pressure of 0.1 Torr, the equilibrium vapor pressure of ozone at 95 K.

4)The superconducting oxide films (YBCO and BSCCO) were successfully fabricated with the ozone generator by MBE method.

REFERENCES

1.Berkley D.D, Goldman A.M, Johnson B.R, Morton J, and Wang T (1989) Rec. Sci. Instrum. 60: 3769

2.Berkley D.D, Johnson B.R, Anand N, Beauchamp K.M, Conroy L.E, Goldman A.M, Maps J, Mauersberger K, Mecartney M.L, Morton J, Tuominen M, and Zhang Y-J (1988) Appl. Phys. Lett. 53: 1973

3.Nakayama Y, Tsukada I, Maeda A, and Uchinokura K (1989) Jpn. J. Appl. Phys. 28: L1809

4.Horvath M, Bilitzky L, and Huttner J (1985), Ozone , Elsevier, New York

5.Hosokawa S, and Ichimura S, Rev. Sci. Instrum. (submitted)

6.Ichimura S, Hosokawa S, Nonaka H, and Arai K, J. Vac. Sci. & Technol. (submitted)

7.Sakai S, Kasai Y, Hosokawa S, and Ichimura S (1990) Proc. of the 3rd Intern. Symp. on Superconductivity (submitted)

Synthesis of $YBa_2Cu_3O_{7-x}$ Thin Films by the Spray-ICP Technique

MITSUO SUZUKI, MASAHIRO KAGAWA, YASUHIKO SYONO, and TOSHIO HIRAI

Institute for Materials Research, Tohoku University, Sendai, 980 Japan

ABSTRACT

$YBa_2Cu_3O_{7-x}$ thin films were synthesized on single crystal MgO (100) by introducing an atomized aqueous solution of appropriate metal nitrates into the inductively coupled rf plasma above 5000 K generated under atmospheric pressure (the spray-ICP technique). c-axis orientated films of 0.8 μ m thick could be synthesized in 20 min run. They were composed of fine crystals of 30 nm in size. A small amount of CuO was present in the films. The as-deposited films had a typical onset temperature of 91 K and zero resisitivity at 85 K.

KEY WORDS: $YBa_2Cu_3O_{7-x}$, thin film, plasma

INTRODUCTION

The inductively coupled rf plasma (ICP) generated under atmospheric pressure offers a contamination-free high temperature reaction zone. The ICP requires large gas consumption for its stabilization. However, it can hold reactive gases such as oxygen and halogens, because the ICP is electrode-less. This feature makes it possible to synthesize ultrafine particles or thin films of oxides by introducing an atomized solution of metal salts into the ICP (the spray-ICP technique) [1]. The fine droplets are completely decomposed in the ICP, and the oxides are formed through the deposition process from the gas phase. Therefore, the spray-ICP technique are regarded as a CVD process. In this method, the ratio of metal concentration can be easily controlled when the mother solution is prepared. The film formation by this method can be performed under atmospheric pressure at larger deposition rates than other techniques using vacuum technologies. These advantages make the spray-ICP technique a candidate method for large scale film production.

Since the discovery of a high Tc oxide superconductor [3], the spray-ICP technique has been applied to the synthesis of high Tc oxide superconductors. The first application was reported on ultrafine particles of $BaPb_{1-x}Bi_xO_3$ [2]. Single phase $La_{1.8}Sr_{0.2}CuO_{4-x}$ films were formed on fused quartz plates, but they revealed no superconductivity [4]. Thick $YBa_2Cu_3O_{7-x}$ films were obtained by post-annealing semiconductive deposits formed on ZrO_2 and Al_2O_3 substrates [5]. Crystallinity and superconductivity of as-deposited $YBa_2Cu_3O_{7-x}$ particles depended on each experimental setup [6-7]. The best sample of post-annealed Bi-Pb-Sr-Ca-Cu-O thin films prepared on polycrystalline MgO had a Tc of 75 K [8]. Recently two papers reported that superconducting YBa2Cu3O7-x films could be synthesized in situ by the spray-ICP technique. The 0.1-0.2 μ m thick films having a typical Tc of 86 K and Jc in excess of 10^5 A/cm^2 at 77 K and zero field could be deposited on single crystal yttria-stabilized zirconia (100) in 15 min of run [9]. The 1-200 μ m thick films were formed on single crystal MgO (100), and an as-deposited film revealed a Tc at 91 K [10].

Previously we described difficulty of the synthesis of $YBa_2Cu_3O_{7-x}$ films by the spray-ICP technique because of the low solubility of their mother metal nitrates when dissolved in water together [4]. However, the study on the synthesis of Bi-Pb-Sr-Ca-Cu-O thin films indicates that a solubility of 0.1-0.3 M is sufficient enough for the film formation, and that the superconducting films are prepared only when the substrates are maintained in a narrow range of distance from the ICP [8]. By taking these into account, we have aimed at the synthesis of $YBa_2Cu_3O_{7-x}$ thin films by the spray-ICP technique.

EXPERIMENTAL

Mother solutions were prepared by dissoving nitrates of Y, Ba and Cu (guararanteed grade; total metal concentration, 0.3 M; Wako Pure Chemicals) with a molar ratio of 1:2:3 in distilled water. They were ultrasonically atomized into fine droplets (droplet size, 1-2 μm; Sowa Kagaku), and were carried into an Ar-ICP (feeding rate, 6 ml/h) through a narrow tipped quartz nozzle (diameter, 1.5 mm) with an Ar carrier gas (1.4 l/min).

The experimental setup is illustrated in Fig. 1. The torch was composed of three coaxial quartz tubes with the nozzle at the center. The Ar-ICP (40 mm in diameter, 160 mm long) was generated with a three-turn work coil (inner diameter, 55 mm) and an rf oscillator (6 MHz-15 kW maximun power, normally run at 6 kW), and was stabilized with an Ar-sheath gas (30 l/min). No plasma gas was used. Single crystal MgO plates (1 mm thick, 5-10 mm wide) of (100) were used as substrates. They were placed on a ceramic plate (3 mm thick) supported with a water-cooled Cu mold.

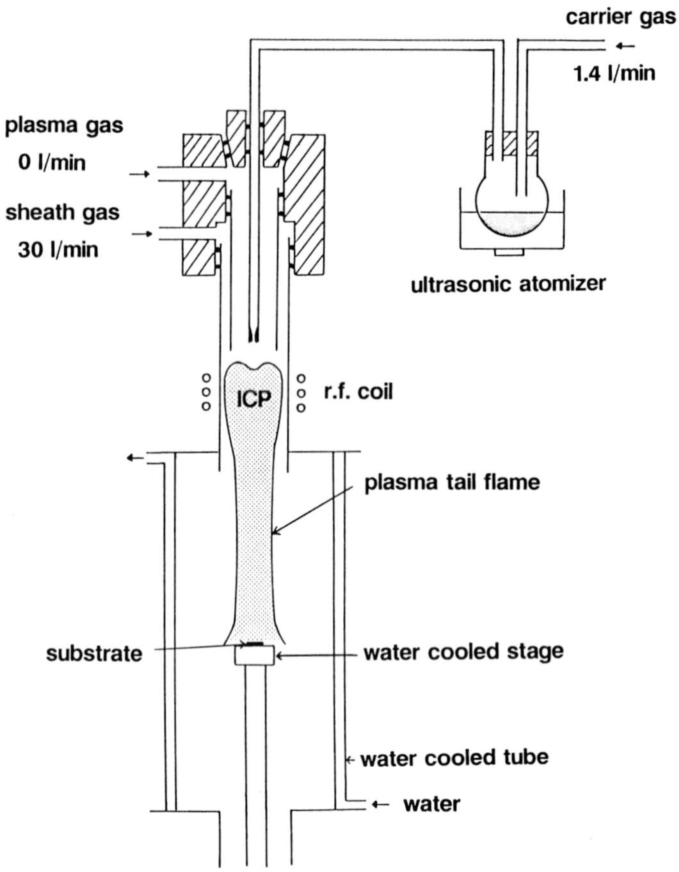

Fig. 1 Experimental setup.

Properties of the films formed by the spray-ICP technique, especially superconductivity, depended on the substrate temperature. However, in the present study it was very difficult to measure with optical methods or thermocouples because of strong emission from the ICP and electromagnetic noise caused by the oscillator. Therefore, x, the distance between the substrate and the center of the work coil was measured at each run, and was utilized as an indication of substrate temperature.

The films were characterized by X-ray diffraction (XRD, CuKα radiation with Ni filter and graphite monochromator, Rigaku Denki) and scanning electron microscopy (field emsission type, Hitachi S-800). Tc's (zero resistance) were determined by measuring resistivities by the four-probe method.

RESULTS AND DISCUSSION

Appropriate x for the synthesis of superconducting films ranged from 500 to 550 mm. In the region of x larger than 550 mm, fine particles were deposited. In some cases films with smooth surface were obtained, but they were not superconducting. Below x=500 mm, the substrate temperature became so high that the evaporation of components occurred, resulting in the formation of brown films. Compared with the synthesis of the Bi system superconductors, the range of appropriate x was narrow. In the previously reported two studies the substrate temperature was monitored by measuring the temperature of substrate holders even during the film formation [9-10]. Probably this monitoring contributed to finding easily the experimental conditions appropriate for the film formation.

An XRD pattern of the as-prepared film formed at x=500 mm is illustrated in Fig. 2. Clearly the film has a c-axis orientation typical of superconductors of this series. The shoulder peak at the right side of (005) can be assigned to (111) of CuO. The presence of CuO in the films comes from the difference in vapor pressure of component oxides. In the case of multi-component oxides, the depoistion of component oxides with high vapor pressure is preceeded by that of oxides with low vapor pressure. Accordingly CuO is likely to deposit partly during the film formation of $YBa_2O_3CuO_{7-x}$, because the vapor pressure of CuO is lower than that of Y_2O_3 and BaO.

Scanning electron micrographs (SEM) of the surface (A) and the fracture surface of the film (B) prepared at x=500 mm are shown in Fig. 3. The film surface consists of projected and faceted crystals of about 0.3 μm in size (marked with a) and dented places (b). The SEM of the fracture surface shows that the film is about 0.8 μm thick, and is composed of double layers. The dented places, (b), correspond to the surface of the first layer formed on the substrate. The first layer is dense while the second one has the island structure. This is interesting from the view of crystal growth. The first layer must also have the island structure at the beginning of deposition. The islands grow large parpendicular to and also along the substrate surface, and finally connect with each other to form the dense film. The second layer is formed by the initiation of new islands on the first layer or further successive growth of particular crystals composing the first layer. This speculation seems to explain well the film structure of double layer.

Energy dispersive spectra obtained at (a) and (b) are shown in Fig. 4. The normalized intensity of Cu to Y is higher at (a) than at (b), indicating that the projected crystals are richer in content of CuO than the first layer. The intensity ratio of Ba to Y is slightly different for (a) and (b), but this difference is within experimental error, because the spectra was obtained for the rough surface.

The as-deposited films had a typical onset temperature of 91 K and zero resistivity at 85 K.

ACKNOWLEDGMENTS

The authors thank Managing Supervisor Dr. T. Fukuda, Dr. T. Kameyama, Mr. T. Tsunoda and Dr. M. Mukaida of National Chemical Laboratory for Industry for kindly permitting them to perform scanning elelctron microscopy and energy dispersive spectrometry and also for valuable discussions.

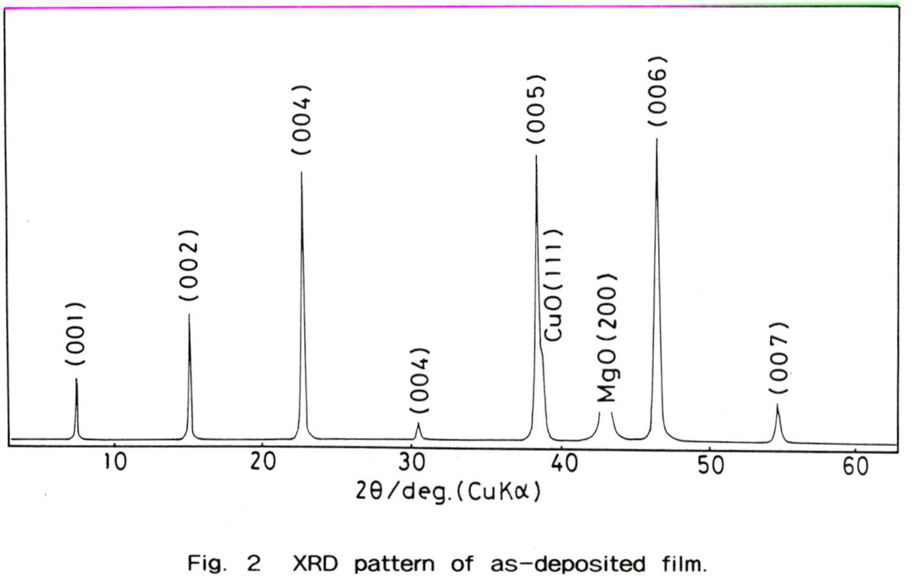

Fig. 2 XRD pattern of as-deposited film.

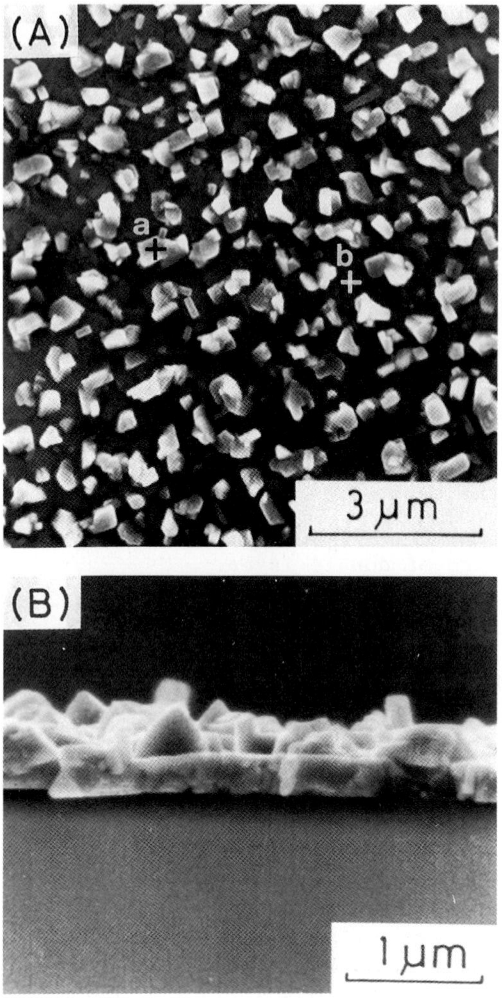

Fig. 3 SEM of film surface (A) and fracture sruface (B).

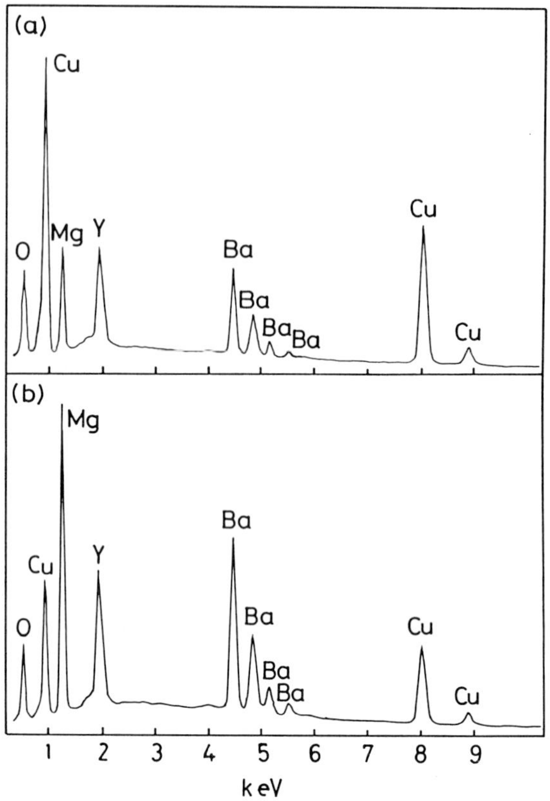

Fig. 4 Energy dispersive spectra obtained at (a) and (b).

REFERENCES

1. Kagawa M. Kikuchi M, Ohno R, Nagae T (1981) J Am Ceram Soc 64: C7-C8
2. Ono T, Kagawa M, Syono Y, Ikebe M, Muto Y (1987) Plasma Chem Plasma Proc 7: 201-209
3. Bednorz JG, Muller KA, (1986) Z Phys B64: 189-193
4. Kagawa M, Suzuki M, Hirai T, Syono Y (1987) In: Akashi K, Kimbara A (eds) 8th international symposium on plasma chemistry. Tokyo, pp 1084-1088
5. Zhu H, Lau YC, Pfender E (1989) In: d'Agostino R (ed) 9th International symposium on plasma chemistry. Italy, pp 1497-1502
6. Zhu H, Lau YC, Pfender E (1989) In: ibid. pp 876-881
7. Tumanov YN, Ivanov AV, Galkin AF, Cucheiko SI (1989) In: ibid. pp 1503-1508.
8. Suzuki M, Kagawa M, Syono Y, Hirai T (1990), under contribution to J Mat Sci
9. Shah A, Patel S, Narumi E, Shaw DT (1990) Appl Phys Lett 57: 1452-1454
10. Wang XW, Zhong HH, Snyder RL (1990) ibid, 57 : 1581-1583

Preparation of YBCO Superconducting Films by MOMBE

K. Endo[1], S. Saya[2], S. Misawa[1], and S. Yoshida[1]

[1] Electrotechnical Laboratory, Tsukuba, 305 Japan
[2] Nippon Institute of Technology, Miyashiro, Saitama, 345 Japan

ABSTRACT

We have fabricated YBCO superconducting thin films on MgO(100) substrates by MOMBE. As source materials, we have used fluorine-free ß-diketonate chelates of Y, Ba and Cu. The films annealed at 940 °C had the c-axis oriented normal to the substrate surface and showed the superconducting properties with Tc(zero resistance) of 45 K.

KEY WORD:YBCO superconducting thin films, MOMBE, MBE, fluorine free ß-diketonate chelates

INTRODUCTION

Metalorganic molecular beam epitaxy (MOMBE) using organometallic compound sources has the advantages of both metalorganic chemical vapor deposition (MOCVD) and molecular beam epitaxy (MBE). In order to develop MOMBE technique for the growth of oxide superconductors, we have carried out MBE using metal chelates as source materials. In conventional MBE, refractory metals, Y and Ba, are used as source materials. So that electron beam evapolators are employed for evaporation of these metals, which restricts the oxygen pressure during growth. The use of metal chelate sources in MBE allows a higher oxygen pressure compared with metal sources, because the volatilizable temperatures of metal chelates are much lower than those of metals. Also, the lower source temperatures make an apparatus simple, where an electron beam evapolator is not necessary.

In the previous study [1], we used $Ba(PPM)_2$ (PPM: pentafloro propanoylpivaloyl methane) as source material for Ba, where PPM ligand contains fluorine. The as-grown films were amorphous and insulating, and contained fluorine. It has been reported that the inclusion of fluorine causes a substantial decrease of superconducting phase [2]. However, high-quality ß-diketonate chelate of Ba without fluorine: $Ba(DPM)_2$ (DPM: dipivaloyl methane) could not be obtained, at that time. Recently, the quality of $Ba(DPM)_2$ has been improved. In this study, we have fabricated YBCO superconducting thin films by MOMBE, using DPM ligand chelates of Y, Ba and Cu as source materials.

SELECTION OF SOURCE MATERIALS

The selection of MO materials holds a key to the MOMBE growth of YBCO films. The knowledge about volatility of these sources is important as basic data for MOMBE. Thermogravimetry (TG) is one of the useful tools to examine volatility. ß-diketonate chelates of Y, Ba and Cu are the candidates for MO source materials. Figure 1(a) shows the structures of the ß-diketone ligands; DPM and PPM. Here, the PPM ligand contains fluorine, while the DPM ligand does not. Figure 1(b) shows the structures of the ß-diketonate chelates of Y, Ba and Cu: $Y(DPM)_3$, $Ba(DPM)_2$ and $Cu(DPM)_2$. It has been pointed out that the Ba chelates have poor volatility among them. Figure 2(a) shows the

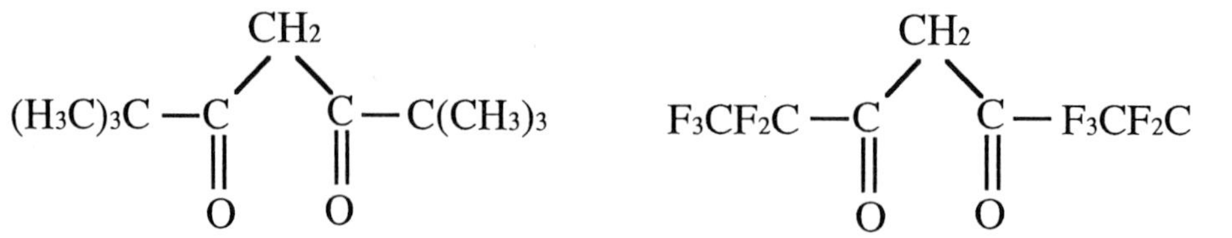

DPM dipivaloyl methane

PPM pentafloro propanoylpivaloyl methane

Fig.1(a) Structural formula of the DPM and PPM ligands.

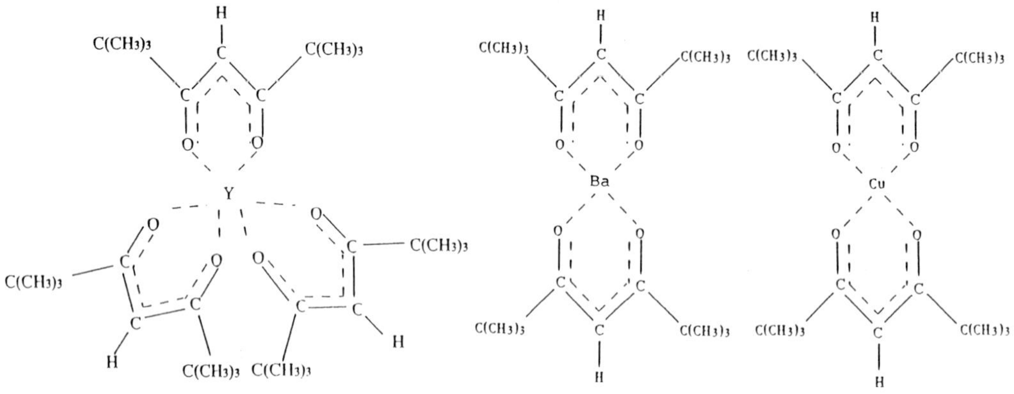

$Y(DPM)_3$ $Ba(DPM)_2$ $Cu(DPM)_2$

Fig.1(b) Structural formula of the source materials.

TG curve for the $Ba(DPM)_2$ powder examined in the previous study. Clearly, multiple steps of the decrease in weight were observed, which suggests that this powder contains some impurities. In contrast, in the TG curve for $Ba(PPM)_2$, only one step of the decrease was observed (Fig.2(b)). Only half amount of $Ba(DPM)_2$ volatilizes in weight even at 500 °C, whereas 90 % of $Ba(PPM)_2$ volatilizes at the same temperature. In the previous study, these results led us to choose $Ba(PPM)_2$ as a source material. However, the as-grown YBCO films by MOMBE had been found to contain fluorine from the PPM ligand. Figure 2(c) shows the TG curve for the improved $Ba(DPM)_2$ powder. In contrast with the previous one, this powder shows a TG curve with single step of the decrease and volatilizes up to 93 % at 400 °C. This result allows us to use $Ba(DPM)_2$ in stead of $Ba(PPM)_2$ as a source material. Also, $Y(DPM)_3$ and $Cu(DPM)_2$ sources volatilized completely up to 100 %.

FILM GROWTH AND CHARACTERIZATION

Figure 3 is a schematic drawing of the MOMBE apparatus used in this study. The chamber was evacuated down to less than 10^{-6} Torr by a turbo molecular pump. As source materials, $Y(DPM)_3$, $Ba(DPM)_2$ and $Cu(DPM)_2$ were used. These sources were loaded into PBN crucibles and were heated at 85, 170 and 56 °C, respectively. MgO(100) crystals were used as substrates, which were mounted on a Mo block with In metal solder and heated by a Pt heater. YBCO films were grown at the substrate temperature of 700 °C for 2 hours. Ozone gas used for the growth was around 10 % in concentration and introduced through a nozzle near the substrates. The flow rate of ozone gas was about 5 sccm. During the deposition, the total pressure in the chamber was 1×10^{-4} Torr.

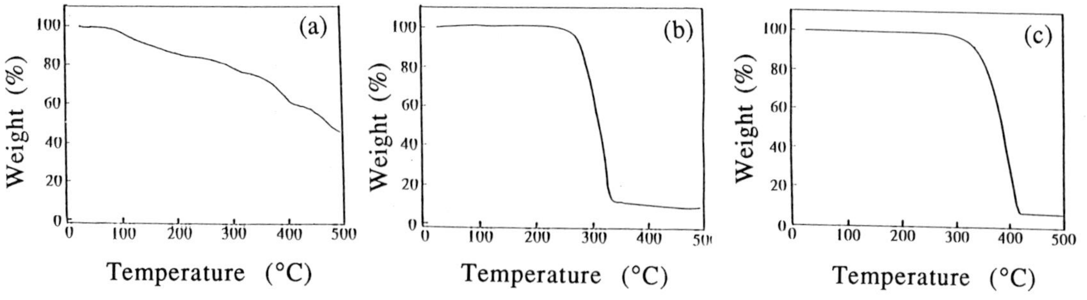

Fig.2 Thermogravimetry of (a) previous $Ba(DPM)_2$, (b) $Ba(PPM)_2$ and (c) improved $Ba(DPM)_2$.

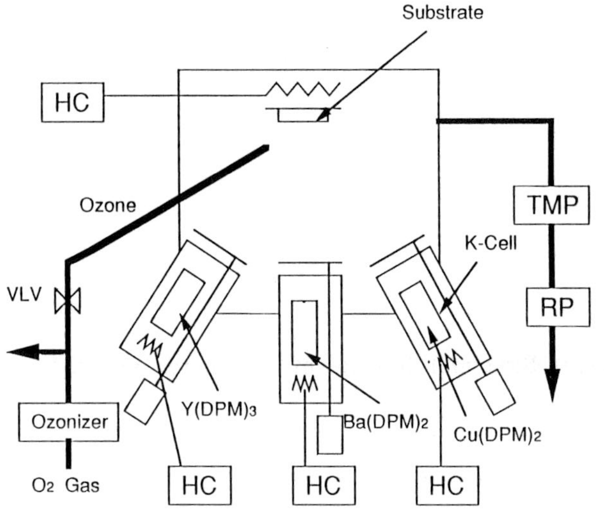

Fig.3 Schematic drawing of MOMBE apparatus. TMP: Turbo Molecular Pump, RP: Rotaly Pump, HC: Heater Controller, VLV: Variable Leak Valve.

The compositions and the structures of the films were examined by inductively coupled plasma spectroscopy (ICP) and X-ray diffraction (XRD), respectively. The incorporation of fluorine and carbon into the films was analyzed by secondary ion mass spectroscopy (SIMS). Resistance measurements were carried out by a standard four probes DC method, where Cu wires were attached with Ag paste to Au electrodes deposited onto the films.

The thicknesses of YBCO films were around 700 Å, and the growth rate was about 6 Å/min. The atomic ratio Y:Ba:Cu in the film was determined to be 1.0:2.2:2.1 by ICP. As-grown films were amorphous and insulating. The films were set in an infrared furnace and annealed at 940 °C for 15 minutes in atmospheric oxygen. An appreciable amount of fluorine and carbon was not detected both in the as-grown films and in annealed ones by SIMS measurement. Figure 4 shows the XRD pattern of an annealed film. All the peaks were assigned as the diffraction lines from the (00n) planes of YBCO. This indicates that the films grow with the c-axis oriented normal to the substrate surface. Figure 5 shows the temperature dependence of the resistance for an annealed film. The superconducting transition was observed at Tc(onset) of 70 K and Tc(zero) of 45 K. These results suggest that the higher substrate temperature is necessary to obtain an as-grown YBCO film showing a superconducting transition.

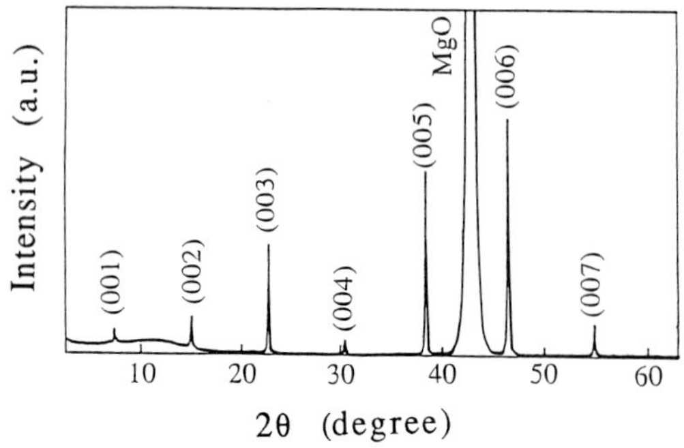

Fig.4 XRD pattern of an annealed film.

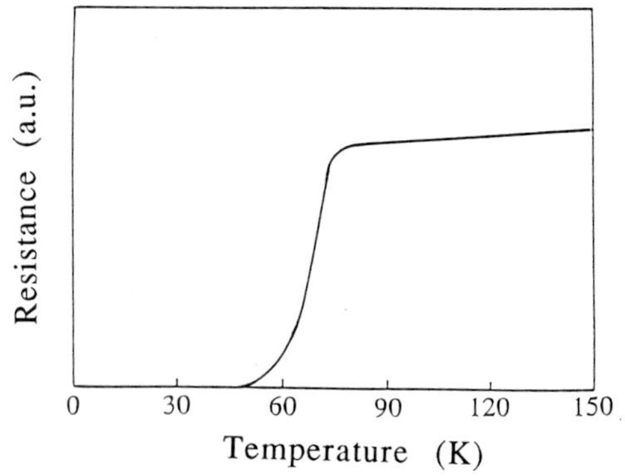

Fig.5 Temperature dependence of electrical resistance of an annealed film.

CONCLUSION

We have fabricated YBCO films on MgO(100) substrates by MOMBE, using the high-quality DPM chelates of Y, Ba and Cu as source materials, which do not contain fluorine. The films annealed at 940 °C show the crystal orientation with c-axis normal to the substrate surface and the superconductivity with Tc(zero) of 45 K.

REFERENCE

1. Endo K, Ikedo Y, Hayashida S, Ishiai J, Nakatsuka K, Misawa S, Yoshida S (1990) In : Ishiguro T, Kajimura K (eds) Advance in superconductivity II. Springer-Verlag, Tokyo, pp805-808

2. Yokoyama Y, Okumura T, Sugise R, Matsubara T, Koshizuka N, Hirabayashi M, Terada N and Ihara H (1988) J Ceram Soc Jpn Inter **96**: 481-484

Reproducible Preparation and Properties of Y-Ba-Cu-O Films by MOCVD

S. MATSUNO, F. UCHIKAWA, S. UTSUNOMIYA, and M. TANAKA*

Materials & Electronic Devices Laboratory, Mitsubishi Electric Corporation, 1-57, Miyashimo 1-chome, Sagamihara, 229 Japan
*Super-GM

ABSTRACT

Y-Ba-Cu-O superconducting films with high Jc over 2×10^6 A/cm^2 at 77 K, 0 T were prepared reproducibly by MOCVD technique. The films had the superconducting transition temperature with zero resistivity Tc of 91.7 ~ 92 K and the critical current density Jc of 1 ~ 2.5×10^4 A/cm^2 at 30.2 T, 77 K. By the observation using transmission electron microscopy, it was confirmed that the films were grown epitaxially on the $SrTiO_3$ (100) substrate.

KEY WORDS: Y-Ba-Cu-O film, MOCVD, high Jc, high magnetic field

INTRODUCTION

Metalorganic chemical vapor deposition (MOCVD) method have some advantages for preparation of high Tc superconductors. This method has high deposition rates [1], compatibility with variously shaped substrates and flexibility for large-scale processing. Accordingly, the superconducting films formed by this method are expected to be used for tape-shaped superconducting wires. Actually, Y-Ba-Cu-O films with excellent high Jc have been prepared by this method [2]. However, a uniformity of superconducting properties for the wires is required in longitudinal direction of conductors. Therefore, a reproducible preparation of the films by MOCVD is strongly demanded. In this paper, we report the reproducible preparation of Y-Ba-Cu-O films with high Jc by MOCVD, from viewpoints of establishment of the stable transport of source gases and reconsideration of the deposition conditions. In addition, we report the result for observation of transmission electron microscopy for the films in this experiment.

EXPERIMENTAL

The MOCVD system in this study was a conventional hot-wall-type system except that a source transport technique with Ba-addition products was used [1]. The deposition conditions are shown in Table 1. $Y(DPM:C_{11}H_{19}O_2)_3$, $Ba(DPM)_2$ and $Cu(DPM)_2$ were adopted as source materials. The substrate temperature was set at 850 °C and the reactor pressure was kept at 10 Torr during the deposition. A $SrTiO_3$ (100) single crystal was used as the substrate. After the deposition for 25 min, the as-deposited films were cooled to room temperature under oxygen flow at pressure of 1 atm. The cooling rate was about 15 °C/min.

The crystal structure of the films was analyzed by an X-ray diffractometer and the surface morphology was examined by scanning electron microscopy (SEM). The film thickness was measured by a stylus thickness gauge. The overall composition and the deposition amount were determined by X-ray spectrometer (XRS) and inductively coupled plasma spectrometer (ICPS). The microscopic composition was observed by electron probe microanalyzer (EPMA). Superconducting properties were measured by a four-probe method. Thickness of the films was 0.25 μm and the deposition rate was 0.6 μm/h. The films were patterned with 0.8 mm wide and 2 mm long. Critical currents were determined in the 1 μV/cm criterion. Magnetic fields up to 30.2 T by the hybrid magnet [3], HM-1a and HM-2 at Tohoku University, were applied to the samples. Observation of micro structure at the interfacial regions between the Y-Ba-Cu-O film and the substrate was preformed by transmission electron microscopy (TEM).

Table 1. Deposition conditions for the Y-Ba-Cu-O films by MOCVD.

Source materials	Y(DPM)$_3$, Cu(DPM)$_2$ Ba(DPM)$_2$-THF addition products
Vaporization temperature	Y: 125 °C, Ba: 210 °C, Cu: 115 °C
Flow rates of carrier gas	8 ~ 12 CCM (Ar), 50 CCM (THF)
Flow rates of reactant gas	200 CCM (O_2)
Substrate temperature	850 °C
Substrate	$SrTiO_3$ (100) (10mm × 10mm)
Reactor pressure	10 Torr
Deposition time	25 min

RESULTS AND DISCUSSION

Figure 1 (a) ~ (c) shows the time dependence of the deposition amount for the each elements. The deposition amount was directly proportional to the deposition time. Therefore, it is indicated that the stable continuous transport of source gases was achieved.

Figure 2 shows the temperature dependence of the resistivity for the films. Those films were prepared at the same deposition conditions in Table 1. It was clearly observed in Fig. 2 that the films had the superconducting transition temperature with zero resistivity Tc of as high as 91.7 ~ 92 K.

Figure 3 shows the SEM photograph of the typical Y-Ba-Cu-O film surface. As shown in Fig. 3, the film had the Y-Ba-Cu-O matrix with smooth surface and numerous small particles which had diameter range of 0.1 ~ 1.5 μm. It was proved by EPMA analysis that the particles over 1 μm in diameter were CuO. The smaller particles are supposed to be CuO, but it was not confirmed because of the resolution of EPMA used. Other films in this experiment had the same surface morphology.

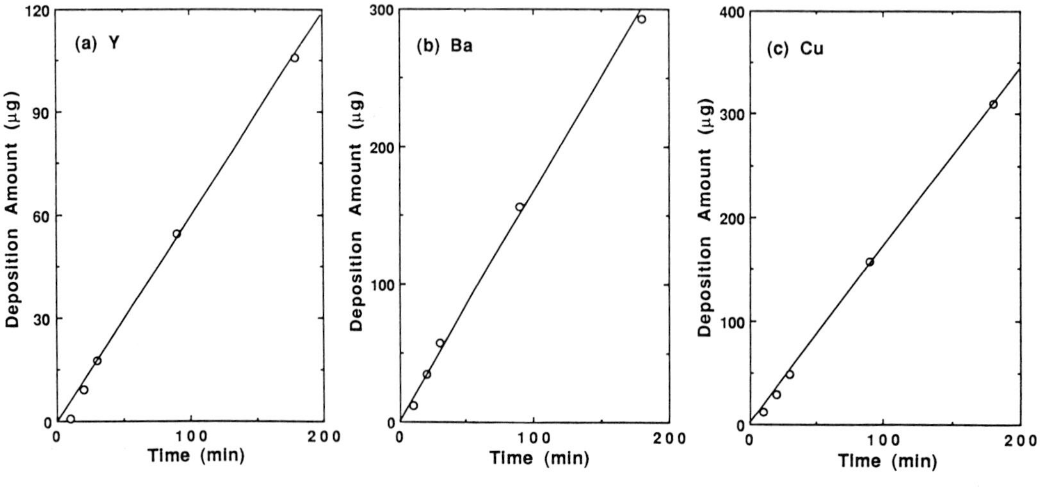

Fig. 1. The time dependence of the deposition amount for the each elements.

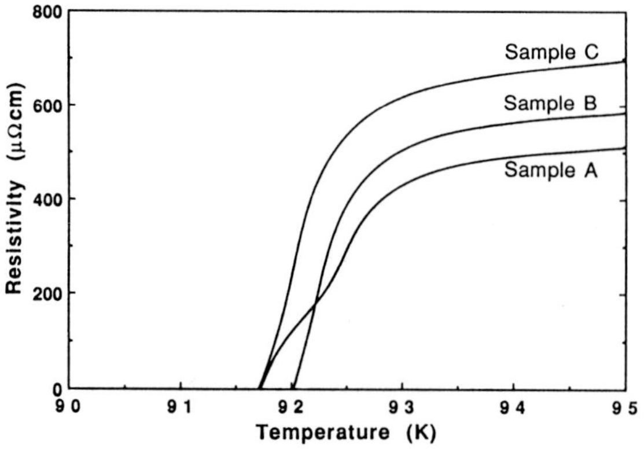

Fig. 2. The temperature dependence of the resistivity for the films.

Fig. 3. The SEM photograph of the typical Y-Ba-Cu-O film surface.

Figure 4 shows the X-ray diffraction pattern of the film on $SrTiO_3$ (100) substrate. The film indicated strong c-axis orientation perpendicular to the substrate surface, although weak CuO peak and Y-Ba-Cu-O (200) peak which showed a-axis orientation were observed. The full width of half maximum intensity in a rocking curve for the peak of Y-Ba-Cu-O (005) were 0.30 ~ 0.33 degree. The (005) spacings for the films measured by precision X-ray diffractometer (Bond method) showed same data of d=2.3330 ±0.0003Å.

Figure 5 shows the critical current (Jc-B) properties for Y-Ba-Cu-O films under high magnetic fields up to 30.2 T. All films had the critical current density Jc of over 2×10^6 A/cm^2 at 0 T and 77 K. When the magnetic field was applied perpendicular to the c-axis of the films, the Jc values decreased with increasing the field. Then, the slope on the magnetic field dependence of the Jc changed at around 20 T. The Jc values appreciably decreased from this point. The Jc-B curve for Sample A is shown to have a excellent property, which had a slow decrease with increasing magnetic fields. The highest Jc at 30.2 T was 2.5×10^4 A/cm^2.

Figure 6 shows the transmission electron micrograph at the interfacial regions between the Y-Ba-Cu-O films and the $SrTiO_3$ substrate. As shown in this figure, Y-Ba-Cu-O (103) planes and $SrTiO_3$ (110) planes were fitted coherently in the interface. However, a dislocation caused by a lattice mismatch between both crystals is observed in this figure at right-hand of the interface.

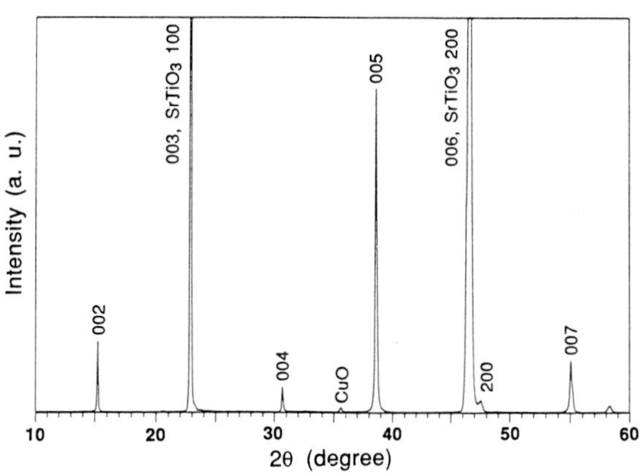

Fig. 4. The X-ray diffraction pattern of the film on $SrTiO_3$ (100) substrate.

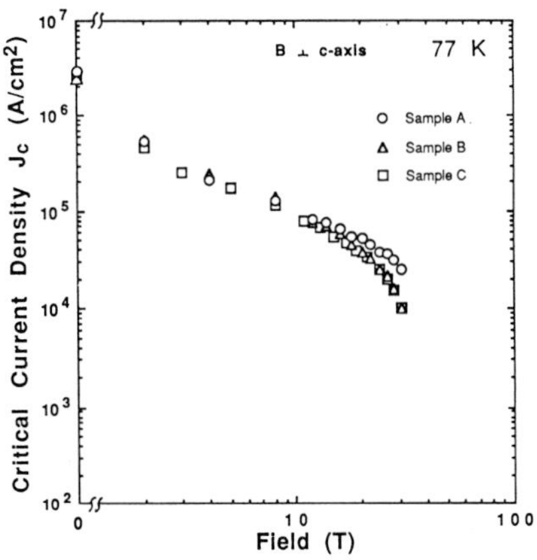

Fig. 5. The magnetic field dependence of the critical current density Jc for Y-Ba-Cu-O films.

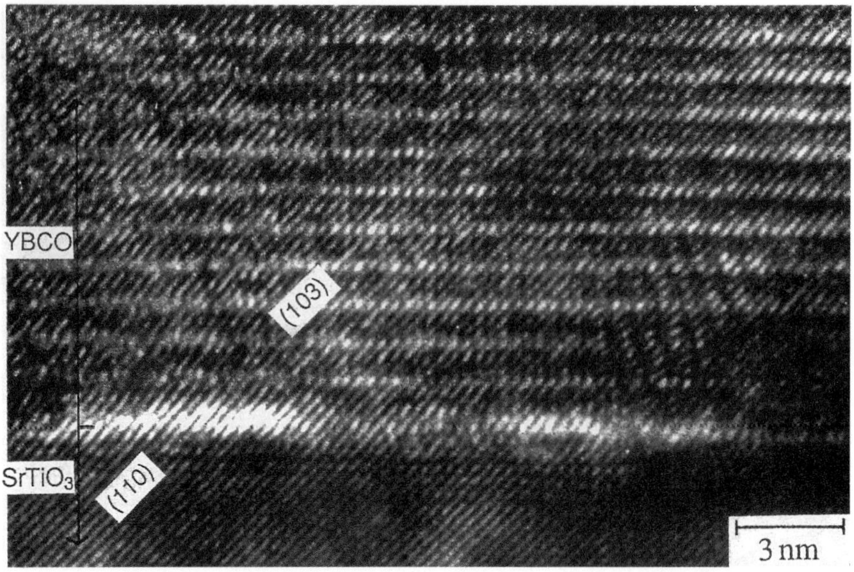

Fig. 6. The transmission electron micrograph in the interfacial regions between the Y-Ba-Cu-O film and $SrTiO_3$ substrate.

CONCLUSIONS

Y-Ba-Cu-O superconducting films with high Jc over 2×10^6 A/cm^2 at 77 K, 0 T were prepared reproducibly by MOCVD technique. The films had the superconducting transition temperature with zero resistivity Tc of 91.7 ~ 92 K and the critical current density Jc of 1 ~ 2.5×10^4 A/cm^2 at 30.2 T, 77 K. The films were grown epitaxially on the $SrTiO_3$ (100) substrate.

ACKNOWLEDGMENTS

This work was performed as a part of" R&D on Superconducting Technology for Electric Power Apparatuses" as a subject of the Engineering Research Association under the Moonlight Project of Agency of Industrial Science and Technology, MITI, being consigned by New Energy and Industrial Technology Development Organization (NEDO).

REFERENCES

[1] S. Matsuno, F. Uchikawa, and K. Yoshizaki, "Preparation of $YBa_2Cu_3O_{7-y}$ Superconducting Films by the Organic Transport-Chemical Vapor Deposition," Advances in Superconductivity II, p. 877 (1990).

[2] S. Matsuno, F. Uchikawa, and K. Yoshizaki, "Y-Ba-Cu-O Superconducting Films with High Jc Values by MOCVD using Ba-Addition Products," Jpn. J. Appl. Phys. Vol. 28, p. L947 (1990).

[3] Y. Muto, Y. Nakagawa, K. Noto, S. Miura, A. Hoshi, K. Watanabe, G. Kido, H. Ichikawa, T. Fujioka, Y. Sato, O. Osaka, and H. Takano, Sci. Rep. RITU A33, p. 221 (1986).

Preparation of Y-Ba-Cu-O Superconducting Films by MOCVD Using Ozone

HIROSHI OHNISHI[1], HIROSHI HARIMA[2], YOSHIHIKO KUSAKABE[1], MINORU KOBAYASHI[1], and KUNIHIDE TACHIBANA[2]

[1] Manufacturing Development Laboratory, Mitsubishi Electric Corporation, Tsukaguchi, Amagasaki, 661 Japan

[2] Department of Electronics and Information Science, Kyoto Institute of Technology, Matsugasaki, Sakyo-ku, Kyoto, 606 Japan

ABSTRACT

As-grown superconducting Y-Ba-Cu-O films were prepared by an MOCVD technique at the substrate temperature below 650°C. As precursors, ß-diketonate complexes were used, and oxygen including ozone was used for oxidation. The highest zero-resistive temperatures (*Tc*) were 84K and 74K prepared at 650°C and 620°C, respectively. The films with *Tc* of 61K were prepared at 620°C, in the nearly stoichiometric composition (Y:Ba:Cu=1:2:3). On the other hand, no zero-resistivity was observed in the films prepared at 580°C. The substrate temperature of about 600°C was the lower threshold to get zero-resistivity in this experiment.

KEY WORDS: YBaCuO, superconductor, MOCVD, as-grown, ozone.

INTRODUCTION

It is obvious that future microelectronic devices will require well-crystallized films of smooth surface and uniform structure which can be prepared at substrate temperature as low as possible. There are some previous reports on the preparation of superconducting oxide films without postannealing by the metalorganic chemical vapor deposition (MOCVD) technique [1-10]. Kanehori *et al.* succeeded in the low-temperature growth of the film with *Tc*=85K at 580°C [5]. Zhao *et al.* also succeeded in low-temperature growth with *Tc*=72K at 570°C [8]. These reports indicate that the MOCVD technique has a potential in low-temperature growth. On the other hand, Zhao *et al* suggested that the future work forcuses on studying the influence of the 1-2-3 metal atomic ratio of the films on superconducting phase formation [8]. Previously, we indicated that an excess of Cu element was required to raise the *Tc*, but this caused the formation of many particles mainly composed of CuO, and the surface smoothness deteriorated [9]. Recently, by introducing ozone into the reactor, we have succeeded in the preparation of the superconducting film with relatively good surface morphology at the nearly stoichiometric composition [10]. In this paper, we will report influences of the substrate temperature on the film character under the condition of using ozone.

EXPERIMENTAL

Figure 1 shows our experimental setup. Dipivaloylmetanato (DPM) chelates of $Y(C_{11}H_{19}O_2)_3$ (=$Y(DPM)_3$), $Ba(C_{11}H_{19}O_2)_2$ (=$Ba(DPM)_2$) and $Cu(C_{11}H_{19}O_2)_2$ (=$Cu(DPM)_2$) were used as precursors for Y, Ba and Cu elements, respectively. They were evaporated in respective source tanks, mixed and led into a reactor by helium carrier gas. The temperatures of the source tanks for $Y(DPM)_3$, $Ba(DPM)_2$ and $Cu(DPM)_2$ were fixed at 100°C, 195°C and 100°C, respectively. To keep the temperature of $Ba(DPM)_2$ tank lower than 220°C was especially important to obtain the reproductivity of Ba concentration because $Ba(DPM)_2$ begins decomposition over 220°C [11]. Oxygen (O_2) gas was fed through an ozonizer to the reactor with a separate gas line. The flow rate of oxygen was fixed at 500 standard cm^3/min. With the present apparatus, the concentration of the ozone in the O_2 was about 1.5%. In this experiment, we changed the flow rate of He carrier gas for controlling the film composition and changed the substrate temperature, as shown in Table 1. The substrate

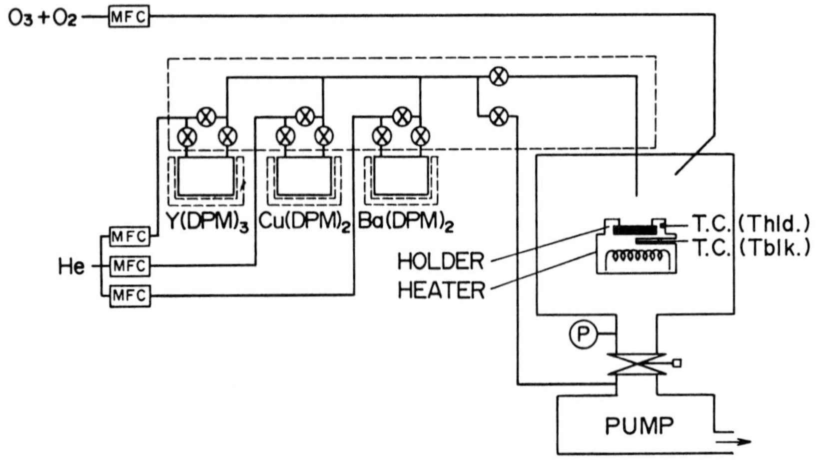

Fig.1 Experimental setup.

temperature was measured by a thermocouple fixed to the surface with ceramic adhesive. Other conditions were fixed as follows. The total gas pressure in the reactor was fixed at 2 Torr. The deposition time was 3 hours. After the deposition, the substrate was cooled to the room temperature in O_2 including 1.5% ozone at a pressure of 2 Torr.
The electrical resistance of the deposited films was measured in vacuum by means of a conventional AC four-probe technique with indium electrodes. The composition was analyzed by an inductively coupled plasma (ICP) method. The crystallinity was analyzed by an X-ray diffraction (XRD). The morphology was observed by scanning electron microscopy (SEM).

Table 1 Experimental Conditions

	$Y(DPM)_3$	$Ba(DPM)_2$	$Cu(DPM)_2$
Flow rate of He carrier gas (cm^3/min)	3-9	50	10-30
Substrate : MgO(100)			
Substrate temperature : 580, 620 and 650°C			

RESULTS AND DISCUSSION

Figure 2 shows the correlation between *Tc* and composition of the films. The marks (○), (◑) and (△) indicate the films prepared at 650°C, 620°C and 580°C, respectively. The Tc value of each film is shown on the right side of the mark. The mark (◈) shows a stoichiometric composition (Y:Ba:Cu=1:2:3). Comparing with our previous results (●) which were prepared at 715°C without ozone [9], it is obvious that the *Tc* value increased greatly in spite of decreasing substrate temperature to 620-650°C. The highest *Tc* values were 74K at 620°C and 84K at 650°C in this experiment. However, we could not obtain the zero-resistivity when we prepared at 580°C. To obtain good surface morphology, the films with stoichiometric composition were best.
Figure 3 shows temperature dependence of resistivity for the samples (a), (b) and (c) in Fig.2. They have almost the same composition. Samples (a) and (b) indicate metallic character in the temperature range above 90K. Although sample (c) shows no zero-resistivity, we can observe the on-set at 78K. This fact indicates that there exists some local superconducting network in this sample, although it is not completed in macroscopic scale (of the order of millimeter) to get zero-resistivity.

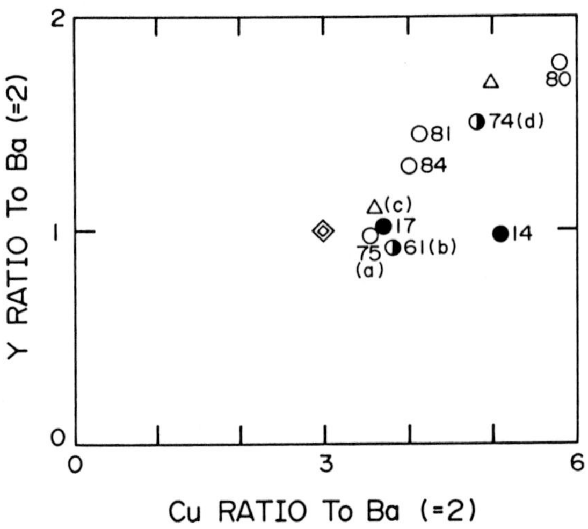

Fig.2 Correlation between *Tc* and film composition which is normalized to Ba=2.

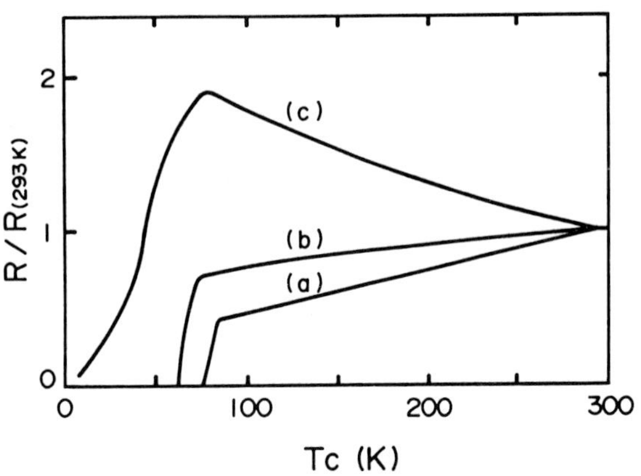

Fig.3 Temperature dependence of resistivity of nearly stoichiometric samples in Fig.2.
Substrate temperatures;
(a)650°C, (b)620°C, (c)580°C

The SEM photograph of sample (b) is shown in Fig.4. A structure with scattered particles is observed in the matrix portion. These particles are inferred to be mainly composed of CuO from our previous results [9]. The diameter of particles is about 0.2um. Comparing with sample (d) having higher Tc (=74K), the number and size of particles have decreased in sample (b), and surface morphology has improved. This is probably due to the preparation condition for nearly stoichiometric composition in sample (b) (see Fig.2).

Figure 5 shows XRD patterns of the samples. Sample (a) is strongly oriented with *c*-axis normal to the substrate. On the contrary, samples (b) and (c) are mainly *a*-axis orientation, although the peak intensity has decreased by lowering the substrate temperature. To see more precisely, we compare the peak intensities of (005) and (200) planes as shown in Fig.6. The *a*-axis oriented portion becomes relatively large by decreasing substrate temperature from 650°C to 620°C. There was no difference in peak intensity ratio between the films prepared at 620°C and at 580°C, but zero-resistivity was not observed on the latter film. This is attributed poorer crystallization at 580°C.

Sample (b) Sample (d)

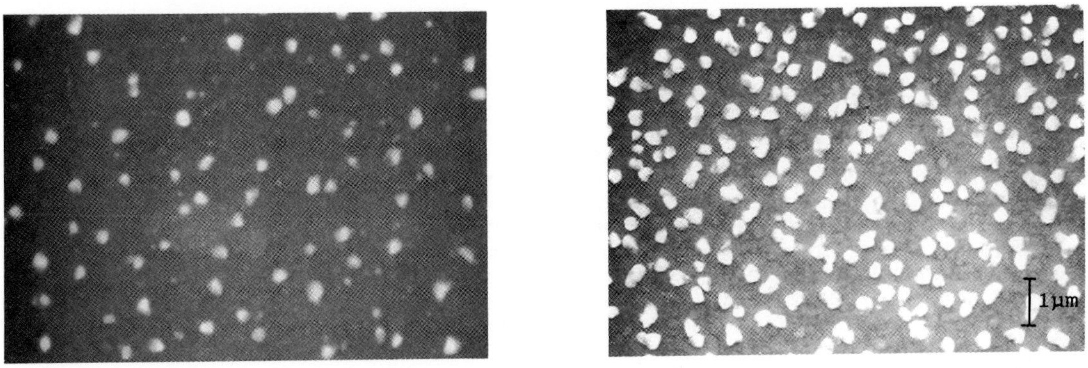

Fig.4 SEM photographs of samples in Fig.2 prepared at 620°C. The *Tc* values were 61K of (b) and 74K of (d).

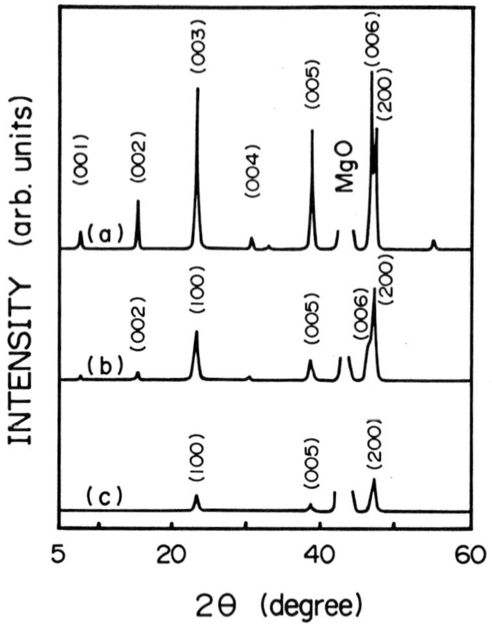

Fig.5 XRD patterns of samples in Fig.2 prepared at various substrate temperature; (a)650°C, (b)620°C, (c)580°C.

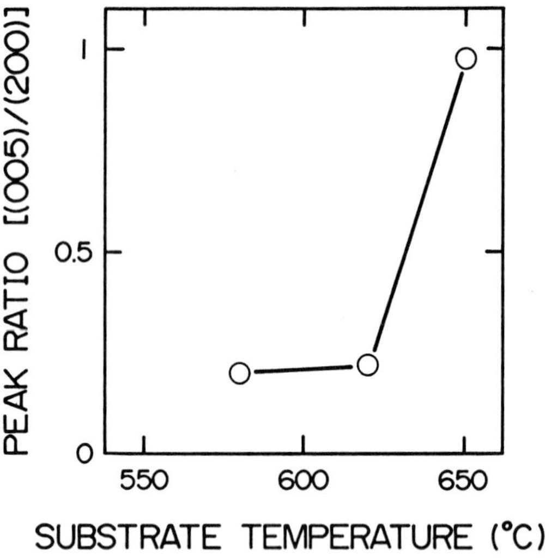

Fig.6 The relationship between substrate temperature and peak intensity ratio (005)/(200).

To conclude, as-grown superconducting films were prepared at the low substrate temperatures of 650°C and 620°C by an MOCVD technique using ozone. However, the temperature of about 600°C was the lower threshold to obtain films with zero-resistivity in our condition. For the preparation of superconducting films at lower temperature, we will try to introduce other techniques to enhance the crystallization together with raising oxidation ability using ozone.

ACKNOWLEDGMENT

The authors would like to thank T.Ogama and K.Yokoyama of Mitsubishi Electric Corp. for performing the ICP analysis.

REFERENCES

1)Abe H, Tsuruoka T, Nakamori T (1988) Jpn. J. Appl. Phys. **27**: L1473-L1475.
2)Yamane H, Kurosawa H, Iwasaki H, Masumoto H, Hirai T, Kobayashi N, Muto Y (1988) Jpn. J. Appl. Phys. **27**: L1275-L1276.
3)Yamane H, Masumoto H, Hirai T, Iwasaki H, Watanabe K, Kobayashi N, Muto Y (1988) Appl. Phys. Lett. **53**: 1548-1550.
4)Tsuruoka T, Kawasaki R, Abe H (1989) Jpn. J. Appl. Phys. **28**: L1800-L1802.
5)Kanehori K, Sugii N, Miyauchi K (1989) Fall MRS Proceedings: M7.172.
6)Kanehori K, Sugii N, Miyauchi K (1989) J. Solid State Chem. **82**: 103-108.
7)Yamane H, Kurosawa H, Hirai T, Watanabe K, Iwasaki H, Kobayashi N, Muto Y (1989) J. Cryst. Growth **98**: 860-866.
8)Zhao J, Noh DW, Chern C, Li YQ, Norris P, Gallois B, Kear B (1990) Appl. Phys. Lett. **56**: 2342-2344.
9)Ohnishi H, Kusakabe Y, Kobayashi M, Hoshinouchi S, Harima H, Tachibana K (1990) Jpn. J. Appl. Phys. **29**: 1070-1075.
10)Ohnishi H, Harima H, Kusakabe Y, Kobayashi M, Hoshinouchi S, Tachibana K (1990) Jpn. J. Appl. Phys. **29**: in press.
11)Harima H, Ohnishi H, Hanaoka K, Tachibana K, Kobayashi M, Hoshinouchi S (1990) Jpn. J. Appl. Phys. **29**: 1932-1938.

Preparation of a-Axis and c-Axis Oriented Y-Ba-Cu-O Superconducting Oxide Films by CVD

Hisanori Yamane[1], Masaharu Hasei[2], Toshio Hirai[1], Kazuo Watanabe[1], Norio Kobayashi[1], Hideyuki Kurosawa[2], and Yoshio Muto[1]

[1] Institute for Materials Research, Tohoku University, 1-1, Katahira 2-chome, Aoba-ku, Sendai, 980 Japan
[2] Riken Co., 810, Kumagaya, Kumagaya, 360 Japan

ABSTRACT

Y-Ba-Cu-O(YBCO) superconducting films were prepared on $SrTiO_3(100)$ at 850°C by chemical vapor deposition using β-diketone metal chelates as sources. The films mainly composed of $YBa_2Cu_3O_x$(123 compound) with c-axis orientation and a-axis orientation were obtained on the substrates on which Ba-O was deposited in advance. The ratio of the a-axis oriented 123 compound and the c-axis oriented 123 compound depended upon the amount of Ba-O. The films having the matrix of the c-axis orientated 123 compound showed superconducting transition temperature defined by zero resistance at 90 K.

KEY WORDS: Y-Ba-Cu-O film, CVD, c-axis orientation, a-axis orientation

INTRODUCTION

A compound in the system of Y-Ba-Cu-O (YBCO), $YBa_2Cu_3O_x$ (123 compound), having CuO layer structure shows anisotropic nature of superconducting properties[1,2]. A problem of the anisotropy for the application of the 123 compound is that critical current density seriously depends on the direction of magnetic fields[3,4]. The superconducting coherence length along a- and b-axis is about 10 Å, whereas the length along the direction perpendicular to the CuO plane (c-axis direction) is about 2 Å[2]. Thus it seems difficult to make Josephson junctions along the c-axis direction[5].

Recently, control of orientation directions for the YBCO films has attracted considerable attention due to the anisotropy of the superconducting properties. Most of the YBCO films prepared by various techniques showed c-axis orientation. Several groups reported that deposition temperature controlled the direction of orientation[5-7]. They obtained a-axis oriented films on $SrTiO_3(100)$ by sputtering at lower deposition temperature and observed c-axis orientation at elevated deposition temperature. The effect of film composition on the ratio of a-axis orientation and c-axis orientation was reported for the YBCO films made by electron beam coevaporation[8]. Kanehori et al. prepared a-axis oriented YBCO films at higher total pressure and c-axis oriented YBCO films at lower total pressure by chemical vapor deposition (CVD)[9]. CVD has advantages of capability for large area deposition and high deposition rate in comparison with physical vapor deposition techniques such as sputtering and electron beam evaporation. The present authors found that a-axis oriented YBCO films were prepared on $SrTiO_3$ (100) by the introduction of a Ba source a few minutes in advance of the introduction of other sources during a CVD process[10].

In the present study, preparation of YBCO films was carried out by CVD using the substrates of $SrTiO_3(100)$ covered with Ba-O in order to understand the mechanism of change in orientation directions and to grasp the possibility of control over the ratio of c-axis orientation and a-axis orientation.

EXPERIMENTAL

CVD sources, 2,2,6,6-tetramethyl-3,5-heptanedionate(thd) of Y, Ba, and Cu, were purchased from Tosoh-Akzo Co.($Y(thd)_3$: Lot No. L8Z127, $Ba(thd)_2$: Lot No. D9Z221, and $Cu(thd)_2$: Lot No. H8Y2127). The CVD reactor used was a vertical

Table 1 Deposition conditions

evaporation temperature	
$Y(thd)_3$	129^oC
$Ba(thd)_2$	242^oC
$Cu(thd)_2$	108^oC
total carrier gas flow rate	
Ar	450 ml/min
oxidation gas flow rate	
O_2	250 ml/min
total pressure	10 Torr
deposition temperature	850^oC
deposition time	30 min
deposition rate	17 nm/min

thd:2,2,6,6-tetramethyl-3,5-heptanedionato

Table 2 Amounts of Ba on the substrates and ratios of a-axis orientation and c-axis orientation.

sample	$m_{Ba}\times10^7$ (mol/cm^2)	α*	Tc (K)
A	0.0	0.004	89.3
B	0.5	0.640	90.8
C	0.6	0.887	89.3
D	1.4	0.991	86.3
E	2.8	0.832	73.5
F	4.3	0.751	66.0

$*\alpha=[I_{ob}(200)/I_r(200)]/[I_{ob}(200)/I_r(200)+I_{ob}(007)/I_r(007)]$

$I_{ob}(200),(007)$: observed XRD intensities; $I_r(200)=12$, $I_r(007)=2$: JCPDS 39-1434[12]

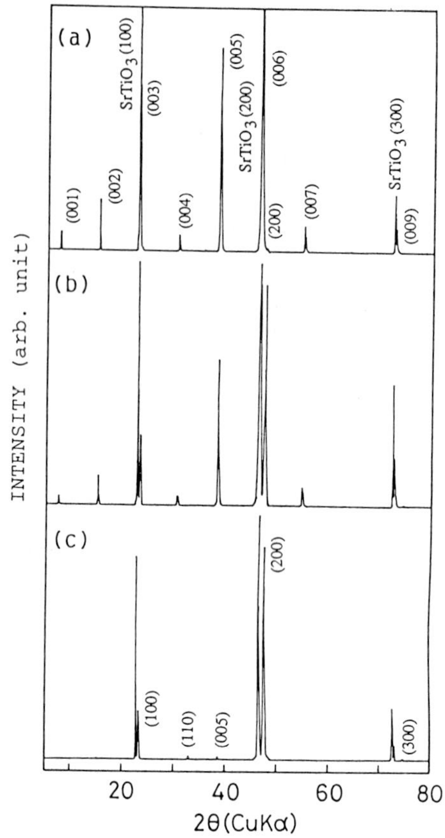

Fig. 1. X-ray diffraction patterns of CVD YBCO films:(a) sample A, (b) sample B, and (c) sample D.

hot-wall type[11]. Substrates were optically polished $SrTiO_3(100)$ single crystals ($5x10x1\ mm^3$).

Various amounts of Ba-O were deposited on $SrTiO_3(100)$ at 850^oC by CVD using $Ba(thd)_2$ evaporated at temperatures from 194° to 234^oC. Evaporated $Ba(thd)_2$ was fed into the reactor with Ar gas at a flow rate of 150 ml/min. O_2 gas (250 ml/min) and Ar gas (300 ml/min) were separately introduced into the reactor. Total pressure in the reactor was constant at 10 Torr. The deposition process continued for a period of 20 min. After the deposition, the deposit was cooled in the reactor under 1 atm of oxygen. The deposits of Ba-O on the substrate had a mirror surface. No peaks were observed in the X-ray diffraction patterns of the deposits. The amounts of Ba-O were represented by moles of Ba (m_{Ba}: mol/cm^2) which were obtained by inductively coupled plasma spectroscopy and X-ray fluorescence analysis.

Table 1 lists the deposition conditions of YBCO films. After the deposition, the YBCO films were cooled in 1 atm of oxygen from a deposition temperature of 850^oC to room temperature in 1 h (in-situ oxygen treatment). The molar ratio of Y:Ba:Cu in the YBCO films was 1:1.9-2.1:3.1-3.6. X-ray diffraction (XRD) patterns of the films were obtained by standard $2\theta-\theta$ method with the CuKα line monochromatized by pyrolytic graphite. The microstructure of film surfaces was observed with a scanning electron microscope (SEM). The resistance of the films was measured by a DC four-probe method having electrodes of Au sputtered on the films.

RESULTS AND DISCUSSION

As listed in Table 2, six samples were prepared on the substrate on which various amounts of Ba-O (m_{Ba}) were deposited in advance.

XRD patterns of sample A, B, and D are shown in Fig. 1(a), (b), and (c), respectively. Prominent peaks of (00l) (l=1,2,3$\cdots$) and a very small peak of (200) were observed in the XRD pattern of sample A which was deposited on the substrate without Ba-O. The peaks in the pattern of sample B (Fig.1(b)) were from the reflections of (100), (200), and (00l). This indicates that large amounts of the a-axis oriented 123 compound are contained in the film with the c-axis oriented 123 compound. In the XRD pattern of sample D(Fig 2(c)), dominant peaks were from (100) and (200) reflections and very small peaks of (005) and (110) were seen. The relatively high peaks of (110) and (103), which were main peaks of the randomly oriented 123 compound, were observed with peaks of (00l), (100), and (200) in the XRD patterns of sample E and F.

The ratio of c-axis orientation and a-axis orientation was evaluated from the equation: $\alpha=[I_{ob}(200)/I_r(200)]/[I_{ob}(200)/I_r(200)+I_{ob}(007)/I_r(007)]$, where $I_{ob}(200)$ and $I_{ob}(007)$ were observed XRD relative intensities and $I_r(200)$ and $I_r(007)$ were the intensities from the sample with random orientation (JCPDS 39-1434[12]). As shown in Table 2, the value of α increased to 0.99 as m_{Ba} increased to 1.4×10^{-7} mol/cm^2(sample D).

Granular precipitates with a size of 2-3 μm were observed for sample A by scanning electron microscopy. The precipitates were considered to be CuO derived from the excess composition from the Y:Ba:Cu=1:2:3 stoichiometric composition[10]. Needle-like precipitates of the a-axis oriented 123 compound were also seen[10]. A SEM photograph of sample B is shown in Fig.2(a). Many rectangular grains aligning two dimensionally are seen with the granular precipitates. The rectangular grains were dense and the part of matrix decreased in sample C. Sample D having the highest α consisted of rectangular grains as shown in Fig. 2(b). The correspondence between the increase of α obtained from XRD data and the change of film morphology revealed that the rectangular grains were the a-axis oriented 123 compound and the matrices were the c-axis oriented 123 compound. Sample E was composed of grains with a size of 1-2 μm. Small rectangular precipitates were seen on the grains. A similar structure was observed in the micrograph of sample F.

On the substrate without the deposit of Ba-O, nucleation probably occurs with c-axis orientation or a-axis orientation, considering the epitaxial relation between the (100) plane of $SrTiO_3$ and the 123 compound. However, the most part of the substrate surface may be covered with the c-axis oriented 123 compound at the early stage of film growth because the crystal growth rate of the 123 compound along the a-b axis is higher than the rate along the c-axis[13].

Two dimensional alignment of the rectangular grains observed in the micrographs of sample B-D means the epitaxial growth of the films on the substrates on which Ba-O was deposited in advance. The thickness of the Ba-O deposit cannot be estimated from m_{Ba} due to uncertainty of the structure of the Ba-O deposit. However, if Ba-O is assumed to be BaO (a=5.539 Å, density:5.992 g/cm^3), the thickness of the deposit is about 40 nm for $m_{Ba}=1.43\times10^{-7}$ mol/cm^2(sample D). On the substrate covered with Ba-O, many nuclei with a-axis orientation may be produced near the substrate surface having epitaxial relation to the (100) plane of $SrTiO_3$ because Cu and Y probably diffuse from the surface of the Ba-O deposit. Nucleation and nuclear growth with random orientation should occur

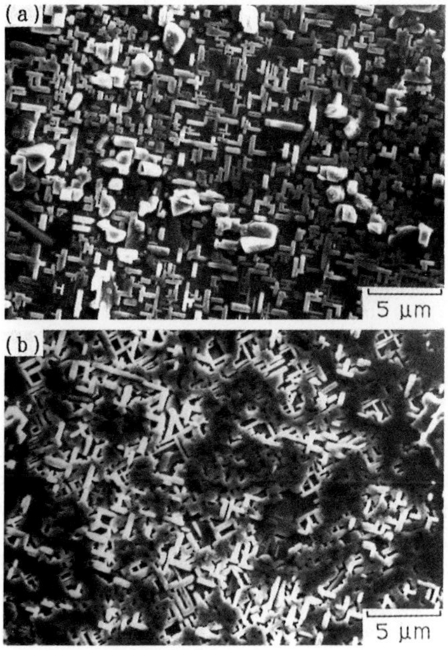

Fig. 2. Scanning electron micrographs of CVD YBCO films: (a) sample B and (b) sample D.

in the thicker deposits of Ba-O (sample E and F). Cross-sectional transmission electron microscope observation is necessary in order to investigate these processes further in detail.

Figure 3 shows temperature dependence of resistance for sample A-F. Superconducting transition temperatures defined by zero resistance (Tc(R=0)) were listed in Table 2. Although all the samples showed metallic behavior at normal state, resistance at room temperature increased with increasing m_{Ba}. Tc(R=0) was around 90 K for sample A, B, and C. Sample D having the highest α showed Tc(R=0) at 87 K. Sample E and F exhibited a broader transition with a tail and had a zero resistivity at 73.5 K and 66 K, respectively.

Almost the same Tc(R=0) of the sample A, B, and C probably attribute to the presence of the c-axis oriented 123 compound matrix. In the case of sample D which included little amount of the c-axis oriented 123 compound, grain boundary between the rectangular grains presumably raised the resistivity at room temperature and broadened the transition width. A lot of grain boundaries among the grains with different orientations may be introduced into sample E and D and reduce Tc(R=0).

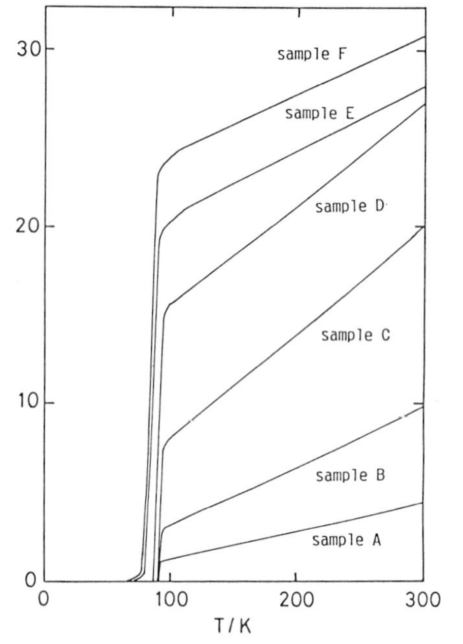

Fig. 3. Resistance versus temperature curves of CVD YBCO films (sample A-F).

To summarize, the c-axis and a-axis oriented YBCO films were prepared on $SrTiO_3$(100) by CVD. The ratio of a-axis orientation and c-axis orientation was controlled with the amount of Ba-O deposited on $SrTiO_3$ in advance of the formation of the YBCO film.

ACKNOWLEDGEMENTS

This work was supported, in part, by a Grant-in-Aid from the Ministry of Education, Science and Culture, Japan.

REFERENCES

1. Worthington TK, Gallagher WJ, Dinger TR (1987) Phys Rev Lett 59: 1160-1163
2. Oh B, Char K, Kent AD, Naito M, Beasley MR, Geballe TH, Hammond RH, Kapitulnik A (1988) Phys Rev B 37: 7861-7864
3. Roas B, Schultz L, Seamann-Ischenko G (1990) Phys Rev Lett 64: 479-482
4. Watanabe K, Awaji S, Kobayashi N, Yamane H, Hirai T, Muto Y (1990) submitted to J Appl Phys
5. Eom CB, Marshall AF, Ladermann SS, Jacowitz RD, Geballe TH (1990) Science in press
6. Asano H, Asahi M, Michikami O (1989) Jpn J Appl Phys 28: L981-L983
7. Linker G, Xi XX, Meyer O, Li Q, Geerk J (1989) Solid State Commun 69: 249-253
8. Oh B, Naito M, Arnason S, Rosenthal P, Barton R, Beasley MR Geballe TH, Hammond RH, Kapitulnik A (1987) 51: 852-854
9. Kanehori K, Sughii N, Fukazawa T, Miyauchi K (1989) Thin Solid Films 182: 265-269
10. Hirai T, Yamane H (1990) J Cryst Growth in press
11. Yamane H, Kurosawa H, Hirai T, Watanabe K, Iwasaki H, Kobayashi N, Muto Y (1989) J Cryst Growth 98: 860-866
12. Powder Diffraction File, JCPDS International Center For Diffraction Data, File No. 39-1434
13. Damento MA, Gshneidner KA Jr, McCallum RW (1987) Appl Phys Lett 51: 690-691

Properties of $YBa_2Cu_3O_y$ Films on $BaSnO_3$ Layers

TOSHIYA MATSUBARA, SHIN-ICHI OHASHI, ATSUYOSHI TAKENAKA, and TAKESHI MORIMOTO

Asahi Glass Co., Ltd., Research Center, 1150, Hazawa-cho, Kanagawa-ku, Yokohama, 221 Japan

ABSTRACT

We have investigated the potentiality of $BaSnO_3$ as the buffer layer between a superconducting Y-Ba-Cu-O layer and a substrate. At first, the reactivities of some barium double oxides with the Y-Ba-Cu-O system were examined at the partial melt region (1050 °C), and $BaSnO_3$ was found to be the most stable. It was then confirmed that $BaSnO_3$ formed a thin and dense layer on a (100)MgO single crystal by a CVD method. Superconducting films of $YBa_2Cu_3O_y$ with critical current densities of up to 3 x 10^5A/cm^2 in zero magnetic field at 77K were prepared on the $BaSnO_3$ layers.

KEY WORDS: superconductor, buffer layer, $BaSnO_3$, CVD, critical current density

INTRODUCTION

It has been shown that the problems caused by weak links at grain boundaries can be overcome and samples with high critical current densities can be prepared by epitaxial growth in thin films [1] and melt-growth in the bulk Y-Ba-Cu-O superconductor[2-5]. The epitaxial growth takes place on single crystals, such as MgO and $SrTiO_3$, in various PVD or CVD processes, while the melt-growth processing can be carried out by heating sintered Y-Ba-Cu-O bulk to the partial melt region (∿1050 °C or above) and continuously cooling in a temperature gradient. As a superconductor is required in a flexible tape or wire form for application to a magnet, the deposition of superconducting thin films without weak links has been tried on flexible substrates, such as polycrystalline YSZ or hastelloy, by the PVD or CVD methods[6-7], but such attempts were unsuccessful because the qualities of the films were strongly affected by those of the substrates. A flexible single crystal substrate should be used but is not available at present. There were also no reports on the preparation of a Y-Ba-Cu-O layer without weak links on a flexible substrate by the melt process because interdiffusion between the Y-Ba-Cu-O layer and the substrate took place at high temperature. Development of a buffer layer to prevent the interdiffusion is desirable.

In our previous work[8], Y-Ba-Cu-O bulk superconductor including small particles of $BaSnO_3$ as pinning centers was synthesized by the unidirectional melt solidification method. In the work, we found that the $BaSnO_3$ particles did not grow or react with any other phases during the melt processing, and there was no appreciable degradation through the addition of $BaSnO_3$ fine particles of superconducting properties of $YBa_2Cu_3O_y$. This indicates that a barium double oxide such as $BaSnO_3$ could be used as the buffer layer if it were formed into a thin and dense layer on a substrate. In this paper, we will report the reactivities of some barium double oxides in the Y-Ba-Cu-O system at high temperature, the capability of $BaSnO_3$ in forming a thin layer on a MgO single crystal and the Jc of the $YBa_2Cu_3O_y$ film prepared on the $BaSnO_3$ layer.

EXPERIMENTAL

Reactivity experiments

Powders of $YBa_2Cu_3O_y$ and Y_2BaCuO_5 were prepared by a conventional solid-state

Table 1 Deposition conditions for $BaSnO_3$ films

Source	Temperature(℃)	Flow(ccm)
$Ba(dpm)_2$	225	40
$Sn(Ph)_4$	160	20
Oxygen Flow		: 40 ccm
Temperature of Substrate		: 830 ℃
Total Gas Pressure		: 3 torr

Table 2 Deposition conditions for Y-Ba-Cu-O films

Source	Temperature(℃)	Flow(ccm)
$Ba(dpm)_2$	225	50
$Y(dpm)_3$	115	50
$Cu(dpm)_2$	110	50
Oxygen Flow		: 140 ccm
Temperature of Substrate		: 830 ℃
Total Gas Pressure		: 20 torr

reaction from stoichiometric mixtures of fine Y_2O_3, $BaCO_3$ and CuO powders. Barium double oxide powders (>99.9%) obtained from High Purity Chemetals Co., Ltd. were used without further treatment. Powders of $YBa_2Cu_3O_y$, Y_2BaCuO_5 and barium double oxides were mixed in a molar ratio of 6:4:10 and pressed into disks of 20 mm in diameter and about 2 mm in thickness. The disks were placed on alumina boats and heated in flowing air at atmospheric pressure according to the temperature profile shown in Fig. 1. For comparison, a disk which consisted of only $YBa_2Cu_3O_y$ and Y_2BaCuO_5 was examined at the same time.

Preparation of films

Both $YBa_2Cu_3O_y$ and $BaSnO_3$ films were prepared by a CVD method. The apparatus consisted of a gas-handling system, four vaporizers, a horizontal reactor of the cold wall type and an evacuation system. The source materials used are tetraphenyltin and the β-diketonate metal chelates. The typical deposition conditions are listed in Table 1 and Table 2. The $YBa_2Cu_3O_y$ films were cooled to room temperature in flowing oxygen gas at atmospheric pressure for 1 hour after deposition. The $YBa_2Cu_3O_y$ films were cut into approximately 1 mm x 1 mm x 10 mm pieces and critical currents were measured in liquid nitrogen by a dc four-probe resistive method. Critical current was defined as the current which generated voltage of 0.2 μV across a 1 mm length of the sample.

RESULTS AND DISCUSSION

Reactivities of barium double oxides

The disks maintained their original shapes after the heat treatment although they bent a little and small amounts of melt escaped from them to the boats. Figure 2 shows the temperature dependence of the ac-susceptibility for the melted disks. Although all samples have onset temperatures of about 90 K, the transition temperature widths of the samples containing $BaSnO_3$ and no barium double oxides are smaller than those of the other samples. Since an increase in width is considered to reflect the degradation of supeı conductivity, it was deduced that some reaction had taken place between $BaZrO_3$ or $BaTiO_3$ and the Y-Ba-Cu-O system. Therefore $BaSnO_3$ is considered to be the most promising material for the buffer layer in the Y-Ba-Cu-O system.

To further confirm this prospect, we used small particles of $BaZrO_3$ as additives instead of those of $BaSnO_3$ to synthesize the Y-Ba-Cu-O bulk specimen

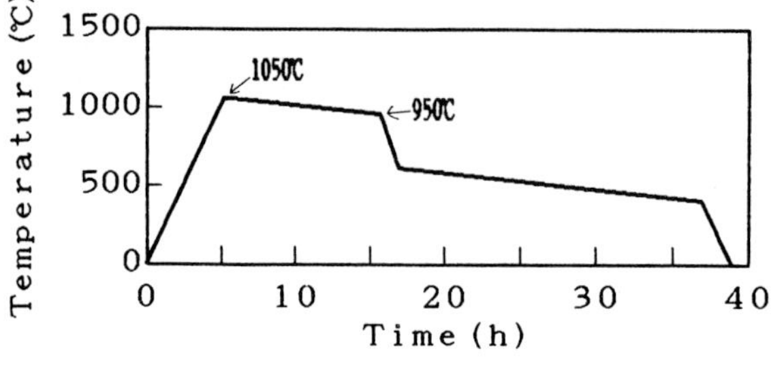

Fig. 1 Temperature profile of the heat treatment.

by the unidirectional melt-solidification method according to the procedure reported before[8]. The particles of $BaZrO_3$ did grow during the melt processing and did not retain their original shapes. This also suggests that some reaction took place between $BaZrO_3$ and the melt.

Thin films of $BaSnO_3$

Transparent colorless films containing Ba and Sn were deposited on MgO single crystals. Figure 3(b) shows an X-ray diffraction pattern for a Ba-Sn-O film on a MgO single crystal. Excepting the peaks for the substrate one strong peak is observed at $2\theta = 44°$, which corresponds to the (200) peak for $BaSnO_3$ powder (Fig.3(a)). The amounts of Ba and Sn in the film were determined by X-ray fluorescence analysis, and the ratio of Ba:Sn was calculated to be 1:1. Thus the Ba-Sn-O film was identified as the $BaSnO_3$ film with (100) orientation. Figure 4 shows a SEM image of a fractured surface of the $BaSnO_3$ film on the MgO substrate. The film appeared to be flat and dense. The value (0.13 μm) of its thickness determined from the image coincided with the value (0.14 μm) which was calculated from the amount of Ba in the film and the density of $BaSnO_3$, 7.238 g/cm^3. Considering that MgO and $BaSnO_3$ both have a cubic structure and nearly the same cell constant of 4.213 Å and 4.1163 Å, respectively, $BaSnO_3$ is epitaxially deposited to form a single crystal-like film on a MgO single crystal by a CVD method.

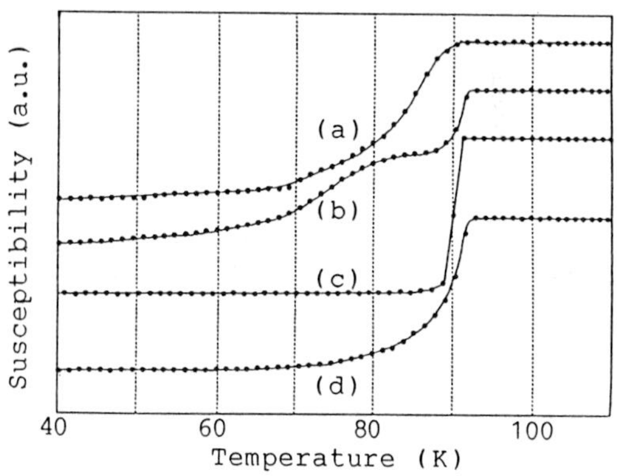

Fig. 2 Temperature dependence of ac-susceptibilities of melted disks.
(a) $YBa_2Cu_3O_y:Y_2BaCuO_5:BaTiO_3$=6: 4:10
(b) $YBa_2Cu_3O_y:Y_2BaCuO_5:BaZrO_3$=6: 4:10
(c) $YBa_2Cu_3O_y:Y_2BaCuO_5:BaSnO_3$=6: 4:10
(d) $YBa_2Cu_3O_y:Y_2BaCuO_5$ =6:14

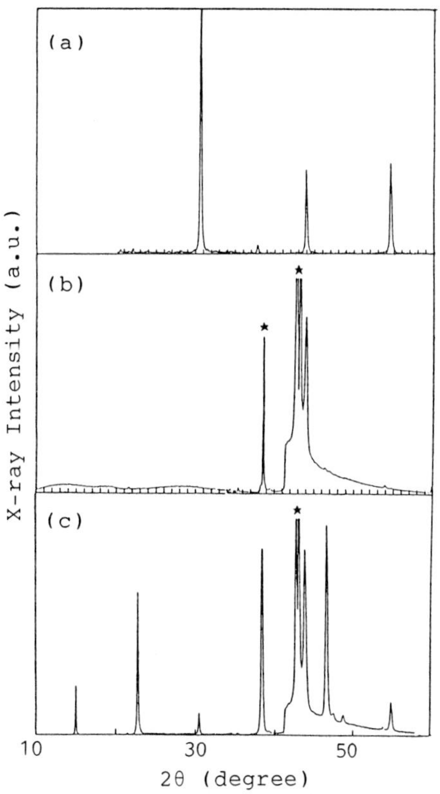

Fig. 3 X-ray diffraction patterns of (a) $BaSnO_3$ powder, (b) a $BaSnO_3$ film on a MgO single crystal and (c) a $YBa_2Cu_3O_y$ film on the $BaSnO_3$ layer. ★ : peaks for the MgO substrate.

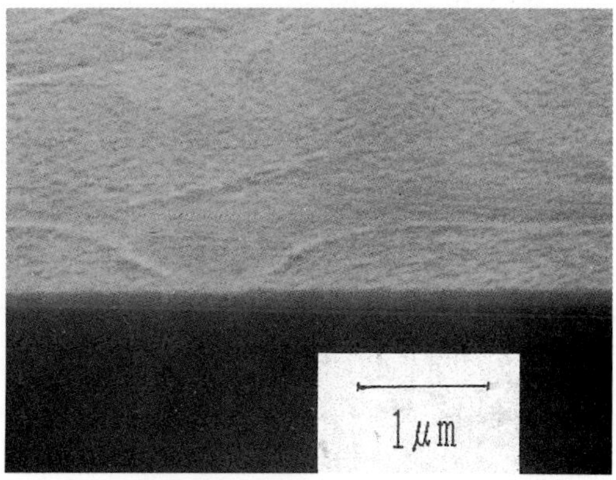

Fig. 4 SEM image of a fractured surface of a $BaSnO_3$ layer on a MgO single crystal.

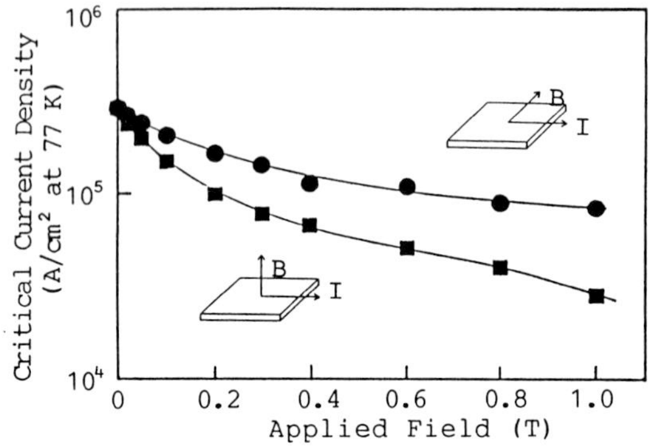

Fig. 5 Magnetic field dependence of critical current density for a $YBa_2Cu_3O_y$ film on a $BaSnO_3$ layer.

Properties of the $YBa_2Cu_3O_y$ films on the $BaSnO_3$ layer

$YBa_2Cu_3O_y$ films with (00ℓ) orientation were successfully prepared on the $BaSnO_3$ layers which had been deposited on MgO crystals (Fig. 3(c)). There was no apparent change in the (200) peak for the $BaSnO_3$ layer before and after the deposition of the $YBa_2Cu_3O_y$ film. Figure 5 shows the magnetic field dependence of the Jc at 77 K for the $YBa_2Cu_3O_y$ film on the $BaSnO_3$ layer. Since the Jc value in zero magnetic field is as high as 3×10^5 A/cm^2 and the field dependence of Jc is much smaller than that for polycrystalline samples, the film may have a single crystal-like structure without significant weak links like those deposited directly on single crystals, such as MgO and $SrTiO_3$. It is well known that the crystal quality of a $YBa_2Cu_3O_y$ film deposited by PVD or CVD is strongly affected by that of a substrate. Therefore it is deduced that the $BaSnO_3$ layer also has a single crystal-like structure.

CONCLUSIONS

$BaSnO_3$ is very stable against the Y-Ba-Cu-O melt at high temperatures (<1050 °C) and capable of forming a thin and dense film by a CVD method. Superconducting $YBa_2Cu_3O_y$ films with the Jc values at 77 K of up to 3×10^5 A/cm^2 in zero magnetic field were prepared on $BaSnO_3$ layers which had been epitaxially deposited on (100)MgO single crystals. These results indicate that $BaSnO_3$ has good potentiality as the buffer layer between a superconducting Y-Ba-Cu-O layer and a substrate.

REFERENCES

1. Chaudhari P, Koch RH, Laibowitz RB, McGuire TR, Gambino RJ (1987) Phy Rev Lett 58: 2684-2686

2. Jin S, Tiefel TH, Sherwood RC, Davis ME, van Dover RB, Kammlott GW, Fastnacht RA, Keith HD (1988) Appl Phys Lett 52: 2074-2076

3. Murakami M, Morita M, Koyama N (1989) Jpn J Appl Phys 28: L1125-L1127

4. Kase J, Shimoyama J, Yanagisawa E, Kondoh S, Matsubara T, Morimoto T, Suzuki M (1990) Jpn J Appl Phys 29: L277-L279

5. Fujimoto J, Murakami M, Gotoh S, Koshizuka N, Oyama T, Shiohara Y, Tanaka S (1989) Proc 2nd Int Sym Superconductivity (ISS'89): 285-288

6. Saitoh J, Fukutomi M, Tananka Y, Asano T, Maeda H, Takahara H (1990) Jpn J Appl Phy 29: L1117-L1119

7. Okuda S, Hayashi N, Takano S, Hitotsuyanagi H, Terai S, Hasegawa K (1989) Proc 2nd Int Sym Superconductivity (ISS'89): 809-812

8. Shimoyama J, Kase J, Kondoh S, Yanagisawa E, Matsubara T, Suzuki M, Morimoto T (1990) Jpn J Appl Phys 29: in press

Excimer Laser-Induced Chemical Vapor Deposition of Metal Oxides Thin Film from β-Diketone Complexes

TAKAHISA USHIDA, KAZUTOSHI HIGASHIYAMA, and IZUMI HIRABAYASHI

Superconductivity Research Laboratory, Nagoya Division, International Superconductivity Technology Center, 4-1, Mutsuno 2-chome, Atsuta-ku, Nagoya, 456 Japan

ABSTRACT

The effect of the UV irradiation (using the KrF(248nm) and XeCl(308nm) excimer laser) on the single source system ($Y(thd)_3$, $Ba(thd)_2$, and $Cu(thd)_2$), the double source system (Y-Ba, Ba-Cu, and Y-Cu) and the triple source system (YBCO) were investigated. The KrF laser irradiation has improved the crystallization in the single source system and also improved both the crystallization and the reactivity of the Ba-Cu and Y-Cu systems in the double source system. These results indicate that the laser irradiation promotes the reactivity of sources, the crystallization and oxidation of films during the deposition.

KEY WORDS: laser-induced CVD, thin film superconductor, excimer laser, β-diketone

INTRODUCTION

Recently, chemical vapor deposition (CVD) process has been reported to prepare high quality YBCO and BSCCO thin films by a number of investigators [1,2]. CVD has high reliability and productivity, and is not in need of low pressure, like sputtering and MBE methods. On the other hand, photoprocessing, such as laser-induced CVD of a variety of metals from the gas phase and laser etching, has proved to be an excellent technology in the production of devices. The laser-induced CVD has received much attention in the research field of amorphous materials, metal films and compound semiconductors, because of its low processing temperature, low contamination and selective film growth. The experiments of monolayer growth of II-VI and III-V semiconductors with visible light irradiation (LALE; laser atomic layer epitaxy) for the surface excitation have been one of the most successful and recently interested examples [3]. However, there has not yet been reported with respect to the preparation of high-Tc superconductor film by laser-induced MOCVD.

In this paper, we have investigated the effect of the UV irradiation (using the KrF(248nm) and XeCl(308nm) excimer laser) on the single source system ($Y(thd)_3$, $Ba(thd)_2$, and $Cu(thd)_2$) (where (thd) represents 2,2,6,6-tetramethyl-3,5-heptanedionate), the double source system (Y-Ba, Ba-Cu, and Y-Cu) and YBCO system expecting the acceleration of deposition by excitation of the gas source materials. The deposited thin films have been investigated on the surface morphology, the microstructure, and the deposition rate.

EXPERIMENTAL

The laser-induced MOCVD apparatus used in this study is shown in Figure 1. The source materials used were β-diketone of $Y(thd)_3$, $Ba(thd)_2$, and $Cu(thd)_2$, which were loaded into individual evaporators equipped with heating mantles at appropriate temperature for each gas source. The source gas lines and reactor were placed in the ovens with the controlled temperature to eliminate the condensation of the source vapor. The reactor was evacuated to less than 1 Pa after the substrate was loaded, and then the gases were introduced. The substrates were heated by ceramics heater. The temperature of the substrate was measured by a thermocouple put on the substrate. The pressure of the reactor was kept at 1.3 kPa during deposition. The details of the experimental condition are shown in Table I. The composition ratio of Y, Ba and Cu of films were controlled by the evaporator temperature and the Ar carrier gas flow rate. The gas flow rates for 2 component operation were adjusted to the atomic ratio of YBCO, i,e., Y : Ba = 1 : 2, Ba : Cu = 2 : 3 and Y : Cu = 1 : 3. After deposition, the films were cooled from the deposition temperature to oven temperature of 240 °C at the rate of 100 °C/min under the oxygen atmosphere of 1 atm. Substrates used were MgO(100) single crystals (10x10x0.5mm). The excimer laser was used as UV light source and operated with KrF(248nm) and XeCl(308nm) at repetition rate of 1~20Hz. Pulse energies were 250mJ and 130mJ, respectively. The laser light was introduced into the reactor through a quartz window and the laser irradiated the substrate directly. The film composition was determined by the inductively coupled plasma chemical analysis (ICP). The structure of the films was examined by the X-ray diffraction (XRD) method. The surface morphology was examined by the scanning electron microscopy (SEM).

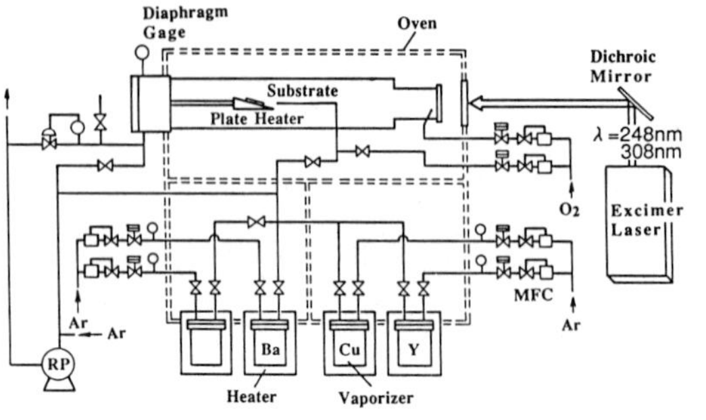

Fig. 1. Apparatus for laser-induced MOCVD.

Table I. Deposition condition.

Vaporizer temperature and Carrier gas (Ar) flow rate :		
$Y(thd)_3$	123~125.8 °C	12.5~27.0 cm^3/min
$Ba(thd)_2$	234~241.5	10.0~25.0
$Cu(thd)_2$	115~120.0	9.0~40.0
Deposition temperature	: 310~850 °C	
Total gas prresure	: 1.3 kPa	
O_2 gas flow rate	: 150 cm^3/min	
Deposition time	: 60 min	
UV light source and Pulse energy :		
KrF excimer laser (248nm)		240 mJ
XeCl excimer laser (308nm)		130 mJ
Repetition rate	: 0~20 Hz	

RESULTS AND DISCUSSION

The UV absorbance of $Y(thd)_3$, $Ba(thd)_2$, and $Cu(thd)_2$ which were dissolved in methanol is shown in Figure 2. In the absorbance of $Cu(thd)_2$, two distinct maxima are observed at 250 and 300 nm. The absorption at about 250nm is assigned to the charge transfer transition from ligand to metal, and that at about 300nm is assigned as $\pi\rightarrow\pi*$ transition of ligand [4]. For similar metal chelate X.Yang and C.Kutal has reported the gas phase spectrum agrees well with that measured in solution [5]. So, we can reasonably assume the spectra in Fig. 2 represent the absorption in the gas phase. The wavelength of KrF and XeCl are 248nm and 308nm, respectively. Although the wavelengths are too long to create ozone from oxygen, if the UV light excites the source materials and accelerates decomposition of them, the lowering of the deposition temperature and the improvement of deposition rate are expected.

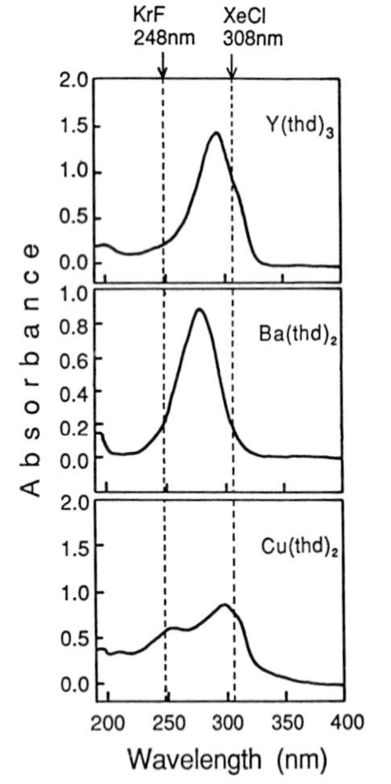

Fig. 2. UV absorbance of $Y(thd)_3$, $Ba(thd)_2$, and $Cu(thd)_2$.

Figure 3 shows deposition rates of the components for the films versus substrate temperature curves of the single source system, where the open circles are for unirradiated samples and the close circles are for KrF laser irradiated ones. The similar results were observed by using XeCl laser. There are no remarkable change in the deposition rates by irradiation in the low temperature range of less than 650°C. It is considered that, since the reaction between β-diketone and oxygen is so fast in the single source system, the decomposition by photo irradiation is far smaller than by oxygen and there is no effect on the deposition rate. At the temperatures higher than 650°C, the deposition rates of irradiated films become smaller than that of unirradiated films. It is considered that the deposited oxides are sputtered by the laser irradiation [6]. The SEM photograph of films with and without KrF laser irradiation are shown in Figure 4. In the Ba film without irradiation, there are many plate-like crystals in the grain boundary, and the composition of film are $Ba(OH)_2$. In the irradiated Ba film, however, the surface is smooth and the composition of film are $Ba(OH)_2$ and BaO_2. In the unirradiated Cu film, there exist mesh-like structures of CuO and Cu_2O. In the irradiated Cu film, a lot of granular crystallites of CuO are isolated and growing randomly. These results indicate the irradiation has promoted the crystallization and oxidation of films for the Ba and Cu films. The improvement of crystallization suggests that the activated oxygen oxidizes the films during deposition. In the Y film with KrF laser irradiation and Y, Ba, and Cu films with XeCl laser irradiation, however, no remarkable effect of irradiation on crystalline morphology is observed. It may be due to the low absorbance of $Y(thd)_3$ for the light of KrF line (248nm) and XeCl line (308nm), and $Ba(thd)_2$ for the light of XeCl line. In the Cu film, KrF laser irradiation was effective, but the XeCl laser irradiation was not effective. It indicates that the absorption at the charge transfer transition from ligand to metal is by far effective than the $\pi\rightarrow\pi*$ transition of ligand. The stationary temperature rise of the films is less than 0.1°C at irradiated power, therefore, these results are photochemical, not thermal. The experimental results indicate that the KrF laser irradiation promotes the crystallization and oxidation for deposited oxides, but there are no remarkable change of the deposition rate.

Figure 5 shows the SEM photographs of Ba-Cu and Y-Cu films without and with KrF laser irradiation, respectively. In the Y-Ba films consist of Y_2O_3, BaY_2O_4, and $Ba_2Y_2O_5$, whereas Y-Ba film KrF laser irradiated at 5Hz consists of Y_2O_3 and $Ba(OH)_2$. From this result, the KrF laser irradiation improve the oxidation of Y and Ba, and the effect of oxidation may be larger than the direct activation of source materials. Therefore, the irradiation impedes the formation of Y-Ba oxides. In the films including Cu are as follows; in the unirradiated Ba-Cu film are CuO and $Ba(OH)_2$, and those in the Y-Cu film are CuO and $Cu_2Y_2O_5$, whereas the films with KrF irradiation at 5Hz

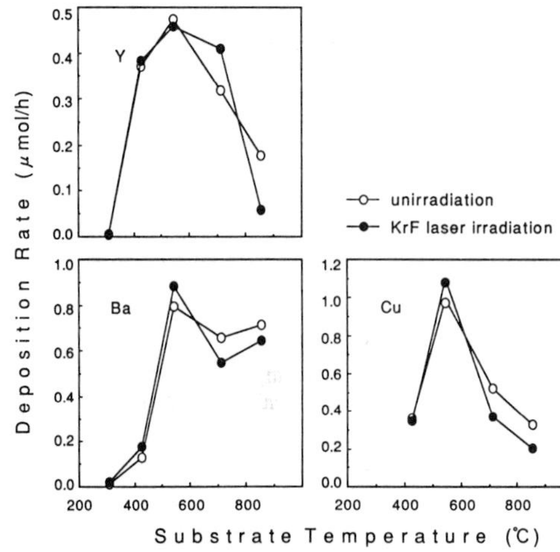

Fig. 3. Deposition rate of the components for the films versus substrate temperature curves of the single source system, where the open circles are for unirradiation and the close circles are for KrF laser irradiation.

consist of only $BaCuO_2$ and $Cu_2Y_2O_5$. The optical absorption of $Cu(thd)_2$ at the charge transfer transition from ligand to metal, is larger comparing with that of $Y(thd)_3$, and $Ba(thd)_2$, (in Fig.2), therefore, the highly excited copper is more active to react with other species and easy to form complex with other metals. On the other hand, in the XeCl laser irradiated films, no remarkable difference are observed for the morphologies and the composition of films. This result indicates that the XeCl laser irradiation neither create activated oxygen nor promote the reactivity of the source materials.

Figure 6 shows the deposition rate of atomic components for the films versus repetition rate curve of the Y-Ba and Ba-Cu films with KrF and XeCl laser irradiation. In the Y-Ba films with KrF laser irradiation, the deposition rates of film irradiated at 1Hz increases than that of unirradiated film. But that of film irradiated at 5Hz decreases slightly comparing to the film irradiated at 1Hz. In the Ba-Cu with KrF irradiation, deposition rates increase with increasing repetition rate, and the similar result is obtained in the Y-Cu film. On the other hand, in the XeCl laser irradiation films, there are no remarkable change of the deposition rate. In the above experiment, the each source gas flow rates are adjusted to the optimal ratio which gives ideal atomic ratio of YBCO, i.e., Y : Ba : Cu = 1 : 2 : 3. Therefore, if the reactions occur after deposition on the substrate, the deposition rate of the atomic components for the films should be Y : Ba = 1 : 2, Ba : Cu = 2 : 3, and Y : Cu = 1 : 3. Actually, in the all films with XeCl laser irradiation and the Y-Cu films with KrF laser irradiation, the deposition rates of the films give ideal atomic ratio of YBCO (1:2:3). However, in the KrF laser irradiation, the deposition rates of Y and Ba are 1 : 1 ~ 1 : 1.6 for the Y-Ba films and that of Ba and Cu are 1 : 1.8 ~ 1 : 3 for the Ba-Cu films, which are obvious deviation from intended deposition rates. It indicates that the source gases including Ba source react in the vapor-phase before

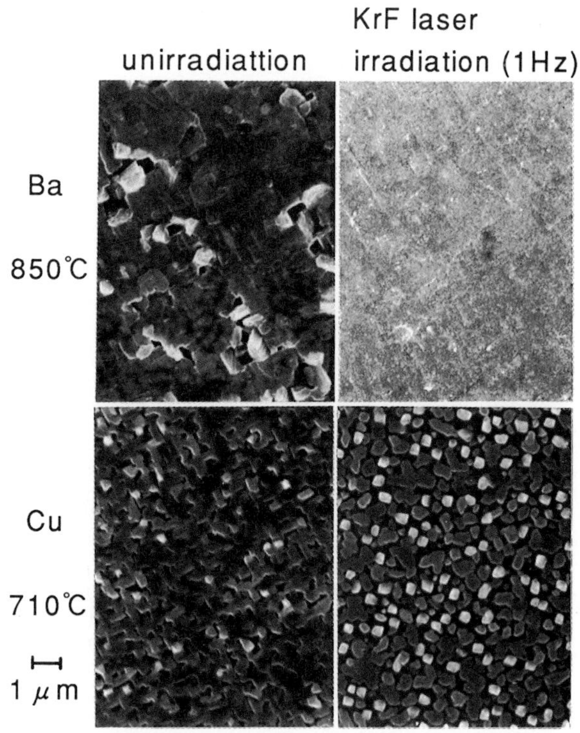

Fig. 4. SEM photograph of Ba and Cu films without and with KrF laser irradiation.

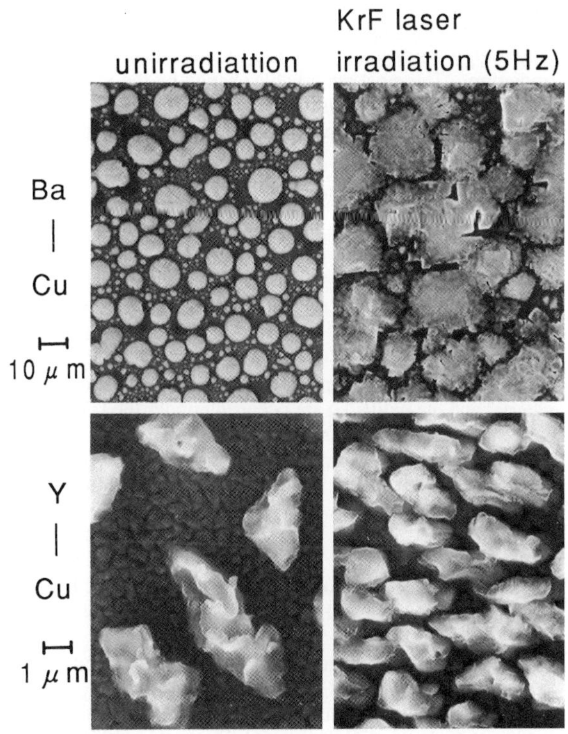

Fig. 5. SEM photograph of Ba-Cu and Y-Cu films without and with KrF laser irradiation at 850°C.

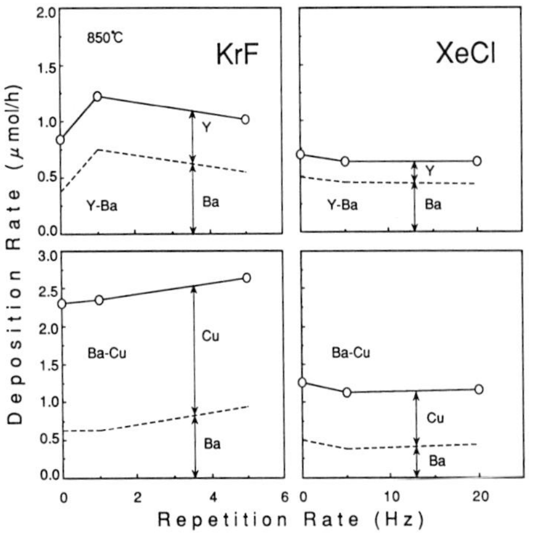

Fig. 6. Deposition rate of the components for the films versus repetition rate curves at 850°C.

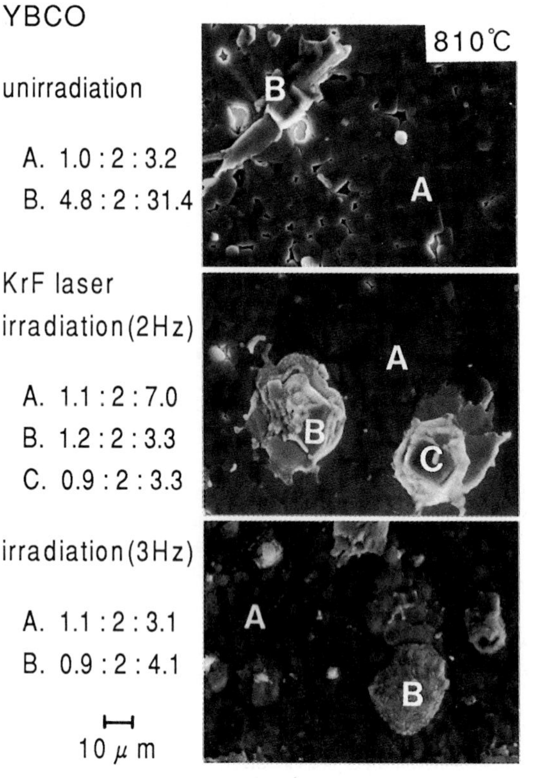

Fig. 7. SEM photograph of YBCO films without and with KrF laser irradiation at 810°C.

depositing on the substrate, and the Ba source is not decomposed by 248 nm light. It is because there is small absorbance as shown in Fig. 2. The results for the double source systems indicate that the KrF laser irradiation promotes the deposition rate and the Cu reactivity with Ba or Y in vapor-phase. It also promotes the crystallization and oxidation because of the uptake of activated oxygen into the compositions on the films during deposition.

Figure 7 shows the SEM photograph of YBCO films without and with KrF laser irradiation. In the unirradiated film, there is obvious grain boundary, but it vanished increasing repetition rate. The film surface morphology irradiated at 3Hz is a rougher than that at 2Hz. It is considered that the surface ablation occurs by the irradiation. The most remarkable change by the irradiation is the composition and the shape of the precipitates (B) in the matrix. The composition ratio of the matrix (A) of each film is about 1:2:3. In the unirradiated film, the crystal (B) is high Cu-rich composition, however, by irradiation the composition ratio approaches to 1:2:3 increasing the repetition rate.

CONCLUSIONS

We investigated the effects of the UV irradiation on the single CVD source system, the double source system, and YBCO system during the deposition. The KrF(248nm) laser irradiation has improved the crystallization and oxidation for some deposited oxide in the single source system and also has improved the crystallization, the oxidation, and the reactivity of Ba-Cu and Y-Cu systems and YBCO system. The improvement of crystallization is also suggested by the uptake of activated oxygen into the films during deposition. These results indicate the decomposition of the Cu source material and the oxidation are occurring under the UV light of 248nm which is resonant to the charge transfer transition from ligand to metal. However, the XeCl(308nm) laser irradiation which is resonant to $\pi \rightarrow \pi^*$ transition of ligand has not improved the crystallization, oxidation, and reactivity.

REFERENCES

1. Oda S, Zama H, Ohtsuka T, Sugiyama K, Hattori T (1989) Jap.J.Appl.Phys. 28(3): 427
2. Yamane H, Kurosawa H, Hirai T, Watanabe K, Iwasaki H, Kobayashi N, Muto Y (1989) Supercond.Sci.Technol. 2: 115
3. Doi A, Aoyagi Y, Namba S (1986) Appl.Phys.Lett. 49: 785
4. Fackler,Jr. JP, Cotton FA, Barnum DW (1963) Inorganic Chem. 2(1): 97
5. Yang X, Kutal C J.Am.Chem.Soc. (1983) 105: 6038
6. Jones CR, Houle FA, Kovac CA, Baum TH (1985) Appl.Phys.Lett. 46(1): 97

Characteristics of YBaCuO Films with Cu-rich Composition Grown by CVD

KAZUTOSHI HIGASHIYAMA, TAKAHISA USHIDA, and IZUMI HIRABAYASHI

Superconductivity Research Laboratory, Nagoya Division, International Superconductivity Technology Center, 4-1, Mutsuno 2-chome, Atsuta-ku, Nagoya, 456 Japan

ABSTRACT

The YBaCuO films with Cu-rich composition were prepared on MgO(100) at 650-750°C by chemical vapor deposition using β-diketone metal chelates. In the films prepared at 650°C and 700°C, the degree of c-axis orientation was enhanced with the increase of Cu composition in the compositional region from Y:Ba:Cu=1:2:3 to 1:2:4, and the films with Cu-rich composition showed higher Tc values than the films with close to stoichiometric one. The films prepared at the higher temperature of 750°C showed no Cu composition dependence on the crystalline orientation and Tc value.

KEY WORDS : chemical vapor deposition, $YBa_2Cu_3O_x$, thin film, crystalline orientation

INTRODUCTION

Chemical vapor deposition (CVD) is extensively used in the industrial applications. Also in the preparation of high-Tc superconducting films, CVD method offers many potentialities i.e., excellent film uniformity, high deposition rate, easily compositional control, and adaptability to deposition in highly oxidizing condition. Therefore, a number of investigators have reported on CVD of the high-Tc superconducting thin films. Some of $YBa_2Cu_3O_x$ films have very high critical current densities even in the strong magnetic fields. Most of the $YBa_2Cu_3O_x$ thin film on (100) perovskite substrates have a mixture of epitaxial grains with a-axis and c-axis perpendicular to the surface [1,2]. A mixture of a-axis and c-axis oriented grains is also observed in the films prepared by CVD [3]. This tendency to form mixed epitaxial grains seriously affects the application of $YBa_2Cu_3O_x$ films because of the anisotropic properties of $YBa_2Cu_3O_x$. On the other hand, Yamane et al. pointed out that the a-axis oriented grains included in the c-axis oriented film might play an important role for a flux pinning [4]. In the same deposition method and same substrate, however, crystalline orientation in $YBa_2Cu_3O_x$ film changes with the details of the growth conditions [5,6]. A detailed understanding of the microstructure and the ability to control the microstructure are key issues for successful application and for understanding for the growth mechanism in CVD as well.

It has been reported that the quantity of a-axis oriented grains in the $YBa_2Cu_3O_x$ films prepared on $SrTiO_3(100)$ and MgO(100) at 850°C changed with chemical composition of Cu [4]. In the $YBa_2Cu_3O_x$ film deposited on MgO(100) at 715°C, crystallinity of the film was improved by supply in a large amount of Cu in excess of the stoichiometric ratio [7]. Recently, we observed that the crystalline orientation in $YBa_2Cu_3O_x$ films prepared on MgO(100) by CVD depended on chemical composition of Cu at the lower substrate temperatures. In this paper, we report on the effects of the excess copper element from stoichiometric composition on crystallographic and superconducting properties of the $YBa_2Cu_3O_x$ films prepared on MgO(100) by CVD.

EXPRIMENTAL

A schematic diagram of the experimental apparatus is shown in Figure 1. A CVD reactor was a horizontal cold-wall type. Substrate temperature was monitored by an inconel sheathed thermocouple set on the substrate with Ag paste. In this apparatus, all of the gas lines, the valves, and the entire reactor were placed in ovens and were heated to temperatures higher than each source temperature to eliminate condensation of the source materials. The source materials used were 2,2,6,6-tetramethyl-3,5-

heptanedione (thd) chelates of Y, Ba, and Cu. The ovens for the Ba line, the Y and Cu lines, and the reactor were kept at 260°C, 180°C, and 250°C, respectively. Argon carrier gas (99.999% purity) was passed over the heated materials in the evaporator. The mixture of the metal chelate vapors was introduced into the reactor through the pipe inlet after premixing with a small amount of oxygen gas. Most of the oxygen gas was separately introduced into the reactor. The films with various Cu composition were prepared on MgO(100) substrates (10x10x0.5mm) at substrate temperatures of 750°C, 700°C, and 650°C. The range of conditions for deposition are summarized in Table 1. The atomic ratio of [Cu]/[Y] in the films was varied in the range of 3 to 4 by controlling the temperature of evaporator and a flow rate of argon gas for $Cu(thd)_2$. The reactor pressure was kept at 10 Torr during the deposition. After each deposition for one hour, the films were cooled to the oven temperature of 250°C at a rate of 100°C/min under the oxygen atmosphere of 1 atm.

The film composition was determined by inductively coupled plasma emission spectroscopy (ICP). The microstructure of the film surface was observed by scanning electron microscope (SEM) with energy dispersive X-ray spectroscopy (EDX). X-ray diffraction (XRD) patterns of the films were obtained by the standard θ-2θ method. The temperature dependence of the electrical resistivity for the films were measured by dc four-probe method.

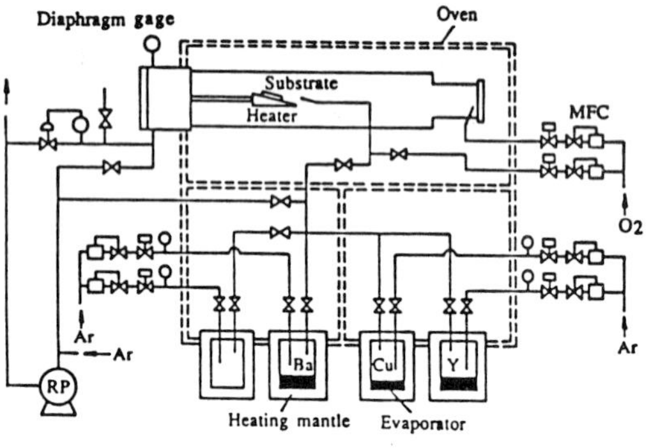

Fig.1 Schematic diagram of CVD apparatus.

Table 1 Deposition conditions.

Evaporator temperature (°C)	
$Y(thd)_3$	125
$Bu(thd)_2$	239
$Cu(thd)_2$	118-120
Argon flow rate (cm^3/min)	
$Y(thd)_3$	29-30
$Bu(thd)_2$	6- 7
$Cu(thd)_2$	20-40
Oxygen flow rate (cm^3/min)	200
Deposition time (min)	60
Reactor pressure (Torr)	10
Substrate temperature (°C)	650-750

RESULTS AND DISCUSSION

The $YBa_2Cu_3O_x$ films prepared at the substrate temperature of 750°C show the high relative intensity peaks of (001) with small peaks of (100), (200) and (110) in the XRD pattern. The relative intensities of these peaks do not depend on the Cu content of the films in the compositional region from Y:Ba:Cu=1:2:3 to 1:2:5. On the other hand, XRD patterns of the films prepared at 700°C and 650°C, vary sensitively with the Cu content, as shown in Figure 2 and 3. The variation of Cu content in the films changes mainly the relative intensities of (001) and (110) peaks. Figure 4 shows the intensity ratios of I(110)/I(005) and I(200)/I(005) as a function of the Cu composition in the range of 3.0 to 4.0. The value of I(110)/I(005) largely depends on the Cu composition. Between the Cu composition of 3 and 4, this value decreases by a factor of 10^3 and 10^2 at 650°C and 700°C, respectively. However, no dependence of Cu composition is shown at 750°C because the intensity of (110) peak is too small even in the film with the stoichiometric composition. On the other hand, the value of I(200)/I(005) is independent of the Cu composition at each temperature, and decreases by a factor of about 10 for each 50°C rise in temperature.

SEM photographs of the films with different Cu composition are shown in Figure 5. The films deposited at 750°C (Figure 5(a),(b)) are composed of a uniform matrix of $YBa_2Cu_3O_x$ with the c-axis oriented perpendicular to the substrate surface, and two types of precipitates being spherical or rectangular in shape. The spherical precipitates are mainly composed of CuO according to analyses of XRD and EDX, and the number of them increases with increasing Cu composition. The rectangular precipitates are regarded as the a-axis oriented $YBa_2Cu_3O_x$ grains corresponding to (100) and (200) peaks in the XRD pattern. The number and the size of the precipitates do not change with Cu composition.

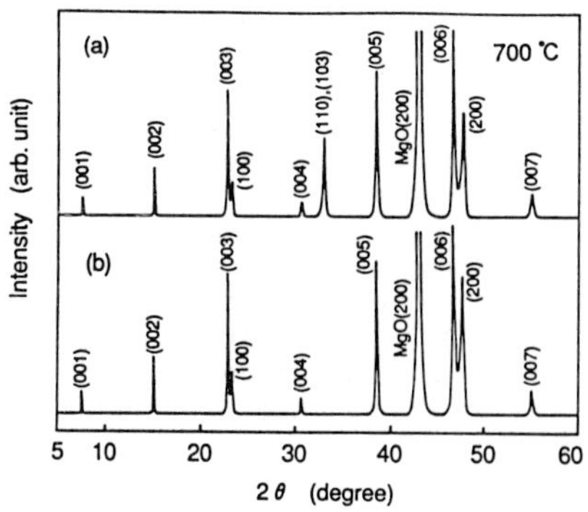

Fig.2 XRD patterns of YBCO films prepared at 700°C. The film compositions are; (a) Y:Ba:Cu=1.1:2:3.2 and (b) 1.2:2:4.0.

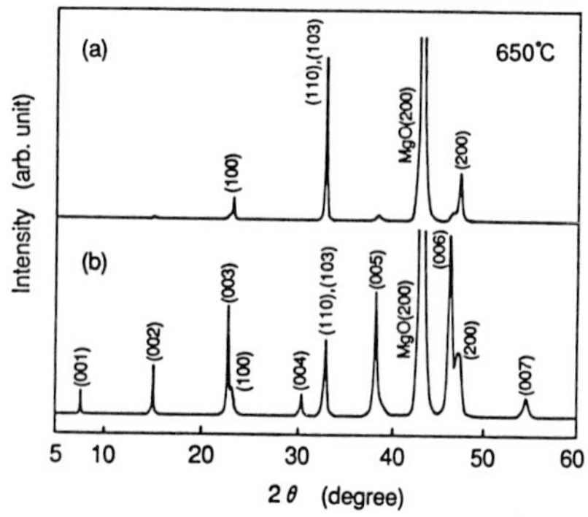

Fig.3 XRD patterns of YBCO films prepared at 650°C. The film compositions are; (a) Y:Ba:Cu=1.0:2:3.2 and (b) 1.1:2:4.0.

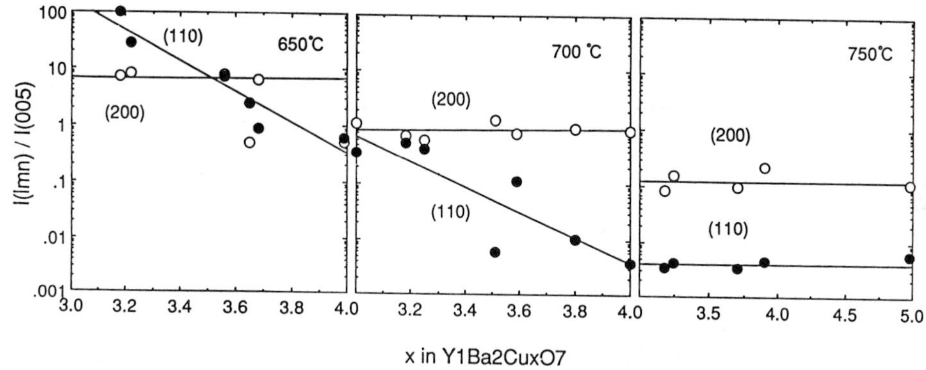

Fig.4 X-ray diffraction intensity ratios of I(110) and I(200) to I(005) for YBCO films as a function of nominal Cu composition.

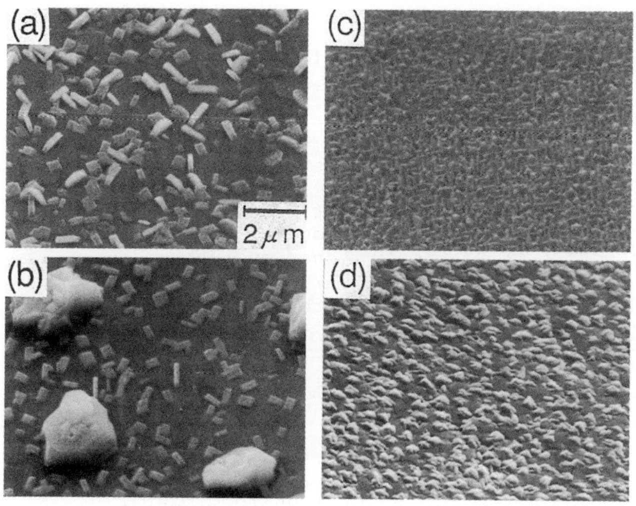

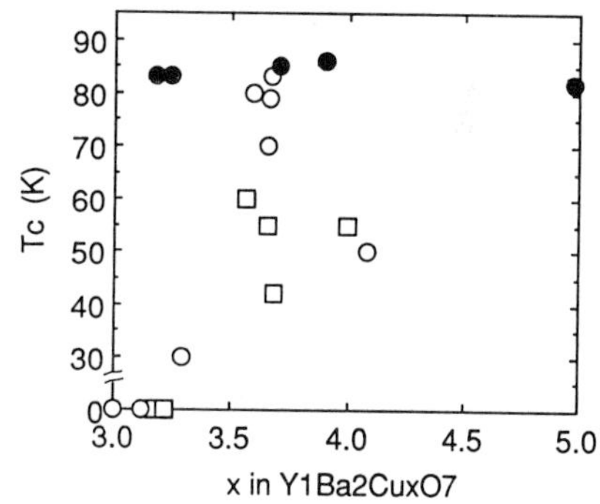

Fig.6 Tc of YBCO films prepared at 650°C(□), 700°C(○), and 750°C(●) as a function of nominal Cu composition.

Fig 5 SEM photographs of YBCO films prepared at 750°C (a),(b) and 650°C (c),(d), respectively. The film compositions are; (a) Y:Ba:Cu=1.1:2:3.2, (b) 1.1:2:3.9, (c) 1.0:2:3.2, and (d) 1.1:2:4.0.

The films prepared at 650°C (Figure 5(c),(d)) show remarkable change in morphology in contrast to the films prepared at 750°C. The film with stoichiometric composition shows the needle like growth patterns and poor c-axis orientation. With increase of Cu composition, spherical precipitates with the size of about 0.1 µm appear in the $YBa_2Cu_3O_x$ matrix, and the matrix becomes smooth and dense. These morphological change is consistent with the results of XRD pattern shown in Figure 4.

Figure 6 shows Tc of the films prepared at 750°C, 700°C, and 650°C as a function of Cu composition. Tc values of the films prepared at 750°C are about 85K independent of Cu composition. For the film prepared at 700°C and 650°C, Tc values are dependent on Cu composition. The films with composition close to stoichiometric one do not show superconductivity, and the films with higher Cu composition than 3.5 show Tc of about 80K for 700°C and about 55K for 650°C.

At the high substrate temperatures, excess amounts of Cu element do not affect Tc value of the film, and it is included as CuO precipitates in the film. In the low temperature region, in which $YBa_2Cu_3O_x$ grows with (100) or (110) orientation, excess amounts of Cu element result in promoting growth of c-axis oriented matrix structure, the film with Cu-rich composition shows higher Tc value than the film with close to stoichiometric composition. Tsuruoka et al. reported that the c-axis oriented $YBa_2Cu_3O_x$ film with Tc value of 65K was prepared at 650°C on MgO(100) by CVD using N_2O as oxygen source [8]. The ratio of [Cu]/[Ba] and [Y]/[Ba] in the films grown by N_2O were larger than those in the films grown by O_2 when other experimental conditions were kept the same. Kanehori reported that the $YBa_2Cu_3O_x$ film with Tc value of 60K was prepared at lower temperature of 515°C on $SrTiO_3$(100) by the plasma-enhanced CVD [9]. Also in this case, Cu composition([Cu]/[Y]+[Ba]) and [Y]/[Ba] ratio of grown films increased with increasing microwave power. In both case, Cu deposition rate was enhanced under the enhanced conditions formed high quality films. These results suggest that activation of Cu precursor may be necessary to form the $YBa_2Cu_3O_x$ structure at low temperature.

In summary, the effects of the excess Cu element from stoichiometric composition on crystallographic and superconducting properties of the $YBa_2Cu_3O_x$ film on MgO(100) were examined. In the XRD patterns of films prepared at 650°C and 700°C, the diffraction intensity ratio of (001) peaks to (110) peak exponentially increases with increase of Cu composition. The films with Cu composition higher than 3.5 show Tc of about 80K for 700°C and about 55K for 650°C. The results show that the supply of excess amount of Cu element is effective to form the highly c-axis oriented $YBa_2Cu_3O_x$ film at the low temperatures.

The authors wish to thank Mr. H. Higa of Nagoya Institute of Technology for helpful discussions.

REFERENCES

1. Eom CB, Sum JZ,Yamamoto K, Marshall AF, Luther KE, Geballe TH, Laderman SS (1989) Appl.Phys.Lett.55:595
2. Hwang DM, Venkatesan T, Chang CC, Nazar L, Wu XD, Inam A, Hegde MS (1989) Appl.Phys.Lett.54:1702
3. Watanabe K, Yamane H, Kurosawa H, Hirai T, Kobayashi N, Iwasaki H, Noto K, Muto Y (1989) Appl.Phys.Lett.54:575
4. Yamane H, Hirai T, Kurosawa H, Suhara A, Watanabe K, Kobayashi N, Iwasaki H, Aoyagi E, Hiraga K, Muto Y (1989) Proceedings of the Second International Symposium on Superconductivity (ISS'89).November 14-17 1989. Tsukuba:767
5. Yamaguchi T, Aoki S, Sadakata N, Kohno O, Osanai H (1989) Appl.Phys. Lett.55:1581
6. Panson AJ, Charles RG, Schomidt DN, Szedon JR, Machiko GJ, Braginski AI (1988) Appl.Phys.Lett.53:1756
7. Ohnishi H, Kusakabe Y, Kobayashi M, Hoshinouchi S, Harima H, Tachibana K (1990) Jpn.J.Appl.Phys.29:1070
8. Tsuruoka T, Kawasaki R, Abe H (1989) Jpn.J.Appl.Phys.Lett.28:1800
9. Kanehori K (1990) Abstracts of the Second ISTEC Workshop on Superconductivity May 28-30 1990. Kagoshima:92

Formation of Y-Ba-Cu-O Ultrathin Films by In-Situ Oxidation of Metallic Components

RYUSUKE KITA, TAKASI HASE, MASATO SASAKI, TADATAKA MORISHITA, and SHOJI TANAKA

Superconductivity Research Laboratory, International Superconductivity Technology Center, 10-13, Shinonome 1-chome, Koto-ku, Tokyo, 135 Japan

ABSTRACT

Formation conditions of each oxide thin film of Y, Ba, Cu and Y-Ba-Cu by in-situ oxidation of the metallic thin films prepared in ultrahigh vacuum have been investigated. A single crystal CuO ultrathin (5nm) film has epitaxially grown on MgO(100) and a Y_2O_3 thin film whose (222) is parallel to the substrate has grown but a Ba oxide film has not. 123 thin film has been formed by in-situ oxidation of metallic thin film of Y-Ba-Cu and the formation condition is discussed.

KEY WORDS : YBCO thin film, CuO thin film, ultrahigh vacuum, in-situ oxidation

INTRODUCTION

It is necessary for preparation of high-Tc oxide superconducting films to understand the growth mechanism of the films under conditions of various temperature and oxygen pressure. And it is also important for the atomic layering growth of a film to investigate the oxidation condition for metallic components of oxide superconductors. We are trying to prepare the films at low temperature by in-situ oxidation of metallic thin films which are free from oxygen contamination. The surface of ultra thin films of metals deposited in ultra high vacuum is very active because it has less adsorption gases. It has been shown that very small amounts of oxygen can reduce reactivity in thin films and multilayers[1][2]. Recently it has been reported that ultra thin films of $Bi_2Sr_2CuO_y$ system is obtained at low temperature by oxidation of metals deposited in ultrahigh vacuum using N_2O gas[3]. For preparing pure Y, Ba layers ultrahigh vacuum is required because they are very active for oxygen. In this paper, we present the formation condition of the each oxide thin films of Y, Ba, Cu and Y-Ba-Cu by in-situ oxidation of the metallic thin films prepared in ultrahigh vacuum.

EXPERIMENTAL

We used a specially designed MBE apparatus which enable the in-situ oxidation of metallic thin films without taking out them from the ultrahigh vacuum chamber. The background pressure is less than $5x10^{-11}$ Torr. The MBE apparatus is characteristic of a in-situ oxidation system, which is composed of the cover for substrates in order to prevent evaporation sources from being oxidized during oxidation of metallic films in a oxygen pressures up to 10^{-1} Torr keeping the growth chamber in ultrahigh vacuum. High purity molecular oxygen gas(99.9999%) is introduced to substrates through the inlet of the cover. The metallic films were deposited on MgO(100) substrates using three electron-beam guns. Reflection high energy electron diffraction (RHEED) patterns are observed in situ. The chemical compositions of oxide films were checked by inductive coupled plasma emission spectroscopy(ICPS).

RESULTS AND DISCUSSION

A. Oxidation of Y, Ba, Cu thin films

We have investigated the in-situ oxidation of Cu thin film deposited in $3x10^{-10}$ Torr first of all. Before deposition of Cu, the surface of MgO was cleaned by heating up to 800°C under $3x10^{-10}$ Torr. The RHEED pattern of MgO is shown in fig. 1(a), which is characteristic of the clean and smooth surface. Cu was deposited on the substrate at 275°C and a deposition rate of 0.017nm/sec. The base pressure during deposition was $3x10^{-10}$ Torr and the thickness of the Cu thin film is 5nm. The RHEED pattern after the deposition is shown in Fig. 1(b). The spots and rings are observed, indicating that the deposited Cu is polycrystalline and the surface has rough surface. After completing of Cu deposition, the substrate was heated up to 550°C(oxidation temperature; *To*) and then oxygen was introduced to the cap. The oxygen pressure(P_{O_2}) near the substrate was $2x10^{-2}$Torr, while the base pressure of the growth chamber was remained less than $3x10^{-10}$ Torr. The film was kept for 10 min under this condition. The thickness of the film resulted was unchanged. Figure 1(c) shows the RHEED pattern of the film after oxidation, where the incident beam is parallel to the [110] direction of the substrate. The streaky pattern indicates that the oxide film is a single crystal epitaxially grown on the substrate and its surface is extremely smooth. The x-ray diffraction(XRD) pattern of the Cu oxide film using Cu Kα radiation is shown in fig. 2. Only the (111) reflection of CuO is observed and full width at half maximum (FWHM) of a rocking curve is 0.8° for the (111) peak.

The Y thin film of 10nm thickness was deposited in the same way. After in-situ oxidation in P_{O_2}=$2x10^{-2}$Torr at *To*=800°C, the film increased in thickness 4 times as large as the metallic Y film, and lost the streaky pattern. The RHEED pattern presents the dim streaks. however the RHEED pattern of the oxide film(40nm) prepared in the condition of *To*=800°C and P_{O_2}=$2x10^{-2}$Torr is not streaky. Only the reflection peak of Y_2O_3 (222) is observed in the XRD pattern of which rocking curve shows 0.82 degrees of FWHM. Ba thin film of 5nm thickness was also deposited. The RHEED pattern of the Ba film is spotty. The spotty RHEED pattern of the Ba film changed into the ring-shaped one after the oxidation of the film(P_{O_2}=$2x10^{-2}$Torr,*To*=400°C). None of

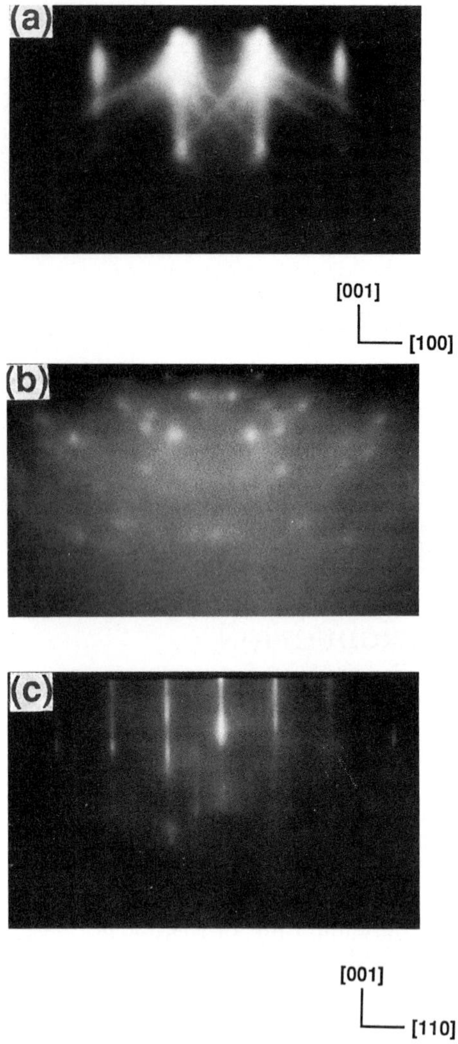

FIG. 1 Reflection high energy electron diffraction patterns for the films. (a) MgO, (b) Cu film, (c) CuO film

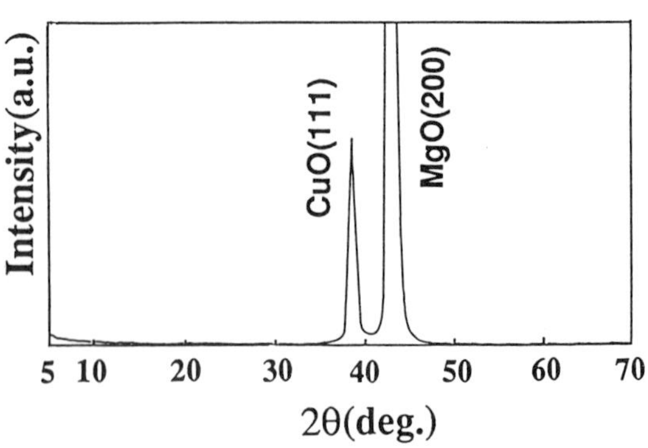

FIG. 2 X-ray diffraction pattern for the CuO thin film.

XRD peaks of BaO or BaO_2 have been observed. More experiments in various condition are in progress.

B. Oxidation of Y-Ba-Cu metallic thin films

It is interesting to apply the results of CuO to the preparation of Y-Ba-Cu-O film. There are two ways for preparing Y-Ba-Cu metallic films. One is co-evaporation and the other is sequential evaporation of constituents.At first, Y, Ba and Cu were simultaneously deposited on MgO at 350°C using electron-beam guns at a background pressure of less than $3x10^{-10}$ Torr. The fluxes of each metal was adjusted to give a 1:2:3=Y:Ba:Cu ratio and the total deposition rate was 0.1nm/sec. The thickness of the metallic film was 10nm. After the deposition of the metals, in-situ oxidation was performed under the following conditions:temperatures *To* from 500°C to 830°C and P_{O_2} from $7x10^{-5}$ Torr to $2x10^{-2}$ Torr to investigate the formation condition of 123 phase. The results are mapped in fig.3. The substrate was kept at 350°C to avoid the desorption of deposited metals when oxygen was introduced. The solid line indicates the critical stability line for the $Y_1Ba_2Cu_3O_{6.0}$ tetragonal phase of the bulk[4]. Only above the line, successful in-situ growth of YBCO thin films have been reported[5].Under the conditions indicated with the closed circles(high P_{O_2}, high *To*), the 123 phase has formed after a hold time of 10min. Fig. 4(a) shows the XRD pattern of the Y-Ba-Cu oxide film prepared under the condition of $P_{O_2}=2x10^{-2}$ Torr, *To*=800°C, showing hat c-axis oriented 123 structure is formed but unidentified oxide phases are contained. The thickness of the obtained oxide film is also increased to 40nm. The rings are observed in the RHEED pattern as shown in fig. 4(b). The region of P_{O_2} and *To* where the 123 phase is realized is below the line of the thermodynamic stability as shown in fig. 3. The reason why the stability line of 123 thin films is probably lower than that of the bulk(shift to lower P_{O_2}) is not clear yet. Under the conditions indicated with the open triangles(low P_{O_2}), the deviation of the chemical composition is severe, and the conditions indicated with the open circles(high P_{O_2}, low *To*) revealed no deviation of the chemical composition but amorphous structure. This corresponds to the tendency that the structure formed in the higher P_{O_2} or lower *To* region has gradually been disordered[5]. It is necessary for growth good

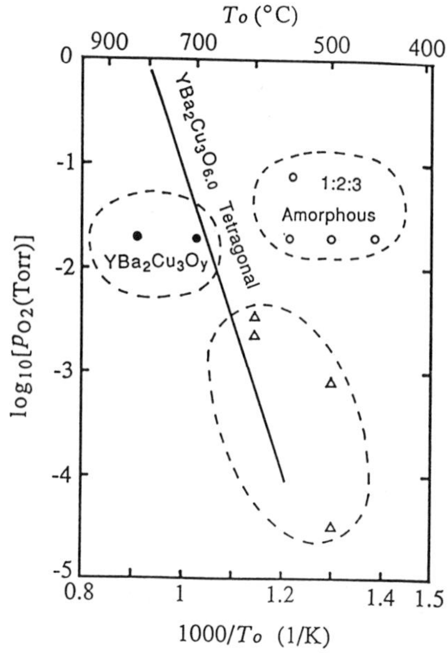

FIG. 3 Oxidation conditions of Y-Ba-Cu metallic thin films. The solid line indicates the critical stability line for the $Y_1Ba_2Cu_3O_{6.0}$ tetragonal phase of the bulk[4].

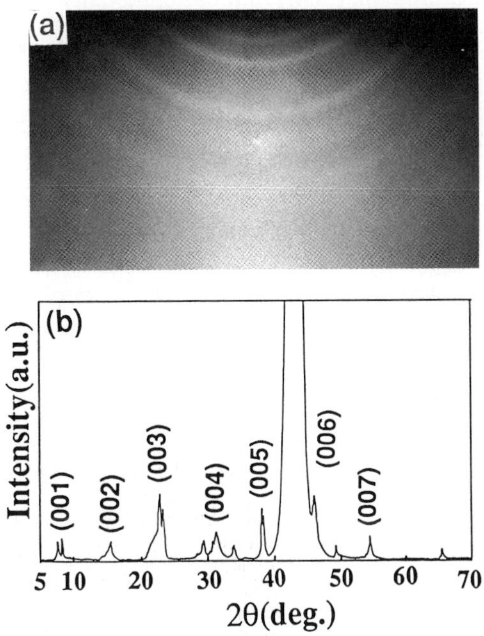

FIG. 4 XRD pattern (a) and RHEED pattern (b) for the YBCO thin film prepared by the in-situ oxidation ($P_{O_2}=2x10^{-2}$ Torr, *To*=800°C).

123 film to oxidize the metallic films along or just below the stability line.

Next Y, Ba and Cu were sequentially deposited (Cu/Ba/Cu/Y/Cu/Ba) on MgO at room temperature.The total thickness of the metallic film is 1.2nm. The RHEED pattern of the layered metallic film shows the same pattern as the Cu layer deposited. The metallic film was in-situ oxidized under the conditions of P_{O_2}=2x10^{-2} Torr and *To* = 300°C. The oxide film is a thickness of 6nm and is insulator. The RHEED pattern of rings is observed in the oxide film as shown in fig. 5(a). Then the film was heated up to 500°C in the same P_{O_2}. The rings pattern becomes clearer as shown in fig. 5(b), but only the peaks of the substrate are observed in XRD. These results indicate that a kind of crystalline oxides are formed even at 300°C. The P_{O_2}-*To* diagram of sequentially deposited metallic films is under investigation.

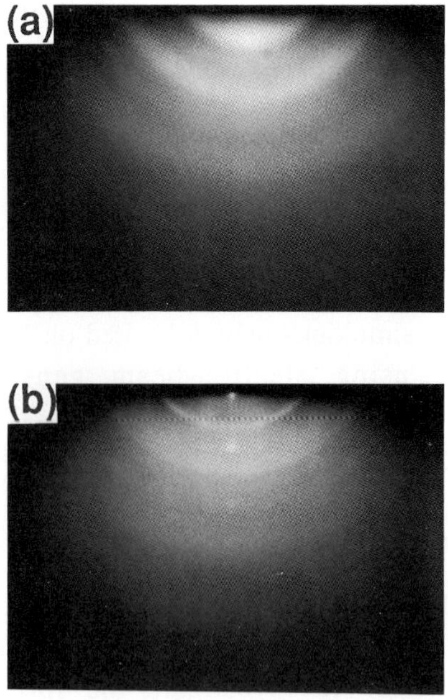

FIG. 5 RHEED patterns for the 123 film prepared by in-situ oxidation of (a)P_{O_2}=2x10^{-2} Torr, *To*=300°C and (b)P_{O_2}=2x10^{-2} Torr, *To*=500°C

CONCLUSION

We have investigated the formation conditions of each oxide thin film of Y, Ba, Cu and Y-Ba-Cu by in-situ oxidation of the metallic thin films prepared in ultrahigh vacuum. A single crystal CuO ultrathin (5nm) film has epitaxially grown on MgO(100). The (111) plane of the CuO film is parallel to the substrate. A Y_2O_3 thin film whose (222) is parallel to the substrate surface has been obtained but a Ba oxide film has not. The 123 thin film has also been formed by in-situ oxidation of the metallic film which is co-evaporated the Y, Ba, and Cu elements in ultrahigh vacuum. The crystalline oxides have been formed at 300°C by sequentially deposition of metallic film following in-situ oxidation .

ACKNOWLEDGEMENT

This work was supported by New Energy and Industrial Technology Development Organization under the management of the R&D of Basic Technology for future Industries.

REFERENCES

1. E. Ma, M-A. Nicolet and M. Nathann (1989) J. Appl. Phys. 65: 2703
2. M. W. Ruckman, L. Jiang and M. Srongin (1990) J. Vac. Sci. Technol. A8: 134
3. S. Watanabe, M. Kawai and T. Hanada (1990) Jpn. J. Appl. Phys. 29: L1111
4. R. Borman and J. Nolting (1989) Appl. Phys. Lett. 54: 2148
5. R. H. Hammond and R. Borman (1989) Physica C 162-164: 703

The Effect of Temperature and Oxygen Partial Pressure on Appearance of 124 Phase in MOCVD Film

HARUMI HAYASHI, KAZUSHI SUGAWARA, and YUH SHIOHARA

Superconductivity Research Laboratory, International Superconductivity Technology Center, 10-13, Shinonome 1-chome, Koto-ku, Tokyo, 135 Japan

ABSTRACT

Recently, we reported that X-ray diffraction peaks due to the $YBa_2Cu_4O_8$ phase were detected[1], when metal organic chemical vapor deposition (MOCVD) films were fabricated at a susceptor temperature of 800°C by controlling the overall composition within certain limits. In this study, efforts were focused on clarifying the effects of oxygen partial pressure and susceptor temperature on appearance of the 124 phase. The 124 phase existed only in films fabricated at susceptor temperatures below 800°C under oxygen partial pressure of 7.5 torr while the overall composition was controlled within certain limits. Oxygen partial pressure was found to affect the appearance of the 124 phase significantly.

KEYWORDS: MOCVD, YBaCuO, 124 phase, superconductor, thin films

INTRODUCTION

The superconducting phase $YBa_2Cu_4O_8$ (124) is superior to the $YBa_2Cu_3O_y$ (123) phase in terms of thermochemical stability of the oxygen content. Previously the 124 phase was discovered as lattice defects in thin film samples[2-5]. Bulk samples were formed by high pressure oxygen annealing[6-8] or ambient pressure oxygen low temperature annealing using alkali carbonate[9,10]. The value of Tc is 80K. However, Miyatake et al.[5] reported that Tc could be increased to 90K by partial substitution of Ca for Y. Thus the 124 has a good potential for practical applications. Thin films of the 124 were fabricated by K. Char et al.[8] by annealing an amorphous film at appropriate temperature under 1 atm oxygen pressure. As for MOCVD, while several groups have fabricated 123 films[11-13], there have been no report of 124 film fabrication. In a previous paper[1], we reported formation of the 124 phase in an MOCVD film. Here we present the effects of oxygen partial pressure and susceptor temperature on appearance of the 124 phase.

EXPERIMENTAL

The MOCVD system used in this study is a conventional horizontal cold-wall type. Deposition conditions are listed in Table 1. The source materials were the ß-diketonate metal chelates: bis-(2,2,6,6-tetramethyl-3,5-heptanedionate)-yttrium $Y(DPM)_3$, $Ba(DPM)_2$ and $Cu(DPM)_2$. The film fabrication process has been described in detail elsewhere[1]. The composition ratios of Y, Ba and Cu in the formed films were analyzed by inductively coupled

Table I. Deposition condition

MO source	Temperature(°C)	Ar flow (sccm)
Y(DPM)3	95-120	50-150
Ba(DPM)2	200-235	50-150
Cu(DPM)2	90-115	50-150
Oxygen flow		:450-2000sccm
Substrate		:MgO(100)
Temperature of susceptor		:750-850°C
Total gas pressure		:10, 20, 40torr
Oxygen partial pressure		:7.5, 17.5, 34torr
Cooling rate		:10°C/min

plasma (ICP) emission spectroscopy. The crystal structure of the films was analyzed by X-ray diffraction (XRD) using Cu-Kα radiation.

RESULTS AND DISCUSSION

Figure 1 shows the XRD patterns of the Y-Ba-Cu-O films fabricated at different susceptor temperatures of (a) 800°C and (b) 750°C. Total pressure (P) and oxygen partial pressure (PO_2) were 10 torr and 7.5 torr, respectively. The overall compositions of the films were found to be (a) Y:Ba:Cu= 0.95:2.0:4.2 and (b) Y:Ba:Cu= 0.79:2.0:3.7. Triangles indicate (001) peaks of the 124 phase, while un marked peaks correspond to the 123 phase. In Figure 1(a), strong (001) peaks of 123 and 124 were detected, but no other peak of 123 or 124 other than (001) was detected. In contrast, in Figure 1(b), (001) peaks of 124 were detected, but (001) peaks of 123 were very weak. Other peaks, which are close to (100) and (200) of both 123 and 124, were detected. Since the lattice constants of the a-axis and b-axis of 124 phase are close to those of the 123 phase, we could not conclude whether the peaks originated from the 123 or the 124 phase. But it can be concluded that it is possible to fabricate films containing the 124 phase at susceptor temperatures below 800°C under oxygen partial pressure of 7.5 torr.

To clarify the effect of the oxygen partial pressure on the appearance of the 124 phase, a number of films were fabricated at different oxygen partial pressures (7.5 torr, 17.5 torr, and 34 torr) at the same temperature (800°C). Figure 2 shows the composition range over which films contained the 124 phase. As the oxygen partial pressure increased, the composition range over which the 124 phase appeared became much larger. From this result, it is clear that the appearance of 124 depends on the oxygen partial pressure. But, so far it has not been succeeded to produce a single phase 124 film. Further studies are needed to determine the conditions for fabrication of single phase 124 films.

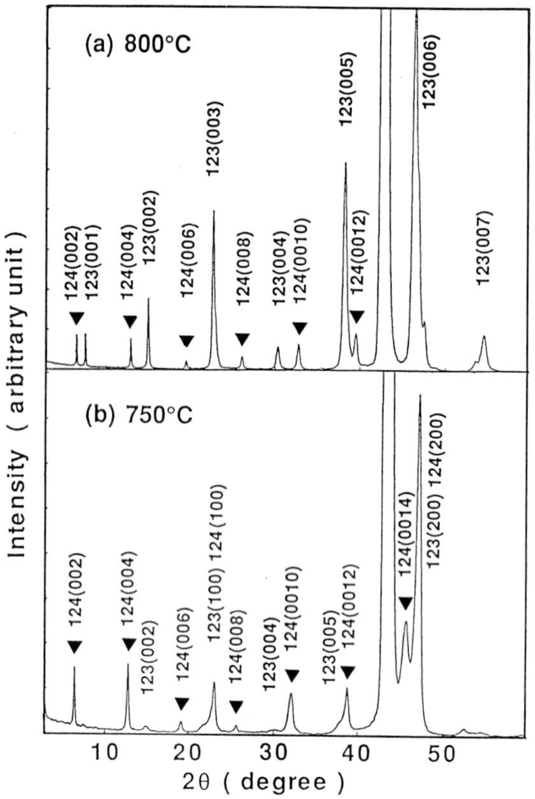

Fig.1 X-ray diffraction pattern of the film fabricated at (a)800°C and (b)750°C.

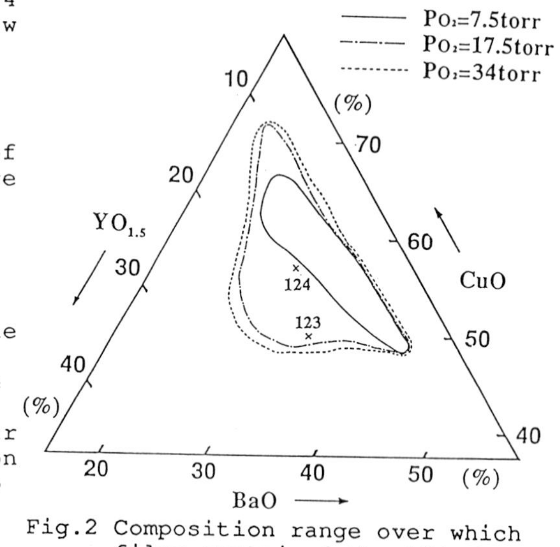

Fig.2 Composition range over which films contained the 124 phase at different oxygen partial pressure (7.5, 17.5, and 34 torr). The susceptor temperature were same(800°C).

CONCLUSION

Y-Ba-Cu-O superconducting films containing the 124 phase were fabricated by MOCVD below 800°C under oxygen partial pressure of 7.5 torr without post annealing. As the oxygen partial pressure increased, the composition range over which the 124 phase appeared became much larger.

ACKNOWLEDGEMENT

The authors express their appreciation to Dr. M. Green and Mr. Y. Yamada for helpful discussions.

REFERENCE

1) H. Hayashi, T. Sugimoto, K. Kikuchi, S. Yuhya, Y. Yamada, M. Yoshida, K. Sugawara and Y. Shiohara: Proc. 2nd Int. Symp. on Superconductivity (ISS'89), Tsukuba, Springer-Verlag Tokyo,(1989), 431
2) A.F.Marshall, R. W. Barton, K. Char, A. Kapitulnik, B. Oh, R. H. Hammond and S. S. Laderman, Phys. Rev. B **37** (1988) 9353
3) M. L. Mandich, A. M. DeSantolo, R. M. Fleming, P. Marsh, S. Nakahara, S. Sunshine, J. Kwo, M. Hong, T. Boone and T. Y. Kometani, Phys. Rev. B **38**,(1988) 5031
4) J. Kwo, M. Hong, R. M. Fleming, A. F. Hebard, M. L. Mandich, A. M. DeSantolo, B. A. Davidson, P. Marsh and W. D. Hobbins, Appl. Phys. Lett. **52** (1988) 1625
5) K. Char, Mark Lee, R. W. Barton, A. F. Marshall, I. Bozovic, R. H. Hammond, M. R. Beasley, T. H. Geballe and A. Kapitulnik, Phys. Rev. B **38** (1988) 834
6) J.Karpinski, E.Kaldis, E, Jilek, S. Rusiecki and B. Bucher, Nature **336** (1988) 660
7) D. E. Morris, J. H. Nickel, J. Y. T. Wei, N. G. Asmer, J. S. Scott, U. M. Scheven, C. T. Hultgren, A. G. Markelz,J. E. Post, P. J. Heaney, D. R. Veblen and R. M. Hazen, Phys. Rev. B**39** (1989) 7347
8) T.Miyatake, K. Yamaguchi. T. Takaka, S. Gotoh, N. Koshizuka and S. Tanaka, Physica C **160** (1989) 541
9) R. J. Cava, J. J. Krajewski, W. F. Peck Jr, B. Batlogg, L. W. Rupp Jr., R. M. Fleming, A. C. W. P. James and P. Marsh, Nature **338** (1989) 328
10) R. J. Cava, J. J. Karajewski, W. F. Peck Jr., B. Batlogg and L. W. Rupp Jr., Physica C **159** (1989) 372
11) H. Abe, T. Turuoka and T. Nakamori: Jpn. J. Appl. Phys. **27** (1988) 1473.
12) H. Yamane, H. Kurosawa, H. Iwasaki, H. Masumoto, T. Hirai, N. Kobayashi and Y. Muto: Jpn. J. Appl. Phys.**27** (1988) 1275
13) K. Shinohara, F. Munakata and M. Yamanaka: Jpn. J. Appl. Phys. **27** (1988) 1683.

Growth and Characterization of a Semiconductor-Superconductor Heterostructure: the $Hg_{1-x}Cd_xTe/YBa_2Cu_3O_{7-x}$ System

JEFFREY T. CHEUNG, ISORIS GERGIS, P. KOBRIN, J. ARIAS, and R.E. DEWAMES

Rockwell International Science Center, Thousand Oaks, CA 91360, USA

ABSTRACT

A semiconductor-superconductor heterostructure has been successfully demonstrated in the $Hg_{1-x}Cd_xTe/YBa_2Cu_3O_{7-x}$ (YBCO) system. YBCO films were deposited on $LaAlO_3$ substrates by Pulsed Laser Deposition and $Hg_{1-x}Cd_xTe$ layers were subsequently grown between 160 °C and 200 °C. Heterostructures grown at 160 °C retain the key electronic properties of both the semiconductor and the superconductor. X-ray diffraction analysis indicates that the $Hg_{1-x}Cd_xTe$ films are oriented with the (111) planes parallel to the substrate. Mesa diodes were fabricated, and rectification in I-V characteristics was observed corresponding to carrier transport across the semiconductor/superconductor interface. For the heterostructure with a $Hg_{0.75}Cd_{0.25}Te$ top layer, the thermal activation barrier to electron transport was found to be 82 meV from the temperature dependence of the zero-bias differential resistance.

KEYWORDS: semiconductor/superconductor heterostructure, super Schottky barrier, $Hg_{1-x}Cd_xTe$

INTRODUCTION

The study of semiconductor/superconductor heterostructures is a subject of fundamental interest. The unique mechanism for electrons transporting across the interface makes this structure one of the most important building blocks in superconductor based electronic devices. There are two major applications: the super Schottky barrier and the SNS junction. A super Schottky barrier is analogous to a normal Schottky barrier in which the metal is replaced by a superconductor. The I-V characteristics are more nonlinear in the super Schottky barrier than the normal Schottky barrier. This property leads to the development of high performance Pb/GaAs millimeter wave detectors and mixers [1]. In the SNS junction, the substitution of the normal metal with a degenerately doped semiconductor with a small electron effective mass and a high mobility can have several advantages. One of them is the longer coherence length which relaxes the gap dimension between the two superconducting electrodes. Subsequently, this makes the planar geometry possible. A semiconductor proximity junction with a planar geometry can be used to switch the supercurrent with high speed and low power consumption by modulating the carrier concentration in the proximity region with a MIS gate. Such a device, known as the Josephson Junction Field Effect Transistor (JOFET), has been recently demonstrated in the Nb/Si/Nb system [2]. The effect of enhancing the coherence length by using narrow bandgap high mobility superconductors in a proximity junction has also been demonstrated with Nb/InAs/Nb [3].

Despite the potential impact of this technology, devices which have been demonstrated to date have been limited to low temperature superconductors. The realization of such devices using high T_c superconductors is hampered by the difficulty in fabricating semiconductor-superconductor heterostructures having a clean and abrupt interfaces while retaining the electronic properties of both materials. The main problem is the incompatibility with their optimal growth conditions and the thermodynamic stability under such conditions. With the advances in technology, high quality superconductor (mostly $YBa_2Cu_3O_x$) thin films can be grown on lattice matching ceramic substrates such as $LaAlO_3$, MgO and $SrTiO_3$ by a number of in-situ techniques at 650 °C in an oxidizing environment. However, this condition, mild by the standard of ceramic substrate, is too severe for semiconductor substrates to tolerate without degradation. Exposing a clean semiconductor to oxygen can immediately form an amorphous native oxide layer which acts as an energy barrier for the electrons transporting across the interface and affects the crystallinity of the subsequently grown superconductor layers. Other deleterious effects such as the decomposition of compound semiconductors at high temperature and the interdiffusion occurring at the interface also impede the progress of this technology. To this date, the direct growth of superconductor layer on the most common semiconductors such as Si and GaAs without a buffer layer has not been successful [4,5]. On GaAs, even the use of a buffer layer has not been successful to produce high quality superconductor films [6].

An alternative approach is to grow a semiconductor layer on a superconductor. The obstacle of this approach is the stability of superconductors in vacuum at the optimal growth temperature as required by in the growth of most semiconductors. Under such conditions, the oxygen affinity of high T_c superconductors is low so that oxygen will outdiffuse from the lattice causing degradation or even a total loss of superconductivity. This rules out most semiconductors including both Si and GaAs to be used in this approach. InAs film has recently grown on a TlBaCaCuO superconducting film by molecular beam migration-enhanced epitaxy at 250 °C [7]. However, even at this temperature, which is lower than the optimal growth temperature for InAs, the superconductor film shows sign of degradation. One solution to circumvent this problem is to use $Hg_{1-x}Cd_xTe$ which has very low growth temperature as the semiconductor material.

In this paper, the results on the structural and electrical properties of $Hg_{1-x}Cd_xTe$/YBCO heterostructures will be presented.

EXPERIMENTAL:

High quality $Hg_{1-x}Cd_xTe$ layers can be grown by molecular beam epitaxy between 160 °C-200 °C which is lowest among all semiconductors. HgCdTe and YBCO growth processes were carried out in separate chambers. The sample transfer after the growth of YBCO films into the HgCdTe grwoth chamber was an ex-situ process. Formation of a native oxide layer on the YBCO film was possible. 2000 Å thick c-axis oriented YBCO films were grown in situ on (100) $LaAlO_3$ substrates by Pulsed Laser Deposition at 700 °C. Films were c-axis oriented with $T_c = 88$ K and $J_c = 3.7\times10^6$ A/cm^2 at 77K. Undoped $Hg_{1-x}Cd_xTe$ layers were grown subsequently by molecular beam epitaxy in a separate chamber at 200 °C, 180 °C and 160 °C with three effusive sources containing Hg, Te and CdTe. The growth rate was approximately 1.5 micron/hr and the thickness of the $Hg_{1-x}Cd_xTe$ layer was about 3 micron. The composition of the $Hg_{1-x}Cd_xTe$ layer was determined from the position of two reflection peaks in the spectral region between 400 nm and 600 nm corresponding to E_1 and $E_1+\Delta_1$ transitions. The uncertainty in the Hg composition determination by this technique is approximately 2%. The YBCO properties both before and after the growth of the HgCdTe layer was determined by the two coil mutual induction technique and X-ray diffraction . Finally, for the $Hg_{0.75}Cd_{0.25}Te$/YBCO heterostructure grown at 160 °C, 1.3×10^{-2} cm by 1.3×10^{-2} cm mesas were delineated and etched in bromine/methanol solution to expose the underlying YBCO layer. The I-V characteristics of the heterostructure were measured as a function of temperature from 300 K to 30 K.

RESULTS AND DISCUSSIONS:

$Hg_{1-x}Cd_xTe$ is a ternary compound semiconductor made of a solid solution of the binary components CdTe and HgTe. It is latticed matched over entire compositional range with a tunable bandgap from 0 eV to 1.6 eV and a wide tunability in other electronic properties as well. In this paper, we grew the semiconductor/superconductor heterostructure by choosing narrow bandgap $Hg_{1-x}Cd_xTe$ which has small electron mass and high mobility. These are desired properties for both the super Schottky barrier as well as the SNS proximity junction.

Fig. 1 shows the x-ray diffraction scan of a $Hg_{0.75}Cd_{0.25}Te/YBCO/LaAlO_3$ heterostructure. The predominant peak at 23.8° corresponds to the (111) reflection of the $Hg_{0.75}Cd_{0.25}Te$ layer. Other peaks are from the substrate and the c-planes of the YBCO film. The position and width of the YBCO (007) peak were measured both before and after the HgCdTe growth to monitor any degradation. C-axis lattice constants increase with the HgCdTe growth temperature. At 160 °C , the c-axis lattice constant was 11.69 Å and did not show any change after the $Hg_{1-x}Cd_xTe$ was grown. It increased to 11.705 Åat 180 °C and 11.725 Å at 200 °C accompanied by peak broadening. The superconductor property was measured by the two coil mutual induction technique shown in Fig. 2. The T_c of the YBCO film before HgCdTe growth was 88 K and the signal reached the 1% level at 86 K. The growth of a HgCdTe layer at 160 °C resulted in negligible change in the superconductor characteristics of the YBCO film. As the growth temperature was increased to 180 °C, the T_c decreased by 1 K and the transition became broader with the 1% signal level at 78 K. At a growth temperature for HgCdTe at 200 °C, the T_c was further lowered to 83 K with a much broader transition. This result suggests that the onset of degradation YBCO in vacuum is between 180 °C and 160 °C. In view of this, we conclude that HgCdTe (or a Hg contained ternary semiconductor compounds such as HgZnTe and HgCdSe) is the most suitable semiconductor system which can be satisfactorily grown under the conditions imposed by the thermal stability of the high T_c superconducting films.

Fig. 3 shows the rectified I-V characteristics at 77 K. The barrier height is determined from the Arhenius plot of the product of the zero-bias differential resistance and the mesa area (R_0A) vs. 1/T shown in Fig. 4. At T> 80 K, the dependence is linear corresponding to a barrier of 82 meV. At lower temperatures, the deviation from linearity may be due to surface leakage. The 82 meV barrier may be a super Schottky barrier or it may be a thin film insulating barrier. An insulating layer may have formed on the YBCO surface as it was exposed to the ambient during transferring from the Pulsed Laser Deposition system to the molecular beam epitaxy apparatus. Studies using in-situ grown HgCdTe layers are underway.

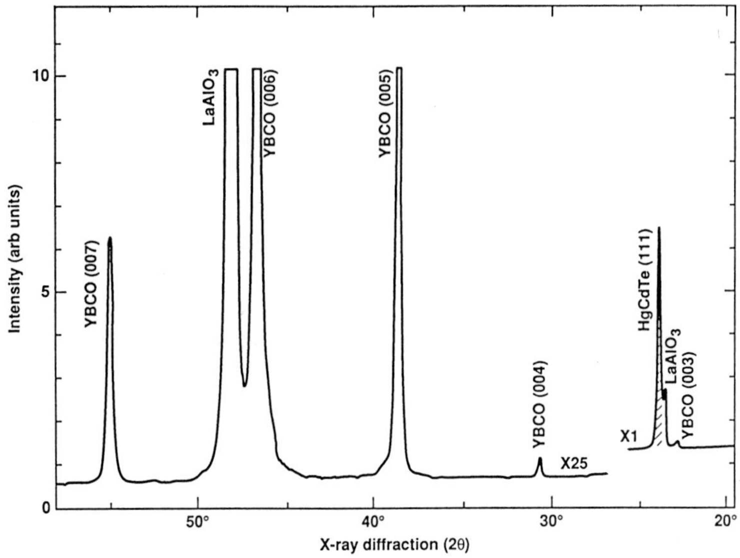

Fig. 1 X-ray diffraction of a HgCdTe/YBCO/LAO heterostructure. The YBCO film is 2000 A thick and the HgCdTe layer is 3 microns thick grown at 160 °C.

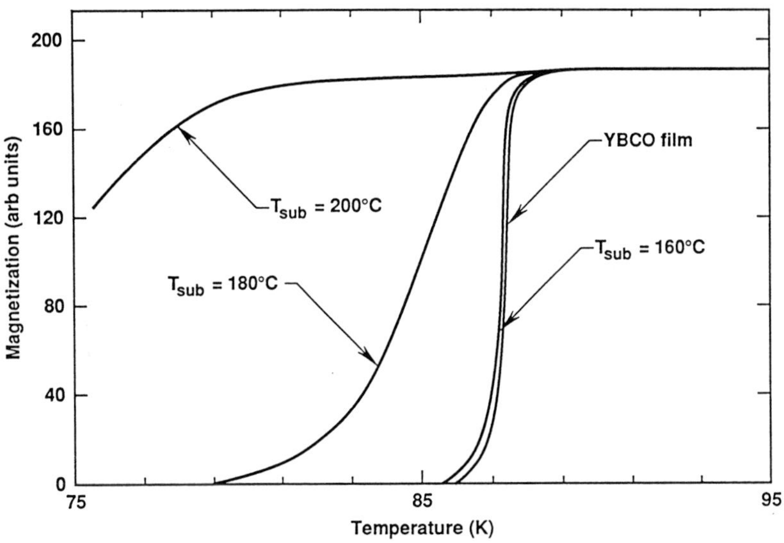

Fig. 2 Two coil mutual inductive measurements on YBCO films and three HgCdTe/YBCO heterostrutures grown at different temperatures. Notice that there is no degradation in the proprerty YBCO film for HgCdTe grown at 160 °C.

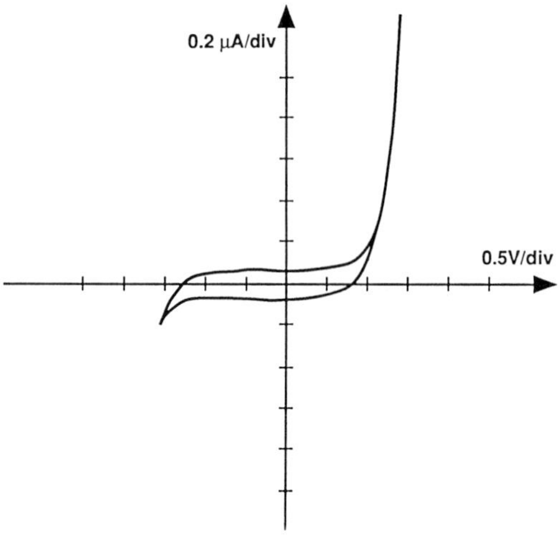

Fig. 3 77 K I-V characteristics of a $Hg_{7.5}Cd_{0.25}Te/YBCO/LaAlO_3$ heterostructure

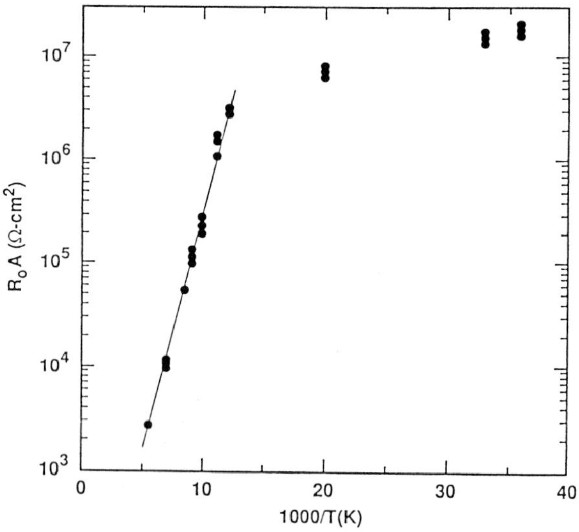

FIg. 4 Temperature dependence of the zero-bias differential resistance of the $Hg_{0.75}Cd_{0.25}Te/YBCO$ heterostructure. The linear portion of the Ahrenius plot gives an activation barrier height of 82 meV.

In summary, we have demonstrated that $HgCdTe/YBCO/LaAlO_3$ heterostructures can be grown at a temperature low enough that the quality of both the superconductor and the semiconductor can be retained. The energy barrier at the interface of the heterostructure is 82 meV.

AKNOWLEDGEMENT:

The authors are grateful to Hitachi Research Laboratory for assisting in the preparation of this manuscript.

REFERENCE

1. A.H. Silver, R.J. Pedersen, M. McMColl, R. L. Dickman, and W.J. Wilson (1981) IEEE Transaction on Magnetics, MAG-17: 698.

2. T. Nishimo, K. Yaki, and U. Kawabe (1990) Hitachi Review 39: 35.

3. H. Takayanagi, and T. Kawahami (1985) Phys. Rev. Lett. 54: 2449.

4. R. M. Silver, A. B. Berzerin, M. Wendman and A. L. deLozanne (1988) Appl. Phys. Lett. 52: 2174.

5. M. R. Rao, E. J. Tarsa, La. A. Samoska, J. H. English, A. C. Gossard, H. Kroemer, P. M. Petroff, and E. L. Hu (1990) Appl. Phys. Lett. 56: 1905.

6. M. Mizuno, M. Miyanchi, K. Setsune, and K. Wasa (1989) Appl. Phys. Lett. 54: 383.

7. M. R. Roa, E. J. Tarsa, H. Kroemer, A. C. Gossard, E. L. Hu, Pm. M. Petroff, W. L. Olser and M. M. Eddy (1990) Appl. Phys. Lett. 55: 490.

Millimeter-Wave Properties of Epitaxial $YBa_2Cu_3O_{7-x}$ Films Grown by Several Methods

P.H. Kobrin, J.T. Cheung, W.W. Ho, N. Glass, J. Lopez, I.S. Gergis, R.E. DeWames, and W.F. Hall

Rockwell International Science Center, Thousand Oaks, CA 91360, USA

ABSTRACT

The complex conductivities of epitaxial $YBa_2Cu_3O_{7-x}$ films grown by pulsed laser deposition, off-axis magnetron sputtering and ion-beam sputtering have been determined at 59 GHz from transmission measurements. The temperature dependence of the real part of the measured conductivities differs significantly from the predictions of the two-fluid and BCS theories but the imaginary parts are in good agreement. The BCS model can be brought into agreement with the measurements by allowing the mean-free path to vary with temperature. Surface resistances of 2 mΩ at 82K and penetration depths, λ_0, of 175 nm have been determined.

KEY WORDS: $YBa_2Cu_3O_{7-x}$, superconductor film, millimeter-wave complex conductivity

The high-frequency properties of high-T_c superconducting films are of interest both for theoretical and technological reasons. The most common measurements are the surface resistance, R_s, and to a lesser extent the penetration depth, λ. The more fundamental parameter, however, is the complex conductivity, $\sigma^* = \sigma_1(\omega,T) - i\sigma_2(\omega,T)$, which is simply related to the complex dielectric constant $\varepsilon^* = i\sigma^*/\varepsilon_0\omega$. A technique to measure the millimeter-wave complex conductivity has been described [1] in which the intensity and phase of millimeter-wave radiation transmitted through a thin superconducting film are measured as a function of temperature. This technique directly measures both the real and imaginary components of the complex conductivity as well as R_s and λ, without any fitting or modeling. This technique has been applied to epitaxial $YBa_2Cu_3O_{7-x}$ (YBCO) films grown on MgO [2] by ion-beam sputtering and post-deposition annealing. However these films had broad superconducting transitions with $T_c(R=0)$ of 55-77K. The unusual behavior measured for σ^* was attributed to mixture effects arising from the broad transition. Similar effects were seen for c-axis oriented Tl-Ba-Ca-Cu-O films and YBCO films on $LaAlO_3$ substrate [3].

In this paper we have extended these measurements to ***in-situ*** YBCO films which are films that do not require high-temperature post-deposition annealing. These films have much narrower superconducting transitions, critical currents over 10^6 A/cm^2 and T_c's of 86-88K. We find that the mixture effects within the superconducting transitions are still present but that the conductivity at lower temperatures is still anomalous.

Three techniques were used to fabricate the five films described in this paper. These techniques include ion-beam sputtering (IBS) and ex-situ annealing at over 900°C in 1 atmosphere of oxygen, off-axis magnetron sputtering (OAMS) and pulsed-laser deposition (PLD). The latter two techniques are ***in-situ*** techniques with post-deposition annealing at 400°C in the growth chamber. In all three techniques the substrate temperature during growth was between 650 - 750°C. X-ray diffraction of all the films showed complete c-axis orientation.

The five films grown on both MgO and $LaAlO_3$ substrates are described in Table 1. Although the c-axis lattice constants and the T_c's of films from these three techniques are similar, the critical currents of the post-annealed IBS films are not as high.

The millimeter-wave transition system has been described in detail previously [1]. Using quasi-optical techniques a near diffraction-limited 59 GHz beam is focussed through the film and the substrate. Because the propagation direction is parallel to the c-axis of these oriented films, the properties that are

TABLE 1 EPITAXIAL YBCO FILMS

Deposition Technique	Substrate	Thickness (nm)	Tc* (K)	c-Constant (Å)	Jc (A/cm^2)
IBS	$LaAlO_3$	100	87	11.70	10^6 @ 73K**
PLD	MgO	100	88	11.70	1.7×10^6 @77K
OAMS	MgO	100	86	11.70	
OAMS	MgO	50	87	11.70	
OAMS	$LaAlO_3$	100	88	11.70	2×10^6 @ 77K**

* Onset of magnetic shielding. ** Measurements on equivalent samples.

measured are those of the a-b plane. Coherent leakage around the sample of -60 dB limits the ultimate sensitivity of the measurement and necessitates the use of very thin films.

The transmission loss and phase change for the 100 nm thick film on MgO grown by OAMS are shown in Fig. 1. In the thin film limit, the maximum change in phase through a superconducting film is 90°. For this film the phase has already changed 80° by 87K. The phase data ends at 87K because there is too much uncertainty in the measurement below that temperature. For all of the 100 nm thick films the phase data was used down to 84 - 87K, while for the 50 nm thick film values down to 75K were used.

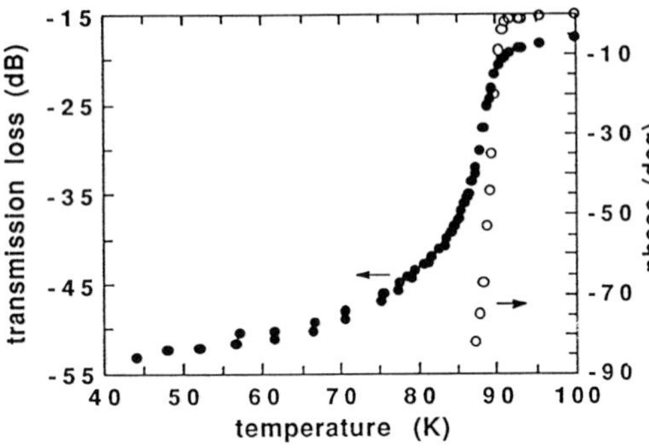

Fig. 1 Temperature dependence of the transmission loss and phase shift for the 100 nm YBCO film on MgO deposited by OAMS.

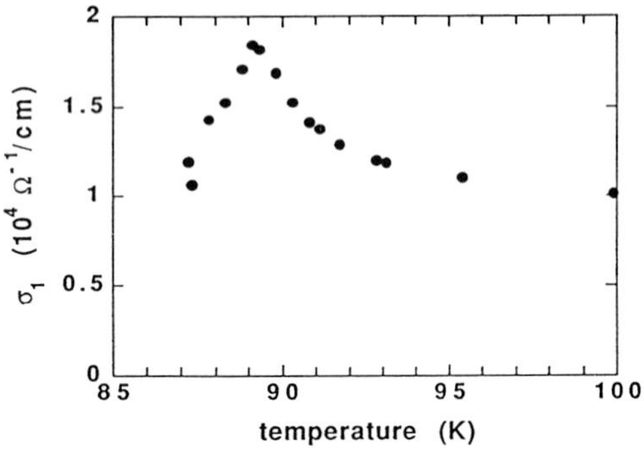

Fig. 2 Real part of the complex conductivity, σ_1 calculated from the data in Fig. 1.

An exact multilayer transmission expression was used to determine σ^* from the transmitted intensity and phase, the film thickness and the substrate properties. Fig. 2 and 3 show the real and imaginary components of σ^* for the 100 nm thick OAMS film on MgO. Changes in σ_1 above ~ 95K are identical to the changes in the dc conductivity. Below that temperature, the conductivity rises to a peak at 89K and then drops to its original value by 87K. The imaginary part of the conductivity, σ_2, is plotted in Fig. 3 as $\sigma_2/\varepsilon_0\omega$ which is the real part of the dielectric constant. The dielectric constant is small and positive above the superconducting transition, but becomes large and negative below the transition. For this sample the real dielectric constant, ε', reaches a value of -2×10^7. When the temperature is $<0.95\ T_c$ (i.e., when the phase is >80°) then $|\sigma_2| >> |\sigma_1|$ and the value of σ_2 is primarily determined by the intensity and is insensitive to the phase. Thus the phase below $0.95\ T_c$ was taken to be equal to the phase change at $0.95\ T_c$ in the calculation of σ_2.

In Fig. 4 are shown the σ_1 and σ_2 results for the PLD film together with theoretical fits using the BCS model for a homogeneous superconductor and the two-fluid model. Both models can be made to fit σ_2, but then the agreement with σ_1 is poor. By adjusting the various parameters it is not possible for either theory to have good agreement with σ_1.

In Fig. 5 are shown the σ_1 data for all of the ***in-situ*** films normalized to σ_n at 100 K. The peak in σ_1 is present in all of the data sets as is the rapid drop at lower temperatures. The lowest temperature at which σ_1 was determined was dependent on how rapidly the transmission was decreasing. The equivalent "peaks" in σ_1 for the lower T_c YBCO films [2] and the higher T_c Tl-Ba-Ca-Cu-O films [3] were broader and sometimes higher.

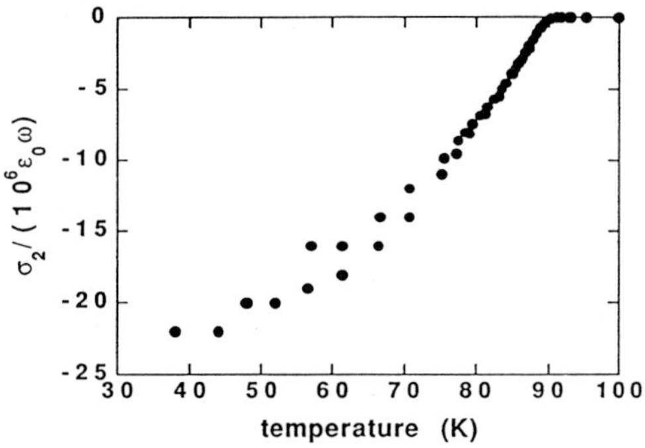

Fig. 3 Imaginary part of the complex conductivity, σ_2 calculated from the data in Fig. 1.

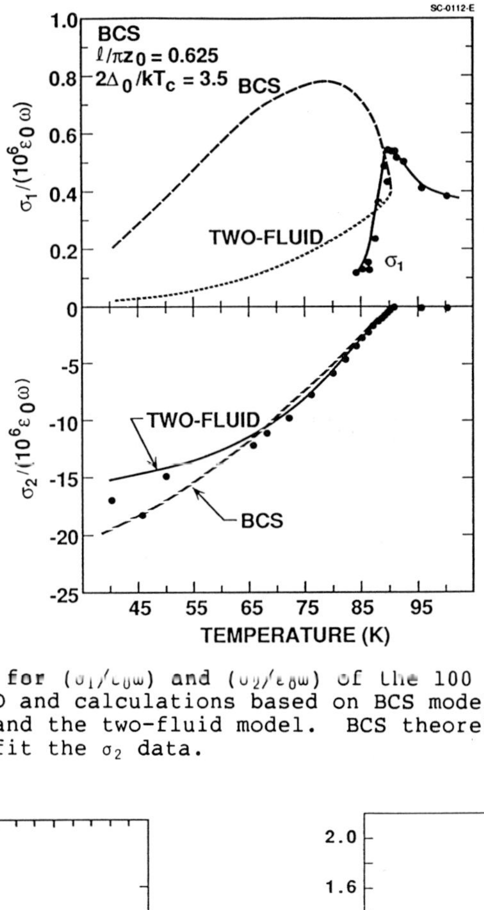

Fig. 4 Measured values for ($\sigma_1/\varepsilon_0\omega$) and ($\sigma_2/\varepsilon_0\omega$) of the 100 nm YBCO film on MgO deposited by PLD and calculations based on BCS model for a homogeneous superconductor and the two-fluid model. BCS theoretical parameters were chosen to best fit the σ_2 data.

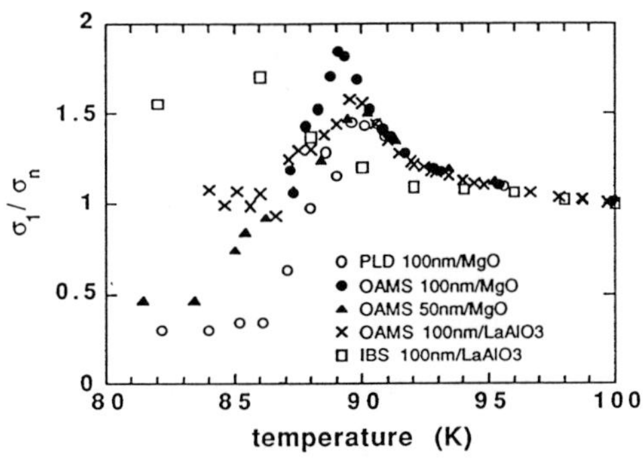

Fig. 5 Comparison of the σ_1/σ_n values measured for four ***in-situ*** YBCO films. All show the narrow peak that is unaccounted by the standard BCS theory.

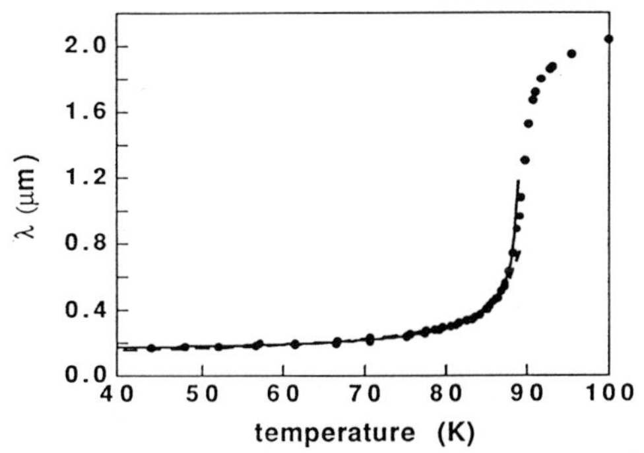

Fig. 6 Penetration depth for the 100 nm YBCO film on MgO deposited by OAMS and calculated curves using the two-fluid and BCS models

An analysis of the 100 nm thick YBCO films on MgO by PLD and OAMS have been made using BCS theory [4]. These studies concluded that:

1. A small spread in T_c (~ 3-4 K) over the grains can cause the enhancement in σ_1 from field enhancement in the normal-conducting regions.

2. Below 88 or 89 K it is impossible to fit the rapid drop in σ_1 using either $2\Delta_0/kT_c = 3.5$ or 8 as long as the mean-free path and the effective mass are held constant.

3. To stay within the BCS framework and to simultaneously fit both parts of σ^* it is necessary to introduce a rapidly increasing mean-free path and a rapidly decreasing effective mass as the temperature is lowered. A rapid rise in the mean-free path has also been postulated from nuclear spin relaxation studies [5].

The surface resistance and the penetration depth are obtained from the complex conductivity using $R_s = \mathrm{Re}[i\omega\mu_0/\sigma^*]^{1/2}$ and $1/\lambda = \mathrm{Im}[i\omega\mu_0\sigma^*]^{1/2}$. Below 0.95 T_c where $|\sigma_2| >> |\sigma_1|$ the surface resistance is proportional to $\sigma_1/\sigma_2^{3/2}$ and the penetration depth is proportional to $\sigma_2^{-1/2}$. At temperatures below where σ_1 could be measured the last value of σ_1 was used. This has virtually no effect on the derived λ, but gives an upper estimate of R_s, provided σ_1 continues to decrease at lower temperatures.

The penetration depth vs. temperature of the 100 nm thick film on MgO grown by OAMS and theoretical fits to the two-fluid and BCS models [6] are shown in Fig. 6. Both theories provide reasonable fits to the data between 45 and 88K; however, the fitting parameters are different. The BCS fit uses a higher T_c (91K vs 89.5K) and a lower λ_0 (0.15 µm vs 0.175 µm). Since λ is dominated by σ_2, not surprisingly they can both be fit by these theories.

The surface resistance of all five films is shown in Fig. 7. At 59 GHz the surface resistance of all the ***in-situ*** films surpasses copper at 87K. The surface resistance of the PLD film reaches 1 mΩ at 77K which, after correcting for frequency, is the lowest surface resistance reported for a high-T_c superconductor. The decrease in the surface resistance is primarily a result of the increase in the magnitude of the dielectric constant.

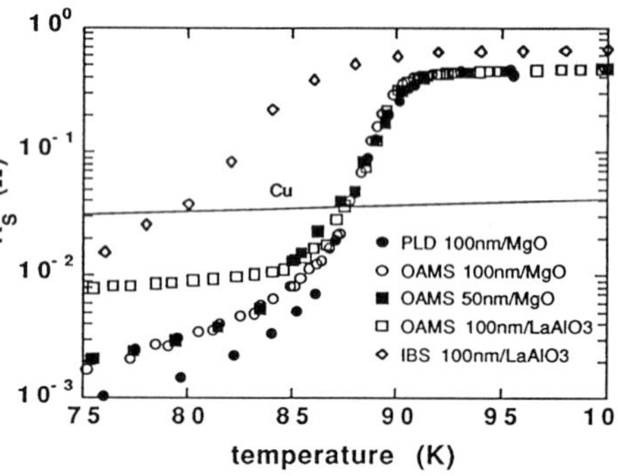

Fig. 7 Surface Resistances for the YBCO films described in Table 1.

In summary, millimeter-wave transmission measurements on several epitaxial YBCO films were used to obtain the temperature dependence of σ^*, R_s and λ at 59 GHz. Both σ_2 and λ are well described by either the two-fluid or BCS theories but σ_1 and R_s are not.

REFERENCES:

1. Ho WW, Hood PJ, Hall WF, Kobrin PH, Harker AB, DeWames RE (1988) Phys. Rev. B 38: 7029
2. Kobrin PH, Ho W, Hall WF, Hood PJ, Gerfis IS, Harker AB, (1990) Phys. Rev. B 42: 6259
3. Kobrin PH, Ho W., Hall WF, Hood PJ, Harker AB, July 23-28, 1989 Materials and Mechanisms of Superconductivity. Stanford, CA.
4. Glass NE, (unpublished)
5. Coffey L (1990) Phys. Rev. Lett. 64: 1071
6. Muhlschelegel B. (1959) Z. Phys. 155: 313

Epitaxial $YBa_2Cu_3O_x/PrBa_2Cu_3O_x/YBa_2Cu_3O_x$ Multilayer Structure by Coevaporation

TETSURO SATOH[1], TSUTOMU YOSHITAKE[1], HITOSHI IGARASHI[1], and HITOSHI HOSHINO[2]

[1] Fundamental Research Laboratories, NEC Corporation, 1-1, Miyazaki 4-chome, Miyamae-ku, Kawasaki, 213 Japan

[2] Resources and Environment Protection Research Laboratories, NEC Corporation, 1-1, Miyazaki 4-chome, Miyamae-ku, Kawasaki, 213 Japan

ABSTRACT

The preparation method and properties of $YBa_2Cu_3O_X/PrBa_2Cu_3O_X/YBa_2Cu_3O_X$ (YBCO/PBCO/YBCO) multilayer structure films were studied for a basis of SIS(or SNS) type superconducting electronic devices. The multilayer structures were prepared *in situ* on (001) $SrTiO_3$ by coevaporation of Y(or Pr), Ba and Cu metals. The detail of their structure, surface morphology and electrical properties is reported.

KEY WORDS: $YBa_2Cu_3O_X$, $PrBa_2Cu_3O_X$, multilayer

INTRODUCTION

Josephson tunnel junctions or weak links are quite essential for many superconducting electronic devices. The artificial junctions or weak links in multilayer structures are favorable from a view point of reproducibility. In the application of high-temperature superconducting(HTSC) materials to electronic devices, it is a key to study the fabrication of multilayer structure such as SIS superconductor-, insulator-, superconductor- multilayer. So far, good HTSC films were prepared as epitaxially grown films. In fabrication of the HTSC multilayer structure, it is favorable to grow each layer epitaxially throughout the growth process in order to attain good electrical properties. The material of barrier layer, therefore, is desired to have a perovskite-like structure.

Several research groups[1-5] have reported the fabrication of epitaxial multilayer structures of the superconducting $YBa_2Cu_3O_X$(YBCO) layers and non-superconducting(semiconductor-like) $PrBa_2Cu_3O_X$(PBCO) layer(s) by sputtering or pulsed laser deposition. There are two interests in YBCO/PBCO multilayer structure. First, the fabrication and properties of YBCO/PBCO superlattices with extremely thin layers down to one unit cell thickness[3-5]. Second, the application of YBCO/PBCO/YBCO multilayer structures to HTSC electronic devices with layer thickness of up to several tens of nanometers[1,2]. PBCO has the same crystal structure as YBCO with a small lattice mismatch. PBCO seems to be a suitable material for the barrier layer of HTSC multilayer structure devices and for the intermediate layers of the superlattices. We were interested in YBCO/PBCO/YBCO multilayer structure as a promising structure for the application to HTSC electronic devices.

We have studied the preparation method of YBCO/PBCO/YBCO multilayer structure by coevaporation. In our evaporation system, good YBCO thin films with the zero resistance superconducting critical temperature (T_C) of 87K have been prepared at the substrate temperature of around 700°C. However, these films had rough surfaces with dispersed precipitates of several hundred nanometers in size. PBCO thin films, which were prepared at the same temperature, had similar surfaces. The films with rough surfaces of this kind are not suitable for the fabrication of the multilayer structures. It was possible to depress the formation of the precipitates by lowering the substrate temperature. Consequently, the multilayer structures were fabricated at about 600°C to obtain flat surfaces. In this condition, however, the crystallinity and the superconducting properties were sacrificed for the flat surfaces.

In this paper, we will report on preparation of YBCO/PBCO/YBCO multilayer structure by coevaporation and some results on structural and electrical studies.

EXPERIMENTAL

The multilayer structures were prepared on (001) $SrTiO_3$ substrates by coevaporation system with three electron-beam guns. Rare earth elements(Y or Pr), Ba and Cu were evaporated separately. The deposition rates were controlled separately by three quartz-crystal-monitor systems. During deposition, the substrates were kept at 600°C. Because flat film surface was necessary for the fabrication of the multilayer structure, the substrate temperature was intended to be kept 50-100°C lower than the optimum film growth temperature. Oxygen gas containing a few percent ozone was produced by passing oxygen gas through an ozone-generator and the gas was directed at the substrate surface during deposition and cooling process. Oxygen partial pressure in the chamber was about 4×10^{-4} Torr. The pressure at the substrate surface was thought to be slightly higher than this value. Typical layer thickness was YBCO(30nm)/PBCO(30nm)/YBCO(30nm).

The composition of single layer films was determined using electron microprobe. The surface of the single layer films and the multilayer structures was observed by secondary electron microscopy(SEM). The structure of the multilayer structures was studied by x-ray diffraction(XRD) with Cu Ka radiation, Rutherford backscattering spectrometry(RBS), secondary ion mass spectrometry(SIMS) and transmission electron microscopy(TEM). The dc-resistivity was measured by a conventional four-probe method.

RESULTS AND DISCUSSION

The surfaces of YBCO and PBCO single layer films and of YBCO/PBCO/YBCO multilayer structure observed by SEM are shown in Fig.1. The surfaces of both YBCO and PBCO single layer films were smooth, except for small particles of a few tens of nanometers. The multilayer structure films had a slightly rougher surface than the single layer films did, and they also had the particles on the surface. The films deposited at about 700°C had precipitates of several hundred nanometers in size, as mentioned above. In deposition at 50-100°C lower substrate temperature, these large precipitates were not formed on the single layer films nor on the multilayer structure films. We believe that the precipitate-free deposition condition revealed by the single layer film growth study were suitable for the multilayer structure fabrications.

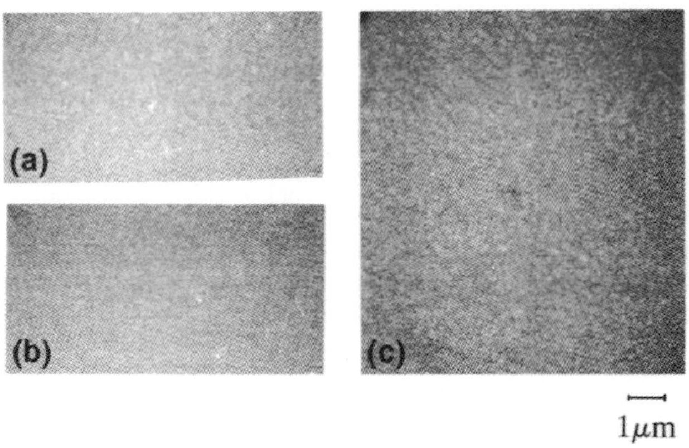

Fig.1. The surfaces of (a)YBCO and (b)PBCO single layer films and of (c)YBCO/PBCO/YBCO multilayer structure observed by SEM.

XRD measurement showed that both YBCO and PBCO single layer films had a preferred orientation with c-axis normal to the substrate surface. A XRD pattern of the YBCO/PBCO/YBCO multilayer structures is shown in Fig.2. The same preferred orientation as that of the single layer films was maintained throughout the multilayer structure. A preliminary TEM observation showed an epitaxial relationship of $[100]_{multilayer}//[100]_{substrate}$. These results showed that the multilayer structure films grew epitaxially on the $SrTiO_3$ substrate as reported[6]. RBS random and [001] aligned measurements were performed to evaluate the crystallinity of the multilayer structures. RBS minimum yields of 40-50% were obtained for integrated Ba+Pr+Y peaks of the multilayer structures. YBCO single layer films deposited at about 700°C had shown RBS integrated minimum yields of less than 20%. These results showed that the multilayer structures had a poor crystallinity in comparison with the YBCO films. The poor crystallinity of the multilayer structure was probably caused by the low substrate temperature during deposition.

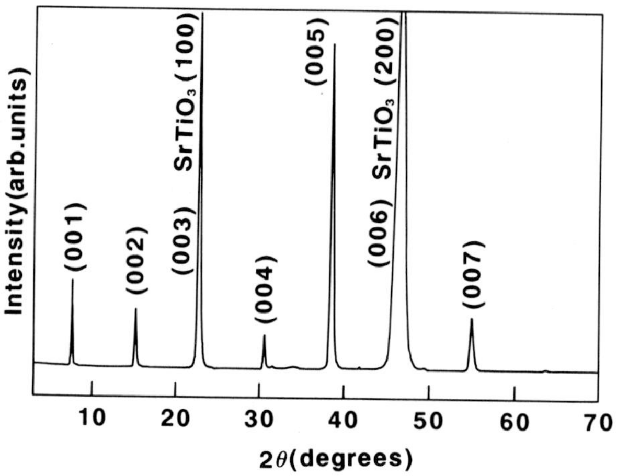

Fig.2. XRD pattern of YBCO/PBCO/YBCO multilayer structure.

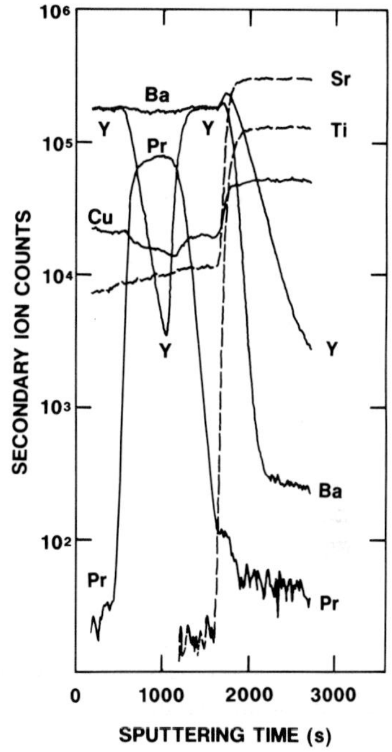

Fig.3. SIMS depth profile of YBCO/PBCO/YBCO multilayer structure.

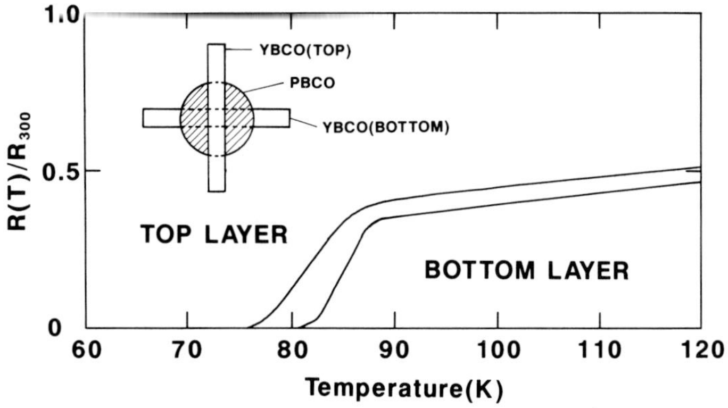

Fig.4. The temperature dependence of resistivity of the top and the bottom YBCO layer in YBCO/PBCO/YBCO multilayer structure. The inset shows the configuration of the multilayer structure for electrical measurement.

The interdiffusion between each layer in YBCO/PBCO/YBCO multilayer structures was investigated by SIMS. The SIMS depth profile(Fig.3) shows that no significant interdiffusion was occurred during deposition and the desired multilayer was fabricated. We believe that the relative unsharpness of the Y profile was artifact of SIMS analysis. The interdiffusion at the interface between each layer was not able to be revealed by SIMS. Further investigation will be performed by cross-sectional TEM.

For electrical measurement of the multilayer structure, the crossover of the multilayer structure shown in Fig.4 was prepared using a metal mask. The strip of the cross was about 15mm long and 2mm wide. The temperature dependence of resistivity of the top and the bottom YBCO layer in YBCO/PBCO/YBCO multilayer structure is shown in Fig.4. Both YBCO layers in the multilayer structure showed superconductivity. T_C of the bottom YBCO layer was about 81K, which was nearly the same as the T_C of YBCO single layer films. However, the T_C of the top YBCO layer was about 5K lower than that of the bottom layer. These results shows that the top layer showed a slight degradation, while the bottom layer did not. The evident causes of the degradation have not been clear. The current-voltage characteristics of the multilayer structures will be measured for the observation of Josephson effect.

In summary, YBCO/PBCO/YBCO multilayer structures were fabricated *in situ* by coevaporation. The multilayer structures had flat surfaces, the preferred orientation with c-axis normal to the substrate surface, and the epitaxial relationship throughout the multilayer structure. However, the crystallinity of the multilayer structures was relatively poor because of low substrate temperature. No significant interdiffusion in the multilayer structures was observed and desired multilayer structures were able to be fabricated. Both of the top and the bottom YBCO layers showed superconductivity. The top YBCO layer had T_C of about 76K, which was about 5K lower than that of the bottom layer. Some degradation of superconducting properties was occurred in the multilayer structures.

REFERENCES

1. Poppe U, Prieto P, Schubert J, Soltner H, Urban K, Buchal Ch (1989) Solid State Commun. 71: 569-572
2. Rogers CT, Inam A, Hedge MS, Dutta B, Wu XD, Venkatesan T (1989) Appl. Phys. Lett. 55: 2032-2034
3. Triscone JM, Fischer O, Brunner O, Antognazza L, Kent AD, Karkut MG (1990) Phys. Rev. Lett. 64: 804-807
4. Li Q, Xi XX, Wu XD, Inam A, Vadlamannati S, Mclean WL, Venkatesan T, Ramesh R, Hwang DM, Martinez JA, Nazar L (1990) Phys. Rev. Lett. 64: 3086-3089
5. Lowndes DH, Norton DP, Budai JD (1990) Phys. Rev. Lett. 65: 1160-1163
6. Jia CL, Kabius B, Soltner H, Poppe U, Urban K, Schubert J, Buchal Ch (1990) Physica C 167: 463-471

6.2 BSCCO and Related Films

Growth of Metastable Phases and Superlattice Structures of Bi-Sr-Ca-Cu-O Compounds by an Atomic Layering MBE Technique

DARRELL G. SCHLOM[1], ANN F. MARSHALL[1], JAMES S. HARRIS, JR.[1], IVAN BOZOVIC[2], and JAMES N. ECKSTEIN[2]

[1] Stanford University, Stanford, CA 94305-4055, USA
[2] Varian Research Center, Palo Alto, CA 94303, USA

ABSTRACT

We have grown Bi-Sr-Ca-Cu-O films by a shuttered, layer-by-layer MBE growth technique. The results demonstrate the ability of this technique to produce smooth, layered, metastable phases, including ordered superlattices, *in situ* using ozone. Both cross-sectional TEM images and a comparison of the observed x-ray diffraction patterns to limiting case simulations show that this growth technique is effective in selecting among nearly energetically degenerate $Bi_2Sr_2Ca_{n-1}Cu_nO_x$ phases and layering them on a half-unit-cell by half-unit-cell basis. Using these techniques, an as-grown superconducting film with T_c^{onset} near 100 K and $T_c\ (\rho = 0)$ of 86 K was achieved.

KEY WORDS: MBE, Bi-Sr-Ca-Cu-O superconductors, TEM, x-ray diffraction

INTRODUCTION

The growth of fully oxidized compounds by molecular beam epitaxy (MBE) poses significant challenges compared to the conventional use of MBE to grow non-oxide semiconductors. To realize the described layered metastable structures, a low-temperature *in situ* growth method is required. In addition, an activated species of oxygen is necessary to bring the constituent elements to the proper oxidation state while maintaining the long mean free path necessary for MBE. Finally, composition control is crucial in order to repeatedly deposit monolayers of the desired composition. This paper describes our progress related to the *in situ* growth of layered $Bi_2Sr_2Ca_{n-1}Cu_nO_x$ compounds by MBE. The *in situ* growth of layered metastable phases and superlattice structures is demonstrated.

EXPERIMENTAL

The Bi-Sr-Ca-Cu-O films were grown in a custom-built MBE machine, which was specifically designed to grow oxide superconductors. Four conventional, resistively heated MBE furnaces were used to generate fluxes of the constituent elements. Four independent shutters, controlled by a computer, were used to supply bursts of the metal constituents to the substrate. The order of the flux bursts was the same as the *c*-axis layering sequence of the desired $Bi_2Sr_2Ca_{n-1}Cu_nO_x$ compound. Oxidation was provided by a purified ozone source. A schematic diagram of this MBE system is shown in Fig. 1. Additional details on the growth and ozone system may be found elsewhere [1-3].

RESULTS AND DISCUSSION

Figure 2 presents θ-2θ x-ray scans of several as-grown films showing *c*-axis oriented 2201, 2212, 2223, 2234, and 2245 phases. Although half unit cells of all of these structures have been seen in regions of bulk compounds [4], single-phase 2234 and 2245 have not been produced by bulk methods. This is believed to be due to the small free energy differences among $Bi_2Sr_2Ca_{n-1}Cu_nO_x$ phases [5]. The nearly single-phase 2234 and 2245 x-ray patterns shown in Fig. 2 indicate the ability of the sequential deposition MBE growth technique to select between nearly energetically degenerate

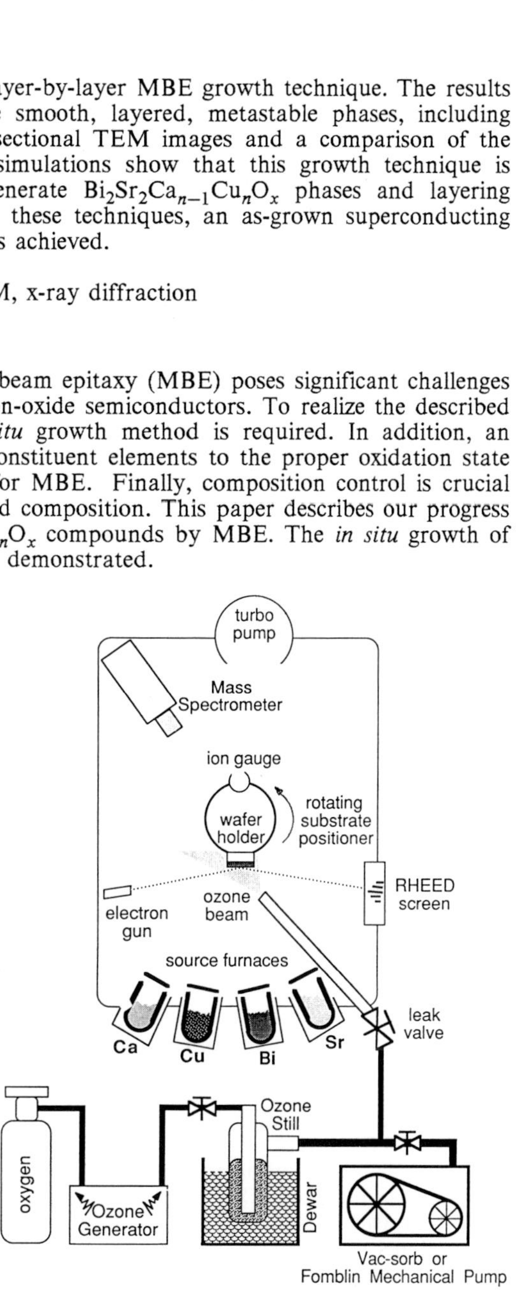

Fig. 1. Schematic diagram of the MBE growth chamber used to grow Bi-Sr-Ca-Cu-O films.

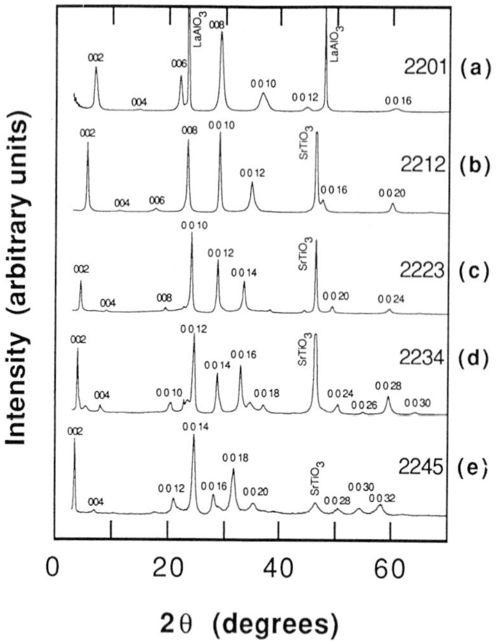

Fig. 2. θ-2θ x-ray scans of as-grown films containing aligned $Bi_2Sr_2Ca_{n-1}Cu_nO_x$ phases.
(a) 2201, rocked 0.5° in omega off alignment to the $LaAlO_3$ {100} substrate.
(b) 2212, rocked 0.5° in omega off alignment to the $SrTiO_3$ {100} substrate.
(c) 2223, rocked 0.5° in omega off alignment to the $SrTiO_3$ {100} substrate.
(d) 2234, rocked 0.5° in omega off alignment to the $SrTiO_3$ {100} substrate.
(e) 2245, rocked 1° in omega off alignment to the $SrTiO_3$ {100} substrate.

structures. Cross-sectional transmission electron microscopy (TEM) has confirmed the layered nature of these films, shown *b*-axis superstructure, and revealed the presence of stacking faults in the films [2]. Atomic force microscopy (AFM) scans over 5000 Å by 5000 Å regions of film surface have revealed a surface roughness of less than 10 Å [6]. Superconducting films with as-grown zero resistance temperatures as high as 86 K have been grown by this shuttered MBE growth method [1].

Out-of-Phase Domains

Since the substrate surface upon which growth is initiated is not perfectly flat, the presence of steps on the substrate might lead to misalignment between adjacent crystalline domains grown on this substrate. However, cross-sectional TEM images show that this does not appear to be a problem for the growth of $Bi_2Sr_2Ca_{n-1}Cu_nO_x$ films by this shuttered MBE technique. Figure 3 shows cross-sectional TEM images of an as-grown film containing 2212 and 2234 layers on a $SrTiO_3$ {100} substrate. This film was formed by the shuttered deposition of ten 2212 half-unit-cell layers on a $SrTiO_3$ substrate beginning with Sr, Cu, Ca, ... and ending with Bi at a substrate temperature of about 650 °C. The film was then *in situ* annealed at 750 °C, cooled to 650 °C for the subsequent growth of fifteen 2234 half-unit-cell layers, and again *in situ* annealed at 750 °C. It was superconducting as-grown with a $T_c\,(\rho = 0)$ of 81 K.

X-ray diffraction (Fig. 2d) as well as electron diffraction and lower-resolution cross-sectional TEM lattice imaging of the outer layers of the as-grown film indicate that the 2234 layering did survive the 750 °C *in situ* anneal. Yet, the 2234 region observed by high-resolution cross-sectional TEM (shown in the upper right of Fig. 3) was rapidly destroyed by beam heating in the TEM as shown by its absence in the lower right image in Fig. 3, taken of the same region shortly after the first image.

The interface between the $SrTiO_3$ substrate and the initial 2212 layer appears to be less abrupt than the interfaces between the subsequent 2212 layers. This may be due to a variation of the position of the $SrTiO_3$/2212 interface through the thickness of the TEM specimen, or to some intermixing between the substrate and the first 2212 half unit cell. Examination of the $SrTiO_3$/2212 film interface suggests that the $(BiO)_2$ layers often lie adjacent to and follow the contours of the stepped $SrTiO_3$ interface. However, far from the interface the $(BiO)_2$ layers are flat and continuous, indicating the absence of out-of-phase domains. The rapid flattening of these layers that occurs over just one or two half unit cells is evident in the magnified view of the stepped region shown on the left side of Fig. 3. Similar observations of the continuity of layers and absence of out-of-phase domains occurring within just a few unit cells of the substrate have also been observed for the growth of $YBa_2Cu_3O_7$ films by laser ablation [7].

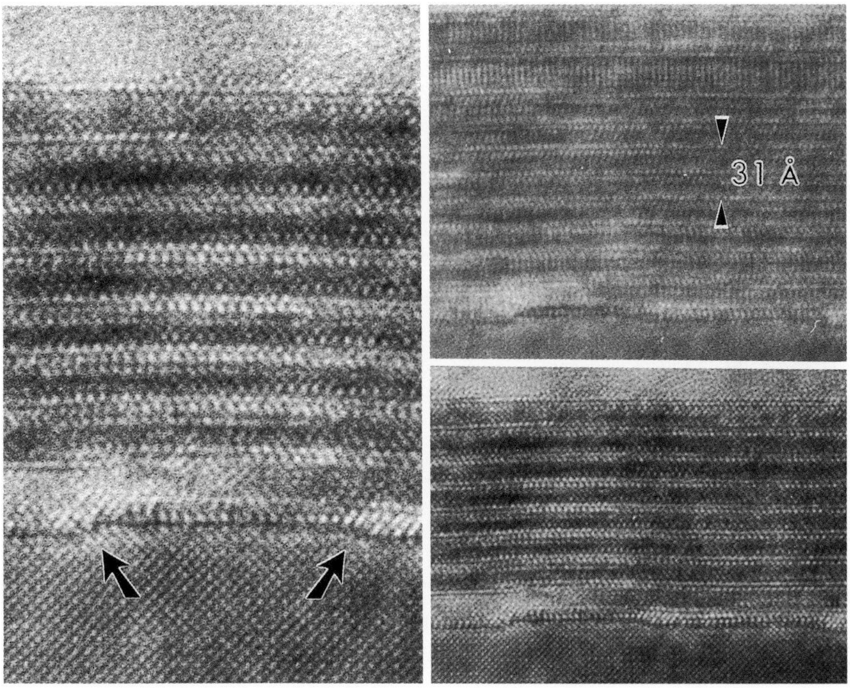

Fig. 3. Cross-sectional TEM images of an as-grown film containing 2212 and 2234 layers grown on top of a stepped region of $SrTiO_3$ {100} substrate. The electron beam is incident along the <100> $SrTiO_3$ zone axis. The image at the left is a close-up of the lower right image and shows a stepped region of the substrate and its influence on the deposited 2212 layers. Note the absence of out-of-phase domains. The upper right image, taken of the same region, shows the beginning of the 2234 layers. These layers were rapidly amorphized during TEM observation. The x-ray diffraction pattern of this as-grown film is shown in Fig. 2d.

Although this shuttered growth began with the strontium layer followed by ten periods of 2212 layer deposition, note that the bismuth layer often appears to be adjacent to the substrate and that only eight 2212 half unit cells are present beneath the 2234 layers in the cross-sectional TEM image. Clearly some rearrangement of the layering has occurred, either during growth or during the *in situ* anneals. The appearance of the $(BiO)_2$ layers adjacent to the $SrTiO_3$ {100} substrate suggests that for growth on $SrTiO_3$ {100} substrates it may be preferable to initiate growth with the bismuth layers.

Ordered Mixtures

Although *in situ* reflection high-energy electron diffraction (RHEED) patterns observed during the growth of $Bi_2Sr_2Ca_{n-1}Cu_nO_x$ phases [8] indicate that this shuttered MBE technique is capable of layering on a half-unit-cell basis, we have directly tested the layering capability by attempting the growth of superlattices of two different, alternating $Bi_2Sr_2Ca_{n-1}Cu_nO_x$ phases [8]. For example, a shuttering sequence corresponding to the deposition of a half unit cell of 2201 followed by a half unit cell of 2223 was repeated 61 times.

The x-ray diffraction pattern of the grown film was compared with two limiting case, simulated x-ray diffraction patterns: an ideal superlattice (perfect layer ordering) and a completely disordered layering consisting of a random stacking sequence of the constituent layers. The theory of Hendricks and Teller [9] was used for the latter simulations. Details of the structural parameters used in these simulations are given elsewhere [2]. The simulated diffraction patterns are less sensitive to errors in the assumed structural parameters (which are extrapolated from the structure of bulk $Bi_2Sr_2Ca_{n-1}Cu_nO_{2n+4}$ phases) at low 2θ angles than at high 2θ angles. A comparison between the observed x-ray diffraction pattern and the two limiting case x-ray diffraction simulations is shown in Fig. 4 for low 2θ values. The 001 peak (at $2\theta = 3.1°$), which would be absent for random layering, is quite evident in the observed diffraction pattern, indicating the presence of repeated ordered layering in this sample. The measured c-axis length (30.75 ± 0.1 Å) is about halfway between that of the 2201 and 2223 phases.

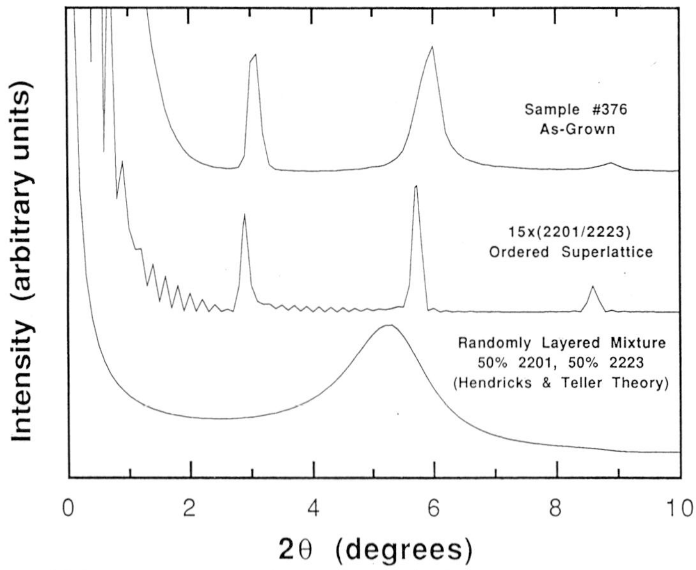

Fig. 4. Observed low angle θ-2θ x-ray scan of an as-grown 2201/2223 superlattice attempt, calculated θ-2θ x-ray intensity pattern for an ordered superlattice consisting of fifteen 2201/2223 periods, and calculated θ-2θ diffracted x-ray intensity pattern for an infinitely thick layered random mixture of 50% 2201 and 50% 2223 layers.

One possible atom rearrangement that could occur with the attempted growth of a 2201/2223 superlattice, distinct from the formation of a randomly ordered mixture of these two phases, is the formation of single-phase 2212. However, since the presence of a glide plane halfway up the 2212 unit cell leads to only even 00ℓ diffraction peaks, the presence of odd 00ℓ diffraction peaks in Fig. 4 rules out this possibility as well and confirms that the shuttered MBE growth method is able to layer $Bi_2Sr_2Ca_{n-1}Cu_nO_x$ phases on a half-unit-cell basis.

Growth of a Metastable 2212-like $Bi_2Sr_2Ca_{n-1}Cu_nO_x$ Phase

The results presented above indicate that for the growth conditions discussed so far, this shuttered MBE growth technique is only capable of controlling the layering on a half-unit-cell basis. Using *in situ* RHEED we previously identified [8] the formation of islands of a $Sr_6Bi_2O_9$-like phase that forms during the first strontium burst of a given half unit cell and persists until the bismuth burst ending the half unit cell. Recently, we have found growth conditions that discourage the formation of islands of this $Sr_6Bi_2O_9$-like phase. Decreasing the ozone pressure during the deposition of the strontium-containing monolayer was found to result in a streaked RHEED pattern throughout the growth of 2201 films. Presumably, the incident flux of oxidant strongly influences the bonding and thus the surface mobility of the adatoms. This technique is analogous to the surface diffusion enhancement method used in the migration enhanced epitaxy (MEE) variant of the MBE technique for the growth of III-V compounds [10].

To test the ability of this growth technique to control film layering on a monolayer basis, we attempted to grow a 2212 structure in which the calcium monolayer was completely replaced by strontium. The target structure might be written $Bi_2Sr_2SrCu_2O_x$ or simply $Bi_2Sr_3Cu_2O_x$. $Bi_2Sr_3Cu_2O_x$ exists as an equilibrium phase in the phase diagram [11], however, its structure is distinct from that of 2212 with strontium on the calcium site [11]. This metastable $Bi_2Sr_3Cu_2O_x$ phase with a 2212-like structure was also recently fabricated by sputtering [12].

The x-ray diffraction pattern of this as-grown film is shown in Fig. 5. Aside from the intense $SrTiO_3$ {100} substrate peaks present, the most intense peaks in the x-ray pattern are due to a *c*-axis oriented 2212-like structure with a *c*-axis length of 31.3 ± 0.3 Å. Some of the other peaks present may be due to the equilibrium polytype of $Bi_2Sr_3Cu_2O_x$, or to 2201 (this sample had a 2201 buffer layer). X-ray diffraction scans of inclined film planes indicate that the *a* and *b*-axes of this metastable 2212-like structure are aligned parallel to the $SrTiO_3$ <110> directions [3]. This sample exhibited zero resistance at 15 K. These experimental results indicate the ability of the shuttered, layer-by-layer MBE technique to customize layering on a monolayer level.

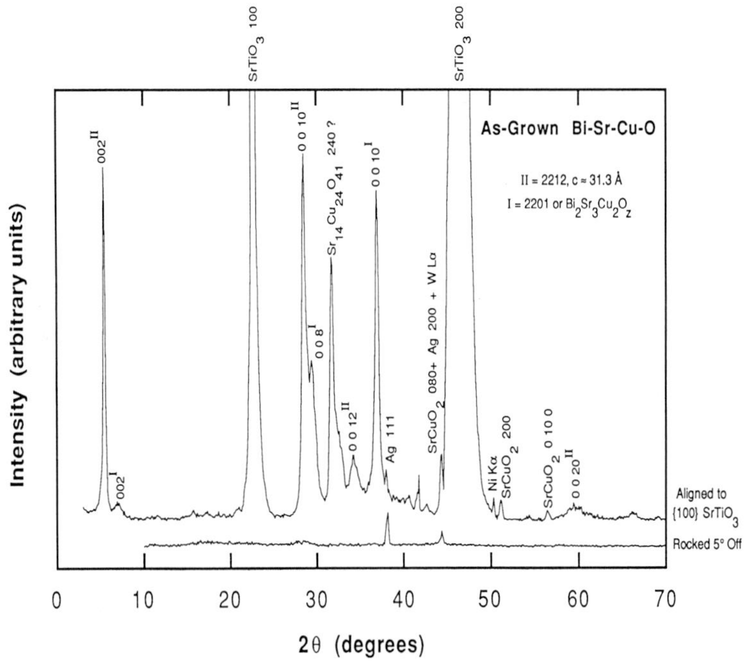

Fig. 5. θ-2θ x-ray diffraction scans of an as-grown film containing a metastable 2212-like $Bi_2Sr_2SrCu_2O_x$ phase. For clarity, the intensity of the scan aligned to the $SrTiO_3$ {100} substrate in omega is offset from the scan rocked 5° in omega off alignment to the $SrTiO_3$ substrate. The 00ℓ peaks due to a 2212-related metastable phase are labeled in addition to those due to other phases.

Single-Crystal Growth on Misoriented $SrTiO_3$ {100}

Although the films discussed above are epitaxial, the presence of reflection twin boundaries or 90° rotation twin boundaries distinguishes these films from true single crystals. However, growth on a misoriented $SrTiO_3$ (001) substrate, misoriented about 3.8° → [110] and 0.9° → [1$\bar{1}$0] as measured by the Laue method, produced the growth of a true single-crystal 2223 film [13].

In situ RHEED patterns indicated the absence of reflection and rotation twin boundaries [13]. X-ray diffraction was also used to confirm the absence of such twin boundaries. The incommensurate superstructure along the *b*-axis of $Bi_2Sr_2Ca_{n-1}Cu_nO_x$ phases causes satellite reflections in x-ray diffraction. In particular the 02$\underline{12}$ reflection of the 2223 phase has satellite peaks on either side of it along the *b** reciprocal space axis, whereas the 20$\underline{12}$ reflection does not have satellite peaks adjacent to it along the *a** reciprocal space axis. Figure 6 shows the observed x-ray diffraction pattern made by aligning the x-ray detector to a satellite reflection of the 02$\underline{12}$ peak and then spinning the ϕ axis of the 4-circle diffractometer around the [001] zone axis of the 2223 film. If the film contained rotation twin boundaries, then this scan would have contained a diffraction peak every 90° ((110) reflection twin boundaries would also lead to x-ray diffraction peaks about every 90° since the *a* and *b*-axes have nearly identical lengths in these $Bi_2Sr_2Ca_{n-1}Cu_nO_x$ materials). Instead, the x-ray diffraction pattern observed contains only two peaks, indicating that the incommensurate superlattice is aligned along only one of the <110>-type $SrTiO_3$ surface directions. Specifically, the x-ray data indicate that the *b*-axis of the film runs up and down the substrate steps and not along the length of the substrate steps.

The RHEED and x-ray diffraction results establish that this film grown on a misoriented $SrTiO_3$ {100} substrate does not contain reflection or rotation twin boundaries. Yet this diffraction data by itself is not sufficient to conclude that the film is a continuous single crystal. Real space crystallographic imaging (e.g., TEM or x-ray topography) is necessary to check for the presence of small angle grain boundaries or amorphous regions in the film. Although this work has not been done, grain boundaries or connected regions having different texture are not apparent by either optical, SEM [1,8], or AFM [6] microscopy. In addition, the TEM images of our twinned $Bi_2Sr_2Ca_{n-1}Cu_nO_x$ films, which were grown under similar growth conditions, do not indicate the presence of boundaries other than twin boundaries in these films, supporting the conclusion that this film grown on a misoriented substrate is a true single crystal.

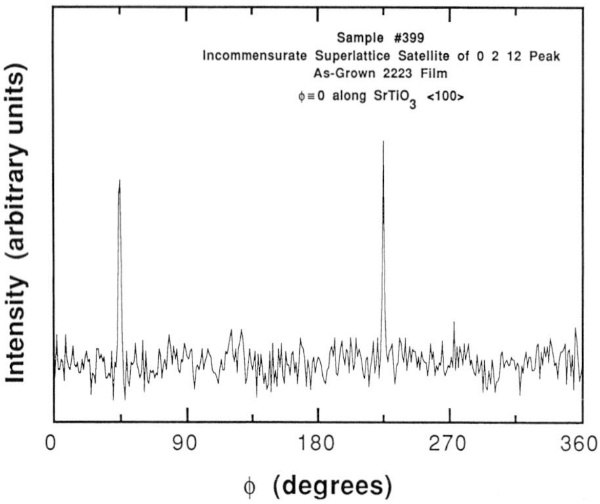

Fig. 6. X-ray diffraction ϕ scan of an incommensurate superlattice satellite peak of the 02$\underline{12}$ peak of a 2223 film grown on a misoriented $SrTiO_3$ {100} substrate. $\phi=0$ was set parallel to $SrTiO_3 < 100 >$. This scan indicates that the *b*-axis of the 2223 film runs up and down the steps of the misoriented substrate (45° and 225° peaks) but not laterally along them (no peak at 135° or 315°), indicating the *absence* of twin boundaries in the film.

CONCLUSIONS

The ability of shuttered MBE to grow smooth, layered, $Bi_2Sr_2Ca_{n-1}Cu_nO_x$ compounds, including ordered superlattices, *in situ* using ozone is demonstrated. This sequential deposition technique is effective in selecting between nearly energetically degenerate $Bi_2Sr_2Ca_{n-1}Cu_nO_x$ phases and layering them on a half-unit-cell by half-unit-cell basis. Using these techniques we achieved an as-grown superconducting film with T_c^{onset} near 100 K and a $T_c(\rho = 0)$ of 86 K. This is a major step in the development of a growth method capable of custom-making layered combinations of oxide compounds. The shuttered MBE growth technique seems quite capable of synthesizing artificially layered epitaxial structures consisting of layers of *oxides* (superconducting oxides, non-superconducting metallic oxides, semiconducting oxides, and insulating oxides), with control of layer thickness and abruptness on an atomic scale.

ACKNOWLEDGEMENTS

We gratefully acknowledge the RBS analysis of S.M. Baumann of Charles Evans & Associates; K. Nelson for the TEM sample preparation; and stimulating interactions with the entire Stanford high T_c thin film group. This work was supported in part by DARPA/ONR under contract number N00014-88-C-0760, Varian Associates research funds, the National Science Foundation Material Research Laboratory Program through the Center for Materials Research at Stanford University, and the Joint Services Electronics program under contract number DAAG29-84-K-0048. DGS acknowledges the support of a Semiconductor Research Corporation fellowship.

REFERENCES

[1]Present address: IBM Research Division, Zurich Research Laboratory, CH-8803 Rüschlikon, Switzerland

[1] Eckstein JN, Bozovic I, von Dessonneck KE, Schlom DG, Harris JS, Baumann SM (1990) Appl. Phys. Lett. 57: 931-933.
[2] Schlom DG, Marshall AF, Sizemore JT, Chen ZJ, Eckstein JN, Bozovic I, von Dessonneck KE, Harris JS, Bravman JC (1990) J. Cryst. Growth 102: 361-375.
[3] Schlom DG (1990) Ph.D thesis. Stanford University.
[4] Torardi CC, Subramanian MA, Calabrese JC, Gopalakrishnan J, Morrissey KJ, Askew TR, Flippen RB, Chowdhry U, Sleight AW (1988) Science 240: 631-633.
[5] Tarascon JM, McKinnon WR, Barboux P, Hwang DM, Bagley BG, Greene LH, Hull GW, LePage Y, Stoffel N, Giroud M (1988) Phys. Rev. B 38: 8885-8892.
[6] Bozovic I, Eckstein JN, Schlom DG, Harris JS (to be published) In: McConnell R, Wolf S (eds) Science and Technology of Thin Film Superconductors II. Plenum Press, New York.
[7] Hwang DM, Venkatesan T, Chang CC, Nazar L, Wu XD, Inam A, Hedge MS (1989) Appl. Phys. Lett. 54: 1702-1704.
[8] Schlom DG, Eckstein JN, Bozovic I, Chen ZJ, Marshall AF, von Dessonneck KE, Harris JS (to be published) In: SPIE proceedings on Growth of Semiconductor Structures and High T_c Thin Films on Semiconductors, and In: SPIE proceedings on High T_c Superconductivity: Thin Films and Applications.
[9] Hendricks S, Teller E (1942) J. Chem. Physics 10: 147-167.
[10] Horikoshi Y, Kawashima M, Yamaguchi H (1986) Jpn. J. Appl. Phys. 25: L868-L870.
[11] Roth RS, Rawn CJ, Burton BP, Beech F (1990) J. Res. NIST. 95: 291-335.
[12] Adachi H, Ichikawa Y, Hirochi K, Matsushima T, Setsune K, Wasa K (1990) Jpn. J. Appl. Phys. 29: L81-L82.
[13] Eckstein JN, Bozovic I, Schlom DG, Harris JS (1990) Appl. Phys. Lett. 57: 1049-1051.

BSCCO Superconducting Films Grown by MOCVD

KAZUHIRO ENDO, KENICHI NAKATSUKA, SHUNJI MISAWA, and SADAFUMI YOSHIDA

Electrotechnical Laboratory, 1-4, Umezono 1-chome, Tsukuba, 305 Japan

ABSTRACT

We have grown Bi-Sr-Ca-Cu-O superconducting films by MOCVD and studied the thickness and growth rate dependences of Tc(zero resistance). We have obtained the as-deposited film showing Tc(zero)=93 K, which is the highest Tc(zero) among as-deposited films by MOCVD reported so far.

KEY WORD: Bi-Sr-Ca-Cu-O superconducting films, MOCVD, high-Tc single phase, as-deposited film, expansion of the c-axis

INTRODUCTION

The Bi-Sr-Ca-Cu-O system attracts much attention for its higher superconducting transition temperature Tc. This oxide system has at least two superconducting phases: the high-Tc phase ($Bi_2Sr_2Ca_2Cu_3O_x$) with Tc around 110 K and the low-Tc phase ($Bi_2Sr_2CaCu_2O_y$) with Tc around 80 K. It was difficult to obtain the high-Tc single phase films. Much effort has been devoted to preparing the as-deposited films with the high-Tc single phase, especially for microelectronics. In the previous study [1], for the first time, we have succeeded in preparing the high-Tc single phase films on MgO substrates without postannealing by a metalorganic chemical vapor deposition (MOCVD). However, Tc(zero resistance) of the films was as low as 74K. On the other hand, it has been reported that Pb doping enhances the high Tc phase formation in the sintering of bulk materials. So, we have carried out in-situ Pb-doping in the growth of Bi-Sr-Ca-Cu-O films by MOCVD and studied its effect on the crystal structures and superconductivity of the films [2]. However, there were little effects of Pb-doping on the rise of Tc(zero), whereas the high-Tc single phase films doped with Pb were formed.

In order to obtain the as-deposited Bi-Sr-Ca-Cu-O films showing the higher Tc(zero), we have grown the films with various thicknesses. We have also grown the films with various growth rates.

EXPERIMENTAL

The source materials were triphenyl bismuth $Bi(C_6H_5)_3$, bis (-2, 2, 6, 6-tetramethyl-3, 5-heptanediono)-strontium $Sr(DPM)_2$, -calcium $Ca(DPM)_2$ and -copper $Cu(DPM)_2$. Each source material was loaded into an individual vaporizer and was heated to an appropriate temperature. These source materials were carried into a quartz reactor with Ar gas. Total flow rate of Ar gas was 770 cc/min. The gas line was heated to the temperatures higher than those of the vaporizers to prevent the condensation of source materials. Oxygen gas was separately introduced into the reactor at a flow rate of 675 cc/min. MgO(100) crystals, 5×10×0.5 mm^3 in size, were used as substrates. They were put on an inconel susceptor and then heated to 800 °C, inductively. The pressure in the reactor was kept at 50 Torr during the film growth. Oxygen partial pressure was 23 Torr. We have grown Bi-Sr-Ca-Cu-O films with the various thicknesses from 200 to 5200 Å at the growth rate of around 100 Å/h. We have also grown the films with the various growth rates between 30 and 500 Å/h.

The compositions and the crystal structures of the films obtained were examined using inductively coupled plasma spectroscopy (ICP) and X-ray diffraction (XRD), respectively. The incorporation of carbon in the films was analyzed by secondary ion mass spectroscopy (SIMS). An appreciable amount of carbon was not detected in the films Resistance measurements were carried out by a standard four-probes DC method, where Cu wires were attached with Ag paste to Au electrodes evaporated onto the films.

RESULTS AND DISCUSSION

Figure 1 shows the dependence of Tc(zero) on the film thickness. It should be noted that Tc(zero) increases steeply with the increase of film thickness, and then looks saturated in the large thickness range. The XRD patterns of these films are classified into two groups. One is the high-Tc single phase pattern: all the XRD peaks were indexed as (00n) of the high-Tc phase, and the other is the high-Tc and low-Tc mixed phases pattern: all the XRD peaks were indexed as (00n) of the high-Tc and low-Tc phases. The thinner films consist of the high-Tc single phase, and the thicker ones consist of the high-Tc and low-Tc mixed phases. Figures 2 and 3 show the XRD patterns of samples A(200 Å thick) and B(5200 Å thick), representative for the thinner and thicker films, respectively. All the films show the crystal orientation with the c-axis normal to the substrate surface.

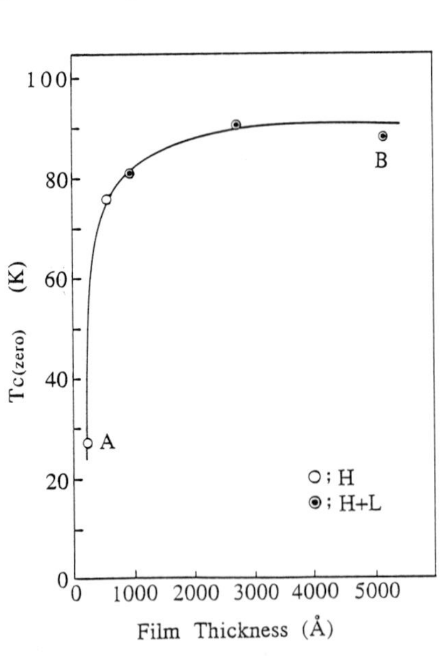

Fig.1. Dependence of Tc(zero) on film thickness. Growth rate is around 100 Å /h; ○ :the film with high-Tc single phase,◉ :the film with high-Tc and low-Tc mixed phases.

We have estimated the lattice constant of c-axis, c, from the XRD patterns. Figure 4 shows the

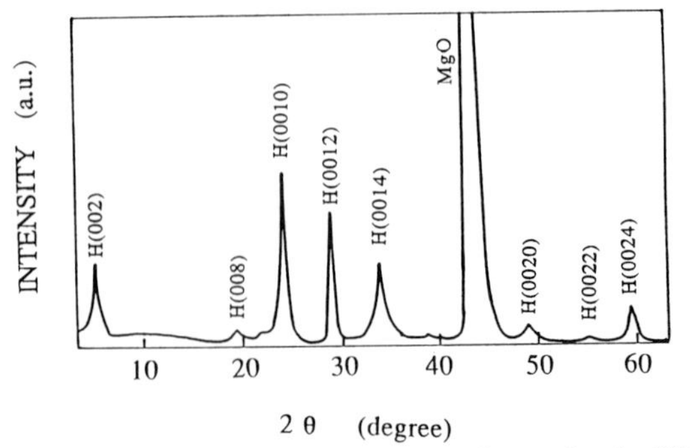

Fig.2. X-ray diffraction pattern of sample A (200 Å thick). H(00n)'s denote the peaks corresponding to the high-Tc phase.

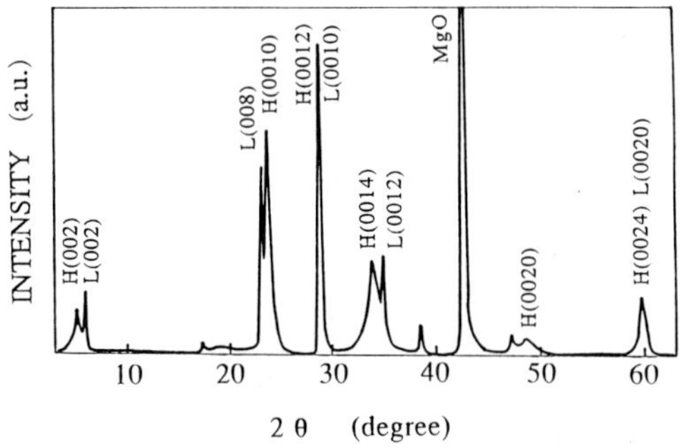

Fig.3. X-ray diffraction pattern of sample B (5200 Å thick). H(00n)'s and L(00n)'s denote the peaks corresponding to the high-Tc and low-Tc phase, respectively.

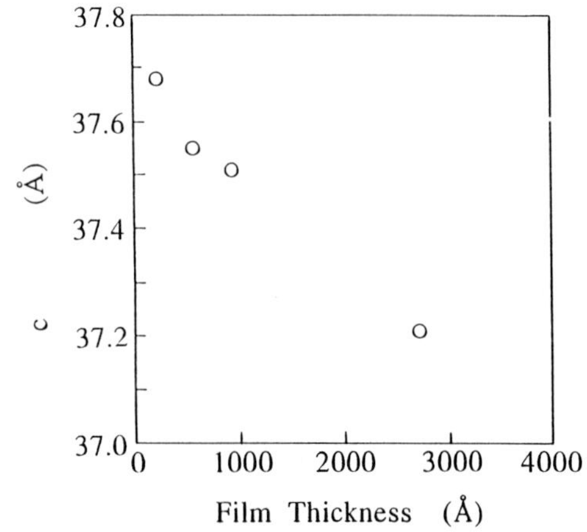

Fig.4. Dependence of lattice constant of c-axis, *c*, on film thickness. Growth rate is around 100 Å/h.

dependence of *c* on film thickness. Clearly, *c* decreases with the increase of film thickness. Such a dependence is considered to be a result of the lattice strain caused by the misfit of in-plane lattice constants between the Bi-Sr-Ca-Cu-O film and the MgO substrate. Namely, the misfit at the interface between the film and the substrate introduces the compressive stress into the film along the interface, because the in-plane lattice constant of the film (the lattice constant of a-axis) is twenty percent larger than that of MgO, although the greater part of the misfit is relaxed by the introduction of so-called misfit dislocation. The origin of the expansion of the c-axis in the films can be attributed to such a compressive stress along the interface. It is presumed that the increase of the film thickness brings about the reduction of the compressive stress, which results in the decrease of *c* with the increase of thickness.

Figure 5 shows the dependence of Tc(zero) on *c*. Tc(zero) increases with the decrease of *c*, and looks saturated to around 90 K. This fact suggests that the use of the substrates with small misfit is favorable to increase Tc(zero).

We have also examined the dependence of Tc(zero) on the growth rate of the films. The result for the films with the thickness of 1800 Å is shown in Fig. 6. It is to be noted that Tc(zero) increases with the decrease of growth rate.

On the basis of these results, we have grown the Bi-Sr-Ca-Cu-O film with the thickness of 1800 Å at the growth rate of 60 Å/h. Figure 7 shows the temperature dependence of the resistivity for the as-deposited film. Tc(zero) was 93 K, which is the highest Tc(zero) among as-deposited films by MOCVD reported so far.

CONCLUSION

We have grown Bi-Sr-Ca-Cu-O superconducting films on MgO(100) by MOCVD. It was found that Tc(zero) increases with the increase of film thickness and with the decrease of

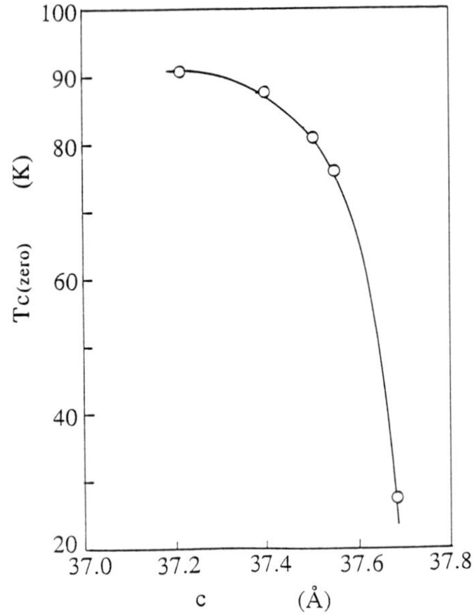

Fig.5. Dependence of Tc(zero) on lattice constant of c-axis, *c*. Growth rate is around 100 Å/h.

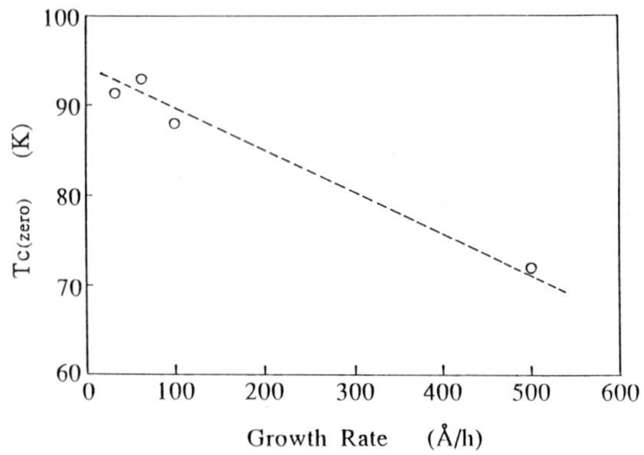

Fig.6. Dependence of Tc(zero) on growth rate. The film thicknesses are around 1800 Å.

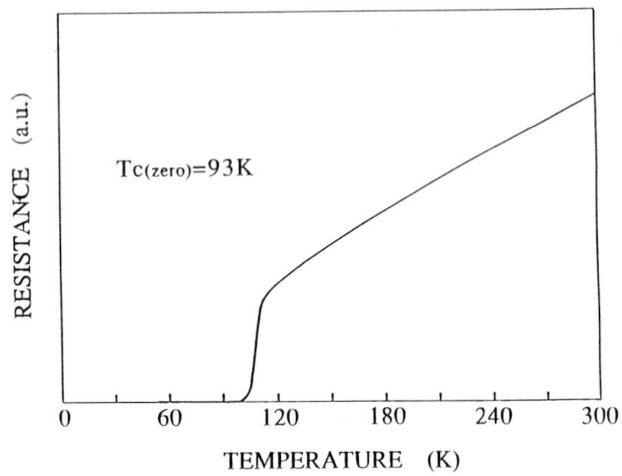

Fig.7. Temperature dependence of the resistance of Bi-Sr-Ca-Cu-O film showing the highest Tc(zero) of 93K (1800 Å thick, 60 Å/h).

explained in terms of lattice strain resulting from the misfit of in-plane lattice constants between Bi-Sr-Ca-Cu-O films and MgO substrates. It was also found that Tc(zero) increases with the decrease of the growth rate. We have succeeded in obtaining the as-deposited Bi-Sr-Ca-Cu-O film showing Tc(zero)=93K.

*On leave of absence from Iwatani Co.,LTD.

REFERENCES

1. Endo K, Hayashida S, Ishiai J, Matsuki Y, Ikedo Y, Misawa S, Yoshida S (1990) Jpn J Appl Phys 29: L294-297
2. Endo K, Hayashida S, Nakatsuka K, Ishiai J, Ikedo Y, Misawa S, Yoshida S (1990) In: Ishiguro T, Kajimura K (eds) Advances in Superconductiviy. Springer, Tokyo, pp 912-914

AS-Deposited Superconducting Bi-Sr-Ca-Cu-O Thin Films with Critical Temperature of 95K

KAZUSHIGE OHBAYASHI[1], MASATOSHI ANMA[2], YOSHIAKI TAKAI[2], and HISAO HAYAKAWA[2]

[1] Research and Development Center, NGK Spark Plug Co., Ltd., 14-18, Takatsuji-cho, Mizuho-ku, Nagoya, 467 Japan

[2] Department of Electrical Engineering, Faculty of Engineering, Nagoya University, Furo-cho, Chikusa-ku, Nagoya, 464 Japan

ABSTRACT

High-Tc superconducting thin films of the Bi-Sr-Ca-Cu-O system were prepared, in situ, on MgO(100) substrates by a three-target rf magnetron sputtering system with a regulated shuttering technique. Where the Bi-Sr-Ca-Cu-O superconducting high-Tc-phase films (700Å) were deposited on a thin Bi-Sr-Cu-O semiconductorlike-phase buffer layer (70Å) under rigorous control of the film composition at 760°C. The superconducting transition temperature of the as-grown film was 115K(Tc_{on}) and 95K(Tc_{zero}). The surface of this film was very smooth and uniform. X-ray diffraction paterns of this film showed a high-Tc-phase with a highly c-axis oriented structure, but small amount of low-Tc-phase and semiconductorlike-phase were obserbed.

KEY WORDS: Bi-Sr-Ca-Cu-O, thin film, as-grown, buffer layer, high-Tc-phase

INTRODUCTION

Bi-Sr-Ca-Cu-O system thin films had been prepared with ex situ heat treatment where the superconducting phases are formed, after deposition of the required stoichiometry of constituent metal atoms. However, using these techniques, it was difficult to obtain a single phase thin film [1]. The compositions of these Bi-Sr-Ca-Cu-O system are represented by $Bi_2Sr_2Ca_{n-1}Cu_nO_x$ (n=1,2 and 3) [2], where n is the number of $Cu\text{-}O_2$ layers with Ca cations between the double Bi-O layers and is related with the superconducting properties of these materials [3,4]. For the purpose of controlling the number of $Cu\text{-}O_2$ layers, the thin film process is a promising technique. Recently, many groups have developed techniques for the as-grown single phase thin films of Bi-Sr-Ca-Cu-O system such as sputtering [5-7], molecular beam epitaxy [8], and coevaporation [9]. These methods succeeded in obtaining films of a highly oriented and single phase without ex situ heat treatment. However, the superconducting transition temperatures of these as-grown films were lower than those of post-annealed films [10]. In particular, the superconducting transition temperature of the film of high-Tc-phase remained under 80K.

In this paper, as-grown superconducting thin films of Bi-Sr-Ca-Cu-O having the highest Tc(zero) as far as the authors know are reported. They were prepared on the Bi-Sr-Cu-O buffer layer by multitarget magnetron sputtering.

EXPERIMENTAL

In order to realize the in situ growth of Bi-Sr-Ca-Cu-O superconducting films, we developed a shuttering equipment [11]. Each individual target was equipped with a shutter. Each shutter was opened or closed independently by an external programmable controller at an arbitrary time. Controlling this shuttering equipment and rf input power, the deposition sequence and the thickness were regulated in the growth direction. The compositions of the targets were Bi_2O_3(4N,powder), $Bi_{2.0}Sr_{2.0}Cu_{1.0}O_x$ (calcined powder) and $Ca_{1.0}Cu_{1.0}O_x$(sintered). The films were grown on polished MgO(100) substrates at the substrate temperature of 760°C. The substrate was fastened to the substrate table which rotated at 24r.p.m., and run periodically just along the circumference which passed through the center of each target. The chamber was evacuated with a cryopump to a base pressure of $7x10^{-4}$ Pa. During growth Ar and O_2 ($Ar:O_2=4:1$) was introduced to the chamber and the pressure was maintained at 1 Pa.

In the first step, the Bi-Sr-Cu-O buffer layer was deposited on the MgO substrate of about 70A using Bi_2O_3 and $Bi_{2.0}Sr_{2.0}Cu_{1.0}O_x$ targets. Following this step, the Bi-Sr-Ca-Cu-O film was deposited. Where the layered structure of the thin films were controlled by the sequence of deposition of the constituent atoms. Here using the shutter control, Bi-Sr-Cu-O blocks and Ca-Cu-O blocks were deposited alternately to promote the growth of the high-Tc-phase Bi-Sr-Ca-Cu-O film. The Bi_2O_3 target was used for the purpose of rigorous control of the Bi content of the films. Detailed sputtering conditions for obtaining Bi-Sr-Cu-O and Bi-Sr-Ca-Cu-O films are listed in Table I. After deposition, air was introduced to the chamber to atmospheric pressure, and the film was cooled down to 300°C in about 20 min inside the chamber. The resistivity of the films was measured as a function of temperature by a standard four-probe method using a pressed indium contact at a current of 10uA from 15K to 280K. The crystal structure of these films was examined by the X-ray diffraction (XRD) method with Cu-Ka radiation.

Table I. Sputtering conditions of Bi-Sr-Cu-O and Bi-Sr-Ca-Cu-O films.

	Target	Bi-Sr-Cu-O	Bi-Sr-Ca-Cu-O
Input power	Bi_2O_3	21W	21W
	$Bi_2Sr_2Cu_1O_x$	50W	50W
	$Ca_1Cu_1O_x$	-	50W
Shutter opening time	Bi_2O_3	600s	50s
	$Bi_2Sr_2Cu_1O_x$	600s	50s
	$Ca_1Cu_1O_x$	-	85s
Deposition sequence	Bi_2O_3	*	3n-2,3n
	$Bi_2Sr_2Cu_1O_x$	*	3n-2,3n
	$Ca_1Cu_1O_x$	-	3n-1
Substrate temperature	760°C		
Gas pressure	1 Pa ($Ar:O_2$ = 4:1)		

(n=layered number : 1,2,3,....40, * : simultaneously sputtered.)

RESULTS AND DISCUSSION

An example of such a effect of the Bi-Sr-Cu-O buffer layer on the superconducting transition temperature of Bi-Sr-Ca-Cu-O thin film is shown in Fig.1. A thin Bi-Sr-Cu-O buffer layer was deposited on a MgO (100) single-crystal substrate, and then, Bi-Sr-Ca-Cu-O film was prepared on the buffer layer. The electrical resistivity of the film is shown in Fig. 1(a). This as-grown film showed an almost linear temperature dependence of the resistivity in the normal state and had a superconducting transition at approximately 115K and reached zero at 95K. The crystallographic properties of the film was characterized by XRD measurement. The XRD analysis identified this film as the high-Tc-phase (indicated by H) and showed a c-axis orientation (Fig. 2(a)). However, small amounts of both low-Tc (indicated by L) and semiconductorlike phases (indicated by S) were observed in the XRD pattern. Since the XRD patterns of the buffer layer itself showed the semiconductorlike-phase, the small peaks of the semiconductorlike-phase in Fig. 2(a) seem to be ascribed to the buffer layer, and the appearance of the low-Tc-phase seems to be a result of the interactive diffusion between the high-Tc-phase and the semiconductor-like-phase.

For contrast, the temperature dependence of the resistivity and XRD pattern for the film without a buffer layer are shown as curves "b" in Figs. 1 and 2, respectively. Typically, the former has a broad resistive transition starting at approximately 110K and the zero-resistivity temperature was about 50K. Nevertheless, the XRD pattern of the film was that of the high-Tc-phase and all the peaks indicated can be indexed as (002n) of an approximately 37.0A unit cell except for the diffraction peaks of the substrate and the peak of CaO at 37.5° and a small amount of the low-Tc-phase. This c-axis lattice constants show a good agreement with the range of measured high-Tc-phase values. On the contrary, in the films with a buffer layer, this CaO peak is not observed or is of weak intensity compared with the CaO peak intensity in the film without a buffer layer. Thus, the main feature of the peaks of XRD patterns does not appreciably differ from each other, but the value of Tc(zero) is remarkably different in the case of having the buffer layer. These results show that the superconducting properties were improved by the presence of the Bi-Sr-Cu-O buffer layer.

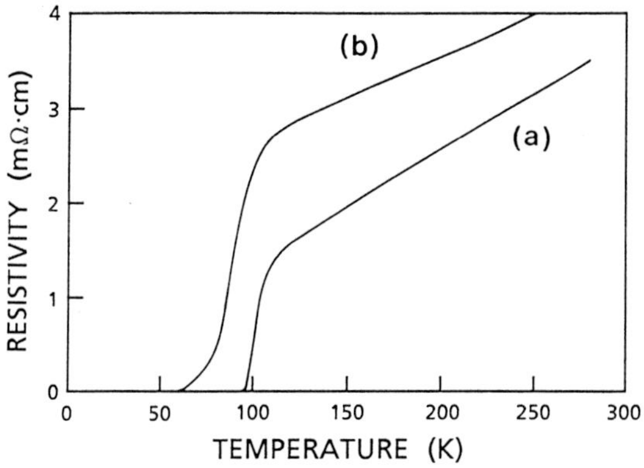

Fig. 1. Temperature dependence of the resistivity for the as-grown films (a) with Bi-Sr-Cu-O buffer layer and (b) without buffer layer.

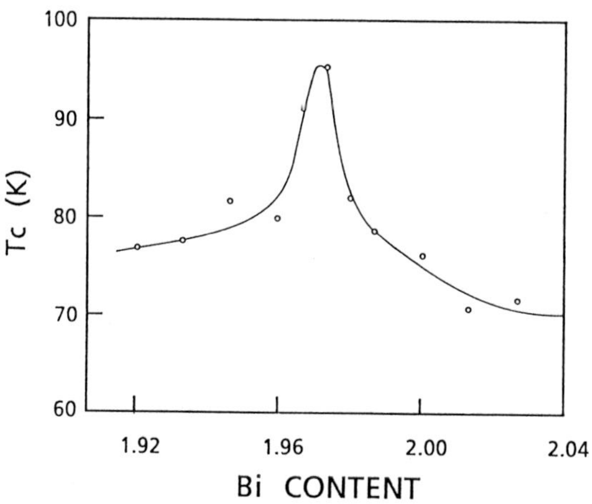

Fig. 3. Bi content dependence of the crytical temperature for the as-grown films with buffer layer. Assuming that the Sr content was 2.00. Bi was supplised with Bi_2O_3 target at the input power range of 17-25W and $Bi_2Sr_2Cu_1O_x$ target at the input power of 50W.

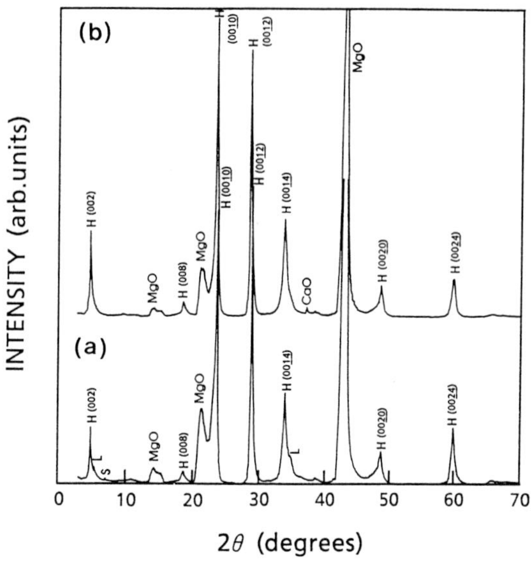

Fig. 2. X-ray diffraction patterns of the films (a) with Bi-Sr-Cu-O buffer layer and (b) without buffer layer. The diffraction lines of the patterns are indexed as the (002n) lines of high-Tc-phase.

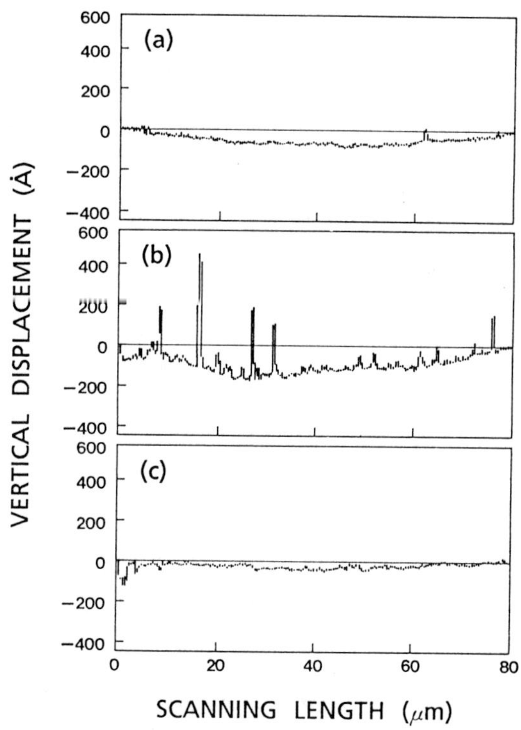

Fig. 4. Measurement of the surface profiles. (a) MgO substrate, (b) the high-Tc-phase film deposited directly cn the MgO substrate and (c) the high-Tc-phase film deposited on the buffer layer.

The semiconductorlike-phase was obtained under a relatively wide range of Bi content ($Bi_xSr_2Cu_1O_y$: x is between 1.68 to 2.22). The compositions were determined by the electron probe microanalysis (EPMA). These films have a smooth surface and a stable c-axis lattice constant. By comparing the high-Tc-phase films which were prepared by changing the Bi content from 1.92 to 2.03 with a step of about 0.015 (assuming that the Sr content was 2.00), we identified that the Bi content in the film affected the critical temperature, the c-axis lattice constant and the surface morphology of the thin films. The Bi content dependence of the Tc(zero) is shown in Fig.3. The values of Tc(zero) varied with the Bi content, and Tc(zero) rose steeply at the Bi content of 1.97; moreover, the film has a smooth surface at this Bi content. These results show that the superconducting properties are quite sensitive to the Bi content, and that the film composition under which the high-Tc-phase can be developed is quite limited.

To examine further, the surface profiles of the substrate and two kinds of films were measured. These results are shown in Fig.4: (a) the surface of a MgO substrate, (b) the surface of higt-Tc-phase film deposited directly on the MgO substrate up to about 700Å and (c) the surface of the high-Tc-phase film with a semiconductorlike-phase buffer layer whose thicknesses were about 700Å and 70Å, respectively. However, on the surface of the high-Tc-phase film which was deposited directly on a MgO substrate, a number of small cornlike particles and pits are observed and the composition of these segregated small particles was identified as CuO by EPMA. On the other hand, the surface of the high-Tc-phase film with a buffer layer is very smooth as same as a surface morphology of MgO substrate. Many small pits, however, are observed on the surface of the film with a buffer layer but no segregation of the constituent elements is observed by EPMA. Thus, the difference of the surface morphology by the presence of the buffer layer is quite apparent.

CONCLUSIONS

We have succeeded in the growth of Bi-Sr-Ca-Cu-O superconducting films by using a layer control technique in the rf magnetron sputtering with a shuttering system. We have also succeeded in controlling the number of $Cu-O_2$ layers. Furthermore, by preparing the Bi-Sr-Cu-O buffer layer on the MgO substrate, we have obtained as-grown superconducting films with zero-resistivity at 95K. X-ray diffraction patterns showed a highly c-axis orientated structure, and these films have very smooth surface. These results show that the superconducting properties of the Bi-Sr-Ca-Cu-O high-Tc-phase are improved by the presence of a Bi-Sr-Cu-O buffer layer. Moerover we suggest that the critical temperature of the high-Tc-phase films are quite sensitive to the Bi content of the films.

REFERENCES

1. K.Ohbayashi,T.Ushida,T.Tunooka,K.Ohya and H.Banno(1989)Proc.1st.Symp.on Superconductivity.Springer-Verlag,Tokyo,pp.551-556
2. E.Takayama-Muromachi,Y.Uchida,A.Ono,F.Izumi,M.Onoda,Y.Matsui, K.Kosuda, S.Takekawa and K.Kato(1988)Jpn.J.Appl.Phys.27:L365
3. Y.Bando,T.Kijima,Y.Kitami,J.Tanaka,F.Izumi and M.Yokoyama(1988)Jpn.J.Appl. Phys.27:L358
4. E.Takayama-Muromachi,Y.Uchida,Y.Matsui,M.Onoda and K.Kato(1988)Jpn.J.Appl. Phys.27:L556
5. K.Kuroda,K.Kojima,M.Tanioku,K.Yokoyama,H.Kuroki and K.Hamanaka(1990)Jpn.J. Appl.Phys.29:1066
6. Y.Hakuraku,D.Miyagi,S.Higo and T.Ogushi(1990)Jpn.J.Appl.Phys.29:L926
7. J.Fujita,T.Tatsumi,T.Yoshitake and H.Igarashi(1989)Appl.Phys.Lett.54:2364
8. Y.Nakayama,H.Ochimizu,A.Maeda,A.Kawazu,K.Uchinokura and S.Tanaka(1989)Jpn. J.Appl.Phys.28:L1217
9. T.Sato,T.Yoshitake,S.Miura,J.Fujita,Y.Kubo and H.Igarashi(1989)Appl.Phys. Lett.55:702
10. K.Ogawa,K.Nakamura.H.Hayakawa,H.Hattori,J.Sato and S.Ikeda(1990)Proc.2nd Int.Symp.on Superconductivity.Springer-Verlag,Tokyo,pp.27-23
11. K.Ohbayashi,S.Suzuki,Y.Takai and H.Hayakawa(1990)Proc.2nd Int.Symp.on Superconductivity.Springer-Verlag,Tokyo,pp.841-844

Epitaxial Growth Study of $Bi_2(Sr, Ca)_3Cu_2O_x$ Films on Cleaved MgO Substrates

J. FUJITA, T. YOSHITAKE, T. SATOH, and H. IGARASHI

Fundamental Research Laboratories, NEC Corporation, 1-1, Miyazaki 4-chome, Miyamae-ku, Kawasaki, 213 Japan

ABSTRACT

In-situ epitaxial growth of $Bi_2(Sr_{0.6}Ca_{0.4})_3Cu_2O_X$ films on cleaved (100)MgO substrates was studied by ion beam sputtering. Epitaxial films having twofold symmetry were successfully grown on cleaved MgO substrates, where the epitaxial relationship of the films on the cleaved MgO substrates was such that the a- and b-axes of the films rotated approximately ±13° from the <100>MgO axis. The early stage of film growth showed the anisotropic growth manner and the step edges tended to be the nucleation site. The step edges played an important role in determining the twofold epitaxial relationship where the b-axis with incommensurate modulation always stayed nearly perpendicular to the step edge directions.

KEY WORDS: Bi based superconductor, thin film, epitaxial growth, cleavage

INTRODUCTION

Bi-based superconductor has an incommensurate modulation with a near-five-time periodicity of the perovskite unit cell along b-axis and an ordered structure with double periodicity along a-axis[1,2,3]. Reflection high energy electron diffraction (RHEED) pattern of in-situ grown films on ordinary polished (100)MgO substrates tended to show a fourfold symmetry pattern[4,5,6,7]. The pattern along <100>MgO directions showed superstructure streaks due to the incommensurate modulation surrounding the [020]main streaks together with superstructure streaks at the half distance between [020]main streaks. This fourfold pattern results from the co-existence of the in-plane 90° rotated domains. Recently, we have succeeded in preparing the $Bi_2(Sr,Ca)_3Cu_2O_X$ (BSCCO) epitaxial films having twofold symmetry by using cleaved MgO substrates. This paper will report the crystallographic analysis of the epitaxial BSCCO films and the manner of the early stage of the film growth on the cleaved MgO substrates.

EXPERIMENTAL

BSCCO epitaxial films were prepared by dual ion beam sputtering method which was described elsewhere[4,5]. Two ion beam sources were equipped to sputter two targets, Bi_2O_3 and $Sr_{1.7}Ca_{1.3}Cu_2O_X$, separately. Films were grown on (100)MgO substrates by co-deposition from the two sputter targets and film composition was controlled by adjusting the sputtering power. Atomic oxygen was also used in order to enhance oxidation during the film growth process[4]. Oxygen partial pressure at the substrate was about $1.3x10^{-3}$torr and substrate temperature range was 640°-750°C. Chemical compositions of the as-grown films were determined by electron probe microanalysis as $Bi_2Sr_{1.7}Ca_{1.3}Cu_{2.1}O_X$. Superconducting transition temperatures of the as- grown films were around 70K.

RESULTS

Typical surface morphology of the cleaved MgO substrate was shown in Fig.1. When a MgO single crystal was cleaved by a sharp knife edge as shown in Fig.2, directions of all step edges tended to run parallel to a <100>MgO direction except for the beginning of the cleavage. This well defined "parallel" region of the cleaved MgO substrate was used in this experiment. Here we define the [100]MgO direction as parallel to the step edges, and the [010]MgO as perpendicular direction to the step edges, namely the up and down direction of the steps. The large step on the cleaved surface indicated by the arrow in Fig.1 had a typical height of several thousand angstroms according to thickness measurement by DEKTAK3030. On the contrary the fine steps which were clearly visible in the figure seemed to be below several ten angstroms in height, that is, below the detection limit of the apparatus.

Fig.1 SEM micrograph of typical surface of the cleaved MgO

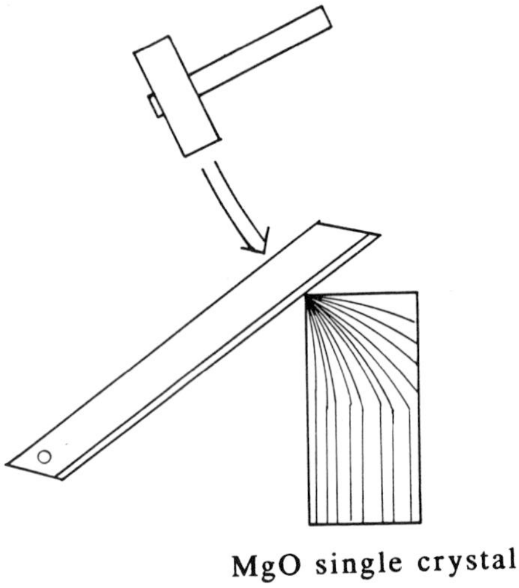

Fig.2 Schematic illustration of the cleavage

Fig.3 shows the RHEED patterns of the as-grown films for a series of beam incidences, where the incident electron beam azimuths are (a)+90° from [010]MgO, (b)+58° from [010]MgO, (c)+13° from [010]MgO, (d)±0° from [010]MgO, (e)-13° from [010]MgO, and (f)-58° from [010]MgO. Two identical RHEED patterns with the incommensurate modulation streaks are observed along the directions deviated by ±13° from the [010]MgO as showed in Fig.3(c) and (e). The tilted streaks with a small curvature indicated by the arrows in Fig.3(c) and (e) seem to be the manifestation of 26° rotated pattern from another equivalent domain. The reciprocal lattice rods are elongated in the incommensurate modulation direction. The crossing of such elongated reciprocal rods and the Ewald sphere will appear as the curved streaks as seen in Fig.3(b),(c),(e), and (f). Thus the existence of the two types of misoriented domains deviating about ±13° from <100>MgO was expected on the cleaved MgO surface.

X-ray diffraction (XRD) measurement was carried out in order to determine the detailed epitaxial relationship by using 4-circle diffractometer in the Bragg-Brentano geometry with Mo Kα radiation. Films were mounted so as to align the c-axis of the film to ϕ axis of the diffractometer. The reflection map in h-k reciprocal plane at l_{MgO}=1.29 was shown in Fig.4, where the hkl indices were defined using the lattice constant, 4.21Å of MgO substrate. The film's (02$\underline{10}$) reflections had the satellite reflections caused by the incommensurate modulation, which correspond to the satellite streaks of the RHEED patterns. The XRD measurement confirmed the unique epitaxial growth manner observed in RHEED. There surely existed the two types of misoriented domains, deviating about ±13° from <100>MgO. The direction of the incommensurate modulation always alined nealy perpendicular to the step edges.

The above epitaxial relationships always reproduced for the film growths on the cleaved MgO substrates. The unique epitaxial relationship was closely related to the fine steps. All fine steps below several ten angstroms were always aligned to the [100]MgO, while the large steps might run with a finite deviation from the [100]MgO direction. This seems to suggest that the growth mechanism of the twofold epitaxial BSCCO films sensitively depend on the existence of the surface steps, especially, the fine steps below several ten angstroms in height. The large lattice misfit between BSCCO and MgO seems to be the reason of the formation of the ±13° misoriented epitaxial growth.

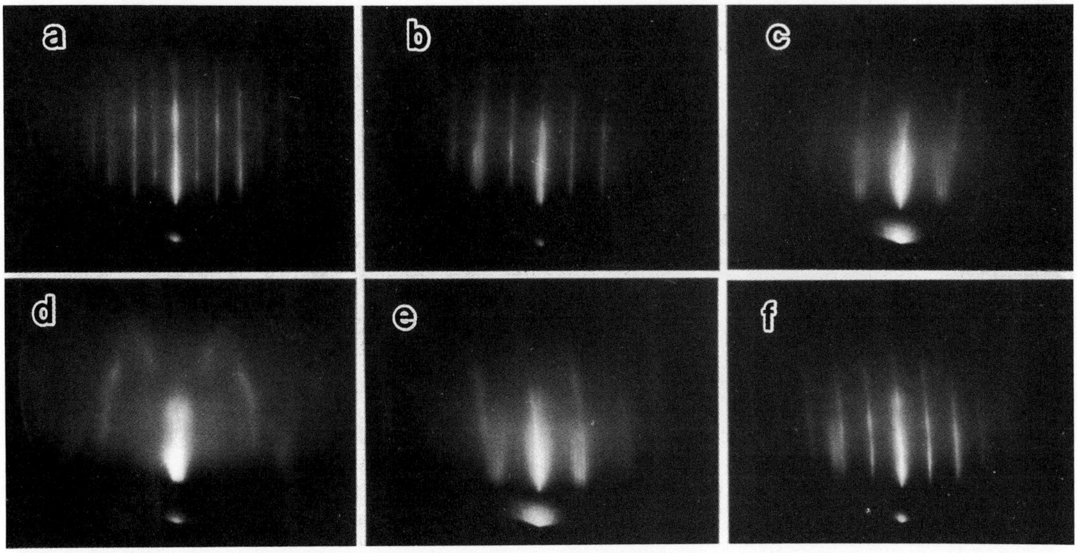

Fig.3. In situ RHEED patterns on cleaved MgO Substrate.
(a) incident electron beam azimuth was +90° from <010>MgO, (b)+58° from <010>MgO, (c)+13° from <010>MgO, (d)±0° from <010>MgO, (e)-13° from <010>MgO, (f)-58° from <010>MgO. Note that the steps are running along <100>MgO, so the incommensurate moderation exists along the up and down direction of terraces.

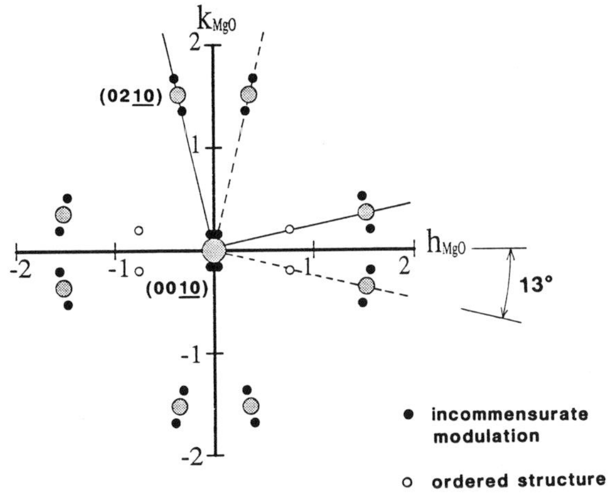

Fig.4. Reflection map in hk reciprocal plane at l=1.29, where the h, k, and l were indexed by using 4.21Å MgO lattice. The center circle was identical to the (0010) reflection, the meshed circles indicated the (0210) reflections.

Characteristic epitaxial growth manners on the cleaved MgO substrate were observed at the early stage film growth. Fig.5 shows the SEM micrographs of the early stage film growth. After the 70Å deposition, the RHEED pattern having two fold symmetry became clearer and the surface was decorated into the characteristic stripe pattern as shown in Fig.5(a). The bright stripes were strongly charged-up regions corresponding to the BSCCO-free portions of the substrate while the step edges tended to become the black lines as the result of the BSCCO nucleation and its growth. The direction of these stripes coincided with that of the fine steps. The stripes appearing on the initial stage partly remained even after the 700Å deposition was performed as shown in

Fig.5(b). These SEM micrographs suggested the anisotropic nature of the film growth where the fine steps on the cleaved MgO surface tended to be the preferential nucleation site and the large film growth rate along the step edge direction. It is noticeable that the epitaxial direction of the incommensurate modulation in the nucleation island already determined so as to avoid the step edges at the early stage of film growth. This film growth manner also supported that the unique direction of the step edges is the origin of the twofold symmetry.

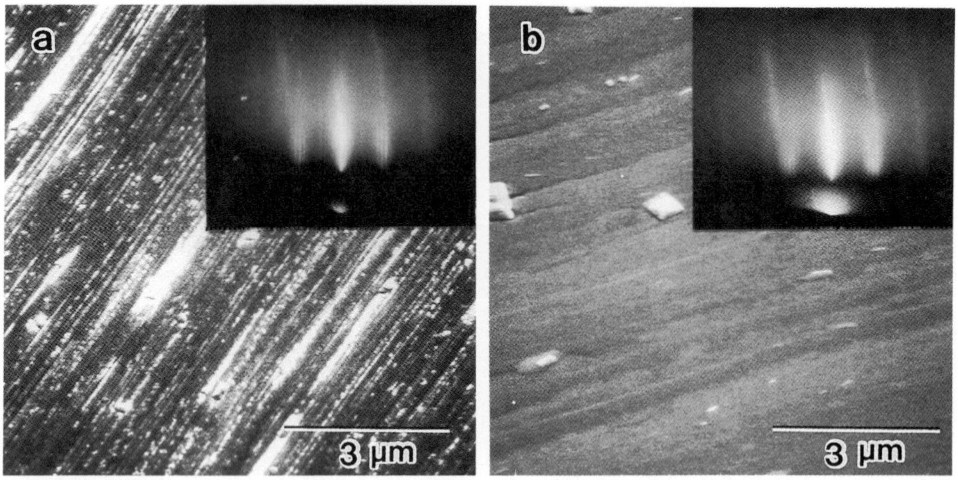

Fig.5. SEM micrographs and RHEED patterns obtained by initial film growth stages. (a) after 70Å deposition, (b) after 700Å deposition.

CONCLUSION

The film growth having twofold symmetry of $Bi_2(Sr,Ca)_3Cu_2O_x$ was performed by using the cleaved MgO substrates. But the films still contained equivalent two type domains with the a-axes deviated about ±13° from the step edge direction. The occurrence of such ±13° deviation is attributed to the large lattice misfit between BSCCO film and MgO. The early stage of film growth showed strong anisotropic growth characteristics and the step edges seemed to be the nucleation sites. The fine steps on the cleaved MgO surface seemed to play an important role in the occurrence of the twofold symmetry.

ACKNOWLEDGMENT

This work was performed under the management of the R & D Association for Future Electron Devices as a part of the R & D of Basic Technology for Future Industries supported by New Energy and Industrial Technology Development Organization.

REFERENCES

1. K. Hiraga, M.Hirabayashi, M.Kikuchi, and Y.Shono (1988) Jpn.J.Appl.Phys. 27: L573.
2. Y.Matsui, H.Maeda, Y.Tanaka, E,Takayama-Muromachi, S.Takekawa, and S.Horiuchi (1988) Jpn.J.Appl.Phys. 27: L827.
3. A.W.Sleight (1988) Science 242: 1519.
4. J.Fujita, T.Tatsumi, T.Yoshitake, and H.Igarashi (1988) Appl.Phys.Lett. 54: 2364.
5. J.Fujita, T.Yoshitake, H.Igarashi, and T.Satoh (1990) Appl.Phys.Lett. 56: 295.
6. J.N.Eckstein, I.Bozovic,K.E.von Dessonneck, D.G.Schlom,J,S,Harris,Jr, and S.M.Baumann. (1989) to be published in Proceedings of Fall MRS meetings, Symposium M:High temperature Superconductors.
7. J.N.Eckstein, I.B.bozovic, D.G.Schlom, and J.S.Harris,Jr, to be appeared in (1990) Appl.Phys.Lett.57: 1049.

Formation of BiSrCaCuO Thin Films with (11n) Orientation on $SrTiO_3$ (110) Substrates by RF Magnetron Sputtering

KEN'ICHI KURODA[1], KAZUYOSHI KOJIMA[1], OSAMU WADA[2], JUNJI TANIMURA[2], MASAMI TANIOKU[1], KAZUO YOKOYAMA[2], and KOICHI HAMANAKA[1]

[1] Central Research Laboratory, [2] Materials and Electronic Devices Laboratory, Mitsubishi Electric Corporation, Tsukaguchi-Honmachi, Amagasaki, 661 Japan

ABSTRACT

As-grown BiSrCaCuO thin films were grown on $SrTiO_3$(110) substrates at 660 °C by consecutively sputtering. XRD measurement showed that the films with a low-T_c phase composition and a high-T_c phase composition had main peaks at 2Θ of 31.0° and 31.9°, respectively. The film with a low-T_c phase composition is considered to mainly consist of a low-T_c phase with (117) orientation.

KEY WORDS: BiSrCaCuO, superconductor, thin films, sputtering, preferential orientation

INTRODUCTION

As-grown superconducting BiSrCaCuO thin films have been formed by various methods, such as RF sputtering [1-8], co-evaporation [9], molecular beam epitaxy[10] and metalorganic chemical vapor deposition [11]. These films were oriented with the *c*-axis perpendicular to the substrate surface, because the films were formed on MgO(100) substrates or $SrTiO_3$(100) substrates. However, films with *c*-axis orientation are considered unsuitable for forming Josephson tunnelling junctions, because the coherence length along the *c*-axis is probably too short to obtain the Josephson effect. Thus, as reported in our previous paper [12], we tried to form as-grown BiSrCaCuO thin films on $SrTiO_3$(110) substrates to obtain films having *c*-axes which are not oriented perpendicular to the substrate surface. We obtained a film with (11n) orientation using single-target sputtering, but the superconducting phase of the film was not clear. Thus, we formed as-grown films on $SrTiO_3$(110) substrates using the consecutive sputtering method, because the film composition could be precisely controlled by this method[4].

In this paper, we report on the crystal structures, surface morphology and superconducting properties of as-grown BiSrCaCuO thin films having a low-T_c phase composition and a high-T_c phase composition.

EXPERIMENTAL

BiSrCaCuO thin films were grown on $SrTiO_3$(110) substrates at 660°C by RF magnetron sputtering (ANELVA SPF-332HS) from three targets. The targets were calcined into a 3 inch diameter disk by Koujundo Kagaku Corp. and had nominal compositions of $Bi_3Sr_2Ca_2Cu_3O_x$, $Bi_2Sr_3Ca_3Cu_3$ and $Bi_2Sr_2Ca_2Cu_{4.5}O_x$. As described in our previous paper [4], it is possible to precisely control the film composition by adjusting the staying times of the substrate over each target. The sputtering gas was neat O_2, and the sputtering pressure was 300mTorr. Discharge power was 100W, and the gap between the target and the substrate was 50mm. The sputtering time was 150min, in which approximately 1200Å-thick films were deposited. After deposition, the films were quickly cooled down to room temperature by introducing air into the chamber.

X-ray diffraction (XRD) and reflection high-energy electron diffraction (RHEED) studies were performed to analyze the crystal structure. The surface morphology was observed by scanning electron microscopy (SEM), and the temperature dependence of resistivity was measured by the AC four-probe method using a lock-in amplifier.

RESULTS AND DISCUSSION

A film with a low-T_c phase composition (film(a)) showed streak-like patterns by the RHEED measurement, as shown in Fig. 1. The RHEED patterns are similar to the previous results from the single-target sputtering [12]. In the figure, streaks consisting of many diffraction spots are observed. The direction of the streaks corresponds to the *c*-axis direction, because the diffraction spots indicate a long period structure. The separation of the diffraction spots corresponds to the length of 15Å. Thus, the film(a) is considered to consist of a low-T_c phase with a half *c*-axis length of 15Å. Also, taking the slope angle of the streaks, it was found that the *c*-axis met the substrate surface at an angle α of about 42°. The plane interval in the direction perpendicular to the substrate surface was 2.8Å, and that in the direction parallel to the *c*-axis was 3.8Å which is equal to that of the (110) surface of the BiSrCaCuO film; the plane interval of the BiSrCaCuO(110) surface (3.8Å) is nearly equal to that of the $SrTiO_3(001)$ surface. The result indicates that film(a) was grown through coordination with the substrate surface; the *a*-axis and the *b*-axis meet at an angle of 45° with the substrate surface. On the other hand, a film with a high-T_c phase composition (film(b)) did not show RHEED patterns. The crystallinity of the film(b) is considered to be bad.

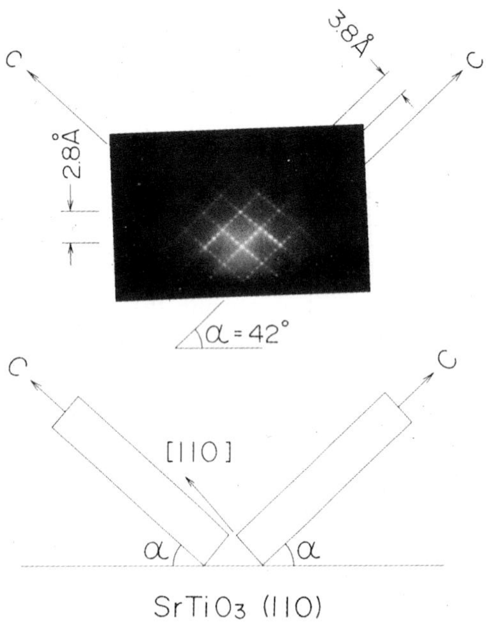

Fig.1. RHEED patterns of BiSrCaCuO film grown on $SrTiO_3(110)$ substrate.

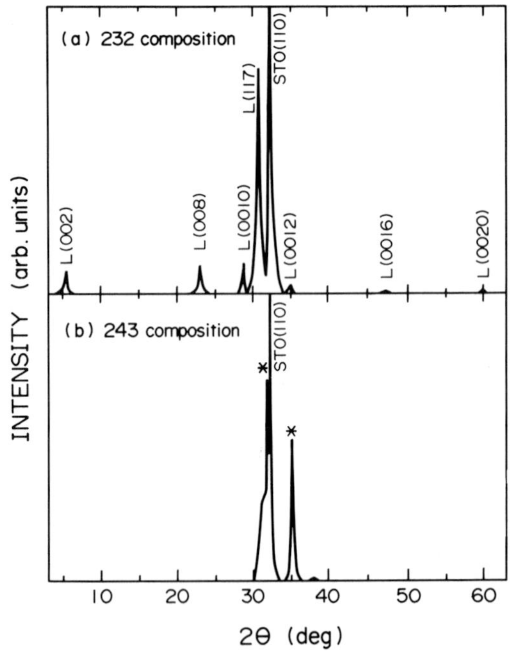

Fig.2 XRD patterns of BiSrCaCuO film grown on $SrTiO_3(110)$ substrate.
(a): low-T_c phase composition,
(b): high-T_c phase composition.

Figures 2 (a) and (b) show the XRD patterns of films(a) and (b), respectively. Film(a) has the main peak at 2Θ of 31.0° and weak (00n)peaks of the low-T_c phase. The plane interval d calculated from the main peak angle is 2.88Å, which agrees with the results of the RHEED measurement. Assuming that the main peak can be assigned to the (117) peak of the low-T_c phase (*c*-axis lattice constant c=30.6Å), 2Θ is calculated to be 31.1° (d=2.87Å) and slope angle α of the *c*-axis to be 41.1°. On the other hand, film(b) has peaks at 31.9° and 35.2°, marked by the asterisks. The (00n) peaks were not observed in film(b). Assuming that the peak at 31.9° can be assigned to the (119) peak of the high-T_c phase (c=37.4Å), 2Θ is calculated to be 31.8°. However, film(b) did not show RHEED patterns, as mentioned above. From these results, we considered film(a) to consist mainly of a low-T_c phase with the (117) orientation, but the crystal structure of the film(b) is not clear.

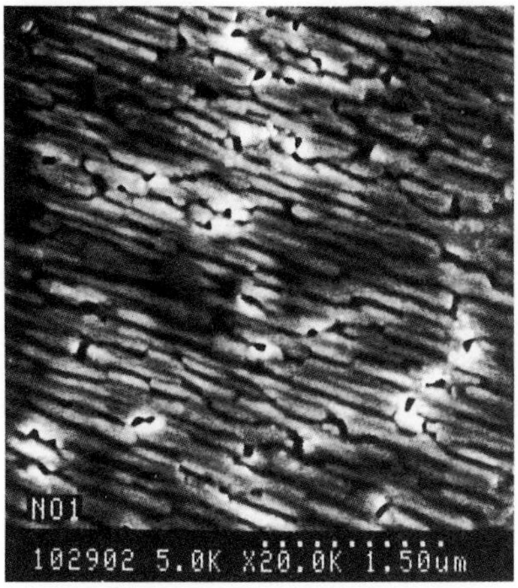

Fig.3. SEM micrograph of the surface of film(a).

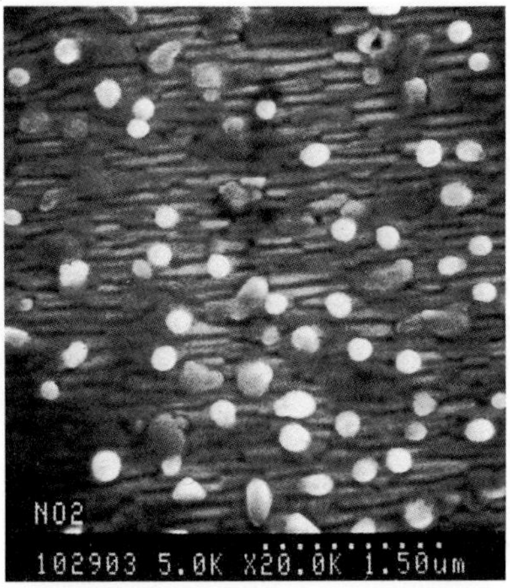

Fig.4. SEM micrograph of the surface of film(b).

Figures 3 and 4 show the surface morphology of films(a) and (b), respectively. In both figures, stripe-like textures are observed. The texture probably corresponds to the film orientation, because the *c*-axis oriented films have smooth surface textures. Particles, which are considered to be CuO, are also observed in film(b).

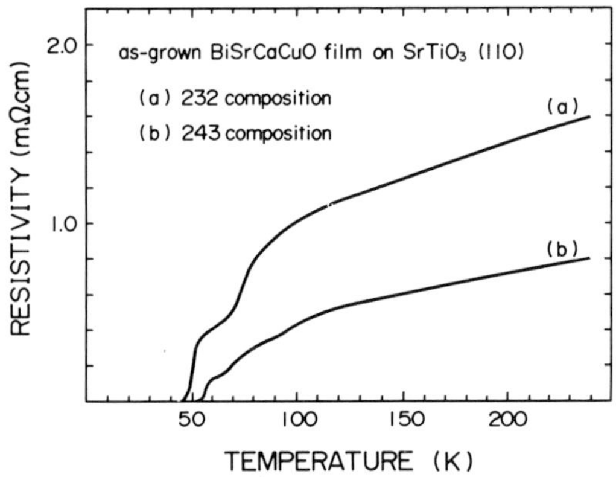

Fig.5 Resistivity vs. temperature curves of the films grown on $SrTiO_3$(110) substrates.

The room-temperature resistivity measured in the parallel direction to the stripe-like texture in Fig.3 was 1.8mΩcm, and that in the longitudinal direction was 3.0mΩcm. On the other hand, the resistivities in the parallel and longitudinal direction to the stripe-like texture in Fig.4 were 0.9mΩcm and 1.9mΩcm, respectively. These values are larger than the room-temperature resistivity (0.5mΩcm) of the *c*-axis oriented film which was formed on a $SrTiO_3$(100) substrate. The temperature dependences of the resistivities of films(a) and (b) are shown in Fig.5. Film(a) showed resistivity drops at about 100K, 80K and 55K. Zero resistivity was obtained at 45K; the resistivity showed the same temperature dependence in both the parallel and longitudinal directions to the stripe-like texture. On the other hand, the resistivity of film(b) dropped at 108K and 60K; zero resistivity occurred at 55K. These results suggest that the films formed on the $SrTiO_3$(110) substrates do not have a single phase.

CONCLUSIONS

The as-grown BiSrCaCuO thin films with a low-T_c phase composition and a high-T_c phase composition were formed at 660°C on the $SrTiO_3$(110) substrates by RF magnetron sputtering. From RHEED observation, the *c*-axis of the film with a low-T_c phase composition was found to meet the substrate surface at an angle of about 42°. It was concluded that the film with the low-T_c phase composition consists mainly of a low-T_c phase with (117) orientation. However, the crystal structure of the film with a high-T_c phase composition was not clear. The surface morphology corresponding to the film orientation was observed. The temperature dependence of the resistivity was different from that of the film formed on the $SrTiO_3$(100) substrate.

REFERENCES

1. Fukutomi M, Machida J, Tanaka Y, Asano T, Yamamoto T and Maeda H (1988) Jpn.J.Appl.Phys. 27: L1484
2. Asano H, Asahi M, Katoh K and Michikami O (1988) Jpn.J.Appl.Phys.27: L1487
3. Miura S, Yoshitake T, Manako T, Miyasaka Y, Shohata N and Satoh T (1989) Appl.Phys.Lett. 55: 1360
4. Kuroda K, Kojima K, Tanioku M, Yokoyama K and Hamanaka K (1989) Jpn.J.Appl.Phys.28: 1586
5. Kuroda K, Kojima K, Tanioku M, Yokoyama K and Hamanaka K (1990) Jpn.J.Appl.Phys. 29: L291
6. Kuroda K, Kojima K, Tanioku M, Yokoyama K, Kuroki H and Hamanaka K (1990) Jpn.J.Appl.Phys. 29: 1066
7. Hakuraku Y, Miyagi D, Higo S and Ogushi T (1990) Jpn.J.Appl.Phys. 29: L926
8. Yamamoto H, Katoh M, Satoh K, Nakamura T and Tanaka M (1990) Physica B 165&166: 1481
9. Satoh T, Yoshitake T, Miura S, Fujita J, Kubo Y and Igarashi H (1989) Appl.Phys.Lett. 55: 70.
10. Schlom D.M, Marshall A.F, Sizemore J.T, Chen Z.J, Eckstein J.N, Bozovic I, von Dessonneck K.E, Harris Jr J.S. and Bravmann J.C (1990) J.Cryst.Growth 102: 361
11. Sugimoto T, Yuhya S, Yamada Y, Hayashi H, Kikuchi K, Yoshida M, Sugawara K and Shiohara Y (1989) *Proc. 2nd Int. Symposium on Superconductivity* (ISS'89). Springer-Verlag, Tokyo, 1990, pp 907-910
12. Kuroda K, Kojima K, Wada O, Tanioku M, Yokoyama K and Hamanaka K (1990) Jpn.J.Appl.Phys. 29: L1816

Preparation of Bi-Sr-Ca-Cu-O Thin Films by Facing Targets Sputtering Method

MUNEYUKI IMAFUKU

R&D Labs-I, Nippon Steel Corporation, 1618, Ida, Nakahara-ku, Kawasaki, 211 Japan

ABSTRACT

Superconducting Bi-Sr-Ca-Cu-O films were prepared onto MgO(100) substrates by facing targets sputtering (FTS) method. Pb-free stoichiometric targets; i.e. Bi:Sr:Ca:Cu=2:2:2:3, were used here. The compositional ratios of the metallic elements in the films were found to be almost the same as those of the targets within the wide range of deposition temperatures except for the slightly decrease of the Bi content at relatively high temperatures. As-deposited films were, however, insulators. After annealing between 820°C and 865°C in air for several hours, I could obtain c-axis oriented films which were the mixture of so-called 2212 and 2223 phases. The superconducting property was sensitive to the annealing condition. The film exhibited a superconducting transition with an onset at 115K and zero resistance at 82K under the appropriate process condition.

KEYWORDS: facing targets sputtering method, compositional deviation, Bi-Sr-Ca-Cu-O, Pb-free, superconducting film

INTRODUCTION

Sputtering deposition methods are widely used techniques for the preparation of high Tc oxide superconducting thin films. Magnetron sputtering deposition is one of the most popular method among them. There exists, however, a dedicate problem in this method. It has been commonly observed that the composition of films prepared by this method deviates considerably from that of target[1,2]. This phenomenon is caused mainly by the selective resputtering of the deposited films by the high-energy negative oxygen ion flux which was emitted from the target during the deposition. It results the poor quality films since the superconducting properties are sensitively influenced by the chemical composition. Therefore, the adjustment of the target composition as to obtain the stoichiometric composition film is very important to produce good quality films. The amount of its compositional compensation depends on the apparatus and the deposition conditions.

Some people have been tried off-axis geometry sputtering[3,4] instead of the adjustment of the target composition. They mounted the substrate as not to face the target to avoid the direct negative ion flux . The facing targets sputtering (FTS) method which was originally developed by Hoshi et al.[5] can be considered as one of the most refined off-axis geometry sputtering methods since the negative ion flux is confined between the two facing targets at the same time that the sputtering plasma was confined with high ionization efficiency. Substrate is set outside of the space between the two targets. Consequently, the compositional deviations caused by the selective resputtering can be reduced to the minimum.

Takagi *et al.* applied FTS method for the preparation of Y-Ba-Cu-O films[6]. They could successfully obtain the superconducting as-deposited films having almost the same compositions of metallic elements as those of the target and having Tc=72K. In this study, I prepared superconducting Bi-Sr-Ca-Cu-O thin films by FTS method. The influences of deposition and annealing conditions on the composition and superconducting property of the films were investigated in detail.

EXPERIMENTAL

I don't mention here the geometry of the FTS system since the details were already described by Takagi *et al.*[6]. A pair of Pb-free stoichiometric oxide targets; i.e. Bi:Sr:Ca:Cu=2:2:2:3, were facing to each other in this study. These targets have the almost same composition of the metallic elements as that of so-called 2223 phase. They worked as the cathodes during the sputtering. MgO(100) single crystals were used for the substrates. The sputtering gas was pre-mixed 75% argon and 25% oxygen. The total sputtering pressure was 5.0 X 10^{-3} Torr. The applied RF power density was 1.4 W/cm^2. The deposition rate was about 40Å/min and the typical thickness of as-deposited film was about 0.5μm.

The substrate temperatures, T_S's, were varied from 70 to 700°C. After the depositions, these films were heated up at the annealing temperatures, Ta's, between 820 and 865°C for several hours (typically, 2 hours) in air. The annealing process was performed with heating rate of 100°C/min and with the quench-cooling from the annealing temperature to room temperature by using the infrared image furnace.

RESULTS AND DISCUSSION

The compositions of the targets and as-deposited films were determined by inductively coupled plasma atomic emission spectroscopy (ICPAES). The structures of the films were characterized by X-ray diffraction (XRD) method with Cu Kα radiation. The temperature dependence of the electric resistivity were measured by dc four-probe method using Ag paste contacts.

The atomic percent of Bi, Sr, Ca and Cu in the targets were proven to be 22.1, 22.0, 23.1 and 32.8, respectively. The compositions of the films are shown in Fig. 1 with various Ts's. We can see from this figure that the compositions of the films were almost same within the accuracy of ±3 atomic %. The composition deviations were fairly small compared with the films prepared by conventional magnetron sputtering method[1,2]. Therefore, we can say that the compositional deviations caused by the selective resputtering could be almost avoided by FTS method. The slightly decrease of the Bi content was also observed at Ts=500°C and 600°C. It means that the composition of the films are governed not only the resputtering rates but also

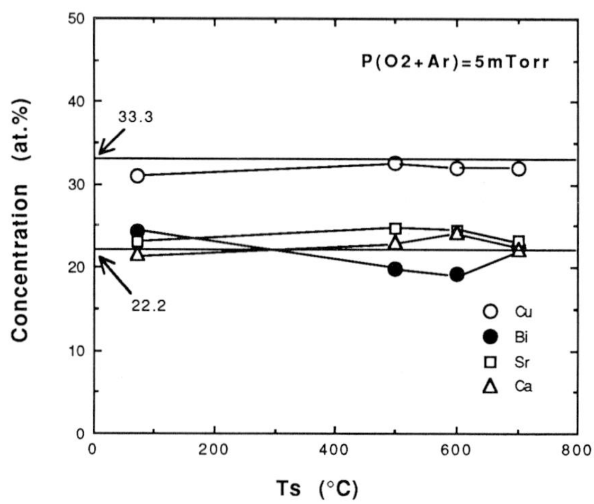

Fig. 1. Dependence of the atomic concentrations of Bi, Sr, Ca and Cu in as-deposited films on Ts.

other thermal properties (for instance, vapor pressures of the oxides of constituents). Altogether, the adhesion coefficient of the Bi or Bi-oxide may slightly decrease with increasing Ts in Bi-Sr-Ca-Cu-O system. The tendency like this was not observed in Y-Ba-Cu-O system[6]. I'm not sure, however, the reason why the content of the Bi recovered at Ts=700°C.

Nevertheless the stoichiometry of the metallic elements was achieved, as-deposited films were all insulators. The film deposited at 70°C was amorphous state and no features were observed by XRD. When Ts ≥ 500°C, only some small diffraction peaks from the oxides of constituents were observed by XRD . After the annealing for several hours between 820 and 865°C, these films were changed into superconducting phase(s). Figures 2(a) and 2(b) show the typical XRD patterns of the films annealed at 850°C for 2 hours which were deposited at Ts=70 °C and 500°C, respectively. These films have almost the same structure and was characterized as the c-axis oriented 2212 phase.

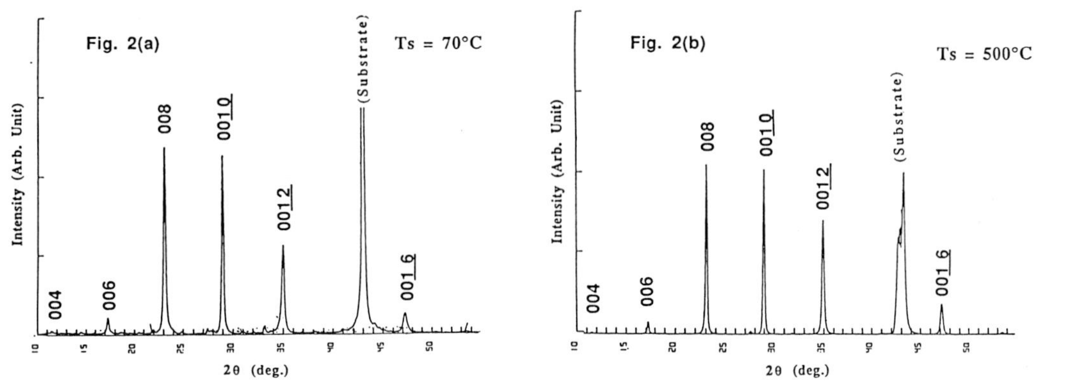

Fig. 2. The XRD patterns of Bi-Sr-Ca-Cu-O films after annealing at 850°C for 2 hours in air. (a): Ts=70°C, (b): Ts=500°C. In these figures, indices are for 2212 phase.

The film deposited at Ts=70°C was cracked and partially peeled off from the substrate after the annealing process. By contrast, the stacking of plate-like grains was observed for the films deposited at Ts ≥ 500°C. Similar grains have been often observed in Bi-Sr-Ca-Cu-O system[7]. Therefore, the measurements of electric resistivity could be performed for the latter films. The temperature dependence of electric resistivity for the film prepared at the condition of Ts=500°C and Ta=850°C is shown in Fig.3. Resistivity drop at 115K was observed and zero-resistance was achieved at 82K. Consequently, this film was considered to be the mixture of 2223 and 2212 phases. The amount of 2223 phase may be so little that it couldn't be distinguished by XRD.

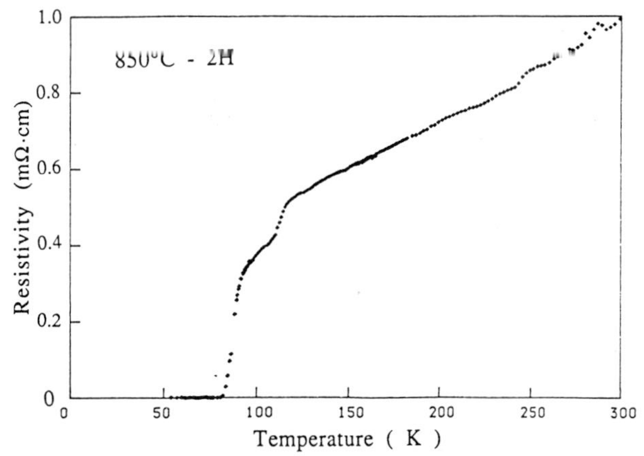

Fig. 3. Temperature dependence of electric resistivity of the film deposited at 500°C and annealed at 850°C for 2 hours.

The relationship between the superconducting property and annealing condition was also studied. Figure 4 shows the dependence of Tc(zero) on Ta for the film deposited at 500°C. We can see from this figure that the zero-resistance temperature, Tc(zero), was very sensitive to Ta. The films having Tc(zero) above the boiling temperature of nitrogen were achieved between 845°C and 860°C. In case of Ta ≥ 865°C, the growth of c-axis oriented 2201 phase besides 2212 phase was observed by XRD and Tc(zero) decreased rapidly with increasing Ta.

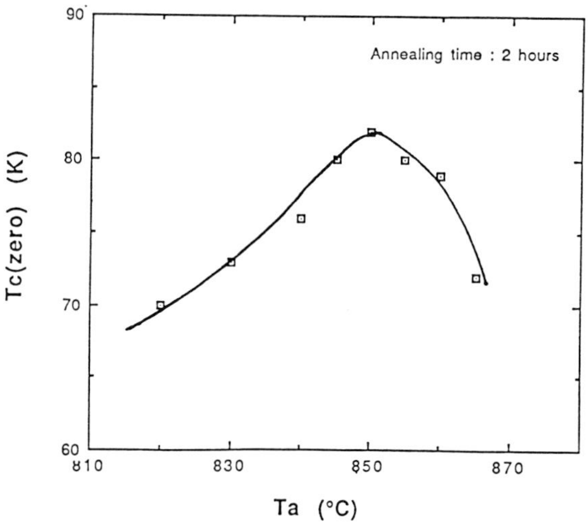

Fig. 4. Dependence of Tc(zero) on Ta. The films were deposited at Ts=500°C and annealed for 2 hours at various Ta's.

CONCLUSIONS

(1) Superconducting Bi-Sr-Ca-Cu-O films were prepared by FTS method. These films have almost the same compositions as those of the targets except for the slightly decrease of Bi content with increasing Ts.
(2) Though as-deposited films were insulators, the mixture of 2212 and 2223 phases were obtained after annealing at around 850°C in air for several hours.
(3) The zero-resistance temperature was very sensitive to the annealing temperature.

REFERENCES

1. Kadin AM and Ballentine TH (1989) *High temperature superconducting films by RF magnetron sputtering*. IEEE Trans. Magn. **25**: 2437-2440.
2. Ishibashi K, Hirata K, Kim K, Hosokawa N, Asamaki T,Kawakita K and Inada T (1989) *Mass spectrometric study in the sputtering of Y-Ba-Cu-O target*. Proc. of 2nd Workshop on High-Temperature Superconducting Electron Devices (R&D Association for Future Electron Devices, Hokkaido Japan)pp.39-44.
3. Terada N, Ihara H, Jo M, Hirabayashi M, Kimura Y, Matsutani K, Hirata K, Ohno E, Sugise R and Kawashima F (1988) *Sputter synthesis of $Ba_2YCu_3O_y$ as-deposited superconducting thin films from stoichiometric target -- a mechanism of compositional deviation and its control*. Jpn J. Appl. Phys. **27**: L639-L642.4.
4. Eom CB, Sun JZ, Yamamoto K, Marshall AF, Luther KE and Geballe H (1989) *In situ grown $YBa_2Cu_3O_{7-\delta}$ thin films from single-target magnetron sputtering*. Appl. Phys. Lett. **55**: 595-597.
5. Hoshi Y, Naoe M and Yamanaka S (1977) *High rate deposition of iron films by sputtering from two facing targets*. J. Appl.Phys. **16**: 1715-1716.
6. Takagi Y, Yamada and Koyama N (1990) *Preparation of thin-film oxide superconductor,$Y_1Ba_2Cu_3O_{7-\delta}$ by facing target sputtering deposition technique*. Proc. of 2nd Int. Symp. on Superconductivity(ISS'89) (Springer-Verlag, Tokyo)pp.777-780.
7. Shah SI, Jones GA and Subramanian MA (1988) *Growth and microstructure of Bi-Sr-Ca-Cu-O thin films*. Appl. Phys. Lett. **53**: 429-431.

Bi-Sr-Ca-Cu-O Thin Films Grown by RF Magnetron Sputtering

NORIKI HAYASHI[1], SHIGERU OKUDA[1], SATOSHI TAKANO[1], DAIJI MINAGUCHI[2], SEIICHIRO TERAI[2], and KIYOSHI HASEGAWA[2]

[1] Osaka Research Laboratories, Sumitomo Electric Industries, 1-3, Shimaya 1-chome, Konohana-ku, Osaka, 554 Japan

[2] Technical Research Center, The Kansai Electric Power Co.

ABSTRACT

Superconducting Bi–Sr–Ca–Cu–O films were prepared on MgO single crystalline substrates by rf magnetron sputtering. To increase Tc Pb doping was carried out by stacking PbO layer. After annealing films the Tc increased over 100 K and the maximum Jc of 4.9×10^5 A/cm^2 at 77.3K was obtained. Magnetic field dependence of the Jc was measured for two cases; magnetic field perpendicular and parallel to the a–b plane. For the former cases, a strong dependence of Jc on magnetic fields was observed. For the latter case the dependence was weaker (<1Tesla) and the maximum Jc at 1 Tesla was 2.2×10^5 at 77.3 K. We also attempted deposition on MgO polycrystalline substrates. The highest Tc of the annealed film was 95 K and the highest Jc was 2700A/cm^2 at 77.3 K and 6.8×10^4 A/cm^2 at 4.2 K.

KEY WORDS: Bi–Sr–Ca–Cu–O, sputtering, thin film, critical current density Jc, PbO doping,

INTRODUCTION

Many processes are being investigated to apply high Tc oxide superconductors to superconducting wires for power transmission lines and high field magnets[1]. We are developing a film process for high Tc oxide superconductors for superconducting wires. Bi–Sr–Ca–Cu–O superconductor discovered by Maeda et al.[2] is one of the most promising system among oxide superconductors. Bi–Sr–Ca–Cu–O superconductors with high Jc (>10^5A/cm^2 at 77.3K) were obtained only for thin films grown with $Bi_2Sr_2Ca_2Cu_3O_x$ (2223) phase by vapor phase deposition[3]. There have been many reports on the fabrication method for Bi–Sr–Ca–Cu–O thin films deposited on single crystalline substrates by vapor phase deposition[4] and few reports on characterization for high Jc films and the films on the polycrystalline substrates. We have prepared films on MgO single crystalline and polycrystalline substrates with c–axis orientation perpendicular to the substrate and measured their superconducting properties.

EXPERIMENTAL

Bi–Sr–Ca–Cu–O films were deposited by rf magnetron sputtering from single target. The target is the complex oxide with a composition of $Bi_4Sr_2Ca_2Cu_3O_x$ to get film composition of a $Bi_2Sr_2Ca_2Cu_3O_x$. The deviation of the film composition from stoichiometry was less than 10%. Single crystalline MgO(110) and poly crystalline MgO were used as substrate. The sputtering was performed in an Ar+20% O_2 atmosphere with a pressure 5–10mTorr. The substrates were kept at 710°C. After the deposition films were cooled down in an oxygen atmosphere. The thickness of the films was 0.4–0.7 μm measured by a profilometer. After taking out from the chamber, PbO layers were stacked on Bi–Sr–Ca–Cu–O films by sputtering without substrate temperature control in another chamber. The thickness of the PbO layers was about 1000A. The films were annealed for 0.5–1 hour at 840~850°C in 10%–20% O_2 atmosphere.
The composition of the films was determined by an ICP emission analyzer. Morphology was observed using scanning electron microscopy (SEM) and transmission electron micrograph(TEM). Film structures were determined by reflected high energy electron diffraction (RHEED) and X–ray diffraction measurements using Cukα radiation. X–ray pole figure measurements were also employed for determining the in–plane orientation of the film. The Schulz geometry was employed for the pole figure measurement. The conventional four probe technique was used for determining the zero resistance critical temperature and the critical current density.

RESULTS AND DISCUSSION

A. Films on single crystalline substrates

As-deposited films were shiny and smooth. X-ray diffraction pattern showed c-axis orientation perpendicular to the substrate with the 2223 phase. The observed broadness of the 2223 phase peak indicated a mixture of the major peak of the 2223 phase and the minor peak of the 2212 phase. Fig.1 shows the temperature dependence of the resistivity for the as-deposited films and the Tc of as-deposited films was around 70K.

To increase Tc Pb doping was carried out by stacking PbO layers on the film and then annealing in 10%–20% oxygen atmosphere at 840–850°C for 0.5–1 hour. Fig.2 shows a X-ray diffraction pattern of the annealed Pb-doping film. This pattern shows a c-axis orientation with two kind of phases ; 2223 and 2212 phase. Even the intensity of the 2212 phase (low Tc phase) was higher, annealed films showed a Tc as high as 100K.

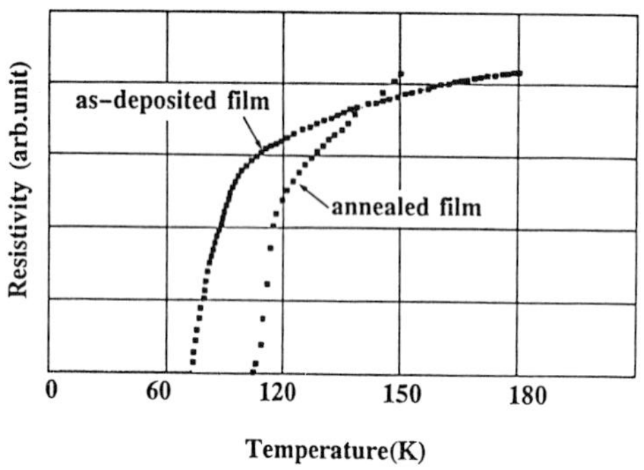

Fig.1 Temperature dependence of the resistivity of BiSrCaCuO films

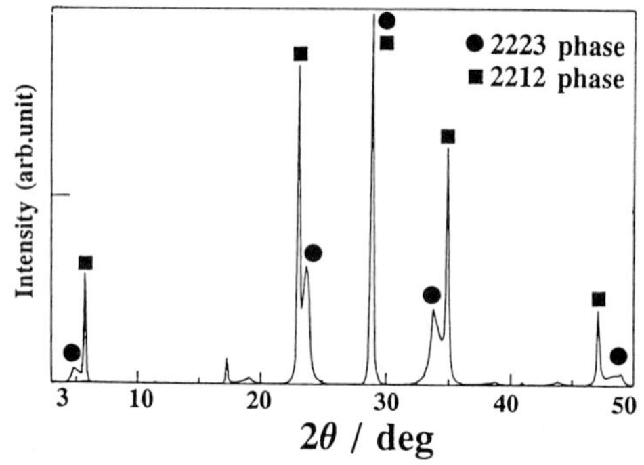

Fig.2 X-ray diffraction pattern of a annealed BiSrCaCuO film

Fig.3 shows the morphology of an annealed film. Flake like grains were piled up layer by layer and the degree of the union of grains was high. This suggests that the path for the superconducting current was not obstructed in the plane of the film. The maximum Jc of the annealed films was 4.9×10^5 A/cm^2 at 77.3 K.

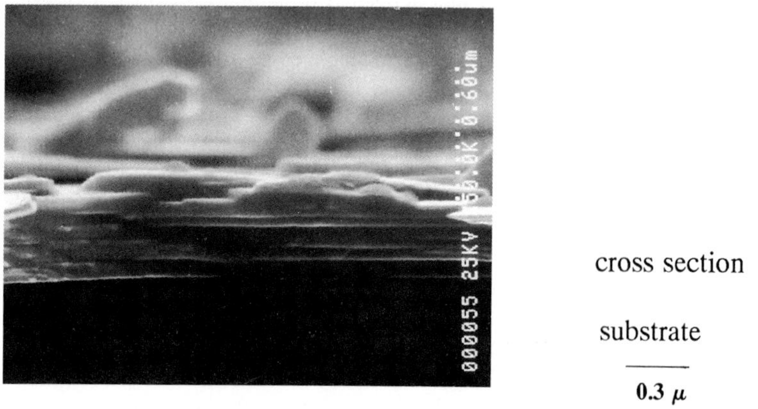

Fig.3 SEM of the cross section of an annealed BiSrCaCuO film

We investigated a-b plane orientation ofannealed films by RHEED ansd X-ray pole figure measurement. Fig.4 shows a RHEED pattern of annealed film. The ring pattern suggests that the surface of the film was oriented randomly in the a-b plane. Fig.5 shows the azimuthal scans of the (115) poles of the 2223 phase. The sample was tilted so that the (115) plane of the 2223 phase reflects the X-ray in the diffractometer plane. The azimuthal distribution of the pole intensity can be interpreted as a statistic of the in-plane orientation for grains of the film. The four major peaks located at every 90° of the pattern reveal the film had an orientation in a–b plane. The difference between the RHEED and the pole figure measurement is considered that the information from RHEED is limited for the extremely thin portion of the surface and the one from the pole figure measurement is for the whole part of the films and we can get average information from the pole figure measurement.

Fig.4 RHEED pattern of annealed film

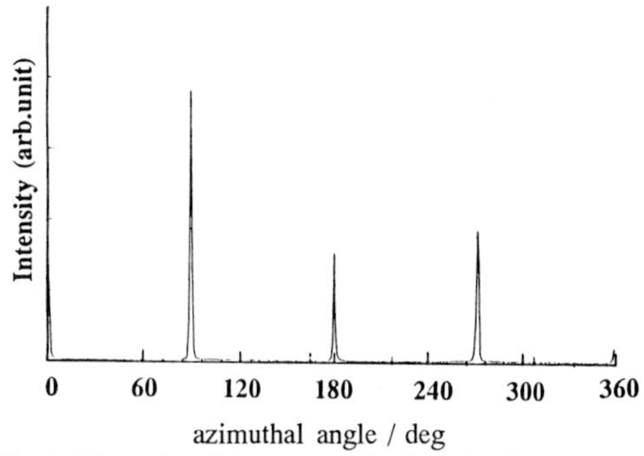

Fig.5 The azimuthal scans of (115) pole of the 2223 phase

The microstructure of the films were investigated by TEM analysis. Fig.6 shows the plane–view TEM of the as–deposited and annealed film. There were some extraneous phase (white portion in the photograph) in the matrix for the as–deposited film. There was not any other phase for the Bi–Sr–Ca–Cu–O system from electron diffraction pattern and the extraneous phase is considered as amorphous or 2212 phase or 2201 phase. Annealed film have much dislocation in the film compared to the as–deposited film and showed an in–plane orientation from the electron diffraction pattern.

(a)

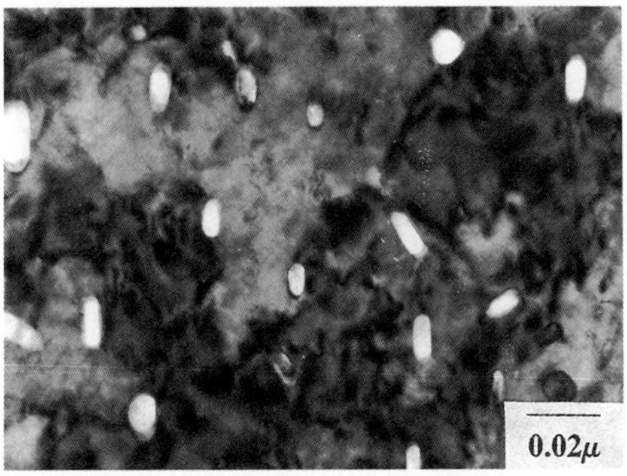

(b)

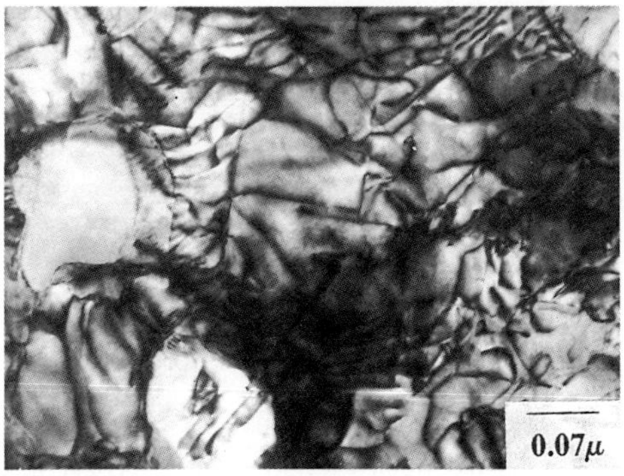

Fig.6 TEM of the plane–view of the BiSrCaCuO films. (a)as–deposited film (b)annealed film

Fig.7 shows the magnetic field dependence of the annealed film on Jc at 77.3 K. A sample with the Jc of $2.4x10^5 A/cm^2$ at 77.3 K and 0 Tesla was used for the measurement. The Jc in the magnetic field perpendicular to the film plane (a–b plane) decreased as magnetic field increased and the Jc at 1 Tesla

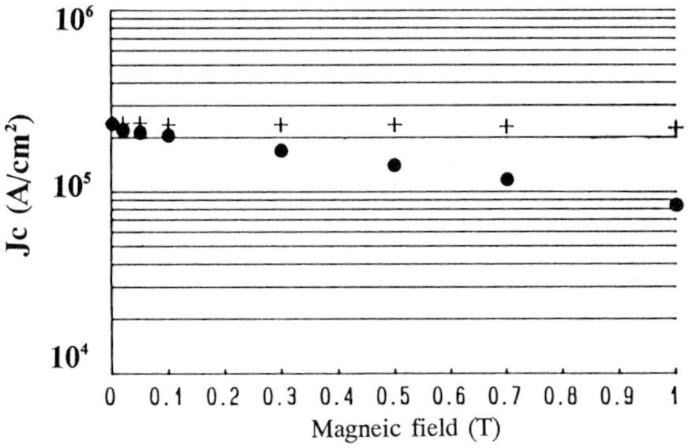

Fig.7 Magnetic field dependence of the annealed film

+ magnetic field parallel to the film plane

● magnetic field perpendicular to the film plane

was $8.4x10^4$ A/cm^2. This tendency of the dependence for magnetic field was different from sample to sample. When magnetic fields were applied parallel to the film plane, the dependence of the Jc was weaker up to 1 Tesla and this tendency was observed by other samples. The Jc at 1 Tesla was $2.2x10^5$ A/cm^2.

B. Films on polycrystalline substrates

We also attempted deposition on polycrystalline substrates and obtained a c–axis orientation perpendicular to the substrate similar to the single crystalline substrate. The post–annealing process was necessary to increase Tc of the films. For the case of the YSZ (yittria–stabilized zirconia) polycrystalline substrates, the Jc of the annealed films at 4.2 K decreased compared to the one of the as–deposited film, though the Tc increased. The elemental yttrium diminishes the superconductor properties reacting with Bi–Sr–Ca–Cu–O system [5]. This suggests that the BiSrCaCuO film reacted with the YSZ substrate during the annealing process. We chose polycrystalline MgO as the substrate. PbO doping and annealing was also carried out as the same way for the single crystal substrates.
Fig.8 shows the dependence of the Jc of the annealed film on the temperature. A Tc of 95 K and a Jc of 2700 A/cm^2 at 77.3 K and $6.8x10^4$ A/cm^2 at 4.2 K was obtained. From observation of the SEM of the annealed films, the surface of the annealed films was slightly rough compared to the films on the single crystalline substrates. We believe that more high Jc would be obtained by the improvement of the morphology of the annealed films.

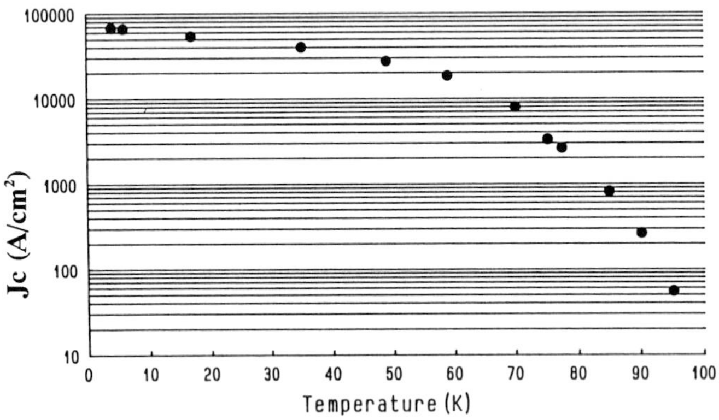

Fig.8 The dependence of the Jc of the annealed film on the temperature

In summary we have prepared Bi–Sr–Ca–Cu–O films on MgO single crystalline and polycrystalline substrates by rf magnetron sputtering. By stacking PbO layers on the films and annealing in the 10–20% oxygen atmosphere, the high Jc (>10^5A/cm^2 at 77.3K) films were obtained on the MgO single crystalline substrates. The films have a c–axis orientation with a mixture of 2223 phase and 2212 phase and an in–plane (a–b plane) orientation. The union of the grains of the films was high and a maximum Jc of $4.9x10^5$ A/cm^2 at 77.3 K was obtained. For the MgO polycrystalline substrate, a Tc of 95 K and a Jc of 2700 A/cm^2 at 77.3 K and $6.8x10^5$ at 4.2 K were obtained.

REFERENCES

1. Nakahara T (1989) In: Kitazawa K, Ishiguro T (eds) in Development of superconducting materials and impacts on the social economy. Advances in superconductivity. Springer–Verlag, Tokyo,pp 725–730
2. Maeda H, Fukutomi Y, Asano T (1988) Jap.J.Appl.Phys. 27: L209
3. Itozaki H, Higaki K, Harada K, Tanaka S, Yazu S, Tada K (1990) Sumitomo Electric Technical Review 29: 41
4. Nakamura K, Sato J, Ogawa K (1990) Jap.J.Appl.Phys. 29: L77
5. Kanai T, Kamo T, Matuda S (1989) Jap.J.Appl.Phys. 28: L551

Effect of High Gravitational Field on Superconducting Oxide Films

MAKOTO OKANO[1], HIROSHI TANAKA[2], KOH AGATSUMA[1], SHUICHIRO FUCHINO[1], and KOJI KAJIMURA[1]

[1] Electrotechnical Laboratory, 1-4, Umezono 1-chome, Tsukuba, 305 Japan
[2] Advanced Materials Research Laboratory, Tosoh Corporation, 2743-1, Hayakawa, Ayase, Kanagawa, 252 Japan

ABSTRACT

The effect of high gravitational field on superconducting properties was investigated on Bi(Pb)SrCaCuO thin films prepared by rf magnetron sputtering on MgO substrates. The gravity acceleration was applied to the samples during oxidation and annealing of the films in a rotating furnace in air. The high gravitational acceleration ranging from 1 G to 3500 G, where G is measured in a unit of 9.8 m/s^2, resulting from centrifugal force was produced by rotating the furnace with high rotational speed. The experimental results have shown a tendency that the superconducting critical temperature increased by applying the high gravitational acceleration.

KEYWORDS: Bi(Pb)SrCaCuO thin film, superconducting critical temperature, rotational furnace, gravity effect

INTRODUCTION

Among a great many works on superconducting oxides with high critical temperature, Tc, hydrostatic pressure effects on the superconducting properties have been investigated expecting to enhance Tc[1-9]. For example, Beille et al. [1] reported that the midpoint resistive transition temperatrue increased with pressure at a rate of +0.2 K/kbar for $Bi_2Sr_2CaCu_2O_{8+x}$ single crystals.

We attempted a new process in which annealing was made under high gravitational field. High gravitational field will produce the same effect as that of the pressure mentioned above. In addition, we expect extra effects such as uniaxial stresses which may change superconducting properties by applying the high gravitational acceleration.

In this paper, we present experimental results on the gravity effect on Tc of Bi(Pb)SrCaCuO thin films prepared by rf magnetron sputtering.

SAMPLE PREPARATION AND TEST APPARATUS

Bi(Pb)SrCaCuO superconducting thin films with thickness of 0.2 μm were prepared by rf magnetron sputtering with a powder-pressed $Bi_{1.8}(Pb_{0.4})Sr_2Ca_{1.9}Cu_{3.3}O_x$ target. Polished MgO(100) single crystals with dimensions of 10 × 10 × 0.5 mm^3 were selected as substrates. A 60 W input power applied to the 125-mm diam. target gave a sputtering rate of about 40 nm/hr. During the sputtering, pure Ar gas with a pressure of 0.5 Pa was introduced. The gravitational acceleration was applied to the samples during oxidation and annealing in a rotating furnace in air. The high gravitational field resulting from centrifugal force was produced by rotating the furnace with high rotational speeds. The samples were annealed at temperatures between 800 ℃ and 860 ℃ for an hour. The applied gravitational acceleration ranged from 1 G to 3500 G, where G is the gravitational acceleration relative to 9.8

m/s². The gravitational acceleration was applied perpendicular to the film surface. The samples cut out of the sputtered substrates and annealed in the small rotational furnace had dimensions of 2.5 × 10 × 0.5 mm³.

Resistivity measurements were made by a standard four-probe method with a current of 0.1 mA. Composition of the sputtered films was determined by an electron probe microanalysis(EPMA). In addition, scanning electron microscope (SEM) not to be shown in this paper and x-ray diffraction(XRD) were used to observe morphology and to determine the crystal structure, respectively.

Figure 1 shows a cross-section of the rotational furnace for applying the gravitational acceleration to the sample. The rotational furnace was made of ceramics. The sample in the holder located 45 mm apart from the center of the rotor was heated up to a required temperature by two ceramic heaters during high speed rotation. The temperature measurment of the sample holder and the input power supply to the heaters were made through a slip ring with six electrodes. Four electrodes were used to measure the temperature using the four-probe method and two were used to supply the controlled heating electric power.

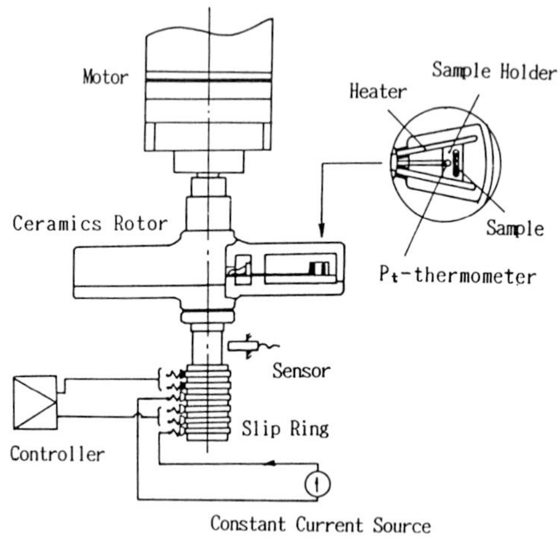

Fig. 1 Cross-section of rotational furnace for applying the gravitational acceleration to samples.

RESULTS AND DISCUSSION

The nominal compositions of prepared films determined by EPMA are indicated in Table 1.

Table 1. Typical compositions of films prepared under various gravitational accelerations.

	annealing conditions		applied gravitational acceleration	compositions of the films				
	T(℃)	t(hr)	g(G)	Bi	Pb	Sr	Ca	Cu
Sample 0	as-sputtered film		–	1.42	0.37	2.00	2.23	3.38
Sample 1	830	1	1	1.48	0	2.00	2.08	3.37
Sample 2	830	1	509	1.39	0	2.00	2.35	3.70
Sample 3	830	1	1923	1.34	0	2.00	2.21	3.78

The zero-resistance temperature, Tc_0, of these crystallized Bi(Pb)SrCaCuO films are shown against the magnitude of applied gravitational acceleration and the annealing temperature in Fig.2(a) and (b), respectively. The results under 1 G were obtained by annealing without rotation using the same furnace. The maximum value of Tc_0 under 1 G was 68 K in the samples annealed at temperatures between 800 ℃ and 860 ℃. Although Tc_0 of the films annealed in the gravitational field are scattered as in the case of 1 G, many films with Tc_0 above 68 K were obtained. In particular, the film annealed at 830 ℃ under the gravitational acceleration of 1923 G showed Tc_0 of 84 K. The annealing temperature which gave higher Tc_0 in this experiment was 830 ℃ as seen in Fig.2(b). These results suggest the existence of optimum condition of annealing temperature and applied gravitational acceleration to obtain higher Tc.

Figure 3 shows the temperature dependence of the resistivity normalized by that at 0 ℃ for samples annealed at 830 ℃ and various gravitational accelerations. These are typical results which were obtained in a series of expriments performed using the samples cut out of the same sputtered lot.

XRD patterns for the samples under 1 G, 509 G and 1923 G were shown in Fig.4(a), (b)and (c), respectively. The result shows that all the films have 2212 phase with the c axis of the superconductors oriented normal to the surface of MgO substrate. It should be noted that the intensities of reflections from the films relative to substrate MgO(200) reflection are stronger for the films prepared under gravitational acceleration than those under 1 G. The correlation of these results and the increase in Tc_0 is under examination.

As described above, Tc_0 of the film annealed under the high gravitational acceleration clearly incresed compared with that of the film under 1G. We presumed that the increase in Tc_0 in this experiment was mainly due to gravity effects, but other causes such as an effect of acceleration change due to vibration occured during rotation of the furnace are possible.

CONCLUSION

We have observed the gravity dependence of Tc for Bi(Pb)SrCaCuO superconducting thin films prepared by rf magnetron sputtering on MgO substrates. The present results have shown a tendency that the superconducting

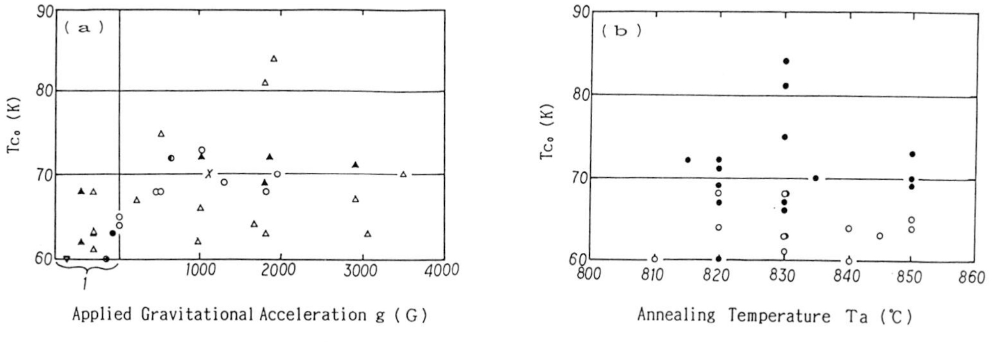

Fig.2(a) Zero-resistance critical temerature as a function of applied gravitational acceleration at various annealed temperatures. (○)850℃, (●)845℃, (◑)840℃, (×)835℃, (△)830℃, (▲)820℃, (◐)815℃, and (▽)810℃. Fig.2(b) Zero-resistance temperature versus annealing temperature under various gravitational accelerations. (○)1G, and (●)high gravitational acceleration.

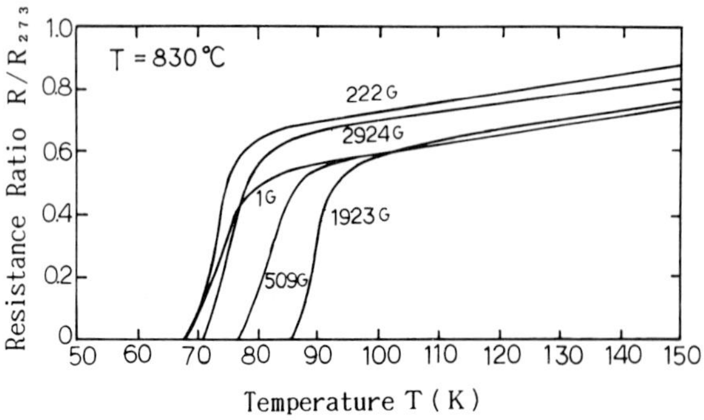

Fig. 3 Temperature dependence of the resistivity normalized by the resistance at 0℃ for the films annealed at 830℃ for an hour under various gravitational accelerations. 1 G, Tc_0=63 K; 222 G, Tc_0=63 K; 509 G, Tc_0=75 K; 1923 G, Tc_0=84 K; 2924 G, Tc_0=67 K.

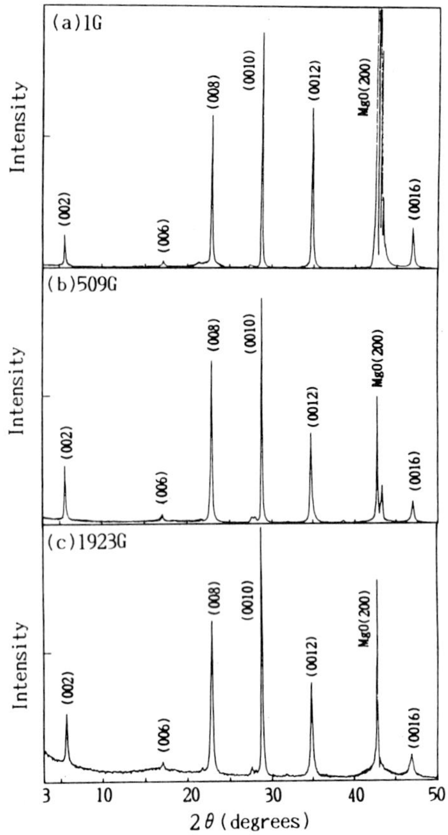

Fig. 4 X-ray diffraction pattern of films prepared on MgO substrates, (a)1 G, (b)509 G and (c)1923 G. Unmarked peaks were not identified.

critical temperature incresed by applying the gravitational acceleration. From now on, we wish to further carry out the experiment at higher gravitational acceleration, such as a few tens of thousands G.

We would like to thank K.Kawaguchi,Y.Nakagawa and K.Endo in the Electrotechnical Laboratory for helpful discussion.

REFERENCES

[1] Beille J, Dupendant H, Laborde O, Lefur Y, Perroux M, Tournier R (1988) Physica C 156: 448-454
[2] Sato M, Onoda M, Shamoto S, Hosoya S, Maruyama Y (1987) Chemistry Letters : 1039-1042
[3] Yoneda T, Mori Y, Akahama Y, Kobayashi M, Kawamura H (1990) Jpn.J.Appl.Phys. 29: L1396-L1398
[4] Murayama C, Mori N, Yomo S, Takagi H, Uchida S, Tokura Y (1989) Nature 339 : 293-294
[5] Yoshida H, Morita H, Noto K, Kaneko T, Fujimori H (1987) Jpn.J.Appl.Phys. 26: L867-L868
[6] Yamada Y, Matsumoto T, Kaieda Y, Mori N (1990) Jpn.J.Appl.Phys. 29: L250-L253
[7] Shun-hui H, Yu-lin Z, Han H, Yong L (1988) Physica C 156: 113-115
[8] Kaldis E, Fischer P, Hewat A.W, Hewat E.A, Karpinski J, Rusiecki S (1990) Physica C 159: 668-680
[9] Hewat A.W, Fischer P, Kaldis E, Karpinski J, Rusiecki S, Jilek E (1990) Physica C 167: 579-590

Preparations of Flat 80K-Phase Bi-Sr-Ca-Cu-O Superconducting Thin Films

HIDEAKI HIGASHI, SATORU KISHIDA, HEIZO TOKUTAKA, KATSUMI NISHIMORI, and NAGANORI ISHIHARA

Department of Electrical and Electronic Engineering, Tottori University, Koyama, Tottori, 680 Japan

ABSTRACT

The Bi-Sr-Ca-Cu-O films were prepared by a rf magnetron sputtering from a single target. The films were deposited using the mixed powder target with the composition of Bi:Sr:Ca:Cu = 2:2:1:0.5 at the substrate temperature of 320～500℃, followed by post-annealing at 865℃ for 4h and at 780℃ for 1h in air. When the film was deposited at the substrate temperature of 380℃ and post-annealed, the film with a very flat surface, covering all over the substrate was obtained. The specimen showed the zero-resistance at ～83K. It was confirmed that the c-axis of the film was highly oriented normal to the substrate surface from X-ray diffraction pattern.

KEY WORD : Bi-system superconducting thin film, rf magnetron sputtering, flat surface, mixed powder target, substrate temperature

1. INTRODUCTION

In order to make devices using the Bi-Sr-Ca-Cu-O superconducting materials [1], it is necessary to make thin films of the materials. From application point of view, the factors such as high critical temperature, high critical current density, surface flatness and interface stability are very important. Up to this time, the films with the most favorable value of critical temperature [2] and critical current density [3] have already been obtained by other workers. However, the films with flat surface have not yet been prepared. In order to prepare the devices with sandwich-type junctions, it is very important to prepare the films with flat surfaces.

In this paper, the optimum preparation conditions of the Bi-Sr-Ca-Cu-O films with flat surface and single phase were investigated by varying the substrate temperatures (T_{sub}) and target compositions.

2. EXPERIMENTAL

The Bi-Sr-Ca-Cu-O thin films were prepared by a rf magnetron sputtering from a single target. The targets with the cation ratios of Bi:Sr:Ca:Cu = 2:2:x:0.5 (x=0.5～1.5) were prepared by mixing the powders of Bi_2O_3, $SrCO_3$, $CaCO_3$ and CuO. Air-cleaved MgO (100) were used as a substrate. The T_{sub} were varied in the range of 320～500℃. The sputtering was carried out in argon and oxygen mixed atmosphere (Ar/O_2=2/1) of 3mTorr. The input rf power was 75W. The distance between the substrate and the target was 60mm. Since the as-deposited thin films were amorphous, they were post-annealed at 865℃ for 4h and at 780℃ for 1h in air. In order to clarify the superconductivity, the crystal structure, the surface morphology and the surface composition of the prepared specimens, resistance-temperature (R-T) characteristics, X-ray diffraction (XRD), X-ray photoelectron spectroscopy (XPS) and scanning electron microscope (SEM) measurements were carried out.

3. RESULTS AND DISCUSSION

3.1 THE DEPENDENCE OF THE SUBSTRATE TEMPERATURE

Figure 1 shows the R-T characteristics of the specimens prepared using the target with the composition of Bi:Sr:Ca:Cu = 2:2:1:0.5. The specimens of (a), (b), (c), (d) and (e) were prepared at the T_{sub} of 340℃, 380℃, 400℃, 450℃ and 500℃, respectively. The specimens of (a) and (e) showed the high resistance in the normal state and the superconducting transition at ~80K. This suggests that the excess components cause the impurities in the films. On the contrary, the specimens of (b), (c) and (d) showed the metallic temperature dependence of the resistance in the normal state, the sharp superconducting transition, and the zero-resistance at ~83K. From the results, the specimens of (b) to (d) have the good superconducting properties.

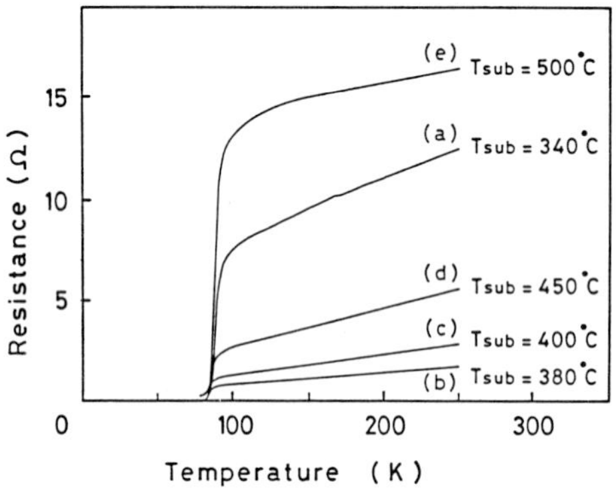

Fig. 1 The R-T characteristics of the Bi-Sr-Ca-Cu-O films prepared at various T_{sub} of 340~500℃.

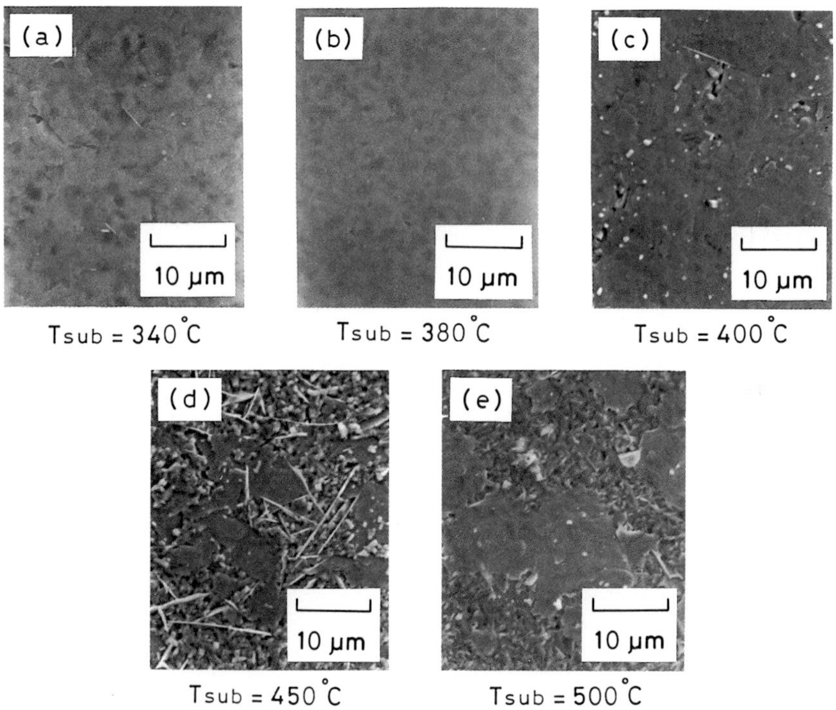

Fig. 2 The SEM photographs of the Bi-Sr-Ca-Cu-O films prepared at various T_{sub} of 340~500℃.

Figure 2 shows the SEM photographs of the specimens of (a) to (e). The specimens of (c) to (e) ($T_{sub} \geq 400$℃) were covered with the plate-like crystals and the whity precipitations were observed on the plate-like crystals as well as at the crystal boundaries. On the contrary, as shown in Fig.2(a) and (b) ($T_{sub} \leq 380$℃), the precipitations were not observed and the plate-like crystals were observed all over the substrate. In the result of the specimen of (b), light and dark areas were observed and any grain boundaries were not observed. The specimen showed the best surface morphology in the prepared specimens of (a) to (e). From the results of XRD patterns, the c-axes of all the films were oriented normal to the substrate surface. The specimen of (b) contained almost no impurities and was the single 80K-phase. The specimens of (c) to (e) ($T_{sub} \geq 400$℃), however, showed that the XRD peaks due to the 80K-phase were weak and the XRD peaks from CuO were strong. This may cause the discontinuity of the plate-like crystals and the existence of the precipitations on the plate-like crystals and at the crystal boundaries, as shown in Fig.2(c) to (e).

We investigated the surface compositions of the as-deposited thin films prepared at the T_{sub} of 320～500℃. Figure 3 shows the relative XPS intensities of the films. From the results, the surface compositions of the specimens vary with the T_{sub}. With increasing the T_{sub}, the Bi relative intensities seem to decrease and the Cu relative intensities seem to increase.

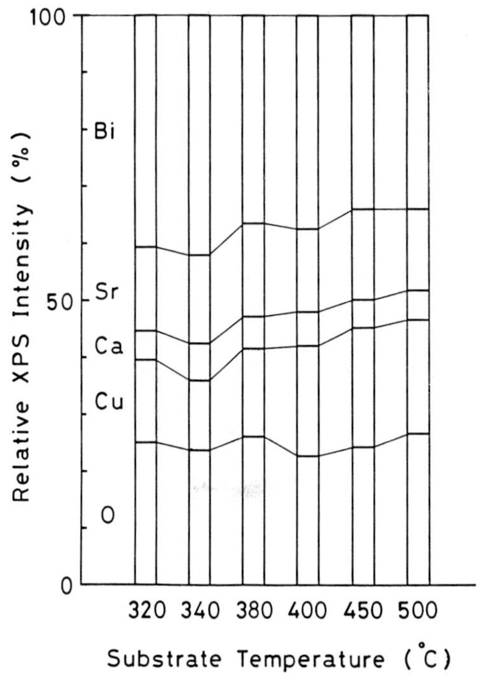

Fig.3 The relative XPS intensities of the as-deposited films prepared at various T_{sub} of 320～500℃.

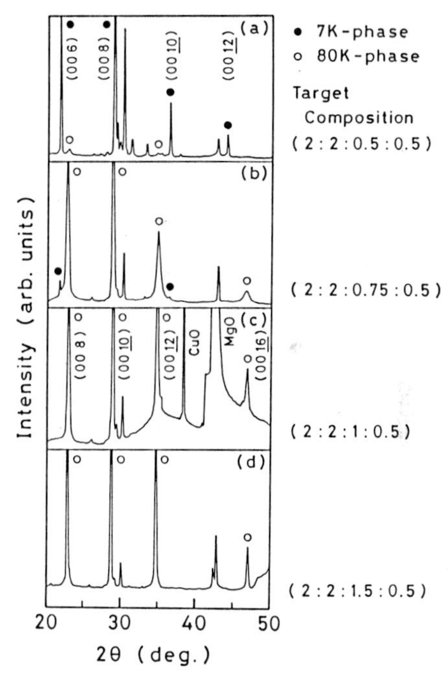

Fig.4 The XRD patterns of the Bi-Sr-Ca-Cu-O films prepared using the targets with the various Ca fractions.

3.2 THE DEPENDENCE OF THE Ca FRACTION IN THE TARGET

As shown in Fig.2(c), on the surface of the specimens prepared at the T_{sub} of 400℃, the whity precipitations were observed. In order to clarify the origins of the chemical species and to obtain the films with more flat surface, we prepared the films using the target compositions with the vaious Ca fractions. The T_{sub} was fixed at 400 ℃ and the Ca fractions in the target were changed from 0.5 to 1.5. The R-T characteristics of the specimens were measured. The specimen prepared at the Ca fraction of 0.5 showed the semiconductor-like temperature dependence of the resistance in the normal state and did not show the zero-resistance. The specimen prepared at the Ca fraction of 1.5 showed a rather high resistance in the normal state and the zero-resistance at ~80K. On the contrary, the specimens prepared at the Ca fractions of 1 (standard sample) and 0.75 showed the metallic temperature dependence of the resistance in the normal state, whose zero-resistances were always above 77K.

Figure 4 shows the XRD patterns of the specimens. The dominant superconducting phases of the specimens vary largely with the Ca fractions in the target compositions: with increasing the Ca fractions, the superconducting phases of the specimens changed from 7K-phase (●) to 80K-phase (○). In the specimen prepared at the Ca fraction of (a) 0.5, the 7K-phase were dominant. This may be caused by the lack of Ca.

Figure 5 shows the SEM photographs of the specimens. As shown in Fig.5(a), the precipitations were observed on the plate-like crystals. The precipitations may be produced by excess Ca, where the 7K-phase was dominant in the specimen (see Fig.4(a)). On the contrary, the specimen prepared at the Ca fraction of (b) 0.75 showed the good surface morphology without any precipitations. As shown in Fig.4(b), the XRD peak due to CuO was not observed in the specimen. However, the specimen included very small amount of 7K-phase.

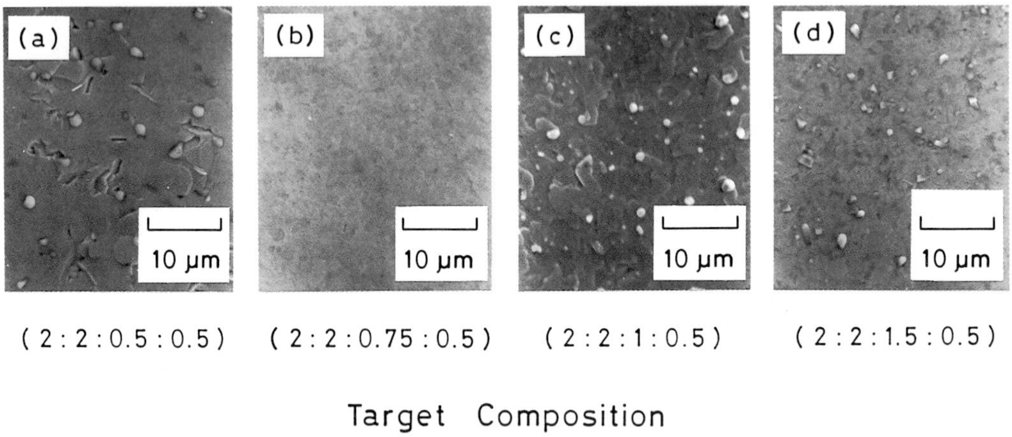

Fig.5 The SEM photographs of the Bi-Sr-Ca-Cu-O films prepared using the targets with the various Ca fractions.

In summary, the Bi-Sr-Ca-Cu-O thin films were prepared by a rf magnetron sputtering using the mixed powder target, and then, post-annealed at 865℃ for 4h and at 780℃ for 1h in air. In order to clarify the optimum preparation conditions of the thin films with flat surface and single phase, we investigated the effects of the T_{sub} during deposition and the Ca fraction in the target on the surface morphology of the thin films. The thin film with the very flat surface without any precipitations was obtained at the T_{sub} of 380℃. We found that the surface compositions of the thin films varied by changing the T_{sub} during deposition. Moreover, we prepared the thin films using the targets of the various Ca fractions. The prepared specimens contained 80K- and 7K-phase. From the results, the precipitations on the plate-like crystals may be the compound containing Ca.

REFERENCES

1. Maeda H, Tanaka Y, Fukutomi M, Asano T (1988) Jpn.J.Appl.Phys.27: L209
2. for example, Hakuraku Y, Higo S, Ogushi T (1989) Appl.Phys.Lett.55: 1569
3. Itozaki H, Higaki K, Harada K, Tanaka S, Fujimori N, Yazu S
Advances in Superconductivity, Nagoya, 1988, ed. Kitazawa K, Ishiguro T (Springer-Verlag, Tokyo, 1989): p.599

(100) Oriented BSCCO Thin Film by RF Magnetron Sputtering

HISAO HATTORI, HIDEO ITOZAKI, and SHUJI YAZU

Itami Research Laboratories, Sumitomo Electric Industries, Ltd., 1-1, Koya-Kita 1-chome, Itami, 664 Japan

ABSTRACT

As-grown superconducting Bi-Sr-Ca-Cu-O thin film with fair (100) orientation was obtained on a MgO(100) substrate. Tc(R=0) of the film was 69K, and the transition width was 5K. Preferred orientation of the film was able to be controlled by the substrate temperature during deposition. Preferred orientation changed from (001) to (100) with decrease in the substrate temperature. This tendency is consistent with the one observed in the $YBa_2Cu_3O_7$ thin film.

KEY WORDS: Bi-Sr-Ca-Cu-O, thin film, sputtering, preferred orientation

INTRODUCTION

One of the unique features of a high Tc superconductor is its anisotropy. Electrical conductivity parallel with the Cu-O planes which are considered as the path of conduction is from two to five orders of magnitude greater than that in the *c*-direction. Superconducting coherence length along the *a-b* plane is also five times or more larger than that of *c*-axis direction. In such an anisotropic material, control of preferred orientation is very important in its application. For example, it has not been reported the observation of Josephson current in a sandwich-type Josephson tunnel junction. Films whose *a-b* plane is perpendicular to the substrate are desirable for this. It is expected that the larger coherence length along the *a- b* plane enables the fabrication of well-behaved Josephson junction. In the fabrication of $YBa_2Cu_3O_7$ (YBCO) thin film, preferred orientation is successively controlled, and (100), (110), and (001) oriented films are prepared by various methods [1-3]. In the Bi-Sr-Ca-Cu-O (BSCCO) system, however, only (001) oriented film has been reported, and preferred orientation of the BSCCO film has not well been investigated. We have examined the relation between preparation conditions and growth of the film using the RF magnetron sputtering method. We have mainly focused on the relation between the preferred orientation of the film and the substrate temperature during deposition. we have achieved as-grown superconducting film which is fairly (100)-oriented.

EXPERIMENTAL

Bi-Sr-Ca-Cu-O thin films were prepared by the 90° off-axis RF magnetron sputtering method. Films were prepared on the MgO (100) substrate heated at 550 to 680°C, using a Bi-Sr-Ca-Cu-O target. The cationic ratio of the target was Bi:Sr:Ca:Cu=1.75:2:1:1.75. The sputtering gas

Table. 1. Deposition condition

target	Bi-Sr-Ca-Cu oxide
target composition	Bi:Sr:Ca:Cu=1.75:2:1:1.75
substrate	MgO(100)
substrate temperature	540-620°C
sputtering gas	$Ar+20\%O_2$
gas pressure	0.05torr
deposition rate	0.06nm/s

was Ar(20%O_2) of 0.05torr. Typical deposition conditions are shown in Table.1. X-ray diffraction (XRD), scanning electron microscopy (SEM), and dc electric resistance measurement were performed on the films to study crystal structure, surface morphology and superconducting properties. Chemical composition of the film was examined by inductive coupled plasma spectroscopy (ICP).

RESULTS AND DISCUSSION

Fairly (100) oriented film with Tc(R=0)=69K has been grown at the substrate temperature of 570C. Figure 1 shows the XRD pattern of the film. As different orientation peaks from (100) are also observed, the film is not perfectly oriented. Most of the diffraction peaks are assigned to (*h*00) and (00*l*) of Bi-Sr-Ca-Cu-O superconductor with a=0.543nm and c=3.79nm. (*h*00) peaks' intensity is very strong compared with (00*l*) peaks.The chemical composition of the film was about $Bi_{1.7}Sr_2Ca_{0.9}Cu_{2.2}O_y$, that is near 2212, but XRD shows the existence of 2223 phase. It is possible to assume that (100) oriented grain is 2212 phase, because the lengths of *a* or *b* axes of the 2212 and 2223 phases are nearly equal to each other.

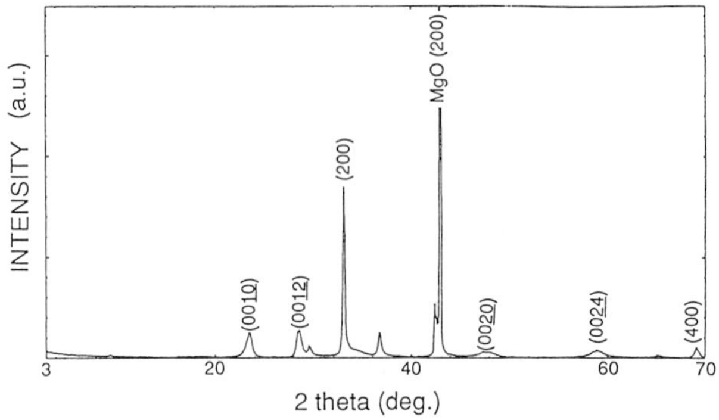

Fig. 1. XRD pattern of (100) oriented thin film. Most of peaks are assigned to (*h*00) or (00*l*) diffraction of BSCCO.

Temperature dependence of the XRD pattern of the film is shown in Fig.2. Most of the peaks are assigned to (*h*00) and (00*l*) of Bi-Sr-Ca-Cu-O superconductor with a=0.543nm and c=3.79nm. With an increase in the substrate temperature, the intensity of (00*l*) peak increases. In other words, by decreasing the substrate temperature from the condition of (001) oriented film well grown, (100) orientation becomes dominant. A similar tendency is also observed in the fabrication of YBCO thin film. A decrease in the substrate temperature results in the crystal orientation of which *c*-axis and substrate become parallel.

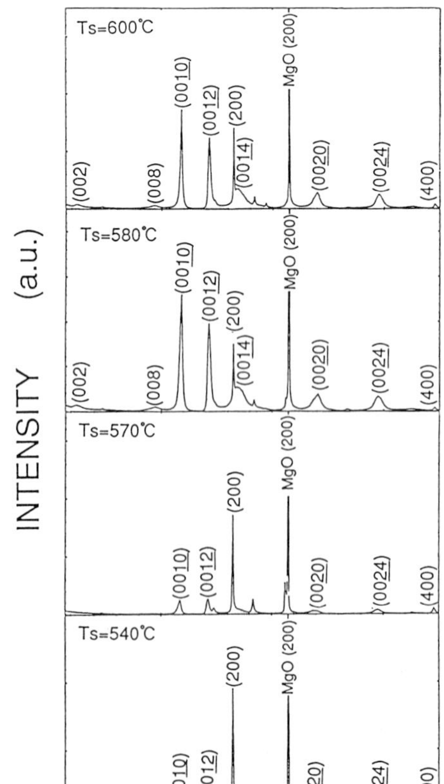

Fig. 2. Substrate temperature dependence of XRD pattern of the films.

Fig. 3. Surface morphology of (100) oriented thin film. This photograph was taken with a 20 degrees tilt from the normal of the surface.

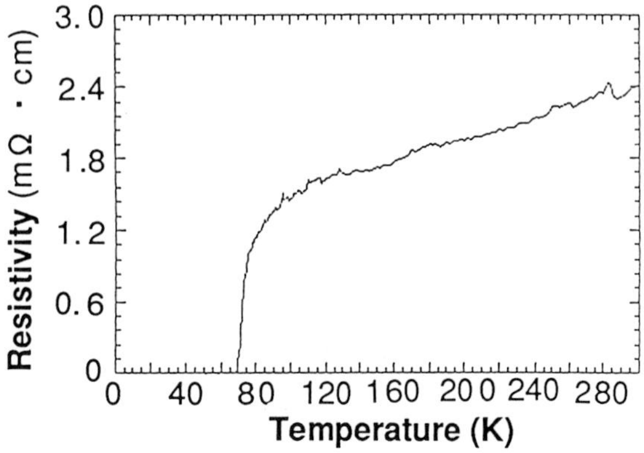

Fig. 4. Temperature dependence of the resistivity of (100) oriented thin film.

Figure 3 shows the surface morphology of the film. This photograph is taken with a 20 degrees tilt from the normal of the surface. The film consists of many grains. Its size is in the range of 200 to 500nm. The shape of each grain is like stacked flakes. The flat planes of the flakes that are perpendicular to the substrate are considered to be *c*-planes of the BSCCO.

Electric properties of the film shown in Fig.1 and 3 were measured. The temperature versus resistivity curve of the film is shown in Fig.4. Its Tc(R=0) is 69K, and the transition width is about 5K. The resistivity is about two times larger than that of the post-annealed (001) oriented film with high critical current density.[4] From the XRD measurement, there exist not only (100) oriented grains but also (001) grains. Therefore, whether this superconducting property originates from (100) oriented grain is not clear.

Temperature dependence of preferred orientation of YBCO and BSCCO is similar. This common feature observed in YBCO and BSCCO systems is considered to be from the anisotropic crystal structure of high Tc superconductor. Temperature dependence of preferred orientation of high Tc superconductor thin film may be understood as follows: At first we assume a matter. That is, growth speed of the crystal along the *a-b* plane is larger enough than that along the *c*-axis. This assumption is reasonable, considering the result of the observation of crystal growth in BSCCO bulk [5]. BSCCO crystal of which shape is like a plate grows planarly. The face of the plate is *a-b* plane. We interpret that in the lower substrate temperature region, atoms reaching to the surface of the substrate can not get enough energy for migration from the substrate. So the atoms on the surface do not spread planarly, and crystal grows perpendicular to the surface. That is (100) or (110) orientation tends to occur. In the higher substrate temperature region, enough energy for migration enables the layered growth of the crystal. That is the film tends to show (001) orientation.

CONCLUSION

We have investigated the controllability of preferred orientation of the BSCCO thin film by RF magnetron sputtering. As-grown superconducting Bi-Sr-Ca-Cu-O thin film with fair (100) orientation has been made on MgO(100). Tc(R=0) of the (100) oriented film is 69K, and the transition width is about 5K. Preferred orientation of the film can be controlled by changing the substrate temperature during deposition. We found fairly (100)-oriented films grew at a rather lower temperature range, between 540 to 570°C, than that in which (100) oriented film preferably grew. This tendency is consistent with the substrate temperature dependence of preferred orientation observed in the YBCO thin film.

REFERENCES

1. Terashima Y, Sagoi M, Kubo K, Mizutani Y, Miura T, Yoshida J, Mizushima K (1989) Jpn. J. Appl. Phys. 28: L653
2. Terashima T, Bando Y, Iijima K, Yamamoto K, Hirata K (1988) Appl. Phys. Lett. 53: 2232
3. Hase T, Izumi H, Ohta K, Suzuki K, Morishita T, Tanaka S (1990) J. Appl. Phys. 68: 374
4. Itozaki H, Higaki K, Harada K, Tanaka S, Fujimori N, Yazu S (1988) In: Kitazawa K, Ishiguro T (eds) in Advances in Superconductivity. Springer-Verlag, Tokyo, pp 599-604
5. Hatano T, Aota K, Ikeda S, Nakamura K, Ogawa K (1988) Jpn. J. Appl. Phys. 27: L2055

Superconducting Multi-layers of $Bi_2Sr_2CaCu_2O_y/Bi_2Sr_2CuO_x$ Grown by RF Magnetron Sputtering

KENTARO SETSUNE, TOMOAKI MATSUSHIMA, YO ICHIKAWA, HIDEAKI ADACHI, and KIYOTAKA WASA

Central Research Laboratories, Matsushita Electric Industrial Co., Ltd., Moriguchi, Osaka, 570 Japan

ABSTRACT

Superconducting $Bi_2Sr_2CaCu_2O_y$(BSCCO) and normal conducting $Bi_2Sr_2CuO_x$(BSCO) thin films were multi-layered using a multi-target RF magnetron sputtering technique. The BSCO films exhibited a semiconductor-like transport property. X-ray diffraction measurements showed that the multi-layers are crystalline with c-axis oriented preferentially perpendicular to the substrate surface and exhibit satellite peaks originated from superlattice structures. Superconductive coupling was studied by varying layer thicknesses of each film. Magnetic field measurements were also done. The same behavior similar to that of YBCO/PrBCO multi-layers was observed in the change of Tc, the broadening of the transition, anisotropy and so on.

KEY WORDS: Multi-layer, Bi system superconductor, Hetero-growth, Proximity

INTRODUCTION

Remarkable progress on thin film growth of high-T_c oxide superconductors have made it possible to fabricate very thin superconducting films and multi-layered structures. The multi-layered film has two dimensional structure and exhibits interesting physical properties when the superconducting films are separated by non-superconducting films with the thickness comparable to the superconducting coherence length. So, it is expected that this multi-layered film structure gives some information about high-Tc superconducting mechanism and its parameters, as high-Tc oxide superconductors are constructed by closely spaced pairs of Cu-O planes and two dimensional crystal structure.

Recently, superconductor/nonsuperconductor multi-layered films such as YBaCuO/ DyBaCuO [1], YBaCuO/PrBaCuO [2-5] and YBaCuO/NdCeCuO [6,7] multi-layers are studied. On their results [2,3], the reduction of the superconducting transition temperature T_c of these multi-layers has been observed for thinning the superconducting film thickness. This reduction was similarly observed in a conventional superconductor such as Nb/Ge multi-layers [8] and discussed in terms of the Josephson coupling including the interlayer coupling and the proximity effect or the other effect such as imperfections of the films. Triscone et al. [3] and Li et al. [4] suggested that this reduction of T_c is related to a reduced coupling between superconducting layers through nonsuperconducting layers in YBaCuO/PrBaCuO multi-layers.

It is well known that Bi-oxide superconductors show higher degree of anisotropy than YBaCuO superconductors which is originated from the strong layered structure of Bi systems. Kanai et al. [9] have studied the superconductivity of $Bi_2Sr_2(Ca_{1-x}Y_x)Cu_2O_8$-based superlattices to evaluate two-dimensional nature in $Bi_2Sr_2CaCu_2O_y$. We have successfully fabricated the S/N/S Josephson junctions of layered BiSrCaCuO/BiSrCuO/BiSrCaCuO [10]. There are two reasons for the choice of BiSrCuO as nonsuperconducting layer materials. One reason is that BSCO lattice parameters of a-axis and b-axis are very close to those of BSCCO. Another reason is that our BSCO films are nonsuperconductors and those temperature dependences of the resistivity show semiconductor-like behavior.

In this paper, the results of the superconductivity of $Bi_2Sr_2CaCu_2O_y$ /$Bi_2Sr_2CuO_x$ (BSCCO/BSCO) multi-layered films which are prepared by RF magnetron sputtering technique are reported and compared with that of YBaCuO-based multi-layers.

SAMPLE PREPARATION

Multi-target magnetron sputtering technique was used for the deposition of BSCCO/BSCO multi-layered films. The BSCCO and the BSCO thin films were sequentially deposited on (100) MgO substrates in situ from the sintered targets of $Bi_{5.7}Sr_{4.2}Ca_{1.7}Cu_{3.2}O_x$ and $Bi_{5.7}Sr_{4.2}Cu_{1.6}O_y$, respectively. In our sputtering chamber, a heated substrate is rotated and faced to each target by a computer controlled stepping motor. During the deposition, the total gas pressure ($Ar+O_2$) of 40 Pa was maintained. The RF power of 50 - 60 W was supplied. The substrate was heated at the temperature of about 700°C. Under these conditions, the deposition rate of 10 - 20 A/min was obtained. The deposition was began with the BSCO layer and ended with the BSCO layer. After the deposition, oxygen gas was introduced into the sputtering chamber and the substrate was quickly cooled down to room temperature for about 30 min. The deposition rate was estimated from total deposition time and total film thickness measurements. Typical thickness of the multi-layers was 1200 A. We have used an X-ray diffractometer and secondary ion mass spectrometry (SIMS) to characterize the structure and compositional modulation of multi-layers, respectively. The resistivity of multi-layers was measured by the standard DC four-probe method using evaporated Au electrodes.

CHARACTERIZATION

Typical trace of X-ray diffractometer is shown for the BSCCO/BSCO multi-layer of (31A/24A) in Fig. 1. This film has the c-axis normal to the substrate surface. In this figure, the satellite peaks which are indicated by the arrows are observed by both sides of the main peaks. The presence of the satellite peaks proves that the modulation is achieved in these structures. The modulation wave length of about 60 A can be calculated from angular positions of the satellite peaks. This value roughly agrees with designed one.

Compositional modulation of multi-layers was measured by SIMS. The depth profile of SIMS for the (90 A/90 A) multi-layered film with total thickness of 2700 A is shown in Fig. 2. The modulation of composition is clearly seen in the signals of Ca and Cu. The modulation wavelength of Ca signal is well consistent with that of Cu signal. The measured modulation wavelength of about 180 A agrees with the designed one. This result also shows that the multi-layers are constructed as we designed. The interdiffusion of atoms seems to be unlikely in our multi-layered films within the SIMS depth resolution.

SUPERCONDUCTING PROPERTIES

Figure 3 shows the temperature dependence of resistivity for BSCCO/BSCO multi-layers with various thicknesses of superconducting BSCCO layers. When the nonsuperconducting BSCO thickness of 100 A is fixed and the BSCCO layer thickness decreases, the transition temperature of multi-layers decreases. The resistance of the (31 A/100 A) multi-layer do not reach to zero even at 4.2 K. But there are some signs of superconducting transition near the transition onset temperature of BSCCO. This can be supposed that superconductivity occurs in BSCCO as thin as 31 A.

The transition temperatures T_c^{on} defined at an inflection point in temperature dependence of resistivity and T_c^{mid} for the multi-layers are shown in Fig. 4 as a function of BSCCO layer thickness. The T_c onset decreases from 85 K to 75 K with decreasing the BSCCO layer thickness from 250 A to 31 A. Generally speaking, the T_c^{mid} is thought to be much affected by imperfections in the films induced by interdiffusion, lattice mismatch between film and substrate and substrate surface roughness. The origin of the reduction of T_c will be discussed late.

Figure 5 shows the BSCO thickness dependences of T_c^{on} and T_c^{mid} of multi-layers with BSCCO layer thicknesses of 31, 62 and 120 A. The BSCO layer thickness was varied from 150 A down to 24 A. It can be seen that the transition temperatures decrease as the BSCO layer thickness increases, and also decreases in case for fixed thickness of the BSCO layer as the BSCCO layer thickness decreases. These tendencies are much clearly seen in the behavior of T_c^{mid}.

As described above, there are many effects to explain the reduction of T_c. In YBaCuO/PrBaCuO multi-layers, a few research groups [3-5] have proposed that the reduction of T_c is related to Josephson coupling such as interlayer coupling between YBaCuO layers through the PrBaCuO layer. In comparison with the experimental data of YBaCuO/PrBaCuO, the reduction of the T_c^{on} in BSCCO/BSCO multi-layers seems to be less than that of

YBaCuO/PrBaCuO multi-layers, but the tendency of reduction is almost the same as that of YBaCuO/PrBaCuO multi-layers. So in our BSCCO/BSCO multi-layers, it seems that the reduction of T_c is also explained by reduced coupling.

Previously, we reported DC and AC Josephson effect in BSCCO/BSCO/BSCCO trilayer structures in which the BSCO layer thickness was 600 - 800 A [10]. In this case, obviously, the nonsuperconducting BSCO layer is rather thick to generate the Josephson tunnel coupling. And in practice, our BSCO films have the resistivity of about one and a half order of magnitude greater than that of BSCCO at low temperature. Thus, observed Josephson effect is most likely caused by the proximity coupling which is induced by the proximity effect. In spite of very short coherence length generally estimated for Bi-oxide superconductors, it is considered that superconducting electrons can adequately diffuse into the BSCO layer, When there is a smooth electric contact between the BSCCO layer and the BSCO layer. And the superconductive coupling is considered to be achieved by the proximity effect through rather thick BSCO barrier layer.

More precise discussions for the reduction of T_c is able to be done on this proximity effect. If the superconducting film thickness becomes comparable to its estimated superconducting coherence length, the proximity effect should generally occur and this effect should saturate for the normal metal layer enough thicker than the proximity coherence length of the normal metal. As shown in Fig. 5, the reduction of the T_c^{mid} for the multi-layers with the BSCCO layer thickness of 31 A seems to saturate for the BSCO layer thickness greater than about 70 A. Similar behavior originated from the proximity effect was observed in Pb/Cu films [11] and Nb/Ge multi-layers [8]. Our result suggests the existence of some effects such as the proximity effect in our multi-layers. This proximity-like effect should induce the coupling effect in BSCCO/BSCO multi-layers. In order to evaluate precisely the influence of such coupling effect in BSCCO/BSCO multi-layers, it is desired to compare the T_c of the multi-layers with that of a corresponding N/S/N single-layer sandwich structure which is designed to have a similar behavior on the reduction of T_c.

SUMMARY

BSCCO/BSCO multi-layers have been prepared using multi-target RF magnetron sputtering technique. It was confirmed that superconductivity occurs in BSCCO films as thin as 31A. The reduction of T_c in BSCCO/BSCO multi-layers is most likely caused by reduced coupling of superconductivity between superconducting layers which is induced by the proximity effect. However this coupling in BSCCO/BSCO multi-layers is weaker than in YBaCuO/PrBaCuO multi-layers. Further study is desired to clear the origin of the reduction of T_c.

ACKNOWLEDGMENTS

We would like to thank Dr. T. Nitta for his support of this work, and Dr. S. Hatta, H. Higashino and K.Mizuno for useful discussions. We thank to K. Tsukamoto of Matsushita Technoresearch Inc., for SIMS measurements.

REFERENCES

1. J.-M. Triscone, M.G. Karkut L. Antognazza, O. Brunner, and O. Fischer (1989) Phys. Rev. Lett. 63: 1016.
2. U. Poppe, P.Prieto, J. Schubert, H. Soltner, K. Urban, and Ch. Buchal (1989) Solid State Commun. 71: 569.
3. J.-M. Triscone, O. Fischer, L. Antognazza, A. D. Kent, and M. G. Karkut (1990) Phys. Rev. Lett. 64: 804.
4. Q. Li, X. X. Xi, X. D. Wu, A. Inam, S. Vadlamannati, W. L. McLean, T. Venkatesan, R. Ramesh, D. M. Hwang, J. A. Martinez, and L. Nazar (1990) Phys. Rev. Lett. 64: 3086.
5. D. H. Lowndes, D. P. Norton, and J. D. Budain (1990) Phys. Rev. Lett. 65: 1160.
6. A. Gupta, R. gross, E. Olsson, A. Segmuller, G. Koren, and C. C. Tsuei (1990) Phys. Rev. Lett. 64: 3191.
7. R. Gross, A. Gupta, E. Olsson, A. Segmuller, and G. Koren (1990) Appl. Phys. Lett. 57: 203.
8. S. T. Ruggiero, T. W. Barbee,Jr., and M. R. Beasley (1982) Phys. Rev. B 26: 4894.
9. M. Kanai, T. Kawai, and S. Kawai (1990) Appl. Phys. Lett. 57: 198.
10. K. Mizuno, K. Higashino, K. Setsune, and K. Wasa (1990) Appl. Phys. Lett. 56: 1469.
11. P. Hilsch (1982) Z. Physik 167: 511.

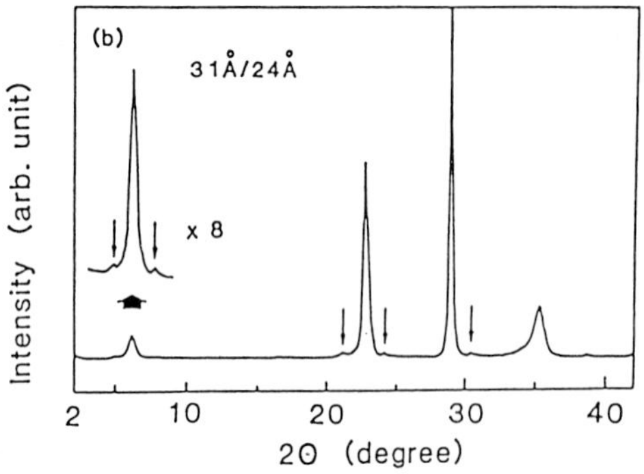

Fig. 1. X-ray diffraction patterns of the multi-layers (31A/24A).
The arrows indicate the satellite peaks.

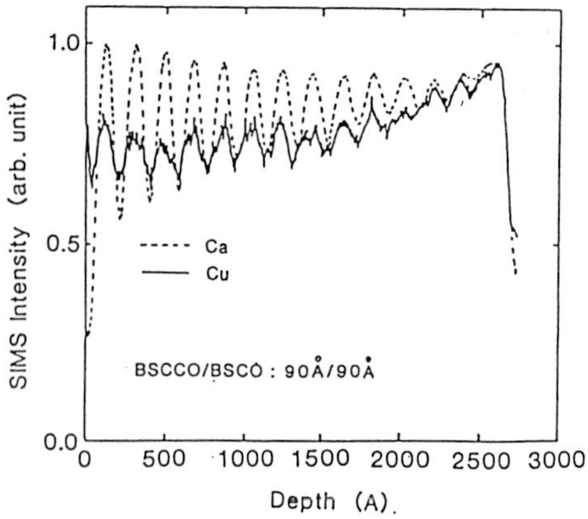

Fig. 2. Depth profile of SIMS for the (90A/90A) multi-layer.

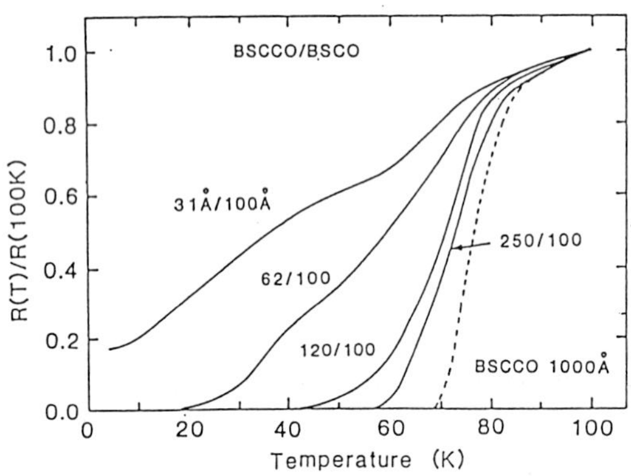

Fig. 3. Temperature dependence of resistivity for the (250A/100A), (120A/100A), (62A/100A), (31A/100A) multi-layers, and the (1000A) single layer BSCCO film.

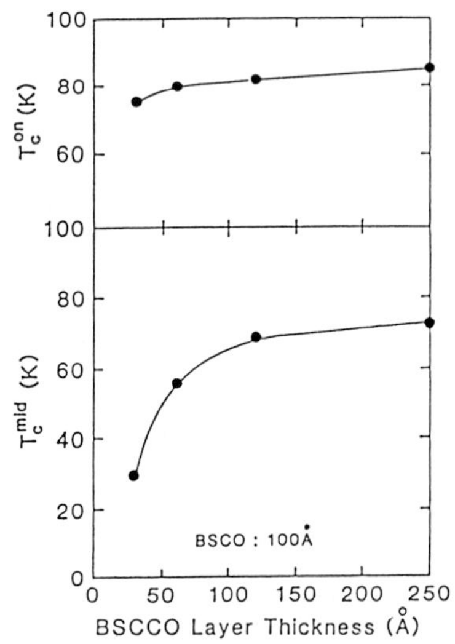

Fig. 4. T_C onset and T_C midpoint as a function for the multi-layers with the constant BSCO of the BSCCO layer thickness

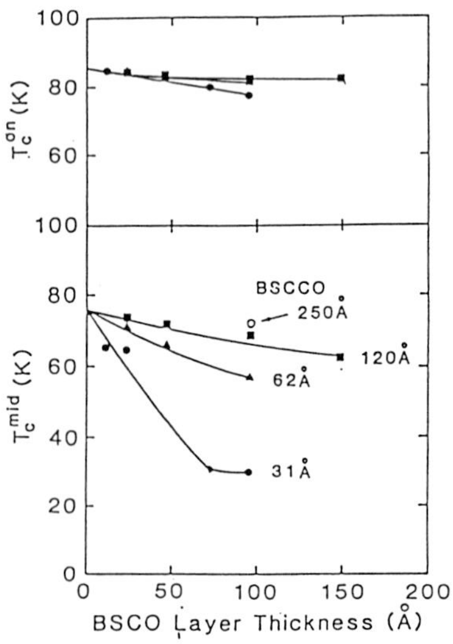

Fig. 5. T_C onset and T_C midpoint as a function of the BSCO layer thickness for the multi-layers with the BSCCO layer thickness of 31A, 62A, and 120A.

The Surface Characterizations of 80K-Phase Bi(Pb)-Sr-Ca-Cu-O Bulk Single Crystals and Thin Films

HIROSHI FUJIMOTO, WATARU FUTO, HIDEAKI HIGASHI, SATORU KISHIDA, HEIZO TOKUTAKA, KATSUMI NISHIMORI, NAGANORI ISHIHARA, KOUJI YAMANO, and SHOICHI KOYAMA

Department of Electrical and Electronic Engineering, Tottori University, Koyama, Tottori, 680 Japan

ABSTRACT

The 80K-phase Bi-system superconducting bulk single crystals and thin films were studied by SEM, LEED, RHEED, AES, XPS and SAM. The clean surfaces of the bulk single crystals were obtained by cleaving them in ultra-high vacuum or heating in vacuum after cleaving in air. The LEED and RHEED patterns from the clean surfaces showed ~1×5 structures. By comparing the results of AES, XPS and SAM from the film surfaces with those from the bulk single crystals, we found that the surfaces of thin films were covered with the impurities formed by post-annealing.

KEY WORDS : bulk single crystal, thin film, surface characterizations, clean surface, 1×5 structure

1. INTRODUCTION

Bi-Sr-Ca-Cu-O (Bi-system) superconductors have different critical temperatures [1-4] and lengths of the c-axis unit cell[1,5,6] as a function of the number of $Cu-O_2$ layers between Bi-O layers. The Bi-system superconductors have also the incommensurate modulation structures which have long range atomic displacement along the b-axis[7]. A Bi-O layer of the (001) cleavage surface also suffers from the structural modulation, almost in the same manner as $(Bi-O)_2$ layers in the bulk structure[8]. The Bi-system superconductors are one of the most interesting materials from a surface-physical point of view. It is important to control the growth process and the surface characterizations of the films in order to fabricate the devices using Bi-system superconductors.

We prepared the 80K-phase Bi(Pb)-Sr-Ca-Cu-O bulk single crystals and Bi-Sr-Ca-Cu-O thin films. In this study, the surface characterizations of the Bi-system bulk single crystals and thin films were studied, where surface morphology was by scanning electron microscope (SEM), surface structure by low-energy electron diffraction (LEED) and reflection high-energy electron diffraction (RHEED), surface composition by Auger electron spectroscopy (AES) and scanning Auger electron microscope (SAM), and the chemical bond natures of the constituent elements by X-ray photoelectron spectroscopy (XPS).

2. EXPERIMENTAL

The 80K-phase Bi(Pb)-Sr-Ca-Cu-O bulk single crystals were prepared using the starting materials with the metal ratio of (Bi,Pb):(Sr,Ca):Cu = 1:1:1 by self-flux method[8]. The bulk single crystals were plate-like and their typically sizes were $\sim 7\times 7mm^2$, and their thicknesses were $0.5\sim 1.0\mu m$. The zero-resistance temperature of bulk single crystal was 63~76K. From the result of X-ray diffraction (XRD), the cleaved surfaces were the a-b face and c-axis unit cell lengths were ~3.08nm.
The 80K-phase Bi-Sr-Ca-Cu-O thin films were prepared on the MgO(001) substrate using a mixed powder target with the metal ratio of Bi:Sr:Ca:Cu = 2:2:1:0.5 by rf magnetron sputtering, and then were post-annealed in air[9]. The thin films were homogenously grown all over the substrate and their average thickness was $\sim 1\mu m$. The zero-resistance temperature of thin films was 80~83K. From the XRD results, the films were c-axis-oriented perpendicular to the substrate and had a single 80K-phase.
In the SEM images, the surfaces obtained by cleaving bulk single crystals in air were very flat, where any crystal boundaries were not observed. The films have also very flat

surfaces. However, several crystal boundaries were only observed at higher magnification of the SEM images.

3. RESULTS AND DISCUSSION

3.1 LEED and RHEED Observations and the Cleaning of Specimens.

Fig.1 shows (a)the LEED pattern from the surface obtained by cleaving a bulk single crystal in ultra-high vacuum (V-C surface), and (b)the RHEED pattern observed in [100] azimuth from the surface obtained by heating a bulk single crystal in vacuum at 300℃ after cleaving in air (A-C & V-H surface). The LEED and RHEED patterns from the V-C surfaces and from the A-C & V-H surfaces showed the very clear diffraction LEED spots and RHEED rods, respectively and the single domain $\sim 1\times 5$ structures which indicate the existence of superstructure along the b-axis direction. The fundamental diffraction spots in the LEED patterns from the V-C surfaces were not clear and spot intensity modulations were observed along the b^*(reciprocal lattice)-axis direction[13]. On the other hand, the fundamental diffraction rods in the RHEED patterns from the A-C & V-H surfaces were clear, but the $(0,1),(0,\bar{1})$ rods intensities were very weak or sometimes were not observed. The calculated lattice parameters were $a = 0.54\sim 0.57$nm, $b = 5a = 2.70\sim 2.85$nm, which were in good agreement with a and b values reported in the reference[7].
The C and Cl impurity signal intensities of the AES spectra from the V-C surfaces and A-C & V-H surfaces(300～330℃) were much smaller than those from the surfaces cleaved in air. This may be due to the contaminations of surface, which are caused by exposing to air and gases generated by chamber baking(150℃,30hours). The impurities were easily removed by cleaving the bulk single crystals in vacuum or heating in vacuum[11].

In order to study the interface of the clean surface with metals, Ag which were used as electrodes was deposited at 300℃ on the clean surfaces. Spotty RHEED patterns from the deposited surfaces were clearly observed. Since the RHEED spotty patterns indicated the transmission type images, Ag islands grew epitaxially on the clean surfaces of bulk single cryatals.

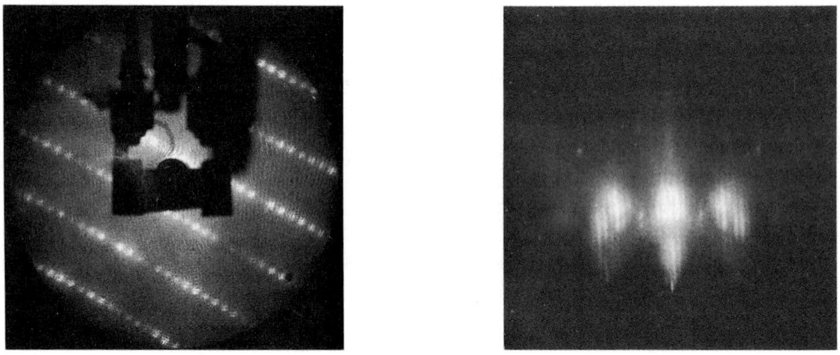

(a) (b)

Fig.1. LEED and RHEED patterns from the clean surfaces of bulk single crystals. (a)The LEED(70eV) pattern from the surface obtained by cleaving the bulk single crystal in ultra-high vacuum. (b)The RHEED(20kV) pattern observed in [100] azimuth from the surface obtained by heating the bulk single crystal in vacuum at 300℃ after cleaving in air.

Fig.2 shows the AES spectra from (a)the film surface heated at 300℃ in vacuum, (b)the film surface heated at 350℃ in vacuum and (c)the V-C surface of the bulk single crystal. The C and Cl impurities from the as-prepared film surface were removed by heating in vacuum at 300～350℃. However, the AES intensities of all the constituent elements from the surfaces heated in vacuum at (a)300℃ and (b)350℃ were different from that from (c)the clean surface of the bulk single crystal. The LEED and RHEED observations were also carried out on the as-prepared film surface and the film surfaces heated in vacuum at 300～350℃, but any diffraction patterns from the surfaces were not observed. The results may indicate that there is not any single cryatals as large as the LEED or RHEED beam spot diameter size (0.1～1.0mm) on the film surface.

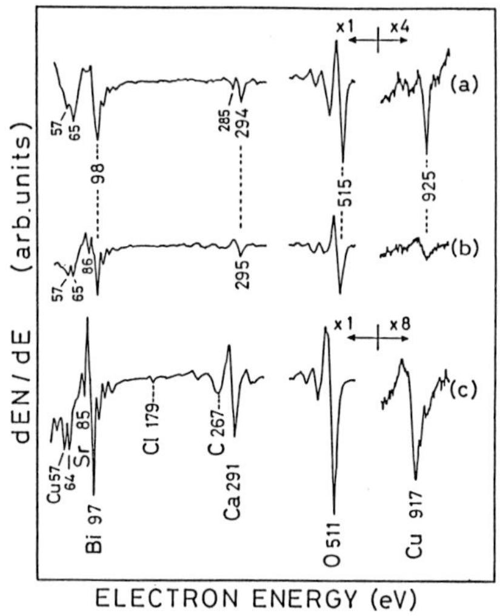

Fig. 2. AES spectra from the surfaces of a film and a bulk single crystal. (a)The spectrum comes from the surface of the film heated at 300℃ in vacuum. Then, (b)the surface was heated at 350℃ in vacuum. (c)The spectrum comes from the surface of the bulk single crystal in ultra-high vacuum.

3.2 The Surface Composition and the Chemical Bond Natures of the Constituent Elements of Specimens.

Fig.3 shows the XPS spectra of Sr-3d core levels. Figs.3 (a),(b) and (c) represent the XPS spectra from the as-prepared film surface, the A-C & V-H surface at 350℃ and the air-polished surface of bulk ceramics, respectively.
In the spectrum from the A-C & V-H surface, clear two peaks at 132.0 and 133.6eV due to Sr-3d 3/2 and 5/2 were observed, as shown Fig.4(b). In the XPS spectrum from the film surface, two peaks at 134.0 and 135.6eV were also observed. However, the energy position of the 133.8eV Sr-3d 3/2 peak from the film surface was higher than that from the bulk single crystal, and the shapes of the spectra from the film surface were also obviously different from those from the bulk ceramics. The spectrum of the bulk ceramics consists of the components due to the Bi-system crystals and the impurity phase of excess constituent elements[12].
Comparing (a) with (b) and (c), the chemical bond nature of Sr from the film surface seems to be different from that of Sr from a 80K-phase Bi(Pb)-Sr-Ca-Cu-O superconductor.
Fig.4 shows the XPS spectra of O-1s core levels. Figs.4 (a),(b),(c) and (d) represent the XPS spectra from the as-prepared surface of film ,the A-C & V-H surface at 350℃, the A-C surface and the air-polished surface of bulk ceramics,respectively.
The XPS spectrum of O-1s core levels from the as-prepared film surface was different from those from the bulk single crystals and bulk ceramics, where three peaks of O_L, O_H and O_n were observed[12].
The XPS spectra of Bi, Sr, Ca and Cu core levels from the A-C surfaces and A-C & V-H surfaces were also measured, and the spectra were sharper and clearer than those from the before-cleaved surfaces. The results were reported previously[12]. The XPS spectra of Bi, Ca and Cu core levels from the as-prepared film surfaces were also measured, and were compared with those from the clean surfaces of bulk single crystals.
From the XPS results of thin films, bulk single crystals and bulk ceramics, we concluded that the film surfaces were covered with oxide impurity layer, which were produced during post-annealing.
From the results of the SAM observations from the film surfaces sputtered by Ar^+ for 40min, we found that there were excess Bi and Sr segregations on the surface and in the bulk of the film.

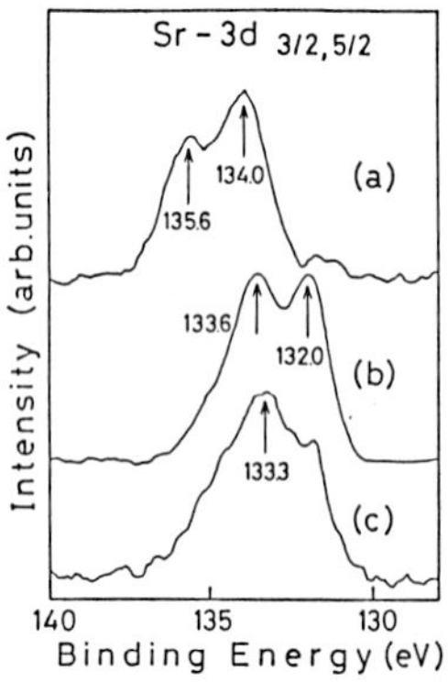

Fig. 3. XPS spectra of Sr-3d core levels of the film, the bulk single crystal and the bulk ceramics. (a) from the surface of the as-prepared film, (b) from the surface of the bulk single crystal heated at 350℃ in vacuum after cleaving in air and (c) from the surface of the bulk ceramics polished in air.

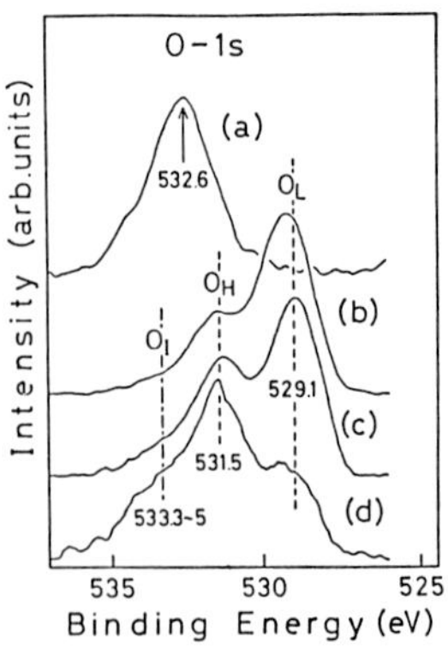

Fig. 4. XPS spectra of O-1s core levels of the film, the bulk single crystal and the bulk ceramics. (a) from the surface of the as-prepared film, (b) from the surface of the bulk single crystal heated at 350℃ in vacuum after cleaving in air, (c) from the surface of the bulk single crystal cleaved in air and (d) from the surface of the bulk ceramics polished in air.

4. SUMMARY

We prepared 80K-phase Bi-system superconducting bulk single crystals and the films, and carried out SEM, LEED, RHEED, AES, XPS and SAM measurements from their surfaces.
The clean surfaces of bulk single crystals were obtained by cleaving them in vacuum and heating in vacuum, and the LEED and RHEED diffraction patterns from the clean surfaces showed ~1×5 structures. In addition, on the cleaned bulk single crystal surface, the island epitaxial-growth of Ag was observed(Ts=300℃).
The SEM image showed that the as-prepared film surface was as flat as the air-cleaved surfaces of bulk single crystals. However, any LEED and RHEED diffraction patterns from the as-prepared surfaces and from the surfaces obtained by heating the as-prepared film at 300~350℃ in vacuum were not observed.
By comparing the AES and XPS results of the films with those of the bulk single crystals, we found that the film surfaces were covered with impurities, which were probably formed under post-annealing.

REFERENCES

1. Akimitsu J, Yamazaki A, Sawa H, and Fujiki H (1987) Jpn.J.Appl.Phys.26:L2080.
2. Subramanian AM, Torardi CC, Calabrese CJ, Gopalakrishnan J, Morrissey JK, Askew RT, Flippen SR, Chowdhry U and Sleight WA (1988) Science 239:1015.
3. Tarascon JM, Page LeY, Barboux P, Bagley GB, Greene HL, Mckinnon RW, Hull WG, Giroud M and Hwang MD (1988) Phys.Rev.B37:9382.
4. Maeda H, Tanaka Y, Fukutomi M and Asano T (1988) Jpn.J.Appl.Phys.27:L209.
5. Kajitani T, Kusaba K, Kikuchi M, Kobayashi N, Syono Y, Williams BT and Hirabayashi M (1988) Jpn.J.Appl.Phys.27:L587.
6. Takayama-Muromachi E, Uchida Y, Matsui Y, Onoda M and Kato K (1988) Jpn.J.Appl.Phys.27:L556.
7. Matsui Y, Maeda H, Tanaka Y and Horiuchi S (1989) Jpn.J.Appl.Phys.28:L946.
8. The preparation conditions will be presented in the session WBP-41 of this ISS,90.
9. The preparation conditions will be presented in the session TFP-44 of this ISS,90.
10. Kishida S, Tokutaka H, Fujimoto H, Nishimori K and Ishihara N (1989) Jpn.J.Appl.Phys.28:L1389.
11. Fujimoto H, Tokutaka H, Kishida S, Nishimori K and Ishihara N (1990) J.Vac.Soc.Jpn.33:p147.
12. Kishida S, Tokutaka H, Nakanishi S, Watanabe Y, Fujimoto H, Nishimori K, Futo W, Ishihara N, Torigoe S and Harada H (1989) Jpn.J.Appl.Phys.28:L949.

Time-Resolved Observation on Substrate Biasing Pulsed Laser Deposition

H. IZUMI, K. OHATA, T. SAWADA, K. SUZUKI, T. MORISHITA, and S. TANAKA

Superconductivity Research Laboratory, International Superconductivity Technology Center,
10-13, Shinonome 1-chome, Koto-ku, Tokyo, 135 Japan

ABSTRACT

The effects of substrate biasing in pulsed laser deposition of superconducting thin films has been investigated by using time-of-flight mass-spectroscopy (TOF-MS). Energy differences between the atomic, monoxides and cluster ions have been observed. While the first ones have an energy of about 200 eV or more, the latter two have only a few eV. The energies of the ions were hardly influenced by the electric field because of the shielding effect of electrons in plume, therefore, this energy difference implies the difference of the generation mechanism. Under the positively biased conditions, the multi-valenced ions (Cu^{2+} and Cu^{3+}) were observed. They should be a one of the key factors in preparing high quality superconducting thin film at lower deposition temperatures[1,2] due to their high internal energies.

KEY WORDS: Pulsed laser deposition, Substrate biasing, Time-of-flight mass-spectroscopy, Multi-valenced ion

INTRODUCTION

Pulsed laser deposition is one of the most promising methods for in-situ preparation of superconducting thin films[3-10]. Several efforts to reduce the substrate temperature have been carried out[11,12] from the point of view of electronics applications. Substrate bias voltages were also found to be significantly effective in the preparation of high quality superconducting $Ba_2Y_1Cu_3O_{7-\delta}$ films at the reduced deposition temperatures by the present authors[1, 2]. Pulsed laser deposition is accompanied with a flash plasma on the target surface, which is usually called a "*plume*", caused by laser irradiation. There are many efforts in laser plume diagnostics using time-resolved optical observation[13-16] and mass-spectroscopy[17, 18] in order to clarify the nature of the laser plume. While Koren et al. did not refer to the proof of the creation of the atomic oxygen[11], Witanachchi et al. also reported the presence of continuous plasma during the deposition[12]. We have revealed by a time-resolved optical observation the presence of the optical emission accompanying the laser plume[1]. We believe that this emission should indicate the occurrence of some activation reaction of ablated species in it, being effective in film formation. There are, however, very few reports concerning the relationship between the creation of activated species and the film formation mechanism at lower deposition temperatures. Therefore, the investigation of the activated states of the species in the laser plume with substrate biasing is of great interest, being relevant to the deposition mechanism. In this study, to investigate the substrate bias voltage effect of charged species on the film growth, mass-spectroscopic studies on the laser plume at the position of substrate set were performed for the study of the behavior of ionic species.

EXPERIMENTAL

ArF (193 nm) excimer laser was used for ablation with a duration of 16 ns. The deposition conditions.were described elsewhere[1]. Since the laser plume involves very complicated mass and energy distribution of the species and it is well known that there are clusters of the mass number up to several thousands in it, we have employed a time-of-flight mass-spectrometry for the mass analysis of the plume. This method is also able to obtain directly the ion energies. As a time-of-flight mass-spectrometer (TOF-MS), we made use of a reflectron type TOF-MS (MSLT-2000: ULVAC JAPAN LTD.), in which the ions introduced from the sampling orifice are turned to the opposite way by the reflection potential and go to the multi channel plate (MCP). This arrangement enables the attainment of a long drift distance of the ions even using a short analyzing tube and enhances the sensitivity in mass and energy separation. Mass-spectroscopic study has been carried out by replacing the substrate holding mechanism of our deposition chamber[1] with the TOF-MS. The sampling orifice was set at the position of the substrate, so we were able to analyze the ions impinging onto the substrate in our setup and determine directly the effects of ions on the film formation process. The identification of ions was carried out by the comparison with the results from the calculation for the each species versus the potential of each target, retarding and reflection. Most of signals has been successfully identified. The energy distribution for the ions was also determined by varying the retarding potential applied to the retarding grid set just behind the sampling orifice. The retarding also has an effect on reducing the noise from the plasma of the laser plume and sharpening the mass spectra. For the better spectra sampling, we usually fixed the target and retarding potentials at +1 and +0.8 kV, respectively. Bias voltages were applied by balancing the potential difference between the target and the grid set in front of the sampling orifice.

RESULTS AND DISCUSSION

Mass analysis of the laser plume is very difficult due to its wide energy and mass distributions of the included species, and the most serious problem is that the laser plume is a kind of plasma in which very complicated reactions take place. Even though the target potential is negative, we observe that the detected ion current is as large as those for the grounded and positive potentials. It seems that the motion of the ablated ions is hardly influenced by the external electric field. By the way, when we measured the transient current between the target and ground during laser ablation, it was observed that very large number of electrons were emitted from the target[19]. This electron emission was suppressed in the case of negative grid potentials against the target and greatly enhanced by positive potentials. There are many in the plume emitted from the target by ablation as many as ions and these electrons are expected to have the ability to shield the inside from the external field due to their large mobility. The mass and energy analysis for the ablated ions, therefore, should be done by extracting ions from the laser plume.

This problem were solved by equipping the retarding grid just behind the sampling orifice. The ions passing through the orifice feel the retarding potential and the low energy ones are repulsed. Only the ions getting over the retarding potential reach the grounded electrode set behind the retarding one and are accelerated again to drift to the detector. Though the sampling orifice is helpful in reducing the obstacle noise from the plume by itself, there still remains an effect from the plasma in the mass spectra as long as a conventional arrangement is used. The retarding circuit was equipped in our TOF-MS just behind the orifice and this device enables us to decrease effectively the obstacle noise coming from the plume, and it serves as a means of sharpening the mass spectra.

We are also able to analyze the ion energies easily by varying the retarding potential. Fig.1 shows the mass spectra of the plume of $Ba_2Y_1Cu_3O_{7-\delta}$ for the various retarding potentials. The target and reflectron potential were fixed at +1.0 kV and +1.3 kV, respectively. The signals from the monoxides of each metal and the cluster ions can be seen in Fig.1 (a) as well as those from the atoms. The clusters observed in this figure are estimated to correspond to mass numbers of several hundreds and

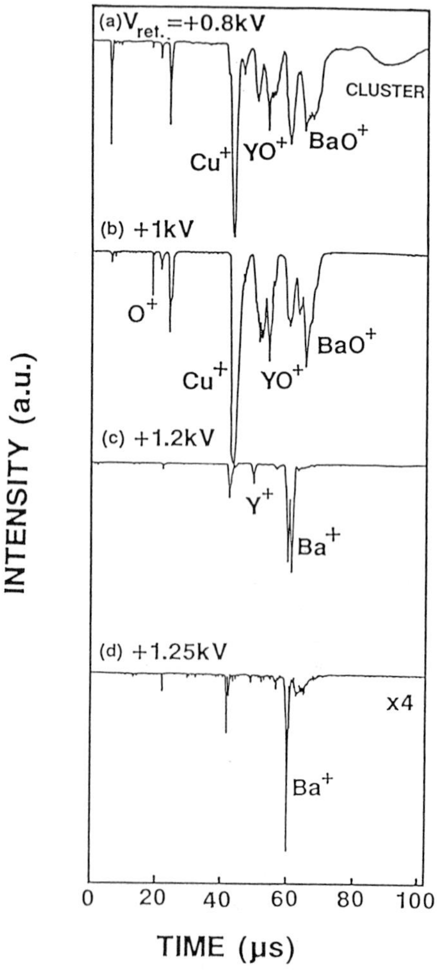

Figure 1 The mass spectra of the plume of the $Ba_2Y_1Cu_3O_{7-\delta}$ target for the retarding potentials of +0.8, +1, +1.2 and +1.25 kV, respectively. The target and reflection potentials were +1 and +1.3 kV, respectively. The pressure around the target was 1.3 Pa and that in the analyzing tube was $2x10^{-4}$ Pa.

heavier ones were not detected. The detailed composition of the clusters is not clear, but there may be some reason for these mass numbers to be dominant. While the signals from monoxide ions have larger intensities than those from atomic ones for Y and Ba, Cu^+ is dominant over CuO^+. This may be due to the characteristics of copper in the $Ba_2Y_1Cu_3O_{7-\delta}$ target and it implies that some additive oxidation process may be useful in preparing high quality films.

Increasing the retarding potential to +1 kV, the signals from cluster ions fall off rapidly (Fig.1 (b)). This indicates that the cluster ions have come into the sampling orifice, which corresponds to the substrate in the case of film deposition, with very low energies. Monoxide ions are also diminished at the higher retarding potential of +1.2 kV. The detail examination has revealed that monoxide ions have energies of the order of a few tens of eV. On the other hand, atomic ions have energies more than 200 eV so that the signals from them still remain, even at the retarding potential of 1.2 kV, as large as those in Fig.1 (a) and (b). At the retarding potential of +1.25 kV, only very weak signals were detected except for Ba^+, so the atomic ions' energies are mainly distributed between 200 to 250 eV but Ba^+ has an energy tail in the energy distribution higher than those of the other metals. The obtained value for the atomic ion energies here are slightly high compared with the previously reported ones which are about 100 eV or less[13, 14, 18]. Further prudent examination must be made to discuss the differences in the values of ion energy, but we have confirmed by another measurement[20] and believe that time-of-flight mass-spectroscopy is one of the most suitable ways to obtain the information about the ion energies directly.

We have reported the value of the order of several eV for the emissive species from a time-resolved optical observation[1] and they were thought to be electrically neutral. Saenger has also reported an energy difference between ions and neutrals[14]. In addition to this, our experiments indicate that neutrals, atomic, monoxide and cluster ions each have different energies in the plume. It is implied by such a wide energy distribution that the ablation mechanism generating these species is very complicated. We suppose that atomic ions with very high energy are generated at the initial stage of the ablation process and there is some mechanism to accelerate ions. On the other hand, monoxide and cluster ions are probably created by plasma reactions in the plume. Since monoxide ions are expected to be directly created by the collision between the high energy atomic ions and oxygen molecules, they should have higher energies than clusters. It is also possible that some of the monoxide ions are created from the decomposition of large clusters. However, we believe that the latter case is not dominant because the various fragments should be observed at the same time and they are not. Clusters are thought to be emitted from the target into the plume with very low energies and they are probably charged in plasma afterward.

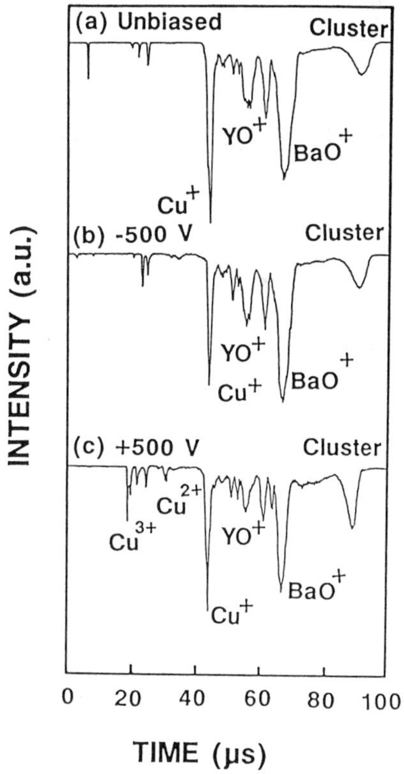

Figure 2 The mass spectra of the plume of $Ba_2Y_1Cu_3O_{7-\delta}$ biasing with 0, -500 and +500 V, respectively. The target, retarding and reflection potentials were +1, +0.8 and +1.3 kV, respectively. The pressure around the target was 1.3 Pa and that in the analyzing tube was $2x10^{-4}$ Pa.

We applied TOF-MS for analysis of the ions in a biased laser ablated plume. Figure 6 shows the results of the plume of $Ba_2Y_1Cu_3O_{7-\delta}$ for the bias voltages of 0, -500 and +500 V, respectively. The retarding and reflectron potentials were fixed at +0.8 and +1.3 kV, respectively. While the detected times for each ion seem not to be influenced by the bias voltages, it should be noted that the multi-valenced ions, Cu^{2+} and Cu^{3+}, are obviously observed in the case of the positive bias voltage. Since the ion energies are expected not to be affected by the external electric field due to plasma shielding, it is natural that the ion energies are hardly influenced in this figure. On the other hand, it is very interesting that positive bias voltages have the effect of generating multi-valenced ions. It has been also reported by the present authors that the electrons emitted from the target to the laser plume are greatly increased when positive bias voltages are applied[19]. When positive bias voltages are applied, the electrons generated by ablation probably are torn off rapidly from the target surface. It can be qualitatively explained that the multi-valenced ions tend to be generated because excess positive charges are left there. It is also expected that the multi-valenced ions would have a higher internal energy than the other ions (for example, ionization potentials for Cu^+ and Cu^{2+} are 7.724 and 20.29 eV, respectively), so they should be helpful for the film formation and be one of the key factors in preparing high quality superconducting thin films at lower deposition temperatures in the biased laser ablation method.

ACNOWLEDGEMENT

A part of this work was supported by New Energy and Industrial Technology Development Organization under the management of the R & D of Basic Technology for Future Industries. We also thank Mr. Kubo from ULVAC JAPAN Ltd. for the assistance in TOF-MS and fruitful discussions.

REFERENCES

1. H. Izumi, K. Ohata, T. Hase, K. Suzuki, T. Morishita and S. Tanaka: to be published in J. Appl. Phys. of Dec. 1.

2. K.Ohata, H. Izumi, T. Hase, K. Suzuki, T. Morishita and S. Tanaka: to be published in Proc. of the Spring Meeting of Mat. Res. Soc., April, 1990, San Francisco, USA.

3. J. Narayan, N. Buinno, R. Singh, O.W. Holland and O Auciello: Appl. Phys. Lett.,**51**, 1845(1987).

4. A. Inam, M.S. Hedge, X.D. Wu, T. Venkatesan, P. England, P.F. Miceli, E.W. Chase, C.C. Chang, J.M. Tarascon and J.B. Watchman: Appl. Phys. Lett.,**53**, 908(1988).

5. W.A. Weimer: Appl. Phys. Lett.,**52**, 2171(1988).

6. S. Deshmuck, E.W. Rothe, G.P. Reck, T. Kushida and Z.G. Wu: Appl. Phys. Lett.,**53**, 2698(1988).

7. G. Koren, A. Gupta, E.A. Giess, A. Segmuller and R.B. Laibowitz: Appl. Phys. Lett.,**54**, 1054(1989).

8. L. Lynds, B.R. Weinberger, G.G. Peterson and H.A. Krasinski: Appl. Phys. Lett.,**52**, 320(1988).

9. H.S. Kwok, P. Mattocks, L. Shi, X.W. Wang, S. Witanachchi, Q.Y. Ying, J.P. Zheng and D.T. Shaw: Appl. Phys. Lett.,**52**, 1825(1988).

10. R.E. Muenchausen, K.M. Hubbard, S. Foltyn, R.C. Estler, N.S. Nogar and C. Jenkins: Appl. Phys. Lett.,**56**, 975(1990).

11. G. Koren, A. Gupta and R.J. Basemann: Appl. Phys. Lett.,**54**, 1920(1989).

12. S. Witanachchi, H.S. Kwok, X.D. Wang and D.T. Shaw: Appl. Phys. Lett.,**53**, 234(1988).

13. E. Dyer, R.D. Greenough, A. Issa and P.H. Key: Appl. Phys. Lett.,**53**, 534(1988).

14. K.L. Saenger: J. Appl. Phys.,**66**, 4435(1989).

15. O. Eryu, K. Murakami, K. Masuda, K. Kasuya and Y. Nishina: Appl. Phys. Lett.,**54**, 2716(1989).

16. O. Auciello, S. Athavale, O.E. Hopkins, M. Sito, A.F. Schreiner and N. Buinno: Appl. Phys. Lett.,**53**, 72(1988).

17. C.H. Becker and J. B. Pallix: J. Appl. Phys.,**64**, 5152(1988).

18. T. Venkatesan, X.D. Wu, A. Inam, Y. Jeon, M. Croft, E.W. Chase, C.C. Chang, J.B. Watchman, R.W. Odom, F.R. di Brozolo and C.A. Magee: Appl. Phys. Lett.,**53**, 1431(1988).

19. H. Izumi, K. Ohata, T. Morishita and S. Tanaka: Proc. of the Appl. Superconductivity Conf., Sept. 24-28, 1990, Snowmass, Colorado, U.S.A.; to be published in IEEE Trans. on Magnetism.

20. H. Izumi, K. Ohata, T. Morishita and S. Tanaka: Proc. of the 4th Ann. Conf. on Superconductivity and Appl., Sept. 18-20, 1990, Buffalo, New York, U.S.A.

Effect of Activated Oxygen on Crystalline Orientation in Bi-Sr-Ca-Cu-O Thin Films

YOSHIKI ISHIZUKA, YOSHIAKI TERASHIMA, and TADAO MIURA

Toshiba Research and Development Center, Komukai Toshiba-cho 1-chome, Saiwai-ku, Kawasaki, 210 Japan

ABSTRACT

Thin films of Bi-Sr-Ca-Cu-O were prepared on $SrTiO_3$ (110) substrates by coevaporation, using a molecular beam epitaxy (MBE) machine with rf oxygen plasma. It has been confirmed that the crystalline orientation of films can be controlled by regulating the activated oxygen flux density. Two orientations could be obtained. These films showed superconducting transitions, regardless of the crystalline orientation.

KEY WORDS: Bi-Sr-Ca-Cu-O, activated oxygen, crystalline orientation

INTRODUCTION

Bi-Sr-Ca-Cu-O superconductors have strong anisotropic properties because of their two-dimensional structures. For example, coherence length perpendicular to the c-axis is much greater than coherence length parallel to the c-axis [1]. Thus, the thin film orientation is of great importance in fabricating multilayer structures, such as tunnel junctions: Films whose a- or b-axes are oriented perpendicularly to the film surfaces are especially desirable.

In Y-Ba-Cu-O superconductors, not only c-axis oriented films but also [100] axis [2-4] [110] axis and [301] axis [5] oriented films were formed by sputtering [2-4] and reactive evaporation [5] by changing the oxygen pressure and/or the substrate temperature. This orientation control could be explained by the lattice matching between a substrate and a film. On the other hand, in Bi-Sr-Ca-Cu-O thin films, only c-axis oriented films have been prepared on $SrTiO_3$ (100) or MgO (100) substrate generally. In the case of MgO (100), there is a large mismatch, 9.1 percent, between a substrate and a film. These facts indicate there is an anisotropic growth mechanism for the crystal. On the contrary, if there is a good lattice matching between a substrate and a film, whose orientation is not c-axis, there is a possibility that the c-axis is not oriented perpendicularly to the film surface. Thus, $SrTiO_3$ (110) was used as a substrate.

From another point of view, thin films of Bi-Sr-Ca-Cu-O are very sensitive to the activated oxygen flux density; the phase formation is controlled by varying the activated oxygen flux density [6]. As an analogy to this fact, the activated oxygen may have an effect on the crystalline orientation.

EXPERIMENTAL

Thin films of Bi-Sr-Ca-Cu-O were prepared on $SrTiO_3$ (110) substrates by coevaporation using a MBE machine [6]. Bi, Sr, Ca, and Cu were evaporated from individual effusion cells. The evaporation rates were measured with a quartz thickness monitor to control the chemical compositions of the films. Plasma, excited by rf (13.56 MHz), was used to oxidize these metals at low oxygen pressure. This oxygen plasma was generated in a quartz tube and supplied to a substrate during deposition. In this experiment, the authors focused on the effect of oxygen on crystalline orientation, especially activated oxygen generated in the plasma. Therefore, the substrate temperature was kept constant, about 800°C, to remove the substrate temperature influence. To investigate the effect of the activated oxygen on crystalline orientation, the activated oxygen flux density was varied by two methods. One was to vary the rf power, from 40 W to 150 W, while the oxygen flow rate was constant. The other was to vary the oxygen flow rate, from 0.45 sccm to 2.3 sccm, while the rf power was constant. The structural properties of films were estimated by X-ray diffraction using Cu Kα radiation. Electrical resistance was measured using an ordinary ac four-probe method. Surface morphology was observed with a scanning electron microscope (SEM).

RESULTS AND DISCUSSION

The X-ray diffraction patterns of the films are shown in Fig. 1. These films were prepared at 800°C substrate temperature and at 40 W rf power. The oxygen flow rate range was between 0.45 sccm and 2.3 sccm. This figure shows that (00k) (k=2,8,10) peaks of the $Bi_2Sr_2CaCu_2O_y$ phase (2212 phase) were observed, when the oxygen flow rate was 1.6 and 2.3 sccm. Other peaks were identified as $SrTiO_3$ (110) and the 2212 phase (117). On the other hand, the peaks identified as $SrTiO_3$ (110) and CuO ($\bar{1}$11) or (002) were only observed when the oxygen flow rates were 0.45 and 0.55 sccm. However, there was a broad peak at about the 2θ of 32 degrees. This peak was identified as the $Bi_2Sr_2Ca_2Cu_3O_y$ phase (2223 phase) (119). In case of the (117) plane of the 2212 phase, parallel to the film surface, or the (119) plane of the 2223 phase, parallel to the film surface, c-axis inclines about 50 degrees from the direction perpendicular to the film surface. As the oxygen flow rate increased, the c-axis tended to be oriented perpendicularly to the film surface. Figure 2 shows the X-ray diffraction patterns for the films prepared at 800°C substrate temperature and at 1.6 sccm oxygen flow rate. The rf power ranged from 40 W to 150 W. This figure shows that, as the rf power increased, (00k) peaks of the 2212 phase became sharp and grew, compared with (117) peak of the 2212 phase. In other words, as the rf power increased, the c-axis tended to be oriented perpendicularly to the film surface. The relation among the crystal phase, the crystalline orientation, the rf power and the oxygen flow rate is shown in Fig. 3. In this figure, the values written under the marks (triangles and squares) show the ratio of the intensity of (00$\underline{10}$) peak for the 2212 phase to that of (117) peak for the 2212 phase. This figure indicates that, as the oxygen flow rate increased, the phase changed from the 2223 phase to the 2212 phase. The c-axis inclination from the direction perpendicular to the film surface changed from about 50 degrees to 0 degrees. In short, as the oxygen flow rate increased, the c-axis tended to be oriented perpendicularly to the film surface. These tendencies meet the case of rf power increase. This fact indicates that activated oxygen effects the crystal phase and the crystalline orientation for films prepared on $SrTiO_3$ (110) substrates.

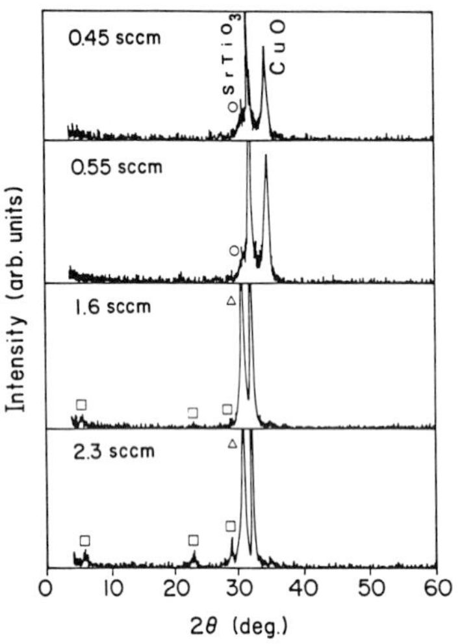

Fig. 1. X-ray diffraction patterns for films prepared under different oxygen flow rate conditions. ○, △ and □ indicate the 2223 phase (119) peak, the 2212 phase (117) peak and the 2212 phase (00k) peaks, respectively.

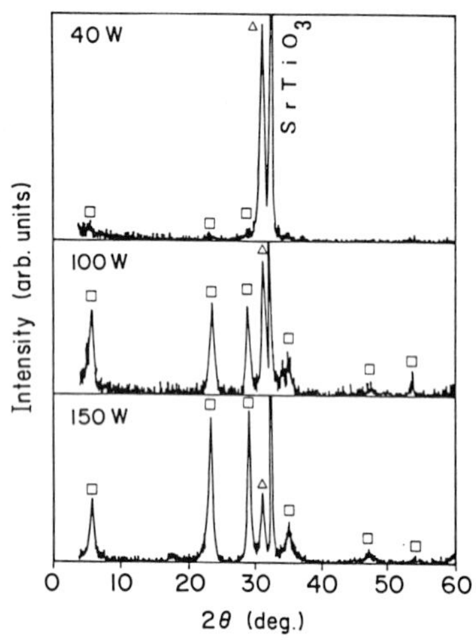

Fig. 2. X-ray diffraction patterns for films prepared under different rf power conditions. △ and □ indicate the 2212 phase (117) peak and the 2212 phase (00k) peaks, respectively.

Figure 4 shows SEM photographs of these films. Figure 4 (a), (b) and (c) correspond to the films prepared under different conditions: Figure 4 (a) shows film prepared at 40 W rf power and 0.55 sccm oxygen flow rate, (b) shows the film prepared at 40 W rf power and 1.6 sccm oxygen flow rate and (c) shows the film prepared at 150 W rf power and 1.6 sccm oxygen flow rate. Therefore, (a) corresponds to the film of (119) plane of the 2223 phase, parallel to the film surface, (b) corresponds to the film of (117) plane of the 2212 phase, parallel to the film surface, and (c) corresponds to the film of c-axis of the 2212 phase oriented perpendicularly to the film surface. They indicate that surface morphology is related to the crystalline orientation. The striped pattern is characteristic of (a). This stripe is parallel to the [001] direction for a $SrTiO_3$ (110) substrate.

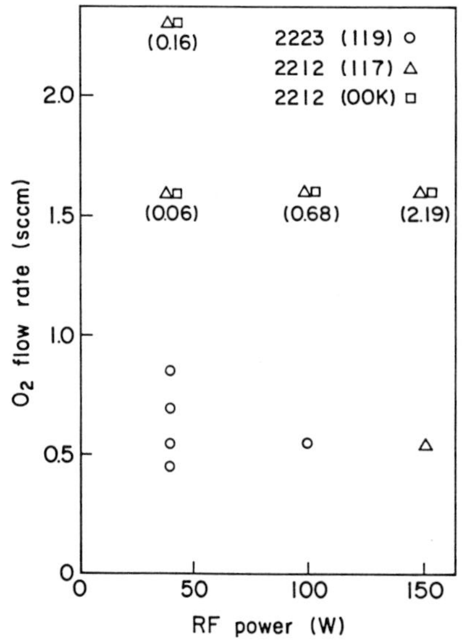

Fig. 3. Relation among phase, orientation, oxygen flow rate, and rf power. The values written under the marks show the ratio of the 2212 phase (00<u>10</u>) peak intensity to that for the 2212 phase (117) peak.

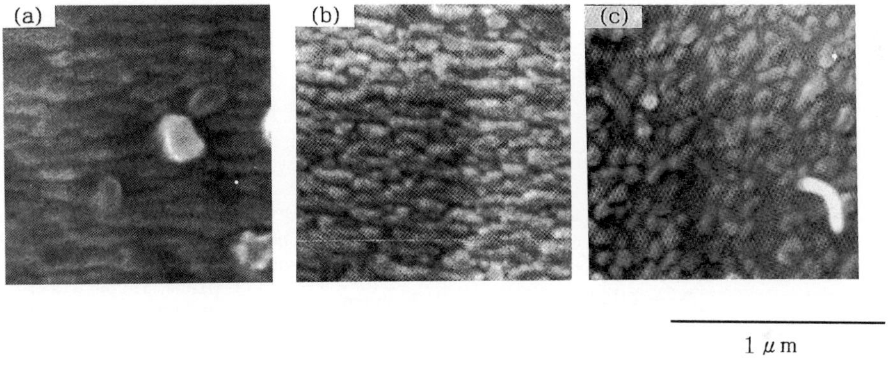

Fig. 4. Scanning electron microscope images. (a) Corresponds to the orientation that (119) plane of the 2223 phase parallel to the film surface. (b) Corresponds to the orientation that (117) plane of the 2212 phase parallel to the film surface. (c) Corresponds to the orientation that c-axis perpendicular to the film surface.

In spite of the differences in these orientations and surface morphologies, all of them showed superconducting transitions. The temperature dependence of the resistance of the film (a), shown in Fig. 5, however, has a characteristic feature. It has two knees at about 100 K and 35 K. The reason for this feature is not clear.

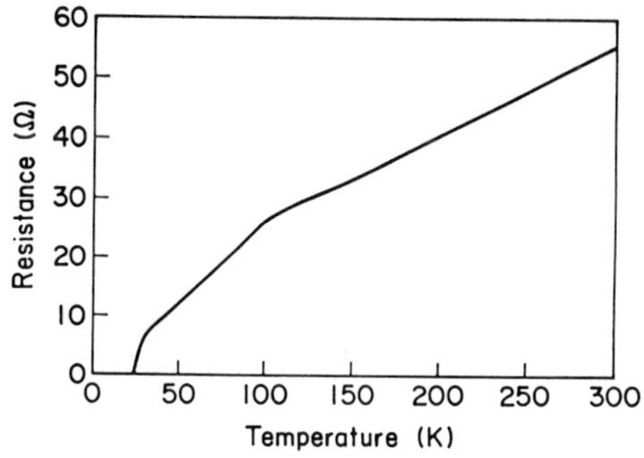

Fig. 5. Temperature dependence of resistance for film (a) in Fig. 4.

The reason for the orientation of (117) plane for the 2212 phase, parallel to the film surface, can be explained by the lattice matching. The $SrTiO_3$ lattice parameter is $a_o=b_o=c_o=3.9$ A, and the 2212 phase lattice parameters are a=b=5.4 A and c=30.6 A. The $5(a_o^2+b_o^2+c_o^2)^{1/2}$ lattice distance in the $[1\bar{1}1]$ direction for $SrTiO_3$ coincides with $3(a^2+b^2+(2c/7)^2)^{1/2}$ in the $[77\bar{2}]$ direction for the 2212 phase with a 3.1 percent mismatch value. On the other hand, the $4(a_o^2+b_o^2+(2c_o)^2)^{1/2}$ lattice distance in the $[\bar{1}12]$ direction for $SrTiO_3$ coincides with $5(a^2+b^2)^{1/2}$ in the $[\bar{1}10]$ direction for the 2212 phase. Geometrical arrangement between a film of the 2212 phase and a substrate of $SrTiO_3$ (110) is shown in Fig. 6. In the case of the orientation of (119) plane for the 2223 phase, parallel to the film surface, a similar discussion can be applied.

From these facts, there are thought to be two film growth mechanism modes of Bi-Sr-Ca-Cu-O. One is the epitaxial mode. The orientations of (117) plane for the 2212 phase, parallel to the film surface, and (119) plane for the 2223 phase, parallel to the film surface, fit this mode. The other mode is the self-completion mode. The c-axis orientation fits this mode. This mode is originated from the anisotropy of the film growth mechanism, which may be caused by the two-dimensional structure. The activated oxygen flux density, varied by the rf power and/or oxygen flow rate, may determine which modes the film growth follows. Therefore, the film orientation is thought to be controlled by regulating the activated oxygen flux density.

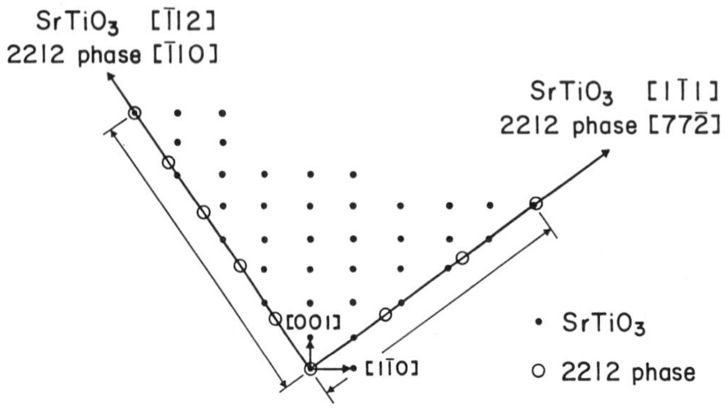

Fig. 6. Geometrical arrangement betweeen a $SrTiO_3$ (110) substrate and the 2212 phase film, whose (117) plane is parallel to the film surface.

In summary, the Bi-Sr-Ca-Cu-O thin film orientation on $SrTiO_3$ (110) substrates is controlled by regulating the activated oxygen flux density. By this method, two orientations could be obtained. One is the orientation that c-axis is oriented perpendicularly to the film surface. The other is the orientation where the c-axis is inclined about 50 degrees from the direction perpendicular to the film surface. The cases where the (117) plane of the 2212 phase is parallel to the film surface and the (119) plane of the 2223 phase is parallel to the film surface meet the latter orientation.

REFERENCES

1. Koike Y, Nakanomyo T, Fukase T (1988) Jap.J.Appl.Phys.27: L841
2. Fujita J, Yoshitake T, Kamijo A, Satoh T, Igarashi H (1988) J.Appl.Phys.64: 1292
3. Asano H, Asahi M, Michikami O (1989) Jap.J.Appl.Phys.28: L981
4. Terashima Y, Sagoi M, Kubo K, Mizutani Y, Miura T, Yoshida J, Mizushima K (1989) Jap.J.Appl.Phys.28: L653
5. Terashima T, Bando Y, Iijima K, Yamamoto K, Hirata K (1988) Appl.Phys.Lett.53: 2232
6. Ishizuka Y, Terashima Y, Miura T (in press) Jap.J.Appl.Phys.

Preparation of Bi-Sr-Ca-Cu-O Films by MBE Using an Oxygen Radical Beam Source

HIROAKI FURUKAWA, SEIICHI TOKUNAGA, and MASAO NAKAO

SANYO Tsukuba Research Center, 2-1, Koyadai, Tsukuba, 305 Japan

ABSTRACT

Bi-Sr-Ca-Cu-O thin films have been grown in situ on (100) MgO substrates using radical beams as a source of reactive oxygen. Oxygen radicals were produced by a RF-excited plasma in a quartz-tube reactor and supplied through fine holes in the beam aperture plate on the substrate. The vacuum pressure was maintained below 4×10^{-6} Torr during growth. Cu deposited on substrates is selectively oxidized by shadowing the beams. The Bi-based film grown at 750 ℃ shows a superconducting transition at 75 K without in-situ annealing. In the x-ray diffraction pattern, the periodical peaks are assigned to (00ℓ) reflections of $Bi_2Sr_2CaCu_2O_y$ as a main phase.

KEY WORDS: Bi-Sr-Ca-Cu-O, MBE, oxygen radical beam, plasma, in-situ growth

INTRODUCTION

Preparation of high temperature superconducting thin films is important for both device applications and fundamental studies. Recently, many groups have studied and developed such techniques for in-situ growth as sputtering, co-eavporation, laser ablation, molecular beam epitaxy (MBE) and metalorganic chemical vapor deposition. In particular, MBE is suitable for producing high T_c films owing to its precise control of layered structures with monoatomic layer scale accuracy as employed presently in semiconductor synthesis. In order to artificially control the layered structures of Bi-based superconductors, the atomically layered epitaxy techniques with shutter control of sources have been developed[1]. One of the crucial problems for cuprate MBE growth is how to oxidize sufficiently the cuprate films in a high vacuum environment using an efficient activated oxygen source. Recently, atomic oxygen (oxygen radicals) [2-4] and pure ozone[1,4,5] are employed as a reactive oxygen source in cuprate MBE growth because the chemical reactivity of these species are much greater than the one of molecular oxygen. In this paper, we describe the in-situ growth of Bi-Sr-Ca-Cu-O films in a conventional MBE apparatus equipped with an oxygen radical beam source. The films were synthesized by sequential deposition techniques[5,6]

EXPERIMENT

Films were prepared by a conventional MBE apparatus (RIBER EVA 32). The growth chamber was pumped by an ion pump and a turbomolecular pump that was added to handle the flow of oxygen species introduced during growth, and has a base pressure of 1.0×10^{-8} Torr without flow of liquid N_2 into cryopanels. Sr,Ca and Cu were evaporated from metal sources and Bi source was oxide. Effusion cells were used to evaporate Sr and Ca. Bi_2O_3 and Cu were heated using electron beam sources. Two feedback circuits used signals from independent quartz-crystal rate monitors to control the rates for electron beam sources. The evaporation rates from each cell were measured using a quartz-crystal-thickness monitor at the substrate position. The substrate holder is made of Inconel, which caused no contamination. The chemical composition was determined by an electron probe x-ray micro analyzer (EPMA).

In order to obtain in-situ growth using MBE techniques, a source of reactive oxygen is necessary. To provide reactive oxygen species, we used beams of activated oxygen species generated by a discharge in the chamber made of a 27-mm-i.d. quartz tube[1]. The discharge was created from inductively-coupled

RF excitation at 13.56 MHz. The RF power is typically a level of 400 W. The activated species escape in a vacuum environment to a substrate via fine holes (0.3 mmφ × 1.5 mm × 37) in a beam aperture plate located about 10 cm from a substrate position. Figure 1 shows a typical emission spectrum for the oxygen plasma in the quartz discharge chamber. The spectrum is very similar to the emission spectrum of thermodynamical non-equilibrium high-temperature (TNH) oxygen plasma[7], implying that many atomic oxygen species are created in the discharge chamber. The ion current of flux from the source was the order of 1 μA. The electrical measurements indicate that the oxygen source provides very little ion current, which decreases with increasing flow rate of oxygen gas at the same RF-power. This result suggests that the majority of the active fraction of total flux is either excited free radicals or otherwise activated molecular species.

During growths the plasma was fed a steady flow of oxygen (5×10^{17} s^{-1}). Taking the production efficiency of activated species by the plasma to be 10 %, the flux of the activated oxygen at the substrate is roughly estimated to be about 1×10^{16} $cm^{-2}s^{-1}$. Background pressure in the growth chamber was maintained below 4×10^{-8} Torr by flow of liquid N_2 into cryopanels. At the such pressure, the mean free path of the metal atoms or molecules is much greater than the source-to-substrate distances. This allows precise control of line-of-sight beams by means of shuttering.

The growths reported here were carried out on (100) MgO substrates that were polished single crystals. The substrates heated at 500 ℃ were cleaned in situ by applying activated oxygen species, while being monitored by the reflection high-energy electron diffraction (RHEED) system. Prior to growth streaky RHEED patterns characteristic of a single-crystal surface were obtained. Growths were done at substrate temperature ranging from 650 to 750 ℃ as measured using optical pyrometer. Typical growth rate were 2500-4000 Å /h. Each source was sequentially supplied to the substrate providing the activated oxygen beams. The sequence was a (-Bi-Sr-Cu-Ca-Cu-Sr-Bi-) pattern[5], which was typically repeated 100 times to grow films. After growth, the deposited films were cooled down below 300 ℃ for about 20 min with turning on the RF power of the oxygen source.

RESULTS AND DISCUSSION

We tried to prepare CuO and Cu films on a same substrate by shadowing the beams from the substrate in order to investigate the directivity of the radical beams. Figure 2 shows the surface of CuO and Cu films on the same MgO substrate. In the region shaded from the activated oxygen beams by a mask that was placed in front of the substrate surface at a distance of 4 mm, a non-oxidized film is formed. In other region, which applied activated oxygen species, an oxidized film is formed. The width of boundary between oxidized and non-oxidized films is about 10-20μm. This result suggests that the beams of activated oxygen species from the oxygen source has sharp directivity.

Sample 1 was synthesized by a 120-cycle deposition and had a 3800-Å thickness. Figure 2 shows temperature dependence of resistivity for as-deposited film and annealed one for the sample 1. The as-deposited film has a broad transition starting at approximately 100 K and reaching zero at 44 K . Such behavior may indicate insufficient oxygen content or existence of many stacking faults.

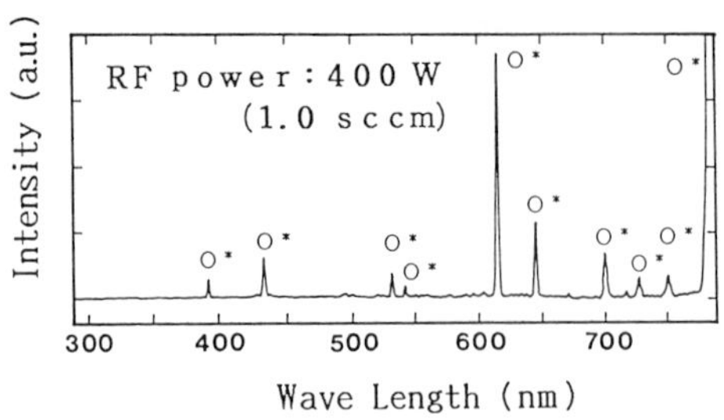

Fig. 1 Typical emission spectrum of the oxygen plasma in the discharge chamber.

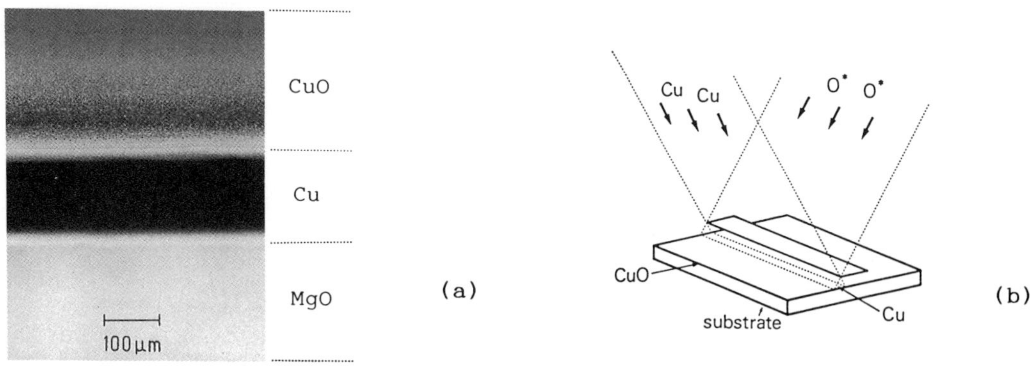

Fig. 2 (a)Film surface of CuO and Cu films on the same MgO substrate. In the region shaded from activated oxygen beams by a mask, as shown in (b), a non-oxidized film is formed.

Figure 4(a) shows an x-ray diffraction pattern from θ -2θ scan of the as-deposited film for the sample 1. Periodical peaks appear in the pattern, but these peaks could not be identified as peaks of Bi-based crystal structures. It should be noted that a peak appears in the low angle region at 3.22 degree. The periodical distance estimated from this peak is 27.4 Å. This result suggests that the film includes the layered structure as long as 27.4 Å, which may correspond to a metastable 2201/2212 phase; the phase consists of a structure alternating 2201 and 2212 half unit cells. When the film was annealed at 840 ℃ for 1 hour in oxygen atmosphere, it showed a superconducting transition around 120 K and a zero resistance at 85 K. The XRD pattern of annealed film is showed Fig. 4(b). The peak at 3.22 degree does not appear. The XRD pattern showed that 2212 phase was formed by annealing of the film and the artificial periodicity of layered structure as long as 27.4 Å disappear in the film.

Films were deposited at substrate temperatures ranging from 650 to 750 ℃ and measured the temperature dependence of resistivity. As shown in an inset of Fig. 5, the superconducting transition temperatures of as-deposited films have a tendency as increasing with increasing the substrate temperature in depositions. The $Bi_{1.92}Sr_{2.22}Ca_{1.72}Cu_{3.00}O_x$ film deposited at 750 ℃ shows a superconducting transition at 75 K with zero resistance. According to analysis for the XRD pattern, the film had 2212 phase as a main phase.

Figure 6 shows XRD patterns of as-deposited films synthesized with different compositions. We have succeeded preparations of films with long c-axis structures. Shutters of sources were operated with the 2212 phase pattern during each deposition. The films with long c-axis structures were formed by increasing opened times of shutters for Ca and Cu sources: chemical composition of Ca and Cu per one shuttering period were increase. The lattice constants of c-axis for

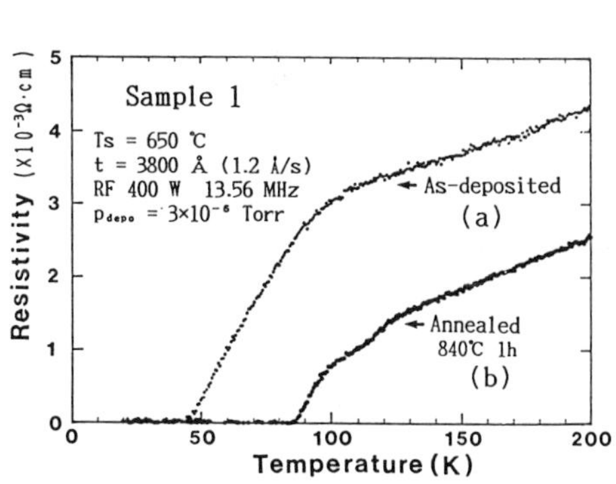

Fig. 3 Temperature dependence of resistivity for an as-deposited $Bi_{1.92}Sr_{2.01}Ca_{1.21}Cu_{2.00}O_x$ film and the same film annealed at 840 ℃ for 1h.

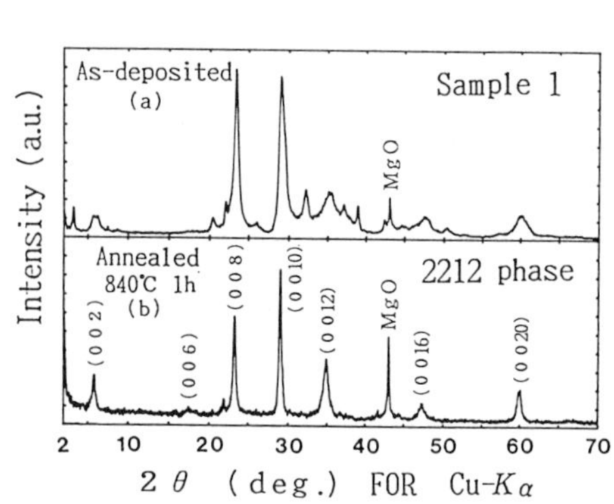

Fig. 4 X-ray diffraction patterns of as-deposited and annealed film for sample 1.

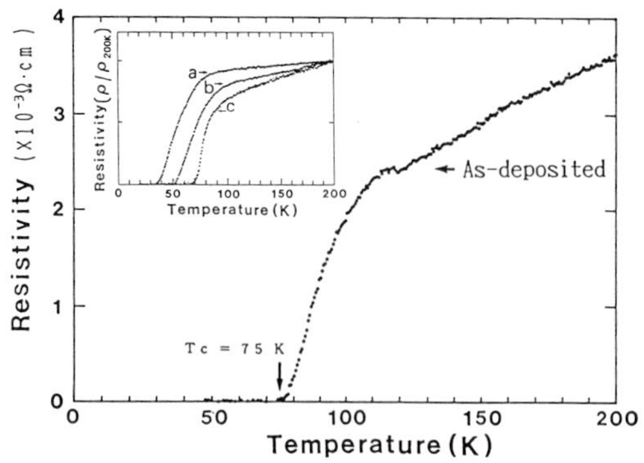

Fig. 5 Temperature dependence of resistivity for an as-deposited film (T_s=750 ℃). Inset shows temperature dependence of resistivity for the Bi based films deposited with different substrate temperature. (T_s=(a)650 ℃, (b)700 ℃ and (c)750 ℃)

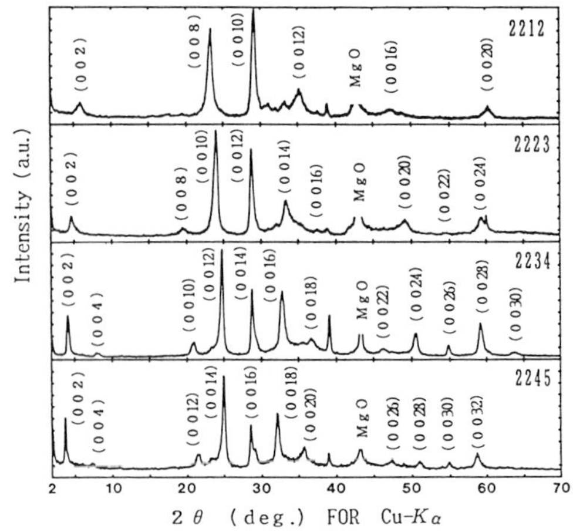

Fig. 6 X-ray diffraction patterns for as-deposited 2212, 2223, 2234 and 2245-like films.

2212, 2223, 2234 and 2245-like films as-grown were 30.6 Å, 37.4 Å, 43.8 Å and 50.4 Å, respectively. 2234 and 2245 phases were not synthesized in bulk samples. The superconducting transition temperatures of 2234 and 2245-like films were 67 K and 25 K, with zero resistance, respectively. However, these T_c's do not seem to be inherent values of the phases.

SUMMARY

Thin films of the Bi-Sr-Ca-Cu-O compound have been grown by MBE using oxygen radical beams as a reactive oxygen source. The as-grown films are superconducting and have shown complete resistive transitions without in-situ annealing. The highest superconducting transition temperature was 75 K with zero resistance. XRD analysis indicated the films to be c-axis oriented perpendicular to the film surfaces. 2234 and 2245-like films were synthesized and showed superconducting transitions. In addition, the activated oxygen beams via beam aperture plate made of quartz have sharp directivity. Therefore, films can be oxidized selectively by shadowing the radical beams from the substrates. In this experiments, however, some problems still remain. The high resolution SEM images show that the micrograins exist on the surface of films. Therefore, the RHEED patterns are spotty. With optimized preparation conditions, smooth epitaxial films could be grown using oxygen radical beams.

REFERENCES

1. Eckstain JN, Bozovic I, von Desonneck KE, Schlom DG, Harris,Jr JS, Baumann SM (1990) Appl Phys Lett 57: 931-933
2. Spah RJ, Hess HF, Stormer HL, White AE, Short KT (1988) Appl Phys Lett 53: 441-443
3. Kwo J, Hong M, Trervor DJ, Fleming RM, White AE, Farrow RC, Kortan AR, Short KT (1988) Appl Phys Lett 53: 2683-2685
4. Eckstein JN, Schlom DG, Hellman ES, von Desonneck KE, Chen ZJ, Webb C, Turner F, Harris,Jr JS, Beasley MR, Geballe TH (1989) J Vac Sci Technol B7: 319-323
5. Berkley DD, Johnson BR, Anand N, Beauchamp KM, Conroy LE, Goldman AM, Maps J, Mauersberger K, Mecartney ML, Morton J, Tuominen M, Zhang Y-J (1988) Appl Phys Lett 53: 1973-1975
6. Nakayama Y, Tsukada I, Maeda A, Uchinokura K (1989) Jpn J Appl Phys 28: L1809-L1811
7. Takagi S, Sekiguchi A, Hosokawa N, Terada N, Jo M, Ihara H (1989) Jpn J Appl Phys 28: L952-L954
8. Sato T, Yoshitake T, Miura S, Fujita J, Kubo Y, Igarashi H (1989) Appl Phys Lett 55: 702-704

Fabrication of Bi-Sr-Ca-Cu-O Thin Films with Layer-by-layer Deposition

K. YOSHIKAWA and N. SASAKI

Advanced Technology Division, Fujitsu Limited, 1015 Kamikodanaka, Nakahara-ku, Kawasaki, 211 Japan

ABSTRACT

We have successfully prepared selectively a single phase as-grown $Bi_2Sr_2Ca_{n-1}Cu_nO_y$ (n = 2, 3, 4) thin films with layer-by-layer deposition using shutter control. The deposition was carried out by reactive evaporation.
As-grown 550 Å thick film formed at 680 °C shows Tc(onset) of 110 K and zero-resistivity at 71 K. Quite smooth surface is obtained. Surface roughness is 20 Å. Precipitate density is decreased to $0.3\mu m^{-2}$ from $25\mu m^{-2}$ of the film by co-evaporation.

KEY WORDS: Bi-Sr-Ca-Cu-O, layer-by-layer, reactive-evaporation, single phase, surface morphology

INTRODUCTION

So many research have been carried out to obtain single phase Bi-Sr-Ca-Cu-O. Pb doping is most popular method for this purpose[1-2]. However, in-situ Pb doping at the temperature above crystallization temperature of 2223 phase is difficult because of high vapor pressure of Pb. Layer-by-layer deposition is promising technique for obtaining single phase Bi-Sr-Ca-Cu-O compared with co-evaporation.
In this study, we report selective fabrication of single phase Bi-Sr-Ca-Cu-O using layer-by-layer deposition.

EXPERIMENTAL

Films were grown on (100) $SrTiO_3$ substrate by reactive evaporation technique [3]. Layer-by-layer deposition were carried out using shutter. Bi_2O_3 and SrF_2 were evaporated from EB guns, and Ca and Cu metals from resistively heated boats. Oxygen was introduced directly onto the substrate. RF plasma was generated between the source and the substrate to enhance oxidation and crvstallization. Growth conditions are summarized in Table 1.

Table 1. Growth Conditions.

Substrate	(100) $SrTiO_3$
Substrate temperature	680 °C
Growth rate	0.2 - 0.3 Å/s
Oxygen pressure	1×10^{-4} Torr
RF power	100 W

Deposition sequence were Bi-Sr-Cu-Ca-Cu-Sr-Bi for the growth of 2212 phase and 2212+2223 phase (described in the next section) , Bi-Sr-Cu-Ca-Cu-Ca-Cu-Sr-Bi for 2223 and 2234 phase. The sequence was repeated 30 times for all films to grow from 400 to 800 Å thick films. After deposition, so-called in-situ annealing was carried out; substrate temperature was kept at 680 °C at the oxygen pressure of 2×10^{-4} Torr for 20 minutes, then cooled down to 500 °C and finally 760 Torr of oxygen were introduced into chamber to cool the film to room temperature.

The crystal structure was studied using X-ray diffraction (XRD). Resistivity was measured with the conventional four-probe method. The surface morphology was examined using High Resolution Scanning Electron Microscopy (HRSEM) and α-step.

RESULTS and DISCUSSION

Figure 1 shows the XRD patterns of the films. Deposition sequence for film A is Bi-Sr-Cu-Ca-Cu-Sr-Bi. Figure 1 shows film A is a single 2212 phase whose lattice constant is 30.9 Å. By increasing the deposition time of Cu and Ca, film C has grown. The lattice constant of film C is 34.1 Å. This value means 2212 and 2223 phases stacked each other in every half unit cell[4]. This phase was also appeared in co-evaporation in our previous work [3].

By increasing the deposition of Cu and Ca in the sequence to Bi-Sr-Cu-Ca-Cu-Ca-Cu-Sr-Bi, film B was obtained. Figure 1 shows film B is single 2223 phase whose lattice constant is 37.4 Å. By increasing the deposition time of Ca and Cu, film D has grown. Figure 1 shows film D is almost single 2234 phase. The lattice constant is 44.3 Å.

Temperature dependence of resistivity are shown in Fig.2. All films show metallic dependence. The best Tc value of as-grown films was Tc(zero)=71 K and Tc(onset)=110K for film C.

Surface morphology of films are investigated, as shown in Fig. 3; these films were not in-situ annealed after deposition. Surface morphology is quite smooth when the number of deposition cycles agrees with the number of half unit cell estimated from 2d / c , where d is film thickness and c is lattice constant.

Figure 4 shows HRSEM photograph of film A. Film thickness is 480 Å. The number of deposition cycles (30) are almost equal to the number of half unit cell (2 x 480 Å / 30.8 Å = 31). The film morphology is quite smooth and featureless (surface roughness is about 20 Å). Precipitate density is decreased to 0.3 μm^{-2} from 25 μm^{-2} of the film by co-evaporation.

CONCLUSION

We have successfully fabricated each single phase of $Bi_2Sr_2Ca_{n-1}Cu_nO_y$ (n = 2, 3, 4) with layer-by-layer deposition using reactive evaporation technique. The best Tc of as-grown films is Tc(onset) = 110 K and Tc(zero) = 71 K. All films show metallic temperature dependence of resistivity.
Surface morphology is quite smooth when the number of deposition cycles agrees with the number of stacked half unit cell. Surface roughness is 20 Å and precipitate density is 0.3 μm^{-2}.

REFERENCES

1. M. Takano, J. Takada, K. Oda, H. Kitaguchi, Y. Ikeda, Y. Tomii, and H. Mazaki (1988) Jpn. J. Appl. Phys.,27: L1041
2. K. Endo, S. Hayasida, K. Nakatsuka, J. Ishiai, Y. Ikedo, S. Misawa, and S. Yoshida, Proceeding of 2nd International Symposium on Superconductivity, November 14-17, 1989, Tsukuba, Japan. pp 911 - 914
3. K. Yoshikawa, T. Satoh, N. Sasaki, and M. Nakano (1989) Proceedings of Material Research Society Fall Meeting, Boston, Massachusetts, (to be published)
4. K. Kojima, D. Schlom, K. Kuroda, M. Tanioku, K. Nakayama, J. Eckstein, and J. Harris (1990) Jpn. J. Appl. Phys.,29: L1638

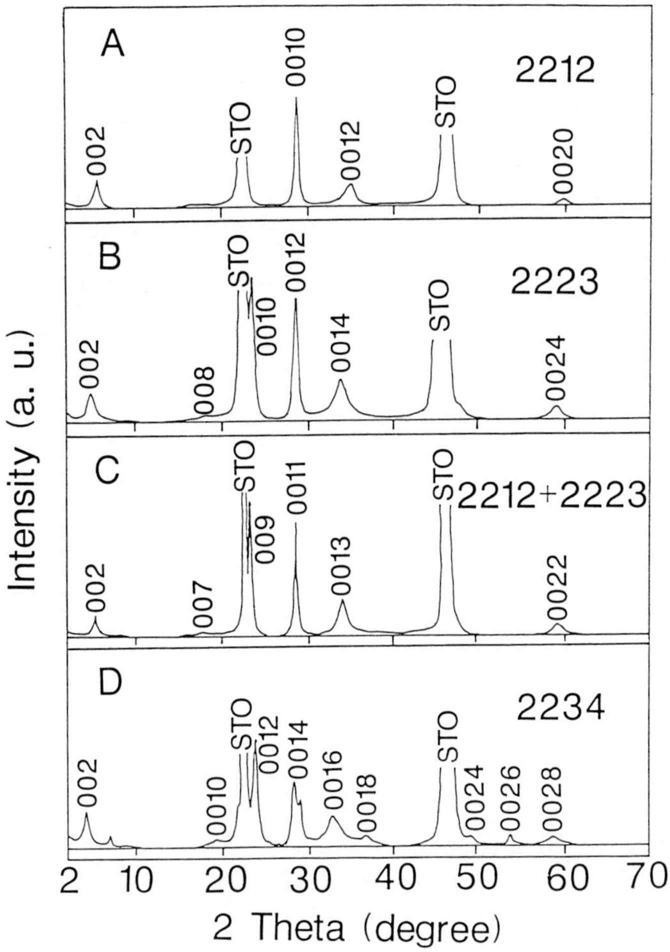

Fig. 1. X-ray diffraction patterns of the as-grown films. 2212, 2223, 2212+2223, 2234 phases are formed (from A to D) by change the deposition sequence and deposition time of Ca and Cu.

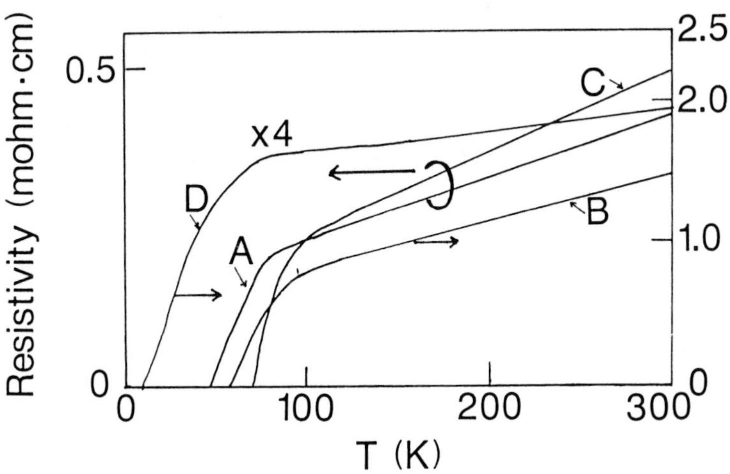

Fig. 2. The temperature dependence of the resistivity.

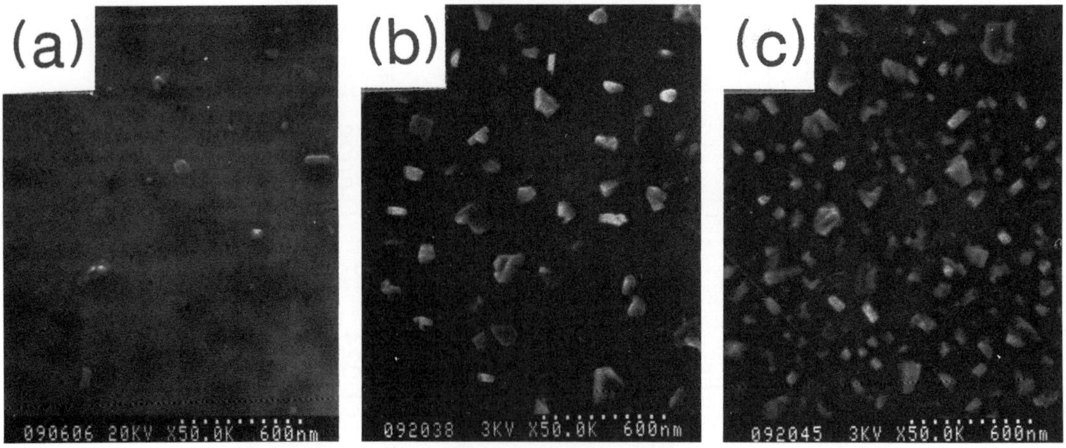

Fig. 3. HRSEM images of 2223 phase films.
(a) 50 cycles of deposition grew 47 half unit cells.
(b) 30 cycles of deposition grew 39 half unit cells.
(c) 20 cycles of deposition grew 44 half unit cells.

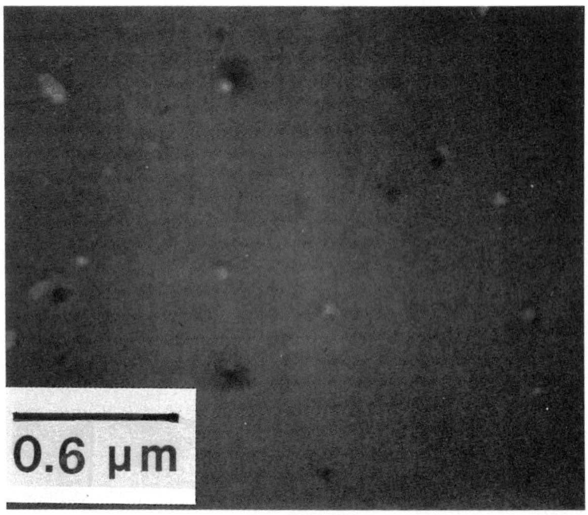

Fig. 4. HRSEM image of film (A). 30 cycles of deposition grew 31 half unit cells.

Growth Characteristics of Bi-Sr-(Ca)-Cu-O Films Prepared by ECR Plasma Assisted Sequential Deposition

M. KAMEI, I. YOSHIDA, K. SUZUKI, T. MORISHITA, and S. TANAKA

Superconductivity Research Laboratory, International Superconductivity Technology Center, 10-13, Shinonome 1-chome, Koto-ku, Tokyo, 135 Japan

ABSTRACT

The growth characteristics of sequentially evaporated Bi-Sr-(Ca)-Cu-O was investigated using in-situ RHEED observation. Particular amorphous like surface structure was observed during the layering of Ca-Cu. The crystallization of Ca-Cu layers is the point to get a $Bi_2Sr_2Ca_2Cu_3Ox$ single phase film.

KEY WORDS: Bi-Sr-Ca-Cu-O, sequential evaporation, in-situ RHEED

INTRODUCTION

Since the discovery of layered perovskite Bi-Sr-Ca-Cu-O with high superconducting transition temperature, much attention has been focused on the thin film synthesis of this material for electronics applications. Almost all the deposition technics have been applied with increasing success [1], [2], [3]. Since Bi-Sr-Ca-Cu-O has many phases with the different number of the Ca-Cu-O layer, sequential evaporation might be one of the most promising methods for tailored growth of each compound. In this letter we report the growth characteristics of sequentially evaporated Bi-Sr-(Ca)-Cu-O films at each atomic layer, using an in-situ RHEED system.

EXPERIMENTAL

A schematic diagram of the apparatus involving the RHEED system is shown in Fig.1. Bi_2O_3 and metallic Cu were heated by electron beams, and metallic Sr and Ca were evaporated using the Knudsen cell. Two

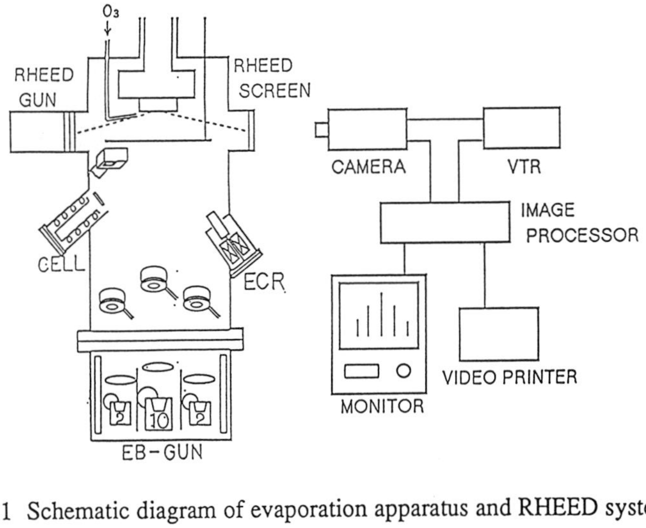

Fig.1 Schematic diagram of evaporation apparatus and RHEED system.

kinds of sensors were used for monitoring an evaporation rate: quartz-crystal thickness sensors for Bi_2O_3 and Cu fluxes and electron impact emission spectroscopy (EIES) sensors [4] for Sr and Ca fluxes. ECR oxygen plasma of a current density of 3μA/cm² near the substrate and 13 wt.% ozone gas were used as activated oxygen sources. The background pressure during deposition was 7.5×10^{-5} Torr.

Independent shutters for each source controlled by a microcomputer were adopted to supply the constituents to the substrate. The deposition was initiated from Bi followed by Sr,Cu,(Ca),etc. in order of atomic layers in the unit cell, ending a period of the sequence with Bi. The shutter timing was estimated from

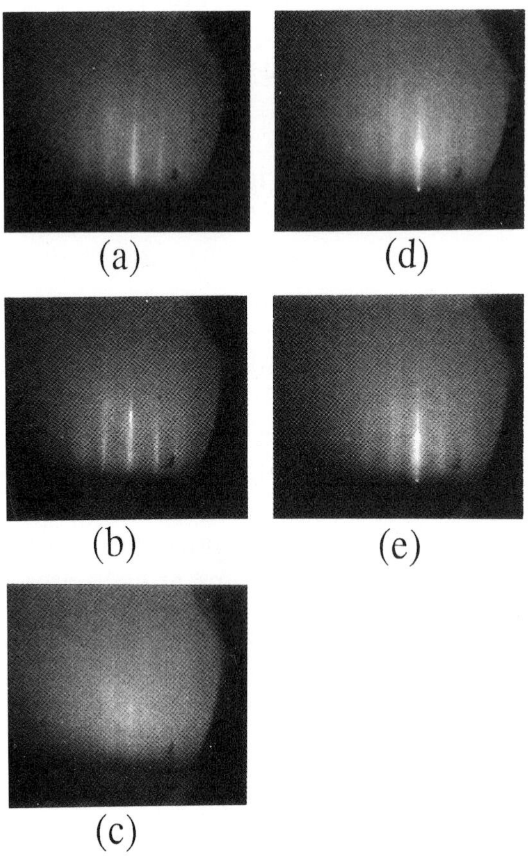

Fig.2 In-situ RHEED patterns observed during sequential deposition of $Bi_2Sr_2Ca_2Cu_3O_x$ along azimuthal direction of <100>.

(a) after the first period of sequence (covered with Bi_2O_3)
(b) after Sr deposition on (a)
(c) after Ca-Cu deposition on (b)
(d) after Sr deposition on (c)
(e) after Bi_2O_3 deposition on (d)

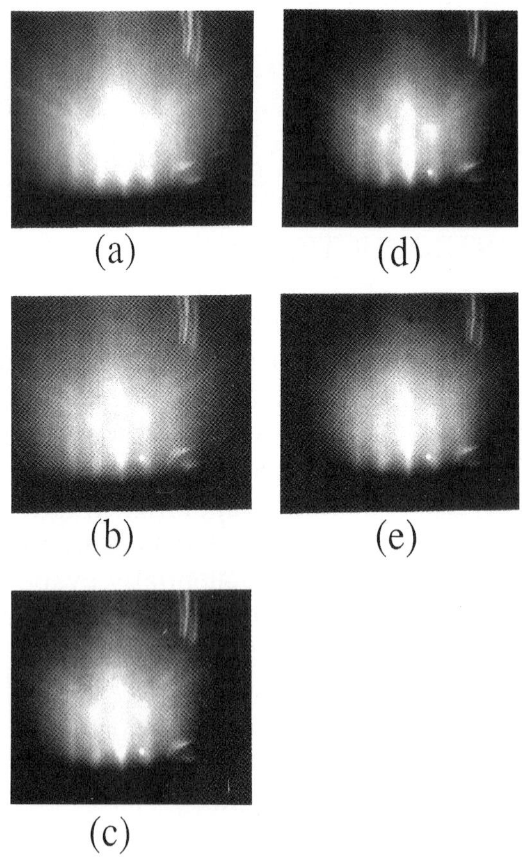

Fig.3 In-situ RHEED patterns observed during sequential deposition of $Bi_2Sr_2Cu_1O_x$ along azimuthal direction of <100>

(a) after the first period of sequence (covered with Bi_2O_3)
(b) after Sr deposition on (a)
(c) after Cu deposition on (b)
(d) after Sr deposition on (c)
(e) after Bi_2O_3 deposition on (d)

the flux density for each element. The sequential deposition corresponding to the chemical composition of $Bi_2Sr_2Ca_2Cu_3Ox$ and $Bi_2Sr_2Cu_1Ox$ were made on MgO (100) substrates at a substrate temperature of 725°C. During growth the crystal structure of films was monitored using RHEED. The films obtained were characterized by x-ray diffraction and inductively coupled plasma spectroscopy.The RHEED system equipped with an electron gun differentially pumped and a video recorder allows for in-situ observations of the nucleation and growth process on a substrate as it is rotating, which enables to homogeneously deposit even during observation of RHEED patterns, and permits to edit the patterns along any azimuth as required after deposition.

RESULTS AND DISCUSSION

The RHEED patterns correspond to the second period of the deposition sequence of $Bi_2Sr_2Ca_2Cu_3Ox$ are shown in Fig.2. During the sequence of Bi and Sr there is no difference in the streak patterns (Fig.2(a),(b)). However, during the layering of Ca-Cu, the streak pattern disappears and a dim pattern suggesting an amorphous structure appears (Fig.2(c)). When the next Sr flux was deposited, the streak pattern gets appeared and is completed during the next Bi flux sequence (Fig.2 (d),(e)). After 50 periods of deposition sequences were made, no streak pattern was observed at any deposition sequence but a spotty pattern correspoding to the segregated CaO could be observed. The crystal structure of this film was investigated using an x-ray diffractometer. The structure of the film shows a mixed phase of $Bi_2Sr_2Ca_1Cu_2Ox$ and $Bi_2Sr_2Cu_1Ox$ although the shuttering sequence was set to form that of $Bi_2Sr_2Ca_2Cu_3Ox$. The weak diffraction peaks of the film compared with that of the substrate suggest that the crystallinity is not so good.

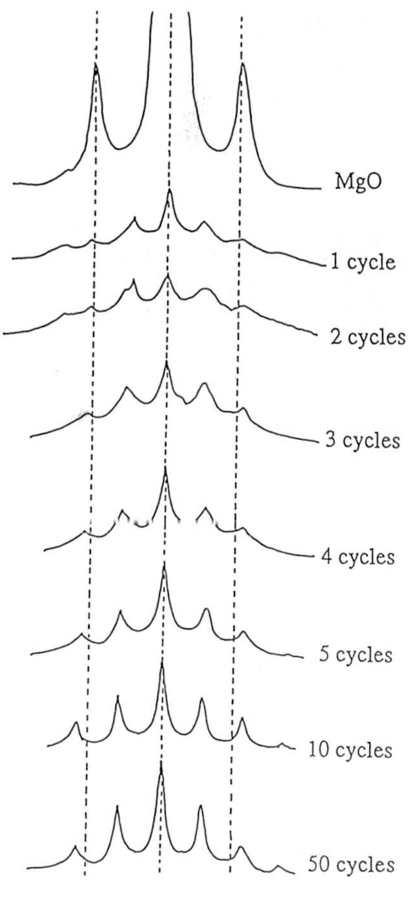

Fig.4 Change of lattice spacing during the growth of $Bi_2Sr_2Cu_1Ox$ on MgO (100) substrate.

For the case of the deposition sequence of $Bi_2Sr_2Cu_1Ox$ (without Ca), the RHEED patterns showed a different character as shown in Fig.3. At any sequence of evaporation, the RHEED pattern keeps the streaky one. Until the fifth period of the sequence, the lattice spacing decreases (the spacing between streaks increases) with thickness, resulting in being constant after that (Fig.4). The film is expanded by the MgO lattice (4.2Å) at the beginning of deposition while recovering to its bulk value (3.8Å). Even after 50 periods of deposition sequences, a clear streak pattern was observed. X-ray diffraction pattern of this film revealed that this film was a single phase film of $Bi_2Sr_2Cu_1Ox$ with rather good crystallinity.

These results suggest that crystallization of Ca and Cu layers is the point in order to get a $Bi_2Sr_2Ca_2Cu_3Ox$ single phase film.

CONCLUSION

The shuttering sequential deposition combined with in-situ RHEED observation has successfully performed to reveal the difference of the crystal growth mechanism between $Bi_2Sr_2Ca_2Cu_3Ox$ and $Bi_2Sr_2Cu_1Ox$ especially at the sequence of Ca. $Bi_2Sr_2Cu_1Ox$ can be epitaxially formed by the sequential atomic layer deposition, while $Bi_2Sr_2Ca_2Cu_3Ox$ cannot complete its crystal structure unless crystallization of Ca and Cu layers is realized.

ACKNOWLEDGEMENT

This work was supported by New Energy and Industrial Technology Development Organization under the management of the R&D of Basic Technology for Future Industries.

REFERENCES

1. Nakayama Y, Ochimizu H, Maeda A, Kawazu A, Uchinokura K, Tanaka S (1989) Jpn. J. Appl. Phys. 28: L1217

2. Fujita J, Tatsumi T, Yoshitake T, Igarashi H (1989) Appl. Phys. Lett. 54(23): 2364

3. Kanai M, Kawai T, Kawai S, Tabata H, (1989) Appl. Phys. Lett. 54:1802

4. Gogol C, Cipro C (1985) Proc. 1st. Int. Symp. Silicon Molecular Beam Epitaxy 857:415

In-Situ Growth of Sequentially Deposited Bi-Sr-Ca-Cu-O Thin Films by Molecular Beam Epitaxy Technique

ICHIRO TSUKADA, YOSHIMI NAKAYAMA*, HIROSHI WATANABE, ICHIRO TERASAKI, ATSUTAKA MAEDA, and KUNIMITSU UCHINOKURA

Department of Applied Physics, The University of Tokyo, Hongo, Bunkyo-ku, Tokyo, 113 Japan

ABSTRACT

Thin Films of $Bi_2Sr_2Ca_{n-1}Cu_nO_y$(n=1~5) were prepared *in situ* by molecular beam epitaxy technique. All films were deposited sequentially by shutter control using pure ozone gas. Films of n=2, 3, 4 and 5 exhibit superconductivity below 65, 40, 75 and 30 K without post annealing process. The resistivity measurements in magnetic fields were performed for the film of n=4. The upper critical field and GL coherence length are estimated.

KEY WORDS: Bi-Sr-Ca-Cu-O Superconductor, Molecular Beam Epitaxy, Thin Film, Upper Critical Field

INTRODUCTION

In recent years, great advancement has been achieved in preparing technique of high-T_c superconducting thin films by molecular beam epitaxy (MBE). In order to oxidize Cu at the pressure of MBE regime (at least below the order of 10^{-3} Pa), it is necessary to provide activated oxygen. Since Berkley *et al.*[1] first reported that pure ozone was very effective for oxidizing $YBa_2Cu_3O_y$, we have applied this technique for preparing Bi-Sr-Ca-Cu-O thin films by co-evaporation technique[2]. After that, we have applied sequential evaporation technique and observed epitaxial growth[3,4], and succeeded in preparing an ultrathin film (thickness≈62 Å) which showed superconductivity below 38 K[5]. There are several structures in Bi-Sr-Ca-Cu-O system which are distinguished by the number of CuO_2 planes n in their half unit cell, and described as $Bi_2Sr_2Ca_{n-1}Cu_nO_y$. So it is possible to synthesize structures with various n, and additionally to construct the stackings of layers with various n. Schlom *et al.* have succeeded in heteroepitaxial growth of $Bi_2Sr_2Ca_{n-1}Cu_nO_y$ [6].

In this paper, we report the results of *in-situ* epitaxial growth of $Bi_2Sr_2Ca_{n-1}Cu_nO_y$ thin films of n=1~5. Films of n>2 showed superconducting transition, and highest T_c was obtained in n=4 films grown on $SrTiO_3$(001) substrates. All films were characterized by x-ray diffraction and temperature dependence of resistivity. Resistivity of n=4 films was measured in magnetic fields in order to estimate the upper critical field H_{c2} and GL coherence length.

SAMPLE PREPARATIONS

All films were prepared in a MBE machine as was previously reported[5]. Our MBE machine mainly consists of growth chamber, preparation chamber and pure ozone providing equipment. We used metallic Bi, Sr, Ca and Cu for sources, which were evaporated by using Knudsen cells. Pure ozone was provided into the growth chamber during growth and cooling down processes. The pressure of ozone gas was monitored by quadrupole mass analyzer.

We used $SrTiO_3$(001) and MgO (001) for substrates. They were mounted on substrate holder made of Inconel by silver paste to be heated uniformly. Substrate temperature was measured by an optical pyrometer. All sources were evaporated sequentially with shutter control. Typical deposition times (approximately corresponding to one atomic layer) are shown in Table I, which are determined by inductively coupled plasma atomic emission spectroscopy (ICP). Typical substrate temperatures were between 720~800°C, and the pressure was maintained around 2×10^{-3} Pa for sufficient oxidation.

* The present address: Department of Industrial Chemistry, The University of Tokyo

Table I Typical deposition time of each sources.

sources	Bi	Sr	Ca	Cu
temp.(°C)	490	530	489	1080
time(sec)	7.8	4.0	4.0	4.7

Table II Chemical compositions of the films with n=2, 4 and 5.

elements	Bi	Sr	Ca	Cu
n=2	1.86	1.92	1.76	2.72
n=4	1.76	1.99	2.73	3.79
n=5	1.98	1.62	4.00	4.74

CHARACTERIZATION OF THE FILMS WITH n=1 ~ 5 PHASES

All films were characterized by x-ray diffraction and temperature dependence of resistivity. The results of x-ray diffraction of the films prepared under the condition that substrate temperature was kept around 720~760°C are shown in Fig. 1[4]. All films are highly c-axis oriented. The c-axis lengths of these films were about 24.7, 30.9, 37.1 and 43.0 Å for n=1, 2, 3 and 4 films, respectively. It shows that the sequential deposition is very useful technique to synthesize structures that we want to obtain. Among these films, n=2, 3 and 4 films showed superconducting transitions at 50, 40 and 33 K, respectively[4,5]. These values are, however, not intrinsic for these phases, so we thought that the deposition conditions were still not good.

After these experiments, we increased substrate temperature up to 780~800°C. Under the new conditions, n=2, 4 and 5 films were synthesized on $SrTiO_3$(001) and MgO (001). Chemical compositions in half unit cell investigated by ICP are shown in Table II. These values show large deviations from the ideal stoichiometric values. Especially, in n=2 phase Ca and Cu were too much contained and the ratio of each element was rather close to those of n=3 phase,but the excess Ca and Cu did not clearly appear in x-ray diffraction patterns as impurity peaks. X-ray diffraction patterns and temperature dependence of the resistivity of these films are shown in Figs. 2 and 3. In Fig. 2(a), each (00l) peak of the film of n=2 becomes sharper than the corresponding one in Fig. 1. A similar result is also observed in the film of n=4 which is shown in Fig. 2(b), in which the c-axis length of this n=4 film is obtained as 43.3 Å. The values of the resistivity at room temperature decreased and the transition temperatures increased up to 65 and 75 K (Figs. 3(a) and (b), respectively). The improvements could be clearly seen also in the transition temperature. Especially the transition width of the n=4 film becomes narrower than 15 K.

The synthesis of the n=5 films was also attempted. The results of x-ray diffraction and temperature dependence of the resistivity of the film on MgO (001) are shown in Figs. 2(c) and 3(c). This film shows superconducting transition at 30 K. The films deposited on $SrTiO_3$(001) were mixed-phase with n=4 and 5. Only the film on MgO (001) was single-phase with n=5. Further experiments to determine the basic growth conditions are still necessary to prepare the films of $n \geq 5$.

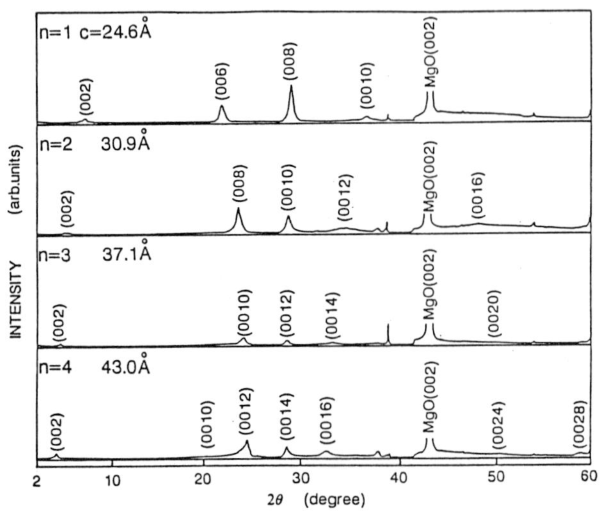

Fig. 1. X-ray diffraction patterns of the films of n=1~4 at early stage of experiments[5].

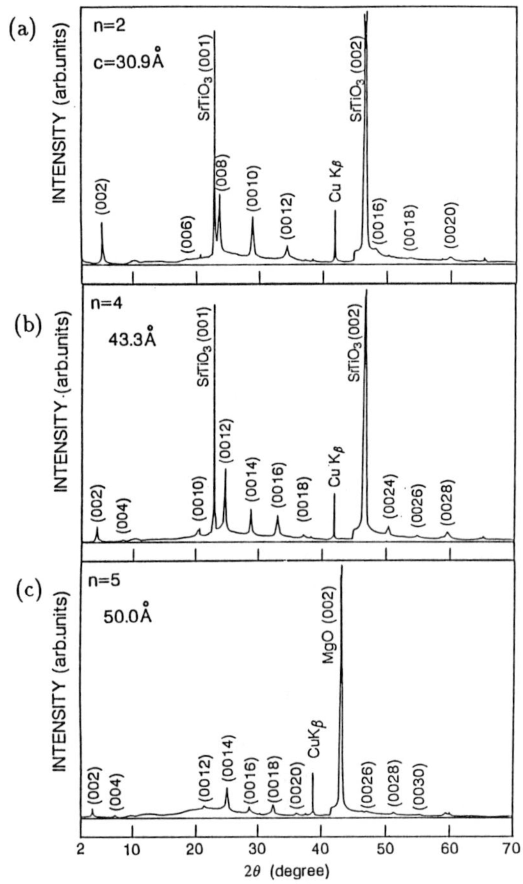

Fig. 2. X-ray diffraction patterns of the films of (a) n=2, (b) n=4 and (c) n=5 deposited under the ozone pressure of 2×10^{-3} Pa and the substrate temperatures of 780~800°C.

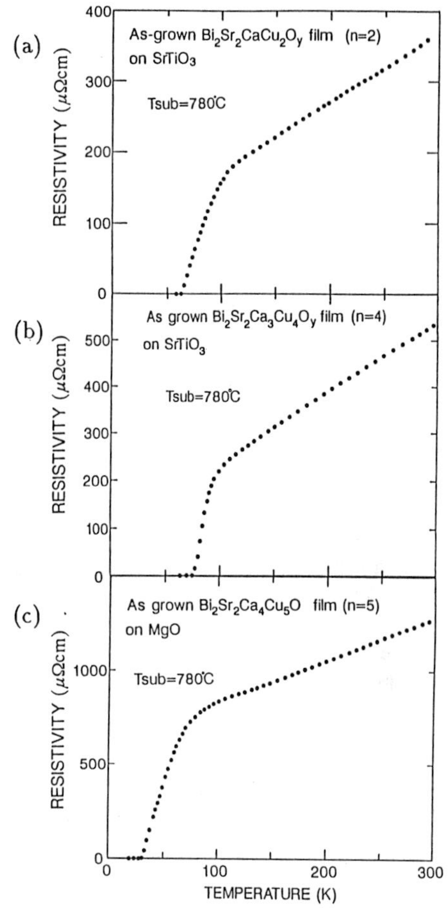

Fig. 3. Temperature dependence of the resistivity of the films of (a) n=2, (b) n=4 and (c) n=5. Samples are the same as the corresponding ones in Fig. 2.

RESISTIVITY OF THE FILM WITH n=4 IN MAGNETIC FIELDS

There arises a problem whether coherence length along the c-axis (ξ_c) changes with the number of CuO_2 planes n, or not. It is well known that the exact estimation of H_{c2} from the resistivity data is impossible for high-T_c oxides, but the resistive transition in magnetic fields still gives us brief information about H_{c2}. So, the resistivity of the n=4 film was measured in magnetic fields below 6 T. Magnetic field was applied to the film parallel and perpendicular to the CuO_2 plane. The experimental results of the broadening are shown Fig. 4. There can be seen a large anisotropy of the resistive transitions as was observed in a single-crystal $Bi_2Sr_2Ca_{n-1}Cu_nO_y$ (n=2: $\xi_{ab}^{GL}(0)$=38 Å and $\xi_c^{GL}(0)$=1.6 Å)[7]. Figure 5 shows transition temperature (determined as the midpoint-temperature) vs applied magnetic field for the film of n=4. According to the WHH theory[8], $H_{c2}^{\perp}(0)$ and $H_{c2}^{\parallel}(0)$ are estimated to be 51 T and 460 T. Then GL coherence lengths $\xi_{ab}^{GL}(0)$ and $\xi_c^{GL}(0)$ can be calculated to be 25.3 Å and 2.8 Å, respectively. Thus, $\xi_c^{GL}(0)$ remains very short with increasing n up to 4.

On the other hand, in the case of $Tl_2Ba_2Ca_{n-1}Cu_nO_y$ system which has almost the same crystal structures as $Bi_2Sr_2Ca_{n-1}Cu_nO_y$ system, $\xi_c^{GL}(0)$ has been estimated to be 6.8 Å in a n=2 single crystal and 10 Å in a n=4 single crystal[9]. In this case, $\xi_c^{GL}(0)$ reaches 10 Å for n=4 phase, which contrasts well with the case of Bi compounds shown above. Although further studies are necessary on many crystals with various qualities, the difference may be due to the difference of the degree of two dimensionality between these two compounds.

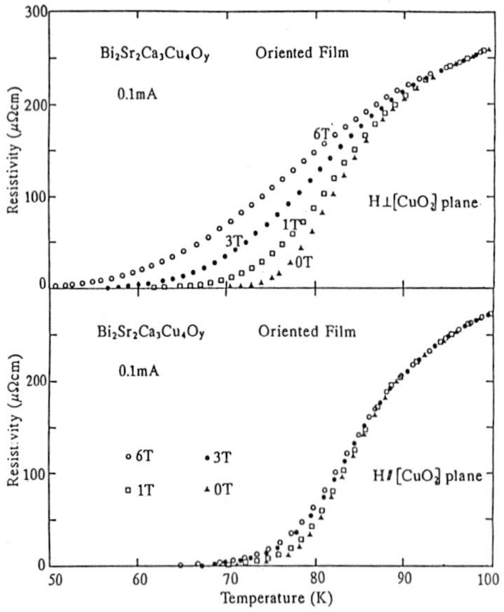

Fig. 4. Temperature dependence of the resistivity of the film of n=4 in magnetic fields. (a) $H \perp CuO_2$ plane, (b) $H \parallel CuO_2$ plane.

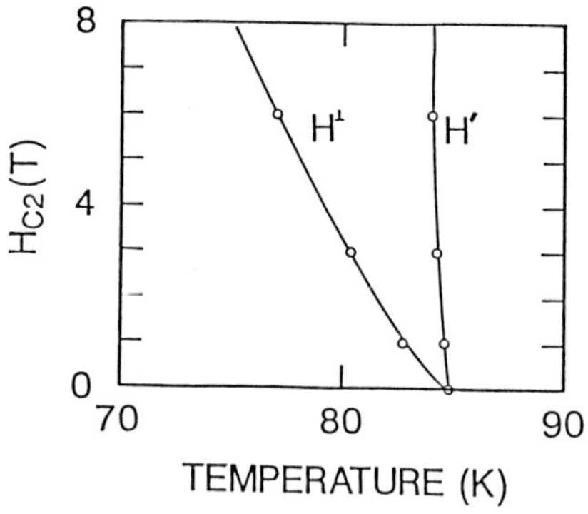

Fig. 5. Transition temperature vs applied magnetic field.

SUMMARY

In conclusion, $Bi_2Sr_2Ca_{n-1}Cu_nO_y$ system superconducting thin films with n=1 to 5 phases were prepared *in situ* by molecular beam epitaxy technique. Films with $n \geq 2$ exhibited superconductivity and one of the n=4 films showed superconducting transition at 75 K. The resistivity of this film was also measured in magnetic fields, and exhibited large anisotropy. GL coherence length $\xi_c^{GL}(0)$ estimated from these experiments suggests that $\xi_c^{GL}(0)$ remained very short with increasing number of CuO_2 planes in $Bi_2Sr_2Ca_{n-1}Cu_nO_y$ system.

ACKNOWLEDGMENTS

The authors would like to thank the members of the Superconductivity Research Laboratory, International Superconductivity Technology Center for the technical assistance in chemical analysis. This work was supported in part by the Grant-in-Aid for Scientific Research from the Ministry of Education, Science and Culture.

REFERENCES

1. Berkley DD, Johnson BR, Anand N, Beauchamp KM, Conroy LE, Goldman AM, Maps J, Mauersberger K, Mecartney ML, Morton J, Tuominen M, Zhang Y-J (1988) Appl. Phys. Lett. 53: 1973-1975
2. Nakayama Y, Ochimizu H, Maeda A, Kawazu A, Uchinokura K, Tanaka S (1989) Jpn. J. Appl. Phys. 28: L1217-L1219
3. Nakayama Y, Tsukada I, Maeda A, Uchinokura K (1989) Jpn. J. Appl. Phys. 28: L1809-L1811
4. Uchinokura K, Nakayama Y, Tsukada I, Terasaki I, Maeda A (1990) To be published in *Proceedings of the Conference on the Science and Technology of Thin-Film Superconductors* Denver Colorado USA April 30-May 4 (New York: Plenum)
5. Tsukada I, Nakayama Y, Terasaki I, Maeda A, Uchinokura K (1990) *Proceedings of the LT-19 Satellite Conference on High Temperature Superconductivity,* in press in Superconductor Science and Technology
6. Schlom DG, Marshall AF, Sizemore JT, Chen ZJ, Eckstein JN, Bozovic I, von Dessonneck KE, Harris Jr. JS, Bravman JC (1990) J. Cryst. Growth 102: 361-375
7. Palstra TTM, Batlogg B, Schneemeyer LF, van Dover RB, Waszczak JV (1988) Phys. Rev. B 38: 5102-5105
8. Werthamer NR, Helfand E, Hohenberg PC (1966) Phys. Rev. 147: 295-302
9. Mukaida H, Kawaguchi K, Nakao M, Kumakura H, Dietderich DR, Togano K (1990) Phys. Rev. B 42: 2659-2661

In-Situ Growth of Bi-Sr-Ca-Cu-O Thin Films by Molecular Beam Epitaxy

SHIGEKI SAKAI, YUJI KASAI, SHUNSUKE HOSOKAWA, and SHINGO ICHIMURA

Electrotechnical Laboratory, 1-4, Umezono 1-chome, Tsukuba, 305 Japan

ABSTRACT

Molecular beam epitaxy technique using an ozone generator as a reactive ozone source is applied to prepare Bi-Sr-Ca-Cu-O thin films. All metal molecular beams are shuttered sequentially to grow films under the control of atomic-layer units. By this way, films with the $Bi_2Sr_2CaCu_2O_x$ and $Bi_2Sr_2Ca_2Cu_3O_x$ phases are prepared. These are superconducting without any post-annealing. Epitaxy is found by observing reflective high energy electron diffraction patterns of a film of the $Bi_2Sr_2CaCu_2O_x$ phase grown on $LaAlO_3$.

KEY WORDS: high temperature superconductor, molecular beam epitaxy, MBE, ozone

INTRODUCTION

The molecular beam epitaxy (MBE) technique has been one of the most important tools for making composition and structure of thin films controlled precisely at the level of atomic layers. This layering capability of MBE seems suitable, not only for preparing copper oxide high temperature superconductors with layered structures, but also for fabricating devices with unique structures such as superlattices or heterojunctions between superconductors and semiconductors. However, copper, being a basic constituent of oxide superconductors, is difficult to be oxidized in a low pressure, and the sticking coefficients of bismuth and thallium are extremely small at high substrate temperature if active oxygen is not supplied. So, special attention for oxygen sources should be paid to accomplishing MBE growth.

In the present paper, we discuss studies on in-situ growth of Bi-Sr-Ca-Cu-O superconductor thin films by MBE having an an ozone generator as a reactive oxygen source.

EXPERIMENTAL

The MBE system for preparing Bi-Sr-Ca-Cu-O superconductor thin films is shown in Fig.1. Bismuth, strontium and calcium are evaporated from effusion cells made of pyrolytic boron nitride, and copper is from an electron beam gun. A new type of ozone generators whose details will be published elsewhere[1] is attached to the MBE machine. The vapor from liquid ozone which is accumulated in a vessel cooled at about 90K is used. The basic idea of this has been known[2], but here a great care is taken in safety and in stability or controllability of ozone flux. A bent stainless tube is introduced from the generator to the MBE chamber. This tube ends at a position near substrates. Furthermore, a quartz tube is extended from this position to a position much closer to the substrates. The nearest distance between the outlet and the substrate is about 5mm, the aperture size of the outlet is about 2mm in diameter and the aperture plane is tilted to the substrates at an angle of about 45 degrees. The total conductance of these connecting tubes is estimated at about 0.01l/s. Films are grown with the control of atomic-layer units, i.e., all metal sources are shuttered sequentially according to the layered structures along the c-axis of $Bi_2Sr_2CaCu_2O_x$ and $Bi_2Sr_2Ca_2Cu_3O_x$. The ozone beam is always irradiated during growth. Substrates are $LaAlO_3(100)$, $SrTiO_3(100)$ and $MgO(100)$. These are soldered by tin on molybdenum sheets. In order to strengthen the solder, titanium, platinum and gold are deposited sequentially both on the surfaces of molybdenum sheets and on the backside of substrates using an ordinary vacuum evaporator. Substrate temperature, Ts, is monitored by an optical pyrometer which measures well calibrated radiations from silicon, soldered by the same way as the substrates are.

RESULTS AND DISCUSSIONS

$Bi_2Sr_2CaCu_2O_x$ (the 2212 phase) thin films were made on wide substrate temperature conditions from 580 ℃ to 770 ℃. All films have the superconducting critical temperature, Tc, of more than 50K without any post-annealing, as shown in Fig.2. Typical growth conditions are listed in Table 1. As far as the present growth conditions are concerned, Tc of films on MgO is small, compared with those on $LaAlO_3$ and $SrTiO_3$. A result of ac susceptibility measurement of a film having Tc of 69K is shown in Fig.3. This is grown on a $SrTiO_3$(100) substrate under the conditions that Ts is 750 ℃, the cycle time of shuttering corresponding to the growth time of the half

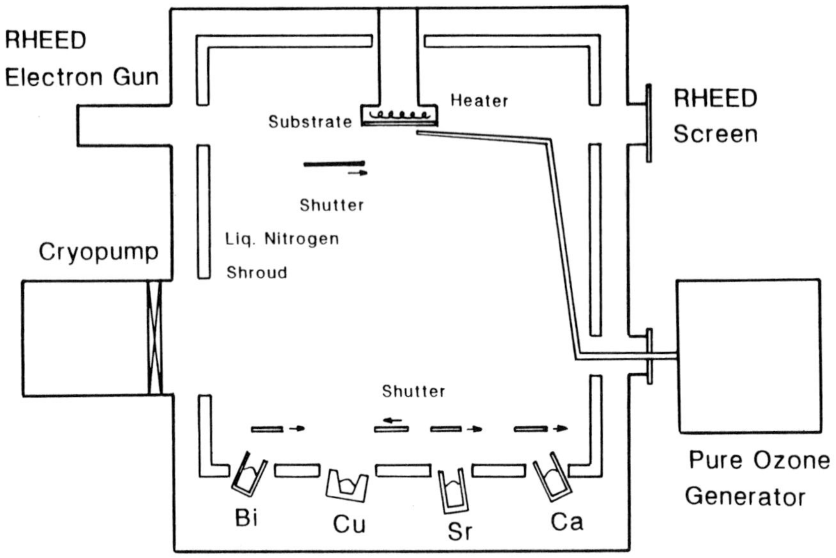

Fig.1 MBE system with an ozone generator for preparing Bi-Sr-Ca-Cu-O superconductor thin films.

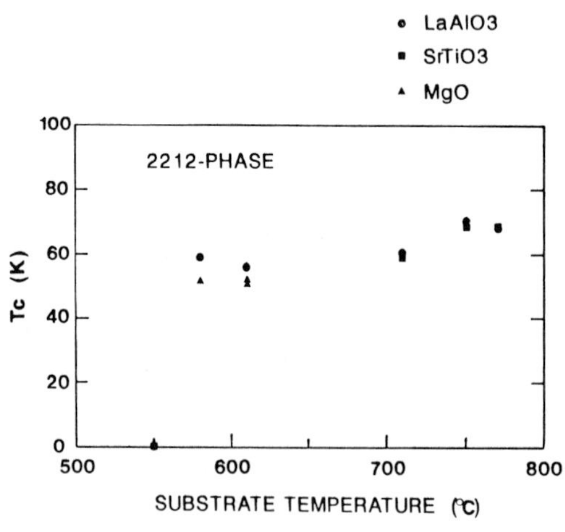

Fig.2 T_c vs. substrate temperature for the growth of $Bi_2Sr_2CaCu_2O_x$ phase.

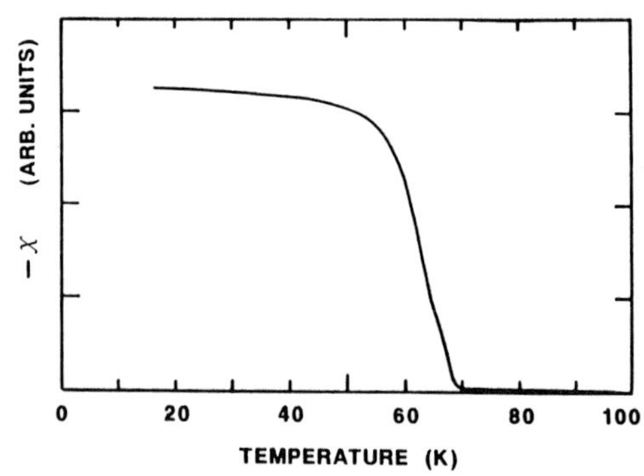

Fig.3 Ac susceptibility χ as a function of temperature. This film referred to as #6 in Table 1 was grown on $SrTiO_3$.

Table 1 Preparation conditions and Tc of $Bi_2Sr_2CaCu_2O_x$ films

sample name	substrate	T_s (℃)	pressure and (temperature) in ozone vessel (Pa, (K))	pressure in growth chamber (Pa)	1 cycle of shuttering growth (s)	total cycle number	T_c (K)
#1	MgO(100)	610	6.7(87.0)	$7x10^{-6}$	59	65	51
#2	$LaAlO_3$(100)	610	11.4(90.0)	$2x10^{-5}$	58	88	56
#3	$LaAlO_3$(100)	580	10.7(90.0)	$1.5x10^{-5}$	57	100	59
#4	$LaAlO_3$(100)	710	21.3(93.0)	$1.1x10^{-4}$	56	100	61
#5	$LaAlO_3$(100)	750	21.5(93.0)	$1.1x10^{-4}$	59	100	70
#6	$SrTiO_3$(100)	750	21.5(93.0)	$1.1x10^{-4}$	59	100	69
#7	$SrTiO_3$(100)	770	21.5(93.0)	$1.1x10^{-4}$	60	90	68

c-axis length is 59s, the total cycle number is 100, the pressure (and the temperature) of the ozone vessel is 21Pa (and 93.0K), and the background pressure in the growth chamber during growth is $1.1x10^{-4}$Pa. Streak patterns by reflective high energy electron diffraction (RHEED) measurement are observed from an early stage of the growth to the last stage, which indicates that the film is epitaxially grown. Figure 4 is the result of the film on $LaAlO_3$ prepared at the same time as the film of Fig.3 is. Figure 4(a) shows the pattern of $LaAlO_3$ before the growth, and (b), (c), (d) show those at the 12th, 24th and 48th cycles during growth, respectively. The streak spacings show that the [110] direction of the 2212 phase coincides with the <100> direction of $LaAlO_3$.

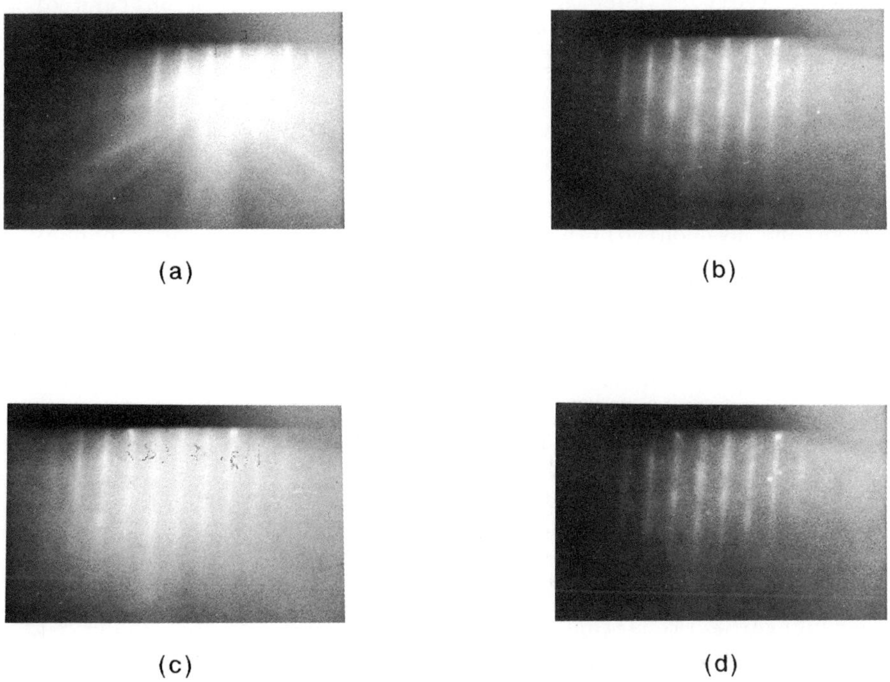

Fig.4 RHEED patterns of a film on $LaAlO_3$ (referred to as #5 in Table 1) during growth. (a) is the pattern of the substrate surface before growth, and (b), (c) and (d) are those at the 12th, 24th and 48th cycles, respectively.

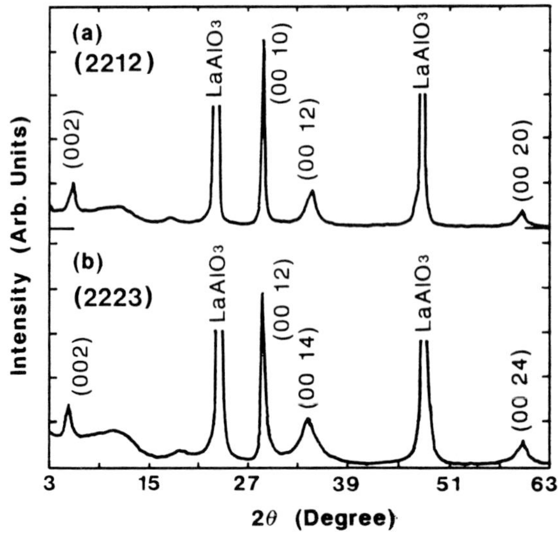

Fig.5 X-ray diffraction patterns of (a) the 2212 phase and (b) the 2223 phase.

$Bi_2Sr_2Ca_2Cu_3O_x$ (the 2223 phase) films are also grown by changing the shuttered conditions from those of the 2212 phase. That is to say, during one cycle, the numbers of the shutter opening of copper and calcium are increased by one, respectively. Figure 5(b) is a X-ray diffraction pattern of the 2223 phase, whereas Fig.5(a) is that of the 2212 phase. At the substrate temperature from 610 ℃ to 750 ℃, the 2223 phase films are prepared. All are superconducting without post-annealing. The critical temperatures are in 42K to 63K.

SUMMARY

Bi-Sr-Cu-Ca-O superconducting thin films have been prepared by MBE using an ozone generator. By controlling metal source shutters, superconducting films with the 2212 and 2223 phases were grown without any post-annealing processes. The 2212 films were made at the substrate temperature from 580 ℃ to 770 ℃. A film made at 750 ℃ was grown epitaxially and its Tc was 69K.

ACKNOWLEDGMENTS

The authors wish to thank H. Kawakami of Mitsubishi Kasei Ltd. for valuable discussions about $LaAlO_3$ substrates, and H. Fujii for his technical assistance at a stage during this work.

REFERENCES

1. Hosokawa S, Ichimura S, submitted to Review of Scientific Instruments
2. Berkley DD, Johnson BR, Anand N, Beanchamp KM, Conroy LE, Goldman AM, Maps J, Mauersberger K, Mecartney ML, Morton J, Tuominen M, Zhang Y-J (1988) Appl Phys Lett 53: 1973-1975

Study on Doping Properties of BSCCO/GaAs Films

SEIICHI TOKUNAGA, HIDEKI KUWAHARA, HIROAKI FURUKAWA, and MASAO NAKAO

SANYO Tsukuba Research Center, 2-1, Koyadai, Tsukuba, 305 Japan

ABSTRACT

We investigated doping properties of $Bi_2Sr_2CaCu_2O_8$/GaAs layered films. Resistivities of the GaAs layers decreased from 10 kΩ·cm to 20 Ω·cm by Si doping. There was no indication that the doping carriers were compensated by deep levels created by oxygen. This suggests that oxygen outdiffusion from the $Bi_2Sr_2CaCu_2O_8$ layer, even though it occurs, has little influence on electrical properties of the $Bi_2Sr_2CaCu_2O_8$/GaAs layered films. To explore the prospect for heteroepitaxy of GaAs on BSCCO, crystal structures of GaAs grown on (001) $Bi_2Sr_2CuO_6$ single crystals and (100) MgO substrates were investigated. On (100) MgO substrates, the (400) peak which results from the lattice matching was observed.

KEY WORDS: heterojunction, BSCCO system, GaAs, dopant, MBE

INTRODUCTION

A hybrid technology of high-T_c oxide superconductor and semiconductor promises an exciting field of electronic applications. If heteroepitaxy of single crystal oxide superconductors and semiconductors is attainable, there will be substantial opportunities for making microelectronic and optoelectronic devices.

In this respect, a progress has been made in growth of $YBa_2Cu_3O_Y$ [1-3] and $Bi_2Sr_2CaCu_2O_8$ [4] thin films on Si substrates. However, these superconducting films deposited on the Si substrates showed poor superconducting properties. The origin of the problems has been that the high temperature process to deposit the high-T_c superconducting films leads to interdiffusion at the interface between the thin films and the Si substrates. Therefore, some intermediate buffer layers are required to prevent the interdiffusion.

In our previous paper [5], we presented an alternative process toward the fabrication of superconductor/semiconductor heterojunction: the direct growth of the GaAs layers on Bi-Sr-Ca-Cu-O superconductor (hereafter referred to as BSCCO) without any buffer layer using MBE techniques. In this process, the growth temperature of GaAs is relatively low as compared with that of the high-T_c oxide superconductors. Therefore no significant interdiffusion between the GaAs layer and the $Bi_2Sr_2CaCu_2O_8$ layer is expected to occur.

For the realization of the BSCCO/GaAs heterojunction, following further investigations are required. In order to explore the prospect for heteroepitaxial growth of GaAs layers on BSCCO, the investigation of the crystal structure of GaAs layers on BSCCO single crystals are needed. Several attempts to fabricate heteroepitaxial GaAs on oxide insulators, such as magnesium aluminate spinel [6] and sapphire [7], have been reported. In addition to the investigation of the crystal structure, the electrical characterization of GaAs layers is required. When GaAs layers are deposited directly on BSCCO, a problem of the outdiffusion of oxygen from BSCCO into the

GaAs layers should be considered. If oxygen outdiffusion occurs, oxygen atoms create deep levels resulting in increase of resistivities of GaAs layers.

In this paper, we report electrical properties of Si-doped GaAs layers on the $Bi_2Sr_2CaCu_2O_8$ polycrystalline thin films to investigate the influence of oxygen interdiffusion in addition to the investigation of the crystal structure of GaAs layers on $Bi_2Sr_2CuO_6$ single crystals. Since the deep levels compensate the increase of carrier concentration caused by Si doping, electrical properties of GaAs layers are influenced by oxygen outdiffusion. Therefore it is possible to reveal the effect of oxygen outdiffusion on electrical properties of GaAs layers by measuring resistivities of the GaAs layers as function of Si concentration.

EXPERIMENT

In order to explore the prospect of the heteroepitaxial growth of GaAs layers on oxide single crystals, GaAs was deposited on the (100) MgO and (001) $Bi_2Sr_2CuO_6$ single crystals. The lattice mismatching between a (100) MgO plane and a (100) GaAs plane (5.4 %) is nearly equal to that between a (001) $Bi_2Sr_2CuO_6$ plane and a (100) GaAs plane (4.5 %). The $Bi_2Sr_2CuO_6$ single crystals were grown by flux method. The growth condition is shown in Table 1. The $Bi_2Sr_2CuO_6$ single crystals were cleaved in a vacuum. The cleaved surface of the $Bi_2Sr_2CuO_6$ single crystal was a (001) surface. The GaAs layers were grown directly on the $Bi_2Sr_2CuO_6$ single crystals. Conventional effusion cells containing As and Ga metals were used as beam sources. The evaporation for each species was set to allow a growth rate of 1 Å /sec for 2 h. The ratio of an As beam flux intensity to a Ga one was the order of 10. The back-ground pressure in the growth chamber remains below 2×10^{-8} Torr during the deposition. The crystal structures of the films were characterized by X-ray diffraction (XRD) with Cu- $K\alpha$ radiation.

Moreover, the $Bi_2Sr_2CaCu_2O_8$ polycrystalline thin films were deposited on (100) MgO substrates by single-target rf magnetron sputtering. The as-deposited films were subsequently annealed in air to form the superconducting phase. The sputtering condition and the post-annealing condition are shown in Table 2. The Si-doped GaAs layers were prepared on the $Bi_2Sr_2CaCu_2O_8$ thin films. To measure the sheet resistances of the Si-doped GaAs layers, a standard four-probe technique was used.

Table 1. Growth condition of the $Bi_2Sr_2CuO_6$ single crystals.

Composition	Bi:Sr:Cu=1:1:1 Bi_2O_3(5N), $SrCO_3$(4N), CuO(4N)
Sintering condition	830℃ 1h in air
Sequence condition	930 ℃ 4 h ↓ -3 ℃ /h 800 ℃ ↓ cooling in the furnace

Table 2. Sputtering and annealing conditions of the $Bi_2Sr_2CaCu_2O_8$ films on (100) MgO.

Target composition	$Bi_3Sr_2Ca_3Cu_6O_x$
Sputtering gas	Pure Ar
Sputtering pressure	30 mTorr
Deposition rate	400 Å /min
Annealing condition	800 ℃ 1 h in air

RESULTS AND DISCUSSION

The RHEED pattern obtained from the GaAs layer on the (100) MgO substrate was spotty. This spotty pattern indicates that the GaAs layer on (100) MgO was a polycrystalline film. Figure 1 gives XRD patterns for the GaAs layers on (100) MgO at the growth temperatures of 430, 450, 470 and 490 ℃, respectively. The (111) and (400) peaks of GaAs are observed. The (400) peak becomes stronger with raising growth temperature. The strong (111) peak is caused by naturally oriented GaAs layers and the (400) peak results from the lattice matching between the MgO (100) plane and the GaAs (100) one.

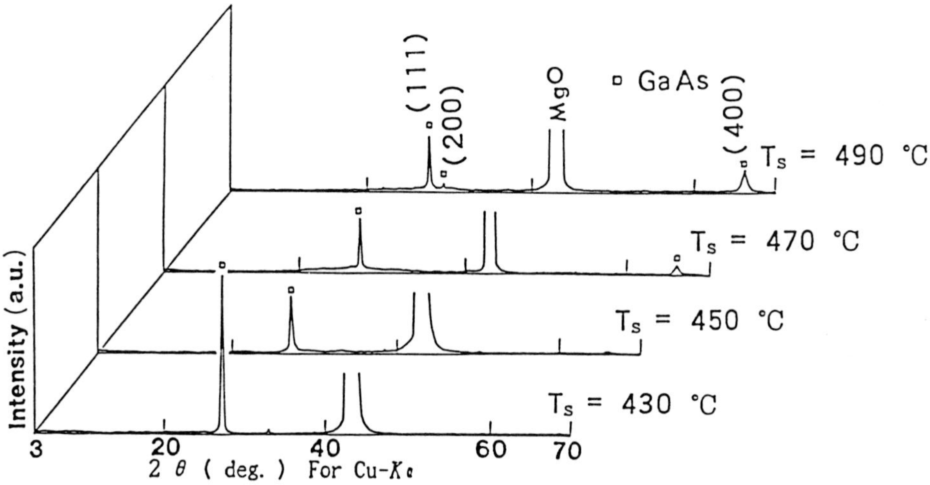

Fig. 1. XRD patterns for the GaAs films on (100) MgO substrates; growth temperatures of the GaAs are 430, 450, 470 and 490 ℃, respectively.

Figure 2 shows XRD patterns of the GaAs layers on the $Bi_2Sr_2CuO_6$ single crystals at the growth temperatures of 440, 450 and 460 ℃, respectively. The peaks of GaAs are most enhanced at 450 ℃. This result is similar to that obtained from the depositions of the GaAs layers on the $Bi_2Sr_2CaCu_2O_8$ polycrystalline thin films. The GaAs layers on the $Bi_2Sr_2CuO_6$ single crystals are strongly (111) oriented. The (400) peak which results from the lattice matching is not observed because of low growth temperature. Since the (001) cleaved surface of the $Bi_2Sr_2CuO_6$ single crystal is smoother than the surface of the $Bi_2Sr_2CaCu_2O_8$ polycrystalline thin film, naturally (111) oriented GaAs layers are grown on the $Bi_2Sr_2CuO_6$ single crystals.

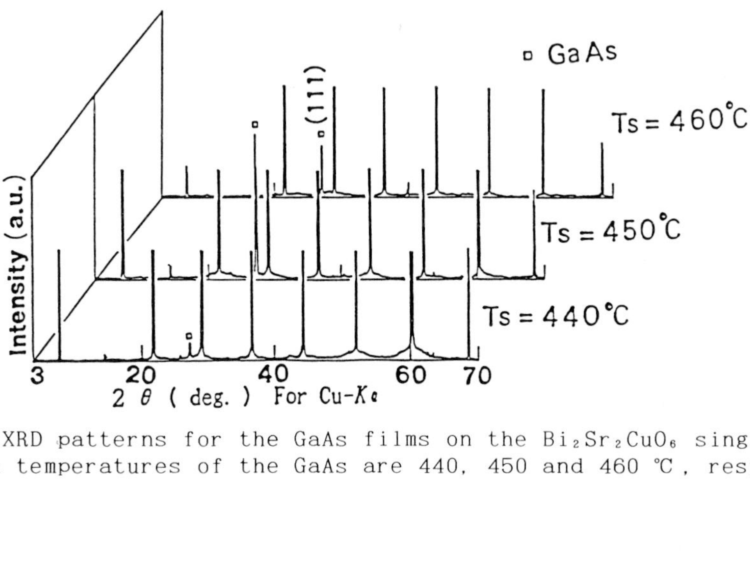

Fig. 2. XRD patterns for the GaAs films on the $Bi_2Sr_2CuO_6$ single crystals; growth temperatures of the GaAs are 440, 450 and 460 ℃, respectively.

Si-doped GaAs layers on the $Bi_2Sr_2CaCu_2O_8$ films showed n-type conductivity. Figure 3 indicates that the resistivities of the GaAs layers are decreasing from 10 kΩ cm to 20 Ω cm with raising doping levels. The carrier concentration could be controlled reproducibly. There is no indication that carrier concentration is compensated by deep levels. This suggests that deep levels created by outdiffusion of oxygen are considered to have little influence on electrical properties of the $Bi_2Sr_2CaCu_2O_8$/GaAs films.

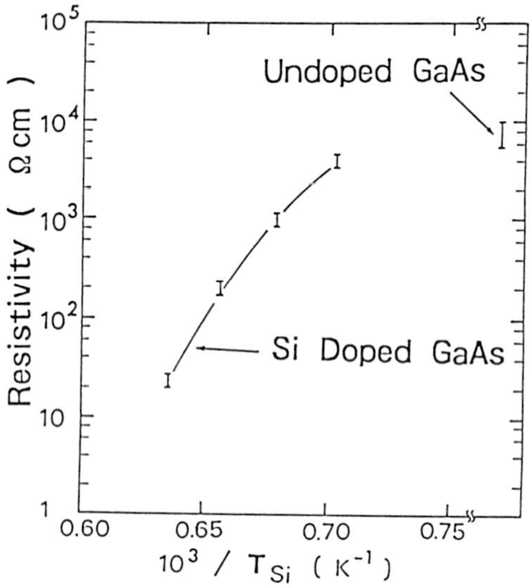

Fig. 3. Resistivity of the GaAs films vs temperature of a conventional effusion cell containing Si; growth temperature of the GaAs is 450 ℃.

CONCLUSIONS

Doping properties of GaAs layers have been reported. It is concluded that oxygen outdiffusion from the BSCCO films has little influence on electrical properties of the GaAs layers on the BSCCO films. The investigation of the crystal structures of the GaAs layers on MgO substrates revealed that the growth temperature plays an important role in determining both the crystallinity and the orientation of the film. The (400) peak which results from lattice matching was observed. At the growth temperature ranging from 440 ℃ to 460 ℃, GaAs films were (111) oriented on (001) $Bi_2Sr_2CuO_6$ single crystals.

REFERENCES

1. Lee WY, Salem J, lee V, Huang T, Savoy R, Deline V, Duran J (1988) Appl Phys Lett 52: 2263-2265
2. Venkatesan T, Chase EW, Wu XD, Inam A, Chang CC, Shokoohi FK (1988) Appl Phys Lett 53: 243-245
3. Berberich P, Take J, Dietsche W, Kinder H (1988) Appl Phys Lett 53: 925-926
4. Hung LS, Agostinelli JA, Paz-Pujalt GR, Mir JM (1988) Appl Phys Lett 53: 2450-2452
5. Kuwahara H, Nakao M (1989) Proc Int Superconductivity Electronics Conf: 28-31
6. Wang CC, Dougherty FC, Zanzucchi PJ, McFarlane SH Ⅲ (1974) J Electrochem Soc 121: 571-582
7. Manasevit HM (1981) Journal of Crystal Growth 55: 1-9

The Role of Grain-Boundaries in the Transport Properties of CVD-Grown Bi-Sr-Ca-Cu-O Films

CHIKAKO TANAKA, TOMOJI NAKAMURA, HIDEKI YAMAWAKI, TAKAFUMI KIMURA, OSAMU UEDA, and MASARU IHARA

Fujitsu Laboratories Ltd., 10-1, Morinosato-Wakamiya, Atsugi, 243-01 Japan

ABSTRACT

The effect on the temperature(T) dependence of electrical resistivity, $\rho(T)$, in the a-b plane of grains/grain-boundaries is studied in Bi-Sr-Ca-Cu-O films grown by CVD. $\rho(T)$ is found to be different in three types of films: (1)$\rho(T)=cT$, $\rho_b(T)/\rho_a(T)\sim 2$, for a single grain, (2)the resistivity perpendicular to the grain-boundaries, $\rho_{\perp GB}(T)=c_1T+c_2/T$ and parallel to the grain-boundaries , $\rho_{//GB}(T)=cT$, in a sample which contains a few grain-boundaries, and (3)$\rho(T)=cT$ and has no anisotropy for many small grains($\geq$ 10000).

KEY WORDS: Bi-Sr-Ca-Cu-O, CVD film, grain, grain-boundary, structural characterization, electrical resistivity

INTRODUCTION

To develop high-Tc superconductor electronic devices, much effort has been spent in fabricating high quality films. In previous work, we have reported the successful CVD-growth of superconducting Bi-Sr-Ca-Cu-O thin films which contain a far lower density of grain boundaries than the usual sputtered films[1]. We now turn our attention to the role of a grain boundary in the superconducting properties of a film; whether it acts as a barrier or a pinning center. In this paper, as a first step toward such a goal, we investigate the influence of grains/grain-boundaries in Bi-Sr-Ca-Cu-O films on the electronic properties in the a-b plain in the normal states.

EXPERIMENTAL PROCEDURES AND STRUCTUAL CHARACTERIZATION

The c-axis oriented Bi-Sr-Ca-Cu-O films are grown on (001)MgO substrate by CVD with a substrate temperature of 740-780°C. In our film growing system, $BiCl_3$, CuBr, CaI_2, and SrI_2 are used as source materials. Helium gas is used as carrier gas and O_2 and/or H_2O are used for oxidizing agents.

As listed in Table 1, the samples are classified into three types of films: type A-consists of large grains, type B-consists of long, narrow grains, and type C-consists of many small grains. By X-ray diffraction we find that the samples contain mainly high-T_c phases (so called 2223). Typical J_c-values for these films are of the order of 10^5-10^6 [A/cm^2] at 10 K.

Table 1

Film type	Growth temp. (°C)	Grain size μm × μm	Morphology of cross pattern
A	780	~300 ×~500	
B	780	~20 ×~200	
C	740	~0.1 ×~0.2	

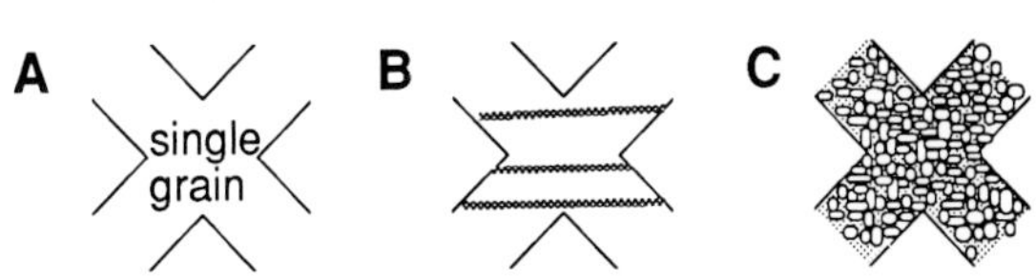

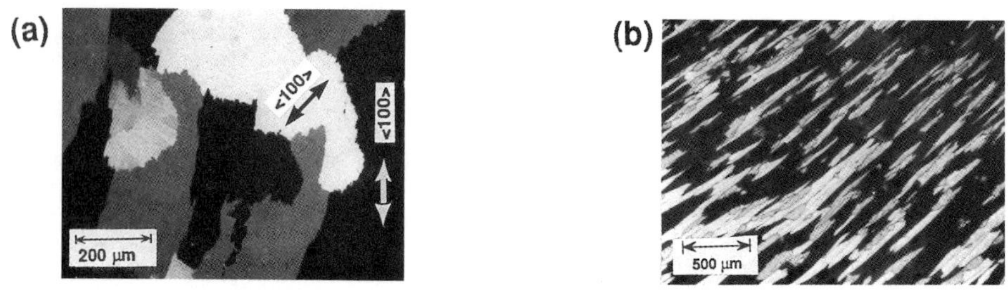

Fig.1 Polarizing micrographs for (a)type A film and (b)type B film.

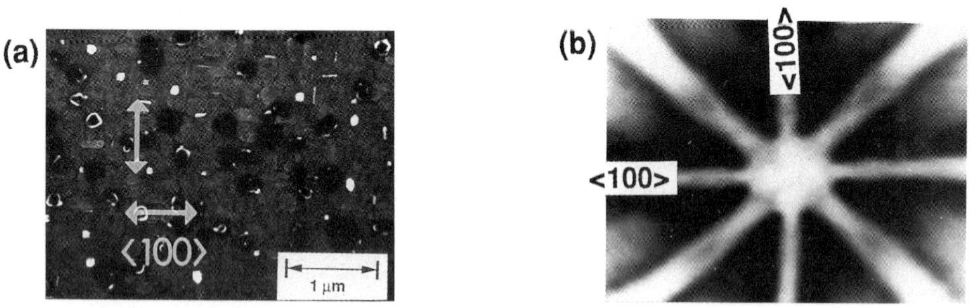

Fig.2 (a)SEM image and (b)ECP for type C film

We developed a method which has advantages in characterization of the grain size and the intra-grain orientation nondestructively using a polarization microscope under the crossed nicols condition[2]. The polarizing micrographs of type A and type B films are shown in Fig.1(a) and (b), respectively. We can observe *black contrast* when the polarized direction of the incident light coincides with the <100> direction(a- or b-axis) of the grain and *white contrast* with the <110> direction. Areas with the same brightness are oriented in the same direction, which is confirmed by the Electron Channelling Pattern(ECP); such a region is defined as a grain. According to Fig.1(a) and (b), the type A film consists of large grains with dimensions of $\sim 300\mu m \times \sim 500\mu m$ and the type B film consists of long, narrow grains with dimensions of $\sim 20\mu m \times \sim 200\mu m$ misaligned about 45° each other

The typical ECP for a (001)Bi-Sr-Ca-Cu-O film is shown in Fig.3(c). The brighter <100> band corresponds to the b-axis, which is determined by Transmission Electron Microscopy(TEM)[2,3]. The SEM image and ECP of the type C film are shown in Fig.2(a) and (b), respectively. This film consists of many small oval-shaped grains with dimensions of $\sim 0.1\mu m \times \sim 0.2\mu m$. As marked in Fig.2(a), these grains orient in the same <100> direction. In the strict sense, the a-axes of some grains lie along the b-axes of others, because the [100] and [010] bands of the ECP for type C film are the same brightness.

To allow variation of the current direction within the same area, the films are fabricated into cross-patterned samples with dimensions of $50\mu m \times 50\mu m$ as shown in Fig.3(a) by using standard photoresist lithography. Film types A, B, and C then yield the following cross-patterned samples: type-A samples consist of a single grain, the type-B samples contain a few grain boundaries, and the type-C samples consist of small grains, as illustrated in Table 1. In Figure 3, the numbers correspond to the electrode terminal numbers. The current is flown as $1 \rightarrow 2 (2 \rightarrow 3)$ and the voltage due to this current flow is measured between 3 and 4(1 and 4).

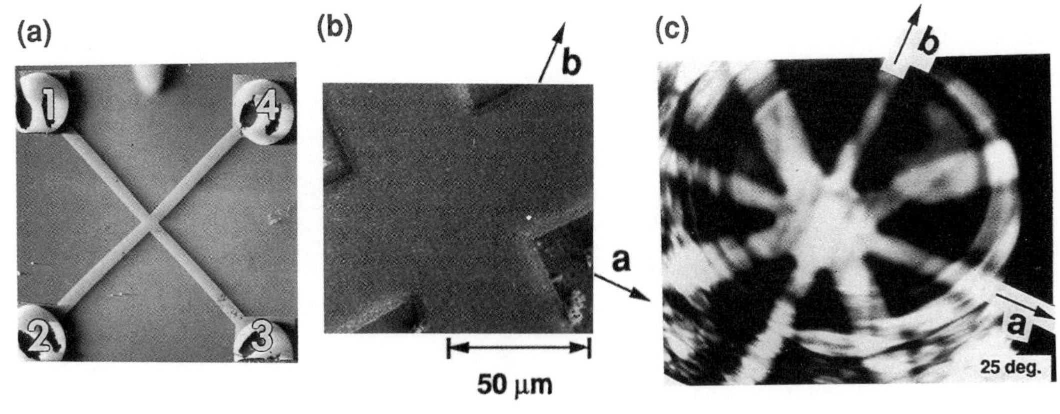

Fig.3 (a) SEM image of cross pattern for resistivity measurement. The numbers in this figure correspond to the electrode terminal number. (b) SEM image of cross pattern and (c) ECP for type A film.

RESULTS AND DISCUSSION

The T dependence of electrical resistivity along the a-axis, $\rho_a(T)$, and b-axis, $\rho_b(T)$ in type-A sample, are shown in Fig.4. Both $\rho_a(T)$ and $\rho_b(T)$ decrease linearly with decreasing temperature over T_c. Measurements in several samples yield values for the resistivity anisotropy of $\rho_b(T)/\rho_a(T)$ ~2. The same results have been reported[4] for single crystals. This anisotropy may arise from modulated structure of Bi-Sr-Ca-Cu-O[5].

The T dependence of resistivity perpendicular to the grain-boundaries, $\rho_{\perp GB}(T)$, and parallel to the grain-boundaries, $\rho_{//GB}(T)$ are shown in Fig.5. $\rho_{//GB}(T)$ shows metallic behavior: $\rho_{//GB}(T)= C+ DT$. On the other hand, $\rho_{\perp GB}(T)$ is semiconductor-like; if we replot $\rho_{\perp GB}(T)T$ versus T^2, the curve becomes a linear relation above T_c. Thus $\rho_{\perp GB}(T)$ can be described by the formula: $\rho_{\perp GB}(T)= A/T + BT$. In addition to the above results, according to Fig.5, T_c becomes higher when current flows along the grain boundaries.

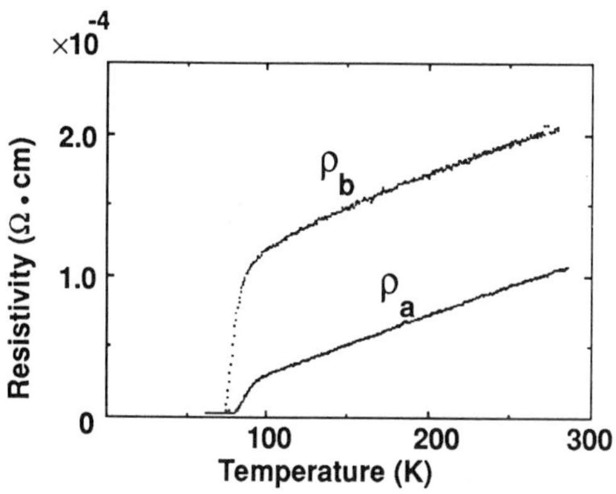

Fig.4 T dependence of resistivity for type-A sample, along a-axis $\rho_a(T)$, and b-axis, $\rho_b(T)$.

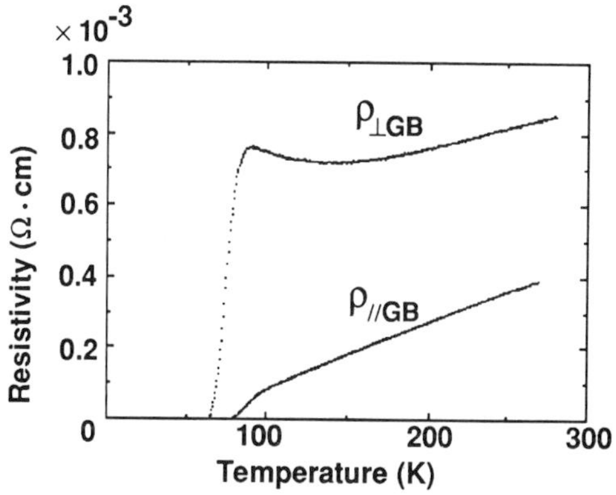

Fig.5 T dependence of resistivity for type-B sample, perpendicular to the grain-boundaries $\rho_{\perp GB}(T)$, and parallel to the grain-boundaries, $\rho_{//GB}(T)$.

We associate our results with the strong anisotropy between the resistivity in the a-b plane, $\rho_{ab}(T)$, and c direction, $\rho_c(T)$ of a single crystal[6]. We confirm that the semiconductor-like behavior of $\rho_{\perp GB}(T)$ comes from the effects of grain boundaries, because all samples without grain boundaries show metallic-behavior of $\rho(T)$[3]. The differing values of $\rho_{\perp GB}$ and $\rho_{//GB}$ from sample to sample seems to depend on the intra-grain orientation[4] rather than on the effects of grain boundaries.

The T dependences of electrical resistivities for type-C sample are shown in Fig.6. Here $\rho_{12}(\rho_{23})$ denotes that the current flows $1\rightarrow2(2\rightarrow3)$. Note that both $\rho_{12}(T)$ and $\rho_{23}(T)$, whose magnitude are almost the same, decrease linearly with decreasing T. The above characteristic behavior of resistivities for type-B sample and C can be attributed to the significant structural differences between these samples; nearest neighbor grains in type-C sample misalign 0° or 90° whereas in type-B sample misalign about 45°.

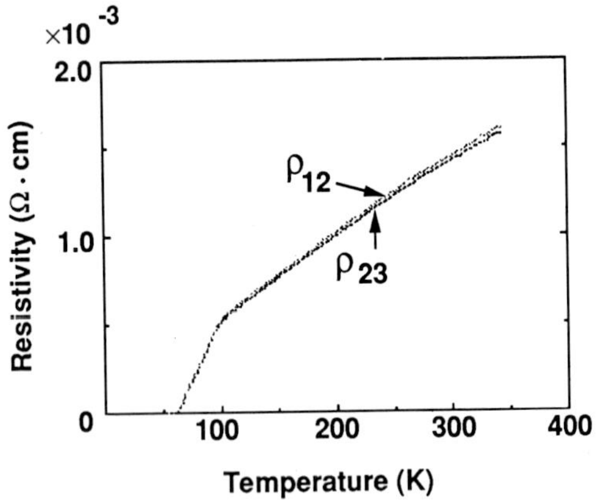

Fig.6 T dependence of resistivity for type-C sample; $\rho_{12}(T)(\rho_{23}(T))$ is the resistivity when the current flows $1\rightarrow2(2\rightarrow3)$.

In summary, we have studied the effect of grains/grain-boundaries on $\rho(T)$ in Bi-Sr-Ca-Cu-O films. $\rho(T)$ is found to be different in three types of films. In type-A sample which consists of a single grain, $\rho(T)=cT$, $\rho_b(T)/\rho_a(T)\sim2$. In type-B sample which includes a few grain-boundaries, $\rho_{\perp GB}(T)=c_1T+c_2/T$ and $\rho_{//GB}(T)=cT$, and $T_{c//GB}>T_{c\perp GB}$. In type-C sample which contains more than 10000 grain-boundaries, $\rho(T)=cT$ and has no anisotropy.

ACKNOWLEDGEMENTS

We would like to thank H. Nakao for his program for resistivity measurement.

REFERENCES

1. Ihara M, Kimura T, Yamawaki H, Ikeda K (1989) IEEE trans. Mag. 25:2470 and Kimura T, Ihara M, Yamawaki H, Ikeda K, Ozeki M Advances in Superconductivity, eds. Kitazawa K and Ishiguro T (Springer-Verlag, Tokyo 1989) p.49.
2. Nakamura T et al. to be published.
3. Tanaka C et al. to be published.
4. Yamaya K, Haga T, Abe Y, Minami F, Takekawa S, Tajima Y, Hidaka Y (1989) Physica C 162-164: 1009, Kadowaki K, Menovsky AA, Franse JJM (1990) Physica B 165-166: 1159.
5. Tarascon JM, et al. (1988) Phys. Rev. B37: 9382.
6. for example, Martin S, Fiory AT, Fleming RM, Schneemeyer LF, Waszczak JV, (1988) Phys. Rev. Lett. 60:2194.

BSCCO/Sapphire Grown by Plasma-Enhanced Halide CVD

HIROSHI NAKAO, TAKAFUMI KIMURA, HIDEKI YAMAWAKI, MASARU IHARA, and MASASHI OZEKI

Fujitsu Laboratories Ltd., 10-1, Morinosato-Wakamiya, Atsugi, 243-01 Japan

ABSTRACT

We grew Bi-Sr-Ca-Cu-O (BSCCO) superconducting film on a $(1\bar{1}02)$ sapphire substrate using plasma-enhanced halide CVD at 580°C. The (2212)-phase was dominant in the film. The c axis of the film was perpendicular to the sapphire $(1\bar{1}02)$ plane. The film started a resistive transition at 110 K and showed zero resistivity at about 70 K. The critical current density of the film was 2×10^6 A/cm^2 at 10 K. These results are almost the same as for BSCCO films on (001) MgO substrates (6×10^6A/cm^2 at 10 K) grown by the same type of CVD.

KEYWORDS: BSCCO, thin film, sapphire, plasma-enhanced halide CVD

INTRODUCTION

Large-area high-quality thin films are necessary for the device application of high-Tc superconductors. Since the discovery of BSCCO [1], many techniques have been tried in obtaining thin films, e. g., evaporation [2], sputtering [3], CVD [4,5], MOCVD [6,7], and laser ablation [8]. Film quality is improving with advances in growth techniques, but film area remains limited by substrate size. The sapphire (α-Al_2O_3) substrate is known for its high quality and large area. Growing BSCCO on sapphire produces large-area high-quality film. We developed a plasma-enhanced halide chemical vapor deposition (PECVD) system enabling low-temperature deposition and grew the superconducting BSCCO film on 3-inch $(1\bar{1}02)$ sapphire substrates without postannealing. This paper details our system and results.

SUBSTRATES

Table 1 compares substrates for BSCCO films. Those usually used for BSCCO film -- MgO and $SrTiO_3$ -- are smaller than sapphire. We studied the crystallinity of (001) MgO and $(1\bar{1}02)$ sapphire using the X-ray topography (Fig. 1). Topographic images were taken of the (200) diffraction of MgO and the (0003) diffraction of sapphire. About 3-mm-diameter grains with network of dislocations are observed in MgO. No granularity is observed in sapphire, however, indicating its much better quality. Sapphire is also chemically stable, and the lattice mismatch between BSCCO and sapphire is the smallest in the three. BSCCO reacts to sapphire in the solid phase at high temperatures above 700°C, however, which means low temperature deposition is required to obtain large-area, high-quality BSCCO thin films on sapphire.

PECVD SYSTEM

Figure 2 diagrams the PECVD system, which has four source cells. Metal-halide sources ($BiCl_3$, SrI_2, CaI_2, and CuI) are heated and vaporized in these cells and carried to the growth region with He gas. Gas concentration in the growth region is controlled by cell temperature and carrier flow rate. Typical cell temperatures were $BiCl_3$ at

Table 1. BSCCO film substrates

Characteristics	MgO	$SrTiO_3$	$(1\bar{1}02)$ Sapphire
Surface lattice constant	a = 4.2 Å	a = 3.9 Å	a' = 5.2 Å b' = 4.75 Å
Misfit with BSCCO $(a_{film}-a_{sub})/a_{film}$	22%	27%	8%
Size	Intermediate	Small	Large
Crystallinity (Fig. 1)	Poor	Poor	Good
Solid phase reaction with BSCCO	Moderate	Moderate	Active

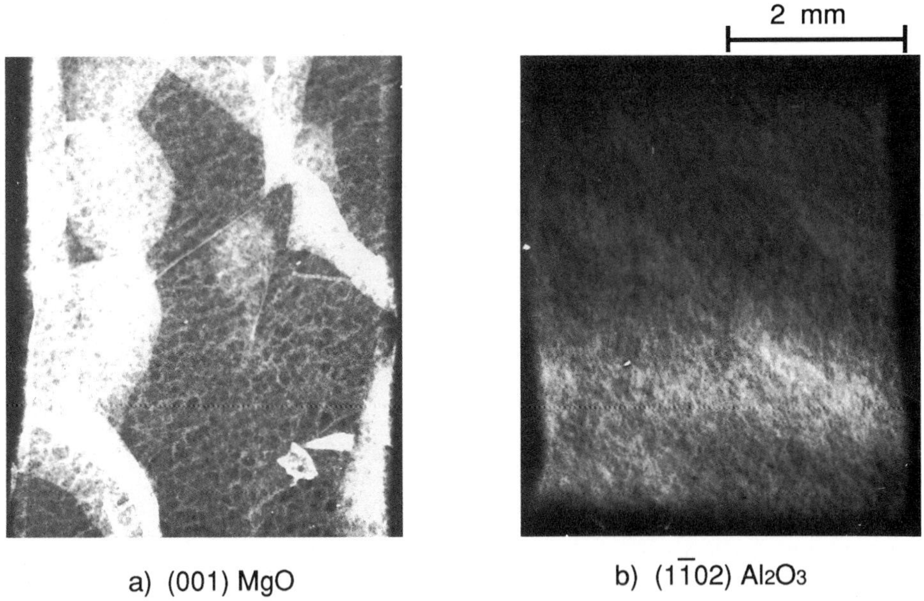

Figure 1 X-ray topographic images of substrates used to study the crystallinity of (001)MgO and (1$\bar{1}$02) sapphire. Images are obtained for the (200) diffraction of MgO and the (0003) diffraction of sapphire. Sapphire dose not show the granularity observed in MgO.

160 to 180°C, SrI_2 at 890 to 910°C, CaI_2 at 870 to 900°C, and CuI at 450 to 500°C. The total flow rate of He carrier gas was 1300 sccm. The growth region has a SiC-coated carbon suscepter 10 cm in diameter heated by a carbon heater and a RF electrode for generating RF plasma. Oxidizing agents are O_2 and H_2O vapor. O_2 is supplied at 100 sccm and H_2O vapor is carried with 100 sccm of He gas. They are supplied to the growth region through the holes in the RF electrode. The total pressure was kept at 0.1 to 1 Torr to supply souce gases to a substrate which is lower temperture than souce sells. RF power was 0.25 W/cm^2. The suscepter is rotated to make film composition uniform. Two to 3-inch-diameter (1$\bar{1}$02) sapphire is used as the substrate. Substrate temperature (T_{sub}) is measured by a thermocouple in the carbon suscepter calibrated by a thermocouple sticked on the sapphire substrate. Substrate temperature varied from 700 to 500°C.

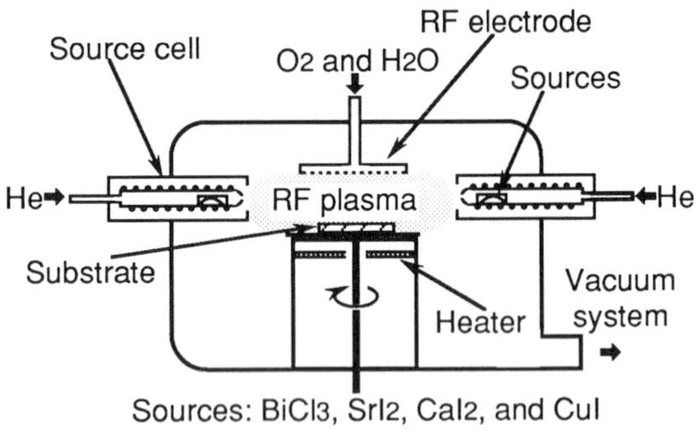

Figure 2 PECVD system having four source cells in which source gases are generated, a rotary suscepter, and a RF electrode for generating RF plasma to enhance film oxidization.

RESULTS AND DISCUSSION

Figure 3 shows the surface morphology for the 700-Å-thick BSCCO/(1$\bar{1}$02) sapphire deposited at 580°C. The film has a flat surface with hillocks. Hillocks were Cu-rich compared to flat parts.

Figure 4 shows X-ray diffraction patterns for the films. Pattern a) suggests that (2212)-phase BSCCO is dominant in the film, and that the c-axis of the film is oriented perpendicular to the sapphire (1$\bar{1}$02) plane. Film deposited without RF plasma has only very small peaks of (2201)-phase BSCCO and the other compounds (Fig. 4-b)). Superconducting BSCCO films become insulating after annealing without plasma in the same oxygen pressure as the growth conditions. RF plasma enhanced the oxidization and crystallization of the film at 580°C without postannealing.

The resistivity of BSCCO/sapphire deposited with RF plasma was about 0.5 mΩ·cm at room temperature and, without RF plasma, was insulating. The conductive film was patterned in a bridge pattern 10 μm width and 20 μm long using photolithography and chemical etching to measure electrical characteristics. Resistivity-temperature characteristics (R-T) were measured at a current density of 1 x 10^4 A/cm^2 (Fig. 5). The R-T curve of the BSCCO/MgO grown by the same type of PECVD system is also plotted (Fig. 5-b). The resistivity of BSCCO/sapphire is metallic above 110 K but the slope of the R-T curve is smaller than that of BSCCO/MgO. The resistive transition starts at around 110 K and zero resistivity occurs at about 70 K, almost the same as for BSCCO/MgO.

Figure 6 shows the critical current density-temperature characteristics (J_c-T). BSCCO/sapphire is shown by the closed circles and BSCCO/MgO by the open ones. BSCCO/sapphire has a rather high J_c of 2 x 10^6 A/cm^2 at 10 K and 2 x 10^3 A/cm^2 at 77 K. These are almost the same as for the BSCCO/MgO (6 x 10^6 A/cm^2 at 10 K), showing sapphire to be a suitable substrate for BSCCO film.

CONCLUSION

We grew BSCCO film on a 3-inch (1$\bar{1}$02) sapphire wafer at 580°C using PECVD. RF-plasma enhancement produced the superconducting thin films at a low temperature without post-annealing. The surface of the film is

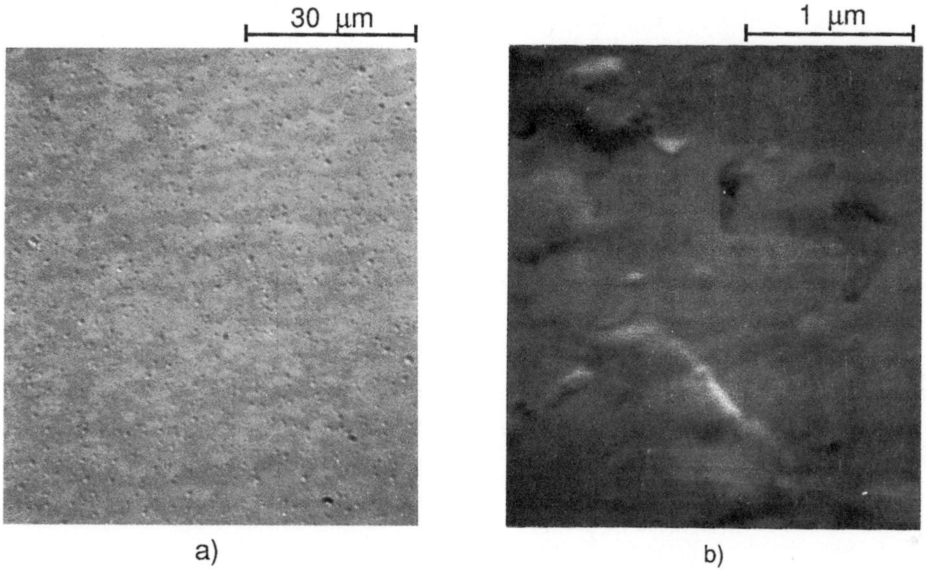

Figure 3 BSCCO/sapphire surface morphology. a) Microscope and b) SEM images. A rather smooth surface, with hillocks, is obtained.

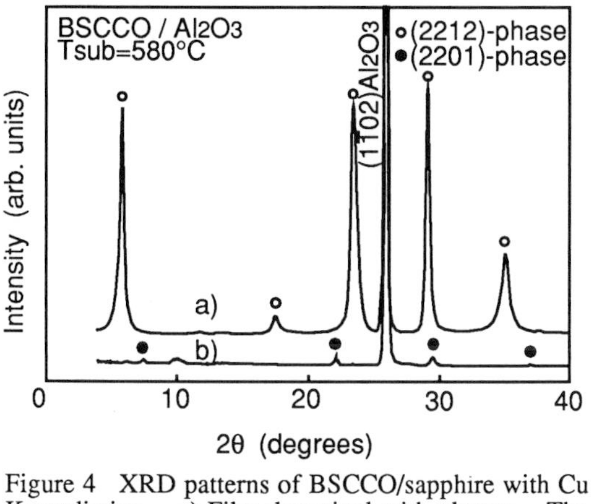

Figure 4 XRD patterns of BSCCO/sapphire with Cu Kα radiation. a) Film deposited with plasma. The (2212)-phase is dominant in the film and the c axis of the film is oriented perpendicular to the sapphire (1 $\bar{1}$02) plane. b) Film deposited without plasma. Film has only very small peaks of (2201)-phase BSCCO.

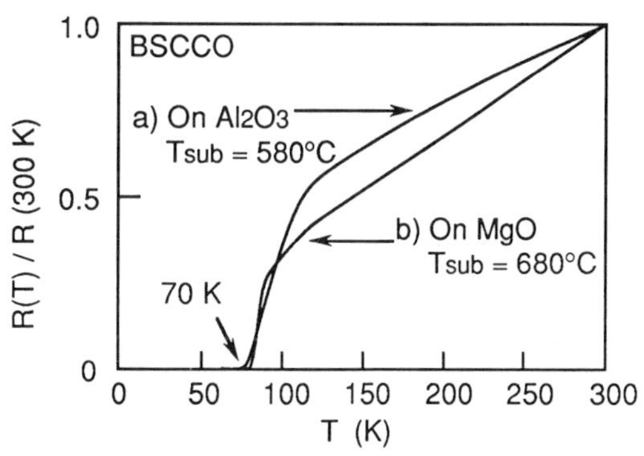

Figure 5 R-T characteristics for a) BSCCO/sapphire and b) BSCCO/MgO. The R-T curves were measured using a 10-μm-wide bridge pattern under a current density of 1 x 10^4 A/cm^2. The resistive transition starts at around 110 K and zero resistivity occurs at about 70 K, almost the same as for BSCCO/MgO.

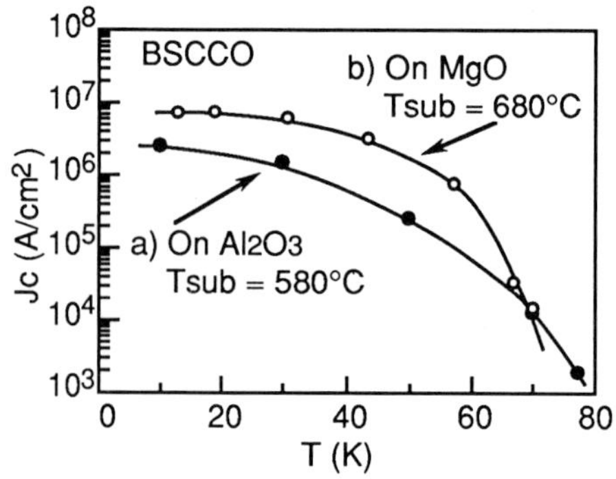

Figure 6 Jc-T characteristics for a) BSCCO/sapphirc and b) BSCCO/MgO. BSCCO/sapphire has a rather high critical current density of 2 x 10^6 A/cm^2 at 10 K.

rather smooth, with hillocks. The critical temperature of the film was about 70 K and critical current density 2 x 10^6 A/cm^2 at 10 K.

ACKNOWLEDGMENT

We thank Drs. S. Komiya, O. Ueda, T. Nakamura, and C. Tanaka for their helpful comments and suggestions regarding crystallinity.

REFERENCES

1. Maeda H, Tanaka Y, Fukutomi M, Asano T (1988) Jpn. J. Appl. Phys. 27: L209-L210
2. Kuroda K, Mukaida M, Yamamoto M, and Miyazawa S (1988) Jpn. J. Appl. Phys. 27: L625-L627
3. Kang JH, Kampwirth RT, Gray KE Marth S, Huff A (1988) Phys. Lett. A 128: 102-104
4. Ihara M, Kimura T (1989) Proc. of ASC88, Aug 1988, San Francisco, IEEE Trans. Magn. 25, 2470-2473
5. Kimura T, Ihara M, Yamawaki H, Ikeda K, Ozeki M (1989) Proc. 1st International Symp. of Superconductivity (ISS'88), Aug 1988, Nagoya In:Kitazawa K, Ishiguro T (eds) in Advances in Superconductivity. Springer-Verlag, Tokyo, pp 495-498
6. Yamane H, Kurosawa H, Hirai T, Iwasaki H, Kobayashi N, Muto Y (1988) Jpn. J. Appl. Phys. 27: L1495-L1497
7. Kanai M, Kawai T, Kawai M, Kawai S (1988) Jpn. J. Appl. Phys. 27: L1293-L1296

Preparation of BSCCO Thin Films by MOCVD Using Purified Source Materials

S. HAYASHIDA, T. NAKAI, I. KOBAYASHI, Y. SUGIMORI, T. TSUYAMA, and S. KOSHIBA

Nippon Sanso Corporation, Technology Division, 4-320, Tsukagoshi, Saiwai-ku, Kawasaki, 210 Japan

ABSTRACT

We synthesized and purified $Bi(C_6H_5)_3$, $Sr(DPM)_2$, $Ca(DPM)_2$ and $Cu(DPM)_2$ as source materials for MOCVD growth of BSCCO thin films, and obtained the purity of 99%. Using these sources, the run-to-run deviation of the deposition rate was reduced to 5%. We grew BSCCO thin films with the low-Tc phase on MgO(100) substrates. The growth temperature and pressure were 750-800℃ and 100 Torr respectively. When the composition of the film is close to the stoichiometry with a small excess of copper, a smooth surface was observed. The XRD pattern of the film has (00n) peaks of the low-Tc phase, and the temperature-vs-resistance curve shows Tc(end)=70K.

KEY WORDS; MOCVD, source material, $Bi_2Sr_2CaCu_2O_x$, surface morphology.

INTRODUCTION

There have been many studies on the superconducting Bi-Sr-Ca-Cu-O (BSCCO) system by thin film fabrication methods; sputtering [1], laser ablation [2], molecular beam epitaxy [3], and metalorganic chemical vapor deposition (MOCVD) [4]. The BSCCO system is a multi-phased superconductor, such as $Bi_2Sr_2Ca_2Cu_3O_x$ with Tc=110K (the high-Tc phase), $Bi_2Sr_2CaCu_2O_x$ with Tc=80K (the low-Tc phase) and $Bi_2Sr_2CuO_x$ with Tc=20K. Most studies were done to investigate growth condition of the high-Tc phase [5,6], but further efforts are needed to obtain the single phase and Tc(end) higher than 100K. There were few studies on growth condition to obtain a smooth surface by MOCVD, which is needed for superconducting junctions.

In this paper, we report the BSCCO thin films with the low-Tc phase grown on MgO(100) substrates by MOCVD. It will be concluded that the stoichiometric composition with a small excess of copper is necessary to obtain a smooth surface. In addition, we discuss significance of the purity of source materials for the repoducibility of the growth rate.

EXPERIMENTAL

The source materials in this study were triphenyl bismuth $Bi(C_6H_5)_3$, dipivaloyl-methanato-strontium $Sr(DPM)_2$, $Ca(DPM)_2$ and $Cu(DPM)_2$. After synthesis and purification, these materials were characterized by the inductively coupled plasma spectroscopy (ICP), the thermogravimetric analysis (TGA) and the elementary analysis.

The growth apparatus (Nippon Sanso Corp. OSC-2000) is composed of 4 argon gas lines, an oxygen line, 4 ovens for source materials, a gas flow control system, a quartz reactor, an RF heating system, a filter and a pumping system. The lines and valves between the outlets of the ovens and the reactor were heated at a constant temperature by a line heater. A susceptor is made of Inconel for the oxidation proof. The vapors of the source materials were carried by argon and mixed with oxygen at the inlet of the reactor. A typical growth condition is shown in Table 1. The growth started more than 20 minites after exhausting gas to the vent line. After growth, films were cooled at the rate of 3-5℃/min in the oxygen atmosphere, and any post-treatment was not done. The growth rate was about 30nm/h, and the thickness of the films was about 500nm.

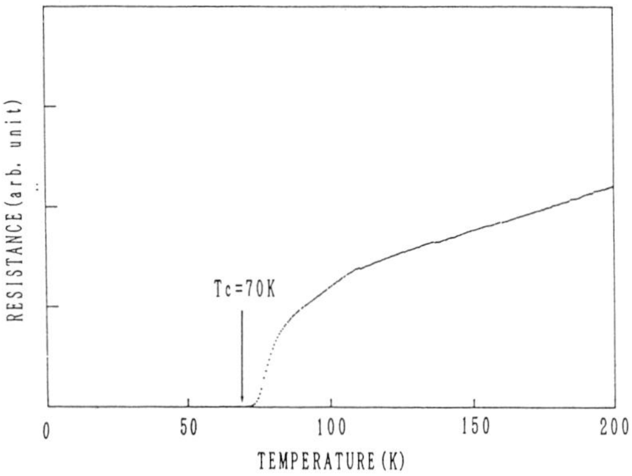

Fig. 4 Temperature-vs-resistance curve of a BSCCO film with the low-Tc phase.

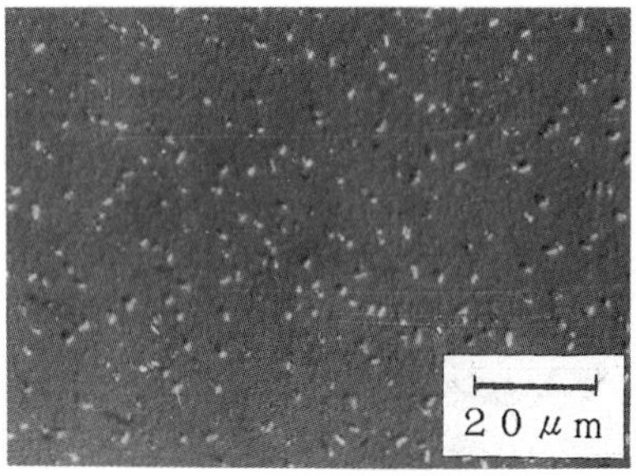

(a)

(b)

Fig. 5 Nomarski micrographs of BSCCO films. The composition of the films was determined by ICP. (a) Bi:Sr:Ca:Cu=1.95:2.00:1.20:2.53 (b) Bi:Sr:Ca:Cu=1.78:2.00:1.06:1.97

Fig. 6 Nomarski micrograph of a BSCCO film with the low-Tc phase.

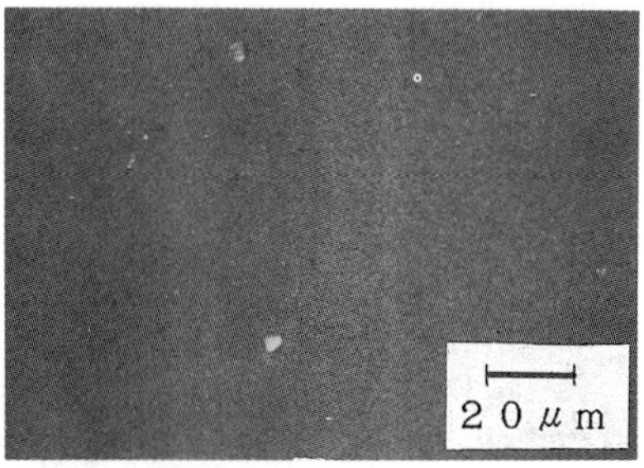

Fig. 7 SEM micrograph of a BSCCO film with the low-Tc phase.

Compositions, crystal structures and superconducting properties of the films were evaluated by the ICP, X-ray diffractometry (XRD) and conventional DC four point probe measurements, respectively. Surface morphology was observed under a Nomarski and a scanning electron microscope (SEM).

RESULTS AND DISCUSSION

Figure 1 shows the TGA curves of the source materials in the argon flow at the heating rate of 10℃/min. Because weight loss around 100℃ caused by loss of impurities and residuals is not or hardly observed, the purity can be estimated to be higher than 99%.

Using the purified materials, we obtained good reproducibility of the deposition rate with the maximum run-to-run deviation of 5% as shown in Figure 2. The deposition rate in the figure was in terms of the weight of strontium in BSCCO films.

We grew films with the low-Tc phase on MgO(100) substrates using the purified source materials. Figure 3 shows a typical XRD pattern of the film. Most of peaks were assigned to (00n) of the the low-Tc phase, which is a c-axis orientaion on substrate. Unknown peaks at $2\theta = 5.3^{\circ}, 21.0^{\circ}, 26.2^{\circ}, 31.4^{\circ}$ were also observed. The temperature-vs-resistance characteristic of the film confirms superconductivity of the low-Tc phase as shown in Figure 4. Tc(onset) is near 80K, and Tc(end) is 70K.

As shown in Figure 5, the morphology of the films observed under a Nomarski microscope is greatly affected by the amount of copper; (a) when copper exceeds stoichiometry, the surface of the film is rather smooth, (b) otherwise the surface is rough. When the composition is close to the stoichiometry within 10% deviation, a specular surface is obtained. The surface structures of the films are shown in Figure 6 and 7. The surface morphology of the films is smooth without any grain boundaries. The density of precipitates in the micrographs is small. The results suggest that stoichiometric composition with a small excess of copper is important to obtain a smooth surface in the case of the low-Tc phase.

CONCLUSION

We studied surface morpholory of BSCCO thin films with the low-Tc phase on MgO(100) by MOCVD. Using purified source materials, we obtained good repodueibility of growth rate. SEM and Nomarski micrographs of the films show smooth surfaces when the composition is close to the stoichiometry with a small excess of copper. The XRD pattern of the film indicates the low-Tc phase with the c-axis orientation, and temperature-vs-resistance curve shows Tc(end)=70K.

REFERENCES

1. Kojima K, Schlom DG, Kuroda K, Tanioku M, Hamanaka K, Eckstein JN, Harris Jr. JS (1990) Superstructure in thin films of Bi-based compounds on MgO. Jpn J Appl Phys 29: L1638-1641
2. Kanai M, Kawai T, Kawai S (1989) Low-temperature formation of multilayered Bi(Pb)-Sr-Ca-Cu-O thin films by successive deposition using laser ablation. Appl Phys Lett 54: 1802-1804
3. Schlom DG, Marshall AF, Sizemore JT, Chen ZJ (1990) J Crstal Growth 102: 361-375
4. Sugimoto T, Yoshida M, Yamaguchi K, Sugawara K, Shiohara Y, Tanaka S (1990) Epitaxial growth of Bi-Sr-Ca-Cu-O superconducting thin films by metalorganic vapor deposition. Appl Phys Lett 57: 928-930
5. Endo K, Hayashida S, Ishiai J, Matsuki Y, Ikedo Y, Misawa S, Yoshida S (1990) Preparation of as-deposited Bi-Sr-Ca-Cu-O films with high-Tc superconducting phase by metalorganic chemical vapor deposition. Jpn J Appl Phys 29: L294-297
6. Ihara M, Yamawaki H, Kimura T, Nakao H (1989) A 110-K phase BiSrCaCuO thin film grown by halide CVD. In: Ishiguro T, Kajimura K (eds), Advances in Superconductivity II. Springer-Verlag, Tokyo Berlin. p.903

Table 1. Typical growth condition

source material	$Bi(C_6H_5)_3$	$Sr(DPM)_2$	$Ca(DPM)_2$	$Cu(DPM)_2$
oven temperature (℃)	100-106	200	196	117
flow rate of Ar (SCCM)	100-125	200-400	30-40	45-50
flow rate of oxygen (SCCM)	1000			
pressure (Torr)	100			
substrate	MgO(100)			
substrate temperature (℃)	750-800			
cooling rate (℃/min)	3-5			
growth rate (nm/h)	30			

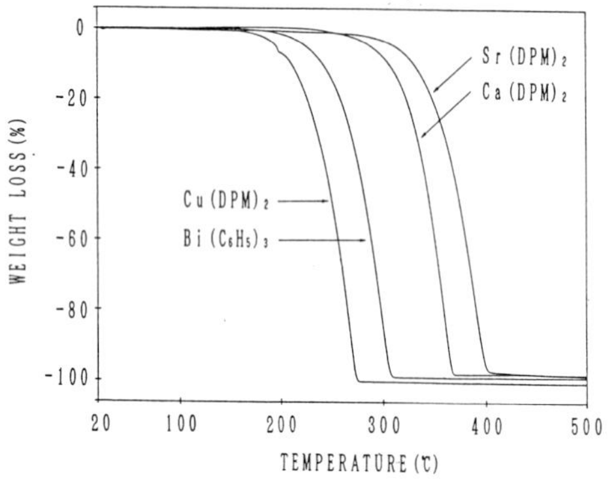

Fig. 1 Thermogravimetric analysis of source materials.

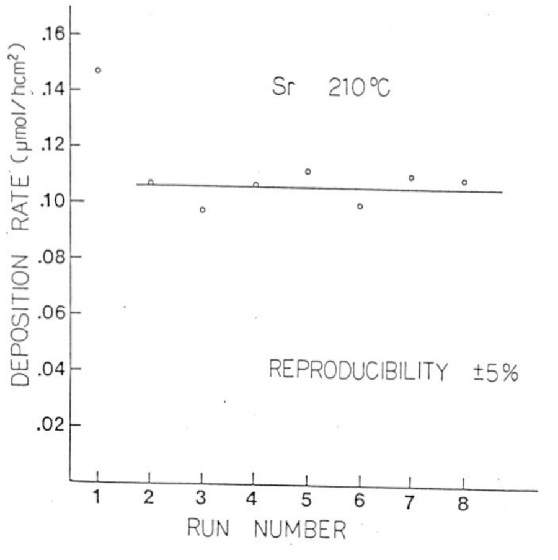

Fig. 2 Run-to-run test of the deposition rate of strontium. The oven of $Sr(DPM)_2$ was heated at 210℃.

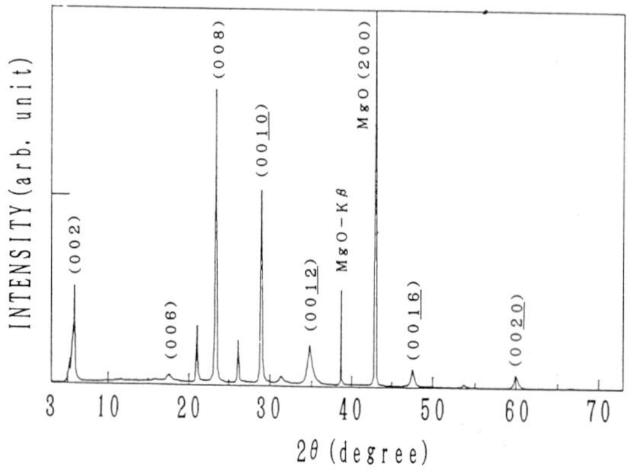

Fig. 3 X-ray diffraction pattern of a BSCCO film with the low-Tc phase.

Preparation of Bi-Sr-Ca-Cu-O Thin Films by Chemical Vapor Deposition

SHIN'YA AOKI[1], TAICHI YAMAGUCHI[1], NOBUYUKI SADAKATA[1], OSAMU KOHNO[1], AKIRA SAJI[2], NOBORU KURODA[2], and HIROSHI YOSHIDA[2]

[1] Fujikura Ltd., 5-1, Kiba 1-chome, Koto-ku, Tokyo, 135 Japan
[2] Chubu Electric Power Co., 20-1, Kita-Sekiyama, Ohdaka-cho, Midori-ku, Nagoya, 459 Japan

ABSTRACT

We report on the Bi-Sr-Ca-Cu-O film formation on MgO(100), $SrTiO_3$(100) and $LaAlO_3$(100) by MOCVD using $Bi(C_6H_5)_3$, $Sr(DPM)_2$, $Ca(DPM)_2$ and $Cu(DPM)_2$ as source materials. The XRD pattern for the films prepared on MgO(100) showed the high Tc phase with the c-axis oriented normal to the substrate surface, whose Tc zero was 75K. On the other hand, the XRD pattern for the films prepared on $SrTiO_3$(100) and $LaAlO_3$(100) showed the mixture of the high Tc and low Tc phases. The Tc zero of 79K on $SrTiO_3$(100) and 82K on $LaAlO_3$(100) were higher than that on MgO(100).

KEY WORDS: Bi-Sr-Ca-Cu-O thin films, MOCVD, oxide superconductor

INTRODUCTION

Since discovery of superconductivity in the Bi-Sr-Ca-Cu-O system [1] , many studies have been made on obtaining the high Tc single phase [2] . Bi-Sr-Ca-Cu-O system has at least two superconducting phases; the high Tc phase ($Bi_2Sr_2Ca_2Cu_3O_x$) with Tc around 110K and the low Tc phase ($Bi_2Sr_2Ca_1Cu_2O_y$) with Tc around 80K.

Characteristics of metalorganic chemical vapor deposition (MOCVD) method compared with sputter and MBE are as follows ; high deposition rate, compatibility with variously shaped substrates, possibility to large scale processing and wide range of oxygen pressure.

It has been reported that in-situ Pb-doping in MOCVD enhances the high Tc phase formation in the Bi system [3] . However, Tc zero of the film was lower than 100 K. In this study, first we report on the preparation of Bi-Sr-Ca-Cu-O films on MgO(100), $SrTiO_3$(100) and $LaAlO_3$(100) substrates by MOCVD and next on the effect of substrates on superconductivity of the films.

EXPERIMENTAL

Figure 1 is a schematic drawing of the MOCVD apparatus used in this study. This system is classified as a hot-wall type MOCVD, where the substrates are heated by an electric furnace. The source materials were triphenyl bismuth $Bi(C_6H_5)_3$, bis(-2,2,2,6,6,-tetramethyl-3,5-heptandione)-strontium $Sr(DPM)_2$, -calcium $Ca(DPM)_2$ and -copper $Cu(DPM)_2$. Figure 2 shows the structure of the source materials. The temperatures of the vessels of $Bi(C_6H_5)_3$, $Sr(DPM)_2$, $Ca(DPM)_2$ and $Cu(DPM)_2$ were 66, 230, 188 and 118℃, respectively. Argon was used as carrier gas. The flow rate of argon gas for $Bi(C_6H_5)_3$, $Sr(DPM)_2$, $Ca(DPM)_2$ and $Cu(DPM)_2$ were 20, 200, 100 and 25 cc/min, respectively. Oxygen gas was separately introduced into the growth chamber at the rate of 200 cc/min. During the deposition, the total pressure in the growth chamber was 10 Torr and the

Sr-, Ca- and Cu- THD chelates and Bi- and Pb- Ph compounds, were investigated by thermogravimetry (TG). Figure 1 shows TG curves of these compounds under 1.33KPa (10mmHg) air.
The temperature dependence of vaporizing velocity of these compounds were clarified from these TG curves and vaporizing temperatures were decided. These vaporizing temperatures of $Bi(Ph)_3$, $Sr(THD)_2$, $Ca(THD)_2$, $Cu(THD)_2$ and $Pb(Ph)_4$ are 115-125°C, 230-240°C, 190-205°C, 100-125°C and 160-180°C, respectively.

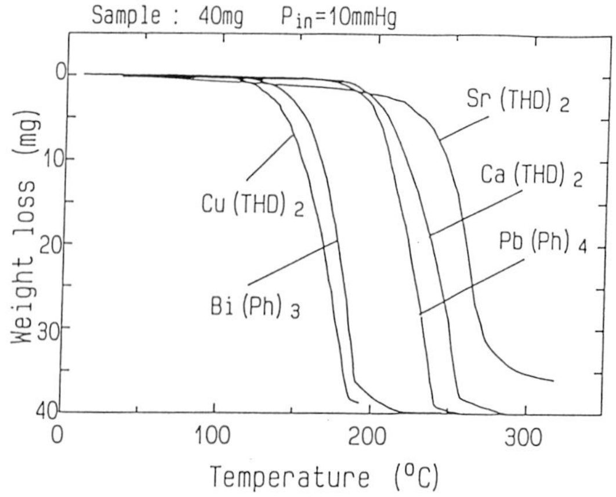

Fig.1 TG curves of source materials of CVD.

Figure 2 shows a schematic diagram of the CVD apparatus used in this experiment.
Source materials in boats were set into each vaporizing cylinder and these were vaporized at prescribed temperatures. The source gases from each vaporizing cylinder were introduced into the reactor chamber with argon gas used as a carrier gas. Oxygen gas for oxydation was separately introduced into the same chamber. The mixed source gases were led to the single crystalline MgO (100) substrates heated at 800-840°C. The total gas pressure was kept under 1.33-2.00KPa (10-15mmHg) during the deposition. After the 30 minutes deposition, films were cooled to room temperature at the rate of -600°C/Hr under 101.3KPa (760mmHg) argon and oxygen atmosphere. The gas ratio of Ar/O_2 is 1.
The thickness of the films were in the range of 2.0 to 2.8 micrometers.

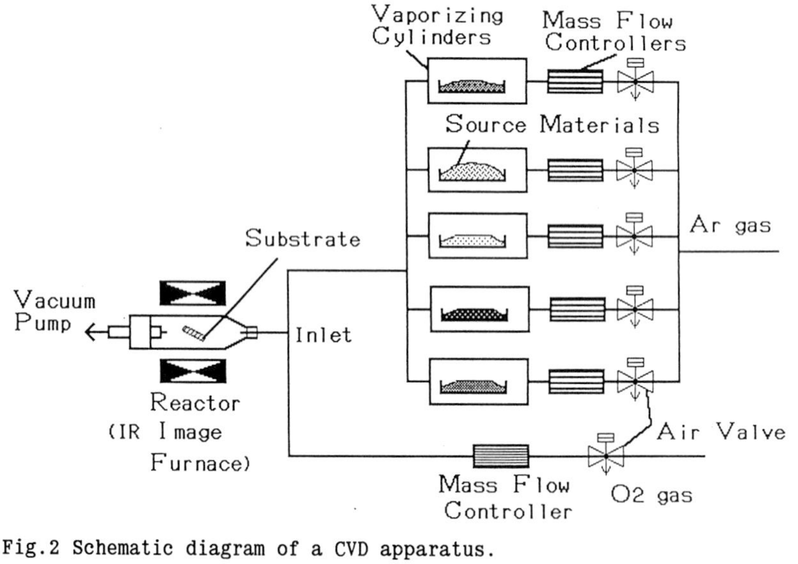

Fig.2 Schematic diagram of a CVD apparatus.

RESULTS AND DISCUSSION

The X-ray diffraction (XRD) patterns of the BSCCO films without Pb on single crystalline MgO (100) are shown in Figure 3. The substrate temperature was 840°C. (a) mainly shows the (00n) peaks of the $Bi_2Sr_2CaCu_2O_x$ (2-2-1-2) phase with weak peaks of the $Bi_2Sr_2CuO_x$ (2-2-0-1) phase. On increasing the contents of Ca and Cu in source gas, the peaks of the $Bi_2Sr_2Ca_2Cu_3O_x$ (2-2-2-3) phase began to appear with the (2-2-1-2) and (2-2-0-1) phases (Figure 3(b)). But the sngle phase of (2-2-2-3) can not be obtained.

The temperature dependence of resistivity of these BSCCO films was measured by a direct current four terminal method with indium electrodes (Figure 4(a)). The film of BSCCO (2-2-1-2) phase showed an onset superconducting transition at 90K, and had a Tc with zero resistivity at 84.5K. On the other hand, the film of BSCCO(2-2-1-2) phase with the (2-2-2-3) phase showed an onset superconducting transition at 110K, but had a Tc at 77K after broad transition. It seems that the electric connection of the (2-2-2-3) phase was incomplete. Critical current density (Jc) at 77.3K, 0 Tesla was measured. The dimension of measuring area was 1.8 mm in width and 2mm in length. The films that have Tc above 77.3K were only BSCCO (2-2-1-2) phase so far. The best BSCCO film, 2.5 micrometer in thickness, had a Jc of $6.4x10^3 A/cm^2$.

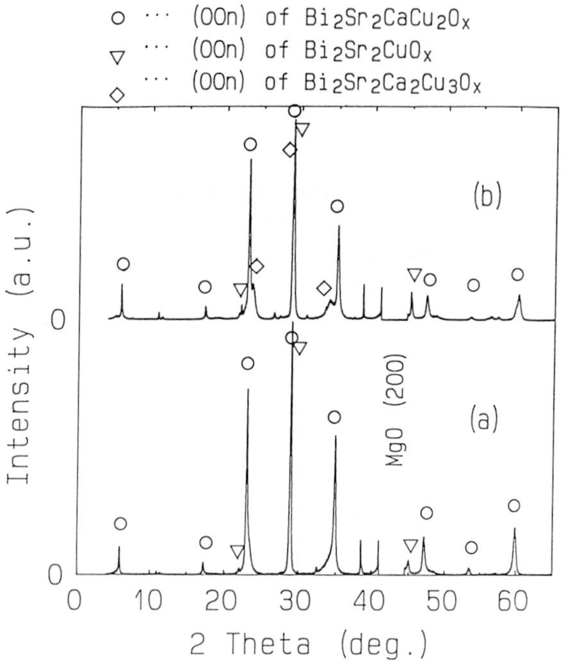

Fig.3 XRD patterns of the BSCCO films on MgO.

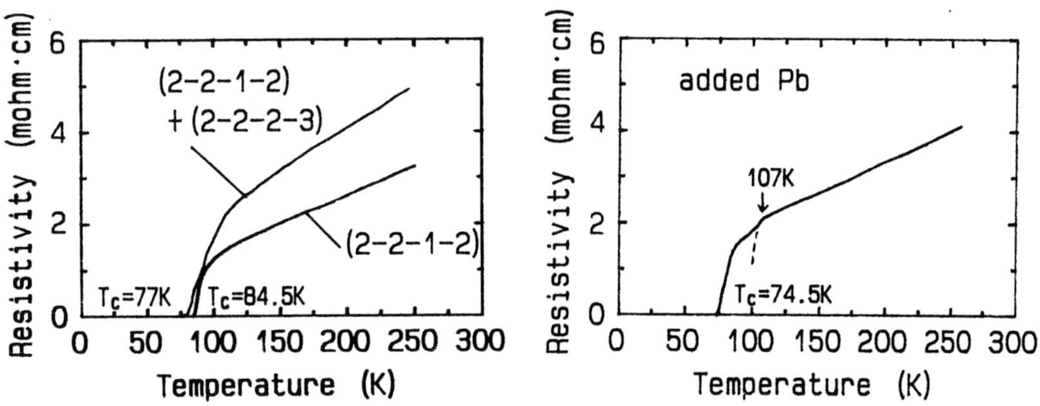

(a) BSCCO without Pb

(b) BSCCO with Pb

Fig.4 Temperature dependence of resistivity of BSCCO films by CVD.

The superconducting BSCCO thin films contained Pb were obtained. The BSCCO films with Pb contain Sr more than the film without Pb, so the Sr content in source gas has to be reduced. But now, according to the XRD analysis, the BSCCO (2-2-2-3) single phase has not been obtained. The temperature dependence of resistivity of this film with Pb was measured (Figure 4(b)). This film had two onset superconducting transitions at 107K and 90K. And Tc with zero resistivity was 74.5K. Compared with the BSCCO film without Pb, the rate of resistivity drop about 110K was sharp in this film.

Figure 5. shows the microstructure of the BSCCO (2-2-1-2) film on MgO (100) observed by scanning electron microscope (SEM). Figure 5 (a) is SEM photograph of the surface of this film. The film consisted of poly-crystalline rectangular grains on the substrate. Figure 5 (b) is SEM of the cross section of this film. Dense structure was observed and voids were rarely recognized in the film. The interface between film and substrate is seems to be strongly adhesive. Thickness of this film was about 2.5 micrometers.

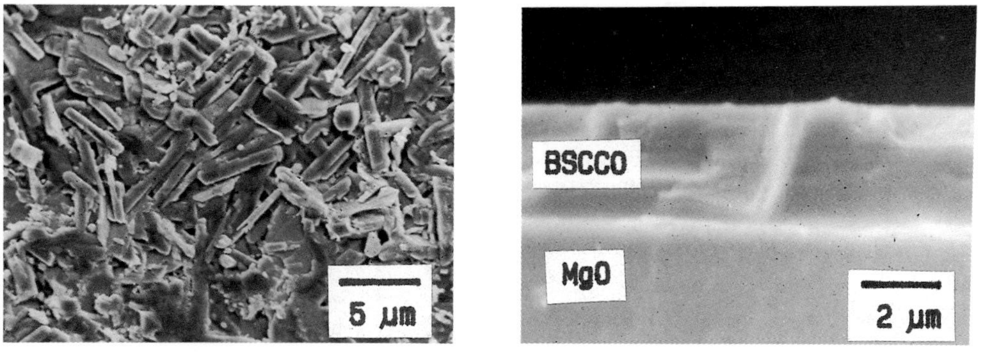

(a) Surface (b) Cross section

Fig.5 SEM photographs of the BSCCO film by CVD.

CONCLUSION

Superconducting (Bi,Pb)-Sr-Ca-Cu-O thin films were prepared by chemical vapor deposition (CVD) method using beta-diketonate chelates and phenyl metals on single crystalline MgO (100) substrates. Superconducting thin films were obtained at 840°C at the high speed growth rate (5 micrometer per hour). The best film on single crystalline MgO had the zero resistivity T_c at 84.5K and critical current density (J_c) of $6.4x10^3$ A/cm^2 at 77.3K, 0T. According to the XRD pattern, the $Bi_2Sr_2CaCu_2O_x$ phase oriented c-axis was observed. In case of adding Pb to the BSCCO film, the $(Bi,Pb)_2Sr_2Ca_2Cu_3O_x$ single phase did not appear in spite of decreasing Sr content in source gas. But the film that had an onset T_c at 107K could be obtained. These BSCCO thin films by CVD was dense and had no void. Synthesis of the higher J_c thin film can be expected.

REFERENCES

1. Noge S, Yamashita T, Wang Z, Matsui T, Kurosawa H, Yamane H and Hirai T, (1989) Jpn. J. Appl. Phys. 28, L1581
2. Irie A, Era M, Yamashita T, Kurosawa H, Yamane H and Hirai T, (1989) Jpn. J. Appl. Phys. 28, L1816
3. Hong M, Liou SH, Kwo J and Davidson B, (1987) Appl. Phys. Lett. 51, 694
4. Yamanishi K, Yasunaga S, Imada K, Sato K and Hashimoto Y, (1988) Mat. Res. Soc. Symp. Proc. 99, 343
5. Wu XD, Dijkkamp D, Ogale SB, Inam A, Chase EW, Miccli PE, Chang CC, Tarascon JM and Venkatesan T, (1987) Appl. Phys. Lett. 51, 861
6. Lackey WJ, Carter WB, Hanigofsky JA, Hill DN, Barefield EK, Neumeier G, O'Brien DF, Shapiro MJ, Thompson JR, Green AJ, Moss,III TS, Jake RA and Efferson KR, (1990) Appl. Phys. Lett. 56, 1175
7. Maeda H, Tanaka Y, Fukutomi M and Asano T, (1988) Jpn. J. Appl. Phys. 27, L209

Preparation and Properties of Bi-Sr-Ca-Cu-O Thin Films by MOCVD

JUNICHI ISHIAI[1], SHINYA YAMADA[1], SHUJI YOSHIZAWA[1], and KAZUHIRO ENDO[2]

[1] Central Research Laboratory, Dowa Mining Co., Ltd., 277-1, Tobuki, Hachioji, Tokyo, 192 Japan
[2] Electrotechnical Laboratory, 1-4, Umezono 1-chome, Tsukuba, 305 Japan

ABSTRACT

We report on the Bi-Sr-Ca-Cu-O film formation on MgO(100), $SrTiO_3$(100) and $LaAlO_3$(100) by MOCVD using $Bi(C_6H_5)_3$, $Sr(DPM)_2$, $Ca(DPM)_2$ and $Cu(DPM)_2$ as source materials. The XRD pattern for the films prepared on MgO(100) showed the high Tc phase with the c-axis oriented normal to the substrate surface, whose Tc zero was 75K. On the other hand, the XRD pattern for the films prepared on $SrTiO_3$(100) and $LaAlO_3$(100) showed the mixture of the high Tc and low Tc phases. The Tc zero of 79K on $SrTiO_3$(100) and 82K on $LaAlO_3$(100) were higher than that on MgO(100).

KEY WORDS: Bi-Sr-Ca-Cu-O thin films, MOCVD, oxide superconductor

INTRODUCTION

Since discovery of superconductivity in the Bi-Sr-Ca-Cu-O system [1] , many studies have been made on obtaining the high Tc single phase [2] . Bi-Sr-Ca-Cu-O system has at least two superconducting phases; the high Tc phase ($Bi_2Sr_2Ca_2Cu_3O_x$) with Tc around 110K and the low Tc phase ($Bi_2Sr_2Ca_1Cu_2O_y$) with Tc around 80K.

Characteristics of metalorganic chemical vapor deposition (MOCVD) method compared with sputter and MBE are as follows ; high deposition rate, compatibility with variously shaped substrates, possibility to large scale processing and wide range of oxygen pressure.

It has been reported that in-situ Pb-doping in MOCVD enhances the high Tc phase formation in the Bi system [3] . However, Tc zero of the film was lower than 100 K. In this study, first we report on the preparation of Bi-Sr-Ca-Cu-O films on MgO(100), $SrTiO_3$(100) and $LaAlO_3$(100) substrates by MOCVD and next on the effect of substrates on superconductivity of the films.

EXPERIMENTAL

Figure 1 is a schematic drawing of the MOCVD apparatus used in this study. This system is classified as a hot-wall type MOCVD, where the substrates are heated by an electric furnace. The source materials were triphenyl bismuth $Bi(C_6H_5)_3$, bis(-2,2,2,6,6,-tetramethyl-3,5-heptandione)-strontium $Sr(DPM)_2$, -calcium $Ca(DPM)_2$ and -copper $Cu(DPM)_2$. Figure 2 shows the structure of the source materials. The temperatures of the vessels of $Bi(C_6H_5)_3$, $Sr(DPM)_2$, $Ca(DPM)_2$ and $Cu(DPM)_2$ were 66, 230, 188 and 118℃, respectively. Argon was used as carrier gas. The flow rate of argon gas for $Bi(C_6H_5)_3$, $Sr(DPM)_2$, $Ca(DPM)_2$ and $Cu(DPM)_2$ were 20, 200, 100 and 25 cc/min, respectively. Oxygen gas was separately introduced into the growth chamber at the rate of 200 cc/min. During the deposition, the total pressure in the growth chamber was 10 Torr and the substrate temperature was 800℃. Single crystals $MgO(100)$, $SrTiO_3(100)$ and $LaAlO_3(100)$ were used as substrates. The composition ratio of Bi, Sr, Ca and Cu in the film was controlled by changing the vessel temperature and the carrier gas flow rate. The films were characterized using X-ray diffraction (XRD), scanning electron microscopy (SEM), electron probe microanalysis (EPMA), auger electron spectroscopy (AES) and resistance measurement. Resistance measurements were carried out by a standard four-probe DC method, where Cu wires were attached with Ag paste.

RESULTS AND DISCUSSION

We have examined the volatility of the source materials using the thermogravimetry. Figure 3 shows the thermogravimetry of the source materials. It was found that $Ca(DPM)_2$ and $Cu(DPM)_2$ volatilize completly up to 100% at 360 and 260℃, respectively, and that $Bi(C_6H_5)_3$ and $Sr(DPM)_2$ volatilize up to 95% at 300 and 390℃, respectively.

We have prepared the films on the substrates of $LaAlO_3(100)$ (sample 1), $SrTiO_3$ (100) (sample 2) and $MgO(100)$ (sample 3). Figure 4 shows the XRD patterns of samples (1), (2) and (3). All the samples showed the preferred orientation of the c-axis normal to the substrate surface. Samples (1) and (2) showed the mixture of the high Tc and low Tc phases. Sample (3) showed the high Tc phase.

Figure 5 shows the temperature dependences of the resistance for the films. All the samples showed the abrupt decrease of the resistance at around 110K. Samples (1), (2) and (3) showed zero resistance at 82, 79 and 75K, respectively.

Figure 6 shows the SEM and the EPMA photographs of the surface of the sample (3). The small precipitates were observed on the surface. EPMA revealed that these precipitates are calcium and copper oxide. The depth profiles of all the samples were obtained by AES. The result showed the diffusion of Mg into the film on $MgO(100)$ substrate, Ti into the film on $SrTiO_3(100)$ substrate and La into the film on $LaAlO_3(100)$ substrate.

CONCLUSION

We have fabricated the Bi-Sr-Ca-Cu-O films on $MgO(100)$, $SrTiO_3(100)$ and $LaAlO_3$ (100) substrates by MOCVD using $Bi(C_6H_5)_3$, $Sr(DPM)_2$, $Ca(DPM)_2$ and $Cu(DPM)_2$ as source materials. The films on $MgO(100)$, $SrTiO_3(100)$ and $LaAlO_3(100)$ substrates showed the superconducting trandition with Tc zero of 75, 79, and 82K, respectively. The film on $MgO(100)$ substrate showed the high Tc phase with the c-axis oriented normal to the substrate surface. In cotrast, the films on $SrTiO_3$ (100) and $LaAlO_3(100)$ substrates showed the mixture of the high Tc and low Tc phases. AES revealed that the substrate elements diffuse into the films.

REFERENCES

1.Maeda H. Tanaka Y. Fukutomi M. Asano T (1988) A new high-Tc superconductor without a rare earth element. Jpn J Appl Phys 27:L209-L210

2.Endo K. Hayashida S. Ishiai J. Matsuki Y. Ikedo Y. Misawa S. Yoshida S (1990) Preparation of As-Deposition Bi-Sr-Ca-Cu-O Films with High Tc Superconducting Phase by Metalorganic Chemical Vapor Deposition. Jpn J Appl Phys 29:L294-L297

3.Endo K. Hayashida S. Ishiai J. Ikedo Y. Misawa S. Yoshida S (1990) In-situ growth of Bi-Sr-Ca-Cu-O films with High-Tc superconducting phase by MOCVD. Advances in Superconductivity II .Springer-Verlag, TOKYO :pp 911-914

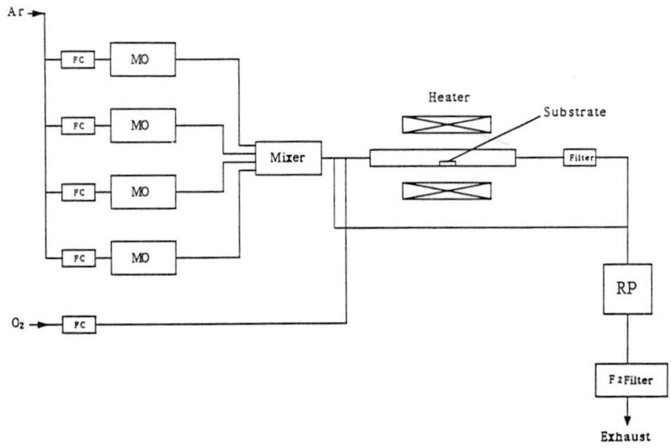

Fig.1 Schematic drawing of MOCVD apparatus.
FC : flow controller.
RP : rotary pump.
MO : vessel for source material.

Source materials

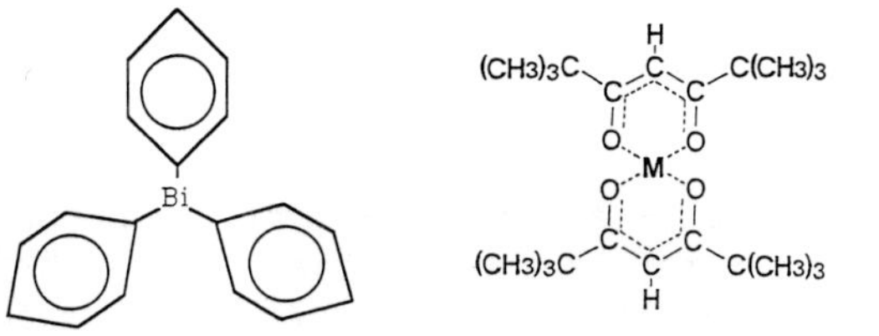

triphenyl bismuth

$Bi(C_6H_5)_3$

bis(dipivaloylmethanato) metal

$M(DPM)_2$

metal : Sr, Ca, Cu

Fig.2 Structure of $Bi(C_6H_5)_3$, $Sr(DPM)_2$, $Ca(DPM)_2$ and $Cu(DPM)_2$.

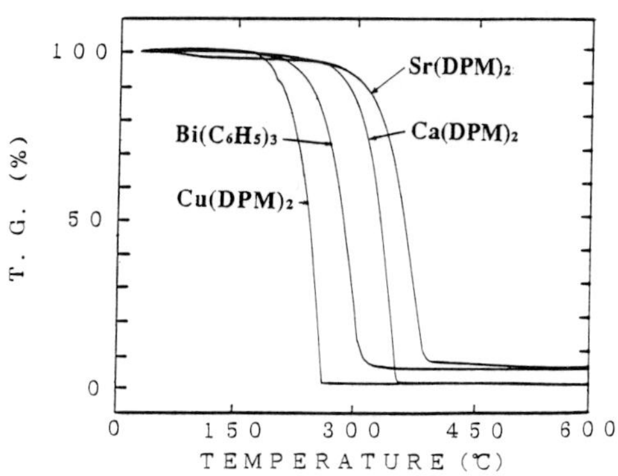

Fig.3 Thermogravimetry of the source materials.

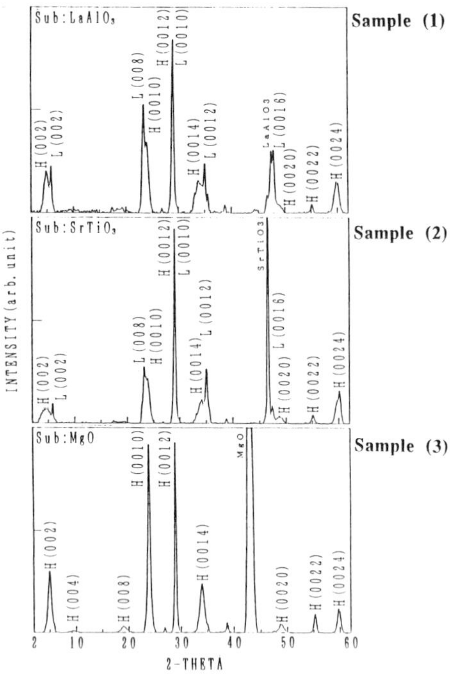

Fig. 4 XRD pattern.

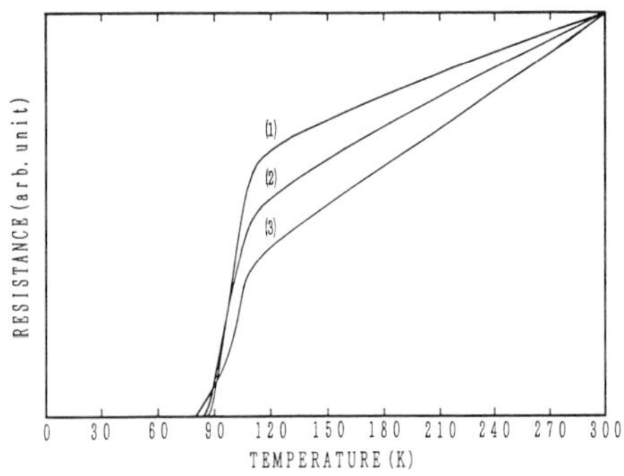

Fig. 5 Temperature dependences of the resistance. Samples (1), (2) and (3) denote the films on MgO(100), SrTiO(100) and LaAlO(100) substrates, respectively.

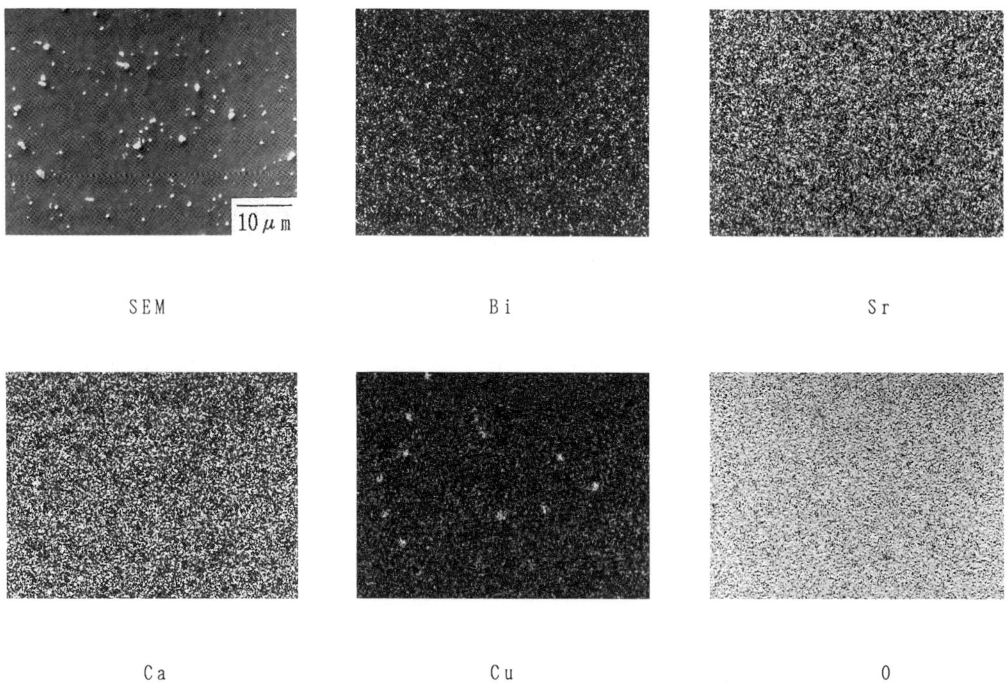

Fig. 6 SEM and EPMA photographs of the surface of the sample (3). The EPMA photographs show the distribution of each element in the film, where each element corresponds to a white dot.

Bi-based Superconducting Films Synthesized by a Combination of Metalorganic Deposition (MOD) and a Diffusion Process

SEIJI YAEGASHI, MARK GREEN, HIROHIKO MURAKAMI, JUNYA NISHINO, and YUH SHIOHARA

Superconductivity Research Laboratory, International Superconductivity Technology Center, 10-13, Shinonome 1-chome, Koto-ku, Tokyo, 135 Japan

ABSTRACT

Bi-based superconducting thin films have been prepared by a combination of metalorganic deposition(MOD) and a diffusion process. The films derived from a layer of Bi/Sr/Ca/Cu=2/2/1/3 and a layer of Ca/Pb=2/1, after firing for only 4 h, yielded a nearly single phase layer of the 2223(high T_c) phase. It was found that Ca_2PbO_4 played a very important role in formation of the 2223 phase in short firing times, acting as a souce of Ca ions to react with the 2212 phase and CuO.

KEYWORD: Bi-Pb-Sr-Ca-Cu-O, superconducting thin film, metalorganic deposition, diffusion process,

INTRODUCTION

Metalorganic deposition (MOD) is one of the techniques for depositing thin film superconductors. A potential advantage of MOD is that it is not a gaseous deposition technique and therefore does not require special complicated equipment. This makes MOD a promising method for fabrication of commercial products under ambient atmosphere. The high T_c phase of Bi-based superconductors(the 2223 phase) generally seems to be difficult to form without long firing times; partial substitution of Pb for Bi has been found to enhance the 2223 phase forming reaction[1] The effect of Pb substitution was clarified to be that the 2223 phase grows from the initially formed low T_c phase(the 2212 phase) and impurity phases such as Ca_2PbO_4 and CuO through a partially melting phase[2]. This suggests that the diffusion process naturally occurs between a liquid phase and a solid phase during firing, even though the precursor was uniformly mixed initially. During a diffusion process, the composition of the liquid phase may be variable.

The diffusion process has been successfully adapted to the synthesis of Y-based and Bi-based superconductors[3,4]. The diffusion process is usually applied by using two kinds of starting materials with different melting points. For Bi-based superconductors, the combinations of Bi_2CaCuO_5 and $Sr_2CaCu_2O_5$, or $Bi_2CaCu_2O_6$ and Sr_2CaCuO_4 have been reported[4]. By partial substitution of Pb for Bi in the low melting point component, formation of the 2223 phase was observed[4].

In this study, a diffusion process was combined with the MOD technique to fabricate Bi-based superconducting thin films. Various combinations of precursor solutions were deposited by MOD(spin coating), and then the films were heated to allow the diffusion process and associated reactions to occur. Formation of the 2223 phase was observed after firing for 4 h. The 2223 phase forming process is discussed from the standpoint of the crystals resulting from different combinations of precursor solutions.

EXPERIMENTAL

Precursor solutions were prepared by mixing carboxylic acid salts in toluene. Bi-2-ethylhexanoate(2EH), Pb-2EH, Sr-2EH, Ca-2EH, and Cu-naphthenate were used as raw materials. Generally, the diffusion reaction proceeds in a combination composed of a high-melting point component and a low-melting point component *via* the formation of a liquid phase. Two pairs of components for the diffusion process to produce Bi-based superconducting films were selected on the basis of one component having a lower melting point than the firing temperature(around 850°C), and the other component having a higher melting point than the firing temperature. The total amount of each component deposited was adjusted so that the overall combination of layers could, if the reaction were to go to completion throughout the entire volume, result in the ideal 2223 phase with some Pb substitution for Bi. Compositions and some characteristics of these precursor solutions are shown in table 1.

Each precursor solution was spin-coated on an MgO(100) single crystal substrate($10x10x1mm^3$) at 2000-3000rpm. The pyrolysis was done by placing the sample directly into a preheated furnace at 600°C and holding for about 30 min in air. The cycle of spin-coating and pyrolysis was repeated from one to several times to control the film thickness and total composition. Pyrolyzed films were heated at a rate of 30°C/min and fired at 845°C for 5 min to 10 h in an atmosphere of O_2:Ar=1:12. To prevent the evaporation of volatile lead compound, a sintered oxide pellet with the compositon of Bi/Pb/Sr/Ca/Cu=0.8/0.2/1/1/1.5 was placed above the film.

The amount of the deposited films was monitored by simply weighing them and by inductively coupled plasma atomic emission spectrometry(ICP). The temperature dependence of the resistivity of the films was measured by a standard four-probe technique. The microstructure was investigated by scanning electron microscopy (SEM) and electron probe microanalysis (EPMA). X-ray diffraction(XRD) was done with Cu(Kα) radiation.

Table 1 Composition, total metal content, viscosity, and melting point of precursor solutions.

Sample Name	Metal ratio	Total metal content(%)	Viscosity (mPas)	Melting point (°C)
BPC	Bi/Pb/Cu=4/1/2	10.0	2.3	780
SCC(i)	Sr/Ca/Cu=8/9/10	3.0	3.0	-
BP	Bi/Pb=4/1	11.4	2.4	755
SCC(ii)	Sr/Ca/Cu=9/10/15	3.1	2.5	-
BPSC	Bi/Pb/Sr/Cu=4/1/4/4	5.1	1.0	-
CC	Ca/Cu=2/1	4.2	5.9	-
BSCC	Bi/Sr/Ca/Cu=2/2/1/3	4.5	1.1	-
CP	Ca/Pb=2/1	2.0	0.5	(820)

RESULTS AND DISCUSSION

Overall composition of the film was adjusted by changing the number of drops used in spin coating. In order to make the 2223 phase, the overall composition nominally equal to 2223, i.e., three drops of BPC and three drops of SCC(i) similar to the third one, was adopted for the experiment. This involved firing the films, all with the same composition, for different lengths of time. The XRD results are shown in Fig. 1. The film fired for 15 min consisted of the 2212 phase, Ca_2PbO_4, a small amount of $Bi_2(Sr,Ca)_2CuO_y$(2201 phase), and CuO. For long firing times, the 2223 phase grew at the expense of the 2212 phase. The film fired for 4 h had the 2223 phase as the dominant

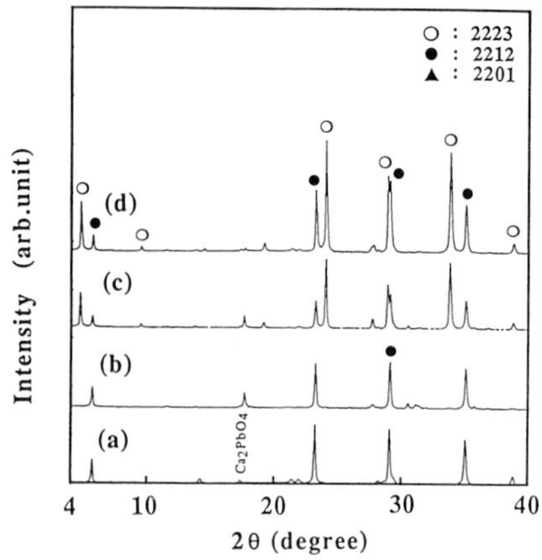

Fig. 1 X-ray diffraction patterns of the films derived from BPC and SCC(i) fired for (a) 15 min, (b) 1 h, (c) 4 h, and (d) 10h.

phase. Figure 2 shows SEM images of the films fired for 15min and 4 h. In the case of the film fired for 15 min, fine rod-like crystals and small island-like crystals can been observed. These crystals were identified by EPMA to be Ca_2PbO_4 and CuO, respcetively. For the film fired for 4 h, these crystals can not be seen. This suggests that the 2223 phase was formed by reaction of the 2212 phase with Ca_2PbO_4 and CuO which were present at the early stage of firing.

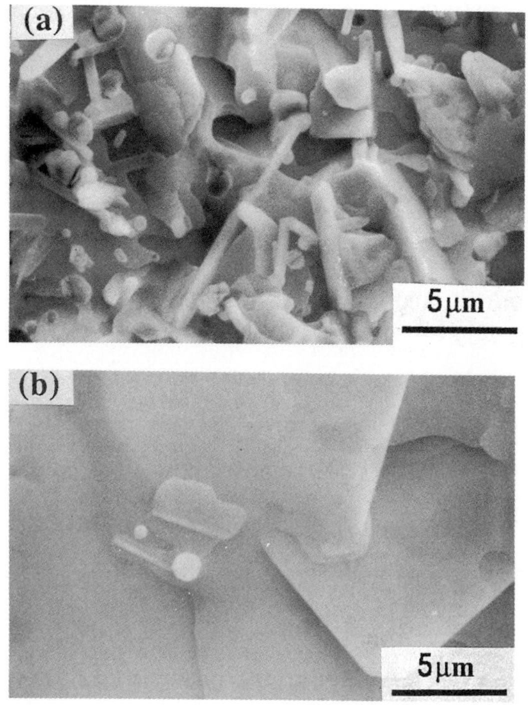

Fig. 2 SEM images of the fims of BPC and SCC(i) fired for (a) 15 min, and (b) 4 h.

In all experiments described from this point, various combinations of precursors were used as the bottom and the top layers, but in all cases, the number of drops used to produce the layers was adjusted in order to keep the overall composition as close as possible to 2223. The objective was to determine which combination(s) would be the most appropriate for producing the 2223 phase. In the case of the combination of BP and SCC(ii) (see Table 1), even after firing for 4 h, the 2223 phase was not produced, although the amount of the 2212 phase, which was present at an early stage of firing, decreased with firing time. According to EPMA, there was no evidence for finely dispersed Ca_2PbO_4 crystals. On the other hand, large needle-like crystals were observed which were identified by EPMA to be Ca_2CuO_3.

Another combination attempted was a bottom layer of bismuth, lead, strontium, and copper, designated BPSC, and a top layer of calcium and copper, designated CC. In this case, even after 10 h, the 2223 phase did not form; the 2212 phase and the 2201 phase persisted as the main phases. This particular result can be explained by the fact that neither layer has a compound with melting point below 900°C, so reactions would be expected to take longer.

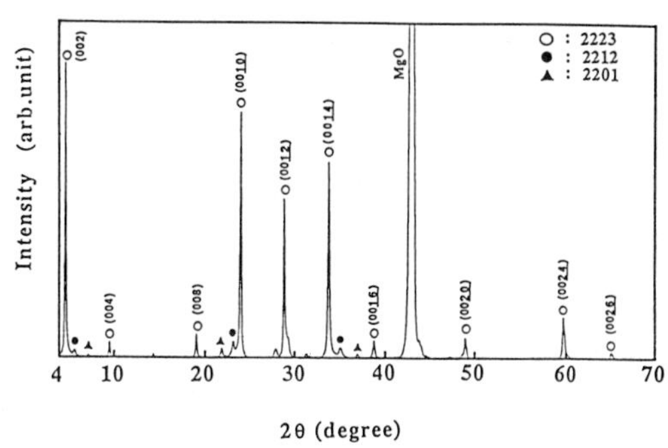

Fig. 3 X-ray diffraction pattern of the film of BSCC and CP fired for 4 h.

In experiments described above, in the one case in which the desired 2223 phase formed, it apparently grew via a reaction involving Ca_2PbO_4 and the 2212 phase. This was true in spite of the fact that the starting compositions of the respective layers were different from these phases. In the other cases, in which 2223 did not form, there was no evidence for finely dispersed Ca_2PbO_4 and CuO crystals at early stage of firing, but huge needle-like crystals were observed. Since it appears that the presence of Ca_2PbO_4, CuO and 2212 in the close proximity to each other is a prerequisite for formation of the 2223 phase, an efficient way to form the 2223 phase would probably be to start with a combination of Ca_2PbO_4 in one layer and 2212+CuO in the other. This possibility was tested by the fourth combination listed in Table 1: BSCC and CP. The film fired at 825°C for 15 min indicated strong peaks for the 2212 phase and weak peaks for Ca_2PbO_4. When the film was fired at 845°C, the 2223 phase was produced relatively quickly, with the amount increasing with firing time. After only 4 h, the 2223 phase was dominant with only trace amounts of the 2212, 2201 and Ca_2PbO_4 phases as shown in Fig. 3. The binary phase diagram of PbO-CaO[5] indicates that a liquid phase appears due to incongruent melting of

Ca_2PbO_4, but another experiment in which only CP was coated, fired at 845°C and quenched could not confirm the phase separation attributed to incongruent melting of Ca_2PbO_4 during firing. EPMA analysis indicated the plate-like crystals to be composed of metals in the ratio of Bi/Pb/Sr/Ca/Cu=0.19/0.034/0.21/0.22/0.35. This result clearly demonstrated that by initially forming Ca_2PbO_4 and the 2212 phase, the subsequent formation of 2223 was enhanced through a solid state reaction. Further support for this point can be obtained by comparing this result with that of Fig. 1, where 2223 also formed at the expense of 2212 and Ca_2PbO_4. In that case, however, Ca_2PbO_4 and 2212 had to form from different starting compositions, so the whole process took longer. After 4 h, as well as after 10 h, the relative amount of the 2223 phase, although dominant, was not as high as in the experiment where the process was begun with the combination of Ca_2PbO_4 in one layer and 2212+CuO in the other.

Table 2 A summary of phase development for each of the starting combinatins at various stage of firing. Region A corresponds to a temperature of about 800°C during the period of temperature increase(30°C/min) ; region B corresponds to the start of the hold period at 845°C ; region C corresponds to the point after holding at 845°C for 4 h.

Region	BPC/SCC(i) [bottom/top]	BP/SCC(ii)	BPSC/CC	BSCC/CP
A	liquid#,DTA 2201	liquid#,DTA 2201	2201	2201
B	liquid* 2201 2212 Ca_2PbO_4	liquid* 2201 2212 Ca_2PbO_4 Ca_2CuO_3	2201 Ca_2CuO_3 CuO	2212 Ca_2PbO_4 CuO
C	2212 2223 Ca_2CuO_3	2212 2201 Ca_2CuO_3	2201 2212 Ca_2CuO_3 CuO	2223

Unless indicated otherwise, the indicated result was obtained in the present work by XRD and/or EPMA.
* Speculation, # ref.5

Table 2 provides a summary of phase development for each of the starting combinations at various stages of firing. Region A corresponds to a temperature of about 800°C during the period of temperature increase(30°C/min); region B corresponds to the start of the hold period at 845°C; region C corresponds to the point after holding at 845°C for 4 h.

In summary, Bi-based superconducting thin films with considerable amount of the 2223 phase could be produced by a combination of MOD and a diffusion process. Ca_2PbO_4 played a very important role to provide the 2212 phase with Ca ions while at the same time preventing the film from forming Ca_2CuO_3. By intentional forming the layer of Ca_2PbO_4, it was found that formation of the 2223 phase could be enhanced.

REFERENCES

1. Takano M, Takada J, Oda K, Kitaguchi H, Miura Y, Ikeda Y, Tomii Y and Mazaki H (1988) Jpn. J. Appl. Phys.27: L1041
2. Nobumasa H, Shimizu K, Kitano Y and Kawai T (1988) Jpn. J. Appl. Phys.27: L1669
3. Tachikawa K, Sadakata N, Sugimoto M, Kohno O (1988) Jpn. J. Appl. Phys.27: L1501
4. Tachikawa K, Watanabe T and Inoue T (1990) Supercond. Sci. Technol. 3:180
5. Roth R S, Negas T and Cook L P (1984) Phase Diagram for Ceramists, ed. G. Smith (The american Ceramic Society, Ohio): Vol.IV, pp94, Fig. 5140

Preparation and Characterization of Ultrathin BSCCO Films by MOCVD

TSUNEMI SUGIMOTO, MASAJI YOSHIDA, KAZUSHI SUGAWARA, and YUH SHIOHARA

Superconductivity Research Laboratory, International Superconductivity Technology Center,
10-13, Shinonome 1-chome, Koto-ku, Tokyo, 135 Japan

ABSTRACT

Ultrathin Bi-Sr-Ca-Cu-O films with high critical transition temperatures were grown on (100) MgO single crystal substrates by MOCVD. The 3.5-nm-thick film close to a single unit cell (3.1 nm) exhibited a complete superconducting resistance transition at 64 K, whereas the superconducting property of much thinner films was drastically deteriorated. From the result of the XRD patterns and SEM observations of ultrathin films, it was found that a 2212 layer was 2-dimensionally grown on the MgO substrate.

KEY WORDS: ultrathin film, Bi-Sr-Ca-Cu-O superconductor, XRD, SEM

INTRODUCTION

The techniques for preparing ultrathin films of high-temperature superconductors are driving forces in the realization of superconductor devices. Moreover, ultrathin films of high-temperature superconductors, especially single-cell-thick films, are also of great importance for the understanding of superconducting mechanism of high-temperature superconductors, because their layered structure and the inter- and/or intra-coupling between neighboring CuO planes play an important role in determining the superconducting properties as well as the basic pairing mechanism. Recently, Y-Ba-Cu-O/M-Ba-Cu-O (M=Pr, Dy) superlattices with single-unit-cell layers were successfully prepared by physical vapor depositions (PVD) including magnetron sputtering [1,2] and laser ablation [3,4], and isolated ultrathin Y-Ba-Cu-O films with the thickness down to 2 nm were also fabricated by magnetron sputtering [5]. On the other hand, metalorganic chemical vapor deposition (MOCVD) is an alternative ultrathin film preparation process because MOCVD has achieved film thickness control with an atomic layer scale in the growth of compound semiconductors.

In this paper, we show the growth of ultrathin Bi-Sr-Ca-Cu-O films close to a single unit cell thickness by MOCVD that exhibit complete superconducting resistive transitions at temperatures as high as 64 K, and also discuss the growth mechanism of Bi-Sr-Ca-Cu-O films prepared by MOCVD.

EXPERIMENTAL

The MOCVD apparatus used in this study consists of a horizontal cold wall quartz reactor, MO feed lines, mass flow controllers, MO source evaporators, and evacuation system. The MO source evaporators were placed in the accurately temperature-controlled ovens and all MO feed lines were also placed in another oven maintained at 200 °C in order to improve the uniformity of line-heatings and to prevent the MO vapors from condensation and/or decomposition. The strontium transport line was wrapped with resistive heating tapes and maintained at 230 °C. The substrate temperature was monitored by a platinum/platinum-rhodium thermocouple placed into the susceptor. The surface temperature was calibrated against the susceptor temperature and monitored 20-30 °C lower than the susceptor temperature. This was achieved by measuring the surface temperature of the CuO thin film on the MgO substrate by an optical pyrometer. A confirmed emissivity value of 0.8 at wavelengths of 0.8-1.1 μm was used for this measurement. In this wavelength range, the transmissivity of the quartz reactor is almost unity.

Volatile β-diketonate complexes of Cu, Sr, Ca with the dipivaloylmethanate ligand (DPM: $C(CH_3)_3COCHCOC(CH_3)$) [that is, $Cu(DPM)_2$, $Sr(DPM)_2$, $Ca(DPM)_2$] were adopted for MO sources. Triphenylbismuth ($Bi(ph)_3$) was employed as the bismuth source because of its superior volatility and chemical stability. Evaporation temperatures of 83-213 °C and argon carrier flow rates of 50 sccm were used to transport the MO sources. The MO vapors and argon carrier gas were mixed with oxygen prier to entering the reactor. A mechanically polished (100) MgO single crystal substrate (10x10x0.5 mm^3) was degreased with acetone and then placed on the inconel alloy susceptor. The ultrathin films were grown at the substrate temperature of 750-850 °C under the total pressure of 10 Torr and the oxygen partial pressure of 8.8 Torr. After the growth process, all MO sources were switched to the vent line and the sample was cooled to 300 °C at the rate of 10 °C/min under the growth conditions. The growth rate employed was as low as sub-namometer/min. The chemical composition of the films analyzed by Inductively coupled plasma emission spectrometry (ICP) was close to $Bi_2Sr_2Ca_2Cu_3O_x$. The average thickness of the films was estimated from the weight of the growth layer on the substrate, using the weight value from measurements by ICP. The theoretical density of a Bi-Sr-Ca-Cu-O superconductor, 6.55 g/cm^3, was adopted as that of the films. The average thickness of ultrathin films estimated by ICP revealed the linear relationship with the growth time from 35 to 2.5 nm and the plotted straight line went through the origin. Moreover, the thickness of thicker films was directly verified by high-resolution scanning electron microscopy (SEM) observation from the cross section of the cleaved samples. The thickness of the films measured by the different techniques was found to be accurate within ±10 %. The ultrathin Bi-Sr-Ca-Cu-O films were characterized by a conventional four-probe resistance measurement, x-ray diffraction (XRD) using Cu-K_α radiation and SEM observation.

RESULTS AND DISCUSSION

Figure 1 shows zero-resistance superconducting transitions of Bi-Sr-Ca-Cu-O films with different thickness between 2.7 and 38 nm. In the films thicker than 10 nm, the zero-resistance superconducting transitions were constant at around 80 K. The zero-resistance superconducting transition as high as 64 K was achieved in the film with the thickness of 3.5 nm close to the c-axis lattice constant of the single unit cell of $Bi_2Sr_2CaCu_2O_x$ (2212) phase (3.1 nm). In the film thinner than the single unit cell of 2212 phase, the temperature dependence of resistivity was metallic with the beginning of a superconducting transition around 85 K, while the zero-resistance superconducting transition has not been achieved. The surface roughness of a mechanically polished (100) MgO substrate is generally in the range from 1 to 2 nm. Therefore, it is suggested that the first half-unit cell layer (c/2=1.5 nm) on the substrate has many defects and a complete superconducting path can not be formed. In view of the surface roughness of the substrate, the transport property of the 3.5-nm-thick film is suggestively determined by the superconducting path of the second half-unit -cell layer, and that of thinner films than 3.5 nm was drastically deteriorated probably because the superconducting path is not continuously connected in the second half-unit-cell layer.

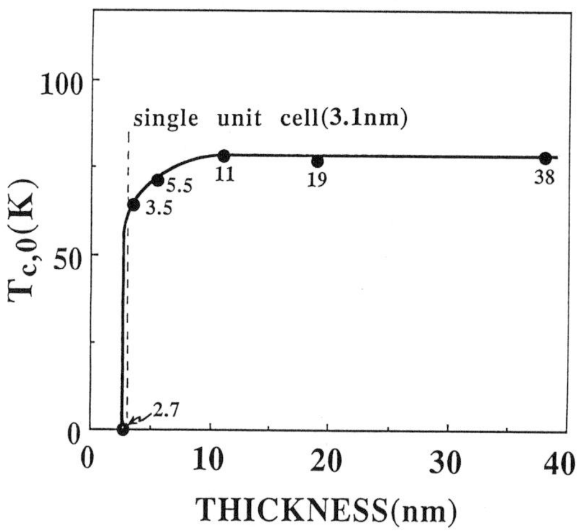

Fig.1 Zero-resistance superconducting transitions of ultrathin films with different thickness between 2.7 and 38 nm.

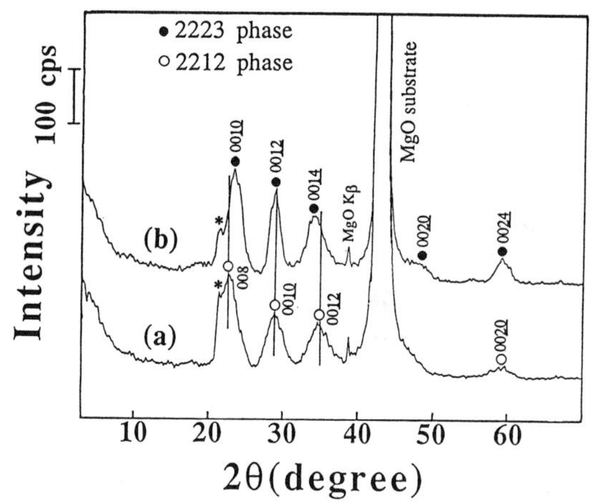

Fig.2 XRD patterns of ultrathin films with the thickness of (a) 3.5 nm and (b) 5.5nm. The asterisks indicate the forbidden reflection from 001 of MgO substrate.

Figure 2 shows XRD patterns from θ-2θ scan of the ultrathin Bi-Sr-Ca-Cu-O films. The c-axis lattice constant of 3.5-nm-thick film was 3.1 nm which is consistent with that of a bulk 2212 phase [6] from its 00*l* reflections as shown in Fig.2 (a). Even in the 2.2-nm-thick film the reflection peaks from 008, 00<u>10</u>, and 00<u>12</u> attributed to 2212 phase were observed while these peaks were much broader due to its thinning. These results suggest that ultrathin Bi-Sr-Ca-Cu-O films on the MgO substrate are free of the stress caused by the large mismatch between the Bi-Sr-Ca-Cu-O layer and the MgO substrate even in the first grown layer, probably by misfit dislocations observed by a cross-sectional transmission electron microscope (TEM) in a thicker film [7]. In the 5.5-nm-thick film the average value of the measured c-axis lattice constant from its 00*l* reflection peaks was 3.7 nm which is in good agreement of a bulk $Bi_2Sr_2Ca_2Cu_3O_x$ (2223) phase while a trace of the 2212 phase was contained since the 00<u>10</u> and 00<u>14</u> reflection peaks were slightly shifted to the side of the 008 and 00<u>10</u> reflection peaks of 2212 phase, respectively.

In order to confirm the growth mechanism (2-dimensional or 3-dimensional growth) , the surface morphology of ultrathin films were inspected by a high-resolution SEM. Figure 3 shows a SEM photograph of the 3.5-nm-thick film. The matrix of the 3.5-nm-thick film found to be continuously smooth, and the 2-dimensional nuclei of the next growth layer were observed on the continuous matrix. It is suggested that the film is 2-dimensionally grown on the MgO substrate . From the result of the XRD pattern of the 3.5-nm-thick film, the 2-dimensional grown layer is a 2212 phase and a 2223 phase is formed by the diffusion of Ca and Cu species to a 2212 grown layer from the gas or solid phase in the thicker films.

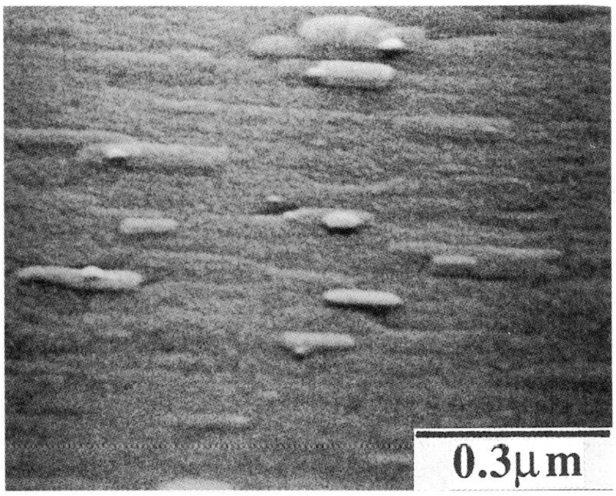

Fig.3 SEM photograph of the 3.5-nm-thick film. The sample was tilted so that the electron beam impinged on it at an incident angle of 70-80 °.

SUMMERY

Ultrathin Bi-Sr-Ca-Cu-O films with high critical transition temperatures were grown on (100) MgO single crystal substrates by MOCVD. The 3.5-nm-thick film close to the c-axis lattice constant of single unit cell (3.1 nm) exhibited a complete superconducting resistance transition at 64 K, whereas the superconducting property of much thinner films was drastically deteriorated. From the result of the XRD patterns and SEM observations of ultrathin films, it was found that a 2212 layer was 2-dimensionally grown on the MgO substrate.

REFERENCE

1.Triscone J-M, Karkut MG, Antognazza L, Brunner O, Fischer Θ (1989) Phys.Rev.Lett. 63: 1016.
2. Triscone J-M, Fischer Θ, Brunner O, Antobnazza L, Kent AD, Karkut MG (1990) Phys.Rev.Lett. 64 : 804.
3. Wu XD, Xi XX, Li Q, Inam A, Dutta B, DiDomenico L, Weiss C, Martinez JA, Wilkens BJ, Schwarz SA, Barner JB, Chang CC, Nazar L, Venkatesan T (1990) Appl.Phys.Lett. 56 : 400.
4. Lowndes DH, Norton DP, Budai JB (1990) Phys.Rev.Lett. 65 : 1160.
5. Xi XX, Geerk J, Linker G, Li Q, Meyer O (1989) Appl.Phys.Lett. 54 : 2367.
6. Matsui Y, Maeda H, Tanaka Y, Horiuchi S (1988) Jpn.J.Appl.Phys. 27 : L372.
7. Sugimoto S, Yoshida M, Yamaguchi Y, Sugawara K, Shiohara Y, Tanaka S (1990) Appl.Phys.Lett. 57 : 928.

Improvement of Critical Current in BiSrCaCuO Micropatterns by Oxygen Plasma Treatment

YASUHIRO NAGAI and KOJI TSURU

NTT Applied Electronics Laboratories, Tokai, Ibaraki, 319-11 Japan

ABSTRACT

The influences of oxygen plasma on critical temperature and critical current density are studied using BiSrCaCuO micropatterns. Although a large critical temperature degradation as well as a slight decrease in a critical current is observed in 1.5-20 μm wide microbridges that have been put into an oxygen plasma, the critical temperatures are recovered by reducing excessive absorbed oxygen. Additionally, these recovered microbridges have higher critical current densities, whose improvement strongly depends on the microbridge width, 5-30 times than those without oxygen plasma treatment.

KEY WORDS: BiSrCaCuO microbridge, Annealed film, Oxygen plasma, Critical current density

INTRODUCTION

Superconducting quantum interference devices (SQUIDs) and microwave components are likely to be one of the earliest applications of high-Tc oxide superconductivity. The high-Tc SQUID might be applied as a magnetic sensor because the grain boundary Josephson fundamentally operates at 77K [1,2] and flux transformers have been intensively developed [3,4]. In the SQUID, it is important to suppress a 1/f noise resulting from flux motion. Thus, some pins are should be introduced into the weak links. In addition, passive microwave components using high-Tc superconducting films, which indicate lower loss and lower dispersion, have been advanced [5]. The lower surface resistance must be maintained even under a relatively high magnetic field. Introducing pins [6] is also effective in keeping the critical current density under the field. Accordingly, we used the oxygen plasma treatment to achieve effective pins, which enable high-Tc microdevices to suppress the 1/f noise and to maintain the lower surface resistance.

This paper reports the effects of oxygen plasma treatment on the critical temperature and critical current density of BiSrCaCuO micropatterns. Reducing excessive absorbed oxygen greatly improves the critical current density and recovers the critical temperature.

EXPERIMENTAL PROCEDURE

1 μm-thick BiSrCaCuO films were prepared on an MgO (100) single crystal substrate by ion beam sputtering. These films were annealed in contact with MgO (100) planes at 880 ℃ for 3hrs [7]. Consequently, the films were made up of a typical 80K-phase with a strong c-axis preferred orientation.

The superconducting films were patterned into 10 μm long and 1.5-20 μm wide microbridges, as shown in Fig. 1, by nitrogen ion beam etching and Ta/resist multilayer mask [8]. After patterning, microbridges with (Fig. 2-a) and without (Fig. 2-b) an Au protective layer of 100 nm were put into the oxygen plasma atmosphere. Consequently, the Au protective layer suppressed the excessive oxygen that was absorbed into the microbridges

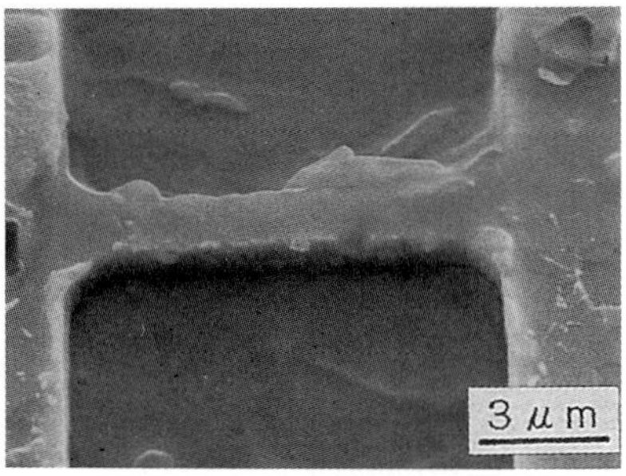

Fig. 1 An example of a patterned BiSrCaCuO microbridge.

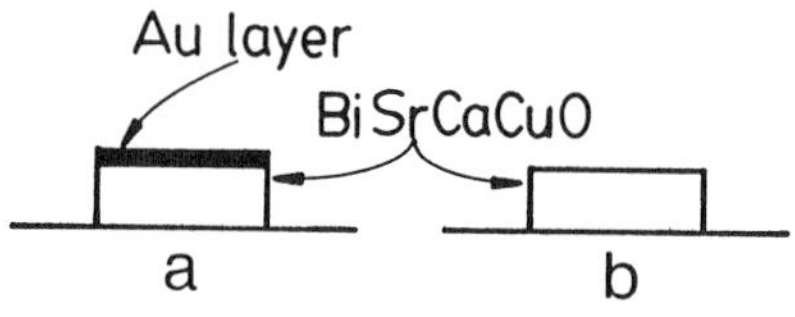

Fig. 2 Cross-section of a microbridge. a and b correspond to microbridges with and without an Au protective layer, respectively.

[9]. Both types of microbridges were put in dry air for several months to slowly reduce the absorbed oxygen (slow reduction process). Then they were annealed at 500 ℃ for 1 min to rapidly reduce the oxygen (rapid reduction process). These oxygen reductions were confirmed from d^9 satellite change around Cu $2p_{3/2}$ spectra, using x-ray photo-electron spectroscopy.

The resistance was measured using the standard d.c. four-probe method, and normalized by the value at 300K. The critical temperature (Tc) was defined as the temperature at zero resistance. The critical current (Ic) was defined as the current at a voltage of 1 μV in the microbridge patterns, and the density (Jc) was normalized by the cross-section.

RESULTS AND DISCUSSION

In order to examine the influences of oxygen plasma on superconducting properties, microbridges such those as shown in Fig. 1 were put into an oxygen plasma atmosphere. First, we'll discuss degradated superconducting properties resulting from oxygen plasma. Figure 3 shows the normalized resistance as a function of temperature. After these patterns were put into oxygen plasma, only microbridges without protection experienced property changes. A large Tc degradation of approximately 10K was confirmed, although

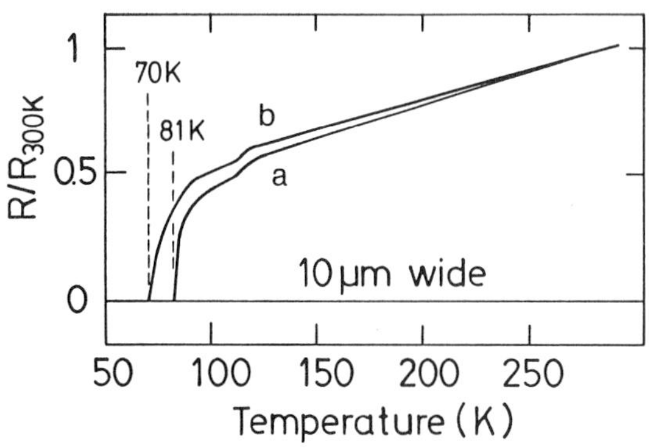

Fig. 3 Normalized resistance as a function of temperature. a and b correspond to microbridges with and without the Au protective layer, respectively. Resistances of a and b are 112 and 90 Ω at 300K, respectively.

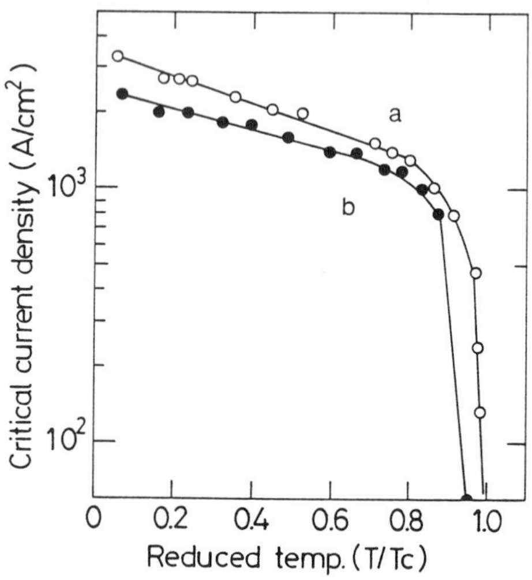

Fig. 4 Critical current density as a function of reduced temperature. a and b correspond to microbridges with and without the Au protective layer, respectively.

resistances at 300K were almost the same. This is because excessive oxygen was absorbed into the film around the bridge [10]. Figure 4 shows the critical current density (Jc) as a function of reduced temperature. At each reduced temperature, the Jc of the microbridge without protection is slightly lower than that of the microbridge with a protective layer. However, the decrease in Jc is not as large as we expected from Tc degradation.

Now, we'll discuss how Tc recovers from reducing the absorbed oxygen. Excessive oxygen absorption induces major degradation in Tc. Figure 5 shows the temperature dependence of the normalized resistance. Although the Tc decreases just after oxygen plasma treatment, the Tc tends to recover through the oxygen reduction process. Table I shows the Tc change at various pattern widths due to two reduction processes. For narrower microbridges, the Tc recovers up to approximately 77K through the slow reduction process. A slight recovery of 2-4K was observed in wider microbridges. However, the rapid reduction process causes a larger increase in Tc even in wider microbridges. Easy escape of excessive oxygen in narrower microbridges, and no Tc change in microbridges with Au protection indicate that Tc recovery corresponds to the reduction of excessive absorbed oxygen.

Table I Critical temperature change due to reduction process

Pattern width	1.5 μm	5 μm	10 μm	20 μm
After oxidation	67.4K	71.2K	69.7K	72.3K
Slow reduction	77.1K	78.2K	73.6K	74.5K
Rapid reduction	77.9K	78K	84.9K	84.3K

Next, we'll discuss large improvements of critical current density (Jc). Figure 6 shows the temperature dependence of the critical current (Ic). At each temperature, the Ic decreases with oxygen plasma treatment. However, the Ic becomes 10 times higher than the degraded microbridge after the slow reduction process, and the Ic increases with the rapid reduction process. Here, no Ic increase occurs in microbridges with Au protection. Figure 7 shows the reduced temperature dependence of Jc at various pattern

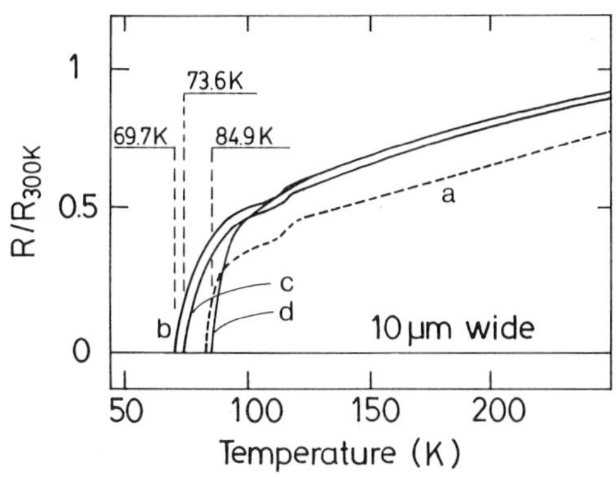

Fig. 5 Normalized resistance as a function of temperature in 10 μm wide microbridges. a and b are the microbridges before and after being putting into oxygen plasma, respectively. c and d are microbridges after slow and rapid reduction process, respectively.

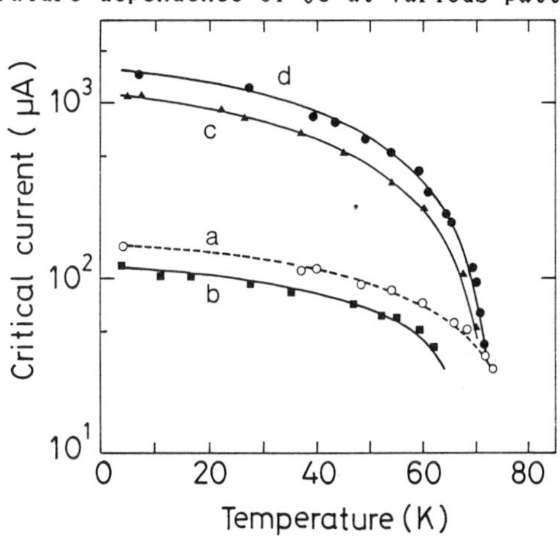

Fig. 6 Temperature dependence of critical current in a 5μm wide microbridge: a and b before and after being put into oxygen plasma, respectively, and c and d after slow and rapid reduction processes, respectively.

widths. The Jc of these microbridges is 5-30 times that of the unpatterned films without oxygen plasma treatment. Additionally, the Jc improvement strongly depends on the pattern width. This phenomenon resembles the recovery of Tc. Thus, great Jc improvement is thought to be strongly related to the reduction process of excessive absorbed oxygen. Although rf-induced steps were obseved in microbridges with Au protection, microbridges with great Jc improvement show no Shapiro steps. Accordingly, excessive oxygen reduction are thought to cause weak links change to the strong connection.

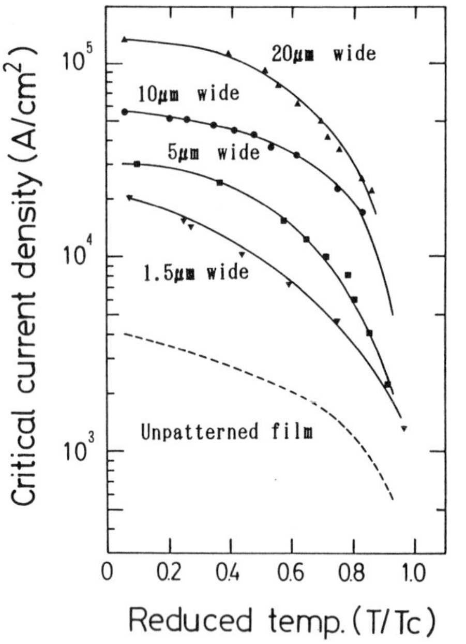

Fig. 7 Reduced temperature dependence of the critical current density at various pattern widths. Broken line indicates the unpatterned film without oxygen plasma treatment.

CONCLUSION

Putting microbridges consisting of a typical 80K-phase into an oxygen plasma caused large critical temperature degradation and a slight decrease in critical current. However, the reduction process of excessive absorbed oxygen greatly improved the critical current and recovered the critical temperature. These phenomena strongly depended on the pattern width of the microbridges. This great improvement in critical current is thought to be useful for high performance of SQUIDs and microwave components.

ACKNOWLEDGEMENTS

The authors wish to thank Y.Koshomoto, O.Ishii and A.Terada for their valuable discussions and encouragement.

REFERENCES

1. R.H.Koch, W.J.Gallagher, B.Bumble and W.Y.Lee (1989) Appl. Phys. Lett., 54: 951
2. Y.Higashino, T.Umezawa, T.Takahashi, K.Mizobuchi and S.Naito (1989) Proc. of ISEC'89, PH-27: 218
3. B.Oh, R.H.Koch, W.J.Gallagher, V.Foglietti, G.Koren and A.Gupta (1990) Appl. Phys. Lett., 56: 2575
4. F.C.Wellstood, J.J.Kingston, and J.Clarke (1990) Appl. Phys. Lett., 56: 2336
5. R.S.Withers and R.W.Ralston (1989) Proc. of IEEE, 77: 1247
6. K.Watanabe, T.Matsushita, N.Kobayashi, H.Kawabe, E.Aoyagi, K.Hiraga, H.Yamane, H.Kurosawa, T.Hirai and Y.Muto (1990) Appl. Phys. Lett., 56: 1490
7. Y.Nagai and K.Tsuru (1989) Jpn. J. Appl. Phys., 28: L1970
8. Y.Nagai and K.Tsuru (1990) Trans. of IEICE, E73: 516
9. Y.Nagai, K.Tsuru, A.Terada and K.Yanagisawa (1990) Abst. of ASC'90, MTP-2.10: 108
10. T.Yoshitake, S.Miura, J.Fujita, N.Shohata, H.Igarashi, T.Satoh, A.Sekiguchi and K.Katoh (1990) Appl. Phys. Lett., 56: 575

6.3 Other Superconducting Films

New Superconducting Thin Films of Pb-Based System Grown by Sputtering

HIDEAKI ADACHI, SEIJI ADACHI, YO ICHIKAWA, KENTARO SETSUNE, and KIYOTAKA WASA

Central Research Laboratories, Matsushita Electric Industrial Co., Ltd., Moriguchi, Osaka, 570 Japan

ABSTRACT

Epitaxial thin films of Pb-based superconductor have been grown in-situ on MgO (100) substrates by rf-magnetron sputtering for the first time. Film structure of Pb-3212 phase could be obtained at lower temperature of about 550°C and at lower oxidizing atmosphere than the preparation of conventional oxide-superconductors. The reducing atmosphere was strongly preferred for the formation of Pb-3212 films and no oxygen was required in the sputtering gas. The film showed superconductive transition at 75 K and zero-resistivity was realized below 60 K without post-heattreatment.

KEYWORDS: Pb-based superconductor, Pb-Sr-Ln-Ca-Cu-O, thin film, sputtering

INTRODUCTION

A variety of Pb-based oxide superconductors have been discovered recently. Structure of the system is composed of Pb/Cu oxide blocking layer which separates perovskite-related layers. The typical compound is $Pb_2Sr_2(Ln,Ca)Cu_3O_8$ which was firstly discovered in the system to have critical temperature (T_c) near 70 K [1-3]. This compound possesses triple PbO-Cu-PbO blocking layer and perovskite-related $Sr_2(Ln,Ca)Cu_2O_6$ layer including double CuO_2 planes, and is generally denoted as Pb-3212. Their sintering temperature is 750°C, which is lower than conventional oxide superconductors, and reducing atmosphere is preferred in the sintering process [4].

Meanwhile, a great development of thin-film technology for oxide high T_c superconductors has been carried out for in-situ formation of superconducting films. These technics need relatively high substrate temperature and special oxidizing methods such as oxygen plasma, activated oxygen, atomic oxygen, ozone and so on. For electronic applications of the superconducting films to combination devices with semiconductors or multilayered junction devices, in-situ preparation of superconducting films at lower temperature with no care of oxidation is expected. From this point of view, the nature of Pb-3212 compound seems to be suitable for the electronic applications, however, thin films of Pb-based superconductors have not been prepared so far. One reason is that Pb-based compounds were not regarded as attractive materials for the electronic applications because of lower T_c than popular 123 compound (Y-system), and the other reason is that the control of volatile Pb content in the film is very difficult. Very recently, single crystals of Pb-3212 compound with zero-resistance T_c of 84 K have been successfully prepared [5], which indicates that the natural T_c of Pb-3212 is near 90 K similar to 123 compound. Since we have previously experienced the preparation of Pb-based ferroelectric thin films with perovskite structure, we have also tried to prepare Pb-based superconducting thin films. Here we describe the preparation and properties of Pb-3212 thin films which could be grown at lower substrate temperature and lower oxidizing atmosphere than conventional oxide-superconducting films.

FILM PREPARATION

In order to prevent the deficiency of volatile Pb in the films, magnetron sputtering with a discharge plasma of high power density was employed as our previous work on thin films of Pb-based ferroelectrics [6]. Typical sputtering conditions are summarized in Table I. The composition of the Pb-Sr-Ln-Ca-Cu-O target was modified so as to make the film composition near Pb/Sr/Ln/Ca/Cu=2/2/0.5/0.5/3. Since the reducing atmosphere is preferred in the preparation of Pb-3212 phase, pure argon or 3%-oxygen mixed argon was used for a sputtering gas. After the deposition, films were cooled to room temperature about 20 min in argon or nitrogen atmosphere and no post-heattreatment was done.

TABLE I. Sputtering conditions.

Target	Pb-Sr-Ln-Ca-Cu-O (Ln=Y and Yb) Pb:Sr:Ln:Ca:Cu=5:2:1-1.5:0.4:2.7-3.3
Substrate	(100) plane of MgO
Substrate temperature	400-650°C
Sputtering gas	Ar (+ 3% O_2)
Gas pressure	0.5 Pa
Rf input power	100 W
Growth rate	12 nm/min

RESULTS AND DISCUSSION

The relation between the substrate temperature and the specification of the films grown in a pure argon atmosphere was summarized in Fig. 1. As is seen in the figure, Pb content in the films decreases with the increase of the substrate temperature, which is similar characteristics as Pb-based ferroelectric films. Thin films with Pb-3212 structure was obtained at the temperature ranged from 500 to 600°C. For the films grown at lower temperature than 450°C, halo pattern originated from amorphous state and other miscellaneous peaks were observed in the X-ray diffraction patterns. While the films grown at higher than 600°C showed significant decrease of Pb content and partially decomposed into several phases. Although it is difficult to evaluate the real surface temperature of substrates, the growth temperature is indeed lower than that for 123 compound about 50°C and that for Bi-based compound about 100°C in our sputtering system.

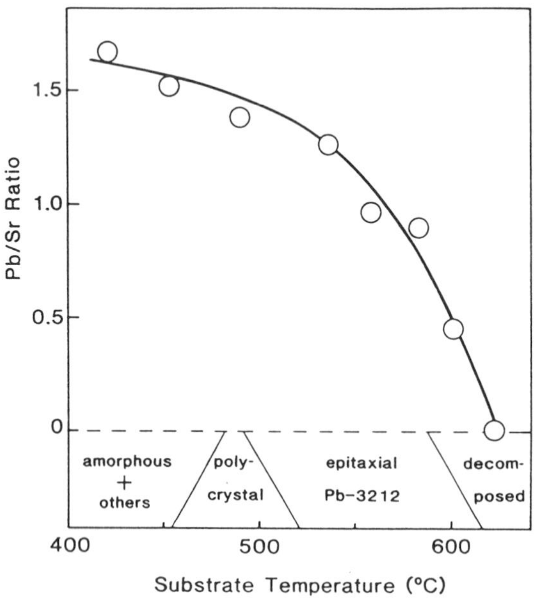

Fig.1. Compositional Pb/Sr ratio and crystalline structure of Pb-Sr-Y-Ca-Cu-O films deposited in pure argon atmosphere at various substrate temperature.

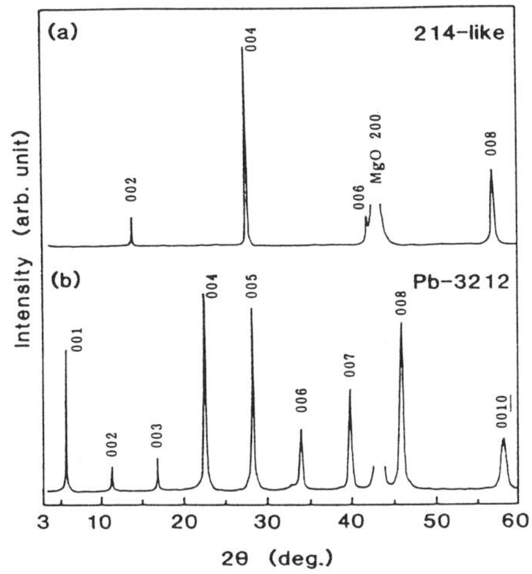

Fig.2. X-ray diffraction patterns of Pb-Sr-Y-Ca-Cu-O films grown in (a)3% oxygen mixed argon and (b) pure argon.

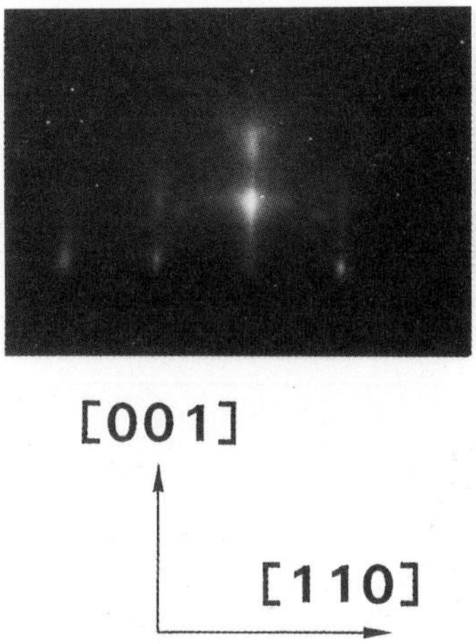

Fig.3. RHEED pattern of Pb-Sr-Yb-Ca-Cu-O film with Pb-3212 phase epitaxially grown on MgO (100) substrate.

The X-ray diffraction patterns of the Pb-Sr-Y-Ca-Cu-O films grown at the substrate temperature of 550°C with different sputtering gas of oxygen (3 %) mixed argon (a) and pure argon (b) are shown in Fig. 2. The films grown in a oxygen mixed gas did not show Pb-3212 structure but showed K_2NiF_4(214 phase)-like structure with c=12.88 Å (Fig.2 (a)), which was sometimes observed as lamellar defects in Pb-Sr-Y-Ca-Cu-O ceramics [7]. The film grown in a pure argon atmosphere showed highly c-axis oriented Pb-3212 structure with c=15.75 Å (Fig.2 (b)). The oxygen only from the target was enough to synthesize Pb-3212 structure in the films. Even if the pure argon was used as a sputtering gas, insufficient presputtering before the deposition sometimes caused the appearance of 214-like structure due to the adsorbed oxygen in the target. It was found that for the preparation of Pb-3212 films by sputtering no oxygen is required in the sputtering gas, neither is high-oxidizing atmosphere. Figure 3 shows the typical reflection high-energy electron diffraction (RHEED) pattern of the Pb-3212 film (Pb-Sr-Yb-Ca-Cu-O). The pattern proved that the film was epitaxially grown on (100) MgO plane with c-axis [001] normal. The streaky patterns maintained in any irradiated position of the film showing that the film has smooth surface uniformly.

Figure 4 shows the electric properties of Pb-3212 films grown at 550°C. The films exhibited metallic behavior with the superconductive transition. As for the Pb-Sr-Yb-Ca-Cu-O film, onset T_c was 75 K and zero-resistivity state was achieved below 60 K. The transport critical-current-density at 50 K was estimated to be 10^4 A/cm^2. Mostly superconducting properties of Pb-Sr-Yb-Ca-Cu-O films was a little better than that of Pb-Sr-Y-Ca-Cu-O films, which is similar characteristics as Pb-Sr-Ln-Ca-Cu-O ceramics [3]. When the deposition temperature was relatively high (near 600°C), the resultant films showed something like melted structure and strong (001) peaks with some other peaks originated from decomposed phases were appeared. Resistivity vs. temperature curve of this Pb-Sr-Yb-Ca-Cu-O film is shown in Fig. 5. Although the resistivity in the normal state was fairly low and the transition was sharp, the T_c was not so high (estimated to be 47 K). Strict control of film composition may be necessary for the realization of higher-T_c films.

Post-heattreatment did not improve the electric properties of the films considerably. Moreover, annealing of the films in air or oxygen at higher temperature than 400°C degraded the superconducting properties of the films, which must be caused by the oxidation of Cu in PbO-Cu-PbO triple blocking layer as shown in previous reports [8]. After that, annealing in a vacuum revived the superconductivity in the films, indicating that the oxygen is put on and put out reversibly similar to 123 films. The T_c of the as-grown Pb-

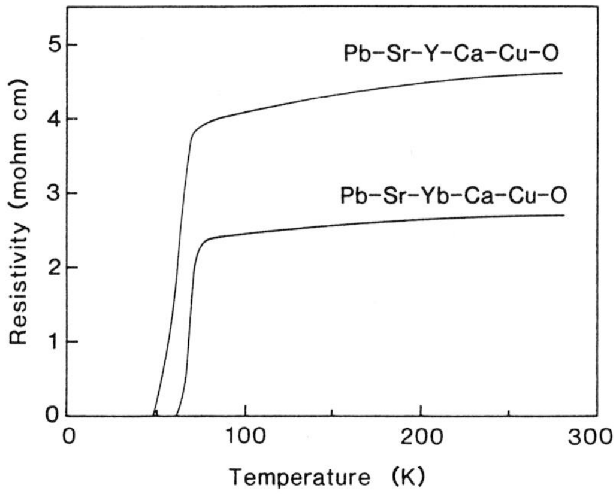

Fig.4. Temperature vs. resistivity curves for Pb-Sr-Ln-Ca-Cu-O (Ln=Y and Yb) films with Pb-3212 phase.

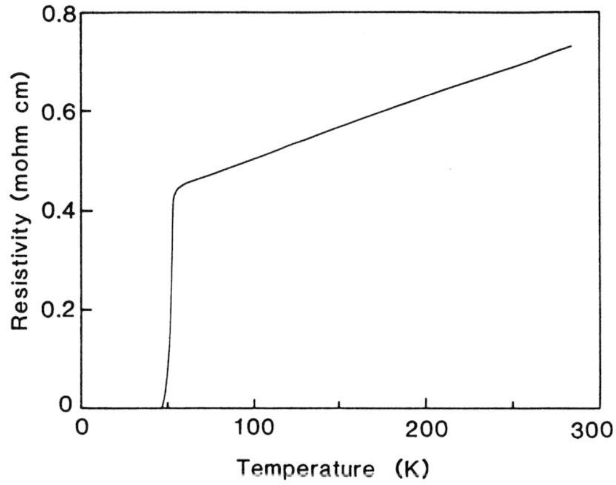

Fig.5. Temperature vs. resistivity curve for Pb-Sr-Yb-Ca-Cu-O film deposited at 590°C.

based film was lower than that of as-grown 123 films. Note that the present study is only a preliminary work for Pb-based films, and further investigation of preparation method will certainly lead to the in-situ formation of excellent superconducting Pb-based films like 123 compounds.

SUMMARY

Thin films of Pb-3212 superconductor were epitaxially grown in-situ at low substrate temperature of 550°C by rf-magnetron sputtering. The film showed superconductive transition at 75 K and zero-resistivity was realized below 60 K without post-heattreatment. The growth temperature was relatively lower than that of conventional oxide-superconducting films. The reducing atmosphere was preferable for the formation of Pb-3212 films and no oxygen gas was required in the sputtering gas. High-vacuum deposition system such as molecular-beam-epitaxy must be suitable for the preparation of the Pb-based superconducting films. The present Pb-3212 film is considered to be one of the most promising materials to make superconducting thin-film devices.

ACKNOWLEDGMENTS

The authors would like to thank Dr. T.Nitta and Dr. K.Kanai for their encouragement and valuable suggestions. They are also indebted to Drs. T.Matsushima, K.Hirochi and S.Hayashi for discussions.

REFERENCES

1. R.J.Cava, B.Batlogg, J.J.Krajewski, L.W.Rupp, L.F.Schneemeyer, T.Siegrist, R.B.vanDover, P.Marsh, W.F.Peck,Jr, P.K.Gallagher, S.H.Glarum, J.H.Marshall, R.C.Farrow, J.V.Waszczak, R.Hull, P.Trevor (1988) Nature 336: 211-214
2. M.A.Subramanian, J.Gopalakrishnan, C.C.Torardi, P.L.Gai, E.D.Boyes, T.R. Askew, R.B.Flippen, W.E.Farneth, A.W.Sleight (1989) Physica C 157: 124-130
3. O.Inoue, S.Adachi, Y.Takahashi, H.Hirano, S.Kawashima (1989) Jpn. J. Appl. Phys. 28: L60-L62
4. M.Masuzawa, T.Noji, Y.Koike, Y.Saito (1989) Jpn. J. Appl. Phys. 28: L1524-L1526
5. J.S.Xue, M.Reedyk, Y.P.Lin, C.V.Stager, J.E.Greedan (1990) Physica C 166: 29-35
6. H.Adachi, T.Mitsuyu, O.Yamazaki, K.Wasa (1986) J. Appl. Phys. 60: 736-741
7. P.Goodman, D.G.Jensen, T.J.White (1989) Physica C 158: 173-177
8. K.Kadowaki, M.J.V.Menken, A.C.Moleman (1989) Physica C 159: 165-172

Superconducting Superlattices of (La, Sr)$_2$CuO$_4$/Sm$_2$CuO$_4$ Prepared by an Excimer Laser Deposition Technique

HITOSHI TABATA[1], TOMOJI KAWAI, and SHICHIO KAWAI

The Institute of Scientific and Industrial Research, Osaka University, 8-1, Mihogaoka, Ibaraki, Osaka, 567 Japan

ABSTRACT

We have grown superlattices of $(La,Sr)_2CuO_4/Sm_2CuO_4$ onto $SrTiO_3$ substrates by an excimer laser deposition technique. The stacking periodicity of $(La,Sr)_2CuO_4/Sm_2CuO_4$ is changed from 1/1(one $(La,Sr)_2CuO_4$ and one Sm_2CuO_4 layer) to 120/120 combinations. The crystal structures and the superconductivities of the superlattices are classified into three regions according to the number of the stacking periodicity: new compounds (2/2-4/4), strained lattices (15/15-60/60) and multi layers (60/60-). In the strained lattice region, the critical temperatures become lower as the number of stacking layers decrease. This phenomenon is explained to be derived from a pressure effect caused by a lattice mismatch between $(La,Sr)_2CuO_4$ and Sm_2CuO_4 layers.

KEY WORDS: oxide superconductor, superlattice, laser deposition, thin film

INTRODUCTION

High Tc oxide superconductors can be classified in terms of well known structural types, such as perovskite, K_2NiF_4, Aurivilius phase and so on. In these copper based oxide superconductors, the common structural feature is CuO_2 sheets. The high Tc superconductivity appears by doping charge carriers into the CuO_2 sheets. Accordingly, any structure of the copper-oxide superconductors can be constructed by inserting various kinds of layers into the infinite stacking of the CuO_2 layers.

Based on this concept, artificially layered new compounds are expected to be synthesized by periodic insertion of the blocking layer into the infinite layers of $CaCuO_2$ which are well known as a parent structure of the superconductors[1]. In fact, the site-selective incorporation of exotic atoms into the Ca or Sr sites with different ionic radii and valence states have been demonstrated with atomic layer scale layer-by-layer deposition technique.[2,3]. Furthermore, the superconducting superlattices with the combination of unit cells, such as $Y_1Ba_2Cu_3O_7/Dy_1Ba_2Cu_3O_7$[4], $Y_1Ba_2Cu_3O_7/(Nd,Ce)_2CuO_4$[5] and $Bi_2Sr_2(Ca_{0.5}Y_{0.5})Cu_2O_8$(semiconductor)/ $Bi_2Sr_2(Ca_{0.85}Y_{0.15})Cu_2O_8$ (superconductor)[6,7] have been synthesized.

In this letter we report in-situ epitaxial growth of $(La,Sr)_2CuO_4/Sm_2CuO_4$ superlattices, having various periodicity from the very short to large periodicity.

EXPERIMENTAL

The films were prepared by a laser deposition technique using ArF excimer laser(193 nm)[2,3]. Thin films were formed on substrates placed at the opposite side of the targets in the presence of O_2/O_3(8%) atmosphere. The substrates used are $SrTiO_3$(100) single crystals. The targets of $(La,Sr)_2CuO_4$ (LSCO) and Sm_2CuO_4 (SCO) for the laser ablation were made by a conventional solid-state reaction with the starting compositions,

[1]On leave from Tech. Inst. Kawasaki Heavy Industries Ltd. 1-1 Kawasaki-cho, Akashi 673 Japan

$La_{1.85}Sr_{0.15}Cu_1O_4$ and $Sm_2Cu_1O_4$, respectively. To produce multilayers, these two targets were set at the center of the vacuum chamber and the laser beam was focused on these targets alternately with the intensity of 1-2 J/cm^2 and the frequency of 15 Hz. The thickness of the thin films was monitored by a quartz crystal oscillating thickness monitor in situ and confirmed by an optical microscope ex-situ. The structures of the films were determined by x-ray diffraction patterns and the superconductivity was measured by a standard four-probe method. The surface morphology was determined by scanning electron microscopy(SEM) and the two dimensional surface structure was studied by reflection high energy electron diffraction(RHEED).

In order to form multilayers, elucidation of the synthetic condition of the thin films using single target is necessary. For the search of the condition in which the thin films of both LSCO and SCO can be formed, each thin film was at first deposited independently under various conditions. Substrate temperature was varied from 550 ℃ to 750 ℃, and oxygen(O_2+8%O_3) partial pressure was varied from 10^{-1} to 10^{-3} Torr. Then we have determined the suitable conditions for the formation of superlattices, and the LSCO and SCO layers were piled up with the cycle of n/m. The n/m means n-layers of $(La,Sr)_2CuO_4$ and m-layers of Sm_2CuO_4 by using LSCO and SCO targets alternately. We have repeated n=m to be 1 - 120 for 20 cycles.

RESULTS AND DISCUSSION

The LSCO thin film can be formed in situ at the substrate temperature of more than 600 °C. The SCO can be made in situ under relatively higher oxygen partial pressure: more than 10^{-2} Torr. We have formed the superlattices under the conditions in which both LSCO and SCO can be crystallized in situ at the substrate temperature of 700 C and the oxygen pressure of $1x10^{-1}$ Torr. The LSCO/SCO superlattices formed here are classified into three regions, according to the number of the stacking periodicity. The first one is the compounds of short periodicity, such as 1/1, 2/2, 3/3, and 4/4. The second one is the strained superlattice which is in the regions of 15/15 to 30/30 periodicity. The third one is the so-called multi-layer of $(La,Sr)_2CuO_4/Sm_2CuO_4$, keeping each original property.

Figure 1 shows the x-ray diffraction patterns and calculated peak position and intensities of the superlattices of LSCO/SCO with the periodicity of (a)1/1, (b)2/2, (c)3/3 and (d)30/30. The pattern of Fig.1(a) is similar to that of the known La(Sr)$SmCuO_4$ having CuO_5 pyramid structure[8]except for the relative intensities of 004 and 006 peaks. In Fig.1(b) and (c), satellite peaks are observed characteristic of superlattices. The larger the periodicity of each layer is, the lower the intensity of the main peak in the x-ray diffraction patterns becomes. Instead, the satellite peaks grow and eventually the main peak separates into two peaks originated from the LSCO and SCO above the periodicity of 15/15(see Fig.1(d)for 30/30).

From the view point of crystal chemistry, the superlattices of 2/2 and 3/3 can be regarded as new compounds having long unit cell by the combination of CuO_6 octahedral, CuO_5 pyramid and CuO_4 sheet in one unit cell(see structures in Fig. 1(b) and (c)). For the case of 3/3 compound, it consists of two octahedral layers, one pyramid layer and two square layers in the ideal structure(see Fig.1(c)). In the x-ray diffraction patterns, however, the relative intensities of 004($2\theta=28°$ in Fig. 1(a)) and 006($2\theta=43°$)peaks are different from that of the calculated pattern. This result may indicate that the mixing of La and Sm occurs at the interface of $(La,Sr)_2CuO_4/Sm_2CuO_4$. Therefore, the perfect pyramid structure is not expected at the interface but a square structure may coexist. Figure 2 shows the x-ray diffraction pattern and calculated pattern of the superlattice with the periodicity of 4/4. Only in this case, the substrate temperature is relatively low(600 °C) and one minute intermission is taken between the stacking of LSCO layer and SCO layer. The pattern (peak position and intensities) of Fig.2(a) is quite similar to calculated one(Fig.2(b)). Furthermore this pattern is different from that of the calculated pattern with the periodicity of 3/5 on the assumption that there is mixing at the interface of LSCO/SCO. The intermission helps the migration of atoms on the film surface, and the low temperature deposition keeps the interface of LSCO/SCO from mixing of La and Sm.

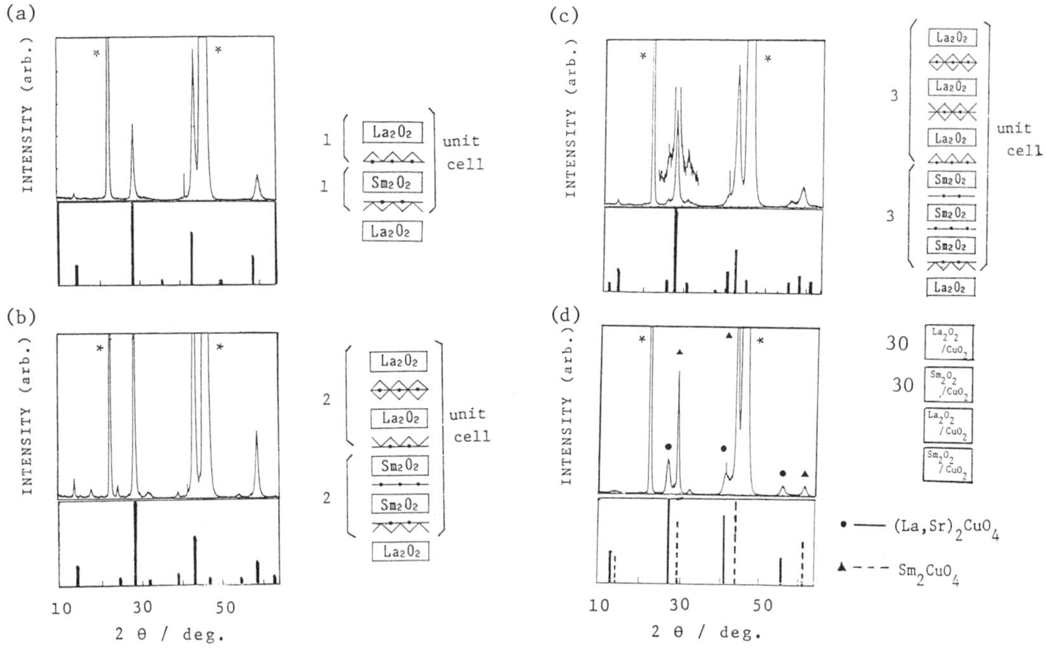

Fig.1 Observed x-ray diffraction patterns of the superlattices and the calculated patterns. The schematic ideal crystal structures are also shown for the stacking periodicity of (a)1/1, (b)2/2, (c)3/3 and (d)30/30.
* : peaks of $SrTiO_3$ substrate.

In the periodicity more than 15/15, the longer the periodic cycle is, the higher the Tc becomes. The superlattices of 15/15, 30/30 and 60/60 show the Tc_{on} of 15K, 25K and 30K (Fig.3), while the LSCO film exhibits the Tc_{on} of 40K and Tc_{zero} of 38K. This phenomenon is explained by a pressure effect due to the strain by the lattice mismatch which is observed in bulk samples [9,10] and also in $Y_1Ba_2Cu_3O_7/(Nd,Ce)_2CuO_4$ system[5]. The a-axis lengths of LSCO and SCO are 3.78Å and 3.91Å, respectively. The CuO_6 octahedral slab of LSCO is forced to expand in the a-direction but the CuO_4 sheet slab of SCO is forced to be shrunken. Accordingly, the bond distance of Cu-O in the CuO_2 sheet of LSCO is expanded for the sake of mismatching the lattice parameters between LSCO and SCO. Fig.3(a) shows the changes of the lattice parameter c against the stacking periodicity. In the range of short periodicity (1/1 to 4/4) the c-axis is defined as the compound of long unit cell, whereas in the region of 15/15 to 60/60, the x-ray peaks due to original LSCO and SCO emerge and the peak position gradually shifts toward the original compounds due to the relaxation of the strain above 60/60 periodicity.

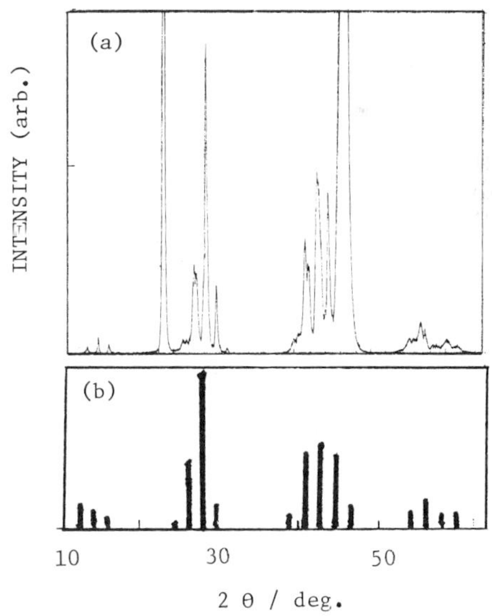

Fig. 2 Obserbed x-ray diffraction pattern of the superlattice and calculated patterns with the periodicity of 4/4.

Above the cycle of 60/60, the superlattice exhibits the similar Tc to the standard LSCO films(Fig. 3(b) for 60/60). Accordingly, this is a normal multilayered thin film having the original properties.

Finally, it should be noted that the SEM images of these superlattices have shown good surface morphology without any particles on the surface. Moreover RHEED pattern shows streaked lines showing epitaxial growth with smooth surface. These results also assure a strict control of the crystal structure of unit cell order.

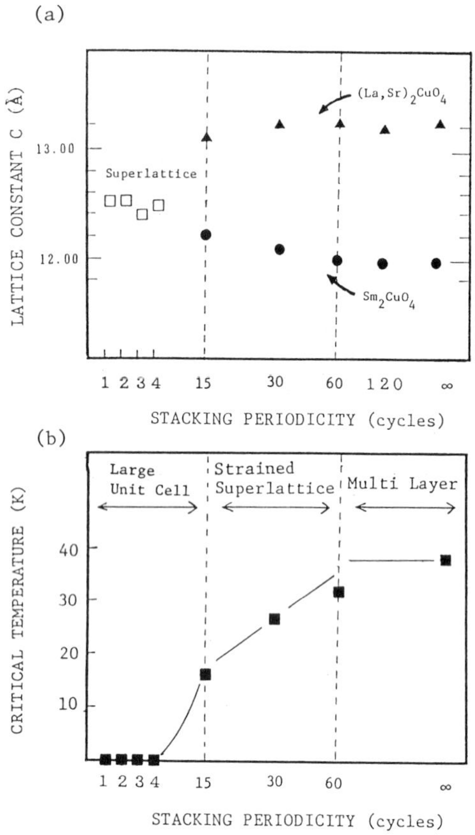

Fig. 3 The relation between stacking periodicity and the lattice constant c (a) and critical temperature (b) of $(La,Sr)_2CuO_4/Sm_2CuO_4$ superlattices.

CONCLUSION

We have formed the superlattices of $(La,Sr)_2CuO_4/Sm_2CuO_4$ by changing the periodicity of each stacking layer. The crystal structures and the superconductivities of the superlattices are classified into three regions according to the number of the stacking periodicity; the new compounds for 2/2 and 4/4, strained lattices for 15/15 to 60/60, multi layers for more than 60/60. In the strained lattice region, the critical temperatures become lower as the number of stacking layers decrease due to the pressure effect to the $(La,Sr)_2CuO_4$ by $SmCuO_4$ layer. The superlattices of 2/2 and 3/3 show the definable main X-ray peaks with satellites and can be regarded as new compounds with large unit cell having the combination of CuO_6 octahedral, CuO_5 pyramid and CuO_2 sheet in one crystallographic unit cell. These structures are interesting in a fundamental concept which leads to the possibilities of creating new superconductors having variety of CuO_2 layered structure.

REFERENCES

1. Siegrist T, M.Zahurak S M, Murphy D W, Roth R S (1988) Nature 334 : 231-232
2. Tabata H, Kawai T, Murata O, Kawai S (1989) Jpn. J. Appl. Phys. 28 : L823-826
3. Tabata H, Kawai T, Murata O, Kawai S (1990) Appl. Phys. Lett. 56 : 1576-1578
4. Triscone J M, Markut M G, Antognazza L, Brunner O, Fischer o (1989) Phys. Rev. Lett. 63: 1016-1019
5. Gupta A, Gross R, Olsson E, Segmuller A, Koren G, Tsuel C C (1990) Phys. Rev. Lett. 64: 3191-3194
6. Kanai M, Kawai T, Kawai S (1990) Appl. Phys. Lett. 57: 198-200
7. Kanai M, Kawai T, Kawai S (in press) Vacuum
8. Sawa H, Suzuki S, Watanabe M, Akimitsu J, Matsubara H, Watanabe H, Uchida S, Kokusho K, Asano H, Izumi F, Takayama-Muromachi E (1989) Nature 337: 347-348
9. Kim H J, Moret R (1988) Physica C, 156: 363-368
10. Yomo S, Murayama C, Takahashi H, Mori N, Kishio K, Kitazawa K, Fueki K (1987) Jpn. J. Appl. Phys. 26 : L603-605

Fabrication by Laser MBE and In Situ Characterization of Layered Cuprates

HIDEOMI KOINUMA, HIROTOSHI NAGATA[1], TAKUYA HASHIMOTO[2], TADASHI TSUKAHARA, SATOSHI GONDA, and MAMORU YOSHIMOTO

Research Laboratory of Engineering Materials, Tokyo Institute of Technology, 4259, Nagatsuta-cho, Yokohama, 227 Japan

ABSTRACT

Epitaxial growth of $(Sr_xCa_{1-x})CuO_y$ layer, an essential constituent of high-*Tc* superconductors, has been investigated by monitoring with RHEED the growing surface of pulsed laser deposition. The RHEED intensity oscillation was detected during the deposition of $(Sr_{0.5}Ca_{0.5})CuO_y$ with NO_2 flushing at a pressure of 10^{-7}Torr. The deposited $(Sr_xCa_{1-x})CuO_y$ layer was analyzed by *in situ* XPS to evaluate the Cu valence and oxygen stoichiometry. The as-grown $SrCuO_y$ (x=1) film was stoichiometric (y≈2.0) and showed a metallic behavior of linearly decreasing resistivities with the temperature decrease from 300 to 4.2K.

KEY WORDS: laser MBE, $(Sr,Ca)CuO_2$, epitaxial growth, RHEED, XPS

INTRODUCTION

New layered structures relating to high-*Tc* superconductors are expected to be made by developing a technology for piling up two-dimensional ceramics layers. According to our three elementary subunits model, high-*Tc* cuprates are fragmented into three types of layers; $(CuO_2)^{2-}$ layer as supercurrent path, coupling or mediating layer sandwiched by two $(CuO_2)^{2-}$ layers, and blocking layer [1]. The coupling layer is an oxygen-free rare earth or alkaline earth plane and therefore positively charged, whereas the (CuO_2) layer should be negatively charged. Combining the $(CuO_2)^{2-}$ and Ca^{2+} layers gives charge neutralized $CaCuO_2$ slab, which can be a growth unit cell.

Atomic scale layer-by-layer growth is one of key technologies for synthesizing ceramics superlattices including possible new high-*Tc* cuprates. We recently reported that ultra-high vacuum (UHV) pulsed laser deposition (laser MBE) was a promising method for layer-by-layer growth of the ceramics thin films [2,3]. In contrast to conventional pulsed laser depositions under relatively high pressures (≈0.1Torr), our laser MBE could deposit cluster-free smooth oxide films [3]. Although UHV is not preferable for oxidizing the metals, this problem can be overcome by flushing a strongly oxidating gas onto the growing film surface. Based on a thermodynamic calculation, we showed that O_3, O^* and NO_2 were far more effective to oxidize Cu into CuO than N_2O, NO or O_2 [1]. Pure O_3 and NO_2 were verified to be useful for forming crystal structures of high-Tc supercondutors in the films as-grown by conventional and laser MBE's [1,4,5].

As described above, $(Sr_xCa_{1-x})CuO_y$ slab can be regarded as a charge neutralized growth unit composed of a $(CuO_2)^{2-}$ layer and a coupling layer of $(Sr_xCa_{1-x})^{2+}$. Therefore, it should be primarily important to establish the technique for the epitaxial growth of this slab as well as to characterize its electrical properties. This paper describes of the epitaxial growth of $(Sr_xCa_{1-x})CuO_y$ films on $SrTiO_3$ by laser MBE.

---

[1]on leave from Sumitomo Cement Co. Ltd., Funabashi-shi, Chiba, Japan.
[2]on leave from University of Tokyo, Bunkyo-ku, Tokyo, Japan.

EXPERIMENTAL

The laser MBE system we used is equipped with in situ RHEED, XPS, OES and QMS and its details were reported in the preceding papers [1,6]. A cleaned $SrTiO_3(001)$ substrate was placed in the laser MBE chamber (ULVAC Co. Ltd.) evacuated to a base-pressure of 2×10^{-9}Torr. Pure NO_2 was flushed through a variable-leak valve onto the substrate. Distance between the substrate and the outlet of gas (1/8" stainless tube) was 20mm. While maintaining the gauge pressure at $1\text{-}3\times10^{-7}$Torr, the substrate was heated up to 700°C (measured by an optical pyrometer). The deposition of $(Sr_xCa_{1-x})CuO_y$ films was carried out by ablating one of the sintered $SrCuO_2$, $(Sr_{0.5}Ca_{0.5})CuO_2$, $(Sr_{0.14}Ca_{0.86})CuO_2$ and, $CaCuO_2$ targets with an ArF pulsed (20ns, 2Hz, 1J/cm^2) excimer laser (LABMDA PHYSIK model LPX-100). Typical deposition rate was 0.1-0.3nm/min. Reflection high energy electron diffraction (RHEED) was used to observe the deposition surface. Beam energy and incidence angle of RHEED were 15keV and 3°, respectively.

After the deposition, the film was cooled in the same atmosphere as that at the period of deposition. Some of the films were transferred to the X-ray photoelectron spectrometer (XPS:JEOL model JPS 80) in vacuo. XPS measurement was carried out using Al-*Kα* radiation to avoid overlapping of Ca *LMM* Auger lines and Cu 2*p* photoelectron lines. The films were also characterized by X-ray diffraction (XRD:MAC SCIENCE model MXP-3) with monochromated Cu-*Kα* radiation and electrical resistivity measurement by the standard four probe method.

RESULTS AND DISCUSSION

$(Sr_xCa_{1-x})CuO_y$ is known to have a tetragonal structure which consists of alternating $(CuO_2)^{2-}$ and $(Sr_xCa_{1-x})^{2+}$ layers. The lattice parameter changes depending on x from a=0.385nm and c=0.316nm (x=0) to a=0.393nm and c=0.343nm (x=1) [7]. $(Sr_{0.5}Ca_{0.5})CuO_2$ (a=0.389nm and c=0.331nm) is expected to grow epitaxially on $SrTiO_3$ (cubic, a=0.390nm) substrate. Figure 1 shows the XRD pattern of a 25nm thick $(Sr_{0.5}Ca_{0.5})CuO_y$ film prepared by the laser MBE. The c-axis length of the film was calculated from the 002 XRD peak to be 0.332nm, agreeing well with 0.331nm reported for the bulk $(Sr_{0.5}Ca_{0.5})CuO_2$ [7]. The full width at half maximum (FWHM) of XRD rocking curve was 0.43° for 002 peak, indicating that the film crystallinity is as good as the substrate (FWHM=0.53° for 002 peak of $SrTiO_3$).

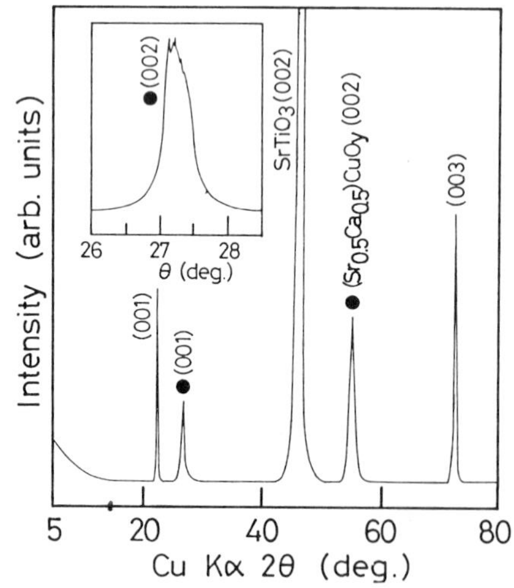

Fig.1 XRD pattern of the $(Sr_{0.5}Ca_{0.5})CuO_y$ film deposited on $SrTiO_3(001)$ substrate. The insertion shows θ scan around the 002 peak of the film.

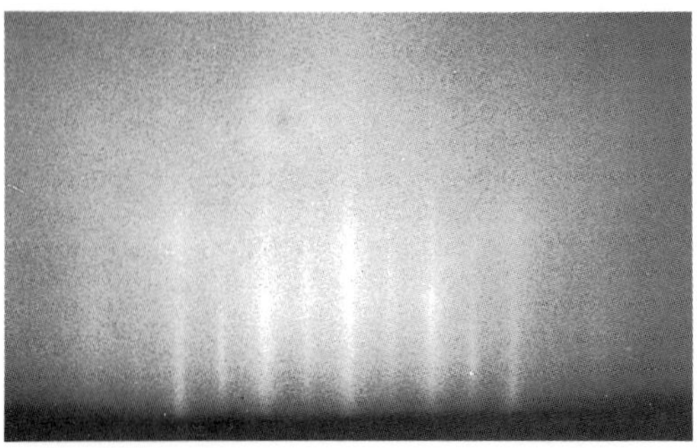

Fig.2 RHEED pattern of the $(Sr_{0.5}Ca_{0.5})CuO_y$ film (12nm thick) deposited on $SrTiO_3(001)$ substrate. Beam azimuth is [010] for both the substrate and the film.

Epitaxial growth of $(Sr_{0.5}Ca_{0.5})CuO_y$ film was verified by the observation of streaky RHEED pattern shown in Fig.2. The pattern is quite similar to that of $SrTiO_3(001)$ substrate. Extra streaks were observed at half positions of 1x1 streaks of the film during the deposition, but they vanished while the film was cooled with NO_2 flushing. With respect to $SrTiO_3(001)$ and $BaTiO_3(001)$ surfaces, Courths reported that high concentration of Ti^{3+}-(O,vacancy) defects gave 2x2 or two mutually perpendicular 2x1 LEED pattern [8]. His result suggests that our as-grown $(Sr_{0.5}Ca_{0.5})CuO_y$ film contains some ordered oxygen defects in its $(CuO_2)^{2-}$ plane and the defects are minimized during the cooling step.

Figure 3 shows a typical XPS spectrum in Cu 2*p* core level region of the $(Sr_{0.5}Ca_{0.5})CuO_y$ film. Satellite structures due to $\underline{2p}3d^9$ ($\underline{2p}$ denotes a Cu 2*p* core hole) configuration are clearly observed, indicating the divalent state of Cu. Intensity ratio of the satellite peak to the main peak ($\underline{2p}3d^{10}\underline{L}$: $\underline{L}$=ligand hole) of 2*p*1/2 was 0.31±0.07 which was close to the values of 0.33 and 0.24 reported for $(La_{0.9}Sr_{0.1})_2CuO_{4-\delta}$ and $YBa_2Cu_3O_{6.69\pm0.03}$, respectively [9]. Thus, the oxygen stoichiometry (y) in the $(Sr_{0.5}Ca_{0.5})CuO_y$ film should be very close to 2.0.

The results on $(Sr_xCa_{1-x})CuO_y$ films with x=1, 0.14 and 0 are summarized in Table I. In addition to the tetragonal structure, $SrCuO_2$ crystal can be orthorhombic (JCPDS No.38-1179) with its *a*=0.3571nm and *c*=0.3912nm lattice parameters close to those of $SrTiO_3$. The deposited $SrCuO_y$ film was predominantly tetragonal judging from the RHEED and XRD patterns, but contained a small amount of orthorhombic phase with its [010] axis oriented perpendicularly to the substrate. Ca-rich (x=0.14, 0) films were not grown epitaxially under the present conditions.

The $SrCuO_y$ and $(Sr_{0.5}Ca_{0.5})CuO_y$ films had black color and very low resistivities in the order of mΩ•cm. Figure 4 shows the temperature dependence of resistivity of 40nm thick $SrCuO_y$ film: the resistivity decreased linearly with the temperature decrease from 300 to 4.2K. On the other hand, the Ca-rich films were insulative.

In order to construct well-defined multilayered structures, it is necessary to evaluate the interface structures on an atomic scale. Spotty RHEED pattern was observed in the initial stage of $(Sr_xCa_{1-x})CuO_y$ growth on $SrTiO_3$, indicating a poor two-dimensionallity of the growth. With the progress of film growth thicker than 10nm, the RHEED pattern changed to streaky and the film growth mode was apparently improved. In this region, two-dimensional epitaxial growth should be dominating, since we observed the RHEED intensity oscillation at the central streak. The oscillation period corresponded well

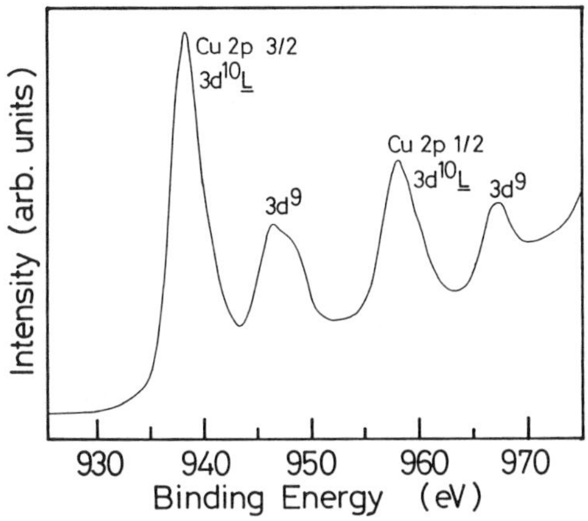

Fig.3 Typical Cu 2*p* XPS spectra of the as-grown $(Sr_{0.5}Ca_{0.5})CuO_y$ film.

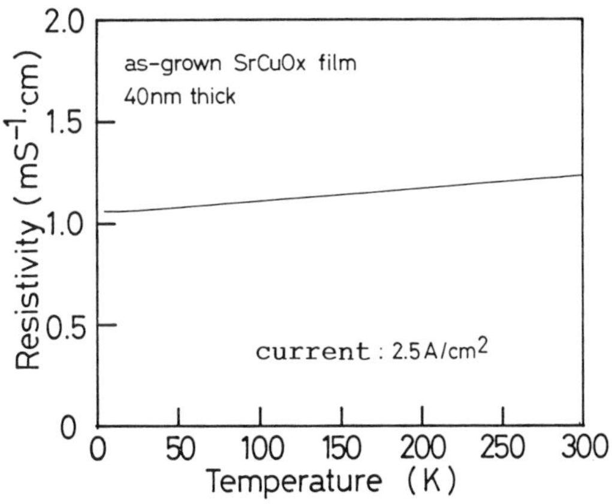

Fig.4 Temperature dependence of resistivity of the as-grown $SrCuO_y$ film.

Table I XPS, RHEED and XRD results on $(Sr_xCa_{1-x})CuO_y$ films deposited on $SrTiO_3(001)$ substrates under various conditions.

Target composition (Sr:Ca):Cu	NO_2 pressure ($\times10^{-7}$Torr)	Substrate temp. (°C)	XPS I_s/I_m[a] Cu2p1/2	RHEED[b] pattern	XRD; $(Sr,Ca)CuO_2$ c-axis length (nm)	rocking curve FWHM of (002) (deg.)
1 : 0 : 1	1	750	0.14	spot	c)	c)
		700	0.38	streak	0.347	0.47
			not measured		0.345	0.32
	5	775	0.54	ring	d)	d)
0.5 : 0.5 : 1	0.1		0.19	spot	unknown peaks	
	1	700	0.24	streak	0.332	0.43
			0.32		0.332	0.39
0.14: 0.86: 1	1	700	0.40	spot	0.325	0.85
0 : 1 : 1	1	700	not measured	spot	(101)//sub. oriented[e]	

[a] XPS peak area ratio of the satellite to the main for Cu 2p 1/2.
[b] Streaky RHEED patterns indicated the film orientation of $(Sr_xCa_{1-x})CuO_y$-(001)//$SrTiO_3$(001) and $(Sr_xCa_{1-x})CuO_y$[010]//$SrTiO_3$[010].
[c] Mixed phases of $Sr_2Cu^{II}O_3$ and $SrCu^{I}{}_2O_2$.
[d] Mixed phases of Sr_2CuO_3 and orthorhombic-$SrCuO_2$.
[e] Only one peak observed at d=0.243nm is assigned to $CaCuO_2$ 101 (d=0.244nm).

to the c-axis lengths (0.34nm, 0.33nm) of $(Sr_xCa_{1-x})CuO_2$ (x=1, 0.5).

At various stages of $SrCuO_y$ film growth on $SrTiO_3$ substrate, the specimen was transferred into the XPS chamber in vacuo to analyze the variation of electronic state in $SrCuO_y$ from the interface. A preliminary result indicates the presence of chemically modulated state within 1nm from the $SrTiO_3$ interface. Beyond this boundary layer, the valence-edge of $SrCuO_y$ film came to be very close to the Fermi level.

ACKNOWLEDGMENTS

The research was supported in part by a Grant-in-Aid for Scientific Research from the Ministry of Education, Science and Culture of Japan.

REFERENCE

1. Koinuma H, Yoshimoto M, Nagata H, Hashimoto T, Tsukahara T, Gonda S, Watanabe S, Kawai M, and Hanada T (1990) Proc. of the 4th Ann. Conf. on Siperconductivity and Applications, Buffalo, NY (to be published)
2. Koinuma H, Nagata H, Tsukahara T, Gonda S, and Yoshimoto M (1990) Ext. Abst. of the 22nd Conf. on Solid State Devices and Materials: 933
3. Koinuma H et al. (unpublished)
4. Schlom DG (1990) Ph.D. Thesis, Stanford Electronics Lab., Stanford Univ.
5. Watanabe S, Kawai M, and Hanada T (1990) Jpn.J.Appl.Phys. 29: L1111
6. Yoshimoto M, Nagata H, Tsukahara T, and Koinuma H (1990) Jpn.J.Appl.Phys.: L1199
7. Takano M, Takeda Y, Okada H, Miyamoto M, and Tsusaka T (1989) Physica C 159: 375
8. Courths R (1980) Phys.Stat.Sol.(b) 100: 135
9. Fujimori A, Takayama-M E, Uchida Y, and Okai B (1987) Phys.Rev.B :8814

7 Application

7.1 Devices

Novel Applications of Tl-Ca-Ba-Cu-O Thin Films to Active and Passive High Frequency Devices

J.S. Martens, D.S. Ginley, T.E. Zipperian, V.M. Hietala, and C.P. Tigges

Sandia National Laboratories, Department 1140, Albuquerque, NM 87185, USA

ABSTRACT

We will present our recent work on the applications of high temperature superconducting thin films to the development of new passive and active circuitry operating up to 40 GHz. We will discuss thin film deposition (including new results on the effect of low temperature anneals) and patterning Tl-based films to linewidths of 3 μm, and contacting. A number of HTS device will be discussed including passive resonators and filters and an active device, the superconducting flux flow transistor (SFFT), that can be used in applications such as amplifiers and phase shifters.

KEY WORDS: Tl superconductor, superconducting devices, microwave applications

INTRODUCTION

In order to develop high frequency applications for HTS materials, three main issues must be addressed. The film process must be stabilized and the resulting films must have low surface resistance at frequencies up into the millimeter wave regime and good surface morphology. Existing device processing technologies must be modified for the Tl-Ca-Ba-Cu-O and other HTS films. This technology must include lithography with resolution of better than a few μm, high quality contacts, and stable non-reactive deposition of dielectrics and normal metal layers compatible with the HTS films. Finally, passive and active device concepts must be found that address applications requirements while fitting the characteristics of these unique materials.

Tl-Ca-Ba-Cu-O FILMS

Due to the volatility of Tl, development of films in the Tl-Ca-Ba-Cu-O system has necessitated ex-situ processing. This processing consists of three main steps: deposition, sintering and annealing. Work to date indicates that the first step is not critical as long as an approximately correct stoichiometry is obtained and the formation of intermetallic phases is prevented. The second step, however, is critical; phase and morphology are established during sintering in air. The third step, a controlled oxygen anneal, optimizes the superconducting properties of the films.

Metallic films are deposited by sequential electron beam evaporation of the pure metals onto the substrate of choice, $LaAlO_3$ for experiments described below, under a slight oxygen overpressure (1-3 x 10^{-5} mbar) to prevent intermetallic phase formation. The evaporations are done in layers starting with Cu and ending with Cu to encapsulate the structure. The as-deposited partially oxidized structures are metallic as expected and show no evidence of superconductivity. Such precursor films are sensitive to oxygen and moisture and consequently are stored in an argon atmosphere dry box while waiting to be sintered and annealed.

Precursor films are oxidized and the Tl-Ca-Ba-Cu-O superconducting phases synthesized during the sintering process. The morphology of the oxidized film is also established at this time. Liquid Tl phases appear to be crucial to the formation of clean, sharp grain boundaries. Films are sintered with the substrate sitting film side up on bulk $Tl_2Ca_2Ba_2Cu_3O_y$ ceramic while a second bulk $Tl_2Ca_2Ba_2Cu_3O_y$ ceramic is located above, but not in contact with, the film. The "sandwich" structure is placed in a 3" Pt crucible with a tight fitting lid. The lid has an approximately 5 mm hole allowing Tl leakage and producing quasi-equalibrium conditions of Tl and oxygen partial pressure. Films are sintered for 16 minutes at 850 °C.

The exact composition of the bulk ceramics used for the Tl source is critical. A narrow window of compositions from $Tl_{2.04}$ to $Tl_{1.96}$ produces the best results. Bulk ceramics that are too Tl rich do not produce uniform growth of superconducting phases on the wafer. Bulk ceramics that are Tl deficient produce good $Tl_2Ca_2Ba_2Cu_3O_y$ phase growth but with random orientation and weak links. The metal ratios in the films as deposited are very close (±5%) to the desired $Tl_2Ca_2Ba_2Cu_3$ composition. Typical anneals are 10 min. (0.3 μm films) at 750 °C followed by a furnace cool. In the best films the effect of this anneal is small with increases in T_c of 2-5 K and increases in J_c of ~10%. Recent work has shown that anneals below 600 °C and as low as 300 °C can substantially improve the superconducting properties of TlCaBaCuO thin films and single crystals. Optimum results are typically obtained at

400 °C in oxygen in a range where DTA and TGA show no significant thermal events. Marginal films (J_c= 10^3-10^5 A/cm^2) can typically be substantially improved while higher quality films (J_c> 10^5 A/cm^2) show less improvement. Best results have shown 10-15K increases in T_c and order-of-magnitude increases in J_c (2 x 10^4 to 2 x 10^5 A/cm^2). Better films show 10-15% increases in T_c and J_c. Work to date indicates that these improvements arise from changes in doping levels due to changes in cationic disorder and oxygen vacancy density. Low temperature annealing potentially provides a mechanism for the precise tailoring of film properties.

The morphology and structure of the films were determined by scanning electron microscopy (SEM) with energy dispersive x-ray analysis (EDS) and by x-ray diffraction. The 0.3 μm films are smooth on a scale ≈ 0.05 μm and are highly dense (> 90%). The grain size varies from 10 to >100 μm, indicative in some cases of an approach to true epitaxial growth. Cleaved films were examined by SEM normal to the substrate surface. In all cases the individual grain thickness is equal to the film thickness and no layering is observed. The lateral and vertical uniformity strongly supports the bulk superconducting nature of the films. Most of the films show complete c-axis orientation perpendicular to the substrate but only partial a-axis orientation in the plane.

PROCESSING DEVICES

All of the circuits to be described are made with standard optical lithography (using a solvent based negative photoresist) and wet etching. A solution of 2 % Br in isopropanol [1] is used for mesa etching with an etch rate of approximately 30 nm/sec. With this process, stable 3 mm linewidths are obtainable. In some applications described below, controlled film thinning is required. For example, a narrow region (e.g. 10 μm by 20 μm) must have its thickness reduced from 300-700 nm to less than 100 nm. This is done with a carefully timed etch of 0.5-0.8% Br in isopropanol.

Dielectric and normal metal levels are also required in some applications. A hard-baked layer of negative photoresist works as an adequate dielectric for many applications (stable and low loss, tan $\delta \approx 10^{-4}$). Other dielectrics are under investigation. Most normal metals are usable for other layers. If the metal is in contact with a Tl-Ca-Ba-Cu-O level, it must not easily oxidize or oxygen may diffuse from the superconductor causing contact problems. The metals used are Au, Ti, Al, W-Si and Ag. The W-Si is used frequently as an oxygen diffusion barrier for Ti and Al resistors. For ohmic contacts, we use annealed silver contacts deposited through a shadow mask. Contact resistances of less than 0.1 Ω have been achieved and do not appear to degrade with time.

PASSIVE DEVICES

A number of different passive microwave components have been developed both for applications (in conjunction with the active devices discussed below) and for use in materials characterization. The first components, the ring resonators, have been built and are used for materials characterization [2]. An all-gold structure is first used to calibrate the system for the surface resistance of gold. To evaluate the HTS material, ring resonators were constructed using a TlCaBaCuO top conductor and a gold ground plane. With the gold surface resistance data and the data from this resonator, it is possible to compute the HTS surface resistance. The method is valid for arbitrary film thicknesses and the structure is easy to measure using a contactless probe (see Fig. 1). For our Tl films, we have measured a surface resistance of about .6 mΩ at 8 GHz and 77K. This can be compared to the theoretical value for gold of about 9 mΩ at 8 GHz and 77K. Modified versions of the Tl ring resonator have been used as frequency-determining networks in oscillators [3].

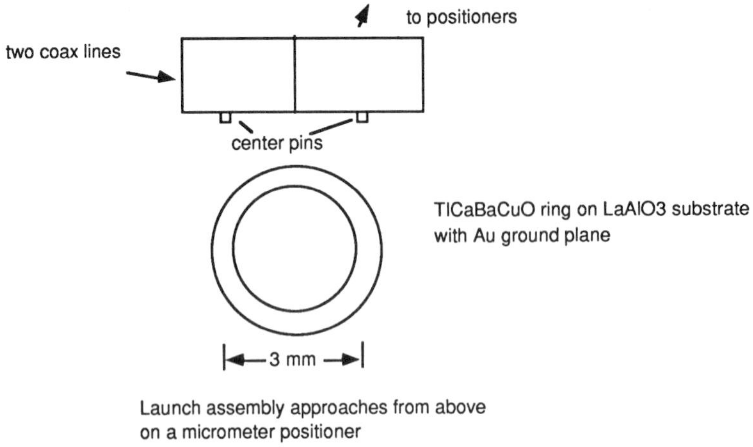

Fig. 1. The ring resonator and probe assembly to measure surface resistance (8.2 GHz).

Confocal resonators have also been used for materials characterization. The structure uses a normal metal mirror focussing a Gaussian beam (mode TEM_{oon}) [4] onto a planar sample as shown in Fig. 2. Because the magnetic fields are strongest at the sample in this configuration, losses in the normal metal are not as important to the total Q as in a cylindrical cavity system. Since this system is quasi-optical, it can be easily extended to much higher frequencies (100 GHz). The surface resistance of TlCaBaCuO films at 77K and 36 GHz has been measured to be ≈.8 mΩ.

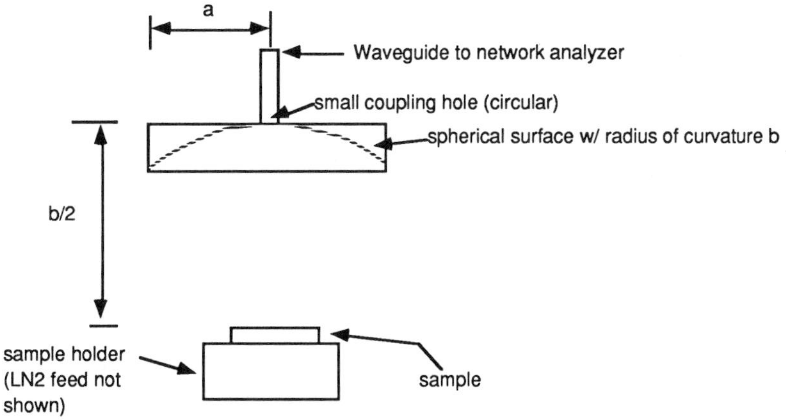

Fig. 2. Experimental arrangement for surface resistance measurements using a confocal resonator. The spherical mirror has a radius of curvature b.

Coplanar waveguide filters in Tl films have been developed. This structure was chosen over the more common microstrip filters[5] because of the planar nature of the active devices used and the desire to avoid requiring double sided films. On a 700 nm Tl film ($LaAlO_3$ substrate) the filter (9.5 GHz cut-off designed) showed a passband loss of .5 dB and matched simulation results to within 1 dB over the range 7 GHz to 11 GHz. The substrate was apparently of sufficient quality that twinning did not disrupt the microwave performance.

ACTIVE DEVICES

The device of most interest to us so far is called the Superconducting Flux Flow Transistor (SFFT) [6]-[9]. It consists of a parallel array of weak superconducting links (separating two unweakened banks of superconductor) and a control line to provide a local magnetic field. An example of this structure is shown in Fig. 3. When the device is biased below the critical current (typically on the order of 1 mA), no flux is admitted into the link system (perfect Meissner state). Above the critical current, flux is admitted in discrete quanta known as vortices[10]. These vortices can move since the bias current generates a Lorentz-type force on the vortices. The vortices are also subject to forces from external magnetic fields (from the control line), viscous damping, pinning forces (which are undesirable in that the average vortex speed is reduced) and surface barriers at the edges of the links which hamper flux entry and exit. The balance of these forces determines the flux motion and hence the terminal voltage. In terms of active device performance, the key principle is the use of an external magnetic field (via the control line) to modulate the flux density in the link system [11],[12] and the resulting flux motion.

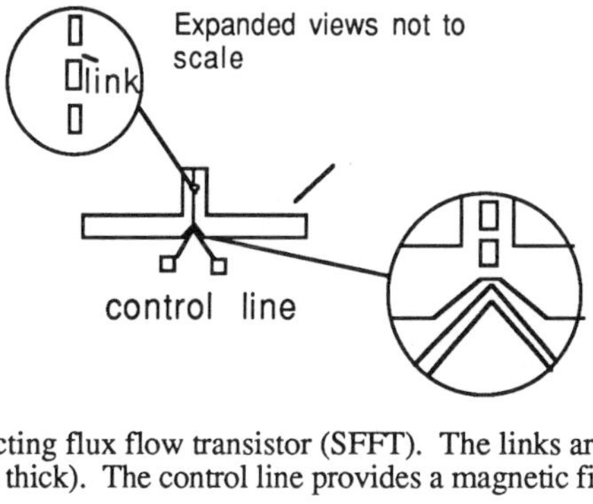

Fig. 3. A layout of the Superconducting flux flow transistor (SFFT). The links are 3-5 μm by 10 μm and are 50-100 nm thick (the banks are 300-700 nm thick). The control line provides a magnetic field that modulates the flux density in the link system.

A typical set of IV curves is shown in Fig.4 The transition between the zero-voltage section and the flux flow section is apparent. By modulating the flux density in the link system, the control field causes a horizontal translation of the flux flow branches (with an accompanying change in critical current).

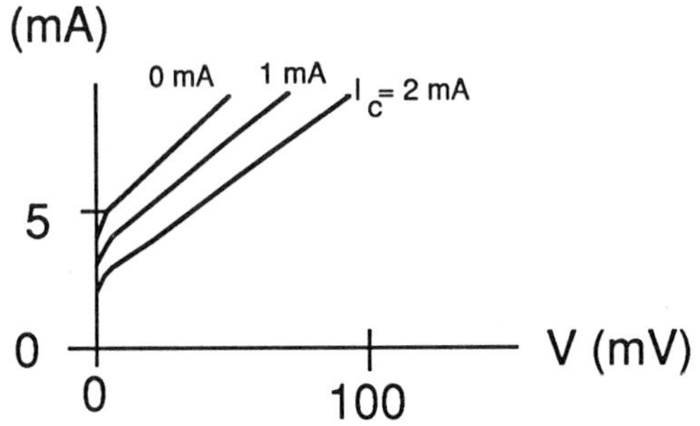

Fig. 4. I-V curves for the device of Fig. 3. Note the flux flow branches and the effects of control current I_c.

The device behavior is closely correlated to the quality of the film remaining in the link region. While increased pinning results in higher critical currents, it also reduces sensitivity to external magnetic fields (and hence gain) and reduces the flux speed and frequency response of the SFFT. A material with a low H_{c1} is desirable in that it is easier to admit flux to the system and increase sensitivity to the control fields. The penetration depth is also important since it strongly influences the surface barriers and hence control-field sensitivity [13]. In addition, the effective penetration depth changes with link thickness, greatly increasing the effect of the film thinning process. Maximum gain occurs for links thinner than a penetration depth but not so thin that the superconductivity is destroyed or the defect density rises enough to induce excessive pinning centers. For a stable film production process, the effect of link thickness must be very reproducible (typical thickness of the active region is about 80 nm). Films with higher J_c tend to have narrower windows of acceptable link thickness but may still be used with good results.

The next step in the development of microwave applications is the creation of an equivalent circuit; one is shown in Fig. 5 [13]. The input impedance is low (a resistance of less than .5 Ω from contacts and surface resistance and an inductance of around 200-300 pH). For applications in a 50 Ω system, impedance matching is necessary. The output is dominated by the transresistance mechanism (a dual of the transconductance mechanism in a FET), an output resistance on the order of 3-5 Ω, and an output inductance that is non-linear for sufficiently large current swings. A typical value of transresistance ($r_m = \Delta V_{out}/\Delta I_{ctl}$) for a TlCaBaCuO device at 77K is 15-20 Ω.

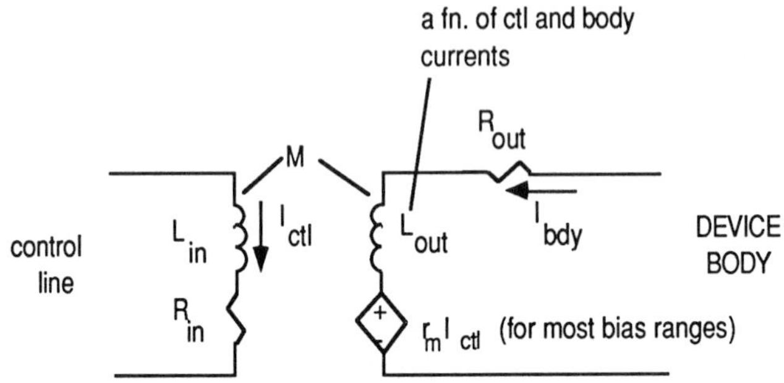

Fig. 5. An equivalent circuit for the device of Fig. 3. For large signal analysis, the output resistance is modeled as part of a non-linear voltage source. The output inductor can also be non-linear.

Two factors affecting device speed are the input and output L/R time constants as can be seen from Fig. 5. By reducing device dimensions, these inductances can be reduced, in turn reducing the time constants below 50 ps in most circuits. Besides the circuit time constants, there is a fundamental device time constant as well. The vortices take a finite amount of time to cross the link system. This delay is the transit time constant. With vortex speeds on the order of 5 x 10^5 m/s (at 77K) [13] and typical link dimensions (3-5 links, each 3 μm in the direction of flux motion) this time constant is about 10-30 ps. As lithographic capabilities improve, this time constant can be reduced further. Device operation has been demonstrated at 35.8 GHz (a smaller device than that described above, used as a mixer) when the net time constant predicts a cut-off of 36.7 GHz [13].

The SFFT can produce net gain in a 50 Ω system [3]. With some impedance matching, one group of devices have the gain shown in Fig. 6. Because of the very low input impedance, very high gain can be achieved with perfect impedance matching. Preliminary noise measurements show a minimum noise figure of about 0.8 dB at 4 GHz [14]. Work continues on broader band amplifiers and on the evaluation of noise and non-linearities.

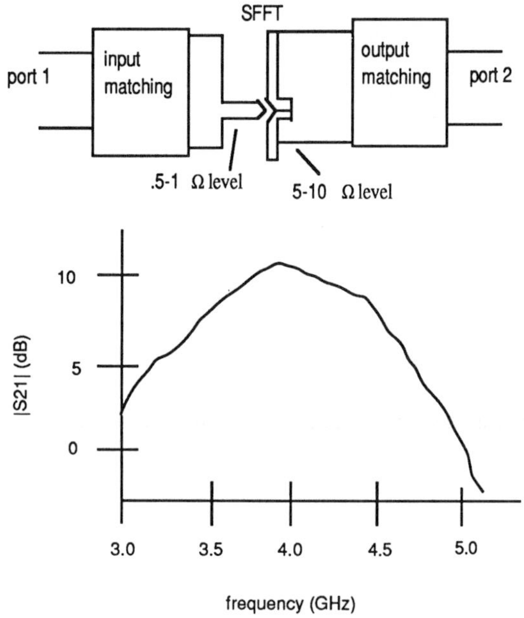

Fig. 6. The upper image is the SFFT embedded in its matching networks (composed of superconducting transmission lines and inductors); the circuit was designed to operate at 4 GHz. The lower image shows amplifier gain in a 50 Ω system. The 3 dB bandwidth was about 1.1 GHz.

The variable output inductance has been used to create a phase shifter. To allow large amounts of linear phase shift (> 2 π radians desired), several devices are used embedded in an artificial transmission line. The individual SFFTs form the inductors of the transmission line and gap capacitors are used to adjust the impedance level to a nominal 50 Ω. At a 4 GHz test frequency, 2π of phase shift was generated by a control current of 3.5 mA as shown in Fig. 7. The corresponding insertion loss was less than 4 dB and the useful bandwidth exceeded 2 GHz.

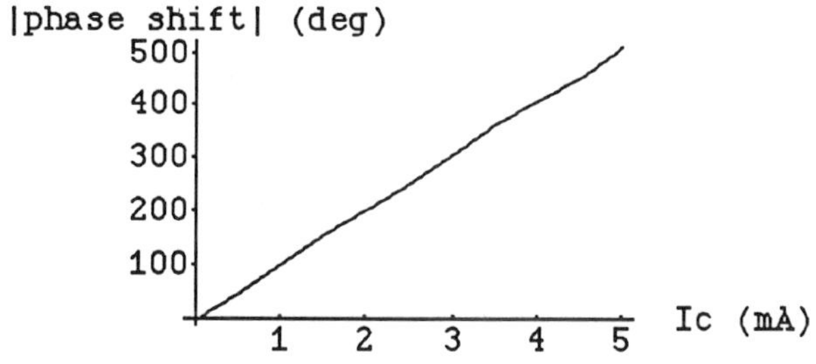

Fig. 7. Phase shifter performance at 4 GHz as a function of control current. Note the linearity and the large amount of phase shift available.

CONCLUSIONS

We have produced TlCaBaCuO films using e-beam evaporation with subsequent sintering and annealing that have critical current densities of at least several hundred kA/cm^2, critical temperatures over 100K and surface resistance lower than that of gold at 77K and 8 GHz. Processing techniques have been developed for making passive microwave elements and a four terminal active device called the superconducting flux flow transistor (SFFT) for microwave applications. The techniques include contacting, dielectric and normal metal layer deposition, and controlled HTS film etching. The SFFT shows promise as a microwave amplifier and as a phase shifter. It has also been used as an active impedance convertor between LTS Josephson junctions and GaAs MESFETs and in prototype digital circuits. The advantages of the device include high speed, potentially low noise and, for some applications, useful impedance levels. Passive resonators also show promise as demonstrated by us and many others.

ACKNOWLEDGEMENTS

Some of this work was performed at the University of Wisconsin-Madison, ECE Department. The authors wish to thank J. B. Beyer, J. E. Nordman, G. K. G. Hohenwarter and S. Whitely for assistance in the course of this research. Thanks also to M. Housel and T. Plut at Sandia National Laboratories for fabrication work. This work was supported in part by a National Science Foundation Graduate fellowship, by the U.S. Department of Energy under contract No. DE-AC04-76P00789, and by DARPA, ONR and SDI under contract No. N0017390WR00172.

REFERENCES

1. Vasquez RP, Stevens HP (1989) JPL invention Report NPO-17712/7216. Jet Propulsion Laboratory. Pasadena, CA.

2. Takemotot J, Oshita F, Fetterman H, Kobrin P, Sovero E (1989), IEEE Trans on MTT 37 : 1650-1652

3. Hohenwarter GKG, Martens JS, Beyer JB, Nordman JE (1991) Resonant Impedance Matching of Abrikosov Vortex Flow Transistors, to be published IEEE Trans. on Mag. 27 (2)

4. Boyd GD, Gordon JP (1961) Bell Sys. Tech. Jour. 40: 489.

5. Hammond RB, Negrete GV, Schmidt MS, Moskowitz MJ, Eddy MM, Strother DD, Skglund DJ (1990) Superconducting Tl-Ca-Ba-Cu-O Thin Film Microstrip Resonator and its Power Handling Performance at 77K, presented at the IEEE MTT-S International Microwave Symposium, Dallas, TX, May 1990

6. Hohenwarter GKG, Martens JS, McGinnis DP, Beyer JB, Nordman JE, Ginley DS (1989) IEEE Trans. on Mag. 25: 954-956

7. Martens JS, Hohenwarter GKG, Beyer JB, Nordman JE, Ginley DS (1989) J. Appl. Phys. 65: 4057-4060

8. Martens JS, Beyer JB, Nordman JE, Hohenwarter GKG, Ginley DS (1989) Proceedings of the 1989 International IEEE MTT-S Symposium: 443-446

9. Martens JS, Beyer JB, Nordman JE, Ginley DS (1990) Micr. and Opt. Tech. Lett. 3: 49-51

10. Likharev KK (1972) Sov. Phys. - JETP 34: 906-912

11. Van Duzer T, Turner CW (1981) Principles of superconductive devices and circuits. Elsevier. New York, chp. 8.

12. Pearl J (1964), Appl. Phys. Lett. 5: 65-66

13. Martens JS (1990) The models and applications of a high frequency three terminal device made of high temperature superconductors, PhD Dissertation. University of Wisconsin-Madison.

14. O'Callaghan JM, Martens JS, Beyer JB, Nordman JE, Ginley DS (1991) Microwave Noise Parameters of a HTS Flux Flow Transistor, to be published IEEE Trans. on Mag. 27 (2)

Surface Impedance Measurements of $YBa_2Cu_3O_{7-x}$ Thin Films in Stripline Resonators

D.E. Oates and Alfredo C. Anderson

Lincoln Laboratory, Massachusetts Institute of Technology, Lexington, MA 02173, USA

ABSTRACT

Microwave surface impedance of thin films of $YBa_2Cu_3O_{7-x}$ deposited by a single-target off-axis magnetron sputtering technique was measured as a function of temperature, frequency and rf magnetic field, H_{rf}, using a stripline resonator. The measurements included determination of the penetration depth λ, equal to 0.167 μm in the best film. The surface resistance R_s was measured over a frequency range of 0.4 to 20 GHz; the temperature range was 4.2 K to T_c, and the H_{rf} ranged from 0 to 225 Oe. The R_s at 4.2 K and 1.5 GHz is 4.5×10^{-6} Ω. We observed that $R_s(H_{rf})$ depends upon the conditions under which the films were deposited. Films deposited at low temperature show a linear dependence on H_{rf} and those deposited at high temperature show a weak dependence upon on H_{rf}. All films show a rapid rise in R_s at high values of H_{rf}. The shape of $R_s(H_{rf})$ is similar to that of Nb and NbN films measured in the same stripline geometry. We propose that $R_s(H_{rf})$ is related to the grain structure of the films. We have measured intermodulation products in these resonators. Films showing weak dependence of R_s on H_{rf} demonstrate smaller intermodulation effects. We discuss the implications for microwave device applications.

INTRODUCTION

Measurements of the rf surface impedance Z_s of superconductors as a function of frequency, temperature, and rf magnetic field are of great interest because they yield information on many fundamental physical parameters of the material such as penetration depth, coherence length, energy gap, and electron mean free path. They are also important for practical applications of superconductors at rf frequencies because the value of the surface resistance determines the losses in the superconductor. $Z_S = R_S + j\omega\mu_o\lambda$, where R_S is the surface resistance, ω is the frequency in rad/s, μ_o the permeability of free space and λ is the penetration depth. Z_S is by definition the surface impedance of a semi-infinite slab of material and in practice implies a material thickness large compared with a penetration depth. Loss occurs at rf frequencies and finite temperatures because the penetration of the magnetic field into the material produces an electric field which couples to the normal unpaired electrons, leading to ohmic losses. At dc the electric field is zero and there are no losses. The R_s of superconductors, however, is usually orders of magnitude smaller than the best normal conductors at microwave frequencies. Because of the low loss, promising applications of superconductors are resonators and filters using miniature planar circuit geometries in the microwave frequency range. Also, since the penetration depth in most superconductors is independent of frequency, delay lines can be made without dispersion which, coupled with low loss, means that high-performance long delay lines and transversal filters are among the potential applications. The discovery of the high-T_c superconductors holds out the hope that these applications can be realized at 77 K and indeed there have been many recent reports of practical microwave devices [1].

Z_s measurements have most often been performed in resonant cavities [2, 3]. With the cavity method, however, the frequency dependence of Z_s is difficult to obtain, and it is difficult to measure R_s with high sensitivity over a wide range of temperatures. Planar transmission line resonators and especially stripline resonators overcome some of the limitations of cavity resonators [4] since R_S can easily be measured as a function of frequency, temperature, and rf magnetic field H_{rf}. Until recently, however, interpretations of stripline measurements were complicated by the nonuniform distribution of currents in the conductors and by the large values of penetration depth of films of the high-T_c materials which in many cases are comparable to the film thickness. Detailed calculations of the current distributions as a function of film thickness and λ for the stripline geometry used in these measurements have recently been completed using full electromagnetic wave methods [5, 6]. With knowledge of the current distribution we can calculate accurately the inductance L of the stripline as a function of penetration depth λ. Then by measuring the resonant frequency, which is proportional to $L^{-0.5}$, as a function of temperature we can measure λ. From knowledge of λ and the current distributions, we can calculate losses in the stripline and thus extract from the measured Q the semi-infinite (intrinsic) R_S. By varying the input power used in the Q measurements we vary the magnitude of the current and can measure R_S as a function of H_{rf} which can be calculated from knowledge of the current distribution. Measurements of Z_S using stripline resonators are reported for thin films of $YBa_2Cu_3O_{7-x}$ deposited on $LaAlO_3$. These reported measurements include determination of R_S as well as λ. We measured the surface resistance as a function of frequency, temperature, and rf magnetic field.

STRIPLINE RESONATORS

Our measurements of Z_S were made using the stripline resonator structure shown in Fig. 1. Stripline is defined as having a center conductor, an upper and lower ground plane, and a symmetric dielectric. The resonator comprises a length of transmission line one-half wavelength long at the fundamental frequency; overtone resonances occur at all multiples of the fundamental frequency. The Q and the resonance frequency of the resonator were measured using transmitted power. The resonator is capacitively coupled to the external circuit by gaps at the two ends of the stripline. The gaps are adjusted so that the resonator is weakly coupled and only small corrections have to be made to

Q_L, the loaded Q, to obtain the unloaded Q. Two different resonators were used for our measurements. In one, the length of line was chosen to yield a fundamental frequency f_0 of 0.4 GHz and in the other f_0 was 1.5 GHz. In both cases the linewidth was 150 μm. This gives a 34-Ω transmission line impedance with 0.5-mm-thick $LaAlO_3$ as the dielectric (ε = 25). The dimensions of the substrates are approximately 2.5 × 1.25 cm for f_0 = 0.4 GHz and 1 × 1 cm for f_0 = 1.5 GHz.

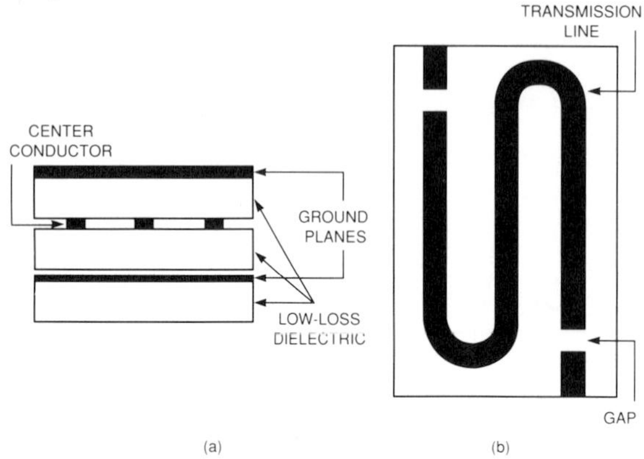

Fig. 1. Schematic (a) cross-sectional and (b) top views of the stripline resonator used in the measurements.

The ultimate sensitivity of R_S measurements using the stripline method is lower than 10^{-7} Ω as has been demonstrated with niobium films at 2 K [7]. This corresponds to a resonator Q of 4×10^6 at 0.6 GHz. A major advantage of the stripline geometry is that radiation losses are low as a result of the presence of upper and lower ground planes. It is important to note that currents flow on both the free surface of the film as well as on the interface between the dielectric substrate and the film. The substrates used in transmission line resonators must have low dielectric losses as discussed in [4]. Because the rf fields penetrate the substrates, tan δ must be less than 1/Q of the resonator.

The resonator is fabricated from three films on three substrates clamped together in a copper package with rf connectors. The measured Z_S contains contributions from all three films so that Z_S is a weighted average, with the center conductor contributing about 80%. The losses are concentrated in the center conductor because the narrow (150 μm) line results in a higher current density in the center conductor than in the ground planes. Each of the resonators we have tested has been fabricated with a set of films deposited under identical conditions, so that the properties are uniform.

Surface impedance measurements have been carried out using resonators fabricated from films of $YBa_2Cu_3O_{7-x}$. Films deposited by electron beam coevaporation of Y, Ba, F_2 and Cu followed by post-annealing [8] have been reported by us in detail previously [4]. The newest results have been obtained with films deposited by in situ off-axis sputtering [9] which we will concentrate on here. Three resonators have been characterized: sample #1 was deposited at 680 °C, #2 at 720 °C, and #3 at 760 °C. As grown, the films are in the tetragonal phase and are converted to the orthorhombic phase after the sputtering has been finished by admitting one atmosphere of O_2 to the deposition chamber and allowing the sample to cool down in the presence of oxygen. The T_c of the films was measured by a 4 point resistance probe and lies between 79 K and 87 K with higher values of T_c for the films deposited at higher temperatures. The critical current density is $J_c = 1.5 \times 10^7$ and 5×10^6 A/cm² at 4.2 K and 50 K, respectively. Film thickness is 3000 Å. The films are highly aligned with the c axis perpendicular to the substrate. The films are granular but we have not yet determined the size of the grains.

PENETRATION DEPTH MEASUREMENTS

The first step in determination of Z_s is the measurement of the penetration depth. Independent determination of λ is necessary also for the extraction of R_s from the measured Q values because the thickness of the films used in the stripline resonator (2 - 3 kÅ) is comparable to λ.

We determined λ by measuring the temperature dependence of the fundamental resonant frequency of the resonator and then by fitting a model of the temperature dependence of λ to the measured data. The resonant frequency is and proportional to $1/L^{0.5}$ where L is the inductance per unit length. L is a function of the geometry of the stripline and λ and can be calculated knowing the current distribution. Figure 2 shows the measured resonant frequency vs T using sputtered film set #3. The solid line is a least-squares fit to the data using the temperature dependence calculated from the two-fluid model and using $\lambda(T = 0) \equiv \lambda_0$ as a free parameter. In the two-fluid model

$$\lambda(T) = \frac{\lambda_0}{\sqrt{1 - \left(\frac{T}{T_c}\right)^4}}$$

T_c was taken to be 86.4 K which gives the best fit and is consistent with the measured resistive transition. The resonant frequency f is given by

$$f(T) = f_0(T_0)\sqrt{\frac{L(\lambda(T_0))}{L(\lambda(T))}}$$

where L(λ) was obtained from the calculated values [5]. The best fit is obtained with λ_0 = 0.167 μm. The fit shows some deviation from the two-fluid model as others have observed [10]. Determination of λ in this manner has an approximate accuracy of 20%. The determination is model dependent. The inferred value of λ is also highly sensitive to the value of T_c chosen. For instance, a shift of 0.1 K in T_c can change λ(0) by 10%.

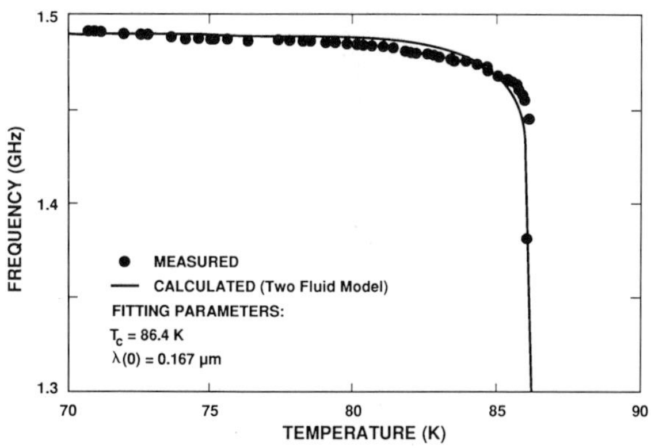

Fig. 2. Measured and calculated values of resonant frequency of the fundamental mode of the resonator vs temperature for sputtered film #3. The calculated values use the two-fluid model. The best fit is obtained with $\lambda(0) = 0.167$ μm and $T_c = 86.4$ K.

TABLE 1

SAMPLE	$\lambda(0)$ (μm)	T_c (K)
Sputtered $YBa_2Cu_3O_{7-X}$ #1	0.316	80
Sputtered $YBa_2Cu_3O_{7-X}$ #2	0.336	80
Sputtered $YBa_2Cu_3O_{7-X}$ #3	0.167	86.4
Sputtered Nb	0.070	9.2

Table 1 shows the measured values of λ for a niobium film which was used as a check of the method as well as the values for the three different sets of sputtered films. The value of λ we obtained for sputtered niobium agrees with previously reported values. The values of λ in the $YBa_2Cu_3O_{7-x}$ films must be understood as an effective penetration depth which is a result of the field penetration in the regions between the grains. We believe that the larger λ of $YBa_2Cu_3O_{7-x}$ in film sets #1 and #2 can be explained by a larger influence of the intergranular material than in film set #3, which we believe was relatively free of intergranular material because of the higher temperature at which it was deposited. The penetration depth of the grains themselves is always smaller than the values quoted here.

MEASUREMENT OF SURFACE RESISTANCE

The R_s (f, T) was determined in the following manner: the unloaded Q of the resonator was measured at low power at each overtone frequency between the fundamental (0.4 or 1.5 GHz depending upon the resonator) and 20 GHz. These measurements were repeated at a series of temperatures between 4.2 K and the transition temperature. To obtain R_s from the Q measurements we used the already determined value of λ for each resonator and the losses which are computed from the distribution of currents. For small λ/t, where t is the conductor thickness, R_S is given by

$$R_S = \Gamma \frac{f}{Q} \qquad (1)$$

where Q is the unloaded Q and Γ is a geometrical factor, which in our case is 0.41 Ω/GHz. When λ/t becomes comparable to unity a correction factor Δ(λ) must be introduced so that Eq. (1) becomes

$$R_S = \frac{\Gamma}{\Delta(\lambda)} \frac{f}{Q}$$

This corrected R_S is the surface resistance which would be measured on a film of thickness t >> λ. Δ(λ) is plotted in [11] and is about 1.5 for λ/t = 0.5 and t = 0.3 μm. This correction factor is slightly larger than that arrived at with the more usual coth (t/2λ) which would apply for uniform transverse current distributions [12].

Figure 3 shows R_s versus temperature at the frequencies given. The correction factor for finite λ, Δ(λ), was computed at each of the measured temperatures. The films produced at higher sputtering temperatures show a sharp transition and approach a residual value within a few degrees of T_c. However the values of R_s at 4 K are quite similar for all three film sets. The two film sets sputtered at higher temperatures differ in transition temperature although the shape of R_s (T) is similar. We believe that the different deposition temperature was the reason for the observed shape, but we note that the sputtering target was changed between #2 and #3 and may have some influence on the results.

Measurements of R_s vs frequency at three different temperatures for film set #3 are shown in Fig. 4. The measurements at 4 K were affected by cavity modes in the resonator package above 10 GHz. Other than this region the general behavior is proportional to f^2 which has been observed and discussed previously by us and others [4, 13]. Many models of R_s yield an f^2 behavior, so this dependence does not support any particular one of these.

The high current density in a stripline resonator enables the measurement of R_S as a function of H_{rf} up to high field values. The total rf current in the resonator center conductor is given by [4]

$$I = \sqrt{\frac{r_v(1-r_v)4Q_cP}{n\pi Z_0}}$$

where r_v is the voltage insertion ratio, related to the insertion loss IL by IL = -20 log r_v, Q_C is the unloaded Q, P is the incident power, Z_o is the characteristic impedance of the line, and n is the resonant mode number. The current density, however, is not uniform but strongly peaked at the edges of the line as shown by the numerical calculations and, for small λ, in the geometry used here the current density at the edges of the line can be more than 100 times greater than the average current density. As λ increases, the current density at the edges is less strongly peaked. The magnetic field at the surface of the conductor is not as strongly peaked at the edges because the field is given by an integral over the current density. Since the magnetic field is the quantity of most interest, we plot in Fig. 5 the magnitude of the magnetic field at the surface of the superconductor as a function of position on the center conductor of the stripline resonator [5]. These calculations were for the case of a 3000-Å-thick film, λ = 2000 Å, and I = 1 A. The magnetic field at the edges is approximately ten times greater than the average field.

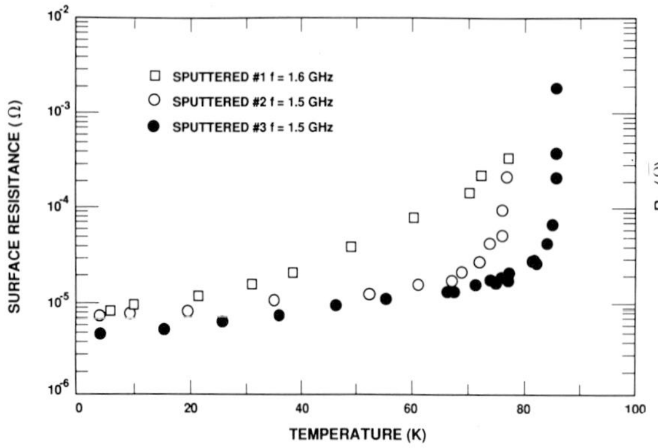

Fig. 3. Surface resistance vs temperature for the three sputtered films. The frequency of each measurement is given.

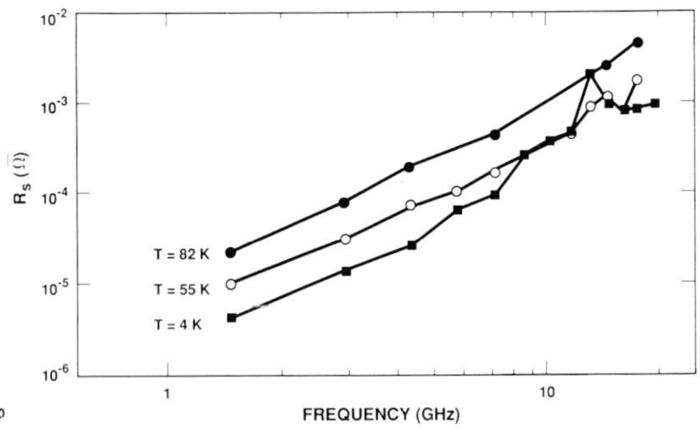

Fig. 4. Surface resistance vs frequency for different temperatures as indicated. These results are for sputtered film #3.

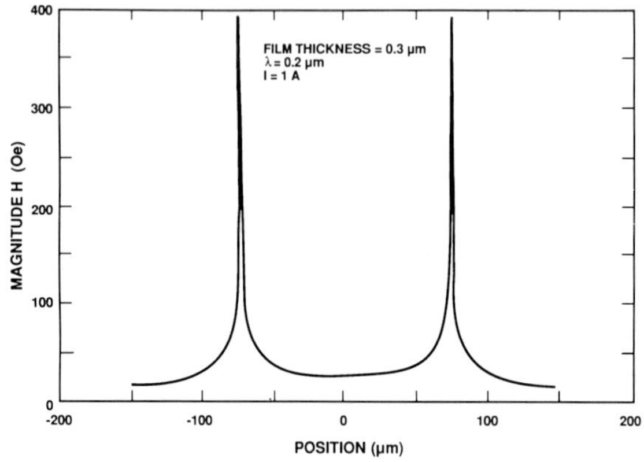

Fig. 5. Plot of the magnitude of the magnetic field at the surface center conductor of the stripline. The center conductor is 150 μm wide.

Figure 6 shows the dependence of R_S on H_{rf} for the three sputtered films and for a postannealed film reported earlier [4]. All were measured at 4 K and at the frequency indicated in the figure caption. $H_{rf,max}$ for sputtered film set #2 is 150 Oe at the point where R_S increases rapidly and for #3 it is 225 Oe. The films sputtered at higher temperatures (#2 and #3) clearly show a greatly reduced dependence of R_S on current and H_{rf}. For both of these films the input power was increased until a sharp increase in R_S was measured, indicating that the critical current or magnetic field was reached. Beyond this point it is meaningless to report values of R_S because the resonance curve is severely distorted and Q is ill defined. Film set #3 shows R_s with little dependence on H_{rf} up to 225 Oe which is to our knowledge the largest reported value of H_{rf} for which R_s remains unchanged.

Figure 7 shows measurements of R_s vs H_{rf} for Nb and NbN [7] at 4 K compared with the results for sputtered film set #3. The Nb films show similar behavior to the best sputtered $YBa_2Cu_3O_{7-X}$ films but the maximum H_{rf} is more than a factor of two greater in Nb. NbN at low field shows a linear dependence on H_{rf}, similar to that of sputtered film #1 and the postannealed films, except that, in NbN, R_s is much smaller and the slope of the R_s vs H_{rf} dependence is smaller. NbN films are composed of small grains and it is believed that the behavior of the NbN films in rf fields results from the penetration of flux in the intergranular regions [4, 7]. Both Nb and NbN show a sharp increase in R_s beyond a maximum H_{rf}, a behavior identical with that of the $YBa_2Cu_3O_{7-X}$ films we have measured.

We have earlier [4] presented the results of $R_s(H_{rf})$ shown in Fig. 6 for the postannealed films and the lowest temperature sputtered film. Both show a linear dependence of R_s on H_{rf} and we proposed that this is due to the penetration of flux at the grain boundaries. In the newer films deposited at higher temperatures there is a very much lower dependence of R_s on H_{rf} which we explain by lowered contributions from the grain boundaries since we expect that the higher temperatures lead to better growth conditions of the films. The sharper transition and flatter behavior in R_s vs T is also consistent with fewer grain boundary effects. It has been suggested [14] that contributions to R_s from the material inside the grains show only weak dependence on temperature within about 10 K of T_c, while the grain boundary material is more strongly temperature dependent. Most high-quality films reported in the literature [2, 3, 13, 15] demonstrate flat R_s vs T curves and sharp transitions.

DISCUSSION/CONCLUSION

The results indicate that the films deposited at the highest substrate temperatures show an improvement in both T_c and in rf properties (both R_s and λ). Further work, however, is needed to establish irrefutably the link between higher deposition temperature and better microwave properties. It is important that the T_c has been raised and that R_s at 77 is low enough to be useful in passive microwave devices such as bandpass filters and high-Q resonators. The most

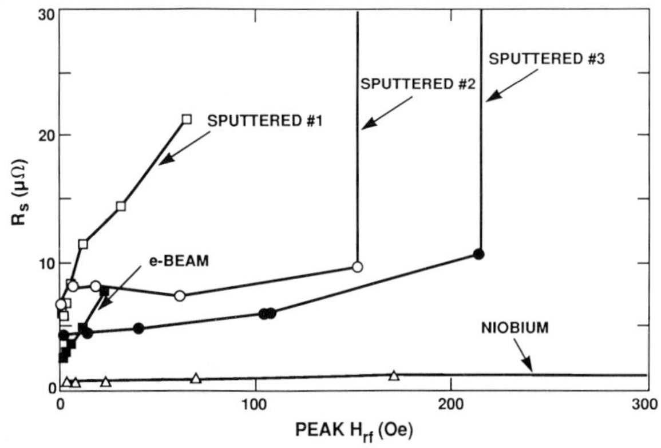

Fig. 6. Surface resistance versus rf magnetic field for four different films as indicated. Lines have been drawn to connect the points. For sputtered films #2 and #3, f = 1.5 GHz. For sputtered and postannealed film #1, f = 1.6 GHz.

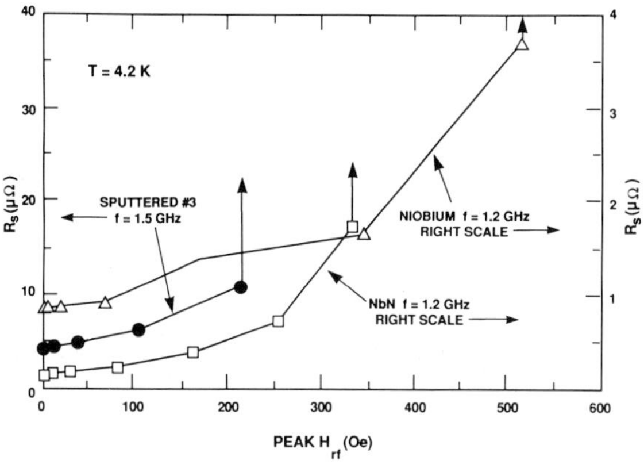

Fig. 7. Comparison of $R_s(H_{rf})$ for $YBa_2Cu_3O_{7-x}$, niobium, NbN films. The Nb and NbN data are multiplied by 10 relative to the $YBa_2Cu_3O_{7-x}$ data and use the scale on the right side. The upward arrows indicate a sharp increase in surface resistance.

important demonstrated improvement in rf properties is the reduction of the dependence of R_s on H_{rf}. It is not enough that the R_s be small at low field because in any practical application the material must be able to carry currents and magnetic fields of some size. We believe that the improvements in the films sputtered at high temperatures result from a reduction in the contribution of the connecting intergranular material to the rf properties as evidenced by the small value of λ and by the small dependence of R_s on H_{rf}. Our previously reported results [4] in postannealed films and in films sputtered at lower temperatures showed a linear dependence on H_{rf} beginning at very low fields which we concluded was due to flux penetration at the grain boundaries. The films produced at higher sputtering temperatures, however, show such low dependence on H_{rf} that we conclude that the contribution to the surface resistance from the intergranular material is greatly reduced. Further work is needed to support this hypothesis. We intend, for instance, to directly measure the grain size and correlate it with the surface impedance.

Measurements of the dc resistance of the sputtered $YBa_2Cu_3O_{7-x}$ films also support the hypothesis that contributions from the grain boundaries are smaller in the films deposited at higher temperatures. The resistivity ρ of the normal state is lower and the curve of $\rho(T)$ extrapolates to zero at zero temperature for the films deposited at higher temperatures. That is, $\rho(T) = cT$, where c is a constant, for $T > T_c$. The $\rho(T)$ of the films deposited at lower temperature does not extrapolate to zero. Figure 8 shows the measurements of $\rho(T)$ for films made at the same time as those of set #1 and #3. Since the thickness of these films was not measured exactly, $\rho(T)$ is shown in arbitrary units and the two curves were normalized by equating the slopes. The lines in the figure show the extrapolation of the curve to zero temperature. When $\rho(T) \rightarrow 0$ as $T \rightarrow 0$ the resistivity of the grains themselves is dominant, because the intragranular material exhibits nearly ideal metallic $\rho(T)$ while the grain boundry material, on the other hand, exhibits nonmetallic behavior and $\rho(T)$ does not fall linearly with temperature. Thus for this limited set of samples $\rho(T) = cT$ correlates with weak dependence of R_s on H_{rf}.

Strong influence from the weak links at the grain boundaries should also correlate with the measurements of λ. It has been proposed [16] that intergranular weak links increase the effective λ. Thus larger λ should be associated with larger dependence of R_s on H_{rf}. We note that sputtered film sets #1 and #2 have similar λ while #2 and #3 have similar $R_s(H_{rf})$. This is not completely understood but does indicate that λ is not the only determinant of R_s.

In applications of the high T_c materials, the dependence of R_s on H_{rf} has important implications. In high Q resonators for frequency control applications, low R_s is the most desirable property. But it is also important that the resonator be operated with a high incident power [17]. One must therefore consider R_s at the operating power and not the values at low power. When R_s depends upon the magnetic field and thus the power carried by the material, nonlinearities become important. These effects can be observed even at relatively low power levels typically found in bandpass filters because the dependence of R_s on H_{rf} extends down to very low values of field as in Fig. 6.

Power dependence of R_S also leads to nonlinearities which generate harmonics and intermodulation products, both of which are important for bandpass filter operation. Figure 9 shows measurements of the intermodulation products in a resonator made from the sputtered film set #1, which showed the linear dependence of R_S on H_{rf}, and for film set #3, which showed very weak dependence on H_{rf}. This was a standard two-tone intermodulation test where two tones of equal power, one at f_1 and the other at f_2, are the input and a frequency of $2f_1 - f_2$ is observed, where f_1, f_2, and $2f_1 - f_2$, are within the passband of the resonator. The third-order intercept, which is the intersection of the extrapolation of slope = 1 connecting the measured outputs at f_1 and slope = 3 line connecting the points measured at $2f_1 - f_2$ is the usual measure of the nonlinearities of a particular device. One sees that the third-order intercept in the $YBa_2Cu_3O_{7-X}$ resonator made with the films of low H_{rf} dependence is approximately 15 dB higher than that of the film which had large H_{rf} dependence. This is a demonstration that lower H_{rf} dependence leads to improved practical performance. Although not shown, a niobium resonator shows better performance that the best $YBa_2Cu_3O_{7-x}$, probably because the critical field is higher.

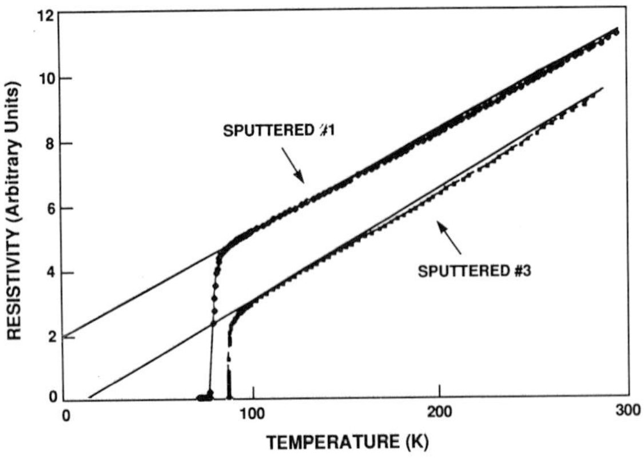

Fig. 8. Resistivity vs temperature for sputtered film set #1 and #3. The straight lines extrapolate the measurements to zero temperature.

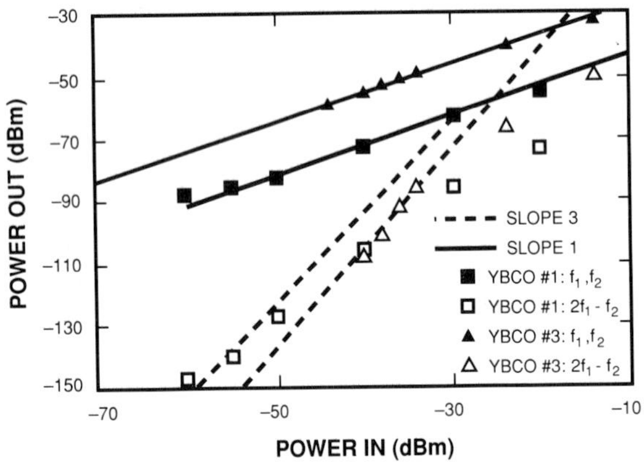

Fig. 9. Intermodulation products in two $YBa_2Cu_3O_{7-x}$ resonators. The lines are drawn to extrapolate the third-order intercept. The third-order intercept is nearly 15 dB higher for the film with weak dependence of R_s on H_{rf}.

The results for the best sputtered film reported here compare very favorably with the best reported values in the literature for R_S of $YBa_2Cu_3O_{7-X}$ thin films. Our results at 77 K and 10 GHz are 7×10^{-4} Ω. The best reported value for an unpatterned film extrapolated from 84 to 10 GHz is [2] 1×10^{-4} Ω. Typical results [14] are 5×10^{-4} Ω. For patterned films, one of the best previously reported values at 77 K and 10 GHz is [13] 1×10^{-3} Ω.

ACKNOWLEDGMENT

This work was conducted under the auspices of the Consortium for Superconducting Electronics with full support by the Defense Advanced Research Projects Agency (Contract MDS972-9-C-0021). We thank A.J. Kelley and R.P. Konieczka for help with device fabrication, and G.L. Fitch for computer programming. We acknowledge J. Halbritter for numerous fruitful discussions concerning R_S.

REFERENCES

1. See for example Lyons WG, Bonetti RR, Williams AE, Mankiewich PM, O'Malley ML, Anderson AC, Withers RS, and Howard RE (in press) Proceedings of 1990 App. Superconductivity Conference
2. Klein N, Müller G, Piel H, Roas B, Schultz L,Klein U, Peiniger M (1989) Appl. Phys. Lett. 54: 757
3. Rubin DL, Green K, Gruschus J, Kirchgessner J, Moffat D, Padamsee H, Sears J, Shu QS, Schneemeyer LF, Waszczak JV (1988) Phys. Rev. B 38: 6538
4. Oates DE, Anderson AC, Mankiewich, PM (1990) J Superconductivity 3: 251
5. Sheen D, Ali S, Oates DE, Withers RS (unpublished)
6. Steinbeck JW, Holtham JH (unpublished)
7. Oates DE (unpublished)
8. Mankiewich PM, Scofield JH, Skocpol WJ, Howard RE, Dayem AH, Good E (1987) Appl. Phys. Lett. 51: 1753
9. Anderson AC, Oates DE, Slattery RL, Lyons WG, Solid State Research Report Lincoln Laboratory MIT 1990:1: 60
10. Anlage SM, Langly BW, Snortland HJ, Eom EB, Geballe TH, Beasley MR (in press) J Superconductivity
11. Oates DE, Anderson AC (in press) Proceedings of 1990 Appl. Superconductivity Conference
12. Mattick, RE (1969) Transmission lines for digital and communication networks. McGraw-Hill. New York
13. Inam A., Wu XD, Nazar L, Hegde MS, Rogers CT, Venkatesan T, Simon RW, Daly K, Padamsee J, Kirchgessner D, Moffat D, Rubin D, Shu SQ, Kalokitis D, Fathy A, Pendrick V, Brown R, Brycki B, Belohoubek E, Drabeck L, Gruner G, Hammond R, Gamble F, Larison BM, Bravman JC (1990) Appl. Phys. Lett. 56: 1178
14. Halbritter J (unpublished)
15. Newman N, Char K, Garrison SM, Barton RW, Taber RC, Emo CB, Geballe TH, Wilkens D (1990) Appl. Phys. Lett. 57: 520
16. Hylton TL, Beasley MR, (1989) Phys. Rev. B 13: 9042
17. Oates DE, Anderson AC, Shin BH (1989) MTT-S Digest: 627

Tl-Ca-Ba-Cu-O Thin Film Passive Microwave Devices

MICHAEL S. SCHMIDT, ROGER FORSE, ROBERT B. HAMMOND, MICHAEL M. EDDY, and WILLIAM L. OLSON

Superconductor Technologies Inc., 460-F Ward Drive, Santa Barbara, CA 93111-2310, USA

ABSTRACT

We report results on several passive microwave devices fabricated and tested using epitaxial thin films of $Tl_2CaBa_2Cu_2O_8$ grown on $LaAlO_3$. We report microstrip resonators with unloaded Q's of 15,000 at 80K and 4.83 GHz, and 1440 at 80K and 33 GHz. We report a superconducting 4.6 GHz bandreject filter with unloaded Q greater than 15,000 when operated at 77K.

INTRODUCTION

The first demonstration of superconductivity above liquid nitrogen temperature, 77K, is now nearly four years past. The microwave performance of the new high temperature materials has improved dramatically during that time.[1,2] In addition, the performance of these films in passive microwave devices has surpassed that of normal metals by one to three orders of magnitude. High Q (up to 100 times higher than can be achieved with normal metals) has been seen in stripline and microstrip thin film resonators at temperatures up to 77K and above.[3,4] In addition, there have been recent reports of superconducting thin film multi-pole bandpass filters with attractive performance up to these high temperatures.[5,6] Finally, these materials offer a number of advantages in microwave antennas.[7] Here we report the development of microstrip resonators at microwave and mm-wave frequencies, and the development of a thin film bandreject filter operating at 77K. In addition, we describe a 2.3 GHz oscillator using a thin film $Tl_2CaBa_2Cu_2O_8$ superconducting resonator and a low noise GaAs FET amplifier, operating at 77K. The design of this oscillator is presented, and the performance of the oscillator using the superconducting and gold resonators are compared.

THIN FILM SYNTHESIS

The films for the microstrip devices were made by our production process for $Tl_2CaBa_2Cu_2O_8$ thin films on $LaAlO_3$. We produce microwave thin films in high yield (>80%). Tl-Ca-Ba-Cu-O thin films were prepared by laser ablation deposition followed by post-deposition thermal processing. The amorphous precursor deposits were prepared by laser ablation because it is an effective method of retaining the stoichiometry of the target in the deposited film. The oxides were ablated at room temperature using a KrF excimer laser (249nm) onto chemically polished $LaAlO_3$ substrates. A laser pulse repetition rate of 10Hz was maintained during the deposition and the deposition time was varied to give the desired film thickness. The amorphous films were heated to high temperature (830-900°C) under controlled thallium and oxygen pressures to prevent excessive loss of Tl_2O_3. Final film thicknesses were ~9000Å. X-ray diffraction showed the films to be oriented with the c-axis of the superconductor perpendicular to the substrate; and, in addition, sharp rocking curves (as sharp as 0.30°) indicated an epitaxial orientation. Epitaxy was substantiated by selected area electron channeling, which showed electron channeling over the whole of the substrate, with the 100 orientation of the superconductor aligned with the 100 of the substrate.

DEVICE FABRICATION

High temperature superconductor thin films were patterned using a wet chemical etch process. The films were rinsed with a series of organic solvents (toluene, acetone, methanol, isopropanol) and dried at 135°C for 30 minutes immediately prior to photoresist deposition. A positive photoresist (Shipley 1713) was spun onto the $1cm^2$ wafer using a spin rate of 5000 rpm for 30 seconds to give a photoresist thickness of 1.3μm. The resist was oven baked at 95°C for 30 minutes to remove casting solvents.

A Kasper Mask Aligner was used to perform the contact lithography. The photoresist layer was exposed for 30 seconds and developed by soaking for 60 seconds in an aqueous hydroxide solution (Shipley MF 319). The etched resist was then examined optically for pinholes/defects. If defects were observed, the photoresist was removed using acetone and the photoresist redeposited. Otherwise, the uncoated film surfaces were chemically removed by etching for 20-30 seconds in a dilute HCl solution (1:150 dilution of concentrated HCl in 18-MΩ deionized water). The film was then rinsed with DI water and dried using dry, hepafiltered nitrogen. Residual resist was removed using acetone. Gold metallization of the film was accomplished using a combined sputtering, lift-off process. Shipley 1713 photoresist was patterned for lift off using the process described above. The exposed HTSC film surfaces were coated with 2000Å of gold using a 1" diameter US sputtering gun. The target Au purity was 99.99%. The sputtering gun was located 3.5" from the substrate and operated using a gun current of 0.5 amps and an Ar partial pressure of 2 x 10^{-2} torr. The gun was presputtered for 15 seconds behind a closed shutter prior to deposition on the substrate. A gold deposition rate of 30Å/sec was typically observed. Liftoff of the undesired Au film and photoresist was performed by soaking the wafer in acetone for 10-20 minutes.

4.83 GHz MICROSTRIP RESONATOR

Figure 1 shows the layout of a linear, end-coupled microstrip resonator developed at STI. This device employed a superconducting ground plane on a separate substrate. The $LaAlO_3$ substrate is 0.020" thick. The fundamental resonance occurs at 4.83 GHz. It was developed for the High Temperature Superconductivity Space Experiment sponsored by the Naval Research Laboratory. The package was designed for ruggedness and repeatability of device performance. The device is permanently bonded in the package. It provides glass-to-metal seals in its SMA connectors and a laser-welded cover. Each finished device was leak checked and temperature cycled several times between room temperature and 77K. The thermal cycling did not affect the device performance.

The packaged microstrip resonator was cooled for measurements in a closed cycle refrigerator. An HP8340 Synthesized Sweeper and HP8757C Network Analyzer were used to make transmission measurements. Various power amplifiers were used to set appropriate signal levels at the device under test and at the test instrumentation ports. Measurements were made on the first resonant peak over a 40 dB dynamic range varying input power from -40 dBm to 0 dBm. Figure 2 is a plot of measured insertion loss at an input power of 0 dBm to the resonator. Loaded Q data were extracted from 3-dB bandwidth measurements on the first resonance. Measurement of insertion loss at resonance allowed calculation of coupling Q and the unloaded Q of the resonator. The measured loaded Q's for the Tl-Ca-Ba-Cu-O microstrip resonator are shown in Figure 3. The measured insertion loss (at -10 dBm input power) was 19 dB.

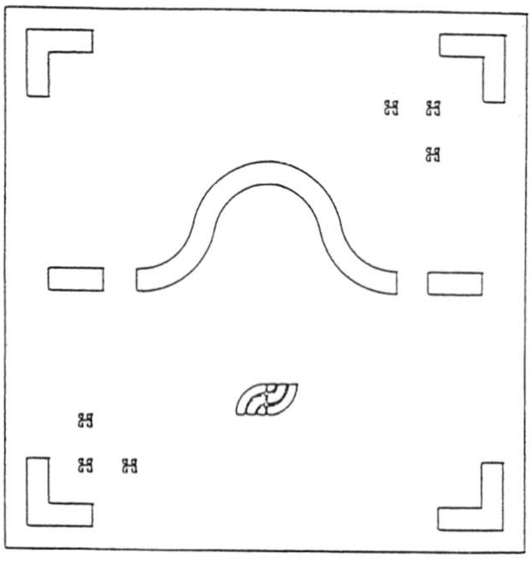

Figure 1
Mask layout for 4.83 GHz microstrip resonator. Angled shapes in the four corners are for aligning the 1 cm substrate with the mask. The mask also includes six small structures for determining Jc in the film.

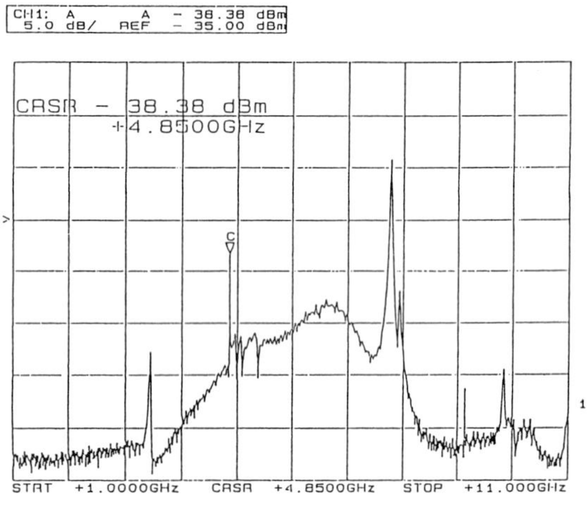

Figure 2
Measured through loss spectrum from 1 to 11 GHz for the 4.83 GHz microstrip resonator.

Figure 2 shows the through loss spectrum (S_{21}) taken at STI on the device. The fundamental resonance is at 4.85 GHz and the first harmonic at 9.1 GHz. The resonances at 3.5 and 7.8 GHz are due to package moding at these frequencies, and are due to the large size of the package. (It accommodates up to three 1 cm substrates.) The package resonances are not in the operating frequency ranges defined in the NRL specification, and do not affect resonator performance. They are sufficiently far away from the resonator fundamental to preclude any coupling to these modes. The device showed excellent Q, more than 40 times higher than gold, and excellent power handling at 77K. Figure 3 contains the final STI test data taken immediately before shipment to NRL and a direct comparison between loaded Q measured at STI and NRL on this device. The agreement is well within experimental error. Similar devices have undergone extensive temperature cycling to 77K and shake testing for space qualification with no change in microwave performance.

33 GHz RING RESONATOR

We developed a resonator for studies above 30 GHz based on a device first developed at NASA Lewis[ref]. The substrate size was 1 cm X 1 cm X 0.025 cm thickness. The layout for the resonator is shown in Figure 4. The resonator is a gap-coupled microstrip ring resonator. (As shown, two complete resonant devices were fabricated on each wafer.) Two resonators were built, one with a superconducting center conductor and Au ground plane, and one with all Au for comparison. Both versions were modeled on the Sonnet Software finite-difference-method electromagnetic modeling software. The ring resonator linewidth is 0.006", with a fundamental resonance at 11 GHz. It is coupled with 0.002" gaps. All measurements were performed in a microwave package with APC 2.4 mm connectors, while the package was held at 77K. The through-loss, S_{21}, spectrum is shown in Figure 5. The data clearly show the three resonances at 11, 22, and 33 GHz. At 33 GHz the unloaded Q was calculated from the measured loaded Q and the magnitude of S_{21} at resonance. As shown in Figure 6 the device showed an insertion loss of 8.6 dB on resonance at 33.6 GHz. The measured loaded Q was 905 and the calculated unloaded Q was 1440. We also measured an identical resonator fabricated in Au and measured the unloaded Q at 550. Using the Sonnet Software program we modeled the performance of this Au resonator and obtained excellent agreement with our measured data. The program predicted unloaded Q of 576 for the Au resonator at 77K.

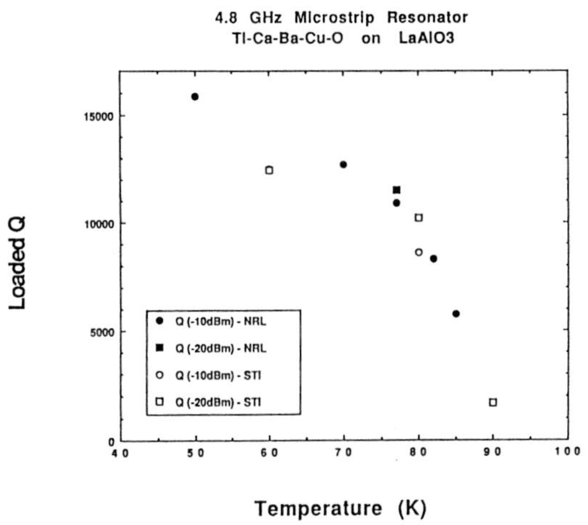

Figure 3
Measured loaded Q on the 4.83 GHz microstrip resonator. Data are taken at two input powers, -10 dBm and -20 dBm, and at two facilities. Solid circles and squares are data from the Naval Research Lab. Open circles and squares are data taken at STI.

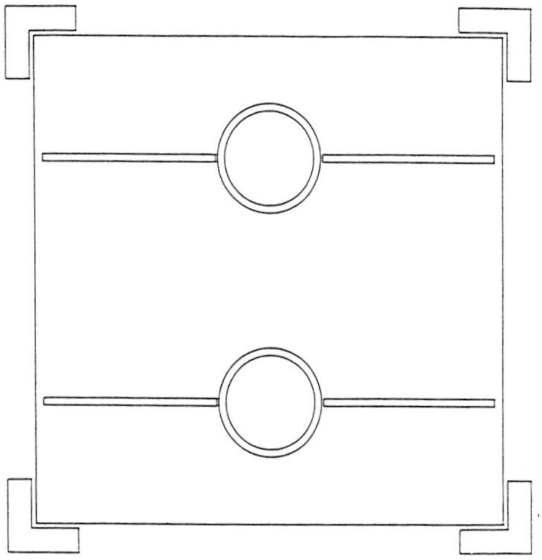

Figure 4
Mask layout for the 33 GHz microstrip resonator. Two complete test structures are formed on a single 1 cm substrate.

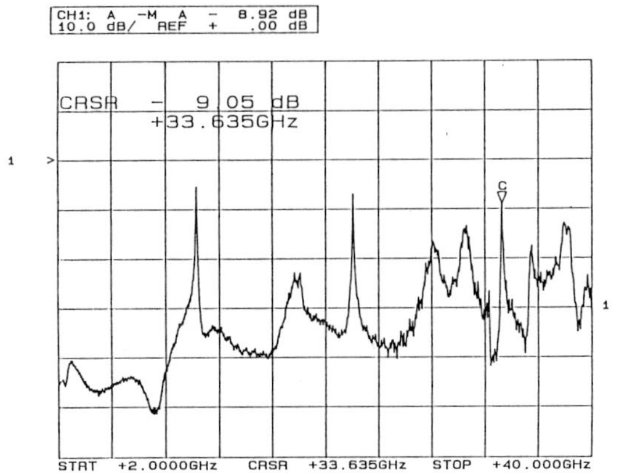

Figure 5
Measure through loss from 2 to 40 GHz on the 33 GHz microstrip resonator. Resonances are seen at 11, 22, and 33 GHz.

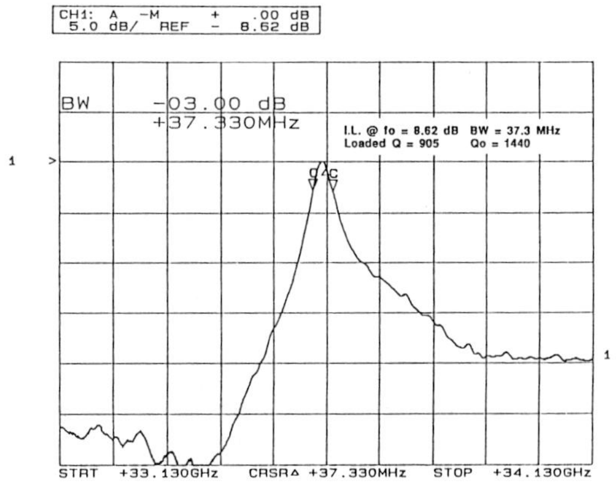

Figure 6
Close up of the through loss in the region of the 33GHz resonance. Loaded Q is 905; unloaded Q is 1440. The measurement temperature is 80K.

4.6 GHz BAND REJECT FILTER

In addition to our work on capacitively coupled resonators (that perform as single-pole bandpass filters), we have also developed inductively coupled resonators that perform as single-pole band reject filters. The device we report here is a single-resonator band-reject filter designed to be as steep as possible while maintaining low loss near the band edges. It is an inductively coupled half-wave-length resonator that provides a reflective short at its resonance, but has no effect off-resonance. The circuit is critically coupled to give it a large shape factor.

As seen in the device layout in Figure 7, the coupling length is a small percentage of the total resonator length and may be adjusted to raise or lower the bandwidth of the filter. It is inductively coupled in the middle of the resonator to make the analysis more accurate than would be possible with end-coupled resonators. The lossless theoretical performance of the filter is shown in Figure 8, and the depth of the null is infinite. The real finite Q of the resonator gives a finite depth and width, which gives a useful measure of microwave loss in the superconducting film. The measured data, shown in Figure 9, shows the good out-of-band performance and the sharp in-band rejection expected. The resonator loaded Q was 12,422 and it provided 17 dB of rejection 1% from the bandedge.

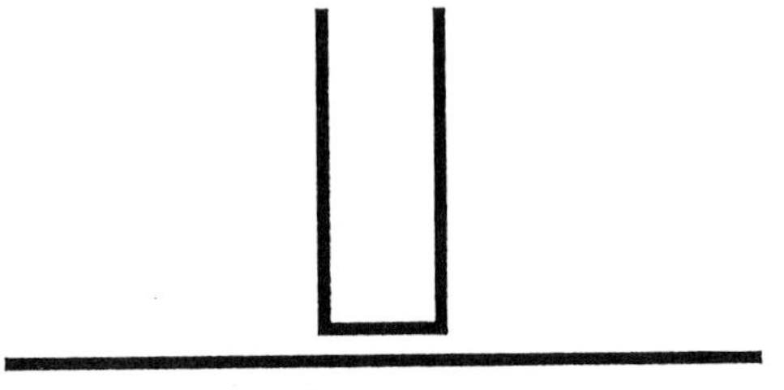

Figure 7
Mask Layout for the 4.6 GHz bandreject filter.

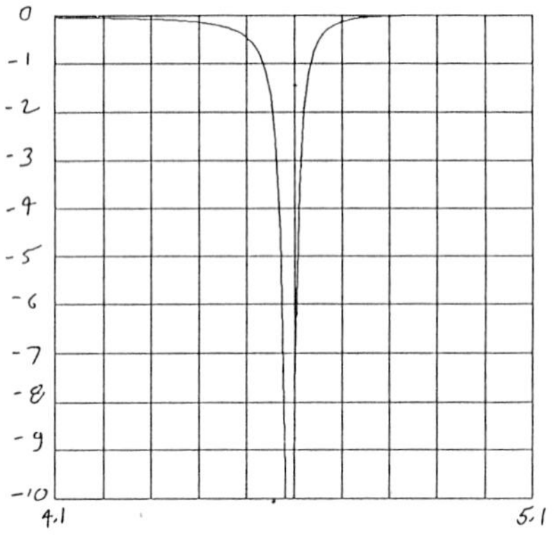

Figure 8
Calculated, ideal response for the 4.6 GHz bandreject filter.

The through line also gives a good measure of the interconnect and general connector losses of the package which is useful in troubleshooting as well as the measurement of the transition temperature and width of the HTSC material. We used this structure to measure loss of our HTSC material and have seen good agreement with the more traditional end-coupled resonator measurements. This topology would be ideal for applications requiring filtering a strong interfering signal near the desired band or rejecting a fixed Local Oscillator from a frequency conversion device.

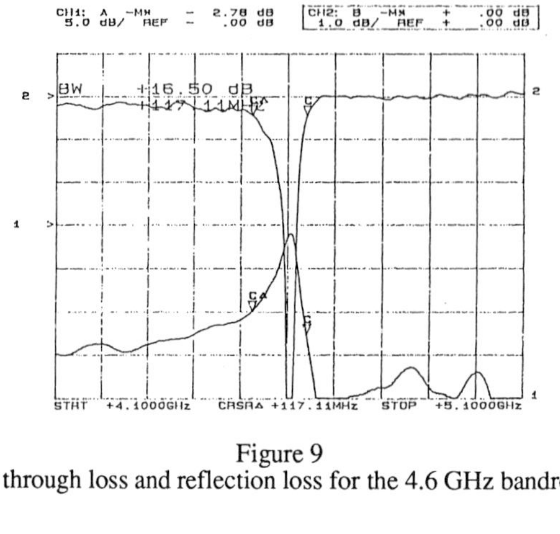

Figure 9
Measured through loss and reflection loss for the 4.6 GHz bandreject filter.

SUMMARY AND CONCLUSIONS

The superconducting passive microwave devices reported here demonstrate that performance 50 times better than gold can be achieved in devices operated at practical power levels at liquid nitrogen temperature, 77K. They demonstrate this in a microstrip configuration, the most thoroughly understood transmission line structure while also allowing easy access to the circuit for tuning. These results show the potential for the application of HTSC films in complex passive microwave structures that can reduce the size and weight and improve performance in satellite communications systems and other high performance microwave systems.

Perhaps the most significant progress shown in our program is that these devices were made with a *production film synthesis process* and fabricated by *production patterning and metallization techniques*. The overall device yield of these processes is higher than 50%. This overall yield permits us to offer these devices on the commercial market. Another significant development in this program has been the rugged and hermetic packaging. The packages use glass to metal seals for microwave transitions and the covers are laser welded. Devices have successfully undergone repeated temperature cycling, as well as shake testing for space qualification with no change in device performance.

ACKNOWLEDGEMENTS

The authors thank the entire technical staff of STI for producing the superconducting thin films and performing the patterning and metallizations needed for the passive microwave devices reported here. In addition, the authors are grateful to Ed Topacio of Avantek for his skillful testing of the resonator and oscillators. We acknowledge the support of the Naval Research Laboratory and the Defense Advanced Research Projects Agency.

REFERENCES

[1] G. Muller, "Microwave Properties of High Temperature Superconductors", Proceedings of the 4th Workshop on RF Superconductivity, KEK Tsukuba-Shi, August, 1989.

[2] R. B. Hammond, G. V. Negrete, L. C. Bourne, D. D. Strother, A. H. Cardona, and M. M. Eddy, "Epitaxial $Tl_2CaBa_2Cu_2O_8$ thin films with low 9.6 GHz surface resistance at high power and above 77K," Applied Physics Letters, Vol. 57, pp. 825-827, 1990.

[3] A.A. Valenzuela and P. Russer, "High Q Coplanar Transmission Line Resonator of $YBa_2Cu_3O_{7-x}$ on MgO," Applied Physics Letters, Volume 55, pp. 1029-1031, Sept., 1989.

[4] R.B. Hammond, G.V. Negrete, M.S. Schmidt, M.J. Moskowitz, M.M. Eddy, D.D. Strother, and D.L. Skoglund, "Superconducting Tl-Ca-Ba-Cu-O Thin Film Microstrip Resonator and its Power Handling Performance at 77K," 1990 IEEE MTT-S International Microwave Symposium Digest, pp. 867-869, 1990.

[5] W. G. Lyons, R. R. Bonetti, A. E. Williams, P. M. Mankiewich, M. L. O'Malley, A. C. Anderson, R. S. Withers, and R. E. Howard, "High-Tc Superconductive Microwave Filters," Proceedings of the Applied Superconductivity Conference, Snowmass, 1990, to be published in IEEE Transactions on Magnetics.

[6] S. H. Talisa, J. Billing, R. Brown, D. C. Buck, M. A. Janocko, C. K. Jones, B. R. McAvoy, C. Moskowitz, J. Talvaccio, and G. R. Wagner, "Microwave Superconducting Filters," Proceedings of the Applied Superconductivity Conference, Snowmass, 1990, to be published in IEEE Transactions on Magnetics.

[7] R. C. Hansen, "Superconducting Antennas," IEEE Transactions on Aerospace and Electronic Systems, Vol. 26, pp. 345-355, 1990.

[8] D. E. Oates, A. C. Anderson, and B. S. Shih, "Superconducting Stripline Resonators and High-Tc Materials," Technical Digest of the IEEE Microwave Symposium, Dallas, pp. 627-630, 1990.

In Situ Fabrication of Reproducible Josephson Tunnel Junctions Employing Co-evaporated YBaCuO Films

IENARI IGUCHI[1], ZHONGMIN WEN[1], HIROAKI OHTAKE[1], MASAKO NASU[1], and KOZO NAKAMURA[2]

[1] Institute of Materials Science, University of Tsukuba, Tsukuba, 305 Japan
[2] Materials Laboratory, Komatsu Ltd., 1200 Manda, Hiratsuka, Kanagawa, 254 Japan

ABSTRACT

The in-situ fabrication of reproducible and controllable YBaCuO(YBCO)/barrier/Pb planar Josephson tunnel junctions utilizing an electron-beam and thermal heater coevaporation chamber equipped with a shadow mask system is presented. As an insulating barrier of tunnel junction, both the artificial and native barriers are used. The tunnel resistance exhibited an exponential behavior against the barrier thickness. The results on the quasiparticle tunneling and Josephson tunneling are shown. The observed Josephson characteristics were appreciably different between the junctions with the c-axis and a(b)-axis oriented YBCO films. The anomalous temperature dependence of Josephson critical current is also reported.

KEY WORDS: YBCO tunnel junction, gap structure, Josephson effect

INTRODUCTION

The fabrication of tunnel junctions using high T_c superconductors is quite important for developing superconducting electronics which may be used at 77K. And the correct information on the gap structure provides the density of states of high T_c superconductors directly related to the superconductivity mechanism. So far the fabrication of high T_c tunnel junctions relied on the accidental formation of a junction in the various junction geometries such as point-contact, break junction, thin film tunnel junction with native barrier etc. The measurements yielded a variety of results [1,2] and the genuine density of states of high T_c superconductors is still unknown. In order to obtain them by means of tunneling spectroscopy, it is necessary to fabricate the tunnel junctions in a more sophisticated way. Quite recently, the reproducible planar YBCO tunnel junctions have been fabricated by an all-in-situ technique in our group [3-6]. The thickness of artificial MgO barrier was typically 2-6nm and the used YBCO films were strongly c-axis oriented. Both the quasiparticle and Josephson tunneling phenomena were observed.

The purpose of this paper is to present the in situ fabrication of YBCO/barrier/Pb Josephson tunnel junctions and their characteristics. Both the c-axis and a(b)-axis oriented YBCO films were used. The tunnel barrier was either an artificial one (MgO, Al_2O_3, Y_2O_3) or a native one. The samples were fabricated by the all-in-situ technique using a coevaporation vacuum chamber equipped with a changeable shadow mask system [3-6].

EXPERIMENTAL

Fig. 1 shows the schematic of a vacuum chamber in which two electron-beam guns were used for Y and Ba and two thermal heaters for Cu and Pb [7]. The distance between the evaporation sources and the substrate was about 70cm and the contact gap between the metal mask and the substrate was a few tenth of mm. For the growth of the c-axis oriented YBCO films, MgO(001), $SrTiO_3$(001), $LaGaO_3$(001), $NdGaO_3$(001) and $LaSrGaO_4$(001) substrates were used. On the other hand, for the growth of the a(b)-axis oriented YBCO films, $LaSrGaO_4$(100) substrates were used. The thickness of YBCO films was 200-250nm.

$LaSrGaO_4$ is a new substrate material with the tetragonal crystal structure of its lattice paramters $\underline{a}$ = 0.3843nm and $\underline{c}$ = 1.2681nm. The $\underline{a}$ parameter of

$LaSrGaO_4$ lies just in between the a and b parameters of YBCO material. The dielectric constant is 25.0 along the c-axis and 24.6 along the a-axis direction. No twin boundary exists in this material. Fig. 2 shows the full width at half-maximum (FWHM) of rocking curves for $LaSrGaO_4$ together with the other materials. The observed FWHM was found to be close to that of Si and better than those of $SrTiO_3$ and $LaGaO_3$.

The YBCO films on $LaSrGaO_4$(100) substrates were deposited at about 600°C. Fig. 3 is the resistivity vs temperature curves for these films. T_c was 85K and the behavior above T_c was quite metallic. From X-ray diffraction method, the presence of small fraction of the c-axis oriented region was recognized.

The MgO and the other oxide barrier materials were evaporated from the Y side electron-beam source with 6 point crucibles. The thickness of insulating barrier was controlled by a calibrated quartz monitor. In case of the growth of native barrier, the in-situ annealing technique of a YBCO film in the oxygen gas at low temperatures(400-500°C) was introduced. The counter-electrode Pb was deposited using a different metal mask.

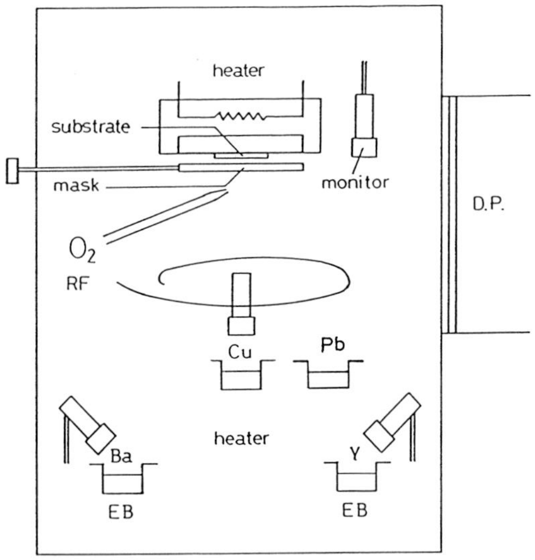

Fig. 1. Schematic of a coevaporation vacuum chamber.

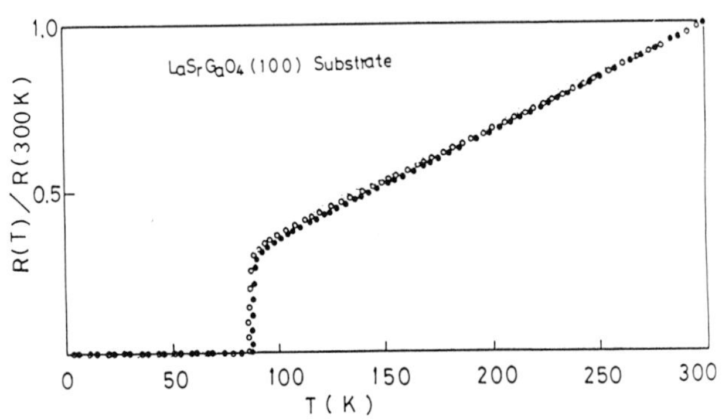

Fig. 3. Resistance vs temperatuare curves for the YBCO films on $LaSrGaO_4$ (100) substrates.

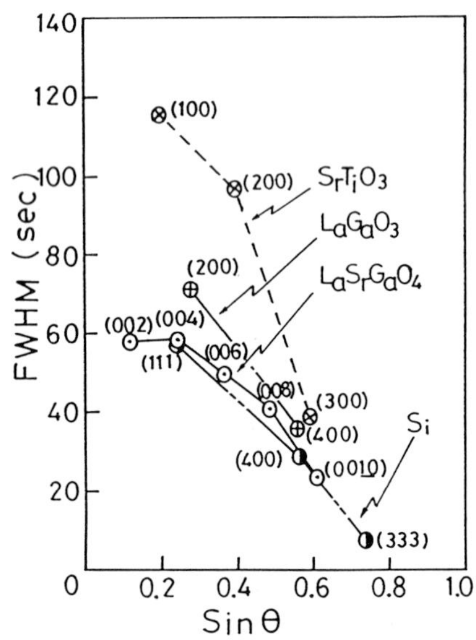

Fig. 2. Full width at half-maximum (FWHM) of rocking curves for various substrate materials. The numbers in the figure indicate the measured crystal planes.

The fabricated tunnel junction had a cross-type geometry with the effective tunneling area of 0.2 x 0.2 mm^2. Fig. 4 depicts the optical photograph of the fabricated junction. The diffusion at the edge was found to be small.

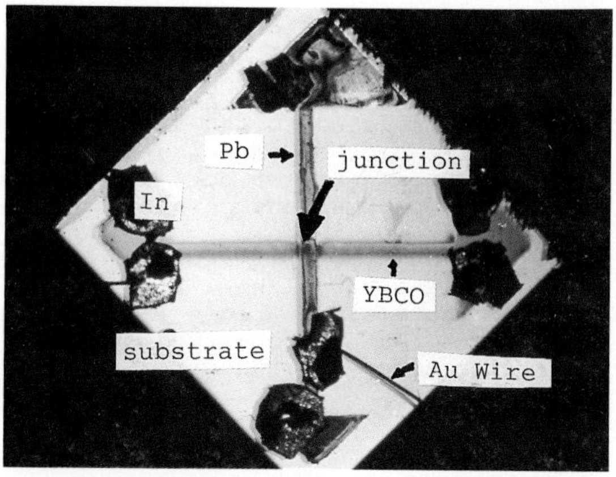

Fig. 4. Optical photograph of the fabricated YBCO/MgO/Pb tunnel junction.

RESULTS AND DICUSSION

The tunnel resistance of a junction with the artificial MgO barrier varied exponentially with the barrier thickness as reported previously [3,6], indicating that the observed data were due to the tunneling processes. For the junctions with the relatively high tunnel resistance ($>0.5\Omega$), the quasiparticle tunneling behavior was observed.

Fig. 5 shows the temperature dependence of the gap structures for the YBCO/MgO/Pb junction with the c-axis oriented YBCO film on the $LaGaO_3$(001) substrate. The tunneling is expected to occur along the c-axis direction of YBCO crystal in this case. The observed gap structures were qualitatively similar among the junctions with the c-axis and a(b)-axis oriented YBCO films, although the gap-edge structure for the latter was more enhanced. The results are rather puzzling since it is natural to expect the different gap structure along the c-axis and a(b)-plane tunneling. Josephson tunneling behavior was, however, found to be quite different between the c-axis and a(b)-axis oriented films.

The YBCO gap opening structure was seen at 18-20mV and the pronounced Pb gap structure was seen below T_c(Pb). Above T_c(Pb), the zero bias anomaly (conductunce dip around zero bias) was observed. The gap structure was significantly broad. The zero bias anomaly was always found for the in-situ fabricated tunnel junctions.

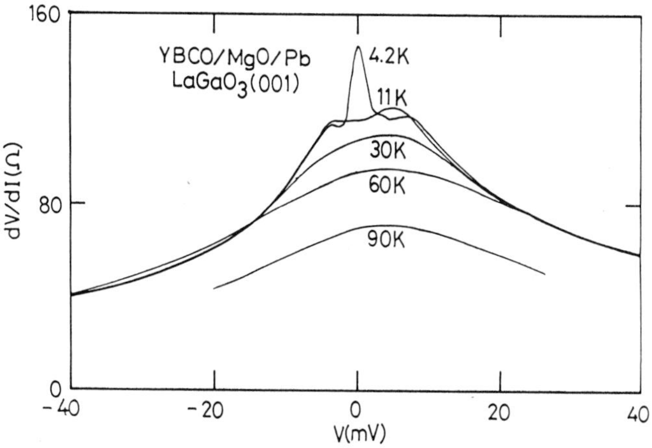

Fig. 5. Differential tunnel resistance as a function of bias voltage for a YBCO/MgO/Pb junction at different temperatures.

The results in Fig. 5 are in a marked contrast to those observed by the other groups utilizing the ex-situ barrier formation technique [8-10]. In the latter case, contrary to Fig. 5, the conductance peak around zero bias voltage was found above T_c(Pb). The reason for this discrepancy is not clear at the present time. It seems, however, that the zero bias anomaly is not likely to be directly related to the 90K superconductivity. We note that the native barrier formation by exposing a YBCO film in air for a few tens of minutes within the framework of our junction-fabrication technique also yielded the conductance peak around zero bias voltage instead of the conductance dip.

In Fig. 6, the tunneling I-V characteristics for the tunnel junctions with different barrier materials (MgO, Al_2O_3, Y_2O_3) of 3nm thick are shown. It is pointed out that the observed characteristics were almost the same irrespective of barrier materials, assuring that the data in Fig. 5 was not due to the barrier effect.

The Josephson effect was observed for the junctions with relatively low tunnel resistance (0.01- 0.2Ω). The observed characteristics were significantly different between the junctions with the c-axis and a(b)-axis oriented YBCO films. In the case of the c-axis tunneling, the I-V curves rather resembled those of S/N/S type junctions. The nonlinear output voltage was only about a few tens of microvolts. In case that the a(b)-axis tunneling was involved, the situation was quite different. Fig. 7 depicts the observed I-V curves at 1.8K for the junction fabricated on a $LaSrGaO_4$(100) substrate. The I-V curve (solid line) was highly nonlinear and a premature switching behavior accompanying the voltage output of about 1mV was seen. Followed by this jump, the S/I/N type Pb gap structure with low leakage current appeared. The I-V curve was nonhysteretic in this case. The dashed line in Fig. 7 corresponds to the I-V cuve of the junction with the c-axis oriented YBCO film at 1.8K. The

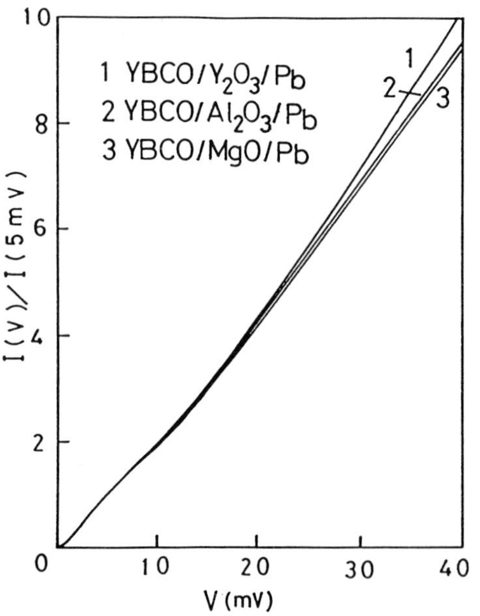

Fig. 6. I-V characteristics of YBCO/barrier/Pb tunnel junctions with different barrier materials at 4.2K. The barrier thicknesses were 3nm for all materials.

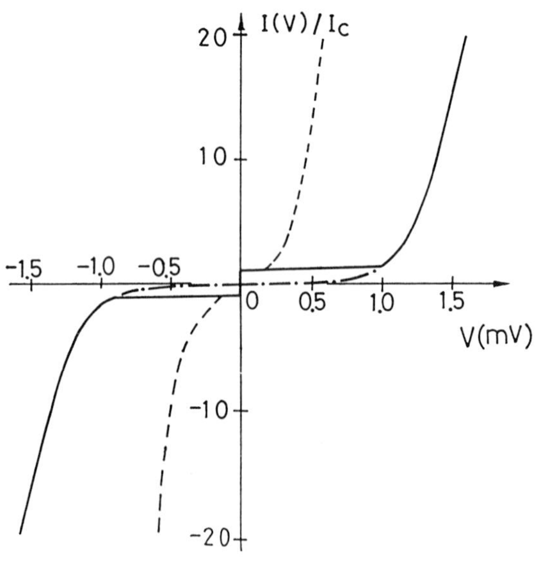

Fig. 7. I-V characteristics of YBCO/ native barrier/Pb Josephson junctions at 1.8K on the $SrTiO_3$(100)substrate (dashed line), on the $LaSrGaO_4$(100) substrate without magnetic field (solid line) and with magnetic field (dash-dotted line). I_c was 0.1mA for the dashed line and 0.18mA for the solid line.

difference between these two curves is evident. Although the YBCO film on the $LaSrGaO_4$(100) substrate contained the small fraction of c-axis oriented region, the observed behavior is considered to arise from the a(b)-axis tunneling contribution since such behavior was never observed in the junctions with the purely c-axis oriented YBCO films. The hysteretic behavior was also observed, but the qualitative feature of the I-V curves was similar to Fig. 7. Josephson current ranged from 0.2mA to 3mA at 4.2K. With applying the magnetic field of about 2mT, Josephson current disappeared. Although the switching behavior was apparently blocked by the Pb gap structure in Fig. 7, it was not directly related to the Pb gap since its temperature dependence was different from that of Pb gap.

The I-V characteristics exhibited a series of Shapiro steps under the microwave irradiation of 9GHz. The Shapiro steps were observable up to the 19th one for the c-axis oriented films. The dependence of current step height for the zeroth and first Shapiro steps on the microwave voltage amplitude almost agreed with those expected from the conventional theory [5].

The dependence of Josephson critical current on the magnetic field showed rather complicated behavior. The periodic Frauhofer-like patterns were observed for the junctions with both the c-axis and a(b)-axis oriented films. They were either symmetrical or asymmetrical around the magnetic-field axis, which seemed not to be related to the YBCO crystal orientation. In Fig. 8, two examples of Fraunhofer-like patterns are shown. Fig. 8(a) was the data for the junction with the c-axis oriented film, whereas, Fig. 8(b) for the junction with the a(b)-axis oriented film. The phenomena seem to depend on the microscopic details of Josephson tunneling behavior in the real junction with some inhomogeneities or imperfectness. The observed period among many samples was from a few times to 8 times longer than the calculated one. The estimated Josephson penetration depth ranged from 0.1mm to about 1mm depending on the Josephson critical current density.

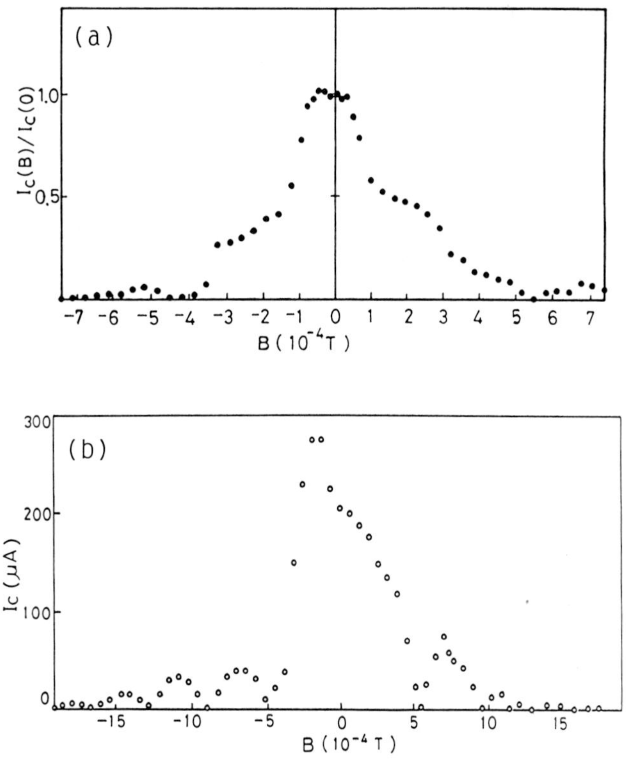

Fig. 8. Dependence of Josephson critical current on the applied magnetic field for YBCO/native barrier/Pb junctions. (a): symmetrical example, (b): asymmetrical example.

Finally, it is very important to point out that the temperature dependence of Josephson critical current was quite anomalous. In Fig. 9, the normalized Josephson critical current against temperature for several junctions are shown. I_c first increased rapidly with reducing temperature below T_c(Pb), then decreased at lower temperatures, forming a broad maximum around 5K [11]. The behavior is hardly expected from the conventional theory. The phenomenon cannot be interpreted at the momoent, but may reflect the special situation of Josephson tunneling between a BCS s-wave superconductor and a high T_c cuprate superconductor with the possible different superconductive mechanism.

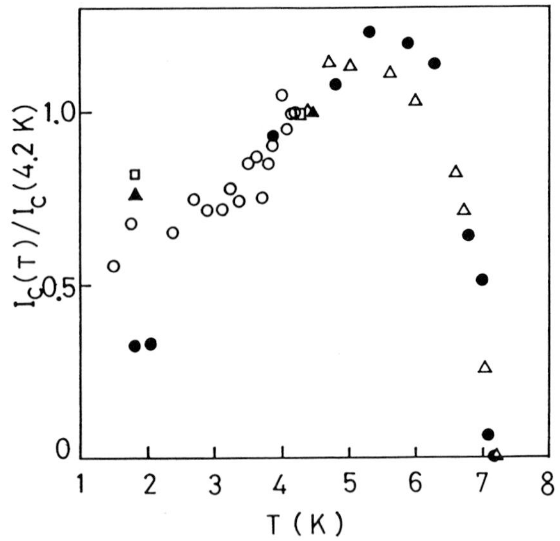

Fig. 9. Temperature dependence of Josephson critical current. The I_c(4.2K) values were 0.35mA(△), 2.2mA(●), 1.0mA(○), 0.22mA(▲) and 0.60mA(□).

In conclusion, we reported the in-situ fabrication of reproducible YBCO/barrier/Pb Josephson tunnel junctions employing a coevaporation vacuum system. The observed characteristics for quasi-particle tunneling was apparently different from that expected from the BCS theory. The gap value was 18-20meV and $2\Delta/k_BT_c \simeq 5$. The zero bias conductance anomaly and the S/I/N type Pb gap structure was observed. The zero bias anomaly seems not to be directly related to the 90K superconductivity phase since it disappeared above 55K. Josephson tunneling was observed for the junctions with low tunnel resistance and the I-V curves were different between the c-axis and a(b)-axis oriented YBCO films. The Shapiro steps under microwave irradiation of 9GHz and the Fraunhofer-like patterns under the applied magnetic field were similarly observed for both type of films. The anomalous temperature dependence of Josephson critical current was also found, which may be quite important for clarifying the superconductivity mechanism of high T_c superconductors.

REFERENCES

1. Gray KE, Hawley ME, Moog ER (1987) Novel Mechanisms of Superconductivity, eds. Wolf SA, Kresin VZ (Plenam, New York, 1987):p.611.
2. Kirtley JR (1990) Intern. Journ. of Mod. Phys. B 4: 201.
3. Furuyama M, Iguchi I, Shirai K, Kusumori T, Ohtake H, Tomura S, Nasu M (1990) Jpn. J. Appl. Phys. 29: L459.
4. Iguchi I, Furuyama M, Kusumori T, Shirai K, Tomura S, Nasu M, Ohtake H (1990) Jpn. J. Appl. Phys. 29: L614.
5. Iguchi I, Wen Z, Furuyama M (1990) Physica B 165&166: 73.
6. Iguchi I (1990) Ext. Abstr. of The Second ISTEC Workshop on Superconductivity, Kagoshima; p.125.
7. Furuyama M, Iguchi I (1989) Ext. Abstr. of 2nd Workshop on High-Temperature Superconducting Electron Devices,Shikabe (R&D Association for Future Electron Devices, Tokyo, 1989): p.105.
8. Geerk J, Xi XX, Linker G (1988) Z. Phys. B 73:329.
9. Gurvitch M, Valles JM, Cucolo AM, Dynes RC, Garno JP, Schneemeyer LF, Waszczak (1989) Phys. Rev. Lett. 63: 1009.
10. Kwo J, Fulton TA, Hong M, Gammel PL (1990) Appl. Phys. Lett. 56:788.
11. Wen Z, Iguchi I, Nakamura K (1990) Appl. Phys. Lett. (submitted).

Electromagnetic Wave Detections by High-T_c Superconducting Active Antennas

T. OHNUMA, T. KUROKO, H. TOMI, M. SHIMEGI, Y. TANAKA, and K. YOSHIDA

Department of Electrical Engineering, Tohoku University, Sendai, 980 Japan

ABSTRACT

Several types of high Tc superconducting active antennas for receiving electromagnetic waves (for examples, the frequency of which is 9.55GHz and 474THz) were confirmed experimentally with a use of the active properties of the high Tc superconductors. The superconducting active antennas consist of thick and thin films on MgO-substrates. Fundamental properties of the superconducting active antennas (the polarization effects of the receiving sensitivity and the directivity of the receiving antenna, the detectivity of terahertz light wave, and the differences of thick and thin film types) were experimentally confirmed. Furthermore, improvements of the radiation efficiency of superconducting folded antennas were shown.

KEYWORDS: Superconducting active antennas, Polarization, Terahertz light waves, Superconducting folded antenna

INTRODUCTION

Electromagnetic wave detections by superconductors have been investigated by several structures, for examples, SIS structures, Josephson structures, et al. High Tc superconducting active antennas in this report, which use active properties of the high Tc superconductors, are receiving antennas of several kinds of electromagnetic waves. The superconducting active antennas consist of thick and thin films of YBaCuO-type on MgO-substrates. The detectivity and directivity of electromagnetic waves by the superconducting active antennas are investigated. Radiation efficiency of a superconducting folded antenna, which consists of a bulk-type superconductor, is also reported experimentally.

EXPERIMENTAL

Superconducting thick films (the thickness ~0.1mm) and bulks (the thickness ~1mm) for antennas were produced by usual methods, and the superconducting thin films (the thickness ~1μm) were produced by DC magnetron plasmas. Those superconductors were remade for several kinds of structures having appropriate structures for antennas. The typical critical temperatures of the superconducting antennas are 78K for thin films, 80K for thick films and 90K for bulk types, respectively.

RESULTS AND DISCUSSIONS

(A) Electromagnetic Detections by Superconducting Active Antennas

(A-1) Polarization Effects of Electromagnetic Wave Detections

Typical superconducting active antenna is of rectangular type as shown in Fig.1. Four electrodes for application of the direct current and detection of the direct resistance are also shown. When electromagnetic waves (f= 9.55GHz) are radiated into the superconducting (SC) active antenna, the resistance (DC voltage) in the SC active antenna is produced. With the measurements of the resistances, one can know the coming electromagnetic waves. For this detectivity of the SC active antenna, the polarization of the radiated electromagnetic waves is very important. The typical experimental results are shown in Figs.1 and 2. In these experiments, the polarization of the electromagnetic wave is fixed. The directions of the electric and magnetic components of the radiated electromagnetic waves are directed as shown in Figs.1 and 2 with respect to the SC active antenna. Figures 1 and 2 show the directive sensitivity of the SC active antenna. The angle $\Theta=0$ indicates for the case that the electromagnetic wave is radiated perpendicular to the plane of the SC active antenna. The differences of the directivities of the SC antennas are clearly shown according to the polarizations of the electromagnetic waves. These experimental results indicate an important fact that the sensitivity of the superconducting active antennas is greatly affected by the direction of the magnetic component of the radiated electromagnetic waves.

(A-2) Structure Effects of the Detectivity

Controlled patterns (line types and meanderline types et al) of the superconducting active antennas make different I-V characteristics of the SC active antennas (I:DC current, V:DC voltage). These controlled I-V characteristics result in the improvement (control) of the detective sensitivity of the SC active antennas. Typical experimental data of the improvement of the detectivity of the SC active antennas are shown in Fig.3. Polarization of the radiated electromagnetic wave is the same as case of Fig.1 The detective angle of the SC antenna is varied. Radiated power P from the horn is indicated in the abscissa. The effective power Pe on the SC antenna area can be estimated as PxA/H (A:antenna area, H:horn area). The value of A/H is 0.9×10^{-3} in this case.

(A-3) Detective Properties of SC Active Antenna of Thin Film Types

The superconducting active antennas in sections of (A-1) and (A-2) are thick films type. The SC active antennas of thin films type, which are produced by DC magnetron sputtering, are described in this section. The typical properties of the SC antennas are shown in Figs.4 and 5. The dependencies of the detectivity of the SC active antennas on the radiated electromagnetic power and the input DC current are shown. The effective power Pe on the SC active antenna area is 0.15×10^{-3}P.

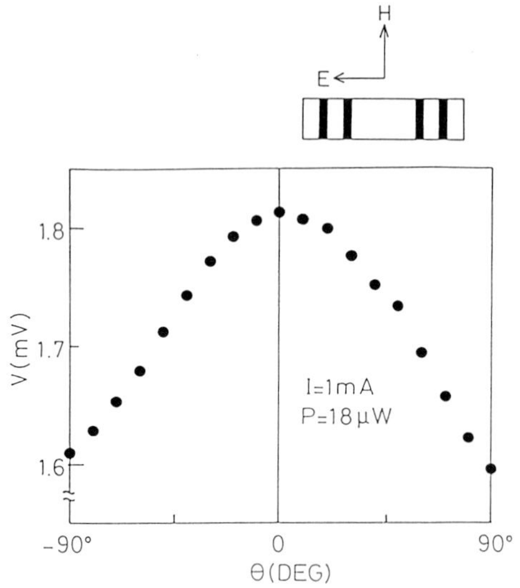

Fig.1. Directive sensitivity of SC active antenna for polarized electromagnetic wave (Polarization-E)

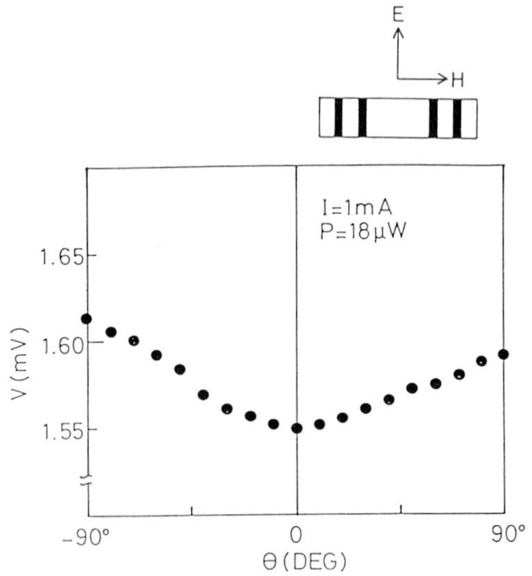

Fig.2 Directive sensitivity of SC active antenna for polarized electromagnetic wave (Polarization-H)

(A-4) Detections of Terahertz Light Waves

Detective properties of Terahertz light waves by the superconducting active antennas are described in this part. As the source, 474THz light waves is used. The typical experimental results on the detectivity of the THz light wave are in Fig.6. The sensitivity of the SC active antenna is shown corresponding to the I-V characteristics, which are controlled by the controlled current patterns of the SC active antennas. The improvement of the sensitivity is obtained by the control of the I-V characteristics.

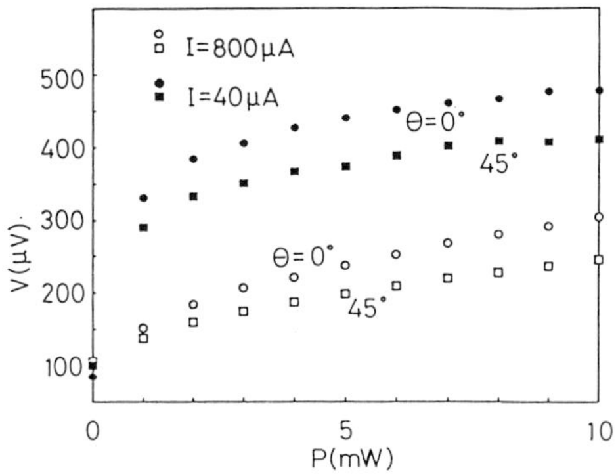

Fig.3 Improved sensitivity of SC active antenna with controlled patterns (solid ones), Pe(for antenna area)=0.9×10^{-3}P

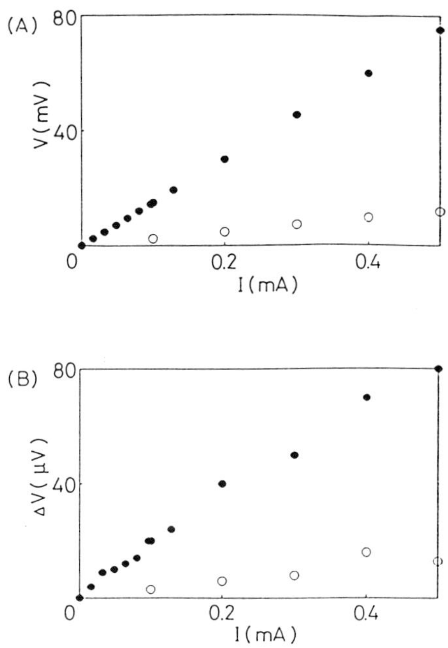

Fig.6 Detective sensitivity (B) of THz light wave corresponding to the controlled I-V characteristics (A)

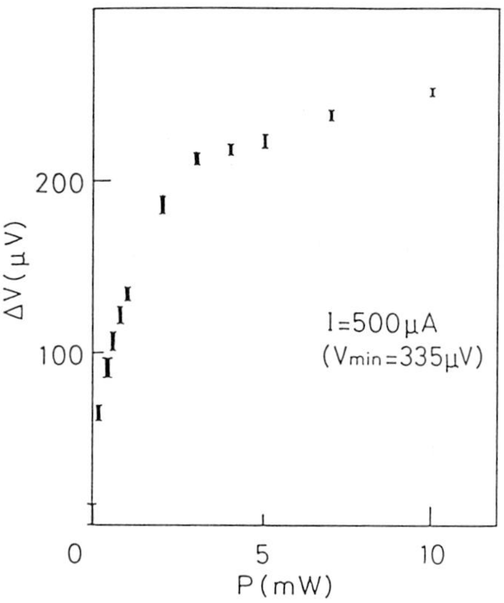

Fig.4 Detectivity of SC active (thin film) antenna, Pe(for antenna area)=0.15×10^{-3}P

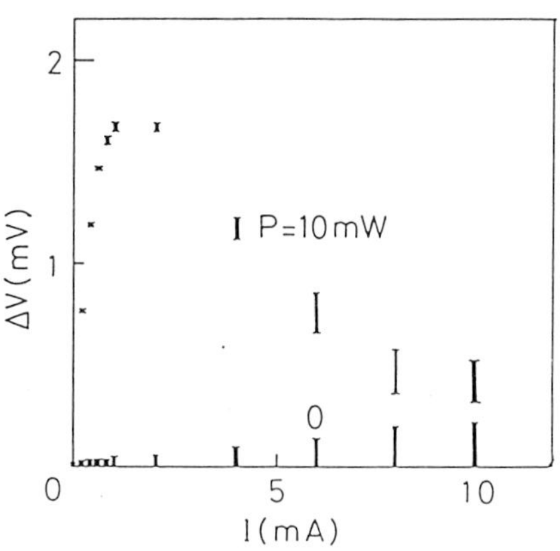

Fig.5 Detectivity of SC active (thin film) antenna for various applied DC current, Pe=0.15×10^{-3}P

(B) Radiations from Superconducting Folded Antenna

Improvements of radiation efficiency of a superconducting folded antenna are investigated experimentally. Schematics of the SC folded antenna is shown in Fig.7. The experimental setup for the radiation system of the SC folded antenna and the detecting system of the radiated electromagnetic wave from the antenna is shown in Fig.8. The stub is used for matching of the input electromagnetic wave to the SC folded antenna. The electric field strength of electromagnetic waves radiated from the SC folded antenna was measured by a spectrum analyzer. Typical experimental data on the radiated electromagnetic waves from the SC folded antenna in a liquid nitrogen are shown in Fig.9 for various frequencies. The improvement of the radiation from the superconducting folded antenna is shown.

CONCLUSIONS

Properties of high Tc superconducting active antennas, i.e. polarization effects and structure effects on electromagnetic wave detections, and detections of Terahertz light waves are confirmed experimentally. Improvements of radiation efficiencies of superconducting folded antennas are also shown experimentally.

REFERENCES

(1). Ohnuma T. (1990) IEICE J73-B-II: 483-484
(2). Ohnuma T. (1990) JJAP 29 : No.11

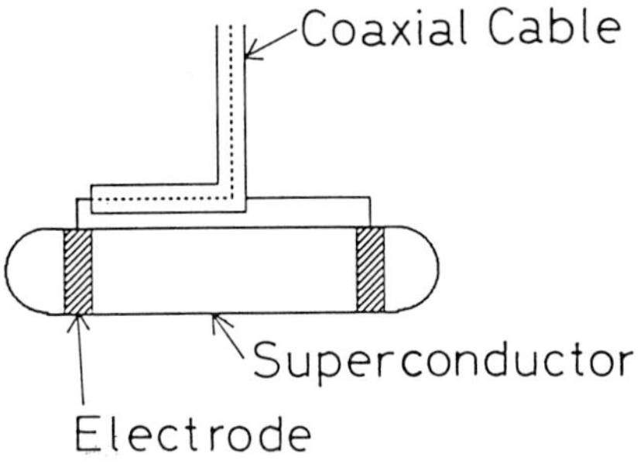

Fig.7 High Tc superconducting folded antenna

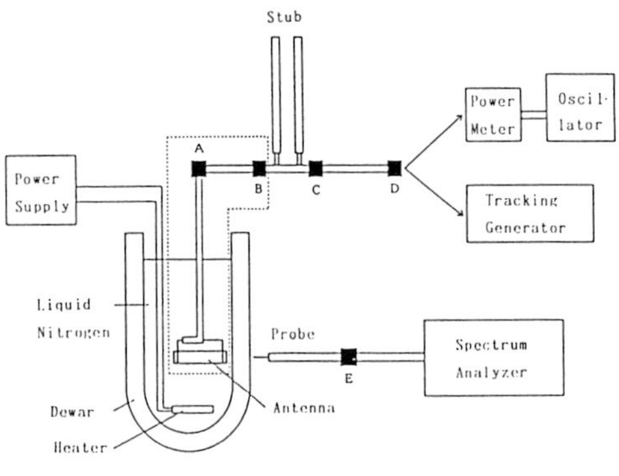

Fig.8 Radiation and detection systems for SC folded antennas

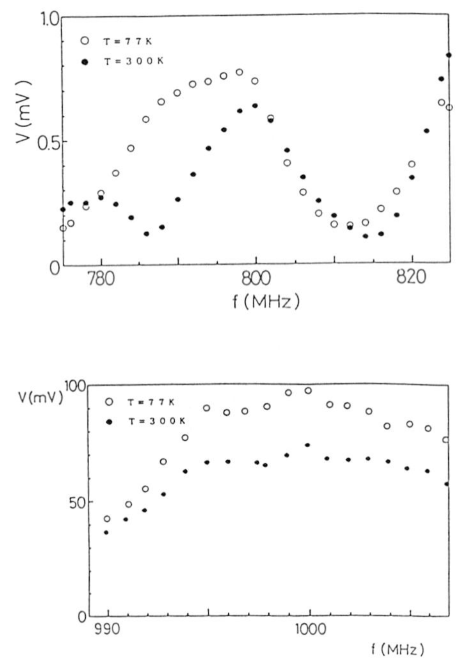

Fig.9 Improved radiation efficiency of SC folded antenna (O) for various frequencies (Tuned freq.= 800MHz, 1000MHz from top to bottom)

Microwave Response of a YBCO Microbridge with Fine Particles and Its Application to a Josephson Mixer

TOSHIMASA HIRANO, YORINOBU YOSHISATO, MINORU TAKAI, KAZUYA NIKI, SHUICHI YOSHIKAWA, and SHOICHI NAKANO

Functional Materials Research Center, SANYO Electric Co., Ltd., 18-13, Hashiridani 1-chome, Hirakata, Osaka, 573 Japan

ABSTRACT

Successful Josephson mixing of microwaves at 20 GHz was carried out up to 60 K using a microbridge prepared of bulk YBCO superconductors for the first time. Though the bridge included many grain boundaries, Shapiro-steps for a single weak link with microwaves were clearly observed. The behavior of induced current-step heights were proven to be similar to those calculated using the RSJ model with a single Josephson junction. The local oscillation wave power required for mixing with a signal wave was less than -28 dBm, which is much lower than that of conventional mixers using a semiconductor diode. Furthermore, it was found that the dynamic resistance, which is necessary for producing highly sensitive, high impedance detectors, increased with the narrower bridge width.

KEY WORDS : Josephson mixing, grain boundary

INTRODUCTION

Since the discovery of high-temperature superconductors with transition temperatures above the boiling point of liquid nitrogen, much research effort has been focused on the fabrication of Josephson junctions operating at 77 K. The Josephson junction is the key device of superconducting electronics. Among the many applications possible, high sensitive microwave detectors are the most promising for Josephson junctions fabricated using high-Tc superconductors. However, the tunnel -type junction using thin oxide superconductors with an insulating barrier necessary for the creation of such a device has not yet been realized owing to the small superconducting coherence length and to degradation of the oxide superconductor at the interface. A clear ac Josephson effect has been observed using point-contact type junctions prepared using a Nb needle and bulk YBCO superconductor.[1)] But in the case of this type of junction, practical applications have been limited because of its poor reproducibility. On the other hand, we previously reported a microwave detector having a long meandering type bridge that uses granular-type bulk YBCO superconductors and operates with high sensitivity (video detection) at temperatures of less than 80 K[2)]. In this paper, we describe Josephson mixing using a bridge-type junction prepared of bulk YBCO superconductors, and discuss the parameter which is necessary for creating a highly sensitive detector.

EXPERIMENTAL

The preparation of bulk YBCO superconductors for the detector was carried out employing the usual solid-state reaction. The bulk grain boundary junctions were prepared using mechanical etching techniques. Details for the fabrication of bridge junctions have been published elsewhere.[3)] Figure 1 shows a photograph of the junction. Figure 2 shows the setup for the microwave mixing experiment. The detector was attached to the cold head in the vacuum space of the cryocooler. The incident signal wave (fs) was irradiated through a horn antenna. The local oscillation (f_L) was irradiated through a coaxial cable in a vacuum space. The I-V characteristics for the bridge junction with

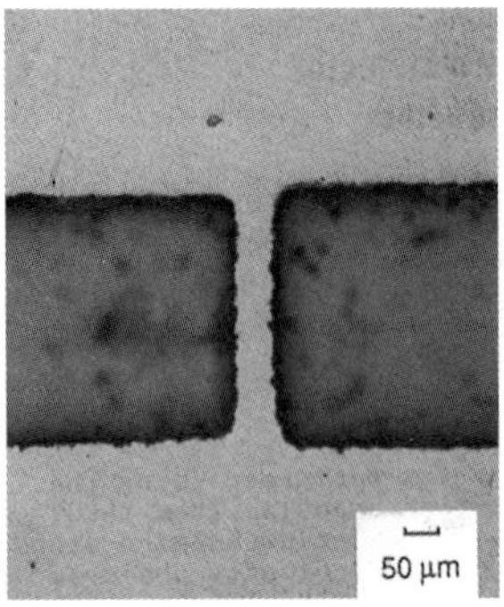

Fig.1. Photograph of a bridge junction with the bulk YBCO superconductor. The thickness of the bridge junction is 50 µm.

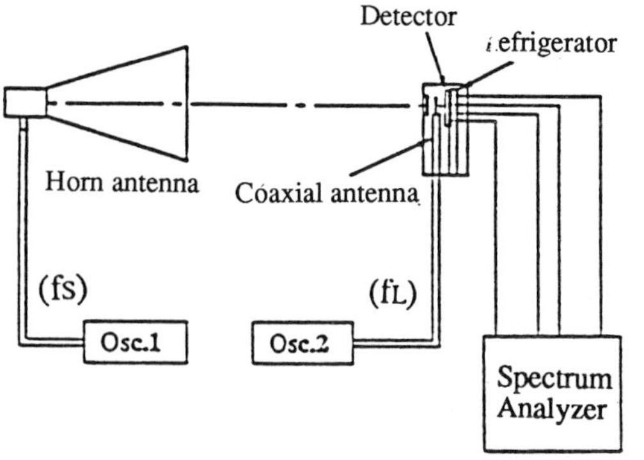

Fig. 2. Schematic diagram of the measuring equipment.

microwave-applied RF power was then measured.

RESULTS AND DISCUSSION

Figure 3 shows the temperature dependence of the critical current (Ic) of the detector without the incident microwave. The critical temperature (Tc) of the detector was 89 K. The solid line in Figure 3 expresses the following equation derived from the Ginzburg-Landau theory for microbridges:

$$Ic=Io(1-T/Tc)^{1.5} \qquad (1)$$

The experimental critical current of the bridge junction agrees with the theoretical value within the calculated range of experimental error. This means that the bridge junctions behave like a typical microbridge junction. Figure 4 shows the I-V characteristics for a bridge junction with 12.15 GHz of RF power applied at 50 K. Clear Shapiro steps were observed. The voltage difference between the steps was 25.1 µV. This value corresponds well with the theoretical vlue of 25.1 µV evaluated from the following equation for a single Josephson junction:

$$V=hf/2e\ , \qquad (2)$$

where f is the RF signal frequency. To investigate I-V characteristics more precisely, we fitted the dependence of the current step heights on the RF current for a simulation based on the Resistively Shunted Junction (RSJ) model.[4] The circuit model for the RSJ is shown in Fig. 5, where the junction capacitance was assumed to be small enough to neglect. The differential equation for the circuit was:

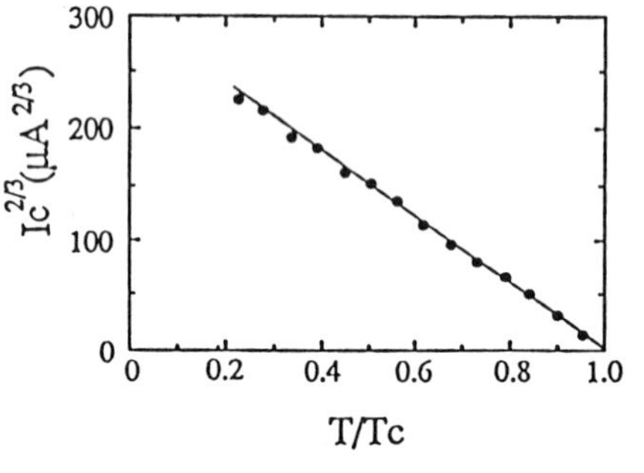

Fig. 3. Dependence of critical current on the temperature.

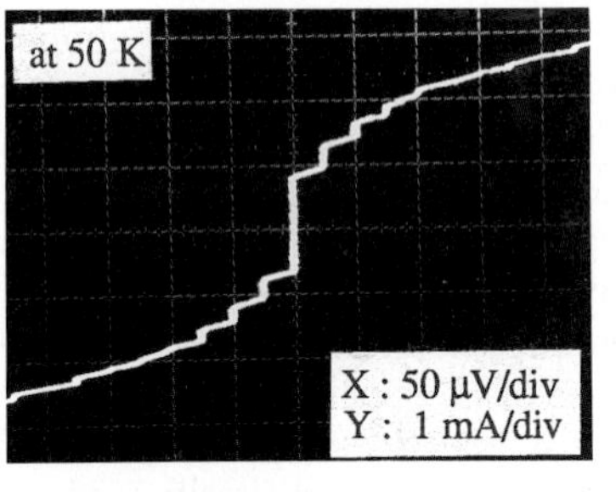

Fig.4. I-V characteristics for a bridge junction with 12.15 GHz applied RF power.

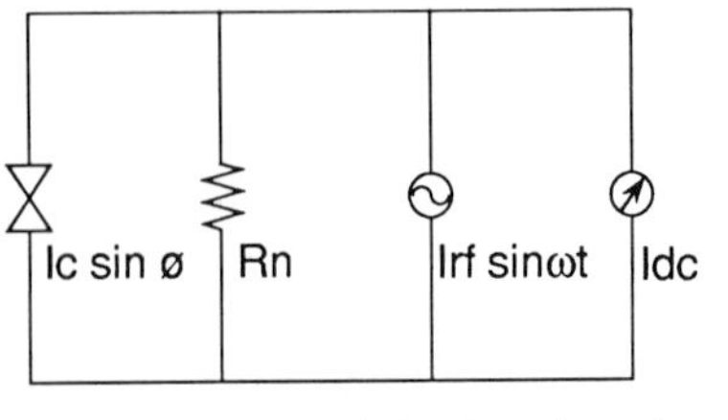

Fig.5. Equivalent circuit for Josephson junction based on the RSJ model.

$$Idc/Ic + (Irf\ /Ic) \sin \Omega\tau = d\phi /d\tau + \sin\phi \ , \qquad (3)$$

where $\Omega = h\omega_r /2eR{\bullet}Ic$ and $\tau = (2eRIc/h)\ t$. Figure 6 shows the dependence of the current step height on the RF current at 25 K. The solid line represents the computer simulation using equation (3), where the dimensionless parameter Ω was assumed to be 0.16. Experimental results fit well with the theory, with the exception of the maximum current (n=0). The first three steps (n = 1, 2 and 3) corresponded exactly to current steps simulated by the RSJ model. The maximum current (n=0) shows an excess current over 1 mA flow compared with the theoretical value. The plotted curve is similar to those simulated. Assuming Ω = 0.16, the value of 0.15 mV for Ic•R was calculated. This value is about one twentieth the 3.25 mV evaluated from the I-V curves. This suggests that superconducting current may increase the apparent maximum current. The large value of Ic•R is necessary to produce a highly sensitive detector. The product of Ic•R was related with an energy gap in the following equation:

$$Ic{\bullet}R = (\pi\Delta/2e)\tanh(\Delta/2kT) \ , \qquad (4)$$

where Δ is the energy gap and k is the Bortzman constant. The value of Ic•R product estimated using equation (4) is considerably larger than the experimental value carried in Table 1.

Fundamental heterodyne mixing with the bridge junction was carried out at a temperature of 22 K. Figure 7 shows the dependence of intermediate frequency (IF) signal on the applied bias voltage, where both signals and local oscillation waves of 19.218 GHz and 19.268 GHz, respectively, were used. The IF signal of 50 MHz frequencies, corresponding to the difference between the signal and local oscillator frequencies, shows the peaks at the nonlinear portions of the I-V curves. The local oscillation wave power required for mixing was less than -28 dBm, much lower than that of conventional mixers using a semiconductor diode. Figure 8 shows the dynamic

Table 1. The experimental value of Ic-Rn for junctions using High-Tc superconductors

Junction type	Institutes	Materials	Ic-Rn (mV)
Bridge	SANYO	Bulk YBCO	3.25 (25 K)
Bridge[5]	Nagaoka University of Technology T. Yamashita et al.	YBCO thin film	0.096 (77 K)
Bridge[6]	University of Caen D. Robbes et al.	Bulk TBCCO	0.4 (77 K)
Point [7] contact	RIKEN Co.	Nb needle and YBCO thin film	6.3 (4.2 K)

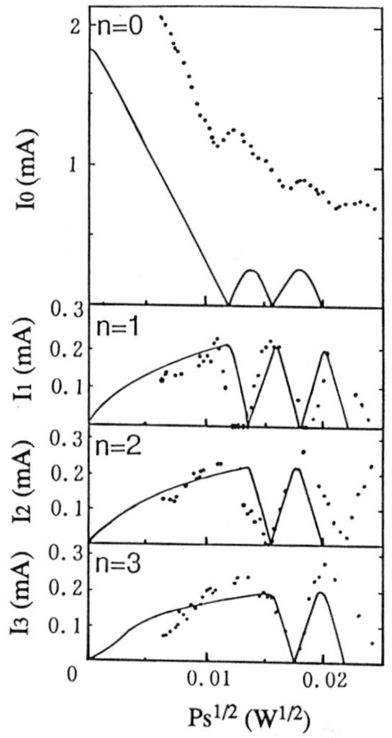

Fig. 6. Dependence of the current step heights on RF power, where the solid line represents the simulation based on the RSJ model.

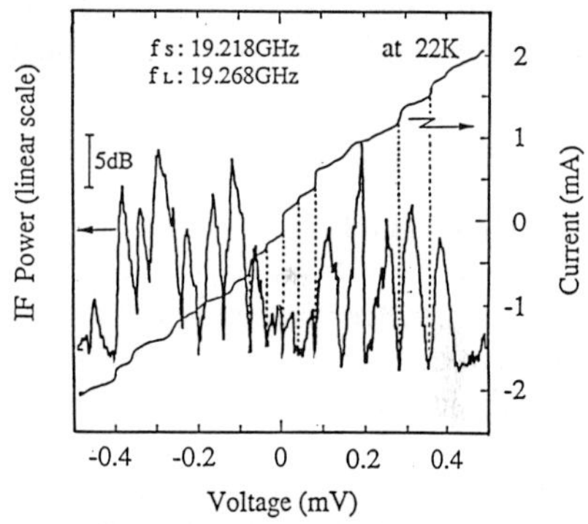

Fig. 7. Dependence of the intermediate frequency (IF) signal on the applied bias voltage where the relative levels of applied microwave of fs and fL power are 35 dBm and 10 dBm respectively.

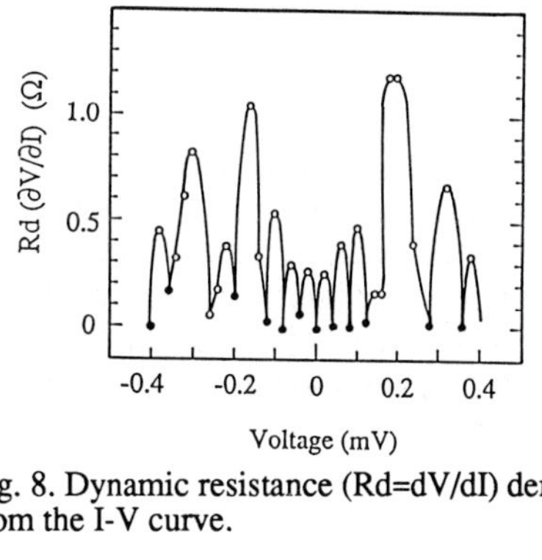

Fig. 8. Dynamic resistance (Rd=dV/dI) derived from the I-V curve.

resistance (Rd=dV/dI) derived from the I-V curve. The Rd obtained in this experiment had values between 0.3 and 1.2 Ω, considerably larger than those of the Josephson weak link using Nb thin-film superconductors. This means that higher output signals can be obtained with an equivalent load resistance. Figure 9 shows the dependence of the IF signal power normalized to that at 39 K. It seems reasonable to conclude that good mixing performance allows operation at or below 60 K, where the conversion efficiency decreased by less than 10 dB of the maximum value. The dotted line shows data for the GdBaCuO bulk-point contact mixer developed by H. K. Olsson et al.[8] The granular -type detector displays better temperature dependence performance than the point-contact type.

Fundamental experiments for the production of large Rd junctions for a highly sensitive microwave detector have been carried out. Figure 10 shows the dependence of Rd on normal resistance (Rn), where Rn was varied as a function of bridge width. A clear increase in Rd with Rn was noted. The maximum Rd value was in the order of about 1.5 Ω, about five times larger than that of normal planar Josephson junctions using Nb thin films. The possibility of a large Rd suggest that the granular-type detector is preferable for creating a good microwave matching system.

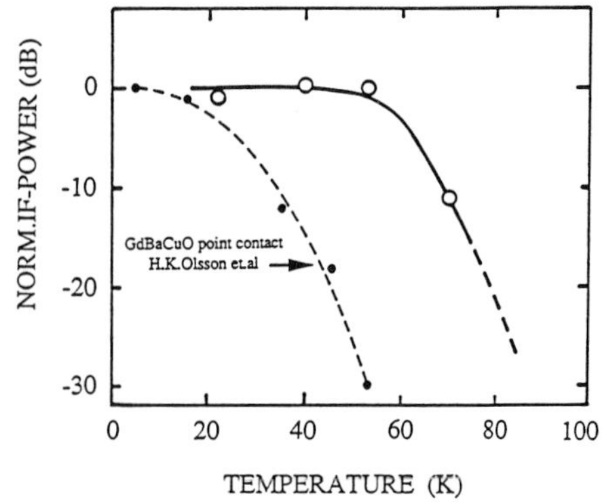

Fig. 9. Dependence of IF signal power normalized to a temperature of 39 K.

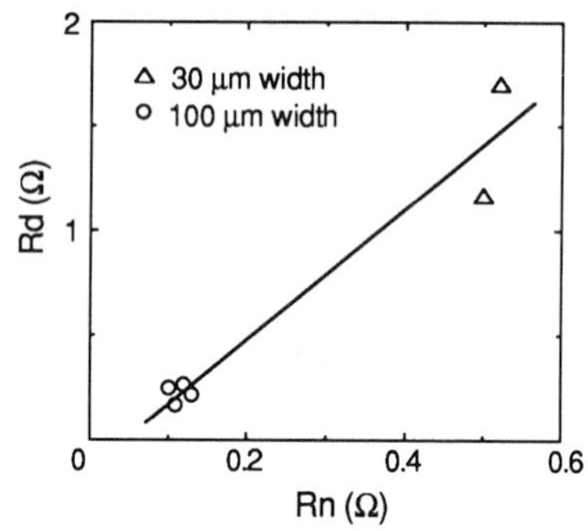

Fig. 10. Dependence of Rd on Rn.

CONCLUSION

In summary, we have investigated the microwave response for a microbridge prepared by a bulk YBCO superconductor. The bridge showed clear Shapiro-steps for a single weak-link junction. The experimental current step heights corresponded well with a simulation using the RSJ model. A small IcRn product of 0.15 mV was estimated from the experimental current step heights. This vlue is about one twentieth the 3.25 mV evaluated from the I-V curves. The possibility of highly sensitive detectors with a large Rd was obtained.

REFERENCES

1. S. Imai, T. Tozawa, K. Sumiya, G. Oya, T. Sugai and N. Mikoshiba : FED HiTcSc-ED Workshop, pp. 255-259 (1988).
2. Y. Yoshisato et al. : Jpn. J. Appl. Phys. 29, 1080 (1990).
3. Y. Yoshisato et al. : Technical Digest of the 8th Sensor Symposium 243 (1989).
4. P. Russer et al. : J. Appl. Phys. 43, 2008 (1972).
5. T. Yamasita, M. Era, S. Noge : Jpn. J. Appl. Phys. 29, 74 (1990).
6. D. Robbes et al. : Appl. Phys. Lett. 54 (1989).
7. N. Sugawara et al. : Extended Abstracts of The 51st Autumn Meeting of JSAP NO. 1:88 (1990).
8. H. K. Olsson et al. : J. Appl. Phys. 62, 4923 (1987).

High Temperature Superconducting Microwave Components

R.G. Dorothy, D.W. Face, W.L. Holstein, C. Wilker, Z-Y. Shen, and D.B. Laubacher

Du Pont Company, P.O. Box 80304, Wilmington, DE 19880-0304, USA

ABSTRACT

We have developed high yield thin film deposition and patterning processes for YBCO ($YBa_2Cu_3O_{7-x}$) and TBCCO ($Tl_2Ba_2CaCu_2O_8$) high temperature superconductors (HTSC) that demonstrate low surface resistance (R_s) at mircowave frequencies (1/10 X to 1/100 X that of copper). These high quality thin films were developed using co-evaporation, *ex situ* sputtering and *in situ* sputtering techniques. We developed and evaluated several passive microwave components based on these YBCO and TBCCO thin films including: resonators, filter, delay lines, and splitter/combiners. HTSC microstipline resonators at 5 GHz provide Q's >40X higher than the same device fabricated with copper thin films and have low power dependence. Delay lines fabricated on 2.5 cm X 2.5 cm substrates have a 11.4 ns delay and an insertion loss of less than 0.05db/ ns at 1.0 GHz . The power dependence of the microwave performance of both YBCO and TBCCO based devices were measured and shown to be low at normally useful power levels.

KEY WORDS: - Microwave Components, Resonator, Delay Line, YBCO, TBCCO

INTRODUCTION

The promise of surface resitivities of HTSC thin films that are one or more orders of magnitude lower than the surface resistivity of copper or noble metal films may allow thin film microwave components to be one of the earliest applications of these superconductors. While many laboratories can routinely produce HTSC thin films with high critical temperature (T_c) and critical current density (J_c), the routine fabrication of thin films with low surface resistivity (R_s) has proven to be a more elusive goal. It appears that many of the parameters that affect R_s such as surface smoothness, surface orientation, grain size and orientation, and surface defect level affect J_c and T_c very little. Hence, more critical processing may be required in order to minimize R_s than are required to maximize J_c and T_c. This study discusses an *in situ* off axis magnetron sputtering process for YBCO, an *ex situ* electron beam coevaporation process for YBCO, and an *ex situ* process for TBCCO that provide high yields of microwave device quality films on acceptable low dielectric loss substrates.

With the availability of films on substrates that have acceptable microwave properties, the challenge becomes to pattern and package the thin films without degradation of R_s. This study presents data on YBCO and TBCCO microstripline resonators and TBCCO coplanar delay lines that have been patterned and packaged without significantly affecting the as deposited R_s. These microwave components have performances 10X to 40X superior to that of some components made with high quality thin film copper. In addition, this study shows that adequately flat power response can be achieved at commercially useful power levels.

FILM FABRICATION

Precursor films of Y, Cu, and BaF_2 have been fabricated on $LaAlO_3$ by coevaporation [1,2]. The films were deposited in a turbo-molecular pumped vacuum chamber with a base pressure of < 8 x 10^{-7} mBar. Y and Cu were electron beam evaporated from separate guns and the rates were controlled Leybold-Heraeus Inficon Sentinel III Electron Impact Emission Sensors (EIES). BaF_2 was thermally evaporated from a covered molybdenum boat and the rate was controlled with a quartz crystal oscillator. Usually, the co-evaporation was carried out in the presence of oxygen at 1 x 10^{-5} mBar. If oxygen is not used during the deposition, the precursor films are very sensitive to exposure to the atmosphere. The deposition rate varied from ~0.5 Å/s to over 1 Å/s. Careful positioning of the guns and thermal source allowed formation of films over a 50 mm diameter wafer, currently the largest $LaAlO_3$ substrate available. Composition ratios as measured by RBS were held to within 1%.

The two stage annealing of the films took place in a quartz-lined tube furnace. The films were taken from room temperature to between 850 °C to 900 °C at a rate of between 10 to 25 °C per minute in either a wet oxygen or dry oxygen atmosphere. The wet oxygen atmosphere was formed by bubbling O_2 through de-ionized water. Then the films were held at temperature for 30 minutes in a wet oxygen atmosphere. During this part of the anneal, the BaF_2 was fully decomposed. Finally, the films were cooled slowly between 0.5 to 1.5 °C/min in dry oxygen until they reached room temperature.

YBCO films were produced in a wide range of thicknesses (1000 Å to 15,000 Å). The microstructure of the films was strongly dependent on the film thickness and anneal conditions. Although the optimum anneal conditions depend upon the substrate material, several broad conclusions remain the same. Correct stoichiometry and higher anneal temperatures favor lower microwave surface resistance. Thicker films (>4000 Å) have a microstructure which consists of a layer of c-axis perpendicular material adjacent to the substrate, and a layer of a-axis perpendicular material with a "basket weave" morphology on the top surface. In thin films, only the c-axis perpendicular material is obtained and the surface is quite smooth. X-ray diffraction indicates that the ratio of c/a axis orientation in 5000 Å thick films annealed at 850 °C is 2 or 3 to 1. X-ray pole figure analysis indicates that the in-plane epitaxy of the films is excellent on this substrate. Both the c-axis perpendicular and the a-axis perpendicular material are well aligned with no evidence of large angle grain misorientations from either pole figure analysis or TEM cross-sections. RBS channeling has shown that higher anneal temperatures and longer anneal times lead to interdiffusion of the substrate with the film. TEM cross-sections show a high-quality interface between $LaAlO_3$ and the superconductor with a perfect lattice after 1 to 2 unit cells.

In situ $Y_1Ba_2Cu_3O_{7-x}$ films were grown by off-axis single-target magnetron sputtering from a stoichiometric $Y_1Ba_2Cu_3O_{7-x}$ target. Sputtering in an off-axis arrangement minimizes re-sputtering of the film by negatively charged oxygen ions, allowing for precursor films to be fabricated with composition close to that of the sputtering target [3,4]. The sputtering atmosphere consisted of 160 mTorr Ar and 40 mTorr O2. Typical substrate temperatures during deposition ranged from 650°C to 750°C. Typical deposition rates were about 500Å/hr. The 1 inch square $LaAlO_3$ substrates were thermally and mechanically anchored to a 9 cm diameter solid nickel block with silver paint. The nickel block rotated continuously during deposition and was heated radiatively from the back-side with quartz halogen lamps. After deposition, the samples were cooled at 15°C/min to room temperature in 200 Torr of oxygen.

TBCCO films were prepared in a two-step process: deposition of precursor films containing Ba, Ca, Cu and O onto $LaAlO_3$, followed by conversion of the precursor films into $Tl_2Ba_2CaCu_2O_8$ [5,6].

Non-thallium containing precursor films were deposited by off-axis rf magnetron sputtering in a load-locked Leybold Heraus L560 deposition system similar to that used for the YBCO films. A 7.5 cm diameter sputtering target pressed from pre-reacted powder of barium copper oxide and calcium copper oxide with nominal composition Ba:Ca:Cu = 2:1:2 was used. Depositions were carried out in 5-50 mtorr Ar at 120 W rf power. Most films were deposited in 5 mtorr Ar at a deposition rate of 1.8 Å/s. Precursor film thickness varied from 0.10-1.00 µm. BaCaCuO precursor films were converted to $Tl_2Ba_2CaCu_2O_8$ by heating in air at 800-900°C for periods of time ranging from a few minutes to an hour in the presence of a powder mixture prepared by mixing $Tl_2Ba_2Ca_2Cu_3O_{10}$ and Tl_2O_3. The entire process was carried out in a covered alumina crucible. Tl_2O volatilizing from the powder mixture is transported through the vapor phase to the BaCaCuO film, with which it reacts to form $Tl_2Ba_2CaCu_2O_8$. Film thickness increased by about 20% during thallination, and final film thickness varied from 0.12-1.2 µm.

MICROWAVE CHARACTERIZATION

Microwave surface resistance (R_S) measurements were carried out using a 20 GHz TE011 end-wall replacement cavity [7]. These measurements were made on 2.54 cm-square samples in a polished copper cavity with a diameter of 2.29 cm. The technique is capable of measuring surface resistance in the range 2 mΩ-10Ω. The best 1.0 µm-thick films have R_S less than 2 Ωr, the resolution of the technique, at 100 K. Figure 1 shows R_S vs. temperature for a typical YBCO film on $LaAlO_3$ with the R_S of copper as a reference and the limit of sensitivity of the 20 GHz cavity indicated. A similar plot of R_S vs. temperature for a TBCCO film on $LaAlO_3$ is shown in Figure 2.

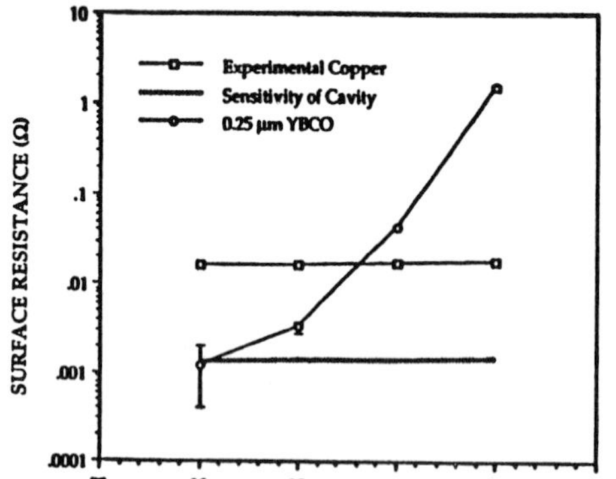

Fig. 1 20 GHz Surface Resistance vs. Temperature for a Co-evaporated Film Measured in a TE_{011} Cavity

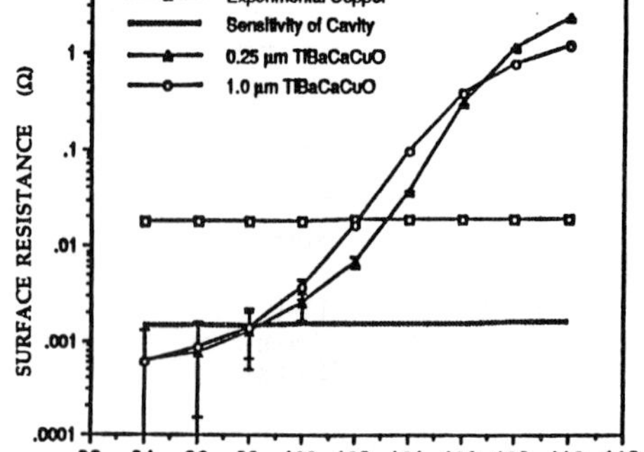

Fig. 2 20 GHz Surface Resistance vs. Temperature for a TBCCO Film Measured in a TE_{011} Cavity

DEVICE FABRICATION

Microwave devices were fabricated using entirely dry processing in order to minimize the degradation of the HTSC films. A two layer photolithography process consisting of 0.5 microns of PMMA under 3 microns of positive photoresist prevented exposure of the films to aqueous-based positive resist developers. The structures defined by the photoresist process were etched using argon ion milling. The resulting patterned lines of HTSC were very sharply defined with process resolution of less than 2 microns. The patterning process was completed by oxygen plasma removal of the remaining photoresits. Gold and silver normal metal contacts to HTSC were defined by either lift-off, or evaporation through etched metal shadow masks. The contacts were then annealed in oxygen at >400°C for one hour. Figures 4 and 5 show the patterns for the active plane of the 5 GHz microstripline resonator and the 11.4 ns coplanar delay line.

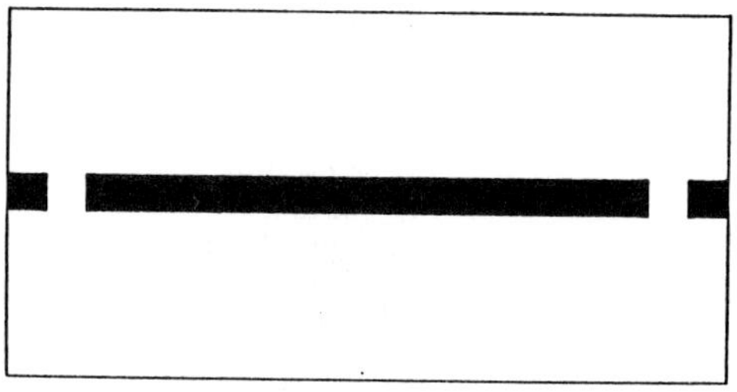

Fig. 3 5GHz Microstrip Resonator Layout (1 cm overall, 0.5 mm line width, 0.5 mm gap)

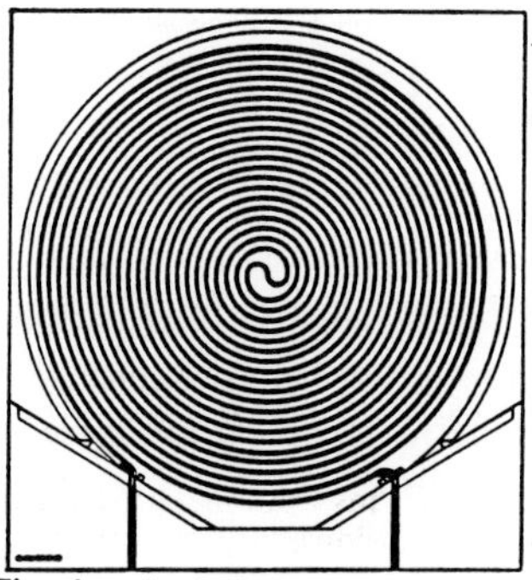

Fig. 4 1 - 5 GHz 11.4 ns Delay Line Layout (25.4 x 25.4 mm)

COMPONENT PERFORMANCE

5GHz Microstipline Resonator

Microstripline resonators with YBCO active and ground planes and TBCCO active and ground planes were fabricated that provide unloaded Qs of up to 17,000 at 5 GHz. Figure 6 shows the response of a YBCO resonator with a loaded Q of 7500 at 77K. Hermetically packaged TBCCO resonators have loaded Qs of >13,000 at 77K. Identical resonators fabricated from high quality evaporated copper 4 μm-thick show a Q of 300 at 77K. The input power dependence of this resonator is relatively flat over the range of -20dBm to +10dBm as shown in Figure 7.

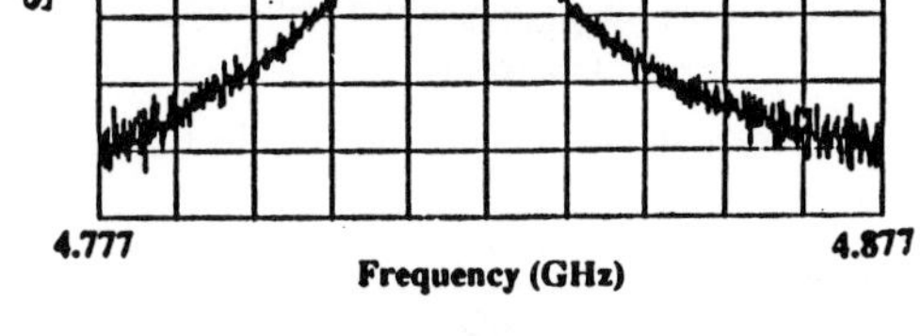

Fig. 5 Output of Half Wave-length Microstrip Resonator Fabricated from YBCO at 77K.

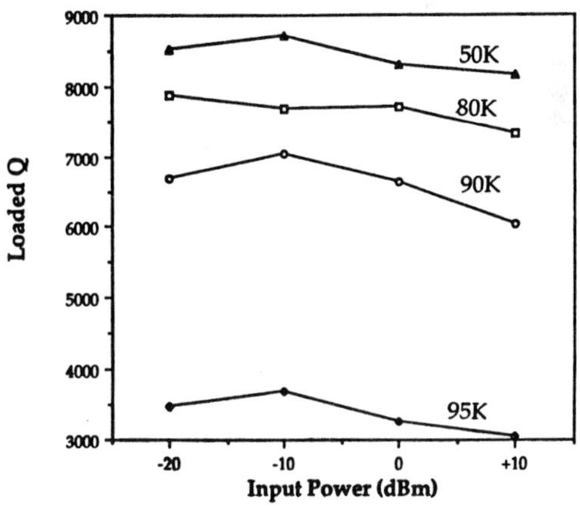

Fig. 6 Loaded Q vs. Input Power for 5 GHz TBCCO Microstipline Resonator

Figure 8 shows the loaded Q for a heavily loaded TBCCO resonator as a function of temperature showing that the Q exceeds that of the copper resonator below 98K over the power range -20dBm to +10dBm.

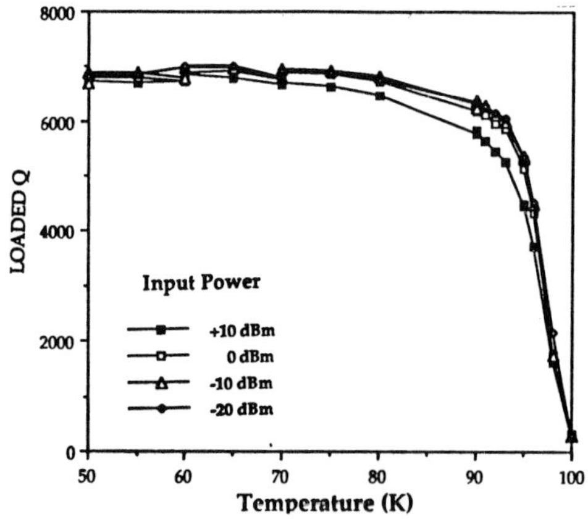

Fig. 7 Loaded Q vs. T for 5GHz TBCCO Microstripline Resonator

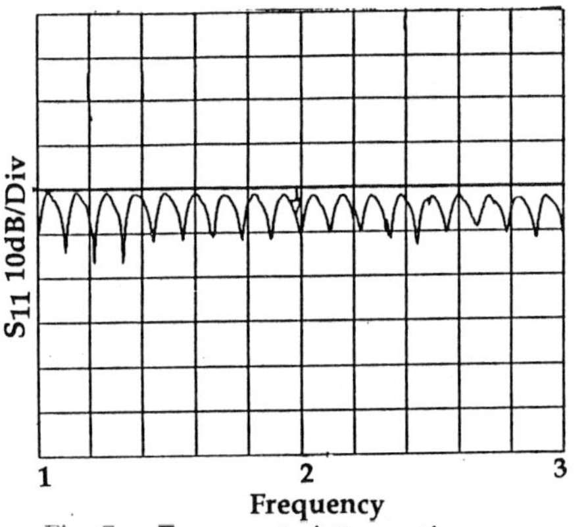

Fig. 8 Resonant Attenuation vs. Frequency for 11.4 ns Coplanar TBCCO Delay Line.

Similar results have been obtained with coevaporated and sputtered YBCO with a corresponding decrease in critical temperature.

Delay Lines

Coplanar lines as long as 11.4 nanoseconds have been fabricated with TBCCO and YBCO on 25.4mm x 25.4mm $LaAlO_3$ substrates that have losses of less than 0.05dBm/nsec. These delay lines provide uniform dispersion free delay characteristics with minimal cross talk to at least 5 GHz with minimal crosstalk. Figure 8 shows the resonance for a TBCCO 11.4 ns delay line from 1-5 GHz.

CONCLUSION

High quality superconducting thin films of YBCO produced by BaF_2 coevaporation and *in situ* off axis magnetron sputtering and *ex situ* processed TBCCO films were fabricated into several building block microwave components. The HTSC microwave resonators and delay lines show a performance advantage of more than 40X when comopared to similar Cu devices. With the demonstration of these preliminary devices, it is now possible to produce more complex HTSC that yield superior overall microwave system performance.

REFERENCES

1. P. M. Mankiewich, *et al.*, "Reproducible technique for fabrication of thin films of high transition temperature superconductor," *Appl. Phys. Lett.*,51, 1753-1755, 1987.

2. F. H. Garzon, *et al.*, "Characterization of Y-Ba-Cu-O superconducting thin films prepared by coevaporation of Y, Cu, and BaF_2." *Appl. Phys. Lett.*, 54, 1365-1367, 1989.

3. C. B. Eom, *et al.*, "*In situ* grown $Y_1Ba_2Cu_3O_{7-x}$ thin films from single-target magnetron sputtering," *Appl. Phys. Lett.*, 55, 595-597, 1989.

4. R. L.Sandstrom, *et al.*, "Reliable single-target sputtering process for high-temperature superconducting films and devices." *Appl. Phys. Lett.*, 53, 444-446, 1988.

5. P. N. Arendt, *et al.*, "Radio frequency surface resistance of Tl-Ba-Ca-Cu-O films on metal and single-crystal substrates." *AIP-Conference Proceedings*, 200, R. L. Stockbauer, S. V. Krishnaswamy, and R. L. Kurtz, Eds., 1989-239-246.

6. S. I. Shah, N. Herron, C. R. Fincher, and W. L. Holstein, "Fabrication of Tl-Ba-Ca-Cu-O films by annealing rf-sputtered Ba-Ca-Cu-O films in thallium oxide vapors," *Appl. Phys. Lett.*, 56, 782-784, 1990.

7. D. W. Cooke, *et al.*, "Microwave characterization of high-temperature superconductors," *Proc. SPIE,* Vol 1187, *Processing of Films for Hight* T_c Superconducting Electronics, *338-347, 1989.*

Acknowledgements:
We wish to acknowledge L. A. Parisi and R. J. Small for their contributions to the film deposition processes, M. S. Warrington and R. D. Nichols for their contributions in patterning the devices described, and C. Wilker, A. L. Matthews, and M. S. Brenner for contributions in film and device characterization.

Microstriplines of Bi-Sr-Ca-Cu-O and Their Microwave Properties

ATSUSHI TANAKA, JASON CRAIN*, NOBUO KAMEHARA, and KOICHI NIWA

Fujitsu Laboratories Ltd., 10-1, Morinosato-Wakamiya, Atsugi, 243-01 Japan

ABSTRACT

Bi-Sr-Ca-Cu-O microstriplines having both a superconducting line and groundplane were reproducibly fabricated on a MgO substrate by rf magnetron sputtering. The samples were 26.4 mm in length and were nearly single phase high-Tc materials. Their microwave properties have been investigated at frequencies up to 20 GHz and the high-frequency loss of the superconducting microstrips were compared with that observed for geometrically equivalent gold lines. The transmission loss of BSCCO stripline showed stronger frequency dependence than that of gold. The effects of synthesis condition and critical current on microwave properties were also investigated.

KEY WORDS: Bi-based superconductor, high-Tc phase, microstripline, transmission loss, rf sputtering

INTRODUCTION

As the quality of thin films of oxide superconductors has improved dramatically since their discovery, research efforts have recently begun to focus on development of possible high-frequency applications for these materials. To date, the majority of investigations into the microwave properties of oxide superconductors have been performed using cavity perturbation techniques and have focused largely on the yttrium-based material (YBCO) [1,2,3,4] and some results of investigations of Tl-Ba-Ca-Cu-0 have also been reported [5,6] Although generally more difficult to synthesize, the high-Tc phase of Bi-Sr-Ca-Cu-O (BSCCO) has several potential advantages over YBCO. In addition to being somewhat more durable than YBCO, its higher Tc implies that less loss due to thermally excited quasiparticles is expected.

In this paper, we report high-frequency measurements on a relatively long microstripline of the high-Tc phase of BSCCO. The samples were prepared on a MgO substrate having both a superconducting line and groundplane. The results are then compared to those obtained for geometrically identical gold microstrips. In addition, the relationship between microstructure and transmission loss is discussed.

SAMPLE SYNTHESIS AND CHARACTERIZATION

Since high-frequency response in superconductors is extremely sensitive to the presence of intergrowths and impurity phases, the synthesis conditions must be carefully monitored and accurately controlled in order to inhibit the formation of such secondary phases as much as possible. Samples were prepared on MgO substrates by RF magnetron sputtering from a composite BSCCO target. Separate Cu-O and PbO targets were also used to independently control Cu content and to dope Pb in appropriate amounts, respectively. The fabrication of the groundplane is described in detail elsewhere [7]. The microstrip was fabricated using a metal mask and was 600 μm wide, 0.85 μm thick and 26.4 mm in

---

* Present address: University of Edinburgh, Department of Physics, Edinburgh, Great Britain EH9 3JZ

length. A sketch of the stripline is shown in Fig. 1. Post-deposition annealing was carefully carried out to form the high-T_c phase. Since Pb evaporates rapidly, the line was covered during the annealing process in an attempt to prevent immediate evaporation of Pb from the stripline. Samples were annealed at 852°C in air for 1 hour. X-ray diffraction results, shown in Fig. 2, indicated that the samples possessed a high degree of c-axis orientation perpendicular to the substrate and that both the line and groundplane were nearly single high-T_c phase material. The DC resistive measurements revealed that the Tc-endpoints of both sides were above 100 K. The average post-anneal composition was determined by Inductively Coupled Plasma Spectroscopy (ICP) to be $Bi_{0.94}Pb_{0.07}Sr_{1.00}Ca_{1.01}Cu_{1.76}O_x$.

MICROWAVE MEASUREMENTS

High-frequency measurements were performed with a Hewlett-Packard model *HP-8510 Network analyzer* operating in a frequency range of 50 MHz to 20 GHz. The microstriplines were mounted on a gold-plated stage and contact to the sample surface was made by silver paint and sliding contacts. The microwave input level was about 3 dBm or 2 mW. The samples were cooled slowly from room temperature and finally immersed in liquid nitrogen bath. The transmission loss was compared with that of geometrically equivalent gold lines at 77 K in order to avoid complications due to dielectric, radiation, and contact losses. In addition to these measurements, transmission losses of several BSCCO samples were compared with each other in order to investigate the effect of microstructure on microwave properties. It was noted that the samples exhibited little or no degradation of their microwave properties even after several thermal cycles.

RESULTS AND DISSCUSSION

The frequency-dependent transmission loss, in dB, from room temperature to 77 K is shown in Fig. 3. The DC resistivity of the sample at room temperature was approximately 240 Ω/cm. Upon cooling, it is observed that the high-frequency loss of this sample decreased by 60 dB at 6 GHz between room temperature and 77 K. We report the loss in dB rather than dB/cm because there is some contribution to the loss from the connectors and contacts meaning that the loss is not due only to the microstripline sample.

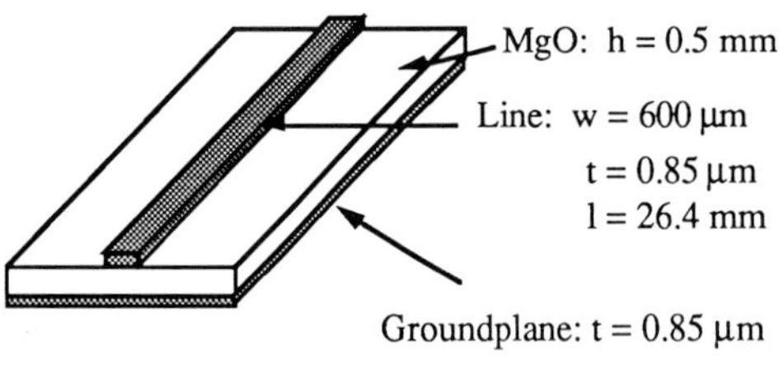

Fig. 1. Sketch of the BSCCO microstripline fabricated on a MgO substrate and having a superconducting groundplane and microstrip.

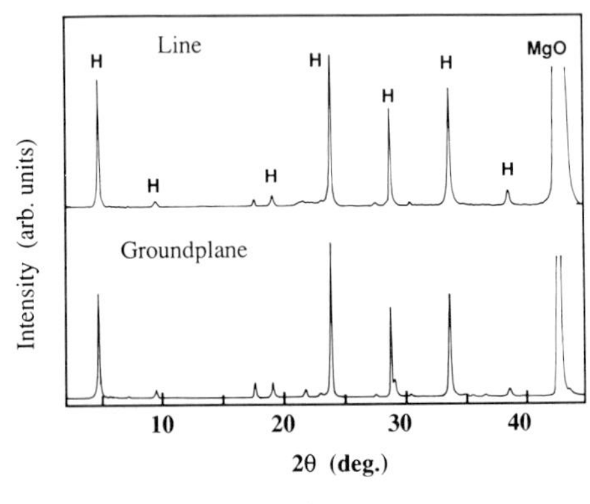

Fig. 2. X-ray diffraction pattern of BSCCO line and groundplane. It is observed that both the line and groundplane are nearly single high-Tc phase material and possess a highly c-axis orientation.

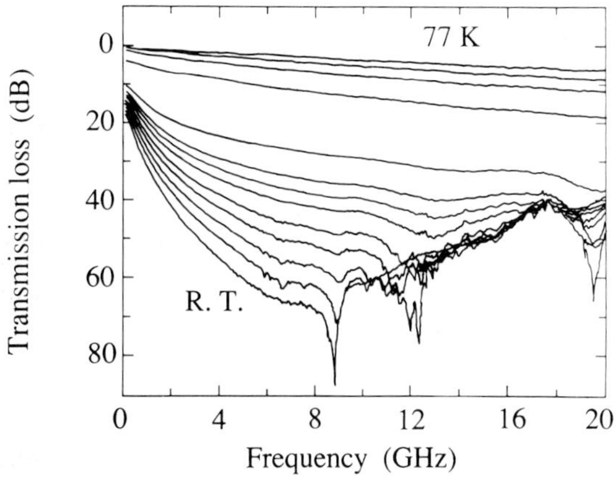

Fig. 3. Transmission loss in dB as a function of frequency for 26.4 mm long BSCCO microstripline from room temperature to 77 K.

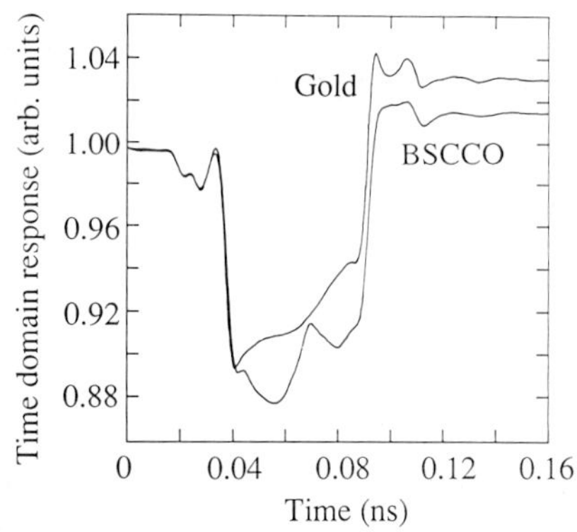

Fig. 4. Time domain response to a step-function input of the BSCCO and that of a geometrically equivalent gold line at 77 K. The similarity in the two traces indicates that the quality of electrical contact was similar.

A plot of the time domain responses to a step-function input of both the gold and superconducting lines are shown in Fig. 4. The similarity of the two time domain responses confirms that the contact quality of the BSCCO line and gold standard are nearly the same. The standing wave ratio (SWR) of the BSCCO microstrip was found to be 2.3 and decreased to 1.16 at 77 K.

In Fig. 5, the transmission losses of the BSCCO striplines and gold samples are compared when they are immersed in liquid nitrogen. Two BSCCO striplines were annealed in different conditions. Both samples are nearly single high-Tc materials and exhibit different microstructure. Sample B in Fig. 5 has better connected grains than sample A. Critical currents (*Jc*) of groundplane are 200 A/cm^2 and 2200 A/cm^2 respectively. Since the groundplane is so large (10x30 mm^2), it is difficult to obtain homogeneous defect-free film. The estimated value of *Jc* is still quite low due to the relatively large (1x6 mm^2) area of groundplane used to determine it and therefor represents a macroscopic property of the film. However, higher Jc is required for less transmission loss. The higher *Jc* sample exhibits superior or equivalent performance to that of the gold stripline for frequencies below roughly 7 GHz. Above this frequency the gold exhibited less loss, however, the difference was only 1 dB at 18 GHz.

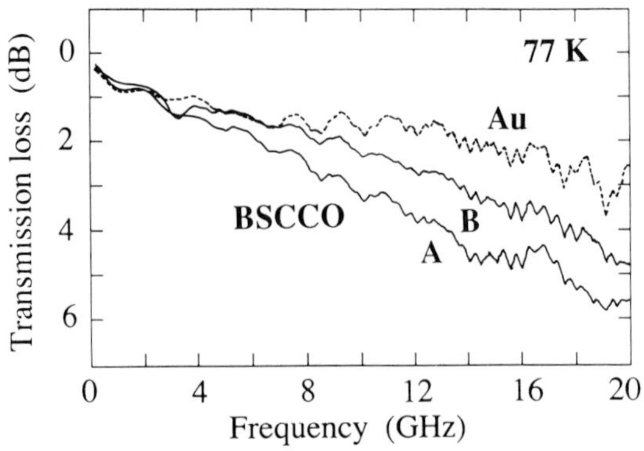

Fig. 5. Transmission loss in dB of two BSCCO microstriplines A and B, prepared under different conditions. Also shown is the loss for a gold film of identical dimension. The *Jc* of samples A and B were 200 A/cm^2 and 2200 A/cm^2 respectively.

According to the Mattis-Bardeen model, the surface resistance of bulk BCS superconductors depends more sensitively on frequency than that of normal metals obeying the classical skin effect [8]. The transmission loss of the superconducting stripline exhibits a stronger frequency dependence than does the gold line as expected; however the influence of grain boundaries cannot be neglected. Since the samples investigated in this work are polycrystalline, it is expected that significant contributions to losses are due to grain boundaries which form a network of weak links through the sample and that the intrinsic high-frequency loss in this material is less than that observed in our samples.

SUMMARY

We have reproducibly fabricated nearly single 110 K phase Bi-Sr-Ca-Cu-O microstrips on MgO substrates by RF magnetron sputtering. The samples have both a superconducting microstrip and groundplane and both possess a high degree of c-axis orientation as well as transition temperatures above 100 K. We report here the results of the first investigations of the high-frequency measurements which indicated that superconducting BSCCO microstrips exhibit superior performance relative to that of gold for low frequencies. Higher Jc BSCCO samples show less loss than lower Jc ones. Further film quality improvement by modifying synthesis condition and techniques is required to fully realize the microwave performance of this potentially promising material.

REFERENCES

1 S. Sridhar, C.A. Shiffman and H. Hamdeh: Phys. Rev. **B36,** 2301, (1987).
2 J.S. Martens and J.B. Beyer, Appl. Phys. Lett. **52** (21) 1822, (1988).
3 A.M. Awasthi, J.P Carini, B. Alavi and G. Grüner, Sol. S. Commun., **67,** 373 (1988).
4 M. Kobayashi, K. Okabe, S. Tanuma, I. Sankawa, M. Sato, T. Konaka and K. Ishihara, IEEE Transactions on Magnetics, **25** No. 2, 997 (1989).
5 H.S. Newman, A.K. Singh, K. Sadanada and M.A. Imam, Appl. Phys. Lett. **54** (4), 389, (1988).
6 L.D. Chang, M.J. Moskowitz, R.B. Hammond, M.M. Eddy, W.L. Olsen, D.D. Casavant, E.J. Smith and M. Robinson, Appl. Phys. Lett. **55** (13), 1357, (1989).
7 A.Tanaka, T. Machi, N. Kamehara and K. Niwa, Appl. Phys. Lett. **54,** 14 1362, (1989).
8 D.C. Mattis and J. Bardeen, Phys. Rev. 111, 412, (1958).

Epitaxial Growth of YBCO/MgO/YBCO Structure

S. TANAKA, H. NAKANISHI, T. MATSUURA, K. HIGAKI, H. ITOZAKI, and S. YAZU

Itami Research Laboratories, Sumitomo Electric Industries, Ltd., 1-1, Koya-kita 1-chome, Itami, 664 Japan

ABSTRACT

A $Y_1Ba_2Cu_3O_{7-y}$(YBCO)/MgO/$Y_1Ba_2Cu_3O_{7-y}$ structure on a MgO(100) substrate was made and its crystallographic structure was investigated by cross-sectional transmission electron microscopy (TEM). A MgO ultra-thin film with (100) orientation as an intermediate layer was obtained on each of the (001) and (100) YBCO thin films at the substrate temperature of about 400°C. From the results of cross-sectional TEM observation of the YBCO/MgO/YBCO structure, it was found that the thickness of the MgO can be reduced to 5nm at a minimum. In the sample with a MgO thickness of 5nm, sharp interfaces were observed without any disrupted portion.

INTRODUCTION

Rapid progress has been made on the preparation of high temperature superconducting (HTSC) films [1-3]. Much attention has been focused on the fabrication of hetero-epitaxial junctions for electronics applications. YBCO/MgO/YBCO,[4-6] YBCO/Y_2O_3/YBCO[7], YBCO/$Pr_1Ba_2Cu_3O_{7-y}$/YBCO[8], and $Bi_2Sr_2Ca_1Cu_2O_y$/$Bi_2Sr_2Cu_1O_y$/$Bi_2Sr_2Ca_1Cu_2O_y$[9] structures have been proposed and investigated. Since there is strong anisotropy in the coherence length of YBCO film, an a- or b-axis oriented YBCO film would be desirable for electronics applications such as a Josephson junction. Kobayashi and Iyori [4] reported successful MgO epitaxial growth on both the (001) and (100) YBCO layers by rf magnetron sputtering. They also reported YBCO(001) epitaxial growth on the YBCO(001) layer via the MgO intermediate layer by the same method. In this paper, we will report on the epitaxial growth of an HTSC SIS structure using a MgO ultra-thin film as an intermediate layer prepared by the vacuum evaporation method.

EXPERIMENTAL

MgO thin film was deposited on a YBCO base layer by a reactive vacuum evaporation system with three electron beam guns [6]. One of the three guns was used for the deposition of the MgO thin film. Base pressure in the chamber was reduced to 10^{-4} Pa with a cryopump. MgO pellets were used as the source. MgO thin films were deposited at several different temperatures from 100 to 600°C. Oxygen gas of 25sccm was introduced near the substrate and the total gas pressure near the evaporation source was maintained at 6.0 x 10^{-2}Pa during the deposition. Oxygen plasma was excited by a ring electrode with rf power of 200W. The film thickness was 5 to 40nm and the growth rate was 0.6nm/min.

A top YBCO layer was sequentially deposited on the base YBCO layer via the MgO intermediate layer by using the three electron beam guns. The specimen was kept in the evaporation chamber during the deposition of the MgO and the top YBCO layer. The interface between the MgO intermediate layer and the top YBCO layer was thereby made by an *in-situ* process. Y, Ba and Cu were used as metallic sources. The composition of the YBCO thin film was adjusted to a 1:2:3 ratio by controlling the evaporation rates of each of the three metal sources. The substrate temperature was kept at 620°C. The film thickness was 290nm and the growth rate was 24nm/min. Other deposition conditions of the YBCO were the same as those of the MgO deposition.

The base YBCO layer which was employed throughout the study in this paper was prepared by an rf magnetron sputtering method. Thus, the interface between the MgO layer and the base YBCO was formed by the so-called ex-situ process. The preparation of the base YBCO film was described in detail in previous papers [2,3]. No post-deposition annealing was performed.

Reflection high-energy electron diffraction (RHEED) using 100kV electrons of 5 degrees grazing incidence and X-ray diffraction using Cu-K_α radiation were used for crystallographic analysis. High-resolution scanning electron microscopy with a field emission electron gun (FE-SEM:HITACHI S-800) was employed for surface observation. For detailed insight into the hetero-epitaxial growth structure, cross-sectional transmission electron microscopy (TEM) with an acceleration voltage of 200kV was used.

RESULTS

Characterization of Base YBCO Layer

Two types of YBCO thin films were prepared as base YBCO layers. One was sputtered at a substrate temperature of 670°C and the other was sputtered at 600°C. Both films were 450nm thick. Figure 1 shows X-ray diffraction patterns for each of the YBCO thin films. In Fig.1(a), sharp (00n) (n=1,2,3...) peaks are observed. This indicates that the thin film consists of the single phase YBCO whose (001) plane is parallel to the MgO(100) substrate. In Fig.1(b), sharp (100) and (200) peaks are observed. This indicates that the (100) plane of the thin film is parallel to the substrate. In this paper, we refer to the YBCO thin films shown in Fig.1(a) as "c-axis oriented films" and the thin films shown in Fig.1(b) as "a-axis oriented films".

MgO/YBCO Epitaxial Growth on MgO Substrate

To form a YBCO/MgO/YBCO structure with smooth and sharp interfaces, it is important to control the orientation of the MgO layer to (100) epitaxial growth. Here, we investigated temperature dependence of morphology and crystallinity for the MgO thin film on the YBCO base layer. MgO ultra-thin films with thicknesses of 10nm were deposited on the c-axis oriented YBCO base layers at several substrate temperatures by the reactive vacuum evaporation system. FE-SEM and RHEED were employed for the investigation of the films. At about 120°C, the RHEED pattern showed weak and broad spots, which showed that the MgO thin film was slightly crystallized. The RHEED patterns at substrate temperatures of 200 and 400°C showed strong spots derived from the (100) growth of the MgO layer as shown in Fig.2. A sample prepared at more than 600°C showed not only spots from (100) growth but also several spots caused by (110) growth. The results are summarized in Fig.3. The surface morphology of the films prepared at substrate temperatures below 400°C is extremely smooth. However, at the substrate temperature of more than 600°C, many meshlike structures, 10 to 100nm in size, appear on the surface. We found that the growth temperature of the MgO layer should be kept at 200 to 400°C in order to obtain (100) growth and a smooth surface for the MgO layer on the c-axis oriented YBCO film.

MgO growth on an a-axis oriented YBCO film was also investigated. The growth temperature was 400°C. The thickness of the MgO layer was 10nm. Figure 4 shows the RHEED pattern for the MgO layer on the a-axis oriented YBCO base film. The direction of the incident electron beam was [010] relative to the substrate. The RHEED pattern shows strong spots derived from (100) growth of the MgO layer, which is almost the same pattern as the MgO layer on the c-axis oriented YBCO film. Therefore, it was found that a (100) epitaxial MgO layer can be grown on an a-axis oriented YBCO film as well as on a c-axis oriented film by the reactive vacuum evaporation method. This result indicates the potential for the growth of heterostructures using a pair of the a-axis oriented YBCO films and a MgO intermediate layer.

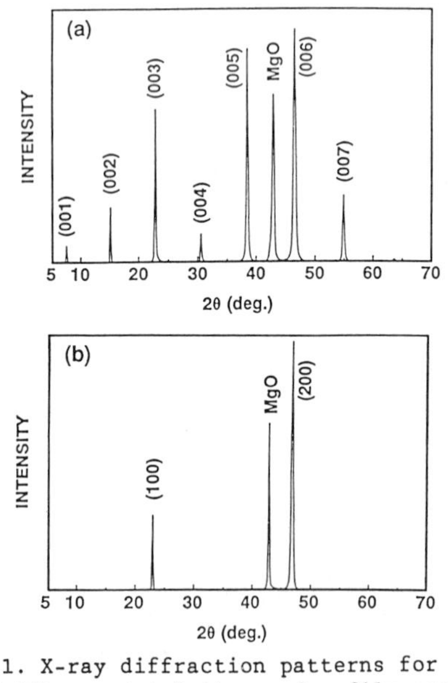

Fig.1. X-ray diffraction patterns for epitaxial YBCO films. (a) Pattern of a film grown at a substrate temperature of 670°C. (b) Pattern of a film grown at 600°C.

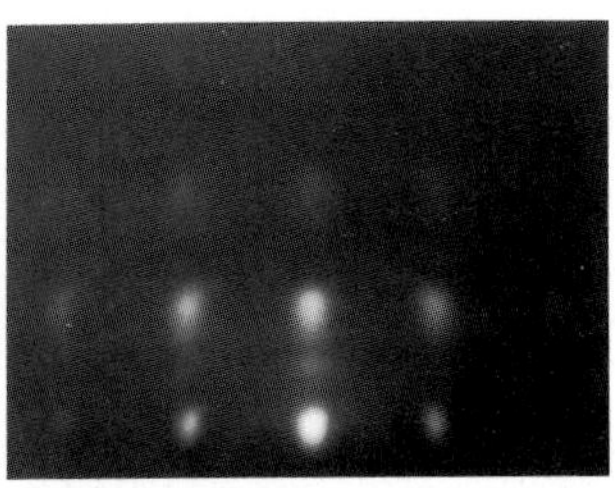

Fig.2. RHEED pattern for 10nm MgO thin film on c-axis oriented base YBCO.

YBCO/MgO/YBCO Structure

As it was demonstrated that a MgO(100) layer can be epitaxially grown on both c- and a-axis oriented YBCO films, epitaxial growth of the top YBCO layer on the base YBCO layer via the MgO intermediate layer was then performed.

Two types of YBCO thin films grown by rf-magnetron sputtering were used as base YBCO layers. A MgO intermediate layer was deposited on each of the c-axis oriented and a-axis oriented films at 400°C by the reactive vacuum evaporation system. A top YBCO layer with a thickness of 290nm was then deposited on each of the MgO/YBCO films at 620°C. To investigate the interface between the MgO intermediate layer and the YBCO layer, cross-sectional TEM observations were performed. Samples were reduced to 50nm in thickness for the observation. Figure 5 shows the results for the c-axis based sample. As shown in this figure, the top YBCO layer is epitaxially grown on the 5nm MgO intermediate layer and the (001) plane of the top layer is parallel to that of the base layer. The top and base YBCO layers were clearly separated by the MgO layer. Figures 6(a) and (b) show the results for the a-axis based samples. The thicknesses of MgO layers were different. When the thickness was 5nm, the top YBCO layer was epitaxially grown on the MgO intermediate layer. The c-axis direction was along the interface and parallel to the c-axis direction of the base layer (Fig.6(a)). When the thickness of the MgO layer was 12nm, the c-axis direction of the top YBCO layer was perpendicular to that of the base YBCO layer (Fig.6(b)).

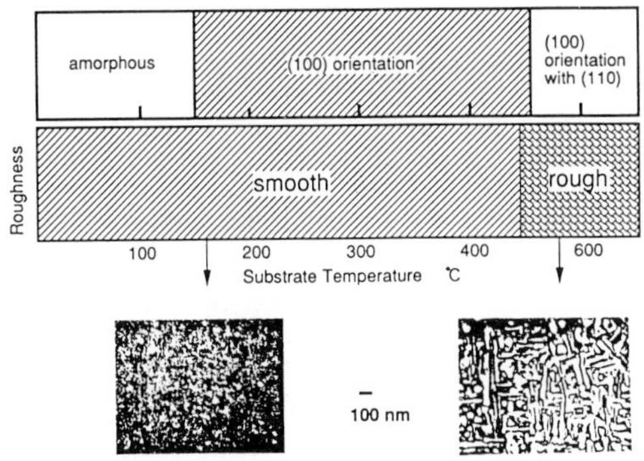

Fig.3. Summary of orientation and surface morphology of MgO layer on c-axis oriented YBCO thin film.

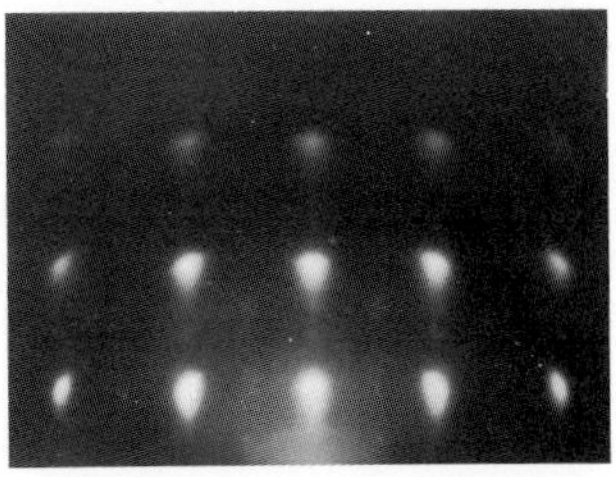

Fig.4. RHEED pattern for 10nm MgO thin film on a-axis oriented base YBCO.

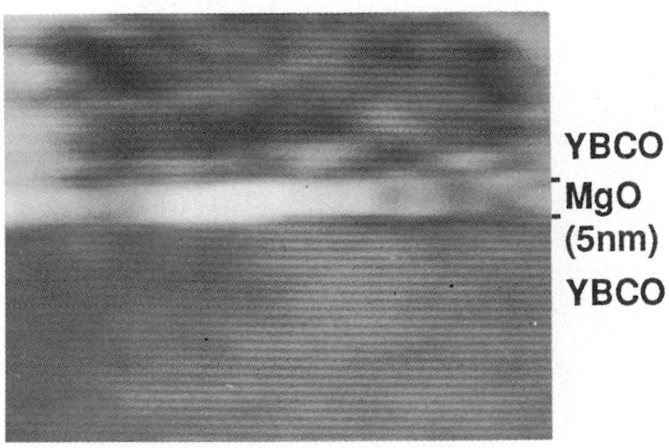

Fig.5. TEM cross-sectional lattice image of trilayered structure YBCO(001)/MgO/YBCO(001).

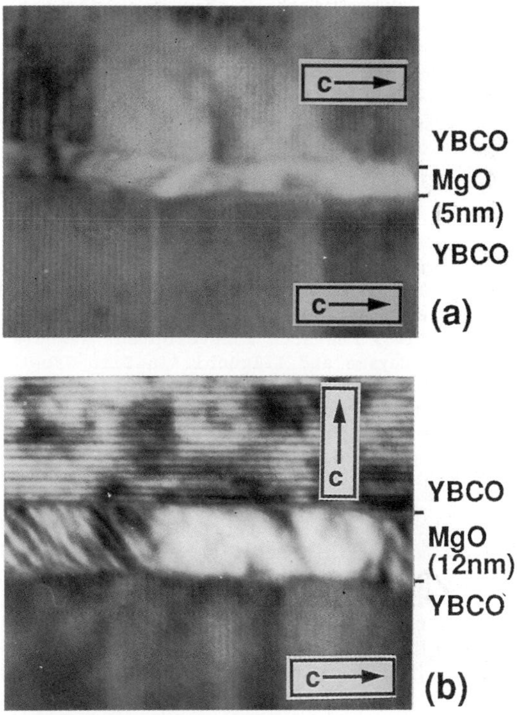

Fig.6. TEM cross-sectional lattice images of trilayered structure:(a) YBCO(100)/MgO/YBCO(100) and (b) YBCO(001)/MgO/YBCO(100).

Slight distortion can be observed on both the MgO-YBCO interfaces (top and base). This distortion seems to be a result of the mismatching of the lattice constant between the MgO and the YBCO. From observation of the cross sections as shown in Figs.6(a) and (b), the thickness of lattice disordering such as the distortion adjacent to the MgO layer seems to be less than 1 or 2nm, which is equal to one or two times the YBCO lattice constant of the c-axis.

DISCUSSION

We had thought that in-situ growth was necessary to form heterostructures with good interfaces using high temperature superconducting thin films. However, we have obtained comparatively sharp interfaces using the ex-situ process, as shown in Fig.5. We think that surface degradation, which may be caused by humidity in the air, may disappear because of the high temperatures of more than 600°C during the deposition of the top layer. Therefore, in-situ growth, which is employed for the fabrication of Nb tunnel junctions[10], may not be necessary in the case of high temperature superconductors. However, we should investigate the interdiffusion of the interface in detail to know whether a sharp interface is formed or not.

CONCLUSIONS

We have investigated the YBCO/MgO/YBCO tri-layered structure. It was confirmed that a MgO(100) ultra-thin film can be grown on a YBCO(100) thin film at about 400°C as well as on a YBCO(001) thin film by the reactive vacuum evaporation technique. From the cross-sectional TEM observation of the YBCO/MgO/YBCO, the following results were obtained. When a YBCO(001) thin film(c-axis oriented) was employed as the base layer, the top YBCO layer was epitaxially grown with (001) orientation via the 5nm intermediate MgO layer at 620°C. However, when a YBCO(100) thin film(a-axis oriented) was used as the base layer, the orientation of the top YBCO layer depended on the thickness of the MgO intermediate layer at the same temperature.

ACKNOWLEDGMENTS

The authors thank T. Nishikawa for the TEM and RHEED observations.

REFERENCES

1. Y.Enomoto, T.Murakami, M.Suzuki and K.Moriwaki "Largely Anisotropic Superconducting Critical Current in Epitaxially Grown $Ba_2YCu_3O_{7-y}$ Thin Film" (1987) *Jpn. J. Appl. Phys.* 26: L1248-1250
2. S.Tanaka and H.Itozaki "High-Jc Superconducting Single Crystalline HoBaCuO Thin Films by Sputtering" (1988) *Jpn. J. Appl. Phys.* 27: L622-624
3. S.Tanaka and H.Itozaki "Crystallinity and Morphology of Superconducting Thin Films Prepared by Sputtering" (1989) *Jpn. J. Appl. Phys.* 28: L441-443
4. T.Kobayashi and M.Iyori "Interface Problems of HTSC SIS Junction" (1989) *Extended Abstracts of 2nd Workshop on High-Temperature Superconducting Electron Devices*, Hokkaido, June, 1989 Research & Development Association for Future Electron Devices, Tokyo: 415-416
5. M.Furuyama and I.Iguchi "In-Situ Growth of $YBa_2Cu_3O_{7-x}$ Thin Films by Reactive Co-Evaporation Technique" (1989) *Extended Abstracts of 2nd Workshop on High-Temperature Superconducting Electron Devices*, Hokkaido, June, 1989 Research & Development Association for Future Electron Devices, Tokyo: 105-108
6. S.Tanaka, H.Nakanishi, K.Higaki and H.Itozaki "Epitaxial Growth of MgO Layer on $Y_1Ba_2Cu_3O_{7-y}$ Thin Film" (1990) *Jpn. J. Appl. Phys.* 29: 1059-1065
7. K.Hirata, K.Yamamoto, K.Iijima, J.Takada, T.Terashima, Y.Bando and H. Mazaki "Tunneling Measurements on Superconductor/Insulator/Superconductor Junctions Using Single-Crystal $YBa_2Cu_3O_{7-x}$ Thin Films", (1990) *Appl. Phys. Lett.* 56: 683-685
8. T.Venkatesan, A.Inam, B.Dutta, R.Ramesh, M.S.Hegde, X.D.Wu, L.Nazar, C.C.Chang, J.B.Barner, D.M.Hwang and C.T.Rogers "Epitaxial $Y_1Ba_2Cu_3O_{7-y}/Y_{1-x}Pr_xBa_2Cu_3O_{7-y}$ Heterostructures" (1990) *Appl. Phys. Lett.* 56: 391-393
9. K.Mizuno, H.Higashino, K.Setsune and K.Wasa "Fabrication of Thin-Film-Type Josephson Junctions Using a Bi-Sr-Ca-Cu-O/Bi-Sr-Cu-O/Bi-Sr-Ca-Cu-O Structure" (1990) *Appl. Phys. Lett.* 56: 1469-1471
10. A.I.Braginski, J.Talvacchio, M.A.Janocko and J.R.Gavaler "Crystalline Oxide Tunnel Barriers Formed by Thermal Oxidation of Aluminum Overlayers on Superconductor Surfaces" (1986) *J. Appl. Phys.* 60: 2058-2064

Electrical Characteristics of a $La_{1.5}Ba_{1.5}Cu_3O_{7-y}$ Thin Film and Its Contact with an $YBa_2Cu_3O_{7-x}$ Superconductor

SHIN'ICHIRO SAITOH, TOKUUMI FUKAZAWA, YOSHINOBU TARUTANI, and KAZUMASA TAKAGI

Central Research Laboratory, Hitachi, Ltd., 1-280, Higashi-Koigakubo, Kokubunji, Tokyo, 185 Japan

ABSTRACT

A $La_{1.5}Ba_{1.5}Cu_3O_{7-y}$ thin film is fabricated for use as a normal layer in an S-N coupling element, and its crystallorgraphic and electrical characteristics are measured. The film does not show superconducting behavior and has a negative coefficient for electrical resistance in relation to temperature. An $YBa_2Cu_3O_{7-x}$-$La_{1.5}Ba_{1.5}Cu_3O_{7-y}$-$YBa_2Cu_3O_{7-x}$ trilayer film is also fabricated by the sequential sputtering method. Epitaxial growth of each layer is confirmed by TEM observation, and a supercurrent is detected through the $La_{1.5}Ba_{1.5}Cu_3O_{7-y}$ layer.

KEY WORDS: $La_{1.5}Ba_{1.5}Cu_3O_{7-y}$ thin film, S-N coupling

INTRODUCTION

A $LaBa_2Cu_3O_{7-y}$ has an isomorphic crystal structure of $YBa_2Cu_3O_{7-x}$, and a critical temperature, T_c, of 90K[1]. The $La_{1+\kappa}Ba_{2-\kappa}Cu_3O_{7-y}$ oxide system also has a wide solid solution range of La atoms at the Ba site. It should be noted that the superconducting characteristics of bulk $La_{1+\kappa}Ba_{2-\kappa}Cu_3O_{7-y}$ have been intensively investigated and their dependence on the La concentration and oxygen content clarified[2]. The T_c of $La_{1+\kappa}Ba_{2-\kappa}Cu_3O_{7-y}$ decreases as La concentration is increased, reaching a normal state up to a k value of 0.5. The oxide is one of the rare 1-2-3 type oxides, which have a very wide solubility range.

On the other hand, the superconducting coupling device (S-N-S) comprising the oxide material is very difficult to fabricate since the superconducting electron density tends to discontinuously decrease at the S-N boundary when the normal layer is made of materials different from those of the oxide superconductor. This may be due to both the formation of an intermediate layer and the short coherence length of oxide superconductors. A $La_{1.5}Ba_{1.5}Cu_3O_{7-y}$ may be suitable for the normal layer in an S-N-S junction due to its having the same crystal structure as $YBa_2Cu_3O_{7-x}$ and almost the same constituent elements.

In this report, a $La_{1.5}Ba_{1.5}Cu_3O_{7-y}$ thin film is fabricated and its crystallographic and electrical characteristics are investigated for use as a normal layer in an S-N-S junction. An $YBa_2Cu_3O_{7-x}$-$La_{1.5}Ba_{1.5}Cu_3O_{7-y}$-$YBa_2Cu_3O_{7-x}$ trilayer is also fabricated and its cross-sectional structure and electrical characteristics are studied. The $La_{1.5}Ba_{1.5}Cu_3O_{7-y}$ structure is discussed based on these experimental results.

EXPERIMENTAL METHODS

The $YBa_2Cu_3O_{7-x}$ and $La_{1.5}Ba_{1.5}Cu_3O_{7-y}$ thin films were formed by an RF magnetron sputtering method. The sputtering gas was Ar+50%O_2 and the total gas pressure was 4Pa. The subtrate temperature during deposition was from

600 to 680℃. The substrate materials used were single crystals of $SrTiO_3$(110) and $SrTiO_3$(100). The composition in the films was measured by electron probe microanalysis (EPMA). A cross section of the trilayer film was observed with a transmission electron microscope (TEM) at an acceleration voltage of 300keV.

RESULTS AND DISCUSSION

The $La_{1.5}Ba_{1.5}Cu_3O_{7-y}$ thin film deposited on a $SrTiO_3$(110) substrate had the isomorphic structure of $YBa_2Cu_3O_{7-x}$, and its (103) plane was parallel to the substrate. An example of the resistivity-temperature (ρ-T) curve for the $La_{1.5}Ba_{1.5}Cu_3O_{7-y}$ thin film is shown in Fig.1. The film did not show superconducting behavior and had a negative coefficient of electrical resistance in relation to temperature. The logρ dependence on $T^{1/4}$ deviated from the linear curve at temperatures higher than 10K.
The reciprocal of the Hall coefficient, $1/eR_H$, and the Hall mobility are shown in Fig.2. The Hall mobility was derived using the Hall coefficient and resistivity values. The value of $1/eR_H$ is almost equivalent to the carrier concentration if the single band model is valid for the $La_{1.5}Ba_{1.5}Cu_3O_{7-y}$ system. The $1/eR_H$ value is greater than $10^{27}/m^3$, and comparable to that of the superconducting $YBa_2Cu_3O_{7-x}$. The temperature dependences of the ρ and Hall mobility indicate that the electrical conduction mechanism in the $La_{1.5}Ba_{1.5}Cu_3O_{7-y}$ thin film does not obey the variable-range hopping model but the one-electron wave function may be overlapping each other at low temperatures[3].
An $YBa_2Cu_3O_{7-x}$-$La_{1.5}Ba_{1.5}Cu_3O_{7-y}$-$YBa_2Cu_3O_{7-x}$ trilayer film was fabricated using a sequential sputtering method. A cross-sectional structure of a trilayer observed by TEM is shown in Fig. 3. The substrate material is $SrTiO_3$(110). The thickness of the $La_{1.5}Ba_{1.5}Cu_3O_{7-y}$ layer is 50nm. The $La_{1.5}Ba_{1.5}Cu_3O_{7-y}$ layer has a uniform thickness distribution and no pinhole in spite of the interface roughness. The c-axis is parallel to the film surface and the [110] direction is perpendicular to the film surface of both $YBa_2Cu_3O_{7-x}$ layers according to the electron diffraction patterns. These results indicate that the crystal growth morphology reaches the upper $YBa_2Cu_3O_{7-x}$ layer.
From Fig. 3(b), can also be seen that the c-axis of both $YBa_2Cu_3O_{7-x}$ layers is parallel to the film plane. Both (110) and (103) planes are mixed in the $La_{1.5}Ba_{1.5}Cu_3O_{7-y}$ layer. The (110) growth morphology may be caused by the

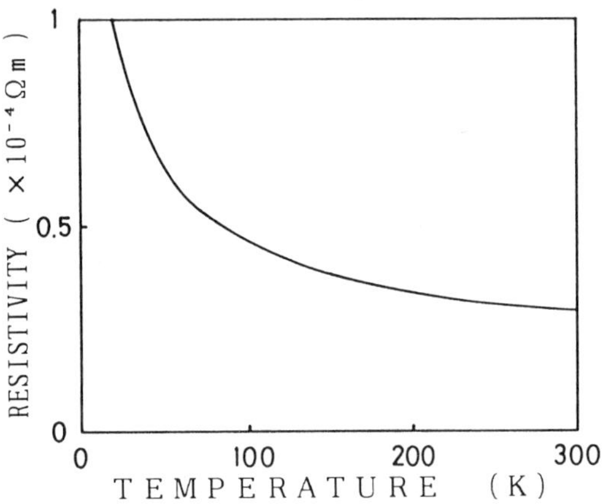

Fig.1 Resistivity-temperature curve of a $La_{1.5}Ba_{1.5}Cu_3O_{7-y}$ thin film.

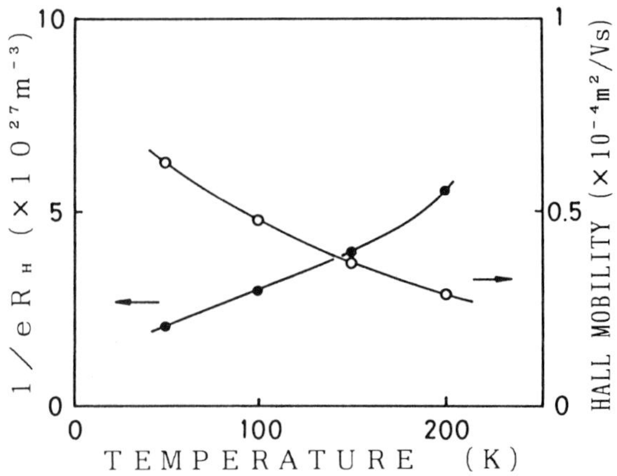

Fig.2 $1/eR_H$ and Hall mobility of a $La_{1.5}Ba_{1.5}Cu_3O_{7-y}$ thin film.

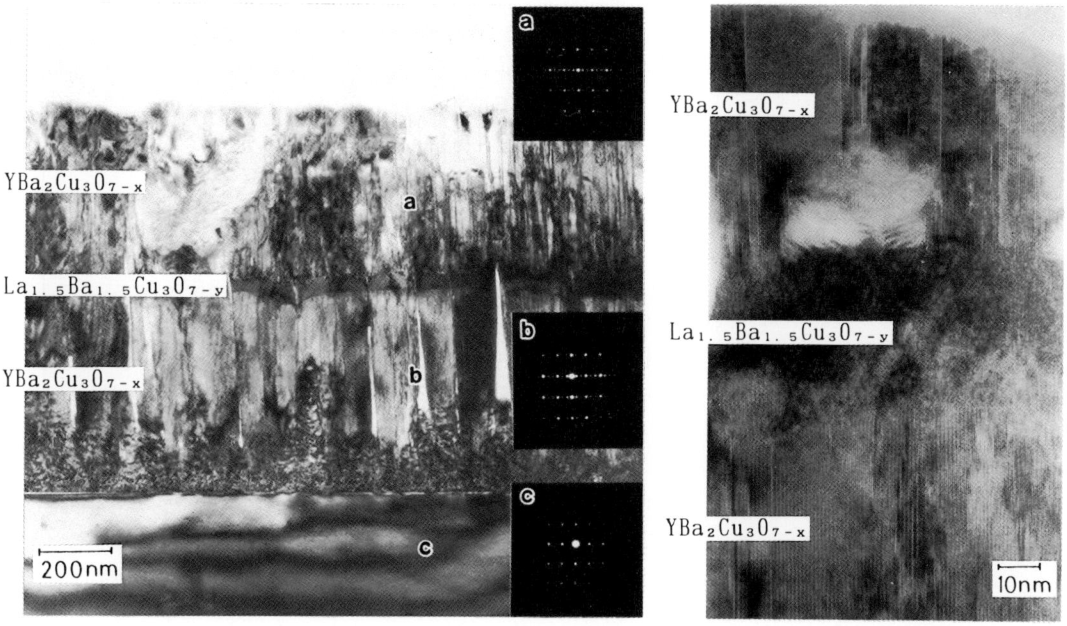

(a) Trilayer (b) $La_{1.5}Ba_{1.5}Cu_3O_{7-y}$ layer

Fig.3 TEM photograph of an $YBa_2Cu_3O_{7-x}$-$La_{1.5}Ba_{1.5}Cu_3O_{7-y}$-$YBa_2Cu_3O_{7-x}$ trilayer.

crystal orientation of the lower $YBa_2Cu_3O_{7-x}$ layer.

The composition distributions of the trilayer film having a $La_{1.5}Ba_{1.5}Cu_3O_{7-y}$ layer with 50nm thick were measured by EPMA along the film thickness direction. The element Y was not detected in the $La_{1.5}Ba_{1.5}Cu_3O_{7-y}$ layer based on the distribution of the characteristic X-ray intensities. In addition, the element La was not detected in the $YBa_2Cu_3O_{7-x}$ layer. The interdiffusion of La and Y did not occur at a detectable level during the trilayer deposition sequence.

An S-N-S junction was fabricated with the trilayer film. The junction area was $0.3 \times 0.3mm^2$. The thickness of the $La_{1.5}Ba_{1.5}Cu_3O_{7-y}$ layer was 800nm. An example of current-voltage characteristics for the S-N-S junction are shown in Fig. 4. The temperatures at which measurements were taken were 4.2K and 25K. A supercurrent was detected up to 25K. The trilayer analysis supported that this supercurrent was not due to any pinhole in the $La_{1.5}Ba_{1.5}Cu_3O_{7-y}$ layer, or the formation of a diffusion layer at the interface. In addition, discontinuity of the $La_{1.5}Ba_{1.5}Cu_3O_{7-y}$ layer was not seen at the lower $YBa_2Cu_3O_{7-x}$ film edges. The crystallographic structure and composition distribution for the trilayer having the thickness of 800nm were almost the same as those for 50nm.

The S-N-S junction was also fabricated on $SrTiO_3$(100) substrate. All three layers had their [001] orientations maintained perpendicular to the film surface. The critical current density of this S-N-S junction was $4 \cdot \times 10^3 A/m^2$ at 4.2K. This value is two orders of magnitude smaller than that of the junction on the $SrTiO_3$(110) substrate. This difference may be due to the strength of the electrical coupling at the interfaces and the electrical characteristics of a $La_{1.5}Ba_{1.5}Cu_3O_{7-y}$ layer owing to the crystal orientation.

The coherence length of a $La_{1.5}Ba_{1.5}Cu_3O_{7-y}$ layer is about 1nm according to the Seto-Van Duzer formula and taking into account the parameter values in Fig. 2[4]. The spacing between two $YBa_2Cu_3O_{7-x}$ electrodes was three orders of

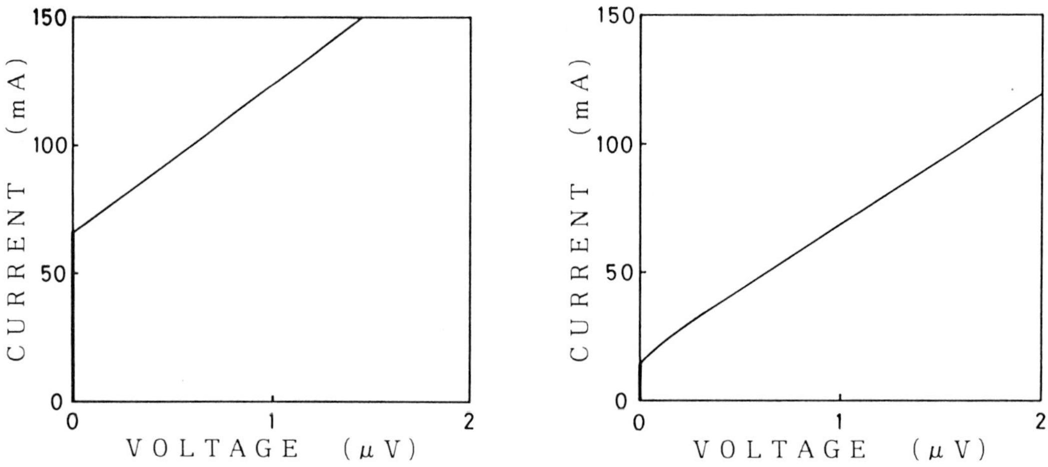

(a) 4.2 K (b) 25 K

Fig.4 Current-voltage characteristics of a $YBa_2Cu_3O_{7-x}$-$La_{1.5}Ba_{1.5}Cu_3O_{7-y}$-$YBa_2Cu_3O_{7-x}$ junction.

magnitude larger than the coherence length. Therefore, a supercurrent could not be detected for the junction in the case of this normal layer thickness due to the proximity effect.

The existence of a supercurrent may originate from the characteristic nature of $La_{1.5}Ba_{1.5}Cu_3O_{7-y}$. A Ba site is only partially substituted by La atoms. The $La_{1.5}Ba_{1.5}Cu_3O_{7-y}$ may be comprised of superconducting domains and nonsuperconducting ones. Such domains can exist when the domain size is larger than coherence length. The $La_{1.5}Ba_{1.5}Cu_3O_{7-y}$ film itself was nonsuperconducting in the film surface direction. However, it may be possible to form a superconducting path through the nonsuperconducting domains with the proximity effect in the film thickness direction. Much more experimental data is needed to fully understand the conduction mechanism of the $La_{1.5}Ba_{1.5}Cu_3O_{7-y}$.

In conclusion, a $La_{1.5}Ba_{1.5}Cu_3O_{7-y}$ thin film was fabricated for use as a coupling layer for an S-N-S junction. The $La_{1.5}Ba_{1.5}Cu_3O_{7-y}$ thin film was nonsuperconducting and had a high carrier concentration that was greater than $10^{27}/m^3$. An $YBa_2Cu_3O_{7-x}$- $La_{1.5}Ba_{1.5}Cu_3O_{7-y}$-$YBa_2Cu_3O_{7-x}$ trilayer showed no interdiffusion of La and Y. The trilayer had epitaxial growth for each layer. Supercurrent was detected through the $La_{1.5}Ba_{1.5}Cu_3O_{7-y}$ layer.

This work was performed under the management of FED (the R & D Association for Future Electron Devices) as a part of the R & D of Basic Technology for Future Industries supported by NEDO (New Energy and Industrial Technology Development Organization). The authors are indebted to Dr. U. Kawabe and Dr. K. Miyauchi for their encouragements through this work.

REFERENCES

1. Segre CU, Dabrowski B, Hinks DG, Zhang K, Jorgensen JD, Beno MA, and Shuller IK (1987), Nature 329: 227
2. Wada T, Suzuki N, Maeda A, Yabe T, Uchinokura K, Uchida S, and Tanaka S (1989), Phys. Rev.B 39, 9126
3. Mott NF (1974), Metal-Insulator transitions, Taylor & Francis, London
4. Seto J and Van Duzer T (1974), Low Temp. Phys (LT-13), 3: 328

Preparation of Tunnel-Type Junctions Consisting of $Bi_2Sr_2CaCu_2O_x$ Single Crystals

K. TAKAHASHI, K. SHIMAOKA, M. KAMINO, Y. YOSHISATO, and S. NAKANO

Functional Materials Research Center, SANYO Electric Co., Ltd., 18-13, Hashiridani 1-chome, Hirakata, Osaka, 537 Japan

ABSTRACT

Quasi-particle injection through the tunnel-type junctions of Nb/MgO/BSCCO has been studied using $Bi_2Sr_2CaCu_2Ox$ single crystals and MgO artificial barriers for the first time, which is necessary for the realization of superconducting devices. Single crystals of $Bi_2Sr_2CaCu_2Ox$ were grown by the self-flux method, then cleaved. MgO and Nb layers were deposited on a cleavege plane of $Bi_2Sr_2CaCu_2Ox$ single crystals by electron beam evaporation to a thickness of 4-7 nm and 200 nm, respectively. The junction area was 0.5 x 0.5 mm^2. Measurement of electrical resistivity showed that the junction maintained a residual resistance of 260 Ω at 5 K. The gap-opening structure was clearly observed at about ± 60 mV in the dI/dV-V characteristics.

KEY WORDS: $Bi_2Sr_2CaCu_2Ox$, self-flux method, single crystal, S/I/S

INTRODUCTION

The discovery of high-Tc superconductors such as Y-Ba-Cu-O [1], Bi-Sr-Ca-Cu-O [2] and Tl-Ba-Ca-Cu-O [3] has generated great impact, leading to the widespread use of these materials for superconductive devices at liquid-nitrogen temperature.

Most superconductive devices need junction structures composed of superconductors (S), insulators (I) and semiconductors (SE). Several kinds of junction structures of S/I/S, S/N/S and S/SE can contribute to the operation of superconductive devices through their non-linear I-V characteristics and high frequency response. Some experimental reports have been written on the fabrication of junction structures such as Y-Ba-Cu-O/Al_2Ox/Nb [4], Y-Ba-Cu-O/Au/Nb [5] and so on.

However, adequate characteristics for the junctions have not been achieved in spite of tremendous efforts. The main difficulty is that the coherent length of high-Tc oxide superconductors is only 2-3 nm at most. It is difficult to reproductively prepare such a thin, smooth layer free of pin-holes between superconductors because of a lack of homogeneity and irregularity in the interfaces. Therefore, the surface morphology of the first superconducting layer, on which the second non-superconducting layer and the third superconducting layer are fabricated, should be exactly controlled for the fabrication of junction structures.

This paper reports on a fabrication process and the characteristics of a Nb/MgO/Bi-Sr-Ca-Cu-O structure using a bulk superconducting substrate of a $Bi_2Sr_2CaCu_2Ox$ single crystal processed by the self-flux method. This substrate is so homogeneous and highly dense that it can be applied as a substrate instead of MgO, $SrTiO_3$ or other conventional substrate materials. It can be considered that the superconductivity of the substrates themselves can simplify the device structure and that the use of the cleavage plane is effective for the solution of interfacial problems in junction structures.

FABRICATION OF $Bi_2Sr_2CaCu_2O_x$ SINGLE CRYSTALS

The single crystals of high-Tc $Bi_2Sr_2CaCu_2O_x$ were grown using a self-flux method. Mixtures of analytical-grade Bi_2O_3, $SrCO_3$, $CaCO_3$ and CuO powders were prepared in cation ratios of 2.2:1.8:1:2. The mixtures were ground using a mortar and pestle, placed in alumina crucibles, heated in a furnace at 950°C for 3 hours, cooled to 800°C at a rate of 2°C/h, then finally cooled to room temperature in the furnace. The reaction product was a solid mass containing single crystals. The single crystals were mechanically isolated by cleavage.

Figure 1 shows a flake of $Bi_2Sr_2CaCu_2O_x$ with typical dimensions on the order of 5x5 mm^2. As shown in Fig. 2, the surface of this single crystal is very smooth compared with that of $Bi_2Sr_2CaCu_2O_x$ thin film processed by the conventional method.

The X-ray diffraction pattern of the facets of $Bi_2Sr_2CaCu_2O_x$ single crystals is shown in Fig. 3. This pattern indicates that the crystal is composed of single-phase $Bi_2Sr_2CaCu_2O_x$, with a facet perpendicular to (001). Furthermore, the chemical composition of single crystals was confirmed to be $Bi_2Sr_2CaCu_2O_x$ using energy dispersive X-ray spectroscopy.

The temperature dependence of resistivity for $Bi_2Sr_2CaCu_2O_x$ single crystals with electrical current perpendicular to the c-axis is shown in Fig. 4. A sharp superconducting transition is observed at 91 K, which is rather higher than previously reported. [6]

Fig. 1 $Bi_2Sr_2CaCu_2O_x$ single crystal

Fig. 2 SEM micrograph of $Bi_2Sr_2CaCu_2O_x$

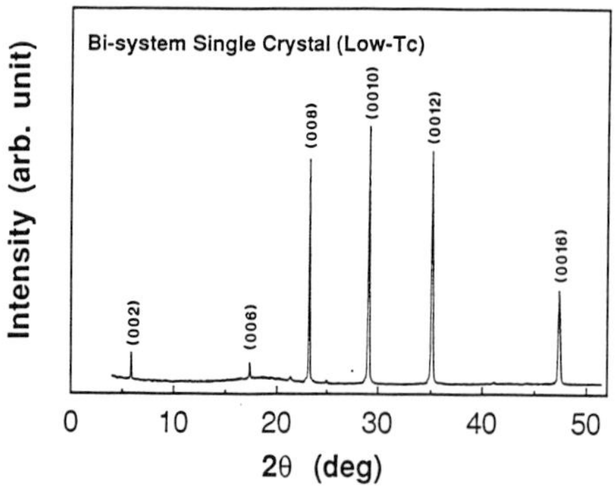

Fig. 3 X-ray diffraction pattern of $Bi_2Sr_2CaCu_2O_x$ single crystal

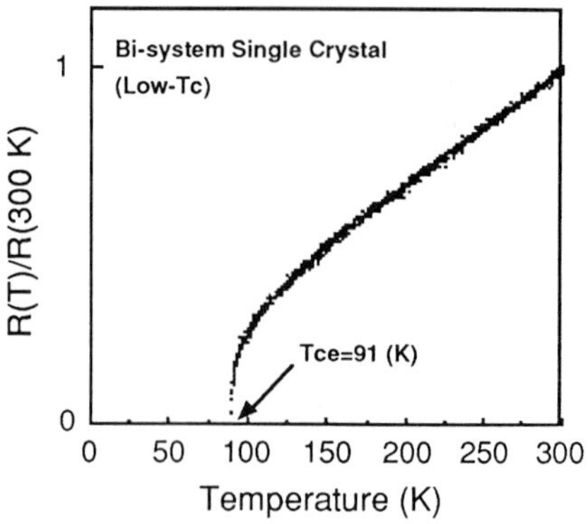

Fig. 4 Temperature dependence of resistivity for $Bi_2Sr_2CaCu_2O_x$ single crystal

FABRICATION OF S/I/S JUNCTION STRUCTURES USING $Bi_2Sr_2CaCu_2O_x$ SINGLE CRYSTALS

Subsequently, S/I/S junction structures were fabricated using $Bi_2Sr_2CaCu_2O_x$ substrates of single crystals as the lower electrode on which an MgO barrier and Nb layer had been deposited with a thickness of 4-7 nm and 200 nm, respectively.

As shown in Fig. 5, first of all, the clean surface of $Bi_2Sr_2CaCu_2O_x$ single crystal was prepared by cleaving it in air. An Au film was fabricated on the cleaved plane as electrodes for electrical measurements. Then, a MgO barrier layer with a thickness of 4-7 nm was deposited onto the cleaved plane at a rate of 1.5-60 nm/min by electron-beam evaporation. A Nb superconducting thin film with a thickness of 200 nm was fabricated on the MgO insulation layer at a rate of 120 nm/min at a substrate temperature of 400°C. Finally, an Au electrode and MgO buffer layer were fabricated to measure electric characteristics through the $Bi_2Sr_2CaCu_2O_x$/MgO/Nb junction structure with a thickness of 500 nm and 300 nm, respectively.

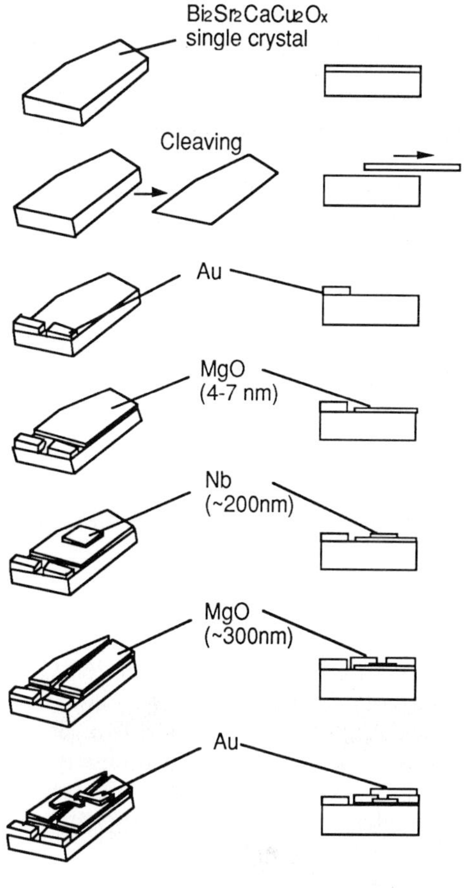

Fig. 5 Fabrication of S/I/S junction structure

Figure 6 shows the I-V characteristics of this structure with a barrier thickness of 7 nm at 5 K. Non-linearity was clearly observed at 20-30 mV. The zero-bias resistance of the junction was about 260 Ω at 5 K.

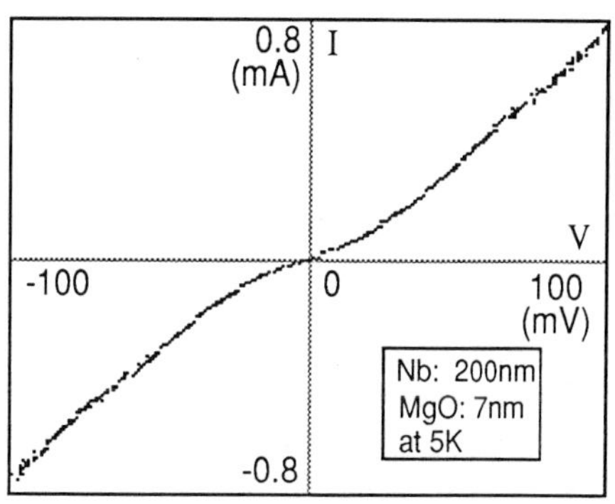

Fig. 6 I-V characteristics of Nb/MgO/BSCCO junction

The tunnel conductance, i.e., dI/dV vs V, was studied as a function of temperature, and the data normalized by that of the normal state are presented in Fig. 7. Information regarding the density of states of the quasiparticles can be deduced by normalizing the dI/dV vs V spectrum of the superconducting state by that of the normal state. The latter is usually obtained by applying a magnetic field higher than the critical field of the specimen. But the upper critical field, H_{c2}, of $Bi_2Sr_2CaCu_2O_x$ is very high [7], and it is very difficult to carry out such measurements. Instead, we used the dI/dV vs V curve at 120 K to normalize the spectra. From this figure, it is obvious that the magnitude of the gap-like portion of the normalized dI/dV vs V curves have only a very weak dependence on temperature, which is different from BCS-type superconductor. The gap-opening structure at about ± 60 mV became slightly apparent at 90 K, which is below the Tc of $Bi_2Sr_2CaCu_2O_x$, 91 K, and was obviously confirmed below 60 K. Furthermore, the anomaly around zero-bias voltage, similar to that of Au/MgO/YBCO reported by Iguchi et al. [8], was observed below 45 K. Although the gap-opening structure at $\pm$ 60 mV seem to be larger than the energy gap previously reported, it can be considered that these structures are strongly involved in the quasi-particle injection into the superconducting $Bi_2Sr_2CaCu_2O_x$ substrate due to the significant change in dI/dV-V characteristics. Further investigation will need the in-situ fabrication and low temperature process in order to eliminate interfacial problems such as adsorption and diffusion.

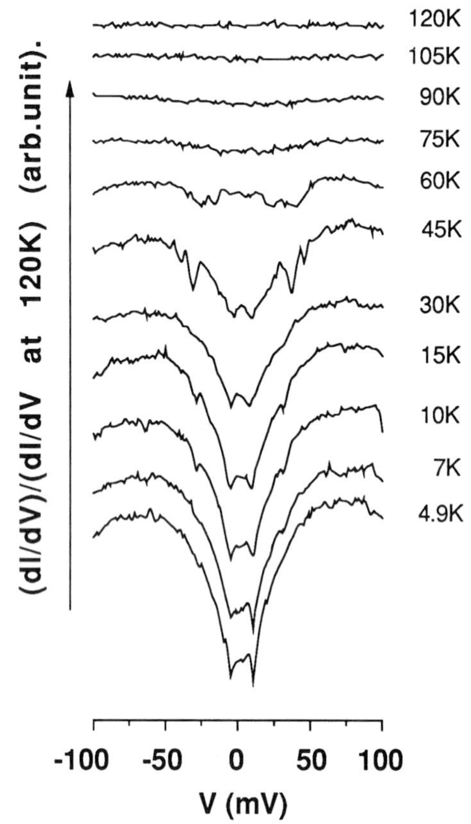

Fig. 7 The differential conductance curves at several temperatures normalized with the differential conductance curve at 120 K.

CONCLUSIONS

Quasi-particle injection through the tunnel-type junctions of Nb/MgO/BSCCO has been studied using the $Bi_2Sr_2CaCu_2O_x$ single crystals grown by the self-flux method. Measurement of electrical resistivity showed that the junction maintained a residual resistance of 260 Ω at 5 K. Non-linearity was observed in the I-V characteristics of this junction. The gap-opening structure at about $\pm$ 60 mV was clearly observed in the normalized dI/dV-V characteristics, corresponding to the appearance of superconductivity in $Bi_2Sr_2CaCu_2O_x$.

ACKNOWLEDGEMENTS

This work was performed under the management of FED (the R & D Association for Future Electron Devices) as a part of the R & D of Basic Technology for Future Industries sponsored by NEDO (New Energy and Industrial Technology Development Organization).

REFERENCES

1. M. K. Wu, J. R. Ashburn, C. J. Trong, P. H. Hor, R. L. Meng, L. Gao, Z. J. Huang, Y.Q. Meng and C. W. Chu: Phys. Rev. Lett., 58, 908 (1987)
2. H. Maeda, Y. Tanaka, M. Fukutomi and T. Asano: Jpn. J. Appl. Phys., 27, L209 (1988)
3. Z. Z. Sheng and A. M. Herman: Nature., 332, 55 (1988)
4. A. Nakayama, A. Inoue, K. Takeuchi and Y. Okabe: Jpn. J. Appl. Phys., 26, L2055 (1987)
5. H. Akoh, F. Shinoki, M. Takahashi and S. Takada., Jpn. J. Appl. Phys., 27, L519 (1988)
6. K. Takahashi, M. Kamino, T. Yokoo, Y. Yoshisato, and S. Nakano: Proc. of the 1st Int. Symp. on Superconductivity, Tukuba., 243 (1989)
7. Y. Hidaka, M. Oda, M. Suzuki, Y. Maeda, Y. Enomoto and T. Murakami: Jpn. J. Appl. Phys. 27 L538 (1988)
8. M. Furuyama, I. Iguchi, K. Shirai, T. Kusumori, H. Ohtaka, S. Tomura, M. Nasu: Jpn. J. Appl. Phys., 29, L459 (1990)

Formation of Tunnel Barriers with Surface Plasma Treatment

TOSHIYUKI MATSUI[1*], MASAYUKI MIYAZAKI[1], TOSHIHIRO KUROSAKA[2], GENSOH MATSUBARA[1], AKIYOSHI NAKAYAMA[2], KAZUO MUKAE[3], and YOICHI OKABE[1]

[1] RCAST. The University of Tokyo, Meguro-ku, Tokyo, 153 Japan
[2] Faculty of Engineering, Kanagawa University, Kanagawa-ku, Yokohama, 221 Japan
[3] Fuji Electric Corp. R&D Ltd., Yokosuka, 240-01 Japan

ABSTRACT

The barriers formed by exposing the YBCO surface to CCl_2F_2 plasma were studied with YBCO-barrier-Au tunnel junctions. Results shows that this barrier has the potential for the Josephson tunnel barrier.

KEY WORD: surface plasma treatment, YBCO, CCl_2F_2, tunnel junction

INTRODUCTION

Various experimental results have been reported about attempts of Josephson tunnel junctions with high critical temperature (high-Tc) superconductors [1-6]. However it seems that perfect Josephson junction has not been reported yet. This is probably due to not only surface problems of high-Tc materials, but also problems of tunnel barriers. For the tunnel barrier, uniformity, no stresses and no damages onto the high-Tc materials during the fabrication process are required. In our recent study, junctions of YBCO(110)-barrier-YBCO(110) layered structure were fabricated using MgO and $SrTiO_3$ for the artificial barriers. However, as the morphology of the barriers depends strongly on that of a base YBCO electrode, the barriers had pinholes in many cases [6]. Moreover after the deposition of these barriers, the lattice constant of the base YBCO surface was changed by stresses with their formation[7]. In these junctions, therefore, only the weak-link type characteristics could be obtained.

On the other hand, in metal-superconductors, two methods to form the Josephson tunnel barriers are known. One is the deposition of insulating material on the base electrode, that is to say "artificial barrier", and the other is a surface reaction of the base electrode with gas such as oxygen. For the latter, good step coverage and controllability of the thickness were reported in Pb-based and Nb-based systems [8-10]. Therefore this technique was tested for the high-Tc superconductors.

In the latter technique for the high-Tc superconductors, barriers formed by exposing the YBCO surface to CF_4 plasma was reported [4]. However, in our recent work, fluorine diffused deeply into YBCO during the CF_4 plasma treatment. This caused the decrease of the superconducting energy gap and Tc of YBCO. It suggests that this is unsuitable material for the tunnel barrier. Therefore, in order to prevent such diffusion, chlorine, whose ion radius is larger than that of fluorine, was tested with YBCO-barrier-Au junctions.

In this paper, we will report the potential of the barriers formed with the surface plasma treatment.

FABRICATION OF JUNCTIONS

The base YBCO electrodes were prepared by an rf magnetron sputtering method at the low gas pressure [11]. By the technique, a high deposition rate, about 12nm/min, and smooth surfaces could be realized in YBCO films. The values of Tc were about 65K. The fabrication procedure of YBCO-barrier-Au junctions is shown in Fig.1. First the oxygen plasma treatment was carried

*on leave from Fuji Electric Corp. R & D Ltd.

out with an oxygen pressure of 10Pa and an rf power density of 140mW/cm^2 for 30min to improve the YBCO surface. Then, to form the barrier, the YBCO surface was exposing to CCl_2F_2 plasma. This plasma treatment was carried out with a gas pressure of 10Pa and an rf power density of 140mW/cm^2. After that Au was deposited on it as a top electrode. Au was selected for the top electrode to avoid a formation of insulating layer with reactions. The thickness of YBCO films was 300nm and of Au was 600nm. The patterning of these junctions were performed by the metal mask process. The junction area was 0.25mm^2.

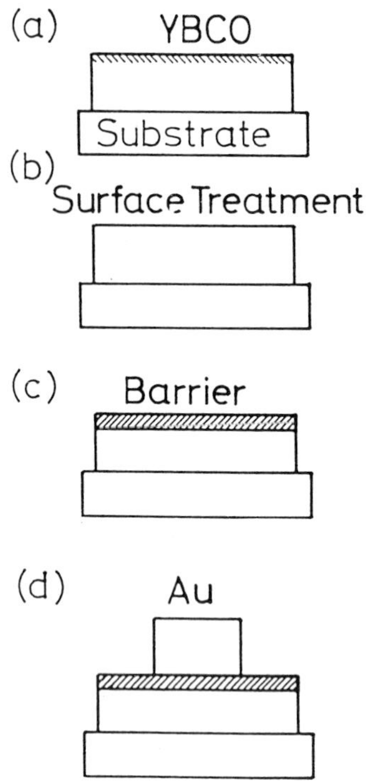

Fig.1 Fabrication procedure of YBCO-barrier-Au junction. (a)Deposition of YBCO film. (b)O_2 plasma treatment to improve the YBCO surface. (c)CCl_2F_2 plasma treatment to form the barrier. (d) Formation of Au for the top electrode.

PROPERTIES OF BARRIER LAYERS

(1) I-V characteristics of YBCO-barrier-Au junctions

In order to obtain the informations of a reaction layer formed by the CCl_2F_2 plasma treatments, the resistance was measured with the conventional four probes method. Figure 2 shows the temperature dependence of the base YBCO and the junction resistance after the junction fabrication. The barrier layer was formed with the CCl_2F_2 plasma treatment for 5min. Although the resistance of the base YBCO decreased with the decrease of the temperature, that of the junction behaved like a semiconductor. The Tc of the base YBCO electrode was not changed after the formation of junction. It suggests that with the CCl_2F_2 plasma treatment, an insulating layer was formed at the surface of YBCO, that is to say, there are no diffusion like the case of the CF_4 treatments. Figure 3 shows the typical I-V characteristic of the junctions at 4.5K. In this figure, c-axis oriented YBCO film was used for the base electrode. On this curve, inflection points around +12mV and -12mV were observed. As the voltages at these points decrease with increasing the temperature as shown in Fig.4, we decide these are corresponding to the superconducting energy gap of the base YBCO. The value of $2\Delta/kTc$, where Δ is the superconducting energy gap of YBCO at 0K and k is a Boltzmann constant, is estimated to 4.6. According to the depth profile by AES (Auger Electron Spectroscopy), the tunnel barriers were formed with chloride as shown in Fig.5. Note that although CCl_2F_2 gas involves fluorine, it was not detected. The thickness of the tunnel barrier was estimated about 5nm for the 5min-treatment.

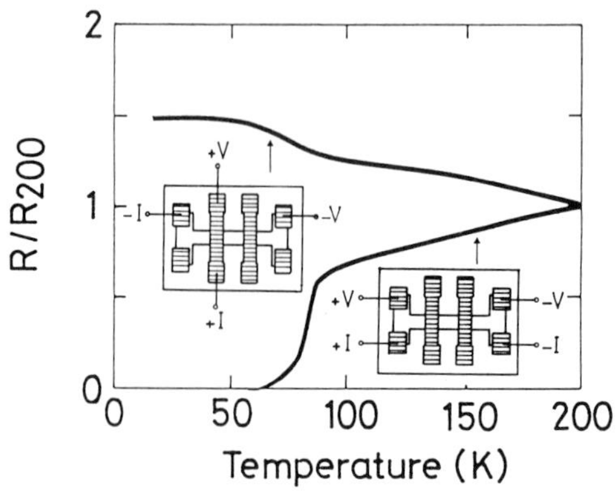

Fig.2 Temperature dependence of the base YBCO and the junction resistance.

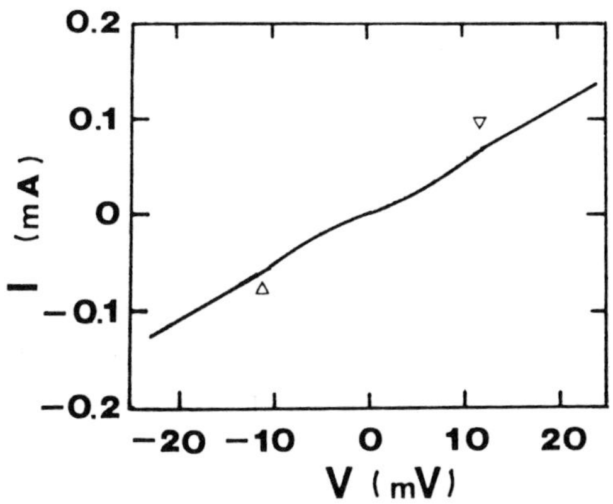

Fig.3 Typical I-V characteristic of YBCO-barrier-Au junction at 4.5K.

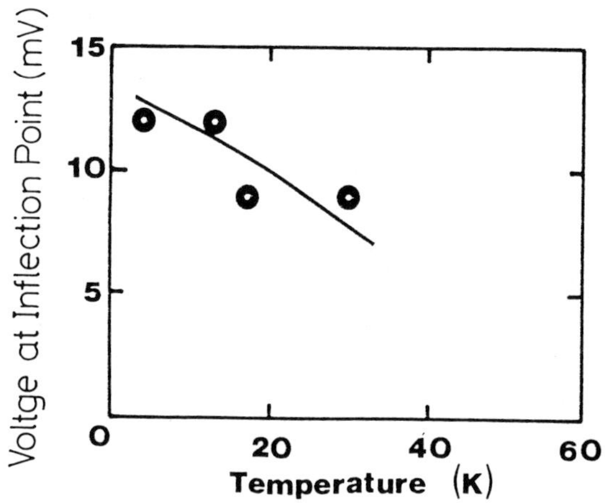

Fig.4 Temperature dependence of the voltage at the inflection point.

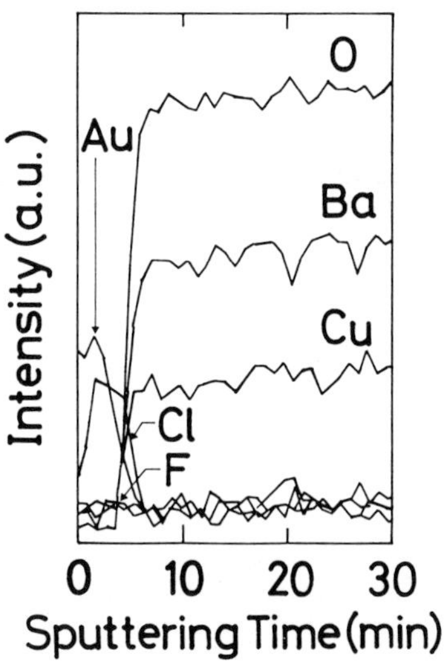

Fig.5 Depth profile of YBCO-barrier-Au junction. The barrier was formed by 5min-treatment.

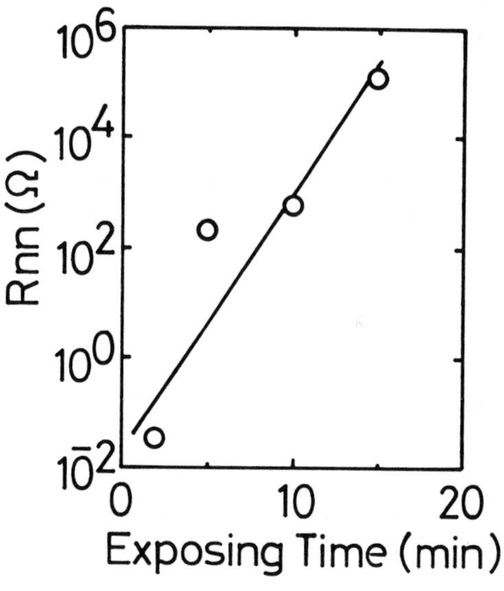

Fig. 6 Characteristic of the junction resistance, Rnn, at 4.5K vs. CCl_2F_2 treatment time.

(2)Junction Resistance

Figure 6 shows the junction resistance, Rnn, of YBCO-barrier-Au as a function of the CCl_2F_2 plasma treatment time. The values of Rnn were defined as the resistances of the linear parts in the I-V characteristics. According to this figure, the junction resistance exponentially increases with the treatment time. This relation was the same when the a- and b-axis oriented films were used for the base YBCO electrode. On the other hand, it is known that in the tunneling process at very small applied voltage, the junction resistance is given as a function of the barrier thickness as following equation:

$$Rnn = \frac{V}{S\ J} = 3x10^{-10} \frac{d}{S\ E_B^{1/2}} \exp(10.25dE_B^{1/2})$$

where J is the tunnel current density in A/cm^2, V is the applied voltage in V, S is the junction area in cm^2, d is the barrier thickness in nm, and E_B is the barrier height in eV. Therefore it shows that the barrier was uniformly formed and its thickness linearly increased with the CCl_2F_2 treatment time. Assuming that the image force is neglected, the barrier height is estimated about 0.15eV.

Hence the CCl_2F_2 plasma treatment has the potential for the formation of the Josephson tunnel barrier.

CONCLUSION

In conclusion, the barriers formed by exposing the YBCO surface to CCl_2F_2 plasma were studied with YBCO-barrier-Au junctions. The superconducting energy gap was observed on the I-V characteristics. Furthermore the junction resistance exponentially increased with the CCl_2F_2 plasma treatment time. Hence this method has the potential for the formation of the Josephson tunnel barrier.

ACKNOWLEDGMENT

One of the authors would like to thank to Mr. K. Tsuda, T.Suzuki, H.Kimura, and Miss Y.Okamura in Fuji Electric Corp. R & D for their help.

The project is partially supported from Special Grant in Aid of Education and Hoso Bunka Fund.

REFERENCES

[1]M.Furuyama, I.Iguchi, K.Shirai, T.Kusumori, H.Ohtake, S.Tomura, M.Nasu (1990)Jap.J.Appl.Phys. 29: ppL614-L617.

[2]K.Harada, S.Tanaka, H.Itozaki and S.Yazu (1990) Jap. J. Appl. Phys. 29:ppL1114-L1116

[3]K.Hirata, K.Yamamoto, K.Iijima, J.Takada, T.Terashima and Y.Bando (1990) Appl. Phys. Lett. 56: pp683-685.

[4]T.Shiota, K.takechi, Y.Takai and H.Hayakawa (1988) In : K.Kitazawa, T.Ishiguro (Eds) Advances in Superconductivity. Springer-Verlag, pp755-759.

[5]T.Matsui, G.Matsubara, A.Nakayama, N.Satoh, K.Mukae, Y.Okabe (1989) In: T.Ishiguro, K.Kajimura (Eds) Advances in Superconductivity 2nd. Springer-Verlag, pp1025-1028.

[6]T.Matsui, K.Takeuchi, A.Nakayama, M.Nagano and Y.Okabe (1989) Proc. Int. Superconductivity Electronics Conf., JSAP, Tokyo: pp222-225.

[7]G.Matsubara, T.Matsui and Y.Okabe (submitted to Jap.J.Appl.Phys.)

[8]J.H.Greiner (1974) J.Appl.Phys. 45 :pp32-37.

[9]P.C.Karulkar and J.E.Nordman (1979) J.Appl.Phys. 50 :pp7051-7059.

[10]A.Matsuda, T.Inamura and H.Yoshikiyo (1980)J.Appl.Phys.51:pp4310-4316.

[11]M.Muroi, Y.Okamura, T.Suzuki, K.Tsuda, M.Nagano and K.Mukae (1990) Jap.J.Appl.Phys.29:pp69-73.

Preparation and Patterning of Bi-System 2-2-1-2 Phase by Ca Diffusion into 2-2-0-1 Single Crystals

Tatsuro Usuki, Ichiro Yasui, Koji Yamano, Maruo Kamino, Yorinobu Yoshisato, and Shoichi Nakano

Functional Materials Research Center, SANYO Electric Co., Ltd., 18-13, Hashiridani 1-chome, Hirakata, Osaka, 573 Japan

ABSTRACT

Preparation of the Bi-system 2212 phase area was carried out by partial recrystallization with Ca-diffusion into 2201 single crystals for the first time. The 2201 single crystal was prepared by a CuO-rich self-flux method. The as-grown crystals were micaceous and typically 5 x 5 mm^2 in size. CaF_2 films were deposited on the *(00l)* cleavage plane of the 2201 crystal to control the Ca composition. Ca was diffused into the 2201 single crystal by annealing. The surface layer was identified as 2212 phase, strongly oriented along the c-axis. The surface layer showed Tc(end) = 85 K. Patterned 2212 phase was fabricated from patterned CaF_2 films. The 2201 host material can be used as the normal barrier of lateral SNS junction along the a-b plane with a long coherent length.

KEY WORDS : Bi-system superconductor, Bi-Sr-Cu-O, CaF_2, Recrystallization, Monolithic

INTRODUCTION

The patterning of films for oxide-superconductors was attempted by several methods.[1-3] In addition these discrete structures, it is necessary to achieve connection and isolation similar to those in semiconductor devices, where isolated regions are constructed by selective doping. Monolithic patterning is one of the necessary processes of function devices. Oxide high-Tc superconductors can alter their electrical properties to form superconductors (S), insulators (I) or semiconductors (SE) through control of their compositions or crystal structures. In Bi-oxide systems there are several phases, such as $Bi_2Sr_2Ca_2Cu_3Ox$ (2223), $Bi_2Sr_2CaCu_2Ox$ (2212) [4] and $Bi_2Sr_2CuO_x$ (2201).[5-6] In these systems, control of the constituent elements, such as Bi, Ca and Sr, seems effective, because their properties are not sensitive to oxygen concentration. Using these Bi-based oxides, an SNS junction structure was reported in the layer structure using the 2212 phase as superconducting electrodes and the 2201 phase as a barrier layer, where each layer was prepared by a sputtering method and the composition of each layer controlled by sputtering with multi-targets of a mixed Bi-based oxide.[7] There have been several studies of 2201 phase single crystals.[8-9] However, there have been no reports of the processes for device structures.

We have reported a new method for the fabrication of high-Tc superconducting films, where a layered structure such as amorphous Bi-O/Sr-Ca-Cu-O film was annealed and yielded high-quality thin films with a high critical temperature.[10] This method suggests that it is possible to make patterning or isolation structures of superconductive devices, especially in the lateral direction, by controlling the composition locally in the films. In this paper, we report a new process for making isolation structures, or patterning on Bi-system single crystals. We have fabricated a BiSrCuO/BiSrCaCuO structure from Ca-F/BiSrCuO layered films. We describe an isolated superconducting BSCCO formation using 2201 single crystals into which Ca has been diffused.

EXPERIMENTAL

Single crystals of $Bi_2Sr_2CuO_x$ (2201 phase) were grown from Cu-O rich melts. Three initial compositions for single crystal growth of Bi: Sr: Cu= 1:1:1, 1:1:2, 1:1:3 were studied. Crystals were prepared by a self flux method. Bi_2O_3 (99.99% purity), $SrCO_3$ (99.99% purity), and CuO

(99.99% purity) powder were mixed and grinded, and those were then placed in an alumina crucible. The growth was carried out in a muffle furnace. The mixed material was heated at 1020°C for 2 hours, then cooled to 800°C at a rate of 2°C/h, after which the furnace was turned off. The as-grown crystals were micaceous. The resulting BSCO crystals were used as host material to form BSCCO.

The guest films were Ca-F films which compensate for the Ca composition. The Ca-F films were deposited on a cleaved surface of as-grown BSCO single crystals by the electron beam evaporation method. The source material for the Ca-F film was CaF_2. The substrate temperature was 100°C and evaporation rate 1 nm/s. Patterning for the Ca-F films was done by a mask deposition method using a metal mask or a lift-off method employing optical photolithography. The deposited films were annealed in a furnace with an air or oxygen atmosphere. The annealing temperature ranged from 800°C to 880°C for 3 hours. Both layered films and unlayered films were then formed to isolate the crystallization resulting from the annealing.

The compositions of the single crystals were measured by ICP (Inductively Coupled Plasma) and EPMA (Electron Probe MicroAnalysis). Depth profiles of the composition were measured by XPS (X-ray Photoelectron Spectroscopy) and SIMS (Secondary Ion Mass Spectroscopy). The structures of films and crystals were analyzed by XRD (X-ray Diffraction) using CuKα. The surface morphology was observed by a photo-microscope or SEM (Scanning Electron Microscope). Electrical properties were measured by the standard direct current four-probe method.

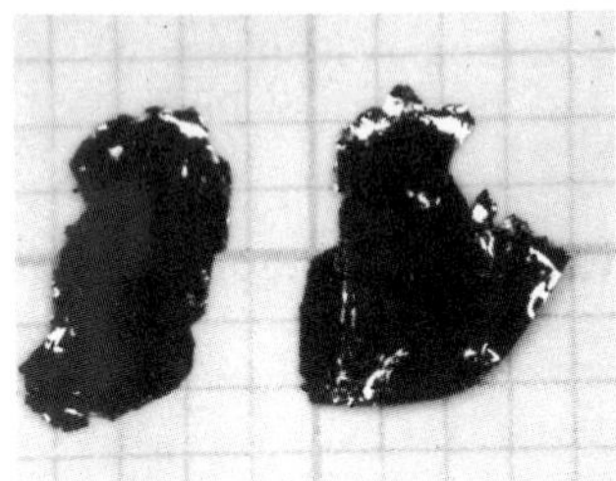

Fig. 1 Typical BSCO single crystal. (Grid: 1mm)

RESULTS AND DISCUSSION

1) BSCO Single Crystals

The single crystals were removed from the matrix. The best initial composition was found to be Bi:Sr:Cu=1:1:1 for growing large crystals. Large crystals were up to 5 x 4 x 0.2 mm^3. A typical BSCO single crystal is shown in Fig. 1. The BSCO single crystals had very shinny surfaces and exhibited a high cleavage along the a-b plane. Figure 2 shows a SEM image of the cleaved surface of the single crystals. The cleaved surface had a good, smooth plane.

Fig. 2 SEM image of cleaved surface of the single crystals.

The X-ray diffraction results indicated the crystals to be single-phase $Bi_2Sr_2CuO_x$ as shown in Fig. 3. It should be noted that the data shown in Fig. 3 were obtained by performing X-ray diffraction on the surface of cleaved

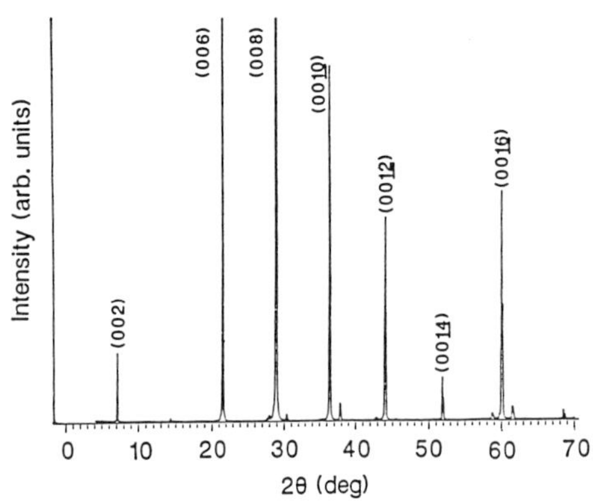

Fig. 3 X-ray diffraction patterns of BSCO single crystal.

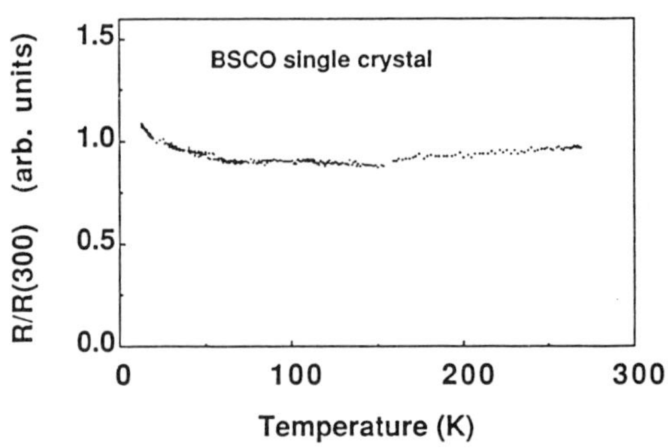

Fig. 4 Temperature dependence of resistivity of the as-grown crystal with a current perpendicular to the surface.

samples. Diffractions from (00l) planes were observed. The lattice parameter was c= 24.0 Å. Figure 4 shows typical data of the temperature dependence of resistivity of the as-grown crystals with a current along the surface. The resistance curves were usually semiconducting and showed no evidence of superconductivity above 4.2 K.

2) Ca diffusion

As mentioned above, it was found that the crystals had good crystallinity and a good cleavage surface. Ca-F films were deposited on the BSCO cleaved surface. After annealing, Ca was induced into the BSCO to form BSCCO. A schematic diagram of this method is shown in Fig 5. Figure 6 shows the X-ray diffraction pattern of Ca-diffused samples from Ca-F with a 300 nm thickness. Annealing condition was 870°C for 30 hours in air. Diffraction peaks were identified as 2212 phase and 2201 phase. Because peaks of 2201 phase were the diffraction from the 2201 single crystal substrate, the 2212-phase layer was found to be fabricated at the surface of the single crystal. The depth profile was measured by XPS and SIMS analysis. Figure 7 shows the depth profile of compositions for the sample with a Ca-F (300 nm)/BSCO structure after annealing measured by XPS. 2212 compositions were kept in a 100 nm layer under the surface. The distribution showed that Ca seemed to diffuse into the BSCO crystal from the 2212 layer. Ca was not found in positions deeper than 700 nm. Analysis by SIMS was consistent and supported the XPS data.

Figure 8 shows the temperature dependence of resistivity of the sample shown in Fig. 6. The data indicated the 2212 phase transition and rather sharp superconducting transitions of ΔTc = 3 K with zero resistance temperature Tc (end)= 85 K. The 2223 phase transition was not observed from the data, which is different from the data of samples using amorphous BSCO films. In our previous report,[11] a BSCCO area was isolated from BSCO film using Ca diffusion into amorphous BSCO films. BSCCO area was formed by annealing for approximately 3 hours. Figure 9 shows the relationship between Ca-F film thickness and critical temperature. Although Tc(onset) was observed in the sample with a 70 nm thickness of Ca-F film, Tc(end) was not observed in the sample with Ca-F film thinner than 100 nm. Figure 10 shows the annealing conditions for the samples with a 300-nm thickness of Ca-F film. BSCO single crystal required a long

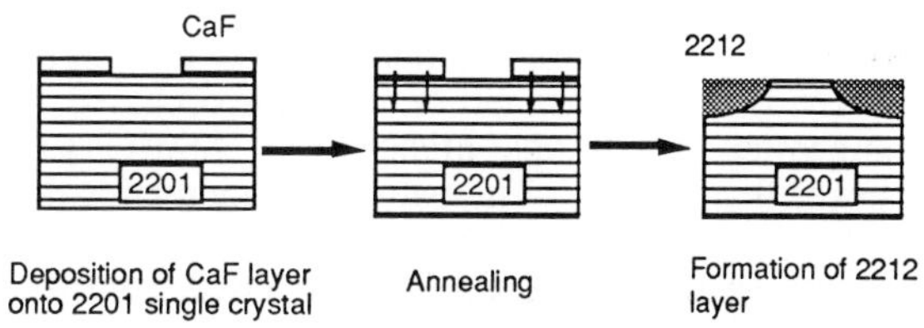

Fig. 5 Fabrication method using single crystals of BSCO.

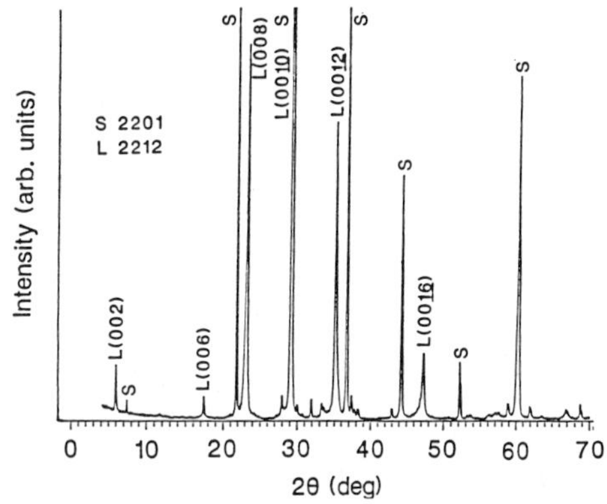

Fig. 6 X-ray diffraction patterns of Ca-diffused samples from Ca-F with a 300-nm thickness.

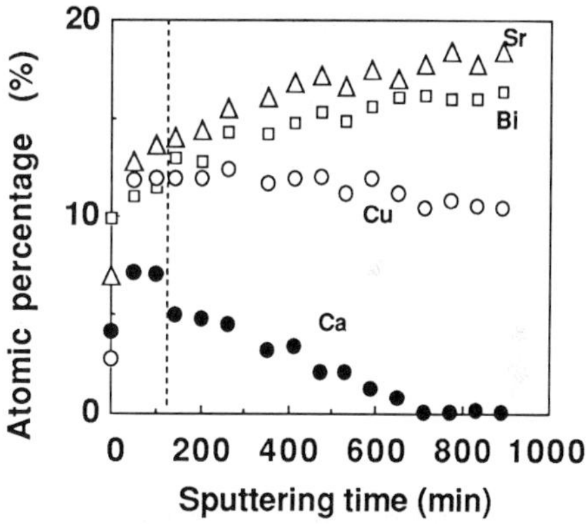

Fig. 7 Depth profile of compositions for the sample with Ca-F(300 nm)/BSCO structure after annealing measured by XPS. Sputtering rate was 1nm.

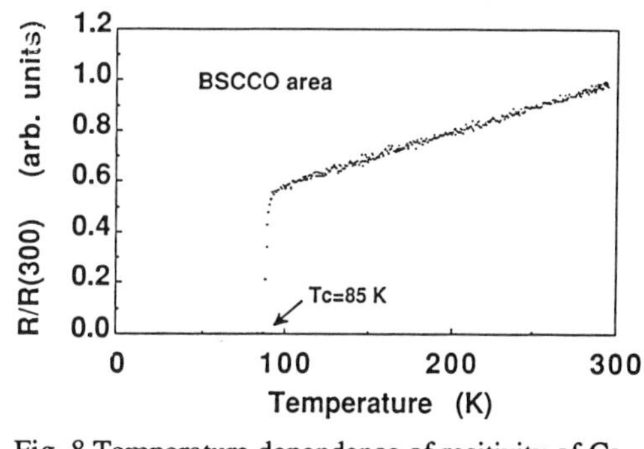

Fig. 8 Temperature dependence of resitivity of Ca diffused samples shown in Fig. 6.

annealing time compared with amorphous BSCO films. As for the annealing temperature, Tc(end) higher than 80 K was not obtained in annealing conditions higher than 880°C and lower than 850 °C. The patterning factor for the diffusion length was defined as the measured length of difference due to annealing from the broadening edge and the patterning line of Ca-F films. In the annealing condition at 870°C for 30 hours, the patterning factor was about 3 μm. It was 50% larger than the value in experiments for BSCO amorphous films,[11] since the annealing time was longer than the experiment for BSCO amorphous films. The patterned structures described above were achieved by a fabrication characterized by thermal equilibrium processes. These processes can be applied to obtain junctions such as S/SE and SIS junctions.

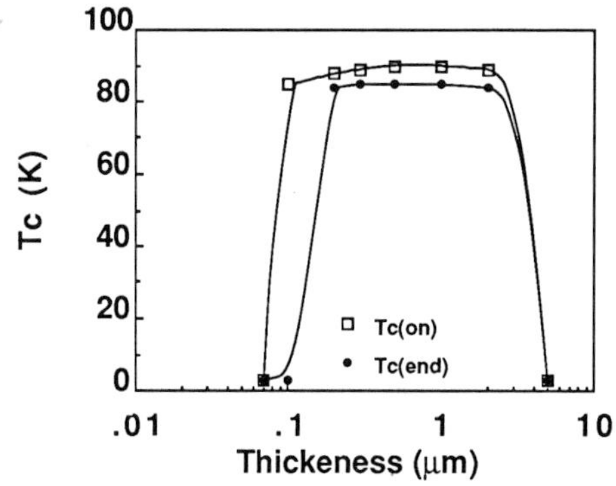

Fig. 9 Relationship between Ca-F film thickness and critical temperature.

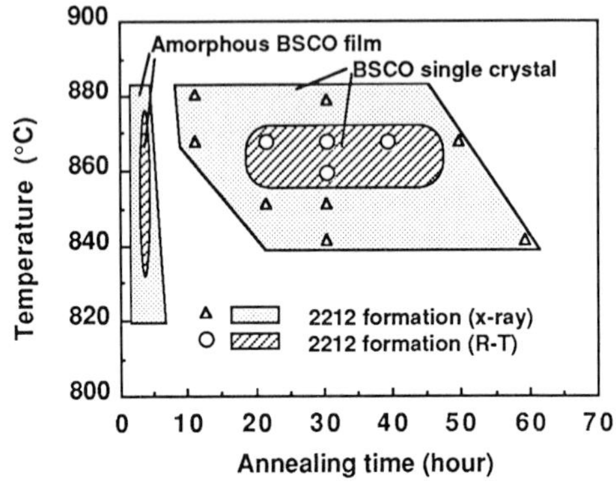

Fig. 10 Annealing conditions for the samples with a 300 nm thickenss of Ca-F films

CONCLUSION

A new kind of monolithic lateral SNS junction structure of Bi-oxide systems has been obtained, where the composition of films was locally controlled in the lateral direction. An isolated BSCCO region was obtained by Ca-diffusion into 2201 single crystals. The BSCO substrate showed normal resistivity above 4.2 K. The BSCCO region was obtained by Ca-diffusion into 2201 single crystals. The BSCO substrate showed normal resistivity above 4.2 K. The BSCCO region indicated 2212 phase with a high orientation perpendicular to the surface of a 2201 single crystal substrate and showed zero resistance at 85 K.

ACKNOWLEDGEMENT

This work was performed under the management of FED (the R&D Association for Future Electron Devices) as a part of the R&D of Basic Technology for Future Industries sponsored by NEDO (New Energy and Industrial Technology Development Organization).

REFERENCES

1. L. R. Harriott, P. A Polakos and C. E. Rice : Appl. Phys. Lett., **55(5)** , 495, (1989)
2. Nir Hess, Leanfl R. Tessler, Uri Dai and Guy Deutscher : Appl. Phys. Lett., **53(8)**, 698, (1989)
3. Francine C. Case : J. Appl. Phys., 67(9), 4365, (1990)
4. H. Maeda, Y. Tanaka, M. Fukutomi and T. Asano : Jpn. J. Phys., **27**, L209, (1988)
5. J. Akimisu, A. Yamazaki, H. Sawa and H. Fujiki : Jpn. J. Appl. Phys., **26**, L208, (1987)
6. C. Michel, M. Hervieu, M. M. Borel, A. Grandin, F. Deslandes, J. Provost and B. Raveau : Z. Phys, B-Condensed Matter, **68**, 421, (1987)
7. K. Mizuno, K. Higashino, K. Setsune and K. Wasa : Appl. Phys. Lett., **56 (15)**, 1469, (1990)
8. E. Sonder, B. C. Chakoumakos and B.C. Sales : Phys. Rev. **B 40**, 6872, (1989)
9. R. Zhifeng, J. G. Wen, J. Yandao and G. Weiyan : J. Crystal Growth, **102**, 349, (1990)
10. T. Usuki, I. Yasui, Y. Yoshisato and S. Nakano : Jpn. J. Appl. Phys., **29**, L1446, (1990)
11. T. Usuki, I. Yasui, Y. Yoshisato and S. Nakano : IEICE Technical Report, **SCE 90-12**, (1990)

Characteristics of $Bi_2Sr_2CaCu_2O_x/Bi_2Sr_2CuO_y/Bi_2Sr_2CaCu_2O_z$ Josephson Junctions

K. MIZUNO, H. HIGASHINO, K. SETSUNE, and K. WASA

Central Research Laboratories, Matsushita Electric Industrial Co., Ltd., Moriguchi, Osaka, 570 Japan

ABSTRACT

We investigated sandwich-type Josephson junctions using a $Bi_2Sr_2CaCu_2O_x/Bi_2Sr_2CuO_y/Bi_2Sr_2CaCu_2O_z$ (BSCCO/BSCO/BSCCO) structure. The barrier layer (BSCO) has a crystalline structure similar to those of the superconducting electrodes (BSCCO). Moreover our BSCO films were normal conductors even at 4.2K. The junctions exhibited S/N/S type I-V characteristics and clear Shapiro steps under microwave radiations. Values of these step heights oscillated according to radiation power. Ic values changed in proportion to $(1-T/Tc)^2$ near Tc.

KEY WORDS: Bi-Sr-Ca-Cu-O, S/N/S, thin film, Josephson junction

INTRODUCTION

Some of the high-Tc superconductors reach a superconducting state over 77.4K. This behavior enables us to construct Josephson junctions or other superconducting devices which can operate in liquefied nitrogen. For practical applications, sandwich-type junctions are preferable than point contacts or micro bridges because their characteristics are stable, reproducible and easily controllable. Several preliminary attempts at building high-Tc oxide superconductor junctions or heterostructures have been reported[1-7]. In these reports, some oxides which have a same crystalline structure were chosen.

Recently we also reported a $Bi_2Sr_2CaCu_2O_x/Bi_2Sr_2CuO_y/Bi_2Sr_2CaCu_2O_z$ structure deposited by rf-magnetron sputtering[8] and observed Josephson effects in a junction structure. These Bi-system oxides have several metastable phases which have nearly the same crystalline structure and contain almost the same elements. On the other hand, their electrical properties, especially those concerning superconductivity, are different. For example, $Bi_2Sr_2CaCu_2O_x$ and $Bi_2Sr_2Ca_2Cu_3O_z$ have Tc of 80K and 115K, respectively. Moreover, deposited $Bi_2Sr_2CuO_y$ films do not exhibit superconductivity even at 4.2K. Using these series of Bi-based oxides may suggest one solution to utilize high-Tc superconductors for electronics.

In this paper, we report on the characteristics of sandwich-type $Bi_2Sr_2CaCu_2O_x/Bi_2Sr_2CuO_y/Bi_2Sr_2CaCu_2O_z$ Josephson junctions especially in reference to temperature dependence of critical currents, on which Bi-based oxides were used for both the superconducting electrodes and for the barrier layer.

EXPERIMENTAL

$Bi_2Sr_2CaCu_2O_x/Bi_2Sr_2CuO_y/Bi_2Sr_2CaCu_2O_z$ multi-layered structures were deposited by in-situ rf magnetron sputtering. (100)MgO single crystals were used as substrates and they were heated up to 650°C during deposition. At this temperature, Ca atoms slightly diffused into the barrier layer from the superconducting electrodes[8]. We, however, suppose that this interdiffusion could not cause degradation in properties of the junctions because remarkable grain growth does not observed in scanning electron microscopy (SEM) images and each layer kept its original crystalline structure. Transition temperatures of the superconducting BSCCO films were about 60K. The barrier layer ($Bi_2Sr_2CuO_y$) of the junctions behaved as normal conductor. The resistivity of the barrier was measured to be about 10^{-1} ohm-cm along the c-axis.

Junctions were fabricated from the BSCCO/BSCO/BSCCO structure using a conventional photo-lithographic method and an Ar ion etching technique. Detailed sputtering conditions and fabrication processes were reported in previous papers[8].

Figures 1 (a) and (b) show SEM images of a BSCCO/BSCO/BSCCO/Pt multi-layer. In these figures, relatively flat surfaces are seen. Moreover flat cleavage planes parallel to the a-b planes of BSCCO or BSCO film are observed. It is because that the both BSCCO and BSCO films were c-axis oriented in X-ray diffraction analyses. A schematic configuration of our junction is shown in fig.2. Rectangular junctions of different area sizes were fabricated on the same substrate.

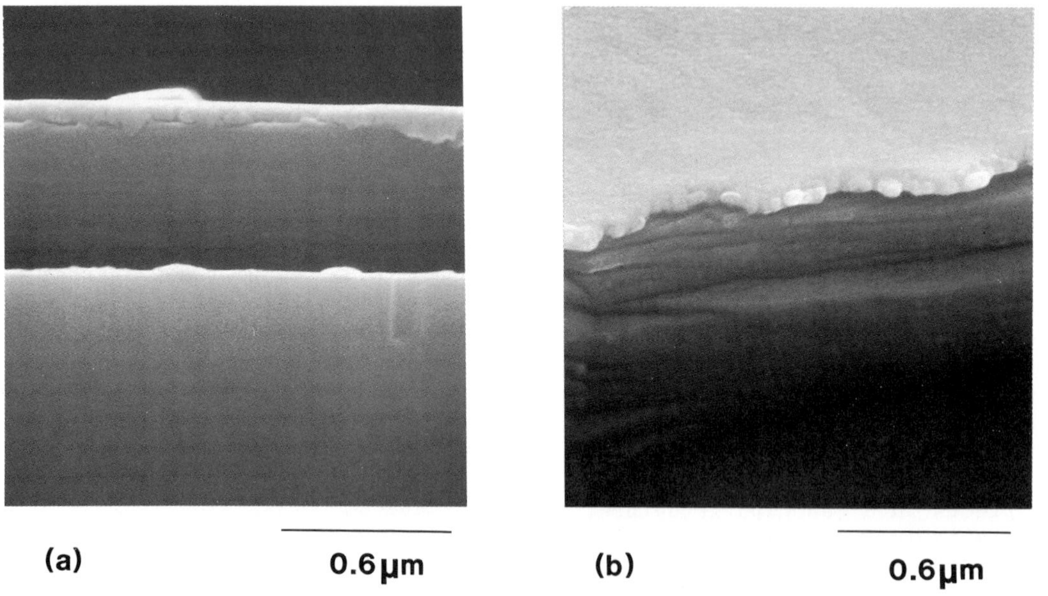

Fig.1 Cross sectional scanning electron microscopy (SEM) images of a BSCCO/BSCO/BSCCO/Pt film.

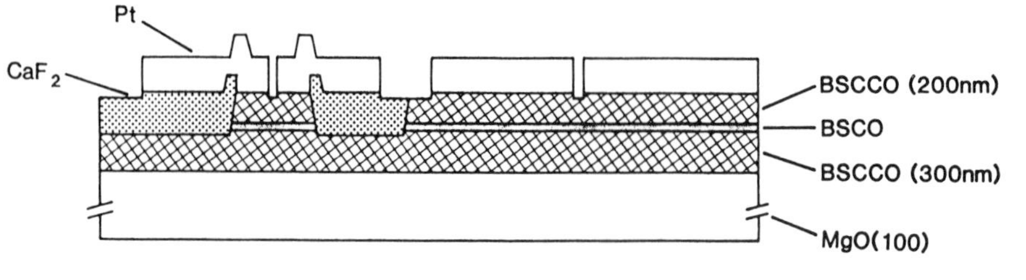

Fig.2 Schematic configuration of a BSCCO/BSCO/BSCCO Josephson junction.

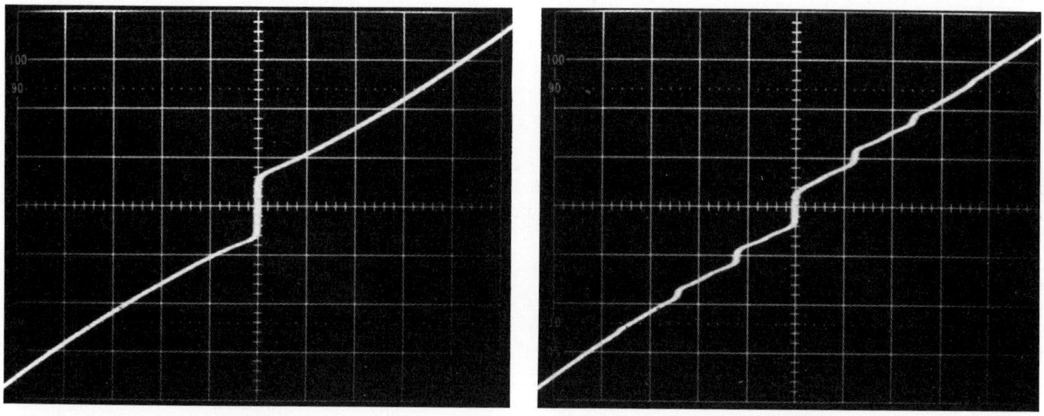

(a) (b)

Fig.3 Current-voltage curves for a BSCCO/BSCO(80nm)/BSCCO junction at 7.1K (200μA/div.vert. and 20μV/div. horiz.); (a) without rf radiation; (b) under 11.9GHz radiation.

Two Pt thin-film contacts were formed on both superconducting electrodes for a set of current lead and a voltage lead. Current-voltage (I-V) characteristics were measured by the ordinal four-probe method.

RESULTS AND DISCUSSION

Figures 3 (a) and (b) show I-V characteristics of a BSCCO/BSCO/BSCCO junction at 7.1K. The junction size was 10X20 μm^2 and the thickness of the barrier layer was 80 nm. The Ic value of fig.3(a) is 120 μA. The IcRn product is to be 0.02 mV at this temperature. This IcRn value became large to be 0.13 mV at 4.2K and its value was the same as those for another junction whose size was 20X40 μm^2 and barrier thickness was 60 nm. Figure 3 (b) shows an I-V characteristic under 11.9 GHz microwave radiation. Clear Shapiro steps are shown and their behavior can be explained qualitatively by numerical calculation results which obtained by the current biased equivalent circuit for the RSJ model.

J. Clarke[9] presented good agreement between experimental and theoretical results on metal-barrier S/N/S junctions. In those results, Ic values changed in proportion to $(1-T/Tc)^2$ at temperatures near Tc and in proportion to $\exp(-T^{1/2})$ at temperatures sufficiently lower than Tc. Figures 4(a) and (b) show the temperature dependence for Ic on a 20x40 μm^2 junction whose barrier layer was 60nm. In fig.4(a), temperature is normalized as $(T/Tc)^{1/2}$ and Ic values are plotted on a logarithmic scale. There is a good liner relationship at lower temperatures between values of log(Ic) and $(T/Tc)^{1/2}$. This means that Ic values change in proportion to $\exp(-T^{1/2})$. Experimental Ic data near Tc are re-plotted in fig.4(b). $(Ic)^{1/2}$ data are plotted on a vertical scale and temperature is normalized as (1-T/Tc) on a horizontal scale. These points also show good liner dependence. These results are consistent with J. Clarke's S/N/S junctions. It is not clear whether high-Tc superconductors can be explained by ordinary BCS theory. But, empirically, it is useful to apply a standard explanation to the behavior of high-Tc superconductors. Concerning our BSCCO/BSCO/BSCCO junctions, the coherence length ξn in a BSCO film must be a most basic physical value. So it is important to estimate it experimentally in further investigations.

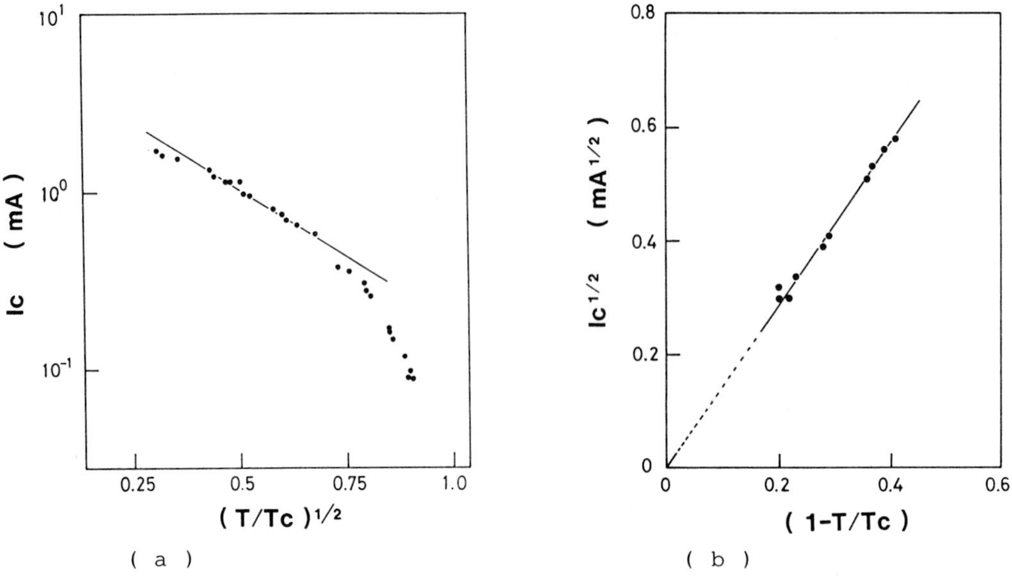

Fig.4 Temperature dependences of critical currents of a BSCCO/BSCO(60nm)/ BSCCO junction; (a) plotted as log(Ic) versus $(T/Tc)^{1/2}$; (b) plotted as $(Ic)^{1/2}$ versus (1-T/Tc) in near Tc.

CONCLUSION

Sandwich type Josephson junctions which were constructed by Bi-based oxide thin films were investigated. Junctions fabricated from BSCCO/BSCO/BSCCO multi-layered structures exhibited non-hysteretic I-V characteristics and also exhibited clear Shapiro steps with microwave radiations. Behavior of the Shapiro steps agreed with theoretical calculations using the RSJ model. Moreover Ic values near Tc changed in proportion to $(1-T/Tc)^2$. We concluded that our junctions operated as S/N/S junctions.

ACKNOWLEDGMENT

The authors wish to thank Dr.T.Nitta for his continuous encouragements.

REFERENCES

1. J.M.Triscone,M.G.Karkut,L.Antogazza,O.Brunner,and Ø.Fischer, Phys.Rev.Lett.63,1016(1989).
2. C.T.Rogers,A.Inam,M.S.Hedge,B.Dutta,X.D.Wu,and T.Venkatesan, Appl.Phys.Lett.55,2032(1989).
3. T.Venkatesan,A.Inam,B.Dutta,R.Ramesh,M.S.Hegde,X.D.Wu,L.Nazar,C.C.Chang, J.B.Barner,D.M.Hwang,and C.T.Rogers,Appl.Phys.Lett.56,391(1990).
4. U.Poppe,P.Prieto,J.Schubert,H.Soltner,K.Urban,and Ch.Buchal Solid State Commun.71,569(1989).
5. J.J.Kingstone,F.C.Wellstood,P.Lerch,A.H.Miklich, and J.Clarke, Appl.Phys.Lett.56,189(1990).
6. A.Gupta,R.Gross,E.Olsson,A.Segmuller,G.Koren,and C.C.Tsuei, Phys.Rev.Lett.64,3191,(1990).
7. K.Harada,S.Tanaka,H.Itozaki,and S.Yazu,Jap.J.appl.phys.29,L1114(1990).
8. K.Mizuno,H.Higashino,K.Setsune,and K.Wasa,Appl.Phys.Lett.56,1469(1990).
9. J.Clarke,Proc.Roy.Soc.London A 308,447(1969).

Proximity Effect in YBaCuO/Al/AlO$_x$/Pb Planar Tunnel Junctions

TAKESHI KUSUMORI, ZHONGMIN WEN, HIROAKI OHTAKE, MAKOTO KAMIYA, and IENARI IGUCHI

Institute of Materials Science, University of Tsukuba, 1-1, Tennoudai 1-chome, Tsukuba, 305 Japan

ABSTRACT

We report the proximity effect of high Tc YBCO(YBaCuO) superconductors using an YBCO/Al/AlOx/Pb(or Au) tunnel-junction geometry. The junctions were fabricated by an in situ deposition technique on YBCO epitaxial films utilizing an electron-beam coevaporation system. The insulating layer was formed by the oxidation of Al surface in oxygen gas atomosphere. The observed characteristics strongly depended on the Al thickness. Without an Al film, a broad YBCO gap structure at about 20mV and a pronounced Pb gap structure around zero bias were observed. The Pb gap structure was broadened as the Al thickness was increased. For the Al thickness above 15nm, the Pb gap structure became completely disappeared.

KEY WORDS: proximity effect, tunneling spectroscopy, YBCO materials, S/N/I/S(N) junction, gap structure

INTRODUCTION

Since the discovery of high Tc oxide superconductors, a variety of researches have been done. The technology for the thin-film fabrication has been developed remarkably and it now became rather easy to fabricate the thin films with high Tc, high current density and good morphology using the various methods. On the other hand, the tunneling spectroscopy which may be a powerful method for investigating the superconducting properties of high Tc materials has been performed to obtain the information on the energy gap, the shape of density of states and the phonon structure using the point contacts, thin film junctions, break juncitons etc. The results were not conclusive in view of the reproducibility and the reliability (1). Quite recently, the fabrication of the reproducible thin-film tunnel junctions by an all-in-situ deposition technique has been reported by our group(2-5). The results yielded the quasiparticle tunneling characteristics with the gap opening structure at about ±20mV and the zero bias anomaly and the Josephson characteristics. To investigate the possible application of high Tc superconductors to electronic devices, it is necessary to fabricate the multiple-layered structures of thin films and study their properties including the tunneling characteristics and the proximity effect. The preliminary studies on the proximity effect were reported by several groups(6-9).
In this paper, we report the investigation of proximity effect using the junctionconfigurations YBCO/Al/AlOx/Au and YBCO/Al/AlOx/Pb where Al and AlOx represent the normal layer and the insulating layer respectively. The junctions were fabricated by the all-in-situ technique utilizing electron-beam and resistive-heater coevaporation chamber equipped with an in-situ metal mask changing system. In this experiments, the YBCO films were grown on the MgO(100) and $SrTiO_3$(100) substrates, hence they had the c-axis orientation perpendicular to the substrate surface. The observed results seem to indicate that the proximity effect is important for the thickness of the normal metal layer less than a few tens of nanometer.

EXPERIMENTAL

The fabrication of YBCO/AL/AlOx/Pb(or Au) tunnel junctions was done by the subsequent deposition of multiple-layered films using the in-situ metal mask changing system. The fabricated junction geometry is shown in Fig. 1. The YBCO and Pb(or Au) strips were 0.2mm wide and 6.0mm long and the normal Al layer had the area of 2X2mm^2. The effective junctionarea was defined by the

overlap of the two strips ($0.2X0.2mm^2$). Because of close contact between the substrate and the mask, little diffusion at the edges was observed.
The YBCO thin films were fabricated by the reactive coevaporation method. Two electron-beam guns were used for Y and Ba evaporation and one resistive heater for Cu evaporation. Near the substrate, oxygen gas of high purity (99.9%) was introduced by a thin nozzle to increase the local oxygen pressure at the substrate. The oxygen background pressure during the deposition was $9X10^{-5}$ Torr. The oxygen plasma was generated by the rf cathode ring, if necessary. The substrate temperature was monitored by a thermocouple embedded in the substrate holder. The evaporation rate of three metal elements was controlled using three individual quartz monitors located closely to each source. The evaporation rate of a YBCO film was monitored by another quartz monitor located near the substrate. Typical rate was about 0.25nm/sec. The deposited YBCO film was annealed in situ during the process of temperature reduction with introducing oxygen gas of 60-200 Torr. The YBCO films exhibited Tc above 85K.

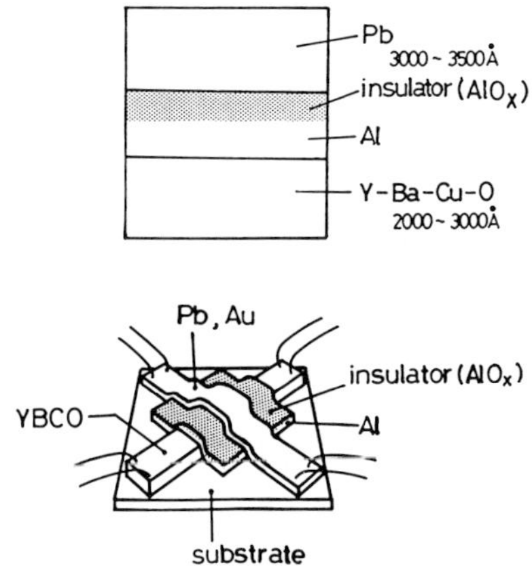

Fig. 1. The fabricated junction geometry.

For the evaporation of Al and Au, the electron-beam guns were used. They were deposited at room temperature. The insulating layer was formed by oxidizing the surface of the Al film for a certain time (typically 10-30mins) using rf glow discharge technique in the oxygen pressure of $2X10^{-4}$ Torr. The rf power was 100W. With this technique of the growth of insulating layer, the junction resistance was typically 10Ω-3kΩ. The Pb film of 0.2mm wide and 300-350nm thick was deposited by a resistive heater using a shadow mask.

RESULTS AND DISCUSSIONS

(1) YBCO/Al/AlOx/Au junction
Fig. 2(b) shows the observed differential tunnel characteristics (dV/dI vs V) at different temperatures together with those of YBCO/MgO/Au (MgO: 4nm) junction without the proximity layer in Fig. 2(a). The thickness of the normal

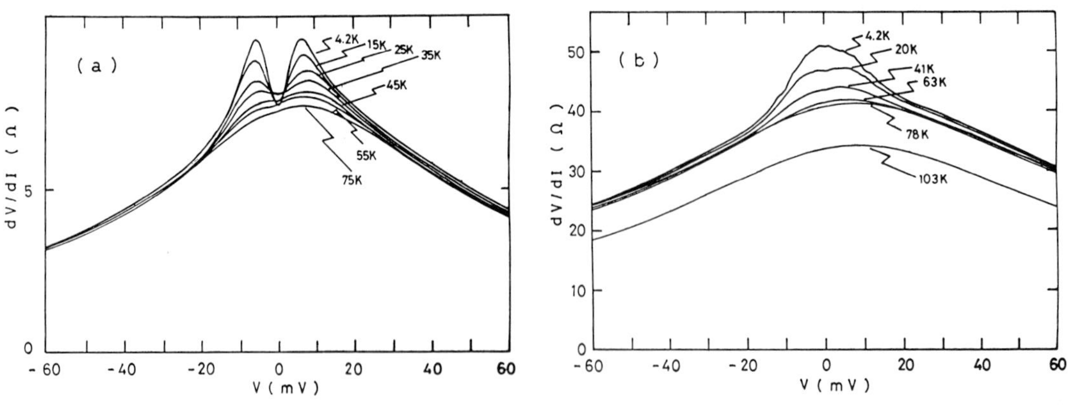

Fig. 2. Differential tunnel characteristics (dV/dI vs V) at different temperatures of YBCO/MgO/Au junction (a) and those of YBCO/Al /AlOx/Au junction (b).

Al layer was 25nm. It is understood that, while the junction without the proximity normal layer had a strong zero bias anomaly, it almost disappeared for a YBCO/Al/AlOx/Au junction, although the gap opening structure at about ±20mV was still present. Fig. 3(a) and (b) show the temperature dependence of the differential tunnel resistance at zero bias voltage for the above two different junctions. Both curves showed a remarkable change at the transition temperature of a YBCO film, indicating the presence of 90K phase region even on the surface of YBCO film. The behavior below Tc was qualitatively different between these two curves. Fig. 3(a) had a upward bending curvature due to the presenceing resistance with reducing temperature. Although the behaviors in Figs. 3(a) and (b) significantly deviated from the ideal BCS behavior, the general tendency of Fig. 3(b) was rather closer to the BCS type.
The result is reasonable in light of the induced superconductivity.

(2) YBCO/Al/AlOx/Pb junction
Fig. 4 shows the differential tunnel characteristics (dV/dI vs V) for a YBCO/Al/AlOx/Pb junction with the Al layer of 7nm thick at different temperatures. The observed behavior resembled those of YBCO/MgO/Pb junctions without the normal layer including the broad gap profile, the zero bias anomaly and the absence of bump at the gap edge. But the Pb gap structure around the zero bias was found to be significantly broadened. Fig. 5 shows the differential tunnel characteristics for YBCO/Al/AlOx/Pb junctions with the normal Al layer of different thicknesses (0-15nm). The gap structure was found to be broadened with increasing the thickness of the normal Al layer.
The Pb gap peak around the zero bias was also broadened and finally disappeared. The phenomena may be consistent with those observed for the YBCO/Al/AlOx/Au junctions. The broadening of Pb gap

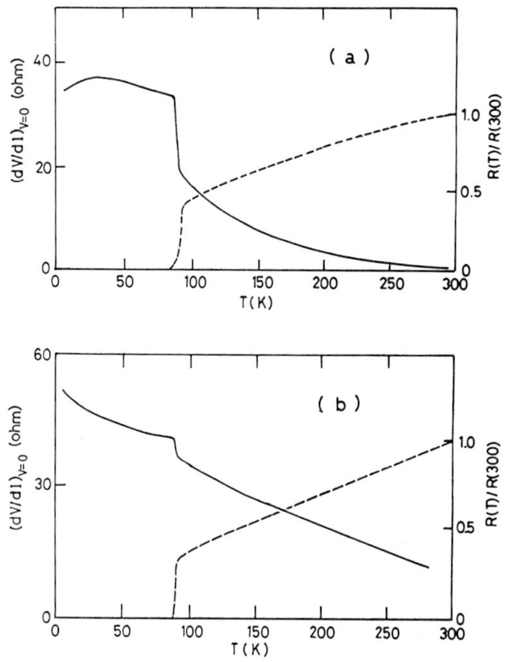

Fig. 3. Temperature dependence of the differential tunnel resistance at zero bias voltage of an YBCO /MgO/Au junction (a) and that of an YBCO/Al/AlOx/Au junction (b).

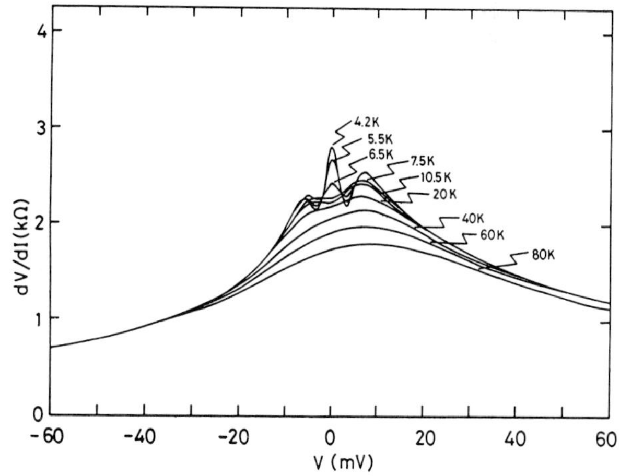

Fig. 4. Differential tunnel characteristics (dV/dI vs V) for a YBCO/Al/AlOx/Pb junction with the Al layer of 7nm thick at different temperatures.

structure means the possible enhancement of Pb gap or the presence of some complicated mechanism of tunneling between Pb and YBCO films. The Pb gap enhancement is, however, not expcted for our junction geometry. The observed behavior was also strange in that the larger gap structure should exhibit a more pronounced peaked structure since the relative peak height was decreased in spite of the broadening of gap width. Hence, the phenomena support the latter idea.

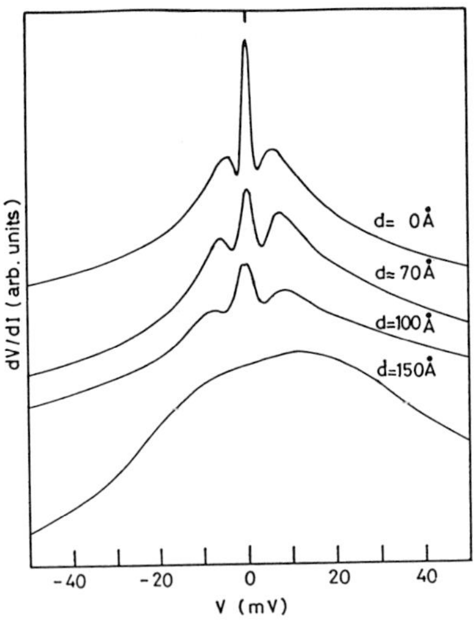

Fig. 5. Differential tunnel characteristics for YBCO/Al/AlOx/Pb junctions with the normal Al layer of different thicknesses (d=0-15nm)

REFERENCES

1. Kirtley JR (1990) Int. J. Mod. Phys. B4 : 201
2. Furuyama M, Iguchi I, Shirai K, Kusumori T, Ohtake H, Tomura S, Nasu M (1990) Jpn. J. Appl. Phys. 29 : L459
3. Iguchi I, Furuyama M, Kusumori T, Shirai K, Tomura S, Nasu M, Ohtake H (1990) Jpn. J. Appl. Phys. 29 : L614
4. Iguchi I, Wen Z, Furuyama M (1990) Physica B 165&166 :73
5. Iguchi I, Wen Z, (1991) IEEE Trans. on MAG, to be published.
6. Jing TW, Wang ZZ, Ong NP (1989) Physica C 162-164 : 1061
7. Greene LH, Barner JB, Feldmann WL, Farrow LA, Miceli PF, Ramesh R, Wilkens BJ, Bagley BG, Tarascon JM, Wernick JH, Giroud M, Rowell JM (1989) Physica C 162-164 : 1069
8. Gijs MAM, Scholten D, van Rooy Th, Gerrits AM (1990) Phys. Rev. B 41 : 11627
9. Fujimaki A, Tamaoki T, Hidaka T, Yanagase M, Shiota T, Takai Y, Hayakawa H (1990) Jpn. J. Appl. Phys. 29 : L1659

Nonbolometric Optical Detection in High-T_c Superconducting Thin Films

KEIICHI TANABE, YOUICHI ENOMOTO, MINORU SUZUKI, SYUGO KUBO, HIDEFUMI ASANO, TSUNEKAZU IWATA, YUKIMICHI TAJIMA, and AKIHIKO YAMAJI

NTT Applied Electronics Laboratories, Nippon Telegraph and Telephone Corporation, Tokai, Ibaraki, 319-11 Japan

ABSTRACT

The electrical responses to infrared radiation of highly oriented thin films of various high-T_c oxide superconductors, $LnBa_2Cu_3O_y$, $La_{2-x}Sr_xCuO_4$, and $Ba_{1-x}K_xBiO_3$, are systematically studied using modulated laser light. All films show clear nonbolometric responses at temperatures below T_c. The strong correlation between the responsivity of the nonbolometric response and the Cooper pair density observed in these materials and $BaPb_{1-x}Bi_xO_3$ confirms that the nonbolometric responses are due to nonequilibrium gap suppression. The present results also suggest that $Ba_{1-x}(K, Rb)_xBiO_3$ is a promising material for use in sensitive detectors which can be operated using a closed-cycle refrigerator.

KEY WORDS: optical detector, high-T_c thin film, nonequilibrium superconductivity

INTRODUCTION

Optical detectors using copper-oxide-based high-T_c superconductors have recently attracted much attention as one of the most promising electronic applications of these materials. The feasibility of transition-edge bolometers including microbolometers has been theoretically established by Richards et al.[1]. There has also been a large number of studies on the optical response of high-T_c superconductor thin films, in particular $YBa_2Cu_3O_y$ films [2-7], with the main hope of developing faster detectors based on nonequilibrium effects which are similar to $BaPb_{1-x}Bi_xO_3$(BPBO) granular thin film detectors [8]. However, no fast response with a responsivity comparable to that of BPBO detectors has been reported. Furthermore, the response mechanism in these materials is still a controversial issue. Only a slow bolometric response was observed in many studies [2,3]. On the other hand, the observation of nonbolometric responses has been claimed by several groups [4-6]. Some of them have also proposed different response mechanisms including the phase slip process [4] and those relating to vortex motions [6,7].

To clarify the response mechanism in high-T_c superconductors and to find a way to develop high-performance optical detectors, we have systematically studied the electrical responses to infrared radiation of highly oriented thin films of various high-T_c oxide superconductors, $LnBa_2Cu_3O_y$, $La_{2-x}Sr_xCuO_4$, and $Ba_{1-x}K_xBiO_3$ [9-11]. These materials show clear nonbolometric responses at temperatures below T_c. A strong correlation between the optical responsivity and the Cooper pair density has also been observed which confirms that the nonbolometric response is caused by nonequilibrium gap suppression. This paper presents some of these recent results as well as the prospect of high-performance infrared detectors based on nonequilibrium superconductivity.

EXPERIMENTAL

$La_{2-x}Sr_xCuO_4$(LSCO) and $LnBa_2Cu_3O_y$(Ln= Eu, Y)(LnBCO) highly oriented thin films were prepared by magnetron sputtering using a fixed composition ceramic target or by reactive evaporation in an rf-activated oxygen atomosphere. $Ba_{1-x}K_xBiO_3$(BKBO) thin films were deposited in an evaporation system equipped with two K-cells for Ba and Bi metal sources, one resistively heated alumina crucible for K, and an ECR type

Table I Properties of highly oriented thin films of various high-T_c superconductors examined in the present study .

film	LSCO(100)	EBCO(001)	YBCO(113)	BKBO(110)
substrate	$SrTiO_3$(100)	MgO(100)	$SrTiO_3$(111)	MgO(100)
thickness (nm)	400	110	100	310
T_c onset (K)	~ 28	~ 88	~ 90	~ 22
T_c zero (K)	19	82	73	15.5
J_c (0 K) (A/cm^2)	3.3×10^5	1.0×10^7	8×10^5	1×10^4

oxygen ion source. The deposition conditions have already been published elsewhere [9-11]. The film properties are summarized in Table I. The thickness of the LSCO and LnBCO films is comparable to the absorption depth ($\sim$ 500 nm for LSCO(100) and $\sim$ 100 nm for the others).

Constrictions with a typical size of 13 $\times 13 \mu$m were fabricated by standard photolithography and ion milling. A semiconductor laser diode (LD) was used as a continuous wave (CW) radiation source (λ = 1.3μm). The laser light was directly modulated with a typical frequency of 50-300 kHz. It was introduced into a cryostat using a single-mode optical fiber and was incident perpendicular to the constrictions. The details of the arrangements for optical response have been described elsewhere [8,9]. The typical peak-to-peak power of the radiation was $2 - 20 \mu$W ($\sim 0.6 - 6$ W/cm^2).

RESULTS AND DISCUSSION

Figure 1 shows the temperature dependence of the radiation-induced peak-to-peak voltage shift (i.e.optical response) observed in the LSCO(100) thin film for various bias current values. For a bias current of 8 mA, the voltage shift shows a small peak near T_c. As the bias current is increased, the peak shifts to lower temperatures well below the transition region and grows larger. The temperature derivative of the resistance (dR/dT) for a bias current of 16 mA is also plotted in the same figure. The difference in temperature for the voltage peak and the dR/dT peak is approximately 10 K. This large difference strongly suggests that most of the observed response is nonbolometric. The maximum responsivity obtained at approximately 5K in this film is 20 V/W. This value is about two orders of magnitude smaller than those for the BPBO detectors [8].

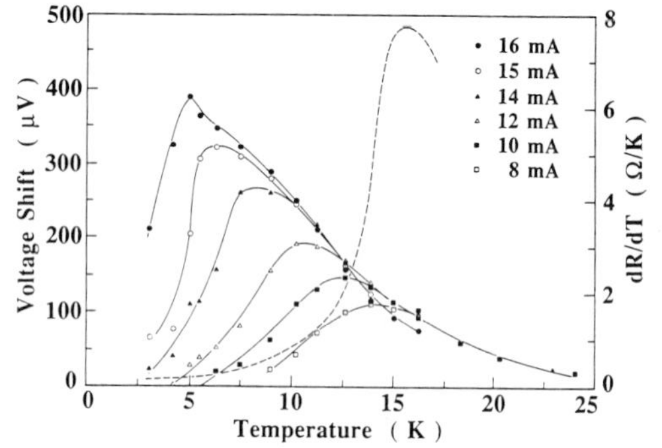

Fig. 1 Temperature dependence of radiation-induced voltage shift (solid lines) and dR/dT (broken line) for $La_{1.76}Sr_{0.24}CuO_y$(100) thin film. The modulation frequency and the radiation power are 300 kHz and 20μW.

Figure 2 illustrates the temperature dependence of both radiation-induced voltage shift and dc voltage

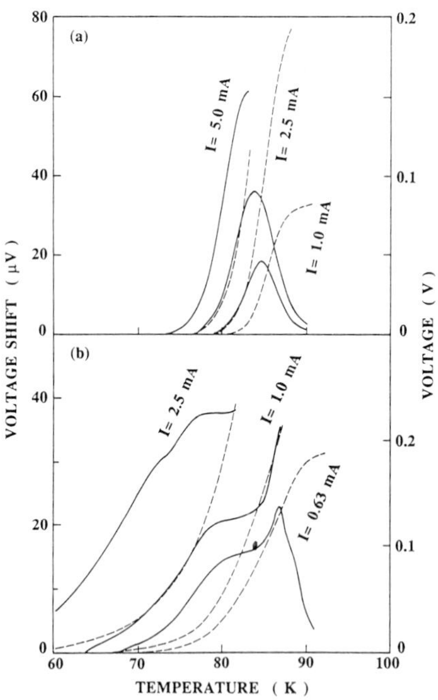

Fig. 2 Temperature dependence of radiation-induced voltage shift (solid lines) and dc voltage (broken lines) for (a) EBCO(001) thin film and (b) YBCO(113) thin film. The modulation frequency and the radiation power are 120 kHz and 20μW.

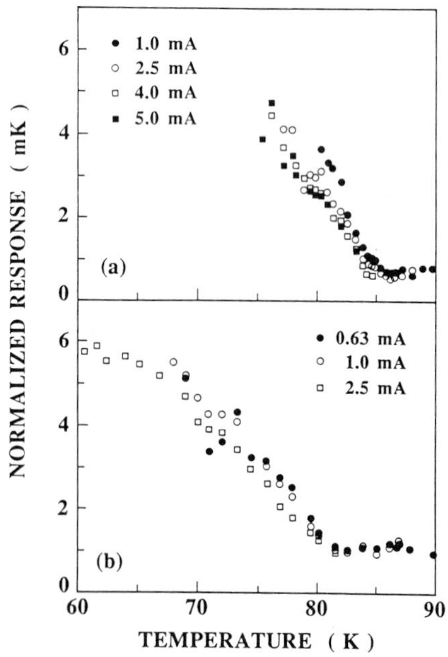

Fig. 3 Temperature dependence of normalized response $\delta V/(dV/dT)$ for (a) EBCO(001) thin film and (b) YBCO(113) thin film.

for LnBCO thin films. EBCO(001) thin film with a high J_c value shows relatively sharp response peaks near the resistive transition. In the YBCO(113) film where the much lower J_c value suggests a greater contribution of the grain boundaries to the transport properties, a shoulder structure at lower temperatures is seen in addition to the sharp peaks. A typical maximum responsivity for both films is 2-3 V/W which is one order of magnitude smaller than the value for LSCO. In Fig. 3, the normalized response $\delta V/(dV/dT)$ for the same films is plotted versus temperature. In spite of the above difference in the response curve, both films show quite similar dependences. At temperatures higher than T_c, the normalized response is almost constant. This indicates that the response in this region is bolometric. The bolometric temperature rises for both films are almost the same ($\sim$ 1mK). On the other hand, as the temperature is decreased below T_c, the normalized response shows a rapid increase which suggests the occurrence of a nonbolometric response. Such a dependence is quite similar to those observed by Zeldov et al. in c-axis oriented YBCO thin films [6]. The ratios of the normalized response at temperatures well below T_c to the bolometric temperature rise for these films are also the same $(4-6)$. These values suggest that the orders of magnitude of the nonbolometric response and the bolometric response are almost equal in LnBCO.

Figure 4 shows the dependence of the voltage shift induced by 2 μW radiation on the bias current for a BKBO thin film. At temperatures well below zero-resistance T_c, the response rapidly rises near the critical current and shows a clear peak. The peak voltage shift increases with a decrease in temperature and reaches a maximum value larger than 600 μV which corresponds to a responsivity of 300 V/W at approximately 13 K. The peak responsivity value at 15 K is still as large as 200 V/W. These values are one or two orders of magnitude larger than those for LSCO and LnBCO. At temperatures below 13 K, however, the response shows multiple peaks and singular dependence curves. The peak height decreases with temperature in this region. Such dependencies are attributed to the occurrence of hysteretic $I-V$ curves which are similar to those observed in BPBO [8]. A rapid increase in the normalized response at temperatures below T_c has also been observed [10].

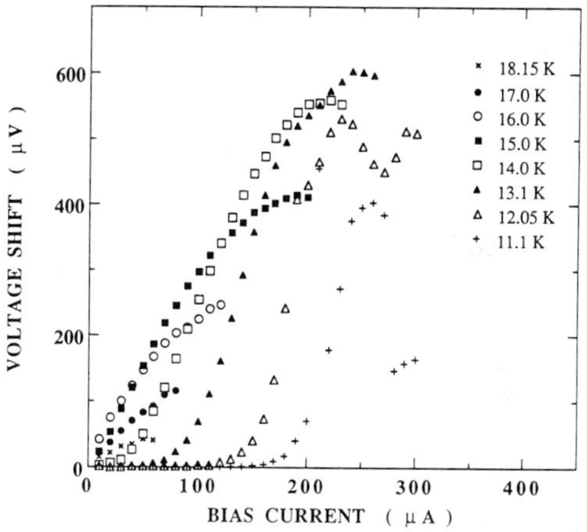

Fig. 4 Dependence of radiation-induced voltage shift on bias current at different temperatures for $Ba_{0.7}K_{0.3}BiO_y(110)$ thin film. The modulation frequency and the radiation power are 50 kHz and 2μW.

The responsivity based on nonequilibrium superconductivity is expressed as

$$r_{NE} = \frac{\partial V}{\partial I_c}\bigg|_{I_B} \frac{\partial I_c}{\partial \Delta}\bigg|_T \frac{\partial \Delta}{\partial n_{qp}}\bigg|_T \frac{\delta n_{qp}}{P} \qquad (1)$$

Here, Δ, n_{qp}, and P are the gap parameter, density of quasiparticles and radiation power, respectively. The magnitude of the responsivity is largely determined by the third and fourth terms of Eq. (1), when the bias current is optimized. As discussed in a recent paper [9], the overall responsivity can be approximately normalized by the density of states at the Fermi level, $N(0)$, multiplied by Δ (that is the Cooper pair density) for the oxide superconductors where the effective quasiparticle recombination time τ_{eff} is possibly dominated by the phonon escape time τ_{es}. Table II summarizes the physical parameters and the observed responsivity values ($r_{obs.}$) for BPBO, BKBO, LSCO, and LnBCO. Although the carrier densities of these materials are quite similar, the $N(0)$ values for LSCO and LnBCO are one or two orders of magnitude larger than those for BPBO and BKBO. In Fig. 5, the observed responsivity is plotted versus $N(0)\Delta$. The responsivity is approximately inversely proportional to the estimated $N(0)\Delta$ value. Such a strong correlation confirms that the nonbolometric responses in these materials are attributable to nonequilibrium gap suppression.

TABLE II Physical parameters as well as observed responsivity values for various oxide superconductors.

	material	BPBO	BKBO	LSCO	YBCO
T_c	(K)	13	30	38	93
n	(10^{21} cm^{-3})	4	6	2	5
$N(0)$	(10^{21} states eV^{-1} cm^{-3}) [a,b]	1.9	4.1	27	102
$N(0)\Delta_0$	(10^{19} cm^{-3})	0.38	1.9	15.6	145
$r_{obs.}$	(V/W)	10^3-10^4	$\sim$300	10-20	1-5

[a] Batlogg B, Cava RJ, Rupp Jr LW, Mujsce AM, Krajewski JJ, Remeika JP, Peck Jr WF, Cooper AS, Espinosa GP (1988) Phys Rev Lett 61: 1670

[b] Fisher RA, Gordon JE, Phillips NE (1988) J. Superconductivity 1: 231

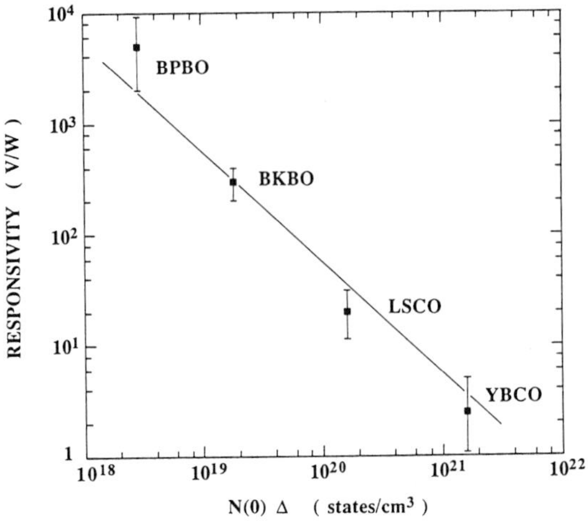

Fig. 5 Observed responsivity versus $N(0)\Delta$ for various oxide superconductors.

CONCLUSIONS

Clear nonbolometric electrical responses to infrared radiation (λ = 1.3μm) are observed in highly oriented thin films of various high-T_c superconductors, $LnBa_2Cu_3O_y$(Ln= Eu, Y), $La_{2-x}Sr_xCuO_4$, and $Ba_{1-x}K_xBiO_3$, at temperatures below T_c. The magnitude of the observed responsivity is largest in BKBO and smallest in LnBCO. In the latter materials, comparable contributions of the nonbolometric and bolometric mechanisms to the response are observed. The responsivity of these materials and BPBO is inversely proportional to the Cooper pair density which confirms that the nonbolometric responses are based on nonequilibrium gap suppression. The present results suggest that high-T_c $LnBa_2Cu_3O_y$ thin films are more suitable for use in a microbolometer than in a nonequilibrium photon detector, although some degree of additional nonbolometric contribution to the responsivity can be expected even in the former case. Recently, a T_c value as high as 37 K has been reported in a bulk Ba_{1-x}(K, Rb)$_x$$BiO_3$ system [12]. Thus, $Ba_{1-x}M_xBiO_3$(M= K, Rb) with a much smaller $N(0)$ than those of copper-oxide-based materials is a promising candidate for fast and sensitive infrared detectors which can be operated using a closed-cycle refrigerator or liquid neon.

REFERENCES

1. Richards PL, Clarke J, Leoni R, Lerch P, Verghese S, Beasley MR, Geballe TH, Hammond RH, Rosenthal P, Spielman SR (1989) Appl Phys Lett 54: 283
   Hu Q, Richards PL (1989) Appl Phys Lett 55: 2444
2. Forrester MG, Gottlieb M, Gavaler JR, Brajinski AI (1988) Appl Phys Lett 53: 1332
3. Brocklesby WS, Monroe D, Leve AFJ, Hong M, Liou SH, Kwo J, Rice CE, Mankiewich PM, Howard RE (1989) Appl Phys Lett 54: 1175
4. Leung M, Broussard PR, Claassen JH, Osofsky M, Wolf SA, Strom U (1987) Appl Phys Lett 51: 2046
5. Frenkel A, Saifi MA, Venkatesan T, England P, Wu XD, Inam A (1989) Appl Phys Lett 54: 1594
6. Zeldov E, Amer NM, Koren G, Gupta A, Gambino RJ, McElfresh (1989) Phys Rev Lett 62: 3093
7. Culbertson JC, Strom U, Wolf SA, Skeath P, West EJ, Burns WK (1989) Phys Rev B 39: 12359
8. Enomoto Y, Murakami T (1986) J Appl Phys 59: 3808
9. Tanabe K, Enomoto Y, Suzuki M, Iwata T, Yamaji A (1990) Jpn J Appl Phys 29: L466
10. Tanabe K, Enomoto Y, Yamaji A (1990) Appl Phys Lett 57: (in press)
11. Tanabe K, Kubo S, Enomoto Y, Asano H, Yamaji A submitted to Jpn J Appl Phys
12. Tseng D, Ruckenstein E (1990) J Mater Res 5: 742

Photoresponse of Bi-Sr-Ca-Cu-O Thin Films

SEISHIRO OHYA[1], SHIRO KARASAWA[1], NAOMI HIDAKA[1], YUKIO KURIHARA[1], and HANNS-ULRICH HABERMEIER[2]

[1] Industrial Research Institute of Kanagawa Prefecture, 3173 Showa-machi, Kanazawa-ku, Yokohama, 236 Japan

[2] Max-Plank-Institut Für Festkörperforschung, Heisenbergstrasse 1, D-7000 Stuttgart 80, Germany

ABSTRACT

The photoresponse of granular Bi-Sr-Ca-Cu-O thin films on $LaAlO_3(10\bar{1}2)$ substrates prepared by excimer laser ablation is reported. I-V characteristics obtained by changing the bias current by several orders, at various temperatures, suggest that the voltage shift by irradiation had the maximum not in the state where the first voltage drop was observed (corresponding to the approach to the critical current), but near the maximum of differential resistance (dV/dI). When chopping frequency increased, the peak of voltage shift decreased. The peak of voltage shift appears to be caused by the transition wherein the entire current path exhibited the normal state with large dissipation.

KEY WORDS: high-T_c oxide superconductor, Bi-Sr-Ca-Cu-O, granular thin films, photoresponse

INTRODUCTION

Soon after the discovery of the high-temperature superconductors, their potential for applications as photodetectors was studied based on the work performed so far using conventional superconductors. Investigation of the photoresponse of superconductors is useful not only for their application but also in understanding the mechanisms of interaction between photons and the superconductor. For the application of superconductors as photodetectors, characteristics such as high sensitivity, high speed, wide range of wavelength, and linearity to input power are desirable. In particular, photodetectors of superconducting Bi-Pb-Ba-O thin films have the characteristics of high sensitivity and high speed[1]. With respect to high-Tc superconductors, several studies on La-Sr-Cu-O and Y-Ba-Cu-O have been reported[2-13]; however, Bi-Sr-Ca-Cu-O was not investigated. Thus we studied the photoresponse of granular Bi-Sr-Ca-Cu-O thin films. Voltage shifts (PR) due to irradiation were measured under various conditions of temperature, bias current, and chopping frequency. There was an optimum temperature and bias current at which the voltage shift was maximum. Comparing photoresponse with I-V characteristics obtained using several orders of bias current, at the bias current giving the peak voltage shift, an abrupt voltage increase occurred in the I-V characteristics. This abrupt voltage increase was found to occur not at the grain boundary but in the grain itself. When chopping frequency increased, the peak of voltage shift decreased. The peak of voltage shift appears to be caused by the transition of the entire current path to a normal state with large dissipation. As a result, response speed was slow although voltage shift was large.

EXPERIMENTAL

In an earlier paper we reported details of our thin film preparation scheme[14]. Ceramic targets with the molar ratio of Bi:Sr:Ca:Cu=3:2:3:4 were prepared from Bi_2O_3, $SrCO_3$, $CaCO_3$, and CuO starting powders. The final sintering was performed in air at 800 °C and the target was furnace-cooled. The target was mounted in a vacuum chamber with a base pressure of 10^{-5} Torr, and irradiated through a quartz window with a KrF excimer laser (Lambda Physik EMG103G, 24ns full width half-maximum, 100mJ/pulse) at a 45° angle of incidence. A quartz lens was used to obtain an energy density greater than $2J/cm^2$ on the target. $LaAlO_3$ single-crystal substrates were grown by SHINKOSHA

CO., LTD. and cut and polished on the $(10\bar{1}2)$ crystallographic face. $LaAlO_3(10\bar{1}2)$ substrates were mounted at a distance of typically 4cm from the pellet surface, close to the normal from the center of the laser spot. The films were deposited with the substrates heated at 600 to 800°C in an atmosphere of 10mTorr of oxygen. The laser repetition rate was 2-5Hz. Film thickness was about 4000-8000Å. For improvement of crystallinity, the films were heated at 900°C for 20 minutes and then annealed at 865°C for 5 hours in flowing oxygen. X-ray diffraction using Cu-Kα showed that the films were c-axis oriented and mainly composed of 2223 and 2212 phases. Platelike grains were observed by SEM. Photoresponse measurements were carried out with the 15mW He-Ne (633nm) laser. The modulation frequency by mechanical chopper was varied in the range of 13Hz to 4kHz. Voltage shift due to irradiation ($0.5W/cm^2$) was measured with a lock-in amplifier. The four-probe technique was used in electric measurements.

RESULTS AND DISCUSSIONS

I-V characteristics at various temperature are shown in Fig. 1. The zero resistance temperature (Tc) of the film was 90K, using 20μA current and a 1μV criterion. At various temperatures, nonlinearity of I-V characteristics was observed. When the temperature was low, nonlinearity of I-V characteristics was more obvious, and therefore the I-V curve could be divided into a nonlinear gradual transition region and an abrupt transition region[15]. At a

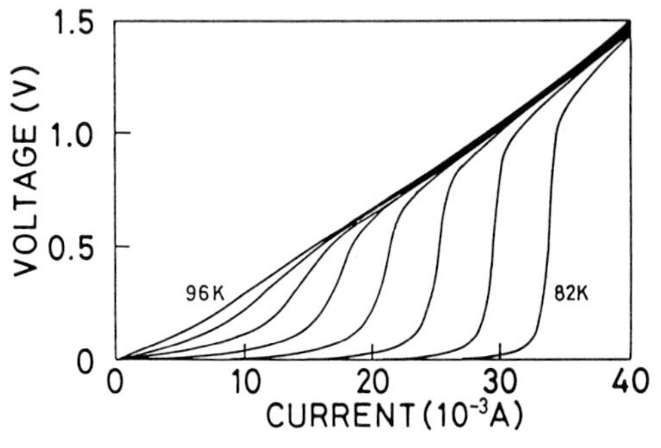

Fig.1 I-V curves. The temperature varies as 82,84,86,88,90,92,94, and 96K from right to left.

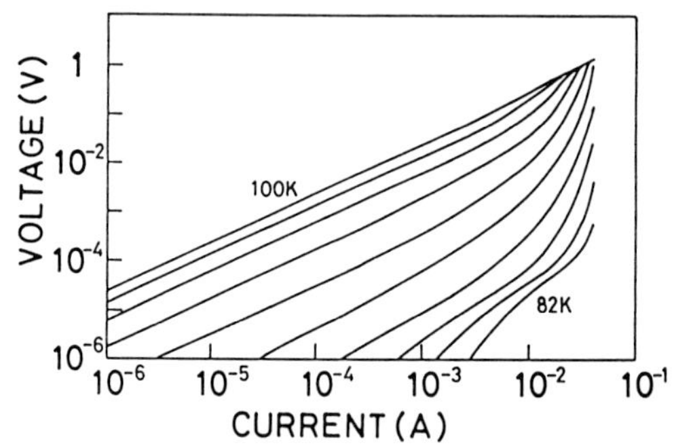

Fig.2 logI-logV curves. The temperature varies as 82,84,86,88,90,92,94,96,98, and 100K from lower right to upper left.

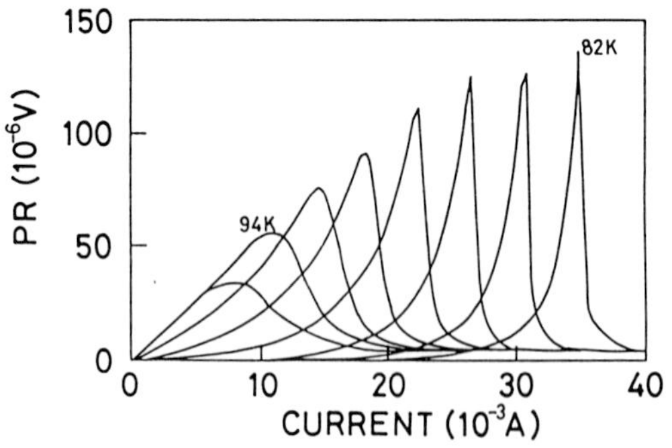

Fig.3 I-PR curves. The temperature varies as 82,84,86,88,90,92,94, and 96K from right to left . The modulation frequency is 1kHz.

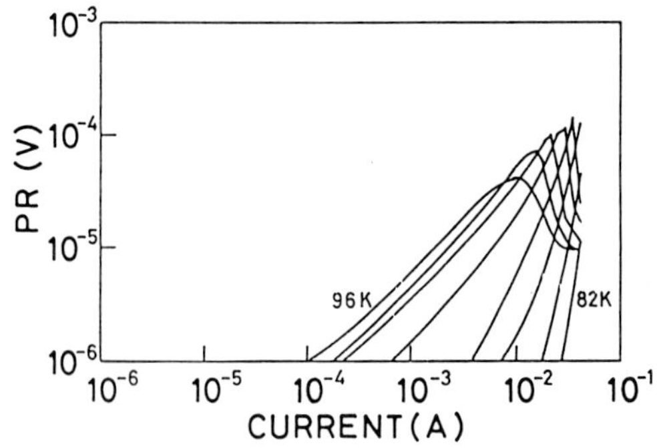

Fig.4 logI-logPR curves. The temperature varies as 82,84,86,88, 90,92,94, and 96K from right to left. The modulation frequency is 1kHz.

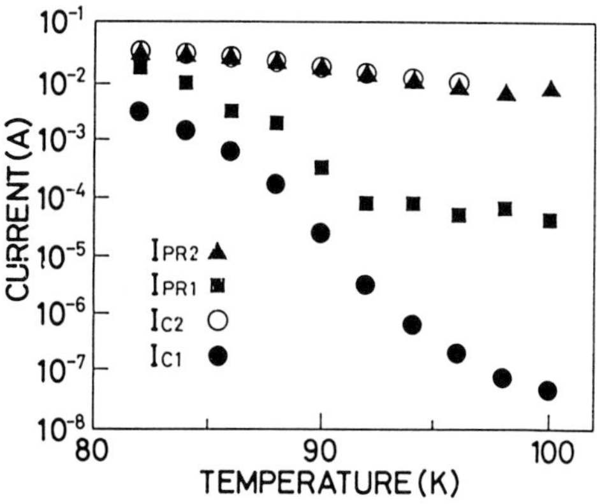

Fig.5 Dependence of Ic1, Ic2, Ipr1, and Ipr2 on temperature.

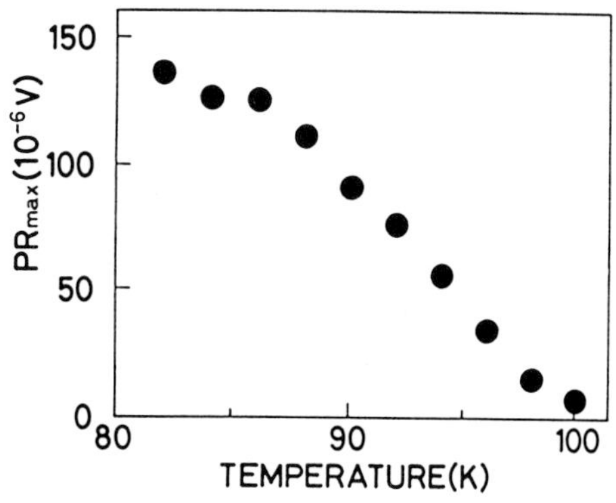

Fig.6 Dependence of the maximum PR on temperature. The modulation frequency is 1kHz.

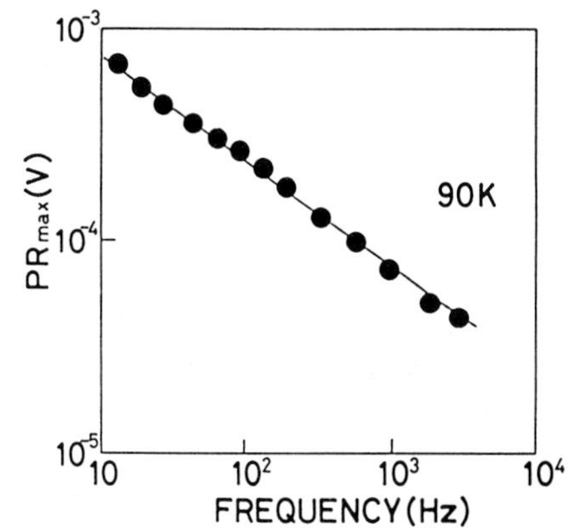

Fig.7 Dependence of the maximum PR at 90K on the modulation frequency.

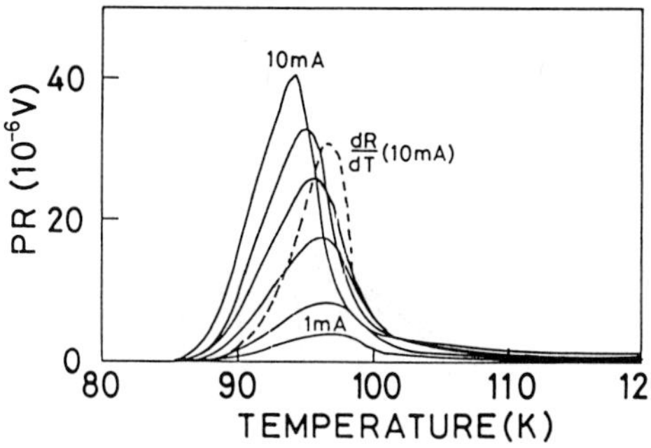

Fig.8 Dependence of PR on temperature. Bias current varies as 1,2,4,6, and 10mA from right to left. Temperature dependence of dR/dT for a bias curret of 10mA is also shown. The modulation frequency is 1kHz.

bias current greater than the current inducing abrupt transition, the I-V curve was ohmic. This resistance agreed with the resistance of the normal state. Above 90K, the transition to the normal state was not abrupt; however, the uppercritical current Ic2 is defined as the current at which the differential resistance (dV/dI) is maximum. Figure 2 shows logI-logV curves. At 92-100K, a low-resistance ohmic state was observed below 1mA. At this current level, regions where superconductivity is weak, such as grain boundaries or some defects, appear to remain in the normal state. At 82-90K, a nonohmic rapid voltage increase was observed. The transition to the normal state around 10mA was observable at various temperatures, as shown in Fig. 1. The lower critical current Ic1 is defined as the current developing 1μV in the I-V curve.

The dependence of photoresponse (PR) on bias current at various temperatures is shown in Fig. 3. The current giving the photoresponse peak is called Ipr2 here. The photo response peak is located in the vicinity of the bias current Ic2 where (dV/dI) derived from I-V characteristics became the maximum, not in the vicinity of bias current Ic1. These two characteristic currents differe by several orders. Furthermore, the peak of voltage shift was also observed when the sample not showing zero resistivity was used. In such a sample, grain boundaries probably remained in the normal state. Figure 4 shows logI-log(PR) curves. The current developing the photoresponse of 1uV is called Ipr1 here. Figure 5 shows Ic1, Ic2, Ipr1, and Ipr2 as a function of temperature. The increase of Ic1 at lower temperature is due to the increase of coupling between the grains. It is noted that Ipr1 depends on Ic1 below 92K but not above 92K. Moreover, Ipr2 is in good agreement with Ic2 above and below 92K. Therefore, these large photoresponses are not caused at grain boundaries but within grains. The dependence of the maximum photoresponse on temperature is shown in Fig. 6. To investigate the dependence of photoresponse on frequency, the chopping frequency was varied in the range of 13Hz to 4kHz. The maximum photoresponse at various frequencies at 90K is shown in Fig. 7. The maximum of photo response decreased when modulation frequency increased. The bias current giving the maximum of photoresponse was almost constant at various frequencies.

In the vicinity of Tc, the slow component of photo response was too large to distinguish the fast component from the slow one. Therefore the dependence of photoresponse on temperature was measured. Figure 8 shows the dependence of photoresponse on temperature at various bias currents and the dR/dT curve at 10mA. When the bias current increased, the peak of PR curves shifted to lower temperature and the maximum of photoresponse increased[7,9,13]. If the photoresponse is bolometric and the temperature rise due to irradiation is constant, it will be proportional to the temperature derivative of the resistance (dR/dT). However, at 10mA bias current, the shift between the peak of photoresponse and the peak of dR/dT was about 2.5K. As a consequence, the existence of nonbolometric response is suggested.

In summary, the I-V measurement suggested that the abrupt voltage increase was not caused at grain boundaries but in the grain itself. The effect of grain boundaries on photoresponse was observed below Tc. In the vicinity of the bias current giving the maximum of (dV/dI), the optical response also had the maximum. This peak of voltage shift by irradiation decreased when chopping frequency increased. Therefore, the peak of voltage shift appears to be caused by the transition of the entire current path to the normal state with large dissipation. As a result, response speed was slow although voltage shift was large. The dependence of photoresponse on temperature at various bias currents also suggests the existence of a nonbolometric component.

REFERENCES

1)Enomoto Y, Murakami T (1989) J.Appl.Phys. 59: 3807

2)Leung M, Broussard PR, Claassen JH, Osofsky M, Wolf SA, Strom U (1987) Appl.Phys.Lett. 51: 2046

3)Enomoto Y, Murakami T, Suzuki M (1988) Physica C 153-155: 1592

4)Richards PL, Clarke J, Leoni R, Lerch Ph, Verghese S, Beasley MR, Geballe TH, Hammond RH, Rosenthal P, Spielman SR (1989) Appl.Phys.Lett. 54: 283

5)Zeldev E, Amer NM, Koren G, Gupta A (1989) Phys.Rev.B39: 9712

6)Brocklesby WS, Monroe D, Levi AFJ, Hong H, Liou SH, Kwo J, Rice CE, Mankiewich PM, Howard RE (1989) Appl.Phys.Lett. 54: 1175

7)Frenkel A, Saifi MA, Venkatesan T, Lin C, Wu XD, Inam A (1989) Appl.Phys.Lett. 54: 1594

8)Culbertson JC, Strom U, Wolf SA, Skeath P, West EJ, Burns WK (1989) Phys.Rev.B39: 12359

9)Zeldev E, Amer NM, Koren G, Gupta A, Gambino RJ, McElfresh MW (1989) Phys.Rev.Lett. 62: 3093

10)Donaldson WR, Kadin AM, Ballentine PH, Sobolewski R (1989) Appl.Phys.Lett. 54: 2470

11)Kwok HS, Zheng JP, Ying QY, Rao R (1989) Appl.Phys.Lett.54: 2473

12)Forrester MG, Gottlieb M, Gavaler JR, Braginski AI (1988) Appl.Phys.Lett. 53: 1332

13)Tanabe K, Enomoto Y, Suzuki M, Iwata T, Yamaji A (1990) Jpn.J.Appl.Phys. 29: L466

14)Ohya S, Kobayashi K, Hirabayashi Y, Kurihara Y, Karasawa S (1989) Jpn.J.Appl.Phys. 28: L978

15)Frenkel A, Clausen E, Chang CC, Venkatesan T, Lin PSD, Wu XD, Inam a, Lalevic B (1989) Appl.Phys.Lett. 55: 911

Optical Responses of Epitaxial $ErBa_2Cu_3O_y$ Superconducting Thin Films in Magnetic Field

KIICHI YOSHIARA, KATSUHIRO IMADA, FUSAOKI UCHIKAWA, and SHIN UTSUNOMIYA

Materials and Electronic Devices Laboratory, Mitsubishi Electric Corporation, 1-57, Miyashimo 1-chome, Sagamihara, Kanagawa, 229 Japan

ABSTRACT

We report the optical response in epitaxial thin films of ErBa2Cu3Oy at wavelength of 0.63 micron in magnetic field. In no magnetic field, these films were shown to have only bolometric response in the resistive transition region. In strong magnetic fields, it shows non-bolometric response depending upon bias currents. These response was supposed to be related to flux-creep.

key words : optical response, ErBa2Cu3Oy thin film, bolometric, non-bolometric, magnetic field

INTRODUCTION

The expectations to device applications of high temperature superconducting materials have increased more and more because of high quality thin films and advanced fabricating techniques. One of the most feasible applications is broadband and sensitive optical sensors in detection of infrared radiation. Optical detection using superconductors has previously been demonstrated in thin films of low temperature superconducting materials, such as Nb, BaPb1-xBiO3 and NbN/BN. The detection mechanism in these detectors can be classified into bolometric or nonequilibrium in origin. The former is simply due to change of resistivity induced by thermal effect. The detector using bolometric effect is simple, but response is slow (~1 ms). On the other hand, the latter is due to quantum effect depending on changes of gap parameter in the weak Boundary Josephson Junction between grains in granular films. This is caused by the distribution change of the quasiparticles due to photoabsorption and recombination. The detection [1-2] using nonequilibrium has a very fast responce determined by the quasiparticle recombination time and has a high S/N ratio by junctions arranged in series. The detectors biased near its critical current could exhibit a voltage change across the Boundary Josephson Junction upon irradiation, with high response time (<1 ns), high detectivity D* ($\sim 10^{11}$cm$(Hz)^{1/2}$/W) and broadband detection. But, very low temperature (6-10K) should be required for operating these detectors.

Since the descovery of high temperature superconducting materials, many groups [3-5] have been reported that optical responses in granular and epitaxial thin films of YBa2Cu3Oy have almost been caused by bolometric effect. These responses have shown a high sensitivity only in resistive transition region. Lately, Zeldov et al. [6] reported from the measurements of V-I characteristic in strong magnetic field that non-bolometric optical response existed even for epitaxial thin films containing no boudary in grain. The proposed mechanism was due to the optically induced flux creep. Furthermore, Culbertson et al. [7] repoted that the destruction of vortex pair by light-irradiation for granular thin films contributed to the photo response.

In this paper, we report the optical response in epitaxial ErBa2Cu3Oy thin film at wavelength of 0. 63 µm in magnetic fields (0-5T). We observed the bolometric and non-bolometric responses depending on magnetic fields.

EXPERIMENTAL

High quality epitaxial thin films of ErBa2Cu3Oy superconducting materials were deposited on single crystal SrTiO3 (100) by I C B (Ionized Cluster Beam) deposition method. Oxidizing gas was supplied into chamber using oxygen gas containing 5 % ozon in a range of 10-4 Torr. The depositing rate of the films was 0.7 A/sec at the substrate temperature of 650 C and the thickness of films was approximately 0.1 µm. The obtained films were epitaxial with the c-axis oriented to the film plane. But, a observation of transmission electron microscopy (TEM) showed that the c-axis oriented films had many micro domains oriented with ab-axis in a size of 10000 square angstrom. A scanning electron microscope photograph of the typical surface in these films are shown in Fig. 1. The surface morphology shows a random mosaic pattern.

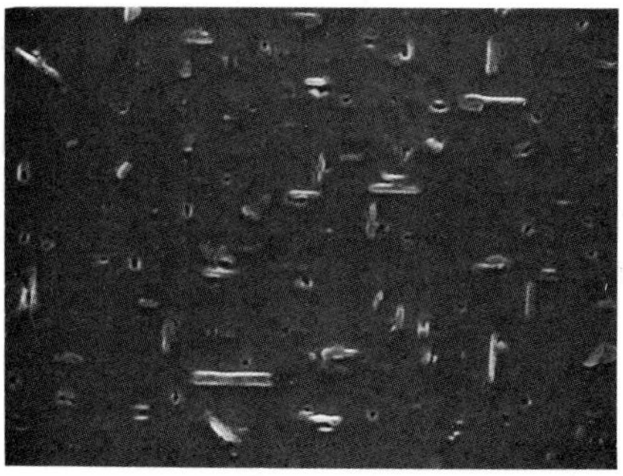

Figure 1 A photograph of scanning electron microscopy (SEM) of c-axis oriented ErBA2Cu3Oy thin films.

The films were patterned to a meanda-line of 20 µm in width and 20 mm in length by a conventional photolithography process. The electrodes of the samples were coated using silver to get low contact resistivity and to avoid light irradiation. Figure 2. shows a sample produced by this manner. The shiny area in the figure is the silver-coated current and voltage electrodes and the narrow dark line at the center is photosensitive superconducting area. The critical temperature Tc and critical current density Jc of these patterned films showed to have 86K and 9*10-5 A/cm2 (at 77K) respectively by four probe technique.

Figure 2 A sample for measurements of optical responce. A shiny area is silver-coated electrodes for current and voltage and a dark narrow area at center is photosensitive line.

The radiation source was a 5 mW He-Ne laser of wavelength 0. 63 micron, and was mechanically chopped at a frequency of 1000 Hz. The single mode optical-fiber (core radius : 6 µm) was used for introduction of the light into cryostat. The chopped light focused by collimater lense was radiated perpendicularly to photosensitive area of the sample. A superconducting magnet (Nb-Ti) was used. The magnet could produce magnetic fields of the range from 0 T to 5T. The magnitude was measured together with optical response by Hall element placed on near sample. The four probe technique was used to measure simultaniously the resistivity of films and the voltage change of optical response.

The bias was applied to the sample at a fixed current. The ac voltage response δV by light irradiation was observed using an oscilloscope or a lock-in amplifier as a function of temperature T and bias current I. Second procedure was to observe the voltage response and resistivity of the sample at the fixed temperature as a function of magnetic field B and bias current I.

RESULTS AND DISCUSSION

Figure 3 shows the temperature dependence of optical response and resistivity of the films at a bias current of 0.1 mA and zero magnetic field. The resistivity shows zero value below 86K and the response voltage exists only above 87.0K. The optical responces were corrected with dR/dT of resistivity of the film. The corrected data was a constant. It may be difficult to compare optical response with dR/dT in the experiment because of a narrow resistive transition temperature (width of Tc = 2K). But, from the results which there is no signal below Tc and the response signal above Tc is due to dR/dT, the observed optical response of epitaxial ErBa2Cu3Oy thin film is thought to show only bolometric in resistive transition regions. The responsivity was 20 V/W and response speed was the order of milisecond. Figure 4-(a) shows the temperature dependence of optical responses with various bias currents at zero magnetic field. From the figure 4-(a), it is found that the response voltage is increasing with bias current and the peak shifts also to low temperature side with increasing bias currents. It is corresponding to the resistivive change in transition region due to high bias currents. Figure 4-(b) shows the bias current dependence of the peak of response voltage at the same conditions. The response voltage is approximately proportional to bias current below 0.3 mA.

Figure 5 shows the temperature dependence of optical response and the resistivity of the same sample at a bias current of 0.1 mA in strong magnetic field of 5T. The resistivity shows zero value below 76K and a broad resistive transition region (about 12K). The response voltage shows also broad signal in a range from 75K to 100K and a peak about 84.5K. In this case, it is found from the figure that there is a response signal below Tc. The corrected data by dR/dT showed changes from a constant value below 70K. This suggests non-bolometric optical response at 5T. But, the response speed was not fast more than 0.1 milisecond order. Though the non-bolometric response which is related to nonequilibrium states of quasiparticles is very fast (~1ns order), the observed response was slow than that. Figure 6 shows the magnetic field dependence of optical response voltages at various bias currents and fixed temperature of 86.0K. By applying high biased current (above 0.8 mA), the response voltage decreased abruptly with increasing magnetic fields in the resistive transition region. The values of magnetic fields at which response voltage changed were 1.58T, 1.16T, 1.0T and 0.71T, respectively, at the biased current of 0.8mA, 1.0mA, 1.2mA and 1.4mA. In the figure, we define the point of magnetic field (Mc) which the optical response voltage decreases abruptly and show the values of bias currents I1~5 and the magnetic fields.

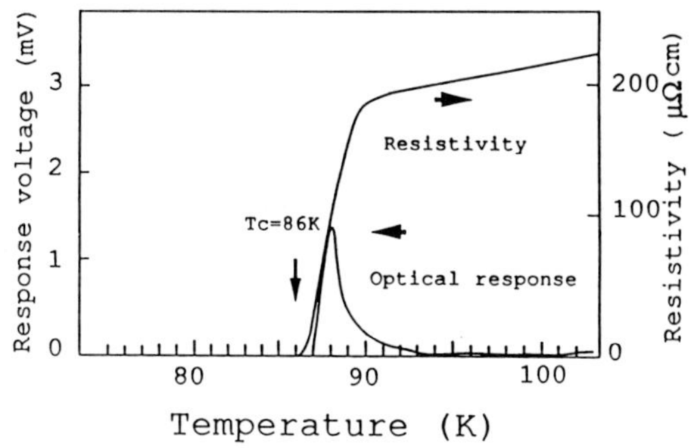

Figure 3 The temperature dependence of the optical responce and the resistivity of films at zero magnetic field.

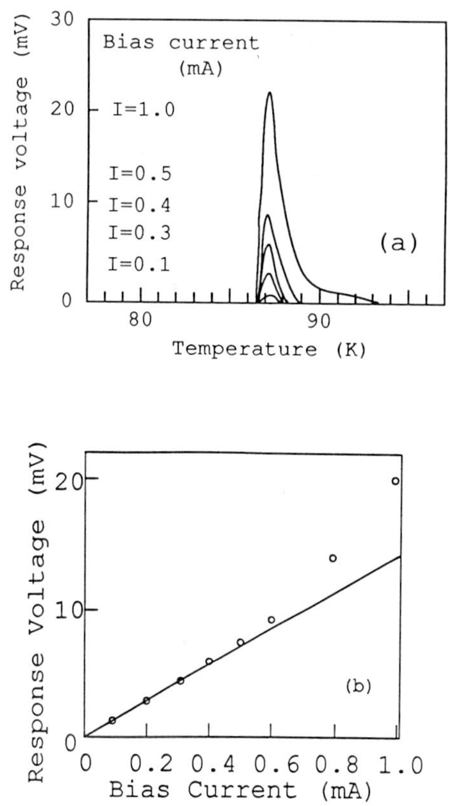

Figure 4 The temperature (a) and bias current (b) dependence of responce voltage at zero magnetic field.

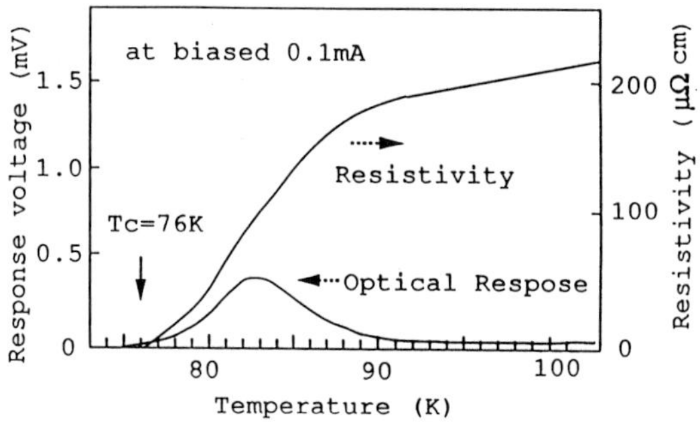

Figure 5 The temperature dependence of optical responce and resistivity of films at a magnetic field of 5T.

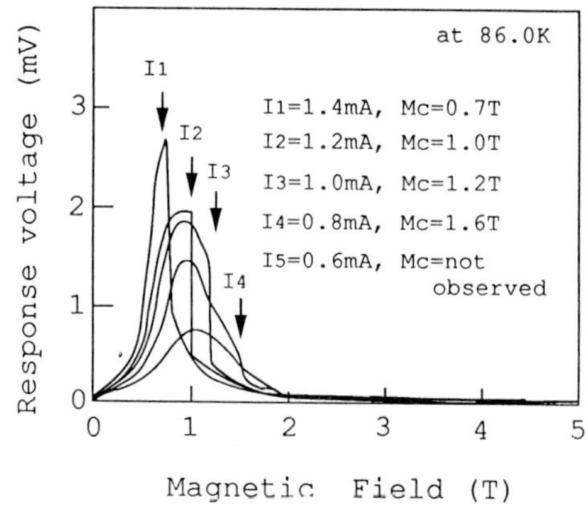

Figure 6 The magnetic field dependence of optical responses at various biased currents.

At the biased current below 0.6mA, there was no abrupt decrease of response voltage with increasing magnetic fields. This suggests that there could be a transition of resistivity depending upon bias currents and magnetic fields. A highly biased superconductor is thought to show the jumps of resistivity caused by the increase of flux flow when magnetic field above critical value is applied. Accordingly, it is thought that the optical responses would suddenly disappear in the resistitive transition region. The sample was broken when the bias current of 1.6 mA was applied at the temperature of 86.0K.

These results may suggest that this response corresponds with the suggestion proposed by Zeldov et al. [6] that flux creeps are induced by optical irradiation. But, we could not explain yet the mechanism of optical response in strong magnetic field and high biased current above Tc.

CONCLUSION

The bolometric response of c-axis oriented ErBa2Cu3Oy superconducting thin film were observed in no magnetic field and the non-bolometric response were observed at high biased currents in magnetic field. This non-bolometric responses were depending upon the bias current and the magnetic field. It may suggest a relationship between the flux-creep and optical irradiations. Furthermore investigation is required to understand the detailed mechanism.

REFERENCE

1. Y. Enomoto, M. Suzuki, and T. Murakami, Jpn. J. Appl. Phys. 23: L333 (1984)
2. M. Leung, U. Strom, J. C. Culbertson, J. H. Claassen, S. A. Wolf, and R. W. Simon, Appl. Phys. Lett. 50: 1691 (1987)
3. M. Leung, P.R. Broussard, J. H. Claassen, M. Osofsky, S. A. Wolf, and U. Stom, Appl. Phys. Lett. 51: 2046 (1987)
4. M. G. Forrester, M. Gottlieb, J. R. Gavaler, and A. I. Braginski, Appl. Phys. Lett. 53: 1332 (1988)
5. W. S. Brocklesby, D. Monroe, A. F. J. Levi, M. Hong, S. H. Liou, J. Kwo, C. E. Rice, P. M. Mankiewich, and R. E. Howard, Appl. Phys. Lett. 54: 1175 (1989)
6. E. Zeldov, N. M. Amer, A. Gupta, R. J. Gambino, and M. W. McElfresh, Phys. Rev. Lett. 62: 3093 (1989)
7. J. C. Culbertson, U. Stom, S. A. Wolf, P. Skeath, E. J. West, and W. K. Burns, Phys. Rev. B39: 12359 (1989)

Superconductive Magnetoresistor Fabricated by Selective Growth of $Y_1Ba_2Cu_3O_{7-x}$ Film Using Electrophoretic Deposition Technique

H. SHINTAKU, H. NOJIMA, M. NAGATA, and M. KOBA

Sharp Corporation, Corporate Research and Development Group, 2613-1, Ichinomoto-cho, Tenri, Nara, 632 Japan

ABSTRACT

A novel patterning method for the improvement in sensitivity of the superconductive magnetoresistor is studied. By use of the electrophoretic deposition technique, $Y_1Ba_2Cu_3O_{7-x}$ film was selectively deposited on the patterned meander-line Cu thin film on YSZ. The meander-line element with a line width of 500 μm and a line length of 250 mm showed a magneto-voltage sensitivity of 57 mV/gauss at 50 K. The sensitivity is about one order higher than that of the previously reported mechanically patterned sensor.

KEY WORDS: superconductive magnetoresistor, magnetic sensor, electrophoretic deposition, selective growth, meander-line

INTRODUCTION

The ceramic superconductive magnetoresistor, which was first proposed by the authors [1,2], has many advantages over the conventional semiconductor magnetoresistive sensor. To date, however, a patterning of the ceramic high Tc superconductor film by chemical etching without degradation of the electric properties has been difficult because grain boundaries, which are essential for our sensor, are damaged during the etching. We already reported that the sensor made of $Y_1Ba_2Cu_3O_{7-x}$ film mechanically patterned to a meander shape showed a magnetic field resolution of 2×10^{-5} gauss/$(Hz)^{1/2}$ at 1 Hz [3]. In this sensor, since the structure of the element is not optimized yet, the realization of a much higher sensitivity can be expected.

In this paper, we report a novel patterning method utilizing a selective growth of electrophoretic deposited $Y_1Ba_2Cu_3O_{7-x}$ film. We succeeded in obtaining a meander-line element with a line width of 500 μm and a line length of 250 mm using this technique. An improvement in magneto-voltage sensitivity of the sensor is described.

EXPERIMENTAL

Selective growth of Y-Ba-Cu-O film was performed using the electrophoretic deposition, which is a promising technique for large-scale as well as electronic device applications [4-6]. The deposition of Y-Ba-Cu-O films was carried out using presintered $Y_1Ba_2Cu_3O_{7-x}$ powder dispersed in distilled acetone. The procedure of the film preparation was described in detail elsewhere [6]. The $Y_1Ba_2Cu_3O_{7-x}$ powder was prepared by a coprecipitation method using solutions of $Y(NO_3)_3$, $Ba(NO_3)_2$ and $Cu(NO_3)_2$ with oxalic acid in cation ratios, Y:Ba:Cu=1:2:3. The $Y_1Ba_2Cu_3O_{7-x}$ powder was dispersed in distilled acetone at a weight ratio of 1.2 %. The substrate on which electrodes were formed was dipped in distilled acetone. A patterned Cu thin films with the thickness of about 2000 Å on yttrium-stabilized zirconia (YSZ) substrate were used as the electrodes. The Cu thin film was prepared by evaporation onto the YSZ surface using electron beam irradiation at a pressure of about 10^{-6} Torr. The Cu film was patterned to a meander-line and co-planar counter-electrode by a chemical wet etching process by use of $FeCl_2$. The line has a width of 500 μm and a length of 250 mm. The distance between the meander-line and the counter-electrode was 100 μm. A mask pattern of Cu thin film prepared on YSZ is shown in Fig.1. A d.c. voltage of 100 V was applied between the electrodes by a constant voltage source. Powder deposition occurred on the meander-line, which was the cathode in this experiment. After the deposition for 20 sec at room temperature, the obtained films were heat-treated in air. The films were kept at 200°C for 10 min to completely evaporate the acetone, and then were heated to 900°C at a heating rate of 17°C/min. The annealing time at 900°C was 3 hours. After that, the films were cooled to 450°C with a rate of 10°C/min. To increase the oxygen content of the films, they were maintained at 450°C for 3 hours. Finally, they were cooled to room temperature at a rate of about 3°C/min. The thickness of the obtained Y-Ba-Cu-O film was about 50 μm. The contacts between the film and the lead wires were made by silver paste. Figure 2 shows the photograph of the element.

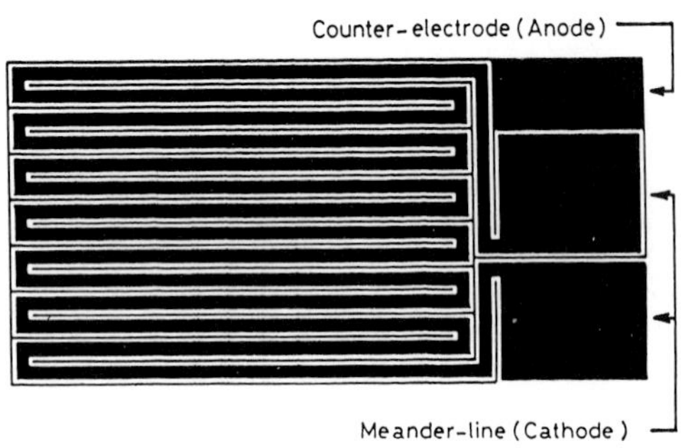

Fig.1. A mask pattern of Cu thin films on YSZ substrate.

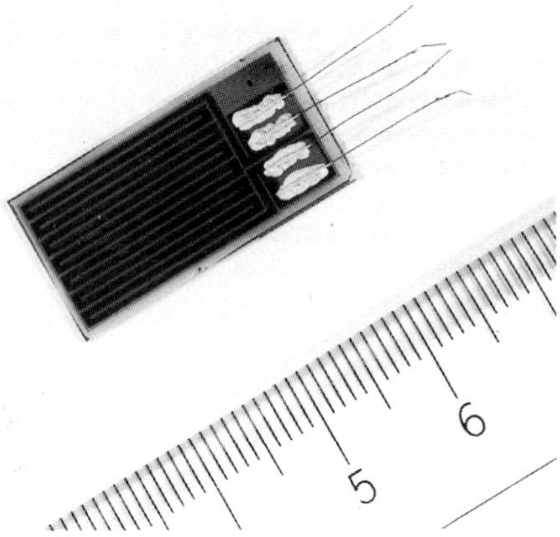

Fig.2. Photograph of the meander-line element

The prepared films were characterized by X-ray diffraction measurement and SEM observation. Electric resistance was measured by standard four-probe techniques. Magnetoresistance measurements were performed at 50 K in a uniform magnetic field generated by the Helmholtz' coil. The instrumental noise level for these resistance measurements was about 10^{-7} V.

RESULTS AND DISCUSSION

Typical X-ray diffraction pattern of the obtained Y-Ba-Cu-O film is shown in Fig.3. The $Y_1Ba_2Cu_3O_{7-x}$ orthorhombic crystal phase and the YSZ phase can be identified. Labeled orientations are indicated to be the $Y_1Ba_2Cu_3O_{7-x}$ phase. Figure 4 shows the SEM image of the Y-Ba-Cu-O film. The size of the grains ranged from 2 μm to 10 μm in diameter.

Temperature dependence of the electric resistance of this meander-line element is shown in Fig.5. The resistance of the element increases with the decrease of the temperature like that of a semiconductor but shows a Tc,onset at 96 K and zero-resistance state at 50 K. In this figure, the temperature dependence of the resistance measured with an application of a magnetic field of 83 gauss is also shown. When the magnetic field is applied, the resistance tail is observed due to the weak couplings of superconductivity at grain boundaries in the film.

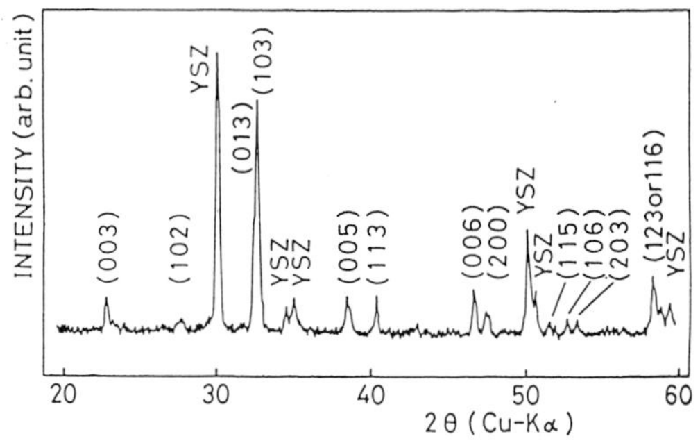

Fig.3. X-ray diffraction pattern of the obtained meander-line element.

Fig.4. Scanning electron micrograph of the obtained Y-Ba-Cu-O film.

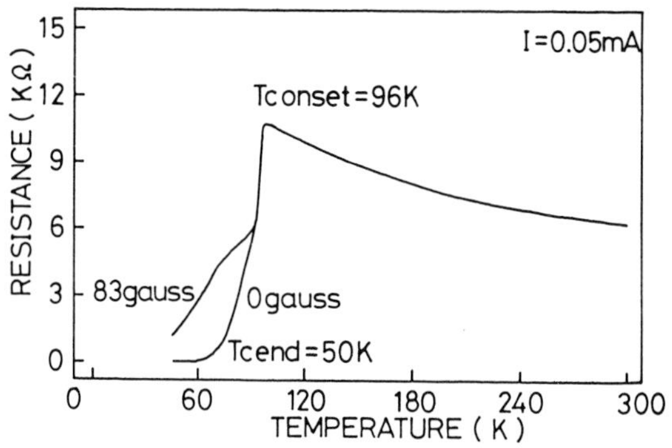

Fig.5. Temperature dependence of the electric resistance of the meander-line element. The electric resistance was measured at zero magnetic field and at a magnetic field of 83 gauss.

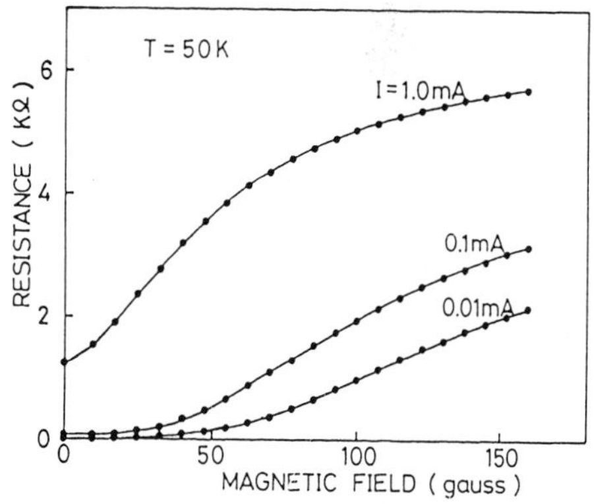

Fig.6. Magnetic field dependence of the resistance of the meander-line element with a d.c. current as a parameter.

Figure 6 shows the magnetic field dependence of the resistance of the element with a d.c. current as a parameter. As shown in this figure, with the increase of the magnetic field, the resistance of the element increases steeply at a low magnetic field. These magnetoresistive characteristics form the basis of the magnetic sensor. The meander-line element indicated the magnetoresistive characteristics similar to the previously reported element [3], which was fabricated using mechanically patterned $Y_1Ba_2Cu_3O_{7-x}$ film prepared by a spray pyrolysis method, but the magnetic sensitivity of this meander-line element was much higher than that of the mechanically patterned element. The magneto-voltage sensitivity of the sensor is defined in this paper as the maximum increase rate of the output voltage with respect to the applied magnetic field, $(\Delta V/\Delta B)max$. Figure 7 shows the magneto-voltage sensitivity of the obtained meander-line element as a function of the current through the element. The magneto-voltage sensitivity of 57 mV/gauss at a d.c. current of 1 mA was obtained. For comparison, the sensitivity for the mechanical patterned element is also shown by open circles in this figure. It is found that the magneto-voltage sensitivity of the electrophoretic selectively deposited meander-line element is about one order higher than that of the previous element. Although the noise measurements are required to clarify the magnetic field resolution of this sensor, the achievement of the patterning of the ceramic high Tc superconductor by the novel selective growth technique shows a potential of the improvement in sensitivity of the superconductive magnetoresistive sensor.

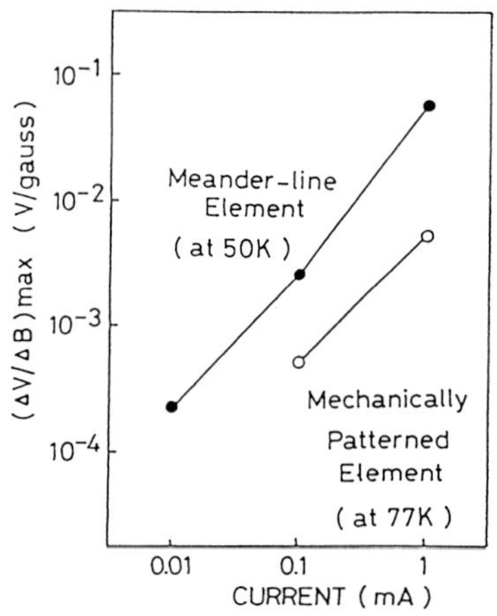

Fig.7. Magneto-voltage sensitivities of the obtained meander-line element (●) and the previously reported mechanically patterned element (○) as a function of the current through the elements.

CONCLUSION

The meander-line element of the superconductive magnetoresistor was fabricated utilizing the selective growth of the electrophoretic deposited $Y_1Ba_2Cu_3O_{7-x}$ on patterned Cu thin film on YSZ. The obtained element has the line width of 500 μm and the length of 250 mm. The element showed the Tc,onset of 96 K and Tc,end of 50 K. The element exhibited the magneto-voltage sensitivity of 57 mV/gauss at 50 K. This sensitivity is about one order higher than that of the previously reported mechanically patterned sensor.

ACKNOWLEDGEMENTS

The authors wish to thank Drs.S.Kataoka and A.Fujisawa for their continuous encouragements throughout this work and for valuable discussions.

REFERENCES

1. Nojima H, Tsuchimoto S, Kataoka S (1988) Jpn.J.Appl.Phys. 27: 746
2. Tsuchimoto S, Kataoka S, Nojima H, Kita R, Nagata M, Shintaku H (1987) : IEDM Tech. Digest (IEEE, New York,1988): 867
3. Shintaku H, Nojima H, Nagata M, Ohno E, Koba M (1989) Int. Symp. on Superconductivity-ISS'89 (Springer-Verlag, Tokyo, 1990): 999
4. Koura N (1988) Denki Kagaku 56: 208 [in Japanese]
5. Chu CT, Dunn B (1989) Appl.Phys.Lett. 55: 492
6. Nojima H, Nagata M, Shintaku H, Koba M (1990) Jpn.J.Appl.Phys. 29: L1655

Current-Injection-Type Three-Terminal Devices Using High-T_c Superconductors

HIDETAKA HIGASHINO, KOICHI MIZUNO, TOMOAKI MATSUSHIMA, KENTARO SETSUNE, and KIYOTAKA WASA

Central Research Laboratories, Matsushita Electric Industrial Co., Ltd., 3-15, Moriguchi, Osaka, 570 Japan

ABSTRACT

Characteristics and new fabrication process of current-injection-type three-terminal devices with low-temperature-grown Bi-Sr-Ca-Cu-O (BSCCO) thin films and Al gate electrodes are described. Degradation mechanism of BSCCO films through the device fabrication process is proved.

KEY WORDS: low-temperature-grown Bi-Sr-Ca-Cu-O film, Al gate electrode, grain boundary degradation

INTRODUCTION

Some kinds of three-terminal devices using high-Tc superconductors have been reported [1-6] up to now. Among them, a current-injection-type device seems to be one of the most promising ones because of its possibility of high speed operation. On the other hand, the high-Tc oxide superconductors of Bi-Sr-Ca-Cu-O (BSCCO) discovered by Maeda et al. [7] have attracted much attention because of their higher Tc over 110 K and of their stability. We have developed this kind of devices using BSCCO thin films and dual Al gate electrodes because of above reasons. We have reported the results of the device using high-temperature annealed BSCCO film, which was dominated by quasi-particle injection effect and exhibited a differential current gain of about 8 [4]. However, degradation of superconductivity was observed through the fabrication process.

In this paper, characteristics of the devices with low-temperature-grown BSCCO films and Al gate electrodes, fabricated by two kinds of methods, by lift-off technique and by Ar ion etching, or in a different gate structure, single or dual, were reported. Furthermore, magnetizations of BSCCO film and of the BSCCO/Al one were also measured in order to examine the degradation. A cause of the degradation will be proved later.

EXPERIMENTAL

A current-injection-type three-terminal device comprises a Dayem-type superconducting bridge of an low-temperature-grown BSCCO film and injection-gate electrodes of Al which are deposited on the bridge. Minimum width and length of the bridge and gate-electrode width are 10 μm each. The dual-gate-type device has separated gate electrodes whose gap spacing is 3 μm, typically, and the gap is arranged around the center of the bridge. The c-axis oriented BSCCO film (260 nm-thick) was deposited on a (100) MgO substrate by RF magnetron sputtering. The sputtering conditions are listed in Table 1 (a). The gate electrodes were fabricated using the lift-off technique in the first experiment. The details of fabrication process is described in [4]. Resistance and critical current values between two banks of the bridge, before and after Al gate formation, were measured by the four-probe method as a function of the device temperature. Variation of the critical-current was measured as a function of the gate-injection current and the power.

In the second experiment, a single-gate device was fabricated by the similar process to the dual-gate one except for the sputtering conditions as listed in Table 1 (b). The thickness of BSCCO film was 330 nm. A different point in a configuration of the device is no gap-spacing of Al gate electrode over the bridge. The electric characteristics between two banks, before and after Al gate formation, were also measured by the four-probe method for comparison.

Table 1. Sputtering and annealing conditions

	(a)	(b)
Target	Bi-Sr-Ca-Cu-O powder	"
Substrate	(100) plane of MgO	"
Sputtering gas	Ar/O_2= 35 / 65	20 / 80
Gas pressure	0.5 Pa	1.1 Pa
RF power	100 W	"
Substrate temperature	650 °C	610 °C
Growth rate	13 nm/min	11 nm/min
In situ anneal	none	slow cooling (O_2 flow x 1 hr)
Film composition	$B_{1.9}Sr_{2.0}Ca_{1.0}Cu_{2.2}O_x$	$B_{2.2}Sr_{2.0}Ca_{1.0}Cu_{2.2}O_y$
Crystal phase	single (2212), c-axis	"
Film thickness	260 nm	330 nm

In order to prove the electric degradation by the Al deposition, a dual-gate device was fabricated using Ar ion etching technique for gate patterning instead of the lift-off, in the third experiment. The fabrication process is shown in the following: The low-temperature-grown BSCCO film (2 mm x 10 mm) was cut out from the same wafer as used in the second experiment. The Dayem-type bridge was patterned from BSCCO film by the Ar ion etching after the first deposition of four Au terminals (2 mm W x 200 nm t, 1.3 mm-spacing) and after the second sequential deposition of Al (30 nm) and Au (60 nm) films. The depositions of metals were carefully carried out to avoid heating up BSCCO film by evaporation. To form a dual-gate electrodes, unnecessary Al/Au was etched off with Ar ions and using a rubber-type photo-resist for a etching mask to make a photo-process damage minimum. After resist removing with O_2 plasma, bridge resistance was measured by the four probe method. The Al/Au dual-gate electrodes were connected with evaporated Al (120 nm) contact electrodes using the lift-off. The variation of the bridge resistance was measured through the process.

For elucidation of grain-boundary degradation by Al deposition, we measured magnetizations of BSCCO films (1.5 mm x 2 mm), with and without evaporated Al (100 nm), using an RF-SQUID magnetometer. The BSCCO films were cut out also from the same wafer as used in the second and third experiments.

RESULTS AND DISCUSSION

In the first experiment, measured resistance variations of the bridge as a function of temperature are shown in Fig. 1; that after the bridge formation, i.e. before Al gate deposition (indicated as "before"), and that after dual-Al gate formation by the lift-off. Tc_0 values were the same before and after the gate formation, while the resistance value at the same temperature was increased after the lift-off. Critical current values were decreased less than two times or more, and at 4.2 K, it was degraded from 2.1 x 10^5 A/cm^2 to 8.0 x 10^4 A/cm^2 (38 % of the former value) after the lift-off. This kind of the degradation has been observed even in annealed samples [4]. Figure 2 shows a critical-current variation as a function of injected gate current which flows from one gate to the other. The curve has a parabola-like shape and a slight asymmetry to the zero gate-current axis. Maximum differential current gains were obtained as 9.4 and 11.5 (reverse gate-current direction) at the gate current of over 50 μA. The gains are larger than that of the annealed sample as about 8 [4]. Impedance between two gate electrodes was about 200 k-ohm and no energy-band gap structure was observed in I-V curves between each gate and the superconductor. The critical current continuously decreases with increasing the gate injection power. This result proves that the device was dominated by simple thermal effect, which is caused by the high gate resistance produced probably by a surface degradation of BSCCO film under the Al gates. The later suspicion will be proved by the second experimental results. The single-gate device after gate formation exhibited no superconductivity but a residual resistance below 45 K, while the bridge

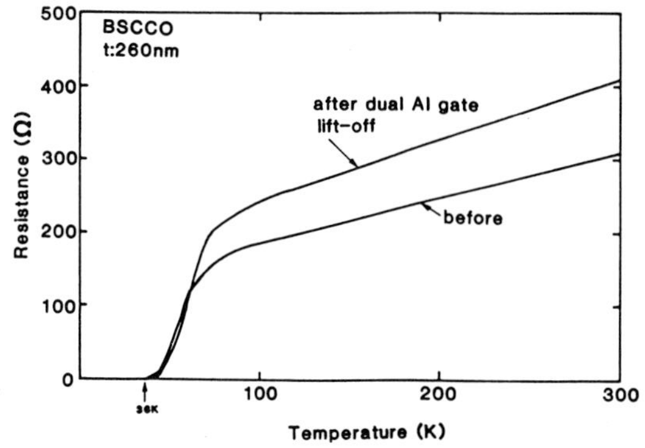

Fig.1. Temperature dependence of resistances of a dual-gate device befor and Al gate lift-off.

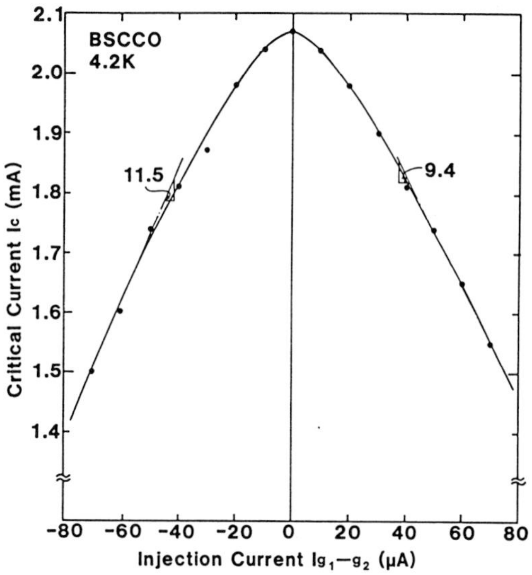

Fig.2. Injection gate current (g_1 -> g_2) dependence of critical current.

before Al deposition did superconductivity below Tc_0 as 45 K. This clearly indicates the degradation of BSCCO film caused by Al deposition.

In the third experiment, three kinds of the temperature-resistance characteristic were measured as shown in Fig. 3; (1) the resistance after probe electrodes deposition and before Al deposition, (2) that after the Al (30 nm) and Au (60nm) deposition and the bridge formation (8 μm W x 10 μm L), and (3) that after Al gate formation (10 μm W, 2 μm spacing) by ion etching and after O_2 plasma ashing. Tc_0 values were 41 K in (1), and 40 K in (3). In (3), the temperature-resistance variation was almost similar to that in (1) except for having about four times greater value and to that in Fig. 1 having about 2.6 times greater, while the bridge in (2) exhibited no superconductivity but normal-metal-like conductivity with a residual resistance of 13 ohm. From these results, it is proved that the degradation was caused by Al deposition and was recovered by removing the Al with Ar ions and by O_2 plasma treatment. Thickness of the degraded layer is considered to be several nm because of low temperature deposition of Al/Au and of short coherence length in BSCCO, while BSCCO thickness was 330 nm. In this case, supercurrent could flow underneath the degraded layer, if there is no grain-boundary degradation. These results suggest us the existence of degradation of grain boundaries of BSCCO film which can be recovered by O_2 plasma treatment.

The normalized magnetization measured with 10 Oe was shown in Fig. 4 as a function of the temperature under the condition of zero field cooling. No significant difference can be fund out between low-temperature-grown BSCCO film (●) and BSCCO/Al (120 nm) film (○). The absolute values of magnetization in the both case were almost the same. This proves that no serious degradation was induced by Al deposition inside the grains of BSCCO film. From above results, we can conclude that the degradation of superconductivity of the devices using low-temperature-grown BSCCO films is caused by Al-gate deposition, which induces the degradation in the grain boundaries.

CONCLUSIONS

Current-injection-type three-terminal devices using low-temperature-grown BSCCO films and Al gate electrodes were fabricated by lift-off technique and by Ar ion etching and were examined. A dual-gate device prepared by the lift-off exhibited higher value of 11.5 in differential current gain than that of

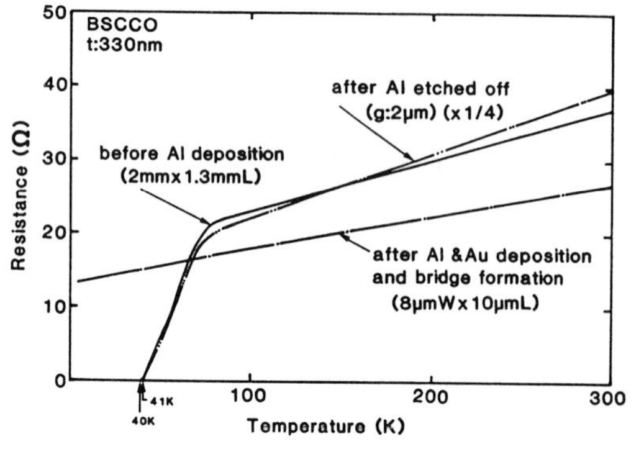

Fig.3. Temperature dependence of resistances for a dual-gate device by Ar ion etching.

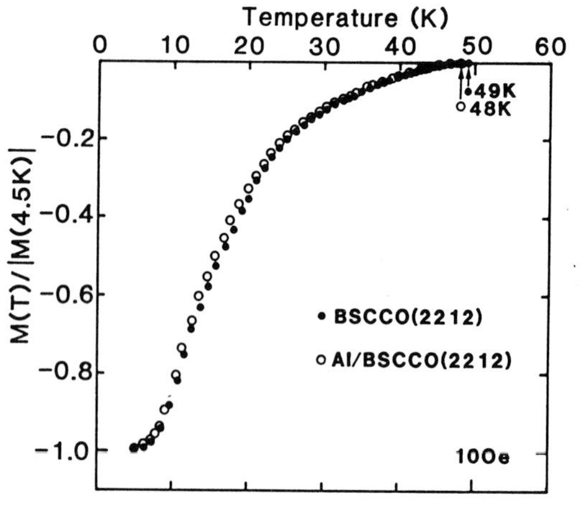

Fig.4. Temperature dependence of normalized magnetization. A filled circle (●) indicates of low-temperature-grown BSCCO film and an open circle (○) does of BSCCO/Al(120 nm) film.

high-temperature-annealed sample [4], however it is proved to be due to simple thermal effect differently from in the high-temperature-annealed sample.

We investigate the cause of degradation in superconductivity observed through the device fabrication process by a single-gate device using the lift-off and by a dual-gate one by Ar ion etching. Magnetizations of BSCCO film and of BSCCO/Al were also measured. From the experimental results, it is proved that the degradation is caused by Al-gate deposition, which induces the degradation in the grain boundaries of BSCCO under the Al metal, and which is recovered by removing Al using Ar ion etching and with O_2 plasma treatment. For high performance devices which have low gate-impedances and accordingly are governed by quasi-particles, it is recommended to use highly epitaxitial BSCCO films having less grain boundaries.

ACKNOWLEDGMENT

The authors would like to thank to the members of superconducting research group for their useful discussions and to Dr. T. Nitta and Dr. K. Kanai of central research laboratories for their encouragements at Matsushita.

REFERENCES

1. T.Kobayashi, K.Hashimoto, U.Kabasawa and M.Tonouchi, IEEE Trans. on Magnetics, **25**, 927 (1989).
2. H.Higashino, A.Enokihara, K.Mizuno, T.Mitsuyu, K.Setsune and K.Wasa, Ext.Abst. of FED HiTcSc-ED Workshop, 267 (1988).
3. H.Higashino, K.Setsune and K.Wasa, Ext.Abst. of ISEC'89, 229 (1989).
4. H.Higashino, A.Enokihara, K.Mizuno, K.Setsune and K.Wasa, Ext.Abst. of FED 2nd Workshop on HiTcSc-ED, 257 (1889).
5. D.F.Moore, Ext.Abst. of FED 2nd Workshop on HiTcSc-ED, 281 (1889).
6. H.Higashino, K.Setsune and K.Wasa, Proceedings of ISS'89, Tsukuba, 981 (1989).
7. H.Maeda, Y.Tanaka, M.Fukutomi and T.Asano, Jpn.J.Appl.Phys., **27**, L209 (1988).

Superconducting Switching Device Using High-T_c Superconductor Thin Film

H. SATOH and O. TSUKAMOTO

Faculty of Engineering, Yokohama National University, 156, Tokiwadai, Hodogaya-ku, Yokohama, 240 Japan

ABSTRACT

We are developing a superconducting current switching device using the super-normal transition of high Tc superconductor thin film. We made an experimental sample using YBaCuO thin film and investigated on the super-normal transition characteristics. We measured turn-off and turn-on times. They were possible to be reduced to less than 2ms. The result suggests that the high Tc superconducting thin film is applicable to the 50/60Hz switching device.

KEY WORDS: superconducting switching device, thin film, electric apparatus

INTRODUCTION

A superconducting current switching device (SCSD) is an important component for superconducting applications to electric power apparatuses. Zero resistivity at "on-state" is an important advantage of the SCSD to conventional semiconductor devices. SCSD can be applied to over-current protection devices for superconducting transformers, fully superconducting generators and superconducting cables, and to a device for limitation of fault currents of electric utility grids. There have been many studies on SCSD using conventional metal superconducting wire [1,2]. Switching of the conventional SCDC is controlled by applying a heat pulse or external magnetic field higher than the critical field to the wire. However, disadvantages of the conventional SCDC are low resistivity at "off-state" and slow turn-on and turn-off speeds (especial the turn-on is slow and usually takes more than a second). High Tc superconductor is expected to greatly improve the performance and economical feasibility of superconducting electrical power apparatuses and the SCDC is not an exception.

We are studying on a superconducting switching device using high Tc superconductor thin film for electric power applications. The switching device is a super-normal transition type and shifted to off-state (normal state) by applying a pulse current larger than the critical current of the superconducting thin film. The switch keeps the off-state by a residual current generating the joule heat and is switched to the on-state (superconducting state) by shutting-off the residual current. The merits of the high-Tc superconducting thin film for the switching device are ① the resistivity at the off-state is high, ② turn-on speed is high, because the surface area of device exposed to liquid nitrogen is large and ③ it is easy to make a large-current-capacity device, because the current density of the high-Tc superconducting thin film is very high.

EXPERIMENT

The Sample to investigate turn-on and turn-off characteristics is YBCO thin film of 2mm × 12mm prepared by EB deposition on a MgO base. The thickness of the film is about 0.3 μ m. A sample arrangement is shown in Fig.1. In the experiment, a bulky current as a thin film superconductor is to be applied to the sample, therefore, a proper preparation has to be done to the current junctions. In the experiment, an indium slug was placed between an electrode of normal metal and the thin film and the electrode was mechanically pressed against the thin film. (Fig.1) Current leads and voltage taps were soldered to the electrodes.

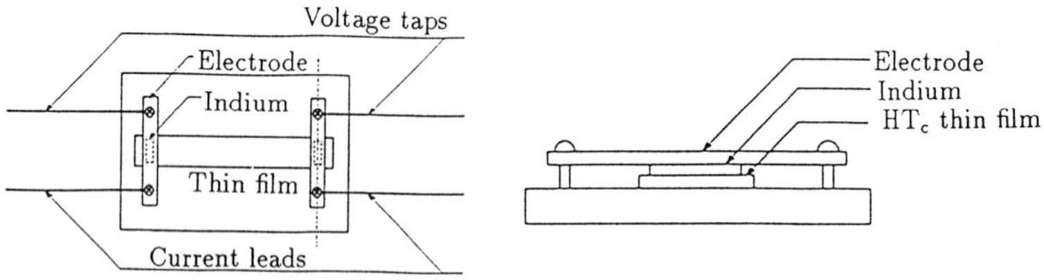

Fig.1 Sample arrangement

The experimental circuit to measure the turn-on and turn-off characteristics is illustrated in Fig.2. The sample current was feed-back controlled to follow the current reference. A patter of the current reference which was generated by a personal computer is shown in Fig.3. The sample was tested in liquid nitrogen. After applying operational current I_1 smaller than the critical current of the sample Ic, a pulse current I_2 larger than Ic was imposed to turn the sample to the off-state (normal state). I_2 and the duration time of the pulse T_2 were changed to investigate the on-off characteristics. After the pulse finishs, the sample current was decreased to I_3. The turn-off characteristics depend on I_1 I_2 and T, and the turn-on characteristics on I_2, I_3 and T. The sample current and voltage traces were recorded by a transient recorder.

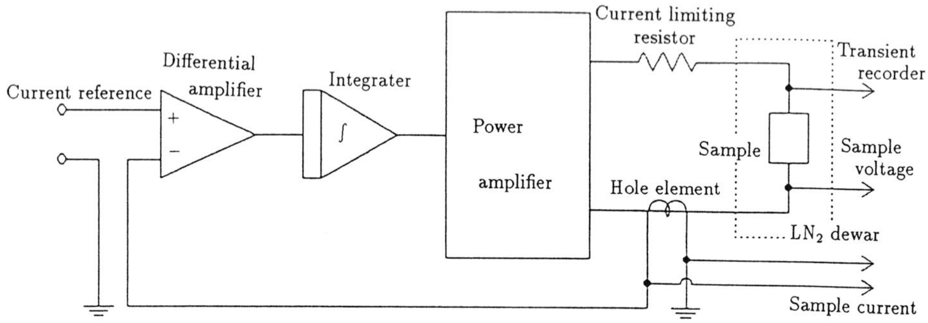

Fig.2 Experimental circuit

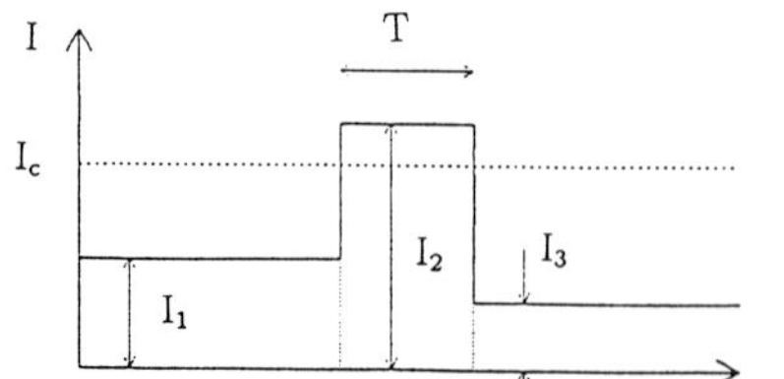

Fig.3 Current reference pattern

EXPERIMENTAL RESULTS

Turn-on and turn-off characteristics of the sample were investigated by changing the sample current following the pattern shown in Fig.3. The data were taken for various sets of I_1,I_2,I_3 and T. The minimum current of the super to normal transition of the sample Ic was 90mA. The figure 4 is the sample voltage and current traces for the case that I_2 was below Ic. The sample voltage was kept zero and the sample was superconducting during the whole current changing sequence. The figure 5 is the traces for the case that I_2 was well over Ic. The voltage appeared at the instant that the pulse current was imposed and kept increasing during the pulse duration. It can be explained as that normal zone was initiated by the current exceeding Ic in some part of the thin film and propagated across the thin film.

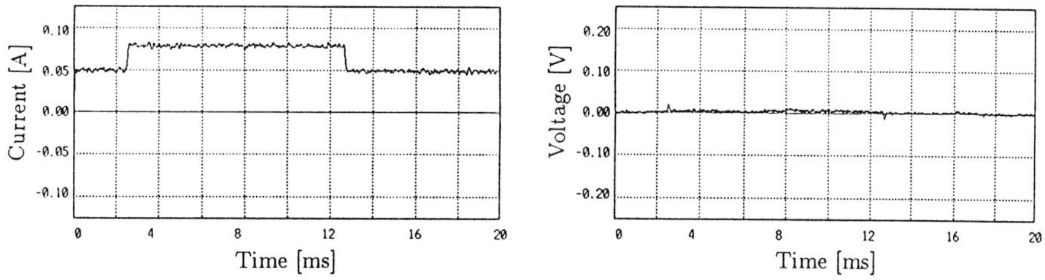

(a) Current trace (b) Voltage trace

Fig.4 Sample current and voltage traces for $I_2 < I_c$
(I_1=50mA, I_2=80mA, I_3=50mA, T=10ms)

In the case of Fig.5, I_2 and the temperature rise of the sample were not so high that the sample recovered the superconducting state almost instantaneously after the current pulse was terminated.

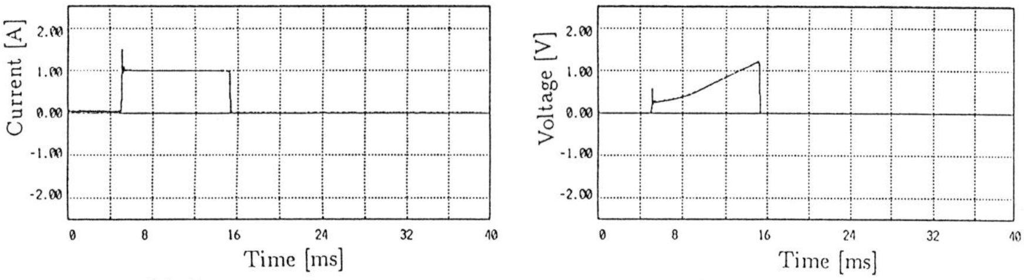

(a) Current trace (b) Voltage trace

Fig.5 Traces for $I_2 > I_c$ with quick recovery of superconducting state
(I_1=50mA, I_2=1.0A, I_3=10mA, T=10ms)

The figure 6 is the traces for higher I_2 where the recovery of the superconducting state was delayed. In this case, the resistivity of the sample became so high that the power amplifier could not sustain the sample current to follow the current reference (see the current drop near the termination of the pulse). The sample temperature rose significantly high and it took a time for the sample to be cooled and recover the superconducting state. The recovery time of the superconducting state depends on the pulse current I_2, the residual current I_3 and the duration time of the pulse T. The recovery time (the turn-on time) t_r is defined as the time necessary for the sample to recover the superconducting state after the termination of the pulse current.

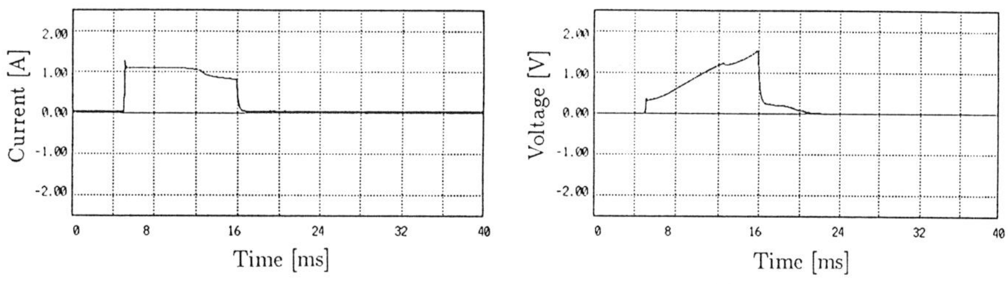

(a) Current trace (b) Voltage trace

Fig.6 Traces for slow recovery of superconducting state
(I_1=50mA, I_2=1.1A, I_3=50mA, T=10ms)

The figure 7 shows the recovery time vs. I_2. For I_2 lower than 0.9A, the sample recovered instantaneously after the finish of the current pulse. However, as I_2 increases higher than 0.9A, t_r increases rapidly.

In the figure 8, t_r is plotted against T for fixed value of I_1,I_2 and I_3. t_r increases almost linearly as T increases beyond about 7ms, which means that excess heating occurs and that the heat is accumulated in the sample for T>7ms.

The figure 9 is for t_r vs. I_3. t_r increases almost linearly as I_3 increases, which suggests that the sample is in a mixed state of super and normal after the finish of the pulse current, while t_r should be proportional to the square of I_3 if the sample is fully normal state.

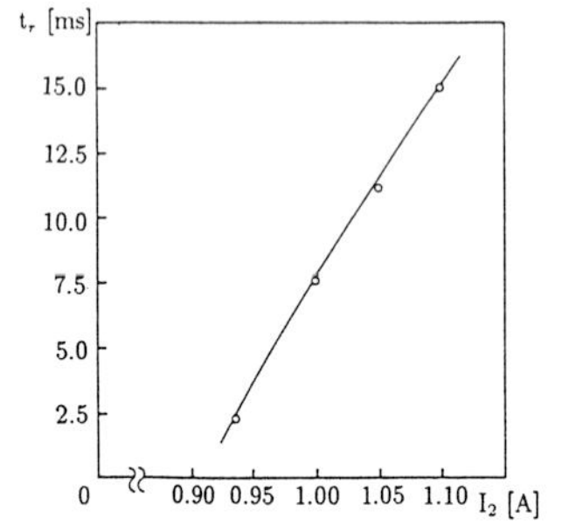

Fig.7 t_r vs. I_2 for fixed values of I_1, I_3 and T (I_1=50mA, I_3=50mA, T=10ms)

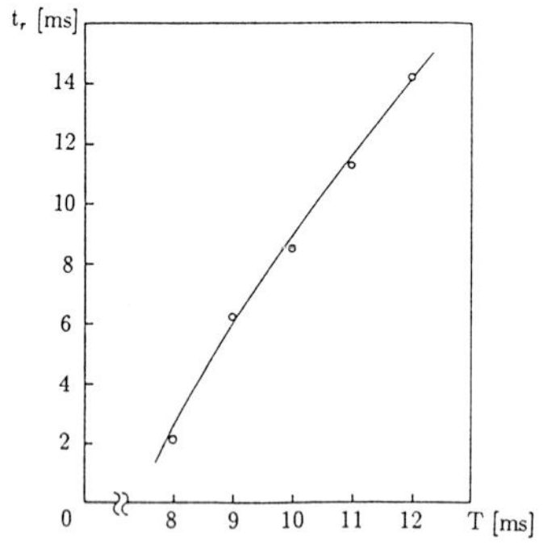

Fig.8 t_r vs. T for fixed values of I_1, I_2 and I_3 (I_1=50mA, I_2=1.0A, I_3=50mA)

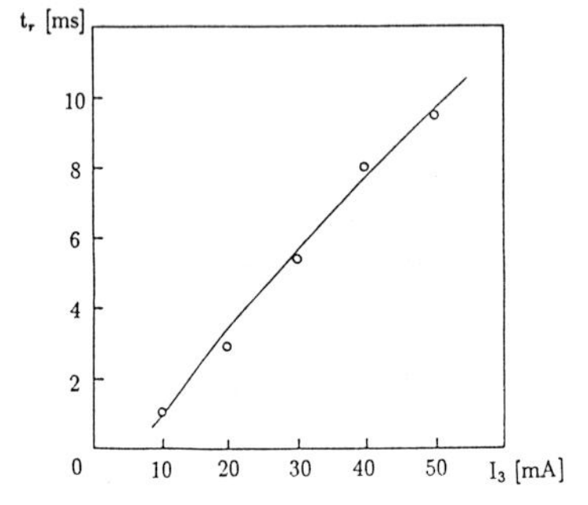

Fig.9 t_r vs. I_3 for fixed values of I_1, I_2 and T (I_1=50mA, I_2=1.0A, T=10ms)

DISCUSSIONS AND CONCLUSIONS

Obviously from the experimental results, if I_2,I_3, and T are properly chosen, the sample can be turned-on in less than 2 msec after the current pulse is removed. The experimental results also show that the sample can be turned-off almost instantaneously by applying the current pulse higher than Ic. Therefore, with a proper arrangement of the thin film and a proper selection of operating conditions, the switching device using high Tc thin film can be applied to 50/60Hz and higher frequency switching. The sample used in this experiment required rather high value of I_2 to get a clear normal transition. We consider that this was due to the non-uniform superconductive characteristic of the sample and that the switching device can have sharper on-off transition, if the characteristic of the thin film is more uniform. Development of high current capacity junction of thin film and an electrode is very important for this kind of switching device.

We acknowledge Dr.Y.Tanaka and Mr.K.Satoh of The Furukawa Electric Co.,Ltd. kindly supplying us the experimental samples.

REFERENCES

1. Herman H.J. ten Kate , Bennie ten Haken, Louis J.M. van de Klundert
   Advances In Cryogenic Engineering, vol 31 1986: 243
2. G.B.J. Mulder, D. ter Avest, H.H.J. ten Kate, H.J.G. Krooshoop and L.J.M. van de Klundert
   IEEE MAGNETICS 1988: 907

Transport Properties of Quasi-Particle in Superconductors Using Monte-Carlo Simulation

TETSUYA YAMAMOTO, HIROSHI SUZUKI, YORINOBU YOSHISATO, and SHOICHI NAKANO

Functional Materials Research Center, SANYO Electric Co., Ltd., Hirakata, Osaka, 573 Japan

ABSTRACT

The transport properties of low energy quasi-particles across the base layer in a superconducting base transistor were investigated by means of Monte-Carlo simulations including the acoustic phonon scattering mechanism. For the isotropic Nb system, a large mean-free path of $L=1.83\times10^{-3}$ (cm) and a diffusion coefficient of $D=2.84\times10^{4}(cm^2/s)$ were obtained. The base layer transit time was estimated to be about 1.0 (ps) for a 1000 Å thin base. We also calculated the quantum mechanical reflection at the base-collector interface using the Bogoliubov equation.

KEY WORDS: transistor, quasi-particle, Monte-Carlo simulation, Bogoliubov equation

INTRODUCTION

There has been considerable research on making an electronic switching device using a superconductor. Superconducting three terminal-devices, which would make it easy to construct complicated circuits, have become of great interest. The low-energy type superconducting base transistor, proposed by D.J.Frank et al.,[1] is expected to realize both high frequency performance and low power consumption. The general structure of this transistor is shown in Fig. 1. The operating principles of the device are similar to that of a conventional bipolar transistor. The characteristics of this transistor were calculated by D.J.Frank et. al.[1] and H.Tamura et. al.[2] In their work, the transport of quasi-particles with energy near the gap (E~Δ) were calculated using a single relaxation-time approximation. However, this approximation is not sufficient for quasi-particles with an energy of several Δ.

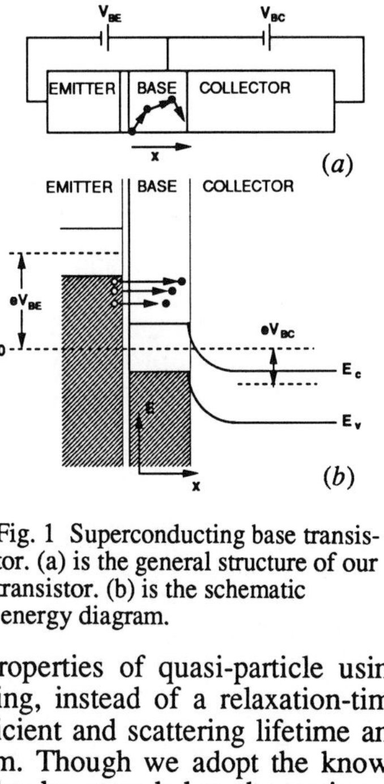

Fig. 1 Superconducting base transistor. (a) is the general structure of our transistor. (b) is the schematic energy diagram.

In this work, we attempt to clarify the characteristics of the superconducting base transistor by investigating the transport properties of quasi-particle using Monte-Carlo simulations, which include acoustic phonon scattering, instead of a relaxation-time approximation. We calculated the mean-free path, diffusion coefficient and scattering lifetime and estimated the base layer transit time for the isotropic Nb system. Though we adopt the known parameters of Nb, the Monte-Carlo method presented here can also be extended to the testing of high-Tc superconductors by modifying the corresponding parameter values. Taking into account the effect of quantum mechanical reflection at the base-collector interface, we also calculated the quasi-particle transmission probability from a normal to a superconducting region through a square potential barrier using the Bogoliubov-de Gennes[3] equation of motion.

MONTE-CARLO SIMULATION IN THE SUPERCONDUCTING BASE LAYER

The transport properties of low energy quasi-particles across the superconducting base layer are described using the following Boltzmann equation based on the BCS theory[4]:

$$\frac{\partial f_k(x,t)}{\partial t}+V_x\frac{\partial f_k(x,t)}{\partial x}=-\sum_{k'}P(k,k')(1-f_{k'})f_k+\sum_{k'}P(k',k)(1-f_k)f_{k'}+\sum_{k'}P_c(k,k')(1-f_k)(1-f_{k'})-\sum_{k'}P_d(k,k')f_kf_{k'} \quad (1)$$

where $f_k(x,t)$ is the quasi-particle distribution function and V_x is the group velocity. $P(k,k')$ is the scattering probability per unit time of a quasi-particle from state k to state k'. $P_c(k,k')$ and $P_d(k,k')$ are the rates of pair-creation and annihilation, respectively. The conceivable scattering mechanisms for quasi-particles are acoustic and optical phonon scattering, impurity scattering, pair-creation and annihilation effect. If the superconductor is a single-crystal, the effects of impurity can be ignored. For low energy quasi-particles, the acoustic phonon scattering is dominant, so that we can neglect the optical phonon scattering. Comparing the quasi-particle lifetimes calculated by Kaplan[5)] et al., we found the rates of pair-creation and annihilation to be less than half of the acoustic phonon scattering rate. Therefore, in our simulation we considered mainly the effects of acoustic phonon scattering.

We can derive the acoustic phonon scattering probability $P(k,q)$ of a quasi-particle from state k to state $k\pm q$ with a first order perturbation approach as follows :

$$P(\vec{k},\vec{q})=\frac{2\pi}{\hbar}|V_q|^2\times\frac{1}{2}\left(1+\frac{\xi_k\xi_{k\pm q}-\Delta^2}{E_kE_{k\pm q}}\right)\times\begin{bmatrix}N_q\\N_q+1\end{bmatrix}\delta\left(E_{k\pm q}-E_k\mp\hbar uq\right) \quad (2)$$

$$E_k=\sqrt{\xi_k^2+\Delta^2}\,,\qquad \xi_k=\frac{\hbar^2k^2}{2m}-\varepsilon_F\,,\qquad N_q=\frac{1}{e^{\hbar uq/kT}-1} \quad (3)$$

where the upper and lower symbols refer to absorption and emission, respectively, and Δ is the gap energy, ε_F the Fermi energy, u the sound velocity, q the phonon wave number and N_q the average phonon number. The square factor that appears in Eq. (2) is reduced to A^2q for the Debey model, where A is a constant characteristic of a material. Integrating Eq. (2) over the polar angle θ between q and k, we derived the following expression for the absorption process :

$$P(\vec{k},q)=\left(\frac{VA^2m}{4\pi\hbar^3}\right)N_qq^2\frac{E_k+\hbar uq}{\sqrt{(E_k+\hbar uq)^2-\Delta^2}}\times\left(1+\frac{\xi_k\sqrt{(E_k+\hbar uq)^2-\Delta^2}-\Delta^2}{E_k(E_k+\hbar uq)}\right) \quad (4)$$

The term which appears in the second bracket in Eq. (4) is called the coherent factor. For low energy quasi-particles, this factor reduces the scattering probability. In the same way, we also obtained similar expression for the emission process. Equation (4) is useful in the Monte-Carlo simulation for the determination of state k' after scattering. We calculated the scattering probability numerically from state k by integrating Eq. (4) over q.

In order to investigate the transport properties, we solved the Boltzmann equation (1) using the Monte-Carlo method[5)]. We carried out the Monte-Carlo simulation under the following conditions : (i) acoustic phonon scattering is dominant, thus the effects of other scattering mechanisms were ignored; (ii) classical statistics were used, and the Pauli exclusion principle neglected; (iii) the gap energy Δ was supposed to be space independent. In our simulations, we dealt with Nb as a superconducting material. Monte-Carlo simulations were carried out using N=100 carriers, N_c=500 collisions for steady-state analysis and N=1000, N_c=50 for transient-state analysis.

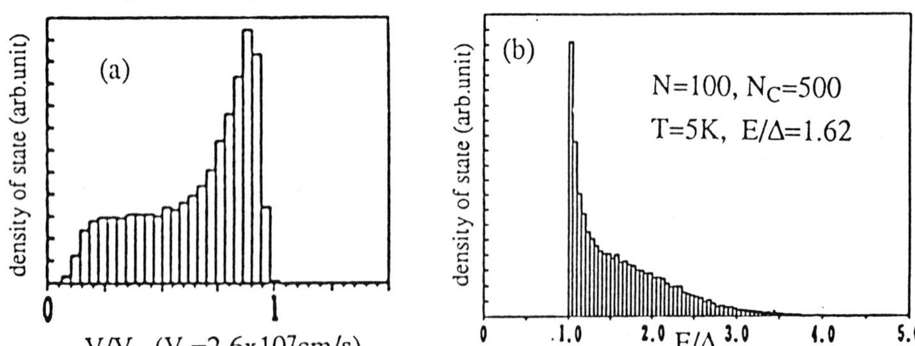

Fig. 2 Histograms of the distribution of the quasi-particle (a) velocity and (b) energy over 100 particles.

Figures 2(a) and (b) show the histograms of the distribution of the injected quasi-particle velocity and energy over 100 carriers at 5(K), respectively. On the basis of this result, we estimated the average velocity V=1.83x10^7(cm/s), average energy E/Δ=1.62, scattering lifetime τ_f=1.57x10^{-10}(s) and mean-free path L=1.83x10^{-3}(cm). The diffusion coefficient D can also be determined from Monte-Carlo simulations through evaluation of the second central moment as a function of time. The results of this steady-state analysis are listed in Table 1. The experimental values of very pure Cu (V_f=1.57x10^8(cm/s)) at 4(K) are also shown for comparison. The values of D and L for Nb are quite large, indicating that the particles are difficult to scatter, as is the case for Cu. This means that the quasi-particles are injected from the emitter across the base layer without scattering. However, it seems that these values are reduced by taking into account of the effects of impurity scattering, pair-creation and annihilation.

We can also simulate time-dependent behavior. Figure 3 shows the distribution of the quasi-particle energy depending on time. The vertical line at x=1000 Å in Fig. 3 indicates the imaginary base-collector boundary. The quasi-particles are injected from the SIS tunneling junction (x=0) to the base layer (x>0) at time t=0. When the thickness of the base layer is W=1000 Å, high energy quasi-particles (E~3Δ) reach the collector region at t=0.5 (ps). After 1.0 (ps) has passed, low energy quasi-particles cross the base layer, and about 50% reach the collector layer. This simulation indicates that the quasi-particles cross the superconducting base layer without scattering. The base layer transit time (W=1000 Å) is estimated to be about 1.0 (ps).

Table 1. Scattering lifetime τ_f, mean-free path L, diffusion coefficient D and average energy E/Δ for Nb at 1(K), 5(K) and for Cu at 4(K).

	Nb(1K)	Nb(5K)	Cu(4K)
τ_f(s)	3.64×10^{-8}	1.57×10^{-10}	2×10^{-9}
L(cm)	2.16×10^{-1}	1.83×10^{-3}	3×10^{-1}
D(cm^2/s)	1.80×10^6	2.84×10^4	1.7×10^7
E/Δ	1.13	1.62	——

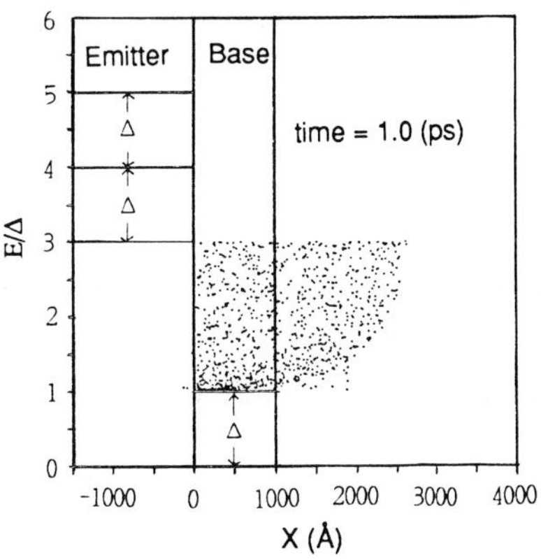

Fig. 3
The distribution of quasi-particle energy after 1.0 (ps). Emitter-base bias is V_{EB} = 4Δ/e. The vertical line at 1000 Å shows the imaginary base-collector boundary.

QUANTUM MECHANICAL REFLECTION AT THE BASE-COLLECTOR INTERFACE

Quantum mechanical reflection (QMR) of low energy quasi-particle at the interface between the base and collector layer is one of the most important factor in evaluating the low-energy superconducting base transistor. In this study, we investigated the effect of the QMR for the normal metal/square-potential-barrier/superconductor geometry (see Fig. 4) using the Bogoliubov equation[3]. Applying the method of A.Griffin and J.Demers[7,8], we can derive an expression for the quasi-particle transmission probability from a normal to a superconducting region through a square potential barrier as follows :

$$\begin{cases} E<\Delta:\ W_{NS}=0 \\ E>\Delta:\ W_{NS}=\begin{cases} (1-B^2)\dfrac{1+(1+B^2)\alpha_2}{1+\alpha_1+\alpha_2+(1-B^2)^2\alpha_1\alpha_2+B^2\sqrt{\alpha_1\alpha_2\beta}}, & E<V_0 \\ (1-B^2)\dfrac{1+(1+B^2)\alpha_2}{1+\tilde{\alpha}_1+\alpha_2+(1-B^2)^2\tilde{\alpha}_1\alpha_2+B^2\sqrt{\tilde{\alpha}_1\alpha_2\tilde{\beta}}}, & E>V_0 \end{cases} \end{cases} \quad (5)$$

$$\alpha_1=\left(\frac{\kappa_+^2+k_F{}^2}{2\kappa_+k_F}\right)^2\sinh^2\kappa_+L,\ \alpha_2=\left(\frac{\kappa_-^2+k_F{}^2}{2\kappa_-k_F}\right)^2\sinh^2\kappa_-L,\ \tilde{\alpha}_1=\left(\frac{\xi^2-k_F{}^2}{2\xi k_F}\right)^2\sin^2\xi L,\ \xi=\sqrt{2m(E-V_0)}/\hbar$$

$$\beta=\left(\frac{\kappa_+^2+\kappa_-^2}{\kappa_+\kappa_-}\right)\sinh(\kappa_+L)\sinh(\kappa_-L)-2\cosh(\kappa_+L)\cosh(\kappa_-L),\ \tilde{\beta}=\left(\frac{\xi^2-\kappa_-^2}{\xi\kappa_-}\right)\sin(\xi L)\sinh(\kappa_-L)+2\cos(\xi L)\cosh(\kappa_-L)$$

where it should be noted, the potential barrier height, V_0, is measured with respect to Fermi energy ε_F. The notation which appears in Eq. (5) is in accordance with Refs. 7 and 8.

In Fig. 5, we give the transmission coefficient W_{NS} as well as the ratio W_{NS}/W_{NN}, where W_{NN} is the usual transmission coefficient in the limit $\Delta \to 0$. We note that $W_{NS}(E<\Delta)=0$. For $E>V_0$, there are oscillations in the transmission probability due to over-the-barrier transmission. For $\alpha>>1$, as shown in Fig. 5, W_{NS}/W_{NN} is always greater than 1. Especially, in the case of low energy quasi-particle W_{NS}/W_{NN} is found to increase considerably. These results suggest that the transmission probability of quasi-particles is larger than that of usual electrons. However, the value of W_{NS} became very small at $E<V_0$, at which point most low energy quasi-particles are reflected by the potential barrier. To achieve high current gain in the transistor action, it is necessary to make the base-collector junction with a very low potential barrier height, nearly equal to the gap energy or ideal ohmic contact, as described by D.J.Frank at el[1].

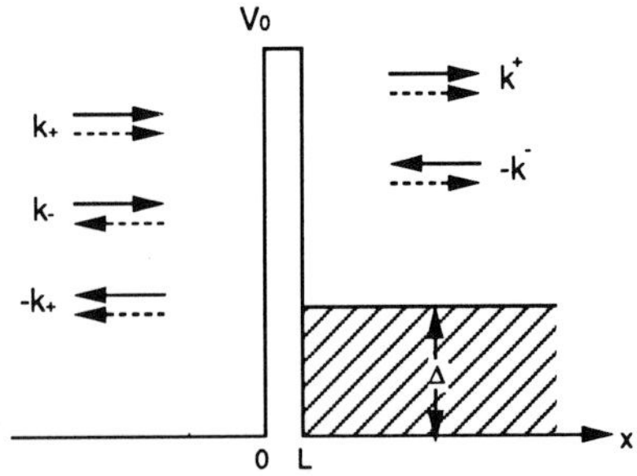

Fig. 4 Normal-metal/square-potential-barrier/superconductor geometry with incident excitation (k+) from the normal side. The solid arrows denote the direction of thc propagation vector while the dashed arrows denote the direction of the group velocity of the various waves.

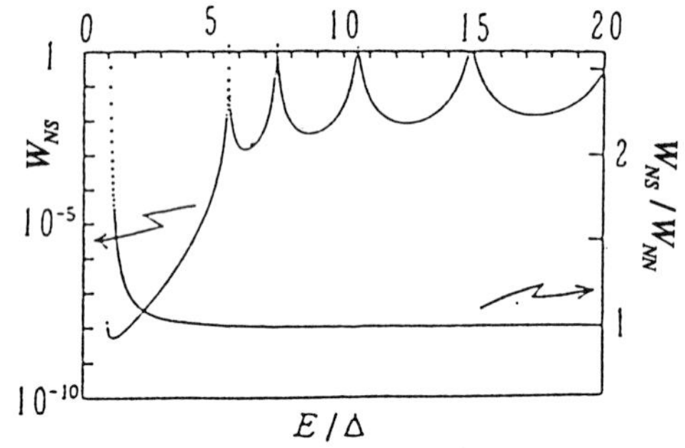

Fig. 5 Quasi-particle current transmission coefficicent W_{NS} and the ratio W_{NS}/W_{NN} as a function of the reduced energy E/Δ. We have taken Δ=1.53 meV, ε_f=5.33 eV, $V_0=5\Delta$ and L=200 Å.

CONCLUSION

We have analyzed the transport properties of quasi-particles in the superconducting base layer by means of Monte-Carlo simulation based on the BCS theory. The scattering lifetime $\tau_f=1.57\times10^{-10}$(s), mean-free path $L=1.83\times10^{-3}$(cm) and diffusion coefficient $D=2.84\times10^{4}$(cm^2/s) were obtained for the isotropic Nb system at 5(K) and transit time was estimated to be about 1.0 (ps) for a 1000 Å thin base. The values of D and L suggest that almost all the quasi-particles injected from the emitter can reach at the base-collector interface. The quasi-particles transmission coefficient at base-collector interface is also calculated using the Bogoluibov equation. It was clarified that a low-barrier-height Schottky contact is required for the collector to realize high current gain in the transistor action.

ACKNOWLEDGEMENT

This work was performed under the management of FED (the R & D Association for Future Electron Devices) as a part of the R & D of Basic Technology for Future Industries sponsored by NEDO (New Energy and Industrial Technology Development Organization).

REFERENCES

1. D.J.Frank,M.J.Brady and A.Davidson : IEEE Trans.Mag. **21**, 721(1988)
2. H.Tamura,N.Fujimaki and S. Hasuo :J.Appl.Phys. **60**, 711(1986)
3. P.G.De Gennes : Superconductivity of metals and alloys, *W.A.Benjamin, N.Y.* (1966)
4. J.Bardeen, G.Rickayzen and L.Tewordt : Phys. Rev. **113**, 982(1959)
5. S.B.Kaplan, C.C.Chi and D.N.Langenberg : Phys. Rev. **B14**, 4854(1976)
6. C.Jacoboni and L.Reggiani : Rev. Mod. Phys. **55**, 645(1983)
7. A.Griffin and J.Demers : Phys. Rev. **B4**, 2202(1971)
8. J.Demens and A.Griffin : Can. J. Phys. **49**, 285(1971)

A Cryogenic Backward Diode

HIROTAKA TAMURA, AKIRA YOSHIDA, and SHINYA HASUO

Fujitsu Laboratories Ltd., 10-1, Morinosato-Wakamiya, Atsugi, 243-01 Japan

ABSTRACT

This paper presents results from experiments with a cryogenic diode we developed which can rectify voltage swings from a few millivolts to a few tens of millivolts. The diode consists of an n-type semiconductor with a high dielectric constant, a metal layer, and an intervening tunneling barrier. As the semiconductor, we chose n-$SrTiO_3$, which has a dielectric constant of $2x10^4$ at around 4.2 K. The tunneling barrier was 6-nm-thick Si, the metal electrode 240-nm-thick Nb, and the $SrTiO_3$ carrier concentration $5x10^{17}$ cm^{-3}. The diode could rectify a voltage swing of a few tens of millivolts. Curvature coefficient γ $(=(d^2I/dV^2)/(dI/dV))$, a measure of nonlinearity, was 380 to 170 V^{-1} for a bias range of 2 to 10 mV.

INTRODUCTION

Low-voltage rectifiers and amplifiers operated with a voltage swing of a few millivolts [1] would be very useful in terms of power consumption and compatibility with conventional Josephson devices. Such a small voltage swing requires operation at cryogenic temperatures because thermal energy kT should be much smaller than a few meV. Electron devices must also have sufficient nonlinearity in current-voltage characteristics to get a low operating voltage.

Ordinary semiconductor devices, however, do not provide a good basis for such low-voltage cryogenic operation because the typical built-in potential in semiconductor devices exceeds 100 mV, producing too high a device impedance at cryogenic temperatures and a bias voltage of a few mV. Josephson junctions have very high nonlinearity at a bias voltage of a few millivolts, but their I-V characteristics are symmetric for applied voltage and do not exhibit the rectification characteristics provided by semiconductor rectifiers such as pn junction and Schottky diodes. What is needed is a new operating principle to enable a low operating voltage at cryogenic temperatures.

This paper presents the design concept and results of experiments with a cryogenic diode we developed which rectifies voltage swings from a few millivolts to a few tens of millivolts at about 4.2 K.

OPERATING PRINCIPLE

Figures 1(a) and 1(b) show schematic band diagrams of the proposed diode, which consists of an n-type semiconductor with a high dielectric constant, a metal layer, and an intervening tunneling barrier. The dielectric constant of the semiconductor depletion layer is assumed to be much higher than that of the barrier in order to make the depletion layer capacitance much higher than that of the barrier. Almost all voltage applied to the diode appears across the barrier, because the continuity of electric displacement makes the electric field in the depletion layer much lower than that in the barrier. From the same reason, almost all bending of the band structure occurs in the barrier, reducing band bending in the semiconductor surface depletion layer.

The effective barrier height for electrons tunneling from the metal to the semiconductor conduction band is the height of the conduction band edge at the barrier/semiconductor interface measured from the metal Fermi level. As shown in Fig. 2, calculation using depletion approximation predicts a zero-bias barrier height of some millivolts, if the semiconductor dielectric constant is around 10^4, the semiconductor carrier concentration 10^{17} cm^{-3} to 10^{18} cm^{-3}, the sum of band bending in the semiconductor and barrier about 1 eV, the barrier 60 Å thick, and the barrier dielectric constant 10.

When a voltage is applied to the diode, the semiconductor band shows a parallel shift because most band bending occurs in the barrier. When the applied voltage is positive on the semiconductor, it reduces the level of the conduction band relative to the metal Fermi level, thus increasing current due to electron tunneling from the metal electrode into the semiconductor conduction band. Negative voltage will make the opposite change, and the tunneling current becomes small. Because the zero-bias effective barrier height is a few millivolts, the diode will rectify a voltage swing of a few millivolts.

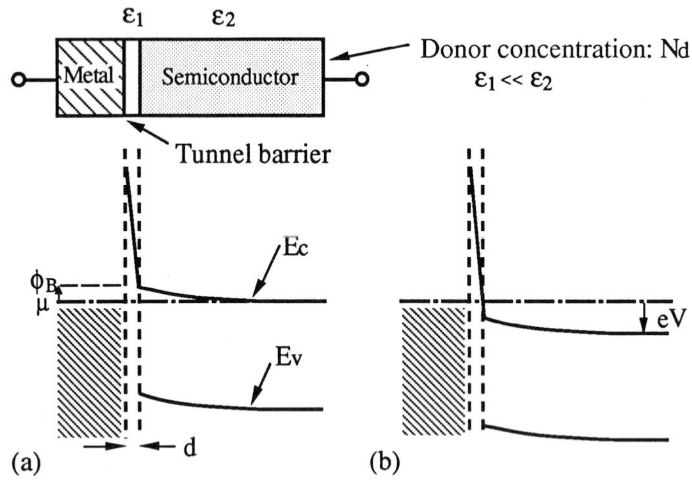

Fig.1. Schematic band diagram of proposed diode, for (a) zero-bias, and (b) positive voltage on semiconductor

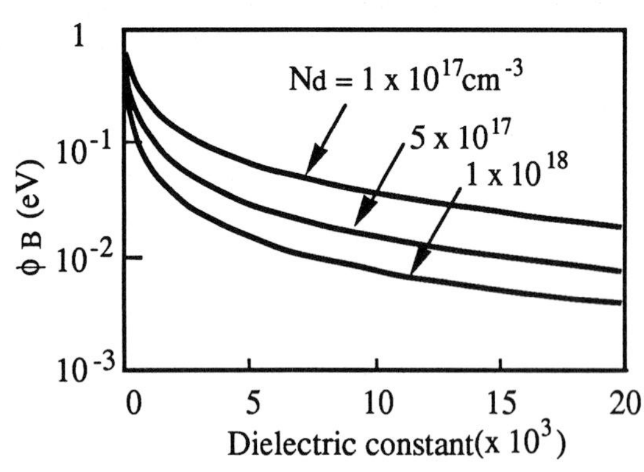

Fig. 2 Effective barrier height versus depletion layer dielectric constant (d=60 Å, ε_1=10, ϕ_B (d=0)=1 eV).

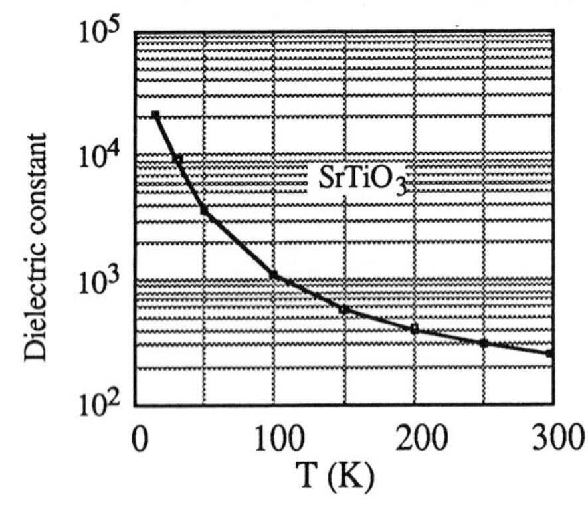

Fig. 3 Temperature dependence of SrTiO3 dielectric constant.

The band diagrams shown in Fig. 1 are basically the same as for Schottky barrier junctions with an insulating interfacial layer. The current-voltage characteristics of such junctions can be expressed as follows [2,3] for an applied voltage about zero.

$$
\begin{aligned}
I &= SA^*T^2\exp\left(-\frac{\phi_B}{kT}\right)\left\{\exp\left(\frac{qV'}{n'kT}\right)-\exp\left[\left(\frac{1}{n'}-1\right)\frac{qV'}{kT}\right]\right\} \\
&= -SA^*T^2\exp\left(-\frac{\phi_B}{kT}\right)\left\{\exp\left(\frac{qV}{nkT}\right)-\exp\left[\left(\frac{1}{n}-1\right)\frac{qV}{kT}\right]\right\}
\end{aligned}
\qquad (1)
$$

Here, S is the diode area, A* the effective Richardson constant including the effect of electron tunneling probability through the barrier, ϕ_B the effective barrier height, T the temperature, V' (=-V) the metal electrode voltage with respect to the semiconductor, n' the ideality factor for the forward bias condition (i.e., forward bias condition if the junction were an *ordinary* n-type Schottky junction), and n the ideality factor for backward operation given by $n^{-1}=1-n'^{-1}$. The value of n' is given by

$$\frac{1}{n'} = 1 - \frac{\partial\phi_B}{\partial(qV')}$$
$$=1 - \left[1 + \left(2\varepsilon_0\varepsilon_s/\left(q^2N_d\right)\right)\left(\varepsilon_B/\left(\varepsilon_s d\right)\right)^2\left(\phi_{B0}-kT\right)\right]^{-1/2} \quad (2)$$

The last expression is obtained by depletion approximation, where N_d is the semiconductor carrier density, ϕ_{B0} the zero bias barrier height in the absence of the tunneling barrier, ε_s the semiconductor dielectric constant, and ε_B the barrier dielectric constant.

If $\varepsilon_s=2\times10^4$, $\varepsilon_B=10$, $N_d=1\times10^{17}$ cm^{-3} to 1×10^{18} cm^{-3}, $d_B=60$ Å, and $\phi_{B0}=1$ eV, n' is about 15 to 130. This means ideality factor n' for forward operation is extremely high. The ideality factor n for backward operation, however, is close to 1, and the diode conducts better in reverse (i.e., semiconductor positive) than forward. Thus the diode should exhibit backward diode characteristics.

RESULTS

We chose n-$SrTiO_3$ as high-dielectric-constant semiconductor. Figure 3 shows the temperature dependence of the dielectric constant of nondoped $SrTiO_3$. The dielectric constant was 250 at room temperature, and rose to 2×10^4 below 10 K. Diodes were made on (110) Nb-doped n-type $SrTiO_3$ substrates. The Nb concentration in substrates was 0.005 wt%, and their electron concentration estimated by Hall measurement was 5×10^{17} cm^{-3}. The carrier concentration showed little temperature dependence between room temperature and 10 K. The carrier mobility measured at 10 K was 7.2×10^3 cm^2/(v·s).

The tunnelling barrier, consisting of 6 nm of Si, and a 240-nm Nb metal electrode were deposited by RF-magnetron sputtering. Films were then patterned by photolithography and reactive ion etching using CF_4+5% O_2 to form 800-μm-diameter dots.

Figure 4 shows the I-V and dI/dV-V characteristics of a diode at 4.2 K. The diode rectified a voltage swing of a few tens of millivolts. When the applied voltage was positive on the n-$SrTiO_3$, forward current flowed as expected, i.e., the diode operated as a backward diode. The ratio of forward and reverse currents was 5 at bias voltages of ±10 mV and 21 at ±20 mV. Curvature coefficient γ (=$(d^2I/dV^2)/(dI/dV)$), a measure of nonlinearity, was 380 to 170 V^{-1} for a bias range of 2 to 10 mV (Fig.5). This value is, to our knowledge, the highest for any rectifying diode in this low bias voltage range.

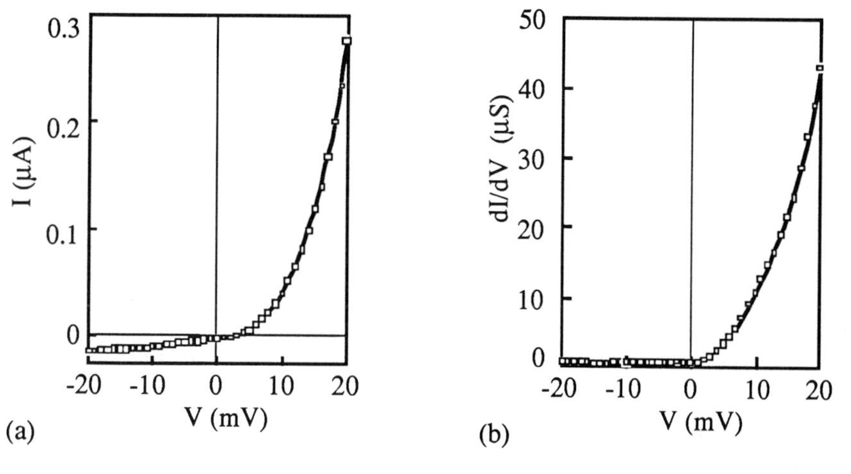

Fig. 4 I-V and dI/dV-V characteristics of diode measured at 4.2 K.
(a) I-V characteristics and (b) dI/dV-V characteristics

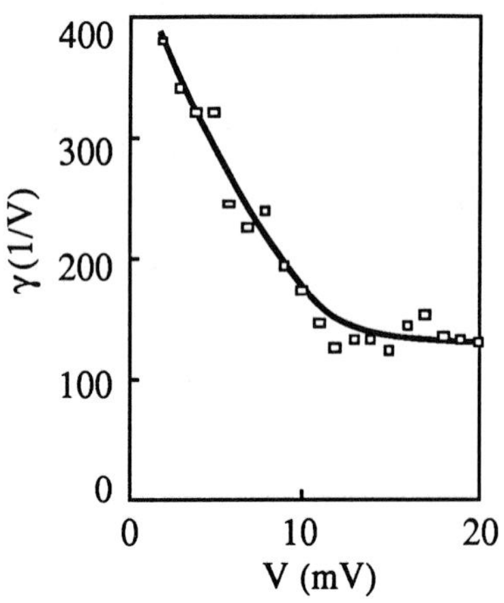

Fig.5 $(d^2I/dV^2)/(dI/dV)$-V characteristics.

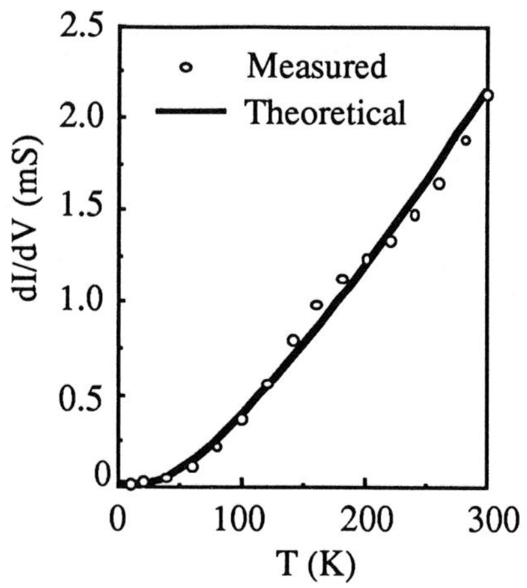

Fig.6 dI/dV vs. temperature characteristics.

Because the dielectric constant of $SrTiO_3$ increases as temperature decreases, zero-bias barrier ϕ_B decreases with decreasing temperature. This compensates somewhat for the temperature dependence of diode conductance, resulting in a dependence much smaller than that of ordinary Schottky diodes. Figure 6 shows a typical conductance versus temperature curve. Results agreed well with the theoretical value estimated from Eq. 1 and the effective barrier height calculated using depletion approximation.

CONCLUSION

We have fabricated diodes which can rectify a voltage swing of a few tens of millivolts. Although the impedance of experimental diodes was too high to be practical, results show that diodes having millivolt operating voltages and high nonlinearity can be made with the proposed structure. The impedance level will be improved by improving materials and process. This structure is applicable to high-Tc superconductors because $SrTiO_3$ is a good substrate for high-Tc materials. The structure can also be used as an emitter and collector of a transistor with a very low operating voltage.

REFERENCES

[1] Tamura H, Yoshida A, Hasuo S, Van Duzer T (1991) IEEE Trans. Magn. **27**, No. 2 (in press).
[2] Rideout VL (1975) Solid-State Electron. **18**: 541-550.
[3] Rideout VL, Crowell CR (1970) Solid-State Electron. **13**: 993-1009.

ACKNOWLEDGMENTS

This work was done under the management of the R&D Association for Future Electron Devices (FED) as a part of the R&D of Basic Technology for Future Industries supported by the New Energy and Industrial Technology Development Organization (NEDO).

Current Transport Characteristics of High-T_c Superconductor Under High Electric Field

RYODO KAWASAKI, TAIJI TSURUOKA, and HITOSHI ABE

Research and Development Group, Oki Electric Inudstry Co., Ltd., 550-5, Higashiasakawa-cho, Hachioji, Tokyo, 193 Japan

ABSTRACT

Current - electric field (I-E) characteristics of BiSrCaCuO and YBaCuO thin film under high electric field in the normal states was investigated. For YBaCuO on (100) $SrTiO_3$, linear I-E characteritics observed when electric field was varied to 800000 V/m. For BiSrCaCuO on (100) MgO, linear I-E characteritic was observed when electric field was varied to 200000 V/m. But nonlinear I-E characteristic was observed when electric field was over 200000 V/m. For our samples, no current saturation was observed.

KEY WORDS: Quasi-particle transport, Current - electric field (I-E) characteristics, YBaCuO thin film, BiSrCaCuO thin film,

INTRODUCTION

We propose Superconducting Base Hot Electron Transistor (SBHET). The SBHET was expected to operate at several hundred GHz. Fig. 1 shows the band diagram of SBHET. The emitter and collector were semiconductor, and the base was high critical temperature (Tc) oxide superconductor. The quasi-particles were injected from emitter into Base and ballistically transport in the base region and arrived at collector electrode. In the estimation of the SBHET performance the quasi-particle transport process in the base was very important. Current - electric field characteristics of YBaCuO thin films in normal state under high electric field has been reported by Osaka University group[1-3]. They observed saturation behavior over E = 200 K V/m.

In this paper, we will report on current - electric field characteristics of our samples i.e. YBaCuO and BiSrCaCuO thin films, to get the information of quasi-particle transport process in the base

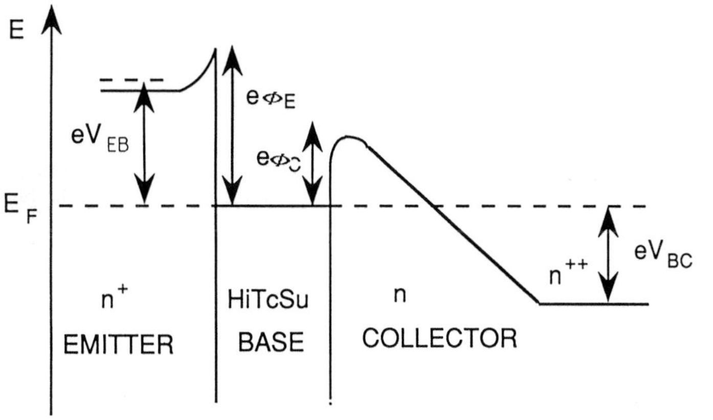

Fig. 1 Energy band diagram of SBHET

SAMPLE PREPARATION

YBaCuO thin films were formed by organometallic chemical vapor deposition (OM-CVD)[4]. Substrates for YBaCuO thin film were used (100) $SrTiO_3$ and (110) $SrTiO_3$ with orientation fraction. BiSrCaCuO thin films of (100) MgO were formed by molecular beam epitaxy (MBE)[5]. Fig. 2 (a) and (b) shows X-ray diffraction pattern and dependence of resistance of temperature of YBaCuO thin films of (100) $SrTiO_3$ and BiSrCaCuO thin film. Fig. 2 (a) indicates (001) oriented YBaCuO thin film on (100) $SrTiO_3$ and Tc=89K. Fig. 2 (b) indicated (001) oriented BiSrCaCuO thin film of (100) MgO and Tc=65K. The reflective high-energy electron diffraction pattern of YBaCuO of (110) $SrTiO_3$ indicates (103) oriented YBaCuO thin film were grown on (110) $SrTiO_3$. The critical temperature of (103) oriented YBaCuO thin film were 85K.

To measure electric conductivity under high voltage, the films was patterned with center narrow current path by photo lithography and dry ethcing. Fig. 3 shows the structure and scanning electron micrograph of a sample after patterning process. The strip width was about 15 um and length was 40 um. The Ag electrode with low contact resistance was formed on the films. For YBaCuO of (110) $SrTiO_3$, two direction paths with right angle each other were made.

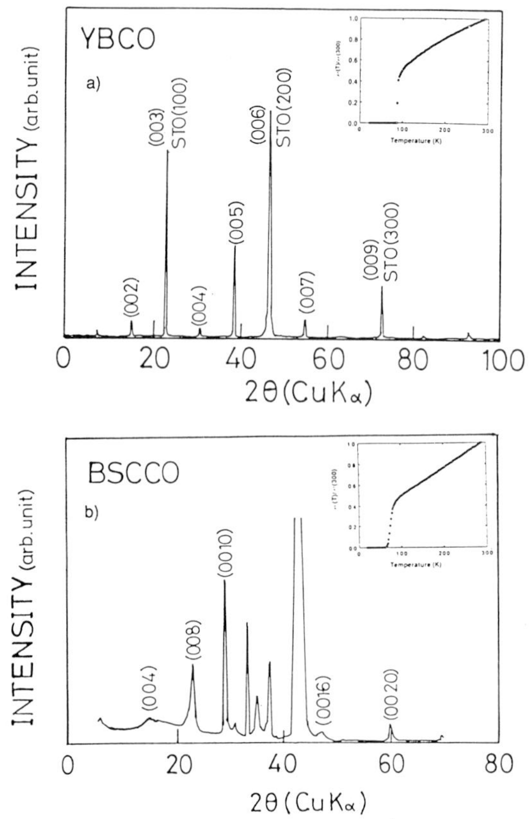

Fig. 2 Characteristics of samples: (a) X-ray diffraction (XRD) pattern and dependence of resistance of temperature of YBaCuO on (100) $SrTiO_3$ and (b) XRD pattern and dependence of resistance of temperature of BiSrCaCuO of (100) MgO.

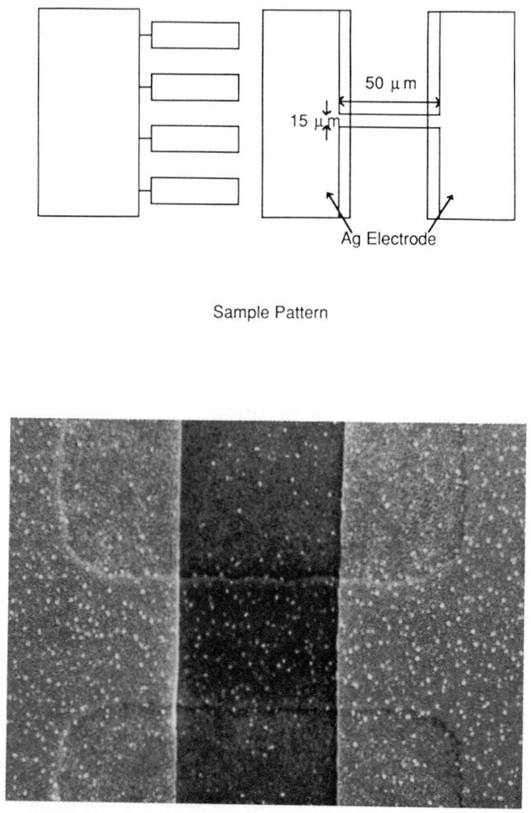

Fig. 3 The structure and scanning electron micrograph of the sample.

MEASUREMENT

Current - electric field characteristics was measured by the typical four probe method. Fig. 4 shows a schematic diagram of experimental setup. As a voltage source, pulse generator was used. Current was converted to voltage by current probe amplifier. Voltage applied on a sample and the output of current probe amplifier were measured by digital storage oscilloscope. The pulse generator generated single pulse. Its pulse duration of voltage was 500 nsec. The observed pulse were kept its rectangular form.

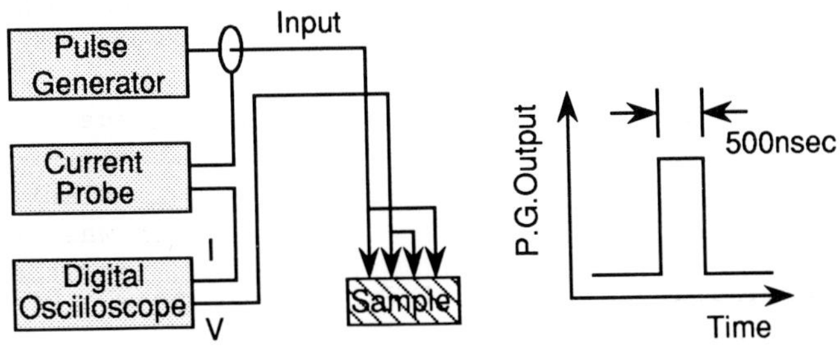

Fig. 4 The schematic diagram of experimental setup.

RESULTS

Fig. 5 show current (I) - electric field (E) characteristics of samples. Fig. 5 (a) shows I - E characteristics of YBaCuO of (100) $SrTiO_3$. The applied electric field was varied about to 800000 V/m. The I - E characteristics of YBaCuO of (100) $SrTiO_3$ obeyed Ohm's law. Fig. 5 (b) shows I - E characteristics of BiSrCaCuO of (100) MgO. The I - E characteristics of BiSrCaCuO of (100) MgO obeyed Ohm's law when electric field was under 200000 V/m. But for over 200000 V/m electric field, non-linear characteristics was observed. Fig. 5 (c) and (d) shows I - E characteristics of YBaCuO of (110) $SrTiO_3$. Low resistance sample shows non-linear I-E characteristics. High resistance sample obeyed Ohm's law. The reason of these non-linearity wasn't clear. For all samples, no current saturation was observed.

SUMMERY

(1) The I - E characteristics of YBaCuO of (100) $SrTiO_3$ obeyed Ohm's law.

(2) The I - E characteristics of BiSrCaCuO of (100) MgO obeyed Ohm's law when electric field was under 200000 V/m. But for over 200000 V/m electric field, non-linear characteristics was observed.

(3) Linearity of I-E curve for YBaCuO on (100) $SrTiO_3$ was kept upto 800000 V/m. No saturation behavior was observed.

ACKNOWLEDGMENT

This work was supported by NEDO under the management of FED.

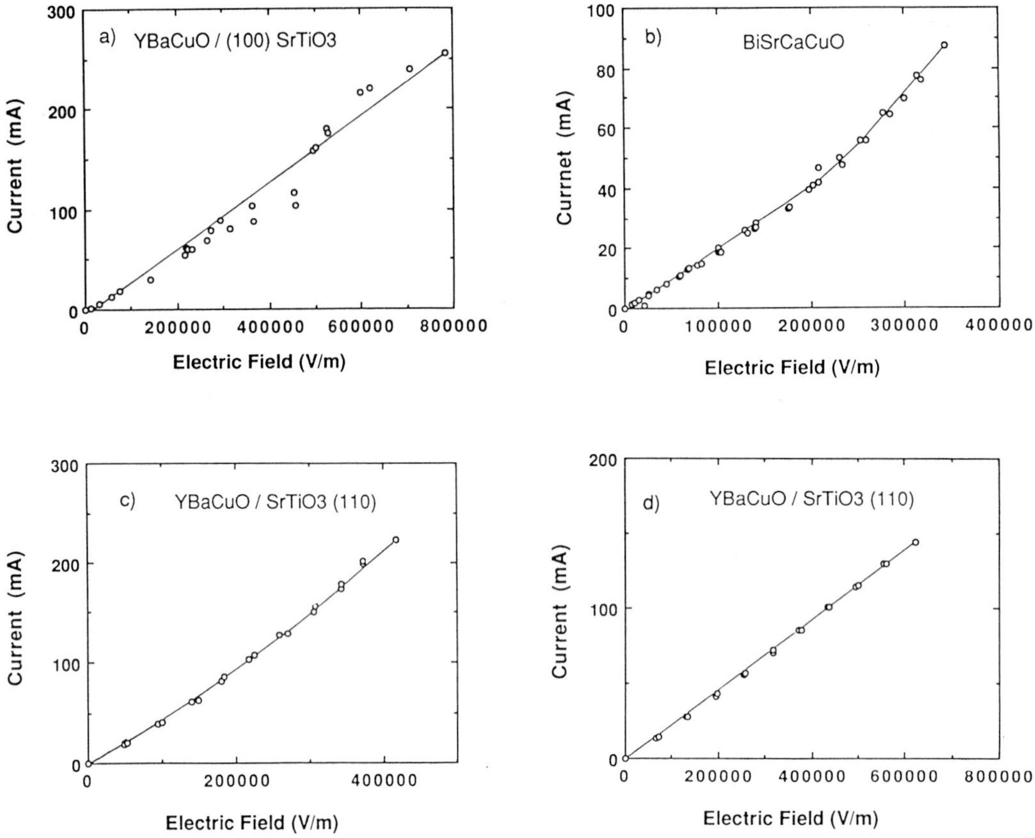

Fig. 5 Current - electric field characteristics of samples: (a) YBaCuO on (100) $SrTiO_3$, (b) BiSrCaCuO on (100) MgO and (c),(d) YBaCuO on (110) $SrTiO_3$.

REFERENCE

[1] U. Kabasawa, K. Asano, M. Iyori, S. Hirata, K. Sakuta, Y. Fujiwara and T. Kobayashi : Jpn. J. Appl. Phys. 29,453(1990)

[2] K. Asano, K. Sakuta and T. Kobayashi: IEICE Technical Report, 90-54,(1990),43,SCE90-16

[3] K. Asano, K. Sakuta and T. Kobayashi: 1990 Fall meeing of the Japan society of applyed Phisics, 26p-ZA-8

[4] M. Ogihara, T. Yamada, T. Tsuruoka and H. Abe: Extended Abstracts of the 22nd (1990 International) Conference on Solid State Device and Materials, Sendai, 1990, 581

[5] H. Abe, T. Nakamori, T. Kanamori and S. Shibata : J. Cryst. Growth 93 (1988) 929.

A Study of Y211/123 Layered Films

MASAAKI NEMOTO, HIROMI MUKAIDA, and MASAO NAKAO

SANYO Tsukuba Research Center, 2-1, Koyadai, Tsukuba, 305 Japan

ABSTRACT

Anisotropic-thermal-expansion measurements of Y_2BaCuO_5 were carried out for the first time. The thermal-expansion coefficients β_a, β_b and β_c (R.T. - 450 °C) are 11.8×10^{-6}, 10.9×10^{-6} and 9.8×10^{-6} K^{-1}, respectively. It was found that there are little differences in thermal expansions between a (001) oriented $YBa_2Cu_3O_{7-\delta}$ layer and a (230) oriented Y_2BaCuO_5 layer below about 450 °C. Moreover, lattice mismatches between these layers are 1.1 - 3.4 %, which are relatively small.

KEY WORDS: $YBa_2Cu_3O_{7-\delta}$, Y_2BaCuO_5, thermal expansion, lattice mismatch

INTRODUCTION

It is very important to fabricate SIS Josephson junctions using the high-T_c oxide superconductors such as $YBa_2Cu_3O_{7-\delta}$ [1] for both fundamental research and device applications. Several attempts to fabricate tunnel junctions using MgO or $SrTiO_3$ as tunnel barriers have been reported [2-3]. Since the high-T_c superconductors have extremely short coherence lengths, major problems of fabricating the tunnel junctions stably are the following: (i) Since tunnel barriers must be very thin, dielectric breakdown is liable to occur. Accordingly, it is necessary that insulators with sufficient dielectric strengths are employed as barrier materials. (ii) Interdiffusion between the superconductor and the insulator layers leads to degrading properties of the layers. (iii) Large differences in thermal expansions between the superconductors and insulators result in formation of microcracks, which cause leaks of currents through the barrier layers or at least reduce the dielectric strengths. (iv) Epitaxial growth of the layers is preferable to obtain flat interfaces. Layered films grown epitaxially, however, cannot necessarily be obtained.

In general, the interface between the superconductor and the insulator becomes broad, when the layered film is annealed. Hence there have been a number of efforts to prepare the layered films without annealing, furthermore at possibly low substrate temperatures. On the other hand, $YBa_2Cu_3O_{7-\delta}$ and Y_2BaCuO_5 have the thermodynamical relation to be segregated according to the equilibrium phase diagram. Therefore, the interface of the layers can be expected to become abrupt, when the layered films are annealed. We have successfully obtained (101) oriented Y_2BaCuO_5 thin films grown epitaxially on MgO (100) [4]. We have also obtained (230) oriented Y_2BaCuO_5 layers on (001) oriented $YBa_2Cu_3O_{7-\delta}$ layers. We reported on effects of annealing on the interface of $YBa_2Cu_3O_{7-\delta}$ and Y_2BaCuO_5 [5]. RBS measurments showed that the interface between $YBa_2Cu_3O_{7-\delta}$ and Y_2BaCuO_5 layers remains abrupt, though the interdiffusion occurs during annealing. Besides, *I–V* characteristics revealed that a dielectric strength of the Y_2BaCuO_5 layer with a thickness of 1000 Å in an $YBa_2Cu_3O_{7-\delta}/Y_2BaCuO_5/Au$ layered film increases from 5 kV/mm to 25 kV/mm as a result of annealing.

In this report, we present results of anisotropic-thermal-expansion measurements of Y_2BaCuO_5 for the first time, and discuss a possibility of the epitaxitial growth of the Y_2BaCuO_5 layer on the $YBa_2Cu_3O_{7-\delta}$ layer.

EXPERIMENT

An Y_2BaCuO_5 powder was produced by a solid-state reaction method. A mixture with a composition of Y_2BaCuO_5 was calcined at 950 °C for 10 hr in air and ground. The powder was pressed into pellets and sintered at 1000 °C for 10 hr. The pellets were ground again. In addition, an $YBa_2Cu_3O_{7-\delta}$ powder was prepared in the usual way. Each resulting powder was confirmed to be a single phase by x-ray diffraction (XRD). Temperature dependences of lattice constants for the powders were measured by XRD over a range of -223 to 593 °C. The XRD measurements were carried out in θ-2θ geometry with utilization of Cu $K\alpha$ radiation (40 kV - 200 mA) from a rotating-anode x-ray generator.

RESULTS AND DISCUSSION

Figure 1 shows an XRD pattern for Y_2BaCuO_5 at R.T.. The lattice parameters were determined from the most intensive twelve peaks using a method of least squares. Figure 2 shows temperature dependences of thermal expansions for $YBa_2Cu_3O_{7-\delta}$ and Y_2BaCuO_5. It was found that the thermal-expansion coefficients for Y_2BaCuO_5 are little anisotropic. In contrast to this, those for $YBa_2Cu_3O_{7-\delta}$ have a large anisotropy. While the thermal expansions for Y_2BaCuO_5 increase almost linearly below 600 °C, those for $YBa_2Cu_3O_{7-\delta}$ increase drastically above about 450 °C because of oxygen deficiency.

Table I shows average values of the thermal-expansion coefficients over a range of R.T.

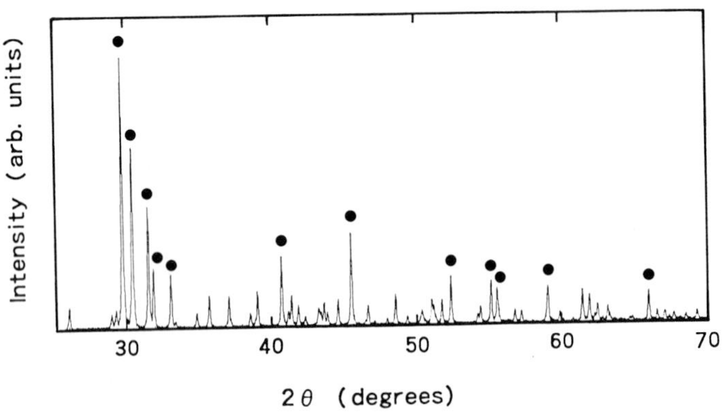

Fig. 1. X-ray diffraction pattern of Y_2BaCuO_5 at R.T.. Peaks marked with ● were used to determine

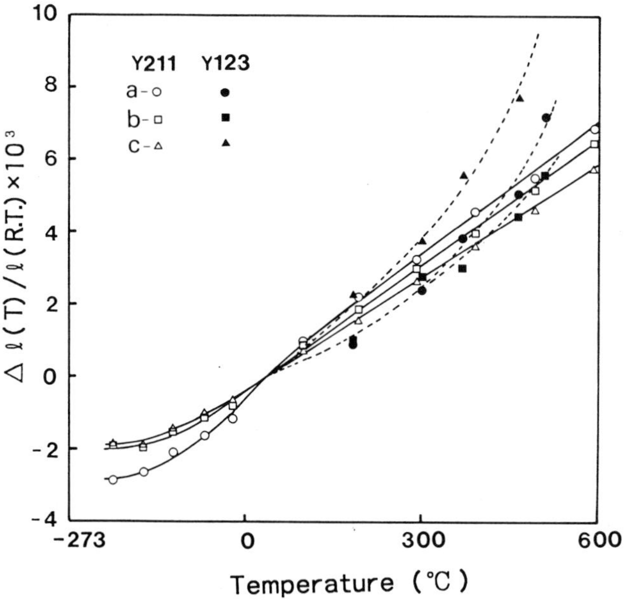

lattice parameters.

Fig. 2. Temperature dependences of thermal expansions for Y_2BaCuO_5 and $YBa_2Cu_3O_{7-\delta}$.

Table I. Thermal-expansion coefficients (R.T.-450 °C) of Y_2BaCuO_5.

	β_a (10^{-6} K^{-1})	β_b (10^{-6} K^{-1})	β_c (10^{-6} K^{-1})
$YBa_2Cu_3O_{7-\delta}$	10.9	11.8	17.0
Y_2BaCuO_5	11.8	10.9	9.8

to 450 °C. The thermal-expansion coefficient along a $[3\bar{2}0]$ azimuth in the (230) plane of Y_2BaCuO_5, $\beta_{[3\bar{2}0]}$ is 11.3×10^{-6} K^{-1}. That along a [001] azimuth, $\beta_{[001]}$ is 9.8×10^{-6} K^{-1}. The thermal-expansion coefficients in the a-b plane of $YBa_2Cu_3O_{7-\delta}$, β_a and β_b are 10.0×10^{-6} and 11.8×10^{-6} K^{-1}, respectively. Consequently, it was turned out that there are little differences in the thermal-expansion coefficients between the (230) oriented Y_2BaCuO_5 layer and the (001) oriented $YBa_2Cu_3O_{7-\delta}$ layer.

Figure 3 shows a relation of crystalline orientations between the $YBa_2Cu_3O_{7-\delta}$ and Y_2BaCuO_5 layers. Here we define $a_{[3\bar{2}0]}$ and $b_{[001]}$ as lattice parameters along the $[3\bar{2}0]$ azimuth and the [001] azimuth in the (230) plane of $YBa_2Cu_3O_{7-\delta}$, respectively. we also define $a_{[1\bar{1}0]}$ and $b_{[110]}$ as the lattice parameters along a $[1\bar{1}0]$ azimuth and a [110] azimuth in the (001) plane of $YBa_2Cu_3O_{7-\delta}$, respectively. The lattice mismatch between $a_{[3\bar{2}0]}$ and six times as large as $a_{[1\bar{1}0]}$ is 1.1 %. The lattice mismatch between $b_{[001]}$ and $b_{[110]}$ is 3.4 %. Thus, the lattice mismatches between the (001) oriented $YBa_2Cu_3O_{7-\delta}$ layer and

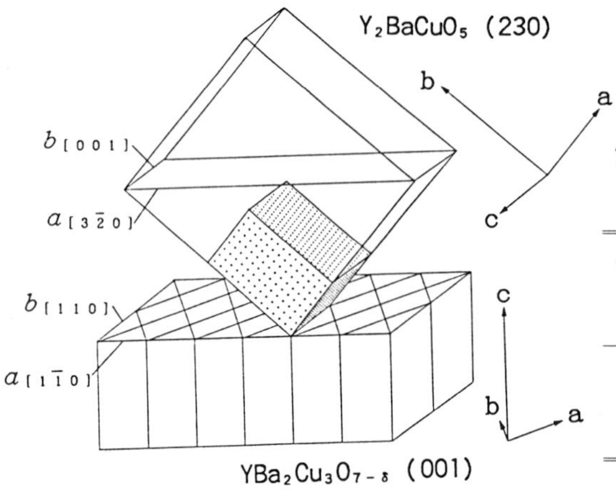

Fig. 3. Relation of crystalline orientations for a (230) oriented Y_2BaCuO_5 layer on a (001) oriented $YBa_2Cu_3O_{7-\delta}$ layer.

Table II. Lattice mismatches between $YBa_2Cu_3O_{7-\delta}$ and Y_2BaCuO_5.

$YBa_2Cu_3O_{7-\delta}$ (001)	Y_2BaCuO_5 (230)	lattice mismatch
$a_{[1\bar{1}0]} \times 6$	$a_{[3\bar{2}0]}$	-1.1%
$b_{[110]}$	$b_{[001]}$	+3.4%

the (230) oriented Y_2BaCuO_5 layer are reratively small. Therefore, it is sufficiently expected that the Y_2BaCuO_5 layer grows epitaxially on the $YBa_2Cu_3O_{7-\delta}$ layer, and vice versa.

CONCLUSIONS

For the $YBa_2Cu_3O_{7-\delta}/Y_2BaCuO_5$ layered film, the matchings of the lattice parameters and the thermal expansions were investigated. It was turned out that when the (230) oriented Y_2BaCuO_5 layer is prepared on the (001) oriented $YBa_2Cu_3O_{7-\delta}$ layer, there are little differences between the layers in the thermal expansions below about 450 °C. Moreover, the lattice mismatches between the layers are small. Consequently, Y_2BaCuO_5 is thought to be suitable for the barrier material.

REFERENCES

1. Wu MK, Ashburn JR, Torng CT, Hor PH, Meng RL, Gao L, Huang ZJ, Wang YQ and Chu CW (1987) Phys. Rev. Lett. 58: 908-910
2. Iguchi I, Furuyama M, Kusumori T, Shirai K, Tomura S, Nasu M, Ohtake H (1990) Jpn. J. Appl. Phys. 29: 614-617
3. Kingston JJ, Wellstood FC, Lerch P, Miklich AH, Clarke J (1990) Appl. Phys. Lett. 56: 189-191
4. Nemoto M, Matsuta Y, Yuasa R, Nakao M (1989) Ext. Abs. ISEC'89: 24-27
5. Nemoto M, Nakao M (1990) Ext. Abs. SSDM'90: 573-576

Chaotic Noise Phenomena in YBCO Thick Film Microbridges

K.N.R. TAYLOR, G. ALVAREZ, A. BAILEY, and G.J. RUSSELL

Advanced Electronic Materials, Physics, University of New South Wales, Kensington, NSW, 2033, Australia

ABSTRACT

Recent observations of the characteristics of microbridges cut into melt processed thick films of YBCO on zirconia substrates have revealed the presence of excessive voltage noise during irradiation with microwaves. The appearance of this noise, and its variation with specimen current suggest strongly that its origin may be associated with chaotic solutions of the resistively shunted junction equation. Measurements have now been made of the microwave power and current dependence of this noise amplitude, and in addition the low frequency spectral dependence of the noise frequency has been determined. The observations are interpreted in terms of the hysteretic junction parameter β and the reduced rf frequency Ω and values for these parameters have been obtained.

KEY WORDS: YBCO, Thick Film, Microbridges, Chaos.

INTRODUCTION

The superconducting behaviour of the ceramic, high T_c materials is dominated by the intergranular contacts which have relatively low current carrying capacity and allow external magnetic fields to penetrate the bulk material as Josephson vortices travelling through the intergranular matrix surrounding the grains. While observations on thin film bicrystals [1] have shown that grain misalignment severely lowers the critical current of the junction, relatively little is known about the behaviour of individual junctions in bulk material. Observations of SQUID like behaviour [2,3] in bulk ceramic material and of typical Josephson features [4,5] in the current-voltage characteristics of solid samples provide some insight into the nature of the intergranular junctions. Similarly the temperature dependence of these characteristics observed by ourselves, and others [6,7] have been used to show that the granular coupling occurs through SNS (or SS'S) junctions. The majority of these observations, however, have been carried out on macroscopic samples which involve a complex three dimensional array involving many contacts even in the variable thickness microbridges carved out of bulk ceramic material.

In our recent work involving the production of highly crystalline, melt processed thick films of YBCO on YSZ [8] the grain sizes are relatively large (<100μm to mm in dimension) and it is relatively straightforward to mechanically define a weak-link bridge so as to include any morphological crystalline feature of interest. As a result of the processing technique, the films contain Y_2BaCuO_5 as well as other insulating phases and these too can be included in the bridge structure. Finally since the film thicknesses can be as little as 5 microns the maximum number of grain boundaries involved in a bridge is necessarily small.

Recent studies of such bridges have shown that with microwave irradiation the Shapiro steps are well defined and follow the predicted Bessel function variation with power level [9]. Further, it is clear from this dependence that the characteristic junction plasma frequency is very much higher than the 10GHz drive frequency [10]. The appearance of Shapiro steps is not uncommon and is generally thought to be the result of the individual junctions locking-on to the microwave drive rather than the bridge itself behaving as a single weak-link due to intergranular coupling. In view of the short coherence length this would not be surprising, however it is clear that some intergranular junction coupling must occur in these materials. In a recent investigation of such junctions, we have observed high levels of voltage noise in these characteristics [11] which we believe may be evidence for chaotic behaviour of the intergranular contacts. A more detailed examination of this behaviour is reported in the following, and origins for the observations are considered.

EXPERIMENTAL

The thick film samples were prepared using our normal melt processing technique [8] to give film thicknesses of approximately 10μm. The microbridges were formed by mechanical abrasion to give dimensions typically 80 x 40 x 10μm^3. The current-voltage characteristics were obtained using conventional four-contact techniques with the junction bias current swept at a constant rate. For observation of the microwave induced transitions, the sample was located parallel to the end face of a shorted X-band waveguide. All characteristics were recorded directly onto an X-Y recorder.

Large bridges in bulk material normally yield characteristics which show a well developed set of Shapiro steps [5], however in the present case, Fig. 1a shows a second type of behaviour in which, with increasing power a voltage noise signal develops on the characteristic extending to lower current-voltage values as the microwave power is further increased. Beyond some microwave power, the Shapiro steps are no longer visible and a clearly defined periodicity appears in the voltage noise. This is observed as a series of horizontal (voltage) steps separated by irregular, apparently random regions for which the average value of ($^{dI}/dv$) changes sign with increasing current. While the general appearance of the irregularities in this staircase characteristic remain the same the details of any particular step vary from run to run, as shown in Fig. 1b. As may be seen from Fig.1c,with increasing microwave power the overall form of the characteristics becomes linear as confirmed by the log-log plot of Fig. 2d which shows that once the staircase structure has developed the characteristic changes from a half-power to a linear relation at the highest power levels used.

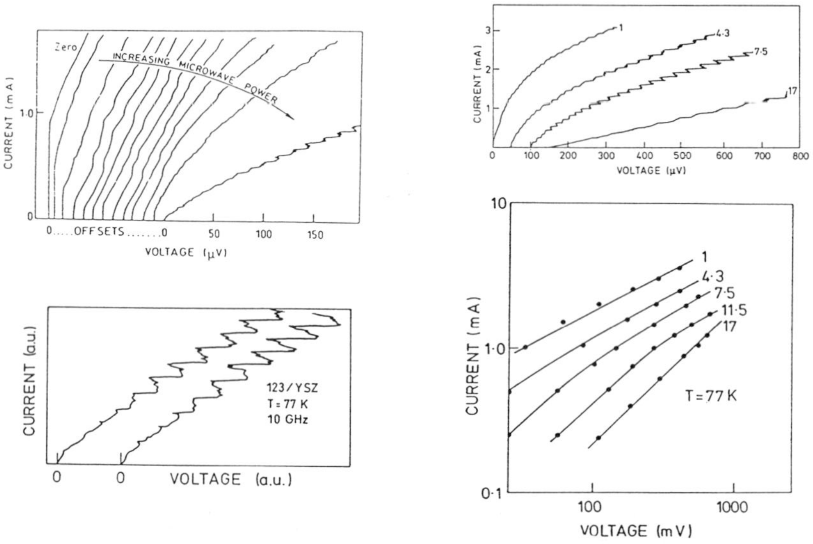

Fig. 1a. The current-voltage characteristics, with 10GHz irradiation for a junction showing excessive voltage noise.

b. The change in the detailed behaviour of the characteristics from run to run, which show variations in both the step position (current) and the noisy region between the steps.

c. The change in the shape of the characteristic at high microwave power level.

d. The overall characteristic i-V relationships showing a change from a half-power to linear dependence.

Measurements of the low frequency voltage noise were made at a fixed frequency using a phase sensitive detection system, and as Fig. 2a shows, with increasing current this is dominated by a series of noise transients which generally coincide with the constant current steps in the staircase characteristic also shown in Fig. 2a for comparison. The amplitude of these transients increases with the bias current at a fixed microwave power and as Fig. 2b shows, this changes from a linear increase at low currents to a square root dependence at higher currents. With increasing microwave power at fixed bias current the noise variation is shown in Fig. 2c and as may be seen, this also follows a change in behaviour from a linear to a half-power law with increasing microwave power.

Preliminary observations of the frequency dependence of the amplitudes of the transient noise signals (at constant current and microwave power) show considerable structure as may be seen from Fig. 3. This spectrum appears to consist of a series of broad peaks separated by 10kHz extending to 55kHz the highest frequency studied.

One final comment on the nature of this noise relates to the detailed change in the phase sensitive detector close to a transient peak (see region indicated in Fig.2c). As may be seen, the output voltage changes sign across the transient implying a phase shift in the noise source of $\cong 180^{\circ}$ during the horizontal voltage step. Clearly, also, there is a gradual phase change in the random signals between the steps.

A second type of noise structure observed in these bridges is shown in Fig. 4 where the Shapiro steps remain well defined but are separated by regions of voltage instability. This also occurs close to zero voltage and the random noise signal appears to be confined by upper and lower envelopes which correspond to different critical bias currents at the beginning and end of each Shapiro step. It is clear that in this case, the amplitudes of the steps follow an oscillatory behaviour decreasing to zero periodically with increasing microwave power. No noise measurements have yet been made on these junctions.

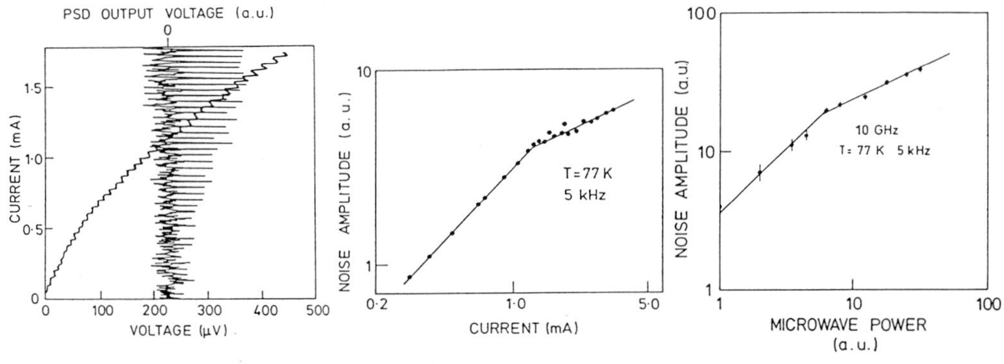

Fig. 2a. The noise signal variation with current at 5kHz overlayed on the i-V characteristic of a typical staircase structure.

b. The variation of peak, noise amplitude with bias current.

c. The microwave power dependence of the noise amplitude at 5kHz.

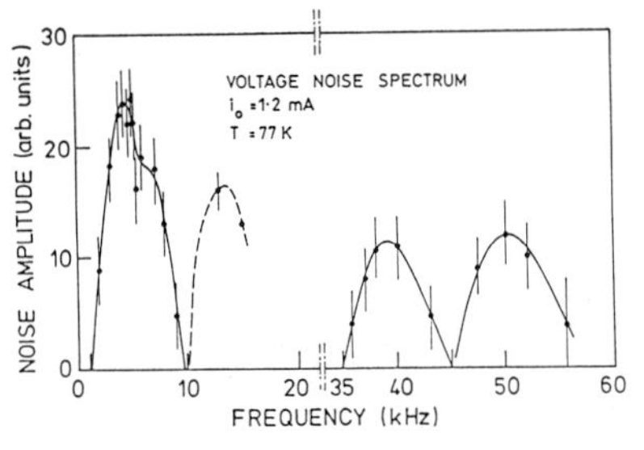

Fig. 3. A preliminary low frequency noise spectrum of the bridge shown in Fig. 2

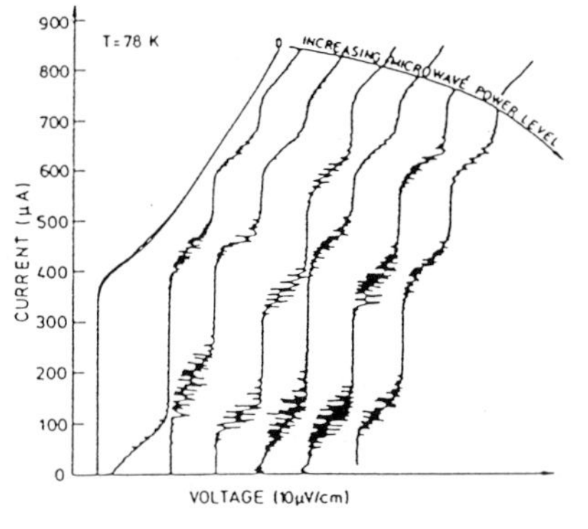

Fig. 4. A second class of characteristic in which the Shapiro steps still dominate to the highest power levels used.

DISCUSSION

The noisy, staircase-like current-voltage characteristics shown in Fig.1 are very similar to those observed in low temperature Josephson junctions [12] and interpreted in terms of chaotic solutions to the resistively shunted junction equation with both dc and rf current bias. Based on the Stewart-McCumber model the dimensionless form of the equation of motion for the junction phase is:

$$\beta \ddot{\phi} + \dot{\phi} + \sin \phi = i_o + i_1 \sin \Omega \tau$$

where $\beta = 2\pi I_c R^2 C/\phi_o$ is the hysteresis parameter and Ω, τ, i_o and i_c are the normalized rf frequency, time, dc and rf bias currents respectively.

The solutions of this equation are dependent upon the two parameters β and Ω and it is possible to use the ($\beta\Omega$) plane to classify the behaviour of particular junctions. Several workers [13,14] have shown that chaotic solutions of this equation exist, which are restricted to a particular region of the ($\beta\Omega$) plane. Fig.5 shows these boundaries and also an estimated location of the present bridge in this plane using values obtained from the observed critical current and normal state resistance (i_c = 1.2mA $R_n = 24\Omega$). Since C is unknown, a range of junction capacitances are shown, which overlap the region for which chaotic solutions are anticipated. The variation in noise amplitude with current and power are not yet understood and we are presently involved in computer simulations in an attempt to reproduce the observed behaviour using the derived values of β and Ω for the bridge.

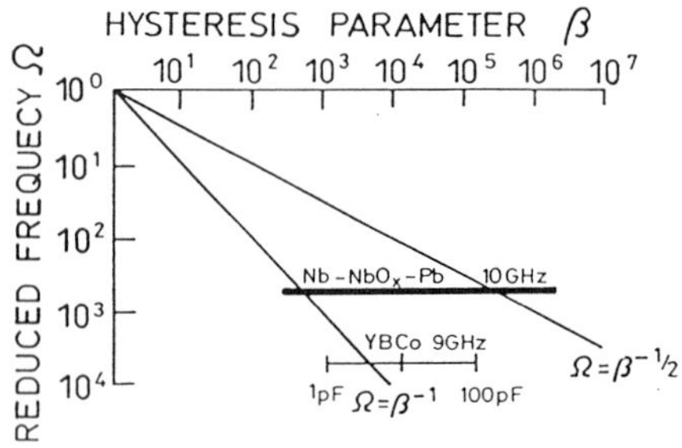

Fig. 5. The (βΩ) plane for a resistively shunted junction showing the boundaries limiting chaotic behaviour and the estimated location of the present sample.

The second type of noise behaviour shown in Fig. 4 is clearly very different from that discussed above, and we suspect that the well defined steps indicate that we are dealing with a number of coupled junctions. The existence of chaotic solutions to the equations describing such systems has been demonstrated by Nerenberg et al [15] and Wu et al [16] for situations in which ac bias is applied to the system. In this behaviour, phase locking and chaotic solutions occur periodically with increasing microwave power. It is also shown that the observed characteristics depend strongly upon the current and voltage of the bridge and further, that strong voltage pulling can occur whenever the junction voltages are the same. While we have not yet been able to simulate the behaviour shown in Fig. 4, we believe that a likely explanation of these unusual characteristics is strongly related to the granular structure of the bridge and involves strong coupling between the individual intergranular contacts. This coupling probably occurs by way of parallel grains which serve to short the various junction pairs and the well defined observations are undoubtedly assisted by the limited number of grains involved in the structure.

CONCLUSIONS

Using microbridge, weak link structures mechanically formed in melt processed thick films of YBCO on YSZ we have observed a range of current-voltage characteristics. Typically, these show well defined Josephson-like critical currents and the presence of Shapiro steps, under microwave irradiation, whose amplitudes follow the anticipated Bessel function behaviour.
In some bridges, whose dimensions are sufficiently small as to involve relatively few intergranular contacts, significant voltage noise appears in the characteristics whose amplitude varies with both dc and rf bias currents. We have observed two types of behaviour for this noise, one of which appears to be similar to the chaotic noise observed in low temperature Josephson junctions and is associated with the intrinsic properties of resistively shunted junctions. In this case, the bridge parameters are in reasonable agreement with the region of the (βΩ) plane in which chaotic behaviour is predicted.

The second type of noise characteristic is quite different, and we suggest that this may be associated with the behaviour of coupled intergranular junctions, in which the coupling is taking place through other grains in parallel to particular junction pairs.

REFERENCES

1. Mannhart J, Chaudhari P, Dimos D, Tsuei C, McGuire T (1988) Phys.Rev.Lett.61: 2476
2. Ketchen MB, Kopley T, H. Ling (1984) Appl. Phys. Lett. 44: 1008
3. de Waele ATAM, Smokers RTM, van der Heijden, RW (1987) Phys. Rev. B35, 8858
4. Aminov BA, Borisovskii W, Brandt NB, Muller R, Thu NM, Ponomarev YG, Sudakova MV, Graboi IE, Kaul AR Tretyakov YB (1988) JETP Lett. 47: 675
5. Alvarez G, Taylor KNR, Russell GJ (1989) Physica C. 162-164 1043
6. Kleiner R, Muller P, Andres K (1989) Physica C 162-164 393
7. Bailey A, Alvarez G, Russell GJ, Taylor KNR (1990) Cryogenics 30: 599
8. Bailey A, Alvarez G, Puzzer T, Matthews DN, Town SL, Russell GJ, Taylor KNR (1990) Aust. J. Phys. 43 347
9. Alvarez G, Bailey A, Russell GJ, Taylor KNR (1990) Physica C. Submitted
10. Russer P (1972) J. Appl. Phys. 43: 2008
11. Alvarez G, Russell GJ, Taylor KNR (1990) Physica C. Accepted
12. Vanneste C, Chi CC, Cronemeyer DC (1985) Phys. Rev. B32 4796 11985.
13. Kautz RL, Monaco R (1985) J. Appl. Phys. 57 875
14. Octavio M. (1984) Phys. Rev. B 29 1231
15. Nerenberg MAH, Blackburn JA, Jillie DW (1980) Phys. Rev. B 21 118
16. Wu BR, Yang, ZJ, Blackburn JA, Smith HJ, Nerenberg MAH (1988) Phys.Rev.B37 3349

7.2 Systems

Recent Progress in Superconducting Generator Program in Japan

TAKASUKE AGETA

Engineering Research Association for Superconductive Generation Equipment and Materials (Super-GM), 14-10, Nishitenma 5-chome, Kita-ku, Osaka, 530 Japan

ABSTRACT

Superconducting generators (SCGs) have advantages of generation efficiency improvement, their reduced sizes and weights, and higher power system stability. Super-GM has been promoting a research and development (R&D) on superconductive generation system since 1988 as a national project. In this R&D, there are four categories - superconducting wires, superconducting generators, refrigeration systems and their total system. In each area, metal and oxide superconducting materials for electric power apparatuses, 70MW class SCGs (model machines), high reliable refrigeration systems, and test methods and operating techniques as well as power system improvement effects are being researched, respectively.

KEY WORDS: superconductivity, superconducting generator, superconductor, oxide superconducting material, refrigerator

INTRODUCTION

The application of superconducting technologies into electric power apparatuses is very important from a viewpoint of energy and resource savings. To cope with the recent steady growth of demand for electricity, capacity enhancement of power source, reduction of power loss and improvement of power system stability are being demanded. SCGs using superconductors as the field windings have the following merits compared with the conventional generators:

(1) Efficiency enhancement (0.5 to 1%)
(2) Reduction in size and weight (both halved)
(3) Improvement of power system stability (power transfer limit can be increased by about 30%) by lowering synchronous reactance (1/2 to 1/5 of that of conventional generators)
(4) Improvement of operation performance by expanding the range of leading power factor operation (leading reactive power equal to rated capacity can be supplied)
(5) Expansion in manufacturing limit of large capacity generators by improved energy density (the maximum capacity of a generator can be increased to about twice)

In Japan, three heavy electric equipment manufacturers studied the basic characteristics using 3MVA to 50MVA SCGs from 1976 to 1985. Thereafter, Agency of Industrial Science and Technology (AIST) of Ministry of International Trade and Industry (MITI) conducted a feasibility study on SCGs for 2 years from fiscal year (FY) 1988 and based on this results, Super-GM was established in September, 1987. As a part of Moonlight Project of AIST of MITI, being commissioned by New Energy and Industrial Technology Development Organization (NEDO), Super-GM has been researching and developing superconductive generation equipment and materials centered on 70MW SCGs.

MASTER PLAN

(1)Main Contents Of R&D

The main R&D contents of this project are shown in Table 1. There are four themes, superconducting wires, SCGs, total system and refrigeration systems. They are being researched and developed for the purpose of application for electric power apparatuses. Especially, in the SCG subject which is the core of this project, 70MW class model machines are being developed as a system, and the reliability will be confirmed by the operating research such as severe rotating tests, and by reflecting the results on the 200MW class pilot machine, we intend to establish the superconducting technology for the application.

Superconducting Wires
In the research of metal superconductors, the development of NbTi and Nb_3Sn superconductors for the generators and the performance improvement of strands for commercial frequency are being researched by wire manufacturers. In the research of oxide superconducting materials, the R&D study of wires, thin films and thick films by means of several synthesizing and crystal orientation methods in order to improve homogenization, compactness, etc. were shared by wire manufacturers, heavy electric equipment manufacturers and Japan Fine Ceramics Center (JFCC).

Table 1. Main contents of R&D

<table>
<tr><th colspan="3">Research Subjects</th><th colspan="4">Main Contents of Research</th><th>Company</th></tr>
<tr><td rowspan="4">Metal super-conductors</td><td rowspan="2">Conductors for generators</td><td>NbTi</td><td colspan="4">• High stability type
• High current density type
• Low AC loss type</td><td>Hitachi Cable
Sumitomo
Furukawa</td></tr>
<tr><td>Nb_3Sn</td><td colspan="4">• Bronze process
• In-situ process</td><td>Furukawa
Fujikura</td></tr>
<tr><td rowspan="2">Performance improvement of strands</td><td>NbTi</td><td colspan="4">• High stability type
• High current density type
• Low AC loss type</td><td>Hitachi Cable
Furukawa
Sumitomo</td></tr>
<tr><td>Nb_3Sn</td><td>• In-situ process
• Bronze process
• Tube process</td><td colspan="2">• Powder metallurgy process
• Internal diffusion process
• External diffusion process</td><td>Fujikura
Furukawa
Showa</td><td>Kobe
Sumitomo
Hitachi Cable</td></tr>
<tr><td rowspan="3">Oxide super-conducting materials</td><td colspan="2">Homogenization and Compactness</td><td colspan="4">• Homogenization and compactness in wire structure
• Homogenization and compactness in large area structure</td><td>Furukawa
Hitachi</td></tr>
<tr><td colspan="2">Grain and Grain boundary control</td><td colspan="4">• Nonequilibrium reaction techniques</td><td>Toshiba</td></tr>
<tr><td colspan="2">Crystal orientation control</td><td colspan="4">• Effects of mechanical stress, etc. on crystal orientation
• Monocrystalline wire processing by CVD, etc
• Monocrystalline wire processing by melt growth
• Methods for synthesizing fine and highly oriented crystallite ceramic wires</td><td>Fujikura
Mitsubishi
Sumitomo
JFCC</td></tr>
<tr><td rowspan="3">Superconducting generators</td><td colspan="2" rowspan="2">Slow response type</td><td>High stability type field winding model</td><td>Multicylindrical rotor model (Single layer type damper)</td><td>Model machine
(Stator)</td><td>Partial stator model</td><td>Hitachi</td></tr>
<tr><td>High current density type field winding model</td><td>Partial rotor model (Squirrel cage type damper)</td><td>Slow response type model machine (Rotor)</td><td>Nb_3Sn field winding model</td><td>Mitsubishi</td></tr>
<tr><td colspan="2">Quick response type</td><td>Low AC loss type field winding model</td><td>Partial rotor model (Three layer type damper)</td><td>Quick response type model machine (Rotor)</td><td>HTC model
Rotating cool-ing model</td><td>Toshiba</td></tr>
<tr><td colspan="3">Total system</td><td colspan="4">Test methods for 70 MW class model machines
Control, protection and operation methods
Power system improvement effects by superconducting generators</td><td>CRIEPI</td></tr>
<tr><td rowspan="2">Refrigeration systems</td><td colspan="2">Conventional type</td><td colspan="4">High reliability of refrigeration system</td><td>Mayekawa</td></tr>
<tr><td colspan="2">Advanced type</td><td colspan="4">Oil-free screw type compressor for 200 MW class pilot machines
Oil-free turbo type compressor for 200 MW class pilot machines</td><td>Mayekawa
IHI</td></tr>
</table>

The slow response type and the quick response type rotors will be developed by three heavy electric equipment manufacturers. The high stability type and the high current density type field windings will be verified by the slow response type rotors, and the low AC loss type field winding will be verified by the quick response type rotor. To achieve the R&D effectively, a company develops common items like a partial stator model or an HTC model, and the company transfers the results technically to two other companies.

Total System
Test methods, power system improvement effects by SCGs, etc. are researched mainly by Central Research Institute of Electric Power Industry (CRIEPI).

Refrigeration Systems
In the conventional type, high reliable methods for 70MW class machines are being researched. In the advanced type, two kinds of oil free compressors, screw type and turbo type, for a 200MW pilot machine are being researched by two refrigerator manufacturers.

(2)Schedule Of R&D Program

Table 2 shows the schedule of R&D program.

In the oxide superconducting materials, research on the methods for synthesizing and wire processing will be finished and the first interim evaluation will be done this year. After that, the improvement of current density and the development of wire manufacture process will be researched. In the metal superconductors, the trial manufacture of conductors for generators will be almost finished and we are planning to establish the conductor manufacturing techniques for 70MW SCGs.

In the SCG development, basic designs of 70MW class model machines are finished and the researches of several element techniques are almost done. We are now manufacturing partial models. After that, we

Table 2. Schedule cf R&D program

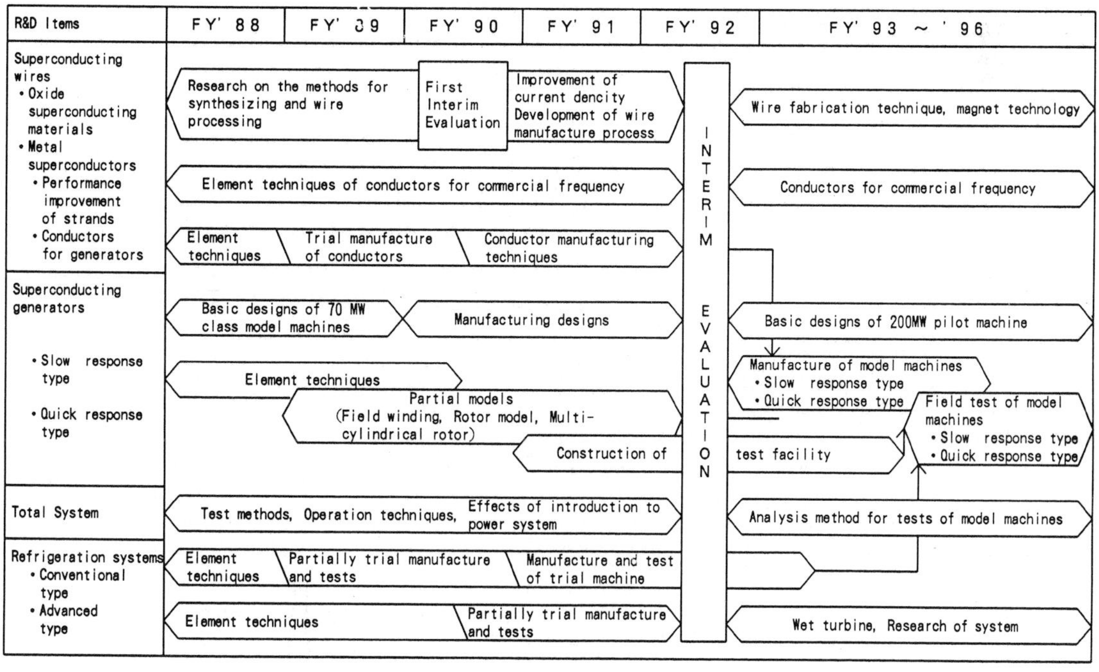

will face the important step of the manufacturing designs and the manufacture of model machines and their field test.

In the research of total system, the test methods, the operation techniques, the effects of introduction to power system are being researched.

As for the refrigeration systems, the study of element techniques like compressors and expansion turbines for conventional type are finished and the partially trial models are being manufactured and tested for the purpose of high reliability. The element techniques of oil free compressors for advanced type are also being researched.

MAIN RESEARCH RESULTS

Table 3 shows the main research results until now.

(1) Superconducting Wires

In the research of the metal superconductors, high stability type, high current density type and low AC loss type superconductors have almost satisfied the development specifications. Figure 1 shows an example of an Al-stabilized conductor for generator field windings, which is superior to Cu-stabilized conductors in thermal and electrical conductivities.
As for the performance improvement of strands, there is a need to make filaments finer in order to make loss smaller in AC machine applications like generator armature windings. Sub-micron diameter filaments of NbTi and Nb_3Sn have been developed.
In the oxide superconducting materials, we have developed superior characteristic wires in critical current density (Jc) and in critical current (Ic). Fig.2 shows Bi based superconducting wires by Laser pedestal method and Doctor blade method, which achieved Ic = 139A and 180A at 77K, 0T, respectively.

Table 3. Main research results

Research subject			Research results
Superconducting wires	Metal	Conductors for generator	• Conductor development satisfying specifications • Element techniques for improvement of stability (Al-stabilized)
		Performance improvement of strands	• Strand development of submicron filament with both NbTi and Nb_3Sn (e.g.: AC loss was lowered at a diameter of 0.05 μm)
	Oxide		• Control factor investigation by various synthesizing methods • Development of high Jc and high Ic superconducting materials CVD method : 6.3×10^6 A/cm² (Y based) Laser pedestal method : 139 A (Bi based) Doctor blade method : 180 A (Bi based)
Superconducting generators	Slow response type Quick response type		• Basic designs of 70 MW class model machines based on element techniques and analyses (slow (A and B) and quick response type) • Selection of candidate structural materials for main components based on various material tests • Element techniques for manufacturing large partial models (field winding models, etc)
Total system			• Test methods of 70 MW class model machines, and control and protect techniques • Introduction effect investigation of SCG into power system
Refrigeration systems	Conventional type		• Element techniques of compressor and expansion turbine (isothermal efficiency 56.3 %) • Techniques for removing impurities of room and low temperature absorbers
	Advanced type		• Thermal strain analysis and manufacturing process investigation of screw rotor • Analyses of stability of turbo shafting, strength of impeller and stability of driver
Independent research			• Impurities (out gas) investigation in the rotor of SCG (Hesper 1) • Basic research on oxide superconductor measuring methods

(2) SCGs

The basic designs of the slow response type and the quick response type 70MW class model machines have finished, including the fundamental dimensions and the electrical and mechanical characteristics, based on element techniques and analysis results. The main component models like damper and flexible disk were manufactured and tested and the main component materials for multi-cylindrical structure were selected. We have verified elemental technologies like manufacturing techniques and operational characteristics toward large partial models using a rotating cryostat model (Fig.3), an HTC model, stator conductor models, etc.

(3) Total System

The test methods of 70MW class model machines and the power system improvement effects by SCGs were investigated. It shows that SCGs are superior in holding the voltage and SCGs improve the power system stability about 5 to 30 % compared with the conventional machines.

(4) Refrigeration Systems

In the research of reliability of the conventional refrigerators, the element techniques of compressor system and the impurity removing techniques were researched. Especially, the performance of the compressor is proved that the isothermal efficiency is 56.3 % (the target is 54.5 %) and the manufacturing techniques for trial model machines were verified.

In the advanced type refrigeration system, the basic analysis of compressor and so on were carried out and the characteristic tests were researched. We are now investigating the reliability by screw type and turbo type oil free compressors.

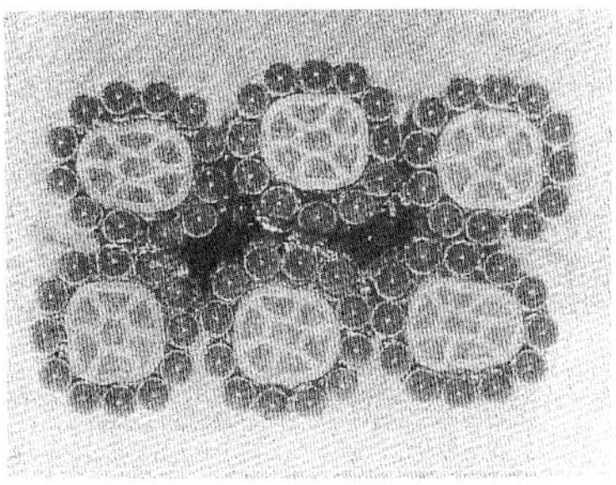

Fig. 1. NbTi Al-stabilized superconducting conductor for SCG field windings

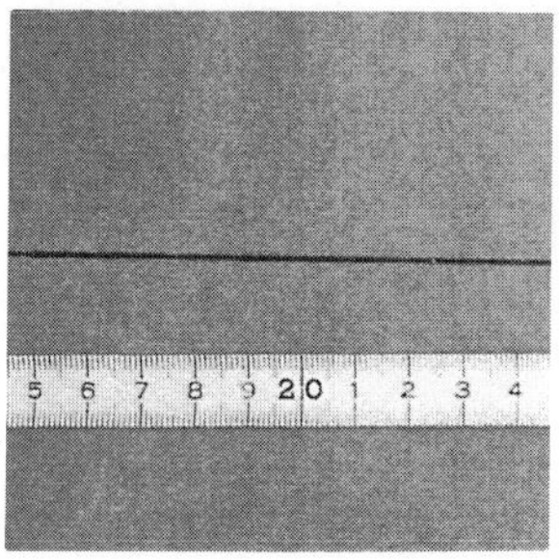

(a) Laser pedestal method (Ic=139A)

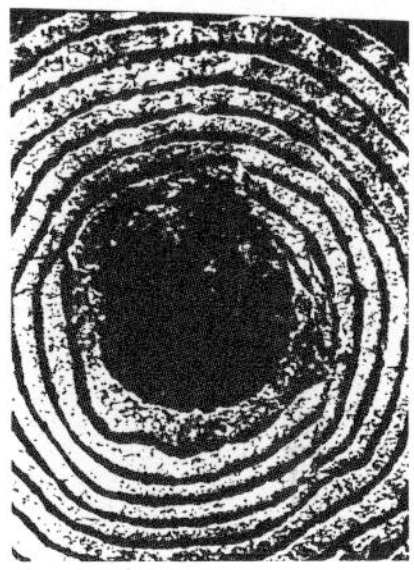

(b) Doctor blade method (Ic=180A)

(at 77K, 0T)

Fig. 2. Bi based superconducting wires

Fig. 3. Rotating cryostat model for examining the characteristics of NbTi superconducting conductor

(3600r/min, Maximum current:3300A, Liquid He cooling)

FUTURE R&D SUBJECTS

Table 4 shows the future R&D subjects.

As for the metal superconducting wires, we give priority to the wires for the 70MW SCGs. Also, we are going to develop lower transient loss AC superconducting wires for the application in AC machines. As for the oxide superconducting materials, we are going to research on higher current density and wire formation for the application to electric power apparatuses.

In the research of SCGs, after the researches on element techniques and partial models, we are planning the manufacture of 70MW model machines, the test and the evaluation.

As for the total system, we would like to establish the field test methods of model machines and the evaluation methods.

As for the refrigeration system, we would like to develop the refrigeration system which can be used during long term continuous operation in electric power business.

CONCLUSION

This research and development of SCGs in Japan is facing the following important steps from now on.

(1) Large partial models and 70MW model machines after the basic designs and the element techniques
(2) The investigation of the test facilities and the establishment of the test organization as well as the investigation of the test methods for the field test of the model machines
(3) The R&D for the purpose of higher performance and higher reliability in the research of AC superconducting wires, oxide superconducting materials and also in the development of the refrigeration systems

We do our best to promote this R&D challengingly and effectively with the cooperation of association members and the advices of AIST, NEDO and universities to establish the superconductive generation technologies.

Table 4. Future R&D subjects

Superconducting wires	• R&D of high performance superconductors for 70 MW class model machines (higher stability, higher Jc and lower AC loss) • Verification of complex properties and reliability, and efficient development of superconductors by utilization of common testing facility • Research on the wire formation by improving synthesis processes for oxide superconducting materials (for application to electric power apparatuses)
Superconducting generators	• Development of manufacturing, fixing, insulating and cooling techniques for superconducting field winding (confirmation of stability and reliability) • Establishment and verification of manufacturing techniques by multi-cylindrical rotor model (efficient development by allotment of subjects) • Verification of manufacturing, fixing, insulating and cooling techniques for air gap armature winding
Total system	• Test methods for verifying the operation performance and reliability of superconducting generators (model machines) • Operation control methods and measuring and diagnosing techniques for superconducting generation system (establishment of testing and evaluation methods and operation techniques)
Refrigeration systems	• R&D of highly reliable refrigeration systems assuring long term continuous operation (conventional type and advanced type - element and system techniques for oil free compressor, etc.)

Current Status of Superconducting Maglev System

JUNJI FUJIE

Technical Development Division, Maglev System Development Division, Railway Technical Research Institute, 8-38, Hikari-cho 2-chome, Kokubunji, Tokyo, 185 Japan

ABSTRACT

In a maglev system utilizing the superconductivity, the superconducting magnets, which correspond to the wheels and motor field in the conventional railway, are required to be reliable, durable, lightweight, compact; permit less heat leak and better stand the vibrational loads. In order to meet these requirements for the maglev system to have higher performance, test runs of MLU002 and electromagnetic vibration tests have been conducted. In consequence, the problems to be resolved have been identified and the countermeasures have been worked out. Now, various improvements such as sidewall levitation coil are steadily being developed. In addition, the introduction of GTO-inverters, the change of coil pitch etc. are being investigated.

KEY WORDS: Maglev, magnetic levitation system, superconducting maget, test line, guideway, superconducting coil, high-speed transportation

INTRODUCTION

In June 1990, it was authorized by the Government to construct an about 43 km long new test line for a superconducting maglev system in Yamanashi Prefecture as a national project. On the test line, test runs with practical versions of maglev vehicles are due to start about three years hence. Things are taking great strides toward the materialization of maglev.

This development aims at completion of an effective, reliable, safe long-distance mass transportation suitable for the 21st century. This being the case, taking into account the results of stationary tests or running tests on the Miyazaki Test Track, a future system answering this purpose is being examined. Superconducting magnet, i.e., vital equipment constituting the system is equivalent to wheels and motors in the conventional railway. Since its reliability and durability are the key factors of a maglev system, the process in which the maglev system has been developed is traced with main emphasis on the superconducting magnet.

VEHICLE AND GUIDEWAY STRUCTURE

The guideway cross-section of the Miyazaki Test Track takes a "U" shape. A superconducting magnet has an I-type structure, under which the coils perform the functions of suspension, guidance and propulsion. On the U-type cross-sectional guideway the loop coils made of aluminium conductor for suspension are laid with the ground coils for propulsion set on both sidewalls along the guideway. The surface of race-track type superconducting coils on board is opposite to the surface of the ground coils for propulsion with the center lines of both coils located at the same level, while the surface of the superconducting coils makes right angles with the surface of the coils for suspension, which generates the suspension forces, that is, the outer edges of the ground coils and the semicircle parts of the race-track coils.

The null-flux connection between the ground coils for propulsion on both sides of the guideway gives the vehicle a guidance function. Superconducting coil and horizontal portion of the ground coil are mainly responsible for guidance force.

SUPERCONDUCTING MAGNET

The superconducting coil for ML500 is an L-type consisting of 2 coils; one for propulsion/guidance and the other for suspension. In order to increase the ratio of load to tare weigth and to obtain light weight, the test vehicles since MLU001 type have adopted the superconducting magnet of a simpler I-type structure which combines the functions of suspension, guidance and propulsion.

Fig. 1 Superconducting Magnet for MLU002

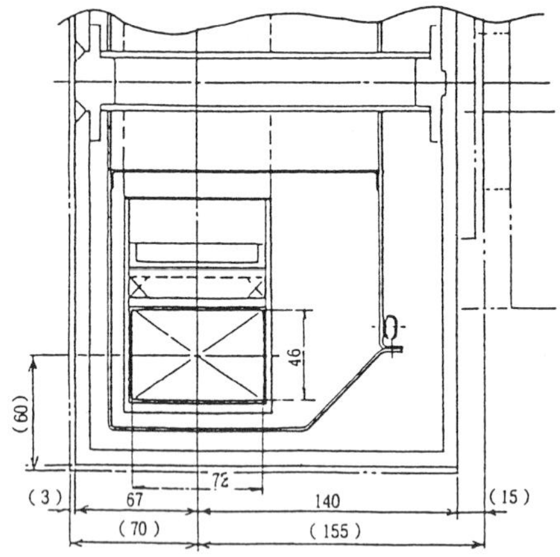

Fig. 2 Typical Cross-section of the Lower Part of a Superconducting Magnet

For MLU001 the load per coil is 1.15 t, whilst for MLU002 it is increased to 1.42 t with a structure making it a step closer to a practical version.

The lower part of the superconducting magnet system, whose typical cross-section is illustrated in Fig. 2, differs from one for MLU001 mainly in shrinkage of cross-sectional area, copper ratio and the inner vessel, which minimizes the dimension between the center of a superconducting coil and the surface of an outer vessel, that is, 70 mm to the side surface of the outer vessel; 60 mm to its bottom surface. The coil consists of wires densely wound and impregnated with epoxy resin. The coil is refrigerated through its surface by means of thermal conduction.

MLU001 has so far logged more than 40000 km without any trouble in the superconducting coil in steady test runs. Meanwhile, MLU002 has met with a trouble due to quenching during a test run. Judging from the test data collected from the test runs and from the electromagnetic exciting test, it is presumable that, when the inner vessel is subjected to an electromagnetic exciting force and the force rises to a certain level, a slight slide occurs between the inner vessel and the superconducting coils, yielding a local heating, which in turn induces quenching.

It is thinkable that too severe design of the inner vessel structure contributes to quenching. So, strength of the inner vessel, current density, exciting force and so on are being reexamined in process of designing a future superconducting magnet. In the meantime, even if quenching occurs within a coil of the superconducting coil for MLU, so long as the thermal shield plate does not happen to be contacted with the inner vessel due to electromagnetic force, there will be no problems about thermal penetration and the superconducting magnet subjected to quenching can be reused, without interrupting test runs.

The off-state resistance value of the persistent current switch is 0.8 ohms for MLU001, whereas it is 50 ohms for MLU002 with exciting speed 5 A/s and magnetically erasing speed 20 A/s. At present, a prototype persistent current switch with an off-state resistance value 100 ohms has already been completed. Higher resistance of the persistent current switch decreases evaporating gas which is generated when magnetically exciting or erasing; and it will realize temporary storage of the gas in a buffer tank built into the truck frame, resulting in a completely closed on-board system independent of the on-ground system.

PROPULSION SYSTEM

An electromagnetic exciting unit has been devised in order to eliminate the effect of harmonic component in the magnetic field, which is induced by the ground coil for propulsion when a maglev vehicle runs, on a superconducting magnet. The test of a prototype superconducting magnet through the exciting unit has revealed that a given exciting frequency generates an abnormally high rate of helium evaporated from the superconducting magnet.

In addition, it has been made clear from the test results that the helium evaporation rate is scarcely increased under a non-exciting state of the superconducting magnet, and that the mechanical vibration state of the inner vessel coincides with the increase of helium evaporation rate. As a result, a model given in Fig. 3 has been supposed and the mechanism of thermal generation has been pursued through this model, revealing the following fact.

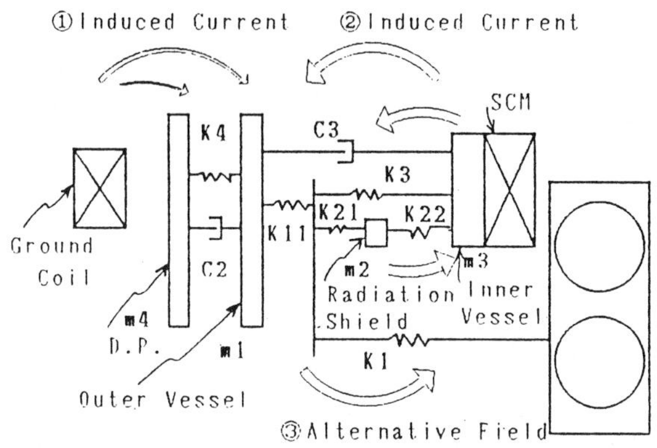

Fig. 3 Vibration Model for Superconducting Magnet

The harmonic component generated from the ground coil vibrates the superconducting coil, which in turn induces vibrations in the outer vessel and in the thermal shield plate both fixed by means of supports. Here, a little fluctuation of a strongly magnetic space generates a large eddy current within the thermal shield plate, etc. This fluctuating magnetic field causes an induced current within the inner vessel, resulting in creation of heat. Therefore, it is desirable to decrease the relative displacement rate. As an effective countermeasure against this, it is thought to support softly the superconducting magnet against the frame, and thereby suppress the vibration within the superconducting magnet or to furnish the magnetic system with any damping element. In accordance with this theory, the related tests have been carried out. However, though it has become possible to lower considerably the thermal generation rate, it has been experimentally confirmed that in principle to decrease harmonics as an exciting source is the most effective at this point of time.

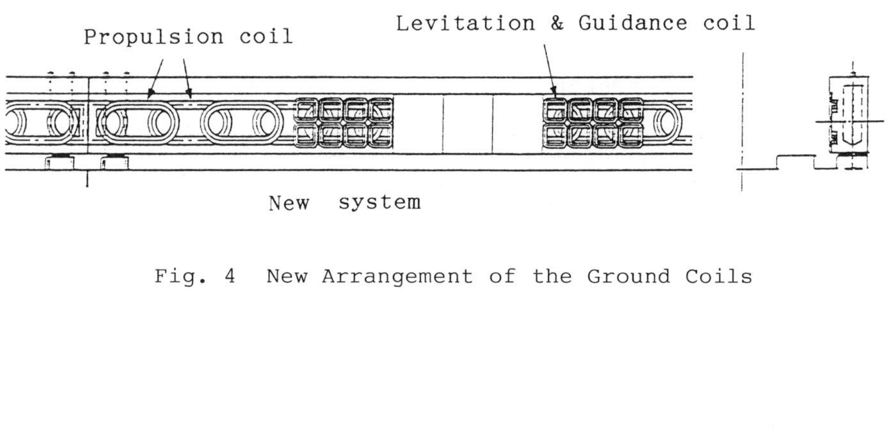

Fig. 4 New Arrangement of the Ground Coils

The propulsion coils on the Miyazaki Test Track are installed in single-layer coil arrangement with U, V and W phases located at a 120° pitch against two poles of the superconducting coils. So, the third harmonic component, which is generated in the propulsion coils, accounts for a high rate of the basic wave. Thus this arrangement presumably generates tremendous heat in a long-distance operation.

As a solution a double-layer arrangement of the propulsion coils (Fig. 4) is considered. However, this arrangement requires two times as many coils as for conventional arrangement. Therefore, a double-layer arrangement with the coils arranged longitudinally has been devised and tested using an exciting test unit. As a result, the effectiveness of this arrangement has been confirmed. Fig. 5 shows a result of calculations about time-dependent fluctuation for y-axial elements of the magnetic fields at the locations in the travelling directions under both the conventional system and the new one. Evidently, the time-dependent fluctuation of the magnetic field for the new system is very small.

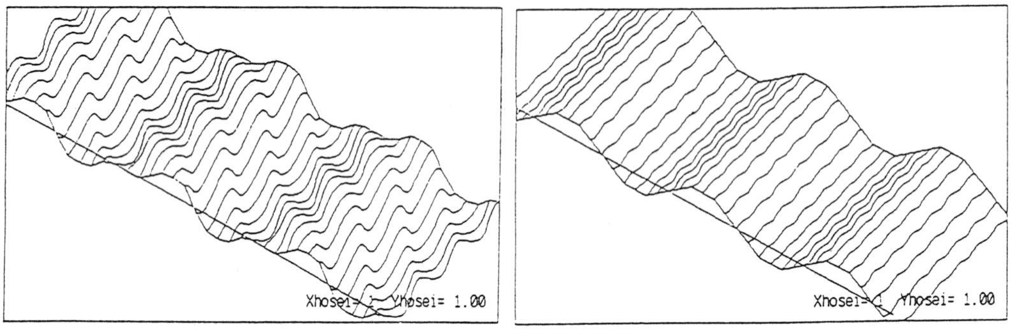

Single layer coil arrangement Double layer coil arrangement

Fig. 5 An Example of Calculation about y-axial Magnetic Field Fluctuation in a Superconducting Magnet due to the Propulsion Coils

LEVITATION/GUIDANCE SYSTEM

For the guidance of a vehicle on the Miyazaki Test Track, a system combining levitation with guidance has been adopted. This system works by the electromagnetic induction action between the propulsion coils on both sides connected via the null-flux connection wires and the superconducting magnets on board of a running vehicle. A maglev train on a future commercial line is assumed to accommodate 1000 passengers. Such a train running at 500 km/h requires about 20 kV for the ground coils. Accordingly, it is necessary to specify high voltage for the null-flux wires connecting the coils on both sides and for the lead portions of the wires, which is structurally and economically disadvantageous.

In 1983, a sidewall levitation system in which the levitation coils are installed on the sidewalls was proposed. After that, in 1988 it was revealed that a null-flux connection of the coils on both sidewalls makes it possible to combine levitation with guidance. The desired performance of the combined system was confirmed as a result of the test runs on a part of the Miyazaki Test Track.

As shown in Fig. 6, this system is set on the sidewalls, where the upper and lower parts of the ground coils consisting of the two parts in two layers are connected to each other in "8" figure and set on the sidewalls. In this system, when the axis of the on-board superconducting coils vertically coincides with the center between the upper and lower parts of the ground coils, an induced current does not flow. When the axis is located in an upper or lower level than the center, a current is indued on the ground coils, which in turn generates a force vertically pulling the superconducting coils back to the center level. Here, at the lower level this pull-back force appears as a levitation force.

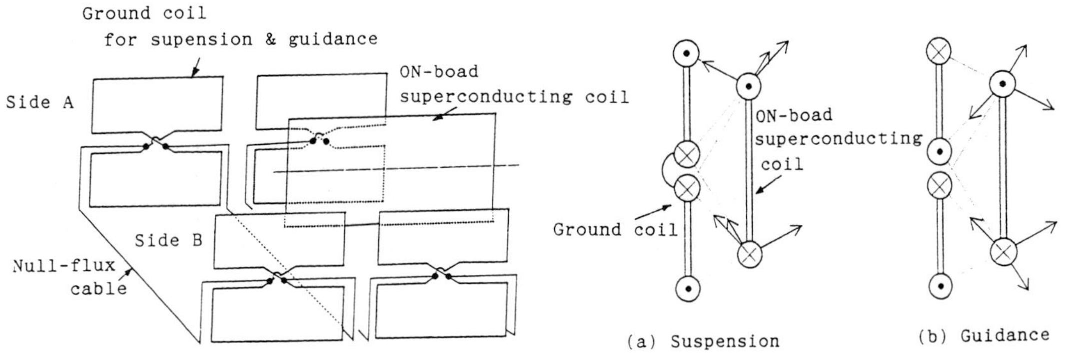

Fig. 6 Sidewall Levitation System

Unlike the conventional system (normal flux levitation system) on the Miyazaki Test Track, in the new system the upside of a superconducting coil also takes part in levitation, resulting in a smaller current in the ground coil and less loss in suspension, which in turn contributes to a large magnetic drag ratio and a small drag peak in a low speed range. When a maglev vehicle runs on wheels in a low speed range with the axes of the superconducting coils and the ground coils keeping the same level, the drag can be made null.

Guidance of a vehicle is done by the null flux connection of the upper parts with the lower ones of the levitation coils on the sidewalls.

Also with the levitation coils like the propulsion coils, there may be a problem of thermal generation due to the harmonic components. The calculation about the time-dependent fluctuation for y-axial components of the magnetic fields is done in both cases: the pitch of suspension coils coincides with that of propulsion coils; the suspension coil pitch is half the propulsion coil one, which is the case of the Miyazaki Test Track. Because in the latter case the third harmonics do not exist and have a small fluctuation of magnetic field and the pitch of the levitation coils is desirably integral number times as long as the propulsion coils, the latter case is found more appropriate for adoption.

In the meantime, as a proposal concerning the sidewall levitation coil system, there is a method in which propulsion, levitation and guidance are simultaneously done by a pair of coils: one on the left sidewall, the other on the right sidewall. Because of various problems mentioned above, this method has not become a theme to be discussed as a system. However, when a magnet made of a high-temperature superconductor available at the liquid nitrogen temperature materializes, the heat generated from fluctuation in the magnetic field will become easy to cope with. Here, the proposed method may possibly become a theme to be examined.

Recently, a GTO (gate turn-off) element with a large capacity has been developed to a stage of applicability to the power sources for maglev to be followed by a plan to introduce GTO inverters. Here, because a higher frequency can be used unlike the case of cycloconverters, it becomes possible to shorten the pole pitch of the superconducting magnets, resulting in an increase of the number of poles. A shortened pitch of the ground coils improves the product of power factor and efficiency for a linear motor with a small fluctuation in a magnetic field, but it decreases a levitation or guidance force per longitudinal length of a superconducting coil, which in turn increases the number of coils. Therefore, at present the appropriate values are being pursued.

Furthermore, the combined propulsion and guidance system on the Miyazaki Test Track will be changed to a system exclusively for propulsion, making it possible to adopt a triple power supply system in which electric power is supplied independently of the propulsion coils on both sides unlike the conventional double power supply system, which contributes to decreasing the power source capacity to three fourths. So, the adoption of this system is being considered.

AN IMAGE OF THE NEW TEST LINE

For the research and development aiming at materialization of a superconducting maglev, goals are set for highspeed, higher traffic capacity, punctuality and economy. Especially as for highspeed, with the aim set at the commercial operation speed of 500 km/h, a safe running at more than 550 km/h will be confirmed on the new test line.

Test vehicles to be operated on the new lines are being designed at present. The tentative technical standards for infrastructures have been established for the assumed maglev vehicles. To be more precise, main technical features are as follows: the minimum radius of curvature 8000 m on the main line; the maximum gradient 40/1000; the minimum distance between track centers 5.8 m.

An image of the vehicles is shown in Fig. 7. The maglev train adopts an articulated truck system in which the superconducting magnets are located at the ends of the vehicles. In order to decrease the aerodynamic draq during running on the line including running in a tunnel, a cross-sectional area of a vehicle body is decreased as far as possible. In addition, an aerodynamic brake making use of aerodynamic drag is to be tested on the new test line for the increased brake efficiency within a highspeed range.

Fig. 4 illustrates a guideway, in which a sidewall levitation/guidance system with a double-layer arrangement of coils mentioned above will be adopted.

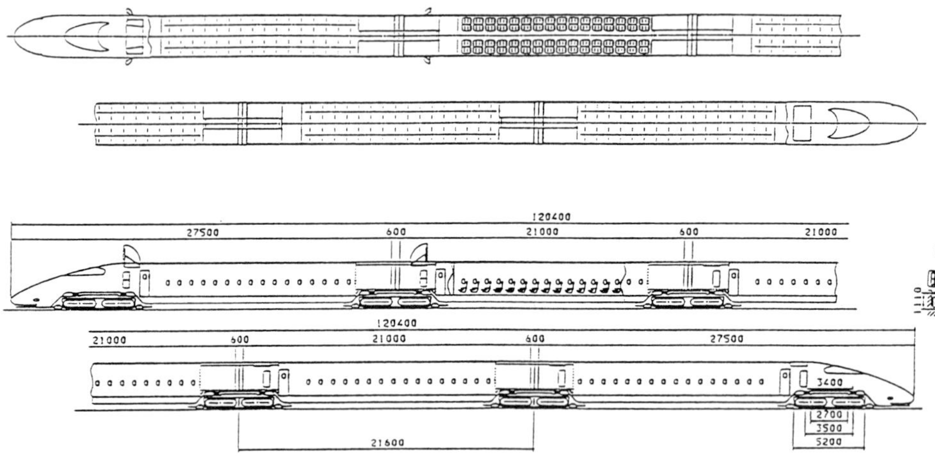

Fig. 7 An Image of a Maglev Vehicle for Commercial Operation

CONCLUDING REMARK

As a mass ground transportation aiming at the commercial operation at a speed of 500 km/h, a superconducting maglev is presumably the most appropriate on account of its stability in running performance. At present, various tests are being carried out to prove reliability and durability of the superconducting magnets playing a key role in this maglev system. In addition, a superconducting magnet more suitable for a commercial version of maglev is being developed. The development for completion of a whole system including superconducting magnets is advancing toward an early completion of the commercial sytem.

REFERENCES

1. Yamaji M, Nakashima H (1988) ISS'88: 72
2. Tsuchijima H, Herai T, Kurihara M, Nakashima H (1988) RTRI Report Vol.2, No. 6: 45
3. Fujiwara S, Fujimoto K (1989) RTRI Report Vol.3, No.11: 17

R&D on Superconducting Electromagnetic Propulsion Ship

S. Takezawa

Japan Ship and Ocean Foundation (SOF), 15-16, Toranomon 1-chome, Minato-ku, Tokyo, 105 Japan

ABSTRACT

In 1985 SOF began R&D of superconducting electromagnetic propulsion ship; the project goal is to develop propulsion equipment utilizing magnetohydrodynamic force as thrust, and to conduct sea trials with a ship furnished with the innovative propulsion equipment. Researches have centered on propulsion system and hull: application of superconducting electromagnets to propulsion equipment and development of an optimum shape of hull. In 1987 our project team succeeded in self-propelling experiment with a 2.6-m long model ship. A manufacturing method for large-scale superconducting coils was established as well. In 1989, based on the research results, the project team designed a 30-m long testcraft with a displacement of 185 t, which will navigate at a speed of 8 kt, carrying up to ten people. The construction of the testcraft is now underway. It will be completed at the end of 1990. Sea trials are slated for 1991.

KEY WORDS

APPLICATION/SYSTEM/COIL

1. INTRODUCTION

The superconducting electromagnetic propulsion (SEMP) ship requires no screw propeller. It is propelled by magnetohydrodynamic force, which is generated by the interaction between an electric current and a magnetic field created by superconducting electromagnets installed inboard when an electric current is applied to the magnetic field. Thrusters have no mechanical power transmission system and are free from noise. Furthermore, the SEMP system is suitable for propelling a ship at high speed. With its many attractive features, the SEMP ship is regarded as a seacraft of the future. Research and development on the SEMP ship, therefore, is being watched with great expectations.

This paper describes the SEMP ship, which is a marine application of superconducting technology, centring on the research and development activities by the Ship and Ocean Foundation.

2. PRINCIPLE AND PROPULSION SYSTEM OF THE SEMP SHIP

The electromagnetic ship propulsion, which utilizes electromagnetic force as thrust, is an application of Fleming's left hand rule, one of the basic rules of electromagnetics: when an electric current flows at a right angle to a magnetic flux, electromagnetic force is generated in the direction at right angles to both the magnetic field and the electric current.

Fig. 1 shows a magnetic field and an electric field crossing each other in sea water, while an electromagnet and electrodes are fixed on a hull. The reaction caused by electromagnetic force in sea water is used as thrust to propel the ship. The SEMP ship employs superconducting electromagnets to create the magnetic field.

The system for propelling a ship by electromagnetic force falls, in terms of the working area of electromagnetic force, into two systems, the external magnetic field system, in which electromagnetic force is generated around the periphery of the ship, and the internal magnetic field system, which produces electromagnetic force inside a duct which runs lengthwise in the hull.

Fig. 2 shows conceptual diagrams of the external magnetic field system and internal magnetic field system. Both systems employ a magnetic field generated by a direct electric current.

In the case of the external magnetic field systems, its effects on the ocean enrivonment cannot be neglected because the magnetic field and electric field are formed around the periphery of the hull. By contrast, the internal magnetic field system limits the working area of the magnetic field and electric field to the part of the duct which runs inside the ship, with the advantage that countermeasures such as magnetic shielding are easier to apply than with the external magnetic field system.

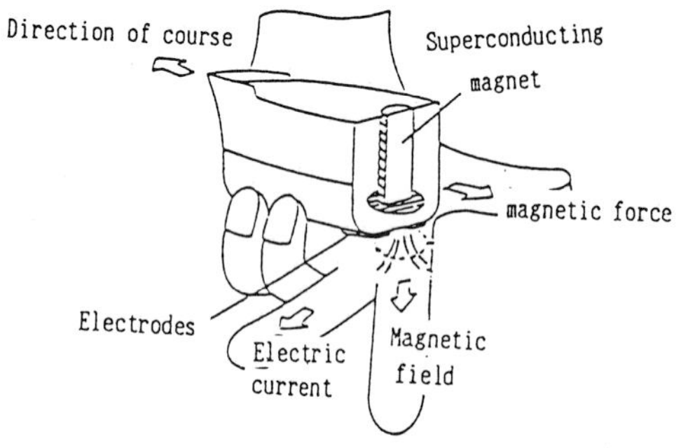

Fig. 1 Principle of the propulsion for electromagnetic propulsion ship

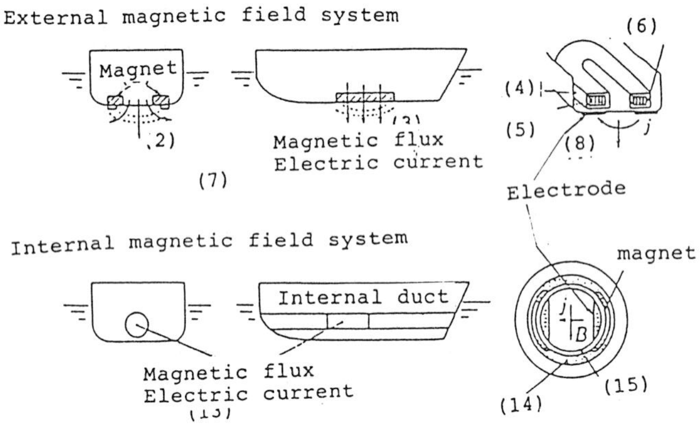

Fig. 2 External or internal magnetic field system

3. OUTLINE OF THE STRUCTURE OF THE SEMP SHIP

It is necessary, when propelling a large object like a ship using electromagnetic force, to make a magnetic field and an electric current of great intensity cross each other orthogonally. In order to generate a powerful magnetic field using an electromagnet it is essential to use a superconducting material for the coil wire, which can reduce electric resistance to zero.

At present, the superconducting material in general use is $N_b \cdot T_i$ alloy which comes into a superconducting state at ultra-low temperatures around -269°C. Liquid helium is used as the cryogen to keep superconducting coils at ultra-low temperatures.

The main equipment which constitute the SEMP system are a superconducting electromagnet, a liquid helium refrigerating machine, and an electric generator to supply electric current to electrodes. To reduce the ship weight and to streamline the system, the superconducting electromagnet is to be operated in a persistent current mode, and excited while the ship is moored. The initial cooling of the superconducting electromagnet and.the supply and recovery of helium will be performed on land.

Fig. 3 shows an example of the concept of the SEMP system based on this structure.

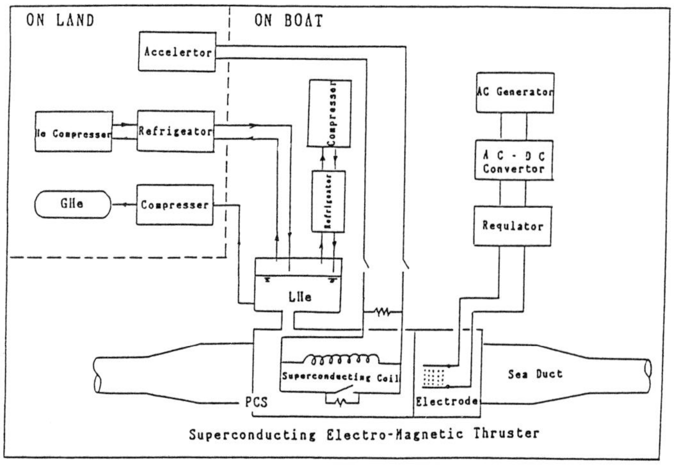

Fig. 3 Conceptual system diagram

4. Research and Development of the SEMP Ship

4-1 Progress of Research and Development

The Japan Foundation of Shipbuilding Advancement, which is the forerunner of the present Ship and Ocean Foundation (SOF), in 1985 established the 'Superconducting Electromagnetic Propulsion Ship Research and Development Committee" (Committee Chairman: Mr. Yohei Sasakawa) to develop superconducting electromagnetic propulsion equipment for ship use. The object of this Committee was to develop the propulsion equipment and to conduct sea trials using a testcraft.
Research and development at present seeks the optimum hull form that is suitable for ships propelled by magnetohydrodynamic force and the application of superconducting coils to the propulsion equipment, as well as measures to improve its performance, namely those related to both hull form and propulsion equipment. In 1988, a self-propelling test of a 2.6 m long model ship equipped with the internal magnetic field-type SEMP system was successfully conducted in the water test tank of SOF's Tsukuba Institute. In the wake of this success, a plan to run a testcraft at sea was developed and construction of the testcraft with a displacement of 185 tons started in 1989. Sea trials with the testcraft are planned for 1991.

4-2 Development of the SEMP Equipment

From extensive studies conducted at the preliminary stage of research and development of the SEMP equipment, $N_b \cdot T_i$ wire was adopted for the superconducting coil for the internal magnetic field-type SEMP system, and the system's basic design was worked out as shown in Fig. 4.

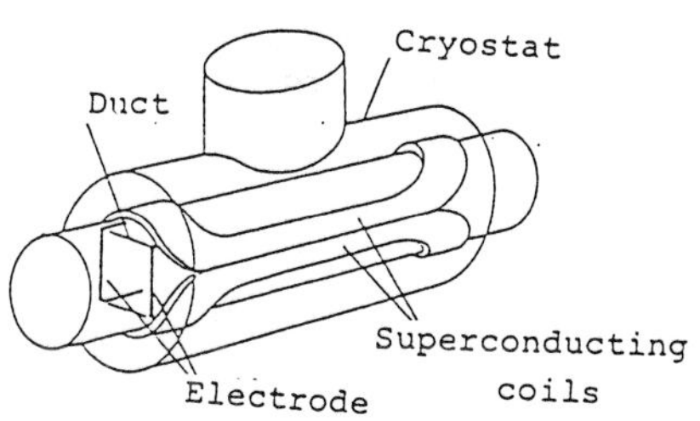

Fig. 4 Basic structure

Because the SEMP equipment of the internal magnetic field type is a novel propulsion equipment, it was necessary to know its performance, characteristics, etc. as propulsion equipment for ship use. Based on the design shown in Fig. 4 which meets the specifications in Table 1, the model SEMP equipment shown in Fig. 5 was constructed.

Table 1 Specification of miniature superconducting electromagnetic thruster

Coil	
Coil type	Dipole type
Operational current	171A
Number of turns	880 x 2
Central magnetic field	1.3T
Superconducting wire	
Superconducting wire material	NbTi/Cu
Inside diameter of a coil	151mm
Total length	534mm
Length of the straight part of a coil	300mm
Cryo	
Bore diameter under an ordinary temperature	100mm
Width	300mm
Length	670mm
Coil cooling system	LHe immersion system

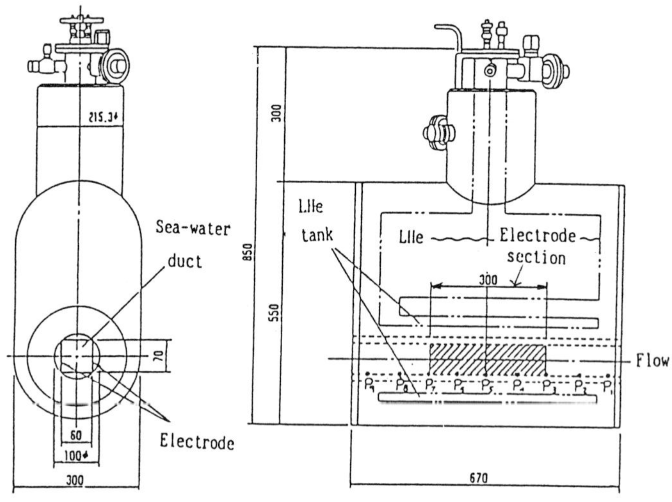

Fig. 5 Miniature superconducting electromagnetic thruster

The performance test was conducted on the propulsion equipment, which was followed by various water tank tests including tests on the mutual effects of the hull and propulsion equipment; the power output of the equipment; and effective thrust.
A 2-m long model ship was used in these water tank tests.
Photo 1 shows the performance test of the equipment and Photo 2 a bollard pull test for the model ship on which the propulsion equipment was loaded.

Photo 1 Self-propelled test

Photo 2 Bollard pull test

Fig. 6 shows an example of the bollard pull test results, which were obtained in the experiment on the propulsion equipment.

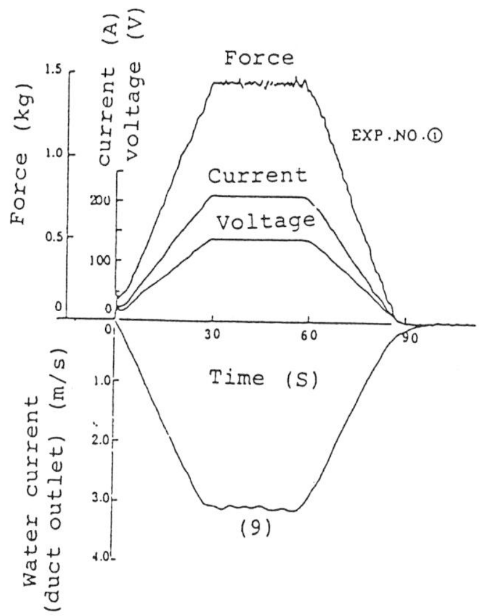

Fig. 6 Result of adjustable speed tests (No. 1)

5. OUTLINE OF THE TESTCRAFT

5-1 Propulsion Equipment

The propulsion equipment to be installed in the testcraft employs, as shown in Fig. 7, six dipole coil units arranged concentrically to reduce the magnetic leakage, thereby avoiding an increase in their weight due to the installation of magnetic shielding. The testcraft, therefore, has no magnetic field shielding. The following specifications were decided with consideration given to the investigations and studies conducted.

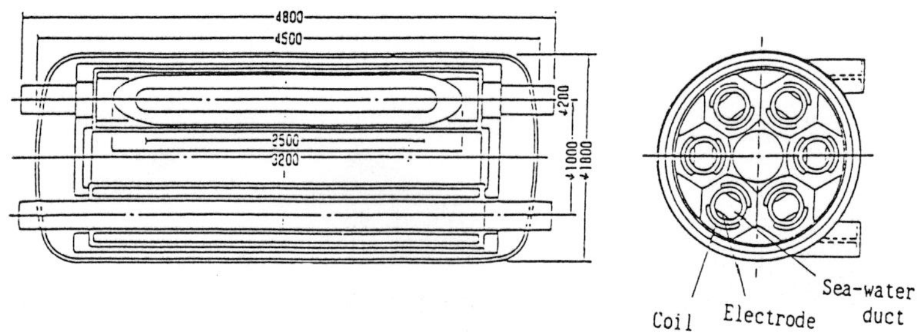

Fig. 7 Basic structure of six dipole units

Number of thrusters	Two
Mode of generating magnetic field	Internal magnetic field type
Type of coil	Dipole type
Effective magnetic field intensity	4 (T)
Effective thrust	4,000 (N) x 2

Listed in Table 2 are the principal particulars of the SEMP equipment designed according to the above basic specifications. Fig. 8 shows its cross-section. Photo 3 shows the exterior of the equipment.

Table 2 Specifications of superconducting magnet for experimental ship

Coil	
Coil type	Dipole type
Operational current	3310A
Number of turns	(125 x 2 layer)x2
Central magnetic field	4.00T
Superconducting wire	
Superconducting wire material	NbTi/Cu
Inside diameter of wound coil	360mm
Total Length	3200mm
Length of straight line part	2500mm
Cryo	
Sea-water duct diameter	260mm
Cryo outside diameter	1800mm
Cryo length	4500mm
Coil cooling system	LHe immersion system

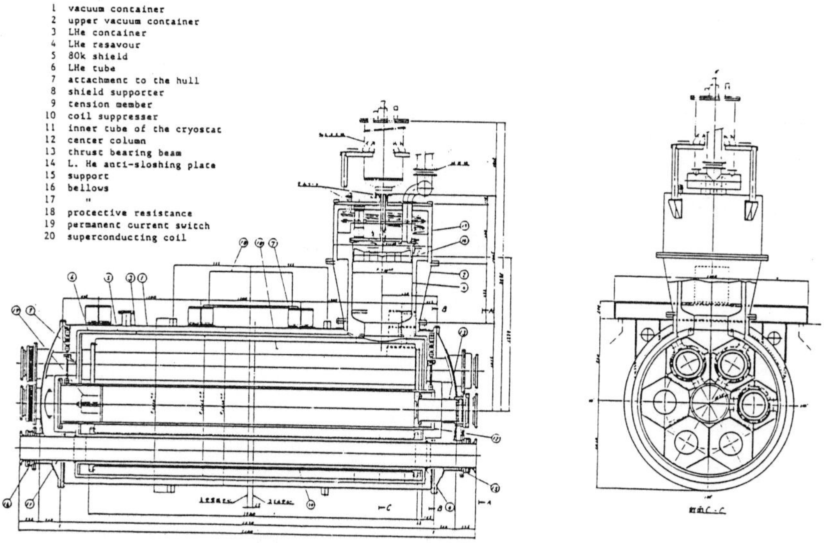

Fig. 8 Thruster for testcraft

Photo 3 Appearance of thruster

5-2 Hull

The SEMP testcraft has as its target a displacement of 185 tons, and a speed of about 8 knots. Much consideration was given not only to the weight of the thrusters and other equipment to be loaded on the ship, but also to their shapes and arrangement. Water tank tests were repeated on the model ship. The form of the hull shown in Fig. 9 was finally selected. Table 3 lists the principal particulars of the testcraft.

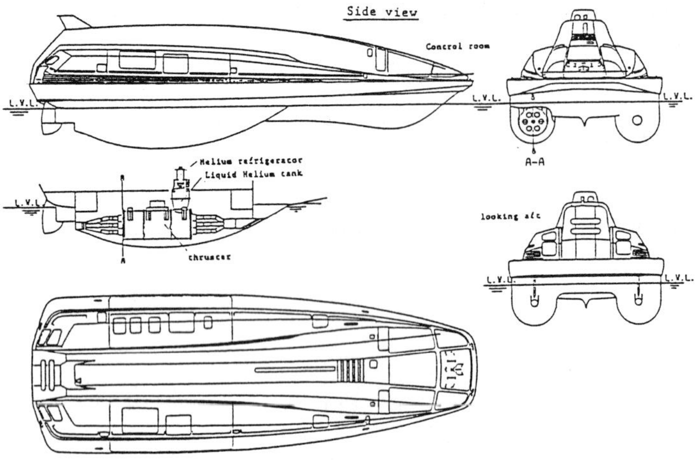

YAMATO－1

Fig. 9 Testcraft

Table 3 Particulars of the experimental ship

Length overall	30 m
Length between perpendiculars	26.4 m
Breadth (molded)	10.39 m
Depth (max)	3.69 m
Draft (max)	2.69 m
Displacement	185 t
(including water in the ducts)	
Speed	about 8kts
(at Lorents force 16,000N)	
Material of the hull	Aluminium
Number of person	10 including 3 crew

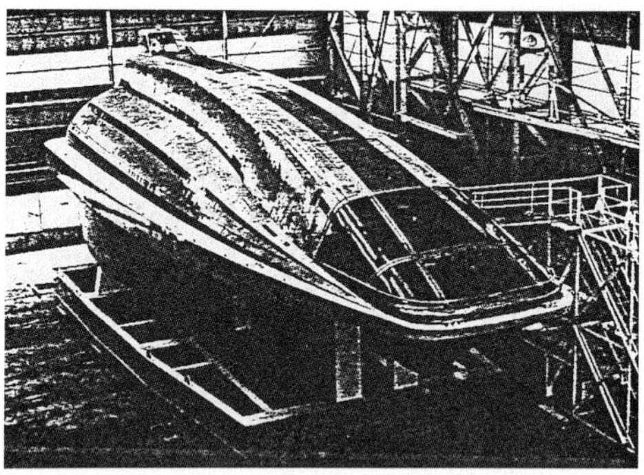

Photo 4 Testcraft

Photo 4 shows the hull of the testcraft under construction. This ship was named "Yamato-1" on 11 July 1989 by Mr. Y. Sasakawa, Chairman of the Research and Development Committee.

The ship's hull has already been completed, and the Yamato-1 is now being fitted with thrusters and other equipment. It is scheduled that the ship will undergo sea trials in the summer of 1991.

6. CONCLUSION

This has been a brief description of our research and development activities for the superconducting electromagnetic propulsion ship, which is propelled by magnetrohydrodynamic force instead of screw propellers. There still remain many technological problems on the development of a superconducting electromagnet to generate a magnetic field of greater intensity, as well as those of the peripheral technologies which should be solved to realize a commercially viable SEMP ship.
Research and development on hightemperature superconducting materials has been progressing rapidly. When a high-performance, hightemperature superconducting electromagnet, which requires no cryogenic technology, is developed, electromagnetic propulsion equipment will become much lighter and its performance will be tremendously improved with the simplification of refrigerating equipment, and a practical superconducting electromagnetic propulsion ship might be realized earlier than we can expect. Because the thrusters and various equipment and machinery used in the ship are required to be lighter and more compact than equipment and machinery for land use, and the service conditions such as temperature, humidity, and pitching and rolling at sea are more demanding. Further research and development is required in the future, but the results will readily propagate to other technical fields with unimaginably favourable effects.

Design Study of 20MWh Solenoidal SMES for Integrated Demonstrative Test Model of Large-Scale SMES

SHIRABE AKITA[1], YOSHINAO SANADA[2], TAKAKAZU SHINTOMI[3], and TOSHIKATSU TANAKA[1]

Research Association of Superconducting Magnetic Energy Storage (RASMES), 9-9, Tokodai 5-chome, Tsukuba, 300-26 Japan

ABSTRACT

To develop SMES technologies, the Research Association of Superconducting Magnetic Energy Storage (RASMES) is conducting design work of large- and small-scale SMES. Recently, a conceptual design of 20MWh solenoidal SMES has been completed as an integrated demonstration test model for 5GWh-class SMES. According to our assessment of SMES technologies, the actual construction of 20MWh solenoidal SMES will improve the technology up to 70 to 80% of the level required to construct large-scale SMES.

KEY WORDS: SMES, solenoid coil, conceptual design

INTRODUCTION

Among the numerous energy storage devices, SMES is the only one which has an extremely high storage efficiency of up to 95% as it utilizes the resistanceless nature of superconductors. When a high Tc superconductor is actually used in SMES, the efficiency will be improved by several percent.

The Research Association of Superconducting Magnetic Energy Storage (RASMES) has been investigating the wide range of possibilities of the application of SMES to electric power systems and other industrial systems. To demonstrate the possible applications of SMES, RASMES has prepared several conceptual designs of solenoidal and toroidal SMES having large and small storage capacities. Since the efficiency of SMES increases with increasing storage capacity, the GWh class of large-scale SMES is one of the goals of technological development. The solenoid coil system requires a lesser amount of superconductor compared to the toroidal coil system for a fixed storage capacity of SMES. RASMES has developed a conceptual design of solenoidal 5GWh SMES with a Nb_3Sn superconductor or a high Tc superconductor. The superconducting coil in the 5GWh SMES has large stored energy capacity and size compared to the superconducting coils which have already been developed in SMES and other superconducting technologies. Thus, RASMES has decided to design a medium-scale solenoidal SMES which will be within the capabilities of state-of-the-art superconducting technologies, and which will have good similarity to 5GWh-class SMES. When this medium-scale SMES is actually constructed, the level of SMES development will be very close to complete because the design is that of an integrated demonstration test model of large-scale SMES.

RESULTS

The stored energy of the medium-scale solenoidal SMES is 20MWh. The basic parameters of this 20MWh solenoidal SMES are shown in Table 1 in

1 Central Research Institute of Electric Power Industry
2 Toshiba Corporation
3 KEK

comparison with those of 5GWh SMES. The coil diameter is almost one-tenth and the height is one-fourth that of the 5GWh model; however, the other parameters are very similar for the two models. The size of the 20MWh coil is good as a demonstration model from the technological point of view because it can be constructed by the same method as is the 5GWh coil.

A 200kA superconductor made of Nb_3Sn and pure aluminum as the stabilizer is embedded into an aluminum alloy mechanical support structure, as shown in Fig. 1. The assembled conductor is wound into a two-layered solenoid coil having 30 ripples, as shown in Fig. 2. These ripples will reduce the thermal contraction stress in a large-scale SMES. Radial magnetic stress on the coil is 3.23MPa, the same level as of 5GWh SMES. The ripples on the coil do not affect the profile of magnetic pressure according to a magnetic field calculation.

A cross section of the 20MWh SMES is shown in Fig. 3. Axial magnetic force is sustained by the aluminum alloy in the conductor. Radial magnetic force is transferred to the rock outside the coil through thermal insulating struts made of alumina-fiber-reinforced plastic (AlFRP). Struts are set inside the coil to support the thermal contraction force and below the coil to support the weight of the coil system. Also, liquid helium tanks are placed under the coil and between the struts to receive liquid helium dumped from the liquid helium vessel around the coil during coil quenching. This procedure keeps the temperature of the coil below room temperature by increasing the quench propagation velocities.

The entire solenoid coil is constructed in a doughnut-shaped cavern excavated in rock. An analysis of the stress in the rock around the coil has shown that an area exists where the stress is tensile during SMES operation. Therefore, the cavern is reinforced by rock bolts and covered with concrete.

The required cooling capacity of the 20MWh SMES is 37W at 4.2K, 607W at 20K, and 4570W at 80K. The helium gas flow through current leads is 18.6g/s. This cooling can be maintained by a conventional refrigerator having 3MW compressors. As the rated charging and discharging capacity is 2 hours, the AC loss of the coil system is ignored for cooling considerations.

The construction is estimated to last approximately four years, excluding exploration and licensing.

ASSESSMENTS

RASMES has assessed the 20MWh solenoidal SMES design from the viewpoint of technological improvements when it is actually constructed. Improvements are shown in percentages of the technological level needed to construct 5GWh-class SMES.

The results are shown in Fig. 4. Almost all of the component technologies are brought up to 70 to 80% of the required level. Some technologies will be fully developed in the 20MWh SMES.

In conclusion, the 20MWh solenoidal SMES designed by RASMES is a good milestone in the development of 5GWh-class SMES technology, and will become an integrated demonstration test model for large-scale SMES.

Table 1. Main parameters of 20MWh solenoidal SMES compared to those of 5GWh SMES.

Stored energy	20 MWh	5GWh
Coil radius	30.0m	250m
Coil height	5.04m	20.04m
Superconductor	Nb_3Sn/Cu	<-
Stabilizer	pure aluminum	<-
Mechanical member	aluminum alloy	<-
Rated current	200kA	500kA
Number of turns	84 turns/2 layers	334 turns/2 layers
Ampere turns	1.68×10^7 AT	1.67×10^8 AT
Number of ripples	30	250
Radius of ripple	inner 6m, outer 2m	<-
Magnetic field at center plane	2.3T	5.9T
Magnetic radial pressure	3.23MPa	3MPa

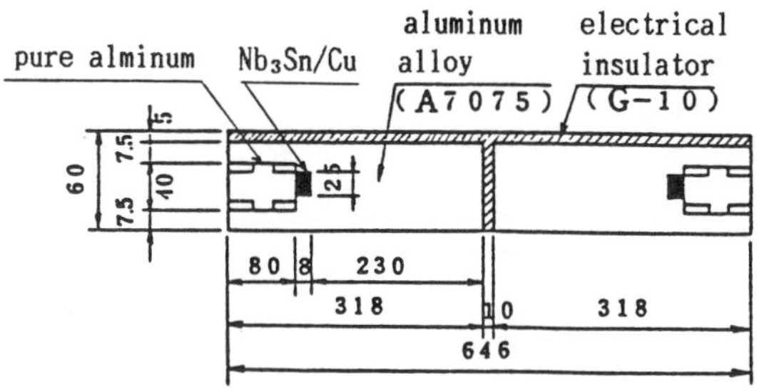

Fig. 1. Cross section of superconducting coil with two 200kA conductors.

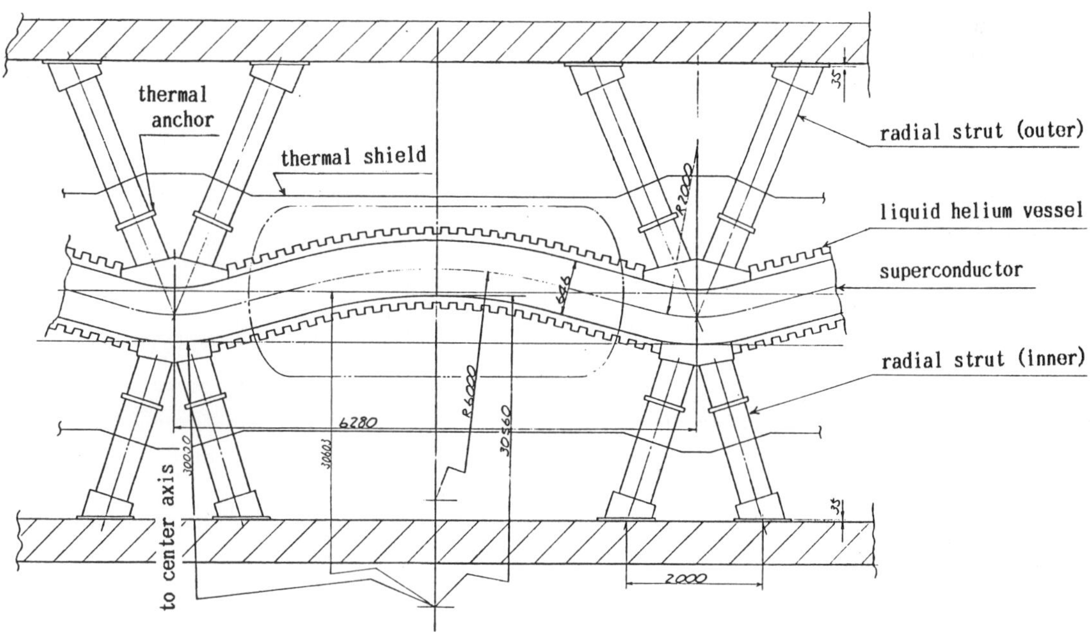

Fig. 2. Horizontal cross section of coil system of 20MWh solenoidal SMES.

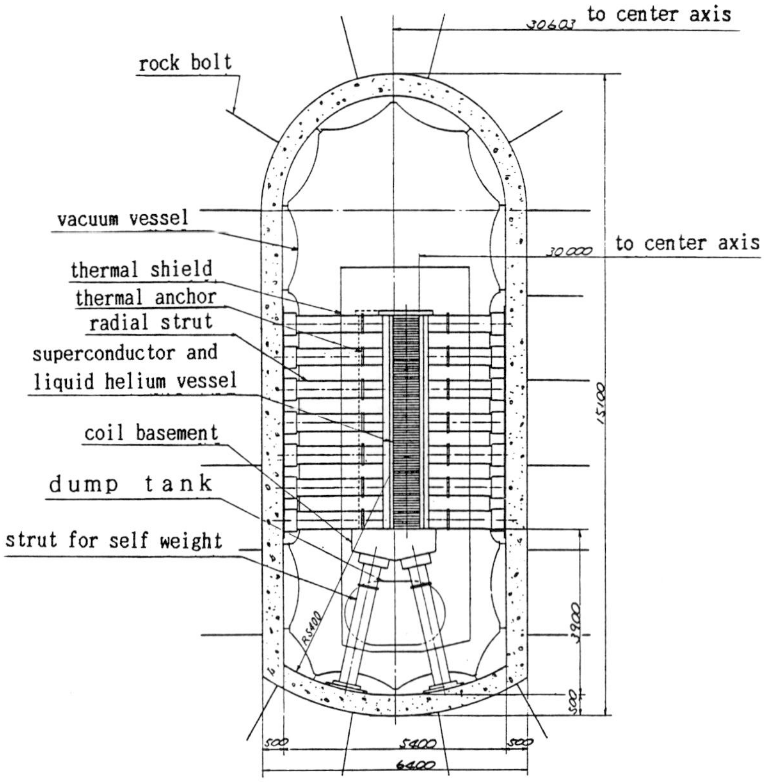

Fig. 3. Vertical cross section of 20MWh solenoidal SMES.

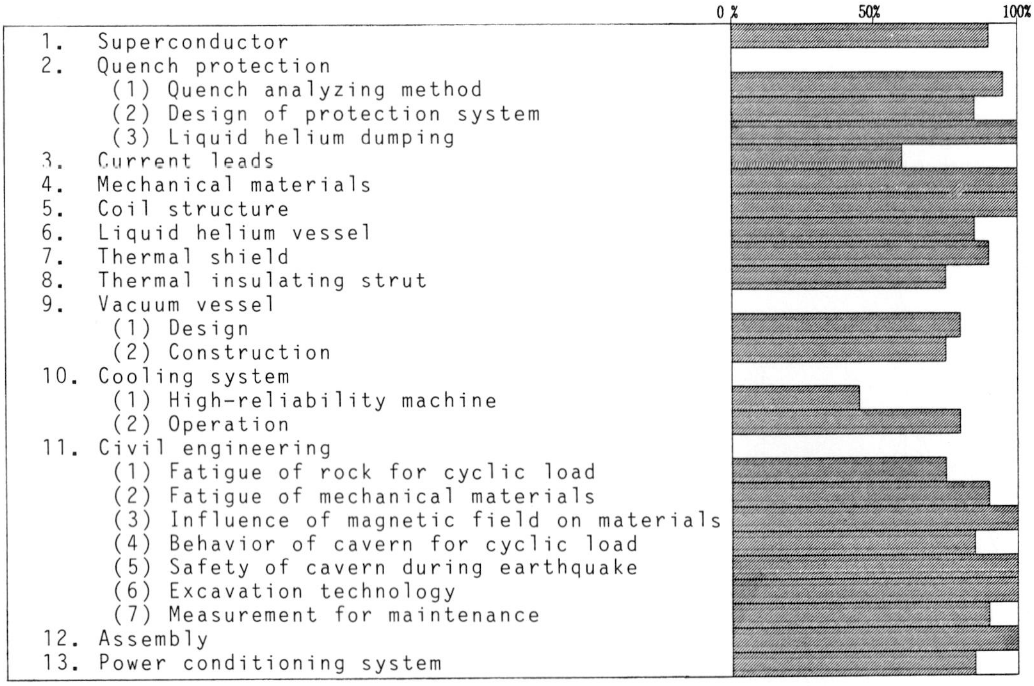

Fig. 4. Improved technological level resulting from the construction of 20MWh SMES compared to the level needed for 5GWh-class SMES.

A Scheme for Development of Superconducting Magnetic Energy Storage System with Function of "Enthalpy Protection"

YUJI KOBAYASHI[1], MISAO JOUSHO[1], MORIHIRO KUBO[2], KATSUHIKO ASANO[2], and YUKIO ISHIGAKI[2]

[1] Tohoku Electric Power Co., Inc., 2-1, Nakayama 7-chome, Aoba-ku, Sendai, 981 Japan
[2] Hitachi Ltd., 1-1, Saiwai-cho 3-chome, Hitachi, Ibaraki, 317 Japan

ABSTRACT

A Superconducting Magnetic Energy Storage (SMES) has many possible applications in an electric utility power system ,such as, power system stabilization, leveling of load fluctuation, substitution for pumped storage power plant etc.
Tohoku Electric Power Company, Inc. and Hitachi Ltd. have started a research of the SMES applications in the electric utility transmission system.
To attain this aim, we have just started the design of the SMES system with the capacity of 1 MJ and a simulated power transmission lines.
In this research, we are going to try a new quench protection system, called "enthalpy protection". If the quench occurs in a part of the magnet, the liquid helium is quickly removed from the magnet , so the entire length of the conductor will go uniformly normal state .

KEY WORDS; SMES, power system, stability, enthalpy protection, superconductivity

INTRODUCTION

For power equipment applied with superconductivity, various introduction effects are expected, such as improvement of equipment efficiency, decrease of size , and reduction of operating costs. Especially , the SMES has such features as high efficiency of power storage , quick response time, simultaneous control for active power(P) and reactive one(Q), etc. The SMES is also expected to achieve wide use according to its scale.

The particular features of the SMES for these experiments are as follows.

1) A concept "Enthalpy Protection " is adopted as quench protection of superconducting coil.
2) A PQ/QI simultaneous control system is adopted .
3) Plural factors for occurrence of disturbance are assumed in the power system stabilization experiment.

OUTLINES

Table 1 gives specifications for major equipments of the 1MJ SMES. Fig. 1 is a diagram of the system. The equipments are composed of superconducting magnet, cryostat, GTO thyristor converter and main control system. The transformer for the converter is also equipped with an AC filter for improving waveform at the secondary side to compose the GTO thyristor bridge. The GTO thyristor bridge outputs a pulse-width-modulation(PWM) wave (550-Hz) according to the control system with α and D . The GTO converter is also equipped with a protection thyristor switch for enthalpy protection at the output side. The system is also equipped with an external protection resister for quench protection and a DC circuit breaker , which are not required in the enthalpy protection system. The superconducting magnet has a rated current of 1000 A and inductance of 2 H, resulting in a storage capacity of 1 MJ. The control system detects active power and reactive one at installed point of the SMES and continuously controls the output of the GTO thyristor converter using signals of the α and D.

SUPERCONDUCTING MAGNET AND CRYOSTAT

Fig. 2 shows the vertical cross section of the superconducting magnet and cryostat. A solenoid-typed superconducting magnet with high energy storage efficiency is used, and 20 layers of double pancake coil are piled up. In each coil interval, spacers are attached vertically and horizontally to maintain cooling channels in order to increase cooling efficiency by liquid He. As a result, the inner diameter of the coil is 360 mm, and the outer dimensions are 814 mm (diameter) by 394 mm (height). The material of the cryostat should be nonmagnetic. To improve reliability for the vacuum seal, SUS304 is used, and the dimensions are 1200 mm (diameter) by 2000 mm (height). The principal features of the superconducting magnet are;

Table 1. Specifications for Principal Equipment in 1 MJ SMES

NO	Equipment name	Item	1 MJ SMES
1	Superconducting coil	Capacity Inductance Rated current	1 MJ 2.0 H 1000 A
2	GTO converter	Capacity Output voltage Output current PWM frequency	25 kW 25 V 1000 A 11 pulses (550 Hz)
3	SMES control unit	control system Various functions	PQ/QI concurrent control Main unit power protection functions Various data processing functions Combined test operation function
4	Simulated electric transmission utility	Generator	20 kVA × 1 unit
5	Others	Quench protection system (Quench detection) Backup	Enthalpy protection (Bridge balance system) Current damping facter di/dt

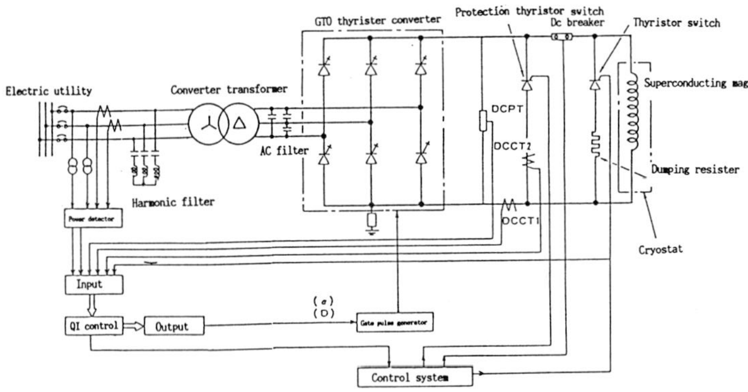

Figure 1 Single-line Diagram for 1 MJ SMES system

1) Adoption of an enthalpy protection system in the superconducting magnet,
2) Design to reduce the AC loss for pulse ,
3) Compact design with high density

The system isolates the coil from the liquid He to raise the temperature of the superconducting magnet uniformly.

Table 2 shows design items of the superconducting wire for reducing AC loss. The superconductor is composed of 14 pieces of element wire approximately 1 mm in diameter. It is composed of a copper block at the center to improve stability on quench. Then the total ratio of superconducting wire to copper is 4.1 to secure stability for pulse use . Each element wire is composed of approximately 10,400 pieces of superconducting material 6 μm in diameter, and the superconducting material is composed of NbTi alloy and copper stabilizing material. As a detection method for coil quench, a bridge balance system is adopted and a detection method using a current damping factor is also adopted as a back up method.

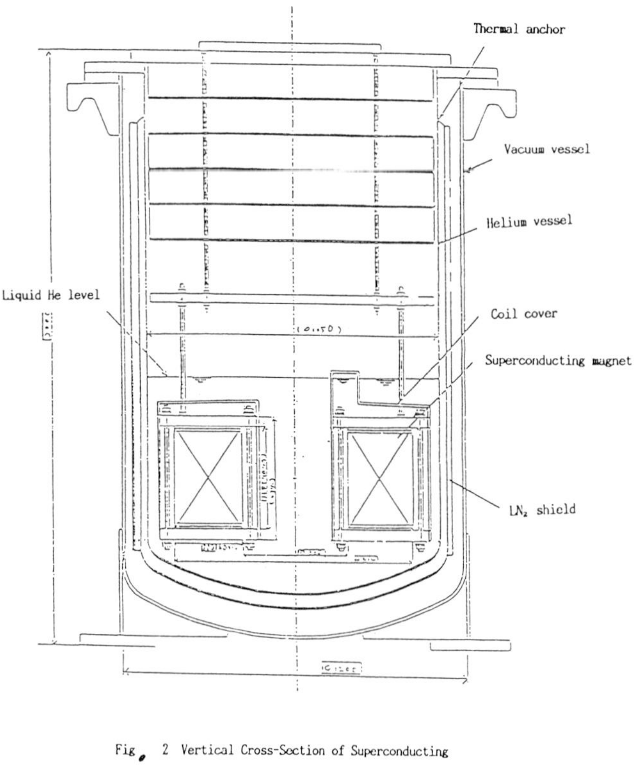

Fig. 2 Vertical Cross-Section of Superconducting Magnet and Cryostat

Table 2 Design Items for Superconducting Wire

	Item		unit	1 MJ SMES			
Superconducting material	Composition			Ti		Nb	
			wt%	47		The remainder	
	Filament diameter		ϕ μm	6.0			
	Number of filaments		pieces	104000			
	Total cross section		mm^2	2.3			
Stabilizing material	Material		-	C1011		C7060	
				(OFC-G)		(Cu-10Ni)	
	Specific resistance at 4.2K		Ωcm	2.2×10^{-8}		-	
	Residual specific resistance		-	80		-	
	Cross-section		mm^2	10.2		1.9	
Complex superconductor	Element wire diameter		ϕ mm	1.0			
	Twist pitch		mm	25			
	Number of strands		pieces	14			
	strands pitch		mm	75			
	Conducter dimensions	T	mm	2.6			
		W	mm	6.4			
	Cu/NbTi (Ratio of copper)		-	4.1			
	Critical current	field	T	5	6	7	7.5
		current	A	4040	3130	2310	1960
	Structure of cross-section			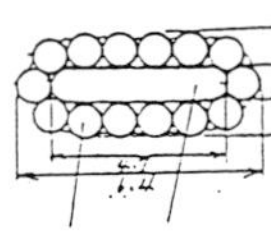 element wire copper			

GTO THYRISTOR CONVERTER AND MAIN CIRCUIT UNIT

The GTO thyristor converter is a 6-phase , single-bridge configuration with a capacity of 25 kW, rated current of 1000 A, and rated voltage of 25 V. The converter is equipped with an AC filter at the secondary side. A device with controllable current of 2000 A and rated voltage of 2500 V is used for the GTO thyristor device to keep reliability for voltage and current.

MAIN CONTROL SYSTEM

The main control system is composed of a microcomputer, AD/DA converter , color monitor and hard disk, etc. This reads the circuit information of the SMES from the main circuit unit and issues references with the α and D to the pulse generation circuit of the GTO converter using high speed arithmetic operation by specified algorithm. The sampling time is 1.6 msec, which can accommodate the high-speed response of the SMES. This unit can handle up to 21 items of trend data for each portion to acquire data of the interior of the SMES.

QI CONTROL PRINCIPLE BY GTO SINGLE BRIDGE

As the SMES is examined as a future device for stabilizing the power system and absorbing load fluctuation, a comprehensive and quick control of active and reactive power is important. To realize this, control of active and reactive power through the GTO converter is fundamental. In this trial manufacture, we examine the system to control direct current concurrently as well as active and reactive power. Since the GTO single-bridge system is, however, a two-variable control system, we adopted a method to decouple the control of active/reactive power and that of reactive power/direct current and to select a control system according to the experimental purposes.

Fig. 3 shows the PQ/QI control principle for SMES through the GTO converter. The control system is composed of an active/reactive power control unit for calculating the α and D and a gate pulse generator. The gate pulse generator is composed of a phase control circuit, pulse-width control circuit, and distribution circuit, and it outputs a gate pulse corresponding to the α and D to the GTO converter.

The α in the gate pulse generator is capable to be ranged from 0° to 360°, and a gate pulse should be output immediately according to a sudden change of the α .

In Fig. 3, the relations between active power P and reactive power Q from the system is expressed as follows;

$$P = Id \cdot Edo \cdot D \cdot \cos\alpha \quad \text{--- (1)}$$
$$Q = Id \cdot Edo \cdot D \cdot \sin\alpha \quad \text{--- (2)}$$

where Id ; DC current
Edo; Maximum no-load DC voltage when $\alpha = 0$
α : Phase-control lag angle
D ; Duty instrumentation signal (average when load is connected with AC power line in one cycle)

We assume that P and Q in formulas (1) and (2) are the reference values Pr and Qr , and that the quotients of P and Q divided by Id and Ed are Po and Qo ,respectively. The α and the D are calculated using the relational expressions that follow.

$$\alpha = \cos^{-1} = \frac{Po}{\sqrt{Po^2+Qo^2}} \qquad (Qo \geqq 0) \cdots\cdots\cdots (3)$$

$$\alpha = 360 - \cos^{-1}\frac{Po}{\sqrt{Po^2+Qo^2}} \qquad (Qo < 0) \cdots\cdots\cdots (4)$$

$$D = \sqrt{Po^2+Qo^2} \qquad \cdots\cdots\cdots (5)$$

If the α calculated from (3) and (4) and the D calculated from (5) are output as pulse-width control reference values to perform phase control and PWM control of GTO converter, P and Q are controlled according to the reference values. Fig. 4 shows the PQ controllable range.

Fig. 5 shows a simulation example for confirming PWM control waveform according to the above principle.

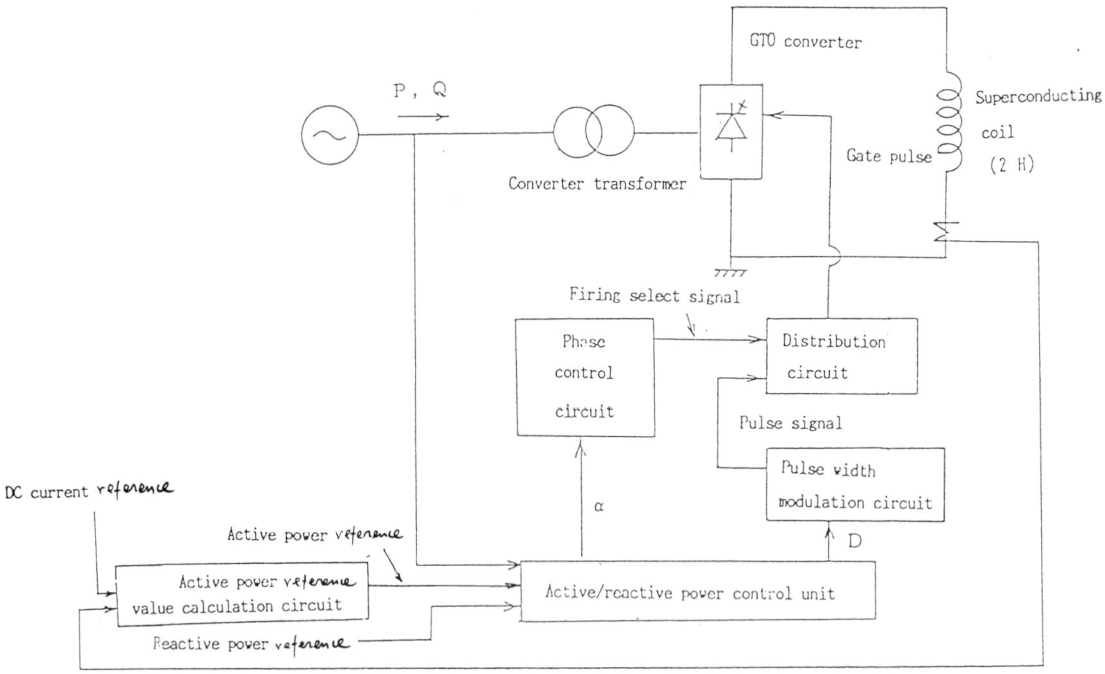

Fig.3 PQ/QI Control Principle through GTO Converter

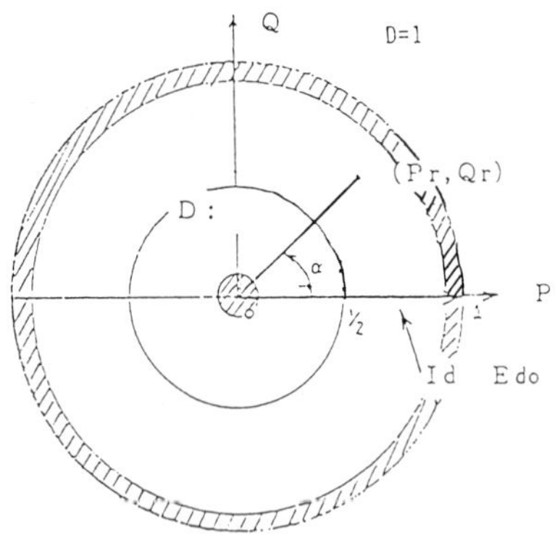

Fig.4 PQ Controllable Range

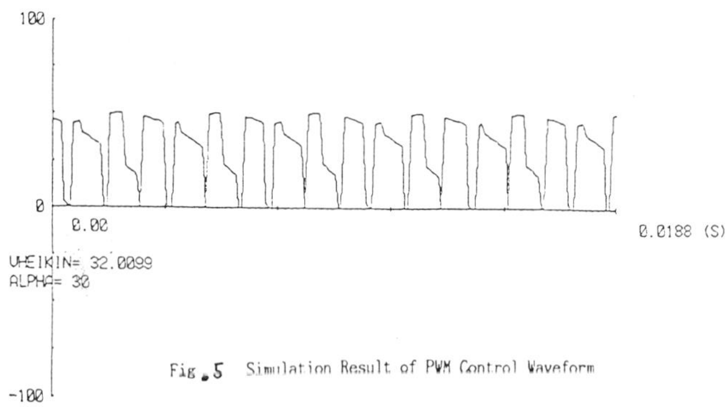

Fig.5 Simulation Result of PWM Control Waveform

COMBINATION TEST OF SMES AND SIMULATED ELECTRIC TRANSMISSION UTILITY

As the simulated electric transmission utility connected with this SMES, a 20-kVA generator simulating a 1000-MVA generator and 440-V line simulating 275-kV transmission lines are designed to simulate 2 lines of 275 kV and 500 km. We perform improvement of system stabilization, countermeasures to voltage fluctuation, leveling of load fluctuation , countermeasures to momentary voltage drop, control of frequency, spinning reserve, etc., independently with this simulated utility. We also perform system control tests regarding multiple factors.

CONCLUSION

We have described the 1 MJ SMES for experiment and the system control test combining with a simulated electric transmission utility. The SMES is currently being examined in various places. We are taking these examinations into account to design a more practical coil structure and control system, and to manufacture the equipment.

Confirmation of SMES Effects for Power System Use

HIDEKI FUJITA

Chubu Electric Power Co., Inc., Electric Power Research & Development Center, 20-1, Kitasekiyama, Ohdaka-cho, Midori-ku, Nagoya, 459 Japan

ABSTRACT

In 1989, Chubu Electric Power Co.,Inc. and Hitachi Ltd. developed a 1 MJ superconducting magnetic energy storage system (SMES) composed of NbTi/CuNi wire for pulsed use (Fig.1). An outline of this system and the results of PQ control experiments were reported at ISS'89.[1] After this symposium, we conducted three series of tests using a simulated power system, confirming the applications as well as several useful effects of SMES. This paper describes these experimental results and confirmed effects.

KEY WORDS: SMES, power system, transmission line, stability, frequency

INTRODUCTION

A power transmission system is a large system composed of huge facilities for supplying the electric energy required by modern society. Owing to the increasing demand for power, needs for a steady power supply and appropriate voltages and frequencies will be enhanced gradually.

For this huge power system, the application of the microphenomenon of superconductivity (in particular, SMES) is considered for energy storage, damping oscillation and supplying pulsed energy which results from zero resistance characteristics.

Concerning the applications of SMES to the power system, its purpose differs according to where it is installed, as shown in Figure 2. Namely, SMES installed near the generator is of advantage to improve the stability of power system, while SMES installed at near the load point stabilizes voltage and active power caused by fluctuating load.

Fig.1 Appearance of SMES

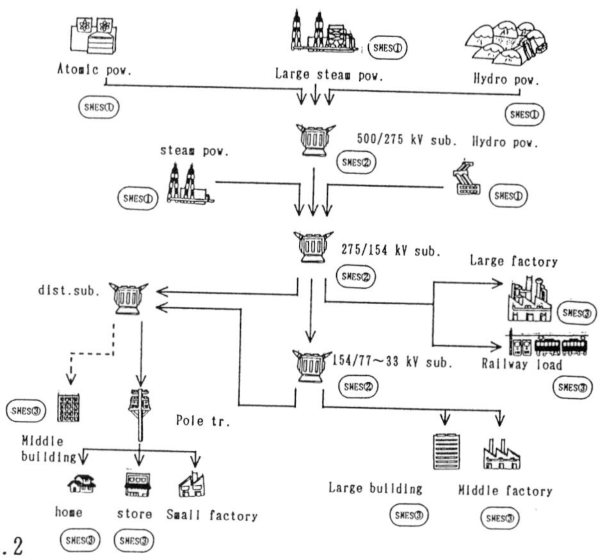

Fig.2 Power system and points of SMES installation.

EXPERIMENTS FOR CONFIRMATION OF SMES EFFECTS

In order to comfirm the effects of SMES, a multifunctional 1MJ-SMES for power system use was manufactured and connected to simulated power systems composed of transmission line coils, four generators and fluctuating loads. Several results of the experiments are presented bellow.

SMES① (installed near power station)

● Improvement of dynamic stability This test aims to stabilize the generator voltage, active output-power occured by slight fault such as 1cct transmission open. A two-generator circuit is used and SMES is set in two configurations as follows. (SMES1# common bus; SMES2# generator bus)

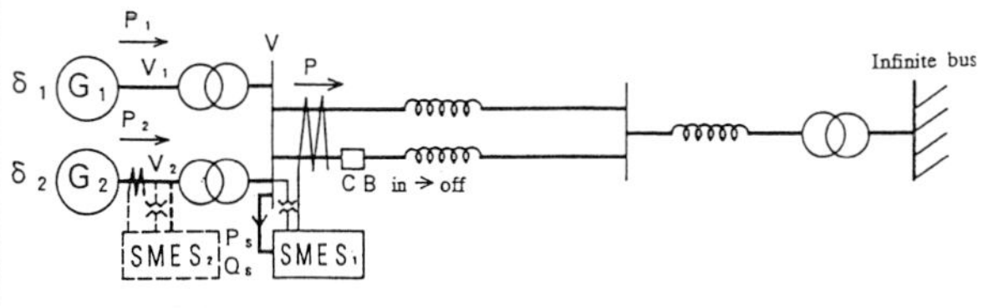

Fig.3 Circuit of experiment (1)

Without SMES1# With SMES1#

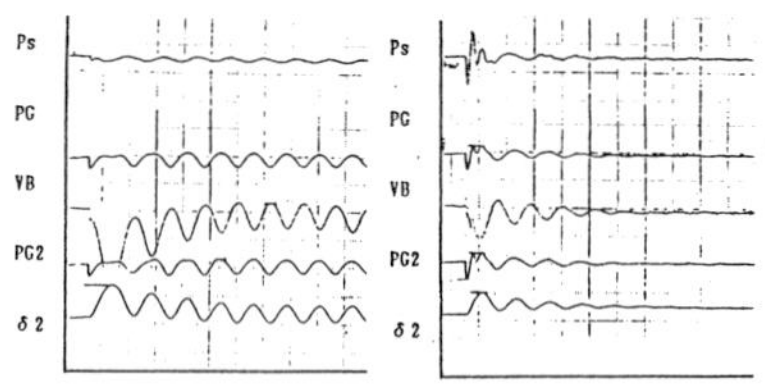

Without SMES2# With SMES2#

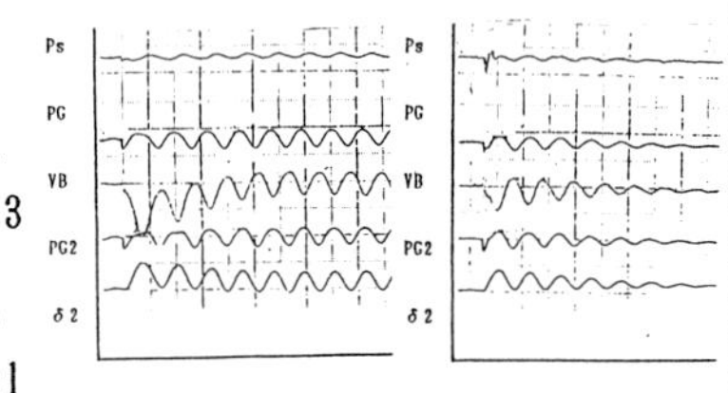

Fig.4 Waveforms#(1)

The results show that SMES can stabilize the oscillation of generators in both cases.

● Improvement of transient stability

This test aims to prevent the generator step-out phenomenon for serious situation such as a 3 phase short circuit near the generator. After the fault rejected, SMES absorbs constant power during a fixed time and applies feedback control while monitoring the actual change in the generatorpower and voltage.

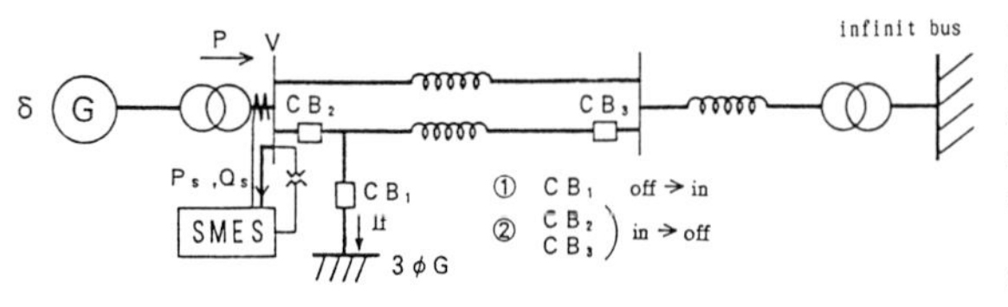

Fig.5 Circuit of experiment (2)

The change of δ shows SMES prevents step-out.

Without SMES With SMES

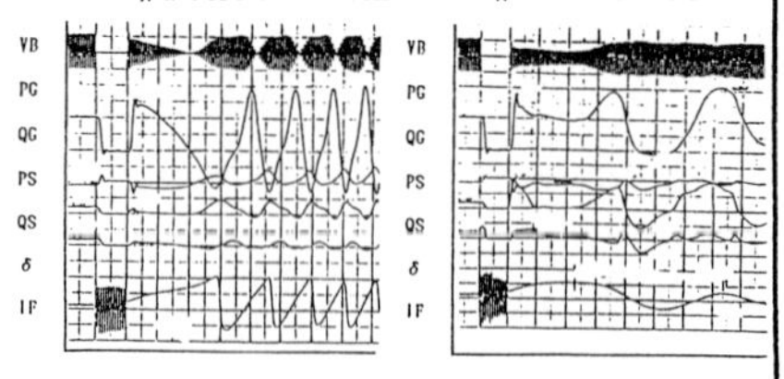

Fig.6 Waveforms#(2)

SMES② (installed near substation)

● Improvement of frequency control This test aims to stabilize the frequency deviation caused by fluctuating load or by cutting off one generator. After detecting the change of frequency

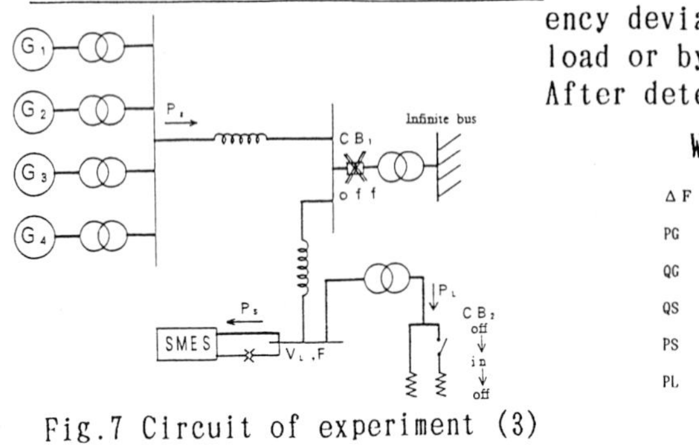

Fig.7 Circuit of experiment (3)

Without SMES With SMES

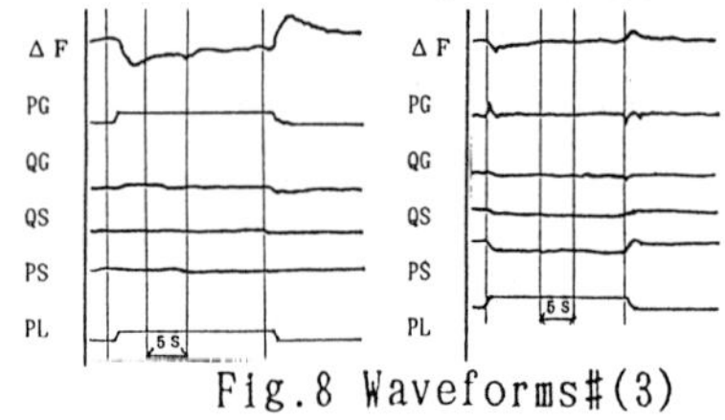

Fig.8 Waveforms#(3)

of the point of SMES installed, SMES controls its active power rapidly. This speed is faster than govenor action of generators. The four-generator circuit is shown in Figure 7. Figure 8 shows the result for the first case; the second case afforded similar results. This experiment demonstrates the SMES effect for frequency control use.

SMES③ (installed near load center)

● Improvement of load change stabilization.

This test aims to stabilize the change of voltage and active power flowing along the transmission line. This test was performed under two conditions: the first one uses a reactor load, and the second uses resistance for fluctuating load. Figure 9 and Figure 10 are for the second case.

As expected, SMES contols the power and stabilizes the change of voltage and power.

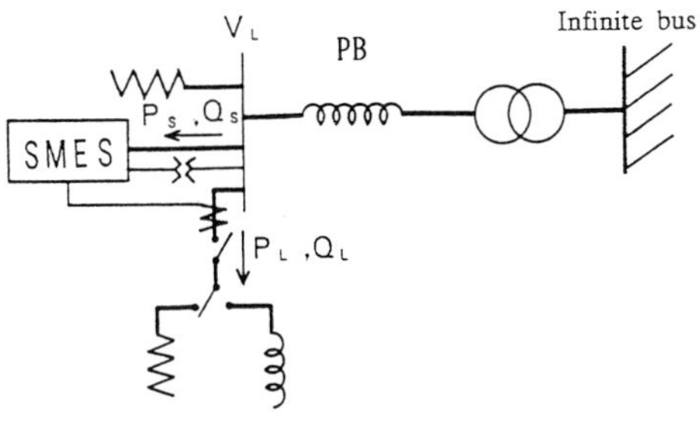

Fig.9 Circuit of experiment (4)

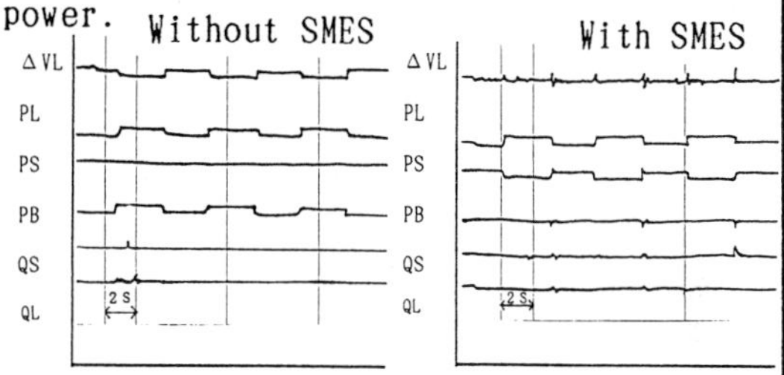

Fig.10 Waveforms#(4)

● Improvement of preventive voltage collapse

This test aims to confirm the prevention of voltage collapse at the end of a long transmission line using resistance load and induction motor load. Figure 11 and Figure 12 show the circuit and the result for the second case. The effectiveness of SMES is confirmed.

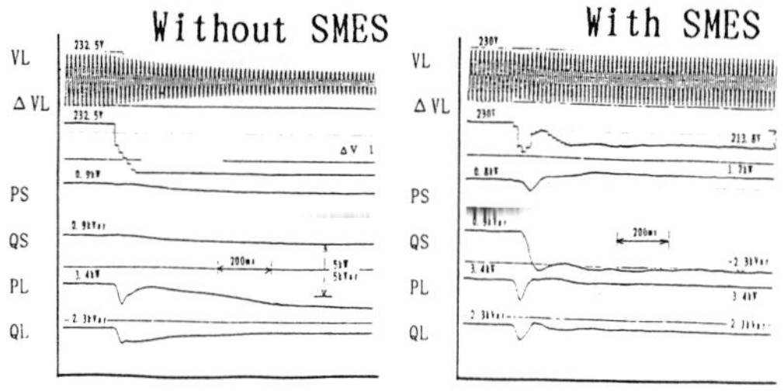

Fig.11 Waveforms#(5)

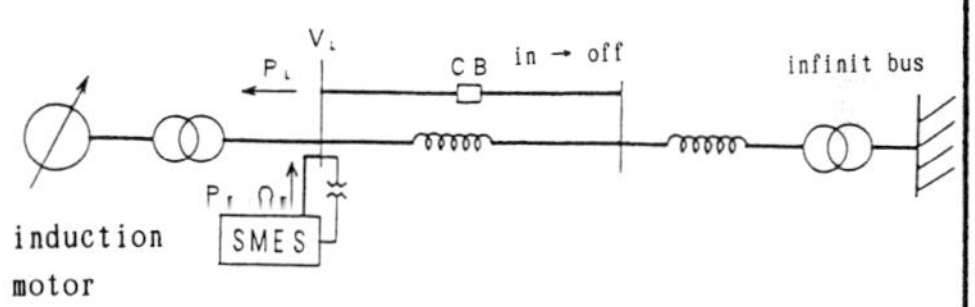

Fig.12 Circuit of experiment (5)

● Effect for a instantaneous voltage drop

This test aims to prevent the drop in voltage caused by a three-phase short circuit at the end of a branching transmission line as shown in Figure 13. Figure 14 shows the effect of SMES. This shows that SMES cannot follow the instantaeous change.

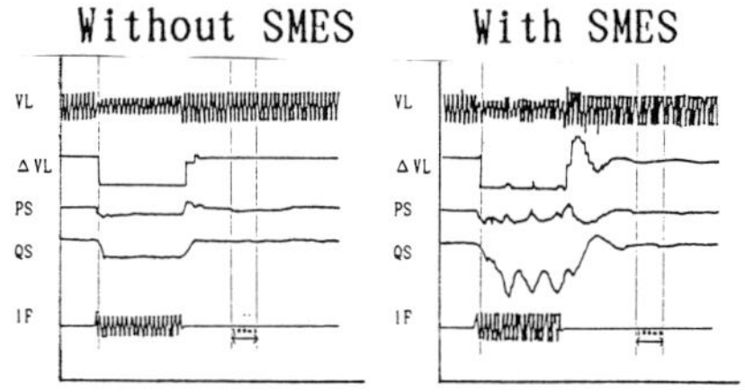

Fig.13 Waveforms#(6)

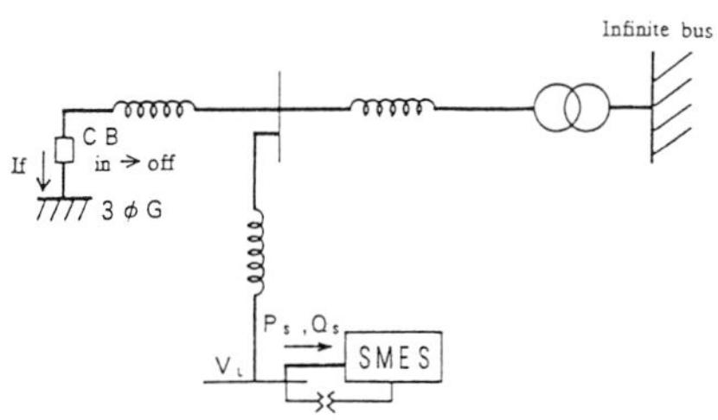

Fig.14 Circuit of experiment (6)

SMES④ (all locations)

○ Improvement of load peak cut This test aims to confirm the basic characteristics for energy storage use. The circuit is shown in Figure 15. Using three steps of fluctuating load, moderate the load curve during one day and the compensatory power generated by SMES show moderate sin-wave functions. Figure 15 shows the power flow along the transmission line whose peak demand is cutted by the generation of SMES. This meens the improvement of load peak cut.

○ Efficiency between charging and discharging This test aims to confirm the basic character for energy storing efficiency by using coil current as the parameter. Figure 16 shows that the efficiency of SMES differes dependin g on the operation condition. The efficiency values are derived from the con dition of constant charging and constant discharging mode. The permanent mod e is included within the two modes. The loss generated at the thyristor conv erter which is related to coil current Ido is not negligible.

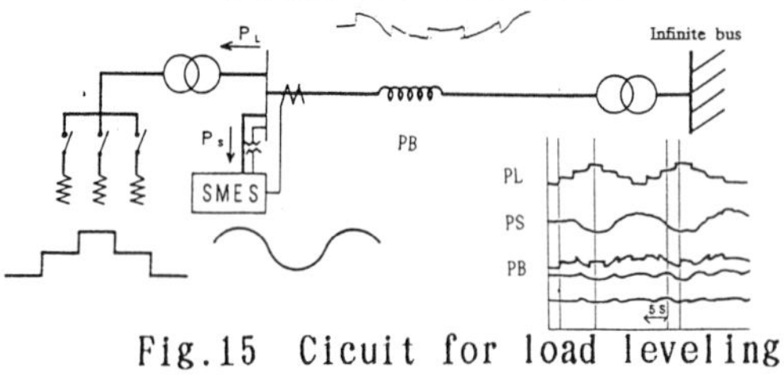

Fig.15 Cicuit for load leveling

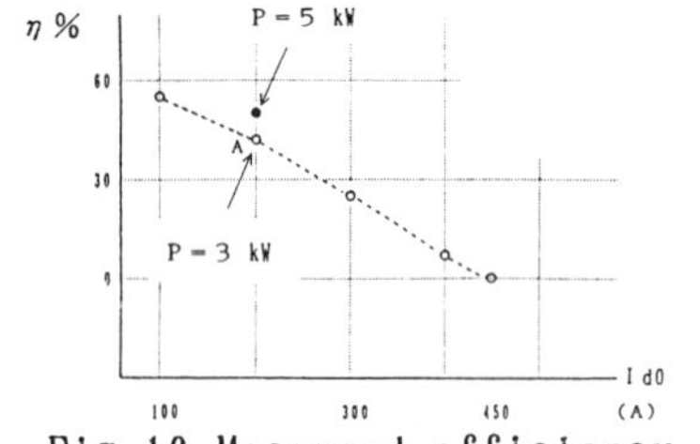

Fig.16 Measured efficiency

CONCLUSION

In conclusion, the confirmed effects of SMES are as follows:
① energy storage in a magnetic field; ② cutoff of peak demand;
③ load leveling effect for fluctuating load; ④ improvement of power system stability; ⑤ frequency control effect in an isolated power system;
⑥ preventive effect for voltage collapse at the end of long transmission line. We confirmed these SMES effects, and they may become very usefull tools for flexbile operation of a power system. However, there are also several disadvantages:

- thyrister loss cannot be neglected, especially for SMES of small size;
- a harmonic filter is necessary, and compensative reactor or capacitance which cancels the reactive power of filter is required;
- the corrective action of SMES is feedback control after detecting a change of voltage or power deviation, so it is not effective against instantanous change.
- the constant values of the control system must be selected properly. If the selection is not appropriate, SMES may became the source of new oscillation.
- a large scale coil is necessary for large amounts of energy storage. But even if the coil can be manufactured, we must systematically consider the design of SMES, including the power converter for scaling up and decreasing thyrister loss, because the GTO-thyrister, a component of the converter, does not have large current or voltage capacity, so the advanced technique of controlling it in parallel or series circuits must be developed.

REFERENCE

1. H.Fujita, Power System Control Experiments Using 1MJ SMES , Proceedings of the 2nd International Symposium on Superconductivity (ISS'89), November 14-17, 1989, Tsukuba, pp.1041-1046.

Economical and Technological Aspect of Small SMES (Superconducting Magnetic Energy Storage)

MASAYOSHI MASUDA

Shonan (Sagami) Institute of Technology, Nishikaigan, Tujidou, Kanagawa, 221 Japan

ABSTRACT

The economical efficiency of small SMES (<10 MWh) for load managing of 30 floors business building by charging cheaper electricity in midnight and to use it in business time, is discussed. The result is suggesting the future commercialization of small SMES for such application.
On the other hand the storage efficiency of small unit is also estimated. Overall efficiency that is considering the losses in keeping cryogenic condition for superconductivity in addition to the loss of energy conversion process, is estimated. It is 75 % for 10 MWh unit. This is not so worth than believed in public.

KEY WORD: Energy storage by superconductivity, economy, storage efficiency

INTRODUCTION

For years, the excellent characteristic features of SMES over than any other energy storages and its good commercialization potential have been demonstrated by studies of many groups. They are University of Wisconsin, LASL, EPRI and recently Bechtel, EBASCO in United states and NEDO(New Energy Development Organization), ENAA(Engineering Advancement Association) and recently RASMES(Research Association of SMES), ISTEC in Japan. The studies, however, have been done for 5,000~10,000 MWh SMES which is able to replace the existing pumped hydro but never done for smaller unit such as 10~20 MWh.
The disadvantages of smaller SMES have been spoken about its poorer efficiency and poorer economy. However the enough knowledges on small SMES have not been available. On the other hand, such small unit is recently spotlighted for applications to an intelligent building and a train business. Thus two issues, the economy and the efficiency of small units, have become the great concern.

STRATEGIC USAGE OF SMALL SMES FOR INTELLIGENT BUILDING

The business buildings which are mainly working in daytime are one of the causes to worse the utility's load condition in a big city. Thus the usage of energy storage for peak shaving of load curve of each building or the business district is a fruitful idea. In addition, the power failure even in short time can be avoided by using SMES. This is an additional benefit of SMES for fully computerized intelligent buildings.
The existing 30 floors business building is picked up as a reference study and the appropriate size of SMES is investigated. Thus the 6.5 MWh,1.4 MW, SMES is considered and the economy is discussed. Table 1 shows the comparison of the cost of electricity and attached equipments between both systems, conventional and SMES equipped. Thus the breakeven capital cost of 6.5 MWh SMES is derived depending on the years of depreciation. It is 1.6 billion yen for 30 years term.
On the other hand the construction cost of 6.5 MWh solenoidal SMES is estimated to be 4.6 billion yen from one of well studied 5,000 MWh unit[1] using a law of $C \propto Es^{2/3}$ where C and Es are the cost and stored energy respectively. The toroidal cost is assumed 2.0 times more expensive than solenoid[2]. Thus the construction cost of 6.5 MWh SMES is estimated as 3 and 6 times more expensive than breakeven for solenoidal and toroidal scheme respectively.
Because of environmental consideration, the toroidal scheme having less leakage magnetic field is favored for densely populated area like business

center. In this sense the economy of small SMES for such purpose is seemed far behind the commercialization front. However the main component of material cost that is the superconducting cable is 45 % in solenoid and 35 % in toroid[3]. Future improvement of productivity of metal superconductor and the use of commercially available High Tc materials enables the great cost reduction of this part.

EFFICIENCIES OF SMALL SMES

The efficiencies of energy storage is defined as follows,

$$\eta = (1- W/Es)100\ \%$$

where W and Es are the losses per cycle and total stored energy respectively. The loss of SMES is divided into two categories, one is of energy conversion and the other is to maintain the cryogenic condition for superconductivity. The latter is specifically associated with SMES technology and is called as the cryogenic loss.
The loss in energy conversion of SMES is small because of simple conversion process between AC and DC. This is performed by a conventional thyristor bridge or a gate triggered thyristor (GTO). In the present paper, only the thyristor bridge is discussed.
The cryogenic loss is generally expressed by a power of refrigerator to liquefy the helium gas evaporated by heat.
The cryogenic losses of full size SMES, 5,000~10,000 MWh, have been investigated in detail[1]. The estimation of cryogenic losses of mall SMES are carried out by scale down of the 5,000 MWh by scaling laws shown as follows.

<u>Conduction through strut</u> : $Es^{2/3}$, <u>Eddy current</u> : Es/t^2

<u>Twist loss</u> : Es/t^2, <u>Magnetic hysteresis in conductor</u> : Es/t

<u>Mechanical hysteresis</u> : Es, <u>Radiation</u> : Es/R (β=constant)

where Es,t and R are the stored energy, decay time and coil radius respectively. The β is the aspect ratio of coil.
Thus the cryogenic losses of various size of SMES are given in table 2.
The persistent switch loss has been ignored in any past studies. The well known magnetic or thermal switch can not be adopted to SMES technology because of low resistance in an open state of such switch. Then only mechanical switch, for instance pneumatically driven by helium gas, is available. The contact resistance of mechanical switch is estimated by separated experiment to be 10^{-6} ohm-cm^2 under the contact pressure of 66 Kg/cm^2. Thus the terms of persistent switch loss in the table are obtained using this data.

DUTY CYCLE AND TERMINAL VOLTAGE OF COIL

In SMES operation, we have many options of duty cycle depending on its application. For example, most typical one is the daily cycle for peak shaving of utility network. The existing pumped hydro charges in midnight during 8~10 hours and discharges in daytime for 8 hours. However for simplification, the charging and discharging time are assumed as given by following equation.

$$T = 3\log Es - 2$$

By this assumption the terminal voltages for nominal operation and peak operation which means the maximum power generation even with 10 % remained energy are obtained.
In old time, the current of superconducting coil was designed by an idea to kept the coil terminal voltage so small that a stored energy could be dumped to the outside resistance as quickly as possible. This led to the low inductance and so the low voltage of coil. However the recent studies of stability of superconducting cables suggest the difficulties of large current diffusion[4]. This favors the high voltage SMES .
The high voltage SMES can also meet the requirement to reduce the cryogenic loss in power lead and thyristor loss of power conditioning unit as shown in later.
The electrical parameters of SMES of various size decided by such consideration are shown in table 3.

LOSSES OF POWER CONDITIONING UNIT

The loss of power conditioning unit is classified into three categories that are of transformer, lead wire and thyristor bridge. The basic equations to estimate such losses are taken from the reference [5].
The reactive losses of convertor transformer of on- and no-load are assumed 0.5 and 0.2 % of nominal power rating. The joule loss of secondary winding of the transformer can not be neglected for such large current DC power source and is assumed 0.2 % of rated power. The loss of lead wire is assumed 0.2 % of rated power. The loss of thyristor bridge depends strongly on the configuration of bridge. The information of high voltage DC transmissions which is connecting Honshu and Hokkaido in Japan, is available to some extent. However the current is too small, 1.2 KA for 300 MW, to compare directly with the SMES.
The number of series connection of thyristor and the forward voltage of thyristor of on-state are assumed 4 and 1 volt respectively.
Thus the losses are given by following equations.

$$W_t(L) = 0.5\,\frac{P}{Es}\,\frac{2t}{3} = \frac{2}{3}\,0.5\ (\%)$$

$$W_t(NL) = 0.2\,\frac{P}{Es}\,(24-2t) = \frac{24-2t}{t}\,0.2\ (\%)$$

$$W_{LW} = \frac{5}{3}\,0.2\,\frac{P}{Es}\,t = \frac{5}{3}\,0.2\ (\%)$$

$$W_{thy} = 8\,\frac{I}{Es}\,t = 8\,\frac{I}{P} = \frac{8}{V} \times 100\ (\%)$$

where $W_t(L)$, $W_t(NL)$, W_{LW} and W_{thy} are the transformer loss of on-load , no-load, lead wire loss and thyristor loss respectively.
From the calculation, it is clear that the main losses in power conditioning unit are independent from the rated power and duty cycle except the no-load transformer loss.
They are summarized in table 3.

EFFICIENCY OF SMES

The efficiency of SMES is estimated by the total loss which is the sum of cryogenic and power conditioning unit loss.
Fig. 1 shows the SMES efficiencies of various size . The dotted line is one of using detachable transformer which can reduce off-load transformer loss. The efficiency of full size unit, 5,000 to 10,000 MWh which can be competing with the existing pumped hydro, is better than 90 % . However the efficiency is coming worse depending on decreasing the size of SMES. It is 75 % and 90 % for 10 and 100 MWh respectively.

CONCLUSION

The economical aspect of small SMES has been discussed for application to the intelligent building . The result shows the economical efficiency of such application of SMES is far from the economical breakeven point. However the possibility of such SMES is thought by the improvement of productivity of metal superconductors or the commercialization of high Tc superconductor.
On the other hand, the storage efficiency of small SMES is not so worth than expected. It is 75 % for 10 MWh unit and it becomes better than 90 % for bigger than 100 MWh unit.

REFERENCES

1. NEDO, NEDO-P-8408 (1984)
2. Wisconsin Superconductive Energy Storage Project, 1:(1974)
3. S.Shoenung and W.Hassenzahl, presented in ASC-90, Snowmass (1990)
4. M.A.Hilal et al (1983) Proc. 9th Symp. on Fusion Technology
5. M.Masuda and T.Shintomi (1977) Cryogenics, 17: 607

Table 1. Comparison of Cost of Electricity Between Conventional and SMES System for Business Building

	conventional	6.5 MWh,1.4MW(SMES)
basic cost(monthly,typical)		
daytime	11.808 Myen(704.5MWh)	9.537 Myen(569.0MWh)
night deduction	1.114 Myen(66.5MWh)	0 (0 MWh)
night	1.200 Myen(204.5MWh)	2.385 Myen(406.5MWh)
sub-total	14.122 Myen(975.5MWh)	11.923 Myen(975.5MWh)
electricity cost		
daytime	10.571 Myen(694.6MWh)	7.955 Myen(522.0MWh)
night deduction	0.984 Myen(58.6MWh)	0 (0 MWh)
night	0.847 Myen(142.6MWh)	2.215 Myen(372.9MWh)
sub-total	12.312 Myen(894.9MWh)	10.170 Myen(894.9MWh)
annual cost	207.702 Myen	156.387 Myen
diesel generator	20.000 Myen	0
CVCF	22.000 Myen	0
SMES	0	1.4(MW)xA Myen
annual total cost	207.702xB+42.000 Myen	156.387xB+1.4xA Myen

A:unit cost of SMES, B:years of depreciation

BECC of SMES : 1.58 billion yen for 30 years depreciation

Table 2. Cryogenic Losses of SMES of Various Size in Refrigerator Load(MW)

Size(MWh)	5,000	100	10
Conduction	2.99	0.22	0.05
Radiation	0.73	0.02	0.002
Hysteresis (conductor)	0.50	0.023	0.01
Hysteresis (mechanical)	0.31	0.006	0.000
Misc.	0.91	0.053	0.012
Power Lead	0.76 (252KA)	0.02 (11.3KA)	0.008 (4.5KA)
Persist.SW.	0.8	0.01	0.001
Total	7.0	0.35	0.083
Daily Loss (MWh)	168.0	8.40	1.98
% of Es	3.4	8.4	19.8

Table 3. Electrical Parameters and Losses of Power Conditioning Unit in % of Stored Energy

Size(MWh)	5,000	100	10
Cycle time(hr)	9	4	1
Output(MW)	555	25	10
Terminal Voltage (KV)	2.2	2.2	2.2
Max. Voltage(KV)	7.0	7.0	7.0
Max.Current(KA)	252	11.3	4.5
Loss of Transf.			
On-load(%)	0.33	0.33	0.33
Off-load(%)	0.00	0.80	4.4
Lead Wire Loss(%)	0.33	0.33	0.33
Thyristor Loss(%)	0.36	0.36	0.36
Total Loss	1.02	1.82	5.42

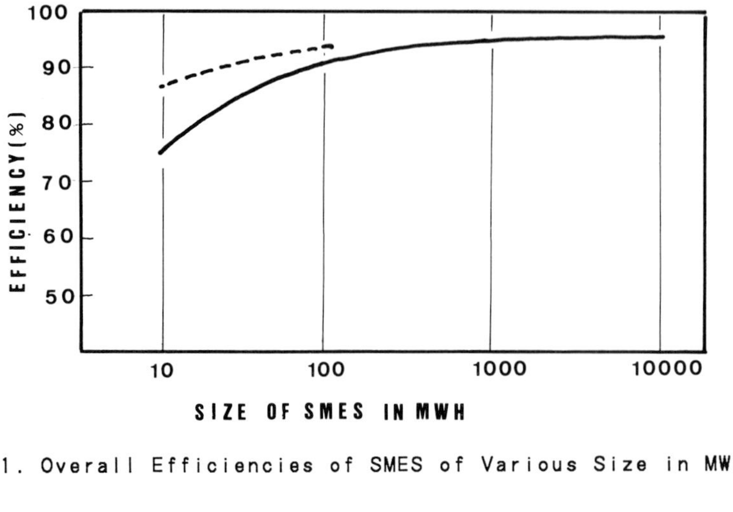

Fig.1. Overall Efficiencies of SMES of Various Size in MWh

New Concept of a Superconducting Fault Current Limiter and Its Evaluation of Quenching Feature

KIYOSHI OKANIWA[1], TSUKUSHI HARA[1], TAKAHIKO YAMAMOTO[1], and HARUHIKO HOSHINO[2]

[1] Engineering Research Center, Tokyo Electric Power Co., Tokyo, 182 Japan
[2] Power and Substation Department, Kandenko, Tokyo, 108 Japan

ABSTRACT

The present paper reviews the concept of a new superconducting fault current limiter with NbTi AC superconducting wire for the suppression of short-circuit currents. This fault current limiter takes full advantage of the quenching feature in the superconductor. On the assumption that the fault current limiter is installed at a distribution substation, parameters to be evaluated in relation to the fault current limiting characteristics of the coils were clarified. Experimental tests were conducted to evaluate 400 V - 100 A class superconducting coils. The results indicated that the superconducting coils exhibit stable characteristics, evidenced by the findings that the quench current remained constant regardless of the load current and short-circuit phase and that the coils did not degrade after repeated operations.

KEY WORDS: AC superconducting wire, fault current limiter, quenching feature, electric power system

INTRODUCTION

Electric power systems are being enhanced and becoming more complex with the increase of power demand, and the occurrence of short-circuit currents in electric power systems is accordingly also on an upward trend. It is therefore a task of critical importance to suppress such fault currents. In view of this situation, we have proposed an approach whereby a fault current limiter which takes full advantage of the phenomenon known as quenching in the superconductor is installed on the secondary bus tie at the distribution substation to suppress short-circuit currents [1]. This fault current limiter employs NbTi AC superconducting coils [2]. Hence, it is important to determine whether the coils meet various requirements and restrictions associated with their use in a fault current limiter, as well as to evaluate the fundamental characteristics of the coils, including AC losses [3].

The present paper first reviews the concept of a new superconducting fault current limiter. Then, taking into consideration the evaluation parameters for coil characteristics required for a fault current limiter, a description is given of tests conducted using 400 V - 100 A class superconducting coils to identify the quenching characteristics of the superconducting coils induced by short-circuit currents.

NEW SUPERCONDUCTING FAULT CURRENT LIMITER AND PATTERN OF SYSTEM APPLICATION

Figures 1 and 2 show a view of a 400 V - 100 A class superconducting coil used in the present test and the principles of operation of the fault current limiter, respectively. As illustrated in the model power system of Fig. 3(a), it was assumed that the fault current limiter would be installed between 6 kV buses [4]. Under these conditions, when a fault occurs in the distribution system, the trigger coil (TC) quenches at current I_q, so that the current transfers to the limiting coil (LC), thereby permitting the fault current to be suppressed, as shown in Fig. 3(b). The use of such fault current limiters allows the bus ties to be used concurrently at all times, thus substantially improving system reliability.

Because the rated current I_o of the fault current limiter must be equal to the rating of the load switch in the bus tie, the following relationship exists:

$$I_o = 2 \times \sqrt{2}\,\text{kA} \qquad (1)$$

The lower limit of I_q is I_o, and its upper limit is imposed as follows because the current flowing in the circuit breaker and transformer shown in Fig. 3(a) must be limited to 20 kA:

$$2 \times \sqrt{2} < I_q < 20 \times \sqrt{2} \times 2/3\text{kA} \qquad (2)$$

Assuming that $K1 = I_q/I_o$, then

$$1.0 < K1 < 6.7 \qquad (3)$$

Thus, K1 is confined within a narrow range. The upper limit of the short-circuit current I_s, theoretically assumed in the absence of any fault current limiter, is determined as follows by the current flowing in the circuit breaker obtained by the calculation of short-circuit capacity:

$$I_q < I_s < 30.2 \times \sqrt{2} \times 2/3\,\text{kA} \qquad (4)$$

Now, assuming that $K2 = I_s/I_q$, then

$$1.0 < K2 < 10.1 \qquad (5)$$

Fig. 1 400 V - 100 A Class Superconducting Coil

LC: Limiting coil (superconductor)
TC: Trigger coil (superconductor)
(Normally, the load current flows in the trigger coil)

$L_{LC} >> L_{TC} \doteqdot 0, I_{C\text{-}LC} > I_{C\text{-}TC}$

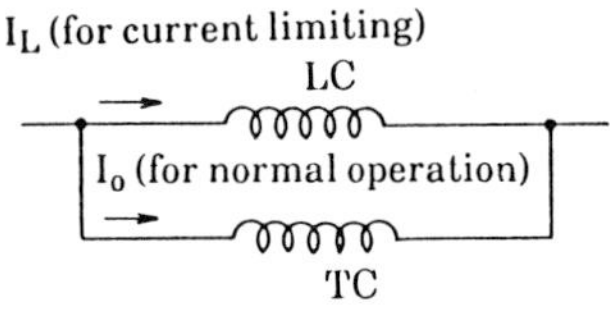

Fig. 2 Principles of Operation of Superconducting Fault Current Limiter

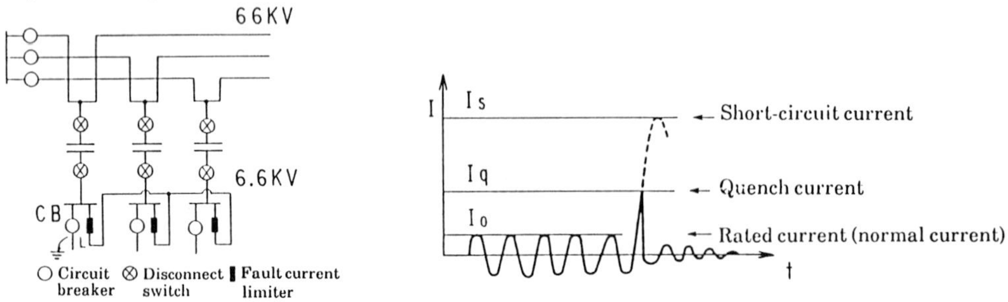

(a) Model System

(b) Current Waveform at a Fault

Fig. 3 Example of Application of Superconducting Fault Current Limiter

EXPERIMENTAL

Taking the above-mentioned restrictions and mode of operation into account for practical application of the system, an experimental evaluation was conducted using a 400 V - 100 A class superconducting coil.

Figure 4 shows the test circuit. On the basis that the quench current I_q in each of the two coils connected in parallel was 60 A (specified value), values of resistance Z, RL, etc. were determined so as to yield the conditions denoted in (3) and (5) above, and an experimental evaluation was conducted on the following parameters:

① Variations in the value of I_q with the magnitudes of I_0 and/or I_s.
② Variations in the value of I_q due to repeated operations and long-term use in normal operation.
③ Variations in the value of I_q and the quench propagation characteristics due to the phases of a short circuit.
④ Quench propagation characteristics in the trigger coil with reference to the resistance developed.

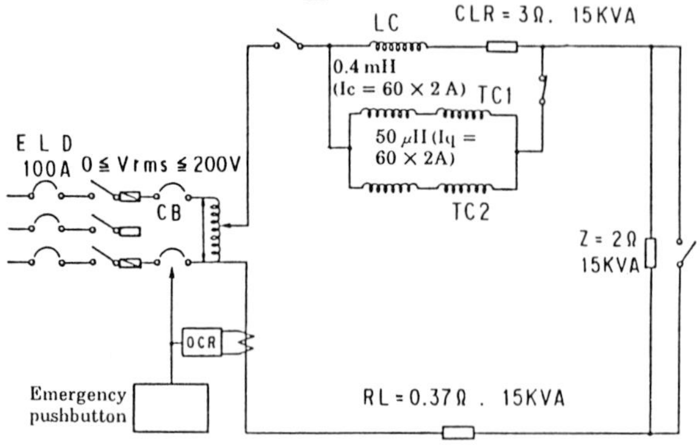

Fig. 4 Test Circuit

RESULTS AND ANALYSIS

Figure 5 shows an example of the measured waveforms of the trigger coil and limiting coil. The current flowing in the trigger coil increased rapidly upon the occurrence of a short circuit, with I_0 = 36.7 A. It was verified that the trigger coil was quenched at 59.1 A, and that the current was transferred to the limiting coil 2 msec after the short circuit, thereby achieving a state where the short-circuit current was suppressed.

Evaluation Parameter ①

Tests were conducted under conditions of $1.1 \leq K1 \leq 4.3$, $1.3 \leq K2 \leq 4.9$, which are more severe than those of (3) and (5) above, to evaluate the variation of the values of I_q. The results are shown in Fig. 6.

The value of I_q increased with increasing values of I_o, showing a tendency toward stabilization at around the specified value. However, values of I_q still lay within a range of 60±6 A, suggesting that the fault current limiter can be sufficiently depended upon for practical applications.

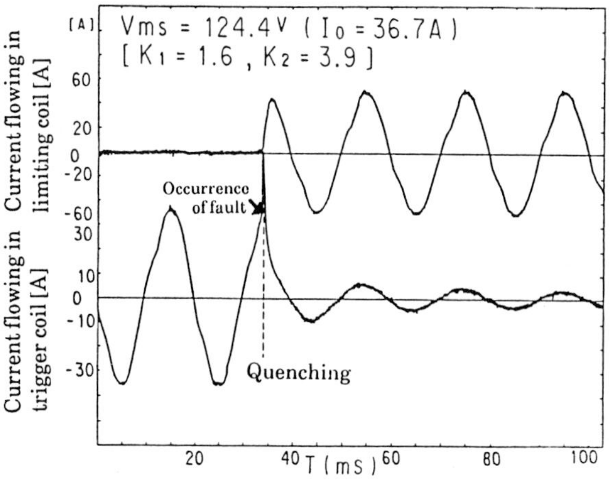

Fig. 5 Example of Measured Waveforms

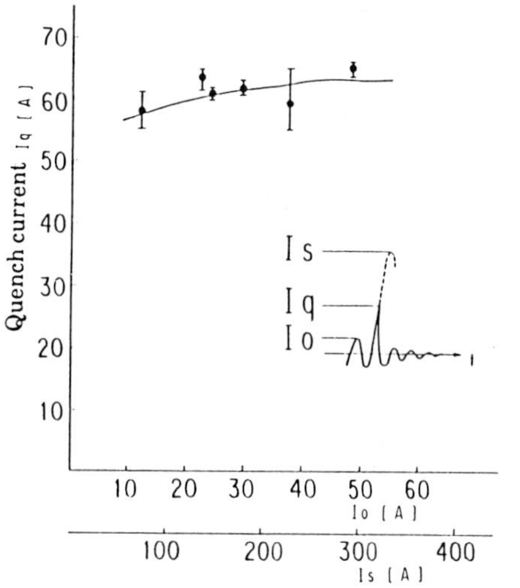

Fig. 6 Quench Current of Trigger Coil

Evaluation parameter ②

When 30 consecutive quenching cycles were conducted at intervals of 30 sec to 1 min with K1 = 1.6 and K2 = 3.9, the following results with respect to the specified quench current value of 60 A were obtained, which demonstrates that virtually constant values of I_q were obtained after repeated operations.

Trigger coil 1: mean value 57.3 A (maximum value 63.5 A; minimum value 54.0 A)

Trigger coil 2: mean value 59.1 A (maximum value 65.0 A, minimum value 55.0 A)

In the case of K1 = 1.4 and K2 = 4.7, the trigger coils were quenched after a maximum of 3 minute normal operation. The results indicate that values of I_q remained constant within a range of ±5 A.

Evaluation parameter ③

Figure 7 shows the quench current I_q as a function of the phase of a short circuit, with K1 = 2.0 and K2 = 3.2, and Fig. 8 indicates the results of investigations for the quench propagation characteristics with reference to the resistance developed, with K1 = 1.3 and K2 = 4.9.

It was found that the quench current was maintained at virtually constant levels regardless of the short-circuit phase and that the resistance developed across the trigger coil after quenching was greater at a phase of 90° than at 45°. This finding may be explained by the supposition that the magnitude of di/dt is greater at a phase of 90°.

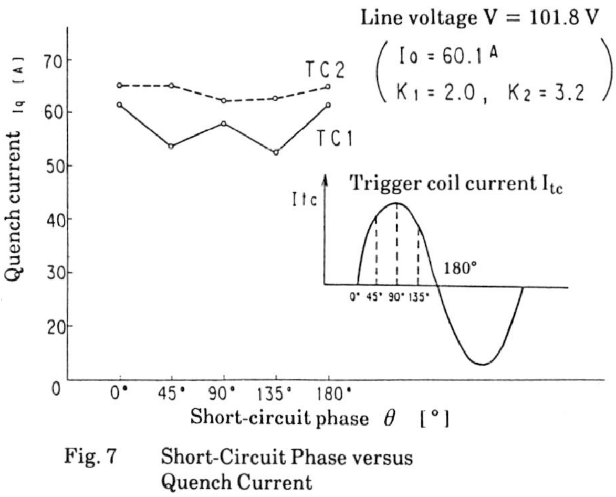

Fig. 7 Short-Circuit Phase versus Quench Current

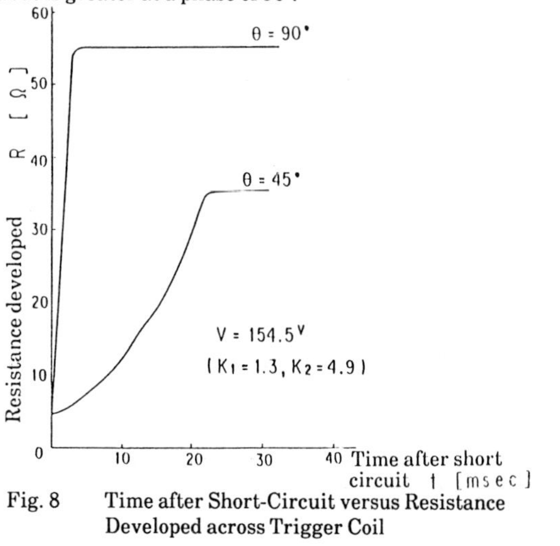

Fig. 8 Time after Short-Circuit versus Resistance Developed across Trigger Coil

Evaluation parameter ④

Figure 9 shows changes in the resistance developed across the trigger coil over time, according to various voltage levels. The range of voltage applied in the present test was extended to a maximum of 154.5 Vrms ($1.1 \leq K1 \leq 4.3$, $1.3 \leq K2 \leq 4.9$). It is clear from this chart that the resistance developed is higher at higher levels of applied voltage and changes more substantially over time. The maximum value of resistance developed was 55 ohms, which is less than 10% of the total resistance of 610 ohms for the coil at 77 K. Consequently, further studies are warranted to determine whether the coil length can be further reduced for more compact design. It is supposed that the resistance developed will be further increased by enhancing the voltage to the rated level.
Figure 10 shows the rate of change in the quench current dI_q/d_t versus quench length. The quench length of the coil increased with an increasing rate of change in the current, with attendant increases in the developed resistance. This finding is presumably due to the phenomenon that AC losses increase with an increasing rate of change in a current, which, in turn, affects quenching.

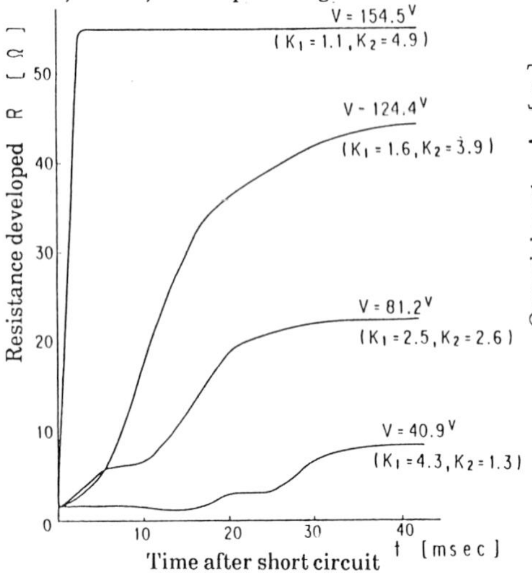

Fig. 9 Resistance Developed across Trigger Coil

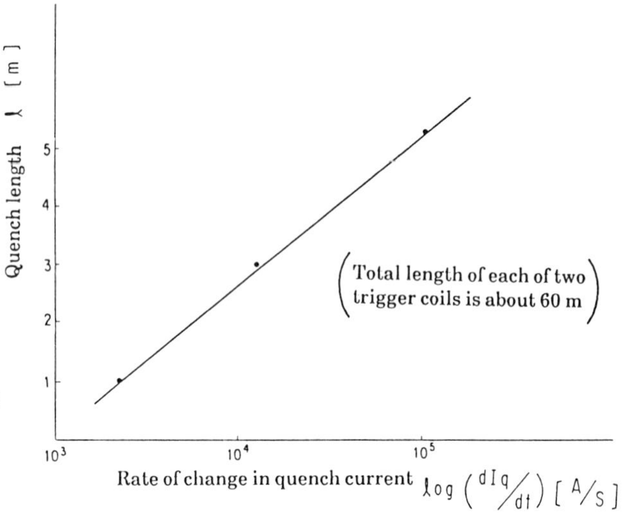

Fig. 10 Rate of Change in Quench Current (after Short Circuit) versus Quench Length

SUMMARY

Using 400 V - 100 A class superconducting coils for a fault current limiter, tests were conducted to evaluate fault current limiting characteristics under simulated conditions that might occur in an actual electric power system. The results of the tests indicated that the fault current limiter did not pose any significant problem in terms of evaluation parameters, such as quench currents supervening on changes in the load current before a short-circuit and in the short-circuit phase and reliability relative to repeated operations, thus exhibiting stable characteristics. The present experimental evaluation demonstrated that a fault current limiter based on the utilization of the quench phenomenon in a superconductor sufficiently fulfills the requirements for use in actual electric power systems with various restrictions, so that it holds promise for future introduction into practical applications.

In the next stage, we plan to undertake the development of an AC superconducting wire with large current capacity and minimal loss in an effort to research and develop a more efficient superconducting fault current limiter. A further evaluation of the quench feature will be conducted using the developed coil. Furthermore, the development of an oxide superconducting power lead and a high-efficiency refrigerating system is required to implement a more compact system design.

ACKNOWLEDGEMENTS

We wish to thank Dr. D. Ito and staff at Toshiba Corporation who fabricated experimental superconducting coils.

REFERENCES

(1) Okaniwa, et al.: IEEJ General Convention, No. 893 (1990-3)
(2) Yamamoto, et al.: IEEJ General Convention, No. 894 (1990-3)
(3) Kasahara, et al.: Central Research Inst. of Electric Power Industry, Report T88075 (1989-5)
(4) Hara, et al.: IEEJ General Convention on Electric Power and Energy, No. 241 (1990-7)

Study of Vertical Transportation System Using Superconducting Liner Motor

H. NAGANO[1], M. KINUGASA[2], T. TOKIZAWA[2], K. HAYAKAWA[3], and K. UESHIRO[3]

[1] University of Toyama, Toyama, 930 Japan
[2] Power Reactor & Nuclear Fuel Development Corporation, Tokyo, Japan
[3] Mitsubishi Heavy Industries Co., Ltd., Kobe, 652 Japan

ABSTRACT

With respect to the transportation into the site deep formation, adaptability of a shaft system is under research and development by means of superconducting linear motor.

In 1988, as basic study (I), a primary experimental equipment of a linear motor was built and an experiment was held in order to verify the principle performing statical floatation test. (ISS.'89 presented before)

In 1990, as basic study (II), an experimental equipment is newly assembled with a three-phase synchronous linear motor and a dynamic experiment is carried out using it.

KEY WORDS : Vertical Transportation System, Superconducting Magnet, Armature Magnet, Three-phase Synchronous motor.

INTRODUCTION

A basic dynamic experiment is performed using the experimental equipment of a vertical shaft which is combined with a three-phase superconducting synchronous linear motor system.

As the result, it is achieved the capsule movement as 2 m/sec up/down speed, 25 kgf (mean) thrust, and ±1mm at stopping accuracy. And also, it is established for the inherent control algorithm of vertical linear motor, the thrust ripple reduction technics, etc.

SPECIFICATION OF EXPERIMENTAL EQUIPMENT

The sketch drawing and the specification of the experimental equipment is shown as Fig.-1 and Table-1.

Mainly, the equipment consists of transfer capsule (cryostat), superconducting magnet, normalconducting armature magnet, power source unit, position detector (control unit), and support structure.

(1) Transfer capsule (cryostat)

The cryostat has a superconducting magnet in its container which is used to cool the magnet using liquid helium. It is double casing type and consists of the vacuum tank outside and helium tank inside. The vacuum tank has the role of heat insulation from the surroundings into the helium tank where the superconducting magnet is placed in immersion cooling condition.

(2) Superconducting magnet

The wire is made of NbTi/Cu and about 5.5 kg weight. The exciting current is 36.4 (A) which induces a magnetomotive force 72,800 (AT) and the central magnetic field intensity 1,200 (Gauss), and also the current density is 100 (A/mm^2). It has a permanent current switch too.

(3) Normalconducting armature magnet

The magnet is made of Cu. The exciting current is AC 30 (A) which corresponds to magnetomotive force 5,040 (AT) and central magnetic field intensity 200 (Gauss). The magnet winding is Star-connection arranged with 120 mm of pitch for the three-phase synchronous linear motor. And the winding is inserted with copper sheet between the each layer of winding which are able to contact to water cooling pipe.

(4) Power source unit

The power source unit consists of three servodrivers. It is able to supply the power synchronized to the capsule position signals which are given from the position detecting system. Additionally, the output current is always controlled to induce a constant thrust force in any position of the capsule.

(5) Position detector

To drive the capsule correctly using a synchronized current by means of the position detector, it is very important in the synchronized linear motorization.
Thus, the requirement is determined as high value as;

- Position detection accuracy > 0.36 mm, within 4 m scope of detection
- Speed of response > 2 m/sec

At the same time, the magnetic-proof is also required in the system.

In view of the above matters, an optical rotary encoder is applied as the position detector. The position detecting system is able to be converted from linear to rotary by use of belt and gear, corresponding to any position of the capsule which is mobile in suspension.

(6) Support

The support is able to accommodate for thirty stacks of the normalconducting amarture magnet sufficiently and has a suspension system on the top.
For the material, it is applied to use SUS304 to prevent from disturbance to the magnetic field of the linear motor.

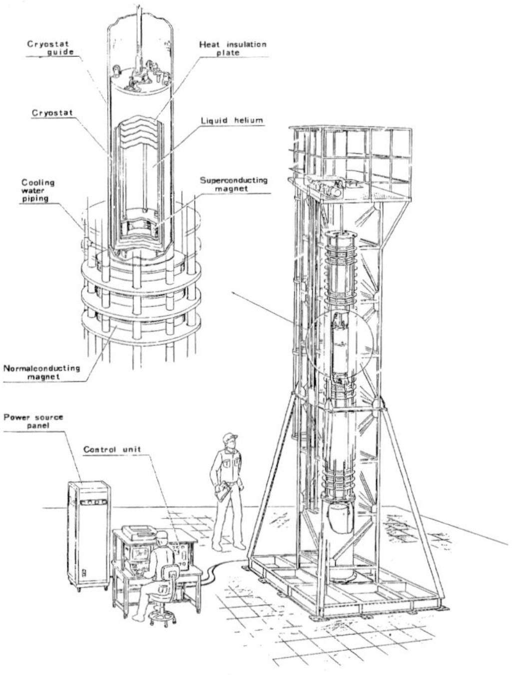

Fig.-1 Sketch drawing of Experimental Equipment

Table-1 Specification of Experimental Equipment

No.	Item	Specification
1.	Capsule (cryostat)	Material : FRP Outline : Outside Dia. 250 mm Inside Dia. 200 mm Hight 660 mm Depth 600 mm LHe quantity : 10 ℓ Weight : 18 kg
2.	Superconducting magnet	Material : NbTi/Cu Outline : Outside Dia. 180 mm Inside Dia. 154 mm Hight 56 mm Current : 36.4 A/72,800 AT Central field intensity : 1,200 Gauss Mean current density : 100 A/mm^2 Weight : 5.5 kg Equipped with permanent current switch
3.	Normalconducting armature magnet	Material : Cu Outline : Outside Dia. 370 mm Inside Dia. 280 mm Hight 40 mm Current : 36 A/5,040 AT Central field intensity : 200 Gauss Weight : 15 kg Equipped with water cooling pipe
4.	Power source unit	Configuration : Control unit 1 set; Power supply to 3 servodrivers for normalconducting armature magnet, and Sequential control for power supply Outline : L570 × W570 × H1,500 mm
		Servodriver Max. output current : 36 A Control type : PWM (Pulse Width Modulation) Brake type : Regenerative control Protection : OC, OV, LV, Overheating, & Primary voltage abnormal Outline : L370 × W223 × H240 mm
5.	Position detector	Position accuracy : ±0.1 mm Scope of detection: about 4,800 mm Speed of response : max. 11.6 m/sec Encorder output : Gray code (absolute)

TEST RESULT

To verify the total performance of the vertical linear motor system combined with the whole components, an integrated experiment is intensively carried out.

(1) Thrust

Thrust test is performed to measure the thrust for the system. The test frame is defined as the superconducting magnet at permanent current mode in the system to measure the thrust resultly induced, using a load cell. The test result is shown in Table-2 and Fig.-2

Table-2 Thrust

	Sine wave	Corrected wave
Relative distance	Thrust f (kg·f)	
0	21.6	23.5
20	26.4	23.7
40	34.0	24.3
60	36.2	26.0
80	31.5	26.7
100	23.9	25.4
120	20.9	23.5
150	29.3	23.5
180	35.5	25.1
MAX - MIN	15.3	3.2

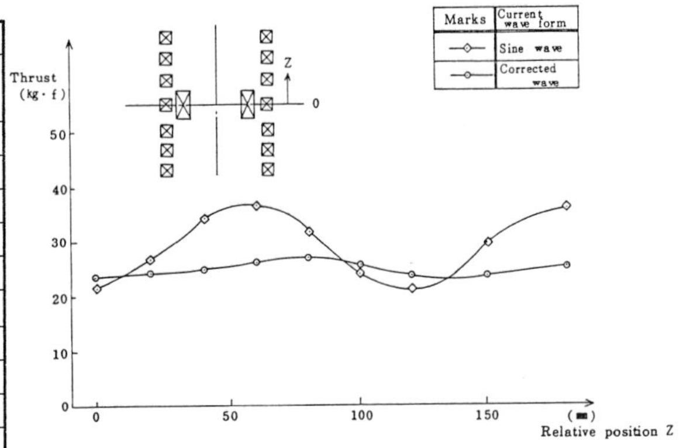

Fig.-2 Thrust

Where; "Corrected wave" means a thrust wave form functionally corrected from the sine wave current which is given by the normalconducting magnet, to make it as less ripple as possible corresponding to any position of the capsule.

(2) Position control

For the position control, the steady-state deviation and the maximum speed are measured under the function*1 given with 2 m elevation height as target position.
(*1 See Fig.-3 "Block diagram of position control")
The test result is shown in Fig.-4.

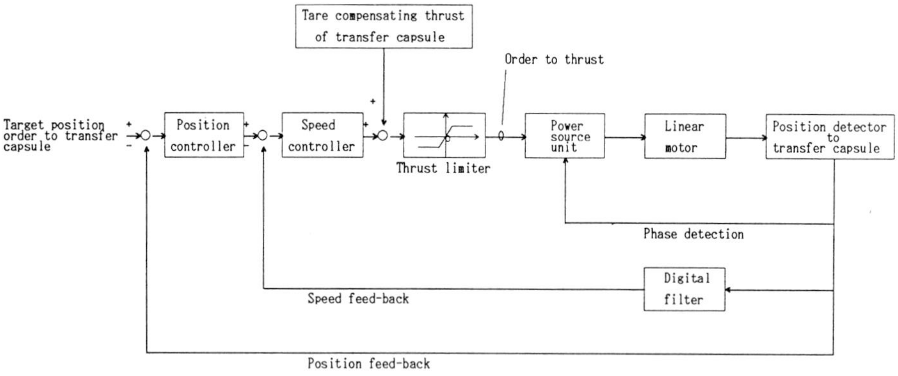

Fig.-3 Block diagram of position control

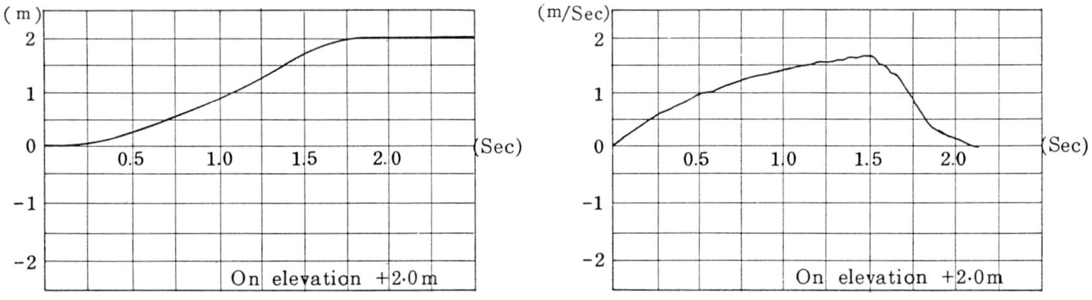

Fig.-4 Result of position control test

CONCLUSION

The following conclusion is attained through the experiment;

- o The transfer capsule has sufficiently 2 m/sec up/down speed.
- o The stopping accuracy of the transfer capsule is ±1 mm.
- o To make the capsule up/down movement smooth, it is effective to reduce the thrust ripple which is caused by the superconducting magnet moving, by means of compensation to the exciting current wave form of the armature magnet.
- o By use of an additional limiter for unbalance thrust in the control algorithm, the capsule up/down speed is able to be made equal.
- o When the capsule comes down*2, it can ensure that any lower armature magnet is prevented from damage caused by free falling of the capsule, by means of the exact position detection.
  (*2 No quench is assumed.)

Consequently, it is reached to the view point that the shaft system with three-phase synchronous linear motor is sufficiently adaptable as well as the other technically area are also expectable to resolve their problems which correspond to scale enlargement, depth increase, safety, and leakage field, etc., through they are still remaining as undeveloped.

Comparison Y-, Bi-, and Tl-compound Bulk Lenses for Relativistic Electron Beams

YOSHIHARU ISHIBASHI, TOMOAKI OSADA, HIDENORI MATSUZAWA, HIROSHI MARUMO, TOMOHIDE YOKOZAWA, YOSHIHIRO KOBAYASHI, MASANORI KAWAMURA, MORIHIRO OHISHI, NOBUO MORIMOTO, TOMIO NAGAKURA, AKIO TOHYAMA, and TETSUYA AKITSU

Faculty of Engineering, Yamanashi University, Kofu, 400 Japan

ABSTRACT

Previously we proposed and demonstrated lenses (Supertrons) for intense electron beams (relativistic electron beams, REBs: 340 keV, 1 kA, duration time of some nanoseconds). In the present paper, the lenses made of Y-, Bi-, and Tl-compound bulks are compared as to their REB-focusing ability in terms of a figure of merit which we proposed. The Tl-compound lens showed the highest value of the figures among the lenses. These results are ascribed to the presumably highest intracurrent density of the Tl-compounds at their operation temperature (about 86 K).

KEYWORDS: Y-compound superconductors, Bi-compound superconducotrs, Tl-compound superconducotrs, lenses, relativistic electron beams (REBs)

INTRODUCTION

Soon after the discovery of the high-T_c superconductor, we proposed and demonstrated lenses (Supertrons) for intense electron beams (relativistic electron beams,REBs) successfully [1-5]. The lenses are one of applications of bulk superconductors, quite different from conventional solenoid coils made of superconductor wires. The principle is as follows: When REBs are injected into small apertures of superconducting tubes, self-magnetic fields of the REBs cannot penetrate into the tubes owing to the Meissner effect. The fields are then confined in a space between the REBs and the wall of the tubes. The enhanced magnetic fields, therefore, focus the REBs. The lenses will guide REBs as optical fibers do the laser light.

In the present paper, we compared the focusing ability of the lenses made of three different high-T_c superconductors (Y-, Bi-, and Tl-compounds). We proposed a figure of merit to evaluate the ability and found that Tl-compound lenses had the highest ability. Their performance was discussed in relation to critical intracurrent of the compounds. These results will give us clues to pulsed power devices of novel type.

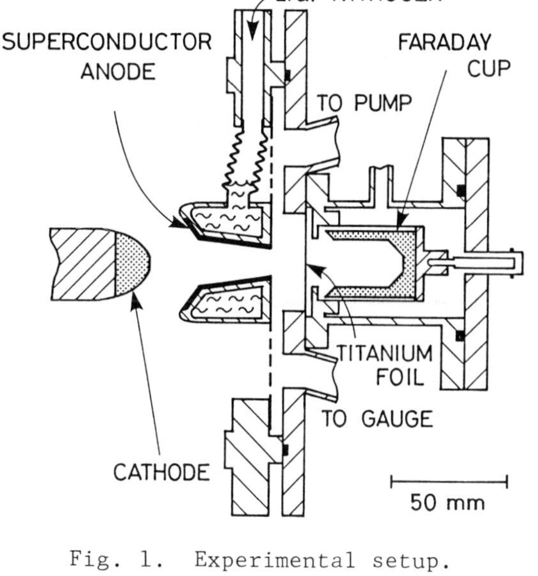

Fig. 1. Experimental setup.

EXPERIMENTAL

Figure 1 shows the experimental setup used. REBs were generated with a diode composed of a carbon cathode and an anode (Supertron). Superconductors pressed to 1 to 1.5 mm in thickness [6] were cemented on the surface of copper-made heat sinks which were cooled with liquid nitrogen [7]. When high-voltage pulses were applied to the diode, intense electron beams (REBs, 340 keV, ~1 kA, some nanoseconds) were field-emitted from the cathode and then were focused with the Supertrons. Radial profiles of the focused REBs were observed with with carbon-made apertures of 2 to 10 mm in diameter and of 1.5 mm in thickness. Neon gases were introduced into the diode chamber to realize charge-neutralization and -multiplication [8].

Figure 2(a) shows radial profiles of focused REBs for Supertrons with 10-mm inner-diameter outlets and with tapered inner surfaces. For comparison, results for the copper-made anodes are also shown. Figure 2(b) shows the normalized radial distributions. The REBs focused have steep slopes as compared with those focused with the cupper-made anodes. To evaluate the performances of the Supertrons, we define a figure of merit as follows:

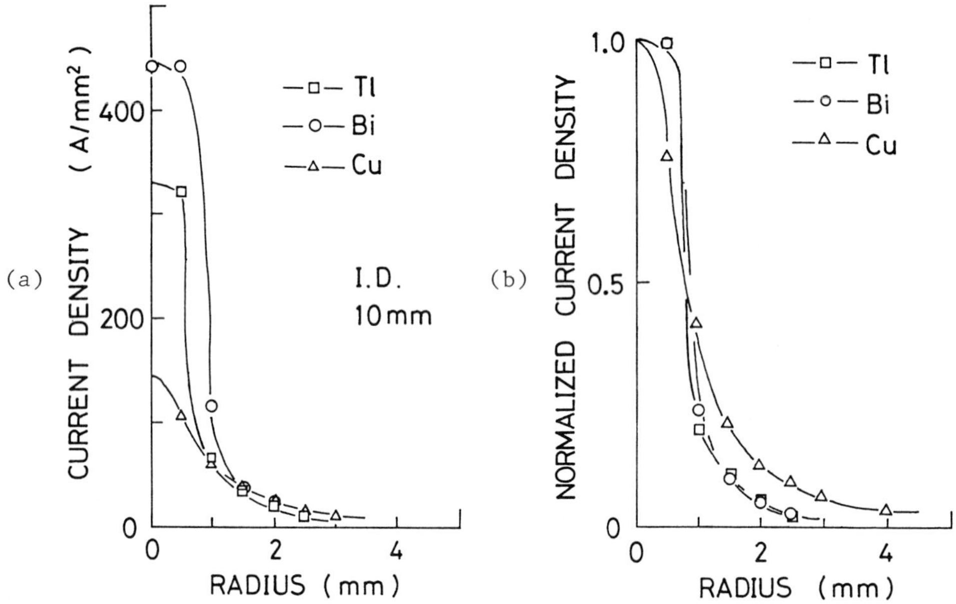

Fig. 2. (a) Focused REBs as a function of radial position. All lenses had a 10-mm-diameter outlet. (b) REBs were normalized with respect to current densities on the axes.

Table 1. Features of lenses tested.

Anode materials	Configurations (diameter in mm)	Total current[A]	$\frac{J_{max}}{J_{mean}}$	$\frac{r_{90}}{r_{10}}$	$\frac{D_I}{2r_f}$	Figure of merit	Critical temperature(k)
Tl	Tapered(11)	850	>36	0.19	12	>82	125
Bi	Tapered(11)	1,800	>19	0.37	6.1	>43	105
	Straight (5, 10, 20)	480 ~3,880	6.7 ~19	0.65 ~0.23	3.8 ~7.4	17 ~32	105
	Straight(10)	1480	3.4	0.63	2.0	4.3	98
Y	Straight(5)	780	3.7	0.73	2.6	7.0	93
	(10)	2640	4.4	0.70	2.3	7.1	93
	(20)	3740	7.0	0.18	4.8	6.0	88
Cu	Tapered(8)	890	7.9	0.17	4.7	6.3	
	Tapered(13)	1,030	5.1	0.24	2.7	3.3	
	Straight(10)	840	6.1	0.20	3.4	4.1	

$$\text{Figure of merit} = J_{max}\, r_{90}\, D_I \,/\, J_{mean}\, r_{10}\, 2r_f$$

where J_{max} and J_{mean} are the current densities on the axes of the lenses and those averaged over their exit area, respectively, r_{90} and r_{10} are the radii at 90 and 10 % of the maximum current densities, respectively, and D_I and $2r_f$ are the inner diameters of the Supertrons at their exits and the radial full widths at half maximum of focused REBs. The factor J_{max}/J_{mean} means the degree of the enhancement in current densities. r_{90}/r_{10} expresses the steepness in radial distribution. The last factor $D_I / 2r_f$ indicates the relative narrowness in beam diameter.

Table I summarizes the experimental results. From the results for Bi-compound, tapered Supertrons are superior to straight ones. Even when we used the same compounds, the Supertrons having higher critical temperatures focused REBs to thinner beams. The figures of merit are plotted in Fig. 3 as a function of temperature margins, which are the temperature differences between the critical temperatures and the operating ones. The open circles indicate the tapered Supertrons, and the triangles show the straight ones. The crosses are for the Supertrons whose critical temperatures were insufficiently high. From Fig. 3, the Supertrons having high critical temperatures show high performances.

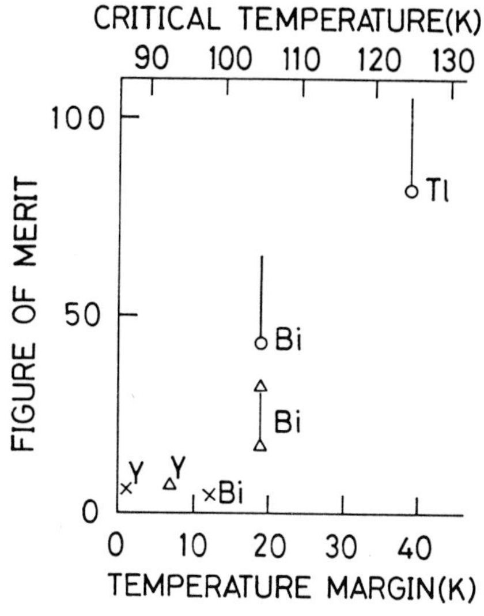

Fig. 3. Figures of merit as a function of temperature margins.

DISCUSSIONS AND CONCLUSIONS

We think that REBs are focused by intracurrents which are induced inside superconductor grains by injected REBs. Supertrons therefore work well when they have higher critical intracurrent densities. There have been many data on intracurrent densities for Y-compounds, while only a few for Bi-compounds [9]. No data are furthermore reported on Tl-compounds. From these facts we think that critical intracurrent densities of the three compounds are higher with increasing critical temperatures when the operating temperatures are limited to ranges higher than liquid nitrogen [9]. Out of the three kinds of the compounds, Bi-compounds seem to be used most favorably because they are stable against water vapor and nontoxic. These conclusions will give us criteria of applying the high-T_c superconductors to devices of novel type.

REFERENCES

1. Ishibashi Y, Osada T, Matsuzawa H, Marumo H, Yokozawa T, Kobayashi Y, Kawamura M, Ohishi M, Morimoto N, Nagakura T, Tohyama A, Akitsu T (1990) Jpn. J. Appl. Phys. 29: L1120
2. Matsuzawa H, Ishibashi Y, Osada T, Irikura K, Okamoto K, Wada H, Mochizuki A, Akitsu T (1990) Jpn. J. Appl. Phys. 29: 785
3. Evans R (1989) Nature 340: 191
4. Matsuzawa H, Osada T, Ishibashi Y, Irikura K, Okamoto K, Mochizuki A, Wada H, Akitsu T (1989) Jpn. J. Appl. Phys. 28: L717
5. Matsuzawa H. Ohmori O, Yamazaki H, Ueno J. Furumizu A, Saito A, Takahashi T, Akitsu T (1989) J. Appl. Phys. 65:2596
6. Osada T, Tohyama Y, Chino Y Matsuzawa H (1989) Rev. Sci. Instrum. 60: 3821
7. Ishibashi Y, Irikura K, Matsuzawa H (1989) J. Vac. Sci. Technol. A 7: 2818
8. Matsuzawa H, Akitsu T (1988) J. Appl. Phys. 63: 4388.
9. Kumakura H, Togano K, Yanagisawa E, Kase J, Maeda H (1990) Jpn. J. Appl. Phys. 29: L1652

Observation of Supertron-Focused Electron Beams with a Streak Camera

TOMOAKI OSADA, YOSHIHARU ISHIBASHI, HIDENORI MATSUZAWA, and TETSUYA AKITSU

Faculty of Engineering, Yamanashi University, Kofu, 400 Japan

ABSTRACT

Relativistic electron beams (REBs, 340 keV, about 1 kA, duration time of 5 ns) were focused in an atmosphere of 0.1-Torr neon gas with Supertrons which are bulk-superconductor-made lenses proposed by the authors. The spatial and temporal behavior of the focused REBs were observed with a conventional camera and with a streak camera (Hamamatsu C1578), respectively. The REBs were, surely, radially focused to diameters of less than 2 mm and seemed to be bunched in their axial direction with a time interval of 3 ns.

KEYWORDS: relativistic electron beams (REBs), high-T_c superconductor lenses, Supertrons, streak camera

INTRODUCTION

As one of novel applications of high-T_c bulk superconductors, we proposed and demonstrated lenses (Supertrons) [1] for relativistic electron beams (REBs, 340 keV, about 1 kA, duration time of 5 ns). The principle of Supertrons is as follows: When intense electron beams are injected into small apertures of superconducting tubes, the self-magnetic fields of the REBs can not penetrate into the tubes owing to the Meissner effect. The fields are thus confined in a space between the REBs and the wall of the tubes. The compressed fields then focus the REBs. These focused REBs will be transported along the axes of the tubes with little loss as optical fibers guide laser light. In previous papers [2-5] we observed spatial profiles of focused REBs with open-shutter photographs and with carbon-made apertures having through-holes at their centers. With these methods we obtained only time-integrated spatial distribution of REBs. In the present paper we try to observe temporal behavior of REBs using a high-speed streak camera (Hamamatsu C1587).

EXPERIMENTAL

Figure 1 shows the arrangement of experimental setup used. REBs were generated with a diode consisting of a velvet-backed carbon-made cathode [6] and a Bi-compound superconductor-made anode (Supertrons). When high-voltage pulses are applied to the diode, REBs are field-emitted from the cathode

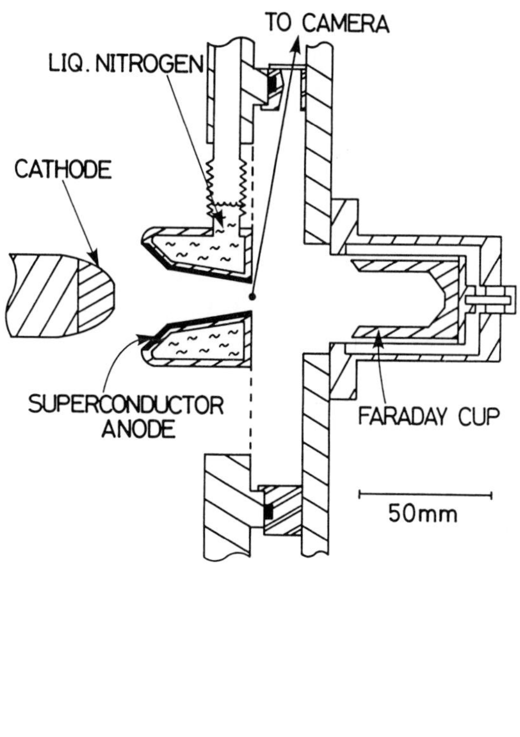

Fig. 1. Experimental arrangement.

and are focused with the Supertron as REBs travel along the axis of the Supertron. Neon gases of 0.1-Torr order were introduced into the diode chamber to neutralize space charge and to multiply REBs generated [7]. The distribution of REBs can therefore be observed with fluorescence emitted from excited neon atoms. Spatial distribution of REBs at the exit of Supertrons was observed through a prism [2], and temporal behavior of the REBs was recorded by the streak camera through a slit 20 mm × 30 μm.

Figure 2(a) is an open-shutter photograph of REBs which were ejected from the exit of Bi-compound Supertrons, and Fig. 2(b) is a contour map of the intensity of fluorescence in Fig. 2(a) [2-3]. REBs were focused to diameters of less than 2 mm. The current densities achieved were as high as 400 A/mm^2. In Fig. 2(a) the REBs diffused downstream 1 cm from the exit. These results indicate that REBs were focused by Supertrons and did not self-pinch.

Figure 3(a) is a waveform of the fluorescence inside the diode chamber, observed with a biplaner-phototube (Hamamatsu R1193U-02). Figure 3(b) is for electrical currents detected with a Faraday cup as in Fig. 1. In this case both ions and secondary electrons (plasma electrons) of low energy were

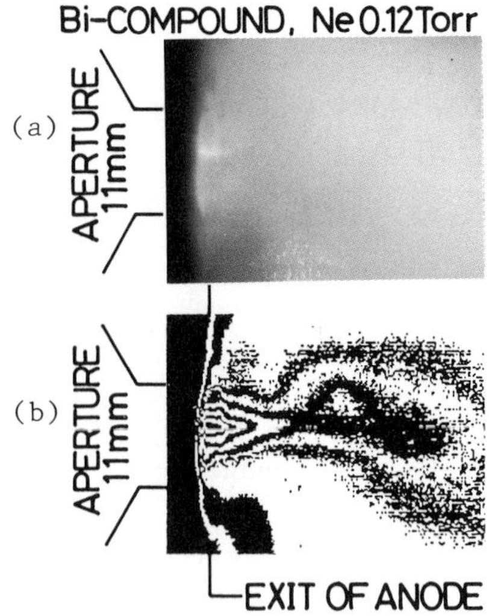

Fig. 2. (a) Open-shutter photograph for Supertron with 10-mm-diameter outlet. (b) Contour map of fluorescence intensity.

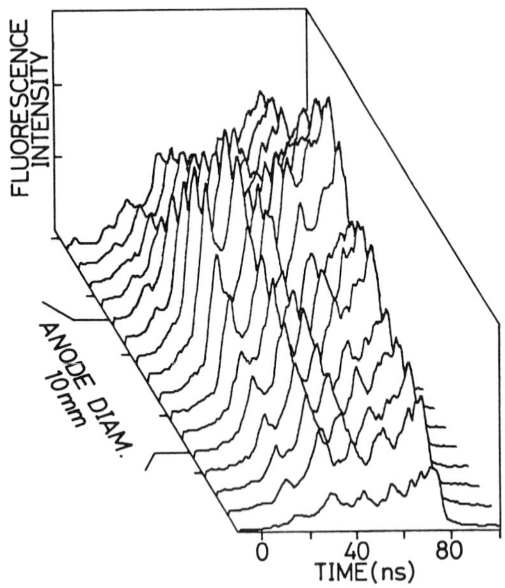

Fig. 4. Temporal behavior of fluorescence emitted from neon atoms at exit of Supertron. Slit of streak camera was 20 mm × 30 μm.

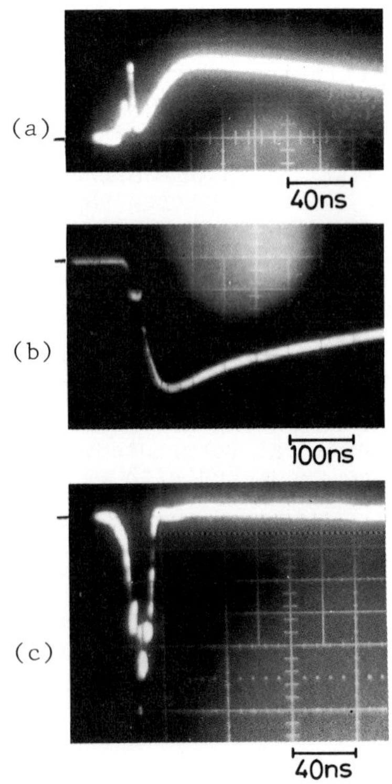

Fig. 3. Temporal behavior of fluorescence at exit of Supertron (a), REBs and secondary electrons (b), and REBs only (c).

were collected in addition to REBs because the Faraday cup was kept in the same atmosphere of Ne gas of 0.1-Torr order as that for the diode chamber. Next, a sheet of 20-μm-thick titanium foil was set in front of the Faraday cup to keep the cup at 10^{-4} Torr or less and to reject secondary charged-particles. Electrons which can pass the titanium foil have kinetic energies higher than 60 keV. We then obtained Fig. 3(c), where secondary charged-particle current was clearly rejected. Each waveform in Fig.3 is for a different shot of operation. The fluorescence in Fig. 3(a) behaves similarly to the net current in Fig. 3(b). Small spikes at the beginning of the waveforms in Figs. 3(a) and (b) correspond to the REBs in Fig. 3(c).

Figure 4 shows temporal and spatial distribution of fluorescence of neon atoms. The horizontal time-scale begins just after REBs were ejected. The intensity of the fluorescence oscillated in axial direction with a period of about 3 ns. In radial direction, however, no distinguished fluctuation is found in amplitude of fluorescence. The oscillation may be induced by axial vibrations of a double layer (plasma sheath) [8]. The double layer represents an electrical capacity and will be formed in regions of the axes of REBs near the corner where the two tapered superconductors are joined (Fig. 1).

CONCLUSIONS

We confirmed optically that Supertrons focused REBs (about 1 kA, 5 ns) to diameters of less than 2 mm and that after-glow plasma which was produced by the REBs oscillated with a period of about 3 ns. The oscillation is possibly due to the time-varying capacity of a double layer. To achieve narrower electron beams we are preparing another Supertrons.

To achieve intrinsic functions of Supertrons in focusing electron beams, we need to use a different electron beam source from the present REBs source. This is because the REBs in the present paper is not monochromatic in kinetic energy [7] and their emittance is not low.

REFERENCES

1. Matsuzawa H, Ohmori O, Yamazaki H, Ueno J, Furumizu A, Saito A, Takahashi T, Akitsu T (1989) J. Appl. Phys. 65: 2596-2603
2. Matsuzawa H, Osada T, Ishibashi Y, Irikura K, Okamoto K, Mochizuki A, Wada H, Akitsu T (1989) Jpn. J. Appl. Phys. 28: L717-719
3. Evans R (1989) Nature 340: 191
4. Matsuzawa H, Ishibashi Y, Osada T, Irikura K, Okamoto K, Wada H, Mochizuki A, Akitsu T (1990) Jpn. J. Appl. Phys. 29: 785-791
5. Ishibashi Y, Osada T, Matsuzawa H, Marumo H, Yokozawa T, Kobayashi Y, Kawamura M, Ohishi M, Morimoto N, Nagakura T, Tohyama A, Akitsu T (1990) Jpn. J. Appl. Phys. 29: L1120-L1123
6. Bekefi G, Hartemann F, Kirkptrick D A (1987) J. Appl. Phys. 62: 1564-1567
7. Matsuzawa H, Akitsu T (1988) J. Appl. Phys. 63: 4388-4391
8. Franklin R N (1976) Plasma Phenomena in Gas Discharges. Clarendon Press. Oxford, pp. 76-80

Author Index

Key Word Index